29 CFR 1926 OSHA Construction Industry Regulations

Including:

Part 1903 - Inspections, Citations, and Proposed Penalties
Part 1904 - Recording and Reporting Occupational Injuries and Illnesses
Part 1910 - General Industry - Selected Topics
Part 1926 - Construction
Letters of Interpretation

Addendum including:

- General Duty Clause
- NAICS Codes
- Sharps Injury Log
- Incidence Rates of Nonfatal Occupational Injuries and Illnesses
- OSHA Forms 300, 300A, and 301
- It's The Law! Mandatory Posting
- Material Approved for Incorp. by Reference
- States with Approved Plans - State Office Directory - Regional and National Offices
- SIC Division Structure
- Safety and Health Management Guidelines
- Multi-Employer Citation Policy
- Most Common Standards Cited for Construction

Updated through July 31, 2012

This publication is also available on CD-Rom 

Changing the Complex Into Compliance®
Mangan Communications, Inc.
http://www.Mancomm.com

315 West Fourth Street
Davenport, Iowa 52801
(563) 323-6245
1-800-MANCOMM
(626-2666)
Fax: (563) 323-0804

Website: http://www.Mancomm.com
E-mail: safetyinfo@mancomm.com

Library of Congress Control Number: 2012944440

ISBN: 1-59959-383-1

RegLogic® . . . a better way

RegLogic® is a graphical approach to help you effortlessly navigate government regulatory information.

MANCOMM® has taken the guesswork out of finding the regulations that keep your workforce safe and compliant. Take advantage of the intuitive format to easily navigate the government's complex standards, saving yourself time and avoiding stress while you solve issues before they become problems.

What makes RegLogic® unique?

▶ Quick Reference

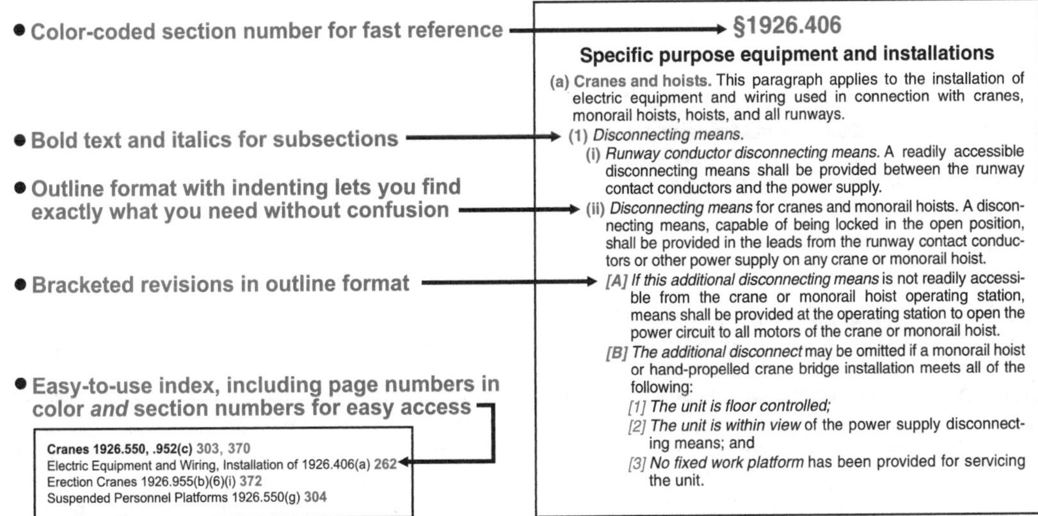

- **Color-coded section number for fast reference**
- **Bold text and italics for subsections**
- **Outline format with indenting lets you find exactly what you need without confusion**
- **Bracketed revisions in outline format**
- **Easy-to-use index, including page numbers in color *and* section numbers for easy access**

Cranes 1926.550, .952(c) 303, 370
Electric Equipment and Wiring, Installation of 1926.406(a) 262
Erection Cranes 1926.955(b)(6)(i) 372
Suspended Personnel Platforms 1926.550(g) 304

§1926.406

Specific purpose equipment and installations

(a) **Cranes and hoists.** This paragraph applies to the installation of electric equipment and wiring used in connection with cranes, monorail hoists, hoists, and all runways.

(1) *Disconnecting means.*

(i) *Runway conductor disconnecting means.* A readily accessible disconnecting means shall be provided between the runway contact conductors and the power supply.

(ii) *Disconnecting means* for cranes and monorail hoists. A disconnecting means, capable of being locked in the open position, shall be provided in the leads from the runway contact conductors or other power supply on any crane or monorail hoist.

[A] If this additional disconnecting means is not readily accessible from the crane or monorail hoist operating station, means shall be provided at the operating station to open the power circuit to all motors of the crane or monorail hoist.

[B] The additional disconnect may be omitted if a monorail hoist or hand-propelled crane bridge installation meets all of the following:

[1] The unit is floor controlled;

[2] The unit is within view of the power supply disconnecting means; and

[3] No fixed work platform has been provided for servicing the unit.

▶ The Graphical Approach

- **Clear, legible text**
- **Enhanced Color Graphics**

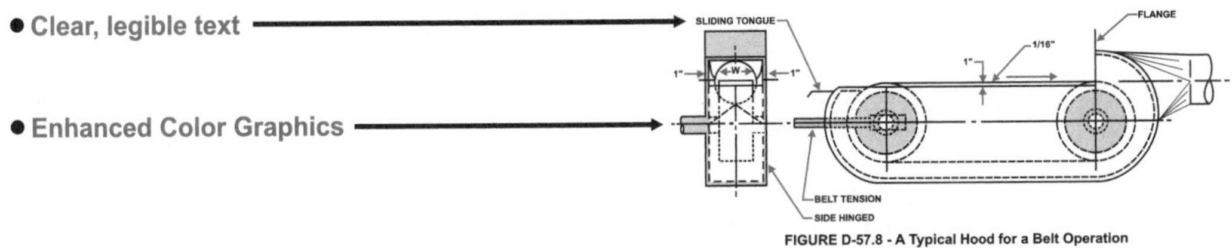

FIGURE D-57.8 - A Typical Hood for a Belt Operation
Entry loss = 0.45 velocity pressure for tapered takeoff.

- **Color-coded headings**
- **Clearly defined lines**

Table F-3

Quantity of LP-Gas Stored	Distance (feet)
500 lbs. or less	0
501 to 6,000 lbs.	10
6,001 to 10,000 lbs.	20
Over 10,000 lbs.	25

RegLogic® is a one-of-a-kind approach that you will find only from MANCOMM®.

Online forms available: Full-size versions of forms in this book are available free of charge at **www.oshacfr.com**.

Recent changes in regulations:

March 26, 2012 (Federal Register Volume 77, No. 58)

In the March 26, 2012 edition of the Federal Register, OSHA modified its Hazard Communication Standard (HCS) to conform to the United Nations' Globally Harmonized System of Classification and Labelling of Chemicals (GHS). OSHA has determined that the modifications will significantly reduce costs and burdens while also improving the quality and consistency of information provided to employers and employees regarding chemical hazards and associated protective measures.

The modifications to the standard include revised criteria for classification of chemical hazards; revised labeling provisions that include requirements for use of standardized signal words, pictograms, hazard statements, and precautionary statements; a specified format for safety data sheets; and related revisions to definitions of terms used in the standard, and requirements for employee training on labels and safety data sheets. OSHA also modified provisions of other standards, including standards for flammable and combustible liquids, process safety management, and most substance-specific health standards, to ensure consistency with the modified HCS requirements. To see a complete list of sections affected and compliance dates, please visit www.mancomm.com

April 18, 2012 (Federal Register Volume 77, No. 75)

In the April 18, 2012 edition of the Federal Register, OSHA corrected its sling standard for construction titled "Rigging Equipment for Material Handling" by removing the rated capacity tables and making minor, nonsubstantive revisions to the regulatory text. This amendment affected section §1926.251, revising paragraphs (b)(5) and (c)(5), and also removed tables H-1 and H-3 through H-19. Table H-2 was redesignated as Table H-1, and Table H-20 was redesignated as Table H-2. This amendment went into effect on April 18, 2012.

June 22, 2012 (Federal Register Volume 77, No. 121)

In the June 22, 2012 edition of the Federal Register, OSHA issued a direct final rule to revise the personal protective equipment (PPE) sections of its general industry, shipyard employment, longshoring, and marine terminals standards regarding requirements for head protection. OSHA is updating the references in its standards to recognize the 2009 edition of the American National Standard for Industrial Head Protection, and is deleting the 1986 edition of that national consensus standard because it is out of date. OSHA also is including the construction industry in this rulemaking to ensure consistency among the Agency's standards. This update becomes effective on September 20, 2012, and affects §1926.6(h)(28)-(30), and §1926.100(b) and (c).

*Editor's note: On July 23, 2012, OSHA published a correction to this update. The correction is included in this edition.

Table of Contents

1903 - Inspections, Citations, and Proposed Penalties

1904 - Recording and Reporting Occupational Injuries and Illnesses

Subpart A - Purpose

Subpart B - Scope

Subpart C - Recordkeeping Forms and Recording Criteria

Subpart D - Other OSHA Injury and Illness Recordkeeping Requirements

Subpart E - Reporting Fatality, Injury and Illness Information to the Government

Subpart P - Excavations

Subpart Q - Concrete and Masonry Construction

Subpart R - Steel Erection

Subpart S - Underground Construction, Caissons, Cofferdams and Compressed Air

Subpart T - Demolition

Subpart U - Blasting and the Use of Explosives

Subpart V - Power Transmission and Distribution

Subpart W - Rollover Protective Structures; Overhead Protection

Subpart X - Stairways and Ladders

Subpart Y - Diving

Subpart Z - Toxic and Hazardous Substances

Subpart AA- BB - Reserved

Subpart CC - Cranes and Derricks in Construction

Subpart DD - Cranes and Derricks Used in Demolition and Underground Construction

Addendum

1903 - Inspections, Citations, and Proposed Penalties

§1903.1
Purpose and scope

The Williams-Steiger Occupational Safety and Health Act of 1970 (84 Stat. 1590 et seq., 29 U.S.C. 651 et seq.) requires, in part, that every employer covered under the Act furnish to his employees employment and a place of employment which are free from recognized hazards that are causing or are likely to cause death or serious physical harm to his employees. The Act also requires that employers comply with occupational safety and health standards promulgated under the Act, and that employees comply with standards, rules, regulations and orders issued under the Act which are applicable to their own actions and conduct. The Act authorizes the Department of Labor to conduct inspections, and to issue citations and proposed penalties for alleged violations. The Act, under section 20(b), also authorizes the Secretary of Health, Education, and Welfare to conduct inspections and to question employers and employees in connection with research and other related activities. The Act contains provisions for adjudication of violations, periods prescribed for the abatement of violations, and proposed penalties by the Occupational Safety and Health Review Commission, if contested by an employer or by an employee or authorized representative of employees, and for judicial review. The purpose of this part 1903 is to prescribe rules and to set forth general policies for enforcement of the inspection, citation, and proposed penalty provisions of the Act. In situations where this part 1903 sets forth general enforcement policies rather than substantive or procedural rules, such policies may be modified in specific circumstances where the Secretary or his designee determines that an alternative course of action would better serve the objectives of the Act. [§1903.1]

§1903.2
Posting of notice; availability of the Act, regulations and applicable standards

(a)(1) *Each employer shall post and keep posted* a notice or notices, to be furnished by the Occupational Safety and Health Administration, U.S. Department of Labor, informing employees of the protections and obligations provided for in the Act, and that for assistance and information, including copies of the Act and of specific safety and health standards, employees should contact the employer or the nearest office of the Department of Labor. Such notice or notices shall be posted by the employer in each establishment in a conspicuous place or places where notices to employees are customarily posted. Each employer shall take steps to insure that such notices are not altered, defaced, or covered by other material. [§1903.2(a)(1)]

(2) *Where a State has an approved poster* informing employees of their protections and obligations as defined in §1952.10 of this chapter, such poster, when posted by employers covered by the State plan, shall constitute compliance with the posting requirements of section 8(c)(1) of the Act. Employers whose operations are not within the issues covered by the State plan must comply with paragraph (a)(1) of this section. [§1903.2(a)(2)]

(3) *Reproductions or facsimiles* of such Federal or State posters shall constitute compliance with the posting requirements of section 8(c)(1) of the Act where such reproductions or facsimiles are at least 8½ inches by 14 inches, and the printing size is at least 10 pt. Whenever the size of the poster increases, the size of the print shall also increase accordingly. The caption or heading on the poster shall be in large type, generally not less than 36 pt. [§1903.2(a)(3)]

(b) **Establishment** means a single physical location where business is conducted or where services or industrial operations are performed. (For example: A factory, mill, store, hotel, restaurant, movie theatre, farm, ranch, bank, sales office, warehouse, or central administrative office.) Where distinctly separate activities are performed at a single physical location (such as contract construction activities from the same physical location as a lumber yard), each activity shall be treated as a separate physical establishment, and a separate notice or notices shall be posted in each such establishment, to the extent that such notices have been furnished by the Occupational Safety and Health Administration, U.S. Department of Labor. Where employers are engaged in activities which are physically dispersed, such as agriculture, construction, transportation, communications, and electric, gas and sanitary services, the notice or notices required by this section shall be posted at the location to which employees report each day. Where employees do not usually work at, or report to, a single establishment, such as longshoremen, traveling salesmen, technicians, engineers, etc., such notice or notices shall be posted at the location from which the employees operate to carry out their activities. In all cases, such notice or notices shall be posted in accordance with the requirements of paragraph (a) of this section. [§1903.2(b)]

(c) **Copies of the Act,** all regulations published in this chapter and all applicable standards will be available at all Area Offices of the Occupational Safety and Health Administration, U.S. Department of Labor. If an employer has obtained copies of these materials, he shall make them available upon request to any employee or his authorized representative for review in the establishment where the employee is employed on the same day the request is made or at the earliest time mutually convenient to the employee or his authorized representative and the employer. [§1903.2(c)]

(d) **Any employer failing to comply with the provisions** of this section shall be subject to citation and penalty in accordance with the provisions of section 17 of the Act. [§1903.2(d)]

[36 FR 17850, Sept. 4, 1971, as amended at 39 FR 39036, Nov. 5, 1974]

§1903.3
Authority for inspection

(a) **Compliance Safety and Health Officers** of the Department of Labor are authorized to enter without delay and at reasonable times any factory, plant, establishment, construction site, or other area, workplace or environment where work is performed by an employee of an employer; to inspect and investigate during regular working hours and at other reasonable times, and within reasonable limits and in a reasonable manner, any such place of employment, and all pertinent conditions, structures, machines, apparatus, devices, equipment and materials therein; to question privately any employer, owner, operator, agent or employee; and to review records required by the Act and regulations published in this chapter, and other records which are directly related to the purpose of the inspection. Representatives of the Secretary of Health, Education, and Welfare are authorized to make inspections and to question employers and employees in order to carry out the functions of the Secretary of Health, Education, and Welfare under the Act. Inspections conducted by Department of Labor Compliance Safety and Health Officers and representatives of the Secretary of Health, Education, and Welfare under section 8 of the Act and pursuant to this part 1903 shall not affect the authority of any State to conduct inspections in accordance with agreements and plans under section 18 of the Act. [§1903.3(a)]

(b) **Prior to inspecting areas containing information** which is classified by an agency of the United States Government in the interest of national security, Compliance Safety and Health Officers shall have obtained the appropriate security clearance. [§1903.3(b)]

§1903.4
Objection to inspection

(a) **Upon a refusal to permit** the Compliance Safety and Health Officer, in exercise of his official duties, to enter without delay and at reasonable times any place of employment or any place therein, to inspect, to review records, or to question any employer, owner, operator, agent, or employee, in accordance with §1903.3 or to permit a representative of employees to accompany the Compliance Safety and Health Officer during the physical inspection of any workplace in accordance with §1903.8, the Safety and Health Officer shall terminate the inspection or confine the inspection to other areas, conditions, structures, machines, apparatus, devices, equipment, materials, records, or interviews concerning which no objection is raised. The Compliance Safety and Health Officer shall endeavor to ascertain the reason for such refusal, and shall immediately report the refusal and the reason therefor to the Area Director. The Area Director shall consult with the Regional Solicitor, who shall take appropriate action, including compulsory process, if necessary. [§1903.4(a)]

(b) **Compulsory process shall be sought** in advance of an attempted inspection or investigation if, in the judgment of the Area Director and the Regional Solicitor, circumstances exist which make such preinspection process desirable or necessary. Some examples of circumstances in which it may be desirable or necessary to seek compulsory process in advance of an attempt to inspect or investigate include (but are not limited to): [§1903.4(b)]

(1) *When the employer's past practice* either implicitly or explicitly puts the Secretary on notice that a warrantless inspection will not be allowed; [§1903.4(b)(1)]

(2) *When an inspection is scheduled* far from the local office and procuring a warrant prior to leaving to conduct the inspection would avoid, in case of refusal of entry, the expenditure of significant time and resources to return to the office, obtain a warrant and return to the worksite; [§1903.4(b)(2)]

(3) *When an inspection includes* the use of special equipment or when the presence of an expert or experts is needed in order to properly conduct the inspection, and procuring a warrant prior to an attempt to inspect would alleviate the difficulties or costs encountered in coordinating the availability of such equipment or expert. [§1903.4(b)(3)]

(c) **With the approval of the Regional Administrator** and the Regional Solicitor, compulsory process may also be obtained by the Area Director or his designee. [§1903.4(c)]

(d) **For purposes of this section,** the term compulsory process shall mean the institution of any appropriate action, including ex parte application for an inspection warrant or its equivalent. Ex parte inspection warrants shall be the preferred form of compulsory process in all circumstances where compulsory process is relied upon to seek entry to a workplace under this section. [§1903.4(d)]

[45 FR 65923, Oct. 3, 1980]

§1903.5
Entry not a waiver

Any permission to enter, inspect, review records, or question any person, shall not imply or be conditioned upon a waiver of any cause of action, citation, or penalty under the Act. Compliance Safety and Health Officers are not authorized to grant any such waiver. [§1903.5]

§1903.6
Advance notice of inspections

(a) **Advance notice of inspections may not be given,** except in the following situations: [§1903.6(a)]

(1) *In cases of apparent imminent danger,* to enable the employer to abate the danger as quickly as possible; [§1903.6(a)(1)]

(2) *In circumstances where the inspection* can most effectively be conducted after regular business hours or where special preparations are necessary for an inspection; [§1903.6(a)(2)]

(3) *Where necessary* to assure the presence of representatives of the employer and employees or the appropriate personnel needed to aid in the inspection; and [§1903.6(a)(3)]

(4) *In other circumstances* where the Area Director determines that the giving of advance notice would enhance the probability of an effective and thorough inspection. [§1903.6(a)(4)]

(b) **In the situations described in paragraph (a) of this section,** advance notice of inspections may be given only if authorized by the Area Director, except that in cases of apparent imminent danger, advance notice may be given by the Compliance Safety and Health Officer without such authorization if the Area Director is not immediately available. When advance notice is given, it shall be the employer's responsibility promptly to notify the authorized representative of employees of the inspection, if the identity of such representative is known to the employer. (See §1903.8(b) as to situations where there is no authorized representative of employees.) Upon the request of the employer, the Compliance Safety and Health Officer will inform the authorized representative of employees of the inspection, provided that the employer furnishes the Compliance Safety and Health Officer with the identity of such representative and with such other information as is necessary to enable him promptly to inform such representative of the inspection. An employer who fails to comply with his obligation under this paragraph promptly to inform the authorized representative of employees of the inspection or to furnish such information as is necessary to enable the Compliance Safety and Health Officer promptly to inform such representative of the inspection, may be subject to citation and penalty under section 17(c) of the Act. Advance notice in any of the situations described in paragraph (a) of this section shall not be given more than 24 hours before the inspection is scheduled to be conducted, except in apparent imminent danger situations and in other unusual circumstances. [§1903.6(b)]

(c) **The Act provides in section 17(f)** that any person who gives advance notice of any inspection to be conducted under the Act, without authority from the Secretary or his designees, shall, upon conviction, be punished by fine of not more than $1,000 or by imprisonment for not more than 6 months, or by both. [§1903.6(c)]

§1903.7
Conduct of inspections

(a) **Subject to the provisions of §1903.3,** inspections shall take place at such times and in such places of employment as the Area Director or the Compliance Safety and Health Officer may direct. At the beginning of an inspection, Compliance Safety and Health Officers shall present their credentials to the owner, operator, or agent in charge at the establishment; explain the nature and purpose of the inspection; and indicate generally the scope of the inspection and the records specified in §1903.3 which they wish to review. However, such designation of records shall not preclude access to additional records specified in §1903.3. [§1903.7(a)]

(b) **Compliance Safety and Health Officers** shall have authority to take environmental samples and to take or obtain photographs related to the purpose of the inspection, employ other reasonable investigative techniques, and question privately any employer, owner, operator, agent or employee of an establishment. (See §1903.9 on trade secrets.) As used herein, the term employ other reasonable investigative techniques includes, but is not limited to, the use of devices to measure employee exposures and the attachment of personal sampling equipment such as dosimeters, pumps, badges and other similar devices to employees in order to monitor their exposures. [§1903.7(b)]

(c) **In taking photographs and samples,** Compliance Safety and Health Officers shall take reasonable precautions to insure that such actions with flash, spark-producing, or other equipment would not be hazardous. Compliance Safety and Health Officers shall comply with all employer safety and health rules and practices at the establishment being inspected, and they shall wear and use appropriate protective clothing and equipment. [§1903.7(c)]

(d) **The conduct of inspections** shall be such as to preclude unreasonable disruption of the operations of the employer's establishment. [§1903.7(d)]

(e) **At the conclusion of an inspection,** the Compliance Safety and Health Officer shall confer with the employer or his representative and informally advise him of any apparent safety or health violations disclosed by the inspection. During such conference, the employer shall be afforded an opportunity to bring to the attention of the Compliance Safety and Health Officer any pertinent information regarding conditions in the workplace. [§1903.7(e)]

(f) **Inspections shall be conducted in accordance** with the requirements of this part. [§1903.7(f)]

[36 FR 17850, Sept. 14, 1971, as amended at 47 FR 6533, Feb. 12, 1982; 47 FR 55481, Dec. 10, 1982]

§1903.8
Representatives of employers and employees

(a) **Compliance Safety and Health Officers** shall be in charge of inspections and questioning of persons. A representative of the employer and a representative authorized by his employees shall be given an opportunity to accompany the Compliance Safety and Health Officer during the physical inspection of any workplace for the purpose of aiding such inspection. A Compliance Safety and Health Officer may permit additional employer representatives and additional representatives authorized by employees to accompany him where he determines that such additional representatives will further aid the inspection. A different employer and employee representative may accompany the Compliance Safety and Health Officer during each different phase of an inspection if this will not interfere with the conduct of the inspection. [§1903.8(a)]

(b) **Compliance Safety and Health Officers** shall have authority to resolve all disputes as to who is the representative authorized by the employer and employees for the purpose of this section. If there is no authorized representative of employees, or if the Compliance Safety and Health Officer is unable to determine with reasonable certainty who is such representative, he shall consult with a reasonable number of employees concerning matters of safety and health in the workplace. [§1903.8(b)]

(c) **The representative(s) authorized by employees** shall be an employee(s) of the employer. However, if in the judgment of the Compliance Safety and Health Officer, good cause has been shown why accompaniment by a third party who is not an employee of the employer (such as an industrial hygienist or a safety engineer) is reasonably necessary to the conduct of an effective and thorough physical inspection of the workplace, such third party may accompany the Compliance Safety and Health Officer during the inspection. [§1903.8(c)]

(d) **Compliance Safety and Health Officers** are authorized to deny the right of accompaniment under this section to any person whose conduct interferes with a fair and orderly inspection. The right of accompaniment in areas containing trade secrets shall be subject to the provisions of §1903.9(d). With regard to information classified by an agency of the U.S. Government in the interest of national security, only persons authorized to have access to such information may accompany a Compliance Safety and Health Officer in areas containing such information. [§1903.8(d)]

§1903.9
Trade secrets

(a) **Section 15 of the Act provides:** "All information reported to or otherwise obtained by the Secretary or his representative in connection with any inspection or proceeding under this Act which contains or which might reveal a trade secret referred to in section 1905 of title 18 of the United States Code shall be considered confidential for the purpose of that section, except that such information may be disclosed to other officers or employees concerned with carrying out this Act or when relevant in any proceeding under this Act. In any such proceeding the Secretary, the Commission, or the court shall issue such orders as may be appropriate to protect the confidentiality of trade secrets." Section 15 of the Act is considered a statute within the meaning of section 552(b)(3) of title 5 of the United States Code, which exempts from the disclosure requirements matters that are "specifically exempted from disclosure by statute." [§1903.9(a)]

(b) **Section 1905 of title 18 of the United States Code provides:** "Whoever, being an officer or employee of the United States or of any department or agency thereof, publishes, divulges, discloses, or makes known in any manner or to any extent not authorized by law any information coming to him in the course of his employment or official duties or by reason of any examination or investigation made by, or return, report or record made to or filed with, such department or agency or officer or employee thereof, which information concerns or relates to the trade secrets, processes, operations, style of work, or apparatus, or to the identity, confidential statistical data, amount or source of any income, profits, losses, or expenditures of any person, firm, partnership, corporation, or association; or permits any income return or copy thereof or any book containing any abstract or particulars thereof to be seen or examined by any person except as provided by law; shall be fined not more than $1,000, or imprisoned not more than 1 year, or both; and shall be removed from office or employment." [§1903.9(b)]

(c) **At the commencement of an inspection,** the employer may identify areas in the establishment which contain or which might reveal a trade secret. If the Compliance Safety and Health Officer has no clear reason to question such identification, information obtained in such areas, including all negatives and prints of photographs, and environmental samples, shall be labeled "confidential — trade secret" and shall not be disclosed except in accordance with the provisions of section 15 of the Act. [§1903.9(c)]

(d) **Upon the request of an employer,** any authorized representative of employees under §1903.8 in an area containing trade secrets shall be an employee in that area or an employee authorized by the employer to enter that area. Where there is no such representative or employee, the Compliance Safety and Health Officer shall consult with a reasonable number of employees who work in that area concerning matters of safety and health. [§1903.9(d)]

§1903.10
Consultation with employees

Compliance Safety and Health Officers may consult with employees concerning matters of occupational safety and health to the extent they deem necessary for the conduct of an effective and thorough inspection. During the course of an inspection, any employee shall be afforded an opportunity to bring any violation of the Act which he has reason to believe exists in the workplace to the attention of the Compliance Safety and Health Officer. [§1903.10]

§1903.11
Complaints by employees

(a) **Any employee or representative of employees** who believe that a violation of the Act exists in any workplace where such employee is employed may request an inspection of such workplace by giving notice of the alleged violation to the Area Director or to a Compliance Safety and Health Officer. Any such notice shall be reduced to writing, shall set forth with reasonable particularity the grounds for the notice, and shall be signed by the employee or representative of employees. A copy shall be provided the employer or his agent by the Area Director or Compliance Safety and Health Officer no later than at the time of inspection, except that, upon the request of the person giving such notice, his name and the names of individual employees referred to therein shall not appear in such copy or on any record published, released, or made available by the Department of Labor. [§1903.11(a)]

(b) **If upon receipt of such notification** the Area Director determines that the complaint meets the requirements set forth in paragraph (a) of this section, and that there are reasonable grounds to believe that the alleged violation exists, he shall cause an inspection to be made as soon as practicable, to determine if such alleged violation exists. Inspections under this section shall not be limited to matters referred to in the complaint. [§1903.11(b)]

(c) **Prior to or during any inspection of a workplace,** any employee or representative of employees employed in such workplace may notify the Compliance Safety and Health Officer, in writing, of any violation of the Act which they have reason to believe exists in such workplace. Any such notice shall comply with the requirements of paragraph (a) of this section. [§1903.11(c)]

(d) **Section 11(c)(1) of the Act provides:** "No person shall discharge or in any manner discriminate against any employee because such employee has filed any complaint or instituted or caused to be instituted any proceeding under or related to this Act or has testified or is about to testify in any such proceeding or because of the exercise by such employee on behalf of himself or others of any right afforded by this Act." [§1903.11(d)]

(Approved by the Office of Management and Budget under control number 1218-0064)

[36 FR 17850, Sept. 4, 1973, as amended at 54 FR 24333, June 7, 1989]

§1903.12
Inspection not warranted; informal review

(a) **If the Area Director determines that an inspection** is not warranted because there are no reasonable grounds to believe that a violation or danger exists with respect to a complaint under §1903.11, he shall notify the complaining party in writing of such determination. The complaining party may obtain review of such determination by submitting a written statement of position with the Assistant Regional Director and, at the same time, providing the employer with a copy of such statement by certified mail. The employer may submit an opposing written statement of position with the Assistant Regional Director and, at the same time, provide the complaining party with a copy of such statement by certified mail. Upon the request of the complaining party or the employer, the Assistant Regional Director, at his discretion, may hold an informal conference in which the complaining party and the employer may orally present their views. After considering all written and oral views presented, the Assistant Regional Director shall affirm, modify, or reverse the determination of the Area Director and furnish the complaining party and the employer and written notification of this decision and the reasons therefor. The decision of the Assistant Regional Director shall be final and not subject to further review. [§1903.12(a)]

(b) **If the Area Director determines that an inspection** is not warranted because the requirements of §1903.11(a) have not been met, he shall notify the complaining party in writing of such determination. Such determination shall be without prejudice to the filing of a new complaint meeting the requirements of §1903.11(a). [§1903.12(b)]

§1903.13

Imminent danger

Whenever and as soon as a Compliance Safety and Health Officer concludes on the basis of an inspection that conditions or practices exist in any place of employment which could reasonably be expected to cause death or serious physical harm immediately or before the imminence of such danger can be eliminated through the enforcement procedures otherwise provided by the Act, he shall inform the affected employees and employers of the danger and that he is recommending a civil action to restrain such conditions or practices and for other appropriate relief in accordance with the provisions of section 13(a) of the Act. Appropriate citations and notices of proposed penalties may be issued with respect to an imminent danger even though, after being informed of such danger by the Compliance Safety and Health Officer, the employer immediately eliminates the imminence of the danger and initiates steps to abate such danger. [§1903.13]

§1903.14

Citations; notices of de minimis violations; policy regarding employee rescue activities

(a) **The Area Director shall review** the inspection report of the Compliance Safety and Health Officer. If, on the basis of the report the Area Director believes that the employer has violated a requirement of section 5 of the Act, of any standard, rule or order promulgated pursuant to section 6 of the Act, or of any substantive rule published in this chapter, he shall, if appropriate, consult with the Regional Solicitor, and he shall issue to the employer either a citation or a notice of de minimis violations which have no direct or immediate relationship to safety or health. An appropriate citation or notice of de minimis violations shall be issued even though after being informed of an alleged violation by the Compliance Safety and Health Officer, the employer immediately abates, or initiates steps to abate, such alleged violation. Any citation or notice of de minimis violations shall be issued with reasonable promptness after termination of the inspection. No citation may be issued under this section after the expiration of 6 months following the occurrence of any alleged violation. [§1903.14(a)]

(b) **Any citation shall describe** with particularity the nature of the alleged violation, including a reference to the provision(s) of the Act, standard, rule, regulation, or order alleged to have been violated. Any citation shall also fix a reasonable time or times for the abatement of the alleged violation. [§1903.14(b)]

(c) **If a citation or notice of de minimis violations is issued** for a violation alleged in a request for inspection under §1903.11(a) or a notification of violation under §1903.11(c), a copy of the citation or notice of de minimis violations shall also be sent to the employee or representative of employees who made such request or notification. [§1903.14(c)]

(d) **After an inspection, if the Area Director determines** that a citation is not warranted with respect to a danger or violation alleged to exist in a request for inspection under §1903.11(a) or a notification of violation under §1903.11(c), the informal review procedures prescribed in §1903.12(a) shall be applicable. After considering all views presented, the Assistant Regional Director shall affirm the determination of the Area Director, order a reinspection, or issue a citation if he believes that the inspection disclosed a violation. The Assistant Regional Director shall furnish the complaining party and the employer with written notification of his determination and the reasons therefor. The determination of the Assistant Regional Director shall be final and not subject to review. [§1903.14(d)]

(e) **Every citation shall state** that the issuance of a citation does not constitute a finding that a violation of the Act has occurred unless there is a failure to contest as provided for in the Act or, if contested, unless the citation is affirmed by the Review Commission. [§1903.14(e)]

(f) **No citation may be issued** to an employer because of a rescue activity undertaken by an employee of that employer with respect to an individual in imminent danger unless: [§1903.14(f)]

(1)(i) *Such employee is designated or assigned* by the employer to have responsibility to perform or assist in rescue operations, and [§1903.14(f)(1)(i)]

(ii) *The employer fails to provide* protection of the safety and health of such employee, including failing to provide appropriate training and rescue equipment; or [§1903.14(f)(1)(ii)]

(2)(i) *Such employee is directed* by the employer to perform rescue activities in the course of carrying out the employee's job duties, and [§1903.14(f)(2)(i)]

(ii) *The employer fails to provide* protection of the safety and health of such employee, including failing to provide appropriate training and rescue equipment; or [§1903.14(f)(2)(ii)]

(3)(i) *Such employee is employed* in a workplace that requires the employee to carry out duties that are directly related to a workplace operation where the likelihood of life-threatening accidents is foreseeable, such as a workplace operation where employees are located in confined spaces or trenches, handle hazardous waste, respond to emergency situations, perform excavations, or perform construction over water; and [§1903.14(f)(3)(i)]

(ii) *Such employee has not been designated* or assigned to perform or assist in rescue operations and voluntarily elects to rescue such an individual; and [§1903.14(f)(3)(ii)]

(iii) *The employer has failed to instruct* employees not designated or assigned to perform or assist in rescue operations of the arrangements for rescue, not to attempt rescue, and of the hazards of attempting rescue without adequate training or equipment. [§1903.14(f)(3)(iii)]

(4) *For purposes of this policy,* the term **imminent danger** means the existence of any condition or practice that could reasonably be expected to cause death or serious physical harm before such condition or practice can be abated. [§1903.14(f)(4)]

[36 FR 17850, Sept. 4, 1971, as amended at 59 FR 66613, Dec. 27, 1994]

§1903.14a

Petitions for modification of abatement date

(a) **An employer may file a petition for modification** of abatement date when he has made a good faith effort to comply with the abatement requirements of a citation, but such abatement has not been completed because of factors beyond his reasonable control. [§1903.14a(a)]

(b) **A petition for modification of abatement date** shall be in writing and shall include the following information: [§1903.14a(b)]

(1) *All steps taken by the employer,* and the dates of such action, in an effort to achieve compliance during the prescribed abatement period. [§1903.14a(b)(1)]

(2) *The specific additional abatement time necessary* in order to achieve compliance. [§1903.14a(b)(2)]

(3) *The reasons such additional time is necessary,* including the unavailability of professional or technical personnel or of materials and equipment, or because necessary construction or alteration of facilities cannot be completed by the original abatement date. [§1903.14a(b)(3)]

(4) *All available interim steps being taken* to safeguard the employees against the cited hazard during the abatement period. [§1903.14a(b)(4)]

(5) *A certification that a copy of the petition* has been posted and, if appropriate, served on the authorized representative of affected employees, in accordance with paragraph (c)(1) of this section and a certification of the date upon which such posting and service was made. [§1903.14a(b)(5)]

(c) **A petition for modification of abatement date** shall be filed with the Area Director of the United States Department of Labor who issued the citation no later than the close of the next working day following the date on which abatement was originally required. A later-filed petition shall be accompanied by the employer's statement of exceptional circumstances explaining the delay. [§1903.14a(c)]

(1) *A copy of such petition shall be posted* in a conspicuous place where all affected employees will have notice thereof or near such location where the violation occurred. The petition shall remain posted for a period of ten (10) working days. Where affected employees are represented by an authorized representative, said representative shall be served with a copy of such petition. [§1903.14a(c)(1)]

(2) *Affected employees or their representatives* may file an objection in writing to such petition with the aforesaid Area Director. Failure to file such objection within ten (10) working days of the date of posting of such petition or of service upon an authorized representative shall constitute a waiver of any further right to object to said petition. [§1903.14a(c)(2)]

(3) *The Secretary or his duly authorized agent* shall have the authority to approve any petition for modification of abatement date filed pursuant to paragraphs (b) and (c) of this section.

Such uncontested petitions shall become final orders pursuant to sections 10 (a) and (c) of the Act. [§1903.14a(c)(3)]

(4) *The Secretary or his authorized representative* shall not exercise his approval power until the expiration of fifteen (15) working days from the date the petition was posted or served pursuant to paragraphs (c)(1) and (2) of this section by the employer. [§1903.14a(c)(4)]

(d) **Where any petition is objected to** by the Secretary or affected employees, the petition, citation, and any objections shall be forwarded to the Commission within three (3) working days after the expiration of the fifteen (15) day period set out in paragraph (c)(4) of this section. [§1903.14a(d)]

[40 FR 6334, Feb. 11, 1975; 40 FR 11351, Mar. 11, 1975]

§1903.15

Proposed penalties

(a) **After, or concurrent with, the issuance of a citation,** and within a reasonable time after the termination of the inspection, the Area Director shall notify the employer by certified mail or by personal service by the Compliance Safety and Health Officer of the proposed penalty under section 17 of the Act, or that no penalty is being proposed. Any notice of proposed penalty shall state that the proposed penalty shall be deemed to be the final order of the Review Commission and not subject to review by any court or agency unless, within 15 working days from the date of receipt of such notice, the employer notifies the Area Director in writing that he intends to contest the citation or the notification of proposed penalty before the Review Commission. [§1903.15(a)]

(b) **The Area Director shall determine the amount** of any proposed penalty, giving due consideration to the appropriateness of the penalty with respect to the size of the business of the employer being charged, the gravity of the violation, the good faith of the employer, and the history of previous violations, in accordance with the provisions of section 17 of the Act. [§1903.15(b)]

(c) **Appropriate penalties may be proposed** with respect to an alleged violation even though after being informed of such alleged violation by the Compliance Safety and Health Officer, the employer immediately abates, or initiates steps to abate, such alleged violation. Penalties shall not be proposed for de minimis violations which have no direct or immediate relationship to safety or health. [§1903.15(c)]

§1903.16

Posting of citations

(a) **Upon receipt of any citation under the Act,** the employer shall immediately post such citation, or a copy thereof, unedited, at or near each place an alleged violation referred to in the citation occurred, except as provided below. Where, because of the nature of the employer's operations, it is not practicable to post the citation at or near each place of alleged violation, such citation shall be posted, unedited, in a prominent place where it will be readily observable by all affected employees. For example, where employers are engaged in activities which are physically dispersed (see §1903.2(b)), the citation may be posted at the location to which employees report each day. Where employees do not primarily work at or report to a single location (see §1903.2(b)), the citation may be posted at the location from which the employees operate to carry out their activities. The employer shall take steps to ensure that the citation is not altered, defaced, or covered by other material. Notices of de minimis violations need not be posted. [§1903.16(a)]

(b) **Each citation, or a copy thereof,** shall remain posted until the violation has been abated, or for 3 working days, whichever is later. The filing by the employer of a notice of intention to contest under §1903.17 shall not affect his posting responsibility under this section unless and until the Review Commission issues a final order vacating the citation. [§1903.16(b)]

(c) **An employer to whom a citation has been issued** may post a notice in the same location where such citation is posted indicating that the citation is being contested before the Review Commission, and such notice may explain the reasons for such contest. The employer may also indicate that specified steps have been taken to abate the violation. [§1903.16(c)]

(d) **Any employer failing to comply with the provisions** of paragraphs (a) and (b) of this section shall be subject to citation and penalty in accordance with the provisions of section 17 of the Act. [§1903.16(d)]

§1903.17

Employer and employee contests before the Review Commission

(a) **Any employer to whom a citation or notice** of proposed penalty has been issued may, under section 10(a) of the Act, notify the Area Director in writing that he intends to contest such citation or proposed penalty before the Review Commission. Such notice of intention to contest shall be postmarked within 15 working days of the receipt by the employer of the notice of proposed penalty. Every notice of intention to contest shall specify whether it is directed to the citation or to the proposed penalty, or both. The Area Director shall immediately transmit such notice to the Review Commission in accordance with the rules of procedure prescribed by the Commission. [§1903.17(a)]

(b) **Any employee or representative of employees** of an employer to whom a citation has been issued may, under section 10(c) of the Act, file a written notice with the Area Director alleging that the period of time fixed in the citation for the abatement of the violation is unreasonable. Such notice shall be postmarked within 15 working days of the receipt by the employer of the notice of proposed penalty or notice that no penalty is being proposed. The Area Director shall immediately transmit such notice to the Review Commission in accordance with the rules of procedure prescribed by the Commission. [§1903.17(b)]

§1903.18

Failure to correct a violation for which a citation has been issued

(a) **If an inspection discloses that an employer** has failed to correct an alleged violation for which a citation has been issued within the period permitted for its correction, the Area Director shall, if appropriate, consult with the Regional Solicitor, and he shall notify the employer by certified mail or by personal service by the Compliance Safety and Health Officer of such failure and of the additional penalty proposed under section 17(d) of the Act by reason of such failure. The period for the correction of a violation for which a citation has been issued shall not begin to run until the entry of a final order of the Review Commission in the case of any review proceedings initiated by the employer in good faith and not solely for delay or avoidance of penalties. [§1903.18(a)]

(b) **Any employer receiving a notification** of failure to correct a violation and of proposed additional penalty may, under section 10(b) of the Act, notify the Area Director in writing that he intends to contest such notification or proposed additional penalty before the Review Commission. Such notice of intention to contest shall be postmarked within 15 working days of the receipt by the employer of the notification of failure to correct a violation and of proposed additional penalty. The Area Director shall immediately transmit such notice to the Review Commission in accordance with the rules of procedure prescribed by the Commission. [§1903.18(b)]

(c) **Each notification of failure to correct a violation** and of proposed additional penalty shall state that it shall be deemed to be the final order of the Review Commission and not subject to review by any court or agency unless, within 15 working days from the date of receipt of such notification, the employer notifies the Area Director in writing that he intends to contest the notification or the proposed additional penalty before the Review Commission. [§1903.18(c)]

§1903.19

Abatement verification

Purpose. OSHA's inspections are intended to result in the abatement of violations of the Occupational Safety and Health Act of 1970 (the OSH Act). This section sets forth the procedures OSHA will use to ensure abatement. These procedures are tailored to the nature of the violation and the employer's abatement actions.[§1903.19]

(a) **Scope and application.** This section applies to employers who receive a citation for a violation of the Occupational Safety and Health Act. [§1903.19(a)]

(b) **Definitions.** [§1903.19(b)]

(1) **Abatement** means action by an employer to comply with a cited standard or regulation or to eliminate a recognized hazard identified by OSHA during an inspection. [§1903.19(b)(1)]

(2) **Abatement date** means: [§1903.19(b)(2)]

(i) *For an uncontested citation item,* the later of: [§1903.19(b)(2)(i)]

1903 Inspections, Citations, and Proposed Penalties

[A] The date in the citation for abatement of the violation; [§1903.19(b)(2)(i)[A]]

[B] The date approved by OSHA or established in litigation as a result of a petition for modification of the abatement date (PMA); or [§1903.19(b)(2)(i)[B]]

[C] The date established in a citation by an informal settlement agreement. [§1903.19(b)(2)(i)[C]]

(ii) *For a contested citation item* for which the Occupational Safety and Health Review Commission (OSHRC) has issued a final order affirming the violation, the later of: [§1903.19(b)(2)(ii)]

[A] The date identified in the final order for abatement; or [§1903.19(b)(2)(ii)[A]]

[B] The date computed by adding the period allowed in the citation for abatement to the final order date; [§1903.19(b)(2)(ii)[B]]

[C] The date established by a formal settlement agreement. [§1903.19(b)(2)(ii)[C]]

(3) **Affected employees** means those employees who are exposed to the hazard(s) identified as violation(s) in a citation. [§1903.19(b)(3)]

(4) **Final order date** means: [§1903.19(b)(4)]

(i) *For an uncontested citation item,* the fifteenth working day after the employer's receipt of the citation; [§1903.19(b)(4)(i)]

(ii) *For a contested citation item:* [§1903.19(b)(4)(ii)]

[A] The thirtieth day after the date on which a decision or order of a commission administrative law judge has been docketed with the commission, unless a member of the commission has directed review; or [§1903.19(b)(4)(ii)[A]]

[B] Where review has been directed, the thirtieth day after the date on which the Commission issues its decision or order disposing of all or pertinent part of a case; or [§1903.19(b)(4)(ii)[B]]

[C] The date on which a federal appeals court issues a decision affirming the violation in a case in which a final order of OSHRC has been stayed. [§1903.19(b)(4)(ii)[C]]

(5) **Movable equipment** means a hand-held or non-hand-held machine or device, powered or unpowered, that is used to do work and is moved within or between worksites. [§1903.19(b)(5)]

(c) Abatement certification. [§1903.19(c)]

(1) *Within 10 calendar days* after the abatement date, the employer must certify to OSHA (the Agency) that each cited violation has been abated, except as provided in paragraph (c)(2) of this section. [§1903.19(c)(1)]

(2) *The employer is not required* to certify abatement if the OSHA Compliance Officer, during the on-site portion of the inspection: [§1903.19(c)(2)]

(i) *Observes, within 24 hours* after a violation is identified, that abatement has occurred; and [§1903.19(c)(2)(i)]

(ii) *Notes in the citation that abatement has occurred.* [§1903.19(c)(2)(ii)]

(3) *The employer's certification* that abatement is complete must include, for each cited violation, in addition to the information required by paragraph (h) of this section, the date and method of abatement and a statement that affected employees and their representatives have been informed of the abatement. [§1903.19(c)(3)]

Note to paragraph (c): Appendix A contains a sample Abatement Certification Letter.

(d) Abatement documentation. [§1903.19(d)]

(1) *The employer must submit to the Agency,* along with the information on abatement certification required by paragraph (c)(3) of this section, documents demonstrating that abatement is complete for each willful or repeat violation and for any serious violation for which the Agency indicates in the citation that such abatement documentation is required. [§1903.19(d)(1)]

(2) *Documents demonstrating* that abatement is complete may include, but are not limited to, evidence of the purchase or repair of equipment, photographic or video evidence of abatement, or other written records. [§1903.19(d)(2)]

(e) Abatement plans. [§1903.19(e)]

(1) *The Agency may require an employer* to submit an abatement plan for each cited violation (except an other-than-serious violation) when the time permitted for abatement is more than 90 calendar days. If an abatement plan is required, the citation must so indicate. [§1903.19(e)(1)]

(2) *The employer must submit an abatement* plan for each cited violation within 25 calendar days from the final order date when the citation indicates that such a plan is required. The abatement plan must identify the violation and the steps to be taken to achieve abatement, including a schedule for completing abatement and, where necessary, how employees will be protected from exposure to the violative condition in the interim until abatement is complete. [§1903.19(e)(2)]

Note to paragraph (e): Appendix B contains a Sample Abatement Plan form.

(f) Progress reports. [§1903.19(f)]

(1) *An employer who is required* to submit an abatement plan may also be required to submit periodic progress reports for each cited violation. The citation must indicate: [§1903.19(f)(1)]

(i) *That periodic progress reports* are required and the citation items for which they are required; [§1903.19(f)(1)(i)]

(ii) *The date on which* an initial progress report must be submitted, which may be no sooner than 30 calendar days after submission of an abatement plan; [§1903.19(f)(1)(ii)]

(iii) *Whether additional progress reports are required; and* [§1903.19(f)(1)(iii)]

(iv) *The date(s) on which* additional progress reports must be submitted. [§1903.19(f)(1)(iv)]

(2) *For each violation,* the progress report must identify, in a single sentence if possible, the action taken to achieve abatement and the date the action was taken. [§1903.19(f)(2)]

Note to paragraph (f): Appendix B contains a Sample Progress Report form.

(g) Employee notification. [§1903.19(g)]

(1) *The employer must inform* affected employees and their representative(s) about abatement activities covered by this section by posting a copy of each document submitted to the Agency or a summary of the document near the place where the violation occurred. [§1903.19(g)(1)]

(2) *Where such posting does not effectively inform* employees and their representatives about abatement activities (for example, for employers who have mobile work operations), the employer must: [§1903.19(g)(2)]

(i) *Post each document* or a summary of the document in a location where it will be readily observable by affected employees and their representatives; or [§1903.19(g)(2)(i)]

(ii) *Take other steps to communicate fully* to affected employees and their representatives about abatement activities. [§1903.19(g)(2)(ii)]

(3) *The employer must inform* employees and their representatives of their right to examine and copy all abatement documents submitted to the Agency. [§1903.19(g)(3)]

(i) *An employee or an employee representative* must submit a request to examine and copy abatement documents within 3 working days of receiving notice that the documents have been submitted. [§1903.19(g)(3)(i)]

(ii) *The employer must comply* with an employee's or employee representative's request to examine and copy abatement documents within 5 working days of receiving the request. [§1903.19(g)(3)(ii)]

(4) *The employer must ensure* that notice to employees and employee representatives is provided at the same time or before the information is provided to the Agency and that abatement documents are: [§1903.19(g)(4)]

(i) *Not altered, defaced, or covered by other material; and* [§1903.19(g)(4)(i)]

(ii) *Remain posted for three working days* after submission to the Agency. [§1903.19(g)(4)(ii)]

(h) Transmitting abatement documents. [§1903.19(h)]

(1) *The employer must include,* in each submission required by this section, the following information: [§1903.19(h)(1)]

(i) *The employer's name and address;* [§1903.19(h)(1)(i)]

(ii) *The inspection number* to which the submission relates; [§1903.19(h)(1)(ii)]

(iii) *The citation and item numbers* to which the submission relates; [§1903.19(h)(1)(iii)]

(iv) *A statement* that the information submitted is accurate; and [§1903.19(h)(1)(iv)]

(v) *The signature* of the employer or the employer's authorized representative. [§1903.19(h)(1)(v)]

(2) *The date of postmark* is the date of submission for mailed documents. For documents transmitted by other means, the date the Agency receives the document is the date of submission. [§1903.19(h)(2)]

(i) **Movable equipment.** [§1903.19(i)]

(1) *For serious, repeat, and willful violations* involving movable equipment, the employer must attach a warning tag or a copy of the citation to the operating controls or to the cited component of equipment that is moved within the worksite or between worksites. [§1903.19(i)(1)]

Note to paragraph (i)(1): Attaching a copy of the citation to the equipment is deemed by OSHA to meet the tagging requirement of paragraph (i)(1) of this section as well as the posting requirement of 29 CFR 1903.16.

(2) *The employer must use a warning* tag that properly warns employees about the nature of the violation involving the equipment and identifies the location of the citation issued. [§1903.19(i)(2)]

Note to paragraph (i)(2): Non-Mandatory Appendix C contains a sample tag that employers may use to meet this requirement.

(3) *If the violation has not already been abated,* a warning tag or copy of the citation must be attached to the equipment: [§1903.19(i)(3)]

(i) *For hand-held equipment,* immediately after the employer receives the citation; or [§1903.19(i)(3)(i)]

(ii) *For non-hand-held equipment,* prior to moving the equipment within or between worksites. [§1903.19(i)(3)(ii)]

(4) *For the construction industry,* a tag that is designed and used in accordance with 29 CFR 1926.20(b)(3) and 29 CFR 1926.200(h) is deemed by OSHA to meet the requirements of this section when the information required by paragraph (i)(2) is included on the tag. [§1903.19(i)(4)]

(5) *The employer must assure* that the tag or copy of the citation attached to movable equipment is not altered, defaced, or covered by other material. [§1903.19(i)(5)]

(6) *The employer must assure* that the tag or copy of the citation attached to movable equipment remains attached until: [§1903.19(i)(6)]

(i) *The violation has been abated* and all abatement verification documents required by this regulation have been submitted to the Agency; [§1903.19(i)(6)(i)]

(ii) *The cited equipment* has been permanently removed from service or is no longer within the employer's control; or [§1903.19(i)(6)(ii)]

(iii) *The Commission issues a final order vacating the citation.* [§1903.19(i)(6)(iii)]

Appendices to §1903.19

Abatement Verification

Note: Appendices A through C provide information and nonmandatory guidelines to assist employers and employees in complying with the appropriate requirements of this section.

§1903.19 Appendix A

Sample Abatement-Certification Letter (non-mandatory)

§1903.19 App A - Sample Abatement-Certification Letter (non-mandatory)

Area Director – Name
U.S. Department of Labor – OSHA

Address of the Area Office (on the citation)

City State Zip Code

Company Name

Company Address

City State Zip Code

The hazard referenced in Inspection Number [insert 9-digit #] ____________ for violation identified as:

Citation #	Item #	Date Corrected	By

I attest that the information contained in this document is accurate.

Signature Title

Typed or Printed Name

*Full-size forms available free of charge at www.oshacfr.com.

§1903.19 Appendix B

Sample Abatement Plan or Progress Report (non-mandatory)

§1903.19 App B - Sample Abatement Plan or Progress Report (non-mandatory)

Area Director – Name
U.S. Department of Labor – OSHA

Address of the Area Office (on the citation)

City State Zip Code

Company Name

Company Address

City State Zip Code

Check One:
Abatement Plan: ☐ Progress Report: ☐ Inspection Number: ____________
Page ______ of ______
Citation Number(s)* ____________ ____________
Item Number(s)* ____________

	Action	Proposed Completion Date (For Abatement Plans Only)	Completion Date (For Progress Reports Only)
1.			
2.			
3.			
4.			
5.			
6.			
7.			
8.			
9.			
10.			
11.			
12.			
13.			
14.			
15.			
16.			
17.			
18.			
19.			
20.			
21.			
22.			
23.			
24.			
25.			
26.			
27.			
28.			
29.			
30.			
31.			
32.			

Date required for final abatement: ____ / ____ / ____
I attest that the information contained in this document is accurate.

Signature

Typed or Printed Name

Name of primary point of contact for questions: (Optional) ____________
Telephone number: (____) ______ - ______ Ext. ______
* Abatement plans or progress reports for more than one citation item may be combined in a single abatement plan or progress report if the abatement actions, proposed completion dates, and actual completion dates (for progress reports only) are the same for each of the citation items.

*Full-size forms available free of charge at www.oshacfr.com.

§1903.19 Appendix C

Sample Warning Tag (non-mandatory)

WARNING:

EQUIPMENT HAZARD CITED BY OSHA

EQUIPMENT CITED:

____

____

____

HAZARD CITED:

____

____

____

FOR DETAILED INFORMATION SEE OSHA CITATION POSTED AT:

____

____

BACKGROUND COLOR - ORANGE
MESSAGE COLOR - BLACK

[62 FR 15337, Mar. 31, 1997]

§1903.20

Informal conferences

At the request of an affected employer, employee, or representative of employees, the Assistant Regional Director may hold an informal conference for the purpose of discussing any issues raised by an inspection, citation, notice of proposed penalty, or notice of intention to contest. The settlement of any issue at such conference shall be subject to the rules of procedure prescribed by the Review Commission. If the conference is requested by the employer, an affected employee or his representative shall be afforded an opportunity to participate, at the discretion of the Assistant Regional Director. If the conference is requested by an employee or representative of employees, the employer shall be afforded an opportunity to participate, at the discretion of the Assistant Regional Director. Any party may be represented by counsel at such conference. No such conference or request for such conference shall operate as a stay of any 15-working-day period for filing a notice of intention to contest as prescribed in §1903.17. [§1903.20]

[36 FR 17850, Sept. 4, 1971. Redesignated at 62 FR 15337, Mar. 31, 1997]

§1903.21

State administration

Nothing in this part 1903 shall preempt the authority of any State to conduct inspections, to initiate enforcement proceedings or otherwise to implement the applicable provisions of State law with respect to State occupational safety and health standards in accordance with agreements and plans under section 18 of the Act and parts 1901 and 1902 of this chapter. [§1903.21)]

[36 FR 17850, Sept. 4, 1971. Redesignated at 62 FR 15337, Mar. 31, 1997]

§1903.22

Definitions

(a) **Act** means the Williams-Steiger Occupational Safety and Health Act of 1970. (84 Stat. 1590 et seq., 29 U.S.C. 651 et seq.) [§1903.22(a)]

(b) The definitions and interpretations contained in section 3 of the Act shall be applicable to such terms when used in this part 1903. [§1903.22(b)]

(c) **Working days** means Mondays through Fridays but shall not include Saturdays, Sundays, or Federal holidays. In computing 15 working days, the day of receipt of any notice shall not be included, and the last day of the 15 working days shall be included. [§1903.22(c)]

(d) **Compliance Safety and Health Officer** means a person authorized by the Occupational Safety and Health Administration, U.S. Department of Labor, to conduct inspections. [§1903.22(d)]

(e) **Area Director** means the employee or officer regularly or temporarily in charge of an Area Office of the Occupational Safety and Health Administration, U.S. Department of Labor, or any other person or persons who are authorized to act for such employee or officer. The latter authorizations may include general delegations of the authority of an Area Director under this part to a Compliance Safety and Health Officer or delegations to such an officer for more limited purposes, such as the exercise of the Area Director's duties under §1903.14(a). The term also includes any employee or officer exercising supervisory responsibilities over an Area Director. A supervisory employee or officer is considered to exercise concurrent authority with the Area Director. [§1903.22(e)]

(f) **Assistant Regional Director** means the employee or officer regularly or temporarily in charge of a Region of the Occupational Safety and Health Administration, U.S. Department of Labor, or any other person or persons who are specifically designated to act for such employee or officer in his absence. The term also includes any employee or officer in the Occupational Safety and Health Administration exercising supervisory responsibilities over the Assistant Regional Director. Such supervisory employee or officer is considered to exercise concurrent authority with the Assistant Regional Director. No delegation of authority under this paragraph shall adversely affect the procedures for independent informal review of investigative determinations prescribed under §1903.12 of this part. [§1903.22(f)]

(g) **Inspection** means any inspection of an employer's factory, plant, establishment, construction site, or other area, workplace or environment where work is performed by an employee of an employer, and includes any inspection conducted pursuant to a complaint filed under §1903.11 (a) and (c), any reinspection, followup inspection, accident investigation or other inspection conducted under section 8(a) of the Act. [§1903.22(g)]

[36 FR 17850, Sept. 4, 1971, as amended at 38 FR 22624, Aug. 23, 1973. Redesignated at 62 FR 15337, Mar. 31, 1997]

Part 1903

Authority for Part 1903

Authority: Sections 8 and 9 of the Occupational Safety and Health Act of 1970 (29 U.S.C. 657, 658); 5 U.S.C. 553; Secretary of Labor's Order No. 1-90 (55 FR 9033) or 6-96 (62 FR 111), as applicable.

Section 1903.7 also issued under 5 U.S.C. 553.

Source: 36 FR 17850, Sept. 4, 1971, unless otherwise noted.

1904 - Recording and Reporting Occupational Injuries and Illnesses

Subpart A - Purpose

§1904.0
Purpose

The purpose of this rule (Part 1904) is to require employers to record and report work-related fatalities, injuries and illnesses. [§1904.0]

Note to §1904.0: Recording or reporting a work-related injury, illness, or fatality does not mean that the employer or employee was at fault, that an OSHA rule has been violated, or that the employee is eligible for workers' compensation or other benefits.

Subpart B - Scope

Note to Subpart B: All employers covered by the Occupational Safety and Health Act (OSH Act) are covered by these Part 1904 regulations. However, most employers do not have to keep OSHA injury and illness records unless OSHA or the Bureau of Labor Statistics (BLS) informs them in writing that they must keep records. For example, employers with 10 or fewer employees and business establishments in certain industry classifications are partially exempt from keeping OSHA injury and illness records.

§1904.1
Partial exemption for employers with 10 or fewer employees

(a) **Basic requirement.** [§1904.1(a)]

(1) *If your company had ten (10)* or fewer employees at all times during the last calendar year, you do not need to keep OSHA injury and illness records unless OSHA or the BLS informs you in writing that you must keep records under §1904.41 or §1904.42. However, as required by §1904.39, all employers covered by the OSH Act must report to OSHA any workplace incident that results in a fatality or the hospitalization of three or more employees. [§1904.1(a)(1)]

(2) *If your company* had more than ten (10) employees at any time during the last calendar year, you must keep OSHA injury and illness records unless your establishment is classified as a partially exempt industry under §1904.2. [§1904.1(a)(2)]

(b) **Implementation.** [§1904.1(b)]

(1) *Is the partial exemption for size based on the size of my entire company or on the size of an individual business establishment?* The partial exemption for size is based on the number of employees in the entire company. [§1904.1(b)(1)]

(2) *How do I determine the size of my company to find out if I qualify for the partial exemption for size?* To determine if you are exempt because of size, you need to determine your company's peak employment during the last calendar year. If you had no more than 10 employees at any time in the last calendar year, your company qualifies for the partial exemption for size. [§1904.1(b)(2)]

§1904.2
Partial exemption for establishments in certain industries

(a) **Basic requirement.** [§1904.2(a)]

(1) *If your business* establishment is classified in a specific low hazard retail, service, finance, insurance or real estate industry listed in Appendix A to this Subpart B, you do not need to keep OSHA injury and illness records unless the government asks you to keep the records under §1904.41 or §1904.42. However, all employers must report to OSHA any workplace incident that results in a fatality or the hospitalization of three or more employees (see §1904.39). [§1904.2(a)(1)]

(2) *If one or more* of your company's establishments are classified in a non-exempt industry, you must keep OSHA injury and illness records for all of such establishments unless your company is partially exempted because of size under §1904.1. [§1904.2(a)(2)]

(b) **Implementation** [§1904.2(b)]

(1) *Does the partial industry classification exemption apply only to business establishments in the retail, services, finance, insurance or real estate industries (SICs 52-89)?* Yes, business establishments classified in agriculture; mining; construction; manufacturing; transportation; communication, electric, gas and sanitary services; or wholesale trade are not eligible for the partial industry classification exemption. [§1904.2(b)(1)]

(2) *Is the partial industry classification exemption based on the industry classification of my entire company or on the classification of individual business establishments operated by my company?* The partial industry classification exemption applies to individual business establishments. If a company has several business establishments engaged in different classes of business activities, some of the company's establishments may be required to keep records, while others may be exempt. [§1904.2(b)(2)]

(3) *How do I determine the Standard Industrial Classification code for my company or for individual establishments?* You determine your Standard Industrial Classification (SIC) code by using the Standard Industrial Classification Manual, Executive Office of the President, Office of Management and Budget. You may contact your nearest OSHA office or State agency for help in determining your SIC. [§1904.2(b)(3)]

§1904.3
Keeping records for more than one agency

If you create records to comply with another government agency's injury and illness recordkeeping requirements, OSHA will consider those records as meeting OSHA's Part 1904 recordkeeping requirements if OSHA accepts the other agency's records under a memorandum of understanding with that agency, or if the other agency's records contain the same information as this Part 1904 requires you to record. You may contact your nearest OSHA office or State agency for help in determining whether your records meet OSHA's requirements.

Appendix A to Subpart B of Part 1904
Partially Exempt Industries (Non-Mandatory)

Employers are not required to keep OSHA injury and illness records for any establishment classified in the following Standard Industrial Classification (SIC) codes, unless they are asked in writing to do so by OSHA, the Bureau of Labor Statistics (BLS), or a state agency operating under the authority of OSHA or the BLS. All employers, including those partially exempted by reason of company size or industry classification, must report to OSHA any workplace incident that results in a fatality or the hospitalization of three or more employees (see §1904.39).

SIC code	Industry description	SIC code	Industry description
525	Hardware Stores	725	Shoe Repair and Shoeshine Parlors
542	Meat and Fish Markets	726	Funeral Service and Crematories
544	Candy, Nut, and Confectionery Stores	729	Miscellaneous Personal Services
545	Dairy Products Stores	731	Advertising Services
546	Retail Bakeries	732	Credit Reporting and Collection Services
549	Miscellaneous Food Stores	733	Mailing, Reproduction, & Stenographic Services
551	New and Used Car Dealers	737	Computer and Data Processing Services
552	Used Car Dealers	738	Miscellaneous Business Services
554	Gasoline Service Stations	764	Reupholstery and Furniture Repair
557	Motorcycle Dealers	78	Motion Picture
56	Apparel and Accessory Stores	791	Dance Studios, Schools, and Halls
573	Radio, Television, & Computer Stores	792	Producers, Orchestras, Entertainers
58	Eating and Drinking Places	793	Bowling Centers
591	Drug Stores and Proprietary Stores	801	Offices & Clinics Of Medical Doctors
592	Liquor Stores	802	Offices and Clinics Of Dentists
594	Miscellaneous Shopping Goods Stores	803	Offices Of Osteopathic
599	Retail Stores, Not Elsewhere Classified	804	Offices Of Other Health Practitioners
60	Depository Institutions (banks & savings institutions)	807	Medical and Dental Laboratories
61	Nondepository	809	Health and Allied Services, Not Elsewhere Classified
62	Security and Commodity Brokers	81	Legal Services

(continued)

SIC code	Industry description	SIC code	Industry description
63	Insurance Carriers	82	Educational Services (schools, colleges, universities and libraries)
64	Insurance Agents, Brokers & Services	832	Individual and Family Services
653	Real Estate Agents and Managers	835	Child Day Care Services
654	Title Abstract Offices	839	Social Services, Not Elsewhere Classified
67	Holding and Other Investment Offices	841	Museums and Art Galleries
722	Photographic Studios, Portrait	86	Membership Organizations
723	Beauty Shops	87	Engineering, Accounting, Research, Management, and Related Services
724	Barber Shops	899	Services, Not Elsewhere Classified

Subpart C - Recordkeeping Forms and Recording Criteria

Note to Subpart C: This Subpart describes the work-related injuries and illnesses that an employer must enter into the OSHA records and explains the OSHA forms that employers must use to record work-related fatalities, injuries, and illnesses.

§1904.4

Recording criteria

(a) **Basic requirement.** Each employer required by this Part to keep records of fatalities, injuries, and illnesses must record each fatality, injury and illness that: [§1904.4(a)]

(1) *Is work-related; and* [§1904.4(a)(1)]

(2) *Is a new case; and* [§1904.4(a)(2)]

(3) *Meets one or more* of the general recording criteria of §1904.7 or the application to specific cases of §1904.8 through §1904.12. [§1904.4(a)(3)]

(b) **Implementation.** [§1904.4(b)]

(1) *What sections of this rule describe recording criteria for recording work-related injuries and illnesses?* The table below indicates which sections of the rule address each topic. [§1904.4(b)(1)]

(i) *Determination of work-relatedness.* See §1904.5. [§1904.4(b)(1)(i)]

(ii) *Determination of a new case.* See §1904.6. [§1904.4(b)(1)(ii)]

(iii) *General recording criteria.* See §1904.7. [§1904.4(b)(1)(iii)]

(iv) *Additional criteria.* (Needlestick and sharps injury cases, tuberculosis cases, hearing loss cases, medical removal cases, and musculoskeletal disorder cases). See §1904.8 through §1904.12. [§1904.4(b)(1)(iv)]

(2) *How do I decide whether a particular injury or illness is recordable?* The decision tree for recording work-related injuries and illnesses below shows the steps involved in making this determination. [§1904.4(b)(2)]

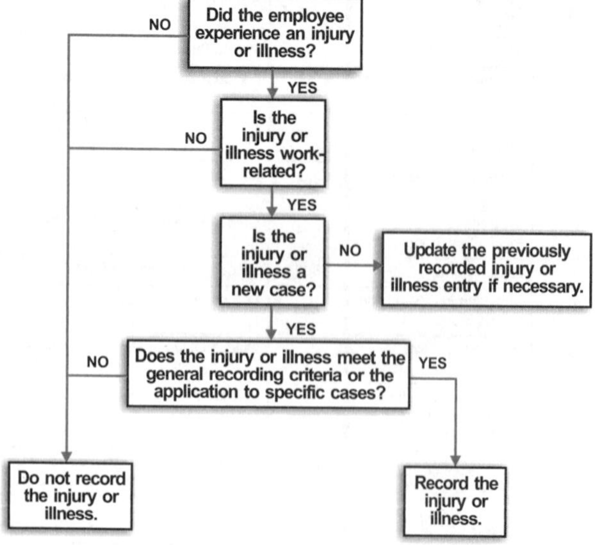

§1904.5

Determination of work-relatedness

(a) **Basic requirement.** You must consider an injury or illness to be work-related if an event or exposure in the work environment either caused or contributed to the resulting condition or significantly aggravated a pre-existing injury or illness. Work-relatedness is presumed for injuries and illnesses resulting from events or exposures occurring in the work environment, unless an exception in §1904.5(b)(2) specifically applies. [§1904.5(a)]

(b) **Implementation.** [§1904.5(b)]

(1) *What is the "work environment"?* OSHA defines the work environment as "the establishment and other locations where one or more employees are working or are present as a condition of their employment. The work environment includes not only physical locations, but also the equipment or materials used by the employee during the course of his or her work." [§1904.5(b)(1)]

(2) *Are there situations where an injury or illness occurs in the work environment and is not considered work-related?* Yes, an injury or illness occurring in the work environment that falls under one of the following exceptions is not work-related, and therefore is not recordable. [§1904.5(b)(2)]

1904.5(b)(2)	You are not required to record injuries and illnesses if . . .
(i)	At the time of the injury or illness, the employee was present in the work environment as a member of the general public rather than as an employee.
(ii)	The injury or illness involves signs or symptoms that surface at work but result solely from a non-work-related event or exposure that occurs outside the work environment.
(iii)	The injury or illness results solely from voluntary participation in a wellness program or in a medical, fitness, or recreational activity such as blood donation, physical examination, flu shot, exercise class, racquetball, or baseball.
(iv)	The injury or illness is solely the result of an employee eating, drinking, or preparing food or drink for personal consumption (whether bought on the employer's premises or brought in). For example, if the employee is injured by choking on a sandwich while in the employer's establishment, the case would not be considered work-related. *Note:* If the employee is made ill by ingesting food contaminated by workplace contaminants (such as lead), or gets food poisoning from food supplied by the employer, the case would be considered work-related.
(v)	The injury or illness is solely the result of an employee doing personal tasks (unrelated to their employment) at the establishment outside of the employee's assigned working hours.
(vi)	The injury or illness is solely the result of personal grooming, self medication for a non-work-related condition, or is intentionally self-inflicted.
(vii)	The injury or illness is caused by a motor vehicle accident and occurs on a company parking lot or company access road while the employee is commuting to or from work.
(viii)	The illness is the common cold or flu. *Note*: Contagious diseases such as tuberculosis, brucellosis, hepatitis A, or plague are considered work-related if the employee is infected at work.
(ix)	The illness is a mental illness. Mental illness will not be considered work-related unless the employee voluntarily provides the employer with an opinion from a physician or other licensed health care professional with appropriate training and experience (psychiatrist, psychologist, psychiatric nurse practitioner, etc.) stating that the employee has a mental illness that is work-related.

(3) *How do I handle a case if it is not obvious whether the precipitating event or exposure occurred in the work environment or occurred away from work?* In these situations, you must evaluate the employee's work duties and environment to decide whether or not one or more events or exposures in the work environment either caused or contributed to the resulting condition or significantly aggravated a pre-existing condition. [§1904.5(b)(3)]

(4) *How do I know if an event or exposure in the work environment "significantly aggravated" a preexisting injury or illness?* A preexisting injury or illness has been significantly aggravated, for purposes of OSHA injury and illness recordkeeping, when an event or exposure in the work environment results in any of the following: [§1904.5(b)(4)]

(i) *Death,* provided that the preexisting injury or illness would likely not have resulted in death but for the occupational event or exposure. [§1904.5(b)(4)(i)]

(ii) *Loss of consciousness,* provided that the preexisting injury or illness would likely not have resulted in loss of consciousness but for the occupational event or exposure. [§1904.5(b)(4)(ii)]

(iii) *One or more days* away from work, or days of restricted work, or days of job transfer that otherwise would not have occurred but for the occupational event or exposure. [§1904.5(b)(4)(iii)]

(iv) *Medical treatment in a case* where no medical treatment was needed for the injury or illness before the workplace event or exposure, or a change in medical treatment was necessitated by the workplace event or exposure. [§1904.5(b)(4)(iv)]

(5) *Which injuries and illnesses are considered pre-existing conditions?* An injury or illness is a preexisting condition if it resulted solely from a non-work-related event or exposure that occured outside the work environment. [§1904.5(b)(5)]

(6) *How do I decide whether an injury or illness is work-related if the employee is on travel status at the time the injury or illness occurs?* Injuries and illnesses that occur while an employee is on travel status are work-related if, at the time of the injury or illness, the employee was engaged in work activities "in the interest of the employer." Examples of such activities include travel to and from customer contacts, conducting job tasks, and entertaining or being entertained to transact, discuss, or promote business (work-related entertainment includes only entertainment activities being engaged in at the direction of the employer).

Injuries or illnesses that occur when the employee is on travel status do not have to be recorded if they meet one of the exceptions listed below. [§1904.5(b)(6)]

1904.5(b)(6)	If the employee has . . .	You may use the following to determine if an injury or illness is work-related
(i)	checked into a hotel or motel for one or more days.	When a traveling employee checks into a hotel, motel, or into another temporary residence, he or she establishes a "home away from home." You must evaluate the employee's activities after he or she checks into the hotel, motel, or other temporary residence for their work-relatedness in the same manner as you evaluate the activities of a non-traveling employee. When the employee checks into the temporary residence, he or she is considered to have left the work environment. When the employee begins work each day, he or she re-enters the work environment. If the employee has established a "home away from home" and is reporting to a fixed worksite each day, you also do not consider injuries or illnesses work-related if they occur while the employee is commuting between the temporary residence and the job location.
(ii)	taken a detour for personal reasons.	Injuries or illnesses are not considered work-related if they occur while the employee is on a personal detour from a reasonably direct route of travel (e.g, has taken a side trip for personal reasons).

(7) *How do I decide if a case is work-related when the employee is working at home?* Injuries and illnesses that occur while an employee is working at home, including work in a home office, will be considered work-related if the injury or illness occurs while the employee is performing work for pay or compensation in the home, and the injury or illness is directly related to the performance of work rather than to the general home environment or setting. For example, if an employee drops a box of work documents and injures his or her foot, the case is considered work-related. If an employee's fingernail is punctured by a needle from a sewing machine used to perform garment work at home, becomes infected and requires medical treatment, the injury is considered work-related. If an employee is injured because he or she trips on the family dog while rushing to answer a work phone call, the case is not considered work-related. If an employee working at home is electrocuted because of faulty home wiring, the injury is not considered work-related. [§1904.5(b)(7)]

§1904.6
Determination of new cases

(a) **Basic requirement.** You must consider an injury or illness to be a "new case" if: [§1904.6(a)]

(1) *The employee has not previously* experienced a recorded injury or illness of the same type that affects the same part of the body, or [§1904.6(a)(1)]

(2) *The employee previously experienced* a recorded injury or illness of the same type that affected the same part of the body but had recovered completely (all signs and symptoms had disappeared) from the previous injury or illness and an event or exposure in the work environment caused the signs or symptoms to reappear. [§1904.6(a)(2)]

(b) **Implementation.** [§1904.6(b)]

(1) *When an employee experiences the signs or symptoms of a chronic work-related illness, do I need to consider each recurrence of signs or symptoms to be a new case?* No, for occupational illnesses where the signs or symptoms may recur or continue in the absence of an exposure in the workplace, the case must only be recorded once. Examples may include occupational cancer, asbestosis, byssinosis and silicosis. [§1904.6(b)(1)]

(2) *When an employee experiences the signs or symptoms of an injury or illness as a result of an event or exposure in the workplace, such as an episode of occupational asthma, must I treat the episode as a new case?* Yes, because the episode or recurrence was caused by an event or exposure in the workplace, the incident must be treated as a new case. [§1904.6(b)(2)]

(3) *May I rely on a physician or other licensed health care professional to determine whether a case is a new case or a recurrence of an old case?* You are not required to seek the advice of a physician or other licensed health care professional. However, if you do seek such advice, you must follow the physician or other licensed health care professional's recommendation about whether the case is a new case or a recurrence. If you receive recommendations from two or more physicians or other licensed health care professionals, you must make a decision as to which recommendation is the most authoritative (best documented, best reasoned, or most authoritative), and record the case based upon that recommendation. [§1904.6(b)(3)]

§1904.7
General recording criteria

(a) **Basic requirement.** You must consider an injury or illness to meet the general recording criteria, and therefore to be recordable, if it results in any of the following: death, days away from work, restricted work or transfer to another job, medical treatment beyond first aid, or loss of consciousness. You must also consider a case to meet the general recording criteria if it involves a significant injury or illness diagnosed by a physician or other licensed health care professional, even if it does not result in death, days away from work, restricted work or job transfer, medical treatment beyond first aid, or loss of consciousness. [§1904.7(a)]

(b) **Implementation.** [§1904.7(b)]

(1) *How do I decide if a case meets one or more of the general recording criteria?* A work-related injury or illness must be recorded if it results in one or more of the following: [§1904.7(b)(1)]

(i) *Death. See §1904.7(b)(2).* [§1904.7(b)(1)(i)]

(ii) *Days away from work. See §1904.7(b)(3).* [§1904.7(b)(1)(ii)]

(iii) *Restricted work or transfer to another job. See §1904.7(b)(4).* [§1904.7(b)(1)(iii)]

(iv) *Medical treatment beyond first aid. See §1904.7(b)(5).* [§1904.7(b)(1)(iv)]

(v) *Loss of consciousness. See §1904.7(b)(6).* [§1904.7(b)(1)(v)]

(vi) *A significant injury or illness* diagnosed by a physician or other licensed health care professional. See §1904.7(b)(7). [§1904.7(b)(1)(vi)]

(2) *How do I record a work-related injury or illness that results in the employee's death?* You must record an injury or illness that results in death by entering a check mark on the OSHA 300 Log in the space for cases resulting in death. You must also report any work-related fatality to OSHA within eight (8) hours, as required by §1904.39. [§1904.7(b)(2)]

(3) *How do I record a work-related injury or illness that results in days away from work?* When an injury or illness involves one or more days away from work, you must record the injury or illness on the OSHA 300 Log with a check mark in the space for cases involving days away and an entry of the number of calendar days away from work in the number of days column. If the employee is out for an extended period of time, you must enter an estimate of the days that the employee will be away, and update the day count when the actual number of days is known. [§1904.7(b)(3)]

(i) *Do I count the day on which the injury occurred or the illness began?* No, you begin counting days away on the day after the injury occurred or the illness began. [§1904.7(b)(3)(i)]

(ii) *How do I record an injury or illness when a physician or other licensed health care professional recommends that the worker stay at home but the employee comes to work anyway?* You must record these injuries and illnesses on the OSHA 300 Log using the check box for cases with days away from work and enter the number of calendar days away recommended by the physician or other licensed health care professional. If a physician or other licensed health care professional recommends days away, you should encourage your employee to follow that recommendation. However, the days away must be recorded whether the injured or ill employee follows the physician or licensed health care professional's recommendation or not. If you receive recommendations from two or more physicians or other licensed health care professionals, you may make a decision as to which recommendation is the most authoritative, and record the case based upon that recommendation. [§1904.7(b)(3)(ii)]

(iii) *How do I handle a case when a physician or other licensed health care professional recommends that the worker return to work but the employee stays at home anyway?* In this situation, you must end the count of days away from work on the date the physician or other licensed health care professional recommends that the employee return to work. [§1904.7(b)(3)(iii)]

(iv) *How do I count weekends, holidays, or other days the employee would not have worked anyway?* You must count the number of calendar days the employee was unable to work as a result of the injury or illness, regardless of whether or not the employee was scheduled to work on those day(s). Weekend days, holidays, vacation days or other days off are included in the total number of days recorded if the employee would not have been able to work on those days because of a work-related injury or illness. [§1904.7(b)(3)(iv)]

(v) *How do I record a case in which a worker is injured or becomes ill on a Friday and reports to work on a Monday, and was not scheduled to work on the weekend?* You need to record this case only if you receive information from a physician or other licensed health care professional indicating that the employee should not have worked, or should have performed only restricted work, during the weekend. If so, you must record the injury or illness as a case with days away from work or restricted work, and enter the day counts, as appropriate. [§1904.7(b)(3)(v)]

(vi) *How do I record a case in which a worker is injured or becomes ill on the day before scheduled time off such as a holiday, a planned vacation, or a temporary plant closing?* You need to record a case of this type only if you receive information from a physician or other licensed health care professional indicating that the employee should not have worked, or should have performed only restricted work, during the scheduled time off. If so, you must record the injury or illness as a case with days away from work or restricted work, and enter the day counts, as appropriate. [§1904.7(b)(3)(vi)]

(vii) *Is there a limit to the number of days away from work I must count?* Yes, you may "cap" the total days away at 180 calendar days. You are not required to keep track of the number of calendar days away from work if the injury or illness resulted in more than 180 calendar days away from work and/or days of job transfer or restriction. In such a case, entering 180 in the total days away column will be considered adequate. [§1904.7(b)(3)(vii)]

(viii) *May I stop counting days if an employee who is away from work because of an injury or illness retires or leaves my company?* Yes, if the employee leaves your company for some reason unrelated to the injury or illness, such as retirement, a plant closing, or to take another job, you may stop counting days away from work or days of restriction/job transfer. If the employee leaves your company because of the injury or illness, you must estimate the total number of days away or days of restriction/job transfer and enter the day count on the 300 Log. [§1904.7(b)(3)(viii)]

(ix) *If a case occurs in one year but results in days away during the next calendar year, do I record the case in both years?* No, you only record the injury or illness once. You must enter the number of calendar days away for the injury or illness on the OSHA 300 Log for the year in which the injury or illness occurred. If the employee is still away from work because of the injury or illness when you prepare the annual summary, estimate the total number of calendar days you expect the employee to be away from work, use this number to calculate the total for the annual summary, and then update the initial log entry later when the day count is known or reaches the 180-day cap. [§1904.7(b)(3)(ix)]

(4) *How do I record a work-related injury or illness that results in restricted work or job transfer?* When an injury or illness involves restricted work or job transfer but does not involve death or days away from work, you must record the injury or illness on the OSHA 300 Log by placing a check mark in the space for job transfer or restriction and an entry of the number of restricted or transferred days in the restricted workdays column. [§1904.7(b)(4)]

(i) *How do I decide if the injury or illness resulted in restricted work?* Restricted work occurs when, as the result of a work-related injury or illness: [§1904.7(b)(4)(i)]

[A] *You keep the employee from performing* one or more of the routine functions of his or her job, or from working the full workday that he or she would otherwise have been scheduled to work; or [§1904.7(b)(4)(i)[A]]

[B] *A physician or other licensed health* care professional recommends that the employee not perform one or more of the routine functions of his or her job, or not work the full workday that he or she would otherwise have been scheduled to work. [§1904.7(b)(4)(i)[B]]

(ii) *What is meant by "routine functions"?* For recordkeeping purposes, an employee's routine functions are those work activities the employee regularly performs at least once per week. [§1904.7(b)(4)(ii)]

(iii) *Do I have to record restricted work or job transfer if it applies only to the day on which the injury occurred or the illness began?* No, you do not have to record restricted work or job transfers if you, or the physician or other licensed health care professional, impose the restriction or transfer only for the day on which the injury occurred or the illness began. [§1904.7(b)(4)(iii)]

(iv) *If you or a physician or other licensed health care professional recommends a work restriction, is the injury or illness automatically recordable as a "restricted work" case?* No, a recommended work restriction is recordable only if it affects one or more of the employee's routine job functions. To determine whether this is the case, you must evaluate the restriction in light of the routine functions of the injured or ill employee's job. If the restriction from you or the physician or other licensed health care professional keeps the employee from performing one or more of his or her routine job functions, or from working the full workday the injured or ill employee would otherwise have worked, the employee's work has been restricted and you must record the case. [§1904.7(b)(4)(iv)]

(v) *How do I record a case where the worker works only for a partial work shift because of a work-related injury or illness?* A partial day of work is recorded as a day of job transfer or restriction for recordkeeping purposes, except for the day on which the injury occurred or the illness began. [§1904.7(b)(4)(v)]

(vi) *If the injured or ill worker produces fewer goods or services than he or she would have produced prior to the injury or illness but otherwise performs all of the routine functions of his or her work, is the case considered a restricted work case?* No, the case is considered restricted work only if the worker does not perform all of the routine functions of his or her job or does not work the full shift that he or she would otherwise have worked. [§1904.7(b)(4)(vi)]

(vii) *How do I handle vague restrictions from a physician or other licensed health care professional, such as that the employee engage only in "light duty" or "take it easy for a week"?* If you are not clear about the physician or other licensed health care professional's recommendation, you may ask that person whether the employee can do all of his or her routine job functions and work all of his or her normally assigned work shift. If the answer to both of these questions is "Yes," then the case does not involve a work restriction and does not have to be recorded as such. If the answer to one or both of these questions is "No," the case involves restricted work and must be recorded as a restricted work case. If you are unable to obtain this additional information from the physician or other licensed health care professional who recommended the restriction, record the injury or illness as a case involving restricted work. [§1904.7(b)(4)(vii)]

(viii) *What do I do if a physician or other licensed health care professional recommends a job restriction meeting OSHA's definition, but the employee does all of his or her routine job functions anyway?* You must record the injury or illness on the OSHA 300 Log as a restricted work case. If a physician or other licensed health care professional recommends a job restriction, you should ensure that the employee complies with that restriction. If you receive recommendations from two or more physicians or other licensed health care professionals, you may make a decision as to which recommendation is the most authoritative, and record the case based upon that recommendation. [§1904.7(b)(4)(viii)]

(ix) *How do I decide if an injury or illness involved a transfer to another job?* If you assign an injured or ill employee to a job other than his or her regular job for part of the day, the case involves transfer to another job. Note: This does not include the day on which the injury or illness occurred. [§1904.7(b)(4)(ix)]

(x) *Are transfers to another job recorded in the same way as restricted work cases?* Yes, both job transfer and restricted work cases are recorded in the same box on the OSHA 300 Log. For example, if you assign, or a physician or other licensed health care professional recommends that you assign, an injured or ill worker to his or her routine job duties for part of the day and to another job for the rest of the day, the injury or illness involves a job transfer. You must record an injury or illness that involves a job transfer by placing a check in the box for job transfer. [§1904.7(b)(4)(x)]

(xi) *How do I count days of job transfer or restriction?* You count days of job transfer or restriction in the same way you count days away from work, using §1904.7(b)(3)(i) to (viii), above. The only difference is that, if you permanently assign the injured or ill employee to a job that has been modified or permanently changed in a manner that eliminates the routine functions the employee was restricted from performing, you may stop the day count when the modification or change is made permanent. You must count at least one day of restricted work or job transfer for such cases. [§1904.7(b)(4)(xi)]

(5) *How do I record an injury or illness that involves medical treatment beyond first aid?* If a work-related injury or illness results in medical treatment beyond first aid, you must record it on the OSHA 300 Log. If the injury or illness did not involve death, one or more days away from work, one or more days of restricted work, or one or more days of job transfer, you enter a check mark in the box for cases where the employee received medical treatment but remained at work and was not transferred or restricted. [§1904.7(b)(5)]

(i) *What is the definition of medical treatment?* **Medical treatment** means the management and care of a patient to combat disease or disorder. For the purposes of Part 1904, medical treatment does not include: [§1904.7(b)(5)(i)]

[A] Visits to a physician or other licensed health care professional solely for observation or counseling; [§1904.7(b)(5)(i)[A]]

[B] The conduct of diagnostic procedures, such as x-rays and blood tests, including the administration of prescription medications used solely for diagnostic purposes (e.g., eye drops to dilate pupils); or [§1904.7(b)(5)(i)[B]]

[C] "First aid" as defined in paragraph (b)(5)(ii) of this section. [§1904.7(b)(5)(i)[C]]

(ii) *What is "first aid"?* For the purposes of Part 1904, **first aid** means the following: [§1904.7(b)(5)(ii)]

[A] Using a non-prescription medication at nonprescription strength (for medications available in both prescription and non-prescription form, a recommendation by a physician or other licensed health care professional to use a non-prescription medication at prescription strength is considered medical treatment for recordkeeping purposes); [§1904.7(b)(5)(ii)[A]]

[B] Administering tetanus immunizations (other immunizations, such as Hepatitis B vaccine or rabies vaccine, are considered medical treatment); [§1904.7(b)(5)(ii)[B]]

[C] Cleaning, flushing or soaking wounds on the surface of the skin; [§1904.7(b)(5)(ii)[C]]

[D] Using wound coverings such as bandages, Band-Aids™, gauze pads, etc.; or using butterfly bandages or Steri-Strips™ (other wound closing devices such as sutures, staples, etc., are considered medical treatment); [§1904.7(b)(5)(ii)[D]]

[E] Using hot or cold therapy; [§1904.7(b)(5)(ii)[E]]

[F] Using any non-rigid means of support, such as elastic bandages, wraps, non-rigid back belts, etc. (devices with rigid stays or other systems designed to immobilize parts of the body are considered medical treatment for recordkeeping purposes); [§1904.7(b)(5)(ii)[F]]

[G] Using temporary immobilization devices while transporting an accident victim (e.g., splints, slings, neck collars, back boards, etc.). [§1904.7(b)(5)(ii)[G]]

[H] Drilling of a fingernail or toenail to relieve pressure, or draining fluid from a blister; [§1904.7(b)(5)(ii)[H]]

[I] Using eye patches; [§1904.7(b)(5)(ii)[I]]

[J] Removing foreign bodies from the eye using only irrigation or a cotton swab; [§1904.7(b)(5)(ii)[J]]

[K] Removing splinters or foreign material from areas other than the eye by irrigation, tweezers, cotton swabs or other simple means; [§1904.7(b)(5)(ii)[K]]

[L] Using finger guards; [§1904.7(b)(5)(ii)[L]]

[M] Using massages (physical therapy or chiropractic treatment are considered medical treatment for recordkeeping purposes); or [§1904.7(b)(5)(ii)[M]]

[N] Drinking fluids for relief of heat stress. [§1904.7(b)(5)(ii)[N]]

(iii) *Are any other procedures included in first aid?* No, this is a complete list of all treatments considered first aid for Part 1904 purposes. [§1904.7(b)(5)(iii)]

(iv) *Does the professional status of the person providing the treatment have any effect on what is considered first aid or medical treatment?* No, OSHA considers the treatments listed in §1904.7(b)(5)(ii) of this Part to be first aid regardless of the professional status of the person providing the treatment. Even when these treatments are provided by a physician or other licensed health care professional, they are considered first aid for the purposes of Part 1904. Similarly, OSHA considers treatment beyond first aid to be medical treatment even when it is provided by someone other than a physician or other licensed health care professional. [§1904.7(b)(5)(iv)]

(v) *What if a physician or other licensed health care professional recommends medical treatment but the employee does not follow the recommendation?* If a physician or other licensed health care professional recommends medical treatment, you should encourage the injured or ill employee to follow that recommendation. However, you must record the case even if the injured or ill employee does not follow the physician or other licensed health care professional's recommendation. [§1904.7(b)(5)(v)]

(6) *Is every work-related injury or illness case involving a loss of consciousness recordable?* Yes, you must record a work-related injury or illness if the worker becomes unconscious, regardless of the length of time the employee remains unconscious. [§1904.7(b)(6)]

(7) *What is a "significant" diagnosed injury or illness that is recordable under the general criteria even if it does not result in death, days away from work, restricted work or job transfer, medical treatment beyond first aid, or loss of consciousness?* Work-related cases involving cancer, chronic irreversible disease, a fractured or cracked bone, or a punctured eardrum must always be recorded under the general criteria at the time of diagnosis by a physician or other licensed health care professional. [§1904.7(b)(7)]

Note to §1904.7: OSHA believes that most significant injuries and illnesses will result in one of the criteria listed in §1904.7(a): death, days away from work, restricted work or job transfer, medical treatment beyond first aid, or loss of consciousness. However, there are some significant injuries, such as a punctured eardrum or a fractured toe or rib, for which neither medical treatment nor work restrictions may be recommended. In addition, there are some significant progressive diseases, such as byssinosis, silicosis, and some types of cancer, for which medical treatment or work restrictions may not be recommended at the time of diagnosis but are likely to be recommended as the disease progresses. OSHA believes that cancer, chronic irreversible diseases, fractured or cracked bones, and punctured eardrums are generally considered significant injuries and illnesses, and must be recorded at the initial diagnosis even if medical treatment or work restrictions are not recommended, or are postponed, in a particular case.

§1904.8
Recording criteria for needlestick and sharps injuries

(a) **Basic requirement.** You must record all work-related needlestick injuries and cuts from sharp objects that are contaminated with another person's blood or other potentially infectious material (as defined by 29 CFR 1910.1030). You must enter the case on the OSHA 300 Log as an injury. To protect the employee's privacy, you may not enter the employee's name on the OSHA 300 Log (see the requirements for privacy cases in paragraphs 1904.29(b)(6) through 1904.29(b)(9)). [§1904.8(a)]

(b) **Implementation.** [§1904.8(b)]

(1) *What does "other potentially infectious material" mean?* The term "other potentially infectious materials" is defined in the OSHA Bloodborne Pathogens standard at §1910.1030(b). These materials include: [§1904.8(b)(1)]

(i) *Human bodily fluids, tissues and organs, and* [§1904.8(b)(1)(i)]

(ii) *Other materials infected with the HIV* or hepatitis B (HBV) virus such as laboratory cultures or tissues from experimental animals. [§1904.8(b)(1)(ii)]

(2) *Does this mean that I must record all cuts, lacerations, punctures, and scratches?* No, you need to record cuts, lacerations, punctures, and scratches only if they are work-related and involve contamination with another person's blood or other potentially infectious material. If the cut, laceration, or scratch involves a clean object, or a contaminant other than blood or other potentially infectious material, you need to record the case only if it meets one or more of the recording criteria in §1904.7. [§1904.8(b)(2)]

(3) *If I record an injury and the employee is later diagnosed with an infectious bloodborne disease, do I need to update the OSHA 300 Log?* Yes, you must update the classification of the case on the OSHA 300 Log if the case results in death, days away from work, restricted work, or job transfer. You must also update the description to identify the infectious disease and change the classification of the case from an injury to an illness. [§1904.8(b)(3)]

(4) *What if one of my employees is splashed or exposed to blood or other potentially infectious material without being cut or scratched? Do I need to record this incident?* You need to record such an incident on the OSHA 300 Log as an illness if: [§1904.8(b)(4)]

(i) *It results in the diagnosis of a bloodborne illness,* such as HIV, hepatitis B, or hepatitis C; or [§1904.8(b)(4)(i)]

(ii) *It meets one or more of the recording criteria in §1904.7.* [§1904.8(b)(4)(ii)]

§1904.9
Recording criteria for cases involving medical removal under OSHA standards

(a) **Basic requirement.** If an employee is medically removed under the medical surveillance requirements of an OSHA standard, you must record the case on the OSHA 300 Log. [§1904.9(a)]

(b) **Implementation.** [§1904.9(b)]

(1) *How do I classify medical removal cases on the OSHA 300 Log?* You must enter each medical removal case on the OSHA 300 Log as either a case involving days away from work or a case involving restricted work activity, depending on how you decide to comply with the medical removal requirement. If the medical removal is the result of a chemical exposure, you must enter the case on the OSHA 300 Log by checking the "poisoning" column. [§1904.9(b)(1)]

(2) *Do all of OSHA's standards have medical removal provisions?* No, some OSHA standards, such as the standards covering bloodborne pathogens and noise, do not have medical removal provisions. Many OSHA standards that cover specific chemical substances have medical removal provisions. These standards include, but are not limited to, lead, cadmium, methylene chloride, formaldehyde, and benzene. [§1904.9(b)(2)]

(3) *Do I have to record a case where I voluntarily removed the employee from exposure before the medical removal criteria in an OSHA standard are met?* No, if the case involves voluntary medical removal before the medical removal levels required by an OSHA standard, you do not need to record the case on the OSHA 300 Log. [§1904.9(b)(3)]

§1904.10
Recording criteria for cases involving occupational hearing loss

(a) **Basic requirement.** If an employee's hearing test (audiogram) reveals that the employee has experienced a work-related Standard Threshold Shift (STS) in hearing in one or both ears, and the employee's total hearing level is 25 decibels (dB) or more above audiometric zero (averaged at 2000, 3000, and 4000 Hz) in the same ear(s) as the STS, you must record the case on the OSHA 300 Log. [§1904.10(a)]

(b) **Implementation.** [§1904.10(b)]

(1) *What is a Standard Threshold Shift?* A Standard Threshold Shift, or STS, is defined in the occupational noise exposure standard at 29 CFR 1910.95(g)(10)(i) as a change in hearing threshold, relative to the baseline audiogram for that employee, of an average of 10 decibels (dB) or more at 2000, 3000, and 4000 hertz (Hz) in one or both ears. [§1904.10(b)(1)]

(2) *How do I evaluate the current audiogram to determine whether an employee has an STS and a 25-dB hearing level?* [§1904.10(b)(2)]

(i) *STS.* If the employee has never previously experienced a recordable hearing loss, you must compare the employee's current audiogram with that employee's baseline audiogram. If the employee has previously experienced a recordable hearing loss, you must compare the employee's current audiogram with the employee's revised baseline audiogram (the audiogram reflecting the employee's previous recordable hearing loss case). [§1904.10(b)(2)(i)]

(ii) *25-dB loss.* Audiometric test results reflect the employee's overall hearing ability in comparison to audiometric zero. Therefore, using the employee's current audiogram, you must use the average hearing level at 2000, 3000, and 4000 Hz to determine whether or not the employee's total hearing level is 25 dB or more. [§1904.10(b)(2)(ii)]

(3) *May I adjust the current audiogram to reflect the effects of aging on hearing?* Yes. When you are determining whether an STS has occurred, you may age adjust the employee's current audiogram results by using Tables F-1 or F-2, as appropriate, in Appendix F of 29 CFR 1910.95. You may not use an age adjustment when determining whether the employee's total hearing level is 25 dB or more above audiometric zero. [§1904.10(b)(3)]

(4) *Do I have to record the hearing loss if I am going to retest the employee's hearing?* No, if you retest the employee's hearing within 30 days of the first test, and the retest does not confirm the recordable STS, you are not required to record the hearing loss case on the OSHA 300 Log. If the retest confirms the recordable STS, you must record the hearing loss illness within seven (7) calendar days of the retest. If subsequent audiometric testing performed under the testing requirements of the §1910.95 noise standard indicates that an STS is not persistent, you may erase or line-out the recorded entry. [§1904.10(b)(4)]

(5) *Are there any special rules for determining whether a hearing loss case is work-related?* No. You must use the rules in §1904.5 to determine if the hearing loss is work-related. If an event or exposure in the work environment either caused or contributed to the hearing loss, or significantly aggravated a pre-existing hearing loss, you must consider the case to be work related. [§1904.10(b)(5)]

(6) *If a physician or other licensed health care professional determines the hearing loss is not work-related, do I still need to record the case?* If a physician or other licensed health care professional determines that the hearing loss is not work-related or has not been significantly aggravated by occupational noise exposure, you are not required to consider the case work-related or to record the case on the OSHA 300 Log. [§1904.10(b)(6)]

(7) *How do I complete the 300 Log for a hearing loss case?* When you enter a recordable hearing loss case on the OSHA 300 Log, you must check the 300 Log column for hearing loss. [§1904.10(b)(7)]

Note: §1904.10(b)(7) is effective beginning January 1, 2004.

[67 FR 44047, July 1, 2002, as amended at 67 FR 77170, Dec. 17, 2002]

§1904.11
Recording criteria for work-related tuberculosis cases

(a) **Basic requirement.** If any of your employees has been occupationally exposed to anyone with a known case of active tuberculosis (TB), and that employee subsequently develops a tuberculosis infection, as evidenced by a positive skin test or diagnosis by a physician or other licensed health care professional, you must record the case on the OSHA 300 Log by checking the "respiratory condition" column. [§1904.11(a)]

(b) **Implementation.** [§1904.11(b)]

(1) *Do I have to record, on the Log, a positive TB skin test result obtained at a pre-employment physical?* No, you do not have to record it because the employee was not occupationally exposed to a known case of active tuberculosis in your workplace. [§1904.11(b)(1)]

(2) *May I line-out or erase a recorded TB case if I obtain evidence that the case was not caused by occupational exposure?* Yes, you may line-out or erase the case from the Log under the following circumstances: [§1904.11(b)(2)]

(i) *The worker is living in a household* with a person who has been diagnosed with active TB; [§1904.11(b)(2)(i)]

(ii) *The Public Health Department* has identified the worker as a contact of an individual with a case of active TB unrelated to the workplace; or [§1904.11(b)(2)(ii)]

(iii) *A medical investigation shows* that the employee's infection was caused by exposure to TB away from work, or proves that the case was not related to the workplace TB exposure. [§1904.11(b)(2)(iii)]

§§1904.13 - 1904.28
[Reserved]

§1904.29
Forms[1]

(a) **Basic requirement.** You must use OSHA 300, 300-A, and 301 forms, or equivalent forms, for recordable injuries and illnesses. The OSHA 300 form is called the Log of Work-Related Injuries and Illnesses, the 300-A is the Summary of Work-Related Injuries and Illnesses, and the OSHA 301 form is called the Injury and Illness Incident Report. [§1904.29(a)]

(b) **Implementation** [§1904.29(b)]

(1) *What do I need to do to complete the OSHA 300 Log?* You must enter information about your business at the top of the OSHA 300 Log, enter a one or two line description for each recordable injury or illness, and summarize this information on the OSHA 300-A at the end of the year. [§1904.29(b)(1)]

(2) *What do I need to do to complete the OSHA 301 Incident Report?* You must complete an OSHA 301 Incident Report form, or an equivalent form, for each recordable injury or illness entered on the OSHA 300 Log. [§1904.29(b)(2)]

(3) *How quickly must each injury or illness be recorded?* You must enter each recordable injury or illness on the OSHA 300 Log and 301 Incident Report within seven (7) calendar days of receiving information that a recordable injury or illness has occurred. [§1904.29(b)(3)]

(4) *What is an equivalent form?* An equivalent form is one that has the same information, is as readable and understandable, and is completed using the same instructions as the OSHA form it replaces. Many employers use an insurance form instead of the OSHA 301 Incident Report, or supplement an insurance form by adding any additional information required by OSHA. [§1904.29(b)(4)]

(5) *May I keep my records on a computer?* Yes, if the computer can produce equivalent forms when they are needed, as described under §5and 1904.40, you may keep your records using the computer system. [§1904.29(b)(5)]

(6) *Are there situations where I do not put the employee's name on the forms for privacy reasons?* Yes, if you have a "privacy concern case," you may not enter the employee's name on the OSHA 300 Log. Instead, enter "privacy case" in the space normally used for the employee's name. This will protect the privacy of the injured or ill employee when another employee, a former employee, or an authorized employee representative is provided access to the OSHA 300 Log under §1904.35(b)(2). You must keep a separate, confidential list of the case numbers and employee names for your privacy concern cases so you can update the cases and provide the information to the government if asked to do so. [§1904.29(b)(6)]

(7) *How do I determine if an injury or illness is a privacy concern case?* You must consider the following injuries or illnesses to be privacy concern cases: [§1904.29(b)(7)]

(i) *An injury or illness to an intimate body part or the reproductive system;* [§1904.29(b)(7)(i)]

(ii) *An injury or illness resulting from a sexual assault;* [§1904.29(b)(7)(ii)]

(iii) *Mental illnesses;* [§1904.29(b)(7)(iii)]

(iv) *HIV infection, hepatitis, or tuberculosis;* [§1904.29(b)(7)(iv)]

(v) *Needlestick injuries and cuts* from sharp objects that are contaminated with another person's blood or other potentially infectious material (see §1904.8 for definitions); and [§1904.29(b)(7)(v)]

(vi) *Other illnesses*, if the employee voluntarily requests that his or her name not be entered on the log. [§1904.29(b)(7)(vi)]

(8) *May I classify any other types of injuries and illnesses as privacy concern cases?* No, this is a complete list of all injuries and illnesses considered privacy concern cases for Part 1904 purposes. [§1904.29(b)(8)]

(9) *If I have removed the employee's name, but still believe that the employee may be identified from the information on the forms, is there anything else that I can do to further protect the employee's privacy?* Yes, if you have a reasonable basis to believe that information describing the privacy concern case may be personally identifiable even though the employee's name has been omitted, you may use discretion in describing the injury or illness on both the OSHA 300 and 301 forms. You must enter enough information to identify the cause of the incident and the general severity of the injury or illness, but you do not need to include details of an intimate or private nature. For example, a sexual assault case could be described as "injury from assault," or an injury to a reproductive organ could be described as "lower abdominal injury." [§1904.29(b)(9)]

(10) *What must I do to protect employee privacy if I wish to provide access to the OSHA Forms 300 and 301 to persons other than government representatives, employees, former employees or authorized representatives?* If you decide to voluntarily disclose the Forms to persons other than government representatives, employees, former employees or authorized representatives (as required by §5and 1904.40), you must remove or hide the employees' names and other personally identifying information, except for the following cases. You may disclose the Forms with personally identifying information only: [§1904.29(b)(10)]

(i) *to an auditor or consultant hired* by the employer to evaluate the safety and health program; [§1904.29(b)(10)(i)]

(ii) *to the extent necessary for processing* a claim for workers' compensation or other insurance benefits; or [§1904.29(b)(10)(ii)]

(iii) *to a public health authority* or law enforcement agency for uses and disclosures for which consent, an authorization, or opportunity to agree or object is not required under Department of Health and Human Services Standards for Privacy of Individually Identifiable Health Information, 45 CFR 164.512. [§1904.29(b)(10)(iii)]

[66 FR 6122, Jan. 19, 2001, as amended at 66 FR 52034, Oct. 12, 2001; 67 FR 77170, Dec. 17, 2002; 68 FR 38607, June 30, 2003]

1. OSHA forms 300, 300A, and 301 are available in the Addendum in the back of this book.

1904

Recording & Reporting Occupational Injuries and Illnesses

Subpart D - Other OSHA Injury and Illness Recordkeeping Requirements

§1904.30
Multiple business establishments

(a) **Basic requirement.** You must keep a separate OSHA 300 Log for each establishment that is expected to be in operation for one year or longer. [§1904.30(a)]

(b) **Implementation** [§1904.30(b)]

(1) *Do I need to keep OSHA injury and illness records for short-term establishments (i.e., establishments that will exist for less than a year)?* Yes, however, you do not have to keep a separate OSHA 300 Log for each such establishment. You may keep one OSHA 300 Log that covers all of your short-term establishments. You may also include the short-term establishments' recordable injuries and illnesses on an OSHA 300 Log that covers short-term establishments for individual company divisions or geographic regions. [§1904.30(b)(1)]

(2) *May I keep the records for all of my establishments at my headquarters location or at some other central location?* Yes, you may keep the records for an establishment at your headquarters or other central location if you can: [§1904.30(b)(2)]

(i) *Transmit information about the injuries* and illnesses from the establishment to the central location within seven (7) calendar days of receiving information that a recordable injury or illness has occurred; and [§1904.30(b)(2)(i)]

(ii) *Produce and send the records* from the central location to the establishment within the time frames required by §1904.35 and §1904.40 when you are required to provide records to a government representative, employees, former employees or employee representatives. [§1904.30(b)(2)(ii)]

(3) *Some of my employees work at several different locations or do not work at any of my establishments at all. How do I record cases for these employees?* You must link each of your employees with one of your establishments, for recordkeeping purposes. You must record the injury and illness on the OSHA 300 Log of the injured or ill employee's establishment, or on an OSHA 300 Log that covers that employee's short-term establishment. [§1904.30(b)(3)]

(4) *How do I record an injury or illness when an employee of one of my establishments is injured or becomes ill while visiting or working at another of my establishments, or while working away from any of my establishments?* If the injury or illness occurs at one of your establishments, you must record the injury or illness on the OSHA 300 Log of the establishment at which the injury or illness occurred. If the employee is injured or becomes ill and is not at one of your establishments, you must record the case on the OSHA 300 Log at the establishment at which the employee normally works. [§1904.30(b)(4)]

§1904.31
Covered employees

(a) **Basic requirement.** You must record on the OSHA 300 Log the recordable injuries and illnesses of all employees on your payroll, whether they are labor, executive, hourly, salary, part-time, seasonal, or migrant workers. You also must record the recordable injuries and illnesses that occur to employees who are not on your payroll if you supervise these employees on a day-to-day basis. If your business is organized as a sole proprietorship or partnership, the owner or partners are not considered employees for recordkeeping purposes. [§1904.31(a)]

(b) **Implementation.** [§1904.31(b)]

(1) *If a self-employed person is injured or becomes ill while doing work at my business, do I need to record the injury or illness?* No, self-employed individuals are not covered by the OSH Act or this regulation. [§1904.31(b)(1)]

(2) *If I obtain employees from a temporary help service, employee leasing service, or personnel supply service, do I have to record an injury or illness occurring to one of those employees?* You must record these injuries and illnesses if you supervise these employees on a day-to-day basis. [§1904.31(b)(2)]

(3) *If an employee in my establishment is a contractor's employee, must I record an injury or illness occurring to that employee?* If the contractor's employee is under the day-to-day supervision of the contractor, the contractor is responsible for recording the injury or illness. If you supervise the contractor employee's work on a day-to-day basis, you must record the injury or illness. [§1904.31(b)(3)]

(4) *Must the personnel supply service, temporary help service, employee leasing service, or contractor also record the injuries or illnesses occurring to temporary, leased or contract employees that I supervise on a day-to-day basis?* No, you and the temporary help service, employee leasing service, personnel supply service, or contractor should coordinate your efforts to make sure that each injury and illness is recorded only once: either on your OSHA 300 Log (if you provide day-to-day supervision) or on the other employer's OSHA 300 Log (if that company provides day-to-day supervision). [§1904.31(b)(4)]

§1904.32
Annual summary

(a) **Basic requirement.** At the end of each calendar year, you must: [§1904.32(a)]

(1) *Review the OSHA 300* Log to verify that the entries are complete and accurate, and correct any deficiencies identified; [§1904.32(a)(1)]

(2) *Create an annual summary* of injuries and illnesses recorded on the OSHA 300 Log; [§1904.32(a)(2)]

(3) *Certify the summary; and* [§1904.32(a)(3)]

(4) *Post the annual summary.* [§1904.32(a)(4)]

(b) **Implementation.** [§1904.32(b)]

(1) *How extensively do I have to review the OSHA 300 Log entries at the end of the year?* You must review the entries as extensively as necessary to make sure that they are complete and correct. [§1904.32(b)(1)]

(2) *How do I complete the annual summary? You must:* [§1904.32(b)(2)]

(i) *Total the columns on the OSHA* 300 Log (if you had no recordable cases, enter zeros for each column total); and [§1904.32(b)(2)(i)]

(ii) *Enter the calendar year* covered, the company's name, establishment name, establishment address, annual average number of employees covered by the OSHA 300 Log, and the total hours worked by all employees covered by the OSHA 300 Log. [§1904.32(b)(2)(ii)]

(iii) *If you are using an equivalent form* other than the OSHA 300-A summary form, as permitted under §1904.6(b)(4), the summary you use must also include the employee access and employer penalty statements found on the OSHA 300-A Summary form. [§1904.32(b)(2)(iii)]

(3) *How do I certify the annual summary?* A company executive must certify that he or she has examined the OSHA 300 Log and that he or she reasonably believes, based on his or her knowledge of the process by which the information was recorded, that the annual summary is correct and complete. [§1904.32(b)(3)]

(4) *Who is considered a company executive?* The company executive who certifies the log must be one of the following persons: [§1904.32(b)(4)]

(i) *An owner of the company* (only if the company is a sole proprietorship or partnership); [§1904.32(b)(4)(i)]

(ii) *An officer of the corporation;* [§1904.32(b)(4)(ii)]

(iii) *The highest ranking company official working at the establishment; or* [§1904.32(b)(4)(iii)]

(iv) *The immediate supervisor of the highest* ranking company official working at the establishment. [§1904.32(b)(4)(iv)]

(5) *How do I post the annual summary?* You must post a copy of the annual summary in each establishment in a conspicuous place or places where notices to employees are customarily posted. You must ensure that the posted annual summary is not altered, defaced or covered by other material. [§1904.32(b)(5)]

(6) *When do I have to post the annual summary?* You must post the summary no later than February 1 of the year following the year covered by the records and keep the posting in place until April 30. [§1904.32(b)(6)]

§1904.33
Retention and updating

(a) **Basic requirement.** You must save the OSHA 300 Log, the privacy case list (if one exists), the annual summary, and the OSHA 301 Incident Report forms for five (5) years following the end of the calendar year that these records cover. [§1904.33(a)]

(b) **Implementation** [§1904.33(b)]

(1) *Do I have to update the OSHA 300 Log during the five-year storage period?* Yes, during the storage period, you must update your stored OSHA 300 Logs to include newly discovered recordable injuries or illnesses and to show any changes that have occurred in the classification of previously recorded injuries and illnesses. If the description or outcome of a case changes, you must remove or line out the original entry and enter the new information. [§1904.33(b)(1)]

(2) *Do I have to update the annual summary?* No, you are not required to update the annual summary, but you may do so if you wish. [§1904.33(b)(2)]

(3) *Do I have to update the OSHA 301 Incident Reports?* No, you are not required to update the OSHA 301 Incident Reports, but you may do so if you wish. [§1904.33(b)(3)]

§1904.34
Change in business ownership

If your business changes ownership, you are responsible for recording and reporting work-related injuries and illnesses only for that period of the year during which you owned the establishment. You must transfer the Part 1904 records to the new owner. The new owner must save all records of the establishment kept by the prior owner, as required by §1904.33 of this Part, but need not update or correct the records of the prior owner. [§1904.34]

§1904.35
Employee involvement

(a) **Basic requirement.** Your employees and their representatives must be involved in the recordkeeping system in several ways. [§1904.35(a)]

(1) *You must inform each employee of how he* or she is to report an injury or illness to you. [§1904.35(a)(1)]

(2) *You must provide limited access* to your injury and illness records for your employees and their representatives. [§1904.35(a)(2)]

(b) **Implementation.** [§1904.35(b)]

(1) *What must I do to make sure that employees report work-related injuries and illnesses to me?* [§1904.35(b)(1)]

(i) *You must set up a way* for employees to report work-related injuries and illnesses promptly; and [§1904.35(b)(1)(i)]

(ii) *You must tell each employee how to report* work-related injuries and illnesses to you. [§1904.35(b)(1)(ii)]

(2) *Do I have to give my employees and their representatives access to the OSHA injury and illness records?* Yes, your employees, former employees, their personal representatives, and their authorized employee representatives have the right to access the OSHA injury and illness records, with some limitations, as discussed below. [§1904.35(b)(2)]

(i) *Who is an authorized employee representative?* An authorized employee representative is an authorized collective bargaining agent of employees. [§1904.35(b)(2)(i)]

(ii) *Who is a "personal representative" of an employee or former employee?* A personal representative is: [§1904.35(b)(2)(ii)]

[A] Any person that the employee or former employee designates as such, in writing; or [§1904.35(b)(2)(ii)[A]]

[B] The legal representative of a deceased or legally incapacitated employee or former employee. [§1904.35(b)(2)(ii)[B]]

(iii) *If an employee or representative asks for access to the OSHA 300 Log, when do I have to provide it?* When an employee, former employee, personal representative, or authorized employee representative asks for copies of your current or stored OSHA 300 Log(s) for an establishment the employee or former employee has worked in, you must give the requester a copy of the relevant OSHA 300 Log(s) by the end of the next business day. [§1904.35(b)(2)(iii)]

(iv) *May I remove the names of the employees or any other information from the OSHA 300 Log before I give copies to an employee, former employee, or employee representative?* No, you must leave the names on the 300 Log. However, to protect the privacy of injured and ill employees, you may not record the employee's name on the OSHA 300 Log for certain "privacy concern cases," as specified in paragraphs 1904.29(b)(6) through 1904.29(b)(9). [§1904.35(b)(2)(iv)]

(v) *If an employee or representative asks for access to the OSHA 301 Incident Report, when do I have to provide it?* [§1904.35(b)(2)(v)]

[A] When an employee, former employee, or personal representative asks for a copy of the OSHA 301 Incident Report describing an injury or illness to that employee or former employee, you must give the requester a copy of the OSHA 301 Incident Report containing that information by the end of the next business day. [§1904.35(b)(2)(v)[A]]

[B] When an authorized employee representative asks for a copies of the OSHA 301 Incident Reports for an establishment where the agent represents employees under a collective bargaining agreement, you must give copies of those forms to the authorized employee representative within 7 calendar days. You are only required to give the authorized employee representative information from the OSHA 301 Incident Report section titled "Tell us about the case." You must remove all other information from the copy of the OSHA 301 Incident Report or the equivalent substitute form that you give to the authorized employee representative. [§1904.35(b)(2)(v)[B]]

(vi) *May I charge for the copies?* No, you may not charge for these copies the first time they are provided. However, if one of the designated persons asks for additional copies, you may assess a reasonable charge for retrieving and copying the records. [§1904.35(b)(2)(vi)]

§1904.36
Prohibition against discrimination

Section 11(c) of the Act prohibits you from discriminating against an employee for reporting a work-related fatality, injury or illness. That provision of the Act also protects the employee who files a safety and health complaint, asks for access to the Part 1904 records, or otherwise exercises any rights afforded by the OSH Act. [§1904.35(a)]

§1904.37
State recordkeeping regulations

(a) **Basic requirement.** Some States operate their own OSHA programs, under the authority of a State Plan approved by OSHA. States operating OSHA-approved State Plans must have occupational injury and illness recording and reporting requirements that are substantially identical to the requirements in this Part (see 29 CFR 1902.3(k), 29 CFR 1952.4 and 29 CFR 1956.10(i)). [§1904.37(a)]

(b) **Implementation.** [§1904.37(b)]

(1) *State-Plan States must have* the same requirements as Federal OSHA for determining which injuries and illnesses are recordable and how they are recorded. [§1904.37(b)(1)]

(2) *For other Part 1904 provisions* (for example, industry exemptions, reporting of fatalities and hospitalizations, record retention, or employee involvement), State-Plan State requirements may be more stringent than or supplemental to the Federal requirements, but because of the unique nature of the national recordkeeping program, States must consult with and obtain approval of any such requirements. [§1904.37(b)(2)]

(3) *Although State and local government employees* are not covered Federally, all State-Plan States must provide coverage, and must develop injury and illness statistics, for these workers. State Plan recording and reporting requirements for State and local government entities may differ from those for the private sector but must meet the requirements of paragraphs 1904.37(b)(1) and (b)(2). [§1904.37(b)(3)]

(4) *A State-Plan State may not issue* a variance to a private sector employer and must recognize all variances issued by Federal OSHA. [§1904.37(b)(4)]

(5) *A State Plan State may only grant* an injury and illness recording and reporting variance to a State or local government employer within the State after obtaining approval to grant the variance from Federal OSHA. [§1904.37(b)(5)]

§1904.38

Variances from the recordkeeping rule

(a) **Basic requirement.** If you wish to keep records in a different manner from the manner prescribed by the Part 1904 regulations, you may submit a variance petition to the Assistant Secretary of Labor for Occupational Safety and Health, U.S. Department of Labor, Washington, DC 20210. You can obtain a variance only if you can show that your alternative recordkeeping system: [§1904.38(a)]

(1) *Collects the same information as this Part requires;* [§1904.38(a)(1)]

(2) *Meets the purposes of the Act; and* [§1904.38(a)(2)]

(3) *Does not interfere with the administration of the Act.* [§1904.38(a)(3)]

(b) **Implementation.** [§1904.38(b)]

(1) *What do I need to include in my variance petition?* You must include the following items in your petition: [§1904.38(b)(1)]

(i) *Your name and address;* [§1904.38(b)(1)(i)]

(ii) *A list of the State(s) where the variance would be used;* [§1904.38(b)(1)(ii)]

(iii) *The address(es) of the business establishment(s) involved;* [§1904.38(b)(1)(iii)]

(iv) *A description of why you are seeking a variance;* [§1904.38(b)(1)(iv)]

(v) *A description of the different recordkeeping procedures you propose to use;* [§1904.38(b)(1)(v)]

(vi) *A description of how your proposed procedures* will collect the same information as would be collected by this Part and achieve the purpose of the Act; and [§1904.38(b)(1)(vi)]

(vii) *A statement that you have informed* your employees of the petition by giving them or their authorized representative a copy of the petition and by posting a statement summarizing the petition in the same way as notices are posted under §1903.2(a). [§1904.38(b)(1)(vii)]

(2) *How will the Assistant Secretary handle my variance petition?* The Assistant Secretary will take the following steps to process your variance petition. [§1904.38(b)(2)]

(i) *The Assistant Secretary will offer* your employees and their authorized representatives an opportunity to submit written data, views, and arguments about your variance petition. [§1904.38(b)(2)(i)]

(ii) *The Assistant Secretary may allow* the public to comment on your variance petition by publishing the petition in the Federal Register. If the petition is published, the notice will establish a public comment period and may include a schedule for a public meeting on the petition. [§1904.38(b)(2)(ii)]

(iii) *After reviewing your variance petition* and any comments from your employees and the public, the Assistant Secretary will decide whether or not your proposed recordkeeping procedures will meet the purposes of the Act, will not otherwise interfere with the Act, and will provide the same information as the Part 1904 regulations provide. If your procedures meet these criteria, the Assistant Secretary may grant the variance subject to such conditions as he or she finds appropriate. [§1904.38(b)(2)(iii)]

(iv) *If the Assistant Secretary* grants your variance petition, OSHA will publish a notice in the Federal Register to announce the variance. The notice will include the practices the variance allows you to use, any conditions that apply, and the reasons for allowing the variance. [§1904.38(b)(2)(iv)]

(3) *If I apply for a variance, may I use my proposed recordkeeping procedures while the Assistant Secretary is processing the variance petition?* No, alternative recordkeeping practices are only allowed after the variance is approved. You must comply with the Part 1904 regulations while the Assistant Secretary is reviewing your variance petition. [§1904.38(b)(3)]

(4) *If I have already been cited by OSHA for not following the Part 1904 regulations, will my variance petition have any effect on the citation and penalty?* No, in addition, the Assistant Secretary may elect not to review your variance petition if it includes an element for which you have been cited and the citation is still under review by a court, an Administrative Law Judge (ALJ), or the OSH Review Commission. [§1904.38(b)(4)]

(5) *If I receive a variance, may the Assistant Secretary revoke the variance at a later date?* Yes, the Assistant Secretary may revoke your variance if he or she has good cause. The procedures revoking a variance will follow the same process as OSHA uses for reviewing variance petitions, as outlined in paragraph 1904.38(b)(2). Except in cases of willfulness or where necessary for public safety, the Assistant Secretary will: [§1904.38(b)(5)]

(i) *Notify you in writing of the facts* or conduct that may warrant revocation of your variance; and [§1904.38(b)(5)(i)]

(ii) *Provide you, your employees,* and authorized employee representatives with an opportunity to participate in the revocation procedures. [§1904.38(b)(5)(ii)]

Subpart E - Reporting Fatality, Injury and Illness Information to the Government

§1904.39

Reporting fatalities and multiple hospitalization incidents to OSHA

(a) **Basic requirement.** Within eight (8) hours after the death of any employee from a work-related incident or the in-patient hospitalization of three or more employees as a result of a work-related incident, you must orally report the fatality/multiple hospitalization by telephone or in person to the Area Office of the Occupational Safety and Health Administration (OSHA), U.S. Department of Labor, that is nearest to the site of the incident. You may also use the OSHA toll-free central telephone number, 1-800-321-OSHA (1-800-321-6742). [§1904.39(a)]

(b) **Implementation.** [§1904.39(b)]

(1) *If the Area Office is closed, may I report the incident by leaving a message on OSHA's answering machine, faxing the area office, or sending an e-mail?* No, if you can't talk to a person at the Area Office, you must report the fatality or multiple hospitalization incident using the 800 number. [§1904.39(b)(1)]

(2) *What information do I need to give to OSHA about the incident?* You must give OSHA the following information for each fatality or multiple hospitalization incident: [§1904.39(b)(2)]

(i) *The establishment name;* [§1904.39(b)(2)(i)]

(ii) *The location of the incident;* [§1904.39(b)(2)(ii)]

(iii) *The time of the incident;* [§1904.39(b)(2)(iii)]

(iv) *The number of fatalities or hospitalized employees;* [§1904.39(b)(2)(iv)]

(v) *The names of any injured employees;* [§1904.39(b)(2)(v)]

(vi) *Your contact person and his or her phone number; and* [§1904.39(b)(2)(vi)]

(vii) *A brief description of the incident.* [§1904.39(b)(2)(vii)]

(3) *Do I have to report every fatality or multiple hospitalization incident resulting from a motor vehicle accident?* No, you do not have to report all of these incidents. If the motor vehicle accident occurs on a public street or highway, and does not occur in a construction work zone, you do not have to report the incident to OSHA. However, these injuries must be recorded on your OSHA injury and illness records, if you are required to keep such records. [§1904.39(b)(3)]

(4) *Do I have to report a fatality or multiple hospitalization incident that occurs on a commercial or public transportation system?* No, you do not have to call OSHA to report a fatality or multiple hospitalization incident if it involves a commercial airplane, train, subway or bus accident. However, these injuries must be recorded on your OSHA injury and illness records, if you are required to keep such records. [§1904.39(b)(4)]

(5) *Do I have to report a fatality caused by a heart attack at work?* Yes, your local OSHA Area Office director will decide whether to investigate the incident, depending on the circumstances of the heart attack. [§1904.39(b)(5)]

(6) *Do I have to report a fatality or hospitalization that occurs long after the incident?* No, you must only report each fatality or multiple hospitalization incident that occurs within thirty (30) days of an incident. [§1904.39(b)(6)]

(7) *What if I don't learn about an incident right away?* If you do not learn of a reportable incident at the time it occurs and the incident would otherwise be reportable under paragraphs (a) and (b) of this section, you must make the report within eight (8) hours of the time the incident is reported to you or to any of your agent(s) or employee(s). [§1904.39(b)(7)]

§1904.40
Providing records to government representatives

(a) **Basic requirement.** When an authorized government representative asks for the records you keep under Part 1904, you must provide copies of the records within four (4) business hours. [§1904.40(a)]

(b) **Implementation.** [§1904.40(b)]

(1) *What government representatives have the right to get copies of my Part 1904 records?* The government representatives authorized to receive the records are: [§1904.40(b)(1)]

(i) *A representative of the Secretary of Labor* conducting an inspection or investigation under the Act; [§1904.40(b)(1)(i)]

(ii) *A representative of the Secretary of Health* and Human Services (including the National Institute for Occupational Safety and Health — NIOSH) conducting an investigation under section 20(b) of the Act, or [§1904.40(b)(1)(ii)]

(iii) *A representative of a State agency* responsible for administering a State plan approved under section 18 of the Act. [§1904.40(b)(1)(iii)]

(2) *Do I have to produce the records within four (4) hours if my records are kept at a location in a different time zone?* OSHA will consider your response to be timely if you give the records to the government representative within four (4) business hours of the request. If you maintain the records at a location in a different time zone, you may use the business hours of the establishment at which the records are located when calculating the deadline. [§1904.40(b)(2)]

§1904.41
Annual OSHA injury and illness survey of ten or more employers

(a) **Basic requirement.** If you receive OSHA's annual survey form, you must fill it out and send it to OSHA or OSHA's designee, as stated on the survey form. You must report the following information for the year described on the form: [§1904.41(a)]

(1) *the number of workers you employed;* [§1904.41(a)(1)]

(2) *the number of hours worked by your employees; and* [§1904.41(a)(2)]

(3) *the requested information from the records that you keep under Part 1904.* [§1904.41(a)(3)]

(b) **Implementation.** [§1904.41(b)]

(1) *Does every employer have to send data to OSHA?* No, each year, OSHA sends injury and illness survey forms to employers in certain industries. In any year, some employers will receive an OSHA survey form and others will not. You do not have to send injury and illness data to OSHA unless you receive a survey form. [§1904.41(b)(1)]

(2) *How quickly do I need to respond to an OSHA survey form?* You must send the survey reports to OSHA, or OSHA's designee, by mail or other means described in the survey form, within 30 calendar days, or by the date stated in the survey form, whichever is later. [§1904.41(b)(2)]

(3) *Do I have to respond to an OSHA survey form if I am normally exempt from keeping OSHA injury and illness records?* Yes, even if you are exempt from keeping injury and illness records under §1904.1 to §1904.3, OSHA may inform you in writing that it will be collecting injury and illness information from you in the following year. If you receive such a letter, you must keep the injury and illness records required by §1904.5 to §1904.15 and make a survey report for the year covered by the survey. [§1904.41(b)(3)]

(4) *Do I have to answer the OSHA survey form if I am located in a State-Plan State?* Yes, all employers who receive survey forms must respond to the survey, even those in State-Plan States. [§1904.41(b)(4)]

(5) *Does this section affect OSHA's authority to inspect my workplace?* No, nothing in this section affects OSHA's statutory authority to investigate conditions related to occupational safety and health. [§1904.41(b)(5)]

§1904.42
Requests from the Bureau of Labor Statistics for data

(a) **Basic requirement.** If you receive a Survey of Occupational Injuries and Illnesses Form from the Bureau of Labor Statistics (BLS), or a BLS designee, you must promptly complete the form and return it following the instructions contained on the survey form. [§1904.42(a)]

(b) **Implementation.** [§1904.42(b)]

(1) *Does every employer have to send data to the BLS?* No, each year, the BLS sends injury and illness survey forms to randomly selected employers and uses the information to create the Nation's occupational injury and illness statistics. In any year, some employers will receive a BLS survey form and others will not. You do not have to send injury and illness data to the BLS unless you receive a survey form. [§1904.42(b)(1)]

(2) *If I get a survey form from the BLS, what do I have to do?* If you receive a Survey of Occupational Injuries and Illnesses Form from the Bureau of Labor Statistics (BLS), or a BLS designee, you must promptly complete the form and return it, following the instructions contained on the survey form. [§1904.42(b)(2)]

(3) *Do I have to respond to a BLS survey form if I am normally exempt from keeping OSHA injury and illness records?* Yes, even if you are exempt from keeping injury and illness records under §1904.1 to §1904.3, the BLS may inform you in writing that it will be collecting injury and illness information from you in the coming year. If you receive such a letter, you must keep the injury and illness records required by §1904.5 to §1904.15 and make a survey report for the year covered by the survey. [§1904.42(b)(3)]

(4) *Do I have to answer the BLS survey form if I am located in a State-Plan State?* Yes, all employers who receive a survey form must respond to the survey, even those in State-Plan States. [§1904.42(b)(4)]

Subpart F - Transition From the Former Rule

§1904.43
Summary and posting of the 2001 data

(a) **Basic requirement.** If you were required to keep OSHA 200 Logs in 2001, you must post a 2000 annual summary from the OSHA 200 Log of occupational injuries and illnesses for each establishment. [§1904.43(a)]

(b) **Implementation.** [§1904.43(b)]

(1) *What do I have to include in the summary?* [§1904.43(b)(1)]

(i) *You must include a copy of the totals from the 2001 OSHA 200 Log and the following information from that form:* [§1904.43(b)(1)(i)]

[A] The calendar year covered; [§1904.43(b)(1)(i)[A]]

[B] Your company name; [§1904.43(b)(1)(i)[B]]

[C] The name and address of the establishment; and [§1904.43(b)(1)(i)[C]]

[D] The certification signature, title and date. [§1904.43(b)(1)(i)[D]]

(ii) *If no injuries or illnesses* occurred at your establishment in 2001, you must enter zeros on the totals line and post the 2001 summary. [§1904.43(b)(1)(ii)]

(2) *When am I required to summarize and post the 2001 information?* [§1904.43(b)(2)]

(i) *You must complete the summary by February 1, 2002; and* [§1904.43(b)(2)(i)]

(ii) *You must post a copy of the summary* in each establishment in a conspicuous place or places where notices to employees are customarily posted. You must ensure that the summary is not altered, defaced or covered by other material. [§1904.43(b)(2)(ii)]

(3) *You must post the 2001 summary* from February 1, 2002 to March 1, 2002. [§1904.43(b)(3)]

§1904.44
Retention and updating of old forms

You must save your copies of the OSHA 200 and 101 forms for five years following the year to which they relate and continue to provide access to the data as though these forms were the OSHA 300 and 301 forms. You are not required to update your old 200 and 101 forms. [§1904.44]

§1904.45
OMB control numbers under the Paperwork Reduction Act

The following sections each contain a collection of information requirement which has been approved by the Office of Management and Budget under the control number listed. [§1904.45]

29 CFR citation	OMB Control No.
1904.4-35	1218-0176
1904.39-41	1218-0176
1904.42	1220-0045
1904.43-44	1218-0176

Subpart G - Definitions

§1904.46
Definitions

The Act. The Act means the Occupational Safety and Health Act of 1970 (29 U.S.C. 651 et seq.). The definitions contained in section 3 of the Act (29 U.S.C. 652) and related interpretations apply to such terms when used in this Part 1904.[§1904.46]

Establishment. An establishment is a single physical location where business is conducted or where services or industrial operations are performed. For activities where employees do not work at a single physical location, such as construction; transportation; communications, electric, gas and sanitary services; and similar operations, the establishment is represented by main or branch offices, terminals, stations, etc. that either supervise such activities or are the base from which personnel carry out these activities. [§1904.46]

(1) **Can one business location include two or more establishments?** Normally, one business location has only one establishment. Under limited conditions, the employer may consider two or more separate businesses that share a single location to be separate establishments. An employer may divide one location into two or more establishments only when: [§1904.46(1)]

(i) *Each of the establishments represents a distinctly separate business;* [§1904.46(1)(i)]

(ii) *Each business is engaged in a different economic activity;* [§1904.46(1)(ii)]

(iii) *No one industry description in the Standard Industrial Classification Manual (1987) applies to the joint activities of the establishments; and* [§1904.46(1)(iii)]

(iv) *Separate reports are routinely* prepared for each establishment on the number of employees, their wages and salaries, sales or receipts, and other business information. For example, if an employer operates a construction company at the same location as a lumber yard, the employer may consider each business to be a separate establishment. [§1904.46(1)(iv)]

(2) **Can an establishment include more than one physical location?** Yes, but only under certain conditions. An employer may combine two or more physical locations into a single establishment only when: [§1904.46(2)]

(i) *The employer operates the locations* as a single business operation under common management; [§1904.46(2)(i)]

(ii) *The locations are all located in close* proximity to each other; and [§1904.46(2)(ii)]

(iii) *The employer keeps one set of business records* for the locations, such as records on the number of employees, their wages and salaries, sales or receipts, and other kinds of business information. For example, one manufacturing establishment might include the main plant, a warehouse a few blocks away, and an administrative services building across the street. [§1904.46(2)(iii)]

(3) **If an employee telecommutes from home, is his or her home considered a separate establishment?** No, for employees who telecommute from home, the employee's home is not a business establishment and a separate 300 Log is not required. Employees who telecommute must be linked to one of your establishments under §1904.30(b)(3). [§1904.46(3)]

Injury or illness. An injury or illness is an abnormal condition or disorder. Injuries include cases such as, but not limited to, a cut, fracture, sprain, or amputation. Illnesses include both acute and chronic illnesses, such as, but not limited to, a skin disease, respiratory disorder, or poisoning. [§1904.46]

(*Note*: Injuries and illnesses are recordable only if they are new, work-related cases that meet one or more of the Part 1904 recording criteria.)

Physician or Other Licensed Health Care Professional. A physician or other licensed health care professional is an individual whose legally permitted scope of practice (i.e., license, registration, or certification) allows him or her to independently perform, or be delegated the responsibility to perform, the activities described by this regulation. [§1904.46]

You. "You" means an employer as defined in Section 3 of the Occupational Safety and Health Act of 1970 (29 U.S.C. 652). [§1904.46]

Part 1904

Authority for Part 1904

Authority: 29 U.S.C. 657, 658, 660, 666, 669, 673, Secretary of Labor, Order No. 3, 2000 (65 FR 50017), and 5 U.S.C. 533.

Source: 66 FR 6122, Jan. 19, 2001, unless otherwise noted.

Selected 1910 General Industry Standards

§1910.12
Construction work

(a) **Standards.** The standards prescribed in part 1926 of this chapter are adopted as occupational safety and health standards under section 6 of the Act and shall apply, according to the provisions thereof, to every employment and place of employment of every employee engaged in construction work. Each employer shall protect the employment and places of employment of each of his employees engaged in construction work by complying with the appropriate standards prescribed in this paragraph.[§1910.12(a)]

(b) **Definition.** For purposes of this section, **Construction work** means work for construction, alteration, and/or repair, including painting and decorating. See discussion of these terms in Sec. 1926.13 of this title. [§1910.12(b)]

(c) **Construction Safety Act distinguished.** This section adopts as occupational safety and health standards under section 6 of the Act the standards which are prescribed in part 1926 of this chapter. Thus, the standards (substantive rules) published in subpart C and the following subparts of part 1926 of this chapter are applied. This section does not incorporate subparts A and B of part 1926 of this chapter. Subparts A and B have pertinence only to the application of section 107 of the Contract Work Hours and Safety Standards Act (the Construction Safety Act). For example, the interpretation of the term "subcontractor" in paragraph (c) of §1926.13 of this chapter is significant in discerning the coverage of the Construction Safety Act and duties thereunder. However, the term "subcontractor" has no significance in the application of the Act, which was enacted under the Commerce Clause and which establishes duties for "employers" which are not dependent for their application upon any contractual relationship with the Federal Government or upon any form of Federal financial assistance.[§1910.12(c)]

(d) **For the purposes of this part,** to the extent that it may not already be included in paragraph (b) of this section, "construction work" includes the erection of new electric transmission and distribution lines and equipment, and the alteration, conversion, and improvement of the existing transmission and distribution lines and equipment.[§1910.12(d)]

§1910.19
Special provisions for air contaminants

(a) **Asbestos, tremolite, anthophyllite, and actinolite dust.** Section 1910.1001 shall apply to the exposure of every employee to asbestos, tremolite, anthophyllite, and actinolite dust in every employment and place of employment covered by Sec. 1910.16, in lieu of any different standard on exposure to asbestos, tremolite, anthophyllite, and actinolite dust which would otherwise be applicable by virtue of any of those sections.[§1910.19(a)]

(b) **Vinyl chloride.** Section 1910.1017 shall apply to the exposure of every employee to vinyl chloride in every employment and place of employment covered by §§1910.12, 1910.13, 1910.14, 1910.15, or 1910.16, in lieu of any different standard on exposure to vinyl chloride which would otherwise be applicable by virtue of any of those sections. [§1910.19(b)]

(c) **Acrylonitrile.** Section 1910.1045 shall apply to the exposure of every employee to acrylonitrile in every employment and place of employment covered by §§1910.12, 1910.13, 1910.14, 1910.15, or Sec. 1910.16, in lieu of any different standard on exposure to acrylonitrile which would otherwise be applicable by virtue of any of those sections. [§1910.19(c)]

(d) **[Reserved]**[§1910.19(d)]

(e) **Inorganic arsenic.** Section 1910.1018 shall apply to the exposure of every employee to inorganic arsenic in every employment covered by §§1910.12, 1910.13, 1910.14, 1910.15, or Sec. 1910.16, in lieu of any different standard on exposure to inorganic arsenic which would otherwise be applicable by virtue of any of those sections.[§1910.19(e)]

(f) **[Reserved]**[§1910.19(f)]

(g) **Lead.** Section 1910.1025 shall apply to the exposure of every employee to lead in every employment and place of employment covered by §§1910.13, 1910.14, 1910.15, and 1910.16, in lieu of any different standard on exposure to lead which would otherwise be applicable by virtue of those sections.[§1910.19(g)]

(h) **Ethylene oxide.** Section 1910.1047 shall apply to the exposure of every employee to ethylene oxide in every employment and place of employment covered by §§1910.12, 1910.13, 1910.14, 1910.15, or 1910.16, in lieu of any different standard on exposure to ethylene oxide which would otherwise be applicable by virtue of those sections. [§1910.19(h)]

(i) **4,4'-Methylenedianiline (MDA).** Section 1910.1050 shall apply to the exposure of every employee to MDA in every employment and place of employment covered by §§1910.13, 1910.14, 1910.15, or 1910.16, in lieu of any different standard on exposure to MDA which would otherwise be applicable by virtue of those sections.[§1910.19(i)]

(j) **Formaldehyde.** Section 1910.1048 shall apply to the exposure of every employee to formaldehyde in every employment and place of employment covered by §§1910.12, 1910.13, 1910.14, 1910.15 or 1910.16 in lieu of any different standard on exposure to formaldehyde which would otherwise be applicable by virtue of those sections. [§1910.19(j)]

(k) **Cadmium.** Section 1910.1027 shall apply to the exposure of every employee to cadmium in every employment and place of employment covered by Sec. 1910.16 in lieu of any different standard on exposures to cadmium that would otherwise be applicable by virtue of those sections.[§1910.19(k)]

(l) **1,3-Butadiene (BD).** Section 1910.1051 shall apply to the exposure of every employee to BD in every employment and place of employment covered by §§1910.12, 1910.13, 1910.14, 1910.15, or 1910.16, in lieu of any different standard on exposure to BD which would otherwise be applicable by virtue of those sections.[§1910.19(l)]

(m) **Methylene chloride (MC).** Section 1910.1052 shall apply to the exposure of every employee to MC in every employment and place of employment covered by Sec. 1910.16 in lieu of any different standard on exposure to MC which would otherwise be applicable by virtue of that section when it is not present in sealed, intact containers. [§1910.19(m)]

§1910.132
General requirements

(a) **Application.** Protective equipment, including personal protective equipment for eyes, face, head, and extremities, protective clothing, respiratory devices, and protective shields and barriers, shall be provided, used, and maintained in a sanitary and reliable condition wherever it is necessary by reason of hazards of processes or environment, chemical hazards, radiological hazards, or mechanical irritants encountered in a manner capable of causing injury or impairment in the function of any part of the body through absorption, inhalation or physical contact.[§1910.132(a)]

(b) **Employee-owned equipment.** Where employees provide their own protective equipment, the employer shall be responsible to assure its adequacy, including proper maintenance, and sanitation of such equipment.[§1910.132(b)]

(c) **Design.** All personal protective equipment shall be of safe design and construction for the work to be performed.[§1910.132(c)]

(d) **Hazard assessment and equipment selection.**[§1910.132(d)]

(1) *The employer shall assess the workplace* to determine if hazards are present, or are likely to be present, which necessitate the use of personal protective equipment (PPE). If such hazards are present, or likely to be present, the employer shall: [§1910.132(d)(1)]

(i) *Select, and have each affected employee use,* the types of PPE that will protect the affected employee from the hazards identified in the hazard assessment;[§1910.132(d)(1)(i)]

(ii) *Communicate selection decisions* to each affected employee; and,[§1910.132(d)(1)(ii)]

(iii) *Select PPE that properly fits each affected employee.* [§1910.132(d)(1)(iii)]

Note: Non-mandatory Appendix B contains an example of procedures that would comply with the requirement for a hazard assessment.

(2) *The employer shall verify* that the required workplace hazard assessment has been performed through a written certification that identifies the workplace evaluated; the person certifying that the evaluation has been performed; the date(s) of the hazard assessment; and, which identifies the document as a certification of hazard assessment.[§1910.132(d)(2)]

(e) **Defective and damaged equipment.** Defective or damaged personal protective equipment shall not be used.[§1910.132(e)]

(f) **Training.**[§1910.132(f)]

(1) *The employer shall provide training* to each employee who is required by this section to use PPE. Each such employee shall be trained to know at least the following:[§1910.132(f)(1)]

(i) *When PPE is necessary;*[§1910.132(f)(1)(i)]

(ii) *What PPE is necessary;*[§1910.132(f)(1)(ii)]

(iii) *How to properly don, doff, adjust, and wear PPE;* [§1910.132(f)(1)(iii)]

(iv) *The limitations of the PPE; and,*[§1910.132(f)(1)(iv)]

(v) *The proper care,* maintenance, useful life and disposal of the PPE.[§1910.132(f)(1)(v)]

Note: This partial selection of Part 1910 "General Industry Standards" is intended for convenience only; some parts may not be applicable to "Construction Standards." A complete Part 1910 CFR, similar to this one, is also available.

(2) *Each affected employee shall demonstrate* an understanding of the training specified in paragraph (f)(1) of this section, and the ability to use PPE properly, before being allowed to perform work requiring the use of PPE.[§1910.132(f)(2)]

(3) *When the employer has reason to believe* that any affected employee who has already been trained does not have the understanding and skill required by paragraph (f)(2) of this section, the employer shall retrain each such employee. Circumstances where retraining is required include, but are not limited to, situations where:[§1910.132(f)(3)]

(i) *Changes in the workplace render previous training obsolete; or*[§1910.132(f)(3)(i)]

(ii) *Changes in the types of PPE* to be used render previous training obsolete; or[§1910.132(f)(3)(ii)]

(iii) *Inadequacies in an affected employee's knowledge* or use of assigned PPE indicate that the employee has not retained the requisite understanding or skill.[§1910.132(f)(3)(iii)]

(g) **Paragraphs (d) and (f)** of this section apply only to §§1910.133, 1910.135, 1919.136, and 1910.138. Paragraphs (d) and (f) of this section do not apply to §§1910.134 and 1910.137.[§1910.132(g)]

(h) **Payment for protective equipment.**[§1910.132(h)]

(1) *Except as provided by paragraphs*(h)(2) through (h)(6) of this section, the protective equipment, including personal protective equipment (PPE), used to comply with this part, shall be provided by the employer at no cost to employees.[§1910.132(h)(1)]

(2) *The employer is not required to pay* for non-specialty safety-toe protective footwear (including steel-toe shoes or steel-toe boots) and nonspecialty prescription safety eyewear, provided that the employer permits such items to be worn off the job-site. [§1910.132(h)(2)]

(3) *When the employer provides metatarsal guards*and allows the employee, at his or her request, to use shoes or boots with built-in metatarsal protection, the employer is not required to reimburse the employee for the shoes or boots.[§1910.132(h)(3)]

(4) *The employer is not required to pay for:*[§1910.132(h)(4)]

(i) *The logging boots required by 29 CFR 1910.266(d)(1)(v);* [§1910.132(h)(4)(i)]

(ii) *Everyday clothing, such as long sleeve shirts,* long pants, street shoes, and normal work boots; or[§1910.132(h)(4)(ii)]

(iii) *Ordinary clothing, skin creams, or other items,* used solely for protection from weather, such as winter coats, jackets, gloves, parkas, rubber boots, hats, raincoats, ordinary sunglasses, and sunscreen.[§1910.132(h)(4)(iii)]

(5) *The employer must pay for replacement PPE,* except when the employee has lost or intentionally damaged the PPE. [§1910.132(h)(5)]

(6) *Where an employee provides adequate protective equipment* he or she owns pursuant to paragraph (b) of this section, the employer may allow the employee to use it and is not required to reimburse the employee for that equipment. The employer shall not require an employee to provide or pay for his or her own PPE, unless the PPE is excepted by paragraphs (h)(2) through (h)(5) of this section.[§1910.132(h)(6)]

(7) *This paragraph (h) shall become effective on February 13, 2008.* Employers must implement the PPE payment requirements no later than May 15, 2008.[§1910.132(h)(7)]

Note to Sec. 1910.132(h): When the provisions of another OSHA standard specify whether or not the employer must pay for specific equipment, the payment provisions of that standard shall prevail.

[39 FR 23502, June 27, 1974, as amended at 59 FR 16334, Apr. 6, 1994; 59 FR 33910, July 1, 1994; 72 FR 64428, Nov. 15, 2007; 76 FR 33606, June 8, 2011]

§1910.134

Respiratory protection

This section applies to General Industry (part 1910), Shipyards (part 1915), Marine Terminals (part 1917), Longshoring (part 1918), and Construction (part 1926).

(a) **Permissible practice.**[§1910.134(a)]

(1) *In the control of those occupational diseases caused* by breathing air contaminated with harmful dusts, fogs, fumes, mists, gases, smokes, sprays, or vapors, the primary objective shall be to prevent atmospheric contamination. This shall be accomplished as far as feasible by accepted engineering control measures (for example, enclosure or confinement of the operation, general and local ventilation, and substitution of less toxic materials). When effective engineering controls are not feasible, or while they are being instituted, appropriate respirators shall be used pursuant to this section.[§1910.134(a)(1)]

(2) *A respirator shall be provided* to each employee when such equipment is necessary to protect the health of such employee. The employer shall provide the respirators which are applicable and suitable for the purpose intended. The employer shall be responsible for the establishment and maintenance of a respiratory protection program, which shall include the requirements outlined in paragraph (c) of this section. The program shall cover each employee required by this section to use a respirator.[§1910.134(a)(2)]

(b) **Definitions.** The following definitions are important terms used in the respiratory protection standard in this section.[§1910.134(b)]

Air-purifying respirator means a respirator with an air-purifying filter, cartridge, or canister that removes specific air contaminants by passing ambient air through the air-purifying element.[§1910.134(b)]

Assigned protection factor (APF) means the workplace level of respiratory protection that a respirator or class of respirators is expected to provide to employees when the employer implements a continuing, effective respiratory protection program as specified by this section.[§1910.134(b)]

Atmosphere-supplying respirator means a respirator that supplies the respirator user with breathing air from a source independent of the ambient atmosphere, and includes supplied-air respirators (SARs) and self-contained breathing apparatus (SCBA) units. [§1910.134(b)]

Canister or cartridge means a container with a filter, sorbent, or catalyst, or combination of these items, which removes specific contaminants from the air passed through the container.[§1910.134(b)]

Demand respirator means an atmosphere-supplying respirator that admits breathing air to the facepiece only when a negative pressure is created inside the facepiece by inhalation.[§1910.134(b)]

Emergency situation means any occurrence such as, but not limited to, equipment failure, rupture of containers, or failure of control equipment that may or does result in an uncontrolled significant release of an airborne contaminant.[§1910.134(b)]

Employee exposure means exposure to a concentration of an airborne contaminant that would occur if the employee were not using respiratory protection.[§1910.134(b)]

End-of-service-life indicator (ESLI) means a system that warns the respirator user of the approach of the end of adequate respiratory protection, for example, that the sorbent is approaching saturation or is no longer effective.[§1910.134(b)]

Escape-only respirator means a respirator intended to be used only for emergency exit.[§1910.134(b)]

Filter or air purifying element means a component used in respirators to remove solid or liquid aerosols from the inspired air. [§1910.134(b)]

Filtering facepiece (dust mask) means a negative pressure particulate respirator with a filter as an integral part of the facepiece or with the entire facepiece composed of the filtering medium. [§1910.134(b)]

Fit factor means a quantitative estimate of the fit of a particular respirator to a specific individual, and typically estimates the ratio of the concentration of a substance in ambient air to its concentration inside the respirator when worn.[§1910.134(b)]

Fit test means the use of a protocol to qualitatively or quantitatively evaluate the fit of a respirator on an individual. (See also Qualitative fit test QLFT and Quantitative fit test QNFT.)[§1910.134(b)]

Helmet means a rigid respiratory inlet covering that also provides head protection against impact and penetration.[§1910.134(b)]

High efficiency particulate air (HEPA) filter means a filter that is at least 99.97% efficient in removing monodisperse particles of 0.3 micrometers in diameter. The equivalent NIOSH 42 CFR 84 particulate filters are the N100, R100, and P100 filters.[§1910.134(b)]

Hood means a respiratory inlet covering that completely covers the head and neck and may also cover portions of the shoulders and torso.[§1910.134(b)]

Immediately dangerous to life or health (IDLH) means an atmosphere that poses an immediate threat to life, would cause irreversible adverse health effects, or would impair an individual's ability to escape from a dangerous atmosphere.[§1910.134(b)]

Interior structural firefighting means the physical activity of fire suppression, rescue or both, inside of buildings or enclosed structures which are involved in a fire situation beyond the incipient stage. (See 29 CFR 1910.155)[§1910.134(b)]

Loose-fitting facepiece means a respiratory inlet covering that is designed to form a partial seal with the face.[§1910.134(b)]

Maximum use concentration (MUC) means the maximum atmospheric concentration of a hazardous substance from which an employee can be expected to be protected when wearing a respirator, and is determined by the assigned protection factor of the respirator or class of respirators and the exposure limit of the hazardous substance. The MUC can be determined mathematically by multiplying the assigned protection factor specified for a respirator by the required OSHA permissible exposure limit, short-term exposure limit, or ceiling limit. When no OSHA exposure limit is available for a hazardous substance, an employer must determine an MUC on the basis of relevant available information and informed professional judgment.[§1910.134(b)]

Negative pressure respirator (tight fitting) means a respirator in which the air pressure inside the facepiece is negative during inhalation with respect to the ambient air pressure outside the respirator.[§1910.134(b)]

Oxygen deficient atmosphere means an atmosphere with an oxygen content below 19.5% by volume.[§1910.134(b)]

Physician or other licensed health care professional (PLHCP) means an individual whose legally permitted scope of practice (i.e., license, registration, or certification) allows him or her to independently provide, or be delegated the responsibility to provide, some or all of the health care services required by paragraph (e) of this section.[§1910.134(b)]

Positive pressure respirator means a respirator in which the pressure inside the respiratory inlet covering exceeds the ambient air pressure outside the respirator.[§1910.134(b)]

Powered air-purifying respirator (PAPR) means an air-purifying respirator that uses a blower to force the ambient air through air-purifying elements to the inlet covering.[§1910.134(b)]

Pressure demand respirator means a positive pressure atmosphere-supplying respirator that admits breathing air to the facepiece when the positive pressure is reduced inside the facepiece by inhalation.[§1910.134(b)]

Qualitative fit test (QLFT) means a pass/fail fit test to assess the adequacy of respirator fit that relies on the individual's response to the test agent.[§1910.134(b)]

Quantitative fit test (QNFT) means an assessment of the adequacy of respirator fit by numerically measuring the amount of leakage into the respirator.[§1910.134(b)]

Respiratory inlet covering means that portion of a respirator that forms the protective barrier between the user's respiratory tract and an air-purifying device or breathing air source, or both. It may be a facepiece, helmet, hood, suit, or a mouthpiece respirator with nose clamp.[§1910.134(b)]

Self-contained breathing apparatus (SCBA) means an atmosphere-supplying respirator for which the breathing air source is designed to be carried by the user.[§1910.134(b)]

Service life means the period of time that a respirator, filter or sorbent, or other respiratory equipment provides adequate protection to the wearer.[§1910.134(b)]

Supplied-air respirator (SAR) or airline respirator means an atmosphere-supplying respirator for which the source of breathing air is not designed to be carried by the user.[§1910.134(b)]

This section means this respiratory protection standard.[§1910.134(b)]

Tight-fitting facepiece means a respiratory inlet covering that forms a complete seal with the face.[§1910.134(b)]

User seal check means an action conducted by the respirator user to determine if the respirator is properly seated to the face. [§1910.134(b)]

(c) **Respiratory protection program.** This paragraph requires the employer to develop and implement a written respiratory protection program with required worksite-specific procedures and elements for required respirator use. The program must be administered by a suitably trained program administrator. In addition, certain program elements may be required for voluntary use to prevent potential hazards associated with the use of the respirator. The Small Entity Compliance Guide contains criteria for the selection of a program administrator and a sample program that meets the requirements of this paragraph. Copies of the Small Entity Compliance Guide will be available on or about April 8, 1998 from the Occupational Safety and Health Administration's Office of Publications, Room N 3101, 200 Constitution Avenue, NW, Washington, DC, 20210 (202-219-4667).[§1910.134(c)]

(1) *In any workplace where respirators are necessary* to protect the health of the employee or whenever respirators are required by the employer, the employer shall establish and implement a written respiratory protection program with worksite-specific procedures. The program shall be updated as necessary to reflect those changes in workplace conditions that affect respirator use. The employer shall include in the program the following provisions of this section, as applicable:[§1910.134(c)(1)]

(i) *Procedures for selecting respirators for use in the workplace;* [§1910.134(c)(1)(i)]

(ii) *Medical evaluations of employees required to use respirators;*[§1910.134(c)(1)(ii)]

(iii) *Fit testing procedures for tight-fitting respirators;* [§1910.134(c)(1)(iii)]

(iv) *Procedures for proper use of respirators* in routine and reasonably foreseeable emergency situations;[§1910.134(c)(1)(iv)]

(v) *Procedures and schedules* for cleaning, disinfecting, storing, inspecting, repairing, discarding, and otherwise maintaining respirators;[§1910.134(c)(1)(v)]

(vi) *Procedures to ensure* adequate air quality, quantity, and flow of breathing air for atmosphere-supplying respirators; [§1910.134(c)(1)(vi)]

(vii) *Training of employees* in the respiratory hazards to which they are potentially exposed during routine and emergency situations;[§1910.134(c)(1)(vii)]

(viii) *Training of employees* in the proper use of respirators, including putting on and removing them, any limitations on their use, and their maintenance; and[§1910.134(c)(1)(viii)]

(ix) *Procedures for regularly evaluating* the effectiveness of the program.[§1910.134(c)(1)(ix)]

(2) *Where respirator use is not required:*[§1910.134(c)(2)]

(i) *An employer may provide respirators* at the request of employees or permit employees to use their own respirators, if the employer determines that such respirator use will not in itself create a hazard. If the employer determines that any voluntary respirator use is permissible, the employer shall provide the respirator users with the information contained in Appendix D to this section ("Information for Employees Using Respirators When Not Required Under the Standard"); and [§1910.134(c)(2)(i)]

(ii) *In addition,* the employer must establish and implement those elements of a written respiratory protection program necessary to ensure that any employee using a respirator voluntarily is medically able to use that respirator, and that the respirator is cleaned, stored, and maintained so that its use does not present a health hazard to the user. [§1910.134(c)(2)(ii)]

Exception: Employers are not required to include in a written respiratory protection program those employees whose only use of respirators involves the voluntary use of filtering facepieces (dust masks).

(3) *The employer shall designate a program administrator* who is qualified by appropriate training or experience that is commensurate with the complexity of the program to administer or oversee the respiratory protection program and conduct the required evaluations of program effectiveness.[§1910.134(c)(3)]

(4) *The employer shall provide respirators,* training, and medical evaluations at no cost to the employee.[§1910.134(c)(4)]

(d) **Selection of respirators.** This paragraph requires the employer to evaluate respiratory hazard(s) in the workplace, identify relevant workplace and user factors, and base respirator selection on these factors. The paragraph also specifies appropriately protective respirators for use in IDLH atmospheres, and limits the selection and use of air-purifying respirators.[§1910.134(d)]

(1) *General requirements.*[§1910.134(d)(1)]

(i) *The employer shall select and provide* an appropriate respirator based on the respiratory hazard(s) to which the worker is exposed and workplace and user factors that affect respirator performance and reliability.[§1910.134(d)(1)(i)]

(ii) *The employer shall select a NIOSH-certified respirator.* The respirator shall be used in compliance with the conditions of its certification.[§1910.134(d)(1)(ii)]

(iii) *The employer shall identify* and evaluate the respiratory hazard(s) in the workplace; this evaluation shall include a reasonable estimate of employee exposures to respiratory hazard(s) and an identification of the contaminant's chemical state and physical form. Where the employer cannot identify or reasonably estimate the employee exposure, the employer shall consider the atmosphere to be IDLH. [§1910.134(d)(1)(iii)]

(iv) *The employer shall select respirators* from a sufficient number of respirator models and sizes so that the respirator is acceptable to, and correctly fits, the user.[§1910.134(d)(1)(iv)]

(2) *Respirators for IDLH atmospheres.*[§1910.134(d)(2)]

(i) *The employer shall provide* the following respirators for employee use in IDLH atmospheres:[§1910.134(d)(2)(i)]

[A] A full facepiece pressure demand SCBA certified by NIOSH for a minimum service life of thirty minutes, or [§1910.134(d)(2)(i)[A]]

[B] A combination full facepiece pressure demand supplied-air respirator (SAR) with auxiliary self-contained air supply.[§1910.134(d)(2)(i)[B]]

(ii) *Respirators provided* only for escape from IDLH atmospheres shall be NIOSH-certified for escape from the atmosphere in which they will be used.[§1910.134(d)(2)(ii)]

(iii) *All oxygen-deficient atmospheres* shall be considered IDLH. [§1910.134(d)(2)(iii)]

Exception: If the employer demonstrates that, under all foreseeable conditions, the oxygen concentration can be maintained within the ranges specified in Table II of this section (i.e., for the altitudes set out in the table), then any atmosphere-supplying respirator may be used.

(3) *Respirators for atmospheres that are not IDLH.*[§1910.134(d)(3)]

(i) *The employer shall provide a respirator* that is adequate to protect the health of the employee and ensure compliance with all other OSHA statutory and regulatory requirements, under routine and reasonably foreseeable emergency situations.[§1910.134(d)(3)(i)]

[A] Assigned Protection Factors (APFs). Employers must use the assigned protection factors listed in Table 1 to select a respirator that meets or exceeds the required level of employee protection. When using a combination respirator (e.g., airline respirators with an air-purifying filter), employers must ensure that the assigned protection factor is appropriate to the mode of operation in which the respirator is being used.[§1910.134(d)(3)(i)[A]]

Table 1 — Assigned Protection Factors [5]

Type of respirator [1,2]	Quarter mask	Half mask	Full facepiece	Helmet/ hood	Loose-fitting facepiece
1. Air-Purifying Respirator	5	[3] 10	50		
2. Powered Air-Purifying Respirator (PAPR)		50	1,000	[4] 25/ 1,000	25
3. Supplied-Air Respirator (SAR) or Airline Respirator					
•Demand mode		10	50		
•Continuous flow mode		50	1,000	[4] 25/ 1,000	25
•Pressure-demand or other positive-pressure mode		50	1,000		
4. Self-Contained Breathing Apparatus (SCBA)					
•Demand mode		10	50	50	
•Pressure-demand or other positive-pressure mode (e.g., open/closed circuit)			10,000	10,000	

Notes:

1 Employers may select respirators assigned for use in higher workplace concentrations of a hazardous substance for use at lower concentrations of that substance, or when required respirator use is independent of concentration.

2 The assigned protection factors in Table 1 are only effective when the employer implements a continuing,effective respirator program as required by this section (29 CFR 1910.134), including training, fit testing,maintenance, and use requirements.

3 This APF category includes filtering facepieces, and half masks with elastomeric facepieces.

4 The employer must have evidence provided by the respirator manufacturer that testing of these respirators demonstrates performance at a level of protection of 1,000 or greater to receive an APF of 1,000. This level of performance can best be demonstrated by performing a WPF or SWPF study or equivalent testing. Absent such testing, all other PAPRs and SARs with helmets/hoods are to be treated as loose-fitting face piece respirators,and receive an APF of 25.

5 These APFs do not apply to respirators used solely for escape. For escape respirators used in association with specific substances covered by 29 CFR 1910 subpart Z, employers must refer to the appropriate substance-specific standards in that subpart. Escape respirators for other IDLH atmospheres are specified by 29 CFR 1910.134 (d)(2)(ii).

[B] Maximum Use Concentration (MUC).[§1910.134(d)(3)(i)[B]]

[1] The employer must select a respirator for employee use that maintains the employee's exposure to the hazardous substance, when measured outside the respirator, at or below the MUC.[§1910.134(d)(3)(i)[B][1]]

[2] Employers must not apply MUCs to conditions that are immediately dangerous to life or health (IDLH); instead, they must use respirators listed for IDLH conditions in paragraph (d)(2) of this standard.[§1910.134(d)(3)(i)[B][2]]

[3] When the calculated MUC exceeds the IDLH level for a hazardous substance, or the performance limits of the cartridge or canister, then employers must set the maximum MUC at that lower limit.[§1910.134(d)(3)(i)[B][3]]

(ii) *The respirator selected* shall be appropriate for the chemical state and physical form of the contaminant.[§1910.134(d)(3)(ii)]

(iii) *For protection against gases and vapors,* the employer shall provide:[§1910.134(d)(3)(iii)]

[A] An atmosphere-supplying respirator, or[§1910.134(d)(3)(iii)[A]]

[B] An air-purifying respirator, provided that: [§1910.134(d)(3)(iii)[B]]

[1] The respirator is equipped with an end-of-service-life indicator (ESLI) certified by NIOSH for the contaminant; or[§1910.134(d)(3)(iii)[B][1]]

[2] If there is no ESLI appropriate for conditions in the employer's workplace, the employer implements a change schedule for canisters and cartridges that is based on objective information or data that will ensure that canisters and cartridges are changed before the end of their service life. The employer shall describe in the respirator program the information and data relied upon and the basis for the canister and cartridge change schedule and the basis for reliance on the data.[§1910.134(d)(3)(iii)[B][2]]

(iv) *For protection against particulates, the employer shall provide:*[§1910.134(d)(3)(iv)]

[A] An atmosphere-supplying respirator; or[§1910.134(d)(3)(iv)[A]]

[B] An air-purifying respirator equipped with a filter certified by NIOSH under 30 CFR part 11 as a high efficiency particulate air (HEPA) filter, or an air-purifying respirator equipped with a filter certified for particulates by NIOSH under 42 CFR part 84; or[§1910.134(d)(3)(iv)[B]]

[C] For contaminants consisting primarily of particles with mass median aerodynamic diameters (MMAD) of at least 2 micrometers, an air-purifying respirator equipped with any filter certified for particulates by NIOSH. [§1910.134(d)(3)(iv)[C]]

Table I — Assigned Protection Factors [Reserved]

Table II

Altitude (ft.)	Oxygen deficient Atmospheres (% O_2) for which the employer may rely on atmosphere-supplying respirators
Less than 3,001	16.0-19.5
3,001-4,000	16.4-19.5
4,001-5,000	17.1-19.5
5,001-6,000	17.8-19.5
6,001-7,000	18.5-19.5
7,001-8,000[1]	19.3-19.5

1 Above 8,000 feet the exception does not apply. Oxygen-enriched breathing air must be supplied above 14,000 feet.

(e) **Medical evaluation.** Using a respirator may place a physiological burden on employees that varies with the type of respirator worn, the job and workplace conditions in which the respirator is used, and the medical status of the employee. Accordingly, this paragraph specifies the minimum requirements for medical evaluation that employers must implement to determine the employee's ability to use a respirator.[§1910.134(e)]

(1) *General.* The employer shall provide a medical evaluation to determine the employee's ability to use a respirator, before the employee is fit tested or required to use the respirator in the workplace. The employer may discontinue an employee's medical evaluations when the employee is no longer required to use a respirator.[§1910.134(e)(1)]

(2) *Medical evaluation procedures.*[§1910.134(e)(2)]

(i) *The employer shall identify* a physician or other licensed health care professional (PLHCP) to perform medical evaluations using a medical questionnaire or an initial medical examination that obtains the same information as the medical questionnaire.[§1910.134(e)(2)(i)]

(ii) *The medical evaluation* shall obtain the information requested by the questionnaire in Sections 1 and 2, Part A of Appendix C of this section.[§1910.134(e)(2)(ii)]

(3) *Follow-up medical examination.*[§1910.134(e)(3)]

(i) *The employer shall ensure* that a follow-up medical examination is provided for an employee who gives a positive response to any question among questions 1 through 8 in Section 2, Part A of Appendix C or whose initial medical examination demonstrates the need for a follow-up medical examination.[§1910.134(e)(3)(i)]

(ii) *The follow-up medical examination* shall include any medical tests, consultations, or diagnostic procedures that the PLHCP deems necessary to make a final determination. [§1910.134(e)(3)(ii)]

(4) *Administration of the medical questionnaire and examinations.* [§1910.134(e)(4)]

(i) *The medical questionnaire and examinations* shall be administered confidentially during the employee's normal working hours or at a time and place convenient to the employee. The medical questionnaire shall be administered in a manner that ensures that the employee understands its content. [§1910.134(e)(4)(i)]

(ii) *The employer shall provide the employee* with an opportunity to discuss the questionnaire and examination results with the PLHCP.[§1910.134(e)(4)(ii)]

(5) *Supplemental information for the PLHCP.*[§1910.134(e)(5)]

(i) *The following information* must be provided to the PLHCP before the PLHCP makes a recommendation concerning an employee's ability to use a respirator:[§1910.134(e)(5)(i)]

[A] The type and weight of the respirator to be used by the employee;[§1910.134(e)(5)(i)[A]]

[B] The duration and frequency of respirator use (including use for rescue and escape);[§1910.134(e)(5)(i)[B]]

[C] The expected physical work effort;[§1910.134(e)(5)(i)[C]]

[D] Additional protective clothing and equipment to be worn; and[§1910.134(e)(5)(i)[D]]

[E] Temperature and humidity extremes that may be encountered.[§1910.134(e)(5)(i)[E]]

(ii) *Any supplemental information* provided previously to the PLHCP regarding an employee need not be provided for a subsequent medical evaluation if the information and the PLHCP remain the same.[§1910.134(e)(5)(ii)]

(iii) *The employer shall provide the PLHCP* with a copy of the written respiratory protection program and a copy of this section.[§1910.134(e)(5)(iii)]

Note to paragraph (e)(5)(iii): When the employer replaces a PLHCP, the employer must ensure that the new PLHCP obtains this information, either by providing the documents directly to the PLHCP or having the documents transferred from the former PLHCP to the new PLHCP. However, OSHA does not expect employers to have employees medically reevaluated solely because a new PLHCP has been selected.

(6) *Medical determination.* In determining the employee's ability to use a respirator, the employer shall:[§1910.134(e)(6)]

(i) *Obtain a written recommendation* regarding the employee's ability to use the respirator from the PLHCP. The recommendation shall provide only the following information: [§1910.134(e)(6)(i)]

[A] Any limitations on respirator use related to the medical condition of the employee, or relating to the workplace conditions in which the respirator will be used, including whether or not the employee is medically able to use the respirator;[§1910.134(e)(6)(i)[A]]

[B] The need, if any, for follow-up medical evaluations; and [§1910.134(e)(6)(i)[B]]

[C] A statement that the PLHCP has provided the employee with a copy of the PLHCP's written recommendation. [§1910.134(e)(6)(i)[C]]

(ii) *If the respirator is a negative pressure respirator* and the PLHCP finds a medical condition that may place the employee's health at increased risk if the respirator is used, the employer shall provide a PAPR if the PLHCP's medical evaluation finds that the employee can use such a respirator; if a subsequent medical evaluation finds that the employee is medically able to use a negative pressure respirator, then the employer is no longer required to provide a PAPR. [§1910.134(e)(6)(ii)]

(7) *Additional medical evaluations.* At a minimum, the employer shall provide additional medical evaluations that comply with the requirements of this section if:[§1910.134(e)(7)]

(i) *An employee reports medical signs or symptoms* that are related to ability to use a respirator;[§1910.134(e)(7)(i)]

(ii) *A PLHCP,* supervisor, or the respirator program administrator informs the employer that an employee needs to be reevaluated;[§1910.134(e)(7)(ii)]

(iii) *Information* from the respiratory protection program, including observations made during fit testing and program evaluation, indicates a need for employee reevaluation; or [§1910.134(e)(7)(iii)]

(iv) *A change occurs in workplace conditions* (e.g., physical work effort, protective clothing, temperature) that may result in a substantial increase in the physiological burden placed on an employee.[§1910.134(e)(7)(iv)]

(f) **Fit testing.** This paragraph requires that, before an employee may be required to use any respirator with a negative or positive pressure tight-fitting facepiece, the employee must be fit tested with the same make, model, style, and size of respirator that will be used. This paragraph specifies the kinds of fit tests allowed, the procedures for conducting them, and how the results of the fit tests must be used.[§1910.134(f)]

(1) *The employer shall ensure* that employees using a tight-fitting facepiece respirator pass an appropriate qualitative fit test (QLFT) or quantitative fit test (QNFT) as stated in this paragraph.[§1910.134(f)(1)]

(2) *The employer shall ensure* that an employee using a tight-fitting facepiece respirator is fit tested prior to initial use of the respirator, whenever a different respirator facepiece (size, style, model or make) is used, and at least annually thereafter.[§1910.134(f)(2)]

(3) *The employer shall conduct* an additional fit test whenever the employee reports, or the employer, PLHCP, supervisor, or program administrator makes visual observations of, changes in the employee's physical condition that could affect respirator fit. Such conditions include, but are not limited to, facial scarring, dental changes, cosmetic surgery, or an obvious change in body weight.[§1910.134(f)(3)]

(4) *If after passing a QLFT or QNFT,* the employee subsequently notifies the employer, program administrator, supervisor, or PLHCP that the fit of the respirator is unacceptable, the employee shall be given a reasonable opportunity to select a different respirator facepiece and to be retested.[§1910.134(f)(4)]

(5) *The fit test shall be administered* using an OSHA-accepted QLFT or QNFT protocol. The OSHA-accepted QLFT and QNFT protocols and procedures are contained in Appendix A of this section.[§1910.134(f)(5)]

(6) *QLFT may only be used* to fit test negative pressure air-purifying respirators that must achieve a fit factor of 100 or less. [§1910.134(f)(6)]

(7) *If the fit factor,* as determined through an OSHA-accepted QNFT protocol, is equal to or greater than 100 for tight-fitting half facepieces, or equal to or greater than 500 for tight-fitting full facepieces, the QNFT has been passed with that respirator. [§1910.134(f)(7)]

(8) *Fit testing* of tight-fitting atmosphere-supplying respirators and tight-fitting powered air-purifying respirators shall be accomplished by performing quantitative or qualitative fit testing in the negative pressure mode, regardless of the mode of operation (negative or positive pressure) that is used for respiratory protection.[§1910.134(f)(8)]

(i) *Qualitative fit testing of these respirators* shall be accomplished by temporarily converting the respirator user's actual facepiece into a negative pressure respirator with appropriate filters, or by using an identical negative pressure air-purifying respirator facepiece with the same sealing surfaces as a surrogate for the atmosphere-supplying or powered air-purifying respirator facepiece.[§1910.134(f)(8)(i)]

(ii) *Quantitative fit testing* of these respirators shall be accomplished by modifying the facepiece to allow sampling inside the facepiece in the breathing zone of the user, midway between the nose and mouth. This requirement shall be accomplished by installing a permanent sampling probe onto a surrogate facepiece, or by using a sampling adapter designed to temporarily provide a means of sampling air from inside the facepiece.[§1910.134(f)(8)(ii)]

(iii) *Any modifications* to the respirator facepiece for fit testing shall be completely removed, and the facepiece restored to NIOSH-approved configuration, before that facepiece can be used in the workplace.[§1910.134(f)(8)(iii)]

(g) **Use of respirators.** This paragraph requires employers to establish and implement procedures for the proper use of respirators. These requirements include prohibiting conditions that may result in facepiece seal leakage, preventing employees from removing respirators in hazardous environments, taking actions to ensure continued effective respirator operation throughout the work shift, and establishing procedures for the use of respirators in IDLH atmospheres or in interior structural firefighting situations. [§1910.134(g)]

(1) *Facepiece seal protection.*[§1910.134(g)(1)]

(i) *The employer* shall not permit respirators with tight-fitting facepieces to be worn by employees who have: [§1910.134(g)(1)(i)]

[A] Facial hair that comes between the sealing surface of the facepiece and the face or that interferes with valve function; or[§1910.134(g)(1)(i)[A]]

[B] Any condition that interferes with the face-to-facepiece seal or valve function.[§1910.134(g)(1)(i)[B]]

(ii) *If an employee wears* corrective glasses or goggles or other personal protective equipment, the employer shall ensure that such equipment is worn in a manner that does not interfere with the seal of the facepiece to the face of the user. [§1910.134(g)(1)(ii)]

(iii) *For all tight-fitting respirators,* the employer shall ensure that employees perform a user seal check each time they put on the respirator using the procedures in Appendix B-1 or procedures recommended by the respirator manufacturer that the employer demonstrates are as effective as those in Appendix B-1 of this section.[§1910.134(g)(1)(iii)]

(2) *Continuing respirator effectiveness.*[§1910.134(g)(2)]

(i) *Appropriate surveillance* shall be maintained of work area conditions and degree of employee exposure or stress. When there is a change in work area conditions or degree of employee exposure or stress that may affect respirator effectiveness, the employer shall reevaluate the continued effectiveness of the respirator.[§1910.134(g)(2)(i)]

(ii) *The employer shall ensure* that employees leave the respirator use area:[§1910.134(g)(2)(ii)]

[A] To wash their faces and respirator facepieces as necessary to prevent eye or skin irritation associated with respirator use; or[§1910.134(g)(2)(ii)[A]]

[B] If they detect vapor or gas breakthrough, changes in breathing resistance, or leakage of the facepiece; or [§1910.134(g)(2)(ii)[B]]

[C] To replace the respirator or the filter, cartridge, or canister elements.[§1910.134(g)(2)(ii)[C]]

(iii) *If the employee* detects vapor or gas breakthrough, changes in breathing resistance, or leakage of the facepiece, the employer must replace or repair the respirator before allowing the employee to return to the work area.[§1910.134(g)(2)(iii)]

(3) *Procedures for IDLH atmospheres.* For all IDLH atmospheres, the employer shall ensure that:[§1910.134(g)(3)]

(i) *One employee or,* when needed, more than one employee is located outside the IDLH atmosphere;[§1910.134(g)(3)(i)]

(ii) *Visual, voice, or signal line communication* is maintained between the employee(s) in the IDLH atmosphere and the employee(s) located outside the IDLH atmosphere; [§1910.134(g)(3)(ii)]

(iii) *The employee(s)* located outside the IDLH atmosphere are trained and equipped to provide effective emergency rescue; [§1910.134(g)(3)(iii)]

(iv) *The employer or designee* is notified before the employee(s) located outside the IDLH atmosphere enter the IDLH atmosphere to provide emergency rescue;[§1910.134(g)(3)(iv)]

(v) *The employer or designee* authorized to do so by the employer, once notified, provides necessary assistance appropriate to the situation;[§1910.134(g)(3)(v)]

(vi) *Employee(s) located* outside the IDLH atmospheres are equipped with:[§1910.134(g)(3)(vi)]

[A] Pressure demand or other positive pressure SCBAs, or a pressure demand or other positive pressure supplied-air respirator with auxiliary SCBA; and either [§1910.134(g)(3)(vi)[A]]

[B] Appropriate retrieval equipment for removing the employee(s) who enter(s) these hazardous atmospheres where retrieval equipment would contribute to the rescue of the employee(s) and would not increase the overall risk resulting from entry; or[§1910.134(g)(3)(vi)[B]]

[C] Equivalent means for rescue where retrieval equipment is not required under paragraph (g)(3)(vi)(B). [§1910.134(g)(3)(vi)[C]]

(4) *Procedures for interior structural firefighting.* In addition to the requirements set forth under paragraph (g)(3), in interior structural fires, the employer shall ensure that:[§1910.134(g)(4)]

(i) *At least two employees* enter the IDLH atmosphere and remain in visual or voice contact with one another at all times;[§1910.134(g)(4)(i)]

(ii) *At least two employees* are located outside the IDLH atmosphere; and[§1910.134(g)(4)(ii)]

(iii) *All employees engaged* in interior structural firefighting use SCBAs.[§1910.134(g)(4)(iii)]

Note 1 to paragraph (g): One of the two individuals located outside the IDLH atmosphere may be assigned to an additional role, such as incident commander in charge of the emergency or safety officer, so long as this individual is able to perform assistance or rescue activities without jeopardizing the safety or health of any firefighter working at the incident.

Note 2 to paragraph (g): Nothing in this section is meant to preclude firefighters from performing emergency rescue activities before an entire team has assembled.

(h) **Maintenance and care of respirators.** This paragraph requires the employer to provide for the cleaning and disinfecting, storage, inspection, and repair of respirators used by employees. [§1910.134(h)]

(1) *Cleaning and disinfecting.* The employer shall provide each respirator user with a respirator that is clean, sanitary, and in good working order. The employer shall ensure that respirators are cleaned and disinfected using the procedures in Appendix B-2 of this section, or procedures recommended by the respirator manufacturer, provided that such procedures are of equivalent effectiveness. The respirators shall be cleaned and disinfected at the following intervals:[§1910.134(h)(1)]

(i) *Respirators issued* for the exclusive use of an employee shall be cleaned and disinfected as often as necessary to be maintained in a sanitary condition;[§1910.134(h)(1)(i)]

(ii) *Respirators issued to more than one employee* shall be cleaned and disinfected before being worn by different individuals;[§1910.134(h)(1)(ii)]

(iii) *Respirators maintained for emergency use* shall be cleaned and disinfected after each use; and[§1910.134(h)(1)(iii)]

(iv) *Respirators used in fit testing and training* shall be cleaned and disinfected after each use.[§1910.134(h)(1)(iv)]

(2) *Storage.* The employer shall ensure that respirators are stored as follows:[§1910.134(h)(2)]

(i) *All respirators shall be stored* to protect them from damage, contamination, dust, sunlight, extreme temperatures, excessive moisture, and damaging chemicals, and they shall be packed or stored to prevent deformation of the facepiece and exhalation valve.[§1910.134(h)(2)(i)]

(ii) *In addition to the requirements* of paragraph (h)(2)(i) of this section, emergency respirators shall be:[§1910.134(h)(2)(ii)]

[A] Kept accessible to the work area;[§1910.134(h)(2)(ii)[A]]

[B] Stored in compartments or in covers that are clearly marked as containing emergency respirators; and [§1910.134(h)(2)(ii)[B]]

[C] Stored in accordance with any applicable manufacturer instructions.[§1910.134(h)(2)(ii)[C]]

(3) *Inspection.*[§1910.134(h)(3)]

(i) *The employer shall ensure* that respirators are inspected as follows:[§1910.134(h)(3)(i)]

[A] All respirators used in routine situations shall be inspected before each use and during cleaning; [§1910.134(h)(3)(i)[A]]

[B] All respirators maintained for use in emergency situations shall be inspected at least monthly and in accordance with the manufacturer's recommendations, and shall be checked for proper function before and after each use; and[§1910.134(h)(3)(i)[B]]

[C] Emergency escape-only respirators shall be inspected before being carried into the workplace for use. [§1910.134(h)(3)(i)[C]]

(ii) *The employer shall ensure* that respirator inspections include the following:[§1910.134(h)(3)(ii)]

[A] A check of respirator function, tightness of connections, and the condition of the various parts including, but not

limited to, the facepiece, head straps, valves, connecting tube, and cartridges, canisters or filters; and [§1910.134(h)(3)(ii)[A]]

[B] A check of elastomeric parts for pliability and signs of deterioration.[§1910.134(h)(3)(ii)[B]]

(iii) *In addition to the requirements* of paragraphs (h)(3)(i) and (ii) of this section, self-contained breathing apparatus shall be inspected monthly. Air and oxygen cylinders shall be maintained in a fully charged state and shall be recharged when the pressure falls to 90% of the manufacturer's recommended pressure level. The employer shall determine that the regulator and warning devices function properly. [§1910.134(h)(3)(iii)]

(iv) *For respirators maintained for emergency use,* the employer shall:[§1910.134(h)(3)(iv)]

[A] Certify the respirator by documenting the date the inspection was performed, the name (or signature) of the person who made the inspection, the findings, required remedial action, and a serial number or other means of identifying the inspected respirator; and[§1910.134(h)(3)(iv)[A]]

[B] Provide this information on a tag or label that is attached to the storage compartment for the respirator, is kept with the respirator, or is included in inspection reports stored as paper or electronic files. This information shall be maintained until replaced following a subsequent certification.[§1910.134(h)(3)(iv)[B]]

(4) *Repairs.* The employer shall ensure that respirators that fail an inspection or are otherwise found to be defective are removed from service, and are discarded or repaired or adjusted in accordance with the following procedures:[§1910.134(h)(4)]

(i) *Repairs or adjustments to respirators* are to be made only by persons appropriately trained to perform such operations and shall use only the respirator manufacturer's NIOSH-approved parts designed for the respirator;[§1910.134(h)(4)(i)]

(ii) *Repairs shall be made* according to the manufacturer's recommendations and specifications for the type and extent of repairs to be performed; and[§1910.134(h)(4)(ii)]

(iii) *Reducing and admission valves,* regulators, and alarms shall be adjusted or repaired only by the manufacturer or a technician trained by the manufacturer.[§1910.134(h)(4)(iii)]

(i) **Breathing air quality and use.** This paragraph requires the employer to provide employees using atmosphere-supplying respirators (supplied-air and SCBA) with breathing gases of high purity.[§1910.134(i)]

(1) *The employer shall ensure* that compressed air, compressed oxygen, liquid air, and liquid oxygen used for respiration accords with the following specifications:[§1910.134(i)(1)]

(i) *Compressed and liquid oxygen* shall meet the United States Pharmacopoeia requirements for medical or breathing oxygen; and[§1910.134(i)(1)(i)]

(ii) *Compressed breathing air* shall meet at least the requirements for Grade D breathing air described in ANSI/Compressed Gas Association Commodity Specification for Air, G-7.1-1989, to include:[§1910.134(i)(1)(ii)]

[A] Oxygen content (v/v) of 19.5-23.5%;[§1910.134(i)(1)(ii)[A]]

[B] Hydrocarbon (condensed) content of 5 milligrams per cubic meter of air or less;[§1910.134(i)(1)(ii)[B]]

[C] Carbon monoxide (CO) content of 10 ppm or less; [§1910.134(i)(1)(ii)[C]]

[D] Carbon dioxide content of 1,000 ppm or less; and [§1910.134(i)(1)(ii)[D]]

[E] Lack of noticeable odor.[§1910.134(i)(1)(ii)[E]]

(2) *The employer shall ensure* that compressed oxygen is not used in atmosphere-supplying respirators that have previously used compressed air.[§1910.134(i)(2)]

(3) *The employer shall ensure* that oxygen concentrations greater than 23.5% are used only in equipment designed for oxygen service or distribution.[§1910.134(i)(3)]

(4) *The employer shall ensure* that cylinders used to supply breathing air to respirators meet the following requirements: [§1910.134(i)(4)]

(i) *Cylinders are tested and maintained* as prescribed in the Shipping Container Specification Regulations of the Department of Transportation (49 CFR part 180);[§1910.134(i)(4)(i)]

(ii) *Cylinders of purchased breathing air* have a certificate of analysis from the supplier that the breathing air meets the requirements for Grade D breathing air; and[§1910.134(i)(4)(ii)]

(iii) *The moisture content in the cylinder* does not exceed a dew point of -50 °F (-45.6 °C) at 1 atmosphere pressure. [§1910.134(i)(4)(iii)]

(5) *The employer shall ensure* that compressors used to supply breathing air to respirators are constructed and situated so as to:[§1910.134(i)(5)]

(i) *Prevent entry of contaminated air into the air-supply system.* [§1910.134(i)(5)(i)]

(ii) *Minimize moisture content* so that the dew point at 1 atmosphere pressure is 10 degrees F (5.56 °C) below the ambient temperature.[§1910.134(i)(5)(ii)]

(iii) *Have suitable* in-line air-purifying sorbent beds and filters to further ensure breathing air quality. Sorbent beds and filters shall be maintained and replaced or refurbished periodically following the manufacturer's instructions.[§1910.134(i)(5)(iii)]

(iv) *Have a tag* containing the most recent change date and the signature of the person authorized by the employer to perform the change. The tag shall be maintained at the compressor.[§1910.134(i)(5)(iv)]

(6) *For compressors that are not oil-lubricated,* the employer shall ensure that carbon monoxide levels in the breathing air do not exceed 10 ppm.[§1910.134(i)(6)]

(7) *For oil-lubricated compressors,* the employer shall use a high-temperature or carbon monoxide alarm, or both, to monitor carbon monoxide levels. If only high-temperature alarms are used, the air supply shall be monitored at intervals sufficient to prevent carbon monoxide in the breathing air from exceeding 10 ppm.[§1910.134(i)(7)]

(8) *The employer shall ensure* that breathing air couplings are incompatible with outlets for nonrespirable worksite air or other gas systems. No asphyxiating substance shall be introduced into breathing air lines.[§1910.134(i)(8)]

(9) *The employer shall use only* the respirator manufacturer's NIOSH-approved breathing-gas containers, marked and maintained in accordance with the Quality Assurance provisions of the NIOSH approval for the SCBA as issued in accordance with the NIOSH respirator-certification standard at 42 CFR part 84. [§1910.134(i)(9)]

(j) **Identification of filters, cartridges, and canisters.** The employer shall ensure that all filters, cartridges and canisters used in the workplace are labeled and color coded with the NIOSH approval label and that the label is not removed and remains legible.[§1910.134(j)]

(k) **Training and information.** This paragraph requires the employer to provide effective training to employees who are required to use respirators. The training must be comprehensive, understandable, and recur annually, and more often if necessary. This paragraph also requires the employer to provide the basic information on respirators in Appendix D of this section to employees who wear respirators when not required by this section or by the employer to do so.[§1910.134(k)]

(1) *The employer shall ensure* that each employee can demonstrate knowledge of at least the following:[§1910.134(k)(1)]

(i) *Why the respirator is necessary* and how improper fit, usage, or maintenance can compromise the protective effect of the respirator;[§1910.134(k)(1)(i)]

(ii) *What the limitations and capabilities of the respirator are;* [§1910.134(k)(1)(ii)]

(iii) *How to use the respirator effectively* in emergency situations, including situations in which the respirator malfunctions;[§1910.134(k)(1)(iii)]

(iv) *How to inspect,* put on and remove, use, and check the seals of the respirator;[§1910.134(k)(1)(iv)]

(v) *What the procedures are* for maintenance and storage of the respirator;[§1910.134(k)(1)(v)]

(vi) *How to recognize medical signs and symptoms* that may limit or prevent the effective use of respirators; and [§1910.134(k)(1)(vi)]

(vii) *The general requirements of this section.*[§1910.134(k)(1)(vii)]

(2) *The training shall be conducted* in a manner that is understandable to the employee.[§1910.134(k)(2)]

(3) *The employer shall provide the training* prior to requiring the employee to use a respirator in the workplace.[§1910.134(k)(3)]

(4) *An employer who is able to demonstrate* that a new employee has received training within the last 12 months that addresses the elements specified in paragraph (k)(1)(i) through (vii) is not required to repeat such training provided that, as required by

paragraph (k)(1), the employee can demonstrate knowledge of those element(s). Previous training not repeated initially by the employer must be provided no later than 12 months from the date of the previous training.[§1910.134(k)(4)]

(5) *Retraining shall be administered annually,* and when the following situations occur:[§1910.134(k)(5)]

(i) *Changes in the workplace* or the type of respirator render previous training obsolete;[§1910.134(k)(5)(i)]

(ii) *Inadequacies in the employee's knowledge* or use of the respirator indicate that the employee has not retained the requisite understanding or skill; or[§1910.134(k)(5)(ii)]

(iii) *Any other situation arises* in which retraining appears necessary to ensure safe respirator use.[§1910.134(k)(5)(iii)]

(6) *The basic advisory information on respirators,* as presented in Appendix D of this section, shall be provided by the employer in any written or oral format, to employees who wear respirators when such use is not required by this section or by the employer.[§1910.134(k)(6)]

(l) **Program evaluation.** This section requires the employer to conduct evaluations of the workplace to ensure that the written respiratory protection program is being properly implemented, and to consult employees to ensure that they are using the respirators properly.[§1910.134(l)]

(1) *The employer shall conduct evaluations* of the workplace as necessary to ensure that the provisions of the current written program are being effectively implemented and that it continues to be effective.[§1910.134(l)(1)]

(2) *The employer shall regularly consult employees* required to use respirators to assess the employees' views on program effectiveness and to identify any problems. Any problems that are identified during this assessment shall be corrected. Factors to be assessed include, but are not limited to:[§1910.134(l)(2)]

(i) *Respirator fit* (including the ability to use the respirator without interfering with effective workplace performance); [§1910.134(l)(2)(i)]

(ii) *Appropriate respirator selection* for the hazards to which the employee is exposed;[§1910.134(l)(2)(ii)]

(iii) *Proper respirator use* under the workplace conditions the employee encounters; and[§1910.134(l)(2)(iii)]

(iv) *Proper respirator maintenance.*[§1910.134(l)(2)(iv)]

(m) **Recordkeeping.** This section requires the employer to establish and retain written information regarding medical evaluations, fit testing, and the respirator program. This information will facilitate employee involvement in the respirator program, assist the employer in auditing the adequacy of the program, and provide a record for compliance determinations by OSHA.[§1910.134(m)]

(1) *Medical evaluation.* Records of medical evaluations required by this section must be retained and made available in accordance with 29 CFR 1910.1020.[§1910.134(m)(1)]

(2) *Fit testing.*[§1910.134(m)(2)]

(i) *The employer shall establish a record* of the qualitative and quantitative fit tests administered to an employee including: [§1910.134(m)(2)(i)]

[A] The name or identification of the employee tested; [§1910.134(m)(2)(i)[A]]

[B] Type of fit test performed;[§1910.134(m)(2)(i)[B]]

[C] Specific make, model, style, and size of respirator tested; [§1910.134(m)(2)(i)[C]]

[D] Date of test; and[§1910.134(m)(2)(i)[D]]

[E] The pass/fail results for QLFTs or the fit factor and strip chart recording or other recording of the test results for QNFTs.[§1910.134(m)(2)(i)[E]]

(ii) *Fit test records shall be retained* for respirator users until the next fit test is administered.[§1910.134(m)(2)(ii)]

(3) *A written copy of the current respirator program* shall be retained by the employer.[§1910.134(m)(3)]

(4) *Written materials required* to be retained under this paragraph shall be made available upon request to affected employees and to the Assistant Secretary or designee for examination and copying.[§1910.134(m)(4)]

(n) **Dates.**[§1910.134(n)]

(1) *Effective date.* Paragraphs (d)(3)(i)(A) and (d)(3)(i)(B) of this section become effective November 22, 2006.[§1910.134(n)(1)]

(o) **Appendices.** Compliance with Appendix A, Appendix B-1, Appendix B-2, Appendix C, and Appendix D to this section are mandatory.[§1910.134(o)]

(1) *Compliance with* Appendix A, Appendix B-1, Appendix B-2, and Appendix C of this section is mandatory.[§1910.134(o)(1)]

(2) *Appendix D of this section* is non-mandatory and is not intended to create any additional obligations not otherwise imposed or to detract from any existing obligations. [§1910.134(o)(2)]

§1910.134 Appendix A

Fit Testing Procedures (Mandatory)

Part I. OSHA-Accepted Fit Test Protocols

A. Fit Testing Procedures — General Requirements.

The employer shall conduct fit testing using the following procedures. The requirements in this appendix apply to all OSHA-accepted fit test methods, both QLFT and QNFT.

1. *The test subject shall be allowed* to pick the most acceptable respirator from a sufficient number of respirator models and sizes so that the respirator is acceptable to, and correctly fits, the user.
2. *Prior to the selection process,* the test subject shall be shown how to put on a respirator, how it should be positioned on the face, how to set strap tension and how to determine an acceptable fit. A mirror shall be available to assist the subject in evaluating the fit and positioning of the respirator. This instruction may not constitute the subject's formal training on respirator use, because it is only a review.
3. *The test subject shall be informed* that he/she is being asked to select the respirator that provides the most acceptable fit. Each respirator represents a different size and shape, and if fitted and used properly, will provide adequate protection.
4. *The test subject shall be instructed* to hold each chosen facepiece up to the face and eliminate those that obviously do not give an acceptable fit.
5. *The more acceptable facepieces* are noted in case the one selected proves unacceptable; the most comfortable mask is donned and worn at least five minutes to assess comfort. Assistance in assessing comfort can be given by discussing the points in the following item A.6. If the test subject is not familiar with using a particular respirator, the test subject shall be directed to don the mask several times and to adjust the straps each time to become adept at setting proper tension on the straps.
6. *Assessment of comfort* shall include a review of the following points with the test subject and allowing the test subject adequate time to determine the comfort of the respirator:
   (a) *Position of the mask on the nose;*
   (b) *Room for eye protection;*
   (c) *Room to talk;*
   (d) *Position of mask on face and cheeks.*
7. *The following criteria shall be used* to help determine the adequacy of the respirator fit:
   (a) *Chin properly placed;*
   (b) *Adequate strap tension, not overly tightened;*
   (c) *Fit across nose bridge;*
   (d) *Respirator of proper size to span distance from nose to chin;*
   (e) *Tendency of respirator to slip;*
   (f) *Self-observation in mirror to evaluate fit and respirator position.*
8. *The test subject shall conduct a user seal check,* either the negative and positive pressure seal checks described in Appendix B-1 of this section or those recommended by the respirator manufacturer which provide equivalent protection to the procedures in Appendix B-1. Before conducting the negative and positive pressure checks, the subject shall be told to seat the mask on the face by moving the head from side-to-side and up and down slowly while taking in a few slow deep breaths. Another facepiece shall be selected and retested if the test subject fails the user seal check tests.
9. *The test shall not be conducted* if there is any hair growth between the skin and the facepiece sealing surface, such as stubble beard growth, beard, mustache or sideburns which cross the respirator sealing surface. Any type of apparel which interferes with a satisfactory fit shall be altered or removed.
10. *If a test subject exhibits difficulty in breathing* during the tests, she or he shall be referred to a physician or other licensed health care professional, as appropriate, to determine whether the test subject can wear a respirator while performing her or his duties.

11. *If the employee finds* the fit of the respirator unacceptable, the test subject shall be given the opportunity to select a different respirator and to be retested.
12. *Exercise regimen.* Prior to the commencement of the fit test, the test subject shall be given a description of the fit test and the test subject's responsibilities during the test procedure. The description of the process shall include a description of the test exercises that the subject will be performing. The respirator to be tested shall be worn for at least 5 minutes before the start of the fit test.
13. *The fit test shall be performed* while the test subject is wearing any applicable safety equipment that may be worn during actual respirator use which could interfere with respirator fit.
14. *Test Exercises.*
   (a) *Employers must perform the following* test exercises for all fit testing methods prescribed in this appendix, except for the CNP quantitative fit testing protocol and the CNP REDON quantitative fit testing protocol. For these two protocols, employers must ensure that the test subjects (i.e., employees) perform the exercise procedure specified in Part I.C.4(b) of this appendix for the CNP quantitative fit testing protocol, or the exercise procedure described in Part I.C.5(b) of this appendix for the CNP REDON quantitative fit-testing protocol. For the remaining fit testing methods, employers must ensure that employees perform the test exercises in the appropriate test environment in the following manner:
      [1] *Normal breathing.* In a normal standing position, without talking, the subject shall breathe normally.
      [2] *Deep breathing.* In a normal standing position, the subject shall breathe slowly and deeply, taking caution so as not to hyperventilate.
      [3] *Turning head side to side.* Standing in place, the subject shall slowly turn his/her head from side to side between the extreme positions on each side. The head shall be held at each extreme momentarily so the subject can inhale at each side.
      [4] *Moving head up and down.* Standing in place, the subject shall slowly move his/her head up and down. The subject shall be instructed to inhale in the up position (i.e., when looking toward the ceiling).
      [5] *Talking.* The subject shall talk out loud slowly and loud enough so as to be heard clearly by the test conductor. The subject can read from a prepared text such as the Rainbow Passage, count backward from 100, or recite a memorized poem or song.

      Rainbow Passage

      When the sunlight strikes raindrops in the air, they act like a prism and form a rainbow. The rainbow is a division of white light into many beautiful colors. These take the shape of a long round arch, with its path high above, and its two ends apparently beyond the horizon. There is, according to legend, a boiling pot of gold at one end. People look, but no one ever finds it. When a man looks for something beyond reach, his friends say he is looking for the pot of gold at the end of the rainbow.
      [6] *Grimace.* The test subject shall grimace by smiling or frowning. (This applies only to QNFT testing; it is not performed for QLFT)
      [7] *Bending over.* The test subject shall bend at the waist as if he/she were to touch his/her toes. Jogging in place shall be substituted for this exercise in those test environments such as shroud type QNFT or QLFT units that do not permit bending over at the waist.
      [8] *Normal breathing.* Same as exercise (1).
   (b) *Each test exercise shall be performed* for one minute except for the grimace exercise which shall be performed for 15 seconds. The test subject shall be questioned by the test conductor regarding the comfort of the respirator upon completion of the protocol. If it has become unacceptable, another model of respirator shall be tried. The respirator shall not be adjusted once the fit test exercises begin. Any adjustment voids the test, and the fit test must be repeated.

B. Qualitative Fit Test (QLFT) Protocols.

1. *General.*
   (a) *The employer shall ensure* that persons administering QLFT are able to prepare test solutions, calibrate equipment and perform tests properly, recognize invalid tests, and ensure that test equipment is in proper working order.
   (b) *The employer shall ensure* that QLFT equipment is kept clean and well maintained so as to operate within the parameters for which it was designed.
2. *Isoamyl Acetate Protocol.*

Note: This protocol is not appropriate to use for the fit testing of particulate respirators. If used to fit test particulate respirators, the respirator must be equipped with an organic vapor filter.

   (a) *Odor Threshold Screening.*
   Odor threshold screening, performed without wearing a respirator, is intended to determine if the individual tested can detect the odor of isoamyl acetate at low levels.
      [1] *Three 1 liter glass jars with metal lids are required.*
      [2] *Odor-free water* (e.g., distilled or spring water) at approximately 25 °C (77 °F) shall be used for the solutions.
      [3] *The isoamyl acetate (IAA)* (also known at isopentyl acetate) stock solution is prepared by adding 1 ml of pure IAA to 800 ml of odor-free water in a 1 liter jar, closing the lid and shaking for 30 seconds. A new solution shall be prepared at least weekly.
      [4] *The screening test* shall be conducted in a room separate from the room used for actual fit testing. The two rooms shall be well-ventilated to prevent the odor of IAA from becoming evident in the general room air where testing takes place.
      [5] *The odor test solution* is prepared in a second jar by placing 0.4 ml of the stock solution into 500 ml of odor-free water using a clean dropper or pipette. The solution shall be shaken for 30 seconds and allowed to stand for two to three minutes so that the IAA concentration above the liquid may reach equilibrium. This solution shall be used for only one day.
      [6] *A test blank shall be prepared* in a third jar by adding 500 cc of odor-free water.
      [7] *The odor test and test blank jar lids* shall be labeled (e.g., 1 and 2) for jar identification. Labels shall be placed on the lids so that they can be peeled off periodically and switched to maintain the integrity of the test.
      [8] *The following instruction* shall be typed on a card and placed on the table in front of the two test jars (i.e., 1 and 2): "The purpose of this test is to determine if you can smell banana oil at a low concentration. The two bottles in front of you contain water. One of these bottles also contains a small amount of banana oil. Be sure the covers are on tight, then shake each bottle for two seconds. Unscrew the lid of each bottle, one at a time, and sniff at the mouth of the bottle. Indicate to the test conductor which bottle contains banana oil."
      [9] *The mixtures used* in the IAA odor detection test shall be prepared in an area separate from where the test is performed, in order to prevent olfactory fatigue in the subject.
      [10] *If the test subject* is unable to correctly identify the jar containing the odor test solution, the IAA qualitative fit test shall not be performed.
      [11] *If the test subject* correctly identifies the jar containing the odor test solution, the test subject may proceed to respirator selection and fit testing.
   (b) *Isoamyl Acetate Fit Test.*
      [1] *The fit test chamber* shall be a clear 55-gallon drum liner suspended inverted over a 2-foot diameter frame so that the top of the chamber is about 6 inches above the test subject's head. If no drum liner is available, a similar chamber shall be constructed using plastic sheeting. The inside top center of the chamber shall have a small hook attached.
      [2] *Each respirator used for the fitting and fit testing* shall be equipped with organic vapor cartridges or offer protection against organic vapors.
      [3] *After selecting, donning, and properly adjusting* a respirator, the test subject shall wear it to the fit testing room. This room shall be separate from the room used for odor threshold screening and respirator selection, and shall be well-ventilated, as by an exhaust fan or lab hood, to prevent general room contamination.
      [4] *A copy of the test exercises* and any prepared text from which the subject is to read shall be taped to the inside of the test chamber.
      [5] *Upon entering the test chamber,* the test subject shall be given a 6-inch by 5-inch piece of paper towel, or

other porous, absorbent, single-ply material, folded in half and wetted with 0.75 ml of pure IAA. The test subject shall hang the wet towel on the hook at the top of the chamber. An IAA test swab or ampule may be substituted for the IAA wetted paper towel provided it has been demonstrated that the alternative IAA source will generate an IAA test atmosphere with a concentration equivalent to that generated by the paper towel method.

[6] *Allow two minutes* for the IAA test concentration to stabilize before starting the fit test exercises. This would be an appropriate time to talk with the test subject; to explain the fit test, the importance of his/her cooperation, and the purpose for the test exercises; or to demonstrate some of the exercises.

[7] *If at any time during the test,* the subject detects the banana-like odor of IAA, the test is failed. The subject shall quickly exit from the test chamber and leave the test area to avoid olfactory fatigue.

[8] *If the test is failed,* the subject shall return to the selection room and remove the respirator. The test subject shall repeat the odor sensitivity test, select and put on another respirator, return to the test area and again begin the fit test procedure described in (b)(1) through (7) above. The process continues until a respirator that fits well has been found. Should the odor sensitivity test be failed, the subject shall wait at least 5 minutes before retesting. Odor sensitivity will usually have returned by this time.

[9] *If the subject passes the test,* the efficiency of the test procedure shall be demonstrated by having the subject break the respirator face seal and take a breath before exiting the chamber.

[10] *When the test subject leaves the chamber,* the subject shall remove the saturated towel and return it to the person conducting the test, so that there is no significant IAA concentration buildup in the chamber during subsequent tests. The used towels shall be kept in a self-sealing plastic bag to keep the test area from being contaminated.

3. *Saccharin Solution Aerosol Protocol.*

The entire screening and testing procedure shall be explained to the test subject prior to the conduct of the screening test.

(a) *Taste threshold screening.* The saccharin taste threshold screening, performed without wearing a respirator, is intended to determine whether the individual being tested can detect the taste of saccharin.

[1] *During threshold screening* as well as during fit testing, subjects shall wear an enclosure about the head and shoulders that is approximately 12 inches in diameter by 14 inches tall with at least the front portion clear and that allows free movements of the head when a respirator is worn. An enclosure substantially similar to the 3M hood assembly, parts #FT 14 and # FT 15 combined, is adequate.

[2] *The test enclosure* shall have a 3/4-inch (1.9 cm) hole in front of the test subject's nose and mouth area to accommodate the nebulizer nozzle.

[3] *The test subject* shall don the test enclosure. Throughout the threshold screening test, the test subject shall breathe through his/her slightly open mouth with tongue extended. The subject is instructed to report when he/she detects a sweet taste.

[4] *Using a DeVilbiss Model 40* Inhalation Medication Nebulizer or equivalent, the test conductor shall spray the threshold check solution into the enclosure. The nozzle is directed away from the nose and mouth of the person. This nebulizer shall be clearly marked to distinguish it from the fit test solution nebulizer.

[5] *The threshold check solution* is prepared by dissolving 0.83 gram of sodium saccharin USP in 100 ml of warm water. It can be prepared by putting 1 ml of the fit test solution (see (b)(5) below) in 100 ml of distilled water.

[6] *To produce the aerosol,* the nebulizer bulb is firmly squeezed so that it collapses completely, then released and allowed to fully expand.

[7] *Ten squeezes are repeated rapidly* and then the test subject is asked whether the saccharin can be tasted. If the test subject reports tasting the sweet taste during the ten squeezes, the screening test is completed. The taste threshold is noted as ten regardless of the number of squeezes actually completed.

[8] *If the first response is negative,* ten more squeezes are repeated rapidly and the test subject is again asked whether the saccharin is tasted. If the test subject reports tasting the sweet taste during the second ten squeezes, the screening test is completed. The taste threshold is noted as twenty regardless of the number of squeezes actually completed.

[9] *If the second response is negative,* ten more squeezes are repeated rapidly and the test subject is again asked whether the saccharin is tasted. If the test subject reports tasting the sweet taste during the third set of ten squeezes, the screening test is completed. The taste threshold is noted as thirty regardless of the number of squeezes actually completed.

[10] *The test conductor* will take note of the number of squeezes required to solicit a taste response.

[11] *If the saccharin is not tasted* after 30 squeezes (step 10), the test subject is unable to taste saccharin and may not perform the saccharin fit test.

Note to paragraph 3(a): If the test subject eats or drinks something sweet before the screening test, he/she may be unable to taste the weak saccharin solution.

[12] *If a taste response is elicited,* the test subject shall be asked to take note of the taste for reference in the fit test.

[13] *Correct use of the nebulizer* means that approximately 1 ml of liquid is used at a time in the nebulizer body.

[14] *The nebulizer shall be* thoroughly rinsed in water, shaken dry, and refilled at least each morning and afternoon or at least every four hours.

(b) *Saccharin solution aerosol fit test procedure.*

[1] *The test subject* may not eat, drink (except plain water), smoke, or chew gum for 15 minutes before the test.

[2] *The fit test uses the same enclosure* described in 3. (a) above.

[3] *The test subject shall don* the enclosure while wearing the respirator selected in section I. A. of this appendix. The respirator shall be properly adjusted and equipped with a particulate filter(s).

[4] *A second DeVilbiss Model 40* Inhalation Medication Nebulizer or equivalent is used to spray the fit test solution into the enclosure. This nebulizer shall be clearly marked to distinguish it from the screening test solution nebulizer.

[5] *The fit test solution is prepared* by adding 83 grams of sodium saccharin to 100 ml of warm water.

[6] *As before,* the test subject shall breathe through the slightly open mouth with tongue extended, and report if he/she tastes the sweet taste of saccharin.

[7] *The nebulizer is inserted* into the hole in the front of the enclosure and an initial concentration of saccharin fit test solution is sprayed into the enclosure using the same number of squeezes (either 10, 20 or 30 squeezes) based on the number of squeezes required to elicit a taste response as noted during the screening test. A minimum of 10 squeezes is required.

[8] *After generating the aerosol,* the test subject shall be instructed to perform the exercises in section I. A. 14. of this appendix.

[9] *Every 30 seconds* the aerosol concentration shall be replenished using one half the original number of squeezes used initially (e.g., 5, 10 or 15).

[10] *The test subject shall indicate* to the test conductor if at any time during the fit test the taste of saccharin is detected. If the test subject does not report tasting the saccharin, the test is passed.

[11] *If the taste of saccharin is detected,* the fit is deemed unsatisfactory and the test is failed. A different respirator shall be tried and the entire test procedure is repeated (taste threshold screening and fit testing).

[12] *Since the nebulizer has a tendency* to clog during use, the test operator must make periodic checks of the nebulizer to ensure that it is not clogged. If clogging is found at the end of the test session, the test is invalid.

4. *Bitrex™ (Denatonium Benzoate)* Solution Aerosol Qualitative Fit Test Protocol.

The Bitrex™ (Denatonium benzoate) solution aerosol QLFT protocol uses the published saccharin test protocol because that protocol is widely accepted. Bitrex is routinely used as a taste aversion agent in household liquids which children should not be drinking and is endorsed by the American Medical Association, the National Safety Council, and the American Association of Poison Control Centers. The entire screening and testing procedure shall be explained to the test subject prior to the conduct of the screening test.

(a) *Taste Threshold Screening.*

The Bitrex taste threshold screening, performed without wearing a respirator, is intended to determine whether the individual being tested can detect the taste of Bitrex.

[1] During threshold screening as well as during fit testing, subjects shall wear an enclosure about the head and shoulders that is approximately 12 inches (30.5 cm) in diameter by 14 inches (35.6 cm) tall. The front portion of the enclosure shall be clear from the respirator and allow free movement of the head when a respirator is worn. An enclosure substantially similar to the 3M hood assembly, parts # FT 14 and # FT 15 combined, is adequate.

[2] The test enclosure shall have a 3/4 inch (1.9 cm) hole in front of the test subject's nose and mouth area to accommodate the nebulizer nozzle.

[3] The test subject shall don the test enclosure. Throughout the threshold screening test, the test subject shall breathe through his or her slightly open mouth with tongue extended. The subject is instructed to report when he/she detects a bitter taste.

[4] Using a DeVilbiss Model 40 Inhalation Medication Nebulizer or equivalent, the test conductor shall spray the Threshold Check Solution into the enclosure. This Nebulizer shall be clearly marked to distinguish it from the fit test solution nebulizer.

[5] The Threshold Check Solution is prepared by adding 13.5 milligrams of Bitrex to 100 ml of 5% salt (NaCl) solution in distilled water.

[6] To produce the aerosol, the nebulizer bulb is firmly squeezed so that the bulb collapses completely, and is then released and allowed to fully expand.

[7] An initial ten squeezes are repeated rapidly and then the test subject is asked whether the Bitrex can be tasted. If the test subject reports tasting the bitter taste during the ten squeezes, the screening test is completed. The taste threshold is noted as ten regardless of the number of squeezes actually completed.

[8] If the first response is negative, ten more squeezes are repeated rapidly and the test subject is again asked whether the Bitrex is tasted. If the test subject reports tasting the bitter taste during the second ten squeezes, the screening test is completed. The taste threshold is noted as twenty regardless of the number of squeezes actually completed.

[9] If the second response is negative, ten more squeezes are repeated rapidly and the test subject is again asked whether the Bitrex is tasted. If the test subject reports tasting the bitter taste during the third set of ten squeezes, the screening test is completed. The taste threshold is noted as thirty regardless of the number of squeezes actually completed.

[10] The test conductor will take note of the number of squeezes required to solicit a taste response.

[11] If the Bitrex is not tasted after 30 squeezes (step 10), the test subject is unable to taste Bitrex and may not perform the Bitrex fit test.

[12] If a taste response is elicited, the test subject shall be asked to take note of the taste for reference in the fit test.

[13] Correct use of the nebulizer means that approximately 1 ml of liquid is used at a time in the nebulizer body.

[14] The nebulizer shall be thoroughly rinsed in water, shaken to dry, and refilled at least each morning and afternoon or at least every four hours.

(b) *Bitrex Solution Aerosol Fit Test Procedure.*

[1] The test subject may not eat, drink (except plain water), smoke, or chew gum for 15 minutes before the test.

[2] The fit test uses the same enclosure as that described in 4. (a) above.

[3] The test subject shall don the enclosure while wearing the respirator selected according to section I. A. of this appendix. The respirator shall be properly adjusted and equipped with any type particulate filter(s).

[4] A second DeVilbiss Model 40 Inhalation Medication Nebulizer or equivalent is used to spray the fit test solution into the enclosure. This nebulizer shall be clearly marked to distinguish it from the screening test solution nebulizer.

[5] The fit test solution is prepared by adding 337.5 mg of Bitrex to 200 ml of a 5% salt (NaCl) solution in warm water.

[6] As before, the test subject shall breathe through his or her slightly open mouth with tongue extended, and be instructed to report if he/she tastes the bitter taste of Bitrex.

[7] The nebulizer is inserted into the hole in the front of the enclosure and an initial concentration of the fit test solution is sprayed into the enclosure using the same number of squeezes (either 10, 20 or 30 squeezes) based on the number of squeezes required to elicit a taste response as noted during the screening test.

[8] After generating the aerosol, the test subject shall be instructed to perform the exercises in section I. A. 14. of this appendix.

[9] Every 30 seconds the aerosol concentration shall be replenished using one half the number of squeezes used initially (e.g., 5, 10 or 15).

[10] The test subject shall indicate to the test conductor if at any time during the fit test the taste of Bitrex is detected. If the test subject does not report tasting the Bitrex, the test is passed.

[11] If the taste of Bitrex is detected, the fit is deemed unsatisfactory and the test is failed. A different respirator shall be tried and the entire test procedure is repeated (taste threshold screening and fit testing).

5. *Irritant Smoke (Stannic Chloride) Protocol.*

This qualitative fit test uses a person's response to the irritating chemicals released in the "smoke" produced by a stannic chloride ventilation smoke tube to detect leakage into the respirator.

(a) *General Requirements and Precautions.*

[1] The respirator to be tested shall be equipped with high efficiency particulate air (HEPA) or P100 series filter(s).

[2] Only stannic chloride smoke tubes shall be used for this protocol.

[3] No form of test enclosure or hood for the test subject shall be used.

[4] The smoke can be irritating to the eyes, lungs, and nasal passages. The test conductor shall take precautions to minimize the test subject's exposure to irritant smoke. Sensitivity varies, and certain individuals may respond to a greater degree to irritant smoke. Care shall be taken when performing the sensitivity screening checks that determine whether the test subject can detect irritant smoke to use only the minimum amount of smoke necessary to elicit a response from the test subject.

[5] The fit test shall be performed in an area with adequate ventilation to prevent exposure of the person conducting the fit test or the build-up of irritant smoke in the general atmosphere.

(b) *Sensitivity Screening Check.*

The person to be tested must demonstrate his or her ability to detect a weak concentration of the irritant smoke.

[1] The test operator shall break both ends of a ventilation smoke tube containing stannic chloride, and attach one end of the smoke tube to a low flow air pump set to deliver 200 milliliters per minute, or an aspirator squeeze bulb. The test operator shall cover the other end of the smoke tube with a short piece of tubing to prevent potential injury from the jagged end of the smoke tube.

[2] The test operator shall advise the test subject that the smoke can be irritating to the eyes, lungs, and nasal passages and instruct the subject to keep his/her eyes closed while the test is performed.

[3] The test subject shall be allowed to smell a weak concentration of the irritant smoke before the respirator is donned to become familiar with its irritating properties and to determine if he/she can detect the irritating properties of the smoke. The test operator shall carefully

direct a small amount of the irritant smoke in the test subject's direction to determine that he/she can detect it.

(c) *Irritant Smoke Fit Test Procedure.*

[1] The person being fit tested shall don the respirator without assistance, and perform the required user seal check(s).

[2] The test subject shall be instructed to keep his/her eyes closed.

[3] The test operator shall direct the stream of irritant smoke from the smoke tube toward the faceseal area of the test subject, using the low flow pump or the squeeze bulb. The test operator shall begin at least 12 inches from the facepiece and move the smoke stream around the whole perimeter of the mask. The operator shall gradually make two more passes around the perimeter of the mask, moving to within six inches of the respirator.

[4] If the person being tested has not had an involuntary response and/or detected the irritant smoke, proceed with the test exercises.

[5] The exercises identified in section I.A. 14. of this appendix shall be performed by the test subject while the respirator seal is being continually challenged by the smoke, directed around the perimeter of the respirator at a distance of six inches.

[6] If the person being fit tested reports detecting the irritant smoke at any time, the test is failed. The person being retested must repeat the entire sensitivity check and fit test procedure.

[7] Each test subject passing the irritant smoke test without evidence of a response (involuntary cough, irritation) shall be given a second sensitivity screening check, with the smoke from the same smoke tube used during the fit test, once the respirator has been removed, to determine whether he/she still reacts to the smoke. Failure to evoke a response shall void the fit test.

[8] If a response is produced during this second sensitivity check, then the fit test is passed.

C. Quantitative Fit Test (QNFT) Protocols.

The following quantitative fit testing procedures have been demonstrated to be acceptable: Quantitative fit testing using a non-hazardous test aerosol (such as corn oil, polyethylene glycol 400 [PEG 400], di-2-ethyl hexyl sebacate [DEHS], or sodium chloride) generated in a test chamber, and employing instrumentation to quantify the fit of the respirator; Quantitative fit testing using ambient aerosol as the test agent and appropriate instrumentation (condensation nuclei counter) to quantify the respirator fit; Quantitative fit testing using controlled negative pressure and appropriate instrumentation to measure the volumetric leak rate of a facepiece to quantify the respirator fit.

1. *General.*

(a) *The employer shall ensure* that persons administering QNFT are able to calibrate equipment and perform tests properly, recognize invalid tests, calculate fit factors properly and ensure that test equipment is in proper working order.

(b) *The employer shall ensure* that QNFT equipment is kept clean, and is maintained and calibrated according to the manufacturer's instructions so as to operate at the parameters for which it was designed.

2. *Generated Aerosol Quantitative Fit Testing Protocol.*

(a) *Apparatus.*

[1] Instrumentation. Aerosol generation, dilution, and measurement systems using particulates (corn oil, polyethylene glycol 400 [PEG 400], di-2-ethyl hexyl sebacate [DEHS] or sodium chloride) as test aerosols shall be used for quantitative fit testing.

[2] Test chamber. The test chamber shall be large enough to permit all test subjects to perform freely all required exercises without disturbing the test agent concentration or the measurement apparatus. The test chamber shall be equipped and constructed so that the test agent is effectively isolated from the ambient air, yet uniform in concentration throughout the chamber.

[3] When testing air-purifying respirators, the normal filter or cartridge element shall be replaced with a high efficiency particulate air (HEPA) or P100 series filter supplied by the same manufacturer.

[4] The sampling instrument shall be selected so that a computer record or strip chart record may be made of the test showing the rise and fall of the test agent concentration with each inspiration and expiration at fit factors of at least 2,000. Integrators or computers that integrate the amount of test agent penetration leakage into the respirator for each exercise may be used provided a record of the readings is made.

[5] The combination of substitute air-purifying elements, test agent and test agent concentration shall be such that the test subject is not exposed in excess of an established exposure limit for the test agent at any time during the testing process, based upon the length of the exposure and the exposure limit duration.

[6] The sampling port on the test specimen respirator shall be placed and constructed so that no leakage occurs around the port (e.g., where the respirator is probed), a free air flow is allowed into the sampling line at all times, and there is no interference with the fit or performance of the respirator. The in-mask sampling device (probe) shall be designed and used so that the air sample is drawn from the breathing zone of the test subject, midway between the nose and mouth and with the probe extending into the facepiece cavity at least 1/4 inch.

[7] The test setup shall permit the person administering the test to observe the test subject inside the chamber during the test.

[8] The equipment generating the test atmosphere shall maintain the concentration of test agent constant to within a 10 percent variation for the duration of the test.

[9] The time lag (interval between an event and the recording of the event on the strip chart or computer or integrator) shall be kept to a minimum. There shall be a clear association between the occurrence of an event and its being recorded.

[10] The sampling line tubing for the test chamber atmosphere and for the respirator sampling port shall be of equal diameter and of the same material. The length of the two lines shall be equal.

[11] The exhaust flow from the test chamber shall pass through an appropriate filter (i.e., high efficiency particulate filter) before release.

[12] When sodium chloride aerosol is used, the relative humidity inside the test chamber shall not exceed 50 percent.

[13] The limitations of instrument detection shall be taken into account when determining the fit factor.

[14] Test respirators shall be maintained in proper working order and be inspected regularly for deficiencies such as cracks or missing valves and gaskets.

(b) *Procedural Requirements.*

[1] When performing the initial user seal check using a positive or negative pressure check, the sampling line shall be crimped closed in order to avoid air pressure leakage during either of these pressure checks.

[2] The use of an abbreviated screening QLFT test is optional. Such a test may be utilized in order to quickly identify poor fitting respirators that passed the positive and/or negative pressure test and reduce the amount of QNFT time. The use of the CNC QNFT instrument in the count mode is another optional method to obtain a quick estimate of fit and eliminate poor fitting respirators before going on to perform a full QNFT.

[3] A reasonably stable test agent concentration shall be measured in the test chamber prior to testing. For canopy or shower curtain types of test units, the determination of the test agent's stability may be established after the test subject has entered the test environment.

[4] Immediately after the subject enters the test chamber, the test agent concentration inside the respirator shall be measured to ensure that the peak penetration does not exceed 5 percent for a half mask or 1 percent for a full facepiece respirator.

[5] A stable test agent concentration shall be obtained prior to the actual start of testing.

[6] Respirator restraining straps shall not be over-tightened for testing. The straps shall be adjusted by the wearer without assistance from other persons to give a reasonably comfortable fit typical of normal use. The respirator shall not be adjusted once the fit test exercises begin.

[7] *The test shall be terminated* whenever any single peak penetration exceeds 5 percent for half masks and 1 percent for full facepiece respirators. The test subject shall be refitted and retested.

[8] *Calculation of fit factors.*

[i] *The fit factor shall be determined* for the quantitative fit test by taking the ratio of the average chamber concentration to the concentration measured inside the respirator for each test exercise except the grimace exercise.

[ii] *The average test chamber concentration* shall be calculated as the arithmetic average of the concentration measured before and after each test (i.e., 7 exercises) or the arithmetic average of the concentration measured before and after each exercise or the true average measured continuously during the respirator sample.

[iii] *The concentration* of the challenge agent inside the respirator shall be determined by one of the following methods:

[A] *Average peak penetration method* means the method of determining test agent penetration into the respirator utilizing a strip chart recorder, integrator, or computer. The agent penetration is determined by an average of the peak heights on the graph or by computer integration, for each exercise except the grimace exercise. Integrators or computers that calculate the actual test agent penetration into the respirator for each exercise will also be considered to meet the requirements of the average peak penetration method.

[B] *Maximum peak penetration method* means the method of determining test agent penetration in the respirator as determined by strip chart recordings of the test. The highest peak penetration for a given exercise is taken to be representative of average penetration into the respirator for that exercise.

[C] *Integration by calculation* of the area under the individual peak for each exercise except the grimace exercise. This includes computerized integration.

[D] *The calculation of the overall fit factor* using individual exercise fit factors involves first converting the exercise fit factors to penetration values, determining the average, and then converting that result back to a fit factor. This procedure is described in the following equation:

$$\text{Overall Fit Factor} = \frac{\text{Number of exercises}}{1/ff_1+1/ff_2+1/ff_3+1/ff_4+1/ff_5+1/ff_7+1/ff_8}$$

Where ff_1, ff_2, ff_3, etc. are the fit factors for exercises 1, 2, 3, etc.

[9] *The test subject* shall not be permitted to wear a half mask or quarter facepiece respirator unless a minimum fit factor of 100 is obtained, or a full facepiece respirator unless a minimum fit factor of 500 is obtained.

[10] *Filters used for quantitative fit testing* shall be replaced whenever increased breathing resistance is encountered, or when the test agent has altered the integrity of the filter media.

3. *Ambient aerosol condensation* nuclei counter (CNC) quantitative fit testing protocol.

The ambient aerosol condensation nuclei counter (CNC) quantitative fit testing (Portacount ™) protocol quantitatively fit tests respirators with the use of a probe. The probed respirator is only used for quantitative fit tests. A probed respirator has a special sampling device, installed on the respirator, that allows the probe to sample the air from inside the mask. A probed respirator is required for each make, style, model, and size that the employer uses and can be obtained from the respirator manufacturer or distributor. The CNC instrument manufacturer, TSI Inc., also provides probe attachments (TSI sampling adapters) that permit fit testing in an employee's own respirator. A minimum fit factor pass level of at least 100 is necessary for a half-mask respirator and a minimum fit factor pass level of at least 500 is required for a full facepiece negative pressure respirator. The entire screening and testing procedure shall be explained to the test subject prior to the conduct of the screening test.

(a) *Portacount Fit Test Requirements.*

[1] *Check the respirator* to make sure the sampling probe and line are properly attached to the facepiece and that the respirator is fitted with a particulate filter capable of preventing significant penetration by the ambient particles used for the fit test (e.g., NIOSH 42 CFR 84 series 100, series 99, or series 95 particulate filter) per manufacturer's instruction.

[2] *Instruct the person to be tested* to don the respirator for five minutes before the fit test starts. This purges the ambient particles trapped inside the respirator and permits the wearer to make certain the respirator is comfortable. This individual shall already have been trained on how to wear the respirator properly.

[3] *Check the following conditions* for the adequacy of the respirator fit: Chin properly placed; Adequate strap tension, not overly tightened; Fit across nose bridge; Respirator of proper size to span distance from nose to chin; Tendency of the respirator to slip; Self-observation in a mirror to evaluate fit and respirator position.

[4] *Have the person wearing the respirator* do a user seal check. If leakage is detected, determine the cause. If leakage is from a poorly fitting facepiece, try another size of the same model respirator, or another model of respirator.

[5] *Follow the manufacturer's instructions* for operating the Portacount and proceed with the test.

[6] *The test subject shall be instructed* to perform the exercises in section I. A. 14. of this appendix.

[7] *After the test exercises,* the test subject shall be questioned by the test conductor regarding the comfort of the respirator upon completion of the protocol. If it has become unacceptable, another model of respirator shall be tried.

(b) *Portacount Test Instrument.*

[1] *The Portacount will automatically stop* and calculate the overall fit factor for the entire set of exercises. The overall fit factor is what counts. The Pass or Fail message will indicate whether or not the test was successful. If the test was a Pass, the fit test is over.

[2] *Since the pass or fail criterion* of the Portacount is user programmable, the test operator shall ensure that the pass or fail criterion meet the requirements for minimum respirator performance in this Appendix.

[3] *A record of the test needs to be kept on file,* assuming the fit test was successful. The record must contain the test subject's name; overall fit factor; make, model, style, and size of respirator used; and date tested.

4. *Controlled negative pressure (CNP) quantitative fit testing protocol.*

The CNP protocol provides an alternative to aerosol fit test methods. The CNP fit test method technology is based on exhausting air from a temporarily sealed respirator facepiece to generate and then maintain a constant negative pressure inside the facepiece. The rate of air exhaust is controlled so that a constant negative pressure is maintained in the respirator during the fit test. The level of pressure is selected to replicate the mean inspiratory pressure that causes leakage into the respirator under normal use conditions. With pressure held constant, air flow out of the respirator is equal to air flow into the respirator. Therefore, measurement of the exhaust stream that is required to hold the pressure in the temporarily sealed respirator constant yields a direct measure of leakage air flow into the respirator. The CNP fit test method measures leak rates through the facepiece as a method for determining the facepiece fit for negative pressure respirators. The CNP instrument manufacturer Occupational Health Dynamics of Birmingham, Alabama also provides attachments (sampling manifolds) that replace the filter cartridges to permit fit testing in an employee's own respirator. To perform the test, the test subject closes his or her mouth and holds his/her breath, after which an air pump removes air from the respirator facepiece at a pre-selected constant pressure. The facepiece fit is expressed as the leak rate through the facepiece, expressed as milliliters per minute. The quality and validity of the CNP fit tests are determined by the degree to which the in-mask pressure tracks the test pressure during the system measurement time of approximately five seconds. Instantaneous feedback in the form of a real-time pressure trace of the in-mask pressure is provided and used to determine test validity and quality. A minimum fit factor pass level of 100 is necessary for a half-mask respirator and a minimum fit factor of at least 500 is required for a full facepiece respirator. The entire screening and testing procedure shall be explained to the test subject prior to the conduct of the screening test.

(a) *CNP Fit Test Requirements.*

[1] The instrument shall have a non-adjustable test pressure of 15.0 mm water pressure.

[2] The CNP system defaults selected for test pressure shall be set at-15 mm of water (-0.58 inches of water) and the modeled inspiratory flow rate shall be 53.8 liters per minute for performing fit tests.

Note: CNP systems have built-in capability to conduct fit testing that is specific to unique work rate, mask, and gender situations that might apply in a specific workplace. Use of system default values, which were selected to represent respirator wear with medium cartridge resistance at a low-moderate work rate, will allow inter-test comparison of the respirator fit.

[3] The individual who conducts the CNP fit testing shall be thoroughly trained to perform the test.

[4] The respirator filter or cartridge needs to be replaced with the CNP test manifold. The inhalation valve downstream from the manifold either needs to be temporarily removed or propped open.

[5] The employer must train the test subject to hold his or her breath for at least 10 seconds.

[6] The test subject must don the test respirator without any assistance from the test administrator who is conducting the CNP fit test. The respirator must not be adjusted once the fit-test exercises begin. Any adjustment voids the test, and the test subject must repeat the fit test.

[7] The QNFT protocol shall be followed according to section I. C. 1. of this appendix with an exception for the CNP test exercises.

(b) *CNP Test Exercises.*

[1] Normal breathing. In a normal standing position, without talking, the subject shall breathe normally for 1 minute. After the normal breathing exercise, the subject needs to hold head straight ahead and hold his or her breath for 10 seconds during the test measurement.

[2] Deep breathing. In a normal standing position, the subject shall breathe slowly and deeply for 1 minute, being careful not to hyperventilate. After the deep breathing exercise, the subject shall hold his or her head straight ahead and hold his or her breath for 10 seconds during test measurement.

[3] Turning head side to side. Standing in place, the subject shall slowly turn his or her head from side to side between the extreme positions on each side for 1 minute. The head shall be held at each extreme momentarily so the subject can inhale at each side. After the turning head side to side exercise, the subject needs to hold head full left and hold his or her breath for 10 seconds during test measurement. Next, the subject needs to hold head full right and hold his or her breath for 10 seconds during test measurement.

[4] Moving head up and down. Standing in place, the subject shall slowly move his or her head up and down for 1 minute. The subject shall be instructed to inhale in the up position (i.e., when looking toward the ceiling). After the moving head up and down exercise, the subject shall hold his or her head full up and hold his or her breath for 10 seconds during test measurement. Next, the subject shall hold his or her head full down and hold his or her breath for 10 seconds during test measurement.

[5] Talking. The subject shall talk out loud slowly and loud enough so as to be heard clearly by the test conductor. The subject can read from a prepared text such as the Rainbow Passage, count backward from 100, or recite a memorized poem or song for 1 minute. After the talking exercise, the subject shall hold his or her head straight ahead and hold his or her breath for 10 seconds during the test measurement.

[6] Grimace. The test subject shall grimace by smiling or frowning for 15 seconds.

[7] Bending Over. The test subject shall bend at the waist as if he or she were to touch his or her toes for 1 minute. Jogging in place shall be substituted for this exercise in those test environments such as shroud-type QNFT units that prohibit bending at the waist. After the bending over exercise, the subject shall hold his or her head straight ahead and hold his or her breath for 10 seconds during the test measurement.

[8] Normal Breathing. The test subject shall remove and re-don the respirator within a one-minute period. Then, in a normal standing position, without talking, the subject shall breathe normally for 1 minute. After the normal breathing exercise, the subject shall hold his or her head straight ahead and hold his or her breath for 10 seconds during the test measurement. After the test exercises, the test subject shall be questioned by the test conductor regarding the comfort of the respirator upon completion of the protocol. If it has become unacceptable, another model of a respirator shall be tried.

(c) *CNP Test Instrument.*

[1] The test instrument must have an effective audio-warning device, or a visual-warning device in the form of a screen tracing, that indicates when the test subject fails to hold his or her breath during the test. The test must be terminated and restarted from the beginning when the test subject fails to hold his or her breath during the test. The test subject then may be refitted and retested.

[2] A record of the test shall be kept on file, assuming the fit test was successful. The record must contain the test subject's name; overall fit factor; make, model, style and size of respirator used; and date tested.

5. *Controlled negative pressure (CNP)* REDON quantitative fit testing protocol.

(a) *When administering this protocol to test subjects,* employers must comply with the requirements specified in paragraphs (a) and (c) of Part I.C.4 of this appendix ("Controlled negative pressure (CNP) quantitative fit testing protocol"), as well as use the test exercises described below in paragraph (b) of this protocol instead of the test exercises specified in paragraph (b) of Part I.C.4 of this appendix.

(b) *Employers must ensure* that each test subject being fit tested using this protocol follows the exercise and measurement procedures, including the order of administration, described below in Table A-1 of this appendix.

Table A-1 — CNP REDON Quantitative Fit Testing Protocol

Exercises[1]	Exercise procedure	Measurement procedure
Facing Forward	Stand and breathe normally, without talking, for 30 seconds	Face forward, while holding breath for 10 seconds
Bending Over	Bend at the waist, as if going to touch his or her toes, for 30 seconds	Face parallel to the floor, while holding breath for 10 seconds
Head Shaking	For about three seconds, shake head back and forth vigorously several times while shouting	Face forward, while holding breath for 10 seconds
REDON 1	Remove the respirator mask, loosen all facepiece straps, and then redon the respirator mask	Face forward, while holding breath for 10 seconds
REDON 2	Remove the respirator mask, loosen all facepiece straps, and then redon the respirator mask again	Face forward, while holding breath for 10 seconds

[1] Exercises are listed in the order in which they are to be administered.

(c) *After completing the test exercises,* the test administrator must question each test subject regarding the comfort of the respirator. When a test subject states that the respirator is unacceptable, the employer must ensure that the test administrator repeats the protocol using another respirator model.

(d) *Employers must determine the overall fit factor* for each test subject by calculating the harmonic mean of the fit testing exercises as follows:

$$\text{Overall Fit Factor} = \frac{N}{[1/ff_1 + 1/ff_2 + \ldots 1/ff_N]}$$

Where:

N = The number of exercises;

FF_1 = The fit factor for the first exercise;

FF_2 = The fit factor for the second exercise; and

FF_N = The fit factor for the nth exercise.

Part II. New Fit Test Protocols

A. **Any person may submit to OSHA** an application for approval of a new fit test protocol. If the application meets the following criteria, OSHA will initiate a rulemaking proceeding under section 6(b)(7) of the OSH Act to determine whether to list the new protocol as an approved protocol in this Appendix A.

B. **The application must include a detailed description** of the proposed new fit test protocol. This application must be supported by either:

1. *A test report prepared* by an independent government research laboratory (e.g., Lawrence Livermore National Laboratory, Los Alamos National Laboratory, the National Institute for Standards

and Technology) stating that the laboratory has tested the protocol and had found it to be accurate and reliable; or

2. *An article that has been published* in a peer-reviewed industrial hygiene journal describing the protocol and explaining how test data support the protocol's accuracy and reliability.

C. If OSHA determines that additional information is required before the Agency commences a rulemaking proceeding under this section, OSHA will so notify the applicant and afford the applicant the opportunity to submit the supplemental information. Initiation of a rulemaking proceeding will be deferred until OSHA has received and evaluated the supplemental information.

§1910.134 Appendix B-1

User Seal Check Procedures (Mandatory)

The individual who uses a tight-fitting respirator is to perform a user seal check to ensure that an adequate seal is achieved each time the respirator is put on. Either the positive and negative pressure checks listed in this appendix, or the respirator manufacturer's recommended user seal check method shall be used. User seal checks are not substitutes for qualitative or quantitative fit tests.

I. Facepiece Positive and/or Negative Pressure Checks.

A. *Positive pressure check.* Close off the exhalation valve and exhale gently into the facepiece. The face fit is considered satisfactory if a slight positive pressure can be built up inside the facepiece without any evidence of outward leakage of air at the seal. For most respirators this method of leak testing requires the wearer to first remove the exhalation valve cover before closing off the exhalation valve and then carefully replacing it after the test.

B. *Negative pressure check.* Close off the inlet opening of the canister or cartridge(s) by covering with the palm of the hand(s) or by replacing the filter seal(s), inhale gently so that the facepiece collapses slightly, and hold the breath for ten seconds. The design of the inlet opening of some cartridges cannot be effectively covered with the palm of the hand. The test can be performed by covering the inlet opening of the cartridge with a thin latex or nitrile glove. If the facepiece remains in its slightly collapsed condition and no inward leakage of air is detected, the tightness of the respirator is considered satisfactory.

II. Manufacturer's Recommended User Seal Check Procedures.

The respirator manufacturer's recommended procedures for performing a user seal check may be used instead of the positive and/or negative pressure check procedures provided that the employer demonstrates that the manufacturer's procedures are equally effective.

§1910.134 Appendix B-2

Respirator Cleaning Procedures (Mandatory)

These procedures are provided for employer use when cleaning respirators. They are general in nature, and the employer as an alternative may use the cleaning recommendations provided by the manufacturer of the respirators used by their employees, provided such procedures are as effective as those listed here in Appendix B-2. Equivalent effectiveness simply means that the procedures used must accomplish the objectives set forth in Appendix B-2, i.e., must ensure that the respirator is properly cleaned and disinfected in a manner that prevents damage to the respirator and does not cause harm to the user.

I. Procedures for Cleaning Respirators.

A. *Remove filters, cartridges, or canisters.* Disassemble facepieces by removing speaking diaphragms, demand and pressure-demand valve assemblies, hoses, or any components recommended by the manufacturer. Discard or repair any defective parts.

B. *Wash components in warm* (43 °C [110 °F] maximum) water with a mild detergent or with a cleaner recommended by the manufacturer. A stiff bristle (not wire) brush may be used to facilitate the removal of dirt.

C. *Rinse components thoroughly* in clean, warm (43 °C [110 °F] maximum), preferably running water. Drain.

D. *When the cleaner used* does not contain a disinfecting agent, respirator components should be immersed for two minutes in one of the following:

1. *Hypochlorite solution (50 ppm of chlorine)* made by adding approximately one milliliter of laundry bleach to one liter of water at 43 °C (110 °F); or,
2. *Aqueous solution of iodine (50 ppm iodine)* made by adding approximately 0.8 milliliters of tincture of iodine (6-8 grams ammonium and/or potassium iodide/100 cc of 45% alcohol) to one liter of water at 43 °C (110 °F); or,
3. *Other commercially available cleansers* of equivalent disinfectant quality when used as directed, if their use is recommended or approved by the respirator manufacturer.

E. *Rinse components thoroughly in clean, warm* (43 °C [110 °F] maximum), preferably running water. Drain. The importance of thorough rinsing cannot be overemphasized. Detergents or disinfectants that dry on facepieces may result in dermatitis. In addition, some disinfectants may cause deterioration of rubber or corrosion of metal parts if not completely removed.

F. *Components should be hand-dried* with a clean lint-free cloth or air-dried.

G. *Reassemble facepiece,* replacing filters, cartridges, and canisters where necessary.

H. *Test the respirator to ensure that all components work properly.*

§1910.134 Appendix C

OSHA Respirator Medical Evaluation Questionnaire (Mandatory)

Appendix C to §1910.134: **OSHA Respirator Medical Evaluation Questionnaire** (Mandatory)

To the employer:
Answers to questions in Section 1, and to question 9 in Section 2 of Part A, do not require a medical examination.

To the employee:
Can you read: ☐Yes ☐No
Your employer must allow you to answer this questionnaire during normal working hours, or at a time and place that is convenient to you. To maintain your confidentiality, your employer or supervisor must not look at or review your answers, and your employer must tell you how to deliver or send this questionnaire to the health care professional who will review it.

Part A. Section 1. (Mandatory)
The following information must be provided by every employee who has been selected to use any type of respirator. (please print)

1. Today's date: ______ / ______ / ______
2. Your name: ____________________
3. Your age (to nearest year): ______ 4. Sex: ☐ M ☐ F 5. Your height: ___ft. ______ in. 6. Your weight: ________ lbs.
7. Your job title: ____________________
8. A phone number where you can be reached by the health care professional who reviews this questionnaire.
   Include Area Code: (________)________ - __________ Ext. __________
9. The best time to phone you at this number:
   ☐ Before ☐ After ☐ Between ______:______ ☐ a.m. ☐ p.m. - ______:______ ☐ a.m. ☐ p.m.
10. Has your employer told you how to contact the health care professional who will review this questionnaire? ☐ Yes ☐ No
11. Check the type of respirator you will use (you can check more than one category):
    a. ☐ N ☐ R ☐ P disposable respirator (filter-mask, non-cartridge type only)
    b. ☐ Other type (for example, half- or full-facepiece type, powered-air purifying, supplied-air, self-contained breathing apparatus)
12. Have you worn a respirator? ☐Yes ☐No If "yes," what type(s): ____________________

*Free forms available free of charge at www.oshacfr.com

Appendix C to §1910.134: **OSHA Respirator Medical Evaluation Questionnaire** (Mandatory)

Part A. Section 2. (Mandatory)
Questions 1 through 9 below must be answered by every employee who has been selected to use any type of respirator. (please check "yes" or "no")

1. **Do you currently smoke tobacco, or have you smoked tobacco in the last month:** ☐ Yes ☐ No
2. **Have you ever had any of the following conditions?**
   a. Seizures (fits): ☐ Yes ☐ No
   b. Diabetes (sugar disease): ☐ Yes ☐ No
   c. Allergic reactions that interfere with your breathing: ☐ Yes ☐ No
   d. Claustrophobia (fear of closed-in places): ☐ Yes ☐ No
   e. Trouble smelling odors: ☐ Yes ☐ No
3. **Have you ever had any of the following pulmonary or lung problems?**
   a. Asbestosis: ☐ Yes ☐ No
   b. Asthma: ☐ Yes ☐ No
   c. Chronic bronchitis: ☐ Yes ☐ No
   d. Emphysema: ☐ Yes ☐ No
   e. Pneumonia: ☐ Yes ☐ No
   f. Tuberculosis: ☐ Yes ☐ No
   g. Silicosis: ☐ Yes ☐ No
   h. Pneumothorax (collapsed lung): ☐ Yes ☐ No
   i. Lung cancer: ☐ Yes ☐ No
   j. Broken ribs: ☐ Yes ☐ No
   k. Any chest injuries or surgeries: ☐ Yes ☐ No
   l. Any other lung problem that you've been told about: ☐ Yes ☐ No
4. **Do you currently have any of the following symptoms of pulmonary or lung illness?**
   a. Shortness of breath: ☐ Yes ☐ No
   b. Shortness of breath when walking fast on level ground or walking up a slight hill or incline: ☐ Yes ☐ No
   c. Shortness of breath when walking with other people at an ordinary pace on level ground: ☐ Yes ☐ No
   d. Have to stop for breath when walking at your own pace on level ground: ☐ Yes ☐ No
   e. Shortness of breath when washing or dressing yourself: ☐ Yes ☐ No
   f. Shortness of breath that interferes with your job: ☐ Yes ☐ No
   g. Coughing that produces phlegm (thick sputum): ☐ Yes ☐ No
   h. Coughing that wakes you early in the morning: ☐ Yes ☐ No
   i. Coughing that occurs mostly when you are lying down: ☐ Yes ☐ No
   j. Coughing up blood in the last month: ☐ Yes ☐ No
   k. Wheezing: ☐ Yes ☐ No
   l. Wheezing that interferes with your job: ☐ Yes ☐ No
   m. Chest pain when you breathe deeply: ☐ Yes ☐ No
   n. Any other symptoms that you think may be related to lung problems: ☐ Yes ☐ No
5. **Have you ever had any of the following cardiovascular or heart problems?**
   a. Heart attack: ☐ Yes ☐ No
   b. Stroke: ☐ Yes ☐ No
   c. Angina: ☐ Yes ☐ No
   d. Heart failure: ☐ Yes ☐ No
   e. Swelling in your legs or feet (not caused by walking): ☐ Yes ☐ No
   f. Heart arrhythmia (heart beating irregularly): ☐ Yes ☐ No
   g. High blood pressure: ☐ Yes ☐ No
   h. Any other heart problem that you've been told about: ☐ Yes ☐ No
6. **Have you ever had any of the following cardiovascular or heart symptoms?**
   a. Frequent pain or tightness in your chest: ☐ Yes ☐ No
   b. Pain or tightness in your chest during physical activity: ☐ Yes ☐ No
   c. Pain or tightness in your chest that interferes with your job: ☐ Yes ☐ No
   d. In the past two years, have you noticed your heart skipping or missing a beat: ☐ Yes ☐ No
   e. Heartburn or indigestion that is not related to eating: ☐ Yes ☐ No
   f. Any other symptoms that you think may be related to heart or circulation problems: ☐ Yes ☐ No
7. **Do you currently take medication for any of the following problems?**
   a. Breathing or lung problems: ☐ Yes ☐ No
   b. Heart trouble: ☐ Yes ☐ No
   c. Blood pressure: ☐ Yes ☐ No
   d. Seizures (fits): ☐ Yes ☐ No
8. **If you've used a respirator, have you ever had any of the following problems?** (If you've never used a respirator, check the following space and go to question 9:) ☐ Never Used
   a. Eye irritation: ☐ Yes ☐ No
   b. Skin allergies or rashes: ☐ Yes ☐ No
   c. Anxiety: ☐ Yes ☐ No
   d. General weakness or fatigue: ☐ Yes ☐ No
   e. Any other problem that interferes with your use of a respirator: ☐ Yes ☐ No
9. **Would you like to talk to the health care professional who will review this questionnaire about your answers to this questionnaire?** ☐ Yes ☐ No

Questions 10 to 15 below must be answered by every employee who has been selected to use either a full-facepiece respirator or a self-contained breathing apparatus (SCBA). For employees who have been selected to use other types of respirators, answering these questions is voluntary.

10. **Have you ever lost vision in either eye (temporarily or permanently):** ☐ Yes ☐ No
11. **Do you currently have any of the following vision problems?**
    a. Wear contact lenses: ☐ Yes ☐ No
    b. Wear glasses: ☐ Yes ☐ No
    c. Color blind: ☐ Yes ☐ No
    d. Any other eye or vision problem: ☐ Yes ☐ No
12. **Have you ever had an injury to your ears, including a broken ear drum:** ☐ Yes ☐ No
13. **Do you currently have any of the following hearing problems?**
    a. Difficulty hearing: ☐ Yes ☐ No
    b. Wear a hearing aid: ☐ Yes ☐ No
    c. Any other hearing or ear problem: ☐ Yes ☐ No
14. **Have you ever had a back injury:** ☐ Yes ☐ No
15. **Do you currently have any of the following musculoskeletal problems?**
    a. Weakness in any of your arms, hands, legs, or feet: ☐ Yes ☐ No
    b. Back pain: ☐ Yes ☐ No
    c. Difficulty fully moving your arms and legs: ☐ Yes ☐ No
    d. Pain or stiffness when you lean forward or backward at the waist: ☐ Yes ☐ No
    e. Difficulty fully moving your head up or down: ☐ Yes ☐ No
    f. Difficulty fully moving your head side to side: ☐ Yes ☐ No
    g. Difficulty bending at your knees: ☐ Yes ☐ No
    h. Difficulty squatting to the ground: ☐ Yes ☐ No
    i. Climbing a flight of stairs or a ladder carrying more than 25 lbs: ☐ Yes ☐ No
    j. Any other muscle or skeletal problem that interferes with using a respirator: ☐ Yes ☐ No

*Free forms available free of charge at www.oshacfr.com

Appendix C to §1910.134: **OSHA Respirator Medical Evaluation Questionnaire** (Mandatory)

Part B.

Any of the following questions, and other questions not listed, may be added to the questionnaire at the discretion of the health care professional who will review the questionnaire.

1. **In your present job, are you working at high altitudes (over 5,000 feet) or in a place that has lower than normal amounts of oxygen:** ☐ Yes ☐ No
   If "Yes", do you have feelings of dizziness, shortness of breath, pounding in your chest, or other symptoms when you're working under these conditions: ☐ Yes ☐ No
2. **At work or at home, have you ever been exposed to hazardous solvents, hazardous airborne chemicals (e.g., gases, fumes, or dust), or have you come into skin contact with hazardous chemicals:** ☐ Yes ☐ No
   If "Yes", name the chemicals if you know them: ______
3. **Have you ever worked with any of the materials, or under any of the conditions, listed below:**
   a. Asbestos: ☐ Yes ☐ No
   b. Silica (e.g., in sandblasting): ☐ Yes ☐ No
   c. Tungsten/cobalt (e.g., grinding or welding this material): ☐ Yes ☐ No
   d. Beryllium: ☐ Yes ☐ No
   e. Aluminum: ☐ Yes ☐ No
   f. Coal (for example, mining): ☐ Yes ☐ No
   g. Iron: ☐ Yes ☐ No
   h. Tin: ☐ Yes ☐ No
   i. Dusty Environments: ☐ Yes ☐ No
   j. Any other hazardous exposures: ☐ Yes ☐ No
   If yes, describe these exposures: ______
4. **List any second jobs or side businesses you have:** ______
5. **List your previous occupations:** ______
6. **List your current and previous hobbies:** ______
7. **Have you been in the military services?:** ☐ Yes ☐ No
   If "Yes", were you exposed to biological or chemical agents (either in training or combat): ☐ Yes ☐ No
8. **Have you ever worked on a HAZMAT team?:** ☐ Yes ☐ No
9. **Other than medications for breathing and lung problems, heart trouble, blood pressure, and seizures mentioned earlier in this questionnaire, are you taking any other medications for any reason (including over-the-counter medications):** ☐ Yes ☐ No
   If "Yes", name the medications, if you know them: ______
10. **Will you be using any of the following items with your respirator(s)?:**
    a. HEPA Filters: ☐ Yes ☐ No
    b. Canisters (for example, gas masks): ☐ Yes ☐ No
    c. Cartridges: ☐ Yes ☐ No
11. **How often are you expected to use the respirator(s)** (check "yes" or "no" for all answers that apply to you)?:
    a. Escape only (no rescue): ☐ Yes ☐ No
    b. Emergency rescue only: ☐ Yes ☐ No
    c. Less than 5 hours per week: ☐ Yes ☐ No
    d. Less than 2 hours per day: ☐ Yes ☐ No
    e. 2 to 4 hours per day: ☐ Yes ☐ No
    f. Over 4 hours per day: ☐ Yes ☐ No
12. **During the period you are using the respirator(s), is your work effort:**
    a. Light (less than 200 kcal per hour): ☐ Yes ☐ No
    If "Yes", how long does this period last during the average shift: ______ hrs. ______ mins.
    Examples of a light work effort are *sitting* while writing, typing, drafting, or performing light assembly work; or *standing* while operating a drill press (1-3 lbs.) or controlling machines.
    b. Moderate (200 to 350 kcal per hour): ☐ Yes ☐ No
    If "Yes", how long does this period last during the average shift: ______ hrs. ______ mins.
    Examples of moderate work effort are *sitting* while nailing or filing; *driving* a truck or bus in urban traffic; *standing* while drilling, nailing, performing assembly work, or transferring a moderate load (about 35 lbs.) at trunk level; *walking* on a level surface about 2 mph or down a 5-degree grade about 3 mph; or pushing a wheelbarrow with a heavy load (about 100 lbs.) on a level surface.
    c. Heavy (above 350 kcal per hour): ☐ Yes ☐ No
    If "Yes", how long does this period last during the average shift: ______ hrs. ______ mins.
    Examples of heavy work are *lifting* a heavy load (about 50 lbs.) from the floor to your waist or shoulder; *working* on a loading dock; *shoveling*; *standing* while bricklaying or chipping castings; *walking* up an 8-degree grade about 2 mph; *climbing* stairs with a heavy load (about 50 lbs.)
13. **Will you be wearing protective clothing and/or equipment (other than the respirator) when you're using your respirator?:** ☐ Yes ☐ No
    If "Yes", describe this protective clothing and/or equipment: ______
14. **Will you be working under hot conditions (temperature exceeding 77 °F)?:** ☐ Yes ☐ No
15. **Will you be working under humid conditions?:** ☐ Yes ☐ No
16. **Describe the work you'll be doing while you're using your respirator(s):** ______
17. **Describe any special or hazardous conditions you might encounter when you're using your respirator(s) (for example, confined spaces, life-threatening gases):** ______
18. **Provide the following information, if you know it, for each toxic substance that you'll be exposed to when you're using your respirator(s):**
    Name of the first toxic substance: ______
    Estimated maximum exposure level per shift: ______ Duration of exposure per shift: ______
    Name of the second toxic substance: ______
    Estimated maximum exposure level per shift: ______ Duration of exposure per shift: ______
    Name of the third toxic substance: ______
    Estimated maximum exposure level per shift: ______ Duration of exposure per shift: ______
    The name of any other toxic substances that you'll be exposed to while using your respirator: ______
19. **Describe any special responsibilities you'll have while using your respirator(s) that may affect the safety and well-being of others (for example, rescue, security):** ______

*Free forms available free of charge at www.oshacfr.com

§1910.134 Appendix D

Information for Employees Using Respirators When Not Required Under the Standard (Mandatory)

Respirators are an effective method of protection against designated hazards when properly selected and worn. Respirator use is encouraged, even when exposures are below the exposure limit, to provide an additional level of comfort and protection for workers. However, if a respirator is used improperly or not kept clean, the respirator itself can become a hazard to the worker. Sometimes, workers may wear respirators to avoid exposures to hazards, even if the amount of hazardous substance does not exceed the limits set by OSHA standards. If your employer provides respirators for your voluntary use, or if you provide your own respirator, you need to take certain precautions to be sure that the respirator itself does not present a hazard.

You should do the following:

1. **Read and heed all instructions** provided by the manufacturer on use, maintenance, cleaning and care, and warnings regarding the respirators limitations.
2. **Choose respirators certified for use**to protect against the contaminant of concern. NIOSH, the National Institute for Occupational Safety and Health of the U.S. Department of Health and Human Services, certifies respirators. A label or statement of certification should appear on the respirator or respirator packaging. It will tell you what the respirator is designed for and how much it will protect you.
3. **Do not wear your respirator** into atmospheres containing contaminants for which your respirator is not designed to protect against. For example, a respirator designed to filter dust particles will not protect you against gases, vapors, or very small solid particles of fumes or smoke.
4. **Keep track of your respirator** so that you do not mistakenly use someone else's respirator.

[63 FR 1270, Jan. 8, 1998; 63 FR 20098, 20099, Apr. 23, 1998, as amended at 69 FR 46993, Aug. 4, 2004; 71 FR 16672, Apr. 3, 2006; 71 FR 50187, Aug. 24, 2006; 73 FR 75584, Dec. 12, 2008; 76 FR 33607, June 8, 2011]

§1910.136

Foot protection

(a) **General requirements.** The employer shall ensure that each affected employee uses protective footwear when working in areas where there is a danger of foot injuries due to falling or rolling objects, or objects piercing the sole, and where such employee's feet are exposed to electrical hazards.[§1910.136(a)]

(b) **Criteria for protective footwear.**[§1910.136(b)]

(1) *Protective footwear must comply* with any of the following consensus standards:[§1910.136(b)(1)]

(i) *ASTM F-2412-2005,* "Standard Test Methods for Foot Protection," and ASTM F-2413-2005, "Standard Specification for Performance Requirements for Protective Footwear," which are incorporated by reference in Sec. 1910.6;[§1910.136(b)(1)(i)]

(ii) *ANSI Z41-1999,* "American National Standard for Personal Protection — Protective Footwear," which is incorporated by reference in Sec. 1910.6; or[§1910.136(b)(1)(ii)]

(iii) *ANSI Z41-1991,* "American National Standard for Personal Protection — Protective Footwear," which is incorporated by reference in Sec. 1910.6.[§1910.136(b)(1)(iii)]

(2) *Protective footwear that the employer* demonstrates is at least as effective as protective footwear that is constructed in accordance with one of the above consensus standards will be deemed to be in compliance with the requirements of this section.[§1910.136(b)(2)]

[59 FR 16362, Apr. 6, 1994; 59 FR 33911, July 1, 1994, as amended at 61 FR 9238, Mar. 7, 1996; 61 FR 19548, May 2, 1996; 61 FR 21228, May 9, 1996; 74 FR 46356, Sept. 9, 2009]

§1910.146

Permit-required confined spaces

(a) **Scope and application.** This section contains requirements for practices and procedures to protect employees in general industry from the hazards of entry into permit-required confined spaces. This section does not apply to agriculture, to construction, or to shipyard employment (Parts 1928, 1926, and 1915 of this chapter, respectively).[§1910.146(a)]

(b) **Definitions.**[§1910.146(b)]

Acceptable entry conditions means the conditions that must exist in a permit space to allow entry and to ensure that employees involved with a permit-required confined space entry can safely enter into and work within the space.[§1910.146(b)]

Attendant means an individual stationed outside one or more permit spaces who monitors the authorized entrants and who performs all attendant's duties assigned in the employer's permit space program.[§1910.146(b)]

Authorized entrant means an employee who is authorized by the employer to enter a permit space.[§1910.146(b)]

Blanking or blinding means the absolute closure of a pipe, line, or duct by the fastening of a solid plate (such as a spectacle blind or a skillet blind) that completely covers the bore and that is capable of withstanding the maximum pressure of the pipe, line, or duct with no leakage beyond the plate.[§1910.146(b)]

Confined space means a space that:[§1910.146(b)]

(1) *Is large enough and so configured* that an employee can bodily enter and perform assigned work; and[§1910.146(b)(1)]

(2) *Has limited or restricted means* for entry or exit (for example, tanks, vessels, silos, storage bins, hoppers, vaults, and pits are spaces that may have limited means of entry.); and [§1910.146(b)(2)]

(3) *Is not designed for continuous employee occupancy.* [§1910.146(b)(3)]

Double block and bleed means the closure of a line, duct, or pipe by closing and locking or tagging two in-line valves and by opening and locking or tagging a drain or vent valve in the line between the two closed valves.[§1910.146(b)]

Emergency means any occurrence (including any failure of hazard control or monitoring equipment) or event internal or external to the permit space that could endanger entrants.[§1910.146(b)]

Engulfment means the surrounding and effective capture of a person by a liquid or finely divided (flowable) solid substance that can be aspirated to cause death by filling or plugging the respiratory system or that can exert enough force on the body to cause death by strangulation, constriction, or crushing.[§1910.146(b)]

Entry means the action by which a person passes through an opening into a permit-required confined space. Entry includes ensuing work activities in that space and is considered to have occurred as soon as any part of the entrant's body breaks the plane of an opening into the space.[§1910.146(b)]

Entry permit (permit) means the written or printed document that is provided by the employer to allow and control entry into a permit space and that contains the information specified in paragraph (f) of this section.[§1910.146(b)]

Entry supervisor means the person (such as the employer, foreman, or crew chief) responsible for determining if acceptable entry conditions are present at a permit space where entry is planned, for authorizing entry and overseeing entry operations, and for terminating entry as required by this section.[§1910.146(b)]

Note: An entry supervisor also may serve as an attendant or as an authorized entrant, as long as that person is trained and equipped as required by this section for each role he or she fills. Also, the duties of entry supervisor may be passed from one individual to another during the course of an entry operation.

Hazardous atmosphere means an atmosphere that may expose employees to the risk of death, incapacitation, impairment of ability to self-rescue (that is, escape unaided from a permit space), injury, or acute illness from one or more of the following causes: [§1910.146(b)]

(1) *Flammable gas, vapor, or mist* in excess of 10 percent of its lower flammable limit (LFL);[§1910.146(b)(1)]

(2) *Airborne combustible dust* at a concentration that meets or exceeds its LFL;[§1910.146(b)(2)]

Note: This concentration may be approximated as a condition in which the dust obscures vision at a distance of 5 feet (1.52 m) or less.

(3) *Atmospheric oxygen concentration* below 19.5 percent or above 23.5 percent;[§1910.146(b)(3)]

(4) *Atmospheric concentration of any substance* for which a dose or a permissible exposure limit is published in Subpart G, Occupational Health and Environmental Control, or in Subpart Z, Toxic and Hazardous Substances, of this part and which could result in employee exposure in excess of its dose or permissible exposure limit;[§1910.146(b)(4)]

Note: An atmospheric concentration of any substance that is not capable of causing death, incapacitation, impairment of ability to self-rescue, injury, or acute illness due to its health effects is not covered by this provision.

(5) *Any other atmospheric condition that is immediately* dangerous to life or health.[§1910.146(b)(5)]

Note: For air contaminants for which OSHA has not determined a dose or permissible exposure limit, other sources of information, such as Material Safety Data Sheets that comply with the Hazard Communication Standard, Sec. 1910.1200 of this part, published information, and internal documents can provide guidance in establishing acceptable atmospheric conditions.

Hot work permit means the employer's written authorization to perform operations (for example, riveting, welding, cutting, burning, and heating) capable of providing a source of ignition.[§1910.146(b)]

Immediately dangerous to life or health (IDLH) means any condition that poses an immediate or delayed threat to life or that would cause irreversible adverse health effects or that would interfere with an individual's ability to escape unaided from a permit space. [§1910.146(b)]

Note: Some materials — hydrogen fluoride gas and cadmium vapor, for example — may produce immediate transient effects that, even if severe, may pass without medical attention, but are followed by sudden, possibly fatal collapse 12-72 hours after exposure. The victim "feels normal" from recovery from transient effects until collapse. Such materials in hazardous quantities are considered to be "immediately" dangerous to life or health.

Inerting means the displacement of the atmosphere in a permit space by a noncombustible gas (such as nitrogen) to such an extent that the resulting atmosphere is noncombustible. [§1910.146(b)]

Note: This procedure produces an IDLH oxygen-deficient atmosphere.

Isolation means the process by which a permit space is removed from service and completely protected against the release of energy and material into the space by such means as: blanking or blinding; misaligning or removing sections of lines, pipes, or ducts; a double block and bleed system; lockout or tagout of all sources of energy; or blocking or disconnecting all mechanical linkages. [§1910.146(b)]

Line breaking means the intentional opening of a pipe, line, or duct that is or has been carrying flammable, corrosive, or toxic material, an inert gas, or any fluid at a volume, pressure, or temperature capable of causing injury.[§1910.146(b)]

Non-permit confined space means a confined space that does not contain or, with respect to atmospheric hazards, have the potential to contain any hazard capable of causing death or serious physical harm.[§1910.146(b)]

Oxygen deficient atmosphere means an atmosphere containing less than 19.5 percent oxygen by volume.[§1910.146(b)]

Oxygen enriched atmosphere means an atmosphere containing more than 23.5 percent oxygen by volume.[§1910.146(b)]

Permit-required confined space (permit space) means a confined space that has one or more of the following characteristics: [§1910.146(b)]

(1) *Contains or has a potential to contain* a hazardous atmosphere;[§1910.146(b)(1)]

(2) *Contains a material that has* the potential for engulfing an entrant;[§1910.146(b)(2)]

(3) *Has an internal configuration* such that an entrant could be trapped or asphyxiated by inwardly converging walls or by a floor which slopes downward and tapers to a smaller cross-section; or[§1910.146(b)(3)]

(4) *Contains any other recognized serious* safety or health hazard. [§1910.146(b)(4)]

Permit-required confined space program (permit space program) means the employer's overall program for controlling, and, where appropriate, for protecting employees from, permit space hazards and for regulating employee entry into permit spaces.[§1910.146(b)]

Permit system means the employer's written procedure for preparing and issuing permits for entry and for returning the permit space to service following termination of entry.[§1910.146(b)]

Prohibited condition means any condition in a permit space that is not allowed by the permit during the period when entry is authorized.[§1910.146(b)]

Rescue service means the personnel designated to rescue employees from permit spaces.[§1910.146(b)]

Retrieval system means the equipment (including a retrieval line, chest or full-body harness, wristlets, if appropriate, and a lifting device or anchor) used for non-entry rescue of persons from permit spaces.[§1910.146(b)]

Testing means the process by which the hazards that may confront entrants of a permit space are identified and evaluated. Testing includes specifying the tests that are to be performed in the permit space.[§1910.146(b)]

Note: Testing enables employers both to devise and implement adequate control measures for the protection of authorized entrants and to determine if acceptable entry conditions are present immediately prior to, and during, entry.

(c) General requirements.[§1910.146(c)]

(1) *The employer shall evaluate the workplace* to determine if any spaces are permit-required confined spaces.[§1910.146(c)(1)]

Note: Proper application of the decision flow chart in appendix A to Sec. 1910.146 would facilitate compliance with this requirement.

(2) *If the workplace contains permit spaces,* the employer shall inform exposed employees, by posting danger signs or by any other equally effective means, of the existence and location of and the danger posed by the permit spaces.[§1910.146(c)(2)]

Note: A sign reading "DANGER — PERMIT-REQUIRED CONFINED SPACE, DO NOT ENTER" or using other similar language would satisfy the requirement for a sign.

(3) *If the employer decides* that its employees will not enter permit spaces, the employer shall take effective measures to prevent its employees from entering the permit spaces and shall comply with paragraphs (c)(1), (c)(2), (c)(6), and (c)(8) of this section. [§1910.146(c)(3)]

(4) *If the employer decides* that its employees will enter permit spaces, the employer shall develop and implement a written permit space program that complies with this section. The written program shall be available for inspection by employees and their authorized representatives.[§1910.146(c)(4)]

(5) *An employer may* use the alternate procedures specified in paragraph (c)(5)(ii) of this section for entering a permit space under the conditions set forth in paragraph (c)(5)(i) of this section.[§1910.146(c)(5)]

(i) *An employer whose employees* enter a permit space need not comply with paragraphs (d) through (f) and (h) through (k) of this section, provided that:[§1910.146(c)(5)(i)]

[A] The employer can demonstrate that the only hazard posed by the permit space is an actual or potential hazardous atmosphere;[§1910.146(c)(5)(i)[A]]

[B] The employer can demonstrate that continuous forced air ventilation alone is sufficient to maintain that permit space safe for entry;[§1910.146(c)(5)(i)[B]]

[C] The employer develops monitoring and inspection data that supports the demonstrations required by paragraphs (c)(5)(i)(A) and (c)(5)(i)(B) of this section; [§1910.146(c)(5)(i)[C]]

[D] If an initial entry of the permit space is necessary to obtain the data required by paragraph (c)(5)(i)(C) of this section, the entry is performed in compliance with paragraphs (d) through (k) of this section;[§1910.146(c)(5)(i)[D]]

[E] The determinations and supporting data required by paragraphs (c)(5)(i)(A), (c)(5)(i)(B), and (c)(5)(i)(C) of this section are documented by the employer and are made available to each employee who enters the permit space under the terms of paragraph (c)(5) of this section or to that employee's authorized representative; and [§1910.146(c)(5)(i)[E]]

[F] Entry into the permit space under the terms of paragraph (c)(5)(i) of this section is performed in accordance with the requirements of paragraph (c)(5)(ii) of this section. [§1910.146(c)(5)(i)[F]]

Note: See paragraph (c)(7) of this section for reclassification of a permit space after all hazards within the space have been eliminated.

(ii) *The following requirements apply* to entry into permit spaces that meet the conditions set forth in paragraph (c)(5)(i) of this section.**[§1910.146(c)(5)(ii)]**

[A] Any conditions making it unsafe to remove an entrance cover shall be eliminated before the cover is removed. **[§1910.146(c)(5)(ii)[A]]**

[B] When entrance covers are removed, the opening shall be promptly guarded by a railing, temporary cover, or other temporary barrier that will prevent an accidental fall through the opening and that will protect each employee working in the space from foreign objects entering the space.**[§1910.146(c)(5)(ii)[B]]**

[C] Before an employee enters the space, the internal atmosphere shall be tested, with a calibrated direct-reading instrument, for oxygen content, for flammable gases and vapors, and for potential toxic air contaminants, in that order. Any employee who enters the space, or that employee's authorized representative, shall be provided an opportunity to observe the pre-entry testing required by this paragraph.**[§1910.146(c)(5)(ii)[C]]**

[D] There may be no hazardous atmosphere within the space whenever any employee is inside the space. **[§1910.146(c)(5)(ii)[D]]**

[E] Continuous forced air ventilation shall be used, as follows: **[§1910.146(c)(5)(ii)[E]]**

[1] An employee may not enter the space until the forced air ventilation has eliminated any hazardous atmosphere;**[§1910.146(c)(5)(ii)[E][1]]**

[2] The forced air ventilation shall be so directed as to ventilate the immediate areas where an employee is or will be present within the space and shall continue until all employees have left the space;**[§1910.146(c)(5)(ii)[E][2]]**

[3] The air supply for the forced air ventilation shall be from a clean source and may not increase the hazards in the space.**[§1910.146(c)(5)(ii)[E][3]]**

[F] The atmosphere within the space shall be periodically tested as necessary to ensure that the continuous forced air ventilation is preventing the accumulation of a hazardous atmosphere. Any employee who enters the space, or that employee's authorized representative, shall be provided with an opportunity to observe the periodic testing required by this paragraph.**[§1910.146(c)(5)(ii)[F]]**

[G] If a hazardous atmosphere is detected during entry: **[§1910.146(c)(5)(ii)[G]]**

[1] Each employee shall leave the space immediately; **[§1910.146(c)(5)(ii)[G][1]]**

[2] The space shall be evaluated to determine how the hazardous atmosphere developed; and **[§1910.146(c)(5)(ii)[G][2]]**

[3] Measures shall be implemented to protect employees from the hazardous atmosphere before any subsequent entry takes place.**[§1910.146(c)(5)(ii)[G][3]]**

[H] The employer shall verify that the space is safe for entry and that the pre-entry measures required by paragraph (c)(5)(ii) of this section have been taken, through a written certification that contains the date, the location of the space, and the signature of the person providing the certification. The certification shall be made before entry and shall be made available to each employee entering the space or to that employee's authorized representative . **[§1910.146(c)(5)(ii)[H]]**

(6) *When there are changes* in the use or configuration of a non-permit confined space that might increase the hazards to entrants, the employer shall reevaluate that space and, if necessary, reclassify it as a permit-required confined space. **[§1910.146(c)(6)]**

(7) *A space classified by the employer* as a permit-required confined space may be reclassified as a non-permit confined space under the following procedures:**[§1910.146(c)(7)]**

(i) *If the permit space* poses no actual or potential atmospheric hazards and if all hazards within the space are eliminated without entry into the space, the permit space may be reclassified as a non-permit confined space for as long as the non-atmospheric hazards remain eliminated.**[§1910.146(c)(7)(i)]**

(ii) *If it is necessary to enter* the permit space to eliminate hazards, such entry shall be performed under paragraphs (d) through (k) of this section. If testing and inspection during that entry demonstrate that the hazards within the permit space have been eliminated, the permit space may be reclassified as a non-permit confined space for as long as the hazards remain eliminated.**[§1910.146(c)(7)(ii)]**

Note: Control of atmospheric hazards through forced air ventilation does not constitute elimination of the hazards. Paragraph (c)(5) covers permit space entry where the employer can demonstrate that forced air ventilation alone will control all hazards in the space.

(iii) *The employer shall document the basis for determining* that all hazards in a permit space have been eliminated, through a certification that contains the date, the location of the space, and the signature of the person making the determination. The certification shall be made available to each employee entering the space or to that employee's authorized representative.**[§1910.146(c)(7)(iii)]**

(iv) *If hazards arise within a permit space* that has been declassified to a non-permit space under paragraph (c)(7) of this section, each employee in the space shall exit the space. The employer shall then reevaluate the space and determine whether it must be reclassified as a permit space, in accordance with other applicable provisions of this section. **[§1910.146(c)(7)(iv)]**

(8) *When an employer (host employer)* arranges to have employees of another employer (contractor) perform work that involves permit space entry, the host employer shall:**[§1910.146(c)(8)]**

(i) *Inform the contractor* that the workplace contains permit spaces and that permit space entry is allowed only through compliance with a permit space program meeting the requirements of this section;**[§1910.146(c)(8)(i)]**

(ii) *Apprise the contractor of the elements,* including the hazards identified and the host employer's experience with the space, that make the space in question a permit space; **[§1910.146(c)(8)(ii)]**

(iii) *Apprise the contractor of any precautions* or procedures that the host employer has implemented for the protection of employees in or near permit spaces where contractor personnel will be working;**[§1910.146(c)(8)(iii)]**

(iv) *Coordinate entry operations* with the contractor, when both host employer personnel and contractor personnel will be working in or near permit spaces, as required by paragraph (d)(11) of this section; and**[§1910.146(c)(8)(iv)]**

(v) *Debrief the contractor at the conclusion* of the entry operations regarding the permit space program followed and regarding any hazards confronted or created in permit spaces during entry operations.**[§1910.146(c)(8)(v)]**

(9) *In addition to complying* with the permit space requirements that apply to all employers, each contractor who is retained to perform permit space entry operations shall:**[§1910.146(c)(9)]**

(i) *Obtain any available information* regarding permit space hazards and entry operations from the host employer; **[§1910.146(c)(9)(i)]**

(ii) *Coordinate entry operations with the host employer,* when both host employer personnel and contractor personnel will be working in or near permit spaces, as required by paragraph (d)(11) of this section; and**[§1910.146(c)(9)(ii)]**

(iii) *Inform the host employer* of the permit space program that the contractor will follow and of any hazards confronted or created in permit spaces, either through a debriefing or during the entry operation.**[§1910.146(c)(9)(iii)]**

(d) Permit-required confined space program (permit space program). Under the permit space program required by paragraph (c)(4) of this section, the employer shall:**[§1910.146(d)]**

(1) *Implement the measures necessary to prevent unauthorized entry;***[§1910.146(d)(1)]**

(2) *Identify and evaluate the hazards* of permit spaces before employees enter them;**[§1910.146(d)(2)]**

(3) *Develop and implement* the means, procedures, and practices necessary for safe permit space entry operations, including, but not limited to, the following:**[§1910.146(d)(3)]**

(i) *Specifying acceptable entry conditions;***[§1910.146(d)(3)(i)]**

(ii) *Providing each authorized entrant* or that employee's authorized representative with the opportunity to observe any monitoring or testing of permit spaces;**[§1910.146(d)(3)(ii)]**

(iii) *Isolating the permit space;***[§1910.146(d)(3)(iii)]**

(iv) *Purging, inerting, flushing, or ventilating* the permit space as necessary to eliminate or control atmospheric hazards; **[§1910.146(d)(3)(iv)]**

(v) *Providing pedestrian, vehicle, or other barriers* as necessary to protect entrants from external hazards; and [§1910.146(d)(3)(v)]

(vi) *Verifying that conditions* in the permit space are acceptable for entry throughout the duration of an authorized entry. [§1910.146(d)(3)(vi)]

(4) *Provide the following equipment* (specified in paragraphs (d)(4)(i) through (d)(4)(ix) of this section) at no cost to employees, maintain that equipment properly, and ensure that employees use that equipment properly:[§1910.146(d)(4)]

(i) *Testing and monitoring equipment* needed to comply with paragraph (d)(5) of this section;[§1910.146(d)(4)(i)]

(ii) *Ventilating equipment* needed to obtain acceptable entry conditions;[§1910.146(d)(4)(ii)]

(iii) *Communications equipment* necessary for compliance with paragraphs (h)(3) and (i)(5) of this section;[§1910.146(d)(4)(iii)]

(iv) *Personal protective equipment* insofar as feasible engineering and work practice controls do not adequately protect employees;[§1910.146(d)(4)(iv)]

(v) *Lighting equipment needed* to enable employees to see well enough to work safely and to exit the space quickly in an emergency;[§1910.146(d)(4)(v)]

(vi) *Barriers and shields as required* by paragraph (d)(3)(v) of this section.[§1910.146(d)(4)(vi)]

(vii) *Equipment, such as ladders,* needed for safe ingress and egress by authorized entrants;[§1910.146(d)(4)(vii)]

(viii) *Rescue and emergency equipment* needed to comply with paragraph (d)(9) of this section, except to the extent that the equipment is provided by rescue services; and [§1910.146(d)(4)(viii)]

(ix) *Any other equipment necessary* for safe entry into and rescue from permit spaces.[§1910.146(d)(4)(ix)]

(5) *Evaluate permit space conditions* as follows when entry operations are conducted:[§1910.146(d)(5)]

(i) *Test conditions in the permit space* to determine if acceptable entry conditions exist before entry is authorized to begin, except that, if isolation of the space is infeasible because the space is large or is part of a continuous system (such as a sewer), pre-entry testing shall be performed to the extent feasible before entry is authorized and, if entry is authorized, entry conditions shall be continuously monitored in the areas where authorized entrants are working; [§1910.146(d)(5)(i)]

(ii) *Test or monitor the permit space* as necessary to determine if acceptable entry conditions are being maintained during the course of entry operations; and[§1910.146(d)(5)(ii)]

(iii) *When testing for atmospheric hazards,* test first for oxygen, then for combustible gases and vapors, and then for toxic gases and vapors.[§1910.146(d)(5)(iii)]

(iv) *Provide each authorized entrant* or that employee's authorized representative an opportunity to observe the pre-entry and any subsequent testing or monitoring of permit spaces; [§1910.146(d)(5)(iv)]

(v) *Reevaluate the permit space* in the presence of any authorized entrant or that employee's authorized representative who requests that the employer conduct such reevaluation because the entrant or representative has reason to believe that the evaluation of that space may not have been adequate;[§1910.146(d)(5)(v)]

(vi) *Immediately provide each authorized entrant* or that employee's authorized representative with the results of any testing conducted in accord with paragraph (d) of this section.[§1910.146(d)(5)(vi)]

Note: Atmospheric testing conducted in accordance with appendix B to Sec. 1910.146 would be considered as satisfying the requirements of this paragraph. For permit space operations in sewers, atmospheric testing conducted in accordance with appendix B, as supplemented by appendix E to Sec. 1910.146, would be considered as satisfying the requirements of this paragraph.

(6) *Provide at least one attendant* outside the permit space into which entry is authorized for the duration of entry operations; [§1910.146(d)(6)]

Note: Attendants may be assigned to monitor more than one permit space provided the duties described in paragraph (i) of this section can be effectively performed for each permit space that is monitored. Likewise, attendants may be stationed at any location outside the permit space to be monitored as long as the duties described in paragraph (i) of this section can be effectively performed for each permit space that is monitored.

(7) *If multiple spaces are to be monitored* by a single attendant, include in the permit program the means and procedures to enable the attendant to respond to an emergency affecting one or more of the permit spaces being monitored without distraction from the attendant's responsibilities under paragraph (i) of this section;[§1910.146(d)(7)]

(8) *Designate the persons* who are to have active roles (as, for example, authorized entrants, attendants, entry supervisors, or persons who test or monitor the atmosphere in a permit space) in entry operations, identify the duties of each such employee, and provide each such employee with the training required by paragraph (g) of this section;[§1910.146(d)(8)]

(9) *Develop and implement procedures* for summoning rescue and emergency services, for rescuing entrants from permit spaces, for providing necessary emergency services to rescued employees, and for preventing unauthorized personnel from attempting a rescue;[§1910.146(d)(9)]

(10) *Develop and implement a system* for the preparation, issuance, use, and cancellation of entry permits as required by this section;[§1910.146(d)(10)]

(11) *Develop and implement procedures* to coordinate entry operations when employees of more than one employer are working simultaneously as authorized entrants in a permit space, so that employees of one employer do not endanger the employees of any other employer;[§1910.146(d)(11)]

(12) *Develop and implement procedures* (such as closing off a permit space and canceling the permit) necessary for concluding the entry after entry operations have been completed; [§1910.146(d)(12)]

(13) *Review entry operations* when the employer has reason to believe that the measures taken under the permit space program may not protect employees and revise the program to correct deficiencies found to exist before subsequent entries are authorized; and[§1910.146(d)(13)]

Note: Examples of circumstances requiring the review of the permit space program are: any unauthorized entry of a permit space, the detection of a permit space hazard not covered by the permit, the detection of a condition prohibited by the permit, the occurrence of an injury or near-miss during entry, a change in the use or configuration of a permit space, and employee complaints about the effectiveness of the program.

(14) *Review the permit space program,* using the canceled permits retained under paragraph (e)(6) of this section within 1 year after each entry and revise the program as necessary, to ensure that employees participating in entry operations are protected from permit space hazards.[§1910.146(d)(14)]

Note: Employers may perform a single annual review covering all entries performed during a 12-month period. If no entry is performed during a 12-month period, no review is necessary.

Appendix C to Sec. 1910.146 presents examples of permit space programs that are considered to comply with the requirements of paragraph (d) of this section.[§1910.146(d)(14)]

(e) **Permit system.**[§1910.146(e)]

(1) *Before entry is authorized,* the employer shall document the completion of measures required by paragraph (d)(3) of this section by preparing an entry permit.[§1910.146(e)(1)]

Note: Appendix D to Sec. 1910.146 presents examples of permits whose elements are considered to comply with the requirements of this section.

(2) *Before entry begins,* the entry supervisor identified on the permit shall sign the entry permit to authorize entry.[§1910.146(e)(2)]

(3) *The completed permit shall be made available* at the time of entry to all authorized entrants or their authorized representatives, by posting it at the entry portal or by any other equally effective means, so that the entrants can confirm that pre-entry preparations have been completed.[§1910.146(e)(3)]

(4) *The duration of the permit may not exceed* the time required to complete the assigned task or job identified on the permit in accordance with paragraph (f)(2) of this section.[§1910.146(e)(4)]

(5) *The entry supervisor shall terminate* entry and cancel the entry permit when:[§1910.146(e)(5)]

(i) *The entry operations* covered by the entry permit have been completed; or[§1910.146(e)(5)(i)]

(ii) *A condition that is not allowed under* the entry permit arises in or near the permit space.[§1910.146(e)(5)(ii)]

(6) *The employer shall retain* each canceled entry permit for at least 1 year to facilitate the review of the permit-required confined space program required by paragraph (d)(14) of this section. Any problems encountered during an entry operation shall be noted on the pertinent permit so that appropriate revisions to the permit space program can be made.[§1910.146(e)(6)]

(f) **Entry permit.** The entry permit that documents compliance with this section and authorizes entry to a permit space shall identify: [§1910.146(f)]

(1) *The permit space to be entered;*[§1910.146(f)(1)]

(2) *The purpose of the entry;*[§1910.146(f)(2)]

(3) *The date and the authorized duration of the entry permit;* [§1910.146(f)(3)]

(4) *The authorized entrants* within the permit space, by name or by such other means (for example, through the use of rosters or tracking systems) as will enable the attendant to determine quickly and accurately, for the duration of the permit, which authorized entrants are inside the permit space;[§1910.146(f)(4)]

Note: This requirement may be met by inserting a reference on the entry permit as to the means used, such as a roster or tracking system, to keep track of the authorized entrants within the permit space.

(5) *The personnel, by name, currently serving as attendants;* [§1910.146(f)(5)]

(6) *The individual, by name,* currently serving as entry supervisor, with a space for the signature or initials of the entry supervisor who originally authorized entry;[§1910.146(f)(6)]

(7) *The hazards of the permit space to be entered;*[§1910.146(f)(7)]

(8) *The measures used* to isolate the permit space and to eliminate or control permit space hazards before entry;[§1910.146(f)(8)]

Note: Those measures can include the lockout or tagging of equipment and procedures for purging, inerting, ventilating, and flushing permit spaces.

(9) *The acceptable entry conditions;*[§1910.146(f)(9)]

(10) *The results of initial and periodic tests* performed under paragraph (d)(5) of this section, accompanied by the names or initials of the testers and by an indication of when the tests were performed;[§1910.146(f)(10)]

(11) *The rescue and emergency services* that can be summoned and the means (such as the equipment to use and the numbers to call) for summoning those services;[§1910.146(f)(11)]

(12) *The communication procedures used* by authorized entrants and attendants to maintain contact during the entry; [§1910.146(f)(12)]

(13) *Equipment,* such as personal protective equipment, testing equipment, communications equipment, alarm systems, and rescue equipment, to be provided for compliance with this section;[§1910.146(f)(13)]

(14) *Any other information* whose inclusion is necessary, given the circumstances of the particular confined space, in order to ensure employee safety; and[§1910.146(f)(14)]

(15) *Any additional permits,* such as for hot work, that have been issued to authorize work in the permit space.[§1910.146(f)(15)]

(g) **Training.**[§1910.146(g)]

(1) *The employer shall provide training* so that all employees whose work is regulated by this section acquire the understanding, knowledge, and skills necessary for the safe performance of the duties assigned under this section.[§1910.146(g)(1)]

(2) *Training shall be provided to each affected employee:* [§1910.146(g)(2)]

(i) *Before the employee* is first assigned duties under this section;[§1910.146(g)(2)(i)]

(ii) *Before there is a change in assigned duties;*[§1910.146(g)(2)(ii)]

(iii) *Whenever there is a change* in permit space operations that presents a hazard about which an employee has not previously been trained;[§1910.146(g)(2)(iii)]

(iv) *Whenever the employer* has reason to believe either that there are deviations from the permit space entry procedures required by paragraph (d)(3) of this section or that there are inadequacies in the employee's knowledge or use of these procedures.[§1910.146(g)(2)(iv)]

(3) *The training shall establish* employee proficiency in the duties required by this section and shall introduce new or revised procedures, as necessary, for compliance with this section. [§1910.146(g)(3)]

(4) *The employer shall certify* that the training required by paragraphs (g)(1) through (g)(3) of this section has been accomplished. The certification shall contain each employee's name, the signatures or initials of the trainers, and the dates of training. The certification shall be available for inspection by employees and their authorized representatives.[§1910.146(g)(4)]

(h) **Duties of authorized entrants.** The employer shall ensure that all authorized entrants:[§1910.146(h)]

(1) *Know the hazards that may be faced* during entry, including information on the mode, signs or symptoms, and consequences of the exposure;[§1910.146(h)(1)]

(2) *Properly use equipment* as required by paragraph (d)(4) of this section;[§1910.146(h)(2)]

(3) *Communicate with the attendant as necessary* to enable the attendant to monitor entrant status and to enable the attendant to alert entrants of the need to evacuate the space as required by paragraph (i)(6) of this section;[§1910.146(h)(3)]

(4) *Alert the attendant whenever:*[§1910.146(h)(4)]

(i) *The entrant recognizes* any warning sign or symptom of exposure to a dangerous situation, or[§1910.146(h)(4)(i)]

(ii) *The entrant detects a prohibited condition; and* [§1910.146(h)(4)(ii)]

(5) *Exit from the permit space as quickly as possible whenever:* [§1910.146(h)(5)]

(i) *An order to evacuate* is given by the attendant or the entry supervisor,[§1910.146(h)(5)(i)]

(ii) *The entrant recognizes* any warning sign or symptom of exposure to a dangerous situation,[§1910.146(h)(5)(ii)]

(iii) *The entrant detects a prohibited condition, or* [§1910.146(h)(5)(iii)]

(iv) *An evacuation alarm is activated.*[§1910.146(h)(5)(iv)]

(i) **Duties of attendants.** The employer shall ensure that each attendant:[§1910.146(i)]

(1) *Knows the hazards that may be faced* during entry, including information on the mode, signs or symptoms, and consequences of the exposure;[§1910.146(i)(1)]

(2) *Is aware of possible behavioral effects* of hazard exposure in authorized entrants;[§1910.146(i)(2)]

(3) *Continuously maintains an accurate count* of authorized entrants in the permit space and ensures that the means used to identify authorized entrants under paragraph (f)(4) of this section accurately identifies who is in the permit space; [§1910.146(i)(3)]

(4) *Remains outside the permit space* during entry operations until relieved by another attendant;[§1910.146(i)(4)]

Note: When the employer's permit entry program allows attendant entry for rescue, attendants may enter a permit space to attempt a rescue if they have been trained and equipped for rescue operations as required by paragraph (k)(1) of this section and if they have been relieved as required by paragraph (i)(4) of this section.

(5) *Communicates with authorized entrants* as necessary to monitor entrant status and to alert entrants of the need to evacuate the space under paragraph (i)(6) of this section;[§1910.146(i)(5)]

(6) *Monitors activities inside and outside* the space to determine if it is safe for entrants to remain in the space and orders the authorized entrants to evacuate the permit space immediately under any of the following conditions;[§1910.146(i)(6)]

(i) *If the attendant detects a prohibited condition;*[§1910.146(i)(6)(i)]

(ii) *If the attendant detects* the behavioral effects of hazard exposure in an authorized entrant;[§1910.146(i)(6)(ii)]

(iii) *If the attendant detects* a situation outside the space that could endanger the authorized entrants; or[§1910.146(i)(6)(iii)]

(iv) *If the attendant cannot* effectively and safely perform all the duties required under paragraph (i) of this section; [§1910.146(i)(6)(iv)]

(7) *Summon rescue and other emergency* services as soon as the attendant determines that authorized entrants may need assistance to escape from permit space hazards;[§1910.146(i)(7)]

(8) *Takes the following actions* when unauthorized persons approach or enter a permit space while entry is underway: [§1910.146(i)(8)]

(i) *Warn the unauthorized persons* that they must stay away from the permit space;[§1910.146(i)(8)(i)]

(ii) *Advise the unauthorized persons* that they must exit immediately if they have entered the permit space; and [§1910.146(i)(8)(ii)]

(iii) *Inform the authorized entrants* and the entry supervisor if unauthorized persons have entered the permit space; [§1910.146(i)(8)(iii)]

(9) *Performs non-entry rescues* as specified by the employer's rescue procedure; and[§1910.146(i)(9)]

(10) *Performs no duties* that might interfere with the attendant's primary duty to monitor and protect the authorized entrants. [§1910.146(i)(10)]

(j) **Duties of entry supervisors.** The employer shall ensure that each entry supervisor:[§1910.146(j)]

(1) *Knows the hazards* that may be faced during entry, including information on the mode, signs or symptoms, and consequences of the exposure;[§1910.146(j)(1)]

(2) *Verifies,* by checking that the appropriate entries have been made on the permit, that all tests specified by the permit have been conducted and that all procedures and equipment speci-

fied by the permit are in place before endorsing the permit and allowing entry to begin;[§1910.146(j)(2)]

(3) *Terminates the entry and cancels the permit* as required by paragraph (e)(5) of this section;[§1910.146(j)(3)]

(4) *Verifies that rescue services* are available and that the means for summoning them are operable;[§1910.146(j)(4)]

(5) *Removes unauthorized individuals* who enter or who attempt to enter the permit space during entry operations; and [§1910.146(j)(5)]

(6) *Determines,* whenever responsibility for a permit space entry operation is transferred and at intervals dictated by the hazards and operations performed within the space, that entry operations remain consistent with terms of the entry permit and that acceptable entry conditions are maintained.[§1910.146(j)(6)]

(k) Rescue and emergency services.[§1910.146(k)]

(1) *An employer who designates* rescue and emergency services, pursuant to paragraph (d)(9) of this section, shall:[§1910.146(k)(1)]

(i) *Evaluate a prospective rescuer's ability* to respond to a rescue summons in a timely manner, considering the hazard(s) identified;[§1910.146(k)(1)(i)]

Note to paragraph (k)(1)(i): What will be considered timely will vary according to the specific hazards involved in each entry. For example, Sec. 1910.134, Respiratory Protection, requires that employers provide a standby person or persons capable of immediate action to rescue employee(s) wearing respiratory protection while in work areas defined as IDLH atmospheres.

(ii) *Evaluate a prospective rescue service's ability,* in terms of proficiency with rescue-related tasks and equipment, to function appropriately while rescuing entrants from the particular permit space or types of permit spaces identified; [§1910.146(k)(1)(ii)]

(iii) *Select a rescue team or service from those evaluated that:* [§1910.146(k)(1)(iii)]

[A] Has the capability to reach the victim(s) within a time frame that is appropriate for the permit space hazard(s) identified;[§1910.146(k)(1)(iii)[A]]

[B] Is equipped for and proficient in performing the needed rescue services;[§1910.146(k)(1)(iii)[B]]

(iv) *Inform each rescue team or service* of the hazards they may confront when called on to perform rescue at the site; and [§1910.146(k)(1)(iv)]

(v) *Provide the rescue team or service* selected with access to all permit spaces from which rescue may be necessary so that the rescue service can develop appropriate rescue plans and practice rescue operations.[§1910.146(k)(1)(v)]

Note to paragraph (k)(1): Non-mandatory appendix F contains examples of criteria which employers can use in evaluating prospective rescuers as required by paragraph (k)(1) of this section.

(2) *An employer whose employees* have been designated to provide permit space rescue and emergency services shall take the following measures:[§1910.146(k)(2)]

(i) *Provide affected employees* with the personal protective equipment (PPE) needed to conduct permit space rescues safely and train affected employees so they are proficient in the use of that PPE, at no cost to those employees; [§1910.146(k)(2)(i)]

(ii) *Train affected employees to perform* assigned rescue duties. The employer must ensure that such employees successfully complete the training required to establish proficiency as an authorized entrant, as provided by paragraphs (g) and (h) of this section;[§1910.146(k)(2)(ii)]

(iii) *Train affected employees* in basic first-aid and cardiopulmonary resuscitation (CPR). The employer shall ensure that at least one member of the rescue team or service holding a current certification in first aid and CPR is available; and [§1910.146(k)(2)(iii)]

(iv) *Ensure that affected employees* practice making permit space rescues at least once every 12 months, by means of simulated rescue operations in which they remove dummies, manikins, or actual persons from the actual permit spaces or from representative permit spaces. Representative permit spaces shall, with respect to opening size, configuration, and accessibility, simulate the types of permit spaces from which rescue is to be performed.[§1910.146(k)(2)(iv)]

(3) *To facilitate non-entry rescue,* retrieval systems or methods shall be used whenever an authorized entrant enters a permit space, unless the retrieval equipment would increase the overall risk of entry or would not contribute to the rescue of the entrant. Retrieval systems shall meet the following requirements.[§1910.146(k)(3)]

(i) *Each authorized entrant shall use* a chest or full body harness, with a retrieval line attached at the center of the entrant's back near shoulder level, above the entrant's head, or at another point which the employer can establish presents a profile small enough for the successful removal of the entrant. Wristlets may be used in lieu of the chest or full body harness if the employer can demonstrate that the use of a chest or full body harness is infeasible or creates a greater hazard and that the use of wristlets is the safest and most effective alternative.[§1910.146(k)(3)(i)]

(ii) *The other end of the retrieval line* shall be attached to a mechanical device or fixed point outside the permit space in such a manner that rescue can begin as soon as the rescuer becomes aware that rescue is necessary. A mechanical device shall be available to retrieve personnel from vertical type permit spaces more than 5 feet (1.52 m) deep. [§1910.146(k)(3)(ii)]

(4) *If an injured entrant is exposed* to a substance for which a Material Safety Data Sheet (MSDS) or other similar written information is required to be kept at the worksite, that MSDS or written information shall be made available to the medical facility treating the exposed entrant.[§1910.146(k)(4)]

(l) Employee participation.[§1910.146(l)]

(1) *Employers shall consult* with affected employees and their authorized representatives on the development and implementation of all aspects of the permit space program required by paragraph (c) of this section.[§1910.146(l)(1)]

(2) *Employers shall make available* to affected employees and their authorized representatives all information required to be developed by this section.[§1910.146(l)(2)]

Appendixes to §1910.146

Permit-required Confined Spaces

Note: Appendixes A through F serve to provide information and non-mandatory guidelines to assist employers and employees in complying with the appropriate requirements of this section.

§1910.146 Appendix A
Permit-Required Confined Space Decision Flow Chart

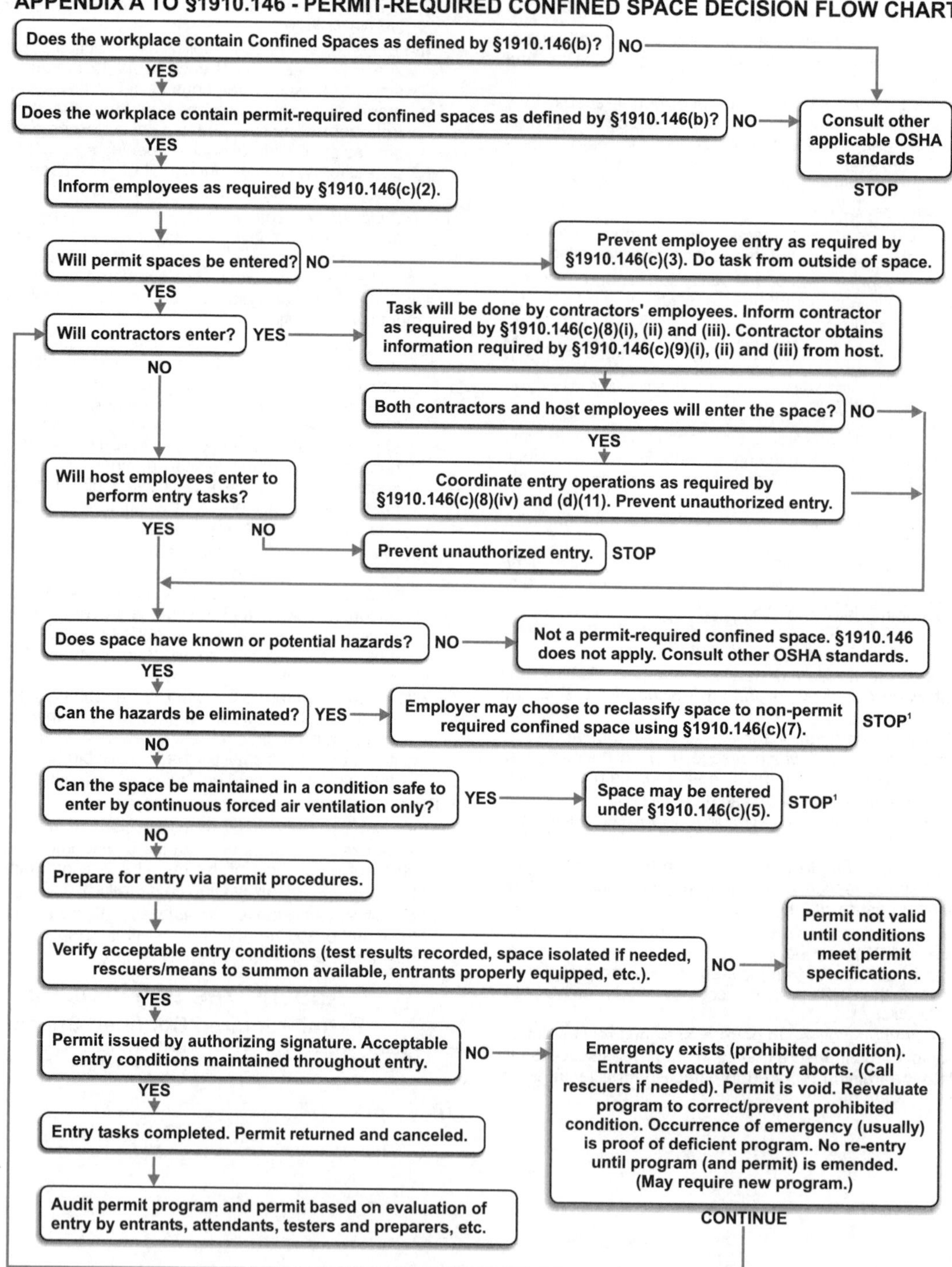

§1910.146 Appendix B
Procedures for Atmospheric Testing

Atmospheric testing is required for two distinct purposes: evaluation of the hazards of the permit space and verification that acceptable entry conditions for entry into that space exist.

(1) **Evaluation testing.** The atmosphere of a confined space should be analyzed using equipment of sufficient sensitivity and specificity to identify and evaluate any hazardous atmospheres that may exist or arise, so that appropriate permit entry procedures can be developed and acceptable entry conditions stipulated for that space. Evaluation and interpretation of these data, and development of the entry procedure, should be done by, or reviewed by, a technically qualified professional (e.g., OSHA consultation service, or certified industrial hygienist, registered safety engineer, certified safety professional, certified marine chemist, etc.) based on evaluation of all serious hazards.

(2) **Verification testing.** The atmosphere of a permit space which may contain a hazardous atmosphere should be tested for residues of all contaminants identified by evaluation testing using permit specified equipment to determine that residual concentrations at the time of testing and entry are within the range of acceptable entry conditions. Results of testing (i.e., actual concentration, etc.) should be recorded on the permit in the space provided adjacent to the stipulated acceptable entry condition.

(3) **Duration of testing.** Measurement of values for each atmospheric parameter should be made for at least the minimum response time of the test instrument specified by the manufacturer.

(4) **Testing stratified atmospheres.** When monitoring for entries involving a descent into atmospheres that may be stratified, the atmospheric envelope should be tested a distance of approximately 4 feet (1.22 m) in the direction of travel and to each side. If a sampling probe is used, the entrant's rate of progress should be slowed to accommodate the sampling speed and detector response.

(5) **Order of testing.** A test for oxygen is performed first because most combustible gas meters are oxygen dependent and will not provide reliable readings in an oxygen deficient atmosphere. Combustible gasses are tested for next because the threat of fire or explosion is both more immediate and more life threatening, in most cases, than exposure to toxic gasses and vapors. If tests for toxic gasses and vapors are necessary, they are performed last.

§1910.146 Appendix C
Examples of Permit-required Confined Space Programs

Example 1.

Workplace. Sewer entry.

Potential hazards. The employees could be exposed to the following:

Engulfment.

Presence of toxic gases. Equal to or more than 10 ppm hydrogen sulfide measured as an 8-hour time-weighted average. If the presence of other toxic contaminants is suspected, specific monitoring programs will be developed.

Presence of explosive/flammable gases. Equal to or greater than 10% of the lower flammable limit (LFL).

Oxygen Deficiency. A concentration of oxygen in the atmosphere equal to or less than 19.5% by volume.

A. Entry Without Permit/Attendant.

Certification. Confined spaces may be entered without the need for a written permit or attendant provided that the space can be maintained in a safe condition for entry by mechanical ventilation alone, as provided in §1910.146(c)(5). All spaces shall be considered permit-required confined spaces until the pre-entry procedures demonstrate otherwise. Any employee required or permitted to pre-check or enter an enclosed/confined space shall have successfully completed, as a minimum, the training as required by the following sections of these procedures. A written copy of operating and rescue procedures as required by these procedures shall be at the work site for the duration of the job. The Confined Space Pre-Entry Check List must be completed by the LEAD WORKER before entry into a confined space. This list verifies completion of items listed below. This check list shall be kept at the job site for duration of the job. If circumstances dictate an interruption in the work, the permit space must be re-evaluated and a new check list must be completed.

Control of atmospheric and engulfment hazards.

Pumps and Lines. All pumps and lines which may reasonably cause contaminants to flow into the space shall be disconnected, blinded and locked out, or effectively isolated by other means to prevent development of dangerous air contamination or engulfment. Not all laterals to sewers or storm drains require blocking. However, where experience or knowledge of industrial use indicates there is a reasonable potential for contamination of air or engulfment into an occupied sewer, then all affected laterals shall be blocked. If blocking and/or isolation requires entry into the space the provisions for entry into a permit-required confined space must be implemented.

Surveillance. The surrounding area shall be surveyed to avoid hazards such as drifting vapors from the tanks, piping, or sewers.

Testing. The atmosphere within the space will be tested to determine whether dangerous air contamination and/or oxygen deficiency exists. Detector tubes, alarm only gas monitors and explosion meters are examples of monitoring equipment that may be used to test permit space atmospheres. Testing shall be performed by the LEAD WORKER who has successfully completed the Gas Detector training for the monitor he will use. The minimum parameters to be monitored are oxygen deficiency, LFL, and hydrogen sulfide concentration. A written record of the pre-entry test results shall be made and kept at the work site for the duration of the job. The supervisor will certify in writing, based upon the results of the pre-entry testing, that all hazards have been eliminated. Affected employees shall be able to review the testing results. The most hazardous conditions shall govern when work is being performed in two adjoining, connecting spaces.

Entry Procedures. If there are no non-atmospheric hazards present and if the pre-entry tests show there is no dangerous air contamination and/or oxygen deficiency within the space and there is no reason to believe that any is likely to develop, entry into and work within may proceed. Continuous testing of the atmosphere in the immediate vicinity of the workers within the space shall be accomplished. The workers will immediately leave the permit space when any of the gas monitor alarm set points are reached as defined. Workers will not return to the area until a SUPERVISOR who has completed the gas detector training has used a direct reading gas detector to evaluate the situation and has determined that it is safe to enter.

Rescue. Arrangements for rescue services are not required where there is no attendant. See the rescue portion of section B., below, for instructions regarding rescue planning where an entry permit is required.

B. Entry Permit Required.

Permits. Confined Space Entry Permit. All spaces shall be considered permit-required confined spaces until the pre-entry procedures demonstrate otherwise. Any employee required or permitted to pre-check or enter a permit-required confined space shall have successfully completed, as a minimum, the training as required by the following sections of these procedures. A written copy of operating and rescue procedures as required by these procedures shall be at the work site for the duration of the job. The Confined Space Entry Permit must be completed before approval can be given to enter a permit-required confined space. This permit verifies completion of items listed below. This permit shall be kept at the job site for the duration of the job. If circumstances cause an interruption in the work or a change in the alarm conditions for which entry was approved, a new Confined Space Entry Permit must be completed.

Control of atmospheric and engulfment hazards.

Surveillance. The surrounding area shall be surveyed to avoid hazards such as drifting vapors from tanks, piping or sewers.

Testing. The confined space atmosphere shall be tested to determine whether dangerous air contamination and/or oxygen deficiency exists. A direct reading gas monitor shall be used. Testing shall be performed by the SUPERVISOR who has successfully completed the gas detector training for the monitor he will use. The minimum parameters to be monitored are oxygen deficiency, LFL and hydrogen sulfide concentration. A written record of the pre-entry test results shall be made and kept at the work site for the duration of the job. Affected employees shall be able to review the testing results. The most hazardous conditions shall govern when work is being performed in two adjoining, connected spaces.

Space Ventilation. Mechanical ventilation systems, where applicable, shall be set at 100% outside air. Where possible, open additional manholes to increase air circulation. Use portable blowers to augment natural circulation if needed. After a suitable ventilating period, repeat the testing. Entry may not begin until testing has demonstrated that the hazardous atmosphere has been eliminated.

Note: This partial selection of Part 1910 "General Industry Standards" is intended for convenience only; some parts may not be applicable to "Construction Standards." A complete Part 1910 CFR, similar to this one, is also available.

Entry Procedures. The following procedure shall be observed under any of the following conditions:

(1) *Testing demonstrates the existence* of dangerous or deficient conditions and additional ventilation cannot reduce concentrations to safe levels;

(2) *The atmosphere tests as safe* but unsafe conditions can reasonably be expected to develop;

(3) *It is not feasible to provide for ready exit* from spaces equipped with automatic fire suppression systems and it is not practical or safe to deactivate such systems; or

(4) *An emergency exists and it is not feasible* to wait for pre-entry procedures to take effect.

All personnel must be trained. A self contained breathing apparatus shall be worn by any person entering the space. At least one worker shall stand by the outside of the space ready to give assistance in case of emergency. The standby worker shall have a self contained breathing apparatus available for immediate use. There shall be at least one additional worker within sight or call of the standby worker. Continuous powered communications shall be maintained between the worker within the confined space and standby personnel.

If at any time there is any questionable action or non-movement by the worker inside, a verbal check will be made. If there is no response, the worker will be moved immediately. Exception: If the worker is disabled due to falling or impact, he/she shall not be removed from the confined space unless there is immediate danger to his/her life. Local fire department rescue personnel shall be notified immediately. The standby worker may only enter the confined space in case of an emergency (wearing the self contained breathing apparatus) and only after being relieved by another worker. Safety belt or harness with attached lifeline shall be used by all workers entering the space with the free end of the line secured outside the entry opening. The standby worker shall attempt to remove a disabled worker via his lifeline before entering the space.

When practical, these spaces shall be entered through side openings— those within 3 1/2 feet (1.07 m) of the bottom. When entry must be through a top opening, the safety belt shall be of the harness type that suspends a person upright and a hoisting device or similar apparatus shall be available for lifting workers out of the space.

In any situation where their use may endanger the worker, use of a hoisting device or safety belt and attached lifeline may be discontinued. When dangerous air contamination is attributable to flammable and/or explosive substances, lighting and electrical equipment shall be Class 1, Division 1 rated per National Electrical Code and no ignition sources shall be introduced into the area.

Continuous gas monitoring shall be performed during all confined space operations. If alarm conditions change adversely, entry personnel shall exit the confined space and a new confined space permit issued.

Rescue. Call the fire department services for rescue. Where immediate hazards to injured personnel are present, workers at the site shall implement emergency procedures to fit the situation.

Example 2.

Workplace. Meat and poultry rendering plants.

Cookers and dryers are either batch or continuous in their operation. Multiple batch cookers are operated in parallel. When one unit of a multiple set is shut down for repairs, means are available to isolate that unit from the others which remain in operation.

Cookers and dryers are horizontal, cylindrical vessels equipped with a center, rotating shaft and agitator paddles or discs. If the inner shell is jacketed, it is usually heated with steam at pressures up to 150 psig (1034.25 kPa). The rotating shaft assembly of the continuous cooker or dryer is also steam heated.

Potential Hazards. The recognized hazards associated with cookers and dryers are the risk that employees could be:

1. *Struck or caught by rotating agitator;*
2. *Engulfed in raw material or hot, recycled fat;*
3. *Burned by steam* from leaks into the cooker/dryer steam jacket or the condenser duct system if steam valves are not properly closed and locked out;
4. *Burned by contact* with hot metal surfaces, such as the agitator shaft assembly, or inner shell of the cooker/dryer;
5. *Heat stress caused by warm atmosphere inside cooker/dryer;*
6. *Slipping and falling on grease in the cooker/dryer;*
7. *Electrically shocked by faulty equipment taken into the cooker/dryer;*
8. *Burned or overcome by fire or products of combustion; or*
9. *Overcome by fumes* generated by welding or cutting done on grease covered surfaces.

Permits. The supervisor in this case is always present at the cooker/dryer or other permit entry confined space when entry is made. The supervisor must follow the pre-entry isolation procedures described in the entry permit in preparing for entry, and ensure that the protective clothing, ventilating equipment and any other equipment required by the permit are at the entry site.

Control of hazards.

Mechanical. Lock out main power switch to agitator motor at main power panel. Affix tag to the lock to inform others that a permit entry confined space entry is in progress.

Engulfment. Close all valves in the raw material blow line. Secure each valve in its closed position using chain and lock. Attach a tag to the valve and chain warning that a permit entry confined space entry is in progress. The same procedure shall be used for securing the fat recycle valve.

Burns and heat stress. Close steam supply valves to jacket and secure with chains and tags. Insert solid blank at flange in cooker vent line to condenser manifold duct system. Vent cooker/dryer by opening access door at discharge end and top center door to allow natural ventilation throughout the entry. If faster cooling is needed, use a portable ventilation fan to increase ventilation. Cooling water may be circulated through the jacket to reduce both outer and inner surface temperatures of cooker/dryers faster. Check air and inner surface temperatures in cooker/dryer to assure they are within acceptable limits before entering, or use proper protective clothing.

Fire and fume hazards. Careful site preparation, such as cleaning the area within 4 inches (10.16 cm) of all welding or torch cutting operations, and proper ventilation are the preferred controls. All welding and cutting operations shall be done in accordance with the requirements of 29 CFR Part 1910, Subpart Q, OSHA's welding standard. Proper ventilation may be achieved by local exhaust ventilation, or the use of portable ventilation fans, or a combination of the two practices.

Electrical shock. Electrical equipment used in cooker/dryers shall be in serviceable condition.

Slips and falls. Remove residual grease before entering cooker/dryer.

Attendant. The supervisor shall be the attendant for employees entering cooker/dryers.

Permit. The permit shall specify how isolation shall be done and any other preparations needed before making entry. This is especially important in parallel arrangements of cooker/dryers so that the entire operation need not be shut down to allow safe entry into one unit.

Rescue. When necessary, the attendant shall call the fire department as previously arranged.

Example 3.

Workplace. Workplaces where tank cars, trucks, and trailers, dry bulk tanks and trailers, railroad tank cars, and similar portable tanks are fabricated or serviced.

A. During fabrication. These tanks and dry-bulk carriers are entered repeatedly throughout the fabrication process. These products are not configured identically, but the manufacturing processes by which they are made are very similar.

Sources of hazards. In addition to the mechanical hazards arising from the risks that an entrant would be injured due to contact with components of the tank or the tools being used, there is also the risk that a worker could be injured by breathing fumes from welding materials or mists or vapors from materials used to coat the tank interior. In addition, many of these vapors and mists are flammable, so the failure to properly ventilate a tank could lead to a fire or explosion.

Control of hazards.

Welding. Local exhaust ventilation shall be used to remove welding fumes once the tank or carrier is completed to the point that workers may enter and exit only through a manhole. (Follow the requirements of 29 CFR 1910, Subpart Q, OSHA's welding standard, at all times.) Welding gas tanks may never be brought into a tank or carrier that is a permit entry confined space.

Application of interior coatings/linings. Atmospheric hazards shall be controlled by forced air ventilation sufficient to keep the atmospheric concentration of flammable materials below 10% of the lower flammable limit (LFL) (or lower explosive limit (LEL), whichever term is used locally). The appropriate respirators are provided and shall be used in addition to providing forced ventilation if the forced ventilation does not maintain acceptable respiratory conditions.

Permits. Because of the repetitive nature of the entries in these operations, an "Area Entry Permit" will be issued for a 1 month period to cover those production areas where tanks are fabricated to the point that entry and exit are made using manholes.

Authorization. Only the area supervisor may authorize an employee to enter a tank within the permit area. The area supervisor must determine that conditions in the tank trailer, dry bulk trailer or truck, etc. meet permit requirements before authorizing entry.

Attendant. The area supervisor shall designate an employee to maintain communication by employer specified means with employees working in tanks to ensure their safety. The attendant may not enter any permit entry confined space to rescue an entrant or for any other reason, unless authorized by the rescue procedure and, and even then, only after calling the rescue team and being relieved by as attendant by another worker.

Communications and observation. Communications between attendant and entrant(s) shall be maintained throughout entry. Methods of communication that may be specified by the permit include voice, voice powered radio, tapping or rapping codes on tank walls, signalling tugs on a rope, and the attendant's observation that work activities such as chipping, grinding, welding, spraying, etc., which require deliberate operator control continue normally. These activities often generate so much noise that the necessary hearing protection makes communication by voice difficult.

Rescue procedures. Acceptable rescue procedures include entry by a team of employee-rescuers, use of public emergency services, and procedures for breaching the tank. The area permit specifies which procedures are available, but the area supervisor makes the final decision based on circumstances. (Certain injuries may make it necessary to breach the tank to remove a person rather than risk additional injury by removal through an existing manhole. However, the supervisor must ensure that no breaching procedure used for rescue would violate terms of the entry permit. For instance, if the tank must be breached by cutting with a torch, the tank surfaces to be cut must be free of volatile or combustible coatings within 4 inches (10.16 cm) of the cutting line and the atmosphere within the tank must be below the LFL.

Retrieval line and harnesses. The retrieval lines and harnesses generally required under this standard are usually impractical for use in tanks because the internal configuration of the tanks and their interior baffles and other structures would prevent rescuers from hauling out injured entrants. However, unless the rescue procedure calls for breaching the tank for rescue, the rescue team shall be trained in the use of retrieval lines and harnesses for removing injured employees through manholes.

B. Repair or service of "used" tanks and bulk trailers.

Sources of hazards. In addition to facing the potential hazards encountered in fabrication or manufacturing, tanks or trailers which have been in service may contain residues of dangerous materials, whether left over from the transportation of hazardous cargoes or generated by chemical or bacterial action on residues of non-hazardous cargoes.

Control of atmospheric hazards. A "used" tank shall be brought into areas where tank entry is authorized only after the tank has been emptied, cleansed (without employee entry) of any residues, and purged of any potential atmospheric hazards.

Welding. In addition to tank cleaning for control of atmospheric hazards, coating and surface materials shall be removed 4 inches (10.16 cm) or more from any surface area where welding or other torch work will be done and care taken that the atmosphere within the tank remains well below the LFL. (Follow the requirements of 29 CFR 1910, Subpart Q, OSHA's welding standard, at all times.)

Permits. An entry permit valid for up to 1 year shall be issued prior to authorization of entry into used tank trailers, dry bulk trailers or trucks. In addition to the pre-entry cleaning requirement, this permit shall require the employee safeguards specified for new tank fabrication or construction permit areas.

Authorization. Only the area supervisor may authorize an employee to enter a tank trailer, dry bulk trailer or truck within the permit area. The area supervisor must determine that the entry permit requirements have been met before authorizing entry.

§1910.146 Appendix D

Sample Permits

Appendix D-1 to §1910.146 - Confined Space Entry Permit

Confined Space Entry Permit:

____/____/____ Date Issued ____:____ ☐ a.m. ☐ p.m. Time Issued ____/____/____ Date Expires ____:____ ☐ a.m. ☐ p.m. Time Expires

Job Site / Space I.D.

Job Supervisor

Equipment To Be Worked On

Work To Be Performed

Stand-By Personnel

1. **Atmospheric Checks:**
   Time: ____:____ ☐ a.m. ☐ p.m.
   Oxygen: ____% Explosive: ____% L.F.L. Toxic: ____PPM
2. Tester's Signature: ____
3. **Source Isolation (No Entry):** N/A Yes No
   Pumps or lines blinded, disconnected, or blocked ☐ ☐ ☐
4. **Ventilation Modification:** N/A Yes No
   Mechanical ☐ ☐ ☐
   Natural Ventilation Only ☐ ☐ ☐
5. **Atmospheric Check After Isolation And Ventilation:**
   Oxygen: ____% > 19.5%
   Explosive: ____% L.F.L. < 10%
   Toxic: ____PPM < 10 PPM H_2S
   Time: ____:____ ☐ a.m. ☐ p.m.
   Tester's Signature: ____
6. **Communication Procedures:**
7. **Rescue Procedures:**
8. **Entry, Standby, and Back Up Persons:** Yes No
   Successfully Completed Required Training? ☐ ☐
   Is It Current? ☐ ☐
9. **Equipment** N/A Yes No
   Direct Reading Gas Monitor - Tested ☐ ☐ ☐
   Safety Harnesses and Lifelines for Entry and Standby Persons ☐ ☐ ☐
   Hoisting Equipment ☐ ☐ ☐
   Powered Communications ☐ ☐ ☐
   SCBA's for Entry and Standby Persons ☐ ☐ ☐
   Protective Clothing ☐ ☐ ☐
   All Electric Equipment Listed
   Class I, Division I, Group D and Non-sparking Tools ☐ ☐ ☐
10. **Periodic Atmospheric Tests:**

Oxygen: ____%	Time: ____:____ ☐ a.m. ☐ p.m.	Oxygen: ____%	Time: ____:____ ☐ a.m. ☐ p.m.
Oxygen: ____%	Time: ____:____ ☐ a.m. ☐ p.m.	Oxygen: ____%	Time: ____:____ ☐ a.m. ☐ p.m.
Explosive: ____%	Time: ____:____ ☐ a.m. ☐ p.m.	Explosive: ____%	Time: ____:____ ☐ a.m. ☐ p.m.
Explosive: ____%	Time: ____:____ ☐ a.m. ☐ p.m.	Explosive: ____%	Time: ____:____ ☐ a.m. ☐ p.m.
Toxic: ____%	Time: ____:____ ☐ a.m. ☐ p.m.	Toxic: ____%	Time: ____:____ ☐ a.m. ☐ p.m.
Toxic: ____%	Time: ____:____ ☐ a.m. ☐ p.m.	Toxic: ____%	Time: ____:____ ☐ a.m. ☐ p.m.

We have reviewed the work authorized by this permit and the information contained herein. Written instructions and safety procedures have been received and are understood. Entry cannot be approved if any squares are marked in the "No" column.

This permit is not valid unless all appropriate items are completed.

Permit Prepared By: (Supervisor) ____ ____

Approved By: (Unit Supervisor) ____ ____

Reviewed By: (Cs Operations Personnel) ____ ____

(Printed Name) (Signature)

This permit to be kept at job site. Return job site copy to Safety Office following job completion.

Copies: White Original (Safety Office) Yellow (Unit Supervisor) Hard (Job Site)

*Full size forms available free of charge at www.oshacfr.com

Appendix D-2 to §1910.146 - Entry Permit

Appendix D-2:
PERMIT VALID FOR 8 HOURS ONLY. ALL PERMIT COPIES REMAIN AT SITE UNTIL JOB COMPLETED.

____/____/____ DATE SITE LOCATION/DESCRIPTION ____

PURPOSE OF ENTRY

SUPERVISOR(S) in charge of crews Type of crew Phone #

COMMUNICATION PROCEDURES

RESCUE PROCEDURES (PHONE NUMBERS AT BOTTOM)

BOLD DENOTES MINIMUM REQUIREMENTS TO BE COMPLETED AND REVIEWED PRIOR TO ENTRY

REQUIREMENTS COMPLETED	DATE	TIME
Lock Out/De-energize/Try-out	____/____/____	____:____ ☐ a.m. ☐ p.m.
Line(s) Broken-Capped-Blank	____/____/____	____:____ ☐ a.m. ☐ p.m.
Purge-Flush and Vent	____/____/____	____:____ ☐ a.m. ☐ p.m.
Ventilation	____/____/____	____:____ ☐ a.m. ☐ p.m.
Secure Area (Post and Flag)	____/____/____	____:____ ☐ a.m. ☐ p.m.
Breathing Apparatus	____/____/____	____:____ ☐ a.m. ☐ p.m.
Resuscitator - Inhalator	____/____/____	____:____ ☐ a.m. ☐ p.m.
Standby Safety Personnel	____/____/____	____:____ ☐ a.m. ☐ p.m.
Full Body Harness w/ "D" ring	____/____/____	____:____ ☐ a.m. ☐ p.m.
Emergency Escape Retrieval Equipment	____/____/____	____:____ ☐ a.m. ☐ p.m.
Lifelines	____/____/____	____:____ ☐ a.m. ☐ p.m.
Fire Extinguishers	____/____/____	____:____ ☐ a.m. ☐ p.m.
Lighting (Explosive Proof)	____/____/____	____:____ ☐ a.m. ☐ p.m.
Protective Clothing	____/____/____	____:____ ☐ a.m. ☐ p.m.
Respirator(s) (Air Purifying)	____/____/____	____:____ ☐ a.m. ☐ p.m.
Burning and Welding Permit	____/____/____	____:____ ☐ a.m. ☐ p.m.

Note: Items that do not apply enter N/A in the blank.

RECORD CONTINUOUS MONITORING RESULTS EVERY 2 HOURS

CONTINUOUS MONITORING** TEST(S) TO BE TAKEN	Permissible Entry Level
PERCENT OF OXYGEN	19.5% TO 23.5%
LOWER FLAMMABLE LIMIT	Under 10%
CARBON MONOXIDE	+35 PPM
Aromatic Hydrocarbon	+ 1 PPM * 5 PPM
Hydrogen Cyanide	(Skin) * 4 PPM
Hydrogen Sulfide	+ 10 PPM * 15 PPM
Sulfur Dioxide	+ 2 PPM * 5 PPM
Ammonia	* 35 PPM

* Short-term exposure limit: Employee can work in the area up to 15 minutes.
\+ 8 hr. Time Weighted Avg.: Employee can work in area 8 hrs (longer with appropriate respiratory protection).

REMARKS:

GAS TESTER NAME & CHECK # INSTRUMENT(S) USED MODEL &/OR TYPE SERIAL &/OR UNIT #

SAFETY STANDBY PERSON IS REQUIRED FOR ALL CONFINED SPACE WORK

SAFETY STANDBY PERSON(S) CHECK # CONFINED SPACE ENTRANT(S) CHECK # CONFINED SPACE ENTRANT(S) CHECK #

SUPERVISOR AUTHORIZATION - ALL CONDITIONS SATISFIED:

DEPARTMENT:

PHONE: (____) ____-____

AMBULANCE 2800 FIRE 2900 SAFETY 4901 GAS COORDINATOR 4529/5387

*Full size forms available free of charge at www.oshacfr.com

§1910.146 Appendix E

Sewer System Entry

Sewer entry differs in three vital respects from other permit entries; first, there rarely exists any way to completely isolate the space (a section of a continuous system) to be entered; second, because isolation is not complete, the atmosphere may suddenly and unpredictably become lethally hazardous (toxic, flammable or explosive) from causes beyond the control of the entrant or employer, and third, experienced sewer workers are especially knowledgeable in entry and work in their permit spaces because of their frequent entries. Unlike other employments where permit space entry is a rare and exceptional event, sewer workers' usual work environment is a permit space.

(1) **Adherence to procedure.** The employer should designate as entrants only employees who are thoroughly trained in the employer's sewer entry procedures and who demonstrate that they follow these entry procedures exactly as prescribed when performing sewer entries.

(2) **Atmospheric monitoring.** Entrants should be trained in the use of, and be equipped with, atmospheric monitoring equipment which sounds an audible alarm, in addition to its visual readout, whenever one of the following conditions are encountered: Oxygen concentration less than 19.5 percent; flammable gas or vapor at 10 percent or more of the lower flammable limit (LFL); or hydrogen sulfide or carbon monoxide at or above 10 ppm or 35 ppm, respectively, measured as an 8-hour time-weighted average. Atmospheric monitoring equipment needs to be calibrated according to the manufacturer's instructions. The oxygen sensor/broad range sensor is best suited for initial use in situations where the actual or potential contaminants have not been identified, because broad range sensors, unlike substance-specific sensors, enable employers to obtain an overall reading of the hydrocarbons (flammables) present in the space. However, such sensors only indicate that a hazardous threshold of a class of chemicals has been exceeded. They do not measure the levels of contamination of specific substances. Therefore, substance-specific devices, which measure the actual levels of specific substances, are best suited for use where actual and potential contaminants have been identified. The measurements obtained with substance-specific devices are of vital importance to the employer when decisions are made concerning the measures necessary to protect entrants (such as ventilation or personal protective equipment) and the setting and attainment of appropriate entry conditions. However, the sewer environment may suddenly and unpredictably change, and the substance-specific devices may not detect the potentially lethal atmospheric hazards which may enter the sewer environment.

Although OSHA considers the information and guidance provided above to be appropriate and useful in most sewer entry situations, the Agency emphasizes that each employer must consider the unique circumstances, including the predictability of the atmosphere, of the sewer permit spaces in the employer's workplace in preparing for entry. Only the employer can decide, based upon his or her knowledge of, and experience with permit spaces in sewer systems, what the best type of testing instrument may be for any specific entry operation.

The selected testing instrument should be carried and used by the entrant in sewer line work to monitor the atmosphere in the entrant's environment, and in advance of the entrant's direction of movement, to warn the entrant of any deterioration in atmospheric conditions. Where several entrants are working together in the same immediate location, one instrument, used by the lead entrant, is acceptable.

(3) **Surge flow and flooding.** Sewer crews should develop and maintain liaison, to the extent possible, with the local weather bureau and fire and emergency services in their area so that sewer work may be delayed or interrupted and entrants withdrawn whenever sewer lines might be suddenly flooded by rain or fire suppression activities, or whenever flammable or other hazardous materials are released into sewers during emergencies by industrial or transportation accidents.

(4) **Special Equipment.** Entry into large bore sewers may require the use of special equipment. Such equipment might include such items as atmosphere monitoring devices with automatic audible alarms, escape self-contained breathing apparatus (ESCBA) with at least 10 minute air supply (or other NIOSH approved self-rescuer), and waterproof flashlights, and may also include boats and rafts, radios and rope stand-offs for pulling around bends and corners as needed.

§1910.146 Appendix F

Rescue Team or Rescue Service Evaluation Criteria (Non-Mandatory)

(1) **This appendix provides guidance** to employers in choosing an appropriate rescue service. It contains criteria that may be used to evaluate the capabilities both of prospective and current rescue teams. Before a rescue team can be trained or chosen, however, a satisfactory permit program, including an analysis of all permit-required confined spaces to identify all potential hazards in those spaces, must be completed. OSHA believes that compliance with all the provisions of §1910.146 will enable employers to conduct permit space operations without recourse to rescue services in nearly all cases. However, experience indicates that circumstances will arise where entrants will need to be rescued from permit spaces. It is therefore important for employers to select rescue services or teams, either on-site or off-site, that are equipped and capable of minimizing harm to both entrants and rescuers if the need arises.

(2) **For all rescue teams or services,** the employer's evaluation should consist of two components: an initial evaluation, in which employers decide whether a potential rescue service or team is adequately trained and equipped to perform permit space rescues of the kind needed at the facility and whether such rescuers can respond in a timely manner, and a performance evaluation, in which employers measure the performance of the team or service during an actual or practice rescue. For example, based on the initial evaluation, an employer may determine that maintaining an on-site rescue team will be more expensive than obtaining the services of an off-site team, without being significantly more effective, and decide to hire a rescue service. During a performance evaluation, the employer could decide, after observing the rescue service perform a practice rescue, that the service's training or preparedness was not adequate to effect a timely or effective rescue at his or her facility and decide to select another rescue service, or to form an internal rescue team.

A. *Initial Evaluation.*

I. *The employer should meet* with the prospective rescue service to facilitate the evaluations required by §1910.146(k)(1)(i) and §1910.146(k)(1)(ii). At a minimum, if an off-site rescue service is being considered, the employer must contact the service to plan and coordinate the evaluations required by the standard. Merely posting the service's number or planning to rely on the 911 emergency phone number to obtain these services at the time of a permit space emergency would not comply with paragraph (k)(1) of the standard.

II. *The capabilities required of a rescue service* vary with the type of permit spaces from which rescue may be necessary and the hazards likely to be encountered in those spaces. Answering the questions below will assist employers in determining whether the rescue service is capable of performing rescues in the permit spaces present at the employer's workplace.

1. *What are the needs of the employer* with regard to response time (time for the rescue service to receive notification, arrive at the scene, and set up and be ready for entry)? For example, if entry is to be made into an IDLH atmosphere, or into a space that can quickly develop an IDLH atmosphere (if ventilation fails or for other reasons), the rescue team or service would need to be standing by at the permit space. On the other hand, if the danger to entrants is restricted to mechanical hazards that would cause injuries (e.g., broken bones, abrasions) a response time of 10 or 15 minutes might be adequate.
2. *How quickly can the rescue team* or service get from its location to the permit spaces from which rescue may be necessary? Relevant factors to consider would include: the location of the rescue team or service relative to the employer's workplace, the quality of roads and highways to be traveled, potential bottlenecks or traffic congestion that might be encountered in transit, the reliability of the rescuer's vehicles, and the training and skill of its drivers.
3. *What is the availability of the rescue service?* Is it unavailable at certain times of the day or in certain situations? What is the likelihood that key personnel of the rescue service might be unavailable at times? If the rescue service becomes unavailable while an entry is underway, does it have the capability of notifying the employer so that the employer can instruct the attendant to abort the entry immediately?

4. *Does the rescue service* meet all the requirements of paragraph (k)(2) of the standard? If not, has it developed a plan that will enable it to meet those requirements in the future? If so, how soon can the plan be implemented?
5. *For off-site services,* is the service willing to perform rescues at the employer's workplace? (An employer may not rely on a rescuer who declines, for whatever reason, to provide rescue services.)
6. *Is an adequate method for communications* between the attendant, employer and prospective rescuer available so that a rescue request can be transmitted to the rescuer without delay? How soon after notification can a prospective rescuer dispatch a rescue team to the entry site?
7. *For rescues into spaces* that may pose significant atmospheric hazards and from which rescue entry, patient packaging and retrieval cannot be safely accomplished in a relatively short time (15-20 minutes), employers should consider using airline respirators (with escape bottles) for the rescuers and to supply rescue air to the patient. If the employer decides to use SCBA, does the prospective rescue service have an ample supply of replacement cylinders and procedures for rescuers to enter and exit (or be retrieved) well within the SCBA's air supply limits?
8. *If the space has a vertical entry* over 5 feet in depth, can the prospective rescue service properly perform entry rescues? Does the service have the technical knowledge and equipment to perform rope work or elevated rescue, if needed?
9. *Does the rescue service* have the necessary skills in medical evaluation, patient packaging and emergency response?
10. *Does the rescue service* have the necessary equipment to perform rescues, or must the equipment be provided by the employer or another source?

B. *Performance Evaluation.*

Rescue services are required by paragraph (k)(2)(iv) of the standard to practice rescues at least once every 12 months, provided that the team or service has not successfully performed a permit space rescue within that time. As part of each practice session, the service should perform a critique of the practice rescue, or have another qualified party perform the critique, so that deficiencies in procedures, equipment, training, or number of personnel can be identified and corrected. The results of the critique, and the corrections made to respond to the deficiencies identified, should be given to the employer to enable it to determine whether the rescue service can quickly be upgraded to meet the employer's rescue needs or whether another service must be selected. The following questions will assist employers and rescue teams and services evaluate their performance.

1. *Have all members of the service* been trained as permit space entrants, at a minimum, including training in the potential hazards of all permit spaces, or of representative permit spaces, from which rescue may be needed? Can team members recognize the signs, symptoms, and consequences of exposure to any hazardous atmospheres that may be present in those permit spaces?
2. *Is every team member* provided with, and properly trained in, the use and need for PPE, such as SCBA or fall arrest equipment, which may be required to perform permit space rescues in the facility? Is every team member properly trained to perform his or her functions and make rescues, and to use any rescue equipment, such as ropes and backboards, that may be needed in a rescue attempt?
3. *Are team members trained* in the first aid and medical skills needed to treat victims overcome or injured by the types of hazards that may be encountered in the permit spaces at the facility?
4. *Do all team members perform their functions* safely and efficiently? Do rescue service personnel focus on their own safety before considering the safety of the victim?
5. *If necessary, can the rescue service properly test the atmosphere to determine if it is IDLH?*
6. *Can the rescue personnel identify information* pertinent to the rescue from entry permits, hot work permits, and MSDSs?
7. *Has the rescue service been informed* of any hazards to personnel that may arise from outside the space, such as those that may be caused by future work near the space?
8. *If necessary,* can the rescue service properly package and retrieve victims from a permit space that has a limited size opening (less than 24 inches (60.9 cm) in diameter), limited internal space, or internal obstacles or hazards?
9. *If necessary, can the rescue service* safely perform an elevated (high angle) rescue?
10. *Does the rescue service have a plan for each of the kinds of permit space rescue operations at the facility?* Is the plan adequate for all types of rescue operations that may be needed at the facility? Teams may practice in representative spaces, or in spaces that are "worst-case" or most restrictive with respect to internal configuration, elevation, and portal size. The following characteristics of a practice space should be considered when deciding whether a space is truly representative of an actual permit space:
   (1) Internal configuration.
   [a] Open — there are no obstacles, barriers, or obstructions within the space. One example is a water tank.
   [b] Obstructed — the permit space contains some type of obstruction that a rescuer would need to maneuver around. An example would be a baffle or mixing blade. Large equipment, such as a ladder or scaffold, brought into a space for work purposes would be considered an obstruction if the positioning or size of the equipment would make rescue more difficult.
   (2) Elevation.
   [a] Elevated — a permit space where the entrance portal or opening is above grade by 4 feet or more. This type of space usually requires knowledge of high angle rescue procedures because of the difficulty in packaging and transporting a patient to the ground from the portal.
   [b] Non-elevated — a permit space with the entrance portal located less than 4 feet above grade. This type of space will allow the rescue team to transport an injured employee normally.
   (3) Portal size.
   [a] Restricted — A portal of 24 inches or less in the least dimension. Portals of this size are too small to allow a rescuer to simply enter the space while using SCBA. The portal size is also too small to allow normal spinal immobilization of an injured employee.
   [b] Unrestricted — A portal of greater than 24 inches in the least dimension. These portals allow relatively free movement into and out of the permit space.
   (4) Space access.
   [a] Horizontal — The portal is located on the side of the permit space. Use of retrieval lines could be difficult.
   [b] Vertical — The portal is located on the top of the permit space, so that rescuers must climb down, or the bottom of the permit space, so that rescuers must climb up to enter the space. Vertical portals may require knowledge of rope techniques, or special patient packaging to safely retrieve a downed entrant.

[58 FR 4549, Jan. 14, 1993; 58 FR 34845, 34846, June 29, 1993, as amended at 59 FR 26114, May 19, 1994; 63 FR 66038, 66039, Dec. 1, 1998; 76 FR 80739, Dec. 27, 2011]

§1910.147

The control of hazardous energy (lockout/tagout)

(a) Scope, application, and purpose[§1910.147(a)]

(1) *Scope.*[§1910.147(a)(1)]

(i) *This standard covers the servicing* and maintenance of machines and equipment in which the unexpected energization or start up of the machines or equipment, or release of stored energy could cause injury to employees This standard establishes minimum performance requirements for the control of such hazardous energy.[§1910.147(a)(1)(i)]

(ii) *This standard does not cover the following:*[§1910.147(a)(1)(ii)]

[A] Construction and agriculture employment; [§1910.147(a)(1)(ii)[A]]

[B] Employment covered by parts 1915, 1917, and 1918 of this title;[§1910.147(a)(1)(ii)[B]]

[C] Installations under the exclusive control of electric utilities for the purpose of power generation, transmission and distribution, including related equipment for communication or metering;[§1910.147(a)(1)(ii)[C]]

[D] Exposure to electrical hazards from work on, near, or with conductors or equipment in electric-utilization installations, which is covered by subpart S of this part; and [§1910.147(a)(1)(ii)[D]]

[E] Oil and gas well drilling and servicing.[§1910.147(a)(1)(ii)[E]]

(2) *Application.*[§1910.147(a)(2)]

(i) *This standard applies* to the control of energy during servicing and/or maintenance of machines and equipment. [§1910.147(a)(2)(i)]

(ii) *Normal production operations* are not covered by this standard (See subpart 0 of this part). Servicing and/or maintenance which takes place during normal production operations is covered by this standard only if:[§1910.147(a)(2)(ii)]

[A] An employee is required to remove or bypass a guard or other safety device; or[§1910.147(a)(2)(ii)[A]]

[B] An employee is required to place any part of his or her body into an area on a machine or piece of equipment where work is actually performed upon the material being processed (point of operation) or where an associated danger zone exists during a machine operating cycle. [§1910.147(a)(2)(ii)[B]]

Note: Exception to paragraph (a)(2)(ii): Minor tool changes and adjustments, and other minor servicing activities, which take place during normal production operations, are not covered by this standard if they are routine, repetitive, and integral to the use of the equipment for production, provided that the work is performed using alternative measures which provide effective protection (See subpart 0 of this part).

(iii) *This standard does not apply to the following.* [§1910.147(a)(2)(iii)]

[A] Work on cord and plug connected electric equipment for which exposure to the hazards of unexpected energization or start up of the equipment is controlled by the unplugging of the equipment from the energy source and by the plug being under the exclusive control of the employee performing the servicing or maintenance. [§1910.147(a)(2)(iii)[A]]

[B] Hot tap operations involving transmission and distribution systems for substances such as gas, steam, water or petroleum products when they are performed on pressurized pipelines, provided that the employer demonstrates that[§1910.147(a)(2)(iii)[B]]

[1] continuity of service is essential;[§1910.147(a)(2)(iii)[B][1]]

[2] shutdown of the system is impractical; and [§1910.147(a)(2)(iii)[B][2]]

[3] documented procedures are followed, and special equipment is used which will provide proven effective protection for employees.[§1910.147(a)(2)(iii)[B][3]]

(3) *Purpose.*[§1910.147(a)(3)]

(i) *This section requires employers* to establish a program and utilize procedures for affixing appropriate lockout devices or tagout devices to energy isolating devices, and to otherwise disable machines or equipment to prevent unexpected energization, start-up or release of stored energy in order to prevent injury to employees.[§1910.147(a)(3)(i)]

(ii) *When other standards in this part require* the use of lockout or tagout, they shall be used and supplemented by the procedural and training requirements of this section. [§1910.147(a)(3)(ii)]

(b) Definitions applicable to this section.[§1910.147(b)]

Affected employee. An employee whose job requires him/her to operate or use a machine or equipment on which servicing or maintenance is being performed under lockout or tagout, or whose job requires him/her to work in an area in which such servicing or maintenance is being performed.[§1910.147(b)]

Authorized employee. A person who locks out or tags out machines or equipment in order to perform servicing or maintenance on that machine or equipment. An affected employee becomes an authorized employee when that employee's duties include performing servicing or maintenance covered under this section.[§1910.147(b)]

Capable of being locked out. An energy isolating device is capable of being locked out if it has a hasp or other means of attachment to which, or through which, a lock can be affixed, or it has a locking mechanism built into it. Other energy isolating devices are capable of being locked out, if lockout can be achieved without the need to dismantle, rebuild, or replace the energy isolating device or permanently alter its energy control capability.[§1910.147(b)]

Energized. Connected to an energy source or containing residual or stored energy.[§1910.147(b)]

Energy isolating device. A mechanical device that physically prevents the transmission or release of energy, including but not limited to the following: A manually operated electrical circuit breaker; a disconnect switch; a manually operated switch by which the conductors of a circuit can be disconnected from all ungrounded supply conductors, and, in addition, no pole can be operated independently; a line valve; a block; and any similar device used to block or isolate energy. Push buttons, selector switches and other control circuit type devices are not energy isolating devices. [§1910.147(b)]

Energy source. Any source of electrical, mechanical, hydraulic, pneumatic, chemical, thermal, or other energy.[§1910.147(b)]

Hot tap. A procedure used in the repair, maintenance and services activities which involves welding on a piece of equipment (pipelines, vessels or tanks) under pressure, in order to install connections or appurtenances. It is commonly used to replace or add sections of pipeline without the interruption of service for air, gas, water, steam, and petrochemical distribution systems.[§1910.147(b)]

Lockout. The placement of a lockout device on an energy isolating device, in accordance with an established procedure, ensuring that the energy isolating device and the equipment being controlled cannot be operated until the lockout device is removed. [§1910.147(b)]

Lockout device. A device that utilizes a positive means such as a lock, either key or combination type, to hold an energy isolating device in a safe position and prevent the energizing of a machine or equipment. Included are blank flanges and bolted slip blinds. [§1910.147(b)]

Normal production operations. The utilization of a machine or equipment to perform its intended production function.

Servicing and/or maintenance. Workplace activities such as constructing, installing, setting up, adjusting, inspecting, modifying, and maintaining and/or servicing machines or equipment. These activities include lubrication, cleaning or unjamming of machines or equipment and making adjustments or tool changes, where the employee may be exposed to the unexpected energization or startup of the equipment or release of hazardous energy. [§1910.147(b)]

Setting up. Any work performed to prepare a machine or equipment to perform its normal production operation.[§1910.147(b)]

Tagout. The placement of a tagout device on an energy isolating device, in accordance with an established procedure, to indicate that the energy isolating device and the equipment being controlled may not be operated until the tagout device is removed. [§1910.147(b)]

Tagout device. A prominent warning device, such as a tag and a means of attachment, which can be securely fastened to an energy isolating device in accordance with an established procedure, to indicate that the energy isolating device and the equipment being controlled may not be operated until the tagout device is removed.[§1910.147(b)]

(c) General.[§1910.147(c)]

(1) *Energy control program.* The employer shall establish a program consisting of energy control procedures, employee training and periodic inspections to ensure that before any employee performs any servicing or maintenance on a machine or equipment where the unexpected energizing, start up or release of stored energy could occur and cause injury, the machine or equipment shall be isolated from the energy source, and rendered inoperative.[§1910.147(c)(1)]

(2) *Lockout/tagout.*[§1910.147(c)(2)]

(i) *If an energy isolating device* is not capable of being locked out, the employer's energy control program under paragraph (c)(1) of this section shall utilize a tagout system. [§1910.147(c)(2)(i)]

(ii) *If an energy isolating device* is capable of being locked out, the employer's energy control program under paragraph (c)(1) of this section shall utilize lockout, unless the employer can demonstrate that the utilization of a tagout system will provide full employee protection as set forth in paragraph (c)(3) of this section.[§1910.147(c)(2)(ii)]

(iii) *After January 2, 1990,* whenever replacement or major repair, renovation or modification of a machine or equipment is performed, and whenever new machines or equipment are installed, energy isolating devices for such machine or equipment shall be designed to accept a lockout device. [§1910.147(c)(2)(iii)]

(3) *Full employee protection.*[§1910.147(c)(3)]

(i) *When a tagout device is used* on an energy isolating device which is capable of being locked out, the tagout device shall be attached at the same location that the lockout device would have been attached, and the employer shall demonstrate that the tagout program will provide a level of safety equivalent to that obtained by using a lockout program. [§1910.147(c)(3)(i)]

(ii) *In demonstrating that a level of safety is achieved* in the tagout program which is equivalent to the level of safety obtained by using a lockout program, the employer shall demonstrate full compliance with all tagout-related provisions of this standard together with such additional elements as are necessary to provide the equivalent safety available from the use of a lockout device. Additional means to be considered as part of the demonstration of full employee protection shall include the implementation of additional safety measures such as the removal of an isolating circuit element, blocking of a controlling switch, opening of an extra disconnecting device, or the removal of a valve handle to reduce the likelihood of inadvertent energization.[§1910.147(c)(3)(ii)]

(4) *Energy control procedure.*[§1910.147(c)(4)]

(i) *Procedures shall be developed,* documented and utilized for the control of potentially hazardous energy when employees are engaged in the activities covered by this section. [§1910.147(c)(4)(i)]

Note: Exception: The employer need not document the required procedure for a particular machine or equipment, when all of the following elements exist:

(1) The machine or equipment has no potential for stored or residual energy or reaccumulation of stored energy after shut down which could endanger employees;[§1910.147(c)(4)(i)]

(2) the machine or equipment has a single energy source which can be readily identified and isolated;[§1910.147(c)(4)(i)]

(3) the isolation and locking out of that energy source will completely deenergize and deactivate the machine or equipment;[§1910.147(c)(4)(i)]

(4) the machine or equipment is isolated from that energy source and locked out during servicing or maintenance; [§1910.147(c)(4)(i)]

(5) a single lockout device will achieve a locked-out condition;[§1910.147(c)(4)(i)]

(6) the lockout device is under the exclusive control of the authorized employee performing the servicing or maintenance;[§1910.147(c)(4)(i)]

(7) the servicing or maintenance does not create hazards for other employees; and[§1910.147(c)(4)(i)]

(8) the employer, in utilizing this exception, has had no accidents involving the unexpected activation or reenergization of the machine or equipment during servicing or maintenance.[§1910.147(c)(4)(i)]

(ii) *The procedures shall clearly and specifically outline* the scope, purpose, authorization, rules, and techniques to be utilized for the control of hazardous energy, and the means to enforce compliance including, but not limited to, the following:[§1910.147(c)(4)(ii)]

[A] A specific statement of the intended use of the procedure; [§1910.147(c)(4)(ii)[A]]

[B] Specific procedural steps for shutting down, isolating, blocking and securing machines or equipment to control hazardous energy;[§1910.147(c)(4)(ii)[B]]

[C] Specific procedural steps for the placement, removal and transfer of lockout devices or tagout devices and the responsibility for them; and[§1910.147(c)(4)(ii)[C]]

[D] Specific requirements for testing a machine or equipment to determine and verify the effectiveness of lockout devices, tagout devices, and other energy control measures.[§1910.147(c)(4)(ii)[D]]

(5) *Protective materials and hardware.*[§1910.147(c)(5)]

(i) *Locks, tags, chains,* wedges, key blocks, adapter pins, self-locking fasteners, or other hardware shall be provided by the employer for isolating, securing or blocking of machines or equipment from energy sources.[§1910.147(c)(5)(i)]

(ii) *Lockout devices and tagout devices* shall be singularly identified; shall be the only devices(s) used for controlling energy; shall not be used for other purposes; and shall meet the following requirements:[§1910.147(c)(5)(ii)]

[A] Durable.[§1910.147(c)(5)(ii)[A]]

[1] Lockout and tagout devices shall be capable of withstanding the environment to which they are exposed for the maximum period of time that exposure is expected. [§1910.147(c)(5)(ii)[A][1]]

[2] Tagout devices shall be constructed and printed so that exposure to weather conditions or wet and damp locations will not cause the tag to deteriorate or the message on the tag to become illegible. [§1910.147(c)(5)(ii)[A][2]]

[3] Tags shall not deteriorate when used in corrosive environments such as areas where acid and alkali chemicals are handled and stored.[§1910.147(c)(5)(ii)[A][3]]

[B] Standardized. Lockout and tagout devices shall be standardized within the facility in at least one of the following criteria: Color; shape; or size; and additionally, in the case of tagout devices, print and format shall be standardized. [§1910.147(c)(5)(ii)[B]]

[C] Substantial.[§1910.147(c)(5)(ii)[C]]

[1] Lockout devices. Lockout devices shall be substantial enough to prevent removal without the use of excessive force or unusual techniques, such as with the use of bolt cutters or other metal cutting tools. [§1910.147(c)(5)(ii)[C][1]]

[2] Tagout devices. Tagout devices, including and their means of attachment, shall be substantial enough to prevent inadvertent or accidental removal. Tagout device attachment means shall be of a non-reusable type, attachable by hand, self-locking, and non-releasable with a minimum unlocking strength of no less than 50 pounds and having the general design and basic characteristics of being at least equivalent to a one-piece, all-environment-tolerant nylon cable tie. [§1910.147(c)(5)(ii)[C][2]]

[D] Identifiable. Lockout devices and tagout devices shall indicate the identity of the employee applying the device(s). [§1910.147(c)(5)(ii)[D]]

(iii) *Tagout devices shall warn* against hazardous conditions if the machine or equipment is energized and shall include a legend such as the following: Do Not Start, Do Not Open, Do Not Close, Do Not Energize, Do Not Operate. [§1910.147(c)(5)(iii)]

(6) *Periodic inspection.*[§1910.147(c)(6)]

(i) *The employer shall conduct a periodic* inspection of the energy control procedure at least annually to ensure that the procedure and the requirements of this standard are being followed.[§1910.147(c)(6)(i)]

[A] The periodic inspection shall be perfomed by an authorized employee other than the ones(s) utilizing the energy control procedure being inspected.[§1910.147(c)(6)(i)[A]]

[B] The periodic inspection shall be conducted to correct any deviations or inadequacies identified.[§1910.147(c)(6)(i)[B]]

[C] Where lockout is used for energy control, the periodic inspection shall include a review, between the inspector and each authorized employee, of that employee's responsibilities under the energy control procedure being inspected.[§1910.147(c)(6)(i)[C]]

[D] Where tagout is used for energy control, the periodic inspection shall include a review, between the inspector and each authorized and affected employee, of that employee's responsibilities under the energy control procedure being inspected, and the elements set forth in paragraph (c)(7)(ii) of this section.[§1910.147(c)(6)(i)[D]]

(ii) *The employer shall certify* that the periodic inspections have been performed. The certification shall identify the machine or equipment on which the energy control procedure was being utilized, the date of the inspection, the employees included in the inspection, and the person performing the inspection.[§1910.147(c)(6)(ii)]

(7) *Training and communication.*[§1910.147(c)(7)]

(i) *The employer shall provide training* to ensure that the purpose and function of the energy control program are understood by employees and that the knowledge and skills required for the safe application, usage, and removal of the energy controls are acquired by employees. The training shall include the following:[§1910.147(c)(7)(i)]

[A] Each authorized employee shall receive training in the recognition of applicable hazardous energy sources, the type and magnitude of the energy available in the workplace, and the methods and means necessary for energy isolation and control.[§1910.147(c)(7)(i)[A]]

[B] Each affected employee shall be instructed in the purpose and use of the energy control procedure. [§1910.147(c)(7)(i)[B]]

[C] All other employees whose work operations are or may be in an area where energy control procedures may be utilized, shall be instructed about the procedure, and about the prohibition relating to attempts to restart or reenergize machines or equipment which are locked out or tagged out.[§1910.147(c)(7)(i)[C]]

(ii) *When tagout systems are used,* employees shall also be trained in the following limitations of tags:[§1910.147(c)(7)(ii)]

[A] Tags are essentially warning devices affixed to energy isolating devices, and do not provide the physical restraint on those devices that is provided by a lock. [§1910.147(c)(7)(ii)[A]]

[B] When a tag is attached to an energy isolating means, it is not to be removed without authorization of the authorized person responsible for it, and it is never to be bypassed, ignored, or otherwise defeated. [§1910.147(c)(7)(ii)[B]]

[C] Tags must be legible and understandable by all authorized employees, affected employees, and all other employees whose work operations are or may be in the area, in order to be effective.[§1910.147(c)(7)(ii)[C]]

[D] Tags and their means of attachment must be made of materials which will withstand the environmental conditions encountered in the workplace.[§1910.147(c)(7)(ii)[D]]

[E] Tags may evoke a false sense of security, and their meaning needs to be understood as part of the overall energy control program.[§1910.147(c)(7)(ii)[E]]

[F] Tags must be securely attached to energy isolating devices so that they cannot be inadvertently or accidentally detached during use.[§1910.147(c)(7)(ii)[F]]

(iii) *Employee retraining.*[§1910.147(c)(7)(iii)]

[A] Retraining shall be provided for all authorized and affected employees whenever there is a change in their job assignments, a change in machines, equipment or processes that present a new hazard, or when there is a change in the energy control procedures. [§1910.147(c)(7)(iii)[A]]

[B] Additional retraining shall also be conducted whenever a periodic inspection under paragraph (c)(6) of this section reveals, or whenever the employer has reason to believe, that there are deviations from or inadequacies in the employee's knowledge or use of the energy control procedures.[§1910.147(c)(7)(iii)[B]]

[C] The retraining shall reestablish employee proficiency and introduce new or revised control methods and procedures, as necessary.[§1910.147(c)(7)(iii)[C]]

(iv) *The employer shall certify* that employee training has been accomplished and is being kept up to date. The certification shall contain each employee's name and dates of training. [§1910.147(c)(7)(iv)]

(8) *Energy isolation.* Lockout or tagout shall be performed only by the authorized employees who are performing the servicing or maintenance.[§1910.147(c)(8)]

(9) *Notification of employees.* Affected employees shall be notified by the employer or authorized employee of the application and removal of lockout devices or tagout devices. Notification shall be given before the controls are applied, and after they are removed from the machine or equipment.[§1910.147(c)(9)]

(d) **Application of control.** The established procedures for the application of energy control (the lockout or tagout procedures) shall cover the following elements and actions and shall be done in the following sequence:[§1910.147(d)]

(1) *Preparation for shutdown.* Before an authorized or affected employee turns off a machine or equipment, the authorized employee shall have knowledge of the type and magnitude of the energy, the hazards of the energy to be controlled, and the method or means to control the energy.[§1910.147(d)(1)]

(2) *Machine or equipment shutdown.* The machine or equipment shall be turned off or shut down using the procedures established for the machine or equipment. An orderly shutdown must be utilized to avoid any additional or increased hazard(s) to employees as a result of the equipment stoppage. [§1910.147(d)(2)]

(3) *Machine or equipment isolation.* All energy isolating devices that are needed to control the energy to the machine or equipment shall be physically located and operated in such a manner as to isolate the machine or equipment from the energy source(s).[§1910.147(d)(3)]

(4) *Lockout or tagout device application.*[§1910.147(d)(4)]

(i) *Lockout or tagout devices* shall be affixed to each energy isolating device by authorized employees.[§1910.147(d)(4)(i)]

(ii) *Lockout devices, where used,* shall be affixed in a manner to that will hold the energy isolating devices in a "safe" or "off" position.[§1910.147(d)(4)(ii)]

(iii) *Tagout devices, where used,* shall be affixed in such a manner as will clearly indicate that the operation or movement of energy isolating devices from the "safe" or "off" position is prohibited.[§1910.147(d)(4)(iii)]

[A] Where tagout devices are used with energy isolating devices designed with the capability of being locked, the tag attachment shall be fastened at the same point at which the lock would have been attached. [§1910.147(d)(4)(iii)[A]]

[B] Where a tag cannot be affixed directly to the energy isolating device, the tag shall be located as close as safely possible to the device, in a position that will be immediately obvious to anyone attempting to operate the device.[§1910.147(d)(4)(iii)[B]]

(5) *Stored energy.*[§1910.147(d)(5)]

(i) *Following the application* of lockout or tagout devices to energy isolating devices, all potentially hazardous stored or residual energy shall be relieved, disconnected, restrained, and otherwise rendered safe.[§1910.147(d)(5)(i)]

(ii) *If there is a possibility* of reaccumulation of stored energy to a hazardous level, verification of isolation shall be continued until the servicing or maintenance is completed, or until the possibility of such accumulation no longer exists. [§1910.147(d)(5)(ii)]

(6) *Verification of isolation.* Prior to starting work on machines or equipment that have been locked out or tagged out, the authorized employee shall verify that isolation and deenergization of the machine or equipment have been accomplished. [§1910.147(d)(6)]

(e) **Release from lockout or tagout.** Before lockout or tagout devices are removed and energy is restored to the machine or equipment, procedures shall be followed and actions taken by the authorized employee(s) to ensure the following:[§1910.147(e)]

(1) *The machine or equipment.* The work area shall be inspected to ensure that nonessential items have been removed and to ensure that machine or equipment components are operationally intact.[§1910.147(e)(1)]

(2) *Employees.*[§1910.147(e)(2)]

(i) *The work area shall be checked* to ensure that all employees have been safely positioned or removed. [§1910.147(e)(2)(i)]

(ii) *After lockout or tagout devices* have been removed and before a machine or equipment is started, affected employees shall be notified that the lockout or tagout device(s) have been removed.[§1910.147(e)(2)(ii)]

(3) *Lockout or tagout devices removal.* Each lockout or tagout device shall be removed from each energy isolating device by the employee who applied the device. Exception to paragraph (e)(3): When the authorized employee who applied the lockout or tagout device is not available to remove it, that device may be removed under the direction of the employer, provided that specific procedures and training for such removal have been developed, documented and incorporated into the employer's energy control program. The employer shall demonstrate that the specific procedure provides equivalent safety to the removal of the device by the authorized employee who applied it. The specific procedure shall include at least the following elements:[§1910.147(e)(3)]

(i) *Verfication by the employer* that the authorized employee who applied the device is not at the facility;[§1910.147(e)(3)(i)]

(ii) *Making all reasonable efforts* to contact the authorized employee to inform him/her that his/her lockout or tagout device has been removed; and[§1910.147(e)(3)(ii)]

(iii) *Ensuring that the authorized employee* has this knowledge before he/she resumes work at that facility.[§1910.147(e)(3)(iii)]

(f) **Additional requirements.**[§1910.147(f)]

(1) *Testing or positioning of machines,* equipment or components thereof. In situations in which lockout or tagout devices must be temporarily removed from the energy isolating device and the machine or equipment energized to test or position the machine, equipment or component thereof, the following sequence of actions shall be followed:[§1910.147(f)(1)]

(i) *Clear the machine or equipment* of tools and materials in accordance with paragraph (e)(1) of this section; [§1910.147(f)(1)(i)]

(ii) *Remove employees from the machine* or equipment area in accordance with paragraph (e)(2) of this section; [§1910.147(f)(1)(ii)]

(iii) *Remove the lockout or tagout devices* as specified in paragraph (e)(3) of this section;[§1910.147(f)(1)(iii)]

(iv) *Energize and proceed with testing or positioning;* [§1910.147(f)(1)(iv)]

(v) *Deenergize all systems* and reapply energy control measures in accordance with paragraph (d) of this section to continue the servicing and/or maintenance.[§1910.147(f)(1)(v)]

(2) *Outside personnel (contractors, etc.).*[§1910.147(f)(2)]

(i) *Whenever outside servicing personnel* are to be engaged in activities covered by the scope and application of this standard, the on-site employer and the outside employer shall inform each other of their respective lockout or tagout procedures.[§1910.147(f)(2)(i)]

(ii) *The on-site employer shall ensure* that his/her employees understand and comply with the restrictions and prohibitions of the outside employer's energy control program. [§1910.147(f)(2)(ii)]

(3) *Group lockout or tagout.*[§1910.147(f)(3)]

(i) *When servicing and/or maintenance is performed* by a crew, craft, department or other group, they shall utilize a procedure which affords the employees a level of protection equivalent to that provided by the implementation of a personal lockout or tagout device.[§1910.147(f)(3)(i)]

(ii) *Group lockout or tagout devices* shall be used in accordance with the procedures required by paragraph (c)(4) of this section including, but not necessarily limited to, the following specific requirements:[§1910.147(f)(3)(ii)]

[A] Primary responsibility is vested in an authorized employee for a set number of employees working under the protection of a group lockout or tagout device (such as an operations lock);[§1910.147(f)(3)(ii)[A]]

[B] Provision for the authorized employee to ascertain the exposure status of individual group members with regard to the lockout or tagout of the machine or equipment; and[§1910.147(f)(3)(ii)[B]]

[C] When more than one crew, craft, department, etc. is involved, assignment of overall job-associated lockout or tagout control responsibility to an authorized employee designated to coordinate affected work forces and ensure continuity of protection; and[§1910.147(f)(3)(ii)[C]]

[D] Each authorized employee shall affix a personal lockout or tagout device to the group lockout device, group lockbox, or comparable mechanism when he or she begins work, and shall remove those devices when he or she stops working on the machine or equipment being serviced or maintained.[§1910.147(f)(3)(ii)[D]]

(4) *Shift or personnel changes.* Specific procedures shall be utilized during shift or personnel changes to ensure the continuity of lockout or tagout protection, including provision for the orderly transfer of lockout or tagout device protection between off-going and oncoming employees, to minimize exposure to hazards from the unexpected energization or start-up of the machine or equipment, or the release of stored energy. [§1910.147(f)(4)]

Note: The following appendix to Sec. 1910.147 services as a non-mandatory guideline to assist employers and employees in complying with the requirements of this section, as well as to provide other helpful information. Nothing in the appendix adds to or detracts from any of the requirements of this section.

§1910.147 Appendix A

Typical Minimal Lockout Procedure

Appendix A to §1910.147 - Typical Minimal Lockout Procedure

General
The following simple lockout procedure is provided to assist employers in developing their procedures so they meet the requirements of this standard. When the energy isolating devices are not lockable, tagout may be used, provided the employer complies with the provisions of the standard which require additional training and more rigorous periodic inspections. When tagout is used and the energy isolating devices are lockable, the employer must provide full employee protection (see paragraph(c)(3)) and additional training and more rigorous periodic inspections are required. For more complex systems, more comprehensive procedures may need to be developed, documented, and utilized.

Lockout Procedure
Lockout procedure for:

____
(Name of Company for single procedure or identification of equipment if multiple procedures are used.)

Purpose
This procedure establishes the minimum requirements for the lockout of energy isolating devices whenever maintenance or servicing is done on machines or equipment. It shall be used to ensure that the machine or equipment is stopped, isolated from all potentially hazardous energy sources and locked out before employees perform any servicing or maintenance where the unexpected energization or start-up of the machine or equipment or release of stored energy could cause injury.

Compliance With This Program
All employees are required to comply with the restrictions and limitations imposed upon them during the use of lockout. The authorized employees are required to perform the lockout in accordance with this procedure. All employees, upon observing a machine or piece of equipment which is locked out to perform servicing or maintenance, shall not attempt to start, energize, or use that machine or equipment.

____
Type of compliance enforcement to be taken for violation of the above.

Sequence of Lockout

(1) Notify all affected employees that servicing or maintenance is required on a machine or equipment and that the machine or equipment must be shut down and locked out to perform the servicing or maintenance.

____
Name(s)/Job Title(s) of affected employees and how to notify.

(2) The authorized employee shall refer to the company procedure to identify the type and magnitude of the energy that the machine or equipment utilizes, shall understand the hazards of the energy, and shall know the methods to control the energy.

____
Type(s) and magnitude(s) of energy, its hazards and the methods to control the energy.

(3) If the machine or equipment is operating, shut it down by the normal stopping procedure (depress the stop button, open switch, close valve, etc.).

____
Type(s) and location(s) of machine or equipment operating controls.

(4) De-activate the energy isolating device(s) so that the machine or equipment is isolated from the energy source(s).

____
Type(s) and location(s) of energy isolating devices.

(5) Lock out the energy isolating device(s) with assigned individual lock(s).

(6) Stored or residual energy (such as that in capacitors, springs, elevated machine members, rotating flywheels, hydraulic systems, and air, gas, steam, or water pressure, etc.) must be dissipated or restrained by methods such as grounding, repositioning, blocking, bleeding down, etc.

____
Type(s) of stored energy – methods to dissipate or restrain.

(7) Ensure that the equipment is disconnected from the energy source(s) by first checking that no personnel are exposed, then verify the isolation of the equipment by operating the push button or other normal operating control(s) or by testing to make certain the equipment will not operate.
Caution: Return operating control(s) to neutral or "off" position after verifying the isolation of the equipment.

____
Method of verifying the isolation of the equipment.

(8) The machine or equipment is now locked out.

Restoring Equipment to Service. When the servicing or maintenance is completed and the machine or equipment is ready to return to normal operating condition, the following steps shall be taken.

(1) Check the machine or equipment and the immediate area around the machine or equipment to ensure that nonessential items have been removed and that the machine or equipment components are operationally intact.
(2) Check the work area to ensure that all employees have been safely positioned or removed from the area.
(3) Verify that the controls are in neutral.
(4) Remove the lockout devices and reenergize the machine or equipment.
Note: The removal of some forms of blocking may require reenergization of the machine before safe removal.
(5) Notify affected employees that the servicing or maintenance is completed and the machine or equipment is ready for use.

*Full size forms available free of charge at www.oshacfr.com

[54 FR 36687, Sept. 1, 1989, as amended at 54 FR 42498, Oct. 17, 1989; 55 FR 38685, 38686, Sept. 20, 1990; 76 FR 24698, May 2, 2011; 76 FR 44265, July 25, 2011]

§1910.178

Powered industrial trucks

(a) **General requirements.** [§1910.178(a)]

(1) *This section contains safety requirements* relating to fire protection, design, maintenance, and use of fork trucks, tractors, platform lift trucks, motorized hand trucks, and other specialized industrial trucks powered by electric motors or internal combustion engines. This section does not apply to compressed air or nonflammable compressed gas-operated industrial trucks, nor to farm vehicles, nor to vehicles intended primarily for earth moving or over-the-road hauling. [§1910.178(a)(1)]

(2) *All new powered industrial trucks acquired* and used by an employer shall meet the design and construction requirements for powered industrial trucks established in the "American National Standard for Powered Industrial Trucks, Part II, ANSI B56.1-1969", which is incorporated by reference as specified in §1910.6, except for vehicles intended primarily for earth moving or over-the-road hauling. [§1910.178(a)(2)]

(3) *Approved trucks shall bear a label* or some other identifying mark indicating approval by the testing laboratory. See paragraph (a)(7) of this section and paragraph 405 of "American National Standard for Powered Industrial Trucks, Part II, ANSI B56.1-1969", which is incorporated by reference in paragraph (a)(2) of this section and which provides that if the powered industrial truck is accepted by a nationally recognized testing laboratory it should be so marked. [§1910.178(a)(3)]

(4) *Modifications and additions which affect capacity* and safe operation shall not be performed by the customer or user without manufacturers prior written approval. Capacity, operation, and maintenance instruction plates, tags, or decals shall be changed accordingly. [§1910.178(a)(4)]

(5) *If the truck is equipped* with front-end attachments other than factory installed attachments, the user shall request that the truck be marked to identify the attachments and show the approximate weight of the truck and attachment combination at maximum elevation with load laterally centered. [§1910.178(a)(5)]

(6) *The user shall see* that all nameplates and markings are in place and are maintained in a legible condition. [§1910.178(a)(6)]

(7) *As used in this section,* the term, **approved truck or approved industrial truck** means a truck that is listed or approved for fire safety purposes for the intended use by a nationally recognized testing laboratory, using nationally recognized testing standards. Refer to §1910.155(c)(3)(iv)(A) for definition of listed, and to §1910.7 for definition of nationally recognized testing laboratory. [§1910.178(a)(7)]

(b) **Designations.** For the purpose of this standard there are eleven different designations of industrial trucks or tractors as follows: D, DS, DY, E, ES, EE, EX, G, GS, LP, and LPS. [§1910.178(b)]

(1) *The D designated units* are units similar to the G units except that they are diesel engine powered instead of gasoline engine powered. [§1910.178(b)(1)]

(2) *The DS designated units* are diesel powered units that are provided with additional safeguards to the exhaust, fuel and electrical systems. They may be used in some locations where a D unit may not be considered suitable. [§1910.178(b)(2)]

(3) *The DY designated units* are diesel powered units that have all the safeguards of the DS units and in addition do not have any electrical equipment including the ignition and are equipped with temperature limitation features. [§1910.178(b)(3)]

‡(4) *The E designated units* are electrically powered units that have minimum acceptable safeguards against inherent fire hazards. [§1910.178(b)(4)]

(5) *The ES designated units* are electrically powered units that, in addition to all of the requirements for the E units, are provided with additional safeguards to the electrical system to prevent emission of hazardous sparks and to limit surface temperatures. They may be used in some locations where the use of an E unit may not be considered suitable. [§1910.178(b)(5)]

(6) *The EE designated units* are electrically powered units that have, in addition to all of the requirements for the E and ES units, the electric motors and all other electrical equipment completely enclosed. In certain locations the EE unit may be used where the use of an E and ES unit may not be considered suitable. [§1910.178(b)(6)]

(7) *The EX designated units* are electrically powered units that differ from the E, ES, or EE units in that the electrical fittings and equipment are so designed, constructed and assembled that the units may be used in certain atmospheres containing flammable vapors or dusts. [§1910.178(b)(7)]

(8) *The G designated units* are gasoline powered units having minimum acceptable safeguards against inherent fire hazards. [§1910.178(b)(8)]

(9) *The GS designated units* are gasoline powered units that are provided with additional safeguards to the exhaust, fuel, and electrical systems. They may be used in some locations where the use of a G unit may not be considered suitable. [§1910.178(b)(9)]

(10) *The LP designated unit* is similar to the G unit except that liquefied petroleum gas is used for fuel instead of gasoline. [§1910.178(b)(10)]

(11) *The LPS designated units* are liquefied petroleum gas powered units that are provided with additional safeguards to the exhaust, fuel, and electrical systems. They may be used in some locations where the use of an LP unit may not be considered suitable. [§1910.178(b)(11)]

(12) *The atmosphere or location* shall have been classified as to whether it is hazardous or nonhazardous prior to the consideration of industrial trucks being used therein and the type of industrial truck required shall be as provided in paragraph (d) of this section for such location. [§1910.178(b)(12)]

(c) **Designated locations.** [§1910.178(c)]

(1) *The industrial trucks specified* under subparagraph (2) of this paragraph are the minimum types required but industrial trucks having greater safeguards may be used if desired. [§1910.178(c)(1)]

(2) *For specific areas of use,* see Table N-1 which tabulates the information contained in this section. References are to the corresponding classification as used in subpart S of this part. [§1910.178(c)(2)]

(i) *Power-operated industrial trucks* shall not be used in atmospheres containing hazardous concentration of acetylene, butadiene, ethylene oxide, hydrogen (or gases or vapors equivalent in hazard to hydrogen, such as manufactured gas), propylene oxide, acetaldehyde, cyclopropane, diethyl ether, ethylene, isoprene, or unsymmetrical dimethyl hydrazine (UDMH). [§1910.178(c)(2)(i)]

(ii)*[a]* *Power-operated industrial trucks* shall not be used in atmospheres containing hazardous concentrations of metal dust, including aluminum, magnesium, and their commercial alloys, other metals of similarly hazardous characteristics, or in atmospheres containing carbon black, coal or coke dust except approved power-operated industrial trucks designated as EX may be used in such atmospheres. [§1910.178(c)(2)(ii)[a]]

[b] *In atmospheres where dust* of magnesium, aluminum or aluminum bronze may be present, fuses, switches, motor controllers, and circuit breakers of trucks shall have enclosures specifically approved for such locations. [§1910.178(c)(2)(ii)[b]]

(iii) *Only approved power-operated industrial trucks* designated as EX may be used in atmospheres containing acetone, acrylonitrile, alcohol, ammonia, benzine, benzol, butane, ethylene dichloride, gasoline, hexane, lacquer solvent vapors, naphtha, natural gas, propane, propylene, styrene, vinyl acetate, vinyl chloride, or xylenes in quantities sufficient to produce explosive or ignitable mixtures and where such concentrations of these gases or vapors exist continuously, intermittently or periodically under normal operating conditions or may exist frequently because of repair, maintenance operations, leakage, breakdown or faulty operation of equipment. [§1910.178(c)(2)(iii)]

(iv) *Power-operated industrial trucks* designated as DY, EE, or EX may be used in locations where volatile flammable liquids or flammable gases are handled, processed or used, but in which the hazardous liquids, vapors or gases will normally be confined within closed containers or closed systems from which they can escape only in case of accidental rupture or breakdown of such containers or systems, or in the case of abnormal operation of equipment; also in locations in which hazardous concentrations of gases or vapors are normally prevented by positive mechanical ventilation but which might become hazardous through failure or abnormal operation of the ventilating equipment; or in locations which are adjacent to Class I, Division 1 locations, and to which hazardous concentrations of gases or vapors might occasionally be communicated unless such communication is prevented by adequate positive-pressure ventilation from a source of clear air, and effective safeguards against ventilation failure are provided. [§1910.178(c)(2)(iv)]

Table N-1 - Summary Table on Use of Industrial Trucks in Various Locations[1]

Classes	Unclassified	Class I locations				Class II locations			Class III locations
Description of classes	Locations not possessing atmospheres as described in other columns.	Locations in which flammable gases or vapors are, or may be, present in the air in quantities sufficient to produce explosive or ignitible mixtures.				Locations which are hazardous because of the presence of combustible dust.			Locations where easily ignitible fibers or flyings are present but not likely to be in suspension in quantities sufficient to produce ignitible mixtures.
Groups in classes	None	A	B	C	D	E	F	G	None
Examples of locations or atmospheres in classes and groups.	Piers and wharves inside and outside general storage, general industrial or commercial properties.	Acetylene	Hydrogen	Ethyl ether	Gasoline Naphtha Alcohols Acetone Lacquer solvent Benzene	Metal dust	Carbon black coal dust, coke dust	Grain dust, flour dust, starch dust, organic dust	Baled waste, cocoa fiber, cotton, excelsior, hemp, istle, jute, kapok, oakum, sisal, Spanish moss, synthetic fibers, tow.

	Unclassified	Class I locations								Class II locations						Class III locations	
		1				2				1			2			1	2
Divisions (nature of hazardous conditions)	None	Above condition exists continuously, intermittently, or periodically under normal operating conditions.				Above condition may occur accidentally as due to a puncture of a storage drum.				Explosive mixture may be present under normal operating conditions, or where failure of equipment may cause the condition to exist simultaneously with arcing or sparking of electrical equipment, or where dusts of an electrically conducting nature may be present.			Explosive mixture not normally present, but where deposits of dust may cause heat rise in electrical equipment, or where such deposits may be ignited by arcs or sparks from electrical equipment.			Locations in which easily ignitable fibers or materials producing combustible flyings are handled, manufactured, or used.	Locations in which easily ignitable fibers are stored or handled (except in the process of manufacture).
Authorized uses of trucks by types in groups of classes and divisions																	
Groups in classes	None	A	B	C	D	A	B	C	D	E	F	G	E	F	G	None	None
Type of truck authorized:																	
Diesel:																	
Type D	D*																
Type DS									DS						DS		DS
Type DY									DY						DY	DY	DY
Electric:																	
Type E	E*																E
Type ES									ES						ES		ES
Type EE									EE						EE	EE	EE
Type EX					EX				EX		EX	EX			EX	EX	EX
Gasoline:																	
Type G	G*																
Type GS									GS						GS		GS
LP-Gas:																	
Type LP	LP*																
Type LPS									LPS						LPS		LPS
Paragraph Ref. in No. 505**.	210.211		201 (a)		203 (a)		209 (a)		204 (a)-(b)		202 (a)	205 (a)		209 (a)	206 (a)-(b)	207 (a)	208 (a)

* Trucks conforming to these types may also be used — see subdivision (c)(2)(x) and (c)(2)(xii) of this section.

** Editor's Note: NFPA 505-1969 Standard for Type Designations, Areas of Use, Maintenance, and Operation of Powered Industrial Trucks.

‡ Editor's Note: This row was added for further clarification.

1. Editor's Note: Please refer to STD 1-11.6A in the Addendum for further clarification of N-1 Table.

(v) *In locations used for the storage* of hazardous liquids in sealed containers or liquefied or compressed gases in containers, approved power-operated industrial trucks designated as DS, ES, GS, or LPS may be used. This classification includes locations where volatile flammable liquids or flammable gases or vapors are used, but which, would become hazardous only in case of an accident or of some unusual operating condition. The quantity of hazardous material that might escape in case of accident, the adequacy of ventilating equipment, the total area involved, and the record of the industry or business with respect to explosions or fires are all factors that should receive consideration in determining whether or not the DS or DY, ES, EE, GS, LPS designated truck possesses sufficient safeguards for the location. Piping without valves, checks, meters and similar devices would not ordinarily be deemed to introduce a hazardous condition even though used for hazardous liquids or gases. Locations used for the storage of hazardous liquids or of liquified or compressed gases in sealed containers would not normally be considered hazardous unless subject to other hazardous conditions also. [§1910.178(c)(2)(v)]

(vi)*[a] Only approved power operated industrial trucks* designated as EX shall be used in atmospheres in which combustible dust is or may be in suspension continuously, intermittently, or periodically under normal operating conditions, in quantities sufficient to produce explosive or ignitable mixtures, or where mechanical failure or abnormal operation of machinery or equipment might cause such mixtures to be produced. [§1910.178(c)(2)(vi)[a]]

[b] The EX classification usually includes the working areas of grain handling and storage plants, room containing grinders or pulverizers, cleaners, graders, scalpers, open conveyors or spouts, open bins or hoppers, mixers, or blenders, automatic or hopper scales, packing machinery, elevator heads and

1910 Selected 1910 General Industry Standards

boots, stock distributors, dust and stock collectors (except all-metal collectors vented to the outside), and all similar dust producing machinery and equipment in grain processing plants, starch plants, sugar pulverizing plants, malting plants, hay grinding plants, and other occupancies of similar nature; coal pulverizing plants (except where the pulverizing equipment is essentially dust tight); all working areas where metal dusts and powders are produced, processed, handled, packed, or stored (except in tight containers); and other similar locations where combustible dust may, under normal operating conditions, be present in the air in quantities sufficient to produce explosive or ignitable mixtures. [§1910.178(c)(2)(vi)[b]]

(vii) *Only approved power-operated industrial trucks* designated as DY, EE, or EX shall be used in atmospheres in which combustible dust will not normally be in suspension in the air or will not be likely to be thrown into suspension by the normal operation of equipment or apparatus in quantities sufficient to produce explosive or ignitable mixtures but where deposits or accumulations of such dust may be ignited by arcs or sparks originating in the truck. [§1910.178(c)(2)(vii)]

(viii) *Only approved power-operated industrial trucks* designated as DY, EE, or EX shall be used in locations which are hazardous because of the presence of easily ignitable fibers or flyings but in which such fibers or flyings are not likely to be in suspension in the air in quantities sufficient to produce ignitable mixtures. [§1910.178(c)(2)(viii)]

(ix) *Only approved power-operated industrial trucks* designated as DS, DY, ES, EE, EX, GS, or LPS shall be used in locations where easily ignitable fibers are stored or handled, including outside storage, but are not being processed or manufactured. Industrial trucks designated as E, which have been previously used in these locations may be continued in use. [§1910.178(c)(2)(ix)]

(x) *On piers and wharves handling general cargo,* any approved power-operated industrial truck designated as Type D, E, G, or LP may be used, or trucks which conform to the requirements for these types may be used. [§1910.178(c)(2)(x)]

(xi) *If storage warehouses and outside storage locations* are hazardous only the approved power-operated industrial truck specified for such locations in this paragraph (c)(2) shall be used. If not classified as hazardous, any approved power-operated industrial truck designated as Type D, E, G, or LP may be used, or trucks which conform to the requirements for these types may be used. [§1910.178(c)(2)(xi)]

(xii) *If general industrial or commercial properties* are hazardous, only approved power-operated industrial trucks specified for such locations in this paragraph (c)(2) shall be used. If not classified as hazardous, any approved power-operated industrial truck designated as Type D, E, G, or LP may be used, or trucks which conform to the requirements of these types may be used. [§1910.178(c)(2)(xii)]

(d) **Converted industrial trucks.** Power-operated industrial trucks that have been originally approved for the use of gasoline for fuel, when converted to the use of liquefied petroleum gas fuel in accordance with paragraph (q) of this section, may be used in those locations where G, GS or LP, and LPS designated trucks have been specified in the preceding paragraphs. [§1910.178(d)]

(e) **Safety guards.** [§1910.178(e)]

(1) *High Lift Rider trucks* shall be fitted with an overhead guard manufactured in accordance with paragraph (a)(2) of this section, unless operating conditions do not permit. [§1910.178(e)(1)]

(2) *If the type of load presents a hazard,* the user shall equip fork trucks with a vertical load backrest extension manufactured in accordance with paragraph (a)(2) of this section. [§1910.178(e)(2)]

(f) **Fuel handling and storage.** [§1910.178(f)]

(1) *The storage and handling of liquid fuels* such as gasoline and diesel fuel shall be in accordance with NFPA Flammable and Combustible Liquids Code (NFPA No. 30-1969), which is incorporated by reference as specified in §1910.6. [§1910.178(f)(1)]

(2) *The storage and handling of liquefied petroleum gas fuel* shall be in accordance with NFPA Storage and Handling of Liquefied Petroleum Gases (NFPA No. 58-1969), which is incorporated by reference as specified in §1910.6. [§1910.178(f)(2)]

(g) **Changing and charging storage batteries.** [§1910.178(g)]

(1) *Battery charging installations* shall be located in areas designated for that purpose. [§1910.178(g)(1)]

(2) *Facilities shall be provided* for flushing and neutralizing spilled electrolyte, for fire protection, for protecting charging apparatus from damage by trucks, and for adequate ventilation for dispersal of fumes from gassing batteries. [§1910.178(g)(2)]

(3) *[Reserved]* [§1910.178(g)(3)]

(4) *A conveyor, overhead hoist,* or equivalent material handling equipment shall be provided for handling batteries. [§1910.178(g)(4)]

(5) *Reinstalled batteries* shall be properly positioned and secured in the truck. [§1910.178(g)(5)]

(6) *A carboy tilter or siphon shall be provided for handling electrolyte.* [§1910.178(g)(6)]

(7) *When charging batteries,* acid shall be poured into water; water shall not be poured into acid. [§1910.178(g)(7)]

(8) *Trucks shall be properly positioned* and brake applied before attempting to change or charge batteries. [§1910.178(g)(8)]

(9) *Care shall be taken to assure* that vent caps are functioning. The battery (or compartment) cover(s) shall be open to dissipate heat. [§1910.178(g)(9)]

(10) *Smoking shall be prohibited in the charging area.* [§1910.178(g)(10)]

(11) *Precautions shall be taken* to prevent open flames, sparks, or electric arcs in battery charging areas. [§1910.178(g)(11)]

(12) *Tools and other metallic objects* shall be kept away from the top of uncovered batteries. [§1910.178(g)(12)]

(h) **Lighting for operating areas.** [§1910.178(h)]

(1) *[Reserved]* [§1910.178(h)(1)]

(2) *Where general lighting* is less than 2 lumens per square foot, auxiliary directional lighting shall be provided on the truck. [§1910.178(h)(2)]

(i) **Control of noxious gases and fumes.** [§1910.178(i)]

(1) *Concentration levels of carbon monoxide gas* created by powered industrial truck operations shall not exceed the levels specified in §1910.1000. [§1910.178(i)(1)]

(j) **Dockboards (bridge plates).** See §1910.30(a). [§1910.178(j)]

(k) **Trucks and railroad cars.** [§1910.178(k)]

(1) *The brakes of highway trucks* shall be set and wheel chocks placed under the rear wheels to prevent the trucks from rolling while they are boarded with powered industrial trucks. [§1910.178(k)(1)]

(2) *Wheel stops or other recognized positive protection* shall be provided to prevent railroad cars from moving during loading or unloading operations. [§1910.178(k)(2)]

(3) *Fixed jacks may be necessary* to support a semitrailer and prevent upending during the loading or unloading when the trailer is not coupled to a tractor. [§1910.178(k)(3)]

(4) *Positive protection shall be provided* to prevent railroad cars from being moved while dockboards or bridge plates are in position. [§1910.178(k)(4)]

(l) **Operator training.** [§1910.178(l)]

(1) *Safe operation.* [§1910.178(l)(1)]

(i) *The employer shall ensure* that each powered industrial truck operator is competent to operate a powered industrial truck safely, as demonstrated by the successful completion of the training and evaluation specified in this paragraph (l). [§1910.178(l)(1)(i)]

(ii) *Prior to permitting an employee to operate* a powered industrial truck (except for training purposes), the employer shall ensure that each operator has successfully completed the training required by this paragraph (l), except as permitted by paragraph (l)(5). [§1910.178(l)(1)(ii)]

(2) *Training program implementation.* [§1910.178(l)(2)]

(i) *Trainees may operate a powered* industrial truck only: [§1910.178(l)(2)(i)]

[A] *Under the direct supervision* of persons who have the knowledge, training, and experience to train operators and evaluate their competence; and [§1910.178(l)(2)(i)[A]]

[B] *Where such operation* does not endanger the trainee or other employees. [§1910.178(l)(2)(i)[B]]

(ii) *Training shall consist of* a combination of formal instruction (e.g., lecture, discussion, interactive computer learning, video tape, written material), practical training (demonstrations performed by the trainer and practical exercises performed by the trainee), and evaluation of the operator's performance in the workplace. [§1910.178(l)(2)(ii)]

(iii) *All operator training and evaluation* shall be conducted by persons who have the knowledge, training, and experience to train powered industrial truck operators and evaluate their competence. [§1910.178(l)(2)(iii)]

(3) *Training program content.* Powered industrial truck operators shall receive initial training in the following topics, except in topics

which the employer can demonstrate are not applicable to safe operation of the truck in the employer's workplace. [§1910.178(l)(3)]

(i) *Truck-related topics:* [§1910.178(l)(3)(i)]

[A] *Operating instructions,* warnings, and precautions for the types of truck the operator will be authorized to operate; [§1910.178(l)(3)(i)[A]]

[B] *Differences between the truck and the automobile;* [§1910.178(l)(3)(i)[B]]

[C] *Truck controls and instrumentation:* where they are located, what they do, and how they work; [§1910.178(l)(3)(i)[C]]

[D] *Engine or motor operation;* [§1910.178(l)(3)(i)[D]]

[E] *Steering and maneuvering;* [§1910.178(l)(3)(i)[E]]

[F] *Visibility (including restrictions due to loading);* [§1910.178(l)(3)(i)[F]]

[G] *Fork and attachment adaptation,* operation, and use limitations; [§1910.178(l)(3)(i)[G]]

[H] *Vehicle capacity;* [§1910.178(l)(3)(i)[H]]

[I] *Vehicle stability;* [§1910.178(l)(3)(i)[I]]

[J] *Any vehicle inspection and maintenance* that the operator will be required to perform; [§1910.178(l)(3)(i)[J]]

[K] *Refueling and/or charging and recharging of batteries;* [§1910.178(l)(3)(i)[K]]

[L] *Operating limitations;* [§1910.178(l)(3)(i)[L]]

[M] *Any other operating instructions,* warnings, or precautions listed in the operator's manual for the types of vehicle that the employee is being trained to operate. [§1910.178(l)(3)(i)[M]]

(ii) *Workplace-related topics:* [§1910.178(l)(3)(ii)]

[A] *Surface conditions where the vehicle will be operated;* [§1910.178(l)(3)(ii)[A]]

[B] *Composition of loads to be carried and load stability;* [§1910.178(l)(3)(ii)[B]]

[C] *Load manipulation, stacking, and unstacking;* [§1910.178(l)(3)(ii)[C]]

[D] *Pedestrian traffic* in areas where the vehicle will be operated; [§1910.178(l)(3)(ii)[D]]

[E] *Narrow aisles and other restricted places* where the vehicle will be operated; [§1910.178(l)(3)(ii)[E]]

[F] *Hazardous (classified) locations* where the vehicle will be operated; [§1910.178(l)(3)(ii)[F]]

[G] *Ramps and other sloped surfaces* that could affect the vehicle's stability; [§1910.178(l)(3)(ii)[G]]

[H] *Closed environments and other areas* where insufficient ventilation or poor vehicle maintenance could cause a buildup of carbon monoxide or diesel exhaust; [§1910.178(l)(3)(ii)[H]]

[I] *Other unique or potentially hazardous* environmental conditions in the workplace that could affect safe operation. [§1910.178(l)(3)(ii)[I]]

(iii) *The requirements of this section.* [§1910.178(l)(3)(iii)]

(4) *Refresher training and evaluation.* [§1910.178(l)(4)]

(i) *Refresher training,* including an evaluation of the effectiveness of that training, shall be conducted as required by paragraph (l)(4)(ii) to ensure that the operator has the knowledge and skills needed to operate the powered industrial truck safely. [§1910.178(l)(4)(i)]

(ii) *Refresher training in relevant topics* shall be provided to the operator when: [§1910.178(l)(4)(ii)]

[A] *The operator has been observed* to operate the vehicle in an unsafe manner; [§1910.178(l)(4)(ii)[A]]

[B] *The operator has been involved* in an accident or near-miss incident; [§1910.178(l)(4)(ii)[B]]

[C] *The operator has received* an evaluation that reveals that the operator is not operating the truck safely; [§1910.178(l)(4)(ii)[C]]

[D] *The operator is assigned to drive* a different type of truck; or [§1910.178(l)(4)(ii)[D]]

[E] *A condition in the workplace* changes in a manner that could affect safe operation of the truck. [§1910.178(l)(4)(ii)[E]]

(iii) *An evaluation* of each powered industrial truck operator's performance shall be conducted at least once every three years. [§1910.178(l)(4)(iii)]

(5) *Avoidance of duplicative training.* If an operator has previously received training in a topic specified in paragraph (l)(3) of this section, and such training is appropriate to the truck and working conditions encountered, additional training in that topic is not required if the operator has been evaluated and found competent to operate the truck safely. [§1910.178(l)(5)]

(6) *Certification.* The employer shall certify that each operator has been trained and evaluated as required by this paragraph (l). The certification shall include the name of the operator, the date of the training, the date of the evaluation, and the identity of the person(s) performing the training or evaluation. [§1910.178(l)(6)]

(7) *Dates.* The employer shall ensure that operators of powered industrial trucks are trained, as appropriate, by the dates shown in the following table. [§1910.178(l)(7)]

If the employee was hired:	The initial training and evaluation of that employee must be completed:
Before December 1, 1999	By December 1, 1999.
After December 1, 1999	Before the employee is assigned to operate a powered industrial truck.

(8) *Appendix A to this section provides* non-mandatory guidance to assist employers in implementing this paragraph (l). This appendix does not add to, alter, or reduce the requirements of this section. [§1910.178(l)(8)]

(m) Truck operations. [§1910.178(m)]

(1) *Trucks shall not be driven* up to anyone standing in front of a bench or other fixed object. [§1910.178(m)(1)]

(2) *No person shall be allowed* to stand or pass under the elevated portion of any truck, whether loaded or empty. [§1910.178(m)(2)]

(3) *Unauthorized personnel shall not be permitted* to ride on powered industrial trucks. A safe place to ride shall be provided where riding of trucks is authorized. [§1910.178(m)(3)]

(4) *The employer shall prohibit* arms or legs from being placed between the uprights of the mast or outside the running lines of the truck. [§1910.178(m)(4)]

(5)(i) *When a powered industrial truck is left unattended,* load engaging means shall be fully lowered, controls shall be neutralized, power shall be shut off, and brakes set. Wheels shall be blocked if the truck is parked on an incline. [§1910.178(m)(5)(i)]

(ii) *A powered industrial truck* is unattended when the operator is 25 ft. or more away from the vehicle which remains in his view, or whenever the operator leaves the vehicle and it is not in his view. [§1910.178(m)(5)(ii)]

(iii) *When the operator of an industrial truck* is dismounted and within 25 ft. of the truck still in his view, the load engaging means shall be fully lowered, controls neutralized, and the brakes set to prevent movement. [§1910.178(m)(5)(iii)]

(6) *A safe distance shall be maintained* from the edge of ramps or platforms while on any elevated dock, or platform or freight car. Trucks shall not be used for opening or closing freight doors. [§1910.178(m)(6)]

(7) *Brakes shall be set and wheel blocks shall be in place* to prevent movement of trucks, trailers, or railroad cars while loading or unloading. Fixed jacks may be necessary to support a semitrailer during loading or unloading when the trailer is not coupled to a tractor. The flooring of trucks, trailers, and railroad cars shall be checked for breaks and weakness before they are driven onto. [§1910.178(m)(7)]

(8) *There shall be sufficient headroom* under overhead installations, lights, pipes, sprinkler system, etc. [§1910.178(m)(8)]

(9) *An overhead guard shall be used* as protection against falling objects. It should be noted that an overhead guard is intended to offer protection from the impact of small packages, boxes, bagged material, etc., representative of the job application, but not to withstand the impact of a falling capacity load. [§1910.178(m)(9)]

(10) *A load backrest extension* shall be used whenever necessary to minimize the possibility of the load or part of it from falling rearward. [§1910.178(m)(10)]

(11) *Only approved industrial trucks shall be used in hazardous locations.* [§1910.178(m)(11)]

(12)-(13) *[Reserved]* [§1910.178(m)(12)-(13)]

(14) *Fire aisles,* access to stairways, and fire equipment shall be kept clear. [§1910.178(m)(14)]

(n) Traveling. [§1910.178(n)]

(1) *All traffic regulations shall be observed,* including authorized plant speed limits. A safe distance shall be maintained approximately three truck lengths from the truck ahead, and the truck shall be kept under control at all times. [§1910.178(n)(1)]

(2) *The right of way shall be yielded* to ambulances, fire trucks, or other vehicles in emergency situations. [§1910.178(n)(2)]

(3) *Other trucks traveling* in the same direction at intersections, blind spots, or other dangerous locations shall not be passed. [§1910.178(n)(3)]

(4) *The driver shall be required* to slow down and sound the horn at cross aisles and other locations where vision is obstructed. If the load being carried obstructs forward view, the driver shall be required to travel with the load trailing. [§1910.178(n)(4)]

(5) *Railroad tracks* shall be crossed diagonally wherever possible. Parking closer than 8 feet from the center of railroad tracks is prohibited. [§1910.178(n)(5)]

(6) *The driver shall be required* to look in the direction of, and keep a clear view of the path of travel. [§1910.178(n)(6)]

(7) *Grades shall be ascended or descended slowly.* [§1910.178(n)(7)]

(i) *When ascending or descending* grades in excess of 10 percent, loaded trucks shall be driven with the load upgrade. [§1910.178(n)(7)(i)]

(ii) *[Reserved]* [§1910.178(n)(7)(ii)]

(iii) *On all grades the load and load engaging means* shall be tilted back if applicable, and raised only as far as necessary to clear the road surface. [§1910.178(n)(7)(iii)]

(8) *Under all travel conditions* the truck shall be operated at a speed that will permit it to be brought to a stop in a safe manner. [§1910.178(n)(8)]

(9) *Stunt driving and horseplay shall not be permitted.* [§1910.178(n)(9)]

(10) *The driver shall be required* to slow down for wet and slippery floors. [§1910.178(n)(10)]

(11) *Dockboard or bridgeplates,* shall be properly secured before they are driven over. Dockboard or bridgeplates shall be driven over carefully and slowly and their rated capacity never exceeded. [§1910.178(n)(11)]

(12) *Elevators shall be approached slowly,* and then entered squarely after the elevator car is properly leveled. Once on the elevator, the controls shall be neutralized, power shut off, and the brakes set. [§1910.178(n)(12)]

(13) *Motorized hand trucks* must enter elevator or other confined areas with load end forward. [§1910.178(n)(13)]

(14) *Running over loose objects* on the roadway surface shall be avoided. [§1910.178(n)(14)]

(15) *While negotiating turns,* speed shall be reduced to a safe level by means of turning the hand steering wheel in a smooth, sweeping motion. Except when maneuvering at a very low speed, the hand steering wheel shall be turned at a moderate, even rate. [§1910.178(n)(15)]

(o) **Loading.** [§1910.178(o)]

(1) *Only stable or safely arranged* loads shall be handled. Caution shall be exercised when handling off-center loads which cannot be centered. [§1910.178(o)(1)]

(2) *Only loads within the rated capacity of the truck shall be handled.* [§1910.178(o)(2)]

(3) *The long or high (including multiple-tiered)* loads which may affect capacity shall be adjusted. [§1910.178(o)(3)]

(4) *Trucks equipped with attachments* shall be operated as partially loaded trucks when not handling a load. [§1910.178(o)(4)]

(5) *A load engaging means* shall be placed under the load as far as possible; the mast shall be carefully tilted backward to stabilize the load. [§1910.178(o)(5)]

(6) *Extreme care shall be used* when tilting the load forward or backward, particularly when high tiering. Tilting forward with load engaging means elevated shall be prohibited except to pick up a load. An elevated load shall not be tilted forward except when the load is in a deposit position over a rack or stack. When stacking or tiering, only enough backward tilt to stabilize the load shall be used. [§1910.178(o)(6)]

(p) **Operation of the truck.** [§1910.178(p)]

(1) *If at any time a powered industrial truck* is found to be in need of repair, defective, or in any way unsafe, the truck shall be taken out of service until it has been restored to safe operating condition. [§1910.178(p)(1)]

(2) *Fuel tanks shall not be filled* while the engine is running. Spillage shall be avoided. [§1910.178(p)(2)]

(3) *Spillage of oil or fuel* shall be carefully washed away or completely evaporated and the fuel tank cap replaced before restarting engine. [§1910.178(p)(3)]

(4) *No truck shall be operated* with a leak in the fuel system until the leak has been corrected. [§1910.178(p)(4)]

(5) *Open flames shall not be used* for checking electrolyte level in storage batteries or gasoline level in fuel tanks. [§1910.178(p)(5)]

(q) **Maintenance of industrial trucks.** [§1910.178(q)]

(1) *Any power-operated industrial truck* not in safe operating condition shall be removed from service. All repairs shall be made by authorized personnel. [§1910.178(q)(1)]

(2) *No repairs shall be made in Class I, II, and III locations.* [§1910.178(q)(2)]

(3) *Those repairs to the fuel and ignition systems* of industrial trucks which involve fire hazards shall be conducted only in locations designated for such repairs. [§1910.178(q)(3)]

(4) *Trucks in need of repairs* to the electrical system shall have the battery disconnected prior to such repairs. [§1910.178(q)(4)]

(5) *All parts of any such industrial truck* requiring replacement shall be replaced only by parts equivalent as to safety with those used in the original design. [§1910.178(q)(5)]

(6) *Industrial trucks shall not be altered* so that the relative positions of the various parts are different from what they were when originally received from the manufacturer, nor shall they be altered either by the addition of extra parts not provided by the manufacturer or by the elimination of any parts, except as provided in paragraph (q)(12) of this section. Additional counterweighting of fork trucks shall not be done unless approved by the truck manufacturer. [§1910.178(q)(6)]

(7) *Industrial trucks shall be examined* before being placed in service, and shall not be placed in service if the examination shows any condition adversely affecting the safety of the vehicle. Such examination shall be made at least daily.

Where industrial trucks are used on a round-the-clock basis, they shall be examined after each shift. Defects when found shall be immediately reported and corrected. [§1910.178(q)(7)]

(8) *Water mufflers shall be filled* daily or as frequently as is necessary to prevent depletion of the supply of water below 75 percent of the filled capacity. Vehicles with mufflers having screens or other parts that may become clogged shall not be operated while such screens or parts are clogged. Any vehicle that emits hazardous sparks or flames from the exhaust system shall immediately be removed from service, and not returned to service until the cause for the emission of such sparks and flames has been eliminated. [§1910.178(q)(8)]

(9) *When the temperature of any part of any truck* is found to be in excess of its normal operating temperature, thus creating a hazardous condition, the vehicle shall be removed from service and not returned to service until the cause for such overheating has been eliminated. [§1910.178(q)(9)]

(10) *Industrial trucks shall be kept* in a clean condition, free of lint, excess oil, and grease. Noncombustible agents should be used for cleaning trucks. Low flash point (below 100 °F.) solvents shall not be used. High flash point (at or above 100 °F.) solvents may be used. Precautions regarding toxicity, ventilation, and fire hazard shall be consonant with the agent or solvent used. [§1910.178(q)(10)]

(11) *[Reserved]* [§1910.178(q)(11)]

(12) *Industrial trucks originally approved* for the use of gasoline for fuel may be converted to liquefied petroleum gas fuel provided the complete conversion results in a truck which embodies the features specified for LP or LPS designated trucks. Such conversion equipment shall be approved. The description of the component parts of this conversion system and the recommended method of installation on specific trucks are contained in the "Listed by Report." [§1910.178(q)(12)]

§1910.178 Appendix A

Stability of Powered Industrial Trucks (Non-mandatory Appendix to Paragraph (l) of This Section)

A-1. Definitions.

The following definitions help to explain the principle of stability:

Center of gravity is the point on an object at which all of the object's weight is concentrated. For symmetrical loads, the center of gravity is at the middle of the load.

Counterweight is the weight that is built into the truck's basic structure and is used to offset the load's weight and to maximize the vehicle's resistance to tipping over.

Fulcrum is the truck's axis of rotation when it tips over.

Grade is the slope of a surface, which is usually measured as the number of feet of rise or fall over a hundred foot horizontal distance (the slope is expressed as a percent).

Lateral stability is a truck's resistance to overturning sideways.

Line of action is an imaginary vertical line through an object's center of gravity.

Load center is the horizontal distance from the load's edge (or the fork's or other attachment's vertical face) to the line of action through the load's center of gravity.

Longitudinal stability is the truck's resistance to overturning forward or rearward.

Moment is the product of the object's weight times the distance from a fixed point (usually the fulcrum). In the case of a powered industrial truck, the distance is measured from the point at which the truck will tip over to the object's line of action. The distance is always measured perpendicular to the line of action.

Track is the distance between the wheels on the same axle of the truck.

Wheelbase is the distance between the centerline of the vehicle's front and rear wheels.

A-2. General.

A-2.1. *Determining the stability* of a powered industrial truck is simple once a few basic principles are understood. There are many factors that contribute to a vehicle's stability: the vehicle's wheelbase, track, and height; the load's weight distribution; and the vehicle's counterweight location (if the vehicle is so equipped).

A-2.2. *The "stability triangle,"* used in most stability discussions, demonstrates stability simply.

A-3. Basic Principles.

A-3.1. *Whether an object is stable* depends on the object's moment at one end of a system being greater than, equal to, or smaller than the object's moment at the system's other end. This principle can be seen in the way a see-saw or teeter-totter works: that is, if the product of the load and distance from the fulcrum (moment) is equal to the moment at the device's other end, the device is balanced and it will not move. However, if there is a greater moment at one end of the device, the device will try to move downward at the end with the greater moment.

A-3.2. *The longitudinal stability* of a counterbalanced powered industrial truck depends on the vehicle's moment and the load's moment. In other words, if the mathematic product of the load moment (the distance from the front wheels, the approximate point at which the vehicle would tip forward) to the load's center of gravity times the load's weight is less than the vehicle's moment, the system is balanced and will not tip forward. However, if the load's moment is greater than the vehicle's moment, the greater load-moment will force the truck to tip forward.

A-4. The Stability Triangle.

A-4.1. *Almost all counterbalanced powered industrial trucks* have a three-point suspension system, that is, the vehicle is supported at three points. This is true even if the vehicle has four wheels. The truck's steer axle is attached to the truck by a pivot pin in the axle's center. When the points are connected with imaginary lines, this three-point support forms a triangle called the stability triangle. Figure 1 depicts the stability triangle.

FIGURE 1

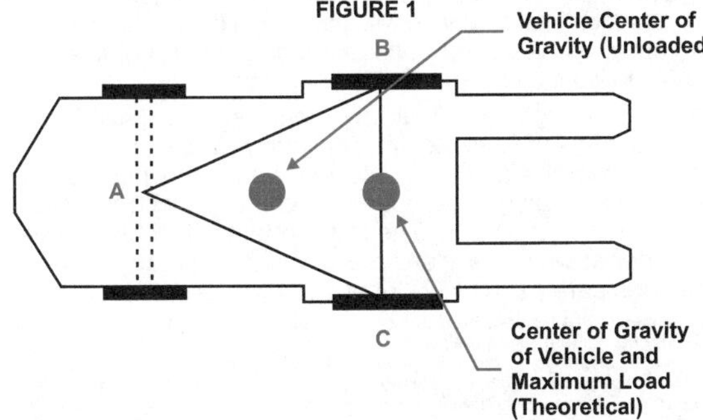

Notes:
1. When the vehicle is loaded, the combined center of gravity (CG) shifts toward line B-C. Theoretically the maximum load will result in the CG at the line B-C. In actual practice, the combined CG should never be at line B-C.
2. The addition of additional counterweight will cause the truck CG to shift toward point A and result in a truck that is less stable laterally.

A-4.2. *When the vehicle's line of action, or load center,* falls within the stability triangle, the vehicle is stable and will not tip over. However, when the vehicle's line of action or the vehicle/load combination falls outside the stability triangle, the vehicle is unstable and may tip over. (See Figure 2.)

FIGURE 2

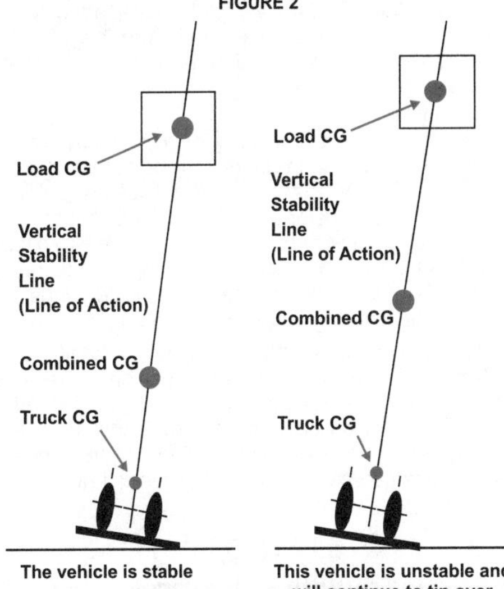

A-5. Longitudinal Stability.

A-5.1. *The axis of rotation* when a truck tips forward is the front wheels' points of contact with the pavement. When a powered industrial truck tips forward, the truck will rotate about this line. When a truck is stable, the vehicle-moment must exceed the load-moment. As long as the vehicle-moment is equal to or exceeds the load-moment, the vehicle will not tip over. On the other hand, if the load moment slightly exceeds the vehicle-moment, the truck will begin to tip forward, thereby causing the rear to lose contact with the floor or ground and resulting in loss of steering control. If the load-moment greatly exceeds the vehicle moment, the truck will tip forward.

A-5.2. *To determine the maximum safe load-moment,* the truck manufacturer normally rates the truck at a maximum load at a given distance from the front face of the forks. The specified distance from the front face of the forks to the line of action of the load is commonly called the load center. Because larger trucks normally handle loads that are physically larger, these vehicles have greater load centers. Trucks with a capacity of 30,000 pounds or less are normally rated at a given load weight at a 24-inch load center. Trucks with a capacity greater than 30,000 pounds are normally rated at a given load weight at a 36- or 48-inch load center. To safely operate the vehicle, the operator should always check the data plate to determine the maximum allowable weight at the rated load center.

A-5.3. *Although the true load-moment distance* is measured from the front wheels, this distance is greater than the distance from the front face of the forks. Calculating the maximum allowable load-moment using the load-center distance always provides a lower load-moment than the truck was designed to handle. When handling unusual loads, such as those that are larger than 48 inches long (the center of gravity is greater than 24 inches) or that have an offset center of gravity, etc., a maximum allowable load-moment should be calculated and used to determine whether a load can be safely handled. For example, if an operator is operating a 3000 pound capacity truck (with a 24-inch load center), the maximum allowable load-moment is 72,000 inch-pounds (3,000 times 24). If a load is 60 inches long (30-inch load center), then the maximum that this load can weigh is 2,400 pounds (72,000 divided by 30).

A-6. Lateral Stability.

A-6.1. *The vehicle's lateral stability* is determined by the line of action's position (a vertical line that passes through the combined vehicle's and load's center of gravity) relative to the stability triangle. When the vehicle is not loaded, the truck's center of gravity location is the only factor to be considered in determining the truck's stability. As long as the line of action of the combined vehicle's and load's center of gravity falls within the stability triangle, the truck is stable and will not tip over. However, if the line of action falls outside the stability triangle, the truck is not stable and may tip over. Refer to Figure 2.

A-6.2. *Factors that affect* the vehicle's lateral stability include the load's placement on the truck, the height of the load above the surface on which the vehicle is operating, and the vehicle's degree of lean.

A-7. Dynamic Stability.

A-7.1. *Up to this point,* the stability of a powered industrial truck has been discussed without considering the dynamic forces that result when the vehicle and load are put into motion. The weight's transfer and the resultant shift in the center of gravity due to the dynamic forces created when the machine is moving, braking, cornering, lifting, tilting, and lowering loads, etc., are important stability considerations.

A-7.2. *When determining whether a load can be safely handled,* the operator should exercise extra caution when handling loads that cause the vehicle to approach its maximum design characteristics. For example, if an operator must handle a maximum load, the load should be carried at the lowest position possible, the truck should be accelerated slowly and evenly, and the forks should be tilted forward cautiously. However, no precise rules can be formulated to cover all of these eventualities.

§1910.332

Training - Electrical

(a) **Scope.** The training requirements contained in this section apply to employees who face a risk of electric shock that is not reduced to a safe level by the electrical installation requirements of §§1910.303 through 1910.308. [§1910.332(a)]

Note: Employees in occupations listed in Table S-4 face such a risk and are required to be trained. Other employees who also may reasonably be expected to face a comparable risk of injury due to electric shock or other electrical hazards must also be trained.

(b) **Content of training.** [§1910.332(b)]

(1) *Practices addressed in this standard.* Employees shall be trained in and familiar with the safety-related work practices required by §§1910.331 through 1910.335 that pertain to their respective job assignments. [§1910.332(b)(1)]

(2) *Additional requirements for unqualified persons.* Employees who are covered by paragraph (a) of this section but who are not qualified persons shall also be trained in and familiar with any electrically related safety practices not specifically addressed by §§1910.331 through 1910.335 but which are necessary for their safety. [§1910.332(b)(2)]

(3) *Additional requirements for qualified persons.* Qualified persons (i.e., those permitted to work on or near exposed energized parts) shall, at a minimum, be trained in and familiar with the following: [§1910.332(b)(3)]

(i) *The skills and techniques necessary* to distinguish exposed live parts from other parts of electric equipment, [§1910.332(b)(3)(i)]

(ii) *The skills and techniques necessary* to determine the nominal voltage of exposed live parts, and [§1910.332(b)(3)(ii)]

(iii) *The clearance distances specified* in §1910.333(c) and the corresponding voltages to which the qualified person will be exposed. [§1910.332(b)(3)(iii)]

Note 1: For the purposes of §§1910.331 through 1910.335, a person must have the training required by paragraph (b)(3) of this section in order to be considered a qualified person.

Note 2: Qualified persons whose work on energized equipment involves either direct contact or contact by means of tools or materials must also have the training needed to meet §1910.333(c)(2).

(c) **Type of training.** The training required by this section shall be of the classroom or on-the-job type. The degree of training provided shall be determined by the risk to the employee. [§1910.332(c)]

Table S-4 — Typical Occupational Categories of Employees Facing a Higher Than Normal Risk of Electrical Accident

Occupation
Blue collar supervisors[1]
Electrical and electronic engineers[1]
Electrical and electronic equipment assemblers[1]
Electrical and electronic technicians[1]
Electricians
Industrial machine operators[1]
Material handling equipment operators[1]
Mechanics and repairers[1]
Painters[1]
Riggers and roustabouts[1]
Stationary engineers[1]
Welders

[1] Workers in these groups do not need to be trained if their work or the work of those they supervise does not bring them or the employees they supervise close enough to exposed parts of electric circuits operating at 50 volts or more to ground for a hazard to exist.

[55 FR 32016, Aug. 6, 1990]

§1910.333

Selection and use of work practices - Electrical

(a) **General.** Safety-related work practices shall be employed to prevent electric shock or other injuries resulting from either direct or indirect electrical contacts, when work is performed near or on equipment or circuits which are or may be energized. The specific safety-related work practices shall be consistent with the nature and extent of the associated electrical hazards. [§1910.333(a)]

(1) *Deenergized parts.* Live parts to which an employee may be exposed shall be deenergized before the employee works on or near them, unless the employer can demonstrate that deenergizing introduces additional or increased hazards or is infeasible due to equipment design or operational limitations. Live parts that operate at less than 50 volts to ground need not be deenergized if there will be no increased exposure to electrical burns or to explosion due to electric arcs. [§1910.333(a)(1)]

Note 1: Examples of increased or additional hazards include interruption of life support equipment, deactivation of emergency alarm systems, shutdown of hazardous location ventilation equipment, or removal of illumination for an area.

Note 2: Examples of work that may be performed on or near energized circuit parts because of infeasibility due to equipment design or operational limitations include testing of electric circuits that can only be performed with the circuit energized and work on circuits that form an integral part of a continuous industrial process in a chemical plant that would otherwise need to be completely shut down in order to permit work on one circuit or piece of equipment.

Note 3: Work on or near deenergized parts is covered by paragraph (b) of this section.

(2) *Energized parts.* If the exposed live parts are not deenergized (i.e., for reasons of increased or additional hazards or infeasibility), other safety-related work practices shall be used to protect employees who may be exposed to the electrical hazards involved. Such work practices shall protect employees against contact with energized circuit parts directly with any part of their body or indirectly through some other conductive object. The work practices that are used shall be suitable for the conditions under which the work is to be performed and for the voltage level of the exposed electric conductors or circuit parts. Specific work practice requirements are detailed in paragraph (c) of this section. [§1910.333(a)(2)]

(b) **Working on or near exposed deenergized parts.** [§1910.333(b)]

(1) *Application.* This paragraph applies to work on exposed deenergized parts or near enough to them to expose the employee to any electrical hazard they present. Conductors and parts of electric equipment that have been deenergized but have not been locked out or tagged in accordance with paragraph (b) of this section shall be treated as energized parts, and paragraph (c) of this section applies to work on or near them. [§1910.333(b)(1)]

(2) *Lockout and tagging.* While any employee is exposed to contact with parts of fixed electric equipment or circuits which have been deenergized, the circuits energizing the parts shall be locked out or tagged or both in accordance with the requirements of this paragraph. The requirements shall be followed in the order in which they are presented (i.e., paragraph (b)(2)(i) first, then paragraph (b)(2)(ii), etc.). [§1910.333(b)(2)]

Note 1: As used in this section, fixed equipment refers to equipment fastened in place or connected by permanent wiring methods.

Note 2: Lockout and tagging procedures that comply with paragraphs (c) through (f) of §1910.147 will also be deemed to comply with paragraph (b)(2) of this section provided that:
(1) *The procedures address* the electrical safety hazards covered by this Subpart; and
(2) *The procedures also incorporate* the requirements of paragraphs (b)(2)(iii)(D) and (b)(2)(iv)(B) of this section.

(i) *Procedures.* The employer shall maintain a written copy of the procedures outlined in paragraph (b)(2) and shall make it available for inspection by employees and by the Assistant Secretary of Labor and his or her authorized representatives. [§1910.333(b)(2)(i)]

Note: The written procedures may be in the form of a copy of paragraph (b) of this section.

(ii) *Deenergizing equipment.* [§1910.333(b)(2)(ii)]

[A] Safe procedures for deenergizing circuits and equipment shall be determined before circuits or equipment are deenergized. [§1910.333(b)(2)(ii)[A]]

[B] The circuits and equipment to be worked on shall be disconnected from all electric energy sources. Control circuit

devices, such as push buttons, selector switches, and interlocks, may not be used as the sole means for deenergizing circuits or equipment. Interlocks for electric equipment may not be used as a substitute for lockout and tagging procedures. [§1910.333(b)(2)(ii)[B]]

[C] *Stored electric energy* which might endanger personnel shall be released. Capacitors shall be discharged and high capacitance elements shall be short-circuited and grounded, if the stored electric energy might endanger personnel. [§1910.333(b)(2)(ii)[C]]

Note: If the capacitors or associated equipment are handled in meeting this requirement, they shall be treated as energized.

[D] *Stored non-electrical energy* in devices that could reenergize electric circuit parts shall be blocked or relieved to the extent that the circuit parts could not be accidentally energized by the device. [§1910.333(b)(2)(ii)[D]]

(iii) *Application of locks and tags.* [§1910.333(b)(2)(iii)]

[A] *A lock and a tag shall be placed* on each disconnecting means used to deenergize circuits and equipment on which work is to be performed, except as provided in paragraphs (b)(2)(iii)(C) and (b)(2)(iii)(E) of this section. The lock shall be attached so as to prevent persons from operating the disconnecting means unless they resort to undue force or the use of tools. [§1910.333(b)(2)(iii)[A]]

[B] *Each tag shall contain a statement* prohibiting unauthorized operation of the disconnecting means and removal of the tag. [§1910.333(b)(2)(iii)[B]]

[C] *If a lock cannot be applied,* or if the employer can demonstrate that tagging procedures will provide a level of safety equivalent to that obtained by the use of a lock, a tag may be used without a lock. [§1910.333(b)(2)(iii)[C]]

[D] *A tag used without a lock,* as permitted by paragraph (b)(2)(iii)(C) of this section, shall be supplemented by at least one additional safety measure that provides a level of safety equivalent to that obtained by the use of a lock. Examples of additional safety measures include the removal of an isolating circuit element, blocking of a controlling switch, or opening of an extra disconnecting device. [§1910.333(b)(2)(iii)[D]]

[E] *A lock may be placed* without a tag only under the following conditions: [§1910.333(b)(2)(iii)[E]]

[1] *Only one circuit or piece of equipment is deenergized, and* [§1910.333(b)(2)(iii)[E][1]]

[2] *The lockout period* does not extend beyond the work shift, and [§1910.333(b)(2)(iii)[E][2]]

[3] *Employees exposed to the hazards* associated with reenergizing the circuit or equipment are familiar with this procedure. [§1910.333(b)(2)(iii)[E][3]]

(iv) *Verification of deenergized condition.* The requirements of this paragraph shall be met before any circuits or equipment can be considered and worked as deenergized. [§1910.333(b)(2)(iv)]

[A] *A qualified person shall operate* the equipment operating controls or otherwise verify that the equipment cannot be restarted. [§1910.333(b)(2)(iv)[A]]

[B] *A qualified person shall use* test equipment to test the circuit elements and electrical parts of equipment to which employees will be exposed and shall verify that the circuit elements and equipment parts are deenergized. The test shall also determine if any energized condition exists as a result of inadvertently induced voltage or unrelated voltage backfeed even though specific parts of the circuit have been deenergized and presumed to be safe. If the circuit to be tested is over 600 volts, nominal, the test equipment shall be checked for proper operation immediately before and immediately after this test. [§1910.333(b)(2)(iv)[B]]

(v) *Reenergizing equipment.* These requirements shall be met, in the order given, before circuits or equipment are reenergized, even temporarily. [§1910.333(b)(2)(v)]

[A] *A qualified person shall conduct tests* and visual inspections, as necessary, to verify that all tools, electrical jumpers, shorts, grounds, and other such devices have been removed, so that the circuits and equipment can be safely energized. [§1910.333(b)(2)(v)[A]]

[B] *Employees exposed to the hazards* associated with reenergizing the circuit or equipment shall be warned to stay clear of circuits and equipment. [§1910.333(b)(2)(v)[B]]

[C] *Each lock and tag shall be removed* by the employee who applied it or under his or her direct supervision. However, if this employee is absent from the workplace, then the lock or tag may be removed by a qualified person designated to perform this task provided that: [§1910.333(b)(2)(v)[C]]

[1] *The employer ensures* that the employee who applied the lock or tag is not available at the workplace, and [§1910.333(b)(2)(v)[C][1]]

[2] *The employer ensures* that the employee is aware that the lock or tag has been removed before he or she resumes work at that workplace. [§1910.333(b)(2)(v)[C][2]]

[D] *There shall be a visual determination* that all employees are clear of the circuits and equipment. [§1910.333(b)(2)(v)[D]]

(c) Working on or near exposed energized parts. [§1910.333(c)]

(1) *Application.* This paragraph applies to work performed on exposed live parts (involving either direct contact or contact by means of tools or materials) or near enough to them for employees to be exposed to any hazard they present. [§1910.333(c)(1)]

(2) *Work on energized equipment.* Only qualified persons may work on electric circuit parts or equipment that have not been deenergized under the procedures of paragraph (b) of this section. Such persons shall be capable of working safely on energized circuits and shall be familiar with the proper use of special precautionary techniques, personal protective equipment, insulating and shielding materials, and insulated tools. [§1910.333(c)(2)]

(3) *Overhead lines.* If work is to be performed near overhead lines, the lines shall be deenergized and grounded, or other protective measures shall be provided before work is started. If the lines are to be deenergized, arrangements shall be made with the person or organization that operates or controls the electric circuits involved to deenergize and ground them. If protective measures, such as guarding, isolating, or insulating are provided, these precautions shall prevent employees from contacting such lines directly with any part of their body or indirectly through conductive materials, tools, or equipment. [§1910.333(c)(3)]

Note: The work practices used by qualified persons installing insulating devices on overhead power transmission or distribution lines are covered by §1910.269 of this Part, not by §§1910.332 through 1910.335 of this Part. Under paragraph (c)(2) of this section, unqualified persons are prohibited from performing this type of work.

(i) *Unqualified persons.* [§1910.333(c)(3)(i)]

[A] *When an unqualified person is working* in an elevated position near overhead lines, the location shall be such that the person and the longest conductive object he or she may contact cannot come closer to any unguarded, energized overhead line than the following distances: [§1910.333(c)(3)(i)[A]]

[1] *For voltages to ground 50kV or below — 10 ft. (305 cm);* [§1910.333(c)(3)(i)[A][1]]

[2] *For voltages to ground* over 50kV — 10 ft. (305 cm) plus 4 in. (10 cm) for every 10kV over 50kV. [§1910.333(c)(3)(i)[A][2]]

[B] *When an unqualified person is working* on the ground in the vicinity of overhead lines, the person may not bring any conductive object closer to unguarded, energized overhead lines than the distances given in paragraph (c)(3)(i)(A) of this section. [§1910.333(c)(3)(i)[B]]

Note: For voltages normally encountered with overhead power lines, objects which do not have an insulating rating for the voltage involved are considered to be conductive.

(ii) *Qualified persons.* When a qualified person is working in the vicinity of overhead lines, whether in an elevated position or on the ground, the person may not approach or take any conductive object without an approved insulating handle closer to exposed energized parts than shown in Table S-5 unless: [§1910.333(c)(3)(ii)]

[A] *The person is insulated from the energized part* (gloves, with sleeves if necessary, rated for the voltage involved are considered to be insulation of the person from the energized part on which work is performed), or [§1910.333(c)(3)(ii)[A]]

[B] *The energized part is insulated* both from all other conductive objects at a different potential and from the person, or [§1910.333(c)(3)(ii)[B]]

[C] *The person is insulated* from all conductive objects at a potential different from that of the energized part. [§1910.333(c)(3)(ii)[C]]

Table S-5 — Approach Distances for Qualified Employees — Alternating Current

Voltage range (phase to phase)	Minimum approach distance
300V and less	Avoid contact
Over 300V, not over 750V	1 ft. 0 in. (30.5 cm)
Over 750V, not over 2kV	1 ft. 6 in. (46 cm)
Over 2kV, not over 15kV	2 ft. 0 in. (61 cm)
Over 15kV, not over 37kV	3 ft. 0 in. (91 cm)
Over 37kV, not over 87.5kV	3 ft. 6 in. (107 cm)
Over 87.5kV, not over 121kV	4 ft. 0 in. (122 cm)
Over 121kV, not over 140kV	4 ft. 6 in. (137 cm)

(iii) *Vehicular and mechanical equipment.* [§1910.333(c)(3)(iii)]

[A] Any vehicle or mechanical equipment capable of having parts of its structure elevated near energized overhead lines shall be operated so that a clearance of 10 ft. (305 cm) is maintained. If the voltage is higher than 50kV, the clearance shall be increased 4 in. (10 cm) for every 10kV over that voltage. However, under any of the following conditions, the clearance may be reduced: [§1910.333(c)(3)(iii)[A]]

[1] If the vehicle is in transit with its structure lowered, the clearance may be reduced to 4 ft. (122 cm). If the voltage is higher than 50kV, the clearance shall be increased 4 in. (10 cm) for every 10kV over that voltage. [§1910.333(c)(3)(iii)[A][1]]

[2] If insulating barriers are installed to prevent contact with the lines, and if the barriers are rated for the voltage of the line being guarded and are not a part of or an attachment to the vehicle or its raised structure, the clearance may be reduced to a distance within the designed working dimensions of the insulating barrier. [§1910.333(c)(3)(iii)[A][2]]

[3] If the equipment is an aerial lift insulated for the voltage involved, and if the work is performed by a qualified person, the clearance (between the uninsulated portion of the aerial lift and the power line) may be reduced to the distance given in Table S-5. [§1910.333(c)(3)(iii)[A][3]]

[B] Employees standing on the ground may not contact the vehicle or mechanical equipment or any of its attachments, unless: [§1910.333(c)(3)(iii)[B]]

[1] The employee is using protective equipment rated for the voltage; or [§1910.333(c)(3)(iii)[B][1]]

[2] The equipment is located so that no uninsulated part of its structure (that portion of the structure that provides a conductive path to employees on the ground) can come closer to the line than permitted in paragraph (c)(3)(iii) of this section. [§1910.333(c)(3)(iii)[B][2]]

[C] If any vehicle or mechanical equipment capable of having parts of its structure elevated near energized overhead lines is intentionally grounded, employees working on the ground near the point of grounding may not stand at the grounding location whenever there is a possibility of overhead line contact. Additional precautions, such as the use of barricades or insulation, shall be taken to protect employees from hazardous ground potentials, depending on earth resistivity and fault currents, which can develop within the first few feet or more outward from the grounding point. [§1910.333(c)(3)(iii)[C]]

(4) *Illumination.* [§1910.333(c)(4)]

(i) *Employees may not enter spaces* containing exposed energized parts, unless illumination is provided that enables the employees to perform the work safely. [§1910.333(c)(4)(i)]

(ii) *Where lack of illumination or an obstruction* precludes observation of the work to be performed, employees may not perform tasks near exposed energized parts. Employees may not reach blindly into areas which may contain energized parts. [§1910.333(c)(4)(ii)]

(5) *Confined or enclosed work spaces.* When an employee works in a confined or enclosed space (such as a manhole or vault) that contains exposed energized parts, the employer shall provide, and the employee shall use, protective shields, protective barriers, or insulating materials as necessary to avoid inadvertent contact with these parts. Doors, hinged panels, and the like shall be secured to prevent their swinging into an employee and causing the employee to contact exposed energized parts. [§1910.333(c)(5)]

(6) *Conductive materials and equipment.* Conductive materials and equipment that are in contact with any part of an employee's body shall be handled in a manner that will prevent them from contacting exposed energized conductors or circuit parts. If an employee must handle long dimensional conductive objects (such as ducts and pipes) in areas with exposed live parts, the employer shall institute work practices (such as the use of insulation, guarding, and material handling techniques) which will minimize the hazard. [§1910.333(c)(6)]

(7) *Portable ladders.* Portable ladders shall have nonconductive siderails if they are used where the employee or the ladder could contact exposed energized parts. [§1910.333(c)(7)]

(8) *Conductive apparel.* Conductive articles of jewelry and clothing (such as watch bands, bracelets, rings, key chains, necklaces, metalized aprons, cloth with conductive thread, or metal headgear) may not be worn if they might contact exposed energized parts. However, such articles may be worn if they are rendered nonconductive by covering, wrapping, or other insulating means. [§1910.333(c)(8)]

(9) *Housekeeping duties.* Where live parts present an electrical contact hazard, employees may not perform housekeeping duties at such close distances to the parts that there is a possibility of contact, unless adequate safeguards (such as insulating equipment or barriers) are provided. Electrically conductive cleaning materials (including conductive solids such as steel wool, metalized cloth, and silicon carbide, as well as conductive liquid solutions) may not be used in proximity to energized parts unless procedures are followed which will prevent electrical contact. [§1910.333(c)(9)]

(10) *Interlocks.* Only a qualified person following the requirements of paragraph (c) of this section may defeat an electrical safety interlock, and then only temporarily while he or she is working on the equipment. The interlock system shall be returned to its operable condition when this work is completed. [§1910.333(c)(10)]

[55 FR 32016, Aug. 6, 1990; 55 FR 42053, Nov. 1, 1990, as amended at 59 FR 4476, Jan. 31, 1994]

GENERAL

§1910.401

Scope and application - Diving

(a) **Scope.**[§1910.401(a)]

(1) *This subpart (standard)* applies to every place of employment within the waters of the United States, or within any State, the District of Columbia, the Commonwealth of Puerto Rico, the Virgin Islands, American Samoa, Guam, the Trust Territory of the Pacific Islands, Wake Island, Johnston Island, the Canal Zone, or within the Outer Continental Shelf lands as defined in the Outer Continental Shelf Lands Act (67 Stat. 462, 43 U.S.C. 1331), where diving and related support operations are performed.[§1910.401(a)(1)]

(2) *This standard applies to diving* and related support operations conducted in connection with all types of work and employments, including general industry, construction, ship repairing, shipbuilding, shipbreaking and longshoring. However, this standard does not apply to any diving operation:[§1910.401(a)(2)]

(i) *Performed solely for instructional purposes,* using open-circuit, compressed-air SCUBA and conducted within the no-decompression limits;[§1910.401(a)(2)(i)]

(ii) *Performed solely* for search, rescue, or related public safety purposes by or under the control of a governmental agency; or[§1910.401(a)(2)(ii)]

(iii) *Governed by 45 CFR part 46* (Protection of Human Subjects, U.S. Department of Health and Human Services) or equivalent rules or regulations established by another federal agency, which regulate research, development, or related purposes involving human subjects.[§1910.401(a)(2)(iii)]

(iv) *Defined as scientific diving* and which is under the direction and control of a diving program containing at least the following elements:[§1910.401(a)(2)(iv)]

[A] Diving safety manual which includes at a minimum: Procedures covering all diving operations specific to the program; procedures for emergency care, including recompression and evacuation; and criteria for diver training and certification.[§1910.401(a)(2)(iv)[A]]

[B] Diving control (safety) board, with the majority of its members being active divers, which shall at a minimum have the authority to: Approve and monitor diving projects; review and revise the diving safety manual; assure compliance with the manual; certify the depths to which a diver has been trained; take disciplinary action for unsafe practices; and, assure adherence to the buddy system (a

diver is accompanied by and is in continuous contact with another diver in the water) for SCUBA diving. [§1910.401(a)(2)(iv)[B]]

(3) *Alternative requirements* for recreational diving instructors and diving guides. Employers of recreational diving instructors and diving guides are not required to comply with the decompression-chamber requirements specified by paragraphs (b)(2) and (c)(3)(iii) of Sec. 1910.423 and paragraph (b)(1) of §1910.426 when they meet all of the following conditions:[§1910.401(a)(3)]

(i) *The instructor or guide* is engaging solely in recreational diving instruction or dive-guiding operations;[§1910.401(a)(3)(i)]

(ii) *The instructor or guide* is diving within the no-decompression limits in these operations;[§1910.401(a)(3)(ii)]

(iii) *The instructor or guide* is using a nitrox breathing-gas mixture consisting of a high percentage of oxygen (more than 22% by volume) mixed with nitrogen;[§1910.401(a)(3)(iii)]

(iv) *The instructor or guide* is using an open-circuit, semi-closed-circuit, or closed-circuit self-contained underwater breathing apparatus (SCUBA); and[§1910.401(a)(3)(iv)]

(v) *The employer of the instructor or guide* is complying with all requirements of Appendix C of this subpart.[§1910.401(a)(3)(v)]

(b) **Application in emergencies.** An employer may deviate from the requirements of this standard to the extent necessary to prevent or minimize a situation which is likely to cause death, serious physical harm, or major environmental damage, provided that the employer:[§1910.401(b)]

(1) *Notifies the Area Director,* Occupational Safety and Health Administration within 48 hours of the onset of the emergency situation indicating the nature of the emergency and extent of the deviation from the prescribed regulations; and[§1910.401(b)(1)]

(2) *Upon request from the Area* Director, submits such information in writing.[§1910.401(b)(2)]

(c) **Employer obligation.** The employer shall be responsible for compliance with:[§1910.401(c)]

(1) *All provisions of this standard of general applicability; and* [§1910.401(c)(1)]

(2) *All requirements pertaining to specific* diving modes to the extent diving operations in such modes are conducted. [§1910.401(c)(2)]

[42 FR 37668, July 22, 1977, as amended at 47 FR 53365, Nov. 26, 1982; 58 FR 35310, June 30, 1993; 69 FR 7363, Feb. 17, 2004]

§1910.402
Definitions - Diving

As used in this standard, the listed terms are defined as follows:

Acfm: Actual cubic feet per minute.[§1910.402]

ASME Code or equivalent: ASME (American Society of Mechanical Engineers) Boiler and Pressure Vessel Code, Section VIII, or an equivalent code which the employer can demonstrate to be equally effective.[§1910.402]

ATA: Atmosphere absolute.[§1910.402]

Bell: An enclosed compartment, pressurized (closed bell) or unpressurized (open bell), which allows the diver to be transported to and from the underwater work area and which may be used as a temporary refuge during diving operations.[§1910.402]

Bottom time: The total elasped time measured in minutes from the time when the diver leaves the surface in descent to the time that the diver begins ascent.[§1910.402]

Bursting pressure: The pressure at which a pressure containment device would fail structurally.[§1910.402]

Cylinder: A pressure vessel for the storage of gases.[§1910.402]

Decompression chamber: A pressure vessel for human occupancy such as a surface decompression chamber, closed bell, or deep diving system used to decompress divers and to treat decompression sickness.[§1910.402]

Decompression sickness: A condition with a variety of symptoms which may result from gas or bubbles in the tissues of divers after pressure reduction.[§1910.402]

Decompression table: A profile or set of profiles of depth-time relationships for ascent rates and breathing mixtures to be followed after a specific depth-time exposure or exposures.[§1910.402]

Dive-guiding operations means leading groups of sports divers, who use an open-circuit, semi-closed-circuit, or closed-circuit self-contained underwater breathing apparatus, to local undersea diving locations for recreational purposes.[§1910.402]

Dive location: A surface or vessel from which a diving operation is conducted.[§1910.402]

Dive-location reserve breathing gas: A supply system of air or mixed-gas (as appropriate) at the dive location which is independent of the primary supply system and sufficient to support divers during the planned decompression.[§1910.402]

Dive team: Divers and support employees involved in a diving operation, including the designated person-in-charge.[§1910.402]

Diver: An employee working in water using underwater apparatus which supplies compressed breathing gas at the ambient pressure. [§1910.402]

Diver-carried reserve breathing gas: A diver-carried supply of air or mixed gas (as appropriate) sufficient under standard operating conditions to allow the diver to reach the surface, or another source of breathing gas, or to be reached by a standby diver.[§1910.402]

Diving mode: A type of diving requiring specific equipment, procedures and techniques (SCUBA, surface-supplied air, or mixed gas). [§1910.402]

Fsw: Feet of seawater (or equivalent static pressure head).[§1910.402]

Heavy gear: Diver-worn deep-sea dress including helmet, breastplate, dry suit, and weighted shoes.[§1910.402]

Hyperbaric conditions: Pressure conditions in excess of surface pressure.[§1910.402]

Inwater stage: A suspended underwater platform which supports a diver in the water.[§1910.402]

Liveboating: The practice of supporting a surfaced-supplied air or mixed gas diver from a vessel which is underway.[§1910.402]

Mixed-gas diving: A diving mode in which the diver is supplied in the water with a breathing gas other than air.[§1910.402]

No-decompression limits: The depth-time limits of the "no-decompression limits and repetitive dive group designation table for no-decompression air dives", U.S. Navy Diving Manual or equivalent limits which the employer can demonstrate to be equally effective. [§1910.402]

Psi(g): Pounds per square inch (gauge).[§1910.402]

Recreational diving instruction means training diving students in the use of recreational diving procedures and the safe operation of diving equipment, including an open-circuit, semi-closed-circuit, or closed-circuit self-contained underwater breathing apparatus, during dives.[§1910.402]

Scientific diving means diving performed solely as a necessary part of a scientific, research, or educational activity by employees whose sole purpose for diving is to perform scientific research tasks. Scientific diving does not include performing any tasks usually associated with commercial diving such as: Placing or removing heavy objects underwater; inspection of pipelines and similar objects; construction; demolition; cutting or welding; or the use of explosives.[§1910.402]

SCUBA diving: A diving mode independent of surface supply in which the diver uses open circuit self-contained underwater breathing apparatus.[§1910.402]

Standby diver: A diver at the dive location available to assist a diver in the water.[§1910.402]

Surface-supplied air diving: A diving mode in which the diver in the water is supplied from the dive location with compressed air for breathing.[§1910.402]

Treatment table: A depth-time and breathing gas profile designed to treat decompression sickness.[§1910.402]

Umbilical: The composite hose bundle between a dive location and a diver or bell, or between a diver and a bell, which supplies the diver or bell with breathing gas, communications, power, or heat as appropriate to the diving mode or conditions, and includes a safety line between the diver and the dive location.[§1910.402]

Volume tank: A pressure vessel connected to the outlet of a compressor and used as an air reservoir.[§1910.402]

Working pressure: The maximum pressure to which a pressure containment device may be exposed under standard operating conditions.[§1910.402]

[42 FR 37668, July 22, 1977, as amended at 47 FR 53365, Nov. 26, 1982; 69 FR 7363, Feb. 17, 2004]

PERSONNEL REQUIREMENTS

§1910.410

Qualifications of dive team - Diving

(a) **General.**[§1910.410(a)]

(1) *Each dive team member* shall have the experience or training necessary to perform assigned tasks in a safe and healthful manner.[§1910.410(a)(1)]

(2) *Each dive team member* shall have experience or training in the following:[§1910.410(a)(2)]

(i) *The use of tools,* equipment and systems relevant to assigned tasks;[§1910.410(a)(2)(i)]

(ii) *Techniques of the assigned diving mode; and* [§1910.410(a)(2)(ii)]

(iii) *Diving operations and emergency procedures.* [§1910.410(a)(2)(iii)]

(3) *All dive team members* shall be trained in cardiopulmonary resuscitation and first aid (American Red Cross standard course or equivalent).[§1910.410(a)(3)]

(4) *Dive team members* who are exposed to or control the exposure of others to hyperbaric conditions shall be trained in diving-related physics and physiology.[§1910.410(a)(4)]

(b) **Assignments.**[§1910.410(b)]

(1) *Each dive team member* shall be assigned tasks in accordance with the employee's experience or training, except that limited additional tasks may be assigned to an employee undergoing training provided that these tasks are performed under the direct supervision of an experienced dive team member. [§1910.410(b)(1)]

(2) *The employer shall not require* a dive team member to be exposed to hyperbaric conditions against the employee's will, except when necessary to complete decompression or treatment procedures.[§1910.410(b)(2)]

(3) *The employer shall not permit* a dive team member to dive or be otherwise exposed to hyperbaric conditions for the duration of any temporary physical impairment or condition which is known to the employer and is likely to affect adversely the safety or health of a dive team member.[§1910.410(b)(3)]

(c) **Designated person-in-charge.**[§1910.410(c)]

(1) *The employer or an employee designated* by the employer shall be at the dive location in charge of all aspects of the diving operation affecting the safety and health of dive team members. [§1910.410(c)(1)]

(2) *The designated person-in-charge* shall have experience and training in the conduct of the assigned diving operation. [§1910.410(c)(2)]

GENERAL OPERATIONS PROCEDURES

§1910.420

Safe practices manual - Diving

(a) **General.** The employer shall develop and maintain a safe practices manual which shall be made available at the dive location to each dive team member.[§1910.420(a)]

(b) **Contents.**[§1910.420(b)]

(1) *The safe practices manual* shall contain a copy of this standard and the employer's policies for implementing the requirements of this standard.[§1910.420(b)(1)]

(2) *For each diving mode engaged in,* the safe practices manual shall include:[§1910.420(b)(2)]

(i) *Safety procedures and checklists for diving operations;* [§1910.420(b)(2)(i)]

(ii) *Assignments and responsibilities of the dive team members;* [§1910.420(b)(2)(ii)]

(iii) *Equipment procedures and checklists; and*[§1910.420(b)(2)(iii)]

(iv) *Emergency procedures* for fire, equipment failure, adverse environmental conditions, and medical illness and injury. [§1910.420(b)(2)(iv)]

[42 FR 37668, July 22, 1977, as amended at 49 FR 18295, Apr. 30, 1984]

§1910.421

Pre-dive procedures - Diving

(a) **General.** The employer shall comply with the following requirements prior to each diving operation, unless otherwise specified. [§1910.421(a)]

(b) **Emergency aid.** A list shall be kept at the dive location of the telephone or call numbers of the following:[§1910.421(b)]

(1) *An operational decompression chamber (if not at the dive location);*[§1910.421(b)(1)]

(2) *Accessible hospitals;*[§1910.421(b)(2)]

(3) *Available physicians;*[§1910.421(b)(3)]

(4) *Available means of transportation; and*[§1910.421(b)(4)]

(5) *The nearest U.S. Coast Guard Rescue Coordination Center.* [§1910.421(b)(5)]

(c) **First aid supplies.**[§1910.421(c)]

(1) *A first aid kit* appropriate for the diving operation and approved by a physician shall be available at the dive location. [§1910.421(c)(1)]

(2) *When used in a decompression chamber or bell,* the first aid kit shall be suitable for use under hyperbaric conditions. [§1910.421(c)(2)]

(3) *In addition to any other first aid supplies,* an American Red Cross standard first aid handbook or equivalent, and a bag-type manual resuscitator with transparent mask and tubing shall be available at the dive location.[§1910.421(c)(3)]

(d) **Planning and assessment.** Planning of a diving operation shall include an assessment of the safety and health aspects of the following:[§1910.421(d)]

(1) *Diving mode;*[§1910.421(d)(1)]

(2) *Surface and underwater conditions and hazards;*[§1910.421(d)(2)]

(3) *Breathing gas supply (including reserves);*[§1910.421(d)(3)]

(4) *Thermal protection;*[§1910.421(d)(4)]

(5) *Diving equipment and systems;*[§1910.421(d)(5)]

(6) *Dive team assignments* and physical fitness of dive team members (including any impairment known to the employer); [§1910.421(d)(6)]

(7) *Repetitive dive designation* or residual inert gas status of dive team members;[§1910.421(d)(7)]

(8) *Decompression and treatment procedures (including altitude corrections); and*[§1910.421(d)(8)]

(9) *Emergency procedures.*[§1910.421(d)(9)]

(e) **Hazardous activities.** To minimize hazards to the dive team, diving operations shall be coordinated with other activities in the vicinity which are likely to interfere with the diving operation. [§1910.421(e)]

(f) **Employee briefing.**[§1910.421(f)]

(1) *Dive team members shall be briefed on:*[§1910.421(f)(1)]

(i) *The tasks to be undertaken;*[§1910.421(f)(1)(i)]

(ii) *Safety procedures for the diving mode;*[§1910.421(f)(1)(ii)]

(iii) *Any unusual hazards* or environmental conditions likely to affect the safety of the diving operation; and[§1910.421(f)(1)(iii)]

(iv) *Any modifications to operating procedures* necessitated by the specific diving operation.[§1910.421(f)(1)(iv)]

(2) *Prior to making individual dive team member assignments,* the employer shall inquire into the dive team member's current state of physical fitness, and indicate to the dive team member the procedure for reporting physical problems or adverse physiological effects during and after the dive.[§1910.421(f)(2)]

(g) **Equipment inspection.** The breathing gas supply system including reserve breathing gas supplies, masks, helmets, thermal protection, and bell handling mechanism (when appropriate) shall be inspected prior to each dive.[§1910.421(g)]

(h) **Warning signal.** When diving from surfaces other than vessels in areas capable of supporting marine traffic, a rigid replica of the international code flag "A" at least one meter in height shall be displayed at the dive location in a manner which allows all-round visibility, and shall be illuminated during night diving operations. [§1910.421(h)]

[42 FR 37668, July 22, 1977, as amended at 47 FR 14706, Apr. 6, 1982; 54 FR 24334, June 7, 1989]

§1910.422

Procedures during dive - Diving

(a) **General.** The employer shall comply with the following requirements which are applicable to each diving operation unless otherwise specified.[§1910.422(a)]

(b) **Water entry and exit.**[§1910.422(b)]

(1) *A means capable of supporting* the diver shall be provided for entering and exiting the water.[§1910.422(b)(1)]

(2) *The means provided for exiting* the water shall extend below the water surface.[§1910.422(b)(2)]

(3) *A means shall be provided to assist* an injured diver from the water or into a bell.[§1910.422(b)(3)]

(c) **Communications.**[§1910.422(c)]

(1) *An operational* two-way voice communication system shall be used between:[§1910.422(c)(1)]

(i) *Each surface-supplied air* or mixed-gas diver and a dive team member at the dive location or bell (when provided or required); and[§1910.422(c)(1)(i)]

(ii) *The bell and the dive location.*[§1910.422(c)(1)(ii)]

(2) *An operational,* two-way communication system shall be available at the dive location to obtain emergency assistance. [§1910.422(c)(2)]

(d) **Decompression tables.** Decompression, repetitive, and no-decompression tables (as appropriate) shall be at the dive location.[§1910.422(d)]

(e) **Dive profiles.** A depth-time profile, including when appropriate any breathing gas changes, shall be maintained for each diver during the dive including decompression.[§1910.422(e)]

(f) **Hand-held power tools and equipment.**[§1910.422(f)]

(1) *Hand-held electrical tools* and equipment shall be de-energized before being placed into or retrieved from the water. [§1910.422(f)(1)]

(2) *Hand-held power tools* shall not be supplied with power from the dive location until requested by the diver.[§1910.422(f)(2)]

(g) **Welding and burning.**[§1910.422(g)]

(1) *A current supply switch* to interrupt the current flow to the welding or burning electrode shall be:[§1910.422(g)(1)]

(i) *Tended by a dive team member* in voice communication with the diver performing the welding or burning; and [§1910.422(g)(1)(i)]

(ii) *Kept in the open position* except when the diver is welding or burning.[§1910.422(g)(1)(ii)]

(2) *The welding machine frame shall be grounded.*[§1910.422(g)(2)]

(3) *Welding and burning cables,* electrode holders, and connections shall be capable of carrying the maximum current required by the work, and shall be properly insulated.[§1910.422(g)(3)]

(4) *Insulated gloves* shall be provided to divers performing welding and burning operations.[§1910.422(g)(4)]

(5) *Prior to welding or burning* on closed compartments, structures or pipes, which contain a flammable vapor or in which a flammable vapor may be generated by the work, they shall be vented, flooded, or purged with a mixture of gases which will not support combustion.[§1910.422(g)(5)]

(h) **Explosives.**[§1910.422(h)]

(1) *Employers shall transport,* store, and use explosives in accordance with this section and the applicable provisions of Sec. 1910.109 and Sec. 1926.912 of Title 29 of the Code of Federal Regulations.[§1910.422(h)(1)]

(2) *Electrical continuity* of explosive circuits shall not be tested until the diver is out of the water.[§1910.422(h)(2)]

(3) *Explosives shall not be detonated while the diver is in the water.* [§1910.422(h)(3)]

(i) **Termination of dive.** The working interval of a dive shall be terminated when:[§1910.422(i)]

(1) *A diver requests termination;*[§1910.422(i)(1)]

(2) *A diver fails to respond* correctly to communications or signals from a dive team member;[§1910.422(i)(2)]

(3) *Communications are lost* and can not be quickly re-established between the diver and a dive team member at the dive location, and between the designated person-in-charge and the person controlling the vessel in liveboating operations; or[§1910.422(i)(3)]

(4) *A diver begins to use* diver-carried reserve breathing gas or the dive-location reserve breathing gas.[§1910.422(i)(4)]

§1910.423

Post-dive procedures - Diving

(a) **General.** The employer shall comply with the following requirements which are applicable after each diving operation, unless otherwise specified.[§1910.423(a)]

(b) **Precautions.**[§1910.423(b)]

(1) *After the completion of any dive, the employer shall:* [§1910.423(b)(1)]

(i) *Check the physical condition of the diver;*[§1910.423(b)(1)(i)]

(ii) *Instruct the diver to report* any physical problems or adverse physiological effects including symptoms of decompression sickness;[§1910.423(b)(1)(ii)]

(iii) *Advise the diver of the location* of a decompression chamber which is ready for use; and[§1910.423(b)(1)(iii)]

(iv) *Alert the diver to the potential hazards of flying after diving.* [§1910.423(b)(1)(iv)]

(2) *For any dive outside the no-decompression limits,* deeper than 100 fsw or using mixed gas as a breathing mixture, the employer shall instruct the diver to remain awake and in the vicinity of the decompression chamber which is at the dive location for at least one hour after the dive (including decompression or treatment as appropriate).[§1910.423(b)(2)]

(c) **Recompression capability.**[§1910.423(c)]

(1) *A decompression chamber* capable of recompressing the diver at the surface to a minimum of 165 fsw (6 ATA) shall be available at the dive location for:[§1910.423(c)(1)]

(i) *Surface-supplied air diving* to depths deeper than 100 fsw and shallower than 220 fsw;[§1910.423(c)(1)(i)]

(ii) *Mixed gas diving* shallower than 300 fsw; or[§1910.423(c)(1)(ii)]

(iii) *Diving outside the no-decompression limits shallower than 300 fsw.*[§1910.423(c)(1)(iii)]

(2) *A decompression chamber* capable of recompressing the diver at the surface to the maximum depth of the dive shall be available at the dive location for dives deeper than 300 fsw. [§1910.423(c)(2)]

(3) *The decompression chamber shall be:*[§1910.423(c)(3)]

(i) *Dual-lock;*[§1910.423(c)(3)(i)]

(ii) *Multiplace; and*[§1910.423(c)(3)(ii)]

(iii) *Located within 5 minutes of the dive location.* [§1910.423(c)(3)(iii)]

(4) *The decompression chamber shall be equipped with:* [§1910.423(c)(4)]

(i) *A pressure gauge* for each pressurized compartment designed for human occupancy;[§1910.423(c)(4)(i)]

(ii) *A built-in-breathing-system* with a minimum of one mask per occupant;[§1910.423(c)(4)(ii)]

(iii) *A two-way voice communication system* between occupants and a dive team member at the dive location; [§1910.423(c)(4)(iii)]

(iv) *A viewport; and*[§1910.423(c)(4)(iv)]

(v) *Illumination capability to light the interior.*[§1910.423(c)(4)(v)]

(5) *Treatment tables,* treatment gas appropriate to the diving mode, and sufficient gas to conduct treatment shall be available at the dive location.[§1910.423(c)(5)]

(6) *A dive team member shall be available* at the dive location during and for at least one hour after the dive to operate the decompression chamber (when required or provided). [§1910.423(c)(6)]

(d) **Record of dive.**[§1910.423(d)]

(1) *The following information* shall be recorded and maintained for each diving operation:[§1910.423(d)(1)]

(i) *Names of dive team members* including designated person-in- charge;[§1910.423(d)(1)(i)]

(ii) *Date, time, and location;*[§1910.423(d)(1)(ii)]

(iii) *Diving modes used;*[§1910.423(d)(1)(iii)]

(iv) *General nature of work performed;*[§1910.423(d)(1)(iv)]

(v) *Approximate underwater and surface conditions* (visibility, water temperature and current); and[§1910.423(d)(1)(v)]

(vi) *Maximum depth and bottom time for each diver.* [§1910.423(d)(1)(vi)]

(2) *For each dive outside the no-decompression limits,* deeper than 100 fsw or using mixed gas, the following additional information shall be recorded and maintained:[§1910.423(d)(2)]

(i) *Depth-time and breathing gas profiles;*[§1910.423(d)(2)(i)]

(ii) *Decompression table designation (including modification); and*[§1910.423(d)(2)(ii)]

(iii) *Elapsed time since last pressure exposure* if less than 24 hours or repetitive dive designation for each diver. [§1910.423(d)(2)(iii)]

(3) *For each dive* in which decompression sickness is suspected or symptoms are evident, the following additional information shall be recorded and maintained:[§1910.423(d)(3)]

(i) *Description of decompression sickness symptoms* (including depth and time of onset); and[§1910.423(d)(3)(i)]

(ii) *Description and results of treatment.*[§1910.423(d)(3)(ii)]

(e) **Decompression procedure assessment.** The employer shall: [§1910.423(e)]

(1) *Investigate and evaluate* each incident of decompression sickness based on the recorded information, consideration of the past performance of decompression table used, and individual susceptibility;[§1910.423(e)(1)]

(2) *Take appropriate corrective action* to reduce the probability of recurrence of decompression sickness; and[§1910.423(e)(2)]

(3) *Prepare a written evaluation* of the decompression procedure assessment, including any corrective action taken, within 45 days of the incident of decompression sickness.[§1910.423(e)(3)]

[42 FR 37668, July 22, 1977, as amended at 49 FR 18295, Apr. 30, 1984]

SPECIFIC OPERATIONS PROCEDURES

§1910.424

SCUBA diving

(a) **General.** Employers engaged in SCUBA diving shall comply with the following requirements, unless otherwise specified.[§1910.424(a)]

(b) **Limits.** SCUBA diving shall not be conducted:[§1910.424(b)]

(1) *At depths deeper than 130 fsw;*[§1910.424(b)(1)]

(2) *At depths deeper than 100 fsw* or outside the no-decompression limits unless a decompression chamber is ready for use; [§1910.424(b)(2)]

(3) *Against currents exceeding one (1) knot unless line-tended; or* [§1910.424(b)(3)]

(4) *In enclosed or physically confining spaces unless line-tended.* [§1910.424(b)(4)]

(c) **Procedures.**[§1910.424(c)]

(1) *A standby diver shall be available while a diver is in the water.* [§1910.424(c)(1)]

(2) *A diver shall be line-tended from the surface,* or accompanied by another diver in the water in continuous visual contact during the diving operations.[§1910.424(c)(2)]

(3) *A diver shall be stationed* at the underwater point of entry when diving is conducted in enclosed or physically confining spaces. [§1910.424(c)(3)]

(4) *A diver-carried reserve breathing gas supply* shall be provided for each diver consisting of:[§1910.424(c)(4)]

(i) *A manual reserve (J valve); or*[§1910.424(c)(4)(i)]

(ii) *An independent reserve cylinder* with a separate regulator or connected to the underwater breathing apparatus. [§1910.424(c)(4)(ii)]

(5) *The valve of the reserve breathing gas supply* shall be in the closed position prior to the dive.[§1910.424(c)(5)]

§1910.425

Surface-supplied air diving

(a) **General.** Employers engaged in surface-supplied air diving shall comply with the following requirements, unless otherwise specified.[§1910.425(a)]

(b) **Limits.**[§1910.425(b)]

(1) *Surface-supplied air diving* shall not be conducted at depths deeper than 190 fsw, except that dives with bottom times of 30 minutes or less may be conducted to depths of 220 fsw. [§1910.425(b)(1)]

(2) *A decompression chamber* shall be ready for use at the dive location for any dive outside the no-decompression limits or deeper than 100 fsw.[§1910.425(b)(2)]

(3) *A bell shall be used* for dives with an inwater decompression time greater than 120 minutes, except when heavy gear is worn or diving is conducted in physically confining spaces. [§1910.425(b)(3)]

(c) **Procedures.**[§1910.425(c)]

(1) *Each diver shall be continuously tended while in the water.* [§1910.425(c)(1)]

(2) *A diver shall be stationed* at the underwater point of entry when diving is conducted in enclosed or physically confining spaces. [§1910.425(c)(2)]

(3) *Each diving operation* shall have a primary breathing gas supply sufficient to support divers for the duration of the planned dive including decompression.[§1910.425(c)(3)]

(4) *For dives deeper than 100 fsw* or outside the no-decompression limits:[§1910.425(c)(4)]

(i) *A separate dive team member shall tend each diver in the water;*[§1910.425(c)(4)(i)]

(ii) *A standby diver shall be available while a diver is in the water;*[§1910.425(c)(4)(ii)]

(iii) *A diver-carried reserve breathing gas supply* shall be provided for each diver except when heavy gear is worn; and [§1910.425(c)(4)(iii)]

(iv) *A dive-location reserve breathing gas supply shall be provided.*[§1910.425(c)(4)(iv)]

(5) *For heavy-gear diving* deeper than 100 fsw or outside the no-decompression limits:[§1910.425(c)(5)]

(i) *An extra breathing gas hose* capable of supplying breathing gas to the diver in the water shall be available to the standby diver.[§1910.425(c)(5)(i)]

(ii) *An inwater stage shall be provided to divers in the water.* [§1910.425(c)(5)(ii)]

(6) *Except when heavy gear is worn* or where physical space does not permit, a diver-carried reserve breathing gas supply shall be provided whenever the diver is prevented by the configuration of the dive area from ascending directly to the surface. [§1910.425(c)(6)]

§1910.426

Mixed-gas diving

(a) **General.** Employers engaged in mixed-gas diving shall comply with the following requirements, unless otherwise specified. [§1910.426(a)]

(b) **Limits.** Mixed-gas diving shall be conducted only when: [§1910.426(b)]

(1) *A decompression chamber is ready for use at the dive location; and*[§1910.426(b)(1)]

(i) *A bell is used* at depths greater than 220 fsw or when the dive involves inwater decompression time of greater than 120 minutes, except when heavy gear is worn or when diving in physically confining spaces; or[§1910.426(b)(1)(i)]

(ii) *A closed bell is used* at depths greater than 300 fsw, except when diving is conducted in physically confining spaces. [§1910.426(b)(1)(ii)]

(c) **Procedures.**[§1910.426(c)]

(1) *A separate dive team member shall tend each diver in the water.*[§1910.426(c)(1)]

(2) *A standby diver shall be available while a diver is in the water.* [§1910.426(c)(2)]

(3) *A diver shall be stationed* at the underwater point of entry when diving is conducted in enclosed or physically confining spaces. [§1910.426(c)(3)]

(4) *Each diving operation* shall have a primary breathing gas supply sufficient to support divers for the duration of the planned dive including decompression.[§1910.426(c)(4)]

(5) *Each diving operation* shall have a dive-location reserve breathing gas supply.[§1910.426(c)(5)]

(6) *When heavy gear is worn:*[§1910.426(c)(6)]

(i) *An extra breathing gas hose* capable of supplying breathing gas to the diver in the water shall be available to the standby diver; and[§1910.426(c)(6)(i)]

(ii) *An inwater stage shall be provided to divers in the water.* [§1910.426(c)(6)(ii)]

(7) *An inwater stage* shall be provided for divers without access to a bell for dives deeper than 100 fsw or outside the no-decompression limits.[§1910.426(c)(7)]

(8) *When a closed bell is used,* one dive team member in the bell shall be available and tend the diver in the water.[§1910.426(c)(8)]

(9) *Except when heavy gear is worn* or where physical space does not permit, a diver-carried reserve breathing gas supply shall be provided for each diver:[§1910.426(c)(9)]

(i) *Diving deeper than 100 fsw* or outside the no-decompression limits; or[§1910.426(c)(9)(i)]

(ii) *Prevented by the configuration* of the dive area from directly ascending to the surface.[§1910.426(c)(9)(ii)]

§1910.427

Liveboating - Diving

(a) **General.** Employers engaged in diving operations involving liveboating shall comply with the following requirements.[§1910.427(a)]

(b) **Limits.** Diving operations involving liveboating shall not be conducted:[§1910.427(b)]

(1) *With an inwater decompression time of greater than 120 minutes;*[§1910.427(b)(1)]

(2) *Using surface-supplied air* at depths deeper than 190 fsw, except that dives with bottom times of 30 minutes or less may be conducted to depths of 220 fsw;[§1910.427(b)(2)]

(3) *Using mixed gas at depths greater than 220 fsw;*[§1910.427(b)(3)]

(4) *In rough seas* which significantly inpede diver mobility or work function; or[§1910.427(b)(4)]

(5) *In other than daylight hours.*[§1910.427(b)(5)]

(c) **Procedures.**[§1910.427(c)]

(1) *The propeller of the vessel* shall be stopped before the diver enters or exits the water.[§1910.427(c)(1)]

(2) *A device shall be used* which minimizes the possibility of entanglement of the diver's hose in the propeller of the vessel. [§1910.427(c)(2)]

(3) *Two-way voice communication* between the designated person-in- charge and the person controlling the vessel shall be available while the diver is in the water.[§1910.427(c)(3)]

(4) *A standby diver* shall be available while a diver is in the water. [§1910.427(c)(4)]

(5) *A diver-carried reserve breathing* gas supply shall be carried by each diver engaged in liveboating operations.[§1910.427(c)(5)]

Equipment Procedures and Requirements

§1910.430

Equipment - Diving

(a) **General.**[§1910.430(a)]

(1) *All employers shall comply* with the following requirements, unless otherwise specified.[§1910.430(a)(1)]

(2) *Each equipment modification, repair,* test, calibration or maintenance service shall be recorded by means of a tagging or logging system, and include the date and nature of work performed, and the name or initials of the person performing the work. [§1910.430(a)(2)]

(b) **Air compressor system.**[§1910.430(b)]

(1) *Compressors used to supply air* to the diver shall be equipped with a volume tank with a check valve on the inlet side, a pressure gauge, a relief valve, and a drain valve.[§1910.430(b)(1)]

(2) *Air compressor intakes* shall be located away from areas containing exhaust or other contaminants.[§1910.430(b)(2)]

(3) *Respirable air supplied to a diver shall not contain:* [§1910.430(b)(3)]

(i) *A level of carbon monoxide (CO) greater than 20 p/m;* [§1910.430(b)(3)(i)]

(ii) *A level of carbon dioxide (CO_2) greater than 1,000 p/m;* [§1910.430(b)(3)(ii)]

(iii) *A level of oil mist greater than 5 milligrams per cubic meter; or*[§1910.430(b)(3)(iii)]

(iv) *A noxious or pronounced odor.*[§1910.430(b)(3)(iv)]

(4) *The output of air compressor systems* shall be tested for air purity every 6 months by means of samples taken at the connection to the distribution system, except that non-oil lubricated compressors need not be tested for oil mist.[§1910.430(b)(4)]

(c) **Breathing gas supply hoses.**[§1910.430(c)]

(1) *Breathing gas supply hoses shall:*[§1910.430(c)(1)]

(i) *Have a working pressure* at least equal to the working pressure of the total breathing gas system;[§1910.430(c)(1)(i)]

(ii) *Have a rated bursting pressure* at least equal to 4 times the working pressure;[§1910.430(c)(1)(ii)]

(iii) *Be tested at least annually* to 1.5 times their working pressure; and[§1910.430(c)(1)(iii)]

(iv) *Have their open ends taped, capped or plugged when not in use.*[§1910.430(c)(1)(iv)]

(2) *Breathing gas supply hose connectors shall:*[§1910.430(c)(2)]

(i) *Be made of corrosion-resistant materials;*[§1910.430(c)(2)(i)]

(ii) *Have a working pressure* at least equal to the working pressure of the hose to which they are attached; and [§1910.430(c)(2)(ii)]

(iii) *Be resistant to accidental disengagement.*[§1910.430(c)(2)(iii)]

(3) *Umbilicals shall:*[§1910.430(c)(3)]

(i) *Be marked in 10-ft. increments* to 100 feet beginning at the diver's end, and in 50 ft. increments thereafter; [§1910.430(c)(3)(i)]

(ii) *Be made of kink-resistant materials; and*[§1910.430(c)(3)(ii)]

(iii) *Have a working pressure* greater than the pressure equivalent to the maximum depth of the dive (relative to the supply source) plus 100 psi.[§1910.430(c)(3)(iii)]

(d) **Buoyancy control.**[§1910.430(d)]

(1) *Helmets or masks* connected directly to the dry suit or other buoyancy-changing equipment shall be equipped with an exhaust valve.[§1910.430(d)(1)]

(2) *A dry suit or other buoyancy-changing equipment* not directly connected to the helmet or mask shall be equipped with an exhaust valve.[§1910.430(d)(2)]

(3) *When used for SCUBA diving,* a buoyancy compensator shall have an inflation source separate from the breathing gas supply.[§1910.430(d)(3)]

(4) *An inflatable flotation device* capable of maintaining the diver at the surface in a face-up position, having a manually activated inflation source independent of the breathing supply, an oral inflation device, and an exhaust valve shall be used for SCUBA diving.[§1910.430(d)(4)]

(e) **Compressed gas cylinders.** Compressed gas cylinders shall: [§1910.430(e)]

(1) *Be designed, constructed and maintained* in accordance with the applicable provisions of 29 CFR 1910.101 and 1910.169 through 1910.171.[§1910.430(e)(1)]

(2) *Be stored in a ventilated area and protected from excessive heat;*[§1910.430(e)(2)]

(3) *Be secured from falling; and*[§1910.430(e)(3)]

(4) *Have shut-off valves* recessed into the cylinder or protected by a cap, except when in use or manifolded, or when used for SCUBA diving.[§1910.430(e)(4)]

(f) **Decompression chambers.**[§1910.430(f)]

(1) *Each decompression chamber* manufactured after the effective date of this standard, shall be built and maintained in accordance with the ASME Code or equivalent.[§1910.430(f)(1)]

(2) *Each decompression chamber* manufactured prior to the effective date of this standard shall be maintained in conformity with the code requirements to which it was built, or equivalent. [§1910.430(f)(2)]

(3) *Each decompression chamber shall be equipped with:* [§1910.430(f)(3)]

(i) *Means to maintain the atmosphere* below a level of 25 percent oxygen by volume;[§1910.430(f)(3)(i)]

(ii) *Mufflers on intake and exhaust lines,* which shall be regularly inspected and maintained;[§1910.430(f)(3)(ii)]

(iii) *Suction guards on exhaust line openings; and* [§1910.430(f)(3)(iii)]

(iv) *A means for extinguishing fire,* and shall be maintained to minimize sources of ignition and combustible material. [§1910.430(f)(3)(iv)]

(g) **Gauges and timekeeping devices.**[§1910.430(g)]

(1) *Gauges indicating diver depth* which can be read at the dive location shall be used for all dives except SCUBA. [§1910.430(g)(1)]

(2) *Each depth gauge* shall be deadweight tested or calibrated against a master reference gauge every 6 months, and when there is a discrepancy greater than two percent (2 percent) of full scale between any two equivalent gauges.[§1910.430(g)(2)]

(3) *A cylinder pressure gauge* capable of being monitored by the diver during the dive shall be worn by each SCUBA diver. [§1910.430(g)(3)]

(4) *A timekeeping device shall be available at each dive location.* [§1910.430(g)(4)]

(h) **Masks and helmets.**[§1910.430(h)]

(1) *Surface-supplied air and mixed-gas masks and helmets shall have:*[§1910.430(h)(1)]

(i) *A non-return valve* at the attachment point between helmet or mask and hose which shall close readily and positively; and[§1910.430(h)(1)(i)]

(ii) *An exhaust valve.*[§1910.430(h)(1)(ii)]

(2) *Surface-supplied air masks* and helmets shall have a minimum ventilation rate capability of 4.5 acfm at any depth at which they are operated or the capability of maintaining the diver's inspired carbon dioxide partial pressure below 0.02 ATA when the diver is producing carbon dioxide at the rate of 1.6 standard liters per minute.[§1910.430(h)(2)]

(i) **Oxygen safety.**[§1910.430(i)]

(1) *Equipment used with oxygen* or mixtures containing over forty percent (40%) by volume oxygen shall be designed for oxygen service.[§1910.430(i)(1)]

(2) *Components (except umbilicals)* exposed to oxygen or mixtures containing over forty percent (40%) by volume oxygen shall be cleaned of flammable materials before use.[§1910.430(i)(2)]

(3) *Oxygen systems over 125 psig* and compressed air systems over 500 psig shall have slow-opening shut-off valves. [§1910.430(i)(3)]

(j) **Weights and harnesses.**[§1910.430(j)]

(1) *Except when heavy gear is worn,* divers shall be equipped with a weight belt or assembly capable of quick release. [§1910.430(j)(1)]

(2) *Except when heavy gear is worn* or in SCUBA diving, each diver shall wear a safety harness with:[§1910.430(j)(2)]

(i) *A positive buckling device;*[§1910.430(j)(2)(i)]

(ii) *An attachment point* for the umbilical to prevent strain on the mask or helmet; and[§1910.430(j)(2)(ii)]

(iii) *A lifting point* to distribute the pull force of the line over the diver's body.[§1910.430(j)(2)(iii)]

[39 FR 23502, June 27, 1974, as amended at 49 FR 18295, Apr. 30, 1984; 51 FR 33033, Sept. 18, 1986]

§1910.440

Recordkeeping requirements - Diving

(a) (1) *[Reserved]*[§1910.440(a)(1)]

(2) *The employer shall record* the occurrence of any diving-related injury or illness which requires any dive team member to be hospitalized for 24 hours or more, specifying the circumstances of the incident and the extent of any injuries or illnesses. [§1910.440(a)(2)]

(b) **Availability of records.**[§1910.440(b)]

(1) *Upon the request* of the Assistant Secretary of Labor for Occupational Safety and Health, or the Director, National Institute for Occupational Safety and Health, Department of Health and Human Services of their designees, the employer shall make available for inspection and copying any record or document required by this standard.[§1910.440(b)(1)]

(2) *Records and documents required by this standard* shall be provided upon request to employees, designated representatives, and the Assistant Secretary in accordance with 29 CFR 1910.1020(a)-(e) and (g)- (i). Safe practices manuals (Sec. 1910.420), depth-time profiles (Sec. 1910.422), recordings of dives (Sec. 1910.423), decompression procedure assessment evaluations (Sec. 1910.423), and records of hospitalizations (Sec. 1910.440) shall be provided in the same manner as employee exposure records or analyses using exposure or medical records. Equipment inspections and testing records which pertain to employees (Sec. 1910.430) shall also be provided upon request to employees and their designated representatives.[§1910.440(b)(2)]

(3) *Records and documents required* by this standard shall be retained by the employer for the following period:[§1910.440(b)(3)]

(i) *[Reserved]*[§1910.440(b)(3)(i)]

(ii) *Safe practices manual (Sec. 1910.420)* — current document only;[§1910.440(b)(3)(ii)]

(iii) *Depth-time profile (Sec. 1910.422)* — until completion of the recording of dive, or until completion of decompression procedure assessment where there has been an incident of decompression sickness;[§1910.440(b)(3)(iii)]

(iv) *Recording of dive (Sec. 1910.423)* — 1 year, except 5 years where there has been an incident of decompression sickness;[§1910.440(b)(3)(iv)]

(v) *Decompression procedure assessment evaluations (Sec. 1910.423)* — 5 years;[§1910.440(b)(3)(v)]

(vi) *Equipment inspections and testing records (Sec. 1910.430)* — current entry or tag, or until equipment is withdrawn from service;[§1910.440(b)(3)(vi)]

(vii) *Records of hospitalizations (Sec. 1910.440)* — 5 years. [§1910.440(b)(3)(vii)]

(4) *The employer shall comply with any additional* requirements set forth at 29 CFR 1910.1020,[§1910.440(b)(4)]

(5) *[Reserved]*[§1910.440(b)(5)]

[42 FR 37668, July 22, 1977, as amended at 45 FR 35281, May 23, 1980; 47 FR 14706, Apr. 6, 1982; 51 FR 34562, Sept. 29, 1986; 61 FR 9242, Mar. 7, 1996; 71 FR 16672, Apr. 3, 2006; 76 FR 33607, June 8, 2011; 76 FR 80740, Dec. 27, 2011]

Subpart T Appendix A

Examples of Conditions Which May Restrict or Limit Exposure to Hyperbaric Conditions

The following disorders may restrict or limit occupational exposure to hyperbaric conditions depending on severity, presence of residual effects, response to therapy, number of occurrences, diving mode, or degree and duration of isolation.

History of seizure disorder other than early febrile convulsions.
Malignancies (active) unless treated and without recurrence for 5 yrs.
Chronic inability to equalize sinus and/or middle ear pressure.
Cystic or cavitary disease of the lungs.
Impaired organ function caused by alcohol or drug use.
Conditions requiring continuous medication for control (e.g., antihistamines, steroids, barbiturates, moodaltering drugs, or insulin).
Meniere's disease.
Hemoglobinopathies.
Obstructive or restrictive lung disease.
Vestibular end organ destruction.
Pneumothorax.
Cardiac abnormalities (e.g., pathological heart block, valvular disease, intraventricular conduction defects other than isolated right bundle branch block, angina pectoris, arrhythmia, coronary artery disease).
Juxta-articular osteonecrosis.

Subpart T Appendix B

Guidelines for Scientific Diving

This appendix contains guidelines that will be used in conjunction with Sec. 1910.401(a)(2)(iv) to determine those scientific diving programs which are exempt from the requirements for commercial diving. The guidelines are as follows:

1. **The Diving Control Board** consists of a majority of active scientific divers and has autonomous and absolute authority over the scientific diving program's operations.
2. **The purpose of the project using scientific diving** is the advancement of science; therefore, information and data resulting from the project are non-proprietary.
3. **The tasks of a scientific diver** are those of an observer and data gatherer. Construction and trouble-shooting tasks traditionally associated with commercial diving are not included within scientific diving.
4. **Scientific divers,** based on the nature of their activities, must use scientific expertise in studying the underwater environment and, therefore, are scientists or scientists in training.

[50 FR 1050, Jan. 9, 1985]

Subpart T Appendix C

Alternative Conditions Under §1910.401(a)(3) for Recreational Diving Instructors and Diving Guides (Mandatory)

Paragraph (a)(3) of §1910.401 specifies that an employer of recreational diving instructors and diving guides (hereafter, "divers" or "employees") who complies with all of the conditions of this appendix need not provide a decompression chamber for these divers as required under §§1910.423(b)(2) or (c)(3) or 1910.426(b)(1).

1. **Equipment Requirements for Rebreathers**

(a) *The employer must ensure* that each employee operates the rebreather (i.e., semi-closed-circuit and closed-circuit self-contained underwater breathing apparatuses (hereafter, "SCUBAs")) according to the rebreather manufacturer's instructions.

(b) *The employer must ensure* that each rebreather has a counterlung that supplies a sufficient volume of breathing gas to their divers to sustain the divers' respiration rates, and contains a baffle system and/ or other moisture separating system that keeps moisture from entering the scrubber.

(c) *The employer must place* a moisture trap in the breathing loop of the rebreather, and ensure that:

(i) *The rebreather manufacturer approves* both the moisture trap and its location in the breathing loop; and

(ii) *Each employee uses* the moisture trap according to the rebreather manufacturer's instructions.

(d) *The employer must ensure* that each rebreather has a continuously functioning moisture sensor, and that:

(i) *The moisture sensor connects* to a visual (e.g., digital, graphic, analog) or auditory (e.g., voice, pure tone) alarm that is readily detectable by the diver under the diving conditions in which the diver operates, and warns the diver of moisture in the breathing loop in sufficient time to terminate the dive and return safely to the surface; and

(ii) *Each diver uses* the moisture sensor according to the rebreather manufacturer's instructions.

(e) *The employer must ensure* that each rebreather contains a continuously functioning CO_2sensor in the breathing loop, and that:

(i) *The rebreather manufacturer approves* the location of the CO_2sensor in the breathing loop;

(ii) *The CO_2sensor* is integrated with an alarm that operates in a visual (e.g., digital, graphic, analog) or auditory (e.g., voice, pure tone) mode that is readily detectable by each diver under the diving conditions in which the diver operates; and

(iii) *The CO_2alarm* remains continuously activated when the inhaled CO_2level reaches and exceeds 0.005 atmospheres absolute (ATA).

(f) *Before each day's diving* operations, and more often when necessary, the employer must calibrate the CO_2sensor according to the sensor manufacturer's instructions, and ensure that:

(i) *The equipment and procedures used* to perform this calibration are accurate to within 10% of a CO_2concentration of 0.005 ATA or less;

(ii) *The equipment and procedures maintain* this accuracy as required by the sensor manufacturer's instructions; and

(iii) *The calibration of the CO_2sensor* is accurate to within 10% of a CO_2concentration of 0.005 ATA or less.

(g) *The employer must replace the CO_2sensor* when it fails to meet the accuracy requirements specified in paragraph 1(f)(iii) of this appendix, and ensure that the replacement CO_2sensor meets the accuracy requirements specified in paragraph 1(f)(iii) of this appendix before placing the rebreather in operation.

(h) *As an alternative to using a continuously functioning CO_2sensor,* the employer may use a schedule for replacing CO_2-sorbent material provided by the rebreather manufacturer. The employer may use such a schedule only when the rebreather manufacturer has developed it according to the canister-testing protocol specified below in Condition 11, and must use the canister within the temperature range for which the manufacturer conducted its scrubber canister tests following that protocol. Variations above or below the range are acceptable only after the manufacturer adds that lower or higher temperature to the protocol.

(i) *When using CO_2-sorbent* replacement schedules, the employer must ensure that each rebreather uses a manufactured (i.e., commercially pre-packed), disposable scrubber cartridge containing a CO_2-sorbent material that:

(i) *Is approved by the rebreather manufacturer;*

(ii) *Removes $CO_2$from* the diver's exhaled gas; and

(iii) *Maintains the CO_2level* in the breathable gas (i.e., the gas that a diver inhales directly from the regulator) below a partial pressure of 0.01 ATA.

(j) *As an alternative to manufactured, disposable scrubber cartridges,* the employer may fill CO_2scrubber cartridges manually with CO_2-sorbent material when:

(i) *The rebreather manufacturer permits* manual filling of scrubber cartridges;

(ii) *The employer fills* the scrubber cartridges according to the rebreather manufacturer's instructions;

(iii) *The employer replaces* the CO_2-sorbent material using a replacement schedule developed under paragraph 1(h) of this appendix; and

(iv) *The employer demonstrates* that manual filling meets the requirements specified in paragraph 1(i) of this appendix.

(k) *The employer must ensure that each rebreather* has an information module that provides:

(i) *A visual (e.g., digital, graphic, analog) or auditory (e.g., voice, pure tone)* display that effectively warns the diver of solenoid failure (when the rebreather uses solenoids) and other electrical weaknesses or failures (e.g., low battery voltage);

(ii) *For a semi-closed circuit rebreather,* a visual display for the partial pressure of CO_2, or deviations above and below a preset CO_2partial pressure of 0.005 ATA; and

(iii) *For a closed-circuit rebreather,* a visual display for: partial pressures of $O_2$and CO_2, or deviations above and below a preset CO_2partial pressure of 0.005 ATA and a preset O_2partial pressure of 1.40 ATA or lower; gas temperature in the breathing loop; and water temperature.

(l) *Before each day's diving operations,* and more often when necessary, the employer must ensure that the electrical power supply and electrical and electronic circuits in each rebreather are operating as required by the rebreather manufacturer's instructions.

2. Special Requirements for Closed-Circuit Rebreathers

(a) *The employer must ensure* that each closed-circuit rebreather uses supply-pressure sensors for the $O_2$and diluent (i.e., air or nitrogen) gases and continuously functioning sensors for detecting temperature in the inhalation side of the gas-loop and the ambient water.

(b) *The employer must ensure that:*

(i) *At least two O_2sensors* are located in the inhalation side of the breathing loop; and

(ii) *The O_2sensors* are: functioning continuously; temperature compensated; and approved by the rebreather manufacturer.

(c) *Before each day's diving operations,* and more often when necessary, the employer must calibrate O_2sensors as required by the sensor manufacturer's instructions. In doing so, the employer must:

(i) *Ensure that the equipment and procedures* used to perform the calibration are accurate to within 1% of the O_2fraction by volume;

(ii) *Maintain this accuracy as required* by the manufacturer of the calibration equipment;

(iii) *Ensure that the sensors are accurate* to within 1% of the O_2fraction by volume;

(iv) *Replace O_2sensors* when they fail to meet the accuracy requirements specified in paragraph 2(c)(iii) of this appendix; and

(v) *Ensure that the replacement O_2sensors* meet the accuracy requirements specified in paragraph 2(c)(iii) of this appendix before placing a rebreather in operation.

(d) *The employer must ensure* that each closed-circuit rebreather has:

(i) *A gas-controller package with electrically operated solenoid O_2-supply valves;*

(ii) *A pressure-activated regulator with a second-stage diluent-gas addition valve;*

(iii) *A manually operated gas-supply bypass valve* to add O_2or diluent gas to the breathing loop; and

(iv) *Separate $O_2$and diluent-gas cylinders* to supply the breathing-gas mixture.

3. O_2Concentration in the Breathing Gas

The employer must ensure that the fraction of O_2in the nitrox breathing-gas mixture:

(a) *Is greater than the fraction of O_2in compressed air* (i.e., exceeds 22% by volume);

(b) *For open-circuit SCUBA,* never exceeds a maximum fraction of breathable $O_2$of 40% by volume or a maximum O_2partial pressure of 1.40 ATA, whichever exposes divers to less O_2; and

(c) *For a rebreather,* never exceeds a maximum O_2partial pressure of 1.40 ATA.

4. Regulating O_2Exposures and Diving Depth

(a) *Regarding O_2exposure,* the employer must:

(i) *Ensure that the exposure of each diver* to partial pressures of O_2between 0.60 and 1.40 ATA does not exceed the 24-hour single-exposure time limits specified either by the 2001 National Oceanic and Atmospheric Administration Diving Manual (the "2001 NOAA Diving Manual"), or by the report entitled "Enriched Air Operations and Resource Guide" published in 1995 by the Professional Association of Diving Instructors (known commonly as the "1995 DSAT Oxygen Exposure Table"); and

(ii) *Determine a diver's O_2-exposure duration* using the diver's maximum O_2exposure (partial pressure of O_2) during the dive and the total dive time (i.e., from the time the diver leaves the surface until the diver returns to the surface).

(b) *Regardless of the diving equipment* used, the employer must ensure that no diver exceeds a depth of 130 feet of sea water ("fsw") or a maximum O_2partial pressure of 1.40 ATA, whichever exposes the diver to less O_2.

5. **Use of No-Decompression Limits**
   (a) *For diving conducted while using nitrox* breathing-gas mixtures, the employer must ensure that each diver remains within the no- decompression limits specified for single and repetitive air diving and published in the 2001 NOAA Diving Manual or the report entitled "Development and Validation of No-Stop Decompression Procedures for Recreational Diving: The DSAT Recreational Dive Planner," published in 1994 by Hamilton Research Ltd. (known commonly as the "1994 DSAT No- Decompression Tables").
   (b) *An employer may permit a diver* to use a dive-decompression computer designed to regulate decompression when the dive-decompression computer uses the no-decompression limits specified in paragraph 5(a) of this appendix, and provides output that reliably represents those limits.
6. **Mixing and Analyzing the Breathing Gas**
   (a) *The employer must ensure that:*
      (i) *Properly trained personnel* mix nitrox-breathing gases, and that nitrogen is the only inert gas used in the breathing-gas mixture; and
      (ii) *When mixing nitrox-breathing gases,* they mix the appropriate breathing gas before delivering the mixture to the breathing-gas cylinders, using the continuous-flow or partial-pressure mixing techniques specified in the 2001 NOAA Diving Manual, or using a filter-membrane system.
   (b) *Before the start of each day's* diving operations, the employer must determine the O_2fraction of the breathing-gas mixture using an O_2analyzer. In doing so, the employer must:
      (i) *Ensure that the O_2analyzer* is accurate to within 1% of the O_2fraction by volume.
      (ii) *Maintain this accuracy* as required by the manufacturer of the analyzer.
   (c) *When the breathing gas is a commercially* supplied nitrox breathing-gas mixture, the employer must ensure that the O_2meets the medical USP specifications (Type I, Quality Verification Level A) or aviator's breathing-oxygen specifications (Type I, Quality Verification Level E) of CGA G-4.3-2000 ("Commodity Specification for Oxygen"). In addition, the commercial supplier must:
      (i) *Determine the O_2fraction* in the breathing-gas mixture using an analytic method that is accurate to within 1% of the O_2fraction by volume;
      (ii) *Make this determination when the mixture* is in the charged tank and after disconnecting the charged tank from the charging apparatus;
      (iii) *Include documentation of the O_2-analysis procedures* and the O_2fraction when delivering the charged tanks to the employer.
   (d) *Before producing nitrox* breathing-gas mixtures using a compressor in which the gas pressure in any system component exceeds 125 pounds per square inch (psi), the:
      (i) *Compressor manufacturer must provide* the employer with documentation that the compressor is suitable for mixing high-pressure air with the highest O_2fraction used in the nitrox breathing-gas mixture when operated according to the manufacturer's operating and maintenance specifications;
      (ii) *Employer must comply with paragraph 6(e)* of this appendix, unless the compressor is rated for O_2service and is oil-less or oil-free; and
      (iii) *Employer must ensure that the compressor* meets the requirements specified in paragraphs (i)(1) and (i)(2) of Sec. 1910.430 whenever the highest O_2fraction used in the mixing process exceeds 40%.
   (e) *Before producing nitrox* breathing-gas mixtures using an oil-lubricated compressor to mix high-pressure air with O_2, and regardless of the gas pressure in any system component, the:
      (i) *Employer must use only* uncontaminated air (i.e., air containing no hydrocarbon particulates) for the nitrox breathing-gas mixture;
      (ii) *Compressor manufacturer must provide* the employer with documentation that the compressor is suitable for mixing the high-pressure air with the highest O_2fraction used in the nitrox breathing-gas mixture when operated according to the manufacturer's operating and maintenance specifications;
      (iii) *Employer must filter* the high-pressure air to produce O_2-compatible air;
      (iv) *The filter-system manufacturer must provide* the employer with documentation that the filter system used for this purpose is suitable for producing O_2-compatible air when operated according to the manufacturer's operating and maintenance specifications; and
      (v) *Employer must continuously monitor* the air downstream from the filter for hydrocarbon contamination.
   (f) *The employer must ensure* that diving equipment using nitrox breathing-gas mixtures or pure O_2under high pressure (i.e., exceeding 125 psi) conforms to the O_2-service requirements specified in paragraphs (i)(1) and (i)(2) of Sec. 1910.430.
7. **Emergency Egress**
   (a) *Regardless of the type of diving equipment* used by a diver (i.e., open-circuit SCUBA or rebreathers), the employer must ensure that the equipment contains (or incorporates) an open-circuit emergency- egress system (a "bail-out" system) in which the second stage of the regulator connects to a separate supply of emergency breathing gas, and the emergency breathing gas consists of air or the same nitrox breathing-gas mixture used during the dive.
   (b) *As an alternative* to the "bail-out" system specified in paragraph 7(a) of this appendix, the employer may use:
      (i) *For open-circuit SCUBA,* an emergency-egress system as specified in Sec. 1910.424(c)(4); or
      (ii) *For a semi-closed-circuit and closed-circuit rebreather,* a system configured so that the second stage of the regulator connects to a reserve supply of emergency breathing gas.
   (c) *The employer must obtain from the rebreather manufacturer* sufficient information to ensure that the bail-out system performs reliably and has sufficient capacity to enable the diver to terminate the dive and return safely to the surface.
8. **Treating Diving-Related Medical Emergencies**
   (a) *Before each day's diving operations, the employer must:*
      (i) *Verify that a hospital,* qualified health-care professionals, and the nearest Coast Guard Coordination Center (or an equivalent rescue service operated by a state, county, or municipal agency) are available to treat diving-related medical emergencies;
      (ii) *Ensure that each dive site* has a means to alert these treatment resources in a timely manner when a diving-related medical emergency occurs; and
      (iii) *Ensure that transportation to a suitable decompression chamber* is readily available when no decompression chamber is at the dive site, and that this transportation can deliver the injured diver to the decompression chamber within four (4) hours travel time from the dive site.
   (b) *The employer must ensure that portable O_2equipment* is available at the dive site to treat injured divers. In doing so, the employer must ensure that:
      (i) *The equipment delivers medical-grade O_2that* meets the requirements for medical USP oxygen (Type I, Quality Verification Level A) of CGA G-4.3-2000 ("Commodity Specification for Oxygen");
      (ii) *The equipment delivers this $O_2$to* a transparent mask that covers the injured diver's nose and mouth; and
      (iii) *Sufficient $O_2$is* available for administration to the injured diver from the time the employer recognizes the symptoms of a diving-related medical emergency until the injured diver reaches a decompression chamber for treatment.
   (c) *Before each day's diving operations,* the employer must:
      (i) *Ensure that at least two attendants,* either employees or non-employees, qualified in first-aid and administering O_2treatment, are available at the dive site to treat diving-related medical emergencies; and
      (ii) *Verify their qualifications for this task.*
9. **Diving Logs and No-Decompression Tables**
   (a) *Before starting each day's diving operations,* the employer must:
      (i) *Designate an employee or a non-employee* to make entries in a diving log; and
      (ii) *Verify that this designee understands* the diving and medical terminology, and proper procedures, for making correct entries in the diving log.
   (b) *The employer must:*
      (i) *Ensure that the diving log* conforms to the requirements specified by paragraph (d) ("Record of dive") of Sec. 1910.423; and
      (ii) *Maintain a record of the dive* according to Sec. 1910.440 ("Recordkeeping requirements").
   (c) *The employer must ensure that a hard-copy* of the no-decompression tables used for the dives (as specified in paragraph

6(a) of this appendix) is readily available at the dive site, whether or not the divers use dive-decompression computers.

10. **Diver Training**
The employer must ensure that each diver receives training that enables the diver to perform work safely and effectively while using open-circuit SCUBAs or rebreathers supplied with nitrox breathing-gas mixtures. Accordingly, each diver must be able to demonstrate the ability to perform critical tasks safely and effectively, including, but not limited to: recognizing the effects of breathing excessive $CO_2$and O_2; taking appropriate action after detecting excessive levels of $CO_2$and O_2; and properly evaluating, operating, and maintaining their diving equipment under the diving conditions they encounter.

11. **Testing Protocol for Determining the CO_2Limits of Rebreather Canisters**

(a) *The employer must ensure* that the rebreather manufacturer has used the following procedures for determining that the CO_2-sorbent material meets the specifications of the sorbent material's manufacturer:

(i) *The North Atlantic Treating Organization CO_2absorbent-activity test;*

(ii) *The RoTap shaker and nested-sieves test;*

(iii) *The Navy Experimental Diving Unit ("NEDU")-derived Schlegel test; and*

(iv) *The NEDU MeshFit software.*

(b) *The employer must ensure* that the rebreather manufacturer has applied the following canister-testing materials, methods, procedures, and statistical analyses:

(i) *Use of a nitrox breathing-gas* mixture that has an O_2fraction maintained at 0.28 (equivalent to 1.4 ATA of O_2at 130 fsw, the maximum O_2concentration permitted at this depth);

(ii) *While operating the rebreather* at a maximum depth of 130 fsw, use of a breathing machine to continuously ventilate the rebreather with breathing gas that is at 100% humidity and warmed to a temperature of 98.6 °F (37 °C) in the heating-humidification chamber;

(iii) *Measurement of the O_2concentration* of the inhalation breathing gas delivered to the mouthpiece;

(iv) *Testing of the canisters* using the three ventilation rates listed in Table I below (with the required breathing-machine tidal volumes and frequencies, and CO_2-injection rates, provided for each ventilation rate):

Table I — Canister Testing Parameters

Ventilation rates (Lpm, ATPS [1])	Breathing machine tidal volumes (L)	Breathing machine frequencies (breaths per min.)	CO2 injection rates (Lpm, STPD [2])
22.5	1.5	15	0.90
40.0	2.0	20	1.35
62.5	2.5	25	2.25

[1] ATPS means ambient temperature and pressure, saturated with water.

[2] STPD means standard temperature and pressure, dry; the standard temperature is 32 degrees F (0 degrees C).

(v) *When using a work rate (i.e.,* breathing-machine tidal volume and frequency) other than the work rates listed in the table above, addition of the appropriate combinations of ventilation rates and CO_2- injection rates;

(vi) *Performance of the CO_2injection* at a constant (steady) and continuous rate during each testing trial;

(vii) *Determination of canister duration* using a minimum of four (4) water temperatures, including 40, 50, 70, and 90 °F (4.4, 10.0, 21.1, and 32.2 °C, respectively);

(viii) *Monitoring of the breathing-gas temperature* at the rebreather mouthpiece (at the "chrome T" connector), and ensuring that this temperature conforms to the temperature of a diver's exhaled breath at the water temperature and ventilation rate used during the testing trial;[1]

(ix) *Implementation of at least eight* (8) testing trials for each combination of temperature and ventilation-CO_2-injection rates (for example, eight testing trials at 40 °F using a ventilation rate of 22.5 Lpm at a CO_2-injection rate of 0.90 Lpm);

(x) *Allowing the water temperature to vary no more than* ± 2.0 °F (± 1.0 degree C) between each of the eight testing trials, and no more than ± 1.0 °F (± 0.5 degree C) within each testing trial;

(xi) *Use of the average temperature* for each set of eight testing trials in the statistical analysis of the testing-trial results, with the testing-trial results being the time taken for the inhaled breathing gas to reach 0.005 ATA of CO_2(i.e., the canister-duration results);

(xii) *Analysis of the canister-duration results* using the repeated- measures statistics described in NEDU Report 2-99;

(xiii) *Specification of the replacement schedule* for the CO_2-sorbent materials in terms of the lower prediction line (or limit) of the 95% confidence interval; and

(xiv) *Derivation of replacement schedules* only by interpolating among, but not by extrapolating beyond, the depth, water temperatures, and exercise levels used during canister testing.

1. NEDU can provide the manufacturer with information on the temperature of a diver's exhaled breath at various water temperatures and ventilation rates, as well as techniques and procedures used to maintain these temperatures during the testing trials.

[69 FR 7363, Feb. 17, 2004]

§1910.1002

Coal tar pitch volatiles; interpretation of term

As used in Sec. 1910.1000 (Table Z-1), coal tar pitch volatiles include the fused polycyclic hydrocarbons which volatilize from the distillation residues of coal, petroleum (excluding asphalt), wood, and other organic matter. Asphalt (CAS 8052-42-4, and CAS 64742-93-4) is not covered under the "coal tar pitch volatiles" standard.

[48 FR 2768, Jan. 21, 1983]

§1910.1003

13 Carcinogens (4-nitrobiphenyl, etc.)

(a) **Scope and application.**[§1910.1003(a)]

(1) *This section applies to any area* in which the 13 carcinogens addressed by this section are manufactured, processed, repackaged, released, handled, or stored, but shall not apply to transshipment in sealed containers, except for the labeling requirements under paragraphs (e)(2), (3) and (4) of this section. The 13 carcinogens are the following:[§1910.1003(a)(1)]

4-Nitrobiphenyl, Chemical Abstracts Service Register Number (CAS No.) 92933;
alpha-Naphthylamine, CAS No. 134327;
methyl chloromethyl ether, CAS No. 107302;
3,3'-Dichlorobenzidine[1] (and its salts) CAS No. 91941;
bis-Chloromethyl ether, CAS No. 542881;
beta-Naphthylamine, CAS No. 91598;
Benzidine, CAS No. 92875;
4-Aminodiphenyl, CAS No. 92671;
Ethyleneimine, CAS No. 151564;
beta-Propiolactone, CAS No. 57578;
2-Acetylaminofluorene, CAS No. 53963;
4-Dimethylaminoazo-benezene, CAS No. 60117; and
N-Nitrosodimethylamine, CAS No. 62759.

(2) *This section shall not apply to the following:*[§1910.1003(a)(2)]

(i) *Solid or liquid mixtures* containing less than 0.1 percent by weight or volume of 4-Nitrobiphenyl; methyl chloromethyl ether; bis-chloromethyl ether; beta-Naphthylamine; benzidine or 4-Aminodiphenyl; and[§1910.1003(a)(2)(i)]

(ii) *Solid or liquid mixtures* containing less than 1.0 percent by weight or volume of alpha-Naphthylamine; 3,3'-Dichlorobenzidine[1] (and its salts); Ethyleneimine; beta-Propiolactone; 2-Acetylaminofluorene; 4-Dimethylaminoazobenzene, or N-Nitrosodimethylamine.[§1910.1003(a)(2)(ii)]

(b) **Definitions.** For the purposes of this section:[§1910.1003(b)]

Absolute filter is one capable of retaining 99.97 percent of a mono disperse aerosol of 0.3 μm particles.

Authorized employee means an employee whose duties require him to be in the regulated area and who has been specifically assigned by the employer.

Clean change room means a room where employees put on clean clothing and/or protective equipment in an environment free of the 13 carcinogens addressed by this section. The clean change room shall be contiguous to and have an entry from a shower room, when the shower room facilities are otherwise required in this section.

Closed system means an operation involving a carcinogen addressed by this section where containment prevents the release of the material into regulated areas, non-regulated areas, or the external environment.

Decontamination means the inactivation of a carcinogen addressed by this section or its safe disposal.

Director means the Director, National Institute for Occupational Safety and Health, or any person directed by him or the Secretary of Health and Human Services to act for the Director.

Disposal means the safe removal of the carcinogens addressed by this section from the work environment.

Emergency means an unforeseen circumstance or set of circumstances resulting in the release of a carcinogen addressed by this section that may result in exposure to or contact with the material.

External environment means any environment external to regulated and nonregulated areas.

Isolated system means a fully enclosed structure other than the vessel of containment of a carcinogen addressed by this section that is impervious to the passage of the material and would prevent the entry of the carcinogen addressed by this section into regulated areas, nonregulated areas, or the external environment, should leakage or spillage from the vessel of containment occur.

Laboratory-type hood is a device enclosed on the three sides and the top and bottom, designed and maintained so as to draw air inward at an average linear face velocity of 150 feet per minute with a minimum of 125 feet per minute; designed, constructed, and maintained in such a way that an operation involving a carcinogen addressed by this section within the hood does not require the insertion of any portion of any employee's body other than his hands and arms.

Nonregulated area means any area under the control of the employer where entry and exit is neither restricted nor controlled.

Open-vessel system means an operation involving a carcinogen addressed by this section in an open vessel that is not in an isolated system, a laboratory-type hood, nor in any other system affording equivalent protection against the entry of the material into regulated areas, non-regulated areas, or the external environment.

Protective clothing means clothing designed to protect an employee against contact with or exposure to a carcinogen addressed by this section.

Regulated area means an area where entry and exit is restricted and controlled.

(c) **Requirements for areas containing a carcinogen addressed by this section.** A regulated area shall be established by an employer where a carcinogen addressed by this section is manufactured, processed, used, repackaged, released, handled or stored. All such areas shall be controlled in accordance with the requirements for the following category or categories describing the operation involved:[§1910.1003(c)]

(1) *Isolated systems.* Employees working with a carcinogen addressed by this section within an isolated system such as a "glove box" shall wash their hands and arms upon completion of the assigned task and before engaging in other activities not associated with the isolated system.[§1910.1003(c)(1)]

(2) *Closed system operation.*[§1910.1003(c)(2)]

(i) *Within regulated areas* where the carcinogens addressed by this section are stored in sealed containers, or contained in a closed system, including piping systems, with any sample ports or openings closed while the carcinogens addressed by this section are contained within, access shall be restricted to authorized employees only.[§1910.1003(c)(2)(i)]

(ii) *Employees exposed to 4-Nitrobiphenyl;* alpha-Naphthylamine; 3,3'-Dichlorobenzidine[1] (and its salts); beta-Naphthylamine; benzidine; 4-Aminodiphenyl; 2-Acetylaminofluorene; 4-Dimethylaminoazo-benzene; and N-Nitrosodimethylamine shall be required to wash hands, forearms, face, and neck upon each exit from the regulated areas, close to the point of exit, and before engaging in other activities.[§1910.1003(c)(2)(ii)]

(3) *Open-vessel system operations.* Open-vessel system operations as defined in paragraph (b)(13) of this section are prohibited.[§1910.1003(c)(3)]

(4) *Transfer from a closed system,* charging or discharging point operations, or otherwise opening a closed system. In operations involving "laboratory-type hoods," or in locations where the carcinogens addressed by this section are contained in an otherwise "closed system," but is transferred, charged, or discharged into other normally closed containers, the provisions of this paragraph shall apply.[§1910.1003(c)(4)]

(i) *Access shall be restricted to authorized employees only.* [§1910.1003(c)(4)(i)]

(ii) *Each operation shall be provided* with continuous local exhaust ventilation so that air movement is always from ordinary work areas to the operation. Exhaust air shall not be discharged to regulated areas, nonregulated areas or the external environment unless decontaminated. Clean makeup air shall be introduced in sufficient volume to maintain the correct operation of the local exhaust system. [§1910.1003(c)(4)(ii)]

(iii) *Employees shall be provided with,* and required to wear, clean, full body protective clothing (smocks, coveralls, or long-sleeved shirt and pants), shoe covers and gloves prior to entering the regulated area.[§1910.1003(c)(4)(iii)]

(iv) *Employers must provide each employee* engaged in handling operations involving the carcinogens 4-Nitrobiphenyl, alpha-Naphthylamine, 3,3'-Dichlorobenzidine (and its salts), beta-Naphthylamine, Benzidine, 4-Aminodiphenyl, 2-Acetylaminofluorene, 4-Dimethylaminoazo-benzene, and N-Nitrosodimethylamine, addressed by this section, with, and ensure that each of these employees wears and uses, a NIOSH-certified air-purifying, half-mask respirator with particulate filters. Employers also must provide each employee engaged in handling operations involving the carcinogens methyl chloromethyl ether, bis-Chloromethyl ether, Ethyleneimine, and beta-Propiolactone, addressed by this section, with, and ensure that each of these employees wears and uses any self-contained breathing apparatus that has a full facepiece and is operated in a pressure-demand or other positive-pressure mode, or any supplied-air respirator that has a full facepiece and is operated in a pressure-demand or other positive-pressure mode in combination with an auxiliary self-contained positive-pressure breathing apparatus. Employers may substitute a respirator affording employees higher levels of protection than these respirators. [§1910.1003(c)(4)(iv)]

(v) *Prior to each exit from a regulated area,* employees shall be required to remove and leave protective clothing and equipment at the point of exit and at the last exit of the day, to place used clothing and equipment in impervious containers at the point of exit for purposes of decontamination or disposal. The contents of such impervious containers shall be identified, as required under paragraph (e) of this section. [§1910.1003(c)(4)(v)]

(vi) *Drinking fountains are prohibited in the regulated area.* [§1910.1003(c)(4)(vi)]

(vii) *Employees shall be required to wash hands,* forearms, face, and neck on each exit from the regulated area, close to the point of exit, and before engaging in other activities and employees exposed to 4-Nitrobiphenyl; alpha-Naphthylamine; 3,3'-dichlorobenzidine (and its salts); beta-Naphthylamine; Benzidine; 4-Aminodiphenyl; 2-Acetylaminofluorene; 4-Dimethylaminoazo-benzene; and N-Nitrosodimethylamine shall be required to shower after the last exit of the day. [§1910.1003(c)(4)(vii)]

(5) *Maintenance and decontamination activities.* In cleanup of leaks of spills, maintenance, or repair operations on contaminated systems or equipment, or any operations involving work in an area where direct contact with a carcinogen addressed by this section could result, each authorized employee entering that area shall:[§1910.1003(c)(5)]

(i) *Be provided with and required to wear clean,* impervious garments, including gloves, boots, and continuous-air supplied hood in accordance with Sec. 1910.134;[§1910.1003(c)(5)(i)]

(ii) *Be decontaminated before removing* the protective garments and hood;[§1910.1003(c)(5)(ii)]

(iii) *Be required to shower upon removing* the protective garments and hood.[§1910.1003(c)(5)(iii)]

(d) **General regulated area requirements**[§1910.1003(d)]

(1) *Respiratory program.* The employer must implement a respiratory protection program in accordance with Sec. 1910.134 (b), (c), (d) (except (d)(1)(iii) and (iv), and (d)(3)), and (e) through (m), which covers each employee required by this section to use a respirator.[§1910.1003(d)(1)]

(2) *Emergencies.* In an emergency, immediate measures including, but not limited to, the requirements of paragraphs (d)(2)(i) through (v) of this section shall be implemented. [§1910.1003(d)(2)]

(i) *The potentially affected area* shall be evacuated as soon as the emergency has been determined.[§1910.1003(d)(2)(i)]

(ii) *Hazardous conditions created* by the emergency shall be eliminated and the potentially affected area shall be decontaminated prior to the resumption of normal operations. [§1910.1003(d)(2)(ii)]

(iii) *Special medical surveillance* by a physician shall be instituted within 24 hours for employees present in the potentially affected area at the time of the emergency.[§1910.1003(d)(2)(iii)]

(iv) *Where an employee has a known contact* with a carcinogen addressed by this section, such employee shall be required to shower as soon as possible, unless contraindicated by physical injuries.[§1910.1003(d)(2)(iv)]

(v) *Emergency deluge showers* and eyewash fountains supplied with running potable water shall be located near, within sight of, and on the same level with locations where a direct exposure to Ethyleneimine or beta-Propiolactone only would be most likely as a result of equipment failure or improper work practice.[§1910.1003(d)(2)(v)]

(3) *Hygiene facilities and practices.*[§1910.1003(d)(3)]

(i) *Storage or consumption of food,* storage or use of containers of beverages, storage or application of cosmetics, smoking, storage of smoking materials, tobacco products or other products for chewing, or the chewing of such products are prohibited in regulated areas.[§1910.1003(d)(3)(i)]

(ii) *Where employees are required by this section to wash,* washing facilities shall be provided in accordance with Sec. 1910.141(d)(1) and (2)(ii) through (vii).[§1910.1003(d)(3)(ii)]

(iii) *Where employees are required by this section to shower,* shower facilities shall be provided in accordance with Sec. 1910.141(d)(3).[§1910.1003(d)(3)(iii)]

(iv) *Where employees wear protective clothing and equipment,* clean change rooms shall be provided for the number of such employees required to change clothes, in accordance with Sec. 1910.141(e).[§1910.1003(d)(3)(iv)]

(v) *Where toilets are in regulated areas,* such toilets shall be in a separate room.[§1910.1003(d)(3)(v)]

(4) *Contamination control.*[§1910.1003(d)(4)]

(i) *Except for outdoor systems,* regulated areas shall be maintained under pressure negative with respect to nonregulated areas. Local exhaust ventilation may be used to satisfy this requirement. Clean makeup air in equal volume shall replace air removed.[§1910.1003(d)(4)(i)]

(ii) *Any equipment,* material, or other item taken into or removed from a regulated area shall be done so in a manner that does not cause contamination in nonregulated areas or the external environment.[§1910.1003(d)(4)(ii)]

(iii) *Decontamination procedures shall be established and implemented* to remove carcinogens addressed by this section from the surfaces of materials, equipment, and the decontamination facility.[§1910.1003(d)(4)(iii)]

(iv) *Dry sweeping and dry mopping are prohibited* for 4-Nitrobiphenyl; alpha-Naphthylamine; 3,3'-Dichlorobenzidine[1] (and its salts); beta-Naphthylamine; Benzidine; 4-Aminodiphenyl; 2-Acetylaminofluorene; 4-Dimethylaminoazo-benzene and N-Nitrosodimethylamine.[§1910.1003(d)(4)(iv)]

(e) Communication of hazards[§1910.1003(e)]

(1) *Hazard communication.*[§1910.1003(e)(1)]

(i) *Chemical manufacturers,* importers, distributors and employers shall comply with all requirements of the Hazard Communication Standard (HCS) (Sec. 1910.1200) for each carcinogen listed in paragraph (e)(1)(iv) of this section. [§1910.1003(e)(1)(i)]

(ii) *In classifying the hazards of carcinogens* listed in paragraph (e)(1)(iv) of this section, at least the hazards listed in paragraph (e)(1)(iv) are to be addressed.[§1910.1003(e)(1)(ii)]

(iii) *Employers shall include the carcinogens* listed in paragraph (e)(1)(iv) of this section in the hazard communication program established to comply with the HCS (Sec. 1910.1200). Employers shall ensure that each employee has access to labels on containers of the carcinogens listed in paragraph (e)(1)(iv) and to safety data sheets, and is trained in accordance with the requirements of HCS and paragraph (e)(4) of this section.[§1910.1003(e)(1)(iii)]

(iv) *List of Carcinogens:*[§1910.1003(e)(1)(iv)]

[A] 4-Nitrobiphenyl: Cancer.[§1910.1003(e)(1)(iv)[A]]

[B] alpha-Naphthylamine: Cancer; skin irritation; and acute toxicity effects.[§1910.1003(e)(1)(iv)[B]]

[C] Methyl chloromethyl ether: Cancer; skin, eye and respiratory effects; acute toxicity effects; and flammability. [§1910.1003(e)(1)(iv)[C]]

[D] 3,3'-Dichlorobenzidine (and its salts): Cancer and skin sensitization.[§1910.1003(e)(1)(iv)[D]]

[E] bis-Chloromethyl ether: Cancer; skin, eye, and respiratory tract effects; acute toxicity effects; and flammability. [§1910.1003(e)(1)(iv)[E]]

[F] beta-Naphthylamine: Cancer and acute toxicity effects. [§1910.1003(e)(1)(iv)[F]]

[G] Benzidine: Cancer and acute toxicity effects. [§1910.1003(e)(1)(iv)[G]]

[H] 4-Aminodiphenyl: Cancer.[§1910.1003(e)(1)(iv)[H]]

[I] Ethyleneimine: Cancer; mutagenicity; skin and eye effects; liver effects; kidney effects; acute toxicity effects; and flammability.[§1910.1003(e)(1)(iv)[I]]

[J] beta-Propiolactone: Cancer; skin irritation; eye effects; and acute toxicity effects.[§1910.1003(e)(1)(iv)[J]]

[K] 2-Acetylaminofluorene: Cancer.[§1910.1003(e)(1)(iv)[K]]

[L] 4-Dimethylaminoazo-benzene: Cancer; skin effects; and respiratory tract irritation.[§1910.1003(e)(1)(iv)[L]]

[M] N-Nitrosodimethylamine: Cancer; liver effects; and acute toxicity effects.[§1910.1003(e)(1)(iv)[M]]

(2) *Signs.*[§1910.1003(e)(2)]

(i) *The employer shall post entrances* to regulated areas with signs bearing the legend: DANGER (CHEMICAL IDENTIFICATION) MAY CAUSE CANCER AUTHORIZED PERSONNEL ONLY[§1910.1003(e)(2)(i)]

(ii) *The employer shall post signs* at entrances to regulated areas containing operations covered in paragraph (c)(5) of this section. The signs shall bear the legend:

DANGER

(CHEMICAL IDENTIFICATION)

MAY CAUSE CANCER

WEAR AIR-SUPPLIED HOODS, IMPERVIOUS SUITS, AND PROTECTIVE EQUIPMENT IN THIS AREA

AUTHORIZED PERSONNEL ONLY

(iii) *Prior to June 1,* 2016, employers may use the following legend in lieu of that specified in paragraph (e)(2)(i) of this section:

CANCER-SUSPECT AGENT

AUTHORIZED PERSONNEL ONLY

(iv) *Prior to June 1,* 2016, employers may use the following legend in lieu of that specified in paragraph (e)(2)(ii) of this section:

CANCER-SUSPECT AGENT EXPOSED IN THIS AREA

IMPERVIOUS SUIT INCLUDING GLOVES, BOOTS, AND AIR-SUPPLIED HOOD REQUIRED AT ALL TIMES

AUTHORIZED PERSONNEL ONLY

(v) *Appropriate signs and instructions* shall be posted at the entrance to, and exit from, regulated areas, informing employees of the procedures that must be followed in entering and leaving a regulated area.[§1910.1003(e)(2)(v)]

(3) *Prohibited statements.* No statement shall appear on or near any required sign, label, or instruction that contradicts or detracts from the effect of any required warning, information, or instruction.[§1910.1003(e)(3)]

(5) *Training and indoctrination.*[§1910.1003(e)(5)]

(i) *Each employee prior to being authorized* to enter a regulated area, shall receive a training and indoctrination program including, but not necessarily limited to:[§1910.1003(e)(5)(i)]

[A] The nature of the carcinogenic hazards of a carcinogen addressed by this section, including local and systemic toxicity;[§1910.1003(e)(5)(i)[A]]

1. Editor's note: The CFR reflects "3,'-Dichlorobenzidine" rather than "3,3'-Dichlorobenzidine". All instances of "3,'-Dichlorobenzidine"of have been replaced with "3,3'-dichlorobenzidine" in this regulation.

1910

Selected 1910 General Industry Standards

[B] The specific nature of the operation involving a carcinogen addressed by this section that could result in exposure;[§1910.1003(e)(5)(i)[B]]

[C] The purpose for and application of the medical surveillance program, including, as appropriate, methods of self-examination;[§1910.1003(e)(5)(i)[C]]

[D] The purpose for and application of decontamination practices and purposes;[§1910.1003(e)(5)(i)[D]]

[E] The purpose for and significance of emergency practices and procedures;[§1910.1003(e)(5)(i)[E]]

[F] The employee's specific role in emergency procedures; [§1910.1003(e)(5)(i)[F]]

[G] Specific information to aid the employee in recognition and evaluation of conditions and situations which may result in the release of a carcinogen addressed by this section;[§1910.1003(e)(5)(i)[G]]

[H] The purpose for and application of specific first aid procedures and practices;[§1910.1003(e)(5)(i)[H]]

[I] A review of this section at the employee's first training and indoctrination program and annually thereafter. [§1910.1003(e)(5)(i)[I]]

(ii) *Specific emergency procedures shall be prescribed,* and posted, and employees shall be familiarized with their terms, and rehearsed in their application.[§1910.1003(e)(5)(ii)]

(iii) *All materials relating to the program* shall be provided upon request to authorized representatives of the Assistant Secretary and the Director.[§1910.1003(e)(5)(iii)]

(f) **[Reserved]**[§1910.1003(f)]

(g) **Medical surveillance.** At no cost to the employee, a program of medical surveillance shall be established and implemented for employees considered for assignment to enter regulated areas, and for authorized employees.[§1910.1003(g)]

(1) *Examinations.*[§1910.1003(g)(1)]

(i) *Before an employee is assigned* to enter a regulated area, a preassignment physical examination by a physician shall be provided. The examination shall include the personal history of the employee, family and occupational background, including genetic and environmental factors.[§1910.1003(g)(1)(i)]

(ii) *Authorized employees shall be provided* periodic physical examinations, not less often than annually, following the preassignment examination.[§1910.1003(g)(1)(ii)]

(iii) *In all physical examinations,* the examining physician shall consider whether there exist conditions of increased risk, including reduced immunological competence, those undergoing treatment with steroids or cytotoxic agents, pregnancy, and cigarette smoking.[§1910.1003(g)(1)(iii)]

(2) *Records.*[§1910.1003(g)(2)]

(i) *Employers of employees examined* pursuant to this paragraph shall cause to be maintained complete and accurate records of all such medical examinations. Records shall be maintained for the duration of the employee's employment. [§1910.1003(g)(2)(i)]

(ii) *Records required by this paragraph* shall be provided upon request to employees, designated representatives, and the Assistant Secretary in accordance with 29 CFR 1910.1020 (a) through (e) and (g) through (i). These records shall also be provided upon request to the Director.[§1910.1003(g)(2)(ii)]

(iii) *Any physician who conducts a medical examination* required by this paragraph shall furnish to the employer a statement of the employee's suitability for employment in the specific exposure.[§1910.1003(g)(2)(iii)]

[61 FR 9242, Mar. 7, 1996, as amended at 63 FR 1286, Jan. 8, 1998; 63 FR 20099, Apr. 23, 1998; 70 FR 1141, Jan. 5, 2005; 71 FR 16672, Apr. 3, 2006; 73 FR 75584, Dec. 2, 2008; 76 FR 33608, June 8, 2011; 76 FR 80740, Dec. 27, 2011; 77 FR 17779, Mar. 26, 2012]

§1910.1018
Inorganic arsenic

(a) **Scope and application.** This section applies to all occupational exposures to inorganic arsenic except that this section does not apply to employee exposures in agriculture or resulting from pesticide application, the treatment of wood with preservatives or the utilization of arsenically preserved wood.[§1910.1018(a)]

(b) **Definitions.**[§1910.1018(b)]

Action level means a concentration of inorganic arsenic of 5 micrograms per cubic meter of air (5 μg/m^3) averaged over any eight (8) hour period.[§1910.1018(b)]

Assistant Secretary means the Assistant Secretary of Labor for Occupational Safety and Health, U.S. Department of Labor, or designee.[§1910.1018(b)]

Authorized person means any person specifically authorized by the employer whose duties require the person to enter a regulated area, or any person entering such an area as a designated representative of employees for the purpose of exercising the right to observe monitoring and measuring procedures under paragraph (e) of this section.[§1910.1018(b)]

Director means the Director, National Institute for Occupational Safety and Health, U.S. Department of Health and Human Services, or designee.[§1910.1018(b)]

Inorganic arsenic means copper aceto-arsenite and all inorganic compounds containing arsenic except arsine, measured as arsenic (As).[§1910.1018(b)]

(c) **Permissible exposure limit.** The employer shall assure that no employee is exposed to inorganic arsenic at concentrations greater than 10 micrograms per cubic meter of air (10 μg/m^3), averaged over any 8-hour period.[§1910.1018(c)]

(d) **[Reserved]**[§1910.1018(d)]

(e) **Exposure monitoring.**[§1910.1018(e)]

(1) *General.*[§1910.1018(e)(1)]

(i) *Determinations of airborne exposure levels* shall be made from air samples that are representative of each employee's exposure to inorganic arsenic over an eight (8) hour period. [§1910.1018(e)(1)(i)]

(ii) *For the purposes of this section,* employee exposure is that exposure which would occur if the employee were not using a respirator.[§1910.1018(e)(1)(ii)]

(iii) *The employer shall collect full shift* (for at least 7 continuous hours) personal samples including at least one sample for each shift for each job classification in each work area. [§1910.1018(e)(1)(iii)]

(2) *Initial monitoring.* Each employer who has a workplace or work operation covered by this standard shall monitor each such workplace and work operation to accurately determine the airborne concentration of inorganic arsenic to which employees may be exposed.[§1910.1018(e)(2)]

(3) *Frequency.*[§1910.1018(e)(3)]

(i) *If the initial monitoring reveals employee exposure* to be below the action level the measurements need not be repeated except as otherwise provided in paragraph (e)(4) of this section.[§1910.1018(e)(3)(i)]

(ii) *If the initial monitoring, required by this section,* or subsequent monitoring reveals employee exposure to be above the permissible exposure limit, the employer shall repeat monitoring at least quarterly.[§1910.1018(e)(3)(ii)]

(iii) *If the initial monitoring, required by this section,* or subsequent monitoring reveals employee exposure to be above the action level and below the permissible exposure limit the employer shall repeat monitoring at least every six months. [§1910.1018(e)(3)(iii)]

(iv) *The employer shall continue monitoring* at the required frequency until at least two consecutive measurements, taken at least seven (7) days apart, are below the action level at which time the employer may discontinue monitoring for that employee until such time as any of the events in paragraph (e)(4) of this section occur.[§1910.1018(e)(3)(iv)]

(4) *Additional monitoring.* Whenever there has been a production, process, control or personal change which may result in new or additional exposure to inorganic arsenic, or whenever the employer has any other reason to suspect a change which may result in new or additional exposures to inorganic arsenic, additional monitoring which complies with paragraph (e) of this section shall be conducted.[§1910.1018(e)(4)]

(5) *Employee notification.*[§1910.1018(e)(5)]

(i) *The employer must,* within 15 working days after the receipt of the results of any monitoring performed under this section, notify each affected employee of these results either individually in writing or by posting the results in an appropriate location that is accessible to affected employees.[§1910.1018(e)(5)(i)]

(ii) *Whenever the results indicate* that the representative employee exposure exceeds the permissible exposure limit, the employer shall include in the written notice a statement that the permissible exposure limit was exceeded and a description of the corrective action taken to reduce exposure to or below the permissible exposure limit.[§1910.1018(e)(5)(ii)]

(6) *Accuracy of measurement.*[§1910.1018(e)(6)]

(i) *The employer shall use a method of monitoring and measurement* which has an accuracy (with a confidence level of 95 percent) of not less than plus or minus 25 percent for concentrations of inorganic arsenic greater than or equal to 10 µg/m^3 .[§1910.1018(e)(6)(i)]

(ii) *The employer shall use a method of monitoring and measurement* which has an accuracy (with confidence level of 95 percent) of not less than plus or minus 35 percent for concentrations of inorganic arsenic greater than 5 µg/m^3 but less than 10 µg/m^3 .[§1910.1018(e)(6)(ii)]

(f) **Regulated area.**[§1910.1018(f)]

(1) *Establishment.* The employer shall establish regulated areas where worker exposures to inorganic arsenic, without regard to the use of respirators, are in excess of the permissible limit. [§1910.1018(f)(1)]

(2) *Demarcation.* Regulated areas shall be demarcated and segregated from the rest of the workplace in any manner that minimizes the number of persons who will be exposed to inorganic arsenic.[§1910.1018(f)(2)]

(3) *Access.* Access to regulated areas shall be limited to authorized persons or to persons otherwise authorized by the Act or regulations issued pursuant thereto to enter such areas. [§1910.1018(f)(3)]

(4) *Provision of respirators.* All persons entering a regulated area shall be supplied with a respirator, selected in accordance with paragraph (h)(2) of this section.[§1910.1018(f)(4)]

(5) *Prohibited activities.* The employer shall assure that in regulated areas, food or beverages are not consumed, smoking products, chewing tobacco and gum are not used and cosmetics are not applied, except that these activities may be conducted in the lunchrooms, change rooms and showers required under paragraph (m) of this section. Drinking water may be consumed in the regulated area.[§1910.1018(f)(5)]

(g) **Methods of compliance.**[§1910.1018(g)]

(1) *Controls.*[§1910.1018(g)(1)]

(i) *The employer shall institute* at the earliest possible time but not later than December 31, 1979, engineering and work practice controls to reduce exposures to or below the permissible exposure limit, except to the extent that the employer can establish that such controls are not feasible. [§1910.1018(g)(1)(i)]

(ii) *Where engineering and work practice controls* are not sufficient to reduce exposures to or below the permissible exposure limit, they shall nonetheless be used to reduce exposures to the lowest levels achievable by these controls and shall be supplemented by the use of respirators in accordance with paragraph (h) of this section and other necessary personal protective equipment. Employee rotation is not required as a control strategy before respiratory protection is instituted.[§1910.1018(g)(1)(ii)]

(2) *Compliance Program.*[§1910.1018(g)(2)]

(i) *The employer shall establish and implement* a written program to reduce exposures to or below the permissible exposure limit by means of engineering and work practice controls.[§1910.1018(g)(2)(i)]

(ii) *Written plans* for these compliance programs shall include at least the following:[§1910.1018(g)(2)(ii)]

[A] A description of each operation in which inorganic arsenic is emitted; e.g. machinery used, material processed, controls in place, crew size, operating procedures and maintenance practices;[§1910.1018(g)(2)(ii)[A]]

[B] Engineering plans and studies used to determine methods selected for controlling exposure to inorganic arsenic; [§1910.1018(g)(2)(ii)[B]]

[C] A report of the technology considered in meeting the permissible exposure limit;[§1910.1018(g)(2)(ii)[C]]

[D] Monitoring data;[§1910.1018(g)(2)(ii)[D]]

[E] A detailed schedule for implementation of the engineering controls and work practices that cannot be implemented immediately and for the adaption and implementation of any additional engineering and work practices necessary to meet the permissible exposure limit;[§1910.1018(g)(2)(ii)[E]]

[F] Whenever the employer will not achieve the permissible exposure limit with engineering controls and work practices by December 31, 1979, the employer shall include in the compliance plan an analysis of the effectiveness of the various controls, shall install engineering controls and institute work practices on the quickest schedule feasible, and shall include in the compliance plan and implement a program to minimize the discomfort and maximize the effectiveness of respirator use; and[§1910.1018(g)(2)(ii)[F]]

[G] Other relevant information.[§1910.1018(g)(2)(ii)[G]]

(iii) *Written plans* for such a program shall be submitted upon request to the Assistant Secretary and the Director, and shall be available at the worksite for examination and copying by the Assistant Secretary, Director, any affected employee or authorized employee representatives.[§1910.1018(g)(2)(iii)]

(iv) *The plans required by this paragraph* must be revised and updated at least annually to reflect the current status of the program.[§1910.1018(g)(2)(iv)]

(h) **Respiratory protection.**[§1910.1018(h)]

(1) *General.* For employees who use respirators required by this section, the employer must provide each employee an appropriate respirator that complies with the requirements of this paragraph. Respirators must be used during:[§1910.1018(h)(1)]

(i) *Periods necessary* to install or implement feasible engineering or work-practice controls.[§1910.1018(h)(1)(i)]

(ii) *Work operations,* such as maintenance and repair activities, for which the employer establishes that engineering and work-practice controls are not feasible.[§1910.1018(h)(1)(ii)]

(iii) *Work operations for which engineering* and work-practice controls are not yet sufficient to reduce employee exposures to or below the permissible exposure limit.[§1910.1018(h)(1)(iii)]

(iv) *Emergencies.*[§1910.1018(h)(1)(iv)]

(2) *Respirator program.*[§1910.1018(h)(2)]

(i) *The employer must implement* a respiratory protection program in accordance with Sec. 1910.134(b) through (d) (except (d)(1)(iii)), and (f) through (m), which covers each employee required by this section to use a respirator. [§1910.1018(h)(2)(i)]

(ii) *If an employee exhibits breathing difficulty* during fit testing or respirator use, they must be examined by a physician trained in pulmonary medicine to determine whether they can use a respirator while performing the required duty. [§1910.1018(h)(2)(ii)]

(3) *Respirator selection.*[§1910.1018(h)(3)]

(i) *Employers must:*[§1910.1018(h)(3)(i)]

[A] Select, and provide to employees, the appropriate respirators specified in paragraph (d)(3)(i)(A) of 29 CFR 1910.134.[§1910.1018(h)(3)(i)[A]]

[B] Ensure that employees do not use half mask respirators for protection against arsenic trichloride because it is absorbed rapidly through the skin.[§1910.1018(h)(3)(i)[B]]

[C] Provide HEPA filters for powered and non-powered air-purifying respirators.[§1910.1018(h)(3)(i)[C]]

[D] Select for employee use:[§1910.1018(h)(3)(i)[D]]

[1] Air-purifying respirators that have a combination HEPA filter with an appropriate gas-sorbent cartridge or canister when the employee's exposure exceeds the permissible exposure level for inorganic arsenic and the relevant limit for other gases.[§1910.1018(h)(3)(i)[D][1]]

[2] Front-or back-mounted gas masks equipped with HEPA filters and acid gas canisters or any full facepiece supplied-air respirators when the inorganic arsenic concentration is at or below 500 mg/m^3 ; and half mask air-purifying respirators equipped with HEPA filters and acid gas cartridges when the inorganic arsenic concentration is at or below 100 µg/m^3 . [§1910.1018(h)(3)(i)[D][2]]

(ii) *Employees required to use respirators may choose,* and the employer must provide, a powered air-purifying respirator if it will provide proper protection. In addition, the employer must provide a combination dust and acid-gas respirator to employees who are exposed to gases over the relevant exposure limits.[§1910.1018(h)(3)(ii)]

(i) **[Reserved]**[§1910.1018(i)]

(j) **Protective work clothing and equipment**[§1910.1018(j)]

(1) *Provision and use.* Where the possibility of skin or eye irritation from inorganic arsenic exists, and for all workers working in regulated areas, the employer shall provide at no cost to the employee and assure that employees use appropriate and

1910 Selected 1910 General Industry Standards

clean protective work clothing and equipment such as, but not limited to:[§1910.1018(j)(1)]

(i) *Coveralls or similar full-body work clothing;*[§1910.1018(j)(1)(i)]

(ii) *Gloves, and shoes or coverlets;*[§1910.1018(j)(1)(ii)]

(iii) *Face shields or vented goggles* when necessary to prevent eye irritation, which comply with the requirements of Sec. 1910.133(a)(2)-(6); and[§1910.1018(j)(1)(iii)]

(iv) *Impervious clothing* for employees subject to exposure to arsenic trichloride.[§1910.1018(j)(1)(iv)]

(2) *Cleaning and replacement.*[§1910.1018(j)(2)]

(i) *The employer shall provide* the protective clothing required in paragraph (j)(1) of this section in a freshly laundered and dry condition at least weekly, and daily if the employee works in areas where exposures are over 100 μg/m^3 of inorganic arsenic or in areas where more frequent washing is needed to prevent skin irritation.[§1910.1018(j)(2)(i)]

(ii) *The employer shall clean,* launder, or dispose of protective clothing required by paragraph (j)(1) of this section. [§1910.1018(j)(2)(ii)]

(iii) *The employer shall repair or replace* the protective clothing and equipment as needed to maintain their effectiveness. [§1910.1018(j)(2)(iii)]

(iv) *The employer shall assure that all protective clothing* is removed at the completion of a work shift only in change rooms prescribed in paragraph (m)(1) of this section. [§1910.1018(j)(2)(iv)]

(v) *The employer shall assure that contaminated* protective clothing which is to be cleaned, laundered, or disposed of, is placed in a closed container in the change-room which prevents dispersion of inorganic arsenic outside the container. [§1910.1018(j)(2)(v)]

(vi) *The employer shall inform in writing* any person who cleans or launders clothing required by this section, of the potentially harmful effects including the carcinogenic effects of exposure to inorganic arsenic.[§1910.1018(j)(2)(vi)]

(vii) *Labels on contaminated protective* clothing and equipment. [§1910.1018(j)(2)(vii)]

[A] The employer shall ensure that the containers of contaminated protective clothing and equipment in the workplace or which are to be removed from the workplace are labeled and that the labels include the following information:
DANGER: CONTAMINATED WITH INORGANIC ARSENIC.
MAY CAUSE CANCER.
DO NOT REMOVE DUST BY BLOWING OR SHAKING.
DISPOSE OF INORGANIC ARSENIC CONTAMINATED WASH WATER IN ACCORDANCE WITH APPLICABLE LOCAL, STATE OR FEDERAL REGULATIONS. [§1910.1018(j)(2)(vii)[A]]

[B] Prior to June 1, 2015, employers may include the following information on containers of protective clothing and equipment in lieu of the labeling requirements in paragraphs (j)(2)(vii) of this section:[§1910.1018(j)(2)(vii)[B]]

CAUTION: Clothing contaminated with inorganic arsenic; do not remove dust by blowing or shaking. Dispose of inorganic arsenic contaminated wash water in accordance with applicable local, State or Federal regulations. [§1910.1018(j)(2)(vii)[B]]

(viii) *The employer shall prohibit the removal* of inorganic arsenic from protective clothing or equipment by blowing or shaking.[§1910.1018(j)(2)(viii)]

(k) Housekeeping.[§1910.1018(k)]

(1) *Surfaces.* All surfaces shall be maintained as free as practicable of accumulations of inorganic arsenic.[§1910.1018(k)(1)]

(2) *Cleaning floors.* Floors and other accessible surfaces contaminated with inorganic arsenic may not be cleaned by the use of compressed air, and shoveling and brushing may be used only where vacuuming or other relevant methods have been tried and found not to be effective.[§1910.1018(k)(2)]

(3) *Vacuuming.* Where vacuuming methods are selected, the vacuums shall be used and emptied in a manner to minimize the reentry of inorganic arsenic into the workplace.[§1910.1018(k)(3)]

(4) *Housekeeping plan.* A written housekeeping and maintenance plan shall be kept which shall list appropriate frequencies for carrying out housekeeping operations, and for cleaning and maintaining dust collection equipment. The plan shall be available for inspection by the Assistant Secretary.[§1910.1018(k)(4)]

(5) *Maintenance of equipment.* Periodic cleaning of dust collection and ventilation equipment and checks of their effectiveness shall be carried out to maintain the effectiveness of the system and a notation kept of the last check of effectiveness and cleaning or maintenance.[§1910.1018(k)(5)]

(l) [Reserved][§1910.1018(l)]

(m) Hygiene facilities and practices.[§1910.1018(m)]

(1) *Change rooms.* The employer shall provide for employees working in regulated areas or subject to the possibility of skin or eye irritation from inorganic arsenic, clean change rooms equipped with storage facilities for street clothes and separate storage facilities for protective clothing and equipment in accordance with 29 CFR 1910.141(e).[§1910.1018(m)(1)]

(2) *Showers.*[§1910.1018(m)(2)]

(i) *The employer shall assure* that employees working in regulated areas or subject to the possibility of skin or eye irritation from inorganic arsenic shower at the end of the work shift. [§1910.1018(m)(2)(i)]

(ii) *The employer shall provide* shower facilities in accordance with Sec. 1910.141(d)(3).[§1910.1018(m)(2)(ii)]

(3) *Lunchrooms.*[§1910.1018(m)(3)]

(i) *The employer shall provide* for employees working in regulated areas, lunchroom facilities which have a temperature controlled, positive pressure, filtered air supply, and which are readily accessible to employees working in regulated areas.[§1910.1018(m)(3)(i)]

(ii) *The employer shall assure* that employees working in the regulated area or subject to the possibility of skin or eye irritation from exposure to inorganic arsenic wash their hands and face prior to eating.[§1910.1018(m)(3)(ii)]

(4) *Lavatories.* The employer shall provide lavatory facilities which comply with Sec. 1910.141(d)(1) and (2).[§1910.1018(m)(4)]

(5) *Vacuuming clothes.* The employer shall provide facilities for employees working in areas where exposure, without regard to the use of respirators, exceeds 100 μg/m^3 to vacuum their protective clothing and clean or change shoes worn in such areas before entering change rooms, lunchrooms or shower rooms required by paragraph (j) of this section and shall assure that such employees use such facilities.[§1910.1018(m)(5)]

(6) *Avoidance of skin irritation.* The employer shall assure that no employee is exposed to skin or eye contact with arsenic trichloride, or to skin or eye contact with liquid or particulate inorganic arsenic which is likely to cause skin or eye irritation. [§1910.1018(m)(6)]

(n) Medical surveillance.[§1910.1018(n)]

(1) *General*[§1910.1018(n)(1)]

(i) *Employees covered.* The employer shall institute a medical surveillance program for the following employees: [§1910.1018(n)(1)(i)]

[A] All employees who are or will be exposed above the action level, without regard to the use of respirators, at least 30 days per year; and[§1910.1018(n)(1)(i)[A]]

[B] All employees who have been exposed above the action level, without regard to respirator use, for 30 days or more per year for a total of 10 years or more of combined employment with the employer or predecessor employers prior to or after the effective date of this standard. The determination of exposures prior to the effective date of this standard shall be based upon prior exposure records, comparison with the first measurements taken after the effective date of this standard, or comparison with records of exposures in areas with similar processes, extent of engineering controls utilized and materials used by that employer.[§1910.1018(n)(1)(i)[B]]

(ii) *Examination by physician.* The employer shall assure that all medical examinations and procedures are performed by or under the supervision of a licensed physician, and shall be provided without cost to the employee, without loss of pay and at a reasonable time and place.[§1910.1018(n)(1)(ii)]

(2) *Initial examinations.* By December 1, 1978, for employees initially covered by the medical provisions of this section, or thereafter at the time of initial assignment to an area where the employee is likely to be exposed over the action level at least 30 days per year, the employer shall provide each affected employee an opportunity for a medical examination, including at least the following elements:[§1910.1018(n)(2)]

(i) *A work history and a medical history* which shall include a smoking history and the presence and degree of respiratory symptoms such as breathlessness, cough, sputum production and wheezing.[§1910.1018(n)(2)(i)]

(ii) *A medical examination* which shall include at least the following:[§1910.1018(n)(2)(ii)]

[A] A standard posterior-anterior chest x-ray; [§1910.1018(n)(2)(ii)[A]]

[B] A nasal and skin examination; and[§1910.1018(n)(2)(ii)[B]]

[C] Other examinations which the physician believes appropriate because of the employees exposure to inorganic arsenic or because of required respirator use. [§1910.1018(n)(2)(ii)[C]]

(3) *Periodic examinations.*[§1910.1018(n)(3)]

(i) *Examinations must be provided* in accordance with this paragraph at least annually.[§1910.1018(n)(3)(i)]

(ii) *Whenever a covered employee* has not taken the examinations specified in paragraphs (n)(2)(i) and (n)(2)(ii) of this section within six (6) months preceding the termination of employment, the employer shall provide such examinations to the employee upon termination of employment. [§1910.1018(n)(3)(ii)]

(4) *Additional examinations.* If the employee for any reason develops signs or symptoms commonly associated with exposure to inorganic arsenic the employer shall provide an appropriate examination and emergency medical treatment.[§1910.1018(n)(4)]

(5) *Information provided to the physician.* The employer shall provide the following information to the examining physician: [§1910.1018(n)(5)]

(i) *A copy of this standard and its appendices;*[§1910.1018(n)(5)(i)]

(ii) *A description of the affected employee's duties* as they relate to the employee's exposure;[§1910.1018(n)(5)(ii)]

(iii) *The employee's representative exposure level* or anticipated exposure level;[§1910.1018(n)(5)(iii)]

(iv) *A description of any personal protective equipment* used or to be used; and[§1910.1018(n)(5)(iv)]

(v) *Information from previous medical examinations* of the affected employee which is not readily available to the examining physician.[§1910.1018(n)(5)(v)]

(6) *Physician's written opinion.*[§1910.1018(n)(6)]

(i) *The employer shall obtain* a written opinion from the examining physician which shall include:[§1910.1018(n)(6)(i)]

[A] The results of the medical examination and tests performed;[§1910.1018(n)(6)(i)[A]]

[B] The physician's opinion as to whether the employee has any detected medical conditions which would place the employee at increased risk of material impairment of the employee's health from exposure to inorganic arsenic; [§1910.1018(n)(6)(i)[B]]

[C] Any recommended limitations upon the employee's exposure to inorganic arsenic or upon the use of protective clothing or equipment such as respirators; and [§1910.1018(n)(6)(i)[C]]

[D] A statement that the employee has been informed by the physician of the results of the medical examination and any medical conditions which require further explanation or treatment.[§1910.1018(n)(6)(i)[D]]

(ii) *The employer shall instruct the physician* not to reveal in the written opinion specific findings or diagnoses unrelated to occupational exposure.[§1910.1018(n)(6)(ii)]

(iii) *The employer shall provide a copy* of the written opinion to the affected employee.[§1910.1018(n)(6)(iii)]

(o) Employee information and training.[§1910.1018(o)]

(1) *Training program.*[§1910.1018(o)(1)]

(i) *The employer shall train each employee* who is subject to exposure to inorganic arsenic above the action level without regard to respirator use, or for whom there is the possibility of skin or eye irritation from inorganic arsenic, in accordance with the requirements of this section. The employer shall institute a training program and ensure employee participation in the program.[§1910.1018(o)(1)(i)]

(ii) *The training program shall be provided* by October 1, 1978, for employees covered by this provision, at the time of initial assignment for those subsequently covered by this provision, and at least annually for other covered employees thereafter; and the employer shall assure that each employee is informed of the following:[§1910.1018(o)(1)(ii)]

[A] The information contained in Appendix A; [§1910.1018(o)(1)(ii)[A]]

[B] The quantity, location, manner of use, storage, sources of exposure, and the specific nature of operations which could result in exposure to inorganic arsenic as well as any necessary protective steps;[§1910.1018(o)(1)(ii)[B]]

[C] The purpose, proper use, and limitation of respirators; [§1910.1018(o)(1)(ii)[C]]

[D] The purpose and a description of the medical surveillance program as required by paragraph (n) of this section; [§1910.1018(o)(1)(ii)[D]]

[E] The engineering controls and work practices associated with the employee's job assignment; and [§1910.1018(o)(1)(ii)[E]]

[F] A review of this standard.[§1910.1018(o)(1)(ii)[F]]

(2) *Access to training materials.*[§1910.1018(o)(2)]

(i) *The employer shall make readily available* to all affected employees a copy of this standard and its appendices. [§1910.1018(o)(2)(i)]

(ii) *The employer shall provide,* upon request, all materials relating to the employee information and training program to the Assistant Secretary and the Director.[§1910.1018(o)(2)(ii)]

(p) Communication of hazards[§1910.1018(p)]

(1) *Hazard communication — General.*[§1910.1018(p)(1)]

(i) *Chemical manufacturers,* importers, distributors and employers shall comply with all requirements of the Hazard Communication Standard (HCS) (Sec. 1910.1200) for inorganic arsenic.[§1910.1018(p)(1)(i)]

(ii) *In classifying the hazards of inorganic* arsenic at least the following hazards are to be addressed: Cancer; liver effects; skin effects; respiratory irritation; nervous system effects; and acute toxicity effects.[§1910.1018(p)(1)(ii)]

(iii) *Employers shall include inorganic* arsenic in the hazard communication program established to comply with the HCS (Sec. 1910.1200). Employers shall ensure that each employee has access to labels on containers of inorganic arsenic and to safety data sheets, and is trained in accordance with the requirements of HCS and paragraph (o) of this section.[§1910.1018(p)(1)(iii)]

(iv) *The employer shall ensure that no statement* appears on or near any sign or label required by this paragraph (p) which contradicts or detracts from the meaning of the required sign or label.[§1910.1018(p)(1)(iv)]

(2) *Signs.*[§1910.1018(p)(2)]

(i) *The employer shall post signs* demarcating regulated areas bearing the legend: DANGER INORGANIC ARSENIC MAY CAUSE CANCER DO NOT EAT, DRINK OR SMOKE WEAR RESPIRATORY PROTECTION IN THIS AREA AUTHORIZED PERSONNEL ONLY[§1910.1018(p)(2)(i)]

(ii) *Prior to June 1,* 2016, employers may use the following legend in lieu of that specified in paragraph (p)(2)(i) of this section: DANGER INORGANIC ARSENIC CANCER HAZARD AUTHORIZED PERSONNEL ONLY NO SMOKING OR EATING RESPIRATOR REQUIRED[§1910.1018(p)(2)(ii)]

(iii) *The employer shall ensure that signs* required by this paragraph (p) are illuminated and cleaned as necessary so that the legend is readily visible.[§1910.1018(p)(2)(iii)]

(3) (i) *Prior to June 1,* 2015, in lieu of the labeling requirements in paragraphs (p)(1)(i) of this section, employers may apply precautionary labels to all shipping and storage containers of inorganic arsenic, and to all products containing inorganic arsenic, bearing the following legend: DANGER CONTAINS INORGANIC ARSENIC CANCER HAZARD HARMFUL IF INHALED OR SWALLOWED USE ONLY WITH ADEQUATE VENTILATION OR RESPIRATORY PROTECTION [§1910.1018(p)(3)(i)]

(ii) *Labels are not required when the inorganic* arsenic in the product is bound in such a manner so as to make unlikely the possibility of airborne exposure to inorganic arsenic. (Possible examples of products not requiring labels are semiconductors, light emitting diodes and glass.)[§1910.1018(p)(3)(ii)]

(q) Recordkeeping.[§1910.1018(q)]

(1) *Exposure monitoring.*[§1910.1018(q)(1)]

(i) *The employer shall establish and maintain* an accurate record of all monitoring required by paragraph (e) of this section.[§1910.1018(q)(1)(i)]

(ii) *This record shall include:*[§1910.1018(q)(1)(ii)]

[A] The date(s), number, duration, location, and results of each of the samples taken, including a description of the sampling procedure used to determine representative employee exposure where applicable;[§1910.1018(q)(1)(ii)[A]]

[B] A description of the sampling and analytical methods used and evidence of their accuracy;[§1910.1018(q)(1)(ii)[B]]

[C] The type of respiratory protective devices worn, if any; [§1910.1018(q)(1)(ii)[C]]

[D] Name, social security number, and job classification of the employees monitored and of all other employees whose exposure the measurement is intended to represent; and[§1910.1018(q)(1)(ii)[D]]

[E] The environmental variables that could affect the measurement of the employee's exposure.[§1910.1018(q)(1)(ii)[E]]

(iii) *The employer shall maintain* these monitoring records for at least 40 years or for the duration of employment plus 20 years, whichever, is longer.[§1910.1018(q)(1)(iii)]

(2) *Medical surveillance.*[§1910.1018(q)(2)]

(i) *The employer shall establish and maintain* an accurate record for each employee subject to medical surveillance as required by paragraph (n) of this section.[§1910.1018(q)(2)(i)]

(ii) *This record shall include:*[§1910.1018(q)(2)(ii)]

[A] The name, social security number, and description of duties of the employee;[§1910.1018(q)(2)(ii)[A]]

[B] A copy of the physician's written opinions; [§1910.1018(q)(2)(ii)[B]]

[C] Results of any exposure monitoring done for that employee and the representative exposure levels supplied to the physician; and[§1910.1018(q)(2)(ii)[C]]

[D] Any employee medical complaints related to exposure to inorganic arsenic.[§1910.1018(q)(2)(ii)[D]]

(iii) *The employer shall in addition keep,* or assure that the examining physician keeps, the following medical records; [§1910.1018(q)(2)(iii)]

[A] A copy of the medical examination results including medical and work history required under paragraph (n) of this section;[§1910.1018(q)(2)(iii)[A]]

[B] A description of the laboratory procedures and a copy of any standards or guidelines used to interpret the test results or references to that information; [§1910.1018(q)(2)(iii)[B]]

[C] The initial X-ray;[§1910.1018(q)(2)(iii)[C]]

[D] The X-rays for the most recent 5 years; and [§1910.1018(q)(2)(iii)[D]]

[E] Any X-rays with a demonstrated abnormality and all subsequent X-rays;[§1910.1018(q)(2)(iii)[E]]

(iv) *The employer shall maintain or assure* that the physician maintains those medical records for at least 40 years, or for the duration of employment plus 20 years whichever is longer.[§1910.1018(q)(2)(iv)]

(3) *Availability.*[§1910.1018(q)(3)]

(i) *The employer shall make available* upon request all records required to be maintained by paragraph (q) of this section to the Assistant Secretary and the Director for examination and copying.[§1910.1018(q)(3)(i)]

(ii) *Records required by this paragraph* shall be provided upon request to employees, designated representatives, and the Assistant Secretary in accordance with 29 CFR 1910.1020 (a) through (e) and (g) through (i).[§1910.1018(q)(3)(ii)]

(4) *Transfer of records.*[§1910.1018(q)(4)]

(i) *Whenever the employer ceases to do business,* the successor employer shall receive and retain all records required to be maintained by this section.[§1910.1018(q)(4)(i)]

(ii) *The employer shall also comply* with any additional requirements involving the transfer of records set in 29 CFR 1910.1020(h).[§1910.1018(q)(4)(ii)]

(r) Observation of monitoring.[§1910.1018(r)]

(1) *Employee observation.* The employer shall provide affected employees or their designated representatives an opportunity to observe any monitoring of employee exposure to inorganic arsenic conducted pursuant to paragraph (e) of this section. [§1910.1018(r)(1)]

(2) *Observation procedures.*[§1910.1018(r)(2)]

(i) *Whenever observation of the monitoring* of employee exposure to inorganic arsenic requires entry into an area where the use of respirators, protective clothing, or equipment is required, the employer shall provide the observer with and assure the use of such respirators, clothing, and such equipment, and shall require the observer to comply with all other applicable safety and health procedures.[§1910.1018(r)(2)(i)]

(ii) *Without interfering with the monitoring,* observers shall be entitled to;[§1910.1018(r)(2)(ii)]

[A] Receive an explanation of the measurement procedures; [§1910.1018(r)(2)(ii)[A]]

[B] Observe all steps related to the monitoring of inorganic arsenic performed at the place of exposure; and [§1910.1018(r)(2)(ii)[B]]

[C] Record the results obtained or receive copies of the results when returned by the laboratory. [§1910.1018(r)(2)(ii)[C]]

(s) Appendices. The information contained in the appendices to this section is not intended by itself, to create any additional obligations not otherwise imposed by this standard nor detract from any existing obligation.[§1910.1018(s)]

§1910.1018 Appendix A

Inorganic Arsenic Substance Information Sheet

I. Substance Identification

A. *Substance.* Inorganic Arsenic.

B. *Definition.* Copper acetoarsenite, arsenic and all inorganic compounds containing arsenic except arsine, measured as arsenic (As).

C. *Permissible Exposure Limit.* 10 micrograms per cubic meter of air as determined as an average over an 8-hour period. No employee may be exposed to any skin or eye contact with arsenic trichloride or to skin or eye contact likely to cause skin or eye irritation.

D. *Regulated Areas.* Only employees authorized by your employer should enter a regulated area.

II. Health Hazard Data

A. *Comments.* The health hazard of inorganic arsenic is high.

B. *Ways in which the chemical affects your body.* Exposure to airborne concentrations of inorganic arsenic may cause lung cancer, and can be a skin irritant. Inorganic arsenic may also affect your body if swallowed. One compound in particular, arsenic trichloride, is especially dangerous because it can be absorbed readily through the skin. Because inorganic arsenic is a poison, you should wash your hands thoroughly prior to eating or smoking.

III. Protective Clothing and Equipment

A. *Respirators.* Respirators will be provided by your employer at no cost to you for routine use if your employer is in the process of implementing engineering and work practice controls or where engineering and work practice controls are not feasible or insufficient. You must wear respirators for non-routine activities or in emergency situations where you are likely to be exposed to levels of inorganic arsenic in excess of the permissible exposure limit. Since how well your respirator fits your face is very important, your employer is required to conduct fit tests to make sure the respirator seals properly when you wear it. These tests are simple and rapid and will be explained to you during training sessions.

B. *Protective clothing.* If you work in a regulated area, your employer is required to provide at no cost to you, and you must wear, appropriate, clean, protective clothing and equipment. The purpose of this equipment is to prevent you from bringing to your home arsenic-contaminated dust and to protect your body from repeated skin contact with inorganic arsenic likely to cause skin irritation. This clothing should include such items as coveralls or similar full-body clothing, gloves, shoes or coverlets, and aprons. Protective equipment should include face shields or vented goggles, where eye irritation may occur.

IV. Hygiene Facilities and Practices

You must not eat, drink, smoke, chew gum or tobacco, or apply cosmetics in the regulated area, except that drinking water is permitted. If you work in a regulated area your employer is required to provide lunchrooms and other areas for these purposes.

If you work in a regulated area, your employer is required to provide showers, washing facilities, and change rooms. You must wash your face, and hands before eating and must shower at the end of the work shift. Do not take used protective clothing out of change rooms without your employer's permission. Your employer is required to provide for laundering or cleaning of your protective clothing.

V. Signs and Labels

Your employer is required to post warning signs and labels for your protection. Signs must be posted in regulated areas. The signs must warn that a cancer hazard is present, that only authorized employees may enter the area, and that no smoking or eating is allowed, and that respirators must be worn.

VI. Medical Examinations

If your exposure to arsenic is over the Action Level (5 µg/m^3) — (including all persons working in regulated areas) at least 30 days per year, or you have been exposed to arsenic for more than 10 years over the Action Level, your employer is required to provide you with a medical examination. The examination shall be every 6 months for employees over 45 years old or with more than 10 years exposure over the Action Level and annually for other covered employees. The medical examination must include a medical history; a chest x-ray; a skin examination and a nasal examination. The examining physician will provide a written opinion to your employer containing the results of your medical exams. You should also receive a copy of this opinion. The physician must not tell your employer any conditions he detects unrelated to occupational exposure to arsenic but must tell you those conditions.

VII. Observation of Monitoring

Your employer is required to monitor your exposure to arsenic and you or your representatives are entitled to observe the monitoring procedure. You are entitled to receive an explanation of the measurement procedure, and to record the results obtained. When the monitoring procedure is taking place in an area where respirators or personal protective clothing and equipment are required to be worn, you must also be provided with and must wear the protective clothing and equipment.

VIII. Access to Records

You or your representative are entitled to records of your exposure to inorganic arsenic and your medical examination records if you request your employer to provide them.

IX. Training and Notification

Additional information on all of these items plus training as to hazards of exposure to inorganic arsenic and the engineering and work practice controls associated with your job will also be provided by your employer. If you are exposed over the permissible exposure limit, your employer must inform you of that fact and the actions he is taking to reduce your exposures.

§1910.1018 Appendix B

Substance Technical Guidelines

Arsenic, Arsenic Trioxide, Arsenic Trichloride (Three Examples)

I. Physical and chemical properties

A. *Arsenic (metal).*

1. *Formula:* As.
2. *Appearance:* Gray metal.
3. *Melting point:* Sublimes without melting at 613 °C.
4. *Specific Gravity:* (H_2O = 1): 5.73.
5. *Solubility in water:* Insoluble.

B. *Arsenic Trioxide.*

1. *Formula:* As_2O_3, (As_4O_6).
2. *Appearance:* White powder.
3. *Melting point:* 315 °C.
4. *Specific Gravity:* (H_2O = 1): 3.74.
5. *Solubility in water:* 3.7 grams in 100cc of water at 20 °C.

C. *Arsenic Trichloride (liquid).*

1. *Formula:* $AsCl_3$.
2. *Appearance:* Colorless or pale yellow liquid.
3. *Melting point:* -8.5 °C.
4. *Boiling point:* 130.2 °C.
5. *Specific Gravity:* (H_2O = 1): 2.16 at 20°C.
6. *Vapor Pressure:* 10mm Hg at 23.5°C.
7. *Solubility in Water:* Decomposes in water.

II. Fire, explosion and reactivity data

A. *Fire:* Arsenic, arsenic Trioxide and Arsenic Trichloride are nonflammable.

B. *Reactivity:*

1. *Conditions Contributing to instability:* Heat.
2. *Incompatibility:* Hydrogen gas can react with inorganic arsenic to form the highly toxic gas arsine.

III. Monitoring and Measurement Procedures

Samples collected should be full shift (at least 7-hour) samples. Sampling should be done using a personal sampling pump at a flow rate of 2 liters per minute. Samples should be collected on 0.8 micrometer pore size membrane filter (37mm diameter). Volatile arsenicals such as arsenic trichloride can be most easily collected in a midget bubbler filled with 15 ml. of 0.1 N NaOH.

The method of sampling and analysis should have an accuracy of not less than ± 25 percent (with a confidence limit of 95 percent) for 10 micrograms per cubic meter of air (10 µg/m^3) and ± 35 percent (with a confidence limit of 95 percent) for concentrations of inorganic arsenic between 5 and 10 µg/m^3.

§1910.1018 Appendix C

Medical Surveillance Guidelines

I. General

Medical examinations are to be provided for all employees exposed to levels of inorganic arsenic above the action level (5 µg/m^3) for at least 30 days per year (which would include among others, all employees, who work in regulated areas). Examinations are also to be provided to all employees who have had 10 years or more exposure above the action level for more than 30 days per year while working for the present or predecessor employer though they may no longer be exposed above the level.

An initial medical examination is to be provided to all such employees by December 1, 1978. In addition, an initial medical examination is to be provided to all employees who are first assigned to areas in which worker exposure will probably exceed 5 µg/m^3 (after the effective date of this standard) at the time of initial assignment. In addition to its immediate diagnostic usefulness, the initial examination will provide a baseline for comparing future test results. The initial examination must include as a minimum the following elements:

(1) *A work and medical history,* including a smoking history, and presence and degree of respiratory symptoms such as breathlessness, cough, sputum production, and wheezing;
(2) *A 14" by 17" posterior-anterior chest X-ray;*
(3) *A nasal and skin examination; and*
(4) *Other examinations* which the physician believes appropriate because of the employee's exposure to inorganic arsenic or because of required respirator use.

Periodic examinations are also to be provided to the employees listed above. The periodic examinations shall be given annually for those covered employees 45 years of age or less with fewer than 10 years employment in areas where employee exposure exceeds the action level (5 µg/m^3). Periodic examinations need not include sputum cytology and only an updated medical history is required.

Periodic examinations for other covered employees, shall be provided every six (6) months. These examinations shall include all tests required in the initial examination, except that the medical history need only be updated.

The examination contents are minimum requirements. Additional tests such as lateral and oblique X-rays or pulmonary function tests may be useful. For workers exposed to three arsenicals which are associated with lymphatic cancer, copper acetoarsenite, potassium arsenite, or sodium arsenite the examination should also include palpation of superficial lymph nodes and complete blood count.

II. Noncarcinogenic Effects

The OSHA standard is based on minimizing risk of exposed workers dying of lung cancer from exposure to inorganic arsenic. It will also minimize skin cancer from such exposures.

The following three sections quoted from "Occupational Diseases: A Guide to Their Recognition", Revised Edition, June 1977, National Institute for Occupational Safety and Health is included to provide information on the nonneoplastic effects of exposure to inorganic arsenic. Such effects should not occur if the OSHA standards are followed.

A. *Local — Trivalent arsenic compounds* are corrosive to the skin. Brief contact has no effect but prolonged contact results in a local hyperemia and later vesicular or pustular eruption. The moist mucous membranes are most sensitive to the irritant action. Conjunctiva, moist and macerated areas of skin, the eyelids, the angles of the ears, nose, mouth, and respiratory mucosa are also vulnerable to the irritant effects. The wrists are common sites of dermatitis, as are the genitalia if personal hygiene is poor. Perforations of the nasal septum may occur. Arsenic trioxide and pentoxide are capable of producing skin sensitization and contact dermatitis. Arsenic is also capable of producing keratoses, especially of the palms and soles.

B. *Systemic — The acute toxic effects* of arsenic are generally seen following ingestion of inorganic arsenical compounds. This rarely occurs in an industrial setting. Symptoms develop within 1/2 to 4 hours following ingestion and are usually characterized by constriction of the throat followed by dysphagia, epigastric pain, vomiting, and watery diarrhea. Blood may appear in vomitus and stools. If the amount ingested is sufficiently high, shock may develop due to severe fluid loss, and death may ensue in 24 hours. If the acute effects are survived, exfoliative dermatitis and peripheral neuritis may develop.

Cases of acute arsenical poisoning due to inhalation are exceedingly rare in industry. When it does occur, respiratory tract symptoms — cough, chest pain, dyspnea — giddiness, headache, and extreme general weakness precede gastrointestinal symptoms. The acute toxic symptoms of trivalent arsenical poisoning are due to severe inflammation of the mucous membranes and greatly increased permeability of the blood capillaries.

Chronic arsenical poisoning due to ingestion is rare and generally confined to patients taking prescribed medications. However, it can be a concomitant of inhaled inorganic arsenic from swallowed sputum and improper eating habits. Symptoms are weight loss, nausea and diarrhea alternating with constipation, pigmentation and eruption of the skin, loss of hair, and peripheral neuritis. Chronic hepatitis and cirrhosis have been described. Polyneuritis may be the salient feature, but more frequently there are numbness and parasthenias of "glove and stocking" distribution. The skin lesions are usually melanotic and keratotic and may occasionally take the form of an intradermal cancer of the squamous cell type, but without infiltrative properties. Horizontal white lines (striations) on the fingernails and toenails are commonly seen in chronic arsenical poisoning and are considered to be a diagnostic accompaniment of arsenical polyneuritis.

Inhalation of inorganic arsenic compounds is the most common cause of chronic poisoning in the industrial situation. This condition is divided into three phases based on signs and symptoms.

First Phase: The worker complains of weakness, loss of appetite, some nausea, occasional vomiting, a sense of heaviness in the stomach, and some diarrhea.

Second Phase: The worker complains of conjunctivitis, a catarrhal state of the mucous membranes of the nose, larynx, and respiratory passage. Coryza, hoarseness, and mild tracheobronchitis may occur. Perforation of the nasal septum is common, and is probably the most typical lesion of the upper respiratory tract in occupational exposure to arsenical dust. Skin lesions, eczematoid and allergic in type, are common.

Third Phase: The worker complains of symptoms of peripheral neuritis, initially of hands and feet, which is essentially sensory. In more severe cases, motor paralyses occur; the first muscles affected are usually the toe extensors and the peronei. In only the most severe cases will paralysis of flexor muscles of the feet or of the extensor muscles of hands occur.

Liver damage from chronic arsenical poisoning is still debated, and as yet the question is unanswered. In cases of chronic and acute arsenical poisoning, toxic effects to the myocardium have been reported based on EKG changes. These findings, however, are now largely discounted and the EKG changes are ascribed to electrolyte disturbances concomitant with arsenicalism. Inhalation of arsenic trioxide and other inorganic arsenical dusts does not give rise to radiological evidence or pneumoconiosis. Arsenic does have a depressant effect upon the bone marrow, with disturbances of both erythropoiesis and myelopoiesis.

Bibliography

Dinman, B. D. 1960. Arsenic; chronic human intoxication. J. Occup. Med. 2:137.

Elkins, H. B. 1959. The Chemistry of Industrial Toxicology, 2nd ed. John Wiley and Sons, New York.

Holmquist, L. 1951. Occupational arsenical dermatitis; a study among employees at a copper-ore smelting works including investigations of skin reactions to contact with arsenic compounds. Acta. Derm. Venereol. (Supp. 26) 31:1.

Pinto, S. S., and C. M. McGill. 1953. Arsenic trioxide exposure in industry. Ind. Med. Surg. 22:281.

Pinto, S. S., and K. W. Nelson. 1976. Arsenic toxicology and industrial exposure. Annu. Rev. Pharmacol. Toxicol. 16:95.

Vallee, B. L., D. D. Ulmer, and W. E. C. Wacker. 1960. Arsenic toxicology and biochemistry. AMA Arch. Indust. Health 21:132.

[39 FR 23502, June 27, 1974, as amended at 43 FR 19624, May 5, 1978; 43 FR 28472, June 30, 1978; 45 FR 35282, May 23, 1980; 54 FR 24334, June 7, 1989; 58 FR 35310, June 30, 1993; 61 FR 5508, Feb. 13, 1996; 61 FR 9245, Mar. 7, 1996; 63 FR 1286, Jan. 8, 1998; 63 FR 33468, June 18, 1998; 70 FR 1141, Jan. 5, 2005; 71 FR 16672, 16673, Apr. 3, 2006; 71 FR 50189, Aug. 24, 2006; 73 FR 75585, Dec. 12, 2008; 76 FR 33608, June 8, 2011; 77 FR 17780, Mar. 26, 2012]

§1910.1020

Access to employee exposure and medical records

(a) Purpose. The purpose of this section is to provide employees and their designated representatives a right of access to relevant exposure and medical records; and to provide representatives of the Assistant Secretary a right of access to these records in order to fulfill responsibilities under the Occupational Safety and Health Act. Access by employees, their representatives, and the Assistant Secretary is necessary to yield both direct and indirect improvements in the detection, treatment, and prevention of occupational disease. Each employer is responsible for assuring compliance with this section, but the activities involved in complying with the access to medical records provisions can be carried out, on behalf of the employer, by the physician or other health care personnel in charge of employee medical records. Except as expressly provided, nothing in this section is intended to affect existing legal and ethical obligations concerning the maintenance and confidentiality of employee medical information, the duty to disclose information to a patient/employee or any other aspect of the medical-care relationship, or affect existing legal obligations concerning the protection of trade secret information.[§1910.1020(a)]

(b) Scope and application.[§1910.1020(b)]

(1) *This section applies to each general* industry, maritime, and construction employer who makes, maintains, contracts for, or has access to employee exposure or medical records, or analyses thereof, pertaining to employees exposed to toxic substances or harmful physical agents.[§1910.1020(b)(1)]

(2) *This section applies to all employee* exposure and medical records, and analyses thereof, of such employees, whether or not the records are mandated by specific occupational safety and health standards.[§1910.1020(b)(2)]

(3) *This section applies to all employee* exposure and medical records, and analyses thereof, made or maintained in any manner, including on an in-house of contractual (e.g., fee-for-service) basis. Each employer shall assure that the preservation and access requirements of this section are complied with regardless of the manner in which the records are made or maintained.[§1910.1020(b)(3)]

(c) Definitions.[§1910.1020(c)]

(1) **Access** means the right and opportunity to examine and copy. [§1910.1020(c)(1)]

(2) **Analysis using exposure or medical records** means any compilation of data or any statistical study based at least in part on information collected from individual employee exposure or medical records or information collected from health insurance claims records, provided that either the analysis has been reported to the employer or no further work is currently being done by the person responsible for preparing the analysis.[§1910.1020(c)(2)]

(3) **Designated representative** means any individual or organization to whom an employee gives written authorization to exercise a right of access. For the purposes of access to employee exposure records and analyses using exposure or medical records, a recognized or certified collective bargaining agent shall be treated automatically as a designated representative without regard to written employee authorization. [§1910.1020(c)(3)]

(4) **Employee** means a current employee, a former employee, or an employee being assigned or transferred to work where there will be exposure to toxic substances or harmful physical agents. In the case of a deceased or legally incapacitated employee, the employee's legal representative may directly exercise all the employee's rights under this section. [§1910.1020(c)(4)]

(5) **Employee exposure record** means a record containing any of the following kinds of information:[§1910.1020(c)(5)]

(i) *Environmental (workplace) monitoring* or measuring of a toxic substance or harmful physical agent, including personal, area, grab, wipe, or other form of sampling, as well as related collection and analytical methodologies, calculations, and other background data relevant to interpretation of the results obtained;[§1910.1020(c)(5)(i)]

(ii) *Biological monitoring results* which directly assess the absorption of a toxic substance or harmful physical agent by body systems (e.g., the level of a chemical in the blood, urine, breath, hair, fingernails, etc) but not including results which assess the biological effect of a substance or agent or which assess an employee's use of alcohol or drugs; [§1910.1020(c)(5)(ii)]

(iii) *Material safety data sheets* indicating that the material may pose a hazard to human health; or[§1910.1020(c)(5)(iii)]

(iv) *In the absence of the above,* a chemical inventory or any other record which reveals where and when used and the identity (e.g., chemical, common, or trade name) of a toxic substance or harmful physical agent.[§1910.1020(c)(5)(iv)]

(6) (i) **Employee medical record** means a record concerning the health status of an employee which is made or maintained by a physician, nurse, or other health care personnel or technician, including:[§1910.1020(c)(6)(i)]

[A] Medical and employment questionnaires or histories (including job description and occupational exposures), [§1910.1020(c)(6)(i)[A]]

[B] The results of medical examinations (pre-employment, pre-assignment, periodic, or episodic) and laboratory tests (including chest and other X-ray examinations taken for the purposes of establishing a base-line or detecting occupational illness, and all biological monitoring not defined as an "employee exposure record"), [§1910.1020(c)(6)(i)[B]]

[C] Medical opinions, diagnoses, progress notes, and recommendations,[§1910.1020(c)(6)(i)[C]]

[D] First aid records,[§1910.1020(c)(6)(i)[D]]

[E] Descriptions of treatments and prescriptions, and [§1910.1020(c)(6)(i)[E]]

[F] Employee medical complaints.[§1910.1020(c)(6)(i)[F]]

(ii) *"Employee medical record"* does not include medical information in the form of:[§1910.1020(c)(6)(ii)]

[A] Physical specimens (e.g., blood or urine samples) which are routinely discarded as a part of normal medical practice; or[§1910.1020(c)(6)(ii)[A]]

[B] Records concerning health insurance claims if maintained separately from the employer's medical program and its records, and not accessible to the employer by employee name or other direct personal identifier (e.g., social security number, payroll number, etc.); or[§1910.1020(c)(6)(ii)[B]]

[C] Records created solely in preparation for litigation which are privileged from discovery under the applicable rules of procedure or evidence; or[§1910.1020(c)(6)(ii)[C]]

[D] Records concerning voluntary employee assistance programs (alcohol, drug abuse, or personal counseling programs) if maintained separately from the employer's medical program and its records.[§1910.1020(c)(6)(ii)[D]]

(7) **Employer** means a current employer, a former employer, or a successor employer.[§1910.1020(c)(7)]

(8) **Exposure or exposed** means that an employee is subjected to a toxic substance or harmful physical agent in the course of employment through any route of entry (inhalation, ingestion, skin contact or absorption, etc.), and includes past exposure and potential (e.g., accidental or possible) exposure, but does not include situations where the employer can demonstrate that the toxic substance or harmful physical agent is not used, handled, stored, generated, or present in the workplace in any manner different from typical non-occupational situations. [§1910.1020(c)(8)]

(9) **Health Professional** means a physician, occupational health nurse, industrial hygienist, toxicologist, or epidemiologist, providing medical or other occupational health services to exposed employees.[§1910.1020(c)(9)]

(10) **Record** means any item, collection, or grouping of information regardless of the form or process by which it is maintained (e.g., paper document, microfiche, microfilm, X-ray film, or automated data processing).[§1910.1020(c)(10)]

(11) **Specific chemical identity** means the chemical name, Chemical Abstracts Service (CAS) Registry Number, or any other information that reveals the precise chemical designation of the substance.[§1910.1020(c)(11)]

(12) (i) **Specific written consent** means a written authorization containing the following:[§1910.1020(c)(12)(i)]

[A] The name and signature of the employee authorizing the release of medical information,[§1910.1020(c)(12)(i)[A]]

[B] The date of the written authorization,[§1910.1020(c)(12)(i)[B]]

[C] The name of the individual or organization that is authorized to release the medical information, [§1910.1020(c)(12)(i)[C]]

[D] The name of the designated representative (individual or organization) that is authorized to receive the released information,[§1910.1020(c)(12)(i)[D]]

[E] A general description of the medical information that is authorized to be released,[§1910.1020(c)(12)(i)[E]]

[F] A general description of the purpose for the release of the medical information, and[§1910.1020(c)(12)(i)[F]]

[G] A date or condition upon which the written authorization will expire (if less than one year).[§1910.1020(c)(12)(i)[G]]

(ii) *A written authorization does not* operate to authorize the release of medical information not in existence on the date of written authorization, unless the release of future information is expressly authorized, and does not operate for more than one year from the date of written authorization. [§1910.1020(c)(12)(ii)]

(iii) *A written authorization may be revoked* in writing prospectively at any time.[§1910.1020(c)(12)(iii)]

(13) **Toxic substance or harmful physical agent** means any chemical substance, biological agent (bacteria, virus, fungus, etc.), or physical stress (noise, heat, cold, vibration, repetitive motion, ionizing and non-ionizing radiation, hypo-or hyperbaric pressure, etc.) which:[§1910.1020(c)(13)]

(i) *Is listed in the latest printed edition* of the National Institute for Occupational Safety and Health (NIOSH) Registry of Toxic Effects of Chemical Substances (RTECS), which is incorporated by reference as specified in Sec. 1910.6; or [§1910.1020(c)(13)(i)]

(ii) *Has yielded positive evidence* of an acute or chronic health hazard in testing conducted by, or known to, the employer; or [§1910.1020(c)(13)(ii)]

(iii) *Is the subject of a material safety data sheet* kept by or known to the employer indicating that the material may pose a hazard to human health.[§1910.1020(c)(13)(iii)]

(14) **Trade secret** means any confidential formula, pattern, process, device, or information or compilation of information that is used in an employer's business and that gives the employer an opportunity to obtain an advantage over competitors who do not know or use it.[§1910.1020(c)(14)]

(d) **Preservation of records.**[§1910.1020(d)]

(1) *Unless a specific occupational safety and health standard* provides a different period of time, each employer shall assure the preservation and retention of records as follows: [§1910.1020(d)(1)]

(i) *Employee medical records.* The medical record for each employee shall be preserved and maintained for at least the duration of employment plus thirty (30) years, except that the following types of records need not be retained for any specified period:[§1910.1020(d)(1)(i)]

[A] Health insurance claims records maintained separately from the employer's medical program and its records, [§1910.1020(d)(1)(i)[A]]

[B] First aid records (not including medical histories) of one-time treatment and subsequent observation of minor scratches, cuts, burns, splinters, and the like which do not involve medical treatment, loss of consciousness, restriction of work or motion, or transfer to another job, if made on-site by a non-physician and if maintained separately from the employer's medical program and its records, and [§1910.1020(d)(1)(i)[B]]

[C] The medical records of employees who have worked for less than (1) year for the employer need not be retained beyond the term of employment if they are provided to the employee upon the termination of employment. [§1910.1020(d)(1)(i)[C]]

(ii) *Employee exposure records.* Each employee exposure record shall be preserved and maintained for at least thirty (30) years, except that:[§1910.1020(d)(1)(ii)]

[A] Background data to environmental (workplace) monitoring or measuring, such as laboratory reports and worksheets, need only be retained for one (1) year as long as the sampling results, the collection methodology (sampling plan), a description of the analytical and mathematical methods used, and a summary of other background data relevant to interpretation of the results obtained, are retained for at least thirty (30) years; and[§1910.1020(d)(1)(ii)[A]]

[B] Material safety data sheets and paragraph (c)(5)(iv) records concerning the identity of a substance or agent need not be retained for any specified period as long as some record of the identity (chemical name if known) of the substance or agent, where it was used, and when it was used is retained for at least thirty (30) years;[1] and [§1910.1020(d)(1)(ii)[B]]

[C] Biological monitoring results designated as exposure records by specific occupational safety and health standards shall be preserved and maintained as required by the specific standard.[§1910.1020(d)(1)(ii)[C]]

(iii) *Analyses using exposure or medical records.* Each analysis using exposure or medial records shall be preserved and maintained for at least thirty (30) years.[§1910.1020(d)(1)(iii)]

(2) *Nothing in this section is intended* to mandate the form, manner, or process by which an employer preserves a record as long as the information contained in the record is preserved and retrievable, except that chest X-ray films shall be preserved in their original state.[§1910.1020(d)(2)]

(e) Access to records.[§1910.1020(e)]

(1) *General.*[§1910.1020(e)(1)]

(i) *Whenever an employee or designated representative* requests access to a record, the employer shall assure that access is provided in a reasonable time, place, and manner. If the employer cannot reasonably provide access to the record within fifteen (15) working days, the employer shall within the fifteen (15) working days apprise the employee or designated representative requesting the record of the reason for the delay and the earliest date when the record can be made available.[§1910.1020(e)(1)(i)]

(ii) *The employer may require of the requester* only such information as should be readily known to the requester and which may be necessary to locate or identify the records being requested (e.g. dates and locations where the employee worked during the time period in question). [§1910.1020(e)(1)(ii)]

(iii) *Whenever an employee or designated representative* requests a copy of a record, the employer shall assure that either:[§1910.1020(e)(1)(iii)]

[A] A copy of the record is provided without cost to the employee or representative,[§1910.1020(e)(1)(iii)[A]]

[B] The necessary mechanical copying facilities (e.g., photocopying) are made available without cost to the employee or representative for copying the record, or [§1910.1020(e)(1)(iii)[B]]

[C] The record is loaned to the employee or representative for a reasonable time to enable a copy to be made. [§1910.1020(e)(1)(iii)[C]]

(iv) *In the case of an original X-ray,* the employer may restrict access to on-site examination or make other suitable arrangements for the temporary loan of the X-ray. [§1910.1020(e)(1)(iv)]

(v) *Whenever a record has been previously provided* without cost to an employee or designated representative, the employer may charge reasonable, non-discriminatory administrative costs (i.e., search and copying expenses but not including overhead expenses) for a request by the employee or designated representative for additional copies of the record, except that[§1910.1020(e)(1)(v)]

[A] An employer shall not charge for an initial request for a copy of new information that has been added to a record which was previously provided; and[§1910.1020(e)(1)(v)[A]]

[B] An employer shall not charge for an initial request by a recognized or certified collective bargaining agent for a copy of an employee exposure record or an analysis using exposure or medical records.[§1910.1020(e)(1)(v)[B]]

(vi) *Nothing in this section is intended* to preclude employees and collective bargaining agents from collectively bargaining to obtain access to information in addition to that available under this section.[§1910.1020(e)(1)(vi)]

(2) *Employee and designated representative access* [§1910.1020(e)(2)]

(i) *Employee exposure records.*[§1910.1020(e)(2)(i)]

[A] Except as limited by paragraph (f) of this section, each employer shall, upon request, assure the access to each employee and designated representative to employee exposure records relevant to the employee. For the purpose of this section, an exposure record relevant to the employee consists of:[§1910.1020(e)(2)(i)[A]]

[1] A record which measures or monitors the amount of a toxic substance or harmful physical agent to which the employee is or has been exposed;[§1910.1020(e)(2)(i)[A][1]]

[2] In the absence of such directly relevant records, such records of other employees with past or present job duties or working conditions related to or similar to those of the employee to the extent necessary to reasonably indicate the amount and nature of the toxic substances or harmful physical agents to which the employee is or has been subjected, and [§1910.1020(e)(2)(i)[A][2]]

[3] Exposure records to the extent necessary to reasonably indicate the amount and nature of the toxic substances or harmful physical agents at workplaces or under working conditions to which the employee is being assigned or transferred.[§1910.1020(e)(2)(i)[A][3]]

[B] Requests by designated representatives for unconsented access to employee exposure records shall be in writing and shall specify with reasonable particularity: [§1910.1020(e)(2)(i)[B]]

[1] The records requested to be disclosed; and [§1910.1020(e)(2)(i)[B][1]]

[2] The occupational health need for gaining access to these records.[§1910.1020(e)(2)(i)[B][2]]

(ii) *Employee medical records.*[§1910.1020(e)(2)(ii)]

[A] Each employer shall, upon request, assure the access of each employee to employee medical records of which the employee is the subject, except as provided in paragraph (e)(2)(ii)(D) of this section.[§1910.1020(e)(2)(ii)[A]]

[B] Each employer shall, upon request, assure the access of each designated representative to the employee medical records of any employee who has given the designated representative specific written consent. Appendix A to this section contains a sample form which may be used to establish specific written consent for access to employee medical records.[§1910.1020(e)(2)(ii)[B]]

[C] Whenever access to employee medical records is requested, a physician representing the employer may recommend that the employee or designated representative:[§1910.1020(e)(2)(ii)[C]]

[1] Consult with the physician for the purposes of reviewing and discussing the records requested, [§1910.1020(e)(2)(ii)[C][1]]

[2] Accept a summary of material facts and opinions in lieu of the records requested, or[§1910.1020(e)(2)(ii)[C][2]]

[3] Accept release of the requested records only to a physician or other designated representative. [§1910.1020(e)(2)(ii)[C][3]]

[D] Whenever an employee requests access to his or her employee medical records, and a physician representing the employer believes that direct employee access to information contained in the records regarding a specific diagnosis of a terminal illness or a psychiatric condition could be detrimental to the employee's health, the employer may inform the employee that access will only be provided to a designated representative of the employee having specific written consent, and deny the employee's request for direct access to this information only. Where a designated representative with specific written consent requests access to information so withheld, the employer shall assure the access of the designated representative to this information, even when it is known that the designated representative will give the information to the employee.[§1910.1020(e)(2)(ii)[D]]

[E] A physician, nurse, or other responsible health care personnel maintaining medical records may delete from requested medical records the identity of a family member, personal friend, or fellow employee who has provided confidential information concerning an employee's health status.[§1910.1020(e)(2)(ii)[E]]

(iii) *Analyses using exposure or medical records.* [§1910.1020(e)(2)(iii)]

1. Material safety data sheets must be kept for those chemicals currently in use that are effected by the Hazard Communication Standard in accordance with 29 CFR 1910.1200(g).

[A] *Each employer*[1] *shall,* upon request, assure the access of each employee and designated representative to each analysis using exposure or medical records concerning the employee's working conditions or workplace. [§1910.1020(e)(2)(iii)[A]]

[B] *Whenever access is requested* to an analysis which reports the contents of employee medical records by either direct identifier (name, address, social security number, payroll number, etc.) or by information which could reasonably be used under the circumstances indirectly to identify specific employees (exact age, height, weight, race, sex, date of initial employment, job title, etc.), the employer shall assure that personal identifiers are removed before access is provided. If the employer can demonstrate that removal of personal identifiers from an analysis is not feasible, access to the personally identifiable portions of the analysis need not be provided. [§1910.1020(e)(2)(iii)[B]]

(3) *OSHA access.*[§1910.1020(e)(3)]

(i) *Each employer shall,* upon request, and without derogation of any rights under the Constitution or the Occupational Safety and Health Act of 1970, 29 U.S.C. 651 et seq., that the employer chooses to exercise, assure the prompt access of representatives of the Assistant Secretary of Labor for Occupational Safety and Health to employee exposure and medical records and to analyses using exposure or medical records. Rules of agency practice and procedure governing OSHA access to employee medical records are contained in 29 CFR 1913.10.[§1910.1020(e)(3)(i)]

(ii) *Whenever OSHA seeks* access to personally identifiable employee medical information by presenting to the employer a written access order pursuant to 29 CFR 1913.10(d), the employer shall prominently post a copy of the written access order and its accompanying cover letter for at least fifteen (15) working days.[§1910.1020(e)(3)(ii)]

(f) Trade secrets.[§1910.1020(f)]

(1) *Except as provided in paragraph (f)(2)* of this section, nothing in this section precludes an employer from deleting from records requested by a health professional, employee, or designated representative any trade secret data which discloses manufacturing processes, or discloses the percentage of a chemical substance in mixture, as long as the health professional, employee, or designated representative is notified that information has been deleted. Whenever deletion of trade secret information substantially impairs evaluation of the place where or the time when exposure to a toxic substance or harmful physical agent occurred, the employer shall provide alternative information which is sufficient to permit the requesting party to identify where and when exposure occurred. [§1910.1020(f)(1)]

(2) *The employer may withhold* the specific chemical identity, including the chemical name and other specific identification of a toxic substance from a disclosable record provided that: [§1910.1020(f)(2)]

(i) *The claim that the information withheld* is a trade secret can be supported;[§1910.1020(f)(2)(i)]

(ii) *All other available information on the properties* and effects of the toxic substance is disclosed;[§1910.1020(f)(2)(ii)]

(iii) *The employer informs the requesting* party that the specific chemical identity is being withheld as a trade secret; and [§1910.1020(f)(2)(iii)]

(iv) *The specific chemical identity* is made available to health professionals, employees and designated representatives in accordance with the specific applicable provisions of this paragraph.[§1910.1020(f)(2)(iv)]

(3) *Where a treating physician or nurse* determines that a medical emergency exists and the specific chemical identity of a toxic substance is necessary for emergency or first-aid treatment, the employer shall immediately disclose the specific chemical identity of a trade secret chemical to the treating physician or nurse, regardless of the existence of a written statement of need or a confidentiality agreement. The employer may require a written statement of need and confidentiality agreement, in accordance with the provisions of paragraphs (f)(4) and (f)(5), as soon as circumstances permit.[§1910.1020(f)(3)]

(4) *In non-emergency situations,* an employer shall, upon request, disclose a specific chemical identity, otherwise permitted to be withheld under paragraph (f)(2) of this section, to a health professional, employee, or designated representative if:[§1910.1020(f)(4)]

(i) *The request is in writing;*[§1910.1020(f)(4)(i)]

(ii) *The request describes with reasonable detail* one or more of the following occupational health needs for the information: [§1910.1020(f)(4)(ii)]

[A] *To assess the hazards of the chemicals* to which employees will be exposed;[§1910.1020(f)(4)(ii)[A]]

[B] *To conduct or assess sampling* of the workplace atmosphere to determine employee exposure levels; [§1910.1020(f)(4)(ii)[B]]

[C] *To conduct pre-assignment* or periodic medical surveillance of exposed employees;[§1910.1020(f)(4)(ii)[C]]

[D] *To provide medical treatment* to exposed employees; [§1910.1020(f)(4)(ii)[D]]

[E] *To select or assess* appropriate personal protective equipment for exposed employees;[§1910.1020(f)(4)(ii)[E]]

[F] *To design or assess* engineering controls or other protective measures for exposed employees; and [§1910.1020(f)(4)(ii)[F]]

[G] *To conduct studies to determine* the health effects of exposure.[§1910.1020(f)(4)(ii)[G]]

(iii) *The request explains in detail* why the disclosure of the specific chemical identity is essential and that, in lieu thereof, the disclosure of the following information would not enable the health professional, employee or designated representative to provide the occupational health services described in paragraph (f)(4)(ii) of this section:[§1910.1020(f)(4)(iii)]

[A] *The properties and effects of the chemical;* [§1910.1020(f)(4)(iii)[A]]

[B] *Measures for controlling workers' exposure* to the chemical;[§1910.1020(f)(4)(iii)[B]]

[C] *Methods of monitoring and analyzing* worker exposure to the chemical; and,[§1910.1020(f)(4)(iii)[C]]

[D] *Methods of diagnosing and treating* harmful exposures to the chemical;[§1910.1020(f)(4)(iii)[D]]

(iv) *The request includes a description* of the procedures to be used to maintain the confidentiality of the disclosed information; and,[§1910.1020(f)(4)(iv)]

(v) *The health professional,* employee, or designated representative and the employer or contractor of the services of the health professional or designated representative agree in a written confidentiality agreement that the health professional, employee or designated representative will not use the trade secret information for any purpose other than the health need(s) asserted and agree not to release the information under any circumstances other than to OSHA, as provided in paragraph (f)(7) of this section, except as authorized by the terms of the agreement or by the employer.[§1910.1020(f)(4)(v)]

(5) *The confidentiality agreement authorized by paragraph (f)(4)(iv)* of this section:[§1910.1020(f)(5)]

(i) *May restrict the use of the information* to the health purposes indicated in the written statement of need;[§1910.1020(f)(5)(i)]

(ii) *May provide for appropriate legal remedies* in the event of a breach of the agreement, including stipulation of a reasonable pre-estimate of likely damages; and,[§1910.1020(f)(5)(ii)]

(iii) *May not include requirements for the posting of a penalty bond.*[§1910.1020(f)(5)(iii)]

(6) *Nothing in this section* is meant to preclude the parties from pursuing non-contractual remedies to the extent permitted by law.[§1910.1020(f)(6)]

(7) *If the health professional,* employee or designated representative receiving the trade secret information decides that there is a need to disclose it to OSHA, the employer who provided the information shall be informed by the health professional prior to, or at the same time as, such disclosure.[§1910.1020(f)(7)]

(8) *If the employer denies* a written request for disclosure of a specific chemical identity, the denial must:[§1910.1020(f)(8)]

(i) *Be provided to the health professional,* employee or designated representative within thirty days of the request; [§1910.1020(f)(8)(i)]

(ii) *Be in writing;*[§1910.1020(f)(8)(ii)]

---

1. Editor's Note: The CFR states "employee" rather than "employer" at 1910.120(e)(2)(iii)[A]. The text is corrected in the regulation appearing above.

(iii) *Include evidence to support* the claim that the specific chemical identity is a trade secret;[§1910.1020(f)(8)(iii)]

(iv) *State the specific reasons* why the request is being denied; and,[§1910.1020(f)(8)(iv)]

(v) *Explain in detail how alternative* information may satisfy the specific medical or occupational health need without revealing the specific chemical identity.[§1910.1020(f)(8)(v)]

(9) *The health professional,* employee, or designated representative whose request for information is denied under paragraph (f)(4) of this section may refer the request and the written denial of the request to OSHA for consideration.[§1910.1020(f)(9)]

(10) *When a heath professional, employee,* or designated representative refers a denial to OSHA under paragraph (f)(9) of this section, OSHA shall consider the evidence to determine if: [§1910.1020(f)(10)]

(i) *The employer has supported* the claim that the specific chemical identity is a trade secret;[§1910.1020(f)(10)(i)]

(ii) *The health professional, employee,* or designated representative has supported the claim that there is a medical or occupational health need for the information; and [§1910.1020(f)(10)(ii)]

(iii) *The health professional,* employee, or designated representative has demonstrated adequate means to protect the confidentiality.[§1910.1020(f)(10)(iii)]

(11) (i) *If OSHA determines* that the specific chemical identity requested under paragraph (f)(4) of this section is not a bona fide trade secret, or that it is a trade secret but the requesting health professional, employee or designated representatives has a legitimate medical or occupational health need for the information, has executed a written confidentiality agreement, and has shown adequate means for complying with the terms of such agreement, the employer will be subject to citation by OSHA.[§1910.1020(f)(11)(i)]

(ii) *If an employer demonstrates* to OSHA that the execution of a confidentiality agreement would not provide sufficient protection against the potential harm from the unauthorized disclosure of a trade secret specific chemical identity, the Assistant Secretary may issue such orders or impose such additional limitations or conditions upon the disclosure of the requested chemical information as may be appropriate to assure that the occupational health needs are met without an undue risk of harm to the employer.[§1910.1020(f)(11)(ii)]

(12) *Notwithstanding the existence* of a trade secret claim, an employer shall, upon request, disclose to the Assistant Secretary any information which this section requires the employer to make available. Where there is a trade secret claim, such claim shall be made no later than at the time the information is provided to the Assistant Secretary so that suitable determinations of trade secret status can be made and the necessary protections can be implemented.[§1910.1020(f)(12)]

(13) *Nothing in this paragraph shall be construed* as requiring the disclosure under any circumstances of process or percentage of mixture information which is trade secret.[§1910.1020(f)(13)]

(g) **Employee information.**[§1910.1020(g)]

(1) *Upon an employee's first* entering into employment, and at least annually thereafter, each employer shall inform current employees covered by this section of the following: [§1910.1020(g)(1)]

(i) *The existence, location, and availability* of any records covered by this section;[§1910.1020(g)(1)(i)]

(ii) *The person responsible for maintaining* and providing access to records; and[§1910.1020(g)(1)(ii)]

(iii) *Each employee's rights of access* to these records. [§1910.1020(g)(1)(iii)]

(2) *Each employer shall keep a copy* of this section and its appendices, and make copies readily available, upon request, to employees. The employer shall also distribute to current employees any informational materials concerning this section which are made available to the employer by the Assistant Secretary of Labor for Occupational Safety and Health. [§1910.1020(g)(2)]

(h) **Transfer of records.**[§1910.1020(h)]

(1) *Whenever an employer is ceasing* to do business, the employer shall transfer all records subject to this section to the successor employer. The successor employer shall receive and maintain these records.[§1910.1020(h)(1)]

(2) *Whenever an employer is ceasing* to do business and there is no successor employer to receive and maintain the records subject to this standard, the employer shall notify affected current employees of their rights of access to records at least three (3) months prior to the cessation of the employer's business.[§1910.1020(h)(2)]

(i) **Appendices.** The information contained in appendices A and B to this section is not intended, by itself, to create any additional obligations not otherwise imposed by this section nor detract from any existing obligation.[§1910.1020(i)]

§1910.1020 Appendix A

Sample Authorization Letter for the Release of Employee Medical Record Information to a Designated Representative (Non-Mandatory)

Appendix A to §1910.1020 - **Sample Authorization Letter For The Release Of Employee Medical Record Information To A Designated Representative (non-mandatory)**

I, (full name of worker/patient) ______________________________,

hereby authorize (individual or organization holding the medical records) ______________________________

to release to (individual or organization authorized to receive the medical information) ______________________________,

the following medical information from my personal medical records: (Describe generally the information desired to be released.)

____

I give my permission for this medical information to be used for the following purpose: ______________________________

____

but I do not give permission for any other use or re-disclosure of this information.

(Note: Several extra lines are provided below so that you can place additional restrictions on this authorization letter if you want to. You may, however, leave these lines blank. On the other hand, you may want to (1) specify a particular expiration date for this letter (if less than one year); (2) describe medical information to be created in the future that you intend to be covered by this authorization letter; or (3) describe portions of the medical information in your records which you do not intend to be released as a result of this letter.)

____

____
Full name of Employee or Legal Representative

______________________________ Date of Signature ____/____/____
Signature of Employee or Legal Representative

* Full-size forms available free of charge at www.oshacfr.com

§1910.1020 Appendix B

Availability of NIOSH Registry of Toxic Effects of Chemical Substances (RTECS) (Non-Mandatory)

The final regulation, 29 CFR 1910.1020[1], applies to all employee exposure and medical records, and analyses thereof, of employees exposed to toxic substances or harmful physical agents (paragraph (b)(2)). The term toxic substance or harmful physical agent is defined by paragraph (c)(13) to encompass chemical substances, biological agents, and physical stresses for which there is evidence of harmful health effects. The regulation uses the latest printed edition of the National Institute for Occupational Safety and Health (NIOSH) Registry of Toxic Effects of Chemical Substances (RTECS) as one of the chief sources of information as to whether evidence of harmful health effects exists. If a substance is listed in the latest printed RTECS, the regulation applies to exposure and medical records (and analyses of these records) relevant to employees exposed to the substance.

It is appropriate to note that the final regulation does not require that employers purchase a copy of RTECS, and many employers need not consult RTECS to ascertain whether their employee exposure or medical records are subject to the rule. Employers who do not cur-

1. Editor's note: The CFR refers in error to "29 CFR 1910.20". This reference has been corrected in the above text.

rently have the latest printed edition of the NIOSH RTECS, however, may desire to obtain a copy. The RTECS is issued in an annual printed edition as mandated by section 20(a)(6) of the Occupational Safety and Health Act (29 U.S.C. 669(a)(6)).

The Introduction to the 1980 printed edition describes the RTECS as follows:

"The 1980 edition of the Registry of Toxic Effects of Chemical Substances, formerly known as the Toxic Substances list, is the ninth revision prepared in compliance with the requirements of Section 20(a)(6) of the Occupational Safety and Health Act of 1970 (Public Law 91-596). The original list was completed on June 28, 1971, and has been updated annually in book format. Beginning in October 1977, quarterly revisions have been provided in microfiche. This edition of the Registry contains 168,096 listings of chemical substances: 45,156 are names of different chemicals with their associated toxicity data and 122,940 are synonyms. This edition includes approximately 5,900 new chemical compounds that did not appear in the 1979 Registry. (p. xi)

"The Registry's purposes are many, and it serves a variety of users. It is a single source document for basic toxicity information and for other data, such as chemical identifiers and information necessary for the preparation of safety directives and hazard evaluations for chemical substances. The various types of toxic effects linked to literature citations provide researchers and occupational health scientists with an introduction to the toxicological literature, making their own review of the toxic hazards of a given substance easier. By presenting data on the lowest reported doses that produce effects by several routes of entry in various species, the Registry furnishes valuable information to those responsible for preparing safety data sheets for chemical substances in the workplace. Chemical and production engineers can use the Registry to identify the hazards which may be associated with chemical intermediates in the development of final products, and thus can more readily select substitutes or alternative processes which may be less hazardous. Some organizations, including health agencies and chemical companies, have included the NIOSH Registry accession numbers with the listing of chemicals in their files to reference toxicity information associated with those chemicals. By including foreign language chemical names, a start has been made toward providing rapid identification of substances produced in other countries. (p. xi)

"In this edition of the Registry, the editors intend to identify "all known toxic substances" which may exist in the environment and to provide pertinent data on the toxic effects from known doses entering an organism by any route described. (p xi)

"It must be reemphasized that the entry of a substance in the Registry does not automatically mean that it must be avoided. A listing does mean, however, that the substance has the documented potential of being harmful if misused, and care must be exercised to prevent tragic consequences. Thus, the Registry lists many substances that are common in everyday life and are in nearly every household in the United States. One can name a variety of such dangerous substances: prescription and non-prescription drugs; food additives; pesticide concentrates, sprays, and dusts; fungicides; herbicides; paints; glazes, dyes; bleaches and other household cleaning agents; alkalies; and various solvents and diluents. The list is extensive because chemicals have become an integral part of our existence."

The RTECS printed edition may be purchased from the Superintendent of Documents, U.S. Government Printing Office (GPO), Washington, DC 20402 (202-783-3238).

Some employers may desire to subscribe to the quarterly update to the RTECS which is published in a microfiche edition. An annual subscription to the quarterly microfiche may be purchased from the GPO (Order the "Microfiche Edition, Registry of Toxic Effects of Chemical Substances"). Both the printed edition and the microfiche edition of RTECS are available for review at many university and public libraries throughout the country. The latest RTECS editions may also be examined at the OSHA Technical Data Center, Room N2439 — Rear, United States Department of Labor, 200 Constitution Avenue, NW., Washington, DC 20210 (202-523-9700), or at any OSHA Regional or Area Office (See, major city telephone directories under United States Government-Labor Department).

[53 FR 38163, Sept. 29, 1988; 53 FR 49981, Dec. 13, 1988, as amended at 54 FR 24333, June 7, 1989; 55 FR 26431, June 28, 1990; 61 FR 9235, Mar. 7, 1996. Redesignated at 61 FR 31430, June 20, 1996, as amended at 71 FR 16673, Apr. 3, 2006; 76 FR 33608, June 8, 2011]

§1910.1028
Benzene

(a) Scope and application.[§1910.1028(a)]

(1) *This section applies to* all occupational exposures to benzene. Chemical Abstracts Service Registry No. 71-43-2, except as provided in paragraphs (a)(2) and (a)(3) of this section. [§1910.1028(a)(1)]

(2) *This section does not apply to:*[§1910.1028(a)(2)]

(i) *The storage, transportation, distribution, dispensing, sale or use* of gasoline, motor fuels, or other fuels containing benzene subsequent to its final discharge from bulk wholesale storage facilities, except that operations where gasoline or motor fuels are dispensed for more than 4 hours per day in an indoor location are covered by this section. [§1910.1028(a)(2)(i)]

(ii) *Loading and unloading operations* at bulk wholesale storage facilities which use vapor control systems for all loading and unloading operations, except for the provisions of 29 CFR 1910.1200 as incorporated into this section and the emergency provisions of paragraphs (g) and (i)(4) of this section. [§1910.1028(a)(2)(ii)]

(iii) *The storage, transportation, distribution or sale* of benzene or liquid mixtures containing more than 0.1 percent benzene in intact containers or in transportation pipelines while sealed in such a manner as to contain benzene vapors or liquid, except for the provisions of 29 CFR 1910.1200 as incorporated into this section and the emergency provisions of paragraphs (g) and (i)(4) of this section.[§1910.1028(a)(2)(iii)]

(iv) *Containers and pipelines carrying mixtures* with less than 0.1 percent benzene and natural gas processing plants processing gas with less than 0.1 percent benzene. [§1910.1028(a)(2)(iv)]

(v) *Work operations where the only exposure* to benzene is from liquid mixtures containing 0.5 percent or less of benzene by volume, or the vapors released from such liquids until September 12, 1988; work operations where the only exposure to benzene is from liquid mixtures containing 0.3 percent or less of benzene by volume or the vapors released from such liquids from September 12, 1988, to September 12, 1989; and work operations where the only exposure to benzene is from liquid mixtures containing 0.1 percent or less of benzene by volume or the vapors released from such liquids after September 12, 1989; except that tire building machine operators using solvents with more than 0.1 percent benzene are covered by paragraph (i) of this section.[§1910.1028(a)(2)(v)]

(vi) *Oil and gas drilling, production and servicing* operations. [§1910.1028(a)(2)(vi)]

(vii) *Coke oven batteries.*[§1910.1028(a)(2)(vii)]

(3) *The cleaning and repair of barges and tankers* which have contained benzene are excluded from paragraph (f) methods of compliance, paragraph (e)(1) exposure monitoring-general, and paragraph (e)(6) accuracy of monitoring. Engineering and work practice controls shall be used to keep exposures below 10 ppm unless it is proven to be not feasible.[§1910.1028(a)(3)]

(b) Definitions.[§1910.1028(b)]

Action level means an airborne concentration of benzene of 0.5 ppm calculated as an 8-hour time-weighted average.

Assistant Secretary means the Assistant Secretary of Labor for Occupational Safety and Health, U.S. Department of Labor, or designee.

Authorized person means any person specifically authorized by the employer whose duties require the person to enter a regulated area, or any person entering such an area as a designated representative of employees for the purpose of exercising the right to observe monitoring and measuring procedures under paragraph (l) of this section, or any other person authorized by the Act or regulations issued under the Act.

Benzene (C_6H_6) (CAS Registry No. 71-43-2) means liquefied or gaseous benzene. It includes benzene contained in liquid mixtures and the benzene vapors released by these liquids. It does not include trace amounts of unreacted benzene contained in solid materials.

Bulk wholesale storage facility means a bulk terminal or bulk plant where fuel is stored prior to its delivery to wholesale customers.

Container means any barrel, bottle, can, cylinder, drum, reaction vessel, storage tank, or the like, but does not include piping systems.

Day means any part of a calendar day.

Director means the Director of the National Institute for Occupational Safety and Health, U.S. Department of Health and Human Services, or designee.

Emergency means any occurrence such as, but not limited to, equipment failure, rupture of containers, or failure of control equipment which may or does result in an unexpected significant release of benzene.

Employee exposure means exposure to airborne benzene which would occur if the employee were not using respiratory protective equipment.

Regulated area means any area where airborne concentrations of benzene exceed or can reasonably be expected to exceed, the permissible exposure limits, either the 8-hour time weighted average exposure of 1 ppm or the short-term exposure limit of 5 ppm for 15 minutes.

Vapor control system means any equipment used for containing the total vapors displaced during the loading of gasoline, motor fuel or other fuel tank trucks and the displacing of these vapors through a vapor processing system or balancing the vapor with the storage tank. This equipment also includes systems containing the vapors displaced from the storage tank during the unloading of the tank truck which balance the vapors back to the tank truck.

(c) Permissible exposure limits (PELs) [§1910.1028(c)]

(1) *Time-weighted average limit (TWA).* The employer shall assure that no employee is exposed to an airborne concentration of benzene in excess of one part of benzene per million parts of air (1 ppm) as an 8-hour time-weighted average. [§1910.1028(c)(1)]

(2) *Short-term exposure limit (STEL).* The employer shall assure that no employee is exposed to an airborne concentration of benzene in excess of five (5) ppm as averaged over any 15 minute period.[§1910.1028(c)(2)]

(d) Regulated areas.[§1910.1028(d)]

(1) *The employer shall establish* a regulated area wherever the airborne concentration of benzene exceeds or can reasonably be expected to exceed the permissible exposure limits, either the 8-hour time weighted average exposure of 1 ppm or the short-term exposure limit of 5 ppm for 15 minutes. [§1910.1028(d)(1)]

(2) *Access* to regulated areas shall be limited to authorized persons.[§1910.1028(d)(2)]

(3) *Regulated areas shall be determined* from the rest of the workplace in any manner that minimizes the number of employees exposed to benzene within the regulated area. [§1910.1028(d)(3)]

(e) Exposure monitoring — [§1910.1028(e)]

(1) *General.*[§1910.1028(e)(1)]

(i) *Determinations of employee exposure* shall be made from breathing zone air samples that are representative of each employee's average exposure to airborne benzene. [§1910.1028(e)(1)(i)]

(ii) *Representative 8-hour TWA* employee exposures shall be determined on the basis of one sample or samples representing the full shift exposure for each job classification in each work area.[§1910.1028(e)(1)(ii)]

(iii) *Determinations of compliance with the STEL* shall be made from 15 minute employee breathing zone samples measured at operations where there is reason to believe exposures are high, such as where tanks are opened, filled, unloaded or gauged; where containers or process equipment are opened and where benzene is used for cleaning or as a solvent in an uncontrolled situation. The employer may use objective data, such as measurements from brief period measuring devices, to determine where STEL monitoring is needed. [§1910.1028(e)(1)(iii)]

(iv) *Except for initial monitoring* as required under paragraph (e)(2) of this section, where the employer can document that one shift will consistently have higher employee exposures for an operation, the employer shall only be required to determine representative employee exposure for that operation during the shift on which the highest exposure is expected. [§1910.1028(e)(1)(iv)]

(2) *Initial monitoring.*[§1910.1028(e)(2)]

(i) *Each employer who has a place* of employment covered under paragraph (a)(1) of this section shall monitor each of these workplaces and work operations to determine accurately the airborne concentrations of benzene to which employees may be exposed.[§1910.1028(e)(2)(i)]

(ii) *The initial monitoring required* under paragraph (e)(2)(i) of this section shall be completed by 60 days after the effective date of this standard or within 30 days of the introduction of benzene into the workplace. Where the employer has monitored within one year prior to the effective date of this standard and the monitoring satisfies all other requirements of this section, the employer may rely on such earlier monitoring results to satisfy the requirements of paragraph (e)(2)(i) of this section.[§1910.1028(e)(2)(ii)]

(3) *Periodic monitoring and monitoring frequency.*[§1910.1028(e)(3)]

(i) *If the monitoring required* by paragraph (e)(2)(i) of this section reveals employee exposure at or above the action level but at or below the TWA, the employer shall repeat such monitoring for each such employee at least every year. [§1910.1028(e)(3)(i)]

(ii) *If the monitoring required* by paragraph (e)(2)(i) of this section reveals employee exposure above the TWA, the employer shall repeat such monitoring for each such employee at least every six (6) months.[§1910.1028(e)(3)(ii)]

(iii) *The employer may alter the monitoring* schedule from every six months to annually for any employee for whom two consecutive measurements taken at least 7 days apart indicate that the employee exposure has decreased to the TWA or below, but is at or above the action level.[§1910.1028(e)(3)(iii)]

(iv) *Monitoring for the STEL shall be repeated* as necessary to evaluate exposures of employees subject to short term exposures.[§1910.1028(e)(3)(iv)]

(4) *Termination of monitoring.*[§1910.1028(e)(4)]

(i) *If the initial monitoring* required by paragraph (e)(2)(i) of this section reveals employee exposure to be below the action level the employer may discontinue the monitoring for that employee, except as otherwise required by paragraph (e)(5) of this section.[§1910.1028(e)(4)(i)]

(ii) *If the periodic monitoring* required by paragraph (e)(3) of this section reveals that employee exposures, as indicated by at least two consecutive measurements taken at least 7 days apart, are below the action level the employer may discontinue the monitoring for that employee, except as otherwise required by paragraph (e)(5).[§1910.1028(e)(4)(ii)]

(5) *Additional monitoring.*[§1910.1028(e)(5)]

(i) *The employer shall institute the exposure* monitoring required under paragraphs (e)(2) and (e)(3) of this section when there has been a change in the production, process, control equipment, personnel or work practices which may result in new or additional exposures to benzene, or when the employer has any reason to suspect a change which may result in new or additional exposures.[§1910.1028(e)(5)(i)]

(ii) *Whenever spills,* leaks, ruptures or other breakdowns occur that may lead to employee exposure, the employer shall monitor (using area or personal sampling) after the cleanup of the spill or repair of the leak, rupture or other breakdown to ensure that exposures have returned to the level that existed prior to the incident.[§1910.1028(e)(5)(ii)]

(6) *Accuracy of monitoring.* Monitoring shall be accurate, to a confidence level of 95 percent, to within plus or minus 25 percent for airborne concentrations of benzene.[§1910.1028(e)(6)]

(7) *Employee notification of monitoring results.*[§1910.1028(e)(7)]

(i) *The employer must,* within 15 working days after the receipt of the results of any monitoring performed under this section, notify each affected employee of these results either individually in writing or by posting the results in an appropriate location that is accessible to employees.[§1910.1028(e)(7)(i)]

(ii) *Whenever the PELs are exceeded,* the written notification required by paragraph (e)(7)(i) of this section shall contain the corrective action being taken by the employer to reduce the employee exposure to or below the PEL, or shall refer to a document available to the employee which states the corrective actions to be taken.[§1910.1028(e)(7)(ii)]

(f) Methods of compliance — [§1910.1028(f)]

(1) *Engineering controls and work practices.*[§1910.1028(f)(1)]

(i) *The employer shall institute engineering* controls and work practices to reduce and maintain employee exposure to benzene at or below the permissible exposure limits, except to the extent that the employer can establish that these controls are not feasible or where the provisions of paragraph (f)(1)(iii) or (g)(1) of this section apply.[§1910.1028(f)(1)(i)]

(ii) *Wherever the feasible engineering* controls and work practices which can be instituted are not sufficient to reduce employee exposure to or below the PELs, the employer shall use them to reduce employee exposure to the lowest levels achievable by these controls and shall supplement them by the use of respiratory protection which complies with the requirements of paragraph (g) of this section. [§1910.1028(f)(1)(ii)]

(iii) *Where the employer can document* that benzene is used in a workplace less than a total of 30 days per year, the employer shall use engineering controls, work practice controls or respiratory protection or any combination of these controls to reduce employee exposure to benzene to or below the PELs, except that employers shall use engineering and work practice controls, if feasible, to reduce exposure to or below 10 ppm as an 8-hour TWA.[§1910.1028(f)(1)(iii)]

(2) *Compliance program.*[§1910.1028(f)(2)]

(i) *When any exposures are over the PEL,* the employer shall establish and implement a written program to reduce employee exposure to or below the PEL primarily by means of engineering and work practice controls, as required by paragraph (f)(1) of this section.[§1910.1028(f)(2)(i)]

(ii) *The written program shall include a schedule* for development and implementation of the engineering and work practice controls. These plans shall be reviewed and revised as appropriate based on the most recent exposure monitoring data, to reflect the current status of the program. [§1910.1028(f)(2)(ii)]

(iii) *Written compliance programs* shall be furnished upon request for examination and copying to the Assistant Secretary, the Director, affected employees and designated employee representatives.[§1910.1028(f)(2)(iii)]

(g) Respiratory protection — [§1910.1028(g)]

(1) *General.* For employees who use respirators required by this section, the employer must provide each employee an appropriate respirator that complies with the requirements of this paragraph. Respirators must be used during:[§1910.1028(g)(1)]

(i) *Periods necessary to install* or implement feasible engineering and work-practice controls.[§1910.1028(g)(1)(i)]

(ii) *Work operations for which* the employer establishes that compliance with either the TWA or STEL through the use of engineering and work-practice controls is not feasible; for example, some maintenance and repair activities, vessel cleaning, or other operations for which engineering and work-practice controls are infeasible because exposures are intermittent and limited in duration.[§1910.1028(g)(1)(ii)]

(iii) *Work operations for which* feasible engineering and work-practice controls are not yet sufficient, or are not required under paragraph (f)(1)(iii) of this section, to reduce employee exposure to or below the PELs.[§1910.1028(g)(1)(iii)]

(iv) *Emergencies.*[§1910.1028(g)(1)(iv)]

(2) *Respirator program.*[§1910.1028(g)(2)]

(i) *The employer must implement* a respiratory protection program in accordance with Sec. 1910.134(b) through (d) (except (d)(1)(iii), (d)(3)(iii)(b)(1) and (2)), and (f) through (m), which covers each employee required by this section to use a respirator.[§1910.1028(g)(2)(i)]

(ii) *For air-purifying respirators,* the employer must replace the air-purifying element at the expiration of its service life or at the beginning of each shift in which such elements are used, whichever comes first.[§1910.1028(g)(2)(ii)]

(iii) *If NIOSH approves* an air-purifying element with an end-of-service-life indicator for benzene, such an element may be used until the indicator shows no further useful life. [§1910.1028(g)(2)(iii)]

(3) *Respirator selection.*[§1910.1028(g)(3)]

(i) *Employers must:*[§1910.1028(g)(3)(i)]

[A] Select, and provide to employees, the appropriate respirators specified in paragraph (d)(3)(i)(A) of 29 CFR 1910.134.[§1910.1028(g)(3)(i)[A]]

[B] Provide employees with any organic vapor gas mask or any self-contained breathing apparatus with a full facepiece to use for escape.[§1910.1028(g)(3)(i)[B]]

[C] Use an organic vapor cartridge or canister with powered and non-powered air-purifying respirators, and a chin-style canister with full facepiece gas masks. [§1910.1028(g)(3)(i)[C]]

[D] Ensure that canisters used with non-powered air-purifying respirators have a minimum service life of four hours when tested at 150 ppm benzene at a flow rate of 64 liters per minute (LPM), a temperature of 25 °C, and a relative humidity of 85%; for canisters used with tight-fitting or loose-fitting powered air-purifying respirators, the flow rates for testing must be 115 LPM and 170 LPM, respectively.[§1910.1028(g)(3)(i)[D]]

(ii) *Any employee who cannot use* a negative-pressure respirator must be allowed to use a respirator with less breathing resistance, such as a powered air-purifying respirator or supplied-air respirator.[§1910.1028(g)(3)(ii)]

(h) Protective clothing and equipment. Personal protective clothing and equipment shall be worn where appropriate to prevent eye contact and limit dermal exposure to liquid benzene. Protective clothing and equipment shall be provided by the employer at no cost to the employee and the employer shall assure its use where appropriate. Eye and face protection shall meet the requirements of 29 CFR 1910.133.[§1910.1028(h)]

(i) Medical surveillance — [§1910.1028(i)]

(1) *General.*[§1910.1028(i)(1)]

(i) *The employer shall make available* a medical surveillance program for employees who are or may be exposed to benzene at or above the action level 30 or more days per year; for employees who are or may be exposed to benzene at or above the PELs 10 or more days per year; for employees who have been exposed to more than 10 ppm of benzene for 30 or more days in a year prior to the effective date of the standard when employed by their current employer; and for employees involved in the tire building operations called tire building machine operators, who use solvents containing greater than 0.1 percent benzene.[§1910.1028(i)(1)(i)]

(ii) *The employer shall assure that all medical examinations* and procedures are performed by or under the supervision of a licensed physician and that all laboratory tests are conducted by an accredited laboratory.[§1910.1028(i)(1)(ii)]

(iii) *The employer shall assure that persons* other than licensed physicians who administer the pulmonary function testing required by this section shall complete a training course in spirometry sponsored by an appropriate governmental, academic or professional institution.[§1910.1028(i)(1)(iii)]

(iv) *The employer shall assure that all examinations* and procedures are provided without cost to the employee and at a reasonable time and place.[§1910.1028(i)(1)(iv)]

(2) *Initial examination.*[§1910.1028(i)(2)]

(i) *Within 60 days* of the effective date of this standard, or before the time of initial assignment, the employer shall provide each employee covered by paragraph (i)(1)(i) of this section with a medical examination including the following elements:[§1910.1028(i)(2)(i)]

[A] A detailed occupational history which includes: [§1910.1028(i)(2)(i)[A]]

[1] Past work exposure to benzene or any other hematological toxins,[§1910.1028(i)(2)(i)[A][1]]

[2] A family history of blood dyscrasias including hematological neoplasms;[§1910.1028(i)(2)(i)[A][2]]

[3] A history of blood dyscrasias including genetic hemoglobin abnormalities, bleeding abnormalities, abnormal function of formed blood elements;[§1910.1028(i)(2)(i)[A][3]]

[4] A history of renal or liver dysfunction; [§1910.1028(i)(2)(i)[A][4]]

[5] A history of medicinal drugs routinely taken; [§1910.1028(i)(2)(i)[A][5]]

[6] A history of previous exposure to ionizing radiation and [§1910.1028(i)(2)(i)[A][6]]

[7] Exposure to marrow toxins outside of the current work situation.[§1910.1028(i)(2)(i)[A][7]]

[B] A complete physical examination.[§1910.1028(i)(2)(i)[B]]

[C] Laboratory tests. A complete blood count including a leukocyte count with differential, a quantitative thrombocyte count, hematocrit, hemoglobin, erythrocyte count and erythrocyte indices (MCV, MCH, MCHC). The results of these tests shall be reviewed by the examining physician. [§1910.1028(i)(2)(i)[C]]

[D] Additional tests as necessary in the opinion of the examining physician, based on alterations to the components of the blood or other signs which may be related to benzene exposure; and[§1910.1028(i)(2)(i)[D]]

[E] For all workers required to wear respirators for at least 30 days a year, the physical examination shall pay special attention to the cardiopulmonary system and shall include a pulmonary function test.[§1910.1028(i)(2)(i)[E]]

(ii) *No initial medical examination* is required to satisfy the requirements of paragraph (i)(2)(i) of this section if adequate records show that the employee has been examined in accordance with the procedures of paragraph (i)(2)(i) of this section within the twelve months prior to the effective date of this standard.[§1910.1028(i)(2)(ii)]

(3) *Periodic examinations.*[§1910.1028(i)(3)]

(i) *The employer shall provide each employee* covered under paragraph (i)(1)(i) of this section with a medical examination annually following the previous examination. These periodic examinations shall incude at least the following elements: [§1910.1028(i)(3)(i)]

[A] A brief history regarding any new exposure to potential marrow toxins, changes in medicinal drug use, and the appearance of physical signs relating to blood disorders: [§1910.1028(i)(3)(i)[A]]

[B] A complete blood count including a leukocyte count with differential, quantitative thrombocyte count, hemoglobin, hematocrit, erythrocyte count and erythrocyte indices (MCV, MCH, MCHC); and[§1910.1028(i)(3)(i)[B]]

[C] Appropriate additional tests as necessary, in the opinion of the examining physician, in consequence of alterations in the components of the blood or other signs which may be related to benzene exposure.[§1910.1028(i)(3)(i)[C]]

(ii) *Where the employee develops signs* and symptoms commonly associated with toxic exposure to benzene, the employer shall provide the employee with an additional medical examination which shall include those elements considered appropriate by the examining physician.[§1910.1028(i)(3)(ii)]

(iii) *For persons required to use respirators* for at least 30 days a year, a pulmonary function test shall be performed every three (3) years. A specific evaluation of the cardiopulmonary system shall be made at the time of the pulmonary function test.[§1910.1028(i)(3)(iii)]

(4) *Emergency examinations.*[§1910.1028(i)(4)]

(i) *In addition to the surveillance required by (i)(1)(i),* if an employee is exposed to benzene in an emergency situation, the employer shall have the employee provide a urine sample at the end of the employee's shift and have a urinary phenol test performed on the sample within 72 hours. The urine specific gravity shall be corrected to 1.024.[§1910.1028(i)(4)(i)]

(ii) *If the result of the urinary* phenol test is below 75 mg phenol/L of urine, no further testing is required.[§1910.1028(i)(4)(ii)]

(iii) *If the result of the urinary* phenol test is equal to or greater than 75 mg phenol/L of urine, the employer shall provide the employee with a complete blood count including an erythrocyte count, leukocyte count with differential and thrombocyte count at monthly intervals for a duration of three (3) months following the emergency exposure.[§1910.1028(i)(4)(iii)]

(iv) *If any of the conditions specified in paragraph (i)(5)(i)* of this section exists, then the further requirements of paragraph (i)(5) of this section shall be met and the employer shall, in addition, provide the employees with periodic examinations if directed by the physician.[§1910.1028(i)(4)(iv)]

(5) *Additional examinations and referrals.*[§1910.1028(i)(5)]

(i) *Where the results of the complete blood count* required for the initial and periodic examinations indicate any of the following abnormal conditions exist, then the blood count shall be repeated within 2 weeks.[§1910.1028(i)(5)(i)]

[A] The hemoglobin level or the hematocrit falls below the normal limit [outside the 95% confidence interval (C.I.)] as determined by the laboratory for the particular geographic area and/or these indices show a persistent downward trend from the individual's pre-exposure norms; provided these findings cannot be explained by other medical reasons.[§1910.1028(i)(5)(i)[A]]

[B] The thrombocyte (platelet) count varies more than 20 percent below the employee's most recent values or falls outside the normal limit (95% C.I.) as determined by the laboratory.[§1910.1028(i)(5)(i)[B]]

[C] The leukocyte count is below 4,000 per mm^3 or there is an abnormal differential count.[§1910.1028(i)(5)(i)[C]]

(ii) *If the abnormality persists,* the examining physician shall refer the employee to a hematologist or an internist for further evaluation unless the physician has good reason to believe such referral is unnecessary. (See Appendix C for examples of conditions where a referral may be unnecessary.) [§1910.1028(i)(5)(ii)]

(iii) *The employer shall provide the hematologist* or internist with the information required to be provided to the physician under paragraph (i)(6) of this section and the medical record required to be maintained by paragraph (k)(2)(ii) of this section.[§1910.1028(i)(5)(iii)]

(iv) *The hematologist's or internist's evaluation* shall include a determination as to the need for additional tests, and the employer shall assure that these tests are provided. [§1910.1028(i)(5)(iv)]

(6) *Information provided to the physician.* The employer shall provide the following information to the examining physician: [§1910.1028(i)(6)]

(i) *A copy of this regulation and its appendices;*[§1910.1028(i)(6)(i)]

(ii) *A description of the affected employee's* duties as they relate to the employee's exposure;[§1910.1028(i)(6)(ii)]

(iii) *The employee's actual or representative* exposure level: [§1910.1028(i)(6)(iii)]

(iv) *A description of any personal protective* equipment used or to be used; and[§1910.1028(i)(6)(iv)]

(v) *Information from previous employment-related* medical examinations of the affected employee which is not otherwise available to the examining physician.[§1910.1028(i)(6)(v)]

(7) *Physician's written opinions.*[§1910.1028(i)(7)]

(i) *For each examination under this section,* the employer shall obtain and provide the employee with a copy of the examining physician's written opinion within 15 days of the examination. The written opinion shall be limited to the following information:[§1910.1028(i)(7)(i)]

[A] The occupationally pertinent results of the medical examination and tests;[§1910.1028(i)(7)(i)[A]]

[B] The physician's opinion concerning whether the employee has any detected medical conditions which would place the employee's health at greater than normal risk of material impairment from exposure to benzene; [§1910.1028(i)(7)(i)[B]]

[C] The physician's recommended limitations upon the employee's exposure to benzene or upon the employee's use of protective clothing or equipment and respirators. [§1910.1028(i)(7)(i)[C]]

[D] A statement that the employee has been informed by the physician of the results of the medical examination and any medical conditions resulting from benzene exposure which require further explanation or treatment. [§1910.1028(i)(7)(i)[D]]

(ii) *The written opinion obtained* by the employer shall not reveal specific records, findings and diagnoses that have no bearing on the employee's ability to work in a benzene-exposed workplace.[§1910.1028(i)(7)(ii)]

(8) *Medical removal plan.*[§1910.1028(i)(8)]

(i) *When a physician makes a referral* to a hematologist/internist as required under paragraph (i)(5)(ii) of this section, the employee shall be removed from areas where exposures may exceed the action level until such time as the physician makes a determination under paragraph (i)(8)(ii) of this section.[§1910.1028(i)(8)(i)]

(ii) *Following the examination and evaluation* by the hematologist/internist, a decision to remove an employee from areas where benzene exposure is above the action level or to allow the employee to return to areas where benzene exposure is above the action level shall be made by the physician in consultation with the hematologist/internist. This decision shall be communicated in writing to the employer and employee. In the case of removal, the physician shall state the required probable duration of removal from occupational exposure to benzene above the action level and the requirements for future medical examinations to review the decision. [§1910.1028(i)(8)(ii)]

(iii) *For any employee who is removed pursuant* to paragraph (i)(8)(ii) of this section, the employer shall provide a follow-up examination. The physician, in consultation with the hematologist/internist, shall make a decision within 6 months of the date the employee was removed as to whether the employee shall be returned to the usual job or whether the employee should be removed permanently.[§1910.1028(i)(8)(iii)]

(iv) *Whenever an employee is temporarily* removed from benzene exposure pursuant to paragraph (i)(8)(i) or (i)(8)(ii) of this section, the employer shall transfer the employee to a comparable job for which the employee is qualified (or can be trained for in a short period) and where benzene exposures are as low as possible, but in no event higher than the action level. The employer shall maintain the employee's current wage rate, seniority and other benefits. If there is no such job available, the employer shall provide medical removal protection benefits until such a job becomes available or for 6 months, whichever comes first.[§1910.1028(i)(8)(iv)]

(v) *Whenever an employee is removed permanently* from benzene exposure based on a physician's recommendation pursuant to paragraph (i)(8)(iii) of this section, the employee shall be given the opportunity to transfer to another position which is available or later becomes available for which the employee is qualified (or can be trained for in a short period) and where benzene exposures are as low as possible but in no event higher than the action level. The employer shall assure that such employee suffers no reduction in current wage rate, seniority or other benefits as a result of the transfer.[§1910.1028(i)(8)(v)]

(9) *Medical removal protection benefits.*[§1910.1028(i)(9)]

(i) *The employer shall provide to an employee* 6 months of medical removal protection benefits immediately following each occasion an employee is removed from exposure to benzene because of hematological findings pursuant to paragraphs (i)(8)(i) and (ii) of this section, unless the employee has been transferred to a comparable job where benzene exposures are below the action level.[§1910.1028(i)(9)(i)]

(ii) *For the purposes of this section,* the requirement that an employer provide medical removal protection benefits means that the employer shall maintain the current wage rate, seniority and other benefits of an employee as though the employee had not been removed.[§1910.1028(i)(9)(ii)]

(iii) *The employer's obligation to provide* medical removal protection benefits to a removed employee shall be reduced to the extent that the employee receives compensation for earnings lost during the period of removal either from a publicly or employer-funded compensation program, or from employment with another employer made possible by virtue of the employee's removal.[§1910.1028(i)(9)(iii)]

(j) Communication of hazards[§1910.1028(j)]

(1) *Hazard communication — general.*[§1910.1028(j)(1)]

(i) *Chemical manufacturers,* importers, distributors and employers shall comply with all requirements of the Hazard Communication Standard (HCS) (Sec. 1910.1200) for benzene. [§1910.1028(j)(1)(i)]

(ii) *In classifying the hazards of benzene* at least the following hazards are to be addressed: Cancer; central nervous system effects; blood effects; aspiration; skin, eye, and respiratory tract irritation; and flammability.[§1910.1028(j)(1)(ii)]

(iii) *Employers shall include benzene* in the hazard communication program established to comply with the HCS (Sec. 1910.1200). Employers shall ensure that each employee has access to labels on containers of benzene and to safety data sheets, and is trained in accordance with the requirements of HCS and paragraph (j)(3) of this section.[§1910.1028(j)(1)(iii)]

(2) *Warning signs and labels.*[§1910.1028(j)(2)]

(i) *The employer shall post signs* at entrances to regulated areas. The signs shall bear the following legend: DANGER BENZENE MAY CAUSE CANCER HIGHLY FLAMMABLE LIQUID AND VAPOR DO NOT SMOKE WEAR RESPIRATORY PROTECTION IN THIS AREA AUTHORIZED PERSONNEL ONLY[§1910.1028(j)(2)(i)]

(ii) *Prior to June 1,* 2016, employers may use the following legend in lieu of that specified in paragraph (j)(2)(i) of this section: DANGER BENZENE CANCER HAZARD FLAMMABLE — NO SMOKING AUTHORIZED PERSONNEL ONLY RESPIRATOR REQUIRED[§1910.1028(j)(2)(ii)]

(iii) *The employer shall ensure that labels* or other appropriate forms of warning are provided for containers of benzene within the workplace. There is no requirement to label pipes. The labels shall comply with the requirements of paragraph (j)(1) of this section and Sec. 1910.1200(f).[§1910.1028(j)(2)(iii)]

(iv) *Prior to June 1,* 2015, employers shall include the following legend or similar language on the labels or other appropriate forms of warning: DANGER CONTAINS BENZENE CANCER HAZARD[§1910.1028(j)(2)(iv)]

(3) *Information and training.*[§1910.1028(j)(3)]

(i) *The employer shall provide employees* with information and training at the time of their initial assignment to a work area where benzene is present. If exposures are above the action level, employees shall be provided with information and training at least annually thereafter.[§1910.1028(j)(3)(i)]

(ii) *The training program shall be in accordance* with the requirements of 29 CFR 1910.1200(h)(1) and (2), and shall include specific information on benzene for each category of information included in that section.[§1910.1028(j)(3)(ii)]

(iii) *In addition to the information required* under 29 CFR 1910.1200, the employer shall:[§1910.1028(j)(3)(iii)]

[A] Provide employees with an explanation of the contents of this section, including Appendices A and B, and indicate to them where the standard is available; and [§1910.1028(j)(3)(iii)[A]]

[B] Describe the medical surveillance program required under paragraph (i) of this section, and explain the information contained in Appendix C.[§1910.1028(j)(3)(iii)[B]]

(k) Recordkeeping — [§1910.1028(k)]

(1) *Exposure measurements.*[§1910.1028(k)(1)]

(i) *The employer shall establish and maintain* an accurate record of all measurements required by paragraph (e) of this section, in accordance with 29 CFR 1910.1020 [1]. [§1910.1028(k)(1)(i)]

(ii) *This record shall include:*[§1910.1028(k)(1)(ii)]

[A] The dates, number, duration, and results of each of the samples taken, including a description of the procedure used to determine representative employee exposures; [§1910.1028(k)(1)(ii)[A]]

[B] A description of the sampling and analytical methods used;[§1910.1028(k)(1)(ii)[B]]

[C] A description of the type of respiratory protective devices worn, if any; and[§1910.1028(k)(1)(ii)[C]]

[D] The name, social security number, job classification and exposure levels of the employee monitored and all other employees whose exposure the measurement is intended to represent.[§1910.1028(k)(1)(ii)[D]]

(iii) *The employer shall maintain this record* for at least 30 years, in accordance with 29 CFR 1910.1020[1].[§1910.1028(k)(1)(iii)]

(2) *Medical surveillance.*[§1910.1028(k)(2)]

(i) *The employer shall establish and maintain* an accurate record for each employee subject to medical surveillance required by paragraph (i) of this section, in accordance with 29 CFR 1910.1020[1].[§1910.1028(k)(2)(i)]

(ii) *This record shall include:*[§1910.1028(k)(2)(ii)]

[A] The name and social security number of the employee; [§1910.1028(k)(2)(ii)[A]]

[B] The employer's copy of the physician's written opinion on the initial, periodic and special examinations, including results of medical examinations and all tests, opinions and recommendations;[§1910.1028(k)(2)(ii)[B]]

[C] Any employee medical complaints related to exposure to benzene;[§1910.1028(k)(2)(ii)[C]]

[D] A copy of the information provided to the physician as required by paragraphs (i)(6)(ii) through (v) of this section; and[§1910.1028(k)(2)(ii)[D]]

[E] A copy of the employee's medical and work history related to exposure to benzene or any other hematologic toxins.[§1910.1028(k)(2)(ii)[E]]

(iii) *The employer shall maintain this record* for at least the duration of employment plus 30 years, in accordance with 29 CFR 1910.1020[1].[§1910.1028(k)(2)(iii)]

(3) *Availability.*[§1910.1028(k)(3)]

(i) *The employer shall assure that all records* required to be maintained by this section shall be made available upon request to the Assistant Secretary and the Director for examination and copying.[§1910.1028(k)(3)(i)]

(ii) *Employee exposure monitoring* records required by this paragraph shall be provided upon request for examination and copying to employees, employee representatives, and the Assistant Secretary in accordance with 29 CFR 1910.1020[1] (a) through (e) and (g) through (i). [§1910.1028(k)(3)(ii)]

(iii) *Employee medical records* required by this paragraph shall be provided upon request for examination and copying, to the subject employee, to anyone having the specific written consent of the subject employee, and to the Assistant Secretary in accordance with 29 CFR 1910.1020[1]. [§1910.1028(k)(3)(iii)]

(4) *Transfer of records.* The employer shall comply with the requirements involving transfer of records as set forth in 29 CFR 1910.1020(h).[§1910.1028(k)(4)]

(l) **Observation of monitoring** — [§1910.1028(l)]

(1) *Employee observation.* The employer shall provide affected employees, or their designated representatives, an opportunity to observe the measuring or monitoring of employee exposure to benzene conducted pursuant to paragraph (e) of this section.[§1910.1028(l)(1)]

(2) *Observation procedures.* When observation of the measuring or monitoring of employee exposure to benzene requires entry into areas where the use of protective clothing and equipment or respirators is required, the employer shall provide the observer with personal protective clothing and equipment or respirators required to be worn by employees working in the area, assure the use of such clothing and equipment or respirators, and require the observer to comply with all other applicable safety and health procedures.[§1910.1028(l)(2)]

(m) **[Reserved]**[§1910.1028(m)]

(n) **Appendices.** The information contained in Appendices A, B, C, and D is not intended, by itself, to create any additional obligations not otherwise imposed or to detract from any existing obligations. [§1910.1028(n)]

§1910.1028 Appendix A

Substance Safety Data Sheet, Benzene

I. Substance Identification

A. *Substance:* Benzene.

B. *Permissible Exposure:* Except as to the use of gasoline, motor fuels and other fuels subsequent to discharge from bulk terminals and other exemptions specified in §1910.1028(a)(2):

1. *Airborne:* The maximum time-weighted average (TWA) exposure limit is 1 part of benzene vapor per million parts of air (1 ppm) for an 8-hour workday and the maximum short-term exposure limit (STEL) is 5 ppm for any 15-minute period.
2. *Dermal:* Eye contact shall be prevented and skin contact with liquid benzene shall be limited.

C. *Appearance and odor:* Benzene is a clear, colorless liquid with a pleasant, sweet odor. The odor of benzene does not provide adequate warning of its hazard.

II. Health Hazard Data

A. *Ways in which benzene affects your health.* Benzene can affect your health if you inhale it, or if it comes in contact with your skin or eyes. Benzene is also harmful if you happen to swallow it.

B. *Effects of overexposure.*

1. *Short-term (acute) overexposure:* If you are overexposed to high concentrations of benzene, well above the levels where its odor is first recognizable, you may feel breathless, irritable, euphoric, or giddy; you may experience irritation in eyes, nose, and respiratory tract. You may develop a headache, feel dizzy, nauseated, or intoxicated. Severe exposures may lead to convulsions and loss of consciousness.
2. *Long-term (chronic) exposure.* Repeated or prolonged exposure to benzene, even at relatively low concentrations, may result in various blood disorders, ranging from anemia to leukemia, an irreversible, fatal disease. Many blood disorders associated with benzene exposure may occur without symptoms.

III. Protective Clothing and Equipment

A. *Respirators.* Respirators are required for those operations in which engineering controls or work practice controls are not feasible to reduce exposure to the permissible level. However, where employers can document that benzene is present in the workplace less than 30 days a year, respirators may be used in lieu of engineering controls. If respirators are worn, they must have joint Mine Safety and Health Administration and the National Institute for Occupational Safety and Health (NIOSH) seal of approval, and cartridge or canisters must be replaced before the end of their service life, or the end of the shift, whichever occurs first. If you experience difficulty breathing while wearing a respirator, you may request a positive pressure respirator from your employer. You must be thoroughly trained to use the assigned respirator, and the training will be provided by your employer.

B. *Protective Clothing.* You must wear appropriate protective clothing (such as boots, gloves, sleeves, aprons, etc.) over any parts of your body that could be exposed to liquid benzene.

C. *Eye and Face Protection.* You must wear splash-proof safety goggles if it is possible that benzene may get into your eyes. In addition, you must wear a face shield if your face could be splashed with benzene liquid.

IV. Emergency and First Aid Procedures

A. *Eye and face exposure.* If benzene is splashed in your eyes, wash it out immediately with large amounts of water. If irritation persists or vision appears to be affected see a doctor as soon as possible.

B. *Skin exposure.* If benzene is spilled on your clothing or skin, remove the contaminated clothing and wash the exposed skin with large amounts of water and soap immediately. Wash contaminated clothing before you wear it again.

C. *Breathing.* If you or any other person breathes in large amounts of benzene, get the exposed person to fresh air at once. Apply artificial respiration if breathing has stopped. Call for medical assistance or a doctor as soon as possible. Never enter any vessel or confined space where the benzene concentration might be high without proper safety equipment and at least one other person present who will stay outside. A life line should be used.

D. *Swallowing.* If benzene has been swallowed and the patient is conscious, do not induce vomiting. Call for medical assistance or a doctor immediately.

V. Medical Requirements

If you are exposed to benzene at a concentration at or above 0.5 ppm as an 8-hour time-weighted average, or have been exposed at or above 10 ppm in the past while employed by your current employer, your employer is required to provide a medical examination and history and laboratory tests within 60 days of the effective date of this standard and annually thereafter. These tests shall be provided without cost to you. In addition, if you are accidentally exposed to benzene (either by ingestion, inhalation, or skin/eye contact) under emergency conditions known or suspected to constitute toxic exposure to benzene, your employer is required to make special laboratory tests available to you.

VI. Observation of Monitoring

Your employer is required to perform measurements that are representative of your exposure to benzene and you or your designated representative are entitled to observe the monitoring procedure. You are entitled to observe the steps taken in the measurement procedure, and to record the results obtained. When the monitoring procedure is taking place in an area where respirators or personal protective clothing and equipment are required to be worn, you or your representative must also be provided with, and must wear the protective clothing and equipment.

VII. Access to Records

You or your representative are entitled to see the records of measurements of your exposure to benzene upon written request to your employer. Your medical examination records can be furnished to yourself, your physician or designated representative upon request by you to your employer.

VIII. Precautions for Safe Use, Handling and Storage

Benzene liquid is highly flammable. It should be stored in tightly closed containers in a cool, well ventilated area. Benzene vapor may form explosive mixtures in air. All sources of ignition must be controlled. Use nonsparking tools when opening or closing benzene containers. Fire extinguishers, where provided, must be readily available. Know where they are located and how to operate them. Smoking is prohibited in areas where benzene is used or stored. Ask your supervisor where benzene is used in your area and for additional plant safety rules.

§1910.1028 Appendix B

Substance Technical Guidelines, Benzene

I. Physical and Chemical Data

A. *Substance identification.*

1. *Synonyms:* Benzol, benzole, coal naphtha, cyclohexatriene, phene, phenyl hydride, pyrobenzol. (Benzin, petroleum benzin and Benzine do not contain benzene).
2. *Formula:* C_6H_6(CAS Registry Number: 71-43-2)

B. *Physical data.*

1. *Boiling Point (760 mm Hg);* 80.1 °C (176 °F)
2. *Specific Gravity (water = 1):* 0.879
3. *Vapor Density (air = 1):* 2.7
4. *Melting Point:* 5.5 °C (42 °F)
5. *Vapor Pressure at 20 °C (68 °F):* 75 mm Hg
6. *Solubility in Water:* .06%
7. *Evaporation Rate (ether = 1):* 2.8
8. *Appearance and Odor:* Clear, colorless liquid with a distinctive sweet odor.

II. Fire, Explosion, and Reactivity Hazard Data

A. *Fire.*

1. *Flash Point (closed* cup): -11 °C (12 °F)
2. *Autoignition Temperature:* 580 °C (1076 °F)
3. *Flammable limits in Air. % by Volume:* Lower: 1.3%, Upper: 7.5%

4. *Extinguishing Media:* Carbon dioxide, dry chemical, or foam.
5. *Special Fire-Fighting procedures:* Do not use solid stream of water, since stream will scatter and spread fire. Fine water spray can be used to keep fire-exposed containers cool.
6. *Unusual fire and explosion hazards:* Benzene is a flammable liquid. Its vapors can form explosive mixtures. All ignition sources must be controlled when benzene is used, handled, or stored. Where liquid or vapor may be released, such areas shall be considered as hazardous locations. Benzene vapors are heavier than air; thus the vapors may travel along the ground and be ignited by open flames or sparks at locations remote from the site at which benzene is handled.
7. *Benzene is classified as a 1 B flammable liquid* for the purpose of conforming to the requirements of 29 CFR 1910.106. A concentration exceeding 3,250 ppm is considered a potential fire explosion hazard. Locations where benzene may be present in quantities sufficient to produce explosive or ignitable mixtures are considered Class I Group D for the purposes of conforming to the requirements of 29 CFR 1910.309.

B. *Reactivity.*
1. *Conditions contributing to instability:* Heat.
2. *Incompatibility:* Heat and oxidizing materials.
3. *Hazardous decomposition products:* Toxic gases and vapors (such as carbon monoxide).

III. Spill and Leak Procedures

A. *Steps to be taken* if the material is released or spilled. As much benzene as possible should be absorbed with suitable materials, such as dry sand or earth. That remaining must be flushed with large amounts of water. Do not flush benzene into a confined space, such as a sewer, because of explosion danger. Remove all ignition sources. Ventilate enclosed places.

B. *Waste disposal method.* Disposal methods must conform to other jurisdictional regulations. If allowed, benzene may be disposed of:
(a) *By absorbing it in dry sand or earth* and disposing in a sanitary landfill;
(b) *if small quantities,* by removing it to a safe location from buildings or other combustible sources, pouring it in dry sand or earth and cautiously igniting it; and
(c) *if large quantities,* by atomizing it in a suitable combustion chamber.

IV. Miscellaneous Precautions

A. *High exposure to benzene* can occur when transferring the liquid from one container to another. Such operations should be well ventilated and good work practices must be established to avoid spills.

B. *Use non-sparking tools* to open benzene containers which are effectively grounded and bonded prior to opening and pouring.

C. *Employers must advise employees* of all plant areas and operations where exposure to benzene could occur. Common operations in which high exposures to benzene may be encountered are: the primary production and utilization of benzene, and transfer of benzene.

§1910.1028 Appendix C

Medical Surveillance Guidelines for Benzene

I. Route of Entry

Inhalation; skin absorption.

II. Toxicology

Benzene is primarily an inhalation hazard. Systemic absorption may cause depression of the hematopoietic system, pancytopenia, aplastic anemia, and leukemia. Inhalation of high concentrations can affect central nervous system function. Aspiration of small amounts of liquid benzene immediately causes pulmonary edema and hemorrhage of pulmonary tissue. There is some absorption through the skin. Absorption may be more rapid in the case of abraded skin, and benzene may be more readily absorbed if it is present in a mixture or as a contaminant in solvents which are readily absorbed. The defatting action of benzene may produce primary irritation due to repeated or prolonged contact with the skin. High concentration are irritating to the eyes and the mucuous membranes of the nose, and respiratory tract.

III. Signs and Symptoms

Direct skin contact with benzene may cause erythema. Repeated or prolonged contact may result in drying, scaling dermatitis, or development of secondary skin infections. In addition, there is benzene absorption through the skin. Local effects of benzene vapor or liquid on the eye are slight. Only at very high concentrations is there any smarting sensation in the eye. Inhalation of high concentrations of benzene may have an initial stimulatory effect on the central nervous system characterized by exhilaration, nervous excitation, and/or giddiness, followed by a period of depression, drowsiness, or fatigue. A sensation of tightness in the chest accompanied by breathlessness may occur and ultimately the victim may lose consciousness. Tremors, convulsions and death may follow from respiratory paralysis or circulatory collapse in a few minutes to several hours following severe exposures.

The detrimental effect on the blood-forming system of prolonged exposure to small quantities of benzene vapor is of extreme importance. The hematopoietic system is the chief target for benzene's toxic effects which are manifested by alterations in the levels of formed elements in the peripheral blood. These effects have occurred at concentrations of benzene which may not cause irritation of mucous membranes, or any unpleasant sensory effects. Early signs and symptoms of benzene morbidity are varied, often not readily noticed and non-specific. Subjective complaints of headache, dizziness, and loss of appetite may precede or follow clinical signs. Rapid pulse and low blood pressure, in addition to a physical appearance of anemia, may accompany a subjective complaint of shortness of breath and excessive tiredness. Bleeding from the nose, gums, or mucous membranes, and the development of purpuric spots (small bruises) may occur as the condition progresses. Clinical evidence of leukopenia, anemia, and thrombocytopenia, singly or in combination, has been frequently reported among the first signs.

Bone marrow may appear normal, aplastic, or hyperplastic, and may not, in all situations, correlate with peripheral blood forming tissues. Because of variations in the susceptibility to benzene morbidity, there is no "typical" blood picture. The onset of effects of prolonged benzene exposure may be delayed for many months or years after the actual exposure has ceased and identification or correlation with benzene exposure must be sought out in the occupational history.

IV. Treatment of Acute Toxic Effects

Remove from exposure immediately. Make sure you are adequately protected and do not risk being overcome by fumes. Give oxygen or artificial resuscitation if indicated. Flush eyes, wash skin if contaminated and remove all contaminated clothing. Symptoms of intoxication may persist following severe exposures. Recovery from mild exposures is usually rapid and complete.

V. Surveillance and Preventive Considerations

A. *General*

The principal effects of benzene exposure which form the basis for this regulation are pathological changes in the hematopoietic system, reflected by changes in the peripheral blood and manifesting clinically as pancytopenia, aplastic anemia, and leukemia. Consequently, the medical surveillance program is designed to observe, on a regular basis, blood indices for early signs of these effects, and although early signs of leukemia are not usually available, emerging diagnostic technology and innovative regimes make consistent surveillance for leukemia, as well as other hematopoietic effects, essential.

Initial examinations are to be provided within 60 days of the effective date of this standard, or at the time of initial assignment, and periodic examinations annually thereafter. There are special provisions for medical tests in the event of hematologic abnormalities or for emergency situations.

The blood values which require referral to a hematologist or internist are noted in the standard in paragraph (i)(5). The standard specifies that blood abnormalities that persist must be referred "unless the physician has good reason to believe such referral is unnecessary" (paragraph (i)(5)). Examples of conditions that could make a referral unnecessary despite abnormal blood limits are iron or folate deficiency, menorrhagia, or blood loss due to some unrelated medical abnormality.

Symptoms and signs of benzene toxicity can be non-specific. Only a detailed history and appropriate investigative procedures will enable a physician to rule out or confirm conditions that place the employee at increased risk. To assist the examining physician with regard to which laboratory tests are necessary and when to refer an employee to the specialist, OSHA has established the following guidelines.

B. *Hematology Guidelines*

A minimum battery of tests is to be performed by strictly standardized methods.

1. *Red cell, white cell, platelet counts, white blood cell differential, hematacrit and red cell indices* must be performed by an accredited laboratory. The normal ranges for the red cell and white cell counts are influenced by altitude, race, and sex, and therefore should be determined by the accredited laboratory in the specific area where the tests are performed.

Either a decline from an absolute normal or an individual's base line to a subnormal value or a rise to a supra-normal value, are indicative of potential toxicity, particularly if all blood parameters decline. The normal total white blood count is approximately 7,200/mm^3 plus or minus 3,000. For cigarette smokers the white count may be higher and the upper range may be 2,000 cells higher than normal for the laboratory. In addition, infection, allergies and some drugs may raise the white cell count. The normal platelet count is approximately 250,000 with a range of 140,000 to 400,000. Counts outside this range should be regarded as possible evidence of benzene toxicity.

Certain abnormalities found through routine screening are of greater significance in the benzene-exposed worker and require prompt consultation with a specialist, namely:

a. *Thrombocytopenia.*

b. *A trend of decreasing white cell, red cell, or platelet indices* in an individual over time is more worrisome than an isolated abnormal finding at one test time. The importance of trend highlights the need to compare an individual's test results to baseline and/or previous periodic tests.

c. *A constellation or pattern of abnormalities* in the different blood indices is of more significance than a single abnormality. A low white count not associated with any abnormalities in other cell indices may be a normal statistical variation, whereas if the low white count is accompanied by decreases in the platelet and/or red cell indices, such a pattern is more likely to be associated with benzene toxicity and merits thorough investigation.

Anemia, leukopenia, macrocytosis or an abnormal differential white blood cell count should alert the physician to further investigate and/or refer the patient if repeat tests confirm the abnormalities. If routine screening detects an abnormality, follow-up tests which may be helpful in establishing the etiology of the abnormality are the peripheral blood smear and the reticulocyte count.

The extreme range of normal for reticulocytes is 0.4 to 2.5 percent of the red cells, the usual range being 0.5 to 1.2 percent of the red cells, but the typical value is in the range of 0.8 to 1.0 percent. A decline in reticulocytes to levels of less than 0.4 percent is to be regarded as possible evidence (unless another specific cause is found) of benzene toxicity requiring accelerated surveillance. An increase in reticulocyte levels to about 2.5 percent may also be consistent with (but is not as characteristic of) benzene toxicity.

2. *An important diagnostic test* is a careful examination of the peripheral blood smear. As with reticulocyte count the smear should be with fresh uncoagulated blood obtained from a needle tip following venipuncture or from a drop of earlobe blood (capillary blood). If necessary, the smear may, under certain limited conditions, be made from a blood sample anticoagulated with EDTA (but never with oxalate or heparin). When the smear is to be prepared from a specimen of venous blood which has been collected by a commercial Vacutainer® type tube containing neutral EDTA, the smear should be made as soon as possible after the venesection. A delay of up to 12 hours is permissible between the drawing of the blood specimen into EDTA and the preparation of the smear if the blood is stored at refrigerator (not freezing) temperature.

3. *The minimum mandatory observations* to be made from the smear are:

a. *The differential white blood cell count.*

b. *Description of abnormalities in the appearance of red cells.*

c. *Description of any abnormalities in the platelets.*

d. *A careful search must be made* throughout of every blood smear for immature white cells such as band forms (in more than normal proportion, i.e., over 10 percent of the total differential count), any number of metamyelocytes, myelocytes or myeloblasts. Any nucleate or multinucleated red blood cells should be reported. Large "giant" platelets or fragments of megakaryocytes must be recognized.

An increase in the proportion of band forms among the neutrophilic granulocytes is an abnormality deserving special mention, for it may represent a change which should be considered as an early warning of benzene toxicity in the absence of other causative factors (most commonly infection). Likewise, the appearance of metamyelocytes, in the absence of another probable cause, is to be considered a possible indication of benzene-induced toxicity.

An upward trend in the number of basophils, which normally do not exceed about 2.0 percent of the total white cells, is to be regarded as possible evidence of benzene toxicity. A rise in the eosinophil count is less specific but also may be suspicious of toxicity if the rises above 6.0 percent of the total white count.

The normal range of monocytes is from 2.0 to 8.0 percent of the total white count with an average of about 5.0 percent. About 20 percent of individuals reported to have mild but persisting abnormalities caused by exposure to benzene show a persistent monocytosis. The findings of a monocyte count which persists at more than 10 to 12 percent of the normal white cell count (when the total count is normal) or persistence of an absolute monocyte count in excess of 800/mm^3 should be regarded as a possible sign of benzene-induced toxicity.

A less frequent but more serious indication of benzene toxicity is the finding in the peripheral blood of the so-called "pseudo" (or acquired) Pelger-Huet anomaly. In this anomaly many, or sometimes the majority, of the neutrophilic granulocytes possess two round nuclear segements — less often one or three round segments — rather than three normally elongated segments. When this anomaly is not hereditary, it is often but not invariably predictive of subsequent leukemia. However, only about two percent of patients who ultimately develop acute myelogenous leukemia show the acquired Pelger-Huet anomaly. Other tests that can be administered to investigate blood abnormalities are discussed below; however, such procedures should be undertaken by the hematologist.

An uncommon sign, which cannot be detected from the smear, but can be elicited by a "sucrose water test" of peripheral blood, is transient paroxysmal nocturnal hemoglobinuria (PNH), which may first occur insidiously during a period of established aplastic anemia, and may be followed within one to a few years by the appearance of rapidly fatal acute myelogenous leukemia. Clinical detection of PNH, which occurs in only one or two percent of those destined to have acute myelogenous leukemia, may be difficult; if the "sucrose water test" is positive, the somewhat more definitive Ham test, also known as the acid-serum hemolysis test, may provide confirmation.

e. *Individuals documented to have* developed acute myelogenous leukemia years after initial exposure to benzene may have progressed through a preliminary phase of hematologic abnormality. In some instances pancytopenia (i.e., a lowering in the counts of all circulating blood cells of bone marrow origin, but not to the extent implied by the term "aplastic anemia") preceded leukemia for many years. Depression of a single blood cell type or platelets may represent a harbinger of aplasia or leukemia. The finding of two or more cytopenias, or pancytopenia in a benzene-exposed individual, must be regarded as highly suspicious of more advanced although still reversible, toxicity. "Pancytopenia" coupled with the appearance of immature cells (myelocytes, myeloblasts, erythroblasts, etc.), with abnormal cells (pseudo Pelger-Huet anomaly, atypical nuclear heterochromatin, etc.), or unexplained elevations of white blood cells must be regarded as evidence of benzene overexposure unless proved otherwise. Many severely aplastic patients manifested the ominous finding of 5-10 percent myeloblasts in the marrow, occasional myeloblasts and myelocytes in the blood and 20-30% monocytes. It is evident that isolated cytopenias, pancytopenias, and even aplastic anemias induced by benzene may be reversible and complete recovery has been reported on cessation of exposure. However, since any of these abnormalities is serious, the employee must immediately be removed from any possible exposure to benzene vapor. Certain tests may substantiate the employee's prospects for progression or regression. One such test would be an examination of the bone marrow, but the decision to perform a bone marrow aspiration or needle biopsy is made by the hematologist.

The findings of basophilic stippling in circulating red blood cells (usually found in 1 to 5% of red cells following marrow injury), and detection in the bone marrow of what are termed "ringed sideroblasts" must be taken seriously, as they have been noted in recent years to be premonitory signs of subsequent leukemia.

Recently peroxidase-staining of circulating or marrow neutrophil granulocytes, employing benzidine dihydrochloride, have revealed the disappearance of, or diminution in, peroxidase in a sizable proportion of the granulocytes, and this has been reported as an early sign of leukemia. However, relatively few patients have been studied to date. Granulocyte granules are normally strongly peroxidase positive. A steady decline in leukocyte alkaline phosphatase has also been reported as suggestive of early acute leukemia. Exposure to benzene may cause an early rise in serum iron, often but not always associated with a fall in the reticulocyte count. Thus, serial measurements of serum iron levels may provide a means of determining whether or not there is a trend representing sustained suppression of erythropoiesis.

Measurement of serum iron, determination of peroxidase and of alkaline phosphatase activity in peripheral granulocytes can be performed in most pathology laboratories. Peroxidase and alkaline phosphatase staining are usually undertaken when the index of suspecion for leukemia is high.

§1910.1028 Appendix D

Sampling and Analytical Methods for Benzene Monitoring and Measurement Procedures

Measurements taken for the purpose of determining employee exposure to benzene are best taken so that the representative average 8-hour exposure may be determined from a single 8-hour sample or two (2) 4-hour samples. Short-time interval samples (or grab samples) may also be used to determine average exposure level if a minimum of five measurements are taken in a random manner over the 8-hour work shift. Random sampling means that any portion of the work shift has the same change of being sampled as any other. The arithmetic average of all such random samples taken on one work shift is an estimate of an employee's average level of exposure for that work shift. Air samples should be taken in the employee's breathing zone (air that would most nearly represent that inhaled by the employee). Sampling and analysis must be performed with procedures meeting the requirements of the standard.

There are a number of methods available for monitoring employee exposures to benzene. The sampling and analysis may be performed by collection of the benzene vapor or charcoal absorption tubes, with subsequent chemical analysis by gas chromatography. Sampling and analysis may also be performed by portable direct reading instruments, real-time continuous monitoring systems, passive dosimeters or other suitable methods. The employer has the obligation of selecting a monitoring method which meets the accuracy and precision requirements of the standard under his unique field conditions. The standard requires that the method of monitoring must have an accuracy, to a 95 percent confidence level, of not less than plus or minus 25 percent for concentrations of benzene greater than or equal to 0.5 ppm.

The OSHA Laboratory modified NIOSH Method S311 and evaluated it at a benzene air concentration of 1 ppm. A procedure for determining the benzene concentration in bulk material samples was also evalauted. This work, reported in OSHA Laboratory Method No. 12, includes the following two analytical procedures:

I. OSHA Method 12 for Air Samples

Analyte: Benzene

Matrix: Air

Procedure: Adsorption on charcoal, desorption with carbon disulfide, analysis by GC.

Detection limit: 0.04 ppm

Recommended air volume and sampling rate: 10L to 0.2 L/min.

1. *Principle of the Method.*
   - 1.1. *A known volume of air* is drawn through a charcoal tube to trap the organic vapors present.
   - 1.2. *The charcoal in the tube is transferred* to a small, stoppered vial, and the anlyte is desorbed with carbon disulfide.
   - 1.3. *An aliquot of the desorbed sample* is injected into a gas chromatograph.
   - 1.4. *The area of the resulting peak* is determined and compared with areas obtained from standards.
2. *Advantages and disadvantages of the method.*
   - 2.1. *The sampling device is small,* portable, and involved no liquids. Interferences are minimal, and most of those which do occur can be eliminated by altering chromatographic conditions. The samples are analyzed by means of a quick, instrumental method.
   - 2.2. *The amount of sample which* can be taken is limited by the number of milligrams that the tube will hold before overloading. When the sample value obtained for the backup section of the charcoal tube exceeds 25 percent of that found on the front section, the possibility of sample loss exists.
3. *Apparatus.*
   - 3.1. *A calibrated personal sampling pump* whose flow can be determined within ± 5 percent at the recommended flow rate.
   - 3.2. *Charcoal tubes:* Glass with both ends flame sealed, 7 cm long with a 6-mm O.D. and a 4-mm I.D., containing 2 sections of 20/40 mesh activated charcoal separated by a 2-mm portion of urethane foam. The activated charcoal is prepared from coconut shells and is fired at 600 °C prior to packing. The adsorbing section contains 100 mg of charcoal, the back-up section 50 mg. A 3-mm portion of urethane foam is placed between the outlet end of the tube and the back-up section. A plug of silanized glass wool is placed in front of the adsorbing section. The pressure drop across the tube must be less than one inch of mercury at a flow rate of 1 liter per minute.
   - 3.3. *Gas chromatograph* equipped with a flame ionization detector.
   - 3.4. *Column (10-ft x 1/8 -in stainless steel)* packed with 80/100 Supelcoport coated with 20 percent SP 2100, 0.1 percent CW 1500.
   - 3.5. *An electronic integrator* or some other suitable method for measuring peak area.
   - 3.6. *Two-milliliter sample vials* with Teflon-lined caps.
   - 3.7. *Microliter syringes:* 10-microliter (10-µL syringe, and other convenient sizes for making standards, 1-µL syringe for sample injections.
   - 3.8. *Pipets:* 1.0 mL delivery pipets
   - 3.9. *Volumetric flasks:* convenient sizes for making standard solutions.
4. *Reagents.*
   - 4.1. *Chromatographic quality carbon disulfide (CS_2).* Most commercially available carbon disulfide contains a trace of benzene which must be removed. It can be removed with the following procedure:

     Heat under reflux for 2 to 3 hours, 500 mL of carbon disulfide, 10 mL concentrated sulfuric acid, and 5 drops of concentrated nitric acid. The benzene is converted to nitrobenzene. The carbon disulfide layer is removed, dried with anhydrous sodium sulfate, and distilled. The recovered carbon disulfide should be benzene free. (It has recently been determined that benzene can also be removed by passing the carbon disulfide through 13x molecular sieve).
   - 4.2. *Benzene,* reagent grade.
   - 4.3. *p-Cymene,* reagent grade, (internal standard).
   - 4.4. *Desorbing reagent.* The desorbing reagent is prepared by adding 0.05 mL of p-cymene per milliliter of carbon disulfide. (The internal standard offers a convenient means correcting analytical response for slight inconsistencies in the size of sample injections. If the external standard technique is preferred, the internal standard can be eliminated).
   - 4.5. *Purified GC grade helium, hydrogen and air.*
5. *Procedure.*
   - 5.1. *Cleaning of equipment.* All glassware used for the laboratory analysis should be properly cleaned and free of organics which could interfere in the analysis.
   - 5.2. *Calibration of personal pumps.* Each pump must be calibrated with a representative charcoal tube in the line.
   - 5.3. *Collection and shipping of samples.*
     - 5.3.1. *Immediately before sampling,* break the ends of the tube to provide an opening at least one-half the internal diameter of the tube (2 mm).
     - 5.3.2. *The smaller section of the charcoal* is used as the backup and should be placed nearest the sampling pump.
     - 5.3.3. *The charcoal tube should be placed* in a vertical position during sampling to minimize channeling through the charcoal.
     - 5.3.4. *Air being sampled should not be passed* through any hose or tubing before entering the charcoal tube.
     - 5.3.5. *A sample size of 10 liters* is recommended. Sample at a flow rate of approximately 0.2 liters per minute. The flow rate should be known with an accuracy of at least ± 5 percent.
     - 5.3.6. *The charcoal tubes should be capped* with the supplied plastic caps immediately after sampling.
     - 5.3.7. *Submit at least one blank tube* (a charcoal tube subjected to the same handling procedures, without having any air drawn through it) with each set of samples.
     - 5.3.8. *Take necessary shipping* and packing precautions to minimize breakage of samples.

5.4. *Analysis of samples.*

5.4.1. *Preparation of samples.* In preparation for analysis, each charcoal tube is scored with a file in front of the first section of charcoal and broken open. The glass wool is removed and discarded. The charcoal in the first (larger) section is transferred to a 2-ml vial. The separating section of foam is removed and discarded; the second section is transferred to another capped vial. These two sections are analyzed separately.

5.4.2. *Desorption of samples.* Prior to analysis, 1.0 mL of desorbing solution is pipetted into each sample container. The desorbing solution consists of 0.05 µL internal standard per mL of carbon disulfide. The sample vials are capped as soon as the solvent is added. Desorption should be done for 30 minutes with occasional shaking.

5.4.3. *GC conditions.* Typical operating conditions for the gas chromatograph are:

1. *30 mL/min* (60 psig) helium carrier gas flow.
2. *30 mL/min* (40 psig) hydrogen gas flow to detector.
3. *240 mL/min* (40 psig) air flow to detector.
4. *150 °C injector* temperature.
5. *250 °C detector* temperature.
6. *100 °C column* temperature.

5.4.4. *Injection size.* 1 µL.

5.4.5. *Measurement of area.* The peak areas are measured by an electronic integrator or some other suitable form of area measurement.

5.4.6. *An internal standard procedure* is used. The integrator is calibrated to report results in ppm for a 10 liter air sample after correction for desorption efficiency.

5.5. *Determination of desorption efficiency.*

5.5.1. *Importance of determination.* The desorption efficiency of a particular compound can vary from one laboratory to another and from one lot of chemical to another. Thus, it is necessary to determine, at least once, the percentage of the specific compound that is removed in the desorption process, provided the same batch of charcoal is used.

5.5.2. *Procedure* for determining desorption efficiency. The reference portion of the charcoal tube is removed. To the remaining portion, amounts representing 0.5X, 1X, and 2X and (X represents target concentration) based on a 10 L air sample are injected into several tubes at each level. Dilutions of benzene with carbon disulfide are made to allow injection of measurable quantities. These tubes are then allowed to equilibrate at least overnight. Following equilibration they are analyzed following the same procedure as the samples. Desorption efficiency is determined by dividing the amount of benzene found by amount spiked on the tube.

6. *Calibration and standards.* A series of standards varying in concentration over the range of interest is prepared and analyzed under the same GC conditions that will be used on the samples. A calibration curve is prepared by plotting concentration (µg/mL) versus peak area.

7. *Calculations.* Benzene air concentration can be calculated from the following equation:

$$mg/m^3 = (A)(B) / (C)(D)$$

Where:
A = µg/mL benzene, obtained from the calibration curve
B = desorption volume (1 mL)
C = Liters of air sampled
D = desorption efficiency

The concentration in mg/m^3 can be converted to ppm (at 25 °C and 760 mm) with following equation:

$$ppm = (mg/m^3)(24.46) / (78.11)$$

Where:
24.46 = molar volume of an ideal gas
25 °C and 760 mm
78.11 = molecular weight of benzene

8. *Backup Data.*

8.1. *Detection limit — Air Samples.*
The detection limit for the analytical procedure is 1.28 ng with a coefficient of variation of 0.023 at this level. This would be equivalent to an air concentration of 0.04 ppm for a 10 L air sample. This amount provided a chromatographic peak that could be identifiable in the presence of possible interferences. The detection limit data were obtained by making 1 µL injections of a 1.283 µg/mL standard.

Injection	Area Count	
1	655.4	
2	617.5	
3	662.0	$\overline{X}$ = 640.2
4	641.1	SD = 14.9
5	636.4	CV = 0.023
6	629.2	

8.2. *Pooled coefficient of variation — Air Samples.*
The pooled coefficient of variation for the analytical procedure was determined by 1 µL replicate injections of analytical standards. The standards were 16.04, 32.08, and 64.16 µg/mL, which are equivalent to 0.5, 1.0, and 2.0 ppm for a 10 L air sample respectively.

Injection	Area Counts		
	0.5 ppm	1.0 ppm	2.0 ppm
1	3996.5	8130.2	16481
2	4059.4	8235.6	16493
3	4052.0	8307.9	16535
4	4027.2	8263.2	16609
5	4046.8	8291.1	16552
6	4137.9	8288.8	16618
$\overline{X}$ =	4053.3	8254.0	16548.3
SD =	47.2	62.5	57.1
CV =	0.0116	0.0076	0.0034
$\overline{CV}$ = 0.008			

8.3. *Storage data — Air Samples*
Samples were generated at 1.03 ppm benzene at 80% relative humidity, 22 °C, and 643 mm. All samples were taken for 50 minutes at 0.2 L/min. Six samples were analyzed immediately and the rest of the samples were divided into two groups by fifteen samples each. One group was stored at refrigerated temperature of -25 °C, and the other group was stored at ambient temperature (approximately 23 °C). These samples were analyzed over a period of fifteen days. The results are tabulated below.

Percent Recovery

Day analyzed	Refrigerated			Ambient		
0	97.4	98.7	98.9	97.4	98.7	98.9
0	97.1	100.6	100.9	97.1	100.6	100.9
2	95.8	96.4	95.4	95.4	96.6	96.9
5	93.9	93.7	92.4	92.4	94.3	94.1
9	93.6	95.5	94.6	95.2	95.6	96.6
13	94.3	95.3	93.7	91.0	95.0	94.6
15	96.8	95.8	94.2	92.9	96.3	95.9

8.4. *Desorption data.*
Samples were prepared by injecting liquid benzene onto the A section of charcoal tubes. Samples were prepared that would be equivalent to 0.5, 1.0, and 2.0 ppm for a 10 L air sample.

Percent Recovery

Sample	0.5 ppm	1.0 ppm	2.0 ppm
1	99.4	98.8	99.5
2	99.5	98.7	99.7
3	99.2	98.6	99.8
4	99.4	99.1	100.0
5	99.2	99.0	99.7
6	99.8	99.1	99.9
$\overline{X}$ =	99.4	98.9	99.8
SD =	0.22	0.21	0.18
CV =	0.0022	0.0021	0.0018
$\overline{X}$ = 99.4			

8.5. *Carbon disulfide.*
Carbon disulfide from a number of sources was analyzed for benzene contamination. The results are given in the following table. The benzene contamiant can be removed with the procedures given in section 4.1.

Sample	µg Benzene/mL	ppm equivalent (for 10 L air sample)
Aldrich Lot 83017	4.20	0.13
Baker Lot 720364	1.01	0.03
Baker Lot 822351	1.01	0.03
Malinkrodt Lot WEMP	1.74	0.05
Malinkrodt Lot WDSJ	5.65	0.18
Malinkrodt Lot WHGA	2.90	0.09
Treated CS_2		

II. **OSHA Laboratory Method No. 12 for Bulk Samples**
Analyte: Benzene.
Matrix: Bulk Samples.
Procedure: Bulk Samples are analyzed directly by high performance liquid chromatography (HPLC).
Detection limits: 0.01% by volume.
1. *Principle of the method.*
   1.1. *An aliquot of the bulk sample* to be analyzed is injected into a liquid chromatograph.
   1.2. *The peak area for benzene* is determined and compared to areas obtained from standards.
2. *Advantages and disadvantages of the method.*
   2.1. *The analytical procedure is quick,* sensitive, and reproducible.
   2.2. *Reanalysis of samples is possible.*
   2.3. *Interferences can be circumvented* by proper selection of HPLC parameters.
   2.4. *Samples must be free of any particulates* that may clog the capillary tubing in the liquid chromatograph. This may require distilling the sample or clarifying with a clarification kit.
3. *Apparatus.*
   3.1. *Liquid chromatograph* equipped with a UV detector.
   3.2. *HPLC Column* that will separate benzene from other components in the bulk sample being analyzed. The column used for validation studies was a Waters uBondapack C18, 30 cm x 3.9 mm.
   3.3. *A clarification kit* to remove any particulates in the bulk if necessary.
   3.4. *A micro-distillation apparatus* to distill any samples if necessary.
   3.5. *An electronic integrator* or some other suitable method of measuring peak areas.
   3.6. *Microliter syringes* — 10 µL syringe and other convenient sizes for making standards. 10 µL syringe for sample injections.
   3.7. *Volumetric flasks,* 5 mL and other convenient sizes for preparing standards and making dilutions.
4. *Reagents.*
   4.1. *Benzene, reagent grade.*
   4.2. *HPLC grade water, methyl alcohol, and isopropyl alcohol.*
5. *Collection and shipment of samples.*
   5.1. *Samples should be transported in glass* containers with Teflon-lined caps.
   5.2. *Samples should not be put in the same* container used for air samples.
6. *Analysis of samples.*
   6.1. *Sample preparation.*
   If necessary, the samples are distilled or clarified. Samples are analyzed undiluted. If the benzene concentration is out of the working range, suitable dilutions are made with isopropyl alcohol.
   6.2. *HPLC conditions*
   The typical operating conditions for the high performance liquid chromatograph are:
   1. *Mobile phase* — Methyl alcohol/water, 50/50
   2. *Analytical wavelength* — 254 nm
   3. *Injection size* — 10 µL
   6.3. *Measurement of peak area* and calibration.
   Peak areas are measured by an integrator or other suitable means. The integrator is calibrated to report results % in benzene by volume.
7. *Calculations.* Since the integrator is programmed to report results in % benzene by volume in an undiluted sample, the following equation is used:

% Benzene by Volume = A x B
Where:
A = % by volume on report
B = Dilution Factor
(B = 1 for undiluted sample)

8. *Backup Data.*
   8.1. *Detection limit — Bulk Samples.*
   The detection limit for the analytical procedure for bulk samples is 0.88 µg, with a coefficient of variation of 0.019 at this level. This amount provided a chromatographic peak that could be identifiable in the presence of possible interferences. The detection limit date were obtained by making 10 µL injections of a 0.10% by volume standard.

Injection	Area Count	
1	45386	
2	44214	
3	43822	$\overline{X}$ = 44040.1
4	44062	SD = 852.5
6	42724	CV = 0.019

   8.2. *Pooled coefficient of variation — Bulk Samples.*
   The pooled coefficient of variation for analytical procedure was determined by 50 µL replicate injections of analytical standards. The standards were 0.01, 0.02, 0.04, 0.10, 1.0, and 2.0% benzene by volume.

Area count (Percent)

Injection No.	0.01	0.02	0.04	0.10	1.0	2.0
1	45386	84737	166097	448497	4395380	9339150
2	44241	84300	170832	441299	4590800	9484900
3	43822	83835	164160	443719	4593200	9557580
4	44062	84381	164445	444842	4642350	9677060
5	44006	83012	168398	442564	4646430	9766240
6	42724	81957	173002	443975	4646260	
$\overline{X}$ =	44040.1	83703.6	167872	444149	4585767	9564986
SD =	852.5	1042.2	3589.8	2459.1	96839.3	166233
CV =	0.0194	0.0125	0.0213	0.0055	0.0211	0.0174
$\overline{CV}$ =	0.017					

[52 FR 34562, Sept. 11, 1987, as amended at 54 FR 24334, June 7, 1989; 61 FR 5508, Feb. 13, 1996; 63 FR 1289, Jan. 8, 1998; 63 FR 20099, Apr. 23, 1998; 70 FR 1142, Jan. 5, 2005; 71 FR 16673, Apr. 3, 2006; 71 FR 50189, Aug. 24, 2006; 73 FR 75585, Dec. 12, 2008; 76 FR 33608, June 8, 2011; 77 FR 17781, Mar. 26, 2012]

§1910.1030
Bloodborne pathogens

(a) **Scope and Application.** This section applies to all occupational exposure to blood or other potentially infectious materials as defined by paragraph (b) of this section.[§1910.1030(a)]

(b) **Definitions.** For purposes of this section, the following shall apply:

Assistant Secretary means the Assistant Secretary of Labor for Occupational Safety and Health, or designated representative.

Blood means human blood, human blood components, and products made from human blood.

Bloodborne Pathogens means pathogenic microorganisms that are present in human blood and can cause disease in humans. These pathogens include, but are not limited to, hepatitis B virus (HBV) and human immunodeficiency virus (HIV).

Clinical Laboratory means a workplace where diagnostic or other screening procedures are performed on blood or other potentially infectious materials.

Contaminated means the presence or the reasonably anticipated presence of blood or other potentially infectious materials on an item or surface.

Contaminated Laundry means laundry which has been soiled with blood or other potentially infectious materials or may contain sharps.

Contaminated Sharps means any contaminated object that can penetrate the skin including, but not limited to, needles, scalpels, broken glass, broken capillary tubes, and exposed ends of dental wires.

Decontamination means the use of physical or chemical means to remove, inactivate, or destroy bloodborne pathogens on a surface or item to the point where they are no longer capable of transmitting infectious particles and the surface or item is rendered safe for handling, use, or disposal.

Director means the Director of the National Institute for Occupational Safety and Health, U.S. Department of Health and Human Services, or designated representative.

Engineering controls means controls (e.g., sharps disposal containers, self-sheathing needles, safer medical devices, such as sharps with engineered sharps injury protections and needleless systems) that isolate or remove the bloodborne pathogens hazard from the workplace.

Exposure Incident means a specific eye, mouth, other mucous membrane, non-intact skin, or parenteral contact with blood or other potentially infectious materials that results from the performance of an employee's duties.

Handwashing facilities means a facility providing an adequate supply of running potable water, soap, and single-use towels or air-drying machines.

Licensed Healthcare Professional is a person whose legally permitted scope of practice allows him or her to independently perform the activities required by paragraph (f) Hepatitis B Vaccination and Post-exposure Evaluation and Follow-up.

HBV means hepatitis B virus.

HIV means human immunodeficiency virus.

Needleless systems means a device that does not use needles for:

(1) *The collection of bodily fluids* or withdrawal of body fluids after initial venous or arterial access is established;

(2) *The administration of medication or fluids; or*

(3) *Any other procedure involving the potential* for occupational exposure to bloodborne pathogens due to percutaneous injuries from contaminated sharps.

Occupational Exposure means reasonably anticipated skin, eye, mucous membrane, or parenteral contact with blood or other potentially infectious materials that may result from the performance of an employee's duties.

Other Potentially Infectious Materials means

(1) *The following human body fluids:* semen, vaginal secretions, cerebrospinal fluid, synovial fluid, pleural fluid, pericardial fluid, peritoneal fluid, amniotic fluid, saliva in dental procedures, any body fluid that is visibly contaminated with blood, and all body fluids in situations where it is difficult or impossible to differentiate between body fluids;

(2) *Any unfixed tissue or organ* (other than intact skin) from a human (living or dead); and

(3) *HIV-containing cell or tissue cultures,* organ cultures, and HIV- or HBV-containing culture medium or other solutions; and blood, organs, or other tissues from experimental animals infected with HIV or HBV.

Parenteral means piercing mucous membranes or the skin barrier through such events as needlesticks, human bites, cuts, and abrasions.

Personal Protective Equipment is specialized clothing or equipment worn by an employee for protection against a hazard. General work clothes (e.g., uniforms, pants, shirts or blouses) not intended to function as protection against a hazard are not considered to be personal protective equipment.

Production Facility means a facility engaged in industrial-scale, large-volume or high concentration production of HIV or HBV.

Regulated Waste means liquid or semi-liquid blood or other potentially infectious materials; contaminated items that would release blood or other potentially infectious materials in a liquid or semi-liquid state if compressed; items that are caked with dried blood or other potentially infectious materials and are capable of releasing these materials during handling; contaminated sharps; and pathological and microbiological wastes containing blood or other potentially infectious materials.

Research Laboratory means a laboratory producing or using research-laboratory-scale amounts of HIV or HBV. Research laboratories may produce high concentrations of HIV or HBV but not in the volume found in production facilities.

Sharps with engineered sharps injury protections means a non-needle sharp or a needle device used for withdrawing body fluids, accessing a vein or artery, or administering medications or other fluids, with a built-in safety feature or mechanism that effectively reduces the risk of an exposure incident.

Source Individual means any individual, living or dead, whose blood or other potentially infectious materials may be a source of occupational exposure to the employee. Examples include, but are not limited to, hospital and clinic patients; clients in institutions for the developmentally disabled; trauma victims; clients of drug and alcohol treatment facilities; residents of hospices and nursing homes; human remains; and individuals who donate or sell blood or blood components.

Sterilize means the use of a physical or chemical procedure to destroy all microbial life including highly resistant bacterial endospores.

Universal Precautions is an approach to infection control. According to the concept of Universal Precautions, all human blood and certain human body fluids are treated as if known to be infectious for HIV, HBV, and other bloodborne pathogens.

Work Practice Controls means controls that reduce the likelihood of exposure by altering the manner in which a task is performed (e.g., prohibiting recapping of needles by a two-handed technique).

(c) Exposure control.[§1910.1030(c)]

(1) *Exposure Control Plan.*[§1910.1030(c)(1)]

(i) *Each employer having an employee(s)* with occupational exposure as defined by paragraph (b) of this section shall establish a written Exposure Control Plan designed to eliminate or minimize employee exposure.[§1910.1030(c)(1)(i)]

(ii) *The Exposure Control Plan* shall contain at least the following elements:[§1910.1030(c)(1)(ii)]

[A] The exposure determination required by paragraph(c)(2), [§1910.1030(c)(1)(ii)[A]]

[B] The schedule and method of implementation for paragraphs (d) Methods of Compliance, (e) HIV and HBV Research Laboratories and Production Facilities, (f) Hepatitis B Vaccination and Post-Exposure Evaluation and Follow-up, (g) Communication of Hazards to Employees, and (h) Recordkeeping, of this standard, and [§1910.1030(c)(1)(ii)[B]]

[C] The procedure for the evaluation of circumstances surrounding exposure incidents as required by paragraph (f)(3)(i) of this standard.[§1910.1030(c)(1)(ii)[C]]

(iii) *Each employer shall ensure that a copy* of the Exposure Control Plan is accessible to employees in accordance with 29 CFR 1910.1020(e)[1].[§1910.1030(c)(1)(iii)]

(iv) *The Exposure Control Plan* shall be reviewed and updated at least annually and whenever necessary to reflect new or modified tasks and procedures which affect occupational exposure and to reflect new or revised employee positions with occupational exposure. The review and update of such plans shall also:[§1910.1030(c)(1)(iv)]

[A] Reflect changes in technology that eliminate or reduce exposure to bloodborne pathogens; and [§1910.1030(c)(1)(iv)[A]]

[B] Document annually consideration and implementation of appropriate commercially available and effective safer medical devices designed to eliminate or minimize occupational exposure.[§1910.1030(c)(1)(iv)[B]]

(v) *An employer,* who is required to establish an Exposure Control Plan shall solicit input from non-managerial employees responsible for direct patient care who are potentially exposed to injuries from contaminated sharps in the identification, evaluation, and selection of effective engineering and work practice controls and shall document the solicitation in the Exposure Control Plan.[§1910.1030(c)(1)(v)]

(vi) *The Exposure Control Plan* shall be made available to the Assistant Secretary and the Director upon request for examination and copying.[§1910.1030(c)(1)(vi)]

(2) *Exposure determination.*[§1910.1030(c)(2)]

(i) *Each employer who has an employee(s)* with occupational exposure as defined by paragraph (b) of this section shall prepare an exposure determination. This exposure determination shall contain the following:[§1910.1030(c)(2)(i)]

[A] A list of all job classifications in which all employees in those job classifications have occupational exposure; [§1910.1030(c)(2)(i)[A]]

[B] A list of job classifications in which some employees have occupational exposure, and[§1910.1030(c)(2)(i)[B]]

[C] A list of all tasks and procedures or groups of closely related task and procedures in which occupational exposure occurs and that are performed by employees in job classifications listed in accordance with the provisions of paragraph (c)(2)(i)(B) of this standard.[§1910.1030(c)(2)(i)[C]]

---

1. Editor's note: The CFR refers in error to "29 CFR 1910.20(e)". This is corrected in the regulation appearing above to refer to "29 CFR 1910.1020(e)".

(ii) *This exposure determination shall be made* without regard to the use of personal protective equipment.[§1910.1030(c)(2)(ii)]

(d) Methods of compliance.[§1910.1030(d)]

(1) *General.* Universal precautions shall be observed to prevent contact with blood or other potentially infectious materials. Under circumstances in which differentiation between body fluid types is difficult or impossible, all body fluids shall be considered potentially infectious materials.[§1910.1030(d)(1)]

(2) *Engineering and work practice controls.*[§1910.1030(d)(2)]

(i) *Engineering and work practice controls* shall be used to eliminate or minimize employee exposure. Where occupational exposure remains after institution of these controls, personal protective equipment shall also be used.[§1910.1030(d)(2)(i)]

(ii) *Engineering controls shall be examined* and maintained or replaced on a regular schedule to ensure their effectiveness. [§1910.1030(d)(2)(ii)]

(iii) *Employers shall provide handwashing facilities* which are readily accessible to employees.[§1910.1030(d)(2)(iii)]

(iv) *When provision of handwashing facilities is not feasible,* the employer shall provide either an appropriate antiseptic hand cleanser in conjunction with clean cloth/paper towels or antiseptic towelettes. When antiseptic hand cleansers or towelettes are used, hands shall be washed with soap and running water as soon as feasible.[§1910.1030(d)(2)(iv)]

(v) *Employers shall ensure that employees* wash their hands immediately or as soon as feasible after removal of gloves or other personal protective equipment.[§1910.1030(d)(2)(v)]

(vi) *Employers shall ensure that employees* wash hands and any other skin with soap and water, or flush mucous membranes with water immediately or as soon as feasible following contact of such body areas with blood or other potentially infectious materials.[§1910.1030(d)(2)(vi)]

(vii) *Contaminated needles and other contaminated sharps* shall not be bent, recapped, or removed except as noted in paragraphs (d)(2)(vii)(A) and (d)(2)(vii)(B) below. Shearing or breaking of contaminated needles is prohibited. [§1910.1030(d)(2)(vii)]

[A] Contaminated needles and other contaminated sharps shall not be bent, recapped or removed unless the employer can demonstrate that no alternative is feasible or that such action is required by a specific medical or dental procedure.[§1910.1030(d)(2)(vii)[A]]

[B] Such bending, recapping or needle removal must be accomplished through the use of a mechanical device or a one-handed technique.[§1910.1030(d)(2)(vii)[B]]

(viii) *Immediately or as soon as possible after use,* contaminated reusable sharps shall be placed in appropriate containers until properly reprocessed. These containers shall be: [§1910.1030(d)(2)(viii)]

[A] Puncture resistant;[§1910.1030(d)(2)(viii)[A]]

[B] Labeled or color-coded in accordance with this standard; [§1910.1030(d)(2)(viii)[B]]

[C] Leakproof on the sides and bottom; and [§1910.1030(d)(2)(viii)[C]]

[D] In accordance with the requirements set forth in paragraph (d)(4)(ii)(E) for reusable sharps. [§1910.1030(d)(2)(viii)[D]]

(ix) *Eating, drinking, smoking,* applying cosmetics or lip balm, and handling contact lenses are prohibited in work areas where there is a reasonable likelihood of occupational exposure.[§1910.1030(d)(2)(ix)]

(x) *Food and drink shall not be kept in refrigerators,* freezers, shelves, cabinets or on countertops or benchtops where blood or other potentially infectious materials are present. [§1910.1030(d)(2)(x)]

(xi) *All procedures involving blood* or other potentially infectious materials shall be performed in such a manner as to minimize splashing, spraying, spattering, and generation of droplets of these substances.[§1910.1030(d)(2)(xi)]

(xii) *Mouth pipetting/suctioning* of blood or other potentially infectious materials is prohibited.[§1910.1030(d)(2)(xii)]

(xiii) *Specimens of blood or other potentially infectious materials* shall be placed in a container which prevents leakage during collection, handling, processing, storage, transport, or shipping.[§1910.1030(d)(2)(xiii)]

[A] The container for storage, transport, or shipping shall be labeled or color-coded according to paragraph (g)(1)(i) and closed prior to being stored, transported, or shipped. When a facility utilizes Universal Precautions in the handling of all specimens, the labeling/color-coding of specimens is not necessary provided containers are recognizable as containing specimens. This exemption only applies while such specimens/containers remain within the facility. Labeling or color-coding in accordance with paragraph (g)(1)(i) is required when such specimens/containers leave the facility.[§1910.1030(d)(2)(xiii)[A]]

[B] If outside contamination of the primary container occurs, the primary container shall be placed within a second container which prevents leakage during handling, processing, storage, transport, or shipping and is labeled or color-coded according to the requirements of this standard. [§1910.1030(d)(2)(xiii)[B]]

[C] If the specimen could puncture the primary container, the primary container shall be placed within a secondary container which is puncture-resistant in addition to the above characteristics.[§1910.1030(d)(2)(xiii)[C]]

(xiv) *Equipment which may become contaminated with blood* or other potentially infectious materials shall be examined prior to servicing or shipping and shall be decontaminated as necessary, unless the employer can demonstrate that decontamination of such equipment or portions of such equipment is not feasible.[§1910.1030(d)(2)(xiv)]

[A] A readily observable label in accordance with paragraph (g)(1)(i)(H) shall be attached to the equipment stating which portions remain contaminated.[§1910.1030(d)(2)(xiv)[A]]

[B] The employer shall ensure that this information is conveyed to all affected employees, the servicing representative, and/or the manufacturer, as appropriate, prior to handling, servicing, or shipping so that appropriate precautions will be taken.[§1910.1030(d)(2)(xiv)[B]]

(3) *Personal protective equipment*[§1910.1030(d)(3)]

(i) *Provision.* When there is occupational exposure, the employer shall provide, at no cost to the employee, appropriate personal protective equipment such as, but not limited to, gloves, gowns, laboratory coats, face shields or masks and eye protection, and mouthpieces, resuscitation bags, pocket masks, or other ventilation devices. Personal protective equipment will be considered "appropriate" only if it does not permit blood or other potentially infectious materials to pass through to or reach the employee's work clothes, street clothes, undergarments, skin, eyes, mouth, or other mucous membranes under normal conditions of use and for the duration of time which the protective equipment will be used. [§1910.1030(d)(3)(i)]

(ii) *Use.* The employer shall ensure that the employee uses appropriate personal protective equipment unless the employer shows that the employee temporarily and briefly declined to use personal protective equipment when, under rare and extraordinary circumstances, it was the employee's professional judgment that in the specific instance its use would have prevented the delivery of health care or public safety services or would have posed an increased hazard to the safety of the worker or co-worker. When the employee makes this judgement, the circumstances shall be investigated and documented in order to determine whether changes can be instituted to prevent such occurrences in the future.[§1910.1030(d)(3)(ii)]

(iii) *Accessibility.* The employer shall ensure that appropriate personal protective equipment in the appropriate sizes is readily accessible at the worksite or is issued to employees. Hypoallergenic gloves, glove liners, powderless gloves, or other similar alternatives shall be readily accessible to those employees who are allergic to the gloves normally provided. [§1910.1030(d)(3)(iii)]

(iv) *Cleaning, Laundering, and Disposal.* The employer shall clean, launder, and dispose of personal protective equipment required by paragraphs (d) and (e) of this standard, at no cost to the employee.[§1910.1030(d)(3)(iv)]

(v) *Repair and Replacement.* The employer shall repair or replace personal protective equipment as needed to maintain its effectiveness, at no cost to the employee. [§1910.1030(d)(3)(v)]

(vi) *If a garment(s) is penetrated* by blood or other potentially infectious materials, the garment(s) shall be removed immediately or as soon as feasible.[§1910.1030(d)(3)(vi)]

(vii) *All personal protective equipment* shall be removed prior to leaving the work area.[§1910.1030(d)(3)(vii)]

(viii) *When personal protective equipment* is removed it shall be placed in an appropriately designated area or container for storage, washing, decontamination or disposal. [§1910.1030(d)(3)(viii)]

(ix) *Gloves.* Gloves shall be worn when it can be reasonably anticipated that the employee may have hand contact with blood, other potentially infectious materials, mucous membranes, and non-intact skin; when performing vascular access procedures except as specified in paragraph (d)(3)(ix)(D); and when handling or touching contaminated items or surfaces.[§1910.1030(d)(3)(ix)]

[A] *Disposable (single use)* gloves such as surgical or examination gloves, shall be replaced as soon as practical when contaminated or as soon as feasible if they are torn, punctured, or when their ability to function as a barrier is compromised.[§1910.1030(d)(3)(ix)[A]]

[B] *Disposable (single use)* gloves shall not be washed or decontaminated for re-use.[§1910.1030(d)(3)(ix)[B]]

[C] *Utility gloves may be decontaminated* for re-use if the integrity of the glove is not compromised. However, they must be discarded if they are cracked, peeling, torn, punctured, or exhibit other signs of deterioration or when their ability to function as a barrier is compromised. [§1910.1030(d)(3)(ix)[C]]

[D] *If an employer* in a volunteer blood donation center judges that routine gloving for all phlebotomies is not necessary then the employer shall:[§1910.1030(d)(3)(ix)[D]]

[1] *Periodically reevaluate this policy;* [§1910.1030(d)(3)(ix)[D][1]]

[2] *Make gloves available* to all employees who wish to use them for phlebotomy;[§1910.1030(d)(3)(ix)[D][2]]

[3] *Not discourage the use of gloves for phlebotomy; and* [§1910.1030(d)(3)(ix)[D][3]]

[4] *Require that gloves be used* for phlebotomy in the following circumstances:[§1910.1030(d)(3)(ix)[D][4]]

[i] *When the employee has cuts,* scratches, or other breaks in his or her skin;[§1910.1030(d)(3)(ix)[D][4][i]]

[ii] *When the employee judges that hand contamination* with blood may occur, for example, when performing phlebotomy on an uncooperative source individual; and[§1910.1030(d)(3)(ix)[D][4][ii]]

[iii] *When the employee is receiving training in phlebotomy.*[§1910.1030(d)(3)(ix)[D][4][iii]]

(x) *Masks, Eye Protection, and Face Shields.* Masks in combination with eye protection devices, such as goggles or glasses with solid side shields, or chin-length face shields, shall be worn whenever splashes, spray, spatter, or droplets of blood or other potentially infectious materials may be generated and eye, nose, or mouth contamination can be reasonably anticipated.[§1910.1030(d)(3)(x)]

(xi) *Gowns, Aprons, and Other Protective Body Clothing.* Appropriate protective clothing such as, but not limited to, gowns, aprons, lab coats, clinic jackets, or similar outer garments shall be worn in occupational exposure situations. The type and characteristics will depend upon the task and degree of exposure anticipated.[§1910.1030(d)(3)(xi)]

(xii) *Surgical caps or hoods and/or shoe covers or boots* shall be worn in instances when gross contamination can reasonably be anticipated (e.g., autopsies, orthopaedic surgery). [§1910.1030(d)(3)(xii)]

(4) *Housekeeping*[§1910.1030(d)(4)]

(i) *General.* Employers shall ensure that the worksite is maintained in a clean and sanitary condition. The employer shall determine and implement an appropriate written schedule for cleaning and method of decontamination based upon the location within the facility, type of surface to be cleaned, type of soil present, and tasks or procedures being performed in the area.[§1910.1030(d)(4)(i)]

(ii) *All equipment and environmental and working surfaces* shall be cleaned and decontaminated after contact with blood or other potentially infectious materials.[§1910.1030(d)(4)(ii)]

[A] *Contaminated work surfaces* shall be decontaminated with an appropriate disinfectant after completion of procedures; immediately or as soon as feasible when surfaces are overtly contaminated or after any spill of blood or other potentially infectious materials; and at the end of the work shift if the surface may have become contaminated since the last cleaning.[§1910.1030(d)(4)(ii)[A]]

[B] *Protective coverings, such as plastic wrap,* aluminum foil, or imperviously-backed absorbent paper used to cover equipment and environmental surfaces, shall be removed and replaced as soon as feasible when they become overtly contaminated or at the end of the workshift if they may have become contaminated during the shift. [§1910.1030(d)(4)(ii)[B]]

[C] *All bins, pails, cans, and similar receptacles* intended for reuse which have a reasonable likelihood for becoming contaminated with blood or other potentially infectious materials shall be inspected and decontaminated on a regularly scheduled basis and cleaned and decontaminated immediately or as soon as feasible upon visible contamination.[§1910.1030(d)(4)(ii)[C]]

[D] *Broken glassware which may be contaminated* shall not be picked up directly with the hands. It shall be cleaned up using mechanical means, such as a brush and dust pan, tongs, or forceps.[§1910.1030(d)(4)(ii)[D]]

[E] *Reusable sharps that are contaminated* with blood or other potentially infectious materials shall not be stored or processed in a manner that requires employees to reach by hand into the containers where these sharps have been placed.[§1910.1030(d)(4)(ii)[E]]

(iii) *Regulated Waste.* [§1910.1030(d)(4)(iii)]

[A] *Contaminated Sharps Discarding and Containment.* [§1910.1030(d)(4)(iii)[A]]

[1] *Contaminated sharps shall be discarded* immediately or as soon as feasible in containers that are: [§1910.1030(d)(4)(iii)[A][1]]

[i] *Closable;* [§1910.1030(d)(4)(iii)[A][1][i]]

[ii] *Puncture resistant;* [§1910.1030(d)(4)(iii)[A][1][ii]]

[iii] *Leakproof on sides and bottom; and* [§1910.1030(d)(4)(iii)[A][1][iii]]

[iv] *Labeled or color-coded in accordance* with paragraph (g)(1)(i) of this standard. [§1910.1030(d)(4)(iii)[A][1][iv]]

[2] *During use,* containers for contaminated sharps shall be: [§1910.1030(d)(4)(iii)[A][2]]

[i] *Easily accessible to personnel* and located as close as is feasible to the immediate area where sharps are used or can be reasonably anticipated to be found (e.g., laundries); [§1910.1030(d)(4)(iii)[A][2][i]]

[ii] *Maintained upright throughout use; and* [§1910.1030(d)(4)(iii)[A][2][ii]]

[iii] *Replaced routinely and not be allowed to overfill.* [§1910.1030(d)(4)(iii)[A][2][iii]]

[3] *When moving containers* of contaminated sharps from the area of use, the containers shall be: [§1910.1030(d)(4)(iii)[A][3]]

[i] *Closed immediately prior to removal or replacement* to prevent spillage or protrusion of contents during handling, storage, transport, or shipping; [§1910.1030(d)(4)(iii)[A][3][i]]

[ii] *Placed in a secondary container* if leakage is possible. The second container shall be: [§1910.1030(d)(4)(iii)[A][3][ii]]

[A] *Closable;* [§1910.1030(d)(4)(iii)[A][3][ii][A]]

[B] *Constructed to contain all contents* and prevent leakage during handling, storage, transport, or shipping; and [§1910.1030(d)(4)(iii)[A][3][ii][B]]

[C] *Labeled or color-coded* according to paragraph (g)(1)(i) of this standard. [§1910.1030(d)(4)(iii)[A][3][ii][C]]

[4] *Reusable containers shall not be opened,* emptied, or cleaned manually or in any other manner which would expose employees to the risk of percutaneous injury. [§1910.1030(d)(4)(iii)[A][4]]

[B] *Other Regulated Waste Containment.* [§1910.1030(d)(4)(iii)[B]]

[1] *Regulated waste shall be placed in containers which are:* [§1910.1030(d)(4)(iii)[B][1]]

[i] *Closable;* [§1910.1030(d)(4)(iii)[B][1][i]]

[ii] *Constructed to contain all contents* and prevent leakage of fluids during handling, storage, transport or shipping; [§1910.1030(d)(4)(iii)[B][1][ii]]

[iii] *Labeled or color-coded in accordance* with paragraph (g)(1)(i) this standard; and [§1910.1030(d)(4)(iii)[B][1][iii]]

[iv] Closed prior to removal to prevent spillage or protrusion of contents during handling, storage, transport, or shipping. [§1910.1030(d)(4)(iii)[B][1][iv]]

[2] If outside contamination of the regulated waste container occurs, it shall be placed in a second container. The second container shall be: [§1910.1030(d)(4)(iii)[B][2]]

[i] Closable; [§1910.1030(d)(4)(iii)[B][2][i]]

[ii] Constructed to contain all contents and prevent leakage of fluids during handling, storage, transport or shipping; [§1910.1030(d)(4)(iii)[B][2][ii]]

[iii] Labeled or color-coded in accordance with paragraph (g)(1)(i) of this standard; and [§1910.1030(d)(4)(iii)[B][2][iii]]

[iv] Closed prior to removal to prevent spillage or protrusion of contents during handling, storage, transport, or shipping. [§1910.1030(d)(4)(iii)[B][2][iv]]

[C] Disposal of all regulated waste shall be in accordance with applicable regulations of the United States, States and Territories, and political subdivisions of States and Territories.[§1910.1030(d)(4)(iii)[C]]

(iv) *Laundry.*[§1910.1030(d)(4)(iv)]

[A] Contaminated laundry shall be handled as little as possible with a minimum of agitation.[§1910.1030(d)(4)(iv)[A]]

[1] Contaminated laundry shall be bagged or containerized at the location where it was used and shall not be sorted or rinsed in the location of use. [§1910.1030(d)(4)(iv)[A][1]]

[2] Contaminated laundry shall be placed and transported in bags or containers labeled or color-coded in accordance with paragraph (g)(1)(i) of this standard. When a facility utilizes Universal Precautions in the handling of all soiled laundry, alternative labeling or color-coding is sufficient if it permits all employees to recognize the containers as requiring compliance with Universal Precautions.[§1910.1030(d)(4)(iv)[A][2]]

[3] Whenever contaminated laundry is wet and presents a reasonable likelihood of soak-through of or leakage from the bag or container, the laundry shall be placed and transported in bags or containers which prevent soak-through and/or leakage of fluids to the exterior. [§1910.1030(d)(4)(iv)[A][3]]

[B] The employer shall ensure that employees who have contact with contaminated laundry wear protective gloves and other appropriate personal protective equipment. [§1910.1030(d)(4)(iv)[B]]

[C] When a facility ships contaminated laundry off-site to a second facility which does not utilize Universal Precautions in the handling of all laundry, the facility generating the contaminated laundry must place such laundry in bags or containers which are labeled or color-coded in accordance with paragraph (g)(1)(i).[§1910.1030(d)(4)(iv)[C]]

(e) HIV and HBV Research Laboratories and Production Facilities.[§1910.1030(e)]

(1) *This paragraph applies to research laboratories* and production facilities engaged in the culture, production, concentration, experimentation, and manipulation of HIV and HBV. It does not apply to clinical or diagnostic laboratories engaged solely in the analysis of blood, tissues, or organs. These requirements apply in addition to the other requirements of the standard.[§1910.1030(e)(1)]

(2) *Research laboratories* and production facilities shall meet the following criteria:[§1910.1030(e)(2)]

(i) *Standard microbiological practices.* All regulated waste shall either be incinerated or decontaminated by a method such as autoclaving known to effectively destroy bloodborne pathogens.[§1910.1030(e)(2)(i)]

(ii) *Special practices.*[§1910.1030(e)(2)(ii)]

[A] Laboratory doors shall be kept closed when work involving HIV or HBV is in progress.[§1910.1030(e)(2)(ii)[A]]

[B] Contaminated materials that are to be decontaminated at a site away from the work area shall be placed in a durable, leakproof, labeled or color-coded container that is closed before being removed from the work area. [§1910.1030(e)(2)(ii)[B]]

[C] Access to the work area shall be limited to authorized persons. Written policies and procedures shall be established whereby only persons who have been advised of the potential biohazard, who meet any specific entry requirements, and who comply with all entry and exit procedures shall be allowed to enter the work areas and animal rooms.[§1910.1030(e)(2)(ii)[C]]

[D] When other potentially infectious materials or infected animals are present in the work area or containment module, a hazard warning sign incorporating the universal biohazard symbol shall be posted on all access doors. The hazard warning sign shall comply with paragraph (g)(1)(ii) of this standard.[§1910.1030(e)(2)(ii)[D]]

[E] All activities involving other potentially infectious materials shall be conducted in biological safety cabinets or other physical-containment devices within the containment module. No work with these other potentially infectious materials shall be conducted on the open bench. [§1910.1030(e)(2)(ii)[E]]

[F] Laboratory coats, gowns, smocks, uniforms, or other appropriate protective clothing shall be used in the work area and animal rooms. Protective clothing shall not be worn outside of the work area and shall be decontaminated before being laundered.[§1910.1030(e)(2)(ii)[F]]

[G] Special care shall be taken to avoid skin contact with other potentially infectious materials. Gloves shall be worn when handling infected animals and when making hand contact with other potentially infectious materials is unavoidable.[§1910.1030(e)(2)(ii)[G]]

[H] Before disposal all waste from work areas and from animal rooms shall either be incinerated or decontaminated by a method such as autoclaving known to effectively destroy bloodborne pathogens.[§1910.1030(e)(2)(ii)[H]]

[I] Vacuum lines shall be protected with liquid disinfectant traps and high-efficiency particulate air (HEPA) filters or filters of equivalent or superior efficiency and which are checked routinely and maintained or replaced as necessary.[§1910.1030(e)(2)(ii)[I]]

[J] Hypodermic needles and syringes shall be used only for parenteral injection and aspiration of fluids from laboratory animals and diaphragm bottles. Only needle-locking syringes or disposable syringe-needle units (i.e., the needle is integral to the syringe) shall be used for the injection or aspiration of other potentially infectious materials. Extreme caution shall be used when handling needles and syringes. A needle shall not be bent, sheared, replaced in the sheath or guard, or removed from the syringe following use. The needle and syringe shall be promptly placed in a puncture-resistant container and autoclaved or decontaminated before reuse or disposal. [§1910.1030(e)(2)(ii)[J]]

[K] All spills shall be immediately contained and cleaned up by appropriate professional staff or others properly trained and equipped to work with potentially concentrated infectious materials.[§1910.1030(e)(2)(ii)[K]]

[L] A spill or accident that results in an exposure incident shall be immediately reported to the laboratory director or other responsible person.[§1910.1030(e)(2)(ii)[L]]

[M] A biosafety manual shall be prepared or adopted and periodically reviewed and updated at least annually or more often if necessary. Personnel shall be advised of potential hazards, shall be required to read instructions on practices and procedures, and shall be required to follow them.[§1910.1030(e)(2)(ii)[M]]

(iii) *Containment equipment.*[§1910.1030(e)(2)(iii)]

[A] Certified biological safety cabinets (Class I, II, or III) or other appropriate combinations of personal protection or physical containment devices, such as special protective clothing, respirators, centrifuge safety cups, sealed centrifuge rotors, and containment caging for animals, shall be used for all activities with other potentially infectious materials that pose a threat of exposure to droplets, splashes, spills, or aerosols.[§1910.1030(e)(2)(iii)[A]]

[B] Biological safety cabinets shall be certified when installed, whenever they are moved and at least annually. [§1910.1030(e)(2)(iii)[B]]

(3) *HIV and HBV research laboratories shall meet the following criteria:*[§1910.1030(e)(3)]

(i) *Each laboratory shall contain a facility* for hand washing and an eye wash facility which is readily available within the work area.[§1910.1030(e)(3)(i)]

(ii) *An autoclave for decontamination* of regulated waste shall be available.[§1910.1030(e)(3)(ii)]

(4) *HIV and HBV production facilities shall meet the following criteria:*[§1910.1030(e)(4)]

(i) *The work areas shall be separated* from areas that are open to unrestricted traffic flow within the building. Passage through two sets of doors shall be the basic requirement for entry into the work area from access corridors or other contiguous areas. Physical separation of the high-containment work area from access corridors or other areas or activities may also be provided by a double-doored clothes-change room (showers may be included), airlock, or other access facility that requires passing through two sets of doors before entering the work area.[§1910.1030(e)(4)(i)]

(ii) *The surfaces of doors, walls, floors and ceilings* in the work area shall be water resistant so that they can be easily cleaned. Penetrations in these surfaces shall be sealed or capable of being sealed to facilitate decontamination. [§1910.1030(e)(4)(ii)]

(iii) *Each work area shall contain a sink* for washing hands and a readily available eye wash facility. The sink shall be foot, elbow, or automatically operated and shall be located near the exit door of the work area.[§1910.1030(e)(4)(iii)]

(iv) *Access doors* to the work area or containment module shall be self-closing.[§1910.1030(e)(4)(iv)]

(v) *An autoclave for decontamination* of regulated waste shall be available within or as near as possible to the work area. [§1910.1030(e)(4)(v)]

(vi) *A ducted exhaust-air ventilation system shall be provided.* This system shall create directional airflow that draws air into the work area through the entry area. The exhaust air shall not be recirculated to any other area of the building, shall be discharged to the outside, and shall be dispersed away from occupied areas and air intakes. The proper direction of the airflow shall be verified (i.e., into the work area). [§1910.1030(e)(4)(vi)]

(5) *Training Requirements.* Additional training requirements for employees in HIV and HBV research laboratories and HIV and HBV production facilities are specified in paragraph (g)(2)(ix). [§1910.1030(e)(5)]

(f) Hepatitis B vaccination and post-exposure evaluation and follow-up.[§1910.1030(f)]

(1) *General.*[§1910.1030(f)(1)]

(i) *The employer shall make available* the hepatitis B vaccine and vaccination series to all employees who have occupational exposure, and post-exposure evaluation and follow-up to all employees who have had an exposure incident. [§1910.1030(f)(1)(i)]

(ii) *The employer shall ensure that all medical evaluations* and procedures including the hepatitis B vaccine and vaccination series and post-exposure evaluation and follow-up, including prophylaxis, are:[§1910.1030(f)(1)(ii)]

[A] Made available at no cost to the employee; [§1910.1030(f)(1)(ii)[A]]

[B] Made available to the employee at a reasonable time and place;[§1910.1030(f)(1)(ii)[B]]

[C] Performed by or under the supervision of a licensed physician or by or under the supervision of another licensed healthcare professional; and[§1910.1030(f)(1)(ii)[C]]

[D] Provided according to recommendations of the U.S. Public Health Service current at the time these evaluations and procedures take place, except as specified by this paragraph (f).[§1910.1030(f)(1)(ii)[D]]

(iii) *The employer shall ensure that all laboratory tests* are conducted by an accredited laboratory at no cost to the employee.[§1910.1030(f)(1)(iii)]

(2) *Hepatitis B Vaccination.*[§1910.1030(f)(2)]

(i) *Hepatitis B vaccination* shall be made available after the employee has received the training required in paragraph (g)(2)(vii)(I) and within 10 working days of initial assignment to all employees who have occupational exposure unless the employee has previously received the complete hepatitis B vaccination series, antibody testing has revealed that the employee is immune, or the vaccine is contraindicated for medical reasons.[§1910.1030(f)(2)(i)]

(ii) *The employer shall not make participation* in a prescreening program a prerequisite for receiving hepatitis B vaccination. [§1910.1030(f)(2)(ii)]

(iii) *If the employee initially declines hepatitis B vaccination* but at a later date while still covered under the standard decides to accept the vaccination, the employer shall make available hepatitis B vaccination at that time.[§1910.1030(f)(2)(iii)]

(iv) *The employer shall assure that employees* who decline to accept hepatitis B vaccination offered by the employer sign the statement in appendix A.[§1910.1030(f)(2)(iv)]

(v) *If a routine booster dose(s) of hepatitis B vaccine* is recommended by the U.S. Public Health Service at a future date, such booster dose(s) shall be made available in accordance with section (f)(1)(ii).[§1910.1030(f)(2)(v)]

(3) *Post-exposure Evaluation and Follow-up.* Following a report of an exposure incident, the employer shall make immediately available to the exposed employee a confidential medical evaluation and follow-up, including at least the following elements:[§1910.1030(f)(3)]

(i) *Documentation of the route(s) of exposure,* and the circumstances under which the exposure incident occurred; [§1910.1030(f)(3)(i)]

(ii) *Identification and documentation of the source individual,* unless the employer can establish that identification is infeasible or prohibited by state or local law;[§1910.1030(f)(3)(ii)]

[A] The source individual's blood shall be tested as soon as feasible and after consent is obtained in order to determine HBV and HIV infectivity. If consent is not obtained, the employer shall establish that legally required consent cannot be obtained. When the source individual's consent is not required by law, the source individual's blood, if available, shall be tested and the results documented. [§1910.1030(f)(3)(ii)[A]]

[B] When the source individual is already known to be infected with HBV or HIV, testing for the source individual's known HBV or HIV status need not be repeated. [§1910.1030(f)(3)(ii)[B]]

[C] Results of the source individual's testing shall be made available to the exposed employee, and the employee shall be informed of applicable laws and regulations concerning disclosure of the identity and infectious status of the source individual.[§1910.1030(f)(3)(ii)[C]]

(iii) *Collection and testing of blood* for HBV and HIV serological status;[§1910.1030(f)(3)(iii)]

[A] The exposed employee's blood shall be collected as soon as feasible and tested after consent is obtained. [§1910.1030(f)(3)(iii)[A]]

[B] If the employee consents to baseline blood collection, but does not give consent at that time for HIV serologic testing, the sample shall be preserved for at least 90 days. If, within 90 days of the exposure incident, the employee elects to have the baseline sample tested, such testing shall be done as soon as feasible.[§1910.1030(f)(3)(iii)[B]]

(iv) *Post-exposure prophylaxis,* when medically indicated, as recommended by the U.S. Public Health Service; [§1910.1030(f)(3)(iv)]

(v) *Counseling; and*[§1910.1030(f)(3)(v)]

(vi) *Evaluation of reported illnesses.*[§1910.1030(f)(3)(vi)]

(4) *Information Provided to the Healthcare Professional.* [§1910.1030(f)(4)]

(i) *The employer shall ensure* that the healthcare professional responsible for the employee's Hepatitis B vaccination is provided a copy of this regulation.[§1910.1030(f)(4)(i)]

(ii) *The employer shall ensure* that the healthcare professional evaluating an employee after an exposure incident is provided the following information:[§1910.1030(f)(4)(ii)]

[A] A copy of this regulation;[§1910.1030(f)(4)(ii)[A]]

[B] A description of the exposed employee's duties as they relate to the exposure incident;[§1910.1030(f)(4)(ii)[B]]

[C] Documentation of the route(s) of exposure and circumstances under which exposure occurred; [§1910.1030(f)(4)(ii)[C]]

[D] Results of the source individual's blood testing, if available; and[§1910.1030(f)(4)(ii)[D]]

[E] All medical records relevant to the appropriate treatment of the employee including vaccination status which are the employer's responsibility to maintain. [§1910.1030(f)(4)(ii)[E]]

(5) *Healthcare Professional's Written Opinion.* The employer shall obtain and provide the employee with a copy of the evaluating

healthcare professional's written opinion within 15 days of the completion of the evaluation.[§1910.1030(f)(5)]

(i) *The healthcare professional's written opinion* for Hepatitis B vaccination shall be limited to whether Hepatitis B vaccination is indicated for an employee, and if the employee has received such vaccination.[§1910.1030(f)(5)(i)]

(ii) *The healthcare professional's written opinion* for post-exposure evaluation and follow-up shall be limited to the following information:[§1910.1030(f)(5)(ii)]

[A] That the employee has been informed of the results of the evaluation; and[§1910.1030(f)(5)(ii)[A]]

[B] That the employee has been told about any medical conditions resulting from exposure to blood or other potentially infectious materials which require further evaluation or treatment. [§1910.1030(f)(5)(ii)[B]]

(iii) *All other findings or diagnoses* shall remain confidential and shall not be included in the written report[§1910.1030(f)(5)(iii)]

(6) *Medical recordkeeping.* Medical records required by this standard shall be maintained in accordance with paragraph (h)(1) of this section.[§1910.1030(f)(6)]

(g) Communication of hazards to employees.[§1910.1030(g)]

(1) *Labels and signs*[§1910.1030(g)(1)]

(i) *Labels.*[§1910.1030(g)(1)(i)]

[A] Warning labels shall be affixed to containers of regulated waste, refrigerators and freezers containing blood or other potentially infectious material; and other containers used to store, transport or ship blood or other potentially infectious materials, except as provided in paragraph (g)(1)(i)(E), (F) and (G).[§1910.1030(g)(1)(i)[A]]

[B] Labels required by this section shall include the following legend:[§1910.1030(g)(1)(i)[B]]

[C] These labels shall be fluorescent orange or orange-red or predominantly so, with lettering and symbols in a contrasting color.[§1910.1030(g)(1)(i)[C]]

[D] Labels shall be affixed as close as feasible to the container by string, wire, adhesive, or other method that prevents their loss or unintentional removal. [§1910.1030(g)(1)(i)[D]]

[E] Red bags or red containers may be substituted for labels. [§1910.1030(g)(1)(i)[E]]

[F] Containers of blood, blood components, or blood products that are labeled as to their contents and have been released for transfusion or other clinical use are exempted from the labeling requirements of paragraph (g). [§1910.1030(g)(1)(i)[F]]

[G] Individual containers of blood or other potentially infectious materials that are placed in a labeled container during storage, transport, shipment or disposal are exempted from the labeling requirement.[§1910.1030(g)(1)(i)[G]]

[H] Labels required for contaminated equipment shall be in accordance with this paragraph and shall also state which portions of the equipment remain contaminated. [§1910.1030(g)(1)(i)[H]]

[I] Regulated waste that has been decontaminated need not be labeled or color-coded.[§1910.1030(g)(1)(i)[I]]

(ii) *Signs.*[§1910.1030(g)(1)(ii)]

[A] The employer shall post signs at the entrance to work areas specified in paragraph (e), HIV and HBV Research Laboratory and Production Facilities, which shall bear the following legend:[§1910.1030(g)(1)(ii)[A]]

(Name of the Infectious Agent)
(Special requirements for entering the area)
(Name, telephone number of the laboratory director or other responsible person.)

[B] These signs shall be fluorescent orange-red or predominantly so, with lettering and symbols in a contrasting color. [§1910.1030(g)(1)(ii)[B]]

(2) *Information and Training.*[§1910.1030(g)(2)]

(i) *The employer shall train* each employee with occupational exposure in accordance with the requirements of this section. Such training must be provided at no cost to the employee and during working hours. The employer shall institute a training program and ensure employee participation in the program.[§1910.1030(g)(2)(i)]

(ii) *Training shall be provided as follows:*[§1910.1030(g)(2)(ii)]

[A] At the time of initial assignment to tasks where occupational exposure may take place;[§1910.1030(g)(2)(ii)[A]]

[B] At least annually thereafter.[§1910.1030(g)(2)(ii)[B]]

(iii) *[Reserved]*[§1910.1030(g)(2)(iii)]

(iv) *Annual training for all employees* shall be provided within one year of their previous training.[§1910.1030(g)(2)(iv)]

(v) *Employers shall provide additional training* when changes such as modification of tasks or procedures or institution of new tasks or procedures affect the employee's occupational exposure. The additional training may be limited to addressing the new exposures created.[§1910.1030(g)(2)(v)]

(vi) *Material appropriate in content* and vocabulary to educational level, literacy, and language of employees shall be used.[§1910.1030(g)(2)(vi)]

(vii) *The training program shall contain* at a minimum the following elements:[§1910.1030(g)(2)(vii)]

[A] An accessible copy of the regulatory text of this standard and an explanation of its contents;[§1910.1030(g)(2)(vii)[A]]

[B] A general explanation of the epidemiology and symptoms of bloodborne diseases;[§1910.1030(g)(2)(vii)[B]]

[C] An explanation of the modes of transmission of bloodborne pathogens;[§1910.1030(g)(2)(vii)[C]]

[D] An explanation of the employer's exposure control plan and the means by which the employee can obtain a copy of the written plan;[§1910.1030(g)(2)(vii)[D]]

[E] An explanation of the appropriate methods for recognizing tasks and other activities that may involve exposure to blood and other potentially infectious materials; [§1910.1030(g)(2)(vii)[E]]

[F] An explanation of the use and limitations of methods that will prevent or reduce exposure including appropriate engineering controls, work practices, and personal protective equipment;[§1910.1030(g)(2)(vii)[F]]

[G] Information on the types, proper use, location, removal, handling, decontamination and disposal of personal protective equipment;[§1910.1030(g)(2)(vii)[G]]

[H] An explanation of the basis for selection of personal protective equipment;[§1910.1030(g)(2)(vii)[H]]

[I] Information on the hepatitis B vaccine, including information on its efficacy, safety, method of administration, the benefits of being vaccinated, and that the vaccine and vaccination will be offered free of charge; [§1910.1030(g)(2)(vii)[I]]

[J] Information on the appropriate actions to take and persons to contact in an emergency involving blood or other potentially infectious materials;[§1910.1030(g)(2)(vii)[J]]

[K] An explanation of the procedure to follow if an exposure incident occurs, including the method of reporting the incident and the medical follow-up that will be made available;[§1910.1030(g)(2)(vii)[K]]

[L] *Information on the post-exposure evaluation* and follow-up that the employer is required to provide for the employee following an exposure incident; [§1910.1030(g)(2)(vii)[L]]

[M] *An explanation of the signs and labels* and/or color coding required by paragraph (g)(1); and[§1910.1030(g)(2)(vii)[M]]

[N] *An opportunity for interactive questions* and answers with the person conducting the training session. [§1910.1030(g)(2)(vii)[N]]

(viii) *The person conducting the training* shall be knowledgeable in the subject matter covered by the elements contained in the training program as it relates to the workplace that the training will address.[§1910.1030(g)(2)(viii)]

(ix) *Additional Initial Training for Employees* in HIV and HBV Laboratories and Production Facilities. Employees in HIV or HBV research laboratories and HIV or HBV production facilities shall receive the following initial training in addition to the above training requirements.[§1910.1030(g)(2)(ix)]

[A] *The employer shall assure that employees demonstrate proficiency* in standard microbiological practices and techniques and in the practices and operations specific to the facility before being allowed to work with HIV or HBV. [§1910.1030(g)(2)(ix)[A]]

[B] *The employer shall assure that employees* have prior experience in the handling of human pathogens or tissue cultures before working with HIV or HBV. [§1910.1030(g)(2)(ix)[B]]

[C] *The employer shall provide a training program* to employees who have no prior experience in handling human pathogens. Initial work activities shall not include the handling of infectious agents. A progression of work activities shall be assigned as techniques are learned and proficiency is developed. The employer shall assure that employees participate in work activities involving infectious agents only after proficiency has been demonstrated.[§1910.1030(g)(2)(ix)[C]]

(h) Recordkeeping.[§1910.1030(h)]

(1) *Medical Records.*[§1910.1030(h)(1)]

(i) *The employer shall establish and maintain* an accurate record for each employee with occupational exposure, in accordance with 29 CFR 1910.1020.[§1910.1030(h)(1)(i)]

(ii) *This record shall include:*[§1910.1030(h)(1)(ii)]

[A] *The name and social security number of the employee;* [§1910.1030(h)(1)(ii)[A]]

[B] *A copy of the employee's hepatitis B vaccination status* including the dates of all the hepatitis B vaccinations and any medical records relative to the employee's ability to receive vaccination as required by paragraph (f)(2); [§1910.1030(h)(1)(ii)[B]]

[C] *A copy of all results of examinations,* medical testing, and follow-up procedures as required by paragraph (f)(3); [§1910.1030(h)(1)(ii)[C]]

[D] *The employer's copy* of the healthcare professional's written opinion as required by paragraph (f)(5); and [§1910.1030(h)(1)(ii)[D]]

[E] *A copy of the information provided* to the healthcare professional as required by paragraphs (f)(4)(ii)(B)(C) and (D).[§1910.1030(h)(1)(ii)[E]]

(iii) *Confidentiality.* The employer shall ensure that employee medical records required by paragraph (h)(1) are: [§1910.1030(h)(1)(iii)]

[A] *Kept confidential; and*[§1910.1030(h)(1)(iii)[A]]

[B] *Not disclosed or reported* without the employee's express written consent to any person within or outside the workplace except as required by this section or as may be required by law.[§1910.1030(h)(1)(iii)[B]]

(iv) *The employer shall maintain the records* required by paragraph (h) for at least the duration of employment plus 30 years in accordance with 29 CFR 1910.1020. [§1910.1030(h)(1)(iv)]

(2) *Training Records.*[§1910.1030(h)(2)]

(i) *Training records shall include the following information:* [§1910.1030(h)(2)(i)]

[A] *The dates of the training sessions;*[§1910.1030(h)(2)(i)[A]]

[B] *The contents or a summary of the training sessions;* [§1910.1030(h)(2)(i)[B]]

[C] *The names and qualifications of persons conducting the training; and*[§1910.1030(h)(2)(i)[C]]

[D] *The names and job titles* of all persons attending the training sessions.[§1910.1030(h)(2)(i)[D]]

(ii) *Training records shall be maintained* for 3 years from the date on which the training occurred.[§1910.1030(h)(2)(ii)]

(3) *Availability.*[§1910.1030(h)(3)]

(i) *The employer shall ensure that all records* required to be maintained by this section shall be made available upon request to the Assistant Secretary and the Director for examination and copying.[§1910.1030(h)(3)(i)]

(ii) *Employee training records* required by this paragraph shall be provided upon request for examination and copying to employees, to employee representatives, to the Director, and to the Assistant Secretary.[§1910.1030(h)(3)(ii)]

(iii) *Employee medical records* required by this paragraph shall be provided upon request for examination and copying to the subject employee, to anyone having written consent of the subject employee, to the Director, and to the Assistant Secretary in accordance with 29 CFR 1910.1020.[§1910.1030(h)(3)(iii)]

(4) *Transfer of Records.* The employer shall comply with the requirements involving transfer of records set forth in 29 CFR 1910.1020(h). [§1910.1030(h)(4)]

(5) *Sharps injury log.*[§1910.1030(h)(5)]

(i) *The employer shall establish and maintain* a sharps injury log for the recording of percutaneous injuries from contaminated sharps. The information in the sharps injury log shall be recorded and maintained in such manner as to protect the confidentiality of the injured employee. The sharps injury log shall contain, at a minimum:[§1910.1030(h)(5)(i)]

[A] *The type and brand of device* involved in the incident, [§1910.1030(h)(5)(i)[A]]

[B] *The department or work area* where the exposure incident occurred, and[§1910.1030(h)(5)(i)[B]]

[C] *An explanation of how the incident occurred.* [§1910.1030(h)(5)(i)[C]]

(ii) *The requirement to establish and maintain* a sharps injury log shall apply to any employer who is required to maintain a log of occupational injuries and illnesses under 29 CFR part 1904.[§1910.1030(h)(5)(ii)]

(iii) *The sharps injury log* shall be maintained for the period required by 29 CFR 1904.33.[§1910.1030(h)(5)(iii)]

(i) Dates[§1910.1030(i)]

(1) *Effective Date.* The standard shall become effective on March 6, 1992.[§1910.1030(i)(1)]

(2) *The Exposure Control Plan* required by paragraph (c) of this section shall be completed on or before May 5, 1992. [§1910.1030(i)(2)]

(3) *Paragraphs (g)(2) Information* and Training and (h) Recordkeeping of this section shall take effect on or before June 4, 1992.[§1910.1030(i)(3)]

(4) *Paragraphs (d)(2) Engineering* and Work Practice Controls, (d)(3) Personal Protective Equipment, (d)(4) Housekeeping, (e) HIV and HBV Research Laboratories and Production Facilities, (f) Hepatitis B Vaccination and Post-Exposure Evaluation and Follow-up, and (g)(1) Labels and Signs of this section, shall take effect July 6, 1992. [§1910.1030(i)(4)]

§1910.1030 Appendix A

Hepatitis B Vaccine Declination (Mandatory)I understand that due to my occupational exposure to blood or other potentially infectious materials I may be at risk of acquiring hepatitis B virus (HBV) infection. I have been given the opportunity to be vaccinated with hepatitis B vaccine, at no charge to myself. However, I decline hepatitis B vaccination at this time. I understand that by declining this vaccine, I continue to be at risk of acquiring hepatitis B, a serious disease. If in the future I continue to have occupational exposure to blood or other potentially infectious materials and I want to be vaccinated with hepatitis B vaccine, I can receive the vaccination series at no charge to me.

[56 FR 64175, Dec. 6, 1991, as amended at 57 FR 12717, Apr. 13, 1992; 57 FR 29206, July 1, 1992; 61 FR 5508, Feb. 13, 1996; 66 FR 5325, Jan. 18, 2001; 71 FR 16672, 16673, Apr. 3, 2006; 73 FR 75586, Dec. 12, 2008; 76 FR 33608, June 8, 2011; 76 FR 80740, Dec. 27, 2011; 77 FR 19934, Apr. 3, 2012]

§1910.1048
Formaldehyde

(a) **Scope and application.** This standard applies to all occupational exposures to formaldehyde, i.e. from formaldehyde gas, its solutions, and materials that release formaldehyde.[§1910.1048(a)]

(b) **Definitions.** For purposes of this standard, the following definitions shall apply:

Action level means a concentration of 0.5 part formaldehyde per million parts of air (0.5 ppm) calculated as an eight (8)-hour time-weighted average (TWA) concentration.

Assistant Secretary means the Assistant Secretary of Labor for the Occupational Safety and Health Administration, U.S. Department of Labor, or designee.

Authorized person means any person required by work duties to be present in regulated areas, or authorized to do so by the employer, by this section, or by the OSH Act of 1970.

Director means the Director of the National Institute for Occupational Safety and Health, U.S. Department of Health and Human Services, or designee.

Emergency is any occurrence, such as but not limited to equipment failure, rupture of containers, or failure of control equipment that results in an uncontrolled release of a significant amount of formaldehyde.

Employee exposure means the exposure to airborne formaldehyde which would occur without corrections for protection provided by any respirator that is in use.

Formaldehyde means the chemical substance, HCHO, Chemical Abstracts Service Registry No. 50-00-0.

(c) **Permissible Exposure Limit (PEL).**[§1910.1048(c)]

(1) *TWA:* The employer shall assure that no employee is exposed to an airborne concentration of formaldehyde which exceeds 0.75 parts formaldehyde per million parts of air (0.75 ppm) as an 8-hour TWA.[§1910.1048(c)(1)]

(2) *Short Term Exposure Limit (STEL):* The employer shall assure that no employee is exposed to an airborne concentration of formaldehyde which exceeds two parts formaldehyde per million parts of air (2 ppm) as a 15-minute STEL.[§1910.1048(c)(2)]

(d) **Exposure monitoring.**[§1910.1048(d)]

(1) *General.*[§1910.1048(d)(1)]

(i) *Each employer who has a workplace* covered by this standard shall monitor employees to determine their exposure to formaldehyde.[§1910.1048(d)(1)(i)]

(ii) *Exception.* Where the employer documents, using objective data, that the presence of formaldehyde or formaldehyde-releasing products in the workplace cannot result in airborne concentrations of formaldehyde that would cause any employee to be exposed at or above the action level or the STEL under foreseeable conditions of use, the employer will not be required to measure employee exposure to formaldehyde.[§1910.1048(d)(1)(ii)]

(iii) *When an employee's exposure is determined* from representative sampling, the measurements used shall be representative of the employee's full shift or short-term exposure to formaldehyde, as appropriate.[§1910.1048(d)(1)(iii)]

(iv) *Representative samples for each job classification* in each work area shall be taken for each shift unless the employer can document with objective data that exposure levels for a given job classification are equivalent for different work shifts.[§1910.1048(d)(1)(iv)]

(2) *Initial monitoring.* The employer shall identify all employees who may be exposed at or above the action level or at or above the STEL and accurately determine the exposure of each employee so identified.[§1910.1048(d)(2)]

(i) *Unless the employer chooses* to measure the exposure of each employee potentially exposed to formaldehyde, the employer shall develop a representative sampling strategy and measure sufficient exposures within each job classification for each workshift to correctly characterize and not underestimate the exposure of any employee within each exposure group.[§1910.1048(d)(2)(i)]

(ii) *The initial monitoring process* shall be repeated each time there is a change in production, equipment, process, personnel, or control measures which may result in new or additional exposure to formaldehyde.[§1910.1048(d)(2)(ii)]

(iii) *If the employer receives reports* of signs or symptoms of respiratory or dermal conditions associated with formaldehyde exposure, the employer shall promptly monitor the affected employee's exposure.[§1910.1048(d)(2)(iii)]

(3) *Periodic monitoring.*[§1910.1048(d)(3)]

(i) *The employer shall periodically measure* and accurately determine exposure to formaldehyde for employees shown by the initial monitoring to be exposed at or above the action level or at or above the STEL.[§1910.1048(d)(3)(i)]

(ii) *If the last monitoring results* reveal employee exposure at or above the action level, the employer shall repeat monitoring of the employees at least every 6 months.[§1910.1048(d)(3)(ii)]

(iii) *If the last monitoring* results reveal employee exposure at or above the STEL, the employer shall repeat monitoring of the employees at least once a year under worst conditions.[§1910.1048(d)(3)(iii)]

(4) *Termination of monitoring.* The employer may discontinue periodic monitoring for employees if results from two consecutive sampling periods taken at least 7 days apart show that employee exposure is below the action level and the STEL. The results must be statistically representative and consistent with the employer's knowledge of the job and work operation.[§1910.1048(d)(4)]

(5) *Accuracy of monitoring.* Monitoring shall be accurate, at the 95 percent confidence level, to within plus or minus 25 percent for airborne concentrations of formaldehyde at the TWA and the STEL and to within plus or minus 35 percent for airborne concentrations of formaldehyde at the action level.[§1910.1048(d)(5)]

(6) *Employee notification* of monitoring results. The employer must, within 15 working days after the receipt of the results of any monitoring performed under this section, notify each affected employee of these results either individually in writing or by posting the results in an appropriate location that is accessible to employees. If employee exposure is above the PEL, affected employees shall be provided with a description of the corrective actions being taken by the employer to decrease exposure.[§1910.1048(d)(6)]

(7) *Observation of monitoring.*[§1910.1048(d)(7)]

(i) *The employer shall provide affected employees* or their designated representatives an opportunity to observe any monitoring of employee exposure to formaldehyde required by this standard.[§1910.1048(d)(7)(i)]

(ii) *When observation of the monitoring of employee* exposure to formaldehyde requires entry into an area where the use of protective clothing or equipment is required, the employer shall provide the clothing and equipment to the observer, require the observer to use such clothing and equipment, and assure that the observer complies with all other applicable safety and health procedures.[§1910.1048(d)(7)(ii)]

(e) **Regulated areas.**[§1910.1048(e)]

(1) *Signs.*[§1910.1048(e)(1)]

(i) *The employer shall establish regulated* areas where the concentration of airborne formaldehyde exceeds either the TWA or the STEL and post all entrances and access ways with signs bearing the following legend: DANGER FORMALDEHYDE MAY CAUSE CANCER CAUSES SKIN, EYE, AND RESPIRATORY IRRITATION AUTHORIZED PERSONNEL ONLY[§1910.1048(e)(1)(i)]

(ii) *Prior to June 1,* 2016, employers may use the following legend in lieu of that specified in paragraph (e)(1)(i) of this section: DANGER FORMALDEHYDE IRRITANT AND POTENTIAL CANCER HAZARD AUTHORIZED PERSONNEL ONLY[§1910.1048(e)(1)(ii)]

(2) *The employer shall limit access* to regulated areas to authorized persons who have been trained to recognize the hazards of formaldehyde.[§1910.1048(e)(2)]

(3) *An employer at a multiemployer worksite* who establishes a regulated area shall communicate the access restrictions and locations of these areas to other employers with work operations at that worksite.[§1910.1048(e)(3)]

(f) **Methods of compliance.**[§1910.1048(f)]

(1) *Engineering controls and work practices.* The employer shall institute engineering and work practice controls to reduce and maintain employee exposures to formaldehyde at or below the TWA and the STEL.[§1910.1048(f)(1)]

(2) *Exception.* Whenever the employer has established that feasible engineering and work practice controls cannot reduce employee exposure to or below either of the PELs, the

employer shall apply these controls to reduce employee exposures to the extent feasible and shall supplement them with respirators which satisfy this standard.[§1910.1048(f)(2)]

(g) Respiratory protection.[§1910.1048(g)]

(1) *General.* For employees who use respirators required by this section, the employer must provide each employee an appropriate respirator that complies with the requirements of this paragraph. Respirators must be used during:[§1910.1048(g)(1)]

(i) *Periods necessary to install* or implement feasible engineering and work-practice controls.[§1910.1048(g)(1)(i)]

(ii) *Work operations,* such as maintenance and repair activities or vessel cleaning, for which the employer establishes that engineering and work-practice controls are not feasible. [§1910.1048(g)(1)(ii)]

(iii) *Work operations for which* feasible engineering and work-practice controls are not yet sufficient to reduce employee exposure to or below the PELs.[§1910.1048(g)(1)(iii)]

(iv) *Emergencies.*[§1910.1048(g)(1)(iv)]

(2) *Respirator program.*[§1910.1048(g)(2)]

(i) *The employer must implement* a respiratory protection program in accordance with Sec. 1910.134(b) through (d) (except (d)(1)(iii), (d)(3)(iii)(b)(1), and (2)), and (f) through (m), which covers each employee required by this section to use a respirator.[§1910.1048(g)(2)(i)]

(ii) *When employees use air-purifying respirators* with chemical cartridges or canisters that do not contain end-of-service-life indicators approved by the National Institute for Occupational Safety and Health, employers must replace these cartridges or canisters as specified by paragraphs (d)(3)(iii)(B)(1) and (B)(2) of 29 CFR 1910.134, or at the end of the workshift, whichever condition occurs first.[§1910.1048(g)(2)(ii)]

(3) *Respirator selection.*[§1910.1048(g)(3)]

(i) *Employers must:*[§1910.1048(g)(3)(i)]

[A] Select, and provide to employees, the appropriate respirators specified in paragraph (d)(3)(i)(A) of 29 CFR 1910.134.[§1910.1048(g)(3)(i)[A]]

[B] Equip each air-purifying, full facepiece respirator with a canister or cartridge approved for protection against formaldehyde.[§1910.1048(g)(3)(i)[B]]

[C] For escape, provide employees with one of the following respirator options: A self-contained breathing apparatus operated in the demand or pressure-demand mode; or a full facepiece respirator having a chin-style, or a front-or back-mounted industrial-size, canister or cartridge approved for protection against formaldehyde. [§1910.1048(g)(3)(i)[C]]

(ii) *Employers may substitute an air-purifying, half mask respirator* for an air-purifying, full facepiece respirator when they equip the half mask respirator with a cartridge approved for protection against formaldehyde and provide the affected employee with effective gas-proof goggles.[§1910.1048(g)(3)(ii)]

(iii) *Employers must provide employees* who have difficulty using negative pressure respirators with powered air-purifying respirators permitted for use under paragraph (g)(3)(i)(A) of this standard and that affords adequate protection against formaldehyde exposures.[§1910.1048(g)(3)(iii)]

(h) Protective equipment and clothing. Employers shall comply with the provisions of 29 CFR 1910.132 and 29 CFR 1910.133. When protective equipment or clothing is provided under these provisions, the employer shall provide these protective devices at no cost to the employee and assure that the employee wears them.[§1910.1048(h)]

(1) *Selection.* The employer shall select protective clothing and equipment based upon the form of formaldehyde to be encountered, the conditions of use, and the hazard to be prevented.[§1910.1048(h)(1)]

(i) *All contact of the eyes and skin with liquids* containing 1 percent or more formaldehyde shall be prevented by the use of chemical protective clothing made of material impervious to formaldehyde and the use of other personal protective equipment, such as goggles and face shields, as appropriate to the operation.[§1910.1048(h)(1)(i)]

(ii) *Contact with irritating or sensitizing materials* shall be prevented to the extent necessary to eliminate the hazard. [§1910.1048(h)(1)(ii)]

(iii) *Where a face shield is worn,* chemical safety goggles are also required if there is a danger of formaldehyde reaching the area of the eye.[§1910.1048(h)(1)(iii)]

(iv) *Full body protection* shall be worn for entry into areas where concentrations exceed 100 ppm and for emergency reentry into areas of unknown concentration.[§1910.1048(h)(1)(iv)]

(2) *Maintenance of protective equipment and clothing.* [§1910.1048(h)(2)]

(i) *The employer shall assure that protective equipment* and clothing that has become contaminated with formaldehyde is cleaned or laundered before its reuse.[§1910.1048(h)(2)(i)]

(ii) *When formaldehyde-contaminated clothing and equipment* is ventilated, the employer shall establish storage areas so that employee exposure is minimized.[§1910.1048(h)(2)(ii)]

[A] Signs. Storage areas for contaminated clothing and equipment shall have signs bearing the following legend:

DANGER

FORMALDEHYDE-CONTAMINATED [CLOTHING] EQUIPMENT

MAY CAUSE CANCER

CAUSES SKIN, EYE AND RESPIRATORY IRRITATION

DO NOT BREATHE VAPOR

DO NOT GET ON SKIN

[B] Labels. The employer shall ensure containers for contaminated clothing and equipment are labeled consistent with the Hazard Communication Standard, Sec. 1910.1200, and shall, as a minimum, include the following:
DANGER
FORMALDEHYDE-CONTAMINATED [CLOTHING] EQUIPMENT
MAY CAUSE CANCER
CAUSES SKIN, EYE, AND RESPIRATORY IRRITATION
DO NOT BREATHE VAPOR DO NOT GET ON SKIN
[§1910.1048(h)(2)(ii)[B]]

[C] Prior to June 1, 2016, employers may use the following legend in lieu of that specified in paragraph (h)(2)(ii)(A) of this section: DANGER FORMALDEHYDE-CONTAMINATED [CLOTHING] EQUIPMENT AVOID INHALATION AND SKIN CONTACT[§1910.1048(h)(2)(ii)[C]]

[D] Prior to June 1, 2015, employers may include the following information on containers of protective clothing and equipment in lieu of the labeling requirements in paragraphs (h)(2)(ii)(B) of this section:
DANGER FORMALDEHYDE-CONTAMINATED [CLOTHING] EQUIPMENT
AVOID INHALATION AND SKIN CONTACT
[§1910.1048(h)(2)(ii)[D]]

(iii) *The employer shall assure* that only persons trained to recognize the hazards of formaldehyde remove the contaminated material from the storage area for purposes of cleaning, laundering, or disposal.[§1910.1048(h)(2)(iii)]

(iv) *The employer shall assure that no employee* takes home equipment or clothing that is contaminated with formaldehyde.[§1910.1048(h)(2)(iv)]

(v) *The employer shall repair or replace* all required protective clothing and equipment for each affected employee as necessary to assure its effectiveness.[§1910.1048(h)(2)(v)]

(vi) *The employer shall inform any person* who launders, cleans, or repairs such clothing or equipment of formaldehyde's potentially harmful effects and of procedures to safely handle the clothing and equipment.[§1910.1048(h)(2)(vi)]

(i) Hygiene protection.[§1910.1048(i)]

(1) *The employer shall provide change rooms,* as described in 29 CFR 1910.141 for employees who are required to change from work clothing into protective clothing to prevent skin contact with formaldehyde.[§1910.1048(i)(1)]

(2) *If employees' skin* may become spashed with solutions containing 1 percent or greater formaldehyde, for example, because of equipment failure or improper work practices, the employer shall provide conveniently located quick drench showers and assure that affected employees use these facilities immediately.[§1910.1048(i)(2)]

(3) *If there is any possibility* that an employee's eyes may be splashed with solutions containing 0.1 percent or greater form-

aldehyde, the employer shall provide acceptable eyewash facilities within the immediate work area for emergency use. [§1910.1048(i)(3)]

(j) **Housekeeping.** For operations involving formaldehyde liquids or gas, the employer shall conduct a program to detect leaks and spills, including regular visual inspections.[§1910.1048(j)]

(1) *Preventative maintenance of equipment,* including surveys for leaks, shall be undertaken at regular intervals.[§1910.1048(j)(1)]

(2) *In work areas where spillage may occur,* the employer shall make provisions to contain the spill, to decontaminate the work area, and to dispose of the waste.[§1910.1048(j)(2)]

(3) *The employer shall assure that all leaks* are repaired and spills are cleaned promptly by employees wearing suitable protective equipment and trained in proper methods for cleanup and decontamination.[§1910.1048(j)(3)]

(4) *Formaldehyde-contaminated waste and debris* resulting from leaks or spills shall be placed for disposal in sealed containers bearing a label warning of formaldehyde's presence and of the hazards associated with formaldehyde. The employer shall ensure that the labels are in accordance with paragraph (m) of this section.[§1910.1048(j)(4)]

(k) **Emergencies.** For each workplace where there is the possibility of an emergency involving formaldehyde, the employer shall assure appropriate procedures are adopted to minimize injury and loss of life. Appropriate procedures shall be implemented in the event of an emergency.[§1910.1048(k)]

(l) **Medical surveillance.**[§1910.1048(l)]

(1) *Employees covered.*[§1910.1048(l)(1)]

(i) *The employer shall institute* medical surveillance programs for all employees exposed to formaldehyde at concentrations at or exceeding the action level or exceeding the STEL. [§1910.1048(l)(1)(i)]

(ii) *The employer shall make medical surveillance available* for employees who develop signs and symptoms of overexposure to formaldehyde and for all employees exposed to formaldehyde in emergencies. When determining whether an employee may be experiencing signs and symptoms of possible overexposure to formaldehyde, the employer may rely on the evidence that signs and symptoms associated with formaldehyde exposure will occur only in exceptional circumstances when airborne exposure is less than 0.1 ppm and when formaldehyde is present in material in concentrations less than 0.1 percent.[§1910.1048(l)(1)(ii)]

(2) *Examination by a physician.* All medical procedures, including administration of medical disease questionnaires, shall be performed by or under the supervision of a licensed physician and shall be provided without cost to the employee, without loss of pay, and at a reasonable time and place.[§1910.1048(l)(2)]

(3) *Medical disease questionnaire.* The employer shall make the following medical surveillance available to employees prior to assignment to a job where formaldehyde exposure is at or above the action level or above the STEL and annually thereafter. The employer shall also make the following medical surveillance available promptly upon determining that an employee is experiencing signs and symptoms indicative of possible overexposure to formaldehyde.[§1910.1048(l)(3)]

(i) *Administration of a medical disease questionnaire,* such as in appendix D, which is designed to elicit information on work history, smoking history, any evidence of eye, nose, or throat irritation; chronic airway problems or hyperreactive airway disease: allergic skin conditions or dermatitis; and upper or lower respiratory problems.[§1910.1048(l)(3)(i)]

(ii) *A determination by the physician,* based on evaluation of the medical disease questionnaire, of whether a medical examination is necessary for employees not required to wear respirators to reduce exposure to formaldehyde.[§1910.1048(l)(3)(ii)]

(4) *Medical examinations.* Medical examinations shall be given to any employee who the physician feels, based on information in the medical disease questionnaire, may be at increased risk from exposure to formaldehyde and at the time of initial assignment and at least annually thereafter to all employees required to wear a respirator to reduce exposure to formaldehyde. The medical examination shall include:[§1910.1048(l)(4)]

(i) *A physical examination with emphasis* on evidence of irritation or sensitization of the skin and respiratory system, shortness of breath, or irritation of the eyes.[§1910.1048(l)(4)(i)]

(ii) *Laboratory examinations for respirator wearers* consisting of baseline and annual pulmonary function tests. As a minimum, these tests shall consist of forced vital capacity (FVC), forced expiratory volume in one second (FEV_1), and forced expiratory flow (FEF).[§1910.1048(l)(4)(ii)]

(iii) *Any other test* which the examining physician deems necessary to complete the written opinion.[§1910.1048(l)(4)(iii)]

(iv) *Counseling of employees* having medical conditions that would be directly or indirectly aggravated by exposure to formaldehyde on the increased risk of impairment of their health.[§1910.1048(l)(4)(iv)]

(5) *Examinations for employees exposed in an emergency.* The employer shall make medical examinations available as soon as possible to all employees who have been exposed to formaldehyde in an emergency.[§1910.1048(l)(5)]

(i) *The examination shall include* a medical and work history with emphasis on any evidence of upper or lower respiratory problems, allergic conditions, skin reaction or hypersensitivity, and any evidence of eye, nose, or throat irritation. [§1910.1048(l)(5)(i)]

(ii) *Other examinations shall consist* of those elements considered appropriate by the examining physician.[§1910.1048(l)(5)(ii)]

(6) *Information provided to the physician.* The employer shall provide the following information to the examining physician: [§1910.1048(l)(6)]

(i) *A copy of this standard and appendix A, C, D, and E*[1]*;* [§1910.1048(l)(6)(i)]

(ii) *A description of the affected employee's job duties* as they relate to the employee's exposure to formaldehyde; [§1910.1048(l)(6)(ii)]

(iii) *The representative exposure level* for the employee's job assignment;[§1910.1048(l)(6)(iii)]

(iv) *Information concerning any personal protective equipment* and respiratory protection used or to be used by the employee; and[§1910.1048(l)(6)(iv)]

(v) *Information from previous medical examinations* of the affected employee within the control of the employer. [§1910.1048(l)(6)(v)]

(vi) *In the event of a nonroutine examination* because of an emergency, the employer shall provide to the physician as soon as possible: A description of how the emergency occurred and the exposure the victim may have received. [§1910.1048(l)(6)(vi)]

[1]**Editor's note: there is not an Appendix E for this document.**

(7) *Physician's written opinion.*[§1910.1048(l)(7)]

(i) *For each examination required under this standard,* the employer shall obtain a written opinion from the examining physician. This written opinion shall contain the results of the medical examination except that it shall not reveal specific findings or diagnoses unrelated to occupational exposure to formaldehyde. The written opinion shall include: [§1910.1048(l)(7)(i)]

[A] The physician's opinion as to whether the employee has any medical condition that would place the employee at an increased risk of material impairment of health from exposure to formaldehyde;[§1910.1048(l)(7)(i)[A]]

[B] Any recommended limitations on the employee's exposure or changes in the use of personal protective equipment, including respirators;[§1910.1048(l)(7)(i)[B]]

[C] A statement that the employee has been informed by the physician of any medical conditions which would be aggravated by exposure to formaldehyde, whether these conditions may have resulted from past formaldehyde exposure or from exposure in an emergency, and whether there is a need for further examination or treatment. [§1910.1048(l)(7)(i)[C]]

(ii) *The employer shall provide for retention* of the results of the medical examination and tests conducted by the physician. [§1910.1048(l)(7)(ii)]

(iii) *The employer shall provide a copy* of the physician's written opinion to the affected employee within 15 days of its receipt. [§1910.1048(l)(7)(iii)]

(8) *Medical removal.*[§1910.1048(l)(8)]

(i) *The provisions of paragraph (l)(8)* apply when an employee reports significant irritation of the mucosa of the eyes or the upper airways, respiratory sensitization, dermal irritation, or dermal sensitization attributed to workplace formaldehyde exposure. Medical removal provisions do not apply in the case of dermal irritation or dermal sensitization when the product suspected of causing the dermal condition contains less than 0.05% formaldehyde.[§1910.1048(l)(8)(i)]

(ii) *An employee's report of signs or symptoms* of possible over-exposure to formaldehyde shall be evaluated by a physician selected by the employer pursuant to paragraph (l)(3). If the physician determines that a medical examination is not necessary under paragraph (l)(3)(ii), there shall be a two-week evaluation and remediation period to permit the employer to ascertain whether the signs or symptoms subside untreated or with the use of creams, gloves, first aid treatment or personal protective equipment. Industrial hygiene measures that limit the employee's exposure to formaldehyde may also be implemented during this period. The employee shall be referred immediately to a physician prior to expiration of the two-week period if the signs or symptoms worsen. Earnings, seniority and benefits may not be altered during the two-week period by virtue of the report.[§1910.1048(l)(8)(ii)]

(iii) *If the signs or symptoms have not subsided* or been remedied by the end of the two-week period, or earlier if signs or symptoms warrant, the employee shall be examined by a physician selected by the employer. The physician shall presume, absent contrary evidence, that observed dermal irritation or dermal sensitization are not attributable to formaldehyde when products to which the affected employee is exposed contain less than 0.1% formaldehyde. [§1910.1048(l)(8)(iii)]

(iv) *Medical examinations shall be conducted* in compliance with the requirements of paragraph (l)(5)(i) and (ii). Additional guidelines for conducting medical exams are contained in appendix C.[§1910.1048(l)(8)(iv)]

(v) *If the physician finds* that significant irritation of the mucosa of the eyes or of the upper airways, respiratory sensitization, dermal irritation, or dermal sensitization result from workplace formaldehyde exposure and recommends restrictions or removal, the employer shall promptly comply with the restrictions or recommendation of removal. In the event of a recommendation of removal, the employer shall remove the effected employee from the current formaldehyde exposure and if possible, transfer the employee to work having no or significantly less exposure to formaldehyde.[§1910.1048(l)(8)(v)]

(vi) *When an employee is removed* pursuant to paragraph (l)(8)(v), the employer shall transfer the employee to comparable work for which the employee is qualified or can be trained in a short period (up to 6 months), where the formaldehyde exposures are as low as possible, but not higher than the action level. The employer shall maintain the employee's current earnings, seniority, and other benefits. If there is no such work available, the employer shall maintain the employee's current earnings, seniority and other benefits until such work becomes available, until the employee is determined to be unable to return to workplace formaldehyde exposure, until the employee is determined to be able to return to the original job status, or for six months, whichever comes first.[§1910.1048(l)(8)(vi)]

(vii) *The employer shall arrange for a follow-up medical examination* to take place within six months after the employee is removed pursuant to this paragraph. This examination shall determine if the employee can return to the original job status, or if the removal is to be permanent. The physician shall make a decision within six months of the date the employee was removed as to whether the employee can be returned to the original job status, or if the removal is to be permanent. [§1910.1048(l)(8)(vii)]

(viii) *An employer's obligation to provide earnings,* seniority and other benefits to a removed employee may be reduced to the extent that the employee receives compensation for earnings lost during the period of removal either from a publicly or employer-funded compensation program or from employment with another employer made possible by virtue of the employee's removal.[§1910.1048(l)(8)(viii)]

(ix) *In making determinations of the formaldehyde content* of materials under this paragraph the employer may rely on objective data.[§1910.1048(l)(8)(ix)]

(9) *Multiple physician review.*[§1910.1048(l)(9)]

(i) *After the employer selects* the initial physician who conducts any medical examination or consultation to determine whether medical removal or restriction is appropriate, the employee may designate a second physician to review any findings, determinations or recommendations of the initial physician and to conduct such examinations, consultations, and laboratory tests as the second physician deems necessary and appropriate to evaluate the effects of formaldehyde exposure and to facilitate this review.[§1910.1048(l)(9)(i)]

(ii) *The employer shall promptly notify* an employee of the right to seek a second medical opinion after each occasion that an initial physician conducts a medical examination or consultation for the purpose of medical removal or restriction. [§1910.1048(l)(9)(ii)]

(iii) *The employer may condition its participation in,* and payment for, the multiple physician review mechanism upon the employee doing the following within fifteen (15) days after receipt of the notification of the right to seek a second medical opinion, or receipt of the initial physician's written opinion, whichever is later;[§1910.1048(l)(9)(iii)]

[A] The employee informs the employer of the intention to seek a second medical opinion, and[§1910.1048(l)(9)(iii)[A]]

[B] The employee initiates steps to make an appointment with a second physician.[§1910.1048(l)(9)(iii)[B]]

(iv) *If the findings,* determinations or recommendations of the second physician differ from those of the initial physician, then the employer and the employee shall assure that efforts are made for the two physicians to resolve the disagreement. If the two physicians are unable to quickly resolve their disagreement, then the employer and the employee through their respective physicians shall designate a third physician who shall be a specialist in the field at issue:[§1910.1048(l)(9)(iv)]

[A] To review the findings, determinations or recommendations of the prior physicians; and[§1910.1048(l)(9)(iv)[A]]

[B] To conduct such examinations, consultations, laboratory tests and discussions with the prior physicians as the third physician deems necessary to resolve the disagreement of the prior physicians.[§1910.1048(l)(9)(iv)[B]]

(v) *In the alternative,* the employer and the employee or authorized employee representative may jointly designate such third physician.[§1910.1048(l)(9)(v)]

(vi) *The employer shall act consistent* with the findings, determinations and recommendations of the third physician, unless the employer and the employee reach an agreement which is otherwise consistent with the recommendations of at least one of the three physicians.[§1910.1048(l)(9)(vi)]

(m) Communication of hazards.[§1910.1048(m)]

(1) *Hazard communication — General.*[§1910.1048(m)(1)]

(i) *Chemical manufacturers,* importers, distributors and employers shall comply with all requirements of the Hazard Communication Standard (HCS) (Sec. 1910.1200) for formaldehyde. [§1910.1048(m)(1)(i)]

(ii) *In classifying the hazards of formaldehyde* at least the following hazards are to be addressed: Cancer; skin and respiratory sensitization; eye, skin and respiratory tract irritation; acute toxicity effects; and flammability.[§1910.1048(m)(1)(ii)]

(iii) *Employers shall include formaldehyde* in the hazard communication program established to comply with the HCS (Sec. 1910.1200). Employers shall ensure that each employee has access to labels on containers of formaldehyde and to safety data sheets, and is trained in accordance with the requirements of HCS and paragraph (n) of this section.[§1910.1048(m)(1)(iii)]

(iv) *Paragraphs (m)(1)(i),* (m)(1)(ii), and (m)(1)(iii) of this section apply to chemicals associated with formaldehyde gas, all mixtures or solutions composed of greater than 0.1 percent formaldehyde, and materials capable of releasing formaldehyde into the air at concentrations reaching or exceeding 0.1 ppm.[§1910.1048(m)(1)(iv)]

(v) *In making the determinations of anticipated* levels of formaldehyde release, the employer may rely on objective data indicating the extent of potential formaldehyde release under reasonably foreseeable conditions of use.[§1910.1048(m)(1)(v)]

(2) (i) *In addition to the requirements in paragraphs* (m)(1) through (m)(1)(iv) of this section, for materials listed in paragraph (m)(1)(iv) capable of releasing formaldehyde at levels above 0.5 ppm, labels shall appropriately address all hazards as defined in paragraph (d) of §1910.1200 and Appendices A and B to Sec. 1910.1200, including cancer and respiratory sensitization, and shall contain the hazard statement "May Cause Cancer."[§1910.1048(m)(2)(i)]

(ii) *As a minimum,* for all materials listed in paragraph (m)(1)(i) and (iv) of this section capable of releasing formaldehyde at levels of 0.1 ppm to 0.5 ppm, labels shall identify that the product contains formaldehyde; list the name and address of the responsible party; and state that physical and health hazard information is readily available from the employer and from safety data sheets.[§1910.1048(m)(2)(ii)]

(iii) *Prior to June 1,* 2015, employers may include the phrase "Potential Cancer Hazard" in lieu of "May Cause Cancer" as specified in paragraph (m)(2)(i) of this section. [§1910.1048(m)(2)(iii)]

(3) *Labels.*[§1910.1048(m)(3)]

(i) *The employer shall assure that hazard warning labels* complying with the requirements of 29 CFR 1910.1200(f) are affixed to all containers of materials listed in paragraph (m)(1)(i), except to the extent that 29 CFR 1910.1200(f) is inconsistent with this paragraph.[§1910.1048(m)(3)(i)]

(ii) *Information on labels.* As a minimum, for all materials listed in paragraph (m)(1)(i) capable of releasing formaldehyde at levels of 0.1 ppm to 0.5 ppm, labels shall identify that the product contains formaldehyde; list the name and address of the responsible party; and state that physical and health hazard information is readily available from the employer and from material safety data sheets.[§1910.1048(m)(3)(ii)]

(iii) *For materials listed in paragraph (m)(1)(i)* capable of releasing formaldehyde at levels above 0.5 ppm, labels shall appropriately address all hazards as defined in 29 CFR 1910.1200(d) and 29 CFR 1910.1200 appendices A and B, including respiratory sensitization, and shall contain the words "Potential Cancer Hazard."[§1910.1048(m)(3)(iii)]

(iv) *In making the determinations of anticipated levels* of formaldehyde release, the employer may rely on objective data indicating the extent of potential formaldehyde release under reasonably foreseeable conditions of use.[§1910.1048(m)(3)(iv)]

(v) *Substitute warning labels.* The employer may use warning labels required by other statutes, regulations, or ordinances which impart the same information as the warning statements required by this paragraph.[§1910.1048(m)(3)(v)]

(4) *Material safety data sheets.*[§1910.1048(m)(4)]

(i) *Any employer who uses formaldehyde-containing* materials listed in paragraph (m)(1)(i) shall comply with the requirements of 29 CFR 1910.1200(g) with regard to the development and updating of material safety data sheets. [§1910.1048(m)(4)(i)]

(ii) *Manufacturers,* importers, and distributors of formaldehyde-containing materials listed in paragraph (m)(1)(i) shall assure that material safety data sheets and updated information are provided to all employers purchasing such materials at the time of the initial shipment and at the time of the first shipment after a material safety data sheet is updated. [§1910.1048(m)(4)(ii)]

(5) *Written hazard communication program.* The employer shall develop, implement, and maintain at the workplace, a written hazard communication program for formaldehyde exposures in the workplace, which at a minimum describes how the requirements specified in this paragraph for labels and other forms of warning and material safety data sheets, and paragraph (n) for employee information and training, will be met. Employers in multi-employer workplaces shall comply with the requirements of 29 CFR 1910.1200(e)(2).[§1910.1048(m)(5)]

(n) Employee information and training.[§1910.1048(n)]

(1) *Participation.* The employer shall assure that all employees who are assigned to workplaces where there is exposure to formaldehyde participate in a training program, except that where the employer can show, using objective data, that employees are not exposed to formaldehyde at or above 0.1 ppm, the employer is not required to provide training. [§1910.1048(n)(1)]

(2) *Frequency.* Employers shall provide such information and training to employees at the time of initial assignment, and whenever a new exposure to formaldehyde is introduced into the work area. The training shall be repeated at least annually. [§1910.1048(n)(2)]

(3) *Training program.* The training program shall be conducted in a manner which the employee is able to understand and shall include:[§1910.1048(n)(3)]

(i) *A discussion of the contents of this regulation* and the contents of the Material Safety Data Sheet.[§1910.1048(n)(3)(i)]

(ii) *The purpose for and a description* of the medical surveillance program required by this standard, including: [§1910.1048(n)(3)(ii)]

[A] A description of the potential health hazards associated with exposure to formaldehyde and a description of the signs and symptoms of exposure to formaldehyde. [§1910.1048(n)(3)(ii)[A]]

[B] Instructions to immediately report to the employer the development of any adverse signs or symptoms that the employee suspects is attributable to formaldehyde exposure.[§1910.1048(n)(3)(ii)[B]]

(iii) *Description of operations in the work area* where formaldehyde is present and an explanation of the safe work practices appropriate for limiting exposure to formaldehyde in each job; [§1910.1048(n)(3)(iii)]

(iv) *The purpose for,* proper use of, and limitations of personal protective clothing and equipment;[§1910.1048(n)(3)(iv)]

(v) *Instructions for the handling of spills,* emergencies, and clean-up procedures;[§1910.1048(n)(3)(v)]

(vi) *An explanation of the importance of engineering* and work practice controls for employee protection and any necessary instruction in the use of these controls; and[§1910.1048(n)(3)(vi)]

(vii) *A review of emergency procedures* including the specific duties or assignments of each employee in the event of an emergency.[§1910.1048(n)(3)(vii)]

(4) *Access to training materials.*[§1910.1048(n)(4)]

(i) *The employer shall inform all affected employees* of the location of written training materials and shall make these materials readily available, without cost, to the affected employees. [§1910.1048(n)(4)(i)]

(ii) *The employer shall provide,* upon request, all training materials relating to the employee training program to the Assistant Secretary and the Director.[§1910.1048(n)(4)(ii)]

(o) Recordkeeping.[§1910.1048(o)]

(1) *Exposure measurements.* The employer shall establish and maintain an accurate record of all measurements taken to monitor employee exposure to formaldehyde. This record shall include:[§1910.1048(o)(1)]

(i) *The date of measurement;*[§1910.1048(o)(1)(i)]

(ii) *The operation being monitored;*[§1910.1048(o)(1)(ii)]

(iii) *The methods of sampling and analysis* and evidence of their accuracy and precision;[§1910.1048(o)(1)(iii)]

(iv) *The number, durations, time, and results of samples taken;* [§1910.1048(o)(1)(iv)]

(v) *The types of protective devices worn; and*[§1910.1048(o)(1)(v)]

(vi) *The names,* job classifications, social security numbers, and exposure estimates of the employees whose exposures are represented by the actual monitoring results. [§1910.1048(o)(1)(vi)]

(2) *Exposure determinations.* Where the employer has determined that no monitoring is required under this standard, the employer shall maintain a record of the objective data relied upon to support the determination that no employee is exposed to formaldehyde at or above the action level. [§1910.1048(o)(2)]

(3) *Medical surveillance.* The employer shall establish and maintain an accurate record for each employee subject to medical surveillance under this standard. This record shall include: [§1910.1048(o)(3)]

(i) *The name and social security number of the employee;* [§1910.1048(o)(3)(i)]

(ii) *The physician's written opinion;*[§1910.1048(o)(3)(ii)]

(iii) *A list of any employee health complaints* that may be related to exposure to formaldehyde; and[§1910.1048(o)(3)(iii)]

(iv) *A copy of the medical examination results,* including medical disease questionnaires and results of any medical tests required by the standard or mandated by the examining physician.[§1910.1048(o)(3)(iv)]

(4) *Respirator fit testing.*[§1910.1048(o)(4)]

(i) *The employer shall establish and maintain* accurate records for employees subject to negative pressure respirator fit testing required by this standard.[§1910.1048(o)(4)(i)]

(ii) *This record shall include:*[§1910.1048(o)(4)(ii)]

[A] A copy of the protocol selected for respirator fit testing. [§1910.1048(o)(4)(ii)[A]]

[B] A copy of the results of any fit testing performed. [§1910.1048(o)(4)(ii)[B]]

[C] The size and manufacturer of the types of respirators available for selection.[§1910.1048(o)(4)(ii)[C]]

[D] The date of the most recent fit testing, the name and social security number of each tested employee, and the respirator type and facepiece selected.[§1910.1048(o)(4)(ii)[D]]

(5) *Record retention.* The employer shall retain records required by this standard for at least the following periods: [§1910.1048(o)(5)]

(i) *Exposure records and determinations* shall be kept for at least 30 years.[§1910.1048(o)(5)(i)]

(ii) *Medical records shall be kept* for the duration of employment plus 30 years.[§1910.1048(o)(5)(ii)]

(iii) *Respirator fit testing records* shall be kept until replaced by a more recent record.[§1910.1048(o)(5)(iii)]

(6) *Availability of records.*[§1910.1048(o)(6)]

(i) *Upon request,* the employer shall make all records maintained as a requirement of this standard available for examination and copying to the Assistant Secretary and the Director.[§1910.1048(o)(6)(i)]

(ii) *The employer shall make employee exposure records,* including estimates made from representative monitoring and available upon request for examination, and copying to the subject employee, or former employee, and employee representatives in accordance with 29 CFR 1910.1020 (a)-(e) and (g)-(i).[§1910.1048(o)(6)(ii)]

(iii) *Employee medical records* required by this standard shall be provided upon request for examination and coying, to the subject employee or former employee or to anyone having the specific written consent of the subject employee or former employee in accordance with 29 CFR 1910.1020 (a)-(e) and (g)-(i).[§1910.1048(o)(6)(iii)]

§1910.1048 Appendix A

Substance Technical Guidelines for Formalin

The following Substance Technical Guideline for Formalin provides information on uninhibited formalin solution (37% formaldehyde, no methanol stabilizer). It is designed to inform employees at the production level of their rights and duties under the formaldehyde standard whether their job title defines them as workers or supervisors. Much of the information provided is general; however, some information is specific for formalin. When employee exposure to formaldehyde is from resins capable of releasing formaldehyde, the resin itself and other impurities or decomposition products may also be toxic, and employers should include this information as well when informing employees of the hazards associated with the materials they handle. The precise hazards associated with exposure to formaldehyde depend both on the form (solid, liquid, or gas) of the material and the concentration of formaldehyde present. For example, 37-50 percent solutions of formaldehyde present a much greater hazard to the skin and eyes from spills or splashes than solutions containing less than 1 percent formaldehyde. Individual Substance Technical Guidelines used by the employer for training employees should be modified to properly give information on the material actually being used.

Substance Identification

Chemical Name: Formaldehyde

Chemical Family: Aldehyde

Chemical Formula: HCHO

Molecular Weight: 30.03

Chemical Abstracts Service Number (CAS Number): 50-00-0

Synonyms: Formalin; Formic Aldehyde; Paraform; Formol; Formalin (Methanol-free); Fyde; Formalith; Methanal; Methyl Aldehyde; Methylene Glycol; Methylene Oxide; Tetraoxymethalene; Oxomethane; Oxymethylene

Components and Contaminants

Percent: 37.0 Formaldehyde

Percent: 63.0 Water

(*Note* — Inhibited solutions contain methanol.)

Other Contaminants: Formic acid (alcohol free)

Exposure Limits:

OSHA TWA — 0.75 ppm

OSHA STEL — 2 ppm

Physical Data

Description: Colorless liquid, pungent odor

Boiling point: 214 °F (101 °C)

Specific Gravity: 1.08 (H_2O = 1 @ 20 °C)

pH: 2.8-4.0

Solubility in Water: Miscible

Solvent Solubility: Soluble in alcohol and acetone

Vapor Density: 1.04 (Air = 1 @ 20 °C)

Odor Threshold: 0.8-1 ppm

Fire and Explosion Hazard

Moderate fire and explosion hazard when exposed to heat or flame.

The flash point of 37% formaldehyde solutions is above normal room temperature, but the explosion range is very wide, from 7 to 73% by volume in air.

Reaction of formaldehyde with nitrogen dioxide, nitromethane, perchloric acid and aniline, or peroxyformic acid yields explosive compounds.

Flash Point: 185 °F (85 °C) closed cup

Lower Explosion Limit: 7%

Upper Explosion Limit: 73%

Autoignition Temperature: 806 °F (430 °C)

Flammability Class (OSHA): III A

Extinguishing Media: Use dry chemical, "alcohol foam", carbon dioxide, or water in flooding amounts as fog. Solid streams may not be effective. Cool fire-exposed containers with water from side until well after fire is out.

Use of water spray to flush spills can also dilute the spill to produce nonflammable mixtures. Water runoff, however, should be contained for treatment.

National Fire Protection Association Section 325M Designation:

Health: 2 — Materials hazardous to health, but areas may be entered with full-faced mask self-contained breathing apparatus which provides eye protection.

Flammability: 2 — Materials which must be moderately heated before ignition will occur. Water spray may be used to extinguish the fire because the material can be cooled below its flash point.

Reactivity: D — Materials which (in themselves) are normally stable even under fire exposure conditions and which are not reactive with water. Normal fire fighting procedures may be used.

Reactivity

Stability: Formaldehyde solutions may self-polymerize to form paraformaldehyde which precipitates.

Incompatibility (Materials to Avoid): Strong oxidizing agents, caustics, strong alkalies, isocyanates, anhydrides, oxides, and inorganic acids. Formaldehyde reacts with hydrochloric acid to form the potent carcinogen, bis-chloromethyl ether. Formaldehyde reacts with nitrogen dioxide, nitromethane, perchloric acid and aniline, or peroxyformic acid to yield explosive compounds. A violent reaction occurs when formaldehyde is mixed with strong oxidizers.

Hazardous Combustion or Decomposition Products: Oxygen from the air can oxidize formaldehyde to formic acid, especially when heated. Formic acid is corrosive.

Health Hazard Data

Acute Effects of Exposure

Ingestion (Swallowing): Liquids containing 10 to 40% formaldehyde cause severe irritation and inflammation of the mouth, throat, and stomach. Severe stomach pains will follow ingestion with possible loss of consciousness and death. Ingestion of dilute formaldehyde solutions (0.03-0.04%) may cause discomfort in the stomach and pharynx.

Inhalation (Breathing): Formaldehyde is highly irritating to the upper respiratory tract and eyes. Concentrations of 0.5 to 2.0 ppm may irritate the eyes, nose, and throat of some individuals. Concentrations of 3 to 5 ppm also cause tearing of the eyes and are intolerable to some persons. Concentrations of 10 to 20 ppm cause difficulty in breathing, burning of the nose and throat, cough, and heavy tearing of the eyes, and 25 to 30 ppm causes severe respiratory tract injury leading to pulmonary edema and pneumonitis. A concentration of 100 ppm is immediately dangerous to life and health. Deaths from accidental exposure to high concentrations of formaldehyde have been reported.

Skin (Dermal): Formalin is a severe skin irritant and a sensitizer. Contact with formalin causes white discoloration, smarting, drying, cracking, and scaling. Prolonged and repeated contact can cause numbness and a hardening or tanning of the skin. Previously exposed persons may react to future exposure with an allergic eczematous dermatitis or hives.

Eye Contact: Formaldehyde solutions splashed in the eye can cause injuries ranging from transient discomfort to severe, permanent corneal clouding and loss of vision. The severity of the effect depends on the concentration of formaldehyde in the solution and whether or not the eyes are flushed with water immediately after the accident.

Note. The perception of formaldehyde by odor and eye irritation becomes less sensitive with time as one adapts to formaldehyde. This can lead to overexposure if a worker is relying on formaldehyde's warning properties to alert him or her to the potential for exposure.

Acute Animal Toxicity:

Oral, rats: LD_{50} = 800 mg/kg

Oral, mouse: LD_{50} = 42 mg/kg

Inhalation, rats: LC_{Lo} = 250 mg/kg

Inhalation, mouse: LC_{Lo} = 900 mg/kg

Inhalation, rats: LC_{50} = 590 mg/kg

Chronic Effects of Exposure

Carcinogenicity: Formaldehyde has the potential to cause cancer in humans. Repeated and prolonged exposure increases the risk. Various animal experiments have conclusively shown formaldehyde to be a carcinogen in rats. In humans, formaldehyde exposure has been associated with cancers of the lung, nasopharynx and oropharynx, and nasal passages.

Mutagenicity: Formaldehyde is genotoxic in several in vitro test systems showing properties of both an initiator and a promoter.

Toxicity: Prolonged or repeated exposure to formaldehyde may result in respiratory impairment. Rats exposed to formaldehyde at 2 ppm developed benign nasal tumors and changes of the cell structure in the nose as well as inflamed mucous membranes of the nose. Structural changes in the epithelial cells in the human nose have also been observed. Some persons have developed asthma or bronchitis following exposure to formaldehyde, most often as the result of an accidental spill involving a single exposure to a high concentration of formaldehyde.

Emergency and First Aid Procedures

Ingestion (Swallowing): If the victim is conscious, dilute, inactivate, or absorb the ingested formaldehyde by giving milk, activated charcoal, or water. Any organic material will inactivate formaldehyde. Keep affected person warm and at rest. Get medical attention immediately. If vomiting occurs, keep head lower than hips.

Inhalation (Breathing): Remove the victim from the exposure area to fresh air immediately. Where the formaldehyde concentration may be very high, each rescuer must put on a self-contained breathing apparatus before attempting to remove the victim, and medical personnel should be informed of the formaldehyde exposure immediately. If breathing has stopped, give artificial respiration. Keep the affected person warm and at rest. Qualified first-aid or medical personnel should administer oxygen, if available, and maintain the patient's airways and blood pressure until the victim can be transported to a medical facility. If exposure results in a highly irritated upper respiratory tract and coughing continues for more than 10 minutes, the worker should be hospitalized for observation and treatment.

Skin Contact: Remove contaminated clothing (including shoes) immediately. Wash the affected area of your body with soap or mild detergent and large amounts of water until no evidence of the chemical remains (at least 15 to 20 minutes). If there are chemical burns, get first aid to cover the area with sterile, dry dressing, and bandages. Get medical attention if you experience appreciable eye or respiratory irritation.

Eye Contact: Wash the eyes immediately with large amounts of water occasionally lifting lower and upper lids, until no evidence of chemical remains (at least 15 to 20 minutes). In case of burns, apply sterile bandages loosely without medication. Get medical attention immediately. If you have experienced appreciable eye irritation from a splash or excessive exposure, you should be referred promptly to an opthamologist for evaluation.

Emergency Procedures

Emergencies: If you work in an area where a large amount of formaldehyde could be released in an accident or from equipment failure, your employer must develop procedures to be followed in event of an emergency. You should be trained in your specific duties in the event of an emergency, and it is important that you clearly understand these duties. Emergency equipment must be accessible and you should be trained to use any equipment that you might need. Formaldehyde contaminated equipment must be cleaned before reuse.

If a spill of appreciable quantity occurs, leave the area quickly unless you have specific emergency duties. Do not touch spilled material. Designated persons may stop the leak and shut off ignition sources if these procedures can be done without risk. Designated persons should isolate the hazard area and deny entry except for necessary people protected by suitable protective clothing and respirators adequate for the exposure. Use water spray to reduce vapors. Do not smoke, and prohibit all flames or flares in the hazard area.

Special Firefighting Procedures: Learn procedures and responsibilities in the event of a fire in your workplace. Become familiar with the appropriate equipment and supplies and their location. In firefighting, withdraw immediately in case of rising sound from venting safety device or any discoloration of storage tank due to fire.

Spill, Leak, and Disposal Procedures

Occupational Spill: For small containers, place the leaking container in a well ventilated area. Take up small spills with absorbent material and place the waste into properly labeled containers for later disposal. For larger spills, dike the spill to minimize contamination and facilitate salvage or disposal. You may be able to neutralize the spill with sodium hydroxide or sodium sulfite. Your employer must comply with EPA rules regarding the clean-up of toxic waste and notify state and local authorities, if required. If the spill is greater than 1,000 lb/day, it is reportable under EPA's Superfund legislation.

Waste Disposal: Your employer must dispose of waste containing formaldehyde in accordance with applicable local, state, and Federal law and in a manner that minimizes exposure of employees at the site and of the clean-up crew.

Monitoring and Measurement Procedures

Monitoring Requirements: If your exposure to formaldehyde exceeds the 0.5 ppm action level or the 2 ppm STEL, your employer must monitor your exposure. Your employer need not measure every exposure if a "high exposure" employee can be identified. This person usually spends the greatest amount of time nearest the process equipment. If you are a "representative employee", you will be asked to wear a sampling device to collect formaldehyde. This device may be a passive badge, a sorbent tube attached to a pump, or an impinger containing liquid. You should perform your work as usual, but inform the person who is conducting the monitoring of any difficulties you are having wearing the device.

Evaluation of 8-hour Exposure: Measurements taken for the purpose of determining time-weighted average (TWA) exposures are best taken with samples covering the full shift. Samples collected must be taken from the employee's breathing zone air.

Short-term Exposure Evaluation: If there are tasks that involve brief but intense exposure to formaldehyde, employee exposure must be measured to assure compliance with the STEL. Sample collections are for brief periods, only 15 minutes, but several samples may be needed to identify the peak exposure.

Monitoring Techniques: OSHA's only requirement for selecting a method for sampling and analysis is that the methods used accurately evaluate the concentration of formaldehyde in employees' breathing zones. Sampling and analysis may be performed by collection of formaldehyde on liquid or solid sorbents with subsequent chemical analysis. Sampling and analysis may also be performed by passive diffusion monitors and short-term exposure may be measured by instruments such as real-time continuous monitoring systems and portable direct reading instruments.

Notification of Results: Your employer must inform you of the results of exposure monitoring representative of your job. You may be informed in writing, but posting the results where you have ready access to them constitutes compliance with the standard.

Protective Equipment and Clothing

[Material impervious to formaldehyde is needed if the employee handles formaldehyde solutions of 1% or more. Other employees may also require protective clothing or equipment to prevent dermatitis.]

Respiratory Protection: Use NIOSH-approved full facepiece negative pressure respirators equipped with approved cartridges or canisters within the use limitations of these devices. (Present restrictions on cartridges and canisters do not permit them to be used for a full workshift.) In all other situations, use positive pressure respirators such as the positive-pressure air purifying respirator or the self-contained breathing apparatus (SCBA). If you use a negative pressure respirator, your employer must provide you with fit testing of the respirator at least once a year.

Protective Gloves: Wear protective (impervious) gloves provided by your employer, at no cost, to prevent contact with formalin. Your employer should select these gloves based on the results of permeation testing and in accordance with the ACGIH Guidelines for Selection of Chemical Protective Clothing.

Eye Protection: If you might be splashed in the eyes with formalin, it is essential that you wear goggles or some other type of complete protection for the eye. You may also need a face shield if your face is likely to be splashed with formalin, but you must not substitute face shields for eye protection. (This section pertains to formaldehyde solutions of 1% or more.)

Other Protective Equipment: You must wear protective (impervious) clothing and equipment provided by your employer at no cost to prevent repeated or prolonged contact with formaldehyde liquids. If you are required to change into whole-body chemical protective clothing, your employer must provide a change room for your privacy and for storage of your normal clothing.

If you are splashed with formaldehyde, use the emergency showers and eyewash fountains provided by your employer immediately to prevent serious injury. Report the incident to your supervisor and obtain necessary medical support.

Entry Into an IDLH Atmosphere

Enter areas where the formaldehyde concentration might be 100 ppm or more only with complete body protection including a self-contained breathing apparatus with a full facepiece operated in a positive pressure mode or a supplied air respirator with full facepiece and operated in a positive pressure mode. This equipment is essential to protect your life and health under such extreme conditions.

Engineering Controls

Ventilation is the most widely applied engineering control method for reducing the concentration of airborne substances in the breathing zones of workers. There are two distinct types of ventilation.

Local Exhaust: Local exhaust ventilation is designed to capture airborne contaminants as near to the point of generation as possible. To protect you, the direction of contaminant flow must always be toward the local exhaust system inlet and away from you.

General (Mechanical): General dilution ventilation involves continuous introduction of fresh air into the workroom to mix with the contaminated air and lower your breathing zone concentration of formaldehyde. Effectiveness depends on the number of air changes per hour. Where devices emitting formaldehyde are spread out over a large area, general dilution ventilation may be the only practical method of control.

Work Practices: Work practices and administrative procedures are an important part of a control system. If you are asked to perform a task in a certain manner to limit your exposure to formaldehyde, it is extremely important that you follow these procedures.

Medical Surveillance

Medical surveillance helps to protect employees' health. You are encouraged strongly to participate in the medical surveillance program.

Your employer must make a medical surveillance program available at no expense to you and at a reasonable time and place if you are exposed to formaldehyde at concentrations above 0.5 ppm as an 8-hour average or 2 ppm over any 15-minute period. You will be offered medical surveillance at the time of your initial assignment and once a year afterward as long as your exposure is at least 0.5 ppm (TWA) or 2 ppm (STEL). Even if your exposure is below these levels, you should inform your employer if you have signs and symptoms that you suspect, through your training, are related to your formaldehyde exposure because you may need medical surveillance to determine if your health is being impaired by your exposure.

The surveillance plan includes:

(a) *A medical disease questionnaire.*

(b) *A physical examination* if the physician determines this is necessary.

If you are required to wear a respirator, your employer must offer you a physical examination and a pulmonary function test every year.

The physician must collect all information needed to determine if you are at increased risk from your exposure to formaldehyde. At the physician's discretion, the medical examination may include other tests, such as a chest x-ray, to make this determination.

After a medical examination the physician will provide your employer with a written opinion which includes any special protective measures recommended and any restrictions on your exposure. The physician must inform you of any medical conditions you have which would be aggravated by exposure to formaldehyde.

All records from your medical examinations, including disease surveys, must be retained at your employer's expense.

Emergencies

If you are exposed to formaldehyde in an emergency and develop signs or symptoms associated with acute toxicity from formaldehyde exposure, your employer must provide you with a medical examination as soon as possible. This medical examination will include all steps necessary to stabilize your health. You may be kept in the hospital for observation if your symptoms are severe to ensure that any delayed effects are recognized and treated.

§1910.1048 Appendix B

Sampling Strategy and Analytical Methods for Formaldehyde

To protect the health of employees, exposure measurements must be unbiased and representative of employee exposure. The proper measurement of employee exposure requires more than a token commitment on the part of the employer. OSHA's mandatory requirements establish a baseline; under the best of circumstances all questions regarding employee exposure will be answered. Many employers, however, will wish to conduct more extensive monitoring before undertaking expensive commitments, such as engineering controls, to assure that the modifications are truly necessary. The following sampling strategy, which was developed at NIOSH by Nelson A. Leidel, Kenneth A. Busch, and Jeremiah R. Lynch and described in NIOSH publication No. 77-173 (Occupational Exposure Sampling Strategy Manual) will assist the employer in developing a strategy for determining the exposure of his or her employees.

There is no one correct way to determine employee exposure. Obviously, measuring the exposure of every employee exposed to formaldehyde will provide the most information on any given day. Where few employees are exposed, this may be a practical solution. For most employers, however, use of the following strategy will give just as much information at less cost.

Exposure data collected on a single day will not automatically guarantee the employer that his or her workplace is always in compliance with the formaldehyde standard. This does not imply, however, that it is impossible for an employer to be sure that his or her worksite is in compliance with the standard. Indeed, a properly designed sampling strategy showing that all employees are exposed below the PELs, at least with a 95 percent certainty, is compelling evidence that the exposure limits are being achieved provided that measurements are conducted using valid sampling strategy and approved analytical methods.

There are two PELs, the TWA concentration and the STEL. Most employers will find that one of these two limits is more critical in the control of their operations, and OSHA expects that the employer will concentrate monitoring efforts on the critical component. If the more difficult exposure is controlled, this information, along with calculations to support the assumptions, should be adequate to show that the other exposure limit is also being achieved.

Sampling Strategy

Determination of the Need for Exposure Measurements

The employer must determine whether employees may be exposed to concentrations in excess of the action level. This determination becomes the first step in an employee exposure monitoring program that minimizes employer sampling burdens while providing adequate employee protection. If employees may be exposed above the action level, the employer must measure exposure. Otherwise, an objective determination that employee exposure is low provides adequate evidence that exposure potential has been examined.

The employer should examine all available relevant information, eg. insurance company and trade association data and information from suppliers or exposure data collected from similar operations. The employer may also use previously-conducted sampling including area monitoring. The employer must make a determination relevant to each operation although this need not be on a separate piece of paper. If the employer can demonstrate conclusively that no employee is exposed above the action level or the STEL through the use of objective data, the employer need proceed no further on employee exposure monitoring until such time that conditions have changed and the determination is no longer valid.

If the employer cannot determine that employee exposure is less than the action level and the STEL, employee exposure monitoring will have to be conducted.

Workplace Material Survey

The primary purpose of a survey of raw material is to determine if formaldehyde is being used in the work environment and if so, the conditions under which formaldehyde is being used.

The first step is to tabulate all situations where formaldehyde is used in a manner such that it may be released into the workplace atmosphere or contaminate the skin. This information should be available through analysis of company records and information on the MSDSs available through provisions of this standard and the Hazard Communication standard.

If there is an indication from materials handling records and accompanying MSDSs that formaldehyde is being used in the following types of processes or work operations, there may be a potential for releasing formaldehyde into the workplace atmosphere:

(1) *Any operation that involves* grinding, sanding, sawing, cutting, crushing, screening, sieving, or any other manipulation of material that generates formaldehyde-bearing dust.

(2) *Any processes* where there have been employee complaints or symptoms indicative of exposure to formaldehyde.

(3) *Any liquid or spray process involving formaldehyde.*

(4) *Any process that uses formaldehyde in preserved tissue.*

(5) *Any process* that involves the heating of a formaldehyde-bearing resin.

Processes and work operations that use formaldehyde in these manners will probably require further investigation at the worksite to determine the extent of employee monitoring that should be conducted.

Workplace Observations

To this point, the only intention has been to provide an indication as to the existence of potentially exposed employees. With this information, a visit to the workplace is needed to observe work operations, to identify potential health hazards, and to determine whether any employees may be exposed to hazardous concentrations of formaldehyde.

In many circumstances, sources of formaldehyde can be identified through the sense of smell. However, this method of detection should be used with caution because of olfactory fatigue.

Employee location in relation to source of formaldehyde is important in determining if an employee may be significantly exposed to formaldehyde. In most instances, the closer a worker is to the source, the higher the probability that a significant exposure will occur.

Other characteristics should be considered. Certain high temperature operations give rise to higher evaporation rates. Locations of open doors and windows provide natural ventilation that tend to dilute formaldehyde emissions. General room ventilation also provides a measure of control.

Calculation of Potential Exposure Concentrations

By knowing the ventilation rate in a workplace and the quantity of formaldehyde generated, the employer may be able to determine by calculation if the PELs might be exceeded. To account for poor mixing of formaldehyde into the entire room, locations of fans and proximity of employees to the work operation, the employer must include a safety factor. If an employee is relatively close to a source, particularly if he or she is located downwind, a safety factor of 100 may be necessary. For other situations, a factor of 10 may be acceptable. If the employer can demonstrate through such calculations that employee exposure does not exceed the action level or the STEL, the employer may use this information as objective data to demonstrate compliance with the standard.

Sampling Strategy

Once the employer determines that there is a possibility of substantial employee exposure to formaldehyde, the employer is obligated to measure employee exposure.

The next step is selection of a maximum risk employee. When there are different processes where employees may be exposed to formaldehyde, a maximum risk employee should be selected for each work operation.

Selection of the maximum risk employee requires professional judgment. The best procedure for selecting the maximum risk employee is to observe employees and select the person closest to the source of formaldehyde. Employee mobility may affect this selection; eg. if the closest employee is mobile in his tasks, he may not be the maximum risk employee. Air movement patterns and differences in work habits will also affect selection of the maximum risk employee.

When many employees perform essentially the same task, a maximum risk employee cannot be selected. In this circumstance, it is necessary to resort to random sampling of the group of workers. The objective is to select a subgroup of adequate size so that there is a high probability that the random sample will contain at least one worker with high exposure if one exists. The number of persons in the group influences the number that need to be sampled to ensure that at least one individual from the highest 10 percent exposure group is contained in the sample. For example, to have 90 percent confidence in the results, if the group size is 10, nine should be sampled; for 50, only 18 need to be sampled.

If measurement shows exposure to formaldehyde at or above the action level or the STEL, the employer needs to identify all other employees who may be exposed at or above the action level or STEL and measure or otherwise accurately characterize the exposure of these employees.

Whether representative monitoring or random sampling are conducted, the purpose remains the same — to determine if the exposure of any employee is above the action level. If the exposure of the most exposed employee is less than the action level and the STEL, regardless of how the employee is identified, then it is reasonable to assume that measurements of exposure of the other employees in that operation would be below the action level and the STEL.

Exposure Measurements

There is no "best" measurement strategy for all situations. Some elements to consider in developing a strategy are:

(1) *Availability and cost of sampling equipment.*

(2) *Availability and cost of analytic facilities.*

(3) *Availability and cost of personnel to take samples.*

(4) *Location of employees and work operations.*

(5) *Intraday and interday variations in the process.*

(6) *Precision and accuracy of sampling and analytic methods, and*

(7) *Number of samples needed.*

Samples taken for determining compliance with the STEL differ from those that measure the TWA concentration in important ways. STEL samples are best taken in a nonrandom fashion using all available knowledge relating to the area, the individual, and the process to obtain samples during periods of maximum expected concentrations. At least three measurements on a shift are generally needed to spot gross errors or mistakes; however, only the highest value represents the STEL.

If an operation remains constant throughout the workshift, a much greater number of samples would need to be taken over the 32 discrete nonoverlapping periods in an 8-hour workshift to verify compliance with a STEL. If employee exposure is truly uniform throughout the workshift, however, an employer in compliance with the 1 ppm TWA would be in compliance with the 2 ppm STEL, and this determination can probably be made using objective data.

Need To Repeat the Monitoring Strategy

Interday and intraday fluctuations in employee exposure are mostly influenced by the physical processes that generate formaldehyde and the work habits of the employee. Hence, in-plant process variations influence the employer's determination of whether or not additional controls need to be imposed. Measurements that employee exposure is low on a day that is not representative of worst conditions may not provide sufficient information to determine whether or not additional engineering controls should be installed to achieve the PELs.

The person responsible for conducting sampling must be aware of systematic changes which will negate the validity of the sampling results. Systematic changes in formaldehyde exposure concentration for an employee can occur due to:

(1) *The employee changing patterns* of movement in the workplace

(2) *Closing of plant doors and windows*

(3) *Changes in ventilation from season to season*

(4) *Decreases in ventilation efficiency* or abrupt failure of engineering control equipment

(5) *Changes in the production process* or work habits of the employee.

Any of these changes, if they may result in additional exposure that reaches the next level of action (i.e. 0.5 or 1.0 ppm as an 8-hr average or 2 ppm over 15 minutes) require the employer to perform additional monitoring to reassess employee exposure.

A number of methods are suitable for measuring employee exposure to formaldehyde or for characterizing emissions within the worksite. The preamble to this standard describes some methods that have been widely used or subjected to validation testing. A detailed analytical procedure derived from the OSHA Method 52 for acrolein and formaldehyde is presented below for informational purposes.

Inclusion of OSHA's method in this appendix in no way implies that it is the only acceptable way to measure employee exposure to formaldehyde. Other methods that are free from significant interferences and that can determine formaldehyde at the permissible exposure limits within ± 25 percent of the "true" value at the 95 percent confidence level are also acceptable. Where applicable, the method should also be capab1e of measuring formaldehyde at the action level to ± 35 percent of the "true" value with a 95 percent confidence level. OSHA encourages employers to choose methods that will be best for their individual needs. The employer must exercise caution, however, in choosing an appropriate method since some techniques suffer from interferences that are likely to be present in workplaces of certain industry sectors where formaldehyde is used.

OSHA's Analytical Laboratory Method

Method No: 52

Matrix: Air

Target Concentration: 1 ppm (1.2 mg/m^3)

Procedures: Air samples are collected by drawing known volumes of air through sampling tubes containing XAD-2 adsorbent which have been coated with 2-(hydroxymethyl) piperidine. The samples are

1910 Selected 1910 General Industry Standards

desorbed with toluene and then analyzed by gas chromatography using a nitrogen selective detector.

Recommended Sampling Rate and Air Volumes: 0.1 L/min and 24 L

*Reliable Quantitation Limit:*16 ppb (20 μg/m^3)

Standard Error of Estimate at the Target Concentration: 7.3%

Status of the Method: A sampling and analytical method that has been subjected to the established evaluation procedures of the Organic Methods Evaluation Branch.

Date: March 1985

1. General Discussion.

1.1 *Background:* The current OSHA method for collecting acrolein vapor recommends the use of activated 13X molecular sieves. The samples must be stored in an ice bath during and after sampling and also they must be analyzed within 48 hours of collection. The current OSHA method for collecting formaldehyde vapor recommends the use of bubblers containing 10% methanol in water as the trapping solution.

This work was undertaken to resolve the sample stability problems associated with acrolein and also to eliminate the need to use bubblers to sample formaldehyde. A goal of this work was to develop and/or to evaluate a common sampling and analytical procedure for acrolein and formaldehyde.

NIOSH has developed independent methodologies for acrolein and formaldehyde which recommend the use of reagent-coated adsorbent tubes to collect the aldehydes as stable derivatives. The formaldehyde sampling tubes contain Chromosorb 102 adsorbent coated with N-benzylethanolamine (BEA) which reacts with formaldehyde vapor to form a stable oxazolidine compound. The acrolein sampling tubes contain XAD-2 adsorbent coated with 2-(hydroxymethyl)piperidine (2-HMP) which reacts with acrolein vapor to form a different, stable oxazolidine derivative. Acrolein does not appear to react with BEA to give a suitable reaction product. Therefore, the formaldehyde procedure cannot provide a common method for both aldehydes. However, formaldehyde does react with 2-HMP to form a very suitable reaction product. It is the quantitative reaction of acrolein and formaldehyde with 2-HMP that provides the basis for this evaluation.

This sampling and analytical procedure is very similar to the method recommended by NIOSH for acrolein. Some changes in the NIOSH methodology were necessary to permit the simultaneous determination of both aldehydes and also to accommodate OSHA laboratory equipment and analytical techniques.

1.2 *Limit-defining parameters:* The analyte air concentrations reported in this method are based on the recommended air volume for each analyte collected separately and a desorption volume of 1 mL. The amounts are presented as acrolein and/or formaldehyde, even though the derivatives are the actual species analyzed.

1.2.1 *Detection limits of the analytical procedure:* The detection limit of the analytical procedure was 386 pg per injection for formaldehyde. This was the amount of analyte which gave a peak whose height was about five times the height of the peak given by the residual formaldehyde derivative in a typical blank front section of the recommended sampling tube.

1.2.2 *Detection limits of the overall procedure:* The detection limits of the overall procedure were 482 ng per sample (16 ppb or 20 μg/m^3 for formaldehyde). This was the amount of analyte spiked on the sampling device which allowed recoveries approximately equal to the detection limit of the analytical procedure.

1.2.3 *Reliable quantitation limits:* The reliable quantitation limit was 482 ng per sample (16 ppb or 20 μg/m^3) for formaldehyde. These were the smallest amounts of analyte which could be quantitated within the limits of a recovery of at least 75% and a precision (± 1.96 SD) of ± 25% or better.

The reliable quantitation limit and detection limits reported in the method are based upon optimization of the instrument for the smallest possible amount of analyte. When the target concentration of an analyte is exceptionally higher than these limits, they may not be attainable at the routine operating parameters.

1.2.4 *Sensitivity:* The sensitivity of the analytical procedure over concentration ranges representing 0.4 to 2 times the target concentration, based on the recommended air volumes, was 7,589 area units per μg/mL for formaldehyde. This value was determined from the slope of the calibration curve. The sensitivity may vary with the particular instrument used in the analysis.

1.2.5 *Recovery:* The recovery of formaldehyde from samples used in an 18-day storage test remained above 92% when the samples were stored at ambient temperature. These values were determined from regression lines which were calculated from the storage data. The recovery of the analyte from the collection device must be at least 75% following storage.

1.2.6 *Precision (analytical method only):* The pooled coefficient of variation obtained from replicate determinations of analytical standards over the range of 0.4 to 2 times the target concentration was 0.0052 for formaldehyde (Section 4.3).

1.2.7 *Precision (overall procedure):* The precision at the 95% confidence level for the ambient temperature storage tests was ± 14.3% for formaldehyde. These values each include an additional ± 5% for sampling error. The overall procedure must provide results at the target concentrations that are ± 25% at the 95% confidence level.

1.2.8 *Reproducibility:* Samples collected from controlled test atmospheres and a draft copy of this procedure were given to a chemist unassociated with this evaluation. The formaldehyde samples were analyzed following 15 days storage. The average recovery was 96.3% and the standard deviation was 1.7%.

1.3 *Advantages:*

1.3.1 *The sampling and analytical procedures* permit the simultaneous determination of acrolein and formaldehyde.

1.3.2 *Samples are stable* following storage at ambient temperature for at least 18 days.

1.4 *Disadvantages:* None.

2. Sampling Procedure.

2.1 *Apparatus:*

2.1.1 *Samples are collected by use* of a personal sampling pump that can be calibrated to within ± 5% of the recommended 0.1 L/min sampling rate with the sampling tube in line.

2.1.2 *Samples are collected with laboratory* prepared sampling tubes. The sampling tube is constructed of silane treated glass and is about 8-cm long. The ID is 4 mm and the OD is 6 mm. One end of the tube is tapered so that a glass wool end plug will hold the contents of the tube in place during sampling. The other end of the sampling tube is open to its full 4-mm ID to facilitate packing of the tube. Both ends of the tube are fire-polished for safety. The tube is packed with a 75-mg backup section, located nearest the tapered end and a 150-mg sampling section of pretreated XAD-2 adsorbent which has been coated with 2-HMP. The two sections of coated adsorbent are separated and retained with small plugs of silanized glass wool. Following packing, the sampling tubes are sealed with two 7/32 inch OD plastic end caps. Instructions for the pretreatment and the coating of XAD-2 adsorbent are presented in Section 4 of this method.

2.1.3 *Sampling tubes,* similar to those recommended in this method, are marketed by Supelco, Inc. These tubes were not available when this work was initiated; therefore, they were not evaluated.

2.2 *Reagents:* None required.

2.3 *Technique:*

2.3.1 *Properly label the sampling tube* before sampling and then remove the plastic end caps.

2.3.2 *Attach the sampling tube* to the pump using a section of flexible plastic tubing such that the large, front section of the sampling tube is exposed directly to the atmosphere. Do not place any tubing ahead of the sampling tube. The sampling tube should be attached in the worker's breathing zone in a vertical manner such that it does not impede work performance.

2.3.3 *After sampling for the appropriate time,* remove the sampling tube from the pump and then seal the tube with plastic end caps.

2.3.4 *Include at least one blank* for each sampling set. The blank should be handled in the same manner as the samples with the exception that air is not drawn through it.

2.3.5 *List any potential interferences on the sample data sheet.*

2.4 *Breakthrough:*

2.4.1 Breakthrough was defined as the relative amount of analyte found on a backup sample in relation to the total amount of analyte collected on the sampling train.

2.4.2 *For formaldehyde collected from test atmospheres* containing 6 times the PEL, the average 5% breakthrough air volume was 41 L. The sampling rate was 0.1 L/min and the average mass of formaldehyde collected was 250 μg.

2.5 *Desorption Efficiency:* No desorption efficiency corrections are necessary to compute air sample results because analytical standards are prepared using coated adsorbent. Desorption efficiencies were determined, however, to investigate the recoveries of the analytes from the sampling device. The average recovery over the range of 0.4 to 2 times the target concentration, based on the recommended air volumes, was 96.2% for formaldehyde. Desorption efficiencies were essentially constant over the ranges studied.

2.6 *Recommended Air Volume and Sampling Rate:*

2.6.1 *The recommended air volume for formaldehyde is 24 L.*

2.6.2 *The recommended sampling rate is 0.1 L/min.*

2.7 *Interferences:*

2.7.1 *Any collected substance that is capable* of reacting 2-HMP and thereby depleting the derivatizing agent is a potential interference. Chemicals which contain a carbonyl group, such as acetone, may be capable of reacting with 2-HMP.

2.7.2 *There are no other known interferences to the sampling method.*

2.8 *Safety Precautions:*

2.8.1 *Attach the sampling equipment* to the worker in such a manner that it well not interfere with work performance or safety.

2.8.2 *Follow all safety practices* that apply to the work area being sampled.

3. Analytical Procedure.

3.1 *Apparatus:*

3.1.1 *A gas chromatograph (GC),* equipped with a nitrogen selective detector. A Hewlett-Packard Model 5840A GC fitted with a nitrogen-phosphorus flame ionization detector (NPD) was used for this evaluation. Injections were performed using a Hewlett-Packard Model 7671A automatic sampler.

3.1.2 *A GC column capable* of resolving the analytes from any interference. A 6 ft x 1/4 in OD (2mm ID) glass GC column containing 10% UCON 50-HB-5100 +2% KOH on 80/100 mesh Chromosorb W-AW was used for the evaluation. Injections were performed on-column.

3.1.3 *Vials, glass 2-mL with Teflon-lined caps.*

3.1.4 *Volumetric flasks, pipets, and syringes* for preparing standards, making dilutions, and performing injections.

3.2 *Reagents:*

3.2.1 *Toluene and dimethylformamide.* Burdick and Jackson solvents were used in this evaluation.

3.2.2 *Helium, hydrogen, and air, GC grade.*

3.2.3 *Formaldehyde, 37%, by weight, in water.* Aldrich Chemical, ACS Reagent Grade formaldehyde was used in this evaluation.

3.2.4 *Amberlite XAD-2 adsorbent* coated with 2-(hydroxymethyl — piperidine (2-HMP), 10% by weight (Section 4).

3.2.5 *Desorbing solution with internal standard.* This solution was prepared by adding 20 μL of dimethylformamide to 100 mL of toluene.

3.3 *Standard preparation:*

3.3.1 *Formaldehyde:* Prepare stock standards by diluting known volumes of 37% formaldehyde solution with methanol. A procedure to determine the formaldehyde content of these standards is presented in Section 4. A standard containing 7.7 mg/mL formaldehyde was prepared by diluting 1 mL of the 37% reagent to 50 mL with methanol.

3.3.2 *It is recommended that analytical standards* be prepared about 16 hours before the air samples are to be analyzed in order to ensure the complete reaction of the analytes with 2-HMP. However, rate studies have shown the reaction to be greater than 95% complete after 4 hours. Therefore, one or two standards can be analyzed after this reduced time if sample results are outside the concentration range of the prepared standards.

3.3.3 *Place 150-mg portions* of coated XAD-2 adsorbent, from the same lot number as used to collect the air samples, into each of several glass 2-mL vials. Seal each vial with a Teflon-lined cap.

3.3.4 *Prepare fresh analytical standards* each day by injecting appropriate amounts of the diluted analyte directly onto 150-mg portions of coated adsorbent. It is permissible to inject both acrolein and formaldehyde on the same adsorbent portion. Allow the standards to stand at room temperature. A standard, approximately the target levels, was prepared by injecting 11 μL of the acrolein and 12 μL of the formaldehyde stock standards onto a single coated XAD-2 adsorbent portion.

3.3.5 *Prepare a sufficient number of standards* to generate the calibration curves. Analytical standard concentrations should bracket sample concentrations. Thus, if samples are not in the concentration range of the prepared standards, additional standards must be prepared to determine detector response.

[1]3.3.7 *Desorb the standards* in the same manner as the samples following the 16-hour reaction time.

3.4 *Sample preparation:*

3.4.1 *Transfer the 150-mg section* of the sampling tube to a 2-mL vial. Place the 75-mg section in a separate vial. If the glass wool plugs contain a significant number of adsorbent beads, place them with the appropriate sampling tube section. Discard the glass wool plugs if they do not contain a significant number of adsorbent beads.

3.4.2 *Add 1 mL of desorbing solution to each vial.*

3.4.3 *Seal the vials with Teflon-lined caps* and then allow them to desorb for one hour. Shake the vials by hand with vigorous force several times during the desorption time.

3.4.4 *Save the used sampling tubes to be cleaned and recycled.*

3.5 *Analysis:*

3.5.1 *GC Conditions*

Column Temperature: Bi-level temperature program —
First level: 100 to 140 °C at 4 °C/min following completion of the first level.
Second level: 140 to 180 °C at 20 °C/min following completion of the first level.
Isothermal period: Hold column at 180 °C until the recorder pen returns to baseline (usually about 25 min after injection).
Injector temperature: 180 °C
Helium flow rate: 30 mL/min (detector response will be reduced if nitrogen is substituted for helium carrier gas).
Injection volume: 0.8 μL
GC column: Six-ft x 1/4-in OD (2 mm ID) glass GC column containing 10% UCON 50-HB-5100+2% KOH on 80/100 Chromosorb W-AW.
NPD conditions:
Hydrogen flow rate: 3 mL/min
Air flow rate: 50 mL/min
Detector temperature: 275 °C

3.5.2 *Chromatogram:* For an example of a typical chromatogram, see Figure 4.11 in OSHA Method 52.

3.5.3 *Use a suitable method,* such as electronic integration, to measure detector response.

3.5.4 *Use an internal standard method* to prepare the calibration curve with several standard solutions of different concentrations. Prepare the calibration curve daily. Program the integrator to report results in μg/mL.

3.5.5 *Bracket sample concentrations with standards.*

3.6 *Interferences (Analytical)*

3.6.1 *Any compound with the same general retention time* as the analytes and which also gives a detector response is a potential interference. Possible interferences should be reported to the laboratory with submitted samples by the industrial hygienist.

3.6.2 *GC parameters* (temperature, column, etc.) may be changed to circumvent interferences.

3.6.3 *A useful means of structure designation is GC/MS.* It is recommended this procedure be used to confirm samples whenever possible.

3.6.4 *The coated adsorbent* usually contains a very small amount of residual formaldehyde derivative (Section 4.8).

3.7 *Calculations:*

3.7.1 *Results are obtained by use of calibration curves.* Calibration curves are prepared by plotting detector response against concentration for each standard. The best line through the data points is determined by curve fitting.

3.7.2 *The concentration,* in μg/mL, for a particular sample is determined by comparing its detector response to the calibration curve. If either of the analytes is found on the backup section, it is added to the amount found on the front section. Blank corrections should be performed before adding the results together.

---

1. Editor's Note: The CFR skips number 3.3.6 in numbering this appendix.

3.7.3 *The acrolein and/or formaldehyde* air concentration can be expressed using the following equation:

$$mg/m^3 = (A)(B) / C$$

where A = μg/mL from 3.7.2, B = desorption volume, and C = L of air sampled.

No desorption efficiency corrections are required.

3.7.4 *The following equation can* be used to convert results in mg/m^3 to ppm.

$$ppm = (mg/m^3)(24.45) / MW$$

where

mg/m^3 =result from 3.7.3,

24.45 = molar volume of an ideal gas at 760 mm Hg and 25 °C,

MW = molecular weight (30.0).

4. Backup Data.

4.1 *Backup data on detection* limits, reliable quantitation limits, sensitivity and precision of the analytical method, breakthrough, desorption efficiency, storage, reproducibility, and generation of test atmospheres are available in OSHA Method 52, developed by the Organics Methods Evaluation Branch, OSHA Analytical Laboratory, Salt Lake City, Utah.

4.2 *Procedure to Coat XAD-2 Adsorbent with 2-HMP:*

4.2.1 *Apparatus:* Soxhlet extraction apparatus, rotary evaporation apparatus, vacuum dessicator, 1-L vacuum flask, 1-L round-bottomed evaporative flask, 1-L Erlenmeyer flask, 250-mL Buchner funnel with a coarse fritted disc, etc.

4.2.2 *Reagents:*

4.2.2.1 Methanol, isooctane, and toluene.

4.2.2.2 2-(Hydroxymethyl)piperidine.

4.2.2.3 Amberlite XAD-2 non-ionic polymeric adsorbent, 20 to 60 mesh, Aldrich Chemical XAD-2 was used in this evaluation.

4.2.3 *Procedure:* Weigh 125 g of crude XAD-2 adsorbent into a 1-L Erlenmeyer flask. Add about 200 mL of water to the flask and then swirl the mixture to wash the adsorbent. Discard any adsorbent that floats to the top of the water and then filter the mixture using a fritted Buchner funnel. Air dry the adsorbent for 2 minutes. Transfer the adsorbent back to the Erlenmeyer flask and then add about 200 mL of methanol to the flask. Swirl and then filter the mixture as before. Transfer the washed adsorbent back to the Erlenmeyer flask and then add about 200 mL of methanol to the flask. Swirl and then filter the mixture as before. Transfer the washed adsorbent to a 1-L round-bottomed evaporative flask, add 13 g of 2-HMP and then 200 mL of methanol, swirl the mixture and then allow it to stand for one hour. Remove the methanol at about 40 °C and reduced pressure using a rotary evaporation apparatus. Transfer the coated adsorbent to a suitable container and store it in a vacuum desiccator at room temperature overnight. Transfer the coated adsorbent to a Soxhlet extractor and then extract the material with toluene for about 24 hours. Discard the contaminated toluene, add methanol in its place and then continue the Soxhlet extraction for an additional 4 hours. Transfer the adsorbent to a weighted 1-L round-bottom evaporative flask and remove the methanol using the rotary evaporation apparatus. Determine the weight of the adsorbent and then add an amount of 2-HMP, which is 10% by weight of the adsorbent. Add 200 mL of methanol and then swirl the mixture. Allow the mixture to stand for one hour. Remove the methanol by rotary evaporation. Transfer the coated adsorbent to a suitable container and store it in a vacuum desiccator until all traces of solvents are gone. Typically, this will take 2-3 days. The coated adsorbent should be protected from contamination. XAD-2 adsorbent treated in this manner will probably not contain residual acrolein derivative. However, this adsorbent will often contain residual formaldehyde derivative levels of about 0.1 μg per 150 mg of adsorbent. If the blank values for a batch of coated adsorbent are too high, then the batch should be returned to the Soxhlet extractor, extracted with toluene again and then recoated. This process can be repeated until the desired blank levels are attained.

The coated adsorbent is now ready to be packed into sampling tubes. The sampling tubes should be stored in a sealed container to prevent contamination. Sampling tubes should be stored in the dark at room temperature. The sampling tubes should be segregated by coated adsorbent lot number. A sufficient amount of each lot number of coated adsorbent should be retained to prepare analytical standards for use with air samples from that lot number.

4.3 *A Procedure to Determine Formaldehyde by Acid Titration:* Standardize the 0.1 N HCl solution using sodium carbonate and methyl orange indicator.

Place 50 mL of 0.1 M sodium sulfite and three drops of thymophthalein indicator into a 250-mL Erlenmeyer flask. Titrate the contents of the flask to a colorless endpoint with 0.1 N HCl (usually one or two drops is sufficient). Transfer 10 mL of the formaldehyde/methanol solution (prepared in 3.3.1) into the same flask and titrate the mixture with 0.1 N HCl, again, to a colorless endpoint. The formaldehyde concentration of the standard may be calculated by the following equation:

$$\text{Formaldehyde, mg/mL} = \frac{\text{acid titer x acid normality x 30.0}}{\text{mL of sample}}$$

This method is based on the quantitative liberation of sodium hydroxide when formaldehyde reacts with sodium sulfite to form the formaldehyde-bisulfite addition product. The volume of sample may be varied depending on the formaldehyde content but the solution to be titrated must contain excess sodium sulfite. Formaldehyde solutions containing substantial amounts of acid or base must be neutralized before analysis.

§1910.1048 Appendix C

Medical Surveillance — Formaldehyde

I. Health Hazards.

The occupational health hazards of formaldehyde are primarily due to its toxic effects after inhalation, after direct contact with the skin or eyes by formaldehyde in liquid or vapor form, and after ingestion.

II. Toxicology.

A. *Acute Effects of Exposure*

1. *Inhalation (breathing):* Formaldehyde is highly irritating to the upper airways. The concentration of formaldehyde that is immediately dangerous to life and health is 100 ppm. Concentrations above 50 ppm can cause severe pulmonary reactions within minutes. These include pulmonary edema, pneumonia, and bronchial irritation which can result in death. Concentrations above 5 ppm readily cause lower airway irritation characterized by cough, chest tightness and wheezing. There is some controversy regarding whether formaldehyde gas is a pulmonary sensitizer which can cause occupational asthma in a previously normal individual. Formaldehyde can produce symptoms of bronchial asthma in humans. The mechanism may be either sensitization of the individual by exposure to formaldehyde or direct irritation by formaldehyde in persons with pre-existing asthma. Upper airway irritation is the most common respiratory effect reported by workers and can occur over a wide range of concentrations, most frequently above 1 ppm. However, airway irritation has occurred in some workers with exposures to formaldehyde as low as 0.1 ppm. Symptoms of upper airway irritation include dry or sore throat, itching and burning sensations of the nose, and nasal congestion. Tolerance to this level of exposure may develop within 1-2 hours. This tolerance can permit workers remaining in an environment of gradually increasing formaldehyde concentrations to be unaware of their increasingly hazardous exposure.
2. *Eye contact:* Concentrations of formaldehyde between 0.05 ppm and 0.5 ppm produce a sensation of irritation in the eyes with burning, itching, redness, and tearing. Increased rate of blinking and eye closure generally protects the eye from damage at these low levels, but these protective mechanisms may interfere with some workers' work abilities. Tolerance can occur in workers continuously exposed to concentrations of formaldehyde in this range. Accidental splash injuries of human eyes to aqueous solutions of formaldehyde (formalin) have resulted in a wide range of ocular injuries including corneal opacities and blindness. The severity of the reactions have been directly dependent on the concentration of formaldehyde in solution and the amount of time lapsed before emergency and medical intervention.
3. *Skin contact:* Exposure to formaldehyde solutions can cause irritation of the skin and allergic contact dermatitis. These skin diseases and disorders can occur at levels well below those encountered by many formaldehyde workers. Symptoms include erythema, edema, and vesiculation or hives. Exposure to liquid formalin or formaldehyde vapor can provoke skin reactions in sensitized individuals even when airborne concentrations of formaldehyde are well below 1 ppm.
4. *Ingestion:* Ingestion of as little as 30 ml of a 37 percent solution of formaldehyde (formalin) can result in death. Gastrointestinal toxicity after ingestion is most severe in the stomach and results in symp-

toms which can include nausea, vomiting, and servere abdominal pain. Diverse damage to other organ systems including the liver, kidney, spleen, pancreas, brain, and central nervous systems can occur from the acute response to ingestion of formaldehyde.

B. *Chronic Effects of Exposure*

Long term exposure to formaldehyde has been shown to be associated with an increased risk of cancer of the nose and accessory sinuses, nasopharyngeal and oropharyngeal cancer, and lung cancer in humans. Animal experiments provide conclusive evidence of a causal relationship between nasal cancer in rats and formaldehyde exposure. Concordant evidence of carcinogenicity includes DNA binding, genotoxicity in short-term tests, and cytotoxic changes in the cells of the target organ suggesting both preneoplastic changes and a dose-rate effect. Formaldehyde is a complete carcinogen and appears to exert an effect on at least two stages of the carcinogenic process.

III. Surveillance considerations.

A. *History*

1. *Medical and occupational history:* Along with its acute irritative effects, formaldehyde can cause allergic sensitization and cancer. One of the goals of the work history should be to elicit information on any prior or additional exposure to formaldehyde in either the occupational or the non-occupational setting.
2. *Respiratory history:* As noted above, formaldehyde has recognized properties as an airway irritant and has been reported by some authors as a cause of occupational asthma. In addition, formaldehyde has been associated with cancer of the entire respiratory system of humans. For these reasons, it is appropriate to include a comprehensive review of the respiratory system in the medical history. Components of this history might include questions regarding dyspnea on exertion, shortness of breath, chronic airway complaints, hyperreactive airway disease, rhinitis, bronchitis, bronchiolitis, asthma, emphysema, respiratory allergic reaction, or other preexisting pulmonary disease.

   In addition, generalized airway hypersensitivity can result from exposures to a single sensitizing agent. The examiner should, therefore, elicit any prior history of exposure to pulmonary irritants, and any short- or long-term effects of that exposure.

   Smoking is known to decrease mucociliary clearance of materials deposited during respiration in the nose and upper airways. This may increase a worker's exposure to inhaled materials such as formaldehyde vapor. In addition, smoking is a potential confounding factor in the investigation of any chronic respiratory disease, including cancer. For these reasons, a complete smoking history should be obtained.
3. *Skin Disorders:* Because of the dermal irritant and sensitizing effects of formaldehyde, a history of skin disorders should be obtained. Such a history might include the existence of skin irritation, previously documented skin sensitivity, and other dermatologic disorders. Previous exposure to formaldehyde and other dermal sensitizers should be recorded.
4. *History of atopic or allergic* diseases: Since formaldehyde can cause allergic sensitization of the skin and airways, it might be useful to identify individuals with prior allergen sensitization. A history of atopic disease and allergies to formaldehyde or any other substances should also be obtained. It is not definitely known at this time whether atopic diseases and allergies to formaldehyde or any other substances should also be obtained. Also it is not definitely known at this time whether atopic individuals have a greater propensity to develop formaldehyde sensitivity than the general population, but identification of these individuals may be useful for ongoing surveillance.
5. *Use of disease questionnaires:* Comparison of the results from previous years with present results provides the best method for detecting a general deterioration in health when toxic signs and symptoms are measured subjectively. In this way recall bias does not affect the results of the analysis. Consequently, OSHA has determined that the findings of the medical and work histories should be kept in a standardized form for comparison of the year-to-year results.

B. *Physical Examination*

1. *Mucosa of eyes and airways:* Because of the irritant effects of formaldehyde, the examining physician should be alert to evidence of this irritation. A speculum examination of the nasal mucosa may be helpful in assessing possible irritation and cytotoxic changes, as may be indirect inspection of the posterior pharynx by mirror.
2. *Pulmonary system:* A conventional respiratory examination, including inspection of the thorax and auscultation and percussion of the lung fields should be performed as part of the periodic medical examination. Although routine pulmonary function testing is only required by the standard once every year for persons who are exposed over the TWA concentration limit, these tests have an obvious value in investigating possible respiratory dysfunction and should be used wherever deemed appropriate by the physician. In cases of alleged formaldehyde-induced airway disease, other possible causes of pulmonary disfunction (including exposures to other substances) should be ruled out. A chest radiograph may be useful in these circumstances. In cases of suspected airway hypersensitivity or allergy, it may be appropriate to use bronchial challenge testing with formaldehyde or methacholine to determine the nature of the disorder. Such testing should be performed by or under the supervision of a physician experienced in the procedures involved.
3. *Skin:* The physician should be alert to evidence of dermal irritation of sensitization, including reddening and inflammation, urticaria, blistering, scaling, formation of skin fissures, or other symptoms. Since the integrity of the skin barrier is compromised by other dermal diseases, the presence of such disease should be noted. Skin sensitivity testing carries with it some risk of inducing sensitivity, and therefore, skin testing for formaldehyde sensitivity should not be used as a routine screening test. Sensitivity testing may be indicated in the investigation of a suspected existing sensitivity. Guidelines for such testing have been prepared by the North American Contact Dermatitis Group.

C. *Additional Examinations or Tests*

The physician may deem it necessary to perform other medical examinations or tests as indicated. The standard provides a mechanism whereby these additional investigations are covered under the standard for occupational exposure to formaldehyde.

D. *Emergencies*

The examination of workers exposed in an emergency should be directed at the organ systems most likely to be affected. Much of the content of the examination will be similar to the periodic examination unless the patient has received a severe acute exposure requiring immediate attention to prevent serious consequences. If a severe overexposure requiring medical intervention or hospitalization has occurred, the physician must be alert to the possibility of delayed symptoms. Followup nonroutine examinations may be necessary to assure the patient's well-being.

E. *Employer Obligations*

The employer is required to provide the physician with the following information: A copy of this standard and appendices A, C, D, and E; a description of the affected employee's duties as they relate to his or her exposure concentration; an estimate of the employee's exposure including duration (e.g. 15 hr/wk, three 8-hour shifts, full-time); a description of any personal protective equipment, including respirators, used by the employee; and the results of any previous medical determinations for the affected employee related to formaldehyde exposure to the extent that this information is within the employer's control.

F. *Physician's Obligations*

The standard requires the employer to obtain a written statement from the physician. This statement must contain the physician's opinion as to whether the employee has any medical condition which would place him or her at increased risk of impaired health from exposure to formaldehyde or use of respirators, as appropriate. The physician must also state his opinion regarding any restrictions that should be placed on the employee's exposure to formaldehyde or upon the use of protective clothing or equipment such as respirators. If the employee wears a respirator as a result of his or her exposure to formaldehyde, the physician's opinion must also contain a statement regarding the suitability of the employee to wear the type of respirator assigned. Finally, the physician must inform the employer that the employee has been told the results of the medical examination and of any medical conditions which require further explanation or treatment. This written opinion is not to contain any information on specific findings or diagnoses unrelated to occupational exposure to formaldehyde.

The purpose in requiring the examining physician to supply the employer with a written opinion is to provide the employer with a medical basis to assist the employer in placing employees initially, in assuring that their health is not being impaired by formaldehyde, and to assess the employee's ability to use any required protective equipment.

§1910.1048 Appendix D
Nonmandatory Medical Disease Questionnaire

Appendix D to §1910.1048
Medical Disease Questionnaire (non-mandatory)

A. IDENTIFICATION

PLANT NAME: ____ DATE: ___–___–___ (MONTH DAY YEAR)

EMPLOYEE NAME: ____

SOCIAL SECURITY NUMBER: ___–___–___ JOB TITLE: ____

BIRTHDATE: ___/___/___ (MONTH DAY YEAR) AGE: ___ SEX: M ☐ F ☐ HEIGHT: ___ FEET ___ INCHES WEIGHT: ___ LBS.

B. MEDICAL HISTORY

1. Have you ever been in the hospital as a patient? ☐ Yes ☐ No
   If yes, what kind of problem were you having? ____
2. Have you ever had any kind of operation? ☐ Yes ☐ No
   If yes, what kind? ____
3. Do you take any kind of medicine regularly? ☐ Yes ☐ No
   If yes, what kind? ____
4. Are you allergic to any drugs, foods, or chemicals? ☐ Yes ☐ No
   If yes, what kind of allergy is it? ____
   What causes the allergy? ____
5. Have you ever been told that you have asthma, hayfever, or sinusitis? ☐ Yes ☐ No
6. Have you ever been told that you have emphysema, bronchitis, or any other respiratory problems? ☐ Yes ☐ No
7. Have you ever been told that you had hepatitis? ☐ Yes ☐ No
8. Have you ever been told that you had cirrhosis? ☐ Yes ☐ No
9. Have you ever been told that you had cancer? ☐ Yes ☐ No
10. Have you ever had arthritis or joint pain? ☐ Yes ☐ No
11. Have you ever been told that you had high blood pressure? ☐ Yes ☐ No
12. Have you ever had a heart attack or heart trouble? ☐ Yes ☐ No

B-1. MEDICAL HISTORY UPDATE

1. Have you been in the hospital as a patient any time within the past year? ☐ Yes ☐ No
   If so, for what condition? ____
2. Have you been under the care of a physician during the past year? ☐ Yes ☐ No
   If so, for what condition? ____
3. Is there any change in your breathing since last year? ☐ Yes ☐ No
   ☐ Better? ☐ Worse? ☐ No change?
   If change, do you know why? ____
4. Is your general health different this year from last year? ☐ Yes ☐ No
   If different, in what way? ____
5. Have you in the past year or are you now taking any medication on a regular basis? ☐ Yes ☐ No
   Name Rx ____
   Condition being treated: ____

C. OCCUPATIONAL HISTORY

1. How long have you worked for your present employer? ____
2. What jobs have you held with this employer? Include job title and length of time in each job. ____
3. In each of these jobs, how many hours a day were you exposed to chemicals? ____
4. What chemicals have you worked with most of the time? ____
5. Have you ever noticed any type of skin rash you feel was related to your work? ☐ Yes ☐ No
6. Have you ever noticed that any kind of chemical makes you cough? ☐ Yes ☐ No Wheeze? ☐ Yes ☐ No
   Become short of breath or cause your chest to become tight? ☐ Yes ☐ No
7. Are you exposed to any dust or chemicals at home? ☐ Yes ☐ No
   If yes, explain: ____
8. In other jobs, have you ever had exposure to:
   Wood dust? ☐ Yes ☐ No Nickel or chromium? ☐ Yes ☐ No Silica (foundry, sand blasting)? ☐ Yes ☐ No
   Arsenic or asbestos? ☐ Yes ☐ No Organic solvents? ☐ Yes ☐ No Urethane foams? ☐ Yes ☐ No

C-1. OCCUPATIONAL HISTORY UPDATE

1. Are you working on the same job this year as you were last year? ☐ Yes ☐ No
   If not, how has your job changed? ____
2. What chemicals are you exposed to on your job. ____
3. How many hours a day are you exposed to chemicals? ____
4. Have you noticed any skin rash within the past year you feel was related to your work? ☐ Yes ☐ No
   If so, explain circumstances: ____
5. Have you noticed that any chemical makes you cough, be short of breath, or wheeze? ☐ Yes ☐ No
   If so, can you identify it? ____

* Full-size forms available free of charge at www.oshacfr.com

Appendix D to §1910.1048
Medical Disease Questionnaire (non-mandatory)
(continued)

D. MISCELLANEOUS

1. Do you smoke? ☐ Yes ☐ No
   If so, how much and for how long? Pipe: ____ Cigars: ____ Cigarettes: ____
2. Do you drink alcohol in any form? ☐ Yes ☐ No
   If so, how much, how long and how often? ____
3. Do you wear glasses or contact lenses? ☐ Yes ☐ No
4. Do you get any physical exercise other than that required to do your job? ☐ Yes ☐ No
   If so, explain: ____
5. Do you have any hobbies or "side jobs" that require you to use chemicals, such as furniture stripping, sand blasting, insulation or manufacture of urethane foam, furniture, etc.? ☐ Yes ☐ No
   If so, please describe, giving type of business or hobby, chemicals used and length of exposures. ____

E. SYMPTOMS QUESTIONNAIRE

1. Do you ever have any shortness of breath? ☐ Yes ☐ No
   If Yes, do you have to rest after climbing several flights of stairs? ☐ Yes ☐ No
   If Yes, if you walk on the level with people your own age, do you walk slower than they do? ☐ Yes ☐ No
   If Yes, if you walk slower than a normal pace, do you have to limit the distance that you walk? ☐ Yes ☐ No
   If Yes, do you have to stop and rest while bathing or dressing? ☐ Yes ☐ No
2. Do you cough as much as three months out of the year? ☐ Yes ☐ No
   If Yes, have you had this cough for more than two years? ☐ Yes ☐ No
   If Yes, do you ever cough anything up from your chest? ☐ Yes ☐ No
3. Do you ever have a feeling of smothering, unable to take a deep breath, or tightness in your chest? ☐ Yes ☐ No
   If Yes, do you notice that this is on any particular day of the week? ☐ Yes ☐ No
   If Yes, what day of the week? ☐ Mon ☐ Tues ☐ Wed ☐ Thurs ☐ Fri ☐ Sat ☐ Sun
   If Yes, do you notice that this occurs at any particular place? ☐ Yes ☐ No
   If Yes, do you notice that this is worse after you have returned to work after being off for several days? ☐ Yes ☐ No
4. Have you ever noticed any wheezing in your chest? ☐ Yes ☐ No
   If Yes, is this only with colds or other infections? ☐ Yes ☐ No
   Is this caused by exposure to any kind of dust or other material? ☐ Yes ☐ No
   If Yes, what kind? ____
5. Have you noticed any burning, tearing, or redness of your eyes when you are at work? ☐ Yes ☐ No
   If so, explain circumstances: ____
6. Have you noticed any sore or burning throat or itchy or burning nose when you are at work? ☐ Yes ☐ No
   If so, explain circumstances: ____
7. Have you noticed any stuffiness or dryness of your nose? ☐ Yes ☐ No
8. Do you ever have swelling of the eyelids or face? ☐ Yes ☐ No
9. Have you ever been jaundiced? ☐ Yes ☐ No
   If Yes, was this accompanied by any pain? ☐ Yes ☐ No
10. Have you ever had a tendency to bruise easily or bleed excessively? ☐ Yes ☐ No
11. Do you have frequent headaches that are not relieved by aspirin or Tylenol? ☐ Yes ☐ No
    If Yes, do they occur at any particular time of the day or week? ☐ Yes ☐ No
    If Yes, when do they occur? ____
12. Do you have frequent episodes of nervousness or irritability? ☐ Yes ☐ No
13. Do you tend to have trouble concentrating or remembering? ☐ Yes ☐ No
14. Do you ever feel dizzy, light-headed, excessively drowsy or like you have been drugged? ☐ Yes ☐ No
15. Does your vision ever become blurred? ☐ Yes ☐ No
16. Do you have numbness or tingling of the hands or feet or other parts of your body? ☐ Yes ☐ No
17. Have you ever had chronic weakness or fatigue? ☐ Yes ☐ No
18. Have you ever had any swelling of your feet or ankles to the point where you could not wear your shoes? ☐ Yes ☐ No
19. Are you bothered by heartburn or indigestion? ☐ Yes ☐ No
20. Do you ever have itching, dryness, or peeling and scaling of the hands? ☐ Yes ☐ No
21. Do you ever have a burning sensation in the hands, or reddening of the skin? ☐ Yes ☐ No
22. Do you ever have cracking or bleeding of the skin on your hands? ☐ Yes ☐ No
23. Are you under a physician's care? ☐ Yes ☐ No
    If Yes, for what are you being treated? ____
24. Do you have any physical complaints today? ☐ Yes ☐ No
    If Yes, explain: ____
25. Do you have other health conditions not covered by these questions? ☐ Yes ☐ No
    If Yes, explain: ____

* Full-size forms available free of charge at www.oshacfr.com

[57 FR 22310, May 27, 1992; 57 FR 27161, June 18, 1992; 61 FR 5508, Feb. 13, 1996; 63 FR 1292, Jan. 8, 1998; 63 FR 20099, Apr. 23, 1998; 70 FR 1143, Jan. 5, 2005; 71 FR 16672, 16673, Apr. 3, 2006; 71 FR 50190, Aug. 24, 2006; 73 FR 75586, Dec. 12, 2008; 77 FR 17784, Mar. 26, 2012]

§1910.1052
Methylene chloride

This occupational health standard establishes requirements for employers to control occupational exposure to methylene chloride (MC). Employees exposed to MC are at increased risk of developing cancer, adverse effects on the heart, central nervous system and liver, and skin or eye irritation. Exposure may occur through inhalation, by absorption through the skin, or through contact with the skin. MC is a solvent which is used in many different types of work activities, such as paint stripping, polyurethane foam manufacturing, and cleaning and degreasing. Under the requirements of paragraph (d) of this section, each covered employer must make an initial determination of each employee's exposure to MC. If the employer determines that employees are exposed below the action level, the only other provisions of this section that apply are that a record must be made of the determination, the employees must receive information and training under paragraph (l) of this section and, where appropriate, employees must be protected from contact with liquid MC under paragraph (h) of this section. The provisions of the MC standard are as follows:

(a) **Scope and application.** This section applies to all occupational exposures to methylene chloride (MC), Chemical Abstracts Service Registry Number 75-09-2, in general industry, construction and shipyard employment.[§1910.1052(a)]

(b) **Definitions.** For the purposes of this section, the following definitions shall apply:

Action level means a concentration of airborne MC of 12.5 parts per million (ppm) calculated as an eight (8)-hour time-weighted average (TWA).

Assistant Secretary means the Assistant Secretary of Labor for Occupational Safety and Health, U.S. Department of Labor, or designee.

Authorized person means any person specifically authorized by the employer and required by work duties to be present in regulated areas, or any person entering such an area as a designated representative of employees for the purpose of exercising the right to observe monitoring and measuring procedures under paragraph (d) of this section, or any other person authorized by the OSH Act or regulations issued under the Act.

Director means the Director of the National Institute for Occupational Safety and Health, U.S. Department of Health and Human Services, or designee.

Emergency means any occurrence, such as, but not limited to, equipment failure, rupture of containers, or failure of control equipment, which results, or is likely to result in an uncontrolled release of MC. If an incidental release of MC can be controlled by employees such as maintenance personnel at the time of release and in accordance with the leak/spill provisions required by paragraph (f) of this section, it is not considered an emergency as defined by this standard.

Employee exposure means exposure to airborne MC which occurs or would occur if the employee were not using respiratory protection.

Methylene chloride (MC) means an organic compound with chemical formula, CH_2Cl_2. Its Chemical Abstracts Service Registry Number is 75-09-2. Its molecular weight is 84.9 g/mole.

Physician or other licensed health care professional is an individual whose legally permitted scope of practice (i.e., license, registration, or certification) allows him or her to independently provide or be delegated the responsibility to provide some or all of the health care services required by paragraph (j) of this section.

Regulated area means an area, demarcated by the employer, where an employee's exposure to airborne concentrations of MC exceeds or can reasonably be expected to exceed either the 8-hour TWA PEL or the STEL.

Symptom means central nervous system effects such as headaches, disorientation, dizziness, fatigue, and decreased attention span; skin effects such as chapping, erythema, cracked skin, or skin burns; and cardiac effects such as chest pain or shortness of breath.

This section means this methylene chloride standard.

(c) **Permissible exposure limits (PELs).**[§1910.1052(c)]

(1) *Eight-hour time-weighted average (TWA) PEL.* The employer shall ensure that no employee is exposed to an airborne concentration of MC in excess of twenty-five parts of MC per million parts of air (25 ppm) as an 8-hour TWA.[§1910.1052(c)(1)]

(2) *Short-term exposure limit (STEL).* The employer shall ensure that no employee is exposed to an airborne concentration of

MC in excess of one hundred and twenty-five parts of MC per million parts of air (125 ppm) as determined over a sampling period of fifteen minutes.[§1910.1052(c)(2)]

(d) **Exposure monitoring.**[§1910.1052(d)]

(1) *Characterization of employee exposure.*[§1910.1052(d)(1)]

(i) *Where MC is present in the workplace,* the employer shall determine each employee's exposure by either: [§1910.1052(d)(1)(i)]

[A] Taking a personal breathing zone air sample of each employee's exposure; or[§1910.1052(d)(1)(i)[A]]

[B] Taking personal breathing zone air samples that are representative of each employee's exposure. [§1910.1052(d)(1)(i)[B]]

(ii) *Representative samples.* The employer may consider personal breathing zone air samples to be representative of employee exposures when they are taken as follows: [§1910.1052(d)(1)(ii)]

[A] 8-hour TWA PEL. The employer has taken one or more personal breathing zone air samples for at least one employee in each job classification in a work area during every work shift, and the employee sampled is expected to have the highest MC exposure.[§1910.1052(d)(1)(ii)[A]]

[B] Short-term exposure limits. The employer has taken one or more personal breathing zone air samples which indicate the highest likely 15-minute exposures during such operations for at least one employee in each job classification in the work area during every work shift, and the employee sampled is expected to have the highest MC exposure.[§1910.1052(d)(1)(ii)[B]]

[C] Exception. Personal breathing zone air samples taken during one work shift may be used to represent employee exposures on other work shifts where the employer can document that the tasks performed and conditions in the workplace are similar across shifts.[§1910.1052(d)(1)(ii)[C]]

(iii) *Accuracy of monitoring.* The employer shall ensure that the methods used to perform exposure monitoring produce results that are accurate to a confidence level of 95 percent, and are:[§1910.1052(d)(1)(iii)]

[A] Within plus or minus 25 percent for airborne concentrations of MC above the 8-hour TWA PEL or the STEL; or [§1910.1052(d)(1)(iii)[A]]

[B] Within plus or minus 35 percent for airborne concentrations of MC at or above the action level but at or below the 8-hour TWA PEL.[§1910.1052(d)(1)(iii)[B]]

(2) *Initial determination.* Each employer whose employees are exposed to MC shall perform initial exposure monitoring to determine each affected employee's exposure, except under the following conditions:[§1910.1052(d)(2)]

(i) *Where objective data demonstrate* that MC cannot be released in the workplace in airborne concentrations at or above the action level or above the STEL. The objective data shall represent the highest MC exposures likely to occur under reasonably foreseeable conditions of processing, use, or handling. The employer shall document the objective data exemption as specified in paragraph (m) of this section; [§1910.1052(d)(2)(i)]

(ii) *Where the employer has performed* exposure monitoring within 12 months prior to April 10, 1997 and that exposure monitoring meets all other requirements of this section, and was conducted under conditions substantially equivalent to existing conditions; or[§1910.1052(d)(2)(ii)]

(iii) *Where employees are exposed to MC* on fewer than 30 days per year (e.g., on a construction site), and the employer has measurements by direct-reading instruments which give immediate results (such as a detector tube) and which provide sufficient information regarding employee exposures to determine what control measures are necessary to reduce exposures to acceptable levels.[§1910.1052(d)(2)(iii)]

(3) *Periodic monitoring.* Where the initial determination shows employee exposures at or above the action level or above the STEL, the employer shall establish an exposure monitoring program for periodic monitoring of employee exposure to MC in accordance with Table 1:[§1910.1052(d)(3)]

Table 1 — Initial Determination Exposure Scenarios and Their Associated Monitoring Frequencies

Exposure scenario	Required monitoring activity
Below the action level and at or below the STEL.	No 8-hour TWA or STEL monitoring required.
Below the action level and above the STEL.	No 8-hour TWA monitoring required; monitor STEL exposures every three months.
At or above the action level, at or below the TWA, and at or below the STEL.	Monitor 8-hour TWA exposures every six months.
At or above the action level, at or below the TWA, and above the STEL.	Monitor 8-hour TWA exposures every six months and monitor STEL exposures every three months.
Above the TWA and at or below the STEL.	Monitor 8-hour TWA exposures every three months. In addition, without regard to the last sentence of the note to paragraph (d)(3), the following employers must monitor STEL exposures every three months until either the date by which they must achieve the 8-hour TWA PEL under paragraph (n) of this section or the date by which they in fact achieve the 8-hour TWA PEL, whichever comes first: employers engaged in polyurethane foam manufacturing; foam fabrication; furniture refinishing; general aviation aircraft stripping; product formulation; use of MC-based adhesives for boat building and repair, recreational vehicle manufacture, van conversion, or upholstery; and use of MC in construction work for restoration and preservation of buildings, painting and paint removal, cabinet making, or floor refinishing and resurfacing.
Above the TWA and above the STEL.	Monitor 8-hour TWA exposures and STEL exposures every three months.

[*Note to paragraph (d)(3):* The employer may decrease the frequency of 8-hour TWA exposure monitoring to every six months when at least two consecutive measurements taken at least seven days apart show exposures to be at or below the 8-hour TWA PEL. The employer may discontinue the periodic 8-hour TWA monitoring for employees where at least two consecutive measurements taken at least seven days apart are below the action level. The employer may discontinue the periodic STEL monitoring for employees where at least two consecutive measurements taken at least 7 days apart are at or below the STEL.]

(4) *Additional monitoring.*[§1910.1052(d)(4)]

(i) *The employer shall perform exposure monitoring* when a change in workplace conditions indicates that employee exposure may have increased. Examples of situations that may require additional monitoring include changes in production, process, control equipment, or work practices, or a leak, rupture, or other breakdown.[§1910.1052(d)(4)(i)]

(ii) *Where exposure monitoring is performed* due to a spill, leak, rupture or equipment breakdown, the employer shall clean-up the MC and perform the appropriate repairs before monitoring.[§1910.1052(d)(4)(ii)]

(5) *Employee notification of monitoring results.*[§1910.1052(d)(5)]

(i) *The employer shall,* within 15 working days after the receipt of the results of any monitoring performed under this section, notify each affected employee of these results in writing, either individually or by posting of results in an appropriate location that is accessible to affected employees. [§1910.1052(d)(5)(i)]

(ii) *Whenever monitoring results* indicate that employee exposure is above the 8-hour TWA PEL or the STEL, the employer shall describe in the written notification the corrective action being taken to reduce employee exposure to or below the 8-hour TWA PEL or STEL and the schedule for completion of this action.[§1910.1052(d)(5)(ii)]

(6) *Observation of monitoring*[§1910.1052(d)(6)]

(i) *Employee observation.* The employer shall provide affected employees or their designated representatives an opportunity to observe any monitoring of employee exposure to MC conducted in accordance with this section.[§1910.1052(d)(6)(i)]

(ii) *Observation procedures.* When observation of the monitoring of employee exposure to MC requires entry into an area where the use of protective clothing or equipment is required, the employer shall provide, at no cost to the observer(s), and the observer(s) shall be required to use such clothing and equipment and shall comply with all other applicable safety and health procedures.[§1910.1052(d)(6)(ii)]

(e) **Regulated areas.**[§1910.1052(e)]

(1) *The employer shall establish a regulated area* wherever an employee's exposure to airborne concentrations of MC exceeds or can reasonably be expected to exceed either the 8-hour TWA PEL or the STEL.[§1910.1052(e)(1)]

(2) *The employer shall limit access* to regulated areas to authorized persons.[§1910.1052(e)(2)]

(3) *The employer shall supply a respirator,* selected in accordance with paragraph (h)(3) of this section, to each person who enters a regulated area and shall require each affected employee to use that respirator whenever MC exposures are likely to exceed the 8-hour TWA PEL or STEL.[§1910.1052(e)(3)]

[*Note to paragraph (e)(3):* An employer who has implemented all feasible engineering, work practice and administrative controls (as required in paragraph (f) of this section), and who has established a regulated area (as required by paragraph (e)(1) of this section) where MC exposure can be reliably predicted to exceed the 8-hour TWA PEL or the STEL only on certain days (for example, because of work or process schedule) would need to have affected employees use respirators in that regulated area only on those days.]

(4) *The employer shall ensure that,* within a regulated area, employees do not engage in non-work activities which may increase dermal or oral MC exposure.[§1910.1052(e)(4)]

(5) *The employer shall ensure* that while employees are wearing respirators, they do not engage in activities (such as taking medication or chewing gum or tobacco) which interfere with respirator seal or performance.[§1910.1052(e)(5)]

(6) *The employer shall demarcate regulated areas* from the rest of the workplace in any manner that adequately establishes and alerts employees to the boundaries of the area and minimizes the number of authorized employees exposed to MC within the regulated area.[§1910.1052(e)(6)]

(7) *An employer at a multi-employer worksite* who establishes a regulated area shall communicate the access restrictions and locations of these areas to all other employers with work operations at that worksite.[§1910.1052(e)(7)]

(f) Methods of compliance.[§1910.1052(f)]

(1) *Engineering and work practice controls.* The employer shall institute and maintain the effectiveness of engineering controls and work practices to reduce employee exposure to or below the PELs except to the extent that the employer can demonstrate that such controls are not feasible. Wherever the feasible engineering controls and work practices which can be instituted are not sufficient to reduce employee exposure to or below the 8-TWA PEL or STEL, the employer shall use them to reduce employee exposure to the lowest levels achievable by these controls and shall supplement them by the use of respiratory protection that complies with the requirements of paragraph (g) of this section.[§1910.1052(f)(1)]

(2) *Prohibition of rotation.* The employer shall not implement a schedule of employee rotation as a means of compliance with the PELs.[§1910.1052(f)(2)]

(3) *Leak and spill detection.*[§1910.1052(f)(3)]

(i) *The employer shall implement procedures* to detect leaks of MC in the workplace. In work areas where spills may occur, the employer shall make provisions to contain any spills and to safely dispose of any MC-contaminated waste materials. [§1910.1052(f)(3)(i)]

(ii) *The employer shall ensure that all incidental leaks* are repaired and that incidental spills are cleaned promptly by employees who use the appropriate personal protective equipment and are trained in proper methods of cleanup. [§1910.1052(f)(3)(ii)]

[*Note to paragraph (f)(3)(ii):* See Appendix A of this section for examples of procedures that satisfy this requirement. Employers covered by this standard may also be subject to the hazardous waste and emergency response provisions contained in 29 CFR 1910.120 (q).]

(g) Respiratory protection.[§1910.1052(g)]

(1) *General.* For employees who use respirators required by this section, the employer must provide each employee an appropriate respirator that complies with the requirements of this paragraph. Respirators must be used during:[§1910.1052(g)(1)]

(i) *Periods when an employee's exposure* to MC exceeds the 8-hour TWA PEL, or STEL (for example, when an employee is using MC in a regulated area).[§1910.1052(g)(1)(i)]

(ii) *Periods necessary to install* or implement feasible engineering and work-practice controls.[§1910.1052(g)(1)(ii)]

(iii) *A few work operations,* such as some maintenance operations and repair activities, for which the employer demonstrates that engineering and work-practice controls are infeasible.[§1910.1052(g)(1)(iii)]

(iv) *Work operations* for which feasible engineering and work-practice controls are not sufficient to reduce employee exposures to or below the PELs.[§1910.1052(g)(1)(iv)]

(v) *Emergencies.*[§1910.1052(g)(1)(v)]

(2) *Respirator program.*[§1910.1052(g)(2)]

(i) *The employer must implement a respiratory protection program* in accordance with §1910.13(b) through (m) (except (d)(1)(iii)), which covers each employee required by this section to use a respirator.[§1910.1052(g)(2)(i)]

(ii) *Employers who provide employees* with gas masks with organic-vapor canisters for the purpose of emergency escape must replace the canisters after any emergency use and before the gas masks are returned to service. [§1910.1052(g)(2)(ii)]

(3) *Respirator selection.* Employers must:[§1910.1052(g)(3)]

(i) *Select,* and provide to employees, the appropriate atmosphere-supplying respirator specified in paragraph (d)(3)(i)(A) of 29 CFR 1910.134; however, employers must not select or use half masks of any type because MC may cause eye irritation or damage.[§1910.1052(g)(3)(i)]

(ii) *For emergency escape,* provide employees with one of the following respirator options: A self-contained breathing apparatus operated in the continuous-flow or pressure-demand mode; or a gas mask with an organic vapor canister. [§1910.1052(g)(3)(ii)]

(4) *Medical evaluation.* Before having an employee use a supplied-air respirator in the negative-pressure mode, or a gas mask with an organic-vapor canister for emergency escape, the employer must:[§1910.1052(g)(4)]

(i) *Have a physician* or other licensed health-care professional (PLHCP) evaluate the employee's ability to use such respiratory protection.[§1910.1052(g)(4)(i)]

(ii) *Ensure that the PLHCP* provides their findings in a written opinion to the employee and the employer.[§1910.1052(g)(4)(ii)]

(h) Protective Work Clothing and Equipment.[§1910.1052(h)]

(1) *Where needed to prevent MC-induced skin or eye irritation,* the employer shall provide clean protective clothing and equipment which is resistant to MC, at no cost to the employee, and shall ensure that each affected employee uses it. Eye and face protection shall meet the requirements of 29 CFR 1910.133 or 29 CFR 1915.153, as applicable. [§1910.1052(h)(1)]

(2) *The employer shall clean,* launder, repair and replace all protective clothing and equipment required by this paragraph as needed to maintain their effectiveness.[§1910.1052(h)(2)]

(3) *The employer shall be responsible* for the safe disposal of such clothing and equipment.[§1910.1052(h)(3)]

[*Note to paragraph (h)(3)*[1]*:* See Appendix A for examples of disposal procedures that will satisfy this requirement.

[1] Editor's Note: The CFR references "(h)(4)", however this section does not exist. This has been corrected in the regulaiton appearing above (from "(h)(4)" to "(h)(3)").

(i) Hygiene facilities.[§1910.1052(i)]

(1) *If it is reasonably foreseeable* that employees' skin may contact solutions containing 0.1 percent or greater MC (for example, through splashes, spills or improper work practices), the employer shall provide conveniently located washing facilities capable of removing the MC, and shall ensure that affected employees use these facilities as needed.[§1910.1052(i)(1)]

(2) *If it is reasonably foreseeable* that an employee's eyes may contact solutions containing 0.1 percent or greater MC (for example through splashes, spills or improper work practices), the employer shall provide appropriate eyewash facilities within the immediate work area for emergency use, and shall ensure that affected employees use those facilities when necessary.[§1910.1052(i)(2)]

(j) Medical surveillance.[§1910.1052(j)]

(1) *Affected employees.* The employer shall make medical surveillance available for employees who are or may be exposed to MC as follows:[§1910.1052(j)(1)]

(i) *At or above the action level* on 30 or more days per year, or above the 8-hour TWA PEL or the STEL on 10 or more days per year;[§1910.1052(j)(1)(i)]

(ii) *Above the 8-TWA PEL or STEL* for any time period where an employee has been identified by a physician or other licensed health care professional as being at risk from cardiac disease or from some other serious MC-related health condition and such employee requests inclusion in the medical surveillance program;[§1910.1052(j)(1)(ii)]

(iii) *During an emergency.*[§1910.1052(j)(1)(iii)]

(2) *Costs.* The employer shall provide all required medical surveillance at no cost to affected employees, without loss of pay and at a reasonable time and place.[§1910.1052(j)(2)]

(3) *Medical personnel.* The employer shall ensure that all medical surveillance procedures are performed by a physician or other licensed health care professional, as defined in paragraph (b) of this section.[§1910.1052(j)(3)]

(4) *Frequency of medical surveillance.* The employer shall make medical surveillance available to each affected employee as follows:[§1910.1052(j)(4)]

(i) *Initial surveillance.* The employer shall provide initial medical surveillance under the schedule provided by paragraph (n)(2)(iii) of this section, or before the time of initial assignment of the employee, whichever is later. The employer need not provide the initial surveillance if medical records show that an affected employee has been provided with medical surveillance that complies with this section within 12 months before April 10, 1997.[§1910.1052(j)(4)(i)]

(ii) *Periodic medical surveillance.* The employer shall update the medical and work history for each affected employee annually. The employer shall provide periodic physical examinations, including appropriate laboratory surveillance, as follows:[§1910.1052(j)(4)(ii)]

[A] For employees 45 years of age or older, within 12 months of the initial surveillance or any subsequent medical surveillance; and[§1910.1052(j)(4)(ii)[A]]

[B] For employees younger than 45 years of age, within 36 months of the initial surveillance or any subsequent medical surveillance.[§1910.1052(j)(4)(ii)[B]]

(iii) *Termination of employment or reassignment.* When an employee leaves the employer's workplace, or is reassigned to an area where exposure to MC is consistently at or below the action level and STEL, medical surveillance shall be made available if six months or more have elapsed since the last medical surveillance.[§1910.1052(j)(4)(iii)]

(iv) *Additional surveillance.* The employer shall provide additional medical surveillance at frequencies other than those listed above when recommended in the written medical opinion. (For example, the physician or other licensed health care professional may determine an examination is warranted in less than 36 months for employees younger than 45 years of age based upon evaluation of the results of the annual medical and work history.)[§1910.1052(j)(4)(iv)]

(5) *Content of medical surveillance*[§1910.1052(j)(5)]

(i) *Medical and work history.* The comprehensive medical and work history shall emphasize neurological symptoms, skin conditions, history of hematologic or liver disease, signs or symptoms suggestive of heart disease (angina, coronary artery disease), risk factors for cardiac disease, MC exposures, and work practices and personal protective equipment used during such exposures.[§1910.1052(j)(5)(i)]

[*Note to paragraph (j)(5)(i):* See Appendix B of this section for an example of a medical and work history format that would satisfy this requirement.]

(ii) *Physical examination.* Where physical examinations are provided as required above, the physician or other licensed health care professional shall accord particular attention to the lungs, cardiovascular system (including blood pressure and pulse), liver, nervous system, and skin. The physician or other licensed health care professional shall determine the extent and nature of the physical examination based on the health status of the employee and analysis of the medical and work history.[§1910.1052(j)(5)(ii)]

(iii) *Laboratory surveillance.* The physician or other licensed health care professional shall determine the extent of any required laboratory surveillance based on the employee's observed health status and the medical and work history. [§1910.1052(j)(5)(iii)]

[*Note to paragraph (j)(5)(iii):* See Appendix B of this section for information regarding medical tests. Laboratory surveillance may include before- and after-shift carboxyhemoglobin determinations, resting ECG, hematocrit, liver function tests and cholesterol levels.]

(iv) *Other information or reports.* The medical surveillance shall also include any other information or reports the physician or other licensed health care professional determines are necessary to assess the employee's health in relation to MC exposure.[§1910.1052(j)(5)(iv)]

(6) *Content of emergency medical surveillance.* The employer shall ensure that medical surveillance made available when an employee has been exposed to MC in emergency situations includes, at a minimum:[§1910.1052(j)(6)]

(i) *Appropriate emergency treatment* and decontamination of the exposed employee;[§1910.1052(j)(6)(i)]

(ii) *Comprehensive physical examination* with special emphasis on the nervous system, cardiovascular system, lungs, liver and skin, including blood pressure and pulse; [§1910.1052(j)(6)(ii)]

(iii) *Updated medical and work history,* as appropriate for the medical condition of the employee; and[§1910.1052(j)(6)(iii)]

(iv) *Laboratory surveillance,* as indicated by the employee's health status.[§1910.1052(j)(6)(iv)]

[*Note to paragraph (j)(6)(iv):* See Appendix B for examples of tests which may be appropriate.]

(7) *Additional examinations and referrals.* Where the physician or other licensed health care professional determines it is necessary, the scope of the medical examination shall be expanded and the appropriate additional medical surveillance, such as referrals for consultation or examination, shall be provided. [§1910.1052(j)(7)]

(8) *Information provided to the physician* or other licensed health care professional. The employer shall provide the following information to a physician or other licensed health care professional who is involved in the diagnosis of MC-induced health effects:[§1910.1052(j)(8)]

(i) *A copy of this section including its applicable appendices;* [§1910.1052(j)(8)(i)]

(ii) *A description of the affected employee's past,* current and anticipated future duties as they relate to the employee's MC exposure;[§1910.1052(j)(8)(ii)]

(iii) *The employee's former or current exposure levels or,* for employees not yet occupationally exposed to MC, the employee's anticipated exposure levels and the frequency and exposure levels anticipated to be associated with emergencies;[§1910.1052(j)(8)(iii)]

(iv) *A description of any personal protective equipment,* such as respirators, used or to be used; and[§1910.1052(j)(8)(iv)]

(v) *Information from previous employment-related medical surveillance* of the affected employee which is not otherwise available to the physician or other licensed health care professional.[§1910.1052(j)(8)(v)]

(9) *Written medical opinions.*[§1910.1052(j)(9)]

(i) *For each physical examination required by this section,* the employer shall ensure that the physician or other licensed health care professional provides to the employer and to the affected employee a written opinion regarding the results of that examination within 15 days of completion of the evaluation of medical and laboratory findings, but not more than 30 days after the examination. The written medical opinion shall be limited to the following information:[§1910.1052(j)(9)(i)]

[A] The physician or other licensed health care professional's opinion concerning whether exposure to MC may contribute to or aggravate the employee's existing cardiac, hepatic, neurological (including stroke) or dermal disease or whether the employee has any other medical condition(s) that would place the employee's health at increased risk of material impairment from exposure to MC.[§1910.1052(j)(9)(i)[A]]

[B] Any recommended limitations upon the employee's exposure to MC, including removal from MC exposure, or upon the employee's use of respirators, protective clothing, or other protective equipment.[§1910.1052(j)(9)(i)[B]]

[C] A statement that the employee has been informed by the physician or other licensed health care professional that MC is a potential occupational carcinogen, of risk factors for heart disease, and the potential for exacerbation of underlying heart disease by exposure to MC through its metabolism to carbon monoxide; and[§1910.1052(j)(9)(i)[C]]

[D] A statement that the employee has been informed by the physician or other licensed health care professional of the results of the medical examination and any medical conditions resulting from MC exposure which require further explanation or treatment.[§1910.1052(j)(9)(i)[D]]

(ii) *The employer shall instruct the physician* or other licensed health care professional not to reveal to the employer, orally or in the written opinion, any specific records, findings, and diagnoses that have no bearing on occupational exposure to MC.[§1910.1052(j)(9)(ii)]

[*Note to paragraph (j)(9)(ii):* The written medical opinion may also include information and opinions generated to comply with other OSHA health standards.]

(10) *Medical presumption.* For purposes of this paragraph (j) of this section, the physician or other licensed health care professional shall presume, unless medical evidence indicates to the

contrary, that a medical condition is unlikely to require medical removal from MC exposure if the employee is not exposed to MC above the 8-hour TWA PEL. If the physician or other licensed health care professional recommends removal for an employee exposed below the 8-hour TWA PEL, the physician or other licensed health care professional shall cite specific medical evidence, sufficient to rebut the presumption that exposure below the 8-hour TWA PEL is unlikely to require removal, to support the recommendation. If such evidence is cited by the physician or other licensed health care professional, the employer must remove the employee. If such evidence is not cited by the physician or other licensed health care professional, the employer is not required to remove the employee.[§1910.1052(j)(10)]

(11) *Medical Removal Protection (MRP).*[§1910.1052(j)(11)]

(i) *Temporary medical removal and return of an employee.* [§1910.1052(j)(11)(i)]

[A] Except as provided in paragraph (j)(10) of this section, when a medical determination recommends removal because the employee's exposure to MC may contribute to or aggravate the employee's existing cardiac, hepatic, neurological (including stroke), or skin disease, the employer must provide medical removal protection benefits to the employee and either:[§1910.1052(j)(11)(i)[A]]

[1] Transfer the employee to comparable work where methylene chloride exposure is below the action level; or[§1910.1052(j)(11)(i)[A][1]]

[2] Remove the employee from MC exposure. [§1910.1052(j)(11)(i)[A][2]]

[B] If comparable work is not available and the employer is able to demonstrate that removal and the costs of extending MRP benefits to an additional employee, considering feasibility in relation to the size of the employer's business and the other requirements of this standard, make further reliance on MRP an inappropriate remedy, the employer may retain the additional employee in the existing job until transfer or removal becomes appropriate, provided: [§1910.1052(j)(11)(i)[B]]

[1] The employer ensures that the employee receives additional medical surveillance, including a physical examination at least every 60 days until transfer or removal occurs; and[§1910.1052(j)(11)(i)[B][1]]

[2] The employer or PLHCP informs the employee of the risk to the employee's health from continued MC exposure.[§1910.1052(j)(11)(i)[B][2]]

[C] The employer shall maintain in effect any job-related protective measures or limitations, other than removal, for as long as a medical determination recommends them to be necessary.[§1910.1052(j)(11)(i)[C]]

(ii) *End of MRP benefits* and return of the employee to former job status.[§1910.1052(j)(11)(ii)]

[A] The employer may cease providing MRP benefits at the earliest of the following:[§1910.1052(j)(11)(ii)[A]]

[1] Six months;[§1910.1052(j)(11)(ii)[A][1]]

[2] Return of the employee to the employee's former job status following receipt of a medical determination concluding that the employee's exposure to MC no longer will aggravate any cardiac, hepatic, neurological (including stroke), or dermal disease; [§1910.1052(j)(11)(ii)[A][2]]

[3] Receipt of a medical determination concluding that the employee can never return to MC exposure. [§1910.1052(j)(11)(ii)[A][3]]

[B] For the purposes of this paragraph (j), the requirement that an employer return an employee to the employee's former job status is not intended to expand upon or restrict any rights an employee has or would have had, absent temporary medical removal, to a specific job classification or position under the terms of a collective bargaining agreement.[§1910.1052(j)(11)(ii)[B]]

(12) *Medical removal protection benefits.*[§1910.1052(j)(12)]

(i) *For purposes of this paragraph (j),* the term medical removal protection benefits means that, for each removal, an employer must maintain for up to six months the earnings, seniority, and other employment rights and benefits of the employee as though the employee had not been removed from MC exposure or transferred to a comparable job. [§1910.1052(j)(12)(i)]

(ii) *During the period of time* that an employee is removed from exposure to MC, the employer may condition the provision of medical removal protection benefits upon the employee's participation in follow-up medical surveillance made available pursuant to this section.[§1910.1052(j)(12)(ii)]

(iii) *If a removed employee* files a workers' compensation claim for a MC-related disability, the employer shall continue the MRP benefits required by this paragraph until either the claim is resolved or the 6-month period for payment f MRP benefits has passed, whichever occurs first. To the extent the employee is entitled to indemnity payments for earnings lost during the period of removal, the employer's obligation to provide medical removal protection benefits to the employee shall be reduced by the amount of such indemnity payments. [§1910.1052(j)(12)(iii)]

(iv) *The employer's obligation* to provide medical removal protection benefits to a removed employee shall be reduced to the extent that the employee receives compensation for earnings lost during the period of removal from either a publicly or an employer-funded compensation program, or receives income from employment with another employer made possible by virtue of the employee's removal. [§1910.1052(j)(12)(iv)]

(13) *Voluntary removal or restriction of an employee.* Where an employer, although not required by this section to do so, removes an employee from exposure to MC or otherwise places any limitation on an employee due to the effects of MC exposure on the employee's medical condition, the employer shall provide medical removal protection benefits to the employee equal to those required by paragraph (j)(12) of this section.[§1910.1052(j)(13)]

(14) *Multiple health care professional review mechanism.* [§1910.1052(j)(14)]

(i) *If the employer* selects the initial physician or licensed health care professional (PLHCP) to conduct any medical examination or consultation provided to an employee under this paragraph (j)(11), the employer shall notify the employee of the right to seek a second medical opinion each time the employer provides the employee with a copy of the written opinion of that PLHCP.[§1910.1052(j)(14)(i)]

(ii) *If the employee* does not agree with the opinion of the employer-selected PLHCP, notifies the employer of that fact, and takes steps to make an appointment with a second PLHCP within 15 days of receiving a copy of the written opinion of the initial PLHCP, the employer shall pay for the PLHCP chosen by the employee to perform at least the following:[§1910.1052(j)(14)(ii)]

[A] Review any findings, determinations or recommendations of the initial PLHCP; and[§1910.1052(j)(14)(ii)[A]]

[B] Conduct such examinations, consultations, and laboratory tests as the PLHCP deems necessary to facilitate this review.[§1910.1052(j)(14)(ii)[B]]

(iii) *If the findings,* determinations or recommendations of the second PLHCP differ from those of the initial PLHCP, then the employer and the employee shall instruct the two health care professionals to resolve the disagreement. [§1910.1052(j)(14)(iii)]

(iv) *If the two health care professionals* are unable to resolve their disagreement within 15 days, then those two health care professionals shall jointly designate a PLHCP who is a specialist in the field at issue. The employer shall pay for the specialist to perform at least the following:[§1910.1052(j)(14)(iv)]

[A] Review the findings, determinations, and recommendations of the first two PLHCPs; and[§1910.1052(j)(14)(iv)[A]]

[B] Conduct such examinations, consultations, laboratory tests and discussions with the prior PLHCPs as the specialist deems necessary to resolve the disagreements of the prior health care professionals.[§1910.1052(j)(14)(iv)[B]]

(v) *The written opinion of the specialist* shall be the definitive medical determination. The employer shall act consistent with the definitive medical determination, unless the employer and employee agree that the written opinion of one of the other two PLHCPs shall be the definitive medical determination.[§1910.1052(j)(14)(v)]

(vi) *The employer and the employee or authorized employee representative* may agree upon the use of any expeditious alternate health care professional determination mechanism in lieu of the multiple health care professional review mechanism provided by this paragraph so long as the alternate mechanism otherwise satisfies the requirements contained in this paragraph.[§1910.1052(j)(14)(vi)]

(k) Hazard communication.[§1910.1052(k)]

(1) *Hazard communication — general.*[§1910.1052(k)(1)]

(i) *Chemical manufacturers,* importers, distributors and employers shall comply with all requirements of the Hazard Communication Standard (HCS) (Sec. 1910.1200) for MC. [§1910.1052(k)(1)(i)]

(ii) *In classifying the hazards of MC* at least the following hazards are to be addressed: Cancer, cardiac effects (including elevation of carboxyhemoglobin), central nervous system effects, liver effects, and skin and eye irritation. [§1910.1052(k)(1)(ii)]

(iii) *Employers shall include MC* in the hazard communication program established to comply with the HCS (Sec. 1910.1200). Employers shall ensure that each employee has access to labels on containers of MC and to safety data sheets, and is trained in accordance with the requirements of HCS and paragraph (l) of this section.[§1910.1052(k)(1)(iii)]

(2) *[Reserved]*[§1910.1052(k)(2)]

(l) Employee information and training.[§1910.1052(l)]

(1) *The employer shall provide information and training* for each affected employee prior to or at the time of initial assignment to a job involving potential exposure to MC.[§1910.1052(l)(1)]

(2) *The employer shall ensure that information and training* is presented in a manner that is understandable to the employees.[§1910.1052(l)(2)]

(3) *In addition to the information required* under the Hazard Communication Standard at 29 CFR 1910.1200, 29 CFR 1915.1200, or 29 CFR 1926.59, as appropiate:[§1910.1052(l)(3)]

(i) *The employer shall inform each affected* employee of the requirements of this section and information available in its appendices, as well as how to access or obtain a copy of it in the workplace;[§1910.1052(l)(3)(i)]

(ii) *Wherever an employee's exposure* to airborne concentrations of MC exceeds or can reasonably be expected to exceed the action level, the employer shall inform each affected employee of the quantity, location, manner of use, release, and storage of MC and the specific operations in the workplace that could result in exposure to MC, particularly noting where exposures may be above the 8-hour TWA PEL or STEL;[§1910.1052(l)(3)(ii)]

(4) *The employer shall train each affected employee* as required under the Hazard Communication standard at 29 CFR 1910.1200, 29 CFR 1915.1200, or 29 CFR 1926.59, as appropiate.[§1910.1052(l)(4)]

(5) *The employer shall re-train each affected employee* as necessary to ensure that each employee exposed above the action level or the STEL maintains the requisite understanding of the principles of safe use and handling of MC in the workplace. [§1910.1052(l)(5)]

(6) *Whenever there are workplace changes,* such as modifications of tasks or procedures or the institution of new tasks or procedures, which increase employee exposure, and where those exposures exceed or can reasonably be expected to exceed the action level, the employer shall update the training as necessary to ensure that each affected employee has the requisite proficiency.[§1910.1052(l)(6)]

(7) *An employer whose employees are exposed to MC* at a multi-employer worksite shall notify the other employers with work operations at that site in accordance with the requirements of the Hazard Communication Standard, 29 CFR 1910.1200, 29 CFR 1915.1200, or 29 CFR 1926.59, as appropiate. [§1910.1052(l)(7)]

(8) *The employer shall provide* to the Assistant Secretary or the Director, upon request, all available materials relating to employee information and training.[§1910.1052(l)(8)]

(m) Recordkeeping.[§1910.1052(m)]

(1) *Objective data.*[§1910.1052(m)(1)]

(i) *Where an employer seeks to demonstrate* that initial monitoring is unnecessary through reasonable reliance on objective data showing that any materials in the workplace containing MC will not release MC at levels which exceed the action level or the STEL under foreseeable conditions of exposure, the employer shall establish and maintain an accurate record of the objective data relied upon in support of the exemption.[§1910.1052(m)(1)(i)]

(ii) *This record shall include at least the following information:* [§1910.1052(m)(1)(ii)]

[A] The MC-containing material in question; [§1910.1052(m)(1)(ii)[A]]

[B] The source of the objective data;[§1910.1052(m)(1)(ii)[B]]

[C] The testing protocol, results of testing, and/or analysis of the material for the release of MC;[§1910.1052(m)(1)(ii)[C]]

[D] A description of the operation exempted under paragraph (d)(2)(i) of this section and how the data support the exemption; and[§1910.1052(m)(1)(ii)[D]]

[E] Other data relevant to the operations, materials, processing, or employee exposures covered by the exemption. [§1910.1052(m)(1)(ii)[E]]

(iii) *The employer shall maintain this record* for the duration of the employer's reliance upon such objective data. [§1910.1052(m)(1)(iii)]

(2) *Exposure measurements.*[§1910.1052(m)(2)]

(i) *The employer shall establish and keep an accurate record* of all measurements taken to monitor employee exposure to MC as prescribed in paragraph (d) of this section. [§1910.1052(m)(2)(i)]

(ii) *Where the employer has 20 or more employees,* this record shall include at least the following information: [§1910.1052(m)(2)(ii)]

[A] The date of measurement for each sample taken; [§1910.1052(m)(2)(ii)[A]]

[B] The operation involving exposure to MC which is being monitored;[§1910.1052(m)(2)(ii)[B]]

[C] Sampling and analytical methods used and evidence of their accuracy;[§1910.1052(m)(2)(ii)[C]]

[D] Number, duration, and results of samples taken; [§1910.1052(m)(2)(ii)[D]]

[E] Type of personal protective equipment, such as respiratory protective devices, worn, if any; and [§1910.1052(m)(2)(ii)[E]]

[F] Name, social security number, job classification and exposure of all of the employees represented by monitoring, indicating which employees were actually monitored. [§1910.1052(m)(2)(ii)[F]]

(iii) *Where the employer has fewer than 20 employees,* the record shall include at least the following information: [§1910.1052(m)(2)(iii)]

[A] The date of measurement for each sample taken; [§1910.1052(m)(2)(iii)[A]]

[B] Number, duration, and results of samples taken; and [§1910.1052(m)(2)(iii)[B]]

[C] Name, social security number, job classification and exposure of all of the employees represented by monitoring, indicating which employees were actually monitored. [§1910.1052(m)(2)(iii)[C]]

(iv) *The employer shall maintain this record* for at least thirty (30) years, in accordance with 29 CFR 1910.1020. [§1910.1052(m)(2)(iv)]

(3) *Medical surveillance.*[§1910.1052(m)(3)]

(i) *The employer shall establish and maintain* an accurate record for each employee subject to medical surveillance under paragraph (j) of this section.[§1910.1052(m)(3)(i)]

(ii) *The record shall include at least the following information:* [§1910.1052(m)(3)(ii)]

[A] The name, social security number and description of the duties of the employee;[§1910.1052(m)(3)(ii)[A]]

[B] Written medical opinions; and[§1910.1052(m)(3)(ii)[B]]

[C] Any employee medical conditions related to exposure to MC.[§1910.1052(m)(3)(ii)[C]]

(iii) *The employer shall ensure that this record* is maintained for the duration of employment plus thirty (30) years, in accordance with 29 CFR 1910.1020.[§1910.1052(m)(3)(iii)]

(4) *Availability.*[§1910.1052(m)(4)]

(i) *The employer, upon written request,* shall make all records required to be maintained by this section available to the Assistant Secretary and the Director for examination and copying in accordance with 29 CFR 1910.1020. [§1910.1052(m)(4)(i)]

[Note to paragraph (m)(4)(i): All records required to be maintained by this section may be kept in the most administratively convenient form (for example, electronic or computer records would satisfy this requirement).]

(ii) *The employer, upon request,* shall make any employee exposure and objective data records required by this section available for examination and copying by affected employees, former employees, and designated representatives in accordance with 29 CFR 1910.1020.[§1910.1052(m)(4)(ii)]

(iii) *The employer, upon request,* shall make employee medical records required to be kept by this section available for examination and copying by the subject employee and by anyone having the specific written consent of the subject employee in accordance with 29 CFR 1910.1020. [§1910.1052(m)(4)(iii)]

(5) *Transfer of records.* The employer shall comply with the requirements concerning transfer of records set forth in 29 CFR 1910.1020(h).[§1910.1052(m)(5)]

(n) **[Reserved]**[§1910.1052(n)]

(o) **Appendices.** The information contained in the appendices does not, by itself, create any additional obligations not otherwise imposed or detract from any existing obligation.[§1910.1052(o)]

[*Note to paragraph (o):* The requirement of 29 CFR 1910.1052(g)(1) to use respiratory protection whenever an employee's exposure to methylene chloride exceeds or can reasonably be expected to exceed the 8-hour TWA PEL is hereby stayed until August 31, 1998 for employers engaged in polyurethane foam manufacturing; foam fabrication; furniture refinishing; general aviation aircraft stripping; formulation of products containing methylene chloride; boat building and repair; recreational vehicle manufacture; van conversion; upholstery; and use of methylene chloride in construction work for restoration and preservation of buildings, painting and paint removal, cabinet making and/or floor refinishing and resurfacing.

The requirement of 29 CFR 1910.1052(f)(1) to implement engineering controls to achieve the 8-hour TWA PEL and STEL is hereby stayed until December 10, 1998 for employers with more than 100 employees engaged in polyurethane foam manufacturing and for employers with more than 20 employees engaged in foam fabrication; furniture refinishing; general aviation aircraft stripping; formulation of products containing methylene chloride; boat building and repair; recreational vehicle manufacture; van conversion; upholstery; and use of methylene chloride in construction work for restoration and preservation of buildings, painting and paint removal, cabinet making and/or floor refinishing and resurfacing.][§1910.1052(o)]

§1910.1052 Appendix A

Substance Safety Data Sheet and Technical Guidelines for Methylene Chloride

I. Substance Identification.

A. *Substance:* Methylene chloride (CH_2Cl_2).

B. *Synonyms:* MC, Dichloromethane (DCM); Methylene dichloride; Methylene bichloride; Methane dichloride; CAS: 75-09-2; NCI-C50102.

C. *Physical data:*

1. *Molecular weight:* 84.9.
2. *Boiling point (760 mm Hg):* 39.8 °C (104 °F).
3. *Specific gravity (water = 1):* 1.3.
4. *Vapor density (air = 1 at boiling point):* 2.9.
5. *Vapor pressure at 20 °C (68 °F):* 350 mm Hg.
6. *Solubility in water, g/100 g water at 20 °C (68 °F)*=1.32.
7. *Appearance and odor:* colorless liquid with a chloroform-like odor.

D. *Uses:*

MC is used as a solvent, especially where high volatility is required. It is a good solvent for oils, fats, waxes, resins, bitumen, rubber and cellulose acetate and is a useful paint stripper and degreaser. It is used in paint removers, in propellant mixtures for aerosol containers, as a solvent for plastics, as a degreasing agent, as an extracting agent in the pharmaceutical industry and as a blowing agent in polyurethane foams. Its solvent property is sometimes increased by mixing with methanol, petroleum naphtha or tetrachloroethylene.

E. *Appearance and odor:*

MC is a clear colorless liquid with a chloroform-like odor. It is slightly soluble in water and completely miscible with most organic solvents.

F. *Permissible exposure:*

Exposure may not exceed 25 parts MC per million parts of air (25 ppm) as an eight-hour time-weighted average (8-hour TWA PEL) or 125 parts of MC per million parts of air (125 ppm) averaged over a 15-minute period (STEL).

II. Health Hazard Data.

A. *MC can affect the body* if it is inhaled or if the liquid comes in contact with the eyes or skin. It can also affect the body if it is swallowed.

B. *Effects of overexposure:*

1. *Short-term Exposure:*

MC is an anesthetic. Inhaling the vapor may cause mental confusion, light-headedness, nausea, vomiting, and headache. Continued exposure may cause increased light-headedness, staggering, unconsciousness, and even death. High vapor concentrations may also cause irritation of the eyes and respiratory tract. Exposure to MC may make the symptoms of angina (chest pains) worse. Skin exposure to liquid MC may cause irritation. If liquid MC remains on the skin, it may cause skin burns. Splashes of the liquid into the eyes may cause irritation.

2. *Long-term (chronic) exposure:*

The best evidence that MC causes cancer is from laboratory studies in which rats, mice and hamsters inhaled MC 6 hours per day, 5 days per week for 2 years. MC exposure produced lung and liver tumors in mice and mammary tumors in rats. No carcinogenic effects of MC were found in hamsters.

There are also some human epidemiological studies which show an association between occupational exposure to MC and increases in biliary (bile duct) cancer and a type of brain cancer. Other epidemiological studies have not observed a relationship between MC exposure and cancer. OSHA interprets these results to mean that there is suggestive (but not absolute) evidence that MC is a human carcinogen.

C. *Reporting signs and symptoms:*

You should inform your employer if you develop any signs or symptoms and suspect that they are caused by exposure to MC.

D. *Warning Properties:*

1. *Odor Threshold:*

Different authors have reported varying odor thresholds for MC. Kirk-Othmer and Sax both reported 25 to 50 ppm; Summer and May both reported 150 ppm; Spector reports 320 ppm. Patty, however, states that since one can become adapted to the odor, MC should not be considered to have adequate warning properties.

2. *Eye Irritation Level:*

Kirk-Othmer reports that "MC vapor is seriously damaging to the eyes." Sax agrees with Kirk-Othmer's statement. The ACGIH Documentation of TLVs states that irritation of the eyes has been observed in workers exposed to concentrations up to 5000 ppm.

3. *Evaluation of Warning Properties:*

Since a wide range of MC odor thresholds are reported (25-320 ppm), and human adaptation to the odor occurs, MC is considered to be a material with poor warning properties.

III. Emergency First Aid Procedures.

In the event of emergency, institute first aid procedures and send for first aid or medical assistance.

A. *Eye and Skin Exposures:*

If there is a potential for liquid MC to come in contact with eye or skin, face shields and skin protective equipment must be provided and used. If liquid MC comes in contact with the eye, get medical attention. Contact lenses should not be worn when working with this chemical.

B. *Breathing:*

If a person breathes in large amounts of MC, move the exposed person to fresh air at once. If breathing has stopped, perform cardiopulmonary resuscitation. Keep the affected person warm and at rest. Get medical attention as soon as possible.

C. *Rescue:*

Move the affected person from the hazardous exposure immediately. If the exposed person has been overcome, notify someone else and put into effect the established emergency rescue procedures. Understand the facility's emergency rescue procedures and know the locations of rescue equipment before the need arises. Do not become a casualty yourself.

IV. Respirators, Protective Clothing, and Eye Protection.

A. *Respirators:*

Good industrial hygiene practices recommend that engineering controls be used to reduce environmental concentrations to the permissible exposure level. However, there are some exceptions where respirators may be used to control exposure. Respirators may be used when engineering and work practice controls are not feasible, when such controls are in the process of being installed, or when these controls fail and need to be supplemented. Respirators may also be used for operations which require entry into tanks or closed vessels, and in emergency situations.

If the use of respirators is necessary, the only respirators permitted are those that have been approved by the Mine Safety and Health Administration (MSHA) or the National Institute for Occupational Safety and Health (NIOSH). Supplied-air respirators are required because air-purifying respirators do not provide adequate respiratory protection against MC.

In addition to respirator selection, a complete written respiratory protection program should be instituted which includes regular training, maintenance, inspection, cleaning, and evaluation. If you can smell MC while wearing a respirator, proceed immediately to fresh air. If you experience difficulty in breathing while wearing a respirator, tell your employer.

B. *Protective Clothing:*
Employees must be provided with and required to use impervious clothing, gloves, face shields (eight-inch minimum), and other appropriate protective clothing necessary to prevent repeated or prolonged skin contact with liquid MC or contact with vessels containing liquid MC. Any clothing which becomes wet with liquid MC should be removed immediately and not reworn until the employer has ensured that the protective clothing is fit for reuse. Contaminated protective clothing should be placed in a regulated area designated by the employer for removal of MC before the clothing is laundered or disposed of. Clothing and equipment should remain in the regulated area until all of the MC contamination has evaporated; clothing and equipment should then be laundered or disposed of as appropriate.

C. *Eye Protection:*
Employees should be provided with and required to use splash-proof safety goggles where liquid MC may contact the eyes.

V. Housekeeping and Hygiene Facilities.
For purposes of complying with 29 CFR 1910.141, the following items should be emphasized:

A. *The workplace should be kept clean,* orderly, and in a sanitary condition. The employer should institute a leak and spill detection program for operations involving liquid MC in order to detect sources of fugitive MC emissions.

B. *Emergency drench showers* and eyewash facilities are recommended. These should be maintained in a sanitary condition. Suitable cleansing agents should also be provided to assure the effective removal of MC from the skin.

C. *Because of the hazardous nature of MC,* contaminated protective clothing should be placed in a regulated area designated by the employer for removal of MC before the clothing is laundered or disposed of.

VI. Precautions for Safe Use, Handling, and Storage.

A. *Fire and Explosion Hazards:*
MC has no flash point in a conventional closed tester, but it forms flammable vapor-air mixtures at approximately 100 °C (212 °F), or higher. It has a lower explosion limit of 12%, and an upper explosion limit of 19% in air. It has an autoignition temperature of 556.1 °C (1033 °F), and a boiling point of 39.8 °C (104 °F). It is heavier than water with a specific gravity of 1.3. It is slightly soluble in water.

B. *Reactivity Hazards:*
Conditions contributing to the instability of MC are heat and moisture. Contact with strong oxidizers, caustics, and chemically active metals such as aluminum or magnesium powder, sodium and potassium may cause fires and explosions.
Special precautions: Liquid MC will attack some forms of plastics, rubber, and coatings.

C. *Toxicity:*
Liquid MC is painful and irritating if splashed in the eyes or if confined on the skin by gloves, clothing, or shoes. Vapors in high concentrations may cause narcosis and death. Prolonged exposure to vapors may cause cancer or exacerbate cardiac disease.

D. *Storage:*
Protect against physical damage. Because of its corrosive properties, and its high vapor pressure, MC should be stored in plain, galvanized or lead lined, mild steel containers in a cool, dry, well ventilated area away from direct sunlight, heat source and acute fire hazards.

E. *Piping Material:*
All piping and valves at the loading or unloading station should be of material that is resistant to MC and should be carefully inspected prior to connection to the transport vehicle and periodically during the operation.

F. *Usual Shipping Containers:*
Glass bottles, 5- and 55-gallon steel drums, tank cars, and tank trucks.

Note: This section addresses MC exposure in marine terminal and longshore employment only where leaking or broken packages allow MC exposure that is not addressed through compliance with 29 CFR parts 1917 and 1918, respectively.

G. *Electrical Equipment:*
Electrical installations in Class I hazardous locations as defined in Article 500 of the National Electrical Code, should be installed according to Article 501 of the code; and electrical equipment should be suitable for use in atmospheres containing MC vapors. See Flammable and Combustible Liquids Code (NFPA No. 325M), Chemical Safety Data Sheet SD-86 (Manufacturing Chemists' Association, Inc.).

H. *Fire Fighting:*
When involved in fire, MC emits highly toxic and irritating fumes such as phosgene, hydrogen chloride and carbon monoxide. Wear breathing apparatus and use water spray to keep fire-exposed containers cool. Water spray may be used to flush spills away from exposures. Extinguishing media are dry chemical, carbon dioxide, foam. For purposes of compliance with 29 CFR 1910.307, locations classified as hazardous due to the presence of MC shall be Class I.

I. *Spills and Leaks:*
Persons not wearing protective equipment and clothing should be restricted from areas of spills or leaks until cleanup has been completed. If MC has spilled or leaked, the following steps should be taken:

1. *Remove all ignition sources.*
2. *Ventilate area of spill or leak.*
3. *Collect for reclamation or absorb in vermiculite,* dry sand, earth, or a similar material.

J. *Methods of Waste Disposal:*
Small spills should be absorbed onto sand and taken to a safe area for atmospheric evaporation. Incineration is the preferred method for disposal of large quantities by mixing with a combustible solvent and spraying into an incinerator equipped with acid scrubbers to remove hydrogen chloride gases formed. Complete combustion will convert carbon monoxide to carbon dioxide. Care should be taken for the presence of phosgene.

K. *You should not keep food,* beverage, or smoking materials, or eat or smoke in regulated areas where MC concentrations are above the permissible exposure limits.

L. *Portable heating units* should not be used in confined areas where MC is used.

M. *Ask your supervisor* where MC is used in your work area and for any additional plant safety and health rules.

VII. Medical Requirements.
Your employer is required to offer you the opportunity to participate in a medical surveillance program if you are exposed to MC at concentrations at or above the action level (12.5 ppm 8-hour TWA) for more than 30 days a year or at concentrations exceeding the PELs (25 ppm 8-hour TWA or 125 ppm 15-minute STEL) for more than 10 days a year. If you are exposed to MC at concentrations over either of the PELs, your employer will also be required to have a physician or other licensed health care professional ensure that you are able to wear the respirator that you are assigned. Your employer must provide all medical examinations relating to your MC exposure at a reasonable time and place and at no cost to you.

VIII. Monitoring and Measurement Procedures.

A. *Exposure above the Permissible* Exposure Limit:

1. *Eight-hour exposure evaluation:* Measurements taken for the purpose of determining employee exposure under this section are best taken with consecutive samples covering the full shift. Air samples must be taken in the employee's breathing zone.
2. *Monitoring techniques:* The sampling and analysis under this section may be performed by collection of the MC vapor on two charcoal adsorption tubes in series or other composition adsorption tubes, with subsequent chemical analysis. Sampling and analysis may also be performed by instruments such as real-time continuous monitoring systems, portable direct reading instruments, or passive dosimeters as long as measurements taken using these methods accurately evaluate the concentration of MC in employees" breathing zones.
OSHA method 80 is an example of a validated method of sampling and analysis of MC. Copies of this method are available from OSHA or can be downloaded from the Internet at http://www.osha.gov. The employer has the obligation of selecting a monitoring method which meets the accuracy and precision requirements of the standard under his or her unique field conditions. The standard requires that the method of monitoring must be accurate, to a 95 percent confidence level, to plus or minus 25 percent for concentrations of MC at or above 25 ppm, and to plus or minus 35 percent for concentrations at or below 25 ppm. In addition to OSHA method 80, there are numerous other methods available for monitoring for MC in the workplace.

B. *Since many of the duties relating to employee exposure* are dependent on the results of measurement procedures, employers must assure that the evaluation of employee exposure is performed by a technically qualified person.

IX. **Observation of Monitoring.**
Your employer is required to perform measurements that are representative of your exposure to MC and you or your designated representative are entitled to observe the monitoring procedure. You are entitled to observe the steps taken in the measurement procedure, and to record the results obtained. When the monitoring procedure is taking place in an area where respirators or personal protective clothing and equipment are required to be worn, you or your representative must also be provided with, and must wear, protective clothing and equipment.

X. **Access to Information.**

A. *Your employer is required* to inform you of the information contained in this Appendix. In addition, your employer must instruct you in the proper work practices for using MC, emergency procedures, and the correct use of protective equipment.

B. *Your employer is required* to determine whether you are being exposed to MC. You or your representative has the right to observe employee measurements and to record the results obtained. Your employer is required to inform you of your exposure. If your employer determines that you are being over exposed, he or she is required to inform you of the actions which are being taken to reduce your exposure to within permissible exposure limits.

C. *Your employer is required* to keep records of your exposures and medical examinations. These records must be kept by the employer for at least thirty (30) years.

D. *Your employer is required* to release your exposure and medical records to you or your representative upon your request.

E. *Your employer*[1] *is required* to provide labels and material safety data sheets (MSDS) for all materials, mixtures or solutions composed of greater than 0.1 percent MC. An example of a label that would satisfy these requirements would be:

DANGER
CONTAINS METHYLENE CHLORIDE
POTENTIAL CANCER HAZARD

May worsen heart disease because methylene chloride is converted to carbon monoxide in the body.

May cause dizziness, headache, irritation of the throat and lungs, loss of consciousness and death at high concentrations (for example, if used in a poorly ventilated room).

Avoid Skin Contact. Contact with liquid causes skin and eye irritation.

XI. **Common Operations and Controls.**
The following list includes some common operations in which exposure to MC may occur and control methods which may be effective in each case:

Operations	Controls
Use as solvent in paint and varnish removers; manufacture of aerosols; cold cleaning and ultrasonic cleaning; and as a solvent in furniture stripping.	General dilution ventilation; local exhaust ventilation; personal protective equipment; substitution.
Use as solvent in vapor degreasing.	Process enclosure; local exhaust ventilation; chilling coils; substitution.
Use as a secondary refrigerant in air conditioning and scientific testing.	General dilution ventilation; local exhaust ventilation; personal protective equipment.

§1910.1052 Appendix B
Medical Surveillance for Methylene Chloride

I. **Primary Route of Entry.**
Inhalation.

II. **Toxicology.**
Methylene Chloride (MC) is primarily an inhalation hazard. The principal acute hazardous effects are the depressant action on the central nervous system, possible cardiac toxicity and possible liver toxicity. The range of CNS effects are from decreased eye/hand coordination and decreased performance in vigilance tasks to narcosis and even death of individuals exposed at very high doses. Cardiac toxicity is due to the metabolism of MC to carbon monoxide, and the effects of carbon monoxide on heart tissue. Carbon monoxide displaces oxygen in the blood, decreases the oxygen available to heart tissue, increasing the risk of damage to the heart, which may result in heart attacks in susceptible individuals. Susceptible individuals include persons with heart disease and those with risk factors for heart disease.

Elevated liver enzymes and irritation to the respiratory passages and eyes have also been reported for both humans and experimental animals exposed to MC vapors.

MC is metabolized to carbon monoxide and carbon dioxide via two separate pathways. Through the first pathway, MC is metabolized to carbon monoxide as an end-product via the P-450 mixed function oxidase pathway located in the microsomal fraction of the cell. This biotransformation of MC to carbon monoxide occurs through the process of microsomal oxidative dechlorination which takes place primarily in the liver. The amount of conversion to carbon monoxide is significant as measured by the concentration of carboxyhemoglobin, up to 12% measured in the blood following occupational exposure of up to 610 ppm. Through the second pathway, MC is metabolized to carbon dioxide as an end product (with formaldehyde and formic acid as metabolic intermediates) via the glutathione dependent enzyme found in the cytosolic fraction of the liver cell. Metabolites along this pathway are believed to be associated with the carcinogenic activity of MC.

MC has been tested for carcinogenicity in several laboratory rodents. These rodent studies indicate that there is clear evidence that MC is carcinogenic to male and female mice and female rats. Based on epidemiologic studies, OSHA has concluded that there is suggestive evidence of increased cancer risk in MC-related worker populations. The epidemiological evidence is consistent with the finding of excess cancer in the experimental animal studies. NIOSH regards MC as a potential occupational carcinogen and the International Agency for Research Cancer (IARC) classifies MC as an animal carcinogen. OSHA considers MC as a suspected human carcinogen.

III. **Medical Signs and Symptoms of Acute Exposure.**
Skin exposure to liquid MC may cause irritation or skin burns. Liquid MC can also be irritating to the eyes. MC is also absorbed through the skin and may contribute to the MC exposure by inhalation.

At high concentrations in air, MC may cause nausea, vomiting, light-headedness, numbness of the extremities, changes in blood enzyme levels, and breathing problems, leading to bronchitis and pulmonary edema, unconsciousness and even death.

At lower concentrations in air, MC may cause irritation to the skin, eye, and respiratory tract and occasionally headache and nausea. Perhaps the greatest problem from exposure to low concentrations of MC is the CNS effects on coordination and alertness that may cause unsafe operations of machinery and equipment, leading to self-injury or accidents.

Low levels and short duration exposures do not seem to produce permanent disability, but chronic exposures to MC have been demonstrated to produce liver toxicity in animals, and therefore, the evidence is suggestive for liver toxicity in humans after chronic exposure.

Chronic exposure to MC may also cause cancer.

IV. **Surveillance and Preventive Considerations.**
As discussed above, MC is classified as a suspect or potential human carcinogen. It is a central nervous system (CNS) depressant and a skin, eye and respiratory tract irritant. At extremely high concentrations, MC has caused liver damage in animals.

MC principally affects the CNS, where it acts as a narcotic. The observation of the symptoms characteristic of CNS depression, along with a physical examination, provides the best detection of early neurological disorders. Since exposure to MC also increases the carboxyhemoglobin level in the blood, ambient carbon monoxide levels would have an additive effect on that carboxyhemoglobin level. Based on such information, a periodic post-shift carboxyhemoglobin test as an index of the presence of carbon monoxide in the blood is recommended, but not required, for medical surveillance.

Based on the animal evidence and three epidemiologic studies previously mentioned, OSHA concludes that MC is a suspect human carcinogen. The medical surveillance program is designed to observe exposed workers on a regular basis. While the medical surveillance program cannot detect MC-induced cancer at a preneoplastic stage, OSHA anticipates that, as in the past, early detection and treatments of cancers leading to enhanced survival rates will continue to evolve.

A. *Medical and Occupational History:*
The medical and occupational work history plays an important role in the initial evaluation of workers exposed to MC. It is therefore extremely important for the examining physician or other licensed health care professional to evaluate the MC-exposed worker carefully and completely and to focus the examination on MC's potentially associated health hazards. The medical evaluation must include an annual detailed work and medical history with special emphasis on cardiac history and neurological symptoms.

An important goal of the medical history is to elicit information from the worker regarding potential signs or symptoms associated with increased levels of carboxyhemoglobin due to the presence of carbon monoxide in the blood. Physicians or other licensed health care professionals should ensure that the smoking history of all MC exposed employees is known. Exposure to MC may cause a

1. Editor's Note: The CFR states "employee" rather than "employer." This is corrected in the regulation appearing above.

significant increase in carboxyhemoglobin level in all exposed persons. However, smokers as well as workers with anemia or heart disease and those concurrently exposed to carbon monoxide are at especially high risk of toxic effects because of an already reduced oxygen carrying capacity of the blood.

A comprehensive or interim medical and work history should also include occurrence of headache, dizziness, fatigue, chest pain, shortness of breath, pain in the limbs, and irritation of the skin and eyes.

In addition, it is important for the physician or other licensed health care professional to become familiar with the operating conditions in which exposure to MC is likely to occur. The physician or other licensed health care professional also must become familiar with the signs and symptoms that may indicate that a worker is receiving otherwise unrecognized and exceptionally high exposure levels of MC.

An example of a medical and work history that would satisfy the requirement for a comprehensive or interim work history is represented by the following:

The following is a list of recommended questions and issues for the self-administered questionnaire for methylene chloride exposure.

Appendix B for §1910.1052

Questionnaire For Methylene Chloride Exposure

I. DEMOGRAPHIC INFORMATION

1. NAME: ____ 2. SOCIAL SECURITY NUMBER: ____—____—____
3. DATE: ____—____—____ (MONTH DAY YEAR) 4. DATE OF BIRTH: ____/____/____ (MONTH DAY YEAR) 5. AGE: ____
6. PRESENT OCCUPATION: ____
7. SEX: ☐ M ☐ F 8. RACE: ____

II. OCCUPATIONAL HISTORY

1. Have you ever worked with methylene chloride, dichloromethane, methylene dichloride, or CH_2Cl_2 (all are different names for the same chemical)? ☐ Yes ☐ No
   Please list which on the occupational history form if you have not already.
2. If you have worked in any of the following industries and have not listed them on the occupational history form, please do so.
   Furniture stripping ☐ Yes ☐ No — Any industry in which you used solvents to clean and degrease equipment or parts ☐ Yes ☐ No
   Polyurethane foam manufacturing ☐ Yes ☐ No — Construction, especially painting and refinishing ☐ Yes ☐ No
   Chemical manufacturing or formulation ☐ Yes ☐ No — Aerosol manufacturing ☐ Yes ☐ No
   Pharmaceutical manufacturing ☐ Yes ☐ No — Any industry in which you used aerosol adhesives ☐ Yes ☐ No
3. If you have not listed hobbies or household projects on the occupational history form, especially furniture refinishing, spray painting, or paint stripping, please do so.

III. MEDICAL HISTORY

A. General

1. Do you consider yourself to be in good health? If no, state reason(s). ____
2. Do you or have you ever had:
   a. ☐ Persistent thirst b. ☐ Frequent urination (three times or more at night) c. ☐ Dermatitis or irritated skin d. ☐ Non-healing wounds
3. What prescription or non-prescription medications do you take, and for what reasons? ____
4. Are you allergic to any medications, and what type of reaction do you have? ____

B. Respiratory

1. Do you have or have you ever had any chest illnesses or diseases? Explain: ____
2. Do you have or have you ever had any of the following: a. ☐ Asthma b. ☐ Wheezing c. ☐ Shortness of breath
3. Have you ever had an abnormal chest X-ray? ☐ Yes ☐ No
   If so, when, where, and what were the findings? ____
4. Have you ever had difficulty using a respirator or breathing apparatus? ☐ Yes ☐ No
   Explain: ____
5. Do any chest or lung diseases run in your family? ☐ Yes ☐ No
   Explain: ____
6. Have you ever smoked cigarettes, cigars, or a pipe? ☐ Yes ☐ No Age started: ____
7. Do you now smoke? ☐ Yes ☐ No
8. If you have stopped smoking completely, how old were you when you stopped? ____
9. On the average of the entire time you smoked, how many packs of cigarettes, cigars, or bowls of tobacco did you smoke per day? ____

C. Cardiovascular

1. Have you ever been diagnosed with any of the following: Which of the following apply to you now or did apply to you at some time in the past, even if the problem is controlled by medication? Please explain any yes answers (i.e., when problem was diagnosed, length of time on medication).
   a. High cholesterol or triglyceride level ☐ Yes ☐ No
   Explain: ____
   b. Hypertension (high blood pressure) ☐ Yes ☐ No
   Explain: ____
   c. Diabetes ☐ Yes ☐ No
   Explain: ____
   d. Family history of heart attack, stroke, or blocked arteries ☐ Yes ☐ No
   Explain: ____
2. Have you ever had chest pain? ☐ Yes ☐ No If so, answer the next five questions.
   a. What was the quality of the pain (i.e., crushing, stabbing, squeezing)? ____
   b. Did the pain go anywhere (i.e., into jaw, left arm)? ____
   c. What brought the pain out? ____
   d. How long did it last? ____
   e. What made the pain go away? ____
3. Have you ever had heart disease, a heart attack, stroke, aneurysm, or blocked arteries anywhere in your body? ☐ Yes ☐ No
   Explain (when, treatment): ____
4. Have you ever had bypass surgery for blocked arteries in your heart or anywhere else? ☐ Yes ☐ No
   Explain: ____
5. Have you ever had any other procedures done to open up a blocked artery (balloon angioplasty, carotid endarterectomy, clot-dissolving drug)? ☐ Yes ☐ No
   Explain: ____

* Full-size forms available free of charge at www.oshacfr.com

Appendix B to §1910.1052

Questionnaire For Methylene Chloride Exposure (Continued)

C. Cardiovascular (continued)

6. Do you have or have you ever had (explain each):
   a. Heart murmur ☐ Yes ☐ No
   Explain: ____
   b. Irregular heartbeat ☐ Yes ☐ No
   Explain: ____
   c. Shortness of breath while lying flat ☐ Yes ☐ No
   Explain: ____
   d. Congestive heart failure ☐ Yes ☐ No
   Explain: ____
   e. Ankle swelling ☐ Yes ☐ No
   Explain: ____
   f. Recurrent pain anywhere below the waist while walking ☐ Yes ☐ No
   Explain: ____
7. Have you ever had an electrocardiogram (EKG)? ☐ Yes ☐ No
   When? DATE: ____—____—____
8. Have you ever had an abnormal EKG? ☐ Yes ☐ No
   If so, when, where, and what were the findings? ____
9. Do any heart diseases, high blood pressure, diabetes, high cholesterol, or high triglycerides run in your family? ☐ Yes ☐ No
   Explain: ____

D. Hepatobiliary and Pancreas

1. Do you now or have you ever drunk alcoholic beverages? ☐ Yes ☐ No Age Started ____ Age Stopped ____
2. Average numbers per week:
   a. Beers: ____, ounces in usual container: ____
   b. Glasses of wine: ____, ounces per glass: ____
   c. Drinks: ____, ounces in usual container: ____
3. Do you have or have you ever had (explain each):
   a. Hepatitis (infectious, autoimmune, drug-induced, or chemical) ☐ Yes ☐ No
   Explain: ____
   b. Jaundice ☐ Yes ☐ No
   Explain: ____
   c. Elevated liver enzymes or elevated bilirubin ☐ Yes ☐ No
   Explain: ____
   d. Liver disease or cancer ☐ Yes ☐ No
   Explain: ____

E. Central Nervous System

1. Do you or have you ever had (explain each):
   a. Headache ☐ Yes ☐ No
   Explain: ____
   b. Dizziness ☐ Yes ☐ No
   Explain: ____
   c. Fainting ☐ Yes ☐ No
   Explain: ____
   d. Loss of consciousness ☐ Yes ☐ No
   Explain: ____
   e. Garbled speech ☐ Yes ☐ No
   Explain: ____
   f. Lack of balance ☐ Yes ☐ No
   Explain: ____
   g. Mental/psychiatric illness ☐ Yes ☐ No
   Explain: ____
   h. Forgetfulness ☐ Yes ☐ No
   Explain: ____

F. Hematologic

1. Do you have, or have you ever had (explain each):
   a. Anemia ☐ Yes ☐ No
   Explain: ____
   b. Sickle cell disease or trait ☐ Yes ☐ No
   Explain: ____
   c. Glucose-6-phosphate dehydrogenase deficiency ☐ Yes ☐ No
   Explain: ____
   d. Bleeding tendency disorder ☐ Yes ☐ No
   Explain: ____
2. If not already mentioned previously, have you ever had a reaction to sulfa drugs or to drugs used to prevent or treat malaria? ☐ Yes ☐ No
   What was the drug? Describe the reaction: ____

* Full-size forms available free of charge at www.oshacfr.com

B. *Physical Examination*

The complete physical examination, when coupled with the medical and occupational history, assists the physician or other licensed health care professional in detecting pre-existing conditions that might place the employee at increased risk, and establishes a baseline for future health monitoring. These examinations should include:

1. *Clinical impressions of the nervous system,* cardiovascular function and pulmonary function, with additional tests conducted where indicated or determined by the examining physician or other licensed health care professional to be necessary.
2. *An evaluation of the advisability of the worker using a respirator,* because the use of certain respirators places an additional burden on the cardiopulmonary system. It is necessary for the attending physician or other licensed health care professional to evaluate the cardiopulmonary function of these workers, in order to inform the employer in a written medical opinion of the worker's ability or fitness to work in an area requiring the use of certain types of respiratory protective equipment. The presence of facial hair or scars that might interfere with the worker's ability to wear certain types of respirators should also be noted during the examination and in the written medical opinion.

   Because of the importance of lung function to workers required to wear certain types of respirators to protect themselves from MC exposure, these workers must receive an assessment of pulmonary function before they begin to wear a negative pressure respirator and at least annually thereafter. The recommended pulmonary function tests include measurement of the employee's forced vital capacity (FVC), forced expiratory volume at one second (FEV1), as well as calculation of the ratios of FEV1 to FVC, and the ratios of measured FVC and measured FEV1 to expected respective values corrected for variation due to age, sex, race, and height. Pulmonary function evaluation must be conducted by a physician or other licensed health care professional experienced in pulmonary function tests.

The following is a summary of the elements of a physical exam which would fulfill the requirements under the MC standard:

Physical Exam

I. *Skin and appendages*
   1. *Irritated or broken skin*
   2. *Jaundice*
   3. *Clubbing cyanosis, edema*
   4. *Capillary refill time*
   5. *Pallor*

II. *Head*
   1. *Facial deformities*
   2. *Scars*
   3. *Hair growth*

III. *Eyes*
   1. *Scleral icterus*
   2. *Corneal arcus*
   3. *Pupillary size and response*
   4. *Fundoscopic exam*

IV. *Chest*
   1. *Standard exam*

V. *Heart*
   1. *Standard exam*
   2. *Jugular vein distension*
   3. *Peripheral pulses*

VI. *Abdomen*
   1. *Liver span*

VII. *Nervous System*
   1. *Complete standard neurologic exam*

VIII. *Laboratory*
   1. *Hemoglobin and hematocrit*
   2. *Alanine aminotransferase (ALT, SGPT)*
   3. *Post-shift carboxyhemoglobin*

IX. *Studies*
   1. *Pulmonary function testing*
   2. *Electrocardiogram*

An evaluation of the oxygen carrying capacity of the blood of employees (for example by measured red blood cell volume) is considered useful, especially for workers acutely exposed to MC. It is also recommended, but not required, that end of shift carboxyhemoglobin levels be determined periodically, and any level above 3% for non-smokers and above 10% for smokers should prompt an investigation of the worker and his workplace. This test is recommended because MC is metabolized to CO, which combines strongly with hemoglobin, resulting in a reduced capacity of the blood to transport oxygen in the body. This is of particular concern for cigarette smokers because they already have a diminished hemoglobin capacity due to the presence of CO in cigarette smoke.

C. *Additional Examinations and Referrals*

1. *Examination by a Specialist*

When a worker examination reveals unexplained symptoms or signs (i.e. in the physical examination or in the laboratory tests), follow-up medical examinations are necessary to assure that MC exposure is not adversely affecting the worker's health. When the examining physician or other licensed health care professional finds it necessary, additional tests should be included to determine the nature of the medical problem and the underlying cause. Where relevant, the worker should be sent to a specialist for further testing and treatment as deemed necessary.

The final rule requires additional investigations to be covered and it also permits physicians or other licensed health care professionals to add appropriate or necessary tests to improve the diagnosis of disease should such tests become available in the future.

2. *Emergencies*

The examination of workers exposed to MC in an emergency should be directed at the organ systems most likely to be affected. If the worker has received a severe acute exposure, hospitalization may be required to assure proper medical intervention. It is not possible to precisely define "severe," but the physician or other licensed health care professional's judgement should not merely rest on hospitalization. If the worker has suffered significant conjunctival, oral, or nasal irritation, respiratory distress, or discomfort, the physician or other licensed health care professional should instigate appropriate follow-up procedures. These include attention to the eyes, lungs and the neurological system. The frequency of follow-up examinations should be determined by the attending physician or other licensed health care professional. This testing permits the early identification essential to proper medical management of such workers.

D. *Employer Obligations*

The employer is required to provide the responsible physician or other licensed health care professional and any specialists involved in a diagnosis with the following information: a copy of the MC standard including relevant appendices, a description of the affected employee's duties as they relate to his or her exposure to MC; an estimate of the employee's exposure including duration (e.g., 15hr/wk, three 8-hour shifts/wk, full time); a description of any personal protective equipment used by the employee, including respirators; and the results of any previous medical determinations for the affected employee related to MC exposure to the extent that this information is within the employer's control.

E. *Physicians' or Other Licensed Health Care Professionals' Obligations*

The standard requires the employer to ensure that the physician or other licensed health care professional provides a written statement to the employee and the employer. This statement should contain the physician's or licensed health care professional's opinion as to whether the employee has any medical condition placing him or her at increased risk of impaired health from exposure to MC or use of respirators, as appropriate. The physician or other licensed health care professional should also state his or her opinion regarding any restrictions that should be placed on the employee's exposure to MC or upon the use of protective clothing or equipment such as respirators. If the employee wears a respirator as a result of his or her exposure to MC, the physician or other licensed health care professional's opinion should also contain a statement regarding the suitability of the employee to wear the type of respirator assigned. Furthermore, the employee should be informed by the physician or other licensed health care professional about the cancer risk of MC and about risk factors for heart disease, and the potential for exacerbation of underlying heart disease by exposure to MC through its metabolism to carbon monoxide. Finally, the physician or other licensed health care professional should inform the employer that the employee has been told the results of the medical examination and of any medical conditions which require further explanation or treatment. This written opinion must not contain any information on specific findings or diagnosis unrelated to employee's occupational exposures.

The purpose in requiring the examining physician or other licensed health care professional to supply the employer with a written opinion is to provide the employer with a medical basis to assist the employer in placing employees initially, in assuring that their health is not being impaired by exposure to MC, and to assess the employee's ability to use any required protective equipment.

§1910.1052 Appendix C

Questions and Answers — Methylene Chloride Control in Furniture Stripping

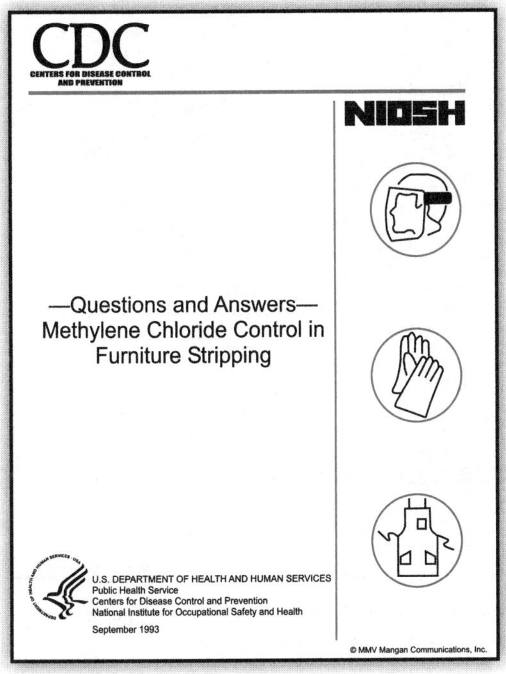

CDC

CENTERS FOR DISEASE CONTROL AND PREVENTION

NIOSH

—Questions and Answers—
Methylene Chloride Control in Furniture Stripping

U.S. DEPARTMENT OF HEALTH AND HUMAN SERVICES
Public Health Service
Centers for Disease Control and Prevention
National Institute for Occupational Safety and Health

September 1993

* Full-size forms available free of charge at www.oshacfr.com

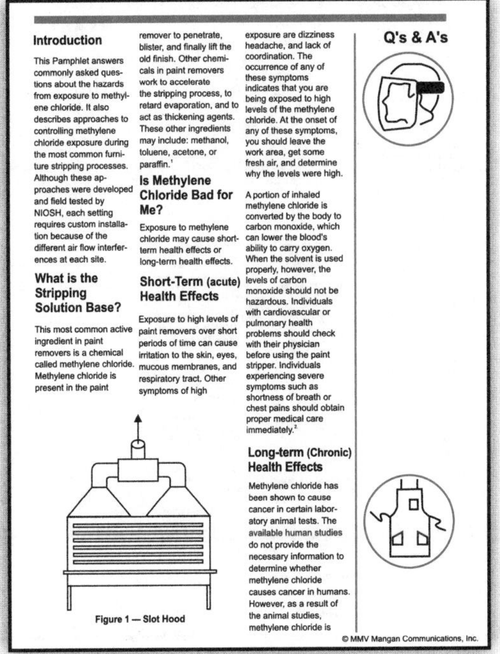

Q's & A's

Introduction

This Pamphlet answers commonly asked questions about the hazards from exposure to methylene chloride. It also describes approaches to controlling methylene chloride exposure during the most common furniture stripping processes. Although these approaches were developed and field tested by NIOSH, each setting requires custom installation because of the different air flow interferences at each site.

What is the Stripping Solution Base?

This most common active ingredient in paint removers is a chemical called methylene chloride. Methylene chloride is present in the paint remover to penetrate, blister, and finally lift the old finish. Other chemicals in paint removers work to accelerate the stripping process, to retard evaporation, and to act as thickening agents. These other ingredients may include: methanol, toluene, acetone, or paraffin.[1]

Is Methylene Chloride Bad for Me?

Exposure to methylene chloride may cause short-term health effects or long-term health effects.

Short-Term (acute) Health Effects

Exposure to high levels of paint removers over short periods of time can cause irritation to the skin, eyes, mucous membranes, and respiratory tract. Other symptoms of high exposure are dizziness headache, and lack of coordination. The occurrence of any of these symptoms indicates that you are being exposed to high levels of the methylene chloride. At the onset of any of these symptoms, you should leave the work area, get some fresh air, and determine why the levels were high.

A portion of inhaled methylene chloride is converted by the body to carbon monoxide, which can lower the blood's ability to carry oxygen. When the solvent is used properly, however, the levels of carbon monoxide should not be hazardous. Individuals with cardiovascular or pulmonary health problems should check with their physician before using the paint stripper. Individuals experiencing severe symptoms such as shortness of breath or chest pains should obtain proper medical care immediately.[2]

Figure 1 — Slot Hood

Long-term (Chronic) Health Effects

Methylene chloride has been shown to cause cancer in certain laboratory animal tests. The available human studies do not provide the necessary information to determine whether methylene chloride causes cancer in humans. However, as a result of the animal studies, methylene chloride is

* Full-size forms available free of charge at www.oshacfr.com

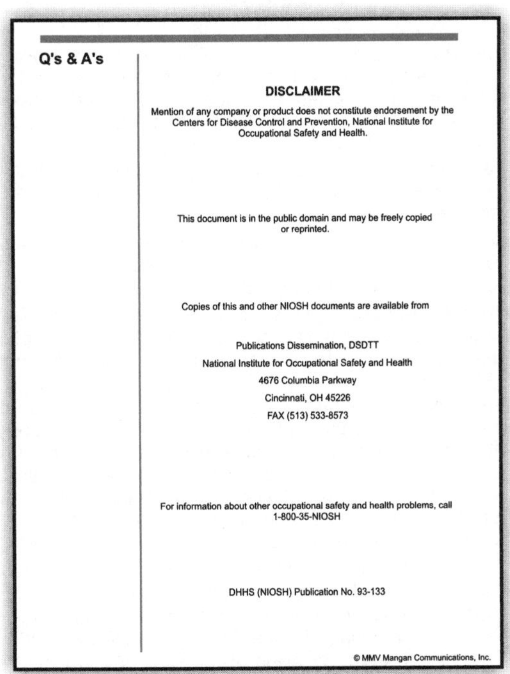

Q's & A's

DISCLAIMER

Mention of any company or product does not constitute endorsement by the Centers for Disease Control and Prevention, National Institute for Occupational Safety and Health.

This document is in the public domain and may be freely copied or reprinted.

Copies of this and other NIOSH documents are available from

Publications Dissemination, DSDTT
National Institute for Occupational Safety and Health
4676 Columbia Parkway
Cincinnati, OH 45226
FAX (513) 533-8573

For information about other occupational safety and health problems, call 1-800-35-NIOSH

DHHS (NIOSH) Publication No. 93-133

* Full-size forms available free of charge at www.oshacfr.com

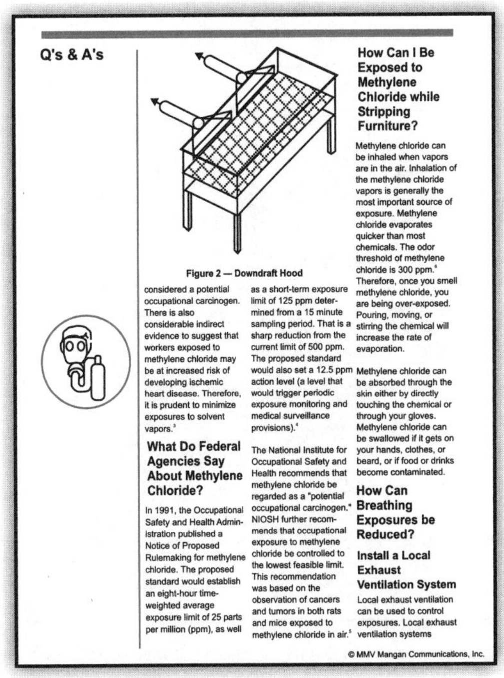

Q's & A's

Figure 2 — Downdraft Hood

considered a potential occupational carcinogen. There is also considerable indirect evidence to suggest that workers exposed to methylene chloride may be at increased risk of developing ischemic heart disease. Therefore, it is prudent to minimize exposures to solvent vapors.[3]

What Do Federal Agencies Say About Methylene Chloride?

In 1991, the Occupational Safety and Health Administration published a Notice of Proposed Rulemaking for methylene chloride. The proposed standard would establish an eight-hour time-weighted average exposure limit of 25 parts per million (ppm), as well as a short-term exposure limit of 125 ppm determined from a 15 minute sampling period. That is a sharp reduction from the current limit of 500 ppm. The proposed standard would also set a 12.5 ppm action level (a level that would trigger periodic exposure monitoring and medical surveillance provisions).[4]

The National Institute for Occupational Safety and Health recommends that methylene chloride be regarded as a "potential occupational carcinogen." NIOSH further recommends that occupational exposure to methylene chloride be controlled to the lowest feasible limit. This recommendation was based on the observation of cancers and tumors in both rats and mice exposed to methylene chloride in air.[5]

How Can I Be Exposed to Methylene Chloride while Stripping Furniture?

Methylene chloride can be inhaled when vapors are in the air. Inhalation of the methylene chloride vapors is generally the most important source of exposure. Methylene chloride evaporates quicker than most chemicals. The odor threshold of methylene chloride is 300 ppm.[6] Therefore, once you smell methylene chloride, you are being over-exposed. Pouring, moving, or stirring the chemical will increase the rate of evaporation.

Methylene chloride can be absorbed through the skin either by directly touching the chemical or through your gloves. Methylene chloride can be swallowed if it gets on your hands, clothes, or beard, or if food or drinks become contaminated.

How Can Breathing Exposures be Reduced?

Install a Local Exhaust Ventilation System

Local exhaust ventilation can be used to control exposures. Local exhaust ventilation systems

* Full-size forms available free of charge at www.oshacfr.com

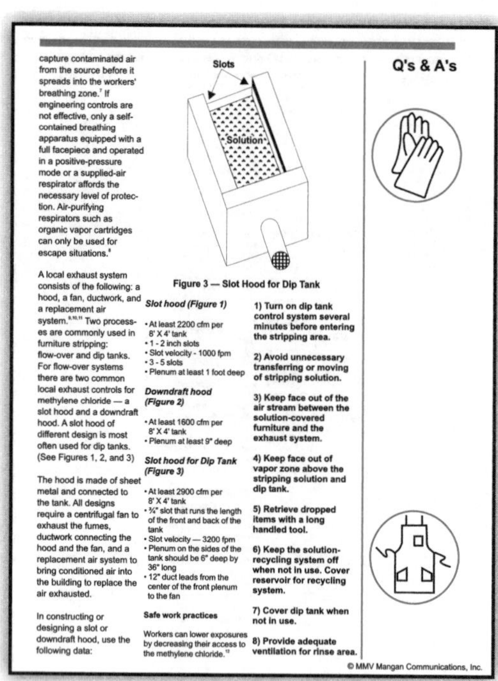

capture contaminated air from the source before it spreads into the workers' breathing zone.[7] If engineering controls are not effective, only a self-contained breathing apparatus equipped with a full facepiece and operated in a positive-pressure mode or a supplied-air respirator affords the necessary level of protection. Air-purifying respirators such as organic vapor cartridges can only be used for escape situations.[8]

A local exhaust system consists of the following: a hood, a fan, ductwork, and a replacement air system.[9,10,11] Two processes are commonly used in furniture stripping: flow-over and dip tanks. For flow-over systems there are two common local exhaust controls for methylene chloride — a slot hood and a downdraft hood. A slot hood of different design is most often used for dip tanks. (See Figures 1, 2, and 3)

The hood is made of sheet metal and connected to the tank. All designs require a centrifugal fan to exhaust the fumes, ductwork connecting the hood and the fan, and a replacement air system to bring conditioned air into the building to replace the air exhausted.

In constructing or designing a slot or downdraft hood, use the following data:

Figure 3 — Slot Hood for Dip Tank

Slot hood (Figure 1)

- At least 2200 cfm per 8' X 4' tank
- 1 - 2 inch slots
- Slot velocity - 1000 fpm
- 3 - 5 slots
- Plenum at least 1 foot deep

Downdraft hood (Figure 2)

- At least 1600 cfm per 8' X 4' tank
- Plenum at least 9" deep

Slot hood for Dip Tank (Figure 3)

- At least 2900 cfm per 8' X 4' tank
- ¾" slot that runs the length of the front and back of the tank
- Slot velocity — 3200 fpm
- Plenum on the sides of the tank should be 6" deep by 36" long
- 12" duct leads from the center of the front plenum to the fan

Safe work practices

Workers can lower exposures by decreasing their access to the methylene chloride.[12]

1) Turn on dip tank control system several minutes before entering the stripping area.

2) Avoid unnecessary transferring or moving of stripping solution.

3) Keep face out of the air stream between the solution-covered furniture and the exhaust system.

4) Keep face out of vapor zone above the stripping solution and dip tank.

5) Retrieve dropped items with a long handled tool.

6) Keep the solution-recycling system off when not in use. Cover reservoir for recycling system.

7) Cover dip tank when not in use.

8) Provide adequate ventilation for rinse area.

Q's & A's

* Full-size forms available free of charge at www.oshacfr.com

REFERENCES

[1] Halogenated Solvents Industry Alliance and Consumer Product Safety Commission [1990]. Stripping Paint from Wood (Pamphlet for consumers on how to strip furniture and precautions to take). Washington DC; Consumer Product Safety Commission.

[2] *Ibid.*

[3] NIOSH [1992]. NIOSH Testimony on Occupational Safety and Health Administration's proposed rule on occupational exposure to methylene chloride, September 21, 1992, OSHA Docket No. H-71. NIOSH policy statements. Cincinnati, OH: U.S. Department of Health and Human Services, Public Health Service, Centers for Disease Control, National Institute for Occupational Safety and Health.

[4] 56 Fed. Reg. 57036 [1991]. Occupational Safety and Health Administration: Proposed rule on occupational exposure to methylene chloride.

[5] NIOSH [1992].

[6] Kirk, R.E. and P.F. Othmer, Eds. [1978]. Encyclopedia of Chemical Technology, 3rd Ed., Vol. 5:690. New York: John Wiley & Sons, Inc.

[7] ACGIH [1988]. Industrial Ventilation: A Manual of Recommended Practice. 20th Ed. Cincinnati, OH: American Conference of Governmental Industrial Hygienists.

[8] NIOSH [1992].

[9] Fairfield, C.L. and A.A. Beasley [1991]. In-depth Survey Report at the Association for Retarded Citizens, Meadowlands, PA. The Control of Methylene Chloride During Furniture Stripping. Cincinnati, OH: U.S. Department of Health and Human Services, Public Health Service, Centers for Disease Control, National Institute for Occupational Safety and Health.

[10] Fairfield, C.L. [1991]. In-depth Survey Report at the J.M. Murray Center, Cortland, NY. The Control of Methylene Chloride During Furniture Stripping. Cincinnati, OH: U.S. Department of Health and Human Services, Public Health Service, Centers for Disease Control, National Institute for Occupational Safety and Health.

[11] Hall, R.M., K.F. Martinez, and P.A. Jensen [1992]. In-depth Survey Report at Tri-County Furniture Stripping and Refinishing, Cincinnati, OH. The Control of Methylene Chloride During Furniture Stripping. Cincinnati, OH: U.S. Department of Health and Human Services, Public Health Service, Centers for Disease Control, National Institute for Occupational Safety and Health.

[12] Fairfield, C.L. and A.A. Beasley [1991]. In-depth Survey Report at the Association for Retarded Citizens, Meadowlands, PA. The Control of Methylene Chloride During Furniture Stripping. Cincinnati, OH: U.S. Department of Health and Human Services, Public Health Service, Centers for Disease Control, National Institute for Occupational Safety and Health.

[13] Roder, M. [1991]. Memorandum of March 11, 1991 from Michael Roder of the Division of Safety Research to Cheryl L. Fairfield of the Division of Physical Sciences and Engineering, National Institute for Occupational Safety and Health, Centers for Disease Control, Public Health Service, U.S. Department of Health and Human Services.

[14] Kwick Kleen Industrial Solvents, Inc. [1981]. Operations Manual, Kwick Kleen Industrial Solvents, Inc., Vincennes, IN.

[15] ACGIH [1988].

[16] *Ibid.*

[17] Gerritsen,, W.B. and C.H. Buschmann [1960]. Phosgene Poisoning Caused by the Use of Chemical Paint Removers containing Methylene Chloride in Ill-Ventilated Rooms Heated by Kerosene Stoves. British Journal of Industrial Medicine 17:187.

U.S. GOVERNMENT PRINTING OFFICE: 1993: 550-147/80026

Q's & A's

Where Should I go for More Information?

The NIOSH 800-number is a toll-free technical information service that provides convenient public access to NIOSH and its information resources. Callers may request information about any aspect of occupational safety and health.

1-800-35-NIOSH
(1-800-356-4674)

* Full-size forms available free of charge at www.oshacfr.com

Q's & A's

How Can Skin Exposures be Reduced?

Skin exposures can be reduced by wearing gloves whenever you are in contact with the stripping solution.[13]

1) Two gloves should be worn. The inner glove should be made from polyethylene/ethylene vinyl alcohol (e.g. Silver Shield®, or 4H®). This material, however, does not provide good physical resistance against tears, so an outer glove made from nitrile or neoprene should be worn.

2) Shoulder-length gloves will be more protective.

3) Change gloves before the break-through time occurs. Rotate several pairs of gloves throughout the day. Let the gloves dry in a warm well ventilated area at least over night before reuse.

4) Keep gloves clean by rinsing often. Keep gloves in good condition. Inspect the gloves before use for pin-holes, cracks, thin spots, and stiffer than normal or sticky surfaces.

5) Wear a face shield or goggles to protect face and eyes.

What Other Problems Occur?

Stripping Solution Temperature

Most manufacturers of stripping solution recommend controlling the solution to a temperature of 70°F. This temperature is required for the wax in the solution to form a vapor barrier on top of the solution to keep the solution from evaporating too quickly. If the temperature is too high, the wax will not form the vapor barrier. If it is too cold, the wax will solidify and separate from the solvent causing increased evaporation. Use a belt heater to heat the solution to the correct temperature. Call your solution manufacturer for the correct temperature for your solution.[14]

Make-Up Air

Air will enter a building in an amount to equal the amount of air exhausted whether or not provision is made for this replacement. If a local exhaust system is added a make-up or replacement air system must be added to replace the air removed. Without a replacement air system, air will enter the building through cracks causing uncontrollable eddy currents. If the building perimeter is tightly sealed, it will prevent the air from entering and severely decrease the amount exhausted from the ventilation system. This will cause the building to be under negative pressure and decrease the performance of the exhaust system.[15]

Dilution Ventilation

With general or dilution ventilation, uncontaminated air is moved through the workroom by means of fans or open windows, which dilutes the pollutants in the air. Dilution ventilation does not provide effective protection to other workers and does not confine the methylene chloride vapors to one area.[16]

Phosgene Poisoning from Use of Kerosene Heaters

Do not use kerosene heaters or other open flame heaters while stripping furniture. Use of kerosene heaters in connection with methylene chloride can create lethal or dangerous concentrations of phosgene. Methylene chloride vapor is mixed with the air used for the combustion of kerosene in kerosene stoves. The vapor thus passes through the flames, coming into close contact with carbon monoxide at high temperatures. Any chlorine formed by decomposition may, under these conditions, react with carbon monoxide and form phosgene.[17]

* Full-size forms available free of charge at www.oshacfr.com

[62 FR 1601, Jan. 10, 1997, as amended at 62 FR 42667, Aug. 8, 1997; 62 FR 54383, Oct. 20, 1997; 62 FR 66277, Dec. 18, 1997; 63 FR 1295, Jan. 8, 1998; 63 FR 20099, Apr. 23, 1998; 63 FR 50729, Sept. 22, 1998; 71 FR 16674, Apr. 3, 2006; 71 FR 50190, Aug. 24, 2006; 73 FR 75587, Dec. 12, 2008; 77 FR 17785, Mar. 26, 2012]

§1910.1096

Ionizing radiation

(a) Definitions applicable to this section.[§1910.1096(a)]

(1) **Radiation** includes alpha rays, beta rays, gamma rays, X-rays, neutrons, high-speed electrons, high-speed protons, and other atomic particles; but such term does not include sound or radio waves, or visible light, or infrared or ultraviolet light. [§1910.1096(a)(1)]

(2) **Radioactive material** means any material which emits, by spontaneous nuclear disintegration, corpuscular or electromagnetic emanations.[§1910.1096(a)(2)]

(3) **Restricted area** means any area access to which is controlled by the employer for purposes of protection of individuals from exposure to radiation or radioactive materials.[§1910.1096(a)(3)]

(4) **Unrestricted area** means any area access to which is not controlled by the employer for purposes of protection of individuals from exposure to radiation or radioactive materials. [§1910.1096(a)(4)]

(5) **Dose** means the quantity of ionizing radiation absorbed, per unit of mass, by the body or by any portion of the body. When the provisions in this section specify a dose during a period of time, the dose is the total quantity of radiation absorbed, per unit of mass, by the body or by any portion of the bodv durina such period of time. Several different units of dose
rent use. Definitions of units used in this section are set forth in paragraphs (a)(6) and (7) of this section.[§1910.1096(a)(5)]

(6) **Rad** means a measure of the dose of any ionizing radiation to body tissues in terms of the energy absorbed per unit of mass of the tissue. One rad is the dose corresponding to the absorption of 100 ergs per gram of tissue (1 millirad (mrad)=0.001 rad).[§1910.1096(a)(6)]

(7) **Rem** means a measure of the dose of any ionizing radiation to body tissue in terms of its estimated biological effect relative to a dose of 1 roentgen (r) of X-rays (1 millirem (mrem)=0.001 rem). The relation of the rem to other dose units depends upon the biological effect under consideration and upon the conditions for irradiation. Each of the following is considered to be equivalent to a dose of 1 rem:[§1910.1096(a)(7)]

(i) *A dose of 1 roentgen* due to X-or gamma radiation; [§1910.1096(a)(7)(i)]

(ii) *A dose of 1 rad* due to X-, gamma, or beta radiation; [§1910.1096(a)(7)(ii)]

(iii) *A dose of 0.1 rad* due to neutrons or high energy protons; [§1910.1096(a)(7)(iii)]

(iv) *A dose of 0.05 rad* due to particles heavier than protons and with sufficient energy to reach the lens of the eye; [§1910.1096(a)(7)(iv)]

(v) *If it is more convenient* to measure the neutron flux, or equivalent, than to determine the neutron dose in rads, as provided in paragraph (a)(7)(iii) of this section, 1 rem of neutron radiation may, for purposes of the provisions in this section be assumed to be equivalent to 14 million neutrons per square centimeter incident upon the body; or, if there is sufficient information to estimate with reasonable accuracy the approximate distribution in energy of the neutrons, the incident number of neutrons per square centimeter equivalent to 1 rem may be estimated from Table G-17:[§1910.1096(a)(7)(v)]

Table G-17 — Neutron Flux Dose Equivalents

Neutron energy (million electron volts (Mev))	Number of neutrons per square centimeter equivalent to a dose of 1 rem (neutrons/cm^2)	Average flux to deliver 100 millirem in 40 hours (neutrons/cm^2 per sec.)
Thermal	970×10^6	670
0.0001	720×10^6	500
0.005	820×10^6	570
0.02	400×10^6	280
0.1	120×10^6	80
0.5	43×10^6	30
1.0	26×10^6	18
2.5	29×10^6	20
5.0	26×10^6	18
7.5	24×10^6	17
10	24×10^6	17
10 to 30	14×10^6	10

(8) *For determining exposures* to X-or gamma rays up to 3 Mev., the dose limits specified in this section may be assumed to be equivalent to the "air dose". For the purpose of this section air dose means that the dose is measured by a properly calibrated appropriate instrument in air at or near the body surface in the region of the highest dosage rate.[§1910.1096(a)(8)]

(b) **Exposure of individuals to radiation in restricted areas.** [§1910.1096(b)]

(1) *Except as provided in paragraph (b)(2)* of this section, no employer shall possess, use, or transfer sources of ionizing radiation in such a manner as to cause any individual in a restricted area to receive in any period of one calendar quarter from sources in the employer's possession or control a dose in excess of the limits specified in Table G-18:[§1910.1096(b)(1)]

Table G-18

	Rems per calendar quarter
Whole body: Head and trunk; active blood-forming organs; lens of eyes; or gonads	1 1/4
Hands and forearms; feet and ankles	18 3/4
Skin of whole body	7 1/2

(2) *An employer may permit* an individual in a restricted area to receive doses to the whole body greater than those permitted under subparagraph (1) of this paragraph, so long as: [§1910.1096(b)(2)]

(i) *During any calendar quarter* the dose to the whole body shall not exceed 3 rems; and[§1910.1096(b)(2)(i)]

(ii) *The dose to the whole body,* when added to the accumulated occupational dose to the whole body, shall not exceed 5 (N-18) rems, where "N" equals the individual's age in years at his last birthday; and[§1910.1096(b)(2)(ii)]

(iii) *The employer maintains* adequate past and current exposure records which show that the addition of such a dose will not cause the individual to exceed the amount authorized in this subparagraph. As used in this subparagraph Dose to the whole body shall be deemed to include any dose to the whole body, gonad, active bloodforming organs, head and trunk, or lens of the eye.[§1910.1096(b)(2)(iii)]

(3) *No employer shall permit* any employee who is under 18 years of age to receive in any period of one calendar quarter a dose in excess of 10 percent of the limits specified in Table G-18. [§1910.1096(b)(3)]

(4) **Calendar quarter** means any 3-month period determined as follows:[§1910.1096(b)(4)]

(i) *The first period of any year* may begin on any date in January: Provided, That the second, third, and fourth periods accordingly begin on the same date in April, July, and October, respectively, and that the fourth period extends into January of the succeeding year, if necessary to complete a 3-month quarter. During the first year of use of this method of determination, the first period for that year shall also include any additional days in January preceding the starting date for the first period; or[§1910.1096(b)(4)(i)]

(ii) *The first period in a calendar year* of 13 complete, consecutive calendar weeks; the second period in a calendar year of 13 complete, consecutive weeks; the third period in a calendar year of 13 complete, consecutive calendar weeks; the fourth period in a calendar year of 13 complete, consecutive calendar weeks. If at the end of a calendar year there are any days not falling within a complete calendar week of that year, such days shall be included within the last complete calendar week of that year. If at the beginning of any calendar year there are days not falling within a complete calendar week of that year, such days shall be included within the last complete calendar week of the previous year; or[§1910.1096(b)(4)(ii)]

(iii) *The four periods in a calendar year* may consist of the first 14 complete, consecutive calendar weeks; the next 12 complete, consecutive calendar weeks, the next 14 complete, consecutive calendar weeks, and the last 12 complete, consecutive calendar weeks. If at the end of a calendar year there are any days not falling within a complete calendar week of that year, such days shall be included (for purposes of this section) within the last complete calendar week of the year. If at the beginning of any calendar year there are days not falling within a complete calendar week of that year, such days shall be included (for purposes of this section) within the last complete week of the previous year. [§1910.1096(b)(4)(iii)]

(c) **Exposure to airborne radioactive material.**[§1910.1096(c)]

(1) *No employer shall possess, use or transport* radioactive material in such a manner as to cause any employee, within a restricted area, to be exposed to airborne radioactive material in an average concentration in excess of the limits specified in Table 1 of appendix B to 10 CFR part 20. The limits given in Table 1 are for exposure to the concentrations specified for 40 hours in any workweek of 7 consecutive days. In any such period where the number of hours of exposure is less than 40, the limits specified in the table may be increased proportionately. In any such period where the number of hours of exposure is greater than 40, the limits specified in the table shall be decreased proportionately.[§1910.1096(c)(1)]

(2) *No employer shall possess, use, or transfer* radioactive material in such a manner as to cause any individual within a restricted area, who is under 18 years of age, to be exposed to airborne radioactive material in an average concentration in excess of the limits specified in Table II of appendix B to 10 CFR part 20. For purposes of this paragraph, concentrations may be averaged over periods not greater than 1 week. [§1910.1096(c)(2)]

(3) **Exposed** as used in this paragraph means that the individual is present in an airborne concentration. No allowance shall be made for the use of protective clothing or equipment, or particle size.[§1910.1096(c)(3)]

(d) **Precautionary procedures and personal monitoring.** [§1910.1096(d)]

(1) *Every employer shall make* such surveys as may be necessary for him to comply with the provisions in this section. Survey means an evaluation of the radiation hazards incident to the production, use, release, disposal, or presence of radioactive materials or other sources of radiation under a specific set of conditions. When appropriate, such evaluation includes a physical survey of the location of materials and equipment, and measurements of levels of radiation or concentrations of radioactive material present.[§1910.1096(d)(1)]

(2) *Every employer shall supply* appropriate personnel monitoring equipment, such as film badges, pocket chambers, pocket dosimeters, or film rings, and shall require the use of such equipment by:[§1910.1096(d)(2)]

(i) *Each employee who enters a restricted* area under such circumstances that he receives, or is likely to receive, a dose in any calendar quarter in excess of 25 percent of the applicable value specified in paragraph (b)(1) of this section; and [§1910.1096(d)(2)(i)]

(ii) *Each employee under 18 years* of age who enters a restricted area under such circumstances that he receives, or is likely to receive, a dose in any calendar quarter in excess of 5 percent of the applicable value specified in paragraph (b)(1) of this section; and[§1910.1096(d)(2)(ii)]

(iii) *Each employee who enters a high radiation area.* [§1910.1096(d)(2)(iii)]

(3) *As used in this section:*[§1910.1096(d)(3)]

(i) **Personnel monitoring equipment** means devices designed to be worn or carried by an individual for the purpose of measuring the dose received (e.g., film badges, pocket chambers, pocket dosimeters, film rings, etc.);[§1910.1096(d)(3)(i)]

(ii) **Radiation area** means any area, accessible to personnel, in which there exists radiation at such levels that a major portion of the body could receive in any 1 hour a dose in excess of 5 millirem, or in any 5 consecutive days a dose in excess of 100 millirem; and[§1910.1096(d)(3)(ii)]

(iii) **High radiation area** means any area, accessible to personnel, in which there exists radiation at such levels that a major portion of the body could receive in any one hour a dose in excess of 100 millirem.[§1910.1096(d)(3)(iii)]

(e) Caution signs, labels, and signals.[§1910.1096(e)]

(1) *General.*[§1910.1096(e)(1)]

(i) *Symbols prescribed by this paragraph* shall use the conventional radiation caution colors (magenta or purple on yellow background). The symbol prescribed by this paragraph is the conventional three-bladed design:[§1910.1096(e)(1)(i)]

RADIATION SYMBOL

1. Cross-hatched area is to be magenta or purple.
2. Background is to be yellow.

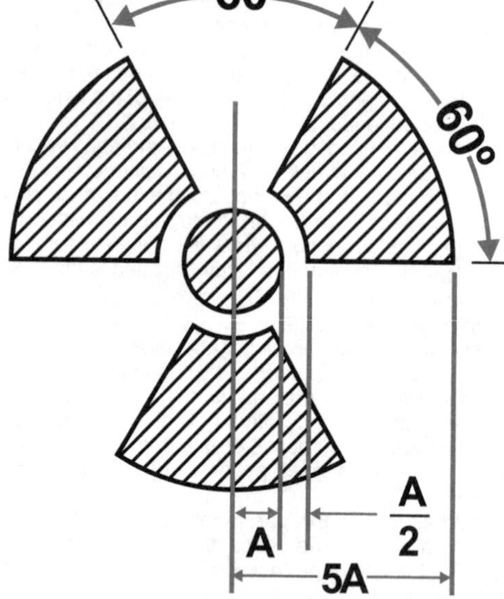

Figure G-10

(ii) *[Reserved]*[§1910.1096(e)(1)(ii)]

(2) *Radiation area.* Each radiation area shall be conspicuously posted with a sign or signs bearing the radiation caution symbol described in subparagraph (1) of this paragraph and the words:[§1910.1096(e)(2)]

CAUTION
RADIATION AREA

(3) *High radiation area.*[§1910.1096(e)(3)]

(i) *Each high radiation area* shall be conspicuously posted with a sign or signs bearing the radiation caution symbol and the words:[§1910.1096(e)(3)(i)]

CAUTION
HIGH RADIATION AREA

(ii) *Each high radiation area* shall be equipped with a control device which shall either cause the level of radiation to be reduced below that at which an individual might receive a dose of 100 millirems in 1 hour upon entry into the area or shall energize a conspicuous visible or audible alarm signal in such a manner that the individual entering and the employer or a supervisor of the activity are made aware of the entry. In the case of a high radiation area established for a period of 30 days or less, such control device is not required.[§1910.1096(e)(3)(ii)]

(4) *Airborne radioactivity area.*[§1910.1096(e)(4)]

(i) *As used in the provisions of this section,* **airborne radioactivity area** means:[§1910.1096(e)(4)(i)]

[a] Any room, enclosure, or operating area in which airborne radioactive materials, composed wholly or partly of radioactive material, exist in concentrations in excess of the amounts specified in column 1 of Table 1 of appendix B to 10 CFR part 20 or[§1910.1096(e)(4)(i)[a]]

[b] Any room, enclosure, or operating area in which airborne radioactive materials exist in concentrations which, averaged over the number of hours in any week during which individuals are in the area, exceed 25 percent of the amounts specified in column 1 of Table 1 of appendix B to 10 CFR part 20.[§1910.1096(e)(4)(i)[b]]

(ii) *Each airborne radioactivity area* shall be conspicuously posted with a sign or signs bearing the radiation caution symbol described in paragraph (e)(1) of this section and the words:[§1910.1096(e)(4)(ii)]

CAUTION
AIRBORNE RADIOACTIVITY AREA

(5) *Additional requirements.*[§1910.1096(e)(5)]

(i) *Each area or room* in which radioactive material is used or stored and which contains any radioactive material (other than natural uranium or thorium) in any amount exceeding 10 times the quantity of such material specified in appendix C to 10 CFR part 20 shall be conspicuously posted with a sign or signs bearing the radiation caution symbol described in paragraph (e)(1) of this section and the words:[§1910.1096(e)(5)(i)]

CAUTION
RADIOACTIVE MATERIALS

(ii) *Each area or room* in which natural uranium or thorium is used or stored in an amount exceeding 100 times the quantity of such material specified in 10 CFR part 20 shall be conspicuously posted with a sign or signs bearing the radiation caution symbol described in paragraph (e)(1) of this section and the words:[§1910.1096(e)(5)(ii)]

CAUTION
RADIOACTIVE MATERIALS

(6) *Containers.*[§1910.1096(e)(6)]

(i) *Each container in which is transported, stored, or used* a quantity of any radioactive material (other than natural uranium or thorium) greater than the quantity of such material specified in appendix C to 10 CFR part 20 shall bear a durable, clearly visible label bearing the radiation caution symbol described in paragraph (e)(1) of this section and the words. [§1910.1096(e)(6)(i)]

CAUTION
RADIOACTIVE MATERIALS

(ii) *Each container in which natural uranium or thorium* is transported, stored, or used in a quantity greater than 10 times the quantity specified in appendix C to 10 CFR part 20 shall bear a durable, clearly visible label bearing the radiation caution symbol described in paragraph (e)(1) of this section and the words:[§1910.1096(e)(6)(ii)]

CAUTION
RADIOACTIVE MATERIALS

(iii) *Notwithstanding the provisions of paragraphs (e)(6)(i) and (ii)* of this section a label shall not be required: [§1910.1096(e)(6)(iii)]

[a] If the concentration of the material in the container does not exceed that specified in column 2 of Table 1 of appendix B to 10 CFR part 20, or[§1910.1096(e)(6)(iii)[a]]

[b] For laboratory containers, such as beakers, flasks, and test tubes, used transiently in laboratory procedures, when the user is present.[§1910.1096(e)(6)(iii)[b]]

(iv) *Where containers are used for storage,* the labels required in this subparagraph shall state also the quantities and kinds of radioactive materials in the containers and the date of measurement of the quantities.[§1910.1096(e)(6)(iv)]

(f) Immediate evacuation warning signal.[§1910.1096(f)]

(1) *Signal characteristics.*[§1910.1096(f)(1)]

(i) *The signal shall be* a midfrequency complex sound wave amplitude modulated at a subsonic frequency. The complex sound wave in free space shall have a fundamental frequency (f_1) between 450 and 500 hertz (Hz) modulated at a subsonic rate between 4 and 5 hertz.[§1910.1096(f)(1)(i)]

(ii) *The signal generator shall not be* less than 75 decibels at every location where an individual may be present whose immediate, rapid, and complete evacuation is essential. [§1910.1096(f)(1)(ii)]

(iii) *A sufficient number of signal units* shall be installed such that the requirements of paragraph (f)(1)(ii) of this section are met at every location where an individual may be present whose immediate, rapid, and complete evacuation is essential.[§1910.1096(f)(1)(iii)]

(iv) *The signal shall be unique in the plant* or facility in which it is installed.[§1910.1096(f)(1)(iv)]

(v) *The minimum duration of the signal* shall be sufficient to insure that all affected persons hear the signal. [§1910.1096(f)(1)(v)]

(vi) *The signal-generating system* shall respond automatically to an initiating event without requiring any human action to sound the signal.[§1910.1096(f)(1)(vi)]

(2) *Design objectives.*[§1910.1096(f)(2)]

(i) *The signal-generating system shall be designed* to incorporate components which enable the system to produce the desired signal each time it is activated within one-half second of activation.[§1910.1096(f)(2)(i)]

(ii) *The signal-generating system shall be provided* with an automatically activated secondary power supply which is adequate to simultaneously power all emergency equipment to which it is connected, if operation during power failure is necessary, except in those systems using batteries as the primary source of power.[§1910.1096(f)(2)(ii)]

(iii) *All components of the signal-generating system* shall be located to provide maximum practicable protection against damage in case of fire, explosion, corrosive atmosphere, or other environmental extremes consistent with adequate system performance.[§1910.1096(f)(2)(iii)]

(iv) *The signal-generating system shall be designed* with the minimum number of components necessary to make it function as intended, and should utilize components which do not require frequent servicing such as lubrication or cleaning. [§1910.1096(f)(2)(iv)]

(v) *Where several activating devices* feed activating information to a central signal generator, failure of any activating device shall not render the signal-generator system inoperable to activating information from the remaining devices. [§1910.1096(f)(2)(v)]

(vi) *The signal-generating system shall be designed* to enhance the probability that alarm occurs only when immediate evacuation is warranted. The number of false alarms shall not be so great that the signal will come to be disregarded and shall be low enough to minimize personal injuries or excessive property damage that might result from such evacuation. [§1910.1096(f)(2)(vi)]

(3) *Testing.*[§1910.1096(f)(3)]

(i) *Initial tests, inspections, and checks* of the signal-generating system shall be made to verify that the fabrication and installation were made in accordance with design plans and specifications and to develop a thorough knowledge of the performance of the system and all components under normal and hostile conditions.[§1910.1096(f)(3)(i)]

(ii) *Once the system has been placed* in service, periodic tests, inspections, and checks shall be made to minimize the possibility of malfunction.[§1910.1096(f)(3)(ii)]

(iii) *Following significant alterations* or revisions to the system, tests and checks similar to the initial installation tests shall be made.[§1910.1096(f)(3)(iii)]

(iv) *Tests shall be designed* to minimize hazards while conducting the tests.[§1910.1096(f)(3)(iv)]

(v) *Prior to normal operation* the signal-generating system shall be checked physically and functionally to assure reliability and to demonstrate accuracy and performance. Specific tests shall include:[§1910.1096(f)(3)(v)]

[a] All power sources.[§1910.1096(f)(3)(v)[a]]

[b] Calibration and calibration stability.[§1910.1096(f)(3)(v)[b]]

[c] Trip levels and stability.[§1910.1096(f)(3)(v)[c]]

[d] Continuity of function with loss and return of required services such as AC or DC power, air pressure, etc. [§1910.1096(f)(3)(v)[d]]

[e] All indicators.[§1910.1096(f)(3)(v)[e]]

[f] Trouble indicator circuits and signals, where used. [§1910.1096(f)(3)(v)[f]]

[g] Air pressure (if used)[§1910.1096(f)(3)(v)[g]]

[h] Determine that sound level of the signal is within the limit of paragraph (f)(1)(ii) of this section at all points that require immediate evacuation.[§1910.1096(f)(3)(v)[h]]

(vi) *In addition to the initial startup* and operating tests, periodic scheduled performance tests and status checks must be made to insure that the system is at all times operating within design limits and capable of the required response. Specific periodic tests or checks or both shall include: [§1910.1096(f)(3)(vi)]

[a] Adequacy of signal activation device.[§1910.1096(f)(3)(vi)[a]]

[b] All power sources.[§1910.1096(f)(3)(vi)[b]]

[c] Function of all alarm circuits and trouble indicator circuits including trip levels.[§1910.1096(f)(3)(vi)[c]]

[d] Air pressure (if used).[§1910.1096(f)(3)(vi)[d]]

[e] Function of entire system including operation without power where required.[§1910.1096(f)(3)(vi)[e]]

[f] Complete operational tests including sounding of the signal and determination that sound levels are adequate. [§1910.1096(f)(3)(vi)[f]]

(vii) *Periodic tests shall be scheduled* on the basis of need, experience, difficulty, and disruption of operations. The entire system should be operationally tested at least quarterly. [§1910.1096(f)(3)(vii)]

(viii) *All employees* whose work may necessitate their presence in an area covered by the signal shall be made familiar with the actual sound of the signal — preferably as it sounds at their work location. Before placing the system into operation, all employees normally working in the area shall be made acquainted with the signal by actual demonstration at their work locations.[§1910.1096(f)(3)(viii)]

(g) Exceptions from posting requirements. Notwithstanding the provisions of paragraph (e) of this section:[§1910.1096(g)]

(1) *A room or area is not required* to be posted with a caution sign because of the presence of a sealed source, provided the radiation level 12 inches from the surface of the source container or housing does not exceed 5 millirem per hour. [§1910.1096(g)(1)]

(2) *Rooms or other areas in onsite medical facilities* are not required to be posted with caution signs because of the presence of patients containing radioactive material, provided that there are personnel in attendance who shall take the precautions necessary to prevent the exposure of any individual to radiation or radioactive material in excess of the limits established in the provisions of this section.[§1910.1096(g)(2)]

(3) *Caution signs are not required* to be posted at areas or rooms containing radioactive materials for periods of less than 8 hours: Provided, That[§1910.1096(g)(3)]

(i) *The materials are constantly attended* during such periods by an individual who shall take the precautions necessary to prevent the exposure of any individual to radiation or radioactive materials in excess of the limits established in the provisions of this section; and[§1910.1096(g)(3)(i)]

(ii) *Such area or room* is subject to the employer's control. [§1910.1096(g)(3)(ii)]

(h) Exemptions for radioactive materials packaged for shipment. Radioactive materials packaged and labeled in accordance with regulations of the Department of Transportation published in 49 CFR Chapter I, are exempt from the labeling and posting requirements of this subpart during shipment, provided that the inside containers are labeled in accordance with the provisions of paragraph (e) of this section.[§1910.1096(h)]

(i) Instruction of personnel, posting.[§1910.1096(i)]

(1) *Employers regulated by* the Nuclear Regulatory Commission shall be governed by 10 CFR part 20 standards. Employers in a State named in paragraph (p)(3) of this section shall be governed by the requirements of the laws and regulations of that State. All other employers shall be regulated by the following: [§1910.1096(i)(1)]

(2) *All individuals working in or frequenting* any portion of a radiation area shall be informed of the occurrence of radioactive materials or of radiation in such portions of the radiation area; shall be instructed in the safety problems associated with exposure to such materials or radiation and in precautions or devices to minimize exposure; shall be instructed in the applicable provisions of this section for the protection of employees from exposure to radiation or radioactive materials; and shall be advised of reports of radiation exposure which employees may request pursuant to the regulations in this section. [§1910.1096(i)(2)]

(3) *Each employer to whom this section applies* shall post a current copy of its provisions and a copy of the operating procedures applicable to the work conspicuously in such locations as to insure that employees working in or frequenting radiation areas will observe these documents on the way to and from their place of employment, or shall keep such documents available for examination of employees upon request. [§1910.1096(i)(3)]

(j) Storage of radioactive materials. Radioactive materials stored in a nonradiation area shall be secured against unauthorized removal from the place of storage.[§1910.1096(j)]

(k) Waste disposal. No employer shall dispose of radioactive material except by transfer to an authorized recipient, or in a manner approved by the Nuclear Regulatory Commission or a State named in paragraph (p)(3) of this section.[§1910.1096(k)]

(l) Notification of incidents.[§1910.1096(l)]

(1) *Immediate notification.* Each employer shall immediately notify the Assistant Secretary of Labor or his duly authorized representative, for employees not protected by the Nuclear Regulatory Commission by means of 10 CFR part 20; paragraph (p)(2) of this section, or the requirements of the laws and regulations of States named in paragraph (p)(3) of this section, by telephone or telegraph of any incident involving radiation which may have caused or threatens to cause:[§1910.1096(l)(1)]

(i) *Exposure of the whole body* of any individual to 25 rems or more of radiation; exposure of the skin of the whole body of any individual to 150 rems or more of radiation; or exposure of the feet, ankles, hands, or forearms of any individual to 375 rems or more of radiation; or[§1910.1096(l)(1)(i)]

(ii) *The release of radioactive material* in concentrations which, if averaged over a period of 24 hours, would exceed 5,000 times the limit specified for such materials in Table II of appendix B to 10 CFR part 20.[§1910.1096(l)(1)(ii)]

(2) *Twenty-four hour notification.* Each employer shall within 24 hours following its occurrence notify the Assistant Secretary of Labor or his duly authorized representative for employees not protected by the Nuclear Regulatory Commission by means of 10 CFR part 20; paragraph (p)(2) of this section, or the requirements of the laws and applicable regulations of States named in paragraph (p)(3) of this section, by telephone or telegraph of any incident involving radiation which may have caused or threatens to cause:[§1910.1096(l)(2)]

(i) *Exposure of the whole body* of any individual to 5 rems or more of radiation; exposure of the skin of the whole body of any individual to 30 rems or more of radiation; or exposure of the feet, ankles, hands, or forearms to 75 rems or more of radiation; or[§1910.1096(l)(2)(i)]

(ii) *[Reserved]*[§1910.1096(l)(2)(ii)]

(m) Reports of overexposure and excessive levels and concentrations.[§1910.1096(m)]

(1) *In addition to any notification* required by paragraph (1) of this section each employer shall make a report in writing within 30 days to the Assistant Secretary of Labor or his duly authorized representative, for employees not protected by the Nuclear Regulatory Commission by means of 10 CFR part 20; or under paragraph (p)(2) of this section, or the requirements of the laws and regulations of States named in paragraph (p)(3) of this section, of each exposure of an individual to radiation or concentrations of radioactive material in excess of any applicable limit in this section. Each report required under this paragraph shall describe the extent of exposure of persons to radiation or to radioactive material; levels of radiation and concentration of radioactive material involved, the cause of the exposure, levels of concentrations; and corrective steps taken or planned to assure against a recurrence.[§1910.1096(m)(1)]

(2) *In any case where an employer is required* pursuant to the provisions of this paragraph to report to the U.S. Department of Labor any exposure of an individual to radiation or to concentrations of radioactive material, the employer shall also notify such individual of the nature and extent of exposure. Such notice shall be in writing and shall contain the following statement: "You should preserve this report for future reference."[§1910.1096(m)(2)]

(n) Records.[§1910.1096(n)]

(1) *Every employer shall maintain* records of the radiation exposure of all employees for whom personnel monitoring is required under paragraph (d) of this section and advise each of his employees of his individual exposure on at least an annual basis.[§1910.1096(n)(1)]

(2) *Every employer shall maintain* records in the same units used in tables in paragraph (b) of this section and appendix B to 10 CFR part 20.[§1910.1096(n)(2)]

(o) Disclosure to former employee of individual employee's record.[§1910.1096(o)]

(1) *At the request of a former employee* an employer shall furnish to the employee a report of the employee's exposure to radiation as shown in records maintained by the employer pursuant to paragraph (n)(1) of this section. Such report shall be furnished within 30 days from the time the request is made, and shall cover each calendar quarter of the individual's employment involving exposure to radiation or such lesser period as may be requested by the employee. The report shall also include the results of any calculations and analysis of radioactive material deposited in the body of the employee. The report shall be in writing and contain the following statement: "You should preserve this report for future reference." [§1910.1096(o)(1)]

(2) *[Reserved]*[§1910.1096(o)(2)]

(p) Nuclear Regulatory Commission licensees — NRC contractors operating NRC plants and facilities — NRC Agreement State licensees or registrants.[§1910.1096(p)]

(1) *Any employer who possesses or uses* source material, byproduct material, or special nuclear material, as defined in the Atomic Energy Act of 1954, as amended, under a license issued by the Nuclear Regulatory Commission and in accordance with the requirements of 10 CFR part 20 shall be deemed to be in compliance with the requirements of this section with respect to such possession and use.[§1910.1096(p)(1)]

(2) *NRC contractors operating NRC plants and facilities:* Any employer who possesses or uses source material, byproduct material, special nuclear material, or other radiation sources under a contract with the Nuclear Regulatory Commission for the operation of NRC plants and facilities and in accordance with the standards, procedures, and other requirements for radiation protection established by the Commission for such contract pursuant to the Atomic Energy Act of 1954 as amended (42 U.S.C. 2011 et seq.), shall be deemed to be in compliance with the requirements of this section with respect to such possession and use.[§1910.1096(p)(2)]

(3) *NRC-agreement State licensees or registrants:*[§1910.1096(p)(3)]

(i) *Atomic Energy Act sources.* Any employer who possesses or uses source material, byproduct material, or special nuclear material, as defined in the Atomic Energy Act of 1954, as amended (42 U.S.C. 2011 et seq.), and has either registered such sources with, or is operating under a license issued by, a State which has an agreement in effect with the Nuclear Regulatory Commission pursuant to section 274(b) (42 U.S.C. 2021(b)) of the Atomic Energy Act of 1954, as amended, and in accordance with the requirements of that State's laws and regulations shall be deemed to be in compliance with the radiation requirements of this section, insofar as his possession and use of such material is concerned, unless the Secretary of Labor, after conference with the Nuclear Regulatory Commission, shall determine that the State's program for control of these radiation sources is incompatible with the requirements of this section. Such agreements currently are in effect only in the States of Alabama, Arkansas, California, Kansas, Kentucky, Florida, Mississippi, New Hampshire, New York, North Carolina, Texas, Tennessee, Oregon, Idaho, Arizona, Colorado, Louisiana,

Nebraska, Washington, Maryland, North Dakota, South Carolina, and Georgia.[§1910.1096(p)(3)(i)]

(ii) *Other sources.* Any employer who possesses or uses radiation sources other than source material, byproduct material, or special nuclear material, as defined in the Atomic Energy Act of 1954, as amended (42 U.S.C. 2011 et seq.), and has either registered such sources with, or is operating under a license issued by a State which has an agreement in effect with the Nuclear Regulatory Commission pursuant to section 274(b) (42 U.S.C. 2021(b)) of the Atomic Energy Act of 1954, as amended, and in accordance with the requirements of that State's laws and regulations shall be deemed to be in compliance with the radiation requirements of this section, insofar as his possession and use of such material is concerned, provided the State's program for control of these radiation sources is the subject of a currently effective determination by the Assistant Secretary of Labor that such program is compatible with the requirements of this section. Such determinations currently are in effect only in the States of Alabama, Arkansas, California, Kansas, Kentucky, Florida, Mississippi, New Hampshire, New York, North Carolina, Texas, Tennessee, Oregon, Idaho, Arizona, Colorado, Louisiana, Nebraska, Washington, Maryland, North Dakota, South Carolina, and Georgia.[§1910.1096(p)(3)(ii)]

[39 FR 23502, June 27, 1974, as amended at 43 FR 49746, Oct. 24, 1978; 43 FR 51759, Nov. 7, 1978; 49 FR 18295, Apr. 30, 1984; 58 FR 35309, June 30, 1993. Redesignated at 61 FR 31430, June 20, 1996]

§1910.1200
Hazard communication

(a) Purpose.[§1910.1200(a)]

(1) *The purpose of this section is to ensure* that the hazards of all chemicals produced or imported are classified, and that information concerning the classified hazards is transmitted to employers and employees. The requirements of this section are intended to be consistent with the provisions of the United Nations Globally Harmonized System of Classification and Labelling of Chemicals (GHS), Revision 3. The transmittal of information is to be accomplished by means of comprehensive hazard communication programs, which are to include container labeling and other forms of warning, safety data sheets and employee training.[§1910.1200(a)(1)]

(2) *This occupational safety and health* standard is intended to address comprehensively the issue of classifying the potential hazards of chemicals, and communicating information concerning hazards and appropriate protective measures to employees, and to preempt any legislative or regulatory enactments of a state, or political subdivision of a state, pertaining to this subject. Classifying the potential hazards of chemicals and communicating information concerning hazards and appropriate protective measures to employees, may include, for example, but is not limited to, provisions for: developing and maintaining a written hazard communication program for the workplace, including lists of hazardous chemicals present; labeling of containers of chemicals in the workplace, as well as of containers of chemicals being shipped to other workplaces; preparation and distribution of safety data sheets to employees and downstream employers; and development and implementation of employee training programs regarding hazards of chemicals and protective measures. Under section 18 of the Act, no state or political subdivision of a state may adopt or enforce any requirement relating to the issue addressed by this Federal standard, except pursuant to a Federally-approved state plan.[§1910.1200(a)(2)]

(b) Scope and application.[§1910.1200(b)]

(1) *This section requires chemical* manufacturers or importers to classify the hazards of chemicals which they produce or import, and all employers to provide information to their employees about the hazardous chemicals to which they are exposed, by means of a hazard communication program, labels and other forms of warning, safety data sheets, and information and training. In addition, this section requires distributors to transmit the required information to employers. (Employers who do not produce or import chemicals need only focus on those parts of this rule that deal with establishing a workplace program and communicating information to their workers.)[§1910.1200(b)(1)]

(2) *This section applies to any chemical* which is known to be present in the workplace in such a manner that employees may be exposed under normal conditions of use or in a foreseeable emergency.[§1910.1200(b)(2)]

(3) *This section applies to laboratories* only as follows:[§1910.1200(b)(3)]

(i) *Employers shall ensure that labels* on incoming containers of hazardous chemicals are not removed or defaced;[§1910.1200(b)(3)(i)]

(ii) *Employers shall maintain any safety data sheets* that are received with incoming shipments of hazardous chemicals, and ensure that they are readily accessible during each workshift to laboratory employees when they are in their work areas;[§1910.1200(b)(3)(ii)]

(iii) *Employers shall ensure that laboratory employees* are provided information and training in accordance with paragraph (h) of this section, except for the location and availability of the written hazard communication program under paragraph (h)(2)(iii) of this section; and,[§1910.1200(b)(3)(iii)]

(iv) *Laboratory employers that ship* hazardous chemicals are considered to be either a chemical manufacturer or a distributor under this rule, and thus must ensure that any containers of hazardous chemicals leaving the laboratory are labeled in accordance with paragraph (f) of this section, and that a safety data sheet is provided to distributors and other employers in accordance with paragraphs (g)(6) and (g)(7) of this section.[§1910.1200(b)(3)(iv)]

(4) *In work operations where employees* only handle chemicals in sealed containers which are not opened under normal conditions of use (such as are found in marine cargo handling, warehousing, or retail sales), this section applies to these operations only as follows:[§1910.1200(b)(4)]

(i) *Employers shall ensure that labels* on incoming containers of hazardous chemicals are not removed or defaced;[§1910.1200(b)(4)(i)]

(ii) *Employers shall maintain copies* of any safety data sheets that are received with incoming shipments of the sealed containers of hazardous chemicals, shall obtain a safety data sheet as soon as possible for sealed containers of hazardous chemicals received without a safety data sheet if an employee requests the safety data sheet, and shall ensure that the safety data sheets are readily accessible during each work shift to employees when they are in their work area(s); and,[§1910.1200(b)(4)(ii)]

(iii) *Employers shall ensure that employees* are provided with information and training in accordance with paragraph (h) of this section (except for the location and availability of the written hazard communication program under paragraph (h)(2)(iii) of this section), to the extent necessary to protect them in the event of a spill or leak of a hazardous chemical from a sealed container.[§1910.1200(b)(4)(iii)]

(5) *This section does not require labeling of the following chemicals:*[§1910.1200(b)(5)]

(i) *Any pesticide* as such term is defined in the Federal Insecticide, Fungicide, and Rodenticide Act (7 U.S.C. 136 et seq.), when subject to the labeling requirements of that Act and labeling regulations issued under that Act by the Environmental Protection Agency;[§1910.1200(b)(5)(i)]

(ii) *Any chemical substance or mixture* as such terms are defined in the Toxic Substances Control Act (15 U.S.C. 2601 et seq.), when subject to the labeling requirements of that Act and labeling regulations issued under that Act by the Environmental Protection Agency.[§1910.1200(b)(5)(ii)]

(iii) *Any food, food additive, color additive, drug,* cosmetic, or medical or veterinary device or product, including materials intended for use as ingredients in such products (e.g. flavors and fragrances), as such terms are defined in the Federal Food, Drug, and Cosmetic Act (21 U.S.C. 301 et seq.) or the Virus-Serum-Toxin Act of 1913 (21 U.S.C. 151 et seq.), and regulations issued under those Acts, when they are subject to the labeling requirements under those Acts by either the Food and Drug Administration or the Department of Agriculture;[§1910.1200(b)(5)(iii)]

(iv) *Any distilled spirits* (beverage alcohols), wine, or malt beverage intended for nonindustrial use, as such terms are defined in the Federal Alcohol Administration Act (27 U.S.C. 201 et seq.) and regulations issued under that Act, when subject to the labeling requirements of that Act and labeling regulations issued under that Act by the Bureau of Alcohol, Tobacco, Firearms and Explosives;[§1910.1200(b)(5)(iv)]

(v) *Any consumer product or hazardous substance* as those terms are defined in the Consumer Product Safety Act (15 U.S.C. 2051 et seq.) and Federal Hazardous Substances Act

(15 U.S.C. 1261 et seq.) respectively, when subject to a consumer product safety standard or labeling requirement of those Acts, or regulations issued under those Acts by the Consumer Product Safety Commission; and, [§1910.1200(b)(5)(v)]

(vi) *Agricultural or vegetable seed* treated with pesticides and labeled in accordance with the Federal Seed Act (7 U.S.C. 1551 et seq.) and the labeling regulations issued under that Act by the Department of Agriculture.[§1910.1200(b)(5)(vi)]

(6) *This section does not apply to:*[§1910.1200(b)(6)]

(i) *Any hazardous waste* as such term is defined by the Solid Waste Disposal Act, as amended by the Resource Conservation and Recovery Act of 1976, as amended (42 U.S.C. 6901 et seq.), when subject to regulations issued under that Act by the Environmental Protection Agency;[§1910.1200(b)(6)(i)]

(ii) *Any hazardous substance as such* term is defined by the Comprehensive Environmental Response, Compensation and Liability Act (CERCLA) (42 U.S.C. 9601 et seq.) when the hazardous substance is the focus of remedial or removal action being conducted under CERCLA in accordance with Environmental Protection Agency regulations. [§1910.1200(b)(6)(ii)]

(iii) *Tobacco or tobacco products;*[§1910.1200(b)(6)(iii)]

(iv) *Wood or wood products,* including lumber which will not be processed, where the chemical manufacturer or importer can establish that the only hazard they pose to employees is the potential for flammability or combustibility (wood or wood products which have been treated with a hazardous chemical covered by this standard, and wood which may be subsequently sawed or cut, generating dust, are not exempted); [§1910.1200(b)(6)(iv)]

(v) *Articles* (as that term is defined in paragraph (c) of this section);[§1910.1200(b)(6)(v)]

(vi) *Food or alcoholic beverages* which are sold, used, or prepared in a retail establishment (such as a grocery store, restaurant, or drinking place), and foods intended for personal consumption by employees while in the workplace; [§1910.1200(b)(6)(vi)]

(vii) *Any drug,* as that term is defined in the Federal Food, Drug, and Cosmetic Act (21 U.S.C. 301 et seq.), when it is in solid, final form for direct administration to the patient (e.g., tablets or pills); drugs which are packaged by the chemical manufacturer for sale to consumers in a retail establishment (e.g., over-the-counter drugs); and drugs intended for personal consumption by employees while in the workplace (e.g., first aid supplies);[§1910.1200(b)(6)(vii)]

(viii) *Cosmetics* which are packaged for sale to consumers in a retail establishment, and cosmetics intended for personal consumption by employees while in the workplace; [§1910.1200(b)(6)(viii)]

(ix) *Any consumer product or hazardous substance,* as those terms are defined in the Consumer Product Safety Act (15 U.S.C. 2051 et seq.) and Federal Hazardous Substances Act (15 U.S.C. 1261 et seq.) respectively, where the employer can show that it is used in the workplace for the purpose intended by the chemical manufacturer or importer of the product, and the use results in a duration and frequency of exposure which is not greater than the range of exposures that could reasonably be experienced by consumers when used for the purpose intended;[§1910.1200(b)(6)(ix)]

(x) *Nuisance particulates* where the chemical manufacturer or importer can establish that they do not pose any physical or health hazard covered under this section;[§1910.1200(b)(6)(x)]

(xi) *Ionizing and nonionizing radiation; and,*[§1910.1200(b)(6)(xi)]

(xii) *Biological hazards.*[§1910.1200(b)(6)(xii)]

(c) Definitions.

Article means a manufactured item other than a fluid or particle:

(i) *which is formed* to a specific shape or design during manufacture;

(ii) *which has end use function(s)* dependent in whole or in part upon its shape or design during end use; and

(iii) *which under normal conditions* of use does not release more than very small quantities, e.g., minute or trace amounts of a hazardous chemical (as determined under paragraph (d) of this section), and does not pose a physical hazard or health risk to employees.

Assistant Secretary means the Assistant Secretary of Labor for Occupational Safety and Health, U.S. Department of Labor, or designee.

Chemical means any substance, or mixture of substances.

Chemical manufacturer means an employer with a workplace where chemical(s) are produced for use or distribution.

Chemical name means the scientific designation of a chemical in accordance with the nomenclature system developed by the International Union of Pure and Applied Chemistry (IUPAC) or the Chemical Abstracts Service (CAS) rules of nomenclature, or a name that will clearly identify the chemical for the purpose of conducting a hazard classification.

Classification means to identify the relevant data regarding the hazards of a chemical; review those data to ascertain the hazards associated with the chemical; and decide whether the chemical will be classified as hazardous according to the definition of hazardous chemical in this section. In addition, classification for health and physical hazards includes the determination of the degree of hazard, where appropriate, by comparing the data with the criteria for health and physical hazards.

Commercial account means an arrangement whereby a retail distributor sells hazardous chemicals to an employer, generally in large quantities over time and/or at costs that are below the regular retail price.

Common name means any designation or identification such as code name, code number, trade name, brand name or generic name used to identify a chemical other than by its chemical name.

Container means any bag, barrel, bottle, box, can, cylinder, drum, reaction vessel, storage tank, or the like that contains a hazardous chemical. For purposes of this section, pipes or piping systems, and engines, fuel tanks, or other operating systems in a vehicle, are not considered to be containers.

Designated representative means any individual or organization to whom an employee gives written authorization to exercise such employee's rights under this section. A recognized or certified collective bargaining agent shall be treated automatically as a designated representative without regard to written employee authorization.

Director means the Director, National Institute for Occupational Safety and Health, U.S. Department of Health and Human Services, or designee.

Distributor means a business, other than a chemical manufacturer or importer, which supplies hazardous chemicals to other distributors or to employers.

Employee means a worker who may be exposed to hazardous chemicals under normal operating conditions or in foreseeable emergencies. Workers such as office workers or bank tellers who encounter hazardous chemicals only in non-routine, isolated instances are not covered.

Employer means a person engaged in a business where chemicals are either used, distributed, or are produced for use or distribution, including a contractor or subcontractor.

Exposure or exposed means that an employee is subjected in the course of employment to a chemical that is a physical or health hazard, and includes potential (e.g. accidental or possible) exposure. "Subjected" in terms of health hazards includes any route of entry (e.g. inhalation, ingestion, skin contact or absorption.)

Foreseeable emergency means any potential occurrence such as, but not limited to, equipment failure, rupture of containers, or failure of control equipment which could result in an uncontrolled release of a hazardous chemical into the workplace.

Hazard category means the division of criteria within each hazard class, e.g., oral acute toxicity and flammable liquids include four hazard categories. These categories compare hazard severity within a hazard class and should not be taken as a comparison of hazard categories more generally.

Hazard class means the nature of the physical or health hazards, e.g., flammable solid, carcinogen, oral acute toxicity.

Hazard not otherwise classified (HNOC) means an adverse physical or health effect identified through evaluation of scientific evidence during the classification process that does not meet the specified criteria for the physical and health hazard classes addressed in this section. This does not extend coverage to adverse physical and health effects for which there is a hazard class addressed in this section, but the effect either falls below the cut-off value/concentration limit of the hazard class or is under a GHS hazard category that has not been adopted by OSHA (e.g., acute toxicity Category 5).

Hazard statement means a statement assigned to a hazard class and category that describes the nature of the hazard(s) of a chemical, including, where appropriate, the degree of hazard.

Hazardous chemical means any chemical which is classified as a physical hazard or a health hazard, a simple asphyxiant, combustible dust, pyrophoric gas, or hazard not otherwise classified.

Health hazard means a chemical which is classified as posing one of the following hazardous effects: acute toxicity (any route of exposure); skin corrosion or irritation; serious eye damage or eye irritation; respiratory or skin sensitization; germ cell mutagenicity; carcinogenicity; reproductive toxicity; specific target organ toxicity (single or repeated exposure); or aspiration hazard. The criteria for determining whether a chemical is classified as a health hazard are detailed in Appendix A to §1910.1200 — Health Hazard Criteria.

Immediate use means that the hazardous chemical will be under the control of and used only by the person who transfers it from a labeled container and only within the work shift in which it is transferred.

Importer means the first business with employees within the Customs Territory of the United States which receives hazardous chemicals produced in other countries for the purpose of supplying them to distributors or employers within the United States.

Label means an appropriate group of written, printed or graphic information elements concerning a hazardous chemical that is affixed to, printed on, or attached to the immediate container of a hazardous chemical, or to the outside packaging.

Label elements means the specified pictogram, hazard statement, signal word and precautionary statement for each hazard class and category.

Mixture means a combination or a solution composed of two or more substances in which they do not react.

Physical hazard means a chemical that is classified as posing one of the following hazardous effects: explosive; flammable (gases, aerosols, liquids, or solids); oxidizer (liquid, solid or gas); self-reactive; pyrophoric (liquid or solid); self-heating; organic peroxide; corrosive to metal; gas under pressure; or in contact with water emits flammable gas. *See* Appendix B to §1910.1200 — Physical Hazard Criteria.

Pictogram means a composition that may include a symbol plus other graphic elements, such as a border, background pattern, or color, that is intended to convey specific information about the hazards of a chemical. Eight pictograms are designated under this standard for application to a hazard category.

Precautionary statement means a phrase that describes recommended measures that should be taken to minimize or prevent adverse effects resulting from exposure to a hazardous chemical, or improper storage or handling.

Produce means to manufacture, process, formulate, blend, extract, generate, emit, or repackage.

Product identifier means the name or number used for a hazardous chemical on a label or in the SDS. It provides a unique means by which the user can identify the chemical. The product identifier used shall permit cross-references to be made among the list of hazardous chemicals required in the written hazard communication program, the label and the SDS.

Pyrophoric gas means a chemical in a gaseous state that will ignite spontaneously in air at a temperature of 130 degrees F (54.4 degrees C) or below.

Responsible party means someone who can provide additional information on the hazardous chemical and appropriate emergency procedures, if necessary.

Safety data sheet (SDS) means written or printed material concerning a hazardous chemical that is prepared in accordance with paragraph (g) of this section.

Signal word means a word used to indicate the relative level of severity of hazard and alert the reader to a potential hazard on the label. The signal words used in this section are "danger" and "warning." "Danger" is used for the more severe hazards, while "warning" is used for the less severe.

Simple asphyxiant means a substance or mixture that displaces oxygen in the ambient atmosphere, and can thus cause oxygen deprivation in those who are exposed, leading to unconsciousness and death.

Specific chemical identity means the chemical name, Chemical Abstracts Service (CAS) Registry Number, or any other information that reveals the precise chemical designation of the substance.

Substance means chemical elements and their compounds in the natural state or obtained by any production process, including any additive necessary to preserve the stability of the product and any impurities deriving from the process used, but excluding any solvent which may be separated without affecting the stability of the substance or changing its composition.

Trade secret means any confidential formula, pattern, process, device, information or compilation of information that is used in an employer's business, and that gives the employer an opportunity to obtain an advantage over competitors who do not know or use it. Appendix E to §1910.1200 — Definition of Trade Secret, sets out the criteria to be used in evaluating trade secrets.

Use means to package, handle, react, emit, extract, generate as a byproduct, or transfer.

Work area means a room or defined space in a workplace where hazardous chemicals are produced or used, and where employees are present.

Workplace means an establishment, job site, or project, at one geographical location containing one or more work areas.

(d) Hazard classification.[§1910.1200(d)]

(1) *Chemical manufacturers and importers* shall evaluate chemicals produced in their workplaces or imported by them to classify the chemicals in accordance with this section. For each chemical, the chemical manufacturer or importer shall determine the hazard classes, and, where appropriate, the category of each class that apply to the chemical being classified. Employers are not required to classify chemicals unless they choose not to rely on the classification performed by the chemical manufacturer or importer for the chemical to satisfy this requirement.[§1910.1200(d)(1)]

(2) *Chemical manufacturers,* importers or employers classifying chemicals shall identify and consider the full range of available scientific literature and other evidence concerning the potential hazards. There is no requirement to test the chemical to determine how to classify its hazards. Appendix A to Sec. 1910.1200 shall be consulted for classification of health hazards, and Appendix B to Sec. 1910.1200 shall be consulted for the classification of physical hazards.[§1910.1200(d)(2)]

(3) *Mixtures.*[§1910.1200(d)(3)]

(i) *Chemical manufacturers,* importers, or employers evaluating chemicals shall follow the procedures described in Appendices A and B to Sec. 1910.1200 to classify the hazards of the chemicals, including determinations regarding when mixtures of the classified chemicals are covered by this section. [§1910.1200(d)(3)(i)]

(ii) *When classifying mixtures they* produce or import, chemical manufacturers and importers of mixtures may rely on the information provided on the current safety data sheets of the individual ingredients, except where the chemical manufacturer or importer knows, or in the exercise of reasonable diligence should know, that the safety data sheet misstates or omits information required by this section.[§1910.1200(d)(3)(ii)]

(4) *Chemical manufacturers, importers and employers* evaluating chemicals shall treat the following sources as establishing that a chemical is a carcinogen or potential carcinogen for hazard communication purposes:[§1910.1200(d)(4)]

(i) *National Toxicology Program* (NTP), Annual Report on Carcinogens (latest edition);[§1910.1200(d)(4)(i)]

(ii) *International Agency for Research on Cancer* (IARC) Monographs (latest editions); or[§1910.1200(d)(4)(ii)]

(iii) *29 CFR part 1910, subpart Z,* Toxic and Hazardous Substances, Occupational Safety and Health Administration. [§1910.1200(d)(4)(iii)]

Note: The Registry of Toxic Effects of Chemical Substances published by the National Institute for Occupational Safety and Health indicates whether a chemical has been found by NTP or IARC to be a potential carcinogen.

(5) *The chemical manufacturer, importer or employer* shall determine the hazards of mixtures of chemicals as follows: [§1910.1200(d)(5)]

(i) *If a mixture has been tested* as a whole to determine its hazards, the results of such testing shall be used to determine whether the mixture is hazardous;[§1910.1200(d)(5)(i)]

(ii) *If a mixture has not been tested* as a whole to determine whether the mixture is a health hazard, the mixture shall be assumed to present the same health hazards as do the components which comprise one percent (by weight or volume) or greater of the mixture, except that the mixture shall be assumed to present a carcinogenic hazard if it contains a component in concentrations of 0.1 percent or greater which is considered to be a carcinogen under paragraph (d)(4) of this section;[§1910.1200(d)(5)(ii)]

(iii) *If a mixture has not been tested* as a whole to determine whether the mixture is a physical hazard, the chemical manufacturer, importer, or employer may use whatever scientifically valid data is available to evaluate the physical hazard potential of the mixture; and,[§1910.1200(d)(5)(iii)]

(iv) *If the chemical manufacturer, importer, or employer* has evidence to indicate that a component present in the mixture in concentrations of less than one percent (or in the case of carcinogens, less than 0.1 percent) could be released in concentrations which would exceed an established OSHA permissible exposure limit or ACGIH Threshold Limit Value, or could present a health risk to employees in those concentrations, the mixture shall be assumed to present the same hazard.[§1910.1200(d)(5)(iv)]

(6) *Chemical manufacturers, importers, or employers* evaluating chemicals shall describe in writing the procedures they use to determine the hazards of the chemical they evaluate. The written procedures are to be made available, upon request, to employees, their designated representatives, the Assistant Secretary and the Director. The written description may be incorporated into the written hazard communication program required under paragraph (e) of this section.[§1910.1200(d)(6)]

(e) Written hazard communication program.[§1910.1200(e)]

(1) *Employers shall develop, implement, and maintain* at each workplace, a written hazard communication program which at least describes how the criteria specified in paragraphs (f), (g), and (h) of this section for labels and other forms of warning, safety data sheets, and employee information and training will be met, and which also includes the following:[§1910.1200(e)(1)]

(i) *A list of the hazardous chemicals* known to be present using a product identifier that is referenced on the appropriate safety data sheet (the list may be compiled for the workplace as a whole or for individual work areas); and,[§1910.1200(e)(1)(i)]

(ii) *The methods the employer will use* to inform employees of the hazards of non-routine tasks (for example, the cleaning of reactor vessels), and the hazards associated with chemicals contained in unlabeled pipes in their work areas. [§1910.1200(e)(1)(ii)]

(2) *Multi-employer workplaces.* Employers who produce, use, or store hazardous chemicals at a workplace in such a way that the employees of other employer(s) may be exposed (for example, employees of a construction contractor working on-site) shall additionally ensure that the hazard communication programs developed and implemented under this paragraph (e) include the following:[§1910.1200(e)(2)]

(i) *The methods the employer will use* to provide the other employer(s) on-site access to safety data sheets for each hazardous chemical the other employer(s)' employees may be exposed to while working;[§1910.1200(e)(2)(i)]

(ii) *The methods the employer will use* to inform the other employer(s) of any precautionary measures that need to be taken to protect employees during the workplace's normal operating conditions and in foreseeable emergencies; and, [§1910.1200(e)(2)(ii)]

(iii) *The methods the employer will use* to inform the other employer(s) of the labeling system used in the workplace. [§1910.1200(e)(2)(iii)]

(3) *The employer may rely* on an existing hazard communication program to comply with these requirements, provided that it meets the criteria established in this paragraph (e). [§1910.1200(e)(3)]

(4) *The employer shall make* the written hazard communication program available, upon request, to employees, their designated representatives, the Assistant Secretary and the Director, in accordance with the requirements of 29 CFR 1910.1020(e).[§1910.1200(e)(4)]

(5) *Where employees must travel* between workplaces during a workshift, i.e., their work is carried out at more than one geographical location, the written hazard communication program may be kept at the primary workplace facility.[§1910.1200(e)(5)]

(f) Labels and other forms of warning[§1910.1200(f)]

(1) *Labels on shipped containers.* The chemical manufacturer, importer, or distributor shall ensure that each container of hazardous chemicals leaving the workplace is labeled, tagged, or marked. Hazards not otherwise classified do not have to be addressed on the container. Where the chemical manufacturer or importer is required to label, tag or mark the following information shall be provided:[§1910.1200(f)(1)]

(i) *Product identifier;*[§1910.1200(f)(1)(i)]

(ii) *Signal word;*[§1910.1200(f)(1)(ii)]

(iii) *Hazard statement(s);*[§1910.1200(f)(1)(iii)]

(iv) *Pictogram(s);*[§1910.1200(f)(1)(iv)]

(v) *Precautionary statement(s); and,*[§1910.1200(f)(1)(v)]

(vi) *Name, address,* and telephone number of the chemical manufacturer, importer, or other responsible party. [§1910.1200(f)(1)(vi)]

(2) *The chemical manufacturer,* importer, or distributor shall ensure that the information provided under paragraphs (f)(1)(i) through (v) of this section is in accordance with Appendix C to Sec. 1910.1200, for each hazard class and associated hazard category for the hazardous chemical, prominently displayed, and in English (other languages may also be included if appropriate).[§1910.1200(f)(2)]

(3) *The chemical manufacturer,* importer, or distributor shall ensure that the information provided under paragraphs (f)(1)(ii) through (iv) of this section is located together on the label, tag, or mark.[§1910.1200(f)(3)]

(4) *Solid materials.*[§1910.1200(f)(4)]

(i) *For solid metal* (such as a steel beam or a metal casting), solid wood, or plastic items that are not exempted as articles due to their downstream use, or shipments of whole grain, the required label may be transmitted to the customer at the time of the initial shipment, and need not be included with subsequent shipments to the same employer unless the information on the label changes;[§1910.1200(f)(4)(i)]

(ii) *The label may be transmitted with the initial* shipment itself, or with the safety data sheet that is to be provided prior to or at the time of the first shipment; and,[§1910.1200(f)(4)(ii)]

(iii) *This exception to requiring labels* on every container of hazardous chemicals is only for the solid material itself, and does not apply to hazardous chemicals used in conjunction with, or known to be present with, the material and to which employees handling the items in transit may be exposed (for example, cutting fluids or pesticides in grains). [§1910.1200(f)(4)(iii)]

(5) *Chemical manufacturers,* importers, or distributors shall ensure that each container of hazardous chemicals leaving the workplace is labeled, tagged, or marked in accordance with this section in a manner which does not conflict with the requirements of the Hazardous Materials Transportation Act (49 U.S.C. 1801 *et seq.*) and regulations issued under that Act by the Department of Transportation.[§1910.1200(f)(5)]

(6) *Workplace labeling.* Except as provided in paragraphs (f)(7) and (f)(8) of this section, the employer shall ensure that each container of hazardous chemicals in the workplace is labeled, tagged or marked with either:[§1910.1200(f)(6)]

(i) *The information specified under* paragraphs (f)(1)(i) through (v) of this section for labels on shipped containers; or, [§1910.1200(f)(6)(i)]

(ii) *Product identifier and words,* pictures, symbols, or combination thereof, which provide at least general information regarding the hazards of the chemicals, and which, in conjunction with the other information immediately available to employees under the hazard communication program, will provide employees with the specific information regarding the physical and health hazards of the hazardous chemical. [§1910.1200(f)(6)(ii)]

(7) *The employer may use signs,* placards, process sheets, batch tickets, operating procedures, or other such written materials in lieu of affixing labels to individual stationary process containers, as long as the alternative method identifies the containers to which it is applicable and conveys the information required by paragraph (f)(6) of this section to be on a label. The employer shall ensure the written materials are readily accessible to the employees in their work area throughout each work shift.[§1910.1200(f)(7)]

(8) *The employer is not required to label* portable containers into which hazardous chemicals are transferred from labeled containers, and which are intended only for the immediate use of the employee who performs the transfer. For purposes of this section, drugs which are dispensed by a pharmacy to a health care provider for direct administration to a patient are exempted from labeling.[§1910.1200(f)(8)]

(9) *The employer shall not remove or deface* existing labels on incoming containers of hazardous chemicals, unless the container is immediately marked with the required information. [§1910.1200(f)(9)]

(10) *The employer shall ensure that workplace* labels or other forms of warning are legible, in English, and prominently displayed on the container, or readily available in the work area throughout each work shift. Employers having employees who speak other languages may add the information in their language to the material presented, as long as the information is presented in English as well.[§1910.1200(f)(10)]

(11) *Chemical manufacturers,* importers, distributors, or employers who become newly aware of any significant information regarding the hazards of a chemical shall revise the labels for the chemical within six months of becoming aware of the new information, and shall ensure that labels on containers of hazardous chemicals shipped after that time contain the new information. If the chemical is not currently produced or imported, the chemical manufacturer, importer, distributor, or employer shall add the information to the label before the chemical is shipped or introduced into the workplace again. [§1910.1200(f)(11)]

(g) Safety data sheets.[§1910.1200(g)]

(1) *Chemical manufacturers and importers* shall obtain or develop a safety data sheet for each hazardous chemical they produce or import. Employers shall have a safety data sheet in the workplace for each hazardous chemical which they use. [§1910.1200(g)(1)]

(2) *The chemical manufacturer or importer* preparing the safety data sheet shall ensure that it is in English (although the employer may maintain copies in other languages as well), and includes at least the following section numbers and headings, and associated information under each heading, in the order listed (*See* Appendix D to §1910.1200 — Safety Data Sheets, for the specific content of each section of the safety data sheet):[§1910.1200(g)(2)]

(i) *Section 1,* Identification;[§1910.1200(g)(2)(i)]

(ii) *Section 2,* Hazard(s) identification;[§1910.1200(g)(2)(ii)]

(iii) *Section 3,* Composition/information on ingredients; [§1910.1200(g)(2)(iii)]

(iv) *Section 4,* First-aid measures;[§1910.1200(g)(2)(iv)]

(v) *Section 5,* Fire-fighting measures;[§1910.1200(g)(2)(v)]

(vi) *Section 6,* Accidental release measures;[§1910.1200(g)(2)(vi)]

(vii) *Section 7,* Handling and storage;[§1910.1200(g)(2)(vii)]

(viii) *Section 8,* Exposure controls/personal protection; [§1910.1200(g)(2)(viii)]

(ix) *Section 9,* Physical and chemical properties; [§1910.1200(g)(2)(ix)]

(x) *Section 10,* Stability and reactivity;[§1910.1200(g)(2)(x)]

(xi) *Section 11,* Toxicological information;[§1910.1200(g)(2)(xi)]

(xii) *Section 12,* Ecological information;[§1910.1200(g)(2)(xii)]

(xiii) *Section 13,* Disposal considerations;[§1910.1200(g)(2)(xiii)]

(xiv) *Section 14,* Transport information;[§1910.1200(g)(2)(xiv)]

(xv) *Section 15,* Regulatory information; and[§1910.1200(g)(2)(xv)]

(xvi) *Section 16,* Other information, including date of preparation or last revision.[§1910.1200(g)(2)(xvi)]

Note 1 to paragraph (g)(2): To be consistent with the GHS, an SDS must also include the headings in paragraphs (g)(2)(xii) through (g)(2)(xv) in order.
Note 2 to paragraph (g)(2): OSHA will not be enforcing information requirements in sections 12 through 15, as these areas are not under its jurisdiction.

(3) *If no relevant information is found* for any sub-heading within a section on the safety data sheet, the chemical manufacturer, importer or employer preparing the safety data sheet shall mark it to indicate that no applicable information was found. [§1910.1200(g)(3)]

(4) *Where complex mixtures have similar hazards and contents* (i.e. the chemical ingredients are essentially the same, but the specific composition varies from mixture to mixture), the chemical manufacturer, importer or employer may prepare one safety data sheet to apply to all of these similar mixtures. [§1910.1200(g)(4)]

(5) *The chemical manufacturer,* importer or employer preparing the safety data sheet shall ensure that the information provided accurately reflects the scientific evidence used in making the hazard classification. If the chemical manufacturer, importer or employer preparing the safety data sheet becomes newly aware of any significant information regarding the hazards of a chemical, or ways to protect against the hazards, this new information shall be added to the safety data sheet within three months. If the chemical is not currently being produced or imported, the chemical manufacturer or importer shall add the information to the safety data sheet before the chemical is introduced into the workplace again.[§1910.1200(g)(5)]

(6) (i) *Chemical manufacturers or importers* shall ensure that distributors and employers are provided an appropriate safety data sheet with their initial shipment, and with the first shipment after a safety data sheet is updated;[§1910.1200(g)(6)(i)]

(ii) *The chemical manufacturer or importer* shall either provide safety data sheets with the shipped containers or send them to the distributor or employer prior to or at the time of the shipment;[§1910.1200(g)(6)(ii)]

(iii) *If the safety data sheet* is not provided with a shipment that has been labeled as a hazardous chemical, the distributor or employer shall obtain one from the chemical manufacturer or importer as soon as possible; and,[§1910.1200(g)(6)(iii)]

(iv) *The chemical manufacturer or importer* shall also provide distributors or employers with a safety data sheet upon request.[§1910.1200(g)(6)(iv)]

(7) (i) *Distributors shall ensure* that safety data sheets, and updated information, are provided to other distributors and employers with their initial shipment and with the first shipment after a safety data sheet is updated;[§1910.1200(g)(7)(i)]

(ii) *The distributor shall either provide* safety data sheets with the shipped containers, or send them to the other distributor or employer prior to or at the time of the shipment; [§1910.1200(g)(7)(ii)]

(iii) *Retail distributors selling hazardous chemicals* to employers having a commercial account shall provide a safety data sheet to such employers upon request, and shall post a sign or otherwise inform them that a safety data sheet is available; [§1910.1200(g)(7)(iii)]

(iv) *Wholesale distributors selling* hazardous chemicals to employers over-the-counter may also provide safety data sheets upon the request of the employer at the time of the over-the-counter purchase, and shall post a sign or otherwise inform such employers that a safety data sheet is available; [§1910.1200(g)(7)(iv)]

(v) *If an employer without a commercial account* purchases a hazardous chemical from a retail distributor not required to have safety data sheets on file (i.e., the retail distributor does not have commercial accounts and does not use the materials), the retail distributor shall provide the employer, upon request, with the name, address, and telephone number of the chemical manufacturer, importer, or distributor from which a safety data sheet can be obtained;[§1910.1200(g)(7)(v)]

(vi) *Wholesale distributors shall also provide* safety data sheets to employers or other distributors upon request; and, [§1910.1200(g)(7)(vi)]

(vii) *Chemical manufacturers, importers, and distributors* need not provide safety data sheets to retail distributors that have informed them that the retail distributor does not sell the product to commercial accounts or open the sealed container to use it in their own workplaces.[§1910.1200(g)(7)(vii)]

(8) *The employer shall maintain in the workplace* copies of the required safety data sheets for each hazardous chemical, and shall ensure that they are readily accessible during each work shift to employees when they are in their work area(s). (Electronic access and other alternatives to maintaining paper copies of the safety data sheets are permitted as long as no barriers to immediate employee access in each workplace are created by such options.)[§1910.1200(g)(8)]

(9) *Where employees must travel between* workplaces during a workshift, i.e., their work is carried out at more than one geographical location, the safety data sheets may be kept at the primary workplace facility. In this situation, the employer shall ensure that employees can immediately obtain the required information in an emergency.[§1910.1200(g)(9)]

(10) *Safety data sheets* may be kept in any form, including operating procedures, and may be designed to cover groups of hazardous chemicals in a work area where it may be more appropriate to address the hazards of a process rather than individual hazardous chemicals. However, the employer shall ensure that in all cases the required information is provided for each hazardous chemical, and is readily accessible during each work shift to employees when they are in their work area(s).[§1910.1200(g)(10)]

(11) *Safety data sheets* shall also be made readily available, upon request, to designated representatives, the Assistant Secretary, and the Director, in accordance with the requirements of Sec. 1910.1020(e).[§1910.1200(g)(11)]

(h) Employee information and training.[§1910.1200(h)]

(1) *Employers shall provide employees* with effective information and training on hazardous chemicals in their work area at the time of their initial assignment, and whenever a new chemical hazard the employees have not previously been trained about is introduced into their work area. Information and training may be designed to cover categories of hazards (e.g., flammability, carcinogenicity) or specific chemicals. Chemical-specific information must always be available through labels and safety data sheets.[§1910.1200(h)(1)]

(2) *Information.* Employees shall be informed of:[§1910.1200(h)(2)]

(i) *The requirements of this section;*[§1910.1200(h)(2)(i)]

(ii) *Any operations in their work area where hazardous chemicals are present; and,*[§1910.1200(h)(2)(ii)]

(iii) *The location and availability* of the written hazard communication program, including the required list(s) of hazardous chemicals, and safety data sheets required by this section. [§1910.1200(h)(2)(iii)]

(3) *Training.* Employee training shall include at least: [§1910.1200(h)(3)]

(i) *Methods and observations that may be used* to detect the presence or release of a hazardous chemical in the work area (such as monitoring conducted by the employer, continuous monitoring devices, visual appearance or odor of hazardous chemicals when being released, etc.); [§1910.1200(h)(3)(i)]

(ii) *The physical,* health, simple asphyxiation, combustible dust, and pyrophoric gas hazards, as well as hazards not otherwise classified, of the chemicals in the work area; [§1910.1200(h)(3)(ii)]

(iii) *The measures employees can take* to protect themselves from these hazards, including specific procedures the employer has implemented to protect employees from exposure to hazardous chemicals, such as appropriate work practices, emergency procedures, and personal protective equipment to be used; and,[§1910.1200(h)(3)(iii)]

(iv) *The details of the hazard communication* program developed by the employer, including an explanation of the labels received on shipped containers and the workplace labeling system used by their employer; the safety data sheet, including the order of information and how employees can obtain and use the appropriate hazard information.[§1910.1200(h)(3)(iv)]

(i) Trade secrets.[§1910.1200(i)]

(1) *The chemical manufacturer,* importer, or employer may withhold the specific chemical identity, including the chemical name, other specific identification of a hazardous chemical, or the exact percentage (concentration) of the substance in a mixture, from the safety data sheet, provided that: [§1910.1200(i)(1)]

(i) *The claim that the information withheld* is a trade secret can be supported;[§1910.1200(i)(1)(i)]

(ii) *Information contained* in the safety data sheet concerning the properties and effects of the hazardous chemical is disclosed;[§1910.1200(i)(1)(ii)]

(iii) *The safety data sheet* indicates that the specific chemical identity and/or percentage of composition is being withheld as a trade secret; and,[§1910.1200(i)(1)(iii)]

(iv) *The specific chemical identity* and percentage is made available to health professionals, employees, and designated representatives in accordance with the applicable provisions of this paragraph (i).[§1910.1200(i)(1)(iv)]

(2) *Where a treating physician or nurse* determines that a medical emergency exists and the specific chemical identity and/or specific percentage of composition of a hazardous chemical is necessary for emergency or first-aid treatment, the chemical manufacturer, importer, or employer shall immediately disclose the specific chemical identity or percentage composition of a trade secret chemical to that treating physician or nurse, regardless of the existence of a written statement of need or a confidentiality agreement. The chemical manufacturer, importer, or employer may require a written statement of need and confidentiality agreement, in accordance with the provisions of paragraphs (i)(3) and (4) of this section, as soon as circumstances permit.[§1910.1200(i)(2)]

(3) *In non-emergency situations,* a chemical manufacturer, importer, or employer shall, upon request, disclose a specific chemical identity or percentage composition, otherwise permitted to be withheld under paragraph (i)(1) of this section, to a health professional (i.e. physician, industrial hygienist, toxicologist, epidemiologist, or occupational health nurse) providing medical or other occupational health services to exposed employee(s), and to employees or designated representatives, if:[§1910.1200(i)(3)]

(i) *The request is in writing;*[§1910.1200(i)(3)(i)]

(ii) *The request describes with reasonable detail* one or more of the following occupational health needs for the information: [§1910.1200(i)(3)(ii)]

[A] To assess the hazards of the chemicals to which employees will be exposed;[§1910.1200(i)(3)(ii)[A]]

[B] To conduct or assess sampling of the workplace atmosphere to determine employee exposure levels; [§1910.1200(i)(3)(ii)[B]]

[C] To conduct pre-assignment or periodic medical surveillance of exposed employees;[§1910.1200(i)(3)(ii)[C]]

[D] To provide medical treatment to exposed employees; [§1910.1200(i)(3)(ii)[D]]

[E] To select or assess appropriate personal protective equipment for exposed employees;[§1910.1200(i)(3)(ii)[E]]

[F] To design or assess engineering controls or other protective measures for exposed employees; and, [§1910.1200(i)(3)(ii)[F]]

[G] To conduct studies to determine the health effects of exposure.[§1910.1200(i)(3)(ii)[G]]

(iii) *The request explains in detail* why the disclosure of the specific chemical identity or percentage composition is essential and that, in lieu thereof, the disclosure of the following information to the health professional, employee, or designated representative, would not satisfy the purposes described in paragraph (i)(3)(ii) of this section:[§1910.1200(i)(3)(iii)]

[A] The properties and effects of the chemical; [§1910.1200(i)(3)(iii)[A]]

[B] Measures for controlling workers' exposure to the chemical;[§1910.1200(i)(3)(iii)[B]]

[C] Methods of monitoring and analyzing worker exposure to the chemical; and,[§1910.1200(i)(3)(iii)[C]]

[D] Methods of diagnosing and treating harmful exposures to the chemical;[§1910.1200(i)(3)(iii)[D]]

(iv) *The request includes a description* of the procedures to be used to maintain the confidentiality of the disclosed information; and,[§1910.1200(i)(3)(iv)]

(v) *The health professional,* and the employer or contractor of the services of the health professional (i.e. downstream employer, labor organization, or individual employee), employee, or designated representative, agree in a written confidentiality agreement that the health professional, employee, or designated representative, will not use the trade secret information for any purpose other than the health need(s) asserted and agree not to release the information under any circumstances other than to OSHA, as provided in paragraph (i)(6) of this section, except as authorized by the terms of the agreement or by the chemical manufacturer, importer, or employer.[§1910.1200(i)(3)(v)]

(4) *The confidentiality agreement* authorized by paragraph (i)(3)(iv) of this section:[§1910.1200(i)(4)]

(i) *May restrict the use of the information* to the health purposes indicated in the written statement of need;[§1910.1200(i)(4)(i)]

(ii) *May provide for appropriate legal remedies* in the event of a breach of the agreement, including stipulation of a reasonable pre-estimate of likely damages; and,[§1910.1200(i)(4)(ii)]

(iii) *May not include requirements for the posting of a penalty bond.*[§1910.1200(i)(4)(iii)]

(5) *Nothing in this standard is meant* to preclude the parties from pursuing non-contractual remedies to the extent permitted by law.[§1910.1200(i)(5)]

(6) *If the health professional, employee, or designated representative* receiving the trade secret information decides that there is a need to disclose it to OSHA, the chemical manufacturer, importer, or employer who provided the information shall be informed by the health professional, employee, or designated representative prior to, or at the same time as, such disclosure.[§1910.1200(i)(6)]

(7) *If the chemical manufacturer,* importer, or employer denies a written request for disclosure of a specific chemical identity or percentage composition, the denial must:[§1910.1200(i)(7)]

(i) *Be provided to the health professional,* employee, or designated representative, within thirty days of the request; [§1910.1200(i)(7)(i)]

(ii) *Be in writing;*[§1910.1200(i)(7)(ii)]

(iii) *Include evidence to support* the claim that the specific chemical identity or percent of composition is a trade secret; [§1910.1200(i)(7)(iii)]

(iv) *State the specific reasons* why the request is being denied; and,[§1910.1200(i)(7)(iv)]

(v) *Explain in detail how alternative* information may satisfy the specific medical or occupational health need without revealing the trade secret.[§1910.1200(i)(7)(v)]

(8) *The health professional, employee, or designated representative* whose request for information is denied under paragraph (i)(3) of this section may refer the request and the written denial of the request to OSHA for consideration.[§1910.1200(i)(8)]

(9) *When a health professional, employee, or designated representative* refers the denial to OSHA under paragraph (i)(8) of this section, OSHA shall consider the evidence to determine if: [§1910.1200(i)(9)]

(i) *The chemical manufacturer,* importer, or employer has supported the claim that the specific chemical identity or percentage composition is a trade secret;[§1910.1200(i)(9)(i)]

(ii) *The health professional, employee, or designated representative* has supported the claim that there is a medical or occupational health need for the information; and, [§1910.1200(i)(9)(ii)]

(iii) *The health professional, employee or designated representative* has demonstrated adequate means to protect the confidentiality.[§1910.1200(i)(9)(iii)]

(10) (i) *If OSHA determines that the specific* chemical identity or percentage composition requested under paragraph (i)(3) of this section is not a "bona fide" trade secret, or that it is a trade secret, but the requesting health professional, employee, or designated representative has a legitimate medical or occupational health need for the information, has executed a written confidentiality agreement, and has shown adequate means to protect the confidentiality of the information, the chemical manufacturer, importer, or employer will be subject to citation by OSHA.[§1910.1200(i)(10)(i)]

(ii) *If a chemical manufacturer,* importer, or employer demonstrates to OSHA that the execution of a confidentiality agreement would not provide sufficient protection against the potential harm from the unauthorized disclosure of a trade secret, the Assistant Secretary may issue such orders or impose such additional limitations or conditions upon the disclosure of the requested chemical information as may be appropriate to assure that the occupational health services are provided without an undue risk of harm to the chemical manufacturer, importer, or employer.[§1910.1200(i)(10)(ii)]

(11) *If a citation for a failure to release* trade secret information is contested by the chemical manufacturer, importer, or employer, the matter will be adjudicated before the Occupational Safety and Health Review Commission in accordance with the Act's enforcement scheme and the applicable Commission rules of procedure. In accordance with the Commission rules, when a chemical manufacturer, importer, or employer continues to withhold the information during the contest, the Administrative Law Judge may review the citation and supporting documentation "in camera" or issue appropriate orders to protect the confidentiality of such matters. [§1910.1200(i)(11)]

(12) *Notwithstanding the existence of a trade secret claim,* a chemical manufacturer, importer, or employer shall, upon request, disclose to the Assistant Secretary any information which this section requires the chemical manufacturer, importer, or employer to make available. Where there is a trade secret claim, such claim shall be made no later than at the time the information is provided to the Assistant Secretary so that suitable determinations of trade secret status can be made and the necessary protections can be implemented. [§1910.1200(i)(12)]

(13) *Nothing in this paragraph shall be construed* as requiring the disclosure under any circumstances of process information which is a trade secret.[§1910.1200(i)(13)]

(j) **Effective dates.**[§1910.1200(j)]

(1) *Employers shall train employees* regarding the new label elements and safety data sheets format by December 1, 2013. [§1910.1200(j)(1)]

(2) *Chemical manufacturers,* importers, distributors, and employers shall be in compliance with all modified provisions of this section no later than June 1, 2015, except:[§1910.1200(j)(2)]

(i) *After December 1,* 2015, the distributor shall not ship containers labeled by the chemical manufacturer or importer unless the label has been modified to comply with paragraph (f)(1) of this section.[§1910.1200(j)(2)(i)]

(ii) *All employers shall,* as necessary, update any alternative workplace labeling used under paragraph (f)(6) of this section, update the hazard communication program required by paragraph (h)(1), and provide any additional employee training in accordance with paragraph (h)(3) for newly identified physical or health hazards no later than June 1, 2016. [§1910.1200(j)(2)(ii)]

(3) *Chemical manufacturers,* importers, distributors, and employers may comply with either Sec. 1910.1200 revised as of October 1, 2011, or the current version of this standard, or both during the transition period.[§1910.1200(j)(3)]

§1910.1200 Appendix A

Health Hazard Criteria (Mandatory)

A.0 GENERAL CLASSIFICATION CONSIDERATIONS

A.0.1 *Classification*

A.0.1.1 *The term "hazard classification"* is used to indicate that only the intrinsic hazardous properties of chemicals are considered. Hazard classification incorporates three steps:

(a) *Identification of relevant data* regarding the hazards of a chemical;

(b) *Subsequent review of those* data to ascertain the hazards associated with the chemical;

(c) *Determination of whether the chemical* will be classified as hazardous and the degree of hazard.

A.0.1.2 *For many hazard classes,* the criteria are semi-quantitative or qualitative and expert judgment is required to interpret the data for classification purposes.

A.0.2 *Available Data,* Test Methods and Test Data Quality

A.0.2.1 *There is no requirement for testing chemicals.*

A.0.2.2 *The criteria for determining health* hazards are test method neutral, i.e., they do not specify particular test methods, as long as the methods are scientifically validated.

A.0.2.3 *The term "scientifically validated"* refers to the process by which the reliability and the relevance of a procedure are established for a particular purpose. Any test that determines hazardous properties, which is conducted according to recognized scientific principles, can be used for purposes of a hazard determination for health hazards. Test conditions need to be standardized so that the results are reproducible with a given substance, and the standardized test yields "valid" data for defining the hazard class of concern.

A.0.2.4 *Existing test data* are acceptable for classifying chemicals, although expert judgment also may be needed for classification purposes.

A.0.2.5 *The effect of a chemical on biological* systems is influenced, by the physico-chemical properties of the substance and/or ingredients of the mixture and the way in which ingredient substances are biologically available. A chemical need not be classified when it can be shown by conclusive experimental data from scientifically validated test methods that the chemical is not biologically available.

A.0.2.6 *For classification purposes,* epidemiological data and experience on the effects of chemicals on humans (e.g., occupational data, data from accident databases) shall be taken into account in the evaluation of human health hazards of a chemical.

A.0.3 *Classification Based on Weight of Evidence*

A.0.3.1 *For some hazard classes,* classification results directly when the data satisfy the criteria. For others, classification of a chemical shall be determined on the basis of the total weight of evidence using expert judgment. This means that all available information bearing on the classification of hazard shall be considered together, including the results of valid *in vitro* tests, relevant animal data, and human experience such as epidemiological and clinical studies and well-documented case reports and observations.

A.0.3.2 *The quality and consistency of the data* shall be considered. Information on chemicals related to the material being classified shall be considered as appropriate, as well as site of action and mechanism or mode of action study results. Both positive and negative results shall be considered together in a single weight-of-evidence determination.

A.0.3.3 *Positive effects which* are consistent with the criteria for classification, whether seen in humans or animals, shall normally justify classification. Where evidence is available from both humans and animals and there is a conflict between the findings, the quality and reliability of the evidence from both sources shall be evaluated in order to resolve the question of classification. Reliable, good quality human data shall generally have precedence over other data. However, even well-designed and conducted epidemiological studies may lack a sufficient number of subjects to detect relatively rare but still significant effects, or to assess potentially confounding factors. Therefore, positive results from well-conducted animal studies are not necessarily negated by the lack of positive human experience but require an assessment of the robustness, quality and statistical power of both the human and animal data.

A.0.3.4 *Route of exposure,* mechanistic information, and metabolism studies are pertinent to determining the relevance of an effect in humans. When such information raises doubt about relevance in humans, a lower classification may be warranted. When there is scientific evidence demonstrating that the mechanism or mode of action is not relevant to humans, the chemical should not be classified.

A.0.3.5 *Both positive and negative* results are considered together in the weight of evidence determination. However, a single positive study performed according to good scientific principles and with statistically and biologically significant positive results may justify classification.

A.0.4 *Considerations for the Classification of Mixtures*

A.0.4.1 *For most hazard classes,* the recommended process of classification of mixtures is based on the following sequence:

(a) *Where test data are available* for the complete mixture, the classification of the mixture will always be based on those data;

(b) *Where test data are not available* for the mixture itself, the bridging principles designated in each health hazard chapter of this appendix shall be considered for classification of the mixture;

(c) *If test data are not available* for the mixture itself, and the available information is not sufficient to allow application of the above-mentioned bridging principles, then the method(s) described in each chapter for estimating the hazards based on the information known will be applied to classify the mixture (e.g., application of cut-off values/concentration limits).

A.0.4.2 *An exception to the above order* or precedence is made for Carcinogenicity, Germ Cell Mutagenicity, and Reproductive Toxicity. For these three hazard classes, mixtures shall be classified based upon information on the ingredient substances, unless on a case-by-case basis, justification can be provided for classifying based upon the mixture as a whole. See chapters A.5, A.6, and A.7 for further information on case-by-case bases.

A.0.4.3 *Use of cut-off values/concentration limits.*

A.0.4.3.1 *When classifying an untested mixture* based on the hazards of its ingredients, cut-off values/concentration limits for the classified ingredients of the mixture are used for several hazard classes. While the adopted cut-off values/concentration limits adequately identify the hazard for most mixtures, there may be some that contain hazardous ingredients at lower concentrations than the specified cut-off values/concentration limits that still pose an identifiable hazard. There may also be cases where the cut-off value/concentration limit is considerably lower than the established non-hazardous level for an ingredient.

A.0.4.3.2 *If the classifier has information* that the hazard of an ingredient will be evident (i.e., it presents a health risk) below the specified cut-off value/concentration limit, the mixture containing that ingredient shall be classified accordingly.

A.0.4.3.3 *In exceptional cases,* conclusive data may demonstrate that the hazard of an ingredient will not be evident (i.e., it does not present a health risk) when present at a level above the specified cut-off value/concentration limit(s). In these cases the mixture may be classified according to those data. The data must exclude the possibility that the ingredient will behave in the mixture in a manner that would increase the hazard over that of the pure substance. Furthermore, the mixture must not contain ingredients that would affect that determination.

A.0.4.4 *Synergistic or antagonistic effects.*

When performing an assessment in accordance with these requirements, the evaluator must take into account all available information about the potential occurrence of synergistic effects among the ingredients of the mixture. Lowering classification of a mixture to a less hazardous category on the basis of antagonistic effects may be done only if the determination is supported by sufficient data.

A.0.5 *Bridging Principles for the Classification* of Mixtures Where Test Data Are Not Available for the Complete Mixture

A.0.5.1 *Where the mixture itself has* not been tested to determine its toxicity, but there are sufficient data on both the individual ingredients and similar tested mixtures to adequately characterize the hazards of the mixture, these data shall be used in accordance with the following bridging principles, subject to any specific provisions for mixtures for each hazard class. These principles ensure that the classification process uses the available data to the greatest extent possible in characterizing the hazards of the mixture.

A.0.5.1.1 Dilution.

For mixtures classified in accordance with A.1 through A.10 of this Appendix, if a tested mixture is diluted with a diluent that has an equivalent or lower toxicity classification than the least toxic original ingredient, and which is not expected to affect the toxicity of other ingredients, then:

(a) *The new diluted mixture shall be classified* as equivalent to the original tested mixture; or

(b) *For classification of acute toxicity* in accordance with A.1 of this Appendix, paragraph A.1.3.6 (the additivity formula) shall be applied.

A.0.5.1.2 Batching.

For mixtures classified in accordance with A.1 through A.10 of this Appendix, the toxicity of a tested production batch of a mixture can be assumed to be substantially equivalent to that of another untested production batch of the same mixture, when produced by or under the control of the same *chemical manufacturer,* unless there is reason to believe there is significant variation such that the toxicity of the untested batch has changed. If the latter occurs, a new classification is necessary.

A.0.5.1.3 Concentration of mixtures.

For mixtures classified in accordance with A.1, A.2, A.3, A.8, A.9, or A.10 of this Appendix, if a tested mixture is classified in Category 1, and the concentration of the ingredients of the tested mixture that are in Category 1 is increased, the resulting untested mixture shall be classified in Category 1.

A.0.5.1.4 Interpolation within one toxicity category.

For mixtures classified in accordance with A.1, A.2, A.3, A.8, A.9, or A.10 of this Appendix, for three mixtures (A, B and C) with identical ingredients, where mixtures A and B have been tested and are in the same toxicity category, and where untested mixture C has the same toxicologically active ingredients as mixtures A and B but has concentrations of toxicologically active ingredients intermediate to the concentrations in mixtures A and B, then mixture C is assumed to be in the same toxicity category as A and B.

A.0.5.1.5 Substantially similar mixtures.

For mixtures classified in accordance with A.1 through A.10 of this Appendix, given the following set of conditions:

(a) *Where there are two mixtures:*

(i) *A + B;*

(ii) *C + B;*

(b) *The concentration of ingredient B* is essentially the same in both mixtures;

(c) *The concentration of ingredient A in mixture* (i) equals that of ingredient C in mixture (ii);

(d) *And data on toxicity for A and C* are available and substantially equivalent; i.e., they are in the same hazard category and are not expected to affect the toxicity of B; then

If mixture (i) or (ii) is already classified based on test data, the other mixture can be assigned the same hazard category.

A.0.5.1.6 Aerosols.

For mixtures classified in accordance with A.1, A.2, A.3, A.4, A.8, or A.9 of this Appendix, an aerosol form of a mixture shall be classified in the same hazard category as the tested, non-aerosolized form of the mixture, provided the added propellant does not affect the toxicity of the mixture when spraying.

A.1 ACUTE TOXICITY

A.1.1 *Definition*

Acute toxicity refers to those adverse effects occurring following oral or dermal administration of a single dose of a substance, or multiple doses given within 24 hours, or an inhalation exposure of 4 hours.

A.1.2 *Classification Criteria for Substances*

A.1.2.1 *Substances can be allocated* to one of four toxicity categories based on acute toxicity by the oral, dermal or inhalation route according to the numeric cut-off criteria as shown in Table A.1.1. Acute toxicity values are expressed as (approximate) LD50 (oral, dermal) or LC50 (inhalation) values or as acute toxicity estimates (ATE). See the footnotes following Table A.1.1 for further explanation on the application of these values.

Table A.1.1 — Acute Toxicity Hazard Categories and Acute Toxicity Estimate (ATE) Values Defining the Respective Categories

Exposure route	Category 1	Category 2	Category 3	Category 4
Oral (mg/kg bodyweight) *see: Note (a), Note (b)*	≤5	>5 and ≤50	>50 and ≤300	>300 and ≤2000.
Dermal (mg/kg bodyweight) *see: Note (a), Note (b)*	≤5	>50 and ≤200	>200 and ≤1000	>1000 and ≤2000.
Inhalation — Gases (ppmV) *see: Note (a), Note (b), Note (c)*	≤100	>100 and ≤500	>500 and ≤2500	>2500 and ≤20000.
Inhalation — Vapors (mg/l) *see: Note (a), Note (b), Note (c), Note (d)*	≤0.5	>0.5 and ≤2.0	>2.0 and ≤10.0	>10.0 and ≤20.0.
Inhalation — Dusts and Mists (mg/l) *see: Note (a), Note (b), Note (c)*	≤0.05	>0.05 and ≤0.5	>0.5 and ≤1.0	>1.0 and ≤5.0.

Note: *Gas concentrations are expressed in parts per million per volume (ppmV).*

Notes to Table A.1.1:

(a) The acute toxicity estimate (ATE) for the classification of a substance is derived using the LD_{50}/LC_{50}Stewardwhere available;

(b) The acute toxicity estimate (ATE) for the classification of a substance or ingredient in a mixture is derived using:

(i) the LD_{50}/$LC_{50}$where available. Otherwise,

(ii) the appropriate conversion value from Table 1.2 that relates to the results of a range test, or

(iii) the appropriate conversion value from Table 1.2 that relates to a classification category;

(c) Inhalation cut-off values in the table are based on 4 hour testing exposures. Conversion of existing inhalation toxicity data which has been generated according to 1 hour exposure is achieved by dividing by a factor of 2 for gases and vapors and 4 for dusts and mists;

(d) For some substances the test atmosphere will be a vapor which consists of a combination of liquid and gaseous phases. For other substances the test atmosphere may consist of a vapor which is nearly all the gaseous phase. In these latter cases, classification is based on ppmV as follows: Category 1 (100 ppmV), Category 2 (500 ppmV), Category 3 (2500 ppmV), Category 4 (20000 ppmV). The terms "dust", "mist" and "vapor" are defined as follows:

(i) Dust: solid particles of a substance or mixture suspended in a gas (usually air);

(ii) Mist: liquid droplets of a substance or mixture suspended in a gas (usually air);

(iii) Vapor: the gaseous form of a substance or mixture released from its liquid or solid state.

A.1.2.3 *The preferred test species* for evaluation of acute toxicity by the oral and inhalation routes is the rat, while the rat or rabbit are preferred for evaluation of acute dermal toxicity. Test data already generated for the classification of chemicals under existing systems should be accepted when reclassifying these chemicals under the harmonized system. When experimental data for acute toxicity are available in several animal species, scientific judgment should be used in selecting the most appropriate LD_{50}value from among scientifically validated tests.

A.1.3 *Classification Criteria for Mixtures*

A.1.3.1 *The approach to classification of mixtures* for acute toxicity is tiered, and is dependent upon the amount of information available for the mixture itself and for its ingredients. The flow chart of Figure A.1.1 indicates the process that must be followed:

Figure A.1.1: Tiered approach to classification of mixtures for acute toxicity

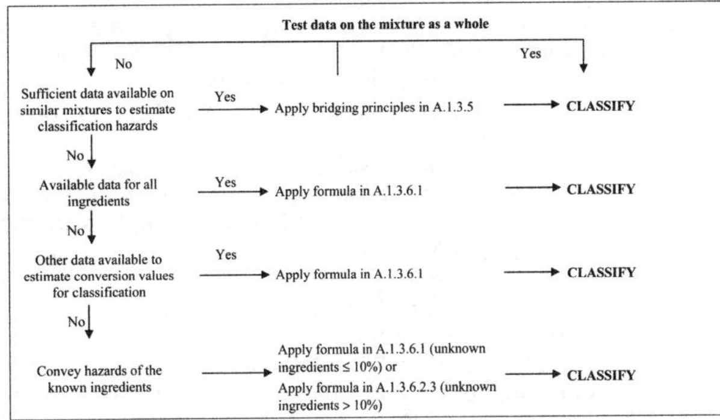

A.1.3.2 *Classification of mixtures for acute* toxicity may be carried out for each route of exposure, but is only required for one route of exposure as long as this route is followed (estimated or tested) for all ingredients and there is no relevant evidence to suggest acute toxicity by multiple routes. When there is relevant evidence of acute toxicity by multiple routes of exposure, classification is to be conducted for all appropriate routes of exposure. All available information shall be considered. The pictogram and signal word used shall reflect the most severe hazard category; and all relevant hazard statements shall be used.

A.1.3.3 *For purposes of classifying the hazards* of mixtures in the tiered approach:

(a) *The "relevant ingredients" of a mixture* are those which are present in concentrations [ge]1% (weight/weight for solids, liquids, dusts, mists and vapors and volume/volume for gases). If there is reason to suspect that an ingredient present at a concentration <1% will affect classification of the mixture for acute toxicity, that ingredient shall also be considered relevant. Consideration of ingredients present at a concentration <1% is particularly important when classifying untested mixtures which contain ingredients that are classified in Category 1 and Category 2;

(b) *Where a classified mixture is used* as an ingredient of another mixture, the actual or derived acute toxicity estimate (ATE) for that mixture is used when calculating the classification of the new mixture using the formulas in A.1.3.6.1 and A.1.3.6.2.4.

(c) *If the converted acute toxicity* point estimates for all ingredients of a mixture are within the same category, then the mixture should be classified in that category.

(d) *When only range data* (or acute toxicity hazard category information) are available for ingredients in a mixture, they may be converted to point estimates in accordance with Table A.1.2 when calculating the classification of the new mixture using the formulas in A.1.3.6.1 and A.1.3.6.2.4.

A.1.3.4 *Classification of Mixtures Where Acute* Toxicity Test Data Are Available for the Complete Mixture

Where the mixture itself has been tested to determine its acute toxicity, it is classified according to the same criteria as those used for substances, presented in Table A.1.1. If test data for the mixture are not available, the procedures presented below must be followed.

A.1.3.5 *Classification of Mixtures Where Acute* Toxicity Test Data Are Not Available for the Complete Mixture: Bridging Principles

A.1.3.5.1 Where the mixture itself has not been tested to determine its acute toxicity, but there are sufficient data on both the individual ingredients and similar tested mixtures to adequately characterize the hazards of the mixture, these data will be used in accordance with the following bridging principles as found in paragraph A.0.5 of this Appendix: Dilution, Batching, Concentration of mixtures, Interpolation within one toxicity category, Substantially similar mixtures, and Aerosols.

A.1.3.6 *Classification of Mixtures Based* on Ingredients of the Mixture (Additivity Formula)

A.1.3.6.1 Data available for all ingredients.

The acute toxicity estimate (ATE) of ingredients is considered as follows:

(a) *Include ingredients with a known* acute toxicity, which fall into any of the acute toxicity categories, or have an oral or dermal LD_{50}greater than 2000 but less than or equal to 5000 mg/kg body weight (or the equivalent dose for inhalation);

(b) *Ignore ingredients that are presumed* not acutely toxic (e.g., water, sugar);

(c) *Ignore ingredients if the data* available are from a limit dose test (at the upper threshold for Category 4 for the appropriate route of exposure as provided in Table A.1.1) and do not show acute toxicity.

Ingredients that fall within the scope of this paragraph are considered to be ingredients with a known acute toxicity estimate (ATE). See note (b) to Table A.1.1 and paragraph A.1.3.3 for appropriate application of available data to the equation below, and paragraph A.1.3.6.2.4.

The ATE of the mixture is determined by calculation from the ATE values for all relevant ingredients according to the following formula below for oral, dermal or inhalation toxicity:

$$\frac{100}{\text{ATEmix}} = \sum_{n} \frac{\text{Ci}}{\text{ATE}_{\text{i}}}$$

Where:

Ci = concentration of ingredient i

n ingredients and i is running from 1 to n

ATEi = acute toxicity estimate of ingredient i.

A.1.3.6.2 Data are not available for one or more ingredients of the mixture.

A.1.3.6.2.1 Where an ATE is not available for an individual ingredient of the mixture, but available information provides a derived conversion value, the formula in A.1.3.6.1 may be applied. This information may include evaluation of:

(a) *Extrapolation between oral,* dermal and inhalation acute toxicity estimates. Such an evaluation requires appropriate pharmacodynamic and pharmacokinetic data;

(b) *Evidence from human exposure* that indicates toxic effects but does not provide lethal dose data;

(c) *Evidence from any other toxicity tests/assays* available on the substance that indicates toxic acute effects but does not necessarily provide lethal dose data; or

(d) *Data from closely analogous* substances using structure/activity relationships.

A.1.3.6.2.2 This approach requires substantial supplemental technical information, and a highly trained and experienced expert, to reliably estimate acute toxicity. If sufficient information is not available to reliably estimate acute toxicity, proceed to the provisions of A.1.3.6.2.3.

A.1.3.6.2.3 In the event that an ingredient with unknown acute toxicity is used in a mixture at a concentration [ge]1%, and the mixture has not been classified based on testing of the mixture as a whole, the mixture cannot be attributed a definitive acute toxicity estimate. In this situation the mixture is classified based on the known ingredients only. (Note: A statement that x percent of the mixture consists of ingredient(s) of unknown toxicity is required on the label and safety data sheet in such cases; see Appendix C to this section, Allocation of Label Elements and Appendix D to this section, Safety Data Sheets.)

Where an ingredient with unknown acute toxicity is used in a mixture at a concentration [ge]1%, and the mixture is not classified based on testing of the mixture as a whole, a statement that X% of the mixture consists of ingredient(s) of unknown acute toxicity is required on the label and safety data sheet in such cases; see Appendix C to this section, Allocation of Label Elements and Appendix D to this section, Safety Data Sheets.)

A.1.3.6.2.4 If the total concentration of the relevant ingredient(s) with unknown acute toxicity is [le]10% then the formula presented in A.1.3.6.1 must be used. If the total concentration of the relevant ingredient(s) with unknown acute toxicity is >10%, the formula presented in A.1.3.6.1 is corrected to adjust for the percentage of the unknown ingredient(s) as follows:

$$\frac{100-(\sum C_{unknown} \text{ if } >10\%)}{ATE_{mix}} = \sum_{n} \frac{Ci}{ATE_i}$$

Table A.1.2 — Conversion From Experimentally Obtained Acute Toxicity Range Values (or Acute Toxicity Hazard Categories) to Acute Toxicity Point Estimates for Use in the Formulas for the Classification of Mixtures

Exposure routes	Classification category or experimentally obtained acute toxicity range estimate	Converted acute toxicity point estimate
Oral (mg/kg bodyweight)	0 <Category 1 ≤5	0.5
	5 <Category 2 ≤50	5
	50 <Category 3 ≤300	100
	300 <Category 4 ≤2000	500
Dermal (mg/kg bodyweight)	0 <Category 1 ≤50	5
	50 <Category 2 ≤200	50
	200 <Category 3 ≤1000	300
	1000 <Category 4 ≤2000	1100
Gases (ppmV)	0 <Category 1 ≤100	10
	100 <Category 2 ≤500	100
	500 <Category 3 ≤2500	700
	2500 <Category 4 ≤20000	4500
Vapors (mg/l)	0 <Category 1 ≤0.5	0.05
	0.5 <Category 2 ≤2.0	0.5
	2.0 <Category 3 ≤10.0	3
	10.0 <Category 4 ≤20.0	11
Dust/mist (mg/l)	0 <Category 1 ≤0.05	0.005
	0.05 <Category 2 ≤0.5	0.05
	0.5 <Category 3 ≤1.0	0.5
	1.0 <Category 4 ≤5.0	1.5

Note: Gas concentrations are expressed in parts per million per volume (ppmV).

A.2 SKIN CORROSION/IRRITATION

A.2.1 *Definitions and General Considerations*

A.2.1.1 *Skin corrosion* is the production of irreversible damage to the skin; namely, visible necrosis through the epidermis and into the dermis, following the application of a test substance for up to 4 hours. Corrosive reactions are typified by ulcers, bleeding, bloody scabs, and, by the end of observation at 14 days, by discoloration due to blanching of the skin, complete areas of alopecia, and scars. Histopathology should be considered to evaluate questionable lesions.

Skin irritation is the production of reversible damage to the skin following the application of a test substance for up to 4 hours.

A.2.1.2 *Skin corrosion/irritation shall be classified* using a tiered approach as detailed in figure A.2.1. Emphasis shall be placed upon existing human data (See A.0.2.6), followed by other sources of information. Classification results directly when the data satisfy the criteria in this section. In case the criteria cannot be directly applied, classification of a substance or a mixture is made on the basis of the total weight of evidence (See A.0.3.1). This means that all available information bearing on the determination of skin corrosion/irritation is considered together, including the results of appropriate scientifically validated in-vitro tests, relevant animal data, and human data such as epidemiological and clinical studies and well-documented case reports and observations.

A.2.2 *Classification Criteria for Substances* Using Animal Test Data

A.2.2.1 *Corrosion*

A.2.2.1.1 A corrosive substance is a chemical that produces destruction of skin tissue, namely, visible necrosis through the epidermis and into the dermis, in at least 1 of 3 tested animals after exposure up to a 4-hour duration. Corrosive reactions are typified by ulcers, bleeding, bloody scabs and, by the end of observation at 14 days, by discoloration due to blanching of the skin, complete areas of alopecia and scars. Histopathology should be considered to discern questionable lesions.

A.2.2.1.2 Three sub-categories of Category 1 are provided in Table A.2.1, all of which shall be regulated as Category 1.

Table A.2.1 — Skin Corrosion Category and Sub-Categories

Category 1: corrosive	Corrosive sub-categories	Corrosive in ≥1 of 3 animals	
		Exposure	Observation
	1A	≤3 min	≤1 h.
	1B	>3 min ≤1 h	≤14 days.
	1C	>1 h ≤4 h	≤14 days.

A.2.2.2 *Irritation*

A.2.2.2.1 A single irritant category (Category 2) is presented in the Table A.2.2. The major criterion for the irritant category is that at least 2 tested animals have a mean score of ≥2.3 ≤4.0.

Table A.2.2 — Skin Irritation Category

	Criteria
Irritant (Category 2)	(1) Mean value of ≥2.3 ≤4.0 for erythema/eschar or for edema in at least 2 of 3 tested animals from gradings at 24, 48 and 72 hours after patch removal or, if reactions are delayed, from grades on 3 consecutive days after the onset of skin reactions; or
	(2) Inflammation that persists to the end of the observation period normally 14 days in at least 2 animals, particularly taking into account alopecia (limited area), hyperkeratosis, hyperplasia, and scaling; or
	(3) In some cases where there is pronounced variability of response among animals, with very definite positive effects related to chemical exposure in a single animal but less than the criteria above.

A.2.2.2.2 Animal irritant responses within a test can be quite variable, as they are with corrosion. A separate irritant criterion accommodates cases when there is a significant irritant response but less than the mean score criterion for a positive test. For example, a substance might be designated as an irritant if at least 1 of 3 tested animals shows a very elevated mean score throughout the study, including lesions persisting at the end of an observation period of normally 14 days. Other responses could also fulfil this criterion. However, it should be ascertained that the responses are the result of chemical exposure. Addition of this criterion increases the sensitivity of the classification system.

A.2.2.2.3 Reversibility of skin lesions is another consideration in evaluating irritant responses. When inflammation persists to the end of the observation period in 2 or more test animals, taking into consideration alopecia (limited area), hyperkeratosis, hyperplasia and scaling, then a chemical should be considered to be an irritant.

A.2.3 *Classification Criteria for Substances* Using Other Data Elements

A.2.3.1 *Existing human and animal* data including information from single or repeated exposure should be the first line of analysis, as they give information directly relevant to effects on the skin. If a substance is highly toxic by the dermal route, a skin corrosion/irritation study may not be practicable since the amount of test substance to be applied would considerably exceed the toxic dose and, consequently, would result in the death of the animals. When observations are made of skin corrosion/irritation in acute toxicity studies and are observed up through the limit dose, these data may be used for classification provided that the dilutions used and species tested are equivalent. *In vitro* alternatives that have been scientifically validated shall be used to make classification decisions. Solid substances (powders) may become corrosive or irritant when moistened or in contact with moist skin or mucous membranes. Likewise, pH extremes like [le]2 and [ge]11.5 may indicate skin effects, especially when associated with significant buffering capacity. Generally, such substances are expected to produce significant effects on the skin. In the absence of any other information, a substance is considered corrosive (Skin Category 1) if it has a pH [le]2 or a pH [ge]11.5. However, if consideration of alkali/acid reserve suggests the substance or mixture may not be corrosive despite the low or high pH value, then further evaluation may be necessary. In some cases enough information may be available from structurally related compounds to make classification decisions.

A.2.3.2 *A tiered approach to the evaluation* of initial information shall be used (Figure A.2.1) recognizing that all elements may not be relevant in certain cases.

A.2.3.3 *The tiered approach explains* how to organize information on a substance and to make a weight-of-evidence decision about hazard assessment and hazard classification.

A.2.3.4 *All the above information that is available* on a substance shall be evaluated. Although information might be gained from the evaluation of single parameters within a tier, there is merit in considering the totality of existing information and making an overall weight of evidence determination. This is especially true when there is information available on some but not all parameters. Emphasis shall be placed upon existing human experience and data, followed by animal experience and testing data, followed by other sources of information, but case-by-case determinations are necessary.

Figure A.2.1: Tiered evaluation of skin corrosion and irritation potential

Step	Parameter		Finding		Conclusion
1a	Existing human or animal data[1] ↓ Not corrosive or no data ↓	→	Skin corrosive	→	Category 1[2]
1b	Existing human or animal data[1] ↓ Not an irritant or no data ↓	→	Skin irritant	→	Category 2[2]
1c	Existing human or animal data[1] ↓ No/Insufficient data ↓	→	Not a skin corrosive or skin irritant	→	Not classified
2:	Other, existing skin data in animals[3] ↓ No/Insufficient data ↓	→	Skin corrosive Skin irritant	→	Category 1[2] Category 2[2]
3:	Existing skin corrosive *ex vivo / in vitro* data[4] ↓ Not corrosive or no data ↓	→	Positive: Skin corrosive	→	Category 1[2]
	Existing skin irritation *ex vivo / in vitro* data[4] ↓ No/Insufficient data ↓	→	Positive: Skin irritant Negative: Not a skin irritant[5]	→	Category 2[2] Not classified
4:	pH-Based assessment (with consideration of buffering capacity of the chemical, or no buffering capacity data)[5] ↓ Not a pH extreme, No pH data or extreme pH with low/no buffering capacity ↓	→	pH ≤ 2 or ≥ 11.5	→	Category 1[2]
5:	Validated Structure/Activity Relationship (SAR) models ↓ No/Insufficient data ↓	→	Skin corrosive Skin irritant	→	Category 1[2] Category 2[2]
6:	Consideration of the total Weight of Evidence[6] ↓ No concern based on consideration of the sum of available data ↓	→	Skin corrosive Skin irritant	→	Category 1[2] Category 2[2]
7:	**Not Classified**	→			Not classified

Notes to Figure A.2.1:

[1] *Evidence of existing human or animal data may be derived from single or repeated exposure(s) in occupational, consumer, transportation, or emergency response scenarios; from ethically-conducted human clinical studies; or from purposely-generated data from animal studies conducted according to scientifically validated test methods (at present, there is no internationally accepted test method for human skin irritation testing).*

[2] *Classify in the appropriate harmonized category, as shown in Tables A.2.1 and A.2.2.*

[3] *Pre-existing animal data (e.g. from an acute dermal toxicity test or a sensitisation test) should be carefully reviewed to determine if sufficient skin corrosion/irritation evidence is available through other, similar information. For example, classification/categorization may be done on the basis of whether a chemical has or has not produced any skin irritation in an acute dermal toxicity test in animals at the limit dose, or produces very toxic effects in an acute dermal toxicity test in animals. In the latter case, the chemical would be classified as being very hazardous by the dermal route for acute toxicity, and it would be moot whether the chemical is also irritating or corrosive on the skin. It should be kept in mind in evaluating acute dermal toxicity information that the reporting of dermal lesions may be incomplete, testing and observations may be made on a species other than the rabbit, and species may differ in sensitivity in their responses.*

[4] *Evidence from studies using scientifically validated protocols with isolated human/animal tissues or other, non-tissue-based, though scientifically validated, protocols should be assessed. Examples of scientifically validated test methods for skin corrosion include OECD TG 430 (Transcutaneous Electrical Resistance Test (TER)), 431 (Human Skin Model Test), and 435 (Membrane Barrier Test Method). OECD TG 439 (Reconstructed Human Epidermis Test Method) is a scientifically validated in vitro test method for skin irritation.*

[5] *Measurement of pH alone may be adequate, but assessment of acid or alkali reserve (buffering capacity) would be preferable. Presently, there is no scientifically validated and internationally accepted method for assessing this parameter.*

[6] *All information that is available on a chemical should be considered and an overall determination made on the total weight of evidence. This is especially true when there is conflict in information available on some parameters. Professional judgment should be exercised in making such a determination.*

A.2.4 *Classification Criteria for Mixtures*

A.2.4.1 *Classification of Mixtures When Data* Are Available for the Complete Mixture

A.2.4.1.1 *The mixture shall be classified using the criteria* for substances (See A.2.3).

A.2.4.2 *Classification of Mixtures When Data* Are Not Available for the Complete Mixture: Bridging Principles

A.2.4.2.1 *Where the mixture itself has* not been tested to determine its skin corrosion/irritation, but there are sufficient data on both the individual ingredients and similar tested mixtures to adequately characterize the hazards of the mixture, these data will be used in accordance with the following bridging principles, as found in paragraph A.0.5 of this Appendix: Dilution, Batching, Concentration of mixtures, Interpolation within one toxicity category, Substantially similar mixtures, and Aerosols.

A.2.4.3 *Classification of Mixtures When Data* Are Available for All Ingredients or Only for Some Ingredients of the Mixture

A.2.4.3.1 *For purposes of classifying the skin* corrosion/irritation hazards of mixtures in the tiered approach:

The "relevant ingredients" of a mixture are those which are present in concentrations >1% (weight/weight for solids, liquids, dusts, mists and vapors and volume/volume for gases.) If the classifier has reason to suspect that an ingredient present at a concentration <1% will affect classification of the mixture for skin corrosion/irritation, that ingredient shall also be considered relevant.

A.2.4.3.2 *In general,* the approach to classification of mixtures as irritant or corrosive to skin when data are available on the ingredients, but not on the mixture as a whole, is based on the theory of additivity, such that each corrosive or irritant ingredient contributes to the overall irritant or corrosive properties of the mixture in proportion to its potency and concentration. A weighting factor of 10 is used for corrosive ingredients when they are present at a concentration below the concentration limit for classification with Category 1, but are at a concentration that will contribute to the classification of the mixture as an irritant. The mixture is classified as corrosive or irritant when the sum of the concentrations of such ingredients exceeds a cut-off value/concentration limit.

A.2.4.3.3 *Table A.2.3 below* provides the cut-off value/concentration limits to be used to determine if the mixture is considered to be an irritant or a corrosive to the skin.

A.2.4.3.4 *Particular care shall be taken* when classifying certain types of chemicals such as acids and bases, inorganic salts, aldehydes, phenols, and surfactants. The approach explained in A.2.4.3.1 and A.2.4.3.2 might not work given that many of such substances are corrosive or irritant at concentrations <1%. For mixtures containing strong acids or bases the pH should be used as classification criteria since pH will be a better indicator of corrosion than the concentration limits of Table A.2.3. A mixture containing corrosive or irritant ingredients that cannot be classified based on the additivity approach shown in Table A.2.3, due to chemical characteristics that make this approach unworkable, should be classified as Skin Category 1 if it contains [ge]1% of a corrosive ingredient and as Skin Category 2 when it contains [ge]3% of an irritant ingredient. Classification of mixtures with ingredients for which the approach in Table A.2.3 does not apply is summarized in Table A.2.4 below.

A.2.4.3.5 *On occasion,* reliable data may show that the skin corrosion/irritation of an ingredient will not be evident when present at a level above the generic concentration cut-off values mentioned in Tables A.2.3 and A.2.4. In these cases the mixture could be classified according to those data (See *Use of cut-off values/concentration limits,* paragraph A.0.4.3 of this Appendix).

A.2.4.3.6 *If there are data showing* that (an) ingredient(s) may be corrosive or irritant at a concentration of <1% (corrosive) or <3% (irritant), the mixture shall be classified accordingly (See *Use of cut-off values/concentration limits,* paragraph A.0.4.3 of this Appendix).

Table A.2.3 — Concentration of Ingredients of a Mixture Classified as Skin Category 1 or 2 That Would Trigger Classification of the Mixture as Hazardous to Skin[1] [Category 1 or 2]

Sum of ingredients classified as:	Concentration triggering classification of a mixture as:	
	Skin corrosive	Skin irritant
	Category 1	Category 2
Skin Category 1	≥5%	≥1% but 5%.
Skin Category 2		≥10%.
(10 x Skin Category 1) + Skin Category 2		≥10%.

1.Editor's Note: The Federal Register cuts off the title for this table after the word "Trigger." The title has been corrected in the above text.

Table A.2.4 — Concentration of Ingredients of a Mixture for Which the Additivity Approach Does Not Apply, That Would Trigger Classification of the Mixture as Hazardous to Skin

Ingredient:	Concentration:	Mixture classified as: Skin
Acid with pH ≤2	≥1%	Category 1.
Base with pH ≥11.5	≥1%	Category 1.
Other corrosive (Category 1) ingredients for which additivity does not apply	≥1%	Category 1.
Other irritant (Category 2) ingredients for which additivity does not apply, including acids and bases	≥3%	Category 2.

A.3 SERIOUS EYE DAMAGE/EYE IRRITATION

A.3.1 *Definitions and General Considerations*

A.3.1.1 *Serious eye damage* is the production of tissue damage in the eye, or serious physical decay of vision, following application of a test substance to the anterior surface of the eye, which is not fully reversible within 21 days of application.

Eye irritation is the production of changes in the eye following the application of test substance to the anterior surface of the eye, which are fully reversible within 21 days of application.

A.3.1.2 *Serious eye damage/eye* irritation shall be classified using a tiered approach as detailed in Figure A.3.1. Emphasis shall be placed upon existing human data (See A.0.2.6), followed by animal data, followed by other sources of information. Classification results directly when the data satisfy the criteria in this section. In case the criteria cannot be directly applied, classification of a substance or a mixture is made on the basis of the total weight of evidence (See A.0.3.1). This means that all available information bearing on the determination of serious eye damage/eye irritation is considered together, including the results of appropriate scientifically validated *in vitro* tests, relevant animal data, and human data such as epidemiological and clinical studies and well-documented case reports and observations.

A.3.2 *Classification Criteria for Substances* Using Animal Test Data

A.3.2.1 *Irreversible effects on the eye/serious* damage to eyes (Category 1).

A single hazard category is provided in Table A.3.1, for substances that have the potential to seriously damage the eyes. Category 1, irreversible effects on the eye, includes the criteria listed below. These observations include animals with grade 4 cornea lesions and other severe reactions (e.g. destruction of cornea) observed at any time during the test, as well as persistent corneal opacity, discoloration of the cornea by a dye substance, adhesion, pannus, and interference with the function of the iris or other effects that impair sight. In this context, persistent lesions are considered those which are not fully reversible within an observation period of normally 21 days. Category 1 also contains substances fulfilling the criteria of corneal opacity [ge]3 and/or iritis >1.5 detected in a Draize eye test with rabbits, because severe lesions like these usually do not reverse within a 21-day observation period.

Table A.3.1 — Irreversible Eye Effects

A substance is classified as Serious Eye Damage Category 1 (irreversible effects on the eye) when it produces:
(a) at least in one tested animal, effects on the cornea, iris or conjunctiva that are not expected to reverse or have not fully reversed within an observation period of normally 21 days; and/or
(b) at least in 2 of 3 tested animals, a positive response of:
(i) corneal opacity ≥3; and/or
(ii) iritis >1.5;
calculated as the mean scores following grading at 24, 48 and 72 hours after instillation of the substance.

A.3.2.2 *Reversible effects on the eye (Category 2).*

A single category is provided in Table A.3.2 for substances that have the potential to induce reversible eye irritation.

Table A.3.2 — Reversible Eye Effects

A substance is classified as Eye irritant Category 2A (irritating to eyes) when it produces in at least in 2 of 3 tested animals a positive response of:
(i) corneal opacity ≥1; and/or
(ii) iritis ≥1; and/or
(iii) conjunctival redness ≥2; and/or
(iv) conjunctival edema (chemosis) ≥2
calculated as the mean scores following grading at 24, 48 and 72 hours after instillation of the substance, and which fully reverses within an observation period of normally 21 days.
An eye irritant is considered mildly irritating to eyes (Category 2B) when the effects listed above are fully reversible within 7 days of observation.

A.3.2.3 *For those chemicals where there* is pronounced variability among animal responses, this information may be taken into account in determining the classification.

A.3.3 *Classification Criteria for Substances* Using Other Data Elements

A.3.3.1 *Existing human and animal* data should be the first line of analysis, as they give information directly relevant to effects on the eye. Possible skin corrosion shall be evaluated prior to consideration of serious eye damage/eye irritation in order to avoid testing for local effects on eyes with skin corrosive substances. *In vitro* alternatives that have been scientifically validated and accepted shall be used to make classification decisions. Likewise, pH extremes like ≤ 2 and ≥ 11.5, may indicate serious eye damage, especially when associated with significant buffering capacity. Generally, such substances are expected to produce significant effects on the eyes. In the absence of any other information, a mixture/substance is considered to cause serious eye damage (Eye Category 1) if it has a pH ≤ 2 or ≥11.5. However, if consideration of acid/alkaline reserve suggests the substance may not have the potential to cause serious eye damage despite the low or high pH value, then further evaluation may be necessary. In some cases enough information may be available from structurally related compounds to make classification decisions.

A.3.3.2 *A tiered approach to the evaluation* of initial information shall be used where applicable, recognizing that all elements may not be relevant in certain cases (Figure A.3.1).

A.3.3.3 *The tiered approach explains* how to organize existing information on a substance and to make a weight-of-evidence decision, where appropriate, about hazard assessment and hazard classification.

A.3.3.4 *All the above information that is available* on a substance shall be evaluated. Although information might be gained from the evaluation of single parameters within a tier, consideration should be given to the totality of existing information and making an overall weight-of-evidence determination. This is especially true when there is conflict in information available on some parameters.

Figure A.3.1 Evaluation strategy for serious eye damage and eye irritation (See also Figure A.2.1)

Step	Parameter	Finding	Conclusion
1a:	Existing human or animal data, eye[1]	→ Serious Eye Damage	→ Category 1[2]
		→ Eye Irritant	→ Category 2[2]
	↓ No/insufficient data or unknown		
1b:	Existing human or animal data, skin corrosion	→ Skin corrosive	→ Category 1[2]
	↓ No/insufficient data or unknown		
1c:	Existing human or animal data, eye[1]	→ Existing data that show that substance does not cause serious eye damage or eye irritation	→ Not Classified
	↓ No/insufficient data		
2:	Other, existing skin/eye data in animals[3]	→ Yes; existing data that show that substance may cause serious eye damage or eye irritation	→ Category1 or Category 2[2]
	↓ No/insufficient data		
3:	Existing *ex vivo / in vitro* data[4]	→ Positive: serious eye damage	→ Category 1[2]
		→ Positive: eye irritant	→ Category 2[2]
	↓ No/insufficient data / negative response		
4:	pH-Based assessment (with consideration of buffering capacity of the chemical, or no buffering capacity data)[5]	→ pH ≤ 2 or ≥11.5	→ Category 1[2]
	↓ Not a pH extreme, no pH data, or extreme pH with low/no buffering capacity		
5:	Validated structure/activity relationship (SAR) models	→ Severe damage to eyes	→ Category 1[2]
		→ Eye irritant	→ Category 2[2]
		→ Skin Corrosive	Category 1[2]
	↓ No/insufficient data		
6:	Consideration of the total weight of evidence[6]	→ Serious eye damage	→ Category 1[2]
		→ Eye irritant	→ Category 2[2]
	↓ No concern based on consideration of the sum of available data		
7:	**Not Classified**		

Notes to Figure A.3.1:

[1] *Evidence of existing human or animal data may be derived from single or repeated exposure(s) in occupational, consumer, transportation, or emergency response scenarios; from ethically-conducted human clinical studies; or from purposely-generated data from animal studies conducted according to scientifically validated test methods. At present, there are no internationally accepted test methods for human skin or eye irritation testing.*

[2] *Classify in the appropriate harmonized category, as shown in Tables A.3.1 and A.3.2.*

[3] *Pre-existing animal data should be carefully reviewed to determine if sufficient skin or eye corrosion/irritation evidence is available through other, similar information.*

[4] *Evidence from studies using scientifically validated protocols with isolated human/animal tissues or other, non-tissue-based, though scientifically validated, protocols should be assessed. Examples of, scientifically validated test methods for identifying eye corrosives and severe irritants (i.e., Serious Eye Damage) include OECD TG 437 (Bovine Corneal Opacity and Permeability (BCOP)) and TG 438 (Isolated Chicken Eye). Positive test results from a scientifically validated in vitro test for skin corrosion would likely also lead to a conclusion to classify as causing Serious Eye Damage.*

[5] *Measurement of pH alone may be adequate, but assessment of acid or alkali reserve (buffering capacity) would be preferable.*

[6] *All information that is available on a chemical should be considered and an overall determination made on the total weight of evidence. This is especially true when there is conflict in information available on some parameters. The weight of evidence including information on skin irritation could lead to classification of eye irritation. It is recognized that not all skin irritants are eye irritants as well. Professional judgment should be exercised in making such a determination.*

A.3.4 *Classification Criteria for Mixtures*

A.3.4.1 *Classification of Mixtures When Data* Are Available for the Complete Mixture

A.3.4.1.1 The mixture will be classified using the criteria for substances.

A.3.4.1.2 Unlike other hazard classes, there are alternative tests available for skin corrosivity of certain types of chemicals that can give an accurate result for classification purposes, as well as being simple and relatively inexpensive to perform. When considering testing of the mixture, chemical manufacturers are encouraged to use a tiered weight of evidence strategy as included in the criteria for classification of substances for skin corrosion and serious eye damage and eye irritation to help ensure an accurate classification, as well as avoid unnecessary animal testing. In the absence of any other information, a mixture is considered to cause serious eye damage (Eye Category 1) if it has a pH ≤ 2 or ≥ 11.5. However, if consideration of acid/alkaline reserve suggests the substance or mixture may not have the potential to cause serious eye damage despite the low or high pH value, then further evaluation may be necessary.

A.3.4.2 *Classification of Mixtures When Data* Are Not Available for the Complete Mixture: Bridging Principles

A.3.4.2.1 Where the mixture itself has not been tested to determine its skin corrosivity or potential to cause serious eye damage or eye irritation, but there are sufficient data on both the individual ingredients and similar tested mixtures to adequately characterize the hazards of the mixture, these data will be used in accordance with the following bridging principles, as found in paragraph A.0.5 of this Appendix: Dilution, Batching, Concentration of mixtures, Interpolation within one toxicity category, Substantially similar mixtures, and Aerosols.

A.3.4.3 *Classification of Mixtures When Data* Are Available for All Ingredients or Only for Some Ingredients of the Mixture

A.3.4.3.1 For purposes of classifying the eye corrosion/irritation hazards of mixtures in the tiered approach:

The "relevant ingredients" of a mixture are those which are present in concentrations >1% (weight/weight for solids, liquids, dusts, mists and vapors and volume/volume for gases). If the classifier has reason to suspect that an ingredient present at a concentration <1% will affect classification of the mixture for eye corrosion/irritation, that ingredient shall also be considered relevant.

A.3.4.3.2 In general, the approach to classification of mixtures as seriously damaging to the eye or eye irritant when data are available on the ingredients, but not on the mixture as a whole, is based on the theory of additivity, such that each corrosive or irritant ingredient contributes to the overall irritant or corrosive properties of the mixture in proportion to its potency and concentration. A weighting factor of 10 is used for corrosive ingredients when they are present at a concentration below the concentration limit for classification with Category 1, but are at a concentration that will contribute to the classification of the mixture as an irritant. The mixture is classified as seriously damaging to the eye or eye irritant when the sum of the concentrations of such ingredients exceeds a threshold cut-off value/concentration limit.

A.3.4.3.3 Table A.3.3 provides the cut-off value/concentration limits to be used to determine if the mixture should be classified as seriously damaging to the eye or an eye irritant.

A.3.4.3.4 Particular care must be taken when classifying certain types of chemicals such as acids and bases, inorganic salts, aldehydes, phenols, and surfactants. The approach explained in A.3.4.3.1 and A.3.4.3.2 might not work given that many of such substances are corrosive or irritant at concentrations <1%. For mixtures containing strong acids or bases, the pH should be used as classification criteria (See A.3.4.1) since pH will be a better indicator of serious eye damage than the concentration limits of Table A.3.3. A mixture containing corrosive or irritant ingredients that cannot be classified based on the additivity approach applied in Table A.3.3 due to chemical characteristics that make this approach unworkable, should be classified as Eye Category 1 if it contains [ge]1% of a corrosive ingredient and as Eye Category 2 when it contains [ge]3% of an irritant ingredient. Classification of mixtures with ingredients for which the approach in Table A.3.3 does not apply is summarized in Table A.3.4.

A.3.4.3.5 On occasion, reliable data may show that the reversible/irreversible eye effects of an ingredient will not be evident when present at a level above the generic cut-off values/concentration limits mentioned in Tables A.3.3 and A.3.4. In these cases the mixture could be classified according to those data (See also A.0.4.3 *Use of cut-off values/concentration limits*"). On occasion, when it is expected that the skin corrosion/irritation or the reversible/irreversible eye effects of an ingredient will not be evident when present at a level above the generic concentration/cut-off levels mentioned in Tables A.3.3 and A.3.4, testing of the mixture may be considered. In those cases, the tiered weight of evidence strategy should be applied as referred to in section A.3.3, Figure A.3.1 and explained in detail in this chapter.

A.3.4.3.6 If there are data showing that (an) ingredient(s) may be corrosive or irritant at a concentration of <1% (corrosive) or <3% (irritant), the mixture should be classified accordingly (See also paragraph A.0.4.3, *Use of cut-off values/concentration limits*).

Table A.3.3 — Concentration of Ingredients of a Mixture Classified as Skin Category 1 and/or Eye Category 1 or 2 That Would Trigger Classification of the Mixtures as Hazardous to the Eye

Sum of ingredients classified as:	Concentration triggering classification of a mixture as:	
	Irreversible eye effects	Reversible eye effects
	Category 1	Category 2
Eye or Skin Category 1	≥3%	≥1% but 3%.
Eye Category 2		≥10%.
(10 x Eye Category 1) + Eye Category 2		≥10%.
Skin Category 1 + Eye Category 1	≥3%	≥1% but 3%.
10 x (Skin Category 1 + Eye Category 1) + Eye Category 2		≥10%.

Note: *A mixture may be classified as Eye Category 2B in cases when all relevant ingredients are classified as Eye Category 2B.*

Table A.3.4 — Concentration of Ingredients of a Mixture for Which the Additivity Approach Does Not Apply, That Would Trigger Classification of the Mixture as Hazardous to the Eye

Ingredient	Concentration	Mixture classified as: Eye
Acid with pH ≤2	≥1%	Category 1.
Base with pH ≥11.5	≥1%	Category 1.
Other corrosive (Category 1) ingredients for which additivity does not apply	≥1%	Category 1.
Other irritant (Category 2) ingredients for which additivity does not apply, including acids and bases	≥3%	Category 2.

A.4 RESPIRATORY OR SKIN SENSITIZATION

A.4.1 *Definitions and General Considerations*

A.4.1.1 **Respiratory sensitizer** means a chemical that will lead to hypersensitivity of the airways following inhalation of the chemical.

Skin sensitizer means a chemical that will lead to an allergic response following skin contact.

A.4.1.2 *For the purpose of this chapter,* sensitization includes two phases: the first phase is induction of specialized immunological memory in an individual by exposure to an allergen. The second phase is elicitation, i.e., production of a cell-mediated or antibody-mediated allergic response by exposure of a sensitized individual to an allergen.

A.4.1.3 *For respiratory sensitization,* the pattern of induction followed by elicitation phases is shared in common with skin sensitization. For skin sensitization, an induction phase is required in which the immune system learns to react; clinical symptoms can then arise when subsequent exposure is sufficient to elicit a visible skin reaction (elicitation phase). As a consequence, predictive tests usually follow this pattern in which there is an induction phase, the response to which is measured by a standardized elicitation phase, typically involving a patch test. The local lymph node assay is the exception, directly measuring the induction response. Evidence of skin sensitization in humans normally is assessed by a diagnostic patch test.

A.4.1.4 *Usually, for both skin* and respiratory sensitization, lower levels are necessary for elicitation than are required for induction.

A.4.1.5 *The hazard class "respiratory* or skin sensitization" is differentiated into:

(a) *Respiratory sensitization; and*

(b) *Skin sensitization.*

A.4.2 *Classification Criteria for Substances*

A.4.2.1 *Respiratory Sensitizers*

A.4.2.1.1 Hazard Categories.

A.4.2.1.1.1 *Effects seen in either* humans or animals will normally justify classification in a weight of evidence approach for respiratory sensitizers. Substances may be allocated to one of the two sub-categories 1A or 1B using a weight of evidence approach in accordance with the criteria given in Table A.4.1 and on the basis of reliable and good quality evidence from human cases or epidemiological studies and/or observations from appropriate studies in experimental animals.

A.4.2.1.1.2 *Where data are not sufficient for sub-categorization,* respiratory sensitizers shall be classified in Category 1.

Table A.4.1 — Hazard Category and Sub-Categories for Respiratory Sensitizers

Category 1	Respiratory sensitizer
	A substance is classified as a respiratory sensitizer.
	(a) if there is evidence in humans that the substance can lead to specific respiratory hypersensitivity and/or
	(b) if there are positive results from an appropriate animal test.[1]
Sub-category 1A	Substances showing a high frequency of occurrence in humans; or a probability of occurrence of a high sensitization rate in humans based on animal or other tests.[1] Severity of reaction may also be considered.
Sub-category 1B	Substances showing a low to moderate frequency of occurrence in humans; or a probability of occurrence of a low to moderate sensitization rate in humans based on animal or other tests.[1] Severity of reaction may also be considered.

1.At this writing, recognized and validated animal models for the testing of respiratory hypersensitivity are not available. Under certain circumstances, data from animal studies may provide valuable information in a weight of evidence assessment.

A.4.2.1.2 Human evidence.

A.4.2.1.2.1 *Evidence that a substance can* lead to specific respiratory hypersensitivity will normally be based on human experience. In this context, hypersensitivity is normally seen as asthma, but other hypersensitivity reactions such as rhinitis/conjunctivitis and alveolitis are also considered. The condition will have the clinical character of an allergic reaction. However, immunological mechanisms do not have to be demonstrated.

A.4.2.1.2.2 *When considering the human evidence,* it is necessary that in addition to the evidence from the cases, the following be taken into account:

(a) *The size of the population exposed;*

(b) *The extent of exposure.*

A.4.2.1.2.3 *The evidence referred to above could be:*

(a) *Clinical history and data* from appropriate lung function tests related to exposure to the substance, confirmed by other supportive evidence which may include:

(i) *In vivo* immunological test (e.g., skin prick test);

(ii) *In vitro* immunological test (e.g., serological analysis);

(iii) *Studies that may indicate other specific* hypersensitivity reactions where immunological mechanisms of action have not been proven, e.g., repeated low-level irritation, pharmacologically mediated effects;

(iv) *A chemical structure related* to substances known to cause respiratory hypersensitivity;

(b) *Data from positive bronchial* challenge tests with the substance conducted according to accepted guidelines for the determination of a specific hypersensitivity reaction.

A.4.2.1.2.4 *Clinical history should include* both medical and occupational history to determine a relationship between exposure to a specific substance and development of respiratory hypersensitivity. Relevant information includes aggravating factors both in the home and workplace, the onset and progress of the disease, family history and medical history of the patient in question. The medical history should also include a note of other allergic or airway disorders from childhood and smoking history.

A.4.2.1.2.5 *The results of positive bronchial* challenge tests are considered to provide sufficient evidence for classification on their own. It is, however, recognized that in practice many of the examinations listed above will already have been carried out.

A.4.2.1.3 Animal studies.

A.4.2.1.3.1 *Data from appropriate animal* studies[2] which may be indicative of the potential of a substance to cause sensitization by inhalation in humans[3] may include:

[2]*At this writing, recognized and validated animal models for the testing of respiratory hypersensitivity are not available. Under certain circumstances, data from animal studies may provide valuable information in a weight of evidence assessment.*

[3] *The mechanisms by which substances induce symptoms of asthma are not yet fully known. For preventive measures, these substances are considered respiratory sensitizers. However, if on the basis of the evidence, it can be demonstrated that these substances induce symptoms of asthma by irritation only in people with bronchial hyperactivity, they should not be considered as respiratory sensitizers.*

(a) *Measurements of Immunoglobulin E* (IgE) and other specific immunological parameters, for example in mice

(b) *Specific pulmonary responses* in guinea pigs.

A.4.2.2 *Skin Sensitizers*

A.4.2.2.1 *Hazard categories.*

A.4.2.2.1.1 *Effects seen in either* humans or animals will normally justify classification in a weight of evidence approach for skin sensitizers. Substances may be allocated to one of the two sub-categories 1A or 1B using a weight of evidence approach in accordance with the criteria given in Table A.4.2 and on the basis of reliable and good quality evidence from human cases or epidemiological studies and/or observations from appropriate studies in experimental animals according to the guidance values provided in A.4.2.2.2.1 and A.4.2.2.3.2 for sub-category 1A and in A.4.2.2.2.2 and A.4.2.2.3.3 for sub-category 1B.

A.4.2.2.1.2 *Where data are not sufficient for sub-categorization,* skin sensitizers shall be classified in Category 1.

Table A.4.2 — Hazard Category and Sub-Categories for Skin Sensitizers

Category 1	Skin sensitizer
	A substance is classified as a skin sensitizer.
	(a) if there is evidence in humans that the substance can lead to sensitization by skin contact in a substantial number of persons, or
	(b) if there are positive results from an appropriate animal test.
Sub-category 1A	Substances showing a high frequency of occurrence in humans and/or a high potency in animals can be presumed to have the potential to produce significant sensitization in humans. Severity of reaction may also be considered.
Sub-category 1B	Substances showing a low to moderate frequency of occurrence in humans and/or a low to moderate potency in animals can be presumed to have the potential to produce sensitization in humans. Severity of reaction may also be considered.

A.4.2.2.2 *Human evidence.*

A.4.2.2.2.1 *Human evidence for sub-category* 1A may include:

(a) *Positive responses at [le]500* µg/cm^2 (Human Repeat Insult Patch Test (HRIPT), Human Maximization Test (HMT) — induction threshold);

(b) *Diagnostic patch test* data where there is a relatively high and substantial incidence of reactions in a defined population in relation to relatively low exposure;

(c) *Other epidemiological evidence where there* is a relatively high and substantial incidence of allergic contact dermatitis in relation to relatively low exposure.

A.4.2.2.2.2 *Human evidence for sub-category* 1B may include:

(a) *Positive responses at >500* µg/cm^2 (HRIPT, HMT — induction threshold);

(b) *Diagnostic patch test* data where there is a relatively low but substantial incidence of reactions in a defined population in relation to relatively high exposure;

(c) *Other epidemiological evidence where there* is a relatively low but substantial incidence of allergic contact dermatitis in relation to relatively high exposure.

A.4.2.2.3 *Animal studies*

A.4.2.2.3.1 *For Category 1,* when an adjuvant type test method for skin sensitization is used, a response of at least 30% of the animals is considered as positive. For a non-adjuvant Guinea pig test method a response of at least 15% of the animals is considered positive. For Category 1, a stimulation index of three or more is considered a positive response in the local lymph node assay.[4]

[4] *Test methods for skin sensitization are described in OECD Guideline 406 (the Guinea Pig Maximization test and the Buehler guinea pig test) and Guideline 429 (Local Lymph Node Assay). Other methods may be used provided that they are scientifically validated. The Mouse Ear Swelling Test (MEST), appears to be a reliable screening test to detect moderate to strong sensitizers, and can be used, in accordance with professional judgment, as a first stage in the assessment of skin sensitization potential.*

A.4.2.2.3.2 *Animal test results* for sub-category 1A can include data with values indicated in Table A.4.3 below:

Table A.4.3 — Animal Test Results for Sub-Category 1A

Assay	Criteria
Local lymph node assay	EC3 value ≤2%.
Guinea pig maximization test	≥30% responding at ≤0.1% intradermal induction dose *or*
	≥60% responding at >0.1% to ≤1% intradermal induction dose.
Buehler assay	≥15% responding at ≤0.2% topical induction dose *or*
	≥60% responding at >0.2% to ≤20% topical induction dose.

Note: *EC3 refers to the estimated concentration of test chemical required to induce a stimulation index of 3 in the local lymph node assay.*

A.4.2.2.3.3 *Animal test results* for sub-category 1B can include data with values indicated in Table A.4.4 below:

Table A.4.4 — Animal Test Results for Sub-Category 1B

Assay	Criteria
Local lymph node assay	EC3 value >2%.
Guinea pig maximization test	≥30% to <60% responding at >0.1% to ≤1% intradermal induction dose or
	≥30% responding at >1% intradermal induction dose.
Buehler assay	≥15% to <60% responding at >0.2% to ≤20% topical induction dose or
	≥15% responding at >20% topical induction dose.

Note: *EC3 refers to the estimated concentration of test chemical required to induce a stimulation index of 3 in the local lymph node assay.*

A.4.2.2.4 *Specific considerations.*

A.4.2.2.4.1 *For classification of a substance,* evidence shall include one or more of the following using a weight of evidence approach:

(a) *Positive data from patch* testing, normally obtained in more than one dermatology clinic;

(b) *Epidemiological studies showing* allergic contact dermatitis caused by the substance. Situations in which a high proportion of those exposed exhibit characteristic symptoms are to be looked at with special concern, even if the number of cases is small;

(c) *Positive data from appropriate* animal studies;

(d) *Positive data from experimental* studies in man (See paragraph A.0.2.6 of this Appendix);

(e) *Well documented episodes* of allergic contact dermatitis, normally obtained in more than one dermatology clinic;

(f) *Severity of reaction.*

A.4.2.2.4.2 *Evidence from animal studies* is usually much more reliable than evidence from human exposure. However, in cases where evidence is available from both sources, and there is conflict between the results, the quality and reliability of the evidence from both sources must be assessed in order to resolve the question of classification on a case-by-case basis. Normally, human data are not generated in controlled experiments with volunteers for the purpose of hazard classification but rather as part of risk assessment to confirm lack of effects seen in animal tests. Consequently, positive human data on skin sensitization are usually derived from case-control or other, less defined studies. Evaluation of human data must, therefore, be carried out with caution as the frequency of cases reflect, in addition to the inherent properties of the substances, factors such as the exposure situation, bioavailability, individual predisposition and preventive measures taken. Negative human data should not normally be used to negate positive results from animal studies. For both animal and human data, consideration should be given to the impact of vehicle.

A.4.2.2.4.3 *If none of the above-mentioned conditions* are met, the substance need not be classified as a skin sensitizer. However, a combination of two or more indicators of skin sensitization, as listed below, may alter the decision. This shall be considered on a case-by-case basis.

(a) *Isolated episodes of allergic* contact dermatitis;

(b) *Epidemiological studies of limited* power, e.g., where chance, bias or confounders have not been ruled out fully with reasonable confidence;

(c) *Data from animal tests,* performed according to existing guidelines, which do not meet the criteria for a positive result described in A.4.2.2.3, but which are sufficiently close to the limit to be considered significant;

(d) *Positive data from non-standard methods;*

(e) *Positive results from close* structural analogues.

A.4.2.2.4.4 *Immunological contact urticaria.*

A.4.2.2.4.4.1 *Substances meeting the criteria* for classification as respiratory sensitizers may, in addition, cause immunological contact urticaria. Consideration shall be given to classifying these substances as skin sensitizers.

A.4.2.2.4.4.2 *Substances which cause* immunological contact urticaria without meeting the criteria for respiratory sensitizers shall be considered for classification as skin sensitizers.

A.4.2.2.4.4.3 *There is no recognized animal* model available to identify substances which cause immunological contact urticaria. Therefore, classification will normally be based on human evidence, similar to that for skin sensitization.

A.4.3 *Classification Criteria for Mixtures*

A.4.3.1 *Classification of Mixtures When Data* Are Available for the Complete Mixture

When reliable and good quality evidence, as described in the criteria for substances, from human experience or appropriate studies in experimental animals, is available for the mixture, then the mixture shall be classified by weight of evidence evaluation of these data. Care must be exercised in evaluating data on mixtures that the dose used does not render the results inconclusive.

A.4.3.2 *Classification of Mixtures When Data* Are Not Available for the Complete Mixture: Bridging Principles

A.4.3.2.1 *Where the mixture itself has* not been tested to determine its sensitizing properties, but there are sufficient data on both the individual ingredients and similar tested mixtures to adequately characterize the hazards of the mixture, these data will be used in accordance with the following agreed bridging principles as found in paragraph A.0.5 of this Appendix: Dilution, Batching, Concentration of mixtures, Interpolation, Substantially similar mixtures, and Aerosols.

A.4.3.3 *Classification of Mixtures When Data* Are Available for All Ingredients or Only for Some Ingredients of the Mixture

The mixture shall be classified as a respiratory or skin sensitizer when at least one ingredient has been classified as a respiratory or skin sensitizer and is present at or above the appropriate cut-off value/concentration limit for the specific endpoint as shown in Table A.4.5.

Table A.4.5 — Cut-Off Values/Concentration Limits of Ingredients of a Mixture Classified as Either Respiratory Sensitizers or Skin Sensitizers That Would Trigger Classification of the Mixture

Ingredient classified as:	Cut-off values/concentration limits triggering classification of a mixture as:		
	Respiratory Sensitizer Category 1		Skin Sensitizer Category 1
	Solid/liquid	Gas	All physical states
Respiratory Sensitizer, Category 1	≥0.1%	≥0.1%	
Respiratory Sensitizer, Sub-category 1A	≥0.1%	≥0.1%	
Respiratory Sensitizer, Sub-category 1B	≥1.0%	≥0.2%	
Skin Sensitizer, Category 1			≥0.1%
Skin Sensitizer, Sub-category 1A			≥0.1%
Skin Sensitizer, Sub-category 1B			≥1.0%

A.5 GERM CELL MUTAGENICITY

A.5.1 *Definitions and General Considerations*

A.5.1.1 *A mutation is defined as a permanent* change in the amount or structure of the genetic material in a cell. The term *mutation* applies both to heritable genetic changes that may be manifested at the phenotypic level and to the underlying DNA modifications when known (including, for example, specific base pair changes and chromosomal translocations). The term *mutagenic* and *mutagen* will be used for agents giving rise to an increased occurrence of mutations in populations of cells and/or organisms.

A.5.1.2 *The more general terms genotoxic* and *genotoxicity* apply to agents or processes which alter the structure, information content, or segregation of DNA, including those which cause DNA damage by interfering with normal replication processes, or which in a non-physiological manner (temporarily) alter its replication. Genotoxicity test results are usually taken as indicators for mutagenic effects.

A.5.1.3 *This hazard class is primarily* concerned with chemicals that may cause mutations in the germ cells of humans that can be transmitted to the progeny. However, mutagenicity/genotoxicity tests *in vitro* and in mammalian somatic cells *in vivo* are also considered in classifying substances and mixtures within this hazard class.

A.5.2 *Classification Criteria for Substances*

A.5.2.1 *The classification system provides* for two different categories of germ cell mutagens to accommodate the weight of evidence available. The two-category system is described in the Figure A.5.1.

Figure A.5.1 — Hazard Categories for Germ Cell Mutagens

CATEGORY 1: Substances known to induce heritable mutations or to be regarded as if they induce heritable mutations in the germ cells of humans.
Category 1A: Substances known to induce heritable mutations in germ cells of humans.
Positive evidence from human epidemiological studies.
Category 1B: Substances which should be regarded as if they induce heritable mutations in the germ cells of humans.
(a) Positive result(s) from *in vivo* heritable germ cell mutagenicity tests in mammals; or
(b) Positive result(s) from *in vivo* somatic cell mutagenicity tests in mammals, in combination with some evidence that the substance has potential to cause mutations to germ cells. This supporting evidence may, for example, be derived from mutagenicity/genotoxicity tests in germ cells *in vivo*, or by demonstrating the ability of the substance or its metabolite(s) to interact with the genetic material of germ cells; or
(c) Positive results from tests showing mutagenic effects in the germ cells of humans, without demonstration of transmission to progeny; for example, an increase in the frequency of aneuploidy in sperm cells of exposed people.
CATEGORY 2: Substances which cause concern for humans owing to the possibility that they may induce heritable mutations in the germ cells of humans.
Positive evidence obtained from experiments in mammals and/or in some cases from *in vitro* experiments, obtained from:
(a) Somatic cell mutagenicity tests *in vivo*, in mammals; or
(b) Other *in vivo* somatic cell genotoxicity tests which are supported by positive results from *in vitro* mutagenicity assays.
Note: *Substances which are positive in in vitro mammalian mutagenicity assays, and which also show chemical structure activity relationship to known germ cell mutagens, should be considered for classification as Category 2 mutagens.*

A.5.2.2 *Specific considerations for classification* of substances as germ cell mutagens:

A.5.2.2.1 *To arrive at a classification,* test results are considered from experiments determining mutagenic and/or genotoxic effects in germ and/or somatic cells of exposed animals. Mutagenic and/or genotoxic effects determined in *in vitro* tests shall also be considered.

A.5.2.2.2 *The system is hazard based,* classifying chemicals on the basis of their intrinsic ability to induce mutations in germ cells. The scheme is, therefore, not meant for the (quantitative) risk assessment of chemical substances.

A.5.2.2.3 Classification for heritable effects in human germ cells is made on the basis of scientifically validated tests. Evaluation of the test results shall be done using expert judgment and all the available evidence shall be weighed for classification.

A.5.2.2.4 The classification of substances shall be based on the total weight of evidence available, using expert judgment. In those instances where a single well-conducted test is used for classification, it shall provide clear and unambiguously positive results. The relevance of the route of exposure used in the study of the substance compared to the route of human exposure should also be taken into account.

A.5.3 *Classification Criteria for Mixtures*[5]

[5] *It should be noted that the classification criteria for health hazards usually include a tiered scheme in which test data available on the complete mixture are considered as the first tier in the evaluation, followed by the applicable bridging principles, and lastly, cut-off values/concentration limits or additivity. However, this approach is not used for Germ Cell Mutagenicity. These criteria for Germ Cell Mutagenicity consider the cut-off values/concentration limits as the primary tier and allow the classification to be modified only on a case-by-case evaluation based on available test data for the mixture as a whole.*

A.5.3.1 *Classification of Mixtures When Data* Are Available for All Ingredients or Only for Some Ingredients of the Mixture

A.5.3.1.1 Classification of mixtures shall be based on the available test data for the individual ingredients of the mixture using cut-off values/concentration limits for the ingredients classified as germ cell mutagens.

A.5.3.1.2 The mixture will be classified as a mutagen when at least one ingredient has been classified as a Category 1A, Category 1B or Category 2 mutagen and is present at or above the appropriate cut-off value/concentration limit as shown in Table A.5.1 below for Category 1 and 2 respectively.

Table A.5.1 — Cut-Off Values/Concentration Limits of Ingredients of a Mixture Classified as Germ Cell Mutagens That Would Trigger Classification of the Mixture

Ingredient classified as:	Cut-off/concentration limits triggering classification of a mixture as:	
	Category 1 mutagen	Category 2 mutagen
Category 1A/B mutagen	≥0.1%	
Category 2 mutagen		≥1.0%

Note: *The cut-off values/concentration limits in the table above apply to solids and liquids (w/w units) as well as gases (v/v units).*

A.5.3.2 *Classification of Mixtures When Data* Are Available for the Mixture Itself

The classification may be modified on a case-by-case basis based on the available test data for the mixture as a whole. In such cases, the test results for the mixture as a whole must be shown to be conclusive taking into account dose and other factors such as duration, observations and analysis (e.g. statistical analysis, test sensitivity) of germ cell mutagenicity test systems.

A.5.3.3 *Classification of Mixtures When Data* Are Not Available for the Complete Mixture: Bridging Principles

A.5.3.3.1 Where the mixture itself has not been tested to determine its germ cell mutagenicity hazard, but there are sufficient data on both the individual ingredients and similar tested mixtures to adequately characterize the hazards of the mixture, these data will be used in accordance with the following bridging principles as found in paragraph A.0.5 of this Appendix: Dilution, Batching, and Substantially similar mixtures.

A.5.4 *Examples of Scientifically Validated Test Methods*

A.5.4.1 *Examples of in vivo* heritable germ cell mutagenicity tests are:

(a) Rodent dominant lethal mutation test (OECD 478)
(b) Mouse heritable translocation assay (OECD 485)
(c) Mouse specific locus test

A.5.4.2 *Examples of in vivo* somatic cell mutagenicity tests are:

(a) Mammalian bone marrow chromosome aberration test (OECD 475)
(b) Mouse spot test (OECD 484)
(c) Mammalian erythrocyte micronucleus test (OECD 474)

A.5.4.3 *Examples of mutagenicity/genotoxicity tests* in germ cells are:

(a) Mutagenicity tests:
(i) Mammalian spermatogonial chromosome aberration test (OECD 483)
(ii) Spermatid micronucleus assay
(b) Genotoxicity tests:
(i) Sister chromatid exchange analysis in spermatogonia
(ii) Unscheduled DNA synthesis test (UDS) in testicular cells

A.5.4.4 *Examples of genotoxicity tests* in somatic cells are:

(a) Liver Unscheduled DNA Synthesis (UDS) *in vivo* (OECD 486)
(b) Mammalian bone marrow Sister Chromatid Exchanges (SCE)

A.5.4.5 *Examples of in vitro* mutagenicity tests are:

(a) In vitro mammalian chromosome aberration test (OECD 473)
(b) In vitro mammalian cell gene mutation test (OECD 476)
(c) Bacterial reverse mutation tests (OECD 471)

A.5.4.6 *As new,* scientifically validated tests arise, these may also be used in the total weight of evidence to be considered.

A.6 CARCINOGENICITY

A.6.1 *Definitions*

Carcinogen means a substance or a mixture of substances which induce cancer or increase its incidence. Substances and mixtures which have induced benign and malignant tumors in well-performed experimental studies on animals are considered also to be presumed or suspected human carcinogens unless there is strong evidence that the mechanism of tumor formation is not relevant for humans.

Classification of a substance or mixture as posing a carcinogenic hazard is based on its inherent properties and does not provide information on the level of the human cancer risk which the use of the substance or mixture may represent.

A.6.2 *Classification Criteria for Substances*[6]

[6] *See Non-mandatory Appendix F Part A for further guidance regarding hazard classification for carcinogenicity. This appendix is consistent with the GHS adn is provided as guidance excerpted from the International Agency for Research on Cancer (IARC) "Monographs on the Evaluation of Carcinogenic Risks to Humans" (2006).*

A.6.2.1 *For the purpose of classification for carcinogenicity,* substances are allocated to one of two categories based on strength of evidence and additional weight of evidence considerations. In certain instances, route-specific classification may be warranted.

Figure A.6.1 — Hazard Categories for Carcinogens

CATEGORY 1: Known or presumed human carcinogens.
The classification of a substance as a Category 1 carcinogen is done on the basis of epidemiological and/or animal data. This classification is further distinguished on the basis of whether the evidence for classification is largely from human data (Category 1A) or from animal data (Category 1B):
Category 1A: Known to have carcinogenic potential for humans. Classification in this category is largely based on human evidence.
Category 1B: Presumed to have carcinogenic potential for humans. Classification in this category is largely based on animal evidence.
The classification of a substance in Category 1A and 1B is based on strength of evidence together with weight of evidence considerations (See paragraph A.6.2.5). Such evidence may be derived from:
— human studies that establish a causal relationship between human exposure to a substance and the development of cancer (known human carcinogen); or
— animal experiments for which there is sufficient evidence to demonstrate animal carcinogenicity (presumed human carcinogen).
In addition, on a case by case basis, scientific judgment may warrant a decision of presumed human carcinogenicity derived from studies showing limited evidence of carcinogenicity in humans together with limited evidence of carcinogenicity in experimental animals.
CATEGORY 2: Suspected human carcinogens.
The classification of a substance in Category 2 is done on the basis of evidence obtained from human and/or animal studies, but which is not sufficiently convincing to place the substance in Category 1A or B. This classification is based on strength of evidence together with weight of evidence considerations (See paragraph A.6.2.5). Such evidence may be from either limited evidence of carcinogenicity in human studies or from limited evidence of carcinogenicity in animal studies.
Other considerations: Where the weight of evidence for the carcinogenicity of a substance does not meet the above criteria, any positive study conducted in accordance with established scientific principles, and which reports statistically significant findings regarding the carcinogenic potential of the substance, must be noted on the safety data sheet.

A.6.2.2 *Classification as a carcinogen is made* on the basis of evidence from reliable and acceptable methods, and is intended to be used for substances which have an intrinsic property to produce such toxic effects. The evaluations are to be based on all existing data, peer-reviewed published studies and additional data accepted by regulatory agencies.

A.6.2.3 *Carcinogen classification* is a one-step, criterion-based process that involves two interrelated determinations: evaluations of strength of evidence and consideration of all other relevant information to place substances with human cancer potential into hazard categories.

A.6.2.4 *Strength of evidence* involves the enumeration of tumors in human and animal studies and determination of their level of statistical significance. Sufficient human evidence demonstrates causality between human exposure and the development of cancer, whereas sufficient evidence in animals shows a causal relationship between the agent and an increased incidence of tumors. Limited evidence in humans is demonstrated by a positive association between exposure and cancer, but a causal relationship cannot be stated. Limited evidence in animals is provided when data suggest a carcinogenic effect, but are less than sufficient. (Guidance on consideration of important factors in the classification of carcinogenicity and a more detailed description of the terms "limited" and "sufficient" have been developed by the International Agency for Research on Cancer (IARC) and are provided in non-mandatory Appendix F).

A.6.2.5 *Weight of evidence:* Beyond the determination of the strength of evidence for carcinogenicity, a number of other factors should be considered that influence the overall likelihood that an agent may pose a carcinogenic hazard in humans. The full list of factors that influence this determination is very lengthy, but some of the important ones are considered here.

A.6.2.5.1 These factors can be viewed as either increasing or decreasing the level of concern for human carcinogenicity. The relative emphasis accorded to each factor depends upon the amount and coherence of evidence bearing on each. Generally there is a requirement for more complete information to decrease than to increase the level of concern. Additional considerations should be used in evaluating the tumor findings and the other factors in a case-by-case manner.

A.6.2.5.2 Some important factors which may be taken into consideration, when assessing the overall level of concern are:

(a) *Tumor type and background incidence;*
(b) *Multisite responses;*
(c) *Progression of lesions to malignancy;*
(d) *Reduced tumor latency;*

Additional factors which may increase or decrease the level of concern include:

(e) *Whether responses are in single or both sexes;*
(f) *Whether responses are in a single* species or several species;
(g) *Structural similarity or not to a substance(s)* for which there is good evidence of carcinogenicity;
(h) *Routes of exposure;*
(i) *Comparison of absorption,* distribution, metabolism and excretion between test animals and humans;
(j) *The possibility of a confounding effect* of excessive toxicity at test doses; and,
(k) *Mode of action and its relevance* for humans, such as mutagenicity, cytotoxicity with growth stimulation, mitogenesis, immunosuppression.

Mutagenicity: It is recognized that genetic events are central in the overall process of cancer development. Therefore evidence of mutagenic activity *in vivo* may indicate that a substance has a potential for carcinogenic effects.

A.6.2.5.3 A substance that has not been tested for carcinogenicity may in certain instances be classified in Category 1A, Category 1B, or Category 2 based on tumor data from a structural analogue together with substantial support from consideration of other important factors such as formation of common significant metabolites, e.g., for benzidine congener dyes.

A.6.2.5.4 The classification should also take into consideration whether or not the substance is absorbed by a given route(s); or whether there are only local tumors at the site of administration for the tested route(s), and adequate testing by other major route(s) show lack of carcinogenicity.

A.6.2.5.5 It is important that whatever is known of the physico-chemical, toxicokinetic and toxicodynamic properties of the substances, as well as any available relevant information on chemical analogues, i.e., structure activity relationship, is taken into consideration when undertaking classification.

A.6.3 *Classification Criteria for Mixtures*[7]

[7] *It should be noted that the classification criteria for health hazards usually include a tiered scheme in which test data available on the complete mixture are considered as the first tier in the evaluation, followed by the applicable bridging principles, and lastly, cut-off values/concentration limit or additivity. However, this approach is not used for Carcinogenicity. These criteria for Carcinogenicity consider the cut-off values/concentration limits as the primary tier and allow the classification to be modified only on a case-by-case evaluation based on available test data for the mixture as a whole.*

A.6.3.1 *The mixture shall be classified as a carcinogen* when at least one ingredient has been classified as a Category 1 or Category 2 carcinogen and is present at or above the appropriate cut-off value/concentration limit as shown in Table A.6.1.

Table A.6.1 — Cut-Off Values/Concentration Limits of Ingredients of a Mixture Classified as Carcinogen That Would Trigger Classification of the Mixture

Ingredient classified as:	Category 1 carcinogen	Category 2 carcinogen
Category 1 carcinogen	≥0.1%	
Category 2 carcinogen		≥0.1% (note 1).

Note: *If a Category 2 carcinogen ingredient is present in the mixture at a concentration between 0.1% and 1%, information is required on the SDS for a product. However, a label warning is optional. If a Category 2 carcinogen ingredient is present in the mixture at a concentration of [ge]1%, both an SDS and a label is required and the information must be included on each.*

A.6.3.2 *Classification of Mixtures When Data* Are Available for the Complete Mixture

A mixture may be classified based on the available test data for the mixture as a whole. In such cases, the test results for the mixture as a whole must be shown to be conclusive taking into account dose and other factors such as duration, observations and analysis (e.g., statistical analysis, test sensitivity) of carcinogenicity test systems.

A.6.3.3 *Classification of Mixtures When Data* Are Not Available for the Complete Mixture: Bridging Principles

Where the mixture itself has not been tested to determine its carcinogenic hazard, but there are sufficient data on both the individual ingredients and similar tested mixtures to adequately characterize the hazards of the mixture, these data will be used in accordance with the following bridging principles as found in paragraph A.0.5 of this Appendix: Dilution; Batching; and Substantially similar mixtures.

A.6.4 *Classification of Carcinogenicity*[8]

[8] *See Non-mandatory Appendix F for further guidance regarding hazard classification for carcinogenicity and how to relate carcinogenicity classification information from IARC and NTP to GHS.*

A.6.4.1 *Chemical manufacturers,* importers and employers evaluating chemicals may treat the following sources as establishing that a substance is a carcinogen or potential carcinogen for hazard communication purposes in lieu of applying the criteria described herein:

A.6.4.1.1 National Toxicology Program (NTP), "Report on Carcinogens" (latest edition);

A.6.4.1.2 International Agency for Research on Cancer (IARC) "Monographs on the Evaluation of Carcinogenic Risks to Humans" (latest editions)

A.6.4.2 *Where OSHA has included* cancer as a health hazard to be considered by classifiers for a chemical covered by 29 CFR part 1910, Subpart Z, Toxic and Hazardous Substances, chemical manufacturers, importers, and employers shall classify the chemical as a carcinogen.

A.7 REPRODUCTIVE TOXICITY

A.7.1 *Definitions and General Considerations*

A.7.1.1 *Reproductive toxicity* includes *adverse effects on sexual function and fertility* in adult males and females, as well as *adverse effects on development of the offspring.* Some reproductive toxic effects cannot be clearly assigned to either impairment of sexual function and fertility or to developmental toxicity. Nonetheless, chemicals with these effects shall be classified as reproductive toxicants.

For classification purposes, the known induction of genetically based inheritable effects in the offspring is addressed in *Germ cell mutagenicity* (See A.5).

A.7.1.2 **Adverse effects on sexual function and fertility** means any effect of chemicals that interferes with reproductive ability or sexual capacity. This includes, but is not limited to, alterations to the female and male reproductive system, adverse effects on onset of puberty, gamete production and transport, reproductive cycle normality, sexual behaviour, fertility, parturition, pregnancy outcomes, premature reproductive senescence, or modifications in other functions that are dependent on the integrity of the reproductive systems.

A.7.1.3 **Adverse effects on development of the offspring** means any effect of chemicals which interferes with normal development of the conceptus either before or after birth, which is induced during pregnancy or results from parental exposure. These effects can be manifested at any point in the life span of the organism. The major manifestations of developmental toxicity include death of the developing organism, structural abnormality, altered growth and functional deficiency.

A.7.1.4 *Adverse effects on or via* lactation are also included in reproductive toxicity, but for classification purposes, such effects are treated separately (See A.7.2.1).

A.7.2 *Classification Criteria for Substances*

A.7.2.1 *For the purpose of classification for reproductive* toxicity, substances shall be classified in one of two categories in accordance with Figure A.7.1(a). Effects on sexual function and fertility, and on development, shall be considered. In addition, effects on or via lactation shall be classified in a separate hazard category in accordance with Figure A.7.1(b).

Figure A.7.1(a) — Hazard Categories for Reproductive Toxicants

CATEGORY 1: Known or presumed human reproductive toxicant.
Substance shall be classified in Category 1 for reproductive toxicity when they are known to have produced an adverse effect on sexual function and fertility or on development in humans or when there is evidence from animal studies, possibly supplemented with other information, to provide a strong presumption that the substance has the capacity to interfere with reproduction in humans. The classification of a substance is further distinguished on the basis of whether the evidence for classification is primarily from human data (Category 1A) or from animal data (Category 1B).
Category 1A: Known human reproductive toxicant.
The classification of a substance in this category is largely based on evidence from humans.
Category 1B: Presumed human reproductive toxicant.
The classification of a substance in this category is largely based on evidence from experimental animals. Data from animal studies shall provide sufficient evidence of an adverse effect on sexual function and fertility or on development in the absence of other toxic effects, or if occurring together with other toxic effects the adverse effect on reproduction is considered not to be a secondary non-specific consequence of other toxic effects. However, when there is mechanistic information that raises doubt about the relevance of the effect for humans, classification in Category 2 may be more appropriate.
CATEGORY 2: Suspected human reproductive toxicant.
Substances shall be classified in Category 2 for reproductive toxicity when there is some evidence from humans or experimental animals, possibly supplemented with other information, of an adverse effect on sexual function and fertility, or on development, in the absence of other toxic effects, or if occurring together with other toxic effects the adverse effect on reproduction is considered not to be a secondary non-specific consequence of the other toxic effects, and where the evidence is not sufficiently convincing to place the substance in Category 1. For instance, deficiencies in the study may make the quality of evidence less convincing, and in view of this, Category 2 would be the more appropriate classification.

Figure A.7.1(b) — Hazard Category for Effects on or Via Lactation

EFFECTS ON OR VIA LACTATION
Effects on or via lactation shall be classified in a separate single category. Chemicals that are absorbed by women and have been shown to interfere with lactation or that may be present (including metabolites) in breast milk in amounts sufficient to cause concern for the health of a breastfed child, shall be classified to indicate this property hazardous to breastfed babies. This classification shall be assigned on the basis of:
(a) absorption, metabolism, distribution and excretion studies that indicate the likelihood the substance would be present in potentially toxic levels in breast milk; and/or
(b) results of one or two generation studies in animals which provide clear evidence of adverse effect in the offspring due to transfer in the milk or adverse effect on the quality of the milk; and/or
(c) human evidence indicating a hazard to babies during the lactation period.

A.7.2.2 *Basis of Classification*

A.7.2.2.1 *Classification is made on the basis* of the criteria, outlined above, an assessment of the total weight of evidence, and the use of expert judgment. Classification as a reproductive toxicant is intended to be used for substances which have an intrinsic, specific property to produce an adverse effect on reproduction and substances should not be so classified if such an effect is produced solely as a non-specific secondary consequence of other toxic effects.

A.7.2.2.2 *In the evaluation of toxic effects* on the developing offspring, it is important to consider the possible influence of maternal toxicity.

A.7.2.2.3 *For human evidence to provide* the primary basis for a Category 1A classification there must be reliable evidence of an adverse effect on reproduction in humans. Evidence used for classification shall be from well conducted epidemiological studies, if available, which include the use of appropriate controls, balanced assessment, and due consideration of bias or confounding factors. Less rigorous data from studies in humans may be sufficient for a Category 1A classification if supplemented with adequate data from studies in experimental animals, but classification in Category 1B may also be considered.

A.7.2.3 *Weight of Evidence*

A.7.2.3.1 *Classification as a reproductive toxicant* is made on the basis of an assessment of the total weight of evidence using expert judgment. This means that all available information that bears on the determination of reproductive toxicity is considered together. Included is information such as epidemiological studies and case reports in humans and specific reproduction studies along with subchronic, chronic and special study results in animals that provide relevant information regarding toxicity to reproductive and related endocrine organs. Evaluation of substances chemically related to the material under study may also be included, particularly when information on the material is scarce. The weight given to the available evidence will be influenced by factors such as the quality of the studies, consistency of results, nature and severity of effects, level of statistical significance for intergroup differences, number of endpoints affected, relevance of route of administration to humans and freedom from bias. Both positive and negative results are considered together in a weight of evidence determination. However, a single, positive study performed according to good scientific principles and with statistically or biologically significant positive results may justify classification (See also A.7.2.2.3).

A.7.2.3.2 *Toxicokinetic studies in animals* and humans, site of action and mechanism or mode of action study results may provide relevant information, which could reduce or increase concerns about the hazard to human health. If it is conclusively demonstrated that the clearly identified mechanism or mode of action has no relevance for humans or when the toxicokinetic differences are so marked that it is certain that the hazardous property will not be expressed in humans then a chemical which produces an adverse effect on reproduction in experimental animals should not be classified.

A.7.2.3.3 *In some reproductive toxicity* studies in experimental animals the only effects recorded may be considered of low or minimal toxicological significance and classification may not necessarily be the outcome. These effects include, for example, small changes in semen parameters or in the incidence of spontaneous defects in the fetus, small changes in the proportions of common fetal variants such as are observed in skeletal examinations, or in fetal weights, or small differences in postnatal developmental assessments.

A.7.2.3.4 *Data from animal studies* shall provide sufficient evidence of specific reproductive toxicity in the absence of other systemic toxic effects. However, if developmental toxicity occurs together with other toxic effects in the dam (mother), the potential influence of the generalized adverse effects should be assessed to the extent possible. The preferred approach is to consider adverse effects in the embryo/fetus first, and then evaluate maternal toxicity, along with any other factors which are likely to have influenced these effects, as part of the weight of evidence. In general, developmental effects that are observed at maternally toxic doses should not be automatically discounted. Discounting developmental effects that are observed at maternally toxic doses can only be done on a case-by-case basis when a causal relationship is established or refuted.

A.7.2.3.5 If appropriate information is available it is important to try to determine whether developmental toxicity is due to a specific maternally mediated mechanism or to a non-specific secondary mechanism, like maternal stress and the disruption of homeostasis. Generally, the presence of maternal toxicity should not be used to negate findings of embryo/fetal effects, unless it can be clearly demonstrated that the effects are secondary non-specific effects. This is especially the case when the effects in the offspring are significant, e.g., irreversible effects such as structural malformations. In some situations it is reasonable to assume that reproductive toxicity is due to a secondary consequence of maternal toxicity and discount the effects, for example if the chemical is so toxic that dams fail to thrive and there is severe inanition; they are incapable of nursing pups; or they are prostrate or dying.

A.7.2.4 *Maternal Toxicity*

A.7.2.4.1 Development of the offspring throughout gestation and during the early postnatal stages can be influenced by toxic effects in the mother either through non-specific mechanisms related to stress and the disruption of maternal homeostasis, or by specific maternally-mediated mechanisms. So, in the interpretation of the developmental outcome to decide classification for developmental effects it is important to consider the possible influence of maternal toxicity. This is a complex issue because of uncertainties surrounding the relationship between maternal toxicity and developmental outcome. Expert judgment and a weight of evidence approach, using all available studies, shall be used to determine the degree of influence to be attributed to maternal toxicity when interpreting the criteria for classification for developmental effects. The adverse effects in the embryo/fetus shall be first considered, and then maternal toxicity, along with any other factors which are likely to have influenced these effects, as weight of evidence, to help reach a conclusion about classification.

A.7.2.4.2 Based on pragmatic observation, it is believed that maternal toxicity may, depending on severity, influence development via non-specific secondary mechanisms, producing effects such as depressed fetal weight, retarded ossification, and possibly resorptions and certain malformations in some strains of certain species. However, the limited numbers of studies which have investigated the relationship between developmental effects and general maternal toxicity have failed to demonstrate a consistent, reproducible relationship across species. Developmental effects which occur even in the presence of maternal toxicity are considered to be evidence of developmental toxicity, unless it can be unequivocally demonstrated on a case by case basis that the developmental effects are secondary to maternal toxicity. Moreover, classification shall be considered where there is a significant toxic effect in the offspring, e.g., irreversible effects such as structural malformations, embryo/fetal lethality, or significant post-natal functional deficiencies.

A.7.2.4.3 Classification shall not automatically be discounted for chemicals that produce developmental toxicity only in association with maternal toxicity, even if a specific maternally-mediated mechanism has been demonstrated. In such a case, classification in Category 2 may be considered more appropriate than Category 1. However, when a chemical is so toxic that maternal death or severe inanition results, or the dams (mothers) are prostrate and incapable of nursing the pups, it is reasonable to assume that developmental toxicity is produced solely as a secondary consequence of maternal toxicity and discount the developmental effects. Classification is not necessarily the outcome in the case of minor developmental changes, e.g., a small reduction in fetal/pup body weight or retardation of ossification when seen in association with maternal toxicity.

A.7.2.4.4 Some of the endpoints used to assess maternal toxicity are provided below. Data on these endpoints, if available, shall be evaluated in light of their statistical or biological significance and dose-response relationship.

(a) *Maternal mortality:* An increased incidence of mortality among the treated dams over the controls shall be considered evidence of maternal toxicity if the increase occurs in a dose-related manner and can be attributed to the systemic toxicity of the test material. Maternal mortality greater than 10% is considered excessive and the data for that dose level shall not normally be considered to need further evaluation.

(b) *Mating index* (Number of animals with seminal plugs or sperm/Number of mated x 100)

(c) *Fertility index* (Number of animals with implants/Number of matings x 100)

(d) *Gestation length* (If allowed to deliver)

(e) *Body weight and body* weight change: Consideration of the maternal body weight change and/or adjusted (corrected) maternal body weight shall be included in the evaluation of maternal toxicity whenever such data are available. The calculation of an adjusted (corrected) mean maternal body weight change, which is the difference between the initial and terminal body weight minus the gravid uterine weight (or alternatively, the sum of the weights of the fetuses), may indicate whether the effect is maternal or intrauterine. In rabbits, the body weight gain may not be a useful indicator of maternal toxicity because of normal fluctuations in body weight during pregnancy.

(f) *Food and water consumption* (if relevant): The observation of a significant decrease in the average food or water consumption in treated dams (mothers) compared to the control group may be useful in evaluating maternal toxicity, particularly when the test material is administered in the diet or drinking water. Changes in food or water consumption must be evaluated in conjunction with maternal body weights when determining if the effects noted are reflective of maternal toxicity or more simply, unpalatability of the test material in feed or water.

(g) *Clinical evaluations* (including clinical signs, markers, and hematology and clinical chemistry studies): The observation of increased incidence of significant clinical signs of toxicity in treated dams (mothers) relative to the control group is useful in evaluating maternal toxicity. If this is to be used as the basis for the assessment of maternal toxicity, the types, incidence, degree and duration of clinical signs shall be reported in the study. Clinical signs of maternal intoxication include, but are not limited to: coma, prostration, hyperactivity, loss of righting reflex, ataxia, or labored breathing.

(h) *Post-mortem data:* Increased incidence and/or severity of post-mortem findings may be indicative of maternal toxicity. This can include gross or microscopic pathological findings or organ weight data, including absolute organ weight, organ-to-body weight ratio, or organ-to-brain weight ratio. When supported by findings of adverse histopathological effects in the affected organ(s), the observation of a significant change in the average weight of suspected target organ(s) of treated dams (mothers), compared to those in the control group, may be considered evidence of maternal toxicity.

A.7.2.5 *Animal and Experimental Data*

A.7.2.5.1 A number of scientifically validated test methods are available, including methods for developmental toxicity testing (e.g., OECD Test Guideline 414, ICH Guideline S5A, 1993), methods for peri- and post-natal toxicity testing (e.g., ICH S5B, 1995), and methods for one or two-generation toxicity testing (e.g., OECD Test Guidelines 415, 416)

A.7.2.5.2 Results obtained from screening tests (e.g., OECD Guidelines 421 — Reproduction/Developmental Toxicity Screening Test, and 422 — Combined Repeated Dose Toxicity Study with Reproduction/Development Toxicity Screening Test) can also be used to justify classification, although the quality of this evidence is less reliable than that obtained through full studies.

A.7.2.5.3 Adverse effects or changes, seen in short- or long-term repeated dose toxicity studies, which are judged likely to impair reproductive function and which occur in the absence of significant generalized toxicity, may be used as a basis for classification, e.g., histopathological changes in the gonads.

A.7.2.5.4 Evidence from in vitro assays, or non-mammalian tests, and from analogous substances using structure-activity relationship (SAR), can contribute to the procedure for classification. In all cases of this nature, expert judgment must be used to assess the adequacy of the data. Inadequate data shall not be used as a primary support for classification.

A.7.2.5.5 *It is preferable that animal studies* are conducted using appropriate routes of administration which relate to the potential route of human exposure. However, in practice, reproductive toxicity studies are commonly conducted using the oral route, and such studies will normally be suitable for evaluating the hazardous properties of the substance with respect to reproductive toxicity. However, if it can be conclusively demonstrated that the clearly identified mechanism or mode of action has no relevance for humans or when the toxicokinetic differences are so marked that it is certain that the hazardous property will not be expressed in humans then a substance which produces an adverse effect on reproduction in experimental animals should not be classified.

A.7.2.5.6 *Studies involving routes* of administration such as intravenous or intraperitoneal injection, which may result in exposure of the reproductive organs to unrealistically high levels of the test substance, or elicit local damage to the reproductive organs, e.g., by irritation, must be interpreted with extreme caution and on their own are not normally the basis for classification.

A.7.2.5.7 *There is general agreement* about the concept of a limit dose, above which the production of an adverse effect may be considered to be outside the criteria which lead to classification. Some test guidelines specify a limit dose, other test guidelines qualify the limit dose with a statement that higher doses may be necessary if anticipated human exposure is sufficiently high that an adequate margin of exposure would not be achieved. Also, due to species differences in toxicokinetics, establishing a specific limit dose may not be adequate for situations where humans are more sensitive than the animal model.

A.7.2.5.8 *In principle,* adverse effects on reproduction seen only at very high dose levels in animal studies (for example doses that induce prostration, severe inappetence, excessive mortality) do not normally lead to classification, unless other information is available, for example, toxicokinetics information indicating that humans may be more susceptible than animals, to suggest that classification is appropriate.

A.7.2.5.9 *However, specification of the actual* "limit dose" will depend upon the test method that has been employed to provide the test results.

A.7.3 *Classification Criteria for Mixtures*[9]

[9] *It should be noted that the classification criteria for health hazards usually include a tiered scheme in which test data available on the complete mixture are considered as the first tier in the evaluation, followed by the applicable bridging principles, and lastly, cut-off values/concentration limits or additivity. However, this approach is not used for Reproductive Toxicity. These criteria for Reproductive Toxicity consider the cut-off values/concentration limits as the primary tier and allow the classification to be modified only on a case-by-case evaluation based on available test data for the mixture as a whole.*

A.7.3.1 *Classification of Mixtures When Data* Are Available for All Ingredients or Only for Some Ingredients of the Mixture

A.7.3.1.1 *The mixture shall be classified as a reproductive* toxicant when at least one ingredient has been classified as a Category 1 or Category 2 reproductive toxicant and is present at or above the appropriate cut-off value/concentration limit specified in Table A.7.1 for Category 1 and 2, respectively.

A.7.3.1.2 *The mixture shall be classified for effects* on or via lactation when at least one ingredient has been classified for effects on or via lactation and is present at or above the appropriate cut-off value/concentration limit specified in Table A.7.1 for the additional category for effects on or via lactation.

Table A.7.1 — Cut-Off Values/Concentration Limits of Ingredients of a Mixture Classified as Reproductive Toxicants or for Effects on or via Lactation That Trigger Classification of the Mixture

Ingredients classified as:	Cut-off values/concentration limits triggering classification of a mixture as:		
	Category 1 reproductive toxicant	Category 2 reproductive toxicant	Additional category for effects on or via lactation
Category 1 reproductive toxicant	≥0.1%		
Category 2 reproductive toxicant		≥0.1%	
Additional category for effects on or via lactation			≥0.1%

A.7.3.2 *Classification of Mixtures When Data* Are Available for the Complete Mixture

Available test data for the mixture as a whole may be used for classification on a case-by-case basis. In such cases, the test results for the mixture as a whole must be shown to be conclusive taking into account dose and other factors such as duration, observations and analysis (e.g., statistical analysis, test sensitivity) of reproduction test systems.

A.7.3.3 *Classification of Mixtures When Data* Are Not Available for the Complete Mixture: Bridging Principles

A.7.3.3.1 *Where the mixture itself has* not been tested to determine its reproductive toxicity, but there are sufficient data on both the individual ingredients and similar tested mixtures to adequately characterize the hazards of the mixture, these data shall be used in accordance with the following bridging principles as found in paragraph A.0.5 of this Appendix: Dilution, Batching, and Substantially similar mixtures.

A.8 SPECIFIC TARGET ORGAN TOXICITY SINGLE EXPOSURE

A.8.1 *Definitions and General Considerations*

A.8.1.1 **Specific target organ toxicity — single exposure, (STOT-SE)** means specific, non-lethal target organ toxicity arising from a single exposure to a chemical. All significant health effects that can impair function, both reversible and irreversible, immediate and/or delayed and not specifically addressed in A.1 to A.7 and A.10 of this Appendix are included. Specific target organ toxicity following repeated exposure is classified in accordance with *SPECIFIC TARGET ORGAN TOXICITY — REPEATED EXPOSURE* (A.9 of this Appendix) and is therefore not included here.

A.8.1.2 *Classification identifies the chemical* as being a specific target organ toxicant and, as such, it presents a potential for adverse health effects in people who are exposed to it.

A.8.1.3 *The adverse health effects* produced by a single exposure include consistent and identifiable toxic effects in humans; or, in experimental animals, toxicologically significant changes which have affected the function or morphology of a tissue/organ, or have produced serious changes to the biochemistry or hematology of the organism, and these changes are relevant for human health. Human data is the primary source of evidence for this hazard class.

A.8.1.4 *Assessment shall take into consideration* not only significant changes in a single organ or biological system but also generalized changes of a less severe nature involving several organs.

A.8.1.5 *Specific target organ* toxicity can occur by any route that is relevant for humans, i.e., principally oral, dermal or inhalation.

A.8.1.6 *The classification criteria for specific* organ systemic toxicity single exposure are organized as criteria for substances Categories 1 and 2 (See A.8.2.1), criteria for substances Category 3 (See A.8.2.2) and criteria for mixtures (See A.8.3). See also Figure A.8.1.

A.8.2 *Classification Criteria for Substances*

A.8.2.1 *Substances of Category 1 and Category 2*

A.8.2.1.1 *Substances shall be classified for immediate* or delayed effects separately, by the use of expert judgment on the basis of the weight of all evidence available, including the use of recommended guidance values (See A.8.2.1.9). Substances shall then be classified in Category 1 or 2, depending upon the nature and severity of the effect(s) observed, in accordance with Figure A.8.1.

Figure A.8.1 — Hazard Categories for Specific Target Organ Toxicity Following Single Exposure

CATEGORY 1: Substances that have produced significant toxicity in humans, or that, on the basis of evidence from studies in experimental animals can be presumed to have the potential to produce significant toxicity in humans following single exposure Substances are classified in Category 1 for STOT-SE on the basis of: (a) reliable and good quality evidence from human cases or epidemiological studies; or (b) observations from appropriate studies in experimental animals in which significant and/or severe toxic effects of relevance to human health were produced at generally low exposure concentrations. Guidance dose/concentration values are provided below (See A.8.2.1.9) to be used as part of weight-of-evidence evaluation.
CATEGORY 2: Substances that, on the basis of evidence from studies in experimental animals, can be presumed to have the potential to be harmful to human health following single exposure Substances are classified in Category 2 for STOT-SE on the basis of observations from appropriate studies in experimental animals in which significant toxic effects, of relevance to human health, were produced at generally moderate exposure concentrations. Guidance dose/concentration values are provided below (See A.8.2.1.9) in order to help in classification. In exceptional cases, human evidence can also be used to place a substance in Category 2 (See A.8.2.1.6).
CATEGORY 3: Transient target organ effects There are target organ effects for which a substance does not meet the criteria to be classified in Categories 1 or 2 indicated above. These are effects which adversely alter human function for a short duration after exposure and from which humans may recover in a reasonable period without leaving significant alteration of structure or function. This category only includes narcotic effects and respiratory tract irritation. Substances are classified specifically for these effects as discussed in A.8.2.2.
Note: The primary target organ/system shall be identified where possible, and where this is not possible, the substance shall be identified as a general toxicant. The data shall be evaluated and, where possible, shall not include secondary effects (e.g., a hepatotoxicant can produce secondary effects in the nervous or gastro-intestinal systems).

A.8.2.1.2 *The relevant route(s) of exposure* by which the classified substance produces damage shall be identified.

A.8.2.1.3 *Classification is determined by expert* judgment, on the basis of the weight of all evidence available including the guidance presented below.

A.8.2.1.4 *Weight of evidence of all available* data, including human incidents, epidemiology, and studies conducted in experimental animals is used to substantiate specific target organ toxic effects that merit classification.

A.8.2.1.5 *The information required to evaluate* specific target organ toxicity comes either from single exposure in humans (e.g., exposure at home, in the workplace or environmentally), or from studies conducted in experimental animals. The standard animal studies in rats or mice that provide this information are acute toxicity studies which can include clinical observations and detailed macroscopic and microscopic examination to enable the toxic effects on target tissues/organs to be identified. Results of acute toxicity studies conducted in other species may also provide relevant information.

A.8.2.1.6 *In exceptional cases,* based on expert judgment, it may be appropriate to place certain substances with human evidence of target organ toxicity in Category 2: (a) when the weight of human evidence is not sufficiently convincing to warrant Category 1 classification, and/or (b) based on the nature and severity of effects. Dose/concentration levels in humans shall not be considered in the classification and any available evidence from animal studies shall be consistent with the Category 2 classification. In other words, if there are also animal data available on the substance that warrant Category 1 classification, the chemical shall be classified as Category 1.

A.8.2.1.7 *Effects considered to support* classification for Category 1 and 2

A.8.2.1.7.1 *Classification is supported by evidence* associating single exposure to the substance with a consistent and identifiable toxic effect.

A.8.2.1.7.2 *Evidence from human experience/incidents* is usually restricted to reports of adverse health consequences, often with uncertainty about exposure conditions, and may not provide the scientific detail that can be obtained from well-conducted studies in experimental animals.

A.8.2.1.7.3 *Evidence from appropriate studies* in experimental animals can furnish much more detail, in the form of clinical observations, and macroscopic and microscopic pathological examination and this can often reveal hazards that may not be life-threatening but could indicate functional impairment. Consequently all available evidence, and evidence relevance to human health, must be taken into consideration in the classification process. Relevant toxic effects in humans and/or animals include, but are not limited to:

(a) *Morbidity resulting from single exposure;*

(b) *Significant functional changes,* more than transient in nature, in the respiratory system, central or peripheral nervous systems, other organs or other organ systems, including signs of central nervous system depression and effects on special senses (e.g., sight, hearing and sense of smell);

(c) *Any consistent and significant adverse* change in clinical biochemistry, hematology, or urinalysis parameters;

(d) *Significant organ damage* that may be noted at necropsy and/or subsequently seen or confirmed at microscopic examination;

(e) *Multi-focal or diffuse necrosis,* fibrosis or granuloma formation in vital organs with regenerative capacity;

(f) *Morphological changes that are potentially* reversible but provide clear evidence of marked organ dysfunction; and,

(g) *Evidence of appreciable cell* death (including cell degeneration and reduced cell number) in vital organs incapable of regeneration.

A.8.2.1.8 *Effects considered not to support* classification for Category 1 and 2

Effects may be seen in humans and/or animals that do not justify classification. Such effects include, but are not limited to:

(a) *Clinical observations or small* changes in bodyweight gain, food consumption or water intake that may have some toxicological importance but that do not, by themselves, indicate "significant" toxicity;

(b) *Small changes in clinical* biochemistry, hematology or urinalysis parameters and/or transient effects, when such changes or effects are of doubtful or of minimal toxicological importance;

(c) *Changes in organ weights* with no evidence of organ dysfunction;

(d) *Adaptive responses that are not considered* toxicologically relevant; and,

(e) *Substance-induced species-specific mechanisms* of toxicity, i.e., demonstrated with reasonable certainty to be not relevant for human health, shall not justify classification.

A.8.2.1.9 *Guidance values to assist* with classification based on the results obtained from studies conducted in experimental animals for Category 1 and 2

A.8.2.1.9.1 *In order to help reach* a decision about whether a substance shall be classified or not, and to what degree it shall be classified (Category 1 vs. Category 2), dose/concentration "guidance values" are provided for consideration of the dose/concentration which has been shown to produce significant health effects. The principal argument for proposing such guidance values is that all chemicals are potentially toxic and there has to be a reasonable dose/concentration above which a degree of toxic effect is acknowledged.

A.8.2.1.9.2 *Thus, in animal studies,* when significant toxic effects are observed that indicate classification, consideration of the dose/concentration at which these effects were seen, in relation to the suggested guidance values, provides useful information to help assess the need to classify (since the toxic effects are a consequence of the hazardous property(ies) and also the dose/concentration).

A.8.2.1.9.3 *The guidance value (C)* ranges for single-dose exposure which has produced a significant non-lethal toxic effect are those applicable to acute toxicity testing, as indicated in Table A.8.1.

Table A.8.1 — Guidance Value Ranges for Single-Dose Exposures

Route of exposure	Units	Guidance value ranges for:		
		Category 1	Category 2	Category 3
Oral (rat)	mg/kg body weight	C ≤300	2000 ≥C >300	Guidance values do not apply.
Dermal (rat or rabbit)	mg/kg body weight	C ≤1,000	2000 ≥C >1,000	
Inhalation (rat) gas	ppmV/4h	C ≤2,500	20,000 ≥C >2,500	
Inhalation (rat) vapor	mg/1/4h	C ≤10	20 ≥C >10	
Inhalation (rat) dust/ mist/fume	mg/l/4h	C ≤1.0	5.0 ≥C >1.0	

A.8.2.1.9.4 The guidance values and ranges mentioned in Table A.8.1 are intended only for guidance purposes, i.e., to be used as part of the weight of evidence approach, and to assist with decisions about classification. They are not intended as strict demarcation values. Guidance values are not provided for Category 3 since this classification is primarily based on human data; animal data may be included in the weight of evidence evaluation.

A.8.2.1.9.5 Thus, it is feasible that a specific profile of toxicity occurs at a dose/concentration below the guidance value, e.g., <2000 mg/kg body weight by the oral route, however the nature of the effect may result in the decision not to classify. Conversely, a specific profile of toxicity may be seen in animal studies occurring at above a guidance value, e.g., [ge]2000 mg/kg body weight by the oral route, and in addition there is supplementary information from other sources, e.g., other single dose studies, or human case experience, which supports a conclusion that, in view of the weight of evidence, classification is the prudent action to take.

A.8.2.1.10 Other considerations

A.8.2.1.10.1 When a substance is characterized only by use of animal data the classification process includes reference to dose/concentration guidance values as one of the elements that contribute to the weight of evidence approach.

A.8.2.1.10.2 When well-substantiated human data are available showing a specific target organ toxic effect that can be reliably attributed to single exposure to a substance, the substance shall be classified. Positive human data, regardless of probable dose, predominates over animal data. Thus, if a substance is unclassified because specific target organ toxicity observed was considered not relevant or significant to humans, if subsequent human incident data become available showing a specific target organ toxic effect, the substance shall be classified.

A.8.2.1.10.3 A substance that has not been tested for specific target organ toxicity shall, where appropriate, be classified on the basis of data from a scientifically validated structure activity relationship and expert judgment-based extrapolation from a structural analogue that has previously been classified together with substantial support from consideration of other important factors such as formation of common significant metabolites.

A.8.2.2 Substances of Category 3

A.8.2.2.1 Criteria for respiratory tract irritation

The criteria for classifying substances as Category 3 for respiratory tract irritation are:

(a) Respiratory irritant effects (characterized by localized redness, edema, pruritis and/or pain) that impair function with symptoms such as cough, pain, choking, and breathing difficulties are included. It is recognized that this evaluation is based primarily on human data;

(b) Subjective human observations supported by objective measurements of clear respiratory tract irritation (RTI) (e.g., electrophysiological responses, biomarkers of inflammation in nasal or bronchoalveolar lavage fluids);

(c) The symptoms observed in humans shall also be typical of those that would be produced in the exposed population rather than being an isolated idiosyncratic reaction or response triggered only in individuals with hypersensitive airways. Ambiguous reports simply of "irritation" should be excluded as this term is commonly used to describe a wide range of sensations including those such as smell, unpleasant taste, a tickling sensation, and dryness, which are outside the scope of classification for respiratory tract irritation;

(d) There are currently no scientifically validated animal tests that deal specifically with RTI; however, useful information may be obtained from the single and repeated inhalation toxicity tests. For example, animal studies may provide useful information in terms of clinical signs of toxicity (dyspnoea, rhinitis etc) and histopathology (e.g., hyperemia, edema, minimal inflammation, thickened mucous layer) which are reversible and may be reflective of the characteristic clinical symptoms described above. Such animal studies can be used as part of weight of evidence evaluation; and,

(e) This special classification will occur only when more severe organ effects including the respiratory system are not observed as those effects would require a higher classification.

A.8.2.2.2 Criteria for narcotic effects

The criteria for classifying substances in Category 3 for narcotic effects are:

(a) Central nervous system depression including narcotic effects in humans such as drowsiness, narcosis, reduced alertness, loss of reflexes, lack of coordination, and vertigo are included. These effects can also be manifested as severe headache or nausea, and can lead to reduced judgment, dizziness, irritability, fatigue, impaired memory function, deficits in perception and coordination, reaction time, or sleepiness; and,

(b) Narcotic effects observed in animal studies may include lethargy, lack of coordination righting reflex, narcosis, and ataxia. If these effects are not transient in nature, then they shall be considered for classification as Category 1 or 2.

A.8.3 Classification Criteria for Mixtures

A.8.3.1 Mixtures are classified using the same criteria as for substances, or alternatively as described below. As with substances, mixtures may be classified for specific target organ toxicity following single exposure, repeated exposure, or both.

A.8.3.2 Classification of Mixtures When Data Are Available for the Complete Mixture

When reliable and good quality evidence from human experience or appropriate studies in experimental animals, as described in the criteria for substances, is available for the mixture, then the mixture shall be classified by weight of evidence evaluation of this data. Care shall be exercised in evaluating data on mixtures, that the dose, duration, observation or analysis, do not render the results inconclusive.

A.8.3.3 Classification of Mixtures When Data Are Not Available for the Complete Mixture: Bridging Principles

A.8.3.3.1 Where the mixture itself has not been tested to determine its specific target organ toxicity, but there are sufficient data on both the individual ingredients and similar tested mixtures to adequately characterize the hazards of the mixture, these data shall be used in accordance with the following bridging principles as found in paragraph A.0.5 of this Appendix: Dilution, Batching, Concentration of mixtures, Interpolation within one toxicity category, Substantially similar mixtures, or Aerosols.

A.8.3.4 Classification of Mixtures When Data Are Available for All Ingredients or Only for Some Ingredients of the Mixture

A.8.3.4.1 Where there is no reliable evidence or test data for the specific mixture itself, and the bridging principles cannot be used to enable classification, then classification of the mixture is based on the classification of the ingredient substances. In this case, the mixture shall be classified as a specific target organ toxicant (specific organ specified), following single exposure, repeated exposure, or both when at least one ingredient has been classified as a Category 1 or Category 2 specific target organ toxicant and is present at or above the appropriate cut-off value/concentration limit specified in Table A.8.2 for Categories 1 and 2, respectively.

Table A.8.2 — Cut-Off Values/Concentration Limits of Ingredients of a Mixture Classified as a Specific Target Organ Toxicant That Would Trigger Classification of the Mixture as Category 1 or 2

Ingredient classified as:	Cut-off values/concentration limits triggering classification of a mixture as:	
	Category 1	Category 2
Category 1 Target organ toxicant	≥1.0%	
Category 2 Target organ toxicant		≥1.0%

A.8.3.4.2 These cut-off values and consequent classifications shall be applied equally and appropriately to both single- and repeated-dose target organ toxicants.

A.8.3.4.3 Mixtures shall be classified for either or both single and repeated dose toxicity independently.

A.8.3.4.4 Care shall be exercised when toxicants affecting more than one organ system are combined that the potentiation or synergistic interactions are considered, because certain substances can cause target organ toxicity at <1% concentration when other ingredients in the mixture are known to potentiate its toxic effect.

A.8.3.4.5 Care shall be exercised when extrapolating the toxicity of a mixture that contains Category 3 ingredient(s). A cut-off value/concentration limit of 20%, considered as an additive of all Category 3 ingredients for each hazard endpoint, is appropriate; however, this cut-off value/concentration limit may be higher or lower depending on the Category 3 ingredient(s) involved and the fact that some effects such as respiratory tract irritation may not occur below a certain concentration while other effects such as narcotic effects may occur below this 20% value. Expert judgment shall be exercised. Respiratory tract irritation and narcotic effects are to be evaluated separately in accordance with the criteria given in A.8.2.2. When conducting classifications for these hazards, the contribution of each ingredient should be considered additive, unless there is evidence that the effects are not additive.

A.9 SPECIFIC TARGET ORGAN TOXICITY REPEATED OR PROLONGED EXPOSURE

A.9.1 *Definitions and general considerations*

A.9.1.1 **Specific target organ toxicity — repeated exposure (STOT-RE)** means specific target organ toxicity arising from repeated exposure to a substance or mixture. All significant health effects that can impair function, both reversible and irreversible, immediate and/or delayed and not specifically addressed in A.1 to A.7 and A.10 of this Appendix are included. Specific target organ toxicity following a single-event exposure is classified in accordance with *SPECIFIC TARGET ORGAN TOXICITY — SINGLE EXPOSURE* (A.8 of this Appendix) and is therefore not included here.

A.9.1.2 *Classification identifies the substance* or mixture as being a specific target organ toxicant and, as such, it may present a potential for adverse health effects in people who are exposed to it.

A.9.1.3 *These adverse health* effects produced by repeated exposure include consistent and identifiable toxic effects in humans, or, in experimental animals, toxicologically significant changes which have affected the function or morphology of a tissue/organ, or have produced serious changes to the biochemistry or hematology of the organism and these changes are relevant for human health. Human data will be the primary source of evidence for this hazard class.

A.9.1.4 *Assessment shall take into consideration* not only significant changes in a single organ or biological system but also generalized changes of a less severe nature involving several organs.

A.9.1.5 *Specific target organ* toxicity can occur by any route that is relevant for humans, e.g., principally oral, dermal or inhalation.

A.9.2 *Classification Criteria for Substances*

A.9.2.1 *Substances shall be classified as STOT-RE* by expert judgment on the basis of the weight of all evidence available, including the use of recommended guidance values which take into account the duration of exposure and the dose/concentration which produced the effect(s), (See A.9.2.9). Substances shall be placed in one of two categories, depending upon the nature and severity of the effect(s) observed, in accordance with Figure A.9.1.

Figure A.9.1 — Hazard Categories for Specific Target Organ Toxicity Following Repeated Exposure

CATEGORY 1: Substances that have produced significant toxicity in humans, or that, on the basis of evidence from studies in experimental animals can be presumed to have the potential to produce significant toxicity in humans following repeated or prolonged exposure
Substances are classified in Category 1 for specific target organ toxicity (repeated exposure) on the basis of:
(a) reliable and good quality evidence from human cases or epidemiological studies; or,
(b) observations from appropriate studies in experimental animals in which significant and/or severe toxic effects, of relevance to human health, were produced at generally low exposure concentrations. Guidance dose/concentration values are provided below (See A.9.2.9) to be used as part of weight-of-evidence evaluation.
CATEGORY 2: Substances that, on the basis of evidence from studies in experimental animals can be presumed to have the potential to be harmful to human health following repeated or prolonged exposure
Substances are classified in Category 2 for specific target organ toxicity (repeated exposure) on the basis of observations from appropriate studies in experimental animals in which significant toxic effects, of relevance to human health, were produced at generally moderate exposure concentrations. Guidance dose/concentration values are provided below (See A.9.2.9) in order to help in classification.
In exceptional cases human evidence can also be used to place a substance in Category 2 (See A.9.2.6).
Note: The primary target organ/system shall be identified where possible, or the substance shall be identified as a general toxicant. The data shall be carefully evaluated and, where possible, shall not include secondary effects (e.g., a hepatotoxicant can produce secondary effects in the nervous or gastro-intestinal systems).

A.9.2.2 *The relevant route of exposure* by which the classified substance produces damage shall be identified.

A.9.2.3 *Classification is determined by expert* judgment, on the basis of the weight of all evidence available including the guidance presented below.

A.9.2.4 *Weight of evidence of all data,* including human incidents, epidemiology, and studies conducted in experimental animals, is used to substantiate specific target organ toxic effects that merit classification.

A.9.2.5 *The information required to evaluate* specific target organ toxicity comes either from repeated exposure in humans, e.g., exposure at home, in the workplace or environmentally, or from studies conducted in experimental animals. The standard animal studies in rats or mice that provide this information are 28 day, 90 day or lifetime studies (up to 2 years) that include hematological, clinico-chemical and detailed macroscopic and microscopic examination to enable the toxic effects on target tissues/organs to be identified. Data from repeat dose studies performed in other species may also be used. Other long-term exposure studies, e.g., for carcinogenicity, neurotoxicity or reproductive toxicity, may also provide evidence of specific target organ toxicity that could be used in the assessment of classification.

A.9.2.6 *In exceptional cases,* based on expert judgment, it may be appropriate to place certain substances with human evidence of specific target organ toxicity in Category 2: (a) when the weight of human evidence is not sufficiently convincing to warrant Category 1 classification, and/or (b) based on the nature and severity of effects. Dose/concentration levels in humans shall not be considered in the classification and any available evidence from animal studies shall be consistent with the Category 2 classification. In other words, if there are also animal data available on the substance that warrant Category 1 classification, the substance shall be classified as Category 1.

A.9.2.7 *Effects Considered To Support Classification*

A.9.2.7.1 Classification is supported by reliable evidence associating repeated exposure to the substance with a consistent and identifiable toxic effect.

A.9.2.7.2 Evidence from human experience/incidents is usually restricted to reports of adverse health consequences, often with uncertainty about exposure conditions, and may not provide the scientific detail that can be obtained from well-conducted studies in experimental animals.

A.9.2.7.3 Evidence from appropriate studies in experimental animals can furnish much more detail, in the form of clinical observations, hematology, clinical chemistry, macroscopic and microscopic pathological examination and this can often reveal hazards that may not be life-threatening but could indicate functional impairment. Consequently all available evidence, and relevance to human health, must be taken into consideration in the classification process.

Relevant toxic effects in humans and/or animals include, but are not limited to:

(a) *Morbidity or death resulting* from repeated or long-term exposure. Morbidity or death may result from repeated exposure, even to relatively low doses/concentrations, due to bioaccumulation of the substance or its metabolites, or due to the overwhelming of the de-toxification process by repeated exposure;

(b) *Significant functional changes* in the central or peripheral nervous systems or other organ systems, including signs of central nervous system depression and effects on special senses (e.g., sight, hearing and sense of smell);

(c) *Any consistent and significant adverse* change in clinical biochemistry, hematology, or urinalysis parameters;

(d) *Significant organ damage* that may be noted at necropsy and/or subsequently seen or confirmed at microscopic examination;

(e) *Multi-focal or diffuse necrosis,* fibrosis or granuloma formation in vital organs with regenerative capacity;

(f) *Morphological changes that are potentially* reversible but provide clear evidence of marked organ dysfunction (e.g., severe fatty change in the liver); and,

(g) *Evidence of appreciable cell* death (including cell degeneration and reduced cell number) in vital organs incapable of regeneration.

A.9.2.8 *Effects Considered Not To Support Classification*

Effects may be seen in humans and/or animals that do not justify classification. Such effects include, but are not limited to:

(a) *Clinical observations or small* changes in bodyweight gain, food consumption or water intake that may have some toxicological importance but that do not, by themselves, indicate "significant" toxicity;

(b) *Small changes in clinical* biochemistry, hematology or urinalysis parameters and/or transient effects, when such changes or effects are of doubtful or of minimal toxicological importance;

(c) *Changes in organ weights* with no evidence of organ dysfunction;

(d) *Adaptive responses that are not considered* toxicologically relevant;

(e) *Substance-induced species-specific mechanisms* of toxicity, i.e., demonstrated with reasonable certainty to be not relevant for human health, shall not justify classification.

A.9.2.9 *Guidance Values To Assist* With Classification Based on the Results Obtained From Studies Conducted in Experimental Animals

A.9.2.9.1 In studies conducted in experimental animals, reliance on observation of effects alone, without reference to the duration of experimental exposure and dose/concentration, omits a fundamental concept of toxicology, i.e., all substances are potentially toxic, and what determines the toxicity is a function of the dose/concentration and the duration of exposure. In most studies conducted in experimental animals the test guidelines use an upper limit dose value.

A.9.2.9.2 In order to help reach a decision about whether a substance shall be classified or not, and to what degree it shall be classified (Category 1 vs. Category 2), dose/concentration "guidance values" are provided in Table A.9.1 for consideration of the dose/concentration which has been shown to produce significant health effects. The principal argument for proposing such guidance values is that all chemicals are potentially toxic and there has to be a reasonable dose/concentration above which a degree of toxic effect is acknowledged. Also, repeated-dose studies conducted in experimental animals are designed to produce toxicity at the highest dose used in order to optimize the test objective and so most studies will reveal some toxic effect at least at this highest dose. What is therefore to be decided is not only what effects have been produced, but also at what dose/concentration they were produced and how relevant is that for humans.

A.9.2.9.3 Thus, in animal studies, when significant toxic effects are observed that indicate classification, consideration of the duration of experimental exposure and the dose/concentration at which these effects were seen, in relation to the suggested guidance values, provides useful information to help assess the need to classify (since the toxic effects are a consequence of the hazardous property(ies) and also the duration of exposure and the dose/concentration).

A.9.2.9.4 The decision to classify at all can be influenced by reference to the dose/concentration guidance values at or below which a significant toxic effect has been observed.

A.9.2.9.5 The guidance values refer to effects seen in a standard 90-day toxicity study conducted in rats. They can be used as a basis to extrapolate equivalent guidance values for toxicity studies of greater or lesser duration, using dose/exposure time extrapolation similar to Haber's rule for inhalation, which states essentially that the effective dose is directly proportional to the exposure concentration and the duration of exposure. The assessment should be done on a case-by-case basis; for example, for a 28-day study the guidance values below would be increased by a factor of three.

A.9.2.9.6 Thus for Category 1 classification, significant toxic effects observed in a 90-day repeated-dose study conducted in experimental animals and seen to occur at or below the (suggested) guidance values (C) as indicated in Table A.9.1 would justify classification:

Table A.9.1 — Guidance Values To Assist in Category 1 Classification
[Applicable to a 90-day study]

Route of exposure	Units	Guidance values (dose/concentration)
Oral (rat)	mg/kg body weight/day	C ≤10.
Dermal (rat or rabbit)	mg/kg body weight/day	C ≤20.
Inhalation (rat) gas	ppmV/6h/day	C ≤50.
Inhalation (rat) vapor	mg/liter/6h/day	C ≤0.2.
Inhalation (rat) dust/mist/fume	mg/liter/6h/day	C ≤0.02.

A.9.2.9.7 For Category 2 classification, significant toxic effects observed in a 90-day repeated-dose study conducted in experimental animals and seen to occur within the (suggested) guidance value ranges as indicated in Table A.9.2 would justify classification:

Table A.9.2 — Guidance Values To Assist in Category 2 Classification
[Applicable to a 90-day study]

Route of exposure	Units	Guidance values (dose/concentration)
Oral (rat)	mg/kg body weight/day	10 <C ≤100.
Dermal (rat or rabbit)	mg/kg body weight/day	20 <C ≤200.
Inhalation (rat) gas	ppmV/6h/day	50 <C ≤250.
Inhalation (rat) vapor	mg/liter/6h/day	0.2 <C ≤1.0.
Inhalation (rat) dust/mist/fume	mg/liter/6h/day	0.02 <C ≤0.2.

A.9.2.9.8 The guidance values and ranges mentioned in A.2.9.9.6 and A.2.9.9.7 are intended only for guidance purposes, i.e., to be used as part of the weight of evidence approach, and to assist with decisions about classification. They are not intended as strict demarcation values.

A.9.2.9.9 Thus, it is possible that a specific profile of toxicity occurs in repeat-dose animal studies at a dose/concentration below the guidance value, e.g., <100 mg/kg body weight/day by the oral route, however the nature of the effect, e.g., nephrotoxicity seen only in male rats of a particular strain known to be susceptible to this effect, may result in the decision not to classify. Conversely, a specific profile of toxicity may be seen in animal studies occurring at above a guidance value, e.g., [ge]100 mg/kg body weight/day by the oral route, and in addition there is supplementary information from other sources, e.g., other long-term administration studies, or human case experience, which supports a conclusion that, in view of the weight of evidence, classification is prudent.

A.9.2.10 *Other Considerations*

A.9.2.10.1 When a substance is characterized only by use of animal data the classification process includes reference to dose/concentration guidance values as one of the elements that contribute to the weight of evidence approach.

A.9.2.10.2 When well-substantiated human data are available showing a specific target organ toxic effect that can be reliably attributed to repeated or prolonged exposure to a substance, the substance shall be classified. Positive human data, regardless of probable dose, predominates over animal data. Thus, if a substance is unclassified because no specific target organ toxicity was seen at or below the dose/concentration guidance value for animal testing, if subsequent human incident data become available showing a specific target organ toxic effect, the substance shall be classified.

A.9.2.10.3 A substance that has not been tested for specific target organ toxicity may in certain instances, where appropriate, be classified on the basis of data from a scientifically validated structure activity relationship and expert judgment-based extrapolation from a structural analogue that has previously been classified together with substantial support from consideration of other important factors such as formation of common significant metabolites.

A.9.3 *Classification Criteria for Mixtures*

A.9.3.1 *Mixtures are classified using the same* criteria as for substances, or alternatively as described below. As with substances, mixtures may be classified for specific target organ toxicity following single exposure, repeated exposure, or both.

A.9.3.2 *Classification of Mixtures When Data* Are Available for the Complete Mixture

When reliable and good quality evidence from human experience or appropriate studies in experimental animals, as described in the criteria for substances, is available for the mixture, then the mixture shall be classified by weight of evidence evaluation of these data. Care shall be exercised in evaluating data on mixtures, that the dose, duration, observation or analysis, do not render the results inconclusive.

A.9.3.3 *Classification of Mixtures When Data* Are Not Available for the Complete Mixture: Bridging Principles

A.9.3.3.1 Where the mixture itself has not been tested to determine its specific target organ toxicity, but there are sufficient data on both the individual ingredients and similar tested mixtures to adequately characterize the hazards of the mixture, these data shall be used in accordance with the following bridging principles as found in paragraph A.0.5 of this Appendix: Dilution; Batching; Concentration of mixtures; Interpolation within one toxicity category; Substantially similar mixtures; and Aerosols.

A.9.3.4 *Classification of Mixtures When Data* Are Available for All Ingredients or Only for Some Ingredients of the Mixture

A.9.3.4.1 Where there is no reliable evidence or test data for the specific mixture itself, and the bridging principles cannot be used to enable classification, then classification of the mixture is based on the classification of the ingredient substances. In this case, the mixture shall be classified as a specific target organ toxicant (specific organ specified), following single exposure, repeated exposure, or both when at least one ingredient has been classified as a Category 1 or Category 2 specific target organ toxicant and is present at or above the appropriate cut-off value/concentration limit specified in Table A.9.3 for Category 1 and 2 respectively.

Table A.9.3 — Cut-Off Value/Concentration Limits of Ingredients of a Mixture Classified as a Specific Target Organ Toxicant That Would Trigger Classification of the Mixture as Category 1 or 2

Ingredient classified as:	Cut-off values/concentration limits triggering classification of a mixture as:	
	Category 1	Category 2
Category 1 Target organ toxicant	≥1.0%	
Category 2 Target organ toxicant		≥1.0%

A.9.3.4.2 These cut-off values and consequent classifications shall be applied equally and appropriately to both single- and repeated-dose target organ toxicants.

A.9.3.4.3 Mixtures shall be classified for either or both single- and repeated-dose toxicity independently.

A.9.3.4.4 Care shall be exercised when toxicants affecting more than one organ system are combined that the potentiation or synergistic interactions are considered, because certain substances can cause specific target organ toxicity at <1% concentration when other ingredients in the mixture are known to potentiate its toxic effect.

A.10 ASPIRATION HAZARD

A.10.1 *Definitions and General and Specific Considerations*

A.10.1.1 **Aspiration** means the entry of a liquid or solid chemical directly through the oral or nasal cavity, or indirectly from vomiting, into the trachea and lower respiratory system.

A.10.1.2 *Aspiration toxicity includes* severe acute effects such as chemical pneumonia, varying degrees of pulmonary injury or death following aspiration.

A.10.1.3 *Aspiration is initiated at the moment* of inspiration, in the time required to take one breath, as the causative material lodges at the crossroad of the upper respiratory and digestive tracts in the laryngopharyngeal region.

A.10.1.4 *Aspiration of a substance or mixture* can occur as it is vomited following ingestion. This may have consequences for labeling, particularly where, due to acute toxicity, a recommendation may be considered to induce vomiting after ingestion. However, if the substance/mixture also presents an aspiration toxicity hazard, the recommendation to induce vomiting may need to be modified.

A.10.1.5 *Specific Considerations*

A.10.1.5.1 The classification criteria refer to kinematic viscosity. The following provides the conversion between dynamic and kinematic viscosity:

$$\frac{\text{Dynamic viscosity (mPa·s)}}{\text{Density (g/cm}^3)} = \text{Kinematic viscosity (mm}^2\text{/s)}$$

A.10.1.5.2 Although the definition of aspiration in A.10.1.1 includes the entry of solids into the respiratory system, classification according to (b) in table A.10.1 for Category 1 is intended to apply to liquid substances and mixtures only.

A.10.1.5.3 Classification of aerosol/mist products.

Aerosol and mist products are usually dispensed in containers such as self-pressurized containers, trigger and pump sprayers. Classification for these products shall be considered if their use may form a pool of product in the mouth, which then may be aspirated. If the mist or aerosol from a pressurized container is fine, a pool may not be formed. On the other hand, if a pressurized container dispenses product in a stream, a pool may be formed that may then be aspirated. Usually, the mist produced by trigger and pump sprayers is coarse and therefore, a pool may be formed that then may be aspirated. When the pump mechanism may be removed and contents are available to be swallowed then the classification of the products should be considered.

A.10.2 *Classification Criteria for Substances*

Table A.10.1 — Criteria for Aspiration Toxicity

Category	Criteria
Category 1: Chemicals known to cause human aspiration toxicity hazards or to be regarded as if they cause human aspiration toxicity hazard	A substance shall be classified in Category 1: (a) If reliable and good quality human evidence indicates that it causes aspiration toxicity (See note); or (b) If it is a hydrocarbon and has a kinematic viscosity [le]20.5 mm²/s, measured at 40 °C.

Note: Examples of substances included in Category 1 are certain hydrocarbons, turpentine and pine oil.

A.10.3 *Classification Criteria for Mixtures*

A.10.3.1 *Classification When Data Are Available* for the Complete Mixture

A mixture shall be classified in Category 1 based on reliable and good quality human evidence.

A.10.3.2 *Classification of Mixtures When Data* Are Not Available for the Complete Mixture: Bridging Principles

A.10.3.2.1 Where the mixture itself has not been tested to determine its aspiration toxicity, but there are sufficient data on both the individual ingredients and similar tested mixtures to adequately characterize the hazard of the mixture, these data shall be used in accordance with the following bridging principles as found in paragraph A.0.5 of this Appendix: Dilution; Batching; Concentration of mixtures; Interpolation within one toxicity category; and Substantially similar mixtures. For application of the dilution bridging principle, the concentration of aspiration toxicants shall not be less than 10%.

A.10.3.3 *Classification of Mixtures When Data* Are Available for All Ingredients or Only for Some Ingredients of the Mixture

A.10.3.3.1 A mixture which contains [ge]10% of an ingredient or ingredients classified in Category 1, and has a kinematic viscosity [le]20.5 mm^2/s, measured at 40 °C, shall be classified in Category 1.

A.10.3.3.2 In the case of a mixture which separates into two or more distinct layers, one of which contains [ge]10% of an ingredient or ingredients classified in Category 1 and has a kinematic viscosity [le]20.5 mm^2/s, measured at 40 °C, then the entire mixture shall be classified in Category 1.

§1910.1200 Appendix B

PHYSICAL CRITERIA (MANDATORY)

B.1 EXPLOSIVES

B.1.1 *Definitions and General Considerations*

B.1.1.1 *An explosive chemical is a solid* or liquid chemical which is in itself capable by chemical reaction of producing gas at such a temperature and pressure and at such a speed as to cause damage to the surroundings. Pyrotechnic chemicals are included even when they do not evolve gases.

A *pyrotechnic chemical* is a chemical designed to produce an effect by heat, light, sound, gas or smoke or a combination of these as the result of non-detonative self-sustaining exothermic chemical reactions.

An *explosive item* is an item containing one or more explosive chemicals.

A *pyrotechnic item* is an item containing one or more pyrotechnic chemicals.

An *unstable explosive* is an explosive which is thermally unstable and/or too sensitive for normal handling, transport, or use.

An *intentional explosive* is a chemical or item which is manufactured with a view to produce a practical explosive or pyrotechnic effect.

B.1.1.2 *The class of explosives comprises:*

(a) *Explosive chemicals;*

(b) *Explosive items,* except devices containing explosive chemicals in such quantity or of such a character that their inadvertent or accidental ignition or initiation shall not cause any effect external to the device either by projection, fire, smoke, heat or loud noise; and

(c) *Chemicals and items not included* under (a) and (b) above which are manufactured with the view to producing a practical explosive or pyrotechnic effect.

B.1.2 *Classification Criteria*

Chemicals and items of this class shall be classified as unstable explosives or shall be assigned to one of the following six divisions depending on the type of hazard they present:

Division 1.1 — Chemicals and items which have a mass explosion hazard (a mass explosion is one which affects almost the entire quantity present virtually instantaneously);

Division 1.2 — Chemicals and items which have a projection hazard but not a mass explosion hazard;

Division 1.3 — Chemicals and items which have a fire hazard and either a minor blast hazard or a minor projection hazard or both, but not a mass explosion hazard:

(i) *Combustion of which gives* rise to considerable radiant heat; or

(ii) *Which burn one* after another, producing minor blast or projection effects or both;

Division 1.4 — Chemicals and items which present no significant hazard: chemicals and items which present only a small hazard in the event of ignition or initiation. The effects are largely confined to the package and no projection of fragments of appreciable size or range is to be expected. An external fire shall not cause virtually instantaneous explosion of almost the entire contents of the package;

Division 1.5 — Very insensitive chemicals which have a mass explosion hazard: chemicals which have a mass explosion hazard but are so insensitive that there is very little probability of initiation or of transition from burning to detonation under normal conditions;

Division 1.6 — Extremely insensitive items which do not have a mass explosion hazard: items which contain only extremely insensitive detonating chemicals and which demonstrate a negligible probability of accidental initiation or propagation.

B.1.3 *Additional Classification Considerations*

B.1.3.1 *Explosives shall be classified as unstable* explosives or shall be assigned to one of the six divisions identified in B.1.2 in accordance with the three step procedure in Part I of the UN ST/SG/AC.10 (incorporated by reference; See Sec. 1910.6). The first step is to ascertain whether the substance or mixture has explosive effects (Test Series 1). The second step is the acceptance procedure (Test Series 2 to 4) and the third step is the assignment to a hazard division (Test Series 5 to 7). The assessment whether a candidate for "ammonium nitrate emulsion or suspension or gel, intermediate for blasting explosives (ANE)" is insensitive enough for inclusion as an oxidizing liquid (See B.13) or an oxidizing solid (See B.14) is determined by Test Series 8 tests.

Note: Classification of solid chemicals shall be based on tests performed on the chemical as presented. If, for example, for the purposes of supply or transport, the same chemical is to be presented in a physical form different from that which was tested and which is considered likely to materially alter its performance in a classification test, classification must be based on testing of the chemical in the new form.

B.1.3.2 *Explosive properties are associated* with the presence of certain chemical groups in a molecule which can react to produce very rapid increases in temperature or pressure. The screening procedure in B.1.3.1 is aimed at identifying the presence of such reactive groups and the potential for rapid energy release. If the screening procedure identifies the chemical as a potential explosive, the acceptance procedure (See section 10.3 of the UN ST/SG/AC.10 (incorporated by reference; See Sec. 1910.6)) is necessary for classification.

Note: Neither a Series 1 type (a) propagation of detonation test nor a Series 2 type (a) test of sensitivity to detonative shock is necessary if the exothermic decomposition energy of organic materials is less than 800 J/g.

B.1.3.3 *If a mixture contains any known* explosives, the acceptance procedure is necessary for classification.

B.1.3.4 *A chemical is not classified as explosive if:*

(a) *There are no chemical groups* associated with explosive properties present in the molecule. Examples of groups which may indicate explosive properties are given in Table A6.1 in Appendix 6 of the UN ST/SG/AC.10 (incorporated by reference; See Sec. 1910.6); or

(b) *The substance contains chemical* groups associated with explosive properties which include oxygen and the calculated oxygen balance is less than - 200.

The oxygen balance is calculated for the chemical reaction:

$C_xH_yO_z$+ [x + (y/4) - (z/2)] O_2[rarr] x. CO_2+ (y/2) H_2O using the formula:

oxygen balance = - 1600 [2x +(y/2) - z]/molecular weight;or

(c) *The organic substance or a homogenous* mixture of organic substances contains chemical groups associated with explosive properties but the exothermic decomposition energy is less than 500 J/g and the onset of exothermic decomposition is below 500 °C (932 °F). The exothermic decomposition energy may be determined using a suitable calorimetric technique; or

(d) *For mixtures of inorganic oxidizing* substances with organic material(s), the concentration of the inorganic oxidizing substance is:

(i) *Less than 15%,* by mass, if the oxidizing substance is assigned to Category 1 or 2;

(ii) *Less than 30%,* by mass, if the oxidizing substance is assigned to Category 3.

B.2 FLAMMABLE GASES

B.2.1 *Definition*

Flammable gas means a gas having a flammable range with air at 20 °C (68 °F) and a standard pressure of 101.3 kPa (14.7 psi).

B.2.2 *Classification Criteria*

A flammable gas shall be classified in one of the two categories for this class in accordance with Table B.2.1:

Table B.2.1 — Criteria for Flammable Gases

Category	Criteria
1	Gases, which at 20 °C (68 °F) and a standard pressure of 101.3 kPa (14.7 psi):
	(a) are ignitable when in a mixture of 13% or less by volume in air; or
	(b) have a flammable range with air of at least 12 percentage points regardless of the lower flammable limit.
2	Gases, other than those of Category 1, which, at 20 °C (68 °F) and a standard pressure of 101.3 kPa (14.7 psi), have a flammable range while mixed in air.

Note: Aerosols should not be classified as flammable gases. See B.3.

B.2.3 *Additional Classification Considerations*

Flammability shall be determined by tests or by calculation in accordance with ISO 10156 (incorporated by reference; See

Sec. 1910.6). Where insufficient data are available to use this method, equivalent validated methods may be used.

B.3 FLAMMABLE AEROSOLS

B.3.1 *Definition*

Aerosol means any non-refillable receptacle containing a gas compressed, liquefied or dissolved under pressure, and fitted with a release device allowing the contents to be ejected as particles in suspension in a gas, or as a foam, paste, powder, liquid or gas.

B.3.2 *Classification Criteria*

B.3.2.1 *Aerosols shall be considered for classification* as flammable if they contain any component which is classified as flammable in accordance with this Appendix, i.e.:

Flammable liquids (See B.6);
Flammable gases (See B.2);
Flammable solids (See B.7).

Note 1: Flammable components do not include pyrophoric, self-heating or water-reactive chemicals.

Note 2: Flammable aerosols do not fall additionally within the scope of flammable gases, flammable liquids, or flammable solids.

B.3.2.2 *A flammable aerosol shall be classified* in one of the two categories for this class in accordance with Table B.3.1.

Table B.3.1 — Criteria for Flammable Aerosols

Category	Criteria
1	Contains ≥85% flammable components and the chemical heat of combustion is ≥30 kJ/g; or
	(a) For spray aerosols, in the ignition distance test, ignition occurs at a distance ≥75 cm (29.5 in), or
	(b) For foam aerosols, in the aerosol foam flammability test. (i) The flame height is ≥20 cm (7.87 in) and the flame duration ≥2 s; or (ii) The flame height is ≥4 cm (1.57 in) and the flame duration ≥7 s.
2	Contains >1% flammable components, or the heat of combustion is ≥20 kJ/g; and
	(a) for spray aerosols, in the ignition distance test, ignition occurs at a distance ≥15 cm (5.9 in), or in the enclosed space ignition test, the (i) Time equivalent is ≤300 s/m^3 ; or (ii) Deflagration density is ≤300 g/m^3 .
	(b) For foam aerosols, in the aerosol foam flammability test, the flame height is ≥4 cm and the flame duration is ≥2 s and it does not meet the criteria for Category 1.

Note: Aerosols not submitted to the flammability classification procedures in this Appendix shall be classified as extremely flammable (Category 1).

B.3.3 *Additional Classification Considerations*

B.3.3.1 *To classify a flammable aerosol,* data on its flammable components, on its chemical heat of combustion and, if applicable, the results of the aerosol foam flammability test (for foam aerosols) and of the ignition distance test and enclosed space test (for spray aerosols) are necessary.

B.3.3.2 *The chemical heat of combustion* ([Delta]Hc), in kilojoules per gram (kJ/g), is the product of the theoretical heat of combustion ([Delta]Hcomb), and a combustion efficiency, usually less than 1.0 (a typical combustion efficiency is 0.95 or 95%).

For a composite aerosol formulation, the chemical heat of combustion is the summation of the weighted heats of combustion for the individual components, as follows:

$$\Delta Hc\ (\text{product}) = \sum_{i}^{n} [\ wi\% \times \Delta Hc(i)]$$

Where:

Where:

ΔHc = chemical heat of combustion (kJ/g);
wi% = mass fraction of component i in the product;
ΔHc(i) = specific heat of combustion (kJ/g) of component i in the product;

The chemical heats of combustion shall be found in literature, calculated or determined by tests (See ASTM D240-02, ISO 13943, Sections 86.1 to 86.3, and NFPA 30B (incorporated by reference; See Sec. 1910.6)).

B.3.3.3 *The Ignition Distance Test,* Enclosed Space Ignition Test and Aerosol Foam Flammability Test shall be performed in accordance with sub-sections 31.4, 31.5 and 31.6 of the of the UN ST/SG/AC.10 (incorporated by reference; See Sec. 1910.6).

B.4 OXIDIZING GASES

B.4.1 *Definition*

Oxidizing gas means any gas which may, generally by providing oxygen, cause or contribute to the combustion of other material more than air does.

Note: "Gases which cause or contribute to the combustion of other material more than air does" means pure gases or gas mixtures with an oxidizing power greater than 23.5% (as determined by a method specified in ISO 10156 or 10156-2 (incorporated by reference, See Sec. 1910.6) or an equivalent testing method.)

B.4.2 *Classification Criteria*

An oxidizing gas shall be classified in a single category for this class in accordance with Table B.4.1:

Table B.4.1 — Criteria for Oxidizing Gases

Category	Criteria
1	Any gas which may, generally by providing oxygen, cause or contribute to the combustion of other material more than air does.

B.4.3 *Additional Classification Considerations*

Classification shall be in accordance with tests or calculation methods as described in ISO 10156 (incorporated by reference; See Sec. 1910.6) and ISO 10156-2 (incorporated by reference; See Sec. 1910.6).

B.5 GASES UNDER PRESSURE

B.5.1 *Definition*

Gases under pressure are gases which are contained in a receptacle at a pressure of 200 kPa (29 psi) (gauge) or more, or which are liquefied or liquefied and refrigerated.

They comprise compressed gases, liquefied gases, dissolved gases and refrigerated liquefied gases.

B.5.2 *Classification Criteria*

Gases under pressure shall be classified in one of four groups in accordance with Table B.5.1:

Table B.5.1 — Criteria for Gases Under Pressure

Group	Criteria
Compressed gas	A gas which when under pressure is entirely gaseous at - 50 °C (- 8 °F), including all gases with a critical temperature[1] ≤ - 50 °C (- 58 °F).
Liquefied gas	A gas which when under pressure is partially liquid at temperatures above - 50 °C (- 58 °F). A distinction is made between:
	(a) High pressure liquefied gas: A gas with a critical temperature[1] between - 50 °C (- 58 °F) and +65 °C (149 °F); and
	(b) Low pressure liquefied gas: A gas with a critical temperature[1] above +65 °C (149 °F).
Refrigerated liquefied gas	A gas which is made partially liquid because of its low temperature.
Dissolved gas	A gas which when under pressure is dissolved in a liquid phase solvent.

[1]*The critical temperature is the temperature above which a pure gas cannot be liquefied, regardless of the degree of compression.*

B.6 FLAMMABLE LIQUIDS

B.6.1 *Definition*

Flammable liquid means a liquid having a flash point of not more than 93 °C (199.4 °F).

Flash point means the minimum temperature at which a liquid gives off vapor in sufficient concentration to form an ignitable mixture with air near the surface of the liquid, as determined by a method identified in Section B.6.3.

B.6.2 *Classification Criteria*

A flammable liquid shall be classified in one of four categories in accordance with Table B.6.1:

Table B.6.1 — Criteria for Flammable Liquids

Category	Criteria
1	Flash point <23 °C (73.4 °F) and initial boiling point ≤35 °C (95 °F).
2	Flash point <23 °C (73.4 °F) and initial boiling point >35 °C (95 °F).
3	Flash point ≥23 °C (73.4 °F) and ≤60 °C (140 °F).
4	Flash point >60 °C (140 °F) and ≤93 °C (199.4 °F).

B.6.3 *Additional Classification Considerations*

The flash point shall be determined in accordance with ASTM D56-05, ASTM D3278, ASTM D3828, ASTM D93-08 (incorporated by reference; See Sec. 1910.6), or any other method specified in GHS Revision 3, Chapter 2.6.

The initial boiling point shall be determined in accordance with ASTM D86-07a or ASTM D1078 (incorporated by reference; See Sec. 1910.6).

B.7 FLAMMABLE SOLIDS

B.7.1 *Definitions*

Flammable solid means a solid which is a readily combustible solid, or which may cause or contribute to fire through friction.

Readily combustible solids are powdered, granular, or pasty chemicals which are dangerous if they can be easily ignited by brief contact with an ignition source, such as a burning match, and if the flame spreads rapidly.

B.7.2 *Classification Criteria*

B.7.2.1 *Powdered, granular or pasty* chemicals shall be classified as flammable solids when the time of burning of one or more of the test runs, performed in accordance with the test method described in the UN ST/SG/AC.10 (incorporated by reference; See Sec. 1910.6), Part III, sub-section 33.2.1, is less than 45 s or the rate of burning is more than 2.2 mm/s (0.0866 in/s).

B.7.2.2 *Powders of metals or metal* alloys shall be classified as flammable solids when they can be ignited and the reaction spreads over the whole length of the sample in 10 min or less.

B.7.2.3 *Solids which may cause* fire through friction shall be classified in this class by analogy with existing entries (e.g., matches) until definitive criteria are established.

B.7.2.4 *A flammable solid shall be classified* in one of the two categories for this class using Method N.1 as described in Part III, sub-section 33.2.1 of the UN ST/SG/AC.10 (incorporated by reference; See Sec. 1910.6), in accordance with Table B.7.1:

Table B.7.1 — Criteria for Flammable Solids

Category	Criteria
1	Burning rate test:
	Chemicals other than metal powders:
	(a) Wetted zone does not stop fire; and
	(b) Burning time <45 s or burning rate >2.2 mm/s.
	Metal powders: Burning time ≤5 min.
2	Burning rate test:
	Chemicals other than metal powders:
	(a) Wetted zone stops the fire for at least 4 min; and
	(b) Burning time <45 s or burning rate >2.2 mm/s.
	Metal powders: Burning time >5 min and ≤10 min.

Note: Classification of solid chemicals shall be based on tests performed on the chemical as presented. If, for example, for the purposes of supply or transport, the same chemical is to be presented in a physical form different from that which was tested and which is considered likely to materially alter its performance in a classification test, classification must be based on testing of the chemical in the new form.

B.8 SELF-REACTIVE CHEMICALS

B.8.1 *Definitions*

Self-reactive chemicals are thermally unstable liquid or solid chemicals liable to undergo a strongly exothermic decomposition even without participation of oxygen (air). This definition excludes chemicals classified under this section as explosives, organic peroxides, oxidizing liquids or oxidizing solids.

A self-reactive chemical is regarded as possessing explosive properties when in laboratory testing the formulation is liable to detonate, to deflagrate rapidly or to show a violent effect when heated under confinement.

B.8.2 *Classification Criteria*

B.8.2.1 *A self-reactive chemical shall be considered* for classification in this class unless:

(a) *It is classified as an explosive according* to B.1 of this appendix;

(b) *It is classified as an oxidizing liquid* or an oxidizing solid according to B.13 or B.14 of this appendix, except that a mixture of oxidizing substances which contains 5% or more of combustible organic substances shall be classified as a self-reactive chemical according to the procedure defined in B.8.2.2;

(c) *It is classified as an organic peroxide* according to B.15 of this appendix;

(d) *Its heat of decomposition is less* than 300 J/g; or

(e) *Its self-accelerating decomposition temperature* (SADT) is greater than 75 °C (167 °F) for a 50 kg (110 lb) package.

B.8.2.2 *Mixtures of oxidizing substances,* meeting the criteria for classification as oxidizing liquids or oxidizing solids, which contain 5% or more of combustible organic substances and which do not meet the criteria mentioned in B.8.2.1 (a), (c), (d) or (e), shall be subjected to the self-reactive chemicals classification procedure in B.8.2.3. Such a mixture showing the properties of a self-reactive chemical type B to F shall be classified as a self-reactive chemical.

B.8.2.3 *Self-reactive chemicals shall be classified* in one of the seven categories of "types A to G" for this class, according to the following principles:

(a) *Any self-reactive chemical which* can detonate or deflagrate rapidly, as packaged, will be defined as self-reactive chemical TYPE A;

(b) *Any self-reactive chemical possessing* explosive properties and which, as packaged, neither detonates nor deflagrates rapidly, but is liable to undergo a thermal explosion in that package will be defined as self-reactive chemical TYPE B;

(c) *Any self-reactive chemical possessing* explosive properties when the chemical as packaged cannot detonate or deflagrate rapidly or undergo a thermal explosion will be defined as self-reactive chemical TYPE C;

(d) *Any self-reactive chemical which* in laboratory testing meets the criteria in (d)(i), (ii), or (iii) will be defined as self-reactive chemical TYPE D:

(i) *Detonates partially,* does not deflagrate rapidly and shows no violent effect when heated under confinement; or

(ii) *Does not detonate at all,* deflagrates slowly and shows no violent effect when heated under confinement; or

(iii) *Does not detonate or deflagrate* at all and shows a medium effect when heated under confinement;

(e) *Any self-reactive chemical which,* in laboratory testing, neither detonates nor deflagrates at all and shows low or no effect when heated under confinement will be defined as self-reactive chemical TYPE E;

(f) *Any self-reactive chemical which,* in laboratory testing, neither detonates in the cavitated state nor deflagrates at all and shows only a low or no effect when heated under confinement as well as low or no explosive power will be defined as self-reactive chemical TYPE F;

(g) *Any self-reactive chemical which,* in laboratory testing, neither detonates in the cavitated state nor deflagrates at all and shows no effect when heated under confinement nor any explosive power, provided that it is thermally stable (self-accelerating decomposition temperature is 60 °C (140 °F) to 75 °C (167 °F) for a 50 kg (110 lb) package), and, for liquid mixtures, a diluent having a boiling point greater than or equal to 150 °C (302 °F) is used for desensitization will be defined as self-reactive chemical TYPE G. If the mixture is not thermally stable or a diluent having a boiling point less than 150 °C (302 °F) is used for desensitization, the mixture shall be defined as self-reactive chemical TYPE F.

B.8.3 *Additional Classification Considerations*

B.8.3.1 *For purposes of classification,* the properties of self-reactive chemicals shall be determined in accordance with test series A to H as described in Part II of the UN ST/SG/AC.10 (incorporated by reference; See Sec. 1910.6).

B.8.3.2 *Self-accelerating decomposition temperature* (SADT) shall be determined in accordance with the UN ST/SG/AC.10, Part II, section 28 (incorporated by reference; See Sec. 1910.6).

B.8.3.3 *The classification procedures for self-reactive* substances and mixtures need not be applied if:

(a) *There are no chemical groups* present in the molecule associated with explosive or self-reactive properties; examples of such groups are given in Tables A6.1 and A6.2 in the Appendix 6 of the UN ST/SG/AC.10 (incorporated by reference; See Sec. 1910.6); or

(b) *For a single organic substance* or a homogeneous mixture of organic substances, the estimated SADT is greater than 75 °C (167 °F) or the exothermic decomposition energy is less than 300 J/g. The onset temperature and decomposition energy may be estimated using a suitable calorimetric technique (See 20.3.3.3 in Part II of the UN

ST/SG/AC.10 (incorporated by reference; See Sec. 1910.6)).

B.9 PYROPHORIC LIQUIDS

B.9.1 *Definition*

Pyrophoric liquid means a liquid which, even in small quantities, is liable to ignite within five minutes after coming into contact with air.

B.9.2 *Classification Criteria*

A pyrophoric liquid shall be classified in a single category for this class using test N.3 in Part III, sub-section 33.3.1.5 of the UN ST/SG/AC.10 (incorporated by reference; See Sec. 1910.6), in accordance with Table B.9.1:

Table B.9.1 — Criteria for Pyrophoric Liquids

Category	Criteria
1	The liquid ignites within 5 min when added to an inert carrier and exposed to air, or it ignites or chars a filter paper on contact with air within 5 min.

B.9.3 *Additional Classification Considerations*

The classification procedure for pyrophoric liquids need not be applied when experience in production or handling shows that the chemical does not ignite spontaneously on coming into contact with air at normal temperatures (i.e., the substance is known to be stable at room temperature for prolonged periods of time (days)).

B.10 PYROPHORIC SOLIDS

B.10.1 *Definition*

Pyrophoric solid means a solid which, even in small quantities, is liable to ignite within five minutes after coming into contact with air.

B.10.2 *Classification Criteria*

A pyrophoric solid shall be classified in a single category for this class using test N.2 in Part III, sub-section 33.3.1.4 of the UN ST/SG/AC.10 (incorporated by reference; See Sec. 1910.6), in accordance with Table B.10.1:

Table B.10.1 — Criteria for Pyrophoric Solids

Category	Criteria
1	The solid ignites within 5 min of coming into contact with air.

Note: Classification of solid chemicals shall be based on tests performed on the chemical as presented. If, for example, for the purposes of supply or transport, the same chemical is to be presented in a physical form different from that which was tested and which is considered likely to materially alter its performance in a classification test, classification must be based on testing of the chemical in the new form.

B.10.3 *Additional Classification Considerations*

The classification procedure for pyrophoric solids need not be applied when experience in production or handling shows that the chemical does not ignite spontaneously on coming into contact with air at normal temperatures (i.e., the chemical is known to be stable at room temperature for prolonged periods of time (days)).

B.11 SELF-HEATING CHEMICALS

B.11.1 *Definition*

A *self-heating chemical* is a solid or liquid chemical, other than a pyrophoric liquid or solid, which, by reaction with air and without energy supply, is liable to self-heat; this chemical differs from a pyrophoric liquid or solid in that it will ignite only when in large amounts (kilograms) and after long periods of time (hours or days).

Note: Self-heating of a substance or mixture is a process where the gradual reaction of that substance or mixture with oxygen (in air) generates heat. If the rate of heat production exceeds the rate of heat loss, then the temperature of the substance or mixture will rise which, after an induction time, may lead to self-ignition and combustion.

B.11.2 *Classification Criteria*

B.11.2.1 *A self-heating chemical shall be classified* in one of the two categories for this class if, in tests performed in accordance with test method N.4 in Part III, sub-section 33.3.1.6 of the UN ST/SG/AC.10 (incorporated by reference; See Sec. 1910.6), the result meets the criteria shown in Table B.11.1.

Table B.11.1 — Criteria for Self-Heating Chemicals

Category	Criteria
1	A positive result is obtained in a test using a 25 mm sample cube at 140 °C (284 °F).
2	A negative result is obtained in a test using a 25 mm cube sample at 140 °C (284 °F), a positive result is obtained in a test using a 100 mm sample cube at 140 °C (284 °F), and:
	(a) The unit volume of the chemical is more than 3 m^3 ; or
	(b) A positive result is obtained in a test using a 100 mm cube sample at 120 °C (248 °F) and the unit volume of the chemical is more than 450 liters; or
	(c) A positive result is obtained in a test using a 100 mm cube sample at 100 °C (212 °F).

B.11.2.2 *Chemicals with a temperature of spontaneous* combustion higher than 50 °C (122 °F) for a volume of 27 m^3 shall not be classified as self-heating chemicals.

B.11.2.3 *Chemicals with a spontaneous ignition* temperature higher than 50 °C (122 °F) for a volume of 450 liters shall not be classified in Category 1 of this class.

B.11.3 *Additional Classification Considerations*

B.11.3.1 *The classification procedure for self-heating* chemicals need not be applied if the results of a screening test can be adequately correlated with the classification test and an appropriate safety margin is applied.

B.11.3.2 *Examples of screening tests are:*

(a) *The Grewer Oven test* (VDI guideline 2263, part 1, 1990, Test methods for the Determination of the Safety Characteristics of Dusts) with an onset temperature 80°K above the reference temperature for a volume of 1 *l;*

(b) *The Bulk Powder Screening* Test (Gibson, N. Harper, D. J. Rogers, R. Evaluation of the fire and explosion risks in drying powders, Plant Operations Progress, 4 (3), 181-189, 1985) with an onset temperature 60°K above the reference temperature for a volume of 1 *l*.

B.12 CHEMICALS WHICH, IN CONTACT WITH WATER, EMIT FLAMMABLE GASES

B.12.1 *Definition*

Chemicals which, in contact with water, emit flammable gases are solid or liquid chemicals which, by interaction with water, are liable to become spontaneously flammable or to give off flammable gases in dangerous quantities.

B.12.2 *Classification Criteria*

B.12.2.1 *A chemical which,* in contact with water, emits flammable gases shall be classified in one of the three categories for this class, using test N.5 in Part III, sub-section 33.4.1.4 of the UN ST/SG/AC.10 (incorporated by reference; See Sec. 1910.6), in accordance with Table B.12.1:

Table B.12.1 — Criteria for Chemicals Which, in Contact With Water, Emit Flammable Gases

Category	Criteria
1	Any chemical which reacts vigorously with water at ambient temperatures and demonstrates generally a tendency for the gas produced to ignite spontaneously, or which reacts readily with water at ambient temperatures such that the rate of evolution of flammable gas is equal to or greater than 10 liters per kilogram of chemical over any one minute.
2	Any chemical which reacts readily with water at ambient temperatures such that the maximum rate of evolution of flammable gas is equal to or greater than 20 liters per kilogram of chemical per hour, and which does not meet the criteria for Category 1.
3	Any chemical which reacts slowly with water at ambient temperatures such that the maximum rate of evolution of flammable gas is equal to or greater than 1 liter per kilogram of chemical per hour, and which does not meet the criteria for Categories 1 and 2.

Note: Classification of solid chemicals shall be based on tests performed on the chemical as presented. If, for example, for the purposes of supply or transport, the same chemical is to be presented in a physical form different from that which was tested and which is considered likely to materially alter its performance in a classification test, classification must be based on testing of the chemical in the new form.

B.12.2.2 *A chemical is classified as a chemical* which, in contact with water emits flammable gases if spontaneous ignition takes place in any step of the test procedure.

B.12.3 *Additional Classification Considerations*

The classification procedure for this class need not be applied if:

The chemical structure of the chemical does not contain metals or metalloids;

Experience in production or handling shows that the chemical does not react with water, (e.g., the chemical is manufactured with water or washed with water); or

The chemical is known to be soluble in water to form a stable mixture.

B.13 OXIDIZING LIQUIDS

B.13.1 *Definition*

Oxidizing liquid means a liquid which, while in itself not necessarily combustible, may, generally by yielding oxygen, cause, or contribute to, the combustion of other material.

B.13.2 *Classification Criteria*

An oxidizing liquid shall be classified in one of the three categories for this class using test O.2 in Part III, sub-section 34.4.2 of the UN ST/SG/AC.10 (incorporated by reference; See Sec. 1910.6), in accordance with Table B.13.1:

Table B.13.1 — Criteria for Oxidizing Liquids

Category	Criteria
1	Any chemical which, in the 1:1 mixture, by mass, of chemical and cellulose tested, spontaneously ignites; or the mean pressure rise time of a 1:1 mixture, by mass, of chemical and cellulose is less than that of a 1:1 mixture, by mass, of 50% perchloric acid and cellulose;
2	Any chemical which, in the 1:1 mixture, by mass, of chemical and cellulose tested, exhibits a mean pressure rise time less than or equal to the mean pressure rise time of a 1:1 mixture, by mass, of 40% aqueous sodium chlorate solution and cellulose; and the criteria for Category 1 are not met;
3	Any chemical which, in the 1:1 mixture, by mass, of chemical and cellulose tested, exhibits a mean pressure rise time less than or equal to the mean pressure rise time of a 1:1 mixture, by mass, of 65% aqueous nitric acid and cellulose; and the criteria for Categories 1 and 2 are not met.

B.13.3 *Additional Classification Considerations*

B.13.3.1 *For organic chemicals,* the classification procedure for this class shall not be applied if:

(a) *The chemical does not contain* oxygen, fluorine or chlorine; or

(b) *The chemical contains oxygen,* fluorine or chlorine and these elements are chemically bonded only to carbon or hydrogen.

B.13.3.2 *For inorganic chemicals,* the classification procedure for this class shall not be applied if the chemical does not contain oxygen or halogen atoms.

B.13.3.3 *In the event of divergence between* test results and known experience in the handling and use of chemicals which shows them to be oxidizing, judgments based on known experience shall take precedence over test results.

B.13.3.4 *In cases where chemicals generate* a pressure rise (too high or too low), caused by chemical reactions not characterizing the oxidizing properties of the chemical, the test described in Part III, sub-section 34.4.2 of the UN ST/SG/AC.10 (incorporated by reference; See Sec. 1910.6) shall be repeated with an inert substance (e.g., diatomite (kieselguhr)) in place of the cellulose in order to clarify the nature of the reaction.

B.14 OXIDIZING SOLIDS

B.14.1 *Definition*

Oxidizing solid means a solid which, while in itself is not necessarily combustible, may, generally by yielding oxygen, cause, or contribute to, the combustion of other material.

B.14.2 *Classification Criteria*

An oxidizing solid shall be classified in one of the three categories for this class using test O.1 in Part III, sub-section 34.4.1 of the UN ST/SG/AC.10 (incorporated by reference; See Sec. 1910.6), in accordance with Table B.14.1:

Table B.14.1 — Criteria for Oxidizing Solids

Category	Criteria
1	Any chemical which, in the 4:1 or 1:1 sample-to-cellulose ratio (by mass) tested, exhibits a mean burning time less than the mean burning time of a 3:2 mixture, by mass, of potassium bromate and cellulose.
2	Any chemical which, in the 4:1 or 1:1 sample-to-cellulose ratio (by mass) tested, exhibits a mean burning time equal to or less than the mean burning time of a 2:3 mixture (by mass) of potassium bromate and cellulose and the criteria for Category 1 are not met.
3	Any chemical which, in the 4:1 or 1:1 sample-to-cellulose ratio (by mass) tested, exhibits a mean burning time equal to or less than the mean burning time of a 3:7 mixture (by mass) of potassium bromate and cellulose and the criteria for Categories 1 and 2 are not met.

Note 1: Some oxidizing solids may present explosion hazards under certain conditions (e.g., when stored in large quantities). For example, some types of ammonium nitrate may give rise to an explosion hazard under extreme conditions and the "Resistance to detonation test" (IMO: Code of Safe Practice for Solid Bulk Cargoes, 2005, Annex 3, Test 5) may be used to assess this hazard. When information indicates that an oxidizing solid may present an explosion hazard, it shall be indicated on the Safety Data Sheet.

Note 2: Classification of solid chemicals shall be based on tests performed on the chemical as presented. If, for example, for the purposes of supply or transport, the same chemical is to be presented in a physical form different from that which was tested and which is considered likely to materially alter its performance in a classification test, classification must be based on testing of the chemical in the new form.

B.14.3 *Additional Classification Considerations*

B.14.3.1 *For organic chemicals,* the classification procedure for this class shall not be applied if:

(a) *The chemical does not contain* oxygen, fluorine or chlorine; or

(b) *The chemical contains oxygen,* fluorine or chlorine and these elements are chemically bonded only to carbon or hydrogen.

B.14.3.2 *For inorganic chemicals,* the classification procedure for this class shall not be applied if the chemical does not contain oxygen or halogen atoms.

B.14.3.3 *In the event of divergence between* test results and known experience in the handling and use of chemicals which shows them to be oxidizing, judgements based on known experience shall take precedence over test results.

B.15 ORGANIC PEROXIDES

B.15.1 *Definition*

B.15.1.1 **Organic peroxide** means a liquid or solid organic chemical which contains the bivalent -0-0- structure and as such is considered a derivative of hydrogen peroxide, where one or both of the hydrogen atoms have been replaced by organic radicals. The term organic peroxide includes organic peroxide mixtures containing at least one organic peroxide. Organic peroxides are thermally unstable chemicals, which may undergo exothermic self-accelerating decomposition. In addition, they may have one or more of the following properties:

(a) *Be liable to explosive decomposition;*

(b) *Burn rapidly;*

(c) *Be sensitive to impact or friction;*

(d) *React dangerously with other substances.*

B.15.1.2 *An organic peroxide is regarded* as possessing explosive properties when in laboratory testing the formulation is liable to detonate, to deflagrate rapidly or to show a violent effect when heated under confinement.

B.15.2 *Classification Criteria*

B.15.2.1 *Any organic peroxide shall be considered* for classification in this class, unless it contains:

(a) *Not more than 1.0%* available oxygen from the organic peroxides when containing not more than 1.0% hydrogen peroxide; or

(b) *Not more than 0.5%* available oxygen from the organic peroxides when containing more than 1.0% but not more than 7.0% hydrogen peroxide.

Note: The available oxygen content (%) of an organic peroxide mixture is given by the formula:

$$16 \times \sum_{i}^{n} \left(\frac{n_i \times c_i}{m_i} \right)$$

Where:

n_i= number of peroxygen groups per molecule of organic peroxide *i*;

c_i= concentration (mass %) of organic peroxide *i*;

m_i= molecular mass of organic peroxide *i*.

B.15.2.2 *Organic peroxides shall be classified* in one of the seven categories of "Types A to G" for this class, according to the following principles:

(a) *Any organic peroxide which,* as packaged, can detonate or deflagrate rapidly shall be defined as organic peroxide TYPE A;

(b) *Any organic peroxide possessing* explosive properties and which, as packaged, neither detonates nor deflagrates rapidly, but is liable to undergo a thermal explosion in that package shall be defined as organic peroxide TYPE B;

(c) *Any organic peroxide possessing* explosive properties when the chemical as packaged cannot detonate or deflagrate rapidly or undergo a thermal explosion shall be defined as organic peroxide TYPE C;

(d) *Any organic peroxide which* in laboratory testing meets the criteria in (d)(i), (ii), or (iii) shall be defined as organic peroxide TYPE D:

(i) *Detonates partially,* does not deflagrate rapidly and shows no violent effect when heated under confinement; or

(ii) *Does not detonate at all,* deflagrates slowly and shows no violent effect when heated under confinement; or

(iii) *Does not detonate or deflagrate* at all and shows a medium effect when heated under confinement;

(e) *Any organic peroxide which,* in laboratory testing, neither detonates nor deflagrates at all and shows low or no effect when heated under confinement shall be defined as organic peroxide TYPE E;

(f) *Any organic peroxide which,* in laboratory testing, neither detonates in the cavitated state nor deflagrates at all and shows only a low or no effect when heated under confinement as well as low or no explosive power shall be defined as organic peroxide TYPE F;

(g) *Any organic peroxide which,* in laboratory testing, neither detonates in the cavitated state nor deflagrates at all and shows no effect when heated under confinement nor any explosive power, provided that it is thermally stable (self-accelerating decomposition temperature is 60 °C (140 °F) or higher for a 50 kg (110 lb) package), and, for liquid mixtures, a diluent having a boiling point of not less than 150 °C (302 °F) is used for desensitization, shall be defined as organic peroxide TYPE G. If the organic peroxide is not thermally stable or a diluent having a boiling point less than 150 °C (302 °F) is used for desensitization, it shall be defined as organic peroxide TYPE F.

B.15.3 *Additional Classification Considerations*

B.15.3.1 *For purposes of classification,* the properties of organic peroxides shall be determined in accordance with test series A to H as described in Part II of the UN ST/SG/AC.10 (incorporated by reference; See Sec. 1910.6).

B.15.3.2 *Self-accelerating decomposition temperature* (SADT) shall be determined in accordance with the UN ST/SG/AC.10 (incorporated by reference; See Sec. 1910.6), Part II, section 28.

B.15.3.3 *Mixtures of organic peroxides* may be classified as the same type of organic peroxide as that of the most dangerous ingredient. However, as two stable ingredients can form a thermally less stable mixture, the SADT of the mixture shall be determined.

B.16 CORROSIVE TO METALS

B.16.1 *Definition*

A ***chemical which is corrosive to metals*** means a chemical which by chemical action will materially damage, or even destroy, metals.

B.16.2 *Classification Criteria*

A chemical which is corrosive to metals shall be classified in a single category for this class, using the test in Part III, sub-section 37.4 of the UN ST/SG/AC.10 (incorporated by reference; See Sec. 1910.6), in accordance with Table B.16.1:

Table B.16.1 — Criteria for Chemicals Corrosive to Metal

Category	Criteria
1	Corrosion rate on either steel or aluminium surfaces exceeding 6.25 mm per year at a test temperature of 55 °C (131 °F) when tested on both materials.

Note: Where an initial test on either steel or aluminium indicates the chemical being tested is corrosive, the follow-up test on the other metal is not necessary.

B.16.3 *Additional Classification Considerations*

The specimen to be used for the test shall be made of the following materials:

For the purposes of testing steel, steel types S235JR+CR (1.0037 resp.St 37-2), S275J2G3+CR (1.0144 resp.St 44-3), ISO 3574, Unified Numbering System (UNS) G 10200, or SAE 1020;

For the purposes of testing aluminium: Non-clad types 7075-T6 or AZ5GU-T6.

§1910.1200 Appendix C

ALLOCATION OF LABEL ELEMENTS (MANDATORY)

C.1 The label for each hazardous chemical shall include the product identifier used on the safety data sheet.

C.1.1 *The labels on shipped containers* shall also include the name, address, and telephone number of the chemical manufacturer, importer, or responsible party.

C.2 The label for each hazardous chemical that is classified shall include the signal word, hazard statement(s), pictogram(s), and precautionary statement(s) specified in C.4 for each hazard class and associated hazard category, except as provided for in C.2.1 through C.2.4.

C.2.1 *Precedence of Hazard Information*

C.2.1.1 *If the signal word "Danger"* is included, the signal word "Warning" shall not appear;

C.2.1.2 *If the skull and crossbones pictogram* is included, the exclamation mark pictogram shall not appear where it is used for acute toxicity;

C.2.1.3 *If the corrosive pictogram is included,* the exclamation mark pictogram shall not appear where it is used for skin or eye irritation;

C.2.1.4 *If the health hazard pictogram* is included for respiratory sensitization, the exclamation mark pictogram shall not appear where it is used for skin sensitization or for skin or eye irritation.

C.2.2 *Hazard Statement Text*

C.2.2.1 *The text of all applicable hazard* statements shall appear on the label, except as otherwise specified. The information in italics shall be included as part of the hazard statement as provided. For example: "causes damage to organs *(state all organs affected)* through prolonged or repeated exposure *(state route of exposure if no other routes of exposure cause the hazard)*". Hazard statements may be combined where appropriate to reduce the information on the label and improve readability, as long as all of the hazards are conveyed as required.

C.2.2.2 *If the chemical manufacturer,* importer, or responsible party can demonstrate that all or part of the hazard statement is inappropriate to a specific substance or mixture, the corresponding statement may be omitted from the label.

C.2.3 *Pictograms*

C.2.3.1 *Pictograms shall be in the shape of a square* set at a point and shall include a black hazard symbol on a white background with a red frame sufficiently wide to be clearly visible. A square red frame set at a point without a hazard symbol is not a pictogram and is not permitted on the label.

C.2.3.2 One of eight standard hazard symbols shall be used in each pictogram. The eight hazard symbols are depicted in Figure C.1. A pictogram using the exclamation mark symbol is presented in Figure C.2, for the purpose of illustration.

Figure C.1 – Hazard Symbols and Classes

Flame	Flame Over Circle	Exclamation Mark	Exploding Bomb
Flammables Self Reactives Pyrophorics Self-heating Emits Flammable Gas Organic Peroxides	Oxidizers	Irritant Dermal Sensitizer Acute Toxicity (harmful) Narcotic Effects Respiratory Tract Irritation	Explosives Self Reactives Organic Peroxides
Corrosion	**Gas Cylinder**	**Health Hazard**	**Skull and Crossbones**
Corrosives	Gases Under Pressure	Carcinogen Respiratory Sensitizer Reproductive Toxicity Target Organ Toxicity Mutagenicity Aspiration Toxicity	Acute Toxicity (severe)

Figure C.2 – Exclamation Mark Pictogram

C.2.3.3 Where a pictogram required by the Department of Transportation under Title 49 of the Code of Federal Regulations appears on a shipped container, the pictogram specified in C.4 for the same hazard shall not appear.

C.2.4 *Precautionary Statement Text*

C.2.4.1 *There are four types* of precautionary statements presented, "prevention," "response," "storage," and "disposal." The core part of the precautionary statement is presented in bold print. This is the text, except as otherwise specified, that shall appear on the label. Where additional information is required, it is indicated in plain text.

C.2.4.2 *When a backslash or diagonal mark* (/) appears in the precautionary statement text, it indicates that a choice has to be made between the separated phrases. In such cases, the chemical manufacturer, importer, or responsible party can choose the most appropriate phrase(s). For example, "Wear protective gloves/protective clothing/eye protection/face protection" could read "wear eye protection".

C.2.4.3 *When three full stops* (* * *) appear in the precautionary statement text, they indicate that all applicable conditions are not listed. For example, in "Use explosion-proof electrical/ventilating/lighting/* * */equipment", the use of "* * *" indicates that other equipment may need to be specified. In such cases, the chemical manufacturer, importer, or responsible party can choose the other conditions to be specified.

C.2.4.4 *When text in italics* is used in a precautionary statement, this indicates specific conditions applying to the use or allocation of the precautionary statement. For example, "Use explosion-proof electrical/ventilating/lighting/* * */equipment" is only required for flammable solids "*if dust clouds can occur*". Text in italics is intended to be an explanatory, conditional note and is not intended to appear on the label.

C.2.4.5 *Where square brackets ([ ])* appear around text in a precautionary statement, this indicates that the text in square brackets is not appropriate in every case and should be used only in certain circumstances. In these cases, conditions for use explaining when the text should be used are provided. For example, one precautionary statement states: "[In case of inadequate ventilation] wear respiratory protection." This statement is given with the condition for use "- text in square brackets may be used if additional information is provided with the chemical at the point of use that explains what type of ventilation would be adequate for safe use". This means that, if additional information is provided with the chemical explaining what type of ventilation would be adequate for safe use, the text in square brackets should be used and the statement would read: "In case of inadequate ventilation wear respiratory protection." However, if the chemical is supplied without such ventilation information, the text in square brackets should not be used, and the precautionary statement should read: "Wear respiratory protection."

C.2.4.6 *Precautionary statements may be combined* or consolidated to save label space and improve readability. For example, "Keep away from heat, sparks and open flame," "Store in a well-ventilated place" and "Keep cool" can be combined to read "Keep away from heat, sparks and open flame and store in a cool, well-ventilated place."

C.2.4.7 *In most cases,* the precautionary statements are independent (e.g., the phrases for explosive hazards do not modify those related to certain health hazards, and products that are classified for both hazard classes shall bear appropriate precautionary statements for both). Where a chemical is classified for a number of hazards, and the precautionary statements are similar, the most stringent shall be included on the label (this will be applicable mainly to preventive measures). An order of precedence may be imposed by the chemical manufacturer, importer or responsible party in situations where phrases concern "Response." Rapid action may be crucial. For example, if a chemical is carcinogenic and acutely toxic, rapid action may be crucial, and first aid measures for acute toxicity will take precedence over those for long-term effects. In addition, medical attention to delayed health effects may be required in cases of incidental exposure, even if not associated with immediate symptoms of intoxication.

C.2.4.8 *If the chemical manufacturer,* importer, or responsible party can demonstrate that a precautionary statement is inappropriate to a specific substance or mixture, the precautionary statement may be omitted from the label.

C.3 Supplementary Hazard Information

C.3.1 *To ensure that non-standardized information* does not lead to unnecessarily wide variation or undermine the required information, supplementary information on the label is limited to when it provides further detail and does not contradict or cast doubt on the validity of the standardized hazard information.

C.3.2 *Where the chemical manufacturer,* importer, or distributor chooses to add supplementary information on the label, the placement of supplemental information shall not impede identification of information required by this section.

C.3.3 *Where an ingredient with unknown acute* toxicity is used in a mixture at a concentration [ge]1%, and the mixture is not classified based on testing of the mixture as a whole, a statement that X% of the mixture consists of ingredient(s) of unknown acute toxicity is required on the label.

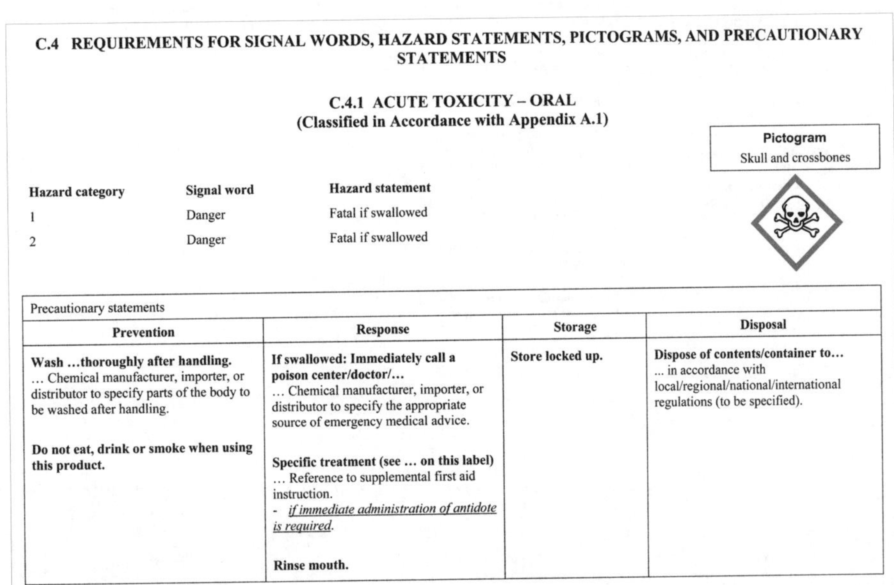

C.4 REQUIREMENTS FOR SIGNAL WORDS, HAZARD STATEMENTS, PICTOGRAMS, AND PRECAUTIONARY STATEMENTS

C.4.1 ACUTE TOXICITY – ORAL
(Classified in Accordance with Appendix A.1)

Hazard category	Signal word	Hazard statement	Pictogram Skull and crossbones
1	Danger	Fatal if swallowed	
2	Danger	Fatal if swallowed	

Precautionary statements

Prevention	Response	Storage	Disposal
Wash …thoroughly after handling. … Chemical manufacturer, importer, or distributor to specify parts of the body to be washed after handling. **Do not eat, drink or smoke when using this product.**	**If swallowed: Immediately call a poison center/doctor/…** … Chemical manufacturer, importer, or distributor to specify the appropriate source of emergency medical advice. **Specific treatment (see … on this label)** … Reference to supplemental first aid instruction. - *if immediate administration of antidote is required.* **Rinse mouth.**	**Store locked up.**	**Dispose of contents/container to…** … in accordance with local/regional/national/international regulations (to be specified).

C.4.1 ACUTE TOXICITY – ORAL (CONTINUED)
(Classified in Accordance with Appendix A.1)

Hazard category	Signal word	Hazard statement
3	Danger	Toxic if swallowed

Pictogram
Skull and crossbones

Precautionary statements			
Prevention	**Response**	**Storage**	**Disposal**
Wash … thoroughly after handling. … Chemical manufacturer, importer, or distributor to specify parts of the body to be washed after handling. **Do not eat, drink or smoke when using this product.**	**If swallowed: Immediately call a poison center/doctor/...** … Chemical manufacturer, importer, or distributor to specify the appropriate source of emergency medical advice. **Specific treatment (see ... on this label)** ... Reference to supplemental first aid instruction. - *if immediate administration of antidote is required*. **Rinse mouth.**	**Store locked up.**	**Dispose of contents/container to...** ... in accordance with local/regional/national/international regulations (to be specified).

C.4.1 ACUTE TOXICITY – ORAL (CONTINUED)
(Classified in Accordance with Appendix A.1)

Hazard category	Signal word	Hazard statement
4	Warning	Harmful if swallowed

Pictogram
Exclamation mark

Precautionary statements			
Prevention	**Response**	**Storage**	**Disposal**
Wash … thoroughly after handling. … Chemical manufacturer, importer, or distributor to specify parts of the body to be washed after handling. **Do not eat, drink or smoke when using this product.**	**If swallowed: Call a poison center/doctor/…/ if you feel unwell.** … Chemical manufacturer, importer, or distributor to specify the appropriate source of emergency medical advice. **Rinse mouth.**		**Dispose of contents/container to...** ... in accordance with local/regional/national/international regulations (to be specified).

C.4.2 ACUTE TOXICITY - DERMAL
(Classified in Accordance with Appendix A.1)

Hazard category	Signal word	Hazard statement
1	Danger	Fatal in contact with skin
2	Danger	Fatal in contact with skin

Pictogram
Skull and crossbones

Precautionary statements			
Prevention	**Response**	**Storage**	**Disposal**
Do not get in eyes, on skin, or on clothing. **Wash … thoroughly after handling.** **…** Chemical manufacturer, importer, or distributor to specify parts of the body to be washed after handling. **Do not eat, drink or smoke when using this product.** **Wear protective gloves/protective clothing.** Chemical manufacturer, importer, or distributor to specify type of equipment.	**If on skin: Wash with plenty of water/...** … Chemical manufacturer, importer, or distributor may specify a cleansing agent if appropriate, or may recommend an alternative agent in exceptional cases if water is clearly inappropriate. **Immediately call a poison center/doctor/...** … Chemical manufacturer, importer, or distributor to specify the appropriate source of emergency medical advice. **Specific treatment (see ... on this label)** ... Reference to supplemental first aid instruction. - *if immediate measures such as specific cleansing agent is advised*. **Take off immediately all contaminated clothing and wash it before reuse.**	**Store locked up.**	**Dispose of contents/container to...** ... in accordance with local/regional/national/international regulations (to be specified).

C.4.2 ACUTE TOXICITY – DERMAL (CONTINUED)
(Classified in Accordance with Appendix A.1)

Hazard category	Signal word	Hazard statement	Pictogram Skull and crossbones
3	Danger	Toxic in contact with skin	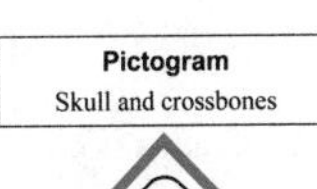

Precautionary statements			
Prevention	**Response**	**Storage**	**Disposal**
Wear protective gloves/protective clothing. Chemical manufacturer, importer, or distributor to specify type of equipment.	**If on skin: Wash with plenty of water/...** ... Chemical manufacturer, importer, or distributor may specify a cleansing agent if appropriate, or may recommend an alternative agent in exceptional cases if water is clearly inappropriate. **Call a poison center/doctor/.../if you feel unwell.** ... Chemical manufacturer, importer, or distributor to specify the appropriate source of emergency medical advice. **Specific treatment (see ... on this label)** ... Reference to supplemental first aid instruction. - *if measures such as specific cleansing agent is advised.* **Take off immediately all contaminated clothing and wash it before reuse.**	**Store locked up.**	**Dispose of contents/container to...** ... in accordance with local/regional/national/international regulations (to be specified).

C.4.2 ACUTE TOXICITY – DERMAL (CONTINUED)
(Classified in Accordance with Appendix A.1)

Hazard category	Signal word	Hazard statement	Pictogram Exclamation mark
4	Warning	Harmful in contact with skin	

Precautionary statements			
Prevention	**Response**	**Storage**	**Disposal**
Wear protective gloves/protective clothing Chemical manufacturer, importer, or distributor to specify type of equipment.	**If on skin: Wash with plenty of water/...** ... Chemical manufacturer, importer, or distributor may specify a cleansing agent if appropriate, or may recommend an alternative agent in exceptional cases if water is clearly inappropriate. **Call a poison center/doctor/.../if you feel unwell.** ... Chemical manufacturer, importer, or distributor to specify the appropriate source of emergency medical advice. **Specific treatment (see ... on this label)** ... Reference to supplemental first aid instruction. - *if measures such as specific cleansing agent is advised.* **Take off contaminated clothing and wash it before reuse.**		**Dispose of contents/container to...** ... in accordance with local/regional/national/international regulations (to be specified).

C.4.3 ACUTE TOXICITY - INHALATION
(Classified in Accordance with Appendix A.1)

Hazard category	Signal word	Hazard statement
1	Danger	Fatal if inhaled
2	Danger	Fatal if inhaled

Pictogram
Skull and crossbones

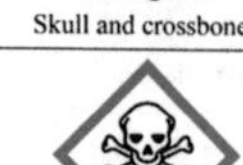

Precautionary statements

Prevention	Response	Storage	Disposal
Do not breathe dust/fume/gas/mist/ vapors/spray. Chemical manufacturer, importer, or distributor to specify applicable conditions. **Use only outdoors or in a well-ventilated area.** **[In case of inadequate ventilation] wear respiratory protection.** Chemical manufacturer, importer, or distributor to specify equipment. - *Text in square brackets may be used if additional information is provided with the chemical at the point of use that explains what type of ventilation would be adequate for safe use.*	**If inhaled: Remove person to fresh air and keep comfortable for breathing.** **Immediately call a poison center/doctor/...** ... Chemical manufacturer, importer, or distributor to specify the appropriate source of emergency medical advice. **Specific treatment is urgent (see ... on this label)** ... Reference to supplemental first aid instruction. - *if immediate administration of antidote is required.*	**Store in a well-ventilated place. Keep container tightly closed.** - *if product is volatile as to generate hazardous atmosphere.* **Store locked up.**	**Dispose of contents/container to...** ... in accordance with local/regional/national/international regulations (to be specified).

C.4.3 ACUTE TOXICITY – INHALATION (CONTINUED)
(Classified in Accordance with Appendix A.1)

Hazard category	Signal word	Hazard statement
3	Danger	Toxic if inhaled

Pictogram
Skull and crossbones

Precautionary statements

Prevention	Response	Storage	Disposal
Avoid breathing dust/fume/gas/mist/ vapors/spray. Chemical manufacturer, importer, or distributor to specify applicable conditions. **Use only outdoors or in a well-ventilated area.**	**If inhaled: Remove person to fresh air and keep comfortable for breathing.** **Call a poison center/doctor/...** ... Chemical manufacturer, importer, or distributor to specify the appropriate source of emergency medical advice. **Specific treatment (see ... on this label)** ... Reference to supplemental first aid instruction. - *if immediate specific measures are required.*	**Store in a well-ventilated place. Keep container tightly closed.** - *if product is volatile so as to generate hazardous atmosphere.* **Store locked up.**	**Dispose of content/container to…** … in accordance with local/regional/national/international regulations (to be specified).

C.4.3 ACUTE TOXICITY – INHALATION (CONTINUED)
(Classified in Accordance with Appendix A.1)

Hazard category	Signal word	Hazard statement
4	Warning	Harmful if inhaled

Pictogram
Exclamation mark

Precautionary statements

Prevention	Response	Storage	Disposal
Avoid breathing dust/fume/gas/mist/ vapors/spray. Chemical manufacturer, importer, or distributor to specify applicable conditions. **Use only outdoors or in a well-ventilated area.**	**If inhaled: Remove person to fresh air and keep comfortable for breathing.** **Call a poison center/doctor/…/if you feel unwell.** … Chemical manufacturer, importer, or distributor to specify the appropriate source of emergency medical advice.		

C.4.4 SKIN CORROSION/IRRITATION
(Classified in Accordance with Appendix A.2)

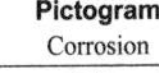

Hazard category	Signal word	Hazard statement
1A to 1C	Danger	Causes severe skin burns and eye damage

Precautionary statements

Prevention	Response	Storage	Disposal
Do not breathe dusts or mists. - _if inhalable particles of dusts or mists may occur during use._ **Wash …thoroughly after handling.** …Chemical manufacturer, importer, or distributor to specify parts of the body to be washed after handling. **Wear protective gloves/protective clothing/eye protection/face protection.** Chemical manufacturer, importer, or distributor to specify type of equipment.	**If swallowed: Rinse mouth. Do NOT induce vomiting.** **If on skin (or hair): Take off immediately all contaminated clothing. Rinse skin with water/shower.** **Wash contaminated clothing before reuse.** **If inhaled: Remove person to fresh air and keep comfortable for breathing.** **Immediately call a poison center/doctor/…** … Chemical manufacturer, importer, or distributor to specify the appropriate source of emergency medical advice. **Specific treatment (see ... on this label)** ... Reference to supplemental first aid instruction. - _Manufacturer, importer, or distributor may specify a cleansing agent if appropriate._ **If in eyes: Rinse cautiously with water for several minutes. Remove contact lenses, if present and easy to do. Continue rinsing.**	**Store locked up.**	**Dispose of contents/container to...** ... in accordance with local/regional/national/international regulations (to be specified).

C.4.4 SKIN CORROSION/IRRITATION (CONTINUED)
(Classified in Accordance with Appendix A.2)

Pictogram
Exclamation mark

Hazard category	Signal word	Hazard statement
2	Warning	Causes skin irritation

Precautionary statements

Prevention	Response	Storage	Disposal
Wash … thoroughly after handling. … Chemical manufacturer, importer, or distributor to specify parts of the body to be washed after handling. **Wear protective gloves.** Chemical manufacturer, importer, or distributor to specify type of equipment.	**If on skin: Wash with plenty of water/...** … Chemical manufacturer, importer, or distributor may specify a cleansing agent if appropriate, or may recommend an alternative agent in exceptional cases if water is clearly inappropriate. **Specific treatment (see ... on this label)** ... Reference to supplemental first aid instruction. - _Manufacturer, importer, or distributor may specify a cleansing agent if appropriate._ **If skin irritation occurs: Get medical advice/attention.** **Take off contaminated clothing and wash it before reuse.**		

C.4.5 EYE DAMAGE/IRRITATION
(Classified in Accordance with Appendix A.3)

Pictogram
Corrosion

Hazard category	Signal word	Hazard statement
1	Danger	Causes serious eye damage

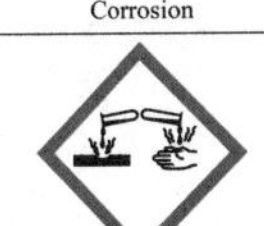

Precautionary statements

Prevention	Response	Storage	Disposal
Wear eye protection/face protection. Chemical manufacturer, importer, or distributor to specify type of equipment.	**If in eyes: Rinse cautiously with water for several minutes. Remove contact lenses, if present and easy to do. Continue rinsing.** **Immediately call a poison center/doctor/…** … Chemical manufacturer, importer, or distributor to specify the appropriate source of emergency medical advice.		

C.4.5 EYE DAMAGE/IRRITATION (CONTINUED)
(Classified in Accordance with Appendix A.3)

Pictogram
Exclamation mark

Hazard category	**Signal word**	**Hazard statement**
2A	Warning	Causes serious eye irritation

Precautionary statements			
Prevention	**Response**	**Storage**	**Disposal**
Wash … thoroughly after handling. … Chemical manufacturer, importer, or distributor to specify parts of the body to be washed after handling. **Wear eye protection/face protection.** Chemical manufacturer, importer, or distributor to specify type of equipment.	**If in eyes: Rinse cautiously with water for several minutes. Remove contact lenses, if present and easy to do. Continue rinsing.** **If eye irritation persists: Get medical advice/attention.**		

C.4.5 EYE DAMAGE/IRRITATION (CONTINUED)
(Classified in Accordance with Appendix A.3)

Pictogram
No Pictogram

Hazard category	**Signal word**	**Hazard statement**
2B	Warning	Causes eye irritation

Precautionary statements			
Prevention	**Response**	**Storage**	**Disposal**
Wash … thoroughly after handling. … Chemical manufacturer, importer, or distributor to specify parts of the body to be washed after handling.	**If in eyes: Rinse cautiously with water for several minutes. Remove contact lenses, if present and easy to do. Continue rinsing.** **If eye irritation persists: Get medical advice/attention.**		

C.4.6 SENSITIZATION - RESPIRATORY
(Classified in Accordance with Appendix A.4)

Pictogram
Health hazard

Hazard category	**Signal word**	**Hazard statement**
1 (including both sub-categories 1A and 1B)	Danger	May cause allergy or asthma symptoms or breathing difficulties if inhaled

Precautionary statements			
Prevention	**Response**	**Storage**	**Disposal**
Avoid breathing dust/fume/gas/mist/ vapors/spray. Chemical manufacturer, importer, or distributor to specify applicable conditions. **[In case of inadequate ventilation] wear respiratory protection.** Chemical manufacturer, importer, or distributor to specify equipment - *Text in square brackets may be used if additional information is provided with the chemical at the point of use that explains what type of ventilation would be adequate for safe use.*	**If inhaled: If breathing is difficult, remove person to fresh air and keep comfortable for breathing.** **If experiencing respiratory symptoms: Call a poison center/doctor/…** … Chemical manufacturer, importer, or distributor to specify the appropriate source of emergency medical advice.		**Dispose of contents/container to...** ... in accordance with local/regional/national/international regulations (to be specified).

C.4.7 SENSITIZATION - SKIN
(Classified in Accordance with Appendix A.4)

Pictogram
Exclamation mark

Hazard category	Signal word	Hazard statement
1 (including both sub-categories 1A and 1B)	Warning	May cause an allergic skin reaction

Precautionary statements

Prevention	Response	Storage	Disposal
Avoid breathing dust/fume/gas/mist/ vapors/spray. Chemical manufacturer, importer, or distributor to specify applicable conditions. **Contaminated work clothing must not be allowed out of the workplace.** **Wear protective gloves.** Chemical manufacturer, importer, or distributor to specify type of equipment.	**If on skin: Wash with plenty of water/...** ... Chemical manufacturer, importer, or distributor may specify a cleansing agent if appropriate, or may recommend an alternative agent in exceptional cases if water is clearly inappropriate. **If skin irritation or rash occurs: Get medical advice/attention.** **Specific treatment (see ... on this label)** ... Reference to supplemental first aid instruction. - *Manufacturer, importer, or distributor may specify a cleansing agent if appropriate.* **Wash contaminated clothing before reuse.**		**Dispose of contents/container to...** ... in accordance with local/regional/national/international regulations (to be specified).

C.4.8 GERM CELL MUTAGENICITY
(Classified in Accordance with Appendix A.5)

Pictogram
Health hazard

Hazard category	Signal word	Hazard statement
1A and 1B	Danger	May cause genetic defects <...>
2	Warning	Suspected of causing genetic defects <...>
		(state route of exposure if no other routes of exposure cause the hazard)

Precautionary statements

Prevention	Response	Storage	Disposal
Obtain special instructions before use. **Do not handle until all safety precautions have been read and understood.** **Wear protective gloves/protective clothing/eye protection/face protection.** Chemical manufacturer, importer, or distributor to specify type of equipment, as required.	**If exposed or concerned: Get medical advice/attention.**	**Store locked up.**	**Dispose of contents/container to...** ... in accordance with local/regional/national/international regulations (to be specified).

C.4.9 CARCINOGENICITY
(Classified in Accordance with Appendix A.6)

Pictogram
Health hazard

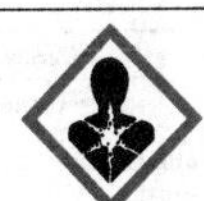

Hazard category	Signal word	Hazard statement
1A and 1B	Danger	May cause cancer <...>
2	Warning	Suspected of causing cancer <...>
		(state route of exposure if no other routes of exposure cause the hazard)

Precautionary statements

Prevention	Response	Storage	Disposal
Obtain special instructions before use. **Do not handle until all safety precautions have been read and understood.** **Wear protective gloves/protective clothing/eye protection/face protection.** Chemical manufacturer, importer, or distributor to specify type of equipment, as required.	**If exposed or concerned: Get medical advice/attention.**	**Store locked up.**	**Dispose of contents/container to...** ... in accordance with local/regional/national/international regulations (to be specified).

Note: If a Category 2 carcinogen ingredient is present in the mixture at a concentration between 0.1% and 1%, information is required on the SDS for a product; however, a label warning is optional. If a Category 2 carcinogen ingredient is present in the mixture at a concentration of ≥ 1%, both an SDS and a label is required and the information must be included on each.

C.4.10 TOXIC TO REPRODUCTION
(Classified in Accordance with Appendix A.7)

Pictogram
Health hazard

Hazard category	Signal word	Hazard statement
1A and 1B	Danger	May damage fertility or the unborn child <...> <<...>>
2	Warning	Suspected of damaging fertility or the unborn child <...> <<...>>
		(state specific effect if known)
		(state route of exposure if no other routes of exposure cause the hazard)

Precautionary statements

Prevention	Response	Storage	Disposal
Obtain special instructions before use. **Do not handle until all safety precautions have been read and understood.** **Wear protective gloves/protective clothing/eye protection/face protection.** Chemical manufacturer, importer, or distributor to specify type of equipment, as required.	**If exposed or concerned: Get medical advice/attention.**	**Store locked up.**	**Dispose of contents/container to...** ... in accordance with local/regional/national/international regulations (to be specified).

C.4.10 TOXIC TO REPRODUCTION (CONTINUED)
(Classified in Accordance with Appendix A.7)
(EFFECTS ON OR VIA LACTATION)

Pictogram
No Pictogram

Hazard category	Signal word	Hazard statement
No designated number	*No signal word*	May cause harm to breast-fed children
(See Table A.7.1 in Appendix A.7)		

Precautionary statements

Prevention	Response	Storage	Disposal
Obtain special instructions before use. **Do not breathe dusts or mists.** - *if inhalable particles of dusts or mists may occur during use.* **Avoid contact during pregnancy/while nursing.** **Wash ... thoroughly after handling.** ...Chemical manufacturer, importer, or distributor to specify parts of the body to be washed after handling. **Do not eat, drink or smoke when using this product.**	**If exposed or concerned: Get medical advice/attention.**		

C.4.11 SPECIFIC TARGET ORGAN TOXICITY (Single Exposure)
(Classified in Accordance with Appendix A.8)

Pictogram
Health hazard

Hazard category	Signal word	Hazard statement
1	Danger	Causes damage to organs <...> <<...>>
		<...> *(or state all organs affected if known)*
		<<...>> *(state route of exposure if no other routes of exposure cause the hazard)*

Precautionary statements

Prevention	Response	Storage	Disposal
Do not breathe dust/fume/gas/mist/ vapors/spray. Chemical manufacturer, importer, or distributor to specify applicable conditions. **Wash ...thoroughly after handling.** ... Chemical manufacturer, importer, or distributor to specify parts of the body to be washed after handling. **Do not eat, drink or smoke when using this product.**	**If exposed: Call a poison center/doctor/...** ... Chemical manufacturer, importer, or distributor to specify the appropriate source of emergency medical advice. **Specific treatment (see ... on this label)** ... Reference to supplemental first aid instruction. - *if immediate measures are required.*	**Store locked up.**	**Dispose of contents/container to...** ... in accordance with local/regional/national/international regulations (to be specified).

C.4.11 SPECIFIC TARGET ORGAN TOXICITY (Single Exposure) (CONTINUED)
(Classified in Accordance with Appendix A.8)

Hazard category	Signal word	Hazard statement	Pictogram Health hazard
2	Warning	May cause damage to organs <...> <<...>> <...> *(or state all organs affected, if known)* <<...>> *(state route of exposure if no other routes of exposure cause the hazard)*	

Precautionary statements

Prevention	Response	Storage	Disposal
Do not breathe dust/fume/gas/mist/ vapors/spray. Chemical manufacturer, importer, or distributor to specify applicable conditions. **Wash … thoroughly after handling.** … Chemical manufacturer, importer, or distributor to specify parts of the body to be washed after handling. **Do not eat, drink or smoke when using this product.**	**If exposed or concerned: Call a poison center/doctor/…** … Chemical manufacturer, importer, or distributor to specify the appropriate source of emergency medical advice.	**Store locked up.**	**Dispose of contents/container to...** ... in accordance with local/regional/national/international regulations (to be specified).

C.4.11 SPECIFIC TARGET ORGAN TOXICITY (Single Exposure) (CONTINUED)
(Classified in Accordance with Appendix A.8)

Hazard category	Signal word	Hazard statement	Pictogram Exclamation mark
3	Warning	May cause respiratory irritation; or May cause drowsiness or dizziness	

Precautionary statements

Prevention	Response	Storage	Disposal
Avoid breathing dust/fume/gas/mist/ vapors/spray. Chemical manufacturer, importer, or distributor to specify applicable conditions. **Use only outdoors or in a well-ventilated area.**	**If inhaled: Remove person to fresh air and keep comfortable for breathing.** **Call a poison center/doctor/…/if you feel unwell.** … Chemical manufacturer, importer, or distributor to specify the appropriate source of emergency medical advice.	**Store in a well-ventilated place. Keep container tightly closed.** - *if product is volatile so as to generate hazardous atmosphere.* **Store locked up.**	**Dispose of contents/container to...** ... in accordance with local/regional/national/international regulations (to be specified).

C.4.12 SPECIFIC TARGET ORGAN TOXICITY (Repeated Exposure)
(Classified in Accordance with Appendix A.9)

Hazard category	Signal word	Hazard statement	Pictogram Health hazard
1	Danger	Causes damage to organs <...> through prolonged or repeated exposure <<...>> <...> *(state all organs affected, if known)* <<...>> *(state route of exposure if no other routes of exposure cause the hazard)*	

Precautionary statements

Prevention	Response	Storage	Disposal
Do not breathe dust/fume/gas/mist/ vapors/spray. Chemical manufacturer, importer, or distributor to specify applicable conditions. **Wash … thoroughly after handling.** …Chemical manufacturer, importer, or distributor to specify parts of the body to be washed after handling. **Do not eat, drink or smoke when using this product.**	**Get medical advice/attention if you feel unwell.**		**Dispose of contents/container to...** ... in accordance with local/regional/national/international regulations (to be specified).

C.4.12 SPECIFIC TARGET ORGAN TOXICITY (Repeated Exposure) (CONTINUED)
(Classified in Accordance with Appendix A.9)

Hazard category	Signal word	Hazard statement	Pictogram Health hazard
2	Warning	May cause damage to organs <...> through prolonged or repeated exposure <<...>> <...> *(state all organs affected, if known)* <<...>> *(state route of exposure if no other routes of exposure cause the hazard)*	

Precautionary statements

Prevention	Response	Storage	Disposal
Do not breathe dust/fume/gas/mist/ vapors/spray. Chemical manufacturer, importer, or distributor to specify applicable conditions.	**Get medical advice/attention if you feel unwell.**		**Dispose of contents/container to...** ... in accordance with local/regional/national/international regulations (to be specified).

C.4.13 ASPIRATION HAZARD
(Classified in Accordance with Appendix A.10)

Hazard category	Signal word	Hazard statement	Pictogram Health hazard
1	Danger	May be fatal if swallowed and enters airways	

Precautionary statements

Prevention	Response	Storage	Disposal
	If swallowed: Immediately call a poison center/doctor/... ... Chemical manufacturer, importer, or distributor to specify the appropriate source of emergency medical advice. **Do NOT induce vomiting.**	**Store locked up.**	**Dispose of contents/container to...** ... in accordance with local/regional/national/international regulations (to be specified).

C.4.14 EXPLOSIVES
(Classified in Accordance with Appendix B.1)

Hazard category	Signal word	Hazard statement	Pictogram Exploding bomb
Unstable explosive	Danger	Unstable explosive	

Precautionary statements

Prevention	Response	Storage	Disposal
Obtain special instructions before use. **Do not handle until all safety precautions have been read and understood.** **Wear personal protective equipment/face protection.** Chemical manufacturer, importer, or distributor to specify type of equipment, as required.	**Explosion risk in case of fire.** **Do NOT fight fire when fire reaches explosives.** **Evacuate area.**	**Store ...** ...in accordance with local/regional/ national/international regulations (to be specified).	**Dispose of contents/container to ...** ...in accordance with local/regional/ national/international regulations (to be specified).

C.4.14 EXPLOSIVES (CONTINUED)
(Classified in Accordance with Appendix B.1)

Hazard category	Signal word	Hazard statement	Pictogram Exploding bomb
Division 1.1	Danger	Explosive; mass explosion hazard	
Division 1.2	Danger	Explosive; severe projection hazard	
Division 1.3	Danger	Explosive; fire, blast or projection hazard	

Precautionary statements			
Prevention	**Response**	**Storage**	**Disposal**
Keep away from heat/sparks/open flames/hot surfaces. - No smoking. Chemical manufacturer, importer, or distributor to specify applicable ignition source(s). **Keep wetted with...** ... Chemical manufacturer, importer, or distributor to specify appropriate material. - *if drying out increases explosion hazard, except as needed for manufacturing or operating processes (e.g., nitrocellulose).* **Ground/bond container and receiving equipment.** - *if the explosive is electrostatically sensitive.* **Do not subject to grinding/shock/.../friction.** ...Chemical manufacturer, importer, or distributor to specify applicable rough handling. **Wear face protection.** Chemical manufacturer, importer, or distributor to specify type of equipment.	**In case of fire: evacuate area.** **Explosion risk in case of fire.** **Do NOT fight fire when fire reaches explosives.**	**Store ...** ...in accordance with local/regional/national/ international regulations (to be specified).	**Dispose of contents/container to ...** ... in accordance with local/ regional/national/ international regulations (to be specified).

Note: Unpackaged explosives or explosives repacked in packagings other than the original or similar packaging shall have the label elements assigned to Division 1.1 unless the hazard is shown to correspond to one of the hazard categories in Appendix B.1, in which case the corresponding symbol, signal word and/or the hazard statement shall be assigned.

C.4.14 EXPLOSIVES (CONTINUED)
(Classified in Accordance with Appendix B.1)

Hazard category	Signal word	Hazard statement	Pictogram Exploding bomb[1]
Division 1.4	Warning	Fire or projection hazard	

Precautionary statements[1]			
Prevention	**Response**	**Storage**	**Disposal**
Keep away from heat/sparks/open flames/hot surfaces. - No smoking. Chemical manufacturer, importer, or distributor to specify applicable ignition source(s). **Ground/bond container and receiving equipment.** - *if the explosive is electrostatically sensitive.* **Do not subject to grinding/shock/.../friction.** Chemical manufacturer, importer, or distributor to specify applicable rough handling. **Wear face protection.** Chemical manufacturer, importer, or distributor to specify type of equipment.	**In case of fire: Evacuate area.** **Explosion risk in case of fire.** - *except if explosives are 1.4S ammunition and components thereof.* **Do NOT fight fire when fire reaches explosives.** **Fight fire with normal precautions from a reasonable distance** - *if explosives are 1.4S ammunition and components thereof.*	**Store ...** ...in accordance with local/regional/ national/internation al regulations (to be specified).	**Dispose of contents/container to...** ... in accordance with local/regional/national/i nternational regulations (to be specified).

Note: Unpackaged explosives or explosives repacked in packagings other than the original or similar packaging shall have the label elements assigned to Division 1.1 unless the hazard is shown to correspond to one of the hazard categories in Appendix B.1, in which case the corresponding symbol, signal word and/or the hazard statement shall be assigned.[1]

[1] *Except no pictogram is required for explosives that are 1.4S small arms ammunition and components thereof. Labels for 1.4S small arms ammunition and components shall include appropriate precautionary statements.*

C.4.14 EXPLOSIVES (CONTINUED)
(Classified in Accordance with Appendix B.1)

Hazard category	Signal word	Hazard statement	Pictogram
Division 1.5	Danger	May mass explode in fire	No pictogram

Prevention	Response	Storage	Disposal
Keep away from heat/sparks/open flames/hot surfaces. - No smoking. Chemical manufacturer, importer, or distributor to specify applicable ignition source(s). **Keep wetted with...** ... Chemical manufacturer, importer, or distributor to specify appropriate material. - *if drying out increases explosion hazard, except as needed for manufacturing or operating processes (e.g., nitrocellulose).* **Ground/bond container and receiving equipment** - *if the explosive is electrostatically sensitive.* **Do not subject to grinding/shock/.../friction.** ...Chemical manufacturer, importer, or distributor to specify applicable rough handling. **Wear face protection.** Chemical manufacturer, importer, or distributor to specify type of equipment.	**In case of fire: Evacuate area.** **Explosion risk in case of fire.** **Do NOT fight fire when fire reaches explosives.**	**Store ...** ...in accordance with local/regional/ national/international regulations (to be specified).	**Dispose of contents/container to ...** ... in accordance with local/regional/ national/international regulations (to be specified).

Note: Unpackaged explosives or explosives repacked in packagings other than the original or similar packaging shall have the label elements assigned to Division 1.1 unless the hazard is shown to correspond to one of the hazard categories in Appendix B.1, in which case the corresponding symbol, signal word and/or the hazard statement shall be assigned.

C.4.14 EXPLOSIVES (CONTINUED)
(Classified in Accordance with Appendix B.1)

Hazard category	Signal word	Hazard statement	Pictogram
Division 1.6	*No signal word*	*No hazard statement*	*No pictogram*

Prevention	Response	Storage	Disposal
None assigned.	None assigned	None assigned	None assigned

Note: Unpackaged explosives or explosives repacked in packagings other than the original or similar packaging shall have the label elements assigned to Division 1.1 unless the hazard is shown to correspond to one of the hazard categories in Appendix B.1, in which case the corresponding symbol, signal word and/or the hazard statement shall be assigned.

C.4.15 FLAMMABLE GASES
(Classified in Accordance with Appendix B.2)

Hazard category	Signal word	Hazard statement	Pictogram
1	Danger	Extremely flammable gas	Flame

Prevention	Response	Storage	Disposal
Keep away from heat/sparks/open flames/hot surfaces. -No smoking. Chemical manufacturer, importer, or distributor to specify applicable ignition source(s).	**Leaking gas fire:** **Do not extinguish, unless leak can be stopped safely.** **Eliminate all ignition sources if safe to do so.**	**Store in well-ventilated place.**	

C.4.15 FLAMMABLE GASES (CONTINUED)
(Classified in Accordance with Appendix B.2)

Hazard category	Signal word	Hazard statement	Pictogram
2	Warning	Flammable gas	*No Pictogram*

Prevention	Response	Storage	Disposal
Keep away from heat/sparks/open flames/hot surfaces. -No smoking. Chemical manufacturer, importer, or distributor to specify applicable ignition sources(s).	**Leaking gas fire: Do not extinguish, unless leak can be stopped safely.** **Eliminate all ignition sources if safe to do so.**	**Store in well-ventilated place.**	

C.4.16 FLAMMABLE AEROSOLS
(Classified in Accordance with Appendix B.3)

Pictogram
Flame

Hazard category	Signal word	Hazard statement
1	Danger	Extremely flammable aerosol
2	Warning	Flammable aerosol

Precautionary statements

Prevention	Response	Storage	Disposal
Keep away from heat/sparks/open flames/hot surfaces. -No smoking. Chemical manufacturer, importer, or distributor to specify applicable ignition sources(s). **Do not spray on an open flame or other ignition source.** **Pressurized container: Do not pierce or burn, even after use.**		**Protect from sunlight. Do not expose to temperatures exceeding 50 °C/122 °F.**	

C.4.17 OXIDIZING GASES
(Classified in Accordance with Appendix B.4)

Pictogram
Flame over circle

Hazard category	Signal word	Hazard statement
1	Danger	May cause or intensify fire; oxidizer

Precautionary statements

Prevention	Response	Storage	Disposal
Keep/Store away from clothing/.../combustible materials. ...Chemical manufacturer, importer, or distributor to specify other incompatible materials. **Keep reduction valves/valves and fittings free from oil and grease.**	**In case of fire: Stop leak if safe to do so.**	**Store in well-ventilated place.**	

C.4.18 GASES UNDER PRESSURE
(Classified in Accordance with Appendix B.5)

Pictogram
Gas cylinder

Hazard category	Signal word	Hazard statement
Compressed gas	Warning	Contains gas under pressure; may explode if heated
Liquefied gas	Warning	Contains gas under pressure; may explode if heated
Dissolved gas	Warning	Contains gas under pressure; may explode if heated

Precautionary statements

Prevention	Response	Storage	Disposal
		Protect from sunlight. Store in a well-ventilated place.	

C.4.18 GASES UNDER PRESSURE (CONTINUED)
(Classified in Accordance with Appendix B.5)

Pictogram
Gas cylinder

Hazard category	Signal word	Hazard statement
Refrigerated liquefied gas	Warning	Contains refrigerated gas; may cause cryogenic burns or injury

Precautionary statements

Prevention	Response	Storage	Disposal
Wear cold insulating gloves/face shield/eye protection.	**Thaw frosted parts with lukewarm water. Do not rub affected area.** **Get immediate medical advice/attention**	**Store in well-ventilated place.**	

C.4.19 FLAMMABLE LIQUIDS
(Classified in Accordance with Appendix B.6)

Hazard category	Signal word	Hazard statement
1	Danger	Extremely flammable liquid and vapor
2	Danger	Highly flammable liquid and vapor
3	Warning	Flammable liquid and vapor

Pictogram
Flame

Precautionary statements			
Prevention	**Response**	**Storage**	**Disposal**
Keep away from heat/sparks/open flames/hot surfaces.– No smoking. Chemical manufacturer, importer, or distributor to specify applicable ignition source(s). **Keep container tightly closed.** **Ground/Bond container and receiving equipment** - *if electrostatically sensitive material is for reloading.* - *if product is volatile so as to generate hazardous atmosphere.* **Use explosion-proof electrical/ventilating/ lighting/.../equipment.** ... Chemical manufacturer, importer, or distributor to specify other equipment. **Use only non-sparking tools.** **Take precautionary measures against static discharge.** **Wear protective gloves/eye protection/face protection** Chemical manufacturer, importer, or distributor to specify type of equipment.	**If on skin (or hair): Take off immediately all contaminated clothing. Rinse skin with water/shower.** **In case of fire: Use ... to extinguish.** ... Chemical manufacturer, importer, or distributor to specify appropriate media. - *if water increases risk.*	**Store in a well-ventilated place. Keep cool.**	**Dispose of contents/container to...** ... in accordance with local/regional/national/ international regulations (to be specified).

C.4.19 FLAMMABLE LIQUIDS (CONTINUED)
(Classified in Accordance with Appendix B.6)

Pictogram
No Pictogram

Hazard category	Signal word	Hazard statement
4	Warning	Combustible liquid

Precautionary statements			
Prevention	**Response**	**Storage**	**Disposal**
Keep away from flames and hot surfaces. – No smoking. **Wear protective gloves/eye protection/face protection** Chemical manufacturer, importer, or distributor to specify type of equipment.	**In case of fire: Use ... to extinguish.** ... Chemical manufacturer, importer, or distributor to specify appropriate media. - *if water increases risk.*	**Store in a well-ventilated place. Keep cool.**	**Dispose of contents/container to...** in accordance with local/regional/ national/international regulations (to be specified).

C.4.20 FLAMMABLE SOLIDS
(Classified in Accordance with Appendix B.7)

Hazard category	Signal word	Hazard statement
1	Danger	Flammable solid
2	Warning	Flammable solid

Pictogram
Flame

Precautionary statements			
Prevention	**Response**	**Storage**	**Disposal**
Keep away from heat/sparks/open flames/hot surfaces. - No smoking. Chemical manufacturer, importer, or distributor to specify applicable ignition source(s). **Ground/Bond container and receiving equipment.** - *if electrostatically sensitive material is for reloading.* **Use explosion-proof electrical/ventilating/ lighting/... /equipment.** ... Chemical manufacturer, importer, or distributor to specify other equipment. - *if dust clouds can occur.* **Wear protective gloves/eye protection/face protection** Chemical manufacturer, importer, or distributor to specify type of equipment.	**In case of fire: Use ... to extinguish** ... Chemical manufacturer, importer, or distributor to specify appropriate media. - *if water increases risk.*		

C.4.21 SELF-REACTIVE SUBSTANCES AND MIXTURES
(Classified in Accordance with Appendix B.8)

Pictogram
Exploding bomb

Hazard category	Signal word	Hazard statement
Type A	Danger	Heating may cause an explosion

Precautionary statements

Prevention	Response	Storage	Disposal
Keep away from heat/sparks/open flames/hot surfaces. - No smoking. Chemical manufacturer, importer, or distributor to specify applicable ignition source(s). **Keep/Store away from clothing/.../combustible materials.** ... Chemical manufacturer, importer, or distributor to specify other incompatible materials. **Keep only in original container.** **Wear protective gloves/eye protection/face protection.** Chemical manufacturer, importer, or distributor to specify type of equipment.	**In case of fire: Use ... to extinguish** ... Chemical manufacturer, importer, or distributor to specify appropriate media. - *if water increases risk.* **In case of fire: Evacuate area. Fight fire remotely due to the risk of explosion.**	**Store in a well-ventilated place. Keep cool.** **Store at temperatures not exceeding ...°C/...°F.** ... Chemical manufacturer, importer, or distributor to specify temperature. **Store away from other materials.**	**Dispose of contents/container to...** ... in accordance with local/regional/national/international regulations (to be specified).

C.4.21 SELF-REACTIVE SUBSTANCES AND MIXTURES (CONTINUED)
(Classified in Accordance with Appendix B.8)

Pictograms
Exploding bomb and flame

Hazard category	Signal word	Hazard statement
Type B	Danger	Heating may cause a fire or explosion

Precautionary statements

Prevention	Response	Storage	Disposal
Keep away from heat/sparks/open flames/hot surfaces. - No smoking. Chemical manufacturer, importer, or distributor to specify applicable ignition source(s). **Keep/Store away from clothing/.../combustible materials.** ... Chemical manufacturer, importer, or distributor to specify other incompatible materials. **Keep only in original container.** **Wear protective gloves/eye protection/face protection.** Chemical manufacturer, importer, or distributor to specify type of equipment.	**In case of fire: Use ... to extinguish.** ... Chemical manufacturer, importer, or distributor to specify appropriate media. - *if water increases risk.* **In case of fire: Evacuate area. Fight fire remotely due to the risk of explosion.**	**Store in a well-ventilated place. Keep cool.** **Store at temperatures not exceeding ...°C/...°F.** ... Chemical manufacturer, importer, or distributor to specify temperature. **Store away from other materials.**	**Dispose of contents/container to...** ...in accordance with local/regional/national/international regulations (to be specified).

C.4.21 SELF-REACTIVE SUBSTANCES AND MIXTURES(CONTINUED)
(Classified in Accordance with Appendix B.8)

Pictogram
Flame

Hazard category	Signal word	Hazard statement
Type C	Danger	Heating may cause a fire
Type D	Danger	Heating may cause a fire
Type E	Warning	Heating may cause a fire
Type F	Warning	Heating may cause a fire

Precautionary statements

Prevention	Response	Storage	Disposal
Keep away from heat/sparks/open flames/hot surfaces. - No smoking. Chemical manufacturer, importer, or distributor to specify applicable ignition source(s). **Keep/Store away from clothing/.../combustible materials.** ...Chemical manufacturer, importer, or distributor to specify other incompatible materials. **Keep only in original container.** **Wear protective gloves/eye protection/face protection.** Chemical manufacturer, importer, or distributor to specify type of equipment.	**In case of fire: Use ... to extinguish** ... Chemical manufacturer, importer, or distributor to specify appropriate media. - *if water increases risk.*	**Store in a well-ventilated place. Keep cool.** **Store at temperatures not exceeding ...°C/...°F.** ...Chemical manufacturer, importer, or distributor to specify temperature. **Store away from other materials.**	**Dispose of contents/container to...** ...in accordance with local/regional/national/international regulations (to be specified).

C.4.22 PYROPHORIC LIQUIDS
(Classified in Accordance with Appendix B.9)

Hazard category	Signal word	Hazard statement
1	Danger	Catches fire spontaneously if exposed to air

Pictogram
Flame

Precautionary statements			
Prevention	**Response**	**Storage**	**Disposal**
Keep away from heat/sparks/open flames/hot surfaces. - No smoking. Chemical manufacturer, importer, or distributor to specify applicable ignition sources(s). **Do not allow contact with air.** **Wear protective gloves/eye protection/face protection.** Chemical manufacturer, importer, or distributor to specify type of equipment.	**If on skin: Immerse in cool water/wrap with wet bandages** **In case of fire: Use ... to extinguish** ... Chemical manufacturer, importer, or distributor to specify appropriate media. - *if water increases risk.*	**Store contents under ...** ... Chemical manufacturer, importer, or distributor to specify appropriate liquid or inert gas.	

C.4.23 PYROPHORIC SOLIDS
(Classified in Accordance with Appendix B.10)

Hazard category	Signal word	Hazard statement
1	Danger	Catches fire spontaneously if exposed to air

Pictogram
Flame

Precautionary statements			
Prevention	**Response**	**Storage**	**Disposal**
Keep away from heat/sparks/open flames/hot surfaces. - No smoking. Chemical manufacturer, importer, or distributor to specify applicable ignition source(s). **Do not allow contact with air.** **Wear protective gloves/eye protection/face protection** Chemical manufacturer, importer, or distributor to specify type of equipment.	**Brush off loose particles from skin. Immerse in cool water/wrap in wet bandages.** **In case of fire: Use ... to extinguish** ... Chemical manufacturer, importer, or distributor to specify appropriate media. - *if water increases risk.*	**Store contents under ...** ...Chemical manufacturer, importer, or distributor to specify appropriate liquid or inert gas.	

C.4.24 SELF-HEATING SUBSTANCES AND MIXTURES
(Classified in Accordance with Appendix B.11)

Hazard category	Signal word	Hazard statement
1	Danger	Self-heating; may catch fire
2	Warning	Self-heating in large quantities; may catch fire

Pictogram
Flame

Precautionary statements			
Prevention	**Response**	**Storage**	**Disposal**
Keep cool. Protect from sunlight. **Wear protective gloves/eye protection/face protection.** Chemical manufacturer, importer, or distributor to specify type of equipment.		**Maintain air gap between stacks/pallets.** **Store bulk masses greater than ... kg/...lbs at temperatures not exceeding ...°C/...°F.** ... Chemical manufacturer, importer, or distributor to specify mass and temperature. **Store away from other materials.**	

C.4.25 SUBSTANCES AND MIXTURES WHICH, IN CONTACT WITH WATER, EMIT FLAMMABLE GASES
(Classified in Accordance with Appendix B.12)

Pictogram
Flame

Hazard category	Signal word	Hazard statement
1	Danger	In contact with water releases flammable gases, which may ignite spontaneously
2	Danger	In contact with water releases flammable gas

Precautionary statements

Prevention	Response	Storage	Disposal
Do not allow contact with water. **Handle under inert gas. Protect from moisture.** **Wear protective gloves/eye protection/face protection.** Chemical manufacturer, importer, or distributor to specify type of equipment.	**Brush off loose particles from skin and immerse in cool water/wrap in wet bandages.** **In case of fire: Use ... to extinguish** ... Chemical manufacturer, importer, or distributor to specify appropriate media. - *if water increases risk.*	**Store in a dry place. Store in a closed container.**	**Dispose of contents/container to...** ...in accordance with local/regional/national/ international regulations (to be specified).

C.4.25 SUBSTANCES AND MIXTURES WHICH, IN CONTACT WITH WATER, EMIT FLAMMABLE GASES (CONTINUED)
(Classified in Accordance with Appendix B.12)

Pictogram
Flame

Hazard category	Signal word	Hazard statement
3	Warning	In contact with water releases flammable gas

Precautionary statements

Prevention	Response	Storage	Disposal
Handle under inert gas. Protect from moisture. **Wear protective gloves/eye protection/face protection.** Chemical manufacturer, importer, or distributor to specify type of equipment.	**In case of fire: Use ... to extinguish.** ... Chemical manufacturer, importer, or distributor to specify appropriate media. - *if water increases risk.*	**Store in a dry place. Store in a closed container.**	**Dispose of contents/container to...** ... in accordance with local/regional/national/international regulations (to be specified).

C.4.26 OXIDIZING LIQUIDS
(Classified in Accordance with Appendix B.13)

Pictogram
Flame over circle

Hazard category	Signal word	Hazard statement
1	Danger	May cause fire or explosion; strong oxidizer

Precautionary statements

Prevention	Response	Storage	Disposal
Keep away from heat. **Keep/Store away from clothing and other combustible materials.** **Take any precaution to avoid mixing with combustibles/...** ... Chemical manufacturer, importer, or distributor to specify other incompatible materials. **Wear protective gloves /eye protection/face protection.** Chemical manufacturer, importer, or distributor to specify type of equipment. **Wear fire/flame resistant/retardant clothing.**	**If on clothing: Rinse immediately contaminated clothing and skin with plenty of water before removing clothes.** **In case of major fire and large quantities: Evacuate area. Fight fire remotely due to the risk of explosion.** **In case of fire: Use ... to extinguish.** ... Chemical manufacturer, importer, or distributor to specify appropriate media. - *if water increases risk.*		**Dispose of contents/container to...** ...in accordance with local/regional/ national/international regulations (to be specified).

C.4.26 OXIDIZING LIQUIDS (CONTINUED)
(Classified in Accordance with Appendix B.13)

Hazard category	Signal word	Hazard statement
2	Danger	May intensify fire; oxidizer
3	Warning	May intensify fire; oxidizer

Pictogram
Flame over circle

Precautionary statements

Prevention	Response	Storage	Disposal
Keep away from heat. **Keep/Store away from clothing/…/combustible materials.** …Chemical manufacturer, importer, or distributor to specify other incompatible materials. **Take any precaution to avoid mixing with combustibles/...** ... Chemical manufacturer, importer, or distributor to specify other incompatible materials. **Wear protective gloves/eye protection/face protection.** Chemical manufacturer, importer, or distributor to specify type of equipment.	**In case of fire: Use ... to extinguish.** ... Chemical manufacturer, importer, or distributor to specify appropriate media. - *if water increases risk.*		**Dispose of contents/container to...** …in accordance with local/regional/ national/international regulations (to be specified).

C.4.27 OXIDIZING SOLIDS
(Classified in Accordance with Appendix B.14)

Hazard category	Signal word	Hazard statement
1	Danger	May cause fire or explosion; strong oxidizer

Pictogram
Flame over circle

Precautionary statements

Prevention	Response	Storage	Disposal
Keep away from heat. **Keep away from clothing and other combustible materials.** **Take any precaution to avoid mixing with combustibles/...** …Chemical manufacturer, importer, or distributor to specify other incompatible materials. **Wear protective gloves/eye protection/face protection.** Chemical manufacturer, importer, or distributor to specify type of equipment. **Wear fire/flame resistant/retardant clothing.**	**If on clothing: Rinse immediately contaminated clothing and skin with plenty of water before removing clothes.** **In case of major fire and large quantities: Evacuate area. Fight fire remotely due to the risk of explosion.** **In case of fire: Use ... to extinguish.** ... Chemical manufacturer, importer, or distributor to specify appropriate media. - *if water increases risk.*		**Dispose of contents/container to...** …in accordance with local/regional/ national/international regulations (to be specified).

C.4.27 OXIDIZING SOLIDS (CONTINUED)
(Classified in Accordance with Appendix B.14)

Hazard category	Signal word	Hazard statement
2	Danger	May intensify fire; oxidizer
3	Warning	May intensify fire; oxidizer

Pictogram
Flame over circle

Precautionary statements

Prevention	Response	Storage	Disposal
Keep away from heat. **Keep/Store away from clothing/…/ combustible materials.** … Chemical manufacturer, importer, or distributor to specify incompatible materials. **Take any precaution to avoid mixing with combustibles/...** …Chemical manufacturer, importer, or distributor to specify other incompatible materials. **Wear protective gloves/eye protection/face protection.** Chemical manufacturer, importer, or distributor to specify type of equipment.	**In case of fire: Use ... to extinguish.** ... Chemical manufacturer, importer, or distributor to specify appropriate media. - *if water increases risk.*		**Dispose of contents/container to...** ... in accordance with local/regional/national/international regulations (to be specified).

C.4.28 ORGANIC PEROXIDES
(Classified in Accordance with Appendix B.15)

Pictogram
Exploding bomb

Hazard category	Signal word	Hazard statement
Type A	Danger	Heating may cause an explosion

Precautionary statements

Prevention	Response	Storage	Disposal
Keep away from heat/sparks/open flames/hot surfaces.- No smoking. Chemical manufacturer, importer, or distributor to specify applicable ignition source(s). **Keep/Store away from clothing/…/combustible materials.** … Chemical manufacturer, importer, or distributor to specify incompatible materials. **Keep only in original container.** **Wear protective gloves/eye protection/face protection.** Chemical manufacturer, importer, or distributor to specify type of equipment.		**Store at temperatures not exceeding …°C/…°F. Keep cool.** … Chemical manufacturer, importer, or distributor to specify temperature. **Protect from sunlight.** **Store away from other materials.**	**Dispose of contents/container to...** … in accordance with local/regional/national/international regulations (to be specified).

C.4.28 ORGANIC PEROXIDES (CONTINUED)
(Classified in Accordance with Appendix B.15)

Pictograms
Exploding bomb and flame

Hazard category	Signal word	Hazard statement
Type B	Danger	Heating may cause a fire or explosion

Precautionary statements

Prevention	Response	Storage	Disposal
Keep away from heat/sparks/open flames/hot surfaces. - No smoking. Chemical manufacturer, importer, or distributor to specify applicable ignition source(s). **Keep /Store away from clothing/…/combustible materials.** … Chemical manufacturer, importer, or distributor to specify incompatible materials. **Keep only in original container.** **Wear protective gloves/eye protection/face protection.** Chemical manufacturer, importer, or distributor to specify type of equipment.		**Store at temperatures not exceeding …°C/…°F. Keep cool.** Chemical manufacturer, importer, or distributor to specify temperature. **Protect from sunlight.** **Store away from other materials.**	**Dispose of contents/container to...** … in accordance with local/regional/national/international regulations (to be specified).

C.4.28 ORGANIC PEROXIDES (CONTINUED)
(Classified in Accordance with Appendix B.15)

Pictogram
Flame

Hazard category	Signal word	Hazard statement
Type C	Danger	Heating may cause a fire
Type D	Danger	Heating may cause a fire
Type E	Warning	Heating may cause a fire
Type F	Warning	Heating may cause a fire

Precautionary statements

Prevention	Response	Storage	Disposal
Keep away from heat/sparks/open flames/hot surfaces. - No smoking. Chemical manufacturer, importer, or distributor to specify applicable ignition source(s). **Keep/Store away from clothing/…/ combustible materials** … Chemical manufacturer, importer, or distributor to specify incompatible materials. **Keep only in original container.** **Wear protective gloves/eyes protection/face protection.** Chemical manufacturer, importer, or distributor to specify type of equipment.		**Store at temperatures not exceeding …°C/…°F. Keep cool.** … Chemical manufacturer, importer, or distributor to specify temperature. **Protect from sunlight.** **Store away from other materials.**	**Dispose of contents/container to...** … in accordance with local/regional/national/international regulations (to be specified).

C.4.29 CORROSIVE TO METALS
(Classified in Accordance with Appendix B.16)

Hazard category	Signal word	Hazard statement	Pictogram
1	Warning	May be corrosive to metals	Corrosion

Precautionary statements			
Prevention	**Response**	**Storage**	**Disposal**
Keep only in original container.	**Absorb spillage to prevent material damage.**	**Store in corrosive resistant/... container with a resistant inner liner.** ... Chemical manufacturer, importer, or distributor to specify other compatible materials.	

C.4.30 Label elements for OSHA defined hazards

Hazard	Signal word	Hazard statement	Pictogram
Pyrophoric Gas	Danger	Catches fire spontaneously if exposed to air	Flame
Simple Asphyxiant	Warning	May displace oxygen and cause rapid suffocation	No Pictogram
Combustible Dust[2]	Warning	May form combustible dust concentrations in air	No Pictogram

[2] The chemical manufacturer or importer shall label chemicals that are shipped in dust form, and present a combustible dust hazard in that form when used downstream, under paragraph (f)(1); 2) the chemical manufacturer or importer shipping chemicals that are in a form that is not yet a dust must provide a label to customers under paragraph (f)(4) if, under normal conditions of use, the chemicals are processed in a downstream workplace in such a way that they present a combustible dust hazard; and 3) the employer shall follow the workplace labeling requirements under paragraph (f)(6) where combustible dust hazards are present.

§1910.1200 Appendix D

Safety Data Sheets (Mandatory)

A safety data sheet (SDS) shall include the information specified in Table D.1 under the section number and heading indicated for sections 1-11 and 16. If no relevant information is found for any given subheading within a section, the SDS shall clearly indicate that no applicable information is available. Sections 12-15 may be included in the SDS, but are not mandatory.

Table D.1 — Minimum Information for an SDS

Heading	Subheading
1. Identification	(a) Product identifier used on the label; (b) Other means of identification; (c) Recommended use of the chemical and restrictions on use; (d) Name, address, and telephone number of the chemical manufacturer, importer, or other responsible party; (e) Emergency phone number.
2. Hazard(s) identification	(a) Classification of the chemical in accordance with paragraph (d) of §1910.1200; (b) Signal word, hazard statement(s), symbol(s) and precautionary statement(s) in accordance with paragraph (f) of §1910.1200. (Hazard symbols may be provided as graphical reproductions in black and white or the name of the symbol, e.g., flame, skull and crossbones); (c) Describe any hazards not otherwise classified that have been identified during the classification process; (d) Where an ingredient with unknown acute toxicity is used in a mixture at a concentration ≥1% and the mixture is not classified based on testing of the mixture as a whole, a statement that X% of the mixture consists of ingredient(s) of unknown acute toxicity is required.

Table D.1 — Minimum Information for an SDS (continued)

Heading	Subheading
3. Composition/ information on ingredients	Except as provided for in paragraph (i) of §1910.1200 on trade secrets:
	For Substances
	(a) Chemical name;
	(b) Common name and synonyms;
	(c) CAS number and other unique identifiers;
	(d) Impurities and stabilizing additives which are themselves classified and which contribute to the classification of the substance.
	For Mixtures
	In addition to the information required for substances:
	(a) The chemical name and concentration (exact percentage) or concentration ranges of all ingredients which are classified as health hazards in accordance with paragraph (d) of §1910.1200 and
	(1) Are present above their cut-off/concentration limits; or
	(2) Present a health risk below the cut-off/concentration limits.
	(b) The concentration (exact percentage) shall be specified unless a trade secret claim is made in accordance with paragraph (i) of §1910.1200, when there is batch-to-batch variability in the production of a mixture, or for a group of substantially similar mixtures (*See* A.0.5.1.2) with similar chemical composition. In these cases, concentration ranges may be used.
	For All Chemicals Where a Trade Secret is Claimed

Table D.1 — Minimum Information for an SDS (continued)

Heading	Subheading
	Where a trade secret is claimed in accordance with paragraph (i) of §1910.1200, a statement that the specific chemical identity and/or exact percentage (concentration) of composition has been withheld as a trade secret is required.
4. First-aid measures	(a) Description of necessary measures, subdivided according to the different routes of exposure, i.e., inhalation, skin and eye contact, and ingestion; (b) Most important symptoms/effects, acute and delayed. (c) Indication of immediate medical attention and special treatment needed, if necessary.
5. Fire-fighting measures	(a) Suitable (and unsuitable) extinguishing media. (b) Specific hazards arising from the chemical (e.g., nature of any hazardous combustion products). (c) Special protective equipment and precautions for fire-fighters.
6. Accidental release measures	(a) Personal precautions, protective equipment, and emergency procedures. (b) Methods and materials for containment and cleaning up.
7. Handling and storage	(a) Precautions for safe handling. (b) Conditions for safe storage, including any incompatibilities.
8. Exposure controls/personal protection	(a) OSHA permissible exposure limit (PEL), American Conference of Governmental Industrial Hygienists (ACGIH) Threshold Limit Value (TLV), and any other exposure limit used or recommended by the chemical manufacturer, importer, or employer preparing the safety data sheet, where available. (b) Appropriate engineering controls. (c) Individual protection measures, such as personal protective equipment.
9. Physical and chemical properties	(a) Appearance (physical state, color, etc.);
	(b) Odor;
	(c) Odor threshold;
	(d) pH;
	(e) Melting point/freezing point;
	(f) Initial boiling point and boiling range;
	(g) Flash point;
	(h) Evaporation rate;
	(i) Flammability (solid, gas);
	(j) Upper/lower flammability or explosive limits;
	(k) Vapor pressure;
	(l) Vapor density;
	(m) Relative density;
	(n) Solubility(ies);
	(o) Partition coefficient: n-octanol/water;
	(p) Auto-ignition temperature;
	(q) Decomposition temperature;
	(r) Viscosity.
10. Stability and reactivity	(a) Reactivity;
	(b) Chemical stability;
	(c) Possibility of hazardous reactions;
	(d) Conditions to avoid (e.g., static discharge, shock, or vibration);
	(e) Incompatible materials;
	(f) Hazardous decomposition products.
11. Toxicological information	Description of the various toxicological (health) effects and the available data used to identify those effects, including:
	(a) Information on the likely routes of exposure (inhalation, ingestion, skin and eye contact);
	(b) Symptoms related to the physical, chemical and toxicological characteristics;
	(c) Delayed and immediate effects and also chronic effects from short- and long-term exposure;
	(d) Numerical measures of toxicity (such as acute toxicity estimates).
	(e) Whether the hazardous chemical is listed in the National Toxicology Program (NTP) Report on Carcinogens (latest edition) or has been found to be a potential carcinogen in the International Agency for Research on Cancer (IARC) Monographs (latest edition), or by OSHA.
12. Ecological information (Non-mandatory)	(a) Ecotoxicity (aquatic and terrestrial, where available); (b) Persistence and degradability; (c) Bioaccumulative potential; (d) Mobility in soil; (e) Other adverse effects (such as hazardous to the ozone layer).
13. Disposal considerations (Non-mandatory)	Description of waste residues and information on their safe handling and methods of disposal, including the disposal of any contaminated packaging.
14. Transport information (Non-mandatory)	(a) UN number;
	(b) UN proper shipping name;
	(c) Transport hazard class(es);
	(d) Packing group, if applicable;
	(e) Environmental hazards (e.g., Marine pollutant (Yes/No));
	(f) Transport in bulk (according to Annex II of MARPOL 73/78 and the IBC Code);
	(g) Special precautions which a user needs to be aware of, or needs to comply with, in connection with transport or conveyance either within or outside their premises.
15. Regulatory information (Non-mandatory)	Safety, health and environmental regulations specific for the product in question.
16. Other information, including date of preparation or last revision	The date of preparation of the SDS or the last change to it.

§1910.1200 Appendix E

Definition of "Trade Secret" (Mandatory)

The following is a reprint of the Restatement of Torts section 757, comment b (1939):

(b) Definition of trade secret.

A trade secret may consist of any formula, pattern, device or compilation of information which is used in one's business, and which gives him an opportunity to obtain an advantage over competitors who do not know or use it. It may be a formula for a chemical compound, a process of manufacturing, treating or preserving materials, a pattern for a machine or other device, or a list of customers. It differs from other secret information in a business (see s759 of the Restatement of Torts which is not included in this Appendix) in that it is not simply information as to single or ephemeral events in the conduct of the business, as, for example, the amount or other terms of a secret bid for a contract or the salary of certain employees, or the security investments made or contemplated, or the date fixed for the announcement of a new policy or for bringing out a new model or the like. A trade secret is a process or device for continuous use in the operations of the business. Generally it relates to the production of goods, as, for example, a machine or formula for the production of an article. It may, however, relate to the sale of goods or to other operations in the business, such as a code for determining discounts, rebates or other concessions in a price list or catalogue, or a list of specialized customers, or a method of bookkeeping or other office management.

Secrecy. The subject matter of a trade secret must be secret. Matters of public knowledge or of general knowledge in an industry cannot be appropriated by one as his secret. Matters which are completely disclosed by the goods which one markets cannot be his secret. Substantially, a trade secret is known only in the particular business in which it is used. It is not requisite that only the proprietor of the business know it. He may, without losing his protection, communicate it to employees involved in its use. He may likewise communicate it to others pledged to secrecy. Others may also know of it independently, as, for example, when they have discovered the process or formula by independent invention and are keeping it secret. Nevertheless, a substantial element of secrecy must exist, so that, except by the use of improper means, there would be difficulty in acquiring the information. An exact definition of a trade secret is not possible. Some factors to be considered in determining whether given information is one's trade secret are:

(1) The extent to which the information is known outside of his business;

(2) the extent to which it is known by employees and others involved in his business;

(3) **the extent of measures taken** by him to guard the secrecy of the information;
(4) **the value of the information** to him and his competitors;
(5) **the amount of effort or money expended** by him in developing the information;
(6) **the ease or difficulty** with which the information could be properly acquired or duplicated by others.

Novelty and prior art. A trade secret may be a device or process which is patentable; but it need not be that. It may be a device or process which is clearly anticipated in the prior art or one which is merely a mechanical improvement that a good mechanic can make. Novelty and invention are not requisite for a trade secret as they are for patentability. These requirements are essential to patentability because a patent protects against unlicensed use of the patented device or process even by one who discovers it properly through independent research. The patent monopoly is a reward to the inventor. But such is not the case with a trade secret. Its protection is not based on a policy of rewarding or otherwise encouraging the development of secret processes or devices. The protection is merely against breach of faith and reprehensible means of learning another's secret. For this limited protection it is not appropriate to require also the kind of novelty and invention which is a requisite of patentability. The nature of the secret is, however, an important factor in determining the kind of relief that is appropriate against one who is subject to liability under the rule stated in this Section. Thus, if the secret consists of a device or process which is a novel invention, one who acquires the secret wrongfully is ordinarily enjoined from further use of it and is required to account for the profits derived from his past use. If, on the other hand, the secret consists of mechanical improvements that a good mechanic can make without resort to the secret, the wrongdoer's liability may be limited to damages, and an injunction against future use of the improvements made with the aid of the secret may be inappropriate.

§1910.1200 Appendix F

Guidance for Hazard Classifications Re: Carcinogenicity (Non-Mandatory)

The mandatory criteria for classification of a chemical for carcinogenicity under HCS (Sec. 1910.1200) are found in Appendix A.6 to this section. This non-mandatory Appendix provides additional guidance on hazard classification for carcinogenicity. Part A of Appendix F includes background guidance provided by GHS based on the Preamble of the International Agency for Research on Cancer (IARC) "Monographs on the Evaluation of Carcinogenic Risks to Humans" (2006). Part B provides IARC classification information. Part C provides background guidance from the National Toxicology Program (NTP) "Report on Carcinogens" (RoC), and Part D is a table that compares GHS carcinogen hazard categories to carcinogen classifications under IARC and NTP, allowing classifiers to be able to use information from IARC and NTP RoC carcinogen classifications to complete their classifications under the GHS, and thus the HCS.

Part A: Background Guidance[1]

[1] *The text of Appendix F, Part A, on the IARC Monographs, is paraphrased from the 2006 Preamble to the "Monographs on the Evaluation of Carcinogenic Risks to Humans"; the Classifier is referred to the full IARC Preamble for the complete text. The text is not part of the agreed GHS text on the harmonized system developed by the OECD Task Force-HCL.*

As noted in Footnote 6 of Appendix A.6. to this section, the GHS includes as guidance for classifiers information taken from the Preamble of the International Agency for Research on Cancer (IARC) "Monographs on the Evaluation of Carcinogenic Risks to Humans" (2006), providing guidance on the evaluation of the strength and evidence of carcinogenic risks to humans. This guidance also discusses some additional considerations in classification and an approach to analysis, rather than hard-and-fast rules. Part A is consistent with Appendix A.6, and should help in evaluating information to determine carcinogenicity.

Carcinogenicity in humans:

The evidence relevant to carcinogenicity from studies in humans is classified into one of the following categories:

(a) *Sufficient evidence of carcinogenicity:* A causal relationship has been established between exposure to the agent and human cancer. That is, a positive relationship has been observed between the exposure and cancer in studies in which chance, bias and confounding could be ruled out with reasonable confidence.

(b) *Limited evidence of carcinogenicity:* A positive association has been observed between exposure to the agent and cancer for which a causal interpretation is considered by the Working Group to be credible, but chance, bias or confounding could not be ruled out with reasonable confidence.

In some instances, the above categories may be used to classify the degree of evidence related to carcinogenicity in specific organs or tissues.

Carcinogenicity in experimental animals:

The evidence relevant to carcinogenicity in experimental animals is classified into one of the following categories:

(a) *Sufficient evidence of carcinogenicity:* A causal relationship has been established between the agent and an increased incidence of malignant neoplasms or of an appropriate combination of benign and malignant neoplasms in two or more species of animals or two or more independent studies in one species carried out at different times or in different laboratories or under different protocols. An increased incidence of tumors in both sexes of a single species in a well-conducted study, ideally conducted under Good Laboratory Practices, can also provide sufficient evidence.

Exceptionally, a single study in one species and sex might be considered to provide sufficient evidence of carcinogenicity when malignant neoplasms occur to an unusual degree with regard to incidence, site, type of tumor or age at onset, or when there are strong findings of tumors at multiple sites.

(b) *Limited evidence of carcinogenicity:* The data suggest a carcinogenic effect but are limited for making a definitive evaluation because, e.g. the evidence of carcinogenicity is restricted to a single experiment; there are unresolved questions regarding the adequacy of the design, conduct or interpretation of the studies; the agent increases the incidence only of benign neoplasms or lesions of uncertain neoplastic potential; or the evidence of carcinogenicity is restricted to studies that demonstrate only promoting activity in a narrow range of tissues or organs.

Guidance on How To Consider Important Factors in Classification of Carcinogenicity (See Reference Section)

The weight of evidence analysis called for in GHS and the HCS (Sec. 1910.1200) is an integrative approach that considers important factors in determining carcinogenic potential along with the strength of evidence analysis. The IPCS *"Conceptual Framework for Evaluating a Mode of Action for Chemical Carcinogenesis"* (2001), International Life Sciences Institute (ILSI) *"Framework for Human Relevance Analysis of Information on Carcinogenic Modes of Action"* (Meek, *et al.*, 2003; Cohen *et al.*, 2003, 2004), and Preamble to the IARC Monographs (2006; Section B.6. (Scientific Review and Evaluation; Evaluation and Rationale)) provide a basis for systematic assessments that may be performed in a consistent fashion. The IPCS also convened a panel in 2004 to further develop and clarify the human relevance framework. However, the above documents are not intended to dictate answers, nor provide lists of criteria to be checked off.

Mode of Action

Various documents on carcinogen assessment all note that mode of action in and of itself, or consideration of comparative metabolism, should be evaluated on a case-by-case basis and are part of an analytic evaluative approach. One must look closely at any mode of action in animal experiments, taking into consideration comparative toxicokinetics/toxicodynamics between the animal test species and humans to determine the relevance of the results to humans. This may lead to the possibility of discounting very specific effects of certain types of substances. Life stage-dependent effects on cellular differentiation may also lead to qualitative differences between animals and humans. Only if a mode of action of tumor development is conclusively determined not to be operative in humans may the carcinogenic evidence for that tumor be discounted. However, a weight of evidence evaluation for a substance calls for any other tumorigenic activity to be evaluated, as well.

Responses in Multiple Animal Experiments

Positive responses in several species add to the weight of evidence that a substance is a carcinogen. Taking into account all of the factors listed in A.6.2.5.2 and more, such chemicals with positive outcomes in two or more species would be provisionally considered to be classified in GHS Category 1B until human relevance of animal results are assessed in their entirety. It should be noted, however, that positive results for one species in at least two independent studies, or a single positive study showing unusually strong evidence of malignancy may also lead to Category 1B.

Responses Are in One Sex or Both Sexes

Any case of gender-specific tumors should be evaluated in light of the total tumorigenic response to the substance observed at other sites (multi-site responses or incidence above background) in determining the carcinogenic potential of the substance.

If tumors are seen only in one sex of an animal species, the mode of action should be carefully evaluated to see if the response is consistent with the postulated mode of action. Effects seen only in one sex in a test species may be less convincing than effects seen in both sexes, unless there is a clear patho-physiological difference consistent with the mode of action to explain the single sex response.

Confounding Effects of Excessive Toxicity or Localized Effects

Tumors occurring only at excessive doses associated with severe toxicity generally have doubtful potential for carcinogenicity in humans. In addition, tumors occurring only at sites of contact and/or only at excessive doses need to be carefully evaluated for human relevance for carcinogenic hazard. For example, forestomach tumors, following administration by gavage of an irritating or corrosive, non-mutagenic chemical, may be of questionable relevance. However, such determinations must be evaluated carefully in justifying the carcinogenic potential for humans; any occurrence of other tumors at distant sites must also be considered.

Tumor Type, Reduced Tumor Latency

Unusual tumor types or tumors occurring with reduced latency may add to the weight of evidence for the carcinogenic potential of a substance, even if the tumors are not statistically significant.

Toxicokinetic behavior is normally assumed to be similar in animals and humans, at least from a qualitative perspective. On the other hand, certain tumor types in animals may be associated with toxicokinetics or toxicodynamics that are unique to the animal species tested and may not be predictive of carcinogenicity in humans. Very few such examples have been agreed internationally. However, one example is the lack of human relevance of kidney tumors in male rats associated with compounds causing α2u-globulin nephropathy (IARC, Scientific Publication N° 147[2]). Even when a particular tumor type may be discounted, expert judgment must be used in assessing the total tumor profile in any animal experiment.

[2] *While most international agencies do not consider kidney tumors coincident with* 2u-globulin nephropathy to be a predictor of risk in humans, this view is not universally held. (See: Doi et al., 2007).

Part B: International Agency for Research on Cancer (IARC)[3]

[3] *Preamble of the International Agency for Research on Cancer (IARC) "Monographs on the Evaluation of Carcinogenic Risks to Humans" (2006).*

IARC Carcinogen Classification Categories:

Group 1: The agent is *carcinogenic to humans*

This category is used when there is *sufficient evidence of carcinogenicity* in humans. Exceptionally, an agent may be placed in this category when evidence of carcinogenicity in humans is less than *sufficient* but there is *sufficient evidence of carcinogenicity* in experimental animals and strong evidence in exposed humans that the agent acts through a relevant mechanism of carcinogenicity.

Group 2:

This category includes agents for which, at one extreme, the degree of evidence of carcinogenicity in humans is almost *sufficient,* as well as those for which, at the other extreme, there are no human data but for which there is evidence of carcinogenicity in experimental animals. Agents are assigned to either Group 2A (*probably carcinogenic to humans*) or Group 2B (*possibly carcinogenic to humans*) on the basis of epidemiological and experimental evidence of carcinogenicity and mechanistic and other relevant data. The terms *probably carcinogenic* and *possibly carcinogenic* have no quantitative significance and are used simply as descriptors of different levels of evidence of human carcinogenicity, with *probably carcinogenic* signifying a higher level of evidence than *possibly carcinogenic.*

Group 2A: The agent is *probably carcinogenic to human.*

This category is used when there is *limited evidence of carcinogenicity* in humans and *sufficient evidence of carcinogenicity* in experimental animals. In some cases, an agent may be classified in this category when there is *inadequate evidence of carcinogenicity* in humans and *sufficient evidence of carcinogenicity* in experimental animals and strong evidence that the carcinogenesis is mediated by a mechanism that also operates in humans. Exceptionally, an agent may be classified in this category solely on the basis of *limited evidence of carcinogenicity* in humans. An agent may be assigned to this category if it clearly belongs, based on mechanistic considerations, to a class of agents for which one or more members have been classified in Group 1 or Group 2A.

Group 2B: The agent is *possibly carcinogenic to humans.*

This category is used for agents for which there is *limited evidence of carcinogenicity* in humans and less than *sufficient evidence of carcinogenicity* in experimental animals. It may also be used when there is *inadequate evidence of carcinogenicity* in humans but there is *sufficient evidence of carcinogenicity* in experimental animals. In some instances, an agent for which there is *inadequate evidence of carcinogenicity* in humans and less than *sufficient evidence of carcinogenicity* in experimental animals together with supporting evidence from mechanistic and other relevant data may be placed in this group. An agent may be classified in this category solely on the basis of strong evidence from mechanistic and other relevant data.

Part C: National Toxicology Program (NTP), "Report on Carcinogens", Background Guidance

NTP Listing Criteria[4] :

[4] *See: http://ntp.niehs.nih.gov/go/15209.*

The criteria for listing an agent, substance, mixture, or exposure circumstance in the Report on Carcinogens (RoC) are as follows:

Known To Be A Human Carcinogen: There is sufficient evidence of carcinogenicity from studies in humans[5] that indicates a causal relationship between exposure to the agent, substance, or mixture, and human cancer.

[5] *This evidence can include traditional cancer epidemiology studies, data from clinical studies, and/or data derived from the study of tissues or cells from humans exposed to the substance in question that can be useful for evaluating whether a relevant cancer mechanism is operating in people.*

Reasonably Anticipated To Be A Human Carcinogen: There is limited evidence of carcinogenicity from studies in humans that indicates that a causal interpretation is credible, but that alternative explanations, such as chance, bias, or confounding factors, could not adequately be excluded,

or

there is sufficient evidence of carcinogenicity from studies in experimental animals that indicates there is an increased incidence of malignant and/or a combination of malignant and benign tumors in multiple species or at multiple tissue sites, or by multiple routes of exposure, or to an unusual degree with regard to incidence, site, or type of tumor, or age at onset,

or

there is less than sufficient evidence of carcinogenicity in humans or laboratory animals; however, the agent, substance, or mixture belongs to a well-defined, structurally-related class of substances whose members are listed in a previous Report on Carcinogens as either known to be a human carcinogen or reasonably anticipated to be a human carcinogen, or there is convincing relevant information that the agent acts through mechanisms indicating it would likely cause cancer in humans.

Conclusions regarding carcinogenicity in humans or experimental animals are based on scientific judgment, with consideration given to all relevant information. Relevant information includes, but is not limited to, dose response, route of exposure, chemical structure, metabolism, pharmacokinetics, sensitive sub-populations, genetic effects, or other data relating to mechanism of action or factors that may be unique to a given substance. For example, there may be substances for which there is evidence of carcinogenicity in laboratory animals, but there are compelling data indicating that the agent acts through mechanisms that do not operate in humans and would therefore not reasonably be anticipated to cause cancer in humans.

Part D: Table Relating Approximate Equivalences Among IARC, NTP RoC, and GHS Carcinogenicity Classifications

The following table may be used to perform hazard classifications for carcinogenicity under the HCS (Sec. 1910.1200). It relates the approximated GHS hazard categories for carcinogenicity to the classifications provided by IARC and NTP, as described in Parts B and C of this Appendix.

Approximate Equivalences Among Carcinogen Classification Schemes

IARC	GHS	NTP RoC
Group 1	Category 1A	Known.
Group 2A	Category 1B	Reasonably Anticipated.
Group 2B	Category 2	(See Note 1).

Note 1:

1. *Limited evidence of carcinogenicity from studies in humans (corresponding to IARC 2A/GHS 1B);*
2. *Sufficient evidence of carcinogenicity from studies in experimental animals (again, essentially corresponding to IARC 2A/GHS 1B);*
3. *Less than sufficient evidence of carcinogenicity in humans or laboratory animals; however:*

c. *The agent, substance, or mixture belongs to a well-defined, structurally-related class of substances whose members are listed in a previous RoC as either "Known" or "Reasonably Anticipated" to be a human carcinogen, or*

1910

Selected 1910 General Industry Standards

d. *There is convincing relevant information that the agent acts through mechanisms indicating it would likely cause cancer in humans.*

References

Cohen, S.M., J. Klaunig, M.E. Meek, R.N. Hill, T. Pastoor, L. Lehman-McKeeman, J. Bucher, D.G. Longfellow, J. Seed, V. Dellarco, P. Fenner-Crisp, and D. Patton. 2004. Evaluating the human relevance of chemically induced animal tumors. *Toxicol. Sci.* 78(2):181-186.

Cohen, S.M., M.E. Meek, J.E. Klaunig, D.E. Patton, P.A. Fenner-Crisp. 2003. The human relevance of information on carcinogenic modes of action: Overview. *Crit. Rev. Toxicol.* 33(6):581-9.

Meek, M.E., J.R. Bucher, S.M. Cohen, V. Dellarco, R.N. Hill, L. Lehman-McKeeman, D.G. Longfellow, T. Pastoor, J. Seed, D.E. Patton. 2003. A framework for human relevance analysis of information on carcinogenic modes of action. *Crit. Rev. Toxicol.* 33(6):591-653.

Sonich-Mullin, C., R. Fielder, J. Wiltse, K. Baetcke, J. Dempsey, P. Fenner-Crisp, D. Grant, M. Hartley, A. Knapp, D. Kroese, I. Mangelsdorf, E. Meek, J.M. Rice, and M. Younes. 2001. The conceptual framework for evaluating a mode of action for chemical carcinogenesis. *Reg. Toxicol. Pharm.* 34:146-152.

International Programme on Chemical Safety Harmonization Group. 2004. Report of the First Meeting of the Cancer Working Group. World Health Organization. Report IPCS/HSC-CWG-1/04. Geneva.

International Agency for Research on Cancer. IARC Monographs on the Evaluation of Carcinogenic Risks to Human. Preambles to Volumes. World Health Organization. Lyon, France.

Cohen, S.M., P.A. Fenner-Crisp, and D.E. Patton. 2003. Special Issue: Cancer Modes of Action and Human Relevance. Critical Reviews in Toxicology, R.O. McClellan, ed., Volume 33/Issue 6. CRC Press.

Capen, C.C., E. Dybing, and J.D. Wilbourn. 1999. Species differences in thyroid, kidney and urinary bladder carcinogenesis. International Agency for Research on Cancer, Scientific Publication N° 147.

Doi, A.M., G. Hill, J. Seely, J.R. Hailey, G. Kissling, and J.R. Buchera. 2007. 2u-Globulin nephropathy and renal tumors in National Toxicology Program studies. *Toxicol. Pathol.* 35:533-540.

§1910.1201

Retention of DOT markings, placards and labels

(a) **Any employer who receives a package** of hazardous material which is required to be marked, labeled or placarded in accordance with the U. S. Department of Transportation's Hazardous Materials Regulations (49 CFR Parts 171 through 180) shall retain those markings, labels and placards on the package until the packaging is sufficiently cleaned of residue and purged of vapors to remove any potential hazards.[§1910.1201(a)]

(b) **Any employer who receives a freight container,** rail freight car, motor vehicle, or transport vehicle that is required to be marked or placarded in accordance with the Hazardous Materials Regulations shall retain those markings and placards on the freight container, rail freight car, motor vehicle or transport vehicle until the hazardous materials which require the marking or placarding are sufficiently removed to prevent any potential hazards.[§1910.1201(b)]

(c) **Markings, placards and labels** shall be maintained in a manner that ensures that they are readily visible.[§1910.1201(c)]

(d) **For non-bulk packages** which will not be reshipped, the provisions of this section are met if a label or other acceptable marking is affixed in accordance with the Hazard Communication Standard (29 CFR 1910.1200).[§1910.1201(d)]

(e) **For the purposes of this section,** the term "hazardous material" and any other terms not defined in this section have the same definition as in the Hazardous Materials Regulations (49 CFR Parts 171 through 180).[§1910.1201(e)]

[59 FR 36700, July 19, 1994]

Subpart A - General

§1926.1
Purpose and scope

(a) **This part sets forth the safety and health standards** promulgated by the Secretary of Labor under section 107 of the Contract Work Hours and Safety Standards Act. The standards are published in subpart C of this part and following subparts. [§1926.1(a)]

(b) **Subpart B of this part contains statements** of general policy and interpretations of section 107 of the Contract Work Hours and Safety Standards Act having general applicability. [§1926.1(b)]

§1926.2
Variances from safety and health standards

(a) **Variances from standards which are,** or may be, published in this part may be granted under the same circumstances whereunder variances may be granted under section 6(b)(A) or 6(d) of the Williams-Steiger Occupational Safety and Health Act of 1970 (29 U.S.C. 65). The procedures for the granting of variances and for related relief under this part are those published in part 1905 of this title. [§1926.2(a)]

(b) **Any requests for variances under this section** shall also be considered requests for variances under the Williams-Steiger Occupational Safety and Health Act of 1970, and any requests for variances under Williams-Steiger Occupational Safety and Health Act with respect to construction safety or health standards shall be considered to be also variances under the Construction Safety Act. Any variance from a construction safety or health standard which is contained in this part and which is incorporated by reference in part 1910 of this title shall be deemed a variance from the standard under both the Construction Safety Act and the Williams-Steiger Occupational Safety and Health Act of 1970. [§1926.2(b)]

§1926.3
Inspections — right of entry

(a) **It shall be a condition of each contract** which is subject to section 107 of the Contract Work Hours and Safety Standards Act that the Secretary of Labor or any authorized representative shall have a right of entry to any site of contract performance for the following purposes: [§1926.3(a)]

(1) *To inspect or investigate the matter* of compliance with the safety and health standards contained in subpart C of this part and following subparts; and [§1926.3(a)(1)]

(2) *To carry out the duties* of the Secretary under section 107(b) of the Act. [§1926.3(a)(2)]

(b) **For the purpose of carrying out his investigative duties** under the Act, the Secretary of Labor may, by agreement, use with or without reimbursement the services, personnel, and facilities of any State or Federal agency. Any agreements with States under this section shall be similar to those provided for under the Walsh-Healey Public Contracts Act under 41 CFR part 50-205. [§1926.3(b)]

§1926.4
Rules of practice for administrative adjudications for enforcement of safety and health standards

(a) **The rules of practice for administrative adjudications** for the enforcement of the safety and health standards contained in subpart C of this part and the following subparts shall be the same as those published in part 6 of this title with respect to safety and health violations of the Service Contract Act of 1965 (69 Stat. 1035), except as provided in paragraph (b) of this section. [§1926.4(a)]

(b) **In the case of debarment,** the findings required by section 107(d) of the Act shall be made by the hearing examiner or the Assistant Secretary of Labor for Occupational Safety and Health, as the case may be. Whenever, as provided in section 107(d)(2), a contractor requests termination of debarment before the end of the 3-year period prescribed in that section, the request shall be filed in writing with the Assistant Secretary of Labor for Occupational Safety and Health who shall publish a notice in the Federal Register that the request has been received and afford interested persons an opportunity to be heard upon the request, and thereafter the provisions of part 6 of this title shall apply with respect to prehearing conferences, hearings and related matters, and decisions and orders. [§1926.4(b)]

§1926.5
OMB control numbers under the Paperwork Reduction Act.

The following sections or paragraphs each contain a collection of information requirement which has been approved by the Office of Management and Budget under the control number listed.

29 CFR citation	OMB control No.
1926.33	1218-0065
1926.50	1218-0093
1926.52	1218-0048
1926.53	1218-0103
1926.59	1218-0072
1926.60	1218-0183
1926.62	1218-0189
1926.64	1218-0200
1926.65	1218-0202
1926.103	1218-0099
1926.200	1218-0132
1926.250	1218-0093
1926.251	1218-0233
1926.403	1218-0130
1926.404	1218-0130
1926.405	1218-0130
1926.407	1218-0130
1926.408	1218-0130
1926.453(a)(2)	1218-0216
1926.502	1218-0197
1926.503	1218-0197
1926.550(a)(1)	1218-0115
1926.550(a)(2)	1218-0115
1926.550(a)(4)	1218-0115
1926.550(a)(6)	1218-0113
1926.550(a)(11)	1218-0054
1926.550(a)(16)	1218-0115
1926.550(b)(2)	1218-0232
1926.550(g)	1218-0151
1926.552	1218-0231
1926.652	1218-0137
1926.703	1218-0095
1926.800	1218-0067
1926.803	1218-0067
1926.900	1218-0217
1926.903	1218-0227
1926.1080	1218-0069
1926.1081	1218-0069
1926.1083	1218-0069
1926.1090	1218-0069
1926.1091	1218-0069
1926.1101	1218-0134
1926.1103	1218-0085
1926.1104	1218-0084
1926.1106	1218-0086
1926.1107	1218-0083
1926.1108	1218-0087
1926.1109	1218-0089
1926.1110	1218-0082
1926.1111	1218-0090
1926.1112	1218-0080
1926.1113	1218-0079
1926.1114	1218-0088
1926.1115	1218-0044
1926.1116	1218-0081

(continued)

29 CFR citation	OMB control No.
1926.1117	1218-0010
1926.1118	1218-0104
1926.1126	1218-0252
1926.1127	1218-0186
1926.1128	1218-0129
1926.1129	1218-0128
1926.1144	1218-0101
1926.1145	1218-0126
1926.1147	1218-0108
1926.1148	1218-0145
1926.1402	1218-0261
1926.1403	1218-0261
1926.1404	1218-0261
1926.1406	1218-0261
1926.1407	1218-0261
1926.1408	1218-0261
1926.1409	1218-0261
1926.1410	1218-0261
1926.1411	1218-0261
1926.1412	1218-0261
1926.1413	1218-0261
1926.1414	1218-0261
1926.1417	1218-0261
1926.1423	1218-0261
1926.1424	1218-0261
1926.1427	1218-0261
1926.1428	1218-0261
1926.1431	1218-0261
1926.1433	1218-0261
1926.1434	1218-0261
1926.1435	1218-0261
1926.1436	1218-0261
1926.1437	1218-0261
1926.1441	1218-0261

[61 FR 5509, Feb. 13, 1996, as amended at 63 FR 3814, Jan. 27, 1998; 63 FR 13340, Mar. 19, 1998; 63 FR 17094, Apr. 8, 1998; 64 FR 18810, Apr. 16, 1999; 71 FR 38086, July 5, 2006; 75 FR 68430, Nov. 8, 2010]

§1926.6
Incorporation by reference

(a) **The standards of agencies of the U.S.** Government, and organizations which are not agencies of the U.S. Government which are incorporated by reference in this part, have the same force and effect as other standards in this part. Only the mandatory provisions (*i.e.,* provisions containing the word "shall" or other mandatory language) of standards incorporated by reference are adopted as standards under the Occupational Safety and Health Act. The locations where these standards may be examined are as follows:[§1926.6(a)]

(1) *Offices of the Occupational Safety* and Health Administration, U.S. Department of Labor, Frances Perkins Building, Washington, DC 20210.[§1926.6(a)(1)]

(2) *The Regional and Field Offices* of the Occupational Safety and Health Administration, which are listed in the U.S. Government Manual.[§1926.6(a)(2)]

(b) **The materials listed in paragraphs** (g) through (ff) of this section are incorporated by reference in the corresponding sections noted as they exist on the date of the approval, and a notice of any change in these materials will be published in the Federal Register. These incorporations by reference were approved by the Director of the Federal Register in accordance with 5 U.S.C. 552(a) and 1 CFR part 51.[§1926.6(b)]

(c) **Copies of standards listed** in this section and issued by private standards organizations are available for purchase from the issuing organizations at the addresses or through the other contact information listed below for these private standards organizations. In addition, these standards are available for inspection at the National Archives and Records Administration (NARA). For information on the availability of these standards at NARA, telephone: 202-741-6030, or go to *http://www.archives.gov/federal[lowbar]register/code[lowbar]of[lowbar]federal[lowbar]regulations/ibr[lowbar]locations.html.* Also, the standards are available for inspection at any Regional Office of the Occupational Safety and Health Administration (OSHA), or at the OSHA Docket Office, U.S. Department of Labor, 200 Constitution Avenue, NW., Room N-2625, Washington, DC 20210; telephone: 202-693-2350 (TTY number: 877-889-5627).[§1926.6(c)]

(d) ***[Reserved.]***[§1926.6(d)]

(e) ***[Reserved.]***[§1926.6(e)]

(f) ***[Reserved.]***[§1926.6(f)]

(g) **The following material is available** for purchase from the American Conference of Governmental Industrial Hygienists (ACGIH), 1330 Kemper Meadow Drive, Cincinnati, OH 45240; telephone: 513-742-6163; fax: 513-742-3355; e-mail: *mail@acgih.org;* Web site: *http://www.acgih.org*.[§1926.6(g)]

(1) *Threshold Limit Values* of Airborne Contaminants for 1970, 1970, IBR approved for Sec. 1926.55(a) and Appendix A of Sec. 1926.55.[§1926.6(g)(1)]

(h) **The following material is available** for purchase from the American National Standards Institute (ANSI), 25 West 43rd Street, Fourth Floor, New York, NY 10036; telephone: 212-642-4900; fax: 212-302-1286; e-mail: *info@ansi.org;* Web site: *http://www.ansi.org/*.[§1926.6(h)]

(1) *ANSI A10.3-1970,* Safety Requirements for Explosive-Actuated Fastening Tools, IBR approved for Sec. 1926.302(e). [§1926.6(h)(1)]

(2) *ANSI A10.4-1963,* Safety Requirements for Workmen's Hoists, IBR approved for Sec. 1926.552(c).[§1926.6(h)(2)]

(3) *ANSI A10.5-1969,* Safety Requirements for Material Hoists, IBR approved for Sec. 1926.552(b).[§1926.6(h)(3)]

(4) *ANSI A11.1-1965 (R1970),* Practice for Industrial Lighting, IBR approved for Sec. 1926.56(b).[§1926.6(h)(4)]

(5) *ANSI A17.1-1965,* Elevators, Dumbwaiters, Escalators, and Moving Walks, IBR approved for Sec. 1926.552(d). [§1926.6(h)(5)]

(6) *ANSI A17.1a-1967,* Elevators, Dumbwaiters, Escalators, and Moving Walks Supplement, IBR approved for Sec. 1926.552(d).[§1926.6(h)(6)]

(7) *ANSI A17.1b-1968,* Elevators, Dumbwaiters, Escalators, and Moving Walks Supplement, IBR approved for Sec. 1926.552(d).[§1926.6(h)(7)]

(8) *ANSI A17.1c-1969,* Elevators, Dumbwaiters, Escalators, and Moving Walks Supplement, IBR approved for Sec. 1926.552(d).[§1926.6(h)(8)]

(9) *ANSI A17.1d-1970,* Elevators, Dumbwaiters, Escalators, and Moving Walks Supplement, IBR approved for Sec. 1926.552(d).[§1926.6(h)(9)]

(10) *ANSI A17.2-1960,* Practice for the Inspection of Elevators (Inspector's Manual), IBR approved for Sec. 1926.552(d). [§1926.6(h)(10)]

(11) *ANSI A17.2a-1965,* Practice for the Inspection of Elevators (Inspector's Manual) Supplement, IBR approved for Sec. 1926.552(d).[§1926.6(h)(11)]

(12) *ANSI A17.2b-1967,* Practice for the Inspection of Elevators (Inspector's Manual) Supplement, IBR approved for Sec. 1926.552(d).[§1926.6(h)(12)]

(13) *ANSI A92.2-1969,* Vehicle Mounted Elevating and Rotating Work Platforms, IBR approved for §§1926.453(a) and 1926.453(b).[§1926.6(h)(13)]

(14) *ANSI B7.1-1970,* Safety Code for the Use, Care, and Protection of Abrasive Wheels, IBR approved for §§1926.57(g), 1926.303(b), 1926.303(c), and 1926.303(d).[§1926.6(h)(14)]

(15) *ANSI B20.1-1957,* Safety Code for Conveyors, Cableways, and Related Equipment, IBR approved for Sec. 1926.555(a). [§1926.6(h)(15)]

(16) *ANSI B56.1-1969,* Safety Standards for Powered Industrial Trucks, IBR approved for Sec. 1926.602(c).[§1926.6(h)(16)]

(17) *ANSI J6.1-1950 (R1971),* Rubber Insulating Line Hose, IBR approved for §1926.951(a).[§1926.6(h)(17)]

(18) *ANSI J6.2-1950 (R1971),* Rubber Insulating Hoods, IBR approved for §1926.951(a).[§1926.6(h)(18)]

(19) *ANSI J6.4-1971,* Rubber Insulating Blankets, IBR approved for §1926.951(a).[§1926.6(h)(19)]

(20) *ANSI J6.5-1971,* Rubber Insulating Sleeves, IBR approved for §1926.951(a).[§1926.6(h)(20)]

(21) *ANSI J6.6-1971,* Rubber Insulating Gloves, IBR approved for §1926.951(a). [§1926.6(h)(21)]

(22) *ANSI J6.7-1935 (R1971),* Rubber Matting for Use Around Electric Apparatus, IBR approved for §1926.951(a). [§1926.6(h)(22)]

(23) *ANSI O1.1-1961,* Safety Code for Woodworking Machinery, IBR approved for Sec. 1926.304(f). [§1926.6(h)(23)]

(24) *ANSI Z35.1-1968,* Specifications for Accident Prevention Signs, IBR approved for Sec. 1926.200(i). [§1926.6(h)(24)]

(25) *ANSI Z35.2-1968,* Specifications for Accident Prevention Tags, IBR approved for Sec. 1926.200(i). [§1926.6(h)(25)]

(26) *ANSI Z49.1-1967,* Safety in Welding and Cutting, IBR approved for Sec. 1926.350(j). [§1926.6(h)(26)]

(27) *ANSI Z87.1-1968,* Practice for Occupational and Educational Eye and Face Protection, IBR approved for Sec. 1926.102(a). [§1926.6(h)(27)]

(28) *ANSI Z89.1-1969,* Safety Requirements for Industrial Head Protection, IBR approved for Sec. 1926.100(b). [§1926.6(h)(28)]

(29) *ANSI Z89.2-1971,* Industrial Protective Helmets for Electrical Workers, Class B, IBR approved for §§1926.100(c) and 1926.951(a). [§1926.6(h)(29)]

(i) *[Reserved.]*[§1926.6(i)]

(j) **The following material is available** for purchase from the American Society for Testing and Materials (ASTM), ASTM International, 100 Barr Harbor Drive, PO Box C700, West Conshohocken, PA, 19428-2959; telephone: 610-832-9585; fax: 610-832-9555; e-mail: *service@astm.org;* Web site: *http://www.astm.org/:*[§1926.6(j)]

(1) *ASTM A370-1968,* Methods and Definitions for Mechanical Testing and Steel Products, IBR approved for Sec. 1926.1001(f). [§1926.6(j)(1)]

(2) *ASTM B117-1964,* 50 Hour Test, IBR approved for Sec. 1926.959(a). [§1926.6(j)(2)]

(3) *ASTM D56-1969,* Standard Method of Test for Flash Point by the Tag Closed Tester, IBR approved for Sec. 1926.155(i). [§1926.6(j)(3)]

(4) *ASTM D93-1969,* Standard Method of Test for Flash Point by the Pensky Martens Closed Tester, IBR approved for Sec. 1926.155(i). [§1926.6(j)(4)]

(5) *ASTM D323-1958 (R1968),* Standard Method of Test for Vapor Pressure of Petroleum Products (Reid Method), IBR approved for Sec. 1926.155(m). [§1926.6(j)(5)]

(k) **The following material is available** for purchase from the American Society of Agricultural and Biological Engineers (ASABE), 2950 Niles Road, St. Joseph, MI 49085; telephone: 269-429-0300; fax: 269-429-3852; e-mail: *hq@asabe.org;* Web site: *http://www.asabe.org/:*[§1926.6(k)]

(1) *ASAE R313.1-1971,* Soil Cone Penetrometer, reaffirmed 1975, IBR approved for Sec. 1926.1002(e). [§1926.6(k)(1)]

(l) **The following material is available** for purchase from the American Society of Mechanical Engineers (ASME), Three Park Avenue, New York, NY 10016; telephone: 1-800-843-2763; fax: 973-882-1717; e-mail: *infocentral@asme.org;* Web site: *http://www.asme.org/:*[§1926.6(l)]

(1) *ASME B30.2-2005,* Overhead and Gantry Cranes (Top Running Bridge, Single or Multiple Girder, Top Running Trolley Hoist), issued Dec. 30, 2005 ("ASME B30.2-2005"), IBR approved for Sec. 1926.1438(b). [§1926.6(l)(1)]

(2) *ASME B30.5-2004,* Mobile and Locomotive Cranes, issued Sept. 27, 2004 ("ASME B30.5-2004"), IBR approved for §§1926.1414(b); 1926.1414(e); 1926.1433(b). [§1926.6(l)(2)]

(3) *ASME B30.7-2001,* Base-Mounted Drum Hoists, issued Jan. 21, 2002 ("ASME B30.7-2001"), IBR approved for Sec. 1926.1436(e). [§1926.6(l)(3)]

(4) *ASME B30.14-2004,* Side Boom Tractors, issued Sept. 20, 2004 ("ASME B30.14-2004"), IBR approved for Sec. 1926.1440(c). [§1926.6(l)(4)]

(5) *ASME Boiler and Pressure* Vessel Code, Section VIII, 1968, IBR approved for §§1926.152(i), 1926.306(a), and 1926.603(a). [§1926.6(l)(5)]

(6) *ASME Power Boilers,* Section I, 1968, IBR approved for Sec. 1926.603(a). [§1926.6(l)(6)]

(m) **The following material is available** for purchase from the American Welding Society (AWS), 550 N.W. LeJeune Road, Miami, Florida 33126; telephone: 1-800-443-9353; Web site: *http://www.aws.org/:*[§1926.6(m)]

(1) *AWS D1.1/D1.1M:2002,* Structural Welding Code — Steel, 18th ed., ANSI approved Aug. 31, 2001 ("AWS D1.1/D1.1M:2002"), IBR approved for Sec. 1926.1436(c). [§1926.6(m)(1)]

(2) *ANSI/AWS D14.3-94,* Specification for Welding Earthmoving and Construction Equipment, ANSI approved Jun. 11, 1993 ("ANSI/AWS D14.3-94"), IBR approved for Sec. 1926.1436(c). [§1926.6(m)(2)]

(n) **The following material is available** for purchase from the British Standards Institution (BSI), 389 Chiswick High Road, London, W4 4AL, United Kingdom; telephone: +44 20 8996 9001; fax: +44 20 8996 7001; e-mail: *cservices@bsigroup.com;* Web site: *http://www.bsigroup.com/:*[§1926.6(n)]

(1) *BS EN 13000:2004,* Cranes — Mobile Cranes, published Jan. 4, 2006 ("BS EN 13000:2004"), IBR approved for Sec. 1926.1433(c). [§1926.6(n)(1)]

(2) *BS EN 14439:2006,* Cranes — Safety — Tower Cranes, published Jan. 31, 2007 ("BS EN 14439:2006"), IBR approved for Sec. 1926.1433(c). [§1926.6(n)(2)]

(o) **The following material is available** for purchase from the Bureau of Reclamation, United States Department of the Interior, 1849 C Street, NW., Washington DC 20240; telephone: 202-208-4501; Web site: *http://www.usbr.gov/:*[§1926.6(o)]

(1) *Safety and Health Regulations* for Construction, Part II, Sept. 1971, IBR approved for Sec. 1926.1000(f). [§1926.6(o)(1)]

(p) **The following material is available** for purchase from the California Department of Industrial Relations, 455 Golden Gate Avenue, San Francisco CA 94102; telephone: (415) 703-5070; e-mail: *info@dir.ca.gov;* Web site: *http://www.dir.ca.gov/:*[§1926.6(p)]

(1) *Construction Safety Orders,* IBR approved for Sec. 1926.1000(f). [§1926.6(p)(1)]

(q) *[Reserved.]*[§1926.6(q)]

(r) *[Reserved.]*[§1926.6(r)]

(s) *[Reserved.]*[§1926.6(s)]

(t) *[Reserved.]*[§1926.6(t)]

(u) **The following material is available** for purchase from the Federal Highway Administration, United States Department of Transportation, 1200 New Jersey Ave., SE., Washington, DC 20590; telephone: 202-366-4000; Web site: *http://www.fhwa.dot.gov/:*[§1926.6(u)]

(1) *Manual on Uniform Traffic* Control Devices, Millennium Edition, Dec. 2000, IBR approved for §§1926.200(g), 1926.201(a), and 1926.202. [§1926.6(u)(1)]

(v) **The following material is available** for purchase from the General Services Administration (GSA), 1800 F Street, NW., Washington, DC 20405; telephone: (202) 501-0800; Web site: *http://www.gsa.gov/:*[§1926.6(v)]

(1) *QQ-P-416, Federal Specification* Plating Cadmium (Electrodeposited), IBR approved for Sec. 1926.104(e). [§1926.6(v)(1)]

(w) **The following material is available** for purchase from the Institute of Makers of Explosives (IME), 1120 19th Street, NW., Suite 310, Washington, DC 20036; telephone: 202-429-9280; fax: 202-429-9280; e-mail: *info@ime.org;* Web site: *http://www.ime.org/:*[§1926.6(w)]

(1) *IME Pub.* No. 2, American Table of Distances for Storage of Explosives, Jun. 5, 1964, IBR approved for Sec. 1926.914(a). [§1926.6(w)(1)]

(2) *IME Pub.* No. 20, Radio Frequency Energy — A Potential Hazard in the Use of Electric Blasting Caps, Mar. 1968, IBR approved for Sec. 1926.900(k). [§1926.6(w)(2)]

(x) **The following material is available** for purchase from the International Organization for Standardization (ISO), 1, ch. de la Voie-Creuse, Case postale 56, CH-1211 Geneva 20, Switzerland; telephone: +41 22 749 01 11; fax: +41 22 733 34 30; Web site: *http://[fnl]www.iso.org/:*[§1926.6(x)]

(1) *ISO 11660-1:2008(E),* Cranes — Access, guards and restraints — Part 1: General, 2d ed., Feb. 15, 2008 ("ISO 11660-1:2008(E)"), IBR approved for Sec. 1926.1423(c). [§1926.6(x)(1)]

(2) *ISO 11660-2:1994(E),* Cranes — Access, guards and restraints — Part 2: Mobile cranes, 1994 ("ISO 11660-2:1994(E)"), IBR approved for Sec. 1926.1423(c). [§1926.6(x)(2)]

(3) *ISO 11660-3:2008(E),* Cranes — Access, guards and restraints — Part 3: Tower cranes, 2d ed., Feb. 15, 2008 ("ISO 11660-3:2008(E)"), IBR approved for Sec. 1926.1423(c). [§1926.6(x)(3)]

(y) **The following material is available** for purchase from the National Fire Protection Association (NFPA), 1 Batterymarch Park, Quincy, MA 02169; telephone: 617-770-3000; fax: 617-770-0700; Web site: *http://www.nfpa.org/*.[§1926.6(y)]

(1) *NFPA 10A-1970,* Maintenance and Use of Portable Fire Extinguishers, IBR approved for Sec. 1926.150(c).[§1926.6(y)(1)]

(2) *NFPA 13-1969,* Standard for the Installation of Sprinkler Systems, IBR approved for Sec. 1926.152(d).[§1926.6(y)(2)]

(3) *NFPA 30-1969,* The Flammable and Combustible Liquids Code, IBR approved for Sec. 1926.152(c).[§1926.6(y)(3)]

(4) *NFPA 80-1970,* Standard for Fire Doors and Windows, Class E or F Openings, IBR approved for Sec. 1926.152(b).[§1926.6(y)(4)]

(5) *NFPA 251-1969,* Standard Methods of Fire Test of Building Construction and Material, IBR approved for §§1926.152(b) and 1926.155(f).[§1926.6(y)(5)]

(6) *NFPA 385-1966,* Standard for Tank Vehicles for Flammable and Combustible Liquids, IBR approved for Sec. 1926.152(g).[§1926.6(y)(6)]

(z) [*Reserved.*][§1926.6(z)]

(aa) **The following material is available** for purchase from the Power Crane and Shovel Association (PCSA), 6737 W. Washington Street, Suite 2400, Milwaukee, WI 53214; telephone: 1-800-369-2310; fax: 414-272-1170; Web site: *http://www.aem.org/CBC/ProdSpec/PCSA/*.[§1926.6(aa)]

(1) *PCSA Std.* No. 1, Mobile Crane and Excavator Standards, 1968, IBR approved for Sec. 1926.602(b).[§1926.6(aa)(1)]

(2) *PCSA Std.* No. 2, Mobile Hydraulic Crane Standards, 1968 ("PCSA Std. No. 2 (1968)"), IBR approved for §§1926.602(b), 1926.1433(a), and 1926.1501(a).[§1926.6(aa)(2)]

(3) *PCSA Std.* No. 3, Mobile Hydraulic Excavator Standards, 1969, IBR approved for Sec. 1926.602(b).[§1926.6(aa)(3)]

(bb) [*Reserved.*][§1926.6(bb)]

(cc) [*Reserved.*][§1926.6(cc)]

(dd) **The following material is available** for purchase from the Society of Automotive Engineers (SAE), 400 Commonwealth Drive, Warrendale, PA 15096; telephone: 1-877-606-7323; fax: 724-776-0790; Web site: *http://www.sae.org/*.[§1926.6(dd)]

(1) *SAE 1970 Handbook,* IBR approved for Sec. 1926.602(b).[§1926.6(dd)(1)]

(2) *SAE 1971 Handbook,* IBR approved for Sec. 1926.1001(h).[§1926.6(dd)(2)]

(3) *SAE J166-1971,* Trucks and Wagons, IBR approved for Sec. 1926.602(a).[§1926.6(dd)(3)]

(4) *SAE J168-1970,* Protective Enclosures — Test Procedures and Performance Requirements, IBR approved for Sec. 1926.1002(a).[§1926.6(dd)(4)]

(5) *SAE J185 (reaf. May 2003),* Access Systems for Off-Road Machines, reaffirmed May 2003 ("SAE J185 (May 1993)"), IBR approved for Sec. 1926.1423(c).[§1926.6(dd)(5)]

(6) *SAE J236-1971,* Self-Propelled Graders, IBR approved for Sec. 1926.602(a).[§1926.6(dd)(6)]

(7) *SAE J237-1971,* Front End Loaders and Dozers, IBR approved for §126.602(a).[§1926.6(dd)(7)]

(8) *SAE J319b-1971,* Self-Propelled Scrapers, IBR approved for Sec. 1926.602(a).[§1926.6(dd)(8)]

(9) *SAE J320a-1971,* Minimum Performance Criteria for Roll-Over Protective Structure for Rubber-Tired, Self-Propelled Scrapers, IBR approved for Sec. 1926.1001(h).[§1926.6(dd)(9)]

(10) *SAE J321a-1970,* Fenders for Pneumatic-Tired Earthmoving Haulage Equipment, IBR approved for Sec. 1926.602(a).[§1926.6(dd)(10)]

(11) *SAE J386-1969,* Seat Belts for Construction Equipment, IBR approved for Sec. 1926.602(a).[§1926.6(dd)(11)]

(12) *SAE J394-1971,* Minimum Performance Criteria for Roll-Over Protective Structure for Rubber-Tired Front End Loaders and Robber-Tired Dozers, IBR approved for Sec. 1926.1001(h).[§1926.6(dd)(12)]

(13) *SAE J395-1971,* Minimum Performance Criteria for Roll-Over Protective Structure for Crawler Tractors and Crawler-Type Loaders, IBR approved for Sec. 1926.1001(h).[§1926.6(dd)(13)]

(14) *SAE J396-1971,* Minimum Performance Criteria for Roll-Over Protective Structure for Motor Graders, IBR approved for Sec. 1926.1001(h).[§1926.6(dd)(14)]

(15) *SAE J397-1969,* Critical Zone Characteristics and Dimensions for Operators of Construction and Industrial Machinery, IBR approved for Sec. 1926.1001(f).[§1926.6(dd)(15)]

(16) *SAE J743a-1964,* Tractor Mounted Side Boom, 1964 ("SAE J743a-1964"), IBR approved for §1926.1501(a).[§1926.6(dd)(16)]

(17) *SAE J959-1966,* Lifting Crane Wire-Rope Strength Factors, 1966 ("SAE J959-1966"), IBR approved for §1926.1501(a).[§1926.6(dd)(17)]

(18) *SAE J987 (rev. Jun. 2003),* Lattice Boom Cranes — Method of Test, revised Jun. 2003 ("SAE J987 (Jun. 2003)"), IBR approved for Sec. 1926.1433(c).[§1926.6(dd)(18)]

(19) *SAE J1063 (rev. Nov. 1993),* Cantilevered Boom Crane Structures — Method of Test, revised Nov. 1993 ("SAE J1063 (Nov. 1993)"), IBR approved for Sec. 1926.1433(c).[§1926.6(dd)(19)]

(ee) **The following material is available** for purchase from the United States Army Corps of Engineers, 441 G Street, NW., Washington, DC 20314; telephone: 202-761-0011; e-mail: *hq-publicaffairs@usace.army.mil;* Web site: *http://www.usace.army.mil/*:[§1926.6(ee)]

(1) *EM-385-1-1, General Safety* Requirements, Mar. 1967, IBR approved for Sec. 1926.1000(f).[§1926.6(ee)(1)]

(ff) **The following material is available** for purchase from standards resellers such as the Document Center Inc., 111 Industrial Road, Suite 9, Belmont, CA 94002; telephone: 650-591-7600; fax: 650-591-7617; e-mail: *info@document-center.com;* Web site: *http://www.document-center.com/*.[§1926.6(ff)]

(1) *ANSI B15.1-1953 (R1958),* Safety Code for Mechanical Power-Transmission Apparatus, revised 1958, IBR approved for §§1926.300(b)(2) and 1926.1501(a).[§1926.6(ff)(1)]

(2) *ANSI B30.2.0-1967,* Safety Code for Overhead and Gantry Cranes, approved May 4, 1967, IBR approved for §1926.1501(d).[§1926.6(ff)(2)]

(3) *ANSI B30.5-1968,* Crawler, Locomotive, and Truck Cranes, approved Dec. 16, 1968, IBR approved for §§1926.1433(a), 1926.1501(a), and 1926.1501(b).[§1926.6(ff)(3)]

(4) *ANSI B30.6-1969,* Safety Code for Derricks, approved Dec. 18, 1967, IBR approved for §1926.1501(e).[§1926.6(ff)(4)]

Federal Register Update

On Friday, June 22, 2012, OSHA amended §1926.6, effective September 20, 2012. The updated text is as follows:

(h) * * *

(28) *American National Standards Institute* (ANSI) Z89.1–2009, American National Standard for Industrial Head Protection, approved January 26, 2009; IBR approved for § 1926.100(b)(1)(i). Copies of ANSI Z89.1–2009 are available for purchase only from the International Safety Equipment Association, 1901 North Moore Street, Arlington, VA 22209–1762; telephone: 703–525–1695; fax: 703–528–2148; Web site: www.safetyequipment.org.

(29) *American National Standards Institute* (ANSI) Z89.1–2003, American National Standard for Industrial Head Protection; IBR approved for § 1926.100(b)(1)(ii). Copies of ANSI Z89.1–2003 are available for purchase only from the International Safety Equipment Association, 1901 North Moore Street, Arlington, VA 22209–1762; telephone: 703–525–1695; fax: 703–528–2148; Web site: www.safetyequipment.org.

(30) *American National Standards Institute* (ANSI) Z89.1–1997, American National Standard for Personnel Protection—Protective Headwear for Industrial Workers—Requirements; IBR approved for § 1926.100(b)(1)(iii). Copies of ANSI Z89.1–1997 are available for purchase only from the International Safety Equipment Association, 1901 North Moore Street, Arlington, VA 22209–1762; telephone: 703–525–1695; fax: 703–528–2148; Web site: www.safetyequipment.org.

[75 FR 48130, Aug. 9, 2010]

Authority: §3704, Contract Work Hours and Safety Standards Act (40 U.S.C. 333); secs. 4, 6, and 8, Occupational Safety and Health Act of 1970 (29 U.S.C. 653, 655, 657); Secretary of Labor's Order No. 12-71 (36 FR 8754), 8-76 (41 FR 25059), 9-83 (48 FR 35736), 6-96 (62 FR 111), or 5-2007 (72 FR 31160) as applicable; and 29 CFR part 1911.

Subpart B - General Interpretations

§1926.10

Scope of subpart

(a) **This subpart contains the general rules** of the Secretary of Labor interpreting and applying the construction safety and health provisions of section 107 of the Contract Work Hours and Safety Standards Act (83 Stat. 96). Section 107 requires as a condition of each contract which is entered into under legislation subject to Reorganization Plan Number 14 of 1950 (64 Stat. 1267), and which is for construction, alteration, and/or repair, including painting and decorating, that no contractor or subcontractor contracting for any part of the contract work shall require any laborer or mechanic employed in the performance of the contract to work in surroundings or under working conditions which are unsanitary, hazardous, or dangerous to his health or safety, as determined under construction safety and health standards promulgated by the Secretary by regulation.[§1926.10(a)]

Authority: §107, Contract Work Hours and Safety Standards Act (Construction Safety Act) (40 U.S.C. 333).

§1926.11

Coverage under section 103 of the act distinguished

(a) **Coverage under section 103.** It is important to note that the coverage of section 107 differs from that for the overtime requirements of the Contract Work Hours and Safety Standards Act. The application of the overtime requirements is governed by section 103, which subject to specific exemptions, includes:[§1926.11(a)]

(1) *Federal contracts requiring* or involving the employment of laborers or mechanics (thus including, but not limited to, contracts for construction), and[§1926.11(a)(1)]

(2) *contracts assisted in whole* or in part by Federal loans, grants, or guarantees under any statute "providing wage standards for such work." The statutes "providing wage standards for such work" include statutes for construction which require the payment of minimum wages in accordance with prevailing wage findings by the Secretary of Labor in accordance with the Davis-Bacon Act. A provision to section 103 excludes from the overtime requirements work where the Federal assistance is only in the form of a loan guarantee or insurance.[§1926.11(a)(2)]

(b) **Coverage under section 107.** To be covered by section 107 of the Contract Work Hours and Safety Standards Act, a contract must be one which[§1926.11(b)]

(1) *is entered into under a statute* that is subject to Reorganization Plan No. 14 of 1950 (64 Stat. 1267); and[§1926.11(b)(1)]

(2) *is for "construction, alteration, and/or repair, including painting and decorating."*[§1926.11(b)(2)]

§1926.12

Reorganization Plan No. 14 of 1950

(a) **General provisions.** Reorganization Plan No. 14 of 1950 relates to the prescribing by the Secretary of Labor of "appropriate standards, regulations, and procedures" with respect to the enforcement of labor standards under Federal and federally assisted contracts which are subject to various statutes subject to the Plan. The rules of the Secretary of Labor implementing the Plan are published in part 5 of this title. Briefly, the statutes subject to the Plan include the Davis-Bacon Act, including its extension to Federal-aid highway legislation subject to 23 U.S.C. 113, and other statutes subject to the Plan by its original terms, statutes by which the Plan is expressly applied, such as the Contract Work Hours Standards Act by virtue of section 104(d) thereof.[§1926.12(a)]

(b) **The Plan.**[§1926.12(b)]

(1) *The statutes subject to* Reorganization Plan No. 14 of 1950 are cited and briefly described in the remaining paragraphs of this section. These descriptions are general in nature and not intended to convey the full scope of the work to be performed under each statute. The individual statutes should be resorted to for a more detailed scope of the work.[§1926.12(b)(1)]

(2) *Federal-Aid Highway Acts.* The provisions codified in 23 U.S.C. 113 apply to the initial construction, reconstruction, or improvement work performed by contractors or subcontractors on highway projects on the Federal-aid systems, the primary and secondary, as well as their extensions in urban areas, and the Interstate System, authorized under the highway laws providing for the expenditure of Federal funds upon the Federal-aid system. As cited in 41 Op. A.G. 488, 496, the Attorney General ruled that the Federal-Aid Highway Acts are subject to Reorganization Plan No. 14 of 1950.[§1926.12(b)(2)]

(3) *National Housing Act* (12 U.S.C. 1713, 1715a, 1715e, 1715k, 1715l(d)(3) and (4), 1715v, 1715w, 1715x, 1743, 1747, 1748, 1748h-2, 1750g, 1715l(h)(1), 1715z(j)(1), 1715z-1, 1715y(d), Subchapter 1x-A and 1x-B, 1715z-7). This act covers construction which is financed with assistance by the Federal Government through programs of loan and mortgage insurance for the following purposes:[§1926.12(b)(3)]

(i) *Rental Housing* — Section 1713 provides mortgage and insurance on rental housing of eight or more units and on mobile-home courts.[§1926.12(b)(3)(i)]

(ii) *Section 1715a* — Repealed.[§1926.12(b)(3)(ii)]

(iii) *Cooperative Housing* — Section 1715e authorizes mortgage insurance on cooperative housing of five or more units as well as supplementary loans for improvement of repair or resale of memberships.[§1926.12(b)(3)(iii)]

(iv) *Urban Renewal Housing* — Section 1715k provides mortgage insurance on single family or multifamily housing in approved urban renewal areas.[§1926.12(b)(3)(iv)]

(v) *Low or Moderate Income Housing* — Section 1715L(d) (3) and (4) insures mortgages on low-cost single family or multifamily housing.[§1926.12(b)(3)(v)]

(vi) *Housing for Elderly* — Section 1715v provides mortgage insurance on rental housing for elderly or handicapped persons.[§1926.12(b)(3)(vi)]

(vii) *Nursing Homes* — Section 1715w authorizes mortgage insurance on nursing home facilities and major equipment.[§1926.12(b)(3)(vii)]

(viii) *Experimental Housing* — Section 1715x provides mortgage insurance on single family or multifamily housing with experimental design of materials.[§1926.12(b)(3)(viii)]

(ix) *War Housing Insurance* — Section 1743 not active.[§1926.12(b)(3)(ix)]

(x) *Yield Insurance* — Section 1747 insures investment returns on multifamily housing.[§1926.12(b)(3)(x)]

(xi) *Armed Services Housing* — Section 1748b to assist in relieving acute shortage and urgent need for family housing at or in areas adjacent to military installations.[§1926.12(b)(3)(xi)]

(xii) *Defense Housing for Impacted Areas* — Section 1748h-2 provides mortgage insurance on single family or multifamily housing for sale or rent primarily to military or civilian personnel of the Armed Services, National Aeronautics and Space Administration, or Atomic Energy Commission.[§1926.12(b)(3)(xii)]

(xiii) *Defense Rental Housing* — Section 1750g provides for mortgage insurance in critical defense housing areas.[§1926.12(b)(3)(xiii)]

(xiv) *Rehabilitation* — Section 1715L (h)(1) provides mortgage insurance for nonprofit organizations to finance the purchase and rehabilitation of deteriorating or substandard housing for subsequent resale to low-income home purchasers. There must be located on the property five or more single family dwellings of detached, semidetached, or row construction.[§1926.12(b)(3)(xiv)]

(xv) *Homeowner Assistance* — Section 1715Z(j)(1) authorizes mortgage insurance to nonprofit organizations or public bodies or agencies executed to finance sale of individual dwellings to lower income individuals or families. Also includes the rehabilitation of such housing if it is deteriorating or substandard for subsequent resale to lower income home purchasers.[§1926.12(b)(3)(xv)]

(xvi) *Rental Housing Assistance* — Section 1715Z-1 authorizes mortgage insurance and interest reduction payments on behalf of owners of rental housing projects designed for occupancy by lower income families. Payments are also authorized for certain State or locally aided projects.[§1926.12(b)(3)(xvi)]

(xvii) *Condominium Housing* — Section 1715y(d) provides mortgage insurance on property purchased for the development of building sites. This includes waterlines and water supply installations, sewer lines and sewage disposal installations, steam, gas, and electrical lines and installations, roads, streets, curbs, gutters, sidewalks, storm drainage facilities, and other installations or work.[§1926.12(b)(3)(xvii)]

(xviii) *Group Medical Practice Facilities* — Subchapter LX-B authorizes mortgage insurance for the financing of construction and equipment, of facilities for group practice of medicine, optometry, or dentistry.[§1926.12(b)(3)(xviii)]

(xix) *Nonprofit Hospitals* — 1715z-7 authorizes mortgage insurance to cover new and rehabilitated hospitals, including initial equipment.[§1926.12(b)(3)(xix)]

(4) *Hospital Survey and Construction Act,* as amended by the Hospital and Medical Facilities Amendments of 1964 (42 U.S.C. 291e). The provisions of this Act cover construction contracts made by State or local authorities or private institutions under Federal grant-in-aid programs for the construction of hospitals and other medical facilities.[§1926.12(b)(4)]

(5) *Federal Airport Act* (49 U.S.C. 1114(b)). The act provides grant- in-aid funds for airport construction limited to general site preparation runways, taxiways, aprons, lighting appurtenant thereto, and fire, rescue, and maintenance buildings. The act excludes construction intended for use as a public parking facility for passenger automobiles and the cost of construction of any part of an airport building except such of those buildings or parts of buildings to house facilities or activities directly related to the safety of persons at the airport.[§1926.12(b)(5)]

(6) *Housing Act of 1949* (42 U.S.C. 1459). Construction contracts awarded by local authorities financed with the assistance of loans and grants from the Federal Government. The construction programs are for slum clearance and urban renewal which includes rehabilitation grants, neighborhood development programs, neighborhood renewal plans, community renewal, demolition projects, and assistance for blighted areas. See the Housing Act of 1964, paragraph (b)(21) of this section, concerning financial assistance for low-rent housing for domestic farm labor.[§1926.12(b)(6)]

(7) *School Survey and Construction Act of 1950* (20 U.S.C. 636). This act provides for a Federal grant-in-aid program to assist in the construction of schools in federally affected areas. [§1926.12(b)(7)]

(8) *Defense Housing & Community Facilities & Services Act of 1951* (42 U.S.C. 1592i). Inactive Program.[§1926.12(b)(8)]

(9) *United States Housing Act of 1937* (42 U.S.C. 1416). This statute covers the construction of low-rent public housing and slum clearance projects awarded by local authorities. These projects are financed with the assistance of loans and grants from the Federal Government. The slum clearance is the demolition and removal of buildings from any slum area to be used for a low-rent housing project.[§1926.12(b)(9)]

(10) *Federal Civil Defense Act of 1950* (50 U.S.C. App. 2281). This act provides for Federal assistance to the several States and their political subdivisions in the field of civil defense which includes procurement, construction, leasing, or renovating of materials and facilities.[§1926.12(b)(10)]

(11) *Delaware River Basin Compact* (sec. 15.1, 75 Stat. 714). This joint resolution creates, by intergovernmental compact between the United States, Delaware, New Jersey, New York, and Pennsylvania, a regional agency for planning, conservation, utilization, development, management and control of the water and related sources of the Delaware River. [§1926.12(b)(11)]

(12) *Cooperative Research Act* (20 U.S.C. 332a(c)). This act provides Federal grants to a university, college, or other appropriate public or nonprofit private agency or institution for part or all of the cost of constructing a facility for research or for research and related purposes. Research and related purposes means research, research training, surveys, or demonstrations in the field of education, or the dissemination of information derived therefrom, or all of such activities, including (but without limitation) experimental schools, except that such term does not include research, research training, surveys, or demonstrations in the field of sectarian instruction or the dissemination of information derived therefrom. Construction includes new buildings, and the acquisition, expansion, remodeling, replacement, and alteration of existing buildings and the equipping of new buildings and existing buildings. [§1926.12(b)(12)]

(13) *Health Professions Educational Assistance Act of 1963* (42 U.S.C. 292d (c)(4), 293a(c)(5)). The provisions of this act provide for grants to assist public and nonprofit medical, dental, and similar schools for the construction, expansion, or renovation of teaching facilities.[§1926.12(b)(13)]

(14) *Mental Retardation Facilities Construction Act* (42 U.S.C. 295(a)(2)(D), 2662(5), 2675(a)(5)). This act authorizes Federal financial assistance in the construction of centers for research on mental retardation and related aspects of human development, of university-affiliated facilities for the mentally retarded and of facilities for the mentally retarded. [§1926.12(b)(14)]

(15) *Community Mental Health Centers Act* (42 U.S.C. 2685(a)(5)). This act authorizes Federal grants for the construction of public and other nonprofit community mental health centers. [§1926.12(b)(15)]

(16) *Higher Education Facilities Act of 1963* (20 U.S.C. 753). This act authorizes the grant or loan of Federal funds to assist public and other nonprofit institutions of higher education in financing the construction, rehabilitation, or improvement of academic and related facilities in undergraduate and graduate schools.[§1926.12(b)(16)]

(17) *Vocational Educational Act of 1963* (20 U.S.C. 35f). This act provides for Federal grants to the various States for construction of area vocational education school facilities. [§1926.12(b)(17)]

(18) *Library Services and Construction Act* (20 U.S.C. 355e(a)(4)). This act provides for Federal assistance to the various States for the construction of public libraries.[§1926.12(b)(18)]

(19) *Urban Mass Transportation Act of 1954* (49 U.S.C. 1609). This act provides for grants and loans to assist States and local public bodies and agencies thereof in financing the acquisition, construction, reconstruction, and improvement of facilities and equipment for use, by operation or lease or otherwise, in mass transportation service in urban areas and in coordinating such service with highway and other transportation in such areas.[§1926.12(b)(19)]

(20) *Economic Opportunity Act of 1964* (42 U.S.C. 2947). This act covers construction which is financed with assistance of the Federal Government for the following purposes:[§1926.12(b)(20)]

(i) *Authorizes Federal assistance* for construction of projects, buildings and works which will provide young men and women in rural and urban residential centers with education, vocational training, and useful work experience (Title I). [§1926.12(b)(20)(i)]

(ii) *Authorizes financial assistance* for construction work planned and carried out at the community level for antipoverty programs (Title II):[§1926.12(b)(20)(ii)]

[A] Authorizes loans to low income rural families by assisting them to acquire or improve real estate or reduce encumbrances or erect improvements thereon, and to participate in cooperative associations and/or to finance nonagricultural enterprises which will enable such families to supplement their income (Title III);[§1926.12(b)(20)(ii)[A]]

[B] Authorizes loans to local cooperative associations furnishing essential processing, purchasing, or marketing services, supplies, or facilities predominantly to low-income rural families (Title III);[§1926.12(b)(20)(ii)[B]]

[C] Authorizes financial assistance to States, political subdivisions of States, public and nonprofit agencies, institutions, organizations, farm associations, or individuals in establishing housing, sanitation, education, and child day-care programs for migrants and other seasonally employed agricultural employees and their families (Title III). [§1926.12(b)(20)(ii)[C]]

(iii) *Authorizes loans or guarantees* loans to small businesses for construction work (Title IV).[§1926.12(b)(20)(iii)]

(iv) *Authorizes the payment of the cost* of experimental, pilot, or demonstration projects to foster State programs providing construction work experience or training for unemployed fathers and needy people (Title V).[§1926.12(b)(20)(iv)]

(21) *Housing Act of 1964* (42 U.S.C. 1486(f); 42 U.S.C. 1452b(e)). Provides financial assistance for low-rent housing for domestic farm labor. The Act further provides for loans, through public or private agencies, where feasible, to owners or tenants of property in urban renewal areas to finance rehabilitation required to conform the property to applicable code requirements or carry out the objectives of the urban renewal plan for the area.[§1926.12(b)(21)]

(22) *The Commercial Fisheries Research and Development Act* of 1964 (16 U.S.C. 779e(b)). This Act authorizes financial assistance to State agencies for construction projects designed for the research and development of the commercial fisheries resources of the Nation.[§1926.12(b)(22)]

(23) *The Nurse Training Act of 1964* (42 U.S.C. 296a(b)(5)). This act provides for grants to assist in the construction of new facilities for collegiate, associate degree, and diploma schools of nursing, or replacement or rehabilitation of existing facilities of such schools.[§1926.12(b)(23)]

(24) *Elementary and Secondary Education Act of 1965* (20 U.S.C. 241i, 848). The purpose of the act is to provide financial assistance to local educational agencies serving areas with concentrations of children from low-income families for construction in connection with the expansion or improvement of their educational programs.[§1926.12(b)(24)]

(25) *Federal Water Pollution Control Act,* as amended by the Water Quality Act of 1965 (3 U.S.C. 466e(g)). Provides for financial assistance to States or municipalities for construction of facilities in connection with the prevention and control of water pollution. This includes projects that will control the discharge into any waters of untreated or inadequately treated sewage.[§1926.12(b)(25)]

(26) *Appalachian Regional Development Act of 1965* (40 U.S.C. App. 402). Authorizes Federal assistance in the construction of an Appalachian development highway system; construction of multicounty demonstration health facilities, hospitals, regional health, diagnostic and treatment centers, and other facilities for health; seal and fill voids in abandoned mines and to rehabilitate strip mine areas; construction of school facilities for vocational education; and to assist in construction of sewage treatment works.[§1926.12(b)(26)]

(27) *National Technical Institute for the Deaf Act* (20 U.S.C. 684(b)(5)). Provides for financial assistance for institutions of higher education for the establishment, construction, including equipment and operation, of a National Institution for the Deaf. [§1926.12(b)(27)]

(28) *Housing Act of 1959* (12 U.S.C. 1701(q)(c)(3)). This act authorizes loans to nonprofit corporations to be used for the construction of housing and related facilities for elderly families. Also, the provisions of the act provide for rehabilitation, alteration, conversion or improvement of existing structures which are otherwise inadequate for proposed dwellings used by such families.[§1926.12(b)(28)]

(29) *College Housing Act of 1950, as amended* (12 U.S.C. 1749a(f)). This act provides for Federal loans to assist educational institutions in providing housing and other educational facilities for students and faculties.[§1926.12(b)(29)]

(30) *Housing and Urban Development Act of 1965* (42 U.S.C. 1500c-3, 3107). This act provides for Federal assistance for the following purposes:[§1926.12(b)(30)]

(i) *Grants to States and local public bodies* to assist in any construction work to be carried out under the open-space land and urban beautification provisions contained therein. It provides for parks and recreation areas, conservation of land and other natural resources, and historical and scenic purposes.[§1926.12(b)(30)(i)]

(ii) *Grants to local public bodies and agencies* to finance specific projects for basic public water facilities (including works for the storage, treatment, purification, and distribution of water), and for basic public sewer facilities (other than "treatment works" as defined in the Federal Water Pollution Control Act).[§1926.12(b)(30)(ii)]

(iii) *Grants to any local public body or agency* to assist in financing neighborhood facilities. These facilities must be necessary for carrying out a program of health, recreational, social, or similar community service and located so as to be available for the use of the area's low or moderate income residents.[§1926.12(b)(30)(iii)]

(31) *National Foundation on the Arts and the Humanities Act of 1965* (20 U.S.C. 954(k)). The act establishes the "National Foundation on the Arts and the Humanities" which may provide matching grants to groups (nonprofit organizations and State and other public organizations) and to individuals engaged in creative and performing arts for the entire range of artistic activity, including construction of necessary facilities. [§1926.12(b)(31)]

(32) *Public Works and Economic Development Act of 1965* (42 U.S.C. 3222). This act provides for Federal assistance for the following purposes:[§1926.12(b)(32)]

(i) *Grants for the acquisition or development of land* or improvements for public works or development facility usage in redevelopment areas. It authorizes loans to assist in financing the purchase or development of land for public works which will assist in the creation of long-term employment opportunities in the area.[§1926.12(b)(32)(i)]

(ii) *Loans for the purchase or development of land and facilities* (including machinery and equipment) for industrial or commercial usage within redevelopment areas; guarantee of loans for working capital made to private borrowers by private lending institutions in connection with direct loan projects; and to contract to pay to, or on behalf of, business entities locating in redevelopment areas, a portion of the interest costs which they incur in financing their expansions from private sources.[§1926.12(b)(32)(ii)]

(iii) *Loans and grants to create economic development centers* within designated county economic development districts. [§1926.12(b)(32)(iii)]

(33) *High-Speed Ground Transportation Study* (40 U.S.C. 1636(b)). This act provides for financial assistance for construction activities in connection with research and development of different forms of high- speed ground transportation and demonstration projects relating to intercity rail passenger service.[§1926.12(b)(33)]

(34) *Heart Disease, Cancer and Stroke Amendments of 1965* (42 U.S.C. 299(b)(4)). This act provides for grants to public or nonprofit private universities, medical schools, research, institutions, hospitals, and other public and nonprofit agencies and institutions, or associations thereof to assist in construction and equipment of facilities in connection with research, training, demonstration of patient care, diagnostic and treatment related to heart disease, cancer, stroke, and other major diseases.[§1926.12(b)(34)]

(35) *Mental Retardation Facilities* and Community Mental Health Centers Construction Act Amendments of 1965 (20 U.S.C. 618(g)). These provisions provide for grants to institutions of higher education for construction of facilities for research or for research and related purposes relating to education for mentally retarded, hard of hearing, deaf, speech impaired, visually handicapped, seriously emotionally disturbed, crippled, or other health impaired children who by reason thereof require special education.[§1926.12(b)(35)]

(36) *Vocational Rehabilitation Act Amendments of 1965* (29 U.S.C. 41a(b)(4)). This act authorizes grants to assist in meeting the costs of construction of public or other nonprofit workshops and rehabilitation facilities.[§1926.12(b)(36)]

(37) *Clean Air and Solid Waste Disposal Acts* (42 U.S.C. 3256). This act provides for financial assistance to public (Federal, State, interstate, or local) authorities, agencies, and institutions, private agencies and institutions, and individuals in the construction of facilities for solid-waste disposal. The term construction includes the installation of initial equipment. [§1926.12(b)(37)]

(38) *Medical Library Assistance Act of 1965* (42 U.S.C. 280b-3(b)(3)). This act provides for grants to public or private nonprofit agencies or institutions for the cost of construction of medical library facilities.[§1926.12(b)(38)]

(39) *Veterans Nursing Home Care Act* (38 U.S.C. 5035(a)(8)). The construction industry health and safety standards do not apply to this act since it is not subject to Reorganization Plan No. 14 of 1950.[§1926.12(b)(39)]

(40) *National Capital Transportation Act of 1965* (40 U.S.C. 682(b)(4)). This act provides for Federal assistance to the National Capital Transportation Agency for construction of a rail rapid transit system and related facilities for the Nation's Capital.[§1926.12(b)(40)]

(41) *Alaska Centennial — 1967 (80* Stat. 82). The program under this legislation has expired.[§1926.12(b)(41)]

(42) *Model Secondary School for the Deaf Act* (80 Stat. 1028). This act provides for funds to establish and operate, including construction and initial equipment of new buildings, expansion, remodeling, and alteration of existing buildings and equipment thereof, a model secondary school for the deaf to serve the residents of the District of Columbia and nearby States.[§1926.12(b)(42)]

(43) *Allied Health Professions Personnel Training Act of 1966* (42 U.S.C. 295h(b)(2)(E)). This act provides for grants to assist in the construction of new facilities for training centers for allied health professions, or replacement or rehabilitation of existing facilities for such centers.[§1926.12(b)(43)]

(44) *Demonstration Cities and Metropolitan Development Act* of 1966 (42 U.S.C. 3310; 12 U.S.C. 1715c; 42 U.S.C. 1416).

This act provides for Federal assistance for the following purposes:[§1926.12(b)(44)]

(i) *Grants to assist in the construction,* rehabilitation, alteration, or repair of residential property only if such residential property is designed for residential use for eight or more families to enable city demonstration agencies to carry out comprehensive city demonstration programs (42 U.S.C. 3310). [§1926.12(b)(44)(i)]

(ii) *Amends the National Housing Act* (12 U.S.C. 1715c) and the Housing Act of 1937 (42 U.S.C. 1416). See these acts for coverage.[§1926.12(b)(44)(ii)]

(45) *Air Quality Act of 1967* (42 U.S.C. 1857j-3). This act provides for Federal assistance to public or nonprofit agencies, institutions, and organizations and to individuals, and contracts with public or private agencies, institutions, or persons for construction of research and development facilities and demonstration plants relating to the application of preventing or controlling discharges into the air of various types of pollutants.[§1926.12(b)(45)]

(46) *Elementary and Secondary Education Amendments of 1967* (Title VII — Bilingual Education Act) (20 U.S.C. 880b-6). This act provides for Federal assistance to local educational agencies or to an institution of higher education applying jointly with a local educational agency for minor remodeling projects in connection with bilingual education programs to meet the special needs of children with limited English- speaking ability in the United States.[§1926.12(b)(46)]

(47) *Vocational Rehabilitation Amendments of 1967* (29 U.S.C. 42a(c)(3)). This act authorizes Federal assistance to any public or nonprofit private agency or organization for the construction of a center for vocational rehabilitation of handicapped individuals who are both deaf and blind which shall be known as the National Center for Deaf-Blind Youths and Adults. Construction includes new buildings and expansion, remodeling, alteration and renovation of existing buildings, and initial equipment of such new, newly acquired, expanded, remodeled, altered, or renovated buildings.[§1926.12(b)(47)]

(48) *National Visitor Center Facilities Act of 1968* (40 U.S.C. 808). This act authorizes agreements and leases with the owner of property in the District of Columbia known as Union Station for the use of all or a part of such property for a national visitor center to be known as the National Visitor Center. The agreements and leases shall provide for such alterations of the Union Station Building as necessary to provide adequate facilities for visitors. They also provide for the construction of a parking facility, including necessary approaches and ramps. [§1926.12(b)(48)]

(49) *Juvenile Delinquency Prevention and Control Act of 1968* (42 U.S.C. 3843). This act provides for Federal grants to State, county, municipal, or other public agency or combination thereof for the construction of facilities to be used in connection with rehabilitation services for the diagnosis, treatment, and rehabilitation of delinquent youths and youths in danger of becoming delinquent.[§1926.12(b)(49)]

(50) *Housing and Urban Development Act of 1968* (including New Communities Act of 1968) (42 U.S.C. 3909). This act provides for Federal assistance for the following purposes: [§1926.12(b)(50)]

(i) *Guarantees, and commitments to guarantee,* the bonds, debentures, notes, and other obligations issued by new community developers to help finance new community development projects.[§1926.12(b)(50)(i)]

(ii) *Amends section 212(a) of the National Housing Act,* adding section 236 for "Rental Housing for Lower Income Families" and section 242 "Mortgage Insurance for Nonprofit Hospitals" thereto.[§1926.12(b)(50)(ii)]

(51) *Public Health Service Act Amendment* (Alcoholic and Narcotic Addict Rehabilitation Amendments of 1968) (42 U.S.C. 2681, et seq.). This act provides for grants to a public and nonprofit private agency or organization for construction projects consisting of any facilities (including post-hospitalization treatment facilities for the prevention and treatment of alcoholism or treatment of narcotic addicts.)[§1926.12(b)(51)]

(52) *Vocational Education Amendments of 1968* (20 U.S.C. 1246). This act provides for grants to States for the construction of area vocational education school facilities. The act further provides grants to public educational agencies, organizations, or institutions for construction of residential schools to provide vocational education for the purpose of demonstrating the feasibility and desirability of such schools. The act still further provides grants to State boards, to colleges and universities, to public educational agencies, organizations or institutions to reduce the cost of borrowing funds for the construction of residential schools and dormitories.[§1926.12(b)(52)]

(53) *Postal Reorganization Act* (39 U.S.C. 410(d)(2)). This Act provides for construction, modification, alteration, repair, and other improvements of postal facilities located in leased buildings.[§1926.12(b)(53)]

(54) *Airport and Airway Development Act of 1970* (Pub. L. 91-258, section 52(b)(7)). This Act provides for Federal financial assistance to States and localities for the construction, improvement, or repair of public airports.[§1926.12(b)(54)]

(55) (i) *Public Law 91-230.* This Act provides for federal financial assistance to institutions of higher learning for the construction of a National Center on Educational Media and Materials for the Handicapped. The program under this statute expires on July 1, 1971. Public Law 91- 230, section 662(1). [§1926.12(b)(55)(i)]

(ii) *Education of the Handicapped Act* (20 U.S.C. 12326, 1404(a)). This Act provides for financial assistance to States for construction, expansion, remodeling, or alteration of facilities for the education of handicapped children at the preschool, elementary school, and secondary school levels. [§1926.12(b)(55)(ii)]

(56) *Housing and Urban Development Act of 1970* (Pub. L. 91-609, section 707(b)). This Act provides for grants to States and local public agencies to help finance the development of open-space or other land in urban areas for open-space uses. This Act becomes effective on July 1, 1971.[§1926.12(b)(56)]

(57) *Developmental Disabilities Services and Facilities Construction Amendments of 1970* (Pub. L. 91-517, section 135(a)(5)). This Act authorizes grants to States for construction of facilities for the provision of services to persons with developmental disabilities who are unable to pay for such services. [§1926.12(b)(57)]

(58) *Rail Passenger Service Act of 1970* (Pub. L. 91-518, section 405(d)). This statute provides that the National Railroad Passenger Corporation may construct physical facilities necessary to intercity rail passenger operations within the basic national rail passenger system designated by the Secretary of Transportation.[§1926.12(b)(58)]

(c) **VA and FHA housing.** In the course of the legislative development of section 107, it was recognized that section 107 would not apply to housing construction for which insurance was issued by the Federal Housing Authority and Veterans' Administration for individual home ownership. Concerning construction under the National Housing Act, Reorganization Plan No. 14 of 1950 applies to construction which is subject to the minimum wage requirements of section 212(a) thereof (12 U.S.C. 1715c).[§1926.12(c)]

§1926.13
Interpretation of statutory terms

(a) **The terms construction,** alteration, and repair used in section 107 of the Act are also used in section 1 of the Davis-Bacon Act (40 U.S.C. 276a), providing minimum wage protection on Federal construction contracts, and section 1 of the Miller Act (40 U.S.C. 270a), providing performance and payment bond protection on Federal construction contracts. Similarly, the terms contractor and subcontractor are used in those statutes, as well as in Copeland (Anti-Kickback) Act (40 U.S.C. 276c) and the Contract Work Hours and Safety Standards Act itself, which apply concurrently with the Miller Act and the Davis-Bacon Act on Federal construction contracts and also apply to most federally assisted construction contracts. The use of the same or identical terms in these statutes which apply concurrently with section 107 of the Act have considerable precedential value in ascertaining the coverage of section 107.[§1926.13(a)]

(b) **It should be noted that section 1** of the Davis-Bacon Act limits minimum wage protection to laborers and mechanics "employed directly" upon the "site of the work." There is no comparable limitation in section 107 of the Act. Section 107 expressly requires as a self- executing condition of each covered contract that no contractor or subcontractor shall require "any laborer or mechanic employed in the performance of the contract to work in surroundings or under working conditions which are unsanitary, hazardous, or dangerous to his health or safety" as these health and safety standards are applied in the rules of the Secretary of Labor. [§1926.13(b)]

(c) The term subcontractor under section 107 is considered to mean a person who agrees to perform any part of the labor or material requirements of a contract for construction, alteration or repair. Cf. MacEvoy Co. v. United States, 322 U.S. 102, 108-9 (1944). A person who undertakes to perform a portion of a contract involving the furnishing of supplies or materials will be considered a "subcontractor" under this part and section 107 if the work in question involves the performance of construction work and is to be performed:[§1926.13(c)]

(1) *Directly on or near the construction site, or*[§1926.13(c)(1)]

(2) *by the employer for the specific project* on a customized basis.[§1926.13(c)(2)]

Thus, a supplier of materials which will become an integral part of the construction is a "subcontractor" if the supplier fabricates or assembles the goods or materials in question specifically for the construction project and the work involved may be said to be construction activity. If the goods or materials in question are ordinarily sold to other customers from regular inventory, the supplier is not a "subcontractor." Generally, the furnishing of prestressed concrete beams and prestressed structural steel would be considered manufacturing; therefore a supplier of such materials would not be considered a "subcontractor." An example of material supplied "for the specific project on a customized basis" as that phrase is used in this section would be ventilating ducts, fabricated in a shop away from the construction job site and specifically cut for the project according to design specifications. On the other hand, if a contractor buys standard size nails from a foundry, the foundry would not be a covered "subcontractor." Ordinarily a contract for the supplying of construction equipment to a contractor would not, in and of itself, be considered a "subcontractor" for purposes of this part.[§1926.13(c)]

§1926.14

Federal contract for "mixed" types of performance

(a) It is the intent of the Congress to provide safety and health protection of Federal, federally financed, or federally assisted construction. See, for example, H. Report No. 91-241, 91st Cong., first session, p. 1 (1969). Thus, it is clear that when a Federal contract calls for mixed types of performance, such as both manufacturing and construction, section 107 would apply to the construction. By its express terms, section 107 applies to a contract which is "for construction, alteration, and/or repair." Such a contract is not required to be exclusively for such services. The application of the section is not limited to contracts which permit an overall characterization as "construction contracts." The text of section 107 is not so limited.[§1926.14(a)]

(b) When the mixed types of performances include both construction and manufacturing, see also Sec. 1926.15(b) concerning the relationship between the Walsh-Healey Public Contracts Act and section 107.[§1926.14(b)]

§1926.15

Relationship to the Service Contract Act; Walsh-Healey Public Contracts Act

(a) A contract for "construction" is one for nonpersonal service. See, e.g., 41 CFR 1-1.208. Section 2(e) of the Service Contract Act of 1965 requires as a condition of every Federal contract (and bid specification therefor) exceeding $2,500, the "principal purpose" of which is to furnish services to the United States through the use of "service employees," that certain safety and health standards be met. See 29 CFR part 1925, which contains the Department rules concerning these standards. Section 7 of the Service Contract Act provides that the Act shall not apply to "any contract of the United States or District of Columbia for construction, alteration, and/or repair, including painting and decorating of public buildings or public works." It is clear from the legislative history of section 107 that no gaps in coverage between the two statutes are intended.[§1926.15(a)]

(b) The Walsh-Healey Public Contracts Act requires that contracts entered into by any Federal agency for the manufacture or furnishing of materials, supplies, articles, and equipment in any amount exceeding $10,000 must contain, among other provisions, a requirement that "no part of such contract will be performed nor will any of the materials, supplies, articles or equipment to be manufactured or furnished under said contract be manufactured or fabricated in any plants, factories, buildings, or surroundings or under working conditions which are unsanitary or hazardous or dangerous to the health and safety of employees engaged in the performance of said contract." The rules of the Secretary concerning these standards are published in 41 CFR part 50-204, and express the Secretary of Labor's interpretation and application of section 1(e) of the Walsh-Healey Public Contracts Act to certain particular working conditions. None of the described working conditions are intended to deal with construction activities, although such activities may conceivably be a part of a contract which is subject to the Walsh-Healey Public Contracts Act. Nevertheless, such activities remain subject to the general statutory duty prescribed by section 1(e). Section 103(b) of the Contract Work Hours and Safety Standards Act provides, among other things, that the Act shall not apply to any work required to be done in accordance with the provisions of the Walsh- Healey Public Contracts Act.[§1926.15(b)]

§1926.16

Rules of construction

(a) The prime contractor and any subcontractors may make their own arrangements with respect to obligations which might be more appropriately treated on a jobsite basis rather than individually. Thus, for example, the prime contractor and his subcontractors may wish to make an express agreement that the prime contractor or one of the subcontractors will provide all required first-aid or toilet facilities, thus relieving the subcontractors from the actual, but not any legal, responsibility (or, as the case may be, relieving the other subcontractors from this responsibility). In no case shall the prime contractor be relieved of overall responsibility for compliance with the requirements of this part for all work to be performed under the contract.[§1926.16(a)]

(b) By contracting for full performance of a contract subject to section 107 of the Act, the prime contractor assumes all obligations prescribed as employer obligations under the standards contained in this part, whether or not he subcontracts any part of the work. [§1926.16(b)]

(c) To the extent that a subcontractor of any tier agrees to perform any part of the contract, he also assumes responsibility for complying with the standards in this part with respect to that part. Thus, the prime contractor assumes the entire responsibility under the contract and the subcontractor assumes responsibility with respect to his portion of the work. With respect to subcontracted work, the prime contractor and any subcontractor or subcontractors shall be deemed to have joint responsibility.[§1926.16(c)]

(d) Where joint responsibility exists, both the prime contractor and his subcontractor or subcontractors, regardless of tier, shall be considered subject to the enforcement provisions of the Act. [§1926.16(d)][

Authority: §107, Contract Work Hours and Safety Standards Act (Construction Safety Act) (40 U.S.C. 333).

Notes

Subpart C - General Safety and Health Provisions

§1926.20

General safety and health provisions

(a) Contractor requirements.[§1926.20(a)]

(1) *Section 107 of the Act requires* that it shall be a condition of each contract which is entered into under legislation subject to Reorganization Plan Number 14 of 1950 (64 Stat. 1267), as defined in Sec. 1926.12, and is for construction, alteration, and/or repair, including painting and decorating, that no contractor or subcontractor for any part of the contract work shall require any laborer or mechanic employed in the performance of the contract to work in surroundings or under working conditions which are unsanitary, hazardous, or dangerous to his health or safety.[§1926.20(a)(1)]

(b) Accident prevention responsibilities.[§1926.20(b)]

(1) *It shall be the responsibility of the employer to initiate* and maintain such programs as may be necessary to comply with this part.[§1926.20(b)(1)]

(2) *Such programs shall provide* for frequent and regular inspections of the job sites, materials, and equipment to be made by competent persons designated by the employers.[§1926.20(b)(2)]

(3) *The use of any machinery,* tool, material, or equipment which is not in compliance with any applicable requirement of this part is prohibited. Such machine, tool, material, or equipment shall either be identified as unsafe by tagging or locking the controls to render them inoperable or shall be physically removed from its place of operation.[§1926.20(b)(3)]

(4) *The employer shall permit only* those employees qualified by training or experience to operate equipment and machinery. [§1926.20(b)(4)]

(c) The standards contained in this part shall apply with respect to employments performed in a workplace in a State, the District of Columbia, the Commonwealth of Puerto Rico, the Virgin Islands, American Samoa, Guam, Trust Territory of the Pacific Islands, Wake Island, Outer Continental Shelf lands defined in the Outer Continental Shelf Lands Act, Johnston Island, and the Canal Zone.[§1926.20(c)]

(d) (1) *If a particular standard* is specifically applicable to a condition, practice, means, method, operation, or process, it shall prevail over any different general standard which might otherwise be applicable to the same condition, practice, means, method, operation, or process.[§1926.20(d)(1)]

(2) *On the other hand,* any standard shall apply according to its terms to any employment and place of employment in any industry, even though particular standards are also prescribed for the industry to the extent that none of such particular standards applies.[§1926.20(d)(2)]

(e) In the event a standard protects on its face a class of persons larger than employees, the standard shall be applicable under this part only to employees and their employment and places of employment.[§1926.20(e)]

(f) Compliance duties owed to each employee.[§1926.20(f)]

(1) *Personal protective equipment.* Standards in this part requiring the employer to provide personal protective equipment (PPE), including respirators and other types of PPE, because of hazards to employees impose a separate compliance duty with respect to each employee covered by the requirement. The employer must provide PPE to each employee required to use the PPE, and each failure to provide PPE to an employee may be considered a separate violation.[§1926.20(f)(1)]

(2) *Training.* Standards in this part requiring training on hazards and related matters, such as standards requiring that employees receive training or that the employer train employees, provide training to employees, or institute or implement a training program, impose a separate compliance duty with respect to each employee covered by the requirement. The employer must train each affected employee in the manner required by the standard, and each failure to train an employee may be considered a separate violation.[§1926.20(f)(2)]

[44 FR 8577, Feb. 9, 1979; 44 FR 20940, Apr. 6, 1979, as amended at 58 FR 35078, June 30, 1993; 73 FR 75588, Dec. 12, 2008]

§1926.21

Safety training and education

(a) General requirements. The Secretary shall, pursuant to section 107(f) of the Act, establish and supervise programs for the education and training of employers and employees in the recognition, avoidance and prevention of unsafe conditions in employments covered by the act.[§1926.21(a)]

(b) Employer responsibility.[§1926.21(b)]

(1) *The employer should avail himself* of the safety and health training programs the Secretary provides.[§1926.21(b)(1)]

(2) *The employer shall instruct each employee* in the recognition and avoidance of unsafe conditions and the regulations applicable to his work environment to control or eliminate any hazards or other exposure to illness or injury.[§1926.21(b)(2)]

(3) *Employees required to handle* or use poisons, caustics, and other harmful substances shall be instructed regarding the safe handling and use, and be made aware of the potential hazards, personal hygiene, and personal protective measures required.[§1926.21(b)(3)]

(4) *In job site areas* where harmful plants or animals are present, employees who may be exposed shall be instructed regarding the potential hazards, and how to avoid injury, and the first aid procedures to be used in the event of injury.[§1926.21(b)(4)]

(5) *Employees required to handle or use* flammable liquids, gases, or toxic materials shall be instructed in the safe handling and use of these materials and made aware of the specific requirements contained in subparts D, F, and other applicable subparts of this part.[§1926.21(b)(5)]

(6) (i) *All employees required to enter* into confined or enclosed spaces shall be instructed as to the nature of the hazards involved, the necessary precautions to be taken, and in the use of protective and emergency equipment required. The employer shall comply with any specific regulations that apply to work in dangerous or potentially dangerous areas. [§1926.21(b)(6)(i)]

(ii) *For purposes of paragraph (b)(6)(i) of this section,* **confined or enclosed space** means any space having a limited means of egress, which is subject to the accumulation of toxic or flammable contaminants or has an oxygen deficient atmosphere. Confined or enclosed spaces include, but are not limited to, storage tanks, process vessels, bins, boilers, ventilation or exhaust ducts, sewers, underground utility vaults, tunnels, pipelines, and open top spaces more than 4 feet in depth such as pits, tubs, vaults, and vessels. [§1926.21(b)(6)(ii)]

§1926.23

First aid and medical attention

First aid services and provisions for medical care shall be made available by the employer for every employee covered by these regulations. Regulations prescribing specific requirements for first aid, medical attention, and emergency facilities are contained in subpart D of this part.

§1926.24

Fire protection and prevention

The employer shall be responsible for the development and maintenance of an effective fire protection and prevention program at the job site throughout all phases of the construction, repair, alteration, or demolition work. The employer shall ensure the availability of the fire protection and suppression equipment required by subpart F of this part.

§1926.25

Housekeeping

(a) During the course of construction, alteration, or repairs, form and scrap lumber with protruding nails, and all other debris, shall be kept cleared from work areas, passageways, and stairs, in and around buildings or other structures.[§1926.25(a)]

(b) Combustible scrap and debris shall be removed at regular intervals during the course of construction. Safe means shall be provided to facilitate such removal.[§1926.25(b)]

(c) Containers shall be provided for the collection and separation of waste, trash, oily and used rags, and other refuse. Containers used for garbage and other oily, flammable, or hazardous wastes, such as caustics, acids, harmful dusts, etc. shall be equipped with covers. Garbage and other waste shall be disposed of at frequent and regular intervals.[§1926.25(c)]

§1926.26
Illumination

Construction areas, aisles, stairs, ramps, runways, corridors, offices, shops, and storage areas where work is in progress shall be lighted with either natural or artificial illumination. The minimum illumination requirements for work areas are contained in subpart D of this part.

§1926.27
Sanitation

Health and sanitation requirements for drinking water are contained in subpart D of this part.

§1926.28
Personal protective equipment

(a) The employer is responsible for requiring the wearing of appropriate personal protective equipment in all operations where there is an exposure to hazardous conditions or where this part indicates the need for using such equipment to reduce the hazards to the employees.[§1926.28(a)]

(b) Regulations governing the use, selection, and maintenance of personal protective and lifesaving equipment are described under subpart E of this part.[§1926.28(b)]

§1926.29
Acceptable certifications

(a) Pressure vessels. Current and valid certification by an insurance company or regulatory authority shall be deemed as acceptable evidence of safe installation, inspection, and testing of pressure vessels provided by the employer.[§1926.29(a)]

(b) Boilers. Boilers provided by the employer shall be deemed to be in compliance with the requirements of this part when evidence of current and valid certification by an insurance company or regulatory authority attesting to the safe installation, inspection, and testing is presented.[§1926.29(b)]

(c) Other requirements. Regulations prescribing specific requirements for other types of pressure vessels and similar equipment are contained in subparts F and O of this part.[§1926.29(c)]

§1926.30
Shipbuilding and ship repairing

(a) General. Shipbuilding, ship repairing, alterations, and maintenance performed on ships under Government contract, except naval ship construction, is work subject to the Act.[§1926.30(a)]

(b) Applicable safety and health standards. For the purpose of work carried out under this section, the safety and health regulations in part 1915 of this title, Shipyard Employment, shall apply. [§1926.30(b)]

[44 FR 8577, Feb. 9, 1979; 44 FR 20940, Apr. 6, 1979, as amended at 61 FR 9249, Mar. 7, 1996]

§1926.31
[Reserved]

§1926.32
Definitions

The following definitions shall apply in the application of the regulations in this part:

(a) **Act** means section 107 of the Contract Work Hours and Safety Standards Act, commonly known as the Construction Safety Act (86 Stat. 96; 40 U.S.C. 333).[§1926.32(a)]

(b) **ANSI** means American National Standards Institute.[§1926.32(b)]

(c) **Approved** means sanctioned, endorsed, accredited, certified, or accepted as satisfactory by a duly constituted and nationally recognized authority or agency.[§1926.32(c)]

(d) **Authorized person** means a person approved or assigned by the employer to perform a specific type of duty or duties or to be at a specific location or locations at the jobsite.[§1926.32(d)]

(e) **Administration** means the Occupational Safety and Health Administration.[§1926.32(e)]

(f) **Competent person** means one who is capable of identifying existing and predictable hazards in the surroundings or working conditions which are unsanitary, hazardous, or dangerous to employees, and who has authorization to take prompt corrective measures to eliminate them.[§1926.32(f)]

(g) **Construction work.** For purposes of this section, Construction work means work for construction, alteration, and/or repair, including painting and decorating.[§1926.32(g)]

(h) **Defect** means any characteristic or condition which tends to weaken or reduce the strength of the tool, object, or structure of which it is a part.[§1926.32(h)]

(i) **Designated person** means "authorized person" as defined in paragraph (d) of this section.[§1926.32(i)]

(j) **Employee** means every laborer or mechanic under the Act regardless of the contractual relationship which may be alleged to exist between the laborer and mechanic and the contractor or subcontractor who engaged him. "Laborer and mechanic" are not defined in the Act, but the identical terms are used in the Davis-Bacon Act (40 U.S.C. 276a), which provides for minimum wage protection on Federal and federally assisted construction contracts. The use of the same term in a statute which often applies concurrently with section 107 of the Act has considerable precedential value in ascertaining the meaning of "laborer and mechanic" as used in the Act. Laborer generally means one who performs manual labor or who labors at an occupation requiring physical strength; mechanic generally means a worker skilled with tools. See 18 Comp. Gen. 341.[§1926.32(j)]

(k) **Employer** means contractor or subcontractor within the meaning of the Act and of this part.[§1926.32(k)]

(l) **Hazardous substance** means a substance which, by reason of being explosive, flammable, poisonous, corrosive, oxidizing, irritating, or otherwise harmful, is likely to cause death or injury. [§1926.32(l)]

(m) **Qualified** means one who, by possession of a recognized degree, certificate, or professional standing, or who by extensive knowledge, training, and experience, has successfully demonstrated his ability to solve or resolve problems relating to the subject matter, the work, or the project.[§1926.32(m)]

(n) **Safety factor** means the ratio of the ultimate breaking strength of a member or piece of material or equipment to the actual working stress or safe load when in use.[§1926.32(n)]

(o) **Secretary** means the Secretary of Labor.[§1926.32(o)]

(p) **SAE** means Society of Automotive Engineers.[§1926.32(p)]

(q) **Shall** means mandatory.[§1926.32(q)]

(r) **Should** means recommended.[§1926.32(r)]

(s) **Suitable** means that which fits, and has the qualities or qualifications to meet a given purpose, occasion, condition, function, or circumstance.[§1926.32(s)]

[44 FR 8577, Feb. 9, 1979; 44 FR 20940, Apr. 6, 1979, as amended at 58 FR 35078, June 30, 1993]

§1926.33
Access to employee exposure and medical records

Note: The requirements applicable to construction work under this section are identical to those set forth at Sec. 1910.1020 of this chapter.

[61 FR 31431, June 20, 1996]

§1926.34
Means of egress

(a) General. In every building or structure exits shall be so arranged and maintained as to provide free and unobstructed egress from all parts of the building or structure at all times when it is occupied. No lock or fastening to prevent free escape from the inside of any building shall be installed except in mental, penal, or corrective institutions where supervisory personnel is continually on duty and effective provisions are made to remove occupants in case of fire or other emergency.[§1926.34(a)]

(b) Exit marking. Exits shall be marked by a readily visible sign. Access to exits shall be marked by readily visible signs in all cases where the exit or way to reach it is not immediately visible to the occupants.[§1926.34(b)]

(c) Maintenance and workmanship. Means of egress shall be continually maintained free of all obstructions or impediments to full instant use in the case of fire or other emergency.[§1926.34(c)]

[58 FR 35083, June 30, 1993]

§1926.35
Employee emergency action plans

(a) Scope and application. This section applies to all emergency action plans required by a particular OSHA standard. The emergency action plan shall be in writing (except as provided in the last sentence of paragraph (e)(3) of this section) and shall cover those designated actions employers and employees must take to ensure employee safety from fire and other emergencies.[§1926.35(a)]

(b) **Elements.** The following elements, at a minimum, shall be included in the plan:[§1926.35(b)]

(1) *Emergency escape procedures* and emergency escape route assignments;[§1926.35(b)(1)]

(2) *Procedures to be followed by employees* who remain to operate critical plant operations before they evacuate;[§1926.35(b)(2)]

(3) *Procedures to account for all employees* after emergency evacuation has been completed;[§1926.35(b)(3)]

(4) *Rescue and medical duties* for those employees who are to perform them;[§1926.35(b)(4)]

(5) *The preferred means of reporting* fires and other emergencies; and[§1926.35(b)(5)]

(6) *Names or regular job titles of persons* or departments who can be contacted for further information or explanation of duties under the plan.[§1926.35(b)(6)]

(c) **Alarm system.**[§1926.35(c)]

(1) *The employer shall establish an employee alarm system* which complies with §1926.159.[§1926.35(c)(1)]

(2) *If the employee alarm system is used* for alerting fire brigade members, or for other purposes, a distinctive signal for each purpose shall be used.[§1926.35(c)(2)]

(d) **Evacuation.** The employer shall establish in the emergency action plan the types of evacuation to be used in emergency circumstances.[§1926.35(d)]

(e) **Training.**[§1926.35(e)]

(1) *Before implementing the emergency action plan,* the employer shall designate and train a sufficient number of persons to assist in the safe and orderly emergency evacuation of employees.[§1926.35(e)(1)]

(2) *The employer shall review the plan* with each employee covered by the plan at the following times:[§1926.35(e)(2)]

(i) *Initially when the plan is developed,*[§1926.35(e)(2)(i)]

(ii) *Whenever the employee's responsibilities* or designated actions under the plan change, and[§1926.35(e)(2)(ii)]

(iii) *Whenever the plan is changed.*[§1926.35(e)(2)(iii)]

(3) *The employer shall review with each employee* upon initial assignment those parts of the plan which the employee must know to protect the employee in the event of an emergency. The written plan shall be kept at the workplace and made available for employee review. For those employers with 10 or fewer employees the plan may be communicated orally to employees and the employer need not maintain a written plan.[§1926.35(e)(3)]

[58 FR 35083, June 30, 1993]

Authority: §107, Contract Work Hours and Safety Standards Act (40 U.S.C. 333); secs. 4, 6, and 8, Occupational Safety and Health Act of 1970 (29 U.S.C. 653, 655, 657); Secretary of Labor's Order No. 12-71 (36 FR 8754), 8-76 (41 FR 25059), 9-83 (48 FR 35736), or 6-96 (62 FR 111), as applicable; and 29 CFR part 1911.

Notes

Subpart D - Occupational Health and Environmental Controls

§1926.50

Medical services and first aid

(a) **The employer shall insure** the availability of medical personnel for advice and consultation on matters of occupational health. [§1926.50(a)]

(b) **Provisions shall be made** prior to commencement of the project for prompt medical attention in case of serious injury. [§1926.50(b)]

(c) **In the absence of an infirmary,** clinic, hospital, or physician, that is reasonably accessible in terms of time and distance to the worksite, which is available for the treatment of injured employees, a person who has a valid certificate in first-aid training from the U.S. Bureau of Mines, the American Red Cross, or equivalent training that can be verified by documentary evidence, shall be available at the worksite to render first aid. [§1926.50(c)]

(d) (1) *First aid supplies* shall be easily accessible when required. [§1926.50(d)(1)]

(2) *The contents of the first aid kit* shall be placed in a weatherproof container with individual sealed packages for each type of item, and shall be checked by the employer before being sent out on each job and at least weekly on each job to ensure that the expended items are replaced. [§1926.50(d)(2)]

(e) **Proper equipment for prompt transportation** of the injured person to a physician or hospital, or a communication system for contacting necessary ambulance service, shall be provided. [§1926.50(e)]

(f) **In areas where 911 is not available,** the telephone numbers of the physicians, hospitals, or ambulances shall be conspicuously posted. [§1926.50(f)]

(g) **Where the eyes or body of any person** may be exposed to injurious corrosive materials, suitable facilities for quick drenching or flushing of the eyes and body shall be provided within the work area for immediate emergency use. [§1926.50(g)]

§1926.50 Appendix A

First aid Kits (Non-Mandatory)

First aid supplies are required to be easily accessible under paragraph Sec. 1926.50(d)(1). An example of the minimal contents of a generic first aid kit is described in American National Standard (ANSI) Z308.1-1978 “Minimum Requirements for Industrial Unit-Type First-aid Kits”. The contents of the kit listed in the ANSI standard should be adequate for small work sites. When larger operations or multiple operations are being conducted at the same location, employers should determine the need for additional first aid kits at the worksite, additional types of first aid equipment and supplies and additional quantities and types of supplies and equipment in the first aid kits.

In a similar fashion, employers who have unique or changing first-aid needs in their workplace may need to enhance their first-aid kits. The employer can use the OSHA 300 log, OSHA 301 log, or other reports to identify these unique problems. Consultation from the local fire/rescue department, appropriate medical professional, or local emergency room may be helpful to employers in these circumstances. By assessing the specific needs of their workplace, employers can ensure that reasonably anticipated supplies are available. Employers should assess the specific needs of their worksite periodically and augment the first aid kit appropriately.

If it is reasonably anticipated employees will be exposed to blood or other potentially infectious materials while using first-aid supplies, employers should provide personal protective equipment (PPE). Appropriate PPE includes gloves, gowns, face shields, masks and eye protection (see “Occupational Exposure to Bloodborne Pathogens”, 29 CFR 1910.1030(d)(3)) (56 FR 64175).

[44 FR 8577, Feb. 9, 1979; 44 FR 20940, Apr. 6, 1979, as amended at 49 FR 18295, Apr. 30, 1984; 58 FR 35084, June 30, 1993; 61 FR 5510, Feb. 13, 1996; 63 FR 33469, June 18, 1998; 76 FR 80740, Dec. 27, 2011]

§1926.51

Sanitation

(a) **Potable water.** [§1926.51(a)]

(1) *An adequate supply of potable* water shall be provided in all places of employment. [§1926.51(a)(1)]

(2) *Portable containers used* to dispense drinking water shall be capable of being tightly closed, and equipped with a tap. Water shall not be dipped from containers. [§1926.51(a)(2)]

(3) *Any container used to distribute* drinking water shall be clearly marked as to the nature of its contents and not used for any other purpose. [§1926.51(a)(3)]

(4) *The common drinking cup is prohibited.* [§1926.51(a)(4)]

(5) *Where single service cups* (to be used but once) are supplied, both a sanitary container for the unused cups and a receptacle for disposing of the used cups shall be provided. [§1926.51(a)(5)]

(6) **Potable water** means water that meets the standards for drinking purposes of the State or local authority having jurisdiction, or water that meets the quality standards prescribed by the U.S. Environmental Protection Agency's National Primary Drinking Water Regulations (40 CFR part 141). [§1926.51(a)(6)]

(b) **Nonpotable water.** [§1926.51(b)]

(1) *Outlets for nonpotable water,* such as water for industrial or firefighting purposes only, shall be identified by signs meeting the requirements of subpart G of this part, to indicate clearly that the water is unsafe and is not to be used for drinking, washing, or cooking purposes. [§1926.51(b)(1)]

(2) *There shall be no cross-connection,* open or potential, between a system furnishing potable water and a system furnishing nonpotable water. [§1926.51(b)(2)]

(c) **Toilets at construction jobsites.** [§1926.51(c)]

(1) *Toilets shall be provided for employees* according to the following table: [§1926.51(c)(1)]

Table D-1

Number of employees	Minimum number of facilities
20 or less	1
20 or more	1 toilet seat and 1 urinal per 40 workers
200 or more	1 toilet seat and 1 urinal per 50 workers

(2) *Under temporary field* conditions, provisions shall be made to assure not less than one toilet facility is available. [§1926.51(c)(2)]

(3) *Job sites,* not provided with a sanitary sewer, shall be provided with one of the following toilet facilities unless prohibited by local codes: [§1926.51(c)(3)]

(i) *Privies (where their* use will not contaminate ground or surface water); [§1926.51(c)(3)(i)]

(ii) *Chemical toilets;* [§1926.51(c)(3)(ii)]

(iii) *Recirculating toilets;* [§1926.51(c)(3)(iii)]

(iv) *Combustion toilets.* [§1926.51(c)(3)(iv)]

(4) *The requirements of this paragraph (c)* for sanitation facilities shall not apply to mobile crews having transportation readily available to nearby toilet facilities. [§1926.51(c)(4)]

(d) **Food handling.** [§1926.51(d)]

(1) *All employees' food service* facilities and operations shall meet the applicable laws, ordinances, and regulations of the jurisdictions in which they are located. [§1926.51(d)(1)]

(2) *All employee food service* facilities and operations shall be carried out in accordance with sound hygienic principles. In all places of employment where all or part of the food service is provided, the food dispensed shall be wholesome, free from spoilage, and shall be processed, prepared, handled, and stored in such a manner as to be protected against contamination. [§1926.51(d)(2)]

(e) **Temporary sleeping quarters.** When temporary sleeping quarters are provided, they shall be heated, ventilated, and lighted. [§1926.51(e)]

(f) **Washing facilities.** [§1926.51(f)]

(1) *The employer shall provide adequate* washing facilities for employees engaged in the application of paints, coating, herbicides, or insecticides, or in other operations where contaminants may be harmful to the employees. Such facilities shall be in near proximity to the worksite and shall be so equipped as to enable employees to remove such substances. [§1926.51(f)(1)]

(2) *General.* Washing facilities shall be maintained in a sanitary condition. [§1926.51(f)(2)]

(3) *Lavatories.* [§1926.51(f)(3)]

(i) *Lavatories shall be made available* in all places of employment. The requirements of this subdivision do not apply to mobile crews or to normally unattended work locations if employees working at these locations have transportation readily available to nearby washing facilities which meet the other requirements of this paragraph. [§1926.51(f)(3)(i)]

(ii) *Each lavatory shall be provided with hot* and cold running water, or tepid running water. [§1926.51(f)(3)(ii)]

(iii) *Hand soap or similar* cleansing agents shall be provided. [§1926.51(f)(3)(iii)]

(iv) *Individual hand towels* or sections thereof, of cloth or paper, air blowers or clean individual sections of continuous cloth toweling, convenient to the lavatories, shall be provided. [§1926.51(f)(3)(iv)]

(4) *Showers.*[§1926.51(f)(4)]

(i) *Whenever showers are required* by a particular standard, the showers shall be provided in accordance with paragraphs (f)(4) (ii) through (v) of this section.[§1926.51(f)(4)(i)]

(ii) *One shower shall be provided* for each 10 employees of each sex, or numerical fraction thereof, who are required to shower during the same shift.[§1926.51(f)(4)(ii)]

(iii) *Body soap or other appropriate* cleansing agents convenient to the showers shall be provided as specified in paragraph (f)(3)(iii) of this section.[§1926.51(f)(4)(iii)]

(iv) *Showers shall be provided with hot* and cold water feeding a common discharge line.[§1926.51(f)(4)(iv)]

(v) *Employees who use showers* shall be provided with individual clean towels.[§1926.51(f)(4)(v)]

(g) **Eating and drinking areas.** No employee shall be allowed to consume food or beverages in a toilet room nor in any area exposed to a toxic material.[§1926.51(g)]

(h) **Vermin control.** Every enclosed workplace shall be so constructed, equipped, and maintained, so far as reasonably practicable, as to prevent the entrance or harborage of rodents, insects, and other vermin. A continuing and effective extermination program shall be instituted where their presence is detected. [§1926.51(h)]

(i) **Change rooms.** Whenever employees are required by a particular standard to wear protective clothing because of the possibility of contamination with toxic materials, change rooms equipped with storage facilities for street clothes and separate storage facilities for the protective clothing shall be provided.[§1926.51(i)]

[44 FR 8577, Feb. 9, 1979; 44 FR 20940, Apr. 6, 1979, as amended at 58 FR 35084, June 30, 1993; 76 FR 33611, June 8, 2011]

§1926.52

Occupational noise exposure

(a) **Protection against the effects** of noise exposure shall be provided when the sound levels exceed those shown in Table D-2 of this section when measured on the A-scale of a standard sound level meter at slow response.[§1926.52(a)]

(b) **When employees are subjected to sound levels** exceeding those listed in Table D-2 of this section, feasible administrative or engineering controls shall be utilized. If such controls fail to reduce sound levels within the levels of the table, personal protective equipment as required in subpart E, shall be provided and used to reduce sound levels within the levels of the table.[§1926.52(b)]

(c) **If the variations in noise level** involve maxima at intervals of 1 second or less, it is to be considered continuous.[§1926.52(c)]

(d) (1) *In all cases where the sound levels* exceed the values shown herein, a continuing, effective hearing conservation program shall be administered.[§1926.52(d)(1)]

Table D-2 — Permissible Noise Exposures

Duration per day, hours	Sound level dBA slow response
8	90
6	92
4	95
3	97
2	100
1 1/2	102
1	105
1/2	110
1/4 or less	115

(2) (i) *When the daily noise exposure* is composed of two or more periods of noise exposure of different levels, their combined effect should be considered, rather than the individual effect of each. Exposure to different levels for various periods of time shall be computed according to the formula set forth in paragraph (d)(2)(ii) of this section.[§1926.52(d)(2)(i)]

(ii) $F_e = (T_1/L_1)+(T_2/L_2)+ \ldots + (T_n/L_n)$

Where:

F_e= The equivalent noise exposure factor.[§1926.52(d)(2)(ii)]

T = The period of noise exposure at any essentially constant level.

L = The duration of the permissible noise exposure at the constant level (from Table D-2).

If the value of F_e exceeds unity (1) the exposure exceeds permissible levels.[§1926.52(d)(2)(ii)]

(iii) *A sample computation showing* an application of the formula in paragraph (d)(2)(ii) of this section is as follows. An employee is exposed at these levels for these periods: [§1926.52(d)(2)(iii)]

110 db A 1/4 hour.

100 db A 1/2 hour.

90 db A 1 1/2 hours.

F_e= (1/4 / 1/2)+(1/2 / 2)+(1 1/2 / 8)

F_e= 0.500+0.25+0.188

F_e= 0.938

Since the value of F_e does not exceed unity, the exposure is within permissible limits.[§1926.52(d)(2)(iii)]

(e) **Exposure to impulsive or impact** noise should not exceed 140 dB peak sound pressure level.[§1926.52(e)]

§1926.53

Ionizing radiation

(a) **In construction and related activities** involving the use of sources of ionizing radiation, the pertinent provisions of the Nuclear Regulatory Commission's Standards for Protection Against Radiation (10 CFR part 20), relating to protection against occupational radiation exposure, shall apply.[§1926.53(a)]

(b) **Any activity which involves** the use of radioactive materials or X-rays, whether or not under license from the Nuclear Regulatory Commission, shall be performed by competent persons specially trained in the proper and safe operation of such equipment. In the case of materials used under Commission license, only persons actually licensed, or competent persons under direction and supervision of the licensee, shall perform such work.[§1926.53(b)]

(c) **- (r) [Reserved]**[§1926.53(c)]

Note: The requirements applicable to construction work under paragraphs (c) through (r) of this section are identical to those set forth at paragraphs (a) through (p) of Sec. 1910.1096 of this chapter.

[44 FR 8577, Feb. 9, 1979; 44 FR 20940, Apr. 6, 1979, as amended at 61 FR 5510, Feb. 13, 1996; 61 FR 31431, June 20, 1996]

§1926.54

Nonionizing radiation

(a) **Only qualified and trained** employees shall be assigned to install, adjust, and operate laser equipment.[§1926.54(a)]

(b) **Proof of qualification of the laser** equipment operator shall be available and in possession of the operator at all times.[§1926.54(b)]

(c) **Employees,** when working in areas in which a potential exposure to direct or reflected laser light greater than 0.005 watts (5 milliwatts) exists, shall be provided with antilaser eye protection devices as specified in subpart E of this part.[§1926.54(c)]

(d) **Areas in which lasers** are used shall be posted with standard laser warning placards.[§1926.54(d)]

(e) **Beam shutters or caps** shall be utilized, or the laser turned off, when laser transmission is not actually required. When the laser is left unattended for a substantial period of time, such as during lunch hour, overnight, or at change of shifts, the laser shall be turned off.[§1926.54(e)]

(f) **Only mechanical or electronic** means shall be used as a detector for guiding the internal alignment of the laser.[§1926.54(f)]

(g) **The laser beam shall not be directed at employees.**[§1926.54(g)]

(h) **When it is raining or snowing,** or when there is dust or fog in the air, the operation of laser systems shall be prohibited where practicable; in any event, employees shall be kept out of range of the area of source and target during such weather conditions. [§1926.54(h)]

(i) **Laser equipment shall bear** a label to indicate maximum output. [§1926.54(i)]

(j) **Employees shall not be exposed** to light intensities above: [§1926.54(j)]

(1) *Direct staring:* 1 micro-watt per square centimeter;[§1926.54(j)(1)]

(2) *Incidental observing:* 1 milliwatt per square centimeter; [§1926.54(j)(2)]

(3) *Diffused reflected light:* 2 1/2 watts per square centimeter. [§1926.54(j)(3)]

(k) **Laser unit in operation** should be set up above the heads of the employees, when possible.[§1926.54(k)]

(l) **Employees shall not be exposed to microwave** power densities in excess of 10 milliwatts per square centimeter.[§1926.54(l)]

§1926.55

Gases, vapors, fumes, dusts, and mists

(a) **Exposure of employees to inhalation,** ingestion, skin absorption, or contact with any material or substance at a concentration above those specified in the "Threshold Limit Values of Airborne Contaminants for 1970" of the American Conference of Governmental Industrial Hygienists, shall be avoided. See Appendix A to this section.[§1926.55(a)]

(b) **To achieve compliance with paragraph (a) of this section,** administrative or engineering controls must first be implemented whenever feasible. When such controls are not feasible to achieve full compliance, protective equipment or other protective measures shall be used to keep the exposure of employees to air contaminants within the limits prescribed in this section. Any equipment and technical measures used for this purpose must first be approved for each particular use by a competent industrial hygienist or other technically qualified person. Whenever respirators are used, their use shall comply with Sec. 1926.103. [§1926.55(b)]

(c) **Paragraphs (a) and (b) of this section** do not apply to the exposure of employees to airborne asbestos, tremolite, anthophyllite, or actinolite dust. Whenever any employee is exposed to airborne asbestos, tremolite, anthophyllite, or actinolite dust, the requirements of §1910.1101 or §1926.58 of this title shall apply. [§1926.55(c)]

(d) **Paragraphs (a) and (b) of this section** do not apply to the exposure of employees to formaldehyde. Whenever any employee is exposed to formaldehyde, the requirements of Sec. 1910.1048 of this title shall apply.[§1926.55(d)]

§1926.55 Appendix A

1970 American Conference of Governmental Industrial Hygienists' Threshold Limit Values of Airborne Contaminants

Threshold Limit Values of Airborne Contaminants for Construction

Substance	CAS. No. [d]	ppm [a]	mg/m^3 [b]	Skin Designation
Abate; see Temephos				
Acetaldehyde	75-07-0	200	360	
Acetic acid	64-19-7	10	25	
Acetic anhydride	108-24-7	5	20	
Acetone	67-64-1	1000	2400	
Acetonitrile	75-05-8	40	70	
2-Acetylaminofluorene; see §1926.1114	53-96-3			
Acetylene	74-86-2	E		
Acetylene dichloride; see 1,2-Dichloroethylene				
Acetylene tetrabromide	79-27-6	1	14	
Acrolein	107-02-8	0.1	0.25	
Acrylamide	79-06-1		0.3	X
Acrylonitrile; see §1926.1145	107-13-1			
Aldrin	309-00-2		0.25	X
Allyl alcohol	107-18-6	2	5	X
Allyl chloride	107-05-1	1	3	
Allyl glycidyl ether (AGE)	106-92-3	(C)10	(C)45	
Allyl propyl disulfide	2179-59-1	2	12	
alpha-Alumina Total dust Respirable fraction	1344-28-1			
Alundum; see alpha-Alumina				
4-Aminodiphenyl; see §1926.1111	92-67-1			
2-Aminoethanol; see Ethanolamine				
2-Aminopyridine	504-29-0	0.5	2	
Ammonia	7664-41-7	50	35	
Ammonium sulfamate Total dust Respirable fraction	7773-06-0		 15 5	
n-Amyl acetate	628-63-7	100	525	

Threshold Limit Values of Airborne Contaminants for Construction (continued)

Substance	CAS. No. [d]	ppm [a]	mg/m^3 [b]	Skin Designation
sec-Amyl acetate	626-38-0	125	650	
Aniline and homologs	62-53-3	5	19	X
Anisidine (o-, p-isomers)	29191-52-4		0.5	X
Antimony and compounds (as Sb)	7440-36-0		0.5	
ANTU (alpha Naphthylthiourea)	86-88-4		0.3	
Argon	7440-37-1	E		
Arsenic, inorganic compounds (as As); see §1926.1118	7440-38-2			
Arsenic, organic compounds (as As)	7440-38-2		0.5	
Arsine	7784-42-1	0.05	0.2	
Asbestos; see §1926.58				
Azinphos-methyl	86-50-0		0.2	X
Barium, soluble compounds (as Ba)	7440-39-3		0.5	
Benzene[g]; see §1926.1128	71-43-2			
Benzidine; see §1926.1110	92-87-5			
p-Benzoquinone; see Quinone				
Benzo(a)pyrene; see Coal tar pitch volatiles				
Benzoyl peroxide	94-36-0		5	
Benzyl chloride	100-44-7	1	5	
Beryllium and beryllium compounds (as Be)	7440-41-7		0.002	
Biphenyl; see Diphenyl				
Bisphenol A; see Diglycidyl ether				
Boron oxide	1303-86-2			
Total dust			15	
Boron tribromide	10294-33-4	1	10	
Boron trifluoride	7637-07-2	(C)1	(C)3	
Bromine	7726-95-6	0.1	0.7	
Bromine pentafluoride	7789-30-2	0.1	0.7	
Bromoform	75-25-2	0.5	5	X
Butadiene (1,3-Butadiene); see 29 CFR 1910.1051; 29 CFR 1910.19(l)	106-99-0	1 ppm/ 5 ppm STEL		
Butanethiol; see Butyl mercaptan				
2-Butanone (Methyl ethyl ketone)	78-93-3	200	590	
2-Butoxyethanol	111-76-2	50	240	X
n-Butyl-acetate	123-86-4	150	710	
sec-Butyl acetate	105-46-4	200	950	
tert-Butyl acetate	540-88-5	200	950	
n-Butyl alcohol	71-36-3	100	300	
sec-Butyl alcohol	78-92-2	150	450	
tert-Butyl alcohol	75-65-0	100	300	
Butylamine	109-73-9	(C)5	(C)15	X
tert-Butyl chromate (as CrO_3); see §1910.1026[n,†]	1189-85-1			
n-Butyl glycidyl ether (BGE)	2426-08-6	50	270	
Butyl mercaptan	109-79-5	0.5	1.5	
p-tert-Butyltoluene	98-51-1	10	60	
Cadmium (as Cd); see §1926.1127	7440-43-9			
Calcium carbonate Total dust Respirable fraction	1317-65-3			
Calcium oxide	1305-78-8		5	
Calcium sulfate Total dust Respirable fraction	7778-18-9		 15 5	
Camphor, synthetic	76-22-2		2	
Carbaryl (Sevin)	63-25-2		5	
Carbon black	1333-86-4		3.5	
Carbon dioxide	124-38-9	5000	9000	
Carbon disulfide	75-15-0	20	60	X
Carbon monoxide	630-08-0	50	55	
Carbon tetrachloride	56-23-5	10	65	X
Cellulose Total dust Respirable fraction	9004-34-6			
Chlordane	57-74-9		0.5	X
Chlorinated camphene	8001-35-2		0.5	X

Threshold Limit Values of Airborne Contaminants for Construction (continued)

Substance	CAS. No. [d]	ppm [a]	mg/m3 [b]	Skin Designation
Chlorinated diphenyl oxide	55720-99-5		0.5	
Chlorine	7782-50-5	1	3	
Chlorine dioxide	10049-04-4	0.1	0.3	
Chlorine trifluoride	7790-91-2	(C)0.1	(C)0.4	
Chloroacetaldehyde	107-20-0	(C)1	(C)3	
a-Chloroacetophenone (Phenacyl chloride)	532-27-4	0.05	0.3	
Chlorobenzene	108-90-7	75	350	
o-Chlorobenzylidene malononitrile	2698-41-1	0.05	0.4	
Chlorobromomethane	74-97-5	200	1050	
2-Chloro-1,3-butadiene; see beta-Chloroprene				
Chlorodiphenyl (42% Chlorine)(PCB)	53469-21-9		1	X
Chlorodiphenyl (54% Chlorine)(PCB)	11097-69-1		0.5	X
1-Chloro-2,3-epoxypropane; see Epichlorohydrin				
2-Chloroethanol; see Ethylene chlorohydrin				
Chloroethylene; see Vinyl chloride				
Chloroform (Trichloromethane)	67-66-3	(C)50	(C)240	
bis(Chloromethyl) ether; see §1926.1108	542-88-1			
Chloromethyl methyl ether; see §1926.1106	107-30-2			
1-Chloro-1-nitropropane	600-25-9	20	100	
Chloropicrin	76-06-2	0.1	0.7	
beta-Chloroprene	126-99-8	25	90	X
Chromium (II) compounds (as Cr)	7440-47-3		0.5	
Chromium (III) compounds (as Cr)	7440-47-3		0.5	
Chromium (VI) compounds; see §1926.1126[o]				
Chromium metal and insol. salts (as Cr)	7440-47-3		1	
Chrysene; see Coal tar pitch volatiles				
Coal tar pitch volatiles (benzene soluble fraction), anthracene, BaP, phenanthrene, acridine, chrysene, pyrene	65996-93-2		0.2	
Cobalt metal, dust, and fume (as Co)	7440-48-4		0.1	
Coke oven emissions; see §1926.1129				
Copper Fume (as Cu) Dusts and mists (as Cu)	7440-50-8		 0.1 1	
Corundum; see Emery				
Cotton dust (raw)			1	
Crag herbicide (Sesone) Total dust Respirable fraction	136-78-7			
Cresol, all isomers	1319-77-3	5	22	X
Crotonaldehyde	123-73-9; 4170-30-3	2	6	
Cumene	98-82-8	50	245	X
Cyanides (as CN)	Varies with compound		5	X
Cyanogen	460-19-5	10		
Cyclohexane	110-82-7	300	1050	
Cyclohexanol	108-93-0	50	200	
Cyclohexanone	108-94-1	50	200	
Cyclohexene	110-83-8	300	1015	
Cyclonite	121-82-4		1.5	X
Cyclopentadiene	542-92-7	75	200	
DDT, see Dichlorodiphenyltrichloroethane				
DDVP, see Dichlorvos				
2,4-D (Dichlorophenoxyacetic acid)	94-75-7		10	
Decaborane	17702-41-9	0.05	0.3	X
Demeton (Systox)	8065-48-3		0.1	X
Diacetone alcohol (4-Hydroxy-4-methyl-2-pentanone)	123-42-2	50	240	
1,2-Diaminoethane; see Ethylenediamine				

Threshold Limit Values of Airborne Contaminants for Construction (continued)

Substance	CAS. No. [d]	ppm [a]	mg/m3 [b]	Skin Designation
Diazomethane	334-88-3	0.2	0.4	
Diborane	19287-45-7	0.1	0.1	
1,2-Dibromo-3-chloropropane (DBCP); see §1926.1144	96-12-8			
1,2-Dibromoethane; see Ethylene dibromide				
Dibutyl phosphate	107-66-4	1	5	
Dibutyl phthalate	84-74-2		5	
Dichloroacetylene	7572-29-4	(C)0.1	(C)0.4	
o-Dichlorobenzene	95-50-1	(C)50	(C)300	
p-Dichlorobenzene	106-46-7	75	450	
3,3'-Dichlorobenzidine; see §1926.1107	91-94-1			
Dichlorodifluoromethane	75-71-8	1000	4950	
1,3-Dichloro-5,5-dimethyl hydantoin	118-52-5		0.2	
Dichlorodiphenyltrichloroethane (DDT)	50-29-3		1	X
1,1-Dichloroethane	75-34-3	100	400	
1,2-Dichloroethane; see Ethylene dichloride				
1,2-Dichloroethylene	540-59-0	200	790	
Dichloroethyl ether	111-44-4	(C)15	(C)90	X
Dichloromethane; see Methylene chloride				
Dichloromonofluoromethane	75-43-4	1000	4200	
1,1-Dichloro-1-nitroethane	594-72-9	(C)10	(C)60	
1,2-Dichloropropane; see Propylene dichloride				
Dichlorotetrafluoroethane	76-14-2	1000	7000	
Dichlorvos (DDVP)	62-73-7		1	X
Dieldrin	60-57-1		0.25	X
Diethylamine	109-89-7	25	75	
2-Diethylaminoethanol	100-37-8	10	50	X
Diethylene triamine	111-40-0	(C)10	(C)42	X
Diethyl ether; see Ethyl ether				
Difluorodibromomethane	75-61-6	100	860	
Diglycidyl ether (DGE)	2238-07-5	(C)0.5	(C)2.8	
Dihydroxybenzene; see Hydroquinone				
Diisobutyl ketone	108-83-8	50	290	
Diisopropylamine	108-18-9	5	20	X
4-Dimethylaminoazobenzene; see §1926.1115	60-11-7			
Dimethoxymethane; see Methylal				
Dimethyl acetamide	127-19-5	10	35	X
Dimethylamine	124-40-3	10	18	
Dimethylaminobenzene; see Xylidine				
Dimethylaniline (N,N-Dimethylaniline)	121-69-7	5	25	X
Dimethylbenzene; see Xylene				
Dimethyl-1,2-dibromo-2,2-dichloroethyl phosphate	300-76-5		3	
Dimethylformamide	68-12-2	10	30	X
2,6-Dimethyl-4-heptanone; see Diisobutyl ketone				
1,1-Dimethylhydrazine	57-14-7	0.5	1	X
Dimethylphthalate	131-11-3		5	
Dimethyl sulfate	77-78-3	1	5	X
Dinitrobenzene (all isomers) (ortho) (meta) (para)	 528-29-0 99-65-0 100-25-4		1	X
Dinitro-o-cresol	534-52-1		0.2	X
Dinitrotoluene	25321-14-6		1.5	X
Dioxane (Diethylene dioxide)	123-91-1	100	360	X
Diphenyl (Biphenyl)	92-52-4	0.2	1	
Diphenylamine	122-39-4		10	
Diphenylmethane diisocyanate; see Methylene bisphenyl isocyanate				
Dipropylene glycol methyl ether	34590-94-8	100	600	X
Di-sec octyl phthalate (Di-(2-ethylhexyl) phthalate)	117-81-7		5	

Threshold Limit Values of Airborne Contaminants for Construction (continued)

Substance	CAS. No. d	ppm a	mg/m3 b	Skin Designation
Emery Total dust Respirable fraction	12415-34-8			
Endosulfan	115-29-7		0.1	X
Endrin	72-20-8		0.1	X
Epichlorohydrin	106-89-8	5	19	X
EPN	2104-64-5		0.5	X
1,2-Epoxypropane; see Propylene oxide				
2,3-Epoxy-1-propanol; see Glycidol				
Ethane	74-84-0	E		
Ethanethiol; see Ethyl mercaptan				
Ethanolamine	141-43-5	3	6	
2-Ethoxyethanol (Cellosolve)	110-80-5	200	740	X
2-Ethoxyethyl acetate (Cellosolve acetate)	111-15-9	100	540	X
Ethyl acetate	141-78-6	400	1400	
Ethyl acrylate	140-88-5	25	100	X
Ethyl alcohol (Ethanol)	64-17-5	1000	1900	
Ethylamine	75-04-7	10	18	
Ethyl amyl ketone (5-Methyl-3-heptanone)	541-85-5	25	130	
Ethyl benzene	100-41-4	100	435	
Ethyl bromide	74-96-4	200	890	
Ethyl butyl ketone (3-Heptanone)	106-35-4	50	230	
Ethyl chloride	75-00-3	1000	2600	
Ethyl ether	60-29-7	400	1200	
Ethyl formate	109-94-4	100	300	
Ethyl mercaptan	75-08-1	0.5	1	
Ethyl silicate	78-10-4	100	850	
Ethylene	74-85-1	E		
Ethylene chlorohydrin	107-07-3	5	16	X
Ethylenediamine	107-15-3	10	25	
Ethylene dibromide	106-93-4	(C)25	(C)190	X
Ethylene dichloride (1,2-Dichloroethane)	107-06-2	50	200	
Ethylene glycol dinitrate	628-96-6	(C)0.2	(C)1	X
Ethylene glycol methyl acetate; see Methyl cellosolve acetate				
Ethyleneimine; see §1926.1112	151-56-4			
Ethylene oxide; see §1926.1147	75-21-8			
Ethylidene chloride; see 1,1-Dichloroethane				
N-Ethylmorpholine	100-74-3	20	94	X
Ferbam Total dust	14484-64-1		15	
Ferrovanadium dust	12604-58-9		1	
Fibrous Glass Total dust Respirable fraction				
Fluorides (as F)	Varies with compound		2.5	
Fluorine	7782-41-4	0.1	0.2	
Fluorotrichloromethane (Trichlorofluoromethane)	75-69-4	1000	5600	
Formaldehyde; see §1926.1148	50-00-0			
Formic acid	64-18-6	5	9	
Furfural	98-01-1	5	20	X
Furfuryl alcohol	98-00-0	50	200	
Gasoline	8006-61-9		A3	
Glycerin (mist) Total dust Respirable fraction	56-81-5			
Glycidol	556-52-5	50	150	
Glycol monoethyl ether; see 2-Ethoxyethanol				
Graphite, natural, respirable dust	7782-42-5	(2)	(2)	(2)
Graphite, synthetic Total dust Respirable Fraction				
Guthion; see Azinphos methyl				

Threshold Limit Values of Airborne Contaminants for Construction (continued)

Substance	CAS. No. d	ppm a	mg/m3 b	Skin Designation
Gypsum Total dust Respirable fraction	13397-24-5			
Hafnium	7440-58-6		0.5	
Helium	7440-59-7	E		
Heptachlor	76-44-8		0.5	X
Heptane (n-Heptane)	142-82-5	500	2000	
Hexachloroethane	67-72-1	1	10	X
Hexachloronaphthalene	1335-87-1		0.2	X
n-Hexane	110-54-3	500	1800	
2-Hexanone (Methyl n-butyl ketone)	591-78-6	100	410	
Hexone (Methyl isobutyl ketone)	108-10-1	100	410	
sec-Hexyl acetate	108-84-9	50	300	
Hydrazine	302-01-2	1	1.3	X
Hydrogen	1333-74-0	E		
Hydrogen bromide	10035-10-6	3	10	
Hydrogen chloride	7647-01-0	(C)5	(C)7	
Hydrogen cyanide	74-90-8	10	11	X
Hydrogen fluoride (as F)	7664-39-3	3	2	
Hydrogen peroxide	7722-84-1	1	1.4	
Hydrogen selenide (as Se)	7783-07-5	0.05	0.2	
Hydrogen sulfide	7783-06-4	10	15	
Hydroquinone	123-31-9		2	
Indene	95-13-6	10	45	
Indium and compounds (as In)	7440-74-6		0.1	
Iodine	7553-56-2	(C)0.1	(C)1	
Iron oxide fume	1309-37-1		10	
Iron salts (soluble) (as Fe)	Varies with compound		1	
Isoamyl acetate	123-92-2	100	525	
Isoamyl alcohol (primary and secondary)	123-51-3	100	360	
Isobutyl acetate	110-19-0	150	700	
Isobutyl alcohol	78-83-1	100	300	
Isophorone	78-59-1	25	140	
Isopropyl acetate	108-21-4	250	950	
Isopropyl alcohol	67-63-0	400	980	
Isopropylamine	75-31-0	5	12	
Isopropyl ether	108-20-3	500	2100	
Isopropyl glycidyl ether (IGE)	4016-14-2	50	240	
Kaolin Total dust Respirable fraction	1332-58-7			
Ketene	463-51-4	0.5	0.9	
Lead, inorganic (as Pb); see §1926.62	7439-92-1			
Limestone Total dust Respirable fraction	1317-65-3			
Lindane	58-89-9		0.5	X
Lithium hydride	7580-67-8		0.025	
LPG (Liquified petroleum gas)	68476-85-7	1000	1800	
Magnesite Total dust Respirable fraction	546-93-0			
Magnesium oxide fume Total particulate	1309-48-4	15		
Malathion Total dust	121-75-5		15	X
Maleic anhydride	108-31-6	0.25		
Manganese compounds (as Mn)	7439-96-5		(C)5	
Manganese fume (as Mn)	7439-96-5		(C)5	
Marble Total dust Respirable fraction	1317-65-3			
Mercury (aryl and inorganic)(as Hg)	7439-97-6		0.1	X
Mercury (organo) alkyl compounds (as Hg)	7439-97-6		0.01	X
Mercury (vapor) (as Hg)	7439-97-6		0.1	X
Mesityl oxide	141-79-7	25	100	
Methane	74-82-8	E		

Threshold Limit Values of Airborne Contaminants for Construction (continued)

Substance	CAS. No. [d]	ppm [a]	mg/m^3 [b]	Skin Designation
Methanethiol; see Methyl mercaptan				
Methoxychlor Total dust	72-43-5		 15	
2-Methoxyethanol (Methyl cellosolve)	109-86-4	25	80	X
2-Methoxyethyl acetate (Methyl cellosolve acetate)	110-49-6	25	120	X
Methyl acetate	79-20-9	200	610	
Methyl acetylene (Propyne)	74-99-7	1000	1650	
Methyl acetylene propadiene mixture (MAPP)		1000	1800	
Methyl acrylate	96-33-3	10	35	X
Methylal (Dimethoxy-methane)	109-87-5	1000	3100	
Methyl alcohol	67-56-1	200	260	
Methylamine	74-89-5	10	12	
Methyl amyl alcohol; see Methyl isobutyl carbinol				
Methyl n-amyl ketone	110-43-0	100	465	
Methyl bromide	74-83-9	(C)20	(C)80	X
Methyl butyl ketone; see 2-Hexanone				
Methyl cellosolve; see 2-Methoxyethanol				
Methyl cellosolve acetate; see 2-Methoxyethyl acetate				
Methylene chloride; see §1910.1052				
Methyl chloroform (1,1,1-Trichloroethane)	71-55-6	350	1900	
Methylcyclohexane	108-87-2	500	2000	
Methylcyclohexanol	25639-42-3	100	470	
o-Methylcyclohexanone	583-60-8	100	460	X
Methylene chloride	75-09-02	500	1740	
Methylenedianiline (MDA)	101-77-9			
Methyl ethyl ketone (MEK); see 2-Butanone				
Methyl formate	107-31-3	100	250	
Methyl hydrazine (Monomethyl hydrazine)	60-34-4	(C)0.2	(C)0.35	X
Methyl iodide	74-88-4	5	28	X
Methyl isoamyl ketone	110-12-3	100	475	
Methyl isobutyl carbinol	108-11-2	25	100	X
Methyl isobutyl ketone; see Hexone				
Methyl isocyanate	624-83-9	0.02	0.05	X
Methyl mercaptan	74-93-1	0.5	1	
Methyl methacrylate	80-62-6	100	410	
Methyl propyl ketone; see 2-Pentanone				
Methyl silicate	681-84-5	(C)5	(C)30	
alpha-Methyl styrene	98-83-9	(C)100	(C)480	
Methylene bisphenyl isocyanate (MDI)	101-68-8	(C)0.02	(C)0.2	
Mica; see Silicates				
Molybdenum (as Mo) Soluble compounds Insoluble compounds Total dust	7439-98-7		 5 15	
Monomethyl aniline	100-61-8	2	9	X
Monomethyl hydrazine; see Methyl hydrazine				
Morpholine	110-91-8	20	70	X
Naphtha (Coal tar)	8030-30-6	100	400	
Naphthalene	91-20-3	10	50	
alpha-Naphthylamine; see §1926.1104	134-32-7			
beta-Naphthylamine; see §1926.1109	91-59-8			
Neon	7440-01-9	E		
Nickel carbonyl (as Ni)	13463-39-3	0.001	0.007	
Nickel, metal and insoluble compounds (as Ni)	7440-02-0		1	
Nickel, soluble compounds (as Ni)	7440-02-0		1	
Nicotine	54-11-5		0.5	X
Nitric acid	7697-37-2	2	5	
Nitric oxide	10102-43-9	25	30	
p-Nitroaniline	100-01-6	1	6	X

Threshold Limit Values of Airborne Contaminants for Construction (continued)

Substance	CAS. No. [d]	ppm [a]	mg/m^3 [b]	Skin Designation
Nitrobenzene	98-95-3	1	5	X
p-Nitrochlorobenzene	100-00-5		1	X
4-Nitrodiphenyl; see §1926.1103	92-93-3			
Nitroethane	79-24-3	100	310	
Nitrogen	7727-37-9	E		
Nitrogen dioxide	10102-44-0	(C)5	(C)9	
Nitrogen trifluoride	7783-54-2	10	29	
Nitroglycerin	55-63-0	(C)0.2	(C)2	X
Nitromethane	75-52-5	100	250	
1-Nitropropane	108-03-2	25	90	
2-Nitropropane	79-46-9	25	90	
N-Nitrosodimethylamine; see §1926.1116	62-79-9			
Nitrotoluene (all isomers) o-isomer m-isomer p-isomer	 88-72-2 99-08-1 99-99-0	5	30	X
Nitrotrichloromethane; see Chloropicrin				
Nitrous oxide	10024-97-2	E		
Octachloronaphthalene	2234-13-1		0.1	X
Octane	111-65-9	400	1900	
Oil mist, mineral	8012-95-1		5	
Osmium tetroxide (as Os)	20816-12-0		0.002	
Oxalic acid	144-62-7		1	
Oxygen difluoride	7783-41-7	0.05	0.1	
Ozone	10028-15-6	0.1	0.2	
Paraquat, respirable dust	4685-14-7; 1910-42-5; 2074-50-2		0.5	X
Parathion	56-38-2		0.1	X
Particulates not otherwise regulated				
Total dust organic and inorganic			15	
PCB; see Chlorodiphenyl (42% and 54% chlorine)				
Pentaborane	19624-22-7	0.005	0.01	
Pentachloronaphthalene	1321-64-8		0.5	X
Pentachlorophenol	87-86-5		0.5	X
Pentaerythritol Total dust Respirable fraction	115-77-5			
Pentane	109-66-0	500	1500	
2-Pentanone (Methyl propyl ketone)	107-87-9	200	700	
Perchloroethylene (Tetrachloroethylene)	127-18-4	100	670	
Perchloromethyl mercaptan	594-42-3	0.1	0.8	
Perchloryl fluoride	7616-94-6	3	13.5	
Petroleum distillates (Naphtha)(Rubber Solvent)			A[3]	
Phenol	108-95-2	5	19	X
p-Phenylene diamine	106-50-3		0.1	X
Phenyl ether, vapor	101-84-8	1	7	
Phenyl ether-biphenyl mixture, vapor		1	7	
Phenylethylene; see Styrene				
Phenyl glycidyl ether (PGE)	122-60-1	10	60	
Phenylhydrazine	100-63-0	5	22	X
Phosdrin (Mevinphos)	7786-34-7		0.1	X
Phosgene (Carbonyl chloride)	75-44-5	0.1	0.4	
Phosphine	7803-51-2	0.3	0.4	
Phosphoric acid	7664-38-2		1	
Phosphorus (yellow)	7723-14-0		0.1	
Phosphorus pentachloride	10026-13-8		1	
Phosphorus pentasulfide	1314-80-3		1	
Phosphorus trichloride	7719-12-2	0.5	3	
Phthalic anhydride	85-44-9	2	12	
Picric acid	88-89-1		0.1	X
Pindone (2-Pivalyl-1,3-indandione)	83-26-1		0.1	
Plaster of Paris Total dust Respirable fraction	26499-65-0			

Threshold Limit Values of Airborne Contaminants for Construction (continued)

Substance	CAS. No. [d]	ppm [a]	mg/m3 [b]	Skin Designation
Platinum (as Pt) Metal Soluble Salts	7440-06-4		0.002	
Polytetrafluoroethylene decomposition products			A2	
Portland cement Total dust Respirable fraction	65997-15-1	5	15	
Propane	74-98-6	E		
Propargyl alcohol	107-19-7	1		X
beta-Propriolactone; see §1926.1113	57-57-8			
n-Propyl acetate	109-60-4	200	840	
n-Propyl alcohol	71-23-8	200	500	
n-Propyl nitrate	627-13-4	25	110	
Propylene dichloride	78-87-5	75	350	
Propylene imine	75-55-8	2	5	X
Propylene oxide	75-56-9	100	240	
Propyne; see Methyl acetylene				
Pyrethrum	8003-34-7		5	
Pyridine	110-86-1	5	15	
Quinone	106-51-4	0.1	0.4	
RDX; see Cyclonite				
Rhodium (as Rh), metal fume and insoluble compounds	7440-16-6		0.1	
Rhodium (as Rh), soluble compounds	7440-16-6		0.001	
Ronnel	299-84-3		10	
Rotenone	83-79-4		5	
Rouge Total dust Respirable fraction				
Selenium compounds (as Se)	7782-49-2		0.2	
Selenium hexafluoride (as Se)	7783-79-1	0.05	0.4	
Silica, amorphous, precipitated and gel	112926-00-8	(2)	(2)	(2)
Silica, amorphous, diatomaceous earth, containing less than 1% crystalline silica	61790-53-2	(2)	(2)	(2)
Silica, crystalline cristobalite, respirable dust	14464-46-1	(2)	(2)	(2)
Silica, crystalline quartz, respirable dust	14808-60-7	(2)	(2)	(2)
Silica, crystalline tripoli (as quartz), respirable dust	1317-95-9	(2)	(2)	(2)
Silica, crystalline tridymite, respirable dust	15468-32-3	(2)	(2)	(2)
Silica, fused, respirable dust	60676-86-0	(2)	(2)	(2)
Silicates (less than 1% crystalline silica) Mica (respirable dust) Soapstone, total dust Soapstone, respirable dust Talc (containing asbestos): use asbestos limit: see §1926.58 Talc (containing no asbestos), respirable dust Tremolite, asbestiform; see §1926.58	12001-26-2 14807-96-6	(2) (2) (2) (2)	(2) (2) (2) (2)	(2) (2) (2) (2)
Silicon carbide Total dust Respirable fraction	409-21-2			
Silver, metal and soluble compounds (as Ag)	7440-22-4		0.01	
Soapstone; see Silicates				
Sodium fluoroacetate	62-74-8		0.05	X
Sodium hydroxide	1310-73-2		2	
Starch Total dust Respirable fraction	9005-25-8			
Stibine	7803-52-3	0.1	0.5	
Stoddard solvent	8052-41-3	200	1150	
Strychnine	57-24-9		0.15	
Styrene	100-42-5	(C)100	(C)420	
Sucrose Total dust Respirable fraction	57-50-1			
Sulfur dioxide	7446-09-5	5	13	

Threshold Limit Values of Airborne Contaminants for Construction (continued)

Substance	CAS. No. [d]	ppm [a]	mg/m3 [b]	Skin Designation
Sulfur hexafluoride	2551-62-4	1000	6000	
Sulfuric acid	7664-93-9		1	
Sulfur monochloride	10025-67-9	1	6	
Sulfur pentafluoride	5714-22-7	0.025	0.25	
Sulfuryl fluoride	2699-79-8	5	20	
Systox; see Demeton				
2,4,5-T (2,4,5-trichlorophenoxyacetic acid)	93-76-5		10	
Talc; see Silicates				
Tantalum, metal and oxide dust	7440-25-7		5	
TEDP (Sulfotep)	3689-24-5		0.2	X
Teflon decomposition products			A2	
Tellurium and compounds (as Te)	13494-80-9		0.1	
Tellurium hexafluoride (as Te)	7783-80-4	0.02	0.2	
Temephos Total dust Respirable fraction	3383-96-8			
TEPP (Tetraethyl pyrophosphate)	107-49-3		0.05	X
Terphenyls	26140-60-3	(C)1	(C)9	
1,1,1,2-Tetrachloro-2,2-difluoroethane	76-11-9	500	4170	
1,1,2,2-Tetrachloro-1,2-difluoroethane	76-12-0	500	4170	
1,1,2,2-Tetrachloroethane	79-34-5	5	35	X
Tetrachloroethylene; see Perchloroethylene				
Tetrachloromethane; see Carbon tetrachloride				
Tetrachloronaphthalene	1335-88-2		2	X
Tetraethyl lead (as Pb)	78-00-2		0.1	X
Tetrahydrofuran	109-99-9	200	590	
Tetramethyl lead, (as Pb)	75-74-1		0.15	X
Tetramethyl succinonitrile	3333-52-6	0.5	3	X
Tetranitromethane	509-14-8	1	8	
Tetryl (2,4,6-Trinitrophenylmethylnitramine)	479-45-8		1.5	X
Thallium, soluble compounds (as Tl)	7440-28-0		0.1	X
Thiram	137-26-8		5	
Tin, inorganic compounds (except oxides) (as Sn)	7440-31-5		2	
Tin, organic compounds (as Sn)	7440-31-5		0.1	
Tin oxide (as Sn) Total dust Respirable fraction	21651-19-4			
Titanium dioxide Total dust	13463-67-7			
Toluene	108-88-3	200	750	
Toluene-2,4-diisocyanate (TDI)	584-84-9	(C)0.02	(C)0.14	
o-Toluidine	95-53-4	5	22	X
Toxaphene; see Chlorinated camphene				
Tremolite; see Silicates				
Tributyl phosphate	126-73-8		5	
1,1,1-Trichloroethane; see Methyl chloroform				
1,1,2-Trichloroethane	79-00-5	10	45	X
Trichloroethylene	79-01-6	100	535	
Trichloromethane; see Chloroform				
Trichloronaphthalene	1321-65-9		5	X
1,2,3-Trichloropropane	96-18-4	50	300	
1,1,2-Trichloro-1,2,2-trifluoroethane	76-13-1	1000	7600	
Triethylamine	121-44-8	25	100	
Trifluorobromomethane	75-63-8	1000	6100	
Trimethyl benzene	25551-13-7	25	120	
2,4,6-Trinitrophenol; see Picric acid				
2,4,6-Trinitrophenylmethylnitramine; see Tetryl				
2,4,6-Trinitrotoluene (TNT)	118-96-7		1.5	X
Triorthocresyl phosphate	78-30-8		0.1	
Triphenyl phosphate	115-86-6		3	

Threshold Limit Values of Airborne Contaminants for Construction (continued)

Substance	CAS. No. [d]	ppm [a]	mg/m^3 [b]	Skin Designation
Tungsten (as W) Insoluble compounds Soluble compounds	7440-33-7		 5 1	
Turpentine	8006-64-2	100	560	
Uranium (as U) Soluble compounds Insoluble compounds	7440-61-1		 0.2 0.2	
Vanadium Respirable dust (as V_2O_5) Fume (as V_2O_5)	1314-62-1		 (C)0.5 (C)0.1	
Vegetable oil mist Total dust Respirable fraction				
Vinyl benzene; see Styrene				
Vinyl chloride; see §1926.1117	75-01-4			
Vinyl cyanide; see Acrylonitrile				
Vinyl toluene	25013-15-4	100	480	
Warfarin	81-81-2		0.1	
Xylenes (o-, m-, p-isomers)	1330-20-7	100	435	
Xylidine	1300-73-8	5	25	X
Yttrium	7440-65-5		1	
Zinc chloride fume	7646-85-7		1	
Zinc oxide fume	1314-13-2		5	
Zinc oxide Total dust Respirable fraction	1314-13-2		 15 5	
Zirconium compounds (as Zr)	7440-67-7		5	

Mineral Dusts

Substance	mppcf [(j)]
SILICA: Crystalline Quartz. Threshold limit calculated from the formula	$\frac{250^{(k)}}{\%SiO_2+5}$
Cristobalite Amorphous, including natural diatomaceous earth	20
SILICATES (less than 1% crystalline silica) Mica	20
Portland Cement	50
Soapstone	20
Talc (non-asbestiform)	20
Talc (fibrous), use asbestos limit	--
Graphite (natural)	15
Inert or Nuisance Particulates:[(m)]: (Inert or Nuisance Dusts includes all mineral, inorganic, and organic dusts as indicated by examples in TLV's Appendix D) Conversion factors mppcf X 35.3 = million particles per cubic meter = particles per c.c.	50 (or 15 mg/m^3 whichever is the smaller) of total dust <1% SiO_2

Footnotes:

[1] [Reserved]

[2] See Mineral Dusts Table.

[3] Use Asbestos Limit § 1926.58.

[4] See §1926.58

* The PELs are 8-hour TWAs unless otherwise noted; a (C) designation denotes a ceiling limit.

** As determined from breathing-zone air samples.

[a] Parts of vapor or gas per million parts of contaminated air by volume at 25 °C and 760 torr.

[b] Milligrams of substance per cubic meter of air. When entry is in this column only, the value is exact; when listed with a ppm entry, it is approximate.

[c] [Reserved]

[d] The CAS number is for information only. Enforcement is based on the substance name. For an entry covering more than one metal compound, measured as the metal, the CAS number for the metal is given–not CAS numbers for the individual compounds.

[e] [Reserved]

[f] [Reserved]

[g] For sectors excluded from §1926.1128 the limit is 10 ppm TWA.

[h] [Reserved]

[i] [Reserved]

[j] Millions of particles per cubic foot of air, based on impinger samples counted by light-field techniques.

[k] The percentage of crystalline silica in the formula is the amount determined from airborne samples, except in those instances in which other methods have been shown to be applicable.

[l] [Reserved]

[m] Covers all organic and inorganic particulates not otherwise regulated. Same as Particulates Not Otherwise Regulated.

[n] If the exposure limit in 1926.1026† is stayed or is not otherwise in effect, the exposure limit is a ceiling of 0.1 mg/m^3.

[o] If the exposure limit in §1926.1026† is stayed or is not otherwise in effect, the exposure limit is 0.1 mg/m^3 (as CrO_3) as an 8-hour TWA.

The 1970 TLV uses letter designations instead of a numerical value as follows:

A[1] [Reserved]

A[2] Polytetrafluoroethylene decomposition products. Because these products decompose in part by hydrolysis in alkaline solution, they can be quantitatively determined in air as fluoride to provide an index of exposure. No TLV is recommended pending determination of the toxicity of the products, but air concentrations should be minimal.

A[3] Gasoline and/or Petroleum Distillates. The composition of these materials varies greatly and thus a single TLV for all types of these materials is no longer applicable. The content of benzene, other aromatics and additives should be determined to arrive at the appropriate TLV.

E Simple asphyxiants. The limiting factor is the available oxygen which shall be at least 19.5% and be within the requirements addressing explosion in Part 1926.

† Editor's Note: The CFR incorrectly references §1926.1126. The entry has been corrected to reflect the Federal Register changes dated June 23, 2006.

[39 FR 22801, June 24, 1974, as amended at 51 FR 37007, Oct. 17, 1986; 52 FR 46312, Dec. 4, 1987; 58 FR 35089, June 30, 1993; 61 FR 9249, 9250, Mar. 7, 1996; 61 FR 56856, Nov. 4, 1996; 62 FR 1619, Jan. 10, 1997; 71 FR 10381, Feb. 28, 2006; 71 FR 36009, June 23, 2006]

§1926.56
Illumination

(a) General. Construction areas, ramps, runways, corridors, offices, shops, and storage areas shall be lighted to not less than the minimum illumination intensities listed in Table D-3 while any work is in progress:**[§1926.56(a)]**

Table D-3 — Minimum Illumination Intensities in Foot-Candles

Foot-candles	Area or operation
5	General construction area lighting.
3	General construction areas, concrete placement, excavation and waste areas, accessways, active storage areas, loading platforms, refueling, and field maintenance areas.
5	Indoors: warehouses, corridors, hallways, and exitways.
5	Tunnels, shafts, and general underground work areas: (Exception: minimum of 10 foot- candles is required at tunnel and shaft heading during drilling, mucking, and scaling. Bureau of Mines approved cap lights shall be acceptable for use in the tunnel heading.)
10	General construction plant and shops (e.g., batch plants, screening plants, mechanical and electrical equipment rooms, carpenter shops, rigging lofts and active storerooms, barracks or living quarters, locker or dressing rooms, mess halls, and indoor toilets and workrooms).
30	First aid stations, infirmaries, and offices.

(b) Other areas. For areas or operations not covered above, refer to the American National Standard A11.1-1965, R1970, Practice for Industrial Lighting, for recommended values of illumination. **[§1926.56(b)]**

§1926.57
Ventilation

(a) General. Whenever hazardous substances such as dusts, fumes, mists, vapors, or gases exist or are produced in the course of construction work, their concentrations shall not exceed the limits specified in Sec. 1926.55(a). When ventilation is used as an engineering control method, the system shall be installed and operated according to the requirements of this section.**[§1926.57(a)]**

(b) Local exhaust ventilation. Local exhaust ventilation when used as described in (a) shall be designed to prevent dispersion into the air of dusts, fumes, mists, vapors, and gases in concentrations causing harmful exposure. Such exhaust systems shall be so designed that dusts, fumes, mists, vapors, or gases are not drawn through the work area of employees.**[§1926.57(b)]**

(c) Design and operation. Exhaust fans, jets, ducts, hoods, separators, and all necessary appurtenances, including refuse receptacles, shall be so designed, constructed, maintained and operated as to ensure the required protection by maintaining a volume and velocity of exhaust air sufficient to gather dusts, fumes, vapors, or gases from said equipment or process, and to convey them to suitable points of safe disposal, thereby preventing their dispersion in harmful quantities into the atmosphere where employees work.**[§1926.57(c)]**

(d) Duration of operations.[§1926.57(d)]

(1) *The exhaust system shall be in operation* continually during all operations which it is designed to serve. If the employee remains in the contaminated zone, the system shall continue to operate after the cessation of said operations, the length of time to depend upon the individual circumstances and effectiveness of the general ventilation system.**[§1926.57(d)(1)]**

(2) *Since dust capable of causing disability is,* according to the best medical opinion, of microscopic size, tending to remain for hours in suspension in still air, it is essential that the exhaust system be continued in operation for a time after the work process or equipment served by the same shall have ceased, in order to ensure the removal of the harmful elements to the required extent. For the same reason, employees wearing respiratory equipment should not remove same immediately until the atmosphere seems clear.[§1926.57(d)(2)]

(e) **Disposal of exhaust materials.** The air outlet from every dust separator, and the dusts, fumes, mists, vapors, or gases collected by an exhaust or ventilating system shall discharge to the outside atmosphere. Collecting systems which return air to work area may be used if concentrations which accumulate in the work area air do not result in harmful exposure to employees. Dust and refuse discharged from an exhaust system shall be disposed of in such a manner that it will not result in harmful exposure to employees. [§1926.57(e)]

(f) **Abrasive blasting**[§1926.57(f)]

(1) *Definitions applicable to this paragraph*[§1926.57(f)(1)]

(i) **Abrasive.** A solid substance used in an abrasive blasting operation.[§1926.57(f)(1)(i)]

(ii) **Abrasive-blasting respirator.** A respirator constructed so that it covers the wearer's head, neck, and shoulders to protect the wearer from rebounding abrasive.[§1926.57(f)(1)(ii)]

(iii) **Blast cleaning barrel.** A complete enclosure which rotates on an axis, or which has an internal moving tread to tumble the parts, in order to expose various surfaces of the parts to the action of an automatic blast spray.[§1926.57(f)(1)(iii)]

(iv) **Blast cleaning room.** A complete enclosure in which blasting operations are performed and where the operator works inside of the room to operate the blasting nozzle and direct the flow of the abrasive material.[§1926.57(f)(1)(iv)]

(v) **Blasting cabinet.** An enclosure where the operator stands outside and operates the blasting nozzle through an opening or openings in the enclosure.[§1926.57(f)(1)(v)]

(vi) **Clean air.** Air of such purity that it will not cause harm or discomfort to an individual if it is inhaled for extended periods of time.[§1926.57(f)(1)(vi)]

(vii) **Dust collector.** A device or combination of devices for separating dust from the air handled by an exhaust ventilation system.[§1926.57(f)(1)(vii)]

(viii) **Exhaust ventilation system.** A system for removing contaminated air from a space, comprising two or more of the following elements [§1926.57(f)(1)(viii)]

[A] enclosure or hood,[§1926.57(f)(1)(viii)[A]]

[B] duct work, [§1926.57(f)(1)(viii)[B]]

[C] dust collecting equipment, [§1926.57(f)(1)(viii)[C]]

[D] exhauster, and [§1926.57(f)(1)(viii)[D]]

[E] discharge stack. [§1926.57(f)(1)(viii)[E]]

(ix) **Particulate-filter respirator.** An air purifying respirator, commonly referred to as a dust or a fume respirator, which removes most of the dust or fume from the air passing through the device.[§1926.57(f)(1)(ix)]

(x) **Respirable dust.** Airborne dust in sizes capable of passing through the upper respiratory system to reach the lower lung passages.[§1926.57(f)(1)(x)]

(xi) **Rotary blast cleaning table.** An enclosure where the pieces to be cleaned are positioned on a rotating table and are passed automatically through a series of blast sprays. [§1926.57(f)(1)(xi)]

(xii) **Abrasive blasting.** The forcible application of an abrasive to a surface by pneumatic pressure, hydraulic pressure, or centrifugal force.[§1926.57(f)(1)(xii)]

(2) *Dust hazards from abrasive blasting.*[§1926.57(f)(2)]

(i) *Abrasives and the surface coatings on the materials blasted* are shattered and pulverized during blasting operations and the dust formed will contain particles of respirable size. The composition and toxicity of the dust from these sources shall be considered in making an evaluation of the potential health hazards.[§1926.57(f)(2)(i)]

(ii) *The concentration of respirable dust* or fume in the breathing zone of the abrasive-blasting operator or any other worker shall be kept below the levels specified in Sec. 1926.55 or other pertinent sections of this part.[§1926.57(f)(2)(ii)]

(iii) *Organic abrasives which are combustible* shall be used only in automatic systems. Where flammable or explosive dust mixtures may be present, the construction of the equipment, including the exhaust system and all electric wiring, shall conform to the requirements of American National Standard Installation of Blower and Exhaust Systems for Dust, Stock, and Vapor Removal or Conveying, Z33.1-1961 (NFPA 91-1961), and subpart S of this part. The blast nozzle shall be bonded and grounded to prevent the build up of static charges. Where flammable or explosive dust mixtures may be present, the abrasive blasting enclosure, the ducts, and the dust collector shall be constructed with loose panels or explosion venting areas, located on sides away from any occupied area, to provide for pressure relief in case of explosion, following the principles set forth in the National Fire Protection Association Explosion Venting Guide. NFPA 68-1954. [§1926.57(f)(2)(iii)]

(3) *Blast-cleaning enclosures.*[§1926.57(f)(3)]

(i) *Blast-cleaning enclosures shall be exhaust ventilated* in such a way that a continuous inward flow of air will be maintained at all openings in the enclosure during the blasting operation. [§1926.57(f)(3)(i)]

[A] All air inlets and access openings shall be baffled or so arranged that by the combination of inward air flow and baffling the escape of abrasive or dust particles into an adjacent work area will be minimized and visible spurts of dust will not be observed.[§1926.57(f)(3)(i)[A]]

[B] The rate of exhaust shall be sufficient to provide prompt clearance of the dust-laden air within the enclosure after the cessation of blasting.[§1926.57(f)(3)(i)[B]]

[C] Before the enclosure is opened, the blast shall be turned off and the exhaust system shall be run for a sufficient period of time to remove the dusty air within the enclosure.[§1926.57(f)(3)(i)[C]]

[D] Safety glass protected by screening shall be used in observation windows, where hard deep-cutting abrasives are used.[§1926.57(f)(3)(i)[D]]

[E] Slit abrasive-resistant baffles shall be installed in multiple sets at all small access openings where dust might escape, and shall be inspected regularly and replaced when needed.[§1926.57(f)(3)(i)[E]]

[1] Doors shall be flanged and tight when closed. [§1926.57(f)(3)(i)[E][1]]

[2] Doors on blast-cleaning rooms shall be operable from both inside and outside, except that where there is a small operator access door, the large work access door may be closed or opened from the outside only. [§1926.57(f)(3)(i)[E][2]]

(4) *Exhaust ventilation systems.*[§1926.57(f)(4)]

(i) *The construction, installation, inspection, and maintenance* of exhaust systems shall conform to the principles and requirements set forth in American National Standard Fundamentals Governing the Design and Operation of Local Exhaust Systems, Z9.2-1960, and ANSI Z33.1-1961. [§1926.57(f)(4)(i)]

[a] When dust leaks are noted, repairs shall be made as soon as possible.[§1926.57(f)(4)(i)[a]]

[b] The static pressure drop at the exhaust ducts leading from the equipment shall be checked when the installation is completed and periodically thereafter to assure continued satisfactory operation. Whenever an appreciable change in the pressure drop indicates a partial blockage, the system shall be cleaned and returned to normal operating condition.[§1926.57(f)(4)(i)[b]]

(ii) *In installations where the abrasive is recirculated,* the exhaust ventilation system for the blasting enclosure shall not be relied upon for the removal of fines from the spent abrasive instead of an abrasive separator. An abrasive separator shall be provided for the purpose.[§1926.57(f)(4)(ii)]

(iii) *The air exhausted from blast-cleaning equipment* shall be discharged through dust collecting equipment. Dust collectors shall be set up so that the accumulated dust can be emptied and removed without contaminating other working areas.[§1926.57(f)(4)(iii)]

(5) *Personal protective equipment.*[§1926.57(f)(5)]

(i) *Employers must use only respirators approved* by NIOSH under 42 CFR part 84 for protecting employees from dusts produced during abrasive-blasting operations.[§1926.57(f)(5)(i)]

(ii) *Abrasive-blasting respirators shall be worn* by all abrasive-blasting operators:[§1926.57(f)(5)(ii)]

[A] When working inside of blast-cleaning rooms, or [§1926.57(f)(5)(ii)[A]]

[B] When using silica sand in manual blasting operations where the nozzle and blast are not physically separated from the operator in an exhaust ventilated enclosure, or [§1926.57(f)(5)(ii)[B]]

[C] Where concentrations of toxic dust dispersed by the abrasive blasting may exceed the limits set in Sec. 1926.55 or other pertinent sections of this part and the nozzle and blast are not physically separated from the operator in an exhaust-ventilated enclosure.[§1926.57(f)(5)(ii)[C]]

(iii) *Properly fitted particulate-filter respirators,* commonly referred to as dust-filter respirators, may be used for short, intermittent, or occasional dust exposures such as cleanup, dumping of dust collectors, or unloading shipments of sand at a receiving point when it is not feasible to control the dust by enclosure, exhaust ventilation, or other means. The respirators used must be approved by NIOSH under 42 CFR part 84 for protection against the specific type of dust encountered.[§1926.57(f)(5)(iii)]

(iv) *A respiratory protection program as defined* and described in Sec. 1926.103, shall be established wherever it is necessary to use respiratory protective equipment.[§1926.57(f)(5)(iv)]

(v) *Operators shall be equipped with heavy canvas* or leather gloves and aprons or equivalent protection to protect them from the impact of abrasives. Safety shoes shall be worn to protect against foot injury where heavy pieces of work are handled.[§1926.57(f)(5)(v)]

[A] Safety shoes shall conform to the requirements of American National Standard for Men's Safety-Toe Footwear, Z41.1-1967.[§1926.57(f)(5)(v)[A]]

[B] Equipment for protection of the eyes and face shall be supplied to the operator when the respirator design does not provide such protection and to any other personnel working in the vicinity of abrasive blasting operations. This equipment shall conform to the requirements of Sec. 1926.102.[§1926.57(f)(5)(v)[B]]

(6) *Air supply and air compressors.* Air for abrasive-blasting respirators must be free of harmful quantities of dusts, mists, or noxious gases, and must meet the requirements for supplied-air quality and use specified in 29 CFR 1910.134(i).[§1926.57(f)(6)]

(7) *Operational procedures and general safety.* Dust shall not be permitted to accumulate on the floor or on ledges outside of an abrasive-blasting enclosure, and dust spills shall be cleaned up promptly. Aisles and walkways shall be kept clear of steel shot or similar abrasive which may create a slipping hazard.[§1926.57(f)(7)]

(8) *Scope.* This paragraph applies to all operations where an abrasive is forcibly applied to a surface by pneumatic or hydraulic pressure, or by centrifugal force. It does not apply to steam blasting, or steam cleaning, or hydraulic cleaning methods where work is done without the aid of abrasives.[§1926.57(f)(8)]

(g) Grinding, polishing, and buffing operations[§1926.57(g)]

(1) *Definitions applicable to this paragraph* — [§1926.57(g)(1)]

(i) **Abrasive cutting-off wheels.** Organic-bonded wheels, the thickness of which is not more than one forty-eighth of their diameter for those up to, and including, 20 inches (50.8 cm) in diameter, and not more than one-sixtieth of their diameter for those larger than 20 inches (50.8 cm) in diameter, used for a multitude of operations variously known as cutting, cutting off, grooving, slotting, coping, and jointing, and the like. The wheels may be "solid" consisting of organic-bonded abrasive material throughout, "steel centered" consisting of a steel disc with a rim of organic-bonded material moulded around the periphery, or of the "inserted tooth" type consisting of a steel disc with organic-bonded abrasive teeth or inserts mechanically secured around the periphery.[§1926.57(g)(1)(i)]

(ii) **Belts.** All power-driven, flexible, coated bands used for grinding, polishing, or buffing purposes.[§1926.57(g)(1)(ii)]

(iii) **Branch pipe.** The part of an exhaust system piping that is connected directly to the hood or enclosure.[§1926.57(g)(1)(iii)]

(iv) **Cradle.** A movable fixture, upon which the part to be ground or polished is placed.[§1926.57(g)(1)(iv)]

(v) **Disc wheels.** All power-driven rotatable discs faced with abrasive materials, artificial or natural, and used for grinding or polishing on the side of the assembled disc. [§1926.57(g)(1)(v)]

(vi) **Entry loss.** The loss in static pressure caused by air flowing into a duct or hood. It is usually expressed in inches of water gauge.[§1926.57(g)(1)(vi)]

(vii) **Exhaust system.** A system consisting of branch pipes connected to hoods or enclosures, one or more header pipes, an exhaust fan, means for separating solid contaminants from the air flowing in the system, and a discharge stack to outside.[§1926.57(g)(1)(vii)]

(viii) **Grinding wheels.** All power-driven rotatable grinding or abrasive wheels, except disc wheels as defined in this standard, consisting of abrasive particles held together by artificial or natural bonds and used for peripheral grinding. [§1926.57(g)(1)(viii)]

(ix) **Header pipe (main pipe).** A pipe into which one or more branch pipes enter and which connects such branch pipes to the remainder of the exhaust system.[§1926.57(g)(1)(ix)]

(x) **Hoods and enclosures.** The partial or complete enclosure around the wheel or disc through which air enters an exhaust system during operation.[§1926.57(g)(1)(x)]

(xi) **Horizontal double-spindle disc grinder.** A grinding machine carrying two power-driven, rotatable, coaxial, horizontal spindles upon the inside ends of which are mounted abrasive disc wheels used for grinding two surfaces simultaneously.[§1926.57(g)(1)(xi)]

(xii) **Horizontal single-spindle disc grinder.** A grinding machine carrying an abrasive disc wheel upon one or both ends of a power-driven, rotatable single horizontal spindle. [§1926.57(g)(1)(xii)]

(xiii) **Polishing and buffing wheels.** All power-driven rotatable wheels composed all or in part of textile fabrics, wood, felt, leather, paper, and may be coated with abrasives on the periphery of the wheel for purposes of polishing, buffing, and light grinding.[§1926.57(g)(1)(xiii)]

(xiv) **Portable grinder.** Any power-driven rotatable grinding, polishing, or buffing wheel mounted in such manner that it may be manually manipulated.[§1926.57(g)(1)(xiv)]

(xv) **Scratch brush wheels.** All power-driven rotatable wheels made from wire or bristles, and used for scratch cleaning and brushing purposes.[§1926.57(g)(1)(xv)]

(xvi) **Swing-frame grinder.** Any power-driven rotatable grinding, polishing, or buffing wheel mounted in such a manner that the wheel with its supporting framework can be manipulated over stationary objects.[§1926.57(g)(1)(xvi)]

(xvii) **Velocity pressure (vp).** The kinetic pressure in the direction of flow necessary to cause a fluid at rest to flow at a given velocity. It is usually expressed in inches of water gauge.[§1926.57(g)(1)(xvii)]

(xviii) **Vertical spindle disc grinder.** A grinding machine having a vertical, rotatable power-driven spindle carrying a horizontal abrasive disc wheel.[§1926.57(g)(1)(xviii)]

(2) *Application.* Wherever dry grinding, dry polishing or buffing is performed, and employee exposure, without regard to the use of respirators, exceeds the permissible exposure limits prescribed in Sec. 1926.55 or other pertinent sections of this part, a local exhaust ventilation system shall be provided and used to maintain employee exposures within the prescribed limits. [§1926.57(g)(2)]

(3) *Hood and branch pipe requirements.*[§1926.57(g)(3)]

(i) *Hoods connected to exhaust systems shall be used,* and such hoods shall be designed, located, and placed so that the dust or dirt particles shall fall or be projected into the hoods in the direction of the air flow. No wheels, discs, straps, or belts shall be operated in such manner and in such direction as to cause the dust and dirt particles to be thrown into the operator's breathing zone.[§1926.57(g)(3)(i)]

(ii) *Grinding wheels on floor stands, pedestals, benches,* and special-purpose grinding machines and abrasive cutting-off wheels shall have not less than the minimum exhaust volumes shown in Table D-57.1 with a recommended minimum duct velocity of 4,500 feet per minute in the branch and 3,500 feet per minute in the main. The entry losses from all hoods except the vertical-spindle disc grinder hood, shall equal 0.65 velocity pressure for a straight takeoff and 0.45 velocity pressure for a tapered takeoff. The entry loss for the vertical-spindle disc grinder hood is shown in figure D-57.1 (following paragraph (g) of this section).[§1926.57(g)(3)(ii)]

Table D-57.1—Grinding and Abrasive Cutting-Off Wheels

Wheel diameter, inches (cm.)	Wheel width, inches (cm.)	Minimum exhaust volume (feet^3 / min.)
To 9 (22.86)	1 1/2 (3.81)	220
Over 9 to 16 (22.86 to 40.64)	2 (5.08)	390
Over 16 to 19 (40.64 to 48.26)	3 (7.62)	500
Over 19 to 24 (48.26 to 60.96)	4 (10.16)	610
Over 24 to 30 (60.96 to 76.2)	5 (12.7)	880
Over 30 to 36 (76.2 to 91.44)	6 (15.24)	1,200

For any wheel wider than wheel diameters shown in Table D-57.1, increase the exhaust volume by the ratio of the new width to the width shown.

Example: If wheel width = 4 1/2 inches (11.43 cm), then 4.5 ÷4 X 610 = 686 (rounded to 690).

(iii) *Scratch-brush wheels and all buffing* and polishing wheels mounted on floor stands, pedestals, benches, or special-purpose machines shall have not less than the minimum exhaust volume shown in Table D-57.2. [§1926.57(g)(3)(iii)]

Table D-57.2—Buffing and Polishing Wheels

Wheel diameter, inches (cm.)	Wheel width, inches cm.	Minimum exhaust volume (feet^3 / min.)
To 9 (22.86)	2 (5.08)	300
Over 9 to 16 (22.86 to 40.64)	3 (7.62)	500
Over 16 to 19 (40.64 to 48.26)	4 (10.16)	610
Over 19 to 24 (48.26 to 60.96)	5 (12.7)	740
Over 24 to 30 (60.96 to 76.2)	6 (15.24)	1,040
Over 30 to 36 (76.2 to 91.44)	6 (15.24)	1,200

(iv) *Grinding wheels or discs for horizontal single-spindle disc grinders* shall be hooded to collect the dust or dirt generated by the grinding operation and the hoods shall be connected to branch pipes having exhaust volumes as shown in Table D-57.3. [§1926.57(g)(3)(iv)]

Table D-57.3—Horizontal Single-Spindle Disc Grinder

Disc diameter, inches (cm.)	Exhaust volume (ft.^{3} / min.)
Up to 12 (30.48)	220
Over 12 to 19 (30.48 to 48.26)	390
Over 19 to 30 (48.26 to 76.2)	610
Over 30 to 36 (76.2 to 91.44)	880

(v) *Grinding wheels or discs* for horizontal double-spindle disc grinders shall have a hood enclosing the grinding chamber and the hood shall be connected to one or more branch pipes having exhaust volumes as shown in Table D-57.4. [§1926.57(g)(3)(v)]

Table D-57.4—Horizontal Double-Spindle Disc Grinder

Disc diameter, inches (cm.)	Exhaust volume (ft.^{3} / min.)
Up to 19 (48.26)	610
Over 19 to 25 (48.26 to 63.5)	880
Over 25 to 30 (63.5 to 76.2)	1,200
Over 30 to 53 (76.2 to 134.62)	1,770
Over 53 to 72 (134.62 to 182.88)	6,280

(vi) *Grinding wheels or discs* for vertical single-spindle disc grinders shall be encircled with hoods to remove the dust generated in the operation. The hoods shall be connected to one or more branch pipes having exhaust volumes as shown in Table D-57.5. [§1926.57(g)(3)(vi)]

Table D-57.5—Vertical Spindle Disc Grinder

Disc diameter, inches (cm.)	One-half or more of disc covered		Disc not covered	
	Number[1]	Exhaust foot^3 / min.	Number[1]	Exhaust foot^3 /min.
Up to 20 (50.8)	1	500	2	780
Over 20 to 30 (50.8 to 76.2)	2	780	2	1,480
Over 30 to 53 (76.2 to 134.62)	2	1,770	4	3,530
Over 53 to 72 (134.62 to 182.88)	2	3,140	5	6,010

[1] Number of exhaust outlets around periphery of hood, or equal distribution provided by other means.

(vii) *Grinding and polishing belts* shall be provided with hoods to remove dust and dirt generated in the operations and the hoods shall be connected to branch pipes having exhaust volumes as shown in Table D-57.6. [§1926.57(g)(3)(vii)]

Table D-57.6—Grinding and Polishing Belts

Belts width, inches (cm.)	Exhaust volume (ft.^{3} / min.)
Up to 3 (7.62)	220
Over 3 to 5 (7.62 to 12.7)	300
Over 5 to 7 (12.7 to 17.78)	390
Over 7 to 9 (17.78 to 22.86)	500
Over 9 to 11 (22.86 to 27.94)	610
Over 11 to 13 (27.94 to 33.02)	740

(viii) *Cradles and swing-frame grinders.* Where cradles are used for handling the parts to be ground, polished, or buffed, requiring large partial enclosures to house the complete operation, a minimum average air velocity of 150 feet per minute shall be maintained over the entire opening of the enclosure. Swing-frame grinders shall also be exhausted in the same manner as provided for cradles. (See fig. D-57.3) [§1926.57(g)(3)(viii)]

(ix) *Where the work is outside the hood,* air volumes must be increased as shown in American Standard Fundamentals Governing the Design and Operation of Local Exhaust Systems, Z9.2-1960 (section 4, exhaust hoods). [§1926.57(g)(3)(ix)]

(4) *Exhaust systems.* [§1926.57(g)(4)]

(i) *Exhaust systems for grinding, polishing, and buffing operations* should be designed in accordance with American Standard Fundamentals Governing the Design and Operation of Local Exhaust Systems, Z9.2-1960. [§1926.57(g)(4)(i)]

(ii) *Exhaust systems for grinding, polishing, and buffing operations* shall be tested in the manner described in American Standard Fundamentals Governing the Design and Operation of Local Exhaust Systems, Z9.2-1960. [§1926.57(g)(4)(ii)]

(iii) *All exhaust systems shall be provided* with suitable dust collectors. [§1926.57(g)(4)(iii)]

(5) *Hood and enclosure design.* [§1926.57(g)(5)]

(i) *[A] It is the dual function of grinding* and abrasive cutting-off wheel hoods to protect the operator from the hazards of bursting wheels as well as to provide a means for the removal of dust and dirt generated. All hoods shall be not less in structural strength than specified in the American National Standard Safety Code for the Use, Care, and Protection of Abrasive Wheels, B7.1-1970. [§1926.57(g)(5)(i)[A]]

[B] Due to the variety of work and types of grinding machines employed, it is necessary to develop hoods adaptable to the particular machine in question, and such hoods shall be located as close as possible to the operation. [§1926.57(g)(5)(i)[B]]

(ii) *Exhaust hoods for floor stands, pedestals, and bench grinders* shall be designed in accordance with figure D-57.2. The adjustable tongue shown in the figure shall be kept in working order and shall be adjusted within one-fourth inch (0.635 cm) of the wheel periphery at all times. [§1926.57(g)(5)(ii)]

(iii) *Swing-frame grinders shall be provided* with exhaust booths as indicated in figure D-57.3. [§1926.57(g)(5)(iii)]

(iv) *Portable grinding operations,* whenever the nature of the work permits, shall be conducted within a partial enclosure. The opening in the enclosure shall be no larger than is actually required in the operation and an average face air velocity of not less than 200 feet per minute shall be maintained. [§1926.57(g)(5)(iv)]

(v) *Hoods for polishing and buffing* and scratch-brush wheels shall be constructed to conform as closely to figure D-57.4 as the nature of the work will permit.[§1926.57(g)(5)(v)]

(vi) *Cradle grinding and polishing operations* shall be performed within a partial enclosure similar to figure D-57.5. The operator shall be positioned outside the working face of the opening of the enclosure. The face opening of the enclosure should not be any greater in area than that actually required for the performance of the operation and the average air velocity into the working face of the enclosure shall not be less than 150 feet per minute.[§1926.57(g)(5)(vi)]

(vii) *Hoods for horizontal single-spindle disc grinders* shall be constructed to conform as closely as possible to the hood shown in figure D-57.6. It is essential that there be a space between the back of the wheel and the hood, and a space around the periphery of the wheel of at least 1 inch (2.54 cm) in order to permit the suction to act around the wheel periphery. The opening on the side of the disc shall be no larger than is required for the grinding operation, but must never be less than twice the area of the branch outlet.[§1926.57(g)(5)(vii)]

(viii) *Horizontal double-spindle disc grinders* shall have a hood encircling the wheels and grinding chamber similar to that illustrated in figure D-57.7. The openings for passing the work into the grinding chamber should be kept as small as possible, but must never be less than twice the area of the branch outlets.[§1926.57(g)(5)(viii)]

(ix) *Vertical-spindle disc grinders* shall be encircled with a hood so constructed that the heavy dust is drawn off a surface of the disc and the lighter dust exhausted through a continuous slot at the top of the hood as shown in figure D-57.1. [§1926.57(g)(5)(ix)]

(x) *Grinding and polishing belt hoods* shall be constructed as close to the operation as possible. The hood should extend almost to the belt, and 1-inch (2.54 cm) wide openings should be provided on either side. Figure D-57.8 shows a typical hood for a belt operation.[§1926.57(g)(5)(x)]

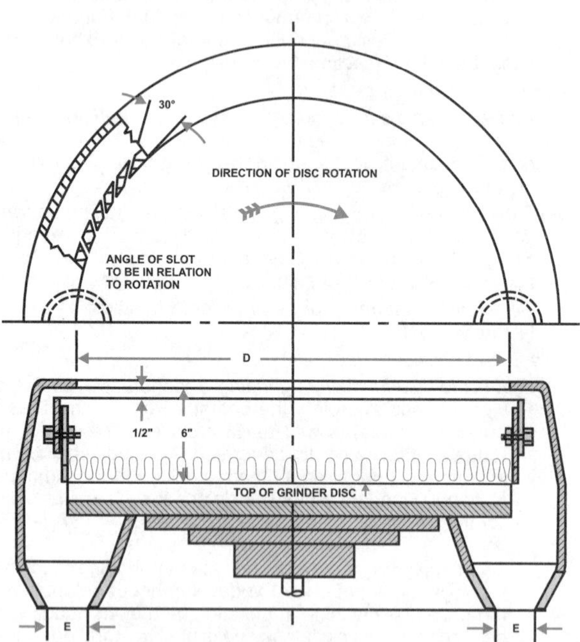

FIGURE D-57.1 - Vertical Spindle Disc Grinder Exhaust Hood and Branch Pipe Connections

Diameter D. inches (cm.)		Exhaust E		Volume exhausted at 4,500 ft./min. ft.³/min.	Note
Min.	Max.	No. pipes	Dia.		
-	20 (50.8)	1	4 1/4 (10.795)	500	When 1/2 or more of the disc can be hooded, use exhaust ducts as shown at left
Over 20 (50.8)	30 (76.2)	2	4 (10.16)	780	
Over 30 (76.2)	72 (182.88)	2	6 (15.24)	1,770	
Over 53 (134.62)	72 (182.88)	2	8 (20.32)	3,140	
-	20 (50.8)	2	4 (10.16)	780	When no hood can be used over disc, use exhaust ducts as shown at left
Over 20 (50.8)	20 (50.8)	2	4 (10.16)	780	
Over 30 (76.2)	30 (76.2)	2	5 1/2 (13.97)	1,480	
Over 53 (134.62)	53 (134.62) 72 (182.88)	4 5	6 (15.24) 7 (17.78)	3,530 6,010	

Entry loss = 1.0 slot velocity pressure + 0.5 branch velocity pressure.
Minimum slot velocity = 2,000 ft/min - 1/2 inch (1.27 cm) slot width.

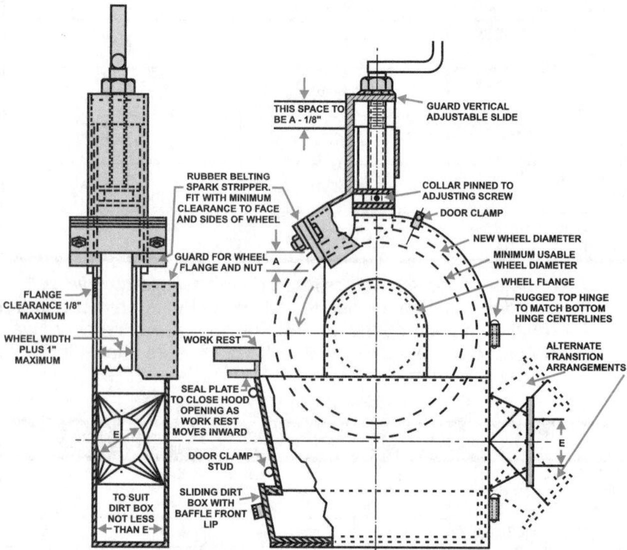

FIGURE D-57.2 - Standard Grinder Hood

Wheel dimension, inches (centimeters)			Exhaust outlet, inches (centimeters) E	Volume of air at 4,500 ft./ min.
Diameter		Width, max.		
Min. = d	Max. = D			
-	9 (22.86)	1 1/2 (3.81)	3	220
Over 9 (22.86)	16 (40.64)	2 (5.08)	4	390
Over 16 (40.64)	19 (48.26)	3 (7.62)	4 1/2	500
Over 19 (48.26)	24 (60.96)	4 (10.16)	5	610
Over 24 (60.96)	30 (76.2)	5 (12.7)	6	880
Over 30 (76.2)	36 (91.44)	6 (15.24)	7	1,200

Entry loss = 0.45 velocity pressure for tapered takeoff 0.65 velocity pressure for straight takeoff.

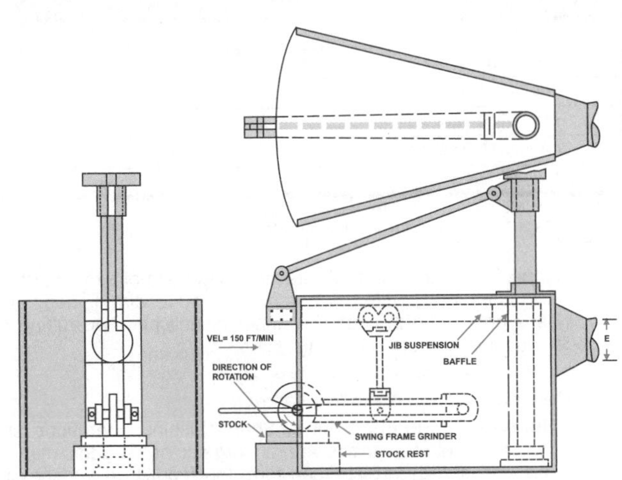

FIGURE D-57.3 - A Method of Applying an Exhaust Enclosure to Swing-Frame Grinders
Note: Baffle to reduce front opening as much as possible.

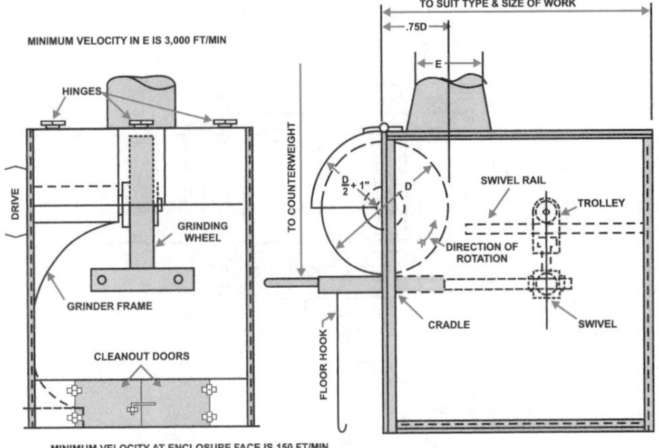

FIGURE D-57.5 - Cradle Polishing or Grinding Enclosure
Entry loss = 0.45 velocity pressure for tapered takeoff.

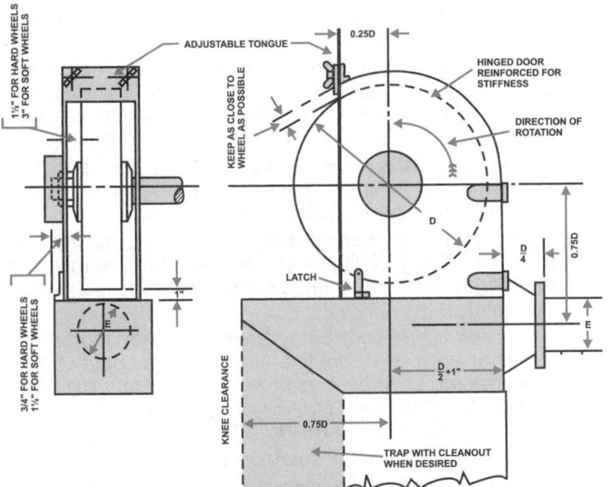

FIGURE D-57.4 - Standard Buffing and Polishing Hood

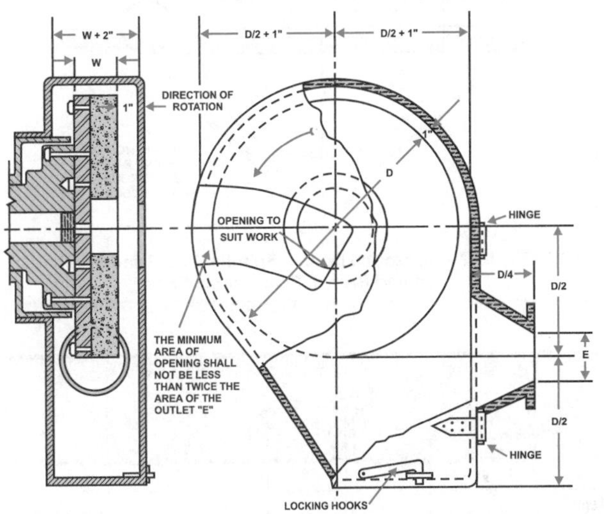

FIGURE D-57.6 - Horizontal Single-Spindle Disc Grinder Exhaust Hood and Branch Pipe Connections

Wheel dimension, inches (centimeters)			Exhaust outlet, inches E	Volume of air at 4,500 ft./min.
Diameter		Width, max		
Min= d	Max. = D			
	9 (22.86)	2 (5.08)	3 1/2 (3.81)	300
Over 9 (22.86)	16 (40.64)	3 (5.08)	4	500
Over 16 (40.64)	19 (48.26)	4 (11.43)	5	610
Over 19 (48.26)	24 (60.96)	5 (12.7)	5 1/2	740
Over 24 (60.96)	30 (76.2)	6 (15.24)	6 1/2	1.040
Over 30 (76.2)	36 (91.44)	6 (15.24)	7	1.200

Entry loss = 0.15 velocity pressure for tapered takeoff; 0.65 velocity pressure for straight takeoff.

Diameter D. inches (cm)		Exhaust E, dia. in. (cm.)	Volume exhausted at 4,500 ft./min. ft.^{3}/ min.
Min.	Max.		
	12 (30.48)	3 (7.6)	220
Over 12 (30.48)	19 (48.26)	4 (10.16)	390
Over 19 (48.26)	30 (76.2)	5 (12.7)	610
Over 30 (76.2)	36 (91.44)	6 (15.24)	880

Note: If grinding wheels are used for disc grinding purposes, hoods must conform to structural strength and materials as described in 9.1.
Entry loss = 0.45 velocity pressure for tapered takeoff.

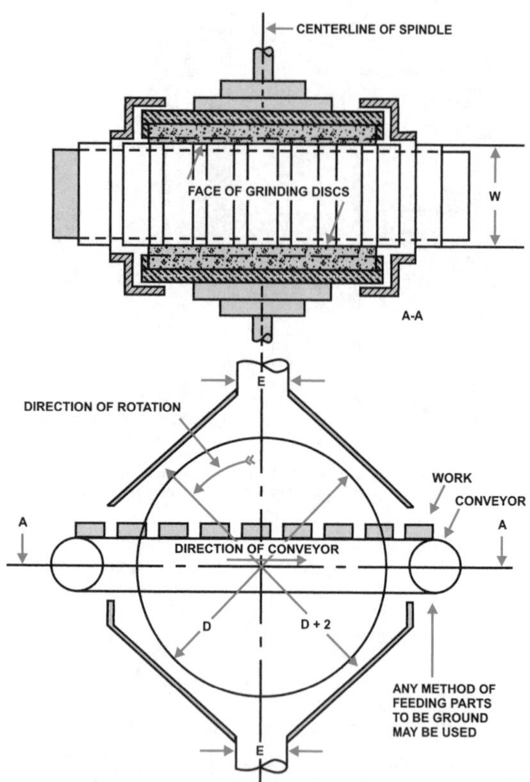

FIGURE D-57.7 - Horizontal Double-Spindle Disc Grinder Exhaust Hood and Branch Pipe Connections

Disc dia. inches (cm.)		Exhaust E		Volume exausted at 4,500 ft./min. ft.^{3}/min.	Note
Min.	Max.	No. pipes	Dia.		
	19 (48.26)	1	5	610	When width "W" permits, exhaust ducts should be as near heaviest grinding as possible.
Over 19 (48.26)	25 (63.5)	1	6	880	
Over 25 (63.5)	30 (76.2)	1	7	1,200	
Over 30 (76.2)	53 (134.62)	2	6	1,770	
Over 53 (134.62)	72 (182.88)	4	8	6,280	

Entry loss = 0.45 velocity pressure for tapered takeoff.

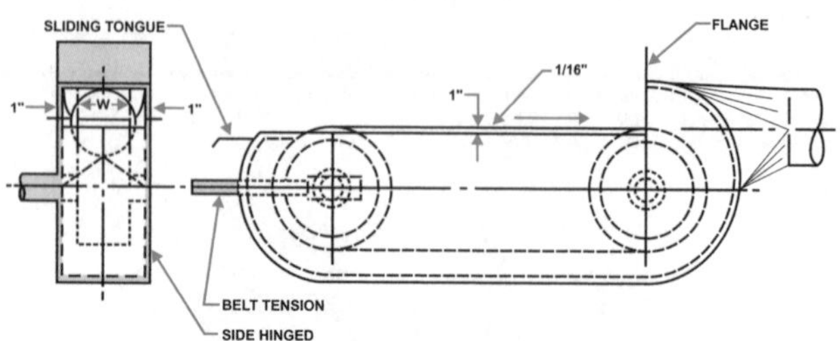

FIGURE D-57.8 - A Typical Hood for a Belt Operation
Entry loss = 0.45 velocity pressure for tapered takeoff.

Belt width W. inches (cm.)	Exhaust volume, ft.^{1}/min.
Up to 3 (7.62)	220
3 to 5 (7.62 to 12.7)	300
5 to 7 (12.7 to 17.78)	390
7 to 9 (17.78 to 22.86)	500
9 to 11 (22.86 to 27.94)	610
11 to 13 (27.94 to 33.02)	740

Minimum duct velocity = 4,500 ft/min branch, 3,500 ft/min main.
Entry loss = 0.45 velocity pressure for tapered takeoff; 0.65 velocity pressure for straight takeoff.

(6) *Scope.* This paragraph (g), prescribes the use of exhaust hood enclosures and systems in removing dust, dirt, fumes, and gases generated through the grinding, polishing, or buffing of ferrous and nonferrous metals.[§1926.57(g)(6)]

(h) **Spray finishing operations**[§1926.57(h)]

(1) *Definitions applicable to this paragraph*[§1926.57(h)(1)]

(i) **Spray-finishing operations.** Spray-finishing operations are employment of methods wherein organic or inorganic materials are utilized in dispersed form for deposit on surfaces to be coated, treated, or cleaned. Such methods of deposit may involve either automatic, manual, or electrostatic deposition but do not include metal spraying or metallizing, dipping, flow coating, roller coating, tumbling, centrifuging, or spray washing and degreasing as conducted in self-contained washing and degreasing machines or systems.[§1926.57(h)(1)(i)]

(ii) **Spray booth.** Spray booths are defined and described in Sec. 1926.66(a). (See sections 103, 104, and 105 of the Standard for Spray Finishing Using Flammable and Combustible Materials, NFPA No. 33-1969).[§1926.57(h)(1)(ii)]

(iii) **Spray room.** A spray room is a room in which spray-finishing operations not conducted in a spray booth are performed separately from other areas.[§1926.57(h)(1)(iii)]

(iv) **Minimum maintained velocity.** Minimum maintained velocity is the velocity of air movement which must be maintained in order to meet minimum specified requirements for health and safety.[§1926.57(h)(1)(iv)]

(2) *Location and application.* Spray booths or spray rooms are to be used to enclose or confine all operations. Spray-finishing operations shall be located as provided in sections 201 through 206 of the Standard for Spray Finishing Using Flammable and Combustible Materials, NFPA No. 33-1969.[§1926.57(h)(2)]

(3) *Design and construction of spray booths.*[§1926.57(h)(3)]

(i) *Spray booths shall be designed* and constructed in accordance with Sec. 1926.66(b) (1) through (4) and (6) through (10) (see sections 301-304 and 306-310 of the Standard for Spray Finishing Using Flammable and Combustible Materials, NFPA No. 33-1969), for general construction specifications. For a more detailed discussion of fundamentals relating to this subject, see ANSI Z9.2-1960[§1926.57(h)(3)(i)]

[A] *Lights, motors, electrical equipment,* and other sources of ignition shall conform to the requirements of Sec. 1926.66(b)(10) and (c). (See section 310 and chapter 4 of the Standard for Spray Finishing Using Flammable and Combustible Materials NFPA No. 33-1969.) [§1926.57(h)(3)(i)[A]]

[B] *In no case shall combustible material* be used in the construction of a spray booth and supply or exhaust duct connected to it.[§1926.57(h)(3)(i)[B]]

(ii) *Unobstructed walkways shall not be less* than 6 1/2 feet (1.976 m) high and shall be maintained clear of obstruction from any work location in the booth to a booth exit or open booth front. In booths where the open front is the only exit, such exits shall be not less than 3 feet (0.912 m) wide. In booths having multiple exits, such exits shall not be less than 2 feet (0.608 m) wide, provided that the maximum distance from the work location to the exit is 25 feet (7.6 m) or less. Where booth exits are provided with doors, such doors shall open outward from the booth.[§1926.57(h)(3)(ii)]

(iii) *Baffles, distribution plates, and dry-type overspray collectors* shall conform to the requirements of Sec. 1926.66(b) (4) and (5). (See sections 304 and 305 of the Standard for Spray Finishing Using Flammable and Combustible Materials, NFPA No. 33-1969.)[§1926.57(h)(3)(iii)]

[A] *Overspray filters shall be installed* and maintained in accordance with the requirements of Sec. 1926.66(b)(5), (see section 305 of the Standard for Spray Finishing Using Flammable and Combustible Materials, NFPA No. 33-1969), and shall only be in a location easily accessible for inspection, cleaning, or replacement.[§1926.57(h)(3)(iii)[A]]

[B] *Where effective means, independent of the overspray filters,* are installed which will result in design air distribution across the booth cross section, it is permissible to operate the booth without the filters in place.[§1926.57(h)(3)(iii)[B]]

(iv) [A] *For wet or water-wash spray booths,* the water-chamber enclosure, within which intimate contact of contaminated air and cleaning water or other cleaning medium is maintained, if made of steel, shall be 18 gage or heavier and adequately protected against corrosion.[§1926.57(h)(3)(iv)[A]]

[B] *Chambers may include* scrubber spray nozzles, headers, troughs, or other devices. Chambers shall be provided with adequate means for creating and maintaining scrubbing action for removal of particulate matter from the exhaust air stream.[§1926.57(h)(3)(iv)[B]]

(v) *Collecting tanks shall be of* welded steel construction or other suitable non-combustible material. If pits are used as collecting tanks, they shall be concrete, masonry, or other material having similar properties.[§1926.57(h)(3)(v)]

[A] *Tanks shall be provided with weirs,* skimmer plates, or screens to prevent sludge and floating paint from entering the pump suction box. Means for automatically maintaining the proper water level shall also be provided. Fresh water inlets shall not be submerged. They shall terminate at least one pipe diameter above the safety overflow level of the tank.[§1926.57(h)(3)(v)[A]]

[B] *Tanks shall be so constructed* as to discourage accumulation of hazardous deposits.[§1926.57(h)(3)(v)[B]]

(vi) *Pump manifolds, risers, and headers* shall be adequately sized to insure sufficient water flow to provide efficient operation of the water chamber.[§1926.57(h)(3)(vi)]

(4) *Design and construction of spray rooms.*[§1926.57(h)(4)]

(i) *Spray rooms, including floors,* shall be constructed of masonry, concrete, or other noncombustible material. [§1926.57(h)(4)(i)]

(ii) *Spray rooms shall have* noncombustible fire doors and shutters.[§1926.57(h)(4)(ii)]

(iii) *Spray rooms shall be adequately ventilated* so that the atmosphere in the breathing zone of the operator shall be maintained in accordance with the requirements of paragraph (h)(6)(ii) of this section.[§1926.57(h)(4)(iii)]

(iv) *Spray rooms used* for production spray-finishing operations shall conform to the requirements for spray booths. [§1926.57(h)(4)(iv)]

(5) *Ventilation.*[§1926.57(h)(5)]

(i) *Ventilation shall be provided* in accordance with provisions of Sec. 1926.66(d) (see chapter 5 of the Standard for Spray Finishing Using Flammable or Combustible Materials, NFPA No. 33- 1969); and in accordance with the following: [§1926.57(h)(5)(i)]

[A] *Where a fan plenum is used* to equalize or control the distribution of exhaust air movement through the booth, it shall be of sufficient strength or rigidity to withstand the differential air pressure or other superficially imposed loads for which the equipment is designed and also to facilitate cleaning. Construction specifications shall be at least equivalent to those of paragraph (h)(5)(iii) of this section.[§1926.57(h)(5)(i)[A]]

[B] *[Reserved]*[§1926.57(h)(5)(i)[B]]

(ii) *Inlet or supply ductwork* used to transport makeup air to spray booths or surrounding areas shall be constructed of noncombustible materials.[§1926.57(h)(5)(ii)]

[A] *If negative pressure exists within inlet ductwork,* all seams and joints shall be sealed if there is a possibility of infiltration of harmful quantities of noxious gases, fumes, or mists from areas through which ductwork passes. [§1926.57(h)(5)(ii)[A]]

[B] *Inlet ductwork shall be sized* in accordance with volume flow requirements and provide design air requirements at the spray booth.[§1926.57(h)(5)(ii)[B]]

[C] *Inlet ductwork shall be adequately* supported throughout its length to sustain at least its own weight plus any negative pressure which is exerted upon it under normal operating conditions.[§1926.57(h)(5)(ii)[C]]

(iii) *[Reserved]*[§1926.57(h)(5)(iii)]

[A] *Exhaust ductwork shall be adequately supported* throughout its length to sustain its weight plus any normal accumulation in interior during normal operating conditions and any negative pressure exerted upon it.[§1926.57(h)(5)(iii)[A]]

[B] *Exhaust ductwork shall be sized* in accordance with good design practice which shall include consideration of fan capacity, length of duct, number of turns and elbows, variation in size, volume, and character of materials being exhausted. See American National Standard Z9.2-1960 for further details and explanation concerning elements of design.[§1926.57(h)(5)(iii)[B]]

[C] *Longitudinal joints in sheet steel ductwork* shall be either lock-seamed, riveted, or welded. For other than steel construction, equivalent securing of joints shall be provided. [§1926.57(h)(5)(iii)[C]]

[D] *Circumferential joints in ductwork* shall be substantially fastened together and lapped in the direction of airflow. At least every fourth joint shall be provided with connecting flanges, bolted together, or of equivalent fastening security.[§1926.57(h)(5)(iii)[D]]

[E] *Inspection or clean-out doors* shall be provided for every 9 to 12 feet (2.736 to 3.648 m) of running length for ducts up to 12 inches (0.304 m) in diameter, but the distance between cleanout doors may be greater for larger pipes. (See 8.3.21 of American National Standard Z9.1-1951.) A clean-out door or doors shall be provided for servicing the fan, and where necessary, a drain shall be provided. [§1926.57(h)(5)(iii)[E]]

[F] *Where ductwork passes through* a combustible roof or wall, the roof or wall shall be protected at the point of penetration by open space or fire-resistive material between the duct and the roof or wall. When ducts pass through firewalls, they shall be provided with automatic fire dampers on both sides of the wall, except that three-eighth-inch steel plates may be used in lieu of automatic fire dampers for ducts not exceeding 18 inches (45.72 cm) in diameter. [§1926.57(h)(5)(iii)[F]]

[G] *Ductwork used for ventilating* any process covered in this standard shall not be connected to ducts ventilating any other process or any chimney or flue used for conveying any products of combustion.[§1926.57(h)(5)(iii)[G]]

(6) *Velocity and air flow requirements.*[§1926.57(h)(6)]

(i) *Except where a spray booth has an adequate air replacement system,* the velocity of air into all openings of a spray booth shall be not less than that specified in Table D-57.7 for the operating conditions specified. An adequate air replacement system is one which introduces replacement air upstream or above the object being sprayed and is so designed that the velocity of air in the booth cross section is not less than that specified in Table D-57.7 when measured upstream or above the object being sprayed.[§1926.57(h)(6)(i)]

Table D-57.7—Minimum Maintained Velocities Into Spray Booths

Operating conditions for objects completely inside booth	Cross-draft, f.p.m.	Airflow velocities, f.p.m.	
		Design	Range
Electrostatic and automatic airless operation contained in booth without operator	Negligible	50 large booth	50-75
		100 small booth	75-125
Air-operated guns, manual or automatic	Up to 50	100 large booth	75-125
		150 small booth	125-175
Air-operated guns, manual or automatic	Up to 100	150 large booth	125-175
		200 small booth	150-250

Notes:
(1) Attention is invited to the fact that the effectiveness of the spray booth is dependent upon the relationship of the depth of the booth to its height and width.
(2) Crossdrafts can be eliminated through proper design and such design should be sought. Crossdrafts in excess of 100fpm (feet per minute) should not be permitted.
(3) Excessive air pressures result in loss of both efficiency and material waste in addition to creating a backlash that may carry overspray and fumes into adjacent work areas.
(4) Booths should be designed with velocities shown in the column headed,"Design." However, booths operating with velocities shown in the column headed ,"Range" are in compliance with this standard.

(ii) *In addition to the requirements in paragraph* (h)(6)(i) of this section the total air volume exhausted through a spray booth shall be such as to dilute solvent vapor to at least 25 percent of the lower explosive limit of the solvent being sprayed. An example of the method of calculating this volume is given below.[§1926.57(h)(6)(ii)]

Example: To determine the lower explosive limits of the most common solvents used in spray finishing, see Table D-57.8. Column 1 gives the number of cubic feet of vapor per gallon of solvent and column 2 gives the lower explosive limit (LEL) in percentage by volume of air. Note that the quantity of solvent will be diminished by the quantity of solids and nonflammables contained in the finish.
To determine the volume of air in cubic feet necessary to dilute the vapor from 1 gallon of solvent to 25 percent of the lower explosive limit, apply the following formula:
Dilution volume required per gallon of solvent=4 (100-LEL)(cubic feet of vapor per gallon)/ LEL
Using toluene as the solvent.
(1) LEL of toluene from Table D-57.8, column 2, is 1.4 percent.
(2) Cubic feet of vapor per gallon from Table D-57.8, column 1, is 30.4 cubic feet per gallon.
(3) Dilution volume required = 4 (100-1.4) 30.4 ÷ 1.4 = 8,564 cubic feet.
(4) To convert to cubic feet per minute of required ventilation, multiply the dilution volume required per gallon of solvent by the number of gallons of solvent evaporated per minute.

Table D-57.8—Lower Explosive Limit of Some Commonly Used Solvents

Solvent	Cubic feet per gallon of vapor of liquid at 70 °F (21.11 °C).	Lower explosive limit inpercent by volume of air at 70 °F (21.11 °C)
Acetone	44.0	2.6
Amyl Acetate (iso)	21.6	[1] 1.0
Amyl Alcohol (n)	29.6	1.2
Amyl Alcohol (iso)	29.6	1.2
Benzene	36.8	[1] 1.4
Butyl Acetate (n)	24.8	1.7
Butyl Alcohol (n)	35.2	1.4
Butyl Cellosolve	24.8	1.1
Cellosolve	33.6	1.8
Cellosolve Acetate	23.2	1.7
Cyclohexanone	31.2	[1] 1.1
1,1 Dichloroethylene	42.4	5.9
1,2 Dichloroethylene	42.4	9.7
Ethyl Acetate	32.8	2.5
Ethyl Alcohol	55.2	4.3
Ethyl Lactate	28.0	[1] 1.5
Methyl Acetate	40.0	3.1
Methyl Alcohol	80.8	7.3
Methyl Cellosolve	40.8	2.5
Methyl Ethyl Ketone	36.0	1.8
Methyl n-Propyl Ketone	30.4	1.5
Naphtha (VM&P) (76° Naphtha)	22.4	0.9
Naphtha (100° Flash) Safety Solvent-Stoddard Solvent	23.2	1.0
Propyl Acetate (n)	27.2	2.8
Propyl Acetate (iso)	28.0	1.1
Propyl Alcohol (n)	44.8	2.1
Propyl Alcohol (iso)	44.0	2.0
Toluene	30.4	1.4
Turpentine	20.8	0.8
Xylene (o)	26.4	1.0

[1] At 212 °F (100 °C).

(iii) *[A] When an operator is in a booth downstream* of the object being sprayed, an air-supplied respirator or other type of respirator approved by NIOSH under 42 CFR Part 84 for the material being sprayed should be used by the operator.[§1926.57(h)(6)(iii)[A]]

[B] Where downdraft booths are provided with doors, such doors shall be closed when spray painting. [§1926.57(h)(6)(iii)[B]]

(7) *Make-up air.*[§1926.57(h)(7)]

(i) *Clean fresh air, free of contamination* from adjacent industrial exhaust systems, chimneys, stacks, or vents, shall be supplied to a spray booth or room in quantities equal to the volume of air exhausted through the spray booth.[§1926.57(h)(7)(i)]

(ii) *Where a spray booth or room* receives make-up air through self-closing doors, dampers, or louvers, they shall be fully open at all times when the booth or room is in use for spraying. The velocity of air through such doors, dampers, or louvers shall not exceed 200 feet per minute. If the fan characteristics are such that the required air flow through the booth will be provided, higher velocities through the doors, dampers, or louvers may be used.[§1926.57(h)(7)(ii)]

(iii) *[A] Where the air supply to a spray booth or room* is filtered, the fan static pressure shall be calculated on the assumption that the filters are dirty to the extent that they require cleaning or replacement.[§1926.57(h)(7)(iii)[A]]

[B] The rating of filters shall be governed by the test data supplied by the manufacturer of the filter. A pressure gage shall be installed to show the pressure drop across the filters. This gage shall be marked to show the pressure drop at which the filters require cleaning or replacement. Filters shall be replaced or cleaned whenever the pressure drop across them becomes excessive or whenever the air flow through the face of the booth falls below that specified in Table D-57.7.[§1926.57(h)(7)(iii)[B]]

(iv)*[A] Means for heating make-up air* to any spray booth or room, before or at the time spraying is normally performed, shall be provided in all places where the outdoor temperature may be expected to remain below 55 °F. (12.77 °C.) for appreciable periods of time during the operation of the booth except where adequate and safe means of radiant heating for all operating personnel affected is provided. The replacement air during the heating seasons shall be maintained at not less than 65 °F. (18.33 °C.) at the point of entry into the spray booth or spray room. When otherwise unheated make-up air would be at a temperature of more than 10 °F. below room temperature, its temperature shall be regulated as provided in section 3.6.3 of ANSI Z9.2-1960.[§1926.57(h)(7)(iv)[A]]

[B] As an alternative to an air replacement system complying with the preceding section, general heating of the building in which the spray room or booth is located may be employed provided that all occupied parts of the building are maintained at not less than 65 °F. (18.33 °C.) when the exhaust system is in operation or the general heating system supplemented by other sources of heat may be employed to meet this requirement.[§1926.57(h)(7)(iv)[B]]

[C] No means of heating make-up air shall be located in a spray booth.[§1926.57(h)(7)(iv)[C]]

[D] Where make-up air is heated by coal or oil, the products of combustion shall not be allowed to mix with the make-up air, and the products of combustion shall be conducted outside the building through a flue terminating at a point remote from all points where make-up air enters the building. [§1926.57(h)(7)(iv)[D]]

[E] Where make-up air is heated by gas, and the products of combustion are not mixed with the make-up air but are conducted through an independent flue to a point outside the building remote from all points where make-up air enters the building, it is not necessary to comply with paragraph (h)(7)(iv)(F) of this section.[§1926.57(h)(7)(iv)[E]]

[F] Where make-up air to any manually operated spray booth or room is heated by gas and the products of combustion are allowed to mix with the supply air, the following precautions must be taken:[§1926.57(h)(7)(iv)[F]]

[1] The gas must have a distinctive and strong enough odor to warn workmen in a spray booth or room of its presence if in an unburned state in the make-up air. [§1926.57(h)(7)(iv)[F][1]]

[2] The maximum rate of gas supply to the make-up air heater burners must not exceed that which would yield in excess of 200 p.p.m. (parts per million) of carbon monoxide or 2,000 p.p.m. of total combustible gases in the mixture if the unburned gas upon the occurrence of flame failure were mixed with all of the make-up air supplied. [§1926.57(h)(7)(iv)[F][2]]

[3] A fan must be provided to deliver the mixture of heated air and products of combustion from the plenum chamber housing the gas burners to the spray booth or room. [§1926.57(h)(7)(iv)[F][3]]

(8) *Scope.* Spray booths or spray rooms are to be used to enclose or confine all spray finishing operations covered by this paragraph (h). This paragraph does not apply to the spraying of the exteriors of buildings, fixed tanks, or similar structures, nor to small portable spraying apparatus not used repeatedly in the same location.[§1926.57(h)(8)]

(i) Open surface tanks[§1926.57(i)]

(1) *General.*[§1926.57(i)(1)]

(i) *This paragraph applies to all operations* involving the immersion of materials in liquids, or in the vapors of such liquids, for the purpose of cleaning or altering the surface or adding to or imparting a finish thereto or changing the character of the materials, and their subsequent removal from the liquid or vapor, draining, and drying. These operations include washing, electroplating, anodizing, pickling, quenching, dying, dipping, tanning, dressing, bleaching, degreasing, alkaline cleaning, stripping, rinsing, digesting, and other similar operations.[§1926.57(i)(1)(i)]

(ii) *Except where specific construction specifications* are prescribed in this section, hoods, ducts, elbows, fans, blowers, and all other exhaust system parts, components, and supports thereof shall be so constructed as to meet conditions of service and to facilitate maintenance and shall conform in construction to the specifications contained in American National Standard Fundamentals Governing the Design and Operation of Local Exhaust Systems, Z9.2-1960. [§1926.57(i)(1)(ii)]

(2) *Classification of open-surface tank operations.*[§1926.57(i)(2)]

(i) *Open-surface tank operations* shall be classified into 16 classes, numbered A-1 to D- 4, inclusive.[§1926.57(i)(2)(i)]

(ii) *Determination of class.* Class is determined by two factors, hazard potential designated by a letter from A to D, inclusive, and rate of gas, vapor, or mist evolution designated by a number from 1 to 4, inclusive (for example, B.3). [§1926.57(i)(2)(ii)]

(iii) *Hazard potential is an index,* on a scale of from A to D, inclusive, of the severity of the hazard associated with the substance contained in the tank because of the toxic, flammable, or explosive nature of the vapor, gas, or mist produced therefrom. The toxic hazard is determined from the concentration, measured in parts by volume of a gas or vapor, per million parts by volume of contaminated air (p.p.m.), or in milligrams of mist per cubic meter of air (mg./m.^{3}), below which ill effects are unlikely to occur to the exposed worker. The concentrations shall be those in Sec. 1926.55 or other pertinent sections of this part.[§1926.57(i)(2)(iii)]

(iv) *The relative fire or explosion hazard* is measured in degrees Fahrenheit in terms of the closed-cup flash point of the substance in the tank. Detailed information on the prevention of fire hazards in dip tanks may be found in Dip Tanks Containing Flammable or Combustible Liquids, NFPA No. 34-1966, National Fire Protection Association. Where the tank contains a mixture of liquids, other than organic solvents, whose effects are additive, the hygienic standard of the most toxic component (for example, the one having the lowest p.p.m. or mg./m.^{3}) shall be used, except where such substance constitutes an insignificantly small fraction of the mixture. For mixtures of organic solvents, their combined effect, rather than that of either individually, shall determine the hazard potential. In the absence of information to the contrary, the effects shall be considered as additive. If the sum of the ratios of the airborne concentration of each contaminant to the toxic concentration of that contaminant exceeds unity, the toxic concentration shall be considered to have been exceeded. (See Note A to paragraph (i)(2)(v) of this section.) [§1926.57(i)(2)(iv)]

(v) *Hazard potential shall be determined from Table D-57.9,* with the value indicating greater hazard being used. When the hazardous material may be either a vapor with a threshhold limit value (TLV) in p.p.m. or a mist with a TLV in mg./m.$_3$, the TLV indicating the greater hazard shall be used (for example, A takes precedence over B or C; B over C; C over D).[§1926.57(i)(2)(v)]

Note A: $(c_1/ TLV_1)+(c_2/ TLV_2)+(c_3/TLV_3)+; . . .(c_N/ TLV_N)^1$
Where:[§1926.57(i)(2)(v)]

c = Concentration measured at the operation in p.p.m. [§1926.57(i)(2)(v)]

Table D-57.9—Determination of Hazard Potential

Hazard potential	Toxicity group		
	Gas or vapor (p.p.m.)	Mist (mg./m^3)	Flash point in degrees F. (C.)
A	0-10	0-0.1	
B	11-100	0.11-1.0	Under 100 (37.77)
C	101-500	1.1-10	100 200(37.77- 93.33)
D	Over 500	Over 10	Over 200 (93.33)

(vi) *Rate of gas, vapor, or mist evolution* is a numerical index, on a scale of from 1 to 4, inclusive, both of the relative capacity of the tank to produce gas, vapor, or mist and of the relative energy with which it is projected or carried upwards from the tank. Rate is evaluated in terms of[§1926.57(i)(2)(vi)]

[a] *The temperature of the liquid in the tank* in degrees Fahrenheit;[§1926.57(i)(2)(vi)[a]]

[b] *The number of degrees Fahrenheit* that this temperature is below the boiling point of the liquid in degrees Fahrenheit;[§1926.57(i)(2)(vi)[b]]

[c] *The relative evaporation of the liquid* in still air at room temperature in an arbitrary scale — fast, medium, slow, or nil; and[§1926.57(i)(2)(vi)[c]]

[d] *The extent that the tank gases or produces mist* in an arbitrary scale — high, medium, low, and nil. (See Table D-57.10, Note 2.) Gassing depends upon electrochemical or mechanical processes, the effects of which have to be individually evaluated for each installation (see Table D-57.10, Note 3).[§1926.57(i)(2)(vi)[d]]

(vii) *Rate of evolution shall be determined* from Table D-57.10. When evaporation and gassing yield different rates, the lowest numerical value shall be used.[§1926.57(i)(2)(vii)]

Table D-57.10—Determination of Rate of Gas, Vapor, or Mist Evolution [1]

Rate	Liquid temperature, °F. (C.)	Degrees below boiling point	Relative evaporation[2]	Gassing[3]
1	Over 200 (93.33)	0-20	Fast	High
2	150-200 (65.55- 93.33)	21-50	Medium	Medium
3	94-149 (34.44-65)	51-100	Slow	Low
4	Under 94 (34.44)	Over 100	Nil	Nil

[1] In certain classes of equipment, specifically vapor degreasers, an internal condenser or vapor level thermostat is used to prevent the vapor from leaving the tank during normal operation. In such cases, rate of vapor evolution from the tank into the workroom is not dependent upon the factors listed in the table, but rather upon abnormalities of operating procedure, such as carryout of vapors from excessively fast action, dragout of liquid by entrainment in parts, contamination of solvent by water and other materials, or improper heat balance. When operating procedure is excellent, effective rate of evolution may be taken as 4. When operating procedure is average, the effective rate of evolution may be taken as 3. When operation is poor, a rate of 2 or 1 is indicated, depending upon observed conditions.

[2] Relative evaporation rate is determined according to the methods described by A. K. Doolittle in Industrial and Engineering Chemistry, vol. 27, p. 1169, (3) where time for 100-percent evaporation is as follows: Fast: 0-3 hours; Medium: 3-12 hours; Slow: 12-50 hours; Nil: more than 50 hours.

[3] **Gassing** means the formation by chemical or electrochemical action of minute bubbles of gas under the surface of the liquid in the tank and is generally limited to aqueous solutions.

(3) *Ventilation.* Where ventilation is used to control potential exposures to workers as defined in paragraph (i)(2)(iii) of this section, it shall be adequate to reduce the concentration of the air contaminant to the degree that a hazard to the worker does not exist. Methods of ventilation are discussed in American National Standard Fundamentals Governing the Design and Operation of Local Exhaust Systems, Z9.2-1960.[§1926.57(i)(3)]

(4) *Control requirements.*[§1926.57(i)(4)]

(i) *Control velocities shall conform to Table D-57.11* in all cases where the flow of air past the breathing or working zone of the operator and into the hoods is undisturbed by local environmental conditions, such as open windows, wall fans, unit heaters, or moving machinery.[§1926.57(i)(4)(i)]

(ii) *All tanks exhausted by means of hoods which*[§1926.57(i)(4)(ii)]

[a] *Project over the entire tank;*[§1926.57(i)(4)(ii)[a]]

[b] *Are fixed in position* in such a location that the head of the workman, in all his normal operating positions while working at the tank, is in front of all hood openings; and [§1926.57(i)(4)(ii)[b]]

[c] *Are completely enclosed* on at least two sides, shall be considered to be exhausted through an enclosing hood. [§1926.57(i)(4)(ii)[c]]

[d] *The quantity of air in cubic feet per minute* necessary to be exhausted through an enclosing hood shall be not less than the product of the control velocity times the net area of all openings in the enclosure through which air can flow into the hood.[§1926.57(i)(4)(ii)[d]]

Table D-57.11—Control Velocities in Feet Per Minute (f.p.m.) for Undisturbed Locations

Class	Enclosing hood		Lateral exhaust[1]	Canopy hood[2]	
	One open side	Two open sides		Three open sides	Four open sides
B-1 and A-2	100	150	150	Do not use	Do not use
A-3[2], B-1, B-2, and C-1	75	100	100	125	175
A-3, C-2, and D-1 [3]	65	90	75	100	150
B-4[2], C-3, and D-2[3]	50	75	50	75	125
A-4, C-4, D-3[3] , and D-4[4]					

[1] See Table D-57.12 for computation of ventilation rate.
[2] Do not use canopy hood for Hazard Potential A processes.
[3] Where complete control of hot water is desired, design as next highest class.
[4] General room ventilation required.

(iii) *All tanks exhausted by means of hoods* which do not project over the entire tank, and in which the direction of air movement into the hood or hoods is substantially horizontal, shall be considered to be laterally exhausted. The quantity of air in cubic feet per minute necessary to be laterally exhausted per square foot of tank area in order to maintain the required control velocity shall be determined from Table D-57.12 for all variations in ratio of tank width (W) to tank length $(L). The total quantity of air in cubic feet per minute required to be exhausted per tank shall be not less than the product of the area of tank surface times the cubic feet per minute per square foot of tank area, determined from Table D-57.12. [§1926.57(i)(4)(iii)]

[a] *For lateral exhaust hoods* over 42 inches (1.06 m) wide, or where it is desirable to reduce the amount of air removed from the workroom, air supply slots or orifices shall be provided along the side or the center of the tank opposite from the exhaust slots. The design of such systems shall meet the following criteria:[§1926.57(i)(4)(iii)[a]]

[1] *The supply air volume* plus the entrained air shall not exceed 50 percent of the exhaust volume. [§1926.57(i)(4)(iii)[a][1]]

[2] *The velocity of the supply airstream* as it reaches the effective control area of the exhaust slot shall be less than the effective velocity over the exhaust slot area. [§1926.57(i)(4)(iii)[a][2]]

Table D-57.12—Minimum Ventilation Rate in Cubic Feet of Air Per Minute Per Square Foot of Tank Area for Lateral Exhaust

Required minimum control velocity, f.p.m. (from Table D-57.11)	C.f.m. per sq. ft. to maintain required minimum velocities at following ratios (tank width (W)/tank length (L))[1,2]				
	0.0-0.09	0.1-0.24	0.25-0.49	0.5-0.99	1.0-2.0
Hood along one side or two parallel sides of tank when one hood is against a wall or baffle.[2] Also for a manifold along tank centerline.[3]					
50	50	60	75	90	100
75	75	90	110	130	150
100	100	125	150	175	200
150	150	190	225	260	300
Hood along one side or two parallel sides of free standing tank not against wall or baffle.					
50	75	90	100	110	125
75	110	130	150	170	190
100	150	175	200	225	250
150	225	260	300	340	375

1. It is not practicable to ventilate across the long dimension of a tank whose ratio W/L exceeds 2.0.
It is undesirable to do so when W/L exceeds 1.0. For circular tanks with lateral exhaust along up to 1/2 the circumference, use W/L=1.0; for over one-half the circumference use W/L=0.5.
2. Baffle is a vertical plate the same length as the tank, and with the top of the plate as high as the tank is wide. If the exhaust hood is on the side of a tank against a building wall or close to it, it is perfectly baffled.
3. Use W/2 as tank width in computing when manifold is along centerline, or when hoods are used on two parallel sides of a tank. **Tank Width** (W) means the effective width over which the hood must pull air to operate (for example, where the hood face is set back from the edge of the tank, this set back must be added in measuring tank width). The surface area of tanks can frequently be reduced and better control obtained (particularly on conveyorized systems) by using covers extending from the upper edges of the slots toward the center of the tank.

[3] *The vertical height* of the receiving exhaust hood, including any baffle, shall not be less than one-quarter the width of the tank.[§1926.57(i)(4)(iii)[a][3]]

[4] *The supply airstream shall not be allowed* to impinge on obstructions between it and the exhaust slot in such a manner as to significantly interfere with the performance of the exhaust hood.[§1926.57(i)(4)(iii)[a][4]]

[5] *Since most failure of push-pull systems* result from excessive supply air volumes and pressures, methods of measuring and adjusting the supply air shall be provided. When satisfactory control has been achieved, the adjustable features of the hood shall be fixed so that they will not be altered.[§1926.57(i)(4)(iii)[a][5]]

(iv) *All tanks exhausted by means of hoods* which project over the entire tank, and which do not conform to the definition of enclosing hoods, shall be considered to be overhead canopy hoods. The quantity of air in cubic feet per minute necessary to be exhausted through a canopy hood shall be not less than the product of the control velocity times the net area of all openings between the bottom edges of the hood and the top edges of the tank.[§1926.57(i)(4)(iv)]

(v) *The rate of vapor evolution* (including steam or products of combustion) from the process shall be estimated. If the rate of vapor evolution is equal to or greater than 10 percent of the calculated exhaust volume required, the exhaust volume shall be increased in equal amount.[§1926.57(i)(4)(v)]

(5) *Spray cleaning and degreasing.* Wherever spraying or other mechanical means are used to disperse a liquid above an open-surface tank, control must be provided for the airborne spray. Such operations shall be enclosed as completely as possible. The inward air velocity into the enclosure shall be sufficient to prevent the discharge of spray into the workroom. Mechanical baffles may be used to help prevent the discharge of spray. Spray painting operations are covered by paragraph (h) of this section.[§1926.57(i)(5)]

(6) *Control means other than ventilation.* Tank covers, foams, beads, chips, or other materials floating on the tank surface so as to confine gases, mists, or vapors to the area under the cover or to the foam, bead, or chip layer; or surface tension depressive agents added to the liquid in the tank to minimize mist formation, or any combination thereof, may all be used as gas, mist, or vapor control means for open- surface tank operations, provided that they effectively reduce the concentrations of hazardous materials in the vicinity of the worker below the limits set in accordance with paragraph (i)(2) of this section. [§1926.57(i)(6)]

(7) *System design.*[§1926.57(i)(7)]

(i) *The equipment for exhausting air* shall have sufficient capacity to produce the flow of air required in each of the hoods and openings of the system.[§1926.57(i)(7)(i)]

(ii) *The capacity required in paragraph (i)(7)(i)* of this section shall be obtained when the airflow producing equipment is operating against the following pressure losses, the sum of which is the static pressure:[§1926.57(i)(7)(ii)]

[a] *Entrance losses into the hood.*[§1926.57(i)(7)(ii)[a]]

[b] *Resistance to airflow* in branch pipe including bends and transformations.[§1926.57(i)(7)(ii)[b]]

[c] *Entrance loss into the main pipe.*[§1926.57(i)(7)(ii)[c]]

[d] *Resistance to airflow* in main pipe including bends and transformations.[§1926.57(i)(7)(ii)[d]]

[e] *Resistance of mechanical equipment;* that is, filters, washers, condensers, absorbers, etc., plus their entrance and exit losses.[§1926.57(i)(7)(ii)[e]]

[f] *Resistance in outlet duct* and discharge stack. [§1926.57(i)(7)(ii)[f]]

(iii) *Two or more operations* shall not be connected to the same exhaust system where either one or the combination of the substances removed may constitute a fire, explosion, or chemical reaction hazard in the duct system. Traps or other devices shall be provided to insure that condensate in ducts does not drain back into any tank.[§1926.57(i)(7)(iii)]

(iv) *The exhaust system,* consisting of hoods, ducts, air mover, and discharge outlet, shall be designed in accordance with American National Standard Fundamentals Governing the Design and Operation of Local Exhaust Systems, Z9.2-1960, or the manual, Industrial Ventilation, published by the American Conference of Governmental Industrial Hygienists 1970. Airflow and pressure loss data provided by the manufacturer of any air cleaning device shall be included in the design calculations.[§1926.57(i)(7)(iv)]

(8) *Operation.* [§1926.57(i)(8)]

(i) *The required airflow shall be maintained* at all times during which gas, mist, or vapor is emitted from the tank, and at all times the tank, the draining, or the drying area is in operation or use. When the system is first installed, the airflow from each hood shall be measured by means of a pitot traverse in the exhaust duct and corrective action taken if the flow is less than that required. When the proper flow is obtained, the hood static pressure shall be measured and recorded. At intervals of not more than 3 months operation, or after a prolonged shutdown period, the hoods and duct system shall be inspected for evidence of corrosion or damage. In any case where the airflow is found to be less than required, it shall be increased to the required value. (Information on airflow and static pressure measurement and calculations may be found in American National Standard Fundamental Governing the Design and Operation of Local Exhaust Systems, Z9.2-1960, or in the manual, Industrial Ventilation, published by the American Conference of Governmental Industrial Hygienists.) [§1926.57(i)(8)(i)]

(ii) *The exhaust system shall discharge* to the outer air in such a manner that the possibility of its effluent entering any building is at a minimum. Recirculation shall only be through a device for contaminant removal which will prevent the creation of a health hazard in the room or area to which the air is recirculated. [§1926.57(i)(8)(ii)]

(iii) *A volume of outside air* in the range of 90 percent to 110 percent of the exhaust volume shall be provided to each room having exhaust hoods. The outside air supply shall enter the workroom in such a manner as not to be detrimental to any exhaust hood. The airflow of the makeup air system shall be measured on installation. Corrective action shall be taken when the airflow is below that required. The makeup air shall be uncontaminated. [§1926.57(i)(8)(iii)]

(9) *Personal protection.* [§1926.57(i)(9)]

(i) *All employees working in and around* open-surface tank operations must be instructed as to the hazards of their respective jobs, and in the personal protection and first aid procedures applicable to these hazards. [§1926.57(i)(9)(i)]

(ii) *All persons required to work* in such a manner that their feet may become wet shall be provided with rubber or other impervious boots or shoes, rubbers, or wooden-soled shoes sufficient to keep feet dry. [§1926.57(i)(9)(ii)]

(iii) *All persons required to handle* work wet with a liquid other than water shall be provided with gloves impervious to such a liquid and of a length sufficient to prevent entrance of liquid into the tops of the gloves. The interior of gloves shall be kept free from corrosive or irritating contaminants. [§1926.57(i)(9)(iii)]

(iv) *All persons required to work* in such a manner that their clothing may become wet shall be provided with such aprons, coats, jackets, sleeves, or other garments made of rubber, or of other materials impervious to liquids other than water, as are required to keep their clothing dry. Aprons shall extend well below the top of boots to prevent liquid splashing into the boots. Provision of dry, clean, cotton clothing along with rubber shoes or short boots and an apron impervious to liquids other than water shall be considered a satisfactory substitute where small parts are cleaned, plated, or acid dipped in open tanks and rapid work is required. [§1926.57(i)(9)(iv)]

(v) *Whenever there is a danger* of splashing, for example, when additions are made manually to the tanks, or when acids and chemicals are removed from the tanks, the employees so engaged shall be required to wear either tight-fitting chemical goggles or an effective face shield. See Sec. 1926.102. [§1926.57(i)(9)(v)]

(vi) *When, during the emergencies* specified in paragraph (i)(11)(v) of this section, employees must be in areas where concentrations of air contaminants are greater than the limits set by paragraph (i)(2)(iii) of this section or oxygen concentrations are less than 19.5 percent, they must use respirators that reduce their exposure to a level below these limits or that provide adequate oxygen. Such respirators must also be provided in marked, quickly-accessible storage compartments built for this purpose when the possibility exists of accidental release of hazardous concentrations of air contaminants. Respirators must be approved by NIOSH under 42 CFR part 84, selected by a competent industrial hygienist or other technically-qualified source, and used in accordance with 29 CFR 1926.103. [§1926.57(i)(9)(vi)]

(vii) *Near each tank containing a liquid* which may burn, irritate, or otherwise be harmful to the skin if splashed upon the worker's body, there shall be a supply of clean cold water. The water pipe (carrying a pressure not exceeding 25 pounds (11.325 kg)) shall be provided with a quick opening valve and at least 48 inches (1.216 m) of hose not smaller than three-fourths inch, so that no time may be lost in washing off liquids from the skin or clothing. Alternatively, deluge showers and eye flushes shall be provided in cases where harmful chemicals may be splashed on parts of the body. [§1926.57(i)(9)(vii)]

(viii) *Operators with sores,* burns, or other skin lesions requiring medical treatment shall not be allowed to work at their regular operations until so authorized by a physician. Any small skin abrasions, cuts, rash, or open sores which are found or reported shall be treated by a properly designated person so that chances of exposures to the chemicals are removed. Workers exposed to chromic acids shall have a periodic examination made of the nostrils and other parts of the body, to detect incipient ulceration. [§1926.57(i)(9)(viii)]

(ix) *Sufficient washing facilities,* including soap, individual towels, and hot water, shall be provided for all persons required to use or handle any liquids which may burn, irritate, or otherwise be harmful to the skin, on the basis of at least one basin (or its equivalent) with a hot water faucet for every 10 employees. See Sec. 1926.51(f). [§1926.57(i)(9)(ix)]

(x) *Locker space or equivalent* clothing storage facilities shall be provided to prevent contamination of street clothing. [§1926.57(i)(9)(x)]

(xi) *First aid facilities specific* to the hazards of the operations conducted shall be readily available. [§1926.57(i)(9)(xi)]

(10) *Special precautions for cyanide.* Dikes or other arrangements shall be provided to prevent the possibility of intermixing of cyanide and acid in the event of tank rupture. [§1926.57(i)(10)]

(11) *Inspection, maintenance, and installation.* [§1926.57(i)(11)]

(i) *Floors and platforms around tanks* shall be prevented from becoming slippery both by original type of construction and by frequent flushing. They shall be firm, sound, and of the design and construction to minimize the possibility of tripping. [§1926.57(i)(11)(i)]

(ii) *Before cleaning the interior of any tank,* the contents shall be drained off, and the cleanout doors shall be opened where provided. All pockets in tanks or pits, where it is possible for hazardous vapors to collect, shall be ventilated and cleared of such vapors. [§1926.57(i)(11)(ii)]

(iii) *Tanks which have been drained* to permit employees to enter for the purposes of cleaning, inspection, or maintenance may contain atmospheres which are hazardous to life or health, through the presence of flammable or toxic air contaminants, or through the absence of sufficient oxygen. Before employees shall be permitted to enter any such tank, appropriate tests of the atmosphere shall be made to determine if the limits set by paragraph (i)(2)(iii) of this section are exceeded, or if the oxygen concentration is less than 19.5 percent. [§1926.57(i)(11)(iii)]

(iv) *If the tests made in accordance with paragraph (i)(11)(iii)* of this section indicate that the atmosphere in the tank is unsafe, before any employee is permitted to enter the tank, the tank shall be ventilated until the hazardous atmosphere is removed, and ventilation shall be continued so as to prevent the occurrence of a hazardous atmosphere as long as an employee is in the tank. [§1926.57(i)(11)(iv)]

(v) *If, in emergencies, such as rescue work,* it is necessary to enter a tank which may contain a hazardous atmosphere, suitable respirators, such as self-contained breathing apparatus; hose mask with blower, if there is a possibility of oxygen deficiency; or a gas mask, selected and operated in accordance with paragraph (i)(9)(vi) of this section, shall be used. If a contaminant in the tank can cause dermatitis, or be absorbed through the skin, the employee entering the tank shall also wear protective clothing. At least one trained standby employee, with suitable respirator, shall be present in the nearest uncontaminated area. The standby employee must be able to communicate with the employee in the tank and be able to haul him out of the tank with a lifeline if necessary. [§1926.57(i)(11)(v)]

(vi) *Maintenance work requiring welding or open flame,* where toxic metal fumes such as cadmium, chromium, or lead may be evolved, shall be done only with sufficient local exhaust ventilation to prevent the creation of a health hazard, or be done with respirators selected and used in accordance with paragraph (i)(9)(vi) of this section. Welding, or the use of open flames near any solvent cleaning equipment shall be permitted only after such equipment has first been thoroughly cleared of solvents and vapors. [§1926.57(i)(11)(vi)]

(12) *Vapor degreasing tanks.* [§1926.57(i)(12)]

(i) *In any vapor degreasing tank equipped* with a condenser or vapor level thermostat, the condenser or thermostat shall keep the level of vapors below the top edge of the tank by a distance at least equal to one-half the tank width, or at least 36 inches (0.912 m), whichever is shorter. [§1926.57(i)(12)(i)]

(ii) *Where gas is used as a fuel* for heating vapor degreasing tanks, the combustion chamber shall be of tight construction, except for such openings as the exhaust flue, and those that are necessary for supplying air for combustion. Flues shall be of corrosion-resistant construction and shall extend to the outer air. If mechanical exhaust is used on this flue, a draft diverter shall be used. Special precautions must be taken to prevent solvent fumes from entering the combustion air of this or any other heater when chlorinated or fluorinated hydrocarbon solvents (for example, trichloroethylene, Freon) are used. [§1926.57(i)(12)(ii)]

(iii) *Heating elements shall be so designed* and maintained that their surface temperature will not cause the solvent or mixture to decompose, break down, or be converted into an excessive quantity of vapor. [§1926.57(i)(12)(iii)]

(iv) *Tanks or machines of more* than 4 square feet (0.368 m^2) of vapor area, used for solvent cleaning or vapor degreasing, shall be equipped with suitable cleanout or sludge doors located near the bottom of each tank or still. These doors shall be so designed and gasketed that there will be no leakage of solvent when they are closed. [§1926.57(i)(12)(iv)]

(13) *Scope.* [§1926.57(i)(13)]

(i) *This paragraph (i) applies to all operations* involving the immersion of materials in liquids, or in the vapors of such liquids, for the purpose of cleaning or altering their surfaces, or adding or imparting a finish thereto, or changing the character of the materials, and their subsequent removal from the liquids or vapors, draining, and drying. Such operations include washing, electroplating, anodizing, pickling, quenching, dyeing, dipping, tanning, dressing, bleaching, degreasing, alkaline cleaning, stripping, rinsing, digesting, and other similar operations, but do not include molten materials handling operations, or surface coating operations. [§1926.57(i)(13)(i)]

(ii) **Molten materials handling operations** means all operations, other than welding, burning, and soldering operations, involving the use, melting, smelting, or pouring of metals, alloys, salts, or other similar substances in the molten state. Such operations also include heat treating baths, descaling baths, die casting stereotyping, galvanizing, tinning, and similar operations. [§1926.57(i)(13)(ii)]

(iii) **Surface coating operations** means all operations involving the application of protective, decorative, adhesive, or strengthening coating or impregnation to one or more surfaces, or into the interstices of any object or material, by means of spraying, spreading, flowing, brushing, roll coating, pouring, cementing, or similar means; and any subsequent draining or drying operations, excluding open-tank operations. [§1926.57(i)(13)(iii)]

[44 FR 8577, Feb. 9, 1979; 44 FR 20940, Apr. 6, 1979, as amended at 58 FR 35099, June 30, 1993; 61 FR 9250, Mar. 3, 1996; 63 FR 1295, Jan. 8, 1998]

§1926.58

[Reserved]

§1926.59

Hazard communication

Note: The requirements applicable to construction work under this section are identical to those set forth at Sec. 1910.1200 of this chapter. (Editor's Note: See 1910 supplement.)

[61 FR 31431, June 20, 1996]

§1926.60

Methylenedianiline

(a) **Scope and application.**[§1926.60(a)]

(1) *This section* applies to all construction work as defined in 29 CFR 1910.12(b), in which there is exposure to MDA, including but not limited to the following:[§1926.60(a)(1)]

(i) *Construction, alteration,* repair, maintenance, or renovation of structures, substrates, or portions thereof, that contain MDA;[§1926.60(a)(1)(i)]

(ii) *Installation or the finishing of surfaces* with products containing MDA;[§1926.60(a)(1)(ii)]

(iii) *MDA spill/emergency cleanup at construction sites; and* [§1926.60(a)(1)(iii)]

(iv) *Transportation, disposal, storage,* or containment of MDA or products containing MDA on the site or location at which construction activities are performed.[§1926.60(a)(1)(iv)]

(2) *Except as provided in paragraphs (a)(7)* and (f)(5) of this section, this section does not apply to the processing, use, and handling of products containing MDA where initial monitoring indicates that the product is not capable of releasing MDA in excess of the action level under the expected conditions of processing, use, and handling which will cause the greatest possible release; and where no "dermal exposure to MDA" can occur.[§1926.60(a)(2)]

(3) *Except as provided in paragraph (a)(7) of this section,* this section does not apply to the processing, use, and handling of products containing MDA where objective data are reasonably relied upon which demonstrate the product is not capable of releasing MDA under the expected conditions of processing, use, and handling which will cause the greatest possible release; and where no "dermal exposure to MDA" can occur. [§1926.60(a)(3)]

(4) *Except as provided in paragraph (a)(7) of this section,* this section does not apply to the storage, transportation, distribution or sale of MDA in intact containers sealed in such a manner as to contain the MDA dusts, vapors, or liquids, except for the provisions of 29 CFR 1910.1200 and paragraph (e) of this section. [§1926.60(a)(4)]

(5) *Except as provided in paragraph (a)(7) of this section,* this section does not apply to materials in any form which contain less than 0.1% MDA by weight or volume.[§1926.60(a)(5)]

(6) *Except as provided in paragraph (a)(7) of this section,* this section does not apply to "finished articles containing MDA." [§1926.60(a)(6)]

(7) *Where products containing MDA* are exempted under paragraphs (a)(2) through (a)(6) of this section, the employer shall maintain records of the initial monitoring results or objective data supporting that exemption and the basis for the employer's reliance on the data, as provided in the recordkeeping provision of paragraph (o) of this section.[§1926.60(a)(7)]

(b) **Definitions.** For the purpose of this section, the following definitions shall apply:[§1926.60(b)]

Action level means a concentration of airborne MDA of 5 ppb as an eight (8)-hour time-weighted average.[§1926.60(b)]

Assistant Secretary means the Assistant Secretary of Labor for Occupational Safety and Health, U.S. Department of Labor, or designee.[§1926.60(b)]

Authorized person means any person specifically authorized by the employer whose duties require the person to enter a regulated area, or any person entering such an area as a designated representative of employees for the purpose of exercising the right to observe monitoring and measuring procedures under paragraph (p) of this section, or any other person authorized by the Act or regulations issued under the Act.[§1926.60(b)]

Container means any barrel, bottle, can, cylinder, drum, reaction vessel, storage tank, commercial packaging or the like, but does not include piping systems.[§1926.60(b)]

Decontamination area means an area outside of but as near as practical to the regulated area, consisting of an equipment storage area, wash area, and clean change area, which is used for the decontamination of workers, materials, and equipment contaminated with MDA.[§1926.60(b)]

Dermal exposure to MDA occurs where employees are engaged in the handling, application or use of mixtures or materials containing MDA, with any of the following non-airborne forms of MDA: [§1926.60(b)]

(i) *Liquid, powdered, granular,* or flaked mixtures containing MDA in concentrations greater than 0.1% by weight or volume; and [§1926.60(b)(i)]

(ii) *Materials other than* "finished articles" containing MDA in concentrations greater than 0.1% by weight or volume. [§1926.60(b)(ii)]

Director means the Director of the National Institute for Occupational Safety and Health, U.S. Department of Health and Human Services, or designee.[§1926.60(b)]

Emergency means any occurrence such as, but not limited to, equipment failure, rupture of containers, or failure of control equipment which results in an unexpected and potentially hazardous release of MDA.[§1926.60(b)]

Employee exposure means exposure to MDA which would occur if the employee were not using respirators or protective work clothing and equipment.[§1926.60(b)]

Finished article containing MDA is defined as a manufactured item: [§1926.60(b)]

(i) *Which is formed to a specific* shape or design during manufacture;[§1926.60(b)(i)]

(ii) *Which has end* use function(s) dependent in whole or part upon its shape or design during end use; and[§1926.60(b)(ii)]

(iii) *Where applicable,* is an item which is fully cured by virtue of having been subjected to the conditions (temperature, time) necessary to complete the desired chemical reaction. [§1926.60(b)(iii)]

Historical monitoring data means monitoring data for construction jobs that meet the following conditions:[§1926.60(b)]

(i) *The data upon which judgments are based* are scientifically sound and were collected using methods that are sufficiently accurate and precise;[§1926.60(b)(i)]

(ii) *The processes and work practices* that were in use when the historical monitoring data were obtained are essentially the same as those to be used during the job for which initial monitoring will not be performed;[§1926.60(b)(ii)]

(iii) *The characteristics of the MDA-containing material* being handled when the historical monitoring data were obtained are the same as those on the job for which initial monitoring will not be performed;[§1926.60(b)(iii)]

(iv) *Environmental conditions prevailing* when the historical monitoring data were obtained are the same as those on the job for which initial monitoring will not be performed; and[§1926.60(b)(iv)]

(v) *Other data relevant to the operations,* materials, processing, or employee exposures covered by the exception are substantially similar. The data must be scientifically sound, the characteristics of the MDA containing material must be similar and the environmental conditions comparable.[§1926.60(b)(v)]

4,4' Methylenedianiline or MDA means the chemical; 4,4'- diaminodiphenylmethane, Chemical Abstract Service Registry number 101- 77-9, in the form of a vapor, liquid, or solid. The definition also includes the salts of MDA.[§1926.60(b)]

Regulated Areas means areas where airborne concentrations of MDA exceed or can reasonably be expected to exceed, the permissible exposure limits, or where "dermal exposure to MDA" can occur.[§1926.60(b)]

STEL means short term exposure limit as determined by any 15-minute sample period.[§1926.60(b)]

(c) **Permissible exposure limits.** The employer shall assure that no employee is exposed to an airborne concentration of MDA in excess of ten parts per billion (10 ppb) as an 8-hour time-weighted average and a STEL of one hundred parts per billion (100 ppb). [§1926.60(c)]

(d) **Communication among employers.** On multi-employer worksites, an employer performing work involving the application of MDA or materials containing MDA for which establishment of one or more regulated areas is required shall inform other employers on the site of the nature of the employer's work with MDA and of the existence of, and requirements pertaining to, regulated areas. [§1926.60(d)]

(e) **Emergency situations.**[§1926.60(e)]

(1) *Written plan.*[§1926.60(e)(1)]

(i) *A written plan for emergency situations* shall be developed for each construction operation where there is a possibility of an emergency. The plan shall include procedures where the employer identifies emergency escape routes for his employees at each construction site before the construction operation begins. Appropriate portions of the plan shall be implemented in the event of an emergency.[§1926.60(e)(1)(i)]

(ii) *The plan shall specifically provide that employees* engaged in correcting emergency conditions shall be equipped with the appropriate personal protective equipment and clothing as required in paragraphs (i) and (j) of this section until the emergency is abated.[§1926.60(e)(1)(ii)]

(iii) *The plan shall specifically include provisions* for alerting and evacuating affected employees as well as the applicable elements prescribed in 29 CFR 1910.38 and 29 CFR 1910.39, "Emergency action plans" and "Fire prevention plans," respectively.[§1926.60(e)(1)(iii)]

(2) *Alerting employees.* Where there is the possibility of employee exposure to MDA due to an emergency, means shall be developed to promptly alert employees who have the potential to be directly exposed. Affected employees not engaged in correcting emergency conditions shall be evacuated immediately in the event that an emergency occurs. Means shall also be developed for alerting other employees who may be exposed as a result of the emergency.[§1926.60(e)(2)]

(f) **Exposure monitoring.**[§1926.60(f)]

(1) *General.*[§1926.60(f)(1)]

(i) *Determinations of employee exposure* shall be made from breathing zone air samples that are representative of each employee's exposure to airborne MDA over an eight (8) hour period. Determination of employee exposure to the STEL shall be made from breathing zone air samples collected over a 15 minute sampling period.[§1926.60(f)(1)(i)]

(ii) *Representative employee exposure* shall be determined on the basis of one or more samples representing full shift exposure for each shift for each job classification in each work area where exposure to MDA may occur.[§1926.60(f)(1)(ii)]

(iii) *Where the employer can document* that exposure levels are equivalent for similar operations in different work shifts, the employer shall only be required to determine representative employee exposure for that operation during one shift. [§1926.60(f)(1)(iii)]

(2) *Initial monitoring.* Each employer who has a workplace or work operation covered by this standard shall perform initial monitoring to determine accurately the airborne concentrations of MDA to which employees may be exposed unless:[§1926.60(f)(2)]

(i) *The employer can demonstrate,* on the basis of objective data, that the MDA-containing product or material being handled cannot cause exposures above the standard's action level, even under worst-case release conditions; or [§1926.60(f)(2)(i)]

(ii) *The employer has historical monitoring* or other data demonstrating that exposures on a particular job will be below the action level.[§1926.60(f)(2)(ii)]

(3) *Periodic monitoring and monitoring frequency.*[§1926.60(f)(3)]

(i) *If the monitoring* required by paragraph (f)(2) of this section reveals employee exposure at or above the action level, but at or below the PELs, the employer shall repeat such monitoring for each such employee at least every six (6) months. [§1926.60(f)(3)(i)]

(ii) *If the monitoring* required by paragraph (f)(2) of this section reveals employee exposure above the PELs, the employer shall repeat such monitoring for each such employee at least every three (3) months.[§1926.60(f)(3)(ii)]

(iii) *Employers who are conducting MDA operations* within a regulated area can forego periodic monitoring if the employees are all wearing supplied-air respirators while working in the regulated area.[§1926.60(f)(3)(iii)]

(iv) *The employer may alter the monitoring schedule* from every three months to every six months for any employee for whom two consecutive measurements taken at least 7 days apart indicate that the employee exposure has decreased to below the PELs but above the action level.[§1926.60(f)(3)(iv)]

(4) *Termination of monitoring.*[§1926.60(f)(4)]

(i) *If the initial monitoring* required by paragraph (f)(2) of this section reveals employee exposure to be below the action level, the employer may discontinue the monitoring for that employee, except as otherwise required by paragraph (f)(5) of this section.[§1926.60(f)(4)(i)]

(ii) *If the periodic monitoring* required by paragraph (f)(3) of this section reveals that employee exposures, as indicated by at least two consecutive measurements taken at least 7 days apart, are below the action level the employer may discontinue the monitoring for that employee, except as otherwise required by paragraph (f)(5) of this section.[§1926.60(f)(4)(ii)]

(5) *Additional monitoring.* The employer shall institute the exposure monitoring required under paragraphs (f)(2) and (f)(3) of this section when there has been a change in production process, chemicals present, control equipment, personnel, or work practices which may result in new or additional exposures to MDA,

or when the employer has any reason to suspect a change which may result in new or additional exposures.[§1926.60(f)(5)]

(6) *Accuracy of monitoring.* Monitoring shall be accurate, to a confidence level of 95 percent, to within plus or minus 25 percent for airborne concentrations of MDA.[§1926.60(f)(6)]

(7) *Employee notification of monitoring results.*[§1926.60(f)(7)]

(i) *The employer must,* as soon as possible but no later than 5 working days after the receipt of the results of any monitoring performed under this section, notify each affected employee of these results either individually in writing or by posting the results in an appropriate location that is accessible to employees.[§1926.60(f)(7)(i)]

(ii) *The written notification* required by paragraph (f)(7)(i) of this section shall contain the corrective action being taken by the employer or any other protective measures which have been implemented to reduce the employee exposure to or below the PELs, wherever the PELs are exceeded.[§1926.60(f)(7)(ii)]

(8) *Visual monitoring.* The employer shall make routine inspections of employee hands, face and forearms potentially exposed to MDA. Other potential dermal exposures reported by the employee must be referred to the appropriate medical personnel for observation. If the employer determines that the employee has been exposed to MDA the employer shall: [§1926.60(f)(8)]

(i) *Determine the source of exposure;*[§1926.60(f)(8)(i)]

(ii) *Implement protective measures to correct the hazard; and* [§1926.60(f)(8)(ii)]

(iii) *Maintain records* of the corrective actions in accordance with paragraph (o) of this section.[§1926.60(f)(8)(iii)]

(g) Regulated areas.[§1926.60(g)]

(1) *Establishment*[§1926.60(g)(1)]

(i) *Airborne exposures.* The employer shall establish regulated areas where airborne concentrations of MDA exceed or can reasonably be expected to exceed, the permissible exposure limits.[§1926.60(g)(1)(i)]

(ii) *Dermal exposures.* Where employees are subject to "dermal exposure to MDA" the employer shall establish those work areas as regulated areas.[§1926.60(g)(1)(ii)]

(2) *Demarcation.* Regulated areas shall be demarcated from the rest of the workplace in a manner that minimizes the number of persons potentially exposed.[§1926.60(g)(2)]

(3) *Access.* Access to regulated areas shall be limited to authorized persons.[§1926.60(g)(3)]

(4) *Personal protective equipment and clothing.* Each person entering a regulated area shall be supplied with, and required to use, the appropriate personal protective clothing and equipment in accordance with paragraphs (i) and (j) of this section. [§1926.60(g)(4)]

(5) *Prohibited activities.* The employer shall ensure that employees do not eat, drink, smoke, chew tobacco or gum, or apply cosmetics in regulated areas.[§1926.60(g)(5)]

(h) Methods of compliance.[§1926.60(h)]

(1) *Engineering controls and work practices and respirators.* [§1926.60(h)(1)]

(i) *The employer shall use* one or any combination of the following control methods to achieve compliance with the permissible exposure limits prescribed by paragraph (c) of this section:[§1926.60(h)(1)(i)]

[A] Local exhaust ventilation equipped with HEPA filter dust collection systems;[§1926.60(h)(1)(i)[A]]

[B] General ventilation systems;[§1926.60(h)(1)(i)[B]]

[C] Use of workpractices; or[§1926.60(h)(1)(i)[C]]

[D] Other engineering controls such as isolation and enclosure that the Assistant Secretary can show to be feasible. [§1926.60(h)(1)(i)[D]]

(ii) *Wherever the feasible engineering controls* and work practices "which can be instituted are not sufficient to reduce employee exposure to or below the PELs, the employer shall use them to reduce employee exposure to the lowest levels achievable by these controls and shall supplement them by the use of respiratory protective devices which comply with the requirements of paragraph (i) of this section. [§1926.60(h)(1)(ii)]

(2) *Special Provisions.* For workers engaged in spray application methods, respiratory protection must be used in addition to feasible engineering controls and work practices to reduce employee exposure to or below the PELs.[§1926.60(h)(2)]

(3) *Prohibitions.* Compressed air shall not be used to remove MDA, unless the compressed air is used in conjunction with an enclosed ventilation system designed to capture the dust cloud created by the compressed air.[§1926.60(h)(3)]

(4) *Employee rotation.* The employer shall not use employee rotation as a means of compliance with the exposure limits prescribed in paragraph (c) of this section.[§1926.60(h)(4)]

(5) *Compliance program.*[§1926.60(h)(5)]

(i) *The employer shall establish and implement* a written program to reduce employee exposure to or below the PELs by means of engineering and work practice controls, as required by paragraph (h)(1) of this section, and by use of respiratory protection where permitted under this section.[§1926.60(h)(5)(i)]

(ii) *Upon request this written program* shall be furnished for examination and copying to the Assistant Secretary, the Director, affected employees and designated employee representatives. The employer shall review and, as necessary, update such plans at least once every 12 months to make certain they reflect the current status of the program. [§1926.60(h)(5)(ii)]

(i) Respiratory protection.[§1926.60(i)]

(1) *General.* For employees who use respirators required by this section, the employer must provide each employee an appropriate respirator that complies with the requirements of this paragraph. Respirators must be used during:[§1926.60(i)(1)]

(i) *Periods necessary to install or implement* feasible engineering and work-practice controls.[§1926.60(i)(1)(i)]

(ii) *Work operations,* such as maintenance and repair activities and spray-application processes, for which engineering and work-practice controls are not feasible.[§1926.60(i)(1)(ii)]

(iii) *Work operations for which feasible engineering* and work-practice controls are not yet sufficient to reduce employee exposure to or below the PELs.[§1926.60(i)(1)(iii)]

(iv) *Emergencies.*[§1926.60(i)(1)(iv)]

(2) *Respirator program.* The employer must implement a respiratory protection program in accordance with Sec. 1910.134 (b) through (d) (except (d)(1)(iii)), and (f) through (m), which covers each employee required by this section to use a respirator. [§1926.60(i)(2)]

(3) *Respirator selection.*[§1926.60(i)(3)]

(i) *Employers must:*[§1926.60(i)(3)(i)]

[A] Select, and provide to employees, the appropriate respirators specified in paragraph (d)(3)(i)(A) of 29 CFR 1910.134.[§1926.60(i)(3)(i)[A]]

[B] Provide HEPA filters for powered and non-powered air-purifying respirators.[§1926.60(i)(3)(i)[B]]

[C] For escape, provide employees with one of the following respirator options: Any self-contained breathing apparatus with a full facepiece or hood operated in the positive-pressure or continuous-flow mode; or a full facepiece air-purifying respirator.[§1926.60(i)(3)(i)[C]]

[D] Provide a combination HEPA filter and organic vapor canister or cartridge with air-purifying respirators when MDA is in liquid form or used as part of a process requiring heat.[§1926.60(i)(3)(i)[D]]

(ii) *An employee* who cannot use a negative-pressure respirator must be given the option of using a positive-pressure respirator, or a supplied-air respirator operated in the continuous-flow or pressure- demand mode.[§1926.60(i)(3)(ii)]

(j) Protective work clothing and equipment.[§1926.60(j)]

(1) *Provision and use.* Where employees are subject to dermal exposure to MDA, where liquids containing MDA can be splashed into the eyes, or where airborne concentrations of MDA are in excess of the PEL, the employer shall provide, at no cost to the employee, and ensure that the employee uses, appropriate protective work clothing and equipment which prevent contact with MDA such as, but not limited to:[§1926.60(j)(1)]

(i) *Aprons, coveralls or other full-body work clothing;* [§1926.60(j)(1)(i)]

(ii) *Gloves, head coverings, and foot coverings; and* [§1926.60(j)(1)(ii)]

(iii) *Face shields, chemical goggles; or*[§1926.60(j)(1)(iii)]

(iv) *Other appropriate protective equipment* which comply with 29 CFR 1910.133.[§1926.60(j)(1)(iv)]

(2) *Removal and storage.*[§1926.60(j)(2)]

(i) *The employer shall ensure that,* at the end of their work shift, employees remove MDA-contaminated protective work clothing and equipment that is not routinely removed throughout the day in change areas provided in accordance with the provisions in paragraph (k) of this section.[§1926.60(j)(2)(i)]

(ii) *The employer shall ensure that,* during their work shift, employees remove all other MDA-contaminated protective work clothing or equipment before leaving a regulated area. [§1926.60(j)(2)(ii)]

(iii) *The employer shall ensure that no employee* takes MDA-contaminated work clothing or equipment out of the decontamination areas, except those employees authorized to do so for the purpose of laundering, maintenance, or disposal. [§1926.60(j)(2)(iii)]

(iv) *MDA-contaminated work clothing or equipment* shall be placed and stored and transported in sealed, impermeable bags, or other closed impermeable containers. [§1926.60(j)(2)(iv)]

(v) *Containers of MDA-contaminated protective work clothing* or equipment which are to be taken out of decontamination areas or the workplace for cleaning, maintenance, or disposal, shall bear labels warning of the hazards of MDA. [§1926.60(j)(2)(v)]

(3) *Cleaning and replacement.*[§1926.60(j)(3)]

(i) *The employer shall provide the employee* with clean protective clothing and equipment. The employer shall ensure that protective work clothing or equipment required by this paragraph is cleaned, laundered, repaired, or replaced at intervals appropriate to maintain its effectiveness.[§1926.60(j)(3)(i)]

(ii) *The employer shall prohibit the removal* of MDA from protective work clothing or equipment by blowing, shaking, or any methods which allow MDA to re-enter the workplace. [§1926.60(j)(3)(ii)]

(iii) *The employer shall ensure that laundering* of MDA-contaminated clothing shall be done so as to prevent the release of MDA in the workplace.[§1926.60(j)(3)(iii)]

(iv) *Any employer who gives MDA-contaminated clothing* to another person for laundering shall inform such person of the requirement to prevent the release of MDA.[§1926.60(j)(3)(iv)]

(v) *The employer shall inform any person* who launders or cleans protective clothing or equipment contaminated with MDA of the potentially harmful effects of exposure. [§1926.60(j)(3)(v)]

(4) *Visual Examination.*[§1926.60(j)(4)]

(i) *The employer shall ensure* that employees' work clothing is examined periodically for rips or tears that may occur during performance of work.[§1926.60(j)(4)(i)]

(ii) *When rips or tears are detected,* the protective equipment or clothing shall be repaired and replaced immediately. [§1926.60(j)(4)(ii)]

(k) Hygiene facilities and practices.[§1926.60(k)]

(1) *General.*[§1926.60(k)(1)]

(i) *The employer shall provide decontamination areas* for employees required to work in regulated areas or required by paragraph (j)(1) of this section to wear protective clothing. Exception: In lieu of the decontamination area requirement specified in paragraph (k)(1)(i) of this section, the employer may permit employees engaged in small scale, short duration operations, to clean their protective clothing or dispose of the protective clothing before such employees leave the area where the work was performed.[§1926.60(k)(1)(i)]

(ii) *Change areas.* The employer shall ensure that change areas are equipped with separate storage facilities for protective clothing and street clothing, in accordance with 29 CFR 1910.141(e).[§1926.60(k)(1)(ii)]

(iii) *Equipment area.* The equipment area shall be supplied with impermeable, labeled bags and containers for the containment and disposal of contaminated protective clothing and equipment.[§1926.60(k)(1)(iii)]

(2) *Shower area.*[§1926.60(k)(2)]

(i) *Where feasible,* shower facilities shall be provided which comply with 29 CFR 1910.141(d)(3) wherever the possibility of employee exposure to airborne levels of MDA in excess of the permissible exposure limit exists.[§1926.60(k)(2)(i)]

(ii) *Where dermal exposure to MDA occurs,* the employer shall ensure that materials spilled or deposited on the skin are removed as soon as possible by methods which do not facilitate the dermal absorption of MDA.[§1926.60(k)(2)(ii)]

(3) *Lunch Areas.*[§1926.60(k)(3)]

(i) *Whenever food or beverages are consumed* at the worksite and employees are exposed to MDA the employer shall provide clean lunch areas were MDA levels are below the action level and where no dermal exposure to MDA can occur. [§1926.60(k)(3)(i)]

(ii) *The employer shall ensure that employees* wash their hands and faces with soap and water prior to eating, drinking, smoking, or applying cosmetics.[§1926.60(k)(3)(ii)]

(iii) *The employer shall ensure that employees* do not enter lunch facilities with contaminated protective work clothing or equipment.[§1926.60(k)(3)(iii)]

(l) Communication of hazards to employees.[§1926.60(l)]

(1) *Hazard communication.* The employer shall include Methylenedianiline (MDA) in the program established to comply with the Hazard Communication Standard (HCS) (Sec. 1910.1200). The employer shall ensure that each employee has access to labels on containers of MDA and safety data sheets, and is trained in accordance with the provisions of HCS and paragraph (l)(3) of this section. The employer shall ensure that at least the following hazards are addressed: Cancer; liver effects; and skin sensitization.[§1926.60(l)(1)]

(2) *Signs and labels* — [§1926.60(l)(2)]

(i) *Signs.*[§1926.60(l)(2)(i)]

[A] The employer shall post and maintain legible signs demarcating regulated areas and entrances or accessways to regulated areas that bear the following legend: [§1926.60(l)(2)(i)[A]]

DANGER
MDA
MAY CAUSE CANCER
CAUSES DAMAGE TO THE LIVER
RESPIRATORY PROTECTION AND PROTECTIVE CLOTHING MAY BE REQUIRED IN THIS AREA
AUTHORIZED PERSONNEL ONLY

[B] Prior to June 1, 2016, employers may use the following legend in lieu of that specified in paragraph (l)(2)(i)(A) of this section:[§1926.60(l)(2)(i)[B]]

DANGER
MDA
MAY CAUSE CANCER
LIVER TOXIN
AUTHORIZED PERSONNEL ONLY
RESPIRATORS AND PROTECTIVE CLOTHING MAY BE REQUIRED TO BE WORN IN THIS AREA

(ii) *Labels.*[§1926.60(l)(2)(ii)]

[A] The employer shall ensure that labels or other appropriate forms of warning are provided for containers of MDA within the workplace. The labels shall comply with the requirements of Sec. 1910.1200(f) and shall include at least the following information for pure MDA and mixtures containing MDA:[§1926.60(l)(2)(ii)[A]]

DANGER
CONTAINS MDA
MAY CAUSE CANCER
CAUSES DAMAGE TO THE LIVER

[B] Prior to June 1, 2015, employers may include the following information workplace labels in lieu of the labeling requirements in paragraph (l)(2)(ii)(A) of this section: [§1926.60(l)(2)(ii)[B]]

[1] For Pure MDA:[§1926.60(l)(2)(ii)[B][1]]

DANGER
CONTAINS MDA
MAY CAUSE CANCER
LIVER TOXIN

[2] For mixtures containing MDA:[§1926.60(l)(2)(ii)[B][2]]

DANGER
CONTAINS MDA
CONTAINS MATERIALS WHICH MAY CAUSE CANCER
LIVER TOXIN

(3) *Information and training.*[§1926.60(l)(3)]

(i) *The employer shall provide employees* with information and training on MDA, in accordance with 29 CFR 1910.1200(h), at the time of initial assignment and at least annually thereafter.[§1926.60(l)(3)(i)]

(ii) *In addition to the information required* under 29 CFR 1910.1200, the employer shall:[§1926.60(l)(3)(ii)]

[A] *Provide an explanation of the contents* of this section, including appendices A and B of this section, and indicate to employees where a copy of the standard is available; [§1926.60(l)(3)(ii)[A]]

[B] *Describe the medical surveillance* program required under paragraph (n) of this section, and explain the information contained in appendix C of this section; and [§1926.60(l)(3)(ii)[B]]

[C] *Describe the medical removal* provision required under paragraph (n) of this section.[§1926.60(l)(3)(ii)[C]]

(4) *Access to training materials.*[§1926.60(l)(4)]

(i) *The employer shall make readily available* to all affected employees, without cost, all written materials relating to the employee training program, including a copy of this regulation.[§1926.60(l)(4)(i)]

(ii) *The employer shall provide* to the Assistant Secretary and the Director, upon request, all information and training materials relating to the employee information and training program.[§1926.60(l)(4)(ii)]

(m) **Housekeeping.**[§1926.60(m)]

(1) *All surfaces shall be maintained* as free as practicable of visible accumulations of MDA.[§1926.60(m)(1)]

(2) *The employer shall institute a program* for detecting MDA leaks, spills, and discharges, including regular visual inspections of operations involving liquid or solid MDA.[§1926.60(m)(2)]

(3) *All leaks shall be repaired* and liquid or dust spills cleaned up promptly.[§1926.60(m)(3)]

(4) *Surfaces contaminated with MDA* may not be cleaned by the use of compressed air.[§1926.60(m)(4)]

(5) *Shoveling*, dry sweeping, and other methods of dry clean-up of MDA may be used where HEPA filtered vacuuming and/or wet cleaning are not feasible or practical.[§1926.60(m)(5)]

(6) *Waste*, scrap, debris, bags, containers, equipment, and clothing contaminated with MDA shall be collected and disposed of in a manner to prevent the re-entry of MDA into the workplace. [§1926.60(m)(6)]

(n) **Medical surveillance.**[§1926.60(n)]

(1) *General.*[§1926.60(n)(1)]

(i) *The employer shall make available* a medical surveillance program for employees exposed to MDA under the following circumstances:[§1926.60(n)(1)(i)]

[A] *Employees exposed at or above* the action level for 30 or more days per year;[§1926.60(n)(1)(i)[A]]

[B] *Employees who are subject* to dermal exposure to MDA for 15 or more days per year;[§1926.60(n)(1)(i)[B]]

[C] *Employees who have been* exposed in an emergency situation;[§1926.60(n)(1)(i)[C]]

[D] *Employees whom the employer,* based on results from compliance with paragraph (f)(8) of this section, has reason to believe are being dermally exposed; and [§1926.60(n)(1)(i)[D]]

[E] *Employees who show signs or symptoms of MDA exposure.*[§1926.60(n)(1)(i)[E]]

(ii) *The employer shall ensure* that all medical examinations and procedures are performed by or under the supervision of a licensed physician at a reasonable time and place, and provided without cost to the employee.[§1926.60(n)(1)(ii)]

(2) *Initial examinations.*[§1926.60(n)(2)]

(i) *Within 150 days of the effective date of this standard,* or before the time of initial assignment, the employer shall provide each employee covered by paragraph (n)(1)(i) of this section with a medical examination including the following elements:[§1926.60(n)(2)(i)]

[A] *A detailed history which includes:*[§1926.60(n)(2)(i)[A]]

[1] *Past work exposure to MDA or any other toxic substances;*[§1926.60(n)(2)(i)[A][1]]

[2] *A history of drugs,* alcohol, tobacco, and medication routinely taken (duration and quantity); and [§1926.60(n)(2)(i)[A][2]]

[3] *A history of dermatitis,* chemical skin sensitization, or previous hepatic disease.[§1926.60(n)(2)(i)[A][3]]

[B] *A physical examination which* includes all routine physical examination parameters, skin examination, and examination for signs of liver disease.[§1926.60(n)(2)(i)[B]]

[C] *Laboratory tests including:*[§1926.60(n)(2)(i)[C]]

[1] *Liver function tests and*[§1926.60(n)(2)(i)[C][1]]

[2] *Urinalysis.*[§1926.60(n)(2)(i)[C][2]]

[D] *Additional tests as necessary in the opinion of the physician.*[§1926.60(n)(2)(i)[D]]

(ii) *No initial medical examination* is required if adequate records show that the employee has been examined in accordance with the requirements of this section within the previous six months prior to the effective date of this standard or prior to the date of initial assignment.[§1926.60(n)(2)(ii)]

(3) *Periodic examinations.*[§1926.60(n)(3)]

(i) *The employer shall provide each employee* covered by this section with a medical examination at least annually following the initial examination. These periodic examinations shall include at least the following elements:[§1926.60(n)(3)(i)]

[A] *A brief history* regarding any new exposure to potential liver toxins, changes in drug, tobacco, and alcohol intake, and the appearance of physical signs relating to the liver, and the skin;[§1926.60(n)(3)(i)[A]]

[B] *The appropriate tests* and examinations including liver function tests and skin examinations; and[§1926.60(n)(3)(i)[B]]

[C] *Appropriate additional tests* or examinations as deemed necessary by the physician.[§1926.60(n)(3)(i)[C]]

(ii) *If in the physician's opinion* the results of liver function tests indicate an abnormality, the employee shall be removed from further MDA exposure in accordance with paragraph (n)(9) of this section. Repeat liver function tests shall be conducted on advice of the physician.[§1926.60(n)(3)(ii)]

(4) *Emergency examinations.* If the employer determines that the employee has been exposed to a potentially hazardous amount of MDA in an emergency situation under paragraph (e) of this section, the employer shall provide medical examinations in accordance with paragraphs (n)(3) (i) and (ii) of this section. If the results of liver function testing indicate an abnormality, the employee shall be removed in accordance with paragraph (n)(9) of this section. Repeat liver function tests shall be conducted on the advice of the physician. If the results of the tests are normal, tests must be repeated two to three weeks from the initial testing. If the results of the second set of tests are normal and on the advice of the physician, no additional testing is required.[§1926.60(n)(4)]

(5) *Additional examinations.* Where the employee develops signs and symptoms associated with exposure to MDA, the employer shall provide the employee with an additional medical examination including liver function tests. Repeat liver function tests shall be conducted on the advice of the physician. If the results of the tests are normal, tests must be repeated two to three weeks from the initial testing. If the results of the second set of tests are normal and on the advice of the physician, no additional testing is required.[§1926.60(n)(5)]

(6) *Multiple physician review mechanism.*[§1926.60(n)(6)]

(i) *If the employer selects* the initial physician who conducts any medical examination or consultation provided to an employee under this section, and the employee has signs or symptoms of occupational exposure to MDA (which could include an abnormal liver function test), and the employee disagrees with the opinion of the examining physician, and this opinion could affect the employee's job status, the employee may designate an appropriate and mutually acceptable second physician:[§1926.60(n)(6)(i)]

[A] *To review any findings,* determinations or recommendations of the initial physician; and[§1926.60(n)(6)(i)[A]]

[B] *To conduct such examinations,* consultations, and laboratory tests as the second physician deems necessary to facilitate this review.[§1926.60(n)(6)(i)[B]]

(ii) *The employer shall promptly notify an employee* of the right to seek a second medical opinion after each occasion that an initial physician conducts a medical examination or consultation pursuant to this section. The employer may condition its participation in, and payment for, the multiple physician review mechanism upon the employee doing the following within fifteen (15) days after receipt of the foregoing notification, or receipt of the initial physician's written opinion, whichever is later:[§1926.60(n)(6)(ii)]

[A] *The employee informing the employer* that he or she intends to seek a second medical opinion, and [§1926.60(n)(6)(ii)[A]]

[B] *The employee initiating steps* to make an appointment with a second physician.[§1926.60(n)(6)(ii)[B]]

(iii) *If the findings, determinations, or recommendations* of the second physician differ from those of the initial physician, then the employer and the employee shall assure that efforts are made for the two physicians to resolve any disagreement.[§1926.60(n)(6)(iii)]

(iv) *If the two physicians* have been unable to quickly resolve their disagreement, then the employer and the employee through their respective physicians shall designate a third physician:[§1926.60(n)(6)(iv)]

[A] *To review any findings,* determinations, or recommendations of the prior physicians; and[§1926.60(n)(6)(iv)[A]]

[B] *To conduct such examinations,* consultations, laboratory tests, and discussions with the prior physicians as the third physician deems necessary to resolve the disagreement of the prior physicians.[§1926.60(n)(6)(iv)[B]]

(v) *The employer shall act consistent with the findings,* determinations, and recommendations of the second physician, unless the employer and the employee reach a mutually acceptable agreement.[§1926.60(n)(6)(v)]

(7) *Information provided to the examining physician.*[§1926.60(n)(7)]

(i) *The employer shall provide* the following information to the examining physician:[§1926.60(n)(7)(i)]

[A] *A copy of this regulation and its appendices;* [§1926.60(n)(7)(i)[A]]

[B] *A description* of the affected employee's duties as they relate to the employee's potential exposure to MDA; [§1926.60(n)(7)(i)[B]]

[C] *The employee's* current actual or representative MDA exposure level;[§1926.60(n)(7)(i)[C]]

[D] *A description* of any personal protective equipment used or to be used; and[§1926.60(n)(7)(i)[D]]

[E] *Information from previous employment* related medical examinations of the affected employee.[§1926.60(n)(7)(i)[E]]

(ii) *The employer shall provide* the foregoing information to a second physician under this section upon request either by the second physician, or by the employee.[§1926.60(n)(7)(ii)]

(8) *Physician's written opinion.*[§1926.60(n)(8)]

(i) *For each examination under this section,* the employer shall obtain, and provide the employee with a copy of, the examining physician's written opinion within 15 days of its receipt. The written opinion shall include the following:[§1926.60(n)(8)(i)]

[A] *The occupationally pertinent results* of the medical examination and tests;[§1926.60(n)(8)(i)[A]]

[B] *The physician's opinion* concerning whether the employee has any detected medical conditions which would place the employee at increased risk of material impairment of health from exposure to MDA;[§1926.60(n)(8)(i)[B]]

[C] *The physician's recommended limitations* upon the employee's exposure to MDA or upon the employee's use of protective clothing or equipment and respirators; and [§1926.60(n)(8)(i)[C]]

[D] *A statement that the employee* has been informed by the physician of the results of the medical examination and any medical conditions resulting from MDA exposure which require further explanation or treatment. [§1926.60(n)(8)(i)[D]]

(ii) *The written opinion obtained* by the employer shall not reveal specific findings or diagnoses unrelated to occupational exposures.[§1926.60(n)(8)(ii)]

(9) *Medical removal*[§1926.60(n)(9)]

(i) *Temporary medical removal of an employee*[§1926.60(n)(9)(i)]

[A] *Temporary removal resulting* from occupational exposure. The employee shall be removed from work environments in which exposure to MDA is at or above the action level or where dermal exposure to MDA may occur, following an initial examination (paragraph (n)(2) of this section), periodic examinations (paragraph (n)(3) of this section), an emergency situation (paragraph (n)(4) of this section), or an additional examination (paragraph (n)(5) of this section) in the following circumstances:[§1926.60(n)(9)(i)[A]]

[1] *When the employee exhibits signs* and/or symptoms indicative of acute exposure to MDA; or [§1926.60(n)(9)(i)[A][1]]

[2] *When the examining physician determines* that an employee's abnormal liver function tests are not associated with MDA exposure but that the abnormalities may be exacerbated as a result of occupational exposure to MDA. [§1926.60(n)(9)(i)[A][2]]

[B] *Temporary removal due to a final medical determination.* [§1926.60(n)(9)(i)[B]]

[1] *The employer shall remove an employee* from work having an exposure to MDA at or above the action level or where the potential for dermal exposure exists on each occasion that a final medical determination results in a medical finding, determination, or opinion that the employee has a detected medical condition which places the employee at increased risk of material impairment to health from exposure to MDA. [§1926.60(n)(9)(i)[B][1]]

[2] *For the purposes of this section,* the phrase "**final medical determination**" shall mean the outcome of the physician review mechanism used pursuant to the medical surveillance provisions of this section. [§1926.60(n)(9)(i)[B][2]]

[3] *Where a final medical determination* results in any recommended special protective measures for an employee, or limitations on an employee's exposure to MDA, the employer shall implement and act consistent with the recommendation.[§1926.60(n)(9)(i)[B][3]]

(ii) *Return of the employee to former job status.*[§1926.60(n)(9)(ii)]

[A] *The employer shall return* an employee to his or her former job status:[§1926.60(n)(9)(ii)[A]]

[1] *When the employee no longer shows* signs or symptoms of exposure to MDA, or upon the advice of the physician.[§1926.60(n)(9)(ii)[A][1]]

[2] *When a subsequent final medical determination* results in a medical finding, determination, or opinion that the employee no longer has a detected medical condition which places the employee at increased risk of material impairment to health from exposure to MDA. [§1926.60(n)(9)(ii)[A][2]]

[B] *For the purposes of this section,* the requirement that an employer return an employee to his or her former job status is not intended to expand upon or restrict any rights an employee has or would have had, absent temporary medical removal, to a specific job classification or position under the terms of a collective bargaining agreement. [§1926.60(n)(9)(ii)[B]]

(iii) *Removal of other employee special protective measure or limitations.* The employer shall remove any limitations placed on an employee or end any special protective measures provided to an employee pursuant to a final medical determination when a subsequent final medical determination indicates that the limitations or special protective measures are no longer necessary.[§1926.60(n)(9)(iii)]

(iv) *Employer options pending a final medical determination.* Where the physician review mechanism used pursuant to the medical surveillance provisions of this section, has not yet resulted in a final medical determination with respect to an employee, the employer shall act as follows:[§1926.60(n)(9)(iv)]

[A] *Removal.* The employer may remove the employee from exposure to MDA, provide special protective measures to the employee, or place limitations upon the employee, consistent with the medical findings, determinations, or recommendations of the physician who has reviewed the employee's health status.[§1926.60(n)(9)(iv)[A]]

[B] *Return.* The employer may return the employee to his or her former job status, and end any special protective measures provided to the employee, consistent with the medical findings, determinations, or recommendations of any of the physicians who have reviewed the employee's health status, with two exceptions:[§1926.60(n)(9)(iv)[B]]

[1] *If the initial removal,* special protection, or limitation of the employee resulted from a final medical determination which differed from the findings, determinations, or recommendations of the initial physician; or [§1926.60(n)(9)(iv)[B][1]]

[2] *The employee has been on removal status* for the preceding six months as a result of exposure to MDA, then the employer shall await a final medical determination. [§1926.60(n)(9)(iv)[B][2]]

(v) *Medical removal protection benefits*[§1926.60(n)(9)(v)]

[A] *Provisions of medical removal protection benefits.* The employer shall provide to an employee up to six (6) months of medical removal protection benefits on each occasion that an employee is removed from exposure to MDA or otherwise limited pursuant to this section. [§1926.60(n)(9)(v)[A]]

[B] Definition of medical removal protection benefits. For the purposes of this section, the requirement that an employer provide medical removal protection benefits means that the employer shall maintain the earnings, seniority, and other employment rights and benefits of an employee as though the employee had not been removed from normal exposure to MDA or otherwise limited. [§1926.60(n)(9)(v)[B]]

[C] Follow-up medical surveillance during the period of employee removal or limitations. During the period of time that an employee is removed from normal exposure to MDA or otherwise limited, the employer may condition the provision of medical removal protection benefits upon the employee's participation in follow-up medical surveillance made available pursuant to this section. [§1926.60(n)(9)(v)[C]]

[D] Workers' compensation claims. If a removed employee files a claim for workers' compensation payments for a MDA-related disability, then the employer shall continue to provide medical removal protection benefits pending disposition of the claim. To the extent that an award is made to the employee for earnings lost during the period of removal, the employer's medical removal protection obligation shall be reduced by such amount. The employer shall receive no credit for workers' compensation payments received by the employee for treatment-related expenses. [§1926.60(n)(9)(v)[D]]

[E] Other credits. The employer's obligation to provide medical removal protection benefits to a removed employee shall be reduced to the extent that the employee receives compensation for earnings lost during the period of removal either from a publicly or employer-funded compensation program, or receives income from employment with any employer made possible by virtue of the employee's removal. [§1926.60(n)(9)(v)[E]]

[F] Employees who do not recover within the 6 months of removal. The employer shall take the following measures with respect to any employee removed from exposure to MDA: [§1926.60(n)(9)(v)[F]]

[1] The employer shall make available to the employee a medical examination pursuant to this section to obtain a final medical determination with respect to the employee; [§1926.60(n)(9)(v)[F][1]]

[2] The employer shall assure that the final medical determination obtained indicates whether or not the employee may be returned to his or her former job status, and, if not, what steps should be taken to protect the employee's health; [§1926.60(n)(9)(v)[F][2]]

[3] Where the final medical determination has not yet been obtained, or once obtained indicates that the employee may not yet be returned to his or her former job status, the employer shall continue to provide medical removal protection benefits to the employee until either the employee is returned to former job status, or a final medical determination is made that the employee is incapable of ever safely returning to his or her former job status; and [§1926.60(n)(9)(v)[F][3]]

[4] Where the employer acts pursuant to a final medical determination which permits the return of the employee to his or her former job status despite what would otherwise be an unacceptable liver function test, later questions concerning removing the employee again shall be decided by a final medical determination. The employer need not automatically remove such an employee pursuant to the MDA removal criteria provided by this section. [§1926.60(n)(9)(v)[F][4]]

(vi) Voluntary removal or restriction of an employee. Where an employer, although not required by this section to do so, removes an employee from exposure to MDA or otherwise places limitations on an employee due to the effects of MDA exposure on the employee's medical condition, the employer shall provide medical removal protection benefits to the employee equal to that required by paragraph (n)(9)(v) of this section. [§1926.60(n)(9)(vi)]

(o) Recordkeeping. [§1926.60(o)]

(1) Objective data for exempted operations. [§1926.60(o)(1)]

(i) Where the employer has relied on objective data that demonstrate that products made from or containing MDA are not capable of releasing MDA or do not present a dermal exposure problem under the expected conditions of processing, use, or handling to exempt such operations from the initial monitoring requirements under paragraph (f)(2) of this section, the employer shall establish and maintain an accurate record of objective data reasonably relied upon in support of the exemption. [§1926.60(o)(1)(i)]

(ii) The record shall include at least the following information: [§1926.60(o)(1)(ii)]

[A] The product qualifying for exemption; [§1926.60(o)(1)(ii)[A]]

[B] The source of the objective data; [§1926.60(o)(1)(ii)[B]]

[C] The testing protocol, results of testing, and/or analysis of the material for the release of MDA; [§1926.60(o)(1)(ii)[C]]

[D] A description of the operation exempted and how the data support the exemption; and [§1926.60(o)(1)(ii)[D]]

[E] Other data relevant to the operations, materials, processing, or employee exposures covered by the exemption. [§1926.60(o)(1)(ii)[E]]

(iii) The employer shall maintain this record for the duration of the employer's reliance upon such objective data. [§1926.60(o)(1)(iii)]

(2) Historical monitoring data. [§1926.60(o)(2)]

(i) Where the employer has relied on historical monitoring data that demonstrate that exposures on a particular job will be below the action level to exempt such operations from the initial monitoring requirements under paragraph (f)(2) of this section, the employer shall establish and maintain an accurate record of historical monitoring data reasonably relied upon in support of the exception. [§1926.60(o)(2)(i)]

(ii) The record shall include information that reflect the following conditions: [§1926.60(o)(2)(ii)]

[A] The data upon which judgments are based are scientifically sound and were collected using methods that are sufficiently accurate and precise; [§1926.60(o)(2)(ii)[A]]

[B] The processes and work practices that were in use when the historical monitoring data were obtained are essentially the same as those to be used during the job for which initial monitoring will not be performed; [§1926.60(o)(2)(ii)[B]]

[C] The characteristics of the MDA-containing material being handled when the historical monitoring data were obtained are the same as those on the job for which initial monitoring will not be performed; [§1926.60(o)(2)(ii)[C]]

[D] Environmental conditions prevailing when the historical monitoring data were obtained are the same as those on the job for which initial monitoring will not be performed; and [§1926.60(o)(2)(ii)[D]]

[E] Other data relevant to the operations, materials, processing, or employee exposures covered by the exception. [§1926.60(o)(2)(ii)[E]]

(iii) The employer shall maintain this record for the duration of the employer's reliance upon such historical monitoring data. [§1926.60(o)(2)(iii)]

(3) The employer may utilize the services of competent organizations such as industry trade associations and employee associations to maintain the records required by this section. [§1926.60(o)(3)]

(4) Exposure measurements. [§1926.60(o)(4)]

(i) The employer shall keep an accurate record of all measurements taken to monitor employee exposure to MDA. [§1926.60(o)(4)(i)]

(ii) This record shall include at least the following information: [§1926.60(o)(4)(ii)]

[A] The date of measurement; [§1926.60(o)(4)(ii)[A]]

[B] The operation involving exposure to MDA; [§1926.60(o)(4)(ii)[B]]

[C] Sampling and analytical methods used and evidence of their accuracy; [§1926.60(o)(4)(ii)[C]]

[D] Number, duration, and results of samples taken; [§1926.60(o)(4)(ii)[D]]

[E] Type of protective devices worn, if any; and [§1926.60(o)(4)(ii)[E]]

[F] Name, social security number, and exposure of the employees whose exposures are represented. [§1926.60(o)(4)(ii)[F]]

(iii) *The employer shall maintain this record* for at least thirty (30) years, in accordance with 29 CFR 1910.33.[§1926.60(o)(4)(iii)]

(5) *Medical surveillance.*[§1926.60(o)(5)]

(i) *The employer shall establish and maintain* an accurate record for each employee subject to medical surveillance by paragraph (n) of this section, in accordance with 29 CFR 1910.33.[§1926.60(o)(5)(i)]

(ii) *The record shall include at least the following information:*[§1926.60(o)(5)(ii)]

[A] The name and social security number of the employee;[§1926.60(o)(5)(ii)[A]]

[B] A copy of the employee's medical examination results, including the medical history, questionnaire responses, results of any tests, and physician's recommendations.[§1926.60(o)(5)(ii)[B]]

[C] Physician's written opinions;[§1926.60(o)(5)(ii)[C]]

[D] Any employee medical complaints related to exposure to MDA; and[§1926.60(o)(5)(ii)[D]]

[E] A copy of the information provided to the physician as required by paragraph (n) of this section.[§1926.60(o)(5)(ii)[E]]

(iii) *The employer shall ensure* that this record is maintained for the duration of employment plus thirty (30) years, in accordance with 29 CFR 1910.33.[§1926.60(o)(5)(iii)]

(iv) *A copy* of the employee's medical removal and return to work status.[§1926.60(o)(5)(iv)]

(6) *Training records.* The employer shall maintain all employee training records for one (1) year beyond the last date of employment.[§1926.60(o)(6)]

(7) *Availability.*[§1926.60(o)(7)]

(i) *The employer,* upon written request, shall make all records required to be maintained by this section available to the Assistant Secretary and the Director for examination and copying.[§1926.60(o)(7)(i)]

(ii) *The employer,* upon request, shall make any exposure records required by paragraphs (f) and (n) of this section available for examination and copying to affected employees, former employees, designated representatives, and the Assistant Secretary, in accordance with 29 CFR 1910.33(a)-(e) and (g)-(i).[§1926.60(o)(7)(ii)]

(iii) *The employer,* upon request, shall make employee medical records required by paragraphs (n) and (o) of this section available for examination and copying to the subject employee, anyone having the specific written consent of the subject employee, and the Assistant Secretary, in accordance with 29 CFR 1910.33.[§1926.60(o)(7)(iii)]

(8) *Transfer of records.* The employer shall comply with the requirements concerning transfer of records set forth in 29 CFR 1910.1020(h).[§1926.60(o)(8)]

(p) **Observation of monitoring.**[§1926.60(p)]

(1) *Employee observation.* The employer shall provide affected employees, or their designated representatives, an opportunity to observe the measuring or monitoring of employee exposure to MDA conducted pursuant to paragraph (f) of this section.[§1926.60(p)(1)]

(2) *Observation procedures.* When observation of the measuring or monitoring of employee exposure to MDA requires entry into areas where the use of protective clothing and equipment or respirators is required, the employer shall provide the observer with personal protective clothing and equipment or respirators required to be worn by employees working in the area, assure the use of such clothing and equipment or respirators, and require the observer to comply with all other applicable safety and health procedures.[§1926.60(p)(2)]

(q) **Appendices.** The information contained in appendices A, B, C, and D of this section is not intended, by itself, to create any additional obligations not otherwise imposed by this standard nor detract from any existing obligation.[§1926.60(q)]

§1926.60Appendix A

Substance Data Sheet, for 4-4' -Methylenedianiline

Note: The requirements applicable to construction work under this Appendix A are identical to those set forth in Appendix A to Sec. 1910.1050 of this chapter.

§1926.60 Appendix B

Substance Technical Guidelines, MDA

Note: The requirements applicable to construction work under this Appendix B are identical to those set forth in Appendix B to Sec. 1910.1050 of this chapter.

§1926.60 Appendix C

Medical Surveillance Guidelines for MDA

Note: The requirements applicable to construction work under this Appendix C are identical to those set forth in Appendix C to Sec. 1910.1050 of this chapter.

§1926.60 Appendix D

Sampling and analytical methods for MDA monitoring and measurement procedures

Note: The requirements applicable to construction work under this Appendix D are identical to those set forth in Appendix D to Sec. 1910.1050 of this chapter. (Editor's Note: See 1910 supplement.)

[57 FR 35681, Aug. 10, 1992, as amended at 57 FR 49649, Nov. 3, 1992; 61 FR 5510, Feb. 13, 1996; 61 FR 31431, June 20, 1996; 63 FR 1296, Jan. 8, 1998; 69 FR 70373, Dec. 6, 2004; 70 FR 1143, Jan. 5, 2005; 71 FR 16674, Apr. 3, 2006; 71 FR 50191, Aug. 24, 2006; 73 FR 75588, Dec. 12, 2008; 76 FR 33611, June 8, 2011; 77 FR 17889, Mar. 26, 2012]

§1926.61

Retention of DOT markings, placards and labels

Note: The requirements applicable to construction work under this section are identical to those set forth at Sec. 1910.1201 of this chapter.

[61 FR 31432, June 20, 1996]

§1926.62

Lead

(a) **Scope.** This section applies to all construction work where an employee may be occupationally exposed to lead. All construction work excluded from coverage in the general industry standard for lead by 29 CFR 1910.1025(a)(2) is covered by this standard. Construction work is defined as work for construction, alteration and/or repair, including painting and decorating. It includes but is not limited to the following:[§1926.62(a)]

(1) *Demolition or salvage of structures* where lead or materials containing lead are present;[§1926.62(a)(1)]

(2) *Removal or encapsulation* of materials containing lead;[§1926.62(a)(2)]

(3) *New construction, alteration, repair, or renovation* of structures, substrates, or portions thereof, that contain lead, or materials containing lead;[§1926.62(a)(3)]

(4) *Installation of products containing lead;*[§1926.62(a)(4)]

(5) *Lead contamination/emergency cleanup;*[§1926.62(a)(5)]

(6) *Transportation, disposal, storage, or containment* of lead or materials containing lead on the site or location at which construction activities are performed, and[§1926.62(a)(6)]

(7) *Maintenance operations* associated with the construction activities described in this paragraph.[§1926.62(a)(7)]

(b) **Definitions.**[§1926.62(b)]

Action level means employee exposure, without regard to the use of respirators, to an airborne concentration of lead of 30 micrograms per cubic meter of air (30 μg/m^3) calculated as an 8-hour time-weighted average (TWA).[§1926.62(b)]

Assistant Secretary means the Assistant Secretary of Labor for Occupational Safety and Health, U.S. Department of Labor, or designee.[§1926.62(b)]

Competent person means one who is capable of identifying existing and predictable lead hazards in the surroundings or working conditions and who has authorization to take prompt corrective measures to eliminate them.[§1926.62(b)]

Director means the Director, National Institute for Occupational Safety and Health (NIOSH), U.S. Department of Health and Human Services, or designee.[§1926.62(b)]

Lead means metallic lead, all inorganic lead compounds, and organic lead soaps. Excluded from this definition are all other organic lead compounds.[§1926.62(b)]

This section means this standard.[§1926.62(b)]

(c) **Permissible exposure limit.**[§1926.62(c)]

(1) *The employer shall assure that no employee is exposed to lead* at concentrations greater than fifty micrograms per cubic meter of air (50 μg/m^3) averaged over an 8- hour period.[§1926.62(c)(1)]

(2) *If an employee is exposed to lead for more than 8 hours* in any work day the employees' allowable exposure, as a time weighted average (TWA) for that day, shall be reduced according to the following formula:[§1926.62(c)(2)]

Allowable employee exposure (in μg/m^3) = 400 divided by hours worked in the day.[§1926.62(c)(2)]

(3) *When respirators are used to limit employee exposure* as required under paragraph (c) of this section and all the requirements of paragraphs (e)(1) and (f) of this section have been met, employee exposure may be considered to be at the level

D

Occupational Health and Environmental Controls

provided by the protection factor of the respirator for those periods the respirator is worn. Those periods may be averaged with exposure levels during periods when respirators are not worn to determine the employee's daily TWA exposure.[§1926.62(c)(3)]

(d) Exposure assessment[§1926.62(d)]

(1) *General.*[§1926.62(d)(1)]

(i) *Each employer who has a workplace* or operation covered by this standard shall initially determine if any employee may be exposed to lead at or above the action level. [§1926.62(d)(1)(i)]

(ii) *For the purposes of paragraph (d) of this section,* employee exposure is that exposure which would occur if the employee were not using a respirator.[§1926.62(d)(1)(ii)]

(iii) *With the exception of monitoring under paragraph (d)(3),* where monitoring is required under this section, the employer shall collect personal samples representative of a full shift including at least one sample for each job classification in each work area either for each shift or for the shift with the highest exposure level.[§1926.62(d)(1)(iii)]

(iv) *Full shift personal* samples shall be representative of the monitored employee's regular, daily exposure to lead. [§1926.62(d)(1)(iv)]

(2) *Protection of employees* during assessment of exposure. [§1926.62(d)(2)]

(i) *With respect to the lead related tasks* listed in paragraph (d)(2)(i) of this section, where lead is present, until the employer performs an employee exposure assessment as required in paragraph (d) of this section and documents that the employee performing any of the listed tasks is not exposed above the PEL, the employer shall treat the employee as if the employee were exposed above the PEL, and not in excess of ten (10) times the PEL, and shall implement employee protective measures prescribed in paragraph (d)(2)(v) of this section. The tasks covered by this requirement are:[§1926.62(d)(2)(i)]

[A] Where lead containing coatings or paint are present: Manual demolition of structures (e.g, dry wall), manual scraping, manual sanding, heat gun applications, and power tool cleaning with dust collection systems; [§1926.62(d)(2)(i)[A]]

[B] Spray painting with lead paint.[§1926.62(d)(2)(i)[B]]

(ii) *In addition, with regard to tasks not listed in paragraph (d)(2)(i),* where the employee has any reason to believe that an employee performing the task may be exposed to lead in excess of the PEL, until the employer performs an employee exposure assessment as required by paragraph (d) of this section and documents that the employee's lead exposure is not above the PEL the employer shall treat the employee as if the employee were exposed above the PEL and shall implememt employee protective measures as prescribed in paragraph (d)(2)(v) of this section.[§1926.62(d)(2)(ii)]

(iii) *With respect to the tasks listed in paragraph (d)(2)(iii) of this section,* where lead is present, until the employer performs an employee exposure assessment as required in paragraph (d) of this section, and documents that the employee performing any of the listed tasks is not exposed in excess of 500 μg/m^3 , the employer shall treat the employee as if the employee were exposed to lead in excess of 500 μg/m^3 and shall implement employee protective measures as prescribed in paragraph (d)(2)(v) of this section. Where the employer does establish that the employee is exposed to levels of lead below 500 μg/m^3 , the employer may provide the exposed employee with the appropriate respirator prescribed for such use at such lower exposures, in accordance with Table 1 of this section. The tasks covered by this requirement are:[§1926.62(d)(2)(iii)]

[A] Using lead containing mortar; lead burning [§1926.62(d)(2)(iii)[A]]

[B] Where lead containing coatings or paint are present: rivet busting; power tool cleaning without dust collection systems; cleanup activities where dry expendable abrasives are used; and abrasive blasting enclosure movement and removal.[§1926.62(d)(2)(iii)[B]]

(iv) *With respect to the tasks listed in paragraph (d)(2)(iv) of this section,* where lead is present, until the employer performs an employee exposure assessment as required in paragraph (d) of this section and documents that the employee performing any of the listed tasks is not exposed to lead in excess of 2,500 μg/m^3 (50xPEL), the employer shall treat the employee as if the employee were exposed to lead in excess of 2,500 μg/m^3 and shall implement employee protective measures as prescribed in paragraph (d)(2)(v) of this section. Where the employer does establish that the employee is exposed to levels of lead below 2,500 μg/m3, the employer may provide the exposed employee with the appropriate respirator prescribed for use at such lower exposures, in accordance with Table I of this section. Interim protection as described in this paragaraph is required where lead containing coatings or paint are present on structures when performing:[§1926.62(d)(2)(iv)]

[A] Abrasive blasting,[§1926.62(d)(2)(iv)[A]]

[B] Welding,[§1926.62(d)(2)(iv)[B]]

[C] Cutting, and[§1926.62(d)(2)(iv)[C]]

[D] Torch burning.[§1926.62(d)(2)(iv)[D]]

(v) *Until the employer performs an employee exposure assessment* as required under paragraph (d) of this section and determines actual employee exposure, the employer shall provide to employees performing the tasks described in paragraphs (d)(2)(i), (d)(2)(ii), (d)(2)(iii), and (d)(2)(iv) of this section with interim protection as follows:[§1926.62(d)(2)(v)]

[A] Appropriate respiratory protection in accordance with paragraph (f) of this section.[§1926.62(d)(2)(v)[A]]

[B] Appropriate personal protective clothing and equipment in accordance with paragraph (g) of this section. [§1926.62(d)(2)(v)[B]]

[C] Change areas in accordance with paragraph (i)(2) of this section.[§1926.62(d)(2)(v)[C]]

[D] Hand washing facilities in accordance with paragraph (i)(5) of this section.[§1926.62(d)(2)(v)[D]]

[E] Biological monitoring in accordance with paragraph (j)(1)(i) of this section, to consist of blood sampling and analysis for lead and zinc protoporphyrin levels, and [§1926.62(d)(2)(v)[E]]

[F] Training as required under paragraph (l)(1)(i) of this section regarding 29 CFR 1926.59, Hazard Communication; training as required under paragraph (1)(2)(iii) of this section, regarding use of respirators; and training in accordance with 29 CFR 1926.21, Safety training and education.[§1926.62(d)(2)(v)[F]]

(3) *Basis of initial determination.*[§1926.62(d)(3)]

(i) *Except as provided under paragraphs (d)(3)(iii) and (d)(3)(iv)* of this section the employer shall monitor employee exposures and shall base initial determinations on the employee exposure monitoring results and any of the following, relevant considerations:[§1926.62(d)(3)(i)]

[A] Any information, observations, or calculations which would indicate employee exposure to lead; [§1926.62(d)(3)(i)[A]]

[B] Any previous measurements of airborne lead; and [§1926.62(d)(3)(i)[B]]

[C] Any employee complaints of symptoms which may be attributable to exposure to lead.[§1926.62(d)(3)(i)[C]]

(ii) *Monitoring for the initial determination where performed* may be limited to a representative sample of the exposed employees who the employer reasonably believes are exposed to the greatest airborne concentrations of lead in the workplace. [§1926.62(d)(3)(ii)]

(iii) *Where the employer has previously monitored for lead exposures,* and the data were obtained within the past 12 months during work operations conducted under workplace conditions closely resembling the processes, type of material, control methods, work practices, and environmental conditions used and prevailing in the employer's current operations, the employer may rely on such earlier monitoring results to satisfy the requirements of paragraphs (d)(3)(i) and (d)(6) of this section if the sampling and analytical methods meet the accuracy and confidence levels of paragraph (d)(10) of this section.[§1926.62(d)(3)(iii)]

(iv) *Where the employer has objective data,* demonstrating that a particular product or material containing lead or a specific process, operation or activity involving lead cannot result in employee exposure to lead at or above the action level during processing, use, or handling, the employer may rely upon such data instead of implementing initial monitoring. [§1926.62(d)(3)(iv)]

[A] The employer shall establish and maintain an accurate record documenting the nature and relevancy of objective data as specified in paragraph (n)(4) of this section, where used in assessing employee exposure in lieu of exposure monitoring.[§1926.62(d)(3)(iv)[A]]

[B] Objective data, as described in paragraph (d)(3)(iv) of this section, is not permitted to be used for exposure assessment in connection with paragraph (d)(2) of this section. [§1926.62(d)(3)(iv)[B]]

(4) *Positive initial determination* and initial monitoring.[§1926.62(d)(4)]

(i) *Where a determination conducted under* paragraphs (d) (1), (2) and (3) of this section shows the possibility of any employee exposure at or above the action level the employer shall conduct monitoring which is representative of the exposure for each employee in the workplace who is exposed to lead.[§1926.62(d)(4)(i)]

(ii) *Where the employer has previously monitored for lead exposure,* and the data were obtained within the past 12 months during work operations conducted under workplace conditions closely resembling the processes, type of material, control methods, work practices, and environmental conditions used and prevailing in the employer's current operations, the employer may rely on such earlier monitoring results to satisfy the requirements of paragraph (d)(4)(i) of this section if the sampling and analytical methods meet the accuracy and confidence levels of paragraph (d)(10) of this section. [§1926.62(d)(4)(ii)]

(5) *Negative initial determination.* Where a determination, conducted under paragraphs (d) (1), (2), and (3) of this section is made that no employee is exposed to airborne concentrations of lead at or above the action level the employer shall make a written record of such determination. The record shall include at least the information specified in paragraph (d)(3)(i) of this section and shall also include the date of determination, location within the worksite, and the name and social security number of each employee monitored.[§1926.62(d)(5)]

(6) *Frequency.*[§1926.62(d)(6)]

(i) *If the initial determination reveals employee exposure* to be below the action level further exposure determination need not be repeated except as otherwise provided in paragraph (d)(7) of this section.[§1926.62(d)(6)(i)]

(ii) *If the initial determination or subsequent determination* reveals employee exposure to be at or above the action level but at or below the PEL the employer shall perform monitoring in accordance with this paragraph at least every 6 months. The employer shall continue monitoring at the required frequency until at least two consecutive measurements, taken at least 7 days apart, are below the action level at which time the employer may discontinue monitoring for that employee except as otherwise provided in paragraph (d)(7) of this section.[§1926.62(d)(6)(ii)]

(iii) *If the initial determination reveals that employee exposure* is above the PEL the employer shall perform monitoring quarterly. The employer shall continue monitoring at the required frequency until at least two consecutive measurements, taken at least 7 days apart, are at or below the PEL but at or above the action level at which time the employer shall repeat monitoring for that employee at the frequency specified in paragraph (d)(6)(ii) of this section, except as otherwise provided in paragraph (d)(7) of this section. The employer shall continue monitoring at the required frequency until at least two consecutive measurements, taken at least 7 days apart, are below the action level at which time the employer may discontinue monitoring for that employee except as otherwise provided in paragraph (d)(7) of this section. [§1926.62(d)(6)(iii)]

(7) *Additional exposure assessments.* Whenever there has been a change of equipment, process, control, personnel or a new task has been initiated that may result in additional employees being exposed to lead at or above the action level or may result in employees already exposed at or above the action level being exposed above the PEL, the employer shall conduct additional monitoring in accordance with this paragraph.[§1926.62(d)(7)]

(8) *Employee notification.*[§1926.62(d)(8)]

(i) *The employer must, as soon as possible* but no later than 5 working days after the receipt of the results of any monitoring performed under this section, notify each affected employee of these results either individually in writing or by posting the results in an appropriate location that is accessible to employees.[§1926.62(d)(8)(i)]

(ii) *Whenever the results indicate* that the representative employee exposure, without regard to respirators, is at or above the PEL the employer shall include in the written notice a statement that the employees exposure was at or above that level and a description of the corrective action taken or to be taken to reduce exposure to below that level. [§1926.62(d)(8)(ii)]

(9) *Accuracy of measurement.* The employer shall use a method of monitoring and analysis which has an accuracy (to a confidence level of 95%) of not less than plus or minus 25 percent for airborne concentrations of lead equal to or greater than 30 $\mu g/m^3$.[§1926.62(d)(9)]

(e) Methods of compliance.[§1926.62(e)]

(1) *Engineering and work practice controls.* The employer shall implement engineering and work practice controls, including administrative controls, to reduce and maintain employee exposure to lead to or below the permissible exposure limit to the extent that such controls are feasible. Wherever all feasible engineering and work practices controls that can be instituted are not sufficient to reduce employee exposure to or below the permissible exposure limit prescribed in paragraph (c) of this section, the employer shall nonetheless use them to reduce employee exposure to the lowest feasible level and shall supplement them by the use of respiratory protection that complies with the requirements of paragraph (f) of this section. [§1926.62(e)(1)]

(2) *Compliance program.*[§1926.62(e)(2)]

(i) *Prior to commencement of the job* each employer shall establish and implement a written compliance program to achieve compliance with paragraph (c) of this section. [§1926.62(e)(2)(i)]

(ii) *Written plans for these* compliance programs shall include at least the following:[§1926.62(e)(2)(ii)]

[A] A description of each activity in which lead is emitted; e.g. equipment used, material involved, controls in place, crew size, employee job responsibilities, operating procedures and maintenance practices;[§1926.62(e)(2)(ii)[A]]

[B] A description of the specific means that will be employed to achieve compliance and, where engineering controls are required engineering plans and studies used to determine methods selected for controlling exposure to lead; [§1926.62(e)(2)(ii)[B]]

[C] A report of the technology considered in meeting the PEL; [§1926.62(e)(2)(ii)[C]]

[D] Air monitoring data which documents the source of lead emissions;[§1926.62(e)(2)(ii)[D]]

[E] A detailed schedule for implementation of the program, including documentation such as copies of purchase orders for equipment, construction contracts, etc.; [§1926.62(e)(2)(ii)[E]]

[F] A work practice program which includes items required under paragraphs (g), (h) and (i) of this section and incorporates other relevant work practices such as those specified in paragraph (e)(5) of this section;[§1926.62(e)(2)(ii)[F]]

[G] An administrative control schedule required by paragraph (e)(4) of this section, if applicable;[§1926.62(e)(2)(ii)[G]]

[H] A description of arrangements made among contractors on multi- contractor sites with respect to informing affected employees of potential exposure to lead and with respect to responsibility for compliance with this section as set-forth in Sec. 1926.16.[§1926.62(e)(2)(ii)[H]]

[I] Other relevant information.[§1926.62(e)(2)(ii)[I]]

(iii) *The compliance program shall provide* for frequent and regular inspections of job sites, materials, and equipment to be made by a competent person.[§1926.62(e)(2)(iii)]

(iv) *Written programs shall be submitted* upon request to any affected employee or authorized employee representatives, to the Assistant Secretary and the Director, and shall be available at the worksite for examination and copying by the Assistant Secretary and the Director.[§1926.62(e)(2)(iv)]

(v) *Written programs must be revised* and updated at least annually to reflect the current status of the program. [§1926.62(e)(2)(v)]

(3) *Mechanical ventilation.* When ventilation is used to control lead exposure, the employer shall evaluate the mechanical perfor-

mance of the system in controlling exposure as necessary to maintain its effectiveness.[§1926.62(e)(3)]

(4) *Administrative controls.* If administrative controls are used as a means of reducing employees TWA exposure to lead, the employer shall establish and implement a job rotation schedule which includes:[§1926.62(e)(4)]

(i) *Name or identification number* of each affected employee; [§1926.62(e)(4)(i)]

(ii) *Duration and exposure levels* at each job or work station where each affected employee is located; and[§1926.62(e)(4)(ii)]

(iii) *Any other information which may be useful* in assessing the reliability of administrative controls to reduce exposure to lead.[§1926.62(e)(4)(iii)]

(5) *The employer shall ensure that,* to the extent relevant, employees follow good work practices such as described in Appendix B of this section.[§1926.62(e)(5)]

(f) **Respiratory protection**[§1926.62(f)]

(1) *General.* For employees who use respirators required by this section, the employer must provide each employee an appropriate respirator that complies with the requirements of this paragraph. Respirators must be used during:[§1926.62(f)(1)]

(i) *Periods when an employee's exposure* to lead exceeds the PEL.[§1926.62(f)(1)(i)]

(ii) *Work operations for which* engineering and work-practice controls are not sufficient to reduce employee exposures to or below the PEL.[§1926.62(f)(1)(ii)]

(iii) *Periods when an employee requests a respirator.* [§1926.62(f)(1)(iii)]

(iv) *Periods when respirators are required* to provide interim protection of employees while they perform the operations specified in paragraph (d)(2) of this section.[§1926.62(f)(1)(iv)]

(2) *Respirator program.*[§1926.62(f)(2)]

(i) *The employer must implement a respiratory protection program* in accordance with 29 CFR 1910.134 (b) through (d) (except (d)(1)(iii)), and (f) through (m), which covers each employee required by this section to use a respirator. [§1926.62(f)(2)(i)]

(ii) *If an employee has breathing difficulty* during fit testing or respirator use, the employer must provide the employee with a medical examination in accordance with paragraph (j)(3)(i)(B) of this section to determine whether or not the employee can use a respirator while performing the required duty.[§1926.62(f)(2)(ii)]

(3) *Respirator selection.*[§1926.62(f)(3)]

(i) *Employers must:*[§1926.62(f)(3)(i)]

[A] Select, and provide to employees, the appropriate respirators specified in paragraph (d)(3)(i)(A) of 29 CFR 1910.134.[§1926.62(f)(3)(i)[A]]

[B] Provide employees with a full facepiece respirator instead of a half mask respirator for protection against lead aerosols that may cause eye or skin irritation at the use concentrations.[§1926.62(f)(3)(i)[B]]

[C] Provide HEPA filters for powered and non-powered air-purifying respirators.[§1926.62(f)(3)(i)[C]]

(ii) *The employer must provide a powered air-purifying respirator* when an employee chooses to use such a respirator and it will provide adequate protection to the employee. [§1926.62(f)(3)(ii)]

(g) **Protective work clothing and equipment**[§1926.62(g)]

(1) *Provision and use.* Where an employee is exposed to lead above the PEL without regard to the use of respirators, where employees are exposed to lead compounds which may cause skin or eye irritation (e.g. lead arsenate, lead azide), and as interim protection for employees performing tasks as specified in paragraph (d)(2) of this section, the employer shall provide at no cost to the employee and assure that the employee uses appropriate protective work clothing and equipment that prevents contamination of the employee and the employee's garments such as, but not limited to:[§1926.62(g)(1)]

(i) *Coveralls or similar full-body work clothing;*[§1926.62(g)(1)(i)]

(ii) *Gloves, hats, and shoes or disposable shoe coverlets; and* [§1926.62(g)(1)(ii)]

(iii) *Face shields, vented goggles,* or other appropriate protective equipment which complies with Sec. 1910.133 of this chapter.[§1926.62(g)(1)(iii)]

(2) *Cleaning and replacement.*[§1926.62(g)(2)]

(i) *The employer shall provide the protective clothing required* in paragraph (g)(1) of this section in a clean and dry condition at least weekly, and daily to employees whose exposure levels without regard to a respirator are over 200 µg/ m^3 of lead as an 8-hour TWA.[§1926.62(g)(2)(i)]

(ii) *The employer shall provide for the cleaning,* laundering, and disposal of protective clothing and equipment required by paragraph (g)(1) of this section.[§1926.62(g)(2)(ii)]

(iii) *The employer shall repair or replace* required protective clothing and equipment as needed to maintain their effectiveness.[§1926.62(g)(2)(iii)]

(iv) *The employer shall assure that all protective clothing* is removed at the completion of a work shift only in change areas provided for that purpose as prescribed in paragraph (i)(2) of this section.[§1926.62(g)(2)(iv)]

(v) *The employer shall assure that contaminated protective clothing* which is to be cleaned, laundered, or disposed of, is placed in a closed container in the change area which prevents dispersion of lead outside the container.[§1926.62(g)(2)(v)]

(vi) *The employer shall inform in writing* any person who cleans or launders protective clothing or equipment of the potentially harmful effects of exposure to lead.[§1926.62(g)(2)(vi)]

(vii) *[A] The employer shall ensure that the containers* of contaminated protective clothing and equipment required by paragraph (g)(2)(v) of this section are labeled as follows: [§1926.62(g)(2)(vii)[A]]

DANGER: CLOTHING AND EQUIPMENT CONTAMINATED WITH LEAD. MAY DAMAGE FERTILITY OR THE UNBORN CHILD. CAUSES DAMAGE TO THE CENTRAL NERVOUS SYSTEM. DO NOT EAT, DRINK OR SMOKE WHEN HANDLING. DO NOT REMOVE DUST BY BLOWING OR SHAKING. DISPOSE OF LEAD CONTAMINATED WASH WATER IN ACCORDANCE WITH APPLICABLE LOCAL, STATE, OR FEDERAL REGULATIONS.

[§1926.62(g)(2)(vii)[A]]

[B] Prior to June 1, 2015, employers may include the following information on bags or containers of contaminated protective clothing and equipment required by paragraph (g)(2)(v) in lieu of the labeling requirements in paragraph (g)(2)(vii)(A) of this section:[§1926.62(g)(2)(vii)[B]]

Caution: Clothing contaminated with lead. Do not remove dust by blowing or shaking. Dispose of lead contaminated wash water in accordance with applicable local, state, or federal regulations.[§1926.62(g)(2)(vii)[B]]

(viii) *The employer shall prohibit the removal* of lead from protective clothing or equipment by blowing, shaking, or any other means which disperses lead into the air. [§1926.62(g)(2)(viii)]

(h) **Housekeeping.**[§1926.62(h)]

(1) *All surfaces shall be maintained as free* as practicable of accumulations of lead.[§1926.62(h)(1)]

(2) *Clean-up of floors and other surfaces* where lead accumulates shall wherever possible, be cleaned by vacuuming or other methods that minimize the likelihood of lead becoming airborne.[§1926.62(h)(2)]

(3) *Shoveling, dry or wet sweeping,* and brushing may be used only where vacuuming or other equally effective methods have been tried and found not to be effective.[§1926.62(h)(3)]

(4) *Where vacuuming methods are selected,* the vacuums shall be equipped with HEPA filters and used and emptied in a manner which minimizes the reentry of lead into the workplace. [§1926.62(h)(4)]

(5) *Compressed air shall not be used* to remove lead from any surface unless the compressed air is used in conjunction with a ventilation system designed to capture the airborne dust created by the compressed air.[§1926.62(h)(5)]

(i) **Hygiene facilities and practices.**[§1926.62(i)]

(1) *The employer shall assure that in areas where employees are exposed to lead* above the PEL without regard to the use of respirators, food or beverage is not present or consumed, tobacco products are not present or used, and cosmetics are not applied.[§1926.62(i)(1)]

(2) *Change areas.*[§1926.62(i)(2)]

(i) *The employer shall provide clean* change areas for employees whose airborne exposure to lead is above the PEL, and as interim protection for employees performing tasks as specified in paragraph (d)(2) of this section, without regard to the use of respirators.[§1926.62(i)(2)(i)]

(ii) *The employer shall assure that change areas are equipped* with separate storage facilities for protective work clothing and equipment and for street clothes which prevent cross-contamination.[§1926.62(i)(2)(ii)]

(iii) *The employer shall assure that employees* do not leave the workplace wearing any protective clothing or equipment that is required to be worn during the work shift.[§1926.62(i)(2)(iii)]

(3) *Showers.*[§1926.62(i)(3)]

(i) *The employer shall provide shower facilities,* where feasible, for use by employees whose airborne exposure to lead is above the PEL.[§1926.62(i)(3)(i)]

(ii) *The employer shall assure,* where shower facilities are available, that employees shower at the end of the work shift and shall provide an adequate supply of cleansing agents and towels for use by affected employees.[§1926.62(i)(3)(ii)]

(4) *Eating facilities.*[§1926.62(i)(4)]

(i) *The employer shall provide lunchroom facilities* or eating areas for employees whose airborne exposure to lead is above the PEL, without regard to the use of respirators. [§1926.62(i)(4)(i)]

(ii) *The employer shall assure that lunchroom facilities* or eating areas are as free as practicable from lead contamination and are readily accessible to employees.[§1926.62(i)(4)(ii)]

(iii) *The employer shall assure that employees* whose airborne exposure to lead is above the PEL, without regard to the use of a respirator, wash their hands and face prior to eating, drinking, smoking or applying cosmetics.[§1926.62(i)(4)(iii)]

(iv) *The employer shall assure that employees* do not enter lunchroom facilities or eating areas with protective work clothing or equipment unless surface lead dust has been removed by vacuuming, downdraft booth, or other cleaning method that limits dispersion of lead dust.[§1926.62(i)(4)(iv)]

(5) *Hand washing facilities.*[§1926.62(i)(5)]

(i) *The employer shall provide adequate handwashing facilities* for use by employees exposed to lead in accordance with 29 CFR 1926.51(f).[§1926.62(i)(5)(i)]

(ii) *Where showers are not provided the employer shall assure* that employees wash their hands and face at the end of the work-shift.[§1926.62(i)(5)(ii)]

(j) **Medical surveillance**[§1926.62(j)]

(1) *General.*[§1926.62(j)(1)]

(i) *The employer shall make available* initial medical surveillance to employees occupationally exposed on any day to lead at or above the action level. Initial medical surveillance consists of biological monitoring in the form of blood sampling and analysis for lead and zinc protoporphyrin levels. [§1926.62(j)(1)(i)]

(ii) *The employer shall institute a medical surveillance program* in accordance with paragraphs (j)(2) and (j)(3) of this section for all employees who are or may be exposed by the employer at or above the action level for more than 30 days in any consecutive 12 months;[§1926.62(j)(1)(ii)]

(iii) *The employer shall assure that all medical examinations* and procedures are performed by or under the supervision of a licensed physician.[§1926.62(j)(1)(iii)]

(iv) *The employer shall make available* the required medical surveillance including multiple physician review under paragraph (j)(3)(iii) without cost to employees and at a reasonable time and place.[§1926.62(j)(1)(iv)]

(2) *Biological monitoring*[§1926.62(j)(2)]

(i) *Blood lead and ZPP level sampling and analysis.* The employer shall make available biological monitoring in the form of blood sampling and analysis for lead and zinc protoporphyrin levels to each employee covered under paragraphs (j)(1)(i) and (ii) of this section on the following schedule:[§1926.62(j)(2)(i)]

[A] For each employee covered under paragraph (j)(1)(ii) of this section, at least every 2 months for the first 6 months and every 6 months thereafter;[§1926.62(j)(2)(i)[A]]

[B] For each employee covered under paragraphs (j)(1) (i) or (ii) of this section whose last blood sampling and analysis indicated a blood lead level at or above 40 µg/dl, at least every two months. This frequency shall continue until two consecutive blood samples and analyses indicate a blood lead level below 40 µg/dl; and[§1926.62(j)(2)(i)[B]]

[C] For each employee who is removed from exposure to lead due to an elevated blood lead level at least monthly during the removal period.[§1926.62(j)(2)(i)[C]]

(ii) *Follow-up blood sampling tests.* Whenever the results of a blood lead level test indicate that an employee's blood lead level is at or above the numerical criterion for medical removal under paragraph (k)(1)(i) of this section, the employer shall provide a second (follow-up) blood sampling test within two weeks after the employer receives the results of the first blood sampling test.[§1926.62(j)(2)(ii)]

(iii) *Accuracy of blood lead level sampling and analysis.* Blood lead level sampling and analysis provided pursuant to this section shall have an accuracy (to a confidence level of 95 percent) within plus or minus 15 percent or 6 µg/dl, whichever is greater, and shall be conducted by a laboratory approved by OSHA.[§1926.62(j)(2)(iii)]

(iv) *Employee notification.*[§1926.62(j)(2)(iv)]

[A] Within five working days after the receipt of biological monitoring results, the employer shall notify each employee in writing of his or her blood lead level; and [§1926.62(j)(2)(iv)[A]]

[B] The employer shall notify each employee whose blood lead level is at or above 40 [mu]g/dl that the standard requires temporary medical removal with Medical Removal Protection benefits when an employee's blood lead level is at or above the numerical criterion for medical removal under paragraph (k)(1)(i) of this section. [§1926.62(j)(2)(iv)[B]]

(3) *Medical examinations and consultations*[§1926.62(j)(3)]

(i) *Frequency.* The employer shall make available medical examinations and consultations to each employee covered under paragraph (j)(1)(ii) of this section on the following schedule:[§1926.62(j)(3)(i)]

[A] At least annually for each employee for whom a blood sampling test conducted at any time during the preceding 12 months indicated a blood lead level at or above 40 µg/dl;[§1926.62(j)(3)(i)[A]]

[B] As soon as possible, upon notification by an employee either that the employee has developed signs or symptoms commonly associated with lead intoxication, that the employee desires medical advice concerning the effects of current or past exposure to lead on the employee's ability to procreate a healthy child, that the employee is pregnant, or that the employee has demonstrated difficulty in breathing during a respirator fitting test or during use; and [§1926.62(j)(3)(i)[B]]

[C] As medically appropriate for each employee either removed from exposure to lead due to a risk of sustaining material impairment to health, or otherwise limited pursuant to a final medical determination.[§1926.62(j)(3)(i)[C]]

(ii) *Content.* The content of medical examinations made available pursuant to paragraph (j)(3)(i)(B)-(C) of this section shall be determined by an examining physician and, if requested by an employee, shall include pregnancy testing or laboratory evaluation of male fertility. Medical examinations made available pursuant to paragraph (j)(3)(i)(A) of this section shall include the following elements:[§1926.62(j)(3)(ii)]

[A] A detailed work history and a medical history, with particular attention to past lead exposure (occupational and non-occupational), personal habits (smoking, hygiene), and past gastrointestinal, hematologic, renal, cardiovascular, reproductive and neurological problems; [§1926.62(j)(3)(ii)[A]]

[B] A thorough physical examination, with particular attention to teeth, gums, hematologic, gastrointestinal, renal, cardiovascular, and neurological systems. Pulmonary status should be evaluated if respiratory protection will be used; [§1926.62(j)(3)(ii)[B]]

[C] A blood pressure measurement;[§1926.62(j)(3)(ii)[C]]

[D] A blood sample and analysis which determines: [§1926.62(j)(3)(ii)[D]]

[1] Blood lead level;[§1926.62(j)(3)(ii)[D][1]]

[2] Hemoglobin and hematocrit determinations, red cell indices, and examination of peripheral smear morphology;[§1926.62(j)(3)(ii)[D][2]]

[3] Zinc protoporphyrin;[§1926.62(j)(3)(ii)[D][3]]

[4] Blood urea nitrogen; and,[§1926.62(j)(3)(ii)[D][4]]

[5] Serum creatinine;[§1926.62(j)(3)(ii)[D][5]]

[E] A routine urinalysis with microscopic examination; and [§1926.62(j)(3)(ii)[E]]

[F] Any laboratory or other test relevant to lead exposure which the examining physician deems necessary by sound medical practice.[§1926.62(j)(3)(ii)[F]]

(iii) *Multiple physician review mechanism.*[§1926.62(j)(3)(iii)]

[A] If the employer selects the initial physician who conducts any medical examination or consultation provided to an employee under this section, the employee may designate a second physician:[§1926.62(j)(3)(iii)[A]]

[1] To review any findings, determinations or recommendations of the initial physician; and[§1926.62(j)(3)(iii)[A][1]]

[2] To conduct such examinations, consultations, and laboratory tests as the second physician deems necessary to facilitate this review.[§1926.62(j)(3)(iii)[A][2]]

[B] The employer shall promptly notify an employee of the right to seek a second medical opinion after each occasion that an initial physician conducts a medical examination or consultation pursuant to this section. The employer may condition its participation in, and payment for, the multiple physician review mechanism upon the employee doing the following within fifteen (15) days after receipt of the foregoing notification, or receipt of the initial physician's written opinion, whichever is later:[§1926.62(j)(3)(iii)[B]]

[1] The employee informing the employer that he or she intends to seek a second medical opinion, and [§1926.62(j)(3)(iii)[B][1]]

[2] The employee initiating steps to make an appointment with a second physician.[§1926.62(j)(3)(iii)[B][2]]

[C] If the findings, determinations or recommendations of the second physician differ from those of the initial physician, then the employer and the employee shall assure that efforts are made for the two physicians to resolve any disagreement.[§1926.62(j)(3)(iii)[C]]

[D] If the two physicians have been unable to quickly resolve their disagreement, then the employer and the employee through their respective physicians shall designate a third physician:[§1926.62(j)(3)(iii)[D]]

[1] To review any findings, determinations or recommendations of the prior physicians; and[§1926.62(j)(3)(iii)[D][1]]

[2] To conduct such examinations, consultations, laboratory tests and discussions with the prior physicians as the third physician deems necessary to resolve the disagreement of the prior physicians.[§1926.62(j)(3)(iii)[D][2]]

[E] The employer shall act consistent with the findings, determinations and recommendations of the third physician, unless the employer and the employee reach an agreement which is otherwise consistent with the recommendations of at least one of the three physicians. [§1926.62(j)(3)(iii)[E]]

(iv) *Information provided to examining* and consulting physicians.[§1926.62(j)(3)(iv)]

[A] The employer shall provide an initial physician conducting a medical examination or consultation under this section with the following information:[§1926.62(j)(3)(iv)[A]]

[1] A copy of this regulation for lead including all Appendices;[§1926.62(j)(3)(iv)[A][1]]

[2] A description of the affected employee's duties as they relate to the employee's exposure;[§1926.62(j)(3)(iv)[A][2]]

[3] The employee's exposure level or anticipated exposure level to lead and to any other toxic substance (if applicable);[§1926.62(j)(3)(iv)[A][3]]

[4] A description of any personal protective equipment used or to be used;[§1926.62(j)(3)(iv)[A][4]]

[5] Prior blood lead determinations; and [§1926.62(j)(3)(iv)[A][5]]

[6] All prior written medical opinions concerning the employee in the employer's possession or control. [§1926.62(j)(3)(iv)[A][6]]

[B] The employer shall provide the foregoing information to a second or third physician conducting a medical examination or consultation under this section upon request either by the second or third physician, or by the employee. [§1926.62(j)(3)(iv)[B]]

(v) *Written medical opinions.*[§1926.62(j)(3)(v)]

[A] The employer shall obtain and furnish the employee with a copy of a written medical opinion from each examining or consulting physician which contains only the following information:[§1926.62(j)(3)(v)[A]]

[1] The physician's opinion as to whether the employee has any detected medical condition which would place the employee at increased risk of material impairment of the employee's health from exposure to lead; [§1926.62(j)(3)(v)[A][1]]

[2] Any recommended special protective measures to be provided to the employee, or limitations to be placed upon the employee's exposure to lead; [§1926.62(j)(3)(v)[A][2]]

[3] Any recommended limitation upon the employee's use of respirators, including a determination of whether the employee can wear a powered air purifying respirator if a physician determines that the employee cannot wear a negative pressure respirator; and[§1926.62(j)(3)(v)[A][3]]

[4] The results of the blood lead determinations. [§1926.62(j)(3)(v)[A][4]]

[B] The employer shall instruct each examining and consulting physician to:[§1926.62(j)(3)(v)[B]]

[1] Not reveal either in the written opinion or orally, or in any other means of communication with the employer, findings, including laboratory results, or diagnoses unrelated to an employee's occupational exposure to lead; and[§1926.62(j)(3)(v)[B][1]]

[2] Advise the employee of any medical condition, occupational or nonoccupational, which dictates further medical examination or treatment.[§1926.62(j)(3)(v)[B][2]]

(vi) *Alternate physician determination mechanisms.* The employer and an employee or authorized employee representative may agree upon the use of any alternate physician determination mechanism in lieu of the multiple physician review mechanism provided by paragraph (j)(3)(iii) of this section so long as the alternate mechanism is as expeditious and protective as the requirements contained in this paragraph.[§1926.62(j)(3)(vi)]

(4) *Chelation.*[§1926.62(j)(4)]

(i) *The employer shall assure that any person* whom he retains, employs, supervises or controls does not engage in prophylactic chelation of any employee at any time.[§1926.62(j)(4)(i)]

(ii) *If therapeutic or diagnostic chelation* is to be performed by any person in paragraph (j)(4)(i) of this section, the employer shall assure that it be done under the supervision of a licensed physician in a clinical setting with thorough and appropriate medical monitoring and that the employee is notified in writing prior to its occurrence.[§1926.62(j)(4)(ii)]

(k) Medical removal protection[§1926.62(k)]

(1) *Temporary medical removal* and return of an employee [§1926.62(k)(1)]

(i) *Temporary removal due to elevated blood lead level.* The employer shall remove an employee from work having an exposure to lead at or above the action level on each occasion that a periodic and a follow-up blood sampling test conducted pursuant to this section indicate that the employee's blood lead level is at or above 50 µg/dl; and,[§1926.62(k)(1)(i)]

(ii) *Temporary removal due* to a final medical determination. [§1926.62(k)(1)(ii)]

[A] The employer shall remove an employee from work having an exposure to lead at or above the action level on each occasion that a final medical determination results in a medical finding, determination, or opinion that the employee has a detected medical condition which places the employee at increased risk of material impairment to health from exposure to lead.[§1926.62(k)(1)(ii)[A]]

[B] For the purposes of this section, the phrase **final medical determination** means the written medical opinion on the employees' health status by the examining physician or, where relevant, the outcome of the multiple physician review mechanism or alternate medical determination mechanism used pursuant to the medical surveillance provisions of this section.[§1926.62(k)(1)(ii)[B]]

[C] Where a final medical determination results in any recommended special protective measures for an employee, or limitations on an employee's exposure to lead, the employer shall implement and act consistent with the recommendation.[§1926.62(k)(1)(ii)[C]]

(iii) *Return of the employee to former job status.*[§1926.62(k)(1)(iii)]

[A] The employer shall return an employee to his or her former job status:[§1926.62(k)(1)(iii)[A]]

[1] For an employee removed due to a blood lead level at or above 50 [mu]g/dl when two consecutive blood sampling tests indicate that the employee's blood lead level is below 40 [mu]g/dl;[§1926.62(k)(1)(iii)[A][1]]

[2] For an employee removed due to a final medical determination, when a subsequent final medical determination results in a medical finding, determination, or opinion that the employee no longer has a detected medical condition which places the employee at increased risk of material impairment to health from exposure to lead. [§1926.62(k)(1)(iii)[A][2]]

[B] For the purposes of this section, the requirement that an employer return an employee to his or her former job status is not intended to expand upon or restrict any rights an employee has or would have had, absent temporary medical removal, to a specific job classification or position under the terms of a collective bargaining agreement. [§1926.62(k)(1)(iii)[B]]

(iv) *Removal of other employee special protective measure or limitations.* The employer shall remove any limitations placed on an employee or end any special protective measures provided to an employee pursuant to a final medical determination when a subsequent final medical determination indicates that the limitations or special protective measures are no longer necessary.[§1926.62(k)(1)(iv)]

(v) *Employer options pending a final medical determination.* Where the multiple physician review mechanism, or alternate medical determination mechanism used pursuant to the medical surveillance provisions of this section, has not yet resulted in a final medical determination with respect to an employee, the employer shall act as follows:[§1926.62(k)(1)(v)]

[A] Removal. The employer may remove the employee from exposure to lead, provide special protective measures to the employee, or place limitations upon the employee, consistent with the medical findings, determinations, or recommendations of any of the physicians who have reviewed the employee's health status.[§1926.62(k)(1)(v)[A]]

[B] Return. The employer may return the employee to his or her former job status, end any special protective measures provided to the employee, and remove any limitations placed upon the employee, consistent with the medical findings, determinations, or recommendations of any of the physicians who have reviewed the employee's health status, with two exceptions.[§1926.62(k)(1)(v)[B]]

[1] If the initial removal, special protection, or limitation of the employee resulted from a final medical determination which differed from the findings, determinations, or recommendations of the initial physician or; [§1926.62(k)(1)(v)[B][1]]

[2] If the employee has been on removal status for the preceding eighteen months due to an elevated blood lead level, then the employer shall await a final medical determination.[§1926.62(k)(1)(v)[B][2]]

(2) *Medical removal protection benefits.*[§1926.62(k)(2)]

(i) *Provision of medical removal protection benefits.* The employer shall provide an employee up to eighteen (18) months of medical removal protection benefits on each occasion that an employee is removed from exposure to lead or otherwise limited pursuant to this section.[§1926.62(k)(2)(i)]

(ii) *Definition of medical removal protection benefits.* For the purposes of this section, the **requirement that an employer provide medical removal protection benefits** means that, as long as the job the employee was removed from continues, the employer shall maintain the total normal earnings, seniority and other employment rights and benefits of an employee, including the employee's right to his or her former job status as though the employee had not been medically removed from the employee's job or otherwise medically limited.[§1926.62(k)(2)(ii)]

(iii) *Follow-up medical surveillance during the period of employee removal or limitation.* During the period of time that an employee is medically removed from his or her job or otherwise medically limited, the employer may condition the provision of medical removal protection benefits upon the employee's participation in follow-up medical surveillance made available pursuant to this section.[§1926.62(k)(2)(iii)]

(iv) *Workers' compensation claims.* If a removed employee files a claim for workers' compensation payments for a lead-related disability, then the employer shall continue to provide medical removal protection benefits pending disposition of the claim. To the extent that an award is made to the employee for earnings lost during the period of removal, the employer's medical removal protection obligation shall be reduced by such amount. The employer shall receive no credit for workers' compensation payments received by the employee for treatment- related expenses.[§1926.62(k)(2)(iv)]

(v) *Other credits.* The employer's obligation to provide medical removal protection benefits to a removed employee shall be reduced to the extent that the employee receives compensation for earnings lost during the period of removal either from a publicly or employer-funded compensation program, or receives income from employment with another employer made possible by virtue of the employee's removal. [§1926.62(k)(2)(v)]

(vi) *Voluntary removal or restriction of an employee.* Where an employer, although not required by this section to do so, removes an employee from exposure to lead or otherwise places limitations on an employee due to the effects of lead exposure on the employee's medical condition, the employer shall provide medical removal protection benefits to the employee equal to that required by paragraph (k)(2) (i) and (ii) of this section.[§1926.62(k)(2)(vi)]

(l) **Communication of hazards.**[§1926.62(l)]

(1) *General.*[§1926.62(l)(1)]

(i) *Hazard communication.* The employer shall include lead in the program established to comply with the Hazard Communication Standard (HCS) (Sec. 1910.1200). The employer shall ensure that each employee has access to labels on containers of lead and safety data sheets, and is trained in accordance with the provisions of HCS and paragraph (l) of this section. The employer shall ensure that at least the following hazards are addressed:[§1926.62(l)(1)(i)]

[A] Reproductive/developmental toxicity;[§1926.62(l)(1)(i)[A]]

[B] Central nervous system effects;[§1926.62(l)(1)(i)[B]]

[C] Kidney effects;[§1926.62(l)(1)(i)[C]]

[D] Blood effects; and[§1926.62(l)(1)(i)[D]]

[E] Acute toxicity effects.[§1926.62(l)(1)(i)[E]]

(ii) *The employer shall train each employee* who is subject to exposure to lead at or above the action level on any day, or who is subject to exposure to lead compounds which may cause skin or eye irritation (e.g., lead arsenate, lead azide), in accordance with the requirements of this section. The employer shall institute a training program and ensure employee participation in the program.[§1926.62(l)(1)(ii)]

(iii) *The employer shall provide the training program* as initial training prior to the time of job assignment or prior to the start up date for this requirement, whichever comes last. [§1926.62(l)(1)(iii)]

(iv) *The employer shall also provide the training program at least annually* for each employee who is subject to lead exposure at or above the action level on any day.[§1926.62(l)(1)(iv)]

(2) *Training program.* The employer shall assure that each employee is trained in the following:[§1926.62(l)(2)]

(i) *The content of this standard and its appendices;* [§1926.62(l)(2)(i)]

(ii) *The specific nature of the operations* which could result in exposure to lead above the action level;[§1926.62(l)(2)(ii)]

(iii) *The purpose,* proper selection, fitting, use, and limitations of respirators;[§1926.62(l)(2)(iii)]

(iv) *The purpose and a description of the medical surveillance program,* and the medical removal protection program including information concerning the adverse health effects associated with excessive exposure to lead (with particular attention to the adverse reproductive effects on both males and females and hazards to the fetus and additional precautions for employees who are pregnant);[§1926.62(l)(2)(iv)]

(v) *The engineering controls and work practices associated with* the employee's job assignment including training of employees to follow relevant good work practices described in Appendix B of this section;[§1926.62(l)(2)(v)]

(vi) *The contents of any compliance plan in effect;* [§1926.62(l)(2)(vi)]

(vii) *Instructions to employees that chelating agents* should not routinely be used to remove lead from their bodies and should not be used at all except under the direction of a licensed physician; and[§1926.62(l)(2)(vii)]

(viii) *The employee's right of access* to records under 29 CFR 1910.1020[1].[§1926.62(l)(2)(viii)]

[1]Editor's Note: The CFR refers in error to §1910.20. This is corrected in the regulation appearing above.

(3) *Access to information and training materials.*[§1926.62(l)(3)]

(i) *The employer shall make readily available* to all affected employees a copy of this standard and its appendices. [§1926.62(l)(3)(i)]

(ii) *The employer shall provide,* upon request, all materials relating to the employee information and training program to affected employees and their designated representatives, and to the Assistant Secretary and the Director. [§1926.62(l)(3)(ii)]

(m) **Signs.**[§1926.62(m)]

(1) *General.*[§1926.62(m)(1)]

(i) *The employer shall post the following* warning signs in each work area where an employee's exposure to lead is above the PEL.[§1926.62(m)(1)(i)]

DANGER
LEAD WORK AREA
MAY DAMAGE FERTILITY OR THE UNBORN CHILD
CAUSES DAMAGE TO THE CENTRAL NERVOUS SYSTEM
DO NOT EAT, DRINK OR SMOKE IN THIS AREA

(ii) *The employer shall ensure that no statement* appears on or near any sign required by this paragraph (m) that contradicts or detracts from the meaning of the required sign. [§1926.62(m)(1)(ii)]

(iii) *The employer shall ensure that signs* required by this paragraph (m) are illuminated and cleaned as necessary so that the legend is readily visible.[§1926.62(m)(1)(iii)]

(iv) *The employer may use signs* required by other statutes, regulations or ordinances in addition to, or in combination with, signs required by this paragraph (m).[§1926.62(m)(1)(iv)]

(v) *Prior to June 1,* 2016, employers may use the following legend in lieu of that specified in paragraph (m)(1)(i) of this section:[§1926.62(m)(1)(v)]

WARNING
LEAD WORK AREA
POISON
NO SMOKING OR EATING

(n) **Recordkeeping.**[§1926.62(n)]

(1) *Exposure assessment.*[§1926.62(n)(1)]

(i) *The employer shall establish and maintain an accurate record* of all monitoring and other data used in conducting employee exposure assessments as required in paragraph (d) of this section.[§1926.62(n)(1)(i)]

(ii) *Exposure monitoring records shall include:*[§1926.62(n)(1)(ii)]

[A] The date(s), number, duration, location and results of each of the samples taken if any, including a description of the sampling procedure used to determine representative employee exposure where applicable;[§1926.62(n)(1)(ii)[A]]

[B] A description of the sampling and analytical methods used and evidence of their accuracy;[§1926.62(n)(1)(ii)[B]]

[C] The type of respiratory protective devices worn, if any; [§1926.62(n)(1)(ii)[C]]

[D] Name, social security number, and job classification of the employee monitored and of all other employees whose exposure the measurement is intended to represent; and[§1926.62(n)(1)(ii)[D]]

[E] The environmental variables that could affect the measurement of employee exposure.[§1926.62(n)(1)(ii)[E]]

(iii) *The employer shall maintain monitoring* and other exposure assessment records in accordance with the provisions of 29 CFR 1910.1020[1].[§1926.62(n)(1)(iii)]

(2) *Medical surveillance.*[§1926.62(n)(2)]

(i) *The employer shall establish and maintain an accurate record* for each employee subject to medical surveillance as required by paragraph (j) of this section.[§1926.62(n)(2)(i)]

(ii) *This record shall include:*[§1926.62(n)(2)(ii)]

[A] The name, social security number, and description of the duties of the employee;[§1926.62(n)(2)(ii)[A]]

[B] A copy of the physician's written opinions; [§1926.62(n)(2)(ii)[B]]

[C] Results of any airborne exposure monitoring done on or for that employee and provided to the physician; and [§1926.62(n)(2)(ii)[C]]

[D] Any employee medical complaints related to exposure to lead.[§1926.62(n)(2)(ii)[D]]

(iii) *The employer shall keep,* or assure that the examining physician keeps, the following medical records:[§1926.62(n)(2)(iii)]

[A] A copy of the medical examination results including medical and work history required under paragraph (j) of this section;[§1926.62(n)(2)(iii)[A]]

[B] A description of the laboratory procedures and a copy of any standards or guidelines used to interpret the test results or references to that information;[§1926.62(n)(2)(iii)[B]]

[C] A copy of the results of biological monitoring. [§1926.62(n)(2)(iii)[C]]

(iv) *The employer shall maintain or assure* that the physician maintains medical records in accordance with the provisions of 29 CFR 1910.1020[2].[§1926.62(n)(2)(iv)]

(3) *Medical removals.*[§1926.62(n)(3)]

(i) *The employer shall establish and maintain an accurate record* for each employee removed from current exposure to lead pursuant to paragraph (k) of this section.[§1926.62(n)(3)(i)]

(ii) *Each record shall include:*[§1926.62(n)(3)(ii)]

[A] The name and social security number of the employee; [§1926.62(n)(3)(ii)[A]]

[B] The date of each occasion that the employee was removed from current exposure to lead as well as the corresponding date on which the employee was returned to his or her former job status;[§1926.62(n)(3)(ii)[B]]

[C] A brief explanation of how each removal was or is being accomplished; and[§1926.62(n)(3)(ii)[C]]

[D] A statement with respect to each removal indicating whether or not the reason for the removal was an elevated blood lead level.[§1926.62(n)(3)(ii)[D]]

(iii) *The employer shall maintain each medical* removal record for at least the duration of an employee's employment. [§1926.62(n)(3)(iii)]

(4) *Objective data for exemption* from requirement for initial monitoring.[§1926.62(n)(4)]

(i) *For purposes of this section,* objective data are information demonstrating that a particular product or material containing lead or a specific process, operation, or activity involving lead cannot release dust or fumes in concentrations at or above the action level under any expected conditions of use. Objective data can be obtained from an industry-wide study or from laboratory product test results from manufacturers of lead containing products or materials. The data the employer uses from an industry-wide survey must be obtained under workplace conditions closely resembling the processes, types of material, control methods, work practices and environmental conditions in the employer's current operations. [§1926.62(n)(4)(i)]

(ii) *The employer shall maintain* the record of the objective data relied upon for at least 30 years.[§1926.62(n)(4)(ii)]

(5) *Availability.* The employer shall make available upon request all records required to be maintained by paragraph (n) of this section to affected employees, former employees, and their designated representatives, and to the Assistant Secretary and the Director for examination and copying.[§1926.62(n)(5)]

(6) *Transfer of records.*[§1926.62(n)(6)]

(i) *Whenever the employer ceases to do business,* the successor employer shall receive and retain all records required to be maintained by paragraph (n) of this section.[§1926.62(n)(6)(i)]

1. Editor's Note: The CFR refers in error to §1910.33. It appears that a reference to §1926.33 (Access to employee exposure and medical records) was intended. As §1926.33 states that the requirements applicable under that section are identical to those set forth at §1910.1020, the regulation appearing above lists §1910.1020.

2. Editor's Note:The CFR refers in error to §1910.20. This is corrected in the regulation appearing above.

(ii) *The employer shall also comply* with any additional requirements involving the transfer of records set forth in 29 CFR 1910.1020(h).[§1926.62(n)(6)(ii)]

(o) **Observation of monitoring.**[§1926.62(o)]

(1) *Employee observation.* The employer shall provide affected employees or their designated representatives an opportunity to observe any monitoring of employee exposure to lead conducted pursuant to paragraph (d) of this section.[§1926.62(o)(1)]

(2) *Observation procedures.*[§1926.62(o)(2)]

(i) *Whenever observation of the monitoring of employee exposure to lead* requires entry into an area where the use of respirators, protective clothing or equipment is required, the employer shall provide the observer with and assure the use of such respirators, clothing and equipment, and shall require the observer to comply with all other applicable safety and health procedures.[§1926.62(o)(2)(i)]

(ii) *Without interfering with the monitoring,* observers shall be entitled to:[§1926.62(o)(2)(ii)]

[A] Receive an explanation of the measurement procedures; [§1926.62(o)(2)(ii)[A]]

[B] Observe all steps related to the monitoring of lead performed at the place of exposure; and[§1926.62(o)(2)(ii)[B]]

[C] Record the results obtained or receive copies of the results when returned by the laboratory.[§1926.62(o)(2)(ii)[C]]

(p) **Appendices.** The information contained in the appendices to this section is not intended by itself, to create any additional obligations not otherwise imposed by this standard nor detract from any existing obligation.[§1926.62(p)]

§1926.62 Appendix A

Substance Data Sheet for Occupational Exposure to Lead

I. SUBSTANCE IDENTIFICATION

A. *Substance:* Pure lead (Pb) is a heavy metal at room temperature and pressure and is a basic chemical element. It can combine with various other substances to form numerous lead compounds.

B. *Compounds covered by the standard:* The word **lead** when used in this interim final standard means elemental lead, all inorganic lead compounds and a class of organic lead compounds called lead soaps. This standard does not apply to other organic lead compounds.

C. *Uses:* Exposure to lead occurs in several different occupations in the construction industry, including demolition or salvage of structures where lead or lead-containing materials are present; removal or encapsulation of lead-containing materials, new construction, alteration, repair, or renovation of structures that contain lead or materials containing lead; installation of products containing lead. In addition, there are construction related activities where exposure to lead may occur, including transportation, disposal, storage, or containment of lead or materials containing lead on construction sites, and maintenance operations associated with construction activities.

D. *Permissible exposure:* The permissible exposure limit (PEL) set by the standard is 50 micrograms of lead per cubic meter of air (50 μg/ m^3), averaged over an 8-hour workday.

E. *Action level:* The interim final standard establishes an action level of 30 micrograms of lead per cubic meter of air (30 μg/ m^3), averaged over an 8-hour workday. The action level triggers several ancillary provisions of the standard such as exposure monitoring, medical surveillance, and training.

II. HEALTH HAZARD DATA

A. *Ways in which lead enters your body.* When absorbed into your body in certain doses, lead is a toxic substance. The object of the lead standard is to prevent absorption of harmful quantities of lead. The standard is intended to protect you not only from the immediate toxic effects of lead, but also from the serious toxic effects that may not become apparent until years of exposure have passed. Lead can be absorbed into your body by inhalation (breathing) and ingestion (eating). Lead (except for certain organic lead compounds not covered by the standard, such as tetraethyl lead) is not absorbed through your skin. When lead is scattered in the air as a dust, fume, or mist it can be inhaled and absorbed through you lungs and upper respiratory tract[1]. Inhalation of airborne lead is generally the most important source of occupational lead absorption. You can also absorb lead through your digestive system if lead gets into your mouth and is swallowed. If you handle food, cigarettes, chewing tobacco, or make-up which have lead on them or handle them with hands contaminated with lead, this will contribute to ingestion. A significant portion of the lead that you inhale or ingest gets into your blood stream. Once in your blood stream, lead is circulated throughout your body and stored in various organs and body tissues. Some of this lead is quickly filtered out of your body and excreted, but some remains in the blood and other tissues. As exposure to lead continues, the amount stored in your body will increase if you are absorbing more lead than your body is excreting. Even though you may not be aware of any immediate symptoms of disease, this lead stored in your tissues can be slowly causing irreversible damage, first to individual cells, then to your organs and whole body systems.

[1]Editor's Note: The CFR has "When lead is scattered in the air as a dust, fume respiratory tract". The text is corrected in the regulation appearing above to match the text in corresponding §1910.1025 Appendix A.

B. *Effects of overexposure to lead*

(1) *Short term (acute) overexposure.* Lead is a potent, systemic poison that serves no known useful function once absorbed by your body. Taken in large enough doses, lead can kill you in a matter of days. A condition affecting the brain called acute encephalopathy may arise which develops quickly to seizures, coma, and death from cardiorespiratory arrest. A short term dose of lead can lead to acute encephalopathy. Short term occupational exposures of this magnitude are highly unusual, but not impossible. Similar forms of encephalopathy may, however, arise from extended, chronic exposure to lower doses of lead. There is no sharp dividing line between rapidly developing acute effects of lead, and chronic effects which take longer to acquire. Lead adversely affects numerous body systems, and causes forms of health impairment and disease which arise after periods of exposure as short as days or as long as several years.

(2) *Long-term (chronic) overexposure.* Chronic overexposure to lead may result in severe damage to your blood-forming, nervous, urinary and reproductive systems. Some common symptoms of chronic overexposure include loss of appetite, metallic taste in the mouth, anxiety, constipation, nausea, pallor, excessive tiredness, weakness, insomnia, headache, nervous irritability, muscle and joint pain or soreness, fine tremors, numbness, dizziness, hyperactivity and colic. In lead colic there may be severe abdominal pain. Damage to the central nervous system in general and the brain (encephalopathy) in particular is one of the most severe forms of lead poisoning. The most severe, often fatal, form of encephalopathy may be preceded by vomiting, a feeling of dullness progressing to drowsiness and stupor, poor memory, restlessness, irritability, tremor, and convulsions. It may arise suddenly with the onset of seizures, followed by coma, and death. There is a tendency for muscular weakness to develop at the same time. This weakness may progress to paralysis often observed as a characteristic "wrist drop" or "foot drop" and is a manifestation of a disease to the nervous system called peripheral neuropathy. Chronic overexposure to lead also results in kidney disease with few, if any, symptoms appearing until extensive and most likely permanent kidney damage has occurred. Routine laboratory tests reveal the presence of this kidney disease only after about two-thirds of kidney function is lost. When overt symptoms of urinary dysfunction arise, it is often too late to correct or prevent worsening conditions, and progression to kidney dialysis or death is possible. Chronic overexposure to lead impairs the reproductive systems of both men and women. Overexposure to lead may result in decreased sex drive, impotence and sterility in men. Lead can alter the structure of sperm cells raising the risk of birth defects. There is evidence of miscarriage and stillbirth in women whose husbands were exposed to lead or who were exposed to lead themselves. Lead exposure also may result in decreased fertility, and abnormal menstrual cycles in women. The course of pregnancy may be adversely affected by exposure to lead since lead crosses the placental barrier and poses risks to developing fetuses. Children born of parents either one of whom were exposed to excess lead levels are more likely to have birth defects, mental retardation, behavioral disorders or die during the first year of childhood. Overexposure to lead also disrupts the blood-forming system resulting in decreased hemoglobin (the substance in the blood that carries oxygen to the cells) and ultimately anemia. Anemia is characterized by weakness, pallor and fatigability as a result of decreased oxygen carrying capacity in the blood.

(3) *Health protection goals of the standard.* Prevention of adverse health effects for most workers from exposure to lead throughout a working lifetime requires that a worker's blood lead level

(BLL, also expressed as PbB) be maintained at or below forty micrograms per deciliter of whole blood (40 µg/dl). The blood lead levels of workers (both male and female workers) who intend to have children should be maintained below 30 µg/dl to minimize adverse reproductive health effects to the parents and to the developing fetus. The measurement of your blood lead level (BLL) is the most useful indicator of the amount of lead being absorbed by your body. Blood lead levels are most often reported in units of milligrams (mg) or micrograms (µg) of lead (1 mg=1000 µg) per 100 grams (100g), 100 milliliters (100 ml) or deciliter (dl) of blood. These three units are essentially the same. Sometime BLLs are expressed in the form of mg% or µg%. This is a shorthand notation for 100g, 100 ml, or dl. (References to BLL measurements in this standard are expressed in the form of µg/dl.)

BLL measurements show the amount of lead circulating in your blood stream, but do not give any information about the amount of lead stored in your various tissues. BLL measurements merely show current absorption of lead, not the effect that lead is having on your body or the effects that past lead exposure may have already caused. Past research into lead-related diseases, however, has focused heavily on associations between BLLs and various diseases. As a result, your BLL is an important indicator of the likelihood that you will gradually acquire a lead- related health impairment or disease.

Once your blood lead level climbs above 40 µg/dl, your risk of disease increases. There is a wide variability of individual response to lead, thus it is difficult to say that a particular BLL in a given person will cause a particular effect. Studies have associated fatal encephalopathy with BLLs as low as 150 µg/dl. Other studies have shown other forms of diseases in some workers with BLLs well below 80 µg/dl. Your BLL is a crucial indicator of the risks to your health, but one other factor is also extremely important. This factor is the length of time you have had elevated BLLs. The longer you have an elevated BLL, the greater the risk that large quantities of lead are being gradually stored in your organs and tissues (body burden). The greater your overall body burden, the greater the chances of substantial permanent damage. The best way to prevent all forms of lead-related impairments and diseases — both short term and long term — is to maintain your BLL below 40 µg/dl. The provisions of the standard are designed with this end in mind.

Your employer has prime responsibility to assure that the provisions of the standard are complied with both by the company and by individual workers. You, as a worker, however, also have a responsibility to assist your employer in complying with the standard. You can play a key role in protecting your own health by learning about the lead hazards and their control, learning what the standard requires, following the standard where it governs your own actions, and seeing that your employer complies with provisions governing his or her actions.

(4) *Reporting signs and symptoms of health problems.* You should immediately notify your employer if you develop signs or symptoms associated with lead poisoning or if you desire medical advice concerning the effects of current or past exposure to lead or your ability to have a healthy child. You should also notify your employer if you have difficulty breathing during a respirator fit test or while wearing a respirator. In each of these cases, your employer must make available to you appropriate medical examinations or consultations. These must be provided at no cost to you and at a reasonable time and place. The standard contains a procedure whereby you can obtain a second opinion by a physician of your choice if your employer selected the initial physician.

§1926.62 Appendix B

Employee Standard Summary

This appendix summarizes key provisions of the interim final standard for lead in construction that you as a worker should become familiar with.

I. Permissible Exposure Limit (PEL) — Paragraph (C)

The standard sets a permissible exposure limit (PEL) of 50 micrograms of lead per cubic meter of air (50 µg/m^3), averaged over an 8-hour workday which is referred to as a time-weighted average (TWA). This is the highest level of lead in air to which you may be permissibly exposed over an 8-hour workday. However, since this is an 8-hour average, short exposures above the PEL are permitted so long as for each 8-hour work day your average exposure does not exceed this level. This interim final standard, however, takes into account the fact that your daily exposure to lead can extend beyond a typical 8-hour workday as the result of overtime or other alterations in your work schedule. To deal with this situation, the standard contains a formula which reduces your permissible exposure when you are exposed more than 8 hours. For example, if you are exposed to lead for 10 hours a day, the maximum permitted average exposure would be 40 µg/m^3 .

II. Exposure Assessment — Paragraph (D)

If lead is present in your workplace in any quantity, your employer is required to make an initial determination of whether any employee's exposure to lead exceeds the action level (30 µg/m^3 averaged over an 8-hour day). Employee exposure is that exposure which would occur if the employee were not using a respirator. This initial determination requires your employer to monitor workers' exposures unless he or she has objective data which can demonstrate conclusively that no employee will be exposed to lead in excess of the action level. Where objective data is used in lieu of actual monitoring the employer must establish and maintain an accurate record, documenting its relevancy in assessing exposure levels for current job conditions. If such objective data is available, the employer need proceed no further on employee exposure assessment until such time that conditions have changed and the determination is no longer valid.

Objective data may be compiled from various sources, e.g., insurance companies and trade associations and information from suppliers or exposure data collected from similar operations. Objective data may also comprise previously-collected sampling data including area monitoring. If it cannot be determined through using objective data that worker exposure is less than the action level, your employer must conduct monitoring or must rely on relevant previous personal sampling, if available. Where monitoring is required for the initial determination, it may be limited to a representative number of employees who are reasonably expected to have the highest exposure levels. If your employer has conducted appropriate air sampling for lead in the past 12 months, he or she may use these results, provided they are applicable to the same employee tasks and exposure conditions and meet the requirements for accuracy as specified in the standard. As with objective data, if such results are relied upon for the initial determination, your employer must establish and maintain a record as to the relevancy of such data to current job conditions.

If there have been any employee complaints of symptoms which may be attributable to exposure to lead or if there is any other information or observations which would indicate employee exposure to lead, this must also be considered as part of the initial determination.

If this initial determination shows that a reasonable possibility exists that any employee may be exposed, without regard to respirators, over the action level, your employer must set up an air monitoring program to determine the exposure level representative of each employee exposed to lead at your workplace. In carrying out this air monitoring program, your employer is not required to monitor the exposure of every employee, but he or she must monitor a representative number of employees and job types. Enough sampling must be done to enable each employee's exposure level to be reasonably represent full shift exposure. In addition, these air samples must be taken under conditions which represent each employee's regular, daily exposure to lead. Sampling performed in the past 12 months may be used to determine exposures above the action level if such sampling was conducted during work activities essentially similar to present work conditions.

The standard lists certain tasks which may likely result in exposures to lead in excess of the PEL and, in some cases, exposures in excess of 50 times the PEL. If you are performing any of these tasks, your employer must provide you with appropriate respiratory protection, protective clothing and equipment, change areas, hand washing facilities, biological monitoring, and training until such time that an exposure assessment is conducted which demonstrates that your exposure level is below the PEL.

If you are exposed to lead and air sampling is performed, your employer is required to notify you in writing within 5 working days of the air monitoring results which represent your exposure. If the results indicate that your exposure exceeds the PEL (without regard to your use of a respirator), then your employer must also notify you of this in writing, and provide you with a description of the corrective action that has been taken or will be taken to reduce your exposure.

Your exposure must be rechecked by monitoring, at least every six months if your exposure is at or over the action level but below the PEL. Your employer may discontinue monitoring for you if 2 consecutive measurements, taken at least 7 days apart, are at or below the action level. Air monitoring must be repeated every 3 months if you are exposed over the PEL. Your employer must continue monitoring for you at this frequency until 2 consecutive measurements, taken at least 7 days apart, are below the PEL but above the action level, at which time your employer must repeat monitoring of your exposure every six months and may discontinue monitoring only after your exposure drops to or below the action level. However, whenever there is a change of equipment, process, control, or personnel or a new type of job is added at your workplace which may result in new or additional exposure to lead, your employer must perform additional monitoring.

III. Methods of Compliance — Paragraph (E)

Your employer is required to assure that no employee is exposed to lead in excess of the PEL as an 8-hour TWA. The interim final standard for lead in construction requires employers to institute engineering and work practice controls including administrative controls to the extent feasible to reduce employee exposure to lead. Where such controls are feasible but not adequate to reduce exposures below the PEL they must be used nonetheless to reduce exposures to the lowest level that can be accomplished by these means and then supplemented with appropriate respiratory protection.

Your employer is required to develop and implement a written compliance program prior to the commencement of any job where employee exposures may reach the PEL as an 8-hour TWA. The interim final standard identifies the various elements that must be included in the plan. For example, employers are required to include a description of operations in which lead is emitted, detailing other relevant information about the operation such as the type of equipment used, the type of material involved, employee job responsibilities, operating procedures and maintenance practices. In addition, your employer's compliance plan must specify the means that will be used to achieve compliance and, where engineering controls are required, include any engineering plans or studies that have been used to select the control methods. If administrative controls involving job rotation are used to reduce employee exposure to lead, the job rotation schedule must be included in the compliance plan. The plan must also detail the type of protective clothing and equipment, including respirators, housekeeping and hygiene practices that will be used to protect you from the adverse effects of exposure to lead.

The written compliance program must be made available, upon request, to affected employees and their designated representatives, the Assistant Secretary and the Director.

Finally, the plan must be reviewed and updated at least every 6 months to assure it reflects the current status in exposure control.

IV. Respiratory Protection — Paragraph (F)

Your employer is required to provide and assure your use of respirators when your exposure to lead is not controlled below the PEL by other means. The employer must pay the cost of the respirator. Whenever you request one, your employer is also required to provide you a respirator even if your air exposure level is not above the PEL. You might desire a respirator when, for example, you have received medical advice that your lead absorption should be decreased. Or, you may intend to have children in the near future, and want to reduce the level of lead in your body to minimize adverse reproductive effects. While respirators are the least satisfactory means of controlling your exposure, they are capable of providing significant protection if properly chosen, fitted, worn, cleaned, maintained, and replaced when they stop providing adequate protection.

Your employer is required to select respirators from the types listed in Table I of the Respiratory Protection section of the standard (Sec. 1926.62 (f)). Any respirator chosen must be approved by the National Institute for Occupational Safety and Health (NIOSH) under the provisions of 42 CFR part 84. This respirator selection table will enable your employer to choose a type of respirator that will give you a proper amount of protection based on your airborne lead exposure. Your employer may select a type of respirator that provides greater protection than that required by the standard; that is, one recommended for a higher concentration of lead than is present in your workplace. For example, a powered air-purifying respirator (PAPR) is much more protective than a typical negative pressure respirator, and may also be more comfortable to wear. A PAPR has a filter, cartridge, or canister to clean the air, and a power source that continuously blows filtered air into your breathing zone. Your employer might make a PAPR available to you to ease the burden of having to wear a respirator for long periods of time. The standard provides that you can obtain a PAPR upon request.

Your employer must also start a Respiratory Protection Program. This program must include written procedures for the proper selection, use, cleaning, storage, and maintenance of respirators.

Your employer must ensure that your respirator facepiece fits properly. Proper fit of a respirator facepiece is critical to your protection from airborne lead. Obtaining a proper fit on each employee may require your employer to make available several different types of respirator masks. To ensure that your respirator fits properly and that facepiece leakage is minimal, your employer must give you either a qualitative or quantitative fit test as specified in Appendix A of the Respiratory Protection standard located at 29 CFR 1910.134.

You must also receive from your employer proper training in the use of respirators. Your employer is required to teach you how to wear a respirator, to know why it is needed, and to understand its limitations.

The standard provides that if your respirator uses filter elements, you must be given an opportunity to change the filter elements whenever an increase in breathing resistance is detected. You also must be permitted to periodically leave your work area to wash your face and respirator facepiece whenever necessary to prevent skin irritation. If you ever have difficulty in breathing during a fit test or while using a respirator, your employer must make a medical examination available to you to determine whether you can safely wear a respirator. The result of this examination may be to give you a positive pressure respirator (which reduces breathing resistance) or to provide alternative means of protection.

V. Protective Work Clothing and Equipment — Paragraph (G)

If you are exposed to lead above the PEL as an 8-hour TWA, without regard to your use of a respirator, or if you are exposed to lead compounds such as lead arsenate or lead azide which can cause skin and eye irritation, your employer must provide you with protective work clothing and equipment appropriate for the hazard. If work clothing is provided, it must be provided in a clean and dry condition at least weekly, and daily if your airborne exposure to lead is greater than 200 μg/m^3 . Appropriate protective work clothing and equipment can include coveralls or similar full-body work clothing, gloves, hats, shoes or disposable shoe coverlets, and face shields or vented goggles. Your employer is required to provide all such equipment at no cost to you. In addition, your employer is responsible for providing repairs and replacement as necessary, and also is responsible for the cleaning, laundering or disposal of protective clothing and equipment.

The interim final standard requires that your employer assure that you follow good work practices when you are working in areas where your exposure to lead may exceed the PEL. With respect to protective clothing and equipment, where appropriate, the following procedures should be observed prior to beginning work:

1. *Change into work clothing* and shoe covers in the clean section of the designated changing areas;
2. *Use work garments* of appropriate protective gear, including respirators before entering the work area; and
3. *Store any clothing not worn* under protective clothing in the designated changing area.

Workers should follow these procedures upon leaving the work area:

1. *HEPA vacuum heavily* contaminated protective work clothing while it is still being worn. At no time may lead be removed from protective clothing by any means which result in uncontrolled dispersal of lead into the air;
2. *Remove shoe covers* and leave them in the work area;
3. *Remove protective clothing* and gear in the dirty area of the designated changing area. Remove protective coveralls by carefully rolling down the garment to reduce exposure to dust.
4. *Remove respirators last; and*
5. *Wash hands and face.*

Workers should follow these procedures upon finishing work for the day (in addition to procedures described above):

1. *Where applicable,* place disposal coveralls and shoe covers with the abatement waste;
2. *Contaminated clothing which* is to be cleaned, laundered or disposed of must be placed in closed containers in the change room.
3. *Clean protective gear,* including respirators, according to standard procedures;
4. *Wash hands and face* again. If showers are available, take a shower and wash hair. If shower facilities are not available at the work site, shower immediately at home and wash hair.

VI. Housekeeping — Paragraph (H)

Your employer must establish a housekeeping program sufficient to maintain all surfaces as free as practicable of accumulations of lead dust. Vacuuming is the preferred method of meeting this requirement, and the use of compressed air to clean floors and other surfaces is generally prohibited unless removal with compressed air is done in conjunction with ventilation systems designed to contain dispersal of the lead dust. Dry or wet sweeping, shoveling, or brushing may not be used except where vacuuming or other equally effective methods have been tried and do not work. Vacuums must be used equipped with a special filter called a high-efficiency particulate air (HEPA) filter and emptied in a manner which minimizes the reentry of lead into the workplace.

VII. Hygiene Facilities and Practices — Paragraph (I)

The standard requires that hand washing facilities be provided where occupational exposure to lead occurs. In addition, change areas, showers (where feasible), and lunchrooms or eating areas are to be made available to workers exposed to lead above the PEL. Your employer must assure that except in these facilities, food and beverage is not present or consumed, tobacco products are not present or used, and cosmetics are not applied, where airborne exposures are above the PEL. Change rooms provided by your employer must be equipped with separate storage facilities for your protective clothing and equipment and street clothes to avoid cross-contamination. After showering, no required protective clothing or equipment worn during the shift may be worn home. It is important that contaminated clothing or equipment be removed in change areas and not be worn home or you will extend your exposure and expose your family since lead from your clothing can accumulate in your house, car, etc.

Lunchrooms or eating areas may not be entered with protective clothing or equipment unless surface dust has been removed by vacuuming, downdraft booth, or other cleaning method. Finally, workers exposed above the PEL must wash both their hands and faces prior to eating, drinking, smoking or applying cosmetics.

All of the facilities and hygiene practices just discussed are essential to minimize additional sources of lead absorption from inhalation or ingestion of lead that may accumulate on you, your clothes, or your possessions. Strict compliance with these provisions can virtually eliminate several sources of lead exposure which significantly contribute to excessive lead absorption.

VIII. Medical Surveillance — Paragraph (J)

The medical surveillance program is part of the standard's comprehensive approach to the prevention of lead-related disease. Its purpose is to supplement the main thrust of the standard which is aimed at minimizing airborne concentrations of lead and sources of ingestion. Only medical surveillance can determine if the other provisions of the standard have effectively protected you as an individual. Compliance with the standard's provision will protect most workers from the adverse effects of lead exposure, but may not be satisfactory to protect individual workers:

(1) *who have high body burdens* of lead acquired over past years;
(2) *who have additional uncontrolled sources of non-occupational lead exposure;*
(3) *who exhibit unusual variations in lead absorption rates; or*
(4) *who have specific non-work related medical conditions* which could be aggravated by lead exposure (e.g., renal disease, anemia). In addition, control systems may fail, or hygiene and respirator programs may be inadequate. Periodic medical surveillance of individual workers will help detect those failures. Medical surveillance will also be important to protect your reproductive ability — regardless of whether you are a man or woman.

All medical surveillance required by the interim final standard must be performed by or under the supervision of a licensed physician. The employer must provide required medical surveillance without cost to employees and at a reasonable time and place. The standard's medical surveillance program has two parts — periodic biological monitoring and medical examinations. Your employer's obligation to offer you medical surveillance is triggered by the results of the air monitoring program. Full medical surveillance must be made available to all employees who are or may be exposed to lead in excess of the action level for more than 30 days a year and whose blood lead level exceeds 40 µg/dl. Initial medical surveillance consisting of blood sampling and analysis for lead and zinc protoporphyrin must be provided to all employees exposed at any time (1 day) above the action level.

Biological monitoring under the standard must be provided at least every 2 months for the first 6 months and every 6 months thereafter until your blood lead level is below 40 µg/dl. A zinc protoporphyrin (ZPP) test is a very useful blood test which measures an adverse metabolic effect of lead on your body and is therefore an indicator of lead toxicity.

If your BLL exceeds 40 µg/dl the monitoring frequency must be increased from every 6 months to at least every 2 months and not reduced until two consecutive BLLs indicate a blood lead level below 40 µg/dl. Each time your BLL is determined to be over 40 µg/dl, your employer must notify you of this in writing within five working days of his or her receipt of the test results. The employer must also inform you that the standard requires temporary medical removal with economic protection when your BLL exceeds 50 µg/dl. (See Discussion of Medical Removal Protection-Paragraph (k).) Anytime your BLL exceeds 50 µg/dl your employer must make available to you within two weeks of receipt of these test results a second follow-up BLL test to confirm your BLL. If the two tests both exceed 50 µg/dl, and you are temporarily removed, then your employer must make successive BLL tests available to you on a monthly basis during the period of your removal.

Medical examinations beyond the initial one must be made available on an annual basis if your blood lead level exceeds 40 µg/dl at any time during the preceding year and you are being exposed above the airborne action level of 30 µg/m^3 for 30 or more days per year. The initial examination will provide information to establish a baseline to which subsequent data can be compared.

An initial medical examination to consist of blood sampling and analysis for lead and zinc protoporphyrin must also be made available (prior to assignment) for each employee being assigned for the first time to an area where the airborne concentration of lead equals or exceeds the action level at any time. In addition, a medical examination or consultation must be made available as soon as possible if you notify your employer that you are experiencing signs or symptoms commonly associated with lead poisoning or that you have difficulty breathing while wearing a respirator or during a respirator fit test. You must also be provided a medical examination or consultation if you notify your employer that you desire medical advice concerning the effects of current or past exposure to lead on your ability to procreate a healthy child.

Finally, appropriate follow-up medical examinations or consultations may also be provided for employees who have been temporarily removed from exposure under the medical removal protection provisions of the standard. (See Part IX, below.)

The standard specifies the minimum content of pre-assignment and annual medical examinations. The content of other types of medical examinations and consultations is left up to the sound discretion of the examining physician. Pre-assignment and annual medical examinations must include

(1) *a detailed work history and medical history;*
(2) *a thorough physical examination,* including an evaluation of your pulmonary status if you will be required to use a respirator;
(3) *a blood pressure measurement; and*
(4) *a series of laboratory tests designed* to check your blood chemistry and your kidney function. In addition, at any time upon your request, a laboratory evaluation of male fertility will be made (microscopic examination of a sperm sample), or a pregnancy test will be given.

The standard does not require that you participate in any of the medical procedures, tests, etc. which your employer is required to make available to you. Medical surveillance can, however, play a very important role in protecting your health. You are strongly encouraged, therefore, to participate in a meaningful fashion. The standard contains a multiple physician review mechanism which will give you a chance to have a physician of your choice directly participate in the medical surveillance program. If you are dissatisfied with an examination by a physician chosen by your employer, you can select a second physician to conduct an independent analysis. The two doctors would attempt to resolve any differences of opinion, and select a third physician to resolve any firm dispute. Generally your employer will choose the physician who conducts medical surveillance under the lead standard- unless you and your employer can agree on the choice of a physician or physicians. Some companies and unions have agreed in advance, for example, to use certain independent medical laboratories or panels of physicians. Any of these arrangements are acceptable so long as required medical surveillance is made available to workers.

The standard requires your employer to provide certain information to a physician to aid in his or her examination of you. This information includes

(1) *the standard and its appendices,*
(2) *a description of your duties as they relate to occupational lead exposure,*
(3) *your exposure level or anticipated exposure level,*
(4) *a description of any personal protective equipment you wear,*
(5) *prior blood lead level results, and*
(6) *prior written medical opinions* concerning you that the employer has. After a medical examination or consultation the physician must prepare a written report which must contain
   (1) *the physician's opinion as to whether you* have any medical condition which places you at increased risk of material impairment to health from exposure to lead,
   (2) *any recommended special protective measures to be provided to you,*
   (3) *any blood lead level determinations, and*
   (4) *any recommended limitation* on your use of respirators. This last element must include a determination of whether you can wear a powered air purifying respirator (PAPR) if you are found unable to wear a negative pressure respirator.

The medical surveillance program of the interim lead standard may at some point in time serve to notify certain workers that they have acquired a disease or other adverse medical condition as a result of occupational lead exposure. If this is true, these workers might have legal rights to compensation from public agencies, their employers, firms that supply hazardous products to their employers, or other persons. Some states have laws, including worker compensation laws, that disallow a worker who learns of a job-related health impairment to sue, unless the worker sues within a short period of time after learning of the impairment. (This period of time may be a matter of months or years.) An attorney can be consulted about these possibilities. It should be stressed that OSHA is in no way trying to either encourage or discourage claims or lawsuits. However, since results of the standard's medical surveillance program can significantly affect the legal remedies of a worker who has acquired a job-related disease or impairment, it is proper for OSHA to make you aware of this.

The medical surveillance section of the standard also contains provisions dealing with chelation. Chelation is the use of certain drugs (administered in pill form or injected into the body) to reduce the amount of lead absorbed in body tissues. Experience accumulated by the medical and scientific communities has largely confirmed the effectiveness of this type of therapy for the treatment of very severe lead poisoning. On the other hand, it has also been established that there can be a long list of extremely harmful side effects associated with the use of chelating agents. The medical community has balanced the advantages and disadvantages resulting from the use of chelating agents in various circumstances and has established when the use of these agents is acceptable. The standard includes these accepted limitations due to a history of abuse of chelation therapy by some lead companies. The most widely used chelating agents are calcium disodium EDTA, (Ca Na2 EDTA), Calcium Disodium Versenate (Versenate), and d-penicillamine (pencillamine or Cupramine).

The standard prohibits "prophylactic chelation" of any employee by any person the employer retains, supervises or controls. **Prophylactic chelation** is the routine use of chelating or similarly acting drugs to prevent elevated blood levels in workers who are occupationally exposed to lead, or the use of these drugs to routinely lower blood lead levels to predesignated concentrations believed to be "safe". It should be emphasized that where an employer takes a worker who has no symptoms of lead poisoning and has chelation carried out by a physician (either inside or outside of a hospital) solely to reduce the worker's blood lead level, that will generally be considered prophylactic chelation. The use of a hospital and a physician does not mean that prophylactic chelation is not being performed. Routine chelation to prevent increased or reduce current blood lead levels is unacceptable whatever the setting.

The standard allows the use of "therapeutic" or "diagnostic" chelation if administered under the supervision of a licensed physician in a clinical setting with thorough and appropriate medical monitoring. Therapeutic chelation responds to severe lead poisoning where there are marked symptoms. Diagnostic chelation involved giving a patient a dose of the drug then collecting all urine excreted for some period of time as an aid to the diagnosis of lead poisoning.

In cases where the examining physician determines that chelation is appropriate, you must be notified in writing of this fact before such treatment. This will inform you of a potentially harmful treatment, and allow you to obtain a second opinion.

IX. Medical Removal Protection — Paragraph (K)

Excessive lead absorption subjects you to increased risk of disease. Medical removal protection (MRP) is a means of protecting you when, for whatever reasons, other methods, such as engineering controls, work practices, and respirators, have failed to provide the protection you need. MRP involves the temporary removal of a worker from his or her regular job to a place of significantly lower exposure without any loss of earnings, seniority, or other employment rights or benefits. The purpose of this program is to cease further lead absorption and allow your body to naturally excrete lead which has previously been absorbed. Temporary medical removal can result from an elevated blood lead level, or a medical opinion. For up to 18 months, or for as long as the job the employee was removed from lasts, protection is provided as a result of either form of removal. The vast majority of removed workers, however, will return to their former jobs long before this eighteen month period expires.

You may also be removed from exposure even if your blood lead level is below 50 µg/dl if a final medical determination indicates that you temporarily need reduced lead exposure for medical reasons. If the physician who is implementing your employers medical program makes a final written opinion recommending your removal or other special protective measures, your employer must implement the physician's recommendation. If you are removed in this manner, you may only be returned when the doctor indicates that it is safe for you to do so.

The standard does not give specific instructions dealing with what an employer must do with a removed worker. Your job assignment upon removal is a matter for you, your employer and your union (if any) to work out consistent with existing procedures for job assignments. Each removal must be accomplished in a manner consistent with existing collective bargaining relationships. Your employer is given broad discretion to implement temporary removals so long as no attempt is made to override existing agreements. Similarly, a removed worker is provided no right to veto an employer's choice which satisfies the standard.

In most cases, employers will likely transfer removed employees to other jobs with sufficiently low lead exposure. Alternatively, a worker's hours may be reduced so that the time weighted average exposure is reduced, or he or she may be temporarily laid off if no other alternative is feasible.

In all of these situations, MRP benefits must be provided during the period of removal — i.e., you continue to receive the same earnings, seniority, and other rights and benefits you would have had if you had not been removed. Earnings includes more than just your base wage; it includes overtime, shift differentials, incentives, and other compensation you would have earned if you had not been removed. During the period of removal you must also be provided with appropriate follow- up medical surveillance. If you were removed because your blood lead level was too high, you must be provided with a monthly blood test. If a medical opinion caused your removal, you must be provided medical tests or examinations that the doctor believes to be appropriate. If you do not participate in this follow up medical surveillance, you may lose your eligibility for MRP benefits.

When you are medically eligible to return to your former job, your employer must return you to your "former job status." This means that you are entitled to the position, wages, benefits, etc., you would have had if you had not been removed. If you would still be in your old job if no removal had occurred that is where you go back. If not, you are returned consistent with whatever job assignment discretion your employer would have had if no removal had occurred. MRP only seeks to maintain your rights, not expand them or diminish them.

If you are removed under MRP and you are also eligible for worker compensation or other compensation for lost wages, your employer's MRP benefits obligation is reduced by the amount that you actually receive from these other sources. This is also true if you obtain other employment during the time you are laid off with MRP benefits.

The standard also covers situations where an employer voluntarily removes a worker from exposure to lead due to the effects of lead on the employee's medical condition, even though the standard does not require removal. In these situations MRP benefits must still be provided as though the standard required removal. Finally, it is important to note that in all cases where removal is required, respirators cannot be used as a substitute. Respirators may be used before removal becomes necessary, but not as an alternative to a transfer to a low exposure job, or to a lay-off with MRP benefits.

X. Employee Information and Training — Paragraph (L)

Your employer is required to provide an information and training program for all employees exposed to lead above the action level or who may suffer skin or eye irritation from lead compounds such as lead arsenate or lead azide. The program must train these employees regarding the specific hazards associated with their work environment, protective measures which can be taken, including the contents of any compliance plan in effect, the danger of lead to their bodies (including their reproductive systems), and their rights under the standard. All employees must be trained prior to initial assignment to areas where there is a possibility of exposure over the action level.

This training program must also be provided at least annually thereafter unless further exposure above the action level will not occur.

XI. Signs — Paragraph (M)

The standard requires that the following warning sign be posted in work areas when the exposure to lead is above the PEL:

DANGER

LEAD WORK AREA

MAY DAMAGE FERTILITY OR THE UNBORN CHILD
CAUSES DAMAGE TO THE CENTRAL NERVOUS SYSTEM
DO NOT EAT, DRINK OR SMOKE IN THIS AREA

Prior to June 1, 2016, employers may use the following legend in lieu of that specified above:

WARNING
LEAD WORK AREA
POISON
NO SMOKING OR EATING

XII. Recordkeeping — Paragraph (N)

Your employer is required to keep all records of exposure monitoring for airborne lead. These records must include the name and job classification of employees measured, details of the sampling and analytical techniques, the results of this sampling, and the type of respiratory protection being worn by the person sampled. Such records are to be retained for at least 30 years. Your employer is also required to keep all records of biological monitoring and medical examination results. These records must include the names of the employees, the physician's written opinion, and a copy of the results of the examination. Medical records must be preserved and maintained for the duration of employment plus 30 years. However, if the employee's duration of employment is less than one year, the employer need not retain that employee's medical records beyond the period of employment if they are provided to the employee upon termination of employment.

Recordkeeping is also required if you are temporarily removed from your job under the medical removal protection program. This record must include your name and social security number, the date of your removal and return, how the removal was or is being accomplished, and whether or not the reason for the removal was an elevated blood lead level. Your employer is required to keep each medical removal record only for as long as the duration of an employee's employment.

The standard requires that if you request to see or copy environmental monitoring, blood lead level monitoring, or medical removal records, they must be made available to you or to a representative that you authorize. Your union also has access to these records. Medical records other than BLL's must also be provided upon request to you, to your physician or to any other person whom you may specifically designate. Your union does not have access to your personal medical records unless you authorize their access.

XIII. Observation of Monitoring — Paragraph (O)

When air monitoring for lead is performed at your workplace as required by this standard, your employer must allow you or someone you designate to act as an observer of the monitoring. Observers are entitled to an explanation of the measurement procedure, and to record the results obtained. Since results will not normally be available at the time of the monitoring, observers are entitled to record or receive the results of the monitoring when returned by the laboratory. Your employer is required to provide the observer with any personal protective devices required to be worn by employees working in the area that is being monitored. The employer must require the observer to wear all such equipment and to comply with all other applicable safety and health procedures.

XIV. For Additional Information

A. *A copy of the interim standard* for lead in construction can be obtained free of charge by calling or writing the OSHA Office of Publications, room N-3101, United States Department of Labor, Washington, DC 20210: Telephone (202) 219-4667.

B. *Additional information about the standard,* its enforcement, and your employer's compliance can be obtained from the nearest OSHA Area Office listed in your telephone directory under United States Government/Department of Labor.

§1926.62 Appendix C

Medical Surveillance Guidelines

Introduction

The primary purpose of the Occupational Safety and Health Act of 1970 is to assure, so far as possible, safe and healthful working conditions for every working man and woman. The interim final occupational health standard for lead in construction is designed to protect workers exposed to inorganic lead including metallic lead, all inorganic lead compounds and organic lead soaps.

Under this interim final standard occupational exposure to inorganic lead is to be limited to 50 μg/m^3 (micrograms per cubic meter) based on an 8 hour time-weighted average (TWA). This permissible exposure limit (PEL) must be achieved through a combination of engineering, work practice and administrative controls to the extent feasible. Where these controls are in place but are found not to reduce employee exposures to or below the PEL, they must be used nonetheless, and supplemented with respirators to meet the 50 μg/m^3 exposure limit.

The standard also provides for a program of biological monitoring for employees exposed to lead above the action level at any time, and additional medical surveillance for all employees exposed to levels of inorganic lead above 30 μg/m^3 (TWA) for more than 30 days per year and whose BLL exceeds 40 μg/dl.

The purpose of this document is to outline the medical surveillance provisions of the interim standard for inorganic lead in construction, and to provide further information to the physician regarding the examination and evaluation of workers exposed to inorganic lead.

Section 1 provides a detailed description of the monitoring procedure including the required frequency of blood testing for exposed workers, provisions for medical removal protection (MRP), the recommended right of the employee to a second medical opinion, and notification and recordkeeping requirements of the employer. A discussion of the requirements for respirator use and respirator monitoring and OSHA's position on prophylactic chelation therapy are also included in this section.

Section 2 discusses the toxic effects and clinical manifestations of lead poisoning and effects of lead intoxication on enzymatic pathways in heme synthesis. The adverse effects on both male and female reproductive capacity and on the fetus are also discussed.

Section 3 outlines the recommended medical evaluation of the worker exposed to inorganic lead, including details of the medical history, physical examination, and recommended laboratory tests, which are based on the toxic effects of lead as discussed in Section 2.

Section 4 provides detailed information concerning the laboratory tests available for the monitoring of exposed workers. Included also is a discussion of the relative value of each test and the limitations and precautions which are necessary in the interpretation of the laboratory results.

I. Medical Surveillance and Monitoring Requirements for Workers Exposed to Inorganic Lead

Under the interim final standard for inorganic lead in the construction industry, initial medical surveillance consisting of biological monitoring to include blood lead and ZPP level determination shall be provided to employees exposed to lead at or above the action level on any one day. In addition, a program of biological monitoring is to be made available to all employees exposed above the action level at any time and additional medical surveillance is to be made available to all employees exposed to lead above 30 μg/m^3 TWA for more than 30 days each year and whose BLL exceeds 40 μg/dl. This program consists of periodic blood sampling and medical evaluation to be performed on a schedule which is defined by previous laboratory results, worker complaints or concerns, and the clinical assessment of the examining physician.

Under this program, the blood lead level (BLL) of all employees who are exposed to lead above 30 μg/m^3 for more than 30 days per year or whose blood lead is above 40 μg/dl but exposed for no more than 30 days per year is to be determined at least every two months for the first six months of exposure and every six months thereafter. The frequency is increased to every two months for employees whose last blood lead level was 40 μg/dl or above. For employees who are removed from exposure to lead due to an elevated blood lead, a new blood lead level must be measured monthly. A zinc protoporphyrin (ZPP) measurement is strongly recommended on each occasion that a blood lead level measurement is made.

An annual medical examination and consultation performed under the guidelines discussed in Section 3 is to be made available to each employee exposed above 30 μg/m^3 for more than 30 days per year for whom a blood test conducted at any time during the preceding 12 months indicated a blood lead level at or above 40 μg/dl. Also, an examination is to be given to all employees prior to their assignment to an area in which airborne lead concentrations reach or exceed the 30 μg/m^3 for more than 30 days per year. In addition, a medical examination must be provided as soon as possible after notification by an employee that the employee has developed signs or symptoms commonly associated with lead intoxication, that the employee desires medical advice regarding lead exposure and the ability to procreate a healthy child, or that the employee has demonstrated difficulty in breathing during a respirator fitting test or during respirator use. An examination is also to be made available to each employee removed from exposure to lead due to a risk of sustaining material impairment to health, or otherwise limited or specially protected pursuant to medical recommendations.

Results of biological monitoring or the recommendations of an examining physician may necessitate removal of an employee from further lead exposure pursuant to the standard's medical removal protection (MRP) program. The object of the MRP program is to provide temporary medical removal to workers either with substantially elevated blood lead levels or otherwise at risk of sustaining material health impairment from continued substantial exposure to lead.

Under the standard's ultimate worker removal criteria, a worker is to be removed from any work having an eight hour TWA exposure to lead of 30 μg/m^3 when his or her blood lead level reaches 50 μg/dl and is confirmed by a second follow-up blood lead level performed within two weeks after the employer receives the results of the first blood sampling test. Return of the employee to his or her job status depends on a worker's blood lead level declining to 40 μg/dl.

As part of the interim standard, the employer is required to notify in writing each employee whose blood lead level exceeds 40 μg/dl. In addition each such employee is to be informed that the standard requires medical removal with MRP benefits, discussed below, when an employee's blood lead level exceeds the above defined limit.

In addition to the above blood lead level criterion, temporary worker removal may also take place as a result of medical determinations and recommendations. Written medical opinions must be prepared after each examination pursuant to the standard. If the examining physician includes a medical finding, determination or opinion that the employee has a medical condition which places the employee at increased risk of material health impairment from exposure to lead, then the employee must be removed from exposure to lead at or above 30 μg/m^3 . Alternatively, if the examining physician recommends special protective measures for an employee (e.g., use of a powered air purifying respirator) or recommends limitations on an employee's exposure to lead, then the employer must implement these recommendations.

Recommendations may be more stringent than the specific provisions of the standard. The examining physician, therefore, is given broad flexibility to tailor special protective procedures to the needs of individual employees. This flexibility extends to the evaluation and management of pregnant workers and male and female workers who are planning to raise children. Based on the history, physical examination, and laboratory studies, the physician might recommend special protective measures or medical removal for an employee who is pregnant or who is planning to conceive a child when, in the physician's judgment, continued exposure to lead at the current job would pose a significant risk. The return of the employee to his or her former job status, or the removal of special protections or limitations, depends upon the examining physician determining that the employee is no longer at increased risk of material impairment or that special measures are no longer needed.

During the period of any form of special protection or removal, the employer must maintain the worker's earnings, seniority, and other employment rights and benefits (as though the worker had not been removed) for a period of up to 18 months or for as long as the job the employee was removed from lasts if less than 18 months. This economic protection will maximize meaningful worker participation in the medical surveillance program, and is appropriate as part of the employer's overall obligation to provide a safe and healthful workplace. The provisions of MRP benefits during the employee's removal period may, however, be conditioned upon participation in medical surveillance.

The lead standard provides for a multiple physician review in cases where the employee wishes a second opinion concerning potential lead poisoning or toxicity. If an employee wishes a second opinion, he or she can make an appointment with a physician of his or her choice. This second physician will review the findings, recommendations or determinations of the first physician and conduct any examinations, consultations or tests deemed necessary in an attempt to make a final medical determination. If the first and second physicians do not agree in their assessment they must try to resolve their differences. If they cannot reach an agreement then they must designate a third physician to resolve the dispute.

The employer must provide examining and consulting physicians with the following specific information: A copy of the lead regulations and all appendices, a description of the employee's duties as related to exposure, the exposure level or anticipated level to lead and any other toxic substances (if applicable), a description of personal protective equipment used, blood lead levels, and all prior written medical opinions regarding the employee in the employer's possession or control. The employer must also obtain from the physician and provide the employee with a written medical opinion containing blood lead levels, the physicians's opinion as to whether the employee is at risk of material impairment to health, any recommended protective measures for the employee if further exposure is permitted, as well as any recommended limitations upon an employee's use of respirators.

Employers must instruct each physician not to reveal to the employer in writing or in any other way his or her findings, laboratory results, or diagnoses which are felt to be unrelated to occupational lead exposure. They must also instruct each physician to advise the employee of any occupationally or non-occupationally related medical condition requiring further treatment or evaluation.

The standard provides for the use of respirators where engineering and other primary controls are not effective. However, the use of respirator protection shall not be used in lieu of temporary medical removal due to elevated blood lead levels or findings that an employee is at risk of material health impairment. This is based on the numerous inadequacies of respirators including skin rash where the facepiece makes contact with the skin, unacceptable stress to breathing in some workers with underlying cardiopulmonary impairment, difficulty in providing adequate fit, the tendency for respirators to create additional hazards by interfering with vision, hearing, and mobility, and the difficulties of assuring the maximum effectiveness of a complicated work practice program involving respirators. Respirators do, however, serve a useful function where engineering and work practice controls are inadequate by providing supplementary, interim, or short- term protection, provided they are properly selected for the environment in which the employee will be working, properly fitted to the employee, maintained and cleaned periodically, and worn by the employee when required.

In its interim final standard on occupational exposure to inorganic lead in the construction industry, OSHA has prohibited prophylactic chelation. Diagnostic and therapeutic chelation are permitted only under the supervision of a licensed physician with appropriate medical monitoring in an acceptable clinical setting. The decision to initiate chelation therapy must be made on an individual basis and take into account the severity of symptoms felt to be a result of lead toxicity along with blood lead levels, ZPP levels, and other laboratory tests as appropriate. EDTA and penicillamine which are the primary chelating agents used in the therapy of occupational lead poisoning have significant potential side effects and their use must be justified on the basis of expected benefits to the worker. Unless frank and severe symptoms are present, therapeutic chelation is not recommended, given the opportunity to remove a worker from exposure and allow the body to naturally excrete accumulated lead. As a diagnostic aid, the chelation mobilization test using CA-EDTA has limited applicability. According to some investigators, the test can differentiate between lead-induced and other nephropathies. The test may also provide an estimation of the mobile fraction of the total body lead burden.

Employers are required to assure that accurate records are maintained on exposure assessment, including environmental monitoring, medical surveillance, and medical removal for each employee. Exposure assessment records must be kept for at least 30 years. Medical surveillance records must be kept for the duration of employment plus 30 years except in cases where the employment was less than one year. If duration of employment is less than one year, the employer need not retain this record beyond the term of employment if the record is provided to the employee upon termination of employment. Medical removal records also must be maintained for the duration of employment. All records required under the standard must be made available upon request to the Assistant Secretary of Labor for Occupational Safety and Health and the Director of the National Institute for Occupational Safety and Health. Employers must also make environmental and biological monitoring and medical removal records available to affected employees and to former employees or their authorized employee representatives. Employees or their specifically designated representatives have access to their entire medical surveillance records.

In addition, the standard requires that the employer inform all workers exposed to lead at or above 30 μg/m^3 of the provisions of the standard and all its appendices, the purpose and description of medical surveillance and provisions for medical removal protection if temporary removal is required. An understanding of the potential health effects of lead exposure by all exposed employees along with full understanding of their rights under the lead standard is essential for an effective monitoring program.

II. Adverse Health Effects of Inorganic Lead

Although the toxicity of lead has been known for 2,000 years, the knowledge of the complex relationship between lead exposure and human response is still being refined. Significant research into the toxic properties of lead continues throughout the world, and it should be anticipated that our understanding of thresholds of effects and margins of safety will be improved in future years. The provisions of the lead standard are founded on two prime medical judgments: First, the prevention of adverse health effects from exposure to lead throughout a working lifetime requires that worker blood lead levels be maintained at or below 40 μg/dl and second, the blood lead levels of workers, male or female, who intend to parent in the near future should be maintained below 30 μg/dl to minimize adverse reproductive health effects to the parents and developing fetus. The adverse effects of lead on reproduction are being actively researched and OSHA encourages the physician to remain abreast of recent developments in the area to best advise pregnant workers or workers planning to conceive children.

The spectrum of health effects caused by lead exposure can be subdivided into five developmental stages: Normal, physiological changes of uncertain significance, pathophysiological changes, overt symptoms (morbidity), and mortality. Within this process there are no sharp distinctions, but rather a continuum of effects. Boundaries between categories overlap due to the wide variation of individual responses and exposures in the working population. OSHA's development of the lead standard focused on pathophysiological changes as well as later stages of disease.

1. *Heme Synthesis Inhibition.* The earliest demonstrated effect of lead involves its ability to inhibit at least two enzymes of the heme synthesis pathway at very low blood levels. Inhibition of delta aminolevulinic acid dehydrase (ALA-D) which catalyzes the conversion of delta-aminolevulinic acid (ALA) to protoporphyrin is observed at a blood lead level below 20 µg/dl. At a blood lead level of 40 µg/ dl, more than 20% of the population would have 70% inhibition of ALA-D. There is an exponential increase in ALA excretion at blood lead levels greater than 40 µg/dl.

   Another enzyme, ferrochelatase, is also inhibited at low blood lead levels. Inhibition of ferrochelatase leads to increased free erythrocyte protoporphyrin (FEP) in the blood which can then bind to zinc to yield zinc protoporphyrin. At a blood lead level of 50 µg/dl or greater, nearly 100% of the population will have an increase in FEP. There is also an exponential relationship between blood lead levels greater than 40 µg/dl and the associated ZPP level, which has led to the development of the ZPP screening test for lead exposure.

   While the significance of these effects is subject to debate, it is OSHA's position that these enzyme disturbances are early stages of a disease process which may eventually result in the clinical symptoms of lead poisoning. Whether or not the effects do progress to the later stages of clinical disease, disruption of these enzyme processes over a working lifetime is considered to be a material impairment of health.

   One of the eventual results of lead-induced inhibition of enzymes in the heme synthesis pathway is anemia which can be asymptomatic if mild but associated with a wide array of symptoms including dizziness, fatigue, and tachycardia when more severe. Studies have indicated that lead levels as low as 50 µg/dl can be associated with a definite decreased hemoglobin, although most cases of lead-induced anemia, as well as shortened red-cell survival times, occur at lead levels exceeding 80 µg/dl. Inhibited hemoglobin synthesis is more common in chronic cases whereas shortened erythrocyte life span is more common in acute cases.

   In lead-induced anemias, there is usually a reticulocytosis along with the presence of basophilic stippling, and ringed sideroblasts, although none of the above are pathognomonic for lead-induced anemia.

2. *Neurological Effects.* Inorganic lead has been found to have toxic effects on both the central and peripheral nervous systems. The earliest stages of lead-induced central nervous system effects first manifest themselves in the form of behavioral disturbances and central nervous system symptoms including irritability, restlessness, insomnia and other sleep disturbances, fatigue, vertigo, headache, poor memory, tremor, depression, and apathy. With more severe exposure, symptoms can progress to drowsiness, stupor, hallucinations, delirium, convulsions and coma.

   The most severe and acute form of lead poisoning which usually follows ingestion or inhalation of large amounts of lead is acute encephalopathy which may arise precipitously with the onset of intractable seizures, coma, cardiorespiratory arrest, and death within 48 hours.

   While there is disagreement about what exposure levels are needed to produce the earliest symptoms, most experts agree that symptoms definitely can occur at blood lead levels of 60 µg/dl whole blood and therefore recommend a 40 µg/dl maximum. The central nervous system effects frequently are not reversible following discontinued exposure or chelation therapy and when improvement does occur, it is almost always only partial.

   The peripheral neuropathy resulting from lead exposure characteristically involves only motor function with minimal sensory damage and has a marked predilection for the extensor muscles of the most active extremity. The peripheral neuropathy can occur with varying degrees of severity. The earliest and mildest form which can be detected in workers with blood lead levels as low as 50 µg/dl is manifested by slowing of motor nerve conduction velocity often without clinical symptoms. With progression of the neuropathy there is development of painless extensor muscle weakness usually involving the extensor muscles of the fingers and hand in the most active upper extremity, followed in severe cases by wrist drop or, much less commonly, foot drop.

   In addition to slowing of nerve conduction, electromyographical studies in patients with blood lead levels greater than 50 µg/dl have demonstrated a decrease in the number of acting motor unit potentials, an increase in the duration of motor unit potentials, and spontaneous pathological activity including fibrillations and fasciculations. Whether these effects occur at levels of 40 µg/dl is undetermined.

   While the peripheral neuropathies can occasionally be reversed with therapy, again such recovery is not assured particularly in the more severe neuropathies and often improvement is only partial. The lack of reversibility is felt to be due in part to segmental demyelination.

3. *Gastrointestinal.* Lead may also affect the gastrointestinal system producing abdominal colic or diffuse abdominal pain, constipation, obstipation, diarrhea, anorexia, nausea and vomiting. Lead colic rarely develops at blood lead levels below 80 µg/dl.

4. *Renal.* Renal toxicity represents one of the most serious health effects of lead poisoning. In the early stages of disease nuclear inclusion bodies can frequently be identified in proximal renal tubular cells. Renal function remains normal and the changes in this stage are probably reversible. With more advanced disease there is progressive interstitial fibrosis and impaired renal function. Eventually extensive interstitial fibrosis ensues with sclerotic glomeruli and dilated and atrophied proximal tubules; all represent end stage kidney disease. Azotemia can be progressive, eventually resulting in frank uremia necessitating dialysis. There is occasionally associated hypertension and hyperuricemia with or without gout.

   Early kidney disease is difficult to detect. The urinalysis is normal in early lead nephropathy and the blood urea nitrogen and serum creatinine increase only when two-thirds of kidney function is lost. Measurement of creatinine clearance can often detect earlier disease as can other methods of measurement of glomerular filtration rate. An abnormal Ca-EDTA mobilization test has been used to differentiate between lead-induced and other nephropathies, but this procedure is not widely accepted. A form of Fanconi syndrome with aminoaciduria, glycosuria, and hyperphosphaturia indicating severe injury to the proximal renal tubules is occasionally seen in children.

5. *Reproductive effects.* Exposure to lead can have serious effects on reproductive function in both males and females. In male workers exposed to lead there can be a decrease in sexual drive, impotence, decreased ability to produce healthy sperm, and sterility. Malformed sperm (teratospermia), decreased number of sperm (hypospermia), and sperm with decreased motility (asthenospermia) can all occur. Teratospermia has been noted at mean blood lead levels of 53 µg/dl and hypospermia and asthenospermia at 41 µg/dl. Furthermore, there appears to be a dose-response relationship for teratospermia in lead exposed workers.

   Women exposed to lead may experience menstrual disturbances including dysmenorrhea, menorrhagia and amenorrhea. Following exposure to lead, women have a higher frequency of sterility, premature births, spontaneous miscarriages, and stillbirths.

   Germ cells can be affected by lead and cause genetic damage in the egg or sperm cells before conception and result in failure to implant, miscarriage, stillbirth, or birth defects.

   Infants of mothers with lead poisoning have a higher mortality during the first year and suffer from lowered birth weights, slower growth, and nervous system disorders.

   Lead can pass through the placental barrier and lead levels in the mother's blood are comparable to concentrations of lead in the umbilical cord at birth. Transplacental passage becomes detectable at 12-14 weeks of gestation and increases until birth.

   There is little direct data on damage to the fetus from exposure to lead but it is generally assumed that the fetus and newborn would be at least as susceptible to neurological damage as young children. Blood lead levels of 50-60 µg/dl in children can cause significant neurobehavioral impairments and there is evidence of hyperactivity at blood levels as low as 25 µg/dl. Given the overall body of literature concerning the adverse health effects of lead in children, OSHA feels that the blood lead level in children should be maintained below 30 µg/dl with a population mean of 15 µg/dl. Blood lead levels in the fetus and newborn likewise should not exceed 30 µg/dl.

   Because of lead's ability to pass through the placental barrier and also because of the demonstrated adverse effects of lead on reproductive function in both the male and female as well as the risk of genetic damage of lead on both the ovum and sperm, OSHA recommends a 30 µg/dl maximum permissible blood lead level in both males and females who wish to bear children.

6. *Other toxic effects.* Debate and research continue on the effects of lead on the human body. Hypertension has frequently been noted in occupationally exposed individuals although it is difficult to

assess whether this is due to lead's adverse effects on the kidney or if some other mechanism is involved. Vascular and electrocardiographic changes have been detected but have not been well characterized. Lead is thought to impair thyroid function and interfere with the pituitary-adrenal axis, but again these effects have not been well defined.

III. Medical Evaluation

The most important principle in evaluating a worker for any occupational disease including lead poisoning is a high index of suspicion on the part of the examining physician. As discussed in Section 2, lead can affect numerous organ systems and produce a wide array of signs and symptoms, most of which are non-specific and subtle in nature at least in the early stages of disease. Unless serious concern for lead toxicity is present, many of the early clues to diagnosis may easily be overlooked.

The crucial initial step in the medical evaluation is recognizing that a worker's employment can result in exposure to lead. The worker will frequently be able to define exposures to lead and lead containing materials but often will not volunteer this information unless specifically asked. In other situations the worker may not know of any exposures to lead but the suspicion might be raised on the part of the physician because of the industry or occupation of the worker. Potential occupational exposure to lead and its compounds occur in many occupations in the construction industry, including demolition and salvaging operations, removal or encapsulation of materials containing lead, construction, alteration, repair or renovation of structures containing lead, transportation, disposal, storage or containment of lead or lead-containing materials on construction sites, and maintenance operations associated with construction activities.

Once the possibility for lead exposure is raised, the focus can then be directed toward eliciting information from the medical history, physical exam, and finally from laboratory data to evaluate the worker for potential lead toxicity.

A complete and detailed work history is important in the initial evaluation. A listing of all previous employment with information on job description, exposure to fumes or dust, known exposures to lead or other toxic substances, a description of any personal protective equipment used, and previous medical surveillance should all be included in the worker's record. Where exposure to lead is suspected, information concerning on-the-job personal hygiene, smoking or eating habits in work areas, laundry procedures, and use of any protective clothing or respiratory protection equipment should be noted. A complete work history is essential in the medical evaluation of a worker with suspected lead toxicity, especially when long term effects such as neurotoxicity and nephrotoxicity are considered.

The medical history is also of fundamental importance and should include a listing of all past and current medical conditions, current medications including proprietary drug intake, previous surgeries and hospitalizations, allergies, smoking history, alcohol consumption, and also non-occupational lead exposures such as hobbies (hunting, riflery). Also known childhood exposures should be elicited. Any previous history of hematological, neurological, gastrointestinal, renal, psychological, gynecological, genetic, or reproductive problems should be specifically noted.

A careful and complete review of systems must be performed to assess both recognized complaints and subtle or slowly acquired symptoms which the worker might not appreciate as being significant. The review of symptoms should include the following:

1. *General*— weight loss fatigue, decreased appetite.
2. *Head, Eyes, Ears, Nose, Throat (HEENT)* — headaches, visual disturbances or decreased visual acuity, hearing deficits or tinnitus, pigmentation of the oral mucosa, or metallic taste in mouth.
3. *Cardio-pulmonary* — shortness of breath, cough, chest pains, palpitations, or orthopnea.
4. *Gastrointestinal* — nausea, vomiting, heartburn, abdominal pain, constipation or diarrhea.
5. *Neurologic* — irritability, insomnia, weakness (fatigue), dizziness, loss of memory, confusion, hallucinations, incoordination, ataxia, decreased strength in hands or feet, disturbances in gait, difficulty in climbing stairs, or seizures.
6. *Hematologic* — pallor, easy fatigability, abnormal blood loss, melena.
7. *Reproductive (male and female and spouse where relevant)* — history of infertility, impotence, loss of libido, abnormal menstrual periods, history of miscarriages, stillbirths, or children with birth defects.
8. *Musculo-skeletal* — muscle and joint pains.

The physical examination should emphasize the neurological, gastrointestinal, and cardiovascular systems. The worker's weight and blood pressure should be recorded and the oral mucosa checked for pigmentation characteristic of a possible Burtonian or lead line on the gingiva. It should be noted, however, that the lead line may not be present even in severe lead poisoning if good oral hygiene is practiced.

The presence of pallor on skin examination may indicate an anemia which, if severe, might also be associated with a tachycardia. If an anemia is suspected, an active search for blood loss should be undertaken including potential blood loss through the gastrointestinal tract.

A complete neurological examination should include an adequate mental status evaluation including a search for behavioral and psychological disturbances, memory testing, evaluation for irritability, insomnia, hallucinations, and mental clouding. Gait and coordination should be examined along with close observation for tremor. A detailed evaluation of peripheral nerve function including careful sensory and motor function testing is warranted. Strength testing particularly of extensor muscle groups of all extremities is of fundamental importance.

Cranial nerve evaluation should also be included in the routine examination.

The abdominal examination should include auscultation for bowel sounds and abdominal bruits and palpation for organomegaly, masses, and diffuse abdominal tenderness.

Cardiovascular examination should evaluate possible early signs of congestive heart failure. Pulmonary status should be addressed particularly if respirator protection is contemplated.

As part of the medical evaluation, the interim lead standard requires the following laboratory studies:

1. *Blood lead level*
2. *Hemoglobin and hematocrit determinations,* red cell indices, and examination of the peripheral blood smear to evaluate red blood cell morphology
3. *Blood urea nitrogen*
4. *Serum creatinine*
5. *Routine urinalysis with microscopic examination.*
6. *A zinc protoporphyrin level.*

In addition to the above, the physician is authorized to order any further laboratory or other tests which he or she deems necessary in accordance with sound medical practice. The evaluation must also include pregnancy testing or laboratory evaluation of male fertility if requested by the employee. Additional tests which are probably not warranted on a routine basis but may be appropriate when blood lead and ZPP levels are equivocal include delta aminolevulinic acid and coproporphyrin concentrations in the urine, and dark-field illumination for detection of basophilic stippling in red blood cells.

If an anemia is detected further studies including a careful examination of the peripheral smear, reticulocyte count, stool for occult blood, serum iron, total iron binding capacity, bilirubin, and, if appropriate, vitamin B12 and folate may be of value in attempting to identify the cause of the anemia.

If a peripheral neuropathy is suspected, nerve conduction studies are warranted both for diagnosis and as a basis to monitor any therapy.

If renal disease is questioned, a 24 hour urine collection for creatinine clearance, protein, and electrolytes may be indicated. Elevated uric acid levels may result from lead-induced renal disease and a serum uric acid level might be performed.

An electrocardiogram and chest x-ray may be obtained as deemed appropriate.

Sophisticated and highly specialized testing should not be done routinely and where indicated should be under the direction of a specialist.

IV. Laboratory Evaluation

The blood lead level at present remains the single most important test to monitor lead exposure and is the test used in the medical surveillance program under the lead standard to guide employee medical removal. The ZPP has several advantages over the blood lead level. Because of its relatively recent development and the lack of extensive data concerning its interpretation, the ZPP currently remains an ancillary test.

This section will discuss the blood lead level and ZPP in detail and will outline their relative advantages and disadvantages. Other blood tests currently available to evaluate lead exposure will also be reviewed.

The blood lead level is a good index of current or recent lead absorption when there is no anemia present and when the worker has not taken any chelating agents. However, blood lead levels along with urinary lead levels do not necessarily indicate the total body burden of lead and are not adequate measures of past exposure. One reason for this is that lead has a high affinity for bone and up to 90% of the body's total lead is deposited there. A very important component of the total lead body burden is lead in soft tissue (liver, kidney, and brain). This fraction of the lead body burden, the biologically active lead, is not entirely reflected by blood lead levels since it is a function of the dynamics of lead absorption, distribution, deposition in bone and excretion. Following discontinuation of exposure to lead, the excess body burden is only slowly mobilized from bone and other relatively stable body stores and excreted. Consequently, a high blood lead level may only represent recent heavy exposure to lead without a significant total body excess and likewise a low blood lead level does not exclude an elevated total body burden of lead.

Also due to its correlation with recent exposures, the blood lead level may vary considerably over short time intervals.

To minimize laboratory error and erroneous results due to contamination, blood specimens must be carefully collected after thorough cleaning of the skin with appropriate methods using lead-free blood containers and analyzed by a reliable laboratory. Under the standard, samples must be analyzed in laboratories which are approved by OSHA. Analysis is to be made using atomic absorption spectrophotometry, anodic stripping voltammetry or any method which meets the accuracy requirements set forth by the standard.

The determination of lead in urine is generally considered a less reliable monitoring technique than analysis of whole blood primarily due to individual variability in urinary excretion capacity as well as the technical difficulty of obtaining accurate 24 hour urine collections. In addition, workers with renal insufficiency, whether due to lead or some other cause, may have decreased lead clearance and consequently urine lead levels may underestimate the true lead burden. Therefore, urine lead levels should not be used as a routine test.

The zinc protoporphyrin test, unlike the blood lead determination, measures an adverse metabolic effect of lead and as such is a better indicator of lead toxicity than the level of blood lead itself. The level of ZPP reflects lead absorption over the preceding 3 to 4 months, and therefore is a better indicator of lead body burden. The ZPP requires more time than the blood lead to read significantly elevated levels; the return to normal after discontinuing lead exposure is also slower. Furthermore, the ZPP test is simpler, faster, and less expensive to perform and no contamination is possible. Many investigators believe it is the most reliable means of monitoring chronic lead absorption.

Zinc protoporphyrin results from the inhibition of the enzyme ferrochelatase which catalyzes the insertion of an iron molecule into the protoporphyrin molecule, which then becomes heme. If iron is not inserted into the molecule then zinc, having a greater affinity for protoporphyrin, takes the place of the iron, forming ZPP.

An elevation in the level of circulating ZPP may occur at blood lead levels as low as 20-30 µg/dl in some workers. Once the blood lead level has reached 40 µg/dl there is more marked rise in the ZPP value from its normal range of less than 100 µg/dl 100 ml. Increases in blood lead levels beyond 40 µg/100 g are associated with exponential increases in ZPP.

Whereas blood lead levels fluctuate over short time spans, ZPP levels remain relatively stable. ZPP is measured directly in red blood cells and is present for the cell's entire 120 day life-span. Therefore, the ZPP level in blood reflects the average ZPP production over the previous 3-4 months and consequently the average lead exposure during that time interval.

It is recommended that a hematocrit be determined whenever a confirmed ZPP of 50 µg/100 ml whole blood is obtained to rule out a significant underlying anemia. If the ZPP is in excess of 100 µg/100 ml and not associated with abnormal elevations in blood lead levels, the laboratory should be checked to be sure that blood leads were determined using atomic absorption spectrophotometry anodic stripping voltammetry, or any method which meets the accuracy requirements set forth by the standard by an OSHA approved laboratory which is experienced in lead level determinations. Repeat periodic blood lead studies should be obtained in all individuals with elevated ZPP levels to be certain that an associated elevated blood lead level has not been missed due to transient fluctuations in blood leads.

ZPP has a characteristic fluorescence spectrum with a peak at 594 nm which is detectable with a hematofluorimeter. The hematofluorimeter is accurate and portable and can provide on-site, instantaneous results for workers who can be frequently tested via a finger prick.

However, careful attention must be given to calibration and quality control procedures. Limited data on blood lead-ZPP correlations and the ZPP levels which are associated with the adverse health effects discussed in Section 2 are the major limitations of the test. Also it is difficult to correlate ZPP levels with environmental exposure and there is some variation of response with age and sex. Nevertheless, the ZPP promises to be an important diagnostic test for the early detection of lead toxicity and its value will increase as more data is collected regarding its relationship to other manifestations of lead poisoning.

Levels of delta-aminolevulinic acid (ALA) in the urine are also used as a measure of lead exposure. Increasing concentrations of ALA are believed to result from the inhibition of the enzyme delta- aminolevulinic acid dehydrase (ALA-D). Although the test is relatively easy to perform, inexpensive, and rapid, the disadvantages include variability in results, the necessity to collect a complete 24 hour urine sample which has a specific gravity greater than 1.010, and also the fact that ALA decomposes in the presence of light.

The pattern of porphyrin excretion in the urine can also be helpful in identifying lead intoxication. With lead poisoning, the urine concentrations of coproporphyrins I and II, porphobilinogen and uroporphyrin I rise. The most important increase, however, is that of coproporphyrin III; levels may exceed 5,000 µg/l in the urine in lead poisoned individuals, but its correlation with blood lead levels and ZPP are not as good as those of ALA. Increases in urinary porphyrins are not diagnostic of lead toxicity and may be seen in porphyria, some liver diseases, and in patients with high reticulocyte counts.

Summary. The Occupational Safety and Health Administration's interim standard for inorganic lead in the construction industry places significant emphasis on the medical surveillance of all workers exposed to levels of inorganic lead above 30 µg/m^3 TWA. The physician has a fundamental role in this surveillance program, and in the operation of the medical removal protection program.

Even with adequate worker education on the adverse health effects of lead and appropriate training in work practices, personal hygiene and other control measures, the physician has a primary responsibility for evaluating potential lead toxicity in the worker. It is only through a careful and detailed medical and work history, a complete physical examination and appropriate laboratory testing that an accurate assessment can be made. Many of the adverse health effects of lead toxicity are either irreversible or only partially reversible and therefore early detection of disease is very important.

This document outlines the medical monitoring program as defined by the occupational safety and health standard for inorganic lead. It reviews the adverse health effects of lead poisoning and describes the important elements of the history and physical examinations as they relate to these adverse effects. Finally, the appropriate laboratory testing for evaluating lead exposure and toxicity is presented.

It is hoped that this review and discussion will give the physician a better understanding of the OSHA standard with the ultimate goal of protecting the health and well-being of the worker exposed to lead under his or her care.

[58 FR 26627, May 4, 1993, as amended at 58 FR 34218, June 24, 1993; 61 FR 5510, Feb. 13, 1996; 63 FR 1296, Jan. 8, 1998; 70 FR 1143, Jan. 5, 2005; 71 FR 16674, Apr. 3, 2006; 71 FR 50191, Aug. 24, 2006; 73 FR 75588, Dec. 12, 2008; 76 FR 33611, June 8, 2011; 76 FR 80741, Dec. 27, 2011; 77 FR 17890, Mar. 26, 2012]

§1926.64

Process safety management of highly hazardous chemicals

Purpose. This section contains requirements for preventing or minimizing the consequences of catastrophic releases of toxic, reactive, flammable, or explosive chemicals. These releases may result in toxic, fire or explosion hazards.[§1926.64]

(a) Application.[§1926.64(a)]

(1) *This section applies to the following:*[§1926.64(a)(1)]

(i) *A process which involves* a chemical at or above the specified threshold quantities listed in Appendix A to this section; [§1926.64(a)(1)(i)]

(ii) *A process which involves* a Category 1 flammable gas (as defined in Sec. 1910.1200(c)) or flammable liquid with a flashpoint below 100 °F (37.8 °C) on site in one location, in a quantity of 10,000 pounds (4535.9 kg) or more except for: [§1926.64(a)(1)(ii)]

[A] Hydrocarbon fuels used solely for workplace consumption as a fuel (e.g., propane used for comfort heating, gasoline for vehicle refueling), if such fuels are not a part of a process containing another highly hazardous chemical covered by this standard;[§1926.64(a)(1)(ii)[A]]

[B] Flammable liquids with a flashpoint below 100 °F (37.8 °C) stored in atmospheric tanks or transferred that are kept below their normal boiling point without benefit of chilling or refrigeration.[§1926.64(a)(1)(ii)[B]]

(2) *This section does not apply to:*[§1926.64(a)(2)]

(i) *Retail facilities;*[§1926.64(a)(2)(i)]

(ii) *Oil or gas well drilling or servicing operations; or,* [§1926.64(a)(2)(ii)]

(iii) *Normally unoccupied remote facilities.*[§1926.64(a)(2)(iii)]

(b) **Definitions.**[§1926.64(b)]

Atmospheric tank means a storage tank which has been designed to operate at pressures from atmospheric through 0.5 p.s.i.g. (pounds per square inch gauge, 3.45 Kpa).[§1926.64(b)]

Boiling point means the boiling point of a liquid at a pressure of 14.7 pounds per square inch absolute (p.s.i.a.) (760 mm.). For the purposes of this section, where an accurate boiling point is unavailable for the material in question, or for mixtures which do not have a constant boiling point, the 10 percent point of a distillation performed in accordance with the Standard Method of Test for Distillation of Petroleum Products, ASTM D-86-62, may be used as the boiling point of the liquid.[§1926.64(b)]

Catastrophic release means a major uncontrolled emission, fire, or explosion, involving one or more highly hazardous chemicals, that presents serious danger to employees in the workplace. [§1926.64(b)]

Facility means the buildings, containers or equipment which contain a process.[§1926.64(b)]

Highly hazardous chemical means a substance possessing toxic, reactive, flammable, or explosive properties and specified by paragraph (a)(1) of this section.[§1926.64(b)]

Hot work means work involving electric or gas welding, cutting, brazing, or similar flame or spark-producing operations.[§1926.64(b)]

Normally unoccupied remote facility means a facility which is operated, maintained or serviced by employees who visit the facility only periodically to check its operation and to perform necessary operating or maintenance tasks. No employees are permanently stationed at the facility. Facilities meeting this definition are not contiguous with, and must be geographically remote from all other buildings, processes or persons.[§1926.64(b)]

Process means any activity involving a highly hazardous chemical including any use, storage, manufacturing, handling, or the on-site movement of such chemicals, or combination of these activities. For purposes of this definition, any group of vessels which are interconnected and separate vessels which are located such that a highly hazardous chemical could be involved in a potential release shall be considered a single process.[§1926.64(b)]

Replacement in kind means a replacement which satisfies the design specification.[§1926.64(b)]

Trade secret means any confidential formula, pattern, process, device, information or compilation of information that is used in an employer's business, and that gives the employer an opportunity to obtain an advantage over competitors who do not know or use it. Appendix D contained in Sec. 1926.59 sets out the criteria to be used in evaluating trade secrets.[§1926.64(b)]

(c) **Employee participation.**[§1926.64(c)]

(1) *Employers shall develop a written plan* of action regarding the implementation of the employee participation required by this paragraph.[§1926.64(c)(1)]

(2) *Employers shall consult with employees* and their representatives on the conduct and development of process hazards analyses and on the development of the other elements of process safety management in this standard.[§1926.64(c)(2)]

(3) *Employers shall provide to employees* and their representatives access to process hazard analyses and to all other information required to be developed under this standard. [§1926.64(c)(3)]

(d) **Process safety information.** In accordance with the schedule set forth in paragraph (e)(1) of this section, the employer shall complete a compilation of written process safety information before conducting any process hazard analysis required by the standard. The compilation of written process safety information is to enable the employer and the employees involved in operating the process to identify and understand the hazards posed by those processes involving highly hazardous chemicals. This process safety information shall include information pertaining to the hazards of the highly hazardous chemicals used or produced by the process, information pertaining to the technology of the process, and information pertaining to the equipment in the process.[§1926.64(d)]

(1) *Information pertaining to the hazards* of the highly hazardous chemicals in the process. This information shall consist of at least the following:[§1926.64(d)(1)]

(i) *Toxicity information;*[§1926.64(d)(1)(i)]

(ii) *Permissible exposure limits;*[§1926.64(d)(1)(ii)]

(iii) *Physical data;*[§1926.64(d)(1)(iii)]

(iv) *Reactivity data:*[§1926.64(d)(1)(iv)]

(v) *Corrosivity data;*[§1926.64(d)(1)(v)]

(vi) *Thermal and chemical stability data; and*[§1926.64(d)(1)(vi)]

(vii) *Hazardous effects of inadvertent* mixing of different materials that could foreseeably occur.[§1926.64(d)(1)(vii)]

Note to paragraph (d)(1): Safety data sheets meeting the requirements of Sec. 1910.1200(g) may be used to comply with this requirement to the extent they contain the information required by this paragraph (d)(1).

(2) *Information pertaining to the technology of the process.* [§1926.64(d)(2)]

(i) *Information concerning the technology* of the process shall include at least the following:[§1926.64(d)(2)(i)]

[A] A block flow diagram or simplified process flow diagram (see Appendix B to this section);[§1926.64(d)(2)(i)[A]]

[B] Process chemistry;[§1926.64(d)(2)(i)[B]]

[C] Maximum intended inventory;[§1926.64(d)(2)(i)[C]]

[D] Safe upper and lower limits for such items as temperatures, pressures, flows or compositions; and, [§1926.64(d)(2)(i)[D]]

[E] An evaluation of the consequences of deviations, including those affecting the safety and health of employees. [§1926.64(d)(2)(i)[E]]

(ii) *Where the original technical information* no longer exists, such information may be developed in conjunction with the process hazard analysis in sufficient detail to support the analysis.[§1926.64(d)(2)(ii)]

(3) *Information pertaining to the equipment* in the process. [§1926.64(d)(3)]

(i) *Information pertaining to the equipment* in the process shall include:[§1926.64(d)(3)(i)]

[A] Materials of construction;[§1926.64(d)(3)(i)[A]]

[B] Piping and instrument diagrams (P&ID's); [§1926.64(d)(3)(i)[B]]

[C] Electrical classification;[§1926.64(d)(3)(i)[C]]

[D] Relief system design and design basis;[§1926.64(d)(3)(i)[D]]

[E] Ventilation system design;[§1926.64(d)(3)(i)[E]]

[F] Design codes and standards employed;[§1926.64(d)(3)(i)[F]]

[G] Material and energy balances for processes built after May 26, 1992; and,[§1926.64(d)(3)(i)[G]]

[H] Safety systems (e.g. interlocks, detection or suppression systems).[§1926.64(d)(3)(i)[H]]

(ii) *The employer shall document that equipment* complies with recognized and generally accepted good engineering practices.[§1926.64(d)(3)(ii)]

(iii) *For existing equipment designed* and constructed in accordance with codes, standards, or practices that are no longer in general use, the employer shall determine and document that the equipment is designed, maintained, inspected, tested, and operating in a safe manner.[§1926.64(d)(3)(iii)]

(e) **Process hazard analysis.**[§1926.64(e)]

(1) *The employer shall perform an initial* process hazard analysis (hazard evaluation) on processes covered by this standard. The process hazard analysis shall be appropriate to the complexity of the process and shall identify, evaluate, and control the hazards involved in the process. Employers shall determine and document the priority order for conducting process hazard analyses based on a rationale which includes such considerations as extent of the process hazards, number of potentially affected employees, age of the process, and operating history of the process. The process hazard analysis shall be conducted as soon as possible, but not later than the following schedule: [§1926.64(e)(1)]

(i) *No less than 25 percent* of the initial process hazards analyses shall be completed by May 26, 1994;[§1926.64(e)(1)(i)]

(ii) *No less than 50 percent* of the initial process hazards analyses shall be completed by May 26, 1995;[§1926.64(e)(1)(ii)]

(iii) *No less than 75 percent* of the initial process hazards analyses shall be completed by May 26, 1996;[§1926.64(e)(1)(iii)]

(iv) *All initial process hazards analyses shall be completed by May 26, 1997.*[§1926.64(e)(1)(iv)]

(v) *Process hazards analyses completed after May 26, 1987* which meet the requirements of this paragraph are acceptable as initial process hazards analyses. These process hazard analyses shall be updated and revalidated, based on their completion date, in accordance with paragraph (e)(6) of this standard.[§1926.64(e)(1)(v)]

(2) *The employer shall use one* or more of the following methodologies that are appropriate to determine and evaluate the hazards of the process being analyzed.[§1926.64(e)(2)]

(i) *What-If;*[§1926.64(e)(2)(i)]

(ii) *Checklist;*[§1926.64(e)(2)(ii)]

(iii) *What-If/Checklist;*[§1926.64(e)(2)(iii)]

(iv) *Hazard and Operability Study (HAZOP);*[§1926.64(e)(2)(iv)]

(v) *Failure Mode and EffectsAnalysis (FMEA);*[§1926.64(e)(2)(v)]

(vi) *Fault-Tree Analysis; or*[§1926.64(e)(2)(vi)]

(vii) *An appropriate equivalent methodology.*[§1926.64(e)(2)(vii)]

(3) *The process hazard analysis shall address:*[§1926.64(e)(3)]

(i) *The hazards of the process;*[§1926.64(e)(3)(i)]

(ii) *The identification of any previous incident* which had a likely potential for catastrophic consequences in the workplace; [§1926.64(e)(3)(ii)]

(iii) *Engineering and administrative controls* applicable to the hazards and their interrelationships such as appropriate application of detection methodologies to provide early warning of releases. (Acceptable detection methods might include process monitoring and control instrumentation with alarms, and detection hardware such as hydrocarbon sensors.); [§1926.64(e)(3)(iii)]

(iv) *Consequences of failure of engineering* and administrative controls;[§1926.64(e)(3)(iv)]

(v) *Facility siting;*[§1926.64(e)(3)(v)]

(vi) *Human factors; and*[§1926.64(e)(3)(vi)]

(vii) *A qualitative evaluation of a range* of the possible safety and health effects of failure of controls on employees in the workplace.[§1926.64(e)(3)(vii)]

(4) *The process hazard analysis* shall be performed by a team with expertise in engineering and process operations, and the team shall include at least one employee who has experience and knowledge specific to the process being evaluated. Also, one member of the team must be knowledgeable in the specific process hazard analysis methodology being used.[§1926.64(e)(4)]

(5) *The employer shall establish a system* to promptly address the team's findings and recommendations; assure that the recommendations are resolved in a timely manner and that the resolution is documented; document what actions are to be taken; complete actions as soon as possible; develop a written schedule of when these actions are to be completed; communicate the actions to operating, maintenance and other employees whose work assignments are in the process and who may be affected by the recommendations or actions.[§1926.64(e)(5)]

(6) *At least every five (5) years* after the completion of the initial process hazard analysis, the process hazard analysis shall be updated and revalidated by a team meeting the requirements in paragraph (e)(4) of this section, to assure that the process hazard analysis is consistent with the current process.[§1926.64(e)(6)]

(7) *Employers shall retain process hazards analyses* and updates or revalidations for each process covered by this section, as well as the documented resolution of recommendations described in paragraph (e)(5) of this section for the life of the process.[§1926.64(e)(7)]

(f) **Operating procedures.**[§1926.64(f)]

(1) *The employer shall develop and implement* written operating procedures that provide clear instructions for safely conducting activities involved in each covered process consistent with the process safety information and shall address at least the following elements.[§1926.64(f)(1)]

(i) *Steps for each operating phase:*[§1926.64(f)(1)(i)]

[A] Initial startup;[§1926.64(f)(1)(i)[A]]

[B] Normal operations;[§1926.64(f)(1)(i)[B]]

[C] Temporary operations;[§1926.64(f)(1)(i)[C]]

[D] Emergency shutdown including the conditions under which emergency shutdown is required, and the assignment of shutdown responsibility to qualified operators to ensure that emergency shutdown is executed in a safe and timely manner.[§1926.64(f)(1)(i)[D]]

[E] Emergency operations;[§1926.64(f)(1)(i)[E]]

[F] Normal shutdown; and,[§1926.64(f)(1)(i)[F]]

[G] Startup following a turnaround,or after an emergency shutdown.[§1926.64(f)(1)(i)[G]]

(ii) *Operating limits:*[§1926.64(f)(1)(ii)]

[A] Consequences of deviation; and[§1926.64(f)(1)(ii)[A]]

[B] Steps required to correct or avoid deviation. [§1926.64(f)(1)(ii)[B]]

(iii) *Safety and health considerations:*[§1926.64(f)(1)(iii)]

[A] Properties of, and hazards presented by, the chemicals used in the process;[§1926.64(f)(1)(iii)[A]]

[B] Precautions necessary to prevent exposure, including engineering controls, administrative controls, and personal protective equipment;[§1926.64(f)(1)(iii)[B]]

[C] Control measures to be taken if physical contact or airborne exposure occurs;[§1926.64(f)(1)(iii)[C]]

[D] Quality control for raw materials and control of hazardous chemical inventory levels; and,[§1926.64(f)(1)(iii)[D]]

[E] Any special or unique hazards.[§1926.64(f)(1)(iii)[E]]

(iv) *Safety systems and their functions.*[§1926.64(f)(1)(iv)]

(2) *Operating procedures* shall be readily accessible to employees who work in or maintain a process.[§1926.64(f)(2)]

(3) *The operating procedures shall be reviewed* as often as necessary to assure that they reflect current operating practice, including changes that result from changes in process chemicals, technology, and equipment, and changes to facilities. The employer shall certify annually that these operating procedures are current and accurate.[§1926.64(f)(3)]

(4) *The employer shall develop and implement* safe work practices to provide for the control of hazards during operations such as lockout/ tagout; confined space entry; opening process equipment or piping; and control over entrance into a facility by maintenance, contractor, laboratory, or other support personnel. These safe work practices shall apply to employees and contractor employees.[§1926.64(f)(4)]

(g) **Training.**[§1926.64(g)]

(1) *Initial training.*[§1926.64(g)(1)]

(i) *Each employee presently involved* in operating a process, and each employee before being involved in operating a newly assigned process, shall be trained in an overview of the process and in the operating procedures as specified in paragraph (f) of this section. The training shall include emphasis on the specific safety and health hazards, emergency operations including shutdown, and safe work practices applicable to the employee's job tasks.[§1926.64(g)(1)(i)]

(ii) *In lieu of initial training* for those employees already involved in operating a process on May 26, 1992, an employer may certify in writing that the employee has the required knowledge, skills, and abilities to safely carry out the duties and responsibilities as specified in the operating procedures. [§1926.64(g)(1)(ii)]

(2) *Refresher training.* Refresher training shall be provided at least every three years, and more often if necessary, to each employee involved in operating a process to assure that the employee understands and adheres to the current operating procedures of the process. The employer, in consultation with the employees involved in operating the process, shall determine the appropriate frequency of refresher training. [§1926.64(g)(2)]

(3) *Training documentation.* The employer shall ascertain that each employee involved in operating a process has received and understood the training required by this paragraph. The employer shall prepare a record which contains the identity of the employee, the date of training, and the means used to verify that the employee understood the training.[§1926.64(g)(3)]

(h) **Contractors.**[§1926.64(h)]

(1) *Application.* This paragraph applies to contractors performing maintenance or repair, turnaround, major renovation, or specialty work on or adjacent to a covered process. It does not apply to contractors providing incidental services which do not influence process safety, such as janitorial work, food and drink services, laundry, delivery or other supply services. [§1926.64(h)(1)]

(2) *Employer responsibilities.*[§1926.64(h)(2)]

(i) *The employer,* when selecting a contractor, shall obtain and evaluate information regarding the contract employer's safety performance and programs.[§1926.64(h)(2)(i)]

(ii) *The employer shall inform* contract employers of the known potential fire, explosion, or toxic release hazards related to the contractor's work and the process.[§1926.64(h)(2)(ii)]

(iii) *The employer shall explain* to contract employers the applicable provisions of the emergency action plan required by paragraph (n) of this section.[§1926.64(h)(2)(iii)]

(iv) *The employer shall develop and implement* safe work practices consistent with paragraph (f)(4) of this section, to control the entrance, presence and exit of contract employers and contract employees in covered process areas. [§1926.64(h)(2)(iv)]

(v) *The employer shall periodically evaluate* the performance of contract employers in fulfilling their obligations as specified in paragraph (h)(3) of this section.[§1926.64(h)(2)(v)]

(vi) *The employer shall maintain a contract* employee injury and illness log related to the contractor's work in process areas. [§1926.64(h)(2)(vi)]

(3) *Contract employer responsibilities.*[§1926.64(h)(3)]

(i) *The contract employer shall assure* that each contract employee is trained in the work practices necessary to safely perform his/her job.[§1926.64(h)(3)(i)]

(ii) *The contract employer shall assure* that each contract employee is instructed in the known potential fire, explosion, or toxic release hazards related to his/her job and the process, and the applicable provisions of the emergency action plan.[§1926.64(h)(3)(ii)]

(iii) *The contract employer shall document* that each contract employee has received and understood the training required by this paragraph. The contract employer shall prepare a record which contains the identity of the contract employee, the date of training, and the means used to verify that the employee understood the training.[§1926.64(h)(3)(iii)]

(iv) *The contract employer shall assure* that each contract employee follows the safety rules of the facility including the safe work practices required by paragraph (f)(4) of this section.[§1926.64(h)(3)(iv)]

(v) *The contract employer shall advise* the employer of any unique hazards presented by the contract employer's work, or of any hazards found by the contract employer's work. [§1926.64(h)(3)(v)]

(i) Pre-startup safety review.[§1926.64(i)]

(1) *The employer shall perform a pre-startup* safety review for new facilities and for modified facilities when the modification is significant enough to require a change in the process safety information.[§1926.64(i)(1)]

(2) *The pre-startup safety review* shall confirm that prior to the introduction of highly hazardous chemicals to a process: [§1926.64(i)(2)]

(i) *Construction and equipment is in accordance* with design specifications;[§1926.64(i)(2)(i)]

(ii) *Safety, operating, maintenance,* and emergency procedures are in place and are adequate;[§1926.64(i)(2)(ii)]

(iii) *For new facilities,* a process hazard analysis has been performed and recommendations have been resolved or implemented before startup; and modified facilities meet the requirements contained in management of change, paragraph (l).[§1926.64(i)(2)(iii)]

(iv) *Training of each employee involved* in operating a process has been completed.[§1926.64(i)(2)(iv)]

(j) Mechanical integrity.[§1926.64(j)]

(1) *Application.* Paragraphs (j)(2) through (j)(6) of this section apply to the following process equipment:[§1926.64(j)(1)]

(i) *Pressure vessels and storage tanks;*[§1926.64(j)(1)(i)]

(ii) *Piping systems* (including piping components such as valves);[§1926.64(j)(1)(ii)]

(iii) *Relief and vent systems and devices;*[§1926.64(j)(1)(iii)]

(iv) *Emergency shutdown systems;*[§1926.64(j)(1)(iv)]

(v) *Controls* (including monitoring devices and sensors, alarms, and interlocks) and,[§1926.64(j)(1)(v)]

(vi) *Pumps.*[§1926.64(j)(1)(vi)]

(2) *Written procedures.* The employer shall establish and implement written procedures to maintain the on-going integrity of process equipment.[§1926.64(j)(2)]

(3) *Training for process maintenance activities.* The employer shall train each employee involved in maintaining the on-going integrity of process equipment in an overview of that process and its hazards and in the procedures applicable to the employee's job tasks to assure that the employee can perform the job tasks in a safe manner.[§1926.64(j)(3)]

(4) *Inspection and testing.*[§1926.64(j)(4)]

(i) *Inspections and tests shall be performed* on process equipment.[§1926.64(j)(4)(i)]

(ii) *Inspection and testing procedures* shall follow recognized and generally accepted good engineering practices. [§1926.64(j)(4)(ii)]

(iii) *The frequency of inspections and tests* of process equipment shall be consistent with applicable manufacturers' recommendations and good engineering practices, and more frequently if determined to be necessary by prior operating experience.[§1926.64(j)(4)(iii)]

(iv) *The employer shall document each inspection* and test that has been performed on process equipment. The documentation shall identify the date of the inspection or test, the name of the person who performed the inspection or test, the serial number or other identifier of the equipment on which the inspection or test was performed, a description of the inspection or test performed, and the results of the inspection or test.[§1926.64(j)(4)(iv)]

(5) *Equipment deficiencies.* The employer shall correct deficiencies in equipment that are outside acceptable limits (defined by the process safety information in paragraph (d) of this section) before further use or in a safe and timely manner when necessary means are taken to assure safe operation.[§1926.64(j)(5)]

(6) *Quality assurance.*[§1926.64(j)(6)]

(i) *In the construction of new plants and equipment,* the employer shall assure that equipment as it is fabricated is suitable for the process application for which they will be used.[§1926.64(j)(6)(i)]

(ii) *Appropriate checks and inspections* shall be performed to assure that equipment is installed properly and consistent with design specifications and the manufacturer's instructions.[§1926.64(j)(6)(ii)]

(iii) *The employer shall assure* that maintenance materials, spare parts and equipment are suitable for the process application for which they will be used.[§1926.64(j)(6)(iii)]

(k) Hot work permit.[§1926.64(k)]

(1) *The employer shall issue* a hot work permit for hot work operations conducted on or near a covered process.[§1926.64(k)(1)]

(2) *The permit shall document* that the fire prevention and protection requirements in 29 CFR 1926.352 have been implemented prior to beginning the hot work operations; it shall indicate the date(s) authorized for hot work; and identify the object on which hot work is to be performed. The permit shall be kept on file until completion of the hot work operations.[§1926.64(k)(2)]

(l) Management of change.[§1926.64(l)]

(1) *The employer shall establish and implement* written procedures to manage changes (except for "replacements in kind") to process chemicals, technology, equipment, and procedures; and, changes to facilities that affect a covered process.[§1926.64(l)(1)]

(2) *The procedures shall assure* that the following considerations are addressed prior to any change:[§1926.64(l)(2)]

(i) *The technical basis for the proposed change;*[§1926.64(l)(2)(i)]

(ii) *Impact of change on safety and health;*[§1926.64(l)(2)(ii)]

(iii) *Modifications to operating procedures;*[§1926.64(l)(2)(iii)]

(iv) *Necessary time period for the change; and,*[§1926.64(l)(2)(iv)]

(v) *Authorization requirements for the proposed change.* [§1926.64(l)(2)(v)]

(3) *Employees involved in operating* a process and maintenance and contract employees whose job tasks will be affected by a change in the process shall be informed of, and trained in, the change prior to start- up of the process or affected part of the process.[§1926.64(l)(3)]

(4) *If a change covered* by this paragraph results in a change in the process safety information required by paragraph (d) of this section, such information shall be updated accordingly. [§1926.64(l)(4)]

(5) *If a change covered* by this paragraph results in a change in the operating procedures or practices required by paragraph (f) of this section, such procedures or practices shall be updated accordingly.[§1926.64(l)(5)]

(m) Incident investigation.[§1926.64(m)]

(1) *The employer shall investigate* each incident which resulted in, or could reasonably have resulted in a catastrophic release of highly hazardous chemical in the workplace.[§1926.64(m)(1)]

(2) *An incident investigation shall be initiated* as promptly as possible, but not later than 48 hours following the incident. [§1926.64(m)(2)]

(3) *An incident investigation team* shall be established and consist of at least one person knowledgeable in the process involved, including a contract employee if the incident involved work of the contractor, and other persons with appropriate knowledge and experience to thoroughly investigate and analyze the incident.[§1926.64(m)(3)]

(4) *A report shall be prepared at the conclusion* of the investigation which includes at a minimum:[§1926.64(m)(4)]

(i) *Date of incident;*[§1926.64(m)(4)(i)]

(ii) *Date investigation began;*[§1926.64(m)(4)(ii)]

(iii) *A description of the incident;*[§1926.64(m)(4)(iii)]

(iv) *The factors that contributed to the incident; and,* [§1926.64(m)(4)(iv)]

(v) *Any recommendations resulting from the investigation.* [§1926.64(m)(4)(v)]

(5) *The employer shall establish a system* to promptly address and resolve the incident report findings and recommendations. Resolutions and corrective actions shall be documented. [§1926.64(m)(5)]

(6) *The report shall be reviewed with all affected* personnel whose job tasks are relevant to the incident findings including contract employees where applicable.[§1926.64(m)(6)]

(7) *Incident investigation reports shall be retained for five years.* [§1926.64(m)(7)]

(n) **Emergency planning and response.** The employer shall establish and implement an emergency action plan for the entire plant in accordance with the provisions of 29 CFR 1926.35(a). In addition, the emergency action plan shall include procedures for handling small releases. Employers covered under this standard may also be subject to the hazardous waste and emergency response provisions contained in 29 CFR 1926.65(a), (p) and (q).[§1926.64(n)]

(o) **Compliance audits.**[§1926.64(o)]

(1) *Employers shall certify that they* have evaluated compliance with the provisions of this section at least every three years to verify that the procedures and practices developed under the standard are adequate and are being followed.[§1926.64(o)(1)]

(2) *The compliance audit shall be conducted* by at least one person knowledgeable in the process.[§1926.64(o)(2)]

(3) *A report of the findings of the audit* shall be developed. [§1926.64(o)(3)]

(4) *The employer shall promptly determine* and document an appropriate response to each of the findings of the compliance audit, and document that deficiencies have been corrected. [§1926.64(o)(4)]

(5) *Employers shall retain the two* (2) most recent compliance audit reports.[§1926.64(o)(5)]

(p) **Trade secrets.**[§1926.64(p)]

(1) *Employers shall make all information* necessary to comply with the section available to those persons responsible for compiling the process safety information (required by paragraph (d) of this section), those assisting in the development of the process hazard analysis (required by paragraph (e) of this section), those responsible for developing the operating procedures (required by paragraph (f) of this section), and those involved in incident investigations (required by paragraph (m) of this section), emergency planning and response (paragraph (n) of this section) and compliance audits (paragraph (o) of this section) without regard to possible trade secret status of such information. [§1926.64(p)(1)]

(2) *Nothing in this paragraph shall preclude* the employer from requiring the persons to whom the information is made available under paragraph (p)(1) of this section to enter into confidentiality agreements not to disclose the information as set forth in 29 CFR 1926.59.[§1926.64(p)(2)]

(3) *Subject to the rules and procedures* set forth in 29 CFR 1926.59(i) (1) through (12), employees and their designated representatives shall have access to trade secret information contained within the process hazard analysis and other documents required to be developed by this standard.[§1926.64(p)(3)]

§1926.64 Appendix A

List of Highly Hazardous Chemicals, Toxics and Reactives (Mandatory)

This Appendix contains a listing of toxic and reactive highly hazardous chemicals which present a potential for a catastrophic event at or above the threshold quantity.

CHEMICAL NAME	CAS[1]	TQ[2]
Acetaldehyde	75-07-0	2500
Acrolein (2-Propenal)	107-02-8	150
Acrylyl Chloride	814-68-6	250
Allyl Chloride	107-05-1	1000
Allylamine	107-11-9	1000
Alkylaluminums	Varies	5000
Ammonia, Anhydrous	7664-41-7	10000
Ammonia solutions (greater than 44% ammonia by weight)	7664-41-7	15000
Ammonium Perchlorate	7790-98-9	500
Ammonium Permanganate	7787-36-2	7500
Arsine (also called Arsenic Hydride)	7784-42-1	100
Bis(Chloromethyl) Ether	542-88-1	100
Boron Trichloride	10294-34-5	2500
Boron Trifluoride	7637-07-2	250
Bromine	7726-95-6	1500
Bromine Chloride	13863-41-7	1500
Bromine Pentafluoride	7789-30-2	2500
Bromine Trifluoride	7787-71-5	15000
3-Bromopropyne (also called Propargyl Bromide)	106-96-7	100
Butyl Hydroperoxide (Tertiary)	75-91-2	5000
Butyl Perbenzoate (Tertiary)	614-45-9	7500
Carbonyl Chloride (see Phosgene)	75-44-5	100
Carbonyl Fluoride*	353-50-4	2500
Cellulose Nitrate (concentration greater than 12.6% nitrogen)	9004-70-0	2500
Chlorine	7782-50-5	1500
Chlorine Dioxide	10049-04-4	1000
Chlorine Pentrafluoride	13637-63-3	1000
Chlorine Trifluoride	7790-91-2	1000
Chlorodiethylaluminum (also called Diethylaluminum Chloride)	96-10-6	5000
1-Chloro-2,4-Dinitrobenzene	97-00-7	5000
Chloromethyl Methyl Ether	107-30-2	500
Chloropicrin	76-06-2	500
Chloropicrin and Methyl Bromide mixture	None	1500
Chloropicrin and Methyl Chloride mixture	None	1500
Cumene Hydroperoxide	80-15-9	5000
Cyanogen	460-19-5	2500
Cyanogen Chloride	506-77-4	500
Cyanuric Fluoride	675-14-9	100
Diacetyl Peroxide (concentration greater than 70%)	110-22-5	5000
Diazomethane	334-88-3	500
Dibenzoyl Peroxide	94-36-0	7500
Diborane	19287-45-7	100
Dibutyl Peroxide (Tertiary)	110-05-4	5000
Dichloro Acetylene	7572-29-4	250
Dichlorosilane	4109-96-0	2500
Diethylzinc	557-20-0	10000
Diisopropyl Peroxydicarbonate	105-64-6	7500
Dilauroyl Peroxide	105-74-8	7500

(continued)

CHEMICAL NAME	CAS[1]	TQ[2]
Dimethyldichlorosilane	75-78-5	1000
Dimethylhydrazine, 1,1-	57-14-7	1000
Dimethylamine, Anhydrous	124-40-3	2500
2,4-Dinitroaniline	97-02-9	5000
Ethyl Methyl Ketone Peroxide (also Methyl Ethyl Ketone Peroxide; concentration greater than 60%)	1338-23-4	5000
Ethyl Nitrite	109-95-5	5000
Ethylamine	75-04-7	7500
Ethylene Fluorohydrin	371-62-0	100
Ethylene Oxide	75-21-8	5000
Ethyleneimine	151-56-4	1000
Fluorine	7782-41-4	1000
Formaldehyde (Formalin)	50-00-0	1000
Furan	110-00-9	500
Hexafluoroacetone	684-16-2	5000
Hydrochloric Acid, Anhydrous	7647-01-0	5000
Hydrofluoric Acid, Anhydrous	7664-39-3	1000
Hydrogen Bromide	10035-10-6	5000
Hydrogen Chloride	7647-01-0	5000
Hydrogen Cyanide, Anhydrous	74-90-8	1000
Hydrogen Fluoride	7664-39-3	1000
Hydrogen Peroxide (52% by weight or greater)	7722-84-1	7500
Hydrogen Selenide	7783-07-5	150
Hydrogen Sulfide	7783-06-4	1500
Hydroxylamine	7803-49-8	2500
Iron, Pentacarbonyl	13463-40-6	250
Isopropylamine	75-31-0	5000
Ketene	463-51-4	100
Methacrylaldehyde	78-85-3	1000
Methacryloyl Chloride	920-46-7	150
Methacryloyloxyethyl Isocyanate	30674-80-7	100
Methyl Acrylonitrile	126-98-7	250
Methylamine, Anhydrous	74-89-5	1000
Methyl Bromide	74-83-9	2500
Methyl Chloride	74-87-3	15000
Methyl Chloroformate	79-22-1	500
Methyl Ethyl Ketone Peroxide (concentration greater than 60%)	1338-23-4	5000
Methyl Fluoroacetate	453-18-9	100
Methyl Fluorosulfate	421-20-5	100
Methyl Hydrazine	60-34-4	100
Methyl Iodide	74-88-4	7500
Methyl Isocyanate	624-83-9	250
Methyl Mercaptan	74-93-1	5000
Methyl Vinyl Ketone	79-84-4	100
Methyltrichlorosilane	75-79-6	500
Nickel Carbonly (Nickel Tetracarbonyl)	13463-39-3	150
Nitric Acid (94.5% by weight or greater)	7697-37-2	500
Nitric Oxide	10102-43-9	250
Nitroaniline (para Nitroaniline)	100-01-6	5000

(continued)

CHEMICAL NAME	CAS[1]	TQ[2]
Nitromethane	75-52-5	2500
Nitrogen Dioxide	10102-44-0	250
Nitrogen Oxides (NO; NO(2);N_2O_4; N_2O_3)	10102-44-0	250
Nitrogen Tetroxide (also called Nitrogen Peroxide).	10544-72-6	250
Nitrogen Trifluoride	7783-54-2	5000
Nitrogen Trioxide	10544-73-7	250
Oleum (65% to 80% by weight; also called Fuming Sulfuric Acid)	8014-94-7	1000
Osmium Tetroxide	20816-12-0	100
Oxygen Difluoride (Fluorine Monoxide)	7783-41-7	100
Ozone	10028-15-6	100
Pentaborane	19624-22-7	100
Peracetic Acid (concentration greater 60% Acetic Acid; also called Peroxyacetic Acid)	79-21-0	1000
Perchloric Acid (concentration greater than 60% by weight)	7601-90-3	5000
Perchloromethyl Mercaptan	594-42-3	150
Perchloryl Fluoride	7616-94-6	5000
Peroxyacetic Acid (concentration greater than 60% Acetic Acid; also called Peracetic Acid)	79-21-0	1000
Phosgene (also called Carbonyl Chloride)	75-44-5	100
Phosphine (Hydrogen Phosphide)	7803-51-2	100
Phosphorus Oxychloride (also called Phosphoryl Chloride)	10025-87-3	1000
Phosphorus Trichloride	7719-12-2	1000
Phosphoryl Chloride (also called Phosphorus Oxychloride)	10025-87-3	1000
Propargyl Bromide	106-96-7	100
Propyl Nitrate	627-3-4	2500
Sarin	107-44-8	100
Selenium Hexafluoride	7783-79-1	1000
Stibine (Antimony Hydride)	7803-52-3	500
Sulfur Dioxide (liquid)	7446-09-5	1000
Sulfur Pentafluoride	5714-22-7	250
Sulfur Tetrafluoride	7783-60-0	250
Sulfur Trioxide (also called Sulfuric Anhydride)	7446-11-9	1000
Sulfuric Anhydride (also called Sulfur Trioxide)	7446-11-9	1000
Tellurium Hexafluoride	7783-80-4	250
Tetrafluoroethylene	116-14-3	5000
Tetrafluorohydrazine	10036-47-2	5000
Tetramethyl Lead	75-74-1	1000
Thionyl Chloride	7719-09-7	250
Trichloro (chloromethyl) Silane	1558-25-4	100
Trichloro (dichlorophenyl) Silane	27137-85-5	2500
Trichlorosilane	10025-78-2	5000
Trifluorochloroethylene	79-38-9	10000
Trimethyoxysilane	2487-90-3	1500

[1] Chemical Abstract Service Numbe.r
[2] Threshold Quantity in Pounds (Amount necessary to be covered by this standard).

§1926.64 Appendix B

Block Flow Diagram and Simplified Process Flow Diagram (Nonmandatory)

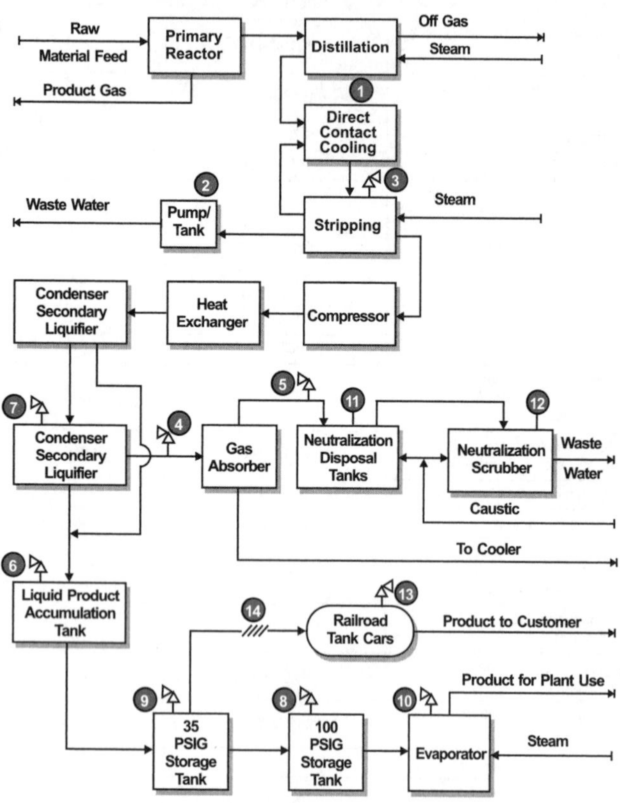

Example of a Block Flow Diagram

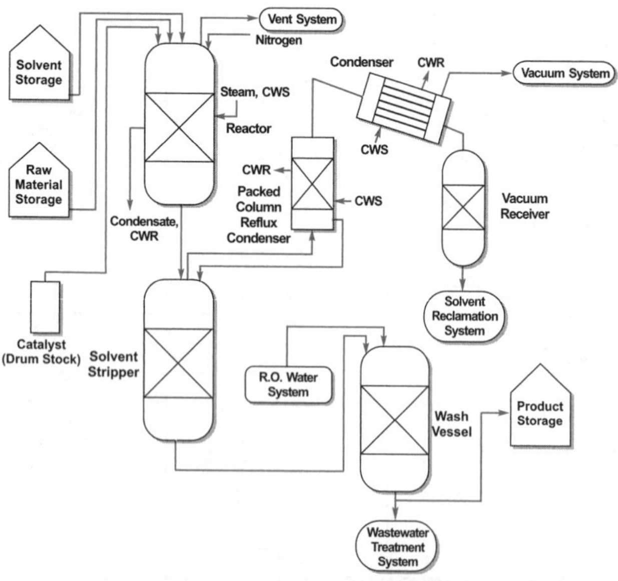

Example of a Process Flow Diagram

§1926.64 Appendix C

Compliance Guidelines and Recommendations for Process Safety Management (Nonmandatory)

This appendix serves as a nonmandatory guideline to assist employers and employees in complying with the requirements of this section, as well as provides other helpful recommendations and information. Examples presented in this appendix are not the only means of achieving the performance goals in the standard. This appendix neither adds nor detracts from the requirements of the standard.

1. **Introduction to Process Safety Management.** The major objective of process safety management of highly hazardous chemicals is to prevent unwanted releases of hazardous chemicals especially into locations which could expose employees and others to serious hazards. An effective process safety management program requires a systematic approach to evaluating the whole process. Using this approach the process design, process technology, operational and maintenance activities and procedures, nonroutine activities and procedures, emergency preparedness plans and procedures, training programs, and other elements which impact the process are all considered in the evaluation. The various lines of defense that have been incorporated into the design and operation of the process to prevent or mitigate the release of hazardous chemicals need to be evaluated and strengthened to assure their effectiveness at each level. Process safety management is the proactive identification, evaluation and mitigation or prevention of chemical releases that could occur as a result of failures in process, procedures or equipment.

   The process safety management standard targets highly hazardous chemicals that have the potential to cause a catastrophic incident. This standard as a whole is to aid employers in their efforts to prevent or mitigate episodic chemical releases that could lead to a catastrophe in the workplace and possibly to the surrounding community. To control these types of hazards, employers need to develop the necessary expertise, experiences, judgment and proactive initiative within their workforce to properly implement and maintain an effective process safety management program as envisioned in the OSHA standard. This OSHA standard is required by the Clean Air Act Amendments as is the Environmental Protection Agency's Risk Management Plan. Employers, who merge the two sets of requirements into their process safety management program, will better assure full compliance with each as well as enhancing their relationship with the local community.

   While OSHA believes process safety management will have a positive effect on the safety of employees in workplaces and also offers other potential benefits to employers (increased productivity), smaller businesses which may have limited resources available to them at this time, might consider alternative avenues of decreasing the risks associated with highly hazardous chemicals at their workplaces. One method which might be considered is the reduction in the inventory of the highly hazardous chemical. This reduction in inventory will result in a reduction of the risk or potential for a catastrophic incident. Also, employers including small employers may be able to establish more efficient inventory control by reducing the quantities of highly hazardous chemicals on site below the established threshold quantities. This reduction can be accomplished by ordering smaller shipments and maintaining the minimum inventory necessary for efficient and safe operation. When reduced inventory is not feasible, then the employer might consider dispersing inventory to several locations on site. Dispersing storage into locations where a release in one location will not cause a release in another location is a practical method to also reduce the risk or potential for catastrophic incidents.

2. **Employee Involvement in Process Safety Management.** Section 304 of the Clean Air Act Amendments states that employers are to consult with their employees and their representatives regarding the employers efforts in the development and implementation of the process safety management program elements and hazard assessments. Section 304 also requires employers to train and educate their employees and to inform affected employees of the findings from incident investigations required by the process safety management program. Many employers, under their safety and health programs, have already established means and methods to keep employees and their representatives informed about relevant safety and health issues and employers may be able to adapt these practices and procedures to meet their obligations under this standard. Employers who have not implemented an occupational safety and health program may wish to form a safety and health committee of employees and management representatives to help the employer meet the obligations specified by this standard. These committees can become a significant ally in helping the employer to implement and maintain an effective process safety management program for all employees.

3. **Process Safety Information.** Complete and accurate written information concerning process chemicals, process technology, and process equipment is essential to an effective process safety management program and to a process hazards analysis. The compiled information will be a necessary resource to a variety of users including the team that will perform the process hazards analysis as required under paragraph (e); those developing the training programs and the operating procedures; contractors whose employees

will be working with the process; those conducting the pre-startup reviews; local emergency preparedness planners; and insurance and enforcement officials.

The information to be compiled about the chemicals, including process intermediates, needs to be comprehensive enough for an accurate assessment of the fire and explosion characteristics, reactivity hazards, the safety and health hazards to workers, and the corrosion and erosion effects on the process equipment and monitoring tools. Current material safety data sheet (MSDS) information can be used to help meet this requirement which must be supplemented with process chemistry information including runaway reaction and over pressure hazards if applicable.

Process technology information will be a part of the process safety information package and it is expected that it will include diagrams of the type shown in Appendix B of this section as well as employer established criteria for maximum inventory levels for process chemicals; limits beyond which would be considered upset conditions; and a qualitative estimate of the consequences or results of deviation that could occur if operating beyond the established process limits. Employers are encouraged to use diagrams which will help users understand the process.

A block flow diagram is used to show the major process equipment and interconnecting process flow lines and show flow rates, stream composition, temperatures, and pressures when necessary for clarity. The block flow diagram is a simplified diagram.

Process flow diagrams are more complex and will show all main flow streams including valves to enhance the understanding of the process, as well as pressures and temperatures on all feed and product lines within all major vessels, in and out of headers and heat exchangers, and points of pressure and temperature control. Also, materials of construction information, pump capacities and pressure heads, compressor horsepower and vessel design pressures and temperatures are shown when necessary for clarity. In addition, major components of control loops are usually shown along with key utilities on process flow diagrams.

Piping and instrument diagrams (P&Ids) may be the more appropriate type of diagrams to show some of the above details and to display the information for the piping designer and engineering staff. The P&IDs are to be used to describe the relationships between equipment and instrumentation as well as other relevant information that will enhance clarity. Computer software programs which do P&Ids or other diagrams useful to the information package, may be used to help meet this requirement.

The information pertaining to process equipment design must be documented. In other words, what were the codes and standards relied on to establish good engineering practice. These codes and standards are published by such organizations as the American Society of Mechanical Engineers, American Petroleum Institute, American National Standards Institute, National Fire Protection Association, American Society for Testing and Materials, National Board of Boiler and Pressure Vessel Inspectors, National Association of Corrosion Engineers, American Society of Exchange Manufacturers Association, and model building code groups.

In addition, various engineering societies issue technical reports which impact process design. For example, the American Institute of Chemical Engineers has published technical reports on topics such as two phase flow for venting devices. This type of technically recognized report would constitute good engineering practice.

For existing equipment designed and constructed many years ago in accordance with the codes and standards available at that time and no longer in general use today, the employer must document which codes and standards were used and that the design and construction along with the testing, inspection and operation are still suitable for the intended use. Where the process technology requires a design which departs from the applicable codes and standards, the employer must document that the design and construction is suitable for the intended purpose.

4. **Process Hazard Analysis.** A process hazard analysis (PHA), sometimes called a process hazard evaluation, is one of the most important elements of the process safety management program. A PHA is an organized and systematic effort to identify and analyze the significance of potential hazards associated with the processing or handling of highly hazardous chemicals. A PHA provides information which will assist employers and employees in making decisions for improving safety and reducing the consequences of unwanted or unplanned releases of hazardous chemicals. A PHA is directed toward analyzing potential causes and consequences of fires, explosions, releases of toxic or flammable chemicals and major spills of hazardous chemicals. The PHA focuses on equipment, instrumentation, utilities, human actions (routine and nonroutine), and external factors that might impact the process. These considerations assist in determining the hazards and potential failure points or failure modes in a process.

The selection of a PHA methodology or technique will be influenced by many factors including the amount of existing knowledge about the process. Is it a process that has been operated for a long period of time with little or no innovation and extensive experience has been generated with its use? Or, is it a new process or one which has been changed frequently by the inclusion of innovative features? Also, the size and complexity of the process will influence the decision as to the appropriate PHA methodology to use. All PHA methodologies are subject to certain limitations. For example, the checklist methodology works well when the process is very stable and no changes are made, but it is not as effective when the process has undergone extensive change. The checklist may miss the most recent changes and consequently the changes would not be evaluated. Another limitation to be considered concerns the assumptions made by the team or analyst. The PHA is dependent on good judgment and the assumptions made during the study need to be documented and understood by the team and reviewer and kept for a future PHA.

The team conducting the PHA need to understand the methodology that is going to be used. A PHA team can vary in size from two people to a number of people with varied operational and technical backgrounds. Some team members may only be a part of the team for a limited time. The team leader needs to be fully knowledgeable in the proper implementation of the PHA methodology that is to be used and should be impartial in the evaluation. The other full or part time team members need to provide the team with expertise in areas such as process technology, process design, operating procedures and practices, including how the work is actually performed, alarms, emergency procedures, instrumentation, maintenance procedures, both routine and nonroutine tasks, including how the tasks are authorized, procurement of parts and supplies, safety and health, and any other relevant subject as the need dictates. At least one team member must be familiar with the process.

The ideal team will have an intimate knowledge of the standards, codes, specifications and regulations applicable to the process being studied. The selected team members need to be compatible and the team leader needs to be able to manage the team and the PHA study. The team needs to be able to work together while benefiting from the expertise of others on the team or outside the team, to resolve issues, and to forge a consensus on the findings of the study and the recommendations.

The application of a PHA to a process may involve the use of different methodologies for various parts of the process. For example, a process involving a series of unit operations of varying sizes, complexities, and ages may use different methodologies and team members for each operation. Then the conclusions can be integrated into one final study and evaluation. A more specific example is the use of a checklist PHA for a standard boiler or heat exchanger and the use of a Hazard and Operability PHA for the overall process. Also, for batch type processes like custom batch operations, a generic PHA of a representative batch may be used where there are only small changes of monomer or other ingredient ratios and the chemistry is documented for the full range and ratio of batch ingredients. Another process that might consider using a generic type of PHA is a gas plant. Often these plants are simply moved from site to site and therefore, a generic PHA may be used for these movable plants. Also, when an employer has several similar size gas plants and no sour gas is being processed at the site, then a generic PHA is feasible as long as the variations of the individual sites are accounted for in the PHA. Finally, when an employer has a large continuous process which has several control rooms for different portions of the process such as for a distillation tower and a blending operation, the employer may wish to do each segment separately and then integrate the final results.

Additionally, small businesses which are covered by this rule, will often have processes that have less storage volume, less capacity, and less complicated than processes at a large facility. Therefore, OSHA would anticipate that the less complex methodologies would be used to meet the process hazard analysis criteria in the standard. These process hazard analyses can be done in less time and with a few people being involved. A less complex process generally means that less data, P&IDs, and process information is needed to perform a process hazard analysis.

Many small businesses have processes that are not unique, such as cold storage lockers or water treatment facilities. Where employer associations have a number of members with such facilities, a generic PHA, evolved from a checklist or what-if questions, could be developed and used by each employer effectively to reflect his/her particular process; this would simplify compliance for them.

When the employer has a number of processes which require a PHA, the employer must set up a priority system of which PHAs to conduct first. A preliminary or gross hazard analysis may be useful in prioritizing the processes that the employer has determined are subject to coverage by the process safety management standard. Consideration should first be given to those processes with the potential of adversely affecting the largest number of employees. This prioritizing should consider the potential severity of a chemical release, the number of potentially affected employees, the operating history of the process such as the frequency of chemical releases, the age of the process and any other relevant factors. These factors would suggest a ranking order and would suggest either using a weighing factor system or a systematic ranking method. The use of a preliminary hazard analysis would assist an employer in determining which process should be of the highest priority and thereby the employer would obtain the greatest improvement in safety at the facility.

Detailed guidance on the content and application of process hazard analysis methodologies is available from the American Institute of Chemical Engineers' Center for Chemical Process Safety (see Appendix D).

5. **Operating Procedures and Practices.** Operating procedures describe tasks to be performed, data to be recorded, operating conditions to be maintained, samples to be collected, and safety and health precautions to be taken. The procedures need to be technically accurate, understandable to employees, and revised periodically to ensure that they reflect current operations. The process safety information package is to be used as a resource to better assure that the operating procedures and practices are consistent with the known hazards of the chemicals in the process and that the operating parameters are accurate. Operating procedures should be reviewed by engineering staff and operating personnel to ensure that they are accurate and provide practical instructions on how to actually carry out job duties safely.

Operating procedures will include specific instructions or details on what steps are to be taken or followed in carrying out the stated procedures. These operating instructions for each procedure should include the applicable safety precautions and should contain appropriate information on safety implications. For example, the operating procedures addressing operating parameters will contain operating instructions about pressure limits, temperature ranges, flow rates, what to do when an upset condition occurs, what alarms and instruments are pertinent if an upset condition occurs, and other subjects. Another example of using operating instructions to properly implement operating procedures is in starting up or shutting down the process. In these cases, different parameters will be required from those of normal operation. These operating instructions need to clearly indicate the distinctions between startup and normal operations such as the appropriate allowances for heating up a unit to reach the normal operating parameters. Also the operating instructions need to describe the proper method for increasing the temperature of the unit until the normal operating temperature parameters are achieved.

Computerized process control systems add complexity to operating instructions. These operating instructions need to describe the logic of the software as well as the relationship between the equipment and the control system; otherwise, it may not be apparent to the operator.

Operating procedures and instructions are important for training operating personnel. The operating procedures are often viewed as the standard operating practices (SOPs) for operations. Control room personnel and operating staff, in general, need to have a full understanding of operating procedures. If workers are not fluent in English then procedures and instructions need to be prepared in a second language understood by the workers. In addition, operating procedures need to be changed when there is a change in the process as a result of the management of change procedures. The consequences of operating procedure changes need to be fully evaluated and the information conveyed to the personnel. For example, mechanical changes to the process made by the maintenance department (like changing a valve from steel to brass or other subtle changes) need to be evaluated to determine if operating procedures and practices also need to be changed. All management of change actions must be coordinated and integrated with current operating procedures and operating personnel must be oriented to the changes in procedures before the change is made. When the process is shutdown in order to make a change, then the operating procedures must be updated before startup of the process.

Training in how to handle upset conditions must be accomplished as well as what operating personnel are to do in emergencies such as when a pump seal fails or a pipeline ruptures. Communication between operating personnel and workers performing work within the process area, such as nonroutine tasks, also must be maintained. The hazards of the tasks are to be conveyed to operating personnel in accordance with established procedures and to those performing the actual tasks. When the work is completed, operating personnel should be informed to provide closure on the job.

6. **Employee Training.** All employees, including maintenance and contractor employees, involved with highly hazardous chemicals need to fully understand the safety and health hazards of the chemicals and processes they work with for the protection of themselves, their fellow employees and the citizens of nearby communities. Training conducted in compliance with 1926.59, the Hazard Communication standard, will help employees to be more knowledgeable about the chemicals they work with as well as familiarize them with reading and understanding MSDS. However, additional training in subjects such as operating procedures and safety work practices, emergency evacuation and response, safety procedures, routine and nonroutine work authorization activities, and other areas pertinent to process safety and health will need to be covered by an employer's training program.

In establishing their training programs, employers must clearly define the employees to be trained and what subjects are to be covered in their training. Employers in setting up their training program will need to clearly establish the goals and objectives they wish to achieve with the training that they provide to their employees. The learning goals or objectives should be written in clear measurable terms before the training begins. These goals and objectives need to be tailored to each of the specific training modules or segments. Employers should describe the important actions and conditions under which the employee will demonstrate competence or knowledge as well as what is acceptable performance.

Hands-on-training where employees are able to use their senses beyond listening, will enhance learning. For example, operating personnel, who will work in a control room or at control panels, would benefit by being trained at a simulated control panel or panels. Upset conditions of various types could be displayed on the simulator, and then the employee could go through the proper operating procedures to bring the simulator panel back to the normal operating parameters. A training environment could be created to help the trainee feel the full reality of the situation but, of course, under controlled conditions. This realistic type of training can be very effective in teaching employees correct procedures while allowing them to also see the consequences of what might happens if they do not follow established operating procedures. Other training techniques using videos or on-the-job training can also be very effective for teaching other job tasks, duties, or other important information. An effective training program will allow the employee to fully participate in the training process and to practice their skill or knowledge.

Employers need to periodically evaluate their training programs to see if the necessary skills, knowledge, and routines are being properly understood and implemented by their trained employees. The means or methods for evaluating the training should be developed along with the training program goals and objectives. Training program evaluation will help employers to determine the amount of training their employees understood, and whether the desired results were obtained. If, after the evaluation, it appears that the trained employees are not at the level of knowledge and skill that was expected, the employer will need to revise the training program, provide retraining, or provide more frequent refresher training sessions until the deficiency is resolved. Those who conducted the training and those who received the training should also be consulted as to how best to improve the training process. If there is a language barrier, the language known to the trainees should be used to reinforce the training messages and information.

Careful consideration must be given to assure that employees including maintenance and contract employees receive current and updated training. For example, if changes are made to a process, impacted employees must be trained in the changes and understand the effects of the changes on their job tasks (e.g., any new operating procedures pertinent to their tasks). Additionally, as already discussed the evaluation of the employee's absorption of training will certainly influence the need for training.

7. **Contractors.** Employers who use contractors to perform work in and around processes that involve highly hazardous chemicals, will need to establish a screening process so that they hire and use contractors who accomplish the desired job tasks without compromising the safety and health of employees at a facility. For contractors, whose safety performance on the job is not known to the hiring employer, the employer will need to obtain information on injury and illness rates and experience and should obtain contractor references. Additionally, the employer must assure that the contractor has the appropriate job skills, knowledge and certifications (such as for pressure vessel welders). Contractor work methods and experi-

ences should be evaluated. For example, does the contractor conducting demolition work swing loads over operating processes or does the contractor avoid such hazards?

Maintaining a site injury and illness log for contractors is another method employers must use to track and maintain current knowledge of work activities involving contract employees working on or adjacent to covered processes. Injury and illness logs of both the employer's employees and contract employees allow an employer to have full knowledge of process injury and illness experience. This log will also contain information which will be of use to those auditing process safety management compliance and those involved in incident investigations.

Contract employees must perform their work safely. Considering that contractors often perform very specialized and potentially hazardous tasks such as confined space entry activities and nonroutine repair activities it is quite important that their activities be controlled while they are working on or near a covered process. A permit system or work authorization system for these activities would also be helpful to all affected employers. The use of a work authorization system keeps an employer informed of contract employee activities, and as a benefit the employer will have better coordination and more management control over the work being performed in the process area. A well run and well maintained process where employee safety is fully recognized will benefit all of those who work in the facility whether they be contract employees or employees of the owner.

8. **Pre-Startup Safety.** For new processes, the employer will find a PHA helpful in improving the design and construction of the process from a reliability and quality point of view. The safe operation of the new process will be enhanced by making use of the PHA recommendations before final installations are completed. P&IDs are to be completed along with having the operating procedures in place and the operating staff trained to run the process before startup. The initial startup procedures and normal operating procedures need to be fully evaluated as part of the pre-startup review to assure a safe transfer into the normal operating mode for meeting the process parameters.

   For existing processes that have been shutdown for turnaround, or modification, etc., the employer must assure that any changes other than "replacement in kind" made to the process during shutdown go through the management of change procedures. P&IDs will need to be updated as necessary, as well as operating procedures and instructions. If the changes made to the process during shutdown are significant and impact the training program, then operating personnel as well as employees engaged in routine and nonroutine work in the process area may need some refresher or additional training in light of the changes. Any incident investigation recommendations, compliance audits or PHA recommendations need to be reviewed as well to see what impacts they may have on the process before beginning the startup.

9. **Mechanical Integrity.** Employers will need to review their maintenance programs and schedules to see if there are areas where "breakdown" maintenance is used rather than an on-going mechanical integrity program. Equipment used to process, store, or handle highly hazardous chemicals needs to be designed, constructed, installed and maintained to minimize the risk of releases of such chemicals. This requires that a mechanical integrity program be in place to assure the continued integrity of process equipment. Elements of a mechanical integrity program include the identification and categorization of equipment and instrumentation, inspections and tests, testing and inspection frequencies, development of maintenance procedures, training of maintenance personnel, the establishment of criteria for acceptable test results, documentation of test and inspection results, and documentation of manufacturer recommendations as to meantime to failure for equipment and instrumentation.

   The first line of defense an employer has available is to operate and maintain the process as designed, and to keep the chemicals contained. This line of defense is backed up by the next line of defense which is the controlled release of chemicals through venting to scrubbers or flares, or to surge or overflow tanks which are designed to receive such chemicals, etc. These lines of defense are the primary lines of defense or means to prevent unwanted releases. The secondary lines of defense would include fixed fire protection systems like sprinklers, water spray, or deluge systems, monitor guns, etc., dikes, designed drainage systems, and other systems which would control or mitigate hazardous chemicals once an unwanted release occurs. These primary and secondary lines of defense are what the mechanical integrity program needs to protect and strengthen these primary and secondary lines of defenses where appropriate.

   The first step of an effective mechanical integrity program is to compile and categorize a list of process equipment and instrumentation for inclusion in the program. This list would include pressure vessels, storage tanks, process piping, relief and vent systems, fire protection system components, emergency shutdown systems and alarms and interlocks and pumps. For the categorization of instrumentation and the listed equipment the employer would prioritize which pieces of equipment require closer scrutiny than others. Meantime to failure of various instrumentation and equipment parts would be known from the manufacturers data or the employer's experience with the parts, which would then influence the inspection and testing frequency and associated procedures. Also, applicable codes and standards such as the National Board Inspection Code, or those from the American Society for Testing and Material, American Petroleum Institute, National Fire Protection Association, American National Standards Institute, American Society of Mechanical Engineers, and other groups, provide information to help establish an effective testing and inspection frequency, as well as appropriate methodologies.

   The applicable codes and standards provide criteria for external inspections for such items as foundation and supports, anchor bolts, concrete or steel supports, guy wires, nozzles and sprinklers, pipe hangers, grounding connections, protective coatings and insulation, and external metal surfaces of piping and vessels, etc. These codes and standards also provide information on methodologies for internal inspection, and a frequency formula based on the corrosion rate of the materials of construction. Also, erosion both internal and external needs to be considered along with corrosion effects for piping and valves. Where the corrosion rate is not known, a maximum inspection frequency is recommended, and methods of developing the corrosion rate are available in the codes. Internal inspections need to cover items such as vessel shell, bottom and head; metallic linings; nonmetallic linings; thickness measurements for vessels and piping; inspection for erosion, corrosion, cracking and bulges; internal equipment like trays, baffles, sensors and screens for erosion, corrosion or cracking and other deficiencies. Some of these inspections may be performed by state or local government inspectors under state and local statutes. However, each employer needs to develop procedures to ensure that tests and inspections are conducted properly and that consistency is maintained even where different employees may be involved. Appropriate training is to be provided to maintenance personnel to ensure that they understand the preventive maintenance program procedures, safe practices, and the proper use and application of special equipment or unique tools that may be required. This training is part of the overall training program called for in the standard.

   A quality assurance system is needed to help ensure that the proper materials of construction are used, that fabrication and inspection procedures are proper, and that installation procedures recognize field installation concerns. The quality assurance program is an essential part of the mechanical integrity program and will help to maintain the primary and secondary lines of defense that have been designed into the process to prevent unwanted chemical releases or those which control or mitigate a release. "As built" drawings, together with certifications of coded vessels and other equipment, and materials of construction need to be verified and retained in the quality assurance documentation. Equipment installation jobs need to be properly inspected in the field for use of proper materials and procedures and to assure that qualified craftsmen are used to do the job. The use of appropriate gaskets, packing, bolts, valves, lubricants and welding rods need to be verified in the field. Also, procedures for installation of safety devices need to be verified, such as the torque on the bolts on ruptured disc installations, uniform torque on flange bolts, proper installation of pump seals, etc. If the quality of parts is a problem, it may be appropriate to conduct audits of the equipment supplier's facilities to better assure proper purchases of required equipment which is suitable for its intended service. Any changes in equipment that may become necessary will need to go through the management of change procedures.

10. **Nonroutine Work Authorizations.** Nonroutine work which is conducted in process areas needs to be controlled by the employer in a consistent manner. The hazards identified involving the work that is to be accomplished must be communicated to those doing the work, but also to those operating personnel whose work could affect the safety of the process. A work authorization notice or permit must have a procedure that describes the steps the maintenance supervisor, contractor representative or other person needs to follow to obtain the necessary clearance to get the job started. The work authorization procedures need to reference and coordinate, as applicable, lockout/tagout procedures, line breaking procedures, confined space entry procedures and hot work authorizations. This procedure also needs to provide clear steps to follow once the job is

completed in order to provide closure for those that need to know the job is now completed and equipment can be returned to normal.

11. **Managing Change.** To properly manage changes to process chemicals, technology, equipment and facilities, one must define what is meant by change. In this process safety management standard, change includes all modifications to equipment, procedures, raw materials and processing conditions other than "replacement in kind." These changes need to be properly managed by identifying and reviewing them prior to implementation of the change. For example, the operating procedures contain the operating parameters (pressure limits, temperature ranges, flow rates, etc.) and the importance of operating within these limits. While the operator must have the flexibility to maintain safe operation within the established parameters, any operation outside of these parameters requires review and approval by a written management of change procedure.

    Management of change covers such as changes in process technology and changes to equipment and instrumentation. Changes in process technology can result from changes in production rates, raw materials, experimentation, equipment unavailability, new equipment, new product development, change in catalyst and changes in operating conditions to improve yield or quality. Equipment changes include among others change in materials of construction, equipment specifications, piping pre-arrangements, experimental equipment, computer program revisions and changes in alarms and interlocks. Employers need to establish means and methods to detect both technical changes and mechanical changes.

    Temporary changes have caused a number of catastrophes over the years, and employers need to establish ways to detect temporary changes as well as those that are permanent. It is important that a time limit for temporary changes be established and monitored since, without control, these changes may tend to become permanent. Temporary changes are subject to the management of change provisions. In addition, the management of change procedures are used to insure that the equipment and procedures are returned to their original or designed conditions at the end of the temporary change. Proper documentation and review of these changes is invaluable in assuring that the safety and health considerations are being incorporated into the operating procedures and the process.

    Employers may wish to develop a form or clearance sheet to facilitate the processing of changes through the management of change procedures. A typical change form may include a description and the purpose of the change, the technical basis for the change, safety and health considerations, documentation of changes for the operating procedures, maintenance procedures, inspection and testing, P&IDs, electrical classification, training and communications, pre-startup inspection, duration if a temporary change, approvals and authorization. Where the impact of the change is minor and well understood, a check list reviewed by an authorized person with proper communication to others who are affected may be sufficient. However, for a more complex or significant design change, a hazard evaluation procedure with approvals by operations, maintenance, and safety departments may be appropriate. Changes in documents such as P&IDs, raw materials, operating procedures, mechanical integrity programs, electrical classifications, etc., need to be noted so that these revisions can be made permanent when the drawings and procedure manuals are updated. Copies of process changes need to be kept in an accessible location to ensure that design changes are available to operating personnel as well as to PHA team members when a PHA is being done or one is being updated.

12. **Investigation of Incidents.** Incident investigation is the process of identifying the underlying causes of incidents and implementing steps to prevent similar events from occurring. The intent of an incident investigation is for employers to learn from past experiences and thus avoid repeating past mistakes. The incidents for which OSHA expects employers to become aware and to investigate are the types of events which result in or could reasonably have resulted in a catastrophic release. Some of the events are sometimes referred to as "near misses," meaning that a serious consequence did not occur, but could have.

    Employers need to develop in-house capability to investigate incidents that occur in their facilities. A team needs to be assembled by the employer and trained in the techniques of investigation including how to conduct interviews of witnesses, needed documentation and report writing. A multi-disciplinary team is better able to gather the facts of the event and to analyze them and develop plausible scenarios as to what happened, and why. Team members should be selected on the basis of their training, knowledge and ability to contribute to a team effort to fully investigate the incident. Employees in the process area where the incident occurred should be consulted, interviewed or made a member of the team. Their knowledge of the events form a significant set of facts about the incident which occurred. The report, its findings and recommendations are to be shared with those who can benefit from the information. The cooperation of employees is essential to an effective incident investigation. The focus of the investigation should be to obtain facts, and not to place blame. The team and the investigation process should clearly deal with all involved individuals in a fair, open and consistent manner.

13. **Emergency Preparedness.** Each employer must address what actions employees are to take when there is an unwanted release of highly hazardous chemicals. Emergency preparedness or the employer's tertiary (third) lines of defense are those that will be relied on along with the secondary lines of defense when the primary lines of defense which are used to prevent an unwanted release fail to stop the release. Employers will need to decide if they want employees to handle and stop small or minor incidental releases. Whether they wish to mobilize the available resources at the plant and have them brought to bear on a more significant release. Or whether employers want their employees to evacuate the danger area and promptly escape to a preplanned safe zone area, and allow the local community emergency response organizations to handle the release. Or whether the employer wants to use some combination of these actions. Employers will need to select how many different emergency preparedness or tertiary lines of defense they plan to have and then develop the necessary plans and procedures, and appropriately train employees in their emergency duties and responsibilities and then implement these lines of defense.

    Employers at a minimum must have an emergency action plan which will facilitate the prompt evacuation of employees when an unwanted release of highly hazardous chemical. This means that the employer will have a plan that will be activated by an alarm system to alert employees when to evacuate and, that employees who are physically impaired, will have the necessary support and assistance to get them to the safe zone as well. The intent of these requirements is to alert and move employees to a safe zone quickly. Delaying alarms or confusing alarms are to be avoided. The use of process control centers or similar process buildings in the process area as safe areas is discouraged. Recent catastrophes have shown that a large life loss has occurred in these structures because of where they have been sited and because they are not necessarily designed to withstand over-pressures from shockwaves resulting from explosions in the process area.

    Unwanted incidental releases of highly hazardous chemicals in the process area must be addressed by the employer as to what actions employees are to take. If the employer wants employees to evacuate the area, then the emergency action plan will be activated. For outdoor processes where wind direction is important for selecting the safe route to a refuge area, the employer should place a wind direction indicator such as a wind sock or pennant at the highest point that can be seen throughout the process area. Employees can move in the direction of cross wind to upwind to gain safe access to the refuge area by knowing the wind direction.

    If the employer wants specific employees in the release area to control or stop the minor emergency or incidental release, these actions must be planned for in advance and procedures developed and implemented. Preplanning for handling incidental releases for minor emergencies in the process area needs to be done, appropriate equipment for the hazards must be provided, and training conducted for those employees who will perform the emergency work before they respond to handle an actual release. The employer's training program, including the Hazard Communication standard training is to address the training needs for employees who are expected to handle incidental or minor releases.

    Preplanning for releases that are more serious than incidental releases is another important line of defense to be used by the employer. When a serious release of a highly hazardous chemical occurs, the employer through preplanning will have determined in advance what actions employees are to take. The evacuation of the immediate release area and other areas as necessary would be accomplished under the emergency action plan. If the employer wishes to use plant personnel such as a fire brigade, spill control team, a hazardous materials team, or use employees to render aid to those in the immediate release area and control or mitigate the incident, these actions are covered by 1926.65, the Hazardous Waste Operations and Emergency Response (HAZWOPER) standard. If outside assistance is necessary, such as through mutual aid agreements between employers or local government emergency response organizations, these emergency responders are also covered by HAZWOPER. The safety and health protections required for emergency responders are the responsibility of their employers and of the on-scene incident commander.

Responders may be working under very hazardous conditions and therefore the objective is to have them competently led by an on-scene incident commander and the commander's staff, properly equipped to do their assigned work safely, and fully trained to carry out their duties safely before they respond to an emergency. Drills, training exercises, or simulations with the local community emergency response planners and responder organizations is one means to obtain better preparedness. This close cooperation and coordination between plant and local community emergency preparedness managers will also aid the employer in complying with the Environmental Protection Agency's Risk Management Plan criteria.

One effective way for medium to large facilities to enhance coordination and communication during emergencies for on plant operations and with local community organizations is for employers to establish and equip an emergency control center. The emergency control center would be sited in a safe zone area so that it could be occupied throughout the duration of an emergency. The center would serve as the major communication link between the on-scene incident commander and plant or corporate management as well as with the local community officials. The communication equipment in the emergency control center should include a network to receive and transmit information by telephone, radio or other means. It is important to have a backup communication network in case of power failure or one communication means fails. The center should also be equipped with the plant layout and community maps, utility drawings including fire water, emergency lighting, appropriate reference materials such as a government agency notification list, company personnel phone list, SARA Title III reports and material safety data sheets, emergency plans and procedures manual, a listing with the location of emergency response equipment, mutual aid information, and access to meteorological or weather condition data and any dispersion modeling data.

14. **Compliance Audits.** Employers need to select a trained individual or assemble a trained team of people to audit the process safety management system and program. A small process or plant may need only one knowledgeable person to conduct an audit. The audit is to include an evaluation of the design and effectiveness of the process safety management system and a field inspection of the safety and health conditions and practices to verify that the employer's systems are effectively implemented. The audit should be conducted or lead by a person knowledgeable in audit techniques and who is impartial towards the facility or area being audited. The essential elements of an audit program include planning, staffing, conducting the audit, evaluation and corrective action, follow-up and documentation.

Planning in advance is essential to the success of the auditing process. Each employer needs to establish the format, staffing, scheduling and verification methods prior to conducting the audit. The format should be designed to provide the lead auditor with a procedure or checklist which details the requirements of each section of the standard. The names of the audit team members should be listed as part of the format as well. The checklist, if properly designed, could serve as the verification sheet which provides the auditor with the necessary information to expedite the review and assure that no requirements of the standard are omitted. This verification sheet format could also identify those elements that will require evaluation or a response to correct deficiencies. This sheet could also be used for developing the follow-up and documentation requirements.

The selection of effective audit team members is critical to the success of the program. Team members should be chosen for their experience, knowledge, and training and should be familiar with the processes and with auditing techniques, practices and procedures. The size of the team will vary depending on the size and complexity of the process under consideration. For a large, complex, highly instrumented plant, it may be desirable to have team members with expertise in process engineering and design, process chemistry, instrumentation and computer controls, electrical hazards and classifications, safety and health disciplines, maintenance, emergency preparedness, warehousing or shipping, and process safety auditing. The team may use part-time members to provide for the depth of expertise required as well as for what is actually done or followed, compared to what is written.

An effective audit includes a review of the relevant documentation and process safety information, inspection of the physical facilities, and interviews with all levels of plant personnel. Utilizing the audit procedure and checklist developed in the preplanning stage, the audit team can systematically analyze compliance with the provisions of the standard and any other corporate policies that are relevant. For example, the audit team will review all aspects of the training program as part of the overall audit. The team will review the written training program for adequacy of content, frequency of training, effectiveness of training in terms of its goals and objectives as well as to how it fits into meeting the standard's requirements, documentation, etc. Through interviews, the team can determine the employee's knowledge and awareness of the safety procedures, duties, rules, emergency response assignments, etc. During the inspection, the team can observe actual practices such as safety and health policies, procedures, and work authorization practices. This approach enables the team to identify deficiencies and determine where corrective actions or improvements are necessary.

An audit is a technique used to gather sufficient facts and information, including statistical information, to verify compliance with standards. Auditors should select as part of their preplanning a sample size sufficient to give a degree of confidence that the audit reflects the level of compliance with the standard. The audit team, through this systematic analysis, should document areas which require corrective action as well as those areas where the process safety management system is effective and working in an effective manner. This provides a record of the audit procedures and findings, and serves as a baseline of operation data for future audits. It will assist future auditors in determining changes or trends from previous audits.

Corrective action is one of the most important parts of the audit. It includes not only addressing the identified deficiencies, but also planning, followup, and documentation. The corrective action process normally begins with a management review of the audit findings. The purpose of this review is to determine what actions are appropriate, and to establish priorities, timetables, resource allocations and requirements and responsibilities. In some cases, corrective action may involve a simple change in procedure or minor maintenance effort to remedy the concern. Management of change procedures need to be used, as appropriate, even for what may seem to be a minor change. Many of the deficiencies can be acted on promptly, while some may require engineering studies or indepth review of actual procedures and practices. There may be instances where no action is necessary and this is a valid response to an audit finding. All actions taken, including an explanation where no action is taken on a finding, needs to be documented as to what was done and why.

It is important to assure that each deficiency identified is addressed, the corrective action to be taken noted, and the audit person or team responsible be properly documented by the employer. To control the corrective action process, the employer should consider the use of a tracking system. This tracking system might include periodic status reports shared with affected levels of management, specific reports such as completion of an engineering study, and a final implementation report to provide closure for audit findings that have been through management of change, if appropriate, and then shared with affected employees and management. This type of tracking system provides the employer with the status of the corrective action. It also provides the documentation required to verify that appropriate corrective actions were taken on deficiencies identified in the audit.

§1926.64 Appendix D

Sources of Further Information (Non-mandatory)

1. Center for Chemical Process Safety, American Institute of Chemical Engineers, 345 East 47th Street, New York, NY 10017, (212) 705-7319.
2. "Guidelines for Hazard Evaluation Procedures," American Institute of Chemical Engineers; 345 East 47th Street, New York, NY 10017.
3. "Guidelines for Technical Management of Chemical Process Safety," Center for Chemical Process Safety of the American Institute of Chemical Engineers; 345 East 47th Street, New York, NY 10017.
4. "Evaluating Process Safety in the Chemical Industry," Chemical Manufacturers Association; 2501 M Street NW, Washington, DC 20037.
5. "Safe Warehousing of Chemicals," Chemical Manufacturers Association; 2501 M Street NW, Washington, DC 20037.
6. "Management of Process Hazards," American Petroleum Institute (API Recommended Practice 750); 1220 L Street, N.W., Washington, DC 20005.

7. "Improving Owner and Contractor Safety Performance," American Petroleum Institute (API Recommended Practice 2220); API, 1220 L Street N.W., Washington, DC 20005.
8. Chemical Manufacturers Association (CMA's Manager Guide), First Edition, September 1991; CMA, 2501 M Street, N.W., Washington, DC 20037.
9. "Improving Construction Safety Performance," Report A-3, The Business Roundtable; The Business Roundtable, 200 Park Avenue, New York, NY 10166. (Report includes criteria to evaluate contractor safety performance and criteria to enhance contractor safety performance).
10. "Recommended Guidelines for Contractor Safety and Health," Texas Chemical Council; Texas Chemical Council, 1402 Nueces Street, Austin, TX 78701-1534.
11. "Loss Prevention in the Process Industries," Volumes I and II; Frank P. Lees, Butterworth; London 1983.
12. "Safety and Health Program Management Guidelines," 1989; U.S. Department of Labor, Occupational Safety and Health Administration.
13. "Safety and Health Guide for the Chemical Industry," 1986, (OSHA 3091); U.S. Department of Labor, Occupational Safety and Health Administration; 200 Constitution Avenue, N.W., Washington, DC 20210.
14. "Review of Emergency Systems," June 1988; U.S. Environmental Protection Agency (EPA), Office of Solid Waste and Emergency Response, Washington, DC 20460.
15. "Technical Guidance for Hazards Analysis, Emergency Planning for Extremely Hazardous Substances," December 1987; U.S. Environmental Protection Agency (EPA), Federal Emergency Management Administration (FEMA) and U.S. Department of Transportation (DOT), Washington, DC 20460.
16. "Accident Investigation...A New Approach," 1983, National Safety Council; 444 North Michigan Avenue, Chicago, IL 60611-3991.
17. "Fire & Explosion Index Hazard Classification Guide," 6th Edition, May 1987, Dow Chemical Company; Midland, Michigan 48674.
18. "Chemical Exposure Index," May 1988, Dow Chemical Company; Midland, Michigan 48674.

[58 FR 35115, June 30, 1993, as amended at 77 FR 17890, Mar. 26, 2012]

§1926.65

Hazardous waste operations and emergency response

(a) Scope, application, and definitions.[§1926.65(a)]

(1) *Scope.* This section covers the following operations, unless the employer can demonstrate that the operation does not involve employee exposure or the reasonable possibility for employee exposure to safety or health hazards:[§1926.65(a)(1)]

(i) *Clean-up operations required by a governmental body,* whether Federal, state, local or other involving hazardous substances that are conducted at uncontrolled hazardous waste sites (including, but not limited to, the EPA's National Priority Site List (NPL), state priority site lists, sites recommended for the EPA NPL, and initial investigations of government identified sites which are conducted before the presence or absence of hazardous substances has been ascertained);[§1926.65(a)(1)(i)]

(ii) *Corrective actions involving clean-up operations* at sites covered by the Resource Conservation and Recovery Act of 1976 (RCRA) as amended (42 U.S.C. 6901 et seq.);[§1926.65(a)(1)(ii)]

(iii) *Voluntary clean-up operations* at sites recognized by Federal, state, local or other governmental bodies as uncontrolled hazardous waste sites;[§1926.65(a)(1)(iii)]

(iv) *Operations involving hazardous wastes* that are conducted at treatment, storage, and disposal (TSD) facilities regulated by 40 CFR parts 264 and 265 pursuant to RCRA; or by agencies under agreement with U.S.E.P.A. to implement RCRA regulations; and[§1926.65(a)(1)(iv)]

(v) *Emergency response operations for releases of,* or substantial threats of releases of, hazardous substances without regard to the location of the hazard.[§1926.65(a)(1)(v)]

(2) *Application.*[§1926.65(a)(2)]

(i) *All requirements of part 1910 and part 1926* of Title 29 of the Code of Federal Regulations apply pursuant to their terms to hazardous waste and emergency response operations whether covered by this section or not. If there is a conflict or overlap, the provision more protective of employee safety and health shall apply without regard to 29 CFR 1926.20(e)(1). [§1926.65(a)(2)(i)]

(ii) *Hazardous substance clean-up operations* within the scope of paragraphs (a)(1)(i) through (a)(1)(iii) of this section must comply with all paragraphs of this section except paragraphs (p) and (q).[§1926.65(a)(2)(ii)]

(iii) *Operations within the scope* of paragraph (a)(1)(iv) of this section must comply only with the requirements of paragraph (p) of this section.[§1926.65(a)(2)(iii)]

Notes and Exceptions:*[A]All provisions of paragraph (p)* of this section cover any treatment, storage or disposal (TSD) operation regulated by 40 CFR parts 264 and 265 or by state law authorized under RCRA, and required to have a permit or interim status from EPA pursuant to 40 CFR 270.1 or from a state agency pursuant to RCRA.[§1926.65(a)(2)(iii)[A]]

[B] Employers who are not required to have a permit or interim status because they are conditionally exempt small quantity generators under 40 CFR 261.5 or are generators who qualify under 40 CFR 262.34 for exemptions from regulation under 40 CFR parts 264, 265 and 270 ("excepted employers") are not covered by paragraphs (p)(1) through (p)(7) of this section. Excepted employers who are required by the EPA or state agency to have their employees engage in emergency response or who direct their employees to engage in emergency response are covered by paragraph (p)(8) of this section, and cannot be exempted by (p)(8)(i) of this section. Excepted employers who are not required to have employees engage in emergency response, who direct their employees to evacuate in the case of such emergencies and who meet the requirements of paragraph (p)(8)(i) of this section are exempt from the balance of paragraph (p)(8) of this section.[§1926.65(a)(2)(iii)[B]]

[C] If an area is used primarily for treatment, storage or disposal, any emergency response operations in that area shall comply with paragraph (p)(8) of this section. In other areas not used primarily for treatment, storage, or disposal, any emergency response operations shall comply with paragraph (q) of this section. Compliance with the requirements of paragraph (q) of this section shall be deemed to be in compliance with the requirements of paragraph (p)(8) of this section.[§1926.65(a)(2)(iii)[C]]

(iv) *Emergency response operations* for releases of, or substantial threats of releases of, hazardous substances which are not covered by paragraphs (a)(1)(i) through (a)(1)(iv) of this section must only comply with the requirements of paragraph (q) of this section.[§1926.65(a)(2)(iv)]

(3) *Definitions* — [§1926.65(a)(3)]

Buddy system means a system of organizing employees into work groups in such a manner that each employee of the work group is designated to be observed by at least one other employee in the work group. The purpose of the buddy system is to provide rapid assistance to employees in the event of an emergency.

Clean-up operation means an operation where hazardous substances are removed, contained, incinerated, neutralized, stabilized, cleared-up, or in any other manner processed or handled with the ultimate goal of making the site safer for people or the environment.

Decontamination means the removal of hazardous substances from employees and their equipment to the extent necessary to preclude the occurrence of foreseeable adverse health affects.

Emergency response or responding to emergencies means a response effort by employees from outside the immediate release area or by other designated responders (i.e., mutual-aid groups, local fire departments, etc.) to an occurrence which results, or is likely to result, in an uncontrolled release of a hazardous substance. Responses to incidental releases of hazardous substances where the substance can be absorbed, neutralized, or otherwise controlled at the time of release by employees in the immediate release area, or by maintenance personnel are not considered to be emergency responses within the scope of this standard. Responses to releases of hazardous substances where there is no potential safety or health hazard (i.e., fire, explosion, or chemical exposure) are not considered to be emergency responses.

Facility means

[A]any building, structure, installation, equipment, pipe or pipeline (including any pipe into a sewer or publicly owned treatment works), well, pit, pond, lagoon, impoundment, ditch, storage container, motor vehicle, rolling stock, or aircraft, or

[B]any site or area where a hazardous substance has been deposited, stored, disposed of, or placed, or otherwise come to be located; but does not include any consumer product in consumer use or any water-borne vessel.

Hazardous materials response (HAZMAT) team means an organized group of employees, designated by the employer, who are expected to perform work to handle and control actual or potential leaks or spills of hazardous substances requiring possible close approach to the substance. The team members perform responses to releases or potential releases of hazardous substances for the purpose of control or stabilization of the incident. A HAZMAT team is not a fire brigade nor is a typical fire brigade a HAZMAT team. A HAZMAT team, however, may be a separate component of a fire brigade or fire department.

Hazardous substance means any substance designated or listed under paragraphs (A) through (D) of this definition, exposure to which results or may result in adverse affects on the health or safety of employees:

[A]Any substance defined under section 101(14) of CERCLA;

[B]Any biological agent and other disease-causing agent which after release into the environment and upon exposure, ingestion, inhalation, or assimilation into any person, either directly from the environment or indirectly by ingestion through food chains, will or may reasonably be anticipated to cause death, disease, behavioral abnormalities, cancer, genetic mutation, physiological malfunctions (including malfunctions in reproduction) or physical deformations in such persons or their offspring;

[C]Any substance listed by the U.S. Department of Transportation as hazardous materials under 49 CFR 172.101 and appendices; and

[D]Hazardous waste as herein defined.

Hazardous waste means

[A]A waste or combination of wastes as defined in 40 CFR 261.3, or

[B]Those substances defined as hazardous wastes in 49 CFR 171.8.

Hazardous waste operation means any operation conducted within the scope of this standard.

Hazardous waste site or Site means any facility or location within the scope of this standard at which hazardous waste operations take place.

Health hazard means a chemical or a pathogen where acute or chronic health effects may occur in exposed employees. It also includes stress due to temperature extremes. The term *health hazard* includes chemicals that are classified in accordance with the Hazard Communication Standard, Sec. 1910.1200, as posing one of the following hazardous effects: acute toxicity (any route of exposure); skin corrosion or irritation; serious eye damage or eye irritation; respiratory or skin sensitization; germ cell mutagenicity; carcinogenicity; reproductive toxicity; specific target organ toxicity (single or repeated exposure); aspiration toxicity or simple asphyxiant. (*See* Appendix A to §1910.1200 — Health Hazard Criteria (Mandatory) for the criteria for determining whether a chemical is classified as a health hazard.)

IDLH or Immediately dangerous to life or health means an atmospheric concentration of any toxic, corrosive or asphyxiant substance that poses an immediate threat to life or would cause irreversible or delayed adverse health effects or would interfere with an individual's ability to escape from a dangerous atmosphere.

Oxygen deficiency means that concentration of oxygen by volume below which atmosphere supplying respiratory protection must be provided. It exists in atmospheres where the percentage of oxygen by volume is less than 19.5 percent oxygen.[

Permissible exposure limit means the exposure, inhalation or dermal permissible exposure limit specified either in Sec. 1926.55, elsewhere in subpart D, or in other pertinent sections of this part.

Published exposure level means the exposure limits published in "NIOSH Recommendations for Occupational Health Standards" dated 1986 incorporated by reference, or if none is specified, the exposure limits published in the standards specified by the American Conference of Governmental Industrial Hygienists in their publication "Threshold Limit Values and Biological Exposure Indices for 1987-88" dated 1987 incorporated by reference.

Post emergency response means that portion of an emergency response performed after the immediate threat of a release has been stabilized or eliminated and clean-up of the site has begun. If post emergency response is performed by an employer's own employees who were part of the initial emergency response, it is considered to be part of the initial response and not post emergency response. However, if a group of an employer's own employees, separate from the group providing initial response, performs the clean-up operation, then the separate group of employees would be considered to be performing post-emergency response and subject to paragraph (q)(11) of this section.

Qualified person means a person with specific training, knowledge and experience in the area for which the person has the responsibility and the authority to control.

Site safety and health supervisor (or official) means the individual located on a hazardous waste site who is responsible to the employer and has the authority and knowledge necessary to implement the site safety and health plan and verify compliance with applicable safety and health requirements.

Small quantity generator means a generator of hazardous wastes who in any calendar month generates no more than 1,000 kilograms (2,205 pounds) of hazardous waste in that month.

Uncontrolled hazardous waste site, means an area identified as an uncontrolled hazardous waste site by a governmental body, whether Federal, state, local or other where an accumulation of hazardous substances creates a threat to the health and safety of individuals or the environment or both. Some sites are found on public lands such as those created by former municipal, county or state landfills where illegal or poorly managed waste disposal has taken place. Other sites are found on private property, often belonging to generators or former generators of hazardous substance wastes. Examples of such sites include, but are not limited to, surface impoundments, landfills, dumps, and tank or drum farms. Normal operations at TSD sites are not covered by this definition.

(b) Safety and health program.[§1926.65(b)]

Note to (b): Safety and health programs developed and implemented to meet other Federal, state, or local regulations are considered acceptable in meeting this requirement if they cover or are modified to cover the topics required in this paragraph. An additional or separate safety and health program is not required by this paragraph.

(1) *General.*[§1926.65(b)(1)]

(i) *Employers shall develop and implement* a written safety and health program for their employees involved in hazardous waste operations. The program shall be designed to identify, evaluate, and control safety and health hazards, and provide for emergency response for hazardous waste operations. [§1926.65(b)(1)(i)]

(ii) *The written safety and health* program shall incorporate the following:[§1926.65(b)(1)(ii)]

[A] An organizational structure;[§1926.65(b)(1)(ii)[A]]

[B] A comprehensive workplan;[§1926.65(b)(1)(ii)[B]]

[C] A site-specific safety and health plan which need not repeat the employer's standard operating procedures required in paragraph (b)(1)(ii)(F) of this section; [§1926.65(b)(1)(ii)[C]]

[D] The safety and health training program;[§1926.65(b)(1)(ii)[D]]

[E] The medical surveillance program;[§1926.65(b)(1)(ii)[E]]

[F] The employer's standard operating procedures for safety and health; and[§1926.65(b)(1)(ii)[F]]

[G] Any necessary interface between general program and site specific activities.[§1926.65(b)(1)(ii)[G]]

(iii) *Site excavation.* Site excavations created during initial site preparation or during hazardous waste operations shall be shored or sloped as appropriate to prevent accidental collapse in accordance with subpart P of 29 CFR part 1926. [§1926.65(b)(1)(iii)]

(iv) *Contractors and sub-contractors.* An employer who retains contractor or sub-contractor services for work in hazardous waste operations shall inform those contractors, sub-contractors, or their representatives of the site emergency response procedures and any potential fire, explosion, health, safety or other hazards of the hazardous waste operation that have been identified by the employer, including those identified in the employer's information program.[§1926.65(b)(1)(iv)]

(v) *Program availability.* The written safety and health program shall be made available to any contractor or subcontractor or their representative who will be involved with the hazardous waste operation; to employees; to employee designated representatives; to OSHA personnel, and to personnel of other Federal, state, or local agencies with regulatory authority over the site.[§1926.65(b)(1)(v)]

(2) *Organizational structure part of the site program*[§1926.65(b)(2)]

(i) *The organizational structure part of the program* shall establish the specific chain of command and specify the overall responsibilities of supervisors and employees. It shall include, at a minimum, the following elements:[§1926.65(b)(2)(i)]

[A] *A general supervisor* who has the responsibility and authority to direct all hazardous waste operations. [§1926.65(b)(2)(i)[A]]

[B] *A site safety and health supervisor* who has the responsibility and authority to develop and implement the site safety and health plan and verify compliance. [§1926.65(b)(2)(i)[B]]

[C] *All other personnel needed* for hazardous waste site operations and emergency response and their general functions and responsibilities.[§1926.65(b)(2)(i)[C]]

[D] *The lines of authority, responsibility, and communication.* [§1926.65(b)(2)(i)[D]]

(ii) *The organizational structure* shall be reviewed and updated as necessary to reflect the current status of waste site operations.[§1926.65(b)(2)(ii)]

(3) *Comprehensive workplan part of the site program.* The comprehensive workplan part of the program shall address the tasks and objectives of the site operations and the logistics and resources required to reach those tasks and objectives. [§1926.65(b)(3)]

(i) *The comprehensive workplan shall address* anticipated clean-up activities as well as normal operating procedures which need not repeat the employer's procedures available elsewhere.[§1926.65(b)(3)(i)]

(ii) *The comprehensive workplan shall define* work tasks and objectives and identify the methods for accomplishing those tasks and objectives.[§1926.65(b)(3)(ii)]

(iii) *The comprehensive workplan shall establish* personnel requirements for implementing the plan.[§1926.65(b)(3)(iii)]

(iv) *The comprehensive workplan shall provide* for the implementation of the training required in paragraph (e) of this section.[§1926.65(b)(3)(iv)]

(v) *The comprehensive workplan shall provide* for the implementation of the required informational programs required in paragraph (i) of this section.[§1926.65(b)(3)(v)]

(vi) *The comprehensive workplan shall provide* for the implementation of the medical surveillance program described in paragraph (f) of this section.[§1926.65(b)(3)(vi)]

(4) *Site-specific safety and health plan part of the program.* [§1926.65(b)(4)]

(i) *General.* The site safety and health plan, which must be kept on site, shall address the safety and health hazards of each phase of site operation and include the requirements and procedures for employee protection.[§1926.65(b)(4)(i)]

(ii) *Elements.* The site safety and health plan, as a minimum, shall address the following:[§1926.65(b)(4)(ii)]

[A] *A safety and health risk* or hazard analysis for each site task and operation found in the workplan. [§1926.65(b)(4)(ii)[A]]

[B] *Employee training assignments* to assure compliance with paragraph (e) of this section.[§1926.65(b)(4)(ii)[B]]

[C] *Personal protective equipment* to be used by employees for each of the site tasks and operations being conducted as required by the personal protective equipment program in paragraph (g)(5) of this section.[§1926.65(b)(4)(ii)[C]]

[D] *Medical surveillance requirements* in accordance with the program in paragraph (f) of this section.[§1926.65(b)(4)(ii)[D]]

[E] *Frequency and types of air monitoring,* personnel monitoring, and environmental sampling techniques and instrumentation to be used, including methods of maintenance and calibration of monitoring and sampling equipment to be used.[§1926.65(b)(4)(ii)[E]]

[F] *Site control measures* in accordance with the site control program required in paragraph (d) of this section. [§1926.65(b)(4)(ii)[F]]

[G] *Decontamination procedures* in accordance with paragraph (k) of this section.[§1926.65(b)(4)(ii)[G]]

[H] *An emergency response plan* meeting the requirements of paragraph (l) of this section for safe and effective responses to emergencies, including the necessary PPE and other equipment.[§1926.65(b)(4)(ii)[H]]

[I] *Confined space entry procedures.*[§1926.65(b)(4)(ii)[I]]

[J] *A spill containment program* meeting the requirements of paragraph (j) of this section.[§1926.65(b)(4)(ii)[J]]

(iii) *Pre-entry briefing.* The site specific safety and health plan shall provide for pre-entry briefings to be held prior to initiating any site activity, and at such other times as necessary to ensure that employees are apprised of the site safety and health plan and that this plan is being followed. The information and data obtained from site characterization and analysis work required in paragraph (c) of this section shall be used to prepare and update the site safety and health plan. [§1926.65(b)(4)(iii)]

(iv) *Effectiveness of site safety and health plan.* Inspections shall be conducted by the site safety and health supervisor or, in the absence of that individual, another individual who is knowledgeable in occupational safety and health, acting on behalf of the employer as necessary to determine the effectiveness of the site safety and health plan. Any deficiencies in the effectiveness of the site safety and health plan shall be corrected by the employer.[§1926.65(b)(4)(iv)]

(c) Site characterization and analysis.[§1926.65(c)]

(1) *General.* Hazardous waste sites shall be evaluated in accordance with this paragraph to identify specific site hazards and to determine the appropriate safety and health control procedures needed to protect employees from the identified hazards. [§1926.65(c)(1)]

(2) *Preliminary evaluation.* A preliminary evaluation of a site's characteristics shall be performed prior to site entry by a qualified person in order to aid in the selection of appropriate employee protection methods prior to site entry. Immediately after initial site entry, a more detailed evaluation of the site's specific characteristics shall be performed by a qualified person in order to further identify existing site hazards and to further aid in the selection of the appropriate engineering controls and personal protective equipment for the tasks to be performed. [§1926.65(c)(2)]

(3) *Hazard identification.* All suspected conditions that may pose inhalation or skin absorption hazards that are immediately dangerous to life or health (IDLH), or other conditions that may cause death or serious harm, shall be identified during the preliminary survey and evaluated during the detailed survey. Examples of such hazards include, but are not limited to, confined space entry, potentially explosive or flammable situations, visible vapor clouds, or areas where biological indicators such as dead animals or vegetation are located.[§1926.65(c)(3)]

(4) *Required information.* The following information to the extent available shall be obtained by the employer prior to allowing employees to enter a site:[§1926.65(c)(4)]

(i) *Location and approximate size of the site.*[§1926.65(c)(4)(i)]

(ii) *Description of the response activity and/or the job task to be performed.*[§1926.65(c)(4)(ii)]

(iii) *Duration of the planned employee activity.*[§1926.65(c)(4)(iii)]

(iv) *Site topography and accessibility by air and roads.* [§1926.65(c)(4)(iv)]

(v) *Safety and health hazards expected at the site.* [§1926.65(c)(4)(v)]

(vi) *Pathways for hazardous substance dispersion.* [§1926.65(c)(4)(vi)]

(vii) *Present status and capabilities* of emergency response teams that would provide assistance to hazardous waste clean-up site employees at the time of an emergency. [§1926.65(c)(4)(vii)]

(viii) *Hazardous substances and health hazards* involved or expected at the site, and their chemical and physical properties.[§1926.65(c)(4)(viii)]

(5) *Personal protective equipment.* Personal protective equipment (PPE) shall be provided and used during initial site entry in accordance with the following requirements:[§1926.65(c)(5)]

(i) *Based upon the results of the preliminary site evaluation,* an ensemble of PPE shall be selected and used during initial site entry which will provide protection to a level of exposure below permissible exposure limits and published exposure levels for known or suspected hazardous substances and health hazards, and which will provide protection against other known and suspected hazards identified during the preliminary site evaluation. If there is no permissible exposure limit or published exposure level, the employer may use other published studies and information as a guide to appropriate personal protective equipment.[§1926.65(c)(5)(i)]

(ii) *If positive-pressure self-contained breathing apparatus* is not used as part of the entry ensemble, and if respiratory protection is warranted by the potential hazards identified during the preliminary site evaluation, an escape self-contained breathing apparatus of at least five minute's duration shall be carried by employees during initial site entry.[§1926.65(c)(5)(ii)]

(iii) *If the preliminary site evaluation does not produce* sufficient information to identify the hazards or suspected hazards of the site, an ensemble providing protection equivalent to Level B PPE shall be provided as minimum protection, and direct reading instruments shall be used as appropriate for identifying IDLH conditions. (See appendix B for a description of Level B hazards and the recommendations for Level B protective equipment.)[§1926.65(c)(5)(iii)]

(iv) *Once the hazards of the site* have been identified, the appropriate PPE shall be selected and used in accordance with paragraph (g) of this section.[§1926.65(c)(5)(iv)]

(6) *Monitoring.* The following monitoring shall be conducted during initial site entry when the site evaluation produces information that shows the potential for ionizing radiation or IDLH conditions, or when the site information is not sufficient reasonably to eliminate these possible conditions:[§1926.65(c)(6)]

(i) *Monitoring with direct reading instruments* for hazardous levels of ionizing radiation.[§1926.65(c)(6)(i)]

(ii) *Monitoring the air* with appropriate direct reading test equipment (i.e., combustible gas meters, detector tubes) for IDLH and other conditions that may cause death or serious harm (combustible or explosive atmospheres, oxygen deficiency, toxic substances).[§1926.65(c)(6)(ii)]

(iii) *Visually observing for signs* of actual or potential IDLH or other dangerous conditions.[§1926.65(c)(6)(iii)]

(iv) *An ongoing air monitoring program* in accordance with paragraph (h) of this section shall be implemented after site characterization has determined the site is safe for the start-up of operations.[§1926.65(c)(6)(iv)]

(7) *Risk identification.* Once the presence and concentrations of specific hazardous substances and health hazards have been established, the risks associated with these substances shall be identified. Employees who will be working on the site shall be informed of any risks that have been identified. In situations covered by the Hazard Communication Standard, 29 CFR 1926.59, training required by that standard need not be duplicated.[§1926.65(c)(7)]

Note to (c)(7): Risks to consider include, but are not limited to:

[a] Exposures exceeding the permissible exposure limits and published exposure levels.

[b] IDLH concentrations.

[c] Potential skin absorption and irritation sources.

[d] Potential eye irritation sources.

[e] Explosion sensitivity and flammability ranges.

[f] Oxygen deficiency.

(8) *Employee notification.* Any information concerning the chemical, physical, and toxicologic properties of each substance known or expected to be present on site that is available to the employer and relevant to the duties an employee is expected to perform shall be made available to the affected employees prior to the commencement of their work activities. The employer may utilize information developed for the hazard communication standard for this purpose.[§1926.65(c)(8)]

(d) Site control.[§1926.65(d)]

(1) *General.* Appropriate site control procedures shall be implemented to control employee exposure to hazardous substances before clean-up work begins.[§1926.65(d)(1)]

(2) *Site control program.* A site control program for protecting employees which is part of the employer's site safety and health program required in paragraph (b) of this section shall be developed during the planning stages of a hazardous waste clean-up operation and modified as necessary as new information becomes available.[§1926.65(d)(2)]

(3) *Elements of the site control program.* The site control program shall, as a minimum, include: A site map; site work zones; the use of a "buddy system"; site communications including alerting means for emergencies; the standard operating procedures or safe work practices; and, identification of the nearest medical assistance. Where these requirements are covered elsewhere they need not be repeated.[§1926.65(d)(3)]

(e) Training.[§1926.65(e)]

(1) *General.*[§1926.65(e)(1)]

(i) *All employees working on site* (such as but not limited to equipment operators, general laborers and others) exposed to hazardous substances, health hazards, or safety hazards and their supervisors and management responsible for the site shall receive training meeting the requirements of this paragraph before they are permitted to engage in hazardous waste operations that could expose them to hazardous substances, safety, or health hazards, and they shall receive review training as specified in this paragraph.[§1926.65(e)(1)(i)]

(ii) *Employees shall not be permitted* to participate in or supervise field activities until they have been trained to a level required by their job function and responsibility. [§1926.65(e)(1)(ii)]

(2) *Elements to be covered.* The training shall thoroughly cover the following:[§1926.65(e)(2)]

(i) *Names of personnel* and alternates responsible for site safety and health;[§1926.65(e)(2)(i)]

(ii) *Safety, health and other hazards present on the site;* [§1926.65(e)(2)(ii)]

(iii) *Use of personal protective equipment;*[§1926.65(e)(2)(iii)]

(iv) *Work practices* by which the employee can minimize risks from hazards;[§1926.65(e)(2)(iv)]

(v) *Safe use of engineering controls and equipment on the site;* [§1926.65(e)(2)(v)]

(vi) *Medical surveillance requirements,* including recognition of symptoms and signs which might indicate overexposure to hazards; and[§1926.65(e)(2)(vi)]

(vii) *The contents of paragraphs (G) through (J)* of the site safety and health plan set forth in paragraph (b)(4)(ii) of this section.[§1926.65(e)(2)(vii)]

(3) *Initial training.*[§1926.65(e)(3)]

(i) *General site workers* (such as equipment operators, general laborers and supervisory personnel) engaged in hazardous substance removal or other activities which expose or potentially expose workers to hazardous substances and health hazards shall receive a minimum of 40 hours of instruction off the site, and a minimum of three days actual field experience under the direct supervision of a trained, experienced supervisor.[§1926.65(e)(3)(i)]

(ii) *Workers on site only occasionally* for a specific limited task (such as, but not limited to, ground water monitoring, land surveying, or geo-physical surveying) and who are unlikely to be exposed over permissible exposure limits and published exposure limits shall receive a minimum of 24 hours of instruction off the site, and the minimum of one day actual field experience under the direct supervision of a trained, experienced supervisor.[§1926.65(e)(3)(ii)]

(iii) *Workers regularly on site* who work in areas which have been monitored and fully characterized indicating that exposures are under permissible exposure limits and published exposure limits where respirators are not necessary, and the characterization indicates that there are no health hazards or the possibility of an emergency developing, shall receive a minimum of 24 hours of instruction off the site and the minimum of one day actual field experience under the direct supervision of a trained, experienced supervisor. [§1926.65(e)(3)(iii)]

(iv) *Workers with 24 hours of training* who are covered by paragraphs (e)(3)(ii) and (e)(3)(iii) of this section, and who become general site workers or who are required to wear respirators, shall have the additional 16 hours and two days of training necessary to total the training specified in paragraph (e)(3)(i).[§1926.65(e)(3)(iv)]

(4) *Management and supervisor training.* On-site management and supervisors directly responsible for, or who supervise employees engaged in, hazardous waste operations shall receive 40 hours initial training, and three days of supervised field experience (the training may be reduced to 24 hours and one day if the only area of their responsibility is employees covered by paragraphs (e)(3)(ii) and (e)(3)(iii)) and at least eight additional hours of specialized training at the time of job assignment on such topics as, but not limited to, the employer's safety and health program and the associated employee training program, personal protective equipment program, spill containment program, and health hazard monitoring procedure and techniques. [§1926.65(e)(4)]

(5) *Qualifications for trainers.* Trainers shall be qualified to instruct employees about the subject matter that is being presented in training. Such trainers shall have satisfactorily completed a training program for teaching the subjects they are expected to teach, or they shall have the academic credentials and instruc-

tional experience necessary for teaching the subjects. Instructors shall demonstrate competent instructional skills and knowledge of the applicable subject matter.[§1926.65(e)(5)]

(6) *Training certification.* Employees and supervisors that have received and successfully completed the training and field experience specified in paragraphs (e)(1) through (e)(4) of this section shall be certified by their instructor or the head instructor and trained supervisor as having successfully completed the necessary training. A written certificate shall be given to each person so certified. Any person who has not been so certified or who does not meet the requirements of paragraph (e)(9) of this section shall be prohibited from engaging in hazardous waste operations.[§1926.65(e)(6)]

(7) *Emergency response.* Employees who are engaged in responding to hazardous emergency situations at hazardous waste clean-up sites that may expose them to hazardous substances shall be trained in how to respond to such expected emergencies.[§1926.65(e)(7)]

(8) *Refresher training.* Employees specified in paragraph (e)(1) of this section, and managers and supervisors specified in paragraph (e)(4) of this section, shall receive eight hours of refresher training annually on the items specified in paragraph (e)(2) and/or (e)(4) of this section, any critique of incidents that have occurred in the past year that can serve as training examples of related work, and other relevant topics.[§1926.65(e)(8)]

(9) *Equivalent training.* Employers who can show by documentation or certification that an employee's work experience and/or training has resulted in training equivalent to that training required in paragraphs (e)(1) through (e)(4) of this section shall not be required to provide the initial training requirements of those paragraphs to such employees and shall provide a copy of the certification or documentation to the employee upon request. However, certified employees or employees with equivalent training new to a site shall receive appropriate, site specific training before site entry and have appropriate supervised field experience at the new site. Equivalent training includes any academic training or the training that existing employees might have already received from actual hazardous waste site work experience.[§1926.65(e)(9)]

(f) **Medical surveillance.**[§1926.65(f)]

(1) *General.* Employers engaged in operations specified in paragraphs (a)(1)(i) through (a)(1)(iv) of this section and not covered by (a)(2)(iii) exceptions and employers of employees specified in paragraph (q)(9) shall institute a medical surveillance program in accordance with this paragraph.[§1926.65(f)(1)]

(2) *Employees covered.* The medical surveillance program shall be instituted by the employer for the following employees: [§1926.65(f)(2)]

(i) *All employees who are or may be exposed* to hazardous substances or health hazards at or above the permissible exposure limits or, if there is no permissible exposure limit, above the published exposure levels for these substances, without regard to the use of respirators, for 30 days or more a year;[§1926.65(f)(2)(i)]

(ii) *All employees who wear a respirator* for 30 days or more a year or as required by Sec. 1926.103;[§1926.65(f)(2)(ii)]

(iii) *All employees who are injured, become ill* or develop signs or symptoms due to possible overexposure involving hazardous substances or health hazards from an emergency response or hazardous waste operation; and[§1926.65(f)(2)(iii)]

(iv) *Members of HAZMAT teams.*[§1926.65(f)(2)(iv)]

(3) *Frequency of medical examinations and consultations.* Medical examinations and consultations shall be made available by the employer to each employee covered under paragraph (f)(2) of this section on the following schedules:[§1926.65(f)(3)]

(i) *For employees covered* under paragraphs (f)(2)(i), (f)(2)(ii), and (f)(2)(iv):[§1926.65(f)(3)(i)]

[A] Prior to assignment;[§1926.65(f)(3)(i)[A]]

[B] At least once every twelve months for each employee covered unless the attending physician believes a longer interval (not greater than biennially) is appropriate; [§1926.65(f)(3)(i)[B]]

[C] At termination of employment or reassignment to an area where the employee would not be covered if the employee has not had an examination within the last six months;[§1926.65(f)(3)(i)[C]]

[D] As soon as possible upon notification by an employee that the employee has developed signs or symptoms indicating possible overexposure to hazardous substances or health hazards, or that the employee has been injured or exposed above the permissible exposure limits or published exposure levels in an emergency situation; [§1926.65(f)(3)(i)[D]]

[E] At more frequent times, if the examining physician determines that an increased frequency of examination is medically necessary.[§1926.65(f)(3)(i)[E]]

(ii) *For employees covered under paragraph (f)(2)(iii)* and for all employees including those of employers covered by paragraph (a)(1)(v) who may have been injured, received a health impairment, developed signs or symptoms which may have resulted from exposure to hazardous substances resulting from an emergency incident, or exposed during an emergency incident to hazardous substances at concentrations above the permissible exposure limits or the published exposure levels without the necessary personal protective equipment being used:[§1926.65(f)(3)(ii)]

[A] As soon as possible following the emergency incident or development of signs or symptoms;[§1926.65(f)(3)(ii)[A]]

[B] At additional times, if the examining physician determines that follow-up examinations or consultations are medically necessary.[§1926.65(f)(3)(ii)[B]]

(4) *Content of medical examinations and consultations.* [§1926.65(f)(4)]

(i) *Medical examinations required by paragraph (f)(3) of this section* shall include a medical and work history (or updated history if one is in the employee's file) with special emphasis on symptoms related to the handling of hazardous substances and health hazards, and to fitness for duty including the ability to wear any required PPE under conditions (i.e., temperature extremes) that may be expected at the work site.[§1926.65(f)(4)(i)]

(ii) *The content of medical examinations* or consultations made available to employees pursuant to paragraph (f) shall be determined by the attending physician. The guidelines in the "Occupational Safety and Health Guidance Manual for Hazardous Waste Site Activities" (See appendix D, Reference #10) should be consulted.[§1926.65(f)(4)(ii)]

(5) *Examination by a physician and costs.* All medical examinations and procedures shall be performed by or under the supervision of a licensed physician, preferably one knowledgeable in occupational medicine, and shall be provided without cost to the employee, without loss of pay, and at a reasonable time and place.[§1926.65(f)(5)]

(6) *Information provided to the physician.* The employer shall provide one copy of this standard and its appendices to the attending physician, and in addition the following for each employee: [§1926.65(f)(6)]

(i) *A description of the employee's duties* as they relate to the employee's exposures.[§1926.65(f)(6)(i)]

(ii) *The employee's exposure levels or anticipated exposure levels.*[§1926.65(f)(6)(ii)]

(iii) *A description of any personal protective equipment used or to be used.*[§1926.65(f)(6)(iii)]

(iv) *Information from previous medical examinations* of the employee which is not readily available to the examining physician.[§1926.65(f)(6)(iv)]

(v) *Information required by Sec. 1926.103.*[§1926.65(f)(6)(v)]

(7) *Physician's written opinion.*[§1926.65(f)(7)]

(i) *The employer shall obtain and furnish* the employee with a copy of a written opinion from the attending physician containing the following:[§1926.65(f)(7)(i)]

[A] The physician's opinion as to whether the employee has any detected medical conditions which would place the employee at increased risk of material impairment of the employee's health from work in hazardous waste operations or emergency response, or from respirator use. [§1926.65(f)(7)(i)[A]]

[B] The physician's recommended limitations upon the employee's assigned work.[§1926.65(f)(7)(i)[B]]

[C] The results of the medical examination and tests if requested by the employee.[§1926.65(f)(7)(i)[C]]

[D] A statement that the employee has been informed by the physician of the results of the medical examination and any medical conditions which require further examination or treatment.[§1926.65(f)(7)(i)[D]]

(ii) *The written opinion obtained* by the employer shall not reveal specific findings or diagnoses unrelated to occupational exposures.[§1926.65(f)(7)(ii)]

(8) *Recordkeeping.*[§1926.65(f)(8)]

(i) *An accurate record of the medical surveillance* required by paragraph (f) of this section shall be retained. This record shall be retained for the period specified and meet the criteria of 29 CFR 1926.33.[§1926.65(f)(8)(i)]

(ii) *The record required in paragraph (f)(8)(i)* of this section shall include at least the following information:[§1926.65(f)(8)(ii)]

[A] The name and social security number of the employee; [§1926.65(f)(8)(ii)[A]]

[B] Physician's written opinions, recommended limitations, and results of examinations and tests;[§1926.65(f)(8)(ii)[B]]

[C] Any employee medical complaints related to exposure to hazardous substances;[§1926.65(f)(8)(ii)[C]]

[D] A copy of the information provided to the examining physician by the employer, with the exception of the standard and its appendices.[§1926.65(f)(8)(ii)[D]]

(g) **Engineering controls, work practices,** and personal protective equipment for employee protection. Engineering controls, work practices, personal protective equipment, or a combination of these shall be implemented in accordance with this paragraph to protect employees from exposure to hazardous substances and safety and health hazards.[§1926.65(g)]

(1) *Engineering controls, work practices* and PPE for substances regulated either in Sec. 1926.55, elsewhere in subpart D, or in other pertinent sections of this part.[§1926.65(g)(1)]

(i) *Engineering controls and work practices* shall be instituted to reduce and maintain employee exposure to or below the permissible exposure limits for substances regulated either in Sec. 1926.55 or other pertinent sections of this part, except to the extent that such controls and practices are not feasible.[§1926.65(g)(1)(i)]

Note to (g)(1)(i): Engineering controls which may be feasible include the use of pressurized cabs or control booths on equipment, and/or the use of remotely operated material handling equipment. Work practices which may be feasible are removing all non-essential employees from potential exposure during opening of drums, wetting down dusty operations and locating employees upwind of possible hazards.

(ii) *Whenever engineering controls and work practices* are not feasible or not required, any reasonable combination of engineering controls, work practices and PPE shall be used to reduce and maintain employee exposures to or below the permissible exposure limits or dose limits for substances regulated either in Sec. 1926.55 or other pertinent sections of this part.[§1926.65(g)(1)(ii)]

(iii) *The employer shall not implement* a schedule of employee rotation as a means of compliance with permissible exposure limits or dose limits except when there is no other feasible way of complying with the airborne or dermal dose limits for ionizing radiation.[§1926.65(g)(1)(iii)]

(iv) *The provisions of subpart D shall be followed.* [§1926.65(g)(1)(iv)]

(2) *Engineering controls, work practices, and PPE for substances not regulated* either in Sec. 1926.55, elsewhere in subpart D, or in other pertinent sections of this part. An appropriate combination of engineering controls, work practices and personal protective equipment shall be used to reduce and maintain employee exposure to or below published exposure levels for hazardous substances and health hazards not regulated either in Sec. 1926.55, elsewhere in subpart D, or in other pertinent sections of this part. The employer may use the published literature and MSDS as a guide in making the employer's determination as to what level of protection the employer believes is appropriate for hazardous substances and health hazards for which there is no permissible exposure limit or published exposure limit.[§1926.65(g)(2)]

(3) *Personal protective equipment selection.*[§1926.65(g)(3)]

(i) *Personal protective equipment (PPE)* shall be selected and used which will protect employees from the hazards and potential hazards they are likely to encounter as identified during the site characterization and analysis.[§1926.65(g)(3)(i)]

(ii) *Personal protective equipment selection* shall be based on an evaluation of the performance characteristics of the PPE relative to the requirements and limitations of the site, the task-specific conditions and duration, and the hazards and potential hazards identified at the site.[§1926.65(g)(3)(ii)]

(iii) *Positive pressure self-contained breathing apparatus,* or positive pressure air-line respirators equipped with an escape air supply, shall be used when chemical exposure levels present will create a substantial possibility of immediate death, immediate serious illness or injury, or impair the ability to escape.[§1926.65(g)(3)(iii)]

(iv) *Totally-encapsulating chemical protective suits* (protection equivalent to Level A protection as recommended in appendix B) shall be used in conditions where skin absorption of a hazardous substance may result in a substantial possibility of immediate death, immediate serious illness or injury, or impair the ability to escape.[§1926.65(g)(3)(iv)]

(v) *The level of protection provided by PPE selection* shall be increased when additional information on site conditions indicates that increased protection is necessary to reduce employee exposures below permissible exposure limits and published exposure levels for hazardous substances and health hazards. (See appendix B for guidance on selecting PPE ensembles.)[§1926.65(g)(3)(v)]

Note to (g)(3): The level of employee protection provided may be decreased when additional information or site conditions show that decreased protection will not result in hazardous exposures to employees.

(vi) *Personal protective equipment* shall be selected and used to meet the requirements of subpart E of this part and additional requirements specified in this section.[§1926.65(g)(3)(vi)]

(4) *Totally-encapsulating chemical protective suits.*[§1926.65(g)(4)]

(i) *Totally-encapsulating suits* shall protect employees from the particular hazards which are identified during site characterization and analysis.[§1926.65(g)(4)(i)]

(ii) *Totally-encapsulating suits shall be capable* of maintaining positive air pressure. (See appendix A for a test method which may be used to evaluate this requirement.) [§1926.65(g)(4)(ii)]

(iii) *Totally-encapsulating suits shall be capable* of preventing inward test gas leakage of more than 0.5 percent. (See appendix A for a test method which may be used to evaluate this requirement.)[§1926.65(g)(4)(iii)]

(5) *Personal protective equipment (PPE) program.* A written personal protective equipment program, which is part of the employer's safety and health program required in paragraph (b) of this section or required in paragraph (p)(1) of this section and which is also a part of the site-specific safety and health plan shall be established. The PPE program shall address the elements listed below. When elements, such as donning and doffing procedures, are provided by the manufacturer of a piece of equipment and are attached to the plan, they need not be rewritten into the plan as long as they adequately address the procedure or element.[§1926.65(g)(5)]

(i) *PPE selection based upon site hazards,*[§1926.65(g)(5)(i)]

(ii) *PPE use and limitations of the equipment,*[§1926.65(g)(5)(ii)]

(iii) *Work mission duration,*[§1926.65(g)(5)(iii)]

(iv) *PPE maintenance and storage,*[§1926.65(g)(5)(iv)]

(v) *PPE decontamination and disposal,*[§1926.65(g)(5)(v)]

(vi) *PPE training and proper fitting,*[§1926.65(g)(5)(vi)]

(vii) *PPE donning and doffing procedures,*[§1926.65(g)(5)(vii)]

(viii) *PPE inspection procedures prior to, during, and after use,* [§1926.65(g)(5)(viii)]

(ix) *Evaluation of the effectiveness of the PPE program, and* [§1926.65(g)(5)(ix)]

(x) *Limitations during temperature extremes,* heat stress, and other appropriate medical considerations.[§1926.65(g)(5)(x)]

(h) **Monitoring.**[§1926.65(h)]

(1) *General.*[§1926.65(h)(1)]

(i) *Monitoring shall be performed in accordance with this paragraph* where there may be a question of employee exposure to hazardous concentrations of hazardous substances in order to assure proper selection of engineering controls, work practices and personal protective equipment so that employees are not exposed to levels which exceed permissible exposure limits, or published exposure levels if there are no permissible exposure limits, for hazardous substances. [§1926.65(h)(1)(i)]

(ii) *Air monitoring shall be used to identify and quantify* airborne levels of hazardous substances and safety and health hazards in order to determine the appropriate level of employee protection needed on site.[§1926.65(h)(1)(ii)]

(2) *Initial entry.* Upon initial entry, representative air monitoring shall be conducted to identify any IDLH condition, exposure over permissible exposure limits or published exposure levels, exposure over a radioactive material's dose limits or other dangerous condition such as

the presence of flammable atmospheres or oxygen-deficient environments.[§1926.65(h)(2)]

(3) *Periodic monitoring.* Periodic monitoring shall be conducted when the possibility of an IDLH condition or flammable atmosphere has developed or when there is indication that exposures may have risen over permissible exposure limits or published exposure levels since prior monitoring. Situations where it shall be considered whether the possibility that exposures have risen are as follows:[§1926.65(h)(3)]

(i) *When work begins on a different portion of the site.* [§1926.65(h)(3)(i)]

(ii) *When contaminants* other than those previously identified are being handled.[§1926.65(h)(3)(ii)]

(iii) *When a different type of operation* is initiated (e.g., drum opening as opposed to exploratory well drilling). [§1926.65(h)(3)(iii)]

(iv) *When employees are handling* leaking drums or containers or working in areas with obvious liquid contamination (e.g., a spill or lagoon).[§1926.65(h)(3)(iv)]

(4) *Monitoring of high-risk employees.* After the actual clean-up phase of any hazardous waste operation commences; for example, when soil, surface water or containers are moved or disturbed; the employer shall monitor those employees likely to have the highest exposures to hazardous substances and health hazards likely to be present above permissible exposure limits or published exposure levels by using personal sampling frequently enough to characterize employee exposures. If the employees likely to have the highest exposure are over permissible exposure limits or published exposure limits, then monitoring shall continue to determine all employees likely to be above those limits. The employer may utilize a representative sampling approach by documenting that the employees and chemicals chosen for monitoring are based on the criteria stated above.[§1926.65(h)(4)]

Note to (h): It is not required to monitor employees engaged in site characterization operations covered by paragraph (c) of this section.

(i) **Informational programs.** Employers shall develop and implement a program, which is part of the employer's safety and health program required in paragraph (b) of this section, to inform employees, contractors, and subcontractors (or their representative) actually engaged in hazardous waste operations of the nature, level and degree of exposure likely as a result of participation in such hazardous waste operations. Employees, contractors and subcontractors working outside of the operations part of a site are not covered by this standard.[§1926.65(i)]

(j) **Handling drums and containers.**[§1926.65(j)]

(1) *General.*[§1926.65(j)(1)]

(i) *Hazardous substances and contaminated soils, liquids,* and other residues shall be handled, transported, labeled, and disposed of in accordance with this paragraph.[§1926.65(j)(1)(i)]

(ii) *Drums and containers used* during the clean-up shall meet the appropriate DOT, OSHA, and EPA regulations for the wastes that they contain.[§1926.65(j)(1)(ii)]

(iii) *When practical, drums and containers* shall be inspected and their integrity shall be assured prior to being moved. Drums or containers that cannot be inspected before being moved because of storage conditions (i.e., buried beneath the earth, stacked behind other drums, stacked several tiers high in a pile, etc.) shall be moved to an accessible location and inspected prior to further handling.[§1926.65(j)(1)(iii)]

(iv) *Unlabelled drums and containers* shall be considered to contain hazardous substances and handled accordingly until the contents are positively identified and labeled. [§1926.65(j)(1)(iv)]

(v) *Site operations shall be organized* to minimize the amount of drum or container movement.[§1926.65(j)(1)(v)]

(vi) *Prior to movement of drums or containers,* all employees exposed to the transfer operation shall be warned of the potential hazards associated with the contents of the drums or containers.[§1926.65(j)(1)(vi)]

(vii) *U.S. Department of Transportation specified salvage drums* or containers and suitable quantities of proper absorbent shall be kept available and used in areas where spills, leaks, or ruptures may occur.[§1926.65(j)(1)(vii)]

(viii) *Where major spills may occur,* a spill containment program, which is part of the employer's safety and health program required in paragraph (b) of this section, shall be implemented to contain and isolate the entire volume of the hazardous substance being transferred.[§1926.65(j)(1)(viii)]

(ix) *Drums and containers* that cannot be moved without rupture, leakage, or spillage shall be emptied into a sound container using a device classified for the material being transferred.[§1926.65(j)(1)(ix)]

(x) *A ground-penetrating system* or other type of detection system or device shall be used to estimate the location and depth of buried drums or containers.[§1926.65(j)(1)(x)]

(xi) *Soil or covering material* shall be removed with caution to prevent drum or container rupture.[§1926.65(j)(1)(xi)]

(xii) *Fire extinguishing equipment* meeting the requirements of subpart F of this part shall be on hand and ready for use to control incipient fires.[§1926.65(j)(1)(xii)]

(2) *Opening drums and containers.* The following procedures shall be followed in areas where drums or containers are being opened:[§1926.65(j)(2)]

(i) *Where an airline respirator system is used,* connections to the source of air supply shall be protected from contamination and the entire system shall be protected from physical damage.[§1926.65(j)(2)(i)]

(ii) *Employees not actually involved* in opening drums or containers shall be kept a safe distance from the drums or containers being opened.[§1926.65(j)(2)(ii)]

(iii) *If employees must work near or adjacent to drums* or containers being opened, a suitable shield that does not interfere with the work operation shall be placed between the employee and the drums or containers being opened to protect the employee in case of accidental explosion. [§1926.65(j)(2)(iii)]

(iv) *Controls for drum or container opening equipment,* monitoring equipment, and fire suppression equipment shall be located behind the explosion-resistant barrier.[§1926.65(j)(2)(iv)]

(v) *When there is a reasonable possibility* of flammable atmospheres being present, material handling equipment and hand tools shall be of the type to prevent sources of ignition. [§1926.65(j)(2)(v)]

(vi) *Drums and containers shall be opened* in such a manner that excess interior pressure will be safely relieved. If pressure can not be relieved from a remote location, appropriate shielding shall be placed between the employee and the drums or containers to reduce the risk of employee injury. [§1926.65(j)(2)(vi)]

(vii) *Employees shall not stand upon or work from drums or containers.*[§1926.65(j)(2)(vii)]

(3) *Material handling equipment.* Material handiing equipment used to transfer drums and containers shall be selected, positioned and operated to minimize sources of ignition related to the equipment from igniting vapors released from ruptured drums or containers.[§1926.65(j)(3)]

(4) *Radioactive wastes.* Drums and containers containing radioactive wastes shall not be handled until such time as their hazard to employees is properly assessed.[§1926.65(j)(4)]

(5) *Shock sensitive wastes.* As a minimum, the following special precautions shall be taken when drums and containers containing or suspected of containing shock-sensitive wastes are handled:[§1926.65(j)(5)]

(i) *All non-essential employees* shall be evacuated from the area of transfer.[§1926.65(j)(5)(i)]

(ii) *Material handling equipment* shall be provided with explosive containment devices or protective shields to protect equipment operators from exploding containers.[§1926.65(j)(5)(ii)]

(iii) *An employee alarm system* capable of being perceived above surrounding light and noise conditions shall be used to signal the commencement and completion of explosive waste handling activities.[§1926.65(j)(5)(iii)]

(iv) *Continuous communications* (i.e., portable radios, hand signals, telephones, as appropriate) shall be maintained between the employee-in-charge of the immediate handling area and both the site safety and health supervisor and the command post until such time as the handling operation is completed. Communication equipment or methods that could cause shock sensitive materials to explode shall not be used. [§1926.65(j)(5)(iv)]

(v) *Drums and containers under pressure,* as evidenced by bulging or swelling, shall not be moved until such time as the cause for excess pressure is determined and appropriate containment procedures have been implemented to protect employees from explosive relief of the drum.[§1926.65(j)(5)(v)]

(vi) *Drums and containers containing* packaged laboratory wastes shall be considered to contain shock-sensitive or explosive materials until they have been characterized. [§1926.65(j)(5)(vi)]

Caution: Shipping of shock sensitive wastes may be prohibited under U.S. Department of Transportation regulations. Employers and their shippers should refer to 49 CFR 173.21 and 173.50.

(6) *Laboratory waste packs.* In addition to the requirements of paragraph (j)(5) of this section, the following precautions shall be taken, as a minimum, in handling laboratory waste packs (lab packs):[§1926.65(j)(6)]

(i) *Lab packs shall be opened* only when necessary and then only by an individual knowledgeable in the inspection, classification, and segregation of the containers within the pack according to the hazards of the wastes.[§1926.65(j)(6)(i)]

(ii) *If crystalline material is noted* on any container, the contents shall be handled as a shock-sensitive waste until the contents are identified.[§1926.65(j)(6)(ii)]

(7) *Sampling of drum and container contents.* Sampling of containers and drums shall be done in accordance with a sampling procedure which is part of the site safety and health plan developed for and available to employees and others at the specific worksite.[§1926.65(j)(7)]

(8) *Shipping and transport.*[§1926.65(j)(8)]

(i) *Drums and containers* shall be identified and classified prior to packaging for shipment.[§1926.65(j)(8)(i)]

(ii) *Drum or container staging areas* shall be kept to the minimum number necessary to identify and classify materials safely and prepare them for transport.[§1926.65(j)(8)(ii)]

(iii) *Staging areas shall be provided* with adequate access and egress routes.[§1926.65(j)(8)(iii)]

(iv) *Bulking of hazardous wastes* shall be permitted only after a thorough characterization of the materials has been completed.[§1926.65(j)(8)(iv)]

(9) *Tank and vault procedures.*[§1926.65(j)(9)]

(i) *Tanks and vaults* containing hazardous substances shall be handled in a manner similar to that for drums and containers, taking into consideration the size of the tank or vault. [§1926.65(j)(9)(i)]

(ii) *Appropriate tank or vault entry procedures* as described in the employer's safety and health plan shall be followed whenever employees must enter a tank or vault. [§1926.65(j)(9)(ii)]

(k) **Decontamination.**[§1926.65(k)]

(1) *General.* Procedures for all phases of decontamination shall be developed and implemented in accordance with this paragraph. [§1926.65(k)(1)]

(2) *Decontamination procedures.*[§1926.65(k)(2)]

(i) *A decontamination procedure shall be developed,* communicated to employees and implemented before any employees or equipment may enter areas on site where potential for exposure to hazardous substances exists.[§1926.65(k)(2)(i)]

(ii) *Standard operating procedures* shall be developed to minimize employee contact with hazardous substances or with equipment that has contacted hazardous substances. [§1926.65(k)(2)(ii)]

(iii) *All employees leaving a contaminated area* shall be appropriately decontaminated; all contaminated clothing and equipment leaving a contaminated area shall be appropriately disposed of or decontaminated.[§1926.65(k)(2)(iii)]

(iv) *Decontamination procedures shall be monitored* by the site safety and health supervisor to determine their effectiveness. When such procedures are found to be ineffective, appropriate steps shall be taken to correct any deficiencies. [§1926.65(k)(2)(iv)]

(3) *Location.* Decontamination shall be performed in geographical areas that will minimize the exposure of uncontaminated employees or equipment to contaminated employees or equipment.[§1926.65(k)(3)]

(4) *Equipment and solvents.* All equipment and solvents used for decontamination shall be decontaminated or disposed of properly.[§1926.65(k)(4)]

(5) *Personal protective clothing and equipment.*[§1926.65(k)(5)]

(i) *Protective clothing and equipment* shall be decontaminated, cleaned, laundered, maintained or replaced as needed to maintain their effectiveness.[§1926.65(k)(5)(i)]

(ii) *Employees whose non-impermeable clothing* becomes wetted with hazardous substances shall immediately remove that clothing and proceed to shower. The clothing shall be disposed of or decontaminated before it is removed from the work zone.[§1926.65(k)(5)(ii)]

(6) *Unauthorized employees.* Unauthorized employees shall not remove protective clothing or equipment from change rooms. [§1926.65(k)(6)]

(7) *Commercial laundries or cleaning establishments.* Commercial laundries or cleaning establishments that decontaminate protective clothing or equipment shall be informed of the potentially harmful effects of exposures to hazardous substances. [§1926.65(k)(7)]

(8) *Showers and change rooms.* Where the decontamination procedure indicates a need for regular showers and change rooms outside of a contaminated area, they shall be provided and meet the requirements of 29 CFR 1910.141. If temperature conditions prevent the effective use of water, then other effective means for cleansing shall be provided and used. [§1926.65(k)(8)]

(l) **Emergency response by employees** at uncontrolled hazardous waste sites.[§1926.65(l)]

(1) *Emergency response plan.*[§1926.65(l)(1)]

(i) *An emergency response plan* shall be developed and implemented by all employers within the scope of paragraphs (a)(1) (i)-(ii) of this section to handle anticipated emergencies prior to the commencement of hazardous waste operations. The plan shall be in writing and available for inspection and copying by employees, their representatives, OSHA personnel and other governmental agencies with relevant responsibilities.[§1926.65(l)(1)(i)]

(ii) *Employers who will evacuate their employees* from the danger area when an emergency occurs, and who do not permit any of their employees to assist in handling the emergency, are exempt from the requirements of this paragraph if they provide an emergency action plan complying with Sec. 1926.35 of this part.[§1926.65(l)(1)(ii)]

(2) *Elements of an emergency response plan.* The employer shall develop an emergency response plan for emergencies which shall address, as a minimum, the following:[§1926.65(l)(2)]

(i) *Pre-emergency planning.*[§1926.65(l)(2)(i)]

(ii) *Personnel roles, lines of authority, and communication.* [§1926.65(l)(2)(ii)]

(iii) *Emergency recognition and prevention.*[§1926.65(l)(2)(iii)]

(iv) *Safe distances and places of refuge.*[§1926.65(l)(2)(iv)]

(v) *Site security and control.*[§1926.65(l)(2)(v)]

(vi) *Evacuation routes and procedures.*[§1926.65(l)(2)(vi)]

(vii) *Decontamination procedures* which are not covered by the site safety and health plan.[§1926.65(l)(2)(vii)]

(viii) *Emergency medical treatment and first aid.*[§1926.65(l)(2)(viii)]

(ix) *Emergency alerting and response procedures.* [§1926.65(l)(2)(ix)]

(x) *Critique of response and follow-up.*[§1926.65(l)(2)(x)]

(xi) *PPE and emergency equipment.*[§1926.65(l)(2)(xi)]

(3) *Procedures for handling emergency incidents.*[§1926.65(l)(3)]

(i) *In addition to the elements* for the emergency response plan required in paragraph (l)(2) of this section, the following elements shall be included for emergency response plans: [§1926.65(l)(3)(i)]

[A] Site topography, layout, and prevailing weather conditions.[§1926.65(l)(3)(i)[A]]

[B] Procedures for reporting incidents to local, state, and federal governmental agencies.[§1926.65(l)(3)(i)[B]]

(ii) *The emergency response plan* shall be a separate section of the Site Safety and Health Plan.[§1926.65(l)(3)(ii)]

(iii) *The emergency response plan* shall be compatible and integrated with the disaster, fire and/or emergency response plans of local, state, and federal agencies.[§1926.65(l)(3)(iii)]

(iv) *The emergency response plan* shall be rehearsed regularly as part of the overall training program for site operations. [§1926.65(l)(3)(iv)]

(v) *The site emergency response plan* shall be reviewed periodically and, as necessary, be amended to keep it current with new or changing site conditions or information.[§1926.65(l)(3)(v)]

(vi) *An employee alarm system* shall be installed in accordance with 29 CFR 1926.159 to notify employees of an emergency situation; to stop work activities if necessary; to lower background noise in order to speed communication; and to begin emergency procedures.[§1926.65(l)(3)(vi)]

(vii) *Based upon the information available* at time of the emergency, the employer shall evaluate the incident and the site response capabilities and proceed with the appropriate steps to implement the site emergency response plan. [§1926.65(l)(3)(vii)]

(m) **Illumination.** Areas accessible to employees shall be lighted to not less than the minimum illumination intensities listed in the following Table D-65.1 while any work is in progress:[§1926.65(m)]

Table D-65.1 — Minimum Illumination Intensities in Foot-Candles

Foot-candles	Area or operations
5	General site areas.
3	Excavation and waste areas, accessways, active storage areas, loading platforms, refueling, and field maintenance areas.
5	Indoors: Warehouses, corridors hallways, and exitways.
5	Tunnels, shafts, and general underground work areas. (Exception: Minimum of 10 foot candles is required at tunnel and shaft heading during drilling mucking, and scaling Mine Safety and Health Administration approved cap lights shall be acceptable for use in the tunnel heading.)
10	General shops (e.g., mechanical and electrical equipment rooms, active storerooms, barracks or living quarters, locker or dressing rooms, dining areas, and indoor toilets and workrooms.)
30	First aid stations, infirmaries, and offices.

(n) **Sanitation at temporary workplaces**[§1926.65(n)]

(1) *Potable water.*[§1926.65(n)(1)]

(i) *An adequate supply* of potable water shall be provided on the site.[§1926.65(n)(1)(i)]

(ii) *Portable containers used* to dispense drinking water shall be capable of being tightly closed, and equipped with a tap. Water shall not be dipped from containers.[§1926.65(n)(1)(ii)]

(iii) *Any container used to distribute* drinking water shall be clearly marked as to the nature of its contents and not used for any other purpose.[§1926.65(n)(1)(iii)]

(iv) *Where single service cups* (to be used but once) are supplied, both a sanitary container for the unused cups and a receptacle for disposing of the used cups shall be provided. [§1926.65(n)(1)(iv)]

(2) *Nonpotable water.*[§1926.65(n)(2)]

(i) *Outlets for nonpotable water,* such as water for firefighting purposes, shall be identified to indicate clearly that the water is unsafe and is not to be used for drinking, washing, or cooking purposes.[§1926.65(n)(2)(i)]

(ii) *There shall be no cross-connection,* open or potential, between a system furnishing potable water and a system furnishing nonpotable water.[§1926.65(n)(2)(ii)]

(3) *Toilet facilities.*[§1926.65(n)(3)]

(i) *Toilets shall be provided for employees* according to the following Table D-65.2.[§1926.65(n)(3)(i)]

Table D-65.2 — Toilet Facilities

Number of employees	Minimum number of facilities
20 or fewer	One.
More than 20, fewer than 200	One toilet seat and one urinal per 40 employees.
More than 200	One toilet seat and one urinal per 50 employees.

(ii) *Under temporary field conditions,* provisions shall be made to assure that at least one toilet facility is available. [§1926.65(n)(3)(ii)]

(iii) *Hazardous waste sites* not provided with a sanitary sewer shall be provided with the following toilet facilities unless prohibited by local codes:[§1926.65(n)(3)(iii)]

[A] Chemical toilets;[§1926.65(n)(3)(iii)[A]]

[B] Recirculating toilets;[§1926.65(n)(3)(iii)[B]]

[C] Combustion toilets; or[§1926.65(n)(3)(iii)[C]]

[D] Flush toilets.[§1926.65(n)(3)(iii)[D]]

(iv) *The requirements of this paragraph* for sanitation facilities shall not apply to mobile crews having transportation readily available to nearby toilet facilities.[§1926.65(n)(3)(iv)]

(v) *Doors entering toilet facilities* shall be provided with entrance locks controlled from inside the facility.[§1926.65(n)(3)(v)]

(4) *Food handling.* All food service facilities and operations for employees shall meet the applicable laws, ordinances, and regulations of the jurisdictions in which they are located.[§1926.65(n)(4)]

(5) *Temporary sleeping quarters.* When temporary sleeping quarters are provided, they shall be heated, ventilated, and lighted. [§1926.65(n)(5)]

(6) *Washing facilities.* The employer shall provide adequate washing facilities for employees engaged in operations where hazardous substances may be harmful to employees. Such facilities shall be in near proximity to the worksite; in areas where exposures are below permissible exposure limits and published exposure levels and which are under the controls of the employer; and shall be so equipped as to enable employees to remove hazardous substances from themselves.[§1926.65(n)(6)]

(7) *Showers and change rooms.* When hazardous waste clean-up or removal operations commence on a site and the duration of the work will require six months or greater time to complete, the employer shall provide showers and change rooms for all employees exposed to hazardous substances and health hazards involved in hazardous waste clean-up or removal operations. [§1926.65(n)(7)]

(i) *Showers shall be provided* and shall meet the requirements of 29 CFR 1926.51(f)(4).[§1926.65(n)(7)(i)]

(ii) *Change rooms shall be provided* and shall meet the requirements of 29 CFR 1926.51(i). Change rooms shall consist of two separate change areas separated by the shower area required in paragraph (n)(7)(i) of this section. One change area, with an exit leading off the worksite, shall provide employees with a clean area where they can remove, store, and put on street clothing. The second area, with an exit to the worksite, shall provide employees with an area where they can put on, remove and store work clothing and personal protective equipment. [§1926.65(n)(7)(ii)]

(iii) *Showers and change rooms* shall be located in areas where exposures are below the permissible exposure limits and published exposure levels. If this cannot be accomplished, then a ventilation system shall be provided that will supply air that is below the permissible exposure limits and published exposure levels.[§1926.65(n)(7)(iii)]

(iv) *Employers shall assure* that employees shower at the end of their work shift and when leaving the hazardous waste site. [§1926.65(n)(7)(iv)]

(o) **New technology programs.**[§1926.65(o)]

(1) *The employer shall develop and implement* procedures for the introduction of effective new technologies and equipment developed for the improved protection of employees working with hazardous waste clean-up operations, and the same shall be implemented as part of the site safety and health program to assure that employee protection is being maintained.[§1926.65(o)(1)]

(2) *New technologies, equipment or control measures* available to the industry, such as the use of foams, absorbents, adsorbents, neutralizers, or other means to suppress the level of air contaminates while excavating the site or for spill control, shall be evaluated by employers or their representatives. Such an evaluation shall be done to determine the effectiveness of the new methods, materials, or equipment before implementing their use on a large scale for enhancing employee protection. Information and data from manufacturers or suppliers may be used as part of the employer's evaluation effort. Such evaluations shall be made available to OSHA upon request.[§1926.65(o)(2)]

(p) **Certain operations conducted under** the Resource Conservation and Recovery Act of 1976 (RCRA). Employers conducting operations at treatment, storage and disposal (TSD) facilities specified in paragraph (a)(1)(iv) of this section shall provide and implement the programs specified in this paragraph. See the "Notes and Exceptions" to paragraph (a)(2)(iii) of this section for employers not covered.)".[§1926.65(p)]

(1) *Safety and health program.* The employer shall develop and implement a written safety and health program for employees involved in hazardous waste operations that shall be available for inspection by employees, their representatives and OSHA personnel. The program shall be designed to identify, evaluate and control safety and health hazards in their facilities for the purpose of employee protection, to provide for emergency response meeting the requirements of paragraph (p)(8) of this section and to address as appropriate site analysis, engineering controls, maximum exposure limits, hazardous waste handling procedures and uses of new technologies.[§1926.65(p)(1)]

(2) *Hazard communication program.* The employer shall implement a hazard communication program meeting the requirements of 29 CFR 1926.59 as part of the employer's safety and program.[§1926.65(p)(2)]

Note to 1926.65: The exemption for hazardous waste provided in Sec. 1926.59 is applicable to this section.

(3) *Medical surveillance program.* The employer shall develop and implement a medical surveillance program meeting the requirements of paragraph (f) of this section.[§1926.65(p)(3)]

(4) *Decontamination program.* The employer shall develop and implement a decontamination procedure meeting the requirements of paragraph (k) of this section.[§1926.65(p)(4)]

(5) *New technology program.* The employer shall develop and implement procedures meeting the requirements of paragraph (o) of this section for introducing new and innovative equipment into the workplace.[§1926.65(p)(5)]

(6) *Material handling program.* Where employees will be handling drums or containers, the employer shall develop and implement procedures meeting the requirements of paragraphs (j)(1) (ii) through (viii) and (xi) of this section, as well as (j)(3) and (j)(8) of this section prior to starting such work.[§1926.65(p)(6)]

(7) *Training program.*[§1926.65(p)(7)]

(i) *New employees.* The employer shall develop and implement a training program, which is part of the employer's safety and health program, for employees exposed to health hazards or hazardous substances at TSD operations to enable the employees to perform their assigned duties and functions in a safe and healthful manner so as not endanger themselves or other employees. The initial training shall be for 24 hours and refresher training shall be for eight hours annually. Employees who have received the initial training required by this paragraph shall be given a written certificate attesting that they have successfully completed the necessary training.[§1926.65(p)(7)(i)]

(ii) *Current employees.* Employers who can show by an employee's previous work experience and/or training that the employee has had training equivalent to the initial training required by this paragraph, shall be considered as meeting the initial training requirements of this paragraph as to that employee. Equivalent training includes the training that existing employees might have already received from actual site work experience. Current employees shall receive eight hours of refresher training annually.[§1926.65(p)(7)(ii)]

(iii) *Trainers.* Trainers who teach initial training shall have satisfactorily completed a training course for teaching the subjects they are expected to teach or they shall have the academic credentials and instruction experience necessary to demonstrate a good command of the subject matter of the courses and competent instructional skills.[§1926.65(p)(7)(iii)]

(8) *Emergency response program.*[§1926.65(p)(8)]

(i) *Emergency response plan.* An emergency response plan shall be developed and implemented by all employers. Such plans need not duplicate any of the subjects fully addressed in the employer's contingency planning required by permits, such as those issued by the U.S. Environmental Protection Agency, provided that the contingency plan is made part of the emergency response plan. The emergency response plan shall be a written portion of the employers safety and health program required in paragraph (p)(1) of this section. Employers who will evacuate their employees from the worksite location when an emergency occurs and who do not permit any of their employees to assist in handling the emergency are exempt from the requirements of paragraph (p)(8) if they provide an emergency action plan complying with Sec. 1926.35 of this part.[§1926.65(p)(8)(i)]

(ii) *Elements of an emergency response plan.* The employer shall develop an emergency response plan for emergencies which shall address, as a minimum, the following areas to the extent that they are not addressed in any specific program required in this paragraph:[§1926.65(p)(8)(ii)]

[A] Pre-emergency planning and coordination with outside parties.[§1926.65(p)(8)(ii)[A]]

[B] Personnel roles, lines of authority, and communication. [§1926.65(p)(8)(ii)[B]]

[C] Emergency recognition and prevention.[§1926.65(p)(8)(ii)[C]]

[D] Safe distances and places of refuge.[§1926.65(p)(8)(ii)[D]]

[E] Site security and control.[§1926.65(p)(8)(ii)[E]]

[F] Evacuation routes and procedures.[§1926.65(p)(8)(ii)[F]]

[G] Decontamination procedures.[§1926.65(p)(8)(ii)[G]]

[H] Emergency medical treatment and first aid. [§1926.65(p)(8)(ii)[H]]

[I] Emergency alerting and response procedures. [§1926.65(p)(8)(ii)[I]]

[J] Critique of response and follow-up.[§1926.65(p)(8)(ii)[J]]

[K] PPE and emergency equipment.[§1926.65(p)(8)(ii)[K]]

(iii) *Training.*[§1926.65(p)(8)(iii)]

[A] Training for emergency response employees shall be completed before they are called upon to perform in real emergencies. Such training shall include the elements of the emergency response plan, standard operating procedures the employer has established for the job, the personal protective equipment to be worn and procedures for handling emergency incidents.[§1926.65(p)(8)(iii)[A]]

Exception #1: An employer need not train all employees to the degree specified if the employer divides the work force in a manner such that a sufficient number of employees who have responsibility to control emergencies have the training specified, and all other employees, who may first respond to an emergency incident, have sufficient awareness training to recognize that an emergency response situation exists and that they are instructed in that case to summon the fully trained employees and not attempt control activities for which they are not trained.

Exception #2: An employer need not train all employees to the degree specified if arrangements have been made in advance for an outside fully-trained emergency response team to respond in a reasonable period and all employees, who may come to the incident first, have sufficient awareness training to recognize that an emergency response situation exists and they have been instructed to call the designated outside fully-trained emergency response team for assistance.

[B] Employee members of TSD facility emergency response organizations shall be trained to a level of competence in the recognition of health and safety hazards to protect themselves and other employees. This would include training in the methods used to minimize the risk from safety and health hazards; in the safe use of control equipment; in the selection and use of appropriate personal protective equipment; in the safe operating procedures to be used at the incident scene; in the techniques of coordination with other employees to minimize risks; in the appropriate response to over exposure from health hazards or injury to themselves and other employees; and in the recognition of subsequent symptoms which may result from over exposures.[§1926.65(p)(8)(iii)[B]]

[C] The employer shall certify that each covered employee has attended and successfully completed the training required in paragraph (p)(8)(iii) of this section, or shall certify the employee's competency at least yearly. The method used to demonstrate competency for certification of training shall be recorded and maintained by the employer.[§1926.65(p)(8)(iii)[C]]

(iv) *Procedures for handling emergency incidents.* [§1926.65(p)(8)(iv)]

[A] In addition to the elements for the emergency response plan required in paragraph (p)(8)(ii) of this section, the following elements shall be included for emergency response plans to the extent that they do not repeat any information already contained in the emergency response plan:[§1926.65(p)(8)(iv)[A]]

[1] Site topography, layout, and prevailing weather conditions.[§1926.65(p)(8)(iv)[A][1]]

[2] Procedures for reporting incidents to local, state, and federal governmental agencies.[§1926.65(p)(8)(iv)[A][2]]

[B] The emergency response plan shall be compatible and integrated with the disaster, fire and/or emergency response plans of local, state, and federal agencies. [§1926.65(p)(8)(iv)[B]]

[C] The emergency response plan shall be rehearsed regularly as part of the overall training program for site operations.[§1926.65(p)(8)(iv)[C]]

[D] The site emergency response plan shall be reviewed periodically and, as necessary, be amended to keep it current with new or changing site conditions or information. [§1926.65(p)(8)(iv)[D]]

[E] An employee alarm system shall be installed in accordance with 29 CFR 1926.159 to notify employees of an emergency situation; to stop work activities if necessary; to lower background noise in order to speed communication; and to begin emergency procedures. [§1926.65(p)(8)(iv)[E]]

[F] Based upon the information available at time of the emergency, the employer shall evaluate the incident and the site response capabilities and proceed with the appropriate steps to implement the site emergency response plan. [§1926.65(p)(8)(iv)[F]]

(q) Emergency response to hazardous substance releases. This paragraph covers employers whose employees are engaged in emergency response no matter where it occurs except that it does not cover employees engaged in operations specified in paragraphs (a)(1)(i) through (a)(1)(iv) of this section. Those emergency response organizations who have developed and implemented programs equivalent to this paragraph for handling releases of hazardous substances pursuant to section 303 of the Superfund Amendments and Reauthorization Act of 1986 (Emergency Planning and Community Right-to-Know Act of 1986, 42 U.S.C. 11003) shall be deemed to have met the requirements of this paragraph.[§1926.65(q)]

(1) *Emergency response plan.* An emergency response plan shall be developed and implemented to handle anticipated emergencies prior to the commencement of emergency response operations. The plan shall be in writing and available for inspection and copying by employees, their representatives and OSHA personnel. Employers who will evacuate their employees from the danger area when an emergency occurs, and who do not permit any of their employees to assist in handling the emergency, are exempt from the requirements of this paragraph if they provide an emergency action plan in accordance with Sec. 1926.35 of this part.[§1926.65(q)(1)]

(2) *Elements of an emergency response plan.* The employer shall develop an emergency response plan for emergencies which shall address, as a minimum, the following to the extent that they are not addressed elsewhere:[§1926.65(q)(2)]

(i) *Pre-emergency planning and coordination with outside parties.*[§1926.65(q)(2)(i)]

(ii) *Personnel roles, lines of authority, training, and communication.*[§1926.65(q)(2)(ii)]

(iii) *Emergency recognition and prevention.*[§1926.65(q)(2)(iii)]

(iv) *Safe distances and places of refuge.*[§1926.65(q)(2)(iv)]

(v) *Site security and control.*[§1926.65(q)(2)(v)]

(vi) *Evacuation routes and procedures.*[§1926.65(q)(2)(vi)]

(vii) *Decontamination.*[§1926.65(q)(2)(vii)]

(viii) *Emergency medical treatment and first aid.*[§1926.65(q)(2)(viii)]

(ix) *Emergency alerting and response procedures.* [§1926.65(q)(2)(ix)]

(x) *Critique of response and follow-up.*[§1926.65(q)(2)(x)]

(xi) *PPE and emergency equipment.*[§1926.65(q)(2)(xi)]

(xii) *Emergency response organizations* may use the local emergency response plan or the state emergency response plan or both, as part of their emergency response plan to avoid duplication. Those items of the emergency response plan that are being properly addressed by the SARA Title III plans may be substituted into their emergency plan or otherwise kept together for the employer and employee's use. [§1926.65(q)(2)(xii)]

(3) *Procedures for handling emergency response.*[§1926.65(q)(3)]

(i) *The senior emergency response official* responding to an emergency shall become the individual in charge of a site-specific Incident Command System (ICS). All emergency responders and their communications shall be coordinated and controlled through the individual in charge of the ICS assisted by the senior official present for each employer. [§1926.65(q)(3)(i)]

Note to (q)(3)(i)[1]: The "senior official" at an emergency response is the most senior official on the site who has the responsibility for controlling the operations at the site. Initially it is the senior officer on the first-due piece of responding emergency apparatus to arrive on the incident scene. As more senior officers arrive (i.e., battalion chief, fire chief, state law enforcement official, site coordinator, etc.) the position is passed up the line of authority which has been previously established.

[1]Editor's Note: The CFR mistakenly labels this note as "(g)(3)(i)." The error is corrected in the above text.

(ii) *The individual in charge of the ICS* shall identify, to the extent possible, all hazardous substances or conditions present and shall address as appropriate site analysis, use of engineering controls, maximum exposure limits, hazardous substance handling procedures, and use of any new technologies.[§1926.65(q)(3)(ii)]

(iii) *Based on the hazardous substances and/or conditions present,* the individual in charge of the ICS shall implement appropriate emergency operations, and assure that the personal protective equipment worn is appropriate for the hazards to be encountered. However, personal protective equipment shall meet, at a minimum, the criteria contained in 29 CFR 1926.97 when worn while performing fire fighting operations beyond the incipient stage for any incident. [§1926.65(q)(3)(iii)]

(iv) *Employees engaged in emergency response* and exposed to hazardous substances presenting an inhalation hazard or potential inhalation hazard shall wear positive pressure self-contained breathing apparatus while engaged in emergency response, until such time that the individual in charge of the ICS determines through the use of air monitoring that a decreased level of respiratory protection will not result in hazardous exposures to employees.[§1926.65(q)(3)(iv)]

(v) *The individual in charge of the ICS* shall limit the number of emergency response personnel at the emergency site, in those areas of potential or actual exposure to incident or site hazards, to those who are actively performing emergency operations. However, operations in hazardous areas shall be performed using the buddy system in groups of two or more. [§1926.65(q)(3)(v)]

(vi) *Back-up personnel shall stand* by with equipment ready to provide assistance or rescue. Advance first aid support personnel, as a minimum, shall also stand by with medical equipment and transportation capability.[§1926.65(q)(3)(vi)]

(vii) *The individual in charge of the ICS* shall designate a safety official, who is knowledgeable in the operations being implemented at the emergency response site, with specific responsibility to identify and evaluate hazards and to provide direction with respect to the safety of operations for the emergency at hand.[§1926.65(q)(3)(vii)]

(viii) *When activities are judged by the safety official* to be an IDLH condition and/or to involve an imminent danger condition, the safety official shall have the authority to alter, suspend, or terminate those activities. The safety official shall immediately inform the individual in charge of the ICS of any actions needed to be taken to correct these hazards at the emergency scene.[§1926.65(q)(3)(viii)]

(ix) *After emergency operations have terminated,* the individual in charge of the ICS shall implement appropriate decontamination procedures.[§1926.65(q)(3)(ix)]

(x) *When deemed necessary for meeting the tasks at hand,* approved self-contained compressed air breathing apparatus may be used with approved cylinders from other approved self-contained compressed air breathing apparatus provided that such cylinders are of the same capacity and pressure rating. All compressed air cylinders used with self-contained breathing apparatus shall meet U.S. Department of Transportation and National Institute for Occupational Safety and Health criteria.[§1926.65(q)(3)(x)]

(4) *Skilled support personnel.* Personnel, not necessarily an employer's own employees, who are skilled in the operation of certain equipment, such as mechanized earth moving or digging equipment or crane and hoisting equipment, and who are needed temporarily to perform immediate emergency support work that cannot reasonably be performed in a timely fashion by an employer's own employees, and who will be or may be exposed to the hazards at an emergency response scene, are not required to meet the training required in this paragraph for the employer's regular employees. However, these personnel shall be given an initial briefing at the site prior to their participation in any emergency response. The initial briefing shall include instruction in the wearing of appropriate personal protective equipment, what chemical hazards are involved, and what duties are to be performed. All other appropriate safety and health precautions provided to the employer's own employees shall be used to assure the safety and health of these personnel.[§1926.65(q)(4)]

(5) *Specialist employees.* Employees who, in the course of their regular job duties, work with and are trained in the hazards of specific hazardous substances, and who will be called upon to provide technical advice or assistance at a hazardous substance release incident to the individual in charge, shall receive training or demonstrate competency in the area of their specialization annually.[§1926.65(q)(5)]

(6) *Training.* Training shall be based on the duties and function to be performed by each responder of an emergency response organization. The skill and knowledge levels required for all new responders, those hired after the effective date of this standard, shall be conveyed to them through training before they are permitted to take part in actual emergency operations on an incident. Employees who participate, or are expected to participate, in emergency response, shall be given training in accordance with the following paragraphs:[§1926.65(q)(6)]

(i) *First responder awareness level.* First responders at the awareness level are individuals who are likely to witness or discover a hazardous substance release and who have been trained to initiate an emergency response sequence by notifying the proper authorities of the release. They would take no further action beyond notifying the authorities of the release. First responders at the awareness level shall have sufficient training or have had sufficient experience to objectively demonstrate competency in the following areas: [§1926.65(q)(6)(i)]

[A] *An understanding of* what hazardous substances are, and the risks associated with them in an incident. [§1926.65(q)(6)(i)[A]]

[B] *An understanding of the potential outcomes* associated with an emergency created when hazardous substances are present. [§1926.65(q)(6)(i)[B]]

[C] *The ability to recognize* the presence of hazardous substances in an emergency. [§1926.65(q)(6)(i)[C]]

[D] *The ability to identify* the hazardous substances, if possible. [§1926.65(q)(6)(i)[D]]

[E] *An understanding of the role* of the first responder awareness individual in the employer's emergency response plan including site security and control and the U.S. Department of Transportation's Emergency Response Guidebook. [§1926.65(q)(6)(i)[E]]

[F] *The ability to realize the need* for additional resources, and to make appropriate notifications to the communication center. [§1926.65(q)(6)(i)[F]]

(ii) *First responder operations level.* First responders at the operations level are individuals who respond to releases or potential releases of hazardous substances as part of the initial response to the site for the purpose of protecting nearby persons, property, or the environment from the effects of the release. They are trained to respond in a defensive fashion without actually trying to stop the release. Their function is to contain the release from a safe distance, keep it from spreading, and prevent exposures. First responders at the operational level shall have received at least eight hours of training or have had sufficient experience to objectively demonstrate competency in the following areas in addition to those listed for the awareness level and the employer shall so certify: [§1926.65(q)(6)(ii)]

[A] *Knowledge of* the basic hazard and risk assessment techniques. [§1926.65(q)(6)(ii)[A]]

[B] *Know how to select and use* proper personal protective equipment provided to the first responder operational level. [§1926.65(q)(6)(ii)[B]]

[C] *An understanding of basic hazardous materials terms.* [§1926.65(q)(6)(ii)[C]]

[D] *Know how to perform* basic control, containment and/or confinement operations within the capabilities of the resources and personal protective equipment available with their unit. [§1926.65(q)(6)(ii)[D]]

[E] *Know how to implement* basic decontamination procedures. [§1926.65(q)(6)(ii)[E]]

[F] *An understanding of* the relevant standard operating procedures and termination procedures. [§1926.65(q)(6)(ii)[F]]

(iii) *Hazardous materials technician.* Hazardous materials technicians are individuals who respond to releases or potential releases for the purpose of stopping the release. They assume a more aggressive role than a first responder at the operations level in that they will approach the point of release in order to plug, patch or otherwise stop the release of a hazardous substance. Hazardous materials technicians shall have received at least 24 hours of training equal to the first responder operations level and in addition have competency in the following areas and the employer shall so certify: [§1926.65(q)(6)(iii)]

[A] *Know how to implement* the employer's emergency response plan. [§1926.65(q)(6)(iii)[A]]

[B] *Know the classification, identification* and verification of known and unknown materials by using field survey instruments and equipment. [§1926.65(q)(6)(iii)[B]]

[C] *Be able to function within* an assigned role in the Incident Command System. [§1926.65(q)(6)(iii)[C]]

[D] *Know how to select and use* proper specialized chemical personal protective equipment provided to the hazardous materials technician. [§1926.65(q)(6)(iii)[D]]

[E] *Understand hazard and risk assessment techniques.* [§1926.65(q)(6)(iii)[E]]

[F] *Be able to perform advance control, containment,* and/or confinement operations within the capabilities of the resources and personal protective equipment available with the unit. [§1926.65(q)(6)(iii)[F]]

[G] *Understand and implement decontamination procedures.* [§1926.65(q)(6)(iii)[G]]

[H] *Understand termination procedures.* [§1926.65(q)(6)(iii)[H]]

[I] *Understand basic* chemical and toxicological terminology and behavior. [§1926.65(q)(6)(iii)[I]]

(iv) *Hazardous materials specialist.* Hazardous materials specialists are individuals who respond with and provide support to hazardous materials technicians. Their duties parallel those of the hazardous materials technician, however, those duties require a more directed or specific knowledge of the various substances they may be called upon to contain. The hazardous materials specialist would also act as the site liaison with Federal, state, local and other government authorities in regards to site activities. Hazardous materials specialists shall have received at least 24 hours of training equal to the technician level and in addition have competency in the following areas and the employer shall so certify: [§1926.65(q)(6)(iv)]

[A] *Know how to implement* the local emergency response plan. [§1926.65(q)(6)(iv)[A]]

[B] *Understand classification, identification and verification* of known and unknown materials by using advanced survey instruments and equipment. [§1926.65(q)(6)(iv)[B]]

[C] *Know of the state emergency response plan.* [§1926.65(q)(6)(iv)[C]]

[D] *Be able to select and use* proper specialized chemical personal protective equipment provided to the hazardous materials specialist. [§1926.65(q)(6)(iv)[D]]

[E] *Understand in-depth hazard and risk techniques.* [§1926.65(q)(6)(iv)[E]]

[F] *Be able to perform specialized control, containment,* and/or confinement operations within the capabilities of the resources and personal protective equipment available. [§1926.65(q)(6)(iv)[F]]

[G] *Be able to determine and implement* decontamination procedures. [§1926.65(q)(6)(iv)[G]]

[H] *Have the ability to develop a site safety and control plan.* [§1926.65(q)(6)(iv)[H]]

[I] *Understand chemical,* radiological and toxicological terminology and behavior. [§1926.65(q)(6)(iv)[I]]

(v) *On scene incident commander.* Incident commanders, who will assume control of the incident scene beyond the first responder awareness level, shall receive at least 24 hours of training equal to the first responder operations level and in addition have competency in the following areas and the employer shall so certify: [§1926.65(q)(6)(v)]

[A] *Know and be able to implement* the employer's incident command system. [§1926.65(q)(6)(v)[A]]

[B] *Know how to implement* the employer's emergency response plan. [§1926.65(q)(6)(v)[B]]

[C] *Know and understand* the hazards and risks associated with employees working in chemical protective clothing. [§1926.65(q)(6)(v)[C]]

[D] *Know how to implement* the local emergency response plan. [§1926.65(q)(6)(v)[D]]

[E] *Know of the state emergency response plan* and of the Federal Regional Response Team. [§1926.65(q)(6)(v)[E]]

[F] *Know and understand* the importance of decontamination procedures. [§1926.65(q)(6)(v)[F]]

(7) *Trainers.* Trainers who teach any of the above training subjects shall have satisfactorily completed a training course for teaching the subjects they are expected to teach, such as the courses offered by the U.S. National Fire Academy, or they shall have the training and/or academic credentials and instructional experience necessary to demonstrate competent instructional skills and a good command of the subject matter of the courses they are to teach. [§1926.65(q)(7)]

(8) *Refresher training.* [§1926.65(q)(8)]

(i) *Those employees who are trained* in accordance with paragraph (q)(6) of this section shall receive annual refresher training of sufficient content and duration to maintain their competencies, or shall demonstrate competency in those areas at least yearly.[§1926.65(q)(8)(i)]

(ii) *A statement shall be made of the training or competency,* and if a statement of competency is made, the employer shall keep a record of the methodology used to demonstrate competency.[§1926.65(q)(8)(ii)]

(9) *Medical surveillance and consultation.*[§1926.65(q)(9)]

(i) *Members of an organized and designated HAZMAT team* and hazardous materials specialists shall receive a baseline physical examination and be provided with medical surveillance as required in paragraph (f) of this section. [§1926.65(q)(9)(i)]

(ii) *Any emergency response employees* who exhibits signs or symptoms which may have resulted from exposure to hazardous substances during the course of an emergency incident, either immediately or subsequently, shall be provided with medical consultation as required in paragraph (f)(3)(ii) of this section.[§1926.65(q)(9)(ii)]

(10) *Chemical protective clothing.* Chemical protective clothing and equipment to be used by organized and designated HAZMAT team members, or to be used by hazardous materials specialists, shall meet the requirements of paragraphs (g) (3) through (5) of this section.[§1926.65(q)(10)]

(11) *Post-emergency response operations.* Upon completion of the emergency response, if it is determined that it is necessary to remove hazardous substances, health hazards, and materials contaminated with them (such as contaminated soil or other elements of the natural environment) from the site of the incident, the employer conducting the clean-up shall comply with one of the following:[§1926.65(q)(11)]

(i) *Meet all of the requirements* of paragraphs (b) through (o) of this section; or[§1926.65(q)(11)(i)]

(ii) *Where the clean-up is done on plant property* using plant or workplace employees, such employees shall have completed the training requirements of the following: 29 CFR 1926.35, 1926.59, and 1926.103, and other appropriate safety and health training made necessary by the tasks that they are expected to be performed such as personal protective equipment and decontamination procedures. All equipment to be used in the performance of the clean-up work shall be in serviceable condition and shall have been inspected prior to use.[§1926.65(q)(11)(ii)]

Appendices to §1926.65

Hazardous Waste Operations and Emergency Response

Note: The following appendices serve as non-mandatory guidelines to assist employees and employers in complying with the appropriate requirements of this section. However §1926.65(g) makes mandatory in certain circumstances the use of Level A and Level B PPE protection.

§1926.65 Appendix A

Personal Protective Equipment Test Methods

This appendix sets forth the non-mandatory examples of tests which may be used to evaluate compliance with §1926.65(g)(4) (ii) and (iii). Other tests and other challenge agents may be used to evaluate compliance.

A. Totally-encapsulating chemical protective suit pressure test

1.0 *Scope*

1.1 *This practice measures the ability* of a gas tight totally-encapsulating chemical protective suit material, seams, and closures to maintain a fixed positive pressure. The results of this practice allow the gas tight integrity of a totally-encapsulating chemical protective suit to be evaluated.

1.2 *Resistance of the suit materials* to permeation, penetration, and degradation by specific hazardous substances is not determined by this test method.

2.0 *Definition of terms*

2.1 **Totally-encapsulated chemical protective suit** (TECP suit) means a full body garment which is constructed of protective clothing materials; covers the wearer's torso, head, arms, legs and respirator; may cover the wearer's hands and feet with tightly attached gloves and boots; completely encloses the wearer and respirator by itself or in combination with the wearer's gloves and boots.

2.2 **Protective clothing material** means any material or combination of materials used in an item of clothing for the purpose of isolating parts of the body from direct contact with a potentially hazardous liquid or gaseous chemicals.

2.3 **Gas tight** means, for the purpose of this test method, the limited flow of a gas under pressure from the inside of a TECP suit to atmosphere at a prescribed pressure and time interval.

3.0 *Summary of test method*

3.1 *The TECP suit is visually inspected* and modified for the test. The test apparatus is attached to the suit to permit inflation to the pre-test suit expansion pressure for removal of suit wrinkles and creases. The pressure is lowered to the test pressure and monitored for three minutes. If the pressure drop is excessive, the TECP suit fails the test and is removed from service. The test is repeated after leak location and repair.

4.0 *Required Supplies*

4.1 *Source of compressed air.*

4.2 *Test apparatus for suit testing,* including a pressure measurement device with a sensitivity of at least 1/4 inch water gauge.

4.3 *Vent valve closure* plugs or sealing tape.

4.4 *Soapy water solution and soft brush.*

4.5 *Stop watch or appropriate timing device.*

5.0 *Safety Precautions*

5.1 *Care shall be taken to provide* the correct pressure safety devices required for the source of compressed air used.

6.0 *Test Procedure*

6.1 *Prior to each test,* the tester shall perform a visual inspection of the suit. Check the suit for seam integrity by visually examining the seams and gently pulling on the seams. Ensure that all air supply lines, fittings, visor, zippers, and valves are secure and show no signs of deterioration.

6.1.1 *Seal off the vent valves* along with any other normal inlet or exhaust points (such as umbilical air line fittings or face piece opening) with tape or other appropriate means (caps, plugs, fixture, etc.). Care should be exercised in the sealing process not to damage any of the suit components.

6.1.2 *Close all closure assemblies.*

6.1.3 *Prepare the suit for inflation* by providing an improvised connection point on the suit for connecting an airline. Attach the pressure test apparatus to the suit to permit suit inflation from a compressed air source equipped with a pressure indicating regulator. The leak tightness of the pressure test apparatus should be tested before and after each test by closing off the end of the tubing attached to the suit and assuring a pressure of three inches water gauge for three minutes can be maintained. If a component is removed for the test, that component shall be replaced and a second test conducted with another component removed to permit a complete test of the ensemble.

6.1.4 *The pre-test expansion pressure* (A) and the suit test pressure (B) shall be supplied by the suit manufacturer, but in no case shall they be less than: (A)=three inches water gauge; and (B)=two inches water gauge. The ending suit pressure (C) shall be no less than 80 percent of the test pressure (B); i.e., the pressure drop shall not exceed 20 percent of the test pressure (B).

6.1.5 *Inflate the suit until the pressure inside* is equal to pressure (A), the pre-test expansion suit pressure. Allow at least one minute to fill out the wrinkles in the suit. Release sufficient air to reduce the suit pressure to pressure (B), the suit test pressure. Begin timing. At the end of three minutes, record the suit pressure as pressure (C), the ending suit pressure. The difference between the suit test pressure and the ending suit test pressure (B-C) shall be defined as the suit pressure drop.

6.1.6 *If the suit pressure drop* is more than 20 percent of the suit test pressure (B) during the three-minute test period, the suit fails the test and shall be removed from service.

7.0 *Retest Procedure*

7.1 *If the suit fails the test check* for leaks by inflating the suit to pressure (A) and brushing or wiping the entire suit (including seams, closures, lens gaskets, glove-to-sleeve joints, etc.) with a mild soap and water solution. Observe the suit for the formation of soap bubbles, which is an indication of a leak. Repair all identified leaks.

7.2 *Retest the TECP suit as outlined in Test procedure 6.0.*

8.0 *Report*

8.1 *Each TECP suit tested* by this practice shall have the following information recorded:

8.1.1 *Unique identification number,* identifying brand name, date of purchase, material of construction, and unique fit features, e.g., special breathing apparatus.

8.1.2 *The actual values* for test pressures (A), (B), and (C) shall be recorded along with the specific observation times. If the ending pressure (C) is less than 80 percent of the test pressure (B), the suit shall be identified as failing the test. When possible, the specific leak location shall be identified in the test records. Retest pressure data shall be recorded as an additional test.

8.1.3 *The source of the test apparatus* used shall be identified and the sensitivity of the pressure gauge shall be recorded.

8.1.4 *Records shall be kept* for each pressure test even if repairs are being made at the test location.

CAUTION: Visually inspect all parts of the suit to be sure they are positioned correctly and secured tightly before putting the suit back into service. Special care should be taken to examine each exhaust valve to make sure it is not blocked. Care should also be exercised to assure that the inside and outside of the suit is completely dry before it is put into storage.

B. Totally-encapsulating chemical protective suit qualitative leak test

1.0 *Scope*

1.1 *This practice semi-qualitatively tests* gas tight totally-encapsulating chemical protective suit integrity by detecting inward leakage of ammonia vapor. Since no modifications are made to the suit to carry out this test, the results from this practice provide a realistic test for the integrity of the entire suit.

1.2 *Resistance of the suit materials* to permeation, penetration, and degradation is not determined by this test method. ASTM test methods are available to test suit materials for these characteristics and the tests are usually conducted by the manufacturers of the suits.

2.0 *Definition of terms*

2.1 **Totally-encapsulated chemical protective suit** (TECP-suit) means a full body garment which is constructed of protective clothing materials; covers the wearer's torso, head, arms, legs and respirator; may cover the wearer's hands and feet with tightly attached gloves and boots; completely encloses the wearer and respirator by itself or in combination with the wearer's gloves, and boots.

2.2 **Protective clothing material** means any material or combination of materials used in an item of clothing for the purpose of isolating parts of the body from direct contact with a potentially hazardous liquid or gaseous chemicals.

2.3 **Gas tight** means, for the purpose of this test method, the limited flow of a gas under pressure from the inside of a TECP suit to atmosphere at a prescribed pressure and time interval.

2.4 **Intrusion Coefficient** means a number expressing the level of protection provided by a gas tight totally-encapsulating chemical protective suit. The intrusion coefficient is calculated by dividing the test room challenge agent concentration by the concentration of challenge agent found inside the suit. The accuracy of the intrusion coefficient is dependent on the challenge agent monitoring methods. The larger the intrusion coefficient the greater the protection provided by the TECP suit.

3.0 *Summary of recommended practice*

3.1 *The volume of concentrated aqueous* ammonia solution (ammonia hydroxide NH_4OH) required to generate the test atmosphere is determined using the directions outlined in 6. The suit is donned by a person wearing the appropriate respiratory equipment (either a positive pressure self-contained breathing apparatus or a positive pressure supplied air respirator) and worn inside the enclosed test room. The concentrated aqueous ammonia solution is taken by the suited individual into the test room and poured into an open plastic pan. A two-minute evaporation period is observed before the test room concentration is measured, using a high range ammonia length of stain detector tube. When the ammonia vapor reaches a concentration of between 1000 and 1200 ppm, the suited individual starts a standardized exercise protocol to stress and flex the suit. After this protocol is completed, the test room concentration is measured again. The suited individual exits the test room and his stand-by person measures the ammonia concentration inside the suit using a low range ammonia length of stain detector tube or other more sensitive ammonia detector. A stand-by person is required to observe the test individual during the test procedure; aid the person in donning and doffing the TECP suit; and monitor the suit interior. The intrusion coefficient of the suit can be calculated by dividing the average test area concentration by the interior suit concentration. A colorimetric ammonia indicator strip of bromophenol blue or equivalent is placed on the inside of the suit face piece lens so that the suited individual is able to detect a color change and know if the suit has a significant leak. If a color change is observed the individual shall leave the test room immediately.

4.0 *Required supplies*

4.1 *A supply of* concentrated aqueous ammonium hydroxide (58% by weight).

4.2 *A supply of* bromophenol/blue indicating paper or equivalent, sensitive to 5-10 ppm ammonia or greater over a two-minute period of exposure. [pH 3.0 (yellow) to pH 4.6 (blue)]

4.3 *A supply of* high range (0.5-10 volume percent) and low range (5-700 ppm) detector tubes for ammonia and the corresponding sampling pump. More sensitive ammonia detectors can be substituted for the low range detector tubes to improve the sensitivity of this practice.

4.4 *A shallow plastic pan* (PVC) at least 12":14":1" and a half pint plastic container (PVC) with tightly closing lid.

4.5 *A graduated cylinder* or other volumetric measuring device of at least 50 milliliters in volume with an accuracy of at least ± 1 milliliters.

5.0 *Safety precautions*

5.1 *Concentrated aqueous ammonium hydroxide, NH_4OH,* is a corrosive volatile liquid requiring eye, skin, and respiratory protection. The person conducting the test shall review the MSDS for aqueous ammonia.

5.2 *Since the established permissible exposure limit* for ammonia is 35 ppm as a 15 minute STEL, only persons wearing a positive pressure self-contained breathing apparatus or a positive pressure supplied air respirator shall be in the chamber. Normally only the person wearing the totally-encapsulating suit will be inside the chamber. A stand-by person shall have a positive pressure self-contained breathing apparatus, or a positive pressure supplied air respirator available to enter the test area should the suited individual need assistance.

5.3 *A method to monitor the suited individual* must be used during this test. Visual contact is the simplest but other methods using communication devices are acceptable.

5.4 *The test room shall be large enough* to allow the exercise protocol to be carried out and then to be ventilated to allow for easy exhaust of the ammonia test atmosphere after the test(s) are completed.

5.5 *Individuals shall be medically screened* for the use of respiratory protection and checked for allergies to ammonia before participating in this test procedure.

6.0 *Test procedure*

6.1.1 *Measure the test area* to the nearest foot and calculate its volume in cubic feet. Multiply the test area volume by 0.2 milliliters of concentrated aqueous ammonia solution per cubic foot of test area volume to determine the approximate volume of concentrated aqueous ammonia required to generate 1000 ppm in the test area.

6.1.2 *Measure this volume from the supply* of concentrated aqueous ammonia and place it into a closed plastic container.

6.1.3 *Place the container,* several high range ammonia detector tubes, and the pump in the clean test pan and locate it near the test area entry door so that the suited individual has easy access to these supplies.

6.2.1 *In a non-contaminated atmosphere,* open a pre-sealed ammonia indicator strip and fasten one end of the strip to the inside of the suit face shield lens where it can be seen by the wearer. Moisten the indicator strip with distilled water. Care shall be taken not to contaminate the detector part of the indicator paper by touching it. A small piece of masking tape or equivalent should be used to attach the indicator strip to the interior of the suit face shield.

6.2.2 *If problems are encountered* with this method of attachment, the indicator strip can be attached to the outside of the respirator face piece lens being used during the test.

6.3 *Don the respiratory protective* device normally used with the suit, and then don the TECP suit to be tested. Check to be sure all openings which are intended to be sealed (zippers, gloves, etc.) are completely sealed. DO NOT, however, plug off any venting valves.

6.4 *Step into the enclosed test* room such as a closet, bathroom, or test booth, equipped with an exhaust fan. No air should be exhausted from the chamber during the test because this will dilute the ammonia challenge concentrations.

6.5 *Open the container with the pre-measured volume* of concentrated aqueous ammonia within the enclosed test room, and pour the liquid into the empty plastic test pan. Wait two minutes to allow for adequate volatilization of the concentrated aqueous ammonia. A small mixing fan can be used near the evaporation pan to increase the evaporation rate of the ammonia solution.

6.6 *After two minutes* a determination of the ammonia concentration within the chamber should be made using the high range colorimetric detector tube. A concentration of 1000 ppm ammonia or greater shall be generated before the exercises are started.

6.7 *To test the integrity of the suit* the following four minute exercise protocol should be followed:

6.7.1 *Raising the arms* above the head with at least 15 raising motions completed in one minute.

6.7.2 *Walking in place* for one minute with at least 15 raising motions of each leg in a one-minute period.

6.7.3 *Touching the toes* with a least 10 complete motions of the arms from above the head to touching of the toes in a one-minute period.

6.7.4 *Knee bends with at least* 10 complete standing and squatting motions in a one-minute period.

6.8 *If at any time during the test* the colorimetric indicating paper should change colors, the test should be stopped and section 6.10 and 6.12 initiated (See §4.2).

6.9 *After completion of the test exercise,* the test area concentration should be measured again using the high range colorimetric detector tube.

6.10 *Exit the test area.*

6.11 *The opening created by the suit zipper* or other appropriate suit penetration should be used to determine the ammonia concentration in the suit with the low range length of stain detector tube or other ammonia monitor. The internal TECP suit air should be sampled far enough from the enclosed test area to prevent a false ammonia reading.

6.12 *After completion of the measurement* of the suit interior ammonia concentration the test is concluded and the suit is doffed and the respirator removed.

6.13 *The ventilating fan for the test room* should be turned on and allowed to run for enough time to remove the ammonia gas. The fan shall be vented to the outside of the building.

6.14 *Any detectable ammonia in the suit interior* (five ppm ammonia (NH_3) or more for the length of stain detector tube) indicates that the suit has failed the test. When other ammonia detectors are used a lower level of detection is possible, and it should be specified as the pass/fail criteria.

6.15 *By following this test method,* an intrusion coefficient of approximately 200 or more can be measured with the suit in a completely operational condition. If the intrusion coefficient is 200 or more, then the suit is suitable for emergency response and field use.

7.0 *Retest procedures*

7.1 *If the suit fails this test,* check for leaks by following the pressure test in test A above.

7.2 *Retest the TECP suit* as outlined in the test procedure 6.0.

8.0 *Report*

8.1 *Each gas tight totally-encapsulating* chemical protective suit tested by this practice shall have the following information recorded.

8.1.1 *Unique identification number,* identifying brand name, date of purchase, material of construction, and unique suit features; e.g., special breathing apparatus.

8.1.2 *General description of test room used for test.*

8.1.3 *Brand name and purchase date* of ammonia detector strips and color change data.

8.1.4 *Brand name, sampling range, and expiration date* of the length of stain ammonia detector tubes. The brand name and model of the sampling pump should also be recorded. If another type of ammonia detector is used, it should be identified along with its minimum detection limit for ammonia.

8.1.5 *Actual test results* shall list the two test area concentrations, their average, the interior suit concentration, and the calculated intrusion coefficient. Retest data shall be recorded as an additional test.

8.2 *The evaluation of the data shall be specified* as "suit passed" or "suit failed," and the date of the test. Any detectable ammonia (five ppm or greater for the length of stain detector tube) in the suit interior indicates the suit has failed this test. When other ammonia detectors are used, a lower level of detection is possible and it should be specified as the pass fail criteria.

CAUTION: Visually inspect all parts of the suit to be sure they are positioned correctly and secured tightly before putting the suit back into service. Special care should be taken to examine each exhaust valve to make sure it is not blocked. Care should also be exercised to assure that the inside and outside of the suit is completely dry before it is put into storage.

§1926.65 Appendix B

General Description and Discussion of the Levels of Protection and Protective Gear

This appendix sets forth information about personal protective equipment (PPE) protection levels which may be used to assist employers in complying with the PPE requirements of this section.

As required by the standard, PPE must be selected which will protect employees from the specific hazards which they are likely to encounter during their work on-site.

Selection of the appropriate PPE is a complex process which should take into consideration a variety of factors. Key factors involved in this process are identification of the hazards, or suspected hazards; their routes of potential hazard to employees (inhalation, skin absorption, ingestion, and eye or skin contact); and the performance of the PPE materials (and seams) in providing a barrier to these hazards. The amount of protection provided by PPE is material-hazard specific. That is, protective equipment materials will protect well against some hazardous substances and poorly, or not at all, against others. In many instances, protective equipment materials cannot be found which will provide continuous protection from the particular hazardous substance. In these cases the breakthrough time of the protective material should exceed the work durations.

Other factors in this selection process to be considered are matching the PPE to the employee's work requirements and task-specific conditions. The durability of PPE materials, such as tear strength and seam strength, should be considered in relation to the employee's tasks. The effects of PPE in relation to heat stress and task duration are a factor in selecting and using PPE. In some cases layers of PPE may be necessary to provide sufficient protection, or to protect expensive PPE inner garments, suits or equipment.

The more that is known about the hazards at the site, the easier the job of PPE selection becomes. As more information about the hazards and conditions at the site becomes available, the site supervisor can make decisions to up-grade or down-grade the level of PPE protection to match the tasks at hand.

The following are guidelines which an employer can use to begin the selection of the appropriate PPE. As noted above, the site information may suggest the use of combinations of PPE selected from the different protection levels (i.e., A, B, C, or D) as being more suitable to the hazards of the work. It should be cautioned that the listing below does not fully address the performance of the specific PPE material in relation to the specific hazards at the job site, and that PPE selection, evaluation and re-selection is an ongoing process until sufficient information about the hazards and PPE performance is obtained.

Part A. Personal protective equipment is divided into four categories based on the degree of protection afforded. (See Part B of this appendix for further explanation of Levels A, B, C, and D hazards.)

I. Level A— To be selected when the greatest level of skin, respiratory, and eye protection is required.

The following constitute Level A equipment; it may be used as appropriate;

1. *Positive pressure,* full face-piece self-contained breathing apparatus (SCBA), or positive pressure supplied air respirator with escape SCBA, approved by the National Institute for Occupational Safety and Health (NIOSH).
2. *Totally-encapsulating chemical-protective suit.*
3. *Coveralls.*[1]
4. *Long underwear.*[1]
5. *Gloves, outer, chemical-resistant.*
6. *Gloves, inner, chemical-resistant.*
7. *Boots, chemical-resistant, steel toe and shank.*
8. *Hard hat (under suit).*[1]

1. Optional, as applicable.

9. *Disposable protective suit, gloves and boots* (depending on suit construction, may be worn over totally-encapsulating suit).

II. **Level B**— The highest level of respiratory protection is necessary but a lesser level of skin protection is needed.

The following constitute Level B equipment; it may be used as appropriate.

1. *Positive pressure,* full-facepiece self-contained breathing apparatus (SCBA), or positive pressure supplied air respirator with escape SCBA (NIOSH approved).
2. *Hooded chemical-resistant clothing* (overalls and long-sleeved jacket; coveralls; one or two-piece chemical-splash suit; disposable chemical-resistant overalls).
3. *Coveralls.* [1]
4. *Gloves, outer, chemical-resistant.*
5. *Gloves, inner, chemical-resistant.*
6. *Boots, outer, chemical-resistant steel toe and shank.*
7. *Boot-covers, outer, chemical-resistant (disposable).* [1]
8. *Hard hat.* [1]
9. *[Reserved]*
10. *Face shield.* [1]

III. **Level C**— The concentration(s) and type(s) of airborne substance(s) is known and the criteria for using air purifying respirators are met.

The following constitute Level C equipment; it may be used as appropriate.

1. *Full-face or half-mask, air purifying respirators (NIOSH approved).*
2. *Hooded chemical-resistant clothing (overalls; two-piece chemical-splash suit; disposable chemical-resistant overalls).*
3. *Coveralls.* [1]
4. *Gloves, outer, chemical-resistant.*
5. *Gloves, inner, chemical-resistant.*
6. *Boots (outer), chemical-resistant steel toe and shank.* [1]
7. *Boot-covers, outer, chemical-resistant (disposable).* [1]
8. *Hard hat.* [1]
9. *Escape mask.* [1]
10. *Face shield.* [1]

IV. **Level D**—A work uniform affording minimal protection, used for nuisance contamination only.

The following constitute Level D equipment; it may be used as appropriate:

1. *Coveralls.*
2. *Gloves.* [1]
3. *Boots/shoes, chemical-resistant steel toe and shank.*
4. *Boots, outer, chemical-resistant (disposable).* [1]
5. *Safety glasses or chemical splash goggles.* [1]
6. *Hard hat.* [1]
7. *Escape mask.* [1]
8. *Face shield.* [1]

Part B. The types of hazards for which levels A, B, C, and D protection are appropriate are described below:

I. **Level A**— Level A protection should be used when:

1. *The hazardous substance has been identified* and requires the highest level of protection for skin, eyes, and the respiratory system based on either the measured (or potential for) high concentration of atmospheric vapors, gases, or particulates; or the site operations and work functions involve a high potential for splash, immersion, or exposure to unexpected vapors, gases, or particulates of materials that are harmful to skin or capable of being absorbed through the skin;
2. *Substances with a high degree* of hazard to the skin are known or suspected to be present, and skin contact is possible; or
3. *Operations are being conducted* in confined, poorly ventilated areas, and the absence of conditions requiring Level A have not yet been determined.

II. **Level B**— Level B protection should be used when:

1. *The type and atmospheric concentration* of substances have been identified and require a high level of respiratory protection, but less skin protection;
2. *The atmosphere contains less than 19.5 percent oxygen; or*
3. *The presence of incompletely identified vapors or gases* is indicated by a direct-reading organic vapor detection instrument, but vapors and gases are not suspected of containing high levels of chemicals harmful to skin or capable of being absorbed through the skin.

Note: This involves atmospheres with IDLH concentrations of specific substances that present severe inhalation hazards and that do not represent a severe skin hazard; or that do not meet the criteria for use of air-purifying respirators.

III. **Level C**— Level C protection should be used when:

1. *The atmospheric contaminants,* liquid splashes, or other direct contact will not adversely affect or be absorbed through any exposed skin;
2. *The types of air contaminants* have been identified, concentrations measured, and an air-purifying respirator is available that can remove the contaminants; and
3. *All criteria for the use of air-purifying respirators are met.*

IV. **Level D**— Level D protection should be used when:

1. *The atmosphere contains no known hazard; and*
2. *Work functions preclude splashes, immersion,* or the potential for unexpected inhalation of or contact with hazardous levels of any chemicals.

Note: As stated before, combinations of personal protective equipment other than those described for Levels A, B, C, and D protection may be more appropriate and may be used to provide the proper level of protection.

As an aid in selecting suitable chemical protective clothing, it should be noted that the National Fire Protection Association (NFPA) has developed standards on chemical protective clothing. The standards that have been adopted by include:

NFPA 1991—Standard on Vapor-Protective Suits for Hazardous Chemical Emergencies (EPA Level A Protective Clothing).

NFPA 1992—Standard on Liquid Splash-Protective Suits for Hazardous Chemical Emergencies (EPA Level B Protective Clothing).

NFPA 1993—Standard on Liquid Splash-Protective Suits for Non-emergency, Non-flammable Hazardous Chemical Situations (EPA Level B Protective Clothing).

These standards apply documentation and performance requirements to the manufacture of chemical protective suits. Chemical protective suits meeting these requirements are labelled as compliant with the appropriate standard. It is recommended that chemical protective suits that meet these standards be used.

§1926.65 Appendix C
Compliance Guidelines

1. **Occupational Safety and Health Program.** Each hazardous waste site clean-up effort will require an occupational safety and health program headed by the site coordinator or the employer's representative. The purpose of the program will be the protection of employees at the site and will be an extension of the employer's overall safety and health program. The program will need to be developed before work begins on the site and implemented as work proceeds as stated in paragraph (b). The program is to facilitate coordination and communication of safety and health issues among personnel responsible for the various activities which will take place at the site. It will provide the overall means for planning and implementing the needed safety and health training and job orientation of employees who will be working at the site. The program will provide the means for identifying and controlling worksite hazards and the means for monitoring program effectiveness. The program will need to cover the responsibilities and authority of the site coordinator or the employer's manager on the site for the safety and health of employees at the site, and the relationships with contractors or support services as to what each employer's safety and health responsibilities are for their employees on the site. Each contractor on the site needs to have its own safety and health program so structured that it will smoothly interface with the program of the site coordinator or principal contractor.

   Also those employers involved with treating, storing or disposal of hazardous waste as covered in paragraph (p) must have implemented a safety and health program for their employees. This program is to include the hazard communication program required in paragraph (p)(1) and the training required in paragraphs (p)(7) and (p)(8) as parts of the employers comprehensive overall safety and health program. This program is to be in writing.

   Each site or workplace safety and health program will need to include the following:

   (1) Policy statements of the line of authority and accountability for implementing the program, the objectives of the program and the role of the site safety and health supervisor or manager and staff;

   (2) Means or methods for the development of procedures for identifying and controlling workplace hazards at the site;

   (3) Means or methods for the development and communication to employees of the various plans, work rules, standard operating procedures and practices that pertain to individual employees and supervisors;

1. Optional, as applicable.

D Occupational Health and Environmental Controls

(4) Means for the training of supervisors and employees to develop the needed skills and knowledge to perform their work in a safe and healthful manner;

(5) Means to anticipate and prepare for emergency situations; and

(6) Means for obtaining information feedback to aid in evaluating the program and for improving the effectiveness of the program.

The management and employees should be trying continually to improve the effectiveness of the program thereby enhancing the protection being afforded those working on the site.

Accidents on the site or workplace should be investigated to provide information on how such occurrences can be avoided in the future. When injuries or illnesses occur on the site or workplace, they will need to be investigated to determine what needs to be done to prevent this incident from occurring again. Such information will need to be used as feedback on the effectiveness of the program and the information turned into positive steps to prevent any reoccurrence. Receipt of employee suggestions or complaints relating to safety and health issues involved with site or workplace activities is also a feedback mechanism that can be used effectively to improve the program and may serve in part as an evaluative tool(s).

For the development and implementation of the program to be the most effective, professional safety and health personnel should be used. Certified Safety Professionals, Board Certified Industrial Hygienists or Registered Professional Safety Engineers are good examples of professional stature for safety and health managers who will administer the employer's program.

2. **Training.** The training programs for employees subject to the requirements of paragraph (e) of this standard should address: the safety and health hazards employees should expect to find on hazardous waste clean-up sites; what control measures or techniques are effective for those hazards; what monitoring procedures are effective in characterizing exposure levels; what makes an effective employer's safety and health program; what a site safety and health plan should include; hands on training with personal protective equipment and clothing they may be expected to use; the contents of the OSHA standard relevant to the employee's duties and function; and, employee's responsibilities under OSHA and other regulations. Supervisors will need training in their responsibilities under the safety and health program and its subject areas such as the spill containment program, the personal protective equipment program, the medical surveillance program, the emergency response plan and other areas.

The training programs for employees subject to the requirements of paragraph (p) of this standard should address: the employers safety and health program elements impacting employees; the hazard communication program; the medical surveillance program; the hazards and the controls for such hazards that employees need to know for their job duties and functions. All require annual refresher training.

The training programs for employees covered by the requirements of paragraph (q) of this standard should address those competencies required for the various levels of response such as: the hazards associated with hazardous substances; hazard identification and awareness; notification of appropriate persons; the need for and use of personal protective equipment including respirators; the decontamination procedures to be used; preplanning activities for hazardous substance incidents including the emergency response plan; company standard operating procedures for hazardous substance emergency responses; the use of the incident command system and other subjects. Hands-on training should be stressed whenever possible. Critiques done after an incident which include an evaluation of what worked and what did not and how could the incident be better handled the next time may be counted as training time.

For hazardous materials specialists (usually members of hazardous materials teams), the training should address the care, use and/or testing of chemical protective clothing including totally encapsulating suits, the medical surveillance program, the standard operating procedures for the hazardous materials team including the use of plugging and patching equipment and other subject areas.

Officers and leaders who may be expected to be in charge at an incident should be fully knowledgeable of their company's incident command system. They should know where and how to obtain additional assistance and be familiar with the local district's emergency response plan and the state emergency response plan.

Specialist employees such as technical experts, medical experts or environmental experts that work with hazardous materials in their regular jobs, who may be sent to the incident scene by the shipper, manufacturer or governmental agency to advise and assist the person in charge of the incident should have training on an annual basis. Their training should include the care and use of personal protective equipment including respirators; knowledge of the incident command system and how they are to relate to it; and those areas needed to keep them current in their respective field as it relates to safety and health involving specific hazardous substances.

Those skilled support personnel, such as employees who work for public works departments or equipment operators who operate bulldozers, sand trucks, backhoes, etc., who may be called to the incident scene to provide emergency support assistance, should have at least a safety and health briefing before entering the area of potential or actual exposure. These skilled support personnel, who have not been a part of the emergency response plan and do not meet the training requirements, should be made aware of the hazards they face and should be provided all necessary protective clothing and equipment required for their tasks.

There are two National Fire Protection Association standards, NFPA 472—"Standard for Professional Competence of Responders to Hazardous Material Incidents" and NFPA 471—"Recommended Practice for Responding to Hazardous Material Incidents", which are excellent resource documents to aid fire departments and other emergency response organizations in developing their training program materials. NFPA 472 provides guidance on the skills and knowledge needed for first responder awareness level, first responder operations level, hazmat technicians, and hazmat specialist. It also offers guidance for the officer corp who will be in charge of hazardous substance incidents.

3. **Decontamination.** Decontamination procedures should be tailored to the specific hazards of the site, and may vary in complexity and number of steps, depending on the level of hazard and the employee's exposure to the hazard. Decontamination procedures and PPE decontamination methods will vary depending upon the specific substance, since one procedure or method may not work for all substances. Evaluation of decontamination methods and procedures should be performed, as necessary, to assure that employees are not exposed to hazards by re-using PPE. References in appendix D may be used for guidance in establishing an effective decontamination program. In addition, the U.S. Coast Guard's Manual, "Policy Guidance for Response to Hazardous Chemical Releases," U.S. Department of Transportation, Washington, DC (COMDTINST M16465.30) is a good reference for establishing an effective decontamination program.

4. **Emergency response plans.** States, along with designated districts within the states, will be developing or have developed local emergency response plans. These state and district plans should be utilized in the emergency response plans called for in the standard. Each employer should assure that its emergency response plan is compatible with the local plan. The major reference being used to aid in developing the state and local district plans is the Hazardous Materials Emergency Planning Guide, NRT-1. The current Emergency Response Guidebook from the U.S. Department of Transportation, CMA's CHEMTREC and the Fire Service Emergency Management Handbook may also be used as resources.

Employers involved with treatment, storage, and disposal facilities for hazardous waste, which have the required contingency plan called for by their permit, would not need to duplicate the same planning elements. Those items of the emergency response plan that are properly addressed in the contingency plan may be substituted into the emergency response plan required in 1926.65 or otherwise kept together for employer and employee use.

5. **Personal protective equipment programs.** The purpose of personal protective clothing and equipment (PPE) is to shield or isolate individuals from the chemical, physical, and biologic hazards that may be encountered at a hazardous substance site.

As discussed in appendix B, no single combination of protective equipment and clothing is capable of protecting against all hazards. Thus PPE should be used in conjunction with other protective methods and its effectiveness evaluated periodically.

The use of PPE can itself create significant worker hazards, such as heat stress, physical and psychological stress, and impaired vision, mobility, and communication. For any given situation, equipment and clothing should be selected that provide an adequate level of protection. However, over-protection, as well as under-protection, can be hazardous and should be avoided where possible.

Two basic objectives of any PPE program should be to protect the wearer from safety and health hazards, and to prevent injury to the wearer from incorrect use and/or malfunction of the PPE. To accomplish these goals, a comprehensive PPE program should include hazard identification, medical monitoring, environmental surveillance, selection, use, maintenance, and decontamination of PPE and its associated training.

The written PPE program should include policy statements, procedures, and guidelines. Copies should be made available to all employees, and a reference copy should be made available at the worksite. Technical data on equipment, maintenance manuals, relevant regulations, and other essential information should also be collected and maintained.

6. **Incident command system (ICS).** Paragraph §1926.65(q)(3)(ii) requires the implementation of an ICS. The ICS is an organized approach to effectively control and manage operations at an emergency incident. The individual in charge of the ICS is the senior official responding to the incident. The ICS is not much different than the "command post" approach used for many years by the fire service. During large complex fires involving several companies and many pieces of apparatus, a command post would be established. This enabled one individual to be in charge of managing the incident, rather than having several officers from different companies making separate, and sometimes conflicting, decisions. The individual in charge of the command post would delegate responsibility for performing various tasks to subordinate officers. Additionally, all communications were routed through the command post to reduce the number of radio transmissions and eliminate confusion. However, strategy, tactics, and all decisions were made by one individual.

   The ICS is a very similar system, except it is implemented for emergency response to all incidents, both large and small, that involve hazardous substances.

   For a small incident, the individual in charge of the ICS may perform many tasks of the ICS. There may not be any, or little, delegation of tasks to subordinates. For example, in response to a small incident, the individual in charge of the ICS, in addition to normal command activities, may become the safety officer and may designate only one employee (with proper equipment) as a back-up to provide assistance if needed. OSHA does recommend, however, that at least two employees be designated as back-up personnel since the assistance needed may include rescue.

   To illustrate the operation of the ICS, the following scenario might develop during a small incident, such as an overturned tank truck with a small leak of flammable liquid.

   The first responding senior officer would implement and take command of the ICS. That person would size-up the incident and determine if additional personnel and apparatus were necessary; would determine what actions to take to control the leak; and, determine the proper level of personal protective equipment. If additional assistance is not needed, the individual in charge of the ICS would implement actions to stop and control the leak using the fewest number of personnel that can effectively accomplish the tasks. The individual in charge of the ICS then would designate himself as the safety officer and two other employees as a back-up in case rescue may become necessary. In this scenario, decontamination procedures would not be necessary.

   A large complex incident may require many employees and difficult, time-consuming efforts to control. In these situations, the individual in charge of the ICS will want to delegate different tasks to subordinates in order to maintain a span of control that will keep the number of subordinates, that are reporting, to a manageable level.

   Delegation of task at large incidents may be by location, where the incident scene is divided into sectors, and subordinate officers coordinate activities within the sector that they have been assigned.

   Delegation of tasks can also be by function. Some of the functions that the individual in charge of the ICS may want to delegate at a large incident are: medical services; evacuation; water supply; resources (equipment, apparatus); media relations; safety; and, site control (integrate activities with police for crowd and traffic control). Also for a large incident, the individual in charge of the ICS will designate several employees as back-up personnel; and a number of safety officers to monitor conditions and recommend safety precautions.

   Therefore, no matter what size or complexity an incident may be, by implementing an ICS there will be one individual in charge who makes the decisions and gives directions; and, all actions, and communications are coordinated through one central point of command. Such a system should reduce confusion, improve safety, organize and coordinate actions, and should facilitate effective management of the incident.

7. **Site Safety and Control Plans.** The safety and security of response personnel and others in the area of an emergency response incident site should be of primary concern to the incident commander. The use of a site safety and control plan could greatly assist those in charge of assuring the safety and health of employees on the site.

   A comprehensive site safety and control plan should include the following: summary analysis of hazards on the site and a risk analysis of those hazards; site map or sketch; site work zones (clean zone, transition or decontamination zone, work or hot zone); use of the buddy system; site communications; command post or command center; standard operating procedures and safe work practices; medical assistance and triage area; hazard monitoring plan (air contaminate monitoring, etc.); decontamination procedures and area; and other relevant areas. This plan should be a part of the employer's emergency response plan or an extension of it to the specific site.

8. **Medical surveillance programs.** Workers handling hazardous substances may be exposed to toxic chemicals, safety hazards, biologic hazards, and radiation. Therefore, a medical surveillance program is essential to assess and monitor workers' health and fitness for employment in hazardous waste operations and during the course of work; to provide emergency and other treatment as needed; and to keep accurate records for future reference.

   The Occupational Safety and Health Guidance Manual for Hazardous Waste Site Activities developed by the National Institute for Occupational Safety and Health (NIOSH), the Occupational Safety and Health Administration (OSHA), the U.S. Coast Guard (USCG), and the Environmental Protection Agency (EPA); October 1985 provides an excellent example of the types of medical testing that should be done as part of a medical surveillance program.

9. **New Technology and Spill Containment Programs.** Where hazardous substances may be released by spilling from a container that will expose employees to the hazards of the materials, the employer will need to implement a program to contain and control the spilled material. Diking and ditching, as well as use of absorbents like diatomaceous earth, are traditional techniques which have proven to be effective over the years. However, in recent years new products have come into the marketplace, the use of which complement and increase the effectiveness of these traditional methods. These new products also provide emergency responders and others with additional tools or agents to use to reduce the hazards of spilled materials.

   These agents can be rapidly applied over a large area and can be uniformly applied or otherwise can be used to build a small dam, thus improving the workers' ability to control spilled material. These application techniques enhance the intimate contact between the agent and the spilled material allowing for the quickest effect by the agent or quickest control of the spilled material. Agents are available to solidify liquid spilled materials, to suppress vapor generation from spilled materials, and to do both. Some special agents, which when applied as recommended by the manufacturer, will react in a controlled manner with the spilled material to neutralize acids or caustics, or greatly reduce the level of hazard of the spilled material.

   There are several modern methods and devices for use by emergency response personnel or others involved with spill control efforts to safely apply spill control agents to control spilled material hazards. These include portable pressurized applicators similar to hand-held portable fire extinguishing devices, and nozzle and hose systems similar to portable fire fighting foam systems which allow the operator to apply the agent without having to come into contact with the spilled material. The operator is able to apply the agent to the spilled material from a remote position.

   The solidification of liquids provides for rapid containment and isolation of hazardous substance spills. By directing the agent at run-off points or at the edges of the spill, the reactant solid will automatically create a barrier to slow or stop the spread of the material. Clean-up of hazardous substances is greatly improved when solidifying agents, acid or caustic neutralizers, or activated carbon adsorbents are used. Properly applied, these agents can totally solidify liquid hazardous substances or neutralize or absorb them, which results in materials which are less hazardous and easier to handle, transport, and dispose of. The concept of spill treatment, to create less hazardous substances, will improve the safety and level of protection of employees working at spill clean-up operations or emergency response operations to spills of hazardous substances.

   The use of vapor suppression agents for volatile hazardous substances, such as flammable liquids and those substances which present an inhalation hazard, is important for protecting workers. The rapid and uniform distribution of the agent over the surface of the spilled material can provide quick vapor knockdown. There are temporary and long-term foam-type agents which are effective on vapors and dusts, and activated carbon adsorption agents which are effective for vapor control and soaking-up of the liquid. The proper use of hose lines or hand-held portable pressurized applicators provides good mobility and permits the worker to deliver the agent from a safe distance without having to step into the untreated spilled material. Some of these systems can be recharged in the field to provide coverage of larger spill areas than the design limits of a single charged applicator unit. Some of the more effective agents can solidify the liquid flammable hazardous substances and at the same time elevate the flashpoint above 140 °F so the resulting substance may be handled as a nonhazardous waste material if it meets the U.S. Environmental Protection Agency's 40 CFR part 261 requirements (See particularly §261.21).

All workers performing hazardous substance spill control work are expected to wear the proper protective clothing and equipment for the materials present and to follow the employer's established standard operating procedures for spill control. All involved workers need to be trained in the established operating procedures; in the use and care of spill control equipment; and in the associated hazards and control of such hazards of spill containment work.

These new tools and agents are the things that employers will want to evaluate as part of their new technology program. The treatment of spills of hazardous substances or wastes at an emergency incident as part of the immediate spill containment and control efforts is sometimes acceptable to EPA and a permit exception is described in 40 CFR 264.1(g)(8) and 265.1(c)(11).

§1926.65 Appendix D

References

The following references may be consulted for further information on the subject of this standard:

1. OSHA Instruction DFO CPL 2.70—January 29, 1986, Special Emphasis Program: Hazardous Waste Sites.
2. OSHA Instruction DFO CPL 2-2.37A—January 29, 1986, Technical Assistance and Guidelines for Superfund and Other Hazardous Waste Site Activities.
3. OSHA Instruction DTS CPL 2.74—January 29, 1986, Hazardous Waste Activity Form, OSHA 175.
4. Hazardous Waste Inspections Reference Manual, U.S. Department of Labor, Occupational Safety and Health Administration, 1986.
5. Memorandum of Understanding Among the National Institute for Occupational Safety and Health, the Occupational Safety and Health Administration, the United States Coast Guard, and the United States Environmental Protection Agency, Guidance for Worker Protection During Hazardous Waste Site Investigations and Clean-up and Hazardous Substance Emergencies. December 18, 1980.
6. National Priorities List, 1st Edition, October 1984; U.S. Environmental Protection Agency, Revised periodically.
7. The Decontamination of Response Personnel, Field Standard Operating Procedures (F.S.O.P.) 7; U.S. Environmental Protection Agency, Office of Emergency and Remedial Response, Hazardous Response Support Division, December 1984.
8. Preparation of a Site Safety Plan, Field Standard Operating Procedures (F.S.O.P.) 9; U.S. Environmental Protection Agency, Office of Emergency and Remedial Response, Hazardous Response Support Division, April 1985.
9. Standard Operating Safety Guidelines; U.S. Environmental Protection Agency, Office of Emergency and Remedial Response, Hazardous Response Support Division, Environmental Response Team; November 1984.
10. Occupational Safety and Health Guidance Manual for Hazardous Waste Site Activities, National Institute for Occupational Safety and Health (NIOSH), Occupational Safety and Health Administration (OSHA), U.S. Coast Guard (USCG), and Environmental Protection Agency (EPA); October 1985.
11. Protecting Health and Safety at Hazardous Waste Sites: An Overview, U.S. Environmental Protection Agency, EPA/625/9-85/006; September 1985.
12. Hazardous Waste Sites and Hazardous Substance Emergencies, NIOSH Worker Bulletin, U.S. Department of Health and Human Services, Public Health Service, Centers for Disease Control, National Institute for Occupational Safety and Health; December 1982.
13. Personal Protective Equipment for Hazardous Materials Incidents: A Selection Guide; U.S. Department of Health and Human Services, Public Health Service, Centers for Disease Control, National Institute for Occupational Safety and Health; October 1984.
14. Fire Service Emergency Management Handbook, International Association of Fire Chiefs Foundation, 101 East Holly Avenue, Unit 10B, Sterling, VA 22170, January 1985.
15. Emergency Response Guidebook, U.S Department of Transportation, Washington, DC, 1987.
16. Report to the Congress on Hazardous Materials Training, Planning and Preparedness, Federal Emergency Management Agency, Washington, DC, July 1986.
17. Workbook for Fire Command, Alan V. Brunacini and J. David Beageron, National Fire Protection Association, Batterymarch Park, Quincy, MA 02269, 1985.
18. Fire Command, Alan V. Brunacini, National Fire Protection Association, Batterymarch Park,, Quincy, MA 02269, 1985.
19. Incident Command System, Fire Protection Publications, Oklahoma State University, Stillwater, OK 74078, 1983.
20. Site Emergency Response Planning, Chemical Manufacturers Association, Washington, DC 20037, 1986.
21. Hazardous Materials Emergency Planning Guide, NRT-1, Environmental Protection Agency, Washington, DC, March 1987.
22. Community Teamwork: Working Together to Promote Hazardous Materials Transportation Safety. U.S. Department of Transportation, Washington, DC, May 1983.
23. Disaster Planning Guide for Business and Industry, Federal Emergency Management Agency, Publication No. FEMA 141, August 1987.

§1926.65 Appendix E

Training Curriculum Guidelines

The following non-mandatory general criteria may be used for assistance in developing site-specific training curriculum used to meet the training requirements of 29 CFR 1926.65(e); 29 CFR 1926.65(p)(7), (p)(8)(iii); and 29 CFR 1926.65(q)(6), (q)(7), and (q)(8). These are generic guidelines and they are not presented as a complete training curriculum for any specific employer. Site-specific training programs must be developed on the basis of a needs assessment of the hazardous waste site, RCRA/TSDF, or emergency response operation in accordance with 29 CFR 1926.65.

It is noted that the legal requirements are set forth in the regulatory text of §1926.65. The guidance set forth here presents a highly effective program that in the areas covered would meet or exceed the regulatory requirements. In addition, other approaches could meet the regulatory requirements.

Suggested General Criteria

Definitions:

Competent means possessing the skills, knowledge, experience, and judgment to perform assigned tasks or activities satisfactorily as determined by the employer.

Demonstration means the showing by actual use of equipment or procedures.

Hands-on training means training in a simulated work environment that permits each student to have experience performing tasks, making decisions, or using equipment appropriate to the job assignment for which the training is being conducted.

Initial training means training required prior to beginning work.

Lecture means an interactive discourse with a class lead by an instructor.

Proficient means meeting a stated level of achievement.

Site-specific means individual training directed to the operations of a specific job site.

Training hours means the number of hours devoted to lecture, learning activities, small group work sessions, demonstration, evaluations, or hands-on experience.

Suggested Core Criteria:

1. *Training facility.* The training facility should have available sufficient resources, equipment, and site locations to perform didactic and hands-on training when appropriate. Training facilities should have sufficient organization, support staff, and services to conduct training in each of the courses offered.
2. *Training Director.* Each training program should be under the direction of a training director who is responsible for the program. The Training Director should have a minimum of two years of employee education experience.
3. *Instructors.* Instructors should be deem competent on the basis of previous documented experience in their area of instruction, successful completion of a "train-the-trainer" program specific to the topics they will teach, and an evaluation of instructional competence by the Training Director.

   Instructors should be required to maintain professional competency by participating in continuing education or professional development programs or by completing successfully an annual refresher course and having an annual review by the Training Director.

The annual review by the Training Director should include observation of an instructor's delivery, a review of those observations with the trainer, and an analysis of any instructor or class evaluations completed by the students during the previous year.

4. *Course materials.* The Training Director should approve all course materials to be used by the training provider. Course materials should be reviewed and updated at least annually. Materials and equipment should be in good working order and maintained properly. All written and audio-visual materials in training curricula should be peer reviewed by technically competent outside reviewers or by a standing advisory committee.
   Reviews should possess expertise in the following disciplines were applicable: occupational health, industrial hygiene and safety, chemical/environmental engineering, employee education, or emergency response. One or more of the peer reviewers should be a employee experienced in the work activities to which the training is directed.
5. *Students.* The program for accepting students should include:
   a. *Assurance that the student is or will be involved* in work where chemical exposures are likely and that the student possesses the skills necessary to perform the work.
   b. *A policy on the necessary medical clearance.*
6. *Ratios.* Student-instructor ratios should not exceed 30 students per instructor. Hands-on activity requiring the use of personal protective equipment should have the following student-instructor ratios. For Level C or Level D personal protective equipment the ratio should be 10 students per instructor. For Level A or Level B personal protective equipment the ratio should be 5 students per instructor.
7. *Proficiency assessment.* Proficiency should be evaluated and documented by the use of a written assessment and a skill demonstration selected and developed by the Training Director and training staff. The assessment and demonstration should evaluate the knowledge and individual skills developed in the course of training. The level of minimum achievement necessary for proficiency shall be specified in writing by the Training Director.
   If a written test is used, there should be a minimum of 50 questions. If a written test is used in combination with a skills demonstration, a minimum of 25 questions should be used. If a skills demonstration is used, the tasks chosen and the means to rate successful completion should be fully documented by the Training Director.
   The content of the written test or of the skill demonstration shall be relevant to the objectives of the course. The written test and skill demonstration should be updated as necessary to reflect changes in the curriculum and any update should be approved by the Training Director.
   The proficiency assessment methods, regardless of the approach or combination of approaches used, should be justified, document and approved by the Training Director.
   The proficiency of those taking the additional courses for supervisors should be evaluated and documented by using proficiency assessment methods acceptable to the Training Director. These proficiency assessment methods must reflect the additional responsibilities borne by supervisory personnel in hazardous waste operations or emergency response.
8. *Course certificate.* Written documentation should be provided to each student who satisfactorily completes the training course. The documentation should include:
   a. *Student's name.*
   b. *Course title.*
   c. *Course date.*
   d. *Statement that the student has successfully completed the course.*
   e. *Name and address of the training provider.*
   f. *An individual identification number for the certificate.*
   g. *List of the levels of personal protective equipment used by the student to complete the course.*
   This documentation may include a certificate and an appropriate wallet-sized laminated card with a photograph of the student and the above information. When such course certificate cards are used, the individual identification number for the training certificate should be shown on the card.
9. *Recordkeeping.* Training providers should maintain records listing the dates courses were presented, the names of the individual course attenders, the names of those students successfully completing each course, and the number of training certificates issued to each successful student. These records should be maintained for a minimum of five years after the date an individual participated in a training program offered by the training provider. These records should be available and provided upon the student's request or as mandated by law.
10. *Program quality control.* The Training Director should conduct or direct an annual written audit of the training program. Program modifications to address deficiencies, if any, should be documented, approved, and implemented by the training provider. The audit and the program modification documents should be maintained at the training facility.

Suggested Program Quality Control Criteria

Factors listed here are suggested criteria for determining the quality and appropriateness of employee health and safety training for hazardous waste operations and emergency response.

A. Training Plan.

Adequacy and appropriateness of the training program's curriculum development, instructor training, distribution of course materials, and direct student training should be considered, including:

1. *The duration of training,* course content, and course schedules/ agendas;
2. *The different training requirements* of the various target populations, as specified in the appropriate generic training curriculum;
3. *The process for the development of curriculum,* which includes appropriate technical input, outside review, evaluation, program pretesting.
4. *The adequate and appropriate inclusion* of hands-on, demonstration, and instruction methods;
5. *Adequate monitoring of student* safety, progress, and performance during the training.

B. Program management, Training Director, staff, and consultants.

Adequacy and appropriateness of staff performance and delivering an effective training program should be considered, including

1. *Demonstration of the training director's leadership* in assuring quality of health and safety training.
2. *Demonstration of the competency of the staff* to meet the demands of delivering high quality hazardous waste employee health and safety training.
3. *Organization charts establishing clear lines of authority.*
4. *Clearly defined staff duties* including the relationship of the training staff to the overall program.
5. *Evidence that the training organizational structure* suits the needs of the training program.
6. *Appropriateness and adequacy* of the training methods used by the instructors.
7. *Sufficiency of the time committed* by the training director and staff to the training program.
8. *Adequacy of the ratio of training staff to students.*
9. *Availability and commitment* of the training program of adequate human and equipment resources in the areas of
   a. *Health effects,*
   b. *Safety,*
   c. *Personal protective equipment (PPE),*
   d. *Operational procedures,*
   e. *Employee protection practices/procedures.*
10. *Appropriateness of management controls.*
11. *Adequacy of the organization* and appropriate resources assigned to assure appropriate training.
12. *In the case of multiple-site training programs,* adequacy of satellite centers management.

C. Training facilities and resources.

Adequacy and appropriateness of the facilities and resources for supporting the training program should be considered, including,

1. *Space and equipment to conduct the training.*
2. *Facilities for representative hands-on training.*
3. *In the case of multiple-site programs,* equipment and facilities at the satellite centers.
4. *Adequacy and appropriateness* of the quality control and evaluations program to account for instructor performance.
5. *Adequacy and appropriateness* of the quality control and evaluation program to ensure appropriate course evaluation, feedback, updating, and corrective action.
6. *Adequacy and appropriateness* of disciplines and expertise being used within the quality control and evaluation program.
7. *Adequacy and appropriateness* of the role of student evaluations to provide feedback for training program improvement.

D. Quality control and evaluation.

Adequacy and appropriateness of quality control and evaluation plans for training programs should be considered, including:

1. *A balanced advisory committee* and/or competent outside reviewers to give overall policy guidance;
2. *Clear and adequate definition* of the composition and active programmatic role of the advisory committee or outside reviewers.
3. *Adequacy of the minutes or reports* of the advisory committee or outside reviewers' meetings or written communication.

4. *Adequacy and appropriateness* of the quality control and evaluations program to account for instructor performance.
5. *Adequacy and appropriateness* of the quality control and evaluation program to ensure appropriate course evaluation, feedback, updating, and corrective action.
6. *Adequacy and appropriateness* of disciplines and expertise being used within the quality control and evaluation program.
7. *Adequacy and appropriateness* of the role of student evaluations to provide feedback for training program improvement.

E. Students.

Adequacy and appropriateness of the program for accepting students should be considered, including:

1. *Assurance that the student* already possess the necessary skills for their job, including necessary documentation.
2. *Appropriateness of methods* the program uses to ensure that recruits are capable of satisfactorily completing training.
3. *Review and compliance with any medical clearance policy.*

F. Institutional Environment and Administrative Support.

The adequacy and appropriateness of the institutional environment and administrative support system for the training program should be considered, including

1. *Adequacy of the institutional commitment* to the employee training program.
2. *Adequacy and appropriateness* of the administrative structure and administrative support.

G. Summary of Evaluation Questions.

Key questions for evaluating the quality and appropriateness of an overall training program should include the following:

1. *Are the program objectives clearly stated?*
2. *Is the program accomplishing its objectives?*
3. *Are appropriate facilities and staff available?*
4. *Is there an appropriate mix of classroom, demonstration, and hands-on training?*
5. *Is the program providing quality employee health and safety training that fully meets the intent of regulatory requirements?*
6. *What are the program's main strengths?*
7. *What are the program's main weaknesses?*
8. *What is recommended to improve the program?*
9. *Are instructors instructing according to their training outlines?*
10. *Is the evaluation tool current and appropriate for the program content?*
11. *Is the course material current and relevant to the target group?*

Suggested Training Curriculum Guidelines

The following training curriculum guidelines are for those operations specifically identified in 29 CFR 1926.65 as requiring training. Issues such as qualifications of instructors, training certification, and similar criteria appropriate to all categories of operations addressed in 1926.65 have been covered in the preceding section and are not re-addressed in each of the generic guidelines. Basic core requirements for training programs that are addressed include:

1. *General Hazardous Waste Operations*
2. *RCRA operations—Treatment, storage, and disposal facilities.*
3. *Emergency Response.*

A. General Hazardous Waste Operations and Site-specific Training

1. *Off-site training.*

Minimum training course content for hazardous waste operations, required by 29 CFR 1926.65(e), should include the following topics or procedures:

a. *Regulatory knowledge.*

(1) *A review of 29 CFR* 1926.65 and the core elements of an occupational safety and health program.
(2) *The content of a medical surveillance program* as outlined in 29 CFR 1926.65(f).
(3) *The content of an effective site safety and health plan* consistent with the requirements of 29 CFR 1926.65(b)(4)(ii).
(4) *Emergency response plan* and procedures as outlined in 29 CFR 1910.38 and 29 CFR 1926.65(l).
(5) *Adequate illumination.*
(6) *Sanitation recommendation and equipment.*
(7) *Review and explanation* of OSHA's hazard-communication standard (29 CFR 1910.1200) and lock-out-tag-out standard (29 CFR 1910.147).
(8) *Review of other applicable standards* including but not limited to those in the construction standards (29 CFR Part 1926).
(9) *Rights and responsibilities* of employers and employees under applicable OSHA and EPA laws.

b. *Technical knowledge.*

(1) *Type of potential exposures* to chemical, biological, and radiological hazards; types of human responses to these hazards and recognition of those responses; principles of toxicology and information about acute and chronic hazards; health and safety considerations of new technology.
(2) *Fundamentals of chemical hazards* including but not limited to vapor pressure, boiling points, flash points, ph, other physical and chemical properties.
(3) *Fire and explosion hazards of chemicals.*
(4) *General safety hazards* such as but not limited to electrical hazards, powered equipment hazards, motor vehicle hazards, walking-working surface hazards, excavation hazards, and hazards associated with working in hot and cold temperature extremes.
(5) *Review and knowledge* of confined space entry procedures in 29 CFR 1910.146.
(6) *Work practices to minimize employee risk from site hazards.*
(7) *Safe use of engineering controls,* equipment, and any new relevant safety technology or safety procedures.
(8) *Review and demonstration of competency* with air sampling and monitoring equipment that may be used in a site monitoring program.
(9) *Container sampling procedures and safeguarding;* general drum and container handling procedures including special requirement for laboratory waste packs, shock-sensitive wastes, and radioactive wastes.
(10) *The elements of a spill control program.*
(11) *Proper use and limitations of material handling equipment.*
(12) *Procedures for safe and healthful* preparation of containers for shipping and transport.
(13) *Methods of communication* including those used while wearing respiratory protection.

c. *Technical skills.*

(1) *Selection, use maintenance, and limitations* of personal protective equipment including the components and procedures for carrying out a respirator program to comply with 29 CFR 1910.134.
(2) *Instruction in decontamination programs* including personnel, equipment, and hardware; hands-on training including level A, B, and C ensembles and appropriate decontamination lines; field activities including the donning and doffing of protective equipment to a level commensurate with the employee's anticipated job function and responsibility and to the degree required by potential hazards.
(3) *Sources for additional hazard information;* exercises using relevant manuals and hazard coding systems.

d. *Additional suggested items.*

(1) *A laminated, dated card or certificate with photo,* denoting limitations and level of protection for which the employee is trained should be issued to those students successfully completing a course.
(2) *Attendance should be required* at all training modules, with successful completion of exercises and a final written or oral examination with at least 50 questions.
(3) *A minimum of one-third of the program* should be devoted to hands-on exercises.
(4) *A curriculum should be established* for the 8-hour refresher training required by 29 CFR 1926.65(e)(8), with delivery of such courses directed toward those areas of previous training that need improvement or reemphasis.
(5) *A curriculum should be established* for the required 8-hour training for supervisors. Demonstrated competency in the skills and knowledge provided in a 40-hour course should be a prerequisite for supervisor training.

2. *Refresher training.*The 8-hour annual refresher training required in 29 CFR 1926.65(e)(8) should be conducted by qualified training providers. Refresher training should include at a minimum the following topics and procedures:

(a) *Review of and retraining on relevant* topics covered in the 40-hour program, as appropriate, using reports by the students on their work experiences.
(b) *Update on developments with respect* to material covered in the 40-hour course.
(c) *Review of changes to pertinent* provisions of EPA or OSHA standards or laws.
(d) *Introduction of additional subject* areas as appropriate.

(e) *Hands-on review of new or altered* PPE or decontamination equipment or procedures. Review of new developments in personal protective equipment.
(f) *Review of newly developed* air and contaminant monitoring equipment.

3. *On-site training.*
   a. *The employer should provide* employees engaged in hazardous waste site activities with information and training prior to initial assignment into their work area, as follows:
      (1) The requirements of the hazard communication program including the location and availability of the written program, required lists of hazardous chemicals, and material safety data sheets.
      (2) Activities and locations in their work area where hazardous substance may be present.
      (3) Methods and observations that may be used to detect the present or release of a hazardous chemical in the work area (such as monitoring conducted by the employer, continuous monitoring devices, visual appearances, or other evidence (sight, sound or smell) of hazardous chemicals being released, and applicable alarms from monitoring devices that record chemical releases.
      (4) The physical and health hazards of substances known or potentially present in the work area.
      (5) The measures employees can take to help protect themselves from work-site hazards, including specific procedures the employer has implemented.
      (6) An explanation of the labeling system and material safety data sheets and how employees can obtain and use appropriate hazard information.
      (7) The elements of the confined space program including special PPE, permits, monitoring requirements, communication procedures, emergency response, and applicable lock-out procedures.
   b. *The employer should provide* hazardous waste employees information and training and should provide a review and access to the site safety and plan as follows:
      (1) Names of personnel and alternate responsible for site safety and health.
      (2) Safety and health hazards present on the site.
      (3) Selection, use, maintenance, and limitations of personal protective equipment specific to the site.
      (4) Work practices by which the employee can minimize risks from hazards.
      (5) Safe use of engineering controls and equipment available on site.
      (6) Safe decontamination procedures established to minimize employee contact with hazardous substances, including:
         [A] Employee decontamination,
         [B] Clothing decontamination, and
         [C] Equipment decontamination.
      (7) Elements of the site emergency response plan, including:
         [A] Pre-emergency planning.
         [B] Personnel roles and lines of authority and communication.
         [C] Emergency recognition and prevention.
         [D] Safe distances and places of refuge.
         [E] Site security and control.
         [F] Evacuation routes and procedures.
         [G] Decontamination procedures not covered by the site safety and health plan.
         [H] Emergency medical treatment and first aid.
         [I] Emergency equipment and procedures for handling emergency incidents.
   c. *The employer should provide* hazardous waste employees information and training on personal protective equipment used at the site, such as the following:
      (1) PPE to be used based upon known or anticipated site hazards.
      (2) PPE limitations of materials and construction; limitations during temperature extremes, heat stress, and other appropriate medical considerations; use and limitations of respirator equipment as well as documentation procedures as outlined in 29 CFR 1910.134.
      (3) PPE inspection procedures prior to, during, and after use.
      (4) PPE donning and doffing procedures.
      (5) PPE decontamination and disposal procedures.
      (6) PPE maintenance and storage.
      (7) Task duration as related to PPE limitations.
   d. *The employer should instruct the employee* about the site medical surveillance program relative to the particular site, including
      (1) Specific medical surveillance programs that have been adapted for the site.
      (2) Specific signs and symptoms related to exposure to hazardous materials on the site.
      (3) The frequency and extent of periodic medical examinations that will be used on the site.
      (4) Maintenance and availability of records.
      (5) Personnel to be contacted and procedures to be followed when signs and symptoms of exposures are recognized.
   e. *The employees will review and discuss* the site safety plan as part of the training program. The location of the site safety plan and all written programs should be discussed with employees including a discussion of the mechanisms for access, review, and references described.

B. **RCRA Operations Training** for Treatment, Storage and Disposal Facilities.

1. *As a minimum,* the training course required in 29 CFR 1926.65 (p) should include the following topics:
   (a) *Review of the applicable paragraphs* of 29 CFR 1926.65 and the elements of the employer's occupational safety and health plan.
   (b) *Review of relevant hazards* such as, but not limited to, chemical, biological, and radiological exposures; fire and explosion hazards; thermal extremes; and physical hazards.
   (c) *General safety hazards* including those associated with electrical hazards, powered equipment hazards, lock-out-tag-out procedures, motor vehicle hazards and walking-working surface hazards.
   (d) *Confined-space hazards and procedures.*
   (e) *Work practices to minimize* employee risk from workplace hazards.
   (f) *Emergency response plan* and procedures including first aid meeting the requirements of paragraph (p)(8).
   (g) *A review of procedures to minimize* exposure to hazardous waste and various type of waste streams, including the materials handling program and spill containment program.
   (h) *A review of hazard communication* programs meeting the requirements of 29 CFR 1910.1200.
   (i) *A review of medical surveillance* programs meeting the requirements of 29 CFR 1926.65(p)(3) including the recognition of signs and symptoms of overexposure to hazardous substance including known synergistic interactions.
   (j) *A review of decontamination programs* and procedures meeting the requirements of 29 CFR 1926.65(p)(4).
   (k) *A review of an employer's requirements* to implement a training program and its elements.
   (l) *A review of the criteria and programs* for proper selection and use of personal protective equipment, including respirators.
   (m) *A review of the applicable appendices* to 29 CFR 1926.65.
   (n) *Principles of toxicology and biological* monitoring as they pertain to occupational health.
   (o) *Rights and responsibilities of employees* and employers under applicable OSHA and EPA laws.
   (p) *Hands-on exercises and demonstrations* of competency with equipment to illustrate the basic equipment principles that may be used during the performance of work duties, including the donning and doffing of PPE.
   (q) *Sources of reference,* efficient use of relevant manuals, and knowledge of hazard coding systems to include information contained in hazardous waste manifests.
   (r) *At least 8 hours of hands-on training.*
   (s) *Training in the job skills* required for an employee's job function and responsibility before they are permitted to participate in or supervise field activities.
2. *The individual employer should provide* hazardous waste employees with information and training prior to an employee's initial assignment into a work area. The training and information should cover the following topics:
   (a) *The Emergency response plan* and procedures including first aid.
   (b) *A review of the employer's* hazardous waste handling procedures including the materials handling program

and elements of the spill containment program, location of spill response kits or equipment, and the names of those trained to respond to releases.

(c) *The hazardous communication program* meeting the requirements of 29 CFR 1910.1200.

(d) *A review of the employer's medical surveillance program* including the recognition of signs and symptoms of exposure to relevant hazardous substance including known synergistic interactions.

(e) *A review of the employer's decontamination* program and procedures.

(f) *An review of the employer's training* program and the parties responsible for that program.

(g) *A review of the employer's personal protective equipment program* including the proper selection and use of PPE based upon specific site hazards.

(h) *All relevant site-specific procedures* addressing potential safety and health hazards. This may include, as appropriate, biological and radiological exposures, fire and explosion hazards, thermal hazards, and physical hazards such as electrical hazards, powered equipment hazards, lock-out-tag-out hazards, motor vehicle hazards, and walking-working surface hazards.

(i) *Safe use engineering controls and equipment on site.*

(j) *Names of personnel and alternates responsible for safety and health.*

C. Emergency response training.

Federal OSHA standards in 29 CFR 1926.65(q) are directed toward private sector emergency responders. Therefore, the guidelines provided in this portion of the appendix are directed toward that employee population. However, they also impact indirectly through State OSHA or USEPA regulations some public sector emergency responders. Therefore, the guidelines provided in this portion of the appendix may be applied to both employee populations.

States with OSHA state plans must cover their employees with regulations at least as effective as the Federal OSHA standards. Public employees in states without approved state OSHA programs covering hazardous waste operations and emergency response are covered by the U.S. EPA under 40 CFR 311, a regulation virtually identical to §1926.65.

Since this is a non-mandatory appendix and therefore not an enforceable standard, OSHA recommends that those employers, employees or volunteers in public sector emergency response organizations outside Federal OSHA jurisdiction consider the following criteria in developing their own training programs. A unified approach to training at the community level between emergency response organizations covered by Federal OSHA and those not covered directly by Federal OSHA can help ensure an effective community response to the release or potential release of hazardous substances in the community.

a. *General considerations.* Emergency response organizations are required to consider the topics listed in §1926.65(q)(6). Emergency response organizations may use some or all of the following topics to supplement those mandatory topics when developing their response training programs. Many of the topics would require an interaction between the response provider and the individuals responsible for the site where the response would be expected.

(1) *Hazard recognition, including:*

[A] Nature of hazardous substances present,

[B] Practical applications of hazard recognition, including presentations on biology, chemistry, and physics.

(2) *Principles of toxicology,* biological monitoring, and risk assessment.

(3) *Safe work practices and general site safety.*

(4) *Engineering controls and hazardous waste operations.*

(5) *Site safety plans and standard operating procedures.*

(6) *Decontamination procedures and practices.*

(7) *Emergency procedures, first aid, and self-rescue.*

(8) *Safe use of field equipment.*

(9) *Storage, handling, use and transportation of hazardous substances.*

(10) *Use, care, and limitations of personal protective equipment.*

(11) *Safe sampling techniques.*

(12) *Rights and responsibilities of employees* under OSHA and other related laws concerning right-to-know, safety and health, compensations and liability.

(13) *Medical monitoring requirements.*

(14) *Community relations.*

b. *Suggested criteria for specific courses.*

(1) *First responder awareness level.*

[A] Review of and demonstration of competency in performing the applicable skills of 29 CFR 1926.65(q).

[B] Hands-on experience with the U.S. Department of Transportation's Emergency Response Guidebook (ERG) and familiarization with OSHA standard 29 CFR 1926.60.

[C] Review of the principles and practices for analyzing an incident to determine both the hazardous substances present and the basic hazard and response information for each hazardous substance present.

[D] Review of procedures for implementing actions consistent with the local emergency response plan, the organization's standard operating procedures, and the current edition of DOT's ERG including emergency notification procedures and follow-up communications.

[E] Review of the expected hazards including fire and explosions hazards, confined space hazards, electrical hazards, powered equipment hazards, motor vehicle hazards, and walking-working surface hazards.

[F] Awareness and knowledge of the competencies for the First Responder at the Awareness Level covered in the National Fire Protection Association's Standard No. 472, Professional Competence of Responders to Hazardous Materials Incidents.

(2) *First responder operations level.*

[A] Review of and demonstration of competency in performing the applicable skills of 29 CFR 1926.65(q).

[B] Hands-on experience with the U.S. Department of Transportation's Emergency Response Guidebook (ERG), manufacturer material safety data sheets, CHEMTREC/CANUTEC, shipper or manufacturer contacts and other relevant sources of information addressing hazardous substance releases. Familiarization with OSHA standard 29 CFR 1926.60.

[C] Review of the principles and practices for analyzing an incident to determine the hazardous substances present, the likely behavior of the hazardous substance and its container, the types of hazardous substance transportation containers and vehicles, the types and selection of the appropriate defensive strategy for containing the release.

[D] Review of procedures for implementing continuing response actions consistent with the local emergency response plan, the organization's standard operating procedures, and the current edition of DOT's ERG including extended emergency notification procedures and follow-up communications.

[E] Review of the principles and practice for proper selection and use of personal protective equipment.

[F] Review of the principles and practice of personnel and equipment decontamination.

[G] Review of the expected hazards including fire and explosions hazards, confined space hazards, electrical hazards, powered equipment hazards, motor vehicle hazards, and walking-working surface hazards.

[H] Awareness and knowledge of the competencies for the First Responder at the Operations Level covered in the National Fire Protection Association's Standard No. 472, Professional Competence of Responders to Hazardous Materials Incidents.

(3) *Hazardous materials technician.*

[A] Review of and demonstration of competency in performing the applicable skills of 29 CFR 1926.65(q).

[B] Hands-on experience with written and electronic information relative to response decision making including but not limited to the U.S. Department of Transportation's Emergency Response Guidebook (ERG), manufacturer material safety data sheets, CHEMTREC/ CANUTEC, shipper or manufacturer contacts, computer data bases and response models, and other relevant sources of information addressing hazardous substance releases. Familiarization with 29 CFR 1926.60.

[C] Review of the principles and practices for analyzing an incident to determine the hazardous substances present, their physical and chemical properties, the likely behavior of the hazardous substance and its container,

the types of hazardous substance transportation containers and vehicles involved in the release, the appropriate strategy for approaching release sites and containing the release.

[D] Review of procedures for implementing continuing response actions consistent with the local emergency response plan, the organization's standard operating procedures, and the current edition of DOT's ERG including extended emergency notification procedures and follow-up communications.

[E] Review of the principles and practice for proper selection and use of personal protective equipment.

[F] Review of the principles and practices of establishing exposure zones, proper decontamination and medical surveillance stations and procedures.

[G] Review of the expected hazards including fire and explosions hazards, confined space hazards, electrical hazards, powered equipment hazards, motor vehicle hazards, and walking-working surface hazards.

[H] Awareness and knowledge of the competencies for the Hazardous Materials Technician covered in the National Fire Protection Association's Standard No. 472, Professional Competence of Responders to Hazardous Materials Incidents.

(4) *Hazardous materials specialist.*

[A] Review of and demonstration of competency in performing the applicable skills of 29 CFR 1926.65(q).

[B] Hands-on experience with retrieval and use of written and electronic information relative to response decision making including but not limited to the U.S. Department of Transportation's Emergency Response Guidebook (ERG), manufacturer material safety data sheets, CHEMTREC/CANUTEC, shipper or manufacturer contacts, computer data bases and response models, and other relevant sources of information addressing hazardous substance releases. Familiarization with 29 CFR 1926.60.

[C] Review of the principles and practices for analyzing an incident to determine the hazardous substances present, their physical and chemical properties, and the likely behavior of the hazardous substance and its container, vessel, or vehicle.

[D] Review of the principles and practices for identification of the types of hazardous substance transportation containers, vessels and vehicles involved in the release; selecting and using the various types of equipment available for plugging or patching transportation containers, vessels or vehicles; organizing and directing the use of multiple teams of hazardous material technicians and selecting the appropriate strategy for approaching release sites and containing or stopping the release.

[E] Review of procedures for implementing continuing response actions consistent with the local emergency response plan, the organization's standard operating procedures, including knowledge of the available public and private response resources, establishment of an incident command post, direction of hazardous material technician teams, and extended emergency notification procedures and follow-up communications.

[F] Review of the principles and practice for proper selection and use of personal protective equipment.

[G] Review of the principles and practices of establishing exposure zones and proper decontamination, monitoring and medical surveillance stations and procedures.

[H] Review of the expected hazards including fire and explosions hazards, confined space hazards, electrical hazards, powered equipment hazards, motor vehicle hazards, and walking-working surface hazards.

[I] Awareness and knowledge of the competencies for the Off-site Specialist Employee covered in the National Fire Protection Association's Standard No. 472, Professional Competence of Responders to Hazardous Materials Incidents.

(5) *Incident commander.*

The incident commander is the individual who, at any one time, is responsible for and in control of the response effort. This individual is the person responsible for the direction and coordination of the response effort. An incident commander's position should be occupied by the most senior, appropriately trained individual present at the response site. Yet, as necessary and appropriate by the level of response provided, the position may be occupied by many individuals during a particular response as the need for greater authority, responsibility, or training increases. It is possible for the first responder at the awareness level to assume the duties of incident commander until a more senior and appropriately trained individual arrives at the response site.

Therefore, any emergency responder expected to perform as an incident commander should be trained to fulfill the obligations of the position at the level of response they will be providing including the following:

[A] Ability to analyze a hazardous substance incident to determine the magnitude of the response problem.

[B] Ability to plan and implement an appropriate response plan within the capabilities of available personnel and equipment.

[C] Ability to implement a response to favorably change the outcome of the incident in a manner consistent with the local emergency response plan and the organization's standard operating procedures.

[D] Ability to evaluate the progress of the emergency response to ensure that the response objectives are being met safely, effectively, and efficiently.

[E] Ability to adjust the response plan to the conditions of the response and to notify higher levels of response when required by the changes to the response plan.

[58 FR 35129, June 30, 1993, as amended at 59 FR 43275, Aug. 22, 1994: 61 FR 5510, Feb. 13, 1996; 77 FR 17890, Mar. 26, 2012]

§1926.66

Criteria for design and construction of spray booths

(a) Definitions applicable to this section[§1926.66(a)]

(1) **Aerated solid powders.** Aerated powders shall mean any powdered material used as a coating material which shall be fluidized within a container by passing air uniformly from below. It is common practice to fluidize such materials to form a fluidized powder bed and then dip the part to be coated into the bed in a manner similar to that used in liquid dipping. Such beds are also used as sources for powder spray operations.[§1926.66(a)(1)]

(2) **Spraying area.** Any area in which dangerous quantities of flammable vapors or mists, or combustible residues, dusts, or deposits are present due to the operation of spraying processes.[§1926.66(a)(2)]

(3) **Spray booth.** A power-ventilated structure provided to enclose or accommodate a spraying operation to confine and limit the escape of spray, vapor, and residue, and to safely conduct or direct them to an exhaust system.[§1926.66(a)(3)]

(4) **Waterwash spray booth.** A spray booth equipped with a water washing system designed to minimize dusts or residues entering exhaust ducts and to permit the recovery of overspray finishing material.[§1926.66(a)(4)]

(5) **Dry spray booth.** A spray booth not equipped with a water washing system as described in paragraph (a)(4) of this section. A dry spray booth may be equipped with[§1926.66(a)(5)]

(i) *Distribution or baffle plates* to promote an even flow of air through the booth or cause the deposit of overspray before it enters the exhaust duct; or[§1926.66(a)(5)(i)]

(ii) *Overspray dry filters to minimize dusts; or*[§1926.66(a)(5)(ii)]

(iii) *Overspray dry filters to minimize* dusts or residues entering exhaust ducts; or[§1926.66(a)(5)(iii)]

(iv) *Overspray dry filter rolls* designed to minimize dusts or residues entering exhaust ducts; or[§1926.66(a)(5)(iv)]

(v) *Where dry powders are being sprayed,* with powder collection systems so arranged in the exhaust to capture oversprayed material.[§1926.66(a)(5)(v)]

(6) **Fluidized bed.** A container holding powder coating material which is aerated from below so as to form an air-supported expanded cloud of such material through which the preheated object to be coated is immersed and transported.[§1926.66(a)(6)]

(7) **Electrostatic fluidized bed.** A container holding powder coating material which is aerated from below so as to form an air-supported expanded cloud of such material which is electrically charged with a charge opposite to the charge of the object to be coated; such object is transported, through the container immediately above the charged and aerated materials in order to be coated.[§1926.66(a)(7)]

(8) **Approved.** Shall mean approved and listed by a nationally recognized testing laboratory.[§1926.66(a)(8)]

(9) **Listed.** See "approved" in paragraph (a)(8) of this section. [§1926.66(a)(9)]

(b) **Spray booths**[§1926.66(b)]

(1) *Construction.* Spray booths shall be substantially constructed of steel, securely and rigidly supported, or of concrete or masonry except that aluminum or other substantial noncombustible material may be used for intermittent or low volume spraying. Spray booths shall be designed to sweep air currents toward the exhaust outlet.[§1926.66(b)(1)]

(2) *Interiors.* The interior surfaces of spray booths shall be smooth and continuous without edges and otherwise designed to prevent pocketing of residues and facilitate cleaning and washing without injury.[§1926.66(b)(2)]

(3) *Floors.* The floor surface of a spray booth and operator's working area, if combustible, shall be covered with noncombustible material of such character as to facilitate the safe cleaning and removal of residues.[§1926.66(b)(3)]

(4) *Distribution or baffle plates.* Distribution or baffle plates, if installed to promote an even flow of air through the booth or cause the deposit of overspray before it enters the exhaust duct, shall be of noncombustible material and readily removable or accessible on both sides for cleaning. Such plates shall not be located in exhaust ducts.[§1926.66(b)(4)]

(5) *Dry type overspray collectors — (exhaust air filters).* In conventional dry type spray booths, overspray dry filters or filter rolls, if installed, shall conform to the following:[§1926.66(b)(5)]

(i) *The spraying operations except electrostatic spraying operations* shall be so designed, installed and maintained that the average air velocity over the open face of the booth (or booth cross section during spraying operations) shall be not less than 100 linear feet per minute. Electrostatic spraying operations may be conducted with an air velocity over the open face of the booth of not less than 60 linear feet per minute, or more, depending on the volume of the finishing material being applied and its flammability and explosion characteristics. Visible gauges or audible alarm or pressure activated devices shall be installed to indicate or insure that the required air velocity is maintained. Filter rolls shall be inspected to insure proper replacement of filter media. [§1926.66(b)(5)(i)]

(ii) *All discarded filter pads* and filter rolls shall be immediately removed to a safe, well-detached location or placed in a water-filled metal container and disposed of at the close of the day's operation unless maintained completely in water. [§1926.66(b)(5)(ii)]

(iii) *The location of filters in a spray booth* shall be so as to not reduce the effective booth enclosure of the articles being sprayed.[§1926.66(b)(5)(iii)]

(iv) *Space within the spray booth* on the downstream and upstream sides of filters shall be protected with approved automatic sprinklers.[§1926.66(b)(5)(iv)]

(v) *Filters or filter rolls* shall not be used when applying a spray material known to be highly susceptible to spontaneous heating and ignition.[§1926.66(b)(5)(v)]

(vi) *Clean filters or filter* rolls shall be noncombustible or of a type having a combustibility not in excess of class 2 filters as listed by Underwriters' Laboratories, Inc. Filters and filter rolls shall not be alternately used for different types of coating materials, where the combination of materials may be conducive to spontaneous ignition.[§1926.66(b)(5)(vi)]

(6) *Frontal area.* Each spray booth having a frontal area larger than 9 square feet shall have a metal deflector or curtain not less than 2 1/2 inches (5.35 cm) deep installed at the upper outer edge of the booth over the opening.[§1926.66(b)(6)]

(7) *Conveyors.* Where conveyors are arranged to carry work into or out of spray booths, the openings therefor shall be as small as practical.[§1926.66(b)(7)]

(8) *Separation of operations.* Each spray booth shall be separated from other operations by not less than 3 feet (0.912 m), or by a greater distance, or by such partition or wall as to reduce the danger from juxtaposition of hazardous operations. See also paragraph (c)(1) of this section.[§1926.66(b)(8)]

(9) *Cleaning.* Spray booths shall be so installed that all portions are readily accessible for cleaning. A clear space of not less than 3 feet (0.912 m) on all sides shall be kept free from storage or combustible construction.[§1926.66(b)(9)]

(10) *Illumination.* When spraying areas are illuminated through glass panels or other transparent materials, only fixed lighting units shall be used as a source of illumination. Panels shall effectively isolate the spraying area from the area in which the lighting unit is located, and shall be of a noncombustible material of such a nature or so protected that breakage will be unlikely. Panels shall be so arranged that normal accumulations of residue on the exposed surface of the panel will not be raised to a dangerous temperature by radiation or conduction from the source of illumination.[§1926.66(b)(10)]

(c) **Electrical and other sources of ignition**[§1926.66(c)]

(1) *Conformance.* All electrical equipment, open flames and other sources of ignition shall conform to the requirements of this paragraph, except as follows:[§1926.66(c)(1)]

(i) *Electrostatic apparatus shall conform* to the requirements of paragraphs (e) and (f) of this section;[§1926.66(c)(1)(i)]

(ii) *Drying, curing, and fusion apparatus* shall conform to the requirements of paragraph (g) of this section;[§1926.66(c)(1)(ii)]

(iii) *[Reserved]*[§1926.66(c)(1)(iii)]

(iv) *Powder coating equipment* shall conform to the requirements of paragraph (c)(1) of this section.[§1926.66(c)(1)(iv)]

(2) *Minimum separation.* There shall be no open flame or spark producing equipment in any spraying area nor within 20 feet (6.08 m) thereof, unless separated by a partition.[§1926.66(c)(2)]

(3) *Hot surfaces.* Space-heating appliances, steampipes, or hot surfaces shall not be located in a spraying area where deposits of combustible residues may readily accumulate.[§1926.66(c)(3)]

(4) *Wiring conformance.* Electrical wiring and equipment shall conform to the provisions of this paragraph and shall otherwise be in accordance with subpart S of this part.[§1926.66(c)(4)]

(5) *Combustible residues, areas.* Unless specifically approved for locations containing both deposits of readily ignitable residue and explosive vapors, there shall be no electrical equipment in any spraying area, whereon deposits of combustible residues may readily accumulate, except wiring in rigid conduit or in boxes or fittings containing no taps, splices, or terminal connections.[§1926.66(c)(5)]

(6) *Wiring type approved.* Electrical wiring and equipment not subject to deposits of combustible residues but located in a spraying area as herein defined shall be of explosion-proof type approved for Class I, group D locations and shall otherwise conform to the provisions of subpart S of this part, for Class I, Division 1, Hazardous Locations. Electrical wiring, motors, and other equipment outside of but within 20 feet (6.08 m) of any spraying area, and not separated therefrom by partitions, shall not produce sparks under normal operating conditions and shall otherwise conform to the provisions of subpart S of this part for Class I, Division 2 Hazardous Locations.[§1926.66(c)(6)]

(7) *Lamps.* Electric lamps outside of, but within 20 feet (6.08 m) of any spraying area, and not separated therefrom by a partition, shall be totally enclosed to prevent the falling of hot particles and shall be protected from mechanical injury by suitable guards or by location.[§1926.66(c)(7)]

(8) *Portable lamps.* Portable electric lamps shall not be used in any spraying area during spraying operations. Portable electric lamps, if used during cleaning or repairing operations, shall be of the type approved for hazardous Class I locations. [§1926.66(c)(8)]

(9) *Grounding.*[§1926.66(c)(9)]

(i) *All metal parts of spray booths,* exhaust ducts, and piping systems conveying flammable or combustible liquids or aerated solids shall be properly electrically grounded in an effective and permanent manner.[§1926.66(c)(9)(i)]

(d) **Ventilation**[§1926.66(d)]

(1) *Conformance.* Ventilating and exhaust systems shall be in accordance with the Standard for Blower and Exhaust Systems for Vapor Removal, NFPA No. 91-1961, where applicable and shall also conform to the provisions of this section.[§1926.66(d)(1)]

(2) *General.* All spraying areas shall be provided with mechanical ventilation adequate to remove flammable vapors, mists, or powders to a safe location and to confine and control combustible residues so that life is not endangered. Mechanical ventilation shall be kept in operation at all times while spraying operations are being conducted and for a sufficient time thereafter to allow vapors from drying coated articles and drying finishing material residue to be exhausted.[§1926.66(d)(2)]

(3) *Independent exhaust.* Each spray booth shall have an independent exhaust duct system discharging to the exterior of the building, except that multiple cabinet spray booths in which identical spray finishing material is used with a combined frontal area of not more than 18 square feet may have a common exhaust. If more than one fan serves one booth, all fans shall be so interconnected that one fan cannot operate without all fans being operated.[§1926.66(d)(3)]

(4) *Fan-rotating element.* The fan-rotating element shall be nonferrous or nonsparking or the casing shall consist of or be lined with such material. There shall be ample clearance between the fan- rotating element and the fan casing to avoid a fire by friction, necessary allowance being made for ordinary expansion and loading to prevent contact between moving parts and the duct or fan housing. Fan blades shall be mounted on a shaft sufficiently heavy to maintain perfect alignment even when the blades of the fan are heavily loaded, the shaft preferably to have bearings outside the duct and booth. All bearings shall be of the self-lubricating type, or lubricated from the outside duct. [§1926.66(d)(4)]

(5) *Electric motors.* Electric motors driving exhaust fans shall not be placed inside booths or ducts. See also paragraph (c) of this section.[§1926.66(d)(5)]

(6) *Belts.* Belts shall not enter the duct or booth unless the belt and pulley within the duct or booth are thoroughly enclosed. [§1926.66(d)(6)]

(7) *Exhaust ducts.* Exhaust ducts shall be constructed of steel and shall be substantially supported. Exhaust ducts without dampers are preferred; however, if dampers are installed, they shall be maintained so that they will be in a full open position at all times the ventilating system is in operation.[§1926.66(d)(7)]

(i) *Exhaust ducts shall be protected* against mechanical damage and have a clearance from unprotected combustible construction or other combustible material of not less than 18 inches (45.72 cm).[§1926.66(d)(7)(i)]

(ii) *If combustible construction* is provided with the following protection applied to all surfaces within 18 inches (45.72 cm), clearances may be reduced to the distances indicated: [§1926.66(d)(7)(ii)]

(a) 28-gage sheet metal on 1/4-inch asbestos mill board.	12 inches (30.48 cm)
(b) 28-gage sheet metal on 1/8-inch asbestos mill board spaced out 1 inch (2.54 cm) on noncombustible spacers.	9 inches (22.86 cm)
(c) 22-gage sheet metal on 1-inch rockwool batts reinforced with wire mesh or the equivalent.	3 inches (7.62 cm)
(d) Where ducts are protected with an approved automatic sprinkler system, properly maintained, the clearance required in paragraph (d)(7)(i) of this section may be reduced to 6 inches (15.24 cm).	

(8) *Discharge clearance.* Unless the spray booth exhaust duct terminal is from a water-wash spray booth, the terminal discharge point shall be not less than 6 feet from any combustible exterior wall or roof nor discharge in the direction of any combustible construction or unprotected opening in any noncombustible exterior wall within 25 feet (7.6 m).[§1926.66(d)(8)]

(9) *Air exhaust.* Air exhaust from spray operations shall not be directed so that it will contaminate makeup air being introduced into the spraying area or other ventilating intakes, nor directed so as to create a nuisance. Air exhausted from spray operations shall not be recirculated.[§1926.66(d)(9)]

(10) *Access doors.* When necessary to facilitate cleaning, exhaust ducts shall be provided with an ample number of access doors. [§1926.66(d)(10)]

(11) *Room intakes.* Air intake openings to rooms containing spray finishing operations shall be adequate for the efficient operation of exhaust fans and shall be so located as to minimize the creation of dead air pockets.[§1926.66(d)(11)]

(12) *Drying spaces.* Freshly sprayed articles shall be dried only in spaces provided with adequate ventilation to prevent the formation of explosive vapors. In the event adequate and reliable ventilation is not provided such drying spaces shall be considered a spraying area.[§1926.66(d)(12)]

(e) Fixed electrostatic apparatus.[§1926.66(e)]

(1) *Conformance.* Where installation and use of electrostatic spraying equipment is used, such installation and use shall conform to all other paragraphs of this section, and shall also conform to the requirements of this paragraph.[§1926.66(e)(1)]

(2) *Type approval.* Electrostatic apparatus and devices used in connection with coating operations shall be of approved types. [§1926.66(e)(2)]

(3) *Location.* Transformers, power packs, control apparatus, and all other electrical portions of the equipment, with the exception of high- voltage grids, electrodes, and electrostatic atomizing heads and their connections, shall be located outside of the spraying area, or shall otherwise conform to the requirements of paragraph (c) of this section.[§1926.66(e)(3)]

(4) *Support.* Electrodes and electrostatic atomizing heads shall be adequately supported in permanent locations and shall be effectively insulated from the ground. Electrodes and electrostatic atomizing heads which are permanently attached to their bases, supports, or reciprocators, shall be deemed to comply with this section. Insulators shall be nonporous and noncombustible.[§1926.66(e)(4)]

(5) *Insulators, grounding.* High-voltage leads to electrodes shall be properly insulated and protected from mechanical injury or exposure to destructive chemicals. Electrostatic atomizing heads shall be effectively and permanently supported on suitable insulators and shall be effectively guarded against accidental contact or grounding. An automatic means shall be provided for grounding the electrode system when it is electrically deenergized for any reason. All insulators shall be kept clean and dry.[§1926.66(e)(5)]

(6) *Safe distance.* A safe distance shall be maintained between goods being painted and electrodes or electrostatic atomizing heads or conductors of at least twice the sparking distance. A suitable sign indicating this safe distance shall be conspicuously posted near the assembly.[§1926.66(e)(6)]

(7) *Conveyors required.* Goods being painted using this process are to be supported on conveyors. The conveyors shall be so arranged as to maintain safe distances between the goods and the electrodes or electrostatic atomizing heads at all times. Any irregularly shaped or other goods subject to possible swinging or movement shall be rigidly supported to prevent such swinging or movement which would reduce the clearance to less than that specified in paragraph (e)(6) of this section.[§1926.66(e)(7)]

(8) *Prohibition.* This process is not acceptable where goods being coated are manipulated by hand. When finishing materials are applied by electrostatic equipment which is manipulated by hand, see paragraph (f) of this section for applicable requirements.[§1926.66(e)(8)]

(9) *Fail-safe controls.* Electrostatic apparatus shall be equipped with automatic controls which will operate without time delay to disconnect the power supply to the high voltage transformer and to signal the operator under any of the following conditions: [§1926.66(e)(9)]

(i) *Stoppage of ventilating fans* or failure of ventilating equipment from any cause.[§1926.66(e)(9)(i)]

(ii) *Stoppage of the conveyor carrying* goods through the high voltage field.[§1926.66(e)(9)(ii)]

(iii) *Occurrence of a ground or of an imminent* ground at any point on the high voltage system.[§1926.66(e)(9)(iii)]

(iv) *Reduction of clearance below* that specified in paragraph (e)(6) of this section.[§1926.66(e)(9)(iv)]

(10) *Guarding.* Adequate booths, fencing, railings, or guards shall be so placed about the equipment that they, either by their location or character or both, assure that a safe isolation of the process is maintained from plant storage or personnel. Such railings, fencing, and guards shall be of conducting material, adequately grounded.[§1926.66(e)(10)]

(11) *Ventilation.* Where electrostatic atomization is used the spraying area shall be so ventilated as to insure safe conditions from a fire and health standpoint.[§1926.66(e)(11)]

(12) *Fire protection.* All areas used for spraying, including the interior of the booth, shall be protected by automatic sprinklers where this protection is available. Where this protection is not available, other approved automatic extinguishing equipment shall be provided.[§1926.66(e)(12)]

(f) Electrostatic hand spraying equipment.[§1926.66(f)]

(1) *Application.* This paragraph shall apply to any equipment using electrostatically charged elements for the atomization and/or, precipitation of materials for coatings on articles, or for other similar purposes in which the atomizing device is hand held and manipulated during the spraying operation.[§1926.66(f)(1)]

(2) *Conformance.* Electrostatic hand spraying equipment shall conform with the other provisions of this section.[§1926.66(f)(2)]

(3) *Equipment approval and specifications.* Electrostatic hand spray apparatus and devices used in connection with coating operations shall be of approved types. The high voltage circuits shall be designed so as to not produce a spark of sufficient intensity to ignite any vapor-air mixtures nor result in appreciable shock hazard upon coming in contact with a grounded object under all normal operating conditions. The electrostatically charged exposed elements of the handgun shall be capable of being energized only by a switch which also controls the coating material supply.[§1926.66(f)(3)]

(4) *Electrical support equipment.* Transformers, powerpacks, control apparatus, and all other electrical portions of the equipment, with the exception of the handgun itself and its connections to the power supply shall be located outside of the spraying area or shall otherwise conform to the requirements of paragraph (c) of this section.[§1926.66(f)(4)]

(5) *Spray gun ground.* The handle of the spraying gun shall be electrically connected to ground by a metallic connection and to be so constructed that the operator in normal operating position is in intimate electrical contact with the grounded handle. [§1926.66(f)(5)]

(6) *Grounding-general.* All electrically conductive objects in the spraying area shall be adequately grounded. This requirement shall apply to paint containers, wash cans, and any other objects or devices in the area. The equipment shall carry a prominent permanently installed warning regarding the necessity for this grounding feature.[§1926.66(f)(6)]

(7) *Maintenance of grounds.* Objects being painted or coated shall be maintained in metallic contact with the conveyor or other grounded support. Hooks shall be regularly cleaned to insure this contact and areas of contact shall be sharp points or knife edges where possible. Points of support of the object shall be concealed from random spray where feasible and where the objects being sprayed are supported from a conveyor, the point of attachment to the conveyor shall be so located as to not collect spray material during normal operation.[§1926.66(f)(7)]

(8) *Interlocks.* The electrical equipment shall be so interlocked with the ventilation of the spraying area that the equipment cannot be operated unless the ventilation fans are in operation. [§1926.66(f)(8)]

(9) *Ventilation.* The spraying operation shall take place within a spray area which is adequately ventilated to remove solvent vapors released from the operation.[§1926.66(f)(9)]

(g) Drying, curing, or fusion apparatus.[§1926.66(g)]

(1) *Conformance.* Drying, curing, or fusion apparatus in connection with spray application of flammable and combustible finishes shall conform to the Standard for Ovens and Furnaces, NFPA 86A-1969, where applicable and shall also conform with the following requirements of this paragraph.[§1926.66(g)(1)]

(2) *Alternate use prohibited.* Spray booths, rooms, or other enclosures used for spraying operations shall not alternately be used for the purpose of drying by any arrangement which will cause a material increase in the surface temperature of the spray booth, room, or enclosure.[§1926.66(g)(2)]

(3) *Adjacent system interlocked.* Except as specifically provided in paragraph (g)(4) of this section, drying, curing, or fusion units utilizing a heating system having open flames or which may produce sparks shall not be installed in a spraying area, but may be installed adjacent thereto when equipped with an interlocked ventilating system arranged to:[§1926.66(g)(3)]

(i) *Thoroughly ventilate* the drying space before the heating system can be started;[§1926.66(g)(3)(i)]

(ii) *Maintain a safe atmosphere* at any source of ignition; [§1926.66(g)(3)(ii)]

(iii) *Automatically shut down* the heating system in the event of failure of the ventilating system.[§1926.66(g)(3)(iii)]

(4) *Alternate use permitted.* Automobile refinishing spray booths or enclosures, otherwise installed and maintained in full conformity with this section, may alternately be used for drying with portable electrical infrared drying apparatus when conforming with the following:[§1926.66(g)(4)]

(i) *Interior (especially floors)* of spray enclosures shall be kept free of overspray deposits.[§1926.66(g)(4)(i)]

(ii) *During spray operations,* the drying apparatus and electrical connections and wiring thereto shall not be located within spray[§1926.66(g)(4)(ii)]

enclosure nor in any other location where spray residues may be deposited thereon.[§1926.66(g)(4)(ii)]

(iii) *The spraying apparatus,* the drying apparatus, and the ventilating system of the spray enclosure shall be equipped with suitable interlocks so arranged that:[§1926.66(g)(4)(iii)]

[a] The spraying apparatus cannot be operated while the drying apparatus is inside the spray enclosure. [§1926.66(g)(4)(iii)[a]]

[b] The spray enclosure will be purged of spray vapors for a period of not less than 3 minutes before the drying apparatus can be energized.[§1926.66(g)(4)(iii)[b]]

[c] The ventilating system will maintain a safe atmosphere within the enclosure during the drying process and the drying apparatus will automatically shut off in the event of failure of the ventilating system.[§1926.66(g)(4)(iii)[c]]

(iv) *All electrical wiring and equipment* of the drying apparatus shall conform with the applicable sections of subpart S of this part. Only equipment of a type approved for Class I, Division 2 hazardous locations shall be located within 18 inches (45.72 cm) of floor level. All metallic parts of the drying apparatus shall be properly electrically bonded and grounded. [§1926.66(g)(4)(iv)]

(v) *The drying apparatus shall contain* a prominently located, permanently attached warning sign indicating that ventilation should be maintained during the drying period and that spraying should not be conducted in the vicinity that spray will deposit on apparatus.[§1926.66(g)(4)(v)]

[58 FR 35149, June 30, 1993]

Subpart E - Personal Protective and Life Saving Equipment

§1926.95

Criteria for personal protective equipment

(a) Application. Protective equipment, including personal protective equipment for eyes, face, head, and extremities, protective clothing, respiratory devices, and protective shields and barriers, shall be provided, used, and maintained in a sanitary and reliable condition wherever it is necessary by reason of hazards of processes or environment, chemical hazards, radiological hazards, or mechanical irritants encountered in a manner capable of causing injury or impairment in the function of any part of the body through absorption, inhalation or physical contact.[§1926.95(a)]

(b) Employee-owned equipment. Where employees provide their own protective equipment, the employer shall be responsible to assure its adequacy, including proper maintenance, and sanitation of such equipment.[§1926.95(b)]

(c) Design. All personal protective equipment shall be of safe design and construction for the work to be performed.[§1926.95(c)]

(d) Payment for protective equipment.[§1926.95(d)]

(1) *Except as provided by paragraphs* (d)(2) through (d)(6) of this section, the protective equipment, including personal protective equipment (PPE), used to comply with this part, shall be provided by the employer at no cost to employees.[§1926.95(d)(1)]

(2) *The employer is not required to pay* for non-specialty safety-toe protective footwear (including steel-toe shoes or steel-toe boots) and non-specialty prescription safety eyewear, provided that the employer permits such items to be worn off the job-site. [§1926.95(d)(2)]

(3) *When the employer provides metatarsal* guards and allows the employee, at his or her request, to use shoes or boots with built-in metatarsal protection, the employer is not required to reimburse the employee for the shoes or boots.[§1926.95(d)(3)]

(4) *The employer is not required to pay for:*[§1926.95(d)(4)]

(i) *Everyday clothing,* such as long-sleeve shirts, long pants, street shoes, and normal work boots; or[§1926.95(d)(4)(i)]

(ii) *Ordinary clothing,* skin creams, or other items, used solely for protection from weather, such as winter coats, jackets, gloves, parkas, rubber boots, hats, raincoats, ordinary sunglasses, and sunscreen.[§1926.95(d)(4)(ii)]

(5) *The employer must pay for replacement PPE,* except when the employee has lost or intentionally damaged the PPE. [§1926.95(d)(5)]

(6) *Where an employee provides adequate* protective equipment he or she owns pursuant to paragraph (b) of this section, the employer may allow the employee to use it and is not required to reimburse the employee for that equipment. The employer shall not require an employee to provide or pay for his or her own PPE, unless the PPE is excepted by paragraphs (d)(2) through (d)(5) of this section.[§1926.95(d)(6)]

(7) *This section shall become effective* on February 13, 2008. Employers must implement the PPE payment requirements no later than May 15, 2008.[§1926.95(d)(7)]

Note to Sec. 1926.95(d): When the provisions of another OSHA standard specify whether or not the employer must pay for specific equipment, the payment provisions of that standard shall prevail.

[58 FR 35152, June 30, 1993, as amended at 72 FR 64429, Nov. 15, 2007]

§1926.96

Occupational foot protection

Safety-toe footwear for employees shall meet the requirements and specifications in American National Standard for Men's Safety-Toe Footwear, Z41.1-1967.

[58 FR 35152, June 30, 1993]

§§1926.97-1926.98

[Reserved]

§1926.100

Head protection

(a) Employees working in areas where there is a possible danger of head injury from impact, or from falling or flying objects, or from electrical shock and burns, shall be protected by protective helmets.[§1926.100(a)]

(b) Helmets for the protection of employees against impact and penetration of falling and flying objects shall meet the specifications contained in American National Standards Institute, Z89.1-1969, Safety Requirements for Industrial Head Protection. [§1926.100(b)]

(c) Helmets for the head protection of employees exposed to high voltage electrical shock and burns shall meet the specifications contained in American National Standards Institute, Z89.2-1971. [§1926.100(c)]

Federal Register Update

On Friday, June 22, ans July 23, 2012, OSHA amended §1926.100, effective September 20, 2012. The updated text is as follows:

(b) Criteria for head protection.The employer must provide each employee with head protection that meets the specifications contained in any of the following consensus standards:

(1) *The employer must provide each employee* with head protection that meets the specifications contained in any of the following consensus standards:

(i) *American National Standards Institute* (ANSI) Z89.1–2009, "American National Standard for Industrial Head Protection," incorporated by reference in § 1926.6;

(ii) *American National Standards Institute* (ANSI) Z89.1–2003, "American National Standard for Industrial Head Protection," incorporated by reference in § 1926.6; or

(iii) *American National Standards Institute* (ANSI) Z89.1–1997, "American National Standard for Personnel Protection—Protective Headwear for Industrial Workers—Requirements," incorporated by reference in § 1926.6.

(2) *The employer must ensure that the head protection* provided for each employee exposed to high-voltage electric shock and burns also meets the specifications contained in Section 9.7 ("Electrical Insulation") of any of the consensus standards identified in paragraph (b)(1) of this section.

(3) *OSHA will deem any head protection device* that the employer demonstrates is at least as effective as a head protection device constructed in accordance with one of the consensus standards identified in paragraph (b)(1) of this section to be in compliance with the requirements of this section.

~~(c)Helmets for the head protection of employees exposed to high voltage electrical shock and burns shall meet the specifications contained in American National Standards Institute, Z89.2-1971~~

§1926.101

Hearing protection

(a) Wherever it is not feasible to reduce the noise levels or duration of exposures to those specified in Table D-2, Permissible Noise Exposures, in Sec. 1926.52, ear protective devices shall be provided and used.[§1926.101(a)]

(b) Ear protective devices inserted in the ear shall be fitted or determined individually by competent persons.[§1926.101(b)]

(c) Plain cotton is not an acceptable protective device. [§1926.101(c)]

§1926.102

Eye and face protection

(a) General.[§1926.102(a)]

(1) *Employees shall be provided with eye and face protection equipment* when machines or operations present potential eye or face injury from physical, chemical, or radiation agents. [§1926.102(a)(1)]

(2) *Eye and face protection equipment* required by this Part shall meet the requirements specified in American National Standards Institute, Z87.1-1968, Practice for Occupational and Educational Eye and Face Protection.[§1926.102(a)(2)]

(3) *Employees whose vision requires the use* of corrective lenses in spectacles, when required by this regulation to wear eye protection, shall be protected by goggles or spectacles of one of the following types:[§1926.102(a)(3)]

(i) *Spectacles whose protective lenses* provide optical correction;[§1926.102(a)(3)(i)]

(ii) *Goggles that can be worn* over corrective spectacles without disturbing the adjustment of the spectacles;[§1926.102(a)(3)(ii)]

(iii) *Goggles that incorporate corrective* lenses mounted behind the protective lenses.[§1926.102(a)(3)(iii)]

(4) *Face and eye protection equipment* shall be kept clean and in good repair. The use of this type equipment with structural or optical defects shall be prohibited.[§1926.102(a)(4)]

(5) *Table E-1 shall be used* as a guide in the selection of face and eye protection for the hazards and operations noted. [§1926.102(a)(5)]

TABLE E-1. Eye and Face Protector Selection Guide

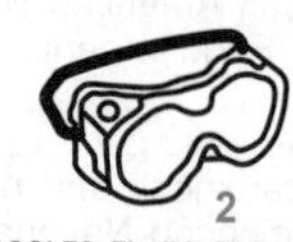

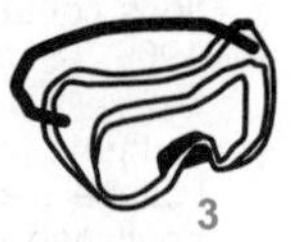

1. GOGGLES, Flexible Fitting, Regular Ventilation
2. GOGGLES, Flexible Fitting, Hooded Ventilation
3. GOGGLES, Cushioned Fitting, Rigid Body

*4. SPECTACLES, Metal Frame, with Sideshields
*5. SPECTACLES, Plastic Frame, with Sideshields
*6 SPECTACLES, Metal-Plastic Frame, with Sideshields

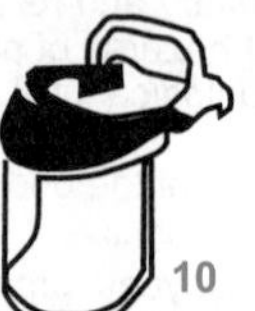

**7. WELDING GOGGLES, Eyecup type, Tinted Lenses (Illustrated)
7A. CHIPPING GOGGLES, Eyecup Type, Clear Safety Lenses (Not Illustrated)
**8. WELDING GOGGLES, Coverspec Type, Tinted Lenses (Illustrated)
8A. CHIPPING GOGGLES, Coverspec Type, Clear Safety Lenses (Not Illustrated)
**9. WELDING GOGGLES, Coverspec Type, Tinted Plate Lens

10. FACE SHIELD (Available with Plastic or Mesh Window)
**11. WELDING HELMETS

* Non-side shield spectacles are available for limited hazard use requiring only frontal protection.
** See Table E-2, in paragraph (b) of this section, Filter Lens Shade Numbers for Protection against Radiant Energy.

Applications		
Operation	Hazards	Recommended protectors: Bold type numbers signify preferred protection
Acetylene—Burning, Acetylene—Cutting, Acetylene—Welding	Sparks, harmful rays, molten metal, flying particles	7, 8, 9
Chemical Handling	Splash, acid burns, fumes	2, 10 (For severe exposure add 10 over 2)
Chipping	Flying particles	1, 3, 4, 5, 6, 7A, 8A
Electric (arc) welding	Sparks, intense rays, molten metal	9, 11, (11 in combination with 4, 5, 6, in tinted lenses, advisable)
Furnace operations	Glare, heat, molten metal	7, 8, 9 (For severe exposure add 10)
Grinding Light	Flying particles	1, 3, 4, 5, 6, 10
Grinding—Heavy	Flying particles	1 , 3, 7A, 8A (For severe exposure add 10)
Laboratory	Chemical splash, glass breakage.	2 (10 when in combination with 4, 5, 6)
Machining	Flying particles	1, 3, 4, 5, 6, 10
Molten metals	Heat, glare, sparks, splash	7, 8, (10 in combination with 4, 5, 6, in tinted lenses)
Spot welding	Flying particles, sparks	1, 3, 4, 5, 6, 10

(6) *Protectors shall meet the following* minimum requirements:[§1926.102(a)(6)]

(i) *They shall provide adequate* protection against the particular hazards for which they are designed.[§1926.102(a)(6)(i)]

(ii) *They shall be reasonably comfortable* when worn under the designated conditions.[§1926.102(a)(6)(ii)]

(iii) *They shall fit snugly* and shall not unduly interfere with the movements of the wearer.[§1926.102(a)(6)(iii)]

(iv) *They shall be durable.*[§1926.102(a)(6)(iv)]

(v) *They shall be capable of being disinfected.*[§1926.102(a)(6)(v)]

(vi) *They shall be easily cleanable.*[§1926.102(a)(6)(vi)]

(7) *Every protector shall be distinctly* marked to facilitate identification only of the manufacturer.[§1926.102(a)(7)]

(8) *When limitations or precautions are indicated* by the manufacturer, they shall be transmitted to the user and care taken to see that such limitations and precautions are strictly observed.[§1926.102(a)(8)]

(b) Protection against radiant energy[§1926.102(b)]

(1) *Selection of shade numbers for welding filter.* Table E-2 shall be used as a guide for the selection of the proper shade numbers of filter lenses or plates used in welding. Shades more dense than those listed may be used to suit the individual's needs.[§1926.102(b)(1)]

Table E-2 — Filter Lens Shade Numbers for Protection Against Radiant Energy

Welding operation	Shade number
Shielded metal-arc welding 1/16-, 3/32-, 1/8-, 5/32- inch diameter electrodes	10
Gas-shielded arc welding (nonferrous) 1/16-, 3/32-, 1/8-, 5/32-inch diameter electrodes	11
Gas-shielded arc welding (ferrous) 1/16-, 3/32-, 1/8-, 5/32-inch diameter electrodes	12
Shielded metal-arc welding 3/16-, 7/32-, 1/4-inch diameter electrodes	12
5/16-, 3/8-inch diameter electrodes	14
Atomic hydrogen welding	10-14
Carbon-arc welding	14
Soldering	2
Torch brazing	3 or 4
Light cutting, up to 1 inch	3 or 4
Medium cutting, 1 inch to 6 inches	4 or 5
Heavy cutting, over 6 inches	5 or 6
Gas welding (light), up to 1/8-inch	4 or 5
Gas welding (medium), 1/8-inch to 1/2-inch	5 or 6
Gas welding (heavy), over 1/2-inch	6 or 8

(2) *Laser protection.*[§1926.102(b)(2)]

(i) *Employees whose occupation* or assignment requires exposure to laser beams shall be furnished suitable laser safety goggles which will protect for the specific wavelength of the laser and be of optical density (O.D.) adequate for the energy involved. Table E-3 lists the maximum power or energy density for which adequate protection is afforded by glasses of optical densities from 5 through 8.[§1926.102(b)(2)(i)]

Table E-3 — Selecting Laser Safety Glass

Intensity, CW maximum power density (watts/cm^2)	Attenuation	
	Optical density (O.D.)	Attenuation factor
10^{-2}	5	10^5
10^{-1}	6	10^6
1.0	7	10^7
10.0	8	10^8

Output levels falling between lines in this table shall require the higher optical density.[§1926.102(b)(2)(i)]

(ii) *All protective goggles shall bear* a label identifying the following data:[§1926.102(b)(2)(ii)]

[a] The laser wavelengths for which use is intended;[§1926.102(b)(2)(ii)[a]]

[b] The optical density of those wavelengths;[§1926.102(b)(2)(ii)[b]]

[c] The visible light transmission.[§1926.102(b)(2)(ii)[c]]

[44 FR 8577, Feb. 9, 1979; 44 FR 20940, Apr. 6, 1979, as amended at 58 FR 35160, June 30, 1993]

§1926.103

Respiratory protection

Note: The requirements applicable to construction work under this section are identical to those set forth at 29 CFR 1910.134 of this chapter.

[63 FR 1297; Jan. 8, 1998]

§1926.104

Safety belts, lifelines, and lanyards

(a) **Lifelines, safety belts, and lanyards** shall be used only for employee safeguarding. Any lifeline, safety belt, or lanyard actually subjected to in-service loading, as distinguished from static load testing, shall be immediately removed from service and shall not be used again for employee safeguarding.[§1926.104(a)]

(b) **Lifelines shall be secured above the point of operation** to an anchorage or structural member capable of supporting a minimum dead weight of 5,400 pounds.[§1926.104(b)]

(c) **Lifelines used on rock-scaling operations,** or in areas where the lifeline may be subjected to cutting or abrasion, shall be a minimum of 7/8 -inch wire core manila rope. For all other lifeline applications, a minimum of 3/4 -inch manila or equivalent, with a minimum breaking strength of 5,400 pounds, shall be used. [§1926.104(c)]

(d) **Safety belt lanyard shall be a minimum of** 1/2 -inch nylon, or equivalent, with a maximum length to provide for a fall of no greater than 6 feet. The rope shall have a nominal breaking strength of 5,400 pounds.[§1926.104(d)]

(e) **All safety belt and lanyard hardware shall be drop forged** or pressed steel, cadmium plated in accordance with type 1, Class B plating specified in Federal Specification QQ-P-416. Surface shall be smooth and free of sharp edges.[§1926.104(e)]

(f) **All safety belt and lanyard hardware, except rivets,** shall be capable of withstanding a tensile loading of 4,000 pounds without cracking, breaking, or taking a permanent deformation.[§1926.104(f)]

§1926.105

Safety nets

(a) **Safety nets shall be provided** when workplaces are more than 25 feet above the ground or water surface, or other surfaces where the use of ladders, scaffolds, catch platforms, temporary floors, safety lines, or safety belts is impractical.[§1926.105(a)]

(b) **Where safety net protection** is required by this part, operations shall not be undertaken until the net is in place and has been tested.[§1926.105(b)]

(c) (1) *Nets shall extend 8 feet* beyond the edge of the work surface where employees are exposed and shall be installed as close under the work surface as practical but in no case more than 25 feet below such work surface. Nets shall be hung with sufficient clearance to prevent user's contact with the surfaces or structures below. Such clearances shall be determined by impact load testing.[§1926.105(c)(1)]

(2) *It is intended that only one* level of nets be required for bridge construction.[§1926.105(c)(2)]

(d) **The mesh size of nets** shall not exceed 6 inches by 6 inches. All new nets shall meet accepted performance standards of 17,500 foot-pounds minimum impact resistance as determined and certified by the manufacturers, and shall bear a label of proof test. Edge ropes shall provide a minimum breaking strength of 5,000 pounds.[§1926.105(d)]

(e) **Forged steel safety** hooks or shackles shall be used to fasten the net to its supports.[§1926.105(e)]

(f) **Connections between net** panels shall develop the full strength of the net.[§1926.105(f)]

§1926.106

Working over or near water

(a) **Employees working over or near water,** where the danger of drowning exists, shall be provided with U.S. Coast Guard-approved life jacket or buoyant work vests.[§1926.106(a)]

(b) **Prior to and after each use, the buoyant work vests** or life preservers shall be inspected for defects which would alter their strength or buoyancy. Defective units shall not be used. [§1926.106(b)]

(c) **Ring buoys with at least 90 feet of line** shall be provided and readily available for emergency rescue operations. Distance between ring buoys shall not exceed 200 feet.[§1926.106(c)]

(d) **At least one lifesaving skiff** shall be immediately available at locations where employees are working over or adjacent to water. [§1926.106(d)]

§1926.107

Definitions applicable to this subpart

(a) **Contaminant** means any material which by reason of its action upon, within, or to a person is likely to cause physical harm. [§1926.107(a)]

(b) **Lanyard** means a rope, suitable for supporting one person. One end is fastened to a safety belt or harness and the other end is secured to a substantial object or a safety line.[§1926.107(b)]

(c) **Lifeline** means a rope, suitable for supporting one person, to which a lanyard or safety belt (or harness) is attached.[§1926.107(c)]

(d) **O.D.** means optical density and refers to the light refractive characteristics of a lens.[§1926.107(d)]

(e) **Radiant energy** means energy that travels outward in all directions from its sources.[§1926.107(e)]

(f) **Safety belt** means a device, usually worn around the waist which, by reason of its attachment to a lanyard and lifeline or a structure, will prevent a worker from falling.[§1926.107(f)]

[44 FR 8577, Feb. 9, 1979]

Authority: Section. 107, Contract Work Hours and Safety Standards Act (Construction Safety Act) (40 U.S.C. 333); Sections. 4, 6, and 8 of the Occupational Safety and Health Act of 1970 (29 U.S.C. 653, 655, 657); Secretary of Labor's Order No. 12-71 (36 FR 8754), 8-76 (41 FR 25059), 9-83 (48 FR 35736), 1-90 (55 FR 9033), 6-96 (62 FR 111), 5-2002 (67 FR 65008), or 5-2007 (72 FR 31160) as applicable; and 29 CFR Part 1911.

Notes

Subpart F - Fire Protection and Prevention

§1926.150
Fire protection

(a) General requirements.[§1926.150(a)]

(1) *The employer shall be responsible for the development* of a fire protection program to be followed throughout all phases of the construction and demolition work, and he shall provide for the firefighting equipment as specified in this subpart. As fire hazards occur, there shall be no delay in providing the necessary equipment.[§1926.150(a)(1)]

(2) *Access to all available firefighting* equipment shall be maintained at all times.[§1926.150(a)(2)]

(3) *All firefighting equipment,* provided by the employer, shall be conspicuously located.[§1926.150(a)(3)]

(4) *All firefighting equipment shall be periodically* inspected and maintained in operating condition. Defective equipment shall be immediately replaced.[§1926.150(a)(4)]

(5) *As warranted by the project,* the employer shall provide a trained and equipped firefighting organization (Fire Brigade) to assure adequate protection to life.[§1926.150(a)(5)]

(b) Water supply.[§1926.150(b)]

(1) *A temporary or permanent water supply,* of sufficient volume, duration, and pressure, required to properly operate the firefighting equipment shall be made available as soon as combustible materials accumulate.[§1926.150(b)(1)]

(2) *Where underground water mains* are to be provided, they shall be installed, completed, and made available for use as soon as practicable.[§1926.150(b)(2)]

(c) Portable firefighting equipment[§1926.150(c)]

(1) *Fire extinguishers and small hose lines.*[§1926.150(c)(1)]

(i) *A fire extinguisher,* rated not less than 2A, shall be provided for each 3,000 square feet of the protected building area, or major fraction thereof. Travel distance from any point of the protected area to the nearest fire extinguisher shall not exceed 100 feet.[§1926.150(c)(1)(i)]

(ii) *One 55-gallon open drum of water* with two fire pails may be substituted for a fire extinguisher having a 2A rating. [§1926.150(c)(1)(ii)]

(iii) *A 1/2 -inch diameter garden-type hose line,* not to exceed 100 feet in length and equipped with a nozzle, may be substituted for a 2A- rated fire extinguisher, providing it is capable of discharging a minimum of 5 gallons per minute with a minimum hose stream range of 30 feet horizontally. The garden-type hose lines shall be mounted on conventional racks or reels. The number and location of hose racks or reels shall be such that at least one hose stream can be applied to all points in the area.[§1926.150(c)(1)(iii)]

(iv) *One or more fire extinguishers,* rated not less than 2A, shall be provided on each floor. In multistory buildings, at least one fire extinguisher shall be located adjacent to stairway. [§1926.150(c)(1)(iv)]

(v) *Extinguishers and water drums,* subject to freezing, shall be protected from freezing.[§1926.150(c)(1)(v)]

(vi) *A fire extinguisher,* rated not less than 10B, shall be provided within 50 feet of wherever more than 5 gallons of flammable or combustible liquids or 5 pounds of flammable gas are being used on the jobsite. This requirement does not apply to the integral fuel tanks of motor vehicles. [§1926.150(c)(1)(vi)]

(vii) *Carbon tetrachloride and other toxic* vaporizing liquid fire extinguishers are prohibited.[§1926.150(c)(1)(vii)]

(viii) *Portable fire extinguishers* shall be inspected periodically and maintained in accordance with Maintenance and Use of Portable Fire Extinguishers, NFPA No. 10A-1970. [§1926.150(c)(1)(viii)]

(ix) *Fire extinguishers which* have been listed or approved by a nationally recognized testing laboratory, shall be used to meet the requirements of this subpart.[§1926.150(c)(1)(ix)]

(x) *Table F-1 may be used* as a guide for selecting the appropriate portable fire extinguishers.[§1926.150(c)(1)(x)]

TABLE F-1 Fire Extinguishers Data

DEPARTMENT OF LABOR · UNITED STATES OF AMERICA ·	WATER TYPE				FOAM	CARBON DIOXIDE	DRY CHEMICAL: SODIUM OR POTASSIUM BICARBONATE		DRY CHEMICAL: MULTI-PURPOSE ABC	
	STORED PRESSURE	CARTRIDGE OPERATED	WATER PUMP TANK	SODA ACID	FOAM	CO_2	CARTRIDGE OPERATED	STORED PRESSURE	STORED PRESSURE	CARTRIDGE OPERATED
CLASS A FIRES WOOD, PAPER, TRASH HAVING GLOWING EMBERS (A – ORDINARY COMBUSTIBLES)	YES	YES	YES	YES	YES	NO (But will control small surface fires)	NO (But will control small surface fires)	NO (But will control small surface fires)	YES	YES
CLASS B FIRES FLAMMABLE LIQUIDS, GASOLINE, OIL, PAINTS, GREASE, ETC. (B – FLAMMABLE LIQUIDS)	NO	NO	NO	NO	YES	YES	YES	YES	YES	YES
CLASS C FIRES ELECTRICAL EQUIPMENT (C – ELECTRICAL EQUIPMENT)	NO	NO	NO	NO	NO	YES	YES	YES	YES	YES
CLASS D FIRES COMBUSTIBLE METALS (D – COMBUSTIBLE METALS)	SPECIAL EXTINGUISHING AGENTS APPROVED BY RECOGNIZED TESTING LABORATORIES									
METHOD OF OPERATION	PULL PIN – SQUEEZE HANDLE	TURN UPSIDE DOWN AND BUMP	PUMP HANDLE	TURN UPSIDE DOWN	TURN UPSIDE DOWN	PULL PIN – SQUEEZE LEVER	RUPTURE CARTRIDGE – SQUEEZE LEVER	PULL PIN – SQUEEZE HANDLE	PULL PIN – SQUEEZE HANDLE	RUPTURE CARTRIDGE – SQUEEZE LEVER
RANGE	30' - 40'	30' - 40'	30' - 40'	30' - 40'	30' - 40'	3' - 8'	5' - 20'	5' - 20'	5' - 20'	5' - 20'
MAINTENANCE	CHECK AIR PRESSURE GAUGE MONTHLY	WEIGH GAS CARTRIDGE – ADD WATER IF REQUIRED ANNUALLY	DISCHARGE AND FILL WITH WATER ANNUALLY	DISCHARGE ANNUALLY – RECHARGE	DISCHARGE ANNUALLY – RECHARGE	WEIGH SEMI-ANNUALLY	WEIGH GAS CARTRIDGE – CHECK CONDITION OF DRY CHEMICAL ANNUALLY	CHECK PRESSURE GAUGE AND CONDITION OF DRY CHEMICAL ANNUALLY	CHECK PRESSURE GAUGE AND CONDITION OF DRY CHEMICAL ANNUALLY	WEIGH GAS CARTRIDGE – CHECK CONDITION OF DRY CHEMICAL ANNUALLY

(2) *Fire hose and connections.*[§1926.150(c)(2)]

(i) *One hundred feet,* or less, of 1 1/2 -inch hose, with a nozzle capable of discharging water at 25 gallons or more per minute, may be substituted for a fire extinguisher rated not more than 2A in the designated area provided that the hose line can reach all points in the area.[§1926.150(c)(2)(i)]

(ii) *If fire hose connections are not compatible* with local firefighting equipment, the contractor shall provide adapters, or equivalent, to permit connections.[§1926.150(c)(2)(ii)]

(iii) *During demolition involving* combustible materials, charged hose lines, supplied by hydrants, water tank trucks with pumps, or equivalent, shall be made available. [§1926.150(c)(2)(iii)]

(d) Fixed firefighting equipment[§1926.150(d)]

(1) *Sprinkler protection.*[§1926.150(d)(1)]

(i) *If the facility being constructed* includes the installation of automatic sprinkler protection, the installation shall closely follow the construction and be placed in service as soon as applicable laws permit following completion of each story. [§1926.150(d)(1)(i)]

(ii) *During demolition or alterations,* existing automatic sprinkler installations shall be retained in service as long as reasonable. The operation of sprinkler control valves shall be permitted only by properly authorized persons. Modification of sprinkler systems to permit alterations or additional demolition should be expedited so that the automatic protection may be returned to service as quickly as possible. Sprinkler control valves shall be checked daily at close of work to ascertain that the protection is in service.[§1926.150(d)(1)(ii)]

(2) *Standpipes.* In all structures in which standpipes are required, or where standpipes exist in structures being altered, they shall be brought up as soon as applicable laws permit, and shall be maintained as construction progresses in such a manner that they are always ready for fire protection use. The standpipes shall be provided with Siamese fire department connections on the outside of the structure, at the street level, which shall be conspicuously marked. There shall be at least one standard hose outlet at each floor.[§1926.150(d)(2)]

(e) Fire alarm devices.[§1926.150(e)]

(1) *An alarm system,* e.g., telephone system, siren, etc., shall be established by the employer whereby employees on the site and the local fire department can be alerted for an emergency. [§1926.150(e)(1)]

(2) *The alarm code and reporting* instructions shall be conspicuously posted at phones and at employee entrances. [§1926.150(e)(2)]

(f) Fire cutoffs.[§1926.150(f)]

(1) *Fire walls and exit stairways,* required for the completed buildings, shall be given construction priority. Fire doors, with automatic closing devices, shall be hung on openings as soon as practicable.[§1926.150(f)(1)]

(2) *Fire cutoffs shall be retained* in buildings undergoing alterations or demolition until operations necessitate their removal. [§1926.150(f)(2)]

[44 FR 8577, Feb. 9, 1979; 44 FR 20940, Apr. 6, 1979, as amended at 58 FR 35162, June 30, 1993; 61 FR 31432, June 20, 1996]

§1926.151
Fire prevention

(a) Ignition hazards.[§1926.151(a)]

(1) *Electrical wiring and equipment for light,* heat, or power purposes shall be installed in compliance with the requirements of subpart K of this part.[§1926.151(a)(1)]

(2) *Internal combustion engine* powered equipment shall be so located that the exhausts are well away from combustible materials. When the exhausts are piped to outside the building under construction, a clearance of at least 6 inches shall be maintained between such piping and combustible material. [§1926.151(a)(2)]

(3) *Smoking shall be prohibited at or in the vicinity* of operations which constitute a fire hazard, and shall be conspicuously posted: "No Smoking or Open Flame."[§1926.151(a)(3)]

(4) *Portable battery powered lighting equipment,* used in connection with the storage, handling, or use of flammable gases or liquids, shall be of the type approved for the hazardous locations.[§1926.151(a)(4)]

(5) *The nozzle of air,* inert gas, and steam lines or hoses, when used in the cleaning or ventilation of tanks and vessels that contain hazardous concentrations of flammable gases or vapors, shall be bonded to the tank or vessel shell. Bonding devices shall not be attached or detached in hazardous concentrations of flammable gases or vapors.[§1926.151(a)(5)]

(b) Temporary buildings.[§1926.151(b)]

(1) *No temporary building shall be erected* where it will adversely affect any means of exit.[§1926.151(b)(1)]

(2) *Temporary buildings,* when located within another building or structure, shall be of either noncombustible construction or of combustible construction having a fire resistance of not less than 1 hour.[§1926.151(b)(2)]

(3) *Temporary buildings,* located other than inside another building and not used for the storage, handling, or use of flammable or combustible liquids, flammable gases, explosives, or blasting agents, or similar hazardous occupancies, shall be located at a distance of not less than 10 feet from another building or structure. Groups of temporary buildings, not exceeding 2,000 square feet in aggregate, shall, for the purposes of this part, be considered a single temporary building. [§1926.151(b)(3)]

(c) Open yard storage.[§1926.151(c)]

(1) *Combustible materials shall be piled* with due regard to the stability of piles and in no case higher than 20 feet. [§1926.151(c)(1)]

(2) *Driveways between and around combustible storage piles* shall be at least 15 feet wide and maintained free from accumulation of rubbish, equipment, or other articles or materials. Driveways shall be so spaced that a maximum grid system unit of 50 feet by 150 feet is produced.[§1926.151(c)(2)]

(3) *The entire storage site* shall be kept free from accumulation of unnecessary combustible materials. Weeds and grass shall be kept down and a regular procedure provided for the periodic cleanup of the entire area.[§1926.151(c)(3)]

(4) *When there is a danger of an underground fire,* that land shall not be used for combustible or flammable storage. [§1926.151(c)(4)]

(5) *Method of piling shall be solid* wherever possible and in orderly and regular piles. No combustible material shall be stored outdoors within 10 feet of a building or structure. [§1926.151(c)(5)]

(6) *Portable fire extinguishing equipment,* suitable for the fire hazard involved, shall be provided at convenient, conspicuously accessible locations in the yard area. Portable fire extinguishers, rated not less than 2A, shall be placed so that maximum travel distance to the nearest unit shall not exceed 100 feet. [§1926.151(c)(6)]

(d) Indoor storage.[§1926.151(d)]

(1) *Storage shall not obstruct, or adversely affect, means of exit.* [§1926.151(d)(1)]

(2) *All materials shall be stored,* handled, and piled with due regard to their fire characteristics.[§1926.151(d)(2)]

(3) *Noncompatible materials,* which may create a fire hazard, shall be segregated by a barrier having a fire resistance of at least 1 hour.[§1926.151(d)(3)]

(4) *Material shall be piled to minimize* the spread of fire internally and to permit convenient access for firefighting. Stable piling shall be maintained at all times. Aisle space shall be maintained to safely accommodate the widest vehicle that may be used within the building for firefighting purposes.[§1926.151(d)(4)]

(5) *Clearance of at least 36 inches* shall be maintained between the top level of the stored material and the sprinkler deflectors. [§1926.151(d)(5)]

(6) *Clearance shall be maintained* around lights and heating units to prevent ignition of combustible materials.[§1926.151(d)(6)]

(7) *A clearance of 24 inches* shall be maintained around the path of travel of fire doors unless a barricade is provided, in which case no clearance is needed. Material shall not be stored within 36 inches of a fire door opening.[§1926.151(d)(7)]

[44 FR 8577, Feb. 9, 1979; 44 FR 20940, Apr. 6, 1979, as amended at 51 FR 25318, July 11, 1986]

§1926.152
Flammable liquids

(a) General requirements.[§1926.152(a)]

(1) *Only approved containers* and portable tanks shall be used for storage and handling of flammable liquids. Approved safety cans or Department of Transportation approved containers shall be used for the handling and use of flammable liquids in

quantities of 5 gallons or less, except that this shall not apply to those flammable liquid materials which are highly viscid (extremely hard to pour), which may be used and handled in original shipping containers. For quantities of one gallon or less, the original container may be used, for storage, use and handling of flammable liquids.[§1926.152(a)(1)]

(2) *Flammable liquids* shall not be stored in areas used for exits, stairways, or normally used for the safe passage of people. [§1926.152(a)(2)]

(b) Indoor storage of flammable liquids.[§1926.152(b)]

(1) *No more than 25 gallons* of flammable liquids shall be stored in a room outside of an approved storage cabinet. For storage of liquefied petroleum gas, see Sec. 1926.153.[§1926.152(b)(1)]

(2) *Quantities of flammable liquid* in excess of 25 gallons shall be stored in an acceptable or approved cabinet meeting the following requirements:[§1926.152(b)(2)]

(i) *Acceptable wooden storage cabinets* shall be constructed in the following manner, or equivalent: The bottom, sides, and top shall be constructed of an exterior grade of plywood at least 1 inch in thickness, which shall not break down or delaminate under standard fire test conditions. All joints shall be rabbeted and shall be fastened in two directions with flathead wood screws. When more than one door is used, there shall be a rabbeted overlap of not less than 1 inch. Steel hinges shall be mounted in such a manner as to not lose their holding capacity due to loosening or burning out of the screws when subjected to fire. Such cabinets shall be painted inside and out with fire retardant paint. [§1926.152(b)(2)(i)]

(ii) *Approved metal storage cabinets will be acceptable.* [§1926.152(b)(2)(ii)]

(iii) *Cabinets shall be labeled in conspicuous* lettering, "Flammable-Keep Away from Open Flames."[§1926.152(b)(2)(iii)]

(3) *Not more than 60* gallons of Category 1, 2 and/or 3 flammable liquids or 120 gallons of Category 4 flammable liquids shall be stored in any one storage cabinet. Not more than three such cabinets may be located in a single storage area. Quantities in excess of this shall be stored in an inside storage room. [§1926.152(b)(3)]

(4) (i) *Inside storage rooms* shall be constructed to meet the required fire-resistive rating for their use. Such construction shall comply with the test specifications set forth in Standard Methods of Fire Test of Building Construction and Material, NFPA 251-1969.[§1926.152(b)(4)(i)]

(ii) *Where an automatic extinguishing system is provided,* the system shall be designed and installed in an approved manner. Openings to other rooms or buildings shall be provided with noncombustible liquid-tight raised sills or ramps at least 4 inches in height, or the floor in the storage area shall be at least 4 inches below the surrounding floor. Openings shall be provided with approved self-closing fire doors. The room shall be liquid-tight where the walls join the floor. A permissible alternate to the sill or ramp is an open-grated trench, inside of the room, which drains to a safe location. Where other portions of the building or other buildings are exposed, windows shall be protected as set forth in the Standard for Fire Doors and Windows, NFPA No. 80-1970, for Class E or F openings. Wood of at least 1-inch nominal thickness may be used for shelving, racks, dunnage, scuffboards, floor overlay, and similar installations.[§1926.152(b)(4)(ii)]

(iii) *Materials which will react* with water and create a fire hazard shall not be stored in the same room with flammable liquids. [§1926.152(b)(4)(iii)]

(iv) *Storage in inside storage rooms shall comply with Table F-2 following:*[§1926.152(b)(4)(iv)]

Table F-2

Fire protection provided	Fire resistance	Maximum size	Total allowable quantities gals./sq. ft./floor area
Yes	2 hrs.	500 sq. ft.	10
No	2 hrs.	500 sq. ft.	4
Yes	1 hr.	150 sq. ft.	5
No	1 hr.	150 sq. ft.	2

Note: Fire protection system shall be sprinkler, water spray, carbon dioxide or other system approved by a nationally recognized testing laboratory for this purpose.

(v) *Electrical wiring and equipment* located in inside storage rooms shall be approved for Class I, Division 1, Hazardous Locations. For definition of Class I, Division 1, Hazardous Locations, see Sec. 1926.449.[§1926.152(b)(4)(v)]

(vi) *Every inside storage* room shall be provided with either a gravity or a mechanical exhausting system. Such system shall commence not more than 12 inches above the floor and be designed to provide for a complete change of air within the room at least 6 times per hour. If a mechanical exhausting system is used, it shall be controlled by a switch located outside of the door. The ventilating equipment and any lighting fixtures shall be operated by the same switch. An electric pilot light shall be installed adjacent to the switch if Category 1, 2, or 3 flammable liquids are dispensed within the room. Where gravity ventilation is provided, the fresh air intake, as well as the exhausting outlet from the room, shall be on the exterior of the building in which the room is located. [§1926.152(b)(4)(vi)]

(vii) *In every inside storage* room there shall be maintained one clear aisle at least 3 feet wide. Containers over 30 gallons capacity shall not be stacked one upon the other. [§1926.152(b)(4)(vii)]

(viii) *Flammable liquids* in excess of that permitted in inside storage rooms shall be stored outside of buildings in accordance with paragraph (c) of this section.[§1926.152(b)(4)(viii)]

(5) *Quantity.* The quantity of flammable liquids kept in the vicinity of spraying operations shall be the minimum required for operations and should ordinarily not exceed a supply for 1 day or one shift. Bulk storage of portable containers of flammable liquids shall be in a separate, constructed building detached from other important buildings or cut off in a standard manner. [§1926.152(b)(5)]

(c) Storage outside buildings.[§1926.152(c)]

(1) *Storage of containers* (not more than 60 gallons each) shall not exceed 1,100 gallons in any one pile or area. Piles or groups of containers shall be separated by a 5-foot clearance. Piles or groups of containers shall not be nearer than 20 feet to a building.[§1926.152(c)(1)]

(2) *Within 200 feet* of each pile of containers, there shall be a 12-foot-wide access way to permit approach of fire control apparatus.[§1926.152(c)(2)]

(3) *The storage area shall be graded* in a manner to divert possible spills away from buildings or other exposures, or shall be surrounded by a curb or earth dike at least 12 inches high. When curbs or dikes are used, provisions shall be made for draining off accumulations of ground or rain water, or spills of flammable liquids. Drains shall terminate at a safe location and shall be accessible to operation under fire conditions. [§1926.152(c)(3)]

(4) *Outdoor portable tank storage:*[§1926.152(c)(4)]

(i) *Portable tanks shall not be nearer* than 20 feet from any building. Two or more portable tanks, grouped together, having a combined capacity in excess of 2,200 gallons, shall be separated by a 5- foot-clear area. Individual portable tanks exceeding 1,100 gallons shall be separated by a 5-foot-clear area.[§1926.152(c)(4)(i)]

(ii) *Within 200 feet* of each portable tank, there shall be a 12-foot-wide access way to permit approach of fire control apparatus.[§1926.152(c)(4)(ii)]

(5) *Storage areas shall be kept* free of weeds, debris, and other combustible material not necessary to the storage. [§1926.152(c)(5)]

(6) *Portable tanks,* not exceeding 660 gallons, shall be provided with emergency venting and other devices, as required by chapters III and IV of NFPA 30-1969, The Flammable and Combustible Liquids Code.[§1926.152(c)(6)]

(7) *Portable tanks,* in excess of 660 gallons, shall have emergency venting and other devices, as required by chapters II and III of The Flammable and Combustible Liquids Code, NFPA 30-1969.[§1926.152(c)(7)]

(d) Fire control for flammable liquid storage.[§1926.152(d)]

(1) *At least one portable* fire extinguisher, having a rating of not less than 20-B units, shall be located outside of, but not more than 10 feet from, the door opening into any room used for storage of more than 60 gallons of flammable liquids. [§1926.152(d)(1)]

(2) *At least one portable* fire extinguisher having a rating of not less than 20-B units shall be located not less than 25 feet, nor more than 75 feet, from any flammable liquid storage area located outside.[§1926.152(d)(2)]

(3) *When sprinklers are provided,* they shall be installed in accordance with the Standard for the Installation of Sprinkler Systems, NFPA 13-1969.[§1926.152(d)(3)]

(4) *At least one portable* fire extinguisher having a rating of not less than 20-B:C units shall be provided on all tank trucks or other vehicles used for transporting and/or dispensing flammable liquids.[§1926.152(d)(4)]

(e) Dispensing liquids.[§1926.152(e)]

(1) *Areas in which flammable liquids* are transferred at one time, in quantities greater than 5 gallons from one tank or container to another tank or container, shall be separated from other operations by 25-feet distance or by construction having a fire resistance of at least 1 hour. Drainage or other means shall be provided to control spills. Adequate natural or mechanical ventilation shall be provided to maintain the concentration of flammable vapor at or below 10 percent of the lower flammable limit.[§1926.152(e)(1)]

(2) *Transfer of Category 1, 2, or 3 flammable liquids* from one container to another shall be done only when containers are electrically interconnected (bonded).[§1926.152(e)(2)]

(3) *Flammable liquids* shall be drawn from or transferred into vessels, containers, or tanks within a building or outside only through a closed piping system, from safety cans, by means of a device drawing through the top, or from a container, or portable tanks, by gravity or pump, through an approved self-closing valve. Transferring by means of air pressure on the container or portable tanks is prohibited.[§1926.152(e)(3)]

(4) *The dispensing units* shall be protected against collision damage.[§1926.152(e)(4)]

(5) *Dispensing devices and nozzles* for Category 1, 2, or 3 flammable liquids shall be of an approved type.[§1926.152(e)(5)]

(f) Handling liquids at point of final use.[§1926.152(f)]

(1) *Category 1, 2, or 3 flammable liquids shall be kept* in closed containers when not actually in use.[§1926.152(f)(1)]

(2) *Leakage or spillage of flammable* liquids shall be disposed of promptly and safely.[§1926.152(f)(2)]

(3) *Category 1, 2, or 3 flammable liquids may be used* only where there are no open flames or other sources of ignition within 50 feet of the operation, unless conditions warrant greater clearance.[§1926.152(f)(3)]

(g) Service and refueling areas.[§1926.152(g)]

(1) *Flammable liquids* shall be stored in approved closed containers, in tanks located underground, or in aboveground portable tanks.[§1926.152(g)(1)]

(2) *The tank trucks shall comply* with the requirements covered in the Standard for Tank Vehicles for Flammable and Combustible Liquids, NFPA No. 385-1966.[§1926.152(g)(2)]

(3) *The dispensing hose shall be an approved type.*[§1926.152(g)(3)]

(4) *The dispensing nozzle shall be an approved* automatic-closing type without a latch-open device.[§1926.152(g)(4)]

(5) *Underground tanks shall not be abandoned.*[§1926.152(g)(5)]

(6) *Clearly identified and easily* accessible switch(es) shall be provided at a location remote from dispensing devices to shut off the power to all dispensing devices in the event of an emergency.[§1926.152(g)(6)]

(7) (i) *Heating equipment of an approved* type may be installed in the lubrication or service area where there is no dispensing or transferring of Category 1, 2, or 3 flammable liquids, provided the bottom of the heating unit is at least 18 inches above the floor and is protected from physical damage. [§1926.152(g)(7)(i)]

(ii) *Heating equipment installed* in lubrication or service areas, where Category 1, 2, or 3 flammable liquids are dispensed, shall be of an approved type for garages, and shall be installed at least 8 feet above the floor.[§1926.152(g)(7)(ii)]

(8) *There shall be no smoking or open* flames in the areas used for fueling, servicing fuel systems for internal combustion engines, receiving or dispensing of flammable liquids. [§1926.152(g)(8)]

(9) *Conspicuous and legible signs* prohibiting smoking shall be posted.[§1926.152(g)(9)]

(10) *The motors of all equipment being fueled* shall be shut off during the fueling operation.[§1926.152(g)(10)]

(11) *Each service or fueling area* shall be provided with at least one fire extinguisher having a rating of not less than 20-B:C located so that an extinguisher will be within 75 feet of each pump, dispenser, underground fill pipe opening, and lubrication or service area.[§1926.152(g)(11)]

(h) Scope. This section applies to the handling, storage, and use of flammable liquids with a flashpoint at or below 199.4 °F (93 °C). This section does not apply to:[§1926.152(h)]

(1) *Bulk transportation of flammable liquids; and*[§1926.152(h)(1)]

(2) *Storage, handling, and use* of fuel oil tanks and containers connected with oil burning equipment.[§1926.152(h)(2)]

(i) Tank storage[§1926.152(i)]

(1) *Design and construction of tanks.*[§1926.152(i)(1)]

(i) *Materials.*[§1926.152(i)(1)(i)]

[A] Tanks shall be built of steel except as provided in paragraphs (i)(1)(i) (B) through (E) of this section. [§1926.152(i)(1)(i)[A]]

[B] Tanks may be built of materials other than steel for installation underground or if required by the properties of the liquid stored. Tanks located above ground or inside buildings shall be of noncombustible construction. [§1926.152(i)(1)(i)[B]]

[C] Tanks built of materials other than steel shall be designed to specifications embodying principles recognized as good engineering design for the material used. [§1926.152(i)(1)(i)[C]]

[D] Unlined concrete tanks may be used for storing flammable liquids having a gravity of 40° API or heavier. Concrete tanks with special lining may be used for other services provided the design is in accordance with sound engineering practice.[§1926.152(i)(1)(i)[D]]

[E] [Reserved][§1926.152(i)(1)(i)[E]]

[F] Special engineering consideration shall be required if the specific gravity of the liquid to be stored exceeds that of water or if the tanks are designed to contain flammable liquids at a liquid temperature below 0 °F. [§1926.152(i)(1)(i)[F]]

(ii) *Fabrication.*[§1926.152(i)(1)(ii)]

[A] [Reserved][§1926.152(i)(1)(ii)[A]]

[B] Metal tanks shall be welded, riveted, and caulked, brazed, or bolted, or constructed by use of a combination of these methods. Filler metal used in brazing shall be nonferrous metal or an alloy having a melting point above 1000 °F. and below that of the metal joined. [§1926.152(i)(1)(ii)[B]]

(iii) *Atmospheric tanks.*[§1926.152(i)(1)(iii)]

[A] Atmospheric tanks shall be built in accordance with acceptable good standards of design. Atmospheric tanks may be built in accordance with:[§1926.152(i)(1)(iii)[A]]

[1] Underwriters' Laboratories, Inc., Subjects No. 142, Standard for Steel Aboveground Tanks for Flammable and Combustible Liquids, 1968; No. 58, Standard for Steel Underground Tanks for Flammable and Combustible Liquids, Fifth Edition, December 1961; or No. 80, Standard for Steel Inside Tanks for Oil-Burner Fuel, September 1963.[§1926.152(i)(1)(iii)[A][1]]

[2] American Petroleum Institute Standards No. 12A, Specification for Oil Storage Tanks with Riveted Shells, Seventh Edition, September 1951, or No. 650, Welded Steel Tanks for Oil Storage, Third Edition, 1966. [§1926.152(i)(1)(iii)[A][2]]

[3] American Petroleum Institute Standards No. 12B, Specification for Bolted Production Tanks, Eleventh Edition, May 1958, and Supplement 1, March 1962; No. 12D, Specification for Large Welded Production Tanks, Seventh Edition, August 1957; or No. 12F, Specification for Small Welded Production Tanks, Fifth Edition, March 1961. Tanks built in accordance with these standards shall be used only as production tanks for storage of crude petroleum in oil-producing areas. [§1926.152(i)(1)(iii)[A][3]]

[B] Tanks designed for underground service not exceeding 2,500 gallons (9,462.5 L) capacity may be used aboveground.[§1926.152(i)(1)(iii)[B]]

[C] Low-pressure tanks and pressure vessels may be used as atmospheric tanks.[§1926.152(i)(1)(iii)[C]]

[D] Atmospheric tanks shall not be used for the storage of a flammable liquid at a temperature at or above its boiling point.[§1926.152(i)(1)(iii)[D]]

(iv) *Low pressure tanks.*[§1926.152(i)(1)(iv)]

[A] *The normal operating pressure* of the tank shall not exceed the design pressure of the tank.[§1926.152(i)(1)(iv)[A]]

[B] *Low-pressure tanks shall be built* in accordance with acceptable standards of design. Low-pressure tanks may be built in accordance with:[§1926.152(i)(1)(iv)[B]]

[1] *American Petroleum Institute* Standard No. 620. Recommended Rules for the Design and Construction of Large, Welded, Low-Pressure Storage Tanks, Third Edition, 1966.[§1926.152(i)(1)(iv)[B][1]]

[2] *The principles of the Code for Unfired Pressure Vessels,* Section VIII of the ASME Boiler and Pressure Vessels Code, 1968.[§1926.152(i)(1)(iv)[B][2]]

[C] *Atmospheric tanks built* according to Underwriters' Laboratories, Inc., requirements in paragraph (i)(1)(iii)(A) of this section and shall be limited to 2.5 p.s.i.g. under emergency venting conditions.[§1926.152(i)(1)(iv)[C]]

This paragraph may be used for operating pressures not exceeding 1 p.s.i.g.[§1926.152(i)(1)(iv)[C]]

[D] *Pressure vessels may be used* as low-pressure tanks. [§1926.152(i)(1)(iv)[D]]

(v) *Pressure vessels.*[§1926.152(i)(1)(v)]

[A] *The normal operating pressure* of the vessel shall not exceed the design pressure of the vessel. [§1926.152(i)(1)(v)[A]]

[B] *Pressure vessels shall be built* in accordance with the Code for Unfired Pressure Vessels, Section VIII of the ASME Boiler and Pressure Vessel Code 1968. [§1926.152(i)(1)(v)[B]]

(vi) *Provisions for internal corrosion.* When tanks are not designed in accordance with the American Petroleum Institute, American Society of Mechanical Engineers, or the Underwriters' Laboratories, Inc.'s, standards, or if corrosion is anticipated beyond that provided for in the design formulas used, additional metal thickness or suitable protective coatings or linings shall be provided to compensate for the corrosion loss expected during the design life of the tank. [§1926.152(i)(1)(vi)]

(2) *Installation of outside aboveground tanks.*[§1926.152(i)(2)]

(i) *[Reserved]*[§1926.152(i)(2)(i)]

(ii) *Spacing (shell-to-shell) between* aboveground tanks. [§1926.152(i)(2)(ii)]

[A] *The distance between any two* flammable liquid storage tanks shall not be less than 3 feet (0.912 m). [§1926.152(i)(2)(ii)[A]]

[B] *Except as provided* in paragraph (i)(2)(ii)(C) of this section, the distance between any two adjacent tanks shall not be less than one- sixth the sum of their diameters. When the diameter of one tank is less than one-half the diameter of the adjacent tank, the distance between the two tanks shall not be less than one-half the diameter of the smaller tank.[§1926.152(i)(2)(ii)[B]]

[C] *Where crude petroleum in conjunction with production facilities* are located in noncongested areas and have capacities not exceeding 126,000 gallons (3,000 barrels), the distance between such tanks shall not be less than 3 feet (0.912 m).[§1926.152(i)(2)(ii)[C]]

[D] *Where unstable flammable liquids* are stored, the distance between such tanks shall not be less than one-half the sum of their diameters.[§1926.152(i)(2)(ii)[D]]

[E] *When tanks are compacted* in three or more rows or in an irregular pattern, greater spacing or other means shall be provided so that inside tanks are accessible for firefighting purposes.[§1926.152(i)(2)(ii)[E]]

[F] *The minimum separation between* a liquefied petroleum gas container and a flammable liquid storage tank shall be 20 feet (6.08 m), except in the case of flammable liquid tanks operating at pressures exceeding 2.5 p.s.i.g. or equipped with emergency venting which will permit pressures to exceed 2.5 p.s.i.g. in which case the provisions of paragraphs (i)(2)(ii) (A) and (B) of this section shall apply. Suitable means shall be taken to prevent the accumulation of flammable liquids under adjacent liquefied petroleum gas containers such as by diversion curbs or grading. When flammable liquid storage tanks are within a diked area, the liquefied petroleum gas containers shall be outside the diked area and at least 10 feet (3.04 m) away from the centerline of the wall of the diked area. The foregoing provisions shall not apply when liquefied petroleum gas containers of 125 gallons (473.125 L) or less capacity are installed adjacent to fuel oil supply tanks of 550 gallons (2,081.75 L) or less capacity. [§1926.152(i)(2)(ii)[F]]

(iii) *[Reserved]*[§1926.152(i)(2)(iii)]

(iv) *Normal venting for aboveground tanks.*[§1926.152(i)(2)(iv)]

[A] *Atmospheric storage tanks* shall be adequately vented to prevent the development of vacuum or pressure sufficient to distort the roof of a cone roof tank or exceeding the design pressure in the case of other atmospheric tanks, as a result of filling or emptying, and atmospheric temperature changes.[§1926.152(i)(2)(iv)[A]]

[B] *Normal vents shall be sized* either in accordance with: [§1926.152(i)(2)(iv)[B]]

[1] *The American Petroleum Institute* Standard 2000 (1968), Venting Atmospheric and Low-Pressure Storage Tanks; or[§1926.152(i)(2)(iv)[B][1]]

[2] *other accepted standard;* or[§1926.152(i)(2)(iv)[B][2]]

[3] *shall be at least as large as the filling* or withdrawal connection, whichever is larger but in no case less than 1 1/4 inch (3.175 cm) nominal inside diameter. [§1926.152(i)(2)(iv)[B][3]]

[C] *Low-pressure tanks and pressure vessels* shall be adequately vented to prevent development of pressure or vacuum, as a result of filling or emptying and atmospheric temperature changes, from exceeding the design pressure of the tank or vessel. Protection shall also be provided to prevent overpressure from any pump discharging into the tank or vessel when the pump discharge pressure can exceed the design pressure of the tank or vessel. [§1926.152(i)(2)(iv)[C]]

[D] *If any tank or pressure vessel* has more than one fill or withdrawal connection and simultaneous filling or withdrawal can be made, the vent size shall be based on the maximum anticipated simultaneous flow. [§1926.152(i)(2)(iv)[D]]

[E] *Unless the vent is designed* to limit the internal pressure 2.5 p.s.i. or less, the outlet of vents and vent drains shall be arranged to discharge in such a manner as to prevent localized overheating of any part of the tank in the event vapors from such vents are ignited.[§1926.152(i)(2)(iv)[E]]

[F] *Tanks and pressure vessels* storing Category 1 flammable liquids shall be equipped with venting devices that shall be normally closed except when venting to pressure or vacuum conditions. Tanks and pressure vessels storing Category 2 flammable liquids, or Category 3 flammable liquids with a flashpoint below 100 °F (37.8 °C), shall be equipped with venting devices that shall be normally closed except when venting under pressure or vacuum conditions, or with approved flame arresters. [§1926.152(i)(2)(iv)[F]]

Exemption: Tanks of 3,000 bbls (barrels) (84 m(3)) capacity or less containing crude petroleum in crude-producing areas; and, outside aboveground atmospheric tanks under 1,000 gallons (3,785 L) capacity containing other than Category 1 flammable liquids may have open vents. (*See* paragraph (i)(2)(vi)(B) of this section.) [§1926.152(i)(2)(iv)[F]]

[G] *Flame arresters or venting* devices required in paragraph (i)(2)(iv)(F) of this section may be omitted for Category 2 flammable liquids or Category 3 flammable liquids with a flashpoint below 100 °F (37.8 °C) where conditions are such that their use may, in case of obstruction, result in tank damage.[§1926.152(i)(2)(iv)[G]]

(v) *Emergency relief venting* for fire exposure for aboveground tanks.[§1926.152(i)(2)(v)]

[A] *Every aboveground storage tank* shall have some form of construction or device that will relieve excessive internal pressure caused by exposure fires.[§1926.152(i)(2)(v)[A]]

[B] *In a vertical tank the construction* referred to in paragraph (i)(2)(v)(A) of this section may take the form of a floating roof, lifter roof, a weak roof-to-shell seam, or other approved pressure relieving construction. The weak roof-to-shell seam shall be constructed to fail preferential to any other seam.[§1926.152(i)(2)(v)[B]]

[C] *Where entire dependence for emergency relief* is placed upon pressure relieving devices, the total venting capacity of both normal and emergency vents shall be enough to prevent rupture of the shell or bottom of the tank if vertical, or of the shell or heads if horizontal. If unstable liquids are stored, the effects of heat or gas resulting from polymerization, decomposition, condensation, or self-reactivity shall be taken into account. The total capacity of both normal and emergency venting devices shall be not less than that derived from Table F-10 except as provided in paragraph (i)(2)(v) (E) or (F) of this section. Such device may be a self-closing manhole cover, or one using long bolts that permit the cover to lift under internal pressure, or an additional or larger relief valve or valves. The wetted area of the tank shall be calculated on the basis of 55 percent of the total exposed area of a sphere or spheroid, 75 percent of the total exposed area of a horizontal tank and the first 30 feet (9.12 m) above grade of the exposed shell area of a vertical tank.[§1926.152(i)(2)(v)[C]]

Table F-10 — Wetted Area Versus Cubic Feet (Meters) Free Air Per Hour [14.7 psia and 60 °F. (15.55 °C)]

Square feet (m^2)	CFH (m^3 H)	Square feet (m^2)	CFH (m^3 H)
20 (1.84)	21,100 (590.8)	350 (32.2)	288,000 (8,064)
30 (2.76)	31,600 (884.8)	400 (36.8)	312,000 (8,736)
40 (3.68)	42,100 (1,178.8)	500 (46)	354,000 (9,912)
50 (4.6)	52,700 (1,475.6)	600 (55.2)	392,000 (10,976)
60 (5.52)	63,200 (1,769.6)	700 (64.4)	428,000 (11,984)
70 (6.44)	73,700 (2,063.6)	800 (73.6)	462,000 (12,936)
80 (7.36)	84,200 (2,357.6)	900 (82.8)	493,000 (13,804)
90 (8.28)	94,800 (2,654.4)	[1] 1,000 (90.2)	524,000 (14,672)
100 (9.2)	105,000 (2,940)	1,200 (110.4)	557,000 (15,596)
120 (11.04)	126,000 (3,528)	1,400 (128.8)	587,000 (16,436)
140 (12.88)	147,000 (4,116)	1,600 (147.2)	614,000 (17,192)
160 (14.72)	168,000 (4,704)	1,800 (165.6)	639,000 (17,892)
180 (16.56)	190,000 (5,320)	2,000 (180.4)	662,000 (18,536)
[1] 200 (18.4)	211,000 (5,908)	2,400 (220.8)	704,000 (19,712)
250 (23)	239,000 (6,692)	2,800 (257.6) and over	742,000 (20,776)
300 (27.6)	265,000 (7,420)		

[D] For tanks and storage vessels designed for pressure over 1 p.s.i.g., the total rate of venting shall be determined in accordance with Table F-10, except that when the exposed wetted area of the surface is greater than 2,800 square feet (257.6 m^2), the total rate of venting shall be calculated by the following formula:[§1926.152(i)(2)(v)[D]]

$CFH = 1{,}107A^{0.82}$

Where:

CFH = Venting requirement, in cubic feet (meters) of free air per hour.

A = Exposed wetted surface, in square feet (m^2).

Note: The foregoing formula is based on $Q=21{,}000A^{0.82}$.

[E] The total emergency relief venting capacity for any specific stable liquid may be determined by the following formula:[§1926.152(i)(2)(v)[E]]

$$V = 1337 \div (L\sqrt{M})$$

V = Cubic feet (meters) of free air per hour from Table F-10.

L = Latent heat of vaporization of specific liquid in B.t.u. per pound.

M = Molecular weight of specific liquids.

[F] The required airflow rate of paragraph (i)(2)(v) (C) or (E) of this section may be multiplied by the appropriate factor listed in the following schedule when protection is provided as indicated. Only one factor may be used for any one tank.[§1926.152(i)(2)(v)[F]]

0.5 for drainage in accordance with paragraph (i)(2)(vii)(B) of this section for tanks over 200 square feet (18.4 m^2) of wetted area.

0.3 for approved water spray.

0.3 for approved insulation.

0.15 for approved water spray with approved insulation.

[G] The outlet of all vents and vent drains on tanks equipped with emergency venting to permit pressures exceeding 2.5 p.s.i.g. shall be arranged to discharge in such a way as to prevent localized overheating of any part of the tank, in the event vapors from such vents are ignited. [§1926.152(i)(2)(v)[G]]

[H] Each commercial tank venting device shall have stamped on it the opening pressure, the pressure at which the valve reaches the full open position, and the flow capacity at the latter pressure, expressed in cubic feet (meters) per hour of air at 60 °F. (15.55 °C) and at a pressure of 14.7 p.s.i.a.[§1926.152(i)(2)(v)[H]]

[I] The flow capacity of tank venting devices 12 inches (30.48 cm) and smaller in nominal pipe size shall be determined by actual test of each type and size of vent. These flow tests may be conducted by the manufacturer if certified by a qualified impartial observer, or may be conducted by an outside agency. The flow capacity of tank venting devices larger than 12 inches (30.48 cm) nominal pipe size, including manhole covers with long bolts or equivalent, may be calculated provided that the opening pressure is actually measured, the rating pressure and corresponding free orifice area are stated, the word "calculated" appears on the nameplate, and the computation is based on a flow coefficient of 0.5 applied to the rated orifice area.[§1926.152(i)(2)(v)[I]]

(vi) *Vent piping for aboveground tanks.*[§1926.152(i)(2)(vi)]

[A] Vent piping shall be constructed in accordance with paragraph (c) of this section.[§1926.152(i)(2)(vi)[A]]

[B] Where vent pipe outlets for tanks storing Category 1 or 2 flammable liquids, or Category 3 flammable liquids with a flashpoint below 100 °F (37.8 °C), are adjacent to buildings or public ways, they shall be located so that the vapors are released at a safe point outside of buildings and not less than 12 feet (3.658 m) above the adjacent ground level. In order to aid their dispersion, vapors shall be discharged upward or horizontally away from closely adjacent walls. Vent outlets shall be located so that flammable vapors will not be trapped by eaves or other obstructions and shall be at least 5 feet (1.52 m) from building openings.[§1926.152(i)(2)(vi)[B]]

[C] When tank vent piping is manifolded, pipe sizes shall be such as to discharge, within the pressure limitations of the system, the vapors they may be required to handle when manifolded tanks are subject to the same fire exposure. [§1926.152(i)(2)(vi)[C]]

(vii) *Drainage,* dikes, and walls for aboveground tanks [§1926.152(i)(2)(vii)]

[A] Drainage and diked areas. The area surrounding a tank or a group of tanks shall be provided with drainage as in paragraph (i)(2)(vii)(B) of this section, or shall be diked as provided in (i)(2)(vii)(C) of this section, to prevent accidental discharge of liquid from endangering adjoining property or reaching waterways.[§1926.152(i)(2)(vii)[A]]

[B] Drainage. Where protection of adjoining property or waterways is by means of a natural or manmade drainage system, such systems shall comply with the following: [§1926.152(i)(2)(vii)[B]]

[1] [Reserved][§1926.152(i)(2)(vii)[B][1]]

[2] The drainage system shall terminate in vacant land or other area or in an impounding basin having a capacity not smaller than that of the largest tank served. This termination area and the route of the drainage system shall be so located that, if the flammable liquids in the drainage system are ignited, the fire will not seriously expose tanks or adjoining property. [§1926.152(i)(2)(vii)[B][2]]

[C] Diked areas. Where protection of adjoining property or waterways is accomplished by retaining the liquid around the tank by means of a dike, the volume of the diked area shall comply with the following requirements: [§1926.152(i)(2)(vii)[C]]

[1] Except as provided in paragraph (i)(2)(vii)(C)(2) of this section, the volumetric capacity of the diked area shall not be less than the greatest amount of liquid that can be released from the largest tank within the diked area, assuming a full tank. The capacity of the diked area enclosing more than one tank shall be calculated by deducting the volume of the tanks other than the largest tank below the height of the dike. [§1926.152(i)(2)(vii)[C][1]]

[2] For a tank or group of tanks with fixed roofs containing crude petroleum with boilover characteristics, the volumetric capacity of the diked area shall be not less than the capacity of the largest tank served by the enclo-

sure, assuming a full tank. The capacity of the diked enclosure shall be calculated by deducting the volume below the height of the dike of all tanks within the enclosure.[§1926.152(i)(2)(vii)[C][2]]

[3] *Walls of the diked area* shall be of earth, steel, concrete or solid masonry designed to be liquidtight and to withstand a full hydrostatic head. Earthen walls 3 feet (0.912 m) or more in height shall have a flat section at the top not less than 2 feet (0.608 m) wide. The slope of an earthen wall shall be consistent with the angle of repose of the material of which the wall is constructed. [§1926.152(i)(2)(vii)[C][3]]

[4] *The walls of the diked area* shall be restricted to an average height of 6 feet (1.824 m) above interior grade. [§1926.152(i)(2)(vii)[C][4]]

[5] *[Reserved]*[§1926.152(i)(2)(vii)[C][5]]

[6] *No loose combustible material,* empty or full drum or barrel, shall be permitted within the diked area. [§1926.152(i)(2)(vii)[C][6]]

(viii) *Tank openings other than* vents for aboveground tanks. [§1926.152(i)(2)(viii)]

[A] *[Reserved]*[§1926.152(i)(2)(viii)[A]]

[B] *[Reserved]*[§1926.152(i)(2)(viii)[B]]

[C] *[Reserved]*[§1926.152(i)(2)(viii)[C]]

[D] *Openings for gaging* shall be provided with a vaportight cap or cover.[§1926.152(i)(2)(viii)[D]]

[E] *For Category 2 flammable* liquids or Category 3 flammable liquids with a flashpoint below 100 °F (37.8 °C), other than crude oils, gasolines, and asphalts, the fill pipe shall be so designed and installed as to minimize the possibility of generating static electricity. A fill pipe entering the top of a tank shall terminate within 6 inches (15.24 cm) of the bottom of the tank and shall be installed to avoid excessive vibration.[§1926.152(i)(2)(viii)[E]]

[F] *Filling and emptying connections* which are made and broken shall be located outside of buildings at a location free from any source of ignition and not less than 5 feet (1.52 m) away from any building opening. Such connection shall be closed and liquidtight when not in use. The connection shall be properly identified.[§1926.152(i)(2)(viii)[F]]

(3) *Installation of underground tanks*[§1926.152(i)(3)]

(i) *Location.* Evacuation for underground storage tanks shall be made with due care to avoid undermining of foundations of existing structures. Underground tanks or tanks under buildings shall be so located with respect to existing building foundations and supports that the loads carried by the latter cannot be transmitted to the tank. The distance from any part of a tank storing Category 1 or 2 flammable liquids, or Category 3 flammable liquids with a flashpoint below 100 °F (37.8 °C), to the nearest wall of any basement or pit shall be not less than 1 foot (0.304 m), and to any property line that may be built upon, not less than 3 feet (0.912 m). The distance from any part of a tank storing Category 3 flammable liquids with a flashpoint at or above 100 °F (37.8 °C) or Category 4 flammable liquids to the nearest wall of any basement, pit or property line shall be not less than 1 foot (0.304 m). [§1926.152(i)(3)(i)]

(ii) *Depth and cover.* Underground tanks shall be set on firm foundations and surrounded with at least 6 inches (15.24 cm) of noncorrosive, inert materials such as clean sand, earth, or gravel well tamped in place. The tank shall be placed in the hole with care since dropping or rolling the tank into the hole can break a weld, puncture or damage the tank, or scrape off the protective coating of coated tanks. Tanks shall be covered with a minimum of 2 feet (0.608 m) of earth, or shall be covered with not less than 1 foot (0.304 m) of earth, on top of which shall be placed a slab of reinforced concrete not less than 4 inches (10.16 cm) thick. When underground tanks are, or are likely to be, subject to traffic, they shall be protected against damage from vehicles passing over them by at least 3 feet (0.912 m) of earth cover, or 18 inches (45.72 cm) of well-tamped earth, plus 6 inches (15.24 cm) of reinforced concrete or 8 inches (20.32 cm) of asphaltic concrete. When asphaltic or reinforced concrete paving is used as part of the protection, it shall extend at least 1 foot (0.304 m) horizontally beyond the outline of the tank in all directions. [§1926.152(i)(3)(ii)]

(iii) *Corrosion protection.* Corrosion protection for the tank and its piping shall be provided by one or more of the following methods:[§1926.152(i)(3)(iii)]

[A] *Use of protective coatings or wrappings;* [§1926.152(i)(3)(iii)[A]]

[B] *Cathodic protection; or,*[§1926.152(i)(3)(iii)[B]]

[C] *Corrosion resistant materials* of construction. [§1926.152(i)(3)(iii)[C]]

(iv) *Vents.*[§1926.152(i)(3)(iv)]

[A] *Location and arrangement of vents* for Category 1 or 2 flammable liquids, or Category 3 flammable liquids with a flashpoint below 100 °F (37.8 °C). Vent pipes from tanks storing Category 1 or 2 flammable liquids, or Category 3 flammable liquids with a flashpoint below 100 °F (37.8 °C), shall be so located that the discharge point is outside of buildings, higher than the fill pipe opening, and not less than 12 feet (3.658 m) above the adjacent ground level. Vent pipes shall discharge only upward in order to disperse vapors. Vent pipes 2 inches (5.08 cm) or less in nominal inside diameter shall not be obstructed by devices that will cause excessive back pressure. Vent pipe outlets shall be so located that flammable vapors will not enter building openings, or be trapped under eaves or other obstructions. If the vent pipe is less than 10 feet (3.04 m) in length, or greater than 2 inches (5.08 cm) in nominal inside diameter, the outlet shall be provided with a vacuum and pressure relief device or there shall be an approved flame arrester located in the vent line at the outlet or within the approved distance from the outlet. [§1926.152(i)(3)(iv)[A]]

[B] *Size of vents.* Each tank shall be vented through piping adequate in size to prevent blow-back of vapor or liquid at the fill opening while the tank is being filled. Vent pipes shall be not less than 1 1/4 inch (3.175 cm) nominal inside diameter.[§1926.152(i)(3)(iv)[B]]

Table F-11 — Vent Line Diameters

Maximum flow GPM (L)	Pipe length[1]		
	50 feet (15.2 m)	100 feet (30.4 m)	200 feet (60.8 m)
	Inches (cm)	Inches (cm)	Inches (cm)
100 (378.5)	1 1/4 (3.175)	1 1/4 (3.175)	1 1/4 (3.175)
200 (757)	1 1/4 (3.175)	1 1/4 (3.175)	1 1/4 (3.175)
300 (1,135.5)	1 1/4 (3.175)	1 1/4 (3.175)	1 1/2 (3.81)
400 (1,514)	1 1/4 (3.175)	1 1/2 (3.81)	2 (5.08)
500 (1,892.5)	1 1/2 (3.81)	1 1/2 (3.81)	2 (5.08)
600 (2,271)	1 1/2 (3.81)	2 (5.08)	2 (5.08)
700 (2,649.5)	2 (5.08)	2 (5.08)	2 (5.08)
800 (3,028)	2 (5.08)	2 (5.08)	3 (7.62)
900 (3,406.5)	2 (5.08)	2 (5.08)	3 (7.62)
1,000 (3,785)	2 (5.08)	2 (5.08)	3 (7.62)

1. Vent lines of 50 ft. (15.2 m), 100 ft. (30.4 m), and 200 ft. (60.8 m) of pipe plus 7 ells.

[C] *Location and arrangement of vents* for Category 3 flammable liquids with a flashpoint at or above 100 °F (37.8 °C) or Category 4 flammable liquids. Vent pipes from tanks storing Category 3 flammable liquids with a flashpoint at or above 100 °F (37.8 °C) or Category 4 flammable liquids shall terminate outside of the building and higher than the fill pipe opening. Vent outlets shall be above normal snow level. They may be fitted with return bends, coarse screens or other devices to minimize ingress of foreign material.[§1926.152(i)(3)(iv)[C]]

[D] *Vent piping shall be constructed* in accordance with paragraph (3)(iv)(C) of this section. Vent pipes shall be so laid as to drain toward the tank without sags or traps in which liquid can collect. They shall be located so that they will not be subjected to physical damage. The tank end of the vent pipe shall enter the tank through the top. [§1926.152(i)(3)(iv)[D]]

[E] *When tank vent piping is manifolded,* pipe sizes shall be such as to discharge, within the pressure limitations of the system, the vapors they may be required to handle when manifolded tanks are filled simultaneously. [§1926.152(i)(3)(iv)[E]]

(v) *Tank openings other than vents.*[§1926.152(i)(3)(v)]

[A] Connections for all tank openings shall be vapor or liquid tight.[§1926.152(i)(3)(v)[A]]

[B] Openings for manual gaging, if independent of the fill pipe, shall be provided with a liquid-tight cap or cover. If inside a building, each such opening shall be protected against liquid overflow and possible vapor release by means of a spring loaded check valve or other approved device.[§1926.152(i)(3)(v)[B]]

[C] Fill and discharge lines shall enter tanks only through the top. Fill lines shall be sloped toward the tank. [§1926.152(i)(3)(v)[C]]

[D] For Category 2 flammable liquids, or Category 3 flammable liquids with a flashpoint below 100 °F (37.8 °C), other than crude oils, gasolines, and asphalts, the fill pipe shall be so designed and installed as to minimize the possibility of generating static electricity by terminating within 6 inches (15.24 cm) of the bottom of the tank. [§1926.152(i)(3)(v)[D]]

[E] Filling and emptying connections which are made and broken shall be located outside of buildings at a location free from any source of ignition and not less than 5 feet (1.52 m) away from any building opening. Such connection shall be closed and liquidtight when not in use. The connection shall be properly identified.[§1926.152(i)(3)(v)[E]]

(4) *Installation of tanks inside of buildings*[§1926.152(i)(4)]

(i) *Location.* Tanks shall not be permitted inside of buildings except as provided in paragraphs (e), (g), (h), or (i) of this section.[§1926.152(i)(4)(i)]

(ii) *Vents.* Vents for tanks inside of buildings shall be as provided in paragraphs (i)(2) (iv), (v), (vi)(B), and (3)(iv) of this section, except that emergency venting by the use of weak roof seams on tanks shall not be permitted. Vents shall discharge vapors outside the buildings.[§1926.152(i)(4)(ii)]

(iii) *Vent piping.* Vent piping shall be constructed in accordance with paragraph (c) of this section.[§1926.152(i)(4)(iii)]

(iv) *Tank openings other than vents.*[§1926.152(i)(4)(iv)]

[A] Connections for all tank openings shall be vapor or liquid-tight. Vents are covered in paragraph (i)(4)(ii) of this section.[§1926.152(i)(4)(iv)[A]]

[B] Each connection to a tank inside of buildings through which liquid can normally flow shall be provided with an internal or an external valve located as close as practical to the shell of the tank. Such valves, when external, and their connections to the tank shall be of steel except when the chemical characteristics of the liquid stored are incompatible with steel. When materials other than steel are necessary, they shall be suitable for the pressures, structural stresses, and temperatures involved, including fire exposures.[§1926.152(i)(4)(iv)[B]]

[C] Flammable liquid tanks located inside of buildings, except in one-story buildings designed and protected for flammable liquid storage, shall be provided with an automatic-closing heat-actuated valve on each withdrawal connection below the liquid level, except for connections used for emergency disposal, to prevent continued flow in the event of fire in the vicinity of the tank. This function may be incorporated in the valve required in paragraph (i)(4)(iv)(B) of this section, and if a separate valve, shall be located adjacent to the valve required in paragraph (i)(4)(iv)(B) of this section.[§1926.152(i)(4)(iv)[C]]

[D] Openings for manual gaging, if independent of the fill pipe (see paragraph (i)(4)(iv)(F) of this section), shall be provided with a vaportight cap or cover. Each such opening shall be protected against liquid overflow and possible vapor release by means of a spring loaded check valve or other approved device.[§1926.152(i)(4)(iv)[D]]

[E] For Category 2 flammable liquids, or Category 3 flammable liquids with a flashpoint below 100 °F (37.8 °C), other than crude oils, gasolines, and asphalts, the fill pipe shall be so designed and installed as to minimize the possibility of generating static electricity by terminating within 6 inches (15.24 cm) of the bottom of the tank. [§1926.152(i)(4)(iv)[E]]

[F] The fill pipe inside of the tank shall be installed to avoid excessive vibration of the pipe.[§1926.152(i)(4)(iv)[F]]

[G] The inlet of the fill pipe shall be located outside of buildings at a location free from any source of ignition and not less than 5 feet (1.52 m) away from any building opening. The inlet of the fill pipe shall be closed and liquidtight when not in use. The fill connection shall be properly identified.[§1926.152(i)(4)(iv)[G]]

[H] Tanks inside buildings shall be equipped with a device, or other means shall be provided, to prevent overflow into the building.[§1926.152(i)(4)(iv)[H]]

(5) *Supports,* foundations, and anchorage for all tank locations [§1926.152(i)(5)]

(i) *General.* Tank supports shall be installed on firm foundations. Tank supports shall be of concrete, masonry, or protected steel. Single wood timber supports (not cribbing) laid horizontally may be used for outside aboveground tanks if not more than 12 inches (30.48 cm) high at their lowest point. [§1926.152(i)(5)(i)]

(ii) *Fire resistance.* Steel supports or exposed piling shall be protected by materials having a fire resistance rating of not less than 2 hours, except that steel saddles need not be protected if less than 12 inches (30.48 cm) high at their lowest point. Water spray protection or its equivalent may be used in lieu of fire-resistive materials to protect supports.[§1926.152(i)(5)(ii)]

(iii) *Spheres.* The design of the supporting structure for tanks such as spheres shall receive special engineering consideration.[§1926.152(i)(5)(iii)]

(iv) *Load distribution.* Every tank shall be so supported as to prevent the excessive concentration of loads on the supporting portion of the shell.[§1926.152(i)(5)(iv)]

(v) *Foundations.* Tanks shall rest on the ground or on foundations made of concrete, masonry, piling, or steel. Tank foundations shall be designed to minimize the possibility of uneven settling of the tank and to minimize corrosion in any part of the tank resting on the foundation.[§1926.152(i)(5)(v)]

(vi) *Flood areas.* Where a tank is located in an area that may be subjected to flooding, the applicable precautions outlined in this subdivision shall be observed.[§1926.152(i)(5)(vi)]

[A] No aboveground vertical storage tank containing a flammable liquid shall be located so that the allowable liquid level within the tank is below the established maximum flood stage, unless the tank is provided with a guiding structure such as described in paragraphs (i)(5)(vi) (M), (N), and (O) of this section.[§1926.152(i)(5)(vi)[A]]

[B] Independent water supply facilities shall be provided at locations where there is no ample and dependable public water supply available for loading partially empty tanks with water.[§1926.152(i)(5)(vi)[B]]

[C] In addition to the preceding requirements, each tank so located that more than 70 percent, but less than 100 percent, of its allowable liquid storage capacity will be submerged at the established maximum flood stage, shall be safeguarded by one of the following methods: Tank shall be raised, or its height shall be increased, until its top extends above the maximum flood stage a distance equivalent to 30 percent or more of its allowable liquid storage capacity: Provided, however, That the submerged part of the tank shall not exceed two and one-half times the diameter. Or, as an alternative to the foregoing, adequate noncombustible structural guides, designed to permit the tank to float vertically without loss of product, shall be provided.[§1926.152(i)(5)(vi)[C]]

[D] Each horizontal tank so located that more than 70 percent of its storage capacity will be submerged at the established flood stage, shall be anchored, attached to a foundation of concrete or of steel and concrete, of sufficient weight to provide adequate load for the tank when filled with flammable liquid and submerged by flood waters to the established flood stage, or adequately secured by other means.[§1926.152(i)(5)(vi)[D]]

[E] [Reserved][§1926.152(i)(5)(vi)[E]]

[F] At locations where there is no ample and dependable water supply, or where filling of underground tanks with liquids is impracticable because of the character of their contents, their use, or for other reasons, each tank shall be safeguarded against movement when empty and submerged by high ground water or flood waters by anchoring, weighting with concrete or other approved solid loading material, or securing by other means. Each such tank shall be so constructed and installed that it will safely resist external pressures due to high ground water or flood waters.[§1926.152(i)(5)(vi)[F]]

[G] At locations where there is an ample and dependable water supply available, underground tanks containing flammable liquids, so installed that more than 70 percent of their storage capacity will be submerged at the maximum flood stage, shall be so anchored, weighted, or secured by other means, as to prevent movement of such tanks when filled with flammable liquids, and submerged by flood waters to the established flood stage. [§1926.152(i)(5)(vi)[G]]

[H] Pipe connections below the allowable liquid level in a tank shall be provided with valves or cocks located as closely as practicable to the tank shell. Such valves and their connections to tanks shall be of steel or other material suitable for use with the liquid being stored. Cast iron shall not be permitted.[§1926.152(i)(5)(vi)[H]]

[I] At locations where an independent water supply is required, it shall be entirely independent of public power and water supply. Independent source of water shall be available when flood waters reach a level not less than 10 feet (3.04 m) below the bottom of the lowest tank on a property.[§1926.152(i)(5)(vi)[I]]

[J] The self-contained power and pumping unit shall be so located or so designed that pumping into tanks may be carried on continuously throughout the rise in flood waters from a level 10 feet (3.04 m) below the lowest tank to the level of the potential flood stage.[§1926.152(i)(5)(vi)[J]]

[K] Capacity of the pumping unit shall be such that the rate of rise of water in all tanks shall be equivalent to the established potential average rate of rise of flood waters at any stage.[§1926.152(i)(5)(vi)[K]]

[L] Each independent pumping unit shall be tested periodically to insure that it is in satisfactory operating condition. [§1926.152(i)(5)(vi)[L]]

[M] Structural guides for holding floating tanks above their foundations shall be so designed that there will be no resistance to the free rise of a tank, and shall be constructed of noncombustible material.[§1926.152(i)(5)(vi)[M]]

[N] The strength of the structure shall be adequate to resist lateral movement of a tank subject to a horizontal force in any direction equivalent to not less than 25 pounds per square foot (1.05 kg m^2) acting on the projected vertical cross-sectional area of the tank.[§1926.152(i)(5)(vi)[N]]

[O] Where tanks are situated on exposed points or bends in a shoreline where swift currents in flood waters will be present, the structures shall be designed to withstand a unit force of not less than 50 pounds per square foot (2.1 kg m^2).[§1926.152(i)(5)(vi)[O]]

[P] The filling of a tank to be protected by water loading shall be started as soon as flood waters reach a dangerous flood stage. The rate of filling shall be at least equal to the rate of rise of the floodwaters (or the established average potential rate of rise).[§1926.152(i)(5)(vi)[P]]

[Q] Sufficient fuel to operate the water pumps shall be available at all times to insure adequate power to fill all tankage with water.[§1926.152(i)(5)(vi)[Q]]

[R] All valves on connecting pipelines shall be closed and locked in closed position when water loading has been completed.[§1926.152(i)(5)(vi)[R]]

[S] Where structural guides are provided for the protection of floating tanks, all rigid connections between tanks and pipelines shall be disconnected and blanked off or blinded before the floodwaters reach the bottom of the tank, unless control valves and their connections to the tank are of a type designed to prevent breakage between the valve and the tank shell.[§1926.152(i)(5)(vi)[S]]

[T] All valves attached to tanks other than those used in connection with water loading operations shall be closed and locked.[§1926.152(i)(5)(vi)[T]]

[U] If a tank is equipped with a swing line, the swing pipe shall be raised to and secured at its highest position. [§1926.152(i)(5)(vi)[U]]

[V] Inspections. The Assistant Secretary or his designated representative shall make periodic inspections of all plants where the storage of flammable liquids is such as to require compliance with the foregoing requirements, in order to assure the following:[§1926.152(i)(5)(vi)[V]]

[1] That all flammable liquid storage tanks are in compliance with these requirements and so maintained. [§1926.152(i)(5)(vi)[V][1]]

[2] That detailed printed instructions of what to do in flood emergencies are properly posted.[§1926.152(i)(5)(vi)[V][2]]

[3] That station operators and other employees depended upon to carry out such instructions are thoroughly informed as to the location and operation of such valves and other equipment necessary to effect these requirements.[§1926.152(i)(5)(vi)[V][3]]

(vii) *Earthquake areas.* In areas subject to earthquakes, the tank supports and connections shall be designed to resist damage as a result of such shocks.[§1926.152(i)(5)(vii)]

(6) *Sources of ignition.* In locations where flammable vapors may be present, precautions shall be taken to prevent ignition by eliminating or controlling sources of ignition. Sources of ignition may include open flames, lightning, smoking, cutting and welding, hot surfaces, frictional heat, sparks (static, electrical, and mechanical), spontaneous ignition, chemical and physical-chemical reactions, and radiant heat.[§1926.152(i)(6)]

(7) *Testing*[§1926.152(i)(7)]

(i) *General.* All tanks, whether shop built or field erected, shall be strength tested before they are placed in service in accordance with the applicable paragraphs of the code under which they were built. The American Society of Mechanical Engineers (ASME) code stamp, American Petroleum Institute (API) monogram, or the label of the Underwriters' Laboratories, Inc., on a tank shall be evidence of compliance with this strength test. Tanks not marked in accordance with the above codes shall be strength tested before they are placed in service in accordance with good engineering principles and reference shall be made to the sections on testing in the codes listed in paragraphs (i)(1) (iii)(A), (iv)(B), or (v)(B) of this section.[§1926.152(i)(7)(i)]

(ii) *Strength.* When the vertical length of the fill and vent pipes is such that when filled with liquid the static head imposed upon the bottom of the tank exceeds 10 pounds per square inch (68.94 kPa), the tank and related piping shall be tested hydrostatically to a pressure equal to the static head thus imposed.[§1926.152(i)(7)(ii)]

(iii) *Tightness.* In addition to the strength test called for in paragraphs (i)(7) (i) and (ii) of this section, all tanks and connections shall be tested for tightness. Except for underground tanks, this tightness test shall be made at operating pressure with air, inert gas, or water prior to placing the tank in service. In the case of field-erected tanks the strength test may be considered to be the test for tank tightness. Underground tanks and piping, before being covered, enclosed, or placed in use, shall be tested for tightness hydrostatically, or with air pressure at not less than 3 pounds per square inch (20.68 kPa) and not more than 5 pounds per square inch (34.47 kPa).[§1926.152(i)(7)(iii)]

(iv) *Repairs.* All leaks or deformations shall be corrected in an acceptable manner before the tank is placed in service. Mechanical caulking is not permitted for correcting leaks in welded tanks except pinhole leaks in the roof. [§1926.152(i)(7)(iv)]

(v) *Derated operations.* Tanks to be operated at pressures below their design pressure may be tested by the applicable provisions of paragraphs (i)(7) (i) or (ii) of this section, based upon the pressure developed under full emergency venting of the tank.[§1926.152(i)(7)(v)]

(j) Piping, valves, and fittings[§1926.152(j)]

(1) *General*[§1926.152(j)(1)]

(i) *Design.* The design (including selection of materials) fabrication, assembly, test, and inspection of piping systems containing flammable liquids shall be suitable for the expected working pressures and structural stresses. Conformity with the applicable provisions of Pressure Piping, ANSI B31 series and the provisions of this paragraph, shall be considered prima facie evidence of compliance with the foregoing provisions.[§1926.152(j)(1)(i)]

(ii) *Exceptions.* This paragraph does not apply to any of the following:[§1926.152(j)(1)(ii)]

[A] Tubing or casing on any oil or gas wells and any piping connected directly thereto.[§1926.152(j)(1)(ii)[A]]

[B] Motor vehicle, aircraft, boat, or portable or stationary engines.[§1926.152(j)(1)(ii)[B]]

[C] *Piping within the scope* of any applicable boiler and pressures vessel code.[§1926.152(j)(1)(ii)[C]]

(iii) *Definitions.* As used in this paragraph, piping systems consist of pipe, tubing, flanges, bolting, gaskets, valves, fittings, the pressure containing parts of other components such as expansion joints and strainers, and devices which serve such purposes as mixing, separating, snubbing, distributing, metering, or controlling flow.[§1926.152(j)(1)(iii)]

(2) *Materials for piping,* valves, and fittings — [§1926.152(j)(2)]

(i) *Required materials.* Materials for piping, valves, or fittings shall be steel, nodular iron, or malleable iron, except as provided in paragraphs (j)(2) (ii), (iii) and (iv) of this section. [§1926.152(j)(2)(i)]

(ii) *Exceptions.* Materials other than steel, nodular iron, or malleable iron may be used underground, or if required by the properties of the flammable liquid handled. Material other than steel, nodular iron, or malleable iron shall be designed to specifications embodying principles recognized as good engineering practices for the material used.[§1926.152(j)(2)(ii)]

(iii) *Linings.* Piping, valves, and fittings may have combustible or noncombustible linings.[§1926.152(j)(2)(iii)]

(iv) *Low-melting materials.* When low-melting point materials such as aluminum and brass or materials that soften on fire exposure such as plastics, or non-ductile materials such as cast iron, are necessary, special consideration shall be given to their behavior on fire exposure. If such materials are used in above ground piping systems or inside buildings, they shall be suitably protected against fire exposure or so located that any spill resulting from the failure of these materials could not unduly expose persons, important buildings or structures or can be readily controlled by remote valves.[§1926.152(j)(2)(iv)]

(3) *Pipe joints.* Joints shall be made liquid tight. Welded or screwed joints or approved connectors shall be used. Threaded joints and connections shall be made up tight with a suitable lubricant or piping compound. Pipe joints dependent upon the friction characteristics of combustible materials for mechanical continuity of piping shall not be used inside buildings. They may be used outside of buildings above or below ground. If used above ground, the piping shall either be secured to prevent disengagement at the fitting or the piping system shall be so designed that any spill resulting from such disengagement could not unduly expose persons, important buildings or structures, and could be readily controlled by remote valves.[§1926.152(j)(3)]

(4) *Supports.* Piping systems shall be substantially supported and protected against physical damage and excessive stresses arising from settlement, vibration, expansion, or contraction. [§1926.152(j)(4)]

(5) *Protection against corrosion.* All piping for flammable liquids, both aboveground and underground, where subject to external corrosion, shall be painted or otherwise protected. [§1926.152(j)(5)]

(6) *Valves.* Piping systems shall contain a sufficient number of valves to operate the system properly and to protect the plant. Piping systems in connection with pumps shall contain a sufficient number of valves to control properly the flow of liquid in normal operation and in the event of physical damage. Each connection to pipelines, by which equipments such as tankcars or tank vehicles discharge liquids by means of pumps into storage tanks, shall be provided with a check valve for automatic protection against backflow if the piping arrangement is such that backflow from the system is possible.[§1926.152(j)(6)]

(7) *Testing.* All piping before being covered, enclosed, or placed in use shall be hydrostatically tested to 150 percent of the maximum anticipated pressure of the system, or pneumatically tested to 110 percent of the maximum anticipated pressure of the system, but not less than 5 pounds per square inch gage at the highest point of the system. This test shall be maintained for a sufficient time to complete visual inspection of all joints and connections, but for at least 10 minutes. [§1926.152(j)(7)]

(k) Marine service stations[§1926.152(k)]

(1) *Dispensing.*[§1926.152(k)(1)]

(i) *The dispensing area shall be located* away from other structures so as to provide room for safe ingress and egress of craft to be fueled. Dispensing units shall in all cases be at least 20 feet (6.08 m) from any activity involving fixed sources of ignition.[§1926.152(k)(1)(i)]

(ii) *Dispensing shall be by approved dispensing* units with or without integral pumps and may be located on open piers, wharves, or floating docks or on shore or on piers of the solid fill type.[§1926.152(k)(1)(ii)]

(iii) *Dispensing nozzles* shall be automatic-closing without a hold- open latch.[§1926.152(k)(1)(iii)]

(2) *Tanks and pumps.*[§1926.152(k)(2)]

(i) *Tanks,* and pumps not integral with the dispensing unit, shall be on shore or on a pier of the solid fill type, except as provided in paragraphs (k)(2) (ii) and (iii) of this section. [§1926.152(k)(2)(i)]

(ii) *Where shore location would* require excessively long supply lines to dispensers, tanks may be installed on a pier provided that applicable portions of paragraph (b) of this section relative to spacing, diking, and piping are complied with and the quantity so stored does not exceed 1,100 gallons (4,163.5 L) aggregate capacity.[§1926.152(k)(2)(ii)]

(iii) *Shore tanks supplying* marine service stations may be located above ground, where rock ledges or high water table make underground tanks impractical.[§1926.152(k)(2)(iii)]

(iv) *Where tanks are at an elevation* which would produce gravity head on the dispensing unit, the tank outlet shall be equipped with a pressure control valve positioned adjacent to and outside the tank block valve specified in §1926.152(c)(8) of this section, so adjusted that liquid cannot flow by gravity from the tank in case of piping or hose failure. [§1926.152(k)(2)(iv)]

(3) *Piping.*[§1926.152(k)(3)]

(i) *Piping between shore tanks* and dispensing units shall be as described in paragraph (k)(2)(iii) of this section, except that, where dispensing is from a floating structure, suitable lengths of oil-resistant flexible hose may be employed between the shore piping and the piping on the floating structure as made necessary by change in water level or shoreline. [§1926.152(k)(3)(i)]

Table F-19 — Electrical Equipment Hazardous Areas — Service Stations

Location	Class I Group D division	Extent of classified area
Underground tank:		
Fill opening	1	Any pit, box, or space below grade level, any part of which is within the Division 1 or 2 classified area.
	2	Up to 18 inches (45.72 cm) above grade level within a horizontal radius of 10 feet (3.04 m) from a loose fill connection and within a horizontal radius of 5 feet (1.52 m) from a tight fill connection.
Vent - Discharging upward	1	Within 3 feet (0.912 m) of open end of vent, extending in all directions.
	2	Area between 3 feet (0.912 m) and 5 feet (1.52 m) of open end of vent, extending in all directions.
Dispenser:		
Pits	1	Any pit, box, or space below grade level, any part of which is within the Division 1 or 2 classified area.
Dispenser enclosure	1	The area 4 feet (1.216 m) vertically above base within the enclosure and 18 inches (45.72 cm) horizontally in all directions.
Outdoor	2	Up to 18 inches (45.72 cm) above grade level within 20 feet (6.08 m) horizontally of any edge of enclosure.
Indoor:		
With mechanical ventilation	2	Up to 18 inches (45.72 cm) above grade or floor level within 20 feet (6.08 m) horizontally of any edge of enclosure.
With gravity ventilation	2	Up to 18 inches (45.72 cm) above grade or floor level within 25 feet (7.6 m) horizontally of any edge of enclosure.

Table F-19 — Electrical Equipment Hazardous Areas — Service Stations (continued)

Location	Class I Group D division	Extent of classified area
Remote pump -Outdoor	1	Any pit, box, or space below grade level if any part is within a horizontal distance of 10 feet (3.04 m) from any edge of pump.
	2	Within 3 feet (0.912 m) of any edge of pump, extending in all directions. Also up to 18 inches (45.72 cm) above grade level within 10 feet (3.04 m) horizontally from any edge of pump.
Remote pump-Indoor	1	Entire area within any pit.
	2	Within 5 feet (1.52 m) of any edge of pump, extending in all directions. Also up to 3 feet (3.04 m) above floor or grade level within 25 feet (6.08 m) horizontally from any edge of pump.
Lubrication or service room	1	Entire area within any pit.
	2	Area up to 18 inches (45.72 cm) above floor or grade level within entire lubrication room.
Dispenser for Class I liquids.	2	Within 3 feet (0.912 m) of any fill or dispensing point, extending in all directions.
Special enclosure inside building per §1910.106(f)(1)(ii)	1	Entire enclosure.
Sales, storage and rest rooms	([1])	If there is any opening to these rooms within the extent of a Division 1 area, the entire room shall be classified as Division 1.

[1] Ordinary.

(ii) *A readily accessible valve* to shut off the supply from shore shall be provided in each pipeline at or near the approach to the pier and at the shore end of each pipeline adjacent to the point where flexible hose is attached.[§1926.152(k)(3)(ii)]

(iii) *Piping shall be located so as to be protected* from physical damage.[§1926.152(k)(3)(iii)]

(iv) *Piping handling Category* 1 or 2 flammable liquids, or Category 3 flammable liquids with a flashpoint below 100 °F (37.8 °C), shall be grounded to control stray currents. [§1926.152(k)(3)(iv)]

(4) *Definition; as used in this section:*[§1926.152(k)(4)]

Marine service station shall mean that portion of a property where flammable liquids used as fuels are stored and dispensed from fixed equipment on shore, piers, wharves, or floating docks into the fuel tanks of self- propelled craft, and shall include all facilities used in connection therewith.[§1926.152(k)(4)]

[44 FR 8577, Feb. 9, 1979; 44 FR 20940, Apr. 6, 1979, as amended at 51 FR 25318, July 11, 1986; 58 FR 35162, June 30, 1993; 63 FR 33469, June 18, 1998; 77 FR 17891, Mar. 26, 2012]

§1926.153
Liquefied petroleum gas (LP-Gas)

(a) Approval of equipment and systems.[§1926.153(a)]

(1) *Each system shall have* containers, valves, connectors, manifold valve assemblies, and regulators of an approved type. [§1926.153(a)(1)]

(2) *All cylinders shall meet* the Department of Transportation specification identification requirements published in 49 CFR Part 178, Shipping Container Specifications.[§1926.153(a)(2)]

(3) *Definition.* As used in this section, Containers — All vessels, such as tanks, cylinders, or drums, used for transportation or storing liquefied petroleum gases.[§1926.153(a)(3)]

(b) Welding on LP-Gas containers. Welding is prohibited on containers.[§1926.153(b)]

(c) Container valves and container accessories.[§1926.153(c)]

(1) *Valves, fittings, and accessories connected directly* to the container, including primary shut off valves, shall have a rated working pressure of at least 250 p.s.i.g. and shall be of material and design suitable for LP-Gas service.[§1926.153(c)(1)]

(2) *Connections to containers, except safety relief connections,* liquid level gauging devices, and plugged openings, shall have shutoff valves located as close to the container as practicable. [§1926.153(c)(2)]

(d) Safety devices.[§1926.153(d)]

(1) *Every container and every vaporizer* shall be provided with one or more approved safety relief valves or devices. These valves shall be arranged to afford free vent to the outer air with discharge not less than 5 feet horizontally away from any opening into a building which is below such discharge. [§1926.153(d)(1)]

(2) *Shutoff valves shall not be installed* between the safety relief device and the container, or the equipment or piping to which the safety relief device is connected, except that a shutoff valve may be used where the arrangement of this valve is such that full required capacity flow through the safety relief device is always afforded.[§1926.153(d)(2)]

(3) *Container safety relief devices* and regulator relief vents shall be located not less than 5 feet in any direction from air openings into sealed combustion system appliances or mechanical ventilation air intakes.[§1926.153(d)(3)]

(e) Dispensing.[§1926.153(e)]

(1) *Filling of fuel containers for trucks or motor vehicles* from bulk storage containers shall be performed not less than 10 feet from the nearest masonry-walled building, or not less than 25 feet from the nearest building or other construction and, in any event, not less than 25 feet from any building opening. [§1926.153(e)(1)]

(2) *Filling of portable containers* or containers mounted on skids from storage containers shall be performed not less than 50 feet from the nearest building.[§1926.153(e)(2)]

(f) Requirements for appliances.[§1926.153(f)]

(1) *LP-Gas consuming appliances* shall be approved types. [§1926.153(f)(1)]

(2) *Any appliance that was originally manufactured* for operation with a gaseous fuel other than LP-Gas, and is in good condition, may be used with LP-Gas only after it is properly converted, adapted, and tested for performance with LP-Gas before the appliance is placed in use.[§1926.153(f)(2)]

(g) Containers and regulating equipment installed outside of buildings or structures. Containers shall be upright upon firm foundations or otherwise firmly secured. The possible effect on the outlet piping of settling shall be guarded against by a flexible connection or special fitting.[§1926.153(g)]

(h) Containers and equipment used inside of buildings or structures.[§1926.153(h)]

(1) *When operational requirements* make portable use of containers necessary, and their location outside of buildings or structures is impracticable, containers and equipment shall be permitted to be used inside of buildings or structures in accordance with paragraphs (h)(2) through (11) of this section. [§1926.153(h)(1)]

(2) **Containers in use** means connected for use.[§1926.153(h)(2)]

(3) *Systems utilizing containers* having a water capacity greater than 2 1/2 pounds (nominal 1 pound LP-Gas capacity) shall be equipped with excess flow valves. Such excess flow valves shall be either integral with the container valves or in the connections to the container valve outlets.[§1926.153(h)(3)]

(4) *Regulators shall be either directly connected* to the container valves or to manifolds connected to the container valves. The regulator shall be suitable for use with LP-Gas. Manifolds and fittings connecting containers to pressure regulator inlets shall be designed for at least 250 p.s.i.g. service pressure. [§1926.153(h)(4)]

(5) *Valves on containers having water capacity* greater than 50 pounds (nominal 20 pounds LP-Gas capacity) shall be protected from damage while in use or storage.[§1926.153(h)(5)]

(6) *Aluminum piping or tubing shall not be used.*[§1926.153(h)(6)]

(7) *Hose shall be designed for a working pressure* of at least 250 p.s.i.g. Design, construction, and performance of hose, and hose connections shall have their suitability determined by listing by a nationally recognized testing agency. The hose length shall be as short as practicable. Hoses shall be long enough to permit compliance with spacing provisions of paragraphs (h)(1) through (13) of this section, without kinking or straining, or causing hose to be so close to a burner as to be damaged by heat.[§1926.153(h)(7)]

(8) *Portable heaters, including salamanders,* shall be equipped with an approved automatic device to shut off the flow of gas to the main burner, and pilot if used, in the event of flame failure. Such heaters, having inputs above 50,000 B.t.u. per hour, shall be equipped with either a pilot, which must be lighted and proved before the main burner can be turned on, or an electrical ignition system.[§1926.153(h)(8)]

Note: The provisions of this subparagraph do not apply to portable heaters under 7,500 B.t.u. per hour input when used with containers having a maximum water capacity of 2 1/2 pounds.

(9) *Container valves, connectors,* regulators, manifolds, piping, and tubing shall not be used as structural supports for heaters.[§1926.153(h)(9)]

(10) *Containers,* regulating equipment, manifolds, pipe, tubing, and hose shall be located to minimize exposure to high temperatures or physical damage.[§1926.153(h)(10)]

(11) *Containers having a water capacity* greater than 2 1/2 pounds (nominal 1 pound LP-Gas capacity) connected for use shall stand on a firm and substantially level surface and, when necessary, shall be secured in an upright position.[§1926.153(h)(11)]

(12) *The maximum water capacity* of individual containers shall be 245 pounds (nominal 100 pounds LP-Gas capacity).[§1926.153(h)(12)]

(13) *For temporary heating,* heaters (other than integral heater-container units) shall be located at least 6 feet from any LP-Gas container. This shall not prohibit the use of heaters specifically designed for attachment to the container or to a supporting standard, provided they are designed and installed so as to prevent direct or radiant heat application from the heater onto the containers. Blower and radiant type heaters shall not be directed toward any LP-Gas container within 20 feet.[§1926.153(h)(13)]

(14) *If two or more heater-container units,* of either the integral or nonintegral type, are located in an unpartitioned area on the same floor, the container or containers of each unit shall be separated from the container or containers of any other unit by at least 20 feet.[§1926.153(h)(14)]

(15) *When heaters are connected to containers* for use in an unpartitioned area on the same floor, the total water capacity of containers, manifolded together for connection to a heater or heaters, shall not be greater than 735 pounds (nominal 300 pounds LP-Gas capacity). Such manifolds shall be separated by at least 20 feet.[§1926.153(h)(15)]

(16) *Storage of containers awaiting use* shall be in accordance with paragraphs (j) and (k) of this section.[§1926.153(h)(16)]

(i) Multiple container systems.[§1926.153(i)]

(1) *Valves in the assembly of multiple container systems* shall be arranged so that replacement of containers can be made without shutting off the flow of gas in the system. This provision is not to be construed as requiring an automatic changeover device.[§1926.153(i)(1)]

(2) *Heaters shall be equipped with an approved* regulator in the supply line between the fuel cylinder and the heater unit. Cylinder connectors shall be provided with an excess flow valve to minimize the flow of gas in the event the fuel line becomes ruptured.[§1926.153(i)(2)]

(3) *Regulators and low-pressure relief devices* shall be rigidly attached to the cylinder valves, cylinders, supporting standards, the building walls, or otherwise rigidly secured, and shall be so installed or protected from the elements.[§1926.153(i)(3)]

(j) Storage of LPG containers. Storage of LPG within buildings is prohibited.[§1926.153(j)]

(k) Storage outside of buildings.[§1926.153(k)]

(1) *Storage outside of buildings,* for containers awaiting use, shall be located from the nearest building or group of buildings, in accordance with the following:[§1926.153(k)(1)]

Table F-3

Quantity of LP-Gas stored	Distance (feet)
500 lbs. or less	0
501 to 6,000 lbs	10
6,001 to 10,000 lbs	20
Over 10,000 lbs	25

(2) *Containers shall be* in a suitable ventilated enclosure or otherwise protected against tampering.[§1926.153(k)(2)]

(l) Fire protection. Storage locations shall be provided with at least one approved portable fire extinguisher having a rating of not less than 20-B:C.[§1926.153(l)]

(m) Systems utilizing containers other than DOT containers[§1926.153(m)]

(1) *Application.* This paragraph applies specifically to systems utilizing storage containers other than those constructed in accordance with DOT specifications. Paragraph (b) of this section applies to this paragraph unless otherwise noted in paragraph (b) of this section.[§1926.153(m)(1)]

(2) *Design pressure and classification* of storage containers. Storage containers shall be designed and classified in accordance with Table F-31.[§1926.153(m)(2)]

Table F-31

Container type	For gases with vapor press. Not to exceed lb. per sq. in. gage at 100 °F. (37.8 °C.)	Minimum design pressure of container, lb. per sq. in. gage	
		1949 and earlier editions of ASME Code (Par. U-68, U-69)	1949 edition of ASME Code (Par. U-200, U-201); 1950, 1952, 1956, 1959, 1962, 1965, and 1968 (Division 1) editions of ASME Code; All editions of API-ASME Code[3]
[1] 80	[1] 80	[1] 80	[1] 100
100	100	100	125
125	125	125	156
150	150	150	187
175	175	175	219
[2] 200	215	200	250

[1] New storage containers of the 80 type have not been authorized since Dec. 31, 1947.

[2] Container type may be increased by increments of 25. The minimum design pressure of containers shall be 100% of the container type designation when constructed under 1949 or earlier editions of the ASME Code (Par. U-68 and U-69). The minimum design pressure of containers shall be 125% of the container type designation when constructed under: (1) the 1949 ASME Code (Par. U-200 and U-201), (2) 1950, 1952, 1956, 1959, 1962, 1965, and 1968 (Division 1) editions of the ASME Code, and (3) all editions of the API-ASME Code.

[3] Construction of containers under the API-ASME Code is not authorized after July 1, 1961.

(3) *Containers with foundations attached* (portable or semiportable b containers with suitable steel "runners" or "skids" and popularly known in the industry as "skid tanks") shall be designed, installed, and used in accordance with these rules subject to the following provisions:[§1926.153(m)(3)]

(i) *If they are to be used at a given general location* for a temporary period not to exceed 6 months they need not have fire- resisting foundations or saddles but shall have adequate ferrous metal supports.[§1926.153(m)(3)(i)]

(ii) *They shall not be located* with the outside bottom of the container shell more than 5 feet (1.52 m) above the surface of the ground unless fire-resisting supports are provided.[§1926.153(m)(3)(ii)]

(iii) *The bottom of the skids shall not be less* than 2 inches (5.08 cm) or more than 12 inches (30.48 cm) below the outside bottom of the container shell.[§1926.153(m)(3)(iii)]

(iv) *Flanges, nozzles, valves, fittings,* and the like, having communication with the interior of the container, shall be protected against physical damage.[§1926.153(m)(3)(iv)]

(v) *When not permanently located on fire-resisting foundations,* piping connections shall be sufficiently flexible to minimize the possibility of breakage or leakage of connections if the container settles, moves, or is otherwise displaced.[§1926.153(m)(3)(v)]

(vi) *Skids, or lugs for attachment of skids,* shall be secured to the container in accordance with the code or rules under which the container is designed and built (with a minimum factor of safety of four) to withstand loading in any direction equal to four times the weight of the container and attachments when filled to the maximum permissible loaded weight.[§1926.153(m)(3)(vi)]

(4) *Field welding where necessary* shall be made only on saddle plates or brackets which were applied by the manufacturer of the tank.[§1926.153(m)(4)]

(n) **When LP-Gas and one or more other gases** are stored or used in the same area, the containers shall be marked to identify their content. Marking shall be in compliance with American National Standard Z48.1- 1954, "Method of Marking Portable Compressed Gas Containers To Identify the Material Contained."[§1926.153(n)]

(o) **Damage from vehicles.** When damage to LP-Gas systems from vehicular traffic is a possibility, precautions against such damage shall be taken.[§1926.153(o)]

[44 FR 8577, Feb. 9, 1979; 44 FR 20940, Apr. 6, 1979, as amended at 58 FR 35170, June 30, 1993]

§1926.154

Temporary heating devices

(a) **Ventilation.**[§1926.154(a)]

(1) *Fresh air shall be supplied* in sufficient quantities to maintain the health and safety of workmen. Where natural means of fresh air supply is inadequate, mechanical ventilation shall be provided.[§1926.154(a)(1)]

(2) *When heaters are used in confined spaces,* special care shall be taken to provide sufficient ventilation in order to ensure proper combustion, maintain the health and safety of workmen, and limit temperature rise in the area.[§1926.154(a)(2)]

(b) **Clearance and mounting.**[§1926.154(b)]

(1) *Temporary heating devices* shall be installed to provide clearance to combustible material not less than the amount shown in Table F-4.[§1926.154(b)(1)]

(2) *Temporary heating devices,* which are listed for installation with lesser clearances than specified in Table F-4, may be installed in accordance with their approval.[§1926.154(b)(2)]

Table F-4

Heating appliances	Minimum clearance, (inches)		
	Sides	Rear	Chimney connector
Room heater, circulating type	12	12	18
Room heater, radiant type	36	36	18

(3) *Heaters not suitable for use* on wood floors shall not be set directly upon them or other combustible materials. When such heaters are used, they shall rest on suitable heat insulating material or at least 1-inch concrete, or equivalent. The insulating material shall extend beyond the heater 2 feet or more in all directions.[§1926.154(b)(3)]

(4) *Heaters used in the vicinity* of combustible tarpaulins, canvas, or similar coverings shall be located at least 10 feet from the coverings. The coverings shall be securely fastened to prevent ignition or upsetting of the heater due to wind action on the covering or other material.[§1926.154(b)(4)]

(c) **Stability.** Heaters, when in use, shall be set horizontally level, unless otherwise permitted by the manufacturer's markings. [§1926.154(c)]

(d) **Solid fuel salamanders.** Solid fuel salamanders are prohibited in buildings and on scaffolds.[§1926.154(d)]

(e) **Oil-fired heaters.**[§1926.154(e)]

(1) *Flammable liquid-fired heaters* shall be equipped with a primary safety control to stop the flow of fuel in the event of flame failure. Barometric or gravity oil feed shall not be considered a primary safety control.[§1926.154(e)(1)]

(2) *Heaters designed for barometric* or gravity oil feed shall be used only with the integral tanks.[§1926.154(e)(2)]

(3) *[Reserved]*[§1926.154(e)(3)]

(4) *Heaters specifically designed and approved* for use with separate supply tanks may be directly connected for gravity feed, or an automatic pump, from a supply tank.[§1926.154(e)(4)]

§1926.155

Definitions applicable to this subpart

(a) **Approved, for the purpose of this subpart,** means equipment that has been listed or approved by a nationally recognized testing laboratory such as Factory Mutual Engineering Corp., or Underwriters' Laboratories, Inc., or Federal agencies such as Bureau of Mines, or U.S. Coast Guard, which issue approvals for such equipment.[§1926.155(a)]

(b) **Closed container** means a container so sealed by means of a lid or other device that neither liquid nor vapor will escape from it at ordinary temperatures.[§1926.155(b)]

(c) **Reserved**[§1926.155(c)]

(d) **Combustion** means any chemical process that involves oxidation sufficient to produce light or heat.[§1926.155(d)]

(e) **Fire brigade** means an organized group of employees that are knowledgeable, trained, and skilled in the safe evacuation of employees during emergency situations and in assisting in fire fighting operations.[§1926.155(e)]

(f) **Fire resistance** means so resistant to fire that, for specified time and under conditions of a standard heat intensity, it will not fail structurally and will not permit the side away from the fire to become hotter than a specified temperature. For purposes of this part, fire resistance shall be determined by the Standard Methods of Fire Tests of Building Construction and Materials, NFPA 251-1969.[§1926.155(f)]

(g) **Flammable** means capable of being easily ignited, burning intensely, or having a rapid rate of flame spread.[§1926.155(g)]

(h) **Flammable liquid** means any liquid having a vapor pressure not exceeding 40 pounds per square inch (absolute) at 100 °F (37.8 °C) and having a flashpoint at or below 199.4 °F (93 °C). Flammable liquids are divided into four categories as follows:[§1926.155(h)]

(1) *Category 1 shall include* liquids having flashpoints below 73.4 °F (23 °C) and having a boiling point at or below 95 °F (35 °C). [§1926.155(h)(1)]

(2) *Category 2 shall include* liquids having flashpoints below 73.4 °F (23 °C) and having a boiling point above 95 °F (35 °C). [§1926.155(h)(2)]

(3) *Category 3 shall include* liquids having flashpoints at or above 73.4 °F (23 °C) and at or below 140 °F (60 °C).[§1926.155(h)(3)]

(4) *Category 4 shall include* liquids having flashpoints above 140 °F (60 °C) and at or below 199.4 °F (93 °C).[§1926.155(h)(4)]

(i) **Flash point of the liquid** means the temperature at which it gives off vapor sufficient to form an ignitable mixture with the air near the surface of the liquid or within the vessel used as determined by appropriate test procedure and apparatus as specified below. [§1926.155(i)]

(1) *The flashpoint of liquids having* a viscosity less than 45 Saybolt Universal Second(s) at 100 °F (37.8 °C) and a flashpoint below 175 °F (79.4 °C) shall be determined in accordance with the Standard Method of Test for Flash Point by the Tag Closed Tester, ASTM D-56-69 (incorporated by reference; *See* Sec. 1926.6), or an equivalent method as defined by Sec. 1910.1200 appendix B.[§1926.155(i)(1)]

(2) *The flashpoints of liquids having* a viscosity of 45 Saybolt Universal Second(s) or more at 175 °F (79.4 °C) or higher shall be determined in accordance with the Standard Method of Test for Flash Point by the Pensky Martens Closed Tester, ASTM D-93-69 (incorporated by reference; *See* Sec. 1926.6), or an equivalent method as defined by Sec. 1910.1200 appendix B.[§1926.155(i)(2)]

(j) **Liquefied petroleum gases, LPG and LP Gas** mean and include any material which is composed predominantly of any of the following hydrocarbons, or mixtures of them, such as propane, propylene, butane (normal butane or iso-butane), and butylenes. [§1926.155(j)]

(k) **Portable tank** means a closed container having a liquid capacity more than 60 U.S. gallons, and not intended for fixed installation. [§1926.155(k)]

(l) **Safety can** means an approved closed container, of not more than 5 gallons capacity, having a flash-arresting screen, spring-closing lid and spout cover and so designed that it will safely relieve internal pressure when subjected to fire exposure.[§1926.155(l)]

(m) **Vapor pressure** means the pressure, measured in pounds per square inch (absolute), exerted by a volatile liquid as determined by the "Standard Method of Test for Vapor Pressure of Petroleum Products (Reid Method)." (ASTM D-323-58).[§1926.155(m)]

[44 FR 8577, Feb. 9, 1979; 44 FR 20940, Apr. 6, 1979, as amended at 77 FR 17894, Mar. 26, 2012]

Authority: Section 107 of the Contract Work Hours and Safety Standards Act (40 U.S.C. 3704); Sections 4, 6, and 8 of the Occupational Safety and Health Act of 1970 (29 U.S.C. 653, 655, 657); Secretary of Labor's Order No. 12-71 (36 FR 8754), 8-76 (41 FR 25059), 9-83 (48 FR 35736),1-90 (55 FR 9033), 6-96 (62 FR 111), 3-2000 (62 FR 50017), 5-2002 (67 FR 650008), 5-2007 (72 FR 31159), 4-2010 (75 FR 55355), or 1-2012 (77 FR 3912), as applicable; and 29 CFR part 1911.

Notes

Subpart G

Signs, Signals, and Barricades

§1926.200

Accident prevention signs and tags

(a) General. Signs and symbols required by this subpart shall be visible at all times when work is being performed, and shall be removed or covered promptly when the hazards no longer exist. [§1926.200(a)]

(b) Danger signs.[§1926.200(b)]

(1) *Danger signs* (see Figure G-1) shall be used only where an immediate hazard exists.[§1926.200(b)(1)]

(2) *Danger signs* shall have red as the predominating color for the upper panel; black outline on the borders; and a white lower panel for additional sign wording.[§1926.200(b)(2)]

(c) Caution signs.[§1926.200(c)]

(1) *Caution signs* (see Figure G-2) shall be used only to warn against potential hazards or to caution against unsafe practices.[§1926.200(c)(1)]

(2) *Caution signs* shall have yellow as the predominating color; black upper panel and borders: yellow lettering of "caution" on the black panel; and the lower yellow panel for additional sign wording. Black lettering shall be used for additional wording. [§1926.200(c)(2)]

FIGURE G-1
Danger

FIGURE G-2
Caution

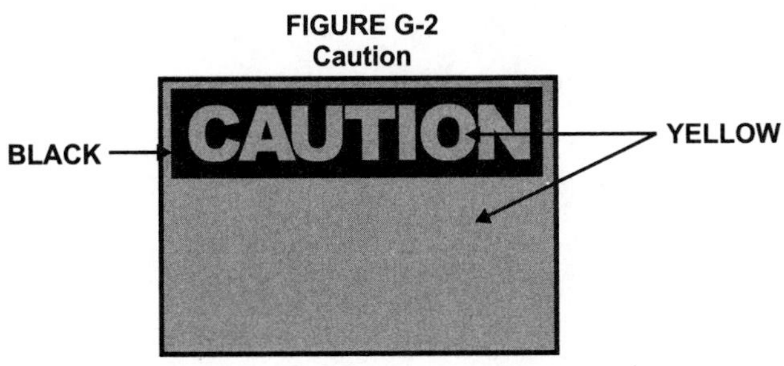

(3) *Standard color of the background* shall be yellow; and the panel, black with yellow letters. Any letters used against the yellow background shall be black. The colors shall be those of opaque glossy samples as specified in Table 1 of American National Standard Z53.1- 1967.[§1926.200(c)(3)]

(d) Exit signs. Exit signs, when required, shall be lettered in legible red letters, not less than 6 inches high, on a white field and the principal stroke of the letters shall be at least three-fourths inch in width.[§1926.200(d)]

(e) Safety instruction signs. Safety instruction signs, when used, shall be white with green upper panel with white letters to convey the principal message. Any additional wording on the sign shall be black letters on the white background.[§1926.200(e)]

(f) Directional signs. Directional signs, other than automotive traffic signs specified in paragraph (g) of this section, shall be white with a black panel and a white directional symbol. Any additional wording on the sign shall be black letters on the white background. [§1926.200(f)]

(g) Traffic signs.[§1926.200(g)]

(1) *Construction areas shall be posted* with legible traffic signs at points of hazard.[§1926.200(g)(1)]

(2) *All traffic control signs or devices* used for protection of construction workers shall conform to Part VI of the Manual of Uniform Traffic Control Devices (AMUTCD"), 1988 Edition, Revision 3, September 3, 1993, FHWA-SA-94-027 or Part VI of the Manual on Uniform Traffic Control Devices, Millennium Edition, December 2000, FHWA, which are incorporated by reference. The Director of the Federal Register approves this incorporation by reference in accordance with 5 U.S.C. 552(a) and 1 CFR part 51.

You may obtain a copy of the Millennium Edition from the following organizations: American Traffic Safety Services Association, 15 Riverside Parkway, Suite 100, Fredericksburg, VA 22406-1022; Telephone: 1-800-231-3475; FAX: (540) 368-1722; www.atssa.com; Institute of Transportation Engineers, 1099 14th Street, NW., Suite 300 West, Washington, DC 20005-3438; FAX: (202) 289-7722; www.ite.org; and American Association of State Highway and Transportation Officials; www.aashto.org; Telephone: 1-800-231-3475; FAX: 1-800-525-5562.

Electronic copies of the MUTCD 2000 are available for downloading at http://mutcd.fhwa.dot.gov/kno-millennium. Electronic copies of the 1988 Edition MUTCD, Revision 3, are available for downloading at http:// www.osha.gov/doc/highway — workzones. Both documents are available for inspection at the OSHA Docket Office, Room N2625, U.S. Department of Labor, 200 Constitution Avenue, NW., Washington, DC 20210 or at the National Archives and Records Administration (NARA).

For information on the availability of this material at NARA, call 202-741-6030, or go to: http://www.archives.gov/federal — register/code — of — federal — regulations/ibr — locations.html. [§1926.200(g)(2)]

(h) Accident prevention tags.[§1926.200(h)]

(1) *Accident prevention tags* shall be used as a temporary means of warning employees of an existing hazard, such as defective tools, equipment, etc. They shall not be used in place of, or as a substitute for, accident prevention signs.[§1926.200(h)(1)]

(2) *Specifications for accident prevention tags* similar to those in Table G-1 shall apply.[§1926.200(h)(2)]

TABLE G-1

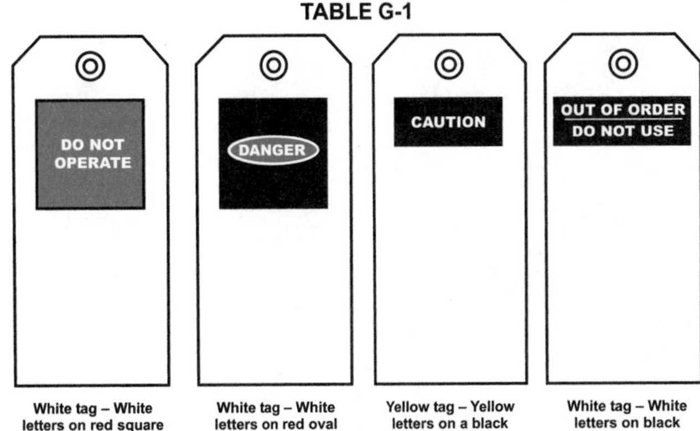

Basic Stock (Background)	Safety Colors (Ink)	Copy Specification (Letters)
White	Red	Do Not Operate
White	Black and Red	Danger
Yellow	Black	Caution
White	Black	Out Of Order Do Not Use

(i) Additional rules. American National Standards Institute (ANSI) Z35.1-1968, Specifications for Accident Prevention Signs, and Z35.2- 1968, Specifications for Accident Prevention Tags, contain rules which are additional to the rules prescribed in this section. The employer shall comply with ANSI Z35.1-1968 and Z35.2-1968 with respect to rules not specifically prescribed in this subpart. [§1926.200(i)]

[44 FR 8577, Feb. 9, 1979; 44 FR 20940, Apr. 6, 1979, as amended at 58 FR 35173, June 30, 1993; 67 FR 57736, Sept. 12, 2002; 69 FR 18803, Apr. 9, 2004]

§1926.201
Signaling

(a) **Flaggers.** Signaling by flaggers and the use of flaggers, including warning garments worn by flaggers shall conform to Part VI of the Manual on Uniform Traffic Control Devices, (1988 Edition, Revision 3 or the Millennium Edition), which are incorporated by reference in Sec. 1926.200(g)(2).[§1926.201(a)]

(b) **Crane and hoist signals.** Regulations for crane and hoist signaling will be found in applicable American National Standards Institute standards.[§1926.201(b)]

[44 FR 8577, Feb. 9, 1979; 44 FR 20940, Apr. 6, 1979, as amended at 67 FR 57736, Sept. 12, 2002]

§1926.202
Barricades

Barricades for protection of employees shall conform to Part VI of the Manual on Uniform Traffic Control Devices (1988 Edition, Revision 3 or Millennium Edition), which are incorporated by reference in Sec. 1926.200(g)(2).

[67 FR 57736, Sept. 12, 2002]

§1926.203
Definitions applicable to this subpart

(a) **Barricade** means an obstruction to deter the passage of persons or vehicles.[§1926.203(a)]

(b) **Signs** are the warnings of hazard, temporarily or permanently affixed or placed, at locations where hazards exist.[§1926.203(b)]

(c) **Signals** are moving signs, provided by workers, such as flaggers, or by devices, such as flashing lights, to warn of possible or existing hazards.[§1926.203(c)]

(d) **Tags** are temporary signs, usually attached to a piece of equipment or part of a structure, to warn of existing or immediate hazards.[§1926.203(d)]

[44 FR 8577, Feb. 9, 1979; 44 FR 20940, Apr. 6, 1979, as amended at 67 FR 57736, Sept. 12, 2002]

Authority: §107, Contract Work Hours and Safety Standards Act (Construction Safety Act) (40 U.S.C. 333); sections 4, 6, 8, Occupational Safety and Health Act of 1970 (29 U.S.C. 653, 655, 657); Secretary of Labor's Order No. 12-71 (36 FR 8754), 8-76 (41 FR 25059), 9-83 (48 FR 35736), or 3-2000 (65 FR 50017) as applicable, 29 CFR part 1911.

Subpart H - Materials Handling, Storage, Use, and Disposal

§1926.250
General requirements for storage

(a) **General.**[§1926.250(a)]

(1) *All materials stored in tiers* shall be stacked, racked, blocked, interlocked, or otherwise secured to prevent sliding, falling or collapse.[§1926.250(a)(1)]

(2) *Maximum safe load limits of floors* within buildings and structures, in pounds per square foot, shall be conspicuously posted in all storage areas, except for floor or slab on grade. Maximum safe loads shall not be exceeded.[§1926.250(a)(2)]

(3) *Aisles and passageways shall be kept* clear to provide for the free and safe movement of material handling equipment or employees. Such areas shall be kept in good repair. [§1926.250(a)(3)]

(4) *When a difference in road or working* levels exist, means such as ramps, blocking, or grading shall be used to ensure the safe movement of vehicles between the two levels.[§1926.250(a)(4)]

(b) **Material storage.**[§1926.250(b)]

(1) *Material stored inside buildings* under construction shall not be placed within 6 feet of any hoistway or inside floor openings, nor within 10 feet of an exterior wall which does not extend above the top of the material stored.[§1926.250(b)(1)]

(2) *Each employee required to work* on stored material in silos, hoppers, tanks, and similar storage areas shall be equipped with personal fall arrest equipment meeting the requirements of subpart M of this part.[§1926.250(b)(2)]

(3) *Noncompatible materials shall be segregated in storage.* [§1926.250(b)(3)]

(4) *Bagged materials shall be stacked* by stepping back the layers and cross-keying the bags at least every 10 bags high. [§1926.250(b)(4)]

(5) *Materials shall not be stored on scaffolds* or runways in excess of supplies needed for immediate operations.[§1926.250(b)(5)]

(6) *Brick stacks shall not be more* than 7 feet in height. When a loose brick stack reaches a height of 4 feet, it shall be tapered back 2 inches in every foot of height above the 4-foot level. [§1926.250(b)(6)]

(7) *When masonry blocks are stacked* higher than 6 feet, the stack shall be tapered back one-half block per tier above the 6-foot level.[§1926.250(b)(7)]

(8) *Lumber:*[§1926.250(b)(8)]

(i) *Used lumber shall have* all nails withdrawn before stacking. [§1926.250(b)(8)(i)]

(ii) *Lumber shall be stacked on level* and solidly supported sills. [§1926.250(b)(8)(ii)]

(iii) *Lumber shall be so stacked as to be stable* and self- supporting.[§1926.250(b)(8)(iii)]

(iv) *Lumber piles shall not exceed* 20 feet in height provided that lumber to be handled manually shall not be stacked more than 16 feet high.[§1926.250(b)(8)(iv)]

(9) *Structural steel,* poles, pipe, bar stock, and other cylindrical materials, unless racked, shall be stacked and blocked so as to prevent spreading or tilting.[§1926.250(b)(9)]

(c) **Housekeeping.** Storage areas shall be kept free from accumulation of materials that constitute hazards from tripping, fire, explosion, or pest harborage. Vegetation control will be exercised when necessary.[§1926.250(c)]

(d) **Dockboards (bridge plates).**[§1926.250(d)]

(1) *Portable and powered dockboards* shall be strong enough to carry the load imposed on them.[§1926.250(d)(1)]

(2) *Portable dockboards shall be secured* in position, either by being anchored or equipped with devices which will prevent their slipping.[§1926.250(d)(2)]

(3) *Handholds, or other effective means,* shall be provided on portable dockboards to permit safe handling.[§1926.250(d)(3)]

(4) *Positive protection shall be provided* to prevent railroad cars from being moved while dockboards or bridge plates are in position.[§1926.250(d)(4)]

[44 FR 8577, Feb. 9, 1979; 44 FR 20940, Apr. 6, 1979, as amended at 49 FR 18295, Apr. 30, 1984; 54 FR 24334, June 7, 1989; 58 FR 35173, June 30, 1993; 59 FR 40729, Aug. 9, 1994; 61 FR 5510, Feb. 13, 1996]

§1926.251
Rigging equipment for material handling

(a) **General.**[§1926.251(a)]

(1) *Rigging equipment for material handling* shall be inspected prior to use on each shift and as necessary during its use to ensure that it is safe. Defective rigging equipment shall be removed from service.[§1926.251(a)(1)]

(2) *Employers must ensure that rigging equipment:*[§1926.251(a)(2)]

(i) *Has permanently affixed* and legible identification markings as prescribed by the manufacturer that indicate the recommended safe working load;[§1926.251(a)(2)(i)]

(ii) *Not be loaded in excess of its recommended* safe working load as prescribed on the identification markings by the manufacturer; and[§1926.251(a)(2)(ii)]

(iii) *Not be used without affixed,* legible identification markings, required by paragraph (a)(2)(i) of this section. [§1926.251(a)(2)(iii)]

(3) *Rigging equipment, when not in use,* shall be removed from the immediate work area so as not to present a hazard to employees.[§1926.251(a)(3)]

(4) *Special custom design grabs, hooks, clamps,* or other lifting accessories, for such units as modular panels, prefabricated structures and similar materials, shall be marked to indicate the safe working loads and shall be proof-tested prior to use to 125 percent of their rated load.[§1926.251(a)(4)]

(5) *Scope.* This section applies to slings used in conjunction with other material handling equipment for the movement of material by hoisting, in employments covered by this part. The types of slings covered are those made from alloy steel chain, wire rope, metal mesh, natural or synthetic fiber rope (conventional three strand construction), and synthetic web (nylon, polyester, and polypropylene).[§1926.251(a)(5)]

(6) *Inspections.* Each day before being used, the sling and all fastenings and attachments shall be inspected for damage or defects by a competent person designated by the employer. Additional inspections shall be performed during sling use, where service conditions warrant. Damaged or defective slings shall be immediately removed from service.[§1926.251(a)(6)]

(b) **Alloy steel chains.**[§1926.251(b)]

(1) *Welded alloy steel chain slings* shall have permanently affixed durable identification stating size, grade, rated capacity, and sling manufacturer.[§1926.251(b)(1)]

(2) *Hooks, rings, oblong links,* pear-shaped links, welded or mechanical coupling links, or other attachments, when used with alloy steel chains, shall have a rated capacity at least equal to that of the chain.[§1926.251(b)(2)]

(3) *Job or shop hooks and links,* or makeshift fasteners, formed from bolts, rods, etc., or other such attachments, shall not be used.[§1926.251(b)(3)]

(4) *Employers must not use alloy* steel-chain slings with loads in excess of the rated capacities (i.e., working load limits) indicated on the sling by permanently affixed and legible identification markings prescribed by the manufacturer.[§1926.251(b)(4)]

(5) *Whenever wear at any point* of any chain link exceeds that shown in Table H-1, the assembly shall be removed from service.[§1926.251(b)(5)]

(6) *Inspections.*[§1926.251(b)(6)]

(i) *In addition to the inspection required* by other paragraphs of this section, a thorough periodic inspection of alloy steel chain slings in use shall be made on a regular basis, to be determined on the basis of[§1926.251(b)(6)(i)]

[A] frequency of sling use;[§1926.251(b)(6)(i)[A]]

[B] severity of service conditions;[§1926.251(b)(6)(i)[B]]

[C] nature of lifts being made; and[§1926.251(b)(6)(i)[C]]

[D] experience gained on the service life of slings used in similar circumstances. Such inspections shall in no event be at intervals greater than once every 12 months. [§1926.251(b)(6)(i)[D]]

(ii) *The employer shall make and maintain* a record of the most recent month in which each alloy steel chain sling was thoroughly inspected, and shall make such record available for examination.[§1926.251(b)(6)(ii)]

(c) **Wire rope.**[§1926.251(c)]

(1) *Employers must not use improved* plow-steel wire rope and wire-rope slings with loads in excess of the rated capacities (i.e., working load limits) indicated on the sling by permanently affixed and legible identification markings prescribed by the manufacturer.[§1926.251(c)(1)]

(2) *Protruding ends of strands* in splices on slings and bridles shall be covered or blunted.[§1926.251(c)(2)]

(3) *Wire rope shall not be secured* by knots, except on haul back lines on scrapers.[§1926.251(c)(3)]

(4) *The following limitations* shall apply to the use of wire rope: [§1926.251(c)(4)]

(i) *An eye splice made in any wire rope* shall have not less than three full tucks. However, this requirement shall not operate to preclude the use of another form of splice or connection which can be shown to be as efficient and which is not otherwise prohibited.[§1926.251(c)(4)(i)]

(ii) *Except for eye splices in the ends of wires* and for endless rope slings, each wire rope used in hoisting or lowering, or in pulling loads, shall consist of one continuous piece without knot or splice.[§1926.251(c)(4)(ii)]

(iii) *Eyes in wire rope bridles,* slings, or bull wires shall not be formed by wire rope clips or knots.[§1926.251(c)(4)(iii)]

(iv) *Wire rope shall not be used* if, in any length of eight diameters, the total number of visible broken wires exceeds 10 percent of the total number of wires, or if the rope shows other signs of excessive wear, corrosion, or defect.[§1926.251(c)(4)(iv)]

(5) *When U-bolt wire rope* clips are used to form eyes, Table H-2 shall be used to determine the number and spacing of clips. [§1926.251(c)(5)]

(i) *When used for eye splices,* the U-bolt shall be applied so that the "U" section is in contact with the dead end of the rope.[§1926.251(c)(5)(i)]

(ii) *Reserved*[§1926.251(c)(5)(ii)]

(6) *Slings shall not be shortened* with knots or bolts or other makeshift devices.[§1926.251(c)(6)]

(7) *Sling legs shall not be kinked.*[§1926.251(c)(7)]

(8) *Slings used in a basket hitch* shall have the loads balanced to prevent slippage.[§1926.251(c)(8)]

(9) *Slings shall be padded or protected* from the sharp edges of their loads.[§1926.251(c)(9)]

(10) *Hands or fingers shall not be placed* between the sling and its load while the sling is being tightened around the load. [§1926.251(c)(10)]

(11) *Shock loading is prohibited.*[§1926.251(c)(11)]

(12) *A sling shall not be pulled* from under a load when the load is resting on the sling.[§1926.251(c)(12)]

(13) *Minimum sling lengths.*[§1926.251(c)(13)]

(i) *Cable laid and 6 x 19 and 6 x 37 slings* shall have a minimum clear length of wire rope 10 times the component rope diameter between splices, sleeves or end fittings.[§1926.251(c)(13)(i)]

(ii) *Braided slings shall have* a minimum clear length of wire rope 40 times the component rope diameter between the loops or end fittings.[§1926.251(c)(13)(ii)]

(iii) *Cable laid grommets,* strand laid grommets and endless slings shall have a minimum circumferential length of 96 times their body diameter.[§1926.251(c)(13)(iii)]

(14) *Safe operating temperatures.* Fiber core wire rope slings of all grades shall be permanently removed from service if they are exposed to temperatures in excess of 200 °F (93.33 °C). When nonfiber core wire rope slings of any grade are used at temperatures above 400 °F (204.44 °C) or below minus 60 °F (15.55 °C), recommendations of the sling manufacturer regarding use at that temperature shall be followed.[§1926.251(c)(14)]

(15) *End attachments.*[§1926.251(c)(15)]

(i) *Welding of end attachments,* except covers to thimbles, shall be performed prior to the assembly of the sling. [§1926.251(c)(15)(i)]

(ii) *All welded end attachments* shall not be used unless proof tested by the manufacturer or equivalent entity at twice their rated capacity prior to initial use. The employer shall retain a certificate of the proof test, and make it available for examination.[§1926.251(c)(15)(ii)]

(16) *Wire rope slings* shall have permanently affixed, legible identification markings stating size, rated capacity for the type(s) of hitch(es) used and the angle upon which it is based, and the number of legs if more than one.[§1926.251(c)(16)]

(d) **Natural rope, and synthetic fiber**[§1926.251(d)]

(1) *Employers must not use natural-* and synthetic-fiber rope slings with loads in excess of the rated capacities (i.e., working load limits) indicated on the sling by permanently affixed and legible identification markings prescribed by the manufacturer. [§1926.251(d)(1)]

(2) *All splices in rope slings* provided by the employer shall be made in accordance with fiber rope manufacturers recommendations.[§1926.251(d)(2)]

(i) *In manila rope,* eye splices shall contain at least three full tucks, and short splices shall contain at least six full tucks (three on each side of the centerline of the splice). [§1926.251(d)(2)(i)]

(ii) *In layed synthetic fiber rope,* eye splices shall contain at least four full tucks, and short splices shall contain at least eight full tucks (four on each side of the centerline of the splice). [§1926.251(d)(2)(ii)]

(iii) *Strand end tails shall not be trimmed short* (flush with the surface of the rope) immediately adjacent to the full tucks. This precaution applies to both eye and short splices and all types of fiber rope. For fiber ropes under 1-inch diameter, the tails shall project at least six rope diameters beyond the last full tuck. For fiber ropes 1- inch diameter and larger, the tails shall project at least 6 inches beyond the last full tuck. In applications where the projecting tails may be objectionable, the tails shall be tapered and spliced into the body of the rope using at least two additional tucks (which will require a tail length of approximately six rope diameters beyond the last full tuck).[§1926.251(d)(2)(iii)]

(iv) *For all eye splices,* the eye shall be sufficiently large to provide an included angle of not greater than 60° at the splice when the eye is placed over the load or support. [§1926.251(d)(2)(iv)]

(v) *Knots shall not be used in lieu of splices.*[§1926.251(d)(2)(v)]

(3) *Safe operating temperatures.* Natural and synthetic fiber rope slings, except for wet frozen slings, may be used in a temperature range from minus 20 °F (-28.88 °C) to plus 180 °F (82.2 °C) without decreasing the working load limit. For operations outside this temperature range and for wet frozen slings, the sling manufacturer's recommendations shall be followed. [§1926.251(d)(3)]

(4) *Splicing.* Spliced fiber rope slings shall not be used unless they have been spliced in accordance with the following minimum requirements and in accordance with any additional recommendations of the manufacturer:[§1926.251(d)(4)]

(i) *In manila rope,* eye splices shall consist of at least three full tucks, and short splices shall consist of at least six full tucks, three on each side of the splice center line.[§1926.251(d)(4)(i)]

(ii) *In synthetic fiber rope,* eye splices shall consist of at least four full tucks, and short splices shall consist of at least eight full tucks, four on each side of the center line.[§1926.251(d)(4)(ii)]

(iii) *Strand end tails shall not be trimmed flush* with the surface of the rope immediately adjacent to the full tucks. This applies to all types of fiber rope and both eye and short splices. For fiber rope under 1 inch (2.54 cm) in diameter, the tail shall project at least six rope diameters beyond the last full tuck. For fiber rope 1 inch (2.54 cm) in diameter and larger, the tail shall project at least 6 inches (15.24 cm) beyond the last full tuck. Where a projecting tail interferes with the use of the sling, the tail shall be tapered and spliced into the body of the rope using at least two additional tucks (which will require a tail length of approximately six rope diameters beyond the last full tuck).[§1926.251(d)(4)(iii)]

(iv) *Fiber rope slings* shall have a minimum clear length of rope between eye splices equal to 10 times the rope diameter. [§1926.251(d)(4)(iv)]

(v) *Knots shall not be used in lieu of splices.*[§1926.251(d)(4)(v)]

(vi) *Clamps not designed* specifically for fiber ropes shall not be used for splicing.[§1926.251(d)(4)(vi)]

(vii) *For all eye splices,* the eye shall be of such size to provide an included angle of not greater than 60 degrees at the splice when the eye is placed over the load or support. [§1926.251(d)(4)(vii)]

(5) *End attachments.* Fiber rope slings shall not be used if end attachments in contact with the rope have sharp edges or projections.[§1926.251(d)(5)]

(6) *Removal from service.* Natural and synthetic fiber rope slings shall be immediately removed from service if any of the following conditions are present:[§1926.251(d)(6)]

(i) *Abnormal wear.*[§1926.251(d)(6)(i)]

(ii) *Powdered fiber between strands.*[§1926.251(d)(6)(ii)]

(iii) *Broken or cut fibers.*[§1926.251(d)(6)(iii)]

(iv) *Variations in the size or roundness of strands.* [§1926.251(d)(6)(iv)]

(v) *Discoloration or rotting.*[§1926.251(d)(6)(v)]

(vi) *Distortion of hardware in the sling.*[§1926.251(d)(6)(vi)]

(7) *Employers must use natural-* and synthetic-fiber rope slings that have permanently affixed and legible identification markings that state the rated capacity for the type(s) of hitch(es) used and the angle upon which it is based, type of fiber material, and the number of legs if more than one.[§1926.251(d)(7)]

(e) **Synthetic webbing (nylon, polyester, and polypropylene).** [§1926.251(e)]

(1) *The employer shall have* each synthetic web sling marked or coded to show:[§1926.251(e)(1)]

(i) *Name or trademark of manufacturer.*[§1926.251(e)(1)(i)]

(ii) *Rated capacities for the type of hitch.*[§1926.251(e)(1)(ii)]

(iii) *Type of material.*[§1926.251(e)(1)(iii)]

(2) *Rated capacity shall not be exceeded.*[§1926.251(e)(2)]

(3) *Webbing.* Synthetic webbing shall be of uniform thickness and width and selvage edges shall not be split from the webbing's width.[§1926.251(e)(3)]

(4) *Fittings.* Fittings shall be:[§1926.251(e)(4)]

(i) *Of a minimum breaking strength* equal to that of the sling; and[§1926.251(e)(4)(i)]

(ii) *Free of all sharp edges* that could in any way damage the webbing.[§1926.251(e)(4)(ii)]

(5) *Attachment of end fittings to webbing* and formation of eyes. Stitching shall be the only method used to attach end fittings to webbing and to form eyes. The thread shall be in an even pattern and contain a sufficient number of stitches to develop the full breaking strength of the sling.[§1926.251(e)(5)]

(6) *Environmental conditions.* When synthetic web slings are used, the following precautions shall be taken:[§1926.251(e)(6)]

(i) *Nylon web slings* shall not be used where fumes, vapors, sprays, mists or liquids of acids or phenolics are present. [§1926.251(e)(6)(i)]

(ii) *Polyester and polypropylene web slings* shall not be used where fumes, vapors, sprays, mists or liquids of caustics are present.[§1926.251(e)(6)(ii)]

(iii) *Web slings with aluminum fittings* shall not be used where fumes, vapors, sprays, mists or liquids of caustics are present.[§1926.251(e)(6)(iii)]

(7) *Safe operating temperatures.* Synthetic web slings of polyester and nylon shall not be used at temperatures in excess of 180 °F (82.2 °C). Polypropylene web slings shall not be used at temperatures in excess of 200 °F (93.33 °C).[§1926.251(e)(7)]

(8) *Removal from service.* Synthetic web slings shall be immediately removed from service if any of the following conditions are present:[§1926.251(e)(8)]

(i) *Acid or caustic burns;*[§1926.251(e)(8)(i)]

(ii) *Melting or charring of any part* of the sling surface; [§1926.251(e)(8)(ii)]

(iii) *Snags,* punctures, tears or cuts;[§1926.251(e)(8)(iii)]

(iv) *Broken or worn stitches; or*[§1926.251(e)(8)(iv)]

(v) *Distortion of fittings.*[§1926.251(e)(8)(v)]

(f) **Shackles and hooks.**[§1926.251(f)]

(1) *Employers must not use shackles* with loads in excess of the rated capacities (i.e., working load limits) indicated on the shackle by permanently affixed and legible identification markings prescribed by the manufacturer.[§1926.251(f)(1)]

(2) *The manufacturer's recommendations shall be followed* in determining the safe working loads of the various sizes and types of specific and identifiable hooks. All hooks for which no applicable manufacturer's recommendations are available shall be tested to twice the intended safe working load before they are initially put into use. The employer shall maintain a record of the dates and results of such tests.[§1926.251(f)(2)]

Table H-1 — Maximum Allowable Wear at any Point of Link

Chain size, inches	Maximum allowable wear, inches
1/4	3/64
3/8	5/64
1/2	7/64
5/8	9/64
3/4	5/32
7/8	11/64
1	3/16
1 1/8	7/32
1 1/4	1/4
1 3/8	9/32
1 1/2	5/16
1 3/4	11/32

Table H-2 — Number and Spacing of U-Bolt Wire Rope Clips

Improved plow steel, rope diameter (inches)	Number of clips		Minimum spacing (inches)
	Drop forged	Other material	
1/2	3	4	3
5/8	3	4	3 3/4
3/4	4	5	4 1/2
7/8	4	5	5 1/4
1	5	6	6
1 1/8	6	6	6 3/4
1 1/4	6	7	7 1/2
1 3/8	7	7	8 1/4
1 1/2	7	8	9

[44 FR 8577, Feb. 9, 1979; 44 FR 20940, Apr. 6, 1979, as amended at 58 FR 35173, June 30, 1993; 76 FR 33611, June 8, 2011; 77 FR 23118, Apr. 18, 2012]

§1926.252
Disposal of waste materials

(a) **Whenever materials are dropped** more than 20 feet to any point lying outside the exterior walls of the building, an enclosed chute of wood, or equivalent material, shall be used. For the purpose of this paragraph, an enclosed chute is a slide, closed in on all sides, through which material is moved from a high place to a lower one. [§1926.252(a)]

(b) **When debris is dropped through holes** in the floor without the use of chutes, the area onto which the material is dropped shall be completely enclosed with barricades not less than 42 inches high and not less than 6 feet back from the projected edge of the opening above. Signs warning of the hazard of falling materials shall be posted at each level. Removal shall not be permitted in this lower area until debris handling ceases above.[§1926.252(b)]

(c) **All scrap lumber,** waste material, and rubbish shall be removed from the immediate work area as the work progresses.[§1926.252(c)]

(d) **Disposal of waste material** or debris by burning shall comply with local fire regulations.[§1926.252(d)]

(e) **All solvent waste,** oily rags, and flammable liquids shall be kept in fire resistant covered containers until removed from worksite. [§1926.252(e)]

Authority: §107, Contract Work Hours and Safety Standards Act (Construction Safety Act) (40 U.S.C. 333); Secs. 4, 6, 8, Occupational Safety and Health Act of 1970 (29 U.S.C. 653, 655, 657); Secretary of Labor's Order No. 12-71 (36 FR 8754), 8-76 (41 FR 25059), 9-83 (48 FR 35736), or 1-90 (50 FR 9033), as applicable. Section 1926.250 also issued under 29 CFR Part 1911.

Notes

Subpart I

Tools Hand and Power

§1926.300

General requirements

(a) Condition of tools. All hand and power tools and similar equipment, whether furnished by the employer or the employee, shall be maintained in a safe condition.[§1926.300(a)]

(b) Guarding.[§1926.300(b)]

(1) *When power operated tools* are designed to accommodate guards, they shall be equipped with such guards when in use. [§1926.300(b)(1)]

(2) *Belts, gears, shafts, pulleys,* sprockets, spindles, drums, fly wheels, chains, or other reciprocating, rotating or moving parts of equipment shall be guarded if such parts are exposed to contact by employees or otherwise create a hazard. Guarding shall meet the requirements as set forth in American National Standards Institute, B15.1-1953 (R1958), Safety Code for Mechanical Power-Transmission Apparatus.[§1926.300(b)(2)]

(3) *Types of guarding.* One or more methods of machine guarding shall be provided to protect the operator and other employees in the machine area from hazards such as those created by point of operation, ingoing nip points, rotating parts, flying chips and sparks. Examples of guarding methods are — barrier guards, two-hand tripping devices, electronic safety devices, etc.[§1926.300(b)(3)]

(4) *Point of operation guarding.*[§1926.300(b)(4)]

(i) *Point of operation is the area on a machine* where work is actually performed upon the material being processed. [§1926.300(b)(4)(i)]

(ii) *The point of operation of machines* whose operation exposes an employee to injury, shall be guarded. The guarding device shall be in conformity with any appropriate standards therefor, or, in the absence of applicable specific standards, shall be so designed and constructed as to prevent the operator from having any part of his body in the danger zone during the operating cycle.[§1926.300(b)(4)(ii)]

(iii) *Special handtools for placing and removing material* shall be such as to permit easy handling of material without the operator placing a hand in the danger zone. Such tools shall not be in lieu of other guarding required by this section, but can only be used to supplement protection provided. [§1926.300(b)(4)(iii)]

(iv) *The following are some of the machines* which usually require point of operation guarding:[§1926.300(b)(4)(iv)]

[a] Guillotine cutters.[§1926.300(b)(4)(iv)[a]]

[b] Shears.[§1926.300(b)(4)(iv)[b]]

[c] Alligator shears.[§1926.300(b)(4)(iv)[c]]

[d] Power presses.[§1926.300(b)(4)(iv)[d]]

[e] Milling machines.[§1926.300(b)(4)(iv)[e]]

[f] Power saws.[§1926.300(b)(4)(iv)[f]]

[g] Jointers.[§1926.300(b)(4)(iv)[g]]

[h] Portable power tools.[§1926.300(b)(4)(iv)[h]]

[i] Forming rolls and calenders.[§1926.300(b)(4)(iv)[i]]

(5) *Exposure of blades.* When the periphery of the blades of a fan is less than 7 feet (2.128 m) above the floor or working level, the blades shall be guarded. The guard shall have openings no larger than 1/2 inch (1.27 cm).[§1926.300(b)(5)]

(6) *Anchoring fixed machinery.* Machines designed for a fixed location shall be securely anchored to prevent walking or moving.[§1926.300(b)(6)]

(7) *Guarding of abrasive wheel machinery — exposure adjustment.* Safety guards of the types described in paragraphs (b) (8) and (9) of this section, where the operator stands in front of the opening, shall be constructed so that the peripheral protecting member can be adjusted to the constantly decreasing diameter of the wheel. The maximum angular exposure above the horizontal plane of the wheel spindle as specified in paragraphs (b) (8) and (9) of this section shall never be exceeded, and the distance between the wheel periphery and the adjustable tongue or the end of the peripheral member at the top shall never exceed 1/4 inch (0.635 cm). (See Figures I-1 through I-6.)[§1926.300(b)(7)]

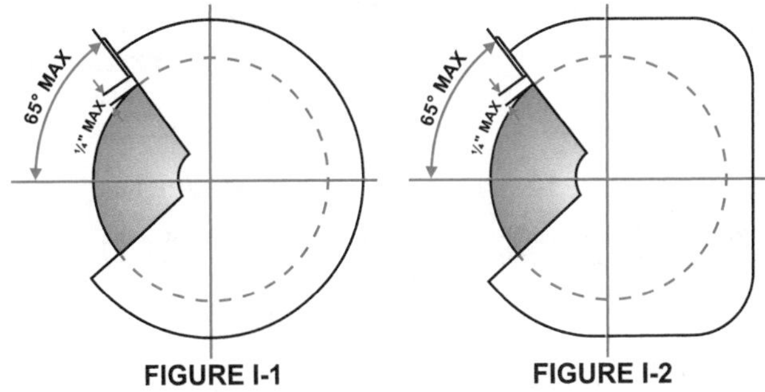

FIGURE I-1 FIGURE I-2

Correct
Showing adjustable tongue giving required angle protection for all sizes of wheel used.

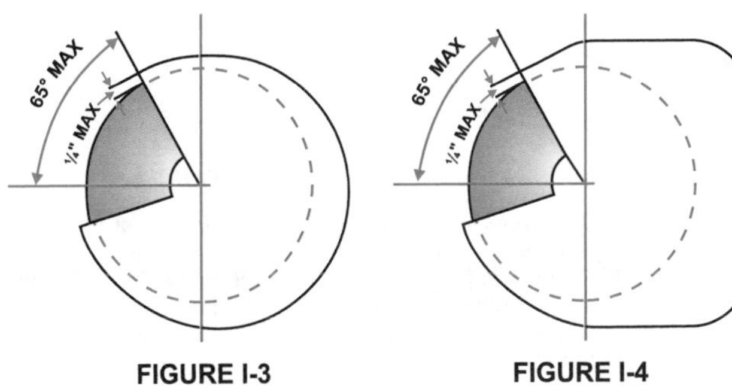

FIGURE I-3 FIGURE I-4

Correct
Showing movable guard with opening small enough to give required protection for the smallest size wheel used.

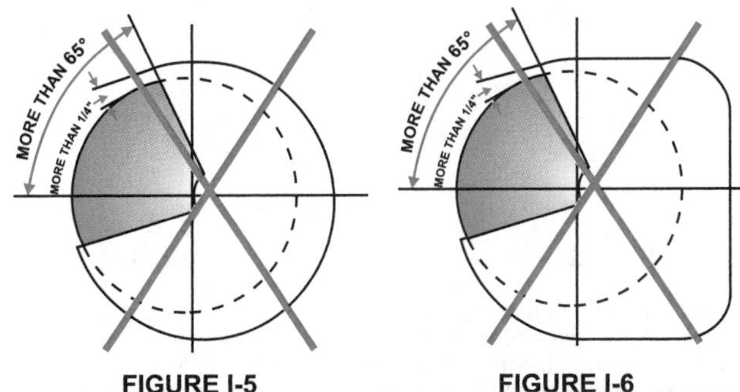

FIGURE I-5 FIGURE I-6

Incorrect
Showing movable guard with size of opening correct for full-size wheel but too large for smaller wheel.

(8) *Bench and floor stands.* The angular exposure of the grinding wheel periphery and sides for safety guards used on machines known as bench and floor stands should not exceed 90° or one-fourth of the periphery. This exposure shall begin at a point not more than 65° above the horizontal plane of the wheel spindle. (See Figures I-7 and I- 8 and paragraph (b)(7) of this section.)[§1926.300(b)(8)]

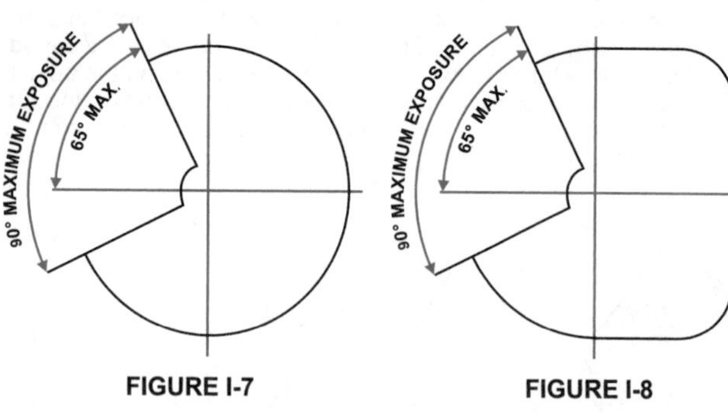

FIGURE I-7 FIGURE I-8

Wherever the nature of the work requires contact with the wheel below the horizontal plane of the spindle, the exposure shall not exceed 125°. (See Figures I-9 and I-10.)

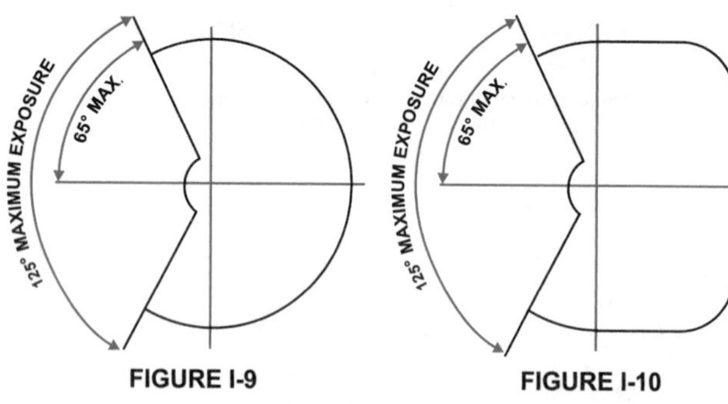

FIGURE I-9 FIGURE I-10

(9) *Cylindrical grinders.* The maximum angular exposure of the grinding wheel periphery and sides for safety guards used on cylindrical grinding machines shall not exceed 180°. This exposure shall begin at a point not more than 65° above the horizontal plane of the wheel spindle. (See Figures I-11 and I-12 and paragraph (b)(7) of this section.)[§1926.300(b)(9)]

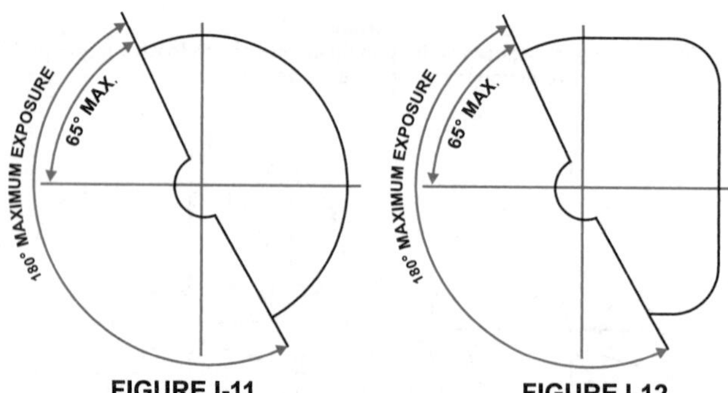

FIGURE I-11 FIGURE I-12

(c) **Personal protective equipment.** Employees using hand and power tools and exposed to the hazard of falling, flying, abrasive, and splashing objects, or exposed to harmful dusts, fumes, mists, vapors, or gases shall be provided with the particular personal protective equipment necessary to protect them from the hazard. All personal protective equipment shall meet the requirements and be maintained according to subparts D and E of this part. [§1926.300(c)]

(d) **Switches.**[§1926.300(d)]

(1) *All hand-held powered platen sanders,* grinders with wheels 2-inch diameter or less, routers, planers, laminate trimmers, nibblers, shears, scroll saws, and jigsaws with blade shanks one-fourth of an inch wide or less may be equipped with only a positive "on-off" control.[§1926.300(d)(1)]

(2) *All hand-held powered drills, tappers, fastener drivers,* horizontal, vertical, and angle grinders with wheels greater than 2 inches in diameter, disc sanders, belt sanders, reciprocating saws, saber saws, and other similar operating powered tools shall be equipped with a momentary contact "on-off" control and may have a lock-on control provided that turnoff can be accomplished by a single motion of the same finger or fingers that turn it on.[§1926.300(d)(2)]

(3) *All other hand-held powered tools,* such as circular saws, chain saws, and percussion tools without positive accessory holding means, shall be equipped with a constant pressure switch that will shut off the power when the pressure is released.[§1926.300(d)(3)]

(4) *The requirements of this paragraph* shall become effective on July 15, 1972.[§1926.300(d)(4)]

(5) *Exception:* This paragraph does not apply to concrete vibrators, concrete breakers, powered tampers, jack hammers, rock drills, and similar hand operated power tools. [§1926.300(d)(5)]

[44 FR 8577, Feb. 9, 1979; 44 FR 20940, Apr. 6, 1979, as amended at 58 FR 35175, June 30, 1993; 61 FR 9250, Mar. 7, 1996]

§1926.301
Hand tools

(a) **Employers shall not issue or permit** the use of unsafe hand tools.[§1926.301(a)]

(b) **Wrenches,** including adjustable, pipe, end, and socket wrenches shall not be used when jaws are sprung to the point that slippage occurs.[§1926.301(b)]

(c) **Impact tools,** such as drift pins, wedges, and chisels, shall be kept free of mushroomed heads.[§1926.301(c)]

(d) **The wooden handles of tools** shall be kept free of splinters or cracks and shall be kept tight in the tool.[§1926.301(d)]

§1926.302
Power-operated hand tools

(a) **Electric power-operated tools.**[§1926.302(a)]

(1) *Electric power operated tools* shall either be of the approved double-insulated type or grounded in accordance with subpart K of this part.[§1926.302(a)(1)]

(2) *The use of electric cords* for hoisting or lowering tools shall not be permitted.[§1926.302(a)(2)]

(b) **Pneumatic power tools.**[§1926.302(b)]

(1) *Pneumatic power tools* shall be secured to the hose or whip by some positive means to prevent the tool from becoming accidentally disconnected.[§1926.302(b)(1)]

(2) *Safety clips or retainers* shall be securely installed and maintained on pneumatic impact (percussion) tools to prevent attachments from being accidentally expelled.[§1926.302(b)(2)]

(3) *All pneumatically driven nailers,* staplers, and other similar equipment provided with automatic fastener feed, which operate at more than 100 p.s.i. pressure at the tool shall have a safety device on the muzzle to prevent the tool from ejecting fasteners, unless the muzzle is in contact with the work surface.[§1926.302(b)(3)]

(4) *Compressed air shall not be used* for cleaning purposes except where reduced to less than 30 p.s.i. and then only with effective chip guarding and personal protective equipment which meets the requirements of subpart E of this part. The 30 p.s.i. requirement does not apply for concrete form, mill scale and similar cleaning purposes.[§1926.302(b)(4)]

(5) *The manufacturer's safe operating pressure* for hoses, pipes, valves, filters, and other fittings shall not be exceeded. [§1926.302(b)(5)]

(6) *The use of hoses for hoisting* or lowering tools shall not be permitted.[§1926.302(b)(6)]

(7) *All hoses exceeding 1/2 -inch inside diameter* shall have a safety device at the source of supply or branch line to reduce pressure in case of hose failure.[§1926.302(b)(7)]

(8) *Airless spray guns* of the type which atomize paints and fluids at high pressures (1,000 pounds or more per square inch) shall be equipped with automatic or visible manual safety devices which will prevent pulling of the trigger to prevent release of the paint or fluid until the safety device is manually released.[§1926.302(b)(8)]

(9) *In lieu of the above,* a diffuser nut which will prevent high pressure, high velocity release, while the nozzle tip is removed, plus a nozzle tip guard which will prevent the tip from coming into contact with the operator, or other equivalent protection, shall be provided.[§1926.302(b)(9)]

(10) *Abrasive blast cleaning nozzles.* The blast cleaning nozzles shall be equipped with an operating valve which must be held open manually. A support shall be provided on which the nozzle may be mounted when it is not in use.[§1926.302(b)(10)]

(c) **Fuel powered tools.**[§1926.302(c)]

(1) *All fuel powered tools* shall be stopped while being refueled, serviced, or maintained, and fuel shall be transported, handled, and stored in accordance with subpart F of this part. [§1926.302(c)(1)]

(2) *When fuel powered tools are used* in enclosed spaces, the applicable requirements for concentrations of toxic gases and use of personal protective equipment, as outlined in subparts D and E of this part, shall apply.[§1926.302(c)(2)]

(d) **Hydraulic power tools.**[§1926.302(d)]

(1) *The fluid used in hydraulic* powered tools shall be fire-resistant fluids approved under Schedule 30 of the U.S. Bureau of Mines, Department of the Interior, and shall retain its operating characteristics at the most extreme temperatures to which it will be exposed.[§1926.302(d)(1)]

(2) *The manufacturer's safe operating pressures* for hoses, valves, pipes, filters, and other fittings shall not be exceeded. [§1926.302(d)(2)]

(e) **Powder-actuated tools.**[§1926.302(e)]

(1) *Only employees who have been trained* in the operation of the particular tool in use shall be allowed to operate a powder-actuated tool.[§1926.302(e)(1)]

(2) *The tool shall be tested each day* before loading to see that safety devices are in proper working condition. The method of testing shall be in accordance with the manufacturer's recommended procedure.[§1926.302(e)(2)]

(3) *Any tool found not in proper working order,* or that develops a defect during use, shall be immediately removed from service and not used until properly repaired.[§1926.302(e)(3)]

(4) *Personal protective equipment* shall be in accordance with subpart E of this part.[§1926.302(e)(4)]

(5) *Tools shall not be loaded until* just prior to the intended firing time. Neither loaded nor empty tools are to be pointed at any employees. Hands shall be kept clear of the open barrel end. [§1926.302(e)(5)]

(6) *Loaded tools shall not be left unattended.*[§1926.302(e)(6)]

(7) *Fasteners shall not be driven* into very hard or brittle materials including, but not limited to, cast iron, glazed tile, surface-hardened steel, glass block, live rock, face brick, or hollow tile. [§1926.302(e)(7)]

(8) *Driving into materials easily penetrated* shall be avoided unless such materials are backed by a substance that will prevent the pin or fastener from passing completely through and creating a flying missile hazard on the other side. [§1926.302(e)(8)]

(9) *No fastener shall be driven* into a spalled area caused by an unsatisfactory fastening.[§1926.302(e)(9)]

(10) *Tools shall not be used* in an explosive or flammable atmosphere.[§1926.302(e)(10)]

(11) *All tools shall be used* with the correct shield, guard, or attachment recommended by the manufacturer.[§1926.302(e)(11)]

(12) *Powder-actuated tools used* by employees shall meet all other applicable requirements of American National Standards Institute, A10.3- 1970, Safety Requirements for Explosive-Actuated Fastening Tools.[§1926.302(e)(12)]

[44 FR 8577, Feb. 9, 1979; 44 FR 20940, Apr. 6, 1979, as amended at 58 FR 35175, June 30, 1993]

§1926.303
Abrasive wheels and tools

(a) **Power.** All grinding machines shall be supplied with sufficient power to maintain the spindle speed at safe levels under all conditions of normal operation.[§1926.303(a)]

(b) **Guarding.**[§1926.303(b)]

(1) *Grinding machines shall be equipped* with safety guards in conformance with the requirements of American National Standards Institute, B7.1-1970, Safety Code for the Use, Care and Protection of Abrasive Wheels, and paragraph (d) of this section.[§1926.303(b)(1)]

(2) *Guard design.* The safety guard shall cover the spindle end, nut, and flange projections. The safety guard shall be mounted so as to maintain proper alignment with the wheel, and the strength of the fastenings shall exceed the strength of the guard, except:[§1926.303(b)(2)]

(i) *Safety guards on all operations* where the work provides a suitable measure of protection to the operator, may be so constructed that the spindle end, nut, and outer flange are exposed; and where the nature of the work is such as to entirely cover the side of the wheel, the side covers of the guard may be omitted; and[§1926.303(b)(2)(i)]

(ii) *The spindle end,* nut, and outer flange may be exposed on machines designed as portable saws.[§1926.303(b)(2)(ii)]

(c) **Use of abrasive wheels.**[§1926.303(c)]

(1) *Floor stand and bench mounted abrasive wheels,* used for external grinding, shall be provided with safety guards (protection hoods). The maximum angular exposure of the grinding wheel periphery and sides shall be not more than 90°, except that when work requires contact with the wheel below the horizontal plane of the spindle, the angular exposure shall not exceed 125°. In either case, the exposure shall begin not more than 65° above the horizontal plane of the spindle. Safety guards shall be strong enough to withstand the effect of a bursting wheel.[§1926.303(c)(1)]

(2) *Floor and bench-mounted grinders* shall be provided with work rests which are rigidly supported and readily adjustable. Such work rests shall be kept at a distance not to exceed one-eighth inch from the surface of the wheel.[§1926.303(c)(2)]

(3) *Cup type wheels used for external grinding* shall be protected by either a revolving cup guard or a band type guard in accordance with the provisions of the American National Standards Institute, B7.1-1970 Safety Code for the Use, Care, and Protection of Abrasive Wheels. All other portable abrasive wheels used for external grinding, shall be provided with safety guards (protection hoods) meeting the requirements of paragraph (c)(5) of this section, except as follows:[§1926.303(c)(3)]

(i) *When the work location makes it impossible,* a wheel equipped with safety flanges, as described in paragraph (c)(6) of this section, shall be used;[§1926.303(c)(3)(i)]

(ii) *When wheels 2 inches or less in diameter* which are securely mounted on the end of a steel mandrel are used. [§1926.303(c)(3)(ii)]

(4) *Portable abrasive wheels used for internal grinding* shall be provided with safety flanges (protection flanges) meeting the requirements of paragraph (c)(6) of this section, except as follows:[§1926.303(c)(4)]

(i) *When wheels 2 inches or less in diameter* which are securely mounted on the end of a steel mandrel are used; [§1926.303(c)(4)(i)]

(ii) *If the wheel is entirely within the work being ground while in use.*[§1926.303(c)(4)(ii)]

(5) *When safety guards are required,* they shall be so mounted as to maintain proper alignment with the wheel, and the guard and its fastenings shall be of sufficient strength to retain fragments of the wheel in case of accidental breakage. The maximum angular exposure of the grinding wheel periphery and sides shall not exceed 180°.[§1926.303(c)(5)]

(6) *When safety flanges are required,* they shall be used only with wheels designed to fit the flanges. Only safety flanges, of a type and design and properly assembled so as to ensure that the pieces of the wheel will be retained in case of accidental breakage, shall be used.[§1926.303(c)(6)]

(7) *All abrasive wheels shall be closely inspected* and ring-tested before mounting to ensure that they are free from cracks or defects.[§1926.303(c)(7)]

(8) *Grinding wheels shall fit* freely on the spindle and shall not be forced on. The spindle nut shall be tightened only enough to hold the wheel in place.[§1926.303(c)(8)]

(9) *All employees using abrasive wheels* shall be protected by eye protection equipment in accordance with the requirements of subpart E of this part, except when adequate eye protection is afforded by eye shields which are permanently attached to the bench or floor stand.[§1926.303(c)(9)]

(d) **Other requirements.** All abrasive wheels and tools used by employees shall meet other applicable requirements of American National Standards Institute, B7.1-1970, Safety Code for the Use, Care and Protection of Abrasive Wheels.[§1926.303(d)]

(e) **Work rests.** On offhand grinding machines, work rests shall be used to support the work. They shall be of rigid construction and designed to be adjustable to compensate for wheel wear. Work rests shall be kept adjusted closely to the wheel with a maximum opening of 1/8 inch (0.3175 cm) to prevent the work from being jammed between the wheel and the rest, which may cause wheel breakage. The work rest shall be securely clamped after each adjustment. The adjustment shall not be made with the wheel in motion.[§1926.303(e)]

[44 FR 8577, Feb. 9, 1979; 44 FR 20940, Apr. 6, 1979, as amended at 58 FR 35175, June 30, 1993]

§1926.304
Woodworking tools

(a) **Disconnect switches.** All fixed power driven woodworking tools shall be provided with a disconnect switch that can either be locked or tagged in the off position.[§1926.304(a)]

(b) **Speeds.** The operating speed shall be etched or otherwise permanently marked on all circular saws over 20 inches in diameter or operating at over 10,000 peripheral feet per minute. Any saw so marked shall not be operated at a speed other than that marked on the blade. When a marked saw is retensioned for a different speed, the marking shall be corrected to show the new speed. [§1926.304(b)]

(c) **Self-feed.** Automatic feeding devices shall be installed on machines whenever the nature of the work will permit. Feeder attachments shall have the feed rolls or other moving parts covered or guarded so as to protect the operator from hazardous points.[§1926.304(c)]

(d) **Guarding.** All portable, power-driven circular saws shall be equipped with guards above and below the base plate or shoe. The upper guard shall cover the saw to the depth of the teeth, except for the minimum arc required to permit the base to be tilted for bevel cuts. The lower guard shall cover the saw to the depth of the teeth, except for the minimum arc required to allow proper retraction and contact with the work. When the tool is withdrawn from the work, the lower guard shall automatically and instantly return to the covering position.[§1926.304(d)]

(e) **Personal protective equipment.** All personal protective equipment provided for use shall conform to subpart E of this part. [§1926.304(e)]

(f) **Other requirements.** All woodworking tools and machinery shall meet other applicable requirements of American National Standards Institute, 01.1-1961, Safety Code for Woodworking Machinery.[§1926.304(f)]

(g) **Radial saws.**[§1926.304(g)]

(1) *The upper hood shall completely enclose the upper portion* of the blade down to a point that will include the end of the saw arbor. The upper hood shall be constructed in such a manner and of such material that it will protect the operator from flying splinters, broken saw teeth, etc., and will deflect sawdust away from the operator. The sides of the lower exposed portion of the blade shall be guarded to the full diameter of the blade by a device that will automatically adjust itself to the thickness of the stock and remain in contact with stock being cut to give maximum protection possible for the operation being performed.[§1926.304(g)(1)]

(h) **Hand-fed crosscut table saws.**[§1926.304(h)]

(1) *Each circular crosscut table saw* shall be guarded by a hood which shall meet all the requirements of paragraph (i)(1) of this section for hoods for circular ripsaws.[§1926.304(h)(1)]

(i) **Hand-fed ripsaws.**[§1926.304(i)]

(1) *Each circular hand-fed ripsaw* shall be guarded by a hood which shall completely enclose that portion of the saw above the table and that portion of the saw above the material being cut. The hood and mounting shall be arranged so that the hood will automatically adjust itself to the thickness of and remain in contact with the material being cut but it shall not offer any considerable resistance to insertion of material to saw or to passage of the material being sawed. The hood shall be made of adequate strength to resist blows and strains incidental to reasonable operation, adjusting, and handling, and shall be so designed as to protect the operator from flying splinters and broken saw teeth. It shall be made of material that is soft enough so that it will be unlikely to cause tooth breakage. The hood shall be so mounted as to insure that its operation will be positive, reliable, and in true alignment with the saw; and the mounting shall be adequate in strength to resist any reasonable side thrust or other force tending to throw it out of line.[§1926.304(i)(1)]

[44 FR 8577, Feb. 9, 1979; 44 FR 20940, Apr. 6, 1979, as amended at 58 FR 35175, June 30, 1993; 61 FR 9251, Mar. 7, 1996]

§1926.305
Jacks — lever and ratchet, screw, and hydraulic

(a) **General requirements.**[§1926.305(a)]

(1) *The manufacturer's rated capacity* shall be legibly marked on all jacks and shall not be exceeded.[§1926.305(a)(1)]

(2) *All jacks shall have a positive stop to prevent overtravel.* [§1926.305(a)(2)]

(b) **[Reserved]**[§1926.305(b)]

(c) **Blocking.** When it is necessary to provide a firm foundation, the base of the jack shall be blocked or cribbed. Where there is a possibility of slippage of the metal cap of the jack, a wood block shall be placed between the cap and the load.[§1926.305(c)]

(d) (1) *Operation and maintenance.*[§1926.305(d)(1)]

(i) *After the load has been raised,* it shall be cribbed, blocked, or otherwise secured at once.[§1926.305(d)(1)(i)]

(ii) *Hydraulic jacks exposed* to freezing temperatures shall be supplied with an adequate antifreeze liquid.[§1926.305(d)(1)(ii)]

(iii) *All jacks shall be properly lubricated at regular intervals.* [§1926.305(d)(1)(iii)]

(iv) *Each jack shall be thoroughly inspected* at times which depend upon the service conditions. Inspections shall be not less frequent than the following:[§1926.305(d)(1)(iv)]

[a] *For constant or intermittent use* at one locality, once every 6 months,[§1926.305(d)(1)(iv)[a]]

[b] *For jacks sent out of shop* for special work, when sent out and when returned,[§1926.305(d)(1)(iv)[b]]

[c] *For a jack subjected* to abnormal load or shock, immediately before and immediately thereafter.[§1926.305(d)(1)(iv)[c]]

(v) *Repair or replacement parts* shall be examined for possible defects.[§1926.305(d)(1)(v)]

(vi) *Jacks which are out of order* shall be tagged accordingly, and shall not be used until repairs are made.[§1926.305(d)(1)(vi)]

[44 FR 8577, Feb. 9, 1979; 44 FR 20940, Apr. 6, 1979, as amended at 55 FR 42328, Oct. 18, 1990; 58 FR 35176, June 30, 1993]

§1926.306
Air receivers

(a) **General requirements.**[§1926.306(a)]

(1) *Application.* This section applies to compressed air receivers, and other equipment used in providing and utilizing compressed air for performing operations such as cleaning, drilling, hoisting, and chipping. On the other hand, however, this section does not deal with the special problems created by using compressed air to convey materials nor the problems created when men work in compressed air as in tunnels and caissons. This section is not intended to apply to compressed air machinery and equipment used on transportation vehicles such as steam railroad cars, electric railway cars, and automotive equipment.[§1926.306(a)(1)]

(2) *New and existing equipment.*[§1926.306(a)(2)]

(i) *All new air receivers installed* after the effective date of these regulations shall be constructed in accordance with the 1968 edition of the A.S.M.E. Boiler and Pressure Vessel Code Section VIII.[§1926.306(a)(2)(i)]

(ii) *All safety valves used* shall be constructed, installed, and maintained in accordance with the A.S.M.E. Boiler and Pressure Vessel Code, Section VIII Edition 1968.[§1926.306(a)(2)(ii)]

(b) **Installation and equipment requirements.**[§1926.306(b)]

(1) *Installation.* Air receivers shall be so installed that all drains, handholes, and manholes therein are easily accessible. Under no circumstances shall an air receiver be buried underground or located in an inaccessible place.[§1926.306(b)(1)]

(2) *Drains and traps.* A drain pipe and valve shall be installed at the lowest point of every air receiver to provide for the removal of accumulated oil and water. Adequate automatic traps may be installed in addition to drain valves. The drain valve on the air receiver shall be opened and the receiver completely drained frequently and at such intervals as to prevent the accumulation of excessive amounts of liquid in the receiver. [§1926.306(b)(2)]

(3) *Gages and valves.*[§1926.306(b)(3)]

(i) *Every air receiver* shall be equipped with an indicating pressure gage (so located as to be readily visible) and with one or more spring-loaded safety valves. The total relieving capacity of such safety valves shall be such as to prevent pressure in the receiver from exceeding the maximum allowable working pressure of the receiver by more than 10 percent. [§1926.306(b)(3)(i)]

(ii) *No valve of any type shall be placed* between the air receiver and its safety valve or valves.[§1926.306(b)(3)(ii)]

(iii) *Safety appliances,* such as safety valves, indicating devices and controlling devices, shall be constructed, located, and installed so that they cannot be readily rendered inoperative by any means, including the elements.[§1926.306(b)(3)(iii)]

(iv) *All safety valves shall be tested* frequently and at regular intervals to determine whether they are in good operating condition.[§1926.306(b)(3)(iv)]

[58 FR 35176, June 30, 1993]

§1926.307
Mechanical power-transmission apparatus

(a) General requirements.[§1926.307(a)]

(1) *This section covers all types* and shapes of power-transmission belts, except the following when operating at two hundred and fifty (250) feet per minute or less:[§1926.307(a)(1)]

(i) *Flat belts 1 inch (2.54 cm) or less in width,*[§1926.307(a)(1)(i)]

(ii) *flat belts 2 inches (5.08 cm)* or less in width which are free from metal lacings or fasteners,[§1926.307(a)(1)(ii)]

(iii) *round belts 1/2 inch (1.27 cm) or less in diameter; and* [§1926.307(a)(1)(iii)]

(iv) *single strand V-belts,* the width of which is thirteen thirty-seconds (13/32) inch or less.[§1926.307(a)(1)(iv)]

(2) *Vertical and inclined belts* (paragraphs (e) (3) and (4) of this section) if not more than 2 1/2 inches (6.35 cm) wide and running at a speed of less than one thousand (1,000) feet per minute, and if free from metal lacings or fastenings may be guarded with a nip-point belt and pulley guard.[§1926.307(a)(2)]

(3) *For the Textile Industry,* because of the presence of excessive deposits of lint, which constitute a serious fire hazard, the sides and face sections only of nip-point belt and pulley guards are required, provided the guard shall extend at least 6 inches (15.24 cm) beyond the rim of the pulley on the in-running and off-running sides of the belt and at least 2 inches (5.08 cm) away from the rim and face of the pulley in all other directions.[§1926.307(a)(3)]

(4) *This section covers the principal* features with which power transmission safeguards shall comply.[§1926.307(a)(4)]

(b) Prime-mover guards.[§1926.307(b)]

(1) *Flywheels.* Flywheels located so that any part is 7 feet (2.128 m) or less above floor or platform shall be guarded in accordance with the requirements of this subparagraph: [§1926.307(b)(1)]

(i) *With an enclosure of sheet,* perforated, or expanded metal, or woven wire;[§1926.307(b)(1)(i)]

(ii) *With guard rails placed* not less than 15 inches (38.1 cm) nor more than 20 inches (50.8 cm) from rim. When flywheel extends into pit or is within 12 inches (30.48 cm) of floor, a standard toeboard shall also be provided;[§1926.307(b)(1)(ii)]

(iii) *When the upper rim of flywheel* protrudes through a working floor, it shall be entirely enclosed or surrounded by a guardrail and toeboard.[§1926.307(b)(1)(iii)]

(iv) *For flywheels with smooth rims* 5 feet (1.52 m) or less in diameter, where the preceding methods cannot be applied, the following may be used: A disk attached to the flywheel in such manner as to cover the spokes of the wheel on the exposed side and present a smooth surface and edge, at the same time providing means for periodic inspection. An open space, not exceeding 4 inches (10.16 cm) in width, may be left between the outside edge of the disk and the rim of the wheel if desired, to facilitate turning the wheel over. Where a disk is used, the keys or other dangerous projections not covered by disk shall be cut off or covered. This subdivision does not apply to flywheels with solid web centers. [§1926.307(b)(1)(iv)]

(v) *Adjustable guard to be used* for starting engine or for running adjustment may be provided at the flywheel of gas or oil engines. A slot opening for jack bar will be permitted. [§1926.307(b)(1)(v)]

(vi) *Wherever flywheels are above* working areas, guards shall be installed having sufficient strength to hold the weight of the flywheel in the event of a shaft or wheel mounting failure. [§1926.307(b)(1)(vi)]

(2) *Cranks and connecting rods.* Cranks and connecting rods, when exposed to contact, shall be guarded in accordance with paragraphs (m) and (n) of this section, or by a guardrail as described in paragraph (o)(5) of this section.[§1926.307(b)(2)]

(3) *Tail rods or extension* piston rods. Tail rods or extension piston rods shall be guarded in accordance with paragraphs (m) and (o) of this section, or by a guardrail on sides and end, with a clearance of not less than 15 (38.1 cm) nor more than 20 inches (50.8 cm) when rod is fully extended.[§1926.307(b)(3)]

(c) Shafting.[§1926.307(c)]

(1) *Installation.*[§1926.307(c)(1)]

(i) *Each continuous line of shafting* shall be secured in position against excessive endwise movement.[§1926.307(c)(1)(i)]

(ii) *Inclined and vertical shafts,* particularly inclined idler shafts, shall be securely held in position against endwise thrust. [§1926.307(c)(1)(ii)]

(2) *Guarding horizontal shafting.*[§1926.307(c)(2)]

(i) *All exposed parts of horizontal* shafting 7 feet (2.128 m) or less from floor or working platform, excepting runways used exclusively for oiling, or running adjustments, shall be protected by a stationary casing enclosing shafting completely or by a trough enclosing sides and top or sides and bottom of shafting as location requires.[§1926.307(c)(2)(i)]

(ii) *Shafting under bench* machines shall be enclosed by a stationary casing, or by a trough at sides and top or sides and bottom, as location requires. The sides of the trough shall come within at least 6 inches (15.24 cm) of the underside of table, or if shafting is located near floor within 6 inches (15.24 cm) of floor. In every case the sides of trough shall extend at least 2 inches (5.08 cm) beyond the shafting or protuberance.[§1926.307(c)(2)(ii)]

(3) *Guarding vertical and inclined* shafting. Vertical and inclined shafting 7 feet (2.128 m) or less from floor or working platform, excepting maintenance runways, shall be enclosed with a stationary casing in accordance with requirements of paragraphs (m) and (o) of this section.[§1926.307(c)(3)]

(4) *Projecting shaft ends.*[§1926.307(c)(4)]

(i) *Projecting shaft ends* shall present a smooth edge and end and shall not project more than one-half the diameter of the shaft unless guarded by nonrotating caps or safety sleeves. [§1926.307(c)(4)(i)]

(ii) *Unused keyways shall be filled up or covered.* [§1926.307(c)(4)(ii)]

(5) *Power-transmission apparatus located* in basements. All mechanical power transmission apparatus located in basements, towers, and rooms used exclusively for power transmission equipment shall be guarded in accordance with this section, except that the requirements for safeguarding belts, pulleys, and shafting need not be complied with when the following requirements are met:[§1926.307(c)(5)]

(i) *The basement,* tower, or room occupied by transmission equipment is locked against unauthorized entrance. [§1926.307(c)(5)(i)]

(ii) *The vertical clearance in passageways* between the floor and power transmission beams, ceiling, or any other objects, is not less than 5 ft. 6 in. (1.672 m).[§1926.307(c)(5)(ii)]

(iii) *The intensity of illumination conforms* to the requirements of ANSI A11.1-1965 (R-1970).[§1926.307(c)(5)(iii)]

(iv) *[Reserved]*[§1926.307(c)(5)(iv)]

(v) *The route followed by the oiler* is protected in such manner as to prevent accident.[§1926.307(c)(5)(v)]

(d) Pulleys.[§1926.307(d)]

(1) *Guarding.* Pulleys, any parts of which are 7 feet (2.128 m) or less from the floor or working platform, shall be guarded in accordance with the standards specified in paragraphs (m) and (o) of this section. Pulleys serving as balance wheels (e.g., punch presses) on which the point of contact between belt and pulley is more than 6 ft. 6 in. (1.976 m) from the floor or platform may be guarded with a disk covering the spokes. [§1926.307(d)(1)]

(2) *Location of pulleys.*[§1926.307(d)(2)]

(i) *Unless the distance to the nearest* fixed pulley, clutch, or hanger exceeds the width of the belt used, a guide shall be provided to prevent the belt from leaving the pulley on the side where insufficient clearance exists.[§1926.307(d)(2)(i)]

(ii) *[Reserved]*[§1926.307(d)(2)(ii)]

(3) *Broken pulleys.* Pulleys with cracks, or pieces broken out of rims, shall not be used.[§1926.307(d)(3)]

(4) *Pulley speeds.* Pulleys intended to operate at rim speed in excess of manufacturers normal recommendations shall be specially designed and carefully balanced for the speed at which they are to operate.[§1926.307(d)(4)]

(e) Belt, rope, and chain drives.[§1926.307(e)]

(1) *Horizontal belts and ropes.*[§1926.307(e)(1)]

(i) *Where both runs of horizontal* belts are 7 feet (2.128 m) or less from the floor level, the guard shall extend to at least 15 inches (38.1 cm) above the belt or to a standard height except that where both runs of a horizontal belt are 42 inches (106.68 cm) or less from the floor, the belt shall be fully enclosed.[§1926.307(e)(1)(i)]

(ii) *In powerplants or power-development rooms,* a guardrail may be used in lieu of the guard required by paragraph (e)(1)(i) of this section.[§1926.307(e)(1)(ii)]

(2) *Overhead horizontal belts.*[§1926.307(e)(2)]

(i) *Overhead horizontal belts,* with lower parts 7 feet (2.128 m) or less from the floor or platform, shall be guarded on sides and bottom in accordance with paragraph (o)(3) of this section.[§1926.307(e)(2)(i)]

(ii) *Horizontal overhead belts* more than 7 feet (2.128 m) above floor or platform shall be guarded for their entire length under the following conditions:[§1926.307(e)(2)(ii)]

[a] If located over passageways or work places and traveling 1,800 feet or more per minute.[§1926.307(e)(2)(ii)[a]]

[b] If center to center distance between pulleys is 10 feet (3.04 m) or more.[§1926.307(e)(2)(ii)[b]]

[c] If belt is 8 inches (20.32 cm) or more in width. [§1926.307(e)(2)(ii)[c]]

(iii) *Where the upper and lower runs* of horizontal belts are so located that passage of persons between them would be possible, the passage shall be either:[§1926.307(e)(2)(iii)]

[a] Completely barred by a guardrail or other barrier in accordance with paragraphs (m) and (o) of this section; or [§1926.307(e)(2)(iii)[a]]

[b] Where passage is regarded as necessary, there shall be a platform over the lower run guarded on either side by a railing completely filled in with wire mesh or other filler, or by a solid barrier. The upper run shall be so guarded as to prevent contact therewith either by the worker or by objects carried by him. In powerplants only the lower run of the belt need be guarded.[§1926.307(e)(2)(iii)[b]]

(iv) *Overhead chain and link* belt drives are governed by the same rules as overhead horizontal belts and shall be guarded in the same manner as belts.[§1926.307(e)(2)(iv)]

(3) *Vertical and inclined belts.*[§1926.307(e)(3)]

(i) *Vertical and inclined belts* shall be enclosed by a guard conforming to standards in paragraphs (m) and (o) of this section.[§1926.307(e)(3)(i)]

(ii) *All guards for inclined belts* shall be arranged in such a manner that a minimum clearance of 7 feet (2.128 m) is maintained between belt and floor at any point outside of guard. [§1926.307(e)(3)(ii)]

(4) *Vertical belts.* Vertical belts running over a lower pulley more than 7 feet (2.128 m) above floor or platform shall be guarded at the bottom in the same manner as horizontal overhead belts, if conditions are as stated in paragraphs (e)(2)(ii) (a) and (c) of this section.[§1926.307(e)(4)]

(5) *Cone-pulley belts.*[§1926.307(e)(5)]

(i) *The cone belt and pulley* shall be equipped with a belt shifter so constructed as to adequately guard the nip point of the belt and pulley. If the frame of the belt shifter does not adequately guard the nip point of the belt and pulley, the nip point shall be further protected by means of a vertical guard placed in front of the pulley and extending at least to the top of the largest step of the cone.[§1926.307(e)(5)(i)]

(ii) *If the belt is of the endless* type or laced with rawhide laces, and a belt shifter is not desired, the belt will be considered guarded if the nip point of the belt and pulley is protected by a nip point guard located in front of the cone extending at least to the top of the largest step of the cone, and formed to show the contour of the cone in order to give the nip point of the belt and pulley the maximum protection.[§1926.307(e)(5)(ii)]

(iii) *If the cone is located* less than 3 feet (0.912 m) from the floor or working platform, the cone pulley and belt shall be guarded to a height of 3 feet (0.912 m) regardless of whether the belt is endless or laced with rawhide.[§1926.307(e)(5)(iii)]

(6) *Belt tighteners.*[§1926.307(e)(6)]

(i) *Suspended counterbalanced tighteners* and all parts thereof shall be of substantial construction and securely fastened; the bearings shall be securely capped. Means must be provided to prevent tightener from falling, in case the belt breaks.[§1926.307(e)(6)(i)]

(ii) *Where suspended counterweights are used* and not guarded by location, they shall be so encased as to prevent accident. [§1926.307(e)(6)(ii)]

(f) **Gears, sprockets, and chains.**[§1926.307(f)]

(1) *Gears.* Gears shall be guarded in accordance with one of the following methods:[§1926.307(f)(1)]

(i) *By a complete enclosure; or*[§1926.307(f)(1)(i)]

(ii) *By a standard guard as described* in paragraph (o) of this section, at least 7 feet (2.128 m) high extending 6 inches (15.24 cm) above the mesh point of the gears; or [§1926.307(f)(1)(ii)]

(iii) *By a band guard covering* the face of gear and having flanges extended inward beyond the root of the teeth on the exposed side or sides. Where any portion of the train of gears guarded by a band guard is less than 6 feet (1.824 m) from the floor a disk guard or a complete enclosure to the height of 6 feet (1.824 m) shall be required.[§1926.307(f)(1)(iii)]

(2) *Hand-operated gears.* Paragraph (f)(1) of this section does not apply to hand-operated gears used only to adjust machine parts and which do not continue to move after hand power is removed. However, the guarding of these gears is highly recommended.[§1926.307(f)(2)]

(3) *Sprockets and chains.* All sprocket wheels and chains shall be enclosed unless they are more than 7 feet (2.128 m) above the floor or platform. Where the drive extends over other machine or working areas, protection against falling shall be provided. This subparagraph does not apply to manually operated sprockets.[§1926.307(f)(3)]

(4) *Openings for oiling.* When frequent oiling must be done, openings with hinged or sliding self-closing covers shall be provided. All points not readily accessible shall have oil feed tubes if lubricant is to be added while machinery is in motion. [§1926.307(f)(4)]

(g) **Guarding friction drives.** The driving point of all friction drives when exposed to contact shall be guarded, all arm or spoke friction drives and all web friction drives with holes in the web shall be entirely enclosed, and all projecting belts on friction drives where exposed to contact shall be guarded.[§1926.307(g)]

(h) **Keys, setscrews, and other projections.**[§1926.307(h)]

(1) *All projecting keys,* setscrews, and other projections in revolving parts shall be removed or made flush or guarded by metal cover. This subparagraph does not apply to keys or setscrews within gear or sprocket casings or other enclosures, nor to keys, setscrews, or oilcups in hubs of pulleys less than 20 inches (50.8 cm) in diameter where they are within the plane of the rim of the pulley.[§1926.307(h)(1)]

(2) *It is recommended,* however, that no projecting setscrews or oilcups be used in any revolving pulley or part of machinery. [§1926.307(h)(2)]

(i) **Collars and couplings.**[§1926.307(i)]

(1) *Collars.* All revolving collars, including split collars, shall be cylindrical, and screws or bolts used in collars shall not project beyond the largest periphery of the collar.[§1926.307(i)(1)]

(2) *Couplings.* Shaft couplings shall be so constructed as to present no hazard from bolts, nuts, setscrews, or revolving surfaces. Bolts, nuts, and setscrews will, however, be permitted where they are covered with safety sleeves or where they are used parallel with the shafting and are countersunk or else do not extend beyond the flange of the coupling.[§1926.307(i)(2)]

(j) **Bearings and facilities for oiling.** All drip cups and pans shall be securely fastened.[§1926.307(j)]

(k) **Guarding of clutches,** cutoff couplings, and clutch pulleys. [§1926.307(k)]

(1) *Guards.* Clutches, cutoff couplings, or clutch pulleys having projecting parts, where such clutches are located 7 feet (2.128 m) or less above the floor or working platform, shall be enclosed by a stationary guard constructed in accordance with this section. A "U" type guard is permissible.[§1926.307(k)(1)]

(2) *Engine rooms.* In engine rooms a guardrail, preferably with toeboard, may be used instead of the guard required by paragraph (k)(1) of this section, provided such a room is occupied only by engine room attendants.[§1926.307(k)(2)]

(l) **Belt shifters,** clutches, shippers, poles, perches, and fasteners. [§1926.307(l)]

(1) *Belt shifters.*[§1926.307(l)(1)]

(i) *Tight and loose pulleys* on all new installations made on or after August 31, 1971, shall be equipped with a permanent belt shifter provided with mechanical means to prevent belt from creeping from loose to tight pulley. It is recommended that old installations be changed to conform to this rule. [§1926.307(l)(1)(i)]

(ii) *Belt shifter and clutch* handles shall be rounded and be located as far as possible from danger of accidental contact, but within easy reach of the operator. Where belt shifters are not directly located over a machine or bench, the handles shall be cut off 6 ft. 6 in. (1.976 m) above floor level. [§1926.307(l)(1)(ii)]

(2) *Belt shippers and shipper* poles. The use of belt poles as substitutes for mechanical shifters is not recommended. [§1926.307(l)(2)]

(3) *Belt perches.* Where loose pulleys or idlers are not practicable, belt perches in form of brackets, rollers, etc., shall be used to keep idle belts away from the shafts. [§1926.307(l)(3)]

(4) *Belt fasteners.* Belts which of necessity must be shifted by hand and belts within 7 feet (2.128 m) of the floor or working platform which are not guarded in accordance with this section shall not be fastened with metal in any case, nor with any other fastening which by construction or wear will constitute an accident hazard. [§1926.307(l)(4)]

(m) Standard guards — general requirements. [§1926.307(m)]

(1) *Materials.* [§1926.307(m)(1)]

(i) *Standard conditions shall be secured* by the use of the following materials. Expanded metal, perforated or solid sheet metal, wire mesh on a frame of angle iron, or iron pipe securely fastened to floor or to frame of machine. [§1926.307(m)(1)(i)]

(ii) *All metal should be free from burrs* and sharp edges. [§1926.307(m)(1)(ii)]

(2) *Methods of manufacture.* [§1926.307(m)(2)]

(i) *Expanded metal,* sheet or perforated metal, and wire mesh shall be securely fastened to frame. [§1926.307(m)(2)(i)]

(n) [Reserved] [§1926.307(n)]

(o) Approved materials. [§1926.307(o)]

(1) *Minimum requirements.* The materials and dimensions specified in this paragraph shall apply to all guards, except horizontal overhead belts, rope, cable, or chain guards more than 7 feet (2.128 m) above floor, or platform. [§1926.307(o)(1)]

(i) *[Reserved]* [§1926.307(o)(1)(i)]

[a] All guards shall be rigidly braced every 3 feet (0.912 m) or fractional part of their height to some fixed part of machinery or building structure. Where guard is exposed to contact with moving equipment additional strength may be necessary. [§1926.307(o)(1)(i)[a]]

(2) *Wood guards.* [§1926.307(o)(2)]

(i) *Wood guards may be used* in the woodworking and chemical industries, in industries where the presence of fumes or where manufacturing conditions would cause the rapid deterioration of metal guards; also in construction work and in locations outdoors where extreme cold or extreme heat make metal guards and railings undesirable. In all other industries, wood guards shall not be used. [§1926.307(o)(2)(i)]

(3) *Guards for horizontal overhead belts.* [§1926.307(o)(3)]

(i) *Guards for horizontal overhead* belts shall run the entire length of the belt and follow the line of the pulley to the ceiling or be carried to the nearest wall, thus enclosing the belt effectively. Where belts are so located as to make it impracticable to carry the guard to wall or ceiling, construction of guard shall be such as to enclose completely the top and bottom runs of belt and the face of pulleys. [§1926.307(o)(3)(i)]

(ii) *[Reserved]* [§1926.307(o)(3)(ii)]

(iii) *Suitable reinforcement shall be provided* for the ceiling rafters or overhead floor beams, where such is necessary, to sustain safely the weight and stress likely to be imposed by the guard. The interior surface of all guards, by which is meant the surface of the guard with which a belt will come in contact, shall be smooth and free from all projections of any character, except where construction demands it; protruding shallow roundhead rivets may be used. Overhead belt guards shall be at least one-quarter wider than belt which they protect, except that this clearance need not in any case exceed 6 inches (15.24 cm) on each side. Overhead rope drive and block and roller-chain-drive guards shall be not less than 6 inches (15.24 cm) wider than the drive on each side. In overhead silent chain-drive guards where the chain is held from lateral displacement on the sprockets, the side clearances required on drives of 20 inch (50.8 cm) centers or under shall be not less than 1/4 inch (0.635 cm) from the nearest moving chain part, and on drives of over 20 inch (50.8 cm) centers a minimum of 1/2 inch (1.27 cm) from the nearest moving chain part. [§1926.307(o)(3)(iii)]

(4) *Guards for horizontal overhead* rope and chain drives. Overhead- rope and chain-drive guard construction shall conform to the rules for overhead-belt guard. [§1926.307(o)(4)]

(5) *Guardrails and toeboards.* [§1926.307(o)(5)]

(i) *Guardrail shall be 42 inches* (106.68 cm) in height, with midrail between top rail and floor. [§1926.307(o)(5)(i)]

(ii) *Posts shall be not more than* 8 feet (2.432 m) apart; they are to be permanent and substantial, smooth, and free from protruding nails, bolts, and splinters. If made of pipe, the post shall be 1 1/4 inches (3.175 cm) inside diameter, or larger. If made of metal shapes or bars, their section shall be equal in strength to that of 1 1/2 (3.81 cm) by 1 1/2 (3.81 cm) by 3/16 inch angle iron. If made of wood, the posts shall be two by four (2 x 4) inches or larger. The upper rail shall be two by four (2 x 4) inches, or two one by four (1 x 4) strips, one at the top and one at the side of posts. The midrail may be one by four (1 x 4) inches or more. Where panels are fitted with expanded metal or wire mesh the middle rails may be omitted. Where guard is exposed to contact with moving equipment, additional strength may be necessary. [§1926.307(o)(5)(ii)]

(iii) *Toeboards shall be 4 inches* (10.16 cm) or more in height, of wood, metal, or of metal grill not exceeding 1 inch (2.54 cm) mesh. [§1926.307(o)(5)(iii)]

(p) Care of equipment. [§1926.307(p)]

(1) *General.* All power-transmission equipment shall be inspected at intervals not exceeding 60 days and be kept in good working condition at all times. [§1926.307(p)(1)]

(2) *Shafting.* [§1926.307(p)(2)]

(i) *Shafting shall be kept in alignment,* free from rust and excess oil or grease. [§1926.307(p)(2)(i)]

(ii) *Where explosives,* explosive dusts, flammable vapors or flammable liquids exist, the hazard of static sparks from shafting shall be carefully considered. [§1926.307(p)(2)(ii)]

(3) *Bearings.* Bearings shall be kept in alignment and properly adjusted. [§1926.307(p)(3)]

(4) *Hangers.* Hangers shall be inspected to make certain that all supporting bolts and screws are tight and that supports of hanger boxes are adjusted properly. [§1926.307(p)(4)]

(5) *Pulleys.* [§1926.307(p)(5)]

(i) *Pulleys shall be kept in proper* alignment to prevent belts from running off. [§1926.307(p)(5)(i)]

(6) *Care of belts.* [§1926.307(p)(6)]

(i) *[Reserved]* [§1926.307(p)(6)(i)]

(ii) *Inspection shall be made of belts,* lacings, and fasteners and such equipment kept in good repair. [§1926.307(p)(6)(ii)]

(7) *Lubrication.* The regular oilers shall wear tight-fitting clothing. Machinery shall be oiled when not in motion, wherever possible. [§1926.307(p)(7)]

[58 FR 35176, June 30, 1993, as amended at 69 FR 31882, June 8, 2004]

Authority: Sections 4, 6, and 8 of the Occupational Safety and Health Act of 1970 (29 U.S.C. 653, 655, 657); Secretary of Labor's Order No. 12-71 (36 FR 8754), 8-76 (41 FR 25059), 9-83 (48 FR 35736), 1-90 (55 FR 9033), or 5-2002 (67 FR 65008), as applicable; and 29 CFR part 1911. Section 1926.307 also issued under 5 U.S.C. 553.

Notes

Subpart J - Welding and Cutting

§1926.350
Gas welding and cutting

(a) **Transporting, moving, and storing compressed gas cylinders.**[§1926.350(a)]

(1) *Valve protection caps* shall be in place and secured. [§1926.350(a)(1)]

(2) *When cylinders are hoisted,* they shall be secured on a cradle, slingboard, or pallet. They shall not be hoisted or transported by means of magnets or choker slings.[§1926.350(a)(2)]

(3) *Cylinders shall be moved by tilting* and rolling them on their bottom edges. They shall not be intentionally dropped, struck, or permitted to strike each other violently.[§1926.350(a)(3)]

(4) *When cylinders are transported* by powered vehicles, they shall be secured in a vertical position.[§1926.350(a)(4)]

(5) *Valve protection caps* shall not be used for lifting cylinders from one vertical position to another. Bars shall not be used under valves or valve protection caps to pry cylinders loose when frozen. Warm, not boiling, water shall be used to thaw cylinders loose.[§1926.350(a)(5)]

(6) *Unless cylinders are firmly secured* on a special carrier intended for this purpose, regulators shall be removed and valve protection caps put in place before cylinders are moved. [§1926.350(a)(6)]

(7) *A suitable cylinder truck,* chain, or other steadying device shall be used to keep cylinders from being knocked over while in use.[§1926.350(a)(7)]

(8) *When work is finished,* when cylinders are empty, or when cylinders are moved at any time, the cylinder valve shall be closed.[§1926.350(a)(8)]

(9) *Compressed gas cylinders* shall be secured in an upright position at all times except, if necessary, for short periods of time while cylinders are actually being hoisted or carried. [§1926.350(a)(9)]

(10) *Oxygen cylinders in storage* shall be separated from fuel-gas cylinders or combustible materials (especially oil or grease), a minimum distance of 20 feet (6.1 m) or by a noncombustible barrier at least 5 feet (1.5 m) high having a fire-resistance rating of at least one-half hour.[§1926.350(a)(10)]

(11) *Inside of buildings, cylinders shall be stored* in a well- protected, well-ventilated, dry location, at least 20 feet (6.1 m) from highly combustible materials such as oil or excelsior. Cylinders should be stored in definitely assigned places away from elevators, stairs, or gangways. Assigned storage places shall be located where cylinders will not be knocked over or damaged by passing or falling objects, or subject to tampering by unauthorized persons. Cylinders shall not be kept in unventilated enclosures such as lockers and cupboards.[§1926.350(a)(11)]

(12) *The in-plant handling,* storage, and utilization of all compressed gases in cylinders, portable tanks, rail tankcars, or motor vehicle cargo tanks shall be in accordance with Compressed Gas Association Pamphlet P-1-1965.[§1926.350(a)(12)]

(b) **Placing cylinders.**[§1926.350(b)]

(1) *Cylinders shall be kept* far enough away from the actual welding or cutting operation so that sparks, hot slag, or flame will not reach them. When this is impractical, fire resistant shields shall be provided.[§1926.350(b)(1)]

(2) *Cylinders shall be placed* where they cannot become part of an electrical circuit. Electrodes shall not be struck against a cylinder to strike an arc.[§1926.350(b)(2)]

(3) *Fuel gas cylinders* shall be placed with valve end up whenever they are in use. They shall not be placed in a location where they would be subject to open flame, hot metal, or other sources of artificial heat.[§1926.350(b)(3)]

(4) *Cylinders containing oxygen* or acetylene or other fuel gas shall not be taken into confined spaces.[§1926.350(b)(4)]

(c) **Treatment of cylinders.**[§1926.350(c)]

(1) *Cylinders,* whether full or empty, shall not be used as rollers or supports.[§1926.350(c)(1)]

(2) *No person other than the gas supplier* shall attempt to mix gases in a cylinder. No one except the owner of the cylinder or person authorized by him, shall refill a cylinder. No one shall use a cylinder's contents for purposes other than those intended by the supplier. All cylinders used shall meet the Department of Transportation requirements published in 49 CFR part 178, subpart C, Specification for Cylinders. [§1926.350(c)(2)]

(3) *No damaged or defective cylinder shall be used.*[§1926.350(c)(3)]

(d) **Use of fuel gas.** The employer shall thoroughly instruct employees in the safe use of fuel gas, as follows:[§1926.350(d)]

(1) *Before a regulator to a cylinder valve is connected,* the valve shall be opened slightly and closed immediately. (This action is generally termed "cracking" and is intended to clear the valve of dust or dirt that might otherwise enter the regulator.) The person cracking the valve shall stand to one side of the outlet, not in front of it. The valve of a fuel gas cylinder shall not be cracked where the gas would reach welding work, sparks, flame, or other possible sources of ignition.[§1926.350(d)(1)]

(2) *The cylinder valve shall always* be opened slowly to prevent damage to the regulator. For quick closing, valves on fuel gas cylinders shall not be opened more than 1 1/2 turns. When a special wrench is required, it shall be left in position on the stem of the valve while the cylinder is in use so that the fuel gas flow can be shut off quickly in case of an emergency. In the case of manifolded or coupled cylinders, at least one such wrench shall always be available for immediate use. Nothing shall be placed on top of a fuel gas cylinder, when in use, which may damage the safety device or interfere with the quick closing of the valve. [§1926.350(d)(2)]

(3) *Fuel gas shall not be used* from cylinders through torches or other devices which are equipped with shutoff valves without reducing the pressure through a suitable regulator attached to the cylinder valve or manifold.[§1926.350(d)(3)]

(4) *Before a regulator is removed* from a cylinder valve, the cylinder valve shall always be closed and the gas released from the regulator.[§1926.350(d)(4)]

(5) *If, when the valve on a fuel gas cylinder is opened,* there is found to be a leak around the valve stem, the valve shall be closed and the gland nut tightened. If this action does not stop the leak, the use of the cylinder shall be discontinued, and it shall be properly tagged and removed from the work area. In the event that fuel gas should leak from the cylinder valve, rather than from the valve stem, and the gas cannot be shut off, the cylinder shall be properly tagged and removed from the work area. If a regulator attached to a cylinder valve will effectively stop a leak through the valve seat, the cylinder need not be removed from the work area.[§1926.350(d)(5)]

(6) *If a leak should develop* at a fuse plug or other safety device, the cylinder shall be removed from the work area.[§1926.350(d)(6)]

(e) **Fuel gas and oxygen manifolds.**[§1926.350(e)]

(1) *Fuel gas and oxygen manifolds* shall bear the name of the substance they contain in letters at least 1- inch high which shall be either painted on the manifold or on a sign permanently attached to it.[§1926.350(e)(1)]

(2) *Fuel gas and oxygen manifolds* shall be placed in safe, well ventilated, and accessible locations. They shall not be located within enclosed spaces.[§1926.350(e)(2)]

(3) *Manifold hose connections,* including both ends of the supply hose that lead to the manifold, shall be such that the hose cannot be interchanged between fuel gas and oxygen manifolds and supply header connections. Adapters shall not be used to permit the interchange of hose. Hose connections shall be kept free of grease and oil.[§1926.350(e)(3)]

(4) *When not in use,* manifold and header hose connections shall be capped.[§1926.350(e)(4)]

(5) *Nothing shall be placed on top* of a manifold, when in use, which will damage the manifold or interfere with the quick closing of the valves.[§1926.350(e)(5)]

(f) **Hose.**[§1926.350(f)]

(1) *Fuel gas hose and oxygen hose* shall be easily distinguishable from each other. The contrast may be made by different colors or by surface characteristics readily distinguishable by the sense of touch. Oxygen and fuel gas hoses shall not be interchangeable. A single hose having more than one gas passage shall not be used.[§1926.350(f)(1)]

(2) *When parallel sections of oxygen and fuel gas hose* are taped together, not more than 4 inches out of 12 inches shall be covered by tape.[§1926.350(f)(2)]

(3) *All hose in use,* carrying acetylene, oxygen, natural or manufactured fuel gas, or any gas or substance which may ignite or enter into combustion, or be in any way harmful to employees, shall be inspected at the beginning of each working shift. Defective hose shall be removed from service.[§1926.350(f)(3)]

(4) *Hose which has been subject to flashback,* or which shows evidence of severe wear or damage, shall be tested to twice the normal pressure to which it is subject, but in no case less than 300 p.s.i. Defective hose, or hose in doubtful condition, shall not be used.[§1926.350(f)(4)]

(5) *Hose couplings shall be of the type* that cannot be unlocked or disconnected by means of a straight pull without rotary motion. [§1926.350(f)(5)]

(6) *Boxes used for the storage* of gas hose shall be ventilated. [§1926.350(f)(6)]

(7) *Hoses, cables, and other equipment* shall be kept clear of passageways, ladders and stairs.[§1926.350(f)(7)]

(g) **Torches.**[§1926.350(g)]

(1) *Clogged torch tip openings* shall be cleaned with suitable cleaning wires, drills, or other devices designed for such purpose.[§1926.350(g)(1)]

(2) *Torches in use shall be inspected* at the beginning of each working shift for leaking shutoff valves, hose couplings, and tip connections. Defective torches shall not be used.[§1926.350(g)(2)]

(3) *Torches shall be lighted by friction lighters* or other approved devices, and not by matches or from hot work.[§1926.350(g)(3)]

(h) **Regulators and gauges.** Oxygen and fuel gas pressure regulators, including their related gauges, shall be in proper working order while in use.[§1926.350(h)]

(i) **Oil and grease hazards.** Oxygen cylinders and fittings shall be kept away from oil or grease. Cylinders, cylinder caps and valves, couplings, regulators, hose, and apparatus shall be kept free from oil or greasy substances and shall not be handled with oily hands or gloves. Oxygen shall not be directed at oily surfaces, greasy clothes, or within a fuel oil or other storage tank or vessel. [§1926.350(i)]

(j) **Additional rules.** For additional details not covered in this subpart, applicable technical portions of American National Standards Institute, Z49.1-1967, Safety in Welding and Cutting, shall apply. [§1926.350(j)]

[44 FR 8577, Feb. 9, 1979; 44 FR 20940, Apr. 6, 1979, as amended at 55 FR 42328, Oct. 18, 1990; 58 FR 35179, June 30, 1993]

§1926.351

Arc welding and cutting

(a) **Manual electrode holders.**[§1926.351(a)]

(1) *Only manual electrode holders* which are specifically designed for arc welding and cutting, and are of a capacity capable of safely handling the maximum rated current required by the electrodes, shall be used.[§1926.351(a)(1)]

(2) *Any current-carrying parts* passing through the portion of the holder which the arc welder or cutter grips in his hand, and the outer surfaces of the jaws of the holder, shall be fully insulated against the maximum voltage encountered to ground. [§1926.351(a)(2)]

(b) **Welding cables and connectors.**[§1926.351(b)]

(1) *All arc welding and cutting cables* shall be of the completely insulated, flexible type, capable of handling the maximum current requirements of the work in progress, taking into account the duty cycle under which the arc welder or cutter is working. [§1926.351(b)(1)]

(2) *Only cable free from repair or splices* for a minimum distance of 10 feet from the cable end to which the electrode holder is connected shall be used, except that cables with standard insulated connectors or with splices whose insulating quality is equal to that of the cable are permitted.[§1926.351(b)(2)]

(3) *When it becomes necessary to connect* or splice lengths of cable one to another, substantial insulated connectors of a capacity at least equivalent to that of the cable shall be used. If connections are effected by means of cable lugs, they shall be securely fastened together to give good electrical contact, and the exposed metal parts of the lugs shall be completely insulated.[§1926.351(b)(3)]

(4) *Cables in need of repair* shall not be used. When a cable, other than the cable lead referred to in paragraph (b)(2) of this section, becomes worn to the extent of exposing bare conductors, the portion thus exposed shall be protected by means of rubber and friction tape or other equivalent insulation.[§1926.351(b)(4)]

(c) **Ground returns and machine grounding.**[§1926.351(c)]

(1) *A ground return cable* shall have a safe current carrying capacity equal to or exceeding the specified maximum output capacity of the arc welding or cutting unit which it services. When a single ground return cable services more than one unit, its safe current-carrying capacity shall equal or exceed the total specified maximum output capacities of all the units which it services.[§1926.351(c)(1)]

(2) *Pipelines containing gases or flammable liquids,* or conduits containing electrical circuits, shall not be used as a ground return. For welding on natural gas pipelines, the technical portions of regulations issued by the Department of Transportation, Office of Pipeline Safety, 49 CFR Part 192, Minimum Federal Safety Standards for Gas Pipelines, shall apply.[§1926.351(c)(2)]

(3) *When a structure or pipeline is employed* as a ground return circuit, it shall be determined that the required electrical contact exists at all joints. The generation of an arc, sparks, or heat at any point shall cause rejection of the structures as a ground circuit.[§1926.351(c)(3)]

(4) *When a structure or pipeline is continuously employed* as a ground return circuit, all joints shall be bonded, and periodic inspections shall be conducted to ensure that no condition of electrolysis or fire hazard exists by virtue of such use. [§1926.351(c)(4)]

(5) *The frames of all arc welding and cutting machines* shall be grounded either through a third wire in the cable containing the circuit conductor or through a separate wire which is grounded at the source of the current. Grounding circuits, other than by means of the structure, shall be checked to ensure that the circuit between the ground and the grounded power conductor has resistance low enough to permit sufficient current to flow to cause the fuse or circuit breaker to interrupt the current. [§1926.351(c)(5)]

(6) *All ground connections shall be inspected* to ensure that they are mechanically strong and electrically adequate for the required current.[§1926.351(c)(6)]

(d) **Operating instructions.** Employers shall instruct employees in the safe means of arc welding and cutting as follows:[§1926.351(d)]

(1) *When electrode holders are to be left unattended,* the electrodes shall be removed and the holders shall be so placed or protected that they cannot make electrical contact with employees or conducting objects.[§1926.351(d)(1)]

(2) *Hot electrode holders* shall not be dipped in water; to do so may expose the arc welder or cutter to electric shock.[§1926.351(d)(2)]

(3) *When the arc welder or cutter* has occasion to leave his work or to stop work for any appreciable length of time, or when the arc welding or cutting machine is to be moved, the power supply switch to the equipment shall be opened.[§1926.351(d)(3)]

(4) *Any faulty or defective equipment* shall be reported to the supervisor.[§1926.351(d)(4)]

(5) *See Sec. 1926.406(c) for additional requirements.*[§1926.351(d)(5)]

(e) **Shielding.** Whenever practicable, all arc welding and cutting operations shall be shielded by noncombustible or flameproof screens which will protect employees and other persons working in the vicinity from the direct rays of the arc.[§1926.351(e)]

[44 FR 8577, Feb. 9, 1979; 44 FR 20940, Apr. 6, 1979, as amended at 51 FR 25318, July 11, 1986]

§1926.352

Fire prevention

(a) **When practical, objects to be welded, cut, or heated** shall be moved to a designated safe location or, if the objects to be welded, cut, or heated cannot be readily moved, all movable fire hazards in the vicinity shall be taken to a safe place, or otherwise protected.[§1926.352(a)]

(b) **If the object to be welded, cut, or heated** cannot be moved and if all the fire hazards cannot be removed, positive means shall be taken to confine the heat, sparks, and slag, and to protect the immovable fire hazards from them.[§1926.352(b)]

(c) **No welding, cutting, or heating** shall be done where the application of flammable paints, or the presence of other flammable compounds, or heavy dust concentrations creates a hazard. [§1926.352(c)]

(d) **Suitable fire extinguishing equipment** shall be immediately available in the work area and shall be maintained in a state of readiness for instant use.[§1926.352(d)]

(e) **When the welding, cutting, or heating operation** is such that normal fire prevention precautions are not sufficient, additional personnel shall be assigned to guard against fire while the actual welding, cutting, or heating operation is being performed, and for a sufficient period of time after completion of the work to ensure that no possibility of fire exists. Such personnel shall be instructed as to the specific anticipated fire hazards and how the firefighting equipment provided is to be used.[§1926.352(e)]

(f) **When welding, cutting, or heating is performed** on walls, floors, and ceilings, since direct penetration of sparks or heat transfer may introduce a fire hazard to an adjacent area, the same precautions shall be taken on the opposite side as are taken on the side on which the welding is being performed.[§1926.352(f)]

(g) **For the elimination of possible fire in enclosed spaces** as a result of gas escaping through leaking or improperly closed torch valves, the gas supply to the torch shall be positively shut off at some point outside the enclosed space whenever the torch is not to be used or whenever the torch is left unattended for a substantial period of time, such as during the lunch period. Overnight and at the change of shifts, the torch and hose shall be removed from the confined space. Open end fuel gas and oxygen hoses shall be immediately removed from enclosed spaces when they are disconnected from the torch or other gas-consuming device. [§1926.352(g)]

(h) **Except when the contents are being removed or transferred,** drums, pails, and other containers which contain or have contained flammable liquids shall be kept closed. Empty containers shall be removed to a safe area apart from hot work operations or open flames.[§1926.352(h)]

(i) **Drums containers, or hollow structures** which have contained toxic or flammable substances shall, before welding, cutting, or heating is undertaken on them, either be filled with water or thoroughly cleaned of such substances and ventilated and tested. For welding, cutting and heating on steel pipelines containing natural gas, the pertinent portions of regulations issued by the Department of Transportation, Office of Pipeline Safety, 49 CFR Part 192, Minimum Federal Safety Standards for Gas Pipelines, shall apply.[§1926.352(i)]

(j) **Before heat is applied to a drum, container, or hollow structure,** a vent or opening shall be provided for the release of any built-up pressure during the application of heat.[§1926.352(j)]

§1926.353

Ventilation and protection in welding, cutting, and heating

(a) **Mechanical ventilation.** For purposes of this section, mechanical ventilation shall meet the following requirements:[§1926.353(a)]

(1) *Mechanical ventilation* shall consist of either general mechanical ventilation systems or local exhaust systems.[§1926.353(a)(1)]

(2) *General mechanical ventilation* shall be of sufficient capacity and so arranged as to produce the number of air changes necessary to maintain welding fumes and smoke within safe limits, as defined in subpart D of this part.[§1926.353(a)(2)]

(3) *Local exhaust ventilation* shall consist of freely movable hoods intended to be placed by the welder or burner as close as practicable to the work. This system shall be of sufficient capacity and so arranged as to remove fumes and smoke at the source and keep the concentration of them in the breathing zone within safe limits as defined in subpart D of this part.[§1926.353(a)(3)]

(4) *Contaminated air exhausted* from a working space shall be discharged into the open air or otherwise clear of the source of intake air.[§1926.353(a)(4)]

(5) *All air replacing that withdrawn* shall be clean and respirable. [§1926.353(a)(5)]

(6) *Oxygen shall not be used for ventilation purposes,* comfort cooling, blowing dust from clothing, or for cleaning the work area.[§1926.353(a)(6)]

(b) **Welding, cutting, and heating in confined spaces.**[§1926.353(b)]

(1) *Except as provided in paragraph (b)(2) of this section,* and paragraph (c)(2) of this section, either general mechanical or local exhaust ventilation meeting the requirements of paragraph (a) of this section shall be provided whenever welding, cutting, or heating is performed in a confined space.[§1926.353(b)(1)]

(2) *When sufficient ventilation cannot be obtained* without blocking the means of access, employees in the confined space shall be protected by air line respirators in accordance with the requirements of subpart E of this part, and an employee on the outside of such a confined space shall be assigned to maintain communication with those working within it and to aid them in an emergency.[§1926.353(b)(2)]

(3) *Lifelines.* Where a welder must enter a confined space through a manhole or other small opening, means shall be provided for quickly removing him in case of emergency. When safety belts and lifelines are used for this purpose they shall be so attached to the welder's body that his body cannot be jammed in a small exit opening. An attendant with a pre-planned rescue procedure shall be stationed outside to observe the welder at all times and be capable of putting rescue operations into effect. [§1926.353(b)(3)]

(c) **Welding,** cutting, or heating of metals of toxic significance. [§1926.353(c)]

(1) *Welding, cutting, or heating in any enclosed spaces* involving the metals specified in this subparagraph shall be performed with either general mechanical or local exhaust ventilation meeting the requirements of paragraph (a) of this section: [§1926.353(c)(1)]

(i) *Zinc-bearing base or filler* metals or metals coated with zinc-bearing materials;[§1926.353(c)(1)(i)]

(ii) *Lead base metals;*[§1926.353(c)(1)(ii)]

(iii) *Cadmium-bearing filler materials;*[§1926.353(c)(1)(iii)]

(iv) *Chromium-bearing metals* or metals coated with chromium-bearing materials.[§1926.353(c)(1)(iv)]

(2) *Welding, cutting, or heating in any enclosed spaces* involving the metals specified in this subparagraph shall be performed with local exhaust ventilation in accordance with the requirements of paragraph (a) of this section, or employees shall be protected by air line respirators in accordance with the requirements of subpart E of this part:[§1926.353(c)(2)]

(i) *Metals containing lead,* other than as an impurity, or metals coated with lead-bearing materials;[§1926.353(c)(2)(i)]

(ii) *Cadmium-bearing or cadmium-coated base metals;* [§1926.353(c)(2)(ii)]

(iii) *Metals coated with mercury-bearing metals;*[§1926.353(c)(2)(iii)]

(iv) *Beryllium-containing base or filler metals.* Because of its high toxicity, work involving beryllium shall be done with both local exhaust ventilation and air line respirators. [§1926.353(c)(2)(iv)]

(3) *Employees performing such operations* in the open air shall be protected by filter-type respirators in accordance with the requirements of subpart E of this part, except that employees performing such operations on beryllium-containing base or filler metals shall be protected by air line respirators in accordance with the requirements of subpart E of this part. [§1926.353(c)(3)]

(4) *Other employees exposed to the same atmosphere* as the welders or burners shall be protected in the same manner as the welder or burner.[§1926.353(c)(4)]

(d) **Inert-gas metal-arc welding.**[§1926.353(d)]

(1) *Since the inert-gas metal-arc welding process* involves the production of ultra-violet radiation of intensities of 5 to 30 times that produced during shielded metal-arc welding, the decomposition of chlorinated solvents by ultraviolet rays, and the liberation of toxic fumes and gases, employees shall not be permitted to engage in, or be exposed to the process until the following special precautions have been taken:[§1926.353(d)(1)]

(i) *The use of chlorinated solvents* shall be kept at least 200 feet, unless shielded, from the exposed arc, and surfaces prepared with chlorinated solvents shall be thoroughly dry before welding is permitted on such surfaces.[§1926.353(d)(1)(i)]

(ii) *Employees in the area not protected from the arc* by screening shall be protected by filter lenses meeting the requirements of subpart E of this part. When two or more welders are exposed to each other's arc, filter lens goggles of a suitable type, meeting the requirements of subpart E of this part, shall be worn under welding helmets. Hand shields to protect the welder against flashes and radiant energy shall be used when either the helmet is lifted or the shield is removed. [§1926.353(d)(1)(ii)]

(iii) *Welders and other employees who are exposed to radiation* shall be suitably protected so that the skin is covered completely to prevent burns and other damage by ultraviolet rays.

Welding helmets and hand shields shall be free of leaks and openings, and free of highly reflective surfaces. [§1926.353(d)(1)(iii)]

(iv) *When inert-gas metal-arc welding* is being performed on stainless steel, the requirements of paragraph (c)(2) of this section shall be met to protect against dangerous concentrations of nitrogen dioxide. [§1926.353(d)(1)(iv)]

(e) **General welding, cutting, and heating.** [§1926.353(e)]

(1) *Welding, cutting, and heating, not involving* conditions or materials described in paragraph (b), (c), or (d) of this section, may normally be done without mechanical ventilation or respiratory protective equipment, but where, because of unusual physical or atmospheric conditions, an unsafe accumulation of contaminants exists, suitable mechanical ventilation or respiratory protective equipment shall be provided. [§1926.353(e)(1)]

(2) *Employees performing any type of welding,* cutting, or heating shall be protected by suitable eye protective equipment in accordance with the requirements of subpart E of this part. [§1926.353(e)(2)]

[44 FR 8577, Feb. 9, 1979; 44 FR 20940, Apr. 6, 1979, as amended at 55 FR 42328, Oct. 18, 1990; 58 FR 35179, June 30, 1993]

§1926.354

Welding, cutting, and heating in way of preservative coatings

(a) **Before welding, cutting, or heating** is commenced on any surface covered by a preservative coating whose flammability is not known, a test shall be made by a competent person to determine its flammability. Preservative coatings shall be considered to be highly flammable when scrapings burn with extreme rapidity. [§1926.354(a)]

(b) **Precautions shall be taken to prevent ignition** of highly flammable hardened preservative coatings. When coatings are determined to be highly flammable, they shall be stripped from the area to be heated to prevent ignition. [§1926.354(b)]

(c) **Protection against toxic preservative coatings:** [§1926.354(c)]

(1) *In enclosed spaces, all surfaces covered* with toxic preservatives shall be stripped of all toxic coatings for a distance of at least 4 inches from the area of heat application, or the employees shall be protected by air line respirators, meeting the requirements of subpart E of this part. [§1926.354(c)(1)]

(2) *In the open air,* employees shall be protected by a respirator, in accordance with requirements of subpart E of this part. [§1926.354(c)(2)]

(d) **The preservative coatings shall be removed** a sufficient distance from the area to be heated to ensure that the temperature of the unstripped metal will not be appreciably raised. Artificial cooling of the metal surrounding the heating area may be used to limit the size of the area required to be cleaned. [§1926.354(d)]

Authority: §107, Contract Work Hours and Safety Standards Act (Construction Safety Act) (40 U.S.C. 333); secs. 4, 6, 8, Occupational Safety and Health Act of 1970 (29 U.S.C. 653, 655, 657); Secretary of Labor's Order No. 12-71 (36 FR 8754), 8-76 (41 FR 25059), or 9-83 (48 FR 35736), as applicable.

Subpart K - Electrical

GENERAL

§1926.400
Introduction

This subpart addresses electrical safety requirements that are necessary for the practical safeguarding of employees involved in construction work and is divided into four major divisions and applicable definitions as follows:

(a) **Installation safety requirements.** Installation safety requirements are contained in §§1926.402 through 1926.408. Included in this category are electric equipment and installations used to provide electric power and light on jobsites.[§1926.400(a)]

(b) **Safety-related work practices.** Safety-related work practices are contained in §§1926.416 and 1926.417. In addition to covering the hazards arising from the use of electricity at jobsites, these regulations also cover the hazards arising from the accidental contact, direct or indirect, by employees with all energized lines, above or below ground, passing through or near the jobsite.[§1926.400(b)]

(c) **Safety-related maintenance and environmental considerations.** Safety-related maintenance and environmental considerations are contained in §§1926.431 and 1926.432.[§1926.400(c)]

(d) **Safety requirements for special equipment.** Safety requirements for special equipment are contained in Sec. 1926.441.[§1926.400(d)]

(e) **Definitions.** Definitions applicable to this subpart are contained in Sec. 1926.449.[§1926.400(e)]

INSTALLATION SAFETY REQUIREMENTS

§1926.401
[Reserved]

§1926.402
Applicability

(a) **Covered.** Sections 1926.402 through 1926.408 contain installation safety requirements for electrical equipment and installations used to provide electric power and light at the jobsite. These sections apply to installations, both temporary and permanent, used on the jobsite; but these sections do not apply to existing permanent installations that were in place before the construction activity commenced.[§1926.402(a)]

Note: If the electrical installation is made in accordance with the National Electrical Code ANSI/NFPA 70-1984, exclusive of Formal Interpretations and Tentative Interim Amendments, it will be deemed to be in compliance with §§1926.403 through 1926.408, except for §§1926.404(b)(1) and 1926.405(a)(2)(ii) (E), (F), (G), and (J).

(b) **Not covered.** Sections 1926.402 through 1926.408 do not cover installations used for the generation, transmission, and distribution of electric energy, including related communication, metering, control, and transformation installations. (However, these regulations do cover portable and vehicle-mounted generators used to provide power for equipment used at the jobsite.) See subpart V of this part for the construction of power distribution and transmission lines.[§1926.402(b)]

§1926.403
General requirements

(a) **Approval.** All electrical conductors and equipment shall be approved.[§1926.403(a)]

(b) **Examination, installation, and use of equipment**[§1926.403(b)]

(1) *Examination.* The employer shall ensure that electrical equipment is free from recognized hazards that are likely to cause death or serious physical harm to employees. Safety of equipment shall be determined on the basis of the following considerations:[§1926.403(b)(1)]

(i) *Suitability for installation and use* in conformity with the provisions of this subpart. Suitability of equipment for an identified purpose may be evidenced by listing, labeling, or certification for that identified purpose.[§1926.403(b)(1)(i)]

(ii) *Mechanical strength and durability,* including, for parts designed to enclose and protect other equipment, the adequacy of the protection thus provided.[§1926.403(b)(1)(ii)]

(iii) *Electrical insulation.*[§1926.403(b)(1)(iii)]

(iv) *Heating effects under conditions of use.*[§1926.403(b)(1)(iv)]

(v) *Arcing effects.*[§1926.403(b)(1)(v)]

(vi) *Classification by type,* size, voltage, current capacity, specific use.[§1926.403(b)(1)(vi)]

(vii) *Other factors which contribute* to the practical safeguarding of employees using or likely to come in contact with the equipment.[§1926.403(b)(1)(vii)]

(2) *Installation and use.* Listed, labeled, or certified equipment shall be installed and used in accordance with instructions included in the listing, labeling, or certification.[§1926.403(b)(2)]

(c) **Interrupting rating.** Equipment intended to break current shall have an interrupting rating at system voltage sufficient for the current that must be interrupted.[§1926.403(c)]

(d) **Mounting and cooling of equipment**[§1926.403(d)]

(1) *Mounting.* Electric equipment shall be firmly secured to the surface on which it is mounted. Wooden plugs driven into holes in masonry, concrete, plaster, or similar materials shall not be used.[§1926.403(d)(1)]

(2) *Cooling.* Electrical equipment which depends upon the natural circulation of air and convection principles for cooling of exposed surfaces shall be installed so that room air flow over such surfaces is not prevented by walls or by adjacent installed equipment. For equipment designed for floor mounting, clearance between top surfaces and adjacent surfaces shall be provided to dissipate rising warm air. Electrical equipment provided with ventilating openings shall be installed so that walls or other obstructions do not prevent the free circulation of air through the equipment.[§1926.403(d)(2)]

(e) **Splices.** Conductors shall be spliced or joined with splicing devices designed for the use or by brazing, welding, or soldering with a fusible metal or alloy. Soldered splices shall first be so spliced or joined as to be mechanically and electrically secure without solder and then soldered. All splices and joints and the free ends of conductors shall be covered with an insulation equivalent to that of the conductors or with an insulating device designed for the purpose.[§1926.403(e)]

(f) **Arcing parts.** Parts of electric equipment which in ordinary operation produce arcs, sparks, flames, or molten metal shall be enclosed or separated and isolated from all combustible material.[§1926.403(f)]

(g) **Marking.** Electrical equipment shall not be used unless the manufacturer's name, trademark, or other descriptive marking by which the organization responsible for the product may be identified is placed on the equipment and unless other markings are provided giving voltage, current, wattage, or other ratings as necessary. The marking shall be of sufficient durability to withstand the environment involved.[§1926.403(g)]

(h) **Identification of disconnecting means and circuits.** Each disconnecting means required by this subpart for motors and appliances shall be legibly marked to indicate its purpose, unless located and arranged so the purpose is evident. Each service, feeder, and branch circuit, at its disconnecting means or overcurrent device, shall be legibly marked to indicate its purpose, unless located and arranged so the purpose is evident. These markings shall be of sufficient durability to withstand the environment involved.[§1926.403(h)]

(i) **600 Volts,** nominal, or less. This paragraph applies to equipment operating at 600 volts, nominal, or less.[§1926.403(i)]

(1) *Working space about electric equipment.* Sufficient access and working space shall be provided and maintained about all electric equipment to permit ready and safe operation and maintenance of such equipment.[§1926.403(i)(1)]

(i) *Working clearances.* Except as required or permitted elsewhere in this subpart, the dimension of the working space in the direction of access to live parts operating at 600 volts or less and likely to require examination, adjustment, servicing, or maintenance while alive shall not be less than indicated in Table K-1. In addition to the dimensions shown in Table K-1, workspace shall not be less than 30 inches (762 mm) wide in front of the electric equipment. Distances shall be measured from the live parts if they are exposed, or from the enclosure front or opening if the live parts are enclosed. Walls constructed of concrete, brick, or tile are considered to be grounded. Working space is not required in back of assemblies such as dead-front switchboards or motor control centers where there are no renewable or adjustable parts such as fuses or switches on the back and where all connections are accessible from locations other than the back.[§1926.403(i)(1)(i)]

Table K-1 - Working Clearances

Nominal voltage to ground	Minimum clear distance for conditions[1]		
	(a)	(b)	(c)
	Feet[2]	Feet[2]	Feet[2]
0-150	3	3	3
151-600	3	3 1/2	4

[1] Conditions (a), (b), and (c) are as follows:

(a) Exposed live parts on one side and no live or grounded parts on the other side of the working space, or exposed live parts on both sides effectively guarded by insulating material. Insulated wire or insulated busbars operating at not over 300 volts are not considered live parts.

(b) Exposed live parts on one side and grounded parts on the other side.

(c) Exposed live parts on both sides of the workspace [not guarded as provided in Condition (a)] with the operator between.

[2] Note: For International System of Units (SI): one foot=0.3048m.

(ii) *Clear spaces.* Working space required by this subpart shall not be used for storage. When normally enclosed live parts are exposed for inspection or servicing, the working space, if in a passageway or general open space, shall be guarded. [§1926.403(i)(1)(ii)]

(iii) *Access and entrance to working space.* At least one entrance shall be provided to give access to the working space about electric equipment. [§1926.403(i)(1)(iii)]

(iv) *Front working space.* Where there are live parts normally exposed on the front of switchboards or motor control centers, the working space in front of such equipment shall not be less than 3 feet (914 mm). [§1926.403(i)(1)(iv)]

(v) *Headroom.* The minimum headroom of working spaces about service equipment, switchboards, panelboards, or motor control centers shall be 6 feet 3 inches (1.91 m). [§1926.403(i)(1)(v)]

(2) *Guarding of live parts.* [§1926.403(i)(2)]

(i) *Except as required or permitted* elsewhere in this subpart, live parts of electric equipment operating at 50 volts or more shall be guarded against accidental contact by cabinets or other forms of enclosures, or by any of the following means: [§1926.403(i)(2)(i)]

[A] By location in a room, vault, or similar enclosure that is accessible only to qualified persons. [§1926.403(i)(2)(i)[A]]

[B] By partitions or screens so arranged that only qualified persons will have access to the space within reach of the live parts. Any openings in such partitions or screens shall be so sized and located that persons are not likely to come into accidental contact with the live parts or to bring conducting objects into contact with them. [§1926.403(i)(2)(i)[B]]

[C] By location on a balcony, gallery, or platform so elevated and arranged as to exclude unqualified persons. [§1926.403(i)(2)(i)[C]]

[D] By elevation of 8 feet (2.44 m) or more above the floor or other working surface and so installed as to exclude unqualified persons. [§1926.403(i)(2)(i)[D]]

(ii) *In locations where electric equipment* would be exposed to physical damage, enclosures or guards shall be so arranged and of such strength as to prevent such damage. [§1926.403(i)(2)(ii)]

(iii) *Entrances to rooms* and other guarded locations containing exposed live parts shall be marked with conspicuous warning signs forbidding unqualified persons to enter. [§1926.403(i)(2)(iii)]

(j) Over 600 volts, nominal [§1926.403(j)]

(1) *General.* Conductors and equipment used on circuits exceeding 600 volts, nominal, shall comply with all applicable provisions of paragraphs (a) through (g) of this section and with the following provisions which supplement or modify those requirements. The provisions of paragraphs (j)(2), (j)(3), and (j)(4) of this section do not apply to equipment on the supply side of the service conductors. [§1926.403(j)(1)]

(2) *Enclosure for electrical installations.* Electrical installations in a vault, room, closet or in an area surrounded by a wall, screen, or fence, access to which is controlled by lock and key or other equivalent means, are considered to be accessible to qualified persons only. A wall, screen, or fence less than 8 feet (2.44 m) in height is not considered adequate to prevent access unless it has other features that provide a degree of isolation equivalent to an 8-foot (2.44-m) fence. The entrances to all buildings, rooms or enclosures containing exposed live parts or exposed conductors operating at over 600 volts, nominal, shall be kept locked or shall be under the observation of a qualified person at all times. [§1926.403(j)(2)]

(i) *Installations accessible to qualified persons only.* Electrical installations having exposed live parts shall be accessible to qualified persons only and shall comply with the applicable provisions of paragraph (j)(3) of this section. [§1926.403(j)(2)(i)]

(ii) *Installations accessible to unqualified persons.* Electrical installations that are open to unqualified persons shall be made with metal-enclosed equipment or shall be enclosed in a vault or in an area, access to which is controlled by a lock. Metal-enclosed switchgear, unit substations, transformers, pull boxes, connection boxes, and other similar associated equipment shall be marked with appropriate caution signs. If equipment is exposed to physical damage from vehicular traffic, guards shall be provided to prevent such damage. Ventilating or similar openings in metal-enclosed equipment shall be designed so that foreign objects inserted through these openings will be deflected from energized parts. [§1926.403(j)(2)(ii)]

(3) *Workspace about equipment.* Sufficient space shall be provided and maintained about electric equipment to permit ready and safe operation and maintenance of such equipment. Where energized parts are exposed, the minimum clear workspace shall not be less than 6 feet 6 inches (1.98 m) high (measured vertically from the floor or platform), or less than 3 feet (914 mm) wide (measured parallel to the equipment). The depth shall be as required in Table K-2. The workspace shall be adequate to permit at least a 90-degree opening of doors or hinged panels. [§1926.403(j)(3)]

(i) *Working space.* The minimum clear working space in front of electric equipment such as switchboards, control panels, switches, circuit breakers, motor controllers, relays, and similar equipment shall not be less than specified in Table K-2 unless otherwise specified in this subpart. Distances shall be measured from the live parts if they are exposed, or from the enclosure front or opening if the live parts are enclosed. However, working space is not required in back of equipment such as deadfront switchboards or control assemblies where there are no renewable or adjustable parts (such as fuses or switches) on the back and where all connections are accessible from locations other than the back. Where rear access is required to work on de- energized parts on the back of enclosed equipment, a minimum working space of 30 inches (762 mm) horizontally shall be provided. [§1926.403(j)(3)(i)]

Table K-2 — Minimum Depth of Clear Working Space in Front of Electric Equipment

Nominal voltage to ground	Conditions[1]		
	(a)	(b)	(c)
	Feet[2]	Feet[2]	Feet[2]
601 to 2,500	3	4	5
2,501 to 9,000	4	5	6
9,001 to 25,000	5	6	9
25,001 to 75 kV	6	8	10
Above 75kV	8	10	12

[1] Conditions (a), (b), and (c) are as follows:

(a) Exposed live parts on one side and no live or grounded parts on the other side of the working space, or exposed live parts on both sides effectively guarded by insulating materials. Insulated wire or insulated busbars operating at not over 300 volts are not considered live parts.

(b) Exposed live parts on one side and grounded parts on the other side. Walls constructed of concrete, brick, or tile are considered to be grounded surfaces.

(c) Exposed live parts on both sides of the workspace [not guarded as provided in Condition (a)] with the operator between.

[2] Note: For SI units: one foot=0.3048 m.

(ii) *Lighting outlets and points of control.* The lighting outlets shall be so arranged that persons changing lamps or making repairs on the lighting system will not be endangered by live parts or other equipment. The points of control shall be so located that persons are not likely to come in contact with any live part or moving part of the equipment while turning on the lights. [§1926.403(j)(3)(ii)]

(iii) *Elevation of unguarded live parts.* Unguarded live parts above working space shall be maintained at elevations not less than specified in Table K-3.[§1926.403(j)(3)(iii)]

Table K-3 — Elevation of Unguarded Energized Parts Above Working Space

Nominal voltage between phases	Minimum elevation
601-7,500	8 feet 6 inches.[1]
7,501-35,000	9 feet.
Over 35kV	9 feet+0.37 inches per kV above 35kV.

[1] Note: For SI units: one inch=25.4 mm; one foot=0.3048 m.

(4) *Entrance and access to workspace.* At least one entrance not less than 24 inches (610 mm) wide and 6 feet 6 inches (1.98 m) high shall be provided to give access to the working space about electric equipment. On switchboard and control panels exceeding 48 inches (1.22 m) in width, there shall be one entrance at each end of such board where practicable. Where bare energized parts at any voltage or insulated energized parts above 600 volts are located adjacent to such entrance, they shall be guarded.[§1926.403(j)(4)]

[51 FR 25318, July 11, 1986, as amended at 61 FR 5510, Feb. 13, 1996]

§1926.404
Wiring design and protection

(a) Use and identification of grounded and grounding conductors[§1926.404(a)]

(1) *Identification of conductors.* A conductor used as a grounded conductor shall be identifiable and distinguishable from all other conductors. A conductor used as an equipment grounding conductor shall be identifiable and distinguishable from all other conductors.[§1926.404(a)(1)]

(2) *Polarity of connections.* No grounded conductor shall be attached to any terminal or lead so as to reverse designated polarity.[§1926.404(a)(2)]

(3) *Use of grounding terminals and devices.* A grounding terminal or grounding-type device on a receptacle, cord connector, or attachment plug shall not be used for purposes other than grounding.[§1926.404(a)(3)]

(b) Branch circuits[§1926.404(b)]

(1) *Ground-fault protection*[§1926.404(b)(1)]

(i) *General.* The employer shall use either ground fault circuit interrupters as specified in paragraph (b)(1)(ii) of this section or an assured equipment grounding conductor program as specified in paragraph (b)(1)(iii) of this section to protect employees on construction sites. These requirements are in addition to any other requirements for equipment grounding conductors.[§1926.404(b)(1)(i)]

(ii) *Ground-fault circuit interrupters.* All 120-volt, single-phase, 15- and 20-ampere receptacle outlets on construction sites, which are not a part of the permanent wiring of the building or structure and which are in use by employees, shall have approved ground-fault circuit interrupters for personnel protection. Receptacles on a two-wire, single-phase portable or vehicle-mounted generator rated not more than 5kW, where the circuit conductors of the generator are insulated from the generator frame and all other grounded surfaces, need not be protected with ground-fault circuit interrupters.[§1926.404(b)(1)(ii)]

(iii) *Assured equipment grounding conductor program.* The employer shall establish and implement an assured equipment grounding conductor program on construction sites covering all cord sets, receptacles which are not a part of the building or structure, and equipment connected by cord and plug which are available for use or used by employees. This program shall comply with the following minimum requirements:[§1926.404(b)(1)(iii)]

[A] A written description of the program, including the specific procedures adopted by the employer, shall be available at the jobsite for inspection and copying by the Assistant Secretary and any affected employee.[§1926.404(b)(1)(iii)[A]]

[B] The employer shall designate one or more competent persons (as defined in Sec. 1926.32(f)) to implement the program.[§1926.404(b)(1)(iii)[B]]

[C] Each cord set, attachment cap, plug and receptacle of cord sets, and any equipment connected by cord and plug, except cord sets and receptacles which are fixed and not exposed to damage, shall be visually inspected before each day's use for external defects, such as deformed or missing pins or insulation damage, and for indications of possible internal damage. Equipment found damaged or defective shall not be used until repaired.[§1926.404(b)(1)(iii)[C]]

[D] The following tests shall be performed on all cord sets, receptacles which are not a part of the permanent wiring of the building or structure, and cord- and plug-connected equipment required to be grounded:[§1926.404(b)(1)(iii)[D]]

[1] All equipment grounding conductors shall be tested for continuity and shall be electrically continuous.[§1926.404(b)(1)(iii)[D][1]]

[2] Each receptacle and attachment cap or plug shall be tested for correct attachment of the equipment grounding conductor. The equipment grounding conductor shall be connected to its proper terminal.[§1926.404(b)(1)(iii)[D][2]]

[E] All required tests shall be performed:[§1926.404(b)(1)(iii)[E]]

[1] Before first use;[§1926.404(b)(1)(iii)[E][1]]

[2] Before equipment is returned to service following any repairs;[§1926.404(b)(1)(iii)[E][2]]

[3] Before equipment is used after any incident which can be reasonably suspected to have caused damage (for example, when a cord set is run over); and [§1926.404(b)(1)(iii)[E][3]]

[4] At intervals not to exceed 3 months, except that cord sets and receptacles which are fixed and not exposed to damage shall be tested at intervals not exceeding 6 months.[§1926.404(b)(1)(iii)[E][4]]

[F] The employer shall not make available or permit the use by employees of any equipment which has not met the requirements of this paragraph (b)(1)(iii) of this section.[§1926.404(b)(1)(iii)[F]]

[G] Tests performed as required in this paragraph shall be recorded. This test record shall identify each receptacle, cord set, and cord- and plug-connected equipment that passed the test and shall indicate the last date it was tested or the interval for which it was tested. This record shall be kept by means of logs, color coding, or other effective means and shall be maintained until replaced by a more current record. The record shall be made available on the jobsite for inspection by the Assistant Secretary and any affected employee.[§1926.404(b)(1)(iii)[G]]

(2) *Outlet devices.* Outlet devices shall have an ampere rating not less than the load to be served and shall comply with the following:[§1926.404(b)(2)]

(i) *Single receptacles.* A single receptacle installed on an individual branch circuit shall have an ampere rating of not less than that of the branch circuit.[§1926.404(b)(2)(i)]

(ii) *Two or more receptacles.* Where connected to a branch circuit supplying two or more receptacles or outlets, receptacle ratings shall conform to the values listed in Table K-4.[§1926.404(b)(2)(ii)]

(iii) *Receptacles used for the connection of motors.* The rating of an attachment plug or receptacle used for cord- and plug-connection of a motor to a branch circuit shall not exceed 15 amperes at 125 volts or 10 amperes at 250 volts if individual overload protection is omitted.[§1926.404(b)(2)(iii)]

Table K-4 — Receptacle Ratings for Various Size Circuits

Circuit rating amperes	Receptacle rating amperes
15	Not over 15
20	15 or 20
30	30
40	40 or 50
50	50

(c) Outside conductors and lamps — [§1926.404(c)]

(1) *600 volts, nominal, or less.* Paragraphs (c)(1)(i) through (c)(1)(iv) of this section apply to branch circuit, feeder, and service conductors rated 600 volts, nominal, or less and run outdoors as open conductors.[§1926.404(c)(1)]

(i) *Conductors on poles.* Conductors supported on poles shall provide a horizontal climbing space not less than the following:[§1926.404(c)(1)(i)]

[A] Power conductors below communication conductors — 30 inches (762 mm) .[§1926.404(c)(1)(i)[A]]

[B] Power conductors alone or above communication conductors: 300 volts or less — 24 inches (610 mm); more than 300 volts — 30 inches (762 mm).[§1926.404(c)(1)(i)[B]]

[C] Communication conductors below power conductors: with power conductors 300 volts or less — 24 inches (610 mm); more than 300 volts — 30 inches (762 mm). [§1926.404(c)(1)(i)[C]]

(ii) *Clearance from ground.* Open conductors shall conform to the following minimum clearances:[§1926.404(c)(1)(ii)]

[A] 10 feet (3.05 m) — above finished grade, sidewalks, or from any platform or projection from which they might be reached.[§1926.404(c)(1)(ii)[A]]

[B] 12 feet (3.66 m) — over areas subject to vehicular traffic other than truck traffic.[§1926.404(c)(1)(ii)[B]]

[C] 15 feet (4.57 m) — over areas other than those specified in paragraph (c)(1)(ii)(D) of this section that are subject to truck traffic.[§1926.404(c)(1)(ii)[C]]

[D] 18 feet (5.49 m) — over public streets, alleys, roads, and driveways.[§1926.404(c)(1)(ii)[D]]

(iii) *Clearance from building openings.* Conductors shall have a clearance of at least 3 feet (914 mm) from windows, doors, fire escapes, or similar locations. Conductors run above the top level of a window are considered to be out of reach from that window and, therefore, do not have to be 3 feet (914 mm) away.[§1926.404(c)(1)(iii)]

(iv) *Clearance over roofs.* Conductors above roof space accessible to employees on foot shall have a clearance from the highest point of the roof surface of not less than 8 feet (2.44 m) vertical clearance for insulated conductors, not less than 10 feet (3.05 m) vertical or diagonal clearance for covered conductors, and not less than 15 feet (4.57 m) for bare conductors, except that:[§1926.404(c)(1)(iv)]

[A] Where the roof space is also accessible to vehicular traffic, the vertical clearance shall not be less than 18 feet (5.49 m), or[§1926.404(c)(1)(iv)[A]]

[B] Where the roof space is not normally accessible to employees on foot, fully insulated conductors shall have a vertical or diagonal clearance of not less than 3 feet (914 mm), or[§1926.404(c)(1)(iv)[B]]

[C] Where the voltage between conductors is 300 volts or less and the roof has a slope of not less than 4 inches (102 mm) in 12 inches (305 mm), the clearance from roofs shall be at least 3 feet (914 mm), or[§1926.404(c)(1)(iv)[C]]

[D] Where the voltage between conductors is 300 volts or less and the conductors do not pass over more than 4 feet (1.22 m) of the overhang portion of the roof and they are terminated at a through-the- roof raceway or support, the clearance from roofs shall be at least 18 inches (457 mm). [§1926.404(c)(1)(iv)[D]]

(2) *Location of outdoor lamps.* Lamps for outdoor lighting shall be located below all live conductors, transformers, or other electric equipment, unless such equipment is controlled by a disconnecting means that can be locked in the open position or unless adequate clearances or other safeguards are provided for relamping operations.[§1926.404(c)(2)]

(d) Services — [§1926.404(d)]

(1) *Disconnecting means* — [§1926.404(d)(1)]

(i) *General.* Means shall be provided to disconnect all conductors in a building or other structure from the service-entrance conductors. The disconnecting means shall plainly indicate whether it is in the open or closed position and shall be installed at a readily accessible location nearest the point of entrance of the service-entrance conductors.[§1926.404(d)(1)(i)]

(ii) *Simultaneous opening of poles.* Each service disconnecting means shall simultaneously disconnect all ungrounded conductors.[§1926.404(d)(1)(ii)]

(2) *Services over 600 volts, nominal.* The following additional requirements apply to services over 600 volts, nominal. [§1926.404(d)(2)]

(i) *Guarding.* Service-entrance conductors installed as open wires shall be guarded to make them accessible only to qualified persons.[§1926.404(d)(2)(i)]

(ii) *Warning signs.* Signs warning of high voltage shall be posted where unauthorized employees might come in contact with live parts.[§1926.404(d)(2)(ii)]

(e) Overcurrent protection — [§1926.404(e)]

(1) *600 volts, nominal, or less.* The following requirements apply to overcurrent protection of circuits rated 600 volts, nominal, or less.[§1926.404(e)(1)]

(i) *Protection of conductors and equipment.* Conductors and equipment shall be protected from overcurrent in accordance with their ability to safely conduct current. Conductors shall have sufficient ampacity to carry the load.[§1926.404(e)(1)(i)]

(ii) *Grounded conductors.* Except for motor-running overload protection, overcurrent devices shall not interrupt the continuity of the grounded conductor unless all conductors of the circuit are opened simultaneously.[§1926.404(e)(1)(ii)]

(iii) *Disconnection of fuses and thermal cutouts.* Except for devices provided for current-limiting on the supply side of the service disconnecting means, all cartridge fuses which are accessible to other than qualified persons and all fuses and thermal cutouts on circuits over 150 volts to ground shall be provided with disconnecting means. This disconnecting means shall be installed so that the fuse or thermal cutout can be disconnected from its supply without disrupting service to equipment and circuits unrelated to those protected by the overcurrent device.[§1926.404(e)(1)(iii)]

(iv) *Location in or on premises.* Overcurrent devices shall be readily accessible. Overcurrent devices shall not be located where they could create an employee safety hazard by being exposed to physical damage or located in the vicinity of easily ignitible material.[§1926.404(e)(1)(iv)]

(v) *Arcing or suddenly moving parts.* Fuses and circuit breakers shall be so located or shielded that employees will not be burned or otherwise injured by their operation. [§1926.404(e)(1)(v)]

(vi) *Circuit breakers*[§1926.404(e)(1)(vi)]

[A] Circuit breakers shall clearly indicate whether they are in the open (off) or closed (on) position.[§1926.404(e)(1)(vi)[A]]

[B] Where circuit breaker handles on switchboards are operated vertically rather than horizontally or rotationally, the up position of the handle shall be the closed (on) position. [§1926.404(e)(1)(vi)[B]]

[C] If used as switches in 120-volt, fluorescent lighting circuits, circuit breakers shall be marked "SWD." [§1926.404(e)(1)(vi)[C]]

(2) *Over 600 volts, nominal.* Feeders and branch circuits over 600 volts, nominal, shall have short-circuit protection.[§1926.404(e)(2)]

(f) Grounding. Paragraphs (f)(1) through (f)(11) of this section contain grounding requirements for systems, circuits, and equipment. [§1926.404(f)]

(1) *Systems to be grounded.* The following systems which supply premises wiring shall be grounded:[§1926.404(f)(1)]

(i) *Three-wire DC systems.* All 3-wire DC systems shall have their neutral conductor grounded.[§1926.404(f)(1)(i)]

(ii) *Two-wire DC systems.* Two-wire DC systems operating at over 50 volts through 300 volts between conductors shall be grounded unless they are rectifier-derived from an AC system complying with paragraphs (f)(1)(iii), (f)(1)(iv), and (f)(1)(v) of this section.[§1926.404(f)(1)(ii)]

(iii) *AC circuits, less than 50 volts.* AC circuits of less than 50 volts shall be grounded if they are installed as overhead conductors outside of buildings or if they are supplied by transformers and the transformer primary supply system is ungrounded or exceeds 150 volts to ground.[§1926.404(f)(1)(iii)]

(iv) *AC systems, 50 volts to 1000 volts.* AC systems of 50 volts to 1000 volts shall be grounded under any of the following conditions, unless exempted by paragraph (f)(1)(v) of this section:[§1926.404(f)(1)(iv)]

[A] If the system can be so grounded that the maximum voltage to ground on the ungrounded conductors does not exceed 150 volts;[§1926.404(f)(1)(iv)[A]]

[B] If the system is nominally rated 480Y/277 volt, 3-phase, 4-wire in which the neutral is used as a circuit conductor; [§1926.404(f)(1)(iv)[B]]

[C] If the system is nominally rated 240/120 volt, 3-phase, 4-wire in which the midpoint of one phase is used as a circuit conductor; or[§1926.404(f)(1)(iv)[C]]

[D] If a service conductor is uninsulated.[§1926.404(f)(1)(iv)[D]]

(v) *Exceptions.* AC systems of 50 volts to 1000 volts are not required to be grounded if the system is separately derived and is supplied by a transformer that has a primary voltage rating less than 1000 volts, provided all of the following conditions are met:[§1926.404(f)(1)(v)]

[A] The system is used exclusively for control circuits, [§1926.404(f)(1)(v)[A]]

[B] The conditions of maintenance and supervision assure that only qualified persons will service the installation, [§1926.404(f)(1)(v)[B]]

[C] Continuity of control power is required, and [§1926.404(f)(1)(v)[C]]

[D] Ground detectors are installed on the control system. [§1926.404(f)(1)(v)[D]]

(2) *Separately derived systems.* Where paragraph (f)(1) of this section requires grounding of wiring systems whose power is derived from generator, transformer, or converter windings and has no direct electrical connection, including a solidly connected grounded circuit conductor, to supply conductors originating in another system, paragraph (f)(5) of this section shall also apply.[§1926.404(f)(2)]

(3) *Portable and vehicle-mounted generators*[§1926.404(f)(3)]

(i) *Portable generators.* Under the following conditions, the frame of a portable generator need not be grounded and may serve as the grounding electrode for a system supplied by the generator:[§1926.404(f)(3)(i)]

[A] The generator supplies only equipment mounted on the generator and/or cord- and plug-connected equipment through receptacles mounted on the generator, and [§1926.404(f)(3)(i)[A]]

[B] The noncurrent-carrying metal parts of equipment and the equipment grounding conductor terminals of the receptacles are bonded to the generator frame.[§1926.404(f)(3)(i)[B]]

(ii) *Vehicle-mounted generators.* Under the following conditions the frame of a vehicle may serve as the grounding electrode for a system supplied by a generator located on the vehicle: [§1926.404(f)(3)(ii)]

[A] The frame of the generator is bonded to the vehicle frame, and[§1926.404(f)(3)(ii)[A]]

[B] The generator supplies only equipment located on the vehicle and/or cord- and plug-connected equipment through receptacles mounted on the vehicle or on the generator, and[§1926.404(f)(3)(ii)[B]]

[C] The noncurrent-carrying metal parts of equipment and the equipment grounding conductor terminals of the receptacles are bonded to the generator frame, and [§1926.404(f)(3)(ii)[C]]

[D] The system complies with all other provisions of this section.[§1926.404(f)(3)(ii)[D]]

(iii) *Neutral conductor bonding.* A neutral conductor shall be bonded to the generator frame if the generator is a component of a separately derived system. No other conductor need be bonded to the generator frame.[§1926.404(f)(3)(iii)]

(4) *Conductors to be grounded.* For AC premises wiring systems the identified conductor shall be grounded.[§1926.404(f)(4)]

(5) *Grounding connections*[§1926.404(f)(5)]

(i) *Grounded system.* For a grounded system, a grounding electrode conductor shall be used to connect both the equipment grounding conductor and the grounded circuit conductor to the grounding electrode. Both the equipment grounding conductor and the grounding electrode conductor shall be connected to the grounded circuit conductor on the supply side of the service disconnecting means, or on the supply side of the system disconnecting means or overcurrent devices if the system is separately derived.[§1926.404(f)(5)(i)]

(ii) *Ungrounded systems.* For an ungrounded service-supplied system, the equipment grounding conductor shall be connected to the grounding electrode conductor at the service equipment. For an ungrounded separately derived system, the equipment grounding conductor shall be connected to the grounding electrode conductor at, or ahead of, the system disconnecting means or overcurrent devices.[§1926.404(f)(5)(ii)]

(6) *Grounding path.* The path to ground from circuits, equipment, and enclosures shall be permanent and continuous. [§1926.404(f)(6)]

(7) *Supports, enclosures, and equipment to be grounded* [§1926.404(f)(7)]

(i) *Supports and enclosures for conductors.* Metal cable trays, metal raceways, and metal enclosures for conductors shall be grounded, except that:[§1926.404(f)(7)(i)]

[A] Metal enclosures such as sleeves that are used to protect cable assemblies from physical damage need not be grounded; and[§1926.404(f)(7)(i)[A]]

[B] Metal enclosures for conductors added to existing installations of open wire, knob-and-tube wiring, and nonmetallic-sheathed cable need not be grounded if all of the following conditions are met:[§1926.404(f)(7)(i)[B]]

[1] Runs are less than 25 feet (7.62 m);[§1926.404(f)(7)(i)[B][1]]

[2] Enclosures are free from probable contact with ground, grounded metal, metal laths, or other conductive materials; and[§1926.404(f)(7)(i)[B][2]]

[3] Enclosures are guarded against employee contact. [§1926.404(f)(7)(i)[B][3]]

(ii) *Service equipment enclosures.* Metal enclosures for service equipment shall be grounded.[§1926.404(f)(7)(ii)]

(iii) *Fixed equipment.* Exposed noncurrent-carrying metal parts of fixed equipment which may become energized shall be grounded under any of the following conditions: [§1926.404(f)(7)(iii)]

[A] If within 8 feet (2.44 m) vertically or 5 feet (1.52 m) horizontally of ground or grounded metal objects and subject to employee contact.[§1926.404(f)(7)(iii)[A]]

[B] If located in a wet or damp location and subject to employee contact.[§1926.404(f)(7)(iii)[B]]

[C] If in electrical contact with metal.[§1926.404(f)(7)(iii)[C]]

[D] If in a hazardous (classified) location.[§1926.404(f)(7)(iii)[D]]

[E] If supplied by a metal-clad, metal-sheathed, or grounded metal raceway wiring method.[§1926.404(f)(7)(iii)[E]]

[F] If equipment operates with any terminal at over 150 volts to ground; however, the following need not be grounded: [§1926.404(f)(7)(iii)[F]]

[1] Enclosures for switches or circuit breakers used for other than service equipment and accessible to qualified persons only;[§1926.404(f)(7)(iii)[F][1]]

[2] Metal frames of electrically heated appliances which are permanently and effectively insulated from ground; and[§1926.404(f)(7)(iii)[F][2]]

[3] The cases of distribution apparatus such as transformers and capacitors mounted on wooden poles at a height exceeding 8 feet (2.44 m) above ground or grade level.[§1926.404(f)(7)(iii)[F][3]]

(iv) *Equipment connected by cord and plug.* Under any of the conditions described in paragraphs (f)(7)(iv)(A) through (f)(7)(iv)(C) of this section, exposed noncurrent-carrying metal parts of cord- and plug-connected equipment which may become energized shall be grounded:[§1926.404(f)(7)(iv)]

[A] If in a hazardous (classified) location (see Sec. 1926.407). [§1926.404(f)(7)(iv)[A]]

[B] If operated at over 150 volts to ground, except for guarded motors and metal frames of electrically heated appliances if the appliance frames are permanently and effectively insulated from ground.[§1926.404(f)(7)(iv)[B]]

[C] If the equipment is one of the types listed in paragraphs (f)(7)(iv)(C)(1) through (f)(7)(iv)(C)(5) of this section. However, even though the equipment may be one of these types, it need not be grounded if it is exempted by paragraph (f)(7)(iv)(C)(6).[§1926.404(f)(7)(iv)[C]]

[1] Hand held motor-operated tools;[§1926.404(f)(7)(iv)[C][1]]

[2] Cord- and plug-connected equipment used in damp or wet locations or by employees standing on the ground or on metal floors or working inside of metal tanks or boilers;[§1926.404(f)(7)(iv)[C][2]]

[3] Portable and mobile X-ray and associated equipment; [§1926.404(f)(7)(iv)[C][3]]

[4] Tools likely to be used in wet and/or conductive locations; and[§1926.404(f)(7)(iv)[C][4]]

[5] Portable hand lamps.[§1926.404(f)(7)(iv)[C][5]]

[6] Tools likely to be used in wet and/or conductive locations need not be grounded if supplied through an isolating transformer with an ungrounded secondary of not over 50 volts. Listed or labeled portable tools and appliances protected by a system of double insulation, or its equivalent, need not be grounded. If such a system is employed, the equipment shall be distinctively marked to indicate that the tool or appliance utilizes a system of double insulation.[§1926.404(f)(7)(iv)[C][6]]

(v) *Nonelectrical equipment.* The metal parts of the following nonelectrical equipment shall be grounded: Frames and tracks of electrically operated cranes; frames of nonelectrically driven elevator cars to which electric conductors are attached; hand-operated metal shifting ropes or cables of electric elevators, and metal partitions, grill work, and similar metal enclosures around equipment of over IkV between conductors.[§1926.404(f)(7)(v)]

(8) *Methods of grounding equipment*[§1926.404(f)(8)]

(i) *With circuit conductors.* Noncurrent-carrying metal parts of fixed equipment, if required to be grounded by this subpart, shall be grounded by an equipment grounding conductor which is contained within the same raceway, cable, or cord, or runs with or encloses the circuit conductors. For DC circuits only, the equipment grounding conductor may be run separately from the circuit conductors.[§1926.404(f)(8)(i)]

(ii) *Grounding conductor.* A conductor used for grounding fixed or movable equipment shall have capacity to conduct safely any fault current which may be imposed on it.[§1926.404(f)(8)(ii)]

(iii) *Equipment considered effectively grounded.* Electric equipment is considered to be effectively grounded if it is secured to, and in electrical contact with, a metal rack or structure that is provided for its support and the metal rack or structure is grounded by the method specified for the noncurrent-carrying metal parts of fixed equipment in paragraph (f)(8)(i) of this section. Metal car frames supported by metal hoisting cables attached to or running over metal sheaves or drums of grounded elevator machines are also considered to be effectively grounded.[§1926.404(f)(8)(iii)]

(9) *Bonding.* If bonding conductors are used to assure electrical continuity, they shall have the capacity to conduct any fault current which may be imposed.[§1926.404(f)(9)]

(10) *Made electrodes.* If made electrodes are used, they shall be free from nonconductive coatings, such as paint or enamel; and, if practicable, they shall be embedded below permanent moisture level. A single electrode consisting of a rod, pipe or plate which has a resistance to ground greater than 25 ohms shall be augmented by one additional electrode installed no closer than 6 feet (1.83 m) to the first electrode.[§1926.404(f)(10)]

(11) *Grounding of systems and circuits* of 1000 volts and over (high voltage).[§1926.404(f)(11)]

(i) *General.* If high voltage systems are grounded, they shall comply with all applicable provisions of paragraphs (f)(1) through (f)(10) of this section as supplemented and modified by this paragraph (f)(11).[§1926.404(f)(11)(i)]

(ii) *Grounding of systems supplying portable or mobile equipment.* Systems supplying portable or mobile high voltage equipment, other than substations installed on a temporary basis, shall comply with the following:[§1926.404(f)(11)(ii)]

[A] Portable and mobile high voltage equipment shall be supplied from a system having its neutral grounded through an impedance. If a delta-connected high voltage system is used to supply the equipment, a system neutral shall be derived.[§1926.404(f)(11)(ii)[A]]

[B] Exposed noncurrent-carrying metal parts of portable and mobile equipment shall be connected by an equipment grounding conductor to the point at which the system neutral impedance is grounded.[§1926.404(f)(11)(ii)[B]]

[C] Ground-fault detection and relaying shall be provided to automatically de-energize any high voltage system component which has developed a ground fault. The continuity of the equipment grounding conductor shall be continuously monitored so as to de-energize automatically the high voltage feeder to the portable equipment upon loss of continuity of the equipment grounding conductor.[§1926.404(f)(11)(ii)[C]]

[D] The grounding electrode to which the portable or mobile equipment system neutral impedance is connected shall be isolated from and separated in the ground by at least 20 feet (6.1 m) from any other system or equipment grounding electrode, and there shall be no direct connection between the grounding electrodes, such as buried pipe, fence or like objects.[§1926.404(f)(11)(ii)[D]]

(iii) *Grounding of equipment.* All noncurrent-carrying metal parts of portable equipment and fixed equipment including their associated fences, housings, enclosures, and supporting structures shall be grounded. However, equipment which is guarded by location and isolated from ground need not be grounded. Additionally, pole-mounted distribution apparatus at a height exceeding 8 feet (2.44 m) above ground or grade level need not be grounded.[§1926.404(f)(11)(iii)]

[51 FR 25318, July 11, 1986, as amended at 54 FR 24334, June 7, 1989; 61 FR 5510, Feb. 13, 1996]

§1926.405

Wiring methods, components, and equipment for general use

(a) **Wiring methods.** The provisions of this paragraph do not apply to conductors which form an integral part of equipment such as motors, controllers, motor control centers and like equipment. [§1926.405(a)]

(1) *General requirements.*[§1926.405(a)(1)]

(i) *Electrical continuity of metal raceways and enclosures.* Metal raceways, cable armor, and other metal enclosures for conductors shall be metallically joined together into a continuous electric conductor and shall be so connected to all boxes, fittings, and cabinets as to provide effective electrical continuity.[§1926.405(a)(1)(i)]

(ii) *Wiring in ducts.* No wiring systems of any type shall be installed in ducts used to transport dust, loose stock or flammable vapors. No wiring system of any type shall be installed in any duct used for vapor removal or in any shaft containing only such ducts.[§1926.405(a)(1)(ii)]

(2) *Temporary wiring.*[§1926.405(a)(2)]

(i) *Scope.* The provisions of paragraph (a)(2) of this section apply to temporary electrical power and lighting wiring methods which may be of a class less than would be required for a permanent installation. Except as specifically modified in paragraph (a)(2) of this section, all other requirements of this subpart for permanent wiring shall apply to temporary wiring installations. Temporary wiring shall be removed immediately upon completion of construction or the purpose for which the wiring was installed.[§1926.405(a)(2)(i)]

(ii) *General requirements for temporary wiring*[§1926.405(a)(2)(ii)]

[A] Feeders shall originate in a distribution center. The conductors shall be run as multiconductor cord or cable assemblies or within raceways; or, where not subject to physical damage, they may be run as open conductors on insulators not more than 10 feet (3.05 m) apart. [§1926.405(a)(2)(ii)[A]]

[B] Branch circuits shall originate in a power outlet or panelboard. Conductors shall be run as multiconductor cord or cable assemblies or open conductors, or shall be run in raceways. All conductors shall be protected by overcurrent devices at their ampacity. Runs of open conductors shall be located where the conductors will not be subject to physical damage, and the conductors shall be fastened at intervals not exceeding 10 feet (3.05 m). No branch-circuit conductors shall be laid on the floor. Each branch circuit that supplies receptacles or fixed equipment shall contain a separate equipment grounding conductor if the branch circuit is run as open conductors. [§1926.405(a)(2)(ii)[B]]

[C] Receptacles shall be of the grounding type. Unless installed in a complete metallic raceway, each branch circuit shall contain a separate equipment grounding conductor, and all receptacles shall be electrically connected to the grounding conductor. Receptacles for uses other than temporary lighting shall not be installed on branch circuits which supply temporary lighting. Receptacles shall not be connected to the same ungrounded conductor of multiwire circuits which supply temporary lighting. [§1926.405(a)(2)(ii)[C]]

[D] Disconnecting switches or plug connectors shall be installed to permit the disconnection of all ungrounded conductors of each temporary circuit.[§1926.405(a)(2)(ii)[D]]

[E] All lamps for general illumination shall be protected from accidental contact or breakage. Metal-case sockets shall be grounded.[§1926.405(a)(2)(ii)[E]]

[F] Temporary lights shall not be suspended by their electric cords unless cords and lights are designed for this means of suspension.[§1926.405(a)(2)(ii)[F]]

[G] Portable electric lighting used in wet and/or other conductive locations, as for example, drums, tanks, and vessels, shall be operated at 12 volts or less. However, 120-volt lights may be used if protected by a ground-fault circuit interrupter.[§1926.405(a)(2)(ii)[G]]

[H] A box shall be used wherever a change is made to a raceway system or a cable system which is metal clad or metal sheathed.[§1926.405(a)(2)(ii)[H]]

[I] Flexible cords and cables shall be protected from damage. Sharp corners and projections shall be avoided. Flexible cords and cables may pass through doorways or other pinch points, if protection is provided to avoid damage.[§1926.405(a)(2)(ii)[I]]

[J] Extension cord sets used with portable electric tools and appliances shall be of three-wire type and shall be designed for hard or extra-hard usage. Flexible cords used with temporary and portable lights shall be designed for hard or extra-hard usage.[§1926.405(a)(2)(ii)[J]]

Note: The National Electrical Code, ANSI/NFPA 70, in Article 400, Table 400-4, lists various types of flexible cords, some of which are noted as being designed for hard or extra-hard usage. Examples of these types of flexible cords include hard service cord (types S, ST, SO, STO) and junior hard service cord (types SJ, SJO, SJT, SJTO).

(iii) *Guarding.* For temporary wiring over 600 volts, nominal, fencing, barriers, or other effective means shall be provided to prevent access of other than authorized and qualified personnel.[§1926.405(a)(2)(iii)]

(b) Cabinets, boxes, and fittings[§1926.405(b)]

(1) *Conductors entering boxes,* cabinets, or fittings. Conductors entering boxes, cabinets, or fittings shall be protected from abrasion, and openings through which conductors enter shall be effectively closed. Unused openings in cabinets, boxes, and fittings shall also be effectively closed.[§1926.405(b)(1)]

(2) *Covers and canopies.* All pull boxes, junction boxes, and fittings shall be provided with covers. If metal covers are used, they shall be grounded. In energized installations each outlet box shall have a cover, faceplate, or fixture canopy. Covers of outlet boxes having holes through which flexible cord pendants pass shall be provided with bushings designed for the purpose or shall have smooth, well-rounded surfaces on which the cords may bear.[§1926.405(b)(2)]

(3) *Pull and junction boxes* for systems over 600 volts, nominal. In addition to other requirements in this section for pull and junction boxes, the following shall apply to these boxes for systems over 600 volts, nominal:[§1926.405(b)(3)]

(i) *Complete enclosure.* Boxes shall provide a complete enclosure for the contained conductors or cables.[§1926.405(b)(3)(i)]

(ii) *Covers.* Boxes shall be closed by covers securely fastened in place. Underground box covers that weigh over 100 pounds (43.6 kg) meet this requirement. Covers for boxes shall be permanently marked "HIGH VOLTAGE." The marking shall be on the outside of the box cover and shall be readily visible and legible.[§1926.405(b)(3)(ii)]

(c) Knife switches. Single-throw knife switches shall be so connected that the blades are dead when the switch is in the open position. Single-throw knife switches shall be so placed that gravity will not tend to close them. Single-throw knife switches approved for use in the inverted position shall be provided with a locking device that will ensure that the blades remain in the open position when so set. Double-throw knife switches may be mounted so that the throw will be either vertical or horizontal. However, if the throw is vertical, a locking device shall be provided to ensure that the blades remain in the open position when so set. [§1926.405(c)]

(d) Switchboards and panelboards. Switchboards that have any exposed live parts shall be located in permanently dry locations and accessible only to qualified persons. Panelboards shall be mounted in cabinets, cutout boxes, or enclosures designed for the purpose and shall be dead front. However, panelboards other than the dead front externally- operable type are permitted where accessible only to qualified persons. Exposed blades of knife switches shall be dead when open.[§1926.405(d)]

(e) Enclosures for damp or wet locations[§1926.405(e)]

(1) *Cabinets, fittings, and boxes.* Cabinets, cutout boxes, fittings, boxes, and panelboard enclosures in damp or wet locations shall be installed so as to prevent moisture or water from entering and accumulating within the enclosures. In wet locations the enclosures shall be weatherproof.[§1926.405(e)(1)]

(2) *Switches and circuit breakers.* Switches, circuit breakers, and switchboards installed in wet locations shall be enclosed in weatherproof enclosures.[§1926.405(e)(2)]

(f) Conductors for general wiring. All conductors used for general wiring shall be insulated unless otherwise permitted in this subpart. The conductor insulation shall be of a type that is suitable for the voltage, operating temperature, and location of use. Insulated conductors shall be distinguishable by appropriate color or other means as being grounded conductors, ungrounded conductors, or equipment grounding conductors.[§1926.405(f)]

(g) Flexible cords and cables.[§1926.405(g)]

(1) *Use of flexible cords and cables.*[§1926.405(g)(1)]

(i) *Permitted uses.* Flexible cords and cables shall be suitable for conditions of use and location. Flexible cords and cables shall be used only for:[§1926.405(g)(1)(i)]

[A] Pendants;[§1926.405(g)(1)(i)[A]]

[B] Wiring of fixtures;[§1926.405(g)(1)(i)[B]]

[C] Connection of portable lamps or appliances; [§1926.405(g)(1)(i)[C]]

[D] Elevator cables;[§1926.405(g)(1)(i)[D]]

[E] Wiring of cranes and hoists;[§1926.405(g)(1)(i)[E]]

[F] Connection of stationary equipment to facilitate their frequent interchange;[§1926.405(g)(1)(i)[F]]

[G] Prevention of the transmission of noise or vibration; or [§1926.405(g)(1)(i)[G]]

[H] Appliances where the fastening means and mechanical connections are designed to permit removal for maintenance and repair.[§1926.405(g)(1)(i)[H]]

(ii) *Attachment plugs for cords.* If used as permitted in paragraphs (g)(1)(i)(C), (g)(1)(i)(F), or (g)(1)(i)(H) of this section, the flexible cord shall be equipped with an attachment plug and shall be energized from a receptacle outlet. [§1926.405(g)(1)(ii)]

(iii) *Prohibited uses.* Unless necessary for a use permitted in paragraph (g)(1)(i) of this section, flexible cords and cables shall not be used:[§1926.405(g)(1)(iii)]

[A] As a substitute for the fixed wiring of a structure; [§1926.405(g)(1)(iii)[A]]

[B] Where run through holes in walls, ceilings, or floors; [§1926.405(g)(1)(iii)[B]]

[C] Where run through doorways, windows, or similar openings, except as permitted in paragraph (a)(2)(ii)(1) of this section;[§1926.405(g)(1)(iii)[C]]

[D] Where attached to building surfaces; or[§1926.405(g)(1)(iii)[D]]

[E] Where concealed behind building walls, ceilings, or floors. [§1926.405(g)(1)(iii)[E]]

(2) *Identification,* splices, and terminations.[§1926.405(g)(2)]

(i) *Identification.* A conductor of a flexible cord or cable that is used as a grounded conductor or an equipment grounding conductor shall be distinguishable from other conductors. [§1926.405(g)(2)(i)]

(ii) *Marking.* Type SJ, SJO, SJT, SJTO, S, SO, ST, and STO cords shall not be used unless durably marked on the surface with the type designation, size, and number of conductors.[§1926.405(g)(2)(ii)]

(iii) *Splices.* Flexible cords shall be used only in continuous lengths without splice or tap. Hard service flexible cords No. 12 or larger may be repaired if spliced so that the splice retains the insulation, outer sheath properties, and usage characteristics of the cord being spliced.[§1926.405(g)(2)(iii)]

(iv) *Strain relief.* Flexible cords shall be connected to devices and fittings so that strain relief is provided which will prevent pull from being directly transmitted to joints or terminal screws.[§1926.405(g)(2)(iv)]

(v) *Cords passing through holes.* Flexible cords and cables shall be protected by bushings or fittings where passing through holes in covers, outlet boxes, or similar enclosures. [§1926.405(g)(2)(v)]

(h) Portable cables over 600 volts, nominal. Multiconductor portable cable for use in supplying power to portable or mobile equipment at over 600 volts, nominal, shall consist of No. 8 or larger conductors employing flexible stranding. Cables operated at over 2000 volts shall be shielded for the purpose of confining the voltage stresses to the insulation. Grounding conductors shall be provided. Connectors for these cables shall be of a locking type with provisions to prevent their opening or closing while energized. Strain relief shall be provided at connections and terminations. Portable cables shall not be operated with splices unless the splices are of the permanent molded, vulcanized, or other equivalent type. Termination enclosures shall be marked with a high voltage hazard warning, and terminations shall be accessible only to authorized and qualified personnel.[§1926.405(h)]

(i) **Fixture wires.**[§1926.405(i)]

(1) *General.* Fixture wires shall be suitable for the voltage, temperature, and location of use. A fixture wire which is used as a grounded conductor shall be identified.[§1926.405(i)(1)]

(2) *Uses permitted.* Fixture wires may be used:[§1926.405(i)(2)]

(i) *For installation in lighting,* fixtures and in similar equipment where enclosed or protected and not subject to bending or twisting in use; or[§1926.405(i)(2)(i)]

(ii) *For connecting lighting fixtures* to the branch-circuit conductors supplying the fixtures.[§1926.405(i)(2)(ii)]

(3) *Uses not permitted.* Fixture wires shall not be used as branch-circuit conductors except as permitted for Class 1 power-limited circuits.[§1926.405(i)(3)]

(j) **Equipment for general use**[§1926.405(j)]

(1) *Lighting fixtures, lampholders, lamps, and receptacles* [§1926.405(j)(1)]

(i) *Live parts.* Fixtures, lampholders, lamps, rosettes, and receptacles shall have no live parts normally exposed to employee contact. However, rosettes and cleat-type lampholders and receptacles located at least 8 feet (2.44 m) above the floor may have exposed parts.[§1926.405(j)(1)(i)]

(ii) *Support.* Fixtures, lampholders, rosettes, and receptacles shall be securely supported. A fixture that weighs more than 6 pounds (2.72 kg) or exceeds 16 inches (406 mm) in any dimension shall not be supported by the screw shell of a lampholder.[§1926.405(j)(1)(ii)]

(iii) *Portable lamps.* Portable lamps shall be wired with flexible cord and an attachment plug of the polarized or grounding type. If the portable lamp uses an Edison-based lampholder, the grounded conductor shall be identified and attached to the screw shell and the identified blade of the attachment plug. In addition, portable handlamps shall comply with the following:[§1926.405(j)(1)(iii)]

[A] Metal shell, paperlined lampholders shall not be used; [§1926.405(j)(1)(iii)[A]]

[B] Handlamps shall be equipped with a handle of molded composition or other insulating material;[§1926.405(j)(1)(iii)[B]]

[C] Handlamps shall be equipped with a substantial guard attached to the lampholder or handle;[§1926.405(j)(1)(iii)[C]]

[D] Metallic guards shall be grounded by the means of an equipment grounding conductor run within the power supply cord.[§1926.405(j)(1)(iii)[D]]

(iv) *Lampholders.* Lampholders of the screw-shell type shall be installed for use as lampholders only. Lampholders installed in wet or damp locations shall be of the weatherproof type. [§1926.405(j)(1)(iv)]

(v) *Fixtures.* Fixtures installed in wet or damp locations shall be identified for the purpose and shall be installed so that water cannot enter or accumulate in wireways, lampholders, or other electrical parts.[§1926.405(j)(1)(v)]

(2) *Receptacles, cord connectors, and attachment plugs (caps)* [§1926.405(j)(2)]

(i) *Configuration.* Receptacles, cord connectors, and attachment plugs shall be constructed so that no receptacle or cord connector will accept an attachment plug with a different voltage or current rating than that for which the device is intended. However, a 20-ampere T-slot receptacle or cord connector may accept a 15-ampere attachment plug of the same voltage rating. Receptacles connected to circuits having different voltages, frequencies, or types of current (ac or dc) on the same premises shall be of such design that the attachment plugs used on these circuits are not interchangeable.[§1926.405(j)(2)(i)]

(ii) *Damp and wet locations.* A receptacle installed in a wet or damp location shall be designed for the location. [§1926.405(j)(2)(ii)]

(3) *Appliances.*[§1926.405(j)(3)]

(i) *Live parts.* Appliances, other than those in which the current-carrying parts at high temperatures are necessarily exposed, shall have no live parts normally exposed to employee contact.[§1926.405(j)(3)(i)]

(ii) *Disconnecting means.* A means shall be provided to disconnect each appliance.[§1926.405(j)(3)(ii)]

(iii) *Rating.* Each appliance shall be marked with its rating in volts and amperes or volts and watts.[§1926.405(j)(3)(iii)]

(4) *Motors.* This paragraph applies to motors, motor circuits, and controllers.[§1926.405(j)(4)]

(i) *In sight from.* If specified that one piece of equipment shall be "in sight from" another piece of equipment, one shall be visible and not more than 50 feet (15.2 m) from the other. [§1926.405(j)(4)(i)]

(ii) *Disconnecting means*[§1926.405(j)(4)(ii)]

[A] A disconnecting means shall be located in sight from the controller location. The controller disconnecting means for motor branch circuits over 600 volts, nominal, may be out of sight of the controller, if the controller is marked with a warning label giving the location and identification of the disconnecting means which is to be locked in the open position.[§1926.405(j)(4)(ii)[A]]

[B] The disconnecting means shall disconnect the motor and the controller from all ungrounded supply conductors and shall be so designed that no pole can be operated independently.[§1926.405(j)(4)(ii)[B]]

[C] If a motor and the driven machinery are not in sight from the controller location, the installation shall comply with one of the following conditions:[§1926.405(j)(4)(ii)[C]]

[1] The controller disconnecting means shall be capable of being locked in the open position.[§1926.405(j)(4)(ii)[C][1]]

[2] A manually operable switch that will disconnect the motor from its source of supply shall be placed in sight from the motor location.[§1926.405(j)(4)(ii)[C][2]]

[D] The disconnecting means shall plainly indicate whether it is in the open (off) or closed (on) position. [§1926.405(j)(4)(ii)[D]]

[E] The disconnecting means shall be readily accessible. If more than one disconnect is provided for the same equipment, only one need be readily accessible. [§1926.405(j)(4)(ii)[E]]

[F] An individual disconnecting means shall be provided for each motor, but a single disconnecting means may be used for a group of motors under any one of the following conditions:[§1926.405(j)(4)(ii)[F]]

[1] If a number of motors drive special parts of a single machine or piece of apparatus, such as a metal or woodworking machine, crane, or hoist; [§1926.405(j)(4)(ii)[F][1]]

[2] If a group of motors is under the protection of one set of branch-circuit protective devices; or [§1926.405(j)(4)(ii)[F][2]]

[3] If a group of motors is in a single room in sight from the location of the disconnecting means. [§1926.405(j)(4)(ii)[F][3]]

(iii) *Motor overload, short-circuit, and ground-fault protection.* Motors, motor-control apparatus, and motor branch-circuit conductors shall be protected against overheating due to motor overloads or failure to start, and against short-circuits or ground faults. These provisions do not require overload protection that will stop a motor where a shutdown is likely to introduce additional or increased hazards, as in the case of fire pumps, or where continued operation of a motor is necessary for a safe shutdown of equipment or process and motor overload sensing devices are connected to a supervised alarm.[§1926.405(j)(4)(iii)]

(iv) *Protection of live parts — all voltages*[§1926.405(j)(4)(iv)]

[A] Stationary motors having commutators, collectors, and brush rigging located inside of motor end brackets and not conductively connected to supply circuits operating at more than 150 volts to ground need not have such parts guarded. Exposed live parts of motors and controllers operating at 50 volts or more between terminals shall be guarded against accidental contact by any of the following:[§1926.405(j)(4)(iv)[A]]

[1] By installation in a room or enclosure that is accessible only to qualified persons;[§1926.405(j)(4)(iv)[A][1]]

[2] By installation on a balcony, gallery, or platform, so elevated and arranged as to exclude unqualified persons; or[§1926.405(j)(4)(iv)[A][2]]

[3] By elevation 8 feet (2.44 m) or more above the floor. [§1926.405(j)(4)(iv)[A][3]]

[B] Where live parts of motors or controllers operating at over 150 volts to ground are guarded against accidental contact only by location, and where adjustment or other attendance may be necessary during the operation of the apparatus, insulating mats or platforms shall be provided so that the attendant cannot readily touch live parts unless standing on the mats or platforms.[§1926.405(j)(4)(iv)[B]]

(5) *Transformers*[§1926.405(j)(5)]

(i) *Application.* The following paragraphs cover the installation of all transformers, except:[§1926.405(j)(5)(i)]

[A] Current transformers;[§1926.405(j)(5)(i)[A]]

[B] Dry-type transformers installed as a component part of other apparatus;[§1926.405(j)(5)(i)[B]]

[C] Transformers which are an integral part of an X-ray, high frequency, or electrostatic-coating apparatus;[§1926.405(j)(5)(i)[C]]

[D] Transformers used with Class 2 and Class 3 circuits, sign and outline lighting, electric discharge lighting, and power-limited fire- protective signaling circuits.[§1926.405(j)(5)(i)[D]]

(ii) *Operating voltage.* The operating voltage of exposed live parts of transformer installations shall be indicated by warning signs or visible markings on the equipment or structure.[§1926.405(j)(5)(ii)]

(iii) *Transformers over 35 kV.* Dry-type, high fire point liquid-insulated, and askarel-insulated transformers installed indoors and rated over 35 kV shall be in a vault.[§1926.405(j)(5)(iii)]

(iv) *Oil-insulated transformers.* If they present a fire hazard to employees, oil-insulated transformers installed indoors shall be in a vault.[§1926.405(j)(5)(iv)]

(v) *Fire protection.* Combustible material, combustible buildings and parts of buildings, fire escapes, and door and window openings shall be safeguarded from fires which may originate in oil-insulated transformers attached to or adjacent to a building or combustible material.[§1926.405(j)(5)(v)]

(vi) *Transformer vaults.* Transformer vaults shall be constructed so as to contain fire and combustible liquids within the vault and to prevent unauthorized access. Locks and latches shall be so arranged that a vault door can be readily opened from the inside.[§1926.405(j)(5)(vi)]

(vii) *Pipes and ducts.* Any pipe or duct system foreign to the vault installation shall not enter or pass through a transformer vault.[§1926.405(j)(5)(vii)]

(viii) *Material storage.* Materials shall not be stored in transformer vaults.[§1926.405(j)(5)(viii)]

(6) *Capacitors.*[§1926.405(j)(6)]

(i) *Drainage of stored charge.* All capacitors, except surge capacitors or capacitors included as a component part of other apparatus, shall be provided with an automatic means of draining the stored charge and maintaining the discharged state after the capacitor is disconnected from its source of supply.[§1926.405(j)(6)(i)]

(ii) *Over 600 volts.* Capacitors rated over 600 volts, nominal, shall comply with the following additional requirements:[§1926.405(j)(6)(ii)]

[A] Isolating or disconnecting switches (with no interrupting rating) shall be interlocked with the load interrupting device or shall be provided with prominently displayed caution signs to prevent switching load current.[§1926.405(j)(6)(ii)[A]]

[B] For series capacitors the proper switching shall be assured by use of at least one of the following:[§1926.405(j)(6)(ii)[B]]

[1] Mechanically sequenced isolating and bypass switches,[§1926.405(j)(6)(ii)[B][1]]

[2] Interlocks, or[§1926.405(j)(6)(ii)[B][2]]

[3] Switching procedure prominently displayed at the switching location.[§1926.405(j)(6)(ii)[B][3]]

[51 FR 25318, July 11, 1986, as amended at 61 FR 5510, Feb. 13, 1996]

§1926.406

Specific purpose equipment and installations

(a) Cranes and hoists. This paragraph applies to the installation of electric equipment and wiring used in connection with cranes, monorail hoists, hoists, and all runways.[§1926.406(a)]

(1) *Disconnecting means.*[§1926.406(a)(1)]

(i) *Runway conductor disconnecting means.* A readily accessible disconnecting means shall be provided between the runway contact conductors and the power supply.[§1926.406(a)(1)(i)]

(ii) *Disconnecting means for cranes and monorail hoists.* A disconnecting means, capable of being locked in the open position, shall be provided in the leads from the runway contact conductors or other power supply on any crane or monorail hoist.[§1926.406(a)(1)(ii)]

[A] If this additional disconnecting means is not readily accessible from the crane or monorail hoist operating station, means shall be provided at the operating station to open the power circuit to all motors of the crane or monorail hoist.[§1926.406(a)(1)(ii)[A]]

[B] The additional disconnect may be omitted if a monorail hoist or hand-propelled crane bridge installation meets all of the following:[§1926.406(a)(1)(ii)[B]]

[1] The unit is floor controlled;[§1926.406(a)(1)(ii)[B][1]]

[2] The unit is within view of the power supply disconnecting means; and[§1926.406(a)(1)(ii)[B][2]]

[3] No fixed work platform has been provided for servicing the unit.[§1926.406(a)(1)(ii)[B][3]]

(2) *Control.* A limit switch or other device shall be provided to prevent the load block from passing the safe upper limit of travel of any hoisting mechanism.[§1926.406(a)(2)]

(3) *Clearance.* The dimension of the working space in the direction of access to live parts which may require examination, adjustment, servicing, or maintenance while alive shall be a minimum of 2 feet 6 inches (762 mm). Where controls are enclosed in cabinets, the door(s) shall open at least 90 degrees or be removable, or the installation shall provide equivalent access.[§1926.406(a)(3)]

(4) *Grounding.* All exposed metal parts of cranes, monorail hoists, hoists and accessories including pendant controls shall be metallically joined together into a continuous electrical conductor so that the entire crane or hoist will be grounded in accordance with Sec. 1926.404(f). Moving parts, other than removable accessories or attachments, having metal-to-metal bearing surfaces shall be considered to be electrically connected to each other through the bearing surfaces for grounding purposes. The trolley frame and bridge frame shall be considered as electrically grounded through the bridge and trolley wheels and its respective tracks unless conditions such as paint or other insulating materials prevent reliable metal-to-metal contact. In this case a separate bonding conductor shall be provided.[§1926.406(a)(4)]

(b) Elevators, escalators, and moving walks[§1926.406(b)]

(1) *Disconnecting means.* Elevators, escalators, and moving walks shall have a single means for disconnecting all ungrounded main power supply conductors for each unit.[§1926.406(b)(1)]

(2) *Control panels.* If control panels are not located in the same space as the drive machine, they shall be located in cabinets with doors or panels capable of being locked closed.[§1926.406(b)(2)]

(c) Electric welders — disconnecting means.[§1926.406(c)]

(1) *Motor-generator, AC transformer,* and DC rectifier arc welders. A disconnecting means shall be provided in the supply circuit for each motor-generator arc welder, and for each AC transformer and DC rectifier arc welder which is not equipped with a disconnect mounted as an integral part of the welder.[§1926.406(c)(1)]

(2) *Resistance welders.* A switch or circuit breaker shall be provided by which each resistance welder and its control equipment can be isolated from the supply circuit. The ampere rating of this disconnecting means shall not be less than the supply conductor ampacity.[§1926.406(c)(2)]

(d) X-Ray equipment.[§1926.406(d)]

(1) *Disconnecting means.*[§1926.406(d)(1)]

(i) *General.* A disconnecting means shall be provided in the supply circuit. The disconnecting means shall be operable from a location readily accessible from the X-ray control. For equipment connected to a 120-volt branch circuit of 30 amperes or less, a grounding-type attachment plug cap and receptacle of proper rating may serve as a disconnecting means.[§1926.406(d)(1)(i)]

(ii) *More than one piece of equipment.* If more than one piece of equipment is operated from the same high-voltage circuit, each piece or each group of equipment as a unit shall be provided with a high-voltage switch or equivalent disconnecting means. This disconnecting means shall be constructed, enclosed, or located so as to avoid contact by employees with its live parts.[§1926.406(d)(1)(ii)]

(2) *Control — Radiographic and fluoroscopic types.* Radiographic and fluoroscopic-type equipment shall be effectively

enclosed or shall have interlocks that deenergize the equipment automatically to prevent ready access to live current-carrying parts.[§1926.406(d)(2)]

§1926.407

Hazardous (classified) locations

(a) **Scope.** This section sets forth requirements for electric equipment and wiring in locations which are classified depending on the properties of the flammable vapors, liquids or gases, or combustible dusts or fibers which may be present therein and the likelihood that a flammable or combustible concentration or quantity is present. Each room, section or area shall be considered individually in determining its classification. These hazardous (classified) locations are assigned six designations as follows:[§1926.407(a)]

Class I, Division 1[§1926.407(a)]

Class I, Division 2[§1926.407(a)]

Class II, Division 1[§1926.407(a)]

Class II, Division 2[§1926.407(a)]

Class III, Division 1[§1926.407(a)]

Class III, Division 2[§1926.407(a)]

For definitions of these locations see Sec. 1926.449. All applicable requirements in this subpart apply to all hazardous (classified) locations, unless modified by provisions of this section. [§1926.407(a)]

(b) **Electrical installations.** Equipment, wiring methods, and installations of equipment in hazardous (classified) locations shall be approved as intrinsically safe or approved for the hazardous (classified) location or safe for the hazardous (classified) location. Requirements for each of these options are as follows:[§1926.407(b)]

(1) *Intrinsically safe.* Equipment and associated wiring approved as intrinsically safe is permitted in any hazardous (classified) location included in its listing or labeling.[§1926.407(b)(1)]

(2) *Approved for the hazardous (classified) location.*[§1926.407(b)(2)]

(i) *General.* Equipment shall be approved not only for the class of location but also for the ignitible or combustible properties of the specific gas, vapor, dust, or fiber that will be present. [§1926.407(b)(2)(i)]

Note: NFPA 70, the National Electrical Code, lists or defines hazardous gases, vapors, and dusts by "Groups" characterized by their ignitible or combustible properties.

(ii) *Marking.* Equipment shall not be used unless it is marked to show the class, group, and operating temperature or temperature range, based on operation in a 40-degree C ambient, for which it is approved. The temperature marking shall not exceed the ignition temperature of the specific gas, vapor, or dust to be encountered. However, the following provisions modify this marking requirement for specific equipment: [§1926.407(b)(2)(ii)]

[A] Equipment of the non-heat-producing type (such as junction boxes, conduit, and fitting) and equipment of the heat-producing type having a maximum temperature of not more than 100 degrees C (212 degrees F) need not have a marked operating temperature or temperature range. [§1926.407(b)(2)(ii)[A]]

[B] Fixed lighting fixtures marked for use only in Class I, Division 2 locations need not be marked to indicate the group. [§1926.407(b)(2)(ii)[B]]

[C] Fixed general-purpose equipment in Class I locations, other than lighting fixtures, which is acceptable for use in Class I, Division 2 locations need not be marked with the class, group, division, or operating temperature. [§1926.407(b)(2)(ii)[C]]

[D] Fixed dust-tight equipment, other than lighting fixtures, which is acceptable for use in Class II, Division 2 and Class III locations need not be marked with the class, group, division, or operating temperature. [§1926.407(b)(2)(ii)[D]]

(3) *Safe for the hazardous (classified) location.* Equipment which is safe for the location shall be of a type and design which the employer demonstrates will provide protection from the hazards arising from the combustibility and flammability of vapors, liquids, gases, dusts, or fibers.[§1926.407(b)(3)]

Note: The National Electrical Code, NFPA 70, contains guidelines for determining the type and design of equipment and installations which will meet this requirement. The guidelines of this document address electric wiring, equipment, and systems installed in hazardous (classified) locations and contain specific provisions for the following: wiring methods, wiring connections, conductor insulation, flexible cords, sealing and drainage, transformers, capacitors, switches, circuit breakers, fuses, motor controllers, receptacles, attachment plugs, meters, relays, instruments, resistors, generators, motors, lighting fixtures, storage battery charging equipment, electric cranes, electric hoists and similar equipment, utilization equipment, signaling systems, alarm systems, remote control systems, local loud speaker and communication systems, ventilation piping, live parts, lightning surge protection, and grounding. Compliance with these guidelines will constitute one means, but not the only means, of compliance with this paragraph.

(c) **Conduits.** All conduits shall be threaded and shall be made wrench-tight. Where it is impractical to make a threaded joint tight, a bonding jumper shall be utilized.[§1926.407(c)]

[51 FR 25318, July 11, 1986, as amended at 61 FR 5510, Feb. 13, 1996]

§1926.408

Special systems

(a) **Systems over 600 volts, nominal.** Paragraphs (a)(1) through (a)(4) of this section contain general requirements for all circuits and equipment operated at over 600 volts.[§1926.408(a)]

(1) *Wiring methods for fixed installations*[§1926.408(a)(1)]

(i) *Above ground.* Above- ground conductors shall be installed in rigid metal conduit, in intermediate metal conduit, in cable trays, in cablebus, in other suitable raceways, or as open runs of metal-clad cable designed for the use and purpose. However, open runs of non-metallic-sheathed cable or of bare conductors or busbars may be installed in locations which are accessible only to qualified persons. Metallic shielding components, such as tapes, wires, or braids for conductors, shall be grounded. Open runs of insulated wires and cables having a bare lead sheath or a braided outer covering shall be supported in a manner designed to prevent physical damage to the braid or sheath.[§1926.408(a)(1)(i)]

(ii) *Installations emerging from the ground.* Conductors emerging from the ground shall be enclosed in raceways. Raceways installed on poles shall be of rigid metal conduit, intermediate metal conduit, PVC schedule 80 or equivalent extending from the ground line up to a point 8 feet (2.44 m) above finished grade. Conductors entering a building shall be protected by an enclosure from the ground line to the point of entrance. Metallic enclosures shall be grounded. [§1926.408(a)(1)(ii)]

(2) *Interrupting and isolating devices.*[§1926.408(a)(2)]

(i) *Circuit breakers.* Circuit breakers located indoors shall consist of metal-enclosed or fire-resistant, cell-mounted units. In locations accessible only to qualified personnel, open mounting of circuit breakers is permitted. A means of indicating the open and closed position of circuit breakers shall be provided.[§1926.408(a)(2)(i)]

(ii) *Fused cutouts.* Fused cutouts installed in buildings or transformer vaults shall be of a type identified for the purpose. They shall be readily accessible for fuse replacement. [§1926.408(a)(2)(ii)]

(iii) *Equipment isolating means.* A means shall be provided to completely isolate equipment for inspection and repairs. Isolating means which are not designed to interrupt the load current of the circuit shall be either interlocked with a circuit interrupter or provided with a sign warning against opening them under load.[§1926.408(a)(2)(iii)]

(3) *Mobile and portable equipment.*[§1926.408(a)(3)]

(i) *Power cable connections to mobile machines.* A metallic enclosure shall be provided on the mobile machine for enclosing the terminals of the power cable. The enclosure shall include provisions for a solid connection for the ground wire(s) terminal to ground effectively the machine frame. The method of cable termination used shall prevent any strain or pull on the cable from stressing the electrical connections. The enclosure shall have provision for locking so only authorized qualified persons may open it and shall be marked with a sign warning of the presence of energized parts. [§1926.408(a)(3)(i)]

(ii) *Guarding live parts.* All energized switching and control parts shall be enclosed in effectively grounded metal cabinets or enclosures. Circuit breakers and protective equipment shall have the operating means projecting through the metal cabinet or enclosure so these units can be reset without locked doors being opened. Enclosures and metal cabinets shall be locked so that only authorized qualified persons have access and shall be marked with a sign warning of the presence of energized parts. Collector ring assemblies on revolving-type machines (shovels, draglines, etc.) shall be guarded. [§1926.408(a)(3)(ii)]

(4) *Tunnel installations.*[§1926.408(a)(4)]

(i) *Application.* The provisions of this paragraph apply to installation and use of high-voltage power distribution and utilization equipment which is associated with tunnels and which is portable and/or mobile, such as substations, trailers, cars, mobile shovels, draglines, hoists, drills, dredges, compressors, pumps, conveyors, and underground excavators. [§1926.408(a)(4)(i)]

(ii) *Conductors.* Conductors in tunnels shall be installed in one or more of the following:[§1926.408(a)(4)(ii)]

[A] Metal conduit or other metal raceway,[§1926.408(a)(4)(ii)[A]]

[B] Type MC cable, or[§1926.408(a)(4)(ii)[B]]

[C] Other suitable multiconductor cable.[§1926.408(a)(4)(ii)[C]]

Conductors shall also be so located or guarded as to protect them from physical damage. Multiconductor portable cable may supply mobile equipment. An equipment grounding conductor shall be run with circuit conductors inside the metal raceway or inside the multiconductor cable jacket. The equipment grounding conductor may be insulated or bare. [§1926.408(a)(4)(ii)[C]]

(iii) *Guarding live parts.* Bare terminals of transformers, switches, motor controllers, and other equipment shall be enclosed to prevent accidental contact with energized parts. Enclosures for use in tunnels shall be drip-proof, weatherproof, or submersible as required by the environmental conditions.[§1926.408(a)(4)(iii)]

(iv) *Disconnecting means.* A disconnecting means that simultaneously opens all ungrounded conductors shall be installed at each transformer or motor location.[§1926.408(a)(4)(iv)]

(v) *Grounding and bonding.* All nonenergized metal parts of electric equipment and metal raceways and cable sheaths shall be grounded and bonded to all metal pipes and rails at the portal and at intervals not exceeding 1000 feet (305 m) throughout the tunnel.[§1926.408(a)(4)(v)]

(b) Class 1, Class 2, and Class 3 remote control, signaling, and power-limited circuits[§1926.408(b)]

(1) *Classification.* Class 1, Class 2, or Class 3 remote control, signaling, or power-limited circuits are characterized by their usage and electrical power limitation which differentiates them from light and power circuits. These circuits are classified in accordance with their respective voltage and power limitations as summarized in paragraphs (b)(1)(i) through (b)(1)(iii) of this section.[§1926.408(b)(1)]

(i) *Class 1 circuits.*[§1926.408(b)(1)(i)]

[A] A Class 1 power-limited circuit is supplied from a source having a rated output of not more than 30 volts and 1000 volt-amperes.[§1926.408(b)(1)(i)[A]]

[B] A Class 1 remote control circuit or a Class 1 signaling circuit has a voltage which does not exceed 600 volts; however, the power output of the source need not be limited. [§1926.408(b)(1)(i)[B]]

(ii) *Class 2 and Class 3 circuits.*[§1926.408(b)(1)(ii)]

[A] Power for Class 2 and Class 3 circuits is limited either inherently (in which no overcurrent protection is required) or by a combination of a power source and overcurrent protection. [§1926.408(b)(1)(ii)[A]]

[B] The maximum circuit voltage is 150 volts AC or DC for a Class 2 inherently limited power source, and 100 volts AC or DC for a Class 3 inherently limited power source.[§1926.408(b)(1)(ii)[B]]

[C] The maximum circuit voltage is 30 volts AC and 60 volts DC for a Class 2 power source limited by overcurrent protection, and 150 volts AC or DC for a Class 3 power source limited by overcurrent protection.[§1926.408(b)(1)(ii)[C]]

(iii) *Application.* The maximum circuit voltages in paragraphs (b)(1)(i) and (b)(1)(ii) of this section apply to sinusoidal AC or continuous DC power sources, and where wet contact occurrence is not likely.[§1926.408(b)(1)(iii)]

(2) *Marking.* A Class 2 or Class 3 power supply unit shall not be used unless it is durably marked where plainly visible to indicate the class of supply and its electrical rating.[§1926.408(b)(2)]

(c) Communications systems.[§1926.408(c)]

(1) *Scope.* These provisions for communication systems apply to such systems as central-station-connected and non-central-station-connected telephone circuits, radio receiving and transmitting equipment, and outside wiring for fire and burglar alarm, and similar central station systems. These installations need not comply with the provisions of §§1926.403 through 1926.408(b), except Sec. 1926.404(c)(1)(ii) and Sec. 1926.407.[§1926.408(c)(1)]

(2) *Protective devices.*[§1926.408(c)(2)]

(i) *Circuits exposed to power conductors.* Communication circuits so located as to be exposed to accidental contact with light or power conductors operating at over 300 volts shall have each circuit so exposed provided with an approved protector.[§1926.408(c)(2)(i)]

(ii) *Antenna lead-ins.* Each conductor of a lead-in from an outdoor antenna shall be provided with an antenna discharge unit or other means that will drain static charges from the antenna system. [§1926.408(c)(2)(ii)]

(3) *Conductor location.*[§1926.408(c)(3)]

(i) *Outside of buildings.*[§1926.408(c)(3)(i)]

[A] Receiving distribution lead-in or aerial-drop cables attached to buildings and lead-in conductors to radio transmitters shall be so installed as to avoid the possibility of accidental contact with electric light or power conductors.[§1926.408(c)(3)(i)[A]]

[B] The clearance between lead-in conductors and any lightning protection conductors shall not be less than 6 feet (1.83 m). [§1926.408(c)(3)(i)[B]]

(ii) *On poles.* Where practicable, communication conductors on poles shall be located below the light or power conductors. Communications conductors shall not be attached to a crossarm that carries light or power conductors.[§1926.408(c)(3)(ii)]

(iii) *Inside of buildings.* Indoor antennas, lead-ins, and other communication conductors attached as open conductors to the inside of buildings shall be located at least 2 inches (50.8 mm) from conductors of any light or power or Class 1 circuits unless a special and equally protective method of conductor separation is employed.[§1926.408(c)(3)(iii)]

(4) *Equipment location.* Outdoor metal structures supporting antennas, as well as self-supporting antennas such as vertical rods or dipole structures, shall be located as far away from overhead conductors of electric light and power circuits of over 150 volts to ground as necessary to avoid the possibility of the antenna or structure falling into or making accidental contact with such circuits. [§1926.408(c)(4)]

(5) *Grounding.*[§1926.408(c)(5)]

(i) *Lead-in conductors.* If exposed to contact with electric light or power conductors, the metal sheath of aerial cables entering buildings shall be grounded or shall be interrupted close to the entrance to the building by an insulating joint or equivalent device. Where protective devices are used, they shall be grounded. [§1926.408(c)(5)(i)]

(ii) *Antenna structures.* Masts and metal structures supporting antennas shall be permanently and effectively grounded without splice or connection in the grounding conductor.[§1926.408(c)(5)(ii)]

(iii) *Equipment enclosures.* Transmitters shall be enclosed in a metal frame or grill or separated from the operating space by a barrier, all metallic parts of which are effectively connected to ground. All external metal handles and controls accessible to the operating personnel shall be effectively grounded. Unpowered equipment and enclosures shall be considered grounded where connected to an attached coaxial cable with an effectively grounded metallic shield.[§1926.408(c)(5)(iii)]

[51 FR 25318, July 11, 1986, as amended at 61 FR 5510, Feb. 13, 1996]

§§1926.409-1926.415
[Reserved]

SAFETY-RELATED WORK PRACTICES

§1926.416
General requirements

(a) Protection of employees.[§1926.416(a)]

(1) *No employer shall permit an employee* to work in such proximity to any part of an electric power circuit that the employee could contact the electric power circuit in the course of work, unless the employee is protected against electric shock by deenergizing the circuit and grounding it or by guarding it effectively by insulation or other means.[§1926.416(a)(1)]

(2) *In work areas where the exact location* of underground electric powerlines is unknown, employees using jack-hammers, bars, or other hand tools which may contact a line shall be provided with insulated protective gloves.[§1926.416(a)(2)]

(3) *Before work is begun* the employer shall ascertain by inquiry or direct observation, or by instruments, whether any part of an energized electric power circuit, exposed or concealed, is so

located that the performance of the work may bring any person, tool, or machine into physical or electrical contact with the electric power circuit. The employer shall post and maintain proper warning signs where such a circuit exists. The employer shall advise employees of the location of such lines, the hazards involved, and the protective measures to be taken. [§1926.416(a)(3)]

(b) **Passageways and open spaces.**[§1926.416(b)]

(1) *Barriers or other means of guarding* shall be provided to ensure that workspace for electrical equipment will not be used as a passageway during periods when energized parts of electrical equipment are exposed.[§1926.416(b)(1)]

(2) *Working spaces,* walkways, and similar locations shall be kept clear of cords so as not to create a hazard to employees. [§1926.416(b)(2)]

(c) **Load ratings.** In existing installations, no changes in circuit protection shall be made to increase the load in excess of the load rating of the circuit wiring.[§1926.416(c)]

(d) **Fuses.** When fuses are installed or removed with one or both terminals energized, special tools insulated for the voltage shall be used.[§1926.416(d)]

(e) **Cords and cables.**[§1926.416(e)]

(1) *Worn or frayed electric cords* or cables shall not be used. [§1926.416(e)(1)]

(2) *Extension cords shall not be fastened* with staples, hung from nails, or suspended by wire.[§1926.416(e)(2)]

[44 FR 8577, Feb. 9, 1979; 44 FR 20940, Apr. 6, 1979, as amended at 55 FR 42328, Oct. 18, 1990; 58 FR 35179, June 30, 1993; 61 FR 9251, Mar. 7, 1996; 61 FR 41738, Aug. 12, 1996]

§1926.417

Lockout and tagging of circuits

(a) **Controls.** Controls that are to be deactivated during the course of work on energized or deenergized equipment or circuits shall be tagged.[§1926.417(a)]

(b) **Equipment and circuits.** Equipment or circuits that are deenergized shall be rendered inoperative and shall have tags attached at all points where such equipment or circuits can be energized. [§1926.417(b)]

(c) **Tags.** Tags shall be placed to identify plainly the equipment or circuits being worked on.[§1926.417(c)]

[44 FR 8577, Feb. 9, 1979; 44 FR 20940, Apr. 6, 1979, as amended at 55 FR 42328, Oct. 18, 1990; 58 FR 35181, June 30, 1993; 61 FR 9251, Mar. 7, 1996; 61 FR 41739, Aug. 12, 1996]]

§§1926.418-1926.430

[Reserved]

SAFETY-RELATED MAINTENANCE AND ENVIRONMENTAL CONSIDERATIONS

§1926.431

Maintenance of equipment

The employer shall ensure that all wiring components and utilization equipment in hazardous locations are maintained in a dust-tight, dust-ignition-proof, or explosion-proof condition, as appropriate. There shall be no loose or missing screws, gaskets, threaded connections, seals, or other impairments to a tight condition.

§1926.432

Environmental deterioration of equipment

(a) **Deteriorating agents**[§1926.432(a)]

(1) *Unless identified for use* in the operating environment, no conductors or equipment shall be located:[§1926.432(a)(1)]

(i) *In damp or wet locations;*[§1926.432(a)(1)(i)]

(ii) *Where exposed to gases, fumes,* vapors, liquids, or other agents having a deteriorating effect on the conductors or equipment; or[§1926.432(a)(1)(ii)]

(iii) *Where exposed to excessive temperatures.*[§1926.432(a)(1)(iii)]

(2) *Control equipment, utilization equipment,* and busways approved for use in dry locations only shall be protected against damage from the weather during building construction. [§1926.432(a)(2)]

(b) **Protection against corrosion.** Metal raceways, cable armor, boxes, cable sheathing, cabinets, elbows, couplings, fittings, supports, and support hardware shall be of materials appropriate for the environment in which they are to be installed.[§1926.432(b)]

§§1926.433-1926.440

[Reserved]

SAFETY REQUIREMENTS FOR SPECIAL EQUIPMENT

§1926.441

Batteries and battery charging

(a) **General requirements.**[§1926.441(a)]

(1) *Batteries of the unsealed type* shall be located in enclosures with outside vents or in well ventilated rooms and shall be arranged so as to prevent the escape of fumes, gases, or electrolyte spray into other areas.[§1926.441(a)(1)]

(2) *Ventilation shall be provided* to ensure diffusion of the gases from the battery and to prevent the accumulation of an explosive mixture.[§1926.441(a)(2)]

(3) *Racks and trays shall be substantial* and shall be treated to make them resistant to the electrolyte.[§1926.441(a)(3)]

(4) *Floors shall be of acid resistant construction* unless protected from acid accumulations.[§1926.441(a)(4)]

(5) *Face shields, aprons, and rubber gloves* shall be provided for workers handling acids or batteries.[§1926.441(a)(5)]

(6) *Facilities for quick drenching* of the eyes and body shall be provided within 25 feet (7.62 m) of battery handling areas. [§1926.441(a)(6)]

(7) *Facilities shall be provided* for flushing and neutralizing spilled electrolyte and for fire protection.[§1926.441(a)(7)]

(b) **Charging.**[§1926.441(b)]

(1) *Battery charging installations* shall be located in areas designated for that purpose.[§1926.441(b)(1)]

(2) *Charging apparatus shall be protected from damage by trucks.*[§1926.441(b)(2)]

(3) *When batteries are being charged,* the vent caps shall be kept in place to avoid electrolyte spray. Vent caps shall be maintained in functioning condition.[§1926.441(b)(3)]

§§1926.442-1926.448

[Reserved]

DEFINITIONS

§1926.449

Definitions applicable to this subpart

The definitions given in this section apply to the terms used in subpart K. The definitions given here for "approved" and "qualified person" apply, instead of the definitions given in Sec. 1926.32, to the use of these terms in subpart K.

Acceptable. An installation or equipment is acceptable to the Assistant Secretary of Labor, and approved within the meaning of this subpart K:[§1926.449]

(a) **If it is accepted,** or certified, or listed, or labeled, or otherwise determined to be safe by a qualified testing laboratory capable of determining the suitability of materials and equipment for installation and use in accordance with this standard; or[§1926.449(a)]

(b) **With respect to an installation or equipment** of a kind which no qualified testing laboratory accepts, certifies, lists, labels, or determines to be safe, if it is inspected or tested by another Federal agency, or by a State, municipal, or other local authority responsible for enforcing occupational safety provisions of the National Electrical Code, and found in compliance with those provisions; or [§1926.449(b)]

(c) **With respect to custom-made equipment** or related installations which are designed, fabricated for, and intended for use by a particular customer, if it is determined to be safe for its intended use by its manufacturer on the basis of test data which the employer keeps and makes available for inspection to the Assistant Secretary and his authorized representatives.[§1926.449(c)]

Accepted. An installation is "accepted" if it has been inspected and found to be safe by a qualified testing laboratory.[§1926.449]

Accessible. (As applied to wiring methods.) Capable of being removed or exposed without damaging the building structure or finish, or not permanently closed in by the structure or finish of the building. (See "concealed" and "exposed.")[§1926.449]

Accessible. (As applied to equipment.) Admitting close approach; not guarded by locked doors, elevation, or other effective means. (See"Readily accessible.")[§1926.449]

Ampacity. The current in amperes a conductor can carry continuously under the conditions of use without exceeding its temperature rating.[§1926.449]

Appliances. Utilization equipment, generally other than industrial, normally built in standardized sizes or types, which is installed or connected as a unit to perform one or more functions.[§1926.449]

Approved. Acceptable to the authority enforcing this subpart. The authority enforcing this subpart is the Assistant Secretary of Labor for Occupational Safety and Health. The definition of "acceptable" indicates what is acceptable to the Assistant Secretary of Labor, and therefore approved within the meaning of this subpart.[§1926.449]

Askarel. A generic term for a group of nonflammable synthetic chlorinated hydrocarbons used as electrical insulating media. Askarels of various compositional types are used. Under arcing conditions the gases produced, while consisting predominantly of noncombustible hydrogen chloride, can include varying amounts of combustible gases depending upon the askarel type.[§1926.449]

Attachment plug (Plug cap)(Cap). A device which, by insertion in a receptacle, establishes connection between the conductors of the attached flexible cord and the conductors connected permanently to the receptacle.[§1926.449]

Automatic. Self-acting, operating by its own mechanism when actuated by some impersonal influence, as for example, a change in current strength, pressure, temperature, or mechanical configuration. [§1926.449]

Bare conductor. See "Conductor."[§1926.449]

Bonding. The permanent joining of metallic parts to form an electrically conductive path which will assure electrical continuity and the capacity to conduct safely any current likely to be imposed.[§1926.449]

Bonding jumper. A reliable conductor to assure the required electrical conductivity between metal parts required to be electrically connected.[§1926.449]

Branch circuit. The circuit conductors between the final overcurrent device protecting the circuit and the outlet(s).[§1926.449]

Building. A structure which stands alone or which is cut off from adjoining structures by fire walls with all openings therein protected by approved fire doors.[§1926.449]

Cabinet. An enclosure designed either for surface or flush mounting, and provided with a frame, mat, or trim in which a swinging door or doors are or may be hung.[§1926.449]

Certified. Equipment is "certified" if it:[§1926.449]

(a) **Has been tested** and found by a qualified testing laboratory to meet applicable test standards or to be safe for use in a specified manner, and[§1926.449(a)]

(b) **Is of a kind whose production** is periodically inspected by a qualified testing laboratory. Certified equipment must bear a label, tag, or other record of certification.[§1926.449(b)]

Circuit breaker.[§1926.449]

(a) **(600 volts nominal, or less.)** A device designed to open and close a circuit by nonautomatic means and to open the circuit automatically on a predetermined overcurrent without injury to itself when properly applied within its rating.[§1926.449(a)]

(b) **(Over 600 volts, nominal.)** A switching device capable of making, carrying, and breaking currents under normal circuit conditions, and also making, carrying for a specified time, and breaking currents under specified abnormal circuit conditions, such as those of short circuit.[§1926.449(b)]

Class I locations. Class I locations are those in which flammable gases or vapors are or may be present in the air in quantities sufficient to produce explosive or ignitible mixtures. Class I locations include the following:[§1926.449]

(a) **Class I, Division 1.**A Class I, Division 1 location is a location: [§1926.449(a)]

(1) *In which ignitible concentrations* of flammable gases or vapors may exist under normal operating conditions; or[§1926.449(a)(1)]

(2) *In which ignitible concentrations* of such gases or vapors may exist frequently because of repair or maintenance operations or because of leakage; or[§1926.449(a)(2)]

(3) *In which breakdown or faulty operation* of equipment or processes might release ignitible concentrations of flammable gases or vapors, and might also cause simultaneous failure of electric equipment.[§1926.449(a)(3)]

Note: This classification usually includes locations where volatile flammable liquids or liquefied flammable gases are transferred from one container to another; interiors of spray booths and areas in the vicinity of spraying and painting operations where volatile flammable solvents are used; locations containing open tanks or vats of volatile flammable liquids; drying rooms or compartments for the evaporation of flammable solvents; inadequately ventilated pump rooms for flammable gas or for volatile flammable liquids; and all other locations where ignitible concentrations of flammable vapors or gases are likely to occur in the course of normal operations.

(b) **Class I, Division 2.** A Class I, Division 2 location is a location: [§1926.449(b)]

(1) *In which volatile flammable liquids or flammable gases* are handled, processed, or used, but in which the hazardous liquids, vapors, or gases will normally be confined within closed containers or closed systems from which they can escape only in case of accidental rupture or breakdown of such containers or systems, or in case of abnormal operation of equipment; or [§1926.449(b)(1)]

(2) *In which ignitible concentrations* of gases or vapors are normally prevented by positive mechanical ventilation, and which might become hazardous through failure or abnormal operations of the ventilating equipment; or[§1926.449(b)(2)]

(3) *That is adjacent to a Class I, Division 1 location,* and to which ignitible concentrations of gases or vapors might occasionally be communicated unless such communication is prevented by adequate positive-pressure ventilation from a source of clean air, and effective safeguards against ventilation failure are provided.[§1926.449(b)(3)]

Note: This classification usually includes locations where volatile flammable liquids or flammable gases or vapors are used, but which would become hazardous only in case of an accident or of some unusual operating condition. The quantity of flammable material that might escape in case of accident, the adequacy of ventilating equipment, the total area involved, and the record of the industry or business with respect to explosions or fires are all factors that merit consideration in determining the classification and extent of each location.

Piping without valves, checks, meters, and similar devices would not ordinarily introduce a hazardous condition even though used for flammable liquids or gases. Locations used for the storage of flammable liquids or of liquefied or compressed gases in sealed containers would not normally be considered hazardous unless also subject to other hazardous conditions.[§1926.449]

Electrical conduits and their associated enclosures separated from process fluids by a single seal or barrier are classed as a Division 2 location if the outside of the conduit and enclosures is a nonhazardous location.[§1926.449]

Class II locations. Class II locations are those that are hazardous because of the presence of combustible dust. Class II locations include the following:[§1926.449]

(a) **Class II, Division 1.**A Class II, Division 1 location is a location: [§1926.449(a)]

(1) *In which combustible dust* is or may be in suspension in the air under normal operating conditions, in quantities sufficient to produce explosive or ignitible mixtures; or[§1926.449(a)(1)]

(2) *Where mechanical failure* or abnormal operation of machinery or equipment might cause such explosive or ignitible mixtures to be produced, and might also provide a source of ignition through simultaneous failure of electric equipment, operation of protection devices, or from other causes, or[§1926.449(a)(2)]

(3) *In which combustible dusts* of an electrically conductive nature may be present.[§1926.449(a)(3)]

Note: Combustible dusts which are electrically nonconductive include dusts produced in the handling and processing of grain and grain products, pulverized sugar and cocoa, dried egg and milk powders, pulverized spices, starch and pastes, potato and woodflour, oil meal from beans and seed, dried hay, and other organic materials which may produce combustible dusts when processed or handled. Dusts containing magnesium or aluminum are particularly hazardous and the use of extreme caution is necessary to avoid ignition and explosion.

(b) **Class II, Division 2.** A Class II, Division 2 location is a location in which:[§1926.449(b)]

(1) *Combustible dust* will not normally be in suspension in the air in quantities sufficient to produce explosive or ignitible mixtures, and dust accumulations are normally insufficient to interfere with the normal operation of electrical equipment or other apparatus; or[§1926.449(b)(1)]

(2) *Dust may be in suspension in the air* as a result of infrequent malfunctioning of handling or processing equipment, and dust accumulations resulting therefrom may be ignitible by abnormal operation or failure of electrical equipment or other apparatus.[§1926.449(b)(2)]

Note: This classification includes locations where dangerous concentrations of suspended dust would not be likely but where dust accumulations might form on or in the vicinity of electric equipment. These areas may contain equipment from which appreciable quantities of dust would escape under abnormal operating conditions or be adjacent to a Class II Division 1 location, as described above, into which an explosive or ignitible concentration of dust may be put into suspension under abnormal operating conditions.

Class III locations. Class III locations are those that are hazardous because of the presence of easily ignitible fibers or flyings but in which such fibers or flyings are not likely to be in suspension in the air in quantities sufficient to produce ignitible mixtures. Class 111 locations include the following:[§1926.449]

(a) **Class III, Division 1.** A Class III, Division 1 location is a location in which easily ignitible fibers or materials producing combustible flyings are handled, manufactured, or used.[§1926.449(a)]

Note: Easily ignitible fibers and flyings include rayon, cotton (including cotton linters and cotton waste), sisal or henequen, istle, jute, hemp, tow, cocoa fiber, oakum, baled waste kapok, Spanish moss, excelsior, sawdust, woodchips, and other material of similar nature.

(b) **Class III, Division 2.** A Class III, Division 2 location is a location in which easily ignitible fibers are stored or handled, except in process of manufacture.[§1926.449(b)]

Collector ring. A collector ring is an assembly of slip rings for transferring electrical energy from a stationary to a rotating member. [§1926.449]

Concealed. Rendered inaccessible by the structure or finish of the building. Wires in concealed raceways are considered concealed, even though they may become accessible by withdrawing them. [See "Accessible. (As applied to wiring methods.)"][§1926.449]

Conductor.[§1926.449]

(a) **Bare.** A conductor having no covering or electrical insulation whatsoever.[§1926.449(a)]

(b) **Covered.** A conductor encased within material of composition or thickness that is not recognized as electrical insulation. [§1926.449(b)]

(c) **Insulated.** A conductor encased within material of composition and thickness that is recognized as electrical insulation. [§1926.449(c)]

Controller. A device or group of devices that serves to govern, in some predetermined manner, the electric power delivered to the apparatus to which it is connected.[§1926.449]

Covered conductor. See "Conductor."[§1926.449]

Cutout. (Over 600 volts, nominal.) An assembly of a fuse support with either a fuseholder, fuse carrier, or disconnecting blade. The fuseholder or fuse carrier may include a conducting element (fuse link), or may act as the disconnecting blade by the inclusion of a nonfusible member.[§1926.449]

Cutout box. An enclosure designed for surface mounting and having swinging doors or covers secured directly to and telescoping with the walls of the box proper. (See "Cabinet.")[§1926.449]

Damp location. See "Location."[§1926.449]

Dead front. Without live parts exposed to a person on the operating side of the equipment.[§1926.449]

Device. A unit of an electrical system which is intended to carry but not utilize electric energy.[§1926.449]

Disconnecting means. A device, or group of devices, or other means by which the conductors of a circuit can be disconnected from their source of supply.[§1926.449]

Disconnecting (or Isolating) switch. (Over 600 volts, nominal.) A mechanical switching device used for isolating a circuit or equipment from a source of power.[§1926.449]

Dry location. See "Location."[§1926.449]

Enclosed. Surrounded by a case, housing, fence or walls which will prevent persons from accidentally contacting energized parts. [§1926.449]

Enclosure. The case or housing of apparatus, or the fence or walls surrounding an installation to prevent personnel from accidentally contacting energized parts, or to protect the equipment from physical damage.[§1926.449]

Equipment. A general term including material, fittings, devices, appliances, fixtures, apparatus, and the like, used as a part of, or in connection with, an electrical installation.[§1926.449]

Equipment grounding conductor. See "Grounding conductor, equipment."[§1926.449]

Explosion-proof apparatus. Apparatus enclosed in a case that is capable of withstanding an explosion of a specified gas or vapor which may occur within it and of preventing the ignition of a specified gas or vapor surrounding the enclosure by sparks, flashes, or explosion of the gas or vapor within, and which operates at such an external temperature that it will not ignite a surrounding flammable atmosphere.[§1926.449]

Exposed. (As applied to live parts.) Capable of being inadvertently touched or approached nearer than a safe distance by a person. It is applied to parts not suitably guarded, isolated, or insulated. (See "Accessible and "Concealed.")[§1926.449]

Exposed. (As applied to wiring methods.) On or attached to the surface or behind panels designed to allow access. [See "Accessible. (As applied to wiring methods.)"][§1926.449]

Exposed. (For the purposes of §1926.408(d), Communications systems.) Where the circuit is in such a position that in case of failure of supports or insulation, contact with another circuit may result. [§1926.449]

Externally operable. Capable of being operated without exposing the operator to contact with live parts.[§1926.449]

Feeder. All circuit conductors between the service equipment, or the generator switchboard of an isolated plant, and the final branch-circuit overcurrent device.[§1926.449]

Festoon lighting. A string of outdoor lights suspended between two points more than 15 feet (4.57 m) apart.[§1926.449]

Fitting. An accessory such as a locknut, bushing, or other part of a wiring system that is intended primarily to perform a mechanical rather than an electrical function.[§1926.449]

Fuse. (Over 600 volts, nominal.) An overcurrent protective device with a circuit opening fusible part that is heated and severed by the passage of overcurrent through it. A fuse comprises all the parts that form a unit capable of performing the prescribed functions. It may or may not be the complete device necessary to connect it into an electrical circuit.[§1926.449]

Ground. A conducting connection, whether intentional or accidental, between an electrical circuit or equipment and the earth, or to some conducting body that serves in place of the earth.[§1926.449]

Grounded. Connected to earth or to some conducting body that serves in place of the earth.[§1926.449]

Grounded, effectively. (Over 600 volts, nominal.) Permanently connected to earth through a ground connection of sufficiently low impedance and having sufficient ampacity that ground fault current which may occur cannot build up to voltages dangerous to personnel. [§1926.449]

Grounded conductor. A system or circuit conductor that is intentionally grounded.[§1926.449]

Grounding conductor. A conductor used to connect equipment or the grounded circuit of a wiring system to a grounding electrode or electrodes.[§1926.449]

Grounding conductor, equipment. The conductor used to connect the noncurrent-carrying metal parts of equipment, raceways, and other enclosures to the system grounded conductor and/or the grounding electrode conductor at the service equipment or at the source of a separately derived system.[§1926.449]

Grounding electrode conductor. The conductor used to connect the grounding electrode to the equipment grounding conductor and/or to the grounded conductor of the circuit at the service equipment or at the source of a separately derived system.[§1926.449]

Ground-fault circuit interrupter. A device for the protection of personnel that functions to deenergize a circuit or portion thereof within an established period of time when a current to ground exceeds some predetermined value that is less than that required to operate the overcurrent protective device of the supply circuit.[§1926.449]

Guarded. Covered, shielded, fenced, enclosed, or otherwise protected by means of suitable covers, casings, barriers, rails, screens, mats, or platforms to remove the likelihood of approach to a point of danger or contact by persons or objects.[§1926.449]

Hoistway. Any shaftway, hatchway, well hole, or other vertical opening or space in which an elevator or dumbwaiter is designed to operate.[§1926.449]

Identified (conductors or terminals). Identified, as used in reference to a conductor or its terminal, means that such conductor or terminal can be recognized as grounded.[§1926.449]

Identified (for the use). Recognized as suitable for the specific purpose, function, use, environment, application, etc. where described as a requirement in this standard. Suitability of equipment for a specific purpose, environment, or application is determined by a qualified testing laboratory where such identification includes labeling or listing. [§1926.449]

Insulated conductor. See "Conductor."[§1926.449]

Interrupter switch. (Over 600 volts, nominal.) A switch capable of making, carrying, and interrupting specified currents.[§1926.449]

Intrinsically safe equipment and associated wiring. Equipment and associated wiring in which any spark or thermal effect, produced either normally or in specified fault conditions, is incapable, under certain prescribed test conditions, of causing ignition of a mixture of flammable or combustible material in air in its most easily ignitible concentration.[§1926.449]

Isolated. Not readily accessible to persons unless special means for access are used.[§1926.449]

Isolated power system. A system comprising an isolating transformer or its equivalent, a line isolation monitor, and its ungrounded circuit conductors.[§1926.449]

Labeled. Equipment or materials to which has been attached a label, symbol or other identifying mark of a qualified testing laboratory which indicates compliance with appropriate standards or performance in a specified manner.[§1926.449]

Lighting outlet. An outlet intended for the direct connection of a lampholder, a lighting fixture, or a pendant cord terminating in a lampholder.[§1926.449]

Listed. Equipment or materials included in a list published by a qualified testing laboratory whose listing states either that the equipment or material meets appropriate standards or has been tested and found suitable for use in a specified manner.[§1926.449]

Location.[§1926.449]

(a) **Damp location.** Partially protected locations under canopies, marquees, roofed open porches, and like locations, and interior locations subject to moderate degrees of moisture, such as some basements.[§1926.449(a)]

(b) **Dry location.** A location not normally subject to dampness or wetness. A location classified as dry may be temporarily subject to dampness or wetness, as in the case of a building under construction.[§1926.449(b)]

(c) **Wet location.** Installations underground or in concrete slabs or masonry in direct contact with the earth, and locations subject to saturation with water or other liquids, such as locations exposed to weather and unprotected.[§1926.449(c)]

Mobile X-ray. X-ray equipment mounted on a permanent base with wheels and/or casters for moving while completely assembled. [§1926.449]

Motor control center. An assembly of one or more enclosed sections having a common power bus and principally containing motor control units.[§1926.449]

Outlet. A point on the wiring system at which current is taken to supply utilization equipment.[§1926.449]

Overcurrent. Any current in excess of the rated current of equipment or the ampacity of a conductor. It may result from overload (see definition), short circuit, or ground fault. A current in excess of rating may be accommodated by certain equipment and conductors for a given set of conditions. Hence the rules for overcurrent protection are specific for particular situations.[§1926.449]

Overload. Operation of equipment in excess of normal, full load rating, or of a conductor in excess of rated ampacity which, when it persists for a sufficient length of time, would cause damage or dangerous overheating. A fault, such as a short circuit or ground fault, is not an overload. (See"Overcurrent.")[§1926.449]

Panelboard. A single panel or group of panel units designed for assembly in the form of a single panel; including buses, automatic overcurrent devices, and with or without switches for the control of light, heat, or power circuits; designed to be placed in a cabinet or cutout box placed in or against a wall or partition and accessible only from the front. (See "Switchboard.")[§1926.449]

Portable X-ray. X-ray equipment designed to be hand-carried. [§1926.449]

Power fuse. (Over 600 volts, nominal.) See "Fuse."[§1926.449]

Power outlet. An enclosed assembly which may include receptacles, circuit breakers, fuseholders, fused switches, buses and watt-hour meter mounting means; intended to serve as a means for distributing power required to operate mobile or temporarily installed equipment. [§1926.449]

Premises wiring system. That interior and exterior wiring, including power, lighting, control, and signal circuit wiring together with all of its associated hardware, fittings, and wiring devices, both permanently and temporarily installed, which extends from the load end of the service drop, or load end of the service lateral conductors to the outlet(s). Such wiring does not include wiring internal to appliances, fixtures, motors, controllers, motor control centers, and similar equipment. [§1926.449]

Qualified person. One familiar with the construction and operation of the equipment and the hazards involved.[§1926.449]

Qualified testing laboratory. A properly equipped and staffed testing laboratory which has capabilities for and which provides the following services:[§1926.449]

(a) Experimental testing for safety of specified items of equipment and materials referred to in this standard to determine compliance with appropriate test standards or performance in a specified manner;[§1926.449(a)]

(b) Inspecting the run of such items of equipment and materials at factories for product evaluation to assure compliance with the test standards;[§1926.449(b)]

(c) Service-value determinations through field inspections to monitor the proper use of labels on products and with authority for recall of the label in the event a hazardous product is installed; [§1926.449(c)]

(d) Employing a controlled procedure for identifying the listed and/or labeled equipment or materials tested; and[§1926.449(d)]

(e) Rendering creditable reports or findings that are objective and without bias of the tests and test methods employed.[§1926.449(e)]

Raceway. A channel designed expressly for holding wires, cables, or busbars, with additional functions as permitted in this subpart. Raceways may be of metal or insulating material, and the term includes rigid metal conduit, rigid nonmetallic conduit, intermediate metal conduit, liquidtight flexible metal conduit, flexible metallic tubing, flexible metal conduit, electrical metallic tubing, underfloor raceways, cellular concrete floor raceways, cellular metal floor raceways, surface raceways, wireways, and busways.[§1926.449]

Readily accessible. Capable of being reached quickly for operation, renewal, or inspections, without requiring those to whom ready access is requisite to climb over or remove obstacles or to resort to portable ladders, chairs, etc. (See "Accessible.")[§1926.449]

Receptacle. A receptacle is a contact device installed at the outlet for the connection of a single attachment plug. A single receptacle is a single contact device with no other contact device on the same yoke. A multiple receptacle is a single device containing two or more receptacles.[§1926.449]

Receptacle outlet. An outlet where one or more receptacles are installed.[§1926.449]

Remote-control circuit. Any electric circuit that controls any other circuit through a relay or an equivalent device.[§1926.449]

Sealable equipment. Equipment enclosed in a case or cabinet that is provided with a means of sealing or locking so that live parts cannot be made accessible without opening the enclosure. The equipment may or may not be operable without opening the enclosure.[§1926.449]

Separately derived system. A premises wiring system whose power is derived from generator, transformer, or converter windings and has no direct electrical connection, including a solidly connected grounded circuit conductor, to supply conductors originating in another system.[§1926.449]

Service. The conductors and equipment for delivering energy from the electricity supply system to the wiring system of the premises served.[§1926.449]

Service conductors. The supply conductors that extend from the street main or from transformers to the service equipment of the premises supplied.[§1926.449]

Service drop. The overhead service conductors from the last pole or other aerial support to and including the splices, if any, connecting to the service-entrance conductors at the building or other structure. [§1926.449]

Service-entrance conductors, overhead system. The service conductors between the terminals of the service equipment and a point usually outside the building, clear of building walls, where joined by tap or splice to the service drop.[§1926.449]

Service-entrance conductors, underground system. The service conductors between the terminals of the service equipment and the point of connection to the service lateral. Where service equipment is located outside the building walls, there may be no service-entrance conductors, or they may be entirely outside the building.[§1926.449]

Service equipment. The necessary equipment, usually consisting of a circuit breaker or switch and fuses, and their accessories, located near the point of entrance of supply conductors to a building or other structure, or an otherwise defined area, and intended to constitute the main control and means of cutoff of the supply.[§1926.449]

Service raceway. The raceway that encloses the service-entrance conductors.[§1926.449]

Signaling circuit. Any electric circuit that energizes signaling equipment.[§1926.449]

Switchboard. A large single panel, frame, or assembly of panels which have switches, buses, instruments, overcurrent and other protective devices mounted on the face or back or both. Switchboards are generally accessible from the rear as well as from the front and are not intended to be installed in cabinets. (See "Panelboard.") [§1926.449]

Switches.[§1926.449]

(a) **General-use switch.** A switch intended for use in general distribution and branch circuits. It is rated in amperes, and it is capable of interrupting its rated current at its rated voltage.[§1926.449(a)]

(b) **General-use snap switch.** A form of general-use switch so constructed that it can be installed in flush device boxes or on outlet box covers, or otherwise used in conjunction with wiring systems recognized by this subpart.[§1926.449(b)]

(c) **Isolating switch.** A switch intended for isolating an electric circuit from the source of power. It has no interrupting rating, and it is intended to be operated only after the circuit has been opened by some other means.[§1926.449(c)]

(d) **Motor-circuit switch.** A switch, rated in horsepower, capable of interrupting the maximum operating overload current of a motor of the same horsepower rating as the switch at the rated voltage. [§1926.449(d)]

Switching devices. (Over 600 volts, nominal.) Devices designed to close and/or open one or more electric circuits. Included in this category are circuit breakers, cutouts, disconnecting (or isolating) switches, disconnecting means, and interrupter switches.[§1926.449]

Transportable X-ray. X-ray equipment installed in a vehicle or that may readily be disassembled for transport in a vehicle.[§1926.449]

Utilization equipment. Utilization equipment means equipment which utilizes electric energy for mechanical, chemical, heating, lighting, or similar useful purpose.[§1926.449]

Utilization system. A utilization system is a system which provides electric power and light for employee workplaces, and includes the premises wiring system and utilization equipment.[§1926.449]

Ventilated. Provided with a means to permit circulation of air sufficient to remove an excess of heat, fumes, or vapors.[§1926.449]

Volatile flammable liquid. A flammable liquid having a flash point below 38 degrees C (100 degrees F) or whose temperature is above its flash point, or a Class II combustible liquid having a vapor pressure not exceeding 40 psia (276 kPa) at 38 °C (100 °F) whose temperature is above its flash point.[§1926.449]

Voltage. (Of a circuit.) The greatest root-mean-square (effective) difference of potential between any two conductors of the circuit concerned.[§1926.449]

Voltage, nominal. A nominal value assigned to a circuit or system for the purpose of conveniently designating its voltage class (as 120/240, 480Y/277, 600, etc.). The actual voltage at which a circuit operates can vary from the nominal within a range that permits satisfactory operation of equipment.[§1926.449]

Voltage to ground. For grounded circuits, the voltage between the given conductor and that point or conductor of the circuit that is grounded; for ungrounded circuits, the greatest voltage between the given conductor and any other conductor of the circuit.[§1926.449]

Watertight. So constructed that moisture will not enter the enclosure. [§1926.449]

Weatherproof. So constructed or protected that exposure to the weather will not interfere with successful operation. Rainproof, raintight, or watertight equipment can fulfill the requirements for weatherproof where varying weather conditions other than wetness, such as snow, ice, dust, or temperature extremes, are not a factor.[§1926.449]

Wet location. See "Location."[§1926.449]

Authority: Sections 6 and 8 of the Occupational Safety and Health Act of 1970 (29 U.S.C. 655 and 657); sec. 107, Contract Work Hours and Safety Standards Act (40 U.S.C. 333); Secretary of Labor's Order No. 9- 83 (48 FR 35736) or 1-90 (55 FR 9033), as applicable; 29 CFR part 1911.

Subpart L - Scaffolds

§1926.450

Scope, application, and definitions applicable to this subpart

(a) **Scope and application.** This subpart applies to all scaffolds used in workplaces covered by this part. It does not apply to crane or derrick suspended personnel platforms. The criteria for aerial lifts are set out exclusively in Sec. 1926.453.[§1926.450(a)]

(b) **Definitions.**

Adjustable suspension scaffold means a suspension scaffold equipped with a hoist(s) that can be operated by an employee(s) on the scaffold.

Bearer (putlog) means a horizontal transverse scaffold member (which may be supported by ledgers or runners) upon which the scaffold platform rests and which joins scaffold uprights, posts, poles, and similar members.

Boatswains' chair means a single-point adjustable suspension scaffold consisting of a seat or sling designed to support one employee in a sitting position.

Body belt (safety belt) means a strap with means both for securing it about the waist and for attaching it to a lanyard, lifeline, or deceleration device.

Body harness means a design of straps which may be secured about the employee in a manner to distribute the fall arrest forces over at least the thighs, pelvis, waist, chest and shoulders, with means for attaching it to other components of a personal fall arrest system.

Brace means a rigid connection that holds one scaffold member in a fixed position with respect to another member, or to a building or structure.

Bricklayers' square scaffold means a supported scaffold composed of framed squares which support a platform.

Carpenters' bracket scaffold means a supported scaffold consisting of a platform supported by brackets attached to building or structural walls.

Catenary scaffold means a suspension scaffold consisting of a platform supported by two essentially horizontal and parallel ropes attached to structural members of a building or other structure. Additional support may be provided by vertical pickups.

Chimney hoist means a multi-point adjustable suspension scaffold used to provide access to work inside chimneys. (See "Multi-point adjustable suspension scaffold".)

Cleat means a structural block used at the end of a platform to prevent the platform from slipping off its supports. Cleats are also used to provide footing on sloped surfaces such as crawling boards.

Competent person means one who is capable of identifying existing and predictable hazards in the surroundings or working conditions which are unsanitary, hazardous, or dangerous to employees, and who has authorization to take prompt corrective measures to eliminate them.

Continuous run scaffold (Run scaffold) means a two- point or multi- point adjustable suspension scaffold constructed using a series of interconnected braced scaffold members or supporting structures erected to form a continuous scaffold.

Coupler means a device for locking together the tubes of a tube and coupler scaffold.

Crawling board (chicken ladder) means a supported scaffold consisting of a plank with cleats spaced and secured to provide footing, for use on sloped surfaces such as roofs.

Deceleration device means any mechanism, such as a rope grab, rip- stitch lanyard, specially-woven lanyard, tearing or deforming lanyard, or automatic self-retracting lifeline lanyard, which dissipates a substantial amount of energy during a fall arrest or limits the energy imposed on an employee during fall arrest.

Double pole (independent pole) scaffold means a supported scaffold consisting of a platform(s) resting on cross beams (bearers) supported by ledgers and a double row of uprights independent of support (except ties, guys, braces) from any structure.

Equivalent means alternative designs, materials or methods to protect against a hazard which the employer can demonstrate will provide an equal or greater degree of safety for employees than the methods, materials or designs specified in the standard.

Exposed power lines means electrical power lines which are accessible to employees and which are not shielded from contact. Such lines do not include extension cords or power tool cords.

Eye or Eye splice means a loop with or without a thimble at the end of a wire rope.

Fabricated decking and planking means manufactured platforms made of wood (including laminated wood, and solid sawn wood planks), metal or other materials.

Fabricated frame scaffold (tubular welded frame scaffold) means a scaffold consisting of a platform(s) supported on fabricated end frames with integral posts, horizontal bearers, and intermediate members.

Failure means load refusal, breakage, or separation of component parts. Load refusal is the point where the ultimate strength is exceeded.

Float (ship) scaffold means a suspension scaffold consisting of a braced platform resting on two parallel bearers and hung from overhead supports by ropes of fixed length.

Form scaffold means a supported scaffold consisting of a platform supported by brackets attached to formwork.

Guardrail system means a vertical barrier, consisting of, but not limited to, toprails, midrails, and posts, erected to prevent employees from falling off a scaffold platform or walkway to lower levels.

Hoist means a manual or power-operated mechanical device to raise or lower a suspended scaffold.

Horse scaffold means a supported scaffold consisting of a platform supported by construction horses (saw horses). Horse scaffolds constructed of metal are sometimes known as trestle scaffolds.

Independent pole scaffold (see "Double pole scaffold").

Interior hung scaffold means a suspension scaffold consisting of a platform suspended from the ceiling or roof structure by fixed length supports.

Ladder jack scaffold means a supported scaffold consisting of a platform resting on brackets attached to ladders.

Ladder stand means a mobile, fixed-size, self-supporting ladder consisting of a wide flat tread ladder in the form of stairs.

Landing means a platform at the end of a flight of stairs.

Large area scaffold means a pole scaffold, tube and coupler scaffold, systems scaffold, or fabricated frame scaffold erected over substantially the entire work area. For example: a scaffold erected over the entire floor area of a room.

Lean-to scaffold means a supported scaffold which is kept erect by tilting it toward and resting it against a building or structure.

Lifeline means a component consisting of a flexible line that connects to an anchorage at one end to hang vertically (vertical lifeline), or that connects to anchorages at both ends to stretch horizontally (horizontal lifeline), and which serves as a means for connecting other components of a personal fall arrest system to the anchorage.

Lower levels means areas below the level where the employee is located and to which an employee can fall. Such areas include, but are not limited to, ground levels, floors, roofs, ramps, runways, excavations, pits, tanks, materials, water, and equipment.

Masons' adjustable supported scaffold (see "Self-contained adjustable scaffold").

Masons' multi-point adjustable suspension scaffold means a continuous run suspension scaffold designed and used for masonry operations.

Maximum intended load means the total load of all persons, equipment, tools, materials, transmitted loads, and other loads reasonably anticipated to be applied to a scaffold or scaffold component at any one time.

Mobile scaffold means a powered or unpowered, portable, caster or wheel-mounted supported scaffold.

Multi-level suspended scaffold means a two-point or multi-point adjustable suspension scaffold with a series of platforms at various levels resting on common stirrups.

Multi-point adjustable suspension scaffold means a suspension scaffold consisting of a platform(s) which is suspended by more than two ropes from overhead supports and equipped with means to raise and lower the platform to desired work levels. Such scaffolds include chimney hoists.

Needle beam scaffold means a platform suspended from needle beams.

Open sides and ends means the edges of a platform that are more than 14 inches (36 cm) away horizontally from a sturdy, continuous, vertical surface (such as a building wall) or a sturdy, continuous horizontal surface (such as a floor), or a point of access. Exception: For plastering and lathing operations the horizontal threshold distance is 18 inches (46 cm).

Outrigger means the structural member of a supported scaffold used to increase the base width of a scaffold in order to provide support for and increased stability of the scaffold.

Outrigger beam (Thrustout) means the structural member of a suspension scaffold or outrigger scaffold which provides support for the scaffold by extending the scaffold point of attachment to a point out and away from the structure or building.

Outrigger scaffold means a supported scaffold consisting of a platform resting on outrigger beams (thrustouts) projecting beyond the wall or face of the building or structure, the inboard ends of which are secured inside the building or structure.

Overhand bricklaying means the process of laying bricks and masonry units such that the surface of the wall to be jointed is on the opposite side of the wall from the mason, requiring the mason to lean over the wall to complete the work. It includes mason tending and electrical installation incorporated into the brick wall during the overhand bricklaying process.

Personal fall arrest system means a system used to arrest an employee's fall. It consists of an anchorage, connectors, a body belt or body harness and may include a lanyard, deceleration device, lifeline, or combinations of these.

Platform means a work surface elevated above lower levels. Platforms can be constructed using individual wood planks, fabricated planks, fabricated decks, and fabricated platforms.

Pole scaffold (see definitions for "Single-pole scaffold" and "Double (independent) pole scaffold").

Power operated hoist means a hoist which is powered by other than human energy.

Pump jack scaffold means a supported scaffold consisting of a platform supported by vertical poles and movable support brackets.

Qualified means one who, by possession of a recognized degree, certificate, or professional standing, or who by extensive knowledge, training, and experience, has successfully demonstrated his/her ability to solve or resolve problems related to the subject matter, the work, or the project.

Rated load means the manufacturer's specified maximum load to be lifted by a hoist or to be applied to a scaffold or scaffold component.

Repair bracket scaffold means a supported scaffold consisting of a platform supported by brackets which are secured in place around the circumference or perimeter of a chimney, stack, tank or other supporting structure by one or more wire ropes placed around the supporting structure.

Roof bracket scaffold means a rooftop supported scaffold consisting of a platform resting on angular-shaped supports.

Runner (ledger or ribbon) means the lengthwise horizontal spacing or bracing member which may support the bearers.

Scaffold means any temporary elevated platform (supported or suspended) and its supporting structure (including points of anchorage), used for supporting employees or materials or both.

Self-contained adjustable scaffold means a combination supported and suspension scaffold consisting of an adjustable platform(s) mounted on an independent supporting frame(s) not a part of the object being worked on, and which is equipped with a means to permit the raising and lowering of the platform(s). Such systems include rolling roof rigs, rolling outrigger systems, and some masons' adjustable supported scaffolds.

Shore scaffold means a supported scaffold which is placed against a building or structure and held in place with props.

Single-point adjustable suspension scaffold means a suspension scaffold consisting of a platform suspended by one rope from an overhead support and equipped with means to permit the movement of the platform to desired work levels.

Single-pole scaffold means a supported scaffold consisting of a platform(s) resting on bearers, the outside ends of which are supported on runners secured to a single row of posts or uprights, and the inner ends of which are supported on or in a structure or building wall.

Stair tower (Scaffold stairway/tower) means a tower comprised of scaffold components and which contains internal stairway units and rest platforms. These towers are used to provide access to scaffold platforms and other elevated points such as floors and roofs.

Stall load means the load at which the prime-mover of a power- operated hoist stalls or the power to the prime-mover is automatically disconnected.

Step, platform, and trestle ladder scaffold means a platform resting directly on the rungs of step ladders or trestle ladders.

Stilts means a pair of poles or similar supports with raised footrests, used to permit walking above the ground or working surface.

Stonesetters' multi-point adjustable suspension scaffold means a continuous run suspension scaffold designed and used for stonesetters' operations.

Supported scaffold means one or more platforms supported by outrigger beams, brackets, poles, legs, uprights, posts, frames, or similar rigid support.

Suspension scaffold means one or more platforms suspended by ropes or other non-rigid means from an overhead structure(s).

System scaffold means a scaffold consisting of posts with fixed connection points that accept runners, bearers, and diagonals that can be interconnected at predetermined levels.

Tank builders' scaffold means a supported scaffold consisting of a platform resting on brackets that are either directly attached to a cylindrical tank or attached to devices that are attached to such a tank.

Top plate bracket scaffold means a scaffold supported by brackets that hook over or are attached to the top of a wall. This type of scaffold is similar to carpenters' bracket scaffolds and form scaffolds and is used in residential construction for setting trusses.

Tube and coupler scaffold means a supported or suspended scaffold consisting of a platform(s) supported by tubing, erected with coupling devices connecting uprights, braces, bearers, and runners.

Tubular welded frame scaffold (see "Fabricated frame scaffold").

Two-point suspension scaffold (swing stage) means a suspension scaffold consisting of a platform supported by hangers (stirrups) suspended by two ropes from overhead supports and equipped with means to permit the raising and lowering of the platform to desired work levels.

Unstable objects means items whose strength, configuration, or lack of stability may allow them to become dislocated and shift and therefore may not properly support the loads imposed on them. Unstable objects do not constitute a safe base support for scaffolds, platforms, or employees. Examples include, but are not limited to, barrels, boxes, loose brick, and concrete blocks.

Vertical pickup means a rope used to support the horizontal rope in catenary scaffolds.

Walkway means a portion of a scaffold platform used only for access and not as a work level.

Window jack scaffold means a platform resting on a bracket or jack which projects through a window opening.

[61 FR 46104, Aug. 30, 1996, as amended at 75 FR 48133, Aug. 9, 2010]

§1926.451
General requirements

This section does not apply to aerial lifts, the criteria for which are set out exclusively in Sec. 1926.453.

(a) Capacity.[§1926.451(a)]

(1) *Except as provided* in paragraphs (a)(2), (a)(3), (a)(4), (a)(5) and (g) of this section, each scaffold and scaffold component shall be capable of supporting, without failure, its own weight and at least 4 times the maximum intended load applied or transmitted to it.[§1926.451(a)(1)]

(2) *Direct connections to roofs and floors,* and counterweights used to balance adjustable suspension scaffolds, shall be capable of resisting at least 4 times the tipping moment imposed by the scaffold operating at the rated load of the hoist, or 1.5 (minimum) times the tipping moment imposed by the scaffold operating at the stall load of the hoist, whichever is greater.[§1926.451(a)(2)]

(3) *Each suspension rope,* including connecting hardware, used on non-adjustable suspension scaffolds shall be capable of supporting, without failure, at least 6 times the maximum intended load applied or transmitted to that rope.[§1926.451(a)(3)]

(4) *Each suspension rope,* including connecting hardware, used on adjustable suspension scaffolds shall be capable of supporting, without failure, at least 6 times the maximum intended load applied or transmitted to that rope with the scaffold operating at either the rated load of the hoist, or 2 (minimum) times the stall load of the hoist, whichever is greater.[§1926.451(a)(4)]

(5) *The stall load of any* scaffold hoist shall not exceed 3 times its rated load.[§1926.451(a)(5)]

(6) *Scaffolds shall be designed by a qualified person* and shall be constructed and loaded in accordance with that design. Non-mandatory Appendix A to this subpart contains examples of criteria that will enable an employer to comply with paragraph (a) of this section.[§1926.451(a)(6)]

(b) Scaffold platform construction.[§1926.451(b)]

(1) *Each platform on all working levels* of scaffolds shall be fully planked or decked between the front uprights and the guardrail supports as follows:[§1926.451(b)(1)]

(i) *Each platform unit* (e.g., scaffold plank, fabricated plank, fabricated deck, or fabricated platform) shall be installed so that the space between adjacent units and the space between the platform and the uprights is no more than 1 inch (2.5 cm) wide, except where the employer can demonstrate that a wider space is necessary (for example, to fit around uprights when side brackets are used to extend the width of the platform).[§1926.451(b)(1)(i)]

(ii) *Where the employer makes the demonstration* provided for in paragraph (b)(1)(i) of this section, the platform shall be planked or decked as fully as possible and the remaining open space between the platform and the uprights shall not exceed 9 1/2 inches (24.1 cm).[§1926.451(b)(1)(ii)]

Exception to paragraph (b)(1): The requirement in paragraph (b)(1) to provide full planking or decking does not apply to platforms used solely as walkways or solely by employees performing scaffold erection or dismantling. In these situations, only the planking that the employer establishes is necessary to provide safe working conditions is required.

(2) *Except as provided in paragraphs* (b)(2)(i) and (b)(2)(ii) of this section, each scaffold platform and walkway shall be at least 18 inches (46 cm) wide.[§1926.451(b)(2)]

(i) *Each ladder jack scaffold,* top plate bracket scaffold, roof bracket scaffold, and pump jack scaffold shall be at least 12 inches (30 cm) wide. There is no minimum width requirement for boatswains' chairs.[§1926.451(b)(2)(i)]

Note to paragraph (b)(2)(i): Pursuant to an administrative stay effective November 29, 1996 and published in the Federal Register on November 25, 1996, the requirement in paragraph (b)(2)(i) that roof bracket scaffolds be at least 12 inches wide is stayed until November 25, 1997 or until rulemaking regarding the minimum width of roof bracket scaffolds has been completed, whichever is later.

(ii) *Where scaffolds must be used* in areas that the employer can demonstrate are so narrow that platforms and walkways cannot be at least 18 inches (46 cm) wide, such platforms and walkways shall be as wide as feasible, and employees on those platforms and walkways shall be protected from fall hazards by the use of guardrails and/or personal fall arrest systems.[§1926.451(b)(2)(ii)]

(3) *Except as provided in paragraphs* (b)(3) (i) and (ii) of this section, the front edge of all platforms shall not be more than 14 inches (36 cm) from the face of the work, unless guardrail systems are erected along the front edge and/or personal fall arrest systems are used in accordance with paragraph (g) of this section to protect employees from falling.[§1926.451(b)(3)]

(i) *The maximum distance from the face* for outrigger scaffolds shall be 3 inches (8 cm);[§1926.451(b)(3)(i)]

(ii) *The maximum distance from the face* for plastering and lathing operations shall be 18 inches (46 cm).[§1926.451(b)(3)(ii)]

(4) *Each end of a platform,* unless cleated or otherwise restrained by hooks or equivalent means, shall extend over the centerline of its support at least 6 inches (15 cm).[§1926.451(b)(4)]

(5) (i) *Each end of a platform 10 feet or less* in length shall not extend over its support more than 12 inches (30 cm) unless the platform is designed and installed so that the cantilevered portion of the platform is able to support employees and/or materials without tipping, or has guardrails which block employee access to the cantilevered end.[§1926.451(b)(5)(i)]

(ii) *Each platform greater than 10 feet in length* shall not extend over its support more than 18 inches (46 cm), unless it is designed and installed so that the cantilevered portion of the platform is able to support employees without tipping, or has guardrails which block employee access to the cantilevered end.[§1926.451(b)(5)(ii)]

(6) *On scaffolds where scaffold planks are abutted* to create a long platform, each abutted end shall rest on a separate support surface. This provision does not preclude the use of common support members, such as "T" sections, to support abutting planks, or hook on platforms designed to rest on common supports.[§1926.451(b)(6)]

(7) *On scaffolds where platforms are overlapped* to create a long platform, the overlap shall occur only over supports, and shall not be less than 12 inches (30 cm) unless the platforms are nailed together or otherwise restrained to prevent movement.[§1926.451(b)(7)]

(8) *At all points of a scaffold* where the platform changes direction, such as turning a corner, any platform that rests on a bearer at an angle other than a right angle shall be laid first, and platforms which rest at right angles over the same bearer shall be laid second, on top of the first platform.[§1926.451(b)(8)]

(9) *Wood platforms shall not be covered* with opaque finishes, except that platform edges may be covered or marked for identification. Platforms may be coated periodically with wood preservatives, fire- retardant finishes, and slip-resistant finishes; however, the coating may not obscure the top or bottom wood surfaces.[§1926.451(b)(9)]

(10) *Scaffold components manufactured* by different manufacturers shall not be intermixed unless the components fit together without force and the scaffold's structural integrity is maintained by the user. Scaffold components manufactured by different manufacturers shall not be modified in order to intermix them unless a competent person determines the resulting scaffold is structurally sound.[§1926.451(b)(10)]

(11) *Scaffold components made* of dissimilar metals shall not be used together unless a competent person has determined that galvanic action will not reduce the strength of any component to a level below that required by paragraph (a)(1) of this section.[§1926.451(b)(11)]

(c) Criteria for supported scaffolds.[§1926.451(c)]

(1) *Supported scaffolds with a height to base width* (including outrigger supports, if used) ratio of more than four to one (4:1) shall be restrained from tipping by guying, tying, bracing, or equivalent means, as follows:[§1926.451(c)(1)]

(i) *Guys, ties, and braces* shall be installed at locations where horizontal members support both inner and outer legs. [§1926.451(c)(1)(i)]

(ii) *Guys, ties, and braces* shall be installed according to the scaffold manufacturer's recommendations or at the closest horizontal member to the 4:1 height and be repeated vertically at locations of horizontal members every 20 feet (6.1 m) or less thereafter for scaffolds 3 feet (0.91 m) wide or less, and every 26 feet (7.9 m) or less thereafter for scaffolds greater than 3 feet (0.91 m) wide. The top guy, tie or brace of completed scaffolds shall be placed no further than the 4:1 height from the top. Such guys, ties and braces shall be installed at each end of the scaffold and at horizontal intervals not to exceed 30 feet (9.1 m) (measured from one end [not both] towards the other).[§1926.451(c)(1)(ii)]

(iii) *Ties, guys, braces, or outriggers* shall be used to prevent the tipping of supported scaffolds in all circumstances where an eccentric load, such as a cantilevered work platform, is applied or is transmitted to the scaffold.[§1926.451(c)(1)(iii)]

(2) *Supported scaffold poles,* legs, posts, frames, and uprights shall bear on base plates and mud sills or other adequate firm foundation.[§1926.451(c)(2)]

(i) *Footings shall be level,* sound, rigid, and capable of supporting the loaded scaffold without settling or displacement. [§1926.451(c)(2)(i)]

(ii) *Unstable objects* shall not be used to support scaffolds or platform units.[§1926.451(c)(2)(ii)]

(iii) *Unstable objects* shall not be used as working platforms. [§1926.451(c)(2)(iii)]

(iv) *Front-end loaders* and similar pieces of equipment shall not be used to support scaffold platforms unless they have been specifically designed by the manufacturer for such use. [§1926.451(c)(2)(iv)]

(v) *Fork-lifts shall not be used to support* scaffold platforms unless the entire platform is attached to the fork and the fork-lift is not moved horizontally while the platform is occupied. [§1926.451(c)(2)(v)]

(3) *Supported scaffold poles,* legs, posts, frames, and uprights shall be plumb and braced to prevent swaying and displacement.[§1926.451(c)(3)]

(d) Criteria for suspension scaffolds.[§1926.451(d)]

(1) *All suspension scaffold support devices,* such as outrigger beams, cornice hooks, parapet clamps, and similar devices, shall rest on surfaces capable of supporting at least 4 times the load imposed on them by the scaffold operating at the rated load of the hoist (or at least 1.5 times the load imposed on them by the scaffold at the stall capacity of the hoist, whichever is greater).[§1926.451(d)(1)]

(2) *Suspension scaffold outrigger beams,* when used, shall be made of structural metal or equivalent strength material, and shall be restrained to prevent movement.[§1926.451(d)(2)]

(3) *The inboard ends of suspension scaffold outrigger beams* shall be stabilized by bolts or other direct connections to the floor or roof deck, or they shall have their inboard ends stabilized by counterweights, except masons' multi-point adjustable suspension scaffold outrigger beams shall not be stabilized by counterweights.[§1926.451(d)(3)]

(i) *Before the scaffold is used,* direct connections shall be evaluated by a competent person who shall confirm, based on the evaluation, that the supporting surfaces are capable of supporting the loads to be imposed. In addition, masons' multi-point adjustable suspension scaffold connections shall be designed by an engineer experienced in such scaffold design.[§1926.451(d)(3)(i)]

(ii) *Counterweights shall be made of non-flowable material.* Sand, gravel and similar materials that can be easily dislocated shall not be used as counterweights.[§1926.451(d)(3)(ii)]

(iii) *Only those items specifically designed* as counterweights shall be used to counterweight scaffold systems. Construction materials such as, but not limited to, masonry units and rolls of roofing felt, shall not be used as counterweights. [§1926.451(d)(3)(iii)]

(iv) *Counterweights shall be secured* by mechanical means to the outrigger beams to prevent accidental displacement. [§1926.451(d)(3)(iv)]

(v) *Counterweights shall not be removed* from an outrigger beam until the scaffold is disassembled.[§1926.451(d)(3)(v)]

(vi) *Outrigger beams which are not stabilized* by bolts or other direct connections to the floor or roof deck shall be secured by tiebacks.[§1926.451(d)(3)(vi)]

(vii) *Tiebacks shall be equivalent in strength to the suspension ropes.*[§1926.451(d)(3)(vii)]

(viii) *Outrigger beams shall be placed* perpendicular to its bearing support (usually the face of the building or structure). However, where the employer can demonstrate that it is not possible to place an outrigger beam perpendicular to the face of the building or structure because of obstructions that cannot be moved, the outrigger beam may be placed at some other angle, provided opposing angle tiebacks are used. [§1926.451(d)(3)(viii)]

(ix) *Tiebacks shall be secured* to a structurally sound anchorage on the building or structure. Sound anchorages include structural members, but do not include standpipes, vents, other piping systems, or electrical conduit.[§1926.451(d)(3)(ix)]

(x) *Tiebacks shall be installed perpendicular* to the face of the building or structure, or opposing angle tiebacks shall be installed. Single tiebacks installed at an angle are prohibited. [§1926.451(d)(3)(x)]

(4) *Suspension scaffold outrigger beams shall be:*[§1926.451(d)(4)]

(i) *Provided with stop bolts* or shackles at both ends; [§1926.451(d)(4)(i)]

(ii) *Securely fastened together* with the flanges turned out when channel iron beams are used in place of I-beams; [§1926.451(d)(4)(ii)]

(iii) *Installed with all bearing supports* perpendicular to the beam center line;[§1926.451(d)(4)(iii)]

(iv) *Set and maintained with the web* in a vertical position; and [§1926.451(d)(4)(iv)]

(v) *When an outrigger beam is used,* the shackle or clevis with which the rope is attached to the outrigger beam shall be placed directly over the center line of the stirrup. [§1926.451(d)(4)(v)]

(5) *Suspension scaffold support* devices such as cornice hooks, roof hooks, roof irons, parapet clamps, or similar devices shall be:[§1926.451(d)(5)]

(i) *Made of steel, wrought iron, or materials of equivalent strength;*[§1926.451(d)(5)(i)]

(ii) *Supported by bearing blocks; and*[§1926.451(d)(5)(ii)]

(iii) *Secured against movement by tiebacks* installed at right angles to the face of the building or structure, or opposing angle tiebacks shall be installed and secured to a structurally sound point of anchorage on the building or structure. Sound points of anchorage include structural members, but do not include standpipes, vents, other piping systems, or electrical conduit.[§1926.451(d)(5)(iii)]

(iv) *Tiebacks shall be equivalent in strength to the hoisting rope.* [§1926.451(d)(5)(iv)]

(6) *When winding drum hoists* are used on a suspension scaffold, they shall contain not less than four wraps of the suspension rope at the lowest point of scaffold travel. When other types of hoists are used, the suspension ropes shall be long enough to allow the scaffold to be lowered to the level below without the rope end passing through the hoist, or the rope end shall be configured or provided with means to prevent the end from passing through the hoist.[§1926.451(d)(6)]

(7) *The use of repaired wire rope* as suspension rope is prohibited.[§1926.451(d)(7)]

(8) *Wire suspension ropes* shall not be joined together except through the use of eye splice thimbles connected with shackles or coverplates and bolts.[§1926.451(d)(8)]

(9) *The load end of wire suspension ropes* shall be equipped with proper size thimbles and secured by eyesplicing or equivalent means.[§1926.451(d)(9)]

(10) *Ropes shall be inspected for defects* by a competent person prior to each workshift and after every occurrence which could affect a rope's integrity. Ropes shall be replaced if any of the following conditions exist:[§1926.451(d)(10)]

(i) *Any physical damage* which impairs the function and strength of the rope.[§1926.451(d)(10)(i)]

(ii) *Kinks that might impair* the tracking or wrapping of rope around the drum(s) or sheave(s).[§1926.451(d)(10)(ii)]

(iii) *Six randomly distributed* broken wires in one rope lay or three broken wires in one strand in one rope lay. [§1926.451(d)(10)(iii)]

(iv) *Abrasion, corrosion,* scrubbing, flattening or peening causing loss of more than one-third of the original diameter of the outside wires.[§1926.451(d)(10)(iv)]

(v) *Heat damage caused* by a torch or any damage caused by contact with electrical wires.[§1926.451(d)(10)(v)]

(vi) *Evidence that the secondary brake* has been activated during an overspeed condition and has engaged the suspension rope.[§1926.451(d)(10)(vi)]

(11) *Swaged attachments or spliced* eyes on wire suspension ropes shall not be used unless they are made by the wire rope manufacturer or a qualified person.[§1926.451(d)(11)]

(12) *When wire rope clips* are used on suspension scaffolds: [§1926.451(d)(12)]

(i) *There shall be a minimum* of 3 wire rope clips installed, with the clips a minimum of 6 rope diameters apart; [§1926.451(d)(12)(i)]

(ii) *Clips shall be installed* according to the manufacturer's recommendations;[§1926.451(d)(12)(ii)]

(iii) *Clips shall be retightened* to the manufacturer's recommendations after the initial loading;[§1926.451(d)(12)(iii)]

(iv) *Clips shall be inspected and retightened* to the manufacturer's recommendations at the start of each workshift thereafter;[§1926.451(d)(12)(iv)]

(v) *U-bolt clips shall not be used* at the point of suspension for any scaffold hoist;[§1926.451(d)(12)(v)]

(vi) *When U-bolt clips are used,* the U-bolt shall be placed over the dead end of the rope, and the saddle shall be placed over the live end of the rope.[§1926.451(d)(12)(vi)]

(13) *Suspension scaffold power-operated hoists* and manual hoists shall be tested by a qualified testing laboratory.[§1926.451(d)(13)]

(14) *Gasoline-powered equipment and hoists* shall not be used on suspension scaffolds.[§1926.451(d)(14)]

(15) *Gears and brakes of power-operated hoists* used on suspension scaffolds shall be enclosed.[§1926.451(d)(15)]

(16) *In addition to the normal operating brake,* suspension scaffold power-operated hoists and manually operated hoists shall have a braking device or locking pawl which engages automatically when a hoist makes either of the following uncontrolled movements: an instantaneous change in momentum or an accelerated overspeed.[§1926.451(d)(16)]

(17) *Manually operated hoists* shall require a positive crank force to descend.[§1926.451(d)(17)]

(18) *Two-point and multi-point suspension scaffolds* shall be tied or otherwise secured to prevent them from swaying, as determined to be necessary based on an evaluation by a competent person. Window cleaners' anchors shall not be used for this purpose.[§1926.451(d)(18)]

(19) *Devices whose sole function* is to provide emergency escape and rescue shall not be used as working platforms. This provision does not preclude the use of systems which are designed to function both as suspension scaffolds and emergency systems.[§1926.451(d)(19)]

(e) **Access.** This paragraph applies to scaffold access for all employees. Access requirements for employees erecting or dismantling supported scaffolds are specifically addressed in paragraph (e)(9) of this section.[§1926.451(e)]

(1) *When scaffold platforms* are more than 2 feet (0.6 m) above or below a point of access, portable ladders, hook-on ladders, attachable ladders, stair towers (scaffold stairways/towers), stairway-type ladders (such as ladder stands), ramps, walkways, integral prefabricated scaffold access, or direct access from another scaffold, structure, personnel hoist, or similar surface shall be used. Crossbraces shall not be used as a means of access.[§1926.451(e)(1)]

(2) *Portable, hook-on, and attachable ladders* (Additional requirements for the proper construction and use of portable ladders are contained in subpart X of this part — Stairways and Ladders):[§1926.451(e)(2)]

(i) *Portable, hook-on, and attachable ladders* shall be positioned so as not to tip the scaffold;[§1926.451(e)(2)(i)]

(ii) *Hook-on and attachable ladders* shall be positioned so that their bottom rung is not more than 24 inches (61 cm) above the scaffold supporting level;[§1926.451(e)(2)(ii)]

(iii) *When hook-on and attachable ladders* are used on a supported scaffold more than 35 feet (10.7 m) high, they shall have rest platforms at 35-foot (10.7 m) maximum vertical intervals.[§1926.451(e)(2)(iii)]

(iv) *Hook-on and attachable ladders* shall be specifically designed for use with the type of scaffold used; [§1926.451(e)(2)(iv)]

(v) *Hook-on and attachable ladders* shall have a minimum rung length of 11 1/2 inches (29 cm); and[§1926.451(e)(2)(v)]

(vi) *Hook-on and attachable ladders* shall have uniformly spaced rungs with a maximum spacing between rungs of 16 3/4 inches.[§1926.451(e)(2)(vi)]

(3) *Stairway-type ladders shall:*[§1926.451(e)(3)]

(i) *Be positioned such that* their bottom step is not more than 24 inches (61 cm) above the scaffold supporting level; [§1926.451(e)(3)(i)]

(ii) *Be provided with rest platforms* at 12 foot (3.7 m) maximum vertical intervals;[§1926.451(e)(3)(ii)]

(iii) *Have a minimum step width* of 16 inches (41 cm), except that mobile scaffold stairway-type ladders shall have a minimum step width of 11 1/2 inches (30 cm); and [§1926.451(e)(3)(iii)]

(iv) *Have slip-resistant treads* on all steps and landings. [§1926.451(e)(3)(iv)]

(4) *Stairtowers (scaffold stairway/towers)* shall be positioned such that their bottom step is not more than 24 inches (61 cm.) above the scaffold supporting level.[§1926.451(e)(4)]

(i) *A stairrail consisting of a toprail* and a midrail shall be provided on each side of each scaffold stairway.[§1926.451(e)(4)(i)]

(ii) *The toprail of each stairrail system* shall also be capable of serving as a handrail, unless a separate handrail is provided. [§1926.451(e)(4)(ii)]

(iii) *Handrails, and toprails* that serve as handrails, shall provide an adequate handhold for employees grasping them to avoid falling.[§1926.451(e)(4)(iii)]

(iv) *Stairrail systems and handrails* shall be surfaced to prevent injury to employees from punctures or lacerations, and to prevent snagging of clothing.[§1926.451(e)(4)(iv)]

(v) *The ends of stairrail systems and handrails* shall be constructed so that they do not constitute a projection hazard. [§1926.451(e)(4)(v)]

(vi) *Handrails, and toprails* that are used as handrails, shall be at least 3 inches (7.6 cm) from other objects.[§1926.451(e)(4)(vi)]

(vii) *Stairrails shall be not less than 28 inches (71 cm)* nor more than 37 inches (94 cm) from the upper surface of the stairrail to the surface of the tread, in line with the face of the riser at the forward edge of the tread.[§1926.451(e)(4)(vii)]

(viii) *A landing platform at least 18 inches (45.7 cm)* wide by at least 18 inches (45.7 cm) long shall be provided at each level.[§1926.451(e)(4)(viii)]

(ix) *Each scaffold stairway* shall be at least 18 inches (45.7 cm) wide between stairrails.[§1926.451(e)(4)(ix)]

(x) *Treads and landings* shall have slip-resistant surfaces. [§1926.451(e)(4)(x)]

(xi) *Stairways shall be installed* between 40 degrees and 60 degrees from the horizontal.[§1926.451(e)(4)(xi)]

(xii) *Guardrails meeting the requirements* of paragraph (g)(4) of this section shall be provided on the open sides and ends of each landing.[§1926.451(e)(4)(xii)]

(xiii) *Riser height shall be uniform,* within 1/4 inch, (0.6 cm) for each flight of stairs. Greater variations in riser height are allowed for the top and bottom steps of the entire system, not for each flight of stairs.[§1926.451(e)(4)(xiii)]

(xiv) *Tread depth shall be uniform,* within 1/4 inch, for each flight of stairs.[§1926.451(e)(4)(xiv)]

(5) *Ramps and walkways.*[§1926.451(e)(5)]

(i) *Ramps and walkways* 6 feet (1.8 m) or more above lower levels shall have guardrail systems which comply with subpart M of this part — Fall Protection;[§1926.451(e)(5)(i)]

(ii) *No ramp or walkway shall be inclined* more than a slope of one (1) vertical to three (3) horizontal (20 degrees above the horizontal).[§1926.451(e)(5)(ii)]

(iii) *If the slope of a ramp* or a walkway is steeper than one (1) vertical in eight (8) horizontal, the ramp or walkway shall have cleats not more than fourteen (14) inches (35 cm) apart which are securely fastened to the planks to provide footing. [§1926.451(e)(5)(iii)]

(6) *Integral prefabricated scaffold access frames shall:* [§1926.451(e)(6)]

(i) *Be specifically designed* and constructed for use as ladder rungs;[§1926.451(e)(6)(i)]

(ii) *Have a rung length of at least 8 inches (20 cm);* [§1926.451(e)(6)(ii)]

(iii) *Not be used as work platforms* when rungs are less than 11 1/2 inches in length, unless each affected employee uses fall protection, or a positioning device, which complies with Sec. 1926.502;[§1926.451(e)(6)(iii)]

(iv) *Be uniformly spaced* within each frame section; [§1926.451(e)(6)(iv)]

(v) *Be provided with rest platforms* at 35-foot (10.7 m) maximum vertical intervals on all supported scaffolds more than 35 feet (10.7 m) high; and[§1926.451(e)(6)(v)]

(vi) *Have a maximum spacing between rungs* of 16 3/4 inches (43 cm). Non-uniform rung spacing caused by joining end frames together is allowed, provided the resulting spacing does not exceed 16 3/4 inches (43 cm).[§1926.451(e)(6)(vi)]

(7) *Steps and rungs of ladder* and stairway type access shall line up vertically with each other between rest platforms. [§1926.451(e)(7)]

(8) *Direct access to or from another surface* shall be used only when the scaffold is not more than 14 inches (36 cm) horizontally and not more than 24 inches (61 cm) vertically from the other surface.[§1926.451(e)(8)]

(9) *Effective September 2, 1997,* access for employees erecting or dismantling supported scaffolds shall be in accordance with the following:[§1926.451(e)(9)]

(i) *The employer shall provide safe means of access* for each employee erecting or dismantling a scaffold where the provision of safe access is feasible and does not create a greater hazard. The employer shall have a competent person determine whether it is feasible or would pose a greater hazard to provide, and have employees use a safe means of access. This determination shall be based on site conditions and the type of scaffold being erected or dismantled.[§1926.451(e)(9)(i)]

(ii) *Hook-on or attachable ladders* shall be installed as soon as scaffold erection has progressed to a point that permits safe installation and use.[§1926.451(e)(9)(ii)]

(iii) *When erecting or dismantling* tubular welded frame scaffolds, (end) frames, with horizontal members that are parallel, level and are not more than 22 inches apart vertically may be used as climbing devices for access, provided they are erected in a manner that creates a usable ladder and provides good hand hold and foot space.[§1926.451(e)(9)(iii)]

(iv) *Cross braces on tubular welded frame scaffolds* shall not be used as a means of access or egress.[§1926.451(e)(9)(iv)]

(f) **Use.**[§1926.451(f)]

(1) *Scaffolds and scaffold components* shall not be loaded in excess of their maximum intended loads or rated capacities, whichever is less.[§1926.451(f)(1)]

(2) *The use of shore or lean-to scaffolds is prohibited.* [§1926.451(f)(2)]

(3) *Scaffolds and scaffold components* shall be inspected for visible defects by a competent person before each work shift, and after any occurrence which could affect a scaffold's structural integrity.[§1926.451(f)(3)]

(4) *Any part of a scaffold damaged or weakened* such that its strength is less than that required by paragraph (a) of this section shall be immediately repaired or replaced, braced to meet those provisions, or removed from service until repaired. [§1926.451(f)(4)]

(5) *Scaffolds shall not be moved horizontally* while employees are on them, unless they have been designed by a registered professional engineer specifically for such movement or, for mobile scaffolds, where the provisions of Sec. 1926.452(w) are followed.[§1926.451(f)(5)]

(6) *The clearance between scaffolds* and power lines shall be as follows: Scaffolds shall not be erected, used, dismantled, altered, or moved such that they or any conductive material handled on them might come closer to exposed and energized power lines than as follows:[§1926.451(f)(6)]

Insulated Lines

Voltage	Minimum distance	Alternatives
Less than 300 volts	3 feet (0.9 m)	
300 volts to 50 kv	10 feet (3.1 m)	
More than 50 kv	10 feet (3.1 m) plus 0.4 inches (1.0 cm) for each 1 kv over 50 kv.	2 times the length of the line insulator but never less than 10 feet (3.1 m).

Uninsulated Lines

Voltage	Minimum distance	Alternatives
Less than 50 kv	10 feet (3.1 m)	
More than 50 kv	10 feet (3.1 m) plus 0.4 inches (1.0 cm) for each 1 kv over 50 kv.	2 times the length of the line insulator, but never less than 10 feet (3.1 m).

Exception to paragraph (f)(6): Scaffolds and materials may be closer to power lines than specified above where such clearance is necessary for performance of work, and only after the utility company, or electrical system operator, has been notified of the need to work closer and the utility company, or electrical system operator, has deenergized the lines, relocated the lines, or installed protective coverings to prevent accidental contact with the lines.

(7) *Scaffolds shall be erected, moved, dismantled,* or altered only under the supervision and direction of a competent person qualified in scaffold erection, moving, dismantling or alteration. Such activities shall be performed only by experienced and trained employees selected for such work by the competent person.[§1926.451(f)(7)]

(8) *Employees shall be prohibited from working* on scaffolds covered with snow, ice, or other slippery material except as necessary for removal of such materials.[§1926.451(f)(8)]

(9) *Where swinging loads are being hoisted* onto or near scaffolds such that the loads might contact the scaffold, tag lines or equivalent measures to control the loads shall be used. [§1926.451(f)(9)]

(10) *Suspension ropes supporting* adjustable suspension scaffolds shall be of a diameter large enough to provide sufficient surface area for the functioning of brake and hoist mechanisms.[§1926.451(f)(10)]

(11) *Suspension ropes shall be shielded* from heat-producing processes. When acids or other corrosive substances are used on a scaffold, the ropes shall be shielded, treated to protect against the corrosive substances, or shall be of a material that will not be damaged by the substance being used. [§1926.451(f)(11)]

(12) *Work on or from scaffolds is prohibited* during storms or high winds unless a competent person has determined that it is safe for employees to be on the scaffold and those employees are protected by a personal fall arrest system or wind screens. Wind screens shall not be used unless the scaffold is secured against the anticipated wind forces imposed.[§1926.451(f)(12)]

(13) *Debris shall not be allowed to accumulate on platforms.* [§1926.451(f)(13)]

(14) *Makeshift devices,* such as but not limited to boxes and barrels, shall not be used on top of scaffold platforms to increase the working level height of employees.[§1926.451(f)(14)]

(15) *Ladders shall not be used on scaffolds* to increase the working level height of employees, except on large area scaffolds where employers have satisfied the following criteria: [§1926.451(f)(15)]

(i) *When the ladder is placed against a structure* which is not a part of the scaffold, the scaffold shall be secured against the sideways thrust exerted by the ladder;[§1926.451(f)(15)(i)]

(ii) *The platform units shall be secured* to the scaffold to prevent their movement;[§1926.451(f)(15)(ii)]

(iii) *The ladder legs shall be on the same platform* or other means shall be provided to stabilize the ladder against unequal platform deflection, and[§1926.451(f)(15)(iii)]

(iv) *The ladder legs shall be secured* to prevent them from slipping or being pushed off the platform.[§1926.451(f)(15)(iv)]

(16) *Platforms shall not deflect more* than 1/60 of the span when loaded.[§1926.451(f)(16)]

(17) *To reduce the possibility of welding* current arcing through the suspension wire rope when performing welding from suspended scaffolds, the following precautions shall be taken, as applicable:[§1926.451(f)(17)]

(i) *An insulated thimble shall be used* to attach each suspension wire rope to its hanging support (such as cornice hook or outrigger). Excess suspension wire rope and any additional independent lines from grounding shall be insulated; [§1926.451(f)(17)(i)]

(ii) *The suspension wire rope* shall be covered with insulating material extending at least 4 feet (1.2 m) above the hoist. If there is a tail line below the hoist, it shall be insulated to prevent contact with the platform. The portion of the tail line that hangs free below the scaffold shall be guided or retained, or both, so that it does not become grounded;[§1926.451(f)(17)(ii)]

(iii) *Each hoist shall be covered* with insulated protective covers; [§1926.451(f)(17)(iii)]

(iv) *In addition to a work lead attachment* required by the welding process, a grounding conductor shall be connected from the scaffold to the structure. The size of this conductor shall be at least the size of the welding process work lead, and this conductor shall not be in series with the welding process or the work piece;[§1926.451(f)(17)(iv)]

(v) *If the scaffold grounding lead* is disconnected at any time, the welding machine shall be shut off; and[§1926.451(f)(17)(v)]

(vi) *An active welding rod* or uninsulated welding lead shall not be allowed to contact the scaffold or its suspension system. [§1926.451(f)(17)(vi)]

(g) **Fall protection.**[§1926.451(g)]

(1) *Each employee on a scaffold* more than 10 feet (3.1 m) above a lower level shall be protected from falling to that lower level. Paragraphs (g)(1) (i) through (vii) of this section establish the types of fall protection to be provided to the employees on each type of scaffold. Paragraph (g)(2) of this section addresses fall protection for scaffold erectors and dismantlers. [§1926.451(g)(1)]

Note to paragraph (g)(1): The fall protection requirements for employees installing suspension scaffold support systems on floors, roofs, and other elevated surfaces are set forth in subpart M of this part.

(i) *Each employee on a boatswains' chair,* catenary scaffold, float scaffold, needle beam scaffold, or ladder jack scaffold shall be protected by a personal fall arrest system; [§1926.451(g)(1)(i)]

(ii) *Each employee on a single-point or two-point* adjustable suspension scaffold shall be protected by both a personal fall arrest system and guardrail system;[§1926.451(g)(1)(ii)]

(iii) *Each employee on a crawling board* (chicken ladder) shall be protected by a personal fall arrest system, a guardrail system (with minimum 200 pound toprail capacity), or by a three-fourth inch (1.9 cm) diameter grabline or equivalent handhold securely fastened beside each crawling board; [§1926.451(g)(1)(iii)]

(iv) *Each employee on a self-contained adjustable scaffold* shall be protected by a guardrail system (with minimum 200 pound toprail capacity) when the platform is supported by the frame structure, and by both a personal fall arrest system and a guardrail system (with minimum 200 pound toprail capacity) when the platform is supported by ropes;[§1926.451(g)(1)(iv)]

(v) *Each employee on a walkway located* within a scaffold shall be protected by a guardrail system (with minimum 200 pound toprail capacity) installed within 9 1/2 inches (24.1 cm) of and along at least one side of the walkway.[§1926.451(g)(1)(v)]

(vi) *Each employee performing* overhand bricklaying operations from a supported scaffold shall be protected from falling from all open sides and ends of the scaffold (except at the side next to the wall being laid) by the use of a personal fall arrest system or guardrail system (with minimum 200 pound toprail capacity).[§1926.451(g)(1)(vi)]

(vii) *For all scaffolds not otherwise specified* in paragraphs (g)(1)(i) through (g)(1)(vi) of this section, each employee shall be protected by the use of personal fall arrest systems or guardrail systems meeting the requirements of paragraph (g)(4) of this section.[§1926.451(g)(1)(vii)]

(2) *Effective September 2, 1997,* the employer shall have a competent person determine the feasibility and safety of providing fall protection for employees erecting or dismantling supported scaffolds. Employers are required to provide fall protection for employees erecting or dismantling supported scaffolds where the installation and use of such protection is feasible and does not create a greater hazard.[§1926.451(g)(2)]

(3) *In addition to meeting the requirements* of Sec. 1926.502(d), personal fall arrest systems used on scaffolds shall be attached by lanyard to a vertical lifeline, horizontal lifeline, or scaffold structural member. Vertical lifelines shall not be used when overhead components, such as overhead protection or additional platform levels, are part of a single-point or two-point adjustable suspension scaffold.[§1926.451(g)(3)]

(i) *When vertical lifelines are used,* they shall be fastened to a fixed safe point of anchorage, shall be independent of the scaffold, and shall be protected from sharp edges and abrasion. Safe points of anchorage include structural members of buildings, but do not include standpipes, vents, other piping systems, electrical conduit, outrigger beams, or counterweights.[§1926.451(g)(3)(i)]

(ii) *When horizontal lifelines are used,* they shall be secured to two or more structural members of the scaffold, or they may be looped around both suspension and independent suspension lines (on scaffolds so equipped) above the hoist and brake attached to the end of the scaffold. Horizontal lifelines shall not be attached only to the suspension ropes. [§1926.451(g)(3)(ii)]

(iii) *When lanyards are connected* to horizontal lifelines or structural members on a single-point or two-point adjustable suspension scaffold, the scaffold shall be equipped with additional independent support lines and automatic locking devices capable of stopping the fall of the scaffold in the event one or both of the suspension ropes fail. The independent support lines shall be equal in number and strength to the suspension ropes.[§1926.451(g)(3)(iii)]

(iv) *Vertical lifelines,* independent support lines, and suspension ropes shall not be attached to each other, nor shall they be attached to or use the same point of anchorage, nor shall they be attached to the same point on the scaffold or personal fall arrest system.[§1926.451(g)(3)(iv)]

(4) *Guardrail systems installed* to meet the requirements of this section shall comply with the following provisions (guardrail systems built in accordance with Appendix A to this subpart will be deemed to meet the requirements of paragraphs (g)(4) (vii), (viii), and (ix) of this section):[§1926.451(g)(4)]

(i) *Guardrail systems shall be installed* along all open sides and ends of platforms. Guardrail systems shall be installed before the scaffold is released for use by employees other than erection/ dismantling crews.[§1926.451(g)(4)(i)]

(ii) *The top edge height of toprails* or equivalent member on supported scaffolds manufactured or placed in service after January 1, 2000 shall be installed between 38 inches (0.97 m) and 45 inches (1.2 m) above the platform surface. The top edge height on supported scaffolds manufactured and placed in service before January 1, 2000, and on all suspended scaffolds where both a guardrail and a personal fall arrest system are required shall be between 36 inches (0.9 m) and 45 inches (1.2 m). When conditions warrant, the height of the top edge may exceed the 45-inch height, provided the guardrail system meets all other criteria of paragraph (g)(4).[§1926.451(g)(4)(ii)]

(iii) *When midrails, screens, mesh,* intermediate vertical members, solid panels, or equivalent structural members are used, they shall be installed between the top edge of the guardrail system and the scaffold platform.[§1926.451(g)(4)(iii)]

(iv) *When midrails are used,* they shall be installed at a height approximately midway between the top edge of the guardrail system and the platform surface.[§1926.451(g)(4)(iv)]

(v) *When screens and mesh are used,* they shall extend from the top edge of the guardrail system to the scaffold platform, and along the entire opening between the supports. [§1926.451(g)(4)(v)]

(vi) *When intermediate members* (such as balusters or additional rails) are used, they shall not be more than 19 inches (48 cm) apart.[§1926.451(g)(4)(vi)]

(vii) *Each toprail or equivalent member* of a guardrail system shall be capable of withstanding, without failure, a force applied in any downward or horizontal direction at any point along its top edge of at least 100 pounds (445 n) for guardrail systems installed on single-point adjustable suspension scaffolds or two-point adjustable suspension scaffolds, and at least 200 pounds (890 n) for guardrail systems installed on all other scaffolds.[§1926.451(g)(4)(vii)]

(viii) *When the loads specified* in paragraph (g)(4)(vii) of this section are applied in a downward direction, the top edge shall not drop below the height above the platform surface that is prescribed in paragraph (g)(4)(ii) of this section. [§1926.451(g)(4)(viii)]

(ix) *Midrails, screens, mesh,* intermediate vertical members, solid panels, and equivalent structural members of a guardrail system shall be capable of withstanding, without failure, a force applied in any downward or horizontal direction at any point along the midrail or other member of at least 75 pounds (333 n) for guardrail systems with a minimum 100 pound toprail capacity, and at least 150 pounds (666 n) for guardrail systems with a minimum 200 pound toprail capacity. [§1926.451(g)(4)(ix)]

(x) *Suspension scaffold hoists* and non-walk-through stirrups may be used as end guardrails, if the space between the hoist or stirrup and the side guardrail or structure does not allow passage of an employee to the end of the scaffold. [§1926.451(g)(4)(x)]

(xi) *Guardrails shall be surfaced* to prevent injury to an employee from punctures or lacerations, and to prevent snagging of clothing.[§1926.451(g)(4)(xi)]

(xii) *The ends of all rails shall not overhang* the terminal posts except when such overhang does not constitute a projection hazard to employees.[§1926.451(g)(4)(xii)]

(xiii) *Steel or plastic banding* shall not be used as a toprail or midrail.[§1926.451(g)(4)(xiii)]

(xiv) *Manila or plastic* (or other synthetic) rope being used for toprails or midrails shall be inspected by a competent person as frequently as necessary to ensure that it continues to meet the strength requirements of paragraph (g) of this section.[§1926.451(g)(4)(xiv)]

(xv) *Crossbracing is acceptable in place of a midrail* when the crossing point of two braces is between 20 inches (0.5 m) and 30 inches (0.8 m) above the work platform or as a toprail when the crossing point of two braces is between 38 inches (0.97 m) and 48 inches (1.3 m) above the work platform. The end points at each upright shall be no more than 48 inches (1.3 m) apart.[§1926.451(g)(4)(xv)]

(h) Falling object protection.[§1926.451(h)]

(1) *In addition to wearing hardhats* each employee on a scaffold shall be provided with additional protection from falling hand tools, debris, and other small objects through the installation of toeboards, screens, or guardrail systems, or through the erection of debris nets, catch platforms, or canopy structures that contain or deflect the falling objects. When the falling objects are too large, heavy or massive to be contained or deflected by any of the above-listed measures, the employer shall place such potential falling objects away from the edge of the surface from which they could fall and shall secure those materials as necessary to prevent their falling.[§1926.451(h)(1)]

(2) *Where there is a danger* of tools, materials, or equipment falling from a scaffold and striking employees below, the following provisions apply:[§1926.451(h)(2)]

(i) *The area below the scaffold* to which objects can fall shall be barricaded, and employees shall not be permitted to enter the hazard area; or[§1926.451(h)(2)(i)]

(ii) *A toeboard shall be erected* along the edge of platforms more than 10 feet (3.1 m) above lower levels for a distance sufficient to protect employees below, except on float (ship) scaffolds where an edging of 3/4 x 1 1/2 inch (2 x 4 cm) wood or equivalent may be used in lieu of toeboards; [§1926.451(h)(2)(ii)]

(iii) *Where tools, materials, or equipment* are piled to a height higher than the top edge of the toeboard, paneling or screening extending from the toeboard or platform to the top of the guardrail shall be erected for a distance sufficient to protect employees below; or[§1926.451(h)(2)(iii)]

(iv) *A guardrail system shall be installed* with openings small enough to prevent passage of potential falling objects; or [§1926.451(h)(2)(iv)]

(v) *A canopy structure,* debris net, or catch platform strong enough to withstand the impact forces of the potential falling objects shall be erected over the employees below. [§1926.451(h)(2)(v)]

(3) *Canopies, when used for falling object protection,* shall comply with the following criteria:[§1926.451(h)(3)]

(i) *Canopies shall be installed between* the falling object hazard and the employees.[§1926.451(h)(3)(i)]

(ii) *When canopies are used on suspension* scaffolds for falling object protection, the scaffold shall be equipped with additional independent support lines equal in number to the number of points supported, and equivalent in strength to the strength of the suspension ropes.[§1926.451(h)(3)(ii)]

(iii) *Independent support lines* and suspension ropes shall not be attached to the same points of anchorage. [§1926.451(h)(3)(iii)]

(4) *Where used,* toeboards shall be:[§1926.451(h)(4)]

(i) *Capable of withstanding, without failure,* a force of at least 50 pounds (222 n) applied in any downward or horizontal direction at any point along the toeboard (toeboards built in accordance with Appendix A to this subpart will be deemed to meet this requirement); and[§1926.451(h)(4)(i)]

(ii) *At least three and one-half inches (9 cm) high* from the top edge of the toeboard to the level of the walking/working surface. Toeboards shall be securely fastened in place at the outermost edge of the platform and have not more than 1/4 inch (0.7 cm) clearance above the walking/working surface. Toeboards shall be solid or with openings not over one inch (2.5 cm) in the greatest dimension.[§1926.451(h)(4)(ii)]

[61 FR 46107, Aug. 30, 1996, as corrected and amended at 61 FR 59831, 59832, Nov. 25, 1996]

§1926.452

Additional requirements applicable to specific types of scaffolds

In addition to the applicable requirements of Sec. 1926.451, the following requirements apply to the specific types of scaffolds indicated. Scaffolds not specifically addressed by Sec. 1926.452, such as but not limited to systems scaffolds, must meet the requirements of Sec. 1926.451.

(a) **Pole scaffolds.**[§1926.452(a)]

(1) *When platforms are being moved to the next level,* the existing platform shall be left undisturbed until the new bearers have been set in place and braced, prior to receiving the new platforms.[§1926.452(a)(1)]

(2) *Crossbracing shall be installed* between the inner and outer sets of poles on double pole scaffolds.[§1926.452(a)(2)]

(3) *Diagonal bracing in both directions* shall be installed across the entire inside face of double-pole scaffolds used to support loads equivalent to a uniformly distributed load of 50 pounds (222 kg) or more per square foot (929 square cm). [§1926.452(a)(3)]

(4) *Diagonal bracing in both directions* shall be installed across the entire outside face of all double- and single-pole scaffolds. [§1926.452(a)(4)]

(5) *Runners and bearers shall be installed on edge.*[§1926.452(a)(5)]

(6) *Bearers shall extend a minimum* of 3 inches (7.6 cm) over the outside edges of runners.[§1926.452(a)(6)]

(7) *Runners shall extend over* a minimum of two poles, and shall be supported by bearing blocks securely attached to the poles.[§1926.452(a)(7)]

(8) *Braces,bearers, and runners shall not be spliced between poles.*[§1926.452(a)(8)]

(9) *Where wooden poles are spliced,* the ends shall be squared and the upper section shall rest squarely on the lower section. Wood splice plates shall be provided on at least two adjacent sides, and shall extend at least 2 feet (0.6 m) on either side of the splice, overlap the abutted ends equally, and have at least the same cross-sectional areas as the pole. Splice plates of other materials of equivalent strength may be used. [§1926.452(a)(9)]

(10) *Pole scaffolds over 60 feet* in height shall be designed by a registered professional engineer, and shall be constructed and loaded in accordance with that design. Non-mandatory Appendix A to this subpart contains examples of criteria that will enable an employer to comply with design and loading requirements for pole scaffolds under 60 feet in height. [§1926.452(a)(10)]

(b) **Tube and coupler scaffolds.**[§1926.452(b)]

(1) *When platforms are being moved to the next level,* the existing platform shall be left undisturbed until the new bearers have been set in place and braced prior to receiving the new platforms.[§1926.452(b)(1)]

(2) *Transverse bracing forming* an "X" across the width of the scaffold shall be installed at the scaffold ends and at least at every third set of posts horizontally (measured from only one end) and every fourth runner vertically. Bracing shall extend diagonally from the inner or outer posts or runners upward to the next outer or inner posts or runners. Building ties shall be installed at the bearer levels between the transverse bracing and shall conform to the requirements of Sec. 1926.451(c)(1). [§1926.452(b)(2)]

(3) *On straight run scaffolds,* longitudinal bracing across the inner and outer rows of posts shall be installed diagonally in both directions, and shall extend from the base of the end posts upward to the top of the scaffold at approximately a 45 degree angle. On scaffolds whose length is greater than their height, such bracing shall be repeated beginning at least at every fifth post. On scaffolds whose length is less than their height, such bracing shall be installed from the base of the end posts upward to the opposite end posts, and then in alternating directions until reaching the top of the scaffold. Bracing shall be installed as close as possible to the intersection of the bearer and post or runner and post.[§1926.452(b)(3)]

(4) *Where conditions preclude the attachment* of bracing to posts, bracing shall be attached to the runners as close to the post as possible.[§1926.452(b)(4)]

(5) *Bearers shall be installed transversely* between posts, and when coupled to the posts, shall have the inboard coupler bear directly on the runner coupler. When the bearers are coupled to the runners, the couplers shall be as close to the posts as possible.[§1926.452(b)(5)]

(6) *Bearers shall extend beyond* the posts and runners, and shall provide full contact with the coupler.[§1926.452(b)(6)]

(7) *Runners shall be installed along* the length of the scaffold, located on both the inside and outside posts at level heights (when tube and coupler guardrails and midrails are used on outside posts, they may be used in lieu of outside runners). [§1926.452(b)(7)]

(8) *Runners shall be interlocked on straight* runs to form continuous lengths, and shall be coupled to each post. The bottom runners and bearers shall be located as close to the base as possible.[§1926.452(b)(8)]

(9) *Couplers shall be of a structural metal,* such as drop-forged steel, malleable iron, or structural grade aluminum. The use of gray cast iron is prohibited.[§1926.452(b)(9)]

(10) *Tube and coupler scaffolds* over 125 feet in height shall be designed by a registered professional engineer, and shall be constructed and loaded in accordance with such design. Non-mandatory Appendix A to this subpart contains examples of criteria that will enable an employer to comply with design and loading requirements for tube and coupler scaffolds under 125 feet in height.[§1926.452(b)(10)]

(c) **Fabricated frame scaffolds** (tubular welded frame scaffolds). [§1926.452(c)]

(1) *When moving platforms to the next level,* the existing platform shall be left undisturbed until the new end frames have been set in place and braced prior to receiving the new platforms.[§1926.452(c)(1)]

(2) *Frames and panels shall be braced by cross,* horizontal, or diagonal braces, or combination thereof, which secure vertical members together laterally. The cross braces shall be of such length as will automatically square and align vertical members so that the erected scaffold is always plumb, level, and square. All brace connections shall be secured.[§1926.452(c)(2)]

(3) *Frames and panels shall be joined* together vertically by coupling or stacking pins or equivalent means.[§1926.452(c)(3)]

(4) *Where uplift can occur* which would displace scaffold end frames or panels, the frames or panels shall be locked together vertically by pins or equivalent means.[§1926.452(c)(4)]

(5) *Brackets used to support* cantilevered loads shall: [§1926.452(c)(5)]

(i) *Be seated with side-brackets parallel* to the frames and end-brackets at 90 degrees to the frames;[§1926.452(c)(5)(i)]

(ii) *Not be bent or twisted from these positions; and* [§1926.452(c)(5)(ii)]

(iii) *Be used only to support personnel,* unless the scaffold has been designed for other loads by a qualified engineer and built to withstand the tipping forces caused by those other loads being placed on the bracket-supported section of the scaffold.[§1926.452(c)(5)(iii)]

(6) *Scaffolds over 125 feet* (38.0 m) in height above their base plates shall be designed by a registered professional engineer, and shall be constructed and loaded in accordance with such design.[§1926.452(c)(6)]

(d) Plasterers', decorators', and large area scaffolds. Scaffolds shall be constructed in accordance with paragraphs (a), (b), or (c) of this section, as appropriate.[§1926.452(d)]

(e) Bricklayers' square scaffolds (squares).[§1926.452(e)]

(1) *Scaffolds made of wood* shall be reinforced with gussets on both sides of each corner.[§1926.452(e)(1)]

(2) *Diagonal braces shall be installed* on all sides of each square. [§1926.452(e)(2)]

(3) *Diagonal braces shall be installed* between squares on the rear and front sides of the scaffold, and shall extend from the bottom of each square to the top of the next square. [§1926.452(e)(3)]

(4) *Scaffolds shall not exceed three* tiers in height, and shall be so constructed and arranged that one square rests directly above the other. The upper tiers shall stand on a continuous row of planks laid across the next lower tier, and shall be nailed down or otherwise secured to prevent displacement.[§1926.452(e)(4)]

(f) Horse scaffolds.[§1926.452(f)]

(1) *Scaffolds shall not be constructed or arranged* more than two tiers or 10 feet (3.0 m) in height, whichever is less. [§1926.452(f)(1)]

(2) *When horses are arranged in tiers,* each horse shall be placed directly over the horse in the tier below.[§1926.452(f)(2)]

(3) *When horses are arranged in tiers,* the legs of each horse shall be nailed down or otherwise secured to prevent displacement.[§1926.452(f)(3)]

(4) *When horses are arranged in tiers,* each tier shall be cross-braced.[§1926.452(f)(4)]

(g) Form scaffolds and carpenters' bracket scaffolds.[§1926.452(g)]

(1) *Each bracket,* except those for wooden bracket-form scaffolds, shall be attached to the supporting formwork or structure by means of one or more of the following: nails; a metal stud attachment device; welding; hooking over a secured structural supporting member, with the form wales either bolted to the form or secured by snap ties or tie bolts extending through the form and securely anchored; or, for carpenters' bracket scaffolds only, by a bolt extending through to the opposite side of the structure's wall.[§1926.452(g)(1)]

(2) *Wooden bracket-form scaffolds* shall be an integral part of the form panel.[§1926.452(g)(2)]

(3) *Folding type metal brackets,* when extended for use, shall be either bolted or secured with a locking-type pin.[§1926.452(g)(3)]

(h) Roof bracket scaffolds.[§1926.452(h)]

(1) *Scaffold brackets shall be constructed* to fit the pitch of the roof and shall provide a level support for the platform. [§1926.452(h)(1)]

(2) *Brackets* (including those provided with pointed metal projections) shall be anchored in place by nails unless it is impractical to use nails. When nails are not used, brackets shall be secured in place with first-grade manila rope of at least three-fourth inch (1.9 cm) diameter, or equivalent.[§1926.452(h)(2)]

(i) Outrigger scaffolds.[§1926.452(i)]

(1) *The inboard end of outrigger beams,* measured from the fulcrum point to the extreme point of anchorage, shall be not less than one and one-half times the outboard end in length. [§1926.452(i)(1)]

(2) *Outrigger beams fabricated* in the shape of an I-beam or channel shall be placed so that the web section is vertical. [§1926.452(i)(2)]

(3) *The fulcrum point of outrigger* beams shall rest on secure bearings at least 6 inches (15.2 cm) in each horizontal dimension.[§1926.452(i)(3)]

(4) *Outrigger beams shall be secured* in place against movement, and shall be securely braced at the fulcrum point against tipping.[§1926.452(i)(4)]

(5) *The inboard ends of outrigger* beams shall be securely anchored either by means of braced struts bearing against sills in contact with the overhead beams or ceiling, or by means of tension members secured to the floor joists underfoot, or by both.[§1926.452(i)(5)]

(6) *The entire supporting structure* shall be securely braced to prevent any horizontal movement.[§1926.452(i)(6)]

(7) *To prevent their displacement,* platform units shall be nailed, bolted, or otherwise secured to outriggers.[§1926.452(i)(7)]

(8) *Scaffolds and scaffold components* shall be designed by a registered professional engineer and shall be constructed and loaded in accordance with such design.[§1926.452(i)(8)]

(j) Pump jack scaffolds.[§1926.452(j)]

(1) *Pump jack brackets,* braces, and accessories shall be fabricated from metal plates and angles. Each pump jack bracket shall have two positive gripping mechanisms to prevent any failure or slippage.[§1926.452(j)(1)]

(2) *Poles shall be secured to the structure* by rigid triangular bracing or equivalent at the bottom, top, and other points as necessary. When the pump jack has to pass bracing already installed, an additional brace shall be installed approximately 4 feet (1.2 m) above the brace to be passed, and shall be left in place until the pump jack has been moved and the original brace reinstalled.[§1926.452(j)(2)]

(3) *When guardrails are used for fall protection,* a workbench may be used as the toprail only if it meets all the requirements in paragraphs (g)(4) (ii), (vii), (viii), and (xiii) of Sec. 1926.451. [§1926.452(j)(3)]

(4) *Work benches shall not be used as scaffold platforms.* [§1926.452(j)(4)]

(5) *When poles are made of wood,* the pole lumber shall be straight- grained, free of shakes, large loose or dead knots, and other defects which might impair strength.[§1926.452(j)(5)]

(6) *When wood poles are constructed* of two continuous lengths, they shall be joined together with the seam parallel to the bracket.[§1926.452(j)(6)]

(7) *When two by fours are spliced* to make a pole, mending plates shall be installed at all splices to develop the full strength of the member.[§1926.452(j)(7)]

(k) Ladder jack scaffolds.[§1926.452(k)]

(1) *Platforms shall not exceed a height of 20 feet (6.1 m).* [§1926.452(k)(1)]

(2) *All ladders used to support* ladder jack scaffolds shall meet the requirements of subpart X of this part — Stairways and Ladders, except that job-made ladders shall not be used to support ladder jack scaffolds.[§1926.452(k)(2)]

(3) *The ladder jack shall be so designed* and constructed that it will bear on the side rails and ladder rungs or on the ladder rungs alone. If bearing on rungs only, the bearing area shall include a length of at least 10 inches (25.4 cm) on each rung.[§1926.452(k)(3)]

(4) *Ladders used to support* ladder jacks shall be placed, fastened, or equipped with devices to prevent slipping.[§1926.452(k)(4)]

(5) *Scaffold platforms shall not be bridged one to another.* [§1926.452(k)(5)]

(l) Window jack scaffolds.[§1926.452(l)]

(1) *Scaffolds shall be securely attached* to the window opening. [§1926.452(l)(1)]

(2) *Scaffolds shall be used only* for the purpose of working at the window opening through which the jack is placed. [§1926.452(l)(2)]

(3) *Window jacks shall not be used* to support planks placed between one window jack and another, or for other elements of scaffolding.[§1926.452(l)(3)]

(m) Crawling boards (chicken ladders).[§1926.452(m)]

(1) *Crawling boards shall extend* from the roof peak to the eaves when used in connection with roof construction, repair, or maintenance.[§1926.452(m)(1)]

(2) *Crawling boards shall be secured* to the roof by ridge hooks or by means that meet equivalent criteria (e.g., strength and durability).[§1926.452(m)(2)]

(n) Step, platform, and trestle ladder scaffolds.[§1926.452(n)]

(1) *Scaffold platforms shall not be placed* any higher than the second highest rung or step of the ladder supporting the platform.[§1926.452(n)(1)]

(2) *All ladders used in conjunction with step,* platform and trestle ladder scaffolds shall meet the pertinent requirements of subpart X of this part — Stairways and Ladders, except that job-made ladders shall not be used to support such scaffolds. [§1926.452(n)(2)]

(3) *Ladders used to support step,* platform, and trestle ladder scaffolds shall be placed, fastened, or equipped with devices to prevent slipping.[§1926.452(n)(3)]

(4) *Scaffolds shall not be bridged one to another.*[§1926.452(n)(4)]

(o) Single-point adjustable suspension scaffolds.[§1926.452(o)]

(1) *When two single-point adjustable* suspension scaffolds are combined to form a two- point adjustable suspension scaffold, the resulting two-point scaffold shall comply with the requirements for two-point adjustable suspension scaffolds in paragraph (p) of this section.[§1926.452(o)(1)]

(2) *The supporting rope between* the scaffold and the suspension device shall be kept vertical unless all of the following conditions are met:[§1926.452(o)(2)]

(i) *The rigging has been* designed by a qualified person, and [§1926.452(o)(2)(i)]

(ii) *The scaffold is accessible to rescuers, and*[§1926.452(o)(2)(ii)]

(iii) *The supporting rope is protected* to ensure that it will not chafe at any point where a change in direction occurs, and [§1926.452(o)(2)(iii)]

(iv) *The scaffold is positioned so that swinging* cannot bring the scaffold into contact with another surface.[§1926.452(o)(2)(iv)]

(3) *Boatswains' chair tackle* shall consist of correct size ball bearings or bushed blocks containing safety hooks and properly "eye- spliced" minimum five-eighth (5/8) inch (1.6 cm) diameter first-grade manila rope, or other rope which will satisfy the criteria (e.g., strength and durability) of manila rope. [§1926.452(o)(3)]

(4) *Boatswains' chair seat* slings shall be reeved through four corner holes in the seat; shall cross each other on the underside of the seat; and shall be rigged so as to prevent slippage which could cause an out-of-level condition.[§1926.452(o)(4)]

(5) *Boatswains' chair seat* slings shall be a minimum of five-eight (5/8) inch (1.6 cm) diameter fiber, synthetic, or other rope which will satisfy the criteria (e.g., strength, slip resistance, durability, etc.) of first grade manila rope.[§1926.452(o)(5)]

(6) *When a heat-producing process such* as gas or arc welding is being conducted, boatswains' chair seat slings shall be a minimum of three-eight (3/8) inch (1.0 cm) wire rope. [§1926.452(o)(6)]

(7) *Non-cross-laminated wood boatswains'* chairs shall be reinforced on their underside by cleats securely fastened to prevent the board from splitting.[§1926.452(o)(7)]

(p) Two-point adjustable suspension scaffolds (swing stages). The following requirements do not apply to two-point adjustable suspension scaffolds used as masons' or stonesetters' scaffolds. Such scaffolds are covered by paragraph (q) of this section. [§1926.452(p)]

(1) *Platforms shall not be more than 36 inches* (0.9 m) wide unless designed by a qualified person to prevent unstable conditions.[§1926.452(p)(1)]

(2) *The platform shall be securely fastened to hangers* (stirrups) by U-bolts or by other means which satisfy the requirements of Sec. 1926.451(a).[§1926.452(p)(2)]

(3) *The blocks for fiber or synthetic ropes* shall consist of at least one double and one single block. The sheaves of all blocks shall fit the size of the rope used.[§1926.452(p)(3)]

(4) *Platforms shall be of the ladder-type,* plank-type, beam-type, or light-metal type. Light metal-type platforms having a rated capacity of 750 pounds or less and platforms 40 feet (12.2 m) or less in length shall be tested and listed by a nationally recognized testing laboratory.[§1926.452(p)(4)]

(5) *Two-point scaffolds shall not be bridged* or otherwise connected one to another during raising and lowering operations unless the bridge connections are articulated (attached), and the hoists properly sized.[§1926.452(p)(5)]

(6) *Passage may be made from one platform* to another only when the platforms are at the same height, are abutting, and walk-through stirrups specifically designed for this purpose are used.[§1926.452(p)(6)]

(q) Multi-point adjustable suspension scaffolds, stonesetters' multi-point adjustable suspension scaffolds, and masons' multi-point adjustable suspension scaffolds.[§1926.452(q)]

(1) *When two or more scaffolds* are used they shall not be bridged one to another unless they are designed to be bridged, the bridge connections are articulated, and the hoists are properly sized.[§1926.452(q)(1)]

(2) *If bridges are not used,* passage may be made from one platform to another only when the platforms are at the same height and are abutting.[§1926.452(q)(2)]

(3) *Scaffolds shall be suspended* from metal outriggers, brackets, wire rope slings, hooks, or means that meet equivalent criteria (e.g., strength, durability).[§1926.452(q)(3)]

(r) Catenary scaffolds.[§1926.452(r)]

(1) *No more than one platform* shall be placed between consecutive vertical pickups, and no more than two platforms shall be used on a catenary scaffold.[§1926.452(r)(1)]

(2) *Platforms supported by wire ropes* shall have hook-shaped stops on each end of the platforms to prevent them from slipping off the wire ropes. These hooks shall be so placed that they will prevent the platform from falling if one of the horizontal wire ropes breaks.[§1926.452(r)(2)]

(3) *Wire ropes shall not be tightened* to the extent that the application of a scaffold load will overstress them.[§1926.452(r)(3)]

(4) *Wire ropes shall be continuous* and without splices between anchors.[§1926.452(r)(4)]

(s) Float (ship) scaffolds.[§1926.452(s)]

(1) *The platform shall be supported* by a minimum of two bearers, each of which shall project a minimum of 6 inches (15.2 cm) beyond the platform on both sides. Each bearer shall be securely fastened to the platform.[§1926.452(s)(1)]

(2) *Rope connections shall be such that the platform cannot shift or slip.*[§1926.452(s)(2)]

(3) *When only two ropes are used with each float:*[§1926.452(s)(3)]

(i) *They shall be arranged so as to provide* four ends which are securely fastened to overhead supports.[§1926.452(s)(3)(i)]

(ii) *Each supporting rope shall be hitched* around one end of the bearer and pass under the platform to the other end of the bearer where it is hitched again, leaving sufficient rope at each end for the supporting ties.[§1926.452(s)(3)(ii)]

(t) Interior hung scaffolds.[§1926.452(t)]

(1) *Scaffolds shall be suspended only* from the roof structure or other structural member such as ceiling beams.[§1926.452(t)(1)]

(2) *Overhead supporting members* (roof structure, ceiling beams, or other structural members) shall be inspected and checked for strength before the scaffold is erected.[§1926.452(t)(2)]

(3) *Suspension ropes and cables* shall be connected to the overhead supporting members by shackles, clips, thimbles, or other means that meet equivalent criteria (e.g., strength, durability).[§1926.452(t)(3)]

(u) Needle beam scaffolds.[§1926.452(u)]

(1) *Scaffold support beams shall be installed on edge.* [§1926.452(u)(1)]

(2) *Ropes or hangers shall be used* for supports, except that one end of a needle beam scaffold may be supported by a permanent structural member.[§1926.452(u)(2)]

(3) *The ropes shall be securely attached* to the needle beams. [§1926.452(u)(3)]

(4) *The support connection shall be arranged* so as to prevent the needle beam from rolling or becoming displaced. [§1926.452(u)(4)]

(5) *Platform units shall be securely* attached to the needle beams by bolts or equivalent means. Cleats and overhang are not considered to be adequate means of attachment. [§1926.452(u)(5)]

(v) Multi-level suspended scaffolds.[§1926.452(v)]

(1) *Scaffolds shall be equipped* with additional independent support lines, equal in number to the number of points supported, and of equivalent strength to the suspension ropes, and rigged to support the scaffold in the event the suspension rope(s) fail. [§1926.452(v)(1)]

(2) *Independent support lines* and suspension ropes shall not be attached to the same points of anchorage.[§1926.452(v)(2)]

(3) *Supports for platforms shall be attached* directly to the support stirrup and not to any other platform.[§1926.452(v)(3)]

(w) Mobile scaffolds.[§1926.452(w)]

(1) *Scaffolds shall be braced by cross,* horizontal, or diagonal braces, or combination thereof, to prevent racking or collapse of the scaffold and to secure vertical members together laterally so as to automatically square and align the vertical members. Scaffolds shall be plumb, level, and squared. All brace connections shall be secured.[§1926.452(w)(1)]

(i) *Scaffolds constructed of tube* and coupler components shall also comply with the requirements of paragraph (b) of this section;[§1926.452(w)(1)(i)]

(ii) *Scaffolds constructed of fabricated* frame components shall also comply with the requirements of paragraph (c) of this section.[§1926.452(w)(1)(ii)]

(2) *Scaffold casters and wheels* shall be locked with positive wheel and/or wheel and swivel locks, or equivalent means, to prevent movement of the scaffold while the scaffold is used in a stationary manner.[§1926.452(w)(2)]

(3) *Manual force used* to move the scaffold shall be applied as close to the base as practicable, but not more than 5 feet (1.5 m) above the supporting surface.[§1926.452(w)(3)]

(4) *Power systems used* to propel mobile scaffolds shall be designed for such use. Forklifts, trucks, similar motor vehicles or add-on motors shall not be used to propel scaffolds unless the scaffold is designed for such propulsion systems. [§1926.452(w)(4)]

(5) *Scaffolds shall be stabilized to prevent* tipping during movement.[§1926.452(w)(5)]

(6) *Employees shall not be allowed to ride* on scaffolds unless the following conditions exist:[§1926.452(w)(6)]

(i) *The surface on which the scaffold* is being moved is within 3 degrees of level, and free of pits, holes, and obstructions; [§1926.452(w)(6)(i)]

(ii) *The height to base width* ratio of the scaffold during movement is two to one or less, unless the scaffold is designed and constructed to meet or exceed nationally recognized stability test requirements such as those listed in paragraph (x) of Appendix A to this subpart (ANSI/SIA A92.5 and A92.6); [§1926.452(w)(6)(ii)]

(iii) *Outrigger frames, when used,* are installed on both sides of the scaffold;[§1926.452(w)(6)(iii)]

(iv) *When power systems are used,* the propelling force is applied directly to the wheels, and does not produce a speed in excess of 1 foot per second (.3 mps); and[§1926.452(w)(6)(iv)]

(v) *No employee is on any part of the scaffold* which extends outward beyond the wheels, casters, or other supports. [§1926.452(w)(6)(v)]

(7) *Platforms shall not extend outward* beyond the base supports of the scaffold unless outrigger frames or equivalent devices are used to ensure stability.[§1926.452(w)(7)]

(8) *Where leveling of the scaffold is necessary,* screw jacks or equivalent means shall be used.[§1926.452(w)(8)]

(9) *Caster stems and wheel stems* shall be pinned or otherwise secured in scaffold legs or adjustment screws.[§1926.452(w)(9)]

(10) *Before a scaffold is moved,* each employee on the scaffold shall be made aware of the move.[§1926.452(w)(10)]

(x) Repair bracket scaffolds.[§1926.452(x)]

(1) *Brackets shall be secured in place* by at least one wire rope at least 1/2 inch (1.27 cm) in diameter.[§1926.452(x)(1)]

(2) *Each bracket shall be attached to the securing wire rope* (or ropes) by a positive locking device capable of preventing the unintentional detachment of the bracket from the rope, or by equivalent means.[§1926.452(x)(2)]

(3) *Each bracket,* at the contact point between the supporting structure and the bottom of the bracket, shall be provided with a shoe (heel block or foot) capable of preventing the lateral movement of the bracket.[§1926.452(x)(3)]

(4) *Platforms shall be secured to the brackets* in a manner that will prevent the separation of the platforms from the brackets and the movement of the platforms or the brackets on a completed scaffold.[§1926.452(x)(4)]

(5) *When a wire rope is placed* around the structure in order to provide a safe anchorage for personal fall arrest systems used by employees erecting or dismantling scaffolds, the wire rope shall meet the requirements of subpart M of this part, but shall be at least 5/16 inch (0.8 cm) in diameter.[§1926.452(x)(5)]

(6) *Each wire rope used* for securing brackets in place or as an anchorage for personal fall arrest systems shall be protected from damage due to contact with edges, corners, protrusions, or other discontinuities of the supporting structure or scaffold components.[§1926.452(x)(6)]

(7) *Tensioning of each wire rope* used for securing brackets in place or as an anchorage for personal fall arrest systems shall be by means of a turnbuckle at least 1 inch (2.54 cm) in diameter, or by equivalent means.[§1926.452(x)(7)]

(8) *Each turnbuckle shall be connected to the other end* of its rope by use of an eyesplice thimble of a size appropriate to the turnbuckle to which it is attached.[§1926.452(x)(8)]

(9) *U-bolt wire rope* clips shall not be used on any wire rope used to secure brackets or to serve as an anchor for personal fall arrest systems.[§1926.452(x)(9)]

(10) *The employer shall ensure that materials* shall not be dropped to the outside of the supporting structure.[§1926.452(x)(10)]

(11) *Scaffold erection shall progress* in only one direction around any structure.[§1926.452(x)(11)]

(y) Stilts. Stilts, when used, shall be used in accordance with the following requirements:[§1926.452(y)]

(1) *An employee may wear stilts* on a scaffold only if it is a large area scaffold.[§1926.452(y)(1)]

(2) *When an employee is using stilts* on a large area scaffold where a guardrail system is used to provide fall protection, the guardrail system shall be increased in height by an amount equal to the height of the stilts being used by the employee. [§1926.452(y)(2)]

(3) *Surfaces on which stilts* are used shall be flat and free of pits, holes and obstructions, such as debris, as well as other tripping and falling hazards.[§1926.452(y)(3)]

(4) *Stilts shall be properly maintained.* Any alteration of the original equipment shall be approved by the manufacturer. [§1926.452(y)(4)]

§1926.453
Aerial lifts

(a) General requirements.[§1926.453(a)]

(1) *Unless otherwise provided in this section,* aerial lifts acquired for use on or after January 22, 1973 shall be designed and constructed in conformance with the applicable requirements of the American National Standards for "Vehicle Mounted Elevating and Rotating Work Platforms," ANSI A92.2-1969, including appendix. Aerial lifts acquired before January 22, 1973 which do not meet the requirements of ANSI A92.2-1969, may not be used after January 1, 1976, unless they shall have been modified so as to conform with the applicable design and construction requirements of ANSI A92.2-1969. Aerial lifts include the following types of vehicle-mounted aerial devices used to elevate personnel to job-sites above ground:[§1926.453(a)(1)]

(i) *Extensible boom platforms;*[§1926.453(a)(1)(i)]

(ii) *Aerial ladders;*[§1926.453(a)(1)(ii)]

(iii) *Articulating boom platforms;*[§1926.453(a)(1)(iii)]

(iv) *Vertical towers; and*[§1926.453(a)(1)(iv)]

(v) *A combination of any such devices.* Aerial equipment may be made of metal, wood, fiberglass reinforced plastic (FRP), or other material; may be powered or manually operated; and are deemed to be aerial lifts whether or not they are capable of rotating about a substantially vertical axis.[§1926.453(a)(1)(v)]

(2) *Aerial lifts may be "field modified"* for uses other than those intended by the manufacturer provided the modification has been certified in writing by the manufacturer or by any other equivalent entity, such as a nationally recognized testing laboratory, to be in conformity with all applicable provisions of ANSI A92.2-1969 and this section and to be at least as safe as the equipment was before modification.[§1926.453(a)(2)]

(b) Specific requirements[§1926.453(b)]

(1) *Ladder trucks and tower trucks.* Aerial ladders shall be secured in the lower traveling position by the locking device on top of the truck cab, and the manually operated device at the base of the ladder before the truck is moved for highway travel.[§1926.453(b)(1)]

(2) *Extensible and articulating boom platforms.*[§1926.453(b)(2)]

(i) *Lift controls shall be tested* each day prior to use to determine that such controls are in safe working condition. [§1926.453(b)(2)(i)]

(ii) *Only authorized persons shall operate an aerial lift.* [§1926.453(b)(2)(ii)]

(iii) *Belting off to an adjacent pole,* structure, or equipment while working from an aerial lift shall not be permitted. [§1926.453(b)(2)(iii)]

(iv) *Employees shall always stand firmly* on the floor of the basket, and shall not sit or climb on the edge of the basket or use planks, ladders, or other devices for a work position. [§1926.453(b)(2)(iv)]

(v) *A body belt shall be worn* and a lanyard attached to the boom or basket when working from an aerial lift. [§1926.453(b)(2)(v)]

Note to paragraph (b)(2)(v): As of January 1, 1998, subpart M of this part (Sec. 1926.502(d)) provides that body belts are not acceptable as part of a personal fall arrest system. The use of a body belt in a tethering system or in a restraint system is acceptable and is regulated under Sec. 1926.502(e).

(vi) *Boom and basket load limits* specified by the manufacturer shall not be exceeded.[§1926.453(b)(2)(vi)]

(vii) *The brakes shall be set and when outriggers are used,* they shall be positioned on pads or a solid surface. Wheel chocks shall be installed before using an aerial lift on an incline, provided they can be safely installed.[§1926.453(b)(2)(vii)]

(viii) *An aerial lift truck* shall not be moved when the boom is elevated in a working position with men in the basket, except for equipment which is specifically designed for this type of operation in accordance with the provisions of paragraphs (a)(1) and (2) of this section.[§1926.453(b)(2)(viii)]

(ix) *Articulating boom and extensible boom platforms,* primarily designed as personnel carriers, shall have both platform (upper) and lower controls. Upper controls shall be in or beside the platform within easy reach of the operator. Lower controls shall provide for overriding the upper controls. Controls shall be plainly marked as to their function. Lower level controls shall not be operated unless permission has been obtained from the employee in the lift, except in case of emergency.[§1926.453(b)(2)(ix)]

(x) *Climbers shall not be worn while performing work from an aerial lift.*[§1926.453(b)(2)(x)]

(xi) *The insulated portion of an aerial lift* shall not be altered in any manner that might reduce its insulating value. [§1926.453(b)(2)(xi)]

(xii) *Before moving an aerial lift for travel,* the boom(s) shall be inspected to see that it is properly cradled and outriggers are in stowed position except as provided in paragraph (b)(2)(viii) of this section.[§1926.453(b)(2)(xii)]

(3) *Electrical tests.* All electrical tests shall conform to the requirements of ANSI A92.2-1969 section 5. However equivalent d.c.; voltage tests may be used in lieu of the a.c. voltage specified in A92.2-1969; d.c. voltage tests which are approved by the equipment manufacturer or equivalent entity shall be considered an equivalent test for the purpose of this paragraph (b)(3).[§1926.453(b)(3)]

(4) *Bursting safety factor.* The provisions of the American National Standards Institute standard ANSI A92.2-1969, section 4.9 Bursting Safety Factor shall apply to all critical hydraulic and pneumatic components. Critical components are those in which a failure would result in a free fall or free rotation of the boom. All noncritical components shall have a bursting safety factor of at least 2 to 1.[§1926.453(b)(4)]

(5) *Welding standards.* All welding shall conform to the following standards as applicable:[§1926.453(b)(5)]

(i) *Standard Qualification Procedure, AWS B3.0-41.* [§1926.453(b)(5)(i)]

(ii) *Recommended Practices for Automotive Welding Design, AWS D8.4- 61.*[§1926.453(b)(5)(ii)]

(iii) *Standard Qualification* of Welding Procedures and Welders for Piping and Tubing, AWS D10.9-69.[§1926.453(b)(5)(iii)]

(iv) *Specifications for Welding Highway and Railway Bridges, AWS D2.0-69.*[§1926.453(b)(5)(iv)]

Note to Sec. 1926.453: Non-mandatory Appendix C to this subpart lists examples of national consensus standards that are considered to provide employee protection equivalent to that provided through the application of ANSI A92.2-1969, where appropriate. This incorporation by reference was approved by the Director of the Federal Register in accordance with 5 U.S.C. 552(a) and 1 CFR part 51. Copies may be obtained from the American National Standards Institute. Copies may be inspected at the Docket Office, Occupational Safety and Health Administration, U.S. Department of Labor, 200 Constitution Avenue, NW., room N2634, Washington, DC or at the National Archives and Records Administration (NARA). For information on the availability of this material at NARA, call 202-741-6030, or go to: http://www.archives.gov/federal_register/code_of_federal_regulations/ibr_locations.html.

[61 FR 46116, Aug. 30, 1996; 61 FR 59832, Nov. 25, 1996, as amended at 69 FR 18803, Apr. 9, 2004]

§1926.454

Training requirements

This section supplements and clarifies the requirements of Sec. 1926.21(b)(2) as these relate to the hazards of work on scaffolds.

(a) **The employer shall have each employee** who performs work while on a scaffold trained by a person qualified in the subject matter to recognize the hazards associated with the type of scaffold being used and to understand the procedures to control or minimize those hazards. The training shall include the following areas, as applicable:[§1926.454(a)]

(1) *The nature of any electrical hazards,* fall hazards and falling object hazards in the work area;[§1926.454(a)(1)]

(2) *The correct procedures* for dealing with electrical hazards and for erecting, maintaining, and disassembling the fall protection systems and falling object protection systems being used; [§1926.454(a)(2)]

(3) *The proper use of the scaffold,* and the proper handling of materials on the scaffold;[§1926.454(a)(3)]

(4) *The maximum intended load* and the load-carrying capacities of the scaffolds used; and[§1926.454(a)(4)]

(5) *Any other pertinent requirements of this subpart.*[§1926.454(a)(5)]

(b) **The employer shall have each employee** who is involved in erecting, disassembling, moving, operating, repairing, maintaining, or inspecting a scaffold trained by a competent person to recognize any hazards associated with the work in question. The training shall include the following topics, as applicable: [§1926.454(b)]

(1) *The nature of scaffold hazards;*[§1926.454(b)(1)]

(2) *The correct procedures for erecting,* disassembling, moving, operating, repairing, inspecting, and maintaining the type of scaffold in question;[§1926.454(b)(2)]

(3) *The design criteria,* maximum intended load-carrying capacity and intended use of the scaffold;[§1926.454(b)(3)]

(4) *Any other pertinent requirements of this subpart.*[§1926.454(b)(4)]

(c) **When the employer has reason to believe** that an employee lacks the skill or understanding needed for safe work involving the erection, use or dismantling of scaffolds, the employer shall retrain each such employee so that the requisite proficiency is regained. Retraining is required in at least the following situations: [§1926.454(c)]

(1) *Where changes at the worksite* present a hazard about which an employee has not been previously trained; or[§1926.454(c)(1)]

(2) *Where changes in the types of scaffolds,* fall protection, falling object protection, or other equipment present a hazard about which an employee has not been previously trained; or [§1926.454(c)(2)]

(3) *Where inadequacies in an affected employee's work* involving scaffolds indicate that the employee has not retained the requisite proficiency.[§1926.454(c)(3)]

Non-Mandatory Appendices
1926 Subpart L Appendix A
Scaffold Specifications (Non-mandatory)

This Appendix provides non-mandatory guidelines to assist employers in complying with the requirements of subpart L of this part. An employer may use these guidelines and tables as a starting point for designing scaffold systems. However, the guidelines do not provide all the information necessary to build a complete system, and the employer is still responsible for designing and assembling these components in such a way that the completed system will meet the requirements of §1926.451(a). Scaffold components which are not selected and loaded in accordance with this Appendix, and components for which no specific guidelines or tables are given in this Appendix (e.g., joints, ties, components for wood pole scaffolds more than 60 feet in height, components for heavy-duty horse scaffolds, components made with other materials, and components with other dimensions, etc.) must be designed and constructed in accordance with the capacity requirements of §1926.451(a), and loaded in accordance with §1926.451(d)(1).

Index to Appendix A for Subpart L

1. *General guidelines and tables.*
2. *Specific guidelines and tables.*
   (a) *Pole scaffolds:*
      Single-pole wood pole scaffolds.
      Independent wood pole scaffolds.
   (b) *Tube and coupler scaffolds.*
   (c) *Fabricated frame scaffolds.*
   (d) *Plasterers',* decorators' and large area scaffolds.
   (e) *Bricklayers' square scaffolds.*
   (f) *Horse scaffolds.*
   (g) *Form scaffolds and carpenters'* bracket scaffolds.
   (h) *Roof bracket scaffolds.*
   (i) *Outrigger scaffolds (one level).*
   (j) *Pump jack scaffolds.*
   (k) *Ladder jack scaffolds.*

(l) *Window jack scaffolds.*
(m) *Crawling boards (chicken ladders).*
(n) *Step,* platform and trestle ladder scaffolds.
(o) *Single-point adjustable suspension scaffolds.*
(p) *Two-point adjustable suspension scaffolds.*
(q)(1) *Stonesetters' multi-point adjustable* suspension scaffolds.
(2) *Masons' multi-point adjustable* suspension scaffolds.
(r) *Catenary scaffolds.*
(s) *Float (ship) scaffolds.*
(t) *Interior hung scaffolds.*
(u) *Needle beam scaffolds.*
(v) *Multi-level suspension scaffolds.*
(w) *Mobile scaffolds.*
(x) *Repair bracket scaffolds.*
(y) *Stilts.*
(z) *Tank builders' scaffolds.*

1. General Guidelines and Tables.

(a) *The following tables, and the tables in Part 2* — Specific guidelines and tables, assume that all load-carrying timber members (except planks) of the scaffold are a minimum of 1,500 lb-f/in^2 (stress grade) construction grade lumber. All dimensions are nominal sizes as provided in the American Softwood Lumber Standards, dated January 1970, except that, where rough sizes are noted, only rough or undressed lumber of the size specified will satisfy minimum requirements.

(b) *Solid sawn wood* used as scaffold planks shall be selected for such use following the grading rules established by a recognized lumber grading association or by an independent lumber grading inspection agency. Such planks shall be identified by the grade stamp of such association or agency. The association or agency and the grading rules under which the wood is graded shall be certified by the Board of Review, American Lumber Standard Committee, as set forth in the American Softwood Lumber Standard of the U.S. Department of Commerce.

(i) *Allowable spans shall be determined* in compliance with the National Design Specification for Wood Construction published by the National Forest Products Association; paragraph 5 of ANSI A10.8-1988 Scaffolding-Safety Requirements published by the American National Standards Institute; or for 2x10 inch (nominal) or 2x9 inch (rough) solid sawn wood planks, as shown in the following table:

Maximum intended nominal load (lb/ft^2)	Maximum permissible span using full thickness undressed lumber (ft.)	Maximum permissible span using nominal thickness lumber (ft)
25	10	8
50	8	6
75	6	

(ii) *The maximum permissible span* for 1 1/4 x 9-inch or wider wood plank of full thickness with a maximum intended load of 50 lb/ft.^{2} shall be 4 feet.

(c) *Fabricated planks and platforms* may be used in lieu of solid sawn wood planks. Maximum spans for such units shall be as recommended by the manufacturer based on the maximum intended load being calculated as follows:

Rated load capacity	Intended load
Light-duty	• 25 pounds per square foot applied uniformly over the entire span area.
Medium-duty	• 50 pounds per square foot applied uniformly over the entire span area.
Heavy-duty	• 75 pounds per square foot applied uniformly over the entire span area.
One-person	• 250 pounds placed at the center of the span (total 250 pounds).
Two-person	• 250 pounds placed 18 inches to the left and right of the center of the span (total 500 pounds).
Three-person	• 250 pounds placed at the center of the span and 250 pounds placed 18 inches to the left and right of the center of the span (total 750 pounds).

Note: Platform units used to make scaffold platforms intended for light-duty use shall be capable of supporting at least 25 pounds per square foot applied uniformly over the entire unit-span area, or a 250-pound point load placed on the unit at the center of the span, whichever load produces the greater shear force.

(d) *Guardrails shall be as follows:*

(i) *Toprails shall be equivalent in strength* to 2 inch by 4 inch lumber; or
1 1/4 inch x 1/8 inch structural angle iron; or
1 inch x .070 inch wall steel tubing; or 1.990 inch x .058 inch wall aluminum tubing.

(ii) *Midrails shall be equivalent in strength* to 1 inch by 6 inch lumber; or
1 1/4 inch x 1 1/4 inch x 1/8 inch structural angle iron; or
1 inch x .070 inch wall steel tubing; or
1.990 inch x .058 inch wall aluminum tubing.

(iii) *Toeboards shall be equivalent in strength* to 1 inch by 4 inch lumber; or
1 1/4 inch x 1 1/4 inch structural angle iron; or
1 inch x .070 inch wall steel tubing; or
1.990 inch x .058 inch wall aluminum tubing.

(iv) *Posts shall be equivalent in strength* to 2 inch by 4 inch lumber; or
1 1/4 inch x 1 1/4 inch x 1/8 structural angle iron; or
1 inch x .070 inch wall steel tubing; or
1.990 inch x .058 inch wall aluminum tubing.

(v) *Distance between posts shall not exceed 8 feet.*

(e) *Overhead protection shall consist* of 2 inch nominal planking laid tight, or 3/4-inch plywood.

(f) *Screen installed between toeboards and midrails or toprails* shall consist of No. 18 gauge U.S. Standard wire one inch mesh.

2. Specific guidelines and tables.

(a) *Pole Scaffolds.*

Single Pole Wood Pole Scaffolds

	Light duty up to 20 feet high	Light duty up to 60 feet high	Medium duty up to 60 feet high	Heavy duty up to 60 feet high
Maximum intended load (lbs./ft.^{2})	25	25	50	75
Poles or uprights	2 X 4 in.	4 X 4 in.	4 X 4 in.	4 X 6 in.
Maximum pole spacing (longitudinal)	6 feet	10 feet	8 feet	6 feet
Maximum pole spacing (transverse)	5 feet	5 feet	5 feet	5 feet
Runners	1 X 4 in.	1 1/4 X 9 in.	2 X 10 in.	2 X 10 in.
Bearers and maximum spacing of bearers:				
3 feet	2 X 4 in.	2 X 4 in.	2 X 10 in. or 3 X 4 in.	2 X 10 in. or 3 X 5 in.
5 feet	2 X 6 in. or 3 X 4 in.	2 X 6 in. or 3 X 4 in. (rough)	2 X 10 in. or 3 X 4 in.	2 X 10 in. or 3 X 5 in.
6 feet			2 X 10 in. or 3 X 4 in.	2 X 10 in. or 3 X 5 in.
8 feet			2 X 10 in. or 3 X 4 in.	
Planking	1 1/4 X 9 in.	2 X 10 in.	2 X 10 in.	2 X 10 in.
Maximum vertical spacing of horizontal members	7 feet	9 feet	7 feet	6 ft. 6 in.
Bracing horizontal	1 X 4 in.	1 X 4 in.	1 X 6 in. or 1 1/4 X 4 in.	2 X 4 in.
Bracing diagonal	1 X 4 in.	1 X 4 in.	1 X 4 in.	2 X 4 in.
Tie-ins	1 X 4 in.	1 X 4 in.	1 X 4 in.	1 X 4 in.

Note: All members except planking are used on edge. All wood bearers shall be reinforced with 3/16 x 2 inch steel strip, or the equivalent, secured to the lower edges for the entire length of the bearer.

Independent Wood Pole Scaffolds

	Light duty up to 20 feet high	Light duty up to 60 feet high	Medium duty up to 60 feet high	Heavy duty up to 60 feet high
Maximum intended load	25 lbs./ft.^{2}	25 lbs./ft.^{2}	50 lbs./ft.^{2}	75 lbs./ft.^{2}
Poles or uprights	2 X 4 in.	4 X 4 in.	4 X 4 in.	4 X 4 in.
Maximum pole spacing (longitudinal)	6 feet	10 feet	8 feet	6 feet
Maximum (transverse)	6 feet	10 feet	8 feet	8 feet
Runners	1 1/4 X 4 in.	1 1/4 X 9 in.	2 X 10 in.	2 X 10 in.
Bearers and maximum spacing of bearers:				
3 feet	2 X 4 in.	2 X 4 in.	2 X 10 in.	2 X 10 in. (rough)
6 feet	2 X 6 in. or 3 X 4 in.	2 X 10 in. (rough) or 3 X 8 in.	2 X 10 in.	2 X 10 in. (rough)
8 feet	2 X 6 in. or 3 X 4 in.	2 X 10 in. (rough) or 3 x 8 in.	2 X 10 in.	
10 feet	2 X 6 in. or 3 X 4 in.	2 X 10 in. (rough) or 3 X 3 in.		
Planking	1 1/4 X 9 in.	2 X 10 in.	2 X 10 in.	2 X 10 in.
Maximum vertical spacing of horizontal members	7 feet	7 feet	6 feet	6 feet
Bracing horizontal	1 X 4 in.	1 X 4 in.	1 X 6 in. or 1 1/4 X 4 in.	2 X 4 in.
Bracing diagonal	1 X 4 in.	1 X 4 in.	1 X 4 in.	2 X 4 in.
Tie-ins	1 X 4 in.	1 X 4 in.	1 X 4 in.	1 X 4 in.

Note: All members except planking are used on edge. All wood bearers shall be reinforced with 3/16 x 2 inch steel strip, or the equivalent, secured to the lower edges for the entire length of the bearer.

(b) *Tube and coupler scaffolds.*

Minimum Size of Members

	Light duty	Medium duty	Heavy duty
Maximum intended load	25 lbs/ft.^{2}	50 lbs/ft.^{2}	75 lbs/ft.^{2}
Posts, runners and braces	Nominal 2 in. (1.90 in.) OD steel tube or pipe	Nominal 2 in. (1.90 in.) OD steel tube or pipe	Nominal 2 in. (1.90 in.) OD steel tube or pipe
Bearers	Nominal 2 in. (1.90 in.) OD steel tube or pipe and a maximum post spacing of 4 ft. X10 ft.	Nominal 2 in. (1.90 in.) OD steel tube or pipe and a maximum post spacing of 4 ft. X 7 ft. or Nominal 2 1/2 in. (2.375 in.). OD steel tube or pipe and a maximum post spacing of 6 ft. X 8 ft. *	Nominal 2 1/2 in. (2.375 in.). OD steel tube or pipe and a maximum post spacing of 6 ft. X 6 ft.
Maximum runner spacing vertically	6 ft. 6 in.	6 ft. 6 in.	6 ft. 6 in.

* Bearers shall be installed in the direction of the shorter dimension.

Note: Longitudinal diagonal bracing shall be installed at an angle of 45° (± 5°).

Maximum Number of Planked Levels

Number of working levels	Maximum number of additional planked levels			Maximum height of scaffold (in feet)
	Light duty	Medium duty	Heavy duty	
1	16	11	6	125
2	11	1	0	125
3	6	0	0	125
4	1	0	0	125

(c) *Fabricated frame scaffolds.* Because of their prefabricated nature, no additional guidelines or tables for these scaffolds are being adopted in this Appendix.

(d) *Plasterers',* decorators', and large area scaffolds. The guidelines for pole scaffolds or tube and coupler scaffolds (Appendix A (a) and (b)) may be applied.

(e) *Bricklayers' square scaffolds.* Maximum intended load: 50 lb/ft.^{2}*

Maximum width: 5 ft.

Maximum height: 5 ft.

Gussets: 1 x 6 in.

Braces: 1 x 8 in.

Legs: 2 x 6 in.

Bearers (horizontal members): 2 x 6 in.

(f) *Horse scaffolds.*

Maximum intended load (light duty): 25 lb/ft.^{2} **

Maximum intended load (medium duty): 50 lb/ft.^{2} **

Horizontal members or bearers:

Light duty: 2 x 4 in.

Medium duty: 3 x 4 in.

Legs: 2 x 4 in.

Longitudinal brace between legs: 1 x 6 in.

Gusset brace at top of legs: 1 x 8 in.

Half diagonal braces: 2 x 4 in.

(g) *Form scaffolds and carpenters'* bracket scaffolds.

(1) *Brackets shall consist of a triangular-shaped* frame made of wood with a cross-section not less than 2 inches by 3 inches, or of 1 1/4 inch x 1 1/4 inch x 1/8 inch structural angle iron.

(2) *Bolts used to attach* brackets to structures shall not be less than 5/8 inches in diameter.

(3) *Maximum bracket spacing* shall be 8 feet on centers.

(4) *No more than two* employees shall occupy any given 8 feet of a bracket or form scaffold at any one time. Tools and materials shall not exceed 75 pounds in addition to the occupancy.

(5) *Wooden figure-four scaffolds:*

Maximum intended load: 25 lb/ft.^{2}

Uprights: 2 x 4 in. or 2 x 6 in.

Bearers (two): 1 x 6 in.

Braces: 1 x 6 in.

Maximum length of bearers (unsupported): 3 ft. 6 in.

(i) *Outrigger bearers shall consist* of two pieces of 1 x 6 inch lumber nailed on opposite sides of the vertical support.

(ii) *Bearers for wood figure-four* brackets shall project not more than 3 feet 6 inches from the outside of the form support, and shall be braced and secured to prevent tipping or turning. The knee or angle brace shall intersect the bearer at least 3 feet from the form at an angle of approximately 45 degrees, and the lower end shall be nailed to a vertical support.

* The squares shall be set not more than 8 feet apart for light duty scaffolds and not more than 5 feet apart for medium duty scaffolds.

** Horses shall be spaced not more than 8 feet apart for light duty loads, and not more than 5 feet apart for medium duty loads.

(6) *Metal bracket scaffolds:*
Maximum intended load: 25 lb/ft.[2]
Uprights: 2 x 4 inch
Bearers: As designed.
Braces: As designed.

(7) *Wood bracket scaffolds:*
Maximum intended load: 25 lb/ft.[2]
Uprights: 2 x 4 in or 2 x 6 in
Bearers: 2 x 6 in
Maximum scaffold width: 3 ft 6 in
Braces: 1 x 6 in

(h) *Roof bracket scaffolds.* No specific guidelines or tables are given.

(i) *Outrigger scaffolds (single* level). No specific guidelines or tables are given.

(j) *Pump jack scaffolds.* Wood poles shall not exceed 30 feet in height. Maximum intended load — 500 lbs between poles; applied at the center of the span. Not more than two employees shall be on a pump jack scaffold at one time between any two supports. When 2 x 4's are spliced together to make a 4 x 4 inch wood pole, they shall be spliced with "10 penny" common nails no more than 12 inches center to center, staggered uniformly from the opposite outside edges.

(k) *Ladder jack scaffolds.* Maximum intended load — 25 lb/ft^2. However, not more than two employees shall occupy any platform at any one time. Maximum span between supports shall be 8 feet.

(l) *Window jack scaffolds.* Not more than one employee shall occupy a window jack scaffold at any one time.

(m) *Crawling boards (chicken* ladders). Crawling boards shall be not less than 10 inches wide and 1 inch thick, with cleats having a minimum 1 x 1 1/2 inch cross-sectional area. The cleats shall be equal in length to the width of the board and spaced at equal intervals not to exceed 24 inches.

(n) *Step,* platform, and trestle ladder scaffolds. No additional guidelines or tables are given.

(o) *Single-point adjustable suspension* scaffolds. Maximum intended load — 250 lbs. Wood seats for boatswains' chairs shall be not less than 1 inch thick if made of non-laminated wood, or 5/8 inches thick if made of marine quality plywood.

(p) *Two-point adjustable suspension scaffolds.*

(1) *In addition to direct connections* to buildings (except window cleaners' anchors) acceptable ways to prevent scaffold sway include angulated roping and static lines. Angulated roping is a system of platform suspension in which the upper wire rope sheaves or suspension points are closer to the plane of the building face than the corresponding attachment points on the platform, thus causing the platform to press against the face of the building. Static lines are separate ropes secured at their top and bottom ends closer to the plane of the building face than the outermost edge of the platform. By drawing the static line taut, the platform is drawn against the face of the building.

(2) *On suspension scaffolds designed* for a working load of 500 pounds, no more than two employees shall be permitted on the scaffold at one time. On suspension scaffolds with a working load of 750 pounds, no more than three employees shall be permitted on the scaffold at one time.

(3) *Ladder-type platforms.* The side stringer shall be of clear straight-grained spruce. The rungs shall be of straight-grained oak, ash, or hickory, at least 1 1/8 inches in diameter, with 7/8 inch tenons mortised into the side stringers at least 7/8 inch. The stringers shall be tied together with tie rods not less than 1/4 inch in diameter, passing through the stringers and riveted up tight against washers on both ends. The flooring strips shall be spaced not more than 5/8 inch apart, except at the side rails where the space may be 1 inch. Ladder-type platforms shall be constructed in accordance with the following table:

Schedule for Ladder-Type Platforms

Length of Platform	12 feet	14 & 16 feet	18 & 20 feet
Side stringers, minimum cross section (finished sizes):			
At ends	1 3/4 x 2 3/4 in.	1 3/4 x 2 3/4 in.	1 3/4 x 3 in..
At middle	1 3/4 x 3 3/4 in.	1 3/4 x 3 3/4 in.	1 3/4 x 4 in.
Reinforcing strip (minimum)	A 1/8 X 7/8 inch steel reinforcing strip shall be attached to the side or underside, full length.		
Rungs	Rungs shall be 1 1/8 inch minimum diameter with at least 7/8 inch in diameter tenons, and the maximum spacing shall be 12 inches to center.		
Tie rods:			
Number (minimum)	3	4	4
Diameter (minimum)	1/4 inch	1/4 inch	1/4 inch
Flooring, minimum finished size	1/2 x 2 3/4 in.	1/2 x 2 3/4 in.	1/2 x 2 3/4 in.

Schedule for Ladder-Type Platforms

Length of Platform	22 & 24 ft.	28 & 30 ft.
Side stringers, minimum cross section (finished sizes):		
At ends	1 3/4 x 3 in.	1 3/4 x 3 1/2 in.
At middle	1 3/4 x 4 1/4 in.	1 3/4 x 5 in.
Reinforcing strip (minimum)	A 1/8 X 7/8 inch steel reinforcing strip shall be attached to the side or underside, full length.	
Rungs	Rungs shall be 1 1/8 inch minimum diameter with at least 7/8 inch in diameter tenons, and the maximum spacing shall be 12 inches to center.	
Tie Rods:		
Number (minimum)	5	6.
Diameter (minimum)	1/4 in.	1/4 in.
Flooring, minimum finished size.	1/2 x 2 3/4 in.	1/2 x 2 3/4 in.

(4) *Plank-Type Platforms.* Plank-type platforms shall be composed of not less than nominal 2 x 8 inch unspliced planks, connected together on the underside with cleats at intervals not exceeding 4 feet, starting 6 inches from each end. A bar or other effective means shall be securely fastened to the platform at each end to prevent the platform from slipping off the hanger. The span between hangers for plank-type platforms shall not exceed 10 feet.

(5) *Beam-Type Platforms.* Beam platforms shall have side stringers of lumber not less than 2 x 6 inches set on edge. The span between hangers shall not exceed 12 feet when beam platforms are used. The flooring shall be supported on 2 x 6 inch cross beams, laid flat and set into the upper edge of the stringers with a snug fit, at intervals of not more than 4 feet, securely nailed to the cross beams. Floorboards shall not be spaced more than 1/2 inch apart.

(q)(1) *Multi-point adjustable suspension scaffolds* and stonesetters' multi-point adjustable suspension scaffolds. No specific guidelines or tables are given for these scaffolds.

(2) *Masons' multi-point adjustable suspension scaffolds.* Maximum intended load — 50 lb/ft^2. Each outrigger beam shall be at least a standard 7 inch, 15.3 pound steel I-beam, at least 15 feet long. Such beams shall not project more than 6 feet 6 inches beyond the bearing point. Where the overhang exceeds 6 feet 6 inches, outrigger beams shall be composed of stronger beams or multiple beams.

(r) *Catenary scaffolds.*

(1) *Maximum intended load — 500 lbs.*

(2) *Not more than two employees* shall be permitted on the scaffold at one time.

(3) *Maximum capacity of come-along shall be 2,000 lbs.*

(4) *Vertical pickups shall be spaced not more than 50 feet apart.*

(5) *Ropes shall be equivalent in strength* to at least 1/2 inch (1.3 cm) diameter improved plow steel wire rope.

(s) *Float (ship) scaffolds.*

(1) *Maximum intended load — 750 lbs.*

(2) *Platforms shall be made of* 3/4 inch plywood, equivalent in rating to American Plywood Association Grade B-B, Group I, Exterior.

(3) *Bearers shall be made from 2 x 4* inch, or 1 x 10 inch rough lumber. They shall be free of knots and other flaws.

(4) *Ropes shall be equivalent in strength* to at least 1 inch (2.5 cm) diameter first grade manila rope.

(t) *Interior hung scaffolds.*
Bearers (use on edge): 2 x 10 in.
Maximum intended load: Maximum span
25 lb/ft.[2]: 10 ft.
50 lb/ft.[2]: 10 ft.
75 lb/ft.[2]: 7 ft.

(u) *Needle beam scaffolds.*

Maximum intended load: 25 lb/ft.^{2}
Beams: 4 x 6 in.
Maximum platform span: 8 ft.
Maximum beam span: 10 ft.

(1) *Ropes shall be attached to the needle beams* by a scaffold hitch or an eye splice. The loose end of the rope shall be tied by a bowline knot or by a round turn and a half hitch.

(2) *Ropes shall be equivalent in strength* to at least 1 inch (2.5 cm) diameter first grade manila rope.

(v) *Multi-level suspension scaffolds.* No additional guidelines or tables are being given for these scaffolds.

(w) *Mobile Scaffolds.* Stability test as described in the ANSI A92 series documents, as appropriate for the type of scaffold, can be used to establish stability for the purpose of §1926.452(w)(6).

(x) *Repair bracket scaffolds.* No additional guidelines or tables are being given for these scaffolds.

(y) *Stilts.* No specific guidelines or tables are given.

(z) *Tank builder's scaffold.*

(1) *The maximum distance between brackets* to which scaffolding and guardrail supports are attached shall be no more than 10 feet 6 inches.

(2) *Not more than three employees* shall occupy a 10 feet 6 inch span of scaffold planking at any time.

(3) *A taut wire or synthetic rope* supported on the scaffold brackets shall be installed at the scaffold plank level between the innermost edge of the scaffold platform and the curved plate structure of the tank shell to serve as a safety line in lieu of an inner guardrail assembly where the space between the scaffold platform and the tank exceeds 12 inches (30.48 cm). In the event the open space on either side of the rope exceeds 12 inches (30.48 cm), a second wire or synthetic rope appropriately placed, or guardrails in accordance with §1926.451(e)(4), shall be installed in order to reduce that open space to less than 12 inches (30.48 cm).

(4) *Scaffold planks* of rough full-dimensioned 2-inch (5.1 cm) x 12-inch (30.5 cm) Douglas Fir or Southern Yellow Pine of Select Structural Grade shall be used. Douglas Fir planks shall have a fiber stress of at least 1900 lb/in^2 (130,929 n/cm^2) and a modulus of elasticity of at least 1,900,000 lb/in^2 (130,929,000 n/cm^2), while Yellow Pine planks shall have a fiber stress of at least 2500 lb/in^2 (172,275 n/cm^2) and a modulus of elasticity of at least 2,000,000 lb/in^2 (137,820,000 n/cm^2).

(5) *Guardrails shall be constructed* of a taut wire or synthetic rope, and shall be supported by angle irons attached to brackets welded to the steel plates. These guardrails shall comply with §1926.451(e)(4). Guardrail supports shall be located at no greater than 10 feet 6 inch intervals.

1926 Subpart L Appendix B

Criteria for Determining the Feasibility of Providing Safe Access and Fall Protection for Scaffold Erectors and Dismantlers (Non-mandatory)

[Reserved]

1926 Subpart L Appendix C

List of National Consensus Standards (Non-mandatory)

ANSI/SIA A92.2-1990 Vehicle-Mounted Elevating and Rotating Aerial Devices

ANSI/SIA A92.3-1990 Manually Propelled Elevating Aerial Platforms

ANSI/SIA A92.5-1990 Boom Supported Elevating Work Platforms

ANSI/SIA A92.6-1990 Self-Propelled Elevating Work Platforms

ANSI/SIA A92.7-1990 Airline Ground Support Vehicle-Mounted Vertical Lift Devices

ANSI/SIA A92.8-1993 Vehicle-Mounted Bridge Inspection and Maintenance Devices

ANSI/SIA A92.9-1993 Mast-Climbing Work Platforms

1926 Subpart L Appendix D

List of Training Topics for Scaffold Erectors and Dismantlers (Non-mandatory)

This Appendix D is provided to serve as a guide to assist employers when evaluating the training needs of employees erecting or dismantling supported scaffolds.

The Agency believes that employees erecting or dismantling scaffolds should be trained in the following topics:

- General Overview of Scaffolding
  - regulations and standards
  - erection/dismantling planning
  - PPE and proper procedures
  - fall protection
  - materials handling
  - access
  - working platforms
  - foundations
  - guys, ties and braces
- Tubular Welded Frame Scaffolds
  - specific regulations and standards
  - components
  - parts inspection
  - erection/dismantling planning
  - guys, ties and braces
  - fall protection
  - general safety
  - access and platforms
  - erection/dismantling procedures
  - rolling scaffold assembly
  - putlogs
- Tube and Clamp Scaffolds
  - specific regulations and standards
  - components
  - parts inspection
  - erection/dismantling planning
  - guys, ties and braces
  - fall protection
  - general safety
  - access and platforms
  - erection/dismantling procedures
  - buttresses, cantilevers, & bridges
- System Scaffolds
  - specific regulations and standards
  - components
  - parts inspection
  - erection/dismantling planning
  - guys, ties and braces
  - fall protection
  - general safety
  - access and platforms
  - erection/dismantling procedures
  - buttresses, cantilevers, & bridges

Scaffold erectors and dismantlers should all receive the general overview, and, in addition, specific training for the type of supported scaffold being erected or dismantled.

1926 Subpart L Appendix E

Drawings and Illustrations (Non-mandatory)

This Appendix provides drawings of particular types of scaffolds and scaffold components, and graphic illustrations of bracing patterns and tie spacing patterns.

This Appendix is intended to provide visual guidance to assist the user in complying with the requirements of subpart L, part 1926.

FIGURE 1. BRACING - TUBE & COUPLER SCAFFOLDS

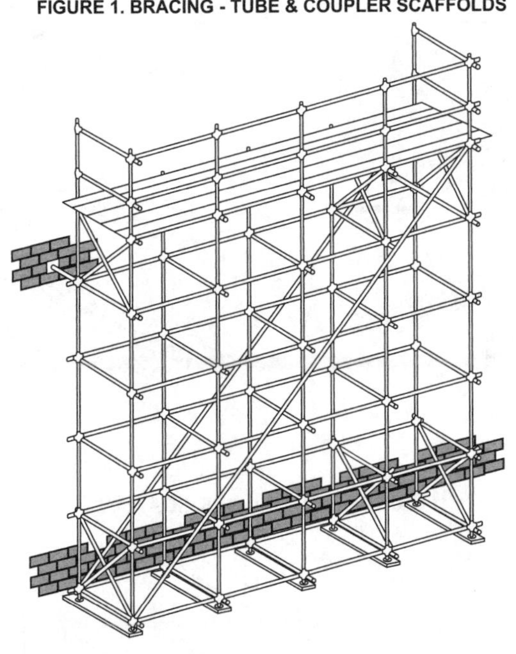

FIGURE 2. SUSPENDED SCAFFOLD PLATFORM WELDING PRECAUTIONS

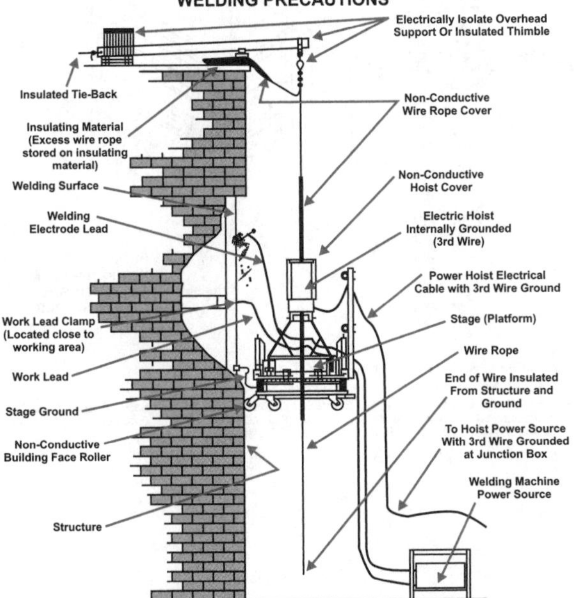

Hoists must be electronically isolated from scaffold.

FIGURE 3. MAXIMUM VERTICAL TIE SPACING WIDER THAN 3'- 0" BASES

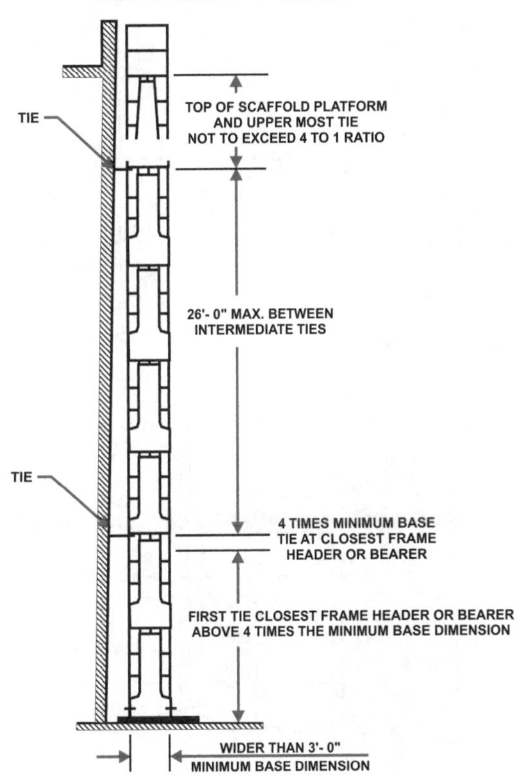

FIGURE 4. MAXIMUM VERTICAL TIE SPACING 3'- 0" AND NARROWER BASES

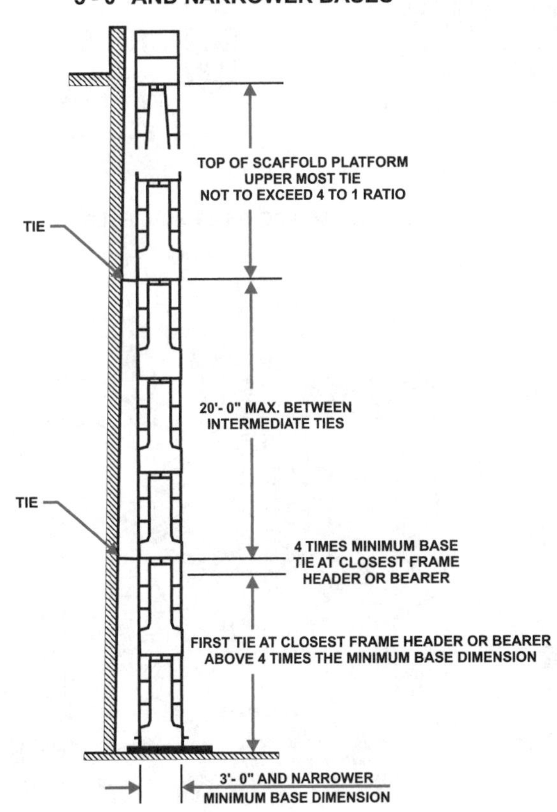

FIGURE 5. SYSTEM SCAFFOLD

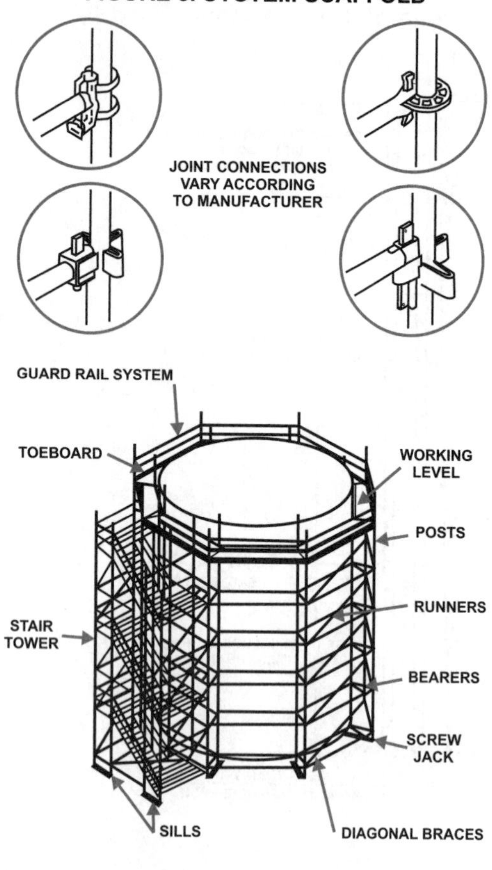

FIGURE 6. GRADE STAMPS

Grade stamp courtesy of Southern Pine Inspection Bureau

Grade stamp courtesy of West Coast Lumber Inspection Bureau

FIGURE 7. TUBE AND COUPLER SCAFFOLD

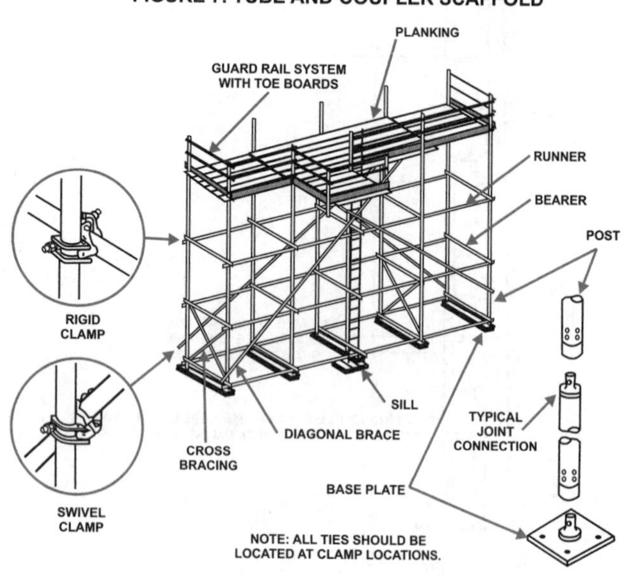

FIGURE 8. SCAFFOLDING WORK SURFACES

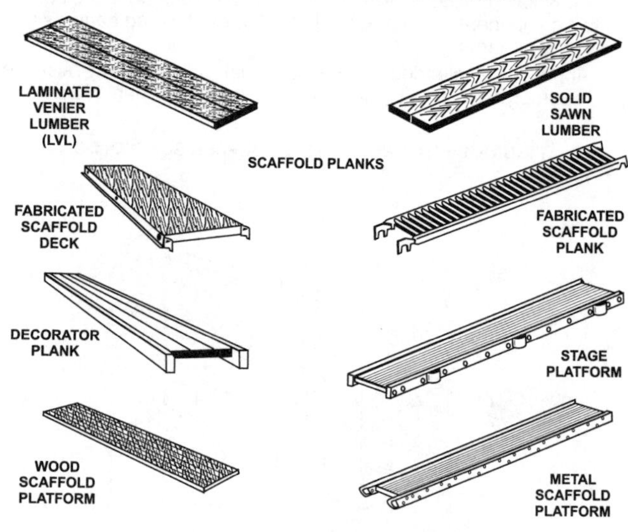

FIGURE 9. OUTRIGGER SCAFFOLD

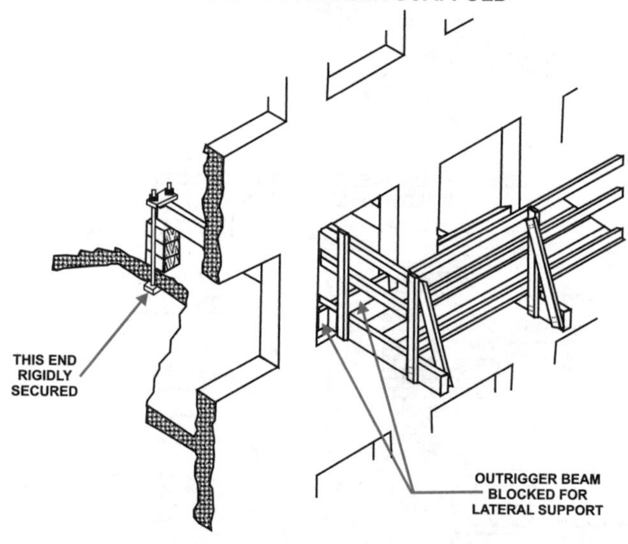

[61 FR 46122, Aug. 30, 1996; 61 FR 59832, Nov. 25, 1996]

Source: 61 FR 46104, Aug. 30, 1996, unless otherwise noted.

Authority: Section 107, Contract Work Hours and Safety Standards Act (Construction Safety Act)(40 U.S.C. 333); Secs. 4, 6, 8, Occupational Safety and Health Act of 1970 (29 U.S.C. 653, 655, 657); Secretary of Labor's Order Nos. 1-90 (55 FR 9033) and 5-2007 (72 FR 31159); and 29 CFR part 1911.

Subpart M - Fall Protection

§1926.500

Scope, application, and definitions applicable to this subpart

(a) Scope and application.[§1926.500(a)]

(1) *This subpart sets forth requirements and criteria* for fall protection in construction workplaces covered under 29 CFR part 1926. Exception: The provisions of this subpart do not apply when employees are making an inspection, investigation, or assessment of workplace conditions prior to the actual start of construction work or after all construction work has been completed.[§1926.500(a)(1)]

(2) *Section 1926.501 sets forth* those workplaces, conditions, operations, and circumstances for which fall protection shall be provided except as follows:[§1926.500(a)(2)]

(i) *Requirements relating to fall protection* for employees working on scaffolds are provided in subpart L of this part. [§1926.500(a)(2)(i)]

(ii) *Requirements relating to fall* protection for employees working on cranes and derricks are provided in subpart CC of this part.[§1926.500(a)(2)(ii)]

(iii) *Fall protection requirements* for employees performing steel erection work (except for towers and tanks) are provided in subpart R of this part.[§1926.500(a)(2)(iii)]

(iv) *Requirements relating to fall protection* for employees working on certain types of equipment used in tunneling operations are provided in subpart S of this part.[§1926.500(a)(2)(iv)]

(v) *Requirements relating to fall protection* for employees engaged in the erection of tanks and communication and broadcast towers are provided in Sec. 1926.105.[§1926.500(a)(2)(v)]

(vi) *Requirements relating to fall protection* for employees engaged in the construction of electric transmission and distribution lines and equipment are provided in subpart V of this part.[§1926.500(a)(2)(vi)]

(vii) *Requirements relating to fall protection* for employees working on stairways and ladders are provided in subpart X of this part.[§1926.500(a)(2)(vii)]

(3) *Section 1926.502 sets forth the requirements* for the installation, construction, and proper use of fall protection required by part 1926, except as follows:[§1926.500(a)(3)]

(i) *Performance requirements for guardrail systems* used on scaffolds and performance requirements for falling object protection used on scaffolds are provided in subpart L of this part.[§1926.500(a)(3)(i)]

(ii) *Performance requirements for stairways,* stairrail systems, and handrails are provided in subpart X of this part. [§1926.500(a)(3)(ii)]

(iii) *Additional performance requirements* for personal climbing equipment, lineman's body belts, safety straps, and lanyards are provided in subpart V of this part.[§1926.500(a)(3)(iii)]

(iv) *Section 1926.502* does not apply to the erection of tanks and communication and broadcast towers. [§1926.500(a)(3)(iv)]

Note: Section 1926.104 sets the criteria for body belts, lanyards and lifelines used for fall protection during tank and communication and broadcast tower erection. Paragraphs (b),(c) and (f) of Sec. 1926.107 provide definitions for the pertinent terms.

(v) *Criteria for steps,* handholds, ladders, and grabrails/guardrails/railings required by subpart CC are provided in subpart CC. Sections 1926.502(a), (c) through (e), and (i) apply to activities covered under subpart CC unless otherwise stated in subpart CC. No other paragraphs of Sec. 1926.502 apply to subpart CC.[§1926.500(a)(3)(v)]

(4) *Section 1926.503 sets* forth requirements for training in the installation and use of fall protection systems, except in relation to steel erection activities and the use of equipment covered by subpart CC.[§1926.500(a)(4)]

(b) Definitions.

Anchorage means a secure point of attachment for lifelines, lanyards or deceleration devices.

Body belt (safety belt) means a strap with means both for securing it about the waist and for attaching it to a lanyard, lifeline, or deceleration device.

Body harness means straps which may be secured about the employee in a manner that will distribute the fall arrest forces over at least the thighs, pelvis, waist, chest and shoulders with means for attaching it to other components of a personal fall arrest system.

Buckle means any device for holding the body belt or body harness closed around the employee's body.

Connector means a device which is used to couple (connect) parts of the personal fall arrest system and positioning device systems together. It may be an independent component of the system, such as a carabiner, or it may be an integral component of part of the system (such as a buckle or dee-ring sewn into a body belt or body harness, or a snap-hook spliced or sewn to a lanyard or self-retracting lanyard).

Controlled access zone (CAZ) means an area in which certain work (e.g., overhand bricklaying) may take place without the use of guardrail systems, personal fall arrest systems, or safety net systems and access to the zone is controlled.

Dangerous equipment means equipment (such as pickling or galvanizing tanks, degreasing units, machinery, electrical equipment, and other units) which, as a result of form or function, may be hazardous to employees who fall onto or into such equipment.

Deceleration device means any mechanism, such as a rope grab, rip- stitch lanyard, specially-woven lanyard, tearing or deforming lanyards, automatic self-retracting lifelines/lanyards, etc., which serves to dissipate a substantial amount of energy during a fall arrest, or otherwise limit the energy imposed on an employee during fall arrest.

Deceleration distance means the additional vertical distance a falling employee travels, excluding lifeline elongation and free fall distance, before stopping, from the point at which the deceleration device begins to operate. It is measured as the distance between the location of an employee's body belt or body harness attachment point at the moment of activation (at the onset of fall arrest forces) of the deceleration device during a fall, and the location of that attachment point after the employee comes to a full stop.

Equivalent means alternative designs, materials, or methods to protect against a hazard which the employer can demonstrate will provide an equal or greater degree of safety for employees than the methods, materials or designs specified in the standard.

Failure means load refusal, breakage, or separation of component parts. Load refusal is the point where the ultimate strength is exceeded.

Free fall means the act of falling before a personal fall arrest system begins to apply force to arrest the fall.

Free fall distance means the vertical displacement of the fall arrest attachment point on the employee's body belt or body harness between onset of the fall and just before the system begins to apply force to arrest the fall. This distance excludes deceleration distance, and lifeline/lanyard elongation, but includes any deceleration device slide distance or self-retracting lifeline/lanyard extension before they operate and fall arrest forces occur.

Guardrail system means a barrier erected to prevent employees from falling to lower levels.

Hole means a gap or void 2 inches (5.1 cm) or more in its least dimension, in a floor, roof, or other walking/working surface.

Infeasible means that it is impossible to perform the construction work using a conventional fall protection system (i.e., guardrail system, safety net system, or personal fall arrest system) or that it is technologically impossible to use any one of these systems to provide fall protection.

Lanyard means a flexible line of rope, wire rope, or strap which generally has a connector at each end for connecting the body belt or body harness to a deceleration device, lifeline, or anchorage.

Leading edge means the edge of a floor, roof, or formwork for a floor or other walking/working surface (such as the deck) which changes location as additional floor, roof, decking, or formwork sections are placed, formed, or constructed. A leading edge is considered to be an "unprotected side and edge" during periods when it is not actively and continuously under construction.

Lifeline means a component consisting of a flexible line for connection to an anchorage at one end to hang vertically (vertical lifeline), or for connection to anchorages at both ends to stretch horizontally (horizontal lifeline), and which serves as a means for connecting other components of a personal fall arrest system to the anchorage.

Low-slope roof means a roof having a slope less than or equal to 4 in 12 (vertical to horizontal).

Lower levels means those areas or surfaces to which an employee can fall. Such areas or surfaces include, but are not limited to, ground levels, floors, platforms, ramps, runways, excavations, pits, tanks, material, water, equipment, structures, or portions thereof.

Mechanical equipment means all motor or human propelled wheeled equipment used for roofing work, except wheelbarrows and mopcarts.

Opening means a gap or void 30 inches (76 cm) or more high and 18 inches (48 cm) or more wide, in a wall or partition, through which employees can fall to a lower level.

Overhand bricklaying and related work means the process of laying bricks and masonry units such that the surface of the wall to be jointed is on the opposite side of the wall from the mason, requiring the mason to lean over the wall to complete the work. Related work includes mason tending and electrical installation incorporated into the brick wall during the overhand bricklaying process.

Personal fall arrest system means a system used to arrest an employee in a fall from a working level. It consists of an anchorage, connectors, a body belt or body harness and may include a lanyard, deceleration device, lifeline, or suitable combinations of these. As of January 1, 1998, the use of a body belt for fall arrest is prohibited.

Positioning device system means a body belt or body harness system rigged to allow an employee to be supported on an elevated vertical surface, such as a wall, and work with both hands free while leaning.

Rope grab means a deceleration device which travels on a lifeline and automatically, by friction, engages the lifeline and locks so as to arrest the fall of an employee. A rope grab usually employs the principle of inertial locking, cam/level locking, or both.

Roof means the exterior surface on the top of a building. This does not include floors or formwork which, because a building has not been completed, temporarily become the top surface of a building.

Roofing work means the hoisting, storage, application, and removal of roofing materials and equipment, including related insulation, sheet metal, and vapor barrier work, but not including the construction of the roof deck.

Safety-monitoring system means a safety system in which a competent person is responsible for recognizing and warning employees of fall hazards.

Self-retracting lifeline/lanyard means a deceleration device containing a drum-wound line which can be slowly extracted from, or retracted onto, the drum under slight tension during normal employee movement, and which, after onset of a fall, automatically locks the drum and arrests the fall.

Snaphook means a connector comprised of a hook-shaped member with a normally closed keeper, or similar arrangement, which may be opened to permit the hook to receive an object and, when released, automatically closes to retain the object. Snaphooks are generally one of two types:

(1) *The locking type with a self-closing,* self-locking keeper which remains closed and locked until unlocked and pressed open for connection or disconnection; or

(2) *The non-locking type with a self-closing keeper* which remains closed until pressed open for connection or disconnection. As of January 1, 1998, the use of a non-locking snaphook as part of personal fall arrest systems and positioning device systems is prohibited.

Steep roof means a roof having a slope greater than 4 in 12 (vertical to horizontal).

Toeboard means a low protective barrier that will prevent the fall of materials and equipment to lower levels and provide protection from falls for personnel.

Unprotected sides and edges means any side or edge (except at entrances to points of access) of a walking/working surface, e.g., floor, roof, ramp, or runway where there is no wall or guardrail system at least 39 inches (1.0 m) high.

Walking/working surface means any surface, whether horizontal or vertical on which an employee walks or works, including, but not limited to, floors, roofs, ramps, bridges, runways, formwork and concrete reinforcing steel but not including ladders, vehicles, or trailers, on which employees must be located in order to perform their job duties.

Warning line system means a barrier erected on a roof to warn employees that they are approaching an unprotected roof side or edge, and which designates an area in which roofing work may take place without the use of guardrail, body belt, or safety net systems to protect employees in the area.

Work area means that portion of a walking/working surface where job duties are being performed.

[59 FR 40730, Aug. 9, 1994, as amended at 60 FR 39255, Aug. 2, 1995; 66 FR 5265, Jan. 18, 2001; 75 FR 48133, Aug. 9, 2010]

Source: 59 FR 40730, Aug. 9, 1994, unless otherwise noted.

§1926.501

Duty to have fall protection

(a) General.[§1926.501(a)]

(1) *This section sets forth requirements* for employers to provide fall protection systems. All fall protection required by this section shall conform to the criteria set forth in Sec. 1926.502 of this subpart.[§1926.501(a)(1)]

(2) *The employer shall determine* if the walking/working surfaces on which its employees are to work have the strength and structural integrity to support employees safely. Employees shall be allowed to work on those surfaces only when the surfaces have the requisite strength and structural integrity. [§1926.501(a)(2)]

(b) (1) *Unprotected sides and edges.* Each employee on a walking/working surface (horizontal and vertical surface) with an unprotected side or edge which is 6 feet (1.8 m) or more above a lower level shall be protected from falling by the use of guardrail systems, safety net systems, or personal fall arrest systems.[§1926.501(b)(1)]

(2) *Leading edges.*[§1926.501(b)(2)]

(i) *Each employee who is constructing* a leading edge 6 feet (1.8 m) or more above lower levels shall be protected from falling by guardrail systems, safety net systems, or personal fall arrest systems. Exception: When the employer can demonstrate that it is infeasible or creates a greater hazard to use these systems, the employer shall develop and implement a fall protection plan which meets the requirements of paragraph (k) of §1926.502.[§1926.501(b)(2)(i)]

Note: There is a presumption that it is feasible and will not create a greater hazard to implement at least one of the above-listed fall protection systems. Accordingly, the employer has the burden of establishing that it is appropriate to implement a fall protection plan which complies with Sec. 1926.502(k) for a particular workplace situation, in lieu of implementing any of those systems.

(ii) *Each employee on a walking/working surface* 6 feet (1.8 m) or more above a lower level where leading edges are under construction, but who is not engaged in the leading edge work, shall be protected from falling by a guardrail system, safety net system, or personal fall arrest system. If a guardrail system is chosen to provide the fall protection, and a controlled access zone has already been established for leading edge work, the control line may be used in lieu of a guardrail along the edge that parallels the leading edge. [§1926.501(b)(2)(ii)]

(3) *Hoist areas.* Each employee in a hoist area shall be protected from falling 6 feet (1.8 m) or more to lower levels by guardrail systems or personal fall arrest systems. If guardrail systems, [or chain, gate, or guardrail] or portions thereof, are removed to facilitate the hoisting operation (e.g., during landing of materials), and an employee must lean through the access opening or out over the edge of the access opening (to receive or guide equipment and materials, for example), that employee shall be protected from fall hazards by a personal fall arrest system.[§1926.501(b)(3)]

(4) *Holes.*[§1926.501(b)(4)]

(i) *Each employee on walking/working surfaces* shall be protected from falling through holes (including skylights) more than 6 feet (1.8 m) above lower levels, by personal fall arrest systems, covers, or guardrail systems erected around such holes.[§1926.501(b)(4)(i)]

(ii) *Each employee on a walking/working surface* shall be protected from tripping in or stepping into or through holes (including skylights) by covers.[§1926.501(b)(4)(ii)]

(iii) *Each employee on a walking/working surface* shall be protected from objects falling through holes (including skylights) by covers.[§1926.501(b)(4)(iii)]

(5) *Formwork and reinforcing steel.* Each employee on the face of formwork or reinforcing steel shall be protected from falling 6 feet (1.8 m) or more to lower levels by personal fall arrest systems, safety net systems, or positioning device systems. [§1926.501(b)(5)]

(6) *Ramps, runways, and other walkways.* Each employee on ramps, runways, and other walkways shall be protected from falling 6 feet (1.8 m) or more to lower levels by guardrail systems.[§1926.501(b)(6)]

(7) *Excavations.*[§1926.501(b)(7)]

(i) *Each employee at the edge of an excavation* 6 feet (1.8 m) or more in depth shall be protected from falling by guardrail systems, fences, or barricades when the excavations are not readily seen because of plant growth or other visual barrier; [§1926.501(b)(7)(i)]

(ii) *Each employee at the edge of a well,* pit, shaft, and similar excavation 6 feet (1.8 m) or more in depth shall be protected from falling by guardrail systems, fences, barricades, or covers. [§1926.501(b)(7)(ii)]

(8) *Dangerous equipment.* [§1926.501(b)(8)]

(i) *Each employee less than 6 feet* (1.8 m) above dangerous equipment shall be protected from falling into or onto the dangerous equipment by guardrail systems or by equipment guards. [§1926.501(b)(8)(i)]

(ii) *Each employee 6 feet* (1.8 m) or more above dangerous equipment shall be protected from fall hazards by guardrail systems, personal fall arrest systems, or safety net systems. [§1926.501(b)(8)(ii)]

(9) *Overhand bricklaying and related work.* [§1926.501(b)(9)]

(i) *Except as otherwise provided* in paragraph (b) of this section, each employee performing overhand bricklaying and related work 6 feet (1.8 m) or more above lower levels, shall be protected from falling by guardrail systems, safety net systems, personal fall arrest systems, or shall work in a controlled access zone. [§1926.501(b)(9)(i)]

(ii) *Each employee reaching more* than 10 inches (25 cm) below the level of the walking/working surface on which they are working, shall be protected from falling by a guardrail system, safety net system, or personal fall arrest system. [§1926.501(b)(9)(ii)]

Note: Bricklaying operations performed on scaffolds are regulated by subpart L — Scaffolds of this part.

(10) *Roofing work on Low-slope roofs.* Except as otherwise provided in paragraph (b) of this section, each employee engaged in roofing activities on low-slope roofs, with unprotected sides and edges 6 feet (1.8 m) or more above lower levels shall be protected from falling by guardrail systems, safety net systems, personal fall arrest systems, or a combination of warning line system and guardrail system, warning line system and safety net system, or warning line system and personal fall arrest system, or warning line system and safety monitoring system. Or, on roofs 50-feet (15.25 m) or less in width (see Appendix A to subpart M of this part), the use of a safety monitoring system alone [i.e. without the warning line system] is permitted. [§1926.501(b)(10)]

(11) *Steep roofs.* Each employee on a steep roof with unprotected sides and edges 6 feet (1.8 m) or more above lower levels shall be protected from falling by guardrail systems with toeboards, safety net systems, or personal fall arrest systems. [§1926.501(b)(11)]

(12) *Precast concrete erection.* Each employee engaged in the erection of precast concrete members (including, but not limited to the erection of wall panels, columns, beams, and floor and roof "tees") and related operations such as grouting of precast concrete members, who is 6 feet (1.8 m) or more above lower levels shall be protected from falling by guardrail systems, safety net systems, or personal fall arrest systems, unless another provision in paragraph (b) of this section provides for an alternative fall protection measure. Exception: When the employer can demonstrate that it is infeasible or creates a greater hazard to use these systems, the employer shall develop and implement a fall protection plan which meets the requirements of paragraph (k) of §1926.502. [§1926.501(b)(12)]

Note: There is a presumption that it is feasible and will not create a greater hazard to implement at least one of the above-listed fall protection systems. Accordingly, the employer has the burden of establishing that it is appropriate to implement a fall protection plan which complies with Sec. 1926.502(k) for a particular workplace situation, in lieu of implementing any of those systems.

(13) *Residential construction.* Each employee engaged in residential construction activities 6 feet (1.8 m) or more above lower levels shall be protected by guardrail systems, safety net system, or personal fall arrest system unless another provision in paragraph (b) of this section provides for an alternative fall protection measure. Exception: When the employer can demonstrate that it is infeasible or creates a greater hazard to use these systems, the employer shall develop and implement a fall protection plan which meets the requirements of paragraph (k) of §1926.502. [§1926.501(b)(13)]

Note: There is a presumption that it is feasible and will not create a greater hazard to implement at least one of the above-listed fall protection systems. Accordingly, the employer has the burden of establishing that it is appropriate to implement a fall protection plan which complies with Sec. 1926.502(k) for a particular workplace situation, in lieu of implementing any of those systems.

(14) *Wall openings.* Each employee working on, at, above, or near wall openings (including those with chutes attached) where the outside bottom edge of the wall opening is 6 feet (1.8 m) or more above lower levels and the inside bottom edge of the wall opening is less than 39 inches (1.0 m) above the walking/working surface, shall be protected from falling by the use of a guardrail system, a safety net system, or a personal fall arrest system. [§1926.501(b)(14)]

(15) *Walking/working surfaces not otherwise addressed.* Except as provided in Sec. 1926.500(a)(2) or in Sec. 1926.501(b)(1) through (b)(14), each employee on a walking/working surface 6 feet (1.8 m) or more above lower levels shall be protected from falling by a guardrail system, safety net system, or personal fall arrest system. [§1926.501(b)(15)]

(c) **Protection from falling objects.** When an employee is exposed to falling objects, the employer shall have each employee wear a hard hat and shall implement one of the following measures: [§1926.501(c)]

(1) *Erect toeboards,* screens, or guardrail systems to prevent objects from falling from higher levels; or, [§1926.501(c)(1)]

(2) *Erect a canopy structure* and keep potential fall objects far enough from the edge of the higher level so that those objects would not go over the edge if they were accidentally displaced; or, [§1926.501(c)(2)]

(3) *Barricade the area to which objects could fall,* prohibit employees from entering the barricaded area, and keep objects that may fall far enough away from the edge of a higher level so that those objects would not go over the edge if they were accidentally displaced. [§1926.501(c)(3)]

§1926.502

Fall protection systems criteria and practices

(a) **General.** [§1926.502(a)]

(1) *Fall protection systems* required by this part shall comply with the applicable provisions of this section. [§1926.502(a)(1)]

(2) *Employers shall provide and install* all fall protection systems required by this subpart for an employee, and shall comply with all other pertinent requirements of this subpart before that employee begins the work that necessitates the fall protection. [§1926.502(a)(2)]

(b) **Guardrail systems.** Guardrail systems and their use shall comply with the following provisions: [§1926.502(b)]

(1) *Top edge height* of top rails, or equivalent guardrail system members, shall be 42 inches (1.1 m) plus or minus 3 inches (8 cm) above the walking/working level. When conditions warrant, the height of the top edge may exceed the 45-inch height, provided the guardrail system meets all other criteria of this paragraph. [§1926.502(b)(1)]

Note: When employees are using stilts, the top edge height of the top rail, or equivalent member, shall be increased an amount equal to the height of the stilts.

(2) *Midrails,* screens, mesh, intermediate vertical members, or equivalent intermediate structural members shall be installed between the top edge of the guardrail system and the walking/working surface when there is no wall or parapet wall at least 21 inches (53 cm) high. [§1926.502(b)(2)]

(i) *Midrails,* when used, shall be installed at a height midway between the top edge of the guardrail system and the walking/working level. [§1926.502(b)(2)(i)]

(ii) *Screens and mesh,* when used, shall extend from the top rail to the walking/working level and along the entire opening between top rail supports. [§1926.502(b)(2)(ii)]

(iii) *Intermediate members* (such as balusters), when used between posts, shall be not more than 19 inches (48 cm) apart. [§1926.502(b)(2)(iii)]

(iv) *Other structural members* (such as additional midrails and architectural panels) shall be installed such that there are no openings in the guardrail system that are more than 19 inches (.5 m) wide. [§1926.502(b)(2)(iv)]

(3) *Guardrail systems shall be capable of withstanding,* without failure, a force of at least 200 pounds (890 N) applied within 2 inches (5.1 cm) of the top edge, in any outward or downward direction, at any point along the top edge. [§1926.502(b)(3)]

(4) *When the 200 pound* (890 N) test load specified in paragraph (b)(3) of this section is applied in a downward direction, the top edge of the guardrail shall not deflect to a height less than 39 inches (1.0 m) above the walking/working level. Guardrail system components selected and constructed in accordance with

the Appendix B to subpart M of this part will be deemed to meet this requirement.[§1926.502(b)(4)]

(5) *Midrails, screens, mesh,* intermediate vertical members, solid panels, and equivalent structural members shall be capable of withstanding, without failure, a force of at least 150 pounds (666 N) applied in any downward or outward direction at any point along the midrail or other member.[§1926.502(b)(5)]

(6) *Guardrail systems shall be so surfaced* as to prevent injury to an employee from punctures or lacerations, and to prevent snagging of clothing.[§1926.502(b)(6)]

(7) *The ends of all top rails* and midrails shall not overhang the terminal posts, except where such overhang does not constitute a projection hazard.[§1926.502(b)(7)]

(8) *Steel banding* and plastic banding shall not be used as top rails or midrails.[§1926.502(b)(8)]

(9) *Top rails and midrails* shall be at least one-quarter inch (0.6 cm) nominal diameter or thickness to prevent cuts and lacerations. If wire rope is used for top rails, it shall be flagged at not more than 6- foot (1.8 m) intervals with high-visibility material. [§1926.502(b)(9)]

(10) *When guardrail systems are used* at hoisting areas, a chain, gate or removable guardrail section shall be placed across the access opening between guardrail sections when hoisting operations are not taking place.[§1926.502(b)(10)]

(11) *When guardrail systems are used* at holes, they shall be erected on all unprotected sides or edges of the hole. [§1926.502(b)(11)]

(12) *When guardrail systems are used* around holes used for the passage of materials, the hole shall have not more than two sides provided with removable guardrail sections to allow the passage of materials. When the hole is not in use, it shall be closed over with a cover, or a guardrail system shall be provided along all unprotected sides or edges.[§1926.502(b)(12)]

(13) *When guardrail systems are used* around holes which are used as points of access (such as ladderways), they shall be provided with a gate, or be so offset that a person cannot walk directly into the hole.[§1926.502(b)(13)]

(14) *Guardrail systems used* on ramps and runways shall be erected along each unprotected side or edge.[§1926.502(b)(14)]

(15) *Manila,* plastic or synthetic rope being used for top rails or midrails shall be inspected as frequently as necessary to ensure that it continues to meet the strength requirements of paragraph (b)(3) of this section.[§1926.502(b)(15)]

(c) **Safety net systems**. Safety net systems and their use shall comply with the following provisions:[§1926.502(c)]

(1) *Safety nets shall be installed* as close as practicable under the walking/working surface on which employees are working, but in no case more than 30 feet (9.1 m) below such level. When nets are used on bridges, the potential fall area from the walking/working surface to the net shall be unobstructed. [§1926.502(c)(1)]

(2) *Safety nets shall extend* outward from the outermost projection of the work surface as follows:[§1926.502(c)(2)]

Vertical distance from working level to horizontal plane of net	Minimum required horizontal distance of outer edge of net from the edge of the working surface
Up to 5 feet	8 feet
More than 5 feet up to 10 feet	10 feet
More than 10 feet	13 feet

(3) *Safety nets shall be installed* with sufficient clearance under them to prevent contact with the surface or structures below when subjected to an impact force equal to the drop test specified in paragraph (c)(4) of this section.[§1926.502(c)(3)]

(4) *Safety nets and their installations* shall be capable of absorbing an impact force equal to that produced by the drop test specified in paragraph (c)(4)(i) of this section.[§1926.502(c)(4)]

(i) *Except as provided in paragraph* (c)(4)(ii) of this section, safety nets and safety net installations shall be drop-tested at the jobsite after initial installation and before being used as a fall protection system, whenever relocated, after major repair, and at 6-month intervals if left in one place. The drop-test shall consist of a 400 pound (180 kg) bag of sand 30 ± 2 inches (76 ± 5 cm) in diameter dropped into the net from the highest walking/working surface at which employees are exposed to fall hazards, but not from less than 42 inches (1.1 m) above that level.[§1926.502(c)(4)(i)]

(ii) *When the employer can demonstrate* that it is unreasonable to perform the drop-test required by paragraph (c)(4)(i) of this section, the employer (or a designated competent person) shall certify that the net and net installation is in compliance with the provisions of paragraphs (c)(3) and (c)(4)(i) of this section by preparing a certification record prior to the net being used as a fall protection system. The certification record must include an identification of the net and net installation for which the certification record is being prepared; the date that it was determined that the identified net and net installation were in compliance with paragraph (c)(3) of this section and the signature of the person making the determination and certification. The most recent certification record for each net and net installation shall be available at the jobsite for inspection.[§1926.502(c)(4)(ii)]

(5) *Defective nets shall not be used.* Safety nets shall be inspected at least once a week for wear, damage, and other deterioration. Defective components shall be removed from service. Safety nets shall also be inspected after any occurrence which could affect the integrity of the safety net system. [§1926.502(c)(5)]

(6) *Materials,* scrap pieces, equipment, and tools which have fallen into the safety net shall be removed as soon as possible from the net and at least before the next work shift. [§1926.502(c)(6)]

(7) *The maximum size* of each safety net mesh opening shall not exceed 36 square inches (230 cm ^{2}) nor be longer than 6 inches (15 cm) on any side, and the opening, measured center-to-center of mesh ropes or webbing, shall not be longer than 6 inches (15 cm). All mesh crossings shall be secured to prevent enlargement of the mesh opening.[§1926.502(c)(7)]

(8) *Each safety net* (or section of it) shall have a border rope for webbing with a minimum breaking strength of 5,000 pounds (22.2 kN).[§1926.502(c)(8)]

(9) *Connections between safety* net panels shall be as strong as integral net components and shall be spaced not more than 6 inches (15 cm) apart.[§1926.502(c)(9)]

(d) **Personal fall arrest systems.** Personal fall arrest systems and their use shall comply with the provisions set forth below. Effective January 1, 1998, body belts are not acceptable as part of a personal fall arrest system. Note: The use of a body belt in a positioning device system is acceptable and is regulated under paragraph (e) of this section.[§1926.502(d)]

(1) *Connectors shall be drop forged,* pressed or formed steel, or made of equivalent materials.[§1926.502(d)(1)]

(2) *Connectors shall have a corrosion-resistant finish,* and all surfaces and edges shall be smooth to prevent damage to interfacing parts of the system.[§1926.502(d)(2)]

(3) *Dee-rings and snaphooks* shall have a minimum tensile strength of 5,000 pounds (22.2 kN).[§1926.502(d)(3)]

(4) *Dee-rings and snaphooks* shall be proof-tested to a minimum tensile load of 3,600 pounds (16 kN) without cracking, breaking, or taking permanent deformation.[§1926.502(d)(4)]

(5) *Snaphooks shall be sized to be compatible* with the member to which they are connected to prevent unintentional disengagement of the snaphook by depression of the snaphook keeper by the connected member, or shall be a locking type snaphook designed and used to prevent disengagement of the snaphook by the contact of the snaphook keeper by the connected member. Effective January 1, 1998, only locking type snaphooks shall be used.[§1926.502(d)(5)]

(6) *Unless the snaphook is a locking type* and designed for the following connections, snaphooks shall not be engaged: [§1926.502(d)(6)]

(i) *directly to webbing, rope or wire rope;*[§1926.502(d)(6)(i)]

(ii) *to each other;*[§1926.502(d)(6)(ii)]

(iii) *to a Dee-ring* to which another snaphook or other connector is attached;[§1926.502(d)(6)(iii)]

(iv) *to a horizontal lifeline;* or[§1926.502(d)(6)(iv)]

(v) *to any object which is incompatibly shaped* or dimensioned in relation to the snaphook such that unintentional disengagement could occur by the connected object being able to depress the snaphook keeper and release itself. [§1926.502(d)(6)(v)]

(7) *On suspended scaffolds* or similar work platforms with horizontal lifelines which may become vertical lifelines, the

devices used to connect to a horizontal lifeline shall be capable of locking in both directions on the lifeline.[§1926.502(d)(7)]

(8) *Horizontal lifelines shall be designed,* installed, and used, under the supervision of a qualified person, as part of a complete personal fall arrest system, which maintains a safety factor of at least two.[§1926.502(d)(8)]

(9) *Lanyards and vertical lifelines* shall have a minimum breaking strength of 5,000 pounds (22.2 kN).[§1926.502(d)(9)]

(10) (i) *Except as provided in paragraph* (d)(10)(ii) of this section, when vertical lifelines are used, each employee shall be attached to a separate lifeline.[§1926.502(d)(10)(i)]

(ii) *During the construction of elevator shafts,* two employees may be attached to the same lifeline in the hoistway, provided both employees are working atop a false car that is equipped with guardrails; the strength of the lifeline is 10,000 pounds [5,000 pounds per employee attached] (44.4 kN); and all other criteria specified in this paragraph for lifelines have been met.[§1926.502(d)(10)(ii)]

(11) *Lifelines shall be protected* against being cut or abraded. [§1926.502(d)(11)]

(12) *Self-retracting lifelines* and lanyards which automatically limit free fall distance to 2 feet (0.61 m) or less shall be capable of sustaining a minimum tensile load of 3,000 pounds (13.3 kN) applied to the device with the lifeline or lanyard in the fully extended position.[§1926.502(d)(12)]

(13) *Self-retracting lifelines* and lanyards which do not limit free fall distance to 2 feet (0.61 m) or less, ripstitch lanyards, and tearing and deforming lanyards shall be capable of sustaining a minimum tensile load of 5,000 pounds (22.2 kN) applied to the device with the lifeline or lanyard in the fully extended position. [§1926.502(d)(13)]

(14) *Ropes and straps (webbing)* used in lanyards, lifelines, and strength components of body belts and body harnesses shall be made from synthetic fibers.[§1926.502(d)(14)]

(15) *Anchorages used for attachment* of personal fall arrest equipment shall be independent of any anchorage being used to support or suspend platforms and capable of supporting at least 5,000 pounds (22.2 kN) per employee attached, or shall be designed, installed, and used as follows:[§1926.502(d)(15)]

(i) *as part of a complete* personal fall arrest system which maintains a safety factor of at least two; and[§1926.502(d)(15)(i)]

(ii) *under the supervision of a qualified person.*[§1926.502(d)(15)(ii)]

(16) *Personal fall arrest systems, when stopping a fall, shall:* [§1926.502(d)(16)]

(i) *limit maximum arresting* force on an employee to 900 pounds (4 kN) when used with a body belt;[§1926.502(d)(16)(i)]

(ii) *limit maximum arresting* force on an employee to 1,800 pounds (8 kN) when used with a body harness; [§1926.502(d)(16)(ii)]

(iii) *be rigged such that an employee* can neither free fall more than 6 feet (1.8 m), nor contact any lower level; [§1926.502(d)(16)(iii)]

(iv) *bring an employee to a complete* stop and limit maximum deceleration distance an employee travels to 3.5 feet (1.07 m); and,[§1926.502(d)(16)(iv)]

(v) *have sufficient strength* to withstand twice the potential impact energy of an employee free falling a distance of 6 feet (1.8 m), or the free fall distance permitted by the system, whichever is less.[§1926.502(d)(16)(v)]

Note: If the personal fall arrest system meets the criteria and protocols contained in Appendix C to subpart M, and if the system is being used by an employee having a combined person and tool weight of less than 310 pounds (140 kg), the system will be considered to be in compliance with the provisions of paragraph (d)(16) of this section. If the system is used by an employee having a combined tool and body weight of 310 pounds (140 kg) or more, then the employer must appropriately modify the criteria and protocols of the Appendix to provide proper protection for such heavier weights, or the system will not be deemed to be in compliance with the requirements of paragraph (d)(16) of this section.

(17) *The attachment point of the body belt* shall be located in the center of the wearer's back. The attachment point of the body harness shall be located in the center of the wearer's back near shoulder level, or above the wearer's head. [§1926.502(d)(17)]

(18) *Body belts, harnesses, and components* shall be used only for employee protection (as part of a personal fall arrest system or positioning device system) and not to hoist materials. [§1926.502(d)(18)]

(19) *Personal fall arrest* systems and components subjected to impact loading shall be immediately removed from service and shall not be used again for employee protection until inspected and determined by a competent person to be undamaged and suitable for reuse.[§1926.502(d)(19)]

(20) *The employer shall provide for prompt* rescue of employees in the event of a fall or shall assure that employees are able to rescue themselves.[§1926.502(d)(20)]

(21) *Personal fall arrest* systems shall be inspected prior to each use for wear, damage and other deterioration, and defective components shall be removed from service.[§1926.502(d)(21)]

(22) *Body belts shall be at least* one and five-eighths (1 5/8) inches (4.1 cm) wide.[§1926.502(d)(22)]

(23) *Personal fall arrest* systems shall not be attached to guardrail systems, nor shall they be attached to hoists except as specified in other subparts of this part.[§1926.502(d)(23)]

(24) *When a personal fall arrest* system is used at hoist areas, it shall be rigged to allow the movement of the employee only as far as the edge of the walking/working surface.[§1926.502(d)(24)]

(e) **Positioning device systems.** Positioning device systems and their use shall conform to the following provisions:[§1926.502(e)]

(1) *Positioning devices shall be rigged* such that an employee cannot free fall more than 2 feet (.6 m).[§1926.502(e)(1)]

(2) *Positioning devices shall be secured* to an anchorage capable of supporting at least twice the potential impact load of an employee's fall or 3,000 pounds (13.3 kN), whichever is greater.[§1926.502(e)(2)]

(3) *Connectors shall be drop forged,* pressed or formed steel, or made of equivalent materials.[§1926.502(e)(3)]

(4) *Connectors shall have a corrosion-resistant finish,* and all surfaces and edges shall be smooth to prevent damage to interfacing parts of this system.[§1926.502(e)(4)]

(5) *Connecting assemblies shall have* a minimum tensile strength of 5,000 pounds (22.2 kN)[§1926.502(e)(5)]

(6) *Dee-rings and snaphooks shall be proof-tested* to a minimum tensile load of 3,600 pounds (16 kN) without cracking, breaking, or taking permanent deformation.[§1926.502(e)(6)]

(7) *Snaphooks shall be sized to be compatible* with the member to which they are connected to prevent unintentional disengagement of the snaphook by depression of the snaphook keeper by the connected member, or shall be a locking type snaphook designed and used to prevent disengagement of the snaphook by the contact of the snaphook keeper by the connected member. As of January 1, 1998, only locking type snaphooks shall be used.[§1926.502(e)(7)]

(8) *Unless the snaphook is a locking type* and designed for the following connections, snaphooks shall not be engaged: [§1926.502(e)(8)]

(i) *directly to webbing, rope or wire rope;*[§1926.502(e)(8)(i)]

(ii) *to each other;*[§1926.502(e)(8)(ii)]

(iii) *to a Dee-ring to which another snaphook or other connector is attached;*[§1926.502(e)(8)(iii)]

(iv) *to a horizontal lifeline; or*[§1926.502(e)(8)(iv)]

(v) *to any object which is incompatibly shaped* or dimensioned in relation to the snaphook such that unintentional disengagement could occur by the connected object being able to depress the snaphook keeper and release itself. [§1926.502(e)(8)(v)]

(9) *Positioning device systems* shall be inspected prior to each use for wear, damage, and other deterioration, and defective components shall be removed from service.[§1926.502(e)(9)]

(10) *Body belts,* harnesses, and components shall be used only for employee protection (as part of a personal fall arrest system or positioning device system) and not to hoist materials. [§1926.502(e)(10)]

(f) **Warning line systems.** Warning line systems [See Sec. 1926.501(b)(10)] and their use shall comply with the following provisions:[§1926.502(f)]

(1) *The warning line shall be erected* around all sides of the roof work area.[§1926.502(f)(1)]

(i) *When mechanical equipment is not being used,* the warning line shall be erected not less than 6 feet (1.8 m) from the roof edge.[§1926.502(f)(1)(i)]

(ii) *When mechanical equipment is being used,* the warning line shall be erected not less than 6 feet (1.8 m) from the roof edge which is parallel to the direction of mechanical equipment operation, and not less than 10 feet (3.1 m) from the roof edge which is perpendicular to the direction of mechanical equipment operation.[§1926.502(f)(1)(ii)]

(iii) *Points of access,* materials handling areas, storage areas, and hoisting areas shall be connected to the work area by an access path formed by two warning lines.[§1926.502(f)(1)(iii)]

(iv) *When the path to a point of access* is not in use, a rope, wire, chain, or other barricade, equivalent in strength and height to the warning line, shall be placed across the path at the point where the path intersects the warning line erected around the work area, or the path shall be offset such that a person cannot walk directly into the work area. [§1926.502(f)(1)(iv)]

(2) *Warning lines shall consist of ropes,* wires, or chains, and supporting stanchions erected as follows:[§1926.502(f)(2)]

(i) *The rope,* wire, or chain shall be flagged at not more than 6-foot (1.8 m) intervals with high-visibility material; [§1926.502(f)(2)(i)]

(ii) *The rope,* wire, or chain shall be rigged and supported in such a way that its lowest point (including sag) is no less than 34 inches (.9 m) from the walking/working surface and its highest point is no more than 39 inches (1.0 m) from the walking/working surface;[§1926.502(f)(2)(ii)]

(iii) *After being erected,* with the rope, wire, or chain attached, stanchions shall be capable of resisting, without tipping over, a force of at least 16 pounds (71 N) applied horizontally against the stanchion, 30 inches (.8 m) above the walking/working surface, perpendicular to the warning line, and in the direction of the floor, roof, or platform edge;[§1926.502(f)(2)(iii)]

(iv) *The rope, wire, or chain* shall have a minimum tensile strength of 500 pounds (2.22 kN), and after being attached to the stanchions, shall be capable of supporting, without breaking, the loads applied to the stanchions as prescribed in paragraph (f)(2)(iii) of this section; and[§1926.502(f)(2)(iv)]

(v) *The line shall be attached at each stanchion* in such a way that pulling on one section of the line between stanchions will not result in slack being taken up in adjacent sections before the stanchion tips over.[§1926.502(f)(2)(v)]

(3) *No employee shall be allowed in the area* between a roof edge and a warning line unless the employee is performing roofing work in that area.[§1926.502(f)(3)]

(4) *Mechanical equipment on roofs* shall be used or stored only in areas where employees are protected by a warning line system, guardrail system, or personal fall arrest system. [§1926.502(f)(4)]

(g) **Controlled access zones.** Controlled access zones [See Sec. 1926.501(b)(9) and Sec. 1926.502(k)] and their use shall conform to the following provisions.[§1926.502(g)]

(1) *When used to control access* to areas where leading edge and other operations are taking place the controlled access zone shall be defined by a control line or by any other means that restricts access.[§1926.502(g)(1)]

(i) *When control lines are used,* they shall be erected not less than 6 feet (1.8 m) nor more than 25 feet (7.7 m) from the unprotected or leading edge, except when erecting precast concrete members.[§1926.502(g)(1)(i)]

(ii) *When erecting precast concrete members,* the control line shall be erected not less than 6 feet (1.8 m) nor more than 60 feet (18 m) or half the length of the member being erected, whichever is less, from the leading edge.[§1926.502(g)(1)(ii)]

(iii) *The control line shall extend* along the entire length of the unprotected or leading edge and shall be approximately parallel to the unprotected or leading edge.[§1926.502(g)(1)(iii)]

(iv) *The control line shall be connected* on each side to a guardrail system or wall.[§1926.502(g)(1)(iv)]

(2) *When used to control access* to areas where overhand bricklaying and related work are taking place:[§1926.502(g)(2)]

(i) *The controlled access zone* shall be defined by a control line erected not less than 10 feet (3.1 m) nor more than 15 feet (4.5 m) from the working edge.[§1926.502(g)(2)(i)]

(ii) *The control line shall extend* for a distance sufficient for the controlled access zone to enclose all employees performing overhand bricklaying and related work at the working edge and shall be approximately parallel to the working edge. [§1926.502(g)(2)(ii)]

(iii) *Additional control lines* shall be erected at each end to enclose the controlled access zone.[§1926.502(g)(2)(iii)]

(iv) *Only employees engaged* in overhand bricklaying or related work shall be permitted in the controlled access zone. [§1926.502(g)(2)(iv)]

(3) *Control lines shall consist of ropes,* wires, tapes, or equivalent materials, and supporting stanchions as follows:[§1926.502(g)(3)]

(i) *Each line shall be flagged* or otherwise clearly marked at not more than 6-foot (1.8 m) intervals with high-visibility material. [§1926.502(g)(3)(i)]

(ii) *Each line shall be rigged and supported* in such a way that its lowest point (including sag) is not less than 39 inches (1 m) from the walking/working surface and its highest point is not more than 45 inches (1.3 m) [50 inches (1.3 m) when overhand bricklaying operations are being performed] from the walking/working surface.[§1926.502(g)(3)(ii)]

(iii) *Each line shall have a minimum* breaking strength of 200 pounds (.88 kN).[§1926.502(g)(3)(iii)]

(4) *On floors and roofs where guardrail systems* are not in place prior to the beginning of overhand bricklaying operations, controlled access zones shall be enlarged, as necessary, to enclose all points of access, material handling areas, and storage areas.[§1926.502(g)(4)]

(5) *On floors and roofs where guardrail systems* are in place, but need to be removed to allow overhand bricklaying work or leading edge work to take place, only that portion of the guardrail necessary to accomplish that day's work shall be removed. [§1926.502(g)(5)]

(h) **Safety monitoring systems.** Safety monitoring systems [See §§1926.501(b)(10) and 1926.502(k)] and their use shall comply with the following provisions:[§1926.502(h)]

(1) *The employer shall designate a competent* person to monitor the safety of other employees and the employer shall ensure that the safety monitor complies with the following requirements:[§1926.502(h)(1)]

(i) *The safety monitor shall be competent* to recognize fall hazards;[§1926.502(h)(1)(i)]

(ii) *The safety monitor shall warn* the employee when it appears that the employee is unaware of a fall hazard or is acting in an unsafe manner;[§1926.502(h)(1)(ii)]

(iii) *The safety monitor shall be on the same* walking/working surface and within visual sighting distance of the employee being monitored;[§1926.502(h)(1)(iii)]

(iv) *The safety monitor shall be close* enough to communicate orally with the employee; and[§1926.502(h)(1)(iv)]

(v) *The safety monitor shall not have* other responsibilities which could take the monitor's attention from the monitoring function.[§1926.502(h)(1)(v)]

(2) *Mechanical equipment shall not be used* or stored in areas where safety monitoring systems are being used to monitor employees engaged in roofing operations on low-slope roofs. [§1926.502(h)(2)]

(3) *No employee,* other than an employee engaged in roofing work [on low-sloped roofs] or an employee covered by a fall protection plan, shall be allowed in an area where an employee is being protected by a safety monitoring system. [§1926.502(h)(3)]

(4) *Each employee working in a controlled* access zone shall be directed to comply promptly with fall hazard warnings from safety monitors.[§1926.502(h)(4)]

(i) **Covers.** Covers for holes in floors, roofs, and other walking/ working surfaces shall meet the following requirements:[§1926.502(i)]

(1) *Covers located in roadways* and vehicular aisles shall be capable of supporting, without failure, at least twice the maximum axle load of the largest vehicle expected to cross over the cover.[§1926.502(i)(1)]

(2) *All other covers shall be capable* of supporting, without failure, at least twice the weight of employees, equipment, and materials that may be imposed on the cover at any one time. [§1926.502(i)(2)]

(3) *All covers shall be secured when installed* so as to prevent accidental displacement by the wind, equipment, or employees.[§1926.502(i)(3)]

(4) *All covers shall be color coded* or they shall be marked with the word "HOLE" or "COVER" to provide warning of the hazard.[§1926.502(i)(4)]

Note: This provision does not apply to cast iron manhole covers or steel grates used on streets or roadways.

(j) **Protection from falling objects.** Falling object protection shall comply with the following provisions:[§1926.502(j)]

(1) *Toeboards,* when used as falling object protection, shall be erected along the edge of the overhead walking/working surface for a distance sufficient to protect employees below. [§1926.502(j)(1)]

(2) *Toeboards shall be capable of withstanding,* without failure, a force of at least 50 pounds (222 N) applied in any downward or outward direction at any point along the toeboard. [§1926.502(j)(2)]

(3) *Toeboards shall be a minimum* of 3 1/2 inches (9 cm) in vertical height from their top edge to the level of the walking/working surface. They shall have not more than 1/4 inch (0.6 cm) clearance above the walking/working surface. They shall be solid or have openings not over 1 inch (2.5 cm) in greatest dimension.[§1926.502(j)(3)]

(4) *Where tools,* equipment, or materials are piled higher than the top edge of a toeboard, paneling or screening shall be erected from the walking/working surface or toeboard to the top of a guardrail system's top rail or midrail, for a distance sufficient to protect employees below.[§1926.502(j)(4)]

(5) *Guardrail systems,* when used as falling object protection, shall have all openings small enough to prevent passage of potential falling objects.[§1926.502(j)(5)]

(6) *During the performance of overhand bricklaying and related work:*[§1926.502(j)(6)]

(i) *No materials or equipment except masonry* and mortar shall be stored within 4 feet (1.2 m) of the working edge. [§1926.502(j)(6)(i)]

(ii) *Excess mortar,* broken or scattered masonry units, and all other materials and debris shall be kept clear from the work area by removal at regular intervals.[§1926.502(j)(6)(ii)]

(7) *During the performance of roofing work:*[§1926.502(j)(7)]

(i) *Materials and equipment shall not be stored* within 6 feet (1.8 m) of a roof edge unless guardrails are erected at the edge. [§1926.502(j)(7)(i)]

(ii) *Materials which are piled,* grouped, or stacked near a roof edge shall be stable and self-supporting.[§1926.502(j)(7)(ii)]

(8) *Canopies,* when used as falling object protection, shall be strong enough to prevent collapse and to prevent penetration by any objects which may fall onto the canopy.[§1926.502(j)(8)]

(k) **Fall protection plan.** This option is available only to employees engaged in leading edge work, precast concrete erection work, or residential construction work [See Sec. 1926.501(b)(2), (b)(12), and (b)(13)] who can demonstrate that it is infeasible or it creates a greater hazard to use conventional fall protection equipment. The fall protection plan must conform to the following provisions.[§1926.502(k)]

(1) *The fall protection plan* shall be prepared by a qualified person and developed specifically for the site where the leading edge work, precast concrete work, or residential construction work is being performed and the plan must be maintained up to date.[§1926.502(k)(1)]

(2) *Any changes to the fall protection plan* shall be approved by a qualified person.[§1926.502(k)(2)]

(3) *A copy of the fall protection plan* with all approved changes shall be maintained at the job site.[§1926.502(k)(3)]

(4) *The implementation of the fall protection plan* shall be under the supervision of a competent person.[§1926.502(k)(4)]

(5) *The fall protection plan* shall document the reasons why the use of conventional fall protection systems (guardrail systems, personal fall arrest systems, or safety nets systems) are infeasible or why their use would create a greater hazard. [§1926.502(k)(5)]

(6) *The fall protection plan* shall include a written discussion of other measures that will be taken to reduce or eliminate the fall hazard for workers who cannot be provided with protection from the conventional fall protection systems. For example, the employer shall discuss the extent to which scaffolds, ladders, or vehicle mounted work platforms can be used to provide a safer working surface and thereby reduce the hazard of falling.[§1926.502(k)(6)]

(7) *The fall protection plan* shall identify each location where conventional fall protection methods cannot be used. These locations shall then be classified as controlled access zones and the employer must comply with the criteria in paragraph (g) of this section.[§1926.502(k)(7)]

(8) *Where no other alternative measure has been implemented,* the employer shall implement a safety monitoring system in conformance with Sec. 1926.502(h).[§1926.502(k)(8)]

(9) *The fall protection plan* must include a statement which provides the name or other method of identification for each employee who is designated to work in controlled access zones. No other employees may enter controlled access zones.[§1926.502(k)(9)]

(10) *In the event an employee falls,* or some other related, serious incident occurs, (e.g., a near miss) the employer shall investigate the circumstances of the fall or other incident to determine if the fall protection plan needs to be changed (e.g. new practices, procedures, or training) and shall implement those changes to prevent similar types of falls or incidents. [§1926.502(k)(10)]

§1926.503
Training requirements.

The following training provisions supplement and clarify the requirements of Sec. 1926.21 regarding the hazards addressed in Subpart M of this part.

(a) **Training Program.**[§1926.503(a)]

(1) *The employer shall provide a training program* for each employee who might be exposed to fall hazards. The program shall enable each employee to recognize the hazards of falling and shall train each employee in the procedures to be followed in order to minimize these hazards.[§1926.503(a)(1)]

(2) *The employer shall assure that each employee* has been trained, as necessary, by a competent person qualified in the following areas:[§1926.503(a)(2)]

(i) *The nature of fall hazards in the work area;*[§1926.503(a)(2)(i)]

(ii) *The correct procedures* for erecting, maintaining, disassembling, and inspecting the fall protection systems to be used; [§1926.503(a)(2)(ii)]

(iii) *The use and operation of guardrail systems,* personal fall arrest systems, safety net systems, warning line systems, safety monitoring systems, controlled access zones, and other protection to be used;[§1926.503(a)(2)(iii)]

(iv) *The role of each employee* in the safety monitoring system when this system is used;[§1926.503(a)(2)(iv)]

(v) *The limitations on the use* of mechanical equipment during the performance of roofing work on low-sloped roofs; [§1926.503(a)(2)(v)]

(vi) *The correct procedures* for the handling and storage of equipment and materials and the erection of overhead protection; and[§1926.503(a)(2)(vi)]

(vii) *The role of employees in fall protection plans;* [§1926.503(a)(2)(vii)]

(viii) *The standards contained in this subpart.*[§1926.503(a)(2)(viii)]

(b) **Certification of training.**[§1926.503(b)]

(1) *The employer shall verify compliance* with paragraph (a) of this section by preparing a written certification record. The written certification record shall contain the name or other identity of the employee trained, the date(s) of the training, and the signature of the person who conducted the training or the signature of the employer. If the employer relies on training conducted by another employer or completed prior to the effective date of this section, the certification record shall indicate the date the employer determined the prior training was adequate rather than the date of actual training.[§1926.503(b)(1)]

(2) *The latest training certification shall be maintained.* [§1926.503(b)(2)]

(c) **Retraining.** When the employer has reason to believe that any affected employee who has already been trained does not have the understanding and skill required by paragraph (a) of this section, the employer shall retrain each such employee. Circumstances where retraining is required include, but are not limited to, situations where:[§1926.503(c)]

(1) *Changes in the workplace render previous training obsolete; or*[§1926.503(c)(1)]

(2) *Changes in the types of fall protection systems* or equipment to be used render previous training obsolete; or[§1926.503(c)(2)]

(3) *Inadequacies in an affected employee's knowledge* or use of fall protection systems or equipment indicate that the employee has not retained the requisite understanding or skill. [§1926.503(c)(3)]

Note: The following appendices to subpart M of this part serve as non-mandatory guidelines to assist employers in complying with the appropriate requirements of subpart M of this part.

1926 Subpart M Appendix A

Determining Roof Widths — Non-mandatory Guidelines for Complying With §1926.501(b)(10)

(1) **This Appendix serves as a guideline** to assist employers complying with the requirements of §1926.501(b)(10). Section 1926.501(b)(10)[1] allows the use of a safety monitoring system alone as a means of providing fall protection during the performance of roofing operations on low-sloped roofs 50 feet (15.25 m) or less in width. Each example in the appendix shows a roof plan or plans and indicates where each roof or roof area is to be measured to determine its width. Section views or elevation views are shown where appropriate. Some examples show "correct" and "incorrect" subdivisions of irregularly shaped roofs divided into smaller, regularly shaped areas. In all examples, the dimension selected to be the width of an area is the lesser of the two primary dimensions of the area, as viewed from above. Example A shows that on a simple rectangular roof, width is the lesser of the two primary overall dimensions. This is also the case with roofs which are sloped toward or away from the roof center, as shown in Example B.

(2) **Many roofs are not simple rectangles.** Such roofs may be broken down into subareas as shown in Example C. The process of dividing a roof area can produce many different configurations. Example C gives the general rule of using dividing lines of minimum length to minimize the size and number of the areas which are potentially less than 50 feet (15.25 m) wide. The intent is to minimize the number of roof areas where safety monitoring systems alone are sufficient protection.

(3) **Roofs which are comprised** of several separate, non-contiguous roof areas, as in Example D, may be considered as a series of individual roofs. Some roofs have penthouses, additional floors, courtyard openings, or similar architectural features; Example E shows how the rule for dividing roofs into subareas is applied to such configurations. Irregular, non-rectangular roofs must be considered on an individual basis, as shown in Example F.

EXAMPLE A. RECTANGULAR SHAPED ROOFS

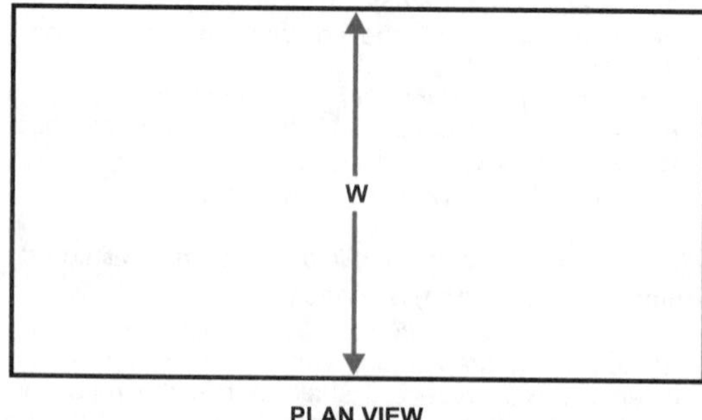

PLAN VIEW

EXAMPLE B. SLOPED RECTANGULAR SHAPED ROOFS

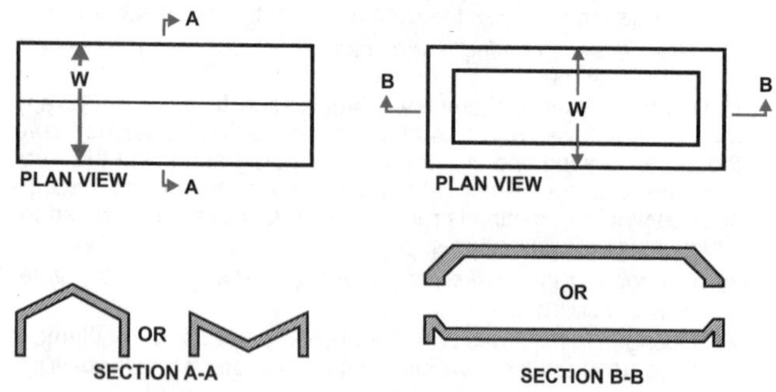

1. Editor's Note: The CFR states ,'1910.501(b)(10)', in this instance. As it appears clear from the context that this was a typographical error, the text is corrected in the regulation appearing above.

EXAMPLE C. IRREGULARLY SHAPED ROOFS WITH RECTANGULAR SHAPED SECTIONS

Such roofs are to be divided into sub-areas by using dividing lines of minimum length to minimize the size and number of the areas which are potentially less than or equal to 50 feet (15.25 meters) in width, in order to limit the size of roof areas where the safety monitoring system alone can be used [1926.502(b)(10)]. Dotted lines are used in the examples to show the location of dividing lines. Ⓦ denotes incorrect measurements of width.

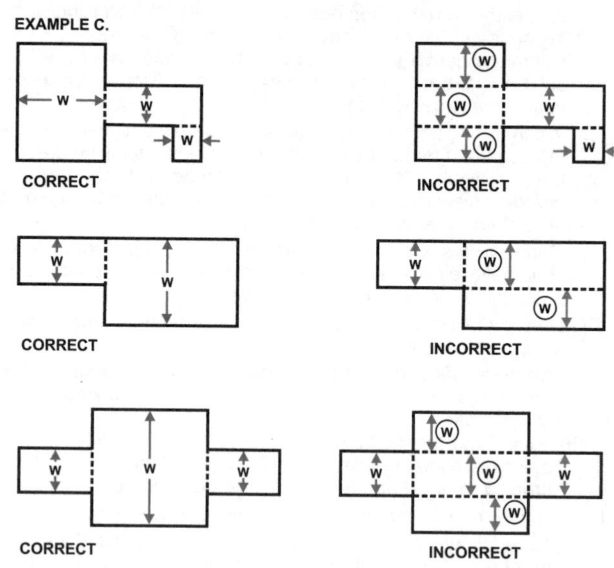

EXAMPLE D. SEPARATE, NON-CONTIGUOUS ROOF AREAS

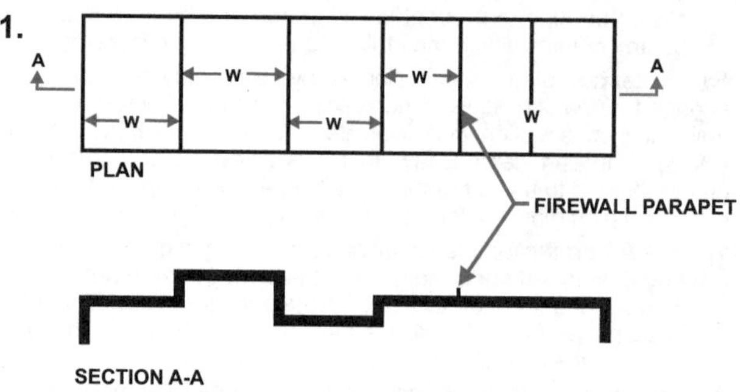

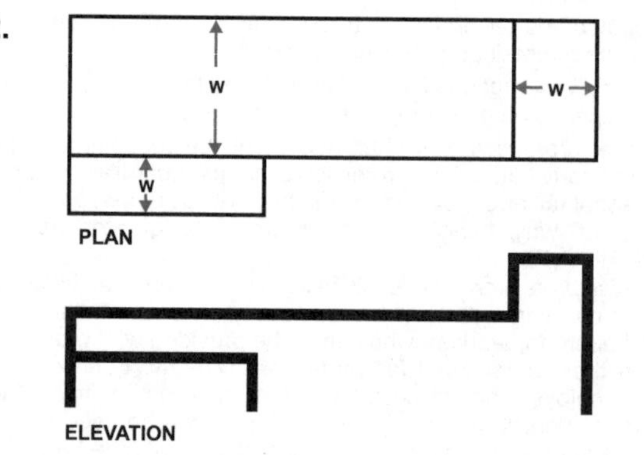

EXAMPLE E.
ROOFS WITH PENTHOUSES, OPEN COURTYARDS, ADDITIONAL FLOORS, ETC.

Such roofs are to be divided into sub-areas by using dividing lines of minimum length to minimize the size and number of the areas which are potentially less than or equal to 50 feet (15.25 meters) in width, in order to limit the size of roof areas where the safety monitoring system alone can be used [1926.502(b)(10)]. Dotted lines are used in the examples to show the location of dividing lines. (W) denotes incorrect measurements of width.

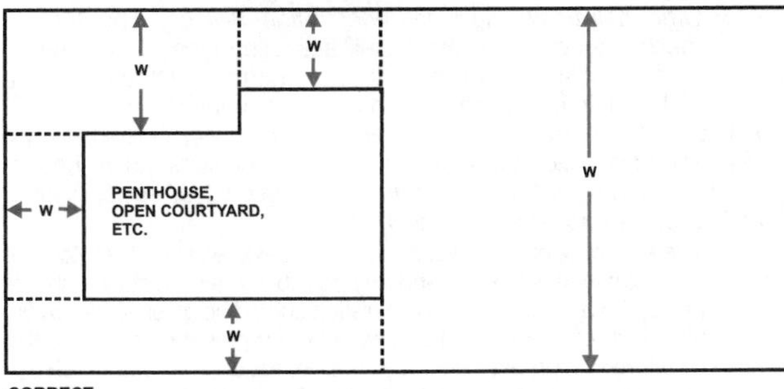

CORRECT

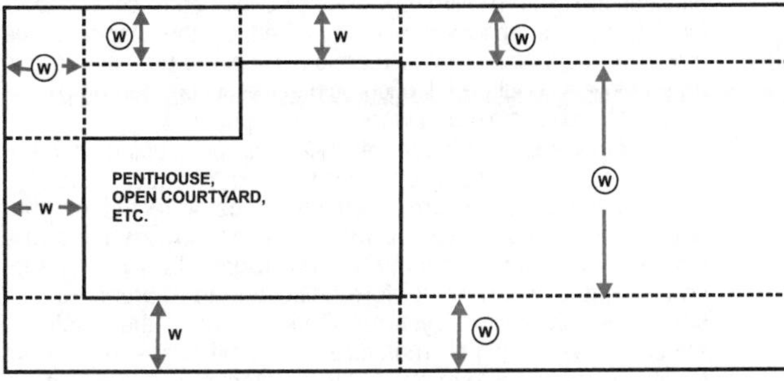

INCORRECT

EXAMPLE F.
IRREGULAR, NON-RECTANGULAR SHAPED ROOFS

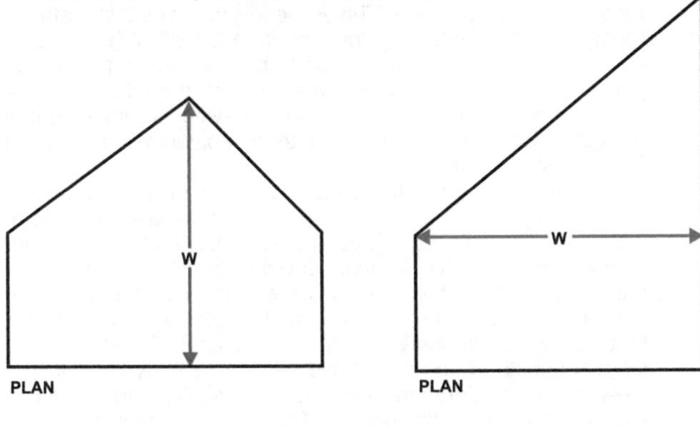

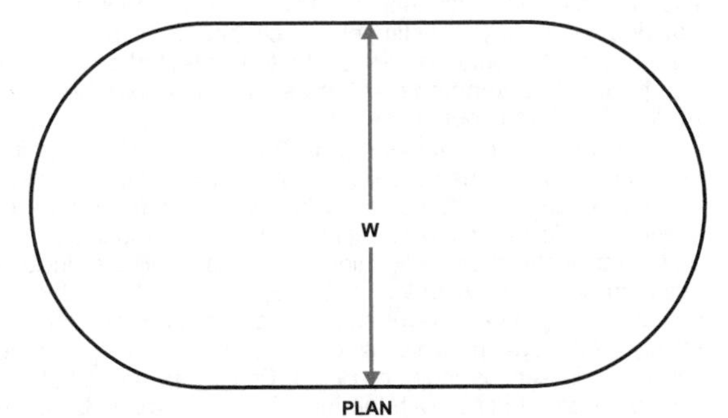

1926 Subpart M Appendix B

Guardrail Systems — Non-Mandatory Guidelines for Complying with §1926.502(b)

The standard requires guardrail systems and components to be designed and built to meet the requirements of §1926.502(b)(3), (4), and (5). This Appendix serves as a non-mandatory guideline to assist employers in complying with these requirements. An employer may use these guidelines as a starting point for designing guardrail systems. However, the guidelines do not provide all the information necessary to build a complete system, and the employer is still responsible for designing and assembling these components in such a way that the completed system will meet the requirements of §1926.502(b)(3), (4), and (5). Components for which no specific guidelines are given in this Appendix (e.g., joints, base connections, components made with other materials, and components with other dimensions) must also be designed and constructed in such a way that the completed system meets the requirements of §1926.502.

(1) **For wood railings:** Wood components shall be minimum 1500 lb-ft/in^2 fiber (stress grade) construction grade lumber; the posts shall be at least 2-inch by 4-inch (5 cm x 10 cm) lumber spaced not more than 8 feet (2.4 m) apart on centers; the top rail shall be at least 2-inch by 4-inch (5 cm x 10 cm) lumber, the intermediate rail shall be at least 1-inch by 6-inch (2.5 cm x 15 cm) lumber. All lumber dimensions are nominal sizes as provided by the American Softwood Lumber Standards, dated January 1970.

(2) **For pipe railings:** posts, top rails, and intermediate railings shall be at least one and one-half inches nominal diameter (schedule 40 pipe) with posts spaced not more than 8 feet (2.4 m) apart on centers.

(3) **For structural steel railings:** posts, top rails, and intermediate rails shall be at least 2-inch by 2-inch (5 cm x 10 cm) by 3/8-inch (1.1 cm) angles, with posts spaced not more than 8 feet (2.4 m) apart on centers.

1926 Subpart M Appendix C

Personal Fall Arrest Systems — Non-Mandatory Guidelines for Complying With §1926.502(d)

I. Test methods for personal fall arrest systems and positioning device systems.

(a) **General.** This appendix serves as a non-mandatory guideline to assist employers comply with the requirements in §1926.502(d). Paragraphs (b), (c), (d) and (e) of this Appendix describe test procedures which may be used to determine compliance with the requirements in §1926.502 (d)(16). As noted in Appendix D of this subpart, the test methods listed here in Appendix C can also be used to assist employers comply with the requirements in §1926.502(e) (3) and (4) for positioning device systems.

(b) **General conditions for all tests** in the Appendix to §1926.502(d).

(1) *Lifelines, lanyards and deceleration devices* should be attached to an anchorage and connected to the body-belt or body harness in the same manner as they would be when used to protect employees.

(2) *The anchorage should be rigid,* and should not have a deflection greater than 0.04 inches (1 mm) when a force of 2,250 pounds (10 kN) is applied.

(3) *The frequency response* of the load measuring instrumentation should be 500 Hz.

(4) *The test weight used* in the strength and force tests should be a rigid, metal, cylindrical or torso-shaped object with a girth of 38 inches plus or minus 4 inches (96 cm plus or minus 10 cm).

(5) *The lanyard or lifeline used* to create the free fall distance should be supplied with the system, or in its absence, the least elastic lanyard or lifeline available to be used with the system.

(6) *The test weight for each test* should be hoisted to the required level and should be quickly released without having any appreciable motion imparted to it.

(7) *The system's performance* should be evaluated taking into account the range of environmental conditions for which it is designed to be used.

(8) *Following the test,* the system need not be capable of further operation.

(c) **Strength test.**

(1) *During the testing of all systems,* a test weight of 300 pounds plus or minus 5 pounds (135 kg plus or minus 2.5 kg) should be used. (See paragraph (b)(4) of this section.)

(2) *The test consists of dropping* the test weight once. A new unused system should be used for each test.

(3) *For lanyard systems,* the lanyard length should be 6 feet plus or minus 2 inches (1.83 m plus or minus 5 cm) as measured from the fixed anchorage to the attachment on the body belt or body harness.

(4) *For rope-grab-type deceleration systems,* the length of the lifeline above the centerline of the grabbing mechanism to the lifeline's anchorage point should not exceed 2 feet (0.61 m).

(5) *For lanyard systems,* for systems with deceleration devices which do not automatically limit free fall distance to 2 feet (0.61 m) or less, and for systems with deceleration devices which have a connection distance in excess of 1 foot (0.3 m) (measured between the centerline of the lifeline and the attachment point to the body belt or harness), the test weight should be rigged to free fall a distance of 7.5 feet (2.3 m) from a point that is 1.5 feet (.46 m) above the anchorage point, to its hanging location (6 feet below the anchorage). The test weight should fall without interference, obstruction, or hitting the floor or ground during the test. In some cases a non-elastic wire lanyard of sufficient length may need to be added to the system (for test purposes) to create the necessary free fall distance.

(6) *For deceleration device systems* with integral lifelines or lanyards which automatically limit free fall distance to 2 feet (0.61 m) or less, the test weight should be rigged to free fall a distance of 4 feet (1.22 m).

(7) *Any weight which detaches* from the belt or harness has failed the strength test.

(d) **Force test.**

(1) *General.* The test consists of dropping the respective test weight once as specified in paragraph (d)(2)(i) or (d)(3)(i) of this section. A new, unused system should be used for each test.

(2) *For lanyard systems.*

(i) *A test weight of 220 pounds* plus or minus 3 pounds (100 kg plus or minus 1.6 kg) should be used. (See paragraph (b)(4) of this appendix).

(ii) *Lanyard length should be 6 feet* plus or minus two inches (1.83 m plus or minus 5 cm) as measured from the fixed anchorage to the attachment on the body belt or body harness.

(iii) *The test weight should fall free* from the anchorage level to its hanging location (a total of 6 feet (1.83 m) free fall distance) without interference, obstruction, or hitting the floor or ground during the test.

(3) *For all other systems.*

(i) *A test weight of 220 pounds* plus or minus 3 pounds (100 kg plus or minus 1.6 kg) should be used. (See paragraph (b)(4) of this appendix)

(ii) *The free fall distance* to be used in the test should be the maximum fall distance physically permitted by the system during normal use conditions, up to a maximum free fall distance for the test weight of 6 feet (1.83 m), except as follows:

[A] *For deceleration systems* which have a connection link or lanyard, the test weight should free fall a distance equal to the connection distance (measured between the centerline of the lifeline and the attachment point to the body belt or harness).

[B] *For deceleration device systems* with integral lifelines or lanyards which automatically limit free fall distance to 2 feet (0.61 m) or less, the test weight should free fall a distance equal to that permitted by the system in normal use. (For example, to test a system with a self-retracting lifeline or lanyard, the test weight should be supported and the system allowed to retract the lifeline or lanyard as it would in normal use. The test weight would then be released and the force and deceleration distance measured).

(4) *A system fails the force test* if the recorded maximum arresting force exceeds 1,260 pounds (5.6 kN) when using a body belt, and/or exceeds 2,520 pounds (11.2 kN) when using a body harness.

(5) *The maximum elongation and deceleration distance* should be recorded during the force test.

(e) *Deceleration device tests*

(1) *General.* The device should be evaluated or tested under the environmental conditions, (such as rain, ice, grease, dirt, type of lifeline, etc.), for which the device is designed.

(2) *Rope-grab-type deceleration devices.*

(i) *Devices should be moved on a lifeline* 1,000 times over the same length of line a distance of not less than 1 foot (30.5 cm), and the mechanism should lock each time.

(ii) *Unless the device is permanently marked* to indicate the type(s) of lifeline which must be used, several types (different diameters and different materials), of lifelines should be used to test the device.

(3) *Other self-activating-type deceleration devices.* The locking mechanisms of other self-activating-type deceleration devices designed for more than one arrest should lock each of 1,000 times as they would in normal service.

II. **Additional non-mandatory guidelines** for personal fall arrest systems. The following information constitutes additional guidelines for use in complying with requirements for a personal fall arrest system.

(a) **Selection and use considerations.**

(1) *The kind of personal fall arrest system selected* should match the particular work situation, and any possible free fall distance should be kept to a minimum. Consideration should be given to the particular work environment. For example, the presence of acids, dirt, moisture, oil, grease, etc., and their effect on the system, should be evaluated. Hot or cold environments may also have an adverse effect on the system. Wire rope should not be used where an electrical hazard is anticipated. As required by the standard, the employer must plan to have means available to promptly rescue an employee should a fall occur, since the suspended employee may not be able to reach a work level independently.

(2) *Where lanyards, connectors, and lifelines* are subject to damage by work operations such as welding, chemical cleaning, and sandblasting, the component should be protected, or other securing systems should be used. The employer should fully evaluate the work conditions and environment (including seasonal weather changes) before selecting the appropriate personal fall protection system. Once in use, the system's effectiveness should be monitored. In some cases, a program for cleaning and maintenance of the system may be necessary.

(b) **Testing considerations.** Before purchasing or putting into use a personal fall arrest system, an employer should obtain from the supplier information about the system based on its performance during testing so that the employer can know if the system meets this standard. Testing should be done using recognized test methods. This Appendix contains test methods recognized for evaluating the performance of fall arrest systems. Not all systems may need to be individually tested; the performance of some systems may be based on data and calculations derived from testing of similar systems, provided that enough information is available to demonstrate similarity of function and design.

(c) **Component compatibility considerations.** Ideally, a personal fall arrest system is designed, tested, and supplied as a complete system. However, it is common practice for lanyards, connectors, lifelines, deceleration devices, body belts and body harnesses to be interchanged since some components wear out before others. The employer and employee should realize that not all components are interchangeable. For instance, a lanyard should not be connected between a body belt (or harness) and a deceleration device of the self-retracting type since this can result in additional free fall for which the system was not designed. Any substitution or change to a personal fall arrest system should be fully evaluated or tested by a competent person to determine that it meets the standard, before the modified system is put in use.

(d) **Employee training considerations.** Thorough employee training in the selection and use of personal fall arrest systems is imperative. Employees must be trained in the safe use of the system. This should include the following: application limits; proper anchoring and tie-off techniques; estimation of free fall distance, including determination of deceleration distance, and total fall distance to prevent striking a lower level; methods of use; and inspection and storage of the system. Careless or improper use of the equipment can result in serious injury or death. Employers and employees should become familiar with the material in this Appendix, as well as manufacturer's recommendations, before a system is used. Of uppermost importance is the reduction in strength caused by certain tie-offs (such as using knots, tying around sharp edges, etc.) and maximum permitted free fall distance. Also, to be stressed are the importance of inspections prior to use, the limitations of the equipment, and unique conditions at the worksite which may be important in determining the type of system to use.

(e) **Instruction considerations.** Employers should obtain comprehensive instructions from the supplier as to the system's proper use and application, including, where applicable:

(1) *The force measured during the sample force test;*

(2) *The maximum elongation* measured for lanyards during the force test;

(3) *The deceleration distance measured* for deceleration devices during the force test;

(4) *Caution statements on critical use limitations;*

(5) *Application limits;*

(6) *Proper hook-up, anchoring and tie-off techniques,* including the proper dee-ring or other attachment point to use on the body belt and harness for fall arrest;

(7) *Proper climbing techniques;*

(8) *Methods of inspection, use, cleaning, and storage; and*

(9) *Specific lifelines which may be used.* This information should be provided to employees during training.

(f) **Rescue considerations.** As required by §1926.502(d)(20), when personal fall arrest systems are used, the employer must assure that employees can be promptly rescued or can rescue themselves should a fall occur. The availability of rescue personnel, ladders or other rescue equipment should be evaluated. In some situations, equipment which allows employees to rescue themselves after the fall has been arrested may be desirable, such as devices which have descent capability.

(g) **Inspection considerations.** As required by §1926.502(d)(21), personal fall arrest systems must be regularly inspected. Any component with any significant defect, such as cuts, tears, abrasions, mold, or undue stretching; alterations or additions which might affect its efficiency; damage due to deterioration; contact with fire, acids, or other corrosives; distorted hooks or faulty hook springs; tongues unfitted to the shoulder of buckles; loose or damaged mountings; non-functioning parts; or wearing or internal deterioration in the ropes must be withdrawn from service immediately, and should be tagged or marked as unusable, or destroyed.

(h) **Tie-off considerations.**

(1) *One of the most important aspects* of personal fall protection systems is fully planning the system before it is put into use. Probably the most overlooked component is planning for suitable anchorage points. Such planning should ideally be done before the structure or building is constructed so that anchorage points can be incorporated during construction for use later for window cleaning or other building maintenance. If properly planned, these anchorage points may be used during construction, as well as afterwards.

(i) *Properly planned anchorages* should be used if they are available. In some cases, anchorages must be installed immediately prior to use. In such cases, a registered professional engineer with experience in designing fall protection systems, or another qualified person with appropriate education and experience should design an anchor point to be installed.

(ii) *In other cases,* the Agency recognizes that there will be a need to devise an anchor point from existing structures. Examples of what might be appropriate anchor points are steel members or I-beams if an acceptable strap is available for the connection (do not use a lanyard with a snaphook clipped onto itself); large eye-bolts made of an appropriate grade steel; guardrails or railings if they have been designed for use as an anchor point; or masonry or wood members only if the attachment point is substantial and precautions have been taken to assure that bolts or other connectors will not pull through. A qualified person should be used to evaluate the suitable of these "make shift" anchorages with a focus on proper strength.

(2) *Employers and employees should at all times* be aware that the strength of a personal fall arrest system is based on its being attached to an anchoring system which does not reduce the strength of the system (such as a properly dimensioned eye-bolt/snap-hook anchorage). Therefore, if a means of attachment is used that will reduce the strength of the system, that component should be replaced by a stronger one, but one that will also maintain the appropriate maximum arrest force characteristics.

(3) *Tie-off using a knot in a rope lanyard* or lifeline (at any location) can reduce the lifeline or lanyard strength by 50 percent or more. Therefore, a stronger lanyard or lifeline should be used to compensate for the weakening effect of the knot, or the lanyard length should be reduced (or the tie-off location raised) to minimize free fall distance, or the lanyard or lifeline should be replaced by one which has an appropriately incorporated connector to eliminate the need for a knot.

(4) *Tie-off of a rope lanyard or lifeline* around an "H" or "I" beam or similar support can reduce its strength as much as 70 percent due to the cutting action of the beam edges. Therefore, use should be made of a webbing lanyard or wire core lifeline around the beam; or the lanyard or lifeline should be protected from the edge; or free fall distance should be greatly minimized.

(5) *Tie-off where the line passes over or around* rough or sharp surfaces reduces strength drastically. Such a tie-off should be avoided or an alternative tie-off rigging should be used. Such alternatives may include use of a snap-hook/dee ring connection, wire rope tie-off, an effective padding of the surfaces, or an abrasion-resistance strap around or over the problem surface.

(6) *Horizontal lifelines may,* depending on their geometry and angle of sag, be subjected to greater loads than the impact load imposed by an attached component. When the angle of horizontal lifeline sag is less than 30 degrees, the impact force imparted to the lifeline by an attached lanyard is greatly amplified. For example, with a sag angle of 15 degrees, the force amplification is about 2:1 and at 5 degrees sag, it is about 6:1. Depending on the angle of sag, and the line's elasticity, the strength of the horizontal lifeline and the anchorages to which it is attached should be increased a number of times over that of the lanyard. Extreme care should be taken in considering a horizontal lifeline for multiple tie-offs. The reason for this is that in multiple tie-offs to a horizontal lifeline, if one employee falls, the movement of the falling employee and the horizontal lifeline during arrest of the fall may cause other employees to fall also. Horizontal lifeline and anchorage strength should be increased for each additional employee to be tied off. For these and other reasons, the design of systems using horizontal lifelines must only be done by qualified persons. Testing of installed lifelines and anchors prior to use is recommended.

(7) *The strength of an eye-bolt is rated* along the axis of the bolt and its strength is greatly reduced if the force is applied at an angle to this axis (in the direction of shear). Also, care should be exercised in selecting the proper diameter of the eye to avoid accidental disengagement of snap-hooks not designed to be compatible for the connection.

(8) *Due to the significant reduction* in the strength of the lifeline/lanyard (in some cases, as much as a 70 percent reduction), the sliding hitch knot (prusik) should not be used for lifeline/lanyard connections except in emergency situations where no other available system is practical. The "one-and-one" sliding hitch knot should never be used because it is unreliable in stopping a fall. The "two-and-two," or "three-and-three" knot (preferable) may be used in emergency situations; however, care should be taken to limit free fall distance to a minimum because of reduced lifeline/lanyard strength.

(i) **Vertical lifeline considerations.** As required by the standard, each employee must have a separate lifeline [except employees engaged in constructing elevator shafts who are permitted to have two employees on one lifeline] when the lifeline is vertical. The reason for this is that in multiple tie-offs to a single lifeline, if one employee falls, the movement of the lifeline during the arrest of the fall may pull other employees' lanyards, causing them to fall as well.

(j) **Snap-hook considerations.**

(1) *Although not required by this standard* for all connections until January 1, 1998, locking snaphooks designed for connection to suitable objects (of sufficient strength) are highly recommended in lieu of the nonlocking type. Locking snaphooks incorporate a positive locking mechanism in addition to the spring loaded keeper, which will not allow the keeper to open under moderate pressure without someone first releasing the mechanism. Such a feature, properly designed, effectively prevents roll-out from occurring.

(2) *As required by §1926.502(d)(6),* the following connections must be avoided (unless properly designed locking snaphooks are used) because they are conditions which can result in roll-out when a nonlocking snaphook is used:

(i) *Direct connection of a snaphook to a horizontal lifeline.*

(ii) *Two (or more) snaphooks connected to one dee-ring.*

(iii) *Two snaphooks connected to each other.*

(iv) *A snaphook connected back on its integral lanyard.*

(v) *A snaphook connected to a webbing loop or webbing lanyard.*

(vi) *Improper dimensions of the dee-ring, rebar,* or other connection point in relation to the snaphook dimensions which would allow the snaphook keeper to be depressed by a turning motion of the snaphook.

(k) **Free fall considerations.** The employer and employee should at all times be aware that a system's maximum arresting force is evaluated under normal use conditions established by the manufacturer, and in no case using a free fall distance in excess of 6 feet (1.8 m). A few extra feet of free fall can significantly increase the arresting force on the employee, possibly to the point of causing injury. Because of this, the free fall distance should be kept at a minimum, and, as required by the standard, in no case greater than 6 feet (1.8 m). To help assure this, the tie-off attachment point to the lifeline or anchor should be located at or above the connection point of the fall arrest equipment to belt or harness. (Since otherwise additional free fall distance is added to the length of the connecting means (i.e. lanyard)). Attaching to the working surface will often result in a free fall greater than 6 feet (1.8 m). For instance, if a 6 foot (1.8 m) lanyard is used, the total free fall distance will be the distance from the working level to the body belt (or harness) attachment point plus the 6 feet (1.8 m) of lanyard length. Another important consideration is that the arresting force which the fall system must withstand also goes up with greater distances of free fall, possibly exceeding the strength of the system.

(l) **Elongation and deceleration distance considerations.** Other factors involved in a proper tie-off are elongation and deceleration distance. During the arresting of a fall, a lanyard will experience a length of stretching or elongation, whereas activation of a deceleration device will result in a certain stopping distance. These distances should be available with the lanyard or device's instructions and must be added to the free fall distance to arrive at the total fall distance before an employee is fully stopped. The additional stopping distance may be very significant if the lanyard or deceleration device is attached near or at the end of a long lifeline, which may itself add considerable distance due to its own elongation. As required by the standard, sufficient distance to allow for all of these factors must also be maintained between the employee and obstructions below, to prevent an injury due to impact before the system fully arrests the fall. In addition, a minimum of 12 feet (3.7 m) of lifeline should be allowed below the securing point of a rope grab type deceleration device, and the end terminated to prevent the device from sliding off the lifeline. Alternatively, the lifeline should extend to the ground or the next working level below. These measures are suggested to prevent the worker from inadvertently moving past the end of the lifeline and having the rope grab become disengaged from the lifeline.

(m) **Obstruction considerations.** The location of the tie-off should also consider the hazard of obstructions in the potential fall path of the employee. Tie-offs which minimize the possibilities of exaggerated swinging should be considered. In addition, when a body belt is used, the employee's body will go through a horizontal position to a jack-knifed position during the arrest of all falls. Thus, obstructions which might interfere with this motion should be avoided or a severe injury could occur.

(n) **Other considerations.** Because of the design of some personal fall arrest systems, additional considerations may be required for proper tie-off. For example, heavy deceleration devices of the self-retracting type should be secured overhead in order to avoid the weight of the device having to be supported by the employee. Also, if self-retracting equipment is connected to a horizontal lifeline, the sag in the lifeline should be minimized to prevent the device from sliding down the lifeline to a position which creates a swing hazard during fall arrest. In all cases, manufacturer's instructions should be followed.

1926 Subpart M Appendix D

Positioning Device Systems — Non-Mandatory Guidelines for Complying With §1926.502(e)

I. **Testing Methods For Positioning Device Systems.** This appendix serves as a non-mandatory guideline to assist employers comply with the requirements for positioning device systems in §1926.502(e). Paragraphs (b), (c), (d) and (e) of Appendix C of subpart M relating to §1926.502(d) — Personal Fall Arrest Systems — set forth test procedures which may be used, along with the procedures listed below, to determine compliance with the requirements for positioning device systems in §1926.502(e) (3) and (4) of subpart M.

(a) **General.**

(1) *Single strap positioning devices* shall have one end attached to a fixed anchorage and the other end connected to a body belt or harness in the same manner as they would be used to protect employees. Double strap positioning devices, similar to window cleaner's belts, shall have one end of the strap attached to a fixed anchorage and the other end shall hang free. The body belt or harness shall be attached to the strap in the same manner as it would be used to protect employees. The two strap ends shall be adjusted to their maximum span.

(2) *The fixed anchorage* shall be rigid, and shall not have a deflection greater than .04 inches (1 mm) when a force of 2,250 pounds (10 kN) is applied.

(3) *During the testing of all systems,* a test weight of 250 pounds plus or minus 3 pounds (113 kg plus or minus 1.6 kg) shall be used. The weight shall be a rigid object with a girth of 38 inches plus or minus 4 inches (96 cm plus or minus 10 cm).

(4) *Each test shall consist of* dropping the specified weight one time without failure of the system being tested. A new system shall be used for each test.

(5) *The test weight for each test* shall be hoisted exactly 4 feet (1.2 m above its "at rest" position), and shall be dropped so as to permit a vertical free fall of 4 feet (1.2 m).

(6) *The test is failed whenever* any breakage or slippage occurs which permits the weight to fall free of the system.

(7) *Following the test,* the system need not be capable of further operation; however, all such incapacities shall be readily apparent.

II. **Inspection Considerations.** As required in §1926.502(e)(5), positioning device systems must be regularly inspected. Any component with any significant defect, such as cuts, tears, abrasions, mold, or undue stretching; alterations or additions which might affect its efficiency; damage due to deterioration; contact with fire, acids, or other corrosives; distorted hooks or faulty hook springs; tongues unfitted to the shoulder of buckles; loose or damaged mountings; non-functioning parts; or wearing or internal deterioration in the ropes must be withdrawn from service immediately, and should be tagged or marked as unusable, or destroyed.

1926 Subpart M Appendix E

Sample Fall Protection Plan — Non-Mandatory Guidelines for Complying With §1926.502(k)

Employers engaged in leading edge work, precast concrete construction work and residential construction work who can demonstrate that it is infeasible or creates a greater hazard to use conventional fall protection systems must develop and follow a fall protection plan. Below are sample fall protection plans developed for precast concrete construction and residential work that could be tailored to be site specific for other precast concrete or residential jobsite. This sample plan can be modified to be used for other work involving leading edge work. The sample plan outlines the elements that must be addressed in any fall protection plan. The reasons outlined in this sample fall protection plan are for illustrative purposes only and are not necessarily a valid, acceptable rationale (unless the conditions at the job site are the same as those covered by these sample plans) for not using conventional fall protection systems for a particular precast concrete or residential construction worksite. However, the sample plans provide guidance to employers on the type of information that is required to be discussed in fall protection plans.

Sample Fall Protection Plans

Fall Protection Plan For Precast/Prestress Concrete Structures

This Fall Protection Plan is specific for the following project:

Location of Job ______________________

Erecting Company ______________________

Date Plan Prepared or Modified ______________________

Plan Prepared By ______________________

Plan Approved By ______________________

Plan Supervised By ______________________

The following Fall Protection Plan is a sample program prepared for the prevention of injuries associated with falls. A Fall Protection Plan must be developed and evaluated on a site by site basis. It is recommended that erectors discuss the written Fall Protection Plan with their OSHA Area Office prior to going on a jobsite.

I. Statement of Company Policy

(Company Name) is dedicated to the protection of its employees from on-the-job injuries. All employees of (Company Name) have the responsibility to work safely on the job. The purpose of this plan is:

(a) *To supplement our standard safety policy* by providing safety standards specifically designed to cover fall protection on this job and;

(b) *to ensure that each employee is trained* and made aware of the safety provisions which are to be implemented by this plan prior to the start of erection.

This Fall Protection Plan addresses the use of other than conventional fall protection at a number of areas on the project, as well as identifying specific activities that require non-conventional means of fall protection. These areas include:

A. *Connecting activity (point of erection).*

B. *Leading edge work.*

C. *Unprotected sides or edge.*

D. *Grouting.*

This plan is designed to enable employers and employees to recognize the fall hazards on this job and to establish the procedures that are to be followed in order to prevent falls to lower levels or through holes and openings in walking/working surfaces. Each employee will be trained in these procedures and strictly adhere to them except when doing so would expose the employee to a greater hazard. If, in the employee's opinion, this is the case, the employee is to notify the foreman of the concern and the concern addressed before proceeding.

Safety policy and procedure on any one project cannot be administered, implemented, monitored and enforced by any one individual. The total objective of a safe, accident free work environment can only be accomplished by a dedicated, concerted effort by every individual involved with the project from management down to the last employee. Each employee must understand their value to the company; the costs of accidents, both monetary, physical, and emotional; the objective of the safety policy and procedures; the safety rules that apply to the safety policy and procedures; and what their individual role is in administering, implementing, monitoring, and compliance of their safety policy and procedures. This allows for a more personal approach to compliance through planning, training, understanding and cooperative effort, rather than by strict enforcement. If for any reason an unsafe act persists, strict enforcement will be implemented.

It is the responsibility of (name of competent person) to implement this Fall Protection Plan. (Name of Competent Person) is responsible for continual observational safety checks of their work operations and to enforce the safety policy and procedures. The foreman also is responsible to correct any unsafe acts or conditions immediately. It is the responsibility of the employee to understand and adhere to the procedures of this plan and to follow the instructions of the foreman. It is also the responsibility of the employee to bring to management's attention any unsafe or hazardous conditions or acts that may cause injury to either themselves or any other employees. Any changes to this Fall Protection Plan must be approved by (name of Qualified Person).

II. Fall Protection Systems to Be Used on This Project

Where conventional fall protection is infeasible or creates a greater hazard at the leading edge and during initial connecting activity, we plan to do this work using a safety monitoring system and expose only a minimum number of employees for the time necessary to actually accomplish the job. The maximum number of workers to be monitored by one safety monitor is six (6). We are designating the following trained employees as designated erectors and they are permitted to enter the controlled access zones and work without the use of conventional fall protection.

Safety monitor:

Designated erector:

Designated erector:

Designated erector:

Designated erector:

Designated erector:

Designated erector:

The safety monitor shall be identified by wearing an orange hard hat. The designated erectors will be identified by one of the following methods:

1. *They will wear a blue colored arm band, or*
2. *They will wear a blue colored hard hat, or*
3. *They will wear a blue colored vest.*

Only individuals with the appropriate experience, skills, and training will be authorized as designated erectors. All employees that will be working as designated erectors under the safety monitoring system shall have been trained and instructed in the following areas:

1. *Recognition of the fall hazards* in the work area (at the leading edge and when making initial connections — point of erection).
2. *Avoidance of fall hazards* using established work practices which have been made known to the employees.
3. *Recognition of unsafe practices* or working conditions that could lead to a fall, such as windy conditions.
4. *The function, use, and operation* of safety monitoring systems, guardrail systems, body belt/harness systems, control zones and other protection to be used.
5. *The correct procedure* for erecting, maintaining, disassembling and inspecting the system(s) to be used.
6. *Knowledge of construction sequence or the erection plan.*

A conference will take place prior to starting work involving all members of the erection crew, crane crew and supervisors of any other concerned contractors. This conference will be conducted by the precast concrete erection supervisor in charge of the project. During the pre-work conference, erection procedures and sequences pertinent to this job will be thoroughly discussed and safety practices to be used throughout the project will be specified. Further, all personnel will be informed that the controlled access zones are off limits to all personnel other than those designated erectors specifically trained to work in that area.

Safety Monitoring System

A safety monitoring system means a fall protection system in which a competent person is responsible for recognizing and warning employees of fall hazards. The duties of the safety monitor are to:

1. *Warn by voice* when approaching the open edge in an unsafe manner.
2. *Warn by voice* if there is a dangerous situation developing which cannot be seen by another person involved with product placement, such as a member getting out of control.
3. *Make the designated erectors* aware they are in a dangerous area.
4. *Be competent in recognizing fall hazards.*
5. *Warn employees* when they appear to be unaware of a fall hazard or are acting in an unsafe manner.
6. *Be on the same walking/working surface* as the monitored employees and within visual sighting distance of the monitored employees.
7. *Be close enough to communicate orally with the employees.*
8. *Not allow other responsibilities to encumber monitoring.* If the safety monitor becomes too encumbered with other responsibilities, the monitor shall:
   (1) *stop the erection process; and*
   (2) *turn over other responsibilities* to a designated erector; or
   (3) *turn over the safety* monitoring function to another designated, competent person. The safety monitoring system shall not be used when the wind is strong enough to cause loads with large surface areas to swing out of radius, or result in loss of control of the load, or when weather conditions cause the walking-working surfaces to become icy or slippery.

M

Fall Protection

Control Zone System

A **controlled access zone** means an area designated and clearly marked, in which leading edge work may take place without the use of guardrail, safety net or personal fall arrest systems to protect the employees in the area. Control zone systems shall comply with the following provisions:

1. *When used to control access to areas* where leading edge and other operations are taking place the controlled access zone shall be defined by a control line or by any other means that restricts access.

   When control lines are used, they shall be erected not less than 6 feet (l.8 m) nor more than 60 feet (18 m) or half the length of the member being erected, whichever is less, from the leading edge.
2. *The control line shall extend* along the entire length of the unprotected or leading edge and shall be approximately parallel to the unprotected or leading edge.
3. *The control line shall be connected* on each side to a guardrail system or wall.
4. *Control lines* shall consist of ropes, wires, tapes, or equivalent materials, and supporting stanchions as follows:
5. *Each line shall be flagged* or otherwise clearly marked at not more than 6-foot (1.8 m) intervals with high-visibility material.
6. *Each line shall be rigged and supported* in such a way that its lowest point (including sag) is not less than 39 inches (1 m) from the walking/working surface and its highest point is not more than 45 inches (1.3 m) from the walking/working surface.
7. *Each line shall have* a minimum breaking strength of 200 pounds (.88 kN).

Holes

All openings greater than 12 in. x 12 in. will have perimeter guarding or covering. All predetermined holes will have the plywood covers made in the precasters' yard and shipped with the member to the jobsite. Prior to cutting holes on the job, proper protection for the hole must be provided to protect the workers. Perimeter guarding or covers will not be removed without the approval of the erection foreman.

Precast concrete column erection through the existing deck requires that many holes be provided through this deck. These are to be covered and protected. Except for the opening being currently used to erect a column, all opening protection is to be left undisturbed. The opening being uncovered to erect a column will become part of the point of erection and will be addressed as part of this Fall Protection Plan. This uncovering is to be done at the erection foreman's direction and will only occur immediately prior to “feeding” the column through the opening. Once the end of the column is through the slab opening, there will no longer exist a fall hazard at this location.

III. Implementation of Fall Protection Plan

The structure being erected is a multistory total precast concrete building consisting of columns, beams, wall panels and hollow core slabs and double tee floor and roof members.

The following is a list of the products and erection situations on this job:

Columns

For columns 10 ft to 36 ft long, employees disconnecting crane hooks from columns will work from a ladder and wear a body belt/harness with lanyard and be tied off when both hands are needed to disconnect. For tying off, a vertical lifeline will be connected to the lifting eye at the top of the column, prior to lifting, to be used with a manually operated or mobile rope grab. For columns too high for the use of a ladder, 36 ft and higher, an added cable will be used to reduce the height of the disconnecting point so that a ladder can be used. This cable will be left in place until a point in erection that it can be removed safely. In some cases, columns will be unhooked from the crane by using an erection tube or shackle with a pull pin which is released from the ground after the column is stabilized.

The column will be adequately connected and/or braced to safely support the weight of a ladder with an employee on it.

Inverted Tee Beams

Employees erecting inverted tee beams, at a height of 6 to 40 ft, will erect the beam, make initial connections, and final alignment from a ladder. If the employee needs to reach over the side of the beam to bar or make an adjustment to the alignment of the beam, they will mount the beam and be tied off to the lifting device in the beam after ensuring the load has been stabilized on its bearing. To disconnect the crane from the beam an employee will stand a ladder against the beam. Because the use of ladders is not practical at heights above 40 ft, beams will be initially placed with the use of tag lines and their final alignment made by a person on a manlift or similar employee positioning systems.

Spandrel Beams

Spandrel beams at the exterior of the building will be aligned as closely as possible with the use of tag lines with the final placement of the spandrel beam made from a ladder at the open end of the structure. A ladder will be used to make the initial connections and a ladder will be used to disconnect the crane. The other end of the beam will be placed by the designated erector from the double tee deck under the observation of the safety monitor.

The beams will be adequately connected and/or braced to safely support the weight of a ladder with an employee on it.

Floor and Roof Members

During installation of the precast concrete floor and/or roof members, the work deck continuously increases in area as more and more units are being erected and positioned. Thus, the unprotected floor/roof perimeter is constantly modified with the leading edge changing location as each member is installed. The fall protection for workers at the leading edge shall be assured by properly constructed and maintained control zone lines not more than 60 ft away from the leading edge supplemented by a safety monitoring system to ensure the safety of all designated erectors working within the area defined by the control zone lines.

The hollow core slabs erected on the masonry portion of the building will be erected and grouted using the safety monitoring system. Grout will be placed in the space between the end of the slab and face shell of the concrete masonry by dumping from a wheelbarrow. The grout in the keyways between the slabs will be dumped from a wheelbarrow and then spread with long handled tools, allowing the worker to stand erect facing toward the unprotected edge and back from any work deck edge.

Whenever possible, the designated erectors will approach the incoming member at the leading edge only after it is below waist height so that the member itself provides protection against falls.

Except for the situations described below, when the arriving floor or roof member is within 2 to 3 inches of its final position, the designated erectors can then proceed to their position of erection at each end of the member under the control of the safety monitor. Crane hooks will be unhooked from double tee members by designated erectors under the direction and supervision of the safety monitor.

Designated erectors, while waiting for the next floor or roof member, will be constantly under the control of the safety monitor for fall protection and are directed to stay a minimum of six (6) ft from the edge. In the event a designated erector must move from one end of a member, which has just been placed at the leading edge, they must first move away from the leading edge a minimum of six (6) ft and then progress to the other end while maintaining the minimum distance of six (6) ft at all times.

Erection of double tees, where conditions require bearing of one end into a closed pocket and the other end on a beam ledge, restricting the tee legs from going directly into the pockets, require special considerations. The tee legs that are to bear in the closed pocket must hang lower than those at the beam bearing. The double tee will be “two-lined” in order to elevate one end higher than the other to allow for the low end to be ducked into the closed pocket using the following procedure.

The double tee will be rigged with a standard four-way spreader off of the main load line. An additional choker will be attached to the married point of the two-legged spreader at the end of the tee that is to be elevated. The double tee will be hoisted with the main load line and swung into a position as close as possible to the tee's final bearing elevation. When the tee is in this position and stabilized, the whip line load block will be lowered to just above the tee deck. At this time, two erectors will walk out on the suspended tee deck at midspan of the tee member and pull the load block to the end of the tee to be elevated and attach the additional choker to the load block. The possibility of entanglement with the crane lines and other obstacles during this two lining process while raising and lowering the crane block on that second line could be hazardous to an encumbered employee. Therefore, the designated erectors will not tie off during any part of this process. While the designated erectors are on the double tee, the safety monitoring system will be used. After attaching the choker, the two erectors then step back on the previously erected tee deck and signal the crane operator to hoist the load with the whip line to the elevation that will allow for enough clearance to let the low end tee legs slide into the pockets when the main load line is lowered. The erector, who is handling the lowered end of the tee at the closed pocket bearing, will step out on the suspended tee. An erection bar will then be placed between the end of the tee leg and the inside face of the pocketed spandrel member. The tee is barred away from the pocketed member to reduce the friction and lateral force against the pocketed member. As the tee is being lowered, the other erector remains on the tee which was previously erected to handle the other end. At this point the tee is slowly lowered by the crane to a point where the tee legs can freely slide into the pockets. The erector working the lowered end of the tee must keep pressure on the bar between the tee and the face of the pocketed spandrel member to very gradually let the tee legs slide into the pocket to its proper bearing dimension. The tee is then slowly lowered into its final erected position.

The designated erector should be allowed onto the suspended double tee, otherwise there is no control over the horizontal movement of the double tee and this movement could knock the spandrel off of its bearing or the column out of plumb. The control necessary to prevent hitting the spandrel can only be done safely from the top of the double tee being erected.

Loadbearing Wall Panels: The erection of the loadbearing wall panels on the elevated decks requires the use of a safety monitor and a controlled access zone that is a minimum of 25 ft and a maximum of 1/2 the length of the wall panels away from the unprotected edge, so that designated erectors can move freely and unencumbered when receiving the panels. Bracing, if required for stability, will be installed by ladder. After the braces are secured, the crane will be disconnected from the wall by using a ladder. The wall to wall connections will also be performed from a ladder.

Non-Loadbearing Panels (Cladding): The locating of survey lines, panel layout and other installation prerequisites (prewelding, etc.) for non-loadbearing panels (cladding) will not commence until floor perimeter and floor openings have been protected. In some areas, it is necessary because of panel configuration to remove the perimeter protection as the cladding is being installed. Removal of perimeter protection will be performed on a bay to bay basis, just ahead of cladding erection to minimize temporarily unprotected floor edges. Those workers within 6 ft of the edge, receiving and positioning the cladding when the perimeter protection is removed shall be tied off.

Detailing

Employees exposed to falls of six (6) feet or more to lower levels, who are not actively engaged in leading edge work or connecting activity, such as welding, bolting, cutting, bracing, guying, patching, painting or other operations, and who are working less than six (6) ft from an unprotected edge will be tied off at all times or guardrails will be installed. Employees engaged in these activities but who are more than six (6) ft from an unprotected edge as defined by the control zone lines, do not require fall protection but a warning line or control lines must be erected to remind employees they are approaching an area where fall protection is required.

IV. **Conventional Fall Protection** Considered for the Point of Erection or Leading Edge Erection Operations

A. *Personal Fall Arrest Systems*

In this particular erection sequence and procedure, personal fall arrest systems requiring body belt/harness systems, lifelines and lanyards will not reduce possible hazards to workers and will create offsetting hazards during their usage at the leading edge of precast/ prestressed concrete construction.

Leading edge erection and initial connections are conducted by employees who are specifically trained to do this type of work and are trained to recognize the fall hazards. The nature of such work normally exposes the employee to the fall hazard for a short period of time and installation of fall protection systems for a short duration is not feasible because it exposes the installers of the system to the same fall hazard, but for a longer period of time.

1. *It is necessary that the employee* be able to move freely without encumbrance in order to guide the sections of precast concrete into their final position without having lifelines attached which will restrict the employee's ability to move about at the point of erection.
2. *A typical procedure requires 2 or more workers* to maneuver around each other as a concrete member is positioned to fit into the structure. If they are each attached to a lifeline, part of their attention must be diverted from their main task of positioning a member weighing several tons to the task of avoiding entanglements of their lifelines or avoiding tripping over lanyards. Therefore, if these workers are attached to lanyards, more fall potential would result than from not using such a device.

   In this specific erection sequence and procedure, retractable lifelines do not solve the problem of two workers becoming tangled. In fact, such a tangle could prevent the lifeline from retracting as the worker moved, thus potentially exposing the worker to a fall greater than 6 ft. Also, a worker crossing over the lifeline of another worker can create a hazard because the movement of one person can unbalance the other. In the event of a fall by one person there is a likelihood that the other person will be caused to fall as well. In addition, if contamination such as grout (during hollow core grouting) enters the retractable housing it can cause excessive wear and damage to the device and could clog the retracting mechanism as the lanyard is dragged across the deck. Obstructing the cable orifice can defeat the device's shock absorbing function, produce cable slack and damage, and adversely affect cable extraction and retraction.
3. *Employees tied to a lifeline* can be trapped and crushed by moving structural members if the employee becomes restrained by the lanyard or retractable lifeline and cannot get out of the path of the moving load.

   The sudden movement of a precast concrete member being raised by a crane can be caused by a number of factors. When this happens, a connector may immediately have to move a considerable distance to avoid injury. If a tied off body belt/harness is being used, the connector could be trapped. Therefore, there is a greater risk of injury if the connector is tied to the structure for this specific erection sequence and procedure.

   When necessary to move away from a retractable device, the worker cannot move at a rate greater than the device locking speed typically 3.5 to 4.5 ft/sec. When moving toward the device it is necessary to move at a rate which does not permit cable slack to build up. This slack may cause cable retraction acceleration and cause a worker to lose their balance by applying a higher than normal jerking force on the body when the cable suddenly becomes taut after building up momentum. This slack can also cause damage to the internal spring-loaded drum, uneven coiling of cable on the drum, and possible cable damage.

   The factors causing sudden movements for this location include:

   (a) *Cranes:*
      (1) *Operator error.*
      (2) *Site conditions (soft* or unstable ground).
      (3) *Mechanical failure.*
      (4) *Structural failure.*
      (5) *Rigging failure.*
      (6) *Crane signal*/radio communication failure.

   (b) *Weather Conditions:*
      (1) *Wind (strong wind /sudden gusting)* — particularly a problem with the large surface areas of precast concrete members.
      (2) *Snow/rain (visibility).*
      (3) *Fog (visibility).*
      (4) *Cold — causing slowed reactions or mechanical problems.*

   (c) *Structure/Product Conditions:*
      (1) *Lifting Eye failure.*
      (2) *Bearing failure or slippage.*
      (3) *Structure shifting.*
      (4) *Bracing failure.*
      (5) *Product failure.*

   (d) *Human Error:*
      (1) *Incorrect tag line procedure.*
      (2) *Tag line hang-up.*
      (3) *Incorrect or misunderstood crane signals.*
      (4) *Misjudged elevation of member.*
      (5) *Misjudged speed of member.*
      (6) *Misjudged angle of member.*
4. *Anchorages or special attachment points* could be cast into the precast concrete members if sufficient preplanning and consideration of erectors' position is done before the members are cast. Any hole or other attachment must be approved by the engineer who designed the member. It is possible that some design restrictions will not allow a member to be weakened by an additional hole; however, it is anticipated that such situations would be the exception, not the rule. Attachment points, other than on the deck surface, will require removal and/or patching. In order to remove and/or patch these points, requires the employee to be exposed to an additional fall hazard at an unprotected perimeter. The fact that attachment points could be available anywhere on the structure does not eliminate the hazards of using these points for tying off as discussed above. A logical point for tying off on double tees would be using the lifting loops, except that they must be cut off to eliminate a tripping hazard at an appropriate time.

5. *Providing attachment at a point* above the walking/working surface would also create fall exposures for employees installing their devices. Final positioning of a precast concrete member requires it to be moved in such a way that it must pass through the area that would be occupied by the lifeline and the lanyards attached to the point above. Resulting entanglements of lifelines and lanyards on a moving member could pull employees from the work surface. Also, the structure is being created and, in most cases, there is no structure above the members being placed.
   (a) *Temporary structural supports,* installed to provide attaching points for lifelines limit the space which is essential for orderly positioning, alignment and placement of the precast concrete members. To keep the lanyards a reasonable and manageable length, lifeline supports would necessarily need to be in proximity to the positioning process. A sudden shift of the precast concrete member being positioned because of wind pressure or crane movement could make it strike the temporary supporting structure, moving it suddenly and causing tied off employees to fall.
   (b) *The time in manhours which would be expended* in placing and maintaining temporary structural supports for lifeline attaching points could exceed the expended manhours involved in placing the precast concrete members. No protection could be provided for the employees erecting the temporary structural supports and these supports would have to be moved for each successive step in the construction process, thus greatly increasing the employee's exposure to the fall hazard.
   (c) *The use of a cable strung horizontally* between two columns to provide tie off lines for erecting or walking a beam for connecting work is not feasible and creates a greater hazard on this multi-story building for the following reasons:
      (1) *If a connector is to use such a line,* it must be installed between the two columns. To perform this installation requires an erector to have more fall exposure time attaching the cable to the columns than would be spent to make the beam to column connection itself.
      (2) *If such a line is to be installed* so that an erector can walk along a beam, it must be overhead or below him. For example, if a connector must walk along a 24 in. wide beam, the presence of a line next to the connector at waist level, attached directly to the columns, would prevent the connector from centering their weight over the beam and balancing themselves. Installing the line above the connector might be possible on the first level of a two-story column; however, the column may extend only a few feet above the floor level at the second level or be flush with the floor level. Attaching the line to the side of the beam could be a solution; however, it would require the connector to attach the lanyard below foot level which would most likely extend a fall farther than 6 ft.
      (3) *When lines are strung over every beam,* it becomes more and more difficult for the crane operator to lower a precast concrete member into position without the member becoming fouled. Should the member become entangled, it could easily dislodge the line from a column. If a worker is tied to it at the time, a fall could be caused.
6. *The ANSI A10.14-1991* American National Standard for Construction and Demolition Operations — Requirements for Safety Belts, Harnesses, Lanyards and Lifelines for Construction and Demolition Use, states that the anchor point of a lanyard or deceleration device should, if possible, be located above the wearer's belt or harness attachment. ANSI A10.14 also states that a suitable anchorage point is one which is located as high as possible to prevent contact with an obstruction below should the worker fall. Most manufacturers also warn in the user's handbook that the safety block/retractable lifeline must be positioned above the D-ring (above the work space of the intended user) and OSHA recommends that fall arrest and restraint equipment be used in accordance with the manufacturer's instructions.

   Attachment of a retractable device to a horizontal cable near floor level or using the inserts in the floor or roof members may result in increased free fall due to the dorsal D-ring of the full-body harness riding higher than the attachment point of the snaphook to the cable or insert (e.g., 6 foot tall worker with a dorsal D-ring at 5 feet above the floor or surface, reduces the working length to only one foot, by placing the anchorage five feet away from the fall hazard). In addition, impact loads may exceed maximum fall arrest forces (MAF) because the fall arrest D-ring would be 4 to 5 feet higher than the safety block/retractable lifeline anchored to the walking-working surface; and the potential for swing hazards is increased.

   Manufacturers also require that workers not work at a level where the point of snaphook attachment to the body harness is above the device because this will increase the free fall distance and the deceleration distance and will cause higher forces on the body in the event of an accidental fall.

   Manufacturers recommend an anchorage for the retractable lifeline which is immovably fixed in space and is independent of the user's support systems. A moveable anchorage is one which can be moved around (such as equipment or wheeled vehicles) or which can deflect substantially under shock loading (such as a horizontal cable or very flexible beam). In the case of a very flexible anchorage, a shock load applied to the anchorage during fall arrest can cause oscillation of the flexible anchorage such that the retractable brake mechanism may undergo one or more cycles of locking/unlocking/locking (ratchet effect) until the anchorage deflection is dampened. Therefore, use of a moveable anchorage involves critical engineering and safety factors and should only be considered after fixed anchorage has been determined to be not feasible.

   Horizontal cables used as an anchorage present an additional hazard due to amplification of the horizontal component of maximum arrest force (of a fall) transmitted to the points where the horizontal cable is attached to the structure. This amplification is due to the angle of sag of a horizontal cable and is most severe for small angles of sag. For a cable sag angle of 2 degrees the horizontal force on the points of cable attachment can be amplified by a factor of 15.

   It is also necessary to install the retractable device vertically overhead to minimize swing falls. If an object is in the worker's swing path (or that of the cable) hazardous situations exist: (1) due to the swing, horizontal speed of the user may be high enough to cause injury when an obstacle in the swing fall path is struck by either the user or the cable; (2) the total vertical fall distance of the user may be much greater than if the user had fallen only vertically without a swing fall path.

   With retractable lines, overconfidence may cause the worker to engage in inappropriate behavior, such as approaching the perimeter of a floor or roof at a distance appreciably greater than the shortest distance between the anchorage point and the leading edge. Though the retractable lifeline may arrest a worker's fall before he or she has fallen a few feet, the lifeline may drag along the edge of the floor or beam and swing the worker like a pendulum until the line has moved to a position where the distance between the anchorage point and floor edge is the shortest distance between those two points. Accompanying this pendulum swing is a lowering of the worker, with the attendant danger that he or she may violently impact the floor or some obstruction below.

   The risk of a cable breaking is increased if a lifeline is dragged sideways across the rough surface or edge of a concrete member at the same moment that the lifeline is being subjected to a maximum impact loading during a fall. The typical 3/16 in. cable in a retractable lifeline has a breaking strength of from 3000 to 3700 lbs.
7. *The competent person,* who can take into account the specialized operations being performed on this project, should determine when and where a designated erector cannot use a personal fall arrest system.

B. *Safety Net Systems*

The nature of this particular precast concrete erection worksite precludes the safe use of safety nets where point of erection or leading edge work must take place.

1. *To install safety nets* in the interior high bay of the single story portion of the building poses rigging attachment problems. Structural members do not exist to which supporting devices for nets can be attached in the area where protection is required. As the erection operation advances, the location of point of erection or leading edge work changes constantly as each member is attached to the structure. Due to this constant change it is not feasible to set net sections and build separate structures to support the nets.

2. *The nature of the erection process* for the precast concrete members is such that an installed net would protect workers as they position and secure only one structural member. After each member is stabilized the net would have to be moved to a new location (this could mean a move of 8 to 10 ft or the possibility of a move to a different level or area of the structure) to protect workers placing the next piece in the construction sequence. The result would be the installation and dismantling of safety nets repeatedly throughout the normal work day. As the time necessary to install a net, test, and remove it is significantly greater than the time necessary to position and secure a precast concrete member, the exposure time for the worker installing the safety net would be far longer than for the workers whom the net is intended to protect. The time exposure repeats itself each time the nets and supporting hardware must be moved laterally or upward to provide protection at the point of erection or leading edge.
3. *Strict interpretation of §1926.502(c)* requires that operations shall not be undertaken until the net is in place and has been tested. With the point of erection constantly changing, the time necessary to install and test a safety net significantly exceeds the time necessary to position and secure the concrete member.
4. *Use of safety nets* on exposed perimeter wall openings and open-sided floors, causes attachment points to be left in architectural concrete which must be patched and filled with matching material after the net supporting hardware is removed. In order to patch these openings, additional numbers of employees must be suspended by swing stages, boatswain chairs or other devices, thereby increasing the amount of fall exposure time to employees.
5. *Installed safety nets* pose an additional hazard at the perimeter of the erected structure where limited space is available in which members can be turned after being lifted from the ground by the crane. There would be a high probability that the member being lifted could become entangled in net hardware, cables, etc.
6. *The use of safety nets* where structural wall panels are being erected would prevent movement of panels to point of installation. To be effective, nets would necessarily have to provide protection across the area where structural supporting wall panels would be set and plumbed before roof units could be placed.
7. *Use of a tower crane* for the erection of the high rise portion of the structure poses a particular hazard in that the crane operator cannot see or judge the proximity of the load in relation to the structure or nets. If the signaler is looking through nets and supporting structural devices while giving instructions to the crane operator, it is not possible to judge precise relationships between the load and the structure itself or to nets and supporting structural devices. This could cause the load to become entangled in the net or hit the structure causing potential damage.

C. *Guardrail Systems*

On this particular worksite, guardrails, barricades, ropes, cables or other perimeter guarding devices or methods on the erection floor will pose problems to safe erection procedures. Typically, a floor or roof is erected by placing 4 to 10 ft. wide structural members next to one another and welding or grouting them together. The perimeter of a floor and roof changes each time a new member is placed into position. It is unreasonable and virtually impossible to erect guardrails and toe boards at the ever changing leading edge of a floor or roof.

1. *To position a member safely* it is necessary to remove all obstructions extending above the floor level near the point of erection. Such a procedure allows workers to swing a new member across the erected surface as necessary to position it properly without worrying about knocking material off of this surface.

   Hollow core slab erection on the masonry wall requires installation of the perimeter protection where the masonry wall has to be constructed. This means the guardrail is installed then subsequently removed to continue the masonry construction. The erector will be exposed to a fall hazard for a longer period of time while installing and removing perimeter protection than while erecting the slabs.

   In hollow core work, as in other precast concrete erection, others are not typically on the work deck until the precast concrete erection is complete. The deck is not complete until the leveling, aligning, and grouting of the joints is done. It is normal practice to keep others off the deck until at least the next day after the installation is complete to allow the grout to harden.
2. *There is no permanent boundary* until all structural members have been placed in the floor or roof. At the leading edge, workers are operating at the temporary edge of the structure as they work to position the next member in the sequence. Compliance with the standard would require a guardrail and toe board be installed along this edge. However, the presence of such a device would prevent a new member from being swung over the erected surface low enough to allow workers to control it safely during the positioning process. Further, these employees would have to work through the guardrail to align the new member and connect it to the structure. The guardrail would not protect an employee who must lean through it to do the necessary work, rather it would hinder the employee to such a degree that a greater hazard is created than if the guardrail were absent.
3. *Guardrail requirements pose* a hazard at the leading edge of installed floor or roof sections by creating the possibility of employees being caught between guardrails and suspended loads. The lack of a clear work area in which to guide the suspended load into position for placement and welding of members into the existing structure creates still further hazards.
4. *Where erection processes require* precast concrete stairways or openings to be installed as an integral part of the overall erection process, it must also be recognized that guardrails or handrails must not project above the surface of the erection floor. Such guardrails should be terminated at the level of the erection floor to avoid placing hazardous obstacles in the path of a member being positioned.

V. Other Fall Protection Measures Considered for This Job

The following is a list and explanation of other fall protection measures available and an explanation of limitations for use on this particular jobsite. If during the course of erecting the building the employee sees an area that could be erected more safely by the use of these fall protection measures, the foreman should be notified.

A. *Scaffolds are not used because:*

1. *The leading edge of the building* is constantly changing and the scaffolding would have to be moved at very frequent intervals. Employees erecting and dismantling the scaffolding would be exposed to fall hazards for a greater length of time than they would by merely erecting the precast concrete member.
2. *A scaffold tower could* interfere with the safe swinging of a load by the crane.
3. *Power lines, terrain and site* do not allow for the safe use of scaffolding.

B. *Vehicle mounted platforms* are not used because:

1. *A vehicle mounted platform* will not reach areas on the deck that are erected over other levels.
2. *The leading edge of the building* is usually over a lower level of the building and this lower level will not support the weight of a vehicle mounted platform.
3. *A vehicle mounted platform* could interfere with the safe swinging of a load by the crane, either by the crane swinging the load over or into the equipment.
4. *Power lines and surrounding* site work do not allow for the safe use of a vehicle mounted platform.

C. *Crane suspended personnel* platforms are not used because:

1. *A second crane close* enough to suspend any employee in the working and erecting area could interfere with the safe swinging of a load by the crane hoisting the product to be erected.
2. *Power lines and surrounding* site work do not allow for the safe use of a second crane on the job.

VI. Enforcement

Constant awareness of and respect for fall hazards, and compliance with all safety rules are considered conditions of employment. The jobsite Superintendent, as well as individuals in the Safety and Personnel Department, reserve the right to issue disciplinary warnings to employees, up to and including termination, for failure to follow the guidelines of this program.

VII. Accident Investigations

All accidents that result in injury to workers, regardless of their nature, shall be investigated and reported. It is an integral part of any safety program that documentation take place as soon as possible so that the cause and means of prevention can be identified to prevent a reoccurrence.

In the event that an employee falls or there is some other related, serious incident occurring, this plan shall be reviewed to determine if additional practices, procedures, or training need to be implemented to prevent similar types of falls or incidents from occurring.

VIII. Changes to Plan

Any changes to the plan will be approved by (name of the qualified person). This plan shall be reviewed by a qualified person as the job progresses to determine if additional practices, procedures or training needs to be implemented by the competent person to improve or provide additional fall protection. Workers shall be notified and trained, if necessary, in the new procedures. A copy of this plan and all approved changes shall be maintained at the jobsite.

Sample Fall Protection Plan for Residential Construction

(Insert Company Name)

This Fall Protection Plan Is Specific For The Following Project:

Location of Job____________________________________

Date Plan Prepared or Modified______________________

Plan Prepared By___________________________________

Plan Approved By __________________________________

Plan Supervised By _________________________________

The following Fall Protection Plan is a sample program prepared for the prevention of injuries associated with falls. A Fall Protection Plan must be developed and evaluated on a site by site basis. It is recommended that builders discuss the written Fall Protection Plan with their OSHA Area Office prior to going on a jobsite.

I. Statement of Company Policy

(Your company name here) is dedicated to the protection of its employees from on-the-job injuries. All employees of (Your company name here) have the responsibility to work safely on the job. The purpose of the plan is to supplement our existing safety and health program and to ensure that every employee who works for (Your company name here) recognizes workplace fall hazards and takes the appropriate measures to address those hazards.

This Fall Protection Plan addresses the use of conventional fall protection at a number of areas on the project, as well as identifies specific activities that require non-conventional means of fall protection. During the construction of residential buildings under 48 feet in height, it is sometimes infeasible or it creates a greater hazard to use conventional fall protection systems at specific areas or for specific tasks. The areas or tasks may include, but are not limited to:

a. *Setting and bracing of roof trusses and rafters;*

b. *Installation of floor sheathing and joists;*

c. *Roof sheathing operations; and*

d. *Erecting exterior walls.*

In these cases, conventional fall protection systems may not be the safest choice for builders. This plan is designed to enable employers and employees to recognize the fall hazards associated with this job and to establish the safest procedures that are to be followed in order to prevent falls to lower levels or through holes and openings in walking/ working surfaces.

Each employee will be trained in these procedures and will strictly adhere to them except when doing so would expose the employee to a greater hazard. If, in the employee's opinion, this is the case, the employee is to notify the competent person of their concern and have the concern addressed before proceeding.

It is the responsibility of (name of competent person) to implement this Fall Protection Plan. Continual observational safety checks of work operations and the enforcement of the safety policy and procedures shall be regularly enforced. The crew supervisor or foreman (insert name) is responsible for correcting any unsafe practices or conditions immediately.

It is the responsibility of the employer to ensure that all employees understand and adhere to the procedures of this plan and to follow the instructions of the crew supervisor. It is also the responsibility of the employee to bring to management's attention any unsafe or hazardous conditions or practices that may cause injury to either themselves or any other employees. Any changes to the Fall Protection Plan must be approved by (name of qualified person).

II. Fall Protection Systems To Be Used on This Job

Installation of roof trusses/rafters, exterior wall erection, roof sheathing, floor sheathing and joist/truss activities will be conducted by employees who are specifically trained to do this type of work and are trained to recognize the fall hazards. The nature of such work normally exposes the employee to the fall hazard for a short period of time. This Plan details how (Your company name here) will minimize these hazards.

Controlled Access Zones

When using the Plan to implement the fall protection options available, workers must be protected through limited access to high hazard locations. Before any non-conventional fall protection systems are used as part of the work plan, a controlled access zone (CAZ) shall be clearly defined by the competent person as an area where a recognized hazard exists. The demarcation of the CAZ shall be communicated by the competent person in a recognized manner, either through signs, wires, tapes, ropes or chains.

(Your company name here) shall take the following steps to ensure that the CAZ is clearly marked or controlled by the competent person:

- All access to the CAZ must be restricted to authorized entrants;
- All workers who are permitted in the CAZ shall be listed in the appropriate sections of the Plan (or be visibly identifiable by the competent person) prior to implementation;
- The competent person shall ensure that all protective elements of the CAZ be implemented prior to the beginning of work.

Installation Procedures for Roof Truss and Rafter Erection

During the erection and bracing of roof trusses/rafters, conventional fall protection may present a greater hazard to workers. On this job, safety nets, guardrails and personal fall arrest systems will not provide adequate fall protection because the nets will cause the walls to collapse, while there are no suitable attachment or anchorage points for guardrails or personal fall arrest systems.

On this job, requiring workers to use a ladder for the entire installation process will cause a greater hazard because the worker must stand on the ladder with his back or side to the front of the ladder. While erecting the truss or rafter the worker will need both hands to maneuver the truss and therefore cannot hold onto the ladder. In addition, ladders cannot be adequately protected from movement while trusses are being maneuvered into place. Many workers may experience additional fatigue because of the increase in overhead work with heavy materials, which can also lead to a greater hazard.

Exterior scaffolds cannot be utilized on this job because the ground, after recent backfilling, cannot support the scaffolding. In most cases, the erection and dismantling of the scaffold would expose workers to a greater fall hazard than erection of the trusses/rafters.

On all walls eight feet or less, workers will install interior scaffolds along the interior wall below the location where the trusses/ rafters will be erected. "Sawhorse" scaffolds constructed of 46 inch sawhorses and 2x10 planks will often allow workers to be elevated high enough to allow for the erection of trusses and rafters without working on the top plate of the wall.

In structures that have walls higher than eight feet and where the use of scaffolds and ladders would create a greater hazard, safe working procedures will be utilized when working on the top plate and will be monitored by the crew supervisor. During all stages of truss/rafter erection the stability of the trusses/rafters will be ensured at all times.

(Your company name here) shall take the following steps to protect workers who are exposed to fall hazards while working from the top plate installing trusses/rafters:

- Only the following trained workers will be allowed to work on the top plate during roof truss or rafter installation:

____

____

____

- Workers shall have no other duties to perform during truss/rafter erection procedures;
- All trusses/rafters will be adequately braced before any worker can use the truss/rafter as a support;
- Workers will remain on the top plate using the previously stabilized truss/rafter as a support while other trusses/ rafters are being erected;
- Workers will leave the area of the secured trusses only when it is necessary to secure another truss/rafter;
- The first two trusses/rafters will be set from ladders leaning on side walls at points where the walls can support the weight of the ladder; and
- A worker will climb onto the interior top plate via a ladder to secure the peaks of the first two trusses/rafters being set.

The workers responsible for detaching trusses from cranes and/or securing trusses at the peaks traditionally are positioned at the peak of the trusses/rafters. There are also situations where workers securing rafters to ridge beams will be positioned on top of the ridge beam.

(Your company name here) shall take the following steps to protect workers who are exposed to fall hazards while securing trusses/rafters at the peak of the trusses/ridge beam:

- Only the following trained workers will be allowed to work at the peak during roof truss or rafter installation:

____

____

____

- Once truss or rafter installation begins, workers not involved in that activity shall not stand or walk below or adjacent to the roof opening or exterior walls in any area where they could be struck by falling objects;
- Workers shall have no other duties than securing/ bracing the trusses/ridge beam;
- Workers positioned at the peaks or in the webs of trusses or on top of the ridge beam shall work from a stable position, either by sitting on a "ridge seat" or other equivalent surface that provides additional stability or by positioning themselves in previously stabilized trusses/rafters and leaning into and reaching through the trusses/rafters;
- Workers shall not remain on or in the peak/ridge any longer than necessary to safely complete the task.

Roof Sheathing Operations

Workers typically install roof sheathing after all trusses/rafters and any permanent truss bracing is in place. Roof structures are unstable until some sheathing is installed, so workers installing roof sheathing cannot be protected from fall hazards by conventional fall protection systems until it is determined that the roofing system can be used as an anchorage point. At that point, employees shall be protected by a personal fall arrest system.

Trusses/rafters are subject to collapse if a worker falls while attached to a single truss with a belt/harness. Nets could also cause collapse, and there is no place to attach guardrails.

All workers will ensure that they have secure footing before they attempt to walk on the sheathing, including cleaning shoes/boots of mud or other slip hazards.

To minimize the time workers must be exposed to a fall hazard, materials will be staged to allow for the quickest installation of sheathing.

(Your company name here) shall take the following steps to protect workers who are exposed to fall hazards while installing roof sheathing:

- Once roof sheathing installation begins, workers not involved in that activity shall not stand or walk below or adjacent to the roof opening or exterior walls in any area where they could be struck by falling objects;
- The competent person shall determine the limits of this area, which shall be clearly communicated to workers prior to placement of the first piece of roof sheathing;
- The competent person may order work on the roof to be suspended for brief periods as necessary to allow other workers to pass through such areas when this would not create a greater hazard;
- Only qualified workers shall install roof sheathing;
- The bottom row of roof sheathing may be installed by workers standing in truss webs;
- After the bottom row of roof sheathing is installed, a slide guard extending the width of the roof shall be securely attached to the roof. Slide guards are to be constructed of no less than nominal 4" height capable of limiting the uncontrolled slide of workers. Workers should install the slide guard while standing in truss webs and leaning over the sheathing;
- Additional rows of roof sheathing may be installed by workers positioned on previously installed rows of sheathing. A slide guard can be used to assist workers in retaining their footing during successive sheathing operations; and
- Additional slide guards shall be securely attached to the roof at intervals not to exceed 13 feet as successive rows of sheathing are installed. For roofs with pitches in excess of 9-in-12, slide guards will be installed at four-foot intervals.
- When wet weather (rain, snow, or sleet) are present, roof sheathing operations shall be suspended unless safe footing can be assured for those workers installing sheathing.
- When strong winds (above 40 miles per hour) are present, roof sheathing operations are to be suspended unless wind breakers are erected.

Installation of Floor Joists and Sheathing

During the installation of floor sheathing/joists (leading edge construction), the following steps shall be taken to protect workers:

- Only the following trained workers will be allowed to install floor joists or sheathing:

____

____

____

- Materials for the operations shall be conveniently staged to allow for easy access to workers;
- The first floor joists or trusses will be rolled into position and secured either from the ground, ladders or sawhorse scaffolds;
- Each successive floor joist or truss will be rolled into place and secured from a platform created from a sheet of plywood laid over the previously secured floor joists or trusses;
- Except for the first row of sheathing which will be installed from ladders or the ground, workers shall work from the established deck; and
- Any workers not assisting in the leading edge construction while leading edges still exist (e.g. cutting the decking for the installers) shall not be permitted within six feet of the leading edge under construction.

Erection of Exterior Walls

During the construction and erection of exterior walls, employers shall take the following steps to protect workers:

- Only the following trained workers will be allowed to erect exterior walls:

____

____

____

- A painted line six feet from the perimeter will be clearly marked prior to any wall erection activities to warn of the approaching unprotected edge;
- Materials for operations shall be conveniently staged to minimize fall hazards; and
- Workers constructing exterior walls shall complete as much cutting of materials and other preparation as possible away from the edge of the deck.

III. Enforcement

Constant awareness of and respect for fall hazards, and compliance with all safety rules are considered conditions of employment. The crew supervisor or foreman, as well as individuals in the Safety and Personnel Department, reserve the right to issue disciplinary warnings to employees, up to and including termination, for failure to follow the guidelines of this program.

IV. Accident Investigations

All accidents that result in injury to workers, regardless of their nature, shall be investigated and reported. It is an integral part of any safety program that documentation take place as soon as possible so that the cause and means of prevention can be identified to prevent a reoccurrence.

In the event that an employee falls or there is some other related, serious incident occurring, this plan shall be reviewed to determine if additional practices, procedures, or training need to be implemented to prevent similar types of falls or incidents from occurring.

V. Changes to Plan

Any changes to the plan will be approved by (name of the qualified person). This plan shall be reviewed by a qualified person as the job progresses to determine if additional practices, procedures or training needs to be implemented by the competent person to improve or provide additional fall protection. Workers shall be notified and trained, if necessary, in the new procedures. A copy of this plan and all approved changes shall be maintained at the jobsite.

[59 FR 40730, Aug. 9, 1994]

Authority: Section 3704 of the Contract Work Hours and Safety Standards Act (Construction Safety Act) (40 U.S.C. 3701); Sections 4, 6 and 8 of the Occupational Safety and Health Act of 1970 (29 U.S.C. 653, 655, 657); Secretary of Labor's Order Nos. 1-90 (55 FR 9033), 6-96 (62 FR 111), 3-2000 (65 FR 50017), and 5-2007 (72 FR 31159); and 29 CFR part 1911.

Notes

Subpart N - Helicopters, Hoists, Elevators, and Conveyors

§1926.550
[Reserved]

§1926.551
Helicopters

(a) **Helicopter regulations.** Helicopter cranes shall be expected to comply with any applicable regulations of the Federal Aviation Administration.[§1926.551(a)]

(b) **Briefing.** Prior to each day's operation a briefing shall be conducted. This briefing shall set forth the plan of operation for the pilot and ground personnel.[§1926.551(b)]

(c) **Slings and tag lines.** Load shall be properly slung. Tag lines shall be of a length that will not permit their being drawn up into rotors. Pressed sleeve, swedged eyes, or equivalent means shall be used for all freely suspended loads to prevent hand splices from spinning open or cable clamps from loosening.[§1926.551(c)]

(d) **Cargo hooks.** All electrically operated cargo hooks shall have the electrical activating device so designed and installed as to prevent inadvertent operation. In addition, these cargo hooks shall be equipped with an emergency mechanical control for releasing the load. The hooks shall be tested prior to each day's operation to determine that the release functions properly, both electrically and mechanically.[§1926.551(d)]

(e) **Personal protective equipment.**[§1926.551(e)]

(1) *Personal protective equipment* for employees receiving the load shall consist of complete eye protection and hard hats secured by chinstraps.[§1926.551(e)(1)]

(2) *Loose-fitting clothing* likely to flap in the downwash, and thus be snagged on hoist line, shall not be worn.[§1926.551(e)(2)]

(f) **Loose gear and objects.** Every practical precaution shall be taken to provide for the protection of the employees from flying objects in the rotor downwash. All loose gear within 100 feet of the place of lifting the load, depositing the load, and all other areas susceptible to rotor downwash shall be secured or removed.[§1926.551(f)]

(g) **Housekeeping.** Good housekeeping shall be maintained in all helicopter loading and unloading areas.[§1926.551(g)]

(h) **Operator responsibility.** The helicopter operator shall be responsible for size, weight, and manner in which loads are connected to the helicopter. If, for any reason, the helicopter operator believes the lift cannot be made safely, the lift shall not be made.[§1926.551(h)]

(i) **Hooking and unhooking loads.** When employees are required to perform work under hovering craft, a safe means of access shall be provided for employees to reach the hoist line hook and engage or disengage cargo slings. Employees shall not perform work under hovering craft except when necessary to hook or unhook loads.[§1926.551(i)]

(j) **Static charge.** Static charge on the suspended load shall be dissipated with a grounding device before ground personnel touch the suspended load, or protective rubber gloves shall be worn by all ground personnel touching the suspended load.[§1926.551(j)]

(k) **Weight limitation.** The weight of an external load shall not exceed the manufacturer's rating.[§1926.551(k)]

(l) **Ground lines.** Hoist wires or other gear, except for pulling lines or conductors that are allowed to "pay out" from a container or roll off a reel, shall not be attached to any fixed ground structure, or allowed to foul on any fixed structure.[§1926.551(l)]

(m) **Visibility.** When visibility is reduced by dust or other conditions, ground personnel shall exercise special caution to keep clear of main and stabilizing rotors. Precautions shall also be taken by the employer to eliminate as far as practical reduced visibility.[§1926.551(m)]

(n) **Signal systems.** Signal systems between aircrew and ground personnel shall be understood and checked in advance of hoisting the load. This applies to either radio or hand signal systems. Hand signals shall be as shown in Figure N-1.[§1926.551(n)]

FIG. N-1 - HELICOPTER HAND SIGNALS

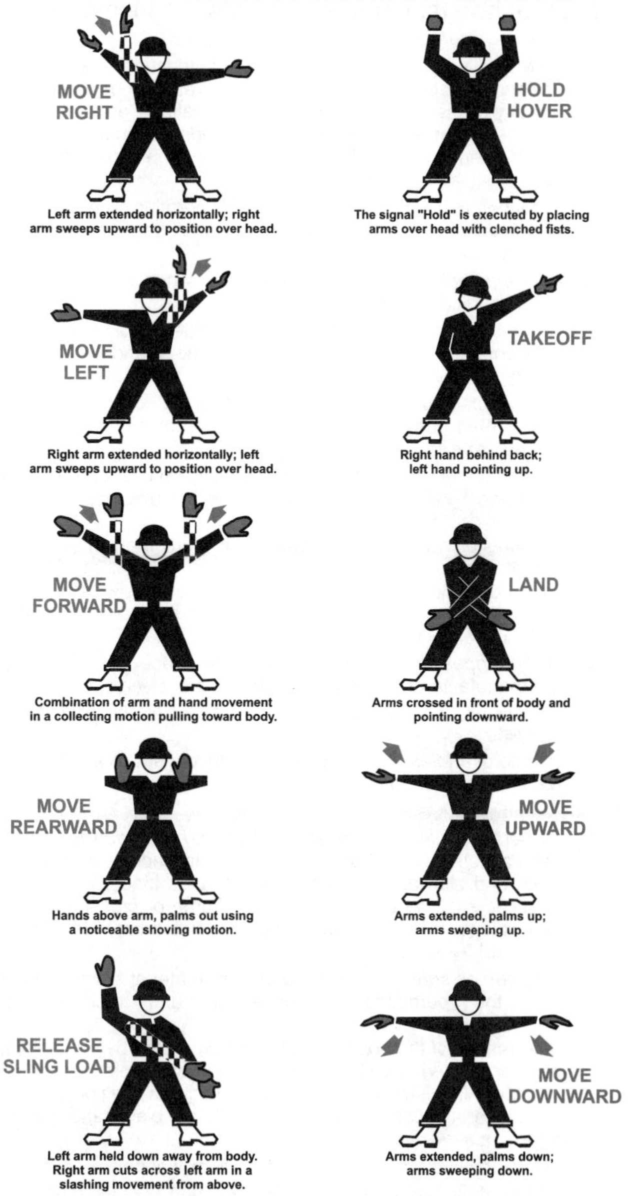

MOVE RIGHT

Left arm extended horizontally; right arm sweeps upward to position over head.

HOLD HOVER

The signal "Hold" is executed by placing arms over head with clenched fists.

MOVE LEFT

Right arm extended horizontally; left arm sweeps upward to position over head.

TAKEOFF

Right hand behind back; left hand pointing up.

MOVE FORWARD

Combination of arm and hand movement in a collecting motion pulling toward body.

LAND

Arms crossed in front of body and pointing downward.

MOVE REARWARD

Hands above arm, palms out using a noticeable shoving motion.

MOVE UPWARD

Arms extended, palms up; arms sweeping up.

RELEASE SLING LOAD

Left arm held down away from body. Right arm cuts across left arm in a slashing movement from above.

MOVE DOWNWARD

Arms extended, palms down; arms sweeping down.

(o) **Approach distance.** No unauthorized person shall be allowed to approach within 50 feet of the helicopter when the rotor blades are turning.[§1926.551(o)]

(p) **Approaching helicopter.** Whenever approaching or leaving a helicopter with blades rotating, all employees shall remain in full view of the pilot and keep in a crouched position. Employees shall avoid the area from the cockpit or cabin rearward unless authorized by the helicopter operator to work there.[§1926.551(p)]

(q) **Personnel.** Sufficient ground personnel shall be provided when required for safe helicopter loading and unloading operations.[§1926.551(q)]

(r) **Communications.** There shall be constant reliable communication between the pilot, and a designated employee of the ground crew who acts as a signalman during the period of loading and unloading. This signalman shall be distinctly recognizable from other ground personnel.[§1926.551(r)]

(s) **Fires.** Open fires shall not be permitted in an area that could result in such fires being spread by the rotor downwash.[§1926.551(s)]

§1926.552

Material hoists, personnel hoists, and elevators

(a) **General requirements.**[§1926.552(a)]

(1) *The employer shall comply* with the manufacturer's specifications and limitations applicable to the operation of all hoists and elevators. Where manufacturer's specifications are not available, the limitations assigned to the equipment shall be based on the determinations of a professional engineer competent in the field.[§1926.552(a)(1)]

(2) *Rated load capacities,* recommended operating speeds, and special hazard warnings or instructions shall be posted on cars and platforms.[§1926.552(a)(2)]

(3) *Wire rope shall be removed* from service when any of the following conditions exists:[§1926.552(a)(3)]

(i) *In hoisting ropes,* six randomly distributed broken wires in one rope lay or three broken wires in one strand in one rope lay;[§1926.552(a)(3)(i)]

(ii) *Abrasion, scrubbing, flattening,* or peening, causing loss of more than one-third of the original diameter of the outside wires;[§1926.552(a)(3)(ii)]

(iii) *Evidence of any heat damage* resulting from a torch or any damage caused by contact with electrical wires; [§1926.552(a)(3)(iii)]

(iv) *Reduction from nominal diameter* of more than three sixty-fourths inch for diameters up to and including three-fourths inch; one-sixteenth inch for diameters seven-eights to 1 1/8 inches; and three thirty-seconds inch for diameters 1 1/4 to 1 1/2 inches.[§1926.552(a)(3)(iv)]

(4) *Hoisting ropes shall be installed* in accordance with the wire rope manufacturers' recommendations.[§1926.552(a)(4)]

(5) *The installation of live booms on hoists is prohibited.* [§1926.552(a)(5)]

(6) *The use of endless belt-type manlifts* on construction shall be prohibited.[§1926.552(a)(6)]

(b) **Material hoists.**[§1926.552(b)]

(1) (i) *Operating rules shall be established and posted* at the operator's station of the hoist. Such rules shall include signal system and allowable line speed for various loads. Rules and notices shall be posted on the car frame or crosshead in a conspicuous location, including the statement "No Riders Allowed."[§1926.552(b)(1)(i)]

(ii) *No person shall be allowed to ride* on material hoists except for the purposes of inspection and maintenance. [§1926.552(b)(1)(ii)]

(2) *All entrances of the hoistways shall be protected* by substantial gates or bars which shall guard the full width of the landing entrance. All hoistway entrance bars and gates shall be painted with diagonal contrasting colors, such as black and yellow stripes.[§1926.552(b)(2)]

(i) *Bars shall be not less than* 2- by 4-inch wooden bars or the equivalent, located 2 feet from the hoistway line. Bars shall be located not less than 36 inches nor more than 42 inches above the floor.[§1926.552(b)(2)(i)]

(ii) *Gates or bars* protecting the entrances to hoistways shall be equipped with a latching device.[§1926.552(b)(2)(ii)]

(3) *Overhead protective covering* of 2-inch planking, 3/4 -inch plywood, or other solid material of equivalent strength, shall be provided on the top of every material hoist cage or platform. [§1926.552(b)(3)]

(4) *The operator's station of a hoisting machine* shall be provided with overhead protection equivalent to tight planking not less than 2 inches thick. The support for the overhead protection shall be of equal strength.[§1926.552(b)(4)]

(5) *Hoist towers may be used* with or without an enclosure on all sides. However, whichever alternative is chosen, the following applicable conditions shall be met:[§1926.552(b)(5)]

(i) *When a hoist tower is enclosed,* it shall be enclosed on all sides for its entire height with a screen enclosure of 1/2 -inch mesh, No. 18 U.S. gauge wire or equivalent, except for landing access.[§1926.552(b)(5)(i)]

(ii) *When a hoist tower is not enclosed,* the hoist platform or car shall be totally enclosed (caged) on all sides for the full height between the floor and the overhead protective covering with 1/2 -inch mesh of No. 14 U.S. gauge wire or equivalent. The hoist platform enclosure shall include the required gates for loading and unloading. A 6-foot high enclosure shall be provided on the unused sides of the hoist tower at ground level.[§1926.552(b)(5)(ii)]

(6) *Car arresting devices* shall be installed to function in case of rope failure.[§1926.552(b)(6)]

(7) *All material hoist towers* shall be designed by a licensed professional engineer.[§1926.552(b)(7)]

(8) *All material hoists* shall conform to the requirements of ANSI A10.5-1969, Safety Requirements for Material Hoists. [§1926.552(b)(8)]

(c) **Personnel hoists.**[§1926.552(c)]

(1) *Hoist towers outside* the structure shall be enclosed for the full height on the side or sides used for entrance and exit to the structure. At the lowest landing, the enclosure on the sides not used for exit or entrance to the structure shall be enclosed to a height of at least 10 feet. Other sides of the tower adjacent to floors or scaffold platforms shall be enclosed to a height of 10 feet above the level of such floors or scaffolds.[§1926.552(c)(1)]

(2) *Towers inside of structures* shall be enclosed on all four sides throughout the full height.[§1926.552(c)(2)]

(3) *Towers shall be anchored to the structure* at intervals not exceeding 25 feet. In addition to tie-ins, a series of guys shall be installed. Where tie-ins are not practical the tower shall be anchored by means of guys made of wire rope at least one-half inch in diameter, securely fastened to anchorage to ensure stability.[§1926.552(c)(3)]

(4) *Hoistway doors or gates* shall be not less than 6 feet 6 inches high and shall be provided with mechanical locks which cannot be operated from the landing side, and shall be accessible only to persons on the car.[§1926.552(c)(4)]

(5) *Cars shall be permanently enclosed* on all sides and the top, except sides used for entrance and exit which have car gates or doors.[§1926.552(c)(5)]

(6) *A door or gate shall be provided* at each entrance to the car which shall protect the full width and height of the car entrance opening.[§1926.552(c)(6)]

(7) *Overhead protective covering* of 2-inch planking, 3/4 -inch plywood or other solid material or equivalent strength shall be provided on the top of every personnel hoist.[§1926.552(c)(7)]

(8) *Doors or gates shall be provided* with electric contacts which do not allow movement of the hoist when door or gate is open. [§1926.552(c)(8)]

(9) *Safeties shall be capable of stopping* and holding the car and rated load when traveling at governor tripping speed. [§1926.552(c)(9)]

(10) *Cars shall be provided with* a capacity and data plate secured in a conspicuous place on the car or crosshead.[§1926.552(c)(10)]

(11) *Internal combustion engines* shall not be permitted for direct drive.[§1926.552(c)(11)]

(12) *Normal and final terminal* stopping devices shall be provided. [§1926.552(c)(12)]

(13) *An emergency stop switch* shall be provided in the car and marked "Stop."[§1926.552(c)(13)]

(14) *Ropes:*[§1926.552(c)(14)]

(i) *The minimum number of hoisting ropes* used shall be three for traction hoists and two for drum-type hoists. [§1926.552(c)(14)(i)]

(ii) *The minimum diameter* of hoisting and counterweight wire ropes shall be 1/2 -inch.[§1926.552(c)(14)(ii)]

(iii) *Safety factors:*[§1926.552(c)(14)(iii)]

Minimum Factors of Safety for Suspension Wire Ropes

Rope speed in feet per minute	Minimum factor of safety
50	7.60
75	7.75
100	7.95
125	8.10
150	8.25
175	8.40
200	8.60
225	8.75
250	8.90
300	9.20
350	9.50
400	9.75

Minimum Factors of Safety for Suspension Wire Ropes (continued)

Rope speed in feet per minute	Minimum factor of safety
450	10.00
500	10.25
550	10.45
600	10.70

(15) *Following assembly and erection of hoists,* and before being put in service, an inspection and test of all functions and safety devices shall be made under the supervision of a competent person. A similar inspection and test is required following major alteration of an existing installation. All hoists shall be inspected and tested at not more than 3-month intervals. The employer shall prepare a certification record which includes the date the inspection and test of all functions and safety devices was performed; the signature of the person who performed the inspection and test; and a serial number, or other identifier, for the hoist that was inspected and tested. The most recent certification record shall be maintained on file. [§1926.552(c)(15)]

(16) *All personnel hoists used by employees* shall be constructed of materials and components which meet the specifications for materials, construction, safety devices, assembly, and structural integrity as stated in the American National Standard A10.4-1963, Safety Requirements for Workmen's Hoists. The requirements of this paragraph (c)(16) do not apply to cantilever type personnel hoists. [§1926.552(c)(16)]

(17)(i) *Personnel hoists used* in bridge tower construction shall be approved by a registered professional engineer and erected under the supervision of a qualified engineer competent in this field.[§1926.552(c)(17)(i)]

(ii) *When a hoist tower is not enclosed,* the hoist platform or car shall be totally enclosed (caged) on all sides for the full height between the floor and the overhead protective covering with 3/4 -inch mesh of No. 14 U.S. gauge wire or equivalent. The hoist platform enclosure shall include the required gates for loading and unloading.[§1926.552(c)(17)(ii)]

(iii) *These hoists shall be inspected* and maintained on a weekly basis. Whenever the hoisting equipment is exposed to winds exceeding 35 miles per hour it shall be inspected and put in operable condition before reuse.[§1926.552(c)(17)(iii)]

(iv) *Wire rope shall be taken* out of service when any of the following conditions exist:[§1926.552(c)(17)(iv)]

[A] *In running ropes,* six randomly distributed broken wires in one lay or three broken wires in one strand in one lay; [§1926.552(c)(17)(iv)[A]]

[B] *Wear of one-third the original diameter* of outside individual wires. Kinking, crushing, bird caging, or any other damage resulting in distortion of the rope structure;[§1926.552(c)(17)(iv)[B]]

[C] *Evidence of any heat damage from any cause;* [§1926.552(c)(17)(iv)[C]]

[D] *Reductions from nominal diameter* of more than three-sixty-fourths inch for diameters to and including three-fourths inch, one-sixteenth inch for diameters seven-eights inch to 1 1/8 inches inclusive, three-thirty-seconds inch for diameters 1 1/4 to 1 1/2 inches inclusive;[§1926.552(c)(17)(iv)[D]]

[E] *In standing ropes,* more than two broken wires in one lay in sections beyond end connections or more than one broken wire at an end connection.[§1926.552(c)(17)(iv)[E]]

(d) **Permanent elevators under the care and custody** of the employer and used by employees for work covered by this Act shall comply with the requirements of American National Standards Institute A17.1-1965 with addenda A17.1a-1967, A17.1b-1968, A17.1c-1969, A17.1d-1970, and inspected in accordance with A17.2-1960 with addenda A17.2a-1965, A17.2b-1967.[§1926.552(d)]

[44 FR 8577, Feb. 9, 1979; 44 FR 20940, Apr. 6, 1979, as amended at 52 FR 36382, Sept. 28, 1987]

§1926.553
Base-mounted drum hoists

(a) **General requirements.**[§1926.553(a)]

(1) *Exposed moving parts* such as gears, projecting screws, setscrews, chain, cables, chain sprockets, and reciprocating or rotating parts, which constitute a hazard, shall be guarded. [§1926.553(a)(1)]

(2) *All controls used during* the normal operation cycle shall be located within easy reach of the operator's station. [§1926.553(a)(2)]

(3) *Electric motor operated hoists shall be provided with:* [§1926.553(a)(3)]

(i) *A device to disconnect all motors* from the line upon power failure and not permit any motor to be restarted until the controller handle is brought to the "off" position;[§1926.553(a)(3)(i)]

(ii) *Where applicable, an overspeed preventive device;* [§1926.553(a)(3)(ii)]

(iii) *A means whereby remotely operated hoists* stop when any control is ineffective.[§1926.553(a)(3)(iii)]

(4) *All base-mounted drum hoists* in use shall meet the applicable requirements for design, construction, installation, testing, inspection, maintenance, and operations, as prescribed by the manufacturer.[§1926.553(a)(4)]

(b) **Specific requirements.** [Reserved][§1926.553(b)]

(c) **This section does not apply** to base-mounted drum hoists used in conjunction with derricks. Base-mounted drum hoists used in conjunction with derricks must conform to Sec. 1926.1436(e). [§1926.553(c)]

[44 FR 8577, Feb. 9, 1979 as amended at 75 FR 48134, Aug. 9, 2010]

§1926.554
Overhead hoists

(a) **General requirements.**[§1926.554(a)]

(1) *The safe working load of the overhead hoist,* as determined by the manufacturer, shall be indicated on the hoist, and this safe working load shall not be exceeded.[§1926.554(a)(1)]

(2) *The supporting structure* to which the hoist is attached shall have a safe working load equal to that of the hoist. [§1926.554(a)(2)]

(3) *The support shall be arranged* so as to provide for free movement of the hoist and shall not restrict the hoist from lining itself up with the load.[§1926.554(a)(3)]

(4) *The hoist shall be installed* only in locations that will permit the operator to stand clear of the load at all times.[§1926.554(a)(4)]

(5) *Air hoists shall be connected* to an air supply of sufficient capacity and pressure to safely operate the hoist. All air hoses supplying air shall be positively connected to prevent their becoming disconnected during use.[§1926.554(a)(5)]

(6) *All overhead hoists in use* shall meet the applicable requirements for construction, design, installation, testing, inspection, maintenance, and operation, as prescribed by the manufacturer.[§1926.554(a)(6)]

(b) **Specific requirements.** [Reserved][§1926.554(b)]

§1926.555
Conveyors

(a) **General requirements.**[§1926.555(a)]

(1) *Means for stopping the motor or engine* shall be provided at the operator's station. Conveyor systems shall be equipped with an audible warning signal to be sounded immediately before starting up the conveyor.[§1926.555(a)(1)]

(2) *If the operator's station is at a remote point,* similar provisions for stopping the motor or engine shall be provided at the motor or engine location.[§1926.555(a)(2)]

(3) *Emergency stop switches* shall be arranged so that the conveyor cannot be started again until the actuating stop switch has been reset to running or "on" position.[§1926.555(a)(3)]

(4) *Screw conveyors shall be guarded* to prevent employee contact with turning flights.[§1926.555(a)(4)]

(5) *Where a conveyor passes over* work areas, aisles, or thoroughfares, suitable guards shall be provided to protect employees required to work below the conveyors.[§1926.555(a)(5)]

(6) *All crossovers, aisles, and passageways* shall be conspicuously marked by suitable signs, as required by subpart G of this part.[§1926.555(a)(6)]

(7) *Conveyors shall be locked out* or otherwise rendered inoperable, and tagged out with a "Do Not Operate" tag during repairs and when operation is hazardous to employees performing maintenance work.[§1926.555(a)(7)]

(8) *All conveyors in use* shall meet the applicable requirements for design, construction, inspection, testing, maintenance, and operation, as prescribed in the ANSI B20.1-1957, Safety Code for Conveyors, Cableways, and Related Equipment. [§1926.555(a)(8)]

Authority: Section 3704 of the Contract Work Hours and Safety Standards Act (40 U.S.C. 3701); Sections 4, 6, and 8 of the Occupational Safety and Health Act of 1970 (29 U.S.C. 653, 655, 657); Secretary of Labor's Order Nos. 12-71 (36 FR 8754), 8-76 (41 FR 25059), or 9-83 (49 FR 35736), and 5-2007 (72 FR 31159).

Notes

Subpart O - Motor Vehicles, Mechanized Equipment, and Marine Operations

§1926.600
Equipment

(a) **General Requirements.** [§1926.600(a)]

(1) *All equipment left unattended at night,* adjacent to a highway in normal use, or adjacent to construction areas where work is in progress, shall have appropriate lights or reflectors, or barricades equipped with appropriate lights or reflectors, to identify the location of the equipment.[§1926.600(a)(1)]

(2) *A safety tire rack, cage, or equivalent protection* shall be provided and used when inflating, mounting, or dismounting tires installed on split rims, or rims equipped with locking rings or similar devices.[§1926.600(a)(2)]

(3) (i) *Heavy machinery, equipment, or parts thereof,* which are suspended or held aloft by use of slings, hoists, or jacks shall be substantially blocked or cribbed to prevent falling or shifting before employees are permitted to work under or between them. Bulldozer and scraper blades, end-loader buckets, dump bodies, and similar equipment, shall be either fully lowered or blocked when being repaired or when not in use. All controls shall be in a neutral position, with the motors stopped and brakes set, unless work being performed requires otherwise.[§1926.600(a)(3)(i)]

(ii) *Whenever the equipment is parked,* the parking brake shall be set. Equipment parked on inclines shall have the wheels chocked and the parking brake set.[§1926.600(a)(3)(ii)]

(4) *The use, care and charging* of all batteries shall conform to the requirements of subpart K of this part.[§1926.600(a)(4)]

(5) *All cab glass shall be safety glass,* or equivalent, that introduces no visible distortion affecting the safe operation of any machine covered by this subpart.[§1926.600(a)(5)]

(6) *All equipment covered by this subpart* shall comply with the following requirements when working or being moved in the vicinity of power lines or energized transmitters, except where electrical distribution and transmission lines have been deenergized and visibly grounded at point of work or where insulating barriers, not a part of or an attachment to the equipment or machinery, have been erected to prevent physical contact with the lines:[§1926.600(a)(6)]

(i) *For lines rated 50* kV or below, minimum clearance between the lines and any part of the crane or load shall be 10 feet; [§1926.600(a)(6)(i)]

(ii) *For lines rated over* 50 kV, minimum clearance between the lines and any part of the crane or load shall be 10 feet plus 0.4 inch for each 1 kV over 50 kV, or twice the length of the line insulator, but never less than 10 feet;[§1926.600(a)(6)(ii)]

(iii) *In transit with no load and boom* lowered, the equipment clearance shall be a minimum of 4 feet for voltages less than 50 kV, and 10 feet for voltages over 50 kV, up to and including 345 kV, and 16 feet for voltages up to and including 750 kV;[§1926.600(a)(6)(iii)]

(iv) *A person shall be designated to observe* clearance of the equipment and give timely warning for all operations where it is difficult for the operator to maintain the desired clearance by visual means;[§1926.600(a)(6)(iv)]

(v) *Cage-type boom guards,* insulating links, or proximity warning devices may be used on cranes, but the use of such devices shall not alter the requirements of any other regulation of this part even if such device is required by law or regulation;[§1926.600(a)(6)(v)]

(vi) *Any overhead wire shall be considered* to be an energized line unless and until the person owning such line or the electrical utility authorities indicate that it is not an energized line and it has been visibly grounded;[§1926.600(a)(6)(vi)]

(vii) *Prior to work near* transmitter towers where an electrical charge can be induced in the equipment or materials being handled, the transmitter shall be de-energized or tests shall be made to determine if electrical charge is induced on the crane. The following precautions shall be taken when necessary to dissipate induced voltages:[§1926.600(a)(6)(vii)]

[A] The equipment shall be provided with an electrical ground directly to the upper rotating structure supporting the boom; and[§1926.600(a)(6)(vii)[A]]

[B] Ground jumper cables shall be attached to materials being handled by boom equipment when electrical charge is induced while working near energized transmitters. Crews shall be provided with nonconductive poles having large alligator clips or other similar protection to attach the ground cable to the load.[§1926.600(a)(6)(vii)[B]]

[C] Combustible and flammable materials shall be removed from the immediate area prior to operations. [§1926.600(a)(6)(vii)[C]]

(7) *Rolling railroad cars.* Derail and/or bumper blocks shall be provided on spur railroad tracks where a rolling car could contact other cars being worked, enter a building, work or traffic area.[§1926.600(a)(7)]

(b) **Specific requirements.** [Reserved][§1926.600(b)]

[44 FR 8577, Feb. 9, 1979; 44 FR 20940, Apr. 6, 1979, as amended at 58 FR 35183, June 30, 1993; 75 FR 48134, Aug. 9, 2010]

§1926.601
Motor vehicles

(a) **Coverage.** Motor vehicles as covered by this part are those vehicles that operate within an off-highway jobsite, not open to public traffic. The requirements of this section do not apply to equipment for which rules are prescribed in Sec. 1926.602.[§1926.601(a)]

(b) **General requirements.**[§1926.601(b)]

(1) *All vehicles shall have a service brake system,* an emergency brake system, and a parking brake system. These systems may use common components, and shall be maintained in operable condition.[§1926.601(b)(1)]

(2) (i) *Whenever visibility conditions* warrant additional light, all vehicles, or combinations of vehicles, in use shall be equipped with at least two headlights and two taillights in operable condition.[§1926.601(b)(2)(i)]

(ii) *All vehicles, or combination of vehicles,* shall have brake lights in operable condition regardless of light conditions. [§1926.601(b)(2)(ii)]

(3) *All vehicles shall be equipped* with an adequate audible warning device at the operator's station and in an operable condition.[§1926.601(b)(3)]

(4) *No employer shall use* any motor vehicle equipment having an obstructed view to the rear unless:[§1926.601(b)(4)]

(i) *The vehicle has a reverse signal alarm* audible above the surrounding noise level or:[§1926.601(b)(4)(i)]

(ii) *The vehicle is backed up* only when an observer signals that it is safe to do so.[§1926.601(b)(4)(ii)]

(5) *All vehicles with cabs shall be equipped* with windshields and powered wipers. Cracked and broken glass shall be replaced. Vehicles operating in areas or under conditions that cause fogging or frosting of the windshields shall be equipped with operable defogging or defrosting devices.[§1926.601(b)(5)]

(6) *All haulage vehicles,* whose pay load is loaded by means of cranes, power shovels, loaders, or similar equipment, shall have a cab shield and/or canopy adequate to protect the operator from shifting or falling materials.[§1926.601(b)(6)]

(7) *Tools and material shall be secured* to prevent movement when transported in the same compartment with employees. [§1926.601(b)(7)]

(8) *Vehicles used to transport employees* shall have seats firmly secured and adequate for the number of employees to be carried.[§1926.601(b)(8)]

(9) *Seat belts and anchorages* meeting the requirements of 49 CFR part 571 (Department of Transportation, Federal Motor Vehicle Safety Standards) shall be installed in all motor vehicles.[§1926.601(b)(9)]

(10) *Trucks with dump bodies* shall be equipped with positive means of support, permanently attached, and capable of being locked in position to prevent accidental lowering of the body while maintenance or inspection work is being done. [§1926.601(b)(10)]

(11) *Operating levers controlling hoisting or dumping devices* on haulage bodies shall be equipped with a latch or other device which will prevent accidental starting or tripping of the mechanism.[§1926.601(b)(11)]

(12) *Trip handles for tailgates* of dump trucks shall be so arranged that, in dumping, the operator will be in the clear. [§1926.601(b)(12)]

(13) (i) *All rubber-tired motor vehicle equipment* manufactured on or after May 1, 1972, shall be equipped with fenders. All rubber-tired motor vehicle equipment manufactured before May 1, 1972, shall be equipped with fenders not later than May 1, 1973.[§1926.601(b)(13)(i)]

(ii) *Mud flaps may be used* in lieu of fenders whenever motor vehicle equipment is not designed for fenders. [§1926.601(b)(13)(ii)]

(14) *All vehicles in use shall be checked at the beginning of each shift* to assure that the following parts, equipment, and accessories are in safe operating condition and free of apparent damage that could cause failure while in use: service brakes, including trailer brake connections; parking system (hand brake); emergency stopping system (brakes); tires; horn; steering mechanism; coupling devices; seat belts; operating controls; and safety devices. All defects shall be corrected before the vehicle is placed in service. These requirements also apply to equipment such as lights, reflectors, windshield wipers, defrosters, fire extinguishers, etc., where such equipment is necessary.[§1926.601(b)(14)]

§1926.602

Material handling equipment

(a) Earthmoving equipment; General.[§1926.602(a)]

(1) *These rules apply* to the following types of earthmoving equipment: scrapers, loaders, crawler or wheel tractors, bulldozers, off-highway trucks, graders, agricultural and industrial tractors, and similar equipment. The promulgation of specific rules for compactors and rubber-tired "skid-steer" equipment is reserved pending consideration of standards currently being developed.[§1926.602(a)(1)]

(2) *Seat belts.*[§1926.602(a)(2)]

(i) *Seat belts shall be provided* on all equipment covered by this section and shall meet the requirements of the Society of Automotive Engineers, J386-1969, Seat Belts for Construction Equipment. Seat belts for agricultural and light industrial tractors shall meet the seat belt requirements of Society of Automotive Engineers J333a-1970, Operator Protection for Agricultural and Light Industrial Tractors.[§1926.602(a)(2)(i)]

(ii) *Seat belts need not be provided* for equipment which is designed only for standup operation.[§1926.602(a)(2)(ii)]

(iii) *Seat belts need not be provided* for equipment which does not have roll-over protective structure (ROPS) or adequate canopy protection.[§1926.602(a)(2)(iii)]

(3) *Access roadways and grades.*[§1926.602(a)(3)]

(i) *No employer shall move or cause to be moved* construction equipment or vehicles upon any access roadway or grade unless the access roadway or grade is constructed and maintained to accommodate safely the movement of the equipment and vehicles involved.[§1926.602(a)(3)(i)]

(ii) *Every emergency access ramp* and berm used by an employer shall be constructed to restrain and control runaway vehicles.[§1926.602(a)(3)(ii)]

(4) *Brakes.* All earthmoving equipment mentioned in this Sec. 1926.602(a) shall have a service braking system capable of stopping and holding the equipment fully loaded, as specified in Society of Automotive Engineers SAE-J237, Loader Dozer-1971, J236, Graders-1971, and J319b, Scrapers-1971. Brake systems for self-propelled rubber-tired off-highway equipment manufactured after January 1, 1972 shall meet the applicable minimum performance criteria set forth in the following Society of Automotive Engineers Recommended Practices: [§1926.602(a)(4)]

Self-Propelled Scrapers	SAE J319b-1971.
Self-Propelled Graders	SAE J236-1971.
Trucks and Wagons	SAE J166-1971.
Front End Loaders and Dozers	SAE J237-1971.

(5) *Fenders.* Pneumatic-tired earth-moving haulage equipment (trucks, scrapers, tractors, and trailing units) whose maximum speed exceeds 15 miles per hour, shall be equipped with fenders on all wheels to meet the requirements of Society of Automotive Engineers SAE J321a-1970, Fenders for Pneumatic-Tired Earthmoving Haulage Equipment. An employer may, of course, at any time seek to show under Sec. 1926.2, that the uncovered wheels present no hazard to personnel from flying materials.[§1926.602(a)(5)]

(6) *Rollover protective structures (ROPS).* See subpart W of this part for requirements for rollover protective structures and overhead protection.[§1926.602(a)(6)]

(7) *Rollover protective structures for off-highway trucks.* The promulgation of standards for rollover protective structures for off- highway trucks is reserved pending further study and development.[§1926.602(a)(7)]

(8) *Specific effective dates — brakes and fenders.*[§1926.602(a)(8)]

(i) *Equipment mentioned in paragraph (a)(4) and (5)* of this section, and manufactured after January 1, 1972, which is used by any employer after that date, shall comply with the applicable rules prescribed therein concerning brakes and fenders. Equipment mentioned in paragraphs (a) (4) and (5) of this section, and manufactured before January 1, 1972, which is used by any employer after that date, shall meet the applicable rules prescribed herein not later than June 30, 1973. It should be noted that, as permitted under Sec. 1926.2, employers may request variations from the applicable brakes and fender standards required by this subpart. Employers wishing to seek variations from the applicable brakes and fenders rules may submit any requests for variations after the publication of this document in the Federal Register. Any statements intending to meet the requirements of §1926.2(b)(4), should specify how the variation would protect the safety of the employees by providing for any compensating restrictions on the operation of equipment. [§1926.602(a)(8)(i)]

(ii) *Notwithstanding the provisions of paragraphs (a)(5) and (a)(8)(i)* of this section, the requirement that fenders be installed on pneumatic-tired earthmoving haulage equipment, is suspended pending reconsideration of the requirement.[§1926.602(a)(8)(ii)]

(9) *Audible alarms.*[§1926.602(a)(9)]

(i) *All bidirectional machines, such as rollers,* compactors, front-end loaders, bulldozers, and similar equipment, shall be equipped with a horn, distinguishable from the surrounding noise level, which shall be operated as needed when the machine is moving in either direction. The horn shall be maintained in an operative condition.[§1926.602(a)(9)(i)]

(ii) *No employer shall permit earthmoving* or compacting equipment which has an obstructed view to the rear to be used in reverse gear unless the equipment has in operation a reverse signal alarm distinguishable from the surrounding noise level or an employee signals that it is safe to do so. [§1926.602(a)(9)(ii)]

(10) *Scissor points.* Scissor points on all front-end loaders, which constitute a hazard to the operator during normal operation, shall be guarded.[§1926.602(a)(10)]

(b) Excavating and other equipment.[§1926.602(b)]

(1) *Tractors covered in paragraph (a) of this section* shall have seat belts as required for the operators when seated in the normal seating arrangement for tractor operation, even though back-hoes, breakers, or other similar attachments are used on these machines for excavating or other work.[§1926.602(b)(1)]

(2) *For the purposes of this subpart* and of subpart N of this part, the nomenclatures and descriptions for measurement of dimensions of machinery and attachments shall be as described in Society of Automotive Engineers 1970 Handbook, pages 1088 through 1103.[§1926.602(b)(2)]

(3) *The safety requirements, ratios, or limitations* applicable to machines or attachment usage covered in Power Crane and Shovel Associations Standards No. 1 and No. 2 of 1968, and No. 3 of 1969, shall be complied with, and shall apply to cranes, machines, and attachments under this part. [§1926.602(b)(3)]

(c) Lifting and hauling equipment (other than equipment covered under subpart N of this part).[§1926.602(c)]

(1) *Industrial trucks* shall meet the requirements of Sec. 1926.600 and the following:[§1926.602(c)(1)]

(i) *Lift trucks, stackers, etc.,* shall have the rated capacity clearly posted on the vehicle so as to be clearly visible to the operator. When auxiliary removable counterweights are provided by the manufacturer, corresponding alternate rated capacities also shall be clearly shown on the vehicle. These ratings shall not be exceeded.[§1926.602(c)(1)(i)]

(ii) *No modifications or additions* which affect the capacity or safe operation of the equipment shall be made without the manufacturer's written approval. If such modifications or changes are made, the capacity, operation, and maintenance instruction plates, tags, or decals shall be changed accordingly. In no case shall the original safety factor of the equipment be reduced.[§1926.602(c)(1)(ii)]

(iii) *If a load is lifted* by two or more trucks working in unison, the proportion of the total load carried by any one truck shall not exceed its capacity.[§1926.602(c)(1)(iii)]

(iv) *Steering or spinner knobs* shall not be attached to the steering wheel unless the steering mechanism is of a type that prevents road reactions from causing the steering handwheel to spin. The steering knob shall be mounted within the periphery of the wheel.[§1926.602(c)(1)(iv)]

(v) *All high lift rider industrial trucks* shall be equipped with overhead guards which meet the configuration and structural requirements as defined in paragraph 421 of American National Standards Institute B56.1-1969, Safety Standards for Powered Industrial Trucks.[§1926.602(c)(1)(v)]

(vi) *All industrial trucks in use* shall meet the applicable requirements of design, construction, stability, inspection, testing, maintenance, and operation, as defined in American National Standards Institute B56.1-1969, Safety Standards for Powered Industrial Trucks.[§1926.602(c)(1)(vi)]

(vii) *Unauthorized personnel shall not be permitted* to ride on powered industrial trucks. A safe place to ride shall be provided where riding of trucks is authorized.[§1926.602(c)(1)(vii)]

(viii) *Whenever a truck is equipped* with vertical only, or vertical and horizontal controls elevatable with the lifting carriage or forks for lifting personnel, the following additional precautions shall be taken for the protection of personnel being elevated.[§1926.602(c)(1)(viii)]

[A] *Use of a safety platform* firmly secured to the lifting carriage and/or forks.[§1926.602(c)(1)(viii)[A]]

[B] *Means shall be provided whereby* personnel on the platform can shut off power to the truck.[§1926.602(c)(1)(viii)[B]]

[C] *Such protection from falling objects* as indicated necessary by the operating conditions shall be provided.[§1926.602(c)(1)(viii)[C]]

(d) Powered industrial truck operator training.[§1926.602(d)]

Note: The requirements applicable to construction work under this paragraph are identical to those set forth at Sec. 1910.178(l) of this chapter.

[44 FR 8577, Feb. 9, 1979; 44 FR 20940, Apr. 6, 1979, as amended at 58 FR 35183, June 30, 1993; 63 FR 66274, Dec. 1, 1998]

§1926.603

Pile driving equipment

(a) General requirements.[§1926.603(a)]

(1) *Boilers and piping systems which are a part of,* or used with, pile driving equipment shall meet the applicable requirements of the American Society of Mechanical Engineers, Power Boilers (section I).[§1926.603(a)(1)]

(2) *All pressure vessels which are a part of,* or used with, pile driving equipment shall meet the applicable requirements of the American Society of Mechanical Engineers, Pressure Vessels (section VIII).[§1926.603(a)(2)]

(3) *Overhead protection, which will not obscure the vision* of the operator and which meets the requirements of subpart N of this part, shall be provided. Protection shall be the equivalent of 2-inch planking or other solid material of equivalent strength.[§1926.603(a)(3)]

(4) *Stop blocks shall be provided* for the leads to prevent the hammer from being raised against the head block.[§1926.603(a)(4)]

(5) *A blocking device, capable of safely supporting* the weight of the hammer, shall be provided for placement in the leads under the hammer at all times while employees are working under the hammer.[§1926.603(a)(5)]

(6) *Guards shall be provided* across the top of the head block to prevent the cable from jumping out of the sheaves.[§1926.603(a)(6)]

(7) *When the leads must be inclined* in the driving of batter piles, provisions shall be made to stabilize the leads.[§1926.603(a)(7)]

(8) *Fixed leads shall be provided with ladder,* and adequate rings, or similar attachment points, so that the loft worker may engage his safety belt lanyard to the leads. If the leads are provided with loft platforms(s), such platform(s) shall be protected by standard guardrails.[§1926.603(a)(8)]

(9) *Steam hose leading to a steam hammer* or jet pipe shall be securely attached to the hammer with an adequate length of at least 1/4 -inch diameter chain or cable to prevent whipping in the event the joint at the hammer is broken. Air hammer hoses shall be provided with the same protection as required for steam lines.[§1926.603(a)(9)]

(10) *Safety chains, or equivalent means,* shall be provided for each hose connection to prevent the line from thrashing around in case the coupling becomes disconnected.[§1926.603(a)(10)]

(11) *Steam line controls* shall consist of two shutoff valves, one of which shall be a quick-acting lever type within easy reach of the hammer operator.[§1926.603(a)(11)]

(12) *Guys, outriggers, thrustouts, or counterbalances* shall be provided as necessary to maintain stability of pile driver rigs.[§1926.603(a)(12)]

(b) Pile driving from barges and floats. Barges or floats supporting pile driving operations shall meet the applicable requirements of Sec. 1926.605.[§1926.603(b)]

(c) Pile driving equipment.[§1926.603(c)]

(1) *Engineers and winchmen* shall accept signals only from the designated signalmen.[§1926.603(c)(1)]

(2) *All employees shall be kept clear* when piling is being hoisted into the leads.[§1926.603(c)(2)]

(3) *When piles are being driven* in an excavated pit, the walls of the pit shall be sloped to the angle of repose or sheet-piled and braced.[§1926.603(c)(3)]

(4) *When steel tube piles are being "blown out",* employees shall be kept well beyond the range of falling materials.[§1926.603(c)(4)]

(5) *When it is necessary* to cut off the tops of driven piles, pile driving operations shall be suspended except where the cutting operations are located at least twice the length of the longest pile from the driver.[§1926.603(c)(5)]

(6) *When driving jacked piles,* all access pits shall be provided with ladders and bulkheaded curbs to prevent material from falling into the pit.[§1926.603(c)(6)]

§1926.604

Site clearing

(a) General requirements.[§1926.604(a)]

(1) *Employees engaged in site clearing* shall be protected from hazards of irritant and toxic plants and suitably instructed in the first aid treatment available.[§1926.604(a)(1)]

(2) *All equipment used in site clearing operations* shall be equipped with rollover guards meeting the requirements of this subpart. In addition, rider-operated equipment shall be equipped with an overhead and rear canopy guard meeting the following requirements:[§1926.604(a)(2)]

(i) *The overhead covering on this canopy structure* shall be of not less than 1/8 -inch steel plate or 1/4 -inch woven wire mesh with openings no greater than 1 inch, or equivalent.[§1926.604(a)(2)(i)]

(ii) *The opening in the rear of the canopy structure* shall be covered with not less than 1/4 -inch woven wire mesh with openings no greater than 1 inch.[§1926.604(a)(2)(ii)]

(b) Specific requirements. [Reserved][§1926.604(b)]

§1926.605

Marine operations and equipment

(a) Material handling operations.[§1926.605(a)]

(1) *Operations fitting the definition* of "material handling" shall be performed in conformance with applicable requirements of part 1918, "Safety and Health Regulations for Longshoring" of this chapter. The term "longshoring operations" means the loading, unloading, moving, or handling of construction materials, equipment and supplies, etc. into, in, on, or out of any vessel from a fixed structure or shore-to-vessel, vessel-to- shore or fixed structure or vessel-to-vessel.[§1926.605(a)(1)]

(b) Access to barges.[§1926.605(b)]

(1) *Ramps for access of vehicles* to or between barges shall be of adequate strength, provided with side boards, well maintained, and properly secured.[§1926.605(b)(1)]

(2) *Unless employees can step safely* to or from the wharf, float, barge, or river towboat, either a ramp, meeting the requirements of paragraph (b)(1) of this section, or a safe walkway, shall be provided.[§1926.605(b)(2)]

(3) *Jacob's ladders* shall be of the double rung or flat tread type. They shall be well maintained and properly secured. [§1926.605(b)(3)]

(4) *A Jacob's ladder* shall either hang without slack from its lashings or be pulled up entirely.[§1926.605(b)(4)]

(5) *When the upper end of the means of access* rests on or is flush with the top of the bulwark, substantial steps properly secured and equipped with at least one substantial hand rail approximately 33 inches in height, shall be provided between the top of the bulwark and the deck.[§1926.605(b)(5)]

(6) *Obstructions shall not be laid on or across the gangway.* [§1926.605(b)(6)]

(7) *The means of access* shall be adequately illuminated for its full length.[§1926.605(b)(7)]

(8) *Unless the structure makes it impossible,* the means of access shall be so located that the load will not pass over employees.[§1926.605(b)(8)]

(c) Working surfaces of barges.[§1926.605(c)]

(1) *Employees shall not be permitted* to walk along the sides of covered lighters or barges with coamings more than 5 feet high, unless there is a 3-foot clear walkway, or a grab rail, or a taut handline is provided.[§1926.605(c)(1)]

(2) *Decks and other working surfaces* shall be maintained in a safe condition.[§1926.605(c)(2)]

(3) *Employees shall not be permitted to pass* fore and aft, over, or around deckloads, unless there is a safe passage. [§1926.605(c)(3)]

(4) *Employees shall not be permitted to walk* over deckloads from rail to coaming unless there is a safe passage. If it is necessary to stand at the outboard or inboard edge of the deckload where less than 24 inches of bulwark, rail, coaming, or other protection exists, all employees shall be provided with a suitable means of protection against falling from the deckload. [§1926.605(c)(4)]

(d) First-aid and lifesaving equipment.[§1926.605(d)]

(1) *Provisions for rendering first aid* and medical assistance shall be in accordance with subpart D of this part.[§1926.605(d)(1)]

(2) *The employer shall ensure* that there is in the vicinity of each barge in use at least one U.S. Coast Guard-approved 30-inch lifering with not less than 90 feet of line attached, and at least one portable or permanent ladder which will reach the top of the apron to the surface of the water. If the above equipment is not available at the pier, the employer shall furnish it during the time that he is working the barge.[§1926.605(d)(2)]

(3) *Employees walking or working* on the unguarded decks of barges shall be protected with U.S. Coast Guard-approved work vests or buoyant vests.[§1926.605(d)(3)]

(e) Commercial diving operations. Commercial diving operations shall be subject to subpart T of part 1910, §§1910.401-1910.441, of this chapter.[§1926.605(e)]

[39 FR 22801, June 24, 1974, as amended at 42 FR 37674, July 22, 1977]

§1926.606
Definitions applicable to this subpart

(a) **Apron** — The area along the waterfront edge of the pier or wharf. [§1926.606(a)]

(b) **Bulwark** — The side of a ship above the upper deck.[§1926.606(b)]

(c) **Coaming** — The raised frame, as around a hatchway in the deck, to keep out water.[§1926.606(c)]

(d) **Jacob's ladder** — A marine ladder of rope or chain with wooden or metal rungs.[§1926.606(d)]

(e) **Rail, for the purpose of Sec. 1926.605,** means a light structure serving as a guard at the outer edge of a ship's deck.[§1926.606(e)]

Authority: Section 107, Construction Work Hours and Safety Standards Act (Construction Safety Act) (40 U.S.C. 333); Secs. 4, 6, 8, Occupational Safety and Health Act of 1970 (29 U.S.C. 653, 655, 657); Secretary of Labor's Order No. 12-71 (36 FR 8754), 8-76 (41 FR 25059), 9-83 (48 FR 35736), 1-90 (55 FR 9033), 6-96 (62 FR 111), or 5-2007 (72 FR 31159), as applicable. Section 1926.602 also issued under 29 CFR part 1911.

Subpart P - Excavations

§1926.650

Scope, application, and definitions applicable to this subpart

(a) **Scope and application.** This subpart applies to all open excavations made in the earth's surface. Excavations are defined to include trenches.[§1926.650(a)]

(b) **Definitions applicable to this subpart.**

Accepted engineering practices means those requirements which are compatible with standards of practice required by a registered professional engineer.

Aluminum Hydraulic Shoring means a pre-engineered shoring system comprised of aluminum hydraulic cylinders (crossbraces) used in conjunction with vertical rails (uprights) or horizontal rails (walers). Such system is designed, specifically to support the sidewalls of an excavation and prevent cave-ins.

Bell-bottom pier hole means a type of shaft or footing excavation, the bottom of which is made larger than the cross section above to form a belled shape.

Benching (Benching system) means a method of protecting employees from cave-ins by excavating the sides of an excavation to form one or a series of horizontal levels or steps, usually with vertical or near- vertical surfaces between levels.

Cave-in means the separation of a mass of soil or rock material from the side of an excavation, or the loss of soil from under a trench shield or support system, and its sudden movement into the excavation, either by falling or sliding, in sufficient quantity so that it could entrap, bury, or otherwise injure and immobilize a person.

Competent person means one who is capable of identifying existing and predictable hazards in the surroundings, or working conditions which are unsanitary, hazardous, or dangerous to employees, and who has authorization to take prompt corrective measures to eliminate them.

Cross braces mean the horizontal members of a shoring system installed perpendicular to the sides of the excavation, the ends of which bear against either uprights or wales.

Excavation means any man-made cut, cavity, trench, or depression in an earth surface, formed by earth removal.

Faces or sides means the vertical or inclined earth surfaces formed as a result of excavation work.

Failure means the breakage, displacement, or permanent deformation of a structural member or connection so as to reduce its structural integrity and its supportive capabilities.

Hazardous atmosphere means an atmosphere which by reason of being explosive, flammable, poisonous, corrosive, oxidizing, irritating, oxygen deficient, toxic, or otherwise harmful, may cause death, illness, or injury.

Kickout means the accidental release or failure of a cross brace.

Protective system means a method of protecting employees from cave- ins, from material that could fall or roll from an excavation face or into an excavation, or from the collapse of adjacent structures. Protective systems include support systems, sloping and benching systems, shield systems, and other systems that provide the necessary protection.

Ramp means an inclined walking or working surface that is used to gain access to one point from another, and is constructed from earth or from structural materials such as steel or wood.

Registered Professional Engineer means a person who is registered as a professional engineer in the state where the work is to be performed. However, a professional engineer, registered in any state is deemed to be a "registered professional engineer" within the meaning of this standard when approving designs for "manufactured protective systems" or "tabulated data" to be used in interstate commerce.

Sheeting means the members of a shoring system that retain the earth in position and in turn are supported by other members of the shoring system.

Shield (Shield system) means a structure that is able to withstand the forces imposed on it by a cave-in and thereby protect employees within the structure. Shields can be permanent structures or can be designed to be portable and moved along as work progresses. Additionally, shields can be either premanufactured or job-built in accordance with Sec. 1926.652 (c)(3) or (c)(4). Shields used in trenches are usually referred to as "trench boxes" or "trench shields."

Shoring (Shoring system) means a structure such as a metal hydraulic, mechanical or timber shoring system that supports the sides of an excavation and which is designed to prevent cave-ins.

Sides. See "Faces."

Sloping (Sloping system) means a method of protecting employees from cave-ins by excavating to form sides of an excavation that are inclined away from the excavation so as to prevent cave-ins. The angle of incline required to prevent a cave-in varies with differences in such factors as the soil type, environmental conditions of exposure, and application of surcharge loads.

Stable rock means natural solid mineral material that can be excavated with vertical sides and will remain intact while exposed. Unstable rock is considered to be stable when the rock material on the side or sides of the excavation is secured against caving-in or movement by rock bolts or by another protective system that has been designed by a registered professional engineer.

Structural ramp means a ramp built of steel or wood, usually used for vehicle access. Ramps made of soil or rock are not considered structural ramps.

Support system means a structure such as underpinning, bracing, or shoring, which provides support to an adjacent structure, underground installation, or the sides of an excavation.

Tabulated data means tables and charts approved by a registered professional engineer and used to design and construct a protective system.

Trench (Trench excavation) means a narrow excavation (in relation to its length) made below the surface of the ground. In general, the depth is greater than the width, but the width of a trench (measured at the bottom) is not greater than 15 feet (4.6 m). If forms or other structures are installed or constructed in an excavation so as to reduce the dimension measured from the forms or structure to the side of the excavation to 15 feet (4.6 m) or less (measured at the bottom of the excavation), the excavation is also considered to be a trench.

Trench box. See "Shield."

Trench shield. See "Shield."

Uprights means the vertical members of a trench shoring system placed in contact with the earth and usually positioned so that individual members do not contact each other. Uprights placed so that individual members are closely spaced, in contact with or interconnected to each other, are often called "sheeting."

Wales means horizontal members of a shoring system placed parallel to the excavation face whose sides bear against the vertical members of the shoring system or earth.

Source: 54 FR 45959, Oct. 31, 1989, unless otherwise noted.

§1926.651

Specific excavation requirements

(a) **Surface encumbrances.** All surface encumbrances that are located so as to create a hazard to employees shall be removed or supported, as necessary, to safeguard employees.[§1926.651(a)]

(b) **Underground installations.**[§1926.651(b)]

(1) *The estimated location of utility installations,* such as sewer, telephone, fuel, electric, water lines, or any other underground installations that reasonably may be expected to be encountered during excavation work, shall be determined prior to opening an excavation.[§1926.651(b)(1)]

(2) *Utility companies or owners* shall be contacted within established or customary local response times, advised of the proposed work, and asked to establish the location of the utility underground installations prior to the start of actual excavation. When utility companies or owners cannot respond to a request to locate underground utility installations within 24 hours (unless a longer period is required by state or local law), or cannot establish the exact location of these installations, the employer may proceed, provided the employer does so with caution, and provided detection equipment or other acceptable means to locate utility installations are used.[§1926.651(b)(2)]

(3) *When excavation operations* approach the estimated location of underground installations, the exact location of the installations shall be determined by safe and acceptable means. [§1926.651(b)(3)]

(4) *While the excavation is open,* underground installations shall be protected, supported or removed as necessary to safeguard employees.[§1926.651(b)(4)]

(c) Access and egress[§1926.651(c)]

(1) *Structural ramps.*[§1926.651(c)(1)]

(i) *Structural ramps that are used solely by employees* as a means of access or egress from excavations shall be designed by a competent person. Structural ramps used for access or egress of equipment shall be designed by a competent person qualified in structural design, and shall be constructed in accordance with the design.[§1926.651(c)(1)(i)]

(ii) *Ramps and runways constructed* of two or more structural members shall have the structural members connected together to prevent displacement.[§1926.651(c)(1)(ii)]

(iii) *Structural members used* for ramps and runways shall be of uniform thickness.[§1926.651(c)(1)(iii)]

(iv) *Cleats or other appropriate means* used to connect runway structural members shall be attached to the bottom of the runway or shall be attached in a manner to prevent tripping.[§1926.651(c)(1)(iv)]

(v) *Structural ramps used* in lieu of steps shall be provided with cleats or other surface treatments on the top surface to prevent slipping.[§1926.651(c)(1)(v)]

(2) *Means of egress from trench excavations.* A stairway, ladder, ramp or other safe means of egress shall be located in trench excavations that are 4 feet (1.22 m) or more in depth so as to require no more than 25 feet (7.62 m) of lateral travel for employees.[§1926.651(c)(2)]

(d) Exposure to vehicular traffic. Employees exposed to public vehicular traffic shall be provided with, and shall wear, warning vests or other suitable garments marked with or made of reflectorized or high- visibility material.[§1926.651(d)]

(e) Exposure to falling loads. No employee shall be permitted underneath loads handled by lifting or digging equipment. Employees shall be required to stand away from any vehicle being loaded or unloaded to avoid being struck by any spillage or falling materials. Operators may remain in the cabs of vehicles being loaded or unloaded when the vehicles are equipped, in accordance with Sec. 1926.601(b)(6), to provide adequate protection for the operator during loading and unloading operations.[§1926.651(e)]

(f) Warning system for mobile equipment. When mobile equipment is operated adjacent to an excavation, or when such equipment is required to approach the edge of an excavation, and the operator does not have a clear and direct view of the edge of the excavation, a warning system shall be utilized such as barricades, hand or mechanical signals, or stop logs. If possible, the grade should be away from the excavation.[§1926.651(f)]

(g) Hazardous atmospheres[§1926.651(g)]

(1) *Testing and controls.* In addition to the requirements set forth in subparts D and E of this part (29 CFR 1926.50-1926.107) to prevent exposure to harmful levels of atmospheric contaminants and to assure acceptable atmospheric conditions, the following requirements shall apply:[§1926.651(g)(1)]

(i) *Where oxygen deficiency* (atmospheres containing less than 19.5 percent oxygen) or a hazardous atmosphere exists or could reasonably be expected to exist, such as in excavations in landfill areas or excavations in areas where hazardous substances are stored nearby, the atmospheres in the excavation shall be tested before employees enter excavations greater than 4 feet (1.22 m) in depth.[§1926.651(g)(1)(i)]

(ii) *Adequate precautions shall be taken* to prevent employee exposure to atmospheres containing less than 19.5 percent oxygen and other hazardous atmospheres. These precautions include providing proper respiratory protection or ventilation in accordance with subparts D and E of this part respectively.[§1926.651(g)(1)(ii)]

(iii) *Adequate precaution shall be taken* such as providing ventilation, to prevent employee exposure to an atmosphere containing a concentration of a flammable gas in excess of 20 percent of the lower flammable limit of the gas.[§1926.651(g)(1)(iii)]

(iv) *When controls are used* that are intended to reduce the level of atmospheric contaminants to acceptable levels, testing shall be conducted as often as necessary to ensure that the atmosphere remains safe.[§1926.651(g)(1)(iv)]

(2) *Emergency rescue equipment.*[§1926.651(g)(2)]

(i) *Emergency rescue equipment, such as breathing apparatus,* a safety harness and line, or a basket stretcher, shall be readily available where hazardous atmospheric conditions exist or may reasonably be expected to develop during work in an excavation. This equipment shall be attended when in use.[§1926.651(g)(2)(i)]

(ii) *Employees entering bell-bottom pier holes,* or other similar deep and confined footing excavations, shall wear a harness with a life- line securely attached to it. The lifeline shall be separate from any line used to handle materials, and shall be individually attended at all times while the employee wearing the lifeline is in the excavation.[§1926.651(g)(2)(ii)]

(h) Protection from hazards associated with water accumulation. [§1926.651(h)]

(1) *Employees shall not work in excavations* in which there is accumulated water, or in excavations in which water is accumulating, unless adequate precautions have been taken to protect employees against the hazards posed by water accumulation. The precautions necessary to protect employees adequately vary with each situation, but could include special support or shield systems to protect from cave-ins, water removal to control the level of accumulating water, or use of a safety harness and lifeline.[§1926.651(h)(1)]

(2) *If water is controlled or prevented* from accumulating by the use of water removal equipment, the water removal equipment and operations shall be monitored by a competent person to ensure proper operation.[§1926.651(h)(2)]

(3) *If excavation work interrupts the natural drainage* of surface water (such as streams), diversion ditches, dikes, or other suitable means shall be used to prevent surface water from entering the excavation and to provide adequate drainage of the area adjacent to the excavation. Excavations subject to runoff from heavy rains will require an inspection by a competent person and compliance with paragraphs (h)(1) and (h)(2) of this section.[§1926.651(h)(3)]

(i) Stability of adjacent structures.[§1926.651(i)]

(1) *Where the stability of adjoining buildings, walls, or other structures* is endangered by excavation operations, support systems such as shoring, bracing, or underpinning shall be provided to ensure the stability of such structures for the protection of employees.[§1926.651(i)(1)]

(2) *Excavation below the level of the base* or footing of any foundation or retaining wall that could be reasonably expected to pose a hazard to employees shall not be permitted except when: [§1926.651(i)(2)]

(i) *A support system, such as underpinning,* is provided to ensure the safety of employees and the stability of the structure; or[§1926.651(i)(2)(i)]

(ii) *The excavation is in stable rock; or*[§1926.651(i)(2)(ii)]

(iii) *A registered professional engineer has approved* the determination that the structure is sufficently removed from the excavation so as to be unaffected by the excavation activity; or[§1926.651(i)(2)(iii)]

(iv) *A registered professional engineer has approved* the determination that such excavation work will not pose a hazard to employees.[§1926.651(i)(2)(iv)]

(3) *Sidewalks, pavements, and appurtenant structure* shall not be undermined unless a support system or another method of protection is provided to protect employees from the possible collapse of such structures.[§1926.651(i)(3)]

(j) Protection of employees from loose rock or soil.[§1926.651(j)]

(1) *Adequate protection shall be provided* to protect employees from loose rock or soil that could pose a hazard by falling or rolling from an excavation face. Such protection shall consist of scaling to remove loose material; installation of protective barricades at intervals as necessary on the face to stop and contain falling material; or other means that provide equivalent protection.[§1926.651(j)(1)]

(2) *Employees shall be protected from excavated or other materials* or equipment that could pose a hazard by falling or rolling into excavations. Protection shall be provided by placing and keeping such materials or equipment at least 2 feet (.61 m) from the edge of excavations, or by the use of retaining devices that are sufficient to prevent materials or equipment from falling or rolling into excavations, or by a combination of both if necessary.[§1926.651(j)(2)]

(k) Inspections.[§1926.651(k)]

(1) *Daily inspections of excavations, the adjacent areas, and protective systems* shall be made by a competent person for evidence of a situation that could result in possible cave-ins, indications of failure of protective systems, hazardous atmospheres, or other hazardous conditions. An inspection

shall be conducted by the competent person prior to the start of work and as needed throughout the shift. Inspections shall also be made after every rainstorm or other hazard increasing occurrence. These inspections are only required when employee exposure can be reasonably anticipated. [§1926.651(k)(1)]

(2) *Where the competent person finds evidence of a situation* that could result in a possible cave-in, indications of failure of protective systems, hazardous atmospheres, or other hazardous conditions, exposed employees shall be removed from the hazardous area until the necessary precautions have been taken to ensure their safety. [§1926.651(k)(2)]

(l) **Walkways shall be provided** where employees or equipment are required or permitted to cross over excavations. Guardrails which comply with Sec. 1926.502(b) shall be provided where walkways are 6 feet (1.8 m) or more above lower levels. [§1926.651(l)]

[54 FR 45959, Oct. 31, 1989, as amended by 59 FR 40730, Aug. 9, 1994]

§1926.652
Requirements for protective systems

(a) **Protection of employees in excavations.** [§1926.652(a)]

(1) *Each employee in an excavation* shall be protected from cave-ins by an adequate protective system designed in accordance with paragraph (b) or (c) of this section except when: [§1926.652(a)(1)]

(i) *Excavations are made entirely in stable rock; or* [§1926.652(a)(1)(i)]

(ii) *Excavations are less than 5 feet (1.52m)* in depth and examination of the ground by a competent person provides no indication of a potential cave-in. [§1926.652(a)(1)(ii)]

(2) *Protective systems* shall have the capacity to resist without failure all loads that are intended or could reasonably be expected to be applied or transmitted to the system. [§1926.652(a)(2)]

(b) **Design of sloping and benching systems.** The slopes and configurations of sloping and benching systems shall be selected and constructed by the employer or his designee and shall be in accordance with the requirements of paragraph (b)(1); or, in the alternative, paragraph (b)(2); or, in the alternative, paragraph (b)(3), or, in the alternative, paragraph (b)(4), as follows: [§1926.652(b)]

(1) *Option (1) — Allowable configurations and slopes.* [§1926.652(b)(1)]

(i) *Excavations shall be sloped at an angle* not steeper than one and one-half horizontal to one vertical (34 degrees measured from the horizontal), unless the employer uses one of the other options listed below. [§1926.652(b)(1)(i)]

(ii) *Slopes specified in paragraph (b)(1)(i) of this section,* shall be excavated to form configurations that are in accordance with the slopes shown for Type C soil in Appendix B to this subpart. [§1926.652(b)(1)(ii)]

(2) *Option (2) — Determination of slopes and configurations* using Appendices A and B. Maximum allowable slopes, and allowable configurations for sloping and benching systems, shall be determined in accordance with the conditions and requirements set forth in appendices A and B to this subpart. [§1926.652(b)(2)]

(3) *Option (3) — Designs using other tabulated data.* [§1926.652(b)(3)]

(i) *Designs of sloping or benching systems* shall be selected from and be in accordance with tabulated data, such as tables and charts. [§1926.652(b)(3)(i)]

(ii) *The tabulated data* shall be in written form and shall include all of the following: [§1926.652(b)(3)(ii)]

[A] Identification of the parameters that affect the selection of a sloping or benching system drawn from such data; [§1926.652(b)(3)(ii)[A]]

[B] Identification of the limits of use of the data, to include the magnitude and configuration of slopes determined to be safe; [§1926.652(b)(3)(ii)[B]]

[C] Explanatory information as may be necessary to aid the user in making a correct selection of a protective system from the data. [§1926.652(b)(3)(ii)[C]]

(iii) *At least one copy of the tabulated data* which identifies the registered professional engineer who approved the data, shall be maintained at the jobsite during construction of the protective system. After that time the data may be stored off the jobsite, but a copy of the data shall be made available to the Secretary upon request. [§1926.652(b)(3)(iii)]

(4) *Option (4) — Design by a registered professional engineer.* [§1926.652(b)(4)]

(i) *Sloping and benching systems* not utilizing Option (1) or Option (2) or Option (3) under paragraph (b) of this section shall be approved by a registered professional engineer. [§1926.652(b)(4)(i)]

(ii) *Designs shall be in written form* and shall include at least the following: [§1926.652(b)(4)(ii)]

[A] The magnitude of the slopes that were determined to be safe for the particular project; [§1926.652(b)(4)(ii)[A]]

[B] The configurations that were determined to be safe for the particular project; and [§1926.652(b)(4)(ii)[B]]

[C] The identity of the registered professional engineer approving the design. [§1926.652(b)(4)(ii)[C]]

(iii) *At least one copy of the design* shall be maintained at the jobsite while the slope is being constructed. After that time the design need not be at the jobsite, but a copy shall be made available to the Secretary upon request. [§1926.652(b)(4)(iii)]

(c) **Design of support systems, shield systems, and other protective systems.** Designs of support systems shield systems, and other protective systems shall be selected and constructed by the employer or his designee and shall be in accordance with the requirements of paragraph (c)(1); or, in the alternative, paragraph (c)(2); or, in the alternative, paragraph (c)(3); or, in the alternative, paragraph (c)(4) as follows: [§1926.652(c)]

(1) *Option (1) — Designs using appendices A, C and D.* Designs for timber shoring in trenches shall be determined in accordance with the conditions and requirements set forth in appendices A and C to this subpart. Designs for aluminum hydraulic shoring shall be in accordance with paragraph (c)(2) of this section, but if manufacturer's tabulated data cannot be utilized, designs shall be in accordance with appendix D. [§1926.652(c)(1)]

(2) *Option (2) — Designs Using Manufacturer's Tabulated Data.* [§1926.652(c)(2)]

(i) *Design of support systems, shield systems,* or other protective systems that are drawn from manufacturer's tabulated data shall be in accordance with all specifications, recommendations, and limitations issued or made by the manufacturer. [§1926.652(c)(2)(i)]

(ii) *Deviation from the specifications, recommendations,* and limitations issued or made by the manufacturer shall only be allowed after the manufacturer issues specific written approval. [§1926.652(c)(2)(ii)]

(iii) *Manufacturer's specifications, recommendations,* and limitations, and manufacturer's approval to deviate from the specifications, recommendations, and limitations shall be in written form at the jobsite during construction of the protective system. After that time this data may be stored off the jobsite, but a copy shall be made available to the Secretary upon request. [§1926.652(c)(2)(iii)]

(3) *Option (3) — Designs using other tabulated data.* [§1926.652(c)(3)]

(i) *Designs of support systems,* shield systems, or other protective systems shall be selected from and be in accordance with tabulated data, such as tables and charts. [§1926.652(c)(3)(i)]

(ii) *The tabulated data* shall be in written form and include all of the following: [§1926.652(c)(3)(ii)]

[A] Identification of the parameters that affect the selection of a protective system drawn from such data; [§1926.652(c)(3)(ii)[A]]

[B] Identification of the limits of use of the data; [§1926.652(c)(3)(ii)[B]]

[C] Explanatory information as may be necessary to aid the user in making a correct selection of a protective system from the data. [§1926.652(c)(3)(ii)[C]]

(iii) *At least one copy of the tabulated data,* which identifies the registered professional engineer who approved the data, shall be maintained at the jobsite during construction of the protective system. After that time the data may be stored off the jobsite, but a copy of the data shall be made available to the Secretary upon request. [§1926.652(c)(3)(iii)]

(4) *Option (4) — Design by a registered professional engineer.* [§1926.652(c)(4)]

(i) *Support systems, shield systems,* and other protective systems not utilizing Option 1, Option 2 or Option 3, above, shall be approved by a registered professional engineer. [§1926.652(c)(4)(i)]

(ii) *Designs shall be in written form* and shall include the following: [§1926.652(c)(4)(ii)]

[A] A plan indicating the sizes, types, and configurations of the materials to be used in the protective system; and [§1926.652(c)(4)(ii)[A]]

[B] The identity of the registered professional engineer approving the design. [§1926.652(c)(4)(ii)[B]]

(iii) *At least one copy of the design* shall be maintained at the jobsite during construction of the protective system. After that time, the design may be stored off the jobsite, but a copy of the design shall be made available to the Secretary upon request. [§1926.652(c)(4)(iii)]

(d) **Materials and equipment.** [§1926.652(d)]

(1) *Materials and equipment* used for protective systems shall be free from damage or defects that might impair their proper function. [§1926.652(d)(1)]

(2) *Manufactured materials and equipment* used for protective systems shall be used and maintained in a manner that is consistent with the recommendations of the manufacturer, and in a manner that will prevent employee exposure to hazards. [§1926.652(d)(2)]

(3) *When material or equipment that is used* for protective systems is damaged, a competent person shall examine the material or equipment and evaluate its suitability for continued use. If the competent person cannot assure the material or equipment is able to support the intended loads or is otherwise suitable for safe use, then such material or equipment shall be removed from service, and shall be evaluated and approved by a registered professional engineer before being returned to service. [§1926.652(d)(3)]

(e) **Installation and removal of support.** [§1926.652(e)]

(1) *General.* [§1926.652(e)(1)]

(i) *Members of support systems* shall be securely connected together to prevent sliding, falling, kickouts, or other predictable failure. [§1926.652(e)(1)(i)]

(ii) *Support systems shall be installed* and removed in a manner that protects employees from cave-ins, structural collapses, or from being struck by members of the support system. [§1926.652(e)(1)(ii)]

(iii) *Individual members of support systems* shall not be subjected to loads exceeding those which those members were designed to withstand. [§1926.652(e)(1)(iii)]

(iv) *Before temporary removal of individual members begins,* additional precautions shall be taken to ensure the safety of employees, such as installing other structural members to carry the loads imposed on the support system. [§1926.652(e)(1)(iv)]

(v) *Removal shall begin at, and progress from,* the bottom of the excavation. Members shall be released slowly so as to note any indication of possible failure of the remaining members of the structure or possible cave-in of the sides of the excavation. [§1926.652(e)(1)(v)]

(vi) *Backfilling shall progress together* with the removal of support systems from excavations. [§1926.652(e)(1)(vi)]

(2) *Additional requirements* for support systems for trench excavations. [§1926.652(e)(2)]

(i) *Excavation of material to a level* no greater than 2 feet (.61 m) below the bottom of the members of a support system shall be permitted, but only if the system is designed to resist the forces calculated for the full depth of the trench, and there are no indications while the trench is open of a possible loss of soil from behind or below the bottom of the support system. [§1926.652(e)(2)(i)]

(ii) *Installation of a support system* shall be closely coordinated with the excavation of trenches. [§1926.652(e)(2)(ii)]

(f) **Sloping and benching systems.** Employees shall not be permitted to work on the faces of sloped or benched excavations at levels above other employees except when employees at the lower levels are adequately protected from the hazard of falling, rolling, or sliding material or equipment. [§1926.652(f)]

(g) **Shield systems.** [§1926.652(g)]

(1) *General.* [§1926.652(g)(1)]

(i) *Shield systems shall not be subjected* to loads exceeding those which the system was designed to withstand. [§1926.652(g)(1)(i)]

(ii) *Shields shall be installed* in a manner to restrict lateral or other hazardous movement of the shield in the event of the application of sudden lateral loads. [§1926.652(g)(1)(ii)]

(iii) *Employees shall be protected* from the hazard of cave-ins when entering or exiting the areas protected by shields. [§1926.652(g)(1)(iii)]

(iv) *Employees shall not be allowed* in shields when shields are being installed, removed, or moved vertically. [§1926.652(g)(1)(iv)]

(2) *Additional requirement for shield systems* used in trench excavations. Excavations of earth material to a level not greater than 2 feet (.61 m) below the bottom of a shield shall be permitted, but only if the shield is designed to resist the forces calculated for the full depth of the trench, and there are no indications while the trench is open of a possible loss of soil from behind or below the bottom of the shield. [§1926.652(g)(2)]

1926 Subpart P
Appendix A
Soil Classification

(a) **Scope and application.**

(1) *Scope.* This appendix describes a method of classifying soil and rock deposits based on site and environmental conditions, and on the structure and composition of the earth deposits. The appendix contains definitions, sets forth requirements, and describes acceptable visual and manual tests for use in classifying soils.

(2) *Application.* This appendix applies when a sloping or benching system is designed in accordance with the requirements set forth in Sec. 1926.652(b)(2) as a method of protection for employees from cave- ins. This appendix also applies when timber shoring for excavations is designed as a method of protection from cave-ins in accordance with appendix C to subpart P of part 1926, and when aluminum hydraulic shoring is designed in accordance with appendix D. This Appendix also applies if other protective systems are designed and selected for use from data prepared in accordance with the requirements set forth in Sec. 1926.652(c), and the use of the data is predicated on the use of the soil classification system set forth in this appendix.

(b) **Definitions.** The definitions and examples given below are based on, in whole or in part, the following: American Society for Testing Materials (ASTM) Standards D653-85 and D2488; The Unified Soils Classification System, The U.S. Department of Agriculture (USDA) Textural Classification Scheme; and The National Bureau of Standards Report BSS-121.

Cemented soil means a soil in which the particles are held together by a chemical agent, such as calcium carbonate, such that a hand-size sample cannot be crushed into powder or individual soil particles by finger pressure.

Cohesive soil means clay (fine grained soil), or soil with a high clay content, which has cohesive strength. Cohesive soil does not crumble, can be excavated with vertical sideslopes, and is plastic when moist. Cohesive soil is hard to break up when dry, and exhibits significant cohesion when submerged. Cohesive soils include clayey silt, sandy clay, silty clay, clay and organic clay.

Dry soil means soil that does not exhibit visible signs of moisture content.

Fissured means a soil material that has a tendency to break along definite planes of fracture with little resistance, or a material that exhibits open cracks, such as tension cracks, in an exposed surface.

Granular soil means gravel, sand, or silt, (coarse grained soil) with little or no clay content. Granular soil has no cohesive strength. Some moist granular soils exhibit apparent cohesion. Granular soil cannot be molded when moist and crumbles easily when dry.

Layered system means two or more distinctly different soil or rock types arranged in layers. Micaceous seams or weakened planes in rock or shale are considered layered.

Moist soil means a condition in which a soil looks and feels damp. Moist cohesive soil can easily be shaped into a ball and rolled into small diameter threads before crumbling. Moist granular soil that contains some cohesive material will exhibit signs of cohesion between particles.

Plastic means a property of a soil which allows the soil to be deformed or molded without cracking, or appreciable volume change.

Saturated soil means a soil in which the voids are filled with water. Saturation does not require flow. Saturation, or near saturation, is necessary for the proper use of instruments such as a pocket penetrometer or sheer vane.

Soil classification system means, for the purpose of this subpart, a method of categorizing soil and rock deposits in a hierarchy of Stable Rock, Type A, Type B, and Type C, in decreasing order of stability. The categories are determined based on an analysis of the properties and performance characteristics of the deposits and the environmental conditions of exposure.

Stable rock means natural solid mineral matter that can be excavated with vertical sides and remain intact while exposed.

Submerged soil means soil which is underwater or is free seeping.

Type A means cohesive soils with an unconfined compressive strength of 1.5 ton per square foot (tsf) (144 kPa) or greater. Examples of cohesive soils are: clay, silty clay, sandy clay, clay loam and, in some cases, silty clay loam and sandy clay loam. Cemented soils such as caliche and hardpan are also considered Type A. However, no soil is Type A if:

(i) *The soil is fissured; or*

(ii) *The soil is subject to vibration* from heavy traffic, pile driving, or similar effects; or

(iii) *The soil has been* previously disturbed; or

(iv) *The soil is part of a sloped,* layered system where the layers dip into the excavation on a slope of four horizontal to one vertical (4H:1V) or greater; or

(v) *The material is subject to other factors* that would require it to be classified as a less stable material.

Type B means:

(i) *Cohesive soil with an unconfined* compressive strength greater than 0.5 tsf (48 kPa) but less than 1.5 tsf (144 kPa); or

(ii) *Granular cohesionless soils* including: angular gravel (similar to crushed rock), silt, silt loam, sandy loam and, in some cases, silty clay loam and sandy clay loam.

(iii) *Previously disturbed soils* except those which would otherwise be classed as Type C soil.

(iv) *Soil that meets the unconfined* compressive strength or cementation requirements for Type A, but is fissured or subject to vibration; or

(v) *Dry rock that is not stable; or*

(vi) *Material that is part of a sloped,* layered system where the layers dip into the excavation on a slope less steep than four horizontal to one vertical (4H:1V), but only if the material would otherwise be classified as Type B.

Type C means:

(i) *Cohesive soil with an unconfined* compressive strength of 0.5 tsf (48 kPa) or less; or

(ii) *Granular soils including* gravel, sand, and loamy sand; or

(iii) *Submerged soil or soil* from which water is freely seeping; or

(iv) *Submerged rock that is not stable, or*

(v) *Material in a sloped,* layered system where the layers dip into the excavation or a slope of four horizontal to one vertical (4H:1V) or steeper.

Unconfined compressive strength means the load per unit area at which a soil will fail in compression. It can be determined by laboratory testing, or estimated in the field using a pocket penetrometer, by thumb penetration tests, and other methods.

Wet soil means soil that contains significantly more moisture than moist soil, but in such a range of values that cohesive material will slump or begin to flow when vibrated. Granular material that would exhibit cohesive properties when moist will lose those cohesive properties when wet.

(c) Requirements.

(1) *Classification of soil and rock deposits.* Each soil and rock deposit shall be classified by a competent person as Stable Rock, Type A, Type B, or Type C in accordance with the definitions set forth in paragraph (b) of this appendix.

(2) *Basis of classification.* The classification of the deposits shall be made based on the results of at least one visual and at least one manual analysis. Such analyses shall be conducted by a competent person using tests described in paragraph (d) below, or in other recognized methods of soil classification and testing such as those adopted by the America Society for Testing Materials, or the U.S. Department of Agriculture textural classification system.

(3) *Visual and manual analyses.* The visual and manual analyses, such as those noted as being acceptable in paragraph (d) of this appendix, shall be designed and conducted to provide sufficient quantitative and qualitative information as may be necessary to identify properly the properties, factors, and conditions affecting the classification of the deposits.

(4) *Layered systems.* In a layered system, the system shall be classified in accordance with its weakest layer. However, each layer may be classified individually where a more stable layer lies under a less stable layer.

(5) *Reclassification.* If, after classifying a deposit, the properties, factors, or conditions affecting its classification change in any way, the changes shall be evaluated by a competent person. The deposit shall be reclassified as necessary to reflect the changed circumstances.

(d) Acceptable visual and manual tests.

(1) *Visual tests.* Visual analysis is conducted to determine qualitative information regarding the excavation site in general, the soil adjacent to the excavation, the soil forming the sides of the open excavation, and the soil taken as samples from excavated material.

(i) *Observe samples of soil* that are excavated and soil in the sides of the excavation. Estimate the range of particle sizes and the relative amounts of the particle sizes. Soil that is primarily composed of fine- grained material is cohesive material. Soil composed primarily of coarse-grained sand or gravel is granular material.

(ii) *Observe soil as it is excavated.* Soil that remains in clumps when excavated is cohesive. Soil that breaks up easily and does not stay in clumps is granular.

(iii) *Observe the side of the opened* excavation and the surface area adjacent to the excavation. Crack-like openings such as tension cracks could indicate fissured material. If chunks of soil spall off a vertical side, the soil could be fissured. Small spalls are evidence of moving ground and are indications of potentially hazardous situations.

(iv) *Observe the area adjacent* to the excavation and the excavation itself for evidence of existing utility and other underground structures, and to identify previously disturbed soil.

(v) *Observe the opened side* of the excavation to identify layered systems. Examine layered systems to identify if the layers slope toward the excavation. Estimate the degree of slope of the layers.

(vi) *Observe the area adjacent* to the excavation and the sides of the opened excavation for evidence of surface water, water seeping from the sides of the excavation, or the location of the level of the water table.

(vii) *Observe the area adjacent* to the excavation and the area within the excavation for sources of vibration that may affect the stability of the excavation face.

(2) *Manual tests.* Manual analysis of soil samples is conducted to determine quantitative as well as qualitative properties of soil and to provide more information in order to classify soil properly.

(i) *Plasticity.* Mold a moist or wet sample of soil into a ball and attempt to roll it into threads as thin as 1/8 -inch in diameter. Cohesive material can be successfully rolled into threads without crumbling. For example, if at least a two inch (50 mm) length of 1/8 - inch thread can be held on one end without tearing, the soil is cohesive.

(ii) *Dry strength.* If the soil is dry and crumbles on its own or with moderate pressure into individual grains or fine powder, it is granular (any combination of gravel, sand, or silt). If the soil is dry and falls into clumps which break up into smaller clumps, but the smaller clumps can only be broken up with difficulty, it may be clay in any combination with gravel, sand or silt. If the dry soil breaks into clumps which do not break up into small clumps and which can only be broken with difficulty, and there is no visual indication the soil is fissured, the soil may be considered unfissured.

(iii) *Thumb penetration.* The thumb penetration test can be used to estimate the unconfined compressive strength of cohesive soils. (This test is based on the thumb penetration test described in American Society for Testing and Materials (ASTM) Standard designation D2488 — "Standard Recommended Practice for Description of Soils (Visual — Manual Procedure).") Type A soils with an unconfined compressive strength of 1.5 tsf can be readily indented by the thumb; however, they can be penetrated by the thumb only with very great effort. Type C soils with an unconfined compressive strength of 0.5 tsf can be easily penetrated several inches by the thumb, and can be molded by light finger pressure. This test should be conducted on an undisturbed soil sample, such as a large clump of spoil, as soon as practicable after excavation to keep to a miminum the effects of exposure to drying influences. If the excavation is later exposed to wetting influences (rain, flooding), the classification of the soil must be changed accordingly.

(iv) *Other strength tests.* Estimates of unconfined compressive strength of soils can also be obtained by use of a pocket penetrometer or by using a hand-operated shearvane.

(v) *Drying test.* The basic purpose of the drying test is to differentiate between cohesive material with fissures, unfissured cohesive material, and granular material. The procedure for the drying test involves drying a sample of soil that is approximately one inch thick (2.54 cm) and six inches (15.24 cm) in diameter until it is thoroughly dry:

[A] *If the sample develops* cracks as it dries, significant fissures are indicated.

[B] *Samples that dry without cracking* are to be broken by hand. If considerable force is necessary to break a sample, the soil has significant cohesive material content. The soil can be classified as a unfissured cohesive material and the unconfined compressive strength should be determined.

[C] *If a sample breaks* easily by hand, it is either a fissured cohesive material or a granular material. To distinguish between the two, pulverize the dried clumps of the sample by hand or by stepping on them. If the clumps do not pulverize easily, the material is cohesive with fissures. If they pulverize easily into very small fragments, the material is granular.

1926 Subpart P Appendix B

Sloping and Benching

(a) **Scope and application.** This appendix contains specifications for sloping and benching when used as methods of protecting employees working in excavations from cave-ins. The requirements of this appendix apply when the design of sloping and benching protective systems is to be performed in accordance with the requirements set forth in §1926.652(b)(2).

(b) **Definitions.**

Actual slope means the slope to which an excavation face is excavated.

Distress means that the soil is in a condition where a cave-in is imminent or is likely to occur. Distress is evidenced by such phenomena as the development of fissures in the face of or adjacent to an open excavation; the subsidence of the edge of an excavation; the slumping of material from the face or the bulging or heaving of material from the bottom of an excavation; the spalling of material from the face of an excavation; and ravelling, i.e., small amounts of material such as pebbles or little clumps of material suddenly separating from the face of an excavation and trickling or rolling down into the excavation.

Maximum allowable slope means the steepest incline of an excavation face that is acceptable for the most favorable site conditions as protection against cave-ins, and is expressed as the ratio of horizontal distance to vertical rise (H:V).

Short term exposure means a period of time less than or equal to 24 hours that an excavation is open.

(c) **Requirements.**

(1) *Soil classification.* Soil and rock deposits shall be classified in accordance with appendix A to subpart P of part 1926.

(2) *Maximum allowable slope.* The maximum allowable slope for a soil or rock deposit shall be determined from Table B-1 of this appendix.

(3) *Actual slope.*

(i) *The actual slope shall not be steeper* than the maximum allowable slope.

(ii) *The actual slope shall be less* steep than the maximum allowable slope, when there are signs of distress. If that situation occurs, the slope shall be cut back to an actual slope which is at least 1/2 horizontal to one vertical (1/2H:1V) less steep than the maximum allowable slope.

(iii) *When surcharge loads from stored* material or equipment, operating equipment, or traffic are present, a competent person shall determine the degree to which the actual slope must be reduced below the maximum allowable slope, and shall assure that such reduction is achieved. Surcharge loads from adjacent structures shall be evaluated in accordance with §1926.651(i).

(4) *Configurations.* Configurations of sloping and benching systems shall be in accordance with Figure B-1.

Table B-1 - Maximum Allowable Slopes

Soil or Rock Type	Maximum Allowable Slopes (H:V)[1] For Excavations Less Than 20 Feet Deep[3]
Stable Rock	Vertical (90°)
Type A[2]	3/4:1 (53°)
Type B	1:1 (45°)
Type C	1 1/2:1 (34°)

[1] Numbers shown in parentheses next to maximum allowable slopes are angles expressed in degrees from the horizontal. Angles have been rounded off.

[2] A short-term maximum allowable slope of 1/2H:1V (63°) is allowed in excavations in Type A soil that are 12 feet (3.67 m) or less in depth. Short-term maximum allowable slopes for excavations greater than 12 feet (3.67 m) in depth shall be 3/4H:1V (53°).

[3] Sloping or benching for excavations greater than 20 feet deep shall be designed by a registered professional engineer.

FIGURE B-1
SLOPE CONFIGURATIONS
(All slopes stated below are in the horizontal to vertical ratio)
B-1.1 EXCAVATIONS MADE IN TYPE A SOIL

1. All simple slope excavations 20 feet or less in depth shall have a maximum allowable slope of 3/4:1

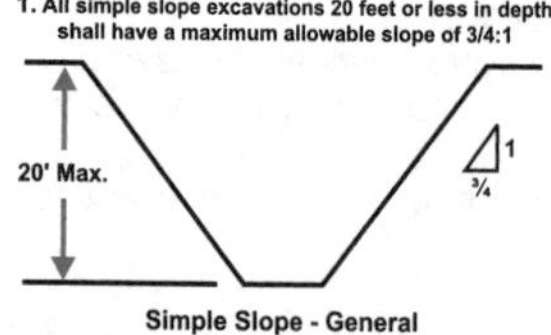

Simple Slope - General

Exception: Simple slope excavations which are open 24 hours or less (short term) and which are 12 feet or less in depth shall have a maximum allowable slope of 1/2:1.

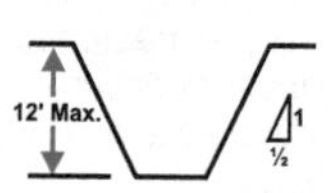

Simple Slope - Short Term

2. All benched excavations 20 feet or less in depth shall have a maximum allowable slope of 3/4:1 and maximum bench dimensions as follows:

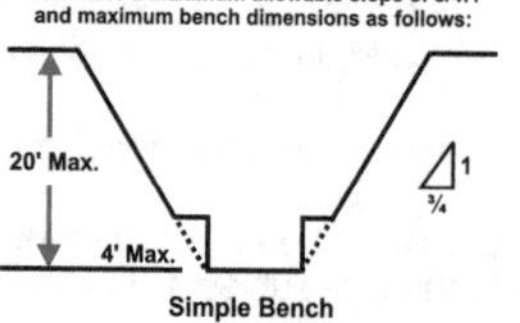

Simple Bench

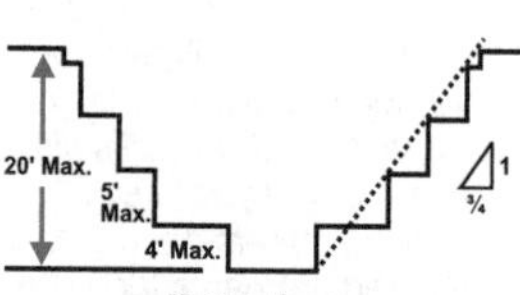

Multiple Bench

3. All excavations 8 feet or less in depth which have unsupported vertically sided lower portions shall have a maximum vertical side of 3 ½ feet.

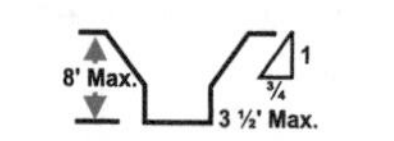

Unsupported Vertically Sided Lower Portion- Maximum 8 Feet in Depth

All excavations more than 8 feet but not more than 12 feet in depth with unsupported vertically sided lower portions shall have a maximum allowable slope of 1:1 and a maximum vertical side of 3 ½ feet.

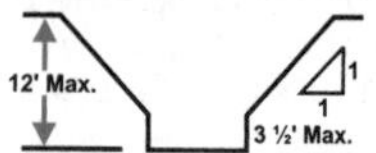

Unsupported Vertically Sided Lower Portion- Maximum 12 Feet in Depth

All excavations 20 feet or less in depth which have vertically sided lower portions that are supported or shielded shall have a maximum allowable slope of 3/4:1. The support or shield system must extend at least 18 inches above the top of the vertical side.

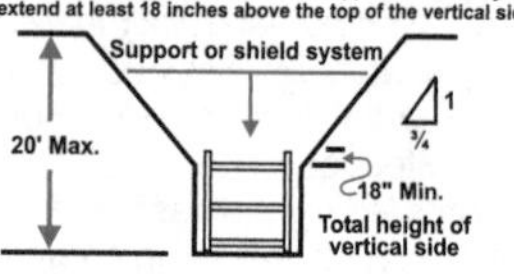

Supported or Shielded Vertically Sided Lower Portion

4. All other simple slope, compound slope, and vertically sided lower portion excavations shall be in accordance with the other options permitted under §1926.652(b).

Editor's Note: In Figure B-1, the CFR contains text that states "2." As it appears clear based on the figure and its context that this was a typographical error, the text is corrected in Figure B-1 appearing above.

FIGURE B-1.2
EXCAVATIONS MADE IN TYPE B SOIL

1. All simple slope excavations 20 feet or less in depth shall have a maximum allowable slope of 1:1.

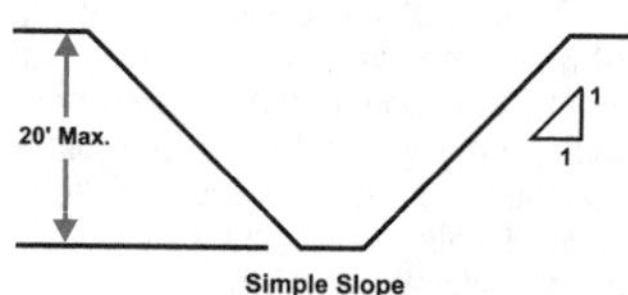

Simple Slope

2. All benched excavations 20 feet or less in depth shall have a maximum allowable slope of 1:1 and maximum bench dimensions as follows:

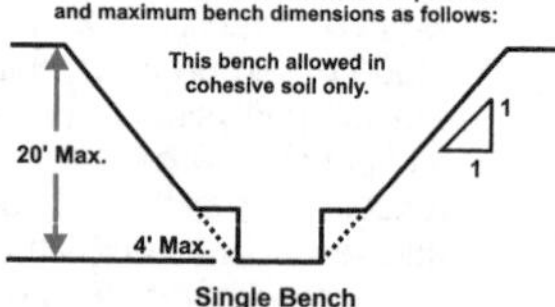

Single Bench

This bench allowed in cohesive soil only.

20' Max.
4' Max.
4' Max.
1
1

Multiple Bench

3. All excavations 20 feet or less in depth which have vertically sided lower portions shall be shielded or supported to a height at least 18 inches above the top of the vertical side. All such excavations shall have a maximum allowable slope of 1:1.

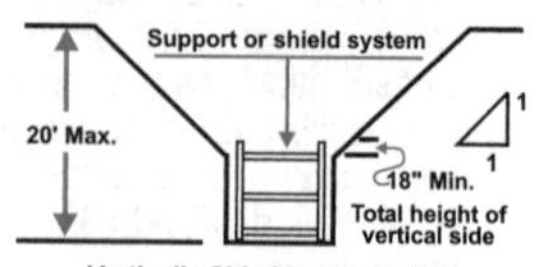

Vertically Sided Lower Portion

4. All other sloped excavations shall be in accordance with the other options permitted in §1926.652(b).

FIGURE B-1.3
EXCAVATIONS MADE IN TYPE C SOIL

1. All simple slope excavations 20 feet or less in depth shall have a maximum allowable slope of 1 ½:1

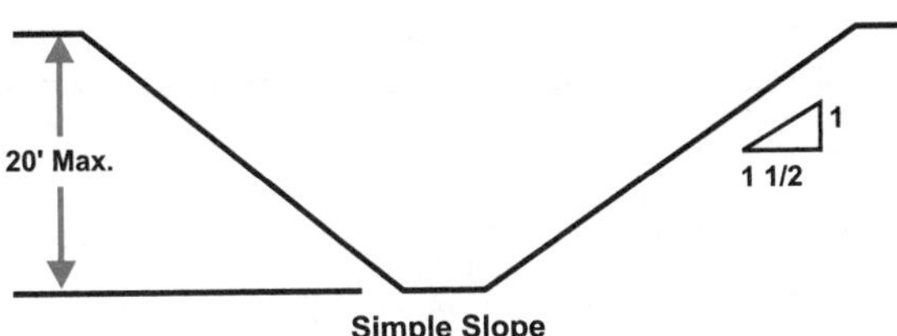

Simple Slope

2. All excavations 20 feet or less in depth which have vertically sided lower portions shall be shielded or supported to a height at least 18 inches above the top of the vertical side. All such excavations shall have a maximum allowable slope of 1 ½:1.

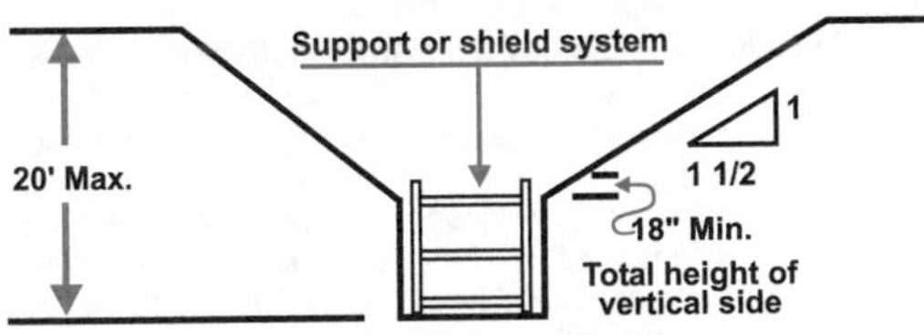

Vertically Sided Lower Portion

3. All other sloped excavations shall be in accordance with the other options permitted in §1926.652(b).

FIGURE B-1.4
EXCAVATIONS MADE IN LAYERED SOILS

1. All excavations 20 feet or less in depth made in layered soils shall have a maximum allowable slope for each layer as set forth below.

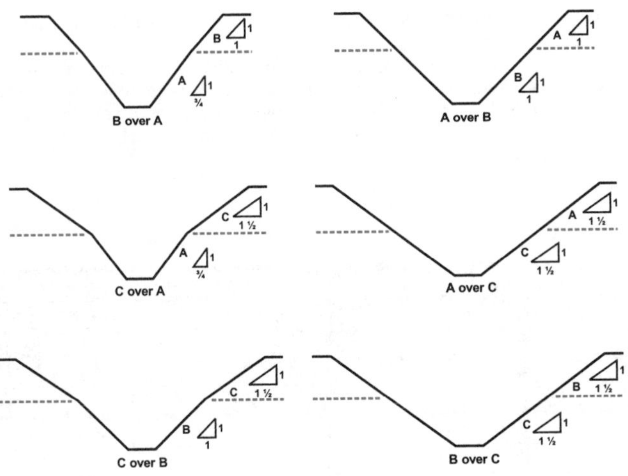

2. All other sloped excavations shall be in accordance with the other options permitted in §1926.652(b).

1926 Subpart P Appendix C

Timber Shoring for Trenches

(a) **Scope.** This appendix contains information that can be used when timber shoring is provided as a method of protection from cave-ins in trenches that do not exceed 20 feet (6.1 m) in depth. This appendix must be used when design of timber shoring protective systems is to be performed in accordance with §1926.652(c)(1). Other timber shoring configurations; other systems of support such as hydraulic and pneumatic systems; and other protective systems such as sloping, benching, shielding, and freezing systems must be designed in accordance with the requirements set forth in §1926.652(b) and §1926.652(c).

(b) **Soil Classification.** In order to use the data presented in this appendix, the soil type or types in which the excavation is made must first be determined using the soil classification method set forth in appendix A of subpart P of this part.

(c) **Presentation of Information.** Information is presented in several forms as follows:

(1) *Information is presented in tabular* form in Tables C-1.1, C-1.2, and C-1.3, and Tables C-2.1, C-2.2 and C-2.3 following paragraph (g) of the appendix. Each table presents the minimum sizes of timber members to use in a shoring system, and each table contains data only for the particular soil type in which the excavation or portion of the excavation is made. The data are arranged to allow the user the flexibility to select from among several acceptable configurations of members based on varying the horizontal spacing of the crossbraces. Stable rock is exempt from shoring requirements and therefore, no data are presented for this condition.

(2) *Information concerning the basis* of the tabular data and the limitations of the data is presented in paragraph (d) of this appendix, and on the tables themselves.

(3) *Information explaining the use* of the tabular data is presented in paragraph (e) of this appendix.

(4) *Information illustrating the use* of the tabular data is presented in paragraph (f) of this appendix.

(5) *Miscellaneous notations regarding* Tables C-1.1 through C-1.3 and Tables C-2.1 through C-2.3 are presented in paragraph (g) of this Appendix.

(d) **Basis and limitations of the data.**

(1) *Dimensions of timber members.*

(i) *The sizes of the timber members* listed in Tables C-1.1 through C-1.3 are taken from the National Bureau of Standards (NBS) report, "Recommended Technical Provisions for Construction Practice in Shoring and Sloping of Trenches and Excavations." In addition, where NBS did not recommend specific sizes of members, member sizes are based on an analysis of the sizes required for use by existing codes and on empirical practice.

(ii) *The required dimensions of the members* listed in Tables C-1.1 through C-1.3 refer to actual dimensions and not nominal dimensions of the timber. Employers wanting to use nominal size shoring are directed to Tables C-2.1 through C-2.3, or have this choice under §1926.652(c)(3), and are referred to The Corps of Engineers, The Bureau of Reclamation or data from other acceptable sources.

(2) *Limitation of application.*

(i) *It is not intended that the timber shoring specification* apply to every situation that may be experienced in the field. These data were developed to apply to the situations that are most commonly experienced in current trenching practice. Shoring systems for use in situations that are not covered by the data in this appendix must be designed as specified in §1926.652(c).

(ii) *When any of the following conditions are present,* the members specified in the tables are not considered adequate. Either an alternate timber shoring system must be designed or another type of protective system designed in accordance with §1926.652.

[A] When loads imposed by structures or by stored material adjacent to the trench weigh in excess of the load imposed by a two-foot soil surcharge. The term "adjacent" as used here means the area within a horizontal distance from the edge of the trench equal to the depth of the trench.

[B] When vertical loads imposed on cross braces exceed a 240-pound gravity load distributed on a one-foot section of the center of the crossbrace.

[C] When surcharge loads are present from equipment weighing in excess of 20,000 pounds.

[D] When only the lower portion of a trench is shored and the remaining portion of the trench is sloped or benched unless: The sloped portion is sloped at an angle less steep than three horizontal to one vertical; or the members are selected from the tables for use at a depth which is determined from the top of the overall trench, and not from the toe of the sloped portion.

(e) **Use of Tables.** The members of the shoring system that are to be selected using this information are the cross braces, the uprights, and the wales, where wales are required. Minimum sizes of members are specified for use in different types of soil. There are six tables of information, two for each soil type. The soil type must first be determined in accordance with the soil classification system described in appendix A to subpart P of part 1926. Using the appropriate table, the selection of the size and spacing of the members is then made. The selection is based on the depth and width of the trench where the members are to be installed and, in most instances, the selection is also based on the horizontal spacing of the crossbraces. Instances where a choice of horizontal spacing of crossbracing is available, the horizontal spacing of the crossbraces must be chosen by the user before the size of any member can be determined. When the soil type, the width and depth of the trench, and the horizontal spacing of the crossbraces are known, the size and vertical spacing of the crossbraces, the size and vertical spacing of the wales, and the size and horizontal spacing of the uprights can be read from the appropriate table.

(f) **Examples to Illustrate the Use** of Tables C-1.1 through C-1.3.

(1) *Example 1.*

A trench dug in Type A soil is 13 feet deep and five feet wide.

From Table C-1.1, for acceptable arrangements of timber can be used.

Arrangement #B1

Space 4x4 crossbraces at six feet horizontally and four feet vertically.

Wales are not required.

Space 3x8 uprights at six feet horizontally. This arrangement is commonly called "skip shoring."

Arrangement #B2

Space 4x6 crossbraces at eight feet horizontally and four feet vertically.

Space 8x8 wales at four feet vertically.

Space 2x6 uprights at four feet horizontally.

Arrangement #B3

Space 6x6 crossbraces at 10 feet horizontally and four feet vertically.

Space 8x10 wales at four feet vertically.

Space 2x6 uprights at five feet horizontally.

Arrangement #B4

Space 6x6 crossbraces at 12 feet horizontally and four feet vertically.

Space 10x10 wales at four feet vertically.

Spaces 3x8 uprights at six feet horizontally.

(2) *Example 2.*

A trench dug in Type B soil in 13 feet deep and five feet wide. From Table C-1.2 three acceptable arrangements of members are listed.

Arrangement #B1

Space 6x6 crossbraces at six feet horizontally and five feet vertically.

Space 8x8 wales at five feet vertically.

Space 2x6 uprights at two feet horizontally.

Arrangement #B2

Space 6x8 crossbraces at eight feet horizontally and five feet vertically.

Space 10x10 wales at five feet vertically.

Space 2x6 uprights at two feet horizontally.

Arrangement #B3

Space 8x8 crossbraces at 10 feet horizontally and five feet vertically.

Space 10x12 wales at five feet vertically.

Space 2x6 uprights at two feet vertically.

(3) *Example 3.*

A trench dug in Type C soil is 13 feet deep and five feet wide.

From Table C-1.3 two acceptable arrangements of members can be used.

Arrangement #B1

Space 8x8 crossbraces at six feet horizontally and five feet vertically.

Space 10x12 wales at five feet vertically.

Position 2x6 uprights as closely together as possible.

If water must be retained use special tongue and groove uprights to form tight sheeting.

Arrangement #B2

Space 8x10 crossbraces at eight feet horizontally and five feet vertically.

Space 12x12 wales at five feet vertically.

Position 2x6 uprights in a close sheeting configuration unless water pressure must be resisted. Tight sheeting must be used where water must be retained.

(4) *Example 4.*

A trench dug in Type C soil is 20 feet deep and 11 feet wide. The size and spacing of members for the section of trench that is over 15 feet in depth is determined using Table C-1.3. Only one arrangement of members is provided.

Space 8x10 crossbraces at six feet horizontally and five feet vertically.

Space 12x12 wales at five feet vertically.

Use 3x6 tight sheeting.

Use of Tables C-2.1 through C-2.3 would follow the same procedures.

(g) **Notes for all Tables.**

1. *Member sizes at spacings* other than indicated are to be determined as specified in §1926.652(c), "Design of Protective Systems."
2. *When conditions are saturated or submerged* use Tight Sheeting. Tight Sheeting refers to the use of specially-edged timber planks (e.g., tongue and groove) at least three inches thick, steel sheet piling, or similar construction that when driven or placed in position provide a tight wall to resist the lateral pressure of water and to prevent the loss of backfill material. Close Sheeting refers to the placement of planks side-by-side allowing as little space as possible between them.
3. *All spacing indicated is measured* center to center.
4. *Wales to be installed with greater* dimension horizontal.
5. *If the vertical distance* from the center of the lowest crossbrace to the bottom of the trench exceeds two and one-half feet, uprights shall be firmly embedded or a mudsill shall be used. Where uprights are embedded, the vertical distance from the center of the lowest crossbrace to the bottom of the trench shall not exceed 36 inches. When mudsills are used, the vertical distance shall not exceed 42 inches. Mudsills are wales that are installed at the toe of the trench side.
6. *Trench jacks may be used* in lieu of or in combination with timber crossbraces.
7. *Placement of crossbraces.* When the vertical spacing of crossbraces is four feet, place the top crossbrace no more than two feet below the top of the trench. When the vertical spacing of crossbraces is five feet, place the top crossbrace no more than 2.5 feet below the top of the trench.

Table C-1.1 Timber Trench Shoring - Minimum Timber Requirements*
Soil Type A $P_a = 25 \times H + 72$ psf (2 ft. Surcharge)

Depth of trench (feet)	Size(actual) and spacing of members**														
	Cross braces							Wales		Uprights					
	Horiz. spacing (feet)	Width of trench (feet)					Vertical spacing (feet)	Size (in.)	Vertical spacing (feet)	Maximum allowable horizontal spacing (feet)					
		Up to 4	Up to 6	Up to 9	Up to 12	Up to 15				Close	4	5	6	8
5 to 10	Up to 6	4 X 4	4 X 4	4 X 6	6 X 6	6 X 6	4	Not req'd	—				2 X 6	
	Up to 8	4 X 4	4 X 4	4 X 6	6 X 6	6 X 6	4	Not req'd	—					2 X 8
	Up to 10	4 X 6	4 X 6	4 X 6	6 X 6	6 X 6	4	8 X 8	4			2 X 6		
	Up to 12	4 X 6	4 X 6	6 X 6	6 X 6	6 X 6	4	8 X 8	4				2 X 6	
10 to 15	Up to 6	4 X 4	4 X 4	4 X 6	6 X 6	6 X 6	4	Not req'd	—				3 X 8	
	Up to 8	4 X 6	4 X 6	6 X 6	6 X 6	6 X 6	4	8 X 8	4		2 X 6			
	Up to 10	6 X 6	6 X 5	6 X 6	6 X 8	6 X 8	4	8 X 10	4			2 X 6		
	Up to 12	6 X 6	6 X 6	6 X 6	6 X 8	6 X 8	4	10 X 10	4				3 X 8	
15 to 20	Up to 6	6 X 6	6 X 6	6 X 6	6 X 8	6 X 8	4	6 X 8	4	3 X 6				
	Up to 8	6 X 6	6 X 6	6 X 6	6 X 8	6 X 8	4	8 X 8	4	3 X 6				
	Up to 10	8 X 8	8 X 8	8 X 8	8 X 8	8 X 10	4	8 X 10	4	3 X 6				
	Up to 12	8 X 8	8 X 8	8 X 8	8 X 8	8 X 10	4	10 X 10	4	3 X 6				
Over 20	See Note 1													

* Mixed oak or equivalent with a bending strength not less than 850 psi.
** Manufactured members of equivalent strength may be substituted for wood.

Table C-1.2 Timber Trench Shoring - Minimum Timber Requirements *
Soil Type B P_a = 45 X H + 72 psf (2 ft. Surcharge)

Depth of trench (feet)	Size (actual) and spacing of members**													
	Cross braces							Wales		Uprights				
	Horiz. spacing (feet)	Width of trench (feet)					Vertical spacing (feet)	Size (in.)	Vertical spacing (feet)	Maximum allowable horizontal spacing (feet)				
		Up to 4	Up to 6	Up to 9	Up to 12	Up to 15				Close	2	3		
5 to 10	Up to 6	4 X 6	4 X 6	6 X 6	6 X 6	6 X 6	5	6 X 8	5			2 X 6		
	Up to 8	6 X 6	6 X 6	6 X 6	6 X 8	6 X 8	5	8 X 10	5			2 X 6		
	Up to 10	6 X 6	6 X 6	6 X 6	6 X 8	6 X 8	5	10 X 10	5			2 X 6		
	See Note 1													
10 to 15	Up to 6	6 X 6	6 X 6	6 X 6	6 X 8	6 X 8	5	8 X 8	5		2 X 6			
	Up to 8	6 X 8	6 X 8	6 X 8	8 X 8	8 X 8	5	10 X 10	5		2 X 6			
	Up to 10	8 X 8	8 X 8	8 X 8	8 X 8	8 X 10	5	10 X 12	5		2 X 6			
	See Note 1													
15 to 20	Up to 6	6 X 8	6 X 8	6 X 8	8 X 8	8 X 8	5	8 X 10	5	3 X 6				
	Up to 8	8 X 8	8 X 8	8 X 8	8 X 8	8 X 10	5	10 X 12	5	3 X 6				
	Up to 10	8 X 10	8 X 10	8 X 10	8 X 10	10 X 10	5	12 X 12	5	3 X 6				
	See Note 1													
Over 20	See Note 1													

* Mixed oak or equivalent with a bending strength not less than 850 psi.
** Manufactured members of equivalent strength may be substituted for wood.

Table C-1.3 Timber Trench Shoring - Minimum Timber Requirements*
Soil Type C P_a = 80 X H + 72 psf (2 ft. Surcharge)

Depth of trench (feet)	Size (actual) and spacing of members**													
	Cross braces							Wales		Uprights				
	Horiz. spacing (feet)	Width of trench (feet)					Vertical spacing (feet)	Size (in.)	Vertical spacing (feet)	Maximum allowable horizontal spacing (feet) (See Note 2)				
		Up to 4	Up to 6	Up to 9	Up to 12	Up to 15				Close				
5 to 10	Up to 6	6X8	6X8	6X8	8X8	8X8	5	8X10	5	2X6				
	Up to 8	8X8	8X8	8X8	8X8	8X10	5	10X12	5	2X6				
	Up to 10	8X10	8X10	8X10	8X10	10X10	5	12X12	5	2X6				
	See Note 1													
10 to 15	Up to 6	8X8	8X8	8X8	8X8	8X10	5	10X12	5	2X6				
	Up to 8	8X10	8X10	8X10	8X10	10X10	5	12X12	5	2X6				
	See Note 1													
	See Note 1													
15 to 20	Up to 6	8X10	8X10	8X10	8X10	10X10	5	12X12	5	3X6				
	See Note 1													
	See Note 1													
	See Note 1													
Over 20	See Note 1													

* Mixed oak or equivalent with a bending strength not less than 850 psi.
** Manufactured members of equivalent strength may be substituted for wood

Table C-2.1 Timber Trench Shoring - Minimum Timber Requirements*
Soil Type A P_a = 25 X H ± 72 psf (2 ft. Surcharge)

Depth of trench (feet)	Size (S4S) and spacing of members**													
	Cross braces							Wales		Uprights				
	Horiz. spacing (feet)	Width of trench (feet)					Vert. spacing (feet)	Size (in.)	Vert. spacing (feet)	Maximum allowable horizontal spacing (feet)				
		Up to 4	Up to 6	Up to 9	Up to 12	Up to 15				Close	4	5	6	8
5 to 10	Up to 6	4 X 4	4 X 4	4 X 4	4 X 4	4 X 6	4	Not req'd	Not req'd				4 X 6	
	Up to 8	4 X 4	4 X 4	4 X 4	4 X 6	4 X 6	4	Not req'd	Not req'd					4 X 8
	Up to 10	4 X 6	4 X 6	4 X 6	6 X 6	6 X 6	4	8 X 8	4			4 X 6		
	Up to 12	4 X 6	4 X 6	4 X 6	6 X 6	6 X 6	4	8 X 8	4				4 X 6	
10 to 15	Up to 6	4 X 4	4 X 4	4 X 4	6 x 6	6 X 6	4	Not req'd	Not req'd				4 X 10	
	Up to 8	4 X 6	4 X 6	4 X 6	6 X 6	6 X 6	4	6 X 8	4		4 X 6			
	Up to 10	6 X 6	6 X 6	6 X 6	6 X 6	6 X 6	4	8 X 8	4			4 X 8		
	Up to 12	6 X 6	6 X 6	6 X 6	6 X 6	6 X 6	4	8 X 10	4		4 X 6		4 X 10	
15 to 20	Up to 6	6 X 6	6 X 6	6 X 6	6 X 6	6 X 6	4	6 X 8	4	3 X 6				
	Up to 8	6 X 6	6 X 6	6 X 6	6 X 6	6 X 6	4	8 X 8	4	3 X 6	4 X 12			
	Up to 10	6 X 6	6 X 6	6 X 6	6 X 6	6 X 8	4	8 X 10	4	3 X 6				
	Up to 12	6 X 6	6 X 6	6 X 6	6 X 8	6 X 8	4	8 X 12	4	3 X 6	4 X 12			
Over 20	See Note 1													

* Douglas fir or equivalent with a bending strength not less than 1500 psi.
** Manufactured members of equivalent strength may be substituted for wood

Table C-2.2 Timber Trench Shoring - Minimum Timber Requirements *
Soil Type B P_a =45 X H + 72 psf (2 ft. Surcharge)

Depth of trench (feet)	Size (S4S) and spacing of members**														
	Cross braces							Wales		Uprights				
	Horiz. spacing (feet)	Width of trench (feet)					Vertical spacing (feet)	Size (in.)	Vertical spacing (feet)	Maximum allowable horizontal spacing (feet)				
		Up to 4	Up to 6	Up to 9	Up to 12	Up to 15				Close	2	3	4	6
5 to 10	Up to 6	4 X 6	4 X 6	4 X 6	6 X 6	6 X 6	5	6 X 8	5			3 X 12 4 X 8		4 X 12
	Up to 8	4 X 6	4 X 6	6 X 6	6 X 6	6 X 6	5	8 X 8	5		3 X 8		4 X 8	
	Up to 10	4 X 6	4 X 6	6 X 6	6 X 6	6 X 8	5	8 X 10	5			4 X 8		
	See Note 1													
10 to 15	Up to 6	6 X 6	6 X 6	6 X 6	6 X 8	6 X 8	5	8 X 8	5	3 X 6	4 X 10			
	Up to 8	6 X 8	6 X 8	6 X 8	8 X 8	8 X 8	5	10 X 10	5	3 X 6	4 X 10			
	Up to 10	6 X 8	6 X 8	8 X 8	8 X 8	8 X 8	5	10 X 12	5	3 X 6	4 X 10			
	See Note 1													
15 to 20	Up to 6	6 X 8	6 X 8	6 X 8	6 X 8	8 X 8	5	8 X 10	5	4 X 6				
	Up to 8	6 X 8	6 X 8	6 X 8	8 X 8	8 X 8	5	10 X 12	5	4 X 6				
	Up to 10	8 X 8	8 X 8	8 X 8	8 X 8	8 X 8	5	12 X 12	5	4 X 6				
	See Note 1													
Over 20	See Note 1													

* Douglas fir or equivalent with a bending strength not less than 1500 psi.
** Manufactured members of equivalent strength may be substituted for wood.

Table C-2.3 Timber Trench Shoring - Minimum Timber Requirements *
Soil Type C P_a = 80 X H + 72 psf (2 ft. Surcharge)

Depth of trench (feet)	Size (S4S) and spacing of members**														
	Cross braces							Wales		Uprights				
	Horiz. spacing (feet)	Width of trench (feet)					Vert. spacing (feet)	Size (in.)	Vert. spacing (feet)	Maximum allowable horizontal spacing (feet)				
		Up to 4	Up to 6	Up to 9	Up to 12	Up to 15				Close				
5 to 10	Up to 6	6 X 6	6 X 6	6 X 6	6 X 6	8 X 8	5	8 X 8	5	3 X 6				
	Up to 8	6 X 6	6 X 6	6 X 6	8 X 8	8 X 8	5	10 X 10	5	3 X 6				
	Up to 10	6 X 6	6 X 6	8 X 8	8 X 8	8 X 8	5	10 X 12	5	3 X 6				
	See Note 1													
10 to 15	Up to 6	6 X 8	6 X 8	6 X 8	8 X 8	8 X 8	5	10 X 10	5	4 X 6				
	Up to 8	8 X 8	8 X 8	8 X 8	8 X 8	8 X 8	5	12 X 12	5	4 X 6				
	See Note 1													
	See Note 1													
15 to 20	Up to 6	8 X 8	8 X 8	8 X 8	8 X 10	8 X 10	5	10 X 12	5	4 X 6				
	See Note 1													
	See Note 1													
	See Note 1													
Over 20	See Note 1													

* Douglas fir or equivalent with a bending strength not less than 1500 psi.
** Manufactured members of equivalent strength may be substituted for wood.

1926 Subpart P Appendix D

Aluminum Hydraulic Shoring for Trenches

(a) **Scope.** This appendix contains information that can be used when aluminum hydraulic shoring is provided as a method of protection against cave-ins in trenches that do not exceed 20 feet (6.1m) in depth. This appendix must be used when design of the aluminum hydraulic protective system cannot be performed in accordance with §1926.652(c)(2).

(b) **Soil Classification.** In order to use data presented in this appendix, the soil type or types in which the excavation is made must first be determined using the soil classification method set forth in appendix A of subpart P of part 1926.

(c) **Presentation of Information.** Information is presented in several forms as follows:

(1) *Information is presented in tabular form* in Tables D-1.1, D-1.2, D-1.3 and D-1.4. Each table presents the maximum vertical and horizontal spacings that may be used with various aluminum member sizes and various hydraulic cylinder sizes. Each table contains data only for the particular soil type in which the excavation or portion of the excavation is made. Tables D-1.1 and D-1.2 are for vertical shores in Types A and B soil. Tables D-1.3 and D1.4 are for horizontal waler systems in Types B and C soil.

(2) *Information concerning the basis* of the tabular data and the limitations of the data is presented in paragraph (d) of this appendix.

(3) *Information explaining the use* of the tabular data is presented in paragraph (e) of this appendix.

(4) *Information illustrating the use* of the tabular data is presented in paragraph (f) of this appendix.

(5) *Miscellaneous notations (footnotes)* regarding Table D-1.1 through D-1.4 are presented in paragraph (g) of this appendix.

(6) *Figures, illustrating typical installations* of hydraulic shoring, are included just prior to the Tables. The illustrations page is entitled "Aluminum Hydraulic Shoring; Typical Installations."

(d) **Basis and limitations of the data.**

(1) *Vertical shore rails* and horizontal wales are those that meet the Section Modulus requirements in the D-1 Tables. Aluminum material is 6061-T6 or material of equivalent strength and properties.

(2) *Hydraulic cylinders specifications.*

(i) *2-inch* cylinders shall be a minimum 2-inch inside diameter with a minimum safe working capacity of no less than 18,000 pounds axial compressive load at maximum extension. Maximum extension is to include full range of cylinder extensions as recommended by product manufacturer.

(ii) *3-inch* cylinders shall be a minimum 3-inch inside diameter with a safe working capacity of not less than 30,000 pounds axial compressive load at extensions as recommended by product manufacturer.

(3) *Limitation of application.*

(i) *It is not intended* that the aluminum hydraulic specification apply to every situation that may be experienced in the field. These data were developed to apply to the situations that are most commonly experienced in current trenching practice. Shoring systems for use in situations that are not covered by the data in this appendix must be otherwise designed as specified in §1926.652(c).

(ii) *When any of the following conditions are present,* the members specified in the Tables are not considered adequate. In this case, an alternative aluminum hydraulic shoring system or other type of protective system must be designed in accordance with §1926.652.

[A] When vertical loads imposed on cross braces exceed a 100 Pound gravity load distributed on a one foot section of the center of the hydraulic cylinder.

[B] When surcharge loads are present from equipment weighing in excess of 20,000 pounds.

[C] When only the lower portion or a trench is shored and the remaining portion of the trench is sloped or benched unless: The sloped portion is sloped at an angle less steep than three horizontal to one vertical; or the members are selected from the tables for use at a depth which is determined from the top of the overall trench, and not from the toe of the sloped portion.

(e) **Use of Tables D-1.1, D-1.2, D-1.3 and D-1.4.** The members of the shoring system that are to be selected using this information are the hydraulic cylinders, and either the vertical shores or the horizontal wales. When a waler system is used the vertical timber sheeting to be used is also selected from these tables. The Tables D-1.1 and D-1.2 for vertical shores are used in Type A and B soils that do not require sheeting. Type B soils that may require sheeting, and Type C soils that always require sheeting are found in the horizontal wale Tables D-1.3 and D-1.4. The soil type must first be determined in accordance with the soil classification system described in appendix A to subpart P of part 1926. Using the appropriate table, the selection of the size and spacing of the members is made. The selection is based on the depth and width of the trench where the members are to be installed. In these tables the vertical spacing is held constant at four feet on center. The tables show the maximum horizontal spacing of cylinders allowed for each size of wale in the waler system tables, and in the vertical shore tables, the hydraulic cylinder horizontal spacing is the same as the vertical shore spacing.

(f) **Example to Illustrate the Use of the Tables:**

(1) *Example 1:* A trench dug in Type A soil is 6 feet deep and 3 feet wide. From Table D-1.1: Find vertical shores and 2 inch diameter cylinders spaced 8 feet on center (o.c.) horizontally and 4 feet on center (o.c.) vertically. (See Figures 1 & 3 for typical installations.)

(2) *Example 2:* A trench is dug in Type B soil that does not require sheeting, 13 feet deep and 5 feet wide. From Table D-1.2: Find vertical shores and 2 inch diameter cylinders spaced 6.5 feet o.c. horizontally and 4 feet o.c. vertically. (See Figures 1 & 3 for typical installations.)

(3) *A trench is dug in Type* B soil that does not require sheeting, but does experience some minor raveling of the trench face. The trench is 16 feet deep and 9 feet wide. From Table D-1.2: Find vertical shores and 2 inch diameter cylinder (with special oversleeves as designated by footnote # B2) spaced 5.5 feet o.c. horizontally and 4 feet o.c. vertically, plywood (per footnote (g)(7) to the D-1 Table) should be used behind the shores. (See Figures 2 & 3 for typical installations.)

(4) *Example 4:* A trench is dug in previously disturbed Type B soil, with characteristics of a Type C soil, and will require sheeting. The trench is 18 feet deep and 12 feet wide. 8 foot horizontal spacing between cylinders is desired for working space. From Table D-1.3: Find horizontal wale with a section modulus of 14.0 spaced at 4 feet o.c. vertically and 3 inch diameter cylinder spaced at 9 feet maximum o.c. horizontally. 3x12 timber sheeting is required at close spacing vertically. (See Figure 4 for typical installation.)

(5) *Example 5:* A trench is dug in Type C soil, 9 feet deep and 4 feet wide. Horizontal cylinder spacing in excess of 6 feet is desired for working space. From Table D-1.4: Find horizontal wale with a section modulus of 7.0 and 2 inch diameter cylinders spaced at 6.5 feet o.c. horizontally. Or, find horizontal wale with a 14.0 section modulus and 3 inch diameter cylinder spaced at 10 feet o.c. horizontally. Both wales are spaced 4 feet o.c. vertically. 3x12 timber sheeting is required at close spacing vertically. (See Figure 4 for typical installation.)

(g) **Footnotes, and general notes,** for Tables D-1.1, D-1.2, D-1.3, and D-1.4.

(1) *For applications other than those* listed in the tables, refer to §1926.652(c)(2) for use of manufacturer's tabulated data. For trench depths in excess of 20 feet, refer to §1926.652(c)(2) and §1926.652(c)(3).

(2) *2 inch diameter* cylinders, at this width, shall have structural steel tube (3.5x3.5x0.1875) oversleeves, or structural oversleeves of manufacturer's specification, extending the full, collapsed length.

(3) *Hydraulic cylinders capacities.*

(i) *2 inch cylinders* shall be a minimum 2-inch inside diameter with a safe working capacity of not less than 18,000 pounds axial compressive load at maximum extension. Maximum extension is to include full range of cylinder extensions as recommended by product manufacturer.

(ii) *3-inch cylinders* shall be a minimum 3-inch inside diameter with a safe work capacity of not less than 30,000 pounds axial compressive load at maximum extension. Maximum extension is to include full range of cylinder extensions as recommended by product manufacturer.

(4) *All spacing indicated is measured center to center.*

(5) *Vertical shoring rails* shall have a minimum section modulus of 0.40 inch.

(6) *When vertical shores are used,* there must be a minimum of three shores spaced equally, horizontally, in a group.

(7) *Plywood shall be 1.125 in.* thick softwood or 0.75 inch. thick, 14 ply, arctic white birch (Finland form). Please note that plywood is not intended as a structural member, but only for prevention of local raveling (sloughing of the trench face) between shores.

(8) *See appendix C for timber specifications.*

(9) *Wales are calculated for simple span conditions.*

(10) *See appendix D, item (d), for basis and limitations of the data.*

ALUMINUM HYDRAULIC SHORING TYPICAL INSTALLATIONS

FIGURE No. 1

Vertical Aluminum Hydraulic Shoring (Spot Bracing)

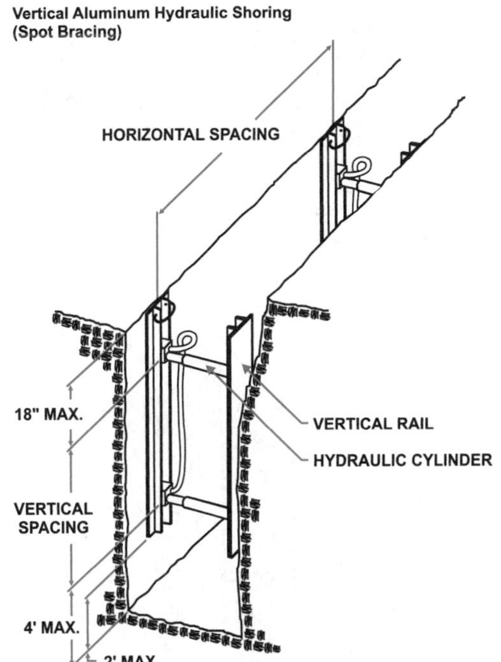

P Excavations

ALUMINUM HYDRAULIC SHORING TYPICAL INSTALLATIONS

FIGURE No. 2

Vertical Aluminum Hydraulic Shoring (With Plywood)

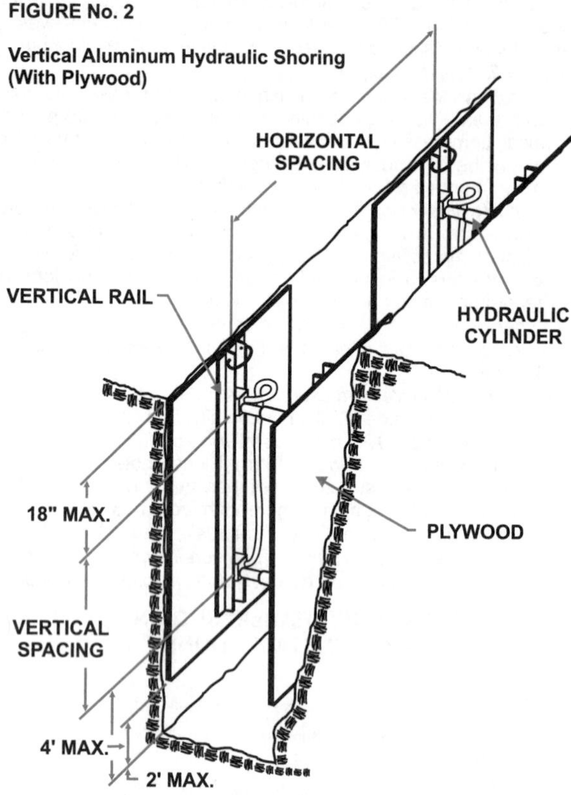

ALUMINUM HYDRAULIC SHORING TYPICAL INSTALLATIONS

FIGURE No. 3

Vertical Aluminum Hydraulic Shoring (Stacked)

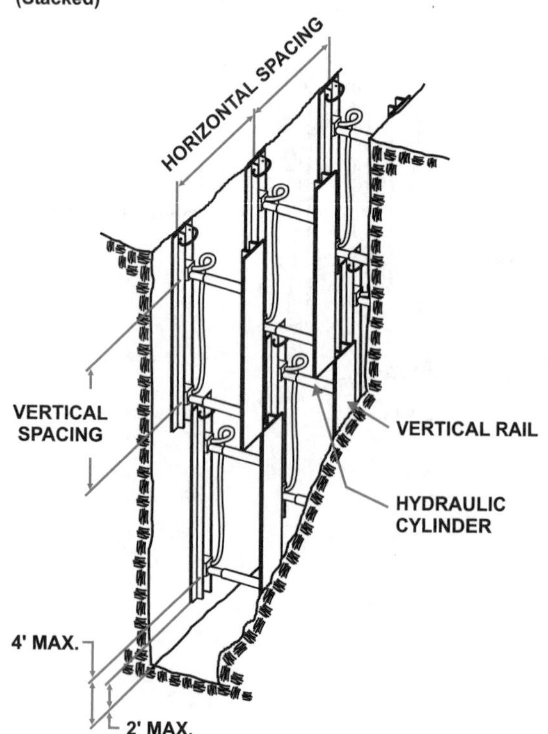

ALUMINUM HYDRAULIC SHORING TYPICAL INSTALLATIONS

FIGURE No. 4

Aluminum Hydraulic Shoring Waler System (Typical)

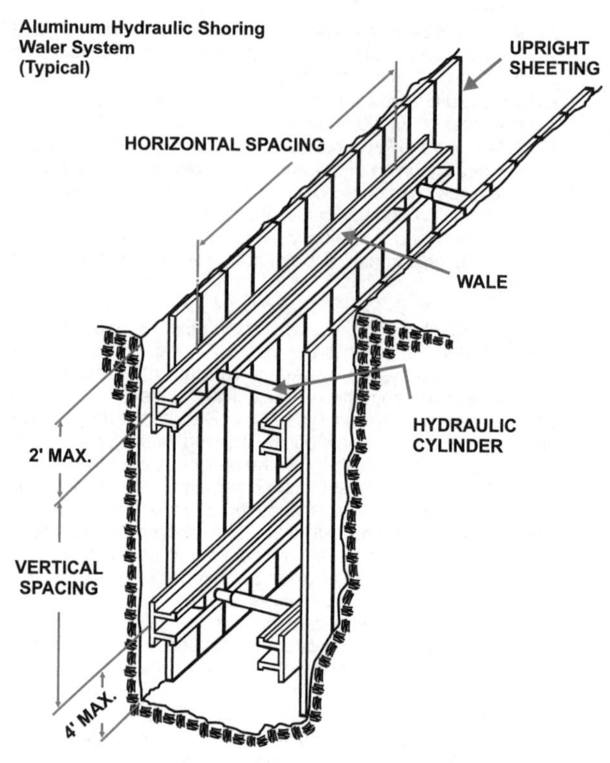

Table D-1.1
Aluminum Hydraulic Shoring
Vertical Shores for Soil Type A

Depth of trench (feet)	Hydraulic cylinders				
	Maximum horizontal spacing (feet)	Maximum vertical spacing (feet)	Width of trench (feet)		
			Up to 8	Over 8 up to 12	Over 12 up to 15
Over 5 up to 10	8	4	2 inch diameter	2 inch diameter (Note 2)	3 inch diameter
Over 10 up to 15	8				
Over 15 up to 20	7				
Over 20	Note 1				

Footnotes to tables, and general notes on hydraulic shoring, are found in Appendix D, Item (g)
Note 1: See Appendix D, Item (g)(1)
Note 2: See Appendix D, Item (g)(2)

Table D-1.2
Aluminum Hydraulic Shoring
Vertical Shores for Soil Type B

Depth of trench (feet)	Hydraulic cylinders				
	Maximum horizontal spacing (feet)	Maximum vertical spacing (feet)	Width of trench (feet)		
			Up to 8	Over 8 up to 12	Over 12 up to 15
Over 5 up to 10	8	4	2 inch diameter	2 inch diameter (Note 2)	3 inch diameter
Over 10 up to 15	6.5				
Over 15 up to 20	5.5				
Over 20	Note 1				

Footnotes to tables, and general notes on hydraulic shoring, are found in Appendix D, Item (g)
Note 1: See Appendix D, Item (g)(1)
Note 2: See Appendix D, Item (g)(2)

Table D-1.3
Aluminum Hydraulic Shoring
Waler Systems for Soil Type B

Depth of trench (feet)	Wales		Hydraulic cylinders						Timber uprights		
			Width of trench (feet)						Max. horiz. spacing (on center)		
	Vertical spacing (feet)	Section modulus* (in.^{3})	Up to 8		Over 8 up to 12		Over 12 up to 15		Solid sheet	2 ft.	3 ft.
			Horiz. spacing	Cylinder diameter	Horiz. spacing	Cylinder diameter	Horiz. spacing	Cylinder diameter			
Over 5 up to 10	4	3.5	8.0	2 in.	8.0	2 in. Note 2	8.0	3 in.			3 X 12
		7.0	9.0	2 in.	9.0	2 in. Note 2	9.0	3 in.			
		14.0	12.0	3 in.	12.0	3 in.	12.0	3 in.			
Over 10 up to 15	4	3.5	6.0	2 in.	6.0	2 in. Note 2	6.0	3 in.		3 X 12	
		7.0	8.0	3 in.	8.0	3 in.	8.0	3 in.			
		14.0	10.0	3 in.	10.0	3 in.	10.0	3 in.			
Over 15 up to 20	4	3.5	5.5	2 in.	5.5	2 in. Note 2	5.5	3 in.	3 X 12		
		7.0	6.0	3 in.	6.0	3 in.	6.0	3 in.			
		14.0	9.0	3 in.	9.0	3 in.	9.0	3 in.			
Over 20	Note 1										

Footnotes to tables, and general notes on hydraulic shoring, are found in Appendix D, Item (g)
Note 1: See Appendix D, Item (g)(1)
Note 2: See Appendix D, Item (g)(2)
* Consult product manufacturer and/or qualified engineer for Section Modulus of available wales

Table D-1.4
Aluminum Hydraulic Shoring
Waler Systems for Soil Type C

Depth of trench (feet)	Wales		Hydraulic cylinders						Timber uprights		
			Width of trench (feet)						Max. horiz. spacing (on center)		
	Vertical spacing (feet)	Section modulus * (in.^{3})	Up to 8		Over 8 up to 12		Over 12 up to 15		Solid sheet	2 ft.	3 ft.
			Horiz. spacing	Cylinder diameter	Horiz. spacing	Cylinder diameter	Horiz. spacing	Cylinder diameter			
Over 5 up to 10	4	3.5	6.0	2 in.	6.0	2 in. Note 2	6.0	3 in.	3 X 12		
		7.0	6.5	2 in.	6.5	2 in. Note 2	6.5	3 in.			
		14.0	10.0	3 in.	10.0	3 in.	10.0	3 in.			
Over 10 up to 15	4	3.5	4.0	2 in.	4.0	2 in. Note 2	4.0	3 in.	3 X 12		
		7.0	5.5	3 in.	5.5	3 in.	5.5	3 in.			
		14.0	8.0	3 in.	8.0	3 in.	8.0	3 in.			
Over 15 up to 20	4	3.5	3.5	2 in.	3.5	2 in. Note 2	3.5	3 in.	3 X 12		
		7.0	5.0	3 in.	5.0	3 in.	5.0	3 in.			
		14.0	6.0	3 in.	6.0	3 in.	6.0	3 in.			
Over 20	Note 1										

Footnotes to tables, and general notes on hydraulic shoring, are found in Appendix D, Item (g)
Note 1: See Appendix D, Item (g)(1)
Note 2: See Appendix D, Item (g)(2)
* Consult product manufacturer and/or qualified engineer for Section Modulus of available wales.

1926 Subpart P
Appendix E
Alternatives to Timber Shoring

Figure E-1. Aluminum Hydraulic Shoring

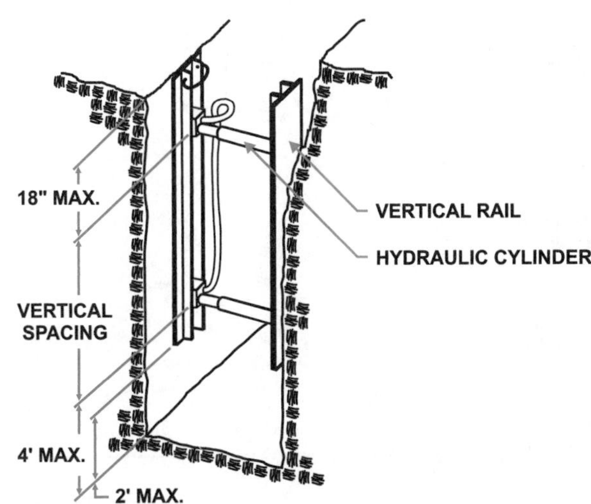

Figure E-2. Pneumatic / Hydraulic Shoring

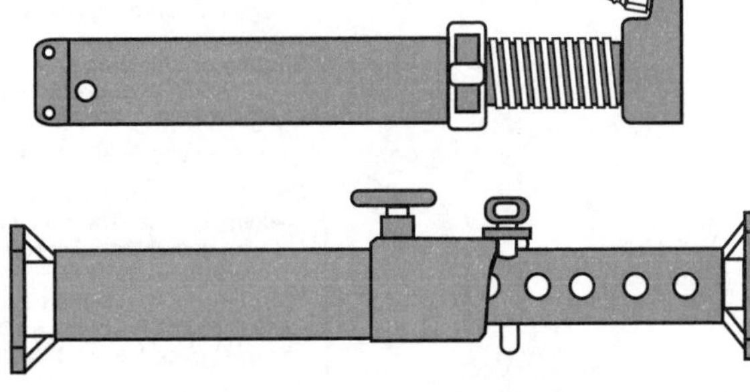

Figure E-3. Trench Jacks (Screw Jacks)

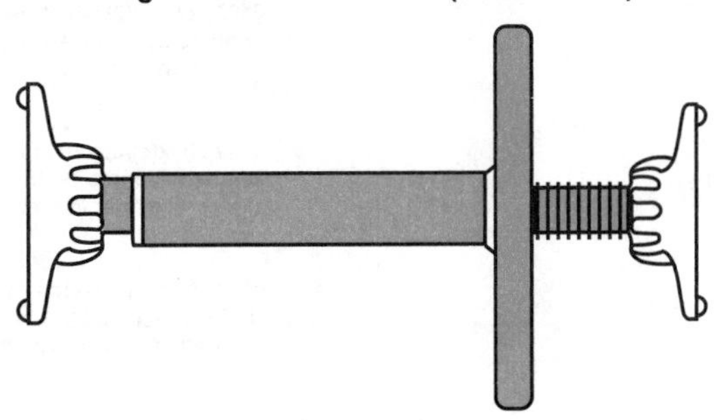

Figure E-4. Trench Shields

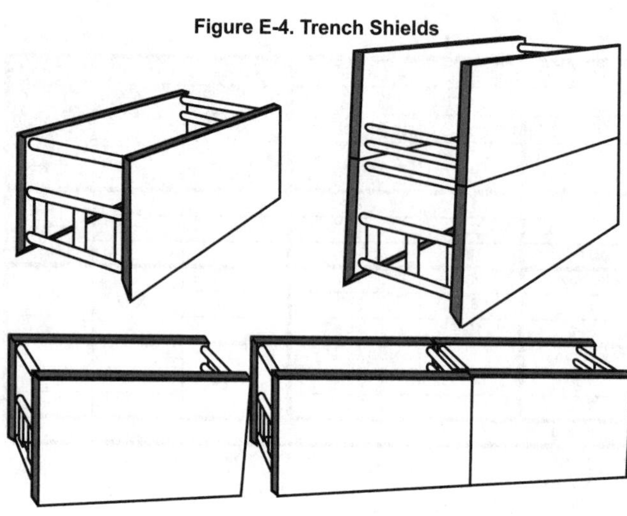

1926 Subpart P Appendix F

Selection of Protective Systems

The following figures are a graphic summary of the requirements contained in subpart P for excavations 20 feet or less in depth. Protective systems for use in excavations more than 20 feet in depth must be designed by a registered professional engineer in accordance with §1926.652 (b) and (c).

Figure 1. Preliminary Decisions

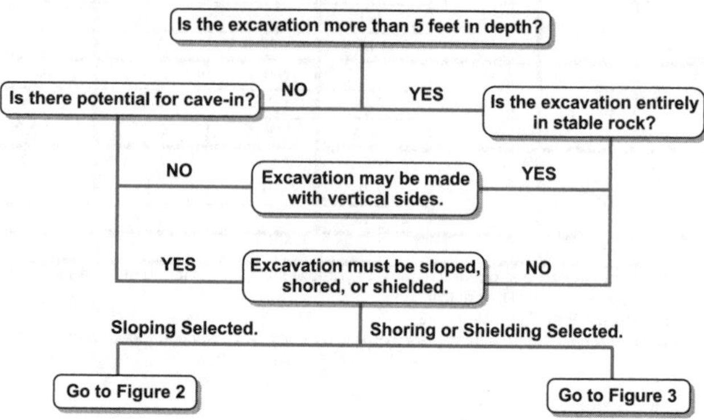

Figure 2. Sloping Options

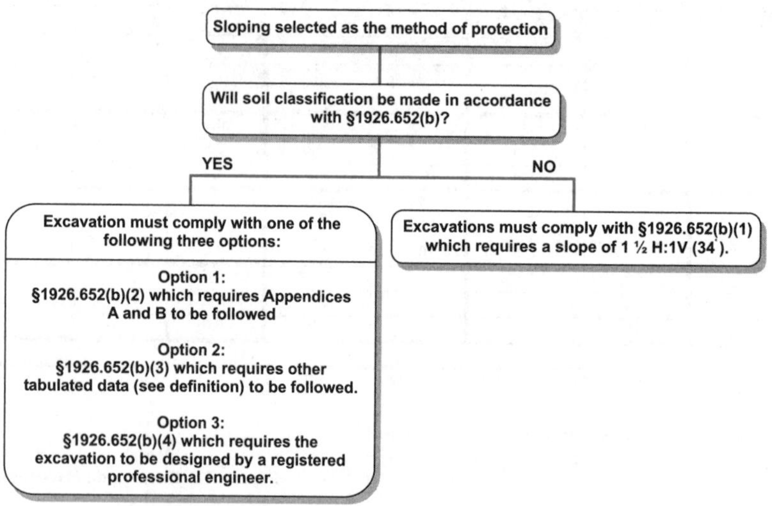

Figure 3. Shoring and Shielding Options

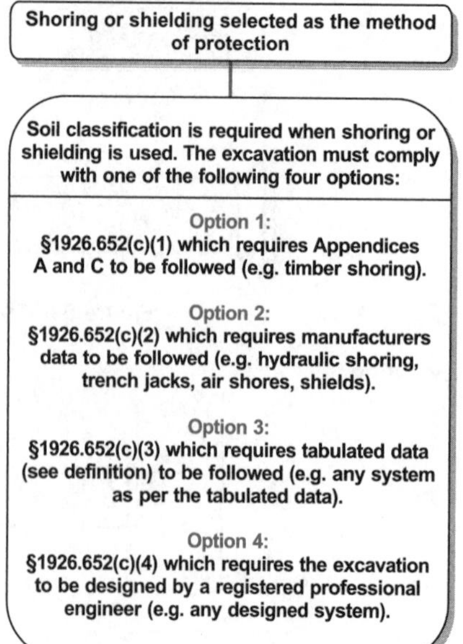

Authority: §107, Contract Worker Hours and Safety Standards Act (Construction Safety Act) (40 U.S.C. 333); Secs. 4, 6, 8, Occupational Safety and Health Act of 1970 (29 U.S.C. 653, 655, 657); Secretary of Labor's Order No. 12-71 (36 FR 8754), 8-76 (41 FR 25059), or 9-83 (48 FR 35736), as applicable, and 29 CFR part 1911.

Subpart Q - Concrete and Masonry Construction

§1926.700

Scope, application, and definitions applicable to this subpart

(a) Scope and application. This subpart sets forth requirements to protect all construction employees from the hazards associated with concrete and masonry construction operations performed in workplaces covered under 29 CFR part 1926. In addition to the requirements in subpart Q, other relevant provisions in parts 1910 and 1926 apply to concrete and masonry construction operations. [§1926.700(a)]

(b) Definitions applicable to this subpart. In addition to the definitions set forth in Sec. 1926.32, the following definitions apply to this subpart. [§1926.700(b)]

(1) **Bull float** means a tool used to spread out and smooth concrete. [§1926.700(b)(1)]

(2) **Formwork** means the total system of support for freshly placed or partially cured concrete, including the mold or sheeting (form) that is in contact with the concrete as well as all supporting members including shores, reshores, hardware, braces, and related hardware. [§1926.700(b)(2)]

(3) **Lift slab** means a method of concrete construction in which floor, and roof slabs are cast on or at ground level and, using jacks, lifted into position. [§1926.700(b)(3)]

(4) **Limited access zone** means an area alongside a masonry wall, which is under construction, and which is clearly demarcated to limit access by employees. [§1926.700(b)(4)]

(5) **Precast concrete** means concrete members (such as walls, panels, slabs, columns, and beams) which have been formed, cast, and cured prior to final placement in a structure. [§1926.700(b)(5)]

(6) **Reshoring** means the construction operation in which shoring equipment (also called reshores or reshoring equipment) is placed, as the original forms and shores are removed, in order to support partially cured concrete and construction loads. [§1926.700(b)(6)]

(7) **Shore** means a supporting member that resists a compressive force imposed by a load. [§1926.700(b)(7)]

(8) **Vertical slip forms** means forms which are jacked vertically during the placement of concrete. [§1926.700(b)(8)]

(9) **Jacking operation** means the task of lifting a slab (or group of slabs) vertically from one location to another (e.g., from the casting location to a temporary (parked) location, or from a temporary location to another temporary location, or to its final location in the structure), during the construction of a building/ structure where the lift-slab process is being used. [§1926.700(b)(9)]

[53 FR 22643, June 16, 1988, as amended at 55 FR 42328, Oct. 18, 1990]

§1926.701

General requirements

(a) Construction loads. No construction loads shall be placed on a concrete structure or portion of a concrete structure unless the employer determines, based on information received from a person who is qualified in structural design, that the structure or portion of the structure is capable of supporting the loads. [§1926.701(a)]

(b) Reinforcing steel. All protruding reinforcing steel, onto and into which employees could fall, shall be guarded to eliminate the hazard of impalement. [§1926.701(b)]

(c) Post-tensioning operations. [§1926.701(c)]

(1) *No employee (except those essential to the post-tensioning operations)* shall be permitted to be behind the jack during tensioning operations. [§1926.701(c)(1)]

(2) *Signs and barriers shall be erected* to limit employee access to the post-tensioning area during tensioning operations. [§1926.701(c)(2)]

(d) Riding concrete buckets. No employee shall be permitted to ride concrete buckets. [§1926.701(d)]

(e) Working under loads. [§1926.701(e)]

(1) *No employee shall be permitted to work* under concrete buckets while buckets are being elevated or lowered into position. [§1926.701(e)(1)]

(2) *To the extent practical,* elevated concrete buckets shall be routed so that no employee, or the fewest number of employees, are exposed to the hazards associated with falling concrete buckets. [§1926.701(e)(2)]

(f) Personal protective equipment. No employee shall be permitted to apply a cement, sand, and water mixture through a pneumatic hose unless the employee is wearing protective head and face equipment. [§1926.701(f)]

[53 FR 22643, June 16, 1988, as amended at 59 FR 40730, Aug. 9, 1994]

§1926.702

Requirements for equipment and tools

(a) Bulk cement storage. [§1926.702(a)]

(1) *Bulk storage bins, containers, and silos shall be equipped with the following:* [§1926.702(a)(1)]

(i) *Conical or tapered bottoms; and* [§1926.702(a)(1)(i)]

(ii) *Mechanical or pneumatic means of starting the flow of material.* [§1926.702(a)(1)(ii)]

(2) *No employee shall be permitted* to enter storage facilities unless the ejection system has been shut down, locked out, and tagged to indicate that the ejection system is not to be operated. [§1926.702(a)(2)]

(b) Concrete mixers. Concrete mixers with one cubic yard (.8m^3) or larger loading skips shall be equipped with the following: [§1926.702(b)]

(1) *A mechanical device to clear the skip of materials; and* [§1926.702(b)(1)]

(2) *Guardrails installed on each side of the skip.* [§1926.702(b)(2)]

(c) Power concrete trowels. Powered and rotating type concrete troweling machines that are manually guided shall be equipped with a control switch that will automatically shut off the power whenever the hands of the operator are removed from the equipment handles. [§1926.702(c)]

(d) Concrete buggies. Concrete buggy handles shall not extend beyond the wheels on either side of the buggy. [§1926.702(d)]

(e) Concrete pumping systems. [§1926.702(e)]

(1) *Concrete pumping systems* using discharge pipes shall be provided with pipe supports designed for 100 percent overload. [§1926.702(e)(1)]

(2) *Compressed air hoses* used on concrete pumping system shall be provided with positive fail-safe joint connectors to prevent separation of sections when pressurized. [§1926.702(e)(2)]

(f) Concrete buckets. [§1926.702(f)]

(1) *Concrete buckets equipped* with hydraulic or pneumatic gates shall have positive safety latches or similar safety devices installed to prevent premature or accidental dumping. [§1926.702(f)(1)]

(2) *Concrete buckets shall be designed* to prevent concrete from hanging up on top and the sides. [§1926.702(f)(2)]

(g) Tremies. Sections of tremies and similar concrete conveyances shall be secured with wire rope (or equivalent materials) in addition to the regular couplings or connections. [§1926.702(g)]

(h) Bull floats. Bull float handles, used where they might contact energized electrical conductors, shall be constructed of nonconductive material or insulated with a nonconductive sheath whose electrical and mechanical characteristics provide the equivalent protection of a handle constructed of nonconductive material. [§1926.702(h)]

(i) Masonry saws. [§1926.702(i)]

(1) *Masonry saws shall be guarded* with a semicircular enclosure over the blade. [§1926.702(i)(1)]

(2) *A method for retaining* blade fragments shall be incorporated in the design of the semicircular enclosure. [§1926.702(i)(2)]

(j) Lockout/Tagout Procedures. [§1926.702(j)]

(1) *No employee shall be permitted* to perform maintenance or repair activity on equipment (such as compressors, mixers, screens or pumps used for concrete and masonry construction activities) where the inadvertent operation of the equipment could occur and cause injury, unless all potentially hazardous energy sources have been locked out and tagged. [§1926.702(j)(1)]

(2) *Tags shall read* Do Not Start or similar language to indicate that the equipment is not to be operated. [§1926.702(j)(2)]

§1926.703

Requirements for cast-in-place concrete

(a) **General requirements for formwork.**[§1926.703(a)]

(1) *Formwork shall be designed, fabricated,* erected, supported, braced and maintained so that it will be capable of supporting without failure all vertical and lateral loads that may reasonably be anticipated to be applied to the formwork. Formwork which is designed, fabricated, erected, supported, braced and maintained in conformance with the Appendix to this section will be deemed to meet the requirements of this paragraph. [§1926.703(a)(1)]

(2) *Drawings or plans, including all revisions,* for the jack layout, formwork (including shoring equipment), working decks, and scaffolds, shall be available at the jobsite.[§1926.703(a)(2)]

(b) **Shoring and reshoring.**[§1926.703(b)]

(1) *All shoring equipment* (including equipment used in reshoring operations) shall be inspected prior to erection to determine that the equipment meets the requirements specified in the formwork drawings.[§1926.703(b)(1)]

(2) *Shoring equipment* found to be damaged such that its strength is reduced to less than that required by Sec. 1926.703(a)(1) shall not be used for shoring.[§1926.703(b)(2)]

(3) *Erected shoring equipment* shall be inspected immediately prior to, during, and immediately after concrete placement. [§1926.703(b)(3)]

(4) *Shoring equipment* that is found to be damaged or weakened after erection, such that its strength is reduced to less than that required by Sec. 1926.703(a)(1), shall be immediately reinforced.[§1926.703(b)(4)]

(5) *The sills for shoring* shall be sound, rigid, and capable of carrying the maximum intended load.[§1926.703(b)(5)]

(6) *All base plates,* shore heads, extension devices, and adjustment screws shall be in firm contact, and secured when necessary, with the foundation and the form.[§1926.703(b)(6)]

(7) *Eccentric loads* on shore heads and similar members shall be prohibited unless these members have been designed for such loading.[§1926.703(b)(7)]

(8) *Whenever single post shores are used* one on top of another (tiered), the employer shall comply with the following specific requirements in addition to the general requirements for formwork:[§1926.703(b)(8)]

(i) *The design of the shoring* shall be prepared by a qualified designer and the erected shoring shall be inspected by an engineer qualified in structural design.[§1926.703(b)(8)(i)]

(ii) *The single post shores shall be vertically aligned.* [§1926.703(b)(8)(ii)]

(iii) *The single post shores shall be spliced to prevent misalignment.*[§1926.703(b)(8)(iii)]

(iv) *The single post shores* shall be adequately braced in two mutually perpendicular directions at the splice level. Each tier shall also be diagonally braced in the same two directions. [§1926.703(b)(8)(iv)]

(9) *Adjustment of single post shores to raise formwork* shall not be made after the placement of concrete.[§1926.703(b)(9)]

(10) *Reshoring shall be erected,* as the original forms and shores are removed, whenever the concrete is required to support loads in excess of its capacity.[§1926.703(b)(10)]

(c) **Vertical slip forms.**[§1926.703(c)]

(1) *The steel rods or pipes on which jacks climb* or by which the forms are lifted shall be — [§1926.703(c)(1)]

(i) *Specifically designed for that purpose; and*[§1926.703(c)(1)(i)]

(ii) *Adequately braced where not encased in concrete.* [§1926.703(c)(1)(ii)]

(2) *Forms shall be designed* to prevent excessive distortion of the structure during the jacking operation.[§1926.703(c)(2)]

(3) *All vertical slip forms* shall be provided with scaffolds or work platforms where employees are required to work or pass. [§1926.703(c)(3)]

(4) *Jacks and vertical supports* shall be positioned in such a manner that the loads do not exceed the rated capacity of the jacks. [§1926.703(c)(4)]

(5) *The jacks or other lifting devices* shall be provided with mechanical dogs or other automatic holding devices to support the slip forms whenever failure of the power supply or lifting mechanism occurs.[§1926.703(c)(5)]

(6) *The form structure* shall be maintained within all design tolerances specified for plumbness during the jacking operation. [§1926.703(c)(6)]

(7) *The predetermined safe rate of lift shall not be exceeded.* [§1926.703(c)(7)]

(d) **Reinforcing steel.**[§1926.703(d)]

(1) *Reinforcing steel* for walls, piers, columns, and similar vertical structures shall be adequately supported to prevent overturning and to prevent collapse.[§1926.703(d)(1)]

(2) *Employers shall take measures* to prevent unrolled wire mesh from recoiling. Such measures may include, but are not limited to, securing each end of the roll or turning over the roll. [§1926.703(d)(2)]

(e) **Removal of formwork.**[§1926.703(e)]

(1) *Forms and shores* (except those used for slabs on grade and slip forms) shall not be removed until the employer determines that the concrete has gained sufficient strength to support its weight and superimposed loads. Such determination shall be based on compliance with one of the following:[§1926.703(e)(1)]

(i) *The plans and specifications* stipulate conditions for removal of forms and shores, and such conditions have been followed, or[§1926.703(e)(1)(i)]

(ii) *The concrete has been properly tested* with an appropriate ASTM standard test method designed to indicate the concrete compressive strength, and the test results indicate that the concrete has gained sufficient strength to support its weight and superimposed loads.[§1926.703(e)(1)(ii)]

(2) *Reshoring shall not be removed* until the concrete being supported has attained adequate strength to support its weight and all loads in place upon it.[§1926.703(e)(2)]

§1926.703(a)(1) Appendix

General Requirements for Formwork

(This Appendix is non-mandatory.)

This appendix serves as a non-mandatory guideline to assist employers in complying with the formwork requirements in Sec. 1926.703(a)(1). Formwork which has been designed, fabricated, erected, braced, supported and maintained in accordance with Sections 6 and 7 of the American National Standard for Construction and Demolition Operations — Concrete and Masonry Work, ANSI A10.9-1983, shall be deemed to be in compliance with the provision of Sec. 1926.703(a)(1).

[53 FR 22643, June 16, 1988, as amended at 61 FR 5510, Feb. 13, 1996]

§1926.704

Requirements for precast concrete

(a) **Precast concrete wall units,** structural framing, and tilt-up wall panels shall be adequately supported to prevent overturning and to prevent collapse until permanent connections are completed. [§1926.704(a)]

(b) **Lifting inserts which are embedded** or otherwise attached to tilt-up precast concrete members shall be capable of supporting at least two times the maximum intended load applied or transmitted to them.[§1926.704(b)]

(c) **Lifting inserts which are embedded** or otherwise attached to precast concrete members, other than the tilt-up members, shall be capable of supporting at least four times the maximum intended load applied or transmitted to them.[§1926.704(c)]

(d) **Lifting hardware shall be capable** of supporting at least five times the maximum intended load applied or transmitted to the lifting hardware.[§1926.704(d)]

(e) **No employee shall be permitted under** precast concrete members being lifted or tilted into position except those employees required for the erection of those members.[§1926.704(e)]

[53 FR 22643, June 16, 1988, as amended at 54 FR 41088, Oct. 5, 1989]

§1926.705

Requirements for lift-slab construction operations

(a) **Lift-slab operations shall be designed and planned** by a registered professional engineer who has experience in lift-slab construction. Such plans and designs shall be implemented by the employer and shall include detailed instructions and sketches indicating the prescribed method of erection. These plans and designs shall also include provisions for ensuring lateral stability of the building/ structure during construction.[§1926.705(a)]

(b) **Jacks/lifting units shall be marked** to indicate their rated capacity as established by the manufacturer.[§1926.705(b)]

(c) **Jacks/lifting units shall not be loaded** beyond their rated capacity as established by the manufacturer.[§1926.705(c)]

(d) **Jacking equipment shall be capable** of supporting at least two and one-half times the load being lifted during jacking operations and the equipment shall not be overloaded. For the purpose of this provision, jacking equipment includes any load bearing component which is used to carry out the lifting operation(s). Such equipment includes, but is not limited, to the following: threaded rods, lifting attachments, lifting nuts, hook-up collars, T-caps, shearheads, columns, and footings.[§1926.705(d)]

(e) **Jacks/lifting units shall be designed and installed** so that they will neither lift nor continue to lift when they are loaded in excess of their rated capacity.[§1926.705(e)]

(f) **Jacks/lifting units shall have a safety device installed** which will cause the jacks/lifting units to support the load in any position in the event any jack/lifting unit malfunctions or loses its lifting ability.[§1926.705(f)]

(g) **Jacking operations shall be synchronized** in such a manner to ensure even and uniform lifting of the slab. During lifting, all points at which the slab is supported shall be kept within 1/2 inch of that needed to maintain the slab in a level position.[§1926.705(g)]

(h) **If leveling is automatically controlled,** a device shall be installed that will stop the operation when the 1/2 inch tolerance set forth in paragraph (g) of this section is exceeded or where there is a malfunction in the jacking (lifting) system.[§1926.705(h)]

(i) **If leveling is maintained by manual controls,** such controls shall be located in a central location and attended by a competent person while lifting is in progress. In addition to meeting the definition in Sec. 1926.32(f), the competent person must be experienced in the lifting operation and with the lifting equipment being used.[§1926.705(i)]

(j) **The maximum number of manually controlled** jacks/lifting units on one slab shall be limited to a number that will permit the operator to maintain the slab level within specified tolerances of paragraph (g) of this section, but in no case shall that number exceed 14.[§1926.705(j)]

(k) (1) *No employee,* except those essential to the jacking operation, shall be permitted in the building/structure while any jacking operation is taking place unless the building/structure has been reinforced sufficiently to ensure its integrity during erection. The phrase "reinforced sufficiently to ensure its integrity" used in this paragraph means that a registered professional engineer, independent of the engineer who designed and planned the lifting operation, has determined from the plans that if there is a loss of support at any jack location, that loss will be confined to that location and the structure as a whole will remain stable.[§1926.705(k)(1)]

(2) *Under no circumstances,* shall any employee who is not essential to the jacking operation be permitted immediately beneath a slab while it is being lifted.[§1926.705(k)(2)]

(3) *For the purpose of paragraph (k) of this section,* a jacking operation begins when a slab or group of slabs is lifted and ends when such slabs are secured (with either temporary connections or permanent connections).[§1926.705(k)(3)]

(4) *Employers who comply* with appendix A to Sec. 1926.705 shall be considered to be in compliance with the provisions of paragraphs (k)(1) through (k)(3) of this section.[§1926.705(k)(4)]

(l) **When making temporary connections to support slabs,** wedges shall be secured by tack welding, or an equivalent method of securing the wedges to prevent them from falling out of position. Lifting rods may not be released until the wedges at that column have been secured.[§1926.705(l)]

(m) **All welding on temporary and permanent connections** shall be performed by a certified welder, familiar with the welding requirements specified in the plans and specifications for the lift-slab operation.[§1926.705(m)]

(n) **Load transfer from jacks/lifting units to building columns** shall not be executed until the welds on the column shear plates (weld blocks) are cooled to air temperature.[§1926.705(n)]

(o) **Jacks/lifting units shall be positively secured** to building columns so that they do not become dislodged or dislocated. [§1926.705(o)]

(p) **Equipment shall be designed and installed** so that the lifting rods cannot slip out of position or the employer shall institute other measures, such as the use of locking or blocking devices, which will provide positive connection between the lifting rods and attachments and will prevent components from disengaging during lifting operations.[§1926.705(p)]

§1926.705 Appendix

Lift-Slab Operations

(This Appendix is non-mandatory.)

In paragraph 1926.705(k), OSHA requires employees to be removed from the building/structure during jacking operations unless an independent registered professional engineer, other than the engineer who designed and planned the lifting operation, has determined that the building/ structure has been sufficiently reinforced to insure the integrity of the building/structure. One method to comply with this provision is for the employer to ensure that continuous bottom steel is provided in every slab and in both directions through every wall or column head area. (Column head area means the distance between lines that are one and one half times the thickness of the slab or drop panel. These lines are located outside opposite faces of the outer edges of the shearhead sections — See Figure 1). The amount of bottom steel shall be established by assuming loss of support at a given lifting jack and then determining the steel necessary to carry, by catenary action over the span between surrounding supports, the slab service dead load plus any service dead and live loads likely to be acting on the slab during jacking. In addition, the surrounding supports must be capable of resisting any additional load transferred to them as a result of the loss of support at the lifting jack considered.

Figure 1. Column Head Area

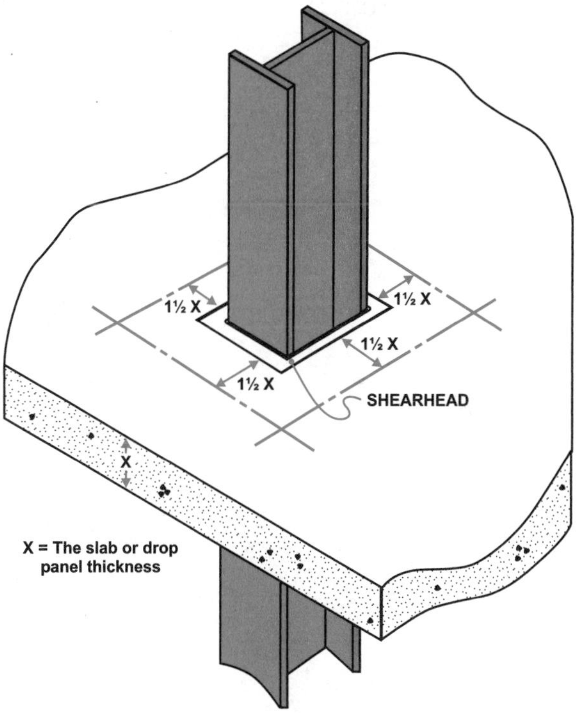

1½ X

1½ X

1½ X

1½ X

SHEARHEAD

X

X = The slab or drop panel thickness

[55 FR 42328, Oct. 18, 1990]

§1926.706

Requirements for masonry construction

(a) **A limited access zone shall be established** whenever a masonry wall is being constructed. The limited access zone shall conform to the following.[§1926.706(a)]

(1) *The limited access zone* shall be established prior to the start of construction of the wall.[§1926.706(a)(1)]

(2) *The limited access zone* shall be equal to the height of the wall to be constructed plus four feet, and shall run the entire length of the wall.[§1926.706(a)(2)]

(3) *The limited access zone* shall be established on the side of the wall which will be unscaffolded.[§1926.706(a)(3)]

(4) *The limited access zone* shall be restricted to entry by employees actively engaged in constructing the wall. No other employees shall be permitted to enter the zone.[§1926.706(a)(4)]

(5) *The limited access zone* shall remain in place until the wall is adequately supported to prevent overturning and to prevent collapse unless the height of wall is over eight feet, in which case, the limited access zone shall remain in place until the requirements of paragraph (b) of this section have been met. [§1926.706(a)(5)]

(b) **All masonry walls over eight feet in height** shall be adequately braced to prevent overturning and to prevent collapse unless the wall is adequately supported so that it will not overturn or collapse. The bracing shall remain in place until permanent supporting elements of the structure are in place.[§1926.706(b)]

Subpart Q Appendix A

References to Subpart Q of Part 1926

(This appendix is non-mandatory.)

The following non-mandatory references provide information which can be helpful in understanding and complying with the requirements contained in subpart Q.

•Accident Prevention Manual for Industrial Operations; Eighth Edition; National Safety Council.

•Building Code Requirements for Reinforced Concrete (ACI 318-83).

•Formwork for Concrete (ACI SP-4).

•Recommended Practice for Concrete Formwork (ACI 347-78).

•Safety Requirements for Concrete and Masonry Work (ANSI A10.9-1983).

•Standard Test Method for Compressive Strength of Cylindrical Concrete Specimens (ASTM C39-86).

•Standard Test Method for Making and Curing Concrete Test Specimens in the Field (ASTM C31-85).

•Standard Test Method for Penetration Resistance of Hardened Concrete (ASTM C803-82).

•Standard Test Method for Compressive Strength of Concrete Cylinders Cast In-Place in Cylindrical Molds (ASTM C873-85).

•Standard Method for Developing Early Age Compressive Test Values and Projecting Later Age Strengths (ASTM C918- 80).

•Recommended Practice for Inspection and Testing Agencies for Concrete, Steel and Bituminous Materials as Used in Construction (ASTM E329-77).

•Method of Making and Curing Concrete Test Specimens in the Laboratory (ASTM C192-88).

•Methods of Obtaining and Testing Drilled Cores and Sawed Beams of Concrete (ASTM C42-87).

•Methods of Securing, Preparing and Testing Specimens from Hardened Lightweight Insulating Concrete for Compressive Strength (ASTM C513-86).

•Test Method for Comprehensive Strength of Lightweight Insulating Concrete (ASTM C495-86).

•Method of Making, Accelerating Curing, and Testing of Concrete Compression Test Specimens (ASTM C684-81).

•Test Method for Compressive Strength of Concrete Using Portions of Beams Broken in Flexure (ASTM C116-68 (1980)).

Authority: §107, Contract Work Hours and Safety Standards Act (Construction Safety Act) (40 U.S.C. 333); Secs. 4, 6 and 8 Occupational Safety and Health Act of 1970 (29 U.S.C. 653, 655, and 657); Secretary of Labor's Order No. 12-71 (36 FR 8754), 8-76 (41 FR 25059), 9-83 (48 FR 35736), or 1-90 (55 FR 9033), as applicable; and 29 CFR part 1911.

Subpart R - Steel Erection

§1926.750
Scope

(a) **This subpart sets forth requirements to protect employees** from the hazards associated with steel erection activities involved in the construction, alteration, and/or repair of single and multi-story buildings, bridges, and other structures where steel erection occurs. The requirements of this subpart apply to employers engaged in steel erection unless otherwise specified. This subpart does not cover electrical transmission towers, communication and broadcast towers, or tanks.[§1926.750(a)]

Note to paragraph (a): Examples of structures where steel erection may occur include but are not limited to the following: Single and multi-story buildings; systems-engineered metal buildings; lift slab/ tilt-up structures; energy exploration structures; energy production, transfer and storage structures and facilities; auditoriums; malls; amphitheaters; stadiums; power plants; mills; chemical process structures; bridges; trestles; overpasses; underpasses; viaducts; aqueducts; aerospace facilities and structures; radar and communication structures; light towers; signage; billboards; scoreboards; conveyor systems; conveyor supports and related framing; stairways; stair towers; fire escapes; draft curtains; fire containment structures; monorails; aerialways; catwalks; curtain walls; window walls; store fronts; elevator fronts; entrances; skylights; metal roofs; industrial structures; hi-bay structures; rail, marine and other transportation structures; sound barriers; water process and water containment structures; air and cable supported structures; space frames; geodesic domes; canopies; racks and rack support structures and frames; platforms; walkways; balconies; atriums; penthouses; car dumpers; stackers/reclaimers; cranes and craneways; bins; hoppers; ovens; furnaces; stacks; amusement park structures and rides; and artistic and monumental structures.

(b) (1) *Steel erection activities include* hoisting, laying out, placing, connecting, welding, burning, guying, bracing, bolting, plumbing and rigging structural steel, steel joists and metal buildings; installing metal decking, curtain walls, window walls, siding systems, miscellaneous metals, ornamental iron and similar materials; and moving point-to-point while performing these activities.[§1926.750(b)(1)]

(2) *The following activities are covered by this subpart* when they occur during and are a part of steel erection activities: rigging, hoisting, laying out, placing, connecting, guying, bracing, dismantling, burning, welding, bolting, grinding, sealing, caulking, and all related activities for construction, alteration and/or repair of materials and assemblies such as structural steel; ferrous metals and alloys; non- ferrous metals and alloys; glass; plastics and synthetic composite materials; structural metal framing and related bracing and assemblies; anchoring devices; structural cabling; cable stays; permanent and temporary bents and towers; falsework for temporary supports of permanent steel members; stone and other non-precast concrete architectural materials mounted on steel frames; safety systems for steel erection; steel and metal joists; metal decking and raceway systems and accessories; metal roofing and accessories; metal siding; bridge flooring; cold formed steel framing; elevator beams; grillage; shelf racks; multi-purpose supports; crane rails and accessories; miscellaneous, architectural and ornamental metals and metal work; ladders; railings; handrails; fences and gates; gratings; trench covers; floor plates; castings; sheet metal fabrications; metal panels and panel wall systems; louvers; column covers; enclosures and pockets; stairs; perforated metals; ornamental iron work, expansion control including bridge expansion joint assemblies; slide bearings; hydraulic structures; fascias; soffit panels; penthouse enclosures; skylights; joint fillers; gaskets; sealants and seals; doors; windows; hardware; detention/ security equipment and doors, windows and hardware; conveying systems; building specialties; building equipment; machinery and plant equipment, furnishings and special construction.[§1926.750(b)(2)]

(c) **The duties of controlling contractors under this subpart** include, but are not limited to, the duties specified in §§1926.752 (a) and (c), 1926.755(b)(2), 1926.759(b), and 1926.760(e). [§1926.750(c)]

§1926.751
Definitions

Anchored bridging means that the steel joist bridging is connected to a bridging terminus point.

Bolted diagonal bridging means diagonal bridging that is bolted to a steel joist or joists.

Bridging clip means a device that is attached to the steel joist to allow the bolting of the bridging to the steel joist.

Bridging terminus point means a wall, a beam, tandem joists (with all bridging installed and a horizontal truss in the plane of the top chord) or other element at an end or intermediate point(s) of a line of bridging that provides an anchor point for the steel joist bridging.

Choker means a wire rope or synthetic fiber rigging assembly that is used to attach a load to a hoisting device.

Cold forming means the process of using press brakes, rolls, or other methods to shape steel into desired cross sections at room temperature.

Column means a load-carrying vertical member that is part of the primary skeletal framing system. Columns do not include posts.

Competent person (also defined in Sec. 1926.32) means one who is capable of identifying existing and predictable hazards in the surroundings or working conditions which are unsanitary, hazardous, or dangerous to employees, and who has authorization to take prompt corrective measures to eliminate them.

Connector means an employee who, working with hoisting equipment, is placing and connecting structural members and/or components.

Constructibility means the ability to erect structural steel members in accordance with subpart R without having to alter the over-all structural design.

Construction load (for joist erection) means any load other than the weight of the employee(s), the joists and the bridging bundle.

Controlled Decking Zone (CDZ) means an area in which certain work (for example, initial installation and placement of metal decking) may take place without the use of guardrail systems, personal fall arrest systems, fall restraint systems, or safety net systems and where access to the zone is controlled.

Controlled load lowering means lowering a load by means of a mechanical hoist drum device that allows a hoisted load to be lowered with maximum control using the gear train or hydraulic components of the hoist mechanism. Controlled load lowering requires the use of the hoist drive motor, rather than the load hoist brake, to lower the load.

Controlling contractor means a prime contractor, general contractor, construction manager or any other legal entity which has the overall responsibility for the construction of the project — its planning, quality and completion.

Critical lift means a lift that (1) exceeds 75 percent of the rated capacity of the crane or derrick, or (2) requires the use of more than one crane or derrick.

Decking hole means a gap or void more than 2 inches (5.1 cm) in its least dimension and less than 12 inches (30.5 cm) in its greatest dimension in a floor, roof or other walking/working surface. Pre- engineered holes in cellular decking (for wires, cables, etc.) are not included in this definition.

Derrick floor means an elevated floor of a building or structure that has been designated to receive hoisted pieces of steel prior to final placement.

Double connection means an attachment method where the connection point is intended for two pieces of steel which share common bolts on either side of a central piece.

Double connection seat means a structural attachment that, during the installation of a double connection, supports the first member while the second member is connected.

Erection bridging means the bolted diagonal bridging that is required to be installed prior to releasing the hoisting cables from the steel joists.

Fall restraint system means a fall protection system that prevents the user from falling any distance. The system is comprised of either a body belt or body harness, along with an anchorage, connectors and other necessary equipment. The other components typically include a lanyard, and may also include a lifeline and other devices.

Final interior perimeter means the perimeter of a large permanent open space within a building such as an atrium or courtyard. This does not include openings for stairways, elevator shafts, etc.

Girt (in systems-engineered metal buildings) means a "Z" or "C" shaped member formed from sheet steel spanning between primary framing and supporting wall material.

Headache ball means a weighted hook that is used to attach loads to the hoist load line of the crane.

Hoisting equipment means commercially manufactured lifting equipment designed to lift and position a load of known weight to a location at some known elevation and horizontal distance from the equipment's center of rotation. "Hoisting equipment" includes but is not limited to cranes, derricks, tower cranes, barge-mounted derricks or cranes, gin poles and gantry hoist systems. A "come-a-long" (a mechanical device, usually consisting of a chain or cable attached at each end, that is used to facilitate movement of materials through leverage) is not considered "hoisting equipment."

Leading edge means the unprotected side and edge of a floor, roof, or formwork for a floor or other walking/working surface (such as deck) which changes location as additional floor, roof, decking or formwork sections are placed, formed or constructed.

Metal decking means a commercially manufactured, structural grade, cold rolled metal panel formed into a series of parallel ribs; for this subpart, this includes metal floor and roof decks, standing seam metal roofs, other metal roof systems and other products such as bar gratings, checker plate, expanded metal panels, and similar products. After installation and proper fastening, these decking materials serve a combination of functions including, but not limited to: a structural element designed in combination with the structure to resist, distribute and transfer loads, stiffen the structure and provide a diaphragm action; a walking/working surface; a form for concrete slabs; a support for roofing systems; and a finished floor or roof.

Multiple lift rigging means a rigging assembly manufactured by wire rope rigging suppliers that facilitates the attachment of up to five independent loads to the hoist rigging of a crane.

Opening means a gap or void 12 inches (30.5 cm) or more in its least dimension in a floor, roof or other walking/working surface. For the purposes of this subpart, skylights and smoke domes that do not meet the strength requirements of Sec. 1926.754(e)(3) shall be regarded as openings.

Permanent floor means a structurally completed floor at any level or elevation (including slab on grade).

Personal fall arrest system means a system used to arrest an employee in a fall from a working level. A personal fall arrest system consists of an anchorage, connectors, a body harness and may include a lanyard, deceleration device, lifeline, or suitable combination of these. The use of a body belt for fall arrest is prohibited.

Positioning device system means a body belt or body harness rigged to allow an employee to be supported on an elevated, vertical surface, such as a wall or column and work with both hands free while leaning.

Post means a structural member with a longitudinal axis that is essentially vertical, that: (1) weighs 300 pounds or less and is axially loaded (a load presses down on the top end), or (2) is not axially loaded, but is laterally restrained by the above member. Posts typically support stair landings, wall framing, mezzanines and other substructures.

Project structural engineer of record means the registered, licensed professional responsible for the design of structural steel framing and whose seal appears on the structural contract documents.

Purlin (in systems-engineered metal buildings) means a "Z" or "C" shaped member formed from sheet steel spanning between primary framing and supporting roof material.

Qualified person (also defined in Sec. 1926.32) means one who, by possession of a recognized degree, certificate, or professional standing, or who by extensive knowledge, training, and experience, has successfully demonstrated the ability to solve or resolve problems relating to the subject matter, the work, or the project.

Safety deck attachment means an initial attachment that is used to secure an initially placed sheet of decking to keep proper alignment and bearing with structural support members.

Shear connector means headed steel studs, steel bars, steel lugs, and similar devices which are attached to a structural member for the purpose of achieving composite action with concrete.

Steel erection means the construction, alteration or repair of steel buildings, bridges and other structures, including the installation of metal decking and all planking used during the process of erection.

Steel joist means an open web, secondary load-carrying member of 144 feet (43.9 m) or less, designed by the manufacturer, used for the support of floors and roofs. This does not include structural steel trusses or cold-formed joists.

Steel joist girder means an open web, primary load-carrying member, designed by the manufacturer, used for the support of floors and roofs. This does not include structural steel trusses.

Steel truss means an open web member designed of structural steel components by the project structural engineer of record. For the purposes of this subpart, a steel truss is considered equivalent to a solid web structural member.

Structural steel means a steel member, or a member made of a substitute material (such as, but not limited to, fiberglass, aluminum or composite members). These members include, but are not limited to, steel joists, joist girders, purlins, columns, beams, trusses, splices, seats, metal decking, girts, and all bridging, and cold formed metal framing which is integrated with the structural steel framing of a building.

Systems-engineered metal building means a metal, field-assembled building system consisting of framing, roof and wall coverings. Typically, many of these components are cold-formed shapes. These individual parts are fabricated in one or more manufacturing facilities and shipped to the job site for assembly into the final structure. The engineering design of the system is normally the responsibility of the systems-engineered metal building manufacturer.

Tank means a container for holding gases, liquids or solids.

Unprotected sides and edges means any side or edge (except at entrances to points of access) of a walking/working surface, for example a, floor, roof, ramp or runway, where there is no wall or guardrail system at least 39 inches (1.0 m) high.

§1926.752

Site layout, site-specific erection plan and construction sequence

(a) **Approval to begin steel erection.** Before authorizing the commencement of steel erection, the controlling contractor shall ensure that the steel erector is provided with the following written notifications:[§1926.752(a)]

(1) *The concrete in the footings, piers and walls* and the mortar in the masonry piers and walls has attained, on the basis of an appropriate ASTM standard test method of field-cured samples, either 75 percent of the intended minimum compressive design strength or sufficient strength to support the loads imposed during steel erection.[§1926.752(a)(1)]

(2) *Any repairs, replacements and modifications* to the anchor bolts were conducted in accordance with Sec. 1926.755(b). [§1926.752(a)(2)]

(b) **Commencement of steel erection.** A steel erection contractor shall not erect steel unless it has received written notification that the concrete in the footings, piers and walls or the mortar in the masonry piers and walls has attained, on the basis of an appropriate ASTM standard test method of field-cured samples, either 75 percent of the intended minimum compressive design strength or sufficient strength to support the loads imposed during steel erection.[§1926.752(b)]

(c) **Site layout.** The controlling contractor shall ensure that the following is provided and maintained:[§1926.752(c)]

(1) *Adequate access roads into and through the site* for the safe delivery and movement of derricks, cranes, trucks, other necessary equipment, and the material to be erected and means and methods for pedestrian and vehicular control. Exception: this requirement does not apply to roads outside of the construction site.[§1926.752(c)(1)]

(2) *A firm, properly graded, drained area,* readily accessible to the work with adequate space for the safe storage of materials and the safe operation of the erector's equipment.[§1926.752(c)(2)]

(d) **Pre-planning of overhead hoisting operations.** All hoisting operations in steel erection shall be pre-planned to ensure that the requirements of Sec. 1926.753(d) are met.[§1926.752(d)]

(e) **Site-specific erection plan.** Where employers elect, due to conditions specific to the site, to develop alternate means and methods that provide employee protection in accordance with Sec. 1926.753(c)(5), Sec. 1926.757(a)(4) or Sec. 1926.757(e)(4), a site- specific erection plan shall be developed by a qualified person and be available at the work site. Guidelines for establishing a site-specific erection plan are contained in Appendix A to this subpart.[§1926.752(e)]

§1926.753

Hoisting and rigging

(a) **All the provisions of subpart CC** apply to hoisting and rigging with the exception of Sec. 1926.1431(a).[§1926.753(a)]

(b) **In addition, paragraphs (c) through (e) of this section** apply regarding the hazards associated with hoisting and rigging. [§1926.753(b)]

(c) **General.**[§1926.753(c)]

(1) *Pre-shift visual inspection of cranes.*[§1926.753(c)(1)]

(i) *Cranes being used in steel erection activities* shall be visually inspected prior to each shift by a competent person; the inspection shall include observation for deficiencies during operation. At a minimum this inspection shall include the following:[§1926.753(c)(1)(i)]

[A] All control mechanisms for maladjustments; [§1926.753(c)(1)(i)[A]]

[B] Control and drive mechanism for excessive wear of components and contamination by lubricants, water or other foreign matter;[§1926.753(c)(1)(i)[B]]

[C] Safety devices, including but not limited to boom angle indicators, boom stops, boom kick out devices, anti-two block devices, and load moment indicators where required;[§1926.753(c)(1)(i)[C]]

[D] Air, hydraulic, and other pressurized lines for deterioration or leakage, particularly those which flex in normal operation;[§1926.753(c)(1)(i)[D]]

[E] Hooks and latches for deformation, chemical damage, cracks, or wear;[§1926.753(c)(1)(i)[E]]

[F] Wire rope reeving for compliance with hoisting equipment manufacturer's specifications;[§1926.753(c)(1)(i)[F]]

[G] Electrical apparatus for malfunctioning, signs of excessive deterioration, dirt, or moisture accumulation; [§1926.753(c)(1)(i)[G]]

[H] Hydraulic system for proper fluid level;[§1926.753(c)(1)(i)[H]]

[I] Tires for proper inflation and condition;[§1926.753(c)(1)(i)[I]]

[J] Ground conditions around the hoisting equipment for proper support, including ground settling under and around outriggers, ground water accumulation, or similar conditions;[§1926.753(c)(1)(i)[J]]

[K] The hoisting equipment for level position; and [§1926.753(c)(1)(i)[K]]

[L] The hoisting equipment for level position after each move and setup.[§1926.753(c)(1)(i)[L]]

(ii) *If any deficiency is identified,* an immediate determination shall be made by the competent person as to whether the deficiency constitutes a hazard.[§1926.753(c)(1)(ii)]

(iii) *If the deficiency is determined to constitute a hazard,* the hoisting equipment shall be removed from service until the deficiency has been corrected.[§1926.753(c)(1)(iii)]

(iv) *The operator shall be responsible for those operations* under the operator's direct control. Whenever there is any doubt as to safety, the operator shall have the authority to stop and refuse to handle loads until safety has been assured.[§1926.753(c)(1)(iv)]

(2) *A qualified rigger* (a rigger who is also a qualified person) shall inspect the rigging prior to each shift in accordance with Sec. 1926.251.[§1926.753(c)(2)]

(3) *The headache ball, hook or load* shall not be used to transport personnel except as provided in paragraph (c)(4) of this section.[§1926.753(c)(3)]

(4) *Cranes or derricks may be used* to hoist employees on a personnel platform when work under this subpart is being conducted, provided that all provisions of Sec. 1926.1431 (except for Sec. 1926.1431(a)) are met.[§1926.753(c)(4)]

(5) *Safety latches on hooks* shall not be deactivated or made inoperable except:[§1926.753(c)(5)]

(i) *When a qualified rigger has determined* that the hoisting and placing of purlins and single joists can be performed more safely by doing so; or[§1926.753(c)(5)(i)]

(ii) *When equivalent protection is provided in a site-specific erection plan.*[§1926.753(c)(5)(ii)]

(d) **Working under loads.**[§1926.753(d)]

(1) *Routes for suspended loads* shall be pre-planned to ensure that no employee is required to work directly below a suspended load except for:[§1926.753(d)(1)]

(i) *Employees engagedin the initial connection of the steel; or* [§1926.753(d)(1)(i)]

(ii) *Employees necessaryfor the hooking or unhooking of the load.*[§1926.753(d)(1)(ii)]

(2) *When working under suspended loads,* the following criteria shall be met:[§1926.753(d)(2)]

(i) *Materials being hoisted* shall be rigged to prevent unintentional displacement;[§1926.753(d)(2)(i)]

(ii) *Hooks with self-closing safety latches* or their equivalent shall be used to prevent components from slipping out of the hook; and[§1926.753(d)(2)(ii)]

(iii) *All loads shall be rigged by a qualified rigger*[§1926.753(d)(2)(iii)]

(e) **Multiple lift rigging procedure.**[§1926.753(e)]

(1) *A multiple lift* shall only be performed if the following criteria are met:[§1926.753(e)(1)]

(i) *A multiple lift rigging assembly is used;*[§1926.753(e)(1)(i)]

(ii) *A maximum of five members are hoisted per lift;* [§1926.753(e)(1)(ii)]

(iii) *Only beams and similar structural members are lifted; and* [§1926.753(e)(1)(iii)]

(iv) *All employees engaged* in the multiple lift have been trained in these procedures in accordance with Sec. 1926.761(c)(1). [§1926.753(e)(1)(iv)]

(v) *No crane is permitted to be used for a multiple lift* where such use is contrary to the manufacturer's specifications and limitations.[§1926.753(e)(1)(v)]

(2) *Components of the multiple lift rigging assembly* shall be specifically designed and assembled with a maximum capacity for total assembly and for each individual attachment point. This capacity, certified by the manufacturer or a qualified rigger, shall be based on the manufacturer's specifications with a 5 to 1 safety factor for all components.[§1926.753(e)(2)]

(3) *The total load shall not exceed:*[§1926.753(e)(3)]

(i) *The rated capacity of the hoisting* equipment specified in the hoisting equipment load charts;[§1926.753(e)(3)(i)]

(ii) *The rigging capacity specified* in the rigging rating chart. [§1926.753(e)(3)(ii)]

(4) *The multiple lift rigging* assembly shall be rigged with members: [§1926.753(e)(4)]

(i) *Attached at their center of gravity and maintained reasonably level;*[§1926.753(e)(4)(i)]

(ii) *Rigged from top down; and*[§1926.753(e)(4)(ii)]

(iii) *Rigged at least 7 feet (2.1 m) apart.*[§1926.753(e)(4)(iii)]

(5) *The members on the multiple lift* rigging assembly shall be set from the bottom up.[§1926.753(e)(5)]

(6) *Controlled load lowering* shall be used whenever the load is over the connectors.[§1926.753(e)(6)]

[66 FR 5265, Jan. 18, 2001, as amended at 75 FR 48134, Aug. 9, 2010]

§1926.754

Structural steel assembly

(a) **Structural stability shall be maintained** at all times during the erection process.[§1926.754(a)]

Note to paragraph (a): Federal Highway Administration (FHWA) regulations incorporate by reference a number of standards, policies, and standard specifications published by the American Association of State Highway and Transportation Officials (AASHTO) and other organizations. (See 23 CFR 625.4). Many of these incorporated provisions may be relevant to maintaining structural stability during the erection process. For instance, as of May 17, 2010, in many cases FHWA requires a Registered Engineer to prepare and seal working drawings for falsework used in highway bridge construction. (See AASHTO Specifications for Highway Bridges, Div. II, §3.2.1, 15th edition, 1992, which FHWA incorporates by reference in 23 CFR 625.4). FHWA also encourages compliance with AASHTO Specifications that the FHWA regulations do not currently incorporate by reference. (See http://www.fhwa.dot.gov/ bridge/lrfd/index.htm.)

(b) **The following additional requirements** shall apply for multi-story structures:[§1926.754(b)]

(1) *The permanent floors shall be installed* as the erection of structural members progresses, and there shall be not more than eight stories between the erection floor and the upper-most permanent floor, except where the structural integrity is maintained as a result of the design.[§1926.754(b)(1)]

(2) *At no time shall there be more than four floors* or 48 feet (14.6 m), whichever is less, of unfinished bolting or welding above the foundation or uppermost permanently secured floor, except where the structural integrity is maintained as a result of the design.[§1926.754(b)(2)]

(3) *A fully planked or decked floor or nets* shall be maintained within two stories or 30 feet (9.1 m), whichever is less, directly under any erection work being performed.[§1926.754(b)(3)]

(c) **Walking/working surfaces** shear connectors and other similar devices[§1926.754(c)]

(1) *Tripping hazards.* Shear connectors (such as headed steel studs, steel bars or steel lugs), reinforcing bars, deformed anchors or threaded studs shall not be attached to the top flanges of beams, joists or beam attachments so that they project vertically from or horizontally across the top flange of the member until after the metal decking, or other walking/working surface, has been installed.[§1926.754(c)(1)]

(2) *Installation of shear connectors* on composite floors, roofs and bridge decks. When shear connectors are used in construction of composite floors, roofs and bridge decks, employees shall lay out and install the shear connectors after the metal decking has been installed, using the metal decking as a working platform. Shear connectors shall not be installed from within a controlled decking zone (CDZ), as specified in §1926.760(c)(8). [§1926.754(c)(2)]

(d) **Plumbing-up.**[§1926.754(d)]

(1) *When deemed necessary* by a competent person, plumbing-up equipment shall be installed in conjunction with the steel erection process to ensure the stability of the structure. [§1926.754(d)(1)]

(2) *When used,* plumbing-up equipment shall be in place and properly installed before the structure is loaded with construction material such as loads of joists, bundles of decking or bundles of bridging.[§1926.754(d)(2)]

(3) *Plumbing-up equipment* shall be removed only with the approval of a competent person.[§1926.754(d)(3)]

(e) **Metal decking**[§1926.754(e)]

(1) *Hoisting, landing and placing of metal decking bundles.* [§1926.754(e)(1)]

(i) *Bundle packaging and strapping* shall not be used for hoisting unless specifically designed for that purpose. [§1926.754(e)(1)(i)]

(ii) *If loose items such as dunnage,* flashing, or other materials are placed on the top of metal decking bundles to be hoisted, such items shall be secured to the bundles.[§1926.754(e)(1)(ii)]

(iii) *Bundles of metal decking* on joists shall be landed in accordance with Sec. 1926.757(e)(4).[§1926.754(e)(1)(iii)]

(iv) *Metal decking bundles* shall be landed on framing members so that enough support is provided to allow the bundles to be unbanded without dislodging the bundles from the supports. [§1926.754(e)(1)(iv)]

(v) *At the end of the shift* or when environmental or jobsite conditions require, metal decking shall be secured against displacement.[§1926.754(e)(1)(v)]

(2) *Roof and floor holes and openings.* Metal decking at roof and floor holes and openings shall be installed as follows: [§1926.754(e)(2)]

(i) *Framed metal deck openings* shall have structural members turned down to allow continuous deck installation except where not allowed by structural design constraints or constructibility.[§1926.754(e)(2)(i)]

(ii) *Roof and floor holes and openings* shall be decked over. Where large size, configuration or other structural design does not allow openings to be decked over (such as elevator shafts, stair wells, etc.) employees shall be protected in accordance with Sec. 1926.760(a)(1).[§1926.754(e)(2)(ii)]

(iii) *Metal decking holes and openings* shall not be cut until immediately prior to being permanently filled with the equipment or structure needed or intended to fulfill its specific use and which meets the strength requirements of paragraph (e)(3) of this section, or shall be immediately covered. [§1926.754(e)(2)(iii)]

(3) *Covering roof and floor openings.*[§1926.754(e)(3)]

(i) *Covers for roof and floor openings* shall be capable of supporting, without failure, twice the weight of the employees, equipment and materials that may be imposed on the cover at any one time.[§1926.754(e)(3)(i)]

(ii) *All covers shall be secured when installed* to prevent accidental displacement by the wind, equipment or employees. [§1926.754(e)(3)(ii)]

(iii) *All covers shall be painted with high-visibility paint* or shall be marked with the word "HOLE" or "COVER" to provide warning of the hazard.[§1926.754(e)(3)(iii)]

(iv) *Smoke dome or skylight fixtures* that have been installed, are not considered covers for the purpose of this section unless they meet the strength requirements of paragraph (e)(3)(i) of this section.[§1926.754(e)(3)(iv)]

(4) *Decking gaps around columns.* Wire mesh, exterior plywood, or equivalent, shall be installed around columns where planks or metal decking do not fit tightly. The materials used must be of sufficient strength to provide fall protection for personnel and prevent objects from falling through.[§1926.754(e)(4)]

(5) *Installation of metal decking.*[§1926.754(e)(5)]

(i) *Except as provided in Sec. 1926.760(c),* metal decking shall be laid tightly and immediately secured upon placement to prevent accidental movement or displacement. [§1926.754(e)(5)(i)]

(ii) *During initial placement,* metal decking panels shall be placed to ensure full support by structural members. [§1926.754(e)(5)(ii)]

(6) *Derrick floors.*[§1926.754(e)(6)]

(i) *A derrick floor shall be fully decked* and/or planked and the steel member connections completed to support the intended floor loading.[§1926.754(e)(6)(i)]

(ii) *Temporary loads placed on a derrick floor* shall be distributed over the underlying support members so as to prevent local overloading of the deck material.[§1926.754(e)(6)(ii)]

[66 FR 5265, Jan. 18, 2001, as amended at 71 FR 2885, Jan. 18, 2006; 71 FR 16674, Apr. 3, 2006; 75 FR 27429, May 17, 2010]

§1926.755
Column anchorage

(a) **General requirements for erection stability.**[§1926.755(a)]

(1) *All columns shall be anchored* by a minimum of 4 anchor rods (anchor bolts).[§1926.755(a)(1)]

(2) *Each column anchor rod (anchor bolt) assembly,* including the column-to-base plate weld and the column foundation, shall be designed to resist a minimum eccentric gravity load of 300 pounds (136.2 kg) located 18 inches (.46m) from the extreme outer face of the column in each direction at the top of the column shaft.[§1926.755(a)(2)]

(3) *Columns shall be set on level* finished floors, pre-grouted leveling plates, leveling nuts, or shim packs which are adequate to transfer the construction loads.[§1926.755(a)(3)]

(4) *All columns shall be evaluated* by a competent person to determine whether guying or bracing is needed; if guying or bracing is needed, it shall be installed.[§1926.755(a)(4)]

(b) **Repair, replacement or field modification of anchor rods (anchor bolts).**[§1926.755(b)]

(1) *Anchor rods (anchor bolts)* shall not be repaired, replaced or field-modified without the approval of the project structural engineer of record.[§1926.755(b)(1)]

(2) *Prior to the erection of a column,* the controlling contractor shall provide written notification to the steel erector if there has been any repair, replacement or modification of the anchor rods (anchor bolts) of that column.[§1926.755(b)(2)]

§1926.756
Beams and columns

(a) **General.**[§1926.756(a)]

(1) *During the final placing of solid web structural members,* the load shall not be released from the hoisting line until the members are secured with at least two bolts per connection, of the same size and strength as shown in the erection drawings, drawn up wrench- tight or the equivalent as specified by the project structural engineer of record, except as specified in paragraph (b) of this section.[§1926.756(a)(1)]

(2) *A competent person shall determine* if more than two bolts are necessary to ensure the stability of cantilevered members; if additional bolts are needed, they shall be installed. [§1926.756(a)(2)]

(b) **Diagonal bracing.** Solid web structural members used as diagonal bracing shall be secured by at least one bolt per connection drawn up wrench-tight or the equivalent as specified by the project structural engineer of record.[§1926.756(b)]

(c) (1) *Double connections at columns* and/or at beam webs over a column. When two structural members on opposite sides of a column web, or a beam web over a column, are connected sharing common connection holes, at least one bolt with its wrench-tight nut shall remain connected to the first member unless a shop-attached or field-attached seat or equivalent connection device is supplied with the member to secure the first member and prevent the column from being displaced (See Appendix H to this subpart for examples of equivalent connection devices).[§1926.756(c)(1)]

(2) *If a seat or equivalent device is used,* the seat (or device) shall be designed to support the load during the double connection process. It shall be adequately bolted or welded to both a supporting member and the first member before the nuts on the shared bolts are removed to make the double connection. [§1926.756(c)(2)]

(d) **Column splices.** Each column splice shall be designed to resist a minimum eccentric gravity load of 300 pounds (136.2 kg) located 18 inches (.46 m) from the extreme outer face of the column in each direction at the top of the column shaft.[§1926.756(d)]

(e) **Perimeter columns.** Perimeter columns shall not be erected unless:[§1926.756(e)]

(1) *The perimeter columns extend* a minimum of 48 inches (1.2 m) above the finished floor to permit installation of perimeter safety cables prior to erection of the next tier, except where constructibility does not allow (see Appendix F to this subpart); [§1926.756(e)(1)]

(2) *The perimeter columns have holes or other devices* in or attached to perimeter columns at 42-45 inches (107-114 cm) above the finished floor and the midpoint between the finished floor and the top cable to permit installation of perimeter safety cables required by Sec. 1926.760(a)(2), except where constructibility does not allow. (See Appendix F to this subpart). [§1926.756(e)(2)]

§1926.757
Open web steel joists

(a) **General.**[§1926.757(a)]

(1) *Except as provided in paragraph (a)(2) of this section,* where steel joists are used and columns are not framed in at least two directions with solid web structural steel members, a steel joist shall be field-bolted at the column to provide lateral stability to the column during erection. For the installation of this joist: [§1926.757(a)(1)]

(i) *A vertical stabilizer plate shall be provided* on each column for steel joists. The plate shall be a minimum of 6 inch by 6 inch (152 mm by 152 mm) and shall extend at least 3 inches (76 mm) below the bottom chord of the joist with a 13/16 inch (21 mm) hole to provide an attachment point for guying or plumbing cables.[§1926.757(a)(1)(i)]

(ii) *The bottom chords* of steel joists at columns shall be stabilized to prevent rotation during erection.[§1926.757(a)(1)(ii)]

(iii) *Hoisting cables shall not be released* until the seat at each end of the steel joist is field-bolted, and each end of the bottom chord is restrained by the column stabilizer plate. [§1926.757(a)(1)(iii)]

(2) *Where constructibility does not allow a steel joist to be installed at the column:*[§1926.757(a)(2)]

(i) *an alternate means of stabilizing joists* shall be installed on both sides near the column and shall:[§1926.757(a)(2)(i)]

[A] provide stability equivalent to paragraph (a)(1) of this section;[§1926.757(a)(2)(i)[A]]

[B] be designed by a qualified person;[§1926.757(a)(2)(i)[B]]

[C] be shop installed; and[§1926.757(a)(2)(i)[C]]

[D] be included in the erection drawings.[§1926.757(a)(2)(i)[D]]

(ii) *hoisting cables shall not be released* until the seat at each end of the steel joist is field-bolted and the joist is stabilized. [§1926.757(a)(2)(ii)]

(3) *Where steel joists at or near columns* span 60 feet (18.3 m) or less, the joist shall be designed with sufficient strength to allow one employee to release the hoisting cable without the need for erection bridging.[§1926.757(a)(3)]

(4) *Where steel joists at or near columns* span more than 60 feet (18.3 m), the joists shall be set in tandem with all bridging installed unless an alternative method of erection, which provides equivalent stability to the steel joist, is designed by a qualified person and is included in the site-specific erection plan. [§1926.757(a)(4)]

(5) *A steel joist or steel joist girder* shall not be placed on any support structure unless such structure is stabilized.[§1926.757(a)(5)]

(6) *When steel joist(s) are landed* on a structure, they shall be secured to prevent unintentional displacement prior to installation.[§1926.757(a)(6)]

(7) *No modification that affects the strength* of a steel joist or steel joist girder shall be made without the approval of the project structural engineer of record.[§1926.757(a)(7)]

(8) *Field-bolted joists.*[§1926.757(a)(8)]

(i) *Except for steel joists that have been pre-assembled into panels,* connections of individual steel joists to steel structures in bays of 40 feet (12.2 m) or more shall be fabricated to allow for field bolting during erection.[§1926.757(a)(8)(i)]

(ii) *These connections shall be field-bolted* unless constructibility does not allow.[§1926.757(a)(8)(ii)]

(9) *Steel joists and steel joist girders* shall not be used as anchorage points for a fall arrest system unless written approval to do so is obtained from a qualified person.[§1926.757(a)(9)]

(10) *A bridging terminus point* shall be established before bridging is installed. (See Appendix C to this subpart.)[§1926.757(a)(10)]

(b) **Attachment of steel joists and steel joist girders.**[§1926.757(b)]

(1) *Each end of "K" series steel joists* shall be attached to the support structure with a minimum of two 1/8 -inch (3 mm) fillet welds 1 inch (25 mm) long or with two 1/2 -inch (13 mm) bolts, or the equivalent.[§1926.757(b)(1)]

(2) *Each end of "LH" and "DLH" series steel joists* and steel joist girders shall be attached to the support structure with a minimum of two 1/4 -inch (6 mm) fillet welds 2 inches (51 mm) long, or with two 3/4 -inch (19 mm) bolts, or the equivalent. [§1926.757(b)(2)]

(3) *Except as provided in paragraph (b)(4) of this section,* each steel joist shall be attached to the support structure, at least at one end on both sides of the seat, immediately upon placement in the final erection position and before additional joists are placed.[§1926.757(b)(3)]

(4) *Panels that have been pre-assembled* from steel joists with bridging shall be attached to the structure at each corner before the hoisting cables are released.[§1926.757(b)(4)]

(c) **Erection of steel joists.**[§1926.757(c)]

(1) *Both sides of the seat* of one end of each steel joist that requires bridging under Tables A and B shall be attached to the support structure before hoisting cables are released. [§1926.757(c)(1)]

(2) *For joists over 60 feet,* both ends of the joist shall be attached as specified in paragraph (b) of this section and the provisions of paragraph (d) of this section met before the hoisting cables are released.[§1926.757(c)(2)]

(3) *On steel joists that do not require erection bridging* under Tables A and B, only one employee shall be allowed on the joist until all bridging is installed and anchored.[§1926.757(c)(3)]

Table A — Erection Bridging for Short Span Joists

Joist	Span	Joist	Span	Joist	Span	Joist	Span
8L1	NM	20K6	36-0	28K6	40-0	18KCS3	NM
10K1	NM	20K7	39-0	28K7	43-0	18KCS4	NM
12K1	23-0	20K9	39-0	28K8	44-0	18KCS5	NM
12K3	NM	20K10	NM	28K9	45-0	20KCS2	36-0
12K5	NM	22K4	34-0	28K10	49-0	20KCS3	39-0
14K1	27-0	22K5	35-0	28K12	53-0	20KCS4	NM
14K3	NM	22K6	36-0	30K7	44-0	20KCS5	NM
14K4	NM	22K7	40-0	30K8	45-0	22KCS2	36-0
14K6	NM	22K9	40-0	30K9	45-0	22KCS3	40-0
16K2	29-0	22K10	40-0	30K10	50-0	22KCS4	NM
16K3	30-0	22K11	40-0	30K11	52-0	22KCS5	NM
16K4	32-0	24K4	36-0	30K12	54-0	24KCS2	39-0
16K5	32-0	24K5	38-0	10KCS1	NM	24KCS3	44-0
16K6	NM	24K6	39-0	10KCS2	NM	24KCS4	NM
16K7	NM	24K7	43-0	10KCS3	NM	24KCS5	NM
16K9	NM	24K8	43-0	12KCS1	NM	26KCS2	39-0
18K3	31-0	24K9	44-0	12KCS2	NM	26KCS3	44-0
18K4	32-0	24K10	NM	12KCS3	NM	26KCS4	NM
18K5	33-0	24K12	NM	14KCS1	NM	26KCS5	NM
18K6	35-0	26K5	38-0	14KCS2	NM	28KCS2	40-0
18K7	NM	26K6	39-0	14KCS3	NM	28KCS3	45-0
18K9	NM	26K7	43-0	16KCS2	NM	28KCS4	53-0
18K10	NM	26K8	44-0	16KCS3	NM	28KCS5	53-0
20K3	32-0	26K9	45-0	16KCS4	NM	30KC53	45-0
20K4	34-0	26K10	49-0	16KCS5	NM	30KCS4	54-0
20K5	34-0	26K12	NM	18KCS2	35-0	30KCS5	54-0

NM=diagonal bolted bridging not mandatory for joists under 40 feet.

Table B — Erection Bridging for Long Span Joists

Joist	Span	Joist	Span	Joist	Span
18LH02	33-0	24LH04	39-0	32LH07	47-0 through 60-0
18LH03	NM	24LH05	40-0	32LH08	55-0 through 60-0
18LH04	NM	24LH06	45-0	32LH09	NM through 60-0
18LH05	NM	24LH07	NM	32LH10	NM through 60-0
18LH06	NM	24LH08	NM	32LH11	NM through 60-0
18LH07	NM	24LH09	NM	32LH12	NM through 60-0
18LH08	NM	24LH10	NM	32LH13	NM through 60-0

Table B — Erection Bridging for Long Span Joists (continued)

Joist	Span	Joist	Span	Joist	Span
18LH09	NM	24LH11	NM	32LH14	NM through 60-0
20LH02	33-0	28LH05	42-0	32LH15	NM through 60-0
20LH03	38-0	28LH06	42-0	36LH07	47-0 through 60-0
20LH04	NM	28LH07	NM	36LH08	47-0 through 60-0
20LH05	NM	28LH08	NM	36LH09	57-0 through 60-0
20LH06	NM	28LH09	NM	36LH10	NM through 60-0
20LH07	NM	28LH10	NM	36LH11	NM through 60-0
20LH08	NM	28LH11	NM	36LH12	NM through 60-0
20LH09	NM	28LH12	NM	36LH13	NM through 60-0
20LH10	NM	28LH13	NM	36LH14	NM through 60-0
24LH03	35-0	32LH06	47-0 through 60-0	36LH15	NM through 60-0

NM = diagonal bolted bridging not mandatory for joists under 40 feet.

(4) *Employees shall not be allowed* on steel joists where the span of the steel joist is equal to or greater than the span shown in Tables A and B except in accordance with Sec. 1926.757(d). [§1926.757(c)(4)]

(5) *When permanent bridging terminus points* cannot be used during erection, additional temporary bridging terminus points are required to provide stability. (See appendix C of this subpart.) [§1926.757(c)(5)]

(d) **Erection bridging.**[§1926.757(d)]

(1) *Where the span of the steel joist* is equal to or greater than the span shown in Tables A and B, the following shall apply: [§1926.757(d)(1)]

(i) *A row of bolted diagonal erection bridging* shall be installed near the midspan of the steel joist;[§1926.757(d)(1)(i)]

(ii) *Hoisting cables shall not be released* until this bolted diagonal erection bridging is installed and anchored; and [§1926.757(d)(1)(ii)]

(iii) *No more than one employee* shall be allowed on these spans until all other bridging is installed and anchored. [§1926.757(d)(1)(iii)]

(2) *Where the span of the steel joist* is over 60 feet (18.3 m) through 100 feet (30.5 m), the following shall apply: [§1926.757(d)(2)]

(i) *All rows of bridging shall be bolted diagonal bridging;* [§1926.757(d)(2)(i)]

(ii) *Two rows of bolted diagonal erection bridging* shall be installed near the third points of the steel joist; [§1926.757(d)(2)(ii)]

(iii) *Hoisting cables shall not be released* until this bolted diagonal erection bridging is installed and anchored; and [§1926.757(d)(2)(iii)]

(iv) *No more than two employees* shall be allowed on these spans until all other bridging is installed and anchored. [§1926.757(d)(2)(iv)]

(3) *Where the span of the steel joist* is over 100 feet (30.5 m) through 144 feet (43.9 m), the following shall apply: [§1926.757(d)(3)]

(i) *All rows of bridging shall be bolted diagonal bridging;* [§1926.757(d)(3)(i)]

(ii) *Hoisting cables shall not be released* until all bridging is installed and anchored; and[§1926.757(d)(3)(ii)]

(iii) *No more than two employees* shall be allowed on these spans until all bridging is installed and anchored. [§1926.757(d)(3)(iii)]

(4) *For steel members spanning* over 144 feet (43.9 m), the erection methods used shall be in accordance with Sec. 1926.756. [§1926.757(d)(4)]

(5) *Where any steel joist specified* in paragraphs (c)(2) and (d)(1), (d)(2), and (d)(3) of this section is a bottom chord bearing joist, a row of bolted diagonal bridging shall be provided near the support(s). This bridging shall be installed and anchored before the hoisting cable(s) is released.[§1926.757(d)(5)]

(6) *When bolted diagonal erection bridging* is required by this section, the following shall apply:[§1926.757(d)(6)]

(i) *The bridging shall be indicated on the erection drawing;* [§1926.757(d)(6)(i)]

(ii) *The erection drawing* shall be the exclusive indicator of the proper placement of this bridging;[§1926.757(d)(6)(ii)]

(iii) *Shop-installed bridging clips,* or functional equivalents, shall be used where the bridging bolts to the steel joists; [§1926.757(d)(6)(iii)]

(iv) *When two pieces of bridging* are attached to the steel joist by a common bolt, the nut that secures the first piece of bridging shall not be removed from the bolt for the attachment of the second; and[§1926.757(d)(6)(iv)]

(v) *Bridging attachments* shall not protrude above the top chord of the steel joist.[§1926.757(d)(6)(v)]

(e) **Landing and placing loads.**[§1926.757(e)]

(1) *During the construction period,* the employer placing a load on steel joists shall ensure that the load is distributed so as not to exceed the carrying capacity of any steel joist.[§1926.757(e)(1)]

(2) *Except for paragraph (e)(4) of this section,* no construction loads are allowed on the steel joists until all bridging is installed and anchored and all joist-bearing ends are attached. [§1926.757(e)(2)]

(3) *The weight of a bundle of joist bridging* shall not exceed a total of 1,000 pounds (454 kg). A bundle of joist bridging shall be placed on a minimum of three steel joists that are secured at one end. The edge of the bridging bundle shall be positioned within 1 foot (.30 m) of the secured end.[§1926.757(e)(3)]

(4) *No bundle of decking may be placed* on steel joists until all bridging has been installed and anchored and all joist bearing ends attached, unless all of the following conditions are met: [§1926.757(e)(4)]

(i) *The employer has first determined* from a qualified person and documented in a site-specific erection plan that the structure or portion of the structure is capable of supporting the load;[§1926.757(e)(4)(i)]

(ii) *The bundle of decking is placed* on a minimum of three steel joists;[§1926.757(e)(4)(ii)]

(iii) *The joists supporting* the bundle of decking are attached at both ends;[§1926.757(e)(4)(iii)]

(iv) *At least one row of bridging is installed and anchored;* [§1926.757(e)(4)(iv)]

(v) *The total weight of the bundle* of decking does not exceed 4,000 pounds (1816 kg); and[§1926.757(e)(4)(v)]

(vi) *Placement of the bundle of decking* shall be in accordance with paragraph (e)(5) of this section.[§1926.757(e)(4)(vi)]

(5) *The edge of the construction load* shall be placed within 1 foot (.30 m) of the bearing surface of the joist end.[§1926.757(e)(5)]

§1926.758
Systems-engineered metal buildings

(a) **All of the requirements of this subpart apply** to the erection of systems-engineered metal buildings except §§1926.755 (column anchorage) and 1926.757 (open web steel joists).[§1926.758(a)]

(b) **Each structural column** shall be anchored by a minimum of four anchor rods (anchor bolts).[§1926.758(b)]

(c) **Rigid frames shall have** 50 percent of their bolts or the number of bolts specified by the manufacturer (whichever is greater) installed and tightened on both sides of the web adjacent to each flange before the hoisting equipment is released.[§1926.758(c)]

(d) **Construction loads shall not be placed** on any structural steel framework unless such framework is safely bolted, welded or otherwise adequately secured.[§1926.758(d)]

(e) **In girt and eave strut-to-frame connections,** when girts or eave struts share common connection holes, at least one bolt with its wrench- tight nut shall remain connected to the first member unless a manufacturer-supplied, field-attached seat or similar connection device is present to secure the first member so that the girt or eave strut is always secured against displacement. [§1926.758(e)]

(f) **Both ends of all steel joists** or cold-formed joists shall be fully bolted and/or welded to the support structure before:[§1926.758(f)]

(1) *Releasing the hoisting cables;*[§1926.758(f)(1)]

(2) *Allowing an employee on the joists; or*[§1926.758(f)(2)]

(3) *Allowing any construction loads on the joists.*[§1926.758(f)(3)]

(g) **Purlins and girts shall not be used** as an anchorage point for a fall arrest system unless written approval is obtained from a qualified person.[§1926.758(g)]

(h) **Purlins may only be used** as a walking/working surface when installing safety systems, after all permanent bridging has been installed and fall protection is provided.[§1926.758(h)]

(i) **Construction loads may be placed** only within a zone that is within 8 feet (2.5 m) of the center-line of the primary support member.[§1926.758(i)]

§1926.759

Falling object protection

(a) **Securing loose items aloft.** All materials, equipment, and tools, which are not in use while aloft, shall be secured against accidental displacement.[§1926.759(a)]

(b) **Protection from falling objects** other than materials being hoisted. The controlling contractor shall bar other construction processes below steel erection unless overhead protection for the employees below is provided.[§1926.759(b)]

§1926.760

Fall protection

(a) **General requirements.**[§1926.760(a)]

(1) *Except as provided by paragraph (a)(3) of this section,* each employee engaged in a steel erection activity who is on a walking/working surface with an unprotected side or edge more than 15 feet (4.6 m) above a lower level shall be protected from fall hazards by guardrail systems, safety net systems, personal fall arrest systems, positioning device systems or fall restraint systems.[§1926.760(a)(1)]

(2) *Perimeter safety cables.* On multi-story structures, perimeter safety cables shall be installed at the final interior and exterior perimeters of the floors as soon as the metal decking has been installed.[§1926.760(a)(2)]

(3) *Connectors and employees working* in controlled decking zones shall be protected from fall hazards as provided in paragraphs (b) and (c) of this section, respectively.[§1926.760(a)(3)]

(b) **Connectors.** Each connector shall:[§1926.760(b)]

(1) *Be protected in accordance with* paragraph (a)(1) of this section from fall hazards of more than two stories or 30 feet (9.1 m) above a lower level, whichever is less;[§1926.760(b)(1)]

(2) *Have completed connector training* in accordance with Sec. 1926.761; and[§1926.760(b)(2)]

(3) *Be provided, at heights over 15* and up to 30 feet above a lower level, with a personal fall arrest system, positioning device system or fall restraint system and wear the equipment necessary to be able to be tied off; or be provided with other means of protection from fall hazards in accordance with paragraph (a)(1) of this section.[§1926.760(b)(3)]

(c) **Controlled Decking Zone (CDZ).** A controlled decking zone may be established in that area of the structure over 15 and up to 30 feet above a lower level where metal decking is initially being installed and forms the leading edge of a work area. In each CDZ, the following shall apply:[§1926.760(c)]

(1) *Each employee working at the leading edge* in a CDZ shall be protected from fall hazards of more than two stories or 30 feet (9.1 m), whichever is less.[§1926.760(c)(1)]

(2) *Access to a CDZ shall be limited* to only those employees engaged in leading edge work.[§1926.760(c)(2)]

(3) *The boundaries of a CDZ shall be designated* and clearly marked. The CDZ shall not be more than 90 feet (27.4 m) wide and 90 (27.4 m) feet deep from any leading edge. The CDZ shall be marked by the use of control lines or the equivalent. Examples of acceptable procedures for demarcating CDZ's can be found in Appendix D to this subpart.[§1926.760(c)(3)]

(4) *Each employee working in a CDZ* shall have completed CDZ training in accordance with Sec. 1926.761.[§1926.760(c)(4)]

(5) *Unsecured decking in a CDZ* shall not exceed 3,000 square feet (914.4 m^2).[§1926.760(c)(5)]

(6) *Safety deck attachments* shall be performed in the CDZ from the leading edge back to the control line and shall have at least two attachments for each metal decking panel.[§1926.760(c)(6)]

(7) *Final deck attachments* and installation of shear connectors shall not be performed in the CDZ.[§1926.760(c)(7)]

(d) **Criteria for fall protection equipment.**[§1926.760(d)]

(1) *Guardrail systems, safety net systems,* personal fall arrest systems, positioning device systems and their components shall conform to the criteria in Sec. 1926.502 (see Appendix G to this subpart).[§1926.760(d)(1)]

(2) *Fall arrest system components* shall be used in fall restraint systems and shall conform to the criteria in Sec. 1926.502 (see Appendix G). Either body belts or body harnesses shall be used in fall restraint systems.[§1926.760(d)(2)]

(3) *Perimeter safety cables* shall meet the criteria for guardrail systems in Sec. 1926.502 (see Appendix G).[§1926.760(d)(3)]

(e) **Custody of fall protection.** Fall protection provided by the steel erector shall remain in the area where steel erection activity has been completed, to be used by other trades, only if the controlling contractor or its authorized representative:[§1926.760(e)]

(1) *Has directed the steel erector* to leave the fall protection in place; and[§1926.760(e)(1)]

(2) *Has inspected and accepted* control and responsibility of the fall protection prior to authorizing persons other than steel erectors to work in the area.[§1926.760(e)(2)]

§1926.761

Training

The following provisions supplement the requirements of Sec. 1926.21 regarding the hazards addressed in this subpart.

(a) **Training personnel.** Training required by this section shall be provided by a qualified person(s).[§1926.761(a)]

(b) **Fall hazard training.** The employer shall train each employee exposed to a fall hazard in accordance with the requirements of this section. The employer shall institute a training program and ensure employee participation in the program.[§1926.761(b)]

(c) **Special training programs.** In addition to the training required in paragraphs (a) and (b) of this section, the employer shall provide special training to employees engaged in the following activities. [§1926.761(c)]

(1) *Multiple lift rigging procedure.* The employer shall ensure that each employee who performs multiple lift rigging has been provided training in the following areas:[§1926.761(c)(1)]

(i) *The nature of the hazards* associated with multiple lifts; and [§1926.761(c)(1)(i)]

(ii) *The proper procedures and equipment* to perform multiple lifts required by Sec. 1926.753(e).[§1926.761(c)(1)(ii)]

(2) *Connector procedures.* The employer shall ensure that each connector has been provided training in the following areas: [§1926.761(c)(2)]

(i) *The nature of the hazards associated with connecting; and* [§1926.761(c)(2)(i)]

(ii) *The establishment,* access, proper connecting techniques and work practices required by Sec. 1926.756(c) and Sec. 1926.760(b).[§1926.761(c)(2)(ii)]

(3) *Controlled Decking Zone Procedures.* Where CDZs are being used, the employer shall assure that each employee has been provided training in the following areas:[§1926.761(c)(3)]

(i) *The nature of the hazards associated* with work within a controlled decking zone; and[§1926.761(c)(3)(i)]

(ii) *The establishment, access,* proper installation techniques and work practices required by Sec. 1926.760(c) and Sec. 1926.754(e).[§1926.761(c)(3)(ii)]

[66 FR 5265, Jan. 18, 2001, as amended at 73 FR 75589, Dec. 12, 2008]

1926 Subpart R Appendix A

Guidelines for Establishing the Components of a Site-specific Erection Plan: Non-mandatory Guidelines for Complying with §1926.752(e)

(a) **General.** This appendix serves as a guideline to assist employers who elect to develop a site-specific erection plan in accordance with §1926.752(e) with alternate means and methods to provide employee protection in accordance with §1926.752(e), §1926.753(c)(5), §1926.757(a)(4) and §1926.757(e)(4).

(b) **Development of a site-specific erection plan.** Pre-construction conference(s) and site inspection(s) are held between the erector and the controlling contractor, and others such as the project engineer and fabricator before the start of steel erection. The purpose of such conference(s) is to develop and review the site-specific erection plan that will meet the requirements of this section.

(c) **Components of a site-specific erection plan.** In developing a site-specific erection plan, a steel erector considers the following elements:

(1) *The sequence of erection activity,* developed in coordination with the controlling contractor, that includes the following:

(i) *Material deliveries:*

(ii) *Material staging and storage; and*

(iii) *Coordination with other trades and construction activities.*

(2) *A description of* the crane and derrick selection and placement procedures, including the following:

(i) *Site preparation;*

(ii) *Path for overhead loads; and*

(iii) *Critical lifts, including rigging supplies and equipment.*

(3) *A description of* steel erection activities and procedures, including the following:

(i) *Stability considerations* requiring temporary bracing and guying;

(ii) *Erection bridging terminus point;*

(iii) *Anchor rod (anchor bolt)* notifications regarding repair, replacement and modifications;

(iv) *Columns and beams (including joists and purlins);*

(v) *Connections;*

(vi) *Decking; and*

(vii) *Ornamental and miscellaneous iron.*

(4) *A description of* the fall protection procedures that will be used to comply with §1926.760.

(5) *A description of* the procedures that will be used to comply with §1926.759.

(6) *A description of* the special procedures required for hazardous non-routine tasks.

(7) *A certification for each employee* who has received training for performing steel erection operations as required by §1926.761.

(8) *A list of the qualified and competent persons.*

(9) *A description of the procedures* that will be utilized in the event of rescue or emergency response.

(d) Other plan information. The plan:

(1) *Includes the identification* of the site and project; and

(2) *Is signed and dated* by the qualified person(s) responsible for its preparation and modification.

1926 Subpart R Appendix B

[Reserved]

1926 Subpart R Appendix C

Illustrations of bridging terminus points: non-mandatory guidelines for complying with §§1926.757(a)(10) and 1926.757(c)(5)

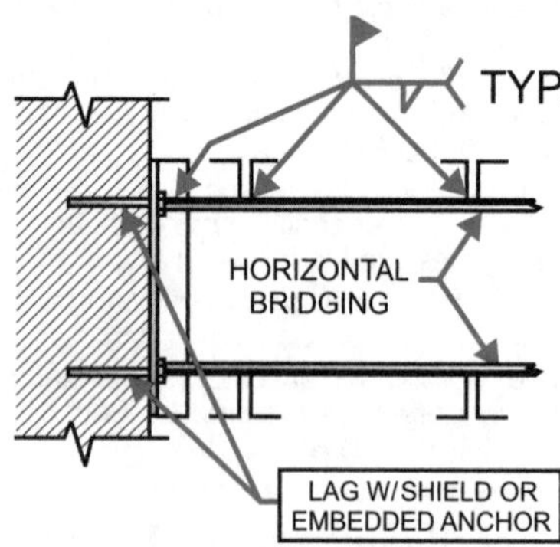

Horizontal Bridging Terminus At Wall

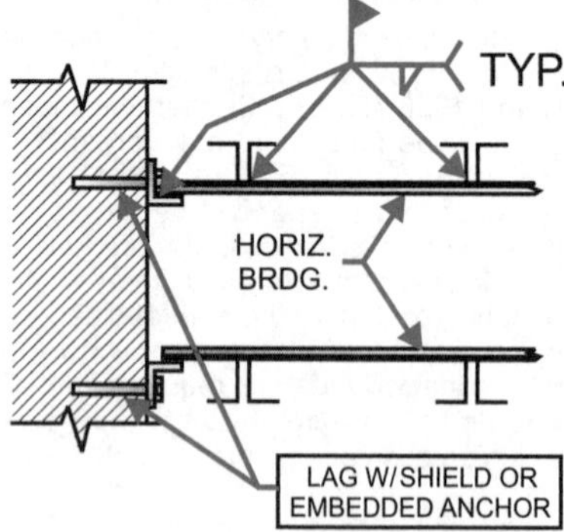

Horizontal Bridging Terminus At Wall

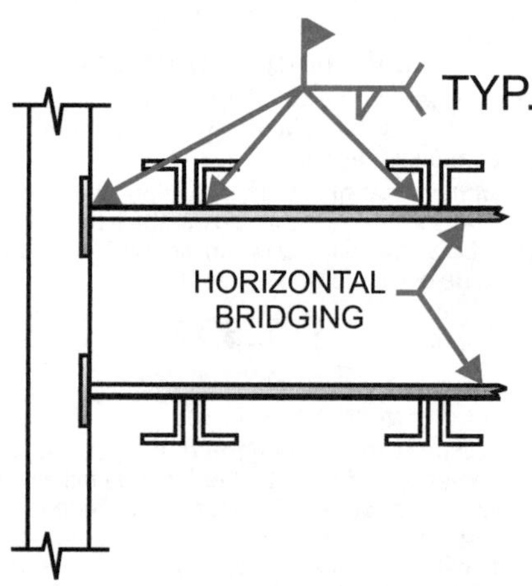

Horizontal Bridging Terminus At Panel Wall

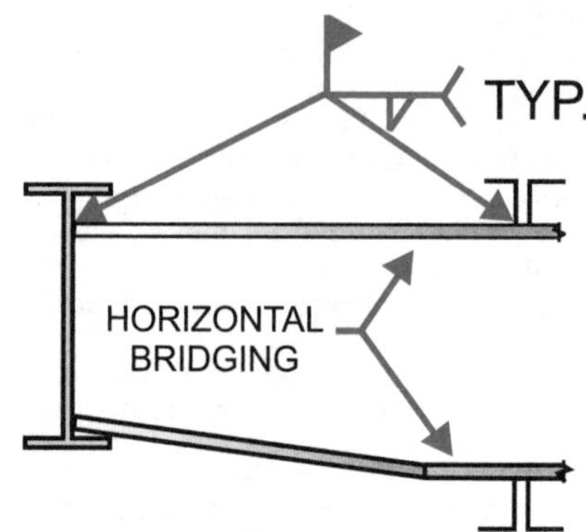

Horizontal Bridging Terminus At Structural Shape

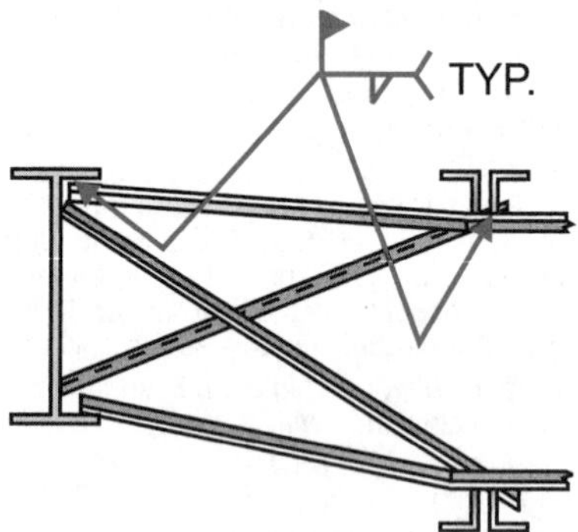

Horizontal Bridging Terminus At Structural Shape With Optional "X-Bridging"

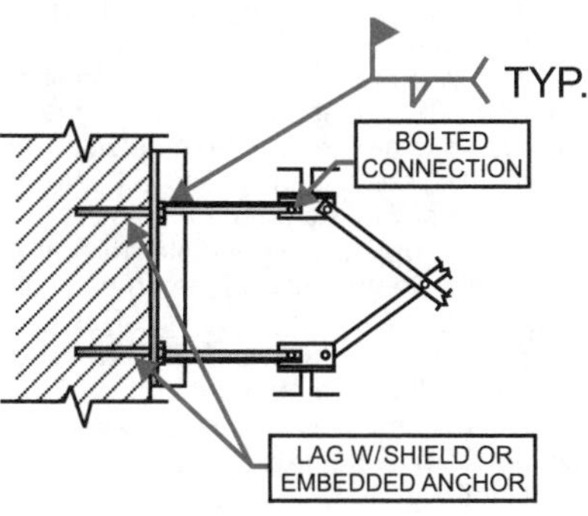

Bolted Diagonal Bridging Terminus At Wall

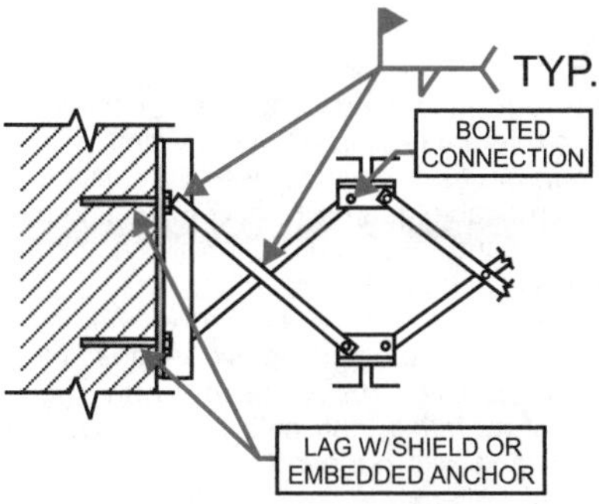

Bolted Diagonal Bridging Terminus At Wall

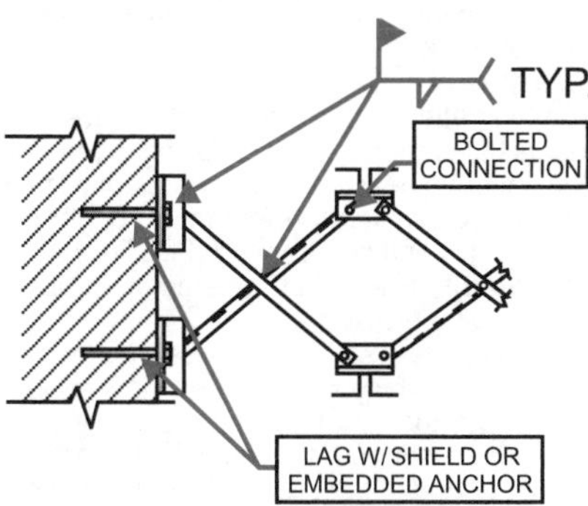

Bolted Diagonal Bridging Terminus At Wall

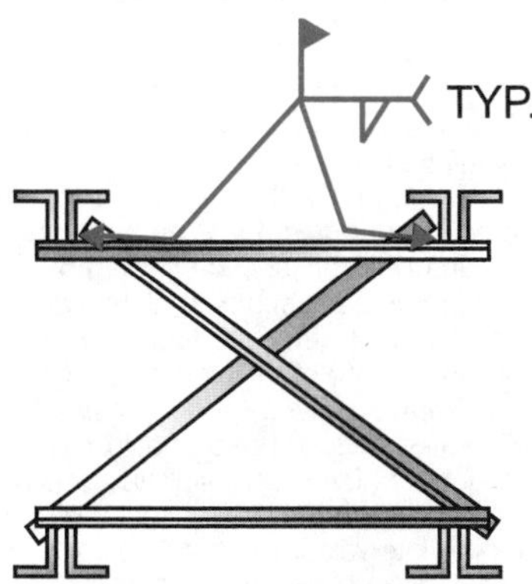

Joists Pair Bridging Terminus Point

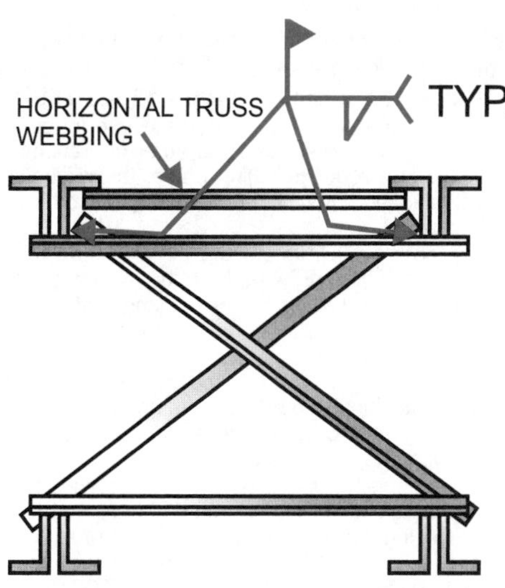

Joists Pair Bridging Terminus Point With Horizontal Truss

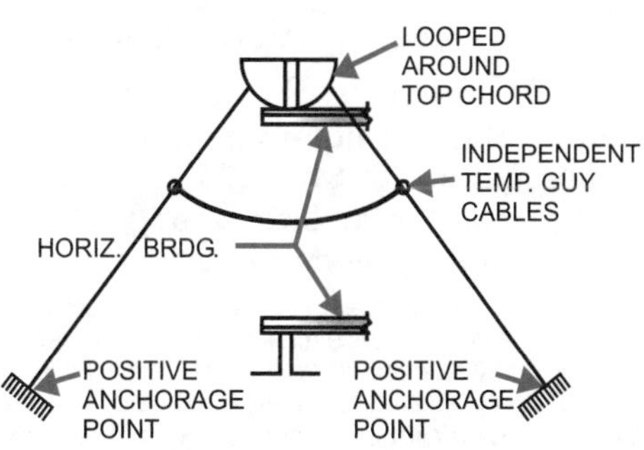

Horizontal Bridging Terminus Point Secured By Temp. Guy Cables

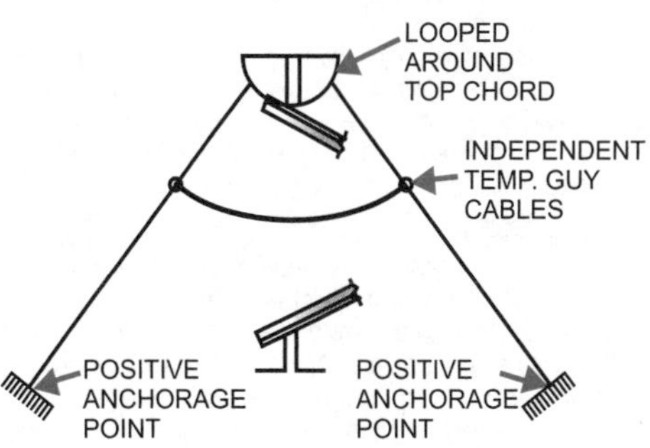

Diagonal Bridging Terminus Point Secured By Temp. Guy Cables

1926 Subpart R Appendix D

Illustration of the use of control lines to demarcate controlled decking zones (CDZs): non-mandatory guidelines for complying with §1926.760(c)(3)

(1) **When used to control access** to areas where leading edge and initial securement of metal deck and other operations connected with leading edge work are taking place, the controlled decking zone (CDZ) is defined by a control line or by any other means that restricts access.

(i) *A control line for a CDZ* is erected not less than 6 feet (1.8 m) nor more than 90 feet (27.4 m) from the leading edge.

(ii) *Control lines extend* along the entire length of the unprotected or leading edge and are approximately parallel to the unprotected or leading edge.

(iii) *Control lines are connected* on each side to a guardrail system, wall, stanchion or other suitable anchorage.

(2) **Control lines consist of** ropes, wires, tapes, or equivalent materials, and supporting stanchions as follows:

(i) *Each line is rigged and supported* in such a way that its lowest point (including sag) is not less than 39 inches (1.0 m) from the walking/working surface and its highest point is not more than 45 inches (1.3 m) from the walking/working surface.

(ii) *Each line has a minimum breaking strength of 200 pounds (90.8 kg).*

1926 Subpart R Appendix E

Training: non-mandatory guidelines for complying with §1926.761

The training requirements of §1926.761 will be deemed to have been met if employees have completed a training course on steel erection, including instruction in the provisions of this standard, that has been approved by the U.S. Department of Labor Bureau of Apprenticeship.

1926 Subpart R Appendix F

Perimeter Columns: non-mandatory guidelines for complying with §1926.756(e) to protect the unprotected side or edge of a walking/working surface

In multi-story structures, when holes in the column web are used for perimeter safety cables, the column splice must be placed sufficiently high so as not to interfere with any attachments to the column necessary for the column splice. Column splices are recommended to be placed at every other or fourth levels as design allows. Column splices at third levels are detrimental to the erection process and should be avoided if possible.

1926 Subpart R Appendix G

§1926.502 (b)-(e) Fall Protection Systems Criteria and Practices

(b) **Guardrail Systems.** Guardrail systems and their use shall comply with the following provisions:

(1) *Top edge height of top rails, or equivalent guardrail system members,* shall be 42 inches (1.1 m) plus or minus 3 inches (8 cm) above the walking/working level. When conditions warrant, the height of the top edge may exceed the 45-inch height, provided the guardrail system meets all other criteria of this paragraph (§1926.502(b)).

Note: When employees are using stilts, the top edge height of the top rail, or equivalent member, shall be increased an amount equal to the height of the stilts.

(2) *Midrails, screens, mesh, intermediate vertical members,* or equivalent intermediate structural members shall be installed between the top edge of the guardrail system and the walking/working surface when there is no wall or parapet wall at least 21 inches (53 cm) high.

(i) *Midrails, when used,* shall be installed at a height midway between the top edge of the guardrail system and the walking/working level.

(ii) *Screens and mesh, when used,* shall extend from the top rail to the walking/working level and along the entire opening between top rail supports.

(iii) *Intermediate members (such as balusters),* when used between posts, shall be not more than 19 inches (48 cm) apart.

(iv) *Other structural members* (such as additional midrails and architectural panels) shall be installed such that there are no openings in the guardrail system that are more than 19 inches (.5 m) wide.

(3) *Guardrail systems shall be capable of withstanding,* without failure, a force of at least 200 pounds (890 N) applied within 2 inches (5.1 cm) of the top edge, in any outward or downward direction, at any point along the top edge.

(4) *When the 200 pound (890* N) test load specified in paragraph (b)(3) of this section (§1926.502) is applied in a downward direction, the top edge of the guardrail shall not deflect to a height less than 39 inches (1.0 m) above the walking/working level. Guardrail system components selected and constructed in accordance with the appendix B to subpart M of this part will be deemed to meet this requirement.

(5) *Midrails,* screens, mesh, intermediate vertical members, solid panels, and equivalent structural members shall be capable of withstanding, without failure, a force of at least 150 pounds. (666 N) applied in any downward or outward direction at any point along the midrail or other member.

(6) *Guardrail systems shall be so surfaced* as to prevent injury to an employee from punctures or lacerations, and to prevent snagging of clothing.

(7) *The ends of all top rails* and midrails shall not overhang the terminal posts, except where such overhang does not constitute a projection hazard.

(8) *Steel banding and plastic* banding shall not be used as top rails or midrails.

(9) *Top rails and midrails* shall be at least one-quarter inch (0.6 cm) nominal diameter or thickness to prevent cuts and lacerations. If wire rope is used for top rails, it shall be flagged at not more than 6-foot intervals with high-visibility material.

(10) *When guardrail systems are used* at hoisting areas, a chain, gate or removable guardrail section shall be placed across the access opening between guardrail sections when hoisting operations are not taking place.

(11) *When guardrail systems are used* at holes, they shall be erected on all unprotected sides or edges of the hole.

(12) *When guardrail systems are used* around holes used for the passage of materials, the hole shall have not more than two sides provided with removable guardrail sections to allow the passage of materials. When the hole is not in use, it shall be closed over with a cover, or a guardrail system shall be provided along all unprotected sides or edges.

(13) *When guardrail systems are used* around holes which are used as points of access (such as ladderways), they shall be provided with a gate, or be so offset that a person cannot walk directly into the hole.

(14) *Guardrail systems used* on ramps and runways shall be erected along each unprotected side or edge.

(15) *Manila, plastic or synthetic rope* being used for top rails or midrails shall be inspected as frequently as necessary to ensure that it continues to meet the strength requirements of paragraph (b)(3) of this section (§1926.502).

(c) **Safety net systems.** Safety net systems and their use shall comply with the following provisions:

(1) *Safety nets shall be installed* as close as practicable under the walking/working surface on which employees are working, but in no case more than 30 feet (9.1 m) below such level. When nets are used on bridges, the potential fall area from the walking/working surface to the net shall be unobstructed.

(2) *Safety nets shall extend* outward from the outermost projection of the work surface as follows:

Vertical distance from working level to horizontal plane of net	Minimum required horizontal distance of outer edge of net from the edge of the working surface
Up to 5 feet	8 feet
More than 5 feet up to 10 feet	10 feet
More than 10 feet	13 feet

(3) *Safety nets shall be installed* with sufficient clearance under them to prevent contact with the surface or structures below when subjected to an impact force equal to the drop test specified in paragraph (4) of this section [§1926.502].

(4) *Safety nets and their installations* shall be capable of absorbing an impact force equal to that produced by the drop test specified in paragraph (c)(4)(i) of this section [§1926.502].

(i) *Except as provided in paragraph (c)(4)(ii)* of this section (§1926.502), safety nets and safety net installations shall be drop-tested at the jobsite after initial installation and before being used as a fall protection system, whenever relocated, after major repair, and at 6-month intervals if left in one place. The drop-test shall consist of a 400 pound (180 kg) bag of sand 30 ± 2 inches (76 ± 5 cm) in diameter dropped into the net from the highest walking/working surface at which employees are exposed to fall hazards, but not from less than 42 inches (1.1 m) above that level.

(ii) *When the employer can demonstrate* that it is unreasonable to perform the drop-test required by paragraph (c)(4)(i) of this section (§1926.502), the employer (or a designated competent person) shall certify that the net and net installation is in compliance with the provisions of paragraphs (c)(3) and (c)(4)(i) of this section (§1926.502) by preparing a certification record prior to the net being used as a fall protection system. The certification record must include an identification of the net and net installation for which the certification record is being prepared; the date that it was determined that the identified net and net installation were in compliance with paragraph (c)(3) of this section (§1926.502) and the signature of the person making the determination and certification. The most recent certification record for each net and net installation shall be available at the jobsite for inspection.

(5) *Defective nets shall not be used.* Safety nets shall be inspected at least once a week for wear, damage, and other deterioration. Defective components shall be removed from service. Safety nets shall also be inspected after any occurrence which could affect the integrity of the safety net system.

(6) *Materials, scrap pieces, equipment,* and tools which have fallen into the safety net shall be removed as soon as possible from the net and at least before the next work shift.

(7) *The maximum size of each safety net mesh opening* shall not exceed 36 square inches (230 cm) nor be longer than 6 inches (15 cm) on any side, and the opening, measured center-to-center of mesh ropes or webbing, shall not be longer than 6 inches (15 cm). All mesh crossings shall be secured to prevent enlargement of the mesh opening.

(8) *Each safety net (or section of it)* shall have a border rope for webbing with a minimum breaking strength of 5,000 pounds (22.2 kN).

(9) *Connections between safety net panels* shall be as strong as integral net components and shall be spaced not more than 6 inches (15 cm) apart.

(d) **Personal fall arrest systems.** Personal fall arrest systems and their use shall comply with the provisions set forth below. Effective January 1, 1998, body belts are not acceptable as part of a personal fall arrest system.

Note: The use of a body belt in a positioning device system is acceptable and is regulated under paragraph (e) of this section (§1926.502).

(1) *Connectors shall be drop forged,* pressed or formed steel, or made of equivalent materials.

(2) *Connectors shall have a corrosion-resistant finish,* and all surfaces and edges shall be smooth to prevent damage to interfacing parts of the system.

(3) *Dee-rings and snaphooks* shall have a minimum tensile strength of 5,000 pounds (22.2 kN).

(4) *Dee-rings and snaphooks* shall be proof-tested to a minimum tensile load of 3,600 pounds (16 kN) without cracking, breaking, or taking permanent deformation.

(5) *Snaphooks shall be sized to be compatible with* the member to which they are connected to prevent unintentional disengagement of the snaphook by depression of the snaphook keeper by the connected member, or shall be a locking type snaphook designed and used to prevent disengagement of the snaphook by the contact of the snaphook keeper by the connected member. Effective January 1, 1998, only locking type snaphooks shall be used.

(6) *Unless the snaphook is a locking type* and designed for the following connections, snaphooks shall not be engaged:

(i) *directly to webbing, rope or wire rope;*

(ii) *to each other;*

(iii) *to a dee-ring to which another snaphook* or other connector is attached;

(iv) *to a horizontal lifeline; or*

(v) *to any object which is incompatibly shaped* or dimensioned in relation to the snaphook such that unintentional disengagement could occur by the connected object being able to depress the snaphook keeper and release itself.

(7) *On suspended scaffolds or similar work platforms* with horizontal lifelines which may become vertical lifelines, the devices used to connect to a horizontal lifeline shall be capable of locking in both directions on the lifeline.

(8) *Horizontal lifelines shall be designed,* installed, and used, under the supervision of a qualified person, as part of a complete personal fall arrest system, which maintains a safety factor of at least two.

(9) *Lanyards and vertical lifelines* shall have a minimum breaking strength of 5,000 pounds (22.2 kN).

(10)(i) *Except as provided in paragraph* (d)(10)(ii) of this section [§1926.502], when vertical lifelines are used, each employee shall be attached to a separate lifeline.

(ii) *During the construction of elevator shafts,* two employees may be attached to the same lifeline in the hoistway, provided both employees are working atop a false car that is equipped with guardrails; the strength of the lifeline is 10,000 pounds [5,000 pounds per employee attached] (44.4 kN); and all other criteria specified in this paragraph for lifelines have been met.

(11) *Lifelines shall be protected against being cut or abraded.*

(12) *Self-retracting lifelines and lanyards* which automatically limit free fall distance to 2 feet (0.61 m) or less shall be capable of sustaining a minimum tensile load of 3,000 pounds (13.3 kN) applied to the device with the lifeline or lanyard in the fully extended position.

(13) *Self-retracting lifelines and lanyards* which do not limit free fall distance to 2 feet (0.61 m) or less, ripstitch lanyards, and tearing and deforming lanyards shall be capable of sustaining a minimum tensile load of 5,000 pounds (22.2 kN) applied to the device with the lifeline or lanyard in the fully extended position.

(14) *Ropes and straps (webbing)* used in lanyards, lifelines, and strength components of body belts and body harnesses shall be made from synthetic fibers.

(15) *Anchorages used for attachment* of personal fall arrest equipment shall be independent of any anchorage being used to support or suspend platforms and capable of supporting at least 5,000 pounds (22.2 kN) per employee attached, or shall be designed, installed, and used as follows:

(i) *as part of a complete personal fall arrest system* which maintains a safety factor of at least two; and

(ii) *under the supervision of a qualified person.*

(16) *Personal fall arrest systems, when stopping a fall, shall:*

(i) *limit maximum arresting force* on an employee to 900 pounds (4 kN) when used with a body belt;

(ii) *limit maximum arresting force* on an employee to 1,800 pounds (8 kN) when used with a body harness;

(iii) *be rigged such that an employee* can neither free fall more than 6 feet (1.8 m), nor contact any lower level;

(iv) *bring an employee to a complete stop* and limit maximum deceleration distance an employee travels to 3.5 feet (1.07 m); and,

(v) *have sufficient strength* to withstand twice the potential impact energy of an employee free falling a distance of 6 feet (1.8 m), or the free fall distance permitted by the system, whichever is less.

Note: If the personal fall arrest system meets the criteria and protocols contained in Appendix C to subpart M, and if the system is being used by an employee having a combined person and tool weight of less than 310 pounds (140 kg), the system will be considered to be in compliance with the provisions of paragraph (d)(16) of this section [§1926.502]. If the system is used by an employee having a combined tool and body weight of 310 pounds (140 kg) or more, then the employer must appropriately modify the criteria and protocols of the Appendix to provide proper protection for such heavier weights, or the system will not be deemed to be in compliance with the requirements of paragraph (d)(16) of this section (§1926.502).

(17) *The attachment point of the body belt* shall be located in the center of the wearer's back. The attachment point of the body harness shall be located in the center of the wearer's back near shoulder level, or above the wearer's head.

(18) *Body belts, harnesses, and components* shall be used only for employee protection (as part of a personal fall arrest system or positioning device system) and not to hoist materials.

(19) *Personal fall arrest systems and components* subjected to impact loading shall be immediately removed from service and shall not be used again for employee protection until inspected and determined by a competent person to be undamaged and suitable for reuse.

(20) *The employer shall provide for prompt rescue* of employees in the event of a fall or shall assure that employees are able to rescue themselves.

(21) *Personal fall arrest systems* shall be inspected prior to each use for wear, damage and other deterioration, and defective components shall be removed from service.

(22) *Body belts shall be at least* one and five-eighths (1 5/8) inches (4.1 cm) wide.

(23) *Personal fall arrest systems* shall not be attached to guardrail systems, nor shall they be attached to hoists except as specified in other subparts of this Part.

(24) *When a personal fall arrest system* is used at hoist areas, it shall be rigged to allow the movement of the employee only as far as the edge of the walking/working surface.

(e) **Positioning device systems.** Positioning device systems and their use shall conform to the following provisions:

(1) *Positioning devices shall be rigged* such that an employee cannot free fall more than 2 feet (.9 m).

(2) *Positioning devices shall be secured* to an anchorage capable of supporting at least twice the potential impact load of an employee's fall or 3,000 pounds (13.3 kN), whichever is greater.
(3) *Connectors shall be drop forged,* pressed or formed steel, or made of equivalent materials.
(4) *Connectors shall have a corrosion-resistant finish,* and all surfaces and edges shall be smooth to prevent damage to interfacing parts of this system.
(5) *Connecting assemblies* shall have a minimum tensile strength of 5,000 pounds (22.2 kN).
(6) *Dee-rings and snaphooks shall be proof-tested* to a minimum tensile load of 3,600 pounds (16 kN) without cracking, breaking, or taking permanent deformation.
(7) *Snaphooks shall be sized to be compatible* with the member to which they are connected to prevent unintentional disengagement of the snaphook by depression of the snaphook keeper by the connected member, or shall be a locking type snaphook designed and used to prevent disengagement of the snaphook by the contact of the snaphook keeper by the connected member. As of January 1, 1998, only locking type snaphooks shall be used.
(8) *Unless the snaphook is a locking type* and designed for the following connections, snaphooks shall not be engaged:
(i) *directly to webbing, rope or wire rope;*
(ii) *to each other;*
(iii) *to a dee-ring to which another snaphook* or other connector is attached;
(iv) *to a horizontal lifeline;* or to depress the snaphook keeper and release itself.
(v) *to any object which is incompatibly* shaped or dimensioned in relation to the snaphook such that unintentional disengagement could occur by the connected object being able to depress the snaphook keeper and release itself.
(9) *Positioning device systems* shall be inspected prior to each use for wear, damage, and other deterioration, and defective components shall be removed from service.
(10) *Body belts, harnesses, and components* shall be used only for employee protection (as part of a personal fall arrest system or positioning device system) and not to hoist materials.

1926 Subpart R Appendix H

Double Connections: Illustration of a Clipped End Connection and a Staggered Connection: Non-Mandatory Guidelines for Complying with §1926.756(c)(1)

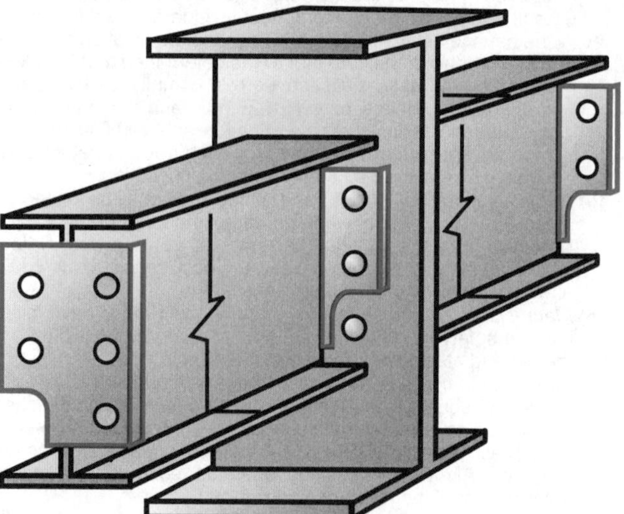

Clipped end connections are connection material on the end of a structural member which has a notch at the bottom and/or top to allow the bolt(s) of the first member placed on the opposite side of the central member to remain in place. The notch(es) fits around the nut or bolt head of the opposing member to allow the second member to be bolted up without removing the bolt(s) holding the first member.

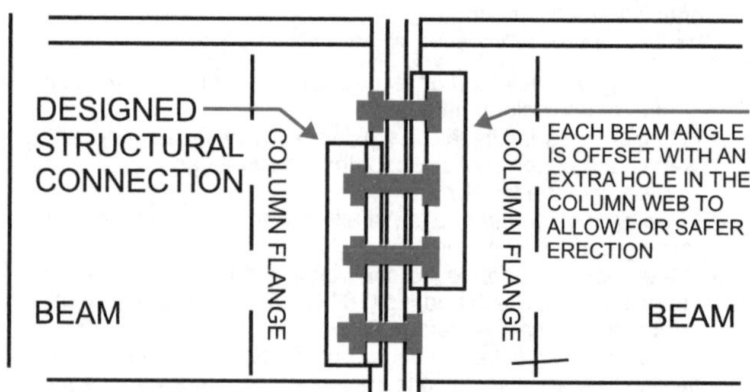

Staggered connections are connection material on a structural member in which all of the bolt holes in the common member web are not shared by the two incoming members in the final connection. The extra hole in the column web allows the erector to maintain at least a one bolt connection at all times while making the double connection.

Authority: Section 3704 of the Contract Work Hours and Safety Standards Act (40 U.S.C. 3701); Sections 4, 6, and 8 of the Occupational Safety and Health Act of 1970 (29 U.S.C. 653, 655, 657); Secretary of Labor's Order Nos. 3-2000 (65 FR 50017), 5-2002 (67 FR 65008), and 5-2007 (72 FR 31159); and 29 CFR part 1911.

Source: 66 FR 5265, Jan. 18, 2001, unless otherwise noted.

Subpart S - Underground Construction, Caissons, Cofferdams and Compressed Air

§1926.800

Underground construction

(a) **Scope and application.**[§1926.800(a)]

(1) *This section applies to the construction* of underground tunnels, shafts, chambers, and passageways. This section also applies to cut-and-cover excavations which are both physically connected to ongoing underground construction operations within the scope of this section, and covered in such a manner as to create conditions characteristic of underground construction. [§1926.800(a)(1)]

(2) *This section does not apply to the following:*[§1926.800(a)(2)]

(i) *Excavation and trenching operations* covered by subpart P of this part, such as foundation operations for above-ground structures that are not physically connected to underground construction operations, and surface excavation; nor [§1926.800(a)(2)(i)]

(ii) *Underground electrical transmission* and distribution lines, as addressed in subpart V of this part.[§1926.800(a)(2)(ii)]

(b) **Access and egress.**[§1926.800(b)]

(1) *The employer shall provide and maintain* safe means of access and egress to all work stations.[§1926.800(b)(1)]

(2) *The employer shall provide access and egress* in such a manner that employees are protected from being struck by excavators, haulage machines, trains and other mobile equipment. [§1926.800(b)(2)]

(3) *The employer shall control access to all openings* to prevent unauthorized entry underground. Unused chutes, manways, or other openings shall be tightly covered, bulkheaded, or fenced off, and shall be posted with warning signs indicating "Keep Out" or similar language. Completed or unused sections of the underground facility shall be barricaded.[§1926.800(b)(3)]

(c) **Check-in/check-out.** The employer shall maintain a check-in/ check-out procedure that will ensure that above-ground personnel can determine an accurate count of the number of persons underground in the event of an emergency. However, this procedure is not required when the construction of underground facilities designed for human occupancy has been sufficiently completed so that the permanent environmental controls are effective, and when the remaining construction activity will not cause any environmental hazard or structural failure within the facilities. [§1926.800(c)]

(d) **Safety instruction.** All employees shall be instructed in the recognition and avoidance of hazards associated with underground construction activities including, where appropriate, the following subjects:[§1926.800(d)]

(1) *Air monitoring;*[§1926.800(d)(1)]

(2) *Ventilation;*[§1926.800(d)(2)]

(3) *Illumination;*[§1926.800(d)(3)]

(4) *Communications;*[§1926.800(d)(4)]

(5) *Flood control;*[§1926.800(d)(5)]

(6) *Mechanical equipment;*[§1926.800(d)(6)]

(7) *Personal protective equipment;*[§1926.800(d)(7)]

(8) *Explosives;*[§1926.800(d)(8)]

(9) *Fire prevention and protection; and*[§1926.800(d)(9)]

(10) *Emergency procedures,* including evacuation plans and check-in/ check-out systems.[§1926.800(d)(10)]

(e) **Notification.**[§1926.800(e)]

(1) *Oncoming shifts shall be informed* of any hazardous occurrences or conditions that have affected or might affect employee safety, including liberation of gas, equipment failures, earth or rock slides, cave-ins, floodings, fires or explosions. [§1926.800(e)(1)]

(2) *The employer shall establish and maintain* direct communications for coordination of activities with other employers whose operations at the jobsite affect or may affect the safety of employees underground.[§1926.800(e)(2)]

(f) **Communications.**[§1926.800(f)]

(1) *When natural unassisted voice communication* is ineffective, a power-assisted means of voice communication shall be used to provide communication between the work face, the bottom of the shaft, and the surface.[§1926.800(f)(1)]

(2) *Two effective means of communication,* at least one of which shall be voice communication, shall be provided in all shafts which are being developed or used either for personnel access or for hoisting. Additional requirements for hoist operator communication are contained in paragraph (t)(3)(xiv) of this section. [§1926.800(f)(2)]

(3) *Powered communication systems* shall operate on an independent power supply, and shall be installed so that the use of or disruption of any one phone or signal location will not disrupt the operation of the system from any other location. [§1926.800(f)(3)]

(4) *Communication systems shall be tested* upon initial entry of each shift to the underground, and as often as necessary at later times, to ensure that they are in working order. [§1926.800(f)(4)]

(5) *Any employee working alone underground* in a hazardous location, who is both out of the range of natural unassisted voice communication and not under observation by other persons, shall be provided with an effective means of obtaining assistance in an emergency.[§1926.800(f)(5)]

(g) **Emergency provisions**[§1926.800(g)]

(1) *Hoisting capability.* When a shaft is used as a means of egress, the employer shall make advance arrangements for power-assisted hoisting capability to be readily available in an emergency, unless the regular hoisting means can continue to function in the event of an electrical power failure at the jobsite. Such hoisting means shall be designed so that the load hoist drum is powered in both directions of rotation and so that the brake is automatically applied upon power release or failure. [§1926.800(g)(1)]

(2) *Self-rescuers.* The employer must provide self-rescuers approved by the National Institute for Occupational Safety and Health under 42 CFR part 84. The respirators must be immediately available to all employees at work stations in underground areas where employees might be trapped by smoke or gas. The selection, issuance, use, and care of respirators must be in accordance with 29 CFR 1926.103.[§1926.800(g)(2)]

(3) *Designated person.* At least one designated person shall be on duty above ground whenever any employee is working underground. This designated person shall be responsible for securing immediate aid and keeping an accurate count of employees underground in case of emergency. The designated person must not be so busy with other responsibilities that the counting function is encumbered.[§1926.800(g)(3)]

(4) *Emergency lighting.* Each employee underground shall have an acceptable portable hand lamp or cap lamp in his or her work area for emergency use, unless natural light or an emergency lighting system provides adequate illumination for escape. [§1926.800(g)(4)]

(5) *Rescue teams.*[§1926.800(g)(5)]

(i) *On jobsites where 25 or more employees* work underground at one time, the employer shall provide (or make arrangements in advance with locally available rescue services to provide) at least two 5-person rescue teams, one on the jobsite or within one-half hour travel time from the entry point, and the other within 2 hours travel time.[§1926.800(g)(5)(i)]

(ii) *On jobsites where less than 25 employees* work underground at one time, the employer shall provide (or make arrangements in advance with locally available rescue services to provide) at least one 5-person rescue team to be either on the jobsite or within one-half hour travel time from the entry point.[§1926.800(g)(5)(ii)]

(iii) *Rescue team members* shall be qualified in rescue procedures, the use and limitations of breathing apparatus, and the use of firefighting equipment. Qualifications shall be reviewed not less than annually.[§1926.800(g)(5)(iii)]

(iv) *On jobsites where flammable or noxious gases* are encountered or anticipated in hazardous quantities, rescue team members shall practice donning and using self-contained breathing apparatus monthly.[§1926.800(g)(5)(iv)]

(v) *The employer shall ensure* that rescue teams are familiar with conditions at the jobsite.[§1926.800(g)(5)(v)]

(h) **Hazardous classifications.**[§1926.800(h)]

(1) *Potentially gassy operations.* Underground construction operations shall be classified as potentially gassy if either: [§1926.800(h)(1)]

(i) *Air monitoring discloses 10 percent or more* of the lower explosive limit for methane or other flammable gases measured at 12 inches (304.8 mm) ± 0.25 inch (6.35 mm) from the roof, face, floor or walls in any underground work area for more than a 24-hour period; or[§1926.800(h)(1)(i)]

(ii) *The history of the geographical area* or geological formation indicates that 10 percent or more of the lower explosive limit for methane or other flammable gases is likely to be encountered in such underground operations.[§1926.800(h)(1)(ii)]

(2) *Gassy operations.* Underground construction operations shall be classified as gassy if:[§1926.800(h)(2)]

(i) *Air monitoring discloses 10 percent or more* of the lower explosive limit for methane or other flammable gases measured at 12 inches (304.8 mm) ± 0.25 inch (6.35 mm) from the roof, face, floor or walls in any underground work area for three consecutive days; or[§1926.800(h)(2)(i)]

(ii) *There has been an ignition of methane* or of other flammable gases emanating from the strata that indicates the presence of such gases; or[§1926.800(h)(2)(ii)]

(iii) *The underground construction operation* is both connected to an underground work area which is currently classified as gassy and is also subject to a continuous course of air containing the flammable gas concentration.[§1926.800(h)(2)(iii)]

(3) *Declassification to potentially gassy operations.* Underground construction gassy operations may be declassified to Potentially Gassy when air monitoring results remain under 10 percent of the lower explosive limit for methane or other flammable gases for three consecutive days.[§1926.800(h)(3)]

(i) Gassy operations-additional requirements.[§1926.800(i)]

(1) *Only acceptable equipment,* maintained in suitable condition, shall be used in gassy operations.[§1926.800(i)(1)]

(2) *Mobile diesel-powered equipment* used in gassy operations shall be either approved in accordance with the requirements of 30 CFR part 36 (formerly Schedule 31) by MSHA, or shall be demonstrated by the employer to be fully equivalent to such MSHA-approved equipment, and shall be operated in accordance with that part.[§1926.800(i)(2)]

(3) *Each entrance to a gassy operation* shall be prominently posted with signs notifying all entrants of the gassy classification.[§1926.800(i)(3)]

(4) *Smoking shall be prohibited* in all gassy operations and the employer shall be responsible for collecting all personal sources of ignition, such as matches and lighters, from all persons entering a gassy operation.[§1926.800(i)(4)]

(5) *A fire watch as described in Sec. 1926.352(e)* shall be maintained when hot work is performed.[§1926.800(i)(5)]

(6) *Once an operation has met the criteria* in paragraph (h)(2) warranting classification as gassy, all operations in the affected area, except the following, shall be discontinued until the operation either is in compliance with all of the gassy operation requirements or has been declassified in accordance with paragraph (h)(3) of this section:[§1926.800(i)(6)]

(i) *Operations related to the control of the gas concentration;* [§1926.800(i)(6)(i)]

(ii) *Installation of new equipment,* or conversion of existing equipment, to comply with this paragraph (i); and [§1926.800(i)(6)(ii)]

(iii) *Installation of above-ground controls for reversing the air flow.*[§1926.800(i)(6)(iii)]

(j) Air quality and monitoring[§1926.800(j)]

(1) *General.* Air quality limits and control requirements for construction are found in Sec. 1926.55, except as modified by this section.[§1926.800(j)(1)]

(i) *[A] The employer shall assign* a competent person who shall perform all air monitoring required by this section. [§1926.800(j)(1)(i)[A]]

[B] Where this paragraph requires monitoring of airborne contaminants "as often as necessary," the competent person shall make a reasonable determination as to which substances to monitor and how frequently to monitor, considering at least the following factors:[§1926.800(j)(1)(i)[B]]

[1] Location of jobsite: Proximity to fuel tanks, sewers, gas lines, old landfills, coal deposits, and swamps; [§1926.800(j)(1)(i)[B][1]]

[2] Geology: Geological studies of the jobsite, particularly involving the soil type and its permeability; [§1926.800(j)(1)(i)[B][2]]

[3] History: Presence of air contaminants in nearby jobsites, changes in levels of substances monitored on the prior shift; and[§1926.800(j)(1)(i)[B][3]]

[4] Work practices and jobsite conditions: The use of diesel engines, use of explosives, use of fuel gas, volume and flow of ventilation, visible atmospheric conditions, decompression of the atmosphere, welding, cutting and hot work, and employees' physical reactions to working underground.[§1926.800(j)(1)(i)[B][4]]

(ii) *[A] The atmosphere in all underground work areas* shall be tested as often as necessary to assure that the atmosphere at normal atmospheric pressure contains at least 19.5 percent oxygen and no more than 22 percent oxygen.[§1926.800(j)(1)(ii)[A]]

[B] Tests for oxygen content shall be made before tests for air contaminants.[§1926.800(j)(1)(ii)[B]]

(iii) *[A] The atmosphere in all underground work areas* shall be tested quantitatively for carbon monoxide, nitrogen dioxide, hydrogen sulfide, and other toxic gases, dusts, vapors, mists, and fumes as often as necessary to ensure that the permissible exposure limits prescribed in Sec. 1926.55 are not exceeded.[§1926.800(j)(1)(iii)[A]]

[B] The atmosphere in all underground work areas shall be tested quantitatively for methane and other flammable gases as often as necessary to determine: [§1926.800(j)(1)(iii)[B]]

[1] Whether action is to be taken under paragraphs (j)(1)(vii), (viii), and (ix), of this section; and [§1926.800(j)(1)(iii)[B][1]]

[2] Whether an operation is to be classified potentially gassy or gassy under paragraph (h) of this section. [§1926.800(j)(1)(iii)[B][2]]

[C] If diesel-engine or gasoline-engine driven ventilating fans or compressors are used, an initial test shall be made of the inlet air of the fan or compressor, with the engines operating, to ensure that the air supply is not contaminated by engine exhaust.[§1926.800(j)(1)(iii)[C]]

[D] Testing shall be performed as often as necessary to ensure that the ventilation requirements of paragraph (k) of this section are met.[§1926.800(j)(1)(iii)[D]]

(iv) *When rapid excavation machines are used,* a continuous flammable gas monitor shall be operated at the face with the sensor(s) placed as high and close to the front of the machine's cutter head as practicable.[§1926.800(j)(1)(iv)]

(v) *[A] Whenever air monitoring* indicates the presence of 5 ppm or more of hydrogen sulfide, a test shall be conducted in the affected underground work area(s), at least at the beginning and midpoint of each shift, until the concentration of hydrogen sulfide has been less than 5 ppm for 3 consecutive days.[§1926.800(j)(1)(v)[A]]

[B] Whenever hydrogen sulfide is detected in an amount exceeding 10 ppm, a continuous sampling and indicating hydrogen sulfide monitor shall be used to monitor the affected work area.[§1926.800(j)(1)(v)[B]]

[C] Employees shall be informed when a concentration of 10 ppm hydrogen sulfide is exceeded.[§1926.800(j)(1)(v)[C]]

[D] The continuous sampling and indicating hydrogen sulfide monitor shall be designed, installed, and maintained to provide a visual and aural alarm when the hydrogen sulfide concentration reaches 20 ppm to signal that additional measures, such as respirator use, increased ventilation, or evacuation, might be necessary to maintain hydrogen sulfide exposure below the permissible exposure limit.[§1926.800(j)(1)(v)[D]]

(vi) *When the competent person determines,* on the basis of air monitoring results or other information, that air contaminants may be present in sufficient quantity to be dangerous to life, the employer shall:[§1926.800(j)(1)(vi)]

[A] Prominently post a notice at all entrances to the underground jobsite to inform all entrants of the hazardous condition; and[§1926.800(j)(1)(vi)[A]]

[B] Ensure that the necessary precautions are taken. [§1926.800(j)(1)(vi)[B]]

(vii) *Whenever five percent or more* of the lower explosive limit for methane or other flammable gases is detected in any underground work area(s) or in the air return, steps shall be taken to increase ventilation air volume or otherwise control the gas concentration, unless the employer is operating in accordance with the potentially gassy or gassy operation requirements. Such additional ventilation controls may be discontinued when gas concentrations are reduced below five percent of the lower explosive limit, but shall be reinstituted whenever the five percent level is exceeded. [§1926.800(j)(1)(vii)]

(viii) *Whenever 10 percent or more* of the lower explosive limit for methane or other flammable gases is detected in the vicinity of welding, cutting, or other hot work, such work shall be suspended until the concentration of such flammable gas is reduced to less than 10 percent of the lower explosive limit.[§1926.800(j)(1)(viii)]

(ix) *Whenever 20 percent or more* of the lower explosive limit for methane or other flammable gases is detected in any underground work area(s) or in the air return:[§1926.800(j)(1)(ix)]

[A] All employees, except those necessary to eliminate the hazard, shall be immediately withdrawn to a safe location above ground; and[§1926.800(j)(1)(ix)[A]]

[B] Electrical power, except for acceptable pumping and ventilation equipment, shall be cut off to the area endangered by the flammable gas until the concentration of such gas is reduced to less than 20 percent of the lower explosive limit.[§1926.800(j)(1)(ix)[B]]

(2) *Additional monitoring* for potentially gassy and gassy operations. Operations which meet the criteria for potentially gassy and gassy operations set forth in paragraph (h) of this section shall be subject to the additional monitoring requirements of this paragraph.[§1926.800(j)(2)]

(i) *A test for oxygen content* shall be conducted in the affected underground work areas and work areas immediately adjacent to such areas at least at the beginning and midpoint of each shift.[§1926.800(j)(2)(i)]

(ii) *When using rapid excavation machines,* continuous automatic flammable gas monitoring equipment shall be used to monitor the air at the heading, on the rib, and in the return air duct. The continuous monitor shall signal the heading, and shut down electric power in the affected underground work area, except for acceptable pumping and ventilation equipment, when 20 percent or more of the lower explosive limit for methane or other flammable gases is encountered. [§1926.800(j)(2)(ii)]

(iii) *A manual flammable gas monitor* shall be used as needed, but at least at the beginning and midpoint of each shift, to ensure that the limits prescribed in paragraphs (h) and (j) are not exceeded. In addition, a manual electrical shut down control shall be provided near the heading.[§1926.800(j)(2)(iii)]

(iv) *Local gas tests* shall be made prior to and continuously during any welding, cutting, or other hot work.[§1926.800(j)(2)(iv)]

(v) *In underground operations* driven by drill-and-blast methods, the air in the affected area shall be tested for flammable gas prior to re- entry after blasting, and continuously when employees are working underground.[§1926.800(j)(2)(v)]

(3) *Recordkeeping.* A record of all air quality tests shall be maintained above ground at the worksite and be made available to the Secretary of Labor upon request. The record shall include the location, date, time, substance and amount monitored. Records of exposures to toxic substances shall be retained in accordance with Sec. 1910.1020[1] of this chapter. All other air quality test records shall be retained until completion of the project.[§1926.800(j)(3)]

[1]**Editor's Note: The CFR refers in error to Sec. 1910.33. The regulation as it appears above is corrected to refer to Sec. 1910.1020, corresponding to the Asbestos standards appearing at Sec. 1910.1001.**

(k) Ventilation.[§1926.800(k)]

(1) (i) *Fresh air shall be supplied* to all underground work areas in sufficient quantities to prevent dangerous or harmful accumulation of dusts, fumes, mists, vapors or gases. [§1926.800(k)(1)(i)]

(ii) *Mechanical ventilation shall be provided* in all underground work areas except when the employer can demonstrate that natural ventilation provides the necessary air quality through sufficient air volume and air flow.[§1926.800(k)(1)(ii)]

(2) *A minimum of 200 cubic feet (5.7 m^3)* of fresh air per minute shall be supplied for each employee underground. [§1926.800(k)(2)]

(3) *The linear velocity of air flow* in the tunnel bore, in shafts, and in all other underground work areas shall be at least 30 feet (9.15 m) per minute where blasting or rock drilling is conducted, or where other conditions likely to produce dust, fumes, mists, vapors, or gases in harmful or explosive quantities are present. [§1926.800(k)(3)]

(4) *The direction of mechanical air flow shall be reversible.* [§1926.800(k)(4)]

(5) *Following blasting,* ventilation systems shall exhaust smoke and fumes to the outside atmosphere before work is resumed in affected areas.[§1926.800(k)(5)]

(6) *Ventilation doors shall be designed* and installed so that they remain closed when in use, regardless of the direction of the air flow.[§1926.800(k)(6)]

(7) *When ventilation has been reduced* to the extent that hazardous levels of methane or flammable gas may have accumulated, a competent person shall test all affected areas after ventilation has been restored and shall determine whether the atmosphere is within flammable limits before any power, other than for acceptable equipment, is restored or work is resumed. [§1926.800(k)(7)]

(8) *Whenever the ventilation system* has been shut down with all employees out of the underground area, only competent persons authorized to test for air contaminants shall be allowed underground until the ventilation has been restored and all affected areas have been tested for air contaminants and declared safe.[§1926.800(k)(8)]

(9) *When drilling rock or concrete,* appropriate dust control measures shall be taken to maintain dust levels within limits set in Sec. 1926.55. Such measures may include, but are not limited to, wet drilling, the use of vacuum collectors, and water mix spray systems.[§1926.800(k)(9)]

(10) (i) *Internal combustion engines,* except diesel-powered engines on mobile equipment, are prohibited underground. [§1926.800(k)(10)(i)]

(ii) *Mobile diesel-powered equipment* used underground in atmospheres other than gassy operations shall be either approved by MSHA in accordance with the provisions of 30 CFR part 32 (formerly Schedule 24), or shall be demonstrated by the employer to be fully equivalent to such MSHA-approved equipment, and shall be operated in accordance with that part. (Each brake horsepower of a diesel engine requires at least 100 cubic feet (28.32 m^3) of air per minute for suitable operation in addition to the air requirements for personnel. Some engines may require a greater amount of air to ensure that the allowable levels of carbon monoxide, nitric oxide, and nitrogen dioxide are not exceeded.) [§1926.800(k)(10)(ii)]

(11) *Potentially gassy or gassy operations* shall have ventilation systems installed which shall:[§1926.800(k)(11)]

(i) *Be constructed of fire-resistant materials; and* [§1926.800(k)(11)(i)]

(ii) *Have acceptable electrical systems, including fan motors.* [§1926.800(k)(11)(ii)]

(12) *Gassy operations shall be provided* with controls located above ground for reversing the air flow of ventilation systems. [§1926.800(k)(12)]

(13) *In potentially gassy or gassy operations,* wherever mine-type ventilation systems using an offset main fan installed on the surface are used, they shall be equipped with explosion-doors or a weak-wall having an area at least equivalent to the cross-sectional area of the airway.[§1926.800(k)(13)]

(l) Illumination.[§1926.800(l)]

(1) *Illumination requirements applicable* to underground construction operations are found in Table D-3 of Sec. 1926.56 of this part.[§1926.800(l)(1)]

(2) *Only acceptable portable lighting equipment* shall be used within 50 feet (15.24 m) of any underground heading during explosives handling.[§1926.800(l)(2)]

(m) Fire prevention and control. Fire prevention and protection requirements applicable to underground construction operations are found in subpart F of this part, except as modified by the following additional standards.[§1926.800(m)]

(1) *Open flames and fires are prohibited* in all underground construction operations except as permitted for welding, cutting and other hot work operations in paragraph (n) of this section. [§1926.800(m)(1)]

(2) (i) *Smoking may be allowed* only in areas free of fire and explosion hazards.[§1926.800(m)(2)(i)]

(ii) *Readily visible signs* prohibiting smoking and open flames shall be posted in areas having fire or explosion hazards. [§1926.800(m)(2)(ii)]

(3) *The employer may store underground* no more than a 24-hour supply of diesel fuel for the underground equipment used at the worksite.[§1926.800(m)(3)]

(4) *The piping of diesel fuel* from the surface to an underground location is permitted only if:[§1926.800(m)(4)]

(i) *Diesel fuel is contained* at the surface in a tank whose maximum capacity is no more than the amount of fuel required to supply for a 24- hour period the equipment serviced by the underground fueling station; and[§1926.800(m)(4)(i)]

(ii) *The surface tank is connected* to the underground fueling station by an acceptable pipe or hose system that is controlled at the surface by a valve, and at the shaft bottom by a hose nozzle; and[§1926.800(m)(4)(ii)]

(iii) *The pipe is empty at all times* except when transferring diesel fuel from the surface tank to a piece of equipment in use underground; and[§1926.800(m)(4)(iii)]

(iv) *Hoisting operations in the shaft* are suspended during refueling operations if the supply piping in the shaft is not protected from damage.[§1926.800(m)(4)(iv)]

(5) (i) *Gasoline shall not be carried, stored, or used underground.* [§1926.800(m)(5)(i)]

(ii) *Acetylene, liquefied petroleum gas,* and Methylacetylene Propadiene Stabilized gas may be used underground only for welding, cutting and other hot work, and only in accordance with subpart J of this part, and paragraphs (j), (k), (m), and (n) of this section.[§1926.800(m)(5)(ii)]

(6) *Oil, grease, and diesel fuel stored underground* shall be kept in tightly sealed containers in fire-resistant areas at least 300 feet (91.44 m) from underground explosive magazines, and at least 100 feet (30.48 m) from shaft stations and steeply inclined passageways. Storage areas shall be positioned or diked so that the contents of ruptured or overturned containers will not flow from the storage area.[§1926.800(m)(6)]

(7) *Flammable or combustible materials* shall not be stored above ground within 100 feet (30.48 m) of any access opening to any underground operation. Where this is not feasible because of space limitations at the jobsite, such materials may be located within the 100-foot limit, provided that:[§1926.800(m)(7)]

(i) *They are located as far as practicable from the opening; and* [§1926.800(m)(7)(i)]

(ii) *Either a fire-resistant barrier* of not less than one-hour rating is placed between the stored material and the opening, or additional precautions are taken which will protect the materials from ignition sources.[§1926.800(m)(7)(ii)]

(8) *Fire-resistant hydraulic fluids* shall be used in hydraulically-actuated underground machinery and equipment unless such equipment is protected by a fire suppression system or by multi-purpose fire extinguisher(s) rated at of sufficient capacity for the type and size of hydraulic equipment involved, but rated at least 4A:40B:C.[§1926.800(m)(8)]

(9) (i) *Electrical installations in underground areas* where oil, grease, or diesel fuel are stored shall be used only for lighting fixtures.[§1926.800(m)(9)(i)]

(ii) *Lighting fixtures in storage areas,* or within 25 feet (7.62 m) of underground areas where oil, grease, or diesel fuel are stored, shall be approved for Class I, Division 2 locations, in accordance with subpart K of this part.[§1926.800(m)(9)(ii)]

(10) *Leaks and spills* of flammable or combustible fluids shall be cleaned up immediately.[§1926.800(m)(10)]

(11) *A fire extinguisher* of at least 4A:40B:C rating or other equivalent extinguishing means shall be provided at the head pulley and at the tail pulley of underground belt conveyors. [§1926.800(m)(11)]

(12) *Any structure located underground* or within 100 feet (30.48 m) of an opening to the underground shall be constructed of material having a fire-resistance rating of at least one hour. [§1926.800(m)(12)]

(n) **Welding, cutting, and other hot work.** In addition to the requirements of subpart J of this part, the following requirements shall apply to underground welding, cutting, and other hot work. [§1926.800(n)]

(1) *No more than the amount of fuel gas and oxygen cylinders* necessary to perform welding, cutting, or other hot work during the next 24-hour period shall be permitted underground. [§1926.800(n)(1)]

(2) *Noncombustible barriers shall be installed* below welding, cutting, or other hot work being done in or over a shaft or raise. [§1926.800(n)(2)]

(o) **Ground support.**[§1926.800(o)]

(1) *Portal areas.* Portal openings and access areas shall be guarded by shoring, fencing, head walls, shotcreting or other equivalent protection to ensure safe access of employees and equipment. Adjacent areas shall be scaled or otherwise secured to prevent loose soil, rock, or fractured materials from endangering the portal and access area.[§1926.800(o)(1)]

(2) *Subsidence areas.* The employer shall ensure ground stability in hazardous subsidence areas by shoring, by filling in, or by erecting barricades and posting warning signs to prevent entry. [§1926.800(o)(2)]

(3) *Underground areas.*[§1926.800(o)(3)]

(i) *[A] A competent person shall inspect* the roof, face, and walls of the work area at the start of each shift and as often as necessary to determine ground stability.[§1926.800(o)(3)(i)[A]]

[B] Competent persons conducting such inspections shall be protected from loose ground by location, ground support or equivalent means.[§1926.800(o)(3)(i)[B]]

(ii) *Ground conditions* along haulageways and travelways shall be inspected as frequently as necessary to ensure safe passage.[§1926.800(o)(3)(ii)]

(iii) *Loose ground* that might be hazardous to employees shall be taken down, scaled or supported.[§1926.800(o)(3)(iii)]

(iv) *[A] Torque wrenches shall be used* wherever bolts that depend on torsionally applied force are used for ground support.[§1926.800(o)(3)(iv)[A]]

[B] A competent person shall determine whether rock bolts meet the necessary torque, and shall determine the testing frequency in light of the bolt system, ground conditions and the distance from vibration sources. [§1926.800(o)(3)(iv)[B]]

(v) *Suitable protection shall be provided* for employees exposed to the hazard of loose ground while installing ground support systems.[§1926.800(o)(3)(v)]

(vi) *Support sets shall be installed* so that the bottoms have sufficient anchorage to prevent ground pressures from dislodging the support base of the sets. Lateral bracing (collar bracing, tie rods, or spreaders) shall be provided between immediately adjacent sets to ensure added stability. [§1926.800(o)(3)(vi)]

(vii) *Damaged or dislodged ground supports* that create a hazardous condition shall be promptly repaired or replaced. When replacing supports, the new supports shall be installed before the damaged supports are removed.[§1926.800(o)(3)(vii)]

(viii) *A shield or other type of support* shall be used to maintain a safe travelway for employees working in dead-end areas ahead of any support replacement operation. [§1926.800(o)(3)(viii)]

(4) *Shafts.*[§1926.800(o)(4)]

(i) *Shafts and wells over 5 feet (1.53 m)* in depth that employees must enter shall be supported by a steel casing, concrete pipe, timber, solid rock or other suitable material. [§1926.800(o)(4)(i)]

(ii) *[A] The full depth of the shaft* shall be supported by casing or bracing except where the shaft penetrates into solid rock having characteristics that will not change as a result of exposure. Where the shaft passes through earth into solid rock, or through solid rock into earth, and where there is potential for shear, the casing or bracing shall extend at least 5 feet (1.53 m) into the solid rock. When the shaft terminates in solid rock, the casing or bracing shall extend to the end of the shaft or 5 feet (1.53 m) into the solid rock, whichever is less.[§1926.800(o)(4)(ii)[A]]

[B] The casing or bracing shall extend 42 inches (1.07 m) plus or minus 3 inches (8 cm) above ground level, except that the minimum casing height may be reduced to 12 inches (0.3 m), provided that a standard railing is installed; that the ground adjacent to the top of the shaft is sloped away from the shaft collar to prevent entry of liquids; and that effective barriers are used to prevent mobile equipment operating near the shaft from jumping over the 12 inch (0.3 m) barrier.[§1926.800(o)(4)(ii)[B]]

(iii) *After blasting operations in shafts,* a competent person shall determine if the walls, ladders, timbers, blocking, or wedges have loosened. If so, necessary repairs shall be made before employees other than those assigned to make the repairs are allowed in or below the affected areas.[§1926.800(o)(4)(iii)]

(p) **Blasting.** This paragraph applies in addition to the requirements for blasting and explosives operations, including handling of misfires, which are found in subpart U of this part.[§1926.800(p)]

(1) *Blasting wires shall be kept clear* of electrical lines, pipes, rails, and other conductive material, excluding earth, to prevent explosives initiation or employee exposure to electric current. [§1926.800(p)(1)]

(2) *Following blasting,* an employee shall not enter a work area until the air quality meets the requirements of paragraph (j) of this section.[§1926.800(p)(2)]

(q) **Drilling.**[§1926.800(q)]

(1) *A competent person shall inspect* all drilling and associated equipment prior to each use. Equipment defects affecting safety shall be corrected before the equipment is used. [§1926.800(q)(1)]

(2) *The drilling area shall be inspected* for hazards before the drilling operation is started.[§1926.800(q)(2)]

(3) *Employees shall not be allowed* on a drill mast while the drill bit is in operation or the drill machine is being moved. [§1926.800(q)(3)]

(4) *When a drill machine is being moved* from one drilling area to another, drill steel, tools, and other equipment shall be secured and the mast shall be placed in a safe position.[§1926.800(q)(4)]

(5) *Receptacles or racks shall be provided* for storing drill steel located on jumbos.[§1926.800(q)(5)]

(6) *Employees working below jumbo decks* shall be warned whenever drilling is about to begin.[§1926.800(q)(6)]

(7) *Drills on columns shall be anchored* firmly before starting drilling, and shall be retightened as necessary thereafter. [§1926.800(q)(7)]

(8) (i) *The employer shall provide* mechanical means on the top deck of a jumbo for lifing unwieldy or heavy material. [§1926.800(q)(8)(i)]

(ii) *When jumbo decks are over 10 feet (3.05 m)* in height, the employer shall install stairs wide enough for two persons. [§1926.800(q)(8)(ii)]

(iii) *Jumbo decks more than 10 feet (3.05 m)* in height shall be equipped with guardrails on all open sides, excluding access openings of platforms, unless an adjacent surface provides equivalent fall protection.[§1926.800(q)(8)(iii)]

(iv) *[A] Only employees assisting the operator* shall be allowed to ride on jumbos, unless the jumbo meets the requirements of paragraph (r)(6)(ii) of this section. [§1926.800(q)(8)(iv)[A]]

[B] Jumbos shall be chocked to prevent movement while employees are working on them.[§1926.800(q)(8)(iv)[B]]

(v) *[A] Walking and working surfaces* of jumbos shall be maintained to prevent the hazards of slipping, tripping and falling.[§1926.800(q)(8)(v)[A]]

[B] Jumbo decks and stair treads shall be designed to be slip- resistent and secured to prevent accidental displacement.[§1926.800(q)(8)(v)[B]]

(9) *Scaling bars shall be available* at scaling operations and shall be maintained in good condition at all times. Blunted or severely worn bars shall not be used.[§1926.800(q)(9)]

(10) (i) *Blasting holes shall not be drilled* through blasted rock (muck) or water.[§1926.800(q)(10)(i)]

(ii) *Employees in a shaft shall be protected* either by location or by suitable barrier(s) if powered mechanical loading equipment is used to remove muck containing unfired explosives. [§1926.800(q)(10)(ii)]

(11) *A caution sign reading "Buried Line,"* or similar wording shall be posted where air lines are buried or otherwise hidden by water or debris.[§1926.800(q)(11)]

(r) **Haulage.**[§1926.800(r)]

(1) (i) *A competent person shall inspect* haulage equipment before each shift.[§1926.800(r)(1)(i)]

(ii) *Equipment defects affecting safety and health* shall be corrected before the equipment is used.[§1926.800(r)(1)(ii)]

(2) *Powered mobile haulage equipment* shall have suitable means of stopping.[§1926.800(r)(2)]

(3) (i) *Power mobile haulage equipment,* including trains, shall have audible warning devices to warn employees to stay clear. The operator shall sound the warning device before moving the equipment and whenever necessary during travel.[§1926.800(r)(3)(i)]

(ii) *The operator shall assure* that lights which are visible to employees at both ends of any mobile equipment, including a train, are turned on whenever the equipment is operating. [§1926.800(r)(3)(ii)]

(4) *In those cabs where glazing is used,* the glass shall be safety glass, or its equivalent, and shall be maintained and cleaned so that vision is not obstructed.[§1926.800(r)(4)]

(5) *Anti-roll back devices or brakes* shall be installed on inclined conveyor drive units to prevent conveyors from inadvertently running in reverse.[§1926.800(r)(5)]

(6) [§1926.800(r)(6)]

(i) [§1926.800(r)(6)(i)]

[A] Employees shall not be permitted to ride a power-driven chain, belt, or bucket conveyor unless the conveyor is specifically designed for the transportation of persons. [§1926.800(r)(6)(i)[A]]

[B] Endless belt-type manlifts are prohibited in underground construction.[§1926.800(r)(6)(i)[B]]

[C] General requirements also applicable to underground construction for use of conveyors in construction are found in Sec. 1926.555 of this part.[§1926.800(r)(6)(i)[C]]

(ii) *No employee shall ride haulage equipment* unless it is equipped with seating for each passenger and protects passengers from being struck, crushed, or caught between other equipment or surfaces. Members of train crews may ride on a locomotive if it is equipped with handholds and nonslip steps or footboards. Requirements applicable to Underground Construction for motor vehicle transportation of employees are found in Sec. 1926.601 of this part.[§1926.800(r)(6)(ii)]

(7) *Powered mobile haulage equipment,* including trains, shall not be left unattended unless the master switch or motor is turned off; operating controls are in neutral or park position; and the brakes are set, or equivalent precautions are taken to prevent rolling.[§1926.800(r)(7)]

(8) *Whenever rails serve* as a return for a trolley circuit, both rails shall be bonded at every joint and crossbonded every 200 feet (60.96 m).[§1926.800(r)(8)]

(9) *When dumping cars by hand,* the car dumps shall have tiedown chains, bumper blocks, or other locking or holding devices to prevent the cars from overturning.[§1926.800(r)(9)]

(10) *Rocker-bottom or bottom-dump cars* shall be equipped with positive locking devices to prevent unintended dumping. [§1926.800(r)(10)]

(11) *Equipment to be hauled* shall be loaded and secured to prevent sliding or dislodgement.[§1926.800(r)(11)]

(12) (i) *Mobile equipment,* including rail-mounted equipment, shall be stopped for manual connecting or service work. [§1926.800(r)(12)(i)]

(ii) *Employees shall not reach* between moving cars during coupling operations.[§1926.800(r)(12)(ii)]

(iii) *Couplings shall not be aligned,* shifted or cleaned on moving cars or locomotives.[§1926.800(r)(12)(iii)]

(13) (i) *Safety chains or other connections* shall be used in addition to couplers to connect man cars or powder cars whenever the locomotive is uphill of the cars.[§1926.800(r)(13)(i)]

(ii) *When the grade exceeds one percent* and there is a potential for runaway cars, safety chains or other connections shall be used in addition to couplers to connect haulage cars or, as an alternative, the locomotive must be downhill of the train. [§1926.800(r)(13)(ii)]

(iii) *Such safety chains or other connections* shall be capable of maintaining connection between cars in the event of either coupler disconnect, failure or breakage.[§1926.800(r)(13)(iii)]

(14) *Parked rail equipment* shall be chocked, blocked, or have brakes set to prevent inadvertent movement.[§1926.800(r)(14)]

(15) *Berms, bumper blocks, safety hooks,* or equivalent means shall be provided to prevent overtravel and overturning of haulage equipment at dumping locations.[§1926.800(r)(15)]

(16) *Bumper blocks or equivalent stopping devices* shall be provided at all track dead ends.[§1926.800(r)(16)]

(17) (i) *Only small handtools,* lunch pails or similar small items may be transported with employees in man-cars, or on top of a locomotive.[§1926.800(r)(17)(i)]

(ii) *When small hand tools* or other small items are carried on top of a locomotive, the top shall be designed or modified to retain them while traveling.[§1926.800(r)(17)(ii)]

(18) (i) *Where switching facilities are available,* occupied personnel-cars shall be pulled, not pushed. If personnel-cars must be pushed and visibility of the track ahead is hampered, then a qualified person shall be stationed in the lead car to give signals to the locomotive operator.[§1926.800(r)(18)(i)]

(ii) *Crew trips shall consist of personnel-loads only.* [§1926.800(r)(18)(ii)]

(s) **Electrical safety.** This paragraph applies in addition to the general requirements for electrical safety which are found in subpart K of this part.[§1926.800(s)]

(1) *Electric power lines* shall be insulated or located away from water lines, telephone lines, air lines, or other conductive materials so that a damaged circuit will not energize the other systems.[§1926.800(s)(1)]

(2) *Lighting circuits shall be located* so that movement of personnel or equipment will not damage the circuits or disrupt service. [§1926.800(s)(2)]

(3) *Oil-filled transformers shall not be used* underground unless they are located in a fire-resistant enclosure suitably vented to the outside and surrounded by a dike to retain the contents of the transformers in the event of rupture.[§1926.800(s)(3)]

(t) **Hoisting unique to underground construction.** Employers must comply with §1926.1501(g) of §1926 subpart DD. Except as modified by this paragraph (t), the following provisions of subpart N of this part apply: Requirements for material hoists are found in §§1926.552(a) and (b) of this part. Requirements for personnel hoists are found in the personnel hoists requirements of §§1926.552(a) and (c) of this part and in the elevator requirement of §§1926.552(a) and (d) of this part.[§1926.800(t)]

(u) **Definitions.**[§1926.800(u)]

Accept — Any device, equipment, or appliance that is either approved by MSHA and maintained in permissible condition, or is listed or labeled for the class and location under subpart K of this part.

Rapid Excavation Machine — Tunnel boring machines, shields, roadheaders, or any other similar excavation machine.

[54 FR 23850, June 2, 1989; 58 FR 35311, June 30, 1993, as amended at 61 FR 5510, Feb. 13, 1996; 63 FR 1297, Jan. 8, 1998; 71 FR 16674, Apr. 3, 2006; 75 FR 48135, Aug. 9, 2010]

§1926.801
Caissons

(a) **Wherever, in caisson work** in which compressed air is used, and the working chamber is less than 11 feet in length, and when such caissons are at any time suspended or hung while work is in progress so that the bottom of the excavation is more than 9 feet below the deck of the working chamber, a shield shall be erected therein for the protection of the employees.[§1926.801(a)]

(b) **Shafts shall be subjected to a hydrostatic** or air-pressure test, at which pressure they shall be tight. The shaft shall be stamped on the outside shell about 12 inches from each flange to show the pressure to which they have been subjected.[§1926.801(b)]

(c) **Whenever a shaft is used, it shall be provided,** where space permits, with a safe, proper, and suitable staircase for its entire length, including landing platforms, not more than 20 feet apart. Where this is impracticable, suitable ladders shall be installed with landing platforms located about 20 feet apart to break the climb. [§1926.801(c)]

(d) **All caissons having a diameter or side** greater than 10 feet shall be provided with a man lock and shaft for the exclusive use of employees.[§1926.801(d)]

(e) **In addition to the gauge in the locks,** an accurate gauge shall be maintained on the outer and inner side of each bulkhead. These gauges shall be accessible at all times and kept in accurate working order.[§1926.801(e)]

(f) **In caisson operations where employees are exposed** to compressed air working environments, the requirements contained in Sec. 1926.803 shall be complied with.[§1926.801(f)]

§1926.802
Cofferdams

(a) **If overtopping of the cofferdam by high waters** is possible, means shall be provided for controlled flooding of the work area. [§1926.802(a)]

(b) **Warning signals for evacuation of employees** in case of emergency shall be developed and posted.[§1926.802(b)]

(c) **Cofferdam walkways, bridges, or ramps** with at least two means of rapid exit shall be provided with guardrails as specified in subpart M of this part.[§1926.802(c)]

(d) **Cofferdams located close to navigable shipping channels** shall be protected from vessels in transit, where possible. [§1926.802(d)]

§1926.803
Compressed air

(a) **General provisions.**[§1926.803(a)]

(1) *There shall be present, at all times,* at least one competent person designated by and representing the employer, who shall be familiar with this subpart in all respects, and responsible for full compliance with these and other applicable subparts. [§1926.803(a)(1)]

(2) *Every employee shall be instructed* in the rules and regulations which concern his safety or the safety of others.[§1926.803(a)(2)]

(b) **Medical attendance, examination, and regulations.**[§1926.803(b)]

(1) *There shall be retained one or more licensed physicians* familiar with and experienced in the physical requirements and the medical aspects of compressed air work and the treatment of decompression illness. He shall be available at all times while work is in progress in order to provide medical supervision of employees employed in compressed air work. He shall himself be physically qualified and be willing to enter a pressurized environment.[§1926.803(b)(1)]

(2) *No employee shall be permitted to enter* a compressed air environment until he has been examined by the physician and reported by him to be physically qualified to engage in such work.[§1926.803(b)(2)]

(3) *In the event an employee is absent from work for 10 days,* or is absent due to sickness or injury, he shall not resume work until he is reexamined by the physician, and his physical condition reported, as provided in this paragraph, to be such as to permit him to work in compressed air.[§1926.803(b)(3)]

(4) *After an employee has been employed continuously* in compressed air for a period designated by the physician, but not to exceed 1 year, he shall be reexamined by the physician to determine if he is still physically qualified to engage in compressed air work.[§1926.803(b)(4)]

(5) *Such physician shall at all times* keep a complete and full record of examinations made by him. The physician shall also keep an accurate record of any decompression illness or other illness or injury incapacitating any employee for work, and of all loss of life that occurs in the operation of a tunnel, caisson, or other compartment in which compressed air is used. [§1926.803(b)(5)]

(6) *Records shall be available for the inspection* of the Secretary or his representatives, and a copy thereof shall be forwarded to OSHA within 48 hours following the occurrence of the accident, death, injury, or decompression illness. It shall state as fully as possible the cause of said death or decompression illness, and the place where the injured or sick employee was taken, and such other relative information as may be required by the Secretary.[§1926.803(b)(6)]

(7) *A fully equipped first aid station* shall be provided at each tunnel project regardless of the number of persons employed. An ambulance or transportation suitable for a litter case shall be at each project.[§1926.803(b)(7)]

(8) *Where tunnels are being excavated from portals* more than 5 road miles apart, a first aid station and transportation facilities shall be provided at each portal.[§1926.803(b)(8)]

(9) *A medical lock shall be established and maintained* in immediate working order whenever air pressure in the working chamber is increased above the normal atmosphere.[§1926.803(b)(9)]

(10) *The medical lock shall:*[§1926.803(b)(10)]

(i) *Have at least 6 feet of clear headroom* at the center, and be subdivided into not less than two compartments; [§1926.803(b)(10)(i)]

(ii) *Be readily accessible* to employees working under compressed air;[§1926.803(b)(10)(ii)]

(iii) *Be kept ready for immediate use* for at least 5 hours subsequent to the emergence of any employee from the working chamber;[§1926.803(b)(10)(iii)]

(iv) *Be properly heated, lighted and ventilated;*[§1926.803(b)(10)(iv)]

(v) *Be maintained in a sanitary condition;*[§1926.803(b)(10)(v)]

(vi) *Have a nonshatterable port* through which the occupant(s) may be kept under constant observation;[§1926.803(b)(10)(vi)]

(vii) *Be designed for a working pressure of 75 p.s.i.g.* [§1926.803(b)(10)(vii)]

(viii) *Be equipped with internal controls* which may be overridden by external controls;[§1926.803(b)(10)(viii)]

(ix) *Be provided with air pressure gauges* to show the air pressure within each compartment to observers inside and outside the medical lock.[§1926.803(b)(10)(ix)]

(x) *Be equipped with a manual type sprinkler system* that can be activated inside the lock or by the outside lock tender. [§1926.803(b)(10)(x)]

(xi) *Be provided with oxygen lines* and fittings leading into external tanks. The lines shall be fitted with check valves to prevent reverse flow. The oxygen system inside the chamber shall be of a closed circuit design and be so designed as to automatically shut off the oxygen supply whenever the fire system is activated.[§1926.803(b)(10)(xi)]

(xii) *Be in constant charge of an attendant* under the direct control of the retained physician. The attendant shall be trained in the use of the lock and suitably instructed regarding steps to be taken in the treatment of employee exhibiting symptoms compatible with a diagnosis of decompression illness; [§1926.803(b)(10)(xii)]

(xiii) *Be adjacent to an adequate emergency medical facility;* [§1926.803(b)(10)(xiii)]

(xiv) *The medical facility shall be equipped* with demand-type oxygen inhalation equipment approved by the U.S. Bureau of Mines;[§1926.803(b)(10)(xiv)]

(xv) *Be capable of being maintained* at a temperature, in use, not to exceed 90 °F. nor be less than 70 °F.; and [§1926.803(b)(10)(xv)]

(xvi) *Be provided with sources of air,* free of oil and carbon monoxide, for normal and emergency use, which are capable of raising the air pressure in the lock from 0 to 75 p.s.i.g. in 5 minutes.[§1926.803(b)(10)(xvi)]

(11) *Identification badges shall be furnished* to all employees, indicating that the wearer is a compressed air worker. A permanent record shall be kept of all identification badges issued. The badge shall give the employee's name, address of the medical lock, the telephone number of the licensed physician for the compressed air project, and contain instructions that in case of emergency of unknown or doubtful cause or illness, the wearer shall be rushed to the medical lock. The badge shall be worn at all times — off the job, as well as on the job.[§1926.803(b)(11)]

(c) Telephone and signal communication.[§1926.803(c)]

(1) *Effective and reliable means of communication,* such as bells, whistles, or telephones, shall be maintained, at all times between all the following locations:[§1926.803(c)(1)]

(i) *The working chamber face;*[§1926.803(c)(1)(i)]

(ii) *The working chamber side* of the man lock near the door; [§1926.803(c)(1)(ii)]

(iii) *The interior of the man lock;*[§1926.803(c)(1)(iii)]

(iv) *Lock attendant's station;*[§1926.803(c)(1)(iv)]

(v) *The compressor plant;*[§1926.803(c)(1)(v)]

(vi) *The first-aid station;*[§1926.803(c)(1)(vi)]

(vii) *The emergency lock (if one is required); and* [§1926.803(c)(1)(vii)]

(viii) *The special decompression chamber (if one is required).* [§1926.803(c)(1)(viii)]

(d) Signs and records.[§1926.803(d)]

(1) *The time of decompression* shall be posted in each man lock as follows:[§1926.803(d)(1)]

Time of Decompression for This Lock[§1926.803(d)(1)]

____ pounds to ____ pounds in ____ minutes.

____ pounds to ____ pounds in ____ minutes.

(Signed by) __________________ (Superintendent)

This form shall be posted in the Man Lock at all times.

(2) *Any code of signals used* shall be conspicuously posted near workplace entrances and such other locations as may be necessary to bring them to the attention of all employees concerned.[§1926.803(d)(2)]

(3) *For each 8-hour shift,* a record of employees employed under air pressure shall be kept by an employee who shall remain outside the lock near the entrance. This record shall show the period each employee spends in the air chamber and the time taken from decompression. A copy shall be submitted to the appointed physician after each shift.[§1926.803(d)(3)]

(e) Compression.[§1926.803(e)]

(1) *Every employee going under air pressure* for the first time shall be instructed on how to avoid excessive discomfort. [§1926.803(e)(1)]

(2) *During the compression of employees,* the pressure shall not be increased to more than 3 p.s.i.g. within the first minute. The pressure shall be held at 3 p.s.i.g. and again at 7 p.s.i.g. sufficiently long to determine if any employees are experiencing discomfort.[§1926.803(e)(2)]

(3) *After the first minute* the pressure shall be raised uniformly and at a rate not to exceed 10 p.s.i. per minute.[§1926.803(e)(3)]

(4) *If any employee complains of discomfort,* the pressure shall be held to determine if the symptoms are relieved. If, after 5 minutes the discomfort does not disappear, the lock attendant shall gradually reduce the pressure until the employee signals that the discomfort has ceased. If he does not indicate that the discomfort has disappeared, the lock attendant shall reduce the pressure to atmospheric and the employee shall be released from the lock.[§1926.803(e)(4)]

(5) *No employee shall be subjected* to pressure exceeding 50 pounds per square inch except in emergency.[§1926.803(e)(5)]

(f) Decompression.[§1926.803(f)]

(1) *Decompression to normal condition* shall be in accordance with the Decompression Tables in Appendix A of this subpart. [§1926.803(f)(1)]

(2) *In the event it is necessary* for an employee to be in compressed air more than once in a 24-hour period, the appointed physician shall be responsible for the establishment of methods and procedures of decompression applicable to repetitive exposures.[§1926.803(f)(2)]

(3) *If decanting is necessary,* the appointed physician shall establish procedures before any employee is permitted to be decompressed by decanting methods. The period of time that the employees spend at atmospheric pressure between the decompression following the shift and recompression shall not exceed 5 minutes.[§1926.803(f)(3)]

(g) Man locks and special decompression chambers[§1926.803(g)]

(1) *Man locks.*[§1926.803(g)(1)]

(i) *Except in emergency,* no employees employed in compressed air shall be permitted to pass from the working chamber to atmospheric pressure until after decompression, in accordance with the procedures in this subpart. [§1926.803(g)(1)(i)]

(ii) *The lock attendant in charge of a man lock* shall be under the direct supervision of the appointed physician. He shall be stationed at the lock controls on the free air side during the period of compression and decompression and shall remain at the lock control station whenever there are men in the working chamber or in the man lock.[§1926.803(g)(1)(ii)]

(iii) *Except where air pressure in the working chamber* is below 12 p.s.i.g., each man lock shall be equipped with automatic controls which, through taped programs, cams, or similar apparatus, shall automatically regulate decompressions. It shall also be equipped with manual controls to permit the lock attendant to override the automatic mechanism in the event of an emergency, as provided in paragraph (g)(1)(viii) of this section.[§1926.803(g)(1)(iii)]

(iv) *A manual control,* which can be used in the event of an emergency, shall be placed inside the man lock. [§1926.803(g)(1)(iv)]

(v) *A clock, thermometer, and continuous recording pressure gauge* with a 4-hour graph shall be installed outside of each man lock and shall be changed prior to each shift's decompression. The chart shall be of sufficient size to register a legible record of variations in pressure within the man lock and shall be visible to the lock attendant. A copy of each graph shall be submitted to the appointed physician after each shift. In addition, a pressure gauge, clock, and thermometer shall also be installed in each man lock. Additional fittings shall be provided so that test gauges may be attached whenever necessary.[§1926.803(g)(1)(v)]

(vi) *Except where air pressure is below 12 p.s.i.g.* and there is no danger of rapid flooding, all caissons having a working area greater than 150 square feet, and each bulkhead in tunnels of 14 feet or more in diameter, or equivalent area, shall have at least two locks in perfect working condition, one of which shall be used exclusively as a man lock, the other, as a materials lock.[§1926.803(g)(1)(vi)]

(vii) *Where only a combination man-and-materials lock* is required, this single lock shall be of sufficient capacity to hold the employees constituting two successive shifts. [§1926.803(g)(1)(vii)]

(viii) *Emergency locks shall be large enough* to hold an entire heading shift and a limit maintained of 12 p.s.i.g. There shall be a chamber available for oxygen decompression therapy to 28 p.s.i.g.[§1926.803(g)(1)(viii)]

(ix) *The man lock shall be large enough* so that those using it are not compelled to be in a cramped position, and shall not have less than 5 feet clear head room at the center and a minimum of 30 cubic feet of air space per occupant. [§1926.803(g)(1)(ix)]

(x) *Locks on caissons shall be so located* that the bottom door shall be not less than 3 feet above the water level surrounding the caisson on the outside. (The water level, where it is affected by tides, is construed to mean high tide.) [§1926.803(g)(1)(x)]

(xi) *In addition to the pressure gauge in the locks,* an accurate pressure gauge shall be maintained on the outer and inner side of each bulkhead. These gauges shall be accessible at all times and shall be kept in accurate working order. [§1926.803(g)(1)(xi)]

(xii) *Man locks shall have an observation port* at least 4 inches in diameter located in such a position that all occupants of the man lock may be observed from the working chamber and from the free air side of the lock.[§1926.803(g)(1)(xii)]

(xiii) *Adequate ventilation in the lock shall be provided.* [§1926.803(g)(1)(xiii)]

(xiv) *Man locks shall be maintained* at a minimum temperature of 70 °F.[§1926.803(g)(1)(xiv)]

(xv) *When locks are not in use* and employees are in the working chamber, lock doors shall be kept open to the working chamber, where practicable.[§1926.803(g)(1)(xv)]

(xvi) *Provision shall be made* to allow for rescue parties to enter the tunnel if the working force is disabled.[§1926.803(g)(1)(xvi)]

(xvii) *A special decompression chamber* of sufficient size to accommodate the entire force of employees being decompressed at the end of a shift shall be provided whenever the regularly established working period requires a total time of decompression exceeding 75 minutes.[§1926.803(g)(1)(xvii)]

(2) *Special decompression chamber.*[§1926.803(g)(2)]

(i) *The headroom in the special decompression chamber* shall be not less than a minimum 7 feet and the cubical content shall provide at least 50 cubic feet of airspace for each employee. For each occupant, there shall be provided 4 square feet of free walking area and 3 square feet of seating space, exclusive of area required for lavatory and toilet facilities. The rated capacity shall be based on the stated minimum space per employee and shall be posted at the chamber entrance. The posted capacity shall not be exceeded, except in case of emergency.[§1926.803(g)(2)(i)]

(ii) *Each special decompression chamber* shall be equipped with the following:[§1926.803(g)(2)(ii)]

[a] *A clock or clocks* suitably placed so that the attendant and the chamber occupants can readily ascertain the time; [§1926.803(g)(2)(ii)[a]]

[b] *Pressure gauges* which will indicate to the attendants and to the chamber occupants the pressure in the chamber; [§1926.803(g)(2)(ii)[b]]

[c] *Valves to enable the attendant* to control the supply and discharge of compressed air into and from the chamber; [§1926.803(g)(2)(ii)[c]]

[d] *Valves and pipes,* in connection with the air supply and exhaust, arranged so that the chamber pressure can be controlled from within and without;[§1926.803(g)(2)(ii)[d]]

[e] *Effective means of oral intercommunication* between the attendant, occupants of the chamber, and the air compressor plant; and[§1926.803(g)(2)(ii)[e]]

[f] *An observation port at the entrance* to permit observation of the chamber occupants.[§1926.803(g)(2)(ii)[f]]

(iii) *Seating facilities in special decompression chambers* shall be so arranged as to permit a normal sitting posture without cramping. Seating space, not less than 18 inches by 24 inches wide, shall be provided per occupant.[§1926.803(g)(2)(iii)]

(iv) *Adequate toilet and washing facilities,* in a screened or enclosed recess, shall be provided. Toilet bowls shall have a built-in protector on the rim so that an air space is created when the seat lid is closed.[§1926.803(g)(2)(iv)]

(v) *Fresh and pure drinking water shall be available.* This may be accomplished by either piping water into the special decompression chamber and providing drinking fountains, or by providing individual canteens, or by some other sanitary means. Community drinking vessels are prohibited. [§1926.803(g)(2)(v)]

(vi) *No refuse or discarded material* of any kind shall be permitted to accumulate, and the chamber shall be kept clean. [§1926.803(g)(2)(vi)]

(vii) *Unless the special decompression chamber* is serving as the man lock to atmospheric pressure, the special decompression chamber shall be situated, where practicable, adjacent to the man lock on the atmospheric pressure side of the bulkhead. A passageway shall be provided, connecting the special chamber with the man lock, to permit employees in the process of decompression to move from the man lock to the special chamber without a reduction in the ambient pressure from that designated for the next stage of decompression. The passageway shall be so arranged as to not interfere with the normal operation of the man lock, nor with the release of the occupants of the special chamber to atmospheric pressure upon the completion of the decompression procedure.[§1926.803(g)(2)(vii)]

(h) Compressor plant and air supply.[§1926.803(h)]

(1) *At all times there shall be a thoroughly experienced,* competent, and reliable person on duty at the air control valves as a gauge tender who shall regulate the pressure in the working areas. During tunneling operations, one gauge tender may regulate the pressure in not more than two headings: Provided, That the gauge and controls are all in one location. In caisson work, there shall be a gauge tender for each caisson. [§1926.803(h)(1)]

(2) *The low air compressor plant* shall be of sufficient capacity to not only permit the work to be done safely, but shall also provide a margin to meet emergencies and repairs.[§1926.803(h)(2)]

(3) *Low air compressor units* shall have at least two independent and separate sources of power supply and each shall be capable of operating the entire low air plant and its accessory systems.[§1926.803(h)(3)]

(4) *The capacity, arrangement, and number of compressors* shall be sufficient to maintain the necessary pressure without overloading the equipment and to assure maintenance of such pressure in the working chamber during periods of breakdown, repair, or emergency.[§1926.803(h)(4)]

(5) *Switching from one independent source of power supply* to the other shall be done periodically to ensure the workability of the apparatus in an emergency.[§1926.803(h)(5)]

(6) *Duplicate low-pressure air feedlines* and regulating valves shall be provided between the source of air supply and a point beyond the locks with one of the lines extending to within 100 feet of the working face.[§1926.803(h)(6)]

(7) *All high- and low-pressure* air supply lines shall be equipped with check valves.[§1926.803(h)(7)]

(8) *Low-pressure air shall be regulated* automatically. In addition, manually operated valves shall be provided for emergency conditions.[§1926.803(h)(8)]

(9) *The air intakes for all air compressors* shall be located at a place where fumes, exhaust, gases, and other air contaminants will be at a minimum.[§1926.803(h)(9)]

(10) *Gauges indicating the pressure* in the working chamber shall be installed in the compressor building, the lock attendant's station, and at the employer's field office.[§1926.803(h)(10)]

(i) Ventilation and air quality.[§1926.803(i)]

(1) *Exhaust valves and exhaust pipes* shall be provided and operated so that the working chamber shall be well ventilated, and there shall be no pockets of dead air. Outlets may be required at intermediate points along the main low-pressure air supply line to the heading to eliminate such pockets of dead air. Ventilating air shall be not less than 30 cubic feet per minute. [§1926.803(i)(1)]

(2) *The air in the workplace shall be analyzed* by the employer not less than once each shift, and records of such tests shall be kept on file at the place where the work is in progress. The test results shall be within the threshold limit values specified in

subpart D of this part, for hazardous gases, and within 10 percent of the lower explosive limit of flammable gases. If these limits are not met, immediate action to correct the situation shall be taken by the employer.[§1926.803(i)(2)]

(3) *The temperature of all working chambers* which are subjected to air pressure shall, by means of after-coolers or other suitable devices, be maintained at a temperature not to exceed 85 °F. [§1926.803(i)(3)]

(4) *Forced ventilation shall be provided* during decompression. During the entire decompression period, forced ventilation through chemical or mechanical air purifying devices that will ensure a source of fresh air shall be provided.[§1926.803(i)(4)]

(5) *Whenever heat-producing machines* (moles, shields) are used in compressed air tunnel operations, a positive means of removing the heat build-up at the heading shall be provided. [§1926.803(i)(5)]

(j) **Electricity.**[§1926.803(j)]

(1) *All lighting in compressed-air chambers* shall be by electricity exclusively, and two independent electric-lighting systems with independent sources of supply shall be used. The emergency source shall be arranged to become automatically operative in the event of failure of the regularly used source.[§1926.803(j)(1)]

(2) *The minimum intensity of light* on any walkway, ladder, stairway, or working level shall be not less than 10 foot-candles, and in all workplaces the lighting shall at all times be such as to enable employees to see clearly.[§1926.803(j)(2)]

(3) *All electrical equipment and wiring for light and power circuits* shall comply with the requirements of subpart K of this part for use in damp, hazardous, high temperature, and compressed air environments.[§1926.803(j)(3)]

(4) *External parts of lighting fixtures* and all other electrical equipment, when within 8 feet of the floor, shall be constructed of noncombustible, nonabsorptive, insulating materials, except that metal may be used if it is effectively grounded. [§1926.803(j)(4)]

(5) *Portable lamps shall be equipped* with noncombustible, nonabsorptive, insulating sockets, approved handles, basket guards, and approved cords.[§1926.803(j)(5)]

(6) *The use of worn or defective* portable and pendant conductors is prohibited.[§1926.803(j)(6)]

(k) **Sanitation.**[§1926.803(k)]

(1) *Sanitary, heated, lighted, and ventilated dressing rooms* and drying rooms shall be provided for all employees engaged in compressed air work. Such rooms shall contain suitable benches and lockers. Bathing accommodations (showers at the ratio of one to 10 employees per shift), equipped with running hot and cold water, and suitable and adequate toilet accommodations, shall be provided. One toilet for each 15 employees, or fractional part thereof, shall be provided.[§1926.803(k)(1)]

(2) *When the toilet bowl is shut by a cover,* there should be an air space so that the bowl or bucket does not implode when pressure is increased.[§1926.803(k)(2)]

(3) *All parts of caissons* and other working compartments shall be kept in a sanitary condition.[§1926.803(k)(3)]

(l) **Fire prevention and protection.**[§1926.803(l)]

(1) *Firefighting equipment shall be available* at all times and shall be maintained in working condition.[§1926.803(l)(1)]

(2) *While welding or flame-cutting* is being done in compressed air, a firewatch with a fire hose or approved extinguisher shall stand by until such operation is completed.[§1926.803(l)(2)]

(3) *Shafts and caissons containing flammable material* of any kind, either above or below ground, shall be provided with a waterline and a fire hose connected thereto, so arranged that all points of the shaft or caisson are within reach of the hose stream. [§1926.803(l)(3)]

(4) *Fire hose shall be at least 1 1/2 inches in nominal diameter;* the water pressure shall at all times be adequate for efficient operation of the type of nozzle used; and the water supply shall be such as to ensure an uninterrupted flow. Fire hose, when not in use, shall be located or guarded to prevent injury thereto. [§1926.803(l)(4)]

(5) *The power house, compressor house, and all buildings* housing ventilating equipment, shall be provided with at least one hose connection in the water line, with a fire hose connected thereto. A fire hose shall be maintained within reach of structures of wood over or near shafts.[§1926.803(l)(5)]

(6) *Tunnels shall be provided with a 2-inch minimum diameter* water line extending into the working chamber and to within 100 feet of the working face. Such line shall have hose outlets with 100 feet of fire hose attached and maintained as follows: One at the working face; one immediately inside of the bulkhead of the working chamber; and one immediately outside such bulkhead. In addition, hose outlets shall be provided at 200-foot intervals throughout the length of the tunnel, and 100 feet of fire hose shall be attached to the outlet nearest to any location where flammable material is being kept or stored or where any flame is being used.[§1926.803(l)(6)]

(7) *In addition to fire hose protection required by this subpart,* on every floor of every building not under compressed air, but used in connection with the compressed air work, there shall be provided at least one approved fire extinguisher of the proper type for the hazard involved. At least two approved fire extinguishers shall be provided in the working chamber as follows: One at the working face and one immediately inside the bulkhead (pressure side). Extinguishers in the working chamber shall use water as the primary extinguishing agent and shall not use any extinguishing agent which could be harmful to the employees in the working chamber. The fire extinguisher shall be protected from damage.[§1926.803(l)(7)]

(8) *Highly combustible materials* shall not be used or stored in the working chamber. Wood, paper, and similar combustible material shall not be used in the working chamber in quantities which could cause a fire hazard. The compressor building shall be constructed of non-combustible material.[§1926.803(l)(8)]

(9) *Man locks shall be equipped* with a manual type fire extinguisher system that can be activated inside the man lock and also by the outside lock attendant. In addition, a fire hose and portable fire extinguisher shall be provided inside and outside the man lock. The portable fire extinguisher shall be the dry chemical type.[§1926.803(l)(9)]

(10) *Equipment, fixtures, and furniture* in man locks and special decompression chambers shall be constructed of noncombustible materials. Bedding, etc., shall be chemically treated so as to be fire resistant.[§1926.803(l)(10)]

(11) *Head frames shall be constructed* of structural steel or open frame-work fireproofed timber. Head houses and other temporary surface buildings or structures within 100 feet of the shaft, caisson, or tunnel opening shall be built of fire-resistant materials.[§1926.803(l)(11)]

(12) *No oil, gasoline, or other combustible material* shall be stored within 100 feet of any shaft, caisson, or tunnel opening, except that oils may be stored in suitable tanks in isolated fireproof buildings, provided such buildings are not less than 50 feet from any shaft, caisson, or tunnel opening, or any building directly connected thereto.[§1926.803(l)(12)]

(13) *Positive means shall be taken* to prevent leaking flammable liquids from flowing into the areas specifically mentioned in the preceding paragraph.[§1926.803(l)(13)]

(14) *All explosives used in connection* with compressed air work shall be selected, stored, transported, and used as specified in subpart U of this part.[§1926.803(l)(14)]

(m) **Bulkheads and safety screens.**[§1926.803(m)]

(1) *Intermediate bulkheads with locks,* or intermediate safety screens or both, are required where there is the danger of rapid flooding.[§1926.803(m)(1)]

(2) *In tunnels 16 feet or more in diameter,* hanging walkways shall be provided from the face to the man lock as high in the tunnel as practicable, with at least 6 feet of head room. Walkways shall be constructed of noncombustible material. Standard railings shall be securely installed throughout the length of all walkways on open sides in accordance with subpart M of this part. Where walkways are ramped under safety screens, the walkway surface shall be skidproofed by cleats or by equivalent means.[§1926.803(m)(2)]

(3) *Bulkheads used to contain compressed air* shall be tested, where practicable, to prove their ability to resist the highest air pressure which may be expected to be used.[§1926.803(m)(3)]

[44 FR 8577, Feb. 9, 1979; 44 FR 20940, Apr. 6, 1979, as amended at 47 FR 14696, 14706, Apr. 6, 1982; 51 FR 25318, July 11, 1986; 61 FR 5510, Feb. 13, 1996]

§1926.804

Definitions applicable to this subpart

(a) **Bulkhead** — An airtight structure separating the working chamber from free air or from another chamber under a lesser pressure than the working pressure.[§1926.804(a)]

(b) **Caisson** — A wood, steel, concrete or reinforced concrete, air- and water-tight chamber in which it is possible for men to work under air pressure greater than atmospheric pressure to excavate material below water level.[§1926.804(b)]

(c) **Decanting** — A method used for decompressing under emergency circumstances. In this procedure, the employees are brought to atmospheric pressure with a very high gas tension in the tissues and then immediately recompressed in a second and separate chamber or lock.[§1926.804(c)]

(d) **Emergency locks** — A lock designed to hold and permit the quick passage of an entire shift of employees.[§1926.804(d)]

(e) **High air** — Air pressure used to supply power to pneumatic tools and devices.[§1926.804(e)]

(f) **Low air** — Air supplied to pressurize working chambers and locks.[§1926.804(f)]

(g) **Man lock** — A chamber through which men pass from one air pressure environment into another.[§1926.804(g)]

(h) **Materials lock** — A chamber through which materials and equipment pass from one air pressure environment into another. [§1926.804(h)]

(i) **Medical lock** — A special chamber in which employees are treated for decompression illness. It may also be used in preemployment physical examinations to determine the adaptability of the prospective employee to changes in pressure.[§1926.804(i)]

(j) **Normal condition** — One during which exposure to compressed air is limited to a single continuous working period followed by a single decompression in any given 24-hour period; the total time of exposure to compressed air during the single continuous working period is not interrupted by exposure to normal atmospheric pressure, and a second exposure to compressed air does not occur until at least 12 consecutive hours of exposure to normal atmospheric pressure has elapsed since the employee has been under pressure.[§1926.804(j)]

(k) **Pressure** — A force acting on a unit area. Usually shown as pounds per square inch. (p.s.i.)[§1926.804(k)]

(l) **Absolute pressure (p.s.i.a.)** — The sum of the atmospheric pressure and gauge pressure (p.s.i.g.).[§1926.804(l)]

(m) **Atmospheric pressure** — The pressure of air at sea level, usually 14.7 p.s.i.a. (1 atmosphere), or 0 p.s.i.g.[§1926.804(m)]

(n) **Gauge pressure (p.s.i.g.)** — Pressure measured by a gauge and indicating the pressure exceeding atmospheric.[§1926.804(n)]

(o) **Safety screen** — An air- and water-tight diaphragm placed across the upper part of a compressed air tunnel between the face and bulkhead, in order to prevent flooding the crown of the tunnel between the safety screen and the bulkhead, thus providing a safe means of refuge and exit from a flooding or flooded tunnel. [§1926.804(o)]

(p) **Special decompression chamber** — A chamber to provide greater comfort of employees when the total decompression time exceeds 75 minutes.[§1926.804(p)]

(q) **Working chamber** — The space or compartment under air pressure in which the work is being done.[§1926.804(q)]

Subpart S Appendix A

Decompression Tables

1. **Explanation.** The decompression tables are computed for working chamber pressures from 0 to 14 pounds, and from 14 to 50 pounds per square inch gauge inclusive by 2-pound increments and for exposure times for each pressure extending from one-half to over 8 hours inclusive. Decompressions will be conducted by two or more stages with a maximum of four stages, the latter for a working chamber pressure of 40 pounds per square inch gauge or over.

Stage 1 consists of a reduction in ambient pressure ranging from 10 to a maximum of 16 pounds per square inch, but in no instance will the pressure be reduced below 4 pounds at the end of stage 1. This reduction in pressure in stage 1 will always take place at a rate not greater than 5 pounds per minute.

Further reduction in pressure will take place during stage 2 and subsequent stages as required at a slower rate, but in no event at a rate greater than 1 pound per minute.

Decompression Table No. 1 indicates in the body of the table the total decompression time in minutes for various combinations of working chamber pressure and exposure time.

Decompression Table No. 2 indicates for the same various combinations of working chamber pressure and exposure time the following:

A. *The number of stages required;*

B. *The reduction in pressure and the terminal* pressure for each required stage;

C. *The time in minutes through* which the reduction in pressure is accomplished for each required stage;

D. *The pressure reduction rate* in minutes per pound for each required stage;

Important Note: The Pressure Reduction in Each Stage is Accomplished at a Uniform Rate. Do Not Interpolate Between Values Shown on the Tables. Use the Next Higher Value of Working Chamber Pressure or Exposure Time Should the Actual Working Chamber Pressure or the Actual Exposure Time, Respectively, Fall Between Those for Which Calculated Values Are Shown in the Body of the Tables.

Examples	Minutes
Example No. 1: 4 hours working period at 20 pounds gauge.	
Decompression Table No. 1: 20 pounds for 4 hours, total decompression time.	43
Decompression Table No. 2: Stage 1: Reduce pressure from 20 pounds to 4 pounds at the uniform rate of 5 pounds per minute. Elapsed time stage 1:16/5	3
Stage 2 (final stage): Reduce pressure at a uniform rate from 4 pounds to 0-pound gage over a period of 40 minutes. Rate — 0.10 pound per minute or 10 minutes per pound. Stage 2 (final) elapsed time:	40
Total time	**43**
Example No. 2: 5-hour working period at 24 pounds gage.	
Decompression Table No. 1: 24 pounds for 5 hours, total decompression time.	117
Decompression Table No. 2: Stage 1: Reduce pressure from 24 pounds to 8 pounds at the uniform rate of 5 pounds per minute. Elapsed time stage 1:16/5	3
Stage 2: Reduce pressure at a uniform rate from 8 pounds to 4 pounds over a period of 4 minutes. Rate, 1 pound per minute. Elapsed time stage 2:	4
Transfer men to special decompression chamber maintaining the 4-pound pressure during the transfer operation.	
Stage 3 (final stage): In the special decompression chamber, reduce the pressure at a uniform rate from 4 pounds to 0-pound gage over a period of 110 minutes. Rate, 0.037 pound per minute or 27.5 minutes per pound. Stage 3 (final) elapsed time:	110
Total time	**117**

Decompression Table No. 1 — Total Decompression Time

Work Pressure p.s.i.g.	Working Period Hours										
	1/2	1	1 1/2	2	3	4	5	6	7	8	Over 8
9-12	3	3	3	3	3	3	3	3	3	3	3
14	6	6	6	6	6	6	6	6	16	16	33
16	7	7	7	7	7	7	17	33	48	48	62
18	7	7	7	8	11	17	48	63	63	73	87
20	7	7	8	15	15	43	63	73	83	103	113
22	9	9	16	24	38	68	93	103	113	128	133
24	11	12	23	27	52	92	117	122	127	137	151
26	13	14	29	34	69	104	126	141	142	142	163
28	15	23	31	41	98	127	143	153	153	165	183
30	17	28	38	62	105	143	165	168	178	188	204
32	19	35	43	85	126	163	178	193	203	213	226
34	21	39	58	98	151	178	195	218	223	233	248
36	24	44	63	113	170	198	223	233	243	253	273
38	28	49	73	128	178	203	223	238	253	263	278
40	31	49	84	143	183	213	233	248	258	278	288
42	37	56	102	144	189	215	245	260	263	268	293
44	43	64	118	154	199	234	254	264	269	269	293
46	44	74	139	171	214	244	269	274	289	299	318
48	51	89	144	189	229	269	299	309	319	319	
50	58	94	164	209	249	279	309	329			

Decompression Table No. 2
[Do not interpolate, use next higher value for conditions not computed]

Working chamber pressure p.s.i.g.	Working period hours	Decompression data					
		Stage no.	Pressure reduction p.s.i.g.		Time in stage min.	Pressure reduction rate min./lb.	Total time decompress minutes
			From	To			
14	1/2	1	14	4	2	0.20	6
		2	4	0	4	1.00	6
	1	1	14	4	2	0.20	6
		2	4	0	4	1.00	6
	1 1/2	1	14	4	2	0.20	6
		2	4	0	4	1.00	6
	2	1	14	4	2	0.20	6
		2	4	0	4	1.00	6
	3	1	14	4	2	0.20	6
		2	4	0	4	1.00	6
	4	1	14	0	2	0.20	6
		2	4	0	4	1.00	6
	5	1	14	4	2	0.20	6
		2	4	0	4	1.00	6
	6	1	14	4	2	0.20	
		2	4	0	4	1.00	6
	7	1	14	4	2	0.20	
		2	4	0	14	3.50	16
	8	1	14	4	2	0.20	
		2	4	0	14	3.50	16
	Over 8	1	14	4	2	0.20	
		2	4	0	30	7.50	32
16	1/2	1	16	4	3	0.20	
		2	4	0	4	1.00	7
	1	1	16	4	3	0.20	7
		2	4	0	4	1.00	7
	1 1/2	1	16	4	3	0.20	
		2	4	0	4	1.00	7
	2	1	16	4	3	0.20	
		2	4	0	4	1.00	7
	3	1	16	4	3	0.20	
		2	4	0	4	1.00	7
	4	1	14	4	3	0.20	
		2	4	0	4	1.00	7
	5	1	14	4	3	0.20	7
		2	4	0	4	3.50	17
	6	1	14	4	3	0.20	
		2	4	0	30	7.50	33
	7	1	14	4	3	0.20	
		2	4	0	45	11.25	48
	8	1	14	4	3	0.20	
		2	4	0	45	11.25	48
	Over 8	1	14	4	3	0.20	
		2	4	0	60	15.00	63
18	1/2	1	18	4	3	0.20	
		2	4	0	4	1.00	7
	1	1	18	4	3	0.20	
		2	4	0	4	1.00	7
	1 1/2	1	18	4	3	0.20	
		2	4	0	4	1.00	7
	2	1	18	4	3	0.20	
		2	4	0	5	1.25	8
	3	1	18	4	3	0.20	
		2	4	0	8	2.00	11

Decompression Table No. 2
[Do not interpolate, use next higher value for conditions not computed] (continued)

Working chamber pressure p.s.i.g.	Working period hours	Decompression data					
		Stage no.	Pressure reduction p.s.i.g.		Time in stage min.	Pressure reduction rate min./lb.	Total time decompress minutes
			From	To			
18	4	1	18	4	3	0.20	
		2	4	0	14	3.50	17
	5	1	18	4	3	0.20	
		2	4	0	45	11.25	48
	6	1	18	4	3	0.20	
		2	4	0	60	15.00	63
	7	1	18	4	3	0.20	
		2	4	0	60	15.00	63
	8	1	18	4	3	0.20	
		2	4	0	70	17.50	73
	Over 8	1	18	4	3	0.20	
		2	4	0	84	21.00	87
20	1/2	1	20	4	3	0.20	
		2	4	0	4	1.00	7
	1	1	20	4	3	0.20	
		2	4	0	4	1.00	7
	1 1/2	1	20	4	3	0.20	
		2	4	0	5	1.25	8
	2	1	20	4	3	0.20	
		2	4	0	12	3.00	15
	3	1	20	4	3	0.20	
		2	4	0	12	3.00	15
	4	1	20	4	3	0.20	
		2	4	0	40	10.00	43
	5	1	20	4	3	0.20	
		2	4	0	60	15.00	63
	6	1	20	4	3	0.20	
		2	4	0	70	17.50	73
	7	1	20	4	3	0.20	
		2	4	0	80	20.00	83
	8	1	20	4	3	0.20	
		2	4	0	100	25.00	103
	Over 8	1	20	4	3	0.20	
		2	4	0	110	27.50	113
22	1/2	1	22	6	3	0.20	
		2	6	0	6	1.00	9
	1	1	22	6	3	0.20	
		2	6	0	6	1.00	9
	1 1/2	1	22	6	3	0.20	
		2	6	0	13	2.20	16
	2	1	22	6	3	0.20	
		2	6	0	21	3.50	24
	3	1	22	6	3	0.20	
		2	6	0	35	5.85	38
	4	1	22	6	3	0.20	
		2	6	0	65	10.83	68
	5	1	22	6	3	0.20	
		2	6	0	90	15.00	93
	6	1	22	6	3	0.20	
		2	4	0	100	16.67	103
	7	1	22	6	3	0.20	
		2	6	0	110	18.35	113
	8	1	22	6	3	0.20	
		2	6	0	125	20.80	128

Decompression Table No. 2
[Do not interpolate, use next higher value for conditions not computed] (continued)

Working chamber pressure p.s.i.g.	Working period hours	Decompression data					
		Stage no.	Pressure reduction p.s.i.g. From	Pressure reduction p.s.i.g. To	Time in stage min.	Pres-sure reduc-tion rate min./lb.	Total time decom-press minutes
	Over 8	1	22	6	3	0.20	
		2	6	0	130	21.70	133
24	1/2	1	24	8	3	0.20	
		2	8	4	4	1.00	
		3	4	0	4	1.00	11
	1	1	24	8	3	0.20	
		2	8	4	4	1.00	
		3	4	0	5	1.25	12
	1 1/2	1	24	8	3	0.20	
		2	8	4	4	1.00	
		3	4	0	16	4.00	23
	2	1	24	8	3	0.20	
		2	8	4	4	1.00	
		3	4	0	20	5.00	27
	3	1	24	8	3	0.20	
		2	8	4	4	1.00	
		3	4	0	45	11.25	52
	4	1	24	8	3	0.20	
		2	8	4	4	1.00	
		3	4	0	85	21.25	92
	5	1	24	8	3	0.20	
		2	8	4	4	1.00	
		3	4	0	110	27.50	117
	6	1	24	8	3	0.20	
		2	8	4	4	1.00	
		3	4	0	115	28.80	122
	7	1	24	8	3	0.20	
		2	8	4	4	1.00	
		3	4	0	120	30.00	127
	8	1	24	8	3	0.20	
		2	8	4	4	1.00	
		3	4	0	130	32.50	137
	Over 8	1	24	8	3	0.20	
		2	8	4	8	2.00	
		3	4	0	140	35.00	151
26	1/2	1	26	10	3	0.20	
		2	10	4	6	1.00	
		3	4	0	4	1.00	13
	1	1	26	10	3	0.20	
		2	10	4	6	1.00	
		3	4	0	5	1.25	14
	1 1/2	1	26	10	3	0.20	
		2	10	4	6	1.00	
		3	4	0	20	5.00	29
	2	1	26	10	3	0.20	
		2	10	4	6	1.00	
		3	4	0	25	6.25	34
	3	1	26	10	3	0.20	
		2	10	4	6	1.00	
		3	4	0	60	15.00	69
	4	1	26	10	3	0.20	
		2	10	4	6	1.00	
		3	4	0	95	23.75	104

Decompression Table No. 2
[Do not interpolate, use next higher value for conditions not computed] (continued)

Working chamber pressure p.s.i.g.	Working period hours	Decompression data					
		Stage no.	Pressure reduction p.s.i.g. From	Pressure reduction p.s.i.g. To	Time in stage min.	Pres-sure reduc-tion rate min./lb.	Total time decom-press minutes
26	5	1	26	10	3	0.20	
		2	10	4	8	1.33	
		3	4	0	115	28.80	126
	6	1	26	10	3	0.20	
		2	10	4	8	1.33	
		3	4	0	130	32.50	141
	7	1	26	10	3	2.20	
		2	10	4	9	1.50	
		3	4	0	130	32.50	142
	8	1	26	10	3	0.20	
		2	10	4	9	1.50	
		3	4	0	130	32.50	142
	Over 8	1	26	10	3	0.20	
		2	10	4	30	5.00	
		3	4	0	30	32.50	163
28	1/2	1	28	12	3	0.20	
		2	12	4	8	1.00	
		3	4	0	4	1.00	15
	1	1	28	12	3	0.20	
		2	12	4	8	1.00	
		3	4	0	12	3.00	23
	1 1/2	1	28	12	3	0.20	
		2	12	4	8	1.00	
		3	4	0	20	5.00	31
	2	1	28	12	3	0.20	
		2	12	4	8	1.00	
		3	4	0	30	7.50	41
	3	1	28	12	3	0.20	
		2	12	4	10	1.25	
		3	4	0	85	21.20	98
	4	1	28	12	3	0.20	
		2	12	4	14	1.75	
		3	4	0	110	27.50	127
	5	1	28	12	3	0.20	
		2	12	4	20	2.50	
		3	4	0	120	30.00	143
	6	1	28	12	3	0.20	
		2	12	4	20	2.50	
		3	4	0	130	32.50	153
	7	1	28	12	3	0.20	
		2	12	4	20	2.50	
		3	4	0	120	32.50	153
	8	1	28	12	3	0.20	
		2	12	4	32	4.00	
		3	4	0	130	32.50	165
	Over 8	1	28	12	3	0.20	
		2	12	4	50	6.25	
		3	4	0	130	32.5	183
30	1/2	1	30	14	3	0.20	
		2	14	4	10	1.00	
		3	4	0	4	1.00	17
	1	1	30	14	3	0.20	
		2	14	4	10	1.00	
		3	4	0	15	3.75	28

Decompression Table No. 2
[Do not interpolate, use next higher value for conditions not computed] (continued)

Working chamber pressure p.s.i.g.	Working period hours	Stage no.	Pressure reduction p.s.i.g. From	Pressure reduction p.s.i.g. To	Time in stage min.	Pressure reduction rate min./lb.	Total time decompress minutes
30	1 1/2	1	30	14	3	0.20	
		2	14	4	10	1.00	
		3	4	0	25	6.25	38
	2	1	30	14	3	0.20	
		2	14	4	14	1.40	
		3	4	0	45	11.25	62
	3	1	30	14	3	0.20	
		2	14	4	17	1.70	
		3	4	0	85	21.20	105
	4	1	30	14	3	0.20	
		2	14	4	30	3.00	
		3	4	0	110	27.50	143
	5	1	30	14	3	0.20	
		2	14	4	35	3.50	
		3	4	0	130	32.50	165
	6	1	30	14	3	0.20	
		2	14	4	35	3.50	
		3	4	0	130	32.50	168
	7	1	30	14	3	0.20	
		2	14	4	45	4.50	
		3	4	0	130	32.50	178
	8	1	30	14	3	0.20	
		2	14	4	55	5.50	
		3	4	0	130	32.50	188
	Over 8	1	30	14	3	0.20	
		2	14	4	71	7.10	
		3	4	0	130	32.50	204
32	1/2	1	32	16	3	0.20	
		2	16	4	12	1.00	
		3	4	0	4	1.00	19
	1	1	32	16	3	0.20	
		2	16	4	12	1.00	
		3	4	0	20	5.00	35
	1 1/2	1	32	16	3	0.20	
		2	16	4	15	1.25	
		3	4	0	25	6.25	43
	2	1	32	16	3	0.20	
		2	16	4	22	1.83	
		3	4	0	60	15.00	85
	3	1	32	16	3	0.20	
		2	16	4	28	2.33	
		3	4	0	95	23.75	126
	4	1	32	16	3	0.20	
		2	16	4	40	3.33	
		3	4	0	120	30.00	163
	5	1	32	16	3	0.20	
		2	16	4	45	3.75	
		3	4	0	130	32.50	178
	6	1	32	16	3	0.20	
		2	16	4	60	5.00	
		3	4	0	130	32.50	193
	7	1	32	16	3	0.20	
		2	16	4	70	5.83	
		3	4	0	130	32.50	203

Decompression Table No. 2
[Do not interpolate, use next higher value for conditions not computed] (continued)

Working chamber pressure p.s.i.g.	Working period hours	Stage no.	Pressure reduction p.s.i.g. From	Pressure reduction p.s.i.g. To	Time in stage min.	Pressure reduction rate min./lb.	Total time decompress minutes
32	8	1	32	16	3	0.20	
		2	16	4	80	6.67	
		3	4	0	130	32.50	213
	Over 8	1	32	16	3	0.20	
		2	16	4	93	7.75	
		3	4	0	130	32.50	226
34	1/2	1	34	18	3	0.20	
		2	18	4	14	1.00	
		3	4	0	4	1.00	21
	1	1	34	18	3	0.20	
		2	18	4	14	1.00	
		3	4	0	22	5.50	39
	1 1/2	1	34	18	3	0.20	
		2	18	4	25	1.80	
		3	4	0	30	7.50	58
	2	1	34	18	3	0.20	
		2	18	4	35	2.50	
		3	4	0	60	15.00	98
	3	1	34	18	3	0.20	
		2	18	4	43	3.10	
		3	4	0	105	26.25	151
	4	1	34	18	3	0.20	
		2	18	4	55	3.93	
		3	4	0	120	30.00	178
	5	1	34	18	3	0.20	
		2	18	4	62	4.43	
		3	4	0	130	32.50	195
	6	1	34	18	3	0.20	
		2	18	4	85	6.07	
		3	4	0	130	32.50	218
	7	1	34	18	3	0.20	
		2	18	4	90	6.43	
		3	4	0	130	32.50	223
	8	1	34	18	3	0.20	
		2	18	4	100	7.15	
		3	4	0	130	32.50	233
	Over 8	1	34	18	3	0.20	
		2	18	4	115	8.23	
		3	4	0	130	32.50	248
36	1/2	1	36	20	3	0.20	
		2	20	4	16	1.00	
		3	4	0	5	1.25	24
	1	1	36	20	3	0.20	
		2	20	4	16	1.00	
		3	4	0	25	6.25	44
	1 1/2	1	36	20	3	0.20	
		2	20	4	30	1.88	
		3	4	0	30	7.50	63
	2	1	36	20	3	0.20	
		2	20	4	40	2.50	
		3	4	0	70	17.50	113
	3	1	36	20	3	0.20	
		2	20	4	52	3.25	
		3	4	0	115	28.75	170

Decompression Table No. 2
[Do not interpolate, use next higher value for conditions not computed] (continued)

Working chamber pressure p.s.i.g.	Working period hours	Stage no.	Pressure reduction p.s.i.g. From	Pressure reduction p.s.i.g. To	Time in stage min.	Pressure reduction rate min./lb.	Total time decompress minutes
36	4	1	36	20	3	0.20	
		2	20	4	65	4.06	
		3	4	0	130	32.50	198
	5	1	36	20	3	0.20	
		2	20	4	90	5.63	
		3	4	0	130	32.50	223
	6	1	36	20	3	0.20	
		2	20	4	100	6.25	
		3	4	0	130	32.50	233
	7	1	36	20	3	0.20	
		2	20	4	110	6.88	
		3	4	0	130	32.50	243
	8	1	36	20	3	0.20	
		2	20	4	120	7.50	
		3	4	0	130	32.50	253
	Over 8	1	36	20	3	0.20	
		2	20	4	140	8.75	
		3	4	0	130	32.50	273
38	1/2	1	38	22	3	0.20	
		2	22	6	16	1.00	
		3	6	0	9	1.50	28
	1	1	38	22	3	0.20	
		2	22	6	16	1.00	
		3	6	0	30	5.00	49
	1 1/2	1	38	22	3	0.20	
		2	22	6	20	1.25	
		3	6	0	50	8.34	73
	2	1	38	22	3	0.20	
		2	22	6	30	1.88	
		3	6	0	95	15.83	128
	3	1	38	22	3	0.20	
		2	22	6	35	2.19	
		3	6	0	140	23.35	178
	4	1	38	22	3	0.20	
		2	22	6	50	3.12	
		3	6	0	150	25.00	203
	5	1	38	22	3	0.20	
		2	22	6	55	3.44	
		3	6	0	165	27.50	223
	6	1	38	22	3	0.20	
		2	22	6	70	4.38	
		3	6	0	165	27.50	238
	7	1	38	22	3	0.20	
		2	22	6	85	5.32	
		3	6	0	165	27.50	253
	8	1	38	22	3	0.20	
		2	22	6	95	5.93	
		3	6	0	165	27.50	263
	Over 8	1	38	22	3	0.20	
		2	22	6	110	6.88	
		3	6	0	165	27.50	278
40	1/2	1	40	24	3	0.20	
		2	24	8	16	1.00	
		3	8	4	4	1.00	
		4	4	0	8	2.00	31

Decompression Table No. 2
[Do not interpolate, use next higher value for conditions not computed] (continued)

Working chamber pressure p.s.i.g.	Working period hours	Stage no.	Pressure reduction p.s.i.g. From	Pressure reduction p.s.i.g. To	Time in stage min.	Pressure reduction rate min./lb.	Total time decompress minutes
40	1	1	40	24	3	0.20	
		2	24	8	16	1.00	
		3	8	4	5	1.25	
		4	4	0	25	6.25	49
	1 1/2	1	40	24	3	0.20	
		2	24	8	16	1.00	
		3	8	4	20	5.00	
		4	4	0	45	11.25	84
	2	1	40	24	3	0.20	
		2	24	8	25	1.56	
		3	8	4	20	5.00	
		4	4	0	95	23.75	143
	3	1	40	24	3	0.20	
		2	24	8	30	1.88	
		3	8	4	30	7.50	
		4	4	0	120	30.00	183
	4	1	40	24	3	0.20	
		2	24	8	45	2.81	
		3	8	4	35	8.75	
		4	4	0	130	32.50	213
	5	1	40	24	3	0.20	
		2	24	8	47	2.94	
		3	8	4	53	13.25	
		4	4	0	130	32.50	233
	6	1	40	24	3	0.20	
		2	24	8	55	3.44	
		3	8	4	60	15.00	
		4	4	0	130	32.50	248
	7	1	40	24	3	0.20	
		2	24	8	65	4.06	
		3	8	4	60	15.00	
		4	4	0	130	32.50	258
	8	1	40	24	3	0.20	
		2	24	8	75	4.70	
		3	8	4	60	15.00	
		4	4	0	130	32.50	268
	Over 8	1	40	24	3	0.20	
		2	24	8	95	5.93	
		3	8	4	60	15.00	
		4	4	0	130	32.50	288
42	1/2	1	42	26	3	0.20	
		2	26	10	16	1.00	
		3	10	4	6	1.00	
		4	4	0	12	3.00	37
	1	1	42	26	3	0.20	
		2	26	10	16	1.00	
		3	10	4	12	2.00	
		4	4	0	25	6.25	56
	1 1/2	1	42	26	3	0.20	
		2	26	10	16	1.00	
		3	10	4	23	3.83	
		4	4	0	60	15.00	102

Decompression Table No. 2
[Do not interpolate, use next higher value for conditions not computed] (continued)

Working chamber pressure p.s.i.g.	Working period hours	Decompression data: Stage no.	Pressure reduction p.s.i.g. From	Pressure reduction p.s.i.g. To	Time in stage min.	Pressure reduction rate min./lb.	Total time decompress minutes
42	2	1	42	26	3	0.20	
		2	26	10	16	1.00	
		3	10	4	30	5.00	
		4	4	0	95	23.75	144
	3	1	42	26	3	0.20	
		2	26	10	16	1.00	
		3	10	4	50	8.34	
		4	4	0	120	30.00	189
	4	1	42	26	3	0.20	
		2	26	10	17	1.06	
		3	10	4	65	10.83	
		4	4	0	130	32.50	215
	5	1	42	26	3	0.20	
		2	26	10	27	1.69	
		3	10	4	85	14.18	
		4	4	0	130	32.50	245
	6	1	42	26	3	0.20	
		2	26	10	27	1.69	
		3	10	4	100	16.67	
		4	4	0	130	32.50	260
	7	1	42	26	3	0.20	
		2	26	10	30	1.88	
		3	10	4	100	16.67	
		4	4	0	130	32.50	263
	8	1	42	26	3	0.20	
		2	26	10	35	2.19	
		3	10	4	100	16.67	
		4	4	0	130	32.50	268
	Over 8	1	42	26	3	0.20	
		2	26	10	60	3.75	
		3	10	4	100	16.67	
		4	4	0	130	32.50	293
44	1/2	1	44	28	3	0.20	
		2	28	12	16	1.00	
		3	12	4	8	1.00	
		4	4	0	16	4.00	43
	1	1	44	28	3	0.20	
		2	28	12	16	1.00	
		3	12	4	20	2.50	
		4	4	0	25	6.25	64
	1 1/2	1	44	28	3	0.20	
		2	28	12	16	1.00	
		3	12	4	27	3.38	
		4	4	0	72	18.00	118
	2	1	44	28	3	0.20	
		2	28	12	16	1.00	
		3	12	4	40	5.00	
		4	4	0	95	23.75	154
	3	1	44	28	3	0.20	
		2	28	12	16	1.00	
		3	12	4	60	7.50	
		4	4	0	120	30.00	199

Decompression Table No. 2
[Do not interpolate, use next higher value for conditions not computed] (continued)

Working chamber pressure p.s.i.g.	Working period hours	Decompression data: Stage no.	Pressure reduction p.s.i.g. From	Pressure reduction p.s.i.g. To	Time in stage min.	Pressure reduction rate min./lb.	Total time decompress minutes
44	4	1	44	28	3	0.20	
		2	28	12	16	1.00	
		3	12	4	85	10.62	
		4	4	0	130	32.50	234
	5	1	44	28	3	0.20	
		2	28	12	16	1.00	
		3	12	4	105	13.13	
		4	4	0	130	32.50	254
	6	1	44	28	3	0.20	
		2	28	12	16	1.00	
		3	12	4	115	14.38	
		4	4	0	130	32.50	264
	7	1	44	28	3	0.20	
		2	28	12	16	1.00	
		3	12	4	120	15.00	
		4	4	0	130	32.50	269
	8	1	44	28	3	0.20	
		2	28	12	16	1.00	
		3	12	4	120	15.00	
		4	4	0	130	32.50	269
	Over 8	1	44	28	3	0.20	
		2	28	12	40	2.50	
		3	12	4	120	15.00	
		4	4	0	130	32.50	293
46	1/2	1	46	30	3	0.20	
		2	30	14	16	1.00	
		3	14	4	10	1.00	
		4	4	0	15	3.75	44
	1	1	46	30	3	0.20	
		2	30	14	16	1.00	
		3	14	4	25	2.50	
		4	4	0	30	7.50	74
	1 1/2	1	46	30	3	0.20	
		2	30	14	16	1.00	
		3	14	4	35	3.50	
		4	4	0	85	21.20	139
	2	1	46	30	3	0.20	
		2	30	14	16	1.00	
		3	14	4	47	4.70	
		4	4	0	105	26.25	171
	3	1	46	30	3	0.20	
		2	30	14	16	1.00	
		3	14	4	65	6.50	
		4	4	0	130	32.50	214
	4	1	46	30	3	0.20	
		2	30	14	16	1.00	
		3	14	4	95	9.50	
		4	4	0	130	32.50	244
	5	1	46	30	3	0.20	
		2	30	14	16	1.00	
		3	14	4	120	12.00	
		4	4	0	130	32.50	269

Decompression Table No. 2
[Do not interpolate, use next higher value for conditions not computed] (continued)

Working chamber pressure p.s.i.g.	Working period hours	Decompression data					
		Stage no.	Pressure reduction p.s.i.g. From	Pressure reduction p.s.i.g. To	Time in stage min.	Pressure reduction rate min./lb.	Total time decompress minutes
46	6	1	46	30	3	0.20	
		2	30	14	16	1.00	
		3	14	4	125	12.50	
		4	4	0	130	32.50	274
	7	1	46	30	3	0.20	
		2	34	14	16	1.00	
		3	10	4	140	14.00	
		4	4	0	130	32.50	289
	8	1	46	30	3	0.20	
		2	30	14	16	1.00	
		3	14	4	150	15.00	
		4	4	0	130	32.50	299
	Over 8	1	46	30	3	0.20	
		2	30	14	25	1.56	
		3	14	4	160	16.00	
		4	4	0	130	32.50	318
48	1/2	1	48	32	3	0.20	
		2	32	16	16	1.00	
		3	16	4	12	1.00	
		4	4	0	20	5.00	51
	1	1	48	32	3	0.20	
		2	32	16	16	1.00	
		3	16	4	35	2.92	
		4	4	0	35	8.75	89
	1 1/2	1	48	32	3	0.20	
		2	32	16	16	1.00	
		3	16	4	45	3.75	
		4	4	0	80	20.00	144
	2	1	48	32	3	0.20	
		2	32	16	16	1.00	
		3	16	4	60	5.00	
		4	4	0	110	27.50	189
	3	1	48	32	3	0.20	
		2	32	16	16	1.00	
		3	16	4	90	7.50	
		4	4	0	120	30.00	229
	4	1	48	32	3	0.20	
		2	32	16	16	1.00	
		3	16	4	120	10.00	
		4	4	0	130	32.50	269
	5	1	48	32	3	0.20	
		2	32	16	16	1.00	
		3	16	4	140	11.67	
		4	4	0	130	32.50	209
	6	1	48	32	3	0.20	
		2	32	16	16	1.00	
		3	16	4	160	13.33	
		4	4	0	130	32.50	309
	7	1	48	32	3	0.20	
		2	32	16	16	1.00	
		3	16	4	170	14.17	
		4	4	0	130	32.50	

Decompression Table No. 2
[Do not interpolate, use next higher value for conditions not computed] (continued)

Working chamber pressure p.s.i.g.	Working period hours	Decompression data					
		Stage no.	Pressure reduction p.s.i.g. From	Pressure reduction p.s.i.g. To	Time in stage min.	Pressure reduction rate min./lb.	Total time decompress minutes
48	8	1	48	32	3	0.20	
		2	32	16	16	1.00	
		3	16	4	170	14.17	
		4	4	0	130	32.50	
50	1/2	1	50	34	3	0.20	
		2	34	18	16	1.00	
		3	18	4	14	1.00	
		4	4	0	25	6.25	58
	1	1	50	34	3	.20	
		2	34	18	16	1.00	
		3	18	4	40	2.86	
		4	4	0	35	8.75	94
	1 1/2	1	50	34	3	0.20	
		2	34	18	16	1.00	
		3	18	4	55	3.93	
		4	4	0	90	22.50	164
	2	1	50	34	3	0.20	
		2	34	18	16	1.00	
		3	18	4	70	5.00	
		4	4	0	120	30.00	209
	3	1	50	34	3	0.20	
		2	34	18	16	1.00	
		3	18	4	100	7.15	
		4	4	0	130	32.50	249
	4	1	50	34	3	0.20	
		2	34	18	16	1.00	
		3	18	4	130	8.58	
		4	4	0	130	32.50	279
	5	1	50	34	3	0.20	
		2	34	18	16	1.00	
		3	18	4	160	11.42	
		4	4	0	130	32.50	309
	6	1	50	34	3	0.20	
		2	34	18	16	1.00	
		3	18	4	180	12.85	
		4	4	0	130	32.50	329

[44 FR 8577, Feb. 9, 1979; 44 FR 20940, Apr. 6. 1979; as amended at 58 FR 35311, June 30, 1993]

Authority: §107, Contract Work Hours and Safety Standards Act (40 U.S.C. 333); secs. 4, 6, and 8 of the Occupational Safety and Health Act of 1970 (29 U.S.C. 653, 655, 657); Secretary of Labor's Orders 12-71 (36 FR 8754), 8-76 (41 FR 25059), 9-83 (48 FR 35736), 1-90 (55 FR 9033), 6-96 (62 FR 111), or 5-2007 (72 FR 31159) as applicable.

Subpart T - Demolition

§1926.850
Preparatory operations

(a) **Prior to permitting employees** to start demolition operations, an engineering survey shall be made, by a competent person, of the structure to determine the condition of the framing, floors, and walls, and possibility of unplanned collapse of any portion of the structure. Any adjacent structure where employees may be exposed shall also be similarly checked. The employer shall have in writing evidence that such a survey has been performed. [§1926.850(a)]

(b) **When employees are required to work** within a structure to be demolished which has been damaged by fire, flood, explosion, or other cause, the walls or floor shall be shored or braced. [§1926.850(b)]

(c) **All electric, gas, water, steam, sewer, and other service lines** shall be shut off, capped, or otherwise controlled, outside the building line before demolition work is started. In each case, any utility company which is involved shall be notified in advance. [§1926.850(c)]

(d) **If it is necessary to maintain any power,** water or other utilities during demolition, such lines shall be temporarily relocated, as necessary, and protected. [§1926.850(d)]

(e) **It shall also be determined** if any type of hazardous chemicals, gases, explosives, flammable materials, or similarly dangerous substances have been used in any pipes, tanks, or other equipment on the property. When the presence of any such substances is apparent or suspected, testing and purging shall be performed and the hazard eliminated before demolition is started. [§1926.850(e)]

(f) **Where a hazard exists** from fragmentation of glass, such hazards shall be removed. [§1926.850(f)]

(g) **Where a hazard exists to employees** falling through wall openings, the opening shall be protected to a height of approximately 42 inches. [§1926.850(g)]

(h) **When debris is dropped through holes in the floor** without the use of chutes, the area onto which the material is dropped shall be completely enclosed with barricades not less than 42 inches high and not less than 6 feet back from the projected edge of the opening above. Signs, warning of the hazard of falling materials, shall be posted at each level. Removal shall not be permitted in this lower area until debris handling ceases above. [§1926.850(h)]

(i) **All floor openings, not used as material drops,** shall be covered over with material substantial enough to support the weight of any load which may be imposed. Such material shall be properly secured to prevent its accidental movement. [§1926.850(i)]

(j) **Except for the cutting of holes in floors for chutes,** holes through which to drop materials, preparation of storage space, and similar necessary preparatory work, the demolition of exterior walls and floor construction shall begin at the top of the structure and proceed downward. Each story of exterior wall and floor construction shall be removed and dropped into the storage space before commencing the removal of exterior walls and floors in the story next below. [§1926.850(j)]

(k) **Employee entrances to multi-story structures** being demolished shall be completely protected by sidewalk sheds or canopies, or both, providing protection from the face of the building for a minimum of 8 feet. All such canopies shall be at least 2 feet wider than the building entrances or openings (1 foot wider on each side thereof), and shall be capable of sustaining a load of 150 pounds per square foot. [§1926.850(k)]

§1926.851
Stairs, passageways, and ladders

(a) **Only those stairways, passageways, and ladders,** designated as means of access to the structure of a building, shall be used. Other access ways shall be entirely closed at all times. [§1926.851(a)]

(b) **All stairs, passageways, ladders** and incidental equipment thereto, which are covered by this section, shall be periodically inspected and maintained in a clean safe condition. [§1926.851(b)]

(c) **In a multistory building, when a stairwell is being used,** it shall be properly illuminated by either natural or artificial means, and completely and substantially covered over at a point not less than two floors below the floor on which work is being performed, and access to the floor where the work is in progress shall be through a properly lighted, protected, and separate passageway. [§1926.851(c)]

§1926.852
Chutes

(a) **No material shall be dropped** to any point lying outside the exterior walls of the structure unless the area is effectively protected. [§1926.852(a)]

(b) **All materials chutes, or sections thereof,** at an angle of more than 45° from the horizontal, shall be entirely enclosed, except for openings equipped with closures at or about floor level for the insertion of materials. The openings shall not exceed 48 inches in height measured along the wall of the chute. At all stories below the top floor, such openings shall be kept closed when not in use. [§1926.852(b)]

(c) **A substantial gate shall be installed in each chute** at or near the discharge end. A competent employee shall be assigned to control the operation of the gate, and the backing and loading of trucks. [§1926.852(c)]

(d) **When operations are not in progress,** the area surrounding the discharge end of a chute shall be securely closed off. [§1926.852(d)]

(e) **Any chute opening, into which workmen dump debris,** shall be protected by a substantial guardrail approximately 42 inches above the floor or other surface on which the men stand to dump the material. Any space between the chute and the edge of openings in the floors through which it passes shall be solidly covered over. [§1926.852(e)]

(f) **Where the material is dumped** from mechanical equipment or wheelbarrows, a securely attached toeboard or bumper, not less than 4 inches thick and 6 inches high, shall be provided at each chute opening. [§1926.852(f)]

(g) **Chutes shall be designed and constructed** of such strength as to eliminate failure due to impact of materials or debris loaded therein. [§1926.852(g)]

§1926.853
Removal of materials through floor openings

Any openings cut in a floor for the disposal of materials shall be no larger in size than 25 percent of the aggregate of the total floor area, unless the lateral supports of the removed flooring remain in place. Floors weakened or otherwise made unsafe by demolition operations shall be shored to carry safely the intended imposed load from demolition operations.

§1926.854
Removal of walls, masonry sections, and chimneys

(a) **Masonry walls, or other sections of masonry,** shall not be permitted to fall upon the floors of the building in such masses as to exceed the safe carrying capacities of the floors. [§1926.854(a)]

(b) **No wall section, which is more than one story in height,** shall be permitted to stand alone without lateral bracing, unless such wall was originally designed and constructed to stand without such lateral support, and is in a condition safe enough to be self-supporting. All walls shall be left in a stable condition at the end of each shift. [§1926.854(b)]

(c) **Employees shall not be permitted** to work on the top of a wall when weather conditions constitute a hazard. [§1926.854(c)]

(d) **Structural or load-supporting members** on any floor shall not be cut or removed until all stories above such a floor have been demolished and removed. This provision shall not prohibit the cutting of floor beams for the disposal of materials or for the installation of equipment, provided that the requirements of §§1926.853 and 1926.855 are met. [§1926.854(d)]

(e) **Floor openings within 10 feet of any wall being demolished** shall be planked solid, except when employees are kept out of the area below. [§1926.854(e)]

(f) **In buildings of "skeleton-steel" construction,** the steel framing may be left in place during the demolition of masonry. Where this is done, all steel beams, girders, and similar structural supports shall be cleared of all loose material as the masonry demolition progresses downward. [§1926.854(f)]

(g) **Walkways or ladders shall be provided** to enable employees to safely reach or leave any scaffold or wall. [§1926.854(g)]

(h) **Walls, which serve as retaining walls** to support earth or adjoining structures, shall not be demolished until such earth has been properly braced or adjoining structures have been properly underpinned. [§1926.854(h)]

(i) **Walls, which are to serve as retaining walls** against which debris will be piled, shall not be so used unless capable of safely supporting the imposed load.[§1926.854(i)]

§1926.855
Manual removal of floors

(a) **Openings cut in a floor shall extend** the full span of the arch between supports.[§1926.855(a)]

(b) **Before demolishing any floor arch,** debris and other material shall be removed from such arch and other adjacent floor area. Planks not less than 2 inches by 10 inches in cross section, full size undressed, shall be provided for, and shall be used by employees to stand on while breaking down floor arches between beams. Such planks shall be so located as to provide a safe support for the workmen should the arch between the beams collapse. The open space between planks shall not exceed 16 inches.[§1926.855(b)]

(c) **Safe walkways, not less than 18 inches wide,** formed of planks not less than 2 inches thick if wood, or of equivalent strength if metal, shall be provided and used by workmen when necessary to enable them to reach any point without walking upon exposed beams.[§1926.855(c)]

(d) **Stringers of ample strength shall be installed** to support the flooring planks, and the ends of such stringers shall be supported by floor beams or girders, and not by floor arches alone. [§1926.855(d)]

(e) **Planks shall be laid together** over solid bearings with the ends overlapping at least 1 foot.[§1926.855(e)]

(f) **When floor arches are being removed,** employees shall not be allowed in the area directly underneath, and such an area shall be barricaded to prevent access to it.[§1926.855(f)]

(g) **Demolition of floor arches shall not be started** until they, and the surrounding floor area for a distance of 20 feet, have been cleared of debris and any other unnecessary materials. [§1926.855(g)]

§1926.856
Removal of walls, floors, and material with equipment

(a) **Mechanical equipment shall not be used** on floors or working surfaces unless such floors or surfaces are of sufficient strength to support the imposed load.[§1926.856(a)]

(b) **Floor openings shall have curbs or stop-logs** to prevent equipment from running over the edge.[§1926.856(b)]

(c) **Mechanical equipment used** shall meet the requirements specified in subparts N and O and §1926.1501 of §1926 subpart DD. [§1926.856(c)]

[44 FR 8577, Feb. 9, 1979 , 75 FR 48135, Aug. 9, 2010]

§1926.857
Storage

(a) **The storage of waste material** and debris on any floor shall not exceed the allowable floor loads.[§1926.857(a)]

(b) **In buildings having wooden floor construction,** the flooring boards may be removed from not more than one floor above grade to provide storage space for debris, provided falling material is not permitted to endanger the stability of the structure. [§1926.857(b)]

(c) **When wood floor beams serve** to brace interior walls or free-standing exterior walls, such beams shall be left in place until other equivalent support can be installed to replace them. [§1926.857(c)]

(d) **Floor arches, to an elevation** of not more than 25 feet above grade, may be removed to provide storage area for debris: Provided, that such removal does not endanger the stability of the structure.[§1926.857(d)]

(e) **Storage space into which material is dumped** shall be blocked off, except for openings necessary for the removal of material. Such openings shall be kept closed at all times when material is not being removed.[§1926.857(e)]

§1926.858
Removal of steel construction

(a) **When floor arches have been removed,** planking in accordance with Sec. 1926.855(b) shall be provided for the workers engaged in razing the steel framing.[§1926.858(a)]

(b) **Cranes, derricks,** and other hoisting equipment used shall meet the requirements specified in §1926.1501 of §1926 subpart DD. [§1926.858(b)]

(c) **Steel construction shall be dismantled** column length by column length, and tier by tier (columns may be in two-story lengths). [§1926.858(c)]

(d) **Any structural member being dismembered** shall not be overstressed.[§1926.858(d)]

[44 FR 8577, Feb. 9, 1979 , 75 FR 48135, Aug. 9, 2010]

§1926.859
Mechanical demolition

(a) **No workers shall be permitted in any area,** which can be adversely affected by demolition operations, when balling or clamming is being performed. Only those workers necessary for the performance of the operations shall be permitted in this area at any other time.[§1926.859(a)]

(b) **The weight of the demolition ball** shall not exceed 50 percent of the crane's rated load, based on the length of the boom and the maximum angle of operation at which the demolition ball will be used, or it shall not exceed 25 percent of the nominal breaking strength of the line by which it is suspended, whichever results in a lesser value.[§1926.859(b)]

(c) **The crane boom and loadline shall be as short as possible.** [§1926.859(c)]

(d) **The ball shall be attached to the loadline** with a swivel-type connection to prevent twisting of the loadline, and shall be attached by positive means in such manner that the weight cannot become accidentally disconnected.[§1926.859(d)]

(e) **When pulling over walls or portions thereof,** all steel members affected shall have been previously cut free.[§1926.859(e)]

(f) **All roof cornices or other such ornamental stonework** shall be removed prior to pulling walls over.[§1926.859(f)]

(g) **During demoliton, continuing inspections by a competent person** shall be made as the work progresses to detect hazards resulting from weakened or deteriorated floors, or walls, or loosened material. No employee shall be permitted to work where such hazards exist until they are corrected by shoring, bracing, or other effective means.[§1926.859(g)]

§1926.860
Selective demolition by explosives

Selective demolition by explosives shall be conducted in accordance with the applicable sections of subpart U of this part.

Authority: §107, Contract Work Hours and Safety Standards Act (40 U.S.C. 333); secs. 4, 6, and 8 of the Occupational Safety and Health Act of 1970 (29 U.S.C. 653, 655, 657); Secretary of Labor's Orders 12-71 (36 FR 8754), 8-76 (41 FR 25059), 9-83 (48 FR 35736), 1-90 (55 FR 9033), 6-96 (62 FR 111), or 5-2007 (72 FR 31159) as applicable.

Subpart U - Blasting and the Use of Explosives

§1926.900

General provisions

(a) **The employer shall permit only** authorized and qualified persons to handle and use explosives.[§1926.900(a)]

(b) **Smoking, firearms, matches, open flame lamps,** and other fires, flame or heat producing devices and sparks shall be prohibited in or near explosive magazines or while explosives are being handled, transported or used.[§1926.900(b)]

(c) **No person shall be allowed** to handle or use explosives while under the influence of intoxicating liquors, narcotics, or other dangerous drugs.[§1926.900(c)]

(d) **All explosives shall be accounted for at all times.** Explosives not being used shall be kept in a locked magazine, unavailable to persons not authorized to handle them. The employer shall maintain an inventory and use record of all explosives. Appropriate authorities shall be notified of any loss, theft, or unauthorized entry into a magazine.[§1926.900(d)]

(e) **No explosives or blasting agents shall be abandoned.** [§1926.900(e)]

(f) **No fire shall be fought** where the fire is in imminent danger of contact with explosives. All employees shall be removed to a safe area and the fire area guarded against intruders.[§1926.900(f)]

(g) **Original containers, or Class II magazines,** shall be used for taking detonators and other explosives from storage magazines to the blasting area.[§1926.900(g)]

(h) **When blasting is done in congested areas** or in proximity to a structure, railway, or highway, or any other installation that may be damaged, the blaster shall take special precautions in the loading, delaying, initiation, and confinement of each blast with mats or other methods so as to control the throw of fragments, and thus prevent bodily injury to employees.[§1926.900(h)]

(i) **Employees authorized to prepare explosive charges** or conduct blasting operations shall use every reasonable precaution including, but not limited to, visual and audible warning signals, flags, or barricades, to ensure employee safety.[§1926.900(i)]

(j) **Insofar as possible, blasting operations above ground** shall be conducted between sunup and sundown.[§1926.900(j)]

(k) **Due precautions shall be taken** to prevent accidental discharge of electric blasting caps from current induced by radar, radio transmitters, lightning, adjacent powerlines, dust storms, or other sources of extraneous electricity. These precautions shall include: [§1926.900(k)]

(1) *Detonators shall be short-circuited* in holes which have been primed and shunted until wired into the blasting circuit. [§1926.900(k)(1)]

(2) *The suspension of all blasting operations* and removal of persons from the blasting area during the approach and progress of an electric storm;[§1926.900(k)(2)]

(3) (i) *The prominent display of adequate signs,* warning against the use of mobile radio transmitters, on all roads within 1,000 feet of blasting operations. Whenever adherence to the 1,000-foot distance would create an operational handicap, a competent person shall be consulted to evaluate the particular situation, and alternative provisions may be made which are adequately designed to prevent any premature firing of electric blasting caps. A description of any such alternatives shall be reduced to writing and shall be certified as meeting the purposes of this subdivision by the competent person consulted. The description shall be maintained at the construction site during the duration of the work, and shall be available for inspection by representatives of the Secretary of Labor.[§1926.900(k)(3)(i)]

(ii) *Specimens of signs which* would meet the requirements of paragraph (k)(3) of this section are the following: [§1926.900(k)(3)(ii)]

About 48" x 48"

About 42" x 36"

TURN OFF
2-WAY
RADIO

(4) *Ensuring that mobile radio transmitters* which are less than 100 feet away from electric blasting caps, in other than original containers, shall be deenergized and effectively locked; [§1926.900(k)(4)]

(5) *Compliance with the recommendations* of The Institute of the Makers of Explosives with regard to blasting in the vicinity of radio transmitters as stipulated in Radio Frequency Energy — A Potential Hazard in the Use of Electric Blasting Caps, IME Publication No. 20, March 1971.[§1926.900(k)(5)]

(l) **Empty boxes and paper and fiber packing materials,** which have previously contained high explosives, shall not be used again for any purpose, but shall be destroyed by burning at an approved location.[§1926.900(l)]

(m) **Explosives, blasting agents, and blasting supplies** that are obviously deteriorated or damaged shall not be used.[§1926.900(m)]

(n) **Delivery and issue of explosives** shall only be made by and to authorized persons and into authorized magazines or approved temporary storage or handling areas.[§1926.900(n)]

(o) **Blasting operations in the proximity** of overhead power lines, communication lines, utility services, or other services and structures shall not be carried on until the operators and/or owners have been notified and measures for safe control have been taken.[§1926.900(o)]

(p) **The use of black powder shall be prohibited.**[§1926.900(p)]

(q) **All loading and firing shall be directed and supervised** by competent persons thoroughly experienced in this field. [§1926.900(q)]

(r) **All blasts shall be fired electrically** with an electric blasting machine or properly designed electric power source, except as provided in Sec. 1926.906 (a) and (r).[§1926.900(r)]

(s) **Buildings used for the mixing of blasting agents** shall conform to the requirements of this section.[§1926.900(s)]

(1) *Buildings shall be of noncombustible construction* or sheet metal on wood studs.[§1926.900(s)(1)]

(2) *Floors in a mixing plant* shall be of concrete or of other nonabsorbent materials.[§1926.900(s)(2)]

(3) *All fuel oil storage facilities* shall be separated from the mixing plant and located in such a manner that in case of tank rupture, the oil will drain away from the mixing plant building.[§1926.900(s)(3)]

(4) *The building shall be well ventilated.*[§1926.900(s)(4)]

(5) *Heating units which do not depend on combustion processes*, when properly designed and located, may be used in the building. All direct sources of heat shall be provided exclusively from units located outside the mixing building.[§1926.900(s)(5)]

(6) *All internal-combustion engines used* for electric power generation shall be located outside the mixing plant building, or shall be properly ventilated and isolated by a firewall. The exhaust systems on all such engines shall be located so any spark emission cannot be a hazard to any materials in or adjacent to the plant.[§1926.900(s)(6)]

(t) **Buildings used for the mixing** of water gels shall conform to the requirements of this subdivision.[§1926.900(t)]

(1) *Buildings shall be of noncombustible construction* or sheet metal on wood studs.[§1926.900(t)(1)]

(2) *Floors in a mixing plant* shall be of concrete or of other nonabsorbent materials.[§1926.900(t)(2)]

(3) *Where fuel oil is used* all fuel oil storage facilities shall be separated from the mixing plant and located in such a manner that in case of tank rupture, the oil will drain away from the mixing plant building.[§1926.900(t)(3)]

(4) *The building shall be well ventilated.*[§1926.900(t)(4)]

(5) *Heating units that do not depend on* combustion processes, when properly designed and located, may be used in the building. All direct sources of heat shall be provided exclusively from units located outside of the mixing building.[§1926.900(t)(5)]

(6) *All internal-combustion engines used* for electric power generation shall be located outside the mixing plant building, or shall be properly ventilated and isolated by a firewall. The exhaust systems on all such engines shall be located so any spark emission cannot be a hazard to any materials in or adjacent to the plant.[§1926.900(t)(6)]

[44 FR 8577, Feb. 9, 1979; 44 FR 20940, Apr. 6, 1979, as amended at 58 FR 35183, June 30, 1993]

§1926.901

Blaster qualifications

(a) **A blaster shall be able to understand** and give written and oral orders.[§1926.901(a)]

(b) **A blaster shall be in good physical condition** and not be addicted to narcotics, intoxicants, or similar types of drugs. [§1926.901(b)]

(c) **A blaster shall be qualified,** by reason of training, knowledge, or experience, in the field of transporting, storing, handling, and use of explosives, and have a working knowledge of State and local laws and regulations which pertain to explosives.[§1926.901(c)]

(d) **Blasters shall be required to furnish** satisfactory evidence of competency in handling explosives and performing in a safe manner the type of blasting that will be required.[§1926.901(d)]

(e) **The blaster shall be knowledgeable and competent** in the use of each type of blasting method used.[§1926.901(e)]

§1926.902

Surface transportation of explosives

(a) **Transportation of explosives** shall meet the provisions of Department of Transportation regulations contained in 46 CFR parts 146- 149, Water Carriers; 49 CFR parts 171-179, Highways and Railways; 49 CFR part 195, Pipelines; and 49 CFR parts 390-397, Motor Carriers.[§1926.902(a)]

(b) **Motor vehicles or conveyances transporting explosives** shall only be driven by, and be in the charge of, a licensed driver who is physically fit. He shall be familiar with the local, State, and Federal regulation governing the transportation of explosives.[§1926.902(b)]

(c) **No person shall smoke, or carry matches** or any other flame-producing device, nor shall firearms or loaded cartridges be carried while in or near a motor vehicle or conveyance transporting explosives.[§1926.902(c)]

(d) **Explosives, blasting agents, and blasting supplies** shall not be transported with other materials or cargoes. Blasting caps (including electric) shall not be transported in the same vehicle with other explosives.[§1926.902(d)]

(e) **Vehicles used for transporting explosives** shall be strong enough to carry the load without difficulty, and shall be in good mechanical condition.[§1926.902(e)]

(f) **When explosives are transported by a vehicle** with an open body, a Class II magazine or original manufacturer's container shall be securely mounted on the bed to contain the cargo. [§1926.902(f)]

(g) **All vehicles used for the transportation of explosives** shall have tight floors and any exposed spark-producing metal on the inside of the body shall be covered with wood, or other nonsparking material, to prevent contact with containers of explosives. [§1926.902(g)]

(h) **Every motor vehicle or conveyance** used for transporting explosives shall be marked or placarded on both sides, the front, and the rear with the word "Explosives" in red letters, not less than 4 inches in height, on white background. In addition to such marking or placarding, the motor vehicle or conveyance may display, in such a manner that it will be readily visible from all directions, a red flag 18 inches by 30 inches, with the word "Explosives" painted, stamped, or sewed thereon, in white letters, at least 6 inches in height.[§1926.902(h)]

(i) **Each vehicle used for transportation** of explosives shall be equipped with a fully charged fire extinguisher, in good condition. An Underwriters Laboratory-approved extinguisher of not less than 10-ABC rating will meet the minimum requirement. The driver shall be trained in the use of the extinguisher on his vehicle. [§1926.902(i)]

(j) **Motor vehicles or conveyances** carrying explosives, blasting agents, or blasting supplies, shall not be taken inside a garage or shop for repairs or servicing.[§1926.902(j)]

(k) **No motor vehicle transporting explosives shall be left unattended.**[§1926.902(k)]

[44 FR 8577, Feb. 9, 1979; 44 FR 20940, Apr. 6, 1979, as amended at 58 FR 35311, June 30, 1993]

§1926.903

Underground transportation of explosives

(a) **All explosives or blasting agents** in transit underground shall be taken to the place of use or storage without delay.[§1926.903(a)]

(b) **The quantity of explosives or blasting** agents taken to an underground loading area shall not exceed the amount estimated to be necessary for the blast.[§1926.903(b)]

(c) **Explosives in transit shall not be left unattended.**[§1926.903(c)]

(d) **The hoist operator shall be notified** before explosives or blasting agents are transported in a shaft conveyance.[§1926.903(d)]

(e) **Trucks used for the transportation** of explosives underground shall have the electrical system checked weekly to detect any failures which may constitute an electrical hazard. A certification record which includes the date of the inspection; the signature of the person who performed the inspection; and a serial number, or other identifier, of the truck inspected shall be prepared and the most recent certification record shall be maintained on file. [§1926.903(e)]

(f) **The installation of auxiliary lights on truck beds,** which are powered by the truck's electrical system, shall be prohibited. [§1926.903(f)]

(g) **Explosives and blasting agents** shall be hoisted, lowered, or conveyed in a powder car. No other materials, supplies, or equipment shall be transported in the same conveyance at the same time.[§1926.903(g)]

(h) **No one, except the operator,** his helper, and the powderman, shall be permitted to ride on a conveyance transporting explosives and blasting agents.[§1926.903(h)]

(i) **No person shall ride in any shaft conveyance** transporting explosives and blasting agents.[§1926.903(i)]

(j) **No explosives or blasting agents** shall be transported on any locomotive. At least two car lengths shall separate the locomotive from the powder car.[§1926.903(j)]

(k) **No explosives or blasting agents** shall be transported on a man haul trip.[§1926.903(k)]

(l) **The car or conveyance** containing explosives or blasting agents shall be pulled, not pushed, whenever possible.[§1926.903(l)]

(m) **The powder car or conveyance** especially built for the purpose of transporting explosives or blasting agents shall bear a reflectorized sign on each side with the word "Explosives" in letters, not less than 4 inches in height; upon a background of sharply contrasting color.[§1926.903(m)]

(n) **Compartments for transporting detonators and explosives** in the same car or conveyance shall be physically separated by a distance of 24 inches or by a solid partition at least 6 inches thick. [§1926.903(n)]

(o) **Detonators and other explosives shall not be transported** at the same time in any shaft conveyance.[§1926.903(o)]

(p) **Explosives, blasting agents, or blasting supplies** shall not be transported with other materials.[§1926.903(p)]

(q) **Explosives or blasting agents,** not in original containers, shall be placed in a suitable container when transported manually. [§1926.903(q)]

(r) **Detonators, primers, and other explosives** shall be carried in separate containers when transported manually.[§1926.903(r)]

[44 FR 8577, Feb. 9, 1979; 44 FR 20940, Apr. 6, 1979, as amended at 52 FR 36382, Sept. 28, 1987]

§1926.904

Storage of explosives and blasting agents

(a) **Explosives and related materials** shall be stored in approved facilities required under the applicable provisions of the Bureau of Alcohol, Tobacco and Firearms regulations contained in 27 CFR part 55, Commerce in Explosives.[§1926.904(a)]

(b) **Blasting caps, electric blasting caps,** detonating primers, and primed cartridges shall not be stored in the same magazine with other explosives or blasting agents.[§1926.904(b)]

(c) **Smoking and open flames shall not be permitted** within 50 feet of explosives and detonator storage magazine.[§1926.904(c)]

(d) **No explosives or blasting agents** shall be permanently stored in any underground operation until the operation has been developed to the point where at least two modes of exit have been provided.[§1926.904(d)]

(e) **Permanent underground storage magazines** shall be at least 300 feet from any shaft, adit, or active underground working area. [§1926.904(e)]

(f) **Permanent underground magazines** containing detonators shall not be located closer than 50 feet to any magazine containing other explosives or blasting agents.[§1926.904(f)]

[44 FR 8577, Feb. 9, 1979; 44 FR 20940, Apr. 6, 1979, as amended at 58 FR 35311, June 30, 1993]

§1926.905

Loading of explosives or blasting agents

(a) **Procedures that permit safe** and efficient loading shall be established before loading is started.[§1926.905(a)]

(b) **All drill holes shall be sufficiently large** to admit freely the insertion of the cartridges of explosives.[§1926.905(b)]

(c) **Tamping shall be done only with wood rods** or plastic tamping poles without exposed metal parts, but nonsparking metal connectors may be used for jointed poles. Violent tamping shall be avoided. The primer shall never be tamped.[§1926.905(c)]

(d) **No holes shall be loaded except those to be fired** in the next round of blasting. After loading, all remaining explosives and detonators shall be immediately returned to an authorized magazine. [§1926.905(d)]

(e) **Drilling shall not be started until** all remaining butts of old holes are examined for unexploded charges, and if any are found, they shall be refired before work proceeds.[§1926.905(e)]

(f) **No person shall be allowed to deepen** drill holes which have contained explosives or blasting agents.[§1926.905(f)]

(g) **No explosives or blasting agents** shall be left unattended at the blast site.[§1926.905(g)]

(h) **Machines and all tools not used for loading explosives** into bore holes shall be removed from the immediate location of holes before explosives are delivered. Equipment shall not be operated within 50 feet of loaded holes.[§1926.905(h)]

(i) **No activity of any nature** other than that which is required for loading holes with explosives shall be permitted in a blast area. [§1926.905(i)]

(j) **Powerlines and portable electric cables** for equipment being used shall be kept a safe distance from explosives or blasting agents being loaded into drill holes. Cables in the proximity of the blast area shall be deenergized and locked out by the blaster. [§1926.905(j)]

(k) **Holes shall be checked prior to loading** to determine depth and conditions. Where a hole has been loaded with explosives but the explosives have failed to detonate, there shall be no drilling within 50 feet of the hole.[§1926.905(k)]

(l) **When loading a long line of holes** with more than one loading crew, the crews shall be separated by practical distance consistent with efficient operation and supervision of crews.[§1926.905(l)]

(m) **No explosive shall be loaded or used underground** in the presence of combustible gases or combustible dusts.[§1926.905(m)]

(n) **No explosives other than those in Fume Class 1,** as set forth by the Institute of Makers of Explosives, shall be used; however, explosives complying with the requirements of Fume Class 2 and Fume Class 3 may be used if adequate ventilation has been provided.[§1926.905(n)]

(o) **All blast holes in open work shall be stemmed** to the collar or to a point which will confine the charge.[§1926.905(o)]

(p) **Warning signs,** indicating a blast area, shall be maintained at all approaches to the blast area. The warning sign lettering shall not be less than 4 inches in height on a contrasting background. [§1926.905(p)]

(q) **A bore hole shall never be sprung** when it is adjacent to or near a hole that is loaded. Flashlight batteries shall not be used for springing holes.[§1926.905(q)]

(r) **Drill holes which have been sprung or chambered,** and which are not water-filled, shall be allowed to cool before explosives are loaded.[§1926.905(r)]

(s) **No loaded holes shall be left unattended or unprotected.** [§1926.905(s)]

(t) **The blaster shall keep an accurate,** up-to-date record of explosives, blasting agents, and blasting supplies used in a blast and shall keep an accurate running inventory of all explosives and blasting agents stored on the operation.[§1926.905(t)]

(u) **When loading blasting agents pneumatically** over electric blasting caps, semiconductive delivery hose shall be used and the equipment shall be bonded and grounded.[§1926.905(u)]

[44 FR 8577, Feb. 9, 1979; 44 FR 20940, Apr. 6, 1979, as amended at 58 FR 35184, June 30, 1993]

§1926.906

Initiation of explosive charges — electric blasting

(a) **Electric blasting caps shall not be used** where sources of extraneous electricity make the use of electric blasting caps dangerous. Blasting cap leg wires shall be kept short-circuited (shunted) until they are connected into the circuit for firing. [§1926.906(a)]

(b) **Before adopting any system of electrical firing,** the blaster shall conduct a thorough survey for extraneous currents, and all dangerous currents shall be eliminated before any holes are loaded.[§1926.906(b)]

(c) **In any single blast using electric blasting caps,** all caps shall be of the same style or function, and of the same manufacture. [§1926.906(c)]

(d) **Electric blasting shall be carried out by using** blasting circuits or power circuits in accordance with the electric blasting cap manufacturer's recommendations, or an approved contractor or his designated representative.[§1926.906(d)]

(e) **When firing a circuit of electric blasting caps,** care must be exercised to ensure that an adequate quantity of delivered current is available, in accordance with the manufacturer's recommendations.[§1926.906(e)]

(f) **Connecting wires and lead wires** shall be insulated single solid wires of sufficient current-carrying capacity.[§1926.906(f)]

(g) **Bus wires shall be solid single wires** of sufficient current-carrying capacity.[§1926.906(g)]

(h) **When firing electrically,** the insulation on all firing lines shall be adequate and in good condition.[§1926.906(h)]

(i) **A power circuit used for firing electric blasting caps** shall not be grounded.[§1926.906(i)]

(j) **In underground operations** when firing from a power circuit, a safety switch shall be placed in the permanent firing line at intervals. This switch shall be made so it can be locked only in the "Off" position and shall be provided with a short-circuiting arrangement of the firing lines to the cap circuit.[§1926.906(j)]

(k) **In underground operations there** shall be a "lightning" gap of at least 5 feet in the firing system ahead of the main firing switch; that is, between this switch and the source of power. This gap shall be bridged by a flexible jumper cord just before firing the blast. [§1926.906(k)]

(l) **When firing from a power circuit,** the firing switch shall be locked in the open or "Off" position at all times, except when firing. It shall be so designed that the firing lines to the cap circuit are automatically short-circuited when the switch is in the "Off" position. Keys to this switch shall be entrusted only to the blaster. [§1926.906(l)]

(m) **Blasting machines shall be in good condition** and the efficiency of the machine shall be tested periodically to make certain that it can deliver power at its rated capacity.[§1926.906(m)]

(n) **When firing with blasting machines,** the connections shall be made as recommended by the manufacturer of the electric blasting caps used.[§1926.906(n)]

(o) **The number of electric blasting caps** connected to a blasting machine shall not be in excess of its rated capacity. Furthermore, in primary blasting, a series circuit shall contain no more caps than the limits recommended by the manufacturer of the electric blasting caps in use.[§1926.906(o)]

(p) **The blaster shall be in charge of the blasting machines,** and no other person shall connect the leading wires to the machine. [§1926.906(p)]

(q) **Blasters, when testing circuits to charged holes,** shall use only blasting galvanometers or other instruments that are specifically designed for this purpose.[§1926.906(q)]

(r) **Whenever the possibility exists** that a leading line or blasting wire might be thrown over a live powerline by the force of an explosion, care shall be taken to see that the total length of wires are kept too short to hit the lines, or that the wires are securely anchored to the ground. If neither of these requirements can be satisfied, a nonelectric system shall be used.[§1926.906(r)]

(s) **In electrical firing,** only the man making leading wire connections shall fire the shot. All connections shall be made from the bore hole back to the source of firing current, and the leading wires shall remain shorted and not be connected to the blasting machine or other source of current until the charge is to be fired. [§1926.906(s)]

(t) **After firing an electric blast from a blasting machine,** the leading wires shall be immediately disconnected from the machine and short-circuited.[§1926.906(t)]

[44 FR 8577, Feb. 9, 1979; 44 FR 20940, Apr. 6, 1979, as amended at 63 FR 33469, June 18, 1998]

§1926.907

Use of safety fuse

(a) **Safety fuse shall only be used** where sources of extraneous electricity make the use of electric blasting caps dangerous. The use of a fuse that has been hammered or injured in any way shall be forbidden.[§1926.907(a)]

(b) **The hanging of a fuse on nails or other projections** which will cause a sharp bend to be formed in the fuse is prohibited. [§1926.907(b)]

(c) **Before capping safety fuse,** a short length shall be cut from the end of the supply reel so as to assure a fresh cut end in each blasting cap.[§1926.907(c)]

(d) **Only a cap crimper of approved design** shall be used for attaching blasting caps to safety fuse. Crimpers shall be kept in good repair and accessible for use.[§1926.907(d)]

(e) **No unused cap or short capped fuse** shall be placed in any hole to be blasted; such unused detonators shall be removed from the working place and destroyed.[§1926.907(e)]

(f) **No fuse shall be capped,** or primers made up, in any magazine or near any possible source of ignition.[§1926.907(f)]

(g) **No one shall be permitted to carry** detonators or primers of any kind on his person.[§1926.907(g)]

(h) **The minimum length of safety fuse** to be used in blasting shall be as required by State law, but shall not be less than 30 inches. [§1926.907(h)]

(i) **At least two men shall be present** when multiple cap and fuse blasting is done by hand lighting methods.[§1926.907(i)]

(j) **Not more than 12 fuses shall be lighted** by each blaster when hand lighting devices are used. However, when two or more safety fuses in a group are lighted as one by means of igniter cord, or other similar fuse-lighting devices, they may be considered as one fuse.[§1926.907(j)]

(k) **The so-called "drop fuse" method** of dropping or pushing a primer or any explosive with a lighted fuse attached is forbidden. [§1926.907(k)]

(l) **Cap and fuse shall not be used** for firing mudcap charges unless charges are separated sufficiently to prevent one charge from dislodging other shots in the blast.[§1926.907(l)]

(m) **When blasting with safety fuses,** consideration shall be given to the length and burning rate of the fuse. Sufficient time, with a margin of safety, shall always be provided for the blaster to reach a place of safety.[§1926.907(m)]

§1926.908

Use of detonating cord

(a) **Care shall be taken to select a detonating cord** consistent with the type and physical condition of the bore hole and stemming and the type of explosives used.[§1926.908(a)]

(b) **Detonating cord shall be handled and used** with the same respect and care given other explosives.[§1926.908(b)]

(c) **The line of detonating cord** extending out of a bore hole or from a charge shall be cut from the supply spool before loading the remainder of the bore hole or placing additional charges. [§1926.908(c)]

(d) **Detonating cord shall be handled and used with care** to avoid damaging or severing the cord during and after loading and hooking-up.[§1926.908(d)]

(e) **Detonating cord connections** shall be competent and positive in accordance with approved and recommended methods. Knot-type or other cord-to-cord connections shall be made only with detonating cord in which the explosive core is dry.[§1926.908(e)]

(f) **All detonating cord trunklines and branchlines** shall be free of loops, sharp kinks, or angles that direct the cord back toward the oncoming line of detonation.[§1926.908(f)]

(g) **All detonating cord connections** shall be inspected before firing the blast.[§1926.908(g)]

(h) **When detonating cord millisecond-delay connectors** or short-interval-delay electric blasting caps are used with detonating cord, the practice shall conform strictly to the manufacturer's recommendations.[§1926.908(h)]

(i) **When connecting a blasting cap** or an electric blasting cap to detonating cord, the cap shall be taped or otherwise attached securely along the side or the end of the detonating cord, with the end of the cap containing the explosive charge pointed in the direction in which the detonation is to proceed.[§1926.908(i)]

(j) **Detonators for firing the trunkline** shall not be brought to the loading area nor attached to the detonating cord until everything else is in readiness for the blast.[§1926.908(j)]

§1926.909

Firing the blast

(a) **A code of blasting signals equivalent to Table U-1,** shall be posted on one or more conspicuous places at the operation, and all employees shall be required to familiarize themselves with the code and conform to it. Danger signs shall be placed at suitable locations.[§1926.909(a)]

(b) **Before a blast is fired,** a loud warning signal shall be given by the blaster in charge, who has made certain that all surplus explosives are in a safe place and all employees, vehicles, and equipment are at a safe distance, or under sufficient cover.[§1926.909(b)]

(c) **Flagmen shall be safely stationed** on highways which pass through the danger zone so as to stop traffic during blasting operations.[§1926.909(c)]

(d) **It shall be the duty of the blaster to fix the time of blasting.** [§1926.909(d)]

(e) **Before firing an underground blast,** warning shall be given, and all possible entries into the blasting area, and any entrances to any working place where a drift, raise, or other opening is about to hole through, shall be carefully guarded. The blaster shall make sure that all employees are out of the blast area before firing a blast.[§1926.909(e)]

Table U-1

WARNING SIGNAL	A 1-minute series of long blasts 5 minutes prior to blast signal.
BLAST SIGNAL	A series of short blasts 1 minute prior to the shot.
ALL CLEAR SIGNAL	A prolonged blast following the inspection of blast area.

§1926.910
Inspection after blasting

(a) **Immediately after the blast has been fired,** the firing line shall be disconnected from the blasting machine, or where power switches are used, they shall be locked open or in the off position.[§1926.910(a)]

(b) **Sufficient time shall be allowed,** not less than 15 minutes in tunnels, for the smoke and fumes to leave the blasted area before returning to the shot. An inspection of the area and the surrounding rubble shall be made by the blaster to determine if all charges have been exploded before employees are allowed to return to the operation, and in tunnels, after the muck pile has been wetted down.[§1926.910(b)]

§1926.911
Misfires

(a) **If a misfire is found,** the blaster shall provide proper safeguards for excluding all employees from the danger zone.[§1926.911(a)]

(b) **No other work shall be done** except that necessary to remove the hazard of the misfire and only those employees necessary to do the work shall remain in the danger zone.[§1926.911(b)]

(c) **No attempt shall be made to extract explosives** from any charged or misfired hole; a new primer shall be put in and the hole reblasted. If refiring of the misfired hole presents a hazard, the explosives may be removed by washing out with water or, where the misfire is under water, blown out with air.[§1926.911(c)]

(d) **If there are any misfires while using cap and fuse,** all employees shall remain away from the charge for at least 1 hour. Misfires shall be handled under the direction of the person in charge of the blasting. All wires shall be carefully traced and a search made for unexploded charges.[§1926.911(d)]

(e) **No drilling, digging, or picking** shall be permitted until all missed holes have been detonated or the authorized representative has approved that work can proceed.[§1926.911(e)]

§1926.912
Underwater blasting

(a) **A blaster shall conduct all blasting operations,** and no shot shall be fired without his approval.[§1926.912(a)]

(b) **Loading tubes and casings of dissimilar metals** shall not be used because of possible electric transient currents from galvanic action of the metals and water.[§1926.912(b)]

(c) **Only water-resistant blasting caps** and detonating cords shall be used for all marine blasting. Loading shall be done through a nonsparking metal loading tube when tube is necessary. [§1926.912(c)]

(d) **No blast shall be fired while** any vessel under way is closer than 1,500 feet to the blasting area. Those on board vessels or craft moored or anchored within 1,500 feet shall be notified before a blast is fired.[§1926.912(d)]

(e) **No blast shall be fired while** any swimming or diving operations are in progress in the vicinity of the blasting area. If such operations are in progress, signals and arrangements shall be agreed upon to assure that no blast shall be fired while any person is in the water.[§1926.912(e)]

(f) **Blasting flags shall be displayed.**[§1926.912(f)]

(g) **The storage and handling of explosives** aboard vessels used in underwater blasting operations shall be according to provisions outlined herein on handling and storing explosives.[§1926.912(g)]

(h) **When more than one charge is placed under water,** a float device shall be attached to an element of each charge in such manner that it will be released by the firing. Misfires shall be handled in accordance with the requirements of Sec. 1926.911.[§1926.912(h)]

§1926.913
Blasting in excavation work under compressed air

(a) **Detonators and explosives** shall not be stored or kept in tunnels, shafts, or caissons. Detonators and explosives for each round shall be taken directly from the magazines to the blasting zone and immediately loaded. Detonators and explosives left over after loading a round shall be removed from the working chamber before the connecting wires are connected up.[§1926.913(a)]

(b) **When detonators or explosives** are brought into an air lock, no employee except the powderman, blaster, lock tender and the employees necessary for carrying, shall be permitted to enter the air lock. No other material, supplies, or equipment shall be locked through with the explosives.[§1926.913(b)]

(c) **Detonators and explosives shall be taken separately** into pressure working chambers.[§1926.913(c)]

(d) **The blaster or powderman shall be responsible** for the receipt, unloading, storage, and on-site transportation of explosives and detonators.[§1926.913(d)]

(e) **All metal pipes, rails, air locks, and steel tunnel lining** shall be electrically bonded together and grounded at or near the portal or shaft, and such pipes and rails shall be cross-bonded together at not less than 1,000-foot intervals throughout the length of the tunnel. In addition, each low air supply pipe shall be grounded at its delivery end.[§1926.913(e)]

(f) **The explosives suitable for use in wet holes** shall be water-resistant and shall be Fume Class 1.[§1926.913(f)]

(g) **When tunnel excavation in rock face** is approaching mixed face, and when tunnel excavation is in mixed face, blasting shall be performed with light charges and with light burden on each hole. Advance drilling shall be performed as tunnel excavation in rock face approaches mixed face, to determine the general nature and extent of rock cover and the remaining distance ahead to soft ground as excavation advances.[§1926.913(g)]

§1926.914
Definitions applicable to this subpart

(a) **American Table of Distances** (also known as Quantity Distance Tables) means American Table of Distances for Storage of Explosives as revised and approved by the Institute of the Makers of Explosives, June 5, 1964.[§1926.914(a)]

(b) **Approved storage facility** — A facility for the storage of explosive materials conforming to the requirements of this part and covered by a license or permit issued under authority of the Bureau of Alcohol, Tobacco and Firearms. (See 27 CFR Part 55)[§1926.914(b)]

(c) **Blast area** — The area in which explosives loading and blasting operations are being conducted.[§1926.914(c)]

(d) **Blaster** — The person or persons authorized to use explosives for blasting purposes and meeting the qualifications contained in Sec. 1926.901.[§1926.914(d)]

(e) **Blasting agent** — A blasting agent is any material or mixture consisting of a fuel and oxidizer used for blasting, but not classified an explosive and in which none of the ingredients is classified as an explosive provided the furnished (mixed) product cannot be detonated with a No. 8 test blasting cap when confined. A common blasting agent presently in use is a mixture of ammonium nitrate (NH_4NO_3) and carbonaceous combustibles, such as fuel oil or coal, and may either be procured, premixed and packaged from explosives companies or mixed in the field.[§1926.914(e)]

(f) **Blasting cap** — A metallic tube closed at one end, containing a charge of one or more detonating compounds, and designed for and capable of detonation from the sparks or flame from a safety fuse inserted and crimped into the open end.[§1926.914(f)]

(g) **Block holing** — The breaking of boulders by firing a charge of explosives that has been loaded in a drill hole.[§1926.914(g)]

(h) **Conveyance** — Any unit for transporting explosives or blasting agents, including but not limited to trucks, trailers, rail cars, barges, and vessels.[§1926.914(h)]

(i) **Detonating cord** — A flexible cord containing a center core of high explosives which when detonated, will have sufficient strength to detonate other cap-sensitive explosives with which it is in contact.[§1926.914(i)]

(j) **Detonator** — Blasting caps, electric blasting caps, delay electric blasting caps, and nonelectric delay blasting caps.[§1926.914(j)]

(k) **Electric blasting cap** — A blasting cap designed for and capable of detonation by means of an electric current.[§1926.914(k)]

(l) **Electric blasting circuitry** — [§1926.914(l)]

(1) **Bus wire.** An expendable wire, used in parallel or series, in parallel circuits, to which are connected the leg wires of electric blasting caps.[§1926.914(l)(1)]

(2) **Connecting wire.** An insulated expendable wire used between electric blasting caps and the leading wires or between the bus wire and the leading wires.[§1926.914(l)(2)]

(3) **Leading wire.** An insulated wire used between the electric power source and the electric blasting cap circuit.[§1926.914(l)(3)]

(4) **Permanent blasting wire.** A permanently mounted insulated wire used between the electric power source and the electric blasting cap circuit.[§1926.914(l)(4)]

(m) **Electric delay blasting caps** — Caps designed to detonate at a predetermined period of time after energy is applied to the ignition system.[§1926.914(m)]

(n) **Explosives** — [§1926.914(n)]

(1) *Any chemical compound, mixture, or device,* the primary or common purpose of which is to function by explosion; that is, with substantially instantaneous release of gas and heat, unless such compound, mixture or device is otherwise specifically classified by the U.S. Department of Transportation. [§1926.914(n)(1)]

(2) *All material which is classified* as Class A, Class B, and Class C Explosives by the U.S. Department of Transportation. [§1926.914(n)(2)]

(3) *Classification of explosives* by the U.S. Department of Transportation is as follows:[§1926.914(n)(3)]

Class A Explosives. Possessing detonating hazard, such as dynamite, nitroglycerin, picric acid, lead azide, fulminate of mercury, black powder, blasting caps, and detonating primers.

Class B Explosives. Possessing flammable hazard, such as propellant explosives, including some smokeless propellants.

Class C Explosives. Include certain types of manufactured articles which contain Class A or Class B explosives, or both, as components, but in restricted quantities.

(o) **Fuse lighters** — Special devices for the purpose of igniting safety fuse.[§1926.914(o)]

(p) **Magazine** — Any building or structure, other than an explosives manufacturing building, used for the storage of explosives. [§1926.914(p)]

(q) **Misfire** — An explosive charge which failed to detonate. [§1926.914(q)]

(r) **Mud-capping** (sometimes known as bulldozing, adobe blasting, or dobying). The blasting of boulders by placing a quantity of explosives against a rock, boulder, or other object without confining the explosives in a drill hole.[§1926.914(r)]

(s) **Nonelectric delay blasting cap** — A blasting cap with an integral delay element in conjunction with and capable of being detonated by a detonation impulse or signal from miniaturized detonating cord.[§1926.914(s)]

(t) **Primary blasting** — The blasting operation by which the original rock formation is dislodged from its natural location.[§1926.914(t)]

(u) **Primer** — A cartridge or container of explosives into which a detonator or detonating cord is inserted or attached.[§1926.914(u)]

(v) **Safety fuse** — A flexible cord containing an internal burning medium by which fire is conveyed at a continuous and uniform rate for the purpose of firing blasting caps.[§1926.914(v)]

(w) **Secondary blasting** — The reduction of oversize material by the use of explosives to the dimension required for handling, including mudcapping and blockholing.[§1926.914(w)]

(x) **Stemming** — A suitable inert incombustible material or device used to confine or separate explosives in a drill hole, or to cover explosives in mud-capping.[§1926.914(x)]

(y) **Springing** — The creation of a pocket in the bottom of a drill hole by the use of a moderate quantity of explosives in order that larger quantities or explosives may be inserted therein.[§1926.914(y)]

(z) **Water gels, or slurry explosives** — A wide variety of materials used for blasting. They all contain substantial proportions of water and high proportions of ammonium nitrate, some of which is in solution in the water. Two broad classes of water gels are: [§1926.914(z)]

(1) *Those which are sensitized* by a material classed as an explosive, such as TNT or smokeless powder, and[§1926.914(z)(1)]

(2) *Those which contain* no ingredient classified as an explosive; these are sensitized with metals such as aluminum or with other fuels. Water gels may be premixed at an explosives plant or mixed at the site immediately before delivery into the bore hole.[§1926.914(z)(2)]

(aa) **Semiconductive hose** — a hose with an electrical resistance high enough to limit flow of stray electric currents to safe levels, yet not so high as to prevent drainage of static electric charges to ground; hose of not more than 2 megohms resistance over its entire length and of not less than 5,000 ohms per foot meets the requirement.[§1926.914(aa)]

[44 FR 8577, Feb. 9, 1979; 44 FR 20940, Apr. 6, 1979, as amended at 58 FR 35184 and 35311, June 30, 1993]

Authority: §107, Contract Work Hours and Safety Standards Act (40 U.S.C. 333); secs. 4, 6, 8, Occupational Safety and Health Act of 1970 (29 U.S.C. 653, 655, 657); Secretary of Labor's Order No. 12-71 (36 FR 8754), 8-76 (41 FR 25059), 9-83 (48 FR 35736), or 6-96 (62 FR 111), as applicable; and 29 CFR part 1911.

Subpart V - Power Transmission and Distribution

§1926.950

General requirements

(a) **Application.** The occupational safety and health standards contained in this subpart V shall apply to the construction of electric transmission and distribution lines and equipment.[§1926.950(a)]

(1) *As used in this subpart V* the term "construction" includes the erection of new electric transmission and distribution lines and equipment, and the alteration, conversion, and improvement of existing electric transmission and distribution lines and equipment.[§1926.950(a)(1)]

(2) *Existing electric transmission and distribution lines* and electrical equipment need not be modified to conform to the requirements of applicable standards in this subpart V, until such work as described in paragraph (a)(1) of this section is to be performed on such lines or equipment.[§1926.950(a)(2)]

(3) *The standards set forth in this subpart V* provide minimum requirements for safety and health. Employers may require adherence to additional standards which are not in conflict with the standards contained in this subpart V.[§1926.950(a)(3)]

(b) **Initial inspections, tests, or determinations.**[§1926.950(b)]

(1) *Existing conditions shall be determined* before starting work, by an inspection or a test. Such conditions shall include, but not be limited to, energized lines and equipment, conditions of poles, and the location of circuits and equipment, including power and communication lines, CATV and fire alarm circuits. [§1926.950(b)(1)]

(2) *Electric equipment and lines* shall be considered energized until determined to be deenergized by tests or other appropriate methods or means.[§1926.950(b)(2)]

(3) *Operating voltage of equipment* and lines shall be determined before working on or near energized parts.[§1926.950(b)(3)]

(c) **Clearances.** The provisions of paragraph (c) (1) or (2) of this section shall be observed.[§1926.950(c)]

(1) *No employee shall be permitted* to approach or take any conductive object without an approved insulating handle closer to exposed energized parts than shown in Table V-1, unless: [§1926.950(c)(1)]

(i) *The employee is insulated or guarded* from the energized part (gloves or gloves with sleeves rated for the voltage involved shall be considered insulation of the employee from the energized part), or[§1926.950(c)(1)(i)]

(ii) *The energized part is insulated or guarded* from him and any other conductive object at a different potential, or [§1926.950(c)(1)(ii)]

(iii) *The employee is isolated, insulated, or guarded* from any other conductive object(s), as during live-line bare-hand work.[§1926.950(c)(1)(iii)]

(2) (i) *The minimum working distance* and minimum clear hot stick distances stated in Table V-1 shall not be violated. The minimum clear hot stick distance is that for the use of live-line tools held by linemen when performing live-line work. [§1926.950(c)(2)(i)]

(ii) *Conductor support tools*, such as link sticks, strain carriers, and insulator cradles, may be used: Provided, That the clear insulation is at least as long as the insulator string or the minimum distance specified in Table V-1 for the operating voltage.[§1926.950(c)(2)(ii)]

Table V-1 — Alternating Current — Minimum Distances

Voltage range (phase to phase) kilovolt)	Minimum working and clear hot stick distance
2.1 to 15	2 ft. 0 in.
15.1 to 35	2 ft. 4 in.
35.1 to 46	2 ft. 6 in.
46.1 to 72.5	3 ft. 0 in.
72.6 to 121	3 ft. 4 in.
138 to 145	3 ft. 6 in.
161 to 169	3 ft. 8 in.
230 to 242	5 ft. 0 in.
345 to 362	[1] 7 ft. 0 in.
500 to 552	[1] 11 ft. 0 in.
700 to 765	[1] 15 ft. 0 in.

[1] Note: For 345-362 kv., 500-552 kv., and 700-765 kv., the minimum working distance and the minimum clear hot stick distance may be reduced provided that such distances are not less than the shortest distance between the energized part and a grounded surface.

(d) **Deenergizing lines and equipment.**[§1926.950(d)]

(1) *When deenergizing lines and equipment* operated in excess of 600 volts, and the means of disconnecting from electric energy is not visibly open or visibly locked out, the provisions of paragraphs (d)(1) (i) through (vii) of this section shall be complied with:[§1926.950(d)(1)]

(i) *The particular section* of line or equipment to be deenergized shall be clearly identified, and it shall be isolated from all sources of voltage.[§1926.950(d)(1)(i)]

(ii) *Notification and assurance* from the designated employee shall be obtained that:[§1926.950(d)(1)(ii)]

[a] All switches and disconnectors through which electric energy may be supplied to the particular section of line or equipment to be worked have been deenergized; [§1926.950(d)(1)(ii)[a]]

[b] All switches and disconnectors are plainly tagged indicating that men are at work;[§1926.950(d)(1)(ii)[b]]

[c] And that where design of such switches and disconnectors permits, they have been rendered inoperable. [§1926.950(d)(1)(ii)[c]]

(iii) *After all designated switches* and disconnectors have been opened, rendered inoperable, and tagged, visual inspection or tests shall be conducted to insure that equipment or lines have been deenergized.[§1926.950(d)(1)(iii)]

(iv) *Protective grounds shall be applied* on the disconnected lines or equipment to be worked on.[§1926.950(d)(1)(iv)]

(v) *Guards or barriers shall be erected* as necessary to adjacent energized lines.[§1926.950(d)(1)(v)]

(vi) *When more than one independent crew requires* the same line or equipment to be deenergized, a prominent tag for each such independent crew shall be placed on the line or equipment by the designated employee in charge. [§1926.950(d)(1)(vi)]

(vii) *Upon completion of work on deenergized lines or equipment,* each designated employee in charge shall determine that all employees in his crew are clear, that protective grounds installed by his crew have been removed, and he shall report to the designated authority that all tags protecting his crew may be removed.[§1926.950(d)(1)(vii)]

(2) *When a crew working on a line or equipment* can clearly see that the means of disconnecting from electric energy are visibly open or visibly locked-out, the provisions of paragraphs (d)(i), and (ii) of this section shall apply:[§1926.950(d)(2)]

(i) *Guards or barriers shall be erected* as necessary to adjacent energized lines.[§1926.950(d)(2)(i)]

(ii) *Upon completion of work* on deenergized lines or equipment, each designated employee in charge shall determine that all employees in his crew are clear, that protective grounds installed by his crew have been removed, and he shall report to the designated authority that all tags protecting his crew may be removed.[§1926.950(d)(2)(ii)]

(e) **Emergency procedures and first aid.**[§1926.950(e)]

(1) *The employer shall provide training* or require that his employees are knowledgeable and proficient in:[§1926.950(e)(1)]

(i) *Procedures involving emergency situations, and* [§1926.950(e)(1)(i)]

(ii) *First-aid fundamentals including resuscitation.* [§1926.950(e)(1)(ii)]

(2) *In lieu of paragraph (e)(1)* of this section the employer may comply with the provisions of Sec. 1926.50(c) regarding first-aid requirements.[§1926.950(e)(2)]

(f) **Night work.** When working at night, spotlights or portable lights for emergency lighting shall be provided as needed to perform the work safely.[§1926.950(f)]

(g) **Work near and over water.** When crews are engaged in work over or near water and when danger of drowning exists, suitable protection shall be provided as stated in Sec. 1926.104, or Sec. 1926.105, or Sec. 1926.106.[§1926.950(g)]

(h) **Sanitation facilities.** The requirements of Sec. 1926.51 of subpart D of this part shall be complied with for sanitation facilities. [§1926.950(h)]

(i) **Hydraulic fluids.** All hydraulic fluids used for the insulated sections of derrick trucks, aerial lifts, and hydraulic tools which are used on or around energized lines and equipment shall be of the insulating type. The requirements for fire resistant fluids of Sec. 1926.302(d)(1) do not apply to hydraulic tools covered by this paragraph.[§1926.950(i)]

§1926.951
Tools and protective equipment

(a) **Protective equipment.**[§1926.951(a)]

(1) (i) *Rubber protective equipment* shall be in accordance with the provisions of the American National Standards Institute (ANSI), ANSI J6 series, as follows:[§1926.951(a)(1)(i)]

Item	Standard
Rubber insulating gloves	J6.6-1971
Rubber matting for use around electric apparatus	J6.7-1935 (R1971)
Rubber insulating blankets	J6.4-1971
Rubber insulating hoods	J6.2-1950 (R1971)
Rubber insulating line hose	J6.1-1950 (R1971)
Rubber insulating sleeves	J6.5-1971

(ii) *Rubber protective equipment* shall be visually inspected prior to use.[§1926.951(a)(1)(ii)]

(iii) *In addition,* an "air" test shall be performed for rubber gloves prior to use.[§1926.951(a)(1)(iii)]

(iv) *Protective equipment of material* other than rubber shall provide equal or better electrical and mechanical protection. [§1926.951(a)(1)(iv)]

(2) *Protective hats shall be in accordance* with the provisions of ANSI Z89.2-1971 Industrial Protective Helmets for Electrical Workers, Class B, and shall be worn at the jobsite by employees who are exposed to the hazards of falling objects, electric shock, or burns.[§1926.951(a)(2)]

(b) **Personal climbing equipment.**[§1926.951(b)]

(1) *Body belts with straps* or lanyards shall be worn to protect employees working at elevated locations on poles, towers, or other structures except where such use creates a greater hazard to the safety of the employees, in which case other safeguards shall be employed.[§1926.951(b)(1)]

(2) *Body belts and safety straps* shall meet the requirements of Sec. 1926.959. In addition to being used as an employee safeguarding item, body belts with approved tool loops may be used for the purpose of holding tools. Body belts shall be free from additional metal hooks and tool loops other than those permitted in Sec. 1926.959.[§1926.951(b)(2)]

(3) *Body belts and straps* shall be inspected before use each day to determine that they are in safe working condition. [§1926.951(b)(3)]

(4) (i) *Lifelines and lanyards shall comply* with the provisions of Sec. 1926.502.[§1926.951(b)(4)(i)]

(ii) *Safety lines are not intended* to be subjected to shock loading and are used for emergency rescue such as lowering a man to the ground. Such safety lines shall be a minimum of one-half-inch diameter and three or four strand first-grade manila or its equivalent in strength (2,650 lb.) and durability. [§1926.951(b)(4)(ii)]

(5) *Defective ropes shall be replaced.*[§1926.951(b)(5)]

(c) **Ladders.**[§1926.951(c)]

(1) *Portable metal or conductive ladders* shall not be used near energized lines or equipment except as may be necessary in specialized work such as in high voltage substations where nonconductive ladders might present a greater hazard than conductive ladders. Conductive or metal ladders shall be prominently marked as conductive and all necessary precautions shall be taken when used in specialized work.[§1926.951(c)(1)]

(2) *Hook or other type ladders used in structures* shall be positively secured to prevent the ladder from being accidentally displaced.[§1926.951(c)(2)]

(d) **Live-line tools.**[§1926.951(d)]

(1) *Only live-line tool poles* having a manufacturer's certification to withstand the following minimum tests shall be used: [§1926.951(d)(1)]

(i) *100,000 volts per foot of length* for 5 minutes when the tool is made of fiberglass; or[§1926.951(d)(1)(i)]

(ii) *75,000 volts per foot of length* for 3 minutes when the tool is made of wood; or[§1926.951(d)(1)(ii)]

(iii) *Other tests equivalent* to paragraph (d) (i) or (ii) of this section as appropriate.[§1926.951(d)(1)(iii)]

(2) *All live-line tools shall be visually inspected* before use each day. Tools to be used shall be wiped clean and if any hazardous defects are indicated such tools shall be removed from service.[§1926.951(d)(2)]

(e) **Measuring tapes or measuring ropes.** Measuring tapes or measuring ropes which are metal or contain conductive strands shall not be used when working on or near energized parts.[§1926.951(e)]

(f) **Handtools.**[§1926.951(f)]

(1) *Switches for all powered hand tools shall comply with Sec. 1926.300(d).*[§1926.951(f)(1)]

(2) *All portable electric handtools shall:*[§1926.951(f)(2)]

(i) *Be equipped with three-wire cord* having the ground wire permanently connected to the tool frame and means for grounding the other end; or[§1926.951(f)(2)(i)]

(ii) *Be of the double insulated type* and permanently labeled as "Double Insulated"; or[§1926.951(f)(2)(ii)]

(iii) *Be connected to the power supply* by means of an isolating transformer, or other isolated power supply.[§1926.951(f)(2)(iii)]

(3) *All hydraulic tools which are used* on or around energized lines or equipment shall use nonconducting hoses having adequate strength for the normal operating pressures. It should be noted that the provisions of Sec. 1926.302(d)(2) shall also apply. [§1926.951(f)(3)]

(4) *All pneumatic tools which are used* on or around energized lines or equipment shall:[§1926.951(f)(4)]

(i) *Have nonconducting hoses* having adequate strength for the normal operating pressures, and[§1926.951(f)(4)(i)]

(ii) *Have an accumulator on the compressor to collect moisture.* [§1926.951(f)(4)(ii)]

[44 FR 8577, Feb. 9, 1979; 44 FR 20940, Apr. 6, 1979, as amended at 59 FR 40730, Aug. 9, 1994]

§1926.952
Mechanical equipment

(a) **General.**[§1926.952(a)]

(1) *Visual inspections shall be made* of the equipment to determine that it is in good condition each day the equipment is to be used.[§1926.952(a)(1)]

(2) *Tests shall be made at the beginning of each shift* during which the equipment is to be used to determine that the brakes and operating systems are in proper working condition. [§1926.952(a)(2)]

(3) *No employer shall use* any motor vehicle equipment having an obstructed view to the rear unless:[§1926.952(a)(3)]

(i) *The vehicle has a reverse signal alarm* audible above the surrounding noise level or:[§1926.952(a)(3)(i)]

(ii) *The vehicle is backed up* only when an observer signals that it is safe to do so.[§1926.952(a)(3)(ii)]

(b) **Aerial lifts.**[§1926.952(b)]

(1) *The provisions of §1926.556,* subpart N of this part, shall apply to the utilization of aerial lifts.[§1926.952(b)(1)]

(2) *When working near energized lines or equipment,* aerial lift trucks shall be grounded or barricaded and considered as energized equipment, or the aerial lift truck shall be insulated for the work being performed.[§1926.952(b)(2)]

(3) *Equipment or material* shall not be passed between a pole or structure and an aerial lift while an employee working from the basket is within reaching distance of energized conductors or equipment that are not covered with insulating protective equipment.[§1926.952(b)(3)]

(c) **Cranes and other lifting equipment.**[§1926.952(c)]

(1) *All equipment shall comply with subparts* CC and O of this part, as applicable.[§1926.952(c)(1)]

(2) *Digger derricks used* for augering holes for poles carrying electric lines, placing and removing poles, or for handling associated materials to be installed or removed from the poles must comply with 29 CFR 1910.269.[§1926.952(c)(2)]

(3) *With the exception of equipment certified* for work on the proper voltage, mechanical equipment shall not be operated closer to any energized line or equipment than the clearances set forth in Sec. 1926.950(c) unless, in addition to the requirements in Sec. 1926.1410:[§1926.952(c)(3)]

(i) *The mechanical equipment is insulated, or*[§1926.952(c)(3)(i)]

(ii) *The mechanical equipment is considered as energized.* [§1926.952(c)(3)(ii)]

Note to paragraph (c)(3): In accordance with 29 CFR 1926.1400(g), compliance with 29 CFR 1910.269(p) will be deemed compliance with §§1926.1407 through 1926.1411, including Sec. 1926.1410.

[44 FR 8577, Feb. 9, 1979 , 75 FR 48135, Aug. 9, 2010]

§1926.953
Material handling

(a) **Unloading.** Prior to unloading steel, poles, cross arms and similar material, the load shall be thoroughly examined to ascertain if the load has shifted, binders or stakes have broken or the load is otherwise hazardous to employees.[§1926.953(a)]

(b) **Pole hauling.**[§1926.953(b)]

(1) *During pole hauling operations,* all loads shall be secured to prevent displacement and a red flag shall be displayed at the trailing end of the longest pole.[§1926.953(b)(1)]

(2) *Precautions shall be exercised* to prevent blocking of roadways or endangering other traffic.[§1926.953(b)(2)]

(3) *When hauling poles* during the hours of darkness, illuminated warning devices shall be attached to the trailing end of the longest pole.[§1926.953(b)(3)]

(c) **Storage.**[§1926.953(c)]

(1) *No materials or equipment shall be stored* under energized bus, energized lines, or near energized equipment, if it is practical to store them elsewhere.[§1926.953(c)(1)]

(2) *When materials or equipment are stored* under energized lines or near energized equipment, applicable clearances shall be maintained as stated in Table V-1; and extraordinary caution shall be exercised when moving materials near such energized equipment.[§1926.953(c)(2)]

(d) **Tag line.** Where hazards to employees exist tag lines or other suitable devices shall be used to control loads being handled by hoisting equipment.[§1926.953(d)]

(e) **Oil filled equipment.** During construction or repair of oil filled equipment the oil may be stored in temporary containers other than those required in Sec. 1926.152, such as pillow tanks. [§1926.953(e)]

(f) **Framing.** During framing operations, employees shall not work under a pole or a structure suspended by a crane, A-frame or similar equipment unless the pole or structure is adequately supported.[§1926.953(f)]

(g) **Attaching the load.** The hoist rope shall not be wrapped around the load. This provision shall not apply to electric construction crews when setting or removing poles.[§1926.953(g)]

§1926.954
Grounding for protection of employees

(a) **General.** All conductors and equipment shall be treated as energized until tested or otherwise determined to be deenergized or until grounded.[§1926.954(a)]

(b) **New construction.** New lines or equipment may be considered deenergized and worked as such where:[§1926.954(b)]

(1) *The lines or equipment are grounded, or*[§1926.954(b)(1)]

(2) *The hazard of induced voltages is not present,* and adequate clearances or other means are implemented to prevent contact with energized lines or equipment and the new lines or equipment.[§1926.954(b)(2)]

(c) **Communication conductors.** Bare wire communication conductors on power poles or structures shall be treated as energized lines unless protected by insulating materials.[§1926.954(c)]

(d) **Voltage testing.** Deenergized conductors and equipment which are to be grounded shall be tested for voltage. Results of this voltage test shall determine the subsequent procedures as required in Sec. 1926.950(d).[§1926.954(d)]

(e) **Attaching grounds.**[§1926.954(e)]

(1) *When attaching grounds,* the ground end shall be attached first, and the other end shall be attached and removed by means of insulated tools or other suitable devices.[§1926.954(e)(1)]

(2) *When removing grounds,* the grounding device shall first be removed from the line or equipment using insulating tools or other suitable devices.[§1926.954(e)(2)]

(f) **Grounds shall be placed between work location** and all sources of energy and as close as practicable to the work location, or grounds shall be placed at the work location. If work is to be performed at more than one location in a line section, the line section must be grounded and short circuited at one location in the line section and the conductor to be worked on shall be grounded at each work location. The minimum distance shown in Table V-1 shall be maintained from ungrounded conductors at the work location. Where the making of a ground is impracticable, or the conditions resulting therefrom would be more hazardous than working on the lines or equipment without grounding, the grounds may be omitted and the line or equipment worked as energized. [§1926.954(f)]

(g) **Testing without grounds.** Grounds may be temporarily removed only when necessary for test purposes and extreme caution shall be exercised during the test procedures.[§1926.954(g)]

(h) **Grounding electrode.** When grounding electrodes are utilized, such electrodes shall have a resistance to ground low enough to remove the danger of harm to personnel or permit prompt operation of protective devices.[§1926.954(h)]

(i) **Grounding to tower.** Grounding to tower shall be made with a tower clamp capable of conducting the anticipated fault current. [§1926.954(i)]

(j) **Ground lead.** A ground lead, to be attached to either a tower ground or driven ground, shall be capable of conducting the anticipated fault current and shall have a minimum conductance of No. 2 AWG copper.[§1926.954(j)]

§1926.955
Overhead lines

(a) **Overhead lines.**[§1926.955(a)]

(1) *When working on or with overhead lines* the provisions of paragraphs (a) (2) through (8) of this section shall be complied with in addition to other applicable provisions of this subpart. [§1926.955(a)(1)]

(2) *Prior to climbing poles, ladders, scaffolds,* or other elevated structures, an inspection shall be made to determine that the structures are capable of sustaining the additional or unbalanced stresses to which they will be subjected.[§1926.955(a)(2)]

(3) *Where poles or structures may be unsafe for climbing,* they shall not be climbed until made safe by guying, bracing, or other adequate means.[§1926.955(a)(3)]

(4) *Before installing or removing wire or cable,* strains to which poles and structures will be subjected shall be considered and necessary action taken to prevent failure of supporting structures.[§1926.955(a)(4)]

(5) (i) *When setting, moving, or removing poles* using cranes, derricks, gin poles, A-frames, or other mechanized equipment near energized lines or equipment, precautions shall be taken to avoid contact with energized lines or equipment, except in bare-hand live-line work, or where barriers or protective devices are used.[§1926.955(a)(5)(i)]

(ii) *Equipment and machinery operating* adjacent to energized lines or equipment shall comply with Sec. 1926.952(c)(2). [§1926.955(a)(5)(ii)]

(6) (i) *Unless using suitable protective equipment* for the voltage involved, employees standing on the ground shall avoid contacting equipment or machinery working adjacent to energized lines or equipment.[§1926.955(a)(6)(i)]

(ii) *Lifting equipment shall be bonded* to an effective ground or it shall be considered energized and barricaded when utilized near energized equipment or lines.[§1926.955(a)(6)(ii)]

(7) *Pole holes shall not be left* unattended or unguarded in areas where employees are currently working.[§1926.955(a)(7)]

(8) *Tag lines shall be* of a nonconductive type when used near

(b) Metal tower construction.[§1926.955(b)]

(1) *When working in unstable material* the excavation for pad- or pile-type footings in excess of 5 feet deep shall be either sloped to the angle of repose as required in Sec. 1926.652 or shored if entry is required. Ladders shall be provided for access to pad- or pile-type footing excavations in excess of 4 feet.[§1926.955(b)(1)]

(2) *When working in unstable material* provision shall be made for cleaning out auger-type footings without requiring an employee to enter the footing unless shoring is used to protect the employee.[§1926.955(b)(2)]

(3) (i) *A designated employee shall be used* in directing mobile equipment adjacent to footing excavations.[§1926.955(b)(3)(i)]

(ii) *No one shall be permitted to remain* in the footing while equipment is being spotted for placement.[§1926.955(b)(3)(ii)]

(iii) *Where necessary to assure* the stability of mobile equipment the location of use for such equipment shall be graded and leveled.[§1926.955(b)(3)(iii)]

(4) (i) *Tower assembly shall be carried out* with a minimum exposure of employees to falling objects when working at two or more levels on a tower.[§1926.955(b)(4)(i)]

(ii) *Guy lines shall be used* as necessary to maintain sections or parts of sections in position and to reduce the possibility of tipping.[§1926.955(b)(4)(ii)]

(iii) *Members and sections being assembled* shall be adequately supported.[§1926.955(b)(4)(iii)]

(5) *When assembling and erecting towers* the provisions of paragraphs (b)(5) (i), (ii) and (iii) of this section shall be complied with:[§1926.955(b)(5)]

(i) *The construction of transmission towers* and the erecting of poles, hoisting machinery, site preparation machinery, and other types of construction machinery shall conform to the applicable requirements of this part.[§1926.955(b)(5)(i)]

(ii) *No one shall be permitted under a tower* which is in the process of erection or assembly, except as may be required to guide and secure the section being set.[§1926.955(b)(5)(ii)]

(iii) *When erecting towers using hoisting equipment* adjacent to energized transmission lines, the lines shall be deenergized when practical. If the lines are not deenergized, extraordinary caution shall be exercised to maintain the minimum clearance distances required by Sec. 1926.950(c), including Table V-1.[§1926.955(b)(5)(iii)]

(6) (i) *Erection cranes shall be set* on firm level foundations and when the cranes are so equipped outriggers shall be used. [§1926.955(b)(6)(i)]

(ii) *Tag lines shall be utilized* to maintain control of tower sections being raised and positioned, except where the use of such lines would create a greater hazard.[§1926.955(b)(6)(ii)]

(iii) *The loadline shall not be detached* from a tower section until the section is adequately secured.[§1926.955(b)(6)(iii)]

(iv) *Except during emergency restoration procedures* erection shall be discontinued in the event of high wind or other adverse weather conditions which would make the work hazardous.[§1926.955(b)(6)(iv)]

(v) *Equipment and rigging* shall be regularly inspected and maintained in safe operating condition.[§1926.955(b)(6)(v)]

(7) *Adequate traffic control* shall be maintained when crossing highways and railways with equipment as required by the provisions of Sec. 1926.200(g) (1) and (2).[§1926.955(b)(7)]

(8) *A designated employee shall be utilized* to determine that required clearance is maintained in moving equipment under or near energized lines.[§1926.955(b)(8)]

(c) Stringing or removing deenergized conductors.[§1926.955(c)]

(1) *When stringing or removing deenergized conductors,* the provisions of paragraphs (c) (2) through (12) of this section shall be complied with.[§1926.955(c)(1)]

(2) *Prior to stringing operations* a briefing shall be held setting forth the plan of operation and specifying the type of equipment to be used, grounding devices and procedures to be followed, crossover methods to be employed, and the clearance authorization required.[§1926.955(c)(2)]

(3) *Where there is a possibility of* the conductor accidentally contacting an energized circuit or receiving a dangerous induced voltage buildup, to further protect the employee from the hazards of the conductor, the conductor being installed or removed shall be grounded or provisions made to insulate or isolate the employee.[§1926.955(c)(3)]

(4) (i) *If the existing line is deenergized,* proper clearance authorization shall be secured and the line grounded on both sides of the crossover or, the line being strung or removed shall be considered and worked as energized.[§1926.955(c)(4)(i)]

(ii) *When crossing over energized conductors* in excess of 600 volts, rope nets or guard structures shall be installed unless provision is made to isolate or insulate the workman or the energized conductor. Where practical the automatic reclosing feature of the circuit interrupting device shall be made inoperative. In addition, the line being strung shall be grounded on either side of the crossover or considered and worked as energized.[§1926.955(c)(4)(ii)]

(5) *Conductors being strung in or removed* shall be kept under positive control by the use of adequate tension reels, guard structures, tielines, or other means to prevent accidental contact with energized circuits.[§1926.955(c)(5)]

(6) *Guard structure members* shall be sound and of adequate dimension and strength, and adequately supported. [§1926.955(c)(6)]

(7) (i) *Catch-off anchors,* rigging, and hoists shall be of ample capacity to prevent loss of the lines.[§1926.955(c)(7)(i)]

(ii) *The manufacturer's load rating* shall not be exceeded for stringing lines, pulling lines, sock connections, and all load-bearing hardware and accessories.[§1926.955(c)(7)(ii)]

(iii) *Pulling lines and accessories* shall be inspected regularly and replaced or repaired when damaged or when dependability is doubtful. The provisions of Sec. 1926.251(c)(4)(ii) (concerning splices) shall not apply.[§1926.955(c)(7)(iii)]

(8) *Conductor grips shall not be used* on wire rope unless designed for this application.[§1926.955(c)(8)]

(9) *While the conductor or pulling line* is being pulled (in motion) employees shall not be permitted directly under overhead operations, nor shall any employee be permitted on the crossarm. [§1926.955(c)(9)]

(10) *A transmission clipping crew* shall have a minimum of two structures clipped in between the crew and the conductor being sagged. When working on bare conductors, clipping and tying crews shall work between grounds at all times. The grounds shall remain intact until the conductors are clipped in, except on dead end structures.[§1926.955(c)(10)]

(11) (i) *Except during emergency restoration procedures,* work from structures shall be discontinued when adverse weather (such as high wind or ice on structures) makes the work hazardous.[§1926.955(c)(11)(i)]

(ii) *Stringing and clipping operations* shall be discontinued during the progress of an electrical storm in the immediate vicinity.[§1926.955(c)(11)(ii)]

(12) (i) *Reel handling equipment,* including pulling and braking machines, shall have ample capacity, operate smoothly, and be leveled and aligned in accordance with the manufacturer's operating instructions.[§1926.955(c)(12)(i)]

(ii) *Reliable communications* between the reel tender and pulling rig operator shall be provided.[§1926.955(c)(12)(ii)]

(iii) *Each pull shall be snubbed* or dead ended at both ends before subsequent pulls.[§1926.955(c)(12)(iii)]

(d) Stringing adjacent to energized lines.[§1926.955(d)]

(1) *Prior to stringing parallel to an existing energized transmission line* a competent determination shall be made to ascertain whether dangerous induced voltage buildups will occur, particularly during switching and ground fault conditions. When there is a possibility that such dangerous induced voltage may exist the employer shall comply with the provisions of paragraphs (d) (2) through (9) of this section in addition to the provisions of paragraph (c) of this Sec. 1926.955, unless the line is worked as energized.[§1926.955(d)(1)]

(2) *When stringing adjacent to energized lines* the tension stringing method or other methods which preclude unintentional contact between the lines being pulled and any employee shall be used.[§1926.955(d)(2)]

(3) *All pulling and tensioning equipment* shall be isolated, insulated, or effectively grounded.[§1926.955(d)(3)]

(4) *A ground shall be installed* between the tensioning reel setup and the first structure in order to ground each bare conductor, subconductor, and overhead ground conductor during stringing operations.[§1926.955(d)(4)]

(5) *During stringing operations,* each bare conductor, subconductor, and overhead ground conductor shall be grounded at the

first tower adjacent to both the tensioning and pulling setup and in increments so that no point is more than 2 miles from a ground.[§1926.955(d)(5)]

(i) *The grounds shall be left in place* until conductor installation is completed.[§1926.955(d)(5)(i)]

(ii) *Such grounds shall be removed* as the last phase of aerial cleanup.[§1926.955(d)(5)(ii)]

(iii) *Except for moving type grounds,* the grounds shall be placed and removed with a hot stick.[§1926.955(d)(5)(iii)]

(6) *Conductors, subconductors,* and overhead ground conductors shall be grounded at all dead-end or catch-off points. [§1926.955(d)(6)]

(7) *A ground shall be located at each side* and within 10 feet of working areas where conductors, subconductors, or overhead ground conductors are being spliced at ground level. The two ends to be spliced shall be bonded to each other. It is recommended that splicing be carried out on either an insulated platform or on a conductive metallic grounding mat bonded to both grounds. When a grounding mat is used, it is recommended that the grounding mat be roped off and an insulated walkway provided for access to the mat.[§1926.955(d)(7)]

(8) (i) *All conductors, subconductors,* and overhead ground conductors shall be bonded to the tower at any isolated tower where it may be necessary to complete work on the transmission line.[§1926.955(d)(8)(i)]

(ii) *Work on dead-end towers* shall require grounding on all deenergized lines.[§1926.955(d)(8)(ii)]

(iii) *Grounds may be removed* as soon as the work is completed: Provided, That the line is not left open circuited at the isolated tower at which work is being completed. [§1926.955(d)(8)(iii)]

(9) *When performing work from the structures,* clipping crews and all others working on conductors, subconductors, or overhead ground conductors shall be protected by individual grounds installed at every work location.[§1926.955(d)(9)]

(e) **Live-line bare-hand work.** In addition to any other applicable standards contained elsewhere in this subpart all live-line bare-hand work shall be performed in accordance with the following requirements:[§1926.955(e)]

(1) *Employees shall be instructed and trained* in the live-line bare-hand technique and the safety requirements pertinent thereto before being permitted to use the technique on energized circuits.[§1926.955(e)(1)]

(2) *Before using the live-line bare-hand technique* on energized high-voltage conductors or parts, a check shall be made of: [§1926.955(e)(2)]

(i) *The voltage rating of the circuit* on which the work is to be performed;[§1926.955(e)(2)(i)]

(ii) *The clearances to ground of lines* and other energized parts on which work is to be performed; and[§1926.955(e)(2)(ii)]

(iii) *The voltage limitations* of the aerial-lift equipment intended to be used.[§1926.955(e)(2)(iii)]

(3) *Only equipment designed,* tested, and intended for live-line bare-hand work shall be used.[§1926.955(e)(3)]

(4) *All work shall be personally supervised* by a person trained and qualified to perform live-line bare-hand work.[§1926.955(e)(4)]

(5) *The automatic reclosing feature* of circuit interrupting devices shall be made inoperative where practical before working on any energized line or equipment.[§1926.955(e)(5)]

(6) *Work shall not be performed* during the progress of an electrical storm in the immediate vicinity.[§1926.955(e)(6)]

(7) *A conductive bucket liner* or other suitable conductive device shall be provided for bonding the insulated aerial device to the energized line or equipment.[§1926.955(e)(7)]

(i) *The employee shall be connected* to the bucket liner by use of conductive shoes, leg clips, or other suitable means. [§1926.955(e)(7)(i)]

(ii) *Where necessary,* adequate electrostatic shielding for the voltage being worked or conductive clothing shall be provided.[§1926.955(e)(7)(ii)]

(8) *Only tools and equipment intended* for live-line bare-hand work shall be used, and such tools and equipment shall be kept clean and dry.[§1926.955(e)(8)]

(9) *Before the boom is elevated,* the outriggers on the aerial truck shall be extended and adjusted to stabilize the truck and the body of the truck shall be bonded to an effective ground, or barricaded and considered as energized equipment.[§1926.955(e)(9)]

(10) *Before moving the aerial lift* into the work position, all controls (ground level and bucket) shall be checked and tested to determine that they are in proper working condition.[§1926.955(e)(10)]

(11) *Arm current tests* shall be made before starting work each day, each time during the day when higher voltage is going to be worked and when changed conditions indicate a need for additional tests. Aerial buckets used for bare-hand live-line work shall be subjected to an arm current test. This test shall consist of placing the bucket in contact with an energized source equal to the voltage to be worked upon for a minimum time of three (3) minutes. the leakage current shall not exceed 1 microampere per kilo-volt of nominal line-to-line voltage. Work operations shall be suspended immediately upon any indication of a malfunction in the equipment.[§1926.955(e)(11)]

(12) *All aerial lifts to be used* for live-line bare-hand work shall have dual controls (lower and upper) as required by paragraph (e)(12) (i) and (ii) of this section.[§1926.955(e)(12)]

(i) *The upper controls shall be* within easy reach of the employee in the basket. If a two basket type lift is used access to the controls shall be within easy reach from either basket.[§1926.955(e)(12)(i)]

(ii) *The lower set of controls* shall be located near base of the boom that will permit over-ride operation of equipment at any time.[§1926.955(e)(12)(ii)]

(13) *Ground level lift control* shall not be operated unless permission has been obtained from the employee in lift, except in case of emergency.[§1926.955(e)(13)]

(14) *Before the employee contacts* the energized part to be worked on, the conductive bucket liner shall be bonded to the energized conductor by means of a positive connection which shall remain attached to the energized conductor until the work on the energized circuit is completed.[§1926.955(e)(14)]

(15) *The minimum clearance distances for live-line bare-hand work* shall be as specified in Table V-2. These minimum clearance distances shall be maintained from all grounded objects and from lines and equipment at a different potential than that to which the insulated aerial device is bonded unless such grounded objects or other lines and equipment are covered by insulated guards. These distances shall be maintained when approaching, leaving, and when bonded to the energized circuit.[§1926.955(e)(15)]

Table V-2 — Minimum Clearance Distances for Live-Line Bare-Hand Work (Alternating Current)

Voltage range (phase-to-phase) kilovolts	Distance in feet and inches for maximum voltage	
	Phase to ground	Phase to phase
2.1 to 15	2'0"	2'0"
15.1 to 35	2'4"	2'4"
35.1 to 46	2'6"	2'6"
46.1 to 72.5	3'0"	3'0"
72.6 to 121	3'4"	4'6"
138 to 145	3'6"	5'0"
161 to 169	3'8"	5'6"
230 to 242	5'0"	8'4"
345 to 362	[1]7'0"	[1]13'4"
500 to 552	[1]11'0"	[1]20'0"
700 to 765	[1]15'0"	[1]31'0"

[1] For 345-362kv., 500-552kv., and 700-765kv., the minimum clearance distance may be reduced provided the distances are not made less than the shortest distance between the energized part and a grounded surface.

(16) *When approaching, leaving, or bonding* to an energized circuit the minimum distances in Table V-2 shall be maintained between all parts of the insulated boom assembly and any grounded parts (including the lower arm or portions of the truck).[§1926.955(e)(16)]

(17) *When positioning the bucket alongside* an energized bushing or insulator string, the minimum line-to-ground clearances of Table V-2 must be maintained between all parts of the bucket and the grounded end of the bushing or insulator string. [§1926.955(e)(17)]

(18) (i) *The use of handlines* between buckets, booms, and the ground is prohibited.[§1926.955(e)(18)(i)]

(ii) *No conductive materials* over 36 inches long shall be placed in the bucket, except for appropriate length jumpers, armor rods, and tools.[§1926.955(e)(18)(ii)]

(iii) *Nonconductive-type handlines may be used* from line to ground when not supported from the bucket. [§1926.955(e)(18)(iii)]

(19) *The bucket and upper insulated boom* shall not be overstressed by attempting to lift or support weights in excess of the manufacturer's rating.[§1926.955(e)(19)]

(20) (i) *A minimum clearance table* (as shown in table V-2) shall be printed on a plate of durable nonconductive material, and mounted in the buckets or its vicinity so as to be visible to the operator of the boom.[§1926.955(e)(20)(i)]

(ii) *It is recommended that* insulated measuring sticks be used to verify clearance distances.[§1926.955(e)(20)(ii)]

§1926.956

Underground lines

(a) **Guarding and ventilating street opening** used for access to underground lines or equipment.[§1926.956(a)]

(1) *Appropriate warning signs* shall be promptly placed when covers of manholes, handholes, or vaults are removed. What is an appropriate warning sign is dependent upon the nature and location of the hazards involved.[§1926.956(a)(1)]

(2) *Before an employee enters a street opening,* such as a manhole or an unvented vault, it shall be promptly protected with a barrier, temporary cover, or other suitable guard.[§1926.956(a)(2)]

(3) *When work is to be performed in a manhole or unvented vault:* [§1926.956(a)(3)]

(i) *No entry shall be permitted unless* forced ventilation is provided or the atmosphere is found to be safe by testing for oxygen deficiency and the presence of explosive gases or fumes;[§1926.956(a)(3)(i)]

(ii) *Where unsafe conditions are detected,* by testing or other means, the work area shall be ventilated and otherwise made safe before entry;[§1926.956(a)(3)(ii)]

(iii) *Provisions shall be made* for an adequate continuous supply of air.[§1926.956(a)(3)(iii)]

(b) **Work in manholes.**[§1926.956(b)]

(1) *While work is being performed in manholes,* an employee shall be available in the immediate vicinity to render emergency assistance as may be required. This shall not preclude the employee in the immediate vicinity from occasionally entering a manhole to provide assistance, other than emergency. This requirement does not preclude a qualified employee, working alone, from entering for brief periods of time, a manhole where energized cables or equipment are in service, for the purpose of inspection, housekeeping, taking readings, or similar work if such work can be performed safely.[§1926.956(b)(1)]

(2) *When open flames must be used* or smoking is permitted in manholes, extra precautions shall be taken to provide adequate ventilation.[§1926.956(b)(2)]

(3) *Before using open flames* in a manhole or excavation in an area where combustible gases or liquids may be present, such as near a gasoline service station, the atmosphere of the manhole or excavation shall be tested and found safe or cleared of the combustible gases or liquids.[§1926.956(b)(3)]

(c) **Trenching and excavating.**[§1926.956(c)]

(1) *During excavation or trenching,* in order to prevent the exposure of employees to the hazards created by damage to dangerous underground facilities, efforts shall be made to determine the location of such facilities and work conducted in a manner designed to avoid damage.[§1926.956(c)(1)]

(2) *Trenching and excavation operations* shall comply with §§1926.651 and 1926.652.[§1926.956(c)(2)]

(3) *When underground facilities are exposed* (electric, gas, water, telephone, etc.) they shall be protected as necessary to avoid damage.[§1926.956(c)(3)]

(4) *Where multiple cables exist in an excavation,* cables other than the one being worked on shall be protected as necessary. [§1926.956(c)(4)]

(5) *When multiple cables exist in an excavation,* the cable to be worked on shall be identified by electrical means unless its identity is obvious by reason of distinctive appearance. [§1926.956(c)(5)]

(6) *Before cutting into a cable or opening a splice,* the cable shall be identified and verified to be the proper cable.[§1926.956(c)(6)]

(7) *When working on buried cable* or on cable in manholes, metallic sheath continuity shall be maintained by bonding across the opening or by equivalent means.[§1926.956(c)(7)]

§1926.957

Construction in energized substations

(a) **Work near energized equipment facilities.**[§1926.957(a)]

(1) *When construction work is performed* in an energized substation, authorization shall be obtained from the designated, authorized person before work is started.[§1926.957(a)(1)]

(2) *When work is to be done* in an energized substation, the following shall be determined:[§1926.957(a)(2)]

(i) *What facilities are energized, and*[§1926.957(a)(2)(i)]

(ii) *What protective equipment and precautions* are necessary for the safety of personnel.[§1926.957(a)(2)(ii)]

(3) *Extraordinary caution shall be exercised* in the handling of busbars, tower steel, materials, and equipment in the vicinity of energized facilities. The requirements set forth in Sec. 1926.950(c), shall be complied with.[§1926.957(a)(3)]

(b) **Deenergized equipment or lines.** When it is necessary to deenergize equipment or lines for protection of employees, the requirements of Sec. 1926.950(d) shall be complied with.[§1926.957(b)]

(c) **Barricades and barriers.**[§1926.957(c)]

(1) *Barricades or barriers shall be installed* to prevent accidental contact with energized lines or equipment.[§1926.957(c)(1)]

(2) *Where appropriate, signs indicating* the hazard shall be posted near the barricade or barrier. These signs shall comply with Sec. 1926.200.[§1926.957(c)(2)]

(d) **Control panels.**[§1926.957(d)]

(1) *Work on or adjacent to energized control* panels shall be performed by designated employees.[§1926.957(d)(1)]

(2) *Precaution shall be taken to prevent* accidental operation of relays or other protective devices due to jarring, vibration, or improper wiring.[§1926.957(d)(2)]

(e) **Mechanized equipment.**[§1926.957(e)]

(1) *Use of vehicles, gin poles, cranes,* and other equipment in restricted or hazardous areas shall at all times be controlled by designated employees.[§1926.957(e)(1)]

(2) *All mobile cranes and derricks* shall be effectively grounded when being moved or operated in close proximity to energized lines or equipment, or the equipment shall be considered energized.[§1926.957(e)(2)]

(3) *Fenders shall not be required* for lowboys used for transporting large electrical equipment, transformers, or breakers. [§1926.957(e)(3)]

(f) **Storage.** The storage requirements of Sec. 1926.953(c) shall be complied with.[§1926.957(f)]

(g) **Substation fences.**[§1926.957(g)]

(1) *When a substation fence must be expanded* or removed for construction purposes, a temporary fence affording similar protection when the site is unattended, shall be provided. Adequate interconnection with ground shall be maintained between temporary fence and permanent fence.[§1926.957(g)(1)]

(2) *All gates to all unattended substations* shall be locked, except when work is in progress.[§1926.957(g)(2)]

(h) **Footing excavation.**[§1926.957(h)]

(1) *Excavation for auger, pad and piling type footings* for structures and towers shall require the same precautions as for metal tower construction (see Sec. 1926.955(b)(1)).[§1926.957(h)(1)]

(2) *No employee shall be permitted to enter* an unsupported auger- type excavation in unstable material for any purpose. Necessary clean- out in such cases shall be accomplished without entry.[§1926.957(h)(2)]

§1926.958

External load helicopters

In all operations performed using a rotorcraft for moving or placing external loads, the provisions of Sec. 1926.551 of subpart N of this part shall be complied with.

§1926.959

Lineman's body belts, safety straps, and lanyards

(a) **General requirements.** The requirements of paragraphs (a) and (b) of this section shall be complied with for all lineman's body belts, safety straps and lanyards acquired for use after the effective date of this subpart.[§1926.959(a)]

(1) *Hardware for lineman's body belts, safety straps,* and lanyards shall be drop forged or pressed steel and have a corrosive resistive finish tested to American Society for Testing and Materials B117-64 (50- hour test). Surfaces shall be smooth and free of sharp edges.[§1926.959(a)(1)]

(2) *All buckles shall withstand* a 2,000-pound tensile test with a maximum permanent deformation no greater than one sixty-fourth inch.[§1926.959(a)(2)]

(3) *D rings shall withstand* a 5,000-pound tensile test without failure. Failure of a D ring shall be considered cracking or breaking.[§1926.959(a)(3)]

(4) *Snaphooks shall withstand* a 5,000-pound tensile test without failure. Failure of a snaphook shall be distortion sufficient to release the keeper.[§1926.959(a)(4)]

(b) **Specific requirements.**[§1926.959(b)]

(1) (i) *All fabric used for safety straps* shall withstand an A.C. dielectric test of not less than 25,000 volts per foot "dry" for 3 minutes, without visible deterioration.[§1926.959(b)(1)(i)]

(ii) *All fabric and leather used* shall be tested for leakage current and shall not exceed 1 milliampere when a potention of 3,000 volts is applied to the electrodes positioned 12 inches apart. [§1926.959(b)(1)(ii)]

(iii) *Direct current tests* may be permitted in lieu of alternating current tests.[§1926.959(b)(1)(iii)]

(2) *The cushion part of the body belt shall:*[§1926.959(b)(2)]

(i) *Contain no exposed rivets on the inside;*[§1926.959(b)(2)(i)]

(ii) *Be at least three (3) inches in width;*[§1926.959(b)(2)(ii)]

(iii) *Be at least five thirty-seconds* (5/32) inch thick, if made of leather; and[§1926.959(b)(2)(iii)]

(iv) *Have pocket tabs that extended* at least 1 1/2 inches down and three (3) inches back of the inside of circle of each D ring for riveting on plier or tool pockets. On shifting D belts, this measurement for pocket tabs shall be taken when the D ring section is centered.[§1926.959(b)(2)(iv)]

(3) *A maximum of four (4) tool loops* shall be so situated on the body belt that four (4) inches of the body belt in the center of the back, measuring from D ring to D ring, shall be free of tool loops, and any other attachments.[§1926.959(b)(3)]

(4) *Suitable copper, steel, or equivalent liners* shall be used around bar of D rings to prevent wear between these members and the leather or fabric enclosing them.[§1926.959(b)(4)]

(5) *All stitching shall be of a minimum* 42-pound weight nylon or equivalent thread and shall be lock stitched. Stitching parallel to an edge shall not be less than three-sixteenths (3/16) inch from edge of narrowest member caught by the thread. The use of cross stitching on leather is prohibited.[§1926.959(b)(5)]

(6) *The keeper of snaphooks shall have* a spring tension that will not allow the keeper to begin to open with a weight of 2 1/2 pounds or less, but the keeper of snaphooks shall begin to open with a weight of four (4) pounds, when the weight is supported on the keeper against the end of the nose.[§1926.959(b)(6)]

(7) *Testing of lineman's safety straps,* body belts and lanyards shall be in accordance with the following procedure: [§1926.959(b)(7)]

(i) *Attach one end of the safety strap* or lanyard to a rigid support, the other end shall be attached to a 250-pound canvas bag of sand:[§1926.959(b)(7)(i)]

(ii) *Allow the 250-pound canvas bag* of sand to free fall 4 feet for (safety strap test) and 6 feet for (lanyard test); in each case stopping the fall of the 250-pound bag:[§1926.959(b)(7)(ii)]

(iii) *Failure of the strap or lanyard shall be indicated* by any breakage, or slippage sufficient to permit the bag to fall free of the strap or lanyard. The entire "body belt assembly" shall be tested using one D ring. A safety strap or lanyard shall be used that is capable of passing the "impact loading test" and attached as required in paragraph (b)(7)(i) of this section. The body belt shall be secured to the 250-pound bag of sand at a point to simulate the waist of a man and allowed to drop as stated in paragraph (b)(7)(ii) of this section. Failure of the body belt shall be indicated by any breakage, or slippage sufficient to permit the bag to fall free of the body belt. [§1926.959(b)(7)(iii)]

§1926.960

Definitions applicable to this subpart

(a) **Alive or live (energized).** The term means electrically connected to a source of potential difference, or electrically charged so as to have a potential significantly different from that of the earth in the vicinity. The term "live" is sometimes used in place of the term "current-carrying," where the intent is clear, to avoid repetition of the longer term.[§1926.960(a)]

(b) **Automatic circuit recloser.** The term means a self-controlled device for automatically interrupting and reclosing an alternating current circuit with a predetermined sequence of opening and reclosing followed by resetting, hold closed, or lockout operation. [§1926.960(b)]

(c) **Barrier.** The term means a physical obstruction which is intended to prevent contact with energized lines or equipment.[§1926.960(c)]

(d) **Barricade.** The term means a physical obstruction such as tapes, screens, or cones intended to warn and limit access to a hazardous area.[§1926.960(d)]

(e) **Bond.** The term means an electrical connection from one conductive element to another for the purpose of minimizing potential differences or providing suitable conductivity for fault current or for mitigation of leakage current and electrolytic action.[§1926.960(e)]

(f) **Bushing.** The term means an insulating structure including a through conductor, or providing a passageway for such a conductor, with provision for mounting on a barrier, conducting or otherwise, for the purpose of insulating the conductor from the barrier and conducting current from one side of the barrier to the other. [§1926.960(f)]

(g) **Cable.** The term means a conductor with insulation, or a stranded conductor with or without insulation and other coverings (single-conductor cable) or a combination of conductors insulated from one another (multiple-conductor cable).[§1926.960(g)]

(h) **Cable sheath.** The term means a protective covering applied to cables.[§1926.960(h)]

Note: A cable sheath may consist of multiple layers of which one or more is conductive.

(i) **Circuit.** The term means a conductor or system of conductors through which an electric current is intended to flow.[§1926.960(i)]

(j) **Communication lines.** The term means the conductors and their supporting or containing structures which are used for public or private signal or communication service, and which operate at potentials not exceeding 400 volts to ground or 750 volts between any two points of the circuit, and the transmitted power of which does not exceed 150 watts. When operating at less than 150 volts no limit is placed on the capacity of the system.[§1926.960(j)]

Note: Telephone, telegraph, railroad signal, data, clock, fire, police-alarm, community television antenna, and other systems conforming with the above are included. Lines used for signaling purposes, but not included under the above definition, are considered as supply lines of the same voltage and are to be so run.

(k) **Conductor.** The term means a material, usually in the form of a wire, cable, or bus bar suitable for carrying an electric current. [§1926.960(k)]

(l) **Conductor shielding.** The term means an envelope which encloses the conductor of a cable and provides an equipotential surface in contact with the cable insulation.[§1926.960(l)]

(m) **Current-carrying part.** The term means a conducting part intended to be connected in an electric circuit to a source of voltage. Non- current-carrying parts are those not intended to be so connected.[§1926.960(m)]

(n) **Dead (deenergized).** The term means free from any electrical connection to a source of potential difference and from electrical charges: Not having a potential difference from that of earth. [§1926.960(n)]

Note: The term is used only with reference to current-carrying parts which are sometimes alive (energized).

(o) **Designated employee.** The term means a qualified person delegated to perform specific duties under the conditions existing. [§1926.960(o)]

(p) **Effectively grounded.** The term means intentionally connected to earth through a ground connection or connections of sufficiently low impedance and having sufficient current-carrying capacity to prevent the buildup of voltages which may result in undue hazard to connected equipment or to persons.[§1926.960(p)]

(q) **Electric line trucks.** The term means a truck used to transport men, tools, and material, and to serve as a traveling workshop for electric power line construction and maintenance work. It is sometimes equipped with a boom and auxiliary equipment for setting poles, digging holes, and elevating material or men.[§1926.960(q)]

(r) **Enclosed.** The term means surrounded by a case, cage, or fence, which will protect the contained equipment and prevent accidental contact of a person with live parts.[§1926.960(r)]

(s) **Equipment.** This is a general term which includes fittings, devices, appliances, fixtures, apparatus, and the like, used as part of, or in connection with, an electrical power transmission and distribution system, or communication systems.[§1926.960(s)]

(t) **Exposed.** The term means not isolated or guarded.[§1926.960(t)]

(u) **Electric supply lines.** The term means those conductors used to transmit electric energy and their necessary supporting or containing structures. Signal lines of more than 400 volts to ground are always supply lines within the meaning of the rules, and those of less than 400 volts to ground may be considered as supply lines, if so run and operated throughout.[§1926.960(u)]

(v) **Guarded.** The term means protected by personnel, covered, fenced, or enclosed by means of suitable casings, barrier rails, screens, mats, platforms, or other suitable devices in accordance with standard barricading techniques designed to prevent dangerous approach or contact by persons or objects.[§1926.960(v)]

Note: Wires, which are insulated but not otherwise protected, are not considered as guarded.

(w) **Ground (Reference).** The term means that conductive body, usually earth, to which an electric potential is referenced.[§1926.960(w)]

(x) **Ground (as a noun).** The term means a conductive connection whether intentional or accidental, by which an electric circuit or equipment is connected to reference ground.[§1926.960(x)]

(y) **Ground (as a verb).** The term means the connecting or establishment of a connection, whether by intention or accident of an electric circuit or equipment to reference ground.[§1926.960(y)]

(z) **Grounding electrode (ground electrode).** The term grounding electrode means a conductor embedded in the earth, used for maintaining ground potential on conductors connected to it, and for dissipating into the earth current conducted to it.[§1926.960(z)]

(aa) **Grounding electrode resistance.** The term means the resistance of the grounding electrode to earth.[§1926.960(aa)]

(bb) **Grounding electrode conductor (grounding conductor).** The term means a conductor used to connect equipment or the grounded circuit of a wiring system to a grounding electrode. [§1926.960(bb)]

(cc) **Grounded conductor.** The term means a system or circuit conductor which is intentionally grounded.[§1926.960(cc)]

(dd) **Grounded system.** The term means a system of conductors in which at least one conductor or point (usually the middle wire, or neutral point of transformer or generator windings) is intentionally grounded, either solidly or through a current-limiting device (not a current- interrupting device).[§1926.960(dd)]

(ee) **Hotline tools and ropes.** The term means those tools and ropes which are especially designed for work on energized high voltage lines and equipment. Insulated aerial equipment especially designed for work on energized high voltage lines and equipment shall be considered hot line.[§1926.960(ee)]

(ff) **Insulated.** The term means separated from other conducting surfaces by a dielectric substance (including air space) offering a high resistance to the passage of current.[§1926.960(ff)]

Note: When any object is said to be insulated, it is understood to be insulated in suitable manner for the conditions to which it is subjected. Otherwise, it is within the purpose of this subpart, uninsulated. Insulating covering of conductors is one means of making the conductor insulated.

(gg) **Insulation (as applied to cable).** The term means that which is relied upon to insulate the conductor from other conductors or conducting parts or from ground.[§1926.960(gg)]

(hh) **Insulation shielding.** The term means an envelope which encloses the insulation of a cable and provides an equipotential surface in contact with cable insulation.[§1926.960(hh)]

(ii) **Isolated.** The term means an object that is not readily accessible to persons unless special means of access are used.[§1926.960(ii)]

(jj) **Manhole.** The term means a subsurface enclosure which personnel may enter and which is used for the purpose of installing, operating, and maintaining equipment and/or cable.[§1926.960(jj)]

(kk) **Pulling tension.** The term means the longitudinal force exerted on a cable during installation.[§1926.960(kk)]

(ll) **Qualified person.** The term means a person who by reason of experience or training is familiar with the operation to be performed and the hazards involved.[§1926.960(ll)]

(mm) **Switch.** The term means a device for opening and closing or changing the connection of a circuit. In these rules, a switch is understood to be manually operable, unless otherwise stated. [§1926.960(mm)]

(nn) **Tag.** The term means a system or method of identifying circuits, systems or equipment for the purpose of alerting persons that the circuit, system or equipment is being worked on.[§1926.960(nn)]

(oo) **Unstable material.** The term means earth material, other than running, that because of its nature or the influence of related conditions, cannot be depended upon to remain in place without extra support, such as would be furnished by a system of shoring. [§1926.960(oo)]

(pp) **Vault.** The term means an enclosure above or below ground which personnel may enter and is used for the purpose of installing, operating, and/or maintaining equipment and/or cable. [§1926.960(pp)]

(qq) **Voltage.** The term means the effective (rms) potential difference between any two conductors or between a conductor and ground. Voltages are expressed in nominal values. The nominal voltage of a system or circuit is the value assigned to a system or circuit of a given voltage class for the purpose of convenient designation. The operating voltage of the system may vary above or below this value.[§1926.960(qq)]

(rr) **Voltage of an effectively grounded circuit.** The term means the voltage between any conductor and ground unless otherwise indicated.[§1926.960(rr)]

(ss) **Voltage of a circuit not effectively grounded.** The term means the voltage between any two conductors. If one circuit is directly connected to and supplied from another circuit of higher voltage (as in the case of an autotransformer), both are considered as of the higher voltage, unless the circuit of lower voltage is effectively grounded, in which case its voltage is not determined by the circuit of higher voltage. Direct connection implies electric connection as distinguished from connection merely through electromagnetic or electrostatic induction.[§1926.960(ss)]

Authority: Section 3704 of the Contract Work Hours and Safety Standards Act (40 U.S.C. 3701); Secs. 4, 6, and 8 of the Occupational Safety and Health Act of 1970 (29 U.S.C. 653, 655, 657); Secretary of Labor's Order Nos. 12-71 (36 FR 8754); 8-76 (41 FR 25059); 9-83 (48 FR 35736, 1-90 (55 FR 9033), and 5-2007 (72 FR 31159). Section 1926.951 also issued under 29 CFR part 1911.

Subpart W - Rollover Protective Structures; Overhead Protection

§1926.1000
Rollover protective structures (ROPS) for material handling equipment

(a) **Coverage.** [§1926.1000(a)]

(1) *This section applies* to the following types of material handling equipment: To all rubber-tired, self-propelled scrapers, rubber-tired front-end loaders, rubber-tired dozers, wheel-type agricultural and industrial tractors, crawler tractors, crawler-type loaders, and motor graders, with or without attachments, that are used in construction work. This requirement does not apply to sideboom pipelaying tractors. [§1926.1000(a)(1)]

(2) *The promulgation of specific standards* for rollover protective structures for compactors and rubber-tired skid-steer equipment is reserved pending consideration of standards currently being developed. [§1926.1000(a)(2)]

(b) **Equipment manufactured on or after September 1, 1972.** Material handling machinery described in paragraph (a) of this section and manufactured on or after September 1, 1972, shall be equipped with rollover protective structures which meet the minimum performance standards prescribed in §1926.1001 and 1926.1002, as applicable. [§1926.1000(b)]

(c) **Equipment manufactured before September 1, 1972.** [§1926.1000(c)]

(1) *All material handling equipment* described in paragraph (a) of this section and manufactured or placed in service (owned or operated by the employer) prior to September 1, 1972, shall be fitted with rollover protective structures no later than the dates listed below: [§1926.1000(c)(1)]

(i) *Machines manufactured on or after* January 1, 1972, shall be fitted no later than April 1, 1973. [§1926.1000(c)(1)(i)]

(ii) *Machines manufactured between* July 1, 1971, and December 31, 1971, shall be fitted no later than July 1, 1973. [§1926.1000(c)(1)(ii)]

(iii) *Machines manufactured between* July 1, 1970, and June 30, 1971, shall be fitted no later than January 1, 1974. [§1926.1000(c)(1)(iii)]

(iv) *Machines manufactured between* July 1, 1969, and June 30, 1970, shall be fitted no later than July 1, 1974. [§1926.1000(c)(1)(iv)]

(v) *Machines manufactured before* July 1, 1969: Reserved pending further study, development, and review. [§1926.1000(c)(1)(v)]

(2) *Rollover protective structures* and supporting attachment shall meet the minimum performance criteria detailed in §1926.1001and 1926.1002, as applicable or shall be designed, fabricated, and installed in a manner which will support, based on the ultimate strength of the metal, at least two times the weight of the prime mover applied at the point of impact. [§1926.1000(c)(2)]

(i) *The design objective shall be to minimize* the likelihood of a complete overturn and thereby minimize the possibility of the operator being crushed as a result of a rollover or upset. [§1926.1000(c)(2)(i)]

(ii) *The design shall provide a vertical clearance* of at least 52 inches from the work deck to the ROPS at the point of ingress or egress. [§1926.1000(c)(2)(ii)]

(d) **Remounting.** ROPS removed for any reason, shall be remounted with equal quality, or better, bolts or welding as required for the original mounting. [§1926.1000(d)]

(e) **Labeling.** Each ROPS shall have the following information permanently affixed to the structure: [§1926.1000(e)]

(1) *Manufacturer or fabricator's name and address;* [§1926.1000(e)(1)]

(2) *ROPS model number, if any;* [§1926.1000(e)(2)]

(3) *Machine make, model, or series number* that the structure is designed to fit. [§1926.1000(e)(3)]

(f) **Machines meeting certain existing** governmental requirements. Any machine in use, equipped with rollover protective structures, shall be deemed in compliance with this section if it meets the rollover protective structure requirements of the State of California, the U.S. Army Corps of Engineers, or the Bureau of Reclamation of the U.S. Department of the Interior in effect on April 5, 1972. The requirements in effect are: [§1926.1000(f)]

(1) *State of California:* Construction Safety Orders, issued by the Department of Industrial Relations pursuant to Division 5, Labor Code, §6312, State of California. [§1926.1000(f)(1)]

(2) *U.S. Army Corps of Engineers:* General Safety Requirements, EM-385-1-1 (March 1967). [§1926.1000(f)(2)]

(3) *Bureau of Reclamation, U.S. Department of the Interior:* Safety and Health Regulations for Construction. Part II (September 1971). [§1926.1000(f)(3)]

§1926.1001
Minimum performance criteria for rollover protective structures for designated scrapers, loaders, dozers, graders, and crawler tractors

(a) **General.** This section prescribes minimum performance criteria for rollover protective structures (ROPS) for rubber-tired self-propelled scrapers; rubber-tired front-end loaders and rubber-tired dozers; crawler tractors, and crawler-type loaders, and motor graders. The vehicle and ROPS as a system shall have the structural characteristics prescribed in paragraph (f) of this section for each type of machine described in this paragraph. [§1926.1001(a)]

(b) **The static laboratory test prescribed herein** will determine the adequacy of the structures used to protect the operator under the following conditions: [§1926.1001(b)]

(1) *For rubber-tired self-propelled scrapers,* rubber-tired front-end loaders, and rubber-tired dozers: Operating between 0 and 10 miles per hour over hard clay where rollover would be limited to a maximum roll angle of 360° down a slope of 30° maximum. [§1926.1001(b)(1)]

(2) *For motor graders:* Operating between 0 and 10 miles per hour over hard clay where rollover would be limited to 360° down a slope of 30° maximum. [§1926.1001(b)(2)]

(3) *For crawler tractors and crawler-type loaders:* Operating between 0 and 10 miles per hour over hard clay where rollover would be limited to a maximum roll angle of 360° down a slope of 45°. [§1926.1001(b)(3)]

(c) **Facilities and apparatus.** [§1926.1001(c)]

(1) *The following material is necessary:* [§1926.1001(c)(1)]

(i) *Material, equipment, and tiedown means* adequate to insure that the ROPS and its vehicle frame absorb the applied energy. [§1926.1001(c)(1)(i)]

(ii) *Equipment necessary to measure* and apply loads to the ROPS. Adequate means to measure deflections and lengths should also be provided. [§1926.1001(c)(1)(ii)]

(iii) *Recommended, but not mandatory,* types of test setups are illustrated in Figure W-1 for all types of equipment to which this section applies; and in Figure W-2 for rubber-tired self-propelled scrapers; Figure W-3 for rubber-tired front-end loaders, rubber-tired dozers, and motor graders; and Figure W-4 for crawler tractors and crawler-type loaders. [§1926.1001(c)(1)(iii)]

(2) *Table W-1 contains a listing* of the required apparatus for all types of equipment described in paragraph (a) of this section. [§1926.1001(c)(2)]

Table W-1

Means to measure	Accuracy
Deflection of ROPS, inches	± 5% of deflection measured.
Vehicle weight, pounds	± 5% of the weight measured.
Force applied to frame, pounds	± 5% of force measured.
Dimensions of critical zone, inches	± 0.5 in.

(d) **Vehicle condition.** The ROPS to be tested must be attached to the vehicle structure in the same manner as it will be attached during vehicle use. A totally assembled vehicle is not required. However, the vehicle structure and frame which support the ROPS must represent the actual vehicle installation. All normally detachable windows, panels, or nonstructural fittings shall be removed so that they do not contribute to the strength of the ROPS. [§1926.1001(d)]

(e) **Test procedure.** The test procedure shall include the following, in the sequence indicated: [§1926.1001(e)]

(1) *Energy absorbing capabilities of ROPS* shall be verified when loaded laterally by incrementally applying a distributed load to the longitudinal outside top member of the ROPS, as shown in Figure W-1, W-2, or W-3, as applicable. The distributed load must be applied so as to result in approximately uniform deflection of the ROPS. The load increments should correspond with approximately 0.5 in. ROPS deflection increment in the direction of the load application, measured at the ROPS top edge. Should the operator's seat be offcenter, the load shall be applied on the offcenter side. For each applied load increment, the total load (lb.) versus corresponding deflection (in.) shall be plotted, and the area under the load -deflection curve shall be calculated. This area is equal to the energy (in.-lb.) absorbed by the ROPS. For a typical load-deflection curve and calculation method, see Figure W-5. [§1926.1001(e)(1)]

Figure W-1. Vertical loading setup for all types of equipment described in §1518.1001(a).

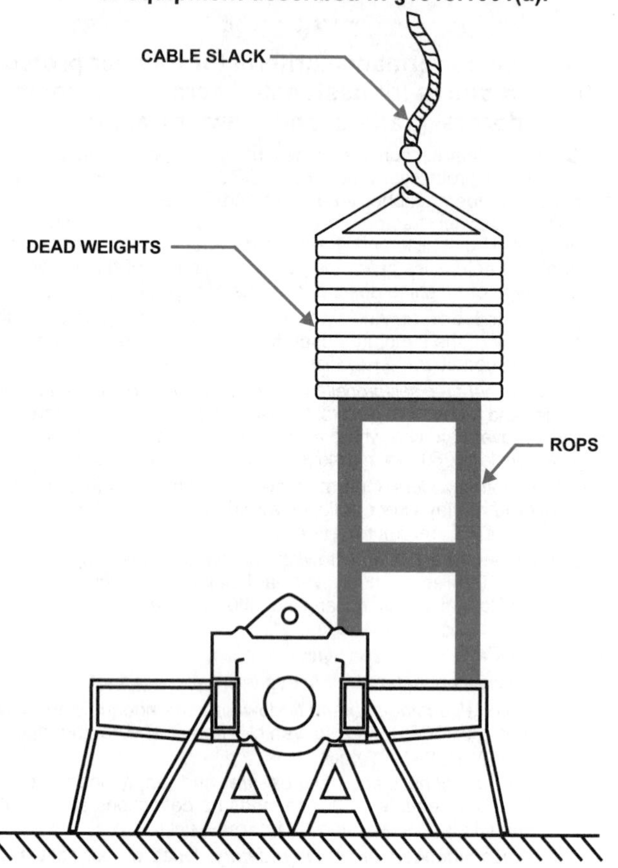

Figure W-2. Test setup for rubber-tired self-propelled scrapers.

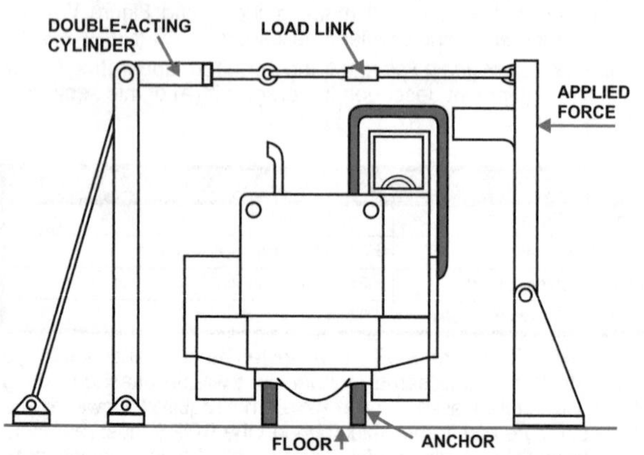

Figure W-3. Test setup for rubber-tired front-end loaders, rubber-tired dozers, and motor graders.

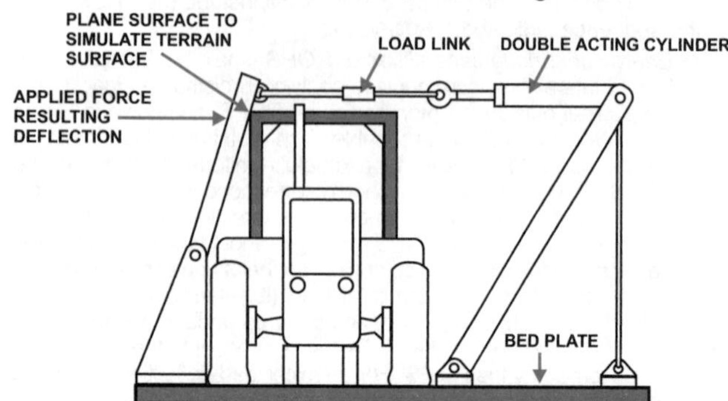

Figure W-4. Side-loading setup for crawler tractors and crawler loaders.

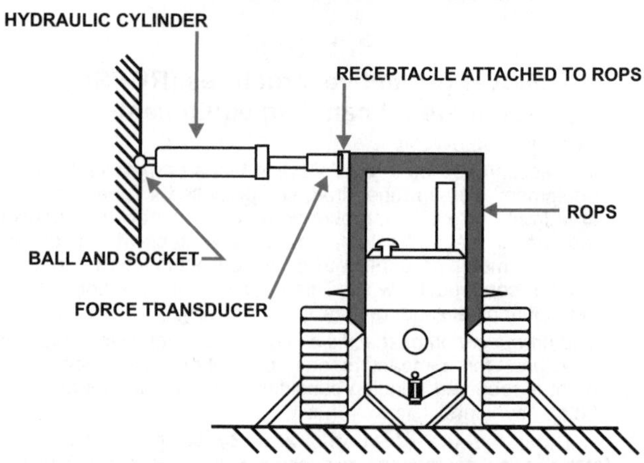

Incremental loading shall be continued until the ROPS has absorbed the amount of energy and the minimum applied load specified under paragraph (f) of this section has been reached or surpassed. [§1926.1001(e)(1)]

(2) *To cover the possibility of the vehicle* coming to rest on its top, the support capability shall be verified by applying a distributed vertical load to the top of the ROPS so as to result in approximately uniform deflection (see Figure W-1). The load magnitude is specified in paragraph (f)(2)(iii) of this section. [§1926.1001(e)(2)]

(3) *The low temperature impact* strength of the material used in the ROPS shall be verified by suitable material tests or material certification (see paragraph (f)(2)(iv) of this section). [§1926.1001(e)(3)]

Figure W-5. Determination of energy area under force deflection curve for all types of ROPS equipment defined in §1926.1001.

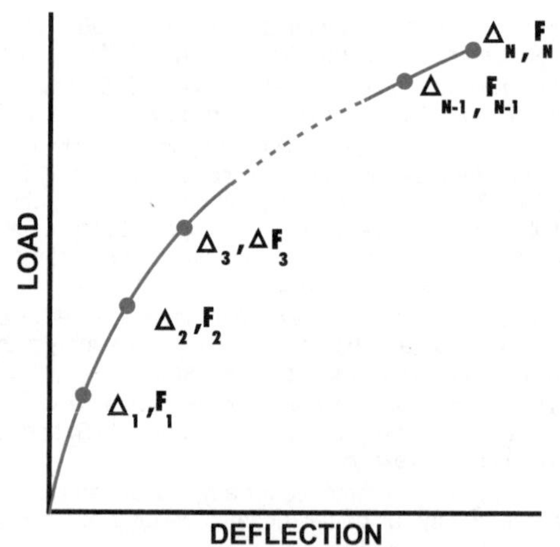

Δ- TOTAL DEFLECTION
F- FORCE APPLIED

$$\text{AREA} = \frac{\Delta_1 F_1}{2} + (\Delta_2 - \Delta_1)\frac{F_1 + F_2}{2} + (\Delta_3 - \Delta_2)\frac{F_2 + F_3}{2} + \cdots + (\Delta_N - \Delta_{N-1})\frac{F_{N-1} + F_N}{2}$$

(f) **Performance requirements.** [§1926.1001(f)]

(1) *General performance requirements.* [§1926.1001(f)(1)]

(i) *No repairs or straightening of any member* shall be carried out between each prescribed test. [§1926.1001(f)(1)(i)]

(ii) *During each test,* no part of the ROPS shall enter the critical zone as detailed in SAE J397 (1969). Deformation of the ROPS shall not allow the plane of the ground to enter this zone. [§1926.1001(f)(1)(ii)]

(2) *Specific performance requirements.* [§1926.1001(f)(2)]

(i) *The energy requirement for purposes* of meeting the requirements of paragraph (e)(1) of this section is to be determined by referring to the plot of the energy versus weight of vehicle (see Figure W-6 for rubber-tired self-propelled scrapers; Figure W-7 for rubber-tired front-end loaders and rubber-tired dozers; Figure W-8 for crawler tractors and crawler-type loaders; and Figure W-9 for motor graders). For purposes of this section, force and weight are measured as pounds (lb.); energy (U) is measured as inch-pounds. [§1926.1001(f)(2)(i)]

FIGURE W-6. Energy absorbed versus vehicle weight.

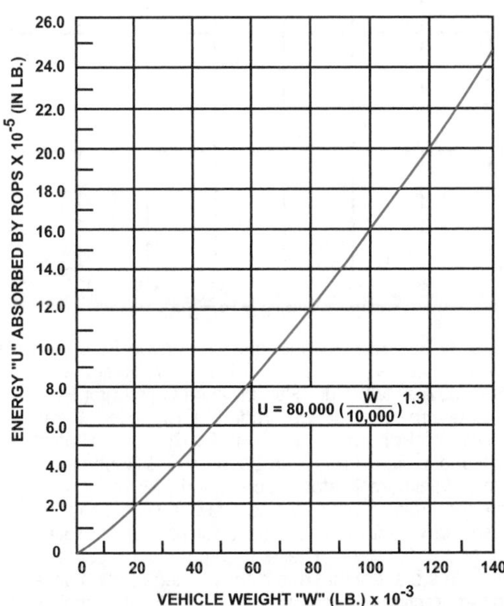

Editor's Note: The CFR states '1 3' in this instance. As it appears clear from the context that this was a typographical error, the text is corrected in the graphic above.

(ii) *The applied load must attain* at least a value which is determined by multiplying the vehicle weight by the corresponding factor shown in Figure W-10 for rubber-tired self-propelled scrapers; in Figure W-11 for rubber-tired front-end loaders and rubber-tired dozers; in Figure W-12 for crawler tractors and crawler-type loaders; and in Figure W-13 for motor graders. [§1926.1001(f)(2)(ii)]

(iii) *The load magnitude for purposes* of compliance with paragraph (e)(2) of this section is equal to the vehicle weight. The test of load magnitude shall only be made after the requirements of paragraph (f)(2)(i) of this section are met. [§1926.1001(f)(2)(iii)]

FIGURE W-7. Energy absorbed versus vehicle weight.

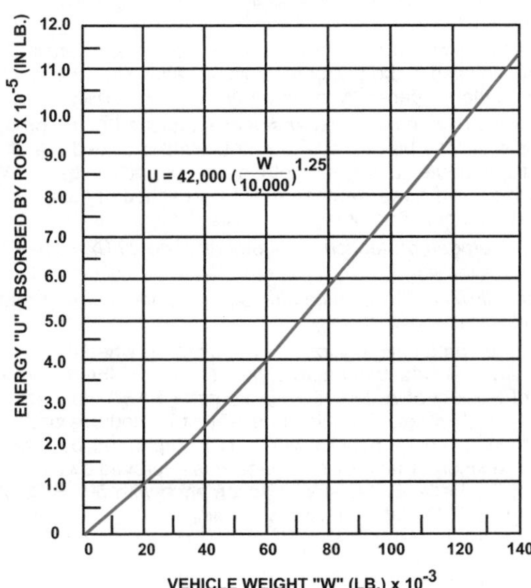

(iv) *Material used in the ROPS* must have the capability of performing at zero degrees Fahrenheit, or exhibit Charpy V notch impact strength of 8 foot-pounds at minus 20 ° Fahrenheit. [§1926.1001(f)(2)(iv)]

This is a standard Charpy specimen as described in American Society of Testing and Materials A 370, Methods and Definitions for Mechanical Testing of Steel Products (available at each Regional Office of the Occupational Safety and Health Administration). The purpose of this requirement is to reduce the tendency of brittle fracture associated with dynamic loading, low temperature operation, and stress raisers which cannot be entirely avoided on welded structures. [§1926.1001(f)(2)(iv)]

FIGURE W-8. Energy absorbed versus vehicle weight.

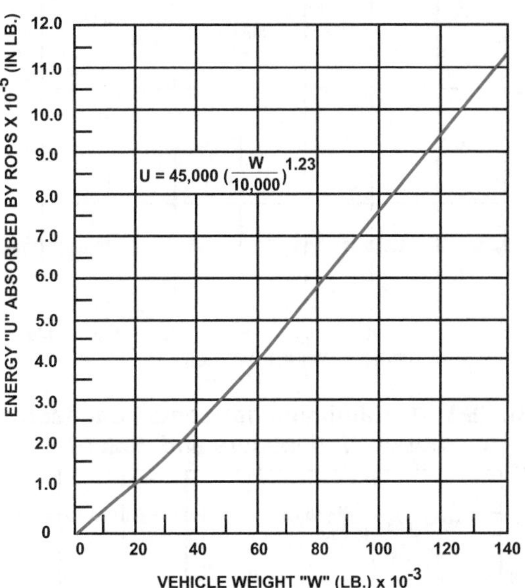

FIGURE W-9. Energy absorbed versus vehicle weight.

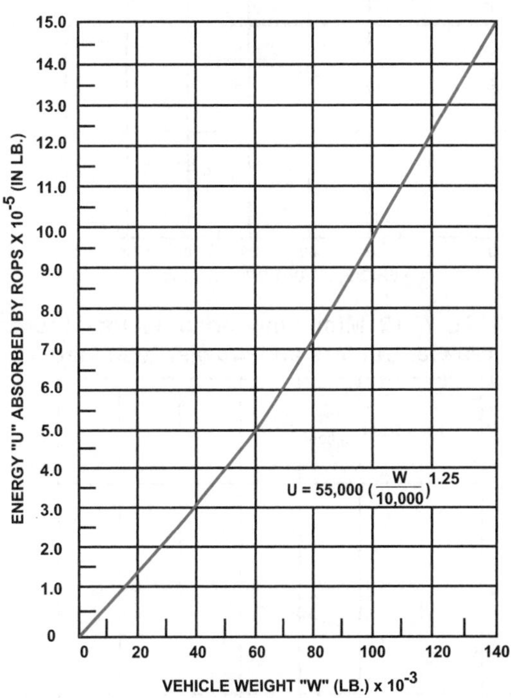

(g) **Definitions.** For purposes of this section, "vehicle weight" means the manufacturer's maximum weight of the prime mover for rubber-tired self-propelled scrapers. For other types of equipment to which this section applies, "vehicle weight" means the manufacturer's maximum recommended weight of the vehicle plus the heaviest attachment. [§1926.1001(g)]

FIGURE W-10. Minimum horizontal load factor for self-propelled scrapers.

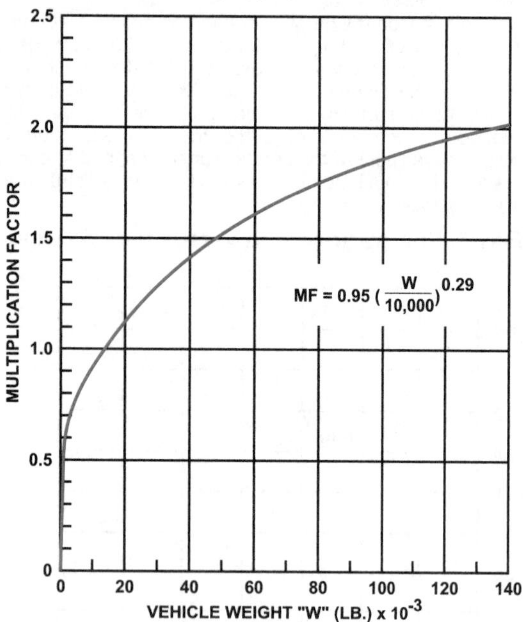

FIGURE W-11. Minimum horizontal load factor for rubber-tired loaders and dozers.

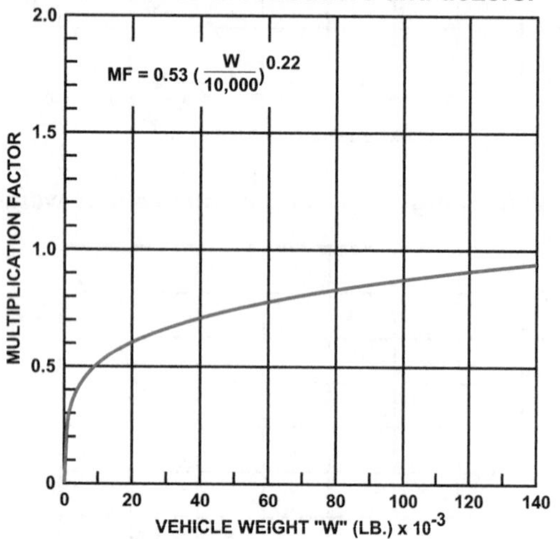

FIGURE W-12. Minimum horizontal load factor for crawler tractors and crawler-type loaders.

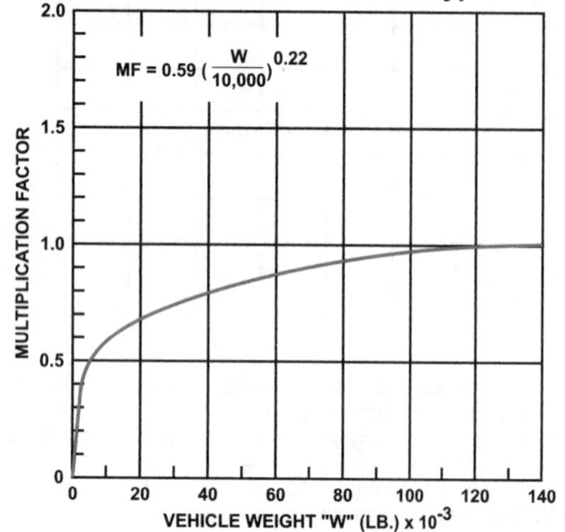

FIGURE W-13. Minimum horizontal load factor for motor graders.

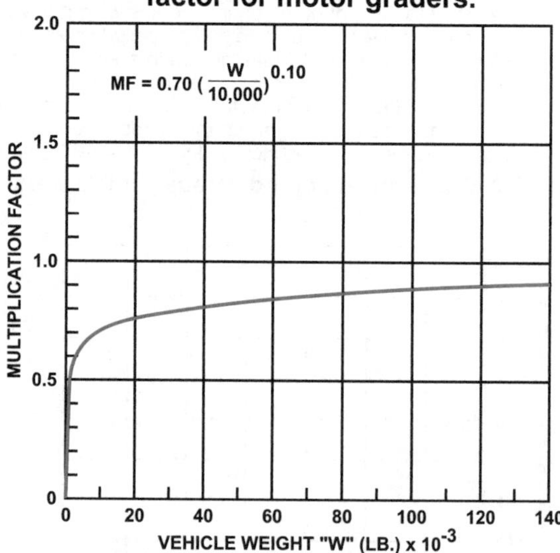

(h) **Source of standard.** This standard is derived from, and restates, the following Society of Automotive Engineers Recommended Practices: SAE J320a, Minimum Performance Criteria for Roll-Over Protective Structure for Rubber-Tired, Self-Propelled Scrapers; SAE J394, Minimum Performance Criteria for Roll-Over Protective Structure for Rubber-Tired Front End Loaders and Rubber-Tired Dozers; SAE J395, Minimum Performance Criteria for Roll-Over Protective Structure for Crawler Tractors and Crawler-Type Loaders; and SAE J396, Minimum Performance Criteria for Roll-Over Protective Structure for Motor Graders. These recommended practices shall be resorted to in the event that questions of interpretation arise. The recommended practices appear in the 1971 SAE Handbook, which may be examined in each of the Regional Offices of the Occupational Safety and Health Administration. [§1926.1001(h)]

§1926.1002

Protective frames (roll-over protective structures, known as ROPS) for wheel-type agricultural and industrial tractors used in construction

(a) **General.** [§1926.1002(a)]

(1) *The purpose of this section* is to set forth requirements for frames used to protect operators of wheel-type agricultural and industrial tractors that will minimize the possibility of operator injury resulting from accidental upsets during normal operation. With respect to agricultural and industrial tractors, the provisions of 29 CFR 1926.1001 and 1926.1003 for rubber-tired dozers and rubber-tired loaders may be used instead of the requirements of this section. [§1926.1002(a)(1)]

(2) *The protective frame* that is the subject of this standard is a structure mounted to the tractor that extends above the operator's seat and conforms generally to Figure W-14. [§1926.1002(a)(2)]

(3) *When an overhead weather shield* is attached to the protective frame, it may be in place during testing, provided that it does not contribute to the strength of the protective frame. When such an overhead weather shield is attached, it must meet the requirements of paragraph (i) of this section. [§1926.1002(a)(3)]

(4) *For overhead protection requirements, see 29 CFR 1926.1003.* [§1926.1002(a)(4)]

(5) *The following provisions address* requirements for protective enclosures. [§1926.1002(a)(5)]

(i) *When protective enclosures are used* on wheel-type agricultural and industrial tractors, they shall meet the requirements of Society of Automotive Engineers ("SAE") standard J168-1970 ("Protective enclosures — test procedures and performance requirements"), which is incorporated by reference. The incorporation by reference was approved by the Director of the Federal Register in accordance with 5 U.S.C. 552(a) and 1 CFR part 51. [§1926.1002(a)(5)(i)]

(ii) *SAE standard J168-1970* appears in the 1971 SAE Handbook, or it may be examined at: any OSHA Regional Office; the OSHA Docket Office, U.S. Department of Labor, 200 Constitution Avenue, NW., Room N-2625, Washington, DC 20210 (telephone: (202) 693-2350 (TTY number: (877) 889-5627)); or the National Archives and Records Administration ("NARA"). (For information on the availability of this material at NARA, telephone (202) 741-6030 or access the NARA Web site at www.archives.gov/federal_register/code_of_federal_regulations/ibr_locations.html.) Copies may be purchased from the Society of Automotive Engineers, 400 Commonwealth Drive, Warrendale, Pennsylvania 15096-0001. [§1926.1002(a)(5)(ii)]

(b) Applicability. The requirements of this section apply to wheel-type agricultural and industrial tractors used in construction work. See paragraph (j) of this section for definitions of agricultural tractors set forth in paragraph (i) of this section. [§1926.1002(b)]

(c) Performance requirements. [§1926.1002(c)]

(1) *Either a laboratory test or a field test* is required to determine the performance requirements set forth in paragraph (i) of this section. [§1926.1002(c)(1)]

(2) *A laboratory test may be either static or dynamic.* The laboratory test must be under conditions of repeatable and controlled loading to permit analysis of the protective frame. [§1926.1002(c)(2)]

(3) *A field-upset test, when used,* shall be conducted under reasonably controlled conditions, both rearward and sideways to verify the effectiveness of the protective frame under actual dynamic conditions. [§1926.1002(c)(3)]

(d) Test procedures — general. [§1926.1002(d)]

(1) *The tractor used* shall be the tractor with the greatest weight on which the protective frame is to be used. [§1926.1002(d)(1)]

(2) *A new protective frame* and mounting connections of the same design shall be used for each test procedure. [§1926.1002(d)(2)]

(3) *Instantaneous and permanent frame deformation* shall be measured and recorded for each segment of the test. [§1926.1002(d)(3)]

(4) *Dimensions relative to the seat* shall be determined with the seat unloaded and adjusted to its highest and most rearward latched position provided for a seated operator. [§1926.1002(d)(4)]

(5) *When the seat is offset,* the frame loading shall be on the side with the least space between the centerline of the seat and the upright. [§1926.1002(d)(5)]

(6) *The low-temperature impact strength* of the material used in the protective structure shall be verified by suitable material tests or material certifications according to 29 CFR 1926.1001(f)(2)(iv). [§1926.1002(d)(6)]

(e) Test procedure for vehicle overturn. [§1926.1002(e)]

(1) *Vehicle weight.* The weight of the tractor, for purposes of this section, includes the protective frame, all fuels, and other components required for normal use of the tractor. Ballast must be added when necessary to achieve a minimum total weight of 130 lb (59 kg) per maximum power-takeoff horsepower at the rated engine speed. The weight of the front end must be at least 33 lb (15 kg) per maximum power-takeoff horsepower. In case power-takeoff horsepower is unavailable, 95 percent of net engine flywheel horsepower shall be used. [§1926.1002(e)(1)]

(2) *Agricultural tractors shall be tested* at the weight set forth in paragraph (e)(1) of this section. [§1926.1002(e)(2)]

(3) *Industrial tractors shall be tested* with items of integral or mounted equipment and ballast that are sold as standard equipment or approved by the vehicle manufacturer for use with the vehicle when the protective frame is expected to provide protection for the operator with such equipment installed. The total vehicle weight and front-end weight as tested shall not be less than the weights established in paragraph (e)(1) of this section. [§1926.1002(e)(3)]

(4) *The following provisions address soil bank test conditions.* [§1926.1002(e)(4)]

(i) *The test shall be conducted* on a dry, firm soil bank as illustrated in Figure W-15. The soil in the impact area shall have an average cone index in the 0-in. to 6-in. (0-mm to 153-mm) layer not less than 150 according to American Society of Agricultural Engineers ("ASAE") recommendation ASAE R313.1-1971 ("Soil cone penetrometer"), as reconfirmed in 1975, which is incorporated by reference. The incorporation by reference was approved by the Director of the Federal Register in accordance with 5 U.S.C. 552(a) and 1 CFR part 51. The path of vehicle travel shall be 12° ± 2° to the top edge of the bank. [§1926.1002(e)(4)(i)]

(ii) *ASAE recommendation ASAE R313.1-1971,* as reconfirmed in 1975, appears in the 1977 Agricultural Engineers Yearbook, or it may be examined at: any OSHA Regional Office; the OSHA Docket Office, U.S. Department of Labor, 200 Constitution Avenue, NW., Room N-2625, Washington, DC 20210 (telephone: (202) 693-2350 (TTY number: (877) 889-5627)); or the National Archives and Records Administration ("NARA"). (For information on the availability of this material at NARA, telephone (202) 741-6030 or access the NARA Web site at www.archives.gov/federal_register/code_of_federal_regulations/ibr_locations.html). Copies may be purchased from the American Society of Agricultural Engineers 2950 Niles Road, St. Joseph, MI 49085. [§1926.1002(e)(4)(ii)]

(5) *The upper edge of the bank* shall be equipped with an 18-in. (457-mm) high ramp as described in Figure W-15 to assist in tipping the vehicle. [§1926.1002(e)(5)]

(6) *The front and rear wheel-tread settings,* when adjustable, shall be at the position nearest to halfway between the minimum and maximum settings obtainable on the vehicle. When only two settings are obtainable, the minimum setting shall be used. [§1926.1002(e)(6)]

(7) *Vehicle overturn test — sideways and rearward.* [§1926.1002(e)(7)]

(i) *The tractor shall be driven under its own power* along the specified path of travel at a minimum speed of 10 mph (16 kph), or maximum vehicle speed when under 10 mph (16 kph), up the ramp as described in paragraph (d)(5) of this section to induce sideways overturn. [§1926.1002(e)(7)(i)]

(ii) *Rear upset shall be induced* by engine power with the tractor operating in gear to obtain 3 to 5 mph (4.8 to 8 kph) at maximum governed engine rpm, preferably by driving forward directly up a minimum slope of two vertical to one horizontal. The engine clutch may be used to aid in inducing the upset. [§1926.1002(e)(7)(ii)]

(f) Other test procedures. When the field-upset test is not used to determine ROPS performance, either the static test or the dynamic test, contained in paragraph (g) or (h) of this section, shall be made. [§1926.1002(f)]

(g) Static test. [§1926.1002(g)]

(1) *Test conditions.* [§1926.1002(g)(1)]

(i) *The laboratory mounting base* shall include that part of the tractor chassis to which the protective frame is attached, including the mounting parts. [§1926.1002(g)(1)(i)]

(ii) *The protective frame shall be instrumented* with the necessary equipment to obtain the required load-deflection data at the locations and directions specified in Figures W-16, W-17, and W-18. [§1926.1002(g)(1)(ii)]

(iii) *The protective frame and mounting connections* shall be instrumented with the necessary recording equipment to obtain the required load-deflection data to be used in calculating FSB (see paragraph (j)(3) of this section). The gauges shall be placed on mounting connections before the installation load is applied. [§1926.1002(g)(1)(iii)]

(2) *Test procedure.* [§1926.1002(g)(2)]

(i) *The side-load application* shall be at the upper extremity of the frame upright at a 90° angle to the centerline of the vehicle. The side load L shall be applied according to Figure W-16. L and D shall be recorded simultaneously. The test shall be stopped when: [§1926.1002(g)(2)(i)]

[A] *The strain energy absorbed* by the frame is equal to the required input energy (E_{is}); [§1926.1002(g)(2)(i)[A]]

[B] *Deflection of the frame* exceeds the allowable deflection; or [§1926.1002(g)(2)(i)[B]]

[C] *The frame load limit* occurs before the allowable deflection is reached in the side load. [§1926.1002(g)(2)(i)[C]]

(ii) *The L-D diagram* (see Figure W-19 for an example) shall be constructed using the data obtained according to paragraph (g)(2)(i) of this section. [§1926.1002(g)(2)(ii)]

(iii) *The modified L_m -D_m diagram* shall be constructed according to paragraph (g)(2)(ii) and Figure W-20 of this section. The strain energy absorbed by the frame (Eu) shall then be determined. [§1926.1002(g)(2)(iii)]

(iv) *E_{is}, FER, and FSB shall be calculated.* [§1926.1002(g)(2)(iv)]

(v) *The test procedure shall be repeated* on the same frame using L (rear input; see Figure W-18) and Eir. Rear-load application shall be distributed uniformly along a maximum projected dimension of 27 in. (686 mm) and a maximum area of 160 sq. in. (1,032 sq. cm) normal to the direction of load application. The load shall be applied to the upper extremity of the frame at the point that is midway between the centerline of the seat and the inside of the frame upright. [§1926.1002(g)(2)(v)]

(h) Dynamic test . [§1926.1002(h)]

(1) *Test conditions.* [§1926.1002(h)(1)]

(i) *The protective frame and tractor* shall meet the requirements of paragraphs (e)(2) or (3) of this section, as appropriate. [§1926.1002(h)(1)(i)]

(ii) *The dynamic loading shall be produced* by using a 4,410-lb (2,000-kg) weight acting as a pendulum. The impact face of the weight shall be 27 ± 1 in. by 27 ± 1in. (686 ± 25 mm by 686 ± 25 mm), and shall be constructed so that its center of gravity is within 1.0 in. (25.4 mm) of its geometric center. The weight shall be suspended from a pivot point 18 to 22 ft. (5.5 to 6.7 m) above the point of impact on the frame, and shall be conveniently and safely adjustable for height (see Figure W-21). [§1926.1002(h)(1)(ii)]

(iii) *For each phase of testing,* the tractor shall be restrained from moving when the dynamic load is applied. The restraining members shall be 0.50- to 0.63-in. (12.5- to 16.0-mm) steel cable, and points for attaching restraining members shall be located an appropriate distance behind the rear axle and in front of the front axle to provide a 15° to 30° angle between the restraining cable and the horizontal. The restraining cables shall either be in the plane in which the center of gravity of the pendulum will swing, or more than one restraining cable shall give a resultant force in this plane (see Figure W-22). [§1926.1002(h)(1)(iii)]

(iv) *The wheel-tread setting shall comply* with the requirements of paragraph (e)(6) of this section. The tires shall have no liquid ballast, and shall be inflated to the maximum operating pressure recommended by the tire manufacturer. With the specified tire inflation, the restraining cables shall be tightened to provide tire deflection of 6 to 8 percent of the nominal tire-section width. After the vehicle is restrained properly, a wooden beam that is 6-in. x 6-in. (150 mm x 150 mm) shall be driven tightly against the appropriate wheels and clamped. For the test to the side, an additional wooden beam shall be placed as a prop against the wheel nearest to the operator's station, and shall be secured to the floor so that when it is positioned against the wheel rim, it is at an angle of 25° to 40° to the horizontal. It shall have a length 20 to 25 times its depth, and a width two to three times its depth (see Figures W-22 and W-23). [§1926.1002(h)(1)(iv)]

(v) *Means shall be provided for indicating* the maximum instantaneous deflection along the line of impact. A simple friction device is illustrated in Figure W-18. [§1926.1002(h)(1)(v)]

(vi) *No repair or adjustments may be carried out during the test.* [§1926.1002(h)(1)(vi)]

(vii) *When any cables, props, or blocking* shift or break during the test, the test shall be repeated. [§1926.1002(h)(1)(vii)]

(2) *Test procedure.* [§1926.1002(h)(2)]

(i) *General.* The frame shall be evaluated by imposing dynamic loading to the rear, followed by a load to the side on the same frame. The pendulum dropped from the height (see the definition of "H" in paragraph (j)(3) of this section) imposes the dynamic load. The position of the pendulum shall be so selected that the initial point of impact on the frame shall be in line with the arc of travel of the center of gravity of the pendulum. A quick-release mechanism should be used but, when used, it shall not influence the attitude of the block. [§1926.1002(h)(2)(i)]

(ii) *Impact at rear.* The tractor shall be restrained properly according to paragraphs (h)(1)(iii) and (h)(1)(iv) of this section. The tractor shall be positioned with respect to the pivot point of the pendulum so that the pendulum is 20° from the vertical prior to impact as shown in Figure W-22. The impact shall be applied to the upper extremity of the frame at the point that is midway between the centerline of the frame and the inside of the frame upright of a new frame. [§1926.1002(h)(2)(ii)]

(iii) *Impact at side.* The blocking and restraining shall conform to paragraphs (h)(1)(iii) and (h)(1)(iv) of this section. The center point of impact shall be that structural member of the protective frame likely to hit the ground first in a sideways accidental upset. The side impact shall be applied to the side opposite that used for rear impact. [§1926.1002(h)(2)(iii)]

(i) Performance requirements. [§1926.1002(i)]

(1) *General.* [§1926.1002(i)(1)]

(i) *The frame, overhead weather shield, fenders,* or other parts in the operator area may be deformed in these tests, but shall not shatter or leave sharp edges exposed to the operator, or violate the dimensions shown in Figures W-16 and W-17, and specified as follows: [§1926.1002(i)(1)(i)]

D = 2 in. (51 mm) inside of the frame upright to the vertical centerline of the seat;

E = 30 in. (762 mm);

F = Not less than 0 in. (0 mm) and not more than 12 in. (305 mm), measured at the centerline of the seat backrest to the crossbar along the line of load application as shown in Figure W-17; and

G = 24 in. (610 mm).

(ii) *The material and design combination* used in the protective structure must be such that the structure can meet all prescribed performance tests at 0 °F (-18 °C) according to 29 CFR 1926.1001(f)(2)(iv). [§1926.1002(i)(1)(ii)]

(2) *Vehicle overturn performance requirements.* The requirements of this paragraph (i) must be met in both side and rear overturns. [§1926.1002(i)(2)]

(3) *Static test performance requirements.* Design factors shall be incorporated in each design to withstand an overturn test as specified by this paragraph (i). The structural requirements will be met generally when FER is greater than 1.0 and FSB is greater than K-1 in both side and rear loadings. [§1926.1002(i)(3)]

(4) *Dynamic test performance requirements.* Design factors shall be incorporated in each design to withstand the overturn test specified by this paragraph (i). The structural requirements will be met generally when the dimensions in this paragraph (i) are used during both side and rear loads. [§1926.1002(i)(4)]

(j) Definitions applicable to this section. [§1926.1002(j)]

(1) **Agricultural tractor** means a wheel-type vehicle of more than 20 engine horsepower, used in construction work, that is designed to furnish the power to pull, propel, or drive implements. (SAE standard J333a-1970 ("Operator protection for wheel-type agricultural and industrial tractors") defines "agricultural tractor" as a "wheel-type vehicle of more than 20 engine horsepower designed to furnish the power to pull, carry, propel, or drive implements that are designed for agricultural usage." Since this part 1926 applies only to construction work, the SAE definition of "agricultural tractor" is adopted for purposes of this subpart.) [§1926.1002(j)(1)]

(2) **Industrial tractor** means that class of wheel-type tractors of more than 20 engine horsepower (other than rubber-tired loaders and dozers described in 29 CFR 1926.1001), used in operations such as landscaping, construction services, loading, digging, grounds keeping, and highway maintenance. [§1926.1002(j)(2)]

(3) *The following symbols,* terms, and explanations apply to this section: [§1926.1002(j)(3)]

E_{is} = Energy input to be absorbed during side loading in ft-lb (E'_{is} in J [joules]);

E_{is} = 723 + 0.4 W ft-lb (E'_{is} = 100 + 0.12 W', J);

E_{ir} = Energy input to be absorbed during rear loading in ft-lb (E'_{ir} in J);

E_{ir} = 0.47 W ft-lb (E'_{ir} = 0.14 W', J);

W = Tractor weight as specified by 29 CFR 1926.1002(e)(1) and (e)(3), in lb (W', kg);

L = Static load, lb (kg);

D = Deflection under L, in. (mm);

L - D = Static load-deflection diagram;

L_m - D_m = Modified static load-deflection diagram (Figure W-20). To account for an increase in strength due to an increase in strain rate, raise L in the plastic range L x K;

K = Increase in yield strength induced by higher rate of loading (1.3 for hot, rolled, low-carbon steel 1010-1030). Low carbon is preferable; however, when higher carbon or other material is used, K must be determined in the laboratory. Refer to Norris, C.H., Hansen, R.J., Holley, M.J., Biggs, J.M., Namyet, S., and Minami, J.V., Structural Design for Dynamic Loads, McGraw-Hill, New York, 1959, p. 3;

L_{max} = Maximum observed static load;

Load Limit = Point on a continuous L - D curve at which the observed static load is 0.8 L_{max} (refer to Figure W-19);

E_u = Strain energy absorbed by the frame, ft-lb (J); area under the L_m - D_m curve;

FER = Factor of energy ratio, FER = E_uE_{is} ; also,

FER = E_uE_{ir} ;

P_b = Maximum observed force in mounting connection under a static load, L, lb (kg);

P_u = Ultimate force capacity of mounting connection, lb (kg);

FSB = Design margin for a mounting connection $(P_u P_b)$-1; and
H = Vertical height of lift of 4,410-lb (2,000-kg) weight, in. (H', mm). The weight shall be pulled back so that the height of its center of gravity above the point of impact is defined as follows:
H = 4.92 + 0.00190 W (H' = 125 + 0.107 W') (see Figure W-24).

(k) **Source of standard.** The standard in this section is derived from, and restates, in part, Society of Automotive Engineers ("SAE") standard J334a-1970 ("Protective frame test procedures and performance requirements"). The SAE standard appears in the 1971 SAE Handbook, which may be examined at any OSHA regional office. [§1926.1002(k)]

[70 FR 76985, Dec. 29, 2005, as amended at 71 FR 41129, July 20, 2006]

§1926.1003

Overhead protection for operators of agricultural and industrial tractors used in construction

(a) **General.** [§1926.1003(a)]

(1) *Purpose.* When overhead protection is provided on wheel-type agricultural and industrial tractors, the overhead protection shall be designed and installed according to the requirements contained in this section. The provisions of 29 CFR 1926.1001 for rubber-tired dozers and rubber-tired loaders may be used instead of the standards contained in this section. The purpose of this standard is to minimize the possibility of operator injury resulting from overhead hazards such as flying and falling objects, and at the same time to minimize the possibility of operator injury from the cover itself in the event of accidental upset. [§1926.1003(a)(1)]

(2) *Applicability.* This standard applies to wheel-type agricultural and industrial tractors used in construction work (see 29 CFR 1926.1002(b) and (j)). In the case of machines to which 29 CFR 1926.604 (relating to site clearing) also applies, the overhead protection may be either the type of protection provided in 29 CFR 1926.604, or the type of protection provided by this section. [§1926.1003(a)(2)]

(b) **Overhead protection.** When overhead protection is installed on wheel-type agricultural or industrial tractors used in construction work, it shall meet the requirements of this paragraph. The overhead protection may be constructed of a solid material. When grid or mesh is used, the largest permissible opening shall be such that the maximum circle that can be inscribed between the elements of the grid or mesh is 1.5 in. (38 mm) in diameter. The overhead protection shall not be installed in such a way as to become a hazard in the case of upset. [§1926.1003(b)]

(c) **Test procedures — general.** [§1926.1003(c)]

(1) *The requirements of 29 CFR 1926.1002(d), (e), and (f) shall be met.* [§1926.1003(c)(1)]

(2) *Static and dynamic rear load application* shall be distributed uniformly along a maximum projected dimension of 27 in. (686 mm), and a maximum area of 160 sq. in. (1,032 sq. cm), normal to the direction of load application. The load shall be applied to the upper extremity of the frame at the point that is midway between the centerline of the seat and the inside of the frame upright. [§1926.1003(c)(2)]

(3) *The static and dynamic side load application* shall be distributed uniformly along a maximum projected dimension of 27 in. (686 mm), and a maximum area of 160 sq. in. (1,032 sq. cm), normal to the direction of load application. The direction of load application is the same as in 29 CFR 1926.1002 (g) and (h). To simulate the characteristics of the structure during an upset, the center of load application may be located from a point 24 in. (610 mm) (K) forward to 12 in. (305 mm) (L) rearward of the front of the seat backrest, to best use the structural strength (see Figure W-25). [§1926.1003(c)(3)]

(d) **Drop test procedures.** [§1926.1003(d)]

(1) *The same frame* shall be subjected to the drop test following either the static or dynamic test. [§1926.1003(d)(1)]

(2) *A solid steel sphere* or material of equivalent spherical dimension weighing 100 lb (45.4 kg) shall be dropped once from a height 10 ft (3.08 m) above the overhead cover. [§1926.1003(d)(2)]

(3) *The point of impact* shall be on the overhead cover at a point within the zone of protection as shown in Figure W-26, which is furthest removed from major structural members. [§1926.1003(d)(3)]

(e) **Crush test procedure.** [§1926.1003(e)]

(1) *The same frame* shall be subjected to the crush test following the drop test and static or dynamic test. [§1926.1003(e)(1)]

(2) *The test load* shall be applied as shown in Figure W-27, with the seat positioned as specified in 29 CFR 1926.1002(d)(4). Loading cylinders shall be mounted pivotally at both ends. Loads applied by each cylinder shall be equal within two percent, and the sum of the loads of the two cylinders shall be two times the tractor weight as set forth in 29 CFR 1926.1002(e)(1). The maximum width of the beam illustrated in Figure W-27 shall be 6 in. (152 mm). [§1926.1003(e)(2)]

(f) **Performance requirements.** [§1926.1003(f)]

(1) *General.* The performance requirements set forth in 29 CFR 1926.1002(i)(2), (3), and (4) shall be met. [§1926.1003(f)(1)]

(2) *Drop test performance requirements.* [§1926.1003(f)(2)]

(i) *Instantaneous deformation due to impact of the sphere* shall not enter the protected zone as illustrated in Figures W-25, W-26, and W-28. [§1926.1003(f)(2)(i)]

(ii) *In addition to the dimensions set forth in 29 CFR 1926.1002(i)(1)(i),* the following dimensions apply to Figure W-28: [§1926.1003(f)(2)(ii)]

H = 17.5 in. (444 mm); and
J = 2 in. (50.8 mm), measured from the outer periphery of the steering wheel.

(3) *Crush test performance requirements.* The protected zone as described in Figure W-28 must not be violated. [§1926.1003(f)(3)]

(g) **Source of standard.** This standard is derived from, and restates, in part, the portions of Society of Automotive Engineers ("SAE") standard J167-1970 ("Protective frame with overhead protection — test procedures and performance requirements"), which pertain to overhead protection requirements. The SAE standard appears in the 1971 SAE Handbook, which may be examined at any OSHA regional office. [§1926.1003(g)]

[70 FR 76985, Dec. 29, 2005]

Appendix A to Subpart W

Figures W-14 through W-28

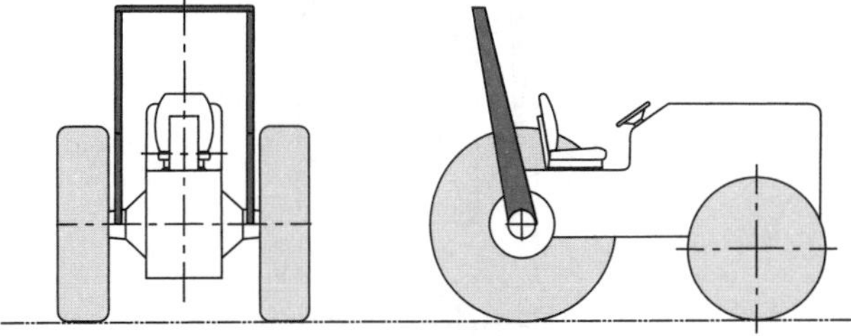

Figure W-14 – Typical Frame Configuration

W

Rollover Protective Structures; Overhead Protection

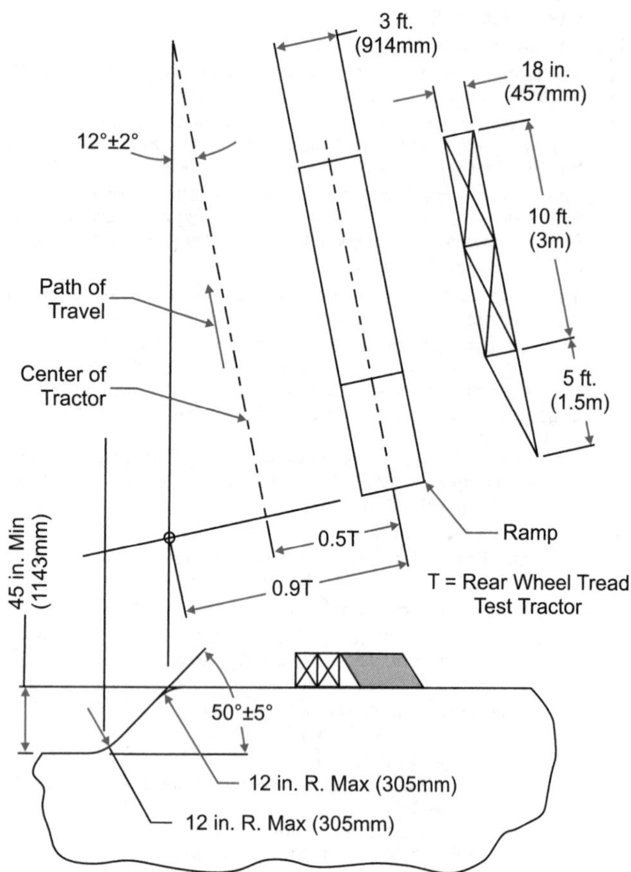

Figure W-15 – Side Overturn Bank And Ramp

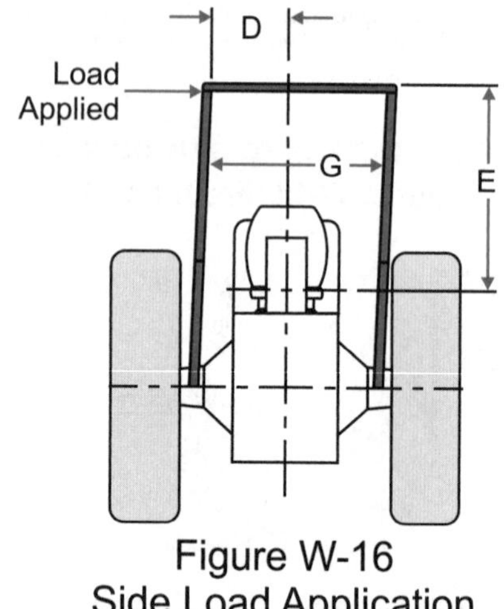

Figure W-16
Side Load Application

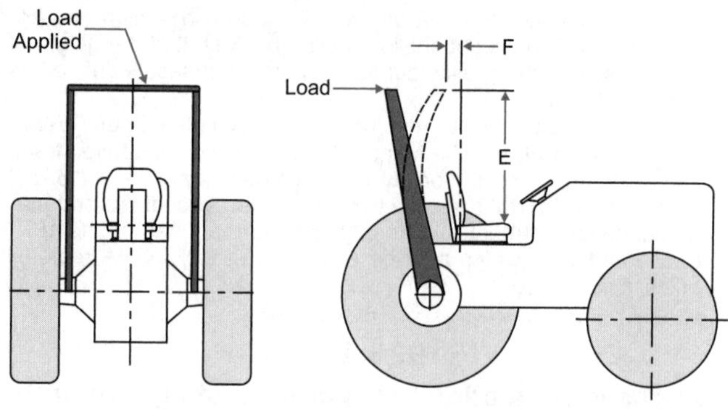

Figure W-17 – Rear Load Application

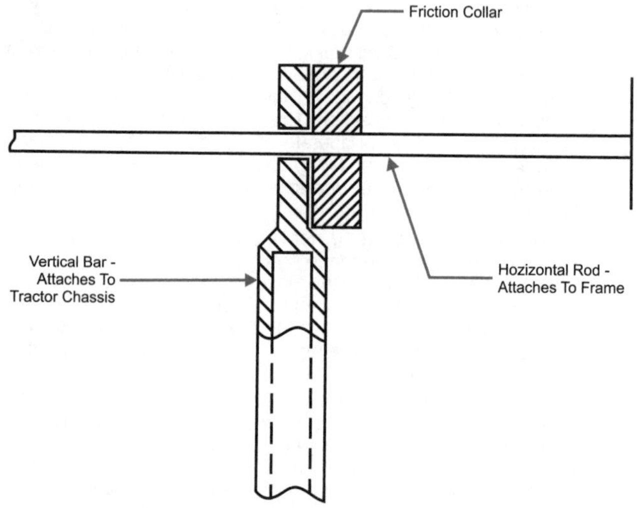

Figure W-18 – Method of Measuring Instantaneous Deflection

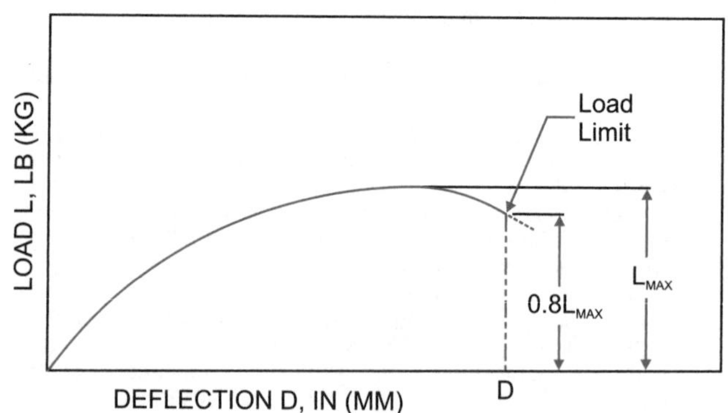

Figure W-19 – Typical L-D Diagram

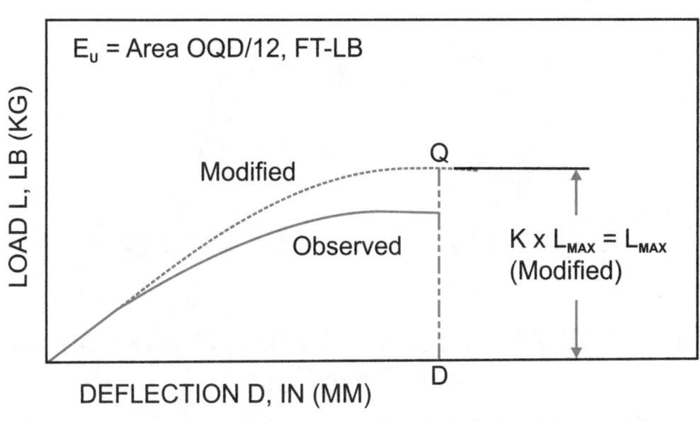

Figure W-20 – Typical Modified L_M-D_M Diagram

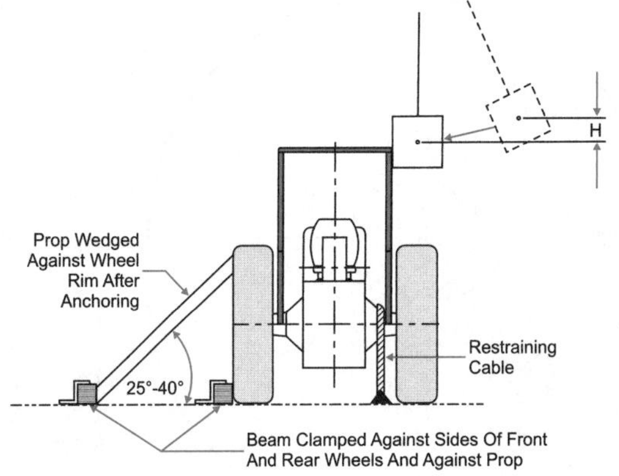

Figure W-23 – Method Of Impact From Side

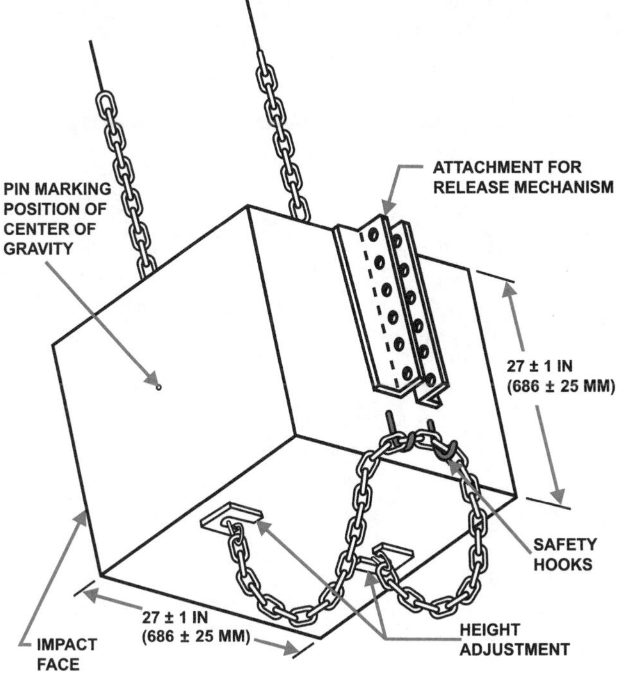

Figure W-21 — Pendulum

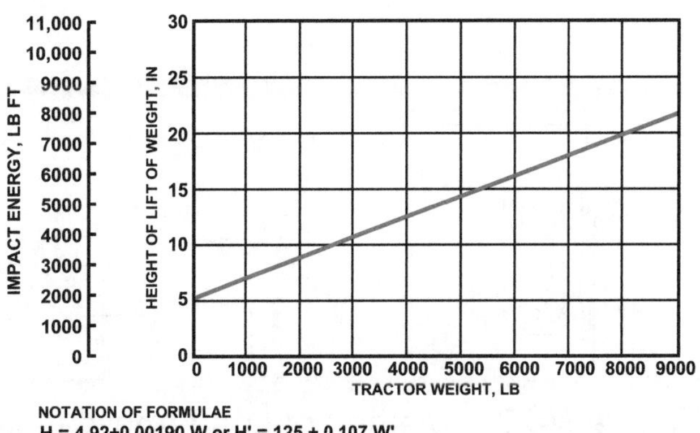

NOTATION OF FORMULAE
H = 4.92+0.00190 W or H' = 125 + 0.107 W'
W = Tractor Weight Specified by 29 CFR 1926.1002(e)(1) and (e)(3) in lbs (W' in kg)

Figure W-24 — Impact Energy and Corresponding Lift Height of 4,410 lb. (2,000 kg) Weight

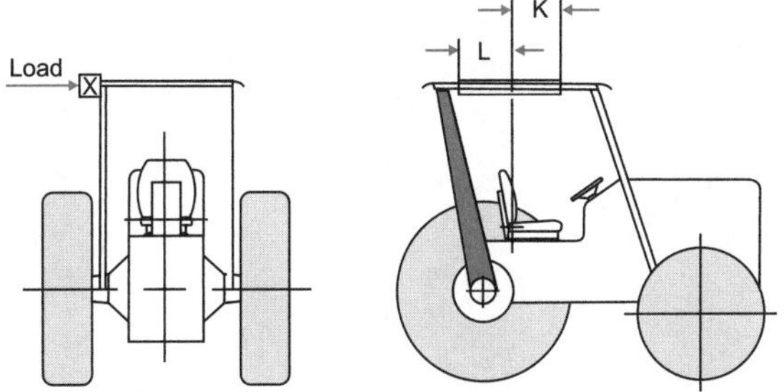

Figure W-25 – Location Of Side Load

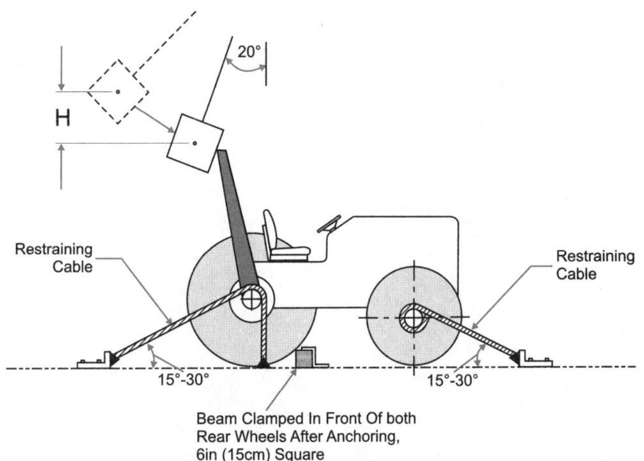

Figure W-22 – Method Of Impact From Rear

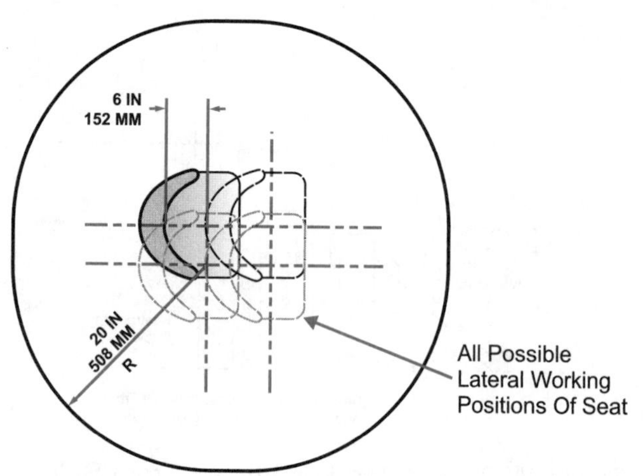

Figure W-26 — Zone of Protection for Drop Test

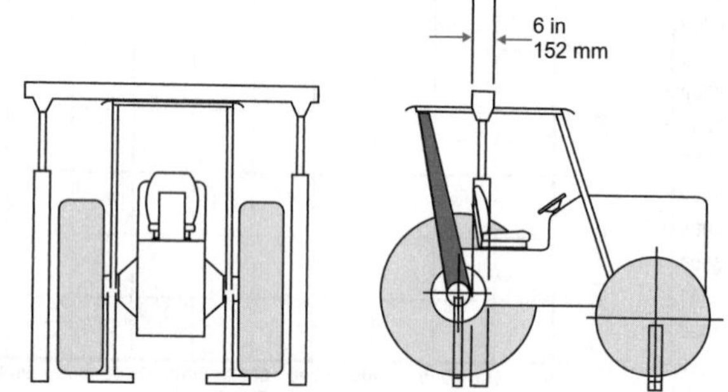

Figure W-27 – Method of Load Application For Crush Test

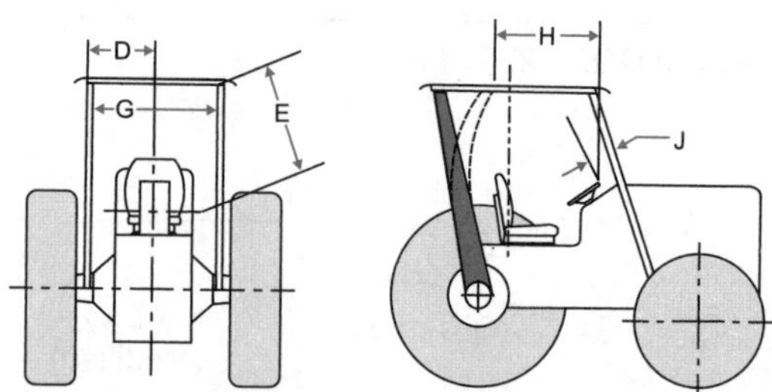

Figure W-28 – Protected Zone During Crush And Drop Test

[70 FR 76985, Dec. 29, 2005, as amended at 71 FR 41130, July 20, 2006]

1926 Subpart W

Authority for 1926 Subpart W

Authority: Section 3704 of the Contract Work Hours and Safety Standards Act (40 U.S.C. 3701); Sections 4, 6, and 8 of the Occupational Safety and Health Act of 1970 (29 U.S.C. 653, 655, 657); and Secretary of Labor's Order No. 12-71 (36 FR 8754), 8-76 (41 FR 25059), 9-83 (48 FR 35736), 1-90 (55 FR 9033), 6-96 (62 FR 111), 3-2000 (65 FR 50017), or 5-2002 (67 FR 65008), as applicable.

Subpart X - Stairways and Ladders

§1926.1050

Scope, application, and definitions applicable to this subpart

(a) **Scope and application.** This subpart applies to all stairways and ladders used in construction, alteration, repair (including painting and decorating), and demolition workplaces covered under 29 CFR part 1926, and also sets forth, in specified circumstances, when ladders and stairways are required to be provided. Additional requirements for ladders used on or with scaffolds are contained in subpart L—Scaffolds. This subpart does not apply to integral components of equipment covered by subpart CC. Subpart CC exclusively sets forth the circumstances when ladders and stairways must be provided on equipment covered by subpart CC. [§1926.1050(a)]

(b) **Definitions.**

Cleat means a ladder crosspiece of rectangular cross section placed on edge upon which a person may step while ascending or descending a ladder. [§1926.1050(b)]

Double-cleat ladder means a ladder similar in construction to a single-cleat ladder, but with a center rail to allow simultaneous two-way traffic for employees ascending or descending. [§1926.1050(b)]

Equivalent means alternative designs, materials, or methods that the employer can demonstrate will provide an equal or greater degree of safety for employees than the method or item specified in the standard. [§1926.1050(b)]

Extension trestle ladder means a self-supporting portable ladder, adjustable in length, consisting of a trestle ladder base and a vertically adjustable extension section, with a suitable means for locking the ladders together. [§1926.1050(b)]

Failure means load refusal, breakage, or separation of component parts. Load refusal is the point where the structural members lose their ability to carry the loads. [§1926.1050(b)]

Fixed ladder means a ladder that cannot be readily moved or carried because it is an integral part of a building or structure. A side-step fixed ladder is a fixed ladder that requires a person getting off at the top to step to the side of the ladder side rails to reach the landing. A through fixed ladder is a fixed ladder that requires a person getting off at the top to step between the side rails of the ladder to reach the landing. [§1926.1050(b)]

Handrail means a rail used to provide employees with a handhold for support. [§1926.1050(b)]

Individual-rung/step ladders means ladders without a side rail or center rail support. Such ladders are made by mounting individual steps or rungs directly to the side or wall of the structure. [§1926.1050(b)]

Job-made ladder means a ladder that is fabricated by employees, typically at the construction site, and is not commercially manufactured. This definition does not apply to any individual-rung/step ladders. [§1926.1050(b)]

Ladder stand. A mobile fixed size self-supporting ladder consisting of a wide flat tread ladder in the form of stairs. The assembly may include handrails. [§1926.1050(b)]

Lower levels means those areas to which an employee can fall from a stairway or ladder. Such areas include ground levels, floors, roofs, ramps, runways, excavations, pits, tanks, material, water, equipment, and similar surfaces. It does not include the surface from which the employee falls. [§1926.1050(b)]

Maximum intended load means the total load of all employees, equipment, tools, materials, transmitted loads, and other loads anticipated to be applied to a ladder component at any one time. [§1926.1050(b)]

Nosing means that portion of a tread projecting beyond the face of the riser immediately below. [§1926.1050(b)]

Point of access means all areas used by employees for work-related passage from one area or level to another. Such open areas include doorways, passageways, stairway openings, studded walls, and various other permanent or temporary openings used for such travel. [§1926.1050(b)]

Portable ladder means a ladder that can be readily moved or carried. [§1926.1050(b)]

Riser height means the vertical distance from the top of a tread to the top of the next higher tread or platform/landing or the distance from the top of a platform/landing to the top of the next higher tread or platform/landing. [§1926.1050(b)]

Side-step fixed ladder. See "Fixed ladder." [§1926.1050(b)]

Single-cleat ladder means a ladder consisting of a pair of side rails, connected together by cleats, rungs, or steps. [§1926.1050(b)]

Single-rail ladder means a portable ladder with rungs, cleats, or steps mounted on a single rail instead of the normal two rails used on most other ladders. [§1926.1050(b)]

Spiral stairway means a series of steps attached to a vertical pole and progressing upward in a winding fashion within a cylindrical space. [§1926.1050(b)]

Stairrail system means a vertical barrier erected along the unprotected sides and edges of a stairway[1] to prevent employees from falling to lower levels. The top surface of a stairrail system may also be a "handrail." [§1926.1050(b)]

Step stool (ladder type) means a self-supporting, foldable, portable ladder, nonadjustable in length, 32 inches or less in overall size, with flat steps and without a pail shelf, designed to be climbed on the ladder top cap as well as all steps. The side rails may continue above the top cap. [§1926.1050(b)]

Through fixed ladder. See "Fixed ladder." [§1926.1050(b)]

Tread depth means the horizontal distance from front to back of a tread (excluding nosing, if any). [§1926.1050(b)]

Unprotected sides and edges means any side or edge (except at entrances to points of access) of a stairway where there is no stairrail system or wall 36 inches (.9 m) or more in height, and any side or edge (except at entrances to points of access) of a stairway landing, or ladder platform where there is no wall or guardrail system 39 inches (1 m) or more in height. [§1926.1050(b)]

[55 FR 47687, Nov. 14, 1990; 56 FR 2585, Jan. 23, 1991, as amended at 58 FR 35184, June 30, 1993]

§1926.1051

General requirements

(a) **A stairway or ladder shall be provided** at all personnel points of access where there is a break in elevation of 19 inches (48 cm) or more, and no ramp, runway, sloped embankment, or personnel hoist is provided. [§1926.1051(a)]

(1) *Employees shall not use* any spiral stairways that will not be a permanent part of the structure on which construction work is being performed. [§1926.1051(a)(1)]

(2) *A double-cleated ladder* or two or more separate ladders shall be provided when ladders are the only mean of access or exit from a working area for 25 or more employees, or when a ladder is to serve simultaneous two-way traffic. [§1926.1051(a)(2)]

(3) *When a building or structure* has only one point of access between levels, that point of access shall be kept clear to permit free passage of employees. When work must be performed or equipment must be used such that free passage at that point of access is restricted, a second point of access shall be provided and used. [§1926.1051(a)(3)]

(4) *When a building or structure* has two or more points of access between levels, at least one point of access shall be kept clear to permit free passage of employees. [§1926.1051(a)(4)]

(b) **Employers shall provide and install** all stairway and ladder fall protection systems required by this subpart and shall comply with all other pertinent requirements of this subpart before employees begin the work that necessitates the installation and use of stairways, ladders, and their respective fall protection systems. [§1926.1051(b)]

§1926.1052

Stairways

(a) **General.** The following requirements apply to all stairways as indicated: [§1926.1052(a)]

(1) *Stairways that will not be a permanent part* of the structure on which construction work is being performed shall have landings of not less than 30 inches (76 cm) in the direction of travel and extend at least 22 inches (56 cm) in width at every 12 feet (3.7 m) or less of vertical rise. [§1926.1052(a)(1)]

(2) *Stairs shall be installed between 30° and 50° from horizontal.* [§1926.1052(a)(2)]

(3) *Riser height and tread depth* shall be uniform within each flight of stairs, including any foundation structure used as one or more treads of the stairs. Variations in riser height or tread depth shall not be over 1/4-inch (0.6 cm) in any stairway system. [§1926.1052(a)(3)]

(4) *Where doors or gates open directly* on a stairway, a platform shall be provided, and the swing of the door shall not reduce the effective width of the platform to less than 20 inches (51 cm). [§1926.1052(a)(4)]

(5) *Metal pan landings and metal pan treads,* when used, shall be secured in place before filling with concrete or other material. [§1926.1052(a)(5)]

(6) *All parts of stairways* shall be free of hazardous projections, such as protruding nails. [§1926.1052(a)(6)]

(7) *Slippery conditions on stairways* shall be eliminated before the stairways are used to reach other levels. [§1926.1052(a)(7)]

(b) **Temporary service.** The following requirements apply to all stairways as indicated: [§1926.1052(b)]

(1) *Except during stairway construction,* foot traffic is prohibited on stairways with pan stairs where the treads and/or landings are to be filled in with concrete or other material at a later date, unless the stairs are temporarily fitted with wood or other solid material at least to the top edge of each pan. Such temporary treads and landings shall be replaced when worn below the level of the top edge of the pan. [§1926.1052(b)(1)]

(2) *Except during stairway construction,* foot traffic is prohibited on skeleton metal stairs where permanent treads and/or landings are to be installed at a later date, unless the stairs are fitted with secured temporary treads and landings long enough to cover the entire tread and/or landing area. [§1926.1052(b)(2)]

(3) *Treads for temporary service* shall be made of wood or other solid material, and shall be installed the full width and depth of the stair. [§1926.1052(b)(3)]

(c) **Stairrails and handrails.** The following requirements apply to all stairways as indicated: [§1926.1052(c)]

(1) *Stairways having four or more risers* or rising more than 30 inches (76 cm), whichever is less, shall be equipped with: [§1926.1052(c)(1)]

(i) *At least one handrail; and* [§1926.1052(c)(1)(i)]

(ii) *One stairrail system along each unprotected side or edge.* [§1926.1052(c)(1)(ii)]

Note: When the top edge of a stairrail system also serves as a handrail, paragraph (c)(7) of this section applies.

(2) *Winding and spiral stairways* shall be equipped with a handrail offset sufficiently to prevent walking on those portions of the stairways where the tread width is less than 6 inches (15 cm). [§1926.1052(c)(2)]

(3) *The height of stairrails shall be as follows:* [§1926.1052(c)(3)]

(i) *Stairrails installed after March 15, 1991,* shall be not less than 36 inches (91.5 cm) from the upper surface of the stairrail system to the surface of the tread, in line with the face of the riser at the forward edge of the tread. [§1926.1052(c)(3)(i)]

(ii) *Stairrails installed before March 15, 1991,* shall be not less than 30 inches (76 cm) nor more than 34 inches (86 cm) from the upper surface of the stairrail system to the surface of the tread, in line with the face of the riser at the forward edge of the tread. [§1926.1052(c)(3)(ii)]

(4) *Midrails, screens, mesh,* intermediate vertical members, or equivalent intermediate structural members, shall be provided between the top rail of the stairrail system and the stairway steps. [§1926.1052(c)(4)]

(i) *Midrails, when used,* shall be located at a height midway between the top edge of the stairrail system and the stairway steps. [§1926.1052(c)(4)(i)]

(ii) *Screens or mesh, when used,* shall extend from the top rail to the stairway step, and along the entire opening between top rail supports. [§1926.1052(c)(4)(ii)]

(iii) *When intermediate vertical members,* such as balusters, are used between posts, they shall be not more than 19 inches (48 cm) apart. [§1926.1052(c)(4)(iii)]

(iv) *Other structural members, when used,* shall be installed such that there are no openings in the stairrail system that are more than 19 inches (48 cm) wide. [§1926.1052(c)(4)(iv)]

(5) *Handrails and the top rails of stairrail systems* shall be capable of withstanding, without failure, a force of at least 200 pounds (890 n) applied within 2 inches (5 cm) of the top edge, in any downward or outward direction, at any point along the top edge. [§1926.1052(c)(5)]

(6) *The height of handrails* shall be not more than 37 inches (94 cm) nor less than 30 inches (76 cm) from the upper surface of the handrail to the surface of the tread, in line with the face of the riser at the forward edge of the tread. [§1926.1052(c)(6)]

(7) *When the top edge of a stairrail system* also serves as a handrail, the height of the top edge shall be not more than 37 inches (94 cm) nor less than 36 inches (91.5 cm) from the upper surface of the stairrail system to the surface of the tread, in line with the face of the riser at the forward edge of the tread. [§1926.1052(c)(7)]

(8) *Stairrail systems and handrails* shall be so surfaced as to prevent injury to employees from punctures or lacerations, and to prevent snagging of clothing. [§1926.1052(c)(8)]

(9) *Handrails shall provide an adequate handhold* for employees grasping them to avoid falling. [§1926.1052(c)(9)]

(10) *The ends of stairrail systems* and handrails shall be constructed so as not to constitute a projection hazard. [§1926.1052(c)(10)]

(11) *Handrails that will not be a permanent part* of the structure being built shall have a minimum clearance of 3 inches (8 cm) between the handrail and walls, stairrail systems, and other objects. [§1926.1052(c)(11)]

(12) *Unprotected sides and edges of stairway landings* shall be provided with guardrail systems. Guardrail system criteria are contained in subpart M of this part. [§1926.1052(c)(12)]

[55 FR 47687, Nov. 14, 1990; 56 FR 2585, Jan. 23, 1991; 56 FR 5061, Feb. 7, 1991; 56 FR 41794, Aug. 23, 1991]

§1926.1053
Ladders

(a) **General.** The following requirements apply to all ladders as indicated, including job-made ladders. [§1926.1053(a)]

(1) *Ladders shall be capable* of supporting the following loads without failure: [§1926.1053(a)(1)]

(i) *Each self-supporting portable ladder:* At least four times the maximum intended load, except that each extra-heavy-duty type 1A metal or plastic ladder shall sustain at least 3.3 times the maximum intended load. The ability of a ladder to sustain the loads indicated in this paragraph shall be determined by applying or transmitting the requisite load to the ladder in a downward vertical direction. Ladders built and tested in conformance with the applicable provisions of appendix A of this subpart will be deemed to meet this requirement. [§1926.1053(a)(1)(i)]

(ii) *Each portable ladder that is not self-supporting:* At least four times the maximum intended load, except that each extra-heavy-duty type 1A metal or plastic ladders shall sustain at least 3.3 times the maximum intended load. The ability of a ladder to sustain the loads indicated in this paragraph shall be determined by applying or transmitting the requisite load to the ladder in a downward vertical direction when the ladder is placed at an angle of 75 1/2 degrees from the horizontal. Ladders built and tested in conformance with the applicable provisions of appendix A will be deemed to meet this requirement. [§1926.1053(a)(1)(ii)]

(iii) *Each fixed ladder:* At least two loads of 250 pounds (114 kg) each, concentrated between any two consecutive attachments (the number and position of additional concentrated loads of 250 pounds (114 kg) each, determined from anticipated usage of the ladder, shall also be included), plus anticipated loads caused by ice buildup, winds, rigging, and impact loads resulting from the use of ladder safety devices. Each step or rung shall be capable of supporting a single concentrated load of at least 250 pounds (114 kg) applied in the middle of the step or rung. Ladders built in conformance with the applicable provisions of appendix A will be deemed to meet this requirement. [§1926.1053(a)(1)(iii)]

(2) *Ladder rungs, cleats, and steps* shall be parallel, level, and uniformly spaced when the ladder is in position for use. [§1926.1053(a)(2)]

(3)(i) *Rungs, cleats, and steps* of portable ladders (except as provided below) and fixed ladders (including individual-rung/step ladders) shall be spaced not less than 10 inches (25 cm) apart, nor more than 14 inches (36 cm) apart, as measured between center lines of the rungs, cleats, and steps. [§1926.1053(a)(3)(i)]

(ii) *Rungs, cleats, and steps* of step stools shall be not less than 8 inches (20 cm) apart, nor more than 12 inches (31 cm) apart, as measured between center lines of the rungs, cleats, and steps. [§1926.1053(a)(3)(ii)]

(iii) *Rungs, cleats, and steps* of the base section of extension trestle ladders shall not be less than 8 inches (20 cm) nor more than 18 inches (46 cm) apart, as measured between center lines of the rungs, cleats, and steps. The rung spacing on the extension section of the extension trestle ladder shall be not less than 6 inches (15 cm) nor more than 12 inches (31 cm), as measured between center lines of the rungs, cleats, and steps. [§1926.1053(a)(3)(iii)]

(4)(i) *The minimum clear distance* between the sides of individual-rung/step ladders and the minimum clear distance between the side rails of other fixed ladders shall be 16 inches (41 cm). [§1926.1053(a)(4)(i)]

(ii) *The minimum clear distance* between side rails for all portable ladders shall be 11 1/2 inches (29 cm). [§1926.1053(a)(4)(ii)]

(5) *The rungs of individual-rung/step ladders* shall be shaped such that employees' feet cannot slide off the end of the rungs. [§1926.1053(a)(5)]

(6)(i) *The rungs and steps of fixed metal ladders* manufactured after March 15, 1991, shall be corrugated, knurled, dimpled, coated with skid-resistant material, or otherwise treated to minimize slipping. [§1926.1053(a)(6)(i)]

(ii) *The rungs and steps of portable metal ladders* shall be corrugated, knurled, dimpled, coated with skid-resistant material, or otherwise treated to minimize slipping. [§1926.1053(a)(6)(ii)]

(7) *Ladders shall not be tied or fastened together* to provide longer sections unless they are specifically designed for such use. [§1926.1053(a)(7)]

(8) *A metal spreader or locking device* shall be provided on each stepladder to hold the front and back sections in an open position when the ladder is being used. [§1926.1053(a)(8)]

(9) *When splicing is required to obtain* a given length of side rail, the resulting side rail must be at least equivalent in strength to a one-piece side rail made of the same material. [§1926.1053(a)(9)]

(10) *Except when portable ladders are used* to gain access to fixed ladders (such as those on utility towers, billboards, and other structures where the bottom of the fixed ladder is elevated to limit access), when two or more separate ladders are used to reach an elevated work area, the ladders shall be offset with a platform or landing between the ladders. (The requirements to have guardrail systems with toeboards for falling object and overhead protection on platforms and landings are set forth in subpart M of this part.) [§1926.1053(a)(10)]

(11) *Ladder components shall be surfaced* so as to prevent injury to an employee from punctures or lacerations, and to prevent snagging of clothing. [§1926.1053(a)(11)]

(12) *Wood ladders shall not be coated* with any opaque covering, except for identification or warning labels which may be placed on one face only of a side rail. [§1926.1053(a)(12)]

(13) *The minimum perpendicular clearance* between fixed ladder rungs, cleats, and steps, and any obstruction behind the ladder shall be 7 inches (18 cm), except in the case of an elevator pit ladder, for which a minimum perpendicular clearance of 4 1/2 inches (11 cm) is required. [§1926.1053(a)(13)]

(14) *The minimum perpendicular clearance* between the center line of fixed ladder rungs, cleats, and steps, and any obstruction on the climbing side of the ladder shall be 30 inches (76 cm), except as provided in paragraph (a)(15) of this section. [§1926.1053(a)(14)]

(15) *When unavoidable obstructions are encountered,* the minimum perpendicular clearance between the centerline of fixed ladder rungs, cleats, and steps, and the obstruction on the climbing side of the ladder may be reduced to 24 inches (61 cm), provided that a deflection device is installed to guide employees around the obstruction. [§1926.1053(a)(15)]

(16) *Through fixed ladders at their point of access/egress* shall have a step-across distance of not less than 7 inches (18 cm) nor more than 12 inches (30 cm) as measured from the centerline of the steps or rungs to the nearest edge of the landing area. If the normal step-across distance exceeds 12 inches (30 cm), a landing platform shall be provided to reduce the distance to the specified limit. [§1926.1053(a)(16)]

(17) *Fixed ladders without cages or wells* shall have a clear width to the nearest permanent object of at least 15 inches (38 cm) on each side of the centerline of the ladder. [§1926.1053(a)(17)]

(18) *Fixed ladders shall be provided* with cages, wells, ladder safety devices, or self-retracting lifelines where the length of climb is less than 24 feet (7.3 m) but the top of the ladder is at a distance greater than 24 feet (7.3 m) above lower levels. [§1926.1053(a)(18)]

(19) *Where the total length of a climb* equals or exceeds 24 feet (7.3 m), fixed ladders shall be equipped with one of the following: [§1926.1053(a)(19)]

(i) *Ladder safety devices; or* [§1926.1053(a)(19)(i)]

(ii) *Self-retracting lifelines,* and rest platforms at intervals not to exceed 150 feet (45.7 m); or [§1926.1053(a)(19)(ii)]

(iii) *A cage or well,* and multiple ladder sections, each ladder section not to exceed 50 feet (15.2 m) in length. Ladder sections shall be offset from adjacent sections, and landing platforms shall be provided at maximum intervals of 50 feet (15.2 m). [§1926.1053(a)(19)(iii)]

(20) *Cages for fixed ladders shall conform to all of the following:* [§1926.1053(a)(20)]

(i) *Horizontal bands shall be fastened* to the side rails of rail ladders, or directly to the structure, building, or equipment for individual-rung ladders; [§1926.1053(a)(20)(i)]

(ii) *Vertical bars shall be on the inside* of the horizontal bands and shall be fastened to them; [§1926.1053(a)(20)(ii)]

(iii) *Cages shall extend* not less than 27 inches (68 cm), or more than 30 inches (76 cm) from the centerline of the step or rung (excluding the flare at the bottom of the cage), and shall not be less than 27 inches (68 cm) in width; [§1926.1053(a)(20)(iii)]

(iv) *The inside of the cage shall be clear of projections;* [§1926.1053(a)(20)(iv)]

(v) *Horizontal bands shall be spaced* not more than 4 feet (1.2 m) on center vertically; [§1926.1053(a)(20)(v)]

(vi) *Vertical bars shall be spaced* at intervals not more than 9 1/2 inches (24 cm) on center horizontally; [§1926.1053(a)(20)(vi)]

(vii) *The bottom of the cage shall be at a level* not less than 7 feet (2.1 m) nor more than 8 feet (2.4 m) above the point of access to the bottom of the ladder. The bottom of the cage shall be flared not less than 4 inches (10 cm) all around within the distance between the bottom horizontal band and the next higher band; [§1926.1053(a)(20)(vii)]

(viii) *The top of the cage shall be a minimum* of 42 inches (1.1 m) above the top of the platform, or the point of access at the top of the ladder, with provision for access to the platform or other point of access. [§1926.1053(a)(20)(viii)]

(21) *Wells for fixed ladders shall conform to all of the following:* [§1926.1053(a)(21)]

(i) *They shall completely encircle the ladder;* [§1926.1053(a)(21)(i)]

(ii) *They shall be free of projections;* [§1926.1053(a)(21)(ii)]

(iii) *Their inside face on the climbing side* of the ladder shall extend not less than 27 inches (68 cm) nor more than 30 inches (76 cm) from the centerline of the step or rung; [§1926.1053(a)(21)(iii)]

(iv) *The inside clear width shall be at least 30 inches (76 cm);* [§1926.1053(a)(21)(iv)]

(v) *The bottom of the wall on the access side* shall start at a level not less than 7 feet (2.1 m) nor more than 8 feet (2.4 m) above the point of access to the bottom of the ladder. [§1926.1053(a)(21)(v)]

(22) *Ladder safety devices,* and related support systems, for fixed ladders shall conform to all of the following: [§1926.1053(a)(22)]

(i) *They shall be capable of withstanding* without failure a drop test consisting of an 18-inch (41 cm) drop of a 500-pound (226 kg) weight; [§1926.1053(a)(22)(i)]

(ii) *They shall permit the employee using the device* to ascend or descend without continually having to hold, push or pull any part of the device, leaving both hands free for climbing; [§1926.1053(a)(22)(ii)]

(iii) *They shall be activated within 2 feet (.61 m)* after a fall occurs, and limit the descending velocity of an employee to 7 feet/sec. (2.1 m/sec.) or less; [§1926.1053(a)(22)(iii)]

(iv) *The connection between the carrier or lifeline* and the point of attachment to the body belt or harness shall not exceed 9 inches (23 cm) in length. [§1926.1053(a)(22)(iv)]

(23) *The mounting of ladder safety devices* for fixed ladders shall conform to the following: [§1926.1053(a)(23)]

(i) *Mountings for rigid carriers* shall be attached at each end of the carrier, with intermediate mountings, as necessary, spaced along the entire length of the carrier, to provide the strength necessary to stop employees' falls. [§1926.1053(a)(23)(i)]

(ii) *Mountings for flexible carriers* shall be attached at each end of the carrier. When the system is exposed to wind, cable guides for flexible carriers shall be installed at a minimum spacing of 25 feet (7.6 m) and maximum spacing of 40 feet (12.2 m) along the entire length of the carrier, to prevent wind damage to the system. [§1926.1053(a)(23)(ii)]

(iii) *The design and installation of mountings* and cable guides shall not reduce the design strength of the ladder. [§1926.1053(a)(23)(iii)]

(24) *The side rails of through or side-step fixed ladders* shall extend 42 inches (1.1 m) above the top of the access level or landing platform served by the ladder. For a parapet ladder, the access level shall be the roof if the parapet is cut to permit passage through the parapet; if the parapet is continuous, the access level shall be the top of the parapet. [§1926.1053(a)(24)]

(25) *For through-fixed-ladder extensions,* the steps or rungs shall be omitted from the extension and the extension of the side rails shall be flared to provide not less than 24 inches (61 cm) nor more than 30 inches (76 cm) clearance between side rails. Where ladder safety devices are provided, the maximum clearance between side rails of the extensions shall not exceed 36 inches (91 cm). [§1926.1053(a)(25)]

(26) *For side-step fixed ladders,* the side rails and the steps or rungs shall be continuous in the extension. [§1926.1053(a)(26)]

(27) *Individual-rung/step ladders,* except those used where their access openings are covered with manhole covers or hatches, shall extend at least 42 inches (1.1 m) above an access level or landing platform either by the continuation of the rung spacings as horizontal grab bars or by providing vertical grab bars that shall have the same lateral spacing as the vertical legs of the rungs. [§1926.1053(a)(27)]

(b) **Use.** The following requirements apply to the use of all ladders, including job-made ladders, except as otherwise indicated: [§1926.1053(b)]

(1) *When portable ladders are used* for access to an upper landing surface, the ladder side rails shall extend at least 3 feet (.9 m) above the upper landing surface to which the ladder is used to gain access; or, when such an extension is not possible because of the ladder's length, then the ladder shall be secured at its top to a rigid support that will not deflect, and a grasping device, such as a grabrail, shall be provided to assist employees in mounting and dismounting the ladder. In no case shall the extension be such that ladder deflection under a load would, by itself, cause the ladder to slip off its support. [§1926.1053(b)(1)]

(2) *Ladders shall be maintained* free of oil, grease, and other slipping hazards. [§1926.1053(b)(2)]

(3) *Ladders shall not be loaded beyond* the maximum intended load for which they were built, nor beyond their manufacturer's rated capacity. [§1926.1053(b)(3)]

(4) *Ladders shall be used only* for the purpose for which they were designed. [§1926.1053(b)(4)]

(5)(i) *Non-self-supporting ladders* shall be used at an angle such that the horizontal distance from the top support to the foot of the ladder is approximately one-quarter of the working length of the ladder (the distance along the ladder between the foot and the top support). [§1926.1053(b)(5)(i)]

(ii) *Wood job-made ladders* with spliced side rails shall be used at an angle such that the horizontal distance is one-eighth the working length of the ladder. [§1926.1053(b)(5)(ii)]

(iii) *Fixed ladders shall be used* at a pitch no greater than 90 degrees from the horizontal, as measured to the back side of the ladder. [§1926.1053(b)(5)(iii)]

(6) *Ladders shall be used only* on stable and level surfaces unless secured to prevent accidental displacement. [§1926.1053(b)(6)]

(7) *Ladders shall not be used* on slippery surfaces unless secured or provided with slip-resistant feet to prevent accidental displacement. Slip-resistant feet shall not be used as a substitute for care in placing, lashing, or holding a ladder that is used upon slippery surfaces including, but not limited to, flat metal or concrete surfaces that are constructed so they cannot be prevented from becoming slippery. [§1926.1053(b)(7)]

(8) *Ladders placed in any location* where they can be displaced by workplace activities or traffic, such as in passageways, doorways, or driveways, shall be secured to prevent accidental displacement, or a barricade shall be used to keep the activities or traffic away from the ladder. [§1926.1053(b)(8)]

(9) *The area around the top and bottom of ladders shall be kept clear.* [§1926.1053(b)(9)]

(10) *The top of a non-self-supporting ladder* shall be placed with the two rails supported equally unless it is equipped with a single support attachment. [§1926.1053(b)(10)]

(11) *Ladders shall not be moved,* shifted, or extended while occupied. [§1926.1053(b)(11)]

(12) *Ladders shall have nonconductive* siderails if they are used where the employee or the ladder could contact exposed energized electrical equipment, except as provided in §1926.951(c)(1) of this part. [§1926.1053(b)(12)]

(13) *The top or top step* of a stepladder shall not be used as a step. [§1926.1053(b)(13)]

(14) *Cross-bracing on the rear section* of stepladders shall not be used for climbing unless the ladders are designed and provided with steps for climbing on both front and rear sections. [§1926.1053(b)(14)]

(15) *Ladders shall be inspected* by a competent person for visible defects on a periodic basis and after any occurrence that could affect their safe use. [§1926.1053(b)(15)]

(16) *Portable ladders with structural defects,* such as, but not limited to, broken or missing rungs, cleats, or steps, broken or split rails, corroded components, or other faulty or defective components, shall either be immediately marked in a manner that readily identifies them as defective, or be tagged with "Do Not Use" or similar language, and shall be withdrawn from service until repaired. [§1926.1053(b)(16)]

(17) *Fixed ladders with structural defects,* such as, but not limited to, broken or missing rungs, cleats, or steps, broken or split rails, or corroded components, shall be withdrawn from service until repaired. The requirement to withdraw a defective ladder from service is satisfied if the ladder is either: [§1926.1053(b)(17)]

(i) *Immediately tagged with* "Do Not Use" or similar language, [§1926.1053(b)(17)(i)]

(ii) *Marked in a manner* that readily identifies it as defective; [§1926.1053(b)(17)(ii)]

(iii) *Or blocked* (such as with a plywood attachment that spans several rungs). [§1926.1053(b)(17)(iii)]

(18) *Ladder repairs shall restore* the ladder to a condition meeting its original design criteria, before the ladder is returned to use. [§1926.1053(b)(18)]

(19) *Single-rail ladders shall not be used.* [§1926.1053(b)(19)]

(20) *When ascending or descending a ladder,* the user shall face the ladder. [§1926.1053(b)(20)]

(21) *Each employee shall use* at least one hand to grasp the ladder when progressing up and/or down the ladder. [§1926.1053(b)(21)]

(22) *An employee shall not carry any object* or load that could cause the employee to lose balance and fall. [§1926.1053(b)(22)]

[55 FR 47687, Nov. 14, 1990; 56 FR 2585, Jan. 23, 1991, as amended at 56 FR 41794, Aug. 23, 1991]

§§1926.1054 - 1926.1059

[Reserved]

§1926.1060

Training requirements

The following training provisions clarify the requirements of §1926.21(b)(2), regarding the hazards addressed in Subpart X.

(a) **The employer shall provide a training program** for each employee using ladders and stairways, as necessary. The program shall enable each employee to recognize hazards related to ladders and stairways, and shall train each employee in the procedures to be followed to minimize these hazards. [§1926.1060(a)]

(1) *The employer shall ensure* that each employee has been trained by a competent person in the following areas, as applicable: [§1926.1060(a)(1)]

(i) *The nature of fall hazards in the work area;* [§1926.1060(a)(1)(i)]

(ii) *The correct procedures* for erecting, maintaining, and disassembling the fall protection systems to be used; [§1926.1060(a)(1)(ii)]

(iii) *The proper construction, use, placement, and care* in handling of all stairways and ladders; [§1926.1060(a)(1)(iii)]

(iv) *The maximum intended* load-carrying capacities of ladders used; and [§1926.1060(a)(1)(iv)]

(v) *The standards contained in this subpart.* [§1926.1060(a)(1)(v)]

(b) **Retraining shall be provided for each employee as necessary** so that the employee maintains the understanding and knowledge acquired through compliance with this section. [§1926.1060(b)]

Appendix A to Subpart X of Part 1926

Ladders

This appendix serves as a non-mandatory guideline to assist employers in complying with the ladder loading and strength requirements of §1926.1053(a)(1). A ladder designed and built in accordance with the applicable national consensus standards, as set forth below, will be considered to meet the requirements of §1926.1053(a)(1):

- Manufactured portable wood ladders: American National Standards Institute (ANSI) A14.1-1982 — American National Standard for Ladders-Portable Wood-Safety Requirements.
- Manufactured portable metal ladders: ANSI A14.2- 1982 — American National Standard for Ladders — Portable Metal-Safety Requirements.
- Manufactured fixed ladders: ANSI A14.3-1984 — American National Standard for Ladders-Fixed-Safety Requirements.
- Job-made ladders: ANSI A14.4-1979 — Safety Requirements for Job-Made Ladders.
- Plastic ladders: ANSI A14.5-1982 — American National Standard for Ladders-Portable Reinforced Plastic-Safety Requirements.

1926 Subpart X

Authority for 1926 Subpart X

Authority: Section 107, Contract Work Hours and Safety Standards Act (Construction Safety Act](40 U.S.C. 333); Secs. 4, 6, 8, Occupational Safety and Health Act of 1970 (29 U.S.C. 653, 655, 657); Secretary of Labor's Order Nos. 1–90 (55 FR 9033), 5–2007 (72 FR 31159); and 29 CFR part 1911.

Subpart Y - Diving

GENERAL

§1926.1071

Scope and application

Note: The requirements applicable to construction work under this section are identical to those set forth at §1910.401 of this chapter. (Editor's Note: See 1910 Supplement.)

[61 FR 31432, June 20, 1996]

§1926.1072

Definitions

Note: The provisions applicable to construction work under this section are identical to those set forth at §1910.402 of this chapter. (Editor's Note: See 1910 Supplement.)

[61 FR 31432, June 20, 1996]

PERSONNEL REQUIREMENTS

§1926.1076

Qualifications of dive team

Note: The requirements applicable to construction work under this section are identical to those set forth at §1910.410 of this chapter. (Editor's Note: See 1910 Supplement.)

[61 FR 31432, June 20, 1996]

GENERAL OPERATIONS PROCEDURES

§1926.1080

Safe practices manual

Note: The requirements applicable to construction work under this section are identical to those set forth at §1910.420 of this chapter. (Editor's Note: See 1910 Supplement.)

[61 FR 31432, June 20, 1996]

§1926.1081

Pre-dive procedures

Note: The requirements applicable to construction work under this section are identical to those set forth at §1910.421 of this chapter. (Editor's Note: See 1910 Supplement.)

[61 FR 31432, June 20, 1996]

§1926.1082

Procedures during dive

Note: The requirements applicable to construction work under this section are identical to those set forth at §1910.422 of this chapter. (Editor's Note: See 1910 Supplement.)

[61 FR 31432, June 20, 1996]

§1926.1083

Post-dive procedures

Note: The requirements applicable to construction work under this section are identical to those set forth at §1910.423 of this chapter. (Editor's Note: See 1910 Supplement.)

[61 FR 31432, June 20, 1996]

SPECIFIC OPERATIONS PROCEDURES

§1926.1084

SCUBA diving

Note: The requirements applicable to construction work under this section are identical to those set forth at §1910.424 of this chapter. (Editor's Note: See 1910 Supplement.)

[61 FR 31432, June 20, 1996]

§1926.1085

Surface-supplied air diving

Note: The requirements applicable to construction work under this section are identical to those set forth at §1910.425 of this chapter. (Editor's Note: See 1910 Supplement.)

[61 FR 31432, June 20, 1996]

§1926.1086

Mixed-gas diving

Note: The requirements applicable to construction work under this section are identical to those set forth at §1910.426 of this chapter. (Editor's Note: See 1910 Supplement.)

[61 FR 31432, June 20, 1996]

§1926.1087

Liveboating

Note: The requirements applicable to construction work under this section are identical to those set forth at §1910.427 of this chapter. (Editor's Note: See 1910 Supplement.)

[61 FR 31432, June 20, 1996]

EQUIPMENT PROCEDURES AND REQUIREMENTS

§1926.1090

Equipment

Note: The requirements applicable to construction work under this section are identical to those set forth at §1910.430 of this chapter. (Editor's Note: See 1910 Supplement.)

[61 FR 31432, June 20, 1996]

RECORDKEEPING

§1926.1091

Recordkeeping requirements

Note: The requirements applicable to construction work under this section are identical to those set forth at §1910.440 of this chapter. (Editor's Note: See 1910 Supplement.)

[61 FR 31432, June 20, 1996]

1926 Subpart Y

Authority for 1926 Subpart Y

Authority: Sections 4, 6, and 8 of the Occupational Safety and Health Act of 1970 (29 U.S.C. 653, 655, 657); §107, Contract Work Hours and Safety Standards Act (the Construction Safety Standards Act) (40 U.S.C. 333); §41, Longshore and Harbor Workers' Compensation Act (33 U.S.C. 941); Secretary of Labor's Order No. 12-71 (36 FR 8754), 8-76 (41 FR 25059), 9-83 (48 FR 35736), 1-90 (55 FR 9033), 3-2000 (65 FR 50017) or 5-2002 (67 FR 65008) as applicable; and 29 CFR part 1911.

Source: 58 FR 35184, June 30, 1993, unless otherwise noted.

1926 Subpart Y Appendix A

Examples of Conditions Which May Restrict or Limit Exposure to Hyperbaric Conditions

*Note:*The requirements applicable to construction work under this Appendix A are identical to those set forth at Appendix A to Subpart T of Part 1910 of this chapter. (Editor's Note: See 1910 Supplement.)

[61 FR 31432, June 20, 1996]

1926 Subpart Y Appendix B

Guidelines for Scientific Diving

*Note:*The requirements applicable to construction work under this Appendix B are identical to those set forth at Appendix B to Subpart T of Part 1910 of this chapter. (Editor's Note: See 1910 Supplement.)

[61 FR 31433, June 20, 1996]

Notes

Subpart Z - Toxic and Hazardous Substances

§§1926.1100 [Reserved]

§1926.1101 Asbestos

(a) Scope and application. This section regulates asbestos exposure in all work as defined in 29 CFR 1910.12(b), including but not limited to the following:[§1926.1101(a)]

(1) *Demolition or salvage of structures where asbestos is present;* [§1926.1101(a)(1)]

(2) *Removal or encapsulation of materials containing asbestos;* [§1926.1101(a)(2)]

(3) *Construction, alteration, repair, maintenance, or renovation* of structures, substrates, or portions thereof, that contain asbestos;[§1926.1101(a)(3)]

(4) *Installation of products containing asbestos;*[§1926.1101(a)(4)]

(5) *Asbestos spill/emergency cleanup; and*[§1926.1101(a)(5)]

(6) *Transportation, disposal, storage,* containment of and housekeeping activities involving asbestos or products containing asbestos, on the site or location at which construction activities are performed.[§1926.1101(a)(6)]

(7) *Coverage under this standard* shall be based on the nature of the work operation involving asbestos exposure.[§1926.1101(a)(7)]

(8) *This section does not apply* to asbestos-containing asphalt roof coatings, cements and mastics.[§1926.1101(a)(8)]

(b) Definitions.[§1926.1101(b)]

Aggressive method means removal or disturbance of building material by sanding, abrading, grinding or other method that breaks, crumbles, or disintegrates intact ACM.[§1926.1101(b)]

Amended water means water to which surfactant (wetting agent) has been added to increase the ability of the liquid to penetrate ACM. [§1926.1101(b)]

Asbestos includes chrysotile, amosite, crocidolite, tremolite asbestos, anthophyllite asbestos, actinolite asbestos, and any of these minerals that has been chemically treated and/or altered. For purposes of this standard, "asbestos" includes PACM, as defined below.[§1926.1101(b)]

Asbestos-containing material (ACM), means any material containing more than one percent asbestos.[§1926.1101(b)]

Assistant Secretary means the Assistant Secretary of Labor for Occupational Safety and Health, U.S. Department of Labor, or designee.[§1926.1101(b)]

Authorized person means any person authorized by the employer and required by work duties to be present in regulated areas. [§1926.1101(b)]

Building/facility owner is the legal entity, including a lessee, which exercises control over management and record keeping functions relating to a building and/or facility in which activities covered by this standard take place.[§1926.1101(b)]

Certified Industrial Hygienist (CIH) means one certified in the practice of industrial hygiene by the American Board of Industrial Hygiene.[§1926.1101(b)]

Class I asbestos work means activities involving the removal of TSI and surfacing ACM and PACM.[§1926.1101(b)]

Class II asbestos work means activities involving the removal of ACM which is not thermal system insulation or surfacing material. This includes, but is not limited to, the removal of asbestos-containing wallboard, floor tile and sheeting, roofing and siding shingles, and construction mastics.[§1926.1101(b)]

Class III asbestos work means repair and maintenance operations, where "ACM", including TSI and surfacing ACM and PACM, is likely to be disturbed.[§1926.1101(b)]

Class IV asbestos work means maintenance and custodial activities during which employees contact but do not disturb ACM or PACM and activities to clean up dust, waste and debris resulting from Class I, II, and III activities.[§1926.1101(b)]

Clean room means an uncontaminated room having facilities for the storage of employees' street clothing and uncontaminated materials and equipment.[§1926.1101(b)]

Closely resemble means that the major workplace conditions which have contributed to the levels of historic asbestos exposure, are no more protective than conditions of the current workplace. [§1926.1101(b)]

Competent person means, in addition to the definition in 29 CFR 1926.32 (f), one who is capable of identifying existing asbestos hazards in the workplace and selecting the appropriate control strategy for asbestos exposure, who has the authority to take prompt corrective measures to eliminate them, as specified in 29 CFR 1926.32(f): in addition, for Class I and Class II work who is specially trained in a training course which meets the criteria of EPA's Model Accreditation Plan (40 CFR part 763) for supervisor, or its equivalent and, for Class III and Class IV work, who is trained in a manner consistent with EPA requirements for training of local education agency maintenance and custodial staff as set forth at 40 CFR 763.92 (a)(2).[§1926.1101(b)]

Critical barrier means one or more layers of plastic sealed over all openings into a work area or any other similarly placed physical barrier sufficient to prevent airborne asbestos in a work area from migrating to an adjacent area.[§1926.1101(b)]

Decontamination area means an enclosed area adjacent and connected to the regulated area and consisting of an equipment room, shower area, and clean room, which is used for the decontamination of workers, materials, and equipment that are contaminated with asbestos.[§1926.1101(b)]

Demolition means the wrecking or taking out of any load-supporting structural member and any related razing, removing, or stripping of asbestos products.[§1926.1101(b)]

Director means the Director, National Institute for Occupational Safety and Health, U.S. Department of Health and Human Services, or designee.[§1926.1101(b)]

Disturbance means activities that disrupt the matrix of ACM or PACM, crumble or pulverize ACM or PACM, or generate visible debris from ACM or PACM. In no event shall the amount of ACM or PACM so disturbed exceed that which can be contained in one glove bag or waste bag which shall not exceed 60 inches in length and width.[§1926.1101(b)]

Employee exposure means that exposure to airborne asbestos that would occur if the employee were not using respiratory protective equipment.[§1926.1101(b)]

Equipment room (change room) means a contaminated room located within the decontamination area that is supplied with impermeable bags or containers for the disposal of contaminated protective clothing and equipment.[§1926.1101(b)]

Fiber means a particulate form of asbestos, 5 micrometers or longer, with a length-to-diameter ratio of at least 3 to 1.[§1926.1101(b)]

Glovebag means not more than a 60x60 inch impervious plastic bag-like enclosure affixed around an asbestos-containing material, with glove-like appendages through which material and tools may be handled.[§1926.1101(b)]

High-efficiency particulate air (HEPA) filter means a filter capable of trapping and retaining at least 99.97 percent of all mono-dispersed particles of 0.3 micrometers in diameter.[§1926.1101(b)]

Homogeneous area means an area of surfacing material or thermal system insulation that is uniform in color and texture.[§1926.1101(b)]

Industrial hygienist means a professional qualified by education, training, and experience to anticipate, recognize, evaluate and develop controls for occupational health hazards.[§1926.1101(b)]

Intact means that the ACM has not crumbled, been pulverized, or otherwise deteriorated so that the asbestos is no longer likely to be bound with its matrix.[§1926.1101(b)]

Modification for purposes of paragraph (g)(6)(ii), means a changed or altered procedure, material or component of a control system, which replaces a procedure, material or component of a required system. Omitting a procedure or component, or reducing or diminishing the stringency or strength of a material or component of the control system is not a "modification" for purposes of paragraph (g)(6) of this section.[§1926.1101(b)]

Negative Initial Exposure Assessment means a demonstration by the employer, which complies with the criteria in paragraph (f)(2)(iii) of this section, that employee exposure during an operation is expected to be consistently below the PELs.[§1926.1101(b)]

PACM means "presumed asbestos containing material".[§1926.1101(b)]

Presumed Asbestos Containing Material means thermal system insulation and surfacing material found in buildings constructed no later than 1980. The designation of a material as "PACM" may be rebutted pursuant to paragraph (k)(5) of this section.[§1926.1101(b)]

Project Designer means a person who has successfully completed the training requirements for an abatement project designer established by 40 U.S.C. 763.90(g).[§1926.1101(b)]

Regulated area means: an area established by the employer to demarcate areas where Class I, II, and III asbestos work is conducted, and any adjoining area where debris and waste from such asbestos work accumulate; and a work area within which airborne concentrations of asbestos, exceed or there is a reasonable possibility they may exceed the permissible exposure limit. Requirements for regulated areas are set out in paragraph (e) of this section.[§1926.1101(b)]

Removal means all operations where ACM and/or PACM is taken out or stripped from structures or substrates, and includes demolition operations.[§1926.1101(b)]

Renovation means the modifying of any existing structure, or portion thereof.[§1926.1101(b)]

Repair means overhauling, rebuilding, reconstructing, or reconditioning of structures or substrates, including encapsulation or other repair of ACM or PACM attached to structures or substrates. [§1926.1101(b)]

Surfacing material means material that is sprayed, troweled-on or otherwise applied to surfaces (such as acoustical plaster on ceilings and fireproofing materials on structural members, or other materials on surfaces for acoustical, fireproofing, and other purposes).[§1926.1101(b)]

Surfacing ACM means surfacing material which contains more than 1% asbestos.[§1926.1101(b)]

Thermal system insulation (TSI) means ACM applied to pipes, fittings, boilers, breeching, tanks, ducts or other structural components to prevent heat loss or gain.[§1926.1101(b)]

Thermal system insulation ACM is thermal system insulation which contains more than 1% asbestos.[§1926.1101(b)]

(c) Permissible exposure limits (PELS)[§1926.1101(c)]

(1) *Time-weighted average limit (TWA).* The employer shall ensure that no employee is exposed to an airborne concentration of asbestos in excess of 0.1 fiber per cubic centimeter of air as an eight (8) hour time-weighted average (TWA), as determined by the method prescribed in Appendix A to this section, or by an equivalent method.[§1926.1101(c)(1)]

(2) *Excursion limit.* The employer shall ensure that no employee is exposed to an airborne concentration of asbestos in excess of 1.0 fiber per cubic centimeter of air (1 f/cc) as averaged over a sampling period of thirty (30) minutes, as determined by the method prescribed in Appendix A to this section, or by an equivalent method.[§1926.1101(c)(2)]

(d) Multi-employer worksites.[§1926.1101(d)]

(1) *On multi-employer worksites,* an employer performing work requiring the establishment of a regulated area shall inform other employers on the site of the nature of the employer's work with asbestos and/or PACM, of the existence of and requirements pertaining to regulated areas, and the measures taken to ensure that employees of such other employers are not exposed to asbestos.[§1926.1101(d)(1)]

(2) *Asbestos hazards at a multi-employer work site* shall be abated by the contractor who created or controls the source of asbestos contamination. For example, if there is a significant breach of an enclosure containing Class I work, the employer responsible for erecting the enclosure shall repair the breach immediately.[§1926.1101(d)(2)]

(3) *In addition,* all employers of employees exposed to asbestos hazards shall comply with applicable protective provisions to protect their employees. For example, if employees working immediately adjacent to a Class I asbestos job are exposed to asbestos due to the inadequate containment of such job, their employer shall either remove the employees from the area until the enclosure breach is repaired; or perform an initial exposure assessment pursuant to (f) of this section.[§1926.1101(d)(3)]

(4) *All employers of employees working adjacent* to regulated areas established by another employer on a multi-employer work-site, shall take steps on a daily basis to ascertain the integrity of the enclosure and/or the effectiveness of the control method relied on by the primary asbestos contractor to assure that asbestos fibers do not migrate to such adjacent areas. [§1926.1101(d)(4)]

(5) *All general contractors on a construction project* which includes work covered by this standard shall be deemed to exercise general supervisory authority over the work covered by this standard, even though the general contractor is not qualified to serve as the asbestos "competent person" as defined by paragraph (b) of this section. As supervisor of the entire project, the general contractor shall ascertain whether the asbestos contractor is in compliance with this standard, and shall require such contractor to come into compliance with this standard when necessary.[§1926.1101(d)(5)]

(e) Regulated areas.[§1926.1101(e)]

(1) *All Class I, II and III asbestos work* shall be conducted within regulated areas. All other operations covered by this standard shall be conducted within a regulated area where airborne concentrations of asbestos exceed, or there is a reasonable possibility they may exceed a PEL. Regulated areas shall comply with the requirements of paragraphs (2), (3),(4) and (5) of this section.[§1926.1101(e)(1)]

(2) *Demarcation.* The regulated area shall be demarcated in any manner that minimizes the number of persons within the area and protects persons outside the area from exposure to airborne asbestos. Where critical barriers or negative pressure enclosures are used, they may demarcate the regulated area. Signs shall be provided and displayed pursuant to the requirements of paragraph (k)(7) of this section.[§1926.1101(e)(2)]

(3) *Access.* Access to regulated areas shall be limited to authorized persons and to persons authorized by the Act or regulations issued pursuant thereto.[§1926.1101(e)(3)]

(4) *Respirators.* All persons entering a regulated area where employees are required pursuant to paragraph (h)(1) of this section to wear respirators shall be supplied with a respirator selected in accordance with paragraph (h)(2) of this section. [§1926.1101(e)(4)]

(5) *Prohibited activities.* The employer shall ensure that employees do not eat, drink, smoke, chew tobacco or gum, or apply cosmetics in the regulated area.[§1926.1101(e)(5)]

(6) *Competent Persons.* The employer shall ensure that all asbestos work performed within regulated areas is supervised by a competent person, as defined in paragraph (b) of this section. The duties of the competent person are set out in paragraph (o) of this section.[§1926.1101(e)(6)]

(f) Exposure assessments and monitoring.[§1926.1101(f)]

(1) *General monitoring criteria.*[§1926.1101(f)(1)]

(i) *Each employer who has a workplace* or work operation where exposure monitoring is required under this section shall perform monitoring to determine accurately the airborne concentrations of asbestos to which employees may be exposed.[§1926.1101(f)(1)(i)]

(ii) *Determinations of employee exposure* shall be made from breathing zone air samples that are representative of the 8-hour TWA and 30-minute short-term exposures of each employee.[§1926.1101(f)(1)(ii)]

(iii) *Representative 8-hour TWA employee exposure* shall be determined on the basis of one or more samples representing full-shift exposure for employees in each work area. Representative 30-minute short-term employee exposures shall be determined on the basis of one or more samples representing 30 minute exposures associated with operations that are most likely to produce exposures above the excursion limit for employees in each work area.[§1926.1101(f)(1)(iii)]

(2) *Initial Exposure Assessment.*[§1926.1101(f)(2)]

(i) *Each employer who has a workplace* or work operation covered by this standard shall ensure that a "competent person" conducts an exposure assessment immediately before or at the initiation of the operation to ascertain expected exposures during that operation or workplace. The assessment must be completed in time to comply with requirements which are triggered by exposure data or the lack of a "negative exposure assessment," and to provide information necessary to assure that all control systems planned are appropriate for that operation and will work properly. [§1926.1101(f)(2)(i)]

(ii) *Basis of Initial Exposure Assessment:* Unless a negative exposure assessment has been made pursuant to paragraph (f)(2)(iii) of this section, the initial exposure assessment shall, if feasible, be based on monitoring conducted pursuant to paragraph (f)(1)(iii) of this section. The assessment shall take into consideration both the monitoring results and all observations, information or calculations which indicate employee exposure to asbestos, including any previous monitoring conducted in the workplace, or of the operations of the employer which indicate the levels of airborne asbestos likely to be encountered on the job. For Class I asbestos work, until the employer conducts exposure monitoring and documents that employees on that job will not be exposed in excess of the PELs, or otherwise makes a negative exposure assessment pursuant to paragraph (f)(2)(iii) of this section, the employer shall presume that employees are exposed in excess of the TWA and excursion limit.[§1926.1101(f)(2)(ii)]

(iii) *Negative Exposure Assessment:* For any one specific asbestos job which will be performed by employees who have been trained in compliance with the standard, the employer may demonstrate that employee exposures will be below the PELs by data which conform to the following criteria;[§1926.1101(f)(2)(iii)]

[A] Objective data demonstrating that the product or material containing asbestos minerals or the activity involving such product or material cannot release airborne fibers in concentrations exceeding the TWA and excursion limit under those work conditions having the greatest potential for releasing asbestos; or[§1926.1101(f)(2)(iii)[A]]

[B] Where the employer has monitored prior asbestos jobs for the PEL and the excursion limit within 12 months of the current or projected job, the monitoring and analysis were performed in compliance with the asbestos standard in effect; and the data were obtained during work operations conducted under workplace conditions "closely resembling" the processes, type of material, control methods, work practices, and environmental conditions used and prevailing in the employer's current operations, the operations were conducted by employees whose training and experience are no more extensive than that of employees performing the current job, and these data show that under the conditions prevailing and which will prevail in the current workplace there is a high degree of certainty that employee exposures will not exceed the TWA and excursion limit; or[§1926.1101(f)(2)(iii)[B]]

[C] The results of initial exposure monitoring of the current job made from breathing zone air samples that are representative of the 8- hour TWA and 30-minute short-term exposures of each employee covering operations which are most likely during the performance of the entire asbestos job to result in exposures over the PELs. [§1926.1101(f)(2)(iii)[C]]

(3) *Periodic monitoring.*[§1926.1101(f)(3)]

(i) *Class I and II operations.* The employer shall conduct daily monitoring that is representative of the exposure of each employee who is assigned to work within a regulated area who is performing Class I or II work, unless the employer pursuant to (f)(2)(iii) of this section, has made a negative exposure assessment for the entire operation.[§1926.1101(f)(3)(i)]

(ii) *All operations under the standard* other than Class I and II operations. The employer shall conduct periodic monitoring of all work where exposures are expected to exceed a PEL, at intervals sufficient to document the validity of the exposure prediction.[§1926.1101(f)(3)(ii)]

(iii) *Exception:* When all employees required to be monitored daily are equipped with supplied-air respirators operated in the pressure demand mode, or other positive pressure mode respirator, the employer may dispense with the daily monitoring required by this paragraph. However, employees performing Class I work using a control method which is not listed in paragraph (g)(4) (i), (ii), or (iii) of this section or using a modification of a listed control method, shall continue to be monitored daily even if they are equipped with supplied-air respirators.[§1926.1101(f)(3)(iii)]

(4) *Termination of monitoring.*[§1926.1101(f)(4)]

(i) *If the periodic monitoring* required by paragraph (f)(3) of this section reveals that employee exposures, as indicated by statistically reliable measurements, are below the permissible exposure limit and excursion limit the employer may discontinue monitoring for those employees whose exposures are represented by such monitoring.[§1926.1101(f)(4)(i)]

(ii) *Additional monitoring.* Notwithstanding the provisions of paragraph (f) (2) and (3), and (f)(4) of this section, the employer shall institute the exposure monitoring required under paragraph (f)(3) of this section whenever there has been a change in process, control equipment, personnel or work practices that may result in new or additional exposures above the permissible exposure limit and/or excursion limit or when the employer has any reason to suspect that a change may result in new or additional exposures above the permissible exposure limit and/or excursion limit. Such additional monitoring is required regardless of whether a "negative exposure assessment" was previously produced for a specific job.[§1926.1101(f)(4)(ii)]

(5) *Employee notification of monitoring results.* The employer must, as soon as possible but no later than 5 working days after the receipt of the results of any monitoring performed under this section, notify each affected employee of these results either individually in writing or by posting the results in an appropriate location that is accessible to employees.[§1926.1101(f)(5)]

(6) *Observation of monitoring.*[§1926.1101(f)(6)]

(i) *The employer shall provide affected employees* and their designated representatives an opportunity to observe any monitoring of employee exposure to asbestos conducted in accordance with this section.[§1926.1101(f)(6)(i)]

(ii) *When observation of the monitoring of employee exposure* to asbestos requires entry into an area where the use of protective clothing or equipment is required, the observer shall be provided with and be required to use such clothing and equipment and shall comply with all other applicable safety and health procedures.[§1926.1101(f)(6)(ii)]

(g) Methods of compliance.[§1926.1101(g)]

(1) *Engineering controls and work practices for all operations* covered by this section. The employer shall use the following engineering controls and work practices in all operations covered by this section, regardless of the levels of exposure:[§1926.1101(g)(1)]

(i) *Vacuum cleaners equipped with HEPA filters* to collect all debris and dust containing ACM and PACM, except as provided in paragraph (g)(8)(ii) of this section in the case of roofing material.[§1926.1101(g)(1)(i)]

(ii) *Wet methods, or wetting agents,* to control employee exposures during asbestos handling, mixing, removal, cutting, application, and cleanup, except where employers demonstrate that the use of wet methods is infeasible due to for example, the creation of electrical hazards, equipment malfunction, and, in roofing, except as provided in paragraph (g)(8)(ii) of this section; and[§1926.1101(g)(1)(ii)]

(iii) *Prompt clean-up and disposal of wastes and debris* contaminated with asbestos in leak-tight containers except in roofing operations, where the procedures specified in paragraph (g)(8)(ii) of this section apply.[§1926.1101(g)(1)(iii)]

(2) *In addition to the requirements of paragraph (g)(1) of this section,* the employer shall use the following control methods to achieve compliance with the TWA permissible exposure limit and excursion limit prescribed by paragraph (c) of this section; [§1926.1101(g)(2)]

(i) *Local exhaust ventilation* equipped with HEPA filter dust collection systems;[§1926.1101(g)(2)(i)]

(ii) *Enclosure or isolation of processes producing asbestos dust;* [§1926.1101(g)(2)(ii)]

(iii) *Ventilation of the regulated area* to move contaminated air away from the breathing zone of employees and toward a filtration or collection device equipped with a HEPA filter; [§1926.1101(g)(2)(iii)]

(iv) *Use of other work practices* and engineering controls that the Assistant Secretary can show to be feasible. [§1926.1101(g)(2)(iv)]

(v) *Wherever the feasible engineering* and work practice controls described above are not sufficient to reduce employee exposure to or below the permissible exposure limit and/or excursion limit prescribed in paragraph (c) of this section, the employer shall use them to reduce employee exposure to the lowest levels attainable by these controls and shall supplement them by the use of respiratory protection that complies with the requirements of paragraph (h) of this section.[§1926.1101(g)(2)(v)]

(3) *Prohibitions.* The following work practices and engineering controls shall not be used for work related to asbestos or for work which disturbs ACM or PACM, regardless of measured levels of asbestos exposure or the results of initial exposure assessments:[§1926.1101(g)(3)]

(i) *High-speed abrasive disc saws* that are not equipped with point of cut ventilator or enclosures with HEPA filtered exhaust air.[§1926.1101(g)(3)(i)]

(ii) *Compressed air used to remove asbestos,* or materials containing asbestos, unless the compressed air is used in conjunction with an enclosed ventilation system designed to capture the dust cloud created by the compressed air. [§1926.1101(g)(3)(ii)]

(iii) *Dry sweeping, shoveling or other dry clean-up* of dust and debris containing ACM and PACM.[§1926.1101(g)(3)(iii)]

(iv) *Employee rotation* as a means of reducing employee exposure to asbestos.[§1926.1101(g)(3)(iv)]

(4) *Class I Requirements.* In addition to the provisions of paragraphs (g) (1) and (2) of this section, the following engineering controls and work practices and procedures shall be used. [§1926.1101(g)(4)]

(i) *All Class I work,* including the installation and operation of the control system shall be supervised by a competent person as defined in paragraph (b) of this section; [§1926.1101(g)(4)(i)]

(ii) *For all Class I jobs* involving the removal of more than 25 linear or 10 square feet of thermal system insulation or surfacing material; for all other Class I jobs, where the employer cannot produce a negative exposure assessment pursuant to paragraph (f)(2)(iii) of this section, or where employees are working in areas adjacent to the regulated area, while the Class I work is being performed, the employer shall use one of the following methods to ensure that airborne asbestos does not migrate from the regulated area:[§1926.1101(g)(4)(ii)]

[A] Critical barriers shall be placed over all the openings to the regulated area, except where activities are performed outdoors; or[§1926.1101(g)(4)(ii)[A]]

[B] The employer shall use another barrier or isolation method which prevents the migration of airborne asbestos from the regulated area, as verified by perimeter area surveillance during each work shift at each boundary of the regulated area, showing no visible asbestos dust; and perimeter area monitoring showing that clearance levels contained in 40 CFR part 763, subpt. E, of the EPA Asbestos in Schools Rule are met, or that perimeter area levels, measured by Phase Contrast Microscopy (PCM) are no more than background levels representing the same area before the asbestos work began. The results of such monitoring shall be made known to the employer no later than 24 hours from the end of the work shift represented by such monitoring. Exception: For work completed outdoors where employees are not working in areas adjacent to the regulated areas, this paragraph (g)(4)(ii) is satisfied when the specific control methods in paragraph (g)(5) of this section are used. [§1926.1101(g)(4)(ii)[B]]

(iii) *For all Class I jobs,* HVAC systems shall be isolated in the regulated area by sealing with a double layer of 6 mil plastic or the equivalent;[§1926.1101(g)(4)(iii)]

(iv) *For all Class I jobs,* impermeable dropcloths shall be placed on surfaces beneath all removal activity;[§1926.1101(g)(4)(iv)]

(v) *For all Class I jobs,* all objects within the regulated area shall be covered with impermeable dropcloths or plastic sheeting which is secured by duct tape or an equivalent. [§1926.1101(g)(4)(v)]

(vi) *For all Class I jobs* where the employer cannot produce a negative exposure assessment, or where exposure monitoring shows that a PEL is exceeded, the employer shall ventilate the regulated area to move contaminated air away from the breathing zone of employees toward a HEPA filtration or collection device.[§1926.1101(g)(4)(vi)]

(5) *Specific control methods for Class I work.* In addition, Class I asbestos work shall be performed using one or more of the following control methods pursuant to the limitations stated below: [§1926.1101(g)(5)]

(i) *Negative Pressure Enclosure (NPE) systems:* NPE systems may be used where the configuration of the work area does not make the erection of the enclosure infeasible, with the following specifications and work practices.[§1926.1101(g)(5)(i)]

[A] Specifications:[§1926.1101(g)(5)(i)[A]]

[1] The negative pressure enclosure (NPE) may be of any configuration,[§1926.1101(g)(5)(i)[A][1]]

[2] At least 4 air changes per hour shall be maintained in the NPE,[§1926.1101(g)(5)(i)[A][2]]

[3] A minimum of -0.02 column inches of water pressure differential, relative to outside pressure, shall be maintained within the NPE as evidenced by manometric measurements,[§1926.1101(g)(5)(i)[A][3]]

[4] The NPE shall be kept under negative pressure throughout the period of its use, and [§1926.1101(g)(5)(i)[A][4]]

[5] Air movement shall be directed away from employees performing asbestos work within the enclosure, and toward a HEPA filtration or a collection device. [§1926.1101(g)(5)(i)[A][5]]

[B] Work Practices:[§1926.1101(g)(5)(i)[B]]

[1] Before beginning work within the enclosure and at the beginning of each shift, the NPE shall be inspected for breaches and smoke-tested for leaks, and any leaks sealed.[§1926.1101(g)(5)(i)[B][1]]

[2] Electrical circuits in the enclosure shall be deactivated, unless equipped with ground-fault circuit interrupters. [§1926.1101(g)(5)(i)[B][2]]

(ii) *Glove bag systems may be used* to remove PACM and/or ACM from straight runs of piping and elbows and other connections with the following specifications and work practices: [§1926.1101(g)(5)(ii)]

[A] Specifications:[§1926.1101(g)(5)(ii)[A]]

[1] Glovebags shall be made of 6 mil thick plastic and shall be seamless at the bottom.[§1926.1101(g)(5)(ii)[A][1]]

[2] Glovebags used on elbows and other connections must be designed for that purpose and used without modifications.[§1926.1101(g)(5)(ii)[A][2]]

[B] Work Practices:[§1926.1101(g)(5)(ii)[B]]

[1] Each glovebag shall be installed so that it completely covers the circumference of pipe or other structure where the work is to be done.[§1926.1101(g)(5)(ii)[B][1]]

[2] Glovebags shall be smoke-tested for leaks and any leaks sealed prior to use.[§1926.1101(g)(5)(ii)[B][2]]

[3] Glovebags may be used only once and may not be moved.[§1926.1101(g)(5)(ii)[B][3]]

[4] Glovebags shall not be used on surfaces whose temperature exceeds 150 °F.[§1926.1101(g)(5)(ii)[B][4]]

[5] Prior to disposal, glovebags shall be collapsed by removing air within them using a HEPA vacuum. [§1926.1101(g)(5)(ii)[B][5]]

[6] Before beginning the operation, loose and friable material adjacent to the glovebag/box operation shall be wrapped and sealed in two layers of six mil plastic or otherwise rendered intact,[§1926.1101(g)(5)(ii)[B][6]]

[7] Where system uses attached waste bag, such bag shall be connected to collection bag using hose or other material which shall withstand pressure of ACM waste and water without losing its integrity: [§1926.1101(g)(5)(ii)[B][7]]

[8] Sliding valve or other device shall separate waste bag from hose to ensure no exposure when waste bag is disconnected:[§1926.1101(g)(5)(ii)[B][8]]

[9] At least two persons shall perform Class I glovebag removal operations.[§1926.1101(g)(5)(ii)[B][9]]

(iii) *Negative Pressure Glove Bag Systems.* Negative pressure glove bag systems may be used to remove ACM or PACM from piping.[§1926.1101(g)(5)(iii)]

[A] Specifications: In addition to specifications for glove bag systems above, negative pressure glove bag systems shall attach HEPA vacuum systems or other devices to bag to prevent collapse during removal. [§1926.1101(g)(5)(iii)[A]]

[B] Work Practices:[§1926.1101(g)(5)(iii)[B]]

[1] The employer shall comply with the work practices for glove bag systems in paragraph (g)(5)(ii)[B][4] of this section.[§1926.1101(g)(5)(iii)[B][1]]

[2] The HEPA vacuum cleaner or other device used to prevent collapse of bag during removal shall run continually during the operation until it is completed at which time the bag shall be collapsed prior to removal of the bag from the pipe.[§1926.1101(g)(5)(iii)[B][2]]

[3] Where a separate waste bag is used along with a collection bag and discarded after one use, the collection bag may be reused if rinsed clean with amended water before reuse.[§1926.1101(g)(5)(iii)[B][3]]

(iv) *Negative Pressure Glove Box Systems:* Negative pressure glove boxes may be used to remove ACM or PACM from pipe runs with the following specifications and work practices.[§1926.1101(g)(5)(iv)]

[A] Specifications:[§1926.1101(g)(5)(iv)[A]]

[1] Glove boxes shall be constructed with rigid sides and made from metal or other material which can withstand

the weight of the ACM and PACM and water used during removal:[§1926.1101(g)(5)(iv)[A][1]]

[2] *A negative pressure generator* shall be used to create negative pressure in the system:[§1926.1101(g)(5)(iv)[A][2]]

[3] *An air filtration unit shall be attached to the box:* [§1926.1101(g)(5)(iv)[A][3]]

[4] *The box shall be fitted with gloved apertures:* [§1926.1101(g)(5)(iv)[A][4]]

[5] *An aperture at the base of the box* shall serve as a bagging outlet for waste ACM and water: [§1926.1101(g)(5)(iv)[A][5]]

[6] *A back-up generator shall be present on site:* [§1926.1101(g)(5)(iv)[A][6]]

[7] *Waste bags shall consist* of 6 mil thick plastic double-bagged before they are filled or plastic thicker than 6 mil.[§1926.1101(g)(5)(iv)[A][7]]

[B] *Work practices:*[§1926.1101(g)(5)(iv)[B]]

[1] *At least two persons shall perform the removal:* [§1926.1101(g)(5)(iv)[B][1]]

[2] *The box shall be smoke-tested* for leaks and any leaks sealed prior to each use.[§1926.1101(g)(5)(iv)[B][2]]

[3] *Loose or damaged ACM* adjacent to the box shall be wrapped and sealed in two layers of 6 mil plastic prior to the job, or otherwise made intact prior to the job. [§1926.1101(g)(5)(iv)[B][3]]

[4] *A HEPA filtration system* shall be used to maintain pressure barrier in box.[§1926.1101(g)(5)(iv)[B][4]]

(v) *Water Spray Process System.* A water spray process system may be used for removal of ACM and PACM from cold line piping if, employees carrying out such process have completed a 40-hour separate training course in its use, in addition to training required for employees performing Class I work. The system shall meet the following specifications and shall be performed by employees using the following work practices.[§1926.1101(g)(5)(v)]

[A] *Specifications:*[§1926.1101(g)(5)(v)[A]]

[1] *Piping shall be surrounded on 3 sides by rigid framing,* [§1926.1101(g)(5)(v)[A][1]]

[2] *A 360 degree water spray,* delivered through nozzles supplied by a high pressure separate water line, shall be formed around the piping.[§1926.1101(g)(5)(v)[A][2]]

[3] *The spray shall collide to form* a fine aerosol which provides a liquid barrier between workers and the ACM and PACM.[§1926.1101(g)(5)(v)[A][3]]

[B] *Work Practices:*[§1926.1101(g)(5)(v)[B]]

[1] *The system shall be run* for at least 10 minutes before removal begins.[§1926.1101(g)(5)(v)[B][1]]

[2] *All removal shall take place within the water barrier.* [§1926.1101(g)(5)(v)[B][2]]

[3] *The system shall be operated* by at least three persons, one of whom shall not perform removal, but shall check equipment, and ensure proper operation of the system.[§1926.1101(g)(5)(v)[B][3]]

[4] *After removal,* the ACM and PACM shall be bagged while still inside the water barrier.[§1926.1101(g)(5)(v)[B][4]]

(vi) *A small walk-in enclosure* which accommodates no more than two persons (mini-enclosure) may be used if the disturbance or removal can be completely contained by the enclosure with the following specifications and work practices. [§1926.1101(g)(5)(vi)]

[A] *Specifications:*[§1926.1101(g)(5)(vi)[A]]

[1] *The fabricated or job-made enclosure* shall be constructed of 6 mil plastic or equivalent: [§1926.1101(g)(5)(vi)[A][1]]

[2] *The enclosure shall be placed* under negative pressure by means of a HEPA filtered vacuum or similar ventilation unit:[§1926.1101(g)(5)(vi)[A][2]]

[B] *Work practices:*[§1926.1101(g)(5)(vi)[B]]

[1] *Before use,* the mini-enclosure shall be inspected for leaks and smoke-tested to detect breaches, and any breaches sealed.[§1926.1101(g)(5)(vi)[B][1]]

[2] *Before reuse,* the interior shall be completely washed with amended water and HEPA-vacuumed. [§1926.1101(g)(5)(vi)[B][2]]

[3] *During use,* air movement shall be directed away from the employee's breathing zone within the mini-enclosure.[§1926.1101(g)(5)(vi)[B][3]]

(6) *Alternative control methods for Class I work.* Class I work may be performed using a control method which is not referenced in paragraph (g)(5) of this section, or which modifies a control method referenced in paragraph (g)(5)of this section, if the following provisions are complied with:[§1926.1101(g)(6)]

(i) *The control method shall enclose,* contain or isolate the processes or source of airborne asbestos dust, or otherwise capture or redirect such dust before it enters the breathing zone of employees.[§1926.1101(g)(6)(i)]

(ii) *A certified industrial hygienist* or licensed professional engineer who is also qualified as a project designer as defined in paragraph (b) of this section, shall evaluate the work area, the projected work practices and the engineering controls and shall certify in writing that the planned control method is adequate to reduce direct and indirect employee exposure to below the PELs under worst-case conditions of use, and that the planned control method will prevent asbestos contamination outside the regulated area, as measured by clearance sampling which meets the requirements of EPA's Asbestos in Schools rule issued under AHERA, or perimeter monitoring which meets the criteria in paragraph (g)(4)(ii)[B] of this section.[§1926.1101(g)(6)(ii)]

[A] *Where the TSI or surfacing material* to be removed is 25 linear or 10 square feet or less , the evaluation required in paragraph (g)(6) of this section may be performed by a "competent person", and may omit consideration of perimeter or clearance monitoring otherwise required. [§1926.1101(g)(6)(ii)[A]]

[B] *The evaluation of employee exposure* required in paragraph (g)(6) of this section, shall include and be based on sampling and analytical data representing employee exposure during the use of such method under worst-case conditions and by employees whose training and experience are equivalent to employees who are to perform the current job.[§1926.1101(g)(6)(ii)[B]]

(7) *Work Practices and Engineering Controls for Class II work.* [§1926.1101(g)(7)]

(i) *All Class II work* shall be supervised by a competent person as defined in paragraph (b) of this section.[§1926.1101(g)(7)(i)]

(ii) *For all indoor Class II jobs,* where the employer has not produced a negative exposure assessment pursuant to paragraph (f)(2)(iii) of this section, or where during the job, changed conditions indicate there may be exposure above the PEL or where the employer does not remove the ACM in a substantially intact state, the employer shall use one of the following methods to ensure that airborne asbestos does not migrate from the regulated area;[§1926.1101(g)(7)(ii)]

[A] *Critical barriers shall be placed* over all openings to the regulated area; or,[§1926.1101(g)(7)(ii)[A]]

[B] *The employer shall use* another barrier or isolation method which prevents the migration of airborne asbestos from the regulated area, as verified by perimeter area monitoring or clearance monitoring which meets the criteria set out in paragraph (g)(4)(ii)[B] of this section. [§1926.1101(g)(7)(ii)[B]]

[C] *Impermeable dropcloths shall be placed* on surfaces beneath all removal activity;[§1926.1101(g)(7)(ii)[C]]

(iii) *[Reserved]*[§1926.1101(g)(7)(iii)]

(iv) *All Class II asbestos work* shall be performed using the work practices and requirements set out above in paragraph (g)(1)(i) through (g)(1)(iii) of this section.[§1926.1101(g)(7)(iv)]

(8) *Additional Controls for Class II work.* Class II asbestos work shall also be performed by complying with the work practices and controls designated for each type of asbestos work to be performed, set out in this paragraph. Where more than one control method may be used for a type of asbestos work, the employer may choose one or a combination of designated control methods. Class II work also may be performed using a method allowed for Class I work, except that glove bags and glove boxes are allowed if they fully enclose the Class II material to be removed.[§1926.1101(g)(8)]

(i) *For removing vinyl and asphalt flooring materials* which contain ACM or for which in buildings constructed no later than 1980, the employer has not verified the absence of ACM pursuant to paragraph (g)(8)(i)[I] of this section. The employer shall ensure that employees comply with the following work practices and that employees are trained in these practices pursuant to paragraph (k)(9):[§1926.1101(g)(8)(i)]

[A] Flooring or its backing shall not be sanded. [§1926.1101(g)(8)(i)[A]]

[B] Vacuums equipped with HEPA filter, disposable dust bag, and metal floor tool (no brush) shall be used to clean floors. [§1926.1101(g)(8)(i)[B]]

[C] Resilient sheeting shall be removed by cutting with wetting of the snip point and wetting during delamination. Rip-up of resilient sheet floor material is prohibited. [§1926.1101(g)(8)(i)[C]]

[D] All scraping of residual adhesive and/or backing shall be performed using wet methods. [§1926.1101(g)(8)(i)[D]]

[E] Dry sweeping is prohibited. [§1926.1101(g)(8)(i)[E]]

[F] Mechanical chipping is prohibited unless performed in a negative pressure enclosure which meets the requirements of paragraph (g)(5)(i) of this section. [§1926.1101(g)(8)(i)[F]]

[G] Tiles shall be removed intact, unless the employer demonstrates that intact removal is not possible. [§1926.1101(g)(8)(i)[G]]

[H] When tiles are heated and can be removed intact, wetting may be omitted. [§1926.1101(g)(8)(i)[H]]

[I] Resilient flooring material including associated mastic and backing shall be assumed to be asbestos-containing unless an industrial hygienist determines that it is asbestos-free using recognized analytical techniques. [§1926.1101(g)(8)(i)[I]]

(ii) *For removing roofing material* which contains ACM the employer shall ensure that the following work practices are followed: [§1926.1101(g)(8)(ii)]

[A] Roofing material shall be removed in an intact state to the extent feasible. [§1926.1101(g)(8)(ii)[A]]

[B] Wet methods shall be used to remove roofing materials that are not intact, or that will be rendered not intact during removal, unless such wet methods are not feasible or will create safety hazards. [§1926.1101(g)(8)(ii)[B]]

[C] Cutting machines shall be continuously misted during use, unless a competent person determines that misting substantially decreases worker safety. [§1926.1101(g)(8)(ii)[C]]

[D] When removing built-up roofs with asbestos-containing roofing felts and an aggregate surface using a power roof cutter, all dust resulting from the cutting operation shall be collected by a HEPA dust collector, or shall be HEPA vacuumed by vacuuming along the cut line. When removing built-up roofs with asbestos-containing roofing felts and a smooth surface using a power roof cutter, the dust resulting from the cutting operation shall be collected either by a HEPA dust collector or HEPA vacuuming along the cut line, or by gently sweeping and then carefully and completely wiping up the still-wet dust and debris left along the cut line. The dust and debris shall be immediately bagged or placed in covered containers. [§1926.1101(g)(8)(ii)[D]]

[E] Asbestos-containing material that has been removed from a roof shall not be dropped or thrown to the ground. Unless the material is carried or passed to the ground by hand, it shall be lowered to the ground via covered, dust-tight chute, crane or hoist: [§1926.1101(g)(8)(ii)[E]]

[1] Any ACM that is not intact shall be lowered to the ground as soon as is practicable, but in any event no later than the end of the work shift. While the material remains on the roof it shall either be kept wet, placed in an impermeable waste bag, or wrapped in plastic sheeting. [§1926.1101(g)(8)(ii)[E][1]]

[2] Intact ACM shall be lowered to the ground as soon as is practicable, but in any event no later than the end of the work shift. [§1926.1101(g)(8)(ii)[E][2]]

[F] Upon being lowered, unwrapped material shall be transferred to a closed receptacle in such manner so as to preclude the dispersion of dust. [§1926.1101(g)(8)(ii)[F]]

[G] Roof level heating and ventilation air intake sources shall be isolated or the ventilation system shall be shut down. [§1926.1101(g)(8)(ii)[G]]

[H] Notwithstanding any other provision of this section, removal or repair of sections of intact roofing less than 25 square feet in area does not require use of wet methods or HEPA vacuuming as long as manual methods which do not render the material non-intact are used to remove the material and no visible dust is created by the removal method used. In determining whether a job involves less than 25 square feet, the employer shall include all removal and repair work performed on the same roof on the same day. [§1926.1101(g)(8)(ii)[H]]

(iii) *When removing cementitious asbestos-containing* siding and shingles or transite panels containing ACM on building exteriors (other than roofs, where paragraph (g)(8)(ii) of this section applies) the employer shall ensure that the following work practices are followed: [§1926.1101(g)(8)(iii)]

[A] Cutting, abrading or breaking siding, shingles, or transite panels, shall be prohibited unless the employer can demonstrate that methods less likely to result in asbestos fiber release cannot be used. [§1926.1101(g)(8)(iii)[A]]

[B] Each panel or shingle shall be sprayed with amended water prior to removal. [§1926.1101(g)(8)(iii)[B]]

[C] Unwrapped or unbagged panels or shingles shall be immediately lowered to the ground via covered dust-tight chute, crane or hoist, or placed in an impervious waste bag or wrapped in plastic sheeting and lowered to the ground no later than the end of the work shift. [§1926.1101(g)(8)(iii)[C]]

[D] Nails shall be cut with flat, sharp instruments. [§1926.1101(g)(8)(iii)[D]]

(iv) *When removing gaskets containing ACM,* the employer shall ensure that the following work practices are followed: [§1926.1101(g)(8)(iv)]

[A] If a gasket is visibly deteriorated and unlikely to be removed intact, removal shall be undertaken within a glovebag as described in paragraph (g)(5)(ii) of this section. [§1926.1101(g)(8)(iv)[A]]

[B] [Reserved] [§1926.1101(g)(8)(iv)[B]]

[C] The gasket shall be immediately placed in a disposal container. [§1926.1101(g)(8)(iv)[C]]

[D] Any scraping to remove residue must be performed wet. [§1926.1101(g)(8)(iv)[D]]

(v) *When performing any other Class II removal* of asbestos containing material for which specific controls have not been listed in paragraph (g)(8)(iv)[A] through [D] of this section, the employer shall ensure that the following work practices are complied with. [§1926.1101(g)(8)(v)]

[A] The material shall be thoroughly wetted with amended water prior to and during its removal. [§1926.1101(g)(8)(v)[A]]

[B] The material shall be removed in an intact state unless the employer demonstrates that intact removal is not possible. [§1926.1101(g)(8)(v)[B]]

[C] Cutting, abrading or breaking the material shall be prohibited unless the employer can demonstrate that methods less likely to result in asbestos fiber release are not feasible. [§1926.1101(g)(8)(v)[C]]

[D] Asbestos-containing material removed, shall be immediately bagged or wrapped, or kept wetted until transferred to a closed receptacle, no later than the end of the work shift. [§1926.1101(g)(8)(v)[D]]

(vi) *Alternative Work Practices and Controls.* Instead of the work practices and controls listed in paragraph (g)(8) (i) through (v) of this section, the employer may use different or modified engineering and work practice controls if the following provisions are complied with. [§1926.1101(g)(8)(vi)]

[A] The employer shall demonstrate by data representing employee exposure during the use of such method under conditions which closely resemble the conditions under which the method is to be used, that employee exposure will not exceed the PELs under any anticipated circumstances. [§1926.1101(g)(8)(vi)[A]]

[B] A competent person shall evaluate the work area, the projected work practices and the engineering controls, and shall certify in writing, that the different or modified controls are adequate to reduce direct and indirect employee exposure to below the PELs under all expected conditions of use and that the method meets the requirements of this standard. The evaluation shall include and be based on data representing employee exposure during the use of such method under conditions which closely resemble the conditions under which the method is to be used for the current job, and by employees whose training and experience are equivalent to employees who are to perform the current job. [§1926.1101(g)(8)(vi)[B]]

(9) *Work Practices and Engineering Controls for Class III asbestos work.* Class III asbestos work shall be conducted using engineering and work practice controls which minimize the exposure to employees performing the asbestos work and to bystander employees.[§1926.1101(g)(9)]

(i) *The work shall be performed using wet methods.* [§1926.1101(g)(9)(i)]

(ii) *To the extent feasible,* the work shall be performed using local exhaust ventilation.[§1926.1101(g)(9)(ii)]

(iii) *Where the disturbance involves drilling,* cutting, abrading, sanding, chipping, breaking, or sawing of thermal system insulation or surfacing material, the employer shall use impermeable dropcloths, and shall isolate the operation using mini-enclosures or glove bag systems pursuant to paragraph (g)(5) of this section or another isolation method. [§1926.1101(g)(9)(iii)]

(iv) *Where the employer does not produce* a "negative exposure assessment" for a job, or where monitoring results show the PEL has been exceeded, the employer shall contain the area using impermeable dropcloths and plastic barriers or their equivalent, or shall isolate the operation using a control system listed in and in compliance with paragraph (g)(5) of this section.[§1926.1101(g)(9)(iv)]

(v) *Employees performing Class III jobs,* which involve the disturbance of thermal system insulation or surfacing material, or where the employer does not produce a "negative exposure assessment" or where monitoring results show a PEL has been exceeded, shall wear respirators which are selected, used and fitted pursuant to provisions of paragraph (h) of this section.[§1926.1101(g)(9)(v)]

(10) *Class IV asbestos work.* Class IV asbestos jobs shall be conducted by employees trained pursuant to the asbestos awareness training program set out in paragraph (k)(9) of this section. In addition, all Class IV jobs shall be conducted in conformity with the requirements set out in paragraph (g)(1) of this section, mandating wet methods, HEPA vacuums, and prompt clean up of debris containing ACM or PACM.[§1926.1101(g)(10)]

(i) *Employees cleaning up debris and waste* in a regulated area where respirators are required shall wear respirators which are selected, used and fitted pursuant to provisions of paragraph (h) of this section.[§1926.1101(g)(10)(i)]

(ii) *Employers of employees who clean up waste and debris in,* and employers in control of, areas where friable thermal system insulation or surfacing material is accessible, shall assume that such waste and debris contain asbestos. [§1926.1101(g)(10)(ii)]

(11) *Alternative methods of compliance* for installation, removal, repair, and maintenance of certain roofing and pipeline coating materials. Notwithstanding any other provision of this section, an employer who complies with all provisions of this paragraph (g)(11) when installing, removing, repairing, or maintaining intact pipeline asphaltic wrap, or roof flashings which contain asbestos fibers encapsulated or coated by bituminous or resinous compounds shall be deemed to be in compliance with this section. If an employer does not comply with all provisions of this paragraph (g)(11) or if during the course of the job the material does not remain intact, the provisions of paragraph (g)(8) of this section apply instead of this paragraph (g)(11). [§1926.1101(g)(11)]

(i) *Before work begins and as needed during the job,* a competent person who is capable of identifying asbestos hazards in the workplace and selecting the appropriate control strategy for asbestos exposure, and who has the authority to take prompt corrective measures to eliminate such hazards, shall conduct an inspection of the worksite and determine that the roofing material is intact and will likely remain intact. [§1926.1101(g)(11)(i)]

(ii) *All employees performing work* covered by this paragraph (g)(11) shall be trained in a training program that meets the requirements of paragraph (k)(9)(viii) of this section. [§1926.1101(g)(11)(ii)]

(iii) *The material shall not be sanded, abraded, or ground.* Manual methods which do not render the material non-intact shall be used.[§1926.1101(g)(11)(iii)]

(iv) *Material that has been removed from a roof* shall not be dropped or thrown to the ground. Unless the material is carried or passed to the ground by hand, it shall be lowered to the ground via covered, dust- tight chute, crane or hoist. All such material shall be removed from the roof as soon as is practicable, but in any event no later than the end of the work shift.[§1926.1101(g)(11)(iv)]

(v) *Where roofing products which have been labeled* as containing asbestos pursuant to paragraph (k)(8) of this section are installed on non-residential roofs during operations covered by this paragraph (g)(11), the employer shall notify the building owner of the presence and location of such materials no later than the end of the job.[§1926.1101(g)(11)(v)]

(vi) *All removal or disturbance of pipeline* asphaltic wrap shall be performed using wet methods.[§1926.1101(g)(11)(vi)]

(h) **Respiratory protection.**[§1926.1101(h)]

(1) *General.* For employees who use respirators required by this section, the employer must provide each employee an appropriate respirator that complies with the requirements of this paragraph. Respirators must be used during:[§1926.1101(h)(1)]

(i) *Class I asbestos work.*[§1926.1101(h)(1)(i)]

(ii) *Class II asbestos work* when ACM is not removed in a substantially intact state.[§1926.1101(h)(1)(ii)]

(iii) *Class II and III asbestos work* that is not performed using wet methods, except for removal of ACM from sloped roofs when a negative- exposure assessment has been conducted and ACM is removed in an intact state.[§1926.1101(h)(1)(iii)]

(iv) *Class II and III asbestos work* for which a negative-exposure assessment has not been conducted.[§1926.1101(h)(1)(iv)]

(v) *Class III asbestos work* when TSI or surfacing ACM or PACM is being disturbed.[§1926.1101(h)(1)(v)]

(vi) *Class IV asbestos work* performed within regulated areas where employees who are performing other work are required to use respirators.[§1926.1101(h)(1)(vi)]

(vii) *Work operations covered* by this section for which employees are exposed above the TWA or excursion limit. [§1926.1101(h)(1)(vii)]

(viii) *Emergencies.*[§1926.1101(h)(1)(viii)]

(2) *Respirator program.*[§1926.1101(h)(2)]

(i) *The employer must implement* a respiratory protection program in accordance with 29 CFR 1910.134 (b) through (d) (except (d)(1)(iii)), and (f) through (m), which covers each employee required by this section to use a respirator. [§1926.1101(h)(2)(i)]

(ii) *No employee shall be assigned to asbestos work* that requires respirator use if, based on their most recent medical examination, the examining physician determines that the employee will be unable to function normally while using a respirator, or that the safety or health of the employee or other employees will be impaired by the employee's respirator use. Such employees must be assigned to another job or given the opportunity to transfer to a different position that they can perform. If such a transfer position is available, it must be with the same employer, in the same geographical area, and with the same seniority, status, rate of pay, and other job benefits the employee had just prior to such transfer.[§1926.1101(h)(2)(ii)]

(3) *Respirator selection.*[§1926.1101(h)(3)]

(i) *Employers must:*[§1926.1101(h)(3)(i)]

[A] Select, and provide to employees, the appropriate respirators specified in paragraph (d)(3)(i)[A] of 29 CFR 1910.134; however, employers must not select or use filtering facepiece respirators for use against asbestos fibers.[§1926.1101(h)(3)(i)[A]]

[B] Provide HEPA filters for powered and non-powered air-purifying respirators.[§1926.1101(h)(3)(i)[B]]

(ii) *Employers must provide an employee* with tight-fitting, powered air-purifying respirator (PAPR) instead of a negative pressure respirator selected according to paragraph (h)(3)(i)(A) of this standard when the employee chooses to use a PAPR and it provides adequate protection to the employee.[§1926.1101(h)(3)(ii)]

(iii) *Employers must provide employees* with an air-purifying half mask respirator, other than a filtering facepiece respirator, whenever the employees perform:[§1926.1101(h)(3)(iii)]

[A] Class II or Class III asbestos work for which no negative exposure assessment is available.[§1926.1101(h)(3)(iii)[A]]

[B] Class III asbestos work involving disturbance of TSI or surfacing ACM or PACM.[§1926.1101(h)(3)(iii)[B]]

(iv) *Employers must provide employees with:*[§1926.1101(h)(3)(iv)]

[A] A tight-fitting powered air-purifying respirator or a full facepiece, supplied-air respirator operated in the pressure-demand mode and equipped with either HEPA egress cartridges or an auxiliary positive-pressure, self-contained breathing apparatus (SCBA) whenever the employees are in a regulated area performing Class I asbestos work for which a negative exposure assessment is not available and the exposure assessment indicates that the exposure level will be at or below 1 f/cc as an 8-hour time-weighted average (TWA).[§1926.1101(h)(3)(iv)[A]]

[B] A full facepiece supplied-air respirator operated in the pressure-demand mode and equipped with an auxiliary positive-pressure SCBA whenever the employees are in a regulated area performing Class I asbestos work for which a negative exposure assessment is not available and the exposure assessment indicates that the exposure level will be above 1 f/cc as an 8-hour TWA. [§1926.1101(h)(3)(iv)[B]]

(i) Protective clothing.[§1926.1101(i)]

(1) *General.* The employer shall provide or require the use of protective clothing, such as coveralls or similar whole-body clothing, head coverings, gloves, and foot coverings for any employee exposed to airborne concentrations of asbestos that exceed the TWA and/or excursion limit prescribed in paragraph (c) of this section, or for which a required negative exposure assessment is not produced, or for any employee performing Class I operations which involve the removal of over 25 linear or 10 square feet of TSI or surfacing ACM and PACM. [§1926.1101(i)(1)]

(2) *Laundering.*[§1926.1101(i)(2)]

(i) *The employer shall ensure* that laundering of contaminated clothing is done so as to prevent the release of airborne asbestos in excess of the TWA or excursion limit prescribed in paragraph (c) of this section.[§1926.1101(i)(2)(i)]

(ii) *Any employer who gives contaminated clothing* to another person for laundering shall inform such person of the requirement in paragraph (i)(2)(i) of this section to effectively prevent the release of airborne asbestos in excess of the TWA and excursion limit prescribed in paragraph (c) of this section.[§1926.1101(i)(2)(ii)]

(3) *Contaminated clothing.* Contaminated clothing shall be transported in sealed impermeable bags, or other closed, impermeable containers, and be labeled in accordance with paragraph (k) of this section.[§1926.1101(i)(3)]

(4) *Inspection of protective clothing.*[§1926.1101(i)(4)]

(i) *The competent person shall examine worksuits* worn by employees at least once per workshift for rips or tears that may occur during performance of work.[§1926.1101(i)(4)(i)]

(ii) *When rips or tears are detected* while an employee is working, rips and tears shall be immediately mended, or the worksuit shall be immediately replaced.[§1926.1101(i)(4)(ii)]

(j) Hygiene facilities and practices for employees.[§1926.1101(j)]

(1) *Requirements for employees* performing Class I asbestos jobs involving over 25 linear or 10 square feet of TSI or surfacing ACM and PACM.[§1926.1101(j)(1)]

(i) *Decontamination areas.* The employer shall establish a decontamination area that is adjacent and connected to the regulated area for the decontamination of such employees. The decontamination area shall consist of an equipment room, shower area, and clean room in series. The employer shall ensure that employees enter and exit the regulated area through the decontamination area.[§1926.1101(j)(1)(i)]

[A] Equipment room. The equipment room shall be supplied with impermeable, labeled bags and containers for the containment and disposal of contaminated protective equipment.[§1926.1101(j)(1)(i)[A]]

[B] Shower area. Shower facilities shall be provided which comply with 29 CFR 1910.141(d)(3), unless the employer can demonstrate that they are not feasible. The showers shall be adjacent both to the equipment room and the clean room, unless the employer can demonstrate that this location is not feasible. Where the employer can demonstrate that it is not feasible to locate the shower between the equipment room and the clean room, or where the work is performed outdoors, the employers shall ensure that employees:[§1926.1101(j)(1)(i)[B]]

[1] Remove asbestos contamination from their worksuits in the equipment room using a HEPA vacuum before proceeding to a shower that is not adjacent to the work area; or[§1926.1101(j)(1)(i)[B][1]]

[2] Remove their contaminated worksuits in the equipment room, then don clean worksuits, and proceed to a shower that is not adjacent to the work area. [§1926.1101(j)(1)(i)[B][2]]

[C] Clean change room. The clean room shall be equipped with a locker or appropriate storage container for each employee's use. When the employer can demonstrate that it is not feasible to provide a clean change area adjacent to the work area or where the work is performed outdoors, the employer may permit employees engaged in Class I asbestos jobs to clean their protective clothing with a portable HEPA-equipped vacuum before such employees leave the regulated area. Following showering, such employees however must then change into street clothing in clean change areas provided by the employer which otherwise meet the requirements of this section.[§1926.1101(j)(1)(i)[C]]

(ii) *Decontamination area entry procedures.* The employer shall ensure that employees:[§1926.1101(j)(1)(ii)]

[A] Enter the decontamination area through the clean room; [§1926.1101(j)(1)(ii)[A]]

[B] Remove and deposit street clothing within a locker provided for their use; and[§1926.1101(j)(1)(ii)[B]]

[C] Put on protective clothing and respiratory protection before leaving the clean room.[§1926.1101(j)(1)(ii)[C]]

[D] Before entering the regulated area, the employer shall ensure that employees pass through the equipment room. [§1926.1101(j)(1)(ii)[D]]

(iii) *Decontamination area exit procedures.* The employer shall ensure that:[§1926.1101(j)(1)(iii)]

[A] Before leaving the regulated area, employees shall remove all gross contamination and debris from their protective clothing.[§1926.1101(j)(1)(iii)[A]]

[B] Employees shall remove their protective clothing in the equipment room and deposit the clothing in labeled impermeable bags or containers.[§1926.1101(j)(1)(iii)[B]]

[C] Employees shall not remove their respirators in the equipment room.[§1926.1101(j)(1)(iii)[C]]

[D] Employees shall shower prior to entering the clean room. [§1926.1101(j)(1)(iii)[D]]

[E] After showering, employees shall enter the clean room before changing into street clothes.[§1926.1101(j)(1)(iii)[E]]

(iv) *Lunch Areas.* Whenever food or beverages are consumed at the worksite where employees are performing Class I asbestos work, the employer shall provide lunch areas in which the airborne concentrations of asbestos are below the permissible exposure limit and/or excursion limit.[§1926.1101(j)(1)(iv)]

(2) *Requirements for Class I work* involving less than 25 linear or 10 square feet of TSI or surfacing ACM and PACM, and for Class II and Class III asbestos work operations where exposures exceed a PEL or where there is no negative exposure assessment produced before the operation.[§1926.1101(j)(2)]

(i) *The employer shall establish an equipment room* or area that is adjacent to the regulated area for the decontamination of employees and their equipment which is contaminated with asbestos which shall consist of an area covered by an impermeable drop cloth on the floor or horizontal working surface. [§1926.1101(j)(2)(i)]

(ii) *The area must be of sufficient size* as to accommodate cleaning of equipment and removing personal protective equipment without spreading contamination beyond the area (as determined by visible accumulations).[§1926.1101(j)(2)(ii)]

(iii) *Work clothing must be cleaned* with a HEPA vacuum before it is removed.[§1926.1101(j)(2)(iii)]

(iv) *All equipment and surfaces of containers* filled with ACM must be cleaned prior to removing them from the equipment room or area.[§1926.1101(j)(2)(iv)]

(v) *The employer shall ensure* that employees enter and exit the regulated area through the equipment room or area. [§1926.1101(j)(2)(v)]

(3) *Requirements for Class IV work.* Employers shall ensure that employees performing Class IV work within a regulated area comply with the hygiene practice required of employees performing work which has a higher classification within that regulated area. Otherwise employers of employees cleaning up debris and material which is TSI or surfacing ACM or identified

as PACM shall provide decontamination facilities for such employees which are required by paragraph (j)(2) of this section.[§1926.1101(j)(3)]

(4) *Smoking in work areas.* The employer shall ensure that employees do not smoke in work areas where they are occupationally exposed to asbestos because of activities in that work area.[§1926.1101(j)(4)]

(k) Communication of hazards.[§1926.1101(k)]

(1) *Hazard communication.*[§1926.1101(k)(1)]

(i) *This section applies to the communication of information* concerning asbestos hazards in construction activities to facilitate compliance with this standard. Most asbestos-related construction activities involve previously installed building materials. Building owners often are the only and/or best sources of information concerning them. Therefore, they, along with employers of potentially exposed employees, are assigned specific information conveying and retention duties under this section. Installed Asbestos Containing Building Material. Employers and building owners shall identify TSI and sprayed or troweled on surfacing materials in buildings as asbestos-containing, unless they determine in compliance with paragraph (k)(5) of this section that the material is not asbestos-containing. Asphalt and vinyl flooring material installed no later than 1980 must also be considered as asbestos containing unless the employer, pursuant to paragraph (g)(8)(i)(I) of this section determines that it is not asbestos- containing. If the employer/building owner has actual knowledge, or should have known through the exercise of due diligence, that other materials are asbestos-containing, they too must be treated as such. When communicating information to employees pursuant to this standard, owners and employers shall identify "PACM" as ACM. Additional requirements relating to communication of asbestos work on multi- employer worksites are set out in paragraph (d) of this section. [§1926.1101(k)(1)(i)]

(ii) *The employer shall include asbestos* in the program established to comply with the Hazard Communication Standard (HCS) (Sec. 1910.1200). The employer shall ensure that each employee has access to labels on containers of asbestos and safety data sheets, and is trained in accordance with the provisions of HCS and paragraphs (k)(9) and (10) of this section. The employer shall provide information on at least the following hazards: Cancer and lung effects. [§1926.1101(k)(1)(ii)]

(2) *Duties of building and facility owners.*[§1926.1101(k)(2)]

(i) *Before work subject to this standard is begun,* building and facility owners shall determine the presence, location, and quantity of ACM and/or PACM at the work site pursuant to paragraph (k)(1)(i) of this section.[§1926.1101(k)(2)(i)]

(ii) *Building and/or facility owners* shall notify the following persons of the presence, location and quantity of ACM or PACM, at the work sites in their buildings and facilities. Notification either shall be in writing, or shall consist of a personal communication between the owner and the person to whom notification must be given or their authorized representatives:[§1926.1101(k)(2)(ii)]

[A] Prospective employers applying or bidding for work whose employees reasonably can be expected to work in or adjacent to areas containing such material; [§1926.1101(k)(2)(ii)[A]]

[B] Employees of the owner who will work in or adjacent to areas containing such material:[§1926.1101(k)(2)(ii)[B]]

[C] On multi-employer worksites, all employers of employees who will be performing work within or adjacent to areas containing such materials;[§1926.1101(k)(2)(ii)[C]]

[D] Tenants who will occupy areas containing such material. [§1926.1101(k)(2)(ii)[D]]

(3) *Duties of employers whose employees perform work* subject to this standard in or adjacent to areas containing ACM and PACM. Building/ facility owners whose employees perform such work shall comply with these provisions to the extent applicable.[§1926.1101(k)(3)]

(i) *Before work in areas containing ACM and PACM is begun;* employers shall identify the presence, location, and quantity of ACM, and/or PACM therein pursuant to paragraph (k)(1)(i) of this section.[§1926.1101(k)(3)(i)]

(ii) *Before work under this standard is performed* employers of employees who will perform such work shall inform the following persons of the location and quantity of ACM and/or PACM present in the area and the precautions to be taken to insure that airborne asbestos is confined to the area. [§1926.1101(k)(3)(ii)]

[A] Owners of the building/facility;[§1926.1101(k)(3)(ii)[A]]

[B] Employees who will perform such work and employers of employees who work and/or will be working in adjacent areas.[§1926.1101(k)(3)(ii)[B]]

(iii) *Within 10 days of the completion of such work,* the employer whose employees have performed work subject to this standard, shall inform the building/facility owner and employers of employees who will be working in the area of the current location and quantity of PACM and/ or ACM remaining in the area and final monitoring results, if any.[§1926.1101(k)(3)(iii)]

(4) *In addition to the above requirements,* all employers who discover ACM and/or PACM on a worksite shall convey information concerning the presence, location and quantity of such newly discovered ACM and/or PACM to the owner and to other employers of employees working at the work site, within 24 hours of the discovery.[§1926.1101(k)(4)]

(5) *Criteria to rebut the designation of installed material as PACM.* [§1926.1101(k)(5)]

(i) *At any time, an employer and/or building owner* may demonstrate, for purposes of this standard, that PACM does not contain asbestos. Building owners and/or employers are not required to communicate information about the presence of building material for which such a demonstration pursuant to the requirements of paragraph (k)(5)(ii) of this section has been made. However, in all such cases, the information, data and analysis supporting the determination that PACM does not contain asbestos, shall be retained pursuant to paragraph (n) of this section.[§1926.1101(k)(5)(i)]

(ii) *An employer or owner may demonstrate* that PACM does not contain more than 1% asbestos by the following: [§1926.1101(k)(5)(ii)]

[A] Having a completed inspection conducted pursuant to the requirements of AHERA (40 CFR Part 763, subpart E) which demonstrates that the material is not ACM; or [§1926.1101(k)(5)(ii)[A]]

[B] Performing tests of the material containing PACM which demonstrate that no ACM is present in the material. Such tests shall include analysis of bulk samples collected in the manner described in 40 CFR 763.86. The tests, evaluation and sample collection shall be conducted by an accredited inspector or by a CIH. Analysis of samples shall be performed by persons or laboratories with proficiency demonstrated by current successful participation in a nationally recognized testing program such as the National Voluntary Laboratory Accreditation Program (NVLAP) or the National Institute for Standards and Technology (NIST) or the Round Robin for bulk samples administered by the American Industrial Hygiene Association (AIHA) or an equivalent nationally-recognized round robin testing program.[§1926.1101(k)(5)(ii)[B]]

(iii) *The employer and/or building owner may demonstrate* that flooring material including associated mastic and backing does not contain asbestos, by a determination of an industrial hygienist based upon recognized analytical techniques showing that the material is not ACM.[§1926.1101(k)(5)(iii)]

(6) *At the entrance to mechanical rooms/areas* in which employees reasonably can be expected to enter and which contain ACM and/or PACM, the building owner shall post signs which identify the material which is present, its location, and appropriate work practices which, if followed, will ensure that ACM and/or PACM will not be disturbed. The employer shall ensure, to the extent feasible, that employees who come in contact with these signs can comprehend them. Means to ensure employee comprehension may include the use of foreign languages, pictographs, graphics, and awareness training.[§1926.1101(k)(6)]

(7) *Signs.*[§1926.1101(k)(7)]

(i) *Warning signs that demarcate the regulated area* shall be provided and displayed at each location where a regulated area is required to be established by paragraph (e) of this section. Signs shall be posted at such a distance from such a location that an employee may read the signs and take necessary protective steps before entering the area marked by the signs.[§1926.1101(k)(7)(i)]

(ii) *[A] The warning signs required* by paragraph (k)(7) of this section shall bear the following information.
DANGER

ASBESTOS MAY CAUSE CANCER
CAUSES DAMAGE TO LUNGS
AUTHORIZED PERSONNEL ONLY[§1926.1101(k)(7)(ii)[A]]

[B] In addition, where the use of respirators and protective clothing is required in the regulated area under this section, the warning signs shall include the following: WEAR RESPIRATORY PROTECTION AND PROTECTIVE CLOTHING IN THIS AREA[§1926.1101(k)(7)(ii)[B]]

[C] Prior to June 1, 2016, employers may use the following legend in lieu of that specified in paragraph (k)(7)(ii)(A) of this section:
DANGER
ASBESTOS
CANCER AND LUNG DISEASE HAZARD
AUTHORIZED PERSONNEL ONLY[§1926.1101(k)(7)(ii)[C]]

[D] Prior to June 1, 2016, employers may use the following legend in lieu of that specified in paragraph (k)(7)(ii)(B) of this section: RESPIRATORS AND PROTECTIVE CLOTHING ARE REQUIRED IN THIS AREA[§1926.1101(k)(7)(ii)[D]]

(iii) *The employer shall ensure that employees* working in and contiguous to regulated areas comprehend the warning signs required to be posted by paragraph (k)(7)(i) of this section. Means to ensure employee comprehension may include the use of foreign languages, pictographs and graphics. [§1926.1101(k)(7)(iii)]

(8) *Labels.*[§1926.1101(k)(8)]

(i) *Labels shall be affixed to all products* containing asbestos and to all containers containing such products, including waste containers. Where feasible, installed asbestos products shall contain a visible label.[§1926.1101(k)(8)(i)]

(ii) *The employer shall ensure that such* labels comply with paragraphs (k) of this section.[§1926.1101(k)(8)(ii)]

(iii) *The employer shall ensure that labels* of bags or containers of protective clothing and equipment, scrap, waste, and debris containing asbestos fibers bear the following information:
DANGER
CONTAINS ASBESTOS FIBERS
MAY CAUSE CANCER CAUSES DAMAGE TO LUNGS
DO NOT BREATHE DUST
AVOID CREATING DUST[§1926.1101(k)(8)(iii)]

(iv) *[A] Prior to June 1,* 2015, employers may include the following information on raw materials, mixtures or labels of bags or containers of protective clothing and equipment, scrap, waste, and debris containing asbestos fibers in lieu of the labeling requirements in paragraphs (k)(8)(ii) and (k)(8)(iii) of this section:
DANGER
CONTAINS ASBESTOS FIBERS
AVOID CREATING DUST
CANCER AND LUNG DISEASE HAZARD [§1926.1101(k)(8)(iv)[A]]

[B] Labels shall also contain a warning statement against breathing asbestos fibers.[§1926.1101(k)(8)(iv)[B]]

(v) *Labels shall contain* a warning statement against breathing asbestos fibers.[§1926.1101(k)(8)(v)]

(vi) *The provisions for labels required* by paragraphs (k)(8)(i) through (k)(8)(iii) of this section do not apply where: [§1926.1101(k)(8)(vi)]

[A] Asbestos fibers have been modified by a bonding agent, coating, binder, or other material, provided that the manufacturer can demonstrate that, during any reasonably foreseeable use, handling, storage, disposal, processing, or transportation, no airborne concentrations of asbestos fibers in excess of the permissible exposure limit and/or excursion limit will be released, or[§1926.1101(k)(8)(vi)[A]]

[B] Asbestos is present in a product in concentrations less than 1.0 percent.[§1926.1101(k)(8)(vi)[B]]

(vii) *When a building owner or employer identifies* previously installed PACM and/or ACM, labels or signs shall be affixed or posted so that employees will be notified of what materials contain PACM and/or ACM. The employer shall attach such labels in areas where they will clearly be noticed by employees who are likely to be exposed, such as at the entrance to mechanical room/areas. Signs required by paragraph (k)(6) of this section may be posted in lieu of labels so long as they contain information required for labelling. The employer shall ensure, to the extent feasible, that employees who come in contact with these signs or labels can comprehend them. Means to ensure employee comprehension may include the use of foreign languages, pictographs, graphics, and awareness training.[§1926.1101(k)(8)(vii)]

(9) *Employee Information and Training.*[§1926.1101(k)(9)]

(i) *The employer shall train each employee* who is likely to be exposed in excess of a PEL, and each employee who performs Class I through IV asbestos operations, in accordance with the requirements of this section. Such training shall be conducted at no cost to the employee. The employer shall institute a training program and ensure employee participation in the program.[§1926.1101(k)(9)(i)]

(ii) *Training shall be provided* prior to or at the time of initial assignment and at least annually thereafter.[§1926.1101(k)(9)(ii)]

(iii) *Training for Class I operations* and for Class II operations that require the use of critical barriers (or equivalent isolation methods) and/or negative pressure enclosures under this section shall be the equivalent in curriculum, training method and length to the EPA Model Accreditation Plan (MAP) asbestos abatement workers training (40 CFR Part 763, subpart E, appendix C).[§1926.1101(k)(9)(iii)]

(iv) *Training for other Class II work.*[§1926.1101(k)(9)(iv)]

[A] For work with asbestos containing roofing materials, flooring materials, siding materials, ceiling tiles, or transite panels, training shall include at a minimum all the elements included in paragraph (k)(9)(viii) of this section and in addition, the specific work practices and engineering controls set forth in paragraph (g) of this section which specifically relate to that category. Such course shall include "hands-on" training and shall take at least 8 hours. [§1926.1101(k)(9)(iv)[A]]

[B] An employee who works with more than one of the categories of material specified in paragraph (k)(9)(iv)[A] of this section shall receive training in the work practices applicable to each category of material that the employee removes and each removal method that the employee uses.[§1926.1101(k)(9)(iv)[B]]

[C] For Class II operations not involving the categories of material specified in paragraph (k)(9)(iv)[A] of this section, training shall be provided which shall include at a minimum all the elements included in paragraph (k)(9)(viii) of this section and in addition, the specific work practices and engineering controls set forth in paragraph (g) of this section which specifically relate to the category of material being removed, and shall include "hands-on" training in the work practices applicable to each category of material that the employee removes and each removal method that the employee uses.[§1926.1101(k)(9)(iv)[C]]

(v) *Training for Class III employees* shall be consistent with EPA requirements for training of local education agency maintenance and custodial staff as set forth at 40 CFR 763.92(a)(2). Such a course shall also include "hands-on" training and shall take at least 16 hours. Exception: For Class III operations for which the competent person determines that the EPA curriculum does not adequately cover the training needed to perform that activity, training shall include as a minimum all the elements included in paragraph (k)(9)(viii) of this section and in addition, the specific work practices and engineering controls set forth in paragraph (g) of this section which specifically relate to that activity, and shall include "hands-on" training in the work practices applicable to each category of material that the employee disturbs. [§1926.1101(k)(9)(v)]

(vi) *Training for employees performing Class IV operations* shall be consistent with EPA requirements for training of local education agency maintenance and custodial staff as set forth at 40 CFR 763.92(a)(1). Such a course shall include available information concerning the locations of thermal system insulation and surfacing ACM/PACM, and asbestos-containing flooring material, or flooring material where the absence of asbestos has not yet been certified; and instruction in recognition of damage, deterioration, and delamination of asbestos containing building materials. Such course shall take at least 2 hours.[§1926.1101(k)(9)(vi)]

(vii) *Training for employees who are likely to be exposed* in excess of the PEL and who are not otherwise required to be trained under paragraph (k)(9)(iii) through (vi) of this section, shall meet the requirements of paragraph (k)(9)(viii) of this section.[§1926.1101(k)(9)(vii)]

(viii) *The training program shall be conducted* in a manner that the employee is able to understand. In addition to the content required by provisions in paragraphs (k)(9)(iii) through (vi) of this section, the employer shall ensure that each such employee is informed of the following:[§1926.1101(k)(9)(viii)]

[A] *Methods of recognizing asbestos,* including the requirement in paragraph (k)(1) of this section to presume that certain building materials contain asbestos; [§1926.1101(k)(9)(viii)[A]]

[B] *The health effects* associated with asbestos exposure; [§1926.1101(k)(9)(viii)[B]]

[C] *The relationship between smoking* and asbestos in producing lung cancer;[§1926.1101(k)(9)(viii)[C]]

[D] *The nature of operations that could result* in exposure to asbestos, the importance of necessary protective controls to minimize exposure including, as applicable, engineering controls, work practices, respirators, housekeeping procedures, hygiene facilities, protective clothing, decontamination procedures, emergency procedures, and waste disposal procedures, and any necessary instruction in the use of these controls and procedures; where Class III and IV work will be or is performed, the contents of EPA 20T-2003, "Managing Asbestos In-Place" July 1990 or its equivalent in content;[§1926.1101(k)(9)(viii)[D]]

[E] *The purpose, proper use, fitting instructions,* and limitations of respirators as required by 29 CFR 1910.134; [§1926.1101(k)(9)(viii)[E]]

[F] *The appropriate work practices* for performing the asbestos job;[§1926.1101(k)(9)(viii)[F]]

[G] *Medical surveillance program requirements;* [§1926.1101(k)(9)(viii)[G]]

[H] *The content of this standard including appendices;* [§1926.1101(k)(9)(viii)[H]]

[I] *The names, addresses and phone numbers* of public health organizations which provide information, materials and/or conduct programs concerning smoking cessation. The employer may distribute the list of such organizations contained in Appendix J to this section, to comply with this requirement; and[§1926.1101(k)(9)(viii)[I]]

[J] *The requirements for posting signs* and affixing labels and the meaning of the required legends for such signs and labels.[§1926.1101(k)(9)(viii)[J]]

(10) *Access to training materials.*[§1926.1101(k)(10)]

(i) *The employer shall make readily available* to affected employees without cost, written materials relating to the employee training program, including a copy of this regulation.[§1926.1101(k)(10)(i)]

(ii) *The employer shall provide* to the Assistant Secretary and the Director, upon request, all information and training materials relating to the employee information and training program.[§1926.1101(k)(10)(ii)]

(iii) *The employer shall inform all employees* concerning the availability of self-help smoking cessation program material. Upon employee request, the employer shall distribute such material, consisting of NIH Publication No, 89-1647, or equivalent self-help material, which is approved or published by a public health organization listed in Appendix J to this section. [§1926.1101(k)(10)(iii)]

(l) Housekeeping[§1926.1101(l)]

(1) *Vacuuming.* Where vacuuming methods are selected, HEPA filtered vacuuming equipment must be used. The equipment shall be used and emptied in a manner that minimizes the reentry of asbestos into the workplace.[§1926.1101(l)(1)]

(2) *Waste disposal.* Asbestos waste, scrap, debris, bags, containers, equipment, and contaminated clothing consigned for disposal shall be collected and disposed of in sealed, labeled, impermeable bags or other closed, labeled, impermeable containers except in roofing operations, where the procedures specified in paragraph (g)(8)(ii) of this section apply. [§1926.1101(l)(2)]

(3) *Care of asbestos-containing flooring material.*[§1926.1101(l)(3)]

(i) *All vinyl and asphalt flooring material* shall be maintained in accordance with this paragraph unless the building/facility owner demonstrates, pursuant to paragraph (g)(8)(i)(I) of this section that the flooring does not contain asbestos. [§1926.1101(l)(3)(i)]

(ii) *Sanding of flooring material is prohibited.*[§1926.1101(l)(3)(ii)]

(iii) *Stripping of finishes shall be conducted* using low abrasion pads at speeds lower than 300 rpm and wet methods. [§1926.1101(l)(3)(iii)]

(iv) *Burnishing or dry buffing may be performed* only on flooring which has sufficient finish so that the pad cannot contact the flooring material.[§1926.1101(l)(3)(iv)]

(4) *Waste and debris and accompanying dust* in an area containing accessible thermal system insulation or surfacing ACM/PACM or visibly deteriorated ACM:[§1926.1101(l)(4)]

(i) *Shall not be dusted or swept dry,* or vacuumed without using a HEPA filter;[§1926.1101(l)(4)(i)]

(ii) *Shall be promptly cleaned up* and disposed of in leak tight containers.[§1926.1101(l)(4)(ii)]

(m) Medical surveillance.[§1926.1101(m)]

(1) *General*[§1926.1101(m)(1)]

(i) *Employees covered.*[§1926.1101(m)(1)(i)]

[A] *The employer shall institute a medical surveillance program* for all employees who for a combined total of 30 or more days per year are engaged in Class I, II and III work or are exposed at or above a permissible exposure limit. For purposes of this paragraph, any day in which a worker engages in Class II or Class III operations or a combination thereof on intact material for one hour or less (taking into account the entire time spent on the removal operation, including cleanup) and, while doing so, adheres fully to the work practices specified in this standard, shall not be counted.[§1926.1101(m)(1)(i)[A]]

[B] *For employees otherwise required* by this standard to wear a negative pressure respirator, employers shall ensure employees are physically able to perform the work and use the equipment. This determination shall be made under the supervision of a physician.[§1926.1101(m)(1)(i)[B]]

(ii) *Examination.*[§1926.1101(m)(1)(ii)]

[A] *The employer shall ensure* that all medical examinations and procedures are performed by or under the supervision of a licensed physician, and are provided at no cost to the employee and at a reasonable time and place. [§1926.1101(m)(1)(ii)[A]]

[B] *Persons other than such licensed physicians* who administer the pulmonary function testing required by this section shall complete a training course in spirometry sponsored by an appropriate academic or professional institution.[§1926.1101(m)(1)(ii)[B]]

(2) *Medical examinations and consultations.*[§1926.1101(m)(2)]

(i) *Frequency.* The employer shall make available medical examinations and consultations to each employee covered under paragraph (m)(1)(i) of this section on the following schedules:[§1926.1101(m)(2)(i)]

[A] *Prior to assignment of the employee* to an area where negative- pressure respirators are worn; [§1926.1101(m)(2)(i)[A]]

[B] *When the employee is assigned to an area* where exposure to asbestos may be at or above the permissible exposure limit for 30 or more days per year, or engage in Class I, II, or III work for a combined total of 30 or more days per year, a medical examination must be given within 10 working days following the thirtieth day of exposure;[§1926.1101(m)(2)(i)[B]]

[C] *And at least annually thereafter.*[§1926.1101(m)(2)(i)[C]]

[D] *If the examining physician determines* that any of the examinations should be provided more frequently than specified, the employer shall provide such examinations to affected employees at the frequencies specified by the physician.[§1926.1101(m)(2)(i)[D]]

[E] *Exception:* No medical examination is required of any employee if adequate records show that the employee has been examined in accordance with this paragraph within the past 1-year period.[§1926.1101(m)(2)(i)[E]]

(ii) *Content.* Medical examinations made available pursuant to paragraphs (m)(2)(i)(A) through (m)(2)(i)(C) of this section shall include:[§1926.1101(m)(2)(ii)]

[A] *A medical and work history* with special emphasis directed to the pulmonary, cardiovascular, and gastrointestinal systems.[§1926.1101(m)(2)(ii)[A]]

[B] *On initial examination,* the standardized questionnaire contained in Part 1 of Appendix D to this section, and, on annual examination, the abbreviated standardized questionnaire contained in Part 2 of Appendix D to this section. [§1926.1101(m)(2)(ii)[B]]

[C] A physical examination directed to the pulmonary and gastrointestinal systems, including a chest roentgenogram to be administered at the discretion of the physician, and pulmonary function tests of forced vital capacity (FVC) and forced expiratory volume at one second (FEV(1)). Interpretation and classification of chest shall be conducted in accordance with Appendix E to this section. [§1926.1101(m)(2)(ii)[C]]

[D] Any other examinations or tests deemed necessary by the examining physician.[§1926.1101(m)(2)(ii)[D]]

(3) *Information provided to the physician.* The employer shall provide the following information to the examining physician: [§1926.1101(m)(3)]

(i) *A copy of this standard and Appendices D, E, and I to this section;*[§1926.1101(m)(3)(i)]

(ii) *A description of the affected employee's duties* as they relate to the employee's exposure;[§1926.1101(m)(3)(ii)]

(iii) *The employee's representative exposure level* or anticipated exposure level;[§1926.1101(m)(3)(iii)]

(iv) *A description of any* personal protective and respiratory equipment used or to be used; and[§1926.1101(m)(3)(iv)]

(v) *Information from previous medical examinations* of the affected employee that is not otherwise available to the examining physician.[§1926.1101(m)(3)(v)]

(4) *Physician's written opinion.*[§1926.1101(m)(4)]

(i) *The employer shall obtain a written opinion* from the examining physician. This written opinion shall contain the results of the medical examination and shall include:[§1926.1101(m)(4)(i)]

[A] The physician's opinion as to whether the employee has any detected medical conditions that would place the employee at an increased risk of material health impairment from exposure to asbestos;[§1926.1101(m)(4)(i)[A]]

[B] Any recommended limitations on the employee or on the use of personal protective equipment such as respirators; and[§1926.1101(m)(4)(i)[B]]

[C] A statement that the employee has been informed by the physician of the results of the medical examination and of any medical conditions that may result from asbestos exposure.[§1926.1101(m)(4)(i)[C]]

[D] A statement that the employee has been informed by the physician of the increased risk of lung cancer attributable to the combined effect of smoking and asbestos exposure.[§1926.1101(m)(4)(i)[D]]

(ii) *The employer shall instruct the physician* not to reveal in the written opinion given to the employer specific findings or diagnoses unrelated to occupational exposure to asbestos. [§1926.1101(m)(4)(ii)]

(iii) *The employer shall provide a copy* of the physician's written opinion to the affected employee within 30 days from its receipt.[§1926.1101(m)(4)(iii)]

(n) Recordkeeping.[§1926.1101(n)]

(1) *Objective data relied on pursuant to paragraph (f) to this section.*[§1926.1101(n)(1)]

(i) *Where the employer has relied on objective data* that demonstrates that products made from or containing asbestos or the activity involving such products or material are not capable of releasing fibers of asbestos in concentrations at or above the permissible exposure limit and/or excursion limit under the expected conditions of processing, use, or handling to satisfy the requirements of paragraph (f), the employer shall establish and maintain an accurate record of objective data reasonably relied upon in support of the exemption.[§1926.1101(n)(1)(i)]

(ii) *The record shall include at least the following information:* [§1926.1101(n)(1)(ii)]

[A] The product qualifying for exemption;[§1926.1101(n)(1)(ii)[A]]

[B] The source of the objective data;[§1926.1101(n)(1)(ii)[B]]

[C] The testing protocol, results of testing, and/or analysis of the material for the release of asbestos; [§1926.1101(n)(1)(ii)[C]]

[D] A description of the operation exempted and how the data support the exemption; and[§1926.1101(n)(1)(ii)[D]]

[E] Other data relevant to the operations, materials, processing, or employee exposures covered by the exemption. [§1926.1101(n)(1)(ii)[E]]

(iii) *The employer shall maintain this record* for the duration of the employer's reliance upon such objective data. [§1926.1101(n)(1)(iii)]

(2) *Exposure measurements.*[§1926.1101(n)(2)]

(i) *The employer shall keep an accurate record* of all measurements taken to monitor employee exposure to asbestos as prescribed in paragraph (f) of this section.[§1926.1101(n)(2)(i)]

Note: The employer may utilize the services of competent organizations such as industry trade associations and employee associations to maintain the records required by this section.

(ii) *This record shall include* at least the following information: [§1926.1101(n)(2)(ii)]

[A] The date of measurement;[§1926.1101(n)(2)(ii)[A]]

[B] The operation involving exposure to asbestos that is being monitored;[§1926.1101(n)(2)(ii)[B]]

[C] Sampling and analytical methods used and evidence of their accuracy;[§1926.1101(n)(2)(ii)[C]]

[D] Number, duration, and results of samples taken; [§1926.1101(n)(2)(ii)[D]]

[E] Type of protective devices worn, if any; and [§1926.1101(n)(2)(ii)[E]]

[F] Name, social security number, and exposure of the employees whose exposures are represented. [§1926.1101(n)(2)(ii)[F]]

(iii) *The employer shall maintain this record* for at least thirty (30) years, in accordance with 29 CFR 1910.1020[1]. [§1926.1101(n)(2)(iii)]

(3) *Medical surveillance.*[§1926.1101(n)(3)]

(i) *The employer shall establish and maintain* an accurate record for each employee subject to medical surveillance by paragraph (m) of this section, in accordance with 29 CFR 1910.1020.[§1926.1101(n)(3)(i)]

(ii) *The record shall include* at least the following information: [§1926.1101(n)(3)(ii)]

[A] The name and social security number of the employee; [§1926.1101(n)(3)(ii)[A]]

[B] A copy of the employee's medical examination results, including the medical history, questionnaire responses, results of any tests, and physician's recommendations. [§1926.1101(n)(3)(ii)[B]]

[C] Physician's written opinions;[§1926.1101(n)(3)(ii)[C]]

[D] Any employee medical complaints related to exposure to asbestos; and[§1926.1101(n)(3)(ii)[D]]

[E] A copy of the information provided to the physician as required by paragraph (m) of this section. [§1926.1101(n)(3)(ii)[E]]

(iii) *The employer shall ensure that this record* is maintained for the duration of employment plus thirty (30) years, in accordance with 29 CFR 1910.1020[1].[§1926.1101(n)(3)(iii)]

(4) *Training records.* The employer shall maintain all employee training records for one (1) year beyond the last date of employment by that employer.[§1926.1101(n)(4)]

(5) *Data to Rebut PACM.* Where the building owner and employer have relied on data to demonstrate that PACM is not asbestos-containing, such data shall be maintained for as long as they are relied upon to rebut the presumption.[§1926.1101(n)(5)]

(6) *Records of required notifications.* Where the building owner has communicated and received information concerning the identification, location and quantity of ACM and PACM, written records of such notifications and their content shall be maintained by the building owner for the duration of ownership and shall be transferred to successive owners of such buildings/facilities. [§1926.1101(n)(6)]

(7) *Availability.*[§1926.1101(n)(7)]

(i) *The employer, upon written request,* shall make all records required to be maintained by this section available to the Assistant Secretary and the Director for examination and copying.[§1926.1101(n)(7)(i)]

(ii) *The employer must comply with the requirements* concerning availability of records set forth in 29 CFR 1910.1020. [§1926.1101(n)(7)(ii)]

---

1. Editor's Note: The CFR refers in error to Sec. 1910.33. The regulation as it appears above is corrected to refer to Sec. 1910.1020, corresponding to the Asbestos standards appearing at Sec. 1910.1001.

(8) *Transfer of records.* The employer must comply with the requirements concerning transfer of records set forth in 29 CFR 1910.1020(h).[§1926.1101(n)(8)]

(o) **Competent person.**[§1926.1101(o)]

(1) *General.* On all construction worksites covered by this standard, the employer shall designate a competent person, having the qualifications and authorities for ensuring worker safety and health required by subpart C, General Safety and Health Provisions for Construction (29 CFR 1926.20 through 1926.32). [§1926.1101(o)(1)]

(2) *Required inspections by the competent person.* Section 1926.20(b)(2) which requires health and safety prevention programs to provide for frequent and regular inspections of the job sites, materials, and equipment to be made by competent persons, is incorporated.[§1926.1101(o)(2)]

(3) *Additional inspections.* In addition, the competent person shall make frequent and regular inspections of the job sites, in order to perform the duties set out below in paragraph (o)(3)(i) of this section. For Class I jobs, on-site inspections shall be made at least once during each work shift, and at any time at employee request. For Class II, III, and IV jobs, on-site inspections shall be made at intervals sufficient to assess whether conditions have changed, and at any reasonable time at employee request.[§1926.1101(o)(3)]

(i) *On all worksites where employees are engaged* in Class I or II asbestos work, the competent person designated in accordance with paragraph (e)(6) of this section shall perform or supervise the following duties, as applicable:[§1926.1101(o)(3)(i)]

[A] Set up the regulated area, enclosure, or other containment;[§1926.1101(o)(3)(i)[A]]

[B] Ensure (by on-site inspection) the integrity of the enclosure or containment;[§1926.1101(o)(3)(i)[B]]

[C] Set up procedures to control entry to and exit from the enclosure and/or area;[§1926.1101(o)(3)(i)[C]]

[D] Supervise all employee exposure monitoring required by this section and ensure that it is conducted as required by paragraph (f) of this section;[§1926.1101(o)(3)(i)[D]]

[E] Ensure that employees working within the enclosure and/or using glove bags wear respirators and protective clothing as required by paragraphs (h) and (i) of this section; [§1926.1101(o)(3)(i)[E]]

[F] Ensure through on-site supervision, that employees set up, use, and remove engineering controls, use work practices and personal protective equipment in compliance with all requirements;[§1926.1101(o)(3)(i)[F]]

[G] Ensure that employees use the hygiene facilities and observe the decontamination procedures specified in paragraph (j) of this section;[§1926.1101(o)(3)(i)[G]]

[H] Ensure that through on-site inspection, engineering controls are functioning properly and employees are using proper work practices; and,[§1926.1101(o)(3)(i)[H]]

[I] Ensure that notification requirement in paragraph (k) of this section are met.[§1926.1101(o)(3)(i)[I]]

(ii) *[Reserved]*[§1926.1101(o)(3)(ii)]

(4) *Training for the competent person.*[§1926.1101(o)(4)]

(i) *For Class I and II asbestos work* the competent person shall be trained in all aspects of asbestos removal and handling, including: abatement, installation, removal and handling; the contents of this standard; the identification of asbestos; removal procedures, where appropriate; and other practices for reducing the hazard. Such training shall be obtained in a comprehensive course for supervisors that meets the criteria of EPA's Model Accreditation Plan (40 CFR part 763, subpart E, Appendix C), such as a course conducted by an EPA-approved or state-approved training provider, certified by EPA or a state, or a course equivalent in stringency, content, and length.[§1926.1101(o)(4)(i)]

(ii) *For Class III and IV asbestos work,* the competent person shall be trained in aspects of asbestos handling appropriate for the nature of the work, to include procedures for setting up glove bags and mini- enclosures, practices for reducing asbestos exposures, use of wet methods, the contents of this standard, and the identification of asbestos. Such training shall include successful completion of a course that is consistent with EPA requirements for training of local education agency maintenance and custodial staff as set forth at 40 CFR 763.92(a)(2), or its equivalent in stringency, content and length. Competent persons for Class III and IV work, may also be trained pursuant to the requirements of paragraph (o)(4)(i) of this section.[§1926.1101(o)(4)(ii)]

(p) **Appendices.**[§1926.1101(p)]

(1) *Appendices A, C, D, and E to this section* are incorporated as part of this section and the contents of these appendices are mandatory.[§1926.1101(p)(1)]

(2) *Appendices B, F, H, I, J, and K to this section* are informational and are not intended to create any additional obligations not otherwise imposed or to detract from any existing obligations. [§1926.1101(p)(2)]

§1926.1101 Appendix A

OSHA Reference Method — Mandatory

This mandatory appendix specifies the procedure for analyzing air samples for asbestos and specifies quality control procedures that must be implemented by laboratories performing the analysis. The sampling and analytical methods described below represent the elements of the available monitoring methods (such as Appendix B of this regulation, the most current version of the OSHA method ID-160, or the most current version of the NIOSH Method 7400). All employers who are required to conduct air monitoring under paragraph (f) of the standard are required to utilize analytical laboratories that use this procedure, or an equivalent method, for collecting and analyzing samples.

Sampling and Analytical Procedure

1. **The sampling medium for air samples** shall be mixed cellulose ester filter membranes. These shall be designated by the manufacturer as suitable for asbestos counting. See below for rejection of blanks.
2. **The preferred collection device** shall be the 25-mm diameter cassette with an open-faced 50-mm electrically conductive extension cowl. The 37-mm cassette may be used if necessary but only if written justification for the need to use the 37-mm filter cassette accompanies the sample results in the employee's exposure monitoring record. Do not reuse or reload cassettes for asbestos sample collection.
3. **An air flow rate** between 0.5 liter/min and 2.5 liters/min shall be selected for the 25/mm cassette. If the 37-mm cassette is used, an air flow rate between 1 liter/min and 2.5 liters/min shall be selected.
4. **Where possible, a sufficient air volume** for each air sample shall be collected to yield between 100 and 1,300 fibers per square millimeter on the membrane filter. If a filter darkens in appearance or if loose dust is seen on the filter, a second sample shall be started.
5. **Ship the samples in a rigid container** with sufficient packing material to prevent dislodging the collected fibers. Packing material that has a high electrostatic charge on its surface (e.g., expanded polystyrene) cannot be used because such material can cause loss of fibers to the sides of the cassette.
6. **Calibrate each personal sampling pump** before and after use with a representative filter cassette installed between the pump and the calibration devices.
7. **Personal samples shall be taken** in the "breathing zone" of the employee (i.e., attached to or near the collar or lapel near the worker's face).
8. **Fiber counts shall be made** by positive phase contrast using a microscope with an 8 to 10 X eyepiece and a 40 to 45 X objective for a total magnification of approximately 400 X and a numerical aperture of 0.65 to 0.75. The microscope shall also be fitted with a green or blue filter.
9. **The microscope shall be fitted** with a Walton-Beckett eyepiece graticule calibrated for a field diameter of 100 micrometers (± 2 micrometers).
10. **The phase-shift detection limit** of the microscope shall be about 3 degrees measured using the HSE phase shift test slide as outlined below.
    a. *Place the test slide* on the microscope stage and center it under the phase objective.
    b. *Bring the blocks of grooved lines into focus.*
    Note: The slide consists of seven sets of grooved lines (ca. 20 grooves to each block) in descending order of visibility from sets 1 to 7, seven being the least visible. The requirements for asbestos counting are that the microscope optics must resolve the grooved lines in set 3 completely, although they may appear somewhat faint, and that the grooved lines in sets 6 and 7 must be invisible. Sets 4 and 5 must be at least partially visible but may vary slightly in visibility between microscopes. A microscope that fails to meet these requirements has either too low or too high a resolution to be used for asbestos counting.
    c. *If the image deteriorates,* clean and adjust the microscope optics. If the problem persists, consult the microscope manufacturer.
11. **Each set of samples taken** will include 10% field blanks or a minimum of 2 field blanks. These blanks must come from the same lot as the filters used for sample collection. The field blank results shall be averaged and subtracted from the analytical results before reporting. A set consists of any sample or group of samples for

which an evaluation for this standard must be made. Any samples represented by a field blank having a fiber count in excess of the detection limit of the method being used shall be rejected.

12. **The samples shall be mounted** by the acetone/triacetin method or a method with an equivalent index of refraction and similar clarity.
13. **Observe the following counting rules.**
    a. *Count only fibers* equal to or longer than 5 micrometers. Measure the length of curved fibers along the curve.
    b. *In the absence of other information,* count all particles as asbestos, that have a length-to-width ratio (aspect ratio) of 3:1 or greater.
    c. *Fibers lying entirely within the boundary* of the Walton-Beckett graticule field shall receive a count of 1. Fibers crossing the boundary once, having one end within the circle, shall receive the count of one half (1/2). Do not count any fiber that crosses the graticule boundary more than once. Reject and do not count any other fibers even though they may be visible outside the graticule area.
    d. *Count bundles of fibers* as one fiber unless individual fibers can be identified by observing both ends of an individual fiber.
    e. *Count enough graticule fields* to yield 100 fibers. Count a minimum of 20 fields; stop counting at 100 fields regardless of fiber count.
14. **Blind recounts shall be conducted at the rate of 10 percent.**

Quality Control Procedures

1. **Intralaboratory program.** Each laboratory and/or each company with more than one microscopist counting slides shall establish a statistically designed quality assurance program involving blind recounts and comparisons between microscopists to monitor the variability of counting by each microscopist and between microscopists. In a company with more than one laboratory, the program shall include all laboratories, and shall also evaluate the laboratory-to-laboratory variability.
2. a. *Interlaboratory program.* Each laboratory analyzing asbestos samples for compliance determination shall implement an interlaboratory quality assurance program that, as a minimum, includes participation of at least two other independent laboratories. Each laboratory shall participate in round robin testing at least once every 6 months with at least all the other laboratories in its interlaboratory quality assurance group. Each laboratory shall submit slides typical of its own workload for use in this program. The round robin shall be designed and results analyzed using appropriate statistical methodology.
    b. *All laboratories should also participate* in a national sample testing scheme such as the Proficiency Analytical Testing Program (PAT), or the Asbestos Registry sponsored by the American Industrial Hygiene Association (AIHA).
3. **All individuals performing asbestos analysis** must have taken the NIOSH course for sampling and evaluating airborne asbestos dust or an equivalent course.
4. **When the use of different microscopes** contributes to differences between counters and laboratories, the effect of the different microscope shall be evaluated and the microscope shall be replaced, as necessary.
5. **Current results of these quality assurance programs** shall be posted in each laboratory to keep the microscopists informed.

§1926.1101 Appendix B

Sampling and Analysis (Non-Mandatory)

Matrix Air:

OSHA Permissible Exposure Limits:

Time Weighted Average	0.1 fiber/cc
Excursion Level (30 minutes)	1.0 fiber/cc

Collection Procedure:

A known volume of air is drawn through a 25-mm diameter cassette containing a mixed-cellulose ester filter. The cassette must be equipped with an electrically conductive 50-mm extension cowl. The sampling time and rate are chosen to give a fiber density of between 100 to 1,300 fibers/mm2 on the filter.

Recommended Sampling Rate 0.5 to 5.0 liters/minute (L/min)

Recommended Air Volumes:

Minimum	25 L
Maximum	2,400 L

Analytical Procedure:

A portion of the sample filter is cleared and prepared for asbestos fiber counting by Phase Contrast Microscopy (PCM) at 400X.

Commercial manufacturers and products mentioned in this method are for descriptive use only and do not constitute endorsements by USDOL-OSHA. Similar products from other sources can be substituted.

1. Introduction

This method describes the collection of airborne asbestos fibers using calibrated sampling pumps with mixed-cellulose ester (MCE) filters and analysis by phase contrast microscopy (PCM). Some terms used are unique to this method and are defined below:

Asbestos: A term for naturally occurring fibrous minerals. Asbestos includes chrysotile, crocidolite, amosite (cummingtonite-grunerite asbestos), tremolite asbestos, actinolite asbestos, anthophyllite asbestos, and any of these minerals that have been chemically treated and/or altered. The precise chemical formulation of each species will vary with the location from which it was mined. Nominal compositions are listed:

Chrysotile..................	$Mg_3Si_2O_5(OH)_4$
Crocidolite.................	$Na_2Fe_3{}^2+Fe_2{}^3+Si_8O_{22}(OH)_2$
Amosite.....................	$(Mg,Fe)_7Si_8O_{22}(OH)_2$
Tremolite-actinolite......	$Ca_2(Mg,Fe)_5Si_8O_{22}(OH)_2$
Anthophyllite..............	$(Mg,Fe)_7Si_8O_{22}(OH)_2$

Asbestos Fiber: A fiber of asbestos which meets the criteria specified below for a fiber.

Aspect Ratio: The ratio of the length of a fiber to it's diameter (e.g. 3:1, 5:1 aspect ratios).

Cleavage Fragments: Mineral particles formed by comminution of minerals, especially those characterized by parallel sides and a moderate aspect ratio (usually less than 20:1).

Detection Limit: The number of fibers necessary to be 95% certain that the result is greater than zero.

Differential Counting: The term applied to the practice of excluding certain kinds of fibers from the fiber count because they do not appear to be asbestos.

Fiber: A particle that is 5 μm or longer, with a length-to-width ratio of 3 to 1 or longer.

Field: The area within the graticule circle that is superimposed on the microscope image.

Set: The samples which are taken, submitted to the laboratory, analyzed, and for which, interim or final result reports are generated.

Tremolite, Anthophyllite, and Actinolite: The non-asbestos form of these minerals which meet the definition of a fiber. It includes any of these minerals that have been chemically treated and/or altered.

Walton-Beckett Graticule: An eyepiece graticule specifically designed for asbestos fiber counting. It consists of a circle with a projected diameter of 100 2 μm (area of about 0.00785 mm^2) with a crosshair having tic-marks at 3-μm intervals in one direction and 5-μm in the orthogonal direction. There are marks around the periphery of the circle to demonstrate the proper sizes and shapes of fibers. This design is reproduced in Figure 1. The disk is placed in one of the microscope eyepieces so that the design is superimposed on the field of view.

1.1. *History*

Early surveys to determine asbestos exposures were conducted using impinger counts of total dust with the counts expressed as million particles per cubic foot. The British Asbestos Research Council recommended filter membrane counting in 1969. In July 1969, the Bureau of Occupational Safety and Health published a filter membrane method for counting asbestos fibers in the United States. This method was refined by NIOSH and published as P & CAM 239. On May 29, 1971, OSHA specified filter membrane sampling with phase contrast counting for evaluation of asbestos exposures at work sites in the United States. The use of this technique was again required by OSHA in 1986. Phase contrast microscopy has continued to be the method of choice for the measurement of occupational exposure to asbestos.

1.2. *Principle*

Air is drawn through a MCE filter to capture airborne asbestos fibers. A wedge shaped portion of the filter is removed, placed on a glass microscope slide and made transparent. A measured area (field) is viewed by PCM. All the fibers meeting defined criteria for asbestos are counted and considered a measure of the airborne asbestos concentration.

1.3. *Advantages and Disadvantages*

There are four main advantages of PCM over other methods:

(1) The technique is specific for fibers. Phase contrast is a fiber counting technique which excludes non-fibrous particles from the analysis.

(2) *The technique is inexpensive and does* not require specialized knowledge to carry out the analysis for total fiber counts.

(3) *The analysis is quick and can* be performed on-site for rapid determination of air concentrations of asbestos fibers.

(4) *The technique has continuity* with historical epidemiological studies so that estimates of expected disease can be inferred from long-term determinations of asbestos exposures.

The main disadvantage of PCM is that it does not positively identify asbestos fibers. Other fibers which are not asbestos may be included in the count unless differential counting is performed. This requires a great deal of experience to adequately differentiate asbestos from non-asbestos fibers. Positive identification of asbestos must be performed by polarized light or electron microscopy techniques. A further disadvantage of PCM is that the smallest visible fibers are about 0.2 μm in diameter while the finest asbestos fibers may be as small as 0.02 μm in diameter. For some exposures, substantially more fibers may be present than are actually counted.

1.4. *Workplace Exposure*

Asbestos is used by the construction industry in such products as shingles, floor tiles, asbestos cement, roofing felts, insulation and acoustical products. Non-construction uses include brakes, clutch facings, paper, paints, plastics, and fabrics. One of the most significant exposures in the workplace is the removal and encapsulation of asbestos in schools, public buildings, and homes. Many workers have the potential to be exposed to asbestos during these operations.

About 95% of the asbestos in commercial use in the United States is chrysotile. Crocidolite and amosite make up most of the remainder. Anthophyllite and tremolite or actinolite are likely to be encountered as contaminants in various industrial products.

1.5. *Physical Properties*

Asbestos fiber possesses a high tensile strength along its axis, is chemically inert, non-combustible, and heat resistant. It has a high electrical resistance and good sound absorbing properties. It can be weaved into cables, fabrics or other textiles, and also matted into asbestos papers, felts, or mats.

2. Range and Detection Limit

2.1. *The ideal counting range* on the filter is 100 to 1,300 fibers/mm^2. With a Walton-Beckett graticule this range is equivalent to 0.8 to 10 fibers/field. Using NIOSH counting statistics, a count of 0.8 fibers/field would give an approximate coefficient of variation (CV) of 0.13.

2.2. *The detection limit for this method* is 4.0 fibers per 100 fields or 5.5 fibers/mm^2. This was determined using an equation to estimate the maximum CV possible at a specific concentration (95% confidence) and a Lower Control Limit of zero. The CV value was then used to determine a corresponding concentration from historical CV vs fiber relationships. As an example:

Lower Control Limit (95% Confidence) = AC — 1.645(CV)(AC)

Where:

AC = Estimate of the airborne fiber concentration (fibers/cc)

Setting the Lower Control Limit=0 and solving for CV:

0 = AC — 1.645(CV)(AC)

CV = 0.61

This value was compared with CV vs. count curves. The count at which CV = 0.61 for Leidel-Busch counting statistics or for an OSHA Salt Lake Technical Center (OSHA-SLTC) CV curve (see Appendix A for further information) was 4.4 fibers or 3.9 fibers per 100 fields, respectively. Although a lower detection limit of 4 fibers per 100 fields is supported by the OSHA-SLTC data, both data sets support the 4.5 fibers per 100 fields value.

3. Method Performance — Precision and Accuracy

Precision is dependent upon the total number of fibers counted and the uniformity of the fiber distribution on the filter. A general rule is to count at least 20 and not more than 100 fields. The count is discontinued when 100 fibers are counted, provided that 20 fields have already been counted. Counting more than 100 fibers results in only a small gain in precision. As the total count drops below 10 fibers, an accelerated loss of precision is noted.

At this time, there is no known method to determine the absolute accuracy of the asbestos analysis. Results of samples prepared through the Proficiency Analytical Testing (PAT) Program and analyzed by the OSHA-SLTC showed no significant bias when compared to PAT reference values. The PAT samples were analyzed from 1987 to 1989 (N=36) and the concentration range was from 120 to 1,300 fibers/mm^2.

4. Interferences

Fibrous substances, if present, may interfere with asbestos analysis. Some common fibers are:

Fiberglass
Anhydrite
Plant fibers
Perlite veins
Gypsum
Some synthetic fibers
Membrane structures
Sponge spicules
Diatoms
Microorganisms
Wollastonite

The use of electron microscopy or optical tests such as polarized light, and dispersion staining may be used to differentiate these materials from asbestos when necessary.

5. Sampling

5.1. *Equipment*

5.1.1. *Sample assembly* (The assembly is shown in Figure 3). Conductive filter holder consisting of a 25-mm diameter, 3-piece cassette having a 50-mm long electrically conductive extension cowl. Backup pad, 25-mm, cellulose. Membrane filter, mixed-cellulose ester (MCE), 25-mm, plain, white, 0.4 to 1.2-μm pore size.

Notes:

(a) DO NOT RE-USE CASSETTES.

(b) Fully conductive cassettes are required to reduce fiber loss to the sides of the cassette due to electrostatic attraction.

(c) Purchase filters which have been selected by the manufacturer for asbestos counting or analyze representative filters for fiber background before use. Discard the filter lot if more than 4 fibers/100 fields are found.

(d) To decrease the possibility of contamination, the sampling system (filter-backup pad-cassette) for asbestos is usually preassembled by the manufacturer.

(e) Other cassettes, such as the Bell-mouth, may be used within the limits of their validation.

5.1.2. *Gel bands for sealing cassettes.*

5.1.3. *Sampling pump.*

Each pump must be a battery operated, self-contained unit small enough to be placed on the monitored employee and not interfere with the work being performed. The pump must be capable of sampling at the collection rate for the required sampling time.

5.1.4. *Flexible tubing, 6-mm bore.*

5.1.5. *Pump calibration.*

Stopwatch and bubble tube/burette or electronic meter.

5.2. *Sampling Procedure*

5.2.1. *Seal the point* where the base and cowl of each cassette meet with a gel band or tape.

5.2.2. *Charge the pumps completely before beginning.*

5.2.3. *Connect each pump* to a calibration cassette with an appropriate length of 6-mm bore plastic tubing. Do not use luer connectors — the type of cassette specified above has built-in adapters.

5.2.4. *Select an appropriate flow* rate for the situation being monitored. The sampling flow rate must be between 0.5 and 5.0 L/min for personal sampling and is commonly set between 1 and 2 L/min. Always choose a flow rate that will not produce overloaded filters.

5.2.5. *Calibrate each sampling pump* before and after sampling with a calibration cassette in-line (Note: This calibration cassette should be from the same lot of cassettes used for sampling). Use a primary standard (e.g. bubble burette) to calibrate each pump. If possible, calibrate at the sampling site.

Note: If sampling site calibration is not possible, environmental influences may affect the flow rate. The extent is dependent on the type of pump used. Consult with the pump manufacturer to determine dependence on environmental influences. If the pump is affected by temperature and pressure changes, correct the flow rate using the formula shown in the section "Sampling Pump Flow Rate Corrections" at the end of this appendix.

5.2.6. *Connect each pump to the base* of each sampling cassette with flexible tubing. Remove the end cap of each cassette and take each air sample open face. Assure that each sample cassette is held open side down in the employee's breathing zone during sampling. The distance from the nose/mouth of the employee to the cassette should be about 10 cm. Secure the cassette on the collar or lapel of the employee using spring clips or other similar devices.

5.2.7. *A suggested minimum air volume* when sampling to determine TWA compliance is 25 L. For Excursion Limit (30 min sampling time) evaluations, a minimum air volume of 48 L is recommended.

5.2.8. *The most significant problem* when sampling for asbestos is overloading the filter with non-asbestos dust. Suggested maximum air sample volumes for specific environments are:

Environment	Air Vol. (L)
Asbestos removal operations (visible dust)	100
Asbestos removal operations (little dust)	240
Office environments	400 to 2,400

CAUTION: Do not overload the filter with dust. High levels of non-fibrous dust particles may obscure fibers on the filter and lower the count or make counting impossible. If more than about 25 to 30% of the field area is obscured with dust, the result may be biased low. Smaller air volumes may be necessary when there is excessive non-asbestos dust in the air.

While sampling, observe the filter with a small flashlight. If there is a visible layer of dust on the filter, stop sampling, remove and seal the cassette, and replace with a new sampling assembly. The total dust loading should not exceed 1 mg.

5.2.9. *Blank samples are used* to determine if any contamination has occurred during sample handling. Prepare two blanks for the first 1 to 20 samples. For sets containing greater than 20 samples, prepare blanks as 10% of the samples. Handle blank samples in the same manner as air samples with one exception: Do not draw any air through the blank samples. Open the blank cassette in the place where the sample cassettes are mounted on the employee. Hold it open for about 30 seconds. Close and seal the cassette appropriately. Store blanks for shipment with the sample cassettes.

5.2.10. *Immediately after sampling,* close and seal each cassette with the base and plastic plugs. Do not touch or puncture the filter membrane as this will invalidate the analysis.

5.2.11. *Attach and secure a sample* seal around each sample cassette in such a way as to assure that the end cap and base plugs cannot be removed without destroying the seal. Tape the ends of the seal together since the seal is not long enough to be wrapped end-to-end. Also wrap tape around the cassette at each joint to keep the seal secure.

5.3. *Sample Shipment*[1]

5.3.1. *Send the samples to the laboratory* with paperwork requesting asbestos analysis. List any known fibrous interferences present during sampling on the paperwork. Also, note the workplace operation(s) sampled.

5.3.2. *Secure and handle the samples* in such that they will not rattle during shipment nor be exposed to static electricity. Do not ship samples in expanded polystyrene peanuts, vermiculite, paper shreds, or excelsior. Tape sample cassettes to sheet bubbles and place in a container that will cushion the samples in such a manner that they will not rattle.

5.3.3. *To avoid the possibility* of sample contamination, always ship bulk samples in separate mailing containers.

6. Analysis

6.1. *Safety Precautions*

6.1.1. *Acetone is extremely flammable* and precautions must be taken not to ignite it. Avoid using large containers or quantities of acetone. Transfer the solvent in a ventilated laboratory hood. Do not use acetone near any open flame. For generation of acetone vapor, use a spark free heat source.

6.1.2. *Any asbestos spills* should be cleaned up immediately to prevent dispersal of fibers. Prudence should be exercised to avoid contamination of laboratory facilities or exposure of personnel to asbestos. Asbestos spills should be cleaned up with wet methods and/or a High Efficiency Particulate-Air (HEPA) filtered vacuum.

CAUTION: Do not use a vacuum without a HEPA filter — It will disperse fine asbestos fibers in the air.

6.2. *Equipment*

6.2.1. *Phase contrast microscope with binocular or trinocular head.*

6.2.2. *Widefield or Huygenian 10X eyepieces* (NOTE: The eyepiece containing the graticule must be a focusing eyepiece. Use a 40X phase objective with a numerical aperture of 0.65 to 0.75).

6.2.3. *Kohler illumination (if possible) with green or blue filter.*

6.2.4. *Walton-Beckett Graticule,* type G-22 with 100 ± 2 μm projected diameter.

6.2.5. *Mechanical stage.* A rotating mechanical stage is convenient for use with polarized light.

6.2.6. *Phase telescope.*

6.2.7. *Stage micrometer with 0.01-mm subdivisions.*

6.2.8. *Phase-shift test slide, mark II* (Available from PTR optics Ltd., and also McCrone).

6.2.9. *Precleaned glass slides,* 25 mmx75 mm. One end can be frosted for convenience in writing sample numbers, etc., or paste-on labels can be used.

6.2.10. *Cover glass # 1 1/2 .*

6.2.11. *Scalpel (#10, curved blade).*

6.2.12. *Fine tipped forceps.*

6.2.13. *Aluminum block for clearing filter* (see Appendix D and Figure 4).

6.2.14. *Automatic adjustable pipette, 100-to 500-μL.*

6.2.15. *Micropipette, 5 μL.*

6.3. *Reagents*

6.3.1. *Acetone (HPLC grade).*

6.3.2. *Triacetin (glycerol triacetate).*

6.3.3. *Lacquer or nail polish.*

6.4. *Standard Preparation*

A way to prepare standard asbestos samples of known concentration has not been developed. It is possible to prepare replicate samples of nearly equal concentration. This has been performed through the PAT program. These asbestos samples are distributed by the AIHA to participating laboratories.

Since only about one-fourth of a 25-mm sample membrane is required for an asbestos count, any PAT sample can serve as a "standard" for replicate counting.

6.5. *Sample Mounting*

Note: See Safety Precautions in Section 6.1. before proceeding. The objective is to produce samples with a smooth (non-grainy) background in a medium with a refractive index of approximately 1.46. The technique below collapses the filter for easier focusing and produces permanent mounts which are useful for quality control and interlaboratory comparison.

An aluminum block or similar device is required for sample preparation.

6.5.1. *Heat the aluminum block* to about 70 °C. The hot block should not be used on any surface that can be damaged by either the heat or from exposure to acetone.

6.5.2. *Ensure that the glass slides* and cover glasses are free of dust and fibers.

6.5.3. *Remove the top plug* to prevent a vacuum when the cassette is opened. Clean the outside of the cassette if necessary. Cut the seal and/or tape on the cassette with a razor blade. Very carefully separate the base from the extension cowl, leaving the filter and backup pad in the base.

6.5.4. *With a rocking motion cut a triangular wedge* from the filter using the scalpel. This wedge should be one-sixth to one-fourth of the filter. Grasp the filter wedge with the forceps on the perimeter of the filter which was clamped between the cassette pieces. DO NOT TOUCH the filter with your finger. Place the filter on the glass slide sample side up. Static electricity will usually keep the filter on the slide until it is cleared.

6.5.5. *Place the tip of the micropipette* containing about 200 μL acetone into the aluminum block. Insert the glass slide into the receiving slot in the aluminum block. Inject the acetone into the block with slow, steady pressure on the plunger while holding the pipette firmly in place. Wait 3 to 5 seconds for the filter to clear, then remove the pipette and slide from the aluminum block.

6.5.6. *Immediately (less than 30 seconds)* place 2.5 to 3.5 μL of triacetin on the filter (NOTE: Waiting longer than 30 seconds will result in increased index of refraction and decreased contrast between the fibers and the preparation. This may also lead to separation of the cover slip from the slide).

6.5.7. *Lower a cover slip* gently onto the filter at a slight angle to reduce the possibility of forming air bubbles. If more than 30 seconds have elapsed between acetone exposure and triacetin application, glue the edges of the cover slip to the slide with lacquer or nail polish.

6.5.8. *If clearing is slow,* warm the slide for 15 min on a hot plate having a surface temperature of about 50 °C to hasten clearing. The top of the hot block can be used if the slide is not heated too long.

6.5.9. *Counting may proceed immediately* after clearing and mounting are completed.

6.6. *Sample Analysis*

Completely align the microscope according to the manufacturer's instructions. Then, align the microscope using the following general alignment routine at the beginning of every counting session and more often if necessary.

6.6.1. *Alignment*

[1] *Clean all optical surfaces.* Even a small amount of dirt can significantly degrade the image.

1. Editor's Note: In this instance, the CFR does not include the subheading '5.3' as it appears in corresponding §1910.1001 Appendix B. The heading and heading title are inserted in the regulation appearing above.

[2] *Rough focus the objective on a sample.*

[3] *Close down the field iris* so that it is visible in the field of view. Focus the image of the iris with the condenser focus. Center the image of the iris in the field of view.

[4] *Install the phase telescope* and focus on the phase rings. Critically center the rings. Misalignment of the rings results in astigmatism which will degrade the image.

[5] *Place the phase-shift test slide* on the microscope stage and focus on the lines. The analyst must see line set 3 and should see at least parts of 4 and 5 but, not see line set 6 or 6. A microscope/ microscopist combination which does not pass this test may not be used.

6.6.2. *Counting Fibers*

[1] *Place the prepared sample slide* on the mechanical stage of the microscope. Position the center of the wedge under the objective lens and focus upon the sample.

[2] *Start counting from one end of the wedge* and progress along a radial line to the other end (count in either direction from perimeter to wedge tip). Select fields randomly, without looking into the eyepieces, by slightly advancing the slide in one direction with the mechanical stage control.

[3] *Continually scan over* a range of focal planes (generally the upper 10 to 15 µm of the filter surface) with the fine focus control during each field count. Spend at least 5 to 15 seconds per field.

[4] *Most samples will contain* asbestos fibers with fiber diameters less than 1 µm. Look carefully for faint fiber images. The small diameter fibers will be very hard to see. However, they are an important contribution to the total count.

[5] *Count only fibers* equal to or longer than 5 µm. Measure the length of curved fibers along the curve.

[6] *Count fibers* which have a length to width ratio of 3:1 or greater.

[7] *Count all the fibers in at least* 20 fields. Continue counting until either 100 fibers are counted or 100 fields have been viewed; whichever occurs first. Count all the fibers in the final field.

[8] *Fibers lying entirely within the boundary* of the Walton-Beckett graticule field shall receive a count of 1. Fibers crossing the boundary once, having one end within the circle shall receive a count of 1/2 . Do not count any fiber that crosses the graticule boundary more than once. Reject and do not count any other fibers even though they may be visible outside the graticule area. If a fiber touches the circle, it is considered to cross the line.

[9] *Count bundles of fibers* as one fiber unless individual fibers can be clearly identified and each individual fiber is clearly not connected to another counted fiber. See Figure 1 for counting conventions.

[10] *Record the number of fibers* in each field in a consistent way such that filter non-uniformity can be assessed.

[11] *Regularly check phase ring alignment.*

[12] *When an agglomerate (mass of material)* covers more than 25% of the field of view, reject the field and select another. Do not include it in the number of fields counted.

[13] *Perform a "blind recount"* of 1 in every 10 filter wedges (slides). Re-label the slides using a person other than the original counter.

6.7. *Fiber Identification*

As previously mentioned in Section 1.3., PCM does not provide positive confirmation of asbestos fibers. Alternate differential counting techniques should be used if discrimination is desirable. Differential counting may include primary discrimination based on morphology, polarized light analysis of fibers, or modification of PCM data by Scanning Electron or Transmission Electron Microscopy.

A great deal of experience is required to routinely and correctly perform differential counting. It is discouraged unless it is legally necessary. Then, only if a fiber is obviously not asbestos should it be excluded from the count. Further discussion of this technique can be found in reference 8.10.

If there is a question whether a fiber is asbestos or not, follow the rule:

"WHEN IN DOUBT, COUNT."

6.8. *Analytical Recommendations — Quality Control System*

6.8.1. *All individuals performing asbestos analysis* must have taken the NIOSH course for sampling and evaluating airborne asbestos or an equivalent course.

6.8.2. *Each laboratory engaged in asbestos counting* shall set up a slide trading arrangement with at least two other laboratories in order to compare performance and eliminate inbreeding of error. The slide exchange occurs at least semiannually. The round robin results shall be posted where all analysts can view individual analyst's results.

6.8.3. *Each laboratory engaged in asbestos counting* shall participate in the Proficiency Analytical Testing Program, the Asbestos Analyst Registry or equivalent.

6.8.4. *Each analyst shall select and count* prepared slides from a "slide bank". These are quality assurance counts. The slide bank shall be prepared using uniformly distributed samples taken from the workload. Fiber densities should cover the entire range routinely analyzed by the laboratory. These slides are counted blind by all counters to establish an original standard deviation. This historical distribution is compared with the quality assurance counts. A counter must have 95% of all quality control samples counted within three standard deviations of the historical mean. This count is then integrated into a new historical mean and standard deviation for the slide.

The analyses done by the counters to establish the slide bank may be used for an interim quality control program if the data are treated in a proper statistical fashion.

7. Calculations

7.1. *Calculate* the estimated airborne asbestos fiber concentration on the filter sample using the following formula:

$$AC = \frac{\left[\left(\frac{FB}{FL}\right) - \left(\frac{BFB}{BFL}\right)\right] \times ECA}{1000 \times FR \times T \times MFA}$$

where:

AC=Airborne fiber concentration

FB=Total number of fibers greater than 5 µm counted

FL=Total number of fields counted on the filter

BFB=Total number of fibers greater than 5 µm counted in the blank

BFL=Total number of fields counted on the blank

ECA=Effective collecting area of filter (385 mm^2 nominal for a 25-mm filter.)

FR=Pump flow rate (L/min)

MFA=Microscope count field area (mm^2). This is 0.00785 mm^2 for a Walton-Beckett Graticule.

T=Sample collection time (min)

1,000=Conversion of L to cc

Note: The collection area of a filter is seldom equal to 385 mm^2. It is appropriate for laboratories to routinely monitor the exact diameter using an inside micrometer. The collection area is calculated according to the formula[1]: Area = $\pi(d/2)^2$

7.2. *Short-Cut Calculation*

Since a given analyst always has the same interpupillary distance, the number of fields per filter for a particular analyst will remain constant for a given size filter. The field size for that analyst is constant (i.e. the analyst is using an assigned microscope and is not changing the reticle).

For example, if the exposed area of the filter is always 385 mm^2 and the size of the field is always 0.00785 mm^2 the number of fields per filter will always be 49,000. In addition it is necessary to convert liters of air to cc. These three constants can then be combined such that ECA/(1,000xMFA)=49. The previous equation simplifies to:

$$AC = \frac{\left(\frac{FB}{FL}\right) - \left(\frac{BFB}{BFL}\right) \times 49}{FR \times T}$$

7.3. *Recount Calculations*

As mentioned in step 13 of Section 6.6.2., a "blind recount" of 10% of the slides is performed. In all cases, differences will be observed between the first and second counts of the same filter wedge. Most of these differences will be due to chance alone, that is, due to the random variability (precision) of the count method. Statistical recount criteria enables one to decide whether observed differences can be explained due to chance

1. Editor's Note: In this instance, the CFR uses the symbol (†) in place of π, as it appears in Calculation 7.1 in corresponding §1910.1001 Appendix B. The corrected equation appears above.

alone or are probably due to systematic differences between analysts, microscopes, or other biasing factors.

The following recount criterion is for a pair of counts that estimate AC in fibers/cc. The criterion is given at the type-I error level. That is, there is 5% maximum risk that we will reject a pair of counts for the reason that one might be biased, when the large observed difference is really due to chance.

Reject a pair of counts if:

$$\left|\sqrt{AC_2} - \sqrt{AC_1}\right| > 2.78 \times \left(\sqrt{AC_{AVG}}\right) \times CV_{FB}$$

Where:

AC_1 =lower estimated airborne fiber concentration

AC_2 =higher estimated airborne fiber concentration

AC_{avg} =average of the two concentration estimates

CV_{FB} =CV for the average of the two concentration estimates

If a pair of counts are rejected by this criterion then, recount the rest of the filters in the submitted set. Apply the test and reject any other pairs failing the test. Rejection shall include a memo to the industrial hygienist stating that the sample failed a statistical test for homogeneity and the true air concentration may be significantly different than the reported value.

7.4. *Reporting Results*

Report results to the industrial hygienist as fibers/cc. Use two significant figures. If multiple analyses are performed on a sample, an average of the results is to be reported unless any of the results can be rejected for cause.

8. References

8.1. *Dreesen, W.C., et al.,* U.S. Public Health Service: A Study of Asbestosis in the Asbestos Textile Industry (Public Health Bulletin No. 241), U.S. Treasury Dept., Washington, DC, 1938.

8.2. *Asbestos Research Council:* The Measurement of Airborne Asbestos Dust by the Membrane Filter Method (Technical Note), Asbestos Research Council, Rockdale, Lancashire, Great Britain, 1969.

8.3. *Bayer, S.G., Zumwalde, R.D., Brown, T.A.,* Equipment and Procedure for Mounting Millipore Filters and Counting Asbestos Fibers by Phase Contrast Microscopy, Bureau of Occupational Health, U.S. Dept. of Health, Education and Welfare, Cincinnati, OH, 1969.

8.4. *NIOSH Manual of Analytical Methods,* 2nd ed., Vol. 1 (DHEW/NIOSH Pub. No. 77-157-A). National Institute for Occupational Safety and Health, Cincinnati, OH, 1977. pp. 239-1 — 239-21.

8.5. *Asbestos, Code of Federal Regulations 29 CFR 1910.1001. 1971.*

8.6. *Occupational Exposure to Asbestos,* Tremolite, Anthophyllite, and Actinolite. Final Rule, Federal Register 51:119 (20 June 1986). pp. 22612-22790.

8.7. *Asbestos, Tremolite, Anthophyllite, and Actinolite,* Code of Federal Regulations 1910.1001. 1988. pp. 711-752.

8.8. *Criteria for a Recommended Standard* — Occupational Exposure to Asbestos (DHEW/NIOSH Pub. No. HSM 72-10267), National Institute for Occupational Safety and Health, NIOSH, Cincinnati, OH, 1972. pp. III-1 — III-24.

8.9. *Leidel,* N.A., Bayer, S.G., Zumwalde, R.D., Busch, K.A., USPHS/ NIOSH Membrane Filter Method for Evaluating Airborne Asbestos Fibers (DHEW/NIOSH Pub. No. 79-127). National Institute for Occupational Safety and Health, Cincinnati, OH, 1979.

8.10. *Dixon,* W.C., Applications of Optical Microscopy in Analysis of Asbestos and Quartz, Analytical Techniques in Occupational Health Chemistry, edited by D.D. Dollberg and A.W. Verstuyft. Wash. DC: American Chemical Society, (ACS Symposium Series 120) 1980. pp. 13-41.

Quality Control

The OSHA asbestos regulations require each laboratory to establish a quality control program. The following is presented as an example of how the OSHA-SLTC constructed its internal CV curve as part of meeting this requirement. Data is from 395 samples collected during OSHA compliance inspections and analyzed from October 1980 through April 1986.

Each sample was counted by 2 to 5 different counters independently of one another. The standard deviation and the CV statistic was calculated for each sample. This data was then plotted on a graph of CV vs. fibers/mm^2. A least squares regression was performed using the following equation:

$$CV = \text{antilog}_{10}\,[A(\log_{10}(x))^2 + B(\log_{10}(x)) + C]$$

where:

x = the number of fibers/mm^2

Application of least squares gave:

A = 0.182205

B = 0.973343

C = 0.327499

Using these values, the equation becomes:

$$CV = \text{antilog}_{10}\,[0.182205(\log_{10}(x))^2 - 0.973343(\log_{10}(x)) + 0.327499]$$

Sampling Pump Flow Rate Corrections

This correction is used if a difference greater than 5% in ambient temperature and/or pressure is noted between calibration and sampling sites and the pump does not compensate for the differences.

$$Q_{act} = Q_{cal} \times \sqrt{\left(\frac{P_{cal}}{P_{act}}\right) \times \left(\frac{T_{act}}{T_{cal}}\right)}$$

Where:

Q_{act} = actual flow rate

Q_{cal} = calibrated flow rate (if a rotameter was used, the rotameter value)

P_{cal} = uncorrected air pressure at calibration

P_{act} = uncorrected air pressure at sampling site

T_{act} = temperature at sampling site (K)

T_{cal} = temperature at calibration (K)

Walton-Beckett Graticule

When ordering the Graticule for asbestos counting, specify the exact disc diameter needed to fit the ocular of the microscope and the diameter (mm) of the circular counting area. Instructions for measuring the dimensions necessary are listed:

(1) *Insert any available graticule* into the focusing eyepiece and focus so that the graticule lines are sharp and clear.

(2) *Align the microscope.*

(3) *Place a stage micrometer* on the microscope object stage and focus the microscope on the graduated lines.

(4) *Measure the magnified grid length,* PL (μm), using the stage micrometer.

(5) *Remove the graticule from the microscope* and measure its actual grid length, AL (mm). This can be accomplished by using a mechanical stage fitted with verniers, or a jeweler's loupe with a direct reading scale.

(6) *Let D=100 μm.* Calculate the circle diameter, d_c (mm), for the Walton-Beckett graticule and specify the diameter when making a purchase:

$$d_c = \frac{AL \times D}{PL}$$

Example: If PL=108 μm, AL=2.93 mm and D=100 μm, then,

$$d_c = \frac{2.93 \times 100}{108} = 2.71\text{mm}$$

(7) *Each eyepiece-objective-reticle combination* on the microscope must be calibrated. Should any of the three be changed (by zoom adjustment, disassembly, replacement, etc.), the combination must be recalibrated. Calibration may change if interpupillary distance is changed.

Measure the field diameter, D (acceptable range: 100 ± 2 μm) with a stage micrometer upon receipt of the graticule from the manufacturer. Determine the field area (mm^2).

Field Area = $\pi(D/2)^2$

If D = 100 μm=0.1 mm, then

Field Area = $\pi(0.1\text{ mm}/2)^2$=0.00785 mm^2

The Graticule is available from: Graticules Ltd., Morley Road, Tonbridge TN9 IRN, Kent, England (Telephone 011-44-732-359061). Also available from PTR Optics Ltd., 145 Newton Street, Waltham, MA 02154 [telephone (617) 891-6000] or McCrone Accessories and Components, 2506 S. Michigan Ave., Chicago, IL 60616 [phone (312)-842-7100]. The graticule is custom made for each microscope.

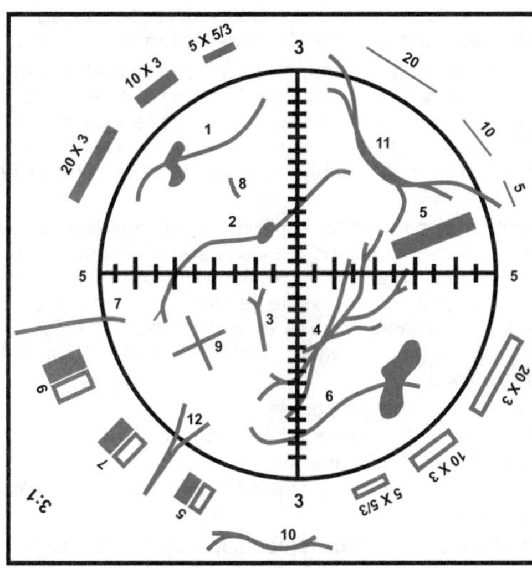

FIGURE 1: Walton-Beckett Graticule with some explanatory fibers.
© MMV Mangan Communications, Inc.

Counts for the Fibers in the Figure

Structure No.	Count	Explanation
1 to 6	1	Single fibers all contained within the Circle.
7	1/2	Fiber crosses circle once.
8	0	Fiber too short.
9	2	Two crossing fibers.
10	0	Fiber outside graticule.
11	0	Fiber crosses graticule twice.
12	1/2	Although split, fiber only crosses once.

§1926.1101 Appendix C
[Reserved]

§1926.1101 Appendix D
Medical Questionnaires; Mandatory

Appendix D to §1926.1101 - Medical Questionnaires - Mandatory

This mandatory appendix contains the medical questionnaires that must be administered to all employees who are exposed to asbestos above the permissible exposure limit, and who will therefore be included in their employer's medical surveillance program. Part 1 of the appendix contains the Initial Medical Questionnaire, which must be obtained for all new hires who will be covered by the medical surveillance requirements. Part 2 includes the abbreviated Periodical Medical Questionnaire, which must be administered to all employees who are provided periodic medical examinations under the medical surveillance provisions of the standard.

Part 1

INITIAL MEDICAL QUESTIONNAIRE:

1. NAME: ______
2. SOCIAL SECURITY NUMBER: ____-____-____
3. CLOCK NUMBER: ______
4. PRESENT OCCUPATION: ______
5. PLANT: ______
6. ADDRESS: ______
7. CITY: ______ ST: ____ ZIP CODE: ______
8. TELEPHONE NUMBER: (____) ____-____ EXT. ____
9. INTERVIEWER: ______
10. DATE: ____/____/____
11. Date of birth: ____/____/____ (Month Day Year)
12. Place of birth: ______
13. Sex: 1. ☐ Male 2. ☐ Female
14. What is your marital status? 1. ☐ Single 2. ☐ Married 3. ☐ Widowed 4. ☐ Separated/Divorced
15. Race: 1. ☐ White 2. ☐ Black 3. ☐ Asian 4. ☐ Hispanic 5. ☐ Indian 6. ☐ Other ______
16. What is the highest grade completed in school? ____ (For example, 12 years is completion of high school)
17. OCCUPATIONAL HISTORY

A. Have you ever worked full time (30 hours per week or more) for 6 months or more? 1. ☐ Yes 2. ☐ No

IF YES TO 17A:

B. Have you ever worked for a year or more in any dusty job? 1. ☐ Yes 2. ☐ No 3. ☐ Does Not Apply

Specify job/industry: ______ Total Years Worked: ____

Was dust exposure: 1. ☐ Mild 2. ☐ Moderate 3. ☐ Severe

C. Have you ever been exposed to gas or chemical fumes in your work? 1. ☐ Yes 2. ☐ No

Specify job/industry: ______ Total Years Worked: ____

Was exposure: 1. ☐ Mild 2. ☐ Moderate 3. ☐ Severe

D. What has been your usual occupation or job - the one you have worked at the longest?

1. Job occupation: ______
2. Number of years employed in this occupation: ____
3. Position/job title: ______
4. Business, field or industry: ______

(Record on lines the years in which you have worked in any of these industries, e.g. 1960-1969)

Have you ever worked:

E. In a mine? ☐ Yes ☐ No ____-____
F. In a quarry? ☐ Yes ☐ No ____-____
G. In a foundry? ☐ Yes ☐ No ____-____
H. In a pottery? ☐ Yes ☐ No ____-____
I. In a cotton, flax, or hemp mill? ☐ Yes ☐ No ____-____
J. With asbestos? ☐ Yes ☐ No ____-____

18. PAST MEDICAL HISTORY

A. Do you consider yourself to be in good health? ☐ Yes ☐ No If "No", state reason: ______
B. Have you any defect of vision? ☐ Yes ☐ No If "Yes", state nature of defect: ______
C. Have you any hearing defect? ☐ Yes ☐ No If "Yes", state nature of defect: ______
D. Are you suffering from or have you ever suffered from:

a. Epilepsy (or fits, seizures, convulsions)? ☐ Yes ☐ No
b. Rheumatic fever? ☐ Yes ☐ No
c. Kidney disease? ☐ Yes ☐ No
d. Bladder disease? ☐ Yes ☐ No
e. Diabetes? ☐ Yes ☐ No
f. Jaundice? ☐ Yes ☐ No

*Full size forms available free of charge at oshacfr.com

Appendix D to §1926.1101 - Medical Questionnaires - Mandatory (Continued)

Part 1 (Continued)

19. CHEST COLDS AND CHEST ILLNESSES:

19A. If you get a cold, does it "usually" go to your chest? (Usually means more than 1/2 the time): 1. ☐ Yes 2. ☐ No 3. ☐ Don't get colds

20A. During the past 3 years, have you had any chest illnesses that have kept you off work, indoors at home, or in bed? 1. ☐ Yes 2. ☐ No IF YES TO 20A:

B. Did you produce phlegm with any of these chest illnesses? 1. ☐ Yes 2. ☐ No 3. ☐ Does Not Apply

C. In the last 3 years, how many such illnesses with (increased) phlegm did you have which lasted a week or more? ______ Number of illnesses ☐ No such illnesses

21. Did you have any lung trouble before the age of 16? 1. ☐ Yes 2. ☐ No

22. Have you ever had any of the following?

1A. Attacks of bronchitis? 1. ☐ Yes 2. ☐ No IF YES TO 1A:

B. Was it confirmed by a doctor? 1. ☐ Yes 2. ☐ No 3. ☐ Does Not Apply

C. At what age was your first attack? ______ Age In Years ☐ Does Not Apply

2A. Pneumonia (include bronchopneumonia)? 1. ☐ Yes 2. ☐ No IF YES TO 2A:

B. Was it confirmed by a doctor? 1. ☐ Yes 2. ☐ No 3. ☐ Does Not Apply

C. At what age did you first have it? ______ Age In Years ☐ Does Not Apply

3A. Hay Fever? 1. ☐ Yes 2. ☐ No IF YES TO 3A:

B. Was it confirmed by a doctor? 1. ☐ Yes 2. ☐ No 3. ☐ Does Not Apply

C. At what age did it start? ______ Age In Years ☐ Does Not Apply

23 A. Have you ever had chronic bronchitis? 1. ☐ Yes 2. ☐ No IF YES TO 23A:

B. Do you still have it? 1. ☐ Yes 2. ☐ No 3. ☐ Does Not Apply

C. Was it confirmed by a doctor? 1. ☐ Yes 2. ☐ No 3. ☐ Does Not Apply

D. At what age did it start? ______ Age In Years ☐ Does Not Apply

24 A. Have you ever had emphysema? 1. ☐ Yes 2. ☐ No IF YES TO 24A:

B. Do you still have it? 1. ☐ Yes 2. ☐ No 3. ☐ Does Not Apply

C. Was it confirmed by a doctor? 1. ☐ Yes 2. ☐ No 3. ☐ Does Not Apply

D. At what age did it start? ______ Age In Years ☐ Does Not Apply

25 A. Have you ever had asthma? 1. ☐ Yes 2. ☐ No IF YES TO 25A:

B. Do you still have it? 1. ☐ Yes 2. ☐ No 3. ☐ Does Not Apply

C. Was it confirmed by a doctor? 1. ☐ Yes 2. ☐ No 3. ☐ Does Not Apply

D. At what age did it start? ______ Age In Years ☐ Does Not Apply

E. If you no longer have it, at what age did it stop? ______ Age Stopped ☐ Does Not Apply

26. Have you ever had:

A. Any other chest illness? 1. ☐ Yes 2. ☐ No If yes, please specify: ______

B. Any chest operations? 1. ☐ Yes 2. ☐ No If yes, please specify: ______

C. Any chest injuries? 1. ☐ Yes 2. ☐ No If yes, please specify: ______

27 A. Has a doctor ever told you that you had heart trouble? 1. ☐ Yes 2. ☐ No IF YES TO 27A:

B. Have you ever had treatment for heart trouble in the past 10 years? 1. ☐ Yes 2. ☐ No 3. ☐ Does Not Apply

28 A. Has a doctor ever told you that you had high blood pressure? 1. ☐ Yes 2. ☐ No IF YES TO 28A:

B. Have you had any treatment for high blood pressure (hypertension) in the past 10 years? 1. ☐ Yes 2. ☐ No 3. ☐ Does Not Apply

29. When did you last have your chest X-rayed? ______ Year

30. Where did you last have your chest X-rayed (if known)? ______

What was the outcome? ______

FAMILY HISTORY

31. Were either of your natural parents ever told by a doctor that they had a chronic lung condition such as:

	FATHER	MOTHER
A. Chronic Bronchitis?	1. ☐ Yes 2. ☐ No 3. ☐ Don't Know	1. ☐ Yes 2. ☐ No 3. ☐ Don't Know
B. Emphysema?	1. ☐ Yes 2. ☐ No 3. ☐ Don't Know	1. ☐ Yes 2. ☐ No 3. ☐ Don't Know
C. Asthma?	1. ☐ Yes 2. ☐ No 3. ☐ Don't Know	1. ☐ Yes 2. ☐ No 3. ☐ Don't Know
D. Lung cancer?	1. ☐ Yes 2. ☐ No 3. ☐ Don't Know	1. ☐ Yes 2. ☐ No 3. ☐ Don't Know
E. Other chest conditions?	1. ☐ Yes 2. ☐ No 3. ☐ Don't Know	1. ☐ Yes 2. ☐ No 3. ☐ Don't Know
F. Is parent currently alive?	1. ☐ Yes 2. ☐ No 3. ☐ Don't Know	1. ☐ Yes 2. ☐ No 3. ☐ Don't Know
G. Please Specify	____ Age if Living ____ Age at Death ☐ Don't Know	____ Age if Living ____ Age at Death ☐ Don't Know
H. Please specify cause of death	______	______

COUGH

32 A. Do you usually have a cough? (Count a cough with first smoke or on first going out of doors. Exclude clearing of throat.) (If No, skip to question 32C.) 1. ☐ Yes 2. ☐ No

B. Do you usually cough as much as 4 to 6 times a day 4 or more days out of the week? 1. ☐ Yes 2. ☐ No

C. Do you usually cough at all on getting up or first thing in the morning? 1. ☐ Yes 2. ☐ No

D. Do you usually cough at all during the rest of the day or at night? 1. ☐ Yes 2. ☐ No

IF YES TO ANY OF ABOVE (32A, B, C, OR D), ANSWER THE FOLLOWING. IF NO TO ALL, CHECK "DOES NOT APPLY" AND SKIP TO PART 2

E. Do you usually cough like this on most days for 3 consecutive months or more during the year? 1. ☐ Yes 2. ☐ No 3. ☐ Does Not Apply

F. For how many years have you had the cough? ______ No. of Years ☐ Does Not Apply

*Full size forms available free of charge at oshacfr.com

Appendix D to §1926.1101 - Medical Questionnaires - Mandatory (Continued)

Part 1 (Continued)

33 A. Do you usually bring up phlegm from your chest? (Count phlegm with the first smoke or on first going out of doors. Exclude phlegm from the nose. Count swallowed phlegm.) If no, skip to 33C. 1. ☐ Yes 2. ☐ No

B. Do you usually bring up phlegm like this as much as twice a day 4 or more days out of the week? 1. ☐ Yes 2. ☐ No

C. Do you usually bring up phlegm at all on getting up or first thing in the morning? 1. ☐ Yes 2. ☐ No

D. Do you usually bring up phlegm at all during the rest of the day or at night? 1. ☐ Yes 2. ☐ No

IF YES TO ANY OF THE ABOVE (33A, B, C, OR D), ANSWER THE FOLLOWING. IF NO TO ALL, CHECK "DOES NOT APPLY" AND SKIP TO 34A.

E. Do you bring up phlegm like this on most days for 3 consecutive months or more during the year? 1. ☐ Yes 2. ☐ No 3. ☐ Does Not Apply

F. For how many years have you had trouble with phlegm? ______ No. of Years ☐ Does Not Apply

EPISODES OF COUGH AND PHLEGM

34 A. Have you had periods or episodes of (increased*) cough and phlegm lasting for 3 weeks or more each year? *(For persons who usually have cough and/or phlegm) 1. ☐ Yes 2. ☐ No IF YES TO 34A:

B. For how long have you had at least 1 such episode per year? ______ No. of Years ☐ Does Not Apply

WHEEZING

35 A. Does your chest ever sound wheezy or whistling:

1. When you have a cold? 1. ☐ Yes 2. ☐ No

2. Occasionally apart from colds? 1. ☐ Yes 2. ☐ No

3. Most days or nights? 1. ☐ Yes 2. ☐ No

IF YES TO 1, 2, or 3 in 35A:

B. For how many years has this been present? ______ No. of Years ☐ Does Not Apply

36 A. Have you ever had an attack of wheezing that has made you feel short of breath? 1. ☐ Yes 2. ☐ No

IF YES TO 36A:

B. How old were you when you had your first such attack? ______ Age in Years ☐ Does Not Apply

C. Have you had 2 or more such episodes? 1. ☐ Yes 2. ☐ No 3. ☐ Does Not Apply

D. Have you ever required medicine or treatment for the(se) attack(s)? 1. ☐ Yes 2. ☐ No 3. ☐ Does Not Apply

BREATHLESSNESS

37. If disabled from walking by any condition other than heart or lung disease, please describe and proceed to question 39A.

Nature of condition(s): ______

38 A. Are you troubled by shortness of breath when hurrying on the level or walking up a slight hill? 1. ☐ Yes 2. ☐ No

IF YES TO 38A:

B. Do you have to walk slower than people of your age on the level because of breathlessness? 1. ☐ Yes 2. ☐ No 3. ☐ Does Not Apply

C. Do you ever have to stop for breath when walking at your own pace on the level? 1. ☐ Yes 2. ☐ No 3. ☐ Does Not Apply

D. Do you ever have to stop for breath after walking about 100 yards (or after a few minutes) on the level? 1. ☐ Yes 2. ☐ No 3. ☐ Does Not Apply

E. Are you too breathless to leave the house or breathless on dressing or climbing one flight of stairs? 1. ☐ Yes 2. ☐ No 3. ☐ Does Not Apply

TOBACCO SMOKING

39 A. Have you ever smoked cigarettes? (No means less than 20 packs of cigarettes or 12 oz. of tobacco in a lifetime or less than 1 cigarette a day for 1 year.) 1. ☐ Yes 2. ☐ No IF YES TO 39A:

B. Do you now smoke cigarettes (as of one month ago) 1. ☐ Yes 2. ☐ No 3. ☐ Does Not Apply

C. How old were you when you first started regular cigarette smoking? ______ Age in Years ☐ Does Not Apply

D. If you have stopped smoking cigarettes completely, how old were you when you stopped? ______ Age Stopped ☐ Still Smoking Cigarettes ☐ Does Not Apply

E. How many cigarettes do you smoke per day now? ______ Cigarettes Per Day ☐ Does Not Apply

F. On the average of the entire time you smoked, how many cigarettes did you smoke per day? ______ Cigarettes Per Day ☐ Does Not Apply

G. Do or did you inhale the cigarette smoke? 1. ☐ Does Not Apply 2. ☐ Not At All 3. ☐ Slightly 4. ☐ Moderately 5. ☐ Deeply

40 A. Have you ever smoked a pipe regularly? (Yes means more than 12 oz. of tobacco in a lifetime.) 1. ☐ Yes 2. ☐ No

IF YES TO 40A:

FOR PERSONS WHO HAVE EVER SMOKED A PIPE

B 1. How old were you when you started to smoke a pipe regularly? ______ Age in Years

2. If you have stopped smoking a pipe completely, how old were you when you stopped? ______ Age Stopped ☐ Still Smoking Pipe ☐ Does Not Apply

C. On the average over the entire time you smoked a pipe, how much pipe tobacco did you smoke per week?

______ Oz. Per Week (a standard pouch of tobacco contains 1.5 oz.) ☐ Does Not Apply

D. How much pipe tobacco are you smoking now? ______ Oz. Per Week ☐ Not Currently Smoking A Pipe

E. Do you or did you inhale the pipe smoke? 1. ☐ Never Smoked 2. ☐ Not At All 3. ☐ Slightly 4. ☐ Moderately 5. ☐ Deeply

41 A. Have you ever smoked cigars regularly? (Yes means more than 1 cigar a week for a year) 1. ☐ Yes 2. ☐ No

IF YES TO 41A:

FOR PERSONS WHO HAVE EVER SMOKED CIGARS

B 1. How old were you when you started smoking cigars regularly? ______ Age in Years

2. If you have stopped smoking cigars completely, how old were you when you stopped? ______ Age Stopped ☐ Still Smoking Cigars ☐ Does Not Apply

C. On the average over the entire time you smoked cigars, how many cigars did you smoke per week? ______ Cigars Per Week ☐ Does Not Apply

D. How many cigars are you smoking per week now? ______ Cigars Per Week ☐ Not Currently Smoking Cigars

E. Do or did you inhale the cigar smoke? 1. ☐ Never Smoked 2. ☐ Not At All 3. ☐ Slightly 4. ☐ Moderately 5. ☐ Deeply

______ Date: ____/____/____

Signature

*Full size forms available free of charge at oshacfr.com

Appendix D to §1926.1101 - **Medical Questionnaires** - Mandatory

Part 2

PERIODIC MEDICAL QUESTIONNAIRE:

1. NAME: ____________________
2. SOCIAL SECURITY NUMBER: ______ — _____ — ________
3. CLOCK NUMBER: ____________________
4. PRESENT OCCUPATION: ____________________
5. PLANT: ____________________
6. ADDRESS: ____________________
7. CITY: ____________________ ST: _____ ZIP CODE: ________
8. TELEPHONE NUMBER: (______) ______ - ________ EXT. ________
9. INTERVIEWER: ____________________
10. DATE: _____ / _____ / _____
11. What is your marital status? 1. ☐ Single 2. ☐ Married 3. ☐ Widowed 4. ☐ Separated/Divorced
12. **OCCUPATIONAL HISTORY**

12A. In the past year, did you work full time (30 hours per week or more) for 6 months or more? 1. ☐ Yes 2. ☐ No

IF YES TO 12A:

12B. In the past year, did you work in a dusty job? 1. ☐ Yes 2. ☐ No 3. ☐ Does Not Apply

12C. Was dust exposure: 1. ☐ Mild 2. ☐ Moderate 3. ☐ Severe

12D. In the past year, were you exposed to gas or chemical fumes in your work? 1. ☐ Yes 2. ☐ No

12E. Was exposure: 1. ☐ Mild 2. ☐ Moderate 3. ☐ Severe

12F. In the past year, what was your:

1. Job/Occupation? ____________________
2. Position/Job Title? ____________________

13. **RECENT MEDICAL HISTORY**

13A. Do you consider yourself to be in good health? 1. ☐ Yes 2. ☐ No

If "No", state reason: ____________________

13B. In the past year, have you developed:

Epilepsy?	☐ Yes	☐ No
Rheumatic Fever?	☐ Yes	☐ No
Kidney Disease?	☐ Yes	☐ No
Bladder Disease?	☐ Yes	☐ No
Diabetes?	☐ Yes	☐ No
Jaundice?	☐ Yes	☐ No
Cancer?	☐ Yes	☐ No

14. **CHEST COLDS AND CHEST ILLNESSES**

14A. If you get a cold, does it "usually" go to your chest? (Usually means more than 1/2 the time) 1. ☐ Yes 2. ☐ No 3. ☐ Don't Get Colds

15A. During the past year, have you had any chest illnesses that have kept you off work, indoors at home, or in bed? 1. ☐ Yes 2. ☐ No 3. ☐ Does Not Apply

IF YES TO 15A:

15B. Did you produce phlegm with any of these chest illnesses? 1. ☐ Yes 2. ☐ No 3. ☐ Does Not Apply

15C. In the past year, how many such illnesses with (increased) phlegm did you have which lasted a week or more? ______ Number of Illnesses ☐ No Such Illnesses

16. **RESPIRATORY SYSTEM**

In the past year have you had:			Further Comment on Positive Answers
Asthma	☐ Yes	☐ No	________
Bronchitis	☐ Yes	☐ No	________
Hay Fever	☐ Yes	☐ No	________
Other Allergies	☐ Yes	☐ No	________
Pneumonia	☐ Yes	☐ No	________
Tuberculosis	☐ Yes	☐ No	________
Chest Surgery	☐ Yes	☐ No	________
Other Lung Problems	☐ Yes	☐ No	________
Heart Disease	☐ Yes	☐ No	________
Do You Have:			
Frequent Colds	☐ Yes	☐ No	________
Chronic Cough	☐ Yes	☐ No	________
Shortness of Breath When Walking or Climbing One Flight of Stairs	☐ Yes	☐ No	________
Do you:			
Wheeze	☐ Yes	☐ No	________
Cough Up Phlegm	☐ Yes	☐ No	________
Smoke Cigarettes	☐ Yes	☐ No	_____ Packs Per Day _____ How Many Years

____________________ Date: _____ / _____ / _____

Signature

*Full size forms available free of charge at oshacfr.com

§1926.1101 Appendix E

Interpretation and Classification of Chest Roentgenograms — Mandatory

(a) **Chest roentgenograms shall be interpreted and classified** in accordance with a professionally accepted classification system and recorded on an interpretation form following the format of the CDC/NIOSH (M) 2.8 form. As a minimum, the content within the bold lines of this form (items 1 through 4) shall be included. This form is not to be submitted to NIOSH.

(b) **Roentgenograms shall be interpreted and classified** only by a B-reader, a board eligible/certified radiologist, or an experienced physician with known expertise in pneumoconioses.[1]

(c) **All interpreters,** whenever interpreting chest roentgenograms made under this section, shall have immediately available for reference a complete set of the ILO-U/C International Classification of Radiographs for Pneumoconioses, 1980.

1926.1101 Appendix F

Work Practices and Engineering Controls for Class I Asbestos Operations (Non-Mandatory)

This is a non-mandatory appendix to the asbestos standards for construction and for shipyards. It describes criteria and procedures for erecting and using negative pressure enclosures for Class I Asbestos Work, when NPEs are used as an allowable control method to comply with paragraph (g)(5)(i) of this section. Many small and variable details are involved in the erection of a negative pressure enclosure. OSHA and most participants in the rulemaking agreed that only the major, more performance oriented criteria should be made mandatory. These criteria are set out in paragraph (g) of this section. In addition, this appendix includes these mandatory specifications and procedures in its guidelines in order to make this appendix coherent and helpful. The mandatory nature of the criteria which appear in the regulatory text is not changed because they are included in this "non-mandatory" appendix.

1. Editor's Note: The CFR states 'known s." Appendix E to §1910.1001, also on Asbestos, states 'known expertise in pneumoconioses', as it appears in the regulation above.

Similarly, the additional criteria and procedures included as guidelines in the appendix, do not become mandatory because mandatory criteria are also included in these comprehensive guidelines.

In addition, none of the criteria, both mandatory and recommended, are meant to specify or imply the need for use of patented or licensed methods or equipment. Recommended specifications included in this attachment should not discourage the use of creative alternatives which can be shown to reliably achieve the objectives of negative-pressure enclosures.

Requirements included in this appendix, cover general provisions to be followed in all asbestos jobs, provisions which must be followed for all Class I asbestos jobs, and provisions governing the construction and testing of negative pressure enclosures. The first category includes the requirement for use of wet methods, HEPA vacuums, and immediate bagging of waste; Class I work must conform to the following provisions:

- oversight by competent person
- use of critical barriers over all openings to work area
- isolation of HVAC systems
- use of impermeable dropcloths and coverage of all objects within regulated areas

In addition, more specific requirements for NPEs include:

- maintenance of -0.02 inches water gauge within enclosure
- manometric measurements
- air movement away from employees performing removal work
- smoke testing or equivalent for detection of leaks and air direction
- deactivation of electrical circuits, if not provided with ground-fault circuit interrupters.

Planning the Project

The standard requires that an exposure assessment be conducted before the asbestos job is begun [Sec. 1926.1101 (f)(1)]. Information needed for that assessment, includes data relating to prior similar jobs, as applied to the specific variables of the current job. The information needed to conduct the assessment will be useful in planning the project, and in complying with any reporting requirements under this standard, when significant changes are being made to a control system listed in the standard, [see also those of USEPA (40 CFR 61, subpart M).

Thus, although the standard does not explicitly require the preparation of a written asbestos removal plan, the usual constituents of such a plan, i.e., a description of the enclosure, the equipment, and the procedures to be used throughout the project, must be determined before the enclosure can be erected. The following information should be included in the planning of the system:

A physical description of the work area;

A description of the approximate amount of material to be removed;

A schedule for turning off and sealing existing ventilation systems;

Personnel hygiene procedures;

A description of personal protective equipment and clothing to be worn by employees;

A description of the local exhaust ventilation systems to be used and how they are to be tested;

A description of work practices to be observed by employees;

An air monitoring plan;

A description of the method to be used to transport waste material; and

The location of the dump site.

Materials and Equipment Necessary for Asbestos Removal

Although individual asbestos removal projects vary in terms of the equipment required to accomplish the removal of the materials, some equipment and materials are common to most asbestos removal operations.

Plastic sheeting used to protect horizontal surfaces, seal HVAC openings or to seal vertical openings and ceilings should have a minimum thickness of 6 mils. Tape or other adhesive used to attach plastic sheeting should be of sufficient adhesive strength to support the weight of the material plus all stresses encountered during the entire duration of the project without becoming detached from the surface.

Other equipment and materials which should be available at the beginning of each project are:

— HEPA Filtered Vacuum is essential for cleaning the work area after the asbestos has been removed. It should have a long hose capable of reaching out-of-the-way places, such as areas above ceiling tiles, behind pipes, etc.

— Portable air ventilation systems installed to provide the negative air pressure and air removal from the enclosure must be equipped with a HEPA filter. The number and capacity of units required to ventilate an enclosure depend on the size of the area to be ventilated. The filters for these systems should be designed in such a manner that they can be replaced when the air flow volume is reduced by the build-up of dust in the filtration material. Pressure monitoring devices with alarms and strip chart recorders attached to each system to indicate the pressure differential and the loss due to dust buildup on the filter are recommended.

— Water sprayers should be used to keep the asbestos material as saturated as possible during removal; the sprayers will provide a fine mist that minimizes the impact of the spray on the material.

— Water used to saturate the asbestos containing material can be amended by adding at least 15 milliliters (1/4 ounce) of wetting agent in 1 liter (1 pint) of water. An example of a wetting agent is a 50/50 mixture of polyoxyethylene ether and polyoxyethylene polyglycol ester.

— Backup power supplies are recommended, especially for ventilation systems.

— Shower and bath water should be with mixed hot and cold water faucets. Water that has been used to clean personnel or equipment should either be filtered or be collected and discarded as asbestos waste. Soap and shampoo should be provided to aid in removing dust from the workers' skin and hair.

— See paragraphs (h) and (i) of this section for appropriate respiratory protection and protective clothing.

— See paragraph (k) of this section for required signs and labels.

Preparing the Work Area

Disabling HVAC Systems: The power to the heating, ventilation, and air conditioning systems that service the restricted area must be deactivated and locked off. All ducts, grills, access ports, windows and vents must be sealed off with two layers of plastic to prevent entrainment of contaminated air.

Operating HVAC Systems in the Restricted Area: If components of a HVAC system located in the restricted area are connected to a system that will service another zone during the project, the portion of the duct in the restricted area must be sealed and pressurized. Necessary precautions include caulking the duct joints, covering all cracks and openings with two layers of sheeting, and pressurizing the duct throughout the duration of the project by restricting the return air flow. The power to the fan supplying the positive pressure should be locked "on" to prevent pressure loss.

Sealing Elevators: If an elevator shaft is located in the restricted area, it should be either shut down or isolated by sealing with two layers of plastic sheeting. The sheeting should provide enough slack to accommodate the pressure changes in the shaft without breaking the air-tight seal.

Removing Mobile Objects: All movable objects should be cleaned and removed from the work area before an enclosure is constructed unless moving the objects creates a hazard. Mobile objects will be assumed to be contaminated and should be either cleaned with amended water and a HEPA vacuum and then removed from the area or wrapped and then disposed of as hazardous waste.

Cleaning and Sealing Surfaces: After cleaning with water and a HEPA vacuum, surfaces of stationary objects should be covered with two layers of plastic sheeting. The sheeting should be secured with duct tape or an equivalent method to provide a tight seal around the object.

Bagging Waste: In addition to the requirement for immediate bagging of waste for disposal, it is further recommended that the waste material be double-bagged and sealed in plastic bags designed for asbestos disposal. The bags should be stored in a waste storage area that can be controlled by the workers conducting the removal. Filters removed from air handling units and rubbish removed from the area are to be bagged and handled as hazardous waste.

Constructing the Enclosure

The enclosure should be constructed to provide an air-tight seal around ducts and openings into existing ventilation systems and around penetrations for electrical conduits, telephone wires, water lines, drain pipes, etc. Enclosures should be both airtight and watertight except for those openings designed to provide entry and/or air flow control.

Size: An enclosure should be the minimum volume to encompass all of the working surfaces yet allow unencumbered movement by the worker(s), provide unrestricted air flow past the worker(s), and ensure walking surfaces can be kept free of tripping hazards.

Shape: The enclosure may be any shape that optimizes the flow of ventilation air past the worker(s).

Structural Integrity: The walls, ceilings and floors must be supported in such a manner that portions of the enclosure will not fall down during normal use.

Openings: It is not necessary that the structure be airtight; openings may be designed to direct air flow. Such openings should be located at a distance from active removal operations. They should be designed to draw air into the enclosure under all anticipated circumstances. In the event that negative pressure is lost, they should be fitted with either HEPA filters to trap dust or automatic trap doors that prevent dust from escaping the enclosure. Openings for exits should be controlled by an airlock or a vestibule.

Barrier Supports: Frames should be constructed to support all unsupported spans of sheeting.

Sheeting: Walls, barriers, ceilings, and floors should be lined with two layers of plastic sheeting having a thickness of at least 6 mil.

Seams: Seams in the sheeting material should be minimized to reduce the possibilities of accidental rips and tears in the adhesive or connections. All seams in the sheeting should overlap, be staggered and not be located at corners or wall-to-floor joints.

Areas Within an Enclosure: Each enclosure consists of a work area, a decontamination area, and waste storage area. The work area where the asbestos removal operations occur should be separated from both the waste storage area and the contamination control area by physical curtains, doors, and/or airflow patterns that force any airborne contamination back into the work area.

See paragraph (j) of this section for requirements for hygiene facilities.

During egress from the work area, each worker should step into the equipment room, clean tools and equipment, and remove gross contamination from clothing by wet cleaning and HEPA vacuuming. Before entering the shower area, foot coverings, head coverings, hand coverings, and coveralls are removed and placed in impervious bags for disposal or cleaning. Airline connections from airline respirators with HEPA disconnects and power cables from powered air-purifying respirators (PAPRs) will be disconnected just prior to entering the shower room.

Establishing Negative Pressure Within the Enclosure

Negative Pressure: Air is to be drawn into the enclosure under all anticipated conditions and exhausted through a HEPA filter for 24 hours a day during the entire duration of the project.

Air Flow Tests: Air flow patterns will be checked before removal operations begin, at least once per operating shift and any time there is a question regarding the integrity of the enclosure. The primary test for air flow is to trace air currents with smoke tubes or other visual methods. Flow checks are made at each opening and at each doorway to demonstrate that air is being drawn into the enclosure and at each worker's position to show that air is being drawn away from the breathing zone.

Monitoring Pressure Within the Enclosure: After the initial air flow patterns have been checked, the static pressure must be monitored within the enclosure. Monitoring may be made using manometers, pressure gauges, or combinations of these devices. It is recommended that they be attached to alarms and strip chart recorders at points identified by the design engineer.

Corrective Actions: If the manometers or pressure gauges demonstrate a reduction in pressure differential below the required level, work should cease and the reason for the change investigated and appropriate changes made. The air flow patterns should be retested before work begins again.

Pressure Differential: The design parameters for static pressure differentials between the inside and outside of enclosures typically range from 0.02 to 0.10 inches of water gauge, depending on conditions. All zones inside the enclosure must have less pressure than the ambient pressure outside of the enclosure (-0.02 inches water gauge differential). Design specifications for the differential vary according to the size, configuration, and shape of the enclosure as well as ambient and mechanical air pressure conditions around the enclosure.

Air Flow Patterns: The flow of air past each worker shall be enhanced by positioning the intakes and exhaust ports to remove contaminated air from the worker's breathing zone, by positioning HEPA vacuum cleaners to draw air from the worker's breathing zone, by forcing relatively uncontaminated air past the worker toward an exhaust port, or by using a combination of methods to reduce the worker's exposure.

Air Handling Unit Exhaust: The exhaust plume from air handling units should be located away from adjacent personnel and intakes for HVAC systems.

Air Flow Volume: The air flow volume (cubic meters per minute) exhausted (removed) from the workplace must exceed the amount of makeup air supplied to the enclosure. The rate of air exhausted from the enclosure should be designed to maintain a negative pressure in the enclosure and air movement past each worker. The volume of air flow removed from the enclosure should replace the volume of the container at every 5 to 15 minutes. Air flow volume will need to be relatively high for large enclosures, enclosures with awkward shapes, enclosures with multiple openings, and operations employing several workers in the enclosure.

Air Flow Velocity: At each opening, the air flow velocity must visibly "drag" air into the enclosure. The velocity of air flow within the enclosure must be adequate to remove airborne contamination from each worker's breathing zone without disturbing the asbestos-containing material on surfaces.

Airlocks: Airlocks are mechanisms on doors and curtains that control the air flow patterns in the doorways. If air flow occurs, the patterns through doorways must be such that the air flows toward the inside of the enclosure. Sometimes vestibules, double doors, or double curtains are used to prevent air movement through the doorways. To use a vestibule, a worker enters a chamber by opening the door or curtain and then closing the entry before opening the exit door or curtain.

Airlocks should be located between the equipment room and shower room, between the shower room and the clean room, and between the waste storage area and the outside of the enclosure. The air flow between adjacent rooms must be checked using smoke tubes or other visual tests to ensure the flow patterns draw air toward the work area without producing eddies.

Monitoring for Airborne Concentrations

In addition to the breathing zone samples taken as outlined in paragraph (f) of this section, samples of air should be taken to demonstrate the integrity of the enclosure, the cleanliness of the clean room and shower area, and the effectiveness of the HEPA filter. If the clean room is shown to be contaminated, the room must be relocated to an uncontaminated area.

Samples taken near the exhaust of portable ventilation systems must be done with care.

General Work Practices

Preventing dust dispersion is the primary means of controlling the spread of asbestos within the enclosure. Whenever practical, the point of removal should be isolated, enclosed, covered, or shielded from the workers in the area. Waste asbestos containing materials must be bagged during or immediately after removal; the material must remain saturated until the waste container is sealed.

Waste material with sharp points or corners must be placed in hard air-tight containers rather than bags.

Whenever possible, large components should be sealed in plastic sheeting and removed intact.

Bags or containers of waste will be moved to the waste holding area, washed, and wrapped in a bag with the appropriate labels.

Cleaning the Work Area

Surfaces within the work area should be kept free of visible dust and debris to the extent feasible. Whenever visible dust appears on surfaces, the surfaces within the enclosure must be cleaned by wiping with a wet sponge, brush, or cloth and then vacuumed with a HEPA vacuum.

All surfaces within the enclosure should be cleaned before the exhaust ventilation system is deactivated and the enclosure is disassembled. An approved encapsulant may be sprayed onto areas after the visible dust has been removed.

§1926.1101 Appendix G

[Reserved]

§1926.1101 Appendix H

Substance Technical Information for Asbestos. Non-Mandatory

I. Substance Identification

A. *Substance:* "Asbestos" is the name of a class of magnesium-silicate minerals that occur in fibrous form. Minerals that are included in this group are chrysotile, crocidolite, amosite, anthophyllite asbestos, tremolite asbestos, and actinolite asbestos.

B. *Asbestos is and was used* in the manufacture of heat-resistant clothing, automotive brake and clutch linings, and a variety of building materials including floor tiles, roofing felts, ceiling tiles, asbestos-cement pipe and sheet, and fire-resistant drywall. Asbestos is also present in pipe and boiler insulation materials and in sprayed-on materials located on beams, in crawlspaces, and between walls.

C. *The potential for an asbestos-containing product* to release breathable fibers depends largely on its degree of friability. Friable means that the material can be crumbled with hand pressure and is therefore likely to emit fibers. The fibrous fluffy sprayed-on materials used for fireproofing, insulation, or sound proofing are considered to be friable, and they readily release airborne fibers if disturbed. Materials such as vinyl-asbestos floor tile or roofing felt are considered non-friable if intact and generally do not emit airborne fibers unless subjected to sanding, sawing and other aggressive operations. Asbestos-cement pipe or sheet can emit airborne fibers if the materials are cut or sawed, or if they are broken.

D. *Permissible exposure:* Exposure to airborne asbestos fibers may not exceed 0.1 fibers per cubic centimeter of air (0.1 f/cc) averaged over the 8-hour workday, and 1 fiber per cubic centimeter of air (1.0 f/ cc) averaged over a 30 minute work period.

II. Health Hazard Data

A. *Asbestos can cause disabling respiratory disease* and various types of cancers if the fibers are inhaled. Inhaling or ingesting fibers from contaminated clothing or skin can also result in these diseases. The symptoms of these diseases generally do not appear for 20 or more years after initial exposure.

B. *Exposure to asbestos has been shown to cause* lung cancer, mesothelioma, and cancer of the stomach and colon. Mesothelioma is a rare cancer of the thin membrane lining of the chest and abdomen. Symptoms of mesothelioma include shortness of breath, pain in the walls of the chest, and/or abdominal pain.

III. Respirators and Protective Clothing

A. *Respirators:* You are required to wear a respirator when performing tasks that result in asbestos exposure that exceeds the permissible exposure limit (PEL) of 0.1 f/cc and when performing certain designated operations. Air-purifying respirators equipped with a high-efficiency particulate air (HEPA) filter can be used where airborne asbestos fiber concentrations do not exceed 1.0 f/cc; otherwise, more protective respirators such as air-supplied, positive-pressure, full facepiece respirators must be used. Disposable respirators or dust masks are not permitted to be used for asbestos work. For effective protection, respirators must fit your face and head snugly. Your employer is required to conduct a fit test when you are first assigned a respirator and every 6 months thereafter. Respirators should not be loosened or removed in work situations where their use is required.

B. *Protective Clothing:* You are required to wear protective clothing in work areas where asbestos fiber concentrations exceed the permissible exposure limit (PEL) of 0.1 f/cc.

IV. Disposal Procedures and Clean-up

A. *Wastes that are generated* by processes where asbestos is present include:

1. *Empty asbestos shipping containers.*
2. *Process wastes such as cuttings, trimmings, or reject materials.*
3. *Housekeeping waste from wet-sweeping or HEPA-vacuuming.*
4. *Asbestos fireproofing or insulating material that is removed from buildings.*
5. *Asbestos-containing building products* removed during building renovation or demolition.
6. *Contaminated disposable protective clothing.*

B. *Empty shipping bags* can be flattened under exhaust hoods and packed into airtight containers for disposal. Empty shipping drums are difficult to clean and should be sealed.

C. *Vacuum bags or disposable paper filters* should not be cleaned, but should be sprayed with a fine water mist and placed into a labeled waste container.

D. *Process waste and housekeeping waste* should be wetted with water or a mixture of water and surfactant prior to packaging in disposable containers.

E. *Asbestos-containing material that is removed* from buildings must be disposed of in leak-tight 6-mil plastic bags, plastic-lined cardboard containers, or plastic-lined metal containers. These wastes, which are removed while wet, should be sealed in containers before they dry out to minimize the release of asbestos fibers during handling.

V. Access to Information

A. *Each year,* your employer is required to inform you of the information contained in this standard and appendices for asbestos. In addition, your employer must instruct you in the proper work practices for handling asbestos-containing materials, and the correct use of protective equipment.

B. *Your employer is required* to determine whether you are being exposed to asbestos. Your employer must treat exposure to thermal system insulation and sprayed-on and troweled-on surfacing material as asbestos exposure, unless results of laboratory analysis show that the material does not contain asbestos. You or your representative has the right to observe employee measurements and to record the results obtained. Your employer is required to inform you of your exposure, and, if you are exposed above the permissible exposure limit, he or she is required to inform you of the actions that are being taken to reduce your exposure to within the permissible limit.

C. *Your employer is required* to keep records of your exposures and medical examinations. These exposure records must be kept for at least thirty (30) years. Medical records must be kept for the period of your employment plus thirty (30) years.

D. *Your employer is required* to release your exposure and medical records to your physician or designated representative upon your written request.

§1926.1101 Appendix I

Medical Surveillance Guidelines for Asbestos, Non-Mandatory

I. Route of Entry

Inhalation, ingestion.

II. Toxicology

Clinical evidence of the adverse effects associated with exposure to asbestos is present in the form of several well-conducted epidemiological studies of occupationally exposed workers, family contacts of workers, and persons living near asbestos mines. These studies have shown a definite association between exposure to asbestos and an increased incidence of lung cancer, pleural and peritoneal mesothelioma, gastrointestinal cancer, and asbestosis. The latter is a disabling fibrotic lung disease that is caused only by exposure to asbestos. Exposure to asbestos has also been associated with an increased incidence of esophageal, kidney, laryngeal, pharyngeal, and buccal cavity cancers. As with other known chronic occupational diseases, disease associated with asbestos generally appears about 20 years following the first occurrence of exposure: There are no known acute effects associated with exposure to asbestos.

Epidemiological studies indicate that the risk of lung cancer among exposed workers who smoke cigarettes is greatly increased over the risk of lung cancer among non-exposed smokers or exposed nonsmokers. These studies suggest that cessation of smoking will reduce the risk of lung cancer for a person exposed to asbestos but will not reduce it to the same level of risk as that existing for an exposed worker who has never smoked.

III. Signs and Symptoms of Exposure-Related Disease

The signs and symptoms of lung cancer or gastrointestinal cancer induced by exposure to asbestos are not unique, except that a chest X-ray of an exposed patient with lung cancer may show pleural plaques, pleural calcification, or pleural fibrosis. Symptoms characteristic of mesothelioma include shortness of breath, pain in the walls of the chest, or abdominal pain. Mesothelioma has a much longer latency period compared with lung cancer (40 years versus 15-20 years), and mesothelioma is therefore more likely to be found among workers who were first exposed to asbestos at an early age. Mesothelioma is always fatal.

Asbestosis is pulmonary tibrosis caused by the accumulation of asbestos fibers in the lungs. Symptoms include shortness of breath, coughing, fatigue, and vague feelings of sickness. When the fibrosis worsens, shortness of breath occurs even at rest. The diagnosis of asbestosis is based on a history of exposure to asbestos, the presence of characteristics radiologic changes, end-inspiratory crackles (rales), and other clinical features of fibrosing lung disease. Pleural plaques and thickening are observed on X-rays taken during the early stages of the disease. Asbestosis is often a progressive disease even in the absence of continued exposure, although this appears to be a highly individualized characteristic. In severe cases, death may be caused by respiratory or cardiac failure.

IV. Surveillance and Preventive Considerations

As noted above, exposure to asbestos has been linked to an increased risk of lung cancer, mesothelioma, gastrointestinal cancer, and asbestosis among occupationally exposed workers. Adequate screening tests to determine an employee's potential for developing serious chronic diseases, such as a cancer, from exposure to asbestos do not presently exist. However, some tests, particularly chest X-rays and pulmonary function tests, may indicate that an employee has been overexposed to asbestos increasing his or her risk of developing exposure related chronic diseases. It is important for the physician to become familiar with the operating conditions in which occupational exposure to asbestos is likely to occur. This is particularly important in evaluating medical and work histories and in conducting physical examinations. When an active employee has been identified as having been overexposed to asbestos measures taken by the employer to eliminate or mitigate further exposure should also lower the risk of serious long-term consequences.

The employer is required to institute a medical surveillance program for all employees who are or will be exposed to asbestos at or above the permissible exposure limit (0.1 fiber per cubic centimeter of air). All examinations and procedures must be performed by or under the supervision of a licensed physician, at a reasonable time and place, and at no cost to the employee.

Although broad latitude is given to the physician in prescribing specific tests to be included in the medical surveillance program, OSHA requires inclusion of the following elements in the routine examination:

(i) *Medical and work histories* with special emphasis directed to symptoms of the respiratory system, cardiovascular system, and digestive tract.

(ii) *Completion of the respiratory disease questionnaire* contained in Appendix D.

(iii) *A physical examination including a chest roentgenogram* and pulmonary function test that includes measurement of the employee's forced vital capacity (FVC) and forced expiratory volume at one second (FEV^1).

(iv) *Any laboratory or other test* that the examining physician deems by sound medical practice to be necessary.

The employer is required to make the prescribed tests available at least annually to those employees covered; more often than specified if recommended by the examining physician; and upon termination of employment.

The employer is required to provide the physician with the following information: A copy of this standard and appendices; a description of the employee's duties as they relate to asbestos exposure; the employee's representative level of exposure to asbestos; a description of any personal protective and respiratory equipment used; and information from previous medical examinations of the affected employee that is not otherwise available to the physician. Making this information available to the physician will aid in the evaluation of the employee's health in relation to assigned duties and fitness to wear personal protective equipment, if required.

The employer is required to obtain a written opinion from the examining physician containing the results of the medical examination; the physician's opinion as to whether the employee has any detected medical conditions that would place the employee at an increased risk of exposure-related disease; any recommended limitations on the employee or on the use of personal protective equipment; and a statement that the employee has been informed by the physician of the results of the medical examination and of any medical conditions related to asbestos exposure that require further explanation or treatment. This written opinion must not reveal specific findings or diagnoses unrelated to exposure to asbestos, and a copy of the opinion must be provided to the affected employee.

§1926.1101 Appendix J

Smoking Cessation Program Information for Asbestos — Non-Mandatory

The following organizations provide smoking cessation information.

1. **The National Cancer Institute** operates a toll-free Cancer Information Service (CIS) with trained personnel to help you. Call 1-800-4-CANCER* to reach the CIS office serving your area, or write: Office of Cancer Communications, National Cancer Institute, National Institutes of Health, Building 31 Room 10A24, Bethesda, Maryland 20892.
2. **American Cancer Society,** 3340 Peachtree Road, N.E., Atlanta, Georgia 30026, (404) 320-3333
   The American Cancer Society (ACS) is a voluntary organization composed of 58 divisions and 3,100 local units. Through "The Great American Smokeout" in November, the annual Cancer Crusade in April, and numerous educational materials, ACS helps people learn about the health hazards of smoking and become successful ex-smokers.
3. **American Heart Association,** 7320 Greenville Avenue, Dallas, Texas 75231, (214) 750-5300
   The American Heart Association (AHA) is a voluntary organization with 130,000 members (physicians, scientists, and laypersons) in 55 state and regional groups. AHA produces a variety of publications and audiovisual materials about the effects of smoking on the heart. AHA also has developed a guidebook for incorporating a weight-control component into smoking cessation programs.
4. **American Lung Association,** 1740 Broadway, New York, New York 10019, (212) 245-8000
   A voluntary organization of 7,500 members (physicians, nurses, and laypersons), the American Lung Association (ALA) conducts numerous public information programs about the health effects of smoking. ALA has 59 state and 85 local units. The organization actively supports legislation and information campaigns for non-smokers' rights and provides help for smokers who want to quit,

for example, through "Freedom From Smoking," a self-help smoking cessation program.

5. **Office on Smoking and Health,** U.S. Department of Health and Human Services, 5600 Fishers Lane, Park Building, Room 110, Rockville, Maryland 20857
The Office on Smoking and Health (OSH) is the Department of Health and Human Services' lead agency in smoking control. OSH has sponsored distribution of publications on smoking-related topics, such as free flyers on relapse after initial quitting, helping a friend or family member quit smoking, the health hazards of smoking, and the effects of parental smoking on teenagers.
*In Hawaii, on Oahu call 524-1234 (call collect from neighboring islands),
Spanish-speaking staff members are available during daytime hours to callers from the following areas: California, Florida, Georgia, Illinois, New Jersey (area code 201), New York, and Texas. Consult your local telephone directory for listings of local chapters.

§1926.1101 Appendix K

Polarized Light Microscopy of Asbestos (Non-Mandatory)

Method number: ID-191

Matrix: Bulk

Collection Procedure: Collect approximately 1 to 2 grams of each type of material and place into separate 20 mL scintillation vials.

Analytical Procedure: A portion of each separate phase is analyzed by gross examination, phase-polar examination, and central stop dispersion microscopy.

Commercial manufacturers and products mentioned in this method are for descriptive use only and do not constitute endorsements by USDOL-OSHA. Similar products from other sources may be substituted.

1. Introduction

This method describes the collection and analysis of asbestos bulk materials by light microscopy techniques including phase-polar illumination and central-stop dispersion microscopy. Some terms unique to asbestos analysis are defined below:

Amphibole: A family of minerals whose crystals are formed by long, thin units which have two thin ribbons of double chain silicate with a brucite ribbon in between. The shape of each unit is similar to an "I beam". Minerals important in asbestos analysis include cummingtonite-grunerite, crocidolite, tremolite-actinolite and anthophyllite.

Asbestos: A term for naturally occurring fibrous minerals. Asbestos includes chrysotile, cummingtonite-grunerite asbestos (amosite), anthophyllite asbestos, tremolite asbestos, crocidolite, actinolite asbestos and any of these minerals which have been chemically treated or altered. The precise chemical formulation of each species varies with the location from which it was mined. Nominal compositions are listed:

Chrysotile..............................$Mg_3\ Si_2\ O_5\ (OH)_4$
Crocidolite (Riebeckite asbestos)......$Na_2\ Fe_3{}^{2}+Fe_2{}^{3}+Si_8\ O_{22}\ (OH)_2$
Cummingtonite-Grunerite asbestos (Amosite).....$(Mg,Fe)_7Si_8O_{22}\ (OH)_2$
Tremolite-Actinolite asbestos..............$Ca_2\ (Mg,Fe)_5\ Si_8\ O_{22}\ (OH)_2$
Anthophyllite asbestos.............$(Mg,Fe)_7\ Si_8\ O_{22}\ (OH)_2$

Asbestos Fiber: A fiber of asbestos meeting the criteria for a fiber. (See section 3.5. of this Appendix)

Aspect Ratio: The ratio of the length of a fiber to its diameter usually defined as "length : width", e.g. 3:1.

Brucite: A sheet mineral with the composition $Mg(OH)_2$.

Central Stop Dispersion Staining (microscope): This is a dark field microscope technique that images particles using only light refracted by the particle, excluding light that travels through the particle unrefracted. This is usually accomplished with a McCrone objective or other arrangement which places a circular stop with apparent aperture equal to the objective aperture in the back focal plane of the microscope.

Cleavage Fragments: Mineral particles formed by the comminution of minerals, especially those characterized by relatively parallel sides and moderate aspect ratio.

Differential Counting: The term applied to the practice of excluding certain kinds of fibers from a phase contrast asbestos count because they are not asbestos.

Fiber: A particle longer than or equal to 5 µm with a length to width ratio greater than or equal to 3:1. This may include cleavage fragments. (see section 3.5 of this appendix).

Phase Contrast: Contrast obtained in the microscope by causing light scattered by small particles to destructively interfere with unscattered light, thereby enhancing the visibility of very small particles and particles with very low intrinsic contrast.

Phase Contrast Microscope: A microscope configured with a phase mask pair to create phase contrast. The technique which uses this is called Phase Contrast Microscopy (PCM).

Phase-Polar Analysis: This is the use of polarized light in a phase contrast microscope. It is used to see the same size fibers that are visible in air filter analysis. Although fibers finer than 1 µm are visible, analysis of these is inferred from analysis of larger bundles that are usually present.

Phase-Polar Microscope: The phase-polar microscope is a phase contrast microscope which has an analyzer, a polarizer, a first order red plate and a rotating phase condenser all in place so that the polarized light image is enhanced by phase contrast.

Sealing Encapsulant: This is a product which can be applied, preferably by spraying, onto an asbestos surface which will seal the surface so that fibers cannot be released.

Serpentine: A mineral family consisting of minerals with the general composition $Mg3(Si_2\ O5(OH)_4$ having the magnesium in brucite layer over a silicate layer. Minerals important in asbestos analysis included in this family are chrysotile, lizardite, antigorite.

1.1. *History*

Light microscopy has been used for well over 100 years for the determination of mineral species. This analysis is carried out using specialized polarizing microscopes as well as bright field microscopes. The identification of minerals is an on-going process with many new minerals described each year. The first recorded use of asbestos was in Finland about 2500 B.C. where the material was used in the mud wattle for the wooden huts the people lived in as well as strengthening for pottery. Adverse health aspects of the mineral were noted nearly 2000 years ago when Pliny the Younger wrote about the poor health of slaves in the asbestos mines. Although known to be injurious for centuries, the first modern references to its toxicity were by the British Labor Inspectorate when it banned asbestos dust from the workplace in 1898. Asbestosis cases were described in the literature after the turn of the century. Cancer was first suspected in the mid 1930's and a causal link to mesothelioma was made in 1965. Because of the public concern for worker and public safety with the use of this material, several different types of analysis were applied to the determination of asbestos content. Light microscopy requires a great deal of experience and craft. Attempts were made to apply less subjective methods to the analysis. X-ray diffraction was partially successful in determining the mineral types but was unable to separate out the fibrous portions from the non-fibrous portions. Also, the minimum detection limit for asbestos analysis by X-ray diffraction (XRD) is about 1%. Differential Thermal Analysis (DTA) was no more successful. These provide useful corroborating information when the presence of asbestos has been shown by microscopy; however, neither can determine the difference between fibrous and non-fibrous minerals when both habits are present. The same is true of Infrared Absorption (IR).

When electron microscopy was applied to asbestos analysis, hundreds of fibers were discovered present too small to be visible in any light microscope. There are two different types of electron microscope used for asbestos analysis: Scanning Electron Microscope (SEM) and Transmission Electron Microscope (TEM). Scanning Electron Microscopy is useful in identifying minerals. The SEM can provide two of the three pieces of information required to identify fibers by electron microscopy: morphology and chemistry. The third is structure as determined by Selected Area Electron Diffraction — SAED which is performed in the TEM. Although the resolution of the SEM is sufficient for very fine fibers to be seen, accuracy of chemical analysis that can be performed on the fibers varies with fiber diameter in fibers of less than 0.2 µm diameter. The TEM is a powerful tool to identify fibers too small to be resolved by light microscopy and should be used in conjunction with this method when necessary. The TEM can provide all three pieces of information required for fiber identification. Most fibers thicker than 1 µm can adequately be defined in the light microscope. The light microscope remains as the best instrument for the determination of mineral type. This is because the minerals under investigation were first described analytically with the light microscope. It is inexpensive and gives positive identification for most samples analyzed. Further, when optical techniques are inadequate, there is ample indication that alternative techniques should be used for complete identification of the sample.

1.2. *Principle*

Minerals consist of atoms that may be arranged in random order or in a regular arrangement. Amorphous materials have atoms in random order while crystalline materials have long range order.

Many materials are transparent to light, at least for small particles or for thin sections. The properties of these materials can be investigated by the effect that the material has on light passing through it. The six asbestos minerals are all crystalline with particular properties that have been identified and cataloged. These six minerals are anisotropic. They have a regular array of atoms, but the arrangement is not the same in all directions. Each major direction of the crystal presents a different regularity. Light photons travelling in each of these main directions will encounter different electrical neighborhoods, affecting the path and time of travel. The techniques outlined in this method use the fact that light traveling through fibers or crystals in different directions will behave differently, but predictably. The behavior of the light as it travels through a crystal can be measured and compared with known or determined values to identify the mineral species. Usually, Polarized Light Microscopy (PLM) is performed with strain-free objectives on a bright-field microscope platform. This would limit the resolution of the microscope to about 0.4 µm. Because OSHA requires the counting and identification of fibers visible in phase contrast, the phase contrast platform is used to visualize the fibers with the polarizing elements added into the light path. Polarized light methods cannot identify fibers finer than about 1 µm in diameter even though they are visible. The finest fibers are usually identified by inference from the presence of larger, identifiable fiber bundles. When fibers are present, but not identifiable by light microscopy, use either SEM or TEM to determine the fiber identity.

1.3. *Advantages and Disadvantages*

The advantages of light microcopy are:

(a) *Basic identification of the materials* was first performed by light microscopy and gross analysis. This provides a large base of published information against which to check analysis and analytical technique.

(b) *The analysis is specific to fibers.* The minerals present can exist in asbestiform, fibrous, prismatic, or massive varieties all at the same time. Therefore, bulk methods of analysis such as X-ray diffraction, IR analysis, DTA, etc. are inappropriate where the material is not known to be fibrous.

(c) *The analysis is quick,* requires little preparation time, and can be performed on-site if a suitably equipped microscope is available.

The disadvantages are:

(a) *Even using phase-polar illumination,* not all the fibers present may be seen. This is a problem for very low asbestos concentrations where agglomerations or large bundles of fibers may not be present to allow identification by inference.

(b) *The method requires a great* degree of sophistication on the part of the microscopist. An analyst is only as useful as his mental catalog of images. Therefore, a microscopist's accuracy is enhanced by experience. The mineralogical training of the analyst is very important. It is the basis on which subjective decisions are made.

(c) *The method uses only* a tiny amount of material for analysis. This may lead to sampling bias and false results (high or low). This is especially true if the sample is severely inhomogeneous.

(d) *Fibers may be bound in a matrix* and not distinguishable as fibers so identification cannot be made.

1.4. *Method Performance*

1.4.1. *This method can be used* for determination of asbestos content from 0 to 100% asbestos. The detection limit has not been adequately determined, although for selected samples, the limit is very low, depending on the number of particles examined. For mostly homogeneous, finely divided samples, with no difficult fibrous interferences, the detection limit is below 1%. For inhomogeneous samples (most samples), the detection limit remains undefined. NIST has conducted proficiency testing of laboratories on a national scale. Although each round is reported statistically with an average, control limits, etc., the results indicate a difficulty in establishing precision especially in the low concentration range. It is suspected that there is significant bias in the low range especially near 1%. EPA tried to remedy this by requiring a mandatory point counting scheme for samples less than 10%. The point counting procedure is tedious, and may introduce significant biases of its own. It has not been incorporated into this method.

1.4.2. *The precision and accuracy* of the quantitation tests performed in this method are unknown. Concentrations are easier to determine in commercial products where asbestos was deliberately added because the amount is usually more than a few percent. An analyst's results can be "calibrated" against the known amounts added by the manufacturer. For geological samples, the degree of homogeneity affects the precision.

1.4.3. *The performance of the method is analyst dependent.* The analyst must choose carefully and not necessarily randomly the portions for analysis to assure that detection of asbestos occurs when it is present. For this reason, the analyst must have adequate training in sample preparation, and experience in the location and identification of asbestos in samples. This is usually accomplished through substantial on-the-job training as well as formal education in mineralogy and microscopy.

1.5. *Interferences*

Any material which is long, thin, and small enough to be viewed under the microscope can be considered an interference for asbestos. There are literally hundreds of interferences in workplaces. The techniques described in this method are normally sufficient to eliminate the interferences. An analyst's success in eliminating the interferences depends on proper training.

Asbestos minerals belong to two mineral families: the serpentines and the amphiboles. In the serpentine family, the only common fibrous mineral is chrysotile. Occasionally, the mineral antigorite occurs in a fibril habit with morphology similar to the amphiboles. The amphibole minerals consist of a score of different minerals of which only five are regulated by federal standard: amosite, crocidolite, anthophyllite asbestos, tremolite asbestos and actinolite asbestos. These are the only amphibole minerals that have been commercially exploited for their fibrous properties; however, the rest can and do occur occasionally in asbestiform habit.

In addition to the related mineral interferences, other minerals common in building material may present a problem for some microscopists: gypsum, anhydrite, brucite, quartz fibers, talc fibers or ribbons, wollastonite, perlite, attapulgite, etc. Other fibrous materials commonly present in workplaces are: fiberglass, mineral wool, ceramic wool, refractory ceramic fibers, kevlar, nomex, synthetic fibers, graphite or carbon fibers, cellulose (paper or wood) fibers, metal fibers, etc.

Matrix embedding material can sometimes be a negative interference. The analyst may not be able to easily extract the fibers from the matrix in order to use the method. Where possible, remove the matrix before the analysis, taking careful note of the loss of weight. Some common matrix materials are: vinyl, rubber, tar, paint, plant fiber, cement, and epoxy. A further negative interference is that the asbestos fibers themselves may be either too small to be seen in Phase contrast Microscopy (PCM) or of a very low fibrous quality, having the appearance of plant fibers. The analyst's ability to deal with these materials increases with experience.

1.6. *Uses and Occupational Exposure*

Asbestos is ubiquitous in the environment. More than 40% of the land area of the United States is composed of minerals which may contain asbestos. Fortunately, the actual formation of great amounts of asbestos is relatively rare. Nonetheless, there are locations in which environmental exposure can be severe such as in the Serpentine Hills of California.

There are thousands of uses for asbestos in industry and the home. Asbestos abatement workers are the most current segment of the population to have occupational exposure to great amounts of asbestos. If the material is undisturbed, there is no exposure. Exposure occurs when the asbestos-containing material is abraded or otherwise disturbed during maintenance operations or some other activity. Approximately 95% of the asbestos in place in the United States is chrysotile.

Amosite and crocidolite make up nearly all the difference. Tremolite and anthophyllite make up a very small percentage. Tremolite is found in extremely small amounts in certain chrysotile deposits. Actinolite exposure is probably greatest from environmental sources, but has been identified in vermiculite containing, sprayed-on insulating materials which may have been certified as asbestos-free.

1.7. *Physical and Chemical Properties*

The nominal chemical compositions for the asbestos minerals were given in Section 1. Compared to cleavage fragments of the same minerals, asbestiform fibers possess a high tensile strength along the fiber axis. They are chemically inert, non-combustible, and heat resistant. Except for chrysotile, they are insoluble in Hydrochloric acid (HCl). Chrysotile is slightly soluble in HCl. Asbestos has high electrical resistance and good sound absorbing characteristics. It can be woven into cables, fabrics or other textiles, or matted into papers, felts, and mats.

1.8. *Toxicology* (This Section is for Information Only and Should Not Be Taken as OSHA Policy)

Possible physiologic results of respiratory exposure to asbestos are mesothelioma of the pleura or peritoneum, interstitial fibrosis, asbestosis, pneumoconiosis, or respiratory cancer. The possible consequences of asbestos exposure are detailed in the NIOSH Criteria Document or in the OSHA Asbestos Standards 29 CFR 1910.1001 and 29 CFR 1926.1101 and 29 CFR 1915.1001.

2. Sampling Procedure

2.1. *Equipment for sampling*

(a) *Tube or cork borer sampling device*

(b) *Knife*

(c) *20 mL scintillation vial or similar vial*

(d) *Sealing encapsulant*

2.2. *Safety Precautions*

Asbestos is a known carcinogen. Take care when sampling. While in an asbestos-containing atmosphere, a properly selected and fit-tested respirator should be worn. Take samples in a manner to cause the least amount of dust. Follow these general guidelines:

(a) *Do not make unnecessary dust.*

(b) *Take only a small amount (1 to 2 g).*

(c) *Tightly close the sample container.*

(d) *Use encapsulant to seal* the spot where the sample was taken, if necessary.

2.3. *Sampling Procedure*

Samples of any suspect material should be taken from an inconspicuous place. Where the material is to remain, seal the sampling wound with an encapsulant to eliminate the potential for exposure from the sample site. Microscopy requires only a few milligrams of material. The amount that will fill a 20 mL scintillation vial is more than adequate. Be sure to collect samples from all layers and phases of material. If possible, make separate samples of each different phase of the material. This will aid in determining the actual hazard. DO NOT USE ENVELOPES, PLASTIC OR PAPER BAGS OF ANY KIND TO COLLECT SAMPLES. The use of plastic bags presents a contamination hazard to laboratory personnel and to other samples. When these containers are opened, a bellows effect blows fibers out of the container onto everything, including the person opening the container.

If a cork-borer type sampler is available, push the tube through the material all the way, so that all layers of material are sampled. Some samplers are intended to be disposable. These should be capped and sent to the laboratory. If a non-disposable cork borer is used, empty the contents into a scintillation vial and send to the laboratory. Vigorously and completely clean the cork borer between samples.

2.4. *Shipment*

Samples packed in glass vials must not touch or they might break in shipment.

(a) *Seal the samples* with a sample seal over the end to guard against tampering and to identify the sample.

(b) *Package the bulk samples* in separate packages from the air samples. They may cross-contaminate each other and will invalidate the results of the air samples.

(c) *Include identifying paperwork* with the samples, but not in contact with the suspected asbestos.

(d) *To maintain sample accountability,* ship the samples by certified mail, overnight express, or hand carry them to the laboratory.

3. Analysis

The analysis of asbestos samples can be divided into two major parts: sample preparation and microscopy. Because of the different asbestos uses that may be encountered by the analyst, each sample may need different preparation steps. The choices are outlined below. There are several different tests that are performed to identify the asbestos species and determine the percentage. They will be explained below.

3.1. *Safety*

(a) *Do not create unnecessary dust.* Handle the samples in HEPA-filter equipped hoods. If samples are received in bags, envelopes or other inappropriate container, open them only in a hood having a face velocity at or greater than 100 fpm. Transfer a small amount to a scintillation vial and only handle the smaller amount.

(b) *Open samples in a hood, never in the open lab area.*

(c) *Index of refraction oils can be toxic.* Take care not to get this material on the skin. Wash immediately with soap and water if this happens.

(d) *Samples that have been heated* in the muffle furnace or the drying oven may be hot. Handle them with tongs until they are cool enough to handle.

(e) *Some of the solvents used,* such as THF (tetrahydrofuran), are toxic and should only be handled in an appropriate fume hood and according to instructions given in the Material Safety Data Sheet (MSDS).

3.2. *Equipment*

(a) *Phase contrast microscope* with 10x, 16x and 40x objectives, 10x wide-field eyepieces, G-22 Walton-Beckett graticule, Whipple disk, polarizer, analyzer and first order red or gypsum plate, 100 Watt illuminator, rotating position condenser with oversize phase rings, central stop dispersion objective, Kohler illumination and a rotating mechanical stage.

(b) *Stereo microscope with reflected light illumination,* transmitted light illumination, polarizer, analyzer and first order red or gypsum plate, and rotating stage.

(c) *Negative pressure hood for the stereo microscope*

(d) *Muffle furnace capable of 600 °C*

(e) *Drying oven capable of 50-150 °C*

(f) *Aluminum specimen pans*

(g) *Tongs for handling samples in the furnace*

(h) *High dispersion index of refraction oils (Special for dispersion staining.)*

n = 1.550
n = 1.585
n = 1.590
n = 1.605
n = 1.620
n = 1.670
n = 1.680
n = 1.690

(i) *A set of index of refraction* oils from about n=1.350 to n=2.000 in n=0.005 increments. (Standard for Becke line analysis.)

(j) *Glass slides with painted* or frosted ends 1x3 inches 1mm (thick, precleaned.)

(k) *Cover Slips 22x22 mm, # 1 1/2*

(l) *Paper clips or dissection needles*

(m) *Hand grinder*

(n) *Scalpel with both # 10 and # 11 blades*

(o) *0.1 molar HCl*

(p) *Decalcifying solution (Baxter* Scientific Products) Ethylenediaminetetraacetic Acid,

Tetrasodium..0.7 g/l
Sodium Potassium Tartrate..................................8.0 mg/liter
Hydrochloric Acid..99.2 g/liter
Sodium Tartrate..0.14 g/liter

(q) *Tetrahydrofuran (THF)*

(r) *Hotplate capable of 60 °C*

(s) *Balance*

(t) *Hacksaw blade*

(u) *Ruby mortar and pestle*

3.3. *Sample Pre-Preparation*

Sample preparation begins with pre-preparation which may include chemical reduction of the matrix, heating the sample to dryness or heating in the muffle furnace. The end result is a sample which has been reduced to a powder that is sufficiently fine to fit under the cover slip. Analyze different phases of samples separately, e.g., tile and the tile mastic should be analyzed separately as the mastic may contain asbestos while the tile may not.

(a) *Wet Samples*

Samples with a high water content will not give the proper dispersion colors and must be dried prior to sample mounting. Remove the lid of the scintillation vial, place the bottle in the drying oven and heat at 100 °C to dryness (usually about 2 h). Samples which are not submitted to the lab in glass must be removed and placed in glass vials or aluminum weighing pans before placing them in the drying oven.

(b) *Samples With Organic Interference — Muffle Furnace*

These may include samples with tar as a matrix, vinyl asbestos tile, or any other organic that can be reduced by heating. Remove the sample from the vial and weigh in a balance to determine the weight of the submitted portion. Place the sample in a muffle furnace at 500 °C for 1 to 2 h or until all obvious organic material has been removed. Retrieve, cool and weigh again to determine the weight loss on ignition. This is necessary to determine the asbestos content of the submitted sample, because the analyst will be looking at a reduced sample.

Note: Heating above 600 °C will cause the sample to undergo a structural change which, given sufficient time, will convert the chrysotile to forsterite. Heating even at lower temperatures for 1 to 2 h may have a measurable effect on the optical properties of the minerals. If the analyst is unsure of what to expect, a sample of standard asbestos should be heated to the same temperature for the same length of time so that it can be examined for the proper interpretation.

(c) Samples With Organic Interference — THF

Vinyl asbestos tile is the most common material treated with this solvent, although, substances containing tar will sometimes yield to this treatment. Select a portion of the material and then grind it up if possible. Weigh the sample and place it in a test tube. Add sufficient THF to dissolve the organic matrix. This is usually about 4 to 5 mL. Remember, THF is highly flammable. Filter the remaining material through a tared silver membrane, dry and weigh to determine how much is left after the solvent extraction. Further process the sample to remove carbonate or mount directly.

(d) Samples With Carbonate Interference

Carbonate material is often found on fibers and sometimes must be removed in order to perform dispersion microscopy. Weigh out a portion of the material and place it in a test tube. Add a sufficient amount of 0.1 M HCl or decalcifying solution in the tube to react all the carbonate as evidenced by gas formation; i.e., when the gas bubbles stop, add a little more solution. If no more gas forms, the reaction is complete. Filter the material out through a tared silver membrane, dry and weigh to determine the weight lost.

3.4. *Sample Preparation*

Samples must be prepared so that accurate determination can be made of the asbestos type and amount present. The following steps are carried out in the low-flow hood (a low-flow hood has less than 50 fpm flow):

(1) If the sample has large lumps, is hard, or cannot be made to lie under a cover slip, the grain size must be reduced. Place a small amount between two slides and grind the material between them or grind a small amount in a clean mortar and pestle. The choice of whether to use an alumina, ruby, or diamond mortar depends on the hardness of the material. Impact damage can alter the asbestos mineral if too much mechanical shock occurs. (Freezer mills can completely destroy the observable crystallinity of asbestos and should not be used). For some samples, a portion of material can be shaved off with a scalpel, ground off with a hand grinder or hack saw blade.

The preparation tools should either be disposable or cleaned thoroughly. Use vigorous scrubbing to loosen the fibers during the washing. Rinse the implements with copious amounts of water and air-dry in a dust-free environment.

(2) If the sample is powder or has been reduced as in (1) above, it is ready to mount. Place a glass slide on a piece of optical tissue and write the identification on the painted or frosted end. Place two drops of index of refraction medium n=1.550 on the slide. (The medium n=1.550 is chosen because it is the matching index for chrysotile. Dip the end of a clean paper-clip or dissecting needle into the droplet of refraction medium on the slide to moisten it. Then dip the probe into the powder sample. Transfer what sticks on the probe to the slide. The material on the end of the probe should have a diameter of about 3 mm for a good mount. If the material is very fine, less sample may be appropriate. For non-powder samples such as fiber mats, forceps should be used to transfer a small amount of material to the slide. Stir the material in the medium on the slide, spreading it out and making the preparation as uniform as possible. Place a cover-slip on the preparation by gently lowering onto the slide and allowing it to fall "trapdoor" fashion on the preparation to push out any bubbles. Press gently on the cover slip to even out the distribution of particulate on the slide. If there is insufficient mounting oil on the slide, one or two drops may be placed near the edge of the coverslip on the slide. Capillary action will draw the necessary amount of liquid into the preparation. Remove excess oil with the point of a laboratory wiper.

Treat at least two different areas of each phase in this fashion. Choose representative areas of the sample. It may be useful to select particular areas or fibers for analysis. This is useful to identify asbestos in severely inhomogeneous samples.

When it is determined that amphiboles may be present, repeat the above process using the appropriate high-dispersion oils until an identification is made or all six asbestos minerals have been ruled out. Note that percent determination must be done in the index medium 1.550 because amphiboles tend to disappear in their matching mediums.

3.5. *Analytical procedure*

Note: This method presumes some knowledge of mineralogy and optical petrography.

The analysis consists of three parts: The determination of whether there is asbestos present, what type is present and the determination of how much is present. The general flow of the analysis is:

(1) Gross examination.

(2) Examination under polarized light on the stereo microscope.

(3) Examination by phase-polar illumination on the compound phase microscope.

(4) Determination of species by dispersion stain. Examination by Becke line analysis may also be used; however, this is usually more cumbersome for asbestos determination.

(5) Difficult samples may need to be analyzed by SEM or TEM, or the results from those techniques combined with light microscopy for a definitive identification.

Identification of a particle as asbestos requires that it be asbestiform. Description of particles should follow the suggestion of Campbell. (Figure 1)

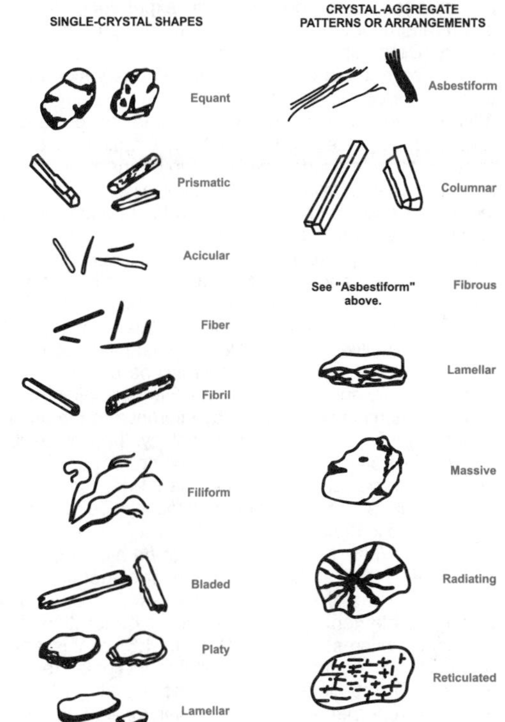

FIGURE 1: Particle definitions showing mineral growth habits. From the U.S. Bureau of Mines

For the purpose of regulation, the mineral must be one of the six minerals covered and must be in the asbestos growth habit. Large specimen samples of asbestos generally have the gross appearance of wood. Fibers are easily parted from it. Asbestos fibers are very long compared with their widths. The fibers have a very high tensile strength as demonstrated by bending without breaking. Asbestos fibers exist in bundles that are easily parted, show longitudinal fine structure and may be tufted at the ends showing "bundle of sticks" morphology. In the microscope some of these properties may not be observable. Amphiboles do not always show striations along their length even when they are asbestos. Neither will they always show tufting. They generally do not show a curved nature except for very long fibers. Asbestos and asbestiform minerals are usually characterized in groups by extremely high aspect ratios (greater than 100:1). While aspect ratio analysis is useful for characterizing populations of fibers, it cannot be used to identify individual fibers of intermediate to short aspect ratio. Observation of many fibers is often necessary to determine whether a sample consists of "cleavage fragments" or of asbestos fibers.

Most cleavage fragments of the asbestos minerals are easily distinguishable from true asbestos fibers. This is because true cleavage fragments usually have larger diameters than 1 µm. Internal structure of particles larger than this usually shows them to have no internal fibrillar structure. In addition, cleavage fragments of the monoclinic amphiboles show inclined extinction under crossed polars with no compensator. Asbestos fibers usually show extinction at zero degrees or ambiguous extinction if any at all. Morphologically, the larger cleavage fragments are obvious by their blunt or stepped ends showing prismatic habit. Also, they tend to be acicular rather than filiform.

Where the particles are less than 1 µm in diameter and have an aspect ratio greater than or equal to 3:1, it is recommended that the sample be analyzed by SEM or TEM if there is any question whether the fibers are cleavage fragments or asbestiform particles. Care must be taken when analyzing by electron microscopy because the interferences are different from those in light microscopy and may structurally be very similar to asbestos. The classic interference is between anthophyllite and biopyribole or intermediate fiber. Use the same morphological clues for electron microscopy as are used for light microscopy, e.g. fibril splitting, internal longitudinal striation, fraying, curvature, etc.

(1) *Gross examination:*

Examine the sample, preferably in the glass vial. Determine the presence of any obvious fibrous component. Estimate a percentage based on previous experience and current observation. Determine whether any pre-preparation is necessary. Determine the number of phases present. This step may be carried out or augmented by observation at 6 to 40x under a stereo microscope.

(2) *After performing any necessary pre-preparation,* prepare slides of each phase as described above. Two preparations of the same phase in the same index medium can be made side-by-side on the same glass for convenience. Examine with the polarizing stereo microscope. Estimate the percentage of asbestos based on the amount of birefringent fiber present.

(3) *Examine the slides on the phase-polar microscopes* at magnifications of 160 and 400x. Note the morphology of the fibers. Long, thin, very straight fibers with little curvature are indicative of fibers from the amphibole family. Curved, wavy fibers are usually indicative of chrysotile. Estimate the percentage of asbestos on the phase-polar microscope under conditions of crossed polars and a gypsum plate. Fibers smaller than 1.0 µm in thickness must be identified by inference to the presence of larger, identifiable fibers and morphology. If no larger fibers are visible, electron microscopy should be performed. At this point, only a tentative identification can be made. Full identification must be made with dispersion microscopy. Details of the tests are included in the appendices.

(4) *Once fibers have been determined to be present,* they must be identified. Adjust the microscope for dispersion mode and observe the fibers. The microscope has a rotating stage, one polarizing element, and a system for generating dark-field dispersion microscopy (see Section 4.6. of this appendix). Align a fiber with its length parallel to the polarizer and note the color of the Becke lines. Rotate the stage to bring the fiber length perpendicular to the polarizer and note the color. Repeat this process for every fiber or fiber bundle examined. The colors must be consistent with the colors generated by standard asbestos reference materials for a positive identification. In n=1.550, amphiboles will generally show a yellow to straw-yellow color indicating that the fiber indices of refraction are higher than the liquid. If long, thin fibers are noted and the colors are yellow, prepare further slides as above in the suggested matching liquids listed below:

Type of asbestos	Index of refraction
Chrysotile	n=1.550.
Amosite	n=1.670 or 1.680.
Crocidolite	n=1.690.
Anthophyllite	n=1.605 and 1.620.
Tremolite	n=1.605 and 1.620.
Actinolite	n=1.620.

Where more than one liquid is suggested, the first is preferred; however, in some cases this liquid will not give good dispersion color. Take care to avoid interferences in the other liquid; e.g., wollastonite in n=1.620 will give the same colors as tremolite. In n=1.605 wollastonite will appear yellow in all directions. Wollastonite may be determined under crossed polars as it will change from blue to yellow as it'is rotated along its fiber axis by tapping on the cover slip. Asbestos minerals will not change in this way.

Determination of the angle of extinction may, when present, aid in the determination of anthophyllite from tremolite. True asbestos fibers usually have 0° extinction or ambiguous extinction, while cleavage fragments have more definite extinction.

Continue analysis until both preparations have been examined and all present species of asbestos are identified. If there are no fibers present, or there is less than 0.1% present, end the analysis with the minimum number of slides (2).

(5) *Some fibers have a coating on them* which makes dispersion microscopy very difficult or impossible. Becke line analysis or electron microscopy may be performed in those cases. Determine the percentage by light microscopy. TEM analysis tends to overestimate the actual percentage present.

(6) *Percentage determination is an estimate* of occluded area, tempered by gross observation. Gross observation information is used to make sure that the high magnification microscopy does not greatly over-or under-estimate the amount of fiber present. This part of the analysis requires a great deal of experience. Satisfactory models for asbestos content analysis have not yet been developed, although some models based on metallurgical grain-size determination have found some utility. Estimation is more easily handled in situations where the grain sizes visible at about 160x are about the same and the sample is relatively homogeneous.

View all of the area under the cover slip to make the percentage determination. View the fields while moving the stage, paying attention to the clumps of material. These are not usually the best areas to perform dispersion microscopy because of the interference from other materials. But, they are the areas most likely to represent the accurate percentage in the sample. Small amounts of asbestos require slower scanning and more frequent analysis of individual fields.

Report the area occluded by asbestos as the concentration. This estimate does not generally take into consideration the difference in density of the different species present in the sample. For most samples this is adequate. Simulation studies with similar materials must be carried out to apply microvisual estimation for that purpose and is beyond the scope of this procedure.

(7) *Where successive concentrations have been made* by chemical or physical means, the amount reported is the percentage of the material in the "as submitted" or original state. The percentage determined by microscopy is multiplied by the fractions remaining after pre-preparation steps to give the percentage in the original sample. For example:

Step 1. 60% remains after heating at 550 °C for 1 h.

Step 2. 30% of the residue of step 1 remains after dissolution of carbonate in 0.1 m HCl.

Step 3. Microvisual estimation determines that 5% of the sample is chrysotile asbestos.

The reported result is:

R = (Microvisual result in percent) x (Fraction remaining after step 2) x (Fraction remaining of original sample after step 1)

R = (5)x(.30)x(.60)=0.9%

(8) *Report the percent and type* of asbestos present. For samples where asbestos was identified, but is less than 1.0%, report "Asbestos present, less than 1.0%." There must have been at least two observed fibers or fiber bundles in the two preparations to be reported as present. For samples where asbestos was not seen, report as "None Detected."

Auxiliary Information

Because of the subjective nature of asbestos analysis, certain concepts and procedures need to be discussed in more depth. This information will help the analyst understand why some of the procedures are carried out the way they are.

4.1. *Light*

Light is electromagnetic energy. It travels from its source in packets called quanta. It is instructive to consider light as a plane wave. The light has a direction of travel. Perpendicular to this and mutually perpendicular to each other, are two vector components. One is the magnetic vector and the other is the electric vector. We shall only be concerned with the electric vector. In this description, the interaction of the vector and the mineral will describe all the observable phenomena. From a light source such a microscope illuminator, light travels in all different direction from the filament.

In any given direction away from the filament, the electric vector is perpendicular to the direction of travel of a light ray. While perpendicular, its orientation is random about the travel axis. If the electric vectors from all the light rays were lined up by passing the light through a filter that would only let light rays with electric vectors oriented in one direction pass, the light would then be POLARIZED.

Polarized light interacts with matter in the direction of the electric vector. This is the polarization direction. Using this property it is possible to use polarized light to probe different materials and identify them by how they interact with light.

The speed of light in a vacuum is a constant at about $2.99x10^8$ m/s. When light travels in different materials such as air, water, minerals or oil, it does not travel at this speed. It travels slower. This slowing is a function of both the material through which the light is traveling and the wavelength or frequency of the light. In general, the more dense the material, the slower the light travels. Also, generally, the higher the frequency, the slower the light will travel. The ratio of the speed of light in a vacuum to that in a material is called the index of refraction (n). It is usually measured at 589 nm (the sodium D line).

If white light (light containing all the visible wavelengths) travels through a material, rays of longer wavelengths will travel faster than those of shorter wavelengths, this separation is called dispersion. Dispersion is used as an identifier of materials as described in Section 4.6.

4.2. *Material Properties*

Materials are either amorphous or crystalline. The difference between these two descriptions depends on the positions of the atoms in them. The atoms in amorphous materials are randomly arranged with no long range order. An example of an amorphous material is glass. The atoms in crystalline materials, on the other hand, are in regular arrays and have long range order. Most of the atoms can be found in highly predictable locations. Examples of crystalline material are salt, gold, and the asbestos minerals.

It is beyond the scope of this method to describe the different types of crystalline materials that can be found, or the full description of the classes into which they can fall. However, some general crystallography is provided below to give a foundation to the procedures described.

With the exception of anthophyllite, all the asbestos minerals belong to the monoclinic crystal type. The unit cell is the basic repeating unit of the crystal and for monoclinic crystals can be described as having three unequal sides, two 90° angles and one angle not equal to 90°. The orthorhombic group, of which anthophyllite is a member has three unequal sides and three 90° angles. The unequal sides are a consequence of the complexity of fitting the different atoms into the unit cell. Although the atoms are in a regular array, that array is not symmetrical in all directions. There is long range order in the three major directions of the crystal. However, the order is different in each of the three directions. This has the effect that the index of refraction is different in each of the three directions. Using polarized light, we can investigate the index of refraction in each of the directions and identify the mineral or material under investigation. The indices α, β, and γ are used to identify the lowest, middle, and highest index of refraction respectively. The x direction, associated with α is called the fast axis. Conversely, the z direction is associated with γ and is the slow direction. Crocidolite has α along the fiber length making it "length-fast". The remainder of the asbestos minerals have the γ axis along the fiber length. They are called "length-slow". This orientation to fiber length is used to aid in the identification of asbestos.

4.3. *Polarized Light Technique*

Polarized light microscopy as described in this section uses the phase-polar microscope described in Section 3.2. A phase contrast microscope is fitted with two polarizing elements, one below and one above the sample. The polarizers have their polarization directions at right angles to each other. Depending on the tests performed, there may be a compensator between these two polarizing elements. A compensator is a piece of mineral with known properties that "compensates" for some deficiency in the optical train. Light emerging from a polarizing element has its electric vector pointing in the polarization direction of the element. The light will not be subsequently transmitted through a second element set at a right angle to the first element. Unless the light is altered as it passes from one element to the other, there is no transmission of light.

4.4. *Angle of Extinction*

Crystals which have different crystal regularity in two or three main directions are said to be anisotropic. They have a different index of refraction in each of the main directions. When such a crystal is inserted between the crossed polars, the field of view is no longer dark but shows the crystal in color. The color depends on the properties of the crystal. The light acts as if it travels through the crystal along the optical axes. If a crystal optical axis were lined up along one of the polarizing directions (either the polarizer or the analyzer) the light would appear to travel only in that direction, and it would blink out or go dark. The difference in degrees between the fiber direction and the angle at which it blinks out is called the angle of extinction. When this angle can be measured, it is useful in identifying the mineral. The procedure for measuring the angle of extinction is to first identify the polarization direction in the microscope. A commercial alignment slide can be used to establish the polarization directions or use anthophyllite or another suitable mineral. This mineral has a zero degree angle of extinction and will go dark to extinction as it aligns with the polarization directions. When a fiber of anthophyllite has gone to extinction, align the eyepiece reticle or graticule with the fiber so that there is a visual cue as to the direction of polarization in the field of view. Tape or otherwise secure the eyepiece in this position so it will not shift.

After the polarization direction has been identified in the field of view, move the particle of interest to the center of the field of view and align it with the polarization direction. For fibers, align the fiber along this direction. Note the angular reading of the rotating stage. Looking at the particle, rotate the stage until the fiber goes dark or "blinks out". Again note the reading of the stage. The difference in the first reading and the second is an angle of extinction.

The angle measured may vary as the orientation of the fiber changes about its long axis. Tables of mineralogical data usually report the maximum angle of extinction. Asbestos forming minerals, when they exhibit an angle of extinction, usually do show an angle of extinction close to the reported maximum, or as appropriate depending on the substitution chemistry.

4.5. *Crossed Polars with Compensator*

When the optical axes of a crystal are not lined up along one of the polarizing directions (either the polarizer or the analyzer) part of the light travels along one axis and part travels along the other visible axis. This is characteristic of birefringent materials.

The color depends on the difference of the two visible indices of refraction and the thickness of the crystal. The maximum difference available is the difference between the α and the γ axes. This maximum difference is usually tabulated as the birefringence of the crystal.

For this test, align the fiber at 45° to the polarization directions in order to maximize the contribution to each of the optical axes. The colors seen are called retardation colors. They arise from the recombination of light which has traveled through the two separate directions of the crystal. One of the rays is retarded behind the other since the light in that direction travels slower. On recombination, some of the colors which make up white light are enhanced by constructive interference and some are suppressed by destructive interference. The result is a color dependent on the difference between the indices and the thickness of the crystal. The proper colors, thicknesses, and retardations are shown on a Michel-Levy chart. The three items, retardation, thickness and birefringence are related by the following relationship:

$R=t(n\gamma-n\alpha)$

R=retardation, t=crystal thickness in µm, and

$n\alpha,\gamma$=indices of refraction.

Examination of the equation for asbestos minerals reveals that the visible colors for almost all common asbestos minerals and fiber sizes are shades of gray and black. The eye is relatively poor at discriminating different shades of gray. It is very good at discriminating different colors. In order to compensate for the low retardation, a compensator is added to the light train between the polarization elements. The compensator used for this test is a gypsum plate of known thickness and birefringence. Such a compensator when oriented at 45° to the polarizer direction, provides a retardation of 530 nm of the 530 nm wavelength color. This enhances the red color and gives the background a characteristic red to red-magenta color. If this "full-wave" compensator is in place when the asbestos preparation is inserted into the light train, the colors seen on the fibers are quite different. Gypsum, like asbestos has a fast axis and a slow axis. When a fiber is aligned with its fast axis in the same direction as the fast axis of the gypsum plate, the ray vibrating in the slow direction is retarded by both the asbestos and the gypsum. This results in a higher retardation than would be present for either of the two minerals. The color seen is a second order blue. When the fiber is rotated 90° using the rotating stage, the slow direction of the fiber is now aligned with the fast direction of the gypsum and the fast direction of the fiber is aligned with the slow direction of the gypsum. Thus, one ray vibrates faster in the fast direction of the gypsum, and slower in the slow direction of the fiber; the other ray will vibrate slower in the slow direction of the gypsum and faster in the fast direction of the fiber. In this case, the effect is subtractive and the color seen is a first order yellow. As long as the fiber thickness does not add appreciably to the color, the same basic colors will be seen for all asbestos types except crocidolite. In crocidolite the colors will be weaker, may be in the opposite directions, and will be altered by the blue absorption color natural to crocidolite. Hundreds of other materials will give the same colors as asbestos, and therefore, this test is not definitive for asbestos. The test is useful in discriminating against fiberglass or other amorphous fibers such as some synthetic fibers. Certain synthetic fibers will show retardation colors different than asbestos; however, there are some forms of polyethylene and aramid which will show morphology and retardation colors similar to asbestos minerals. This test must be supplemented with a positive identification test when birefringent fibers are present which can not be excluded by morphology. This test is relatively ineffective for use on fibers less than 1 μm in diameter. For positive confirmation TEM or SEM should be used if no larger bundles or fibers are visible.

4.6. *Dispersion Staining*

Dispersion microscopy or dispersion staining is the method of choice for the identification of asbestos in bulk materials. Becke line analysis is used by some laboratories and yields the same results as does dispersion staining for asbestos and can be used in lieu of dispersion staining. Dispersion staining is performed on the same platform as the phase-polar analysis with the analyzer and compensator removed. One polarizing element remains to define the direction of the light so that the different indices of refraction of the fibers may be separately determined. Dispersion microscopy is a dark-field technique when used for asbestos. Particles are imaged with scattered light. Light which is unscattered is blocked from reaching the eye either by the back field image mask in a McCrone objective or a back field image mask in the phase condenser. The most convenient method is to use the rotating phase condenser to move an oversized phase ring into place. The ideal size for this ring is for the central disk to be just larger than the objective entry aperture as viewed in the back focal plane. The larger the disk, the less scattered light reaches the eye. This will have the effect of diminishing the intensity of dispersion color and will shift the actual color seen. The colors seen vary even on microscopes from the same manufacturer. This is due to the different bands of wavelength exclusion by different mask sizes. The mask may either reside in the condenser or in the objective back focal plane. It is imperative that the analyst determine by experimentation with asbestos standards what the appropriate colors should be for each asbestos type. The colors depend also on the temperature of the preparation and the exact chemistry of the asbestos. Therefore, some slight differences from the standards should be allowed. This is not a serious problem for commercial asbestos uses. This technique is used for identification of the indices of refraction for fibers by recognition of color. There is no direct numerical readout of the index of refraction. Correlation of color to actual index of refraction is possible by referral to published conversion tables. This is not necessary for the analysis of asbestos. Recognition of appropriate colors along with the proper morphology are deemed sufficient to identify the commercial asbestos minerals. Other techniques including SEM, TEM, and XRD may be required to provide additional information in order to identify other types of asbestos.

Make a preparation in the suspected matching high dispersion oil, e.g., n=1.550 for chrysotile. Perform the preliminary tests to determine whether the fibers are birefringent or not. Take note of the morphological character. Wavy fibers are indicative of chrysotile while long, straight, thin, frayed fibers are indicative of amphibole asbestos. This can aid in the selection of the appropriate matching oil. The microscope is set up and the polarization direction is noted as in Section 4.4. Align a fiber with the polarization direction. Note the color. This is the color parallel to the polarizer. Then rotate the fiber rotating the stage 90° so that the polarization direction is across the fiber. This is the perpendicular position. Again note the color. Both colors must be consistent with standard asbestos minerals in the correct direction for a positive identification of asbestos. If only one of the colors is correct while the other is not, the identification is not positive. If the colors in both directions are bluish-white, the analyst has chosen a matching index oil which is higher than the correct matching oil, e.g. the analyst has used n=1.620 where chrysotile is present. The next lower oil (Section 3.5.) should be used to prepare another specimen. If the color in both directions is yellow-white to straw-yellow-white, this indicates that the index of the oil is lower than the index of the fiber, e.g. the preparation is in n=1.550 while anthophyllite is present. Select the next higher oil (Section 3.5.) and prepare another slide. Continue in this fashion until a positive identification of all asbestos species present has been made or all possible asbestos species have been ruled out by negative results in this test. Certain plant fibers can have similar dispersion colors as asbestos. Take care to note and evaluate the morphology of the fibers or remove the plant fibers in pre-preparation. Coating material on the fibers such as carbonate or vinyl may destroy the dispersion color. Usually, there will be some outcropping of fiber which will show the colors sufficient for identification. When this is not the case, treat the sample as described in Section 3.3. and then perform dispersion staining. Some samples will yield to Becke line analysis if they are coated or electron microscopy can be used for identification.

5. References

5.1. *Crane, D.T.,* Asbestos in Air, OSHA method ID160, Revised November 1992.

5.2. *Ford, W.E.,* Dana's Textbook of Mineralogy; Fourth Ed.; John Wiley and Son, New York, 1950, p. vii.

5.3. *Selikoff, I.J., Lee, D.H.K.,* Asbestos and Disease, Academic Press, New York, 1978, pp. 3,20.

5.4. *Women Inspectors of Factories.* Annual Report for 1898, H.M. Statistical Office, London, p. 170 (1898).

5.5. *Selikoff,.I.J.,* Lee, D.H.K., Asbestos and Disease, Academic Press, New York, 1978, pp. 26,30.

5.6. *Campbell, W.J., et al,* Selected Silicate Minerals and Their Asbestiform Varieties, United States Department of the Interior, Bureau of Mines, Information Circular 8751, 1977.

5.7. *Asbestos, Code of Federal Regulations,* 29 CFR 1910.1001 and 29 CFR 1926.58.

5.8. *National Emission Standards* for Hazardous Air Pollutants; Asbestos NESHAP Revision, Federal Register, Vol. 55, No. 224, 20 November 1990, p. 48410.

5.9. *Ross, M.* The Asbestos Minerals: Definitions, Description, Modes of Formation, Physical and Chemical Properties and Health Risk to the Mining Community, Nation Bureau of Standards Special Publication, Washington, DC, 1977.

5.10. *Lilis, R.,* Fibrous Zeolites and Endemic Mesothelioma in Cappadocia, Turkey, J. Occ Medicine, 1981, 23,(8),548-550.

5.11. *Occupational Exposure to Asbestos — 1972,* U.S. Department of Health Education and Welfare, Public Health Service, Center for Disease Control, National Institute for Occupational Safety and Health, HSM-72- 10267.

5.12. *Campbell,W.J., et al,* Relationship of Mineral Habit to Size Characteristics for Tremolite Fragments and Fibers, United States Department of the Interior, Bureau of Mines, Information Circular 8367, 1979.

5.13. *Mefford, D.,* DCM Laboratory, Denver, private communication, July 1987.

5.14. *Deer, W.A., Howie, R.A., Zussman, J.,* Rock Forming Minerals, Longman, Thetford, UK, 1974.

5.15. *Kerr, P.F.,* Optical Mineralogy; Third Ed. McGraw-Hill, New York, 1959.

5.16. *Veblen, D.R. (Ed.),* Amphiboles and Other Hydrous Pyriboles —Mineralogy, Reviews in Mineralogy, Vol 9A, Michigan, 1982, pp 1-102.

5.17. *Dixon, W.C.,* Applications of Optical Microscopy in the Analysis of Asbestos and Quartz, ACS Symposium Series, No. 120, Analytical Techniques in Occupational Health Chemistry, 1979.

5.18. *Polarized Light Microscopy,* McCrone Research Institute, Chicago, 1976.

5.19. *Asbestos Identification,* McCrone Research Institute, G & G printers, Chicago, 1987.

5.20. *McCrone, W.C.,* Calculation of Refractive Indices from Dispersion Staining Data, The Microscope, No 37, Chicago, 1989.

5.21. *Levadie, B. (Ed.),* Asbestos and Other Health Related Silicates, ASTM Technical Publication 834, ASTM, Philadelphia 1982.

5.22. *Steel, E. and Wylie, A., Riordan, P.H. (Ed.),* Mineralogical Characteristics of Asbestos, Geology of Asbestos Deposits, pp. 93-101, SME-AIME, 1981.

5.23. *Zussman, J.,* The Mineralogy of Asbestos, Asbestos: Properties, Applications and Hazards, pp. 45-67 Wiley, 1979.

[51 FR 22756, June 20, 1986, as amended at 52 FR 17755, May 12, 1987; 53 FR 27346, July 20, 1988; 55 FR 3732, Feb. 5, 1990; 57 FR 24331, June 8, 1992. Redesignated at 59 FR 41131, Aug. 10, 1994 and further amended at 59 FR 41144, 41150, 41152, and 41153, Aug. 10, 1994; 60 FR 33345, June 28, 1995; 60 FR 33995, June 29, 1995; 60 FR 36044, July 13, 1995; 60 FR 50412, 50413, Sept. 29, 1995; 61 FR 5510, Feb. 13, 1996; 61 FR 43458, Aug. 23, 1996; 63 FR 1298, Jan. 8, 1998; 63 FR 20099, Apr. 23, 1998; 63 FR 35138, June 29, 1998; 70 FR 1143, Jan. 5, 2005; 71 FR 16674, 16675, Apr. 3, 2006; 71 FR 50191, Aug. 24, 2006]

§1926.1102

Coal tar pitch volatiles; interpretation of term

Note: The requirements applicable to construction work under this section are identical to those set forth at Sec. 1910.1002 of this chapter. (Editor's Note: See 1910 Supplement.)

[61 FR 31433, June 20, 1996]

§1926.1103

13 carcinogens (4-Nitrobiphenyl, etc.)

Note: The requirements applicable to construction work under this section are identical to those set forth at Sec. 1910.1003 of this chapter. (Editor's Note: See 1910 Supplement.)

[61 FR 31433, June 20, 1996]

§1926.1104

alpha-Naphthylamine

Note: The requirements applicable to construction work under this section are identical to those set forth at Sec. 1910.1003 of this chapter. (Editor's Note: See 1910 Supplement.)

[61 FR 31433, June 20, 1996]

§1926.1105

[Reserved]

§1926.1106

Methyl chloromethyl ether

Note: The requirements applicable to construction work under this section are identical to those set forth at Sec. 1910.1003 of this chapter. (Editor's Note: See 1910 Supplement.)

[61 FR 31433, June 20, 1996]

§1926.1107

3,3'-Dichlorobenzidiene (and its salts)

Note: The requirements applicable to construction work under this section are identical to those set forth at Sec. 1910.1003 of this chapter. (Editor's Note: See 1910 Supplement.)

[61 FR 31433, June 20, 1996]

§1926.1108

bis-Chloromethyl ether

Note: The requirements applicable to construction work under this section are identical to those set forth at Sec. 1910.1003 of this chapter. (Editor's Note: See 1910 Supplement.)

[61 FR 31433, June 20, 1996]

§1926.1109

beta-Naphthylamine

Note: The requirements applicable to construction work under this section are identical to those set forth at Sec. 1910.1003 of this chapter. (Editor's Note: See 1910 Supplement.)

[61 FR 31433, June 20, 1996]

§1926.1110

Benzidine

Note: The requirements applicable to construction work under this section are identical to those set forth at Sec. 1910.1003 of this chapter. (Editor's Note: See 1910 Supplement.)

[61 FR 31433, June 20, 1996]

§1926.1111

4-Aminodiphenyl

Note: The requirements applicable to construction work under this section are identical to those set forth at Sec. 1910.1003 of this chapter. (Editor's Note: See 1910 Supplement.)

[61 FR 31433, June 20, 1996]

§1926.1112

Ethyleneimine

Note: The requirements applicable to construction work under this section are identical to those set forth at Sec. 1910.1003 of this chapter. (Editor's Note: See 1910 Supplement.)

[61 FR 31433, June 20, 1996]

§1926.1113

beta-Propiolactone

Note: The requirements applicable to construction work under this section are identical to those set forth at Sec. 1910.1003 of this chapter. (Editor's Note: See 1910 Supplement.)

[61 FR 31433, June 20, 1996]

§1926.1114

2-Acetylaminofluorene

Note: The requirements applicable to construction work under this section are identical to those set forth at Sec. 1910.1003 of this chapter. (Editor's Note: See 1910 Supplement.)

[61 FR 31433, June 20, 1996]

§1926.1115

4-Dimethylaminoazobenzene

Note: The requirements applicable to construction work under this section are identical to those set forth at Sec. 1910.1003 of this chapter. (Editor's Note: See 1910 Supplement.)

[61 FR 31433, June 20, 1996]

§1926.1116

N-Nitrosodimethylamine

Note: The requirements applicable to construction work under this section are identical to those set forth at Sec. 1910.1003 of this chapter. (Editor's Note: See 1910 Supplement.)

[61 FR 31433, June 20, 1996]

§1926.1117

Vinyl chloride

Note: The requirements applicable to construction work under this section are identical to those set forth at Sec. 1910.1017 of this chapter. (Editor's Note: See 1910 Supplement.)

[61 FR 31433, June 20, 1996]

§1926.1118

Inorganic arsenic

Note: The requirements applicable to construction work under this section are identical to those set forth at Sec. 1910.1018 of this chapter. (Editor's Note: See 1910 Supplement.)

[61 FR 31433, June 20, 1996]

§1926.1126

Chromium (VI)

(a) **Scope.**[§1926.1126(a)]

(1) *This standard applies to* occupational exposures to chromium (VI) in all forms and compounds in construction, except: [§1926.1126(a)(1)]

(2) *Exposures that occur in the application* of pesticides regulated by the Environmental Protection Agency or another Federal government agency (e.g., the treatment of wood with preservatives);[§1926.1126(a)(2)]

(3) *Exposures to portland cement; or*[§1926.1126(a)(3)]

(4) *Where the employer has objective data* demonstrating that a material containing chromium or a specific process, operation, or activity involving chromium cannot release dusts, fumes, or mists of chromium (VI) in concentrations at or above 0.5 μgm/m^3 as an 8- hour time-weighted average (TWA) under any expected conditions of use.[§1926.1126(a)(4)]

(b) **Definitions.** For the purposes of this section the following definitions apply:[§1926.1126(b)]

Action level means a concentration of airborne chromium (VI) of 2.5 micrograms per cubic meter of air (2.5 μgm/m^3) calculated as an 8-hour time-weighted average (TWA).[§1926.1126(b)]

Assistant Secretary means the Assistant Secretary of Labor for Occupational Safety and Health, U.S. Department of Labor, or designee.[§1926.1126(b)]

Chromium (VI) [hexavalent chromium or Cr(VI)] means chromium with a valence of positive six, in any form and in any compound. [§1926.1126(b)]

Director means the Director of the National Institute for Occupational Safety and Health (NIOSH), U.S. Department of Health and Human Services, or designee.[§1926.1126(b)]

Emergency means any occurrence that results, or is likely to result, in an uncontrolled release of chromium (VI). If an incidental release of chromium (VI) can be controlled at the time of release by employees in the immediate release area, or by maintenance personnel, it is not an emergency.[§1926.1126(b)]

Employee exposure means the exposure to airborne chromium (VI) that would occur if the employee were not using a respirator. [§1926.1126(b)]

High-efficiency particulate air [HEPA] filter means a filter that is at least 99.97 percent efficient in removing mono-dispersed particles of 0.3 micrometers in diameter or larger.[§1926.1126(b)]

Historical monitoring data means data from chromium (VI) monitoring conducted prior to May 30, 2006, obtained during work operations conducted under workplace conditions closely resembling the processes, types of material, control methods, work practices, and environmental conditions in the employer's current operations.[§1926.1126(b)]

Objective data means information such as air monitoring data from industry-wide surveys or calculations based on the composition or chemical and physical properties of a substance demonstrating the employee exposure to chromium (VI) associated with a particular product or material or a specific process, operation, or activity. The data must reflect workplace conditions closely resembling the processes, types of material, control methods, work practices, and environmental conditions in the employer's current operations. [§1926.1126(b)]

Physician or other licensed health care professional [PLHCP] is an individual whose legally permitted scope of practice (i.e., license, registration, or certification) allows him or her to independently provide or be delegated the responsibility to provide some or all of the particular health care services required by paragraph (i) of this section.[§1926.1126(b)]

This section means this Sec. 1926.1126 chromium (VI) standard. [§1926.1126(b)]

(c) **Permissible exposure limit (PEL).** The employer shall ensure that no employee is exposed to an airborne concentration of chromium (VI) in excess of 5 micrograms per cubic meter of air (5 μgm/m^3), calculated as an 8-hour time-weighted average (TWA). [§1926.1126(c)]

(d) **Exposure determination**[§1926.1126(d)]

(1) *General.* Each employer who has a workplace or work operation covered by this section shall determine the 8-hour TWA exposure for each employee exposed to chromium (VI). This determination shall be made in accordance with either paragraph (d)(2) or paragraph (d)(3) of this section.[§1926.1126(d)(1)]

(2) *Scheduled monitoring option.*[§1926.1126(d)(2)]

(i) *The employer shall perform initial monitoring* to determine the 8-hour TWA exposure for each employee on the basis of a sufficient number of personal breathing zone air samples to accurately characterize full shift exposure on each shift, for each job classification, in each work area. Where an employer does representative sampling instead of sampling all employees in order to meet this requirement, the employer shall sample the employee(s) expected to have the highest chromium (VI) exposures.[§1926.1126(d)(2)(i)]

(ii) *If initial monitoring indicates* that employee exposures are below the action level, the employer may discontinue monitoring for those employees whose exposures are represented by such monitoring.[§1926.1126(d)(2)(ii)]

(iii) *If monitoring reveals* employee exposures to be at or above the action level, the employer shall perform periodic monitoring at least every six months.[§1926.1126(d)(2)(iii)]

(iv) *If monitoring reveals* employee exposures to be above the PEL, the employer shall perform periodic monitoring at least every three months.[§1926.1126(d)(2)(iv)]

(v) *If periodic monitoring indicates* that employee exposures are below the action level, and the result is confirmed by the result of another monitoring taken at least seven days later, the employer may discontinue the monitoring for those employees whose exposures are represented by such monitoring.[§1926.1126(d)(2)(v)]

(vi) *The employer shall perform additional monitoring* when there has been any change in the production process, raw materials, equipment, personnel, work practices, or control methods that may result in new or additional exposures to chromium (VI), or when the employer has any reason to believe that new or additional exposures have occurred. [§1926.1126(d)(2)(vi)]

(3) *Performance-oriented option.* The employer shall determine the 8- hour TWA exposure for each employee on the basis of any combination of air monitoring data, historical monitoring data, or objective data sufficient to accurately characterize employee exposure to chromium (VI).[§1926.1126(d)(3)]

(4) *Employee notification of determination results.*[§1926.1126(d)(4)]

(i) *Within 5 work* days after making an exposure determination in accordance with paragraph (d)(2) or paragraph (d)(3) of this section, the employer shall individually notify each affected employee in writing of the results of that determination or post the results in an appropriate location accessible to all affected employees.[§1926.1126(d)(4)(i)]

(ii) *Whenever the exposure determination indicates* that employee exposure is above the PEL, the employer shall describe in the written notification the corrective action being taken to reduce employee exposure to or below the PEL. [§1926.1126(d)(4)(ii)]

(5) *Accuracy of measurement.* Where air monitoring is performed to comply with the requirements of this section, the employer shall use a method of monitoring and analysis that can measure chromium (VI) to within an accuracy of plus or minus 25 percent (± 25%) and can produce accurate measurements to within a statistical confidence level of 95 percent for airborne concentrations at or above the action level.[§1926.1126(d)(5)]

(6) *Observation of monitoring.*[§1926.1126(d)(6)]

(i) *Where air monitoring is performed* to comply with the requirements of this section, the employer shall provide affected employees or their designated representatives an opportunity to observe any monitoring of employee exposure to chromium (VI).[§1926.1126(d)(6)(i)]

(ii) *When observation of monitoring requires entry* into an area where the use of protective clothing or equipment is required, the employer shall provide the observer with clothing and equipment and shall assure that the observer uses such clothing and equipment and complies with all other applicable safety and health procedures.[§1926.1126(d)(6)(ii)]

(e) **Methods of compliance.**[§1926.1126(e)]

(1) *Engineering and work practice controls.*[§1926.1126(e)(1)]

(i) *Except as permitted in paragraph (e)(1)(ii) of this section,* the employer shall use engineering and work practice controls to reduce and maintain employee exposure to chromium (VI) to or below the PEL unless the employer can demonstrate that such controls are not feasible. Wherever feasible engineering and work practice controls are not sufficient to reduce employee exposure to or below the PEL, the employer shall use them to reduce employee exposure to the lowest levels achievable, and shall supplement them by the use of respiratory protection that complies with the requirements of paragraph (f) of this section.[§1926.1126(e)(1)(i)]

(ii) *Where the employer can demonstrate* that a process or task does not result in any employee exposure to chromium (VI) above the PEL for 30 or more days per year (12 consecutive months), the requirement to implement engineering and work practice controls to achieve the PEL does not apply to that process or task.[§1926.1126(e)(1)(ii)]

(2) *Prohibition of rotation.* The employer shall not rotate employees to different jobs to achieve compliance with the PEL. [§1926.1126(e)(2)]

(f) **Respiratory protection**[§1926.1126(f)]

(1) *General.* Where respiratory protection is required by this section, the employer must provide each employee an appropriate respirator that complies with the requirements of this paragraph. Respiratory protection is required during:[§1926.1126(f)(1)]

(i) *Periods necessary to install* or implement feasible engineering and work practice controls;[§1926.1126(f)(1)(i)]

(ii) *Work operations,* such as maintenance and repair activities, for which engineering and work practice controls are not feasible;[§1926.1126(f)(1)(ii)]

(iii) *Work operations for which* an employer has implemented all feasible engineering and work practice controls and such controls are not sufficient to reduce exposures to or below the PEL;[§1926.1126(f)(1)(iii)]

(iv) *Work operations where employees* are exposed above the PEL for fewer than 30 days per year, and the employer has elected not to implement engineering and work practice controls to achieve the PEL; or[§1926.1126(f)(1)(iv)]

(v) *Emergencies.*[§1926.1126(f)(1)(v)]

(2) *Respiratory protection program.* Where respirator use is required by this section, the employer shall institute a respiratory protection program in accordance with 29 CFR 1910.134, which covers each employee required to use a respirator. [§1926.1126(f)(2)]

(g) **Protective work clothing and equipment**[§1926.1126(g)]

(1) *Provision and use.* Where a hazard is present or is likely to be present from skin or eye contact with chromium (VI), the employer shall provide appropriate personal protective clothing and equipment at no cost to employees, and shall ensure that employees use such clothing and equipment.[§1926.1126(g)(1)]

(2) *Removal and storage.*[§1926.1126(g)(2)]

(i) *The employer shall ensure* that employees remove all protective clothing and equipment contaminated with chromium (VI) at the end of the work shift or at the completion of their tasks involving chromium (VI) exposure, whichever comes first.[§1926.1126(g)(2)(i)]

(ii) *The employer shall ensure* that no employee removes chromium (VI)-contaminated protective clothing or equipment from the workplace, except for those employees whose job it is to launder, clean, maintain, or dispose of such clothing or equipment.[§1926.1126(g)(2)(ii)]

(iii) *When contaminated protective clothing* or equipment is removed for laundering, cleaning, maintenance, or disposal, the employer shall ensure that it is stored and transported in sealed, impermeable bags or other closed, impermeable containers.[§1926.1126(g)(2)(iii)]

(iv) *The employer shall ensure that bags* or containers of contaminated protective clothing or equipment that are removed from change rooms for laundering, cleaning, maintenance, or disposal shall be labeled in accordance with the requirements of the Hazard Communication Standard, Sec. 1910.1200.[§1926.1126(g)(2)(iv)]

(3) *Cleaning and replacement.*[§1926.1126(g)(3)]

(i) *The employer shall clean,* launder, repair and replace all protective clothing and equipment required by this section as needed to maintain its effectiveness.[§1926.1126(g)(3)(i)]

(ii) *The employer shall prohibit the removal* of chromium (VI) from protective clothing and equipment by blowing, shaking, or any other means that disperses chromium (VI) into the air or onto an employee's body.[§1926.1126(g)(3)(ii)]

(iii) *The employer shall inform any person* who launders or cleans protective clothing or equipment contaminated with chromium (VI) of the potentially harmful effects of exposure to chromium (VI) and that the clothing and equipment should be laundered or cleaned in a manner that minimizes skin or eye contact with chromium (VI) and effectively prevents the release of airborne chromium (VI) in excess of the PEL. [§1926.1126(g)(3)(iii)]

(h) **Hygiene areas and practices**[§1926.1126(h)]

(1) *General.* Where protective clothing and equipment is required, the employer shall provide change rooms in conformance with 29 CFR 1926.51 Where skin contact with chromium (VI) occurs, the employer shall provide washing facilities in conformance with 29 CFR 1926.51. Eating and drinking areas provided by the employer shall also be in conformance with Sec. 1926.51. [§1926.1126(h)(1)]

(2) *Change rooms.* The employer shall assure that change rooms are equipped with separate storage facilities for protective clothing and equipment and for street clothes, and that these facilities prevent cross-contamination.[§1926.1126(h)(2)]

(3) *Washing facilities.*[§1926.1126(h)(3)]

(i) *The employer shall provide* readily accessible washing facilities capable of removing chromium (VI) from the skin, and shall ensure that affected employees use these facilities when necessary.[§1926.1126(h)(3)(i)]

(ii) *The employer shall ensure* that employees who have skin contact with chromium (VI) wash their hands and faces at the end of the work shift and prior to eating, drinking, smoking, chewing tobacco or gum, applying cosmetics, or using the toilet.[§1926.1126(h)(3)(ii)]

(4) *Eating and drinking areas.*[§1926.1126(h)(4)]

(i) *Whenever the employer allows employees* to consume food or beverages at a worksite where chromium (VI) is present, the employer shall ensure that eating and drinking areas and surfaces are maintained as free as practicable of chromium (VI).[§1926.1126(h)(4)(i)]

(ii) *The employer shall ensure* that employees do not enter eating and drinking areas with protective work clothing or equipment unless surface chromium (VI) has been removed from the clothing and equipment by methods that do not disperse chromium (VI) into the air or onto an employee's body. [§1926.1126(h)(4)(ii)]

(5) *Prohibited activities.* The employer shall ensure that employees do not eat, drink, smoke, chew tobacco or gum, or apply cosmetics in areas where skin or eye contact with chromium (VI) occurs; or carry the products associated with these activities, or store such products in these areas.[§1926.1126(h)(5)]

(i) **Medical surveillance**[§1926.1126(i)]

(1) *General.*[§1926.1126(i)(1)]

(i) *The employer shall make medical surveillance* available at no cost to the employee, and at a reasonable time and place, for all employees:[§1926.1126(i)(1)(i)]

[A] Who are or may be occupationally exposed to chromium (VI) at or above the action level for 30 or more days a year;[§1926.1126(i)(1)(i)[A]]

[B] Experiencing signs or symptoms of the adverse health effects associated with chromium (VI) exposure; or [§1926.1126(i)(1)(i)[B]]

[C] Exposed in an emergency.[§1926.1126(i)(1)(i)[C]]

(ii) *The employer shall assure* that all medical examinations and procedures required by this section are performed by or under the supervision of a PLHCP.[§1926.1126(i)(1)(ii)]

(2) *Frequency.* The employer shall provide a medical examination: [§1926.1126(i)(2)]

(i) *Within 30 days after initial assignment,* unless the employee has received a chromium (VI) related medical examination that meets the requirements of this paragraph within the last twelve months;[§1926.1126(i)(2)(i)]

(ii) *Annually;*[§1926.1126(i)(2)(ii)]

(iii) *Within 30 days* after a PLHCP's written medical opinion recommends an additional examination;[§1926.1126(i)(2)(iii)]

(iv) *Whenever an employee shows signs or symptoms* of the adverse health effects associated with chromium (VI) exposure;[§1926.1126(i)(2)(iv)]

(v) *Within 30 days after exposure* during an emergency which results in an uncontrolled release of chromium (VI); or [§1926.1126(i)(2)(v)]

(vi) *At the termination of employment,* unless the last examination that satisfied the requirements of paragraph (i) of this section was less than six months prior to the date of termination.[§1926.1126(i)(2)(vi)]

(3) *Contents of examination.* A medical examination consists of: [§1926.1126(i)(3)]

(i) *A medical and work history,* with emphasis on: past, present, and anticipated future exposure to chromium (VI); any history of respiratory system dysfunction; any history of asthma, dermatitis, skin ulceration, or nasal septum perforation; and smoking status and history;[§1926.1126(i)(3)(i)]

(ii) *A physical examination* of the skin and respiratory tract; and [§1926.1126(i)(3)(ii)]

(iii) *Any additional tests* deemed appropriate by the examining PLHCP.[§1926.1126(i)(3)(iii)]

(4) *Information provided to the PLHCP.* The employer shall ensure that the examining PLHCP has a copy of this standard, and shall provide the following information:[§1926.1126(i)(4)]

(i) *A description of the affected employee's* former, current, and anticipated duties as they relate to the employee's occupational exposure to chromium (VI);[§1926.1126(i)(4)(i)]

(ii) *The employee's former,* current, and anticipated levels of occupational exposure to chromium (VI);[§1926.1126(i)(4)(ii)]

(iii) *A description of any personal protective equipment* used or to be used by the employee, including when and for how long the employee has used that equipment; and[§1926.1126(i)(4)(iii)]

(iv) *Information from records* of employment-related medical examinations previously provided to the affected employee, currently within the control of the employer.[§1926.1126(i)(4)(iv)]

(5) *PLHCP's written medical opinion.*[§1926.1126(i)(5)]

(i) *The employer shall obtain* a written medical opinion from the PLHCP, within 30 days for each medical examination performed on each employee, which contains:[§1926.1126(i)(5)(i)]

[A] The PLHCP's opinion as to whether the employee has any detected medical condition(s) that would place the employee at increased risk of material impairment to health from further exposure to chromium (VI); [§1926.1126(i)(5)(i)[A]]

[B] Any recommended limitations upon the employee's exposure to chromium (VI) or upon the use of personal protective equipment such as respirators;[§1926.1126(i)(5)(i)[B]]

[C] A statement that the PLHCP has explained to the employee the results of the medical examination, including any medical conditions related to chromium (VI) exposure that require further evaluation or treatment, and any special provisions for use of protective clothing or equipment.[§1926.1126(i)(5)(i)[C]]

(ii) *The PLHCP shall not reveal* to the employer specific findings or diagnoses unrelated to occupational exposure to chromium (VI).[§1926.1126(i)(5)(ii)]

(iii) *The employer shall provide a copy* of the PLHCP's written medical opinion to the examined employee within two weeks after receiving it.[§1926.1126(i)(5)(iii)]

(j) **Communication of chromium (VI) hazards to employees.** [§1926.1126(j)]

(1) *Hazard communication.* The employer shall include chromium (VI) in the program established to comply with the Hazard Communication Standard (HCS) (Sec. 1910.1200). The employer shall ensure that each employee has access to labels on containers of chromium and safety data sheets, and is trained in accordance with the provisions of Sec. 1910.1200 and paragraph (j)(2) of this section. The employer shall provide information on at least the following hazards: Cancer; eye irritation; and skin sensitization.[§1926.1126(j)(1)]

(2) *Employee information and training.*[§1926.1126(j)(2)]

(i) *The employer shall ensure* that each employee can demonstrate knowledge of at least the following:[§1926.1126(j)(2)(i)]

[A] The contents of this section; and[§1926.1126(j)(2)(i)[A]]

[B] The purpose and a description of the medical surveillance program required by paragraph (i) of this section. [§1926.1126(j)(2)(i)[B]]

(ii) *The employer shall make a copy* of this section readily available without cost to all affected employees.[§1926.1126(j)(2)(ii)]

(k) **Recordkeeping**[§1926.1126(k)]

(1) *Air monitoring data.*[§1926.1126(k)(1)]

(i) *The employer shall maintain* an accurate record of all air monitoring conducted to comply with the requirements of this section.[§1926.1126(k)(1)(i)]

(ii) *This record shall include* at least the following information: [§1926.1126(k)(1)(ii)]

[A] The date of measurement for each sample taken; [§1926.1126(k)(1)(ii)[A]]

[B] The operation involving exposure to chromium (VI) that is being monitored;[§1926.1126(k)(1)(ii)[B]]

[C] Sampling and analytical methods used and evidence of their accuracy;[§1926.1126(k)(1)(ii)[C]]

[D] Number, duration, and the results of samples taken; [§1926.1126(k)(1)(ii)[D]]

[E] Type of personal protective equipment, such as respirators worn; and[§1926.1126(k)(1)(ii)[E]]

[F] Name, social security number, and job classification of all employees represented by the monitoring, indicating which employees were actually monitored. [§1926.1126(k)(1)(ii)[F]]

(iii) *The employer shall ensure* that exposure records are maintained and made available in accordance with 29 CFR 1910.1020.[§1926.1126(k)(1)(iii)]

(2) *Historical monitoring data.*[§1926.1126(k)(2)]

(i) *Where the employer has relied* on historical monitoring data to determine exposure to chromium (VI), the employer shall establish and maintain an accurate record of the historical monitoring data relied upon.[§1926.1126(k)(2)(i)]

(ii) *The record shall include* information that reflects the following conditions:[§1926.1126(k)(2)(ii)]

[A] The data were collected using methods that meet the accuracy requirements of paragraph (d)(5) of this section; [§1926.1126(k)(2)(ii)[A]]

[B] The processes and work practices that were in use when the historical monitoring data were obtained are essentially the same as those to be used during the job for which exposure is being determined;[§1926.1126(k)(2)(ii)[B]]

[C] The characteristics of the chromium (VI) containing material being handled when the historical monitoring data were obtained are the same as those on the job for which exposure is being determined;[§1926.1126(k)(2)(ii)[C]]

[D] Environmental conditions prevailing when the historical monitoring data were obtained are the same as those on the job for which exposure is being determined; and [§1926.1126(k)(2)(ii)[D]]

[E] Other data relevant to the operations, materials, processing, or employee exposures covered by the exception. [§1926.1126(k)(2)(ii)[E]]

(iii) *The employer shall ensure* that historical exposure records are maintained and made available in accordance with 29 CFR 1910.1020.[§1926.1126(k)(2)(iii)]

(3) *Objective data.*[§1926.1126(k)(3)]

(i) *The employer shall maintain* an accurate record of all objective data relied upon to comply with the requirements of this section.[§1926.1126(k)(3)(i)]

(ii) *This record shall include* at least the following information: [§1926.1126(k)(3)(ii)]

[A] The chromium containing material in question; [§1926.1126(k)(3)(ii)[A]]

[B] The source of the objective data,[§1926.1126(k)(3)(ii)[B]]

[C] The testing protocol and results of testing, or analysis of the material for the release of chromium (VI); [§1926.1126(k)(3)(ii)[C]]

[D] A description of the process, operation, or activity and how the data support the determination; and [§1926.1126(k)(3)(ii)[D]]

[E] Other data relevant to the process, operation, activity, material, or employee exposures.[§1926.1126(k)(3)(ii)[E]]

(iii) *The employer shall ensure* that objective data are maintained and made available in accordance with 29 CFR 1910.1020.[§1926.1126(k)(3)(iii)]

(4) *Medical surveillance.*[§1926.1126(k)(4)]

(i) *The employer shall establish and maintain* an accurate record for each employee covered by medical surveillance under paragraph (i) of this section.[§1926.1126(k)(4)(i)]

(ii) *The record shall include* the following information about the employee:[§1926.1126(k)(4)(ii)]

[A] Name and social security number;[§1926.1126(k)(4)(ii)[A]]

[B] A copy of the PLHCP's written opinions; [§1926.1126(k)(4)(ii)[B]]

[C] A copy of the information provided to the PLHCP as required by paragraph (i)(4) of this section. [§1926.1126(k)(4)(ii)[C]]

(iii) *The employer shall ensure* that medical records are maintained and made available in accordance with 29 CFR 1910.1020.[§1926.1126(k)(4)(iii)]

(l) **Dates.**[§1926.1126(l)]

(1) *For employers with 20 or more employees,* all obligations of this section, except engineering controls required by paragraph (e) of this section, commence November 27, 2006. [§1926.1126(l)(1)]

(2) *For employers with 19 or fewer employees,* all obligations of this section, except engineering controls required by paragraph (e) of this section, commence May 30, 2007.[§1926.1126(l)(2)]

(3) *For all employers,* engineering controls required by paragraph (e) of this section shall be implemented no later than May 31, 2010.[§1926.1126(l)(3)]

[71 FR 10382, Feb. 28, 2006, as amended at 73 FR 75589, Dec. 12, 2008; 75 FR 12686, Mar. 17, 2010; 77 FR 17895, Mar. 26, 2012]

§1926.1127
Cadmium

(a) **Scope.** This standard applies to all occupational exposures to cadmium and cadmium compounds, in all forms, in all construction work where an employee may potentially be exposed to cadmium. Construction work is defined as work involving construction, alteration and/or repair, including but not limited to the following:[§1926.1127(a)]

(1) *Wrecking, demolition or salvage* of structures where cadmium or materials containing cadmium are present;[§1926.1127(a)(1)]

(2) *Use of cadmium containing-paints* and cutting, brazing, burning, grinding or welding on surfaces that were painted with cadmium-containing paints;[§1926.1127(a)(2)]

(3) *Construction, alteration, repair, maintenance,* or renovation of structures, substrates, or portions thereof, that contain cadmium, or materials containing cadmium;[§1926.1127(a)(3)]

(4) *Cadmium welding;* cutting and welding cadmium-plated steel; brazing or welding with cadmium alloys;[§1926.1127(a)(4)]

(5) *Installation of products containing cadmium;*[§1926.1127(a)(5)]

(6) *Electrical grounding with cadmium welding,* or electrical work using cadmium-coated conduit;[§1926.1127(a)(6)]

(7) *Maintaining or retrofitting cadmium-coated equipment;* [§1926.1127(a)(7)]

(8) *Cadmium contamination/emergency cleanup; and*[§1926.1127(a)(8)]

(9) *Transportation, disposal, storage,* or containment of cadmium or materials containing cadmium on the site or location at which construction activities are performed.[§1926.1127(a)(9)]

(b) **Definitions.**[§1926.1127(b)]

Action level (AL) is defined as an airborne concentration of cadmium of 2.5 micrograms per cubic meter of air (2.5 µg/m^3), calculated as an 8-hour time-weighted average (TWA).[§1926.1127(b)]

Assistant Secretary means the Assistant Secretary of Labor for Occupational Safety and Health, U.S. Department of Labor, or designee. [§1926.1127(b)]

Authorized person means any person authorized by the employer and required by work duties to be present in regulated areas or any person authorized by the OSH Act or regulations issued under it to be in regulated areas.[§1926.1127(b)]

Competent person, in accordance with 29 CFR 1926.32(f), means a person designated by the employer to act on the employer's behalf who is capable of identifying existing and potential cadmium hazards in the workplace and the proper methods to control them in order to protect workers, and has the authority necessary to take prompt corrective measures to eliminate or control such hazards. The duties of a competent person include at least the following: Determining prior to the performance of work whether cadmium is present in the workplace; establishing, where necessary, regulated areas and assuring that access to and from those areas is limited to authorized employees; assuring the adequacy of any employee exposure monitoring required by this standard; assuring that all employees exposed to air cadmium levels above the PEL wear appropriate personal protective equipment and are trained in the use of appropriate methods of exposure control; assuring that proper hygiene facilities are provided and that workers are trained to use those facilities; and assuring that the engineering controls required by this standard are implemented, maintained in proper operating condition, and functioning properly.[§1926.1127(b)]

Director means the Director of the National Institute for Occupational Safety and Health (NIOSH), U.S. Department of Health and Human Services, or designee.[§1926.1127(b)]

Employee exposure and similar language referring to the air cadmium level to which an employee is exposed means the exposure to airborne cadmium that would occur if the employee were not using respiratory protective equipment.[§1926.1127(b)]

Final medical determination is the written medical opinion of the employee's health status by the examining physician under paragraphs (l)(3)-(12) of this section or, if multiple physician review under paragraph (l)(13) of this section or the alternative physician determination under paragraph (l)(14) of this section is invoked, it is the final, written medical finding, recommendation or determination that emerges from that process.[§1926.1127(b)]

High-efficiency Particulate Air [HEPA] filter means a filter capable of trapping and retaining at least 99.97 percent of mono-dispersed particles of 0.3 micrometers in diameter.[§1926.1127(b)]

Regulated area means an area demarcated by the employer where an employee's exposure to airborne concentrations of cadmium exceeds, or can reasonably be expected to exceed the permissible exposure limit (PEL).[§1926.1127(b)]

This section means this cadmium standard.[§1926.1127(b)]

(c) Permissible Exposure Limit (PEL). The employer shall assure that no employee is exposed to an airborne concentration of cadmium in excess of five micrograms per cubic meter of air ($5\ \mu g/m^3$), calculated as an eight-hour time-weighted average exposure (TWA).[§1926.1127(c)]

(d) Exposure Monitoring[§1926.1127(d)]

(1) *General.*[§1926.1127(d)(1)]

(i) *Prior to the performance of any construction work* where employees may be potentially exposed to cadmium, the employer shall establish the applicability of this standard by determining whether cadmium is present in the workplace and whether there is the possibility that employee exposures will be at or above the action level. The employer shall designate a competent person who shall make this determination. Investigation and material testing techniques shall be used, as appropriate, in the determination. Investigation shall include a review of relevant plans, past reports, material safety data sheets, and other available records, and consultations with the property owner and discussions with appropriate individuals and agencies.[§1926.1127(d)(1)(i)]

(ii) *Where cadmium has been determined to be present* in the workplace, and it has been determined that there is a possibility the employee's exposure will be at or above the action level, the competent person shall identify employees potentially exposed to cadmium at or above the action level.[§1926.1127(d)(1)(ii)]

(iii) *Determinations of employee exposure* shall be made from breathing-zone air samples that reflect the monitored employee's regular, daily 8-hour TWA exposure to cadmium. [§1926.1127(d)(1)(iii)]

(iv) *Eight-hour TWA exposures* shall be determined for each employee on the basis of one or more personal breathing-zone air samples reflecting full shift exposure on each shift, for each job classification, in each work area. Where several employees perform the same job tasks, in the same job classification, on the same shift, in the same work area, and the length, duration, and level of cadmium exposures are similar, an employer may sample a representative fraction of the employees instead of all employees in order to meet this requirement. In representative sampling, the employer shall sample the employee(s) expected to have the highest cadmium exposures.[§1926.1127(d)(1)(iv)]

(2) *Specific.*[§1926.1127(d)(2)]

(i) *Initial monitoring.* Except as provided for in paragraph (d)(2)(iii) of this section, where a determination conducted under paragraph (d)(1)(i) of this section shows the possibility of employee exposure to cadmium at or above the action level, the employer shall conduct exposure monitoring as soon as practicable that is representative of the exposure for each employee in the workplace who is or may be exposed to cadmium at or above the action level.[§1926.1127(d)(2)(i)]

(ii) *In addition,* if the employee periodically performs tasks that may expose the employee to a higher concentration of airborne cadmium, the employee shall be monitored while performing those tasks.[§1926.1127(d)(2)(ii)]

(iii) *Where the employer has objective data,* as defined in paragraph (n)(2) of this section, demonstrating that employee exposure to cadmium will not exceed airborne concentrations at or above the action level under the expected conditions of processing, use, or handling, the employer may rely upon such data instead of implementing initial monitoring.[§1926.1127(d)(2)(iii)]

(iv) *Where a determination* conducted under paragraphs (d)(1) or (d)(2) of this section is made that a potentially exposed employee is not exposed to airborne concentrations of cadmium at or above the action level, the employer shall make a written record of such determination. The record shall include at least the monitoring data developed under paragraphs (d)(2)(i)-(iii) of this section, where applicable, and shall also include the date of determination, and the name and social security number of each employee. [§1926.1127(d)(2)(iv)]

(3) *Monitoring frequency (periodic monitoring).*[§1926.1127(d)(3)]

(i) *If the initial monitoring or periodic monitoring* reveals employee exposures to be at or above the action level, the employer shall monitor at a frequency and pattern needed to assure that the monitoring results reflect with reasonable accuracy the employee's typical exposure levels, given the variability in the tasks performed, work practices, and environmental conditions on the job site, and to assure the adequacy of respiratory selection and the effectiveness of engineering and work practice controls.[§1926.1127(d)(3)(i)]

(ii) *If the initial monitoring or the periodic monitoring* indicates that employee exposures are below the action level and that result is confirmed by the results of another monitoring taken at least seven days later, the employer may discontinue the monitoring for those employees whose exposures are represented by such monitoring.[§1926.1127(d)(3)(ii)]

(4) *Additional monitoring.* The employer also shall institute the exposure monitoring required under paragraphs (d)(2)(i) and (d)(3) of this section whenever there has been a change in the raw materials, equipment, personnel, work practices, or finished products that may result in additional employees being exposed to cadmium at or above the action level or in employees already exposed to cadmium at or above the action level being exposed above the PEL, or whenever the employer or competent person has any reason to suspect that any other change might result in such further exposure. [§1926.1127(d)(4)]

(5) *Employee notification of monitoring results.*[§1926.1127(d)(5)]

(i) *The employer must,* as soon as possible but no later than 5 working days after the receipt of the results of any monitoring performed under this section, notify each affected employee of these results either individually in writing or by posting the results in an appropriate location that is accessible to employees. [§1926.1127(d)(5)(i)]

(ii) *Wherever monitoring results indicate* that employee exposure exceeds the PEL, the employer shall include in the written notice a statement that the PEL has been exceeded and a description of the corrective action being taken by the employer to reduce employee exposure to or below the PEL.[§1926.1127(d)(5)(ii)]

(6) *Accuracy of measurement.* The employer shall use a method of monitoring and analysis that has an accuracy of not less than plus or minus 25 percent (± 25%), with a confidence level of 95 percent, for airborne concentrations of cadmium at or above the action level and the permissible exposure limit.[§1926.1127(d)(6)]

(e) Regulated areas[§1926.1127(e)]

(1) *Establishment.* The employer shall establish a regulated area wherever an employee's exposure to airborne concentrations of cadmium is, or can reasonably be expected to be in excess of the permissible exposure limit (PEL).[§1926.1127(e)(1)]

(2) *Demarcation.* Regulated areas shall be demarcated from the rest of the workplace in any manner that adequately establishes and alerts employees of the boundaries of the regulated area, including employees who are or may be incidentally in the regulated areas, and that protects persons outside the area from exposure to airborne concentrations of cadmium in excess of the PEL. [§1926.1127(e)(2)]

(3) *Access.* Access to regulated areas shall be limited to authorized persons.[§1926.1127(e)(3)]

(4) *Provision of respirators.* Each person entering a regulated area shall be supplied with and required to use a respirator, selected in accordance with paragraph (g)(2) of this section.[§1926.1127(e)(4)]

(5) *Prohibited activities.* The employer shall assure that employees do not eat, drink, smoke, chew tobacco or gum, or apply cosmetics in regulated areas, or carry the products associated with any of these activities into regulated areas or store such products in those areas. [§1926.1127(e)(5)]

(f) **Methods of compliance**[§1926.1127(f)]

(1) *Compliance hierarchy.*[§1926.1127(f)(1)]

(i) *Except as specified* in paragraph (f)(1)(ii) of this section, the employer shall implement engineering and work practice controls to reduce and maintain employee exposure to cadmium at or below the PEL, except to the extent that the employer can demonstrate that such controls are not feasible.[§1926.1127(f)(1)(i)]

(ii) *The requirement to implement engineering controls* to achieve the PEL does not apply where the employer demonstrates the following:[§1926.1127(f)(1)(ii)]

[A] The employee is only intermittently exposed; and [§1926.1127(f)(1)(ii)[A]]

[B] The employee is not exposed above the PEL on 30 or more days per year (12 consecutive months).[§1926.1127(f)(1)(ii)[B]]

(iii) *Wherever engineering and work practice controls* are not sufficient to reduce employee exposure to or below the PEL, the employer nonetheless shall implement such controls to reduce exposures to the lowest levels achievable. The employer shall supplement such controls with respiratory protection that complies with the requirements of paragraph (g) of this section and the PEL.[§1926.1127(f)(1)(iii)]

(iv) *The employer shall not use* employee rotation as a method of compliance.[§1926.1127(f)(1)(iv)]

(2) *Specific operations*[§1926.1127(f)(2)]

(i) *Abrasive blasting.* Abrasive blasting on cadmium or cadmium-containing materials shall be conducted in a manner that will provide adequate protection.[§1926.1127(f)(2)(i)]

(ii) *Heating cadmium and cadmium-containing materials.* Welding, cutting, and other forms of heating of cadmium or cadmium-containing materials shall be conducted in accordance with the requirements of 29 CFR 1926.353 and 29 CFR 1926.354, where applicable.[§1926.1127(f)(2)(ii)]

(3) *Prohibitions.*[§1926.1127(f)(3)]

(i) *High speed abrasive disc saws* and similar abrasive power equipment shall not be used for work on cadmium or cadmium-containing materials unless they are equipped with appropriate engineering controls to minimize emissions, if the exposure levels are above the PEL.[§1926.1127(f)(3)(i)]

(ii) *Materials containing cadmium* shall not be applied by spray methods, if exposures are above the PEL, unless employees are protected with supplied-air respirators with full facepiece, hood, helmet, suit, operated in positive pressure mode and measures are instituted to limit overspray and prevent contamination of adjacent areas.[§1926.1127(f)(3)(ii)]

(4) *Mechanical ventilation.*[§1926.1127(f)(4)]

(i) *When ventilation is used to control exposure,* measurements that demonstrate the effectiveness of the system in controlling exposure, such as capture velocity, duct velocity, or static pressure shall be made as necessary to maintain its effectiveness. [§1926.1127(f)(4)(i)]

(ii) *Measurements of the system's effectiveness* in controlling exposure shall be made as necessary within five working days of any change in production, process, or control that might result in a significant increase in employee exposure to cadmium. [§1926.1127(f)(4)(ii)]

(iii) *Recirculation of air.* If air from exhaust ventilation is recirculated into the workplace, the system shall have a high efficiency filter and be monitored to assure effectiveness.[§1926.1127(f)(4)(iii)]

(iv) *Procedures shall be developed* and implemented to minimize employee exposure to cadmium when maintenance of ventilation systems and changing of filters is being conducted. [§1926.1127(f)(4)(iv)]

(5) *Compliance program.*[§1926.1127(f)(5)]

(i) *Where employee exposure to cadmium* exceeds the PEL and the employer is required under paragraph (f)(1) of this section to implement controls to comply with the PEL, prior to the commencement of the job the employer shall establish and implement a written compliance program to reduce employee exposure to or below the PEL. To the extent that engineering and work practice controls cannot reduce exposures to or below the PEL, the employer shall include in the written compliance program the use of appropriate respiratory protection to achieve compliance with the PEL.[§1926.1127(f)(5)(i)]

(ii) *Written compliance programs* shall be reviewed and updated as often and as promptly as necessary to reflect significant changes in the employer's compliance status or significant changes in the lowest air cadmium level that is technologically feasible. [§1926.1127(f)(5)(ii)]

(iii) *A competent person* shall review the comprehensive compliance program initially and after each change.[§1926.1127(f)(5)(iii)]

(iv) *Written compliance programs* shall be provided upon request for examination and copying to the Assistant Secretary, the Director, affected employees, and designated employee representatives. [§1926.1127(f)(5)(iv)]

(g) **Respirator protection**[§1926.1127(g)]

(1) *General.* For employees who use respirators required by this section, the employer must provide each employee an appropriate respirator that complies with the requirements of this paragraph. Respirators must be used during:[§1926.1127(g)(1)]

(i) *Periods necessary to install* or implement feasible engineering and work-practice controls when employee exposures exceed the PEL.[§1926.1127(g)(1)(i)]

(ii) *Maintenance and repair activities,* and brief or intermittent work operations, for which employee exposures exceed the PEL and engineering and work-practice controls are not feasible or are not required.[§1926.1127(g)(1)(ii)]

(iii) *Work operations in the regulated areas* specified in paragraph (e) of this section.[§1926.1127(g)(1)(iii)]

(iv) *Work operations for which the employer* has implemented all feasible engineering and work-practice controls, and such controls are not sufficient to reduce employee exposures to or below the PEL.[§1926.1127(g)(1)(iv)]

(v) *Work operations for which an employee,* who is exposed to cadmium at or above the action level, requests a respirator. [§1926.1127(g)(1)(v)]

(vi) *Work operations for which engineering controls* are not required by paragraph (f)(1)(ii) of this section to reduce employee exposures that exceed the PEL.[§1926.1127(g)(1)(vi)]

(vii) *Emergencies.*[§1926.1127(g)(1)(vii)]

(2) *Respirator program.*[§1926.1127(g)(2)]

(i) *The employer must implement* a respiratory protection program in accordance with 29 CFR 1910.134 (b) through (d) (except (d)(1)(iii)), and (f) through (m), which covers each employee required by this section to use a respirator.[§1926.1127(g)(2)(i)]

(ii) *If an employee exhibits breathing difficulty* during fit testing or respirator use, the employer must provide the employee with a medical examination in accordance with paragraph (l)(6)(ii) of this section to determine if the employee can use a respirator while performing the required duties.[§1926.1127(g)(2)(ii)]

(iii) *No employee must use a respirator when,* based on their most recent medical examination, the examining physician determines that the employee will be unable to continue to function normally while using a respirator. If the physician determines the employee must be limited in, or removed from, their current job because of the employee's inability to use a respirator, the job limitation or removal must be conducted in accordance with paragraphs (l) (11) and (12) of this section.[§1926.1127(g)(2)(iii)]

(3) *Respirator selection.*[§1926.1127(g)(3)]

(i) *Employers must:*[§1926.1127(g)(3)(i)]

[A] Select, and provide to employees, the appropriate respirators specified in paragraph (d)(3)(i)(A) of 29 CFR 1910.134. [§1926.1127(g)(3)(i)[A]]

[B] Provide employees with full facepiece respirators when they experience eye irritation.[§1926.1127(g)(3)(i)[B]]

[C] Provide HEPA filters for powered and non-powered air-purifying respirators.[§1926.1127(g)(3)(i)[C]]

(ii) *The employer must provide* a powered air-purifying respirator instead of a negative-pressure respirator when an employee entitled to a respirator chooses to use this type of respirator and such a respirator will provide adequate protection to the employee. [§1926.1127(g)(3)(ii)]

(h) **Emergency situations.** The employer shall develop and implement a written plan for dealing with emergency situations involving substantial releases of airborne cadmium. The plan shall include provisions for the use of appropriate respirators and personal protective equipment. In addition, employees not essential to correcting the emergency situation shall be restricted from the area and normal operations halted in that area until the emergency is abated.[§1926.1127(h)]

(i) **Protective work clothing and equipment**[§1926.1127(i)]

(1) *Provision and use.* If an employee is exposed to airborne cadmium above the PEL or where skin or eye irritation is associated with cadmium exposure at any level, the employer shall provide at no cost to the employee, and assure that the employee uses, appropriate protective work clothing and equipment that prevents contamination of the employee and the employee's garments. Protective work clothing and equipment includes, but is not limited to:[§1926.1127(i)(1)]

(i) *Coveralls or similar full-body work clothing;*[§1926.1127(i)(1)(i)]

(ii) *Gloves, head coverings, and boots or foot coverings; and* [§1926.1127(i)(1)(ii)]

(iii) *Face shields, vented goggles,* or other appropriate protective equipment that complies with 29 CFR 1910.133.[§1926.1127(i)(1)(iii)]

(2) *Removal and storage.*[§1926.1127(i)(2)]

(i) *The employer shall assure* that employees remove all protective clothing and equipment contaminated with cadmium at the completion of the work shift and do so only in change rooms provided in accordance with paragraph (j)(1) of this section.[§1926.1127(i)(2)(i)]

(ii) *The employer shall assure* that no employee takes cadmium-contaminated protective clothing or equipment from the workplace, except for employees authorized to do so for purposes of laundering, cleaning, maintaining, or disposing of cadmium-contaminated protective clothing and equipment at an appropriate location or facility away from the workplace.[§1926.1127(i)(2)(ii)]

(iii) *The employer shall assure* that contaminated protective clothing and equipment, when removed for laundering, cleaning, maintenance, or disposal, is placed and stored in sealed, impermeable bags or other closed, impermeable containers that are designed to prevent dispersion of cadmium dust.[§1926.1127(i)(2)(iii)]

(iv) *The employer shall ensure that containers* of contaminated protective clothing and equipment that are to be taken out of the change rooms or the workplace for laundering, cleaning, maintenance or disposal shall bear labels in accordance with paragraph (m)(3)(ii) of this section.[§1926.1127(i)(2)(iv)]

(3) *Cleaning, replacement, and disposal.*[§1926.1127(i)(3)]

(i) *The employer shall provide* the protective clothing and equipment required by paragraph (i)(1) of this section in a clean and dry condition as often as necessary to maintain its effectiveness, but in any event at least weekly. The employer is responsible for cleaning and laundering the protective clothing and equipment required by this paragraph to maintain its effectiveness and is also responsible for disposing of such clothing and equipment. [§1926.1127(i)(3)(i)]

(ii) *The employer also is responsible for repairing or replacing* required protective clothing and equipment as needed to maintain its effectiveness. When rips or tears are detected while an employee is working they shall be immediately mended, or the worksuit shall be immediately replaced.[§1926.1127(i)(3)(ii)]

(iii) *The employer shall prohibit* the removal of cadmium from protective clothing and equipment by blowing, shaking, or any other means that disperses cadmium into the air.[§1926.1127(i)(3)(iii)]

(iv) *The employer shall assure* that any laundering of contaminated clothing or cleaning of contaminated equipment in the workplace is done in a manner that prevents the release of airborne cadmium in excess of the permissible exposure limit prescribed in paragraph (c) of this section.[§1926.1127(i)(3)(iv)]

(v) *The employer shall inform any person* who launders or cleans protective clothing or equipment contaminated with cadmium of the potentially harmful effects of exposure to cadmium, and that the clothing and equipment should be laundered or cleaned in a manner to effectively prevent the release of airborne cadmium in excess of the PEL.[§1926.1127(i)(3)(v)]

(j) **Hygiene areas and practices**[§1926.1127(j)]

(1) *General.* For employees whose airborne exposure to cadmium is above the PEL, the employer shall provide clean change rooms, handwashing facilities, showers, and lunchroom facilities that comply with 29 CFR 1926.51.[§1926.1127(j)(1)]

(2) *Change rooms.* The employer shall assure that change rooms are equipped with separate storage facilities for street clothes and for protective clothing and equipment, which are designed to prevent dispersion of cadmium and contamination of the employee's street clothes.[§1926.1127(j)(2)]

(3) *Showers and handwashing facilities.*[§1926.1127(j)(3)]

(i) *The employer shall assure* that employees whose airborne exposure to cadmium is above the PEL shower during the end of the work shift.[§1926.1127(j)(3)(i)]

(ii) *The employer shall assure* that employees who are exposed to cadmium above the PEL wash their hands and faces prior to eating, drinking, smoking, chewing tobacco or gum, or applying cosmetics.[§1926.1127(j)(3)(ii)]

(4) *Lunchroom facilities.*[§1926.1127(j)(4)]

(i) *The employer shall assure* that the lunchroom facilities are readily accessible to employees, that tables for eating are maintained free of cadmium, and that no employee in a lunchroom facility is exposed at any time to cadmium at or above a concentration of 2.5 $\mu g/m^3$.[§1926.1127(j)(4)(i)]

(ii) *The employer shall assure* that employees do not enter lunchroom facilities with protective work clothing or equipment unless surface cadmium has been removed from the clothing and equipment by HEPA vacuuming or some other method that removes cadmium dust without dispersing it.[§1926.1127(j)(4)(ii)]

(k) **Housekeeping.**[§1926.1127(k)]

(1) *All surfaces shall be maintained* as free as practicable of accumulations of cadmium.[§1926.1127(k)(1)]

(2) *All spills and sudden releases* of material containing cadmium shall be cleaned up as soon as possible.[§1926.1127(k)(2)]

(3) *Surfaces contaminated with cadmium shall,* wherever possible, be cleaned by vacuuming or other methods that minimize the likelihood of cadmium becoming airborne.[§1926.1127(k)(3)]

(4) *HEPA-filtered vacuuming equipment* or equally effective filtration methods shall be used for vacuuming. The equipment shall be used and emptied in a manner that minimizes the reentry of cadmium into the workplace.[§1926.1127(k)(4)]

(5) *Shoveling, dry or wet sweeping, and brushing* may be used only where vacuuming or other methods that minimize the likelihood of cadmium becoming airborne have been tried and found not to be effective.[§1926.1127(k)(5)]

(6) *Compressed air shall not be used* to remove cadmium from any surface unless the compressed air is used in conjunction with a ventilation system designed to capture the dust cloud created by the compressed air.[§1926.1127(k)(6)]

(7) *Waste, scrap,* debris, bags, and containers, personal protective equipment and clothing contaminated with cadmium and consigned for disposal shall be collected and disposed of in sealed impermeable bags or other closed, impermeable containers. These bags and containers shall be labeled in accordance with paragraph (m)(3)(ii) of this section.[§1926.1127(k)(7)]

(l) **Medical Surveillance**[§1926.1127(l)]

(1) *General*[§1926.1127(l)(1)]

(i) *Scope*[§1926.1127(l)(1)(i)]

[A] Currently exposed— The employer shall institute a medical surveillance program for all employees who are or may be exposed at or above the action level and all employees who perform the following tasks, operations or jobs: Electrical grounding with cadmium welding; cutting, brazing, burning, grinding or welding on surfaces that were painted with cadmium- containing paints; electrical work using cadmium-coated conduit; use of cadmium containing paints; cutting and welding cadmium-plated steel; brazing or welding with cadmium alloys; fusing of reinforced steel by cadmium welding; maintaining or retrofitting cadmium-coated equipment; and, wrecking and demolition where cadmium is present. A medical surveillance program will not be required if the employer demonstrates that the employee:[§1926.1127(l)(1)(i)[A]]

[1] Is not currently exposed by the employer to airborne concentrations of cadmium at or above the action level on 30 or more days per year (twelve consecutive months); and, [§1926.1127(l)(1)(i)[A][1]]

[2] Is not currently exposed by the employer in those tasks on 30 or more days per year (twelve consecutive months). [§1926.1127(l)(1)(i)[A][2]]

[B] Previously exposed— The employer shall also institute a medical surveillance program for all employees who might previously have been exposed to cadmium by the employer prior to the effective date of this standard in tasks specified under paragraph (l)(1)(i)(A) of this section, unless the employer demonstrates that the employee did not in the years prior to the effective date of this section work in those tasks for the employer with exposure to cadmium for an aggregated total of more than 12 months.[§1926.1127(l)(1)(i)[B]]

(ii) *To determine an employee's fitness* for using a respirator, the employer shall provide the limited medical examination specified in paragraph (l)(6) of this section.[§1926.1127(l)(1)(ii)]

(iii) *The employer shall assure* that all medical examinations and procedures required by this section are performed by or under the supervision of a licensed physician, who has read and is familiar with the health effects section of appendix A to this section, the regulatory text of this section, the protocol for sample handling and lab selection in appendix F to this section, and the questionnaire of appendix D to this section.[§1926.1127(l)(1)(iii)]

(iv) *The employer shall provide* the medical surveillance required by this section, including multiple physician review under paragraph (l)(13) of this section without cost to employees, and at a time and place that is reasonable and convenient to employees. [§1926.1127(l)(1)(iv)]

(v) *The employer shall assure* that the collecting and handling of biological samples of cadmium in urine (CdU), cadmium in blood (CdB), and beta-2 microglobulin in urine (β_2-M) taken from employees under this section is done in a manner that assures their reliability and that analysis of biological samples of cadmium in urine (CdU), cadmium in blood (CdB), and beta-2 microglobulin in urine (β_2-M) taken from employees under this section is performed in laboratories with demonstrated proficiency to perform the particular analysis. (See appendix F to this section.) [§1926.1127(l)(1)(v)]

(2) *Initial Examination.*[§1926.1127(l)(2)]

(i) *For employees covered by medical surveillance* under paragraph (l)(1)(i) of this section, the employer shall provide an initial medical examination. The examination shall be provided to those employees within 30 days after initial assignment to a job with exposure to cadmium or no later than 90 days after the effective date of this section, whichever date is later.[§1926.1127(l)(2)(i)]

(ii) *The initial medical examination shall include:*[§1926.1127(l)(2)(ii)]

[A] *A detailed medical and work history,* with emphasis on: Past, present, and anticipated future exposure to cadmium; any history of renal, cardiovascular, respiratory, hematopoietic, reproductive, and/or musculo-skeletal system dysfunction; current usage of medication with potential nephrotoxic side-effects; and smoking history and current status; and[§1926.1127(l)(2)(ii)[A]]

[B] *Biological monitoring that includes the following tests:* [§1926.1127(l)(2)(ii)[B]]

[1] *Cadmium in urine (CdU),* standardized to grams of creatinine (g/ Cr);[§1926.1127(l)(2)(ii)[B][1]]

[2] *Beta-2 microglobulin in urine (□$_2$-M),* standardized to grams of creatinine (g/Cr), with pH specified, as described in Appendix F to this section; and[§1926.1127(l)(2)(ii)[B][2]]

[3] *Cadmium in blood (CdB),* standardized to liters of whole blood (lwb).[§1926.1127(l)(2)(ii)[B][3]]

(iii) *Recent Examination:* An initial examination is not required to be provided if adequate records show that the employee has been examined in accordance with the requirements of paragraph (l)(2)(ii) of this section within the past 12 months. In that case, such records shall be maintained as part of the employee's medical record and the prior exam shall be treated as if it were an initial examination for the purposes of paragraphs (l)(3) and (4) of this section.[§1926.1127(l)(2)(iii)]

(3) *Actions triggered by initial biological monitoring.*[§1926.1127(l)(3)]

(i) *If the results of the biological monitoring tests* in the initial examination show the employee's CdU level to be at or below 3 µg/g Cr, β_2-M level to be at or below 300 µg/g Cr and CdB level to be at or below 5 µg/lwb, then:[§1926.1127(l)(3)(i)]

[A] *For employees who are subject to* medical surveillance under paragraphs (l)(1)(i)(A) of this section because of current or anticipated exposure to cadmium, the employer shall provide the minimum level of periodic medical surveillance in accordance with the requirements in paragraph (l)(4)(i) of this section; and[§1926.1127(l)(3)(i)[A]]

[B] *For employees who are subject to* medical surveillance under paragraph (l)(1)(i)(B) of this section because of prior but not current exposure, the employer shall provide biological monitoring for CdU, B_2-M, and CdB one year after the initial biological monitoring and then the employer shall comply with the requirements of paragraph (l)(4)(vi) of this section. [§1926.1127(l)(3)(i)[B]]

(ii) *For all employees who are subject to* medical surveillance under paragraph (l)(1)(i) of this section, if the results of the initial biological monitoring tests show the level of CdU to exceed 3 µg/g Cr, the level of β_2-M to be in excess of 300 µg/g Cr, or the level of CdB to be in excess of 5 µg/lwb, the employer shall: [§1926.1127(l)(3)(ii)]

[A] *Within two weeks* after receipt of biological monitoring results, reassess the employee's occupational exposure to cadmium as follows:[§1926.1127(l)(3)(ii)[A]]

[1] *Reassess the employee's work practices and personal hygiene;*[§1926.1127(l)(3)(ii)[A][1]]

[2] *Reevaluate the employee's respirator use,* if any, and the respirator program;[§1926.1127(l)(3)(ii)[A][2]]

[3] *Review the hygiene facilities;*[§1926.1127(l)(3)(ii)[A][3]]

[4] *Reevaluate the maintenance* and effectiveness of the relevant engineering controls;[§1926.1127(l)(3)(ii)[A][4]]

[5] *Assess the employee's smoking history and status;* [§1926.1127(l)(3)(ii)[A][5]]

[B] *Within 30 days after the exposure reassessment,* specified in paragraph (l)(3)(ii)(A) of this section, take reasonable steps to correct any deficiencies found in the reassessment that may be responsible for the employee's excess exposure to cadmium; and,[§1926.1127(l)(3)(ii)[B]]

[C] *Within 90 days after receipt of biological monitoring results,* provide a full medical examination to the employee in accordance with the requirements of paragraph (l)(4)(ii) of this section. After completing the medical examination, the examining physician shall determine in a written medical opinion whether to medically remove the employee. If the physician determines that medical removal is not necessary, then until the employee's CdU level falls to or below 3 µg/g Cr, β_2-M level falls to or below 300 µg/g Cr and CdB level falls to or below 5 µg/lwb, the employer shall:[§1926.1127(l)(3)(ii)[C]]

[1] *Provide biological monitoring* in accordance with paragraph (l)(2)(ii)(B) of this section on a semiannual basis; and [§1926.1127(l)(3)(ii)[C][1]]

[2] *Provide annual medical examinations* in accordance with paragraph (l)(4)(ii) of this section.[§1926.1127(l)(3)(ii)[C][2]]

(iii) *For all employees who are subject to medical surveillance* under paragraph (l)(1)(i) of this section, if the results of the initial biological monitoring tests show the level of CdU to be in excess of 15 µg/g Cr, or the level of CdB to be in excess of 15 µg/lwb, or the level of β_2-M to be in excess of 1,500 µg/g Cr, the employer shall comply with the requirements of paragraphs (l)(3)(ii)(A)-(B) of this section. Within 90 days after receipt of biological monitoring results, the employer shall provide a full medical examination to the employee in accordance with the requirements of paragraph (l)(4)(ii) of this section. After completing the medical examination, the examining physician shall determine in a written medical opinion whether to medically remove the employee. However, if the initial biological monitoring results and the biological monitoring results obtained during the medical examination both show that: CdU exceeds 15 µg/g Cr; or CdB exceeds 15 µg/lwb; or β_2-M exceeds 1500 µg/g Cr, and in addition CdU exceeds 3 µg/g Cr or CdB exceeds 5 µg/liter of whole blood, then the physician shall medically remove the employee from exposure to cadmium at or above the action level. If the second set of biological monitoring results obtained during the medical examination does not show that a mandatory removal trigger level has been exceeded, then the employee is not required to be removed by the mandatory provisions of this paragraph. If the employee is not required to be removed by the mandatory provisions of this paragraph or by the physician's determination, then until the employee's CdU level falls to or below 3 µg/g Cr, β_2-M level falls to or below 300 µg/g Cr and CdB level falls to or below 5 µg/lwb, the employer shall: [§1926.1127(l)(3)(iii)]

[A] *Periodically reassess* the employee's occupational exposure to cadmium;[§1926.1127(l)(3)(iii)[A]]

[B] *Provide biological monitoring* in accordance with paragraph (l)(2)(ii)(B) of this section on a quarterly basis; and [§1926.1127(l)(3)(iii)[B]]

[C] *Provide semiannual medical examinations* in accordance with paragraph (l)(4)(ii) of this section.[§1926.1127(l)(3)(iii)[C]]

(iv) *For all employees to whom medical surveillance is provided,* beginning on January 1, 1999, and in lieu of paragraph (l)(3)(iii) of this section, whenever the results of initial biological monitoring tests show the employee's CdU level to be in excess of 7 µg/g Cr, or β_2-M level to be in excess of 750 µg/g Cr, or CdB level to be in excess of 10 µg/lwb, the employer shall comply with the requirements of paragraphs (l)(3)(ii)(A)-(B) of this section. Within 90 days after receipt of biological monitoring results, the employer shall provide a full medical examination to the employee in accordance with the requirements of paragraph (l)(4)(ii) of this section. After completing the medical examination, the examining physician shall determine in a written medical opinion whether to medically remove the employee. However, if the initial biological monitoring results and the biological monitoring results obtained during the medical examination both show that: CdU exceeds 7 µg/g Cr; or CdB exceeds 10 µg/lwb; or β_2-M exceeds 750 µg/g Cr, and in addition CdU exceeds 3 µg/g Cr or CdB exceeds 5 µg/liter of whole blood, then the physician shall medically remove the employee from exposure to cadmium at or above the action level. If the second set of biological monitoring results obtained during the medical examination does not show that a mandatory removal trigger level has been exceeded, then the employee is not required to be removed by the mandatory provisions of this paragraph. If the employee is not required to be removed by the mandatory provisions of this paragraph or by the physician's determination, then until the employee's CdU level falls to or below 3 µg/g Cr, β_2-M level falls to or below 300 µg/g Cr and CdB level falls to or below 5 µg/lwb, the employer shall: [§1926.1127(l)(3)(iv)]

[A] Periodically reassess the employee's occupational exposure to cadmium;[§1926.1127(l)(3)(iv)[A]]

[B] Provide biological monitoring in accordance with paragraph (l)(2)(ii)(B) of this section on a quarterly basis; and [§1926.1127(l)(3)(iv)[B]]

[C] Provide semiannual medical examinations in accordance with paragraph (l)(4)(ii) of this section.[§1926.1127(l)(3)(iv)[C]]

(4) *Periodic medical surveillance.*[§1926.1127(l)(4)]

(i) *For each employee who is covered by medical surveillance* under paragraph (l)(1)(i)(A) of this section because of current or anticipated exposure to cadmium, the employer shall provide at least the minimum level of periodic medical surveillance, which consists of periodic medical examinations and periodic biological monitoring. A periodic medical examination shall be provided within one year after the initial examination required by paragraph (l)(2) of this section and thereafter at least biennially. Biological sampling shall be provided at least annually either as part of a periodic medical examination or separately as periodic biological monitoring.[§1926.1127(l)(4)(i)]

(ii) *The periodic medical examination shall include:*[§1926.1127(l)(4)(ii)]

[A] A detailed medical and work history, or update thereof, with emphasis on: Past, present and anticipated future exposure to cadmium; smoking history and current status; reproductive history; current use of medications with potential nephrotoxic side-effects; any history of renal, cardiovascular, respiratory, hematopoietic, and/or musculo- skeletal system dysfunction; and as part of the medical and work history, for employees who wear respirators, questions 3-11 and 25-32 in appendix D to this section;[§1926.1127(l)(4)(ii)[A]]

[B] A complete physical examination with emphasis on: blood pressure, the respiratory system, and the urinary system; [§1926.1127(l)(4)(ii)[B]]

[C] A 14 inch by 17 inch, or a reasonably standard sized posterior- anterior chest X-ray (after the initial X-ray, the frequency of chest X- rays is to be determined by the examining physician);[§1926.1127(l)(4)(ii)[C]]

[D] Pulmonary function tests, including forced vital capacity (FVC) and forced expiratory volume at 1 second (FEV1); [§1926.1127(l)(4)(ii)[D]]

[E] Biological monitoring, as required in paragraph (l)(2)(ii)(B) of this section;[§1926.1127(l)(4)(ii)[E]]

[F] Blood analysis, in addition to the analysis required under paragraph (l)(2)(ii)(B) of this section, including blood urea nitrogen, complete blood count, and serum creatinine; [§1926.1127(l)(4)(ii)[F]]

[G] Urinalysis, in addition to the analysis required under paragraph (l)(2)(ii)(B) of this section, including the determination of albumin, glucose, and total and low molecular weight proteins; [§1926.1127(l)(4)(ii)[G]]

[H] For males over 40 years old, prostate palpation, or other at least as effective diagnostic test(s), and;[§1926.1127(l)(4)(ii)[H]]

[I] Any additional tests or procedures deemed appropriate by the examining physician.[§1926.1127(l)(4)(ii)[I]]

(iii) *Periodic biological monitoring* shall be provided in accordance with paragraph (l)(2)(ii)(B) of this section.[§1926.1127(l)(4)(iii)]

(iv) *If the results of periodic biological monitoring* or the results of biological monitoring performed as part of the periodic medical examination show the level of the employee's CdU, β_2-M, or CdB to be in excess of the levels specified in paragraphs (l)(3)(ii) or (iii) of this section; or beginning on January 1, 1999, in excess of the levels specified in paragraphs (l)(3)(ii) or (iv), the employer shall take the appropriate actions specified in paragraphs (l)(3)(ii)- (iv) of this section, respectively.[§1926.1127(l)(4)(iv)]

(v) *For previously exposed employees* under paragraph (l)(1)(i)(B) of this section:[§1926.1127(l)(4)(v)]

[A] If the employee's levels of CdU did not exceed 3 µg/g Cr, CdB did not exceed 5 µg/lwb, and β_2-M did not exceed 300 µg/g Cr in the initial biological monitoring tests, and if the results of the followup biological monitoring required by paragraph (l)(3)(i)(B) of this section one year after the initial examination confirm the previous results, the employer may discontinue all periodic medical surveillance for that employee. [§1926.1127(l)(4)(v)[A]]

[B] If the initial biological monitoring results for CdU, CdB, or β_2-M were in excess of the levels specified in paragraph (l)(3)(i) of this section, but subsequent biological monitoring results required by paragraph (l)(3)(ii)- (iv) of this section show that the employee's CdU levels no longer exceed 3 µg/g Cr, CdB levels no longer exceed 5 µg/lwb, and β_2-M levels no longer exceed 300 µg/g Cr, the employer shall provide biological monitoring for CdU, CdB, and β_2-M one year after these most recent biological monitoring results. If the results of the followup biological monitoring specified in this paragraph, confirm the previous results, the employer may discontinue all periodic medical surveillance for that employee. [§1926.1127(l)(4)(v)[B]]

[C] However, if the results of the follow-up tests specified in paragraph (l)(4)(v)(A) or (B) of this section indicate that the level of the employee's CdU, β_2-M, or CdB exceeds these same levels, the employer is required to provide annual medical examinations in accordance with the provisions of paragraph (l)(4)(ii) of this section until the results of biological monitoring are consistently below these levels or the examining physician determines in a written medical opinion that further medical surveillance is not required to protect the employee's health. [§1926.1127(l)(4)(v)[C]]

(vi) *A routine, biennial medical examination* is not required to be provided in accordance with paragraphs (l)(3)(i) and (l)(4) of this section if adequate medical records show that the employee has been examined in accordance with the requirements of paragraph (l)(4)(ii) of this section within the past 12 months. In that case, such records shall be maintained by the employer as part of the employee's medical record, and the next routine, periodic medical examination shall be made available to the employee within two years of the previous examination.[§1926.1127(l)(4)(vi)]

(5) *Actions triggered by medical examinations.*[§1926.1127(l)(5)]

(i) *If the results of a medical examination carried out* in accordance with this section indicate any laboratory or clinical finding consistent with cadmium toxicity that does not require employer action under paragraphs (l)(2), (3) or (4) of this section, the employer shall take the following steps and continue to take them until the physician determines that they are no longer necessary. [§1926.1127(l)(5)(i)]

[A] Periodically reassess: The employee's work practices and personal hygiene; the employee's respirator use, if any; the employee's smoking history and status; the respiratory protection program; the hygiene facilities; the maintenance and effectiveness of the relevant engineering controls; and take all reasonable steps to correct the deficiencies found in the reassessment that may be responsible for the employee's excess exposure to cadmium.[§1926.1127(l)(5)(i)[A]]

[B] Provide semi-annual medical reexaminations to evaluate the abnormal clinical sign(s) of cadmium toxicity until the results are normal or the employee is medically removed; and [§1926.1127(l)(5)(i)[B]]

[C] Where the results of tests for total proteins in urine are abnormal, provide a more detailed medical evaluation of the toxic effects of cadmium on the employee's renal system. [§1926.1127(l)(5)(i)[C]]

(6) *Examination for respirator use.*[§1926.1127(l)(6)]

(i) *To determine an employee's fitness for respirator use,* the employer shall provide a medical examination that includes the elements specified in paragraph (l)(6)(i)(A)-(D) of this section. This examination shall be provided prior to the employee's being assigned to a job that requires the use of a respirator or no later than 90 days after this section goes into effect, whichever date is later, to any employee without a medical examination within the preceding 12 months that satisfies the requirements of this paragraph.[§1926.1127(l)(6)(i)]

[A] A detailed medical and work history, or update thereof, with emphasis on: past exposure to cadmium; smoking history and current status; any history of renal, cardiovascular, respiratory, hematopoietic, and/or musculo-skeletal system dysfunction; a description of the job for which the respirator is required; and questions 3-11 and 25-32 in appendix D;[§1926.1127(l)(6)(i)[A]]

[B] A blood pressure test;[§1926.1127(l)(6)(i)[B]]

[C] Biological monitoring of the employee's levels of CdU, CdB and β_2-M in accordance with the requirements of paragraph (l)(2)(ii)(B) of this section, unless such results already have been obtained within the twelve months; and [§1926.1127(l)(6)(i)[C]]

[D] Any other test or precedure that the examining physician deems appropriate.[§1926.1127(l)(6)(i)[D]]

(ii) *After reviewing all the information* obtained from the medical examination required in paragraph (l)(6)(i) of this section, the physician shall determine whether the employee is fit to wear a respirator.[§1926.1127(l)(6)(ii)]

(iii) *Whenever an employee has exhibited difficulty* in breathing during a respirator fit test or during use of a respirator, the employer, as soon as possible, shall provide the employee with a periodic medical examination in accordance with paragraph (l)(4)(ii) of this section to determine the employee's fitness to wear a respirator. [§1926.1127(l)(6)(iii)]

(iv) *Where the results of the examination required* under paragraphs (l)(6)(i), (ii), or (iii) of this section are abnormal, medical limitation or prohibition of respirator use shall be considered. If the employee is allowed to wear a respirator, the employee's ability to continue to do so shall be periodically evaluated by a physician. [§1926.1127(l)(6)(iv)]

(7) *Emergency Examinations.*[§1926.1127(l)(7)]

(i) *In addition to the medical surveillance* required in paragraphs (l)(2)-(6) of this section, the employer shall provide a medical examination as soon as possible to any employee who may have been acutely exposed to cadmium because of an emergency. [§1926.1127(l)(7)(i)]

(ii) *The examination shall include* the requirements of paragraph (l)(4)(ii), of this section, with emphasis on the respiratory system, other organ systems considered appropriate by the examining physician, and symptoms of acute overexposure, as identified in paragraphs II(B)(1)-(2) and IV of appendix A of this section. [§1926.1127(l)(7)(ii)]

(8) *Termination of employment examination.*[§1926.1127(l)(8)]

(i) *At termination of employment,* the employer shall provide a medical examination in accordance with paragraph (l)(4)(ii) of this section, including a chest X-ray where necessary, to any employee to whom at any prior time the employer was required to provide medical surveillance under paragraph (l)(1)(i) or (l)(7) of this section. However, if the last examination satisfied the requirements of paragraph (l)(4)(ii) of this section and was less than six months prior to the date of termination, no further examination is required unless otherwise specified in paragraph (l)(3) or (l)(5) of this section;[§1926.1127(l)(8)(i)]

(ii) *In addition, if the employer has discontinued* all periodic medical surveillance under paragraph (l)(4)(v) of this section, no termination of employment medical examination is required. [§1926.1127(l)(8)(ii)]

(9) *Information provided to the physician.* The employer shall provide the following information to the examining physician:[§1926.1127(l)(9)]

(i) *A copy of this standard and appendices;*[§1926.1127(l)(9)(i)]

(ii) *A description of the affected employee's* former, current, and anticipated duties as they relate to the employee's occupational exposure to cadmium;[§1926.1127(l)(9)(ii)]

(iii) *The employee's former, current,* and anticipated future levels of occupational exposure to cadmium;[§1926.1127(l)(9)(iii)]

(iv) *A description of any personal protective equipment,* including respirators, used or to be used by the employee, including when and for how long the employee has used that equipment; and [§1926.1127(l)(9)(iv)]

(v) *Relevant results of previous biological monitoring and medical examinations.*[§1926.1127(l)(9)(v)]

(10) *Physician's written medical opinion.*[§1926.1127(l)(10)]

(i) *The employer shall promptly obtain* a written, medical opinion from the examining physician for each medical examination performed on each employee. This written opinion shall contain: [§1926.1127(l)(10)(i)]

[A] The physician's diagnosis for the employee; [§1926.1127(l)(10)(i)[A]]

[B] The physician's opinion as to whether the employee has any detected medical condition(s) that would place the employee at increased risk of material impairment to health from further exposure to cadmium, including any indications of potential cadmium toxicity;[§1926.1127(l)(10)(i)[B]]

[C] The results of any biological or other testing or related evaluations that directly assess the employee's absorption of cadmium;[§1926.1127(l)(10)(i)[C]]

[D] Any recommended removal from, or limitation on the activities or duties of the employee or on the employee's use of personal protective equipment, such as respirators; [§1926.1127(l)(10)(i)[D]]

[E] A statement that the physician has clearly and carefully explained to the employee the results of the medical examination, including all biological monitoring results and any medical conditions related to cadmium exposure that require further evaluation or treatment, and any limitation on the employee's diet or use of medications.[§1926.1127(l)(10)(i)[E]]

(ii) *The employer shall promptly obtain a copy of the results* of any biological monitoring provided by an employer to an employee independently of a medical examination under paragraphs (l)(2) and (l)(4) of this section, and, in lieu of a written medical opinion, an explanation sheet explaining those results.[§1926.1127(l)(10)(ii)]

(iii) *The employer shall instruct the physician* not to reveal orally or in the written medical opinion given to the employer specific findings or diagnoses unrelated to occupational exposure to cadmium. [§1926.1127(l)(10)(iii)]

(11) *Medical Removal Protection (MRP)*[§1926.1127(l)(11)]

(i) *General.*[§1926.1127(l)(11)(i)]

[A] The employer shall temporarily remove an employee from work where there is excess exposure to cadmium on each occasion that medical removal is required under paragraphs (l)(3), (l)(4), or (l)(6) of this section and on each occasion that a physician determines in a written medical opinion that the employee should be removed from such exposure. The physician's determination may be based on biological monitoring results, inability to wear a respirator, evidence of illness, other signs or symptoms of cadmium-related dysfunction or disease, or any other reason deemed medically sufficient by the physician.[§1926.1127(l)(11)(i)[A]]

[B] The employer shall medically remove an employee in accordance with paragraph (l)(11) of this section regardless of whether at the time of removal a job is available into which the removed employee may be transferred.[§1926.1127(l)(11)(i)[B]]

[C] Whenever an employee is medically removed under paragraph (l)(11) of this section, the employer shall transfer the removed employee to a job where the exposure to cadmium is within the permissible levels specified in that paragraph as soon as one becomes available.[§1926.1127(l)(11)(i)[C]]

[D] For any employee who is medically removed under the provisions of paragraph (l)(11)(i) of this section, the employer shall provide follow-up medical examinations semi-annually until, in a written medical opinion, the examining physician determines that either the employee may be returned to his/her former job status or the employee must be permanently removed from excess cadmium exposure.[§1926.1127(l)(11)(i)[D]]

[E] The employer may not return an employee who has been medically removed for any reason to his/her former job status until a physician determines in a written medical opinion that continued medical removal is no longer necessary to protect the employee's health.[§1926.1127(l)(11)(i)[E]]

(ii) *Where an employee is found unfit to wear a respirator* under paragraph (l)(6)(ii) of this section, the employer shall remove the employee from work where exposure to cadmium is above the PEL.[§1926.1127(l)(11)(ii)]

(iii) *Where removal is based upon any reason* other than the employee's inability to wear a respirator, the employer shall remove the employee from work where exposure to cadmium is at or above the action level.[§1926.1127(l)(11)(iii)]

(iv) *Except as specified in paragraph (l)(11)(v) of this section,* no employee who was removed because his/her level of CdU, CdB and/or β_2-M exceeded the trigger levels in paragraph (l)(3) or (l)(4) of this section may be returned to work with exposure to cadmium at or above the action level until the employee's levels of CdU fall to or below 3 µg/g Cr, CdB fall to or below 5 µg/lwb, and β_2-M fall to or below 300 µg/g Cr.[§1926.1127(l)(11)(iv)]

(v) *However, when in the examining physician's opinion* continued exposure to cadmium will not pose an increased risk to the employee's health and there are special circumstances that make continued medical removal an inappropriate remedy, the physician shall fully discuss these matters with the employee, and then in a written determination may return a worker to his/her former job status despite what would otherwise be unacceptably high biological monitoring results. Thereafter and until such time as the employee's biological monitoring results have decreased to levels where he/she could have been returned to his/her former job status, the returned employee shall continue medical surveillance as if he/she were still on medical removal. Until such time, the employee is no longer subject to mandatory medical removal. Subsequent questions regarding the employee's medical removal shall be decided solely by a final medical determination. [§1926.1127(l)(11)(v)]

(vi) *Where an employer,* although not required by this section to do so, removes an employee from exposure to cadmium or otherwise places limitations on an employee due to the effects of cadmium exposure on the employee's medical condition, the employer shall provide the same medical removal protection benefits to that employee under paragraph (l)(12) of this section as would have been provided had the removal been required under paragraph (l)(11) of this section.[§1926.1127(l)(11)(vi)]

(12) *Medical removal protection benefits.*[§1926.1127(l)(12)]

(i) *The employer shall provide medical removal protection benefits* to an employee for up to a maximum of 18 months each time, and while the employee is temporarily medically removed under paragraph (l)(11) of this section.[§1926.1127(l)(12)(i)]

(ii) *For purposes of this section,* the requirement that the employer provide medical removal protection benefits means that the employer shall maintain the total normal earnings, seniority, and all other employee rights and benefits of the removed employee, including the employee's right to his/her former job status, as if the employee had not been removed from the employee's job or otherwise medically limited.[§1926.1127(l)(12)(ii)]

(iii) *Where, after 18 months on medical removal* because of elevated biological monitoring results, the employee's monitoring results have not declined to a low enough level to permit the employee to be returned to his/her former job status: [§1926.1127(l)(12)(iii)]

[A] The employer shall make available to the employee a medical examination pursuant to this section in order to obtain a final medical determination as to whether the employee may be returned to his/her former job status or must be permanently removed from excess cadmium exposure; and [§1926.1127(l)(12)(iii)[A]]

[B] The employer shall assure that the final medical determination indicates whether the employee may be returned to his/her former job status and what steps, if any, should be taken to protect the employee's health;[§1926.1127(l)(12)(iii)[B]]

(iv) *The employer may condition the provision* of medical removal protection benefits upon the employee's participation in medical surveillance provided in accordance with this section. [§1926.1127(l)(12)(iv)]

(13) *Multiple physician review.*[§1926.1127(l)(13)]

(i) *If the employer selects the initial physician* to conduct any medical examination or consultation provided to an employee under this section, the employee may designate a second physician to: [§1926.1127(l)(13)(i)]

[A] Review any findings, determinations, or recommendations of the initial physician; and[§1926.1127(l)(13)(i)[A]]

[B] Conduct such examinations, consultations, and laboratory tests as the second physician deems necessary to facilitate this review.[§1926.1127(l)(13)(i)[B]]

(ii) *The employer shall promptly notify an employee* of the right to seek a second medical opinion after each occasion that an initial physician provided by the employer conducts a medical examination or consultation pursuant to this section. The employer may condition its participation in, and payment for, multiple physician review upon the employee doing the following within fifteen (15) days after receipt of this notice, or receipt of the initial physician's written opinion, whichever is later:[§1926.1127(l)(13)(ii)]

[A] Informing the employer that he or she intends to seek a medical opinion; and[§1926.1127(l)(13)(ii)[A]]

[B] Initiating steps to make an appointment with a second physician.[§1926.1127(l)(13)(ii)[B]]

(iii) *If the findings, determinations, or recommendations* of the second physician differ from those of the initial physician, then the employer and the employee shall assure that efforts are made for the two physicians to resolve any disagreement.[§1926.1127(l)(13)(iii)]

(iv) *If the two physicians have been unable* to quickly resolve their disagreement, then the employer and the employee, through their respective physicians, shall designate a third physician to: [§1926.1127(l)(13)(iv)]

[A] Review any findings, determinations, or recommendations of the other two physicians; and[§1926.1127(l)(13)(iv)[A]]

[B] Conduct such examinations, consultations, laboratory tests, and discussions with the other two physicians as the third physician deems necessary to resolve the disagreement among them.[§1926.1127(l)(13)(iv)[B]]

(v) *The employer shall act consistently* with the findings, determinations, and recommendations of the third physician, unless the employer and the employee reach an agreement that is consistent with the recommendations of at least one of the other two physicians.[§1926.1127(l)(13)(v)]

(14) *Alternate physician determination.* The employer and an employee or designated employee representative may agree upon the use of any alternate form of physician determination in lieu of the multiple physician review provided by paragraph (l)(13) of this section, so long as the alternative is expeditious and at least as protective of the employee.[§1926.1127(l)(14)]

(15) *Information the employer must provide the employee.* [§1926.1127(l)(15)]

(i) *The employer shall provide a copy* of the physician's written medical opinion to the examined employee within five working days after receipt thereof.[§1926.1127(l)(15)(i)]

(ii) *The employer shall provide the employee* with a copy of the employee's biological monitoring results and an explanation sheet explaining the results within five working days after receipt thereof.[§1926.1127(l)(15)(ii)]

(iii) *Within 30 days after a request by an employee,* the employer shall provide the employee with the information the employer is required to provide the examining physician under paragraph (l)(9) of this section.[§1926.1127(l)(15)(iii)]

(16) *Reporting.* In addition to other medical events that are required to be reported on the OSHA Form No. 200, the employer shall report any abnormal condition or disorder caused by occupational exposure to cadmium associated with employment as specified in Chapter (V)(E) of the Reporting Guidelines for Occupational Injuries and Illnesses.[§1926.1127(l)(16)]

(m) Communication of cadmium hazards to employees[§1926.1127(m)]

(1) *Hazard communication.* The employer shall include cadmium in the program established to comply with the Hazard Communication Standard (HCS) (Sec. 1910.1200). The employer shall ensure that each employee has access to labels on containers of cadmium and safety data sheets, and is trained in accordance with the provisions of HCS and paragraph (m)(4) of this section. The employer shall provide information on at least the following hazards: Cancer; lung effects; kidney effects; and acute toxicity effects.[§1926.1127(m)(1)]

(2) *Warning signs.*[§1926.1127(m)(2)]

(i) *Warning signs shall be provided* and displayed in regulated areas. In addition, warning signs shall be posted at all approaches to regulated areas so that an employee may read the signs and take necessary protective steps before entering the area.[§1926.1127(m)(2)(i)]

(ii) *Warning signs required* by paragraph (m)(2)(i) of this section shall bear the following legend: DANGER CADMIUM MAY CAUSE CANCER CAUSES DAMAGE TO LUNGS AND KIDNEYS WEAR RESPIRATORY PROTECTION IN THIS AREA AUTHORIZED PERSONNEL ONLY[§1926.1127(m)(2)(ii)]

(iii) *The employer shall ensure that signs* required by this paragraph (m)(2) are illuminated, cleaned, and maintained as necessary so that the legend is readily visible.[§1926.1127(m)(2)(iii)]

(iv) *Prior to June 1,* 2016, employers may use the following legend in lieu of that specified in paragraph (m)(2)(ii) of this section: DANGER CADMIUM CANCER HAZARD CAN CAUSE LUNG AND KIDNEY DISEASE AUTHORIZED PERSONNEL ONLY RESPIRATORS REQUIRED IN THIS AREA[§1926.1127(m)(2)(iv)]

(3) *Warning labels.*[§1926.1127(m)(3)]

(i) *Shipping and storage containers* containing cadmium or cadmium compounds shall bear appropriate warning labels, as specified in paragraph (m)(1) of this section.[§1926.1127(m)(3)(i)]

(ii) *The warning labels for containers* of cadmium-contaminated protective clothing, equipment, waste, scrap, or debris shall include at least the following information: DANGER CONTAINS CADMIUM MAY CAUSE CANCER CAUSES DAMAGE TO LUNGS AND KIDNEYS AVOID CREATING DUST[§1926.1127(m)(3)(ii)]

(iii) *Where feasible,* installed cadmium products shall have a visible label or other indication that cadmium is present. [§1926.1127(m)(3)(iii)]

(iv) *Prior to June 1,* 2015, employers may include the following information on shipping and storage containers containing cadmium, cadmium compounds, or cadmium-contaminated clothing, equipment, waste, scrap, or debris in lieu of the labeling requirements specified in paragraphs (m)(3)(i) and (m)(3)(ii) of this section: DANGER CONTAINS CADMIUM CANCER HAZARD AVOID CREATING DUST CAN CAUSE LUNG AND KIDNEY DISEASE [§1926.1127(m)(3)(iv)]

(4) *Employee information and training.*[§1926.1127(m)(4)]

(i) *The employer shall train each employee* who is potentially exposed to cadmium in accordance with the requirements of this section. The employer shall institute a training program, ensure employee participation in the program, and maintain a record of the contents of the training program.[§1926.1127(m)(4)(i)]

(ii) *Training shall be provided prior to* or at the time of initial assignment to a job involving potential exposure to cadmium and at least annually thereafter.[§1926.1127(m)(4)(ii)]

(iii) *The employer shall make the training program* understandable to the employee and shall assure that each employee is informed of the following:[§1926.1127(m)(4)(iii)]

[A] *The health hazards associated* with cadmium exposure, with special attention to the information incorporated in appendix A to this section;[§1926.1127(m)(4)(iii)[A]]

[B] *The quantity, location, manner of use,* release, and storage of cadmium in the workplace and the specific nature of operations that could result in exposure to cadmium, especially exposures above the PEL;[§1926.1127(m)(4)(iii)[B]]

[C] *The engineering controls* and work practices associated with the employee's job assignment;[§1926.1127(m)(4)(iii)[C]]

[D] *The measures employees can take* to protect themselves from exposure to cadmium, including modification of such habits as smoking and personal hygiene, and specific procedures the employer has implemented to protect employees from exposure to cadmium such as appropriate work practices, emergency procedures, and the provision of personal protective equipment;[§1926.1127(m)(4)(iii)[D]]

[E] *The purpose, proper selection,* fitting, proper use, and limitations of respirators and protective clothing;[§1926.1127(m)(4)(iii)[E]]

[F] *The purpose and a description* of the medical surveillance program required by paragraph (l) of this section; [§1926.1127(m)(4)(iii)[F]]

[G] *The contents of this section and its appendices, and,* [§1926.1127(m)(4)(iii)[G]]

[H] *The employee's rights of access* to records under §1926.33(g)(1) and (2).[§1926.1127(m)(4)(iii)[H]]

(iv) *Additional access to information* and training program and materials.[§1926.1127(m)(4)(iv)]

[A] *The employer shall make a copy* of this section and its appendices readily available to all affected employees and shall provide a copy without cost if requested.[§1926.1127(m)(4)(iv)[A]]

[B] *Upon request, the employer shall provide* to the Assistant Secretary or the Director all materials relating to the employee information and the training program.[§1926.1127(m)(4)(iv)[B]]

(5) *Multi-employer workplace.* In a multi-employer workplace, an employer who produces, uses, or stores cadmium in a manner that may expose employees of other employers to cadmium shall notify those employers of the potential hazard in accordance with paragraph (e) of the hazard communication standard for construction, 29 CFR 1926.59.[§1926.1127(m)(5)]

(n) Recordkeeping[§1926.1127(n)]

(1) *Exposure monitoring.*[§1926.1127(n)(1)]

(i) *The employer shall establish and keep* an accurate record of all air monitoring for cadmium in the workplace.[§1926.1127(n)(1)(i)]

(ii) *This record shall include* at least the following information: [§1926.1127(n)(1)(ii)]

[A] *The monitoring date, shift, duration,* air volume, and results in terms of an 8-hour TWA of each sample taken, and if cadmium is not detected, the detection level;[§1926.1127(n)(1)(ii)[A]]

[B] *The name, social security number, and job classification* of all employees monitored and of all other employees whose exposures the monitoring result is intended to represent, including, where applicable, a description of how it was determined that the employee's monitoring result could be taken to represent other employee's exposures;[§1926.1127(n)(1)(ii)[B]]

[C] *A description* of the sampling and analytical methods used and evidence of their accuracy;[§1926.1127(n)(1)(ii)[C]]

[D] *The type of respiratory protective device,* if any, worn by the monitored employee and by any other employee whose exposure the monitoring result is intended to represent; [§1926.1127(n)(1)(ii)[D]]

[E] *A notation of any other conditions* that might have affected the monitoring results.[§1926.1127(n)(1)(ii)[E]]

[F] *Any exposure monitoring* or objective data that were used and the levels.[§1926.1127(n)(1)(ii)[F]]

(iii) *The employer shall maintain this record* for at least thirty (30) years, in accordance with Sec. 1926.33 of this part. [§1926.1127(n)(1)(iii)]

(iv) *The employer shall also provide a copy* of the results of an employee's air monitoring prescribed in paragraph (d) of this section to an industry trade association and to the employee's union, if any, or, if either of such associations or unions do not exist, to another comparable organization that is competent to maintain such records and is reasonably accessible to employers and employees in the industry.[§1926.1127(n)(1)(iv)]

(2) *Objective data for exemption from requirement for initial monitoring.* [§1926.1127(n)(2)]

(i) *For purposes of this section,* objective data are information demonstrating that a particular product or material containing cadmium or a specific process, operation, or activity involving cadmium cannot release dust or fumes in concentrations at or above the action level even under the worst-case release conditions. Objective data can be obtained from an industry-wide study or from laboratory product test results from manufacturers of cadmium-containing products or materials. The data the employer uses from an industry-wide survey must be obtained under workplace conditions closely resembling the processes, types of material, control methods, work practices and environmental conditions in the employer's current operations.[§1926.1127(n)(2)(i)]

(ii) *The employer shall maintain* the record for at least 30 years of the objective data relied upon.[§1926.1127(n)(2)(ii)]

(3) *Medical surveillance.*[§1926.1127(n)(3)]

(i) *The employer shall establish and maintain* an accurate record for each employee covered by medical surveillance under paragraph (l)(1)(i) of this section.[§1926.1127(n)(3)(i)]

(ii) *The record shall include* at least the following information about the employee:[§1926.1127(n)(3)(ii)]

[A] *Name, social security number, and description of duties;* [§1926.1127(n)(3)(ii)[A]]

[B] *A copy of the physician's written opinions* and of the explanation sheets for biological monitoring results;[§1926.1127(n)(3)(ii)[B]]

[C] *A copy of the medical history, and the results* of any physical examination and all test results that are required to be provided by this section, including biological tests, X-rays, pulmonary function tests, etc., or that have been obtained to further evaluate any condition that might be related to cadmium exposure;[§1926.1127(n)(3)(ii)[C]]

[D] *The employee's medical symptoms* that might be related to exposure to cadmium; and[§1926.1127(n)(3)(ii)[D]]

[E] *A copy of the information provided* to the physician as required by paragraph (l)(9) of this section.[§1926.1127(n)(3)(ii)[E]]

(iii) *The employer shall assure that this record* is maintained for the duration of employment plus thirty (30) years, in accordance with Sec. 1926.33 of this part.[§1926.1127(n)(3)(iii)]

(iv) *At the employee's request,* the employer shall promptly provide a copy of the employee's medical record, or update as appropriate, to a medical doctor or a union specified by the employee. [§1926.1127(n)(3)(iv)]

(4) *Availability.*[§1926.1127(n)(4)]

(i) *Except as otherwise provided* for in this section, access to all records required to be maintained by paragraphs (n)(1) through (3) of this section shall be in accordance with the provisions of [fnl]29 CFR 1910.1020.[§1926.1127(n)(4)(i)]

(ii) *Within 15 days after a request,* the employer shall make an employee's medical records required to be kept by paragraph (n)(3) of this section available for examination and copying to the subject employee, to designated representatives, to anyone having the specific written consent of the subject employee, and after the employee's death or incapacitation, to the employee's family members.[§1926.1127(n)(4)(ii)]

(5) *Transfer of records.* Whenever an employer ceases to do business and there is no successor employer or designated organization to receive and retain records for the prescribed period, the employer shall comply with the requirements concerning transfer of records set forth in §1926.33 (h) of this part.[§1926.1127(n)(5)]

(o) Observation of monitoring[§1926.1127(o)]

(1) *Employee observation.* The employer shall provide affected employees or their designated representatives an opportunity to observe any monitoring of employee exposure to cadmium.[§1926.1127(o)(1)]

(2) *Observation procedures.* When observation of monitoring requires entry into an area where the use of protective clothing or equipment is required, the employer shall provide the observer with that clothing and equipment and shall assure that the observer uses such clothing and equipment and complies with all other applicable safety and health procedures.[§1926.1127(o)(2)]

(p) [Reserved][§1926.1127(p)]

(q) Appendices. Except where portions of appendices A, B, D, E, and F to this section are expressly incorporated in requirements of this section, these appendices are purely informational and are not intended to create any additional obligations not otherwise imposed or to detract from any existing obligations.[§1926.1127(q)]

§1926.1127 Appendix A

Substance Safety Data Sheet

Note: The requirements applicable to construction work under this Appendix A are identical to those set forth in Appendix A to §1910.1027 of this chapter.

§1926.1127 Appendix B

Substance Technical Guidelines for Cadmium

Note: The requirements applicable to construction work under this Appendix B are identical to those set forth in Appendix B to §1910.1027 of this chapter.

§1926.1127 Appendix C

[Reserved]

§1926.1127 Appendix D

Occupational Health History Interview With Reference to Cadmium Exposure

Note: The requirements applicable to construction work under this Appendix D are identical to those set forth in Appendix D to §1910.1027 of this chapter.

§1926.1127 Appendix E

Cadmium in Workplace Atmospheres

Note: The requirements applicable to construction work under this Appendix E are identical to those set forth in Appendix E to §1910.1027 of this chapter.

§1926.1127 Appendix F

Nonmandatory Protocol for Biological Monitoring

Note: The requirements applicable to construction work under this Appendix F are identical to those set forth in Appendix F to §1910.1027 of this chapter.

[57 FR 42452, Sept. 14, 1992, as amended at 57 FR 49272, Oct. 30, 1992; 58 FR 21787, Apr. 23, 1993. Redesignated and amended at 59 FR 215, Jan. 3, 1994; 61 FR 5510, Feb. 13, 1996; 61 FR 31433, 31434, June 20, 1996; 63 FR 1298, Jan. 8, 1998; 70 FR 1144, Jan. 5, 2005; 71 FR 16675, Apr. 3, 2006; 71 FR 50192, Aug. 24, 2006]

§1926.1128

Benzene

Note: The requirements applicable to construction work under this section are identical to those set forth at §1910.1028 of this chapter. (Editor's Note: See 1910 Supplement.)

[61 FR 31434, June 20, 1996]

§1926.1129

Coke oven emissions

Note: The requirements applicable to construction work under this section are identical to those set forth at §1910.1029 of this chapter.

[61 FR 31434, June 20, 1996]

§1926.1144

1,2-dibromo-3-chloropropane

Note: The requirements applicable to construction work under this section are identical to those set forth at §1910.1044 of this chapter.

[61 FR 31434, June 20, 1996]

§1926.1145

Acrylonitrile

Note: The requirements applicable to construction work under this section are identical to those set forth at §1910.1045 of this chapter.

[61 FR 31434, June 20, 1996]

§1926.1147

Ethylene oxide

Note: The requirements applicable to construction work under this section are identical to those set forth at §1910.1047 of this chapter.

[61 FR 31434, June 20, 1996]

§1926.1148

Formaldehyde

Note: The requirements applicable to construction work under this section are identical to those set forth at §1910.1048 of this chapter. (Editor's Note: See 1910 Supplement.)

[61 FR 31434, June 20, 1996]

§1926.1152

Methylene chloride

Note: The requirements applicable to construction employment under this section are identical to those set forth at 29 CFR 1910.1052. (Editor's Note: See 1910 Supplement.)

1926 Subpart Z

Authority for 1926 Subpart Z

Authority: Section 3704 of the Contract Work Hours and Safety Standards Act (40 U.S.C. 3701 et seq.); Sections 4, 6, and 8 of the Occupational Safety and Health Act of 1970 (29 U.S.C. 653, 655, 657); Secretary of Labor's Orders 12–71 (36 FR 8754), 8–76 (41 FR 25059), 9–83 (48 FR 35736), 1–90 (55 FR 9033), 6–96 (62 FR 111), 3–2000 (62 FR 50017), or 5–2002 (67 FR 65008) as applicable; and 29 CFR part 11. Section 1926.1102 of 29 CFR not issued under 29 U.S.C. 655 or 29 CFR part 1911; also issued under 5 U.S.C. 553.

1926 Subpart Z Appendix A

Designations for General Industry Standards Incorporated Into Body of Construction Standards

New Designations for General Industry Standards Incorporated Into Body of Construction Standards

1926 Designations for Applicable 1910 Standards

New section number and/or paragraph	Source section number and/or paragraph
1926.20(c)	1910.5(a)
[Do.] (d)	[Do.] (c)
[Do.] (e)	[Do.] (d)
1926.32(g)	1910.12(b)
1926.33	1910.1020
1926.34(a)	1910.36(b)(4)
[Do.] (b)	1910.37(q)(1)
[Do.] (c)	[Do.] (k)(2)
1926.35	1910.38(a)
1926.50(g)	1910.151(c)
1926.51(a)(6)	1910.141(a)(2)(v)
[Do.] (d)(2)	[Do.] (h)
[Do.] (f)(2)-(4)	[Do.] (d)(1)-(3)
[Do.] (g)	[Do.] (g)(2)
[Do.] (h)	[Do.] (a)(5)
[Do.] (i)	[Do.] (e)
1926.53(c)-(r)	1910.96
1926.57(f)-(i)	1910.94
1926.64	1910.119
1926.65	1910.120
1926.66(a)	1910.107(a)
[Do.] (b)	[Do.] (b)(1)-(10)
[Do.] (c)-(d)	[Do.] (c)-(d)
[Do.] (e)-(g)	[Do.] (h)-(j)
1926.95	1910.132
1926.96	1910.136
1926.97(a)-(e)	1910.156(e)
[Do.] (f)-(h)	[Do.] Subpart L App. E
1926.98	1910.156(f)
1926.102(a)(6)	1910.133(a)(2)
[Do.] (7)	[Do.] (4)
[Do.] (8)	[Do.] (5)
1926.103(d)	1910.134(a)
[Do.] (e)	[Do.] (b)
[Do.] (f)-(i)	[Do.] (d)-(g)
1926.150(c)(1)(xi)	1910.157(g)(1)
[Do.] (xii)	[Do.] (g)(2)
[Do.] (xiii)	[Do.] (c)(4)
[Do.] (xiv)	[Do.] (e)(3)
1926.152(b)(5)	1910.107(e)(2)

1926 Designations for Applicable 1910 Standards (continued)

New section number and/or paragraph	Source section number and/or paragraph
[Do.] (h)	1910.106(j)
[Do.] (i)	[Do.] (b)
[Do.] (j)	[Do.] (c)
[Do.] (k)(1)-(3)	[Do.] (g)(4)
[Do.] (k)(4)	[Do.] (a)(22)
1926.153(a)(3)	1910.110(a)(4)
[Do.] (m)(1)	[Do.] (d)(1)
[Do.] (2)	[Do.] (d)(2)
[Do.] (3)	[Do.] (d)(7)(vii)
[Do.] (4)	[Do.] (d)(7)(viii)
[Do.] (n)	[Do.] (b)(5)(iii)
[Do.] (o)	[Do.] (d)(10)
1926.156	1910.160
1926.157	1910.162
1926.158	1910.164
1926.159	1910.165
1926.200(c)(3)	1910.145(d)(4)
1926.250(c)	1910.176(c)
[Do.] (d)(1)-(4)	1910.30(a)(1),(2),(4)-(5)
1926.251(a)(5)	1910.184(a)
[Do.] (a)(6)	[Do.] (d)
[Do.] (b)(6)(i)-(ii)	[Do.] (e)(3)(i)-(ii)
[Do.] (c)(6)-(7)	[Do.] (c)(2)-(3)
[Do.] (c)(8)	[Do.] (c)(5)
[Do.] (c)(9)	[Do.] (c)(7)
[Do.] (c)(10)-(12)	[Do.] (c)(10)-(12)
[Do.] (c)(13)-(15)	[Do.] (f)(2)-(4)
[Do.] (d)(3)-(6)	[Do.] (h)(2)-(5)
[Do.] (e)(3)-(5)	[Do.] (i)(2)-(4)
[Do.] (e)(6)-(7)	[Do.] (i)(6)-(7)
[Do.] (e)(8)	[Do.] (i)(9)
1926.300(b)(3)	1910.212(a)(1)
[Do.] (4)	[Do.] (a)(3)
[Do.] (5)	[Do.] (a)(5)
[Do.] (6)	[Do.] (b)
[Do.] (7)	1910.215(b)(9)
[Do.] (8)-(9)	[Do.] (b)(3)-(4)
1926.302(b)(10)	1910.244(b)
1926.303(b)(2)	1910.215(a)(2)
[Do.] (e)	[Do.] (4)
1926.304(g)	1910.213(h)(1)
[Do.] (h)	[Do.] (d)(1)
[Do.] (i)	[Do.] (c)(1)
1926.305(d)(1)	1910.244(a)(2)(iii)-(viii)
1926.306	1910.169
1926.307	1910.219
1926.350(a)(10)	1910.253(b)(4)(iii)
[Do.] (11)	[Do.] (2)(ii)
[Do.] (12)	1910.101(b)
1926.353(b)(3)	1910.252(b)(4)(iv)
1926.416(a)(4)	1910.333(c)(2)
[Do.] (f)(1)	[Do.] (c)(10)
[Do.] (2)	1910.334(a)(1)
[Do.] (3)	[Do.] (a)(2)(iii)
[Do.] (4)	[Do.] (a)(5)
[Do.] (5)-(6)	[Do.] (b)(1)-(2)
[Do.] (7)-(9)	[Do.] (c)(1)-(3)

1926 Designations for Applicable 1910 Standards (continued)

New section number and/or paragraph	Source section number and/or paragraph
[Do.] (10)	[Do.] (d)
1926.417(d)	1910.333(b)(2)
1926.451(a)(22)	1910.28(a)(15)
[Do.] (23)	[Do.] (18)
[Do.] (24)	[Do.] (20)
1926.453(a)	1910.29(a)
[Do.] (b)	[Do.] (c)
1926.600(a)(7)	1910.176(f)
1926.602(c)(1)(vii)	1910.178(m)(3)
[Do.] (viii)	[Do.] (12)
1926.900(s)	1910.109(g)(2)(ii)
[Do.] (t)	[Do.] (h)(3)(ii)
1926.905(u)	[Do.] (e)(3)(iii)
1926.914(aa)	[Do.] (a)(12)
1926.1050(b)	1910.21(g)(9)
1926.1071	1910.401
1926.1072	1910.402
1926.1076	1910.410
1926.1080	1910.420
1926.1081	1910.421
1926.1082	1910.422
1926.1083	1910.423
1926.1084	1910.424
1926.1085	1910.425
1926.1086	1910.426
1926.1087	1910.427
1926.1090	1910.430
1926.1091	1910.440
1926.1092	1910.441
1926.1102	1910.1002
1926.1103	1910.1003
1926.1104	1910.1004
1926.1105	1910.1005
1926.1106	1910.1006
1926.1107	1910.1007
1926.1108	1910.1008
1926.1109	1910.1009
1926.1110	1910.1010
1926.1111	1910.1011
1926.1112	1910.1012
1926.1113	1910.1013
1926.1114	1910.1014
1926.1115	1910.1015
1926.1116	1910.1016
1926.1117	1910.1017
1926.1118	1910.1018
1926.1128	1910.1028
1926.1129	1910.1029
1926.1144	1910.1044
1926.1145	1910.1045
1926.1147	1910.1047
1926.1148	1910.1048

[58 FR 35305, June 30, 1993, as amended at 61 FR 9255, Mar. 7, 1996]

Authority: Section 107 of the Contract Work Hours and Safety Standards Act (40 U.S.C. 3704); Sections 4, 6, and 8 of the Occupational Safety and Health Act of 1970 (29 U.S.C. 653, 655, 657); Secretary of Labor's Order No. 12-71 (36 FR 8754), 8-76 (41 FR 25059), 9-83 (48 FR 35736),1-90 (55 FR 9033), 6-96 (62 FR 111), 3-2000 (62 FR 50017), 5-2002 (67 FR 650008), 5-2007 (72 FR 31159), 4-2010 (75 FR 55355), or 1-2012 (77 FR 3912), as applicable; and 29 CFR part 1911.

Subpart AA- BB
Reserved

Subpart CC
Cranes and Derricks in Construction

§1926.1400
Scope

(a) **This standard applies to power-operated** equipment, when used in construction, that can hoist, lower and horizontally move a suspended load. Such equipment includes, but is not limited to: Articulating cranes (such as knuckle-boom cranes); crawler cranes; floating cranes; cranes on barges; locomotive cranes; mobile cranes (such as wheel-mounted, rough-terrain, all-terrain, commercial truck-mounted, and boom truck cranes); multi-purpose machines when configured to hoist and lower (by means of a winch or hook) and horizontally move a suspended load; industrial cranes (such as carry-deck cranes); dedicated pile drivers; service/mechanic trucks with a hoisting device; a crane on a monorail; tower cranes (such as a fixed jib, *i.e.,* "hammerhead boom"), luffing boom and self-erecting); pedestal cranes; portal cranes; overhead and gantry cranes; straddle cranes; sideboom cranes; derricks; and variations of such equipment. However, items listed in paragraph (c) of this section are excluded from the scope of this standard. [§1926.1400(a)]

(b) **Attachments.** This standard applies to equipment included in paragraph (a) of this section when used with attachments. Such attachments, whether crane-attached or suspended include, but are not limited to: Hooks, magnets, grapples, clamshell buckets, orange peel buckets, concrete buckets, drag lines, personnel platforms, augers or drills and pile driving equipment. [§1926.1400(b)]

(c) **Exclusions.** This subpart does not cover: [§1926.1400(c)]

(1) *Machinery included in paragraph* (a) of this section while it has been converted or adapted for a non-hoisting/lifting use. Such conversions/adaptations include, but are not limited to, power shovels, excavators and concrete pumps. [§1926.1400(c)(1)]

(2) *Power shovels,* excavators, wheel loaders, backhoes, loader backhoes, track loaders. This machinery is also excluded when used with chains, slings or other rigging to lift suspended loads. [§1926.1400(c)(2)]

(3) *Automotive wreckers and tow* trucks when used to clear wrecks and haul vehicles. [§1926.1400(c)(3)]

(4) *Digger derricks when used* for augering holes for poles carrying electric and telecommunication lines, placing and removing the poles, and for handling associated materials to be installed on or removed from the poles. Digger derricks used in work subject to 29 CFR part 1926, subpart V, must comply with 29 CFR 1910.269. Digger derricks used in construction work for telecommunication service (as defined at 29 CFR 1910.268(s)(40)) must comply with 29 CFR 1910.268. [§1926.1400(c)(4)]

(5) *Machinery originally designed* as vehicle-mounted aerial devices (for lifting personnel) and self-propelled elevating work platforms. [§1926.1400(c)(5)]

(6) *Telescopic/hydraulic gantry systems.* [§1926.1400(c)(6)]

(7) *Stacker cranes.* [§1926.1400(c)(7)]

(8) *Powered industrial trucks* (forklifts), except when configured to hoist and lower (by means of a winch or hook) and horizontally move a suspended load. [§1926.1400(c)(8)]

(9) *Mechanic's truck with a hoisting* device when used in activities related to equipment maintenance and repair. [§1926.1400(c)(9)]

(10) *Machinery that hoists by using a come-a-long or chainfall.* [§1926.1400(c)(10)]

(11) *Dedicated drilling rigs.* [§1926.1400(c)(11)]

(12) *Gin poles when used* for the erection of communication towers. [§1926.1400(c)(12)]

(13) *Tree trimming and tree removal work.* [§1926.1400(c)(13)]

(14) *Anchor handling or dredge-related* operations with a vessel or barge using an affixed A-frame. [§1926.1400(c)(14)]

(15) *Roustabouts.* [§1926.1400(c)(15)]

(16) *Helicopter cranes.* [§1926.1400(c)(16)]

(17) *Material Delivery* [§1926.1400(c)(17)]

(i) *Articulating/knuckle-boom truck cranes* that deliver material to a construction site when used to transfer materials from the truck crane to the ground, without arranging the materials in a particular sequence for hoisting. [§1926.1400(c)(17)(i)]

(ii) *Articulating/knuckle-boom truck cranes* that deliver material to a construction site when the crane is used to transfer building supply sheet goods or building supply packaged materials from the truck crane onto a structure, using a fork/cradle at the end of the boom, but only when the truck crane is equipped with a properly functioning automatic overload prevention device. Such sheet goods or packaged materials include, but are not limited to: Sheets of sheet rock, sheets of plywood, bags of cement, sheets or packages of roofing shingles, and rolls of roofing felt. [§1926.1400(c)(17)(ii)]

(iii) *This exclusion does not apply when:* [§1926.1400(c)(17)(iii)]

[A] The articulating/knuckle-boom crane is used to hold, support or stabilize the material to facilitate a construction activity, such as holding material in place while it is attached to the structure; [§1926.1400(c)(17)(iii)[A]]

[B] The material being handled by the articulating/knuckle-boom crane is a prefabricated component. Such prefabricated components include, but are not limited to: Precast concrete members or panels, roof trusses (wooden, cold-formed metal, steel, or other material), prefabricated building sections such as, but not limited to: Floor panels, wall panels, roof panels, roof structures, or similar items; [§1926.1400(c)(17)(iii)[B]]

[C] The material being handled by the crane is a structural steel member (for example, steel joists, beams, columns, steel decking (bundled or unbundled) or a component of a systems-engineered metal building (as defined in 29 CFR 1926 subpart R). [§1926.1400(c)(17)(iii)[C]]

[D] The activity is not specifically excluded under §1400(c)(17)(i) and (ii). [§1926.1400(c)(17)(iii)[D]]

(d) **All sections of this subpart CC** apply to the equipment covered by this standard unless specified otherwise. [§1926.1400(d)]

(e) **The duties of controlling entities** under this subpart include, but are not limited to, the duties specified in Sec. 1926.1402(c), Sec. 1926.1402(e) and Sec. 1926.1424(b). [§1926.1400(e)]

(f) **Where provisions of this standard direct** an operator, crewmember, or other employee to take certain actions, the employer must establish, effectively communicate to the relevant persons, and enforce, work rules to ensure compliance with such provisions. [§1926.1400(f)]

(g) **For work covered by subpart** V of this part, compliance with 29 CFR §1910.269(p) is deemed compliance with §§1926.1407 through 1926.1411. [§1926.1400(g)]

(h) **Section 1926.1402 does** not apply to cranes designed for use on railroad tracks, when used on railroad tracks that are part of the general railroad system of transportation that is regulated pursuant to the Federal Railroad Administration under 49 CFR part 213, and that comply with applicable Federal Railroad Administration requirements. *See* Sec. 1926.1402(f). [§1926.1400(h)]

§1926.1401
Definitions

A/D director (Assembly/Disassembly director) means an individual who meets this subpart's requirements for an A/D director, irrespective of the person's formal job title or whether the person is non-management or management personnel.

Articulating crane means a crane whose boom consists of a series of folding, pin connected structural members, typically manipulated to extend or retract by power from hydraulic cylinders.

Assembly/Disassembly means the assembly and/or disassembly of equipment covered under this standard. With regard to tower cranes, "erecting and climbing" replaces the term "assembly," and "dismantling" replaces the term "disassembly." Regardless of whether the crane is initially erected to its full height or is climbed in stages, the process of increasing the height of the crane is an erection process.

Assist crane means a crane used to assist in assembling or disassembling a crane.

Attachments means any device that expands the range of tasks that can be done by the equipment. Examples include, but are not limited to: An auger, drill, magnet, pile-driver, and boom-attached personnel platform.

Audible signal means a signal made by a distinct sound or series of sounds. Examples include, but are not limited to, sounds made by a bell, horn, or whistle.

Blocking (also referred to as "cribbing") is wood or other material used to support equipment or a component and distribute loads to the ground. It is typically used to support lattice boom sections during assembly/disassembly and under outrigger and stabilizer floats.

Boatswain's chair means a single-point adjustable suspension scaffold consisting of a seat or sling (which may be incorporated into a full body harness) designed to support one employee in a sitting position.

Bogie means "travel bogie," which is defined below.

Boom (equipment other than tower crane) means an inclined spar, strut, or other long structural member which supports the upper hoisting tackle on a crane or derrick. Typically, the length and vertical angle of the boom can be varied to achieve increased height or height and reach when lifting loads. Booms can usually be grouped into general categories of hydraulically extendible, cantilevered type, latticed section, cable supported type or articulating type.

Boom (tower cranes): On tower cranes, if the "boom" (*i.e.,* principal horizontal structure) is fixed, it is referred to as a jib; if it is moveable up and down, it is referred to as a boom.

Boom angle indicator means a device which measures the angle of the boom relative to horizontal.

Boom hoist limiting device includes boom hoist disengaging device, boom hoist shut-off, boom hoist disconnect, boom hoist hydraulic relief, boom hoist kick-outs, automatic boom stop device, or derricking limiter. This type of device disengages boom hoist power when the boom reaches a predetermined operating angle. It also sets brakes or closes valves to prevent the boom from lowering after power is disengaged.

Boom length indicator indicates the length of the permanent part of the boom (such as ruled markings on the boom) or, as in some computerized systems, the length of the boom with extensions/attachments.

Boom stop includes boom stops, (belly straps with struts/standoff), telescoping boom stops, attachment boom stops, and backstops. These devices restrict the boom from moving above a certain maximum angle and toppling over backward.

Boom suspension system means a system of pendants, running ropes, sheaves, and other hardware which supports the boom tip and controls the boom angle.

Builder means the builder/constructor of equipment.

Center of gravity: The center of gravity of any object is the point in the object around which its weight is evenly distributed. If you could put a support under that point, you could balance the object on the support.

Certified welder means a welder who meets nationally recognized certification requirements applicable to the task being performed.

Climbing means the process in which a tower crane is raised to a new working height, either by adding additional tower sections to the top of the crane (top climbing), or by a system in which the entire crane is raised inside the structure (inside climbing).

Come-a-long means a mechanical device typically consisting of a chain or cable attached at each end that is used to facilitate movement of materials through leverage.

Competent person means one who is capable of identifying existing and predictable hazards in the surroundings or working conditions which are unsanitary, hazardous, or dangerous to employees, and who has authorization to take prompt corrective measures to eliminate them.

Controlled load lowering means lowering a load by means of a mechanical hoist drum device that allows a hoisted load to be lowered with maximum control using the gear train or hydraulic components of the hoist mechanism. Controlled load lowering requires the use of the hoist drive motor, rather than the load hoist brake, to lower the load.

Controlling entity means an employer that is a prime contractor, general contractor, construction manager or any other legal entity which has the overall responsibility for the construction of the project — its planning, quality and completion.

Counterweight means a weight used to supplement the weight of equipment in providing stability for lifting loads by counterbalancing those loads.

Crane/derrick includes all equipment covered by this subpart.

Crawler crane means equipment that has a type of base mounting which incorporates a continuous belt of sprocket driven track.

Crossover points means locations on a wire rope which is spooled on a drum where one layer of rope climbs up on and crosses over the previous layer. This takes place at each flange of the drum as the rope is spooled onto the drum, reaches the flange, and begins to wrap back in the opposite direction.

Dedicated channel means a line of communication assigned by the employer who controls the communication system to only one signal person and crane/derrick or to a coordinated group of cranes/derricks/signal person(s).

Dedicated pile-driver is a machine that is designed to function exclusively as a pile-driver. These machines typically have the ability to both hoist the material that will be pile-driven and to pile-drive that material.

Dedicated spotter (power lines): To be considered a dedicated spotter, the requirements of Sec. 1926.1428 (Signal person qualifications) must be met and his/her sole responsibility is to watch the separation between the power line and the equipment, load line and load (including rigging and lifting accessories), and ensure through communication with the operator that the applicable minimum approach distance is not breached.

Directly under the load means a part or all of an employee is directly beneath the load.

Dismantling includes partial dismantling (such as dismantling to shorten a boom or substitute a different component).

Drum rotation indicator means a device on a crane or hoist which indicates in which direction and at what relative speed a particular hoist drum is turning.

Electrical contact occurs when a person, object, or equipment makes contact or comes in close proximity with an energized conductor or equipment that allows the passage of current.

Employer-made equipment means floating cranes/derricks designed and built by an employer for the employer's own use.

Encroachment is where any part of the crane, load line or load (including rigging and lifting accessories) breaches a minimum clearance distance that this subpart requires to be maintained from a power line.

Equipment means equipment covered by this subpart.

Equipment criteria means instructions, recommendations, limitations and specifications.

Fall protection equipment means guardrail systems, safety net systems, personal fall arrest systems, positioning device systems or fall restraint systems.

Fall restraint system means a fall protection system that prevents the user from falling any distance. The system is comprised of either a body belt or body harness, along with an anchorage, connectors and other necessary equipment. The other components typically include a lanyard, and may also include a lifeline and other devices.

Fall zone means the area (including but not limited to the area directly beneath the load) in which it is reasonably foreseeable that partially or completely suspended materials could fall in the event of an accident.

Flange points are points of contact between rope and drum flange where the rope changes layers.

Floating cranes/derricks means equipment designed by the manufacturer (or employer) for marine use by permanent attachment to a barge, pontoons, vessel or other means of flotation.

For example means "one example, although there are others."

Free fall (of the load line) means that only the brake is used to regulate the descent of the load line (the drive mechanism is not used to drive the load down faster or retard its lowering).

Free surface effect is the uncontrolled transverse movement of liquids in compartments which reduce a vessel's transverse stability.

Hoist means a mechanical device for lifting and lowering loads by winding a line onto or off a drum.

Hoisting is the act of raising, lowering or otherwise moving a load in the air with equipment covered by this standard. As used in this standard, "hoisting" can be done by means other than wire rope/hoist drum equipment.

Include/including means "including, but not limited to."

Insulating link/device means an insulating device listed, labeled, or accepted by a Nationally Recognized Testing Laboratory in accordance with 29 CFR 1910.7.

Jib stop (also referred to as a jib backstop), is the same type of device as a boom stop but is for a fixed or luffing jib.

Land crane/derrick is equipment not originally designed by the manufacturer for marine use by permanent attachment to barges, pontoons, vessels, or other means of floatation.

List means the angle of inclination about the longitudinal axis of a barge, pontoons, vessel or other means of floatation.

Load refers to the object(s) being hoisted and/or the weight of the object(s); both uses refer to the object(s) and the load-attaching equipment, such as, the load block, ropes, slings, shackles, and any other ancillary attachment.

Load moment (or rated capacity) indicator means a system which aids the equipment operator by sensing (directly or indirectly) the overturning moment on the equipment, *i.e.,* load multiplied by radius. It compares this lifting condition to the equipment's rated capacity, and indicates to the operator the percentage of capacity at which the equipment is working. Lights, bells, or buzzers may be incorporated as a warning of an approaching overload condition.

Load moment (or rated capacity) limiter means a system which aids the equipment operator by sensing (directly or indirectly) the overturning moment on the equipment, *i.e.*, load multiplied by radius. It compares this lifting condition to the equipment's rated capacity, and when the rated capacity is reached, it shuts off power to those equipment functions which can increase the severity of loading on the equipment, *e.g.*, hoisting, telescoping out, or luffing out. Typically, those functions which decrease the severity of loading on the equipment remain operational, *e.g.*, lowering, telescoping in, or luffing in.

Locomotive crane means a crane mounted on a base or car equipped for travel on a railroad track.

Luffing jib limiting device is similar to a boom hoist limiting device, except that it limits the movement of the luffing jib.

Marine hoisted personnel transfer device means a device, such as a "transfer net," that is designed to protect the employees being hoisted during a marine transfer and to facilitate rapid entry into and exit from the device. Such devices do not include boatswain's chairs when hoisted by equipment covered by this standard.

Marine worksite means a construction worksite located in, on or above the water.

Mobile crane means a lifting device incorporating a cable suspended latticed boom or hydraulic telescopic boom designed to be moved between operating locations by transport over the road.

Moving point-to-point means the times during which an employee is in the process of going to or from a work station.

Multi-purpose machine means a machine that is designed to be configured in various ways, at least one of which allows it to hoist (by means of a winch or hook) and horizontally move a suspended load. For example, a machine that can rotate and can be configured with removable forks/tongs (for use as a forklift) or with a winch pack, jib (with a hook at the end) or jib used in conjunction with a winch. When configured with the forks/tongs, it is not covered by this subpart. When configured with a winch pack, jib (with a hook at the end) or jib used in conjunction with a winch, it is covered by this subpart.

Nationally recognized accrediting agency is an organization that, due to its independence and expertise, is widely recognized as competent to accredit testing organizations. Examples of such accrediting agencies include, but are not limited to, the National Commission for Certifying Agencies and the American National Standards Institute.

Nonconductive means that, because of the nature and condition of the materials used, and the conditions of use (including environmental conditions and condition of the material), the object in question has the property of not becoming energized (that is, it has high dielectric properties offering a high resistance to the passage of current under the conditions of use).

Operational aids are devices that assist the operator in the safe operation of the crane by providing information or automatically taking control of a crane function. These include, but are not limited to, the devices listed in Sec. 1926.1416 ("listed operational aids").

Operational controls means levers, switches, pedals and other devices for controlling equipment operation.

Operator means a person who is operating the equipment.

Overhead and gantry cranes includes overhead/bridge cranes, semigantry, cantilever gantry, wall cranes, storage bridge cranes, launching gantry cranes, and similar equipment, irrespective of whether it travels on tracks, wheels, or other means.

Paragraph refers to a paragraph in the same section of this subpart that the word "paragraph" is used, unless otherwise specified.

Pendants includes both wire and bar types. Wire type: A fixed length of wire rope with mechanical fittings at both ends for pinning segments of wire rope together. Bar type: Instead of wire rope, a bar is used. Pendants are typically used in a latticed boom crane system to easily change the length of the boom suspension system without completely changing the rope on the drum when the boom length is increased or decreased.

Personal fall arrest system means a system used to arrest an employee in a fall from a working level. It consists of an anchorage, connectors, a body harness and may include a lanyard, deceleration device, lifeline, or suitable combination of these.

Portal crane is a type of crane consisting of a rotating upperstructure, hoist machinery, and boom mounted on top of a structural gantry which may be fixed in one location or have travel capability. The gantry legs or columns usually have portal openings in between to allow passage of traffic beneath the gantry.

Power lines means electric transmission and distribution lines.

Procedures include, but are not limited to: Instructions, diagrams, recommendations, warnings, specifications, protocols and limitations.

Proximity alarm is a device that provides a warning of proximity to a power line and that has been listed, labeled, or accepted by a Nationally Recognized Testing Laboratory in accordance with 29 CFR 1910.7.

Qualified evaluator (not a third party) means a person employed by the signal person's employer who has demonstrated that he/she is competent in accurately assessing whether individuals meet the Qualification Requirements in this subpart for a signal person.

Qualified evaluator (third party) means an entity that, due to its independence and expertise, has demonstrated that it is competent in accurately assessing whether individuals meet the Qualification Requirements in this subpart for a signal person.

Qualified person means a person who, by possession of a recognized degree, certificate, or professional standing, or who by extensive knowledge, training and experience, successfully demonstrated the ability to solve/resolve problems relating to the subject matter, the work, or the project.

Qualified rigger is a rigger who meets the criteria for a qualified person.

Range control limit device is a device that can be set by an equipment operator to limit movement of the boom or jib tip to a plane or multiple planes.

Range control warning device is a device that can be set by an equipment operator to warn that the boom or jib tip is at a plane or multiple planes.

Rated capacity means the maximum working load permitted by the manufacturer under specified working conditions. Such working conditions typically include a specific combination of factors such as equipment configuration, radii, boom length, and other parameters of use.

Rated capacity indicator: See load moment indicator.

Rated capacity limiter: See load moment limiter.

Repetitive pickup points refer to, when operating on a short cycle operation, the rope being used on a single layer and being spooled repetitively over a short portion of the drum.

Running wire rope means a wire rope that moves over sheaves or drums.

Runway means a firm, level surface designed, prepared and designated as a path of travel for the weight and configuration of the crane being used to lift and travel with the crane suspended platform. An existing surface may be used as long as it meets these criteria.

Section means a section of this subpart, unless otherwise specified.

Sideboom crane means a track-type or wheel-type tractor having a boom mounted on the side of the tractor, used for lifting, lowering or transporting a load suspended on the load hook. The boom or hook can be lifted or lowered in a vertical direction only.

Special hazard warnings means warnings of site-specific hazards (for example, proximity of power lines).

Stability (flotation device) means the tendency of a barge, pontoons, vessel or other means of flotation to return to an upright position after having been inclined by an external force.

Standard Method means the protocol in Appendix A of this subpart for hand signals.

Such as means "such as, but not limited to."

Superstructure: See Upperworks.

Tagline means a rope (usually fiber) attached to a lifted load for purposes of controlling load spinning and pendular motions or used to stabilize a bucket or magnet during material handling operations.

Tender means an individual responsible for monitoring and communicating with a diver.

Tilt up or tilt down operation means raising/lowering a load from the horizontal to vertical or vertical to horizontal.

Tower crane is a type of lifting structure which utilizes a vertical mast or tower to support a working boom (jib) in an elevated position. Loads are suspended from the working boom. While the working boom may be of the fixed type (horizontal or angled) or have luffing capability, it can always rotate to swing loads, either by rotating on the top of the tower (top slewing) or by the rotation of the tower (bottom slewing). The tower base may be fixed in one location or ballasted and moveable between locations. Mobile cranes that are configured with luffing jib and/or tower attachments are not considered tower cranes under this section.

Travel bogie (tower cranes) is an assembly of two or more axles arranged to permit vertical wheel displacement and equalize the loading on the wheels.

Trim means angle of inclination about the transverse axis of a barge, pontoons, vessel or other means of floatation.

Two blocking means a condition in which a component that is uppermost on the hoist line such as the load block, hook block, overhaul ball, or similar component, comes in contact with the boom tip, fixed upper block or similar component. This binds the system and continued application of power can cause failure of the hoist rope or other component.

Unavailable procedures means procedures that are no longer available from the manufacturer, or have never been available, from the manufacturer.

Upperstructure: See Upperworks.

Upperworks means the revolving frame of equipment on which the operating machinery (and many cases the engine) are mounted along with the operator's cab. The counterweight is typically supported on the rear of the upperstructure and the boom or other front end attachment is mounted on the front.

Up to means "up to and including."

Wire rope means a flexible rope constructed by laying steel wires into various patterns of multi-wired strands around a core system to produce a helically wound rope.

§1926.1402
Ground conditions

(a) **Definitions.** [§1926.1402(a)]

(1) **"Ground conditions"** means the ability of the ground to support the equipment (including slope, compaction, and firmness). [§1926.1402(a)(1)]

(2) **"Supporting materials"** means blocking, mats, cribbing, marsh buggies (in marshes/wetlands), or similar supporting materials or devices. [§1926.1402(a)(2)]

(b) **The equipment must not be assembled or used** unless ground conditions are firm, drained, and graded to a sufficient extent so that, in conjunction (if necessary) with the use of supporting materials, the equipment manufacturer's specifications for adequate support and degree of level of the equipment are met. The requirement for the ground to be drained does not apply to marshes/wetlands. [§1926.1402(b)]

(c) **The controlling entity must:** [§1926.1402(c)]

(1) *Ensure that ground preparations* necessary to meet the requirements in paragraph (b) of this section are provided. [§1926.1402(c)(1)]

(2) *Inform the user of the equipment* and the operator of the location of hazards beneath the equipment set-up area (such as voids, tanks, utilities) if those hazards are identified in documents (such as site drawings, as-built drawings, and soil analyses) that are in the possession of the controlling entity (whether at the site or off-site) or the hazards are otherwise known to that controlling entity. [§1926.1402(c)(2)]

(d) **If there is no controlling entity** for the project, the requirement in paragraph (c)(1) of this section must be met by the employer that has authority at the site to make or arrange for ground preparations needed to meet paragraph (b) of this section. [§1926.1402(d)]

(e) **If the A/D director or the operator** determines that ground conditions do not meet the requirements in paragraph (b) of this section, that person's employer must have a discussion with the controlling entity regarding the ground preparations that are needed so that, with the use of suitable supporting materials/devices (if necessary), the requirements in paragraph (b) of this section can be met. [§1926.1402(e)]

(f) **This section does not apply** to cranes designed for use on railroad tracks when used on railroad tracks that are part of the general railroad system of transportation that is regulated pursuant to the Federal Railroad Administration under 49 CFR part 213 and that comply with applicable Federal Railroad Administration requirements. [§1926.1402(f)]

§1926.1403
Assembly/Disassembly — selection of manufacturer or employer procedures

When assembling or disassembling equipment (or attachments), the employer must comply with all applicable manufacturer prohibitions and must comply with either:

(a) **Manufacturer procedures applicable** to assembly and disassembly, or [§1926.1403(a)]

(b) **Employer procedures for assembly** and disassembly. Employer procedures may be used only where the employer can demonstrate that the procedures used meet the requirements in Sec. 1926.1406. Note: The employer must follow manufacturer procedures when an employer uses synthetic slings during assembly or disassembly rigging. (*See* Sec. 1926.1404(r).) [§1926.1403(b)]

§1926.1404
Assembly/Disassembly — general requirements (applies to all assembly and disassembly operations)

(a) **Supervision — competent-qualified person.** [§1926.1404(a)]

(1) *Assembly/disassembly must be directed by a person* who meets the criteria for both a competent person and a qualified person, or by a competent person who is assisted by one or more qualified persons ("A/D director"). [§1926.1404(a)(1)]

(2) *Where the assembly/disassembly is being performed by only* one person, that person must meet the criteria for both a competent person and a qualified person. For purposes of this standard, that person is considered the A/D director. [§1926.1404(a)(2)]

(b) **Knowledge of procedures.** The A/D director must understand the applicable assembly/disassembly procedures. [§1926.1404(b)]

(c) **Review of procedures.** The A/D director must review the applicable assembly/disassembly procedures immediately prior to the commencement of assembly/disassembly unless the A/D director understands the procedures and has applied them to the same type and configuration of equipment (including accessories, if any). [§1926.1404(c)]

(d) **Crew instructions.** [§1926.1404(d)]

(1) *Before commencing assembly/disassembly* operations, the A/D director must ensure that the crew members understand all of the following: [§1926.1404(d)(1)]

(i) *Their tasks.* [§1926.1404(d)(1)(i)]

(ii) *The hazards associated with their tasks.* [§1926.1404(d)(1)(ii)]

(iii) *The hazardous positions/locations that they need to avoid.* [§1926.1404(d)(1)(iii)]

(2) *During assembly/disassembly operations,* before a crew member takes on a different task, or when adding new personnel during the operations, the requirements in paragraphs (d)(1)(i) through (d)(1)(iii) of this section must be met. [§1926.1404(d)(2)]

(e) **Protecting assembly/disassembly crew members out of operator view.** [§1926.1404(e)]

(1) *Before a crew member* goes to a location that is out of view of the operator and is either in, on, or under the equipment, or near the equipment (or load) where the crew member could be injured by movement of the equipment (or load), the crew member must inform the operator that he/she is going to that location. [§1926.1404(e)(1)]

(2) *Where the operator knows that a crew* member went to a location covered by paragraph (e)(1) of this section, the operator must not move any part of the equipment (or load) until the operator is informed in accordance with a pre-arranged system of communication that the crew member is in a safe position. [§1926.1404(e)(2)]

(f) **Working under the boom, jib or other components.** [§1926.1404(f)]

(1) *When pins (or similar devices)* are being removed, employees must not be under the boom, jib, or other components, except where the requirements of paragraph (f)(2) of this section are met. [§1926.1404(f)(1)]

(2) *Exception.* Where the employer demonstrates that site constraints require one or more employees to be under the boom, jib, or other components when pins (or similar devices) are being removed, the A/D director must implement procedures that minimize the risk of unintended dangerous movement and minimize the duration and extent of exposure under the boom. (*See* Non-mandatory Appendix B of this subpart for an example.) [§1926.1404(f)(2)]

(g) **Capacity limits.** During all phases of assembly/disassembly, rated capacity limits for loads imposed on the equipment, equipment components (including rigging), lifting lugs and equipment accessories, must not be exceeded for the equipment being assembled/disassembled. [§1926.1404(g)]

(h) **Addressing specific hazards.** The A/D director supervising the assembly/disassembly operation must address the hazards associated with the operation, which include: [§1926.1404(h)]

(1) *Site and ground bearing conditions.* Site and ground conditions must be adequate for safe assembly/disassembly operations and to support the equipment during assembly/disassembly (*see* Sec. 1926.1402 for ground condition requirements). [§1926.1404(h)(1)]

(2) *Blocking material.* The size, amount, condition and method of stacking the blocking must be sufficient to sustain the loads and maintain stability. [§1926.1404(h)(2)]

(3) *Proper location of blocking.* When used to support lattice booms or components, blocking must be appropriately placed to: [§1926.1404(h)(3)]

(i) *Protect the structural integrity* of the equipment, and [§1926.1404(h)(3)(i)]

(ii) *Prevent dangerous movement and collapse.* [§1926.1404(h)(3)(ii)]

(4) *Verifying assist crane loads.* When using an assist crane, the loads that will be imposed on the assist crane at each phase of assembly/disassembly must be verified in accordance with Sec. 1926.1417(o)(3) before assembly/disassembly begins. [§1926.1404(h)(4)]

(5) *Boom and jib pick points.* The point(s) of attachment of rigging to a boom (or boom sections or jib or jib sections) must be suitable for preventing structural damage and facilitating safe handling of these components. [§1926.1404(h)(5)]

(6) *Center of gravity.* [§1926.1404(h)(6)]

(i) *The center of gravity of the load* must be identified if that is necessary for the method used for maintaining stability. [§1926.1404(h)(6)(i)]

(ii) *Where there is insufficient information* to accurately identify the center of gravity, measures designed to prevent unintended dangerous movement resulting from an inaccurate identification of the center of gravity must be used. (*See* Non-mandatory Appendix B of this subpart for an example.) [§1926.1404(h)(6)(ii)]

(7) *Stability upon pin removal.* The boom sections, boom suspension systems (such as gantry A-frames and jib struts), and components must be rigged or supported to maintain stability upon the removal of the pins. [§1926.1404(h)(7)]

(8) *Snagging.* Suspension ropes and pendants must not be allowed to catch on the boom or jib connection pins or cotter pins (including keepers and locking pins). [§1926.1404(h)(8)]

(9) *Struck by counterweights.* The potential for unintended movement from inadequately supported counterweights and from hoisting counterweights. [§1926.1404(h)(9)]

(10) *Boom hoist brake failure.* Each time reliance is to be placed on the boom hoist brake to prevent boom movement during assembly/disassembly, the brake must be tested prior to such reliance to determine if it is sufficient to prevent boom movement. If it is not sufficient, a boom hoist pawl, other locking device/back-up braking device, or another method of preventing dangerous movement of the boom (such as blocking or using an assist crane) from a boom hoist brake failure must be used. [§1926.1404(h)(10)]

(11) *Loss of backward stability.* Backward stability before swinging the upperworks, travel, and when attaching or removing equipment components. [§1926.1404(h)(11)]

(12) *Wind speed and weather.* The effect of wind speed and weather on the equipment. [§1926.1404(h)(12)]

(i) ***[Reserved.]*** [§1926.1404(i)]

(j) **Cantilevered boom sections.** Manufacturer limitations on the maximum amount of boom supported only by cantilevering must not be exceeded. Where these are unavailable, a registered professional engineer familiar with the type of equipment involved must determine in writing this limitation, which must not be exceeded. [§1926.1404(j)]

(k) **Weight of components.** The weight of each of the components must be readily available. [§1926.1404(k)]

(l) ***[Reserved.]*** [§1926.1404(l)]

(m) **Components and configuration.** [§1926.1404(m)]

(1) *The selection of components,* and configuration of the equipment, that affect the capacity or safe operation of the equipment must be in accordance with: [§1926.1404(m)(1)]

(i) *Manufacturer instructions,* prohibitions, limitations, and specifications. Where these are unavailable, a registered professional engineer familiar with the type of equipment involved must approve, in writing, the selection and configuration of components; or [§1926.1404(m)(1)(i)]

(ii) *Approved modifications that meet* the requirements of Sec. 1926.1434 (Equipment modifications). [§1926.1404(m)(1)(ii)]

(2) *Post-assembly inspection.* Upon completion of assembly, the equipment must be inspected to ensure compliance with paragraph (m)(1) of this section (*see* Sec. 1926.1412(c) for post-assembly inspection requirements). [§1926.1404(m)(2)]

(n) ***[Reserved.]*** [§1926.1404(n)]

(o) **Shipping pins.** Reusable shipping pins, straps, links, and similar equipment must be removed. Once they are removed they must either be stowed or otherwise stored so that they do not present a falling object hazard. [§1926.1404(o)]

(p) **Pile driving.** Equipment used for pile driving must not have a jib attached during pile driving operations. [§1926.1404(p)]

(q) **Outriggers and Stabilizers.** When the load to be handled and the operating radius require the use of outriggers or stabilizers, or at any time when outriggers or stabilizers are used, all of the following requirements must be met (except as otherwise indicated): [§1926.1404(q)]

(1) *The outriggers or stabilizers must be either* fully extended or, if manufacturer procedures permit, deployed as specified in the load chart. [§1926.1404(q)(1)]

(2) *The outriggers must be set to remove* the equipment weight from the wheels, except for locomotive cranes (*see* paragraph (q)(6) of this section for use of outriggers on locomotive cranes). This provision does not apply to stabilizers. [§1926.1404(q)(2)]

(3) *When outrigger floats are used,* they must be attached to the outriggers. When stabilizer floats are used, they must be attached to the stabilizers. [§1926.1404(q)(3)]

(4) *Each outrigger or stabilizer must be visible* to the operator or to a signal person during extension and setting. [§1926.1404(q)(4)]

(5) *Outrigger and stabilizer blocking must:* [§1926.1404(q)(5)]

(i) *Meet the requirements in paragraphs* (h)(2) and (h)(3) of this section. [§1926.1404(q)(5)(i)]

(ii) *Be placed only under* the outrigger or stabilizer float/pad of the jack or, where the outrigger or stabilizer is designed without a jack, under the outer bearing surface of the extended outrigger or stabilizer beam. [§1926.1404(q)(5)(ii)]

(6) *For locomotive cranes,* when using outriggers or stabilizers to handle loads, the manufacturer's procedures must be followed. When lifting loads without using outriggers or stabilizers, the manufacturer's procedures must be met regarding truck wedges or screws. [§1926.1404(q)(6)]

(r) **Rigging.** In addition to following the requirements in 29 CFR 1926.251 and other requirements in this and other standards applicable to rigging, when rigging is used for assembly/disassembly, the employer must ensure that: [§1926.1404(r)]

(1) *The rigging work is done* by a qualified rigger. [§1926.1404(r)(1)]

(2) *Synthetic slings are protected* from: Abrasive, sharp or acute edges, and configurations that could cause a reduction of the sling's rated capacity, such as distortion or localized compression. Note: Requirements for the protection of wire rope slings are contained in 29 CFR 1926.251(c)(9). [§1926.1404(r)(2)]

(3) *When synthetic slings are used,* the synthetic sling manufacturer's instructions, limitations, specifications and recommendations must be followed. [§1926.1404(r)(3)]

§1926.1405

Disassembly — additional requirements for dismantling of booms and jibs (applies to both the use of manufacturer procedures and employer procedures)

Dismantling (including dismantling for changing the length of) booms and jibs.

(a) **None of the pins in the pendants** are to be removed (partly or completely) when the pendants are in tension. [§1926.1405(a)]

(b) **None of the pins (top or bottom)** on boom sections located between the pendant attachment points and the crane/derrick body are to be removed (partly or completely) when the pendants are in tension. [§1926.1405(b)]

(c) **None of the pins (top or bottom)** on boom sections located between the uppermost boom section and the crane/derrick body are to be removed (partly or completely) when the boom is being supported by the uppermost boom section resting on the ground (or other support). [§1926.1405(c)]

(d) **None of the top pins** on boom sections located on the cantilevered portion of the boom being removed (the portion being removed ahead of the pendant attachment points) are to be removed (partly or completely) until the cantilevered section to be removed is fully supported. [§1926.1405(d)]

§1926.1406

Assembly/Disassembly — employer procedures — general requirements

(a) **When using employer procedures instead** of manufacturer procedures for assembly/disassembly, the employer must ensure that the procedures: [§1926.1406(a)]

(1) *Prevent unintended dangerous* movement, and prevent collapse, of any part of the equipment. [§1926.1406(a)(1)]

(2) *Provide adequate support* and stability of all parts of the equipment. [§1926.1406(a)(2)]

(3) *Position employees involved* in the assembly/disassembly operation so that their exposure to unintended movement or collapse of part or all of the equipment is minimized. [§1926.1406(a)(3)]

(b) **Qualified person.** Employer procedures must be developed by a qualified person. [§1926.1406(b)]

§1926.1407

Power line safety (up to 350 kV) — assembly and disassembly

(a) **Before assembling or disassembling** equipment, the employer must determine if any part of the equipment, load line, or load (including rigging and lifting accessories) could get, in the direction or area of assembly/disassembly, closer than 20 feet to a power line during the assembly/disassembly process. If so, the employer must meet the requirements in Option (1), Option (2), or Option (3) of this section, as follows: [§1926.1407(a)]

(1) *Option (1) — Deenergize and ground.* Confirm from the utility owner/operator that the power line has been deenergized and visibly grounded at the worksite. [§1926.1407(a)(1)]

(2) *Option (2) — 20 foot clearance.* Ensure that no part of the equipment, load line or load (including rigging and lifting accessories), gets closer than 20 feet to the power line by implementing the measures specified in paragraph (b) of this section. [§1926.1407(a)(2)]

(3) *Option (3) — Table A clearance.* [§1926.1407(a)(3)]

(i) *Determine the line's voltage* and the minimum clearance distance permitted under Table A (*see* Sec. 1926.1408). [§1926.1407(a)(3)(i)]

(ii) *Determine if any part of the equipment,* load line, or load (including rigging and lifting accessories), could get closer than the minimum clearance distance to the power line permitted under Table A (*see* Sec. 1926.1408). If so, then the employer must follow the requirements in paragraph (b) of this section to ensure that no part of the equipment, load line, or load (including rigging and lifting accessories), gets closer to the line than the minimum clearance distance. [§1926.1407(a)(3)(ii)]

(b) **Preventing encroachment/electrocution.** Where encroachment precautions are required under Option (2), or Option (3) of this section, all of the following requirements must be met: [§1926.1407(b)]

(1) *Conduct a planning meeting* with the Assembly/Disassembly director (A/D director), operator, assembly/disassembly crew and the other workers who will be in the assembly/disassembly area to review the location of the power line(s) and the steps that will be implemented to prevent encroachment/electrocution. [§1926.1407(b)(1)]

(2) *If tag lines are used,* they must be nonconductive. [§1926.1407(b)(2)]

(3) *At least one of the following* additional measures must be in place. The measure selected from this list must be effective in preventing encroachment. [§1926.1407(b)(3)]

The additional measures are: [§1926.1407(b)(3)]

(i) *Use a dedicated spotter* who is in continuous contact with the equipment operator. The dedicated spotter must: [§1926.1407(b)(3)(i)]

[A] Be equipped with a visual aid to assist in identifying the minimum clearance distance. Examples of a visual aid include, but are not limited to: A clearly visible line painted on the ground; a clearly visible line of stanchions; a set of clearly visible line-of-sight landmarks (such as a fence post behind the dedicated spotter and a building corner ahead of the dedicated spotter). [§1926.1407(b)(3)(i)[A]]

[B] Be positioned to effectively gauge the clearance distance. [§1926.1407(b)(3)(i)[B]]

[C] Where necessary, use equipment that enables the dedicated spotter to communicate directly with the operator. [§1926.1407(b)(3)(i)[C]]

[D] Give timely information to the operator so that the required clearance distance can be maintained. [§1926.1407(b)(3)(i)[D]]

(ii) *A proximity alarm set* to give the operator sufficient warning to prevent encroachment. [§1926.1407(b)(3)(ii)]

(iii) *A device that automatically warns* the operator when to stop movement, such as a range control warning device. Such a device must be set to give the operator sufficient warning to prevent encroachment. [§1926.1407(b)(3)(iii)]

(iv) *A device that automatically limits* range of movement, set to prevent encroachment. [§1926.1407(b)(3)(iv)]

(v) *An elevated warning line,* barricade, or line of signs, in view of the operator, equipped with flags or similar high-visibility markings. [§1926.1407(b)(3)(v)]

(c) **Assembly/disassembly below power lines prohibited.** No part of a crane/derrick, load line, or load (including rigging and lifting accessories), whether partially or fully assembled, is allowed below a power line unless the employer has confirmed that the utility owner/operator has deenergized and (at the worksite) visibly grounded the power line. [§1926.1407(c)]

(d) **Assembly/disassembly inside Table A clearance prohibited.** No part of a crane/derrick, load line, or load (including rigging and lifting accessories), whether partially or fully assembled, is allowed closer than the minimum approach distance under Table A (*see* Sec. 1926.1408) to a power line unless the employer has confirmed that the utility owner/operator has deenergized and (at the worksite) visibly grounded the power line. [§1926.1407(d)]

(e) **Voltage information.** Where Option (3) of this section is used, the utility owner/operator of the power lines must provide the requested voltage information within two working days of the employer's request. [§1926.1407(e)]

(f) **Power lines presumed energized.** The employer must assume that all power lines are energized unless the utility owner/operator confirms that the power line has been and continues to be deenergized and visibly grounded at the worksite. [§1926.1407(f)]

(g) **Posting of electrocution warnings.** There must be at least one electrocution hazard warning conspicuously posted in the cab so that it is in view of the operator and (except for overhead gantry and tower cranes) at least two on the outside of the equipment. [§1926.1407(g)]

§1926.1408

Power line safety (up to 350 kV) — equipment operations

(a) **Hazard assessments and precautions inside the work zone.** Before beginning equipment operations, the employer must: [§1926.1408(a)]

(1) *Identify the work zone by either:* [§1926.1408(a)(1)]

(i) *Demarcating boundaries* (such as with flags, or a device such as a range limit device or range control warning device) and prohibiting the operator from operating the equipment past those boundaries, or [§1926.1408(a)(1)(i)]

(ii) *Defining the work zone* as the area 360 degrees around the equipment, up to the equipment's maximum working radius. [§1926.1408(a)(1)(ii)]

(2) *Determine if any part of the equipment,* load line or load (including rigging and lifting accessories), if operated up to the equipment's maximum working radius in the work zone, could get closer than 20 feet to a power line. If so, the employer must meet the requirements in Option (1), Option (2), or Option (3) of this section, as follows: [§1926.1408(a)(2)]

(i) *Option (1) — Deenergize and ground.* Confirm from the utility owner/operator that the power line has been deenergized and visibly grounded at the worksite. [§1926.1408(a)(2)(i)]

(ii) *Option (2) — 20 foot clearance.* Ensure that no part of the equipment, load line, or load (including rigging and lifting accessories), gets closer than 20 feet to the power line by implementing the measures specified in paragraph (b) of this section. [§1926.1408(a)(2)(ii)]

(iii) *Option (3) — Table A clearance.* [§1926.1408(a)(2)(iii)]

[A] Determine the line's voltage and the minimum approach distance permitted under Table A (*see* Sec. 1926.1408). [§1926.1408(a)(2)(iii)[A]]

[B] Determine if any part of the equipment, load line or load (including rigging and lifting accessories), while operating up to the equipment's maximum working radius in the work zone, could get closer than the minimum approach distance of the power line permitted under Table A (*see* Sec. 1926.1408). If so, then the employer must follow the requirements in paragraph (b) of this section to ensure that no part of the equipment, load line, or load (including rigging and lifting accessories), gets closer to the line than the minimum approach distance. [§1926.1408(a)(2)(iii)[B]]

(b) **Preventing encroachment/electrocution.** Where encroachment precautions are required under Option (2) or Option (3) of this section, all of the following requirements must be met: [§1926.1408(b)]

(1) *Conduct a planning meeting* with the operator and the other workers who will be in the area of the equipment or load to review the location of the power line(s), and the steps that will be implemented to prevent encroachment/electrocution. [§1926.1408(b)(1)]

(2) *If tag lines are used,* they must be non-conductive. [§1926.1408(b)(2)]

(3) *Erect and maintain an elevated* warning line, barricade, or line of signs, in view of the operator, equipped with flags or similar high-visibility markings, at 20 feet from the power line (if using Option (2) of this section) or at the minimum approach distance under Table A (*see* Sec. 1926.1408) (if using Option (3) of this section). If the operator is unable to see the elevated warning line, a dedicated spotter must be used as described in Sec. 1926.1408(b)(4)(ii) in addition to implementing one of the measures described in §§1926.1408(b)(4)(i), (iii), (iv) and (v). [§1926.1408(b)(3)]

(4) *Implement at least one* of the following measures: [§1926.1408(b)(4)]

(i) *A proximity alarm set* to give the operator sufficient warning to prevent encroachment. [§1926.1408(b)(4)(i)]

(ii) *A dedicated spotter who is in continuous* contact with the operator. Where this measure is selected, the dedicated spotter must: [§1926.1408(b)(4)(ii)]

[A] Be equipped with a visual aid to assist in identifying the minimum clearance distance. Examples of a visual aid include, but are not limited to: A clearly visible line painted on the ground; a clearly visible line of stanchions; a set of clearly visible line-of-sight landmarks (such as a fence post behind the dedicated spotter and a building corner ahead of the dedicated spotter). [§1926.1408(b)(4)(ii)[A]]

[B] Be positioned to effectively gauge the clearance distance. [§1926.1408(b)(4)(ii)[B]]

[C] Where necessary, use equipment that enables the dedicated spotter to communicate directly with the operator. [§1926.1408(b)(4)(ii)[C]]

[D] Give timely information to the operator so that the required clearance distance can be maintained. [§1926.1408(b)(4)(ii)[D]]

(iii) *A device that automatically warns* the operator when to stop movement, such as a range control warning device. Such a device must be set to give the operator sufficient warning to prevent encroachment. [§1926.1408(b)(4)(iii)]

(iv) *A device that automatically limits* range of movement, set to prevent encroachment. [§1926.1408(b)(4)(iv)]

(v) *An insulating link/device,* as defined in Sec. 1926.1401, installed at a point between the end of the load line (or below) and the load. [§1926.1408(b)(4)(v)]

(5) *The requirements of paragraph (b)(4)* of this section do not apply to work covered by subpart V of this part. [§1926.1408(b)(5)]

(c) **Voltage information.** Where Option (3) of this section is used, the utility owner/operator of the power lines must provide the requested voltage information within two working days of the employer's request. [§1926.1408(c)]

(d) **Operations below power lines.** [§1926.1408(d)]

(1) *No part of the equipment,* load line, or load (including rigging and lifting accessories) is allowed below a power line unless the employer has confirmed that the utility owner/operator has deenergized and (at the worksite) visibly grounded the power line, except where one of the exceptions in paragraph (d)(2) of this section applies. [§1926.1408(d)(1)]

(2) *Exceptions.* Paragraph (d)(1) of this section is inapplicable where the employer demonstrates that one of the following applies: [§1926.1408(d)(2)]

(i) *The work is covered by subpart* V of this part. [§1926.1408(d)(2)(i)]

(ii) *For equipment with non-extensible booms:* The uppermost part of the equipment, with the boom at true vertical, would be more than 20 feet below the plane of the power line or more than the Table A of this section minimum clearance distance below the plane of the power line. [§1926.1408(d)(2)(ii)]

(iii) *For equipment with articulating or extensible* booms: The uppermost part of the equipment, with the boom in the fully extended position, at true vertical, would be more than 20 feet below the plane of the power line or more than the Table A of this section minimum clearance distance below the plane of the power line. [§1926.1408(d)(2)(iii)]

(iv) *The employer demonstrates that compliance* with paragraph (d)(1) of this section is infeasible and meets the requirements of Sec. 1926.1410. [§1926.1408(d)(2)(iv)]

(e) **Power lines presumed energized.** The employer must assume that all power lines are energized unless the utility owner/operator confirms that the power line has been and continues to be deenergized and visibly grounded at the worksite. [§1926.1408(e)]

(f) **When working near transmitter/communication** towers where the equipment is close enough for an electrical charge to be induced in the equipment or materials being handled, the transmitter must be deenergized or the following precautions must be taken: [§1926.1408(f)]

(1) *The equipment must be provided with an electrical ground.* [§1926.1408(f)(1)]

(2) *If tag lines are used,* they must be non-conductive. [§1926.1408(f)(2)]

(g) **Training.** [§1926.1408(g)]

(1) *The employer must train each operator* and crew member assigned to work with the equipment on all of the following: [§1926.1408(g)(1)]

(i) *The procedures to be followed in the event* of electrical contact with a power line. Such training must include: [§1926.1408(g)(1)(i)]

[A] Information regarding the danger of electrocution from the operator simultaneously touching the equipment and the ground. [§1926.1408(g)(1)(i)[A]]

[B] The importance to the operator's safety of remaining inside the cab except where there is an imminent danger of fire, explosion, or other emergency that necessitates leaving the cab. [§1926.1408(g)(1)(i)[B]]

[C] The safest means of evacuating from equipment that may be energized. [§1926.1408(g)(1)(i)[C]]

[D] The danger of the potentially energized zone around the equipment (step potential). [§1926.1408(g)(1)(i)[D]]

[E] The need for crew in the area to avoid approaching or touching the equipment and the load. [§1926.1408(g)(1)(i)[E]]

[F] Safe clearance distance from power lines. [§1926.1408(g)(1)(i)[F]]

(ii) *Power lines are presumed* to be energized unless the utility owner/operator confirms that the power line has been and continues to be deenergized and visibly grounded at the worksite. [§1926.1408(g)(1)(ii)]

(iii) *Power lines are presumed* to be uninsulated unless the utility owner/operator or a registered engineer who is a qualified person with respect to electrical power transmission and distribution confirms that a line is insulated. [§1926.1408(g)(1)(iii)]

(iv) *The limitations of an insulating link/device,* proximity alarm, and range control (and similar) device, if used. [§1926.1408(g)(1)(iv)]

(v) *The procedures to be followed to properly* ground equipment and the limitations of grounding. [§1926.1408(g)(1)(v)]

(2) *Employees working as dedicated* spotters must be trained to enable them to effectively perform their task, including training on the applicable requirements of this section. [§1926.1408(g)(2)]

(3) *Training under this section* must be administered in accordance with Sec. 1926.1430(g). [§1926.1408(g)(3)]

(h) **Devices originally designed** by the manufacturer for use as: A safety device (*see* Sec. 1926.1415), operational aid, or a means to prevent power line contact or electrocution, when used to comply with this section, must meet the manufacturer's procedures for use and conditions of use. [§1926.1408(h)]

Table A — Minimum Clearance Distances

Voltage (nominal, kV, alternating current)	Minimum clearance distance (feet)
up to 50	10
over 50 to 200	15
over 200 to 350	20
over 350 to 500	25
over 500 to 750	35
over 750 to 1,000	45
over 1,000	(as established by the utility owner/operator or registered professional engineer who is a qualified person with respect to electrical power transmission and distribution).

Note: The value that follows "to" is up to and includes that value. For example, over 50 to 200 means up to and including 200kV.

§1926.1409
Power line safety (over 350 kV)

The requirements of Sec. 1926.1407 and Sec. 1926.1408 apply to power lines over 350 kV except:

(a) **For power lines at or below** 1000 kV, wherever the distance "20 feet" is specified, the distance "50 feet" must be substituted; and [§1926.1409(a)]

(b) **For power lines over** 1000 kV, the minimum clearance distance must be established by the utility owner/operator or registered professional engineer who is a qualified person with respect to electrical power transmission and distribution. [§1926.1409(b)]

§1926.1410
Power line safety (all voltages) — equipment operations closer than the Table A zone

Equipment operations in which any part of the equipment, load line, or load (including rigging and lifting accessories) is closer than the minimum approach distance under Table A of Sec. 1926.1408 to an energized power line is prohibited, except where the employer demonstrates that all of the following requirements are met:

(a) **The employer determines that it is infeasible** to do the work without breaching the minimum approach distance under Table A of Sec. 1926.1408. [§1926.1410(a)]

(b) **The employer determines that,** after consultation with the utility owner/operator, it is infeasible to deenergize and ground the power line or relocate the power line. [§1926.1410(b)]

(c) **Minimum clearance distance.** [§1926.1410(c)]

(1) *The power line owner/operator* or registered professional engineer who is a qualified person with respect to electrical power transmission and distribution determines the minimum clearance distance that must be maintained to prevent electrical contact in light of the on-site conditions. The factors that must be considered in making this determination include, but are not limited to: Conditions affecting atmospheric conductivity; time necessary to bring the equipment, load line, and load (including rigging and lifting accessories) to a complete stop; wind conditions; degree of sway in the power line; lighting conditions, and other conditions affecting the ability to prevent electrical contact. [§1926.1410(c)(1)]

(2) *Paragraph (c)(1) of this section* does not apply to work covered by subpart V of this part; instead, for such work, the minimum clearance distances specified in Sec. 1926.950 Table V-1 apply. Employers engaged in subpart V work are permitted to work closer than the distances in Sec. 1926.950 Table V-1 where both the requirements of this section and Sec. 1926.952(c)(3)(i) or (ii) are met. [§1926.1410(c)(2)]

(d) **A planning meeting with the employer** and utility owner/operator (or registered professional engineer who is a qualified person with respect to electrical power transmission and distribution) is held to determine the procedures that will be followed to prevent electrical contact and electrocution. At a minimum these procedures must include: [§1926.1410(d)]

(1) *If the power line is equipped* with a device that automatically reenergizes the circuit in the event of a power line contact, before the work begins, the automatic reclosing feature of the circuit interrupting device must be made inoperative if the design of the device permits. [§1926.1410(d)(1)]

(2) *A dedicated spotter who is in continuous* contact with the operator. The dedicated spotter must: [§1926.1410(d)(2)]

(i) *Be equipped with a visual aid* to assist in identifying the minimum clearance distance. Examples of a visual aid include, but are not limited to: A line painted on the ground; a clearly visible line of stanchions; a set of clearly visible line-of-sight landmarks (such as a fence post behind the dedicated spotter and a building corner ahead of the dedicated spotter). [§1926.1410(d)(2)(i)]

(ii) *Be positioned to effectively gauge* the clearance distance. [§1926.1410(d)(2)(ii)]

(iii) *Where necessary,* use equipment that enables the dedicated spotter to communicate directly with the operator. [§1926.1410(d)(2)(iii)]

(iv) *Give timely information* to the operator so that the required clearance distance can be maintained. [§1926.1410(d)(2)(iv)]

(3) *An elevated warning line,* or barricade (not attached to the crane), in view of the operator (either directly or through video equipment), equipped with flags or similar high-visibility markings, to prevent electrical contact. However, this provision does not apply to work covered by subpart V of this part. [§1926.1410(d)(3)]

(4) *Insulating link/device.* [§1926.1410(d)(4)]

(i) *An insulating link/device installed* at a point between the end of the load line (or below) and the load. [§1926.1410(d)(4)(i)]

(ii) *For work covered by subpart* V of this part, the requirement in paragraph (d)(4)(i) of this section applies only when working inside the Sec. 1926.950 Table V-1 clearance distances. [§1926.1410(d)(4)(ii)]

(iii) *For work covered by subpart* V of this part involving operations where use of an insulating link/device is infeasible, the requirements of Sec. 1910.269(p)(4)(iii)(B) or (C) may be substituted for the requirement in (d)(4)(i) of this section. [§1926.1410(d)(4)(iii)]

(iv) *Until November 8,* 2011, the following procedure may be substituted for the requirement in paragraph (d)(4)(i) of this section: All employees, excluding equipment operators located on the equipment, who may come in contact with the equipment, the load line, or the load must be insulated or guarded from the equipment, the load line, and the load. Insulating gloves rated for the voltage involved are adequate insulation for the purposes of this paragraph. [§1926.1410(d)(4)(iv)]

(v) *Until November 8,* 2013, the following procedure may be substituted for the requirement in (d)(4)(i) of this section: [§1926.1410(d)(4)(v)]

[A] The employer must use a link/device manufactured on or before November 8, 2011, that meets the definition of an insulating link/device, except that it has not been approved by a Nationally Recognized Testing Laboratory, and that is maintained and used in accordance with manufacturer requirements and recommendations, and is installed at a point between the end of the load line (or below) and the load; and [§1926.1410(d)(4)(v)[A]]

[B] All employees, excluding equipment operators located on the equipment, who may come in contact with the equipment, the load line, or the load must be insulated or guarded from the equipment, the load line, and the load through an additional means other than the device described in paragraph (d)(4)(v)(A) of this section. Insulating gloves rated for the voltage involved are adequate additional means of protection for the purposes of this paragraph. [§1926.1410(d)(4)(v)[B]]

(5) *Nonconductive rigging if the rigging* may be within the Table A of Sec. 1926.1408 distance during the operation. [§1926.1410(d)(5)]

(6) *If the equipment is equipped with a device* that automatically limits range of movement, it must be used and set to prevent any part of the equipment, load line, or load (including rigging and lifting accessories) from breaching the minimum approach distance established under paragraph (c) of this section. [§1926.1410(d)(6)]

(7) *If a tag line is used,* it must be of the nonconductive type. [§1926.1410(d)(7)]

(8) *Barricades forming a perimeter* at least 10 feet away from the equipment to prevent unauthorized personnel from entering the work area. In areas where obstacles prevent the barricade from being at least 10 feet away, the barricade must be as far from the equipment as feasible. [§1926.1410(d)(8)]

(9) *Workers other than the operator* must be prohibited from touching the load line above the insulating link/device and crane. Operators remotely operating the equipment from the ground must use either wireless controls that isolate the operator from the equipment or insulating mats that insulate the operator from the ground. [§1926.1410(d)(9)]

(10) *Only personnel essential* to the operation are permitted to be in the area of the crane and load. [§1926.1410(d)(10)]

(11) *The equipment must be properly grounded.* [§1926.1410(d)(11)]

(12) *Insulating line hose* or cover-up must be installed by the utility owner/operator except where such devices are unavailable for the line voltages involved. [§1926.1410(d)(12)]

(e) **The procedures developed to comply** with paragraph (d) of this section are documented and immediately available on-site. [§1926.1410(e)]

(f) **The equipment user and utility** owner/operator (or registered professional engineer) meet with the equipment operator and the other workers who will be in the area of the equipment or load to review the procedures that will be implemented to prevent breaching the minimum approach distance established in paragraph (c) of this section and prevent electrocution. [§1926.1410(f)]

(g) **The procedures developed to comply** with paragraph (d) of this section are implemented. [§1926.1410(g)]

(h) **The utility owner/operator** (or registered professional engineer) and all employers of employees involved in the work must identify one person who will direct the implementation of the procedures. The person identified in accordance with this paragraph must direct the implementation of the procedures and must have the authority to stop work at any time to ensure safety. [§1926.1410(h)]

(i) ***[Reserved.]*** [§1926.1410(i)]

(j) **If a problem occurs implementing** the procedures being used to comply with paragraph (d) of this section, or indicating that those procedures are inadequate to prevent electrocution, the employer must safely stop operations and either develop new procedures to comply with paragraph (d) of this section or have the utility owner/operator deenergize and visibly ground or relocate the power line before resuming work. [§1926.1410(j)]

(k) **Devices originally designed** by the manufacturer for use as a safety device (*see* Sec. 1926.1415), operational aid, or a means to prevent power line contact or electrocution, when used to comply with this section, must comply with the manufacturer's procedures for use and conditions of use. [§1926.1410(k)]

(l) ***[Reserved.]*** [§1926.1410(l)]

(m) **The employer must train each operator** and crew member assigned to work with the equipment in accordance with Sec. 1926.1408(g). [§1926.1410(m)]

§1926.1411

Power line safety — while traveling under or near power lines with no load

(a) **This section establishes procedures** and criteria that must be met for equipment traveling under or near a power line on a construction site with no load. Equipment traveling on a construction site with a load is governed by §§1926.1408, 1926.1409 or 1926.1410, whichever is appropriate, and Sec. 1926.1417(u). [§1926.1411(a)]

(b) **The employer must ensure that:** [§1926.1411(b)]

(1) *The boom/mast and boom/mast support* system are lowered sufficiently to meet the requirements of this paragraph. [§1926.1411(b)(1)]

(2) *The clearances specified in Table* T of this section are maintained. [§1926.1411(b)(2)]

(3) *The effects of speed and terrain* on equipment movement (including movement of the boom/mast) are considered so that those effects do not cause the minimum clearance distances specified in Table T of this section to be breached. [§1926.1411(b)(3)]

(4) *Dedicated spotter.* If any part of the equipment while traveling will get closer than 20 feet to the power line, the employer must ensure that a dedicated spotter who is in continuous contact with the driver/operator is used. The dedicated spotter must: [§1926.1411(b)(4)]

(i) *Be positioned to effectively gauge* the clearance distance. [§1926.1411(b)(4)(i)]

(ii) *Where necessary,* use equipment that enables the dedicated spotter to communicate directly with the operator. [§1926.1411(b)(4)(ii)]

(iii) *Give timely information* to the operator so that the required clearance distance can be maintained. [§1926.1411(b)(4)(iii)]

(5) *Additional precautions for traveling in poor visibility.* When traveling at night, or in conditions of poor visibility, in addition to the measures specified in paragraphs (b)(1) through (4) of this section, the employer must ensure that: [§1926.1411(b)(5)]

(i) *The power lines are illuminated* or another means of identifying the location of the lines is used. [§1926.1411(b)(5)(i)]

(ii) *A safe path of travel* is identified and used. [§1926.1411(b)(5)(ii)]

Table T — Minimum Clearance Distances While Traveling With No Load

Voltage (nominal, kV, alternating current)	While traveling — minimum clearance distance (feet)
up to 0.75	4
over .75 to 50	6
over 50 to 345	10
over 345 to 750	16
Over 750 to 1,000	20
Over 1,000	(as established by the utility owner/operator or registered professional engineer who is a qualified person with respect to electrical power transmission and distribution).

§1926.1412

Inspections

(a) **Modified equipment.** [§1926.1412(a)]

(1) *Equipment that has had* modifications or additions which affect the safe operation of the equipment (such as modifications or additions involving a safety device or operational aid, critical part of a control system, power plant, braking system, load-sustaining structural components, load hook, or in-use operating mechanism) or capacity must be inspected by a qualified person after such modifications/additions have been completed, prior to initial use. The inspection must meet all of the following requirements: [§1926.1412(a)(1)]

(i) *The inspection must assure that the modifications* or additions have been done in accordance with the approval obtained pursuant to Sec. 1926.1434 (Equipment modifications). [§1926.1412(a)(1)(i)]

(ii) *The inspection must include functional* testing of the equipment. [§1926.1412(a)(1)(ii)]

(2) *Equipment must not be used until* an inspection under this paragraph demonstrates that the requirements of paragraph (a)(1)(i) of this section have been met. [§1926.1412(a)(2)]

(b) **Repaired/adjusted equipment.** [§1926.1412(b)]

(1) *Equipment that has had* a repair or adjustment that relates to safe operation (such as: A repair or adjustment to a safety device or operator aid, or to a critical part of a control system, power plant, braking system, load-sustaining structural components, load hook, or in-use operating mechanism), must be inspected by a qualified person after such a repair or adjustment has been completed, prior to initial use. The inspection must meet all of the following requirements: [§1926.1412(b)(1)]

(i) *The qualified person must determine* if the repair/adjustment meets manufacturer equipment criteria (where applicable and available). [§1926.1412(b)(1)(i)]

(ii) *Where manufacturer equipment criteria* are unavailable or inapplicable, the qualified person must: [§1926.1412(b)(1)(ii)]

[A] Determine if a registered professional engineer (RPE) is needed to develop criteria for the repair/adjustment. If an RPE is not needed, the employer must ensure that the criteria are developed by the qualified person. If an RPE is needed, the employer must ensure that they are developed by an RPE. [§1926.1412(b)(1)(ii)[A]]

[B] Determine if the repair/adjustment meets the criteria developed in accordance with paragraph (b)(1)(ii)(A) of this section. [§1926.1412(b)(1)(ii)[B]]

(iii) *The inspection must include functional* testing of the repaired/adjusted parts and other components that may be affected by the repair/adjustment. [§1926.1412(b)(1)(iii)]

(4)[1] *Equipment must not be used until* an inspection under this paragraph demonstrates that the repair/adjustment meets the requirements of paragraph (b)(1)(i) of this section (or, where applicable, paragraph (b)(1)(ii) of this section). [§1926.1412(b)(4)]

(c) **Post-assembly.** [§1926.1412(c)]

(1) *Upon completion of assembly,* the equipment must be inspected by a qualified person to assure that it is configured in accordance with manufacturer equipment criteria. [§1926.1412(c)(1)]

(2) *Where manufacturer equipment criteria* are unavailable, a qualified person must: [§1926.1412(c)(2)]

(i) *Determine if a registered professional* engineer (RPE) familiar with the type of equipment involved is needed to develop criteria for the equipment configuration. If an RPE is not needed, the employer must ensure that the criteria are developed by the qualified person. If an RPE is needed, the employer must ensure that they are developed by an RPE. [§1926.1412(c)(2)(i)]

(ii) *Determine if the equipment meets* the criteria developed in accordance with paragraph (c)(2)(i) of this section. [§1926.1412(c)(2)(ii)]

(3) *Equipment must not be used until* an inspection under this paragraph demonstrates that the equipment is configured in accordance with the applicable criteria. [§1926.1412(c)(3)]

(d) **Each shift.** [§1926.1412(d)]

(1) *A competent person must begin* a visual inspection prior to each shift the equipment will be used, which must be completed before or during that shift. The inspection must consist of observation for apparent deficiencies. Taking apart equipment components and booming down is not required as part of this inspection unless the results of the visual inspection or trial operation indicate that further investigation necessitating taking apart equipment components or booming down is needed. Determinations made in conducting the inspection must be reassessed in light of observations made during operation. At a minimum the inspection must include all of the following: [§1926.1412(d)(1)]

(i) *Control mechanisms for maladjustments* interfering with proper operation. [§1926.1412(d)(1)(i)]

(ii) *Control and drive mechanisms* for apparent excessive wear of components and contamination by lubricants, water or other foreign matter. [§1926.1412(d)(1)(ii)]

(iii) *Air, hydraulic,* and other pressurized lines for deterioration or leakage, particularly those which flex in normal operation. [§1926.1412(d)(1)(iii)]

(iv) *Hydraulic system for proper fluid level.* [§1926.1412(d)(1)(iv)]

(v) *Hooks and latches for deformation,* cracks, excessive wear, or damage such as from chemicals or heat. [§1926.1412(d)(1)(v)]

(vi) *Wire rope reeving* for compliance with the manufacturer's specifications. [§1926.1412(d)(1)(vi)]

(vii) *Wire rope,* in accordance with Sec. 1926.1413(a). [§1926.1412(d)(1)(vii)]

(viii) *Electrical apparatus for malfunctioning,* signs of apparent excessive deterioration, dirt or moisture accumulation. [§1926.1412(d)(1)(viii)]

(ix) *Tires (when in use)* for proper inflation and condition. [§1926.1412(d)(1)(ix)]

(x) *Ground conditions around* the equipment for proper support, including ground settling under and around outriggers/stabilizers and supporting foundations, ground water accumulation, or similar conditions. This paragraph does not apply to the inspection of ground conditions for railroad tracks and their underlying support when the railroad tracks are part of the general railroad system of transportation that is regulated pursuant to the Federal Railroad Administration under 49 CFR part 213. [§1926.1412(d)(1)(x)]

(xi) *The equipment for level position* within the tolerances specified by the equipment manufacturer's recommendations, both before each shift and after each move and setup. [§1926.1412(d)(1)(xi)]

(xii) *Operator cab windows* for significant cracks, breaks, or other deficiencies that would hamper the operator's view. [§1926.1412(d)(1)(xii)]

(xiii) *Rails, rail stops,* rail clamps and supporting surfaces when the equipment has rail traveling. This paragraph does not apply to the inspection of rails, rail stops, rail clamps and supporting surfaces when the railroad tracks are part of the general railroad system of transportation that is regulated pursuant to the Federal Railroad Administration under 49 CFR part 213. [§1926.1412(d)(1)(xiii)]

(xiv) *Safety devices and operational* aids for proper operation. [§1926.1412(d)(1)(xiv)]

(2) *If any deficiency in paragraphs (d)(1)(i)* through (xiii) of this section (or in additional inspection items required to be checked for specific types of equipment in accordance with other sections of this standard) is identified, an immediate determination must be made by the competent person as to whether the deficiency constitutes a safety hazard. If the deficiency is determined to constitute a safety hazard, the equipment must be taken out of service until it has been corrected. *See* Sec. 1926.1417. [§1926.1412(d)(2)]

(3) *If any deficiency in paragraph (d)(1)(xiv)* of this section (safety devices/operational aids) is identified, the action specified in Sec. 1926.1415 and Sec. 1926.1416 must be taken prior to using the equipment. [§1926.1412(d)(3)]

(e) **Monthly.** [§1926.1412(e)]

(1) *Each month the equipment is in service* it must be inspected in accordance with paragraph (d) of this section (each shift). [§1926.1412(e)(1)]

(2) *Equipment must not be used until* an inspection under this paragraph demonstrates that no corrective action under paragraphs (d)(2) and (3) of this section is required. [§1926.1412(e)(2)]

(3) *Documentation.* [§1926.1412(e)(3)]

(i) *The following information must be documented* and maintained by the employer that conducts the inspection: [§1926.1412(e)(3)(i)]

[A] *The items checked and the results* of the inspection. [§1926.1412(e)(3)(i)[A]]

[B] *The name and signature of the person* who conducted the inspection and the date. [§1926.1412(e)(3)(i)[B]]

(ii) *This document must be retained for a minimum* of three months. [§1926.1412(e)(3)(ii)]

(f) **Annual/comprehensive.** [§1926.1412(f)]

(1) *At least every 12* months the equipment must be inspected by a qualified person in accordance with paragraph (d) of this section (each shift) except that the corrective action set forth in paragraphs (f)(4), (f)(5), and (f)(6) of this section must apply in place of the corrective action required by paragraphs (d)(2) and (d)(3) of this section. [§1926.1412(f)(1)]

(2) *In addition,* at least every 12 months, the equipment must be inspected by a qualified person. Disassembly is required, as necessary, to complete the inspection. The equipment must be inspected for all of the following: [§1926.1412(f)(2)]

(i) *Equipment structure* (including the boom and, if equipped, the jib): [§1926.1412(f)(2)(i)]

[A] *Structural members:* Deformed, cracked, or significantly corroded. [§1926.1412(f)(2)(i)[A]]

[B] *Bolts, rivets and other fasteners:* loose, failed or significantly corroded. [§1926.1412(f)(2)(i)[B]]

[C] *Welds for cracks.* [§1926.1412(f)(2)(i)[C]]

(ii) *Sheaves and drums for cracks* or significant wear. [§1926.1412(f)(2)(ii)]

(iii) *Parts such as pins,* bearings, shafts, gears, rollers and locking devices for distortion, cracks or significant wear. [§1926.1412(f)(2)(iii)]

(iv) *Brake and clutch system* parts, linings, pawls and ratchets for excessive wear. [§1926.1412(f)(2)(iv)]

(v) *Safety devices and operational* aids for proper operation (including significant inaccuracies). [§1926.1412(f)(2)(v)]

(vi) *Gasoline, diesel,* electric, or other power plants for safety-related problems (such as leaking exhaust and emergency shut-down feature) and conditions, and proper operation. [§1926.1412(f)(2)(vi)]

(vii) *Chains and chain drive* sprockets for excessive wear of sprockets and excessive chain stretch. [§1926.1412(f)(2)(vii)]

(viii) *Travel steering,* brakes, and locking devices, for proper operation. [§1926.1412(f)(2)(viii)]

(ix) *Tires for damage or excessive wear.* [§1926.1412(f)(2)(ix)]

1. Editor's Note: This paragraph was mislabeled as paragraph (4) in the August 9, 2010 Federal Register. There is no content missing.

(x) *Hydraulic, pneumatic and other pressurized* hoses, fittings and tubing, as follows: [§1926.1412(f)(2)(x)]

[A] Flexible hose or its junction with the fittings for indications of leaks. [§1926.1412(f)(2)(x)[A]]

[B] Threaded or clamped joints for leaks. [§1926.1412(f)(2)(x)[B]]

[C] Outer covering of the hose for blistering, abnormal deformation or other signs of failure/impending failure. [§1926.1412(f)(2)(x)[C]]

[D] Outer surface of a hose, rigid tube, or fitting for indications of excessive abrasion or scrubbing. [§1926.1412(f)(2)(x)[D]]

(xi) *Hydraulic and pneumatic pumps* and motors, as follows: [§1926.1412(f)(2)(xi)]

[A] Performance indicators: Unusual noises or vibration, low operating speed, excessive heating of the fluid, low pressure. [§1926.1412(f)(2)(xi)[A]]

[B] Loose bolts or fasteners. [§1926.1412(f)(2)(xi)[B]]

[C] Shaft seals and joints between pump sections for leaks. [§1926.1412(f)(2)(xi)[C]]

(xii) *Hydraulic and pneumatic valves, as follows:* [§1926.1412(f)(2)(xii)]

[A] Spools: Sticking, improper return to neutral, and leaks. [§1926.1412(f)(2)(xii)[A]]

[B] Leaks. [§1926.1412(f)(2)(xii)[B]]

[C] Valve housing cracks. [§1926.1412(f)(2)(xii)[C]]

[D] Relief valves: Failure to reach correct pressure (if there is a manufacturer procedure for checking pressure, it must be followed). [§1926.1412(f)(2)(xii)[D]]

(xiii) *Hydraulic and pneumatic cylinders, as follows:* [§1926.1412(f)(2)(xiii)]

[A] Drifting caused by fluid leaking across the piston. [§1926.1412(f)(2)(xiii)[A]]

[B] Rod seals and welded joints for leaks. [§1926.1412(f)(2)(xiii)[B]]

[C] Cylinder rods for scores, nicks, or dents. [§1926.1412(f)(2)(xiii)[C]]

[D] Case (barrel) for significant dents. [§1926.1412(f)(2)(xiii)[D]]

[E] Rod eyes and connecting joints: Loose or deformed. [§1926.1412(f)(2)(xiii)[E]]

(xiv) *Outrigger or stabilizer pads/floats* for excessive wear or cracks. [§1926.1412(f)(2)(xiv)]

(xv) *Slider pads for excessive* wear or cracks. [§1926.1412(f)(2)(xv)]

(xvi) *Electrical components and wiring* for cracked or split insulation and loose or corroded terminations. [§1926.1412(f)(2)(xvi)]

(xvii) *Warning labels and decals* originally supplied with the equipment by the manufacturer or otherwise required under this standard: Missing or unreadable. [§1926.1412(f)(2)(xvii)]

(xviii) *Originally equipped operator* seat (or equivalent): Missing. [§1926.1412(f)(2)(xviii)]

(xix) *Operator seat: Unserviceable.* [§1926.1412(f)(2)(xix)]

(xx) *Originally equipped steps,* ladders, handrails, guards: Missing. [§1926.1412(f)(2)(xx)]

(xxi) *Steps, ladders,* handrails, guards: In unusable/unsafe condition. [§1926.1412(f)(2)(xxi)]

(3) *This inspection must include functional* testing to determine that the equipment as configured in the inspection is functioning properly. [§1926.1412(f)(3)]

(4) *If any deficiency is identified,* an immediate determination must be made by the qualified person as to whether the deficiency constitutes a safety hazard or, though not yet a safety hazard, needs to be monitored in the monthly inspections. [§1926.1412(f)(4)]

(5) *If the qualified person determines* that a deficiency is a safety hazard, the equipment must be taken out of service until it has been corrected, except when temporary alternative measures are implemented as specified in Sec. 1926.1416(d) or Sec. 1926.1435(e). *See* Sec. 1926.1417. [§1926.1412(f)(5)]

(6) *If the qualified person determines* that, though not presently a safety hazard, the deficiency needs to be monitored, the employer must ensure that the deficiency is checked in the monthly inspections. [§1926.1412(f)(6)]

(7) *Documentation of annual/comprehensive inspection.* The following information must be documented, maintained, and retained for a minimum of 12 months, by the employer that conducts the inspection: [§1926.1412(f)(7)]

(i) *The items checked and the results* of the inspection. [§1926.1412(f)(7)(i)]

(ii) *The name and signature of the person* who conducted the inspection and the date. [§1926.1412(f)(7)(ii)]

(g) **Severe service.** Where the severity of use/conditions is such that there is a reasonable probability of damage or excessive wear (such as loading that may have exceeded rated capacity, shock loading that may have exceeded rated capacity, prolonged exposure to a corrosive atmosphere), the employer must stop using the equipment and a qualified person must: [§1926.1412(g)]

(1) *Inspect the equipment for structural* damage to determine if the equipment can continue to be used safely. [§1926.1412(g)(1)]

(2) *In light of the use/conditions determine* whether any items/conditions listed in paragraph (f) of this section need to be inspected; if so, the qualified person must inspect those items/conditions. [§1926.1412(g)(2)]

(3) *If a deficiency is found,* the employer must follow the requirements in paragraphs (f)(4) through (6) of this section. [§1926.1412(g)(3)]

(h) **Equipment not in regular use.** Equipment that has been idle for 3 months or more must be inspected by a qualified person in accordance with the requirements of paragraph (e) (Monthly) of this section before initial use. [§1926.1412(h)]

(i) *[Reserved.]* [§1926.1412(i)]

(j) **Any part of a manufacturer's procedures** regarding inspections that relate to safe operation (such as to a safety device or operational aid, critical part of a control system, power plant, braking system, load-sustaining structural components, load hook, or in-use operating mechanism) that is more comprehensive or has a more frequent schedule of inspection than the requirements of this section must be followed. [§1926.1412(j)]

(k) **All documents produced under** this section must be available, during the applicable document retention period, to all persons who conduct inspections under this section. [§1926.1412(k)]

§1926.1413

Wire rope — inspection

(a) **Shift inspection.** [§1926.1413(a)]

(1) *A competent person must begin* a visual inspection prior to each shift the equipment is used, which must be completed before or during that shift. The inspection must consist of observation of wire ropes (running and standing) that are likely to be in use during the shift for apparent deficiencies, including those listed in paragraph (a)(2) of this section. Untwisting (opening) of wire rope or booming down is not required as part of this inspection. [§1926.1413(a)(1)]

(2) *Apparent deficiencies.* [§1926.1413(a)(2)]

(i) *Category I.* Apparent deficiencies in this category include the following: [§1926.1413(a)(2)(i)]

[A] Significant distortion of the wire rope structure such as kinking, crushing, unstranding, birdcaging, signs of core failure or steel core protrusion between the outer strands. [§1926.1413(a)(2)(i)[A]]

[B] Significant corrosion. [§1926.1413(a)(2)(i)[B]]

[C] Electric arc damage (from a source other than power lines) or heat damage. [§1926.1413(a)(2)(i)[C]]

[D] Improperly applied end connections. [§1926.1413(a)(2)(i)[D]]

[E] Significantly corroded, cracked, bent, or worn end connections (such as from severe service). [§1926.1413(a)(2)(i)[E]]

(ii) *Category II.* Apparent deficiencies in this category are: [§1926.1413(a)(2)(ii)]

[A] Visible broken wires, as follows: [§1926.1413(a)(2)(ii)[A]]

[1] In running wire ropes: Six randomly distributed broken wires in one rope lay or three broken wires in one strand in one rope lay, where a rope lay is the length along the rope in which one strand makes a complete revolution around the rope. [§1926.1413(a)(2)(ii)[A][1]]

[2] In rotation resistant ropes: Two randomly distributed broken wires in six rope diameters or four randomly distributed broken wires in 30 rope diameters. [§1926.1413(a)(2)(ii)[A][2]]

[3] In pendants or standing wire ropes: More than two broken wires in one rope lay located in rope beyond end connections and/or more than one broken wire in a rope lay located at an end connection. [§1926.1413(a)(2)(ii)[A][3]]

[B] A diameter reduction of more than 5% from nominal diameter. [§1926.1413(a)(2)(ii)[B]]

(iii) *Category III.* Apparent deficiencies in this category include the following: [§1926.1413(a)(2)(iii)]

[A] In rotation resistant wire rope, core protrusion or other distortion indicating core failure. [§1926.1413(a)(2)(iii)[A]]

[B] *Prior electrical contact* with a power line. [§1926.1413(a)(2)(iii)[B]]

[C] *A broken strand.* [§1926.1413(a)(2)(iii)[C]]

(3) *Critical review items.* The competent person must give particular attention to all of the following: [§1926.1413(a)(3)]

(i) *Rotation resistant wire rope in use.* [§1926.1413(a)(3)(i)]

(ii) *Wire rope being used* for boom hoists and luffing hoists, particularly at reverse bends. [§1926.1413(a)(3)(ii)]

(iii) *Wire rope at flange* points, crossover points and repetitive pickup points on drums. [§1926.1413(a)(3)(iii)]

(iv) *Wire rope at or near terminal ends.* [§1926.1413(a)(3)(iv)]

(v) *Wire rope in contact* with saddles, equalizer sheaves or other sheaves where rope travel is limited. [§1926.1413(a)(3)(v)]

(4) *Removal from service.* [§1926.1413(a)(4)]

(i) *If a deficiency in Category I* (*see* paragraph (a)(2)(i) of this section) is identified, an immediate determination must be made by the competent person as to whether the deficiency constitutes a safety hazard. If the deficiency is determined to constitute a safety hazard, operations involving use of the wire rope in question must be prohibited until: [§1926.1413(a)(4)(i)]

[A] *The wire rope is replaced* (*see* Sec. 1926.1417), or [§1926.1413(a)(4)(i)[A]]

[B] *If the deficiency is localized,* the problem is corrected by severing the wire rope in two; the undamaged portion may continue to be used. Joining lengths of wire rope by splicing is prohibited. If a rope is shortened under this paragraph, the employer must ensure that the drum will still have two wraps of wire when the load and/or boom is in its lowest position. [§1926.1413(a)(4)(i)[B]]

(ii) *If a deficiency in Category II* (*see* paragraph (a)(2)(ii) of this section) is identified, operations involving use of the wire rope in question must be prohibited until: [§1926.1413(a)(4)(ii)]

[A] *The employer complies with the wire* rope manufacturer's established criterion for removal from service or a different criterion that the wire rope manufacturer has approved in writing for that specific wire rope (*see* Sec. 1926.1417), [§1926.1413(a)(4)(ii)[A]]

[B] *The wire rope is replaced* (*see* Sec. 1926.1417), or [§1926.1413(a)(4)(ii)[B]]

[C] *If the deficiency is localized,* the problem is corrected by severing the wire rope in two; the undamaged portion may continue to be used. Joining lengths of wire rope by splicing is prohibited. If a rope is shortened under this paragraph, the employer must ensure that the drum will still have two wraps of wire when the load and/or boom is in its lowest position. [§1926.1413(a)(4)(ii)[C]]

(iii) *If a deficiency in Category III* is identified, operations involving use of the wire rope in question must be prohibited until: [§1926.1413(a)(4)(iii)]

[A] *The wire rope is replaced* (*see* Sec. 1926.1417), or [§1926.1413(a)(4)(iii)[A]]

[B] *If the deficiency* (other than power line contact) is localized, the problem is corrected by severing the wire rope in two; the undamaged portion may continue to be used. Joining lengths of wire rope by splicing is prohibited. Repair of wire rope that contacted an energized power line is also prohibited. If a rope is shortened under this paragraph, the employer must ensure that the drum will still have two wraps of wire when the load and/or boom is in its lowest position. [§1926.1413(a)(4)(iii)[B]]

(iv) *Where a wire rope is required* to be removed from service under this section, either the equipment (as a whole) or the hoist with that wire rope must be tagged-out, in accordance with Sec. 1926.1417(f)(1), until the wire rope is repaired or replaced. [§1926.1413(a)(4)(iv)]

(b) **Monthly inspection.** [§1926.1413(b)]

(1) *Each month an inspection must be conducted* in accordance with paragraph (a) (shift inspection) of this section. [§1926.1413(b)(1)]

(2) *The inspection must include any deficiencies* that the qualified person who conducts the annual inspection determines under paragraph (c)(3)(ii) of this section must be monitored. [§1926.1413(b)(2)]

(3) *Wire ropes on equipment* must not be used until an inspection under this paragraph demonstrates that no corrective action under paragraph (a)(4) of this section is required. [§1926.1413(b)(3)]

(4) *The inspection must be documented according* to Sec. 1926.1412(e)(3) (monthly inspection documentation). [§1926.1413(b)(4)]

(c) **Annual/comprehensive.** [§1926.1413(c)]

(1) *At least every 12* months, wire ropes in use on equipment must be inspected by a qualified person in accordance with paragraph (a) of this section (shift inspection). [§1926.1413(c)(1)]

(2) *In addition,* at least every 12 months, the wire ropes in use on equipment must be inspected by a qualified person, as follows: [§1926.1413(c)(2)]

(i) *The inspection must be for deficiencies of the types* listed in paragraph (a)(2) of this section. [§1926.1413(c)(2)(i)]

(ii) *The inspection must be complete and thorough,* covering the surface of the entire length of the wire ropes, with particular attention given to all of the following: [§1926.1413(c)(2)(ii)]

[A] *Critical review items* listed in paragraph (a)(3) of this section. [§1926.1413(c)(2)(ii)[A]]

[B] *Those sections that are normally* hidden during shift and monthly inspections. [§1926.1413(c)(2)(ii)[B]]

[C] *Wire rope subject* to reverse bends. [§1926.1413(c)(2)(ii)[C]]

[D] *Wire rope passing over sheaves.* [§1926.1413(c)(2)(ii)[D]]

(iii) *Exception:* In the event an inspection under paragraph (c)(2) of this section is not feasible due to existing set-up and configuration of the equipment (such as where an assist crane is needed) or due to site conditions (such as a dense urban setting), such inspections must be conducted as soon as it becomes feasible, but no longer than an additional 6 months for running ropes and, for standing ropes, at the time of disassembly. [§1926.1413(c)(2)(iii)]

(3) *If a deficiency is identified,* an immediate determination must be made by the qualified person as to whether the deficiency constitutes a safety hazard. [§1926.1413(c)(3)]

(i) *If the deficiency is determined to constitute* a safety hazard, operations involving use of the wire rope in question must be prohibited until: [§1926.1413(c)(3)(i)]

[A] *The wire rope is replaced* (*see* Sec. 1926.1417), or [§1926.1413(c)(3)(i)[A]]

[B] *If the deficiency is localized,* the problem is corrected by severing the wire rope in two; the undamaged portion may continue to be used. Joining lengths of wire rope by splicing is prohibited. If a rope is shortened under this paragraph, the employer must ensure that the drum will still have two wraps of wire when the load and/or boom is in its lowest position. [§1926.1413(c)(3)(i)[B]]

(ii) *If the qualified person determines* that, though not presently a safety hazard, the deficiency needs to be monitored, the employer must ensure that the deficiency is checked in the monthly inspections. [§1926.1413(c)(3)(ii)]

(4) *The inspection must be documented according* to Sec. 1926.1412(f)(7) (annual/comprehensive inspection documentation). [§1926.1413(c)(4)]

(d) **Rope lubricants that are of the type** that hinder inspection must not be used. [§1926.1413(d)]

(e) **All documents produced under** this section must be available, during the applicable document retention period, to all persons who conduct inspections under this section. [§1926.1413(e)]

§1926.1414

Wire rope — selection and installation criteria

(a) **Original equipment wire** rope and replacement wire rope must be selected and installed in accordance with the requirements of this section. Selection of replacement wire rope must be in accordance with the recommendations of the wire rope manufacturer, the equipment manufacturer, or a qualified person. [§1926.1414(a)]

(b) **Wire rope design criteria:** Wire rope (other than rotation resistant rope) must comply with either Option (1) or Option (2) of this section, as follows: [§1926.1414(b)]

(1) *Option (1).* Wire rope must comply with section 5-1.7.1 of ASME B30.5-2004 (incorporated by reference, *see* Sec. 1926.6) except that section's paragraph (c) must not apply. [§1926.1414(b)(1)]

(2) *Option (2).* Wire rope must be designed to have, in relation to the equipment's rated capacity, a sufficient minimum breaking force and design factor so that compliance with the applicable inspection provisions in Sec. 1926.1413 will be an effective means of preventing sudden rope failure. [§1926.1414(b)(2)]

(c) **Wire rope must be compatible** with the safe functioning of the equipment. [§1926.1414(c)]

(d) **Boom hoist reeving.** [§1926.1414(d)]

(1) *Fiber core ropes* must not be used for boom hoist reeving, except for derricks. [§1926.1414(d)(1)]

(2) *Rotation resistant ropes* must be used for boom hoist reeving only where the requirements of paragraph (e)(4)(ii) of this section are met. [§1926.1414(d)(2)]

(e) **Rotation resistant ropes.** [§1926.1414(e)]

(1) *Definitions.* [§1926.1414(e)(1)]

(i) *Type I rotation resistant wire rope ("Type I").* Type I rotation resistant rope is stranded rope constructed to have little or no tendency to rotate or, if guided, transmits little or no torque. It has at least 15 outer strands and comprises an assembly of at least three layers of strands laid helically over a center in two operations. The direction of lay of the outer strands is opposite to that of the underlying layer. [§1926.1414(e)(1)(i)]

(ii) *Type II rotation resistant wire rope ("Type II").* Type II rotation resistant rope is stranded rope constructed to have significant resistance to rotation. It has at least 10 outer strands and comprises an assembly of two or more layers of strands laid helically over a center in two or three operations. The direction of lay of the outer strands is opposite to that of the underlying layer. [§1926.1414(e)(1)(ii)]

(iii) *Type III rotation resistant wire rope ("Type III").* Type III rotation resistant rope is stranded rope constructed to have limited resistance to rotation. It has no more than nine outer strands, and comprises an assembly of two layers of strands laid helically over a center in two operations. The direction of lay of the outer strands is opposite to that of the underlying layer. [§1926.1414(e)(1)(iii)]

(2) *Requirements.* [§1926.1414(e)(2)]

(i) *Types II and III* with an operating design factor of less than 5 must not be used for duty cycle or repetitive lifts. [§1926.1414(e)(2)(i)]

(ii) *Rotation resistant ropes* (including Types I, II and III) must have an operating design factor of no less than 3.5. [§1926.1414(e)(2)(ii)]

(iii) *Type I must have* an operating design factor of no less than 5, except where the wire rope manufacturer and the equipment manufacturer approves the design factor, in writing. [§1926.1414(e)(2)(iii)]

(iv) *Types II and III* must have an operating design factor of no less than 5, except where the requirements of paragraph (e)(3) of this section are met. [§1926.1414(e)(2)(iv)]

(3) *When Types II and III* with an operating design factor of less than 5 are used (for non-duty cycle, non-repetitive lifts), the following requirements must be met for each lifting operation: [§1926.1414(e)(3)]

(i) *A qualified person must inspect* the rope in accordance with Sec. 1926.1413(a). The rope must be used only if the qualified person determines that there are no deficiencies constituting a hazard. In making this determination, more than one broken wire in any one rope lay must be considered a hazard. [§1926.1414(e)(3)(i)]

(ii) *Operations must be conducted in such* a manner and at such speeds as to minimize dynamic effects. [§1926.1414(e)(3)(ii)]

(iii) *Each lift made under* Sec. 1926.1414(e)(3) must be recorded in the monthly and annual inspection documents. Such prior uses must be considered by the qualified person in determining whether to use the rope again. [§1926.1414(e)(3)(iii)]

(4) *Additional requirements for rotation resistant ropes for boom hoist reeving.* [§1926.1414(e)(4)]

(i) *Rotation resistant ropes* must not be used for boom hoist reeving, except where the requirements of paragraph (e)(4)(ii) of this section are met. [§1926.1414(e)(4)(i)]

(ii) *Rotation resistant ropes* may be used as boom hoist reeving when load hoists are used as boom hoists for attachments such as luffing attachments or boom and mast attachment systems. Under these conditions, all of the following requirements must be met: [§1926.1414(e)(4)(ii)]

[A] The drum must provide a first layer rope pitch diameter of not less than 18 times the nominal diameter of the rope used. [§1926.1414(e)(4)(ii)[A]]

[B] The requirements in Sec. 1926.1426(a) (irrespective of the date of manufacture of the equipment), and Sec. 1926.1426(b). [§1926.1414(e)(4)(ii)[B]]

[C] The requirements in ASME B30.5-2004 sections 5-1.3.2(a), (a)(2) through (a)(4), (b) and (d) (incorporated by reference, *see* Sec. 1926.6) except that the minimum pitch diameter for sheaves used in multiple rope reeving is 18 times the nominal diameter of the rope used (instead of the value of 16 specified in section 5-1.3.2(d)). [§1926.1414(e)(4)(ii)[C]]

[D] All sheaves used in the boom hoist reeving system must have a rope pitch diameter of not less than 18 times the nominal diameter of the rope used. [§1926.1414(e)(4)(ii)[D]]

[E] The operating design factor for the boom hoist reeving system must be not less than five. [§1926.1414(e)(4)(ii)[E]]

[F] The operating design factor for these ropes must be the total minimum breaking force of all parts of rope in the system divided by the load imposed on the rope system when supporting the static weights of the structure and the load within the equipment's rated capacity. [§1926.1414(e)(4)(ii)[F]]

[G] When provided, a power-controlled lowering system must be capable of handling rated capacities and speeds as specified by the manufacturer. [§1926.1414(e)(4)(ii)[G]]

(f) **Wire rope clips** used in conjunction with wedge sockets must be attached to the unloaded dead end of the rope only, except that the use of devices specifically designed for dead-ending rope in a wedge socket is permitted. [§1926.1414(f)]

(g) **Socketing must be done in the manner** specified by the manufacturer of the wire rope or fitting. [§1926.1414(g)]

(h) **Prior to cutting a wire** rope, seizings must be placed on each side of the point to be cut. The length and number of seizings must be in accordance with the wire rope manufacturer's instructions. [§1926.1414(h)]

§1926.1415
Safety devices

(a) **Safety devices.** The following safety devices are required on all equipment covered by this subpart, unless otherwise specified: [§1926.1415(a)]

(1) *Crane level indicator.* [§1926.1415(a)(1)]

(i) *The equipment must have a crane* level indicator that is either built into the equipment or is available on the equipment. [§1926.1415(a)(1)(i)]

(ii) *If a built-in crane level* indicator is not working properly, it must be tagged-out or removed. If a removable crane level indicator is not working properly, it must be removed. [§1926.1415(a)(1)(ii)]

(iii) *This requirement does not apply* to portal cranes, derricks, floating cranes/derricks and land cranes/derricks on barges, pontoons, vessels or other means of flotation. [§1926.1415(a)(1)(iii)]

(2) *Boom stops,* except for derricks and hydraulic booms. [§1926.1415(a)(2)]

(3) *Jib stops (if a jib is attached),* except for derricks. [§1926.1415(a)(3)]

(4) *Equipment with foot pedal* brakes must have locks. [§1926.1415(a)(4)]

(5) *Hydraulic outrigger jacks* and hydraulic stabilizer jacks must have an integral holding device/check valve. [§1926.1415(a)(5)]

(6) *Equipment on rails must have* rail clamps and rail stops, except for portal cranes. [§1926.1415(a)(6)]

(7) *Horn* [§1926.1415(a)(7)]

(i) *The equipment must have a horn* that is either built into the equipment or is on the equipment and immediately available to the operator. [§1926.1415(a)(7)(i)]

(ii) *If a built-in horn is not working* properly, it must be tagged-out or removed. If a removable horn is not working properly, it must be removed. [§1926.1415(a)(7)(ii)]

(b) **Proper operation required.** Operations must not begin unless all of the devices listed in this section are in proper working order. If a device stops working properly during operations, the operator must safely stop operations. If any of the devices listed in this section are not in proper working order, the equipment must be taken out of service and operations must not resume until the device is again working properly. *See* §1926.1417 (Operation). Alternative measures are not permitted to be used. [§1926.1415(b)]

§1926.1416
Operational aids

(a) **The devices listed in this section** ("listed operational aids") are required on all equipment covered by this subpart, unless otherwise specified. [§1926.1416(a)]

(1) *The requirements in paragraphs (e)(1),* (e)(2), and (e)(3) of this section do not apply to articulating cranes. [§1926.1416(a)(1)]

(2) *The requirements in paragraphs (d)(3),* (e)(1), and (e)(4) of this section apply only to those digger derricks manufactured after November 8, 2011. [§1926.1416(a)(2)]

(b) **Operations must not begin unless** the listed operational aids are in proper working order, except where an operational aid is being repaired the employer uses the specified temporary alternative measures. The time periods permitted for repairing defective operational aids are specified in paragraphs (d) and (e) of this section. More protective alternative measures specified by the crane/derrick manufacturer, if any, must be followed. [§1926.1416(b)]

(c) **If a listed operational aid** stops working properly during operations, the operator must safely stop operations until the temporary alternative measures are implemented or the device is again working properly. If a replacement part is no longer available, the use of a substitute device that performs the same type of function is permitted and is not considered a modification under Sec. 1926.1434. [§1926.1416(c)]

(d) **Category I operational aids and alternative measures.** Operational aids listed in this paragraph that are not working properly must be repaired no later than 7 calendar days after the deficiency occurs. *Exception:* If the employer documents that it has ordered the necessary parts within 7 calendar days of the occurrence of the deficiency, the repair must be completed within 7 calendar days of receipt of the parts. *See* Sec. 1926.1417(j) for additional requirements. [§1926.1416(d)]

(1) *Boom hoist limiting device.* [§1926.1416(d)(1)]

(i) *For equipment manufactured after* December 16, 1969, a boom hoist limiting device is required. *Temporary alternative measures (use at least one).* One or more of the following methods must be used: [§1926.1416(d)(1)(i)]

[A] Use a boom angle indicator. [§1926.1416(d)(1)(i)[A]]

[B] Clearly mark the boom hoist cable (so that it can easily be seen by the operator) at a point that will give the operator sufficient time to stop the hoist to keep the boom within the minimum allowable radius. In addition, install mirrors or remote video cameras and displays if necessary for the operator to see the mark. [§1926.1416(d)(1)(i)[B]]

[C] Clearly mark the boom hoist cable (so that it can easily be seen by a spotter) at a point that will give the spotter sufficient time to signal the operator and have the operator stop the hoist to keep the boom within the minimum allowable radius. [§1926.1416(d)(1)(i)[C]]

(ii) *If the equipment was manufactured on or before* December 16, 1969, and is not equipped with a boom hoist limiting device, at least one of the measures in paragraphs (d)(1)(i)(A) through (C) of this section must be used. [§1926.1416(d)(1)(ii)]

(2) *Luffing jib limiting device.* Equipment with a luffing jib must have a luffing jib limiting device. Temporary alternative measures are the same as in paragraph (d)(1)(i) of this section, except to limit the movement of the luffing jib rather than the boom hoist. [§1926.1416(d)(2)]

(3) *Anti two-blocking device.* [§1926.1416(d)(3)]

(i) *Telescopic boom cranes* manufactured after February 28, 1992, must be equipped with a device which automatically prevents damage from contact between the load block, overhaul ball, or similar component, and the boom tip (or fixed upper block or similar component). The device(s) must prevent such damage at all points where two-blocking could occur.

Temporary alternative measures: Clearly mark the cable (so that it can easily be seen by the operator) at a point that will give the operator sufficient time to stop the hoist to prevent two-blocking, and use a spotter when extending the boom. [§1926.1416(d)(3)(i)]

(ii) *Lattice boom cranes.* [§1926.1416(d)(3)(ii)]

[A] Lattice boom cranes manufactured after Feb 28, 1992, must be equipped with a device that either automatically prevents damage and load failure from contact between the load block, overhaul ball, or similar component, and the boom tip (or fixed upper block or similar component), or warns the operator in time for the operator to prevent two-blocking. The device must prevent such damage/failure or provide adequate warning for all points where two-blocking could occur. [§1926.1416(d)(3)(ii)[A]]

[B] Lattice boom cranes and derricks manufactured after November 8, 2011 must be equipped with a device which automatically prevents damage and load failure from contact between the load block, overhaul ball, or similar component, and the boom tip (or fixed upper block or similar component). The device(s) must prevent such damage/failure at all points where two-blocking could occur. [§1926.1416(d)(3)(ii)[B]]

[C] Exception. The requirements in paragraphs (d)(3)(ii)(A) and (B) of this section do not apply to such lattice boom equipment when used for dragline, clamshell (grapple), magnet, drop ball, container handling, concrete bucket, marine operations that do not involve hoisting personnel, and pile driving work. [§1926.1416(d)(3)(ii)[C]]

[D] Temporary alternative measures. Clearly mark the cable (so that it can easily be seen by the operator) at a point that will give the operator sufficient time to stop the hoist to prevent two-blocking, or use a spotter. [§1926.1416(d)(3)(ii)[D]]

(iii) *Articulating cranes manufactured* after December 31, 1999, that are equipped with a load hoist must be equipped with a device that automatically prevents damage from contact between the load block, overhaul ball, or similar component, and the boom tip (or fixed upper block or similar component). The device must prevent such damage at all points where two-blocking could occur. *Temporary alternative measures:* When two-blocking could only occur with movement of the load hoist, clearly mark the cable (so that it can easily be seen by the operator) at a point that will give the operator sufficient time to stop the hoist to prevent two-blocking, or use a spotter. When two-blocking could occur without movement of the load hoist, clearly mark the cable (so that it can easily be seen by the operator) at a point that will give the operator sufficient time to stop the hoist to prevent two-blocking, and use a spotter when extending the boom. [§1926.1416(d)(3)(iii)]

(e) **Category II operational aids and alternative measures.** Operational aids listed in this paragraph that are not working properly must be repaired no later than 30 calendar days after the deficiency occurs. *Exception:* If the employer documents that it has ordered the necessary parts within 7 calendar days of the occurrence of the deficiency, and the part is not received in time to complete the repair in 30 calendar days, the repair must be completed within 7 calendar days of receipt of the parts. *See* Sec. 1926.1417(j) for additional requirements. [§1926.1416(e)]

(1) *Boom angle or radius indicator.* The equipment must have a boom angle or radius indicator readable from the operator's station. *Temporary alternative measures:* Radii or boom angle must be determined by measuring the radii or boom angle with a measuring device. [§1926.1416(e)(1)]

(2) *Jib angle indicator* if the equipment has a luffing jib. *Temporary alternative measures:* Radii or jib angle must be determined by ascertaining the main boom angle and then measuring the radii or jib angle with a measuring device. [§1926.1416(e)(2)]

(3) *Boom length indicator* if the equipment has a telescopic boom, except where the rated capacity is independent of the boom length. *Temporary alternative measures.* One or more of the following methods must be used: [§1926.1416(e)(3)]

(i) *Mark the boom with measured* marks to calculate boom length, [§1926.1416(e)(3)(i)]

(ii) *Calculate boom length* from boom angle and radius measurements, [§1926.1416(e)(3)(ii)]

(iii) *Measure the boom with a measuring device.* [§1926.1416(e)(3)(iii)]

(4) *Load weighing and similar devices.* [§1926.1416(e)(4)]

(i) *Equipment* (other than derricks and articulating cranes) manufactured after March 29, 2003 with a rated capacity over 6,000 pounds must have at least one of the following: load weighing device, load moment (or rated capacity) indicator, or load moment (or rated capacity) limiter. *Temporary alternative measures:* The weight of the load must be determined from a source recognized by the industry (such as the load's manufacturer) or by a calculation method recognized by the industry (such as calculating a steel beam from measured dimensions and a known per foot weight). This information must be provided to the operator prior to the lift. [§1926.1416(e)(4)(i)]

(ii) *Articulating cranes manufactured* after November 8, 2011 must have at least one of the following: automatic overload prevention device, load weighing device, load moment (or rated capacity) indicator, or load moment (rated capacity) limiter. *Temporary alternative measures:* The weight of the load must be determined from a source recognized by the industry (such as the load's manufacturer) or by a calculation method recognized by the industry (such as calculating a steel beam from measured dimensions and a known per foot weight). This information must be provided to the operator prior to the lift. [§1926.1416(e)(4)(ii)]

(5) *The following devices are required* on equipment manufactured after November 8, 2011: [§1926.1416(e)(5)]

(i) *Outrigger/stabilizer position* (horizontal beam extension) sensor/monitor if the equipment has outriggers or stabilizers. *Temporary alternative measures:* The operator must verify that the position of the outriggers or stabilizers is correct (in accordance with manufacturer procedures) before beginning operations requiring outrigger or stabilizer deployment. [§1926.1416(e)(5)(i)]

(ii) *Hoist drum rotation* indicator if the equipment has a hoist drum not visible from the operator's station. *Temporary alternative measures:* Mark the drum to indicate the rotation of the drum. In addition, install mirrors or remote video cameras and displays if necessary for the operator to see the mark. [§1926.1416(e)(5)(ii)]

§1926.1417 Operation

(a) **The employer must comply with all manufacturer** procedures applicable to the operational functions of equipment, including its use with attachments. [§1926.1417(a)]

(b) **Unavailable operation procedures.** [§1926.1417(b)]

(1) *Where the manufacturer procedures are unavailable,* the employer must develop and ensure compliance with all procedures necessary for the safe operation of the equipment and attachments. [§1926.1417(b)(1)]

(2) *Procedures for the operational controls* must be developed by a qualified person. [§1926.1417(b)(2)]

(3) *Procedures related to the capacity* of the equipment must be developed and signed by a registered professional engineer familiar with the equipment. [§1926.1417(b)(3)]

(c) **Accessibility of procedures.** [§1926.1417(c)]

(1) *The procedures applicable to the operation* of the equipment, including rated capacities (load charts), recommended operating speeds, special hazard warnings, instructions, and operator's manual, must be readily available in the cab at all times for use by the operator. [§1926.1417(c)(1)]

(2) *Where rated capacities are available* in the cab only in electronic form: In the event of a failure which makes the rated capacities inaccessible, the operator must immediately cease operations or follow safe shut-down procedures until the rated capacities (in electronic or other form) are available. [§1926.1417(c)(2)]

(d) **The operator must not engage in any practice** or activity that diverts his/her attention while actually engaged in operating the equipment, such as the use of cellular phones (other than when used for signal communications). [§1926.1417(d)]

(e) **Leaving the equipment unattended.** [§1926.1417(e)]

(1) *The operator must not leave the controls* while the load is suspended, except where all of the following are met: [§1926.1417(e)(1)]

(i) *The operator remains adjacent* to the equipment and is not engaged in any other duties. [§1926.1417(e)(1)(i)]

(ii) *The load is to be held suspended* for a period of time exceeding normal lifting operations. [§1926.1417(e)(1)(ii)]

(iii) *The competent person determines* that it is safe to do so and implements measures necessary to restrain the boom hoist and telescoping, load, swing, and outrigger or stabilizer functions. [§1926.1417(e)(1)(iii)]

(iv) *Barricades or caution lines,* and notices, are erected to prevent all employees from entering the fall zone. No employees, including those listed in §§1926.1425(b)(1) through (3), Sec. 1926.1425(d) or Sec. 1926.1425(e), are permitted in the fall zone. [§1926.1417(e)(1)(iv)]

(2) *The provisions in Sec. 1926.1417(e)(1) do* not apply to working gear (such as slings, spreader bars, ladders, and welding machines) where the weight of the working gear is negligible relative to the lifting capacity of the equipment as positioned, and the working gear is suspended over an area other than an entrance or exit. [§1926.1417(e)(2)]

(f) **Tag-out.** [§1926.1417(f)]

(1) *Tagging out of service equipment/functions.* Where the employer has taken the equipment out of service, a tag must be placed in the cab stating that the equipment is out of service and is not to be used. Where the employer has taken a function(s) out of service, a tag must be placed in a conspicuous position stating that the function is out of service and is not to be used. [§1926.1417(f)(1)]

(2) *Response to "do not operate"/tag-out signs.* [§1926.1417(f)(2)]

(i) *If there is a warning* (tag-out or maintenance/do not operate) sign on the equipment or starting control, the operator must not activate the switch or start the equipment until the sign has been removed by a person authorized to remove it, or until the operator has verified that: [§1926.1417(f)(2)(i)]

[A] No one is servicing, working on, or otherwise in a dangerous position on the machine. [§1926.1417(f)(2)(i)[A]]

[B] The equipment has been repaired and is working properly. [§1926.1417(f)(2)(i)[B]]

(ii) *If there is a warning* (tag-out or maintenance/do not operate) sign on any other switch or control, the operator must not activate that switch or control until the sign has been removed by a person authorized to remove it, or until the operator has verified that the requirements in paragraphs (f)(2)(i)(A) and (B) of this section have been met. [§1926.1417(f)(2)(ii)]

(g) **Before starting the engine,** the operator must verify that all controls are in the proper starting position and that all personnel are in the clear. [§1926.1417(g)]

(h) **Storm warning.** When a local storm warning has been issued, the competent person must determine whether it is necessary to implement manufacturer recommendations for securing the equipment. [§1926.1417(h)]

(i) *[Reserved.]* [§1926.1417(i)]

(j) **If equipment adjustments or repairs are necessary:** [§1926.1417(j)]

(1) *The operator must,* in writing, promptly inform the person designated by the employer to receive such information and, where there are successive shifts, to the next operator; and [§1926.1417(j)(1)]

(2) *The employer must notify all affected* employees, at the beginning of each shift, of the necessary adjustments or repairs and all alternative measures. [§1926.1417(j)(2)]

(k) **Safety devices and operational** aids must not be used as a substitute for the exercise of professional judgment by the operator. [§1926.1417(k)]

(l) *[Reserved.]* [§1926.1417(l)]

(m) **If the competent person determines** that there is a slack rope condition requiring re-spooling of the rope, it must be verified (before starting to lift) that the rope is seated on the drum and in the sheaves as the slack is removed. [§1926.1417(m)]

(n) **The competent person must adjust** the equipment and/or operations to address the effect of wind, ice, and snow on equipment stability and rated capacity. [§1926.1417(n)]

(o) **Compliance with rated capacity.** [§1926.1417(o)]

(1) *The equipment must not be operated in excess* of its rated capacity. [§1926.1417(o)(1)]

(2) *The operator must not be required to operate* the equipment in a manner that would violate paragraph (o)(1) of this section. [§1926.1417(o)(2)]

(3) *Load weight.* The operator must verify that the load is within the rated capacity of the equipment by at least one of the following methods: [§1926.1417(o)(3)]

(i) *The weight of the load must be determined* from a source recognized by the industry (such as the load's manufacturer), or by a calculation method recognized by the industry (such as calculating a steel beam from measured dimensions and a known per foot weight), or by other equally reliable means. In addition, when requested by the operator, this information must be provided to the operator prior to the lift; or [§1926.1417(o)(3)(i)]

(ii) *The operator must begin hoisting* the load to determine, using a load weighing device, load moment indicator, rated capacity indicator, or rated capacity limiter, if it exceeds 75 percent of the maximum rated capacity at the longest radius that will be used during the lift operation. If it does, the operator must not proceed with the lift until he/she verifies the weight of the load in accordance with paragraph (o)(3)(i) of this section. [§1926.1417(o)(3)(ii)]

(p) **The boom or other parts of the equipment** must not contact any obstruction. [§1926.1417(p)]

(q) **The equipment must not be used to drag** or pull loads sideways. [§1926.1417(q)]

(r) **On wheel-mounted equipment,** no loads must be lifted over the front area, except as permitted by the manufacturer. [§1926.1417(r)]

(s) **The operator must test the brakes** each time a load that is 90% or more of the maximum line pull is handled by lifting the load a few inches and applying the brakes. In duty cycle and repetitive lifts where each lift is 90% or more of the maximum line pull, this requirement applies to the first lift but not to successive lifts. [§1926.1417(s)]

(t) **Neither the load nor** the boom must be lowered below the point where less than two full wraps of rope remain on their respective drums. [§1926.1417(t)]

(u) **Traveling with a load.** [§1926.1417(u)]

(1) *Traveling with a load is prohibited* if the practice is prohibited by the manufacturer. [§1926.1417(u)(1)]

(2) *Where traveling with a load,* the employer must ensure that: [§1926.1417(u)(2)]

(i) *A competent person supervises* the operation, determines if it is necessary to reduce rated capacity, and makes determinations regarding load position, boom location, ground support, travel route, overhead obstructions, and speed of movement necessary to ensure safety. [§1926.1417(u)(2)(i)]

(ii) *The determinations of the competent person* required in paragraph (u)(2)(i) of this section are implemented. [§1926.1417(u)(2)(ii)]

(iii) *For equipment with tires,* tire pressure specified by the manufacturer is maintained. [§1926.1417(u)(2)(iii)]

(v) **Rotational speed of the equipment** must be such that the load does not swing out beyond the radius at which it can be controlled. [§1926.1417(v)]

(w) **A tag or restraint line** must be used if necessary to prevent rotation of the load that would be hazardous. [§1926.1417(w)]

(x) **The brakes must be adjusted in accordance** with manufacturer procedures to prevent unintended movement. [§1926.1417(x)]

(y) **The operator must obey a stop** (or emergency stop) signal, irrespective of who gives it. [§1926.1417(y)]

(z) **Swinging locomotive cranes.** A locomotive crane must not be swung into a position where railway cars on an adjacent track could strike it, until it is determined that cars are not being moved on the adjacent track and that proper flag protection has been established. [§1926.1417(z)]

(aa) **Counterweight/ballast.** [§1926.1417(aa)]

(1) *The following applies to equipment* other than tower cranes: [§1926.1417(aa)(1)]

(i) *Equipment must not be operated without* the counterweight or ballast in place as specified by the manufacturer. [§1926.1417(aa)(1)(i)]

(ii) *The maximum counterweight or ballast* specified by the manufacturer for the equipment must not be exceeded. [§1926.1417(aa)(1)(ii)]

(2) *Counterweight/ballast requirements for tower* cranes are specified in Sec. 1926.1435(b)(8). [§1926.1417(aa)(2)]

§1926.1418
Authority to stop operation

Whenever there is a concern as to safety, the operator must have the authority to stop and refuse to handle loads until a qualified person has determined that safety has been assured.

§1926.1419
Signals — general requirements

(a) **A signal person must be provided** in each of the following situations: [§1926.1419(a)]

(1) *The point of operation,* meaning the load travel or the area near or at load placement, is not in full view of the operator. [§1926.1419(a)(1)]

(2) *When the equipment is traveling,* the view in the direction of travel is obstructed. [§1926.1419(a)(2)]

(3) *Due to site specific* safety concerns, either the operator or the person handling the load determines that it is necessary. [§1926.1419(a)(3)]

(b) **Types of signals.** Signals to operators must be by hand, voice, audible, or new signals. [§1926.1419(b)]

(c) **Hand signals.** [§1926.1419(c)]

(1) *When using hand signals,* the Standard Method must be used (*see* Appendix A of this subpart). *Exception:* Where use of the Standard Method for hand signals is infeasible, or where an operation or use of an attachment is not covered in the Standard Method, non-standard hand signals may be used in accordance with paragraph (c)(2) of this section. [§1926.1419(c)(1)]

(2) *Non-standard hand signals.* When using non-standard hand signals, the signal person, operator, and lift director (where there is one) must contact each other prior to the operation and agree on the non-standard hand signals that will be used. [§1926.1419(c)(2)]

(d) **New signals.** Signals other than hand, voice, or audible signals may be used where the employer demonstrates that: [§1926.1419(d)]

(1) *The new signals provide at least* equally effective communication as voice, audible, or Standard Method hand signals, or [§1926.1419(d)(1)]

(2) *The new signals comply with a national* consensus standard that provides at least equally effective communication as voice, audible, or Standard Method hand signals. [§1926.1419(d)(2)]

(e) **Suitability.** The signals used (hand, voice, audible, or new), and means of transmitting the signals to the operator (such as direct line of sight, video, radio, *etc.*), must be appropriate for the site conditions. [§1926.1419(e)]

(f) **During operations requiring** signals, the ability to transmit signals between the operator and signal person must be maintained. If that ability is interrupted at any time, the operator must safely stop operations requiring signals until it is reestablished and a proper signal is given and understood. [§1926.1419(f)]

(g) **If the operator becomes aware** of a safety problem and needs to communicate with the signal person, the operator must safely stop operations. Operations must not resume until the operator and signal person agree that the problem has been resolved. [§1926.1419(g)]

(h) **Only one person** may give signals to a crane/derrick at a time, except in circumstances covered by paragraph (j) of this section. [§1926.1419(h)]

(i) *[Reserved.]* [§1926.1419(i)]

(j) **Anyone who becomes aware** of a safety problem must alert the operator or signal person by giving the stop or emergency stop signal. (Note: Sec. 1926.1417(y) requires the operator to obey a stop or emergency stop signal). [§1926.1419(j)]

(k) **All directions given to the operator** by the signal person must be given from the operator's direction perspective. [§1926.1419(k)]

(l) *[Reserved.]* [§1926.1419(l)]

(m) **Communication with multiple cranes/derricks.** Where a signal person(s) is in communication with more than one crane/derrick, a system must be used for identifying the crane/derrick each signal is for, as follows: [§1926.1419(m)]

(1) *for each signal,* prior to giving the function/direction, the signal person must identify the crane/derrick the signal is for, or [§1926.1419(m)(1)]

(2) *must use an equally effective* method of identifying which crane/derrick the signal is for. [§1926.1419(m)(2)]

§1926.1420
Signals — radio, telephone or other electronic transmission of signals

(a) **The device(s) used to transmit** signals must be tested on site before beginning operations to ensure that the signal transmission is effective, clear, and reliable. [§1926.1420(a)]

(b) **Signal transmission must be through** a dedicated channel, except: [§1926.1420(b)]

(1) *Multiple cranes/derricks and one* or more signal persons may share a dedicated channel for the purpose of coordinating operations. [§1926.1420(b)(1)]

(2) *Where a crane is being operated on or adjacent* to railroad tracks, and the actions of the crane operator need to be coordinated with the movement of other equipment or trains on the same or adjacent tracks. [§1926.1420(b)(2)]

(c) **The operator's reception of signals** must be by a hands-free system. [§1926.1420(c)]

§1926.1421
Signals — voice signals — additional requirements

(a) **Prior to beginning operations,** the operator, signal person and lift director (if there is one), must contact each other and agree on the voice signals that will be used. Once the voice signals are agreed upon, these workers need not meet again to discuss voice signals unless another worker is added or substituted, there is confusion about the voice signals, or a voice signal is to be changed. [§1926.1421(a)]

(b) **Each voice signal must contain** the following three elements, given in the following order: function (such as hoist, boom, *etc.*), direction; distance and/or speed; function, stop command. [§1926.1421(b)]

(c) **The operator,** signal person and lift director (if there is one), must be able to effectively communicate in the language used. [§1926.1421(c)]

§1926.1422
Signals — hand signal chart

Hand signal charts must be either posted on the equipment or conspicuously posted in the vicinity of the hoisting operations.

§1926.1423
Fall protection

(a) **Application.** [§1926.1423(a)]

(1) *Paragraphs (b),* (c)(3), (e) and (f) of this section apply to all equipment covered by this subpart except tower cranes. [§1926.1423(a)(1)]

(2) *Paragraphs (c)(1),* (c)(2), (d), (g), (j) and (k) of this section apply to all equipment covered by this subpart. [§1926.1423(a)(2)]

(3) *Paragraphs (c)(4) and (h)* of this section apply only to tower cranes. [§1926.1423(a)(3)]

(b) **Boom walkways.** [§1926.1423(b)]

(1) *Equipment manufactured after* November 8, 2011 with lattice booms must be equipped with walkways on the boom(s) if the vertical profile of the boom (from cord centerline to cord centerline) is 6 or more feet. [§1926.1423(b)(1)]

(2) *Boom walkway criteria.* [§1926.1423(b)(2)]

(i) *The walkways must be at least 12 inches wide.* [§1926.1423(b)(2)(i)]

(ii) *Guardrails, railings and other permanent* fall protection attachments along walkways are: [§1926.1423(b)(2)(ii)]

[A] Not required. [§1926.1423(b)(2)(ii)[A]]

[B] Prohibited on booms supported by pendant ropes or bars if the guardrails/railings/attachments could be snagged by the ropes or bars. [§1926.1423(b)(2)(ii)[B]]

[C] Prohibited if of the removable type (designed to be installed and removed each time the boom is assembled/disassembled). [§1926.1423(b)(2)(ii)[C]]

[D] Where not prohibited, guardrails or railings may be of any height up to, but not more than, 45 inches. [§1926.1423(b)(2)(ii)[D]]

(c) **Steps, handholds, ladders, grabrails, guardrails and railings.** [§1926.1423(c)]

(1) *Section 1926.502(b) does* not apply to equipment covered by this subpart. [§1926.1423(c)(1)]

(2) *The employer must maintain in good* condition originally-equipped steps, handholds, ladders and guardrails/railings/grabrails. [§1926.1423(c)(2)]

(3) *Equipment manufactured after* November 8, 2011 must be equipped so as to provide safe access and egress between the ground and the operator work station(s), including the forward and rear positions, by the provision of devices such as steps, handholds, ladders, and guardrails/railings/grabrails. These devices must meet the following criteria: [§1926.1423(c)(3)]

(i) *Steps, handholds,* ladders and guardrails/railings/grabrails must meet the criteria of SAE J185 (May 2003) (incorporated by reference, *see* Sec. 1926.6) or ISO 11660-2:1994(E) (incorporated by reference, *see* Sec. 1926.6) except where infeasible. [§1926.1423(c)(3)(i)]

(ii) *Walking/stepping surfaces,* except for crawler treads, must have slip-resistant features/properties (such as diamond plate metal, strategically placed grip tape, expanded metal, or slip-resistant paint). [§1926.1423(c)(3)(ii)]

(4) *Tower cranes manufactured* after November 8, 2011 must be equipped so as to provide safe access and egress between the ground and the cab, machinery platforms, and tower (mast), by the provision of devices such as steps, handholds, ladders, and guardrails/railings/grabrails. These devices must meet the following criteria: [§1926.1423(c)(4)]

(i) *Steps, handholds,* ladders, and guardrails/railings/grabrails must meet the criteria of ISO 11660-1:2008(E) (incorporated by reference, *see* Sec. 1926.6) and ISO 11660-3:2008(E) (incorporated by reference, *see* Sec. 1926.6) or SAE J185 (May 2003) (incorporated by reference, *see* Sec. 1926.6) except where infeasible. [§1926.1423(c)(4)(i)]

(ii) *Walking/stepping surfaces must have* slip-resistant features/properties (such as diamond plate metal, strategically placed grip tape, expanded metal, or slip-resistant paint). [§1926.1423(c)(4)(ii)]

(d) **Personal fall arrest and fall restraint systems.** Personal fall arrest system components must be used in personal fall arrest and fall restraint systems and must conform to the criteria in Sec. 1926.502(d) except that Sec. 1926.502(d)(15) does not apply to components used in personal fall arrest and fall restraint systems. Either body belts or body harnesses must be used in personal fall arrest and fall restraint systems. [§1926.1423(d)]

(e) **For non-assembly/disassembly work,** the employer must provide and ensure the use of fall protection equipment for employees who are on a walking/working surface with an unprotected side or edge more than 6 feet above a lower level as follows: [§1926.1423(e)]

(1) *When moving point-to-point:* [§1926.1423(e)(1)]

(i) *On non-lattice booms* (whether horizontal or not horizontal). [§1926.1423(e)(1)(i)]

(ii) *On lattice booms that are not horizontal.* [§1926.1423(e)(1)(ii)]

(iii) *On horizontal lattice booms* where the fall distance is 15 feet or more. [§1926.1423(e)(1)(iii)]

(2) *While at a work station* on any part of the equipment (including the boom, of any type), except when the employee is at or near draw-works (when the equipment is running), in the cab, or on the deck. [§1926.1423(e)(2)]

(f) **For assembly/disassembly work,** the employer must provide and ensure the use of fall protection equipment for employees who are on a walking/working surface with an unprotected side or edge more than 15 feet above a lower level, except when the employee is at or near draw-works (when the equipment is running), in the cab, or on the deck. [§1926.1423(f)]

(g) **Anchorage criteria.** [§1926.1423(g)]

(1) *Sections 1926.502(d)(15) and 1926.502(e)(2)* apply to equipment covered by this subpart only to the extent delineated in paragraph (g)(2) of this section. [§1926.1423(g)(1)]

(2) *Anchorages for personal fall arrest and positioning device systems.* [§1926.1423(g)(2)]

(i) *Personal fall arrest* systems must be anchored to any apparently substantial part of the equipment unless a competent person, from a visual inspection, without an engineering analysis, would conclude that the criteria in Sec. 1926.502(d)(15) would not be met. [§1926.1423(g)(2)(i)]

(ii) *Positioning device systems* must be anchored to any apparently substantial part of the equipment unless a competent person, from a visual inspection, without an engineering analysis, would conclude that the criteria in Sec. 1926.502(e)(2) would not be met. [§1926.1423(g)(2)(ii)]

(iii) *Attachable anchor devices* (portable anchor devices that are attached to the equipment) must meet the anchorage criteria in Sec. 1926.502(d)(15) for personal fall arrest systems and Sec. 1926.502(e)(2) for positioning device systems. [§1926.1423(g)(2)(iii)]

(3) *Anchorages for fall restraint systems.* Fall restraint systems must be anchored to any part of the equipment that is capable of withstanding twice the maximum load that an employee may impose on it during reasonably anticipated conditions of use. [§1926.1423(g)(3)]

(h) **Tower cranes.** [§1926.1423(h)]

(1) *For work other than erecting,* climbing, and dismantling, the employer must provide and ensure the use of fall protection equipment for employees who are on a walking/working surface with an unprotected side or edge more than 6 feet above a lower level, except when the employee is at or near draw-works (when the equipment is running), in the cab, or on the deck. [§1926.1423(h)(1)]

(2) *For erecting,* climbing, and dismantling work, the employer must provide and ensure the use of fall protection equipment for employees who are on a walking/working surface with an unprotected side or edge more than 15 feet above a lower level. [§1926.1423(h)(2)]

(i) ***[Reserved.]*** [§1926.1423(i)]

(j) **Anchoring to the load line.** A personal fall arrest system is permitted to be anchored to the crane/derrick's hook (or other part of the load line) where all of the following requirements are met: [§1926.1423(j)]

(1) *A qualified person has* determined that the set-up and rated capacity of the crane/derrick (including the hook, load line and rigging) meets or exceeds the requirements in Sec. 1926.502(d)(15). [§1926.1423(j)(1)]

(2) *The equipment operator must be at the work* site and informed that the equipment is being used for this purpose. [§1926.1423(j)(2)]

(3) *No load is suspended from the load* line when the personal fall arrest system is anchored to the crane/derrick's hook (or other part of the load line). [§1926.1423(j)(3)]

(k) **Training.** The employer must train each employee who may be exposed to fall hazards while on, or hoisted by, equipment covered by this subpart on all of the following: [§1926.1423(k)]

(1) *the requirements in this subpart that address* fall protection. [§1926.1423(k)(1)]

(2) *the applicable requirements in §§1926.500 and 1926.502.* [§1926.1423(k)(2)]

§1926.1424

Work area control

(a) **Swing radius hazards.** [§1926.1424(a)]

(1) *The requirements in paragraph (a)(2)* of this section apply where there are accessible areas in which the equipment's rotating superstructure (whether permanently or temporarily mounted) poses a reasonably foreseeable risk of: [§1926.1424(a)(1)]

(i) *Striking and injuring an employee; or* [§1926.1424(a)(1)(i)]

(ii) *Pinching/crushing an employee against* another part of the equipment or another object. [§1926.1424(a)(1)(ii)]

(2) *To prevent employees from entering* these hazard areas, the employer must: [§1926.1424(a)(2)]

(i) *Train each employee assigned* to work on or near the equipment ("authorized personnel") in how to recognize struck-by and pinch/crush hazard areas posed by the rotating superstructure. [§1926.1424(a)(2)(i)]

(ii) *Erect and maintain control* lines, warning lines, railings or similar barriers to mark the boundaries of the hazard areas. *Exception:* When the employer can demonstrate that it is neither feasible to erect such barriers on the ground nor on the equipment, the hazard areas must be clearly marked by a combination of warning signs (such as "Danger — Swing/ Crush Zone") and high visibility markings on the equipment that identify the hazard areas. In addition, the employer must train each employee to understand what these markings signify. [§1926.1424(a)(2)(ii)]

(3) *Protecting employees in the hazard area.* [§1926.1424(a)(3)]

(i) *Before an employee goes* to a location in the hazard area that is out of view of the operator, the employee (or someone instructed by the employee) must ensure that the operator is informed that he/she is going to that location. [§1926.1424(a)(3)(i)]

(ii) *Where the operator knows that an employee* went to a location covered by paragraph (a)(1) of this section, the operator must not rotate the superstructure until the operator is informed in accordance with a pre-arranged system of communication that the employee is in a safe position. [§1926.1424(a)(3)(ii)]

(b) **Where any part of a crane/derrick is within** the working radius of another crane/derrick, the controlling entity must institute a system to coordinate operations. If there is no controlling entity, the employer (if there is only one employer operating the multiple pieces of equipment), or employers, must institute such a system. [§1926.1424(b)]

§1926.1425

Keeping clear of the load

(a) **Where available,** hoisting routes that minimize the exposure of employees to hoisted loads must be used, to the extent consistent with public safety. [§1926.1425(a)]

(b) **While the operator is not moving** a suspended load, no employee must be within the fall zone, except for employees: [§1926.1425(b)]

(1) *Engaged in hooking,* unhooking or guiding a load; [§1926.1425(b)(1)]

(2) *Engaged in the initial attachment* of the load to a component or structure; or [§1926.1425(b)(2)]

(3) *Operating a concrete hopper* or concrete bucket. [§1926.1425(b)(3)]

(c) **When employees are engaged in hooking,** unhooking, or guiding the load, or in the initial connection of a load to a component or structure and are within the fall zone, all of the following criteria must be met: [§1926.1425(c)]

(1) *The materials being hoisted must be rigged* to prevent unintentional displacement. [§1926.1425(c)(1)]

(2) *Hooks with self-closing latches* or their equivalent must be used. *Exception:* "J" hooks are permitted to be used for setting wooden trusses. [§1926.1425(c)(2)]

(3) *The materials must be rigged by a qualified rigger.* [§1926.1425(c)(3)]

(d) **Receiving a load.** Only employees needed to receive a load are permitted to be within the fall zone when a load is being landed. [§1926.1425(d)]

(e) **During a tilt-up or tilt-down operation:** [§1926.1425(e)]

(1) *No employee must be directly under the load.* [§1926.1425(e)(1)]

(2) *Only employees essential* to the operation are permitted in the fall zone (but not directly under the load). An employee is essential to the operation if the employee is conducting one of the following operations and the employer can demonstrate it is infeasible for the employee to perform that operation from outside the fall zone: (1) Physically guide the load; (2) closely monitor and give instructions regarding the load's movement; or (3) either detach it from or initially attach it to another component or structure (such as, but not limited to, making an initial connection or installing bracing). [§1926.1425(e)(2)]

Note: Boom free fall is prohibited when an employee is in the fall zone of the boom or load, and load line free fall is prohibited when an employee is directly under the load; *see* Sec. 1926.1426.

§1926.1426

Free fall and controlled load lowering

(a) **Boom free fall prohibitions.** [§1926.1426(a)]

(1) *The use of equipment in which* the boom is designed to free fall (live boom) is prohibited in each of the following circumstances: [§1926.1426(a)(1)]

(i) *An employee is in the fall zone* of the boom or load. [§1926.1426(a)(1)(i)]

(ii) *An employee is being hoisted.* [§1926.1426(a)(1)(ii)]

(iii) *The load or boom is directly* over a power line, or over any part of the area extending the Table A of Sec. 1926.1408 clearance distance to each side of the power line; or any part of the area extending the Table A clearance distance to each side of the power line is within the radius of vertical travel of the boom or the load. [§1926.1426(a)(1)(iii)]

(iv) *The load is over a shaft,* except where there are no employees in the shaft. [§1926.1426(a)(1)(iv)]

(v) *The load is over a cofferdam,* except where there are no employees in the fall zone of the boom or the load. [§1926.1426(a)(1)(v)]

(vi) *Lifting operations are taking* place in a refinery or tank farm. [§1926.1426(a)(1)(vi)]

(2) *The use of equipment in which* the boom is designed to free fall (live boom) is permitted only where none of the circumstances listed in paragraph (a)(1) of this section are present and: [§1926.1426(a)(2)]

(i) *The equipment was manufactured prior* to October 31, 1984; or [§1926.1426(a)(2)(i)]

(ii) *The equipment is a floating crane/derrick* or a land crane/derrick on a vessel/flotation device. [§1926.1426(a)(2)(ii)]

(b) **Preventing boom free fall.** Where the use of equipment with a boom that is designed to free fall (live boom) is prohibited, the boom hoist must have a secondary mechanism or device designed to prevent the boom from falling in the event the primary system used to hold or regulate the boom hoist fails, as follows: [§1926.1426(b)]

(1) *Friction drums must have:* [§1926.1426(b)(1)]

(i) *A friction clutch and,* in addition, a braking device, to allow for controlled boom lowering. [§1926.1426(b)(1)(i)]

(ii) *A secondary braking or locking* device, which is manually or automatically engaged, to back-up the primary brake while the boom is held (such as a secondary friction brake or a ratchet and pawl device). [§1926.1426(b)(1)(ii)]

(2) *Hydraulic drums must have* an integrally mounted holding device or internal static brake to prevent boom hoist movement in the event of hydraulic failure. [§1926.1426(b)(2)]

(3) *Neither clutches nor* hydraulic motors must be considered brake or locking devices for purposes of this subpart. [§1926.1426(b)(3)]

(4) *Hydraulic boom cylinders* must have an integrally mounted holding device. [§1926.1426(b)(4)]

(c) **Preventing uncontrolled retraction.** Hydraulic telescoping booms must have an integrally mounted holding device to prevent the boom from retracting in the event of hydraulic failure. [§1926.1426(c)]

(d) **Load line free fall.** In each of the following circumstances, controlled load lowering is required and free fall of the load line hoist is prohibited: [§1926.1426(d)]

(1) *An employee is directly under the load.* [§1926.1426(d)(1)]

(2) *An employee is being hoisted.* [§1926.1426(d)(2)]

(3) *The load is directly over* a power line, or over any part of the area extending the Table A of Sec. 1926.1408 clearance distance to each side of the power line; or any part of the area extending the Table A of Sec. 1926.1408 clearance distance to each side of the power line is within the radius of vertical travel of the load. [§1926.1426(d)(3)]

(4) *The load is over a shaft.* [§1926.1426(d)(4)]

(5) *The load is over a cofferdam,* except where there are no employees in the fall zone of the load. [§1926.1426(d)(5)]

§1926.1427

Operator qualification and certification

(a) **The employer must ensure that,** prior to operating any equipment covered under subpart CC, the person is operating the equipment during a training period in accordance with paragraph (f) of this section, or the operator is qualified or certified to operate the equipment in accordance with the following: [§1926.1427(a)]

(1) *When a non-military government entity* issues operator licenses for equipment covered under subpart CC, and that government licensing program meets the requirements of paragraphs (e)(2) and (j) of this section, the equipment operator must either be: [§1926.1427(a)(1)]

(i) *Licensed by that government entity* for operation of equipment within that entity's jurisdiction; or [§1926.1427(a)(1)(i)]

(ii) *qualified in compliance with paragraph* (d) of this section. [§1926.1427(a)(1)(ii)]

(2) *Where paragraph (a)(1) of this section* is not applicable, the certification or qualification must comply with one of the options in paragraphs (b) through (d) of this section. [§1926.1427(a)(2)]

(3) *Exceptions:* Operator qualification or certification under this section is not required for operators of derricks (*see* Sec. 1926.1436), sideboom cranes (*see* Sec. 1926.1440), or equipment with a maximum manufacturer-rated hoisting/lifting capacity of 2,000 pounds or less (*see* Sec. 1926.1441). [§1926.1427(a)(3)]

(4) *Whenever operator qualification* or certification is required under Sec. 1926.1427, the employer must provide the qualification or certification at no cost to operators who are employed by the employer on November 8, 2010. [§1926.1427(a)(4)]

(b) **Option (1):** Certification by an accredited crane operator testing organization. [§1926.1427(b)]

(1) *For a testing organization to be considered* accredited to certify operators under this subpart, it must: [§1926.1427(b)(1)]

(i) *Be accredited by a nationally recognized* accrediting agency based on that agency's determination that industry recognized criteria for written testing materials, practical examinations, test administration, grading, facilities/equipment and personnel have been met. [§1926.1427(b)(1)(i)]

(ii) *Administer written and practical tests that:* [§1926.1427(b)(1)(ii)]

[A] *Assess the operator applicant* regarding, at a minimum, the knowledge and skills listed in paragraphs (j)(1) and (2) of this section. [§1926.1427(b)(1)(ii)[A]]

[B] *Provide different levels* of certification based on equipment capacity and type. [§1926.1427(b)(1)(ii)[B]]

(iii) *Have procedures for operators* to re-apply and be re-tested in the event an operator applicant fails a test or is decertified. [§1926.1427(b)(1)(iii)]

(iv) *Have testing procedures* for re-certification designed to ensure that the operator continues to meet the technical knowledge and skills requirements in paragraphs (j)(1) and (2) of this section. [§1926.1427(b)(1)(iv)]

(v) *Have its accreditation reviewed* by the nationally recognized accrediting agency at least every three years. [§1926.1427(b)(1)(v)]

(2) *An operator will be deemed qualified* to operate a particular piece of equipment if the operator is certified under paragraph (b) of this section for that type and capacity of equipment or for higher-capacity equipment of that type. If no accredited testing agency offers certification examinations for a particular type and/or capacity of equipment, an operator will be deemed qualified to operate that equipment if the operator has been certified for the type/capacity that is most similar to that equipment and for which a certification examination is available. The operator's certificate must state the type/capacity of equipment for which the operator is certified. [§1926.1427(b)(2)]

(3) *A certification issued under* this option is portable and meets the requirements of paragraph (a)(2) of this section. [§1926.1427(b)(3)]

(4) *A certification issued under* this paragraph is valid for 5 years. [§1926.1427(b)(4)]

(c) **Option (2):** Qualification by an audited employer program. The employer's qualification of its employee must meet the following requirements: [§1926.1427(c)]

(1) *The written and practical tests* must be either: [§1926.1427(c)(1)]

(i) *Developed by an accredited crane* operator testing organization (*see* paragraph (b) of this section); or [§1926.1427(c)(1)(i)]

(ii) *Approved by an auditor in accordance* with the following requirements: [§1926.1427(c)(1)(ii)]

[A] *The auditor is certified to evaluate* such tests by an accredited crane operator testing organization (*see* paragraph (b) of this section). [§1926.1427(c)(1)(ii)[A]]

[B] *The auditor is not an employee of the employer.* [§1926.1427(c)(1)(ii)[B]]

[C] *The approval must be based on the auditor's* determination that the written and practical tests meet nationally recognized test development criteria and are valid and reliable in assessing the operator applicants regarding, at a minimum, the knowledge and skills listed in paragraphs (j)(1) and (2) of this section. [§1926.1427(c)(1)(ii)[C]]

[D] *The audit must be conducted in accordance* with nationally recognized auditing standards. [§1926.1427(c)(1)(ii)[D]]

(2) *Administration of tests.* [§1926.1427(c)(2)]

(i) *The written and practical tests* must be administered under circumstances approved by the auditor as meeting nationally recognized test administration standards. [§1926.1427(c)(2)(i)]

(ii) *The auditor must be certified to evaluate* the administration of the written and practical tests by an accredited crane operator testing organization (*see* paragraph (b) of this section). [§1926.1427(c)(2)(ii)]

(iii) *The auditor must not be an employee of the employer.* [§1926.1427(c)(2)(iii)]

(iv) *The audit must be conducted in accordance* with nationally recognized auditing standards. [§1926.1427(c)(2)(iv)]

(3) *The employer program must be audited* within 3 months of the beginning of the program and at least every 3 years thereafter. [§1926.1427(c)(3)]

(4) *The employer program must have* testing procedures for re-qualification designed to ensure that the operator continues to meet the technical knowledge and skills requirements in paragraphs (j)(1) and (2) of this section. The re-qualification procedures must be audited in accordance with paragraphs (c)(1) and (2) of this section. [§1926.1427(c)(4)]

(5) *Deficiencies.* If the auditor determines that there is a significant deficiency ("deficiency") in the program, the employer must ensure that: [§1926.1427(c)(5)]

(i) *No operator is qualified until* the auditor confirms that the deficiency has been corrected. [§1926.1427(c)(5)(i)]

(ii) *The program is audited again* within 180 days of the confirmation that the deficiency was corrected. [§1926.1427(c)(5)(ii)]

(iii) *The auditor files a documented* report of the deficiency to the appropriate Regional Office of the Occupational Safety and Health Administration within 15 days of the auditor's determination that there is a deficiency. [§1926.1427(c)(5)(iii)]

(iv) *Records of the audits of the employer's* program are maintained by the auditor for three years and are made available by the auditor to the Secretary of Labor or the Secretary's designated representative upon request. [§1926.1427(c)(5)(iv)]

(6) *A qualification under this paragraph is:* [§1926.1427(c)(6)]

(i) *Not portable.* Such a qualification meets the requirements of paragraph (a) of this section only where the operator is employed by (and operating the equipment for) the employer that issued the qualification. [§1926.1427(c)(6)(i)]

(ii) *Valid for 5 years.* [§1926.1427(c)(6)(ii)]

(d) **Option (3):** Qualification by the U.S. military. [§1926.1427(d)]

(1) *For purposes of this section,* an operator who is an employee of the U.S. military is considered qualified if he/she has a current operator qualification issued by the U.S. military for operation of the equipment. An employee of the U.S. military is a Federal employee of the Department of Defense or Armed Forces and does not include employees of private contractors. [§1926.1427(d)(1)]

(2) *A qualification under this paragraph is:* [§1926.1427(d)(2)]

(i) *Not portable.* Such a qualification meets the requirements of paragraph (a) of this section only where the operator is employed by (and operating the equipment for) the employer that issued the qualification. [§1926.1427(d)(2)(i)]

(ii) *Valid for the period of time* stipulated by the issuing entity. [§1926.1427(d)(2)(ii)]

(e) **Option (4):** Licensing by a government entity. [§1926.1427(e)]

(1) *For purposes of this section,* a government licensing department/office that issues operator licenses for operating equipment covered by this standard is considered a government accredited crane operator testing organization if the criteria in paragraph (e)(2) of this section are met. [§1926.1427(e)(1)]

(2) *Licensing criteria.* [§1926.1427(e)(2)]

(i) *The requirements for obtaining the license* include an assessment, by written and practical tests, of the operator applicant regarding, at a minimum, the knowledge and skills listed in paragraphs (j)(1) and (2) of this section. [§1926.1427(e)(2)(i)]

(ii) *The testing meets industry* recognized criteria for written testing materials, practical examinations, test administration, grading, facilities/equipment and personnel. [§1926.1427(e)(2)(ii)]

(iii) *The government authority that oversees* the licensing department/office, has determined that the requirements in paragraphs (e)(2)(i) and (ii) of this section have been met. [§1926.1427(e)(2)(iii)]

(iv) *The licensing department/office has* testing procedures for re-licensing designed to ensure that the operator continues to meet the technical knowledge and skills requirements in paragraphs (j)(1) and (2) of this section. [§1926.1427(e)(2)(iv)]

(3) *A license issued by a government* accredited crane operator testing organization that meets the requirements of this option: [§1926.1427(e)(3)]

(i) *Meets the operator qualification* requirements of this section for operation of equipment only within the jurisdiction of the government entity. [§1926.1427(e)(3)(i)]

(ii) *Is valid for the period of time* stipulated by the licensing department/office, but no longer than 5 years. [§1926.1427(e)(3)(ii)]

(f) **Pre-qualification/certification training period.** An employee who is not qualified or certified under this section is permitted to operate equipment only as an operator-in-training and only where the requirements of this paragraph are met. [§1926.1427(f)]

(1) *The employer must provide each operator-in-training* with sufficient training prior to operating the equipment to enable the operator-in-training to operate the equipment safely under limitations established by this section (including continuous monitoring) and any additional limitations established by the employer. [§1926.1427(f)(1)]

(2) *The tasks performed by the operator-in-training* while operating the equipment must be within the operator-in-training's ability. [§1926.1427(f)(2)]

(3) *Trainer.* While operating the equipment, the operator-in-training must be continuously monitored by an individual ("operator's trainer") who meets all of the following requirements: [§1926.1427(f)(3)]

(i) *The operator's trainer is an employee* or agent of the operator-in-training's employer. [§1926.1427(f)(3)(i)]

(ii) *The operator's trainer is either* a certified operator under this section, or has passed the written portion of a certification test under one of the options in paragraphs (b) through (e) of this section, and is familiar with the proper use of the equipment's controls. [§1926.1427(f)(3)(ii)]

(iii) *While monitoring the operator-in-training,* the operator's trainer performs no tasks that detract from the trainer's ability to monitor the operator-in-training. [§1926.1427(f)(3)(iii)]

(iv) *For equipment other than tower* cranes: The operator's trainer and the operator-in-training must be in direct line of sight of each other. In addition, they must communicate verbally or by hand signals. For tower cranes: The operator's trainer and the operator-in-training must be in direct communication with each other. [§1926.1427(f)(3)(iv)]

(4) *Continuous monitoring.* The operator-in-training must be monitored by the operator's trainer at all times, except for short breaks where all of the following are met: [§1926.1427(f)(4)]

(i) *The break lasts no longer* than 15 minutes and there is no more than one break per hour. [§1926.1427(f)(4)(i)]

(ii) *Immediately prior to the break* the operator's trainer informs the operator-in-training of the specific tasks that the operator-in-training is to perform and limitations to which he/she must adhere during the operator trainer's break. [§1926.1427(f)(4)(ii)]

(iii) *The specific tasks that the operator-in-training* will perform during the operator trainer's break are within the operator-in-training's abilities. [§1926.1427(f)(4)(iii)]

(5) *The operator-in-training must not operate the equipment* in any of the following circumstances unless the exception stated in paragraph (f)(5)(v) of this section is applicable: [§1926.1427(f)(5)]

(i) *If any part of the equipment,* load line or load (including rigging and lifting accessories), if operated up to the equipment's maximum working radius in the work zone (*see* Sec. 1926.1408(a)(1)), could get within 20 feet of a power line that is up to 350 kV, or within 50 feet of a power line that is over 350 kV. [§1926.1427(f)(5)(i)]

(ii) *If the equipment is used to hoist personnel.* [§1926.1427(f)(5)(ii)]

(iii) *In multiple-equipment lifts.* [§1926.1427(f)(5)(iii)]

(iv) *If the equipment is used over* a shaft, cofferdam, or in a tank farm. [§1926.1427(f)(5)(iv)]

(v) *In multiple-lift rigging operations,* except where the operator's trainer determines that the operator-in-training skills are sufficient for this high-skill work. [§1926.1427(f)(5)(v)]

(g) **Under this section,** a testing entity is permitted to provide training as well as testing services as long as the criteria of the applicable accrediting agency (in the option selected) for an organization providing both services are met. [§1926.1427(g)]

(h) **Language and Literacy Requirements.** [§1926.1427(h)]

(1) *Tests under this section* may be administered verbally, with answers given verbally, where the operator candidate: [§1926.1427(h)(1)]

(i) *Passes a written demonstration* of literacy relevant to the work. [§1926.1427(h)(1)(i)]

(ii) *Demonstrates the ability to use* the type of written manufacturer procedures applicable to the class/type of equipment for which the candidate is seeking certification. [§1926.1427(h)(1)(ii)]

(2) *Tests under this section* may be administered in any language the operator candidate understands, and the operator's certificate must note the language in which the test was given. The operator is qualified under paragraph (b)(2) of this section to operate equipment that is furnished with materials required by this subpart that are written in the language of the certification. The operator may only operate equipment furnished with such materials. [§1926.1427(h)(2)]

(i) ***[Reserved.]*** [§1926.1427(i)]

(j) **Certification criteria.** Qualifications and certifications must be based, at a minimum, on the following: [§1926.1427(j)]

(1) *A determination through a written test that:* [§1926.1427(j)(1)]

(i) *The individual knows the information* necessary for safe operation of the specific type of equipment the individual will operate, including all of the following: [§1926.1427(j)(1)(i)]

[A] The controls and operational/performance characteristics. [§1926.1427(j)(1)(i)[A]]

[B] Use of, and the ability to calculate (manually or with a calculator), load/capacity information on a variety of configurations of the equipment. [§1926.1427(j)(1)(i)[B]]

[C] Procedures for preventing and responding to power line contact. [§1926.1427(j)(1)(i)[C]]

[D] Technical knowledge similar to the subject matter criteria listed in Appendix C of this subpart applicable to the specific type of equipment the individual will operate. Use of the Appendix C criteria meets the requirements of this provision. [§1926.1427(j)(1)(i)[D]]

[E] Technical knowledge applicable to: [§1926.1427(j)(1)(i)[E]]

[1] The suitability of the supporting ground and surface to handle expected loads. [§1926.1427(j)(1)(i)[E][1]]

[2] Site hazards. [§1926.1427(j)(1)(i)[E][2]]

[3] Site access. [§1926.1427(j)(1)(i)[E][3]]

[F] This subpart, including applicable incorporated materials. [§1926.1427(j)(1)(i)[F]]

(ii) *The individual is able to read* and locate relevant information in the equipment manual and other materials containing information referred to in paragraph (j)(1)(i) of this section. [§1926.1427(j)(1)(ii)]

(2) *A determination through a practical* test that the individual has the skills necessary for safe operation of the equipment, including the following: [§1926.1427(j)(2)]

(i) *Ability to recognize,* from visual and auditory observation, the items listed in Sec. 1926.1412(d) (shift inspection). [§1926.1427(j)(2)(i)]

(ii) *Operational and maneuvering skills.* [§1926.1427(j)(2)(ii)]

(iii) *Application of load chart information.* [§1926.1427(j)(2)(iii)]

(iv) *Application of safe shut-down* and securing procedures. [§1926.1427(j)(2)(iv)]

(k) **Phase-in.** [§1926.1427(k)]

(1) *The provisions of this section are applicable* November 8, 2010, except for paragraphs (a)(2) and (f) which are applicable November 10, 2014. [§1926.1427(k)(1)]

(2) *When Sec. 1926.1427(a)(1) is not applicable,* all of the requirements in paragraphs (k)(2)(i) and (ii) of this section apply until November 10, 2014: [§1926.1427(k)(2)]

(i) *The employer must ensure that operators* of equipment covered by this standard are competent to operate the equipment safely. [§1926.1427(k)(2)(i)]

(ii) *Where an employee assigned to operate* machinery does not have the required knowledge or ability to operate the equipment safely, the employer must train that employee prior to operating the equipment. The employer must ensure that each operator is evaluated to confirm that he/she understands the information provided in the training. [§1926.1427(k)(2)(ii)]

§1926.1428
Signal person qualifications

(a) **The employer of the signal person** must ensure that each signal person meets the Qualification Requirements (paragraph (c) of this section) prior to giving any signals. This requirement must be met by using either Option (1) or Option (2) of this section. [§1926.1428(a)]

(1) *Option (1) — Third party qualified evaluator.* The signal person has documentation from a third party qualified evaluator (*see* Qualified Evaluator (third party), Sec. 1926.1401 for definition) showing that the signal person meets the Qualification Requirements (*see* paragraph (c) of this section). [§1926.1428(a)(1)]

(2) *Option (2) — Employer's qualified evaluator.* The employer's qualified (*see* Qualified Evaluator (not a third party), Sec. 1926.1401 for definition) evaluator assesses the individual and determines that the individual meets the Qualification Requirements (*see* paragraph (c) of this section) and provides documentation of that determination. An assessment by an employer's qualified evaluator under this option is not portable — other employers are not permitted to use it to meet the requirements of this section. [§1926.1428(a)(2)]

(3) *The employer must make the documentation* for whichever option is used available at the site while the signal person is employed by the employer. The documentation must specify each type of signaling (*e.g.* hand signals, radio signals, *etc.*) for which the signal person meets the requirements of paragraph (c) of this section. [§1926.1428(a)(3)]

(b) **If subsequent actions by the signal** person indicate that the individual does not meet the Qualification Requirements (*see* paragraph (c) of this section), the employer must not allow the individual to continue working as a signal person until re-training is provided and a re-assessment is made in accordance with paragraph (a) of this section that confirms that the individual meets the Qualification Requirements. [§1926.1428(b)]

(c) **Qualification Requirements.** Each signal person must: [§1926.1428(c)]

(1) *Know and understand the type* of signals used. If hand signals are used, the signal person must know and understand the Standard Method for hand signals. [§1926.1428(c)(1)]

(2) *Be competent in the application of the type* of signals used. [§1926.1428(c)(2)]

(3) *Have a basic understanding* of equipment operation and limitations, including the crane dynamics involved in swinging and stopping loads and boom deflection from hoisting loads. [§1926.1428(c)(3)]

(4) *Know and understand the relevant* requirements of Sec. 1926.1419 through Sec. 1926.1422 and Sec. 1926.1428. [§1926.1428(c)(4)]

(5) *Demonstrate that he/she meets* the requirements in paragraphs (c)(1) through (4) of this section through an oral or written test, and through a practical test. [§1926.1428(c)(5)]

§1926.1429
Qualifications of maintenance & repair employees

(a) **Maintenance, inspection and repair** personnel are permitted to operate the equipment only where all of the following requirements are met: [§1926.1429(a)]

(1) *The operation is limited to those* functions necessary to perform maintenance, inspect the equipment, or verify its performance. [§1926.1429(a)(1)]

(2) *The personnel either:* [§1926.1429(a)(2)]

(i) *Operate the equipment under* the direct supervision of an operator who meets the requirements of Sec. 1926.1427 (Operator qualification and certification); or [§1926.1429(a)(2)(i)]

(ii) *Are familiar with the operation,* limitations, characteristics and hazards associated with the type of equipment. [§1926.1429(a)(2)(ii)]

(b) **Maintenance and repair personnel** must meet the definition of a qualified person with respect to the equipment and maintenance/repair tasks performed. [§1926.1429(b)]

§1926.1430
Training

The employer must provide training as follows:

(a) **Overhead powerlines.** The employer must train each employee specified in Sec. 1926.1408(g) and Sec. 1926.1410(m) in the topics listed in Sec. 1926.1408(g). [§1926.1430(a)]

(b) **Signal persons.** The employer must train each employee who will be assigned to work as a signal persons who does not meet the requirements of Sec. 1926.1428(c) in the areas addressed in that paragraph. [§1926.1430(b)]

(c) **Operators.** [§1926.1430(c)]

(1) *Operators-in-Training for equipment where certification or qualification is required by this subpart.* The employer must train each operator-in-training in the areas addressed in Sec. 1926.1427(j). The employer must provide re-training if the operator-in-training does not pass a qualification or certification test. [§1926.1430(c)(1)]

(2) *Transitional Period.* During the four-year phase-in period for operator certification or qualification, as provided in Sec. 1926.1427(k), employers must train each operator who has not yet been certified or qualified in the areas addressed in Sec. 1926.1427(j). [§1926.1430(c)(2)]

(3) *Operators excepted from the requirements of Sec. 1926.1427.* The employer must train each operator excepted under Sec. 1926.1427(a) from the requirements of Sec. 1926.1427 on the safe operation of the equipment the operator will be using. [§1926.1430(c)(3)]

(4) *The employer must train each operator* of the equipment covered by this subpart in the following practices: [§1926.1430(c)(4)]

(i) *On friction equipment,* whenever moving a boom off a support, first raise the boom a short distance (sufficient to take the load of the boom) to determine if the boom hoist brake needs to be adjusted. On other types of equipment with a boom, the same practice is applicable, except that typically there is no means of adjusting the brake; if the brake does not hold, a repair is necessary. *See* Sec. 1926.1417(f) and (j) for additional requirements. [§1926.1430(c)(4)(i)]

(ii) *Where available,* the manufacturer's emergency procedures for halting unintended equipment movement. [§1926.1430(c)(4)(ii)]

(d) **Competent persons and qualified persons.** The employer must train each competent person and each qualified person regarding the requirements of this subpart applicable to their respective roles. [§1926.1430(d)]

(e) **Crush/pinch points.** The employer must train each employee who works with the equipment to keep clear of holes, and crush/pinch points and the hazards addressed in Sec. 1926.1424 (Work area control). [§1926.1430(e)]

(f) **Tag-out.** The employer must train each operator and each additional employee authorized to start/energize equipment or operate equipment controls (such as maintenance and repair employees), in the tag-out and start-up procedures in §§1926.1417(f) and (g). [§1926.1430(f)]

(g) **Training administration.** [§1926.1430(g)]

(1) *The employer must evaluate each employee* required to be trained under this subpart to confirm that the employee understands the information provided in the training. [§1926.1430(g)(1)]

(2) *The employer must provide refresher* training in relevant topics for each employee when, based on the conduct of the employee or an evaluation of the employee's knowledge, there is an indication that retraining is necessary. [§1926.1430(g)(2)]

(3) *Whenever training is required* under subpart CC, the employer must provide the training at no cost to the employee. [§1926.1430(g)(3)]

§1926.1431
Hoisting personnel

The requirements of this section are supplemental to the other requirements in this subpart and apply when one or more employees are hoisted.

(a) **The use of equipment to hoist** employees is prohibited except where the employer demonstrates that the erection, use, and dismantling of conventional means of reaching the work area, such as a personnel hoist, ladder, stairway, aerial lift, elevating work platform, or scaffold, would be more hazardous, or is not possible because of the project's structural design or worksite conditions. This paragraph does not apply to work covered by subpart R (Steel Erection) of this part. [§1926.1431(a)]

(b) **Use of personnel platform.** [§1926.1431(b)]

(1) *When using equipment to hoist employees,* the employees must be in a personnel platform that meets the requirements of paragraph (e) of this section. [§1926.1431(b)(1)]

(2) *Exceptions:* A personnel platform is not required for hoisting employees: [§1926.1431(b)(2)]

(i) *Into and out of drill shafts that are up* to and including 8 feet in diameter (*see* paragraph (o) of this section for requirements for hoisting these employees). [§1926.1431(b)(2)(i)]

(ii) *In pile driving operations* (*see* paragraph (p) of this section for requirements for hoisting these employees). [§1926.1431(b)(2)(ii)]

(iii) *Solely for transfer to or from a marine* worksite in a marine-hoisted personnel transfer device (*see* paragraph (r) of this section for requirements for hoisting these employees). [§1926.1431(b)(2)(iii)]

(iv) *In storage-tank (steel or concrete),* shaft and chimney operations (*see* paragraph (s) of this section for requirements for hoisting these employees). [§1926.1431(b)(2)(iv)]

(c) **Equipment set-up.** [§1926.1431(c)]

(1) *The equipment must be uniformly level,* within one percent of level grade, and located on footing that a qualified person has determined to be sufficiently firm and stable. [§1926.1431(c)(1)]

(2) *Equipment with outriggers or stabilizers* must have them all extended and locked. The amount of extension must be the same for all outriggers and stabilizers and in accordance with manufacturer procedures and load charts. [§1926.1431(c)(2)]

(d) **Equipment criteria.** [§1926.1431(d)]

(1) *Capacity: Use of suspended personnel platforms.* The total load (with the platform loaded, including the hook, load line and rigging) must not exceed 50 percent of the rated capacity for the radius and configuration of the equipment, except during proof testing. [§1926.1431(d)(1)]

(2) *Capacity: Use of boom-attached personnel platforms.* The total weight of the loaded personnel platform must not exceed 50 percent of the rated capacity for the radius and configuration of the equipment (except during proof testing). [§1926.1431(d)(2)]

(3) *Capacity: Hoisting personnel without a personnel platform.* When hoisting personnel without a personnel platform pursuant to paragraph (b)(2) of this section, the total load (including the hook, load line, rigging and any other equipment that imposes a load) must not exceed 50 percent of the rated capacity for the radius and configuration of the equipment, except during proof testing. [§1926.1431(d)(3)]

(4) *When the occupied personnel platform* is in a stationary working position, the load and boom hoist brakes, swing brakes, and operator actuated secondary braking and locking features (such as pawls or dogs) or automatic secondary brakes must be engaged. [§1926.1431(d)(4)]

(5) *Devices.* [§1926.1431(d)(5)]

(i) *Equipment* (except for derricks and articulating cranes) with a variable angle boom must be equipped with all of the following: [§1926.1431(d)(5)(i)]

[A] *A boom angle indicator,* readily visible to the operator, and [§1926.1431(d)(5)(i)[A]]

[B] *A boom hoist limiting device.* [§1926.1431(d)(5)(i)[B]]

(ii) *Articulating cranes must be equipped* with a properly functioning automatic overload protection device. [§1926.1431(d)(5)(ii)]

(iii) *Equipment with a luffing jib* must be equipped with: [§1926.1431(d)(5)(iii)]

[A] *A jib angle indicator,* readily visible to the operator, and. [§1926.1431(d)(5)(iii)[A]]

[B] *A jib hoist limiting device.* [§1926.1431(d)(5)(iii)[B]]

(iv) *Equipment with telescoping booms* must be equipped with a device to indicate the boom's extended length clearly to the operator, or must have measuring marks on the boom. [§1926.1431(d)(5)(iv)]

(v) *Anti two-block.* A device which automatically prevents damage and load failure from contact between the load block, overhaul ball, or similar component, and the boom tip (or fixed upper block or similar component) must be used. The device(s) must prevent such damage/failure at all points where two-blocking could occur. *Exception:* This device is not required when hoisting personnel in pile driving operations. Instead, paragraph (p)(2) of this section specifies how to prevent two-blocking during such operations. [§1926.1431(d)(5)(v)]

(vi) *Controlled load lowering.* The load line hoist drum must have a system, other than the load line hoist brake, which regulates the lowering rate of speed of the hoist mechanism. This system or device must be used when hoisting personnel. [§1926.1431(d)(5)(vi)]

Note: Free fall of the load line hoist is prohibited (*see* Sec. 1926.1426(d); the use of equipment in which the boom hoist mechanism can free fall is also prohibited (*see* Sec. 1926.1426(a)(1).

(vii) *Proper operation required.* Personnel hoisting operations must not begin unless the devices listed in this section are in proper working order. If a device stops working properly during such operations, the operator must safely stop operations. Personnel hoisting operations must not resume until the device is again working properly. Alternative measures are not permitted. (*See* Sec. 1926.1417 for tag-out and related requirements.) [§1926.1431(d)(5)(vii)]

(6) *Direct attachment of a personnel* platform to a luffing jib is prohibited. [§1926.1431(d)(6)]

(e) **Personnel platform criteria.** [§1926.1431(e)]

(1) *A qualified person familiar* with structural design must design the personnel platform and attachment/suspension system used for hoisting personnel. [§1926.1431(e)(1)]

(2) *The system used to connect* the personnel platform to the equipment must allow the platform to remain within 10 degrees of level, regardless of boom angle. [§1926.1431(e)(2)]

(3) *The suspension system must be designed* to minimize tipping of the platform due to movement of employees occupying the platform. [§1926.1431(e)(3)]

(4) *The personnel platform itself* (excluding the guardrail system and personal fall arrest system anchorages), must be capable of supporting, without failure, its own weight and at least five times the maximum intended load. [§1926.1431(e)(4)]

(5) *All welding of the personnel platform* and its components must be performed by a certified welder familiar with the weld grades, types and material specified in the platform design. [§1926.1431(e)(5)]

(6) *The personnel platform must be equipped* with a guardrail system which meets the requirements of subpart M of this part, and must be enclosed at least from the toeboard to mid-rail with either solid construction material or expanded metal having openings no greater than 1/2 inch (1.27 cm). Points to which personal fall arrest systems are attached must meet the anchorage requirements in subpart M of this part. [§1926.1431(e)(6)]

(7) *A grab rail must be installed* inside the entire perimeter of the personnel platform except for access gates/doors. [§1926.1431(e)(7)]

(8) *Access gates/doors.* If installed, access gates/doors of all types (including swinging, sliding, folding, or other types) must: [§1926.1431(e)(8)]

(i) *Not swing outward.* If due to the size of the personnel platform, such as a 1-person platform, it is infeasible for the door to swing inward and allow safe entry for the platform occupant, then the access gate/door may swing outward. [§1926.1431(e)(8)(i)]

(ii) *Be equipped with a device that prevents* accidental opening. [§1926.1431(e)(8)(ii)]

(9) *Headroom must be sufficient to allow* employees to stand upright in the platform. [§1926.1431(e)(9)]

(10) *In addition to the use of hard* hats, employees must be protected by overhead protection on the personnel platform when employees are exposed to falling objects. The platform overhead protection must not obscure the view of the operator or platform occupants (such as wire mesh that has up to 1/2 inch openings), unless full protection is necessary. [§1926.1431(e)(10)]

(11) *All edges exposed to employee* contact must be smooth enough to prevent injury. [§1926.1431(e)(11)]

(12) *The weight of the platform and its rated* capacity must be conspicuously posted on the platform with a plate or other permanent marking. [§1926.1431(e)(12)]

(f) **Personnel platform loading.** [§1926.1431(f)]

(1) *The personnel platform must not be loaded* in excess of its rated capacity. [§1926.1431(f)(1)]

(2) *Use.* [§1926.1431(f)(2)]

(i) *Personnel platforms must be used* only for employees, their tools, and the materials necessary to do their work. Platforms must not be used to hoist materials or tools when not hoisting personnel. [§1926.1431(f)(2)(i)]

(ii) *Exception:* Materials and tools to be used during the lift, if secured and distributed in accordance with paragraph (f)(3) of this section may be in the platform for trial lifts. [§1926.1431(f)(2)(ii)]

(3) *Materials and tools must be:* [§1926.1431(f)(3)]

(i) *Secured to prevent displacement.* [§1926.1431(f)(3)(i)]

(ii) *Evenly distributed within* the confines of the platform while it is suspended. [§1926.1431(f)(3)(ii)]

(4) *The number of employees occupying* the personnel platform must not exceed the maximum number the platform was designed to hold or the number required to perform the work, whichever is less. [§1926.1431(f)(4)]

(g) **Attachment and rigging.** [§1926.1431(g)]

(1) *Hooks and other detachable devices.* [§1926.1431(g)(1)]

(i) *Hooks used in the connection* between the hoist line and the personnel platform (including hooks on overhaul ball assemblies, lower load blocks, bridle legs, or other attachment assemblies or components) must be: [§1926.1431(g)(1)(i)]

[A] Of a type that can be closed and locked, eliminating the throat opening. [§1926.1431(g)(1)(i)[A]]

[B] Closed and locked when attached. [§1926.1431(g)(1)(i)[B]]

(ii) *Shackles used in place* of hooks must be of the alloy anchor type, with either: [§1926.1431(g)(1)(ii)]

[A] A bolt, nut and retaining pin, in place; or [§1926.1431(g)(1)(ii)[A]]

[B] Of the screw type, with the screw pin secured from accidental removal. [§1926.1431(g)(1)(ii)[B]]

(iii) *Where other detachable devices are used,* they must be of the type that can be closed and locked to the same extent as the devices addressed in paragraphs (g)(1)(i) and (ii) of this section. Such devices must be closed and locked when attached. [§1926.1431(g)(1)(iii)]

(2) *Rope bridle.* When a rope bridle is used to suspend the personnel platform, each bridle leg must be connected to a master link or shackle (*see* paragraph (g)(1) of this section) in a manner that ensures that the load is evenly divided among the bridle legs. [§1926.1431(g)(2)]

(3) *Rigging hardware* (including wire rope, shackles, rings, master links, and other rigging hardware) and hooks must be capable of supporting, without failure, at least five times the maximum intended load applied or transmitted to that component. Where rotation resistant rope is used, the slings must be capable of supporting without failure at least ten times the maximum intended load. [§1926.1431(g)(3)]

(4) *Eyes in wire rope* slings must be fabricated with thimbles. [§1926.1431(g)(4)]

(5) *Bridles and associated rigging* for suspending the personnel platform must be used only for the platform and the necessary employees, their tools and materials necessary to do their work. The bridles and associated rigging must not have been used for any purpose other than hoisting personnel. [§1926.1431(g)(5)]

(h) **Trial lift and inspection.** [§1926.1431(h)]

(1) *A trial lift with the unoccupied* personnel platform loaded at least to the anticipated liftweight must be made from ground level, or any other location where employees will enter the platform, to each location at which the platform is to be hoisted and positioned. Where there is more than one location to be reached from a single set-up position, either individual trial lifts for each location, or a single trial lift, in which the platform is moved sequentially to each location, must be performed; the method selected must be the same as the method that will be used to hoist the personnel. [§1926.1431(h)(1)]

(2) *The trial lift must be performed* immediately prior to each shift in which personnel will be hoisted. In addition, the trial lift must be repeated prior to hoisting employees in each of the following circumstances: [§1926.1431(h)(2)]

(i) *The equipment is moved and set* up in a new location or returned to a previously used location. [§1926.1431(h)(2)(i)]

(ii) *The lift route is changed,* unless the competent person determines that the new route presents no new factors affecting safety. [§1926.1431(h)(2)(ii)]

(3) *The competent person must determine that:* [§1926.1431(h)(3)]

(i) *Safety devices and operational* aids required by this section are activated and functioning properly. Other safety devices and operational aids must meet the requirements of Sec. 1926.1415 and Sec. 1926.1416. [§1926.1431(h)(3)(i)]

(ii) *Nothing interferes with the equipment* or the personnel platform in the course of the trial lift. [§1926.1431(h)(3)(ii)]

(iii) *The lift will not exceed 50* percent of the equipment's rated capacity at any time during the lift. [§1926.1431(h)(3)(iii)]

(iv) *The load radius to be used* during the lift has been accurately determined. [§1926.1431(h)(3)(iv)]

(4) *Immediately after the trial* lift, the competent person must: [§1926.1431(h)(4)]

(i) *Conduct a visual inspection* of the equipment, base support or ground, and personnel platform, to determine whether the trial lift has exposed any defect or problem or produced any adverse effect. [§1926.1431(h)(4)(i)]

(ii) *Confirm that,* upon the completion of the trial lift process, the test weight has been removed. [§1926.1431(h)(4)(ii)]

(5) *Immediately prior to each lift:* [§1926.1431(h)(5)]

(i) *The platform must be hoisted a few* inches with the personnel and materials/tools on board and inspected by a competent person to ensure that it is secure and properly balanced. [§1926.1431(h)(5)(i)]

(ii) *The following conditions must be determined* by a competent person to exist before the lift of personnel proceeds: [§1926.1431(h)(5)(ii)]

[A] Hoist ropes must be free of deficiencies in accordance with Sec. 1926.1413(a). [§1926.1431(h)(5)(ii)[A]]

[B] *Multiple part lines* must not be twisted around each other. [§1926.1431(h)(5)(ii)[B]]

[C] *The primary attachment must be centered* over the platform. [§1926.1431(h)(5)(ii)[C]]

[D] *If the load rope is slack,* the hoisting system must be inspected to ensure that all ropes are properly seated on drums and in sheaves. [§1926.1431(h)(5)(ii)[D]]

(6) *Any condition found during* the trial lift and subsequent inspection(s) that fails to meet a requirement of this standard or otherwise creates a safety hazard must be corrected before hoisting personnel. (*See* Sec. 1926.1417 for tag-out and related requirements.) [§1926.1431(h)(6)]

(i) **[*Reserved.*]** [§1926.1431(i)]

(j) **Proof testing.** [§1926.1431(j)]

(1) *At each jobsite,* prior to hoisting employees on the personnel platform, and after any repair or modification, the platform and rigging must be proof tested to 125 percent of the platform's rated capacity. The proof test may be done concurrently with the trial lift. [§1926.1431(j)(1)]

(2) *The platform must be lowered by controlled* load lowering, braked, and held in a suspended position for a minimum of five minutes with the test load evenly distributed on the platform. [§1926.1431(j)(2)]

(3) *After proof testing,* a competent person must inspect the platform and rigging to determine if the test has been passed. If any deficiencies are found that pose a safety hazard, the platform and rigging must not be used to hoist personnel unless the deficiencies are corrected, the test is repeated, and a competent person determines that the test has been passed. (*See* Sec. 1926.1417 for tag-out and related requirements.) [§1926.1431(j)(3)]

(4) *Personnel hoisting must not be conducted* until the competent person determines that the platform and rigging have successfully passed the proof test. [§1926.1431(j)(4)]

(k) **Work practices.** [§1926.1431(k)]

(1) *Hoisting of the personnel platform* must be performed in a slow, controlled, cautious manner, with no sudden movements of the equipment or the platform. [§1926.1431(k)(1)]

(2) *Platform occupants must:* [§1926.1431(k)(2)]

(i) *Keep all parts of the body* inside the platform during raising, lowering, and horizontal movement. This provision does not apply to an occupant of the platform when necessary to position the platform or while performing the duties of a signal person. [§1926.1431(k)(2)(i)]

(ii) *Not stand,* sit on, or work from the top or intermediate rail or toeboard, or use any other means/device to raise their working height. [§1926.1431(k)(2)(ii)]

(iii) *Not pull the platform out of plumb* in relation to the hoisting equipment. [§1926.1431(k)(2)(iii)]

(3) *Before employees exit* or enter a hoisted personnel platform that is not landed, the platform must be secured to the structure where the work is to be performed, unless the employer can demonstrate that securing to the structure would create a greater hazard. [§1926.1431(k)(3)]

(4) *If the platform is tied to the structure,* the operator must not move the platform until the operator receives confirmation that it is freely suspended. [§1926.1431(k)(4)]

(5) *Tag lines must be used* when necessary to control the platform. [§1926.1431(k)(5)]

(6) *Platforms without controls.* Where the platform is not equipped with controls, the equipment operator must remain at the equipment controls, on site, and in view of the equipment, at all times while the platform is occupied. [§1926.1431(k)(6)]

(7) *Platforms with controls.* Where the platform is equipped with controls, all of the following must be met at all times while the platform is occupied: [§1926.1431(k)(7)]

(i) *The occupant using the controls in the platform* must be a qualified person with respect to their use, including the safe limitations of the equipment and hazards associated with its operation. [§1926.1431(k)(7)(i)]

(ii) *The equipment operator must be at a set* of equipment controls that include boom and swing functions of the equipment, and must be on site and in view of the equipment. [§1926.1431(k)(7)(ii)]

(iii) *The platform operating manual* must be in the platform or on the equipment. [§1926.1431(k)(7)(iii)]

(8) *Environmental conditions.* [§1926.1431(k)(8)]

(i) *Wind.* When wind speed (sustained or gusts) exceeds 20 mph at the personnel platform, a qualified person must determine if, in light of the wind conditions, it is not safe to lift personnel. If it is not, the lifting operation must not begin (or, if already in progress, must be terminated). [§1926.1431(k)(8)(i)]

(ii) *Other weather and environmental conditions.* A qualified person must determine if, in light of indications of dangerous weather conditions, or other impending or existing danger, it is not safe to lift personnel. If it is not, the lifting operation must not begin (or, if already in progress, must be terminated). [§1926.1431(k)(8)(ii)]

(9) *Employees being hoisted must remain* in direct communication with the signal person (where used), or the operator. [§1926.1431(k)(9)]

(10) *Fall protection.* [§1926.1431(k)(10)]

(i) *Except over water,* employees occupying the personnel platform must be provided and use a personal fall arrest system. The system must be attached to a structural member within the personnel platform. When working over or near water, the requirements of Sec. 1926.106 apply. [§1926.1431(k)(10)(i)]

(ii) *The fall arrest system,* including the attachment point (anchorage) used to comply with paragraph (i) of this section, must meet the requirements in Sec. 1926.502. [§1926.1431(k)(10)(ii)]

(11) *Other load lines.* [§1926.1431(k)(11)]

(i) *No lifts must be made on any other of the equipment's* load lines while personnel are being hoisted, except in pile driving operations. [§1926.1431(k)(11)(i)]

(ii) *Factory-produced boom-mounted personnel platforms that incorporate a winch as original equipment.* Loads are permitted to be hoisted by such a winch while employees occupy the personnel platform only where the load on the winch line does not exceed 500 pounds and does not exceed the rated capacity of the winch and platform. [§1926.1431(k)(11)(ii)]

(12) *Traveling — equipment other than derricks.* [§1926.1431(k)(12)]

(i) *Hoisting of employees while* the equipment is traveling is prohibited, except for: [§1926.1431(k)(12)(i)]

[A] *Equipment that travels on fixed rails; or* [§1926.1431(k)(12)(i)[A]]

[B] *Where the employer demonstrates that there* is no less hazardous way to perform the work. [§1926.1431(k)(12)(i)[B]]

[C] *This exception does not apply* to rubber-tired equipment. [§1926.1431(k)(12)(i)[C]]

(ii) *Where employees are hoisted while* the equipment is traveling, all of the following criteria must be met: [§1926.1431(k)(12)(ii)]

[A] *Equipment travel must be restricted* to a fixed track or runway. [§1926.1431(k)(12)(ii)[A]]

[B] *Where a runway is used,* it must be a firm, level surface designed, prepared and designated as a path of travel for the weight and configuration of the equipment being used to lift and travel with the personnel platform. An existing surface may be used as long as it meets these criteria. [§1926.1431(k)(12)(ii)[B]]

[C] *Equipment travel must be limited* to boom length. [§1926.1431(k)(12)(ii)[C]]

[D] *The boom must be parallel to the direction* of travel, except where it is safer to do otherwise. [§1926.1431(k)(12)(ii)[D]]

[E] *A complete trial run* must be performed to test the route of travel before employees are allowed to occupy the platform. This trial run can be performed at the same time as the trial lift required by paragraph (h) of this section which tests the lift route. [§1926.1431(k)(12)(ii)[E]]

(13) *Traveling — derricks.* Derricks are prohibited from traveling while personnel are hoisted. [§1926.1431(k)(13)]

(l) **[*Reserved.*]** [§1926.1431(l)]

(m) **Pre-lift meeting.** A pre-lift meeting must be: [§1926.1431(m)]

(1) *Held to review the applicable* requirements of this section and the procedures that will be followed. [§1926.1431(m)(1)]

(2) *Attended by the equipment operator,* signal person (if used for the lift), employees to be hoisted, and the person responsible for the task to be performed. [§1926.1431(m)(2)]

(3) *Held prior to the trial* lift at each new work location, and must be repeated for any employees newly assigned to the operation. [§1926.1431(m)(3)]

(n) **Hoisting personnel near power lines.** Hoisting personnel within 20 feet of a power line that is up to 350 kV, and hoisting personnel within 50 feet of a power line that is over 350 kV, is prohibited, except for work covered by subpart V of this part (Power Transmission and Distribution). [§1926.1431(n)]

(o) **Hoisting personnel in drill shafts.** When hoisting employees into and out of drill shafts that are up to and including 8 feet in diameter, all of the following requirements must be met: [§1926.1431(o)]

(1) *The employee must be in either a personnel* platform or on a boatswain's chair. [§1926.1431(o)(1)]

(2) *If using a personnel platform,* paragraphs (a) through (n) of this section apply. [§1926.1431(o)(2)]

(3) *If using a boatswain's chair:* [§1926.1431(o)(3)]

(i) *The following paragraphs of this section* apply: (a), (c), (d)(1), (d)(3), (d)(4), (e)(1), (e)(2), (e)(3), (f)(1), (f)(2)(i), (f)(3)(i), (g), (h), (k)(1), (k)(6), (k)(8), (k)(9), (k)(11)(i), (m), (n). Where the terms "personnel platform" or "platform" are used in these paragraphs, substitute them with "boatswain's chair." [§1926.1431(o)(3)(i)]

(ii) *A signal person must be stationed* at the shaft opening. [§1926.1431(o)(3)(ii)]

(iii) *The employee must be hoisted in a slow,* controlled descent and ascent. [§1926.1431(o)(3)(iii)]

(iv) *The employee must use personal* fall protection equipment, including a full body harness, attached independent of the crane/derrick. [§1926.1431(o)(3)(iv)]

(v) *The fall protection equipment* must meet the applicable requirements in Sec. 1926.502. [§1926.1431(o)(3)(v)]

(vi) *The boatswain's chair itself* (excluding the personal fall arrest system anchorages), must be capable of supporting, without failure, its own weight and at least five times the maximum intended load. [§1926.1431(o)(3)(vi)]

(vii) *No more than one* person must be hoisted at a time. [§1926.1431(o)(3)(vii)]

(p) **Hoisting personnel for pile driving operations.** When hoisting an employee in pile driving operations, the following requirements must be met: [§1926.1431(p)]

(1) *The employee must be in a personnel platform* or boatswain's chair. [§1926.1431(p)(1)]

(2) *For lattice boom cranes:* Clearly mark the cable (so that it can easily be seen by the operator) at a point that will give the operator sufficient time to stop the hoist to prevent two-blocking, or use a spotter who is in direct communication with the operator to inform the operator when this point is reached. For telescopic boom cranes: Clearly mark the cable (so that it can be easily seen by the operator) at a point that will give the operator sufficient time to stop the hoist to prevent two-blocking, and use a spotter who is in direct communication with the operator to inform the operator when this point is reached. [§1926.1431(p)(2)]

(3) *If using a personnel platform,* paragraphs (b) through (n) of this section apply. [§1926.1431(p)(3)]

(4) *If using a boatswain's chair:* [§1926.1431(p)(4)]

(i) *The following paragraphs of this section* apply: (a), (c), (d)(1), (d)(3), (d)(4), (e)(1), (e)(2), (e)(3), (f)(1), (f)(2)(i), (f)(3)(i), (g), (h), (j), (k)(1), (k)(6), (k)(8), (k)(9), (k)(11)(i), (m), and (n). Where the terms "personnel platform" or "platform" are used in these paragraphs, substitute them with "boatswains chair." [§1926.1431(p)(4)(i)]

(ii) *The employee must be hoisted in a slow,* controlled descent and ascent. [§1926.1431(p)(4)(ii)]

(iii) *The employee must use personal* fall protection equipment, including a full body harness, independently attached to the lower load block or overhaul ball. [§1926.1431(p)(4)(iii)]

(iv) *The fall protection equipment* must meet the applicable requirements in Sec. 1926.502. [§1926.1431(p)(4)(iv)]

(v) *The boatswain's chair itself* (excluding the personal fall arrest system anchorages), must be capable of supporting, without failure, its own weight and at least five times the maximum intended load. [§1926.1431(p)(4)(v)]

(vi) *No more than one* person must be hoisted at a time. [§1926.1431(p)(4)(vi)]

(q) *[Reserved.]* [§1926.1431(q)]

(r) **Hoisting personnel for marine transfer.** When hoisting employees solely for transfer to or from a marine worksite, the following requirements must be met: [§1926.1431(r)]

(1) *The employee must be in either a personnel* platform or a marine-hoisted personnel transfer device. [§1926.1431(r)(1)]

(2) *If using a personnel platform,* paragraphs (a) through (n) of this section apply. [§1926.1431(r)(2)]

(3) *If using a marine-hoisted personnel transfer device:* [§1926.1431(r)(3)]

(i) *The following paragraphs of this section* apply: (a), (c)(2), (d)(1), (d)(3), (d)(4), (e)(1) through (5), (e)(12), (f)(1), (g), (h), (j), (k)(1), (k)(8), (k)(9), (k)(10)(ii), (k)(11)(i), (k)(12), (m), and (n). Where the terms "personnel platform" or "platform" are used in these paragraphs, substitute them with "marine-hoisted personnel transfer device." [§1926.1431(r)(3)(i)]

(ii) *The transfer device must be used* only for transferring workers. [§1926.1431(r)(3)(ii)]

(iii) *The number of workers occupying* the transfer device must not exceed the maximum number it was designed to hold. [§1926.1431(r)(3)(iii)]

(iv) *Each employee must wear a U.S.* Coast Guard personal flotation device approved for industrial use. [§1926.1431(r)(3)(iv)]

(s) **Hoisting personnel for storage-tank (steel or concrete), shaft and chimney operations.** When hoisting an employee in storage tank (steel or concrete), shaft and chimney operations, the following requirements must be met: [§1926.1431(s)]

(1) *The employee must be in a personnel platform* except when the employer can demonstrate that use of a personnel platform is infeasible; in such a case, a boatswain's chair must be used. [§1926.1431(s)(1)]

(2) *If using a personnel platform,* paragraphs (a) through (n) of this section apply. [§1926.1431(s)(2)]

(3) *If using a boatswain's chair:* [§1926.1431(s)(3)]

(i) *The following paragraphs of this section* apply: (a), (c), (d)(1), (d)(3), (d)(4), (e)(1), (e)(2), (e)(3), (f)(1), (f)(2)(i), (f)(3)(i), (g), (h), (k)(1), (k)(6), (k)(8), (k)(9), (k)(11)(i), (m), (n). Where the terms "personnel platform" or "platform" are used in these paragraphs, substitute them with "boatswains chair." [§1926.1431(s)(3)(i)]

(ii) *The employee must be hoisted in a slow,* controlled descent and ascent. [§1926.1431(s)(3)(ii)]

(iii) *The employee must use personal* fall protection equipment, including a full body harness, attached independent of the crane/derrick. When there is no adequate structure for attachment of personal fall arrest equipment as required in Sec. 1926.502(d)(15), the attachment must be to the lower load block or overhaul ball. [§1926.1431(s)(3)(iii)]

(iv) *The fall protection equipment* must meet the applicable requirements in Sec. 1926.502. [§1926.1431(s)(3)(iv)]

(v) *The boatswain's chair itself* (excluding the personal fall arrest system anchorages), must be capable of supporting, without failure, its own weight and at least five times the maximum intended load. [§1926.1431(s)(3)(v)]

(vi) *No more than one* person must be hoisted at a time. [§1926.1431(s)(3)(vi)]

§1926.1432
Multiple-crane/derrick lifts — supplemental requirements

(a) **Plan development.** Before beginning a crane/derrick operation in which more than one crane/derrick will be supporting the load, the operation must be planned. The planning must meet the following requirements: [§1926.1432(a)]

(1) *The plan must be developed by a qualified person.* [§1926.1432(a)(1)]

(2) *The plan must be designed to ensure* that the requirements of this subpart are met. [§1926.1432(a)(2)]

(3) *Where the qualified person determines* that engineering expertise is needed for the planning, the employer must ensure that it is provided. [§1926.1432(a)(3)]

(b) **Plan implementation.** [§1926.1432(b)]

(1) *The multiple-crane/derrick lift must be directed* by a person who meets the criteria for both a competent person and a qualified person, or by a competent person who is assisted by one or more qualified persons (lift director). [§1926.1432(b)(1)]

(2) *The lift director must review* the plan in a meeting with all workers who will be involved with the operation. [§1926.1432(b)(2)]

§1926.1433
Design, construction and testing

The following requirements apply to equipment that has a manufacturer-rated hoisting/lifting capacity of more than 2,000 pounds.

(a) **Crawler, truck and locomotive** cranes manufactured prior to November 8, 2010 must meet the applicable requirements for design, construction, and testing as prescribed in ANSI B30.5-1968 (incorporated by reference, *see* Sec. 1926.6), PCSA Std. No. 2 (1968) (incorporated by reference, *see* Sec. 1926.6), the requirements in paragraph (b) of this section, or the applicable DIN standards that were in effect at the time of manufacture. [§1926.1433(a)]

(b) **Mobile** (including crawler and truck) and locomotive cranes manufactured on or after November 8, 2010 must meet the following portions of ASME B30.5-2004 (incorporated by reference, *see* Sec. 1926.6) as applicable: [§1926.1433(b)]

(1) *In section 5-1.1.1* ("Load Ratings — Where Stability Governs Lifting Performance"), paragraphs (a) — (d) (including subparagraphs). [§1926.1433(b)(1)]

(2) *In section 5-1.1.2* ("Load Ratings — Where Structural Competence Governs Lifting Performance"), paragraph (b). [§1926.1433(b)(2)]

(3) *Section 5-1.2* ("Stability (Backward and Forward)"). [§1926.1433(b)(3)]

(4) *In section 5-1.3.1* ("Boom Hoist Mechanism"), paragraphs (a), (b)(1) and (b)(2), except that when using rotation resistant rope, §1926.1414(c)(4)(ii)(A) applies. [§1926.1433(b)(4)]

(5) *In section 5-1.3.2* ("Load Hoist Mechanism"), paragraphs (a)(2) through (a)(4) (including subparagraphs), (b) (including subparagraphs), (c) (first sentence only) and (d). [§1926.1433(b)(5)]

(6) *Section 5-1.3.3 ("Telescoping Boom").* [§1926.1433(b)(6)]

(7) *Section 5-1.4 ("Swing Mechanism").* [§1926.1433(b)(7)]

(8) *In section 5-1.5 ("Crane Travel"),* all provisions except 5-1.5.3(d). [§1926.1433(b)(8)]

(9) *In section 5-1.6 ("Controls"),* all provisions except 5-1.6.1 (c). [§1926.1433(b)(9)]

(10) *Section 5-1.7.4 ("Sheaves").* [§1926.1433(b)(10)]

(11) *Section 5-1.7.5 ("Sheave sizes").* [§1926.1433(b)(11)]

(12) *In section 5-1.9.1 ("Booms"), paragraph (f).* [§1926.1433(b)(12)]

(13) *Section 5-1.9.3 ("Outriggers").* [§1926.1433(b)(13)]

(14) *Section 5-1.9.4* ("Locomotive Crane Equipment"). [§1926.1433(b)(14)]

(15) *Section 5-1.9.7* ("Clutch and Brake Protection"). [§1926.1433(b)(15)]

(16) *In section 5-1.9.11* ("Miscellaneous equipment"), paragraphs (a), (c), (e), and (f). [§1926.1433(b)(16)]

(c) **Prototype testing:** mobile (including crawler and truck) and locomotive cranes manufactured on or after November 8, 2010 must meet the prototype testing requirements in Test Option A or Test Option B of this section. Tower cranes manufactured on or after November 8, 2010 must meet the prototype testing requirements in BS EN 14439:2006 (incorporated by reference, *see* Sec. 1926.6). [§1926.1433(c)]

Note: Prototype testing of crawler, locomotive and truck cranes manufactured prior to November 8, 2010 must conform to paragraph (a) of this section.

(1) *Test Option A.* [§1926.1433(c)(1)]

(i) *The following applies to equipment* with cantilevered booms (such as hydraulic boom cranes): All the tests listed in SAE J1063 (Nov. 1993) Table 1 (incorporated by reference, *see* Sec. 1926.6) must be performed to load all critical structural elements to their respective limits. All the strength margins listed in SAE J1063 (Nov. 1993) Table 2 (incorporated by reference, *see* Sec. 1926.6) must be met. [§1926.1433(c)(1)(i)]

(ii) *The following applies to equipment* with pendant supported lattice booms: All the tests listed in SAE J987 (Jun. 2003) Table 1 (incorporated by reference, *see* Sec. 1926.6) must be performed to load all critical structural elements to their respective limits. All the strength margins listed in SAE J987 (Jun. 2003) Table 2 (incorporated by reference, *see* Sec. 1926.6) must be met. [§1926.1433(c)(1)(ii)]

(2) *Test Option B.* The testing and verification requirements of BS EN 13000:2004 (incorporated by reference, *see* Sec. 1926.6) must be met. In applying BS EN 13000:2004, the following additional requirements must be met: [§1926.1433(c)(2)]

(i) *The following applies to equipment* with cantilevered booms (such as hydraulic boom cranes): The analysis methodology (computer modeling) must demonstrate that all load cases listed in SAE J1063 (Nov. 1993) (incorporated by reference, *see* Sec. 1926.6) meet the strength margins listed in SAE J1063 (Nov. 1993) Table 2. [§1926.1433(c)(2)(i)]

(ii) *The following applies to equipment* with pendant supported lattice booms: The analysis methodology (computer modeling) must demonstrate that all load cases listed in SAE J987 (Jun. 2003) (incorporated by reference, *see* Sec. 1926.6) meet the strength margins listed in SAE J987 (Jun. 2003) Table 2. [§1926.1433(c)(2)(ii)]

(iii) *Analysis verification.* The physical testing requirements under SAE J1063 (Nov. 1993) (incorporated by reference, *see* Sec. 1926.6) and SAE J987 (Jun. 2003) (incorporated by reference, *see* Sec. 1926.6) must be met unless the reliability of the analysis methodology (computer modeling) has been demonstrated by a documented history of verification through strain gauge measuring or strain gauge measuring in combination with other physical testing. [§1926.1433(c)(2)(iii)]

(d) **All equipment covered by this subpart** must meet the following requirements: [§1926.1433(d)]

(1) *Rated capacity and related information.* The information available in the cab (*see* Sec. 1926.1417(c)) regarding "rated capacity" and related information must include, at a minimum, the following information: [§1926.1433(d)(1)]

(i) *A complete range of the manufacturer's* equipment rated capacities, as follows: [§1926.1433(d)(1)(i)]

[A] At all manufacturer approved operating radii, boom angles, work areas, boom lengths and configurations, jib lengths and angles (or offset). [§1926.1433(d)(1)(i)[A]]

[B] Alternate ratings for use and nonuse of option equipment which affects rated capacities, such as outriggers, stabilizers, and extra counterweights. [§1926.1433(d)(1)(i)[B]]

(ii) *A work area chart* for which capacities are listed in the load chart. (Note: An example of this type of chart is in ASME B30.5-2004, section 5-1.1.3, Figure 11). [§1926.1433(d)(1)(ii)]

(iii) *The work area figure* and load chart must clearly indicate the areas where no load is to be handled. [§1926.1433(d)(1)(iii)]

(iv) *Recommended reeving for the hoist* lines must be shown. [§1926.1433(d)(1)(iv)]

(v) *Recommended parts of hoist* reeving, size, and type of wire rope for various equipment loads. [§1926.1433(d)(1)(v)]

(vi) *Recommended boom hoist* reeving diagram, where applicable; size, type and length of wire rope. [§1926.1433(d)(1)(vi)]

(vii) *Tire pressure (where applicable).* [§1926.1433(d)(1)(vii)]

(viii) *Caution or warnings relative* to limitations on equipment and operating procedures, including an indication of the least stable direction. [§1926.1433(d)(1)(viii)]

(ix) *Position of the gantry and requirements* for intermediate boom suspension (where applicable). [§1926.1433(d)(1)(ix)]

(x) *Instructions for boom erection* and conditions under which the boom, or boom and jib combinations, may be raised or lowered. [§1926.1433(d)(1)(x)]

(xi) *Whether the hoist holding* mechanism is automatically or manually controlled, whether free fall is available, or any combination of these. [§1926.1433(d)(1)(xi)]

(xii) *The maximum telescopic travel* length of each boom telescopic section. [§1926.1433(d)(1)(xii)]

(xiii) *Whether sections are telescoped* manually or with power. [§1926.1433(d)(1)(xiii)]

(xiv) *The sequence and procedure for extending* and retracting the telescopic boom section. [§1926.1433(d)(1)(xiv)]

(xv) *Maximum loads permitted* during the boom extending operation, and any limiting conditions or cautions. [§1926.1433(d)(1)(xv)]

(xvi) *Hydraulic relief valve* settings specified by the manufacturer. [§1926.1433(d)(1)(xvi)]

(2) *Load hooks* (including latched and unlatched types), ball assemblies and load blocks must be of sufficient weight to overhaul the line from the highest hook position for boom or boom and jib lengths and the number of parts of the line in use. [§1926.1433(d)(2)]

(3) *Hook and ball assemblies* and load blocks must be marked with their rated capacity and weight. [§1926.1433(d)(3)]

(4) *Latching hooks.* [§1926.1433(d)(4)]

(i) *Hooks must be equipped with latches,* except where the requirements of paragraph (d)(4)(ii) of this section are met. [§1926.1433(d)(4)(i)]

(ii) *Hooks without latches,* or with latches removed or disabled, must not be used unless: [§1926.1433(d)(4)(ii)]

[A] A qualified person has determined that it is safer to hoist and place the load without latches (or with the latches removed/ tied-back). [§1926.1433(d)(4)(ii)[A]]

[B] Routes for the loads are pre-planned to ensure that no employee is required to work in the fall zone except for employees necessary for the hooking or unhooking of the load. [§1926.1433(d)(4)(ii)[B]]

(iii) *The latch must close the throat* opening and be designed to retain slings or other lifting devices/accessories in the hook when the rigging apparatus is slack. [§1926.1433(d)(4)(iii)]

(5) *Posted warnings.* Posted warnings required by this subpart as well as those originally supplied with the equipment by the manufacturer must be maintained in legible condition. [§1926.1433(d)(5)]

(6) *An accessible fire extinguisher* must be on the equipment. [§1926.1433(d)(6)]

(7) *Cabs.* Equipment with cabs must meet the following requirements: [§1926.1433(d)(7)]

(i) *Cabs must be designed with a form* of adjustable ventilation and method for clearing the windshield for maintaining visibility and air circulation. Examples of means for adjustable ventilation include air conditioner or window that can be opened (for ventilation and air circulation); examples of means for maintaining visibility include heater (for preventing windshield icing), defroster, fan, windshield wiper. [§1926.1433(d)(7)(i)]

(ii) *Cab doors (swinging, sliding)* must be designed to prevent inadvertent opening or closing while traveling or operating the machine. Swinging doors adjacent to the operator must open outward. Sliding operator doors must open rearward. [§1926.1433(d)(7)(ii)]

(iii) *Windows.* [§1926.1433(d)(7)(iii)]

[A] The cab must have windows in front and on both sides of the operator. Forward vertical visibility must be sufficient to give the operator a view of the boom point at all times. [§1926.1433(d)(7)(iii)[A]]

[B] Windows may have sections designed to be opened or readily removed. Windows with sections designed to be opened must be designed so that they can be secured to prevent inadvertent closure. [§1926.1433(d)(7)(iii)[B]]

[C] Windows must be of safety glass or material with similar optical and safety properties, that introduce no visible distortion or otherwise obscure visibility that interferes with the safe operation of the equipment. [§1926.1433(d)(7)(iii)[C]]

(iv) *A clear passageway must be provided* from the operator's station to an exit door on the operator's side. [§1926.1433(d)(7)(iv)]

(v) *Areas of the cab roof* that serve as a workstation for rigging, maintenance or other equipment-related tasks must be capable of supporting 250 pounds without permanent distortion. [§1926.1433(d)(7)(v)]

(8) *Belts, gears,* shafts, pulleys, sprockets, spindles, drums, fly wheels, chains, and other parts or components that reciprocate, rotate or otherwise move must be guarded where contact by employees (except for maintenance and repair employees) is possible in the performance of normal duties. [§1926.1433(d)(8)]

(9) *All exhaust pipes,* turbochargers, and charge air coolers must be insulated or guarded where contact by employees (except for maintenance and repair employees) is possible in the performance of normal duties. [§1926.1433(d)(9)]

(10) *Hydraulic and pneumatic lines* must be protected from damage to the extent feasible. [§1926.1433(d)(10)]

(11) *The equipment must be designed so that exhaust* fumes are not discharged in the cab and are discharged in a direction away from the operator. [§1926.1433(d)(11)]

(12) *Friction mechanisms.* Where friction mechanisms (such as brakes and clutches) are used to control the boom hoist or load line hoist, they must be: [§1926.1433(d)(12)]

(i) *Of a size and thermal capacity* sufficient to control all rated loads with the minimum recommended reeving. [§1926.1433(d)(12)(i)]

(ii) *Adjustable to permit compensation* for lining wear to maintain proper operation. [§1926.1433(d)(12)(ii)]

(13) *Hydraulic load hoists.* Hydraulic drums must have an integrally mounted holding device or internal static brake to prevent load hoist movement in the event of hydraulic failure. [§1926.1433(d)(13)]

(e) **The employer's obligations under** paragraphs (a) through (c) and (d)(7) through (13) of this section are met where the equipment has not changed (except in accordance with Sec. 1926.1434 (Equipment modifications)) and it can refer to documentation from the manufacturer showing that the equipment has been designed, constructed and tested in accordance with those paragraphs. [§1926.1433(e)]

§1926.1434
Equipment modifications

(a) **Modifications or additions which** affect the capacity or safe operation of the equipment are prohibited except where the requirements of paragraphs (a)(1), (a)(2), (a)(3), (a)(4), or (a)(5) of this section are met. [§1926.1434(a)]

(1) *Manufacturer review and approval.* [§1926.1434(a)(1)]

(i) *The manufacturer approves the modifications/additions in writing.* [§1926.1434(a)(1)(i)]

(ii) *The load charts,* procedures, instruction manuals and instruction plates/tags/decals are modified as necessary to accord with the modification/addition. [§1926.1434(a)(1)(ii)]

(iii) *The original safety factor* of the equipment is not reduced. [§1926.1434(a)(1)(iii)]

(2) *Manufacturer refusal to review request.* The manufacturer is provided a detailed description of the proposed modification/addition, is asked to approve the modification/addition, but it declines to review the technical merits of the proposal or fails, within 30 days, to acknowledge the request or initiate the review, and all of the following are met: [§1926.1434(a)(2)]

(i) *A registered professional engineer* who is a qualified person with respect to the equipment involved: [§1926.1434(a)(2)(i)]

[A] Approves the modification/addition and specifies the equipment configurations to which that approval applies, and [§1926.1434(a)(2)(i)[A]]

[B] Modifies load charts, procedures, instruction manuals and instruction plates/tags/decals as necessary to accord with the modification/addition. [§1926.1434(a)(2)(i)[B]]

(ii) *The original safety factor* of the equipment is not reduced. [§1926.1434(a)(2)(ii)]

(3) *Unavailable manufacturer.* The manufacturer is unavailable and the requirements of paragraphs (a)(2)(i) and (ii) of this section are met. [§1926.1434(a)(3)]

(4) *Manufacturer does not complete the review within 120 days of the request.* The manufacturer is provided a detailed description of the proposed modification/addition, is asked to approve the modification/addition, agrees to review the technical merits of the proposal, but fails to complete the review of the proposal within 120 days of the date it was provided the detailed description of the proposed modification/addition, and the requirements of paragraphs (a)(2)(i) and (ii) of this section are met. [§1926.1434(a)(4)]

(5) *Multiple manufacturers of equipment designed for use on marine work sites.* The equipment is designed for marine work sites, contains major structural components from more than one manufacturer, and the requirements of paragraphs (a)(2)(i) and (ii) of this section are met. [§1926.1434(a)(5)]

(b) **Modifications or additions which** affect the capacity or safe operation of the equipment are prohibited where the manufacturer, after a review of the technical safety merits of the proposed modification/ addition, rejects the proposal and explains the reasons for the rejection in a written response. If the manufacturer rejects the proposal but does not explain the reasons for the rejection in writing, the employer may treat this as a manufacturer refusal to review the request under paragraph (a)(2) of this section. [§1926.1434(b)]

(c) **The provisions in paragraphs (a)** and (b) of this section do not apply to modifications made or approved by the U.S. military. [§1926.1434(c)]

§1926.1435
Tower cranes

(a) **This section contains supplemental** requirements for tower cranes; all sections of this subpart apply to tower cranes unless specified otherwise. [§1926.1435(a)]

(b) Erecting, climbing and dismantling. [§1926.1435(b)]

(1) *Section 1926.1403* (Assembly/Disassembly — selection of manufacturer or employer procedures), Sec. 1926.1404 (Assembly/Disassembly — general requirements (applies to all assembly and disassembly operations)), Sec. 1926.1405 (Disassembly — additional requirements for dismantling of booms and jibs (applies to both the use of manufacturer procedures and employer procedures)), and Sec. 1926.1406 (Assembly/Disassembly — employer procedures — general requirements), apply to tower cranes (except as otherwise specified), except that the term "assembly/disassembly" is replaced by "erecting, climbing and dismantling," and the term "disassembly" is replaced by "dismantling." [§1926.1435(b)(1)]

(2) *Dangerous areas (self-erecting tower cranes).* In addition to the requirements in Sec. 1926.1404(e), for self-erecting tower cranes, the following applies: Employees must not be in or under the tower, jib, or rotating portion of the crane during erecting, climbing and dismantling operations until the crane is secured in a locked position and the competent person in charge indicates it is safe to enter this area, unless the manufacturer's instructions direct otherwise and only the necessary personnel are permitted in this area. [§1926.1435(b)(2)]

(3) *Foundations and structural supports.* Tower crane foundations and structural supports (including both the portions of the structure used for support and the means of attachment) must be designed by the manufacturer or a registered professional engineer. [§1926.1435(b)(3)]

(4) *Addressing specific hazards.* The requirements in Sec. 1926.1404(h)(1) through (9) apply. In addition, the A/D director must address the following: [§1926.1435(b)(4)]

(i) *Foundations and structural supports.* The A/D director must determine that tower crane foundations and structural supports are installed in accordance with their design. [§1926.1435(b)(4)(i)]

(ii) *Loss of backward stability.* Backward stability before swinging self erecting cranes or cranes on traveling or static undercarriages. [§1926.1435(b)(4)(ii)]

(iii) *Wind speed.* Wind must not exceed the speed recommended by the manufacturer or, where manufacturer does not specify this information, the speed determined by a qualified person. [§1926.1435(b)(4)(iii)]

(5) *Plumb tolerance.* Towers must be erected plumb to the manufacturer's tolerance and verified by a qualified person. Where the manufacturer does not specify plumb tolerance, the crane tower must be plumb to a tolerance of at least 1:500 (approximately 1 inch in 40 feet). [§1926.1435(b)(5)]

(6) *Multiple tower crane jobsites.* On jobsites where more than one fixed jib (hammerhead) tower crane is installed, the cranes must be located such that no crane can come in contact with the structure of another crane. Cranes are permitted to pass over one another. [§1926.1435(b)(6)]

(7) *Climbing procedures.* Prior to, and during, all climbing procedures (including inside climbing and top climbing), the employer must: [§1926.1435(b)(7)]

(i) *Comply with all manufacturer prohibitions.* [§1926.1435(b)(7)(i)]

(ii) *Have a registered professional* engineer verify that the host structure is strong enough to sustain the forces imposed through the braces, brace anchorages and supporting floors. [§1926.1435(b)(7)(ii)]

(8) *Counterweight/ballast.* [§1926.1435(b)(8)]

(i) *Equipment must not be erected,* dismantled or operated without the amount and position of counterweight and/or ballast in place as specified by the manufacturer or a registered professional engineer familiar with the equipment. [§1926.1435(b)(8)(i)]

(ii) *The maximum counterweight and/or* ballast specified by the manufacturer or registered professional engineer familiar with the equipment must not be exceeded. [§1926.1435(b)(8)(ii)]

(c) Signs. The size and location of signs installed on tower cranes must be in accordance with manufacturer specifications. Where these are unavailable, a registered professional engineer familiar with the type of equipment involved must approve in writing the size and location of any signs. [§1926.1435(c)]

(d) Safety devices. [§1926.1435(d)]

(1) *Section 1926.1415 does* not apply to tower cranes. [§1926.1435(d)(1)]

(2) *The following safety devices* are required on all tower cranes unless otherwise specified: [§1926.1435(d)(2)]

(i) *Boom stops on luffing* boom type tower cranes. [§1926.1435(d)(2)(i)]

(ii) *Jib stops on luffing* boom type tower cranes if equipped with a jib attachment. [§1926.1435(d)(2)(ii)]

(iii) *Travel rail end* stops at both ends of travel rail. [§1926.1435(d)(2)(iii)]

(iv) *Travel rail clamps* on all travel bogies. [§1926.1435(d)(2)(iv)]

(v) *Integrally mounted check* valves on all load supporting hydraulic cylinders. [§1926.1435(d)(2)(v)]

(vi) *Hydraulic system pressure* limiting device. [§1926.1435(d)(2)(vi)]

(vii) *The following brakes,* which must automatically set in the event of pressure loss or power failure, are required: [§1926.1435(d)(2)(vii)]

[A] A hoist brake on all hoists. [§1926.1435(d)(2)(vii)[A]]

[B] Swing brake. [§1926.1435(d)(2)(vii)[B]]

[C] Trolley brake. [§1926.1435(d)(2)(vii)[C]]

[D] Rail travel brake. [§1926.1435(d)(2)(vii)[D]]

(viii) *Deadman control or forced* neutral return control (hand) levers. [§1926.1435(d)(2)(viii)]

(ix) *Emergency stop switch* at the operator's station. [§1926.1435(d)(2)(ix)]

(x) *Trolley end stops* must be provided at both ends of travel of the trolley. [§1926.1435(d)(2)(x)]

(3) *Proper operation required.* Operations must not begin unless the devices listed in this section are in proper working order. If a device stops working properly during operations, the operator must safely stop operations. The equipment must be taken out of service, and operations must not resume until the device is again working properly. *See* Sec. 1926.1417(f). Alternative measures are not permitted to be used. [§1926.1435(d)(3)]

(e) Operational aids. [§1926.1435(e)]

(1) *Section 1926.1416 does* not apply to tower cranes. [§1926.1435(e)(1)]

(2) *The devices listed in this section* ("operational aids") are required on all tower cranes covered by this subpart, unless otherwise specified. [§1926.1435(e)(2)]

(3) *Operations must not begin unless* the operational aids are in proper working order, except where the employer meets the specified temporary alternative measures. More protective alternative measures specified by the tower crane manufacturer, if any, must be followed. *See* Sec. 1926.1417(j) for additional requirements. [§1926.1435(e)(3)]

(4) *If an operational aid stops* working properly during operations, the operator must safely stop operations until the temporary alternative measures are implemented or the device is again working properly. If a replacement part is no longer available, the use of a substitute device that performs the same type of function is permitted and is not considered a modification under Sec. 1926.1434. [§1926.1435(e)(4)]

(5) *Category I operational aids and alternative measures.* Operational aids listed in this paragraph that are not working properly must be repaired no later than 7 calendar days after the deficiency occurs. *Exception:* If the employer documents that it has ordered the necessary parts within 7 calendar days of the occurrence of the deficiency, the repair must be completed within 7 calendar days of receipt of the parts. [§1926.1435(e)(5)]

(i) *Trolley travel limiting device.* The travel of the trolley must be restricted at both ends of the jib by a trolley travel limiting device to prevent the trolley from running into the trolley end stops. *Temporary alternative measures:* [§1926.1435(e)(5)(i)]

[A] Option A. The trolley rope must be marked (so it can be seen by the operator) at a point that will give the operator sufficient time to stop the trolley prior to the end stops. [§1926.1435(e)(5)(i)[A]]

[B] Option B. A spotter who is in direct communication with the operator must be used when operations are conducted within 10 feet of the outer or inner trolley end stops. [§1926.1435(e)(5)(i)[B]]

(ii) *Boom hoist limiting device.* The range of the boom must be limited at the minimum and maximum radius. *Temporary alternative measures:* Clearly mark the cable (so it can be seen by the operator) at a point that will give the operator sufficient time to stop the boom hoist within the minimum and maximum boom radius, or use a spotter who is in direct communication with the operator to inform the operator when this point is reached. [§1926.1435(e)(5)(ii)]

(iii) *Anti two-blocking device.* The tower crane must be equipped with a device which automatically prevents damage from contact between the load block, overhaul ball, or similar component, and the boom tip (or fixed upper block or similar component). The device(s) must prevent such damage at all points where two-blocking could occur. *Temporary alternative measures:* Clearly mark the cable (so it can be seen by the operator) at a point that will give the operator sufficient time to stop the hoist to prevent two-blocking, or use a spotter who is in direct communication with the operator to inform the operator when this point is reached. [§1926.1435(e)(5)(iii)]

(iv) *Hoist drum lower limiting device.* Tower cranes manufactured after November 8, 2011 must be equipped with a device that prevents the last 2 wraps of hoist cable from being spooled off the drum. *Temporary alternative measures:* Mark the cable (so it can be seen by the operator) at a point that will give the operator sufficient time to stop the hoist prior to last 2 wraps of hoist cable being spooled off the drum, or use a spotter who is in direct communication with the operator to inform the operator when this point is reached [§1926.1435(e)(5)(iv)]

(v) *Load moment limiting device.* The tower crane must have a device that prevents moment overloading. *Temporary alternative measures:* A radius indicating device must be used (if the tower crane is not equipped with a radius indicating device, the radius must be measured to ensure the load is within the rated capacity of the crane). In addition, the weight of the load must be determined from a source recognized by the industry (such as the load's manufacturer), or by a calculation method recognized by the industry (such as calculating a steel beam from measured dimensions and a known per foot weight), or by other equally reliable means. This information must be provided to the operator prior to the lift. [§1926.1435(e)(5)(v)]

(vi) *Hoist line pull limiting device.* The capacity of the hoist must be limited to prevent overloading, including each individual gear ratio if equipped with a multiple speed hoist transmission. *Temporary alternative measures:* The operator must ensure that the weight of the load does not exceed the capacity of the hoist (including for each individual gear ratio if equipped with a multiple speed hoist transmission). [§1926.1435(e)(5)(vi)]

(vii) *Rail travel limiting device.* The travel distance in each direction must be limited to prevent the travel bogies from running into the end stops or buffers. *Temporary alternative measures:* A spotter who is in direct communication with the operator must be used when operations are conducted within 10 feet of either end of the travel rail end stops; the spotter must inform the operator of the distance of the travel bogies from the end stops or buffers. [§1926.1435(e)(5)(vii)]

(viii) *Boom hoist drum positive locking device and control.* The boom hoist drum must be equipped with a control that will enable the operator to positively lock the boom hoist drum from the cab. *Temporary alternative measures:* The device must be manually set when required if an electric, hydraulic or automatic control is not functioning. [§1926.1435(e)(5)(viii)]

(6) *Category II operational aids and alternative measures.* Operational aids listed in this paragraph that are not working properly must be repaired no later than 30 calendar days after the deficiency occurs. *Exception:* If the employer documents that it has ordered the necessary parts within 7 calendar days of the occurrence of the deficiency, and the part is not received in time to complete the repair in 30 calendar days, the repair must be completed within 7 calendar days of receipt of the parts. [§1926.1435(e)(6)]

(i) *Boom angle or hook radius indicator.* [§1926.1435(e)(6)(i)]

[A] *Luffing boom tower* cranes must have a boom angle indicator readable from the operator's station. [§1926.1435(e)(6)(i)[A]]

[B] *Hammerhead tower cranes* manufactured after November 8, 2011 must have a hook radius indicator readable from the operator's station. [§1926.1435(e)(6)(i)[B]]

[C] *Temporary alternative measures:* Hook radii or boom angle must be determined by measuring the hook radii or boom angle with a measuring device. [§1926.1435(e)(6)(i)[C]]

(ii) *Trolley travel deceleration device.* The trolley speed must be automatically reduced prior to the trolley reaching the end limit in both directions. *Temporary alternative measure:* The employer must post a notice in the cab of the crane notifying the operator that the trolley travel deceleration device is malfunctioning and instructing the operator to take special care to reduce the trolley speed when approaching the trolley end limits. [§1926.1435(e)(6)(ii)]

(iii) *Boom hoist deceleration device.* The boom speed must be automatically reduced prior to the boom reaching the minimum or maximum radius limit. *Temporary alternative measure:* The employer must post a notice in the cab of the crane notifying the operator that the boom hoist deceleration device is malfunctioning and instructing the operator to take special care to reduce the boom speed when approaching the minimum or maximum radius limits. [§1926.1435(e)(6)(iii)]

(iv) *Load hoist deceleration device.* The load speed must be automatically reduced prior to the hoist reaching the upper limit. *Temporary alternative measure:* The employer must post a notice in the cab of the crane notifying the operator that the load hoist deceleration device is malfunctioning and instructing the operator to take special care to reduce the load speed when approaching the upper limits. [§1926.1435(e)(6)(iv)]

(v) *Wind speed indicator.* A device must be provided to display the wind speed and must be mounted above the upper rotating structure on tower cranes. On self erecting cranes, it must be mounted at or above the jib level. *Temporary alternative measures:* Use of wind speed information from a properly functioning indicating device on another tower crane on the same site, or a qualified person estimates the wind speed. [§1926.1435(e)(6)(v)]

(vi) *Load indicating device.* Cranes manufactured after November 8, 2011 must have a device that displays the magnitude of the load on the hook. Displays that are part of load moment limiting devices that display the load on the hook meet this requirement. *Temporary alternative measures:* The weight of the load must be determined from a source recognized by the industry (such as the load's manufacturer), or by a calculation method recognized by the industry (such as calculating a steel beam from measured dimensions and a known per foot weight), or by other equally reliable means. This information must be provided to the operator prior to the lift. [§1926.1435(e)(6)(vi)]

(f) Inspections. [§1926.1435(f)]

(1) *Section 1926.1412 (Inspections)* applies to tower cranes, except that the term "assembly" is replaced by "erection." Section 1926.1413 (Wire rope — inspection) applies to tower cranes. [§1926.1435(f)(1)]

(2) *Pre-erection inspection.* Before each crane component is erected, it must be inspected by a qualified person for damage or excessive wear. [§1926.1435(f)(2)]

(i) *The qualified person must pay* particular attention to components that will be difficult to inspect thoroughly during shift inspections. [§1926.1435(f)(2)(i)]

(ii) *If the qualified person determines* that a component is damaged or worn to the extent that it would create a safety hazard if used on the crane, that component must not be erected on the crane unless it is repaired and, upon reinspection by the qualified person, found to no longer create a safety hazard. [§1926.1435(f)(2)(ii)]

(iii) *If the qualified person determines* that, though not presently a safety hazard, the component needs to be monitored, the employer must ensure that the component is checked in the monthly inspections. Any such determination must be documented, and the documentation must be available to any individual who conducts a monthly inspection. [§1926.1435(f)(2)(iii)]

(3) *Post-erection inspection.* In addition to the requirements in Sec. 1926.1412(c), the following requirements must be met: [§1926.1435(f)(3)]

(i) *A load test using certified* weights, or scaled weights using a certified scale with a current certificate of calibration, must be conducted after each erection. [§1926.1435(f)(3)(i)]

(ii) *The load test must be conducted* in accordance with the manufacturer's instructions when available. Where these instructions are unavailable, the test must be conducted in accordance with written load test procedures developed by a registered professional engineer familiar with the type of equipment involved. [§1926.1435(f)(3)(ii)]

(4) *Monthly.* The following additional items must be included: [§1926.1435(f)(4)]

(i) *Tower (mast) bolts* and other structural bolts (for loose or dislodged condition) from the base of the tower crane up or, if the crane is tied to or braced by the structure, those above the upper-most brace support. [§1926.1435(f)(4)(i)]

(ii) *The upper-most tie-in,* braces, floor supports and floor wedges where the tower crane is supported by the structure, for loose or dislodged components. [§1926.1435(f)(4)(ii)]

(5) *Annual.* In addition to the items that must be inspected under Sec. 1926.1412(f), all turntable and tower bolts must be inspected for proper condition and torque. [§1926.1435(f)(5)]

§1926.1436
Derricks

(a) **This section contains supplemental** requirements for derricks, whether temporarily or permanently mounted; all sections of this subpart apply to derricks unless specified otherwise. A derrick is powered equipment consisting of a mast or equivalent member that is held at or near the end by guys or braces, with or without a boom, and its hoisting mechanism. The mast/equivalent member and/or the load is moved by the hoisting mechanism (typically base-mounted) and operating ropes. Derricks include: A-frame, basket, breast, Chicago boom, gin pole (except gin poles used for erection of communication towers), guy, shearleg, stiffleg, and variations of such equipment. [§1926.1436(a)]

(b) **Operation — procedures.** [§1926.1436(b)]

(1) *Section 1926.1417 (Operation)* applies except for Sec. 1926.1417(c) (Accessibility of procedures). [§1926.1436(b)(1)]

(2) *Load chart contents.* Load charts must contain at least the following information: [§1926.1436(b)(2)]

(i) *Rated capacity at corresponding* ranges of boom angle or operating radii. [§1926.1436(b)(2)(i)]

(ii) *Specific lengths of components* to which the rated capacities apply. [§1926.1436(b)(2)(ii)]

(iii) *Required parts for hoist reeving.* [§1926.1436(b)(2)(iii)]

(iv) *Size and construction of rope* must be included on the load chart or in the operating manual. [§1926.1436(b)(2)(iv)]

(3) *Load chart location.* [§1926.1436(b)(3)]

(i) *Permanent installations.* For permanently installed derricks with fixed lengths of boom, guy, and mast, a load chart must be posted where it is visible to personnel responsible for the operation of the equipment. [§1926.1436(b)(3)(i)]

(ii) *Non-permanent installations.* For derricks that are not permanently installed, the load chart must be readily available at the job site to personnel responsible for the operation of the equipment. [§1926.1436(b)(3)(ii)]

(c) **Construction.** [§1926.1436(c)]

(1) *General requirements.* [§1926.1436(c)(1)]

(i) *Derricks must be constructed to meet* all stresses imposed on members and components when installed and operated in accordance with the manufacturer's/builder's procedures and within its rated capacity. [§1926.1436(c)(1)(i)]

(ii) *Welding of load sustaining* members must conform to recommended practices in ANSI/AWS D14.3-94 (incorporated by reference, *see* Sec. 1926.6) or AWS D1.1/D1.1M:2002 (incorporated by reference, *see* Sec. 1926.6). [§1926.1436(c)(1)(ii)]

(2) *Guy derricks.* [§1926.1436(c)(2)]

(i) *The minimum number of guys* must be 6, with equal spacing, except where a qualified person or derrick manufacturer approves variations from these requirements and revises the rated capacity to compensate for such variations. [§1926.1436(c)(2)(i)]

(ii) *Guy derricks must not be used* unless the employer has the following guy information from the manufacturer or a qualified person, when not available from the manufacturer: [§1926.1436(c)(2)(ii)]

[A] *The number of guys.* [§1926.1436(c)(2)(ii)[A]]

[B] *The spacing around the mast.* [§1926.1436(c)(2)(ii)[B]]

[C] *The size,* grade, and construction of rope to be used for each guy. [§1926.1436(c)(2)(ii)[C]]

(iii) *For guy derricks manufactured* after December 18, 1970, in addition to the information required in paragraph (c)(2)(ii) of this section, the employer must have the following guy information from the manufacturer or a qualified person, when not available from the manufacturer: [§1926.1436(c)(2)(iii)]

[A] *The amount of initial sag or tension.* [§1926.1436(c)(2)(iii)[A]]

[B] *The amount of tension in guy* line rope at anchor. [§1926.1436(c)(2)(iii)[B]]

(iv) *The mast base must permit* the mast to rotate freely with allowance for slight tilting of the mast caused by guy slack. [§1926.1436(c)(2)(iv)]

(v) *The mast cap must:* [§1926.1436(c)(2)(v)]

[A] *Permit the mast to rotate freely.* [§1926.1436(c)(2)(v)[A]]

[B] *Withstand tilting and cramping* caused by the guy loads. [§1926.1436(c)(2)(v)[B]]

[C] *Be secured to the mast to prevent* disengagement during erection. [§1926.1436(c)(2)(v)[C]]

[D] *Be provided with means for attaching guy ropes.* [§1926.1436(c)(2)(v)[D]]

(3) *Stiffleg derricks.* [§1926.1436(c)(3)]

(i) *The mast must be supported in the vertical* position by at least two stifflegs; one end of each must be connected to the top of the mast and the other end securely anchored. [§1926.1436(c)(3)(i)]

(ii) *The stifflegs must be capable of withstanding* the loads imposed at any point of operation within the load chart range. [§1926.1436(c)(3)(ii)]

(iii) *The mast base must:* [§1926.1436(c)(3)(iii)]

[A] *Permit the mast to rotate* freely (when necessary). [§1926.1436(c)(3)(iii)[A]]

[B] *Permit deflection of the mast* without binding. [§1926.1436(c)(3)(iii)[B]]

(iv) *The mast must be prevented from lifting* out of its socket when the mast is in tension. [§1926.1436(c)(3)(iv)]

(v) *The stiffleg connecting member* at the top of the mast must: [§1926.1436(c)(3)(v)]

[A] *Permit the mast to rotate* freely (when necessary). [§1926.1436(c)(3)(v)[A]]

[B] *Withstand the loads imposed* by the action of the stifflegs. [§1926.1436(c)(3)(v)[B]]

[C] *Be secured so as to oppose separating forces.* [§1926.1436(c)(3)(v)[C]]

(4) *Gin pole derricks.* [§1926.1436(c)(4)]

(i) *Guy lines must be sized* and spaced so as to make the gin pole stable in both boomed and vertical positions. *Exception:* Where the size and/or spacing of guy lines do not result in the gin pole being stable in both boomed and vertical positions, the employer must ensure that the derrick is not used in an unstable position. [§1926.1436(c)(4)(i)]

(ii) *The base of the gin pole* must permit movement of the pole (when necessary). [§1926.1436(c)(4)(ii)]

(iii) *The gin pole must be anchored* at the base against horizontal forces (when such forces are present). [§1926.1436(c)(4)(iii)]

(5) *Chicago boom derricks.* The fittings for stepping the boom and for attaching the topping lift must be arranged to: [§1926.1436(c)(5)]

(i) *Permit the derrick to swing* at all permitted operating radii and mounting heights between fittings. [§1926.1436(c)(5)(i)]

(ii) *Accommodate attachment to the upright* member of the host structure. [§1926.1436(c)(5)(ii)]

(iii) *Withstand the forces applied* when configured and operated in accordance with the manufacturer's/builder's procedures and within its rated capacity. [§1926.1436(c)(5)(iii)]

(iv) *Prevent the boom or topping* lift from lifting out under tensile forces. [§1926.1436(c)(5)(iv)]

(d) **Anchoring and guying.** [§1926.1436(d)]

(1) *Load anchoring data* developed by the manufacturer or a qualified person must be used. [§1926.1436(d)(1)]

(2) *Guy derricks.* [§1926.1436(d)(2)]

(i) *The mast base must be anchored.* [§1926.1436(d)(2)(i)]

(ii) *The guys must be secured to the ground* or other firm anchorage. [§1926.1436(d)(2)(ii)]

(iii) *The anchorage and guying must be designed* to withstand maximum horizontal and vertical forces encountered when operating within rated capacity with the particular guy slope and spacing specified for the application. [§1926.1436(d)(2)(iii)]

(3) *Stiffleg derricks.* [§1926.1436(d)(3)]

(i) *The mast base and stifflegs* must be anchored. [§1926.1436(d)(3)(i)]

(ii) *The mast base and stifflegs* must be designed to withstand maximum horizontal and vertical forces encountered when operating within rated capacity with the particular stiffleg spacing and slope specified for the application. [§1926.1436(d)(3)(ii)]

(e) Swingers and hoists. [§1926.1436(e)]

(1) *The boom,* swinger mechanisms and hoists must be suitable for the derrick work intended and must be anchored to prevent displacement from the imposed loads. [§1926.1436(e)(1)]

(2) *Hoists.* [§1926.1436(e)(2)]

(i) *Base mounted drum* hoists must meet the requirements in the following sections of ASME B30.7-2001 (incorporated by reference, *see* Sec. 1926.6): [§1926.1436(e)(2)(i)]

[A] Sections 7-1.1 ("Load ratings and markings"). [§1926.1436(e)(2)(i)[A]]

[B] Section 7-1.2 ("Construction"), except: 7-1.2.13 ("Operator's cab"); 7-1.2.15 ("Fire extinguishers"). [§1926.1436(e)(2)(i)[B]]

[C] Section 7-1.3 ("Installation"). [§1926.1436(e)(2)(i)[C]]

[D] Applicable terms in section 7-0.2 ("Definitions"). [§1926.1436(e)(2)(i)[D]]

(ii) *Load tests for new hoists.* The employer must ensure that new hoists are load tested to a minimum of 110% of rated capacity, but not more than 125% of rated capacity, unless otherwise recommended by the manufacturer. This requirement is met where the manufacturer has conducted this testing. [§1926.1436(e)(2)(ii)]

(iii) *Repaired or modified hoists.* Hoists that have had repairs, modifications or additions affecting their capacity or safe operation must be evaluated by a qualified person to determine if a load test is necessary. If it is, load testing must be conducted in accordance with paragraphs (e)(2)(ii) and (iv) of this section. [§1926.1436(e)(2)(iii)]

(iv) *Load test procedure.* Load tests required by paragraphs (e)(2)(ii) or (e)(2)(iii) of this section must be conducted as follows: [§1926.1436(e)(2)(iv)]

[A] The test load must be hoisted a vertical distance to assure that the load is supported by the hoist and held by the hoist brake(s). [§1926.1436(e)(2)(iv)[A]]

[B] The test load must be lowered, stopped and held with the brake(s). [§1926.1436(e)(2)(iv)[B]]

[C] The hoist must not be used unless a competent person determines that the test has been passed. [§1926.1436(e)(2)(iv)[C]]

(f) Operational aids. [§1926.1436(f)]

(1) *Section 1926.1416* (Operational aids) applies, except for Sec. 1926.1416(d)(1) (Boom hoist limiting device), Sec. 1926.1416(e)(1) (Boom angle or radius indicator), and Sec. 1926.1416(e)(4) (Load weighing and similar devices). [§1926.1436(f)(1)]

(2) *Boom angle aid.* A boom angle indicator is not required but if the derrick is not equipped with a functioning one, the employer must ensure that either: [§1926.1436(f)(2)]

(i) *The boom hoist cable* must be marked with caution and stop marks. The stop marks must correspond to maximum and minimum allowable boom angles. The caution and stop marks must be in view of the operator, or a spotter who is in direct communication with the operator; or [§1926.1436(f)(2)(i)]

(ii) *An electronic or other device that signals* the operator in time to prevent the boom from moving past its maximum and minimum angles, or automatically prevents such movement, is used. [§1926.1436(f)(2)(ii)]

(3) *Load weight/capacity devices.* [§1926.1436(f)(3)]

(i) *Derricks manufactured more* than one year after November 8, 2010 with a maximum rated capacity over 6,000 pounds must have at least one of the following: load weighing device, load moment indicator, rated capacity indicator, or rated capacity limiter. *Temporary alternative measures:* The weight of the load must be determined from a source recognized by the industry (such as the load's manufacturer), or by a calculation method recognized by the industry (such as calculating a steel beam from measured dimensions and a known per foot weight), or by other equally reliable means. This information must be provided to the operator prior to the lift. *See* Sec. 1926.1417(j) for additional requirements. [§1926.1436(f)(3)(i)]

(ii) *A load weight/capacity device* that is not working properly must be repaired no later than 30 days after the deficiency occurs. *Exception:* If the employer documents that it has ordered the necessary parts within 7 days of the occurrence of the deficiency, and the part is not received in time to complete the repair in 30 days, the repair must be completed within 7 days of receipt of the parts. [§1926.1436(f)(3)(ii)]

(g) Post-assembly approval and testing — new or reinstalled derricks. [§1926.1436(g)]

(1) *Anchorages.* [§1926.1436(g)(1)]

(i) *Anchorages, including the structure* to which the derrick is attached (if applicable), must be approved by a qualified person. [§1926.1436(g)(1)(i)]

(ii) *If using a rock or hairpin anchorage,* the qualified person must determine if any special testing of the anchorage is needed. If so, it must be tested accordingly. [§1926.1436(g)(1)(ii)]

(2) *Functional test.* Prior to initial use, new or reinstalled derricks must be tested by a competent person with no hook load to verify proper operation. This test must include: [§1926.1436(g)(2)]

(i) *Lifting and lowering the hook(s)* through the full range of hook travel. [§1926.1436(g)(2)(i)]

(ii) *Raising and lowering the boom* through the full range of boom travel. [§1926.1436(g)(2)(ii)]

(iii) *Swinging in each direction through* the full range of swing. [§1926.1436(g)(2)(iii)]

(iv) *Actuating the anti two-block* and boom hoist limit devices (if provided). [§1926.1436(g)(2)(iv)]

(v) *Actuating locking,* limiting and indicating devices (if provided). [§1926.1436(g)(2)(v)]

(3) *Load test.* Prior to initial use, new or reinstalled derricks must be load tested by a competent person. The test load must meet the following requirements: [§1926.1436(g)(3)]

(i) *Test loads must be at least* 100% and no more than 110% of the rated capacity, unless otherwise recommended by the manufacturer or qualified person, but in no event must the test load be less than the maximum anticipated load. [§1926.1436(g)(3)(i)]

(ii) *The test must consist of:* [§1926.1436(g)(3)(ii)]

[A] Hoisting the test load a few inches and holding to verify that the load is supported by the derrick and held by the hoist brake(s). [§1926.1436(g)(3)(ii)[A]]

[B] Swinging the derrick, if applicable, the full range of its swing, at the maximum allowable working radius for the test load. [§1926.1436(g)(3)(ii)[B]]

[C] Booming the derrick up and down within the allowable working radius for the test load. [§1926.1436(g)(3)(ii)[C]]

[D] Lowering, stopping and holding the load with the brake(s). [§1926.1436(g)(3)(ii)[D]]

(iii) *The derrick must not be used unless* the competent person determines that the test has been passed. [§1926.1436(g)(3)(iii)]

(4) *Documentation.* Tests conducted under this paragraph must be documented. The document must contain the date, test results and the name of the tester. The document must be retained until the derrick is re-tested or dismantled, whichever occurs first. All such documents must be available, during the applicable document retention period, to all persons who conduct inspections in accordance with Sec. 1926.1412. [§1926.1436(g)(4)]

(h) Load testing repaired or modified derricks. Derricks that have had repairs, modifications or additions affecting the derrick's capacity or safe operation must be evaluated by a qualified person to determine if a load test is necessary. If it is, load testing must be conducted and documented in accordance with paragraph (g) of this section. [§1926.1436(h)]

(i) ***[Reserved.]*** [§1926.1436(i)]

(j) Power failure procedures. If power fails during operations, the derrick operator must safely stop operations. This must include: [§1926.1436(j)]

(1) *Setting all brakes or locking devices.* [§1926.1436(j)(1)]

(2) *Moving all clutch and other power* controls to the off position. [§1926.1436(j)(2)]

(k) Use of winch heads. [§1926.1436(k)]

(1) *Ropes must not be handled on a winch* head without the knowledge of the operator. [§1926.1436(k)(1)]

(2) *While a winch head* is being used, the operator must be within reach of the power unit control lever. [§1926.1436(k)(2)]

(l) ***[Reserved.]*** [§1926.1436(l)]

(m) Securing the boom. [§1926.1436(m)]

(1) *When the boom is being held in a fixed* position, dogs, pawls, or other positive holding mechanisms on the boom hoist must be engaged. [§1926.1436(m)(1)]

(2) *When taken out of service for 30* days or more, the boom must be secured by one of the following methods: [§1926.1436(m)(2)]

(i) *Laid down.* [§1926.1436(m)(2)(i)]

(ii) *Secured to a stationary member,* as nearly under the head as possible, by attachment of a sling to the load block. [§1926.1436(m)(2)(ii)]

(iii) *For guy derricks,* lifted to a vertical position and secured to the mast. [§1926.1436(m)(2)(iii)]

(iv) *For stiffleg derricks,* secured against the stiffleg. [§1926.1436(m)(2)(iv)]

(n) **The process of jumping the derrick** must be supervised by the A/D director. [§1926.1436(n)]

(o) **Derrick operations must be supervised** by a competent person. [§1926.1436(o)]

(p) **Inspections.** In addition to the requirements in Sec. 1926.1412, the following additional items must be included in the inspections: [§1926.1436(p)]

(1) *Daily:* Guys for proper tension. [§1926.1436(p)(1)]

(2) *Annual.* [§1926.1436(p)(2)]

(i) *Gudgeon pin for cracks,* wear, and distortion. [§1926.1436(p)(2)(i)]

(ii) *Foundation supports for continued* ability to sustain the imposed loads. [§1926.1436(p)(2)(ii)]

(q) **Qualification and Training.** The employer must train each operator of a derrick on the safe operation of equipment the individual will operate. Section 1926.1427 of this subpart (Operator qualification and certification) does not apply. [§1926.1436(q)]

§1926.1437

Floating cranes/derricks and land cranes/derricks on barges

(a) **This section contains supplemental requirements** for floating cranes/derricks and land cranes/derricks on barges, pontoons, vessels or other means of flotation (i.e., vessel/flotation device). The sections of this subpart apply to floating cranes/derricks and land cranes/derricks on barges, pontoons, vessels or other means of flotation, unless specified otherwise. The requirements of this section do not apply when using jacked barges when the jacks are deployed to the river, lake, or sea bed and the barge is fully supported by the jacks. [§1926.1437(a)]

(b) **General requirements.** The requirements in paragraphs (c) through (k) of this section apply to both floating cranes/derricks and land cranes/derricks on barges, pontoons, vessels or other means of flotation. [§1926.1437(b)]

(c) **Work area control.** [§1926.1437(c)]

(1) *The requirements of §1926.1424* (Work area control) apply, except for §1926.1424(a)(2)(ii). [§1926.1437(c)(1)]

(2) *The employer must either:* [§1926.1437(c)(2)]

(i) *Erect and maintain control lines,* warning lines, railings or similar barriers to mark the boundaries of the hazard areas; or [§1926.1437(c)(2)(i)]

(ii) *Clearly mark the hazard areas* by a combination of warning signs (such as, "Danger – Swing/Crush Zone") and high visibility markings on the equipment that identify the hazard areas. In addition, the employer must train each employee to understand what these markings signify. [§1926.1437(c)(2)(ii)]

(d) **Keeping clear of the load.** Section 1926.1425 does not apply. [§1926.1437(d)]

(e) **Additional safety devices.** In addition to the safety devices listed in §1926.1415, the following safety devices are required: [§1926.1437(e)]

(1) *Barge, pontoon,* vessel or other means of flotation list and trim device. The safety device must be located in the cab or, when there is no cab, at the operator's station. [§1926.1437(e)(1)]

(2) *Positive equipment house lock.* [§1926.1437(c)(2)]

(3) *Wind speed and direction indicator.* A competent person must determine if wind is a factor that needs to be considered; if wind needs to be considered, a wind speed and direction indicator must be used. [§1926.1437(c)(3)]

(f) **Operational aids.** [§1926.1437(f)]

(1) *An anti two-block device* is required only when hoisting personnel or hoisting over an occupied cofferdam or shaft. [§1926.1437(f)(1)]

(2) *Section 1926.1416(e)(4)* (Load weighing and similar devices) does not apply to dragline, clamshell (grapple), magnet, drop ball, container handling, concrete bucket, and pile driving work performed under this section. [§1926.1437(f)(2)]

(g) **Accessibility of procedures** applicable to equipment operation. If the crane/derrick has a cab, the requirements of §1926.1417(c) apply. If the crane/derrick does not have a cab, the employer must ensure that: [§1926.1437(g)]

(1) *Rated capacities (load charts)* are posted at the operator's station. If the operator's station is moveable (such as with pendant-controlled equipment), the load charts are posted on the equipment. [§1926.1437(g)(1)]

(2) *Procedures applicable to the operation* of the equipment (other than load charts), recommended operating speeds, special hazard warnings, instructions and operators manual, must be readily available on board the vessel/flotation device. [§1926.1437(g)(2)]

(h) **Inspections.** In addition to meeting the requirements of §1926.1412 for inspecting the crane/derrick, the employer must inspect the barge, pontoons, vessel or other means of flotation used to support a floating crane/derrick or land crane/derrick, and ensure that: [§1926.1437(h)]

(1) *Shift.* For each shift inspection, the means used to secure/attach the equipment to the vessel/flotation device is in proper condition, including wear, corrosion, loose or missing fasteners, defective welds, and (when applicable) insufficient tension. [§1926.1437(h)(1)]

(2) *Monthly.* For each monthly inspection: [§1926.1437(h)(2)]

(i) *The means used to secure/attach* the equipment to the vessel/flotation device is in proper condition, including inspection for wear, corrosion, and, when applicable, insufficient tension. [§1926.1437(h)(2)(i)]

(ii) *The vessel/flotation device* is not taking on water. [§1926.1437(h)(2)(ii)]

(iii) *The deckload* is properly secured. [§1926.1437(h)(2)(iii)]

(iv) *The vessel/flotation device* is watertight based on the condition of the chain lockers, storage, fuel compartments, and hatches. [§1926.1437(h)(2)(iv)]

(v) *The firefighting and lifesaving equipment* is in place and functional. [§1926.1437(h)(2)(iv)]

(3) *The shift and monthly inspections* are conducted by a competent person, and: [§1926.1437(h)(3)]

(i) *If any deficiency is identified,* an immediate determination is made by a qualified person whether the deficiency constitutes a hazard. [§1926.1437(h)(3)(i)]

(ii) *If the deficiency is determined* to constitute a hazard, the vessel/flotation device is removed from service until the deficiency has been corrected. [§1926.1437(h)(3)(ii)]

(4) *Annual:* external vessel/flotation device inspection. For each annual inspection: [§1926.1437(h)(4)]

(i) *The external portion of the barge,* pontoons, vessel or other means of flotation used is inspected annually by a qualified person who has expertise with respect to vessels/flotation devices and that the inspection includes the following items: [§1926.1437(h)(4)(i)]

[A] *The items identified* in paragraphs (h)(1) (Shift) and (h)(2) (Monthly) of this section. [§1926.1437(h)(4)(i)[A]]

[B] *Cleats, bitts, chocks,* fenders, capstans, ladders, and stanchions, for significant corrosion, wear, deterioration, or deformation that could impair the function of these items. [§1926.1437(h)(4)(i)[B]]

[C] *External evidence of leaks* and structural damage; evidence of leaks and damage below the waterline may be determined through internal inspection of the vessel/flotation device. [§1926.1437(h)(4)(i)[C]]

[D] *Four-corner draft readings.* [§1926.1437(h)(4)(i)[D]]

[E] *Firefighting equipment* for serviceability. [§1926.1437(h)(4)(i)[E]]

(ii) *Rescue skiffs,* lifelines, work vests, life preservers and ring buoys are inspected for proper condition. [§1926.1437(h)(4)(ii)]

(iii) *If any deficiency is identified,* an immediate determination is made by the qualified person whether the deficiency constitutes a hazard or, though not yet a hazard, needs to be monitored in the monthly inspections. [§1926.1437(h)(4)(iii)]

[A] *If the qualified person determines* that the deficiency constitutes a hazard, the vessel/flotation device is removed from service until it has been corrected. See requirements in §1926.1417(f). [§1926.1437(h)(4)(iii)[A]]

[B] *If the qualified person determines that,* though not presently a hazard, the deficiency needs to be monitored, the deficiency is checked in the monthly inspections. [§1926.1437(h)(4)(iii)[B]]

(5) *Four-year:* internal vessel/flotation device inspection. For each four-year inspection: [§1926.1437(h)(5)]

(i) *A marine engineer,* marine architect, licensed surveyor, or other qualified person who has expertise with respect to vessels/flotation devices surveys the internal portion of the barge, pontoons, vessel, or other means of flotation. [§1926.1437(h)(5)(i)]

(ii) *If the surveyor identifies a deficiency,* an immediate determination is made by the surveyor as to whether the deficiency constitutes a hazard or, though not yet a hazard, needs to be monitored in the monthly or annual inspections, as appropriate. [§1926.1437(h)(5)(ii)]

[A] *If the surveyor determines* that the deficiency constitutes a hazard, the vessel/flotation device is removed from service until it has been corrected. [§1926.1437(h)(5)(ii)[A]]

[B] *If the surveyor determines that,* though not presently a hazard, the deficiency needs to be monitored, the deficiency is checked in the monthly or annual inspections, as appropriate. [§1926.1437(h)(5)(ii)[B]]

(6) *Documentation.* The monthly and annual inspections required in paragraphs (h)(2) and (h)(4) of this section are documented in accordance with §1926.1412 (e)(3) and 1926.1412(f)(7), respectively, and that the four-year inspection required in paragraph (h)(5) of this section is documented in accordance with §1926.1412(f)(7), except that the documentation for that inspection must be retained for a minimum of 4 years. All such documents must be made available, during the applicable document retention period, to all persons who conduct inspections in accordance with §1926.1412. [§1926.1437(h)(6)]

(i) **[Reserved.]**

(j) **Working with a diver.** The employer must meet the following additional requirements when working with a diver in the water: [§1926.1437(j)]

(1) *If a crane/derrick is used to get a diver* into and out of the water, it must not be used for any other purpose until the diver is back on board. When used for more than one diver, it must not be used for any other purpose until all divers are back on board. [§1926.1437(j)(1)]

(2) *The operator must remain at the controls* of the crane/derrick at all times. [§1926.1437(j)(2)]

(3) *In addition to the requirements* in §1926.1419 through 1926.1422 (Signals), either: [§1926.1437(j)(3)]

(i) *A clear line of sight* must be maintained between the operator and tender; or [§1926.1437(j)(3)(i)]

(ii) *The signals between the operator* and tender must be transmitted electronically. [§1926.1437(j)(3)(ii)]

(4) *The means used to secure* the crane/derrick to the vessel/flotation device (see paragraph (n)(5) of this section) must not allow any amount of shifting in any direction. [§1926.1437(j)(4)]

(k) **Manufacturer's specifications and limitations.** [§1926.1437(k)]

(1) *The employer must ensure that the barge,* pontoons, vessel, or other means of flotation must be capable of withstanding imposed environmental, operational and in-transit loads when used in accordance with the manufacturer's specifications and limitations. [§1926.1437(k)(1)]

(2) *The employer must ensure* that the manufacturer's specifications and limitations with respect to environmental, operational, and in-transit loads for a barge, pontoon, vessel, or other means of flotation are not exceeded or violated. [§1926.1437(k)(2)]

(3) *When the manufacturer's specifications* and limitations are unavailable, the employer must ensure that the specifications and limitations established by a qualified person with respect to environmental, operational and in-transit loads for the barge, pontoons, vessel, or other means of flotation are not exceeded or violated. [§1926.1437(k)(3)]

(l) **[Reserved.]**

(m) **Floating cranes/derricks.** For equipment designed by the manufacturer (or employer) for marine use by permanent attachment to barges, pontoons, vessels or other means of flotation: [§1926.1437(m)]

(1) *Load charts.* [§1926.1437(m)(1)]

(i) *The employer must not exceed* the manufacturer load charts applicable to operations on water. When using these charts, the employer must comply with all parameters and limitations (such as dynamic and environmental parameters) applicable to the use of the charts. [§1926.1437(m)(1)(i)]

(ii) *The employer must ensure* that load charts take into consideration a minimum wind speed of 40 miles per hour. [§1926.1437(m)(1)(ii)]

(2) *The employer must ensure* that the requirements for maximum allowable list and maximum allowable trim as specified in Table M1 of this section are met. [§1926.1437(m)(2)]

Table M1 - Equipment designed for marine use by permanent attachment (other than derricks):

Rated Capacity	Maximum Allowable List	Maximum Allowable Trim
25 tons or less	5 degrees	5 degrees
25 tons or less	5 degrees	5 degrees
Over 25 tons	7 degrees	7 degrees
Derricks designed for marine use by permanent attachment:		
Any rated capacity	10 degrees	10 degrees

(3) *The employer must ensure that the equipment* is stable under the conditions specified in Tables M2 and M3 of this section. (Note: Freeboard is the vertical distance between the water line and the main deck of the vessel.) [§1926.1437(m)(3)]

Table M2

Operated at	Wind speed	Minimum freeboard
Rated capacity	60 mph	2 ft
Rated capacity plus 25%	60 mph	1 ft
High boom, no load	60 mph	2 ft

Table M3- For backward stability of the boom:

Operated at	Wind speed
High boom, no load, full back list (least stable condition)	90 mph

(4) *If the equipment is employer-made,* it must not be used unless the employer has documents demonstrating that the load charts and applicable parameters for use meet the requirements of paragraphs (m)(1) through (3) of this section. Such documents must be signed by a registered professional engineer who is a qualified person with respect to the design of this type of equipment (including the means of flotation). [§1926.1437(m)(4)]

(5) *The employer must ensure that the barge,* pontoons, vessel or other means of flotation used: [§1926.1437(m)(5)]

(i) *Are structurally sufficient to withstand the static* and dynamic loads of the crane/derrick when operating at the crane/derrick's maximum rated capacity with all planned and actual deck loads and ballasted compartments. [§1926.1437(m)(5)(i)]

(ii) *Have a subdivided hull* with one or more longitudinal watertight bulkheads for reducing the free-surface effect. [§1926.1437(m)(5)(ii)]

(iii) *Have access to void compartments* to allow for inspection and pumping. [§1926.1437(m)(5)(iii)]

(n) **Land cranes/derricks.** For land cranes/derricks used on barges, pontoons, vessels or other means of flotation, the employer must ensure that: [§1926.1437(n)]

(1) *The rated capacity of the equipment* (including but not limited to modification of load charts) applicable for use on land is reduced to: [§1926.1437(n)(1)]

(i) *Account for increased loading* from list, trim, wave action, and wind. [§1926.1437(n)(1)(i)]

(ii) *Be applicable to a specified location(s)* on the specific barge, pontoons, vessel or other means of flotation that will be used, under the environmental conditions expected and encountered. [§1926.1437(n)(1)(ii)]

(iii) *The conditions required* in paragraphs (n)(3) and (n)(4) of this section are met. [§1926.1437(n)(1)(iii)]

(2) *The rated capacity modification required* in paragraph (n)(1) of this section is performed by the equipment manufacturer, or a qualified person who has expertise with respect to both land crane/derrick capacity and the stability of vessels/flotation devices. [§1926.1437(n)(2)]

(3) *For list and trim.* [§1926.1437(n)(3)]

(i) *The maximum allowable list* and the maximum allowable trim for the barge, pontoon, vessel or other means of flotation must not exceed the amount necessary to ensure that the conditions in paragraph (n)(4) of this section are met. In addition, the maximum allowable list and the maximum allowable trim does not exceed the least of the following: 5 degrees, the amount specified by the crane/derrick manufacturer, or, when, an amount is not so specified, the amount specified by the qualified person. [§1926.1437(n)(3)(i)]

(ii) *The maximum allowable list* and the maximum allowable trim for the land crane/derrick does not exceed the amount specified by the crane/derrick manufacturer, or, when, an amount is not so specified, the amount specified by the qualified person. [§1926.1437(n)(3)(ii)]

(4) *For the following conditions:* [§1926.1437(n)(4)]

(i) *All deck surfaces of the barge,* pontoons, vessel or other means of flotation used are above water. [§1926.1437(n)(4)(i)]

(ii) *The entire bottom area* of the barge, pontoons, vessel or other means of flotation used is submerged. [§1926.1437(n)(4)(ii)]

(5) *Physical attachment,* corralling, rails system and centerline cable system meet the requirements in Option (1), Option (2), Option (3), or Option (4) of this section, and that whichever option is used also meets the requirements of paragraph (n)(5)(v) of this section. [§1926.1437(n)(5)]

(i) *Option (1) – Physical attachment.* The crane/derrick is physically attached to the barge, pontoons, vessel or other means of flotation. Methods of physical attachment include crossed-cable systems attached to the crane/derrick and vessel/flotation device, bolting or welding the crane/derrick to the vessel/flotation device, strapping the crane/derrick to the vessel/flotation device with chains, or other methods of physical attachment. [§1926.1437(n)(5)(i)]

(ii) *Option (2) – Corralling.* The crane/derrick is prevented from shifting by installing barricade restraints (i.e., a corralling system). Employers must ensure that corralling systems do not allow the equipment to shift by any amount of shifting in any direction. [§1926.1437(n)(5)(ii)]

(iii) *Option (3) – Rails.* The crane/derrick must be prevented from shifting by being mounted on a rail system. Employers must ensure that rail clamps and rail stops are used unless the system is designed to prevent movement during operation by other means. [§1926.1437(n)(5)(iii)]

(iv) *Option (4) – Centerline cable system.* The crane/derrick is prevented from shifting by being mounted to a wire rope system. The employer must ensure that the wire rope system meets the following requirements: [§1926.1437(n)(5)(iv)]

[A] The wire rope and attachments are of sufficient size and strength to support the side load of crane/derrick. [§1926.1437(n)(5)(iv)[A]]

[B] The wire rope is attached physically to the vessel/flotation device. [§1926.1437(n)(5)(iv)[B]]

[C] The wire rope is attached to the crane/derrick by appropriate attachment methods (such as shackles or sheaves) on the undercarriage, and that the method used will allow the crew to secure the crane/derrick from movement during operation and to move the crane/derrick longitudinally along the vessel/flotation device for repositioning. [§1926.1437(n)(5)(iv)[C]]

[D] Means are installed to prevent the crane/derrick from passing the forward or aft end of the wire rope attachments. [§1926.1437(n)(5)(iv)[D]]

[E] The crane/derrick is secured from movement during operation. [§1926.1437(n)(5)(iv)[E]]

(v) *The systems/means used to comply* with Option (1), Option (2), Option (3), or Option (4) of this section are designed by a marine engineer, registered professional engineer familiar with floating crane/derrick design, or qualified person familiar with floating crane/derrick design. [§1926.1437(n)(5)(v)]

(6) *Exception.* For mobile auxiliary cranes used on the deck of a floating crane/derrick, the requirement specified by paragraph (n)(5) of this section to use Option (1), Option (2), Option (3), or Option (4) does not apply when the employer demonstrates implementation of a plan and procedures that meet the following requirements: [§1926.1437(n)(6)]

(i) *A marine engineer or registered professional engineer* familiar with floating crane/derrick design develops and signs a written plan for the use of the mobile auxiliary crane. [§1926.1437(n)(6)(i)]

(ii) *The plan is designed* so that the applicable requirements of this section are met despite the position, travel, operation, and lack of physical attachment (or corralling, use of rails or cable system) of the mobile auxiliary crane. [§1926.1437(n)(6)(ii)]

(iii) *The plan specifies the areas* of the deck where the mobile auxiliary crane is permitted to be positioned, travel, and operate, and the parameters and limitations of such movements and operation. [§1926.1437(n)(6)(iii)]

(iv) *The deck is marked to identify* the permitted areas for positioning, travel, and operation. [§1926.1437(n)(6)(iv)]

(v) *The plan specifies* the dynamic and environmental conditions that must be present for use of the plan. [§1926.1437(n)(6)(v)]

(vi) *If the dynamic and environmental conditions* in paragraph (n)(6)(v) of this section are exceeded, the mobile auxiliary crane is attached physically or corralled in accordance with Option (1), Option (2) or Option (4) of paragraph (n)(5) of this section. [§1926.1437(n)(6)(vi)]

(7) *The barge, pontoons,* vessel or other means of flotation used: [§1926.1437(n)(7)]

(i) *Are structurally sufficient to withstand* the static and dynamic loads of the crane/derrick when operating at the crane/derrick's maximum rated capacity with all anticipated deck loads and ballasted compartments. [§1926.1437(n)(7)(i)]

(ii) *Have a subdivided hull* with one or more longitudinal watertight bulkheads for reducing the free surface effect. [§1926.1437(n)(7)(ii)]

(iii) *Have access to void compartments* to allow for inspection and pumping. [§1926.1437(n)(7)(iii)]

§1926.1438
Overhead & gantry cranes

(a) **Permanently installed overhead and gantry cranes.** The requirements of Sec. 1910.179, except for Sec. 1910.179(b)(1), and not the requirements of this subpart CC, apply to the following equipment when used in construction and permanently installed in a facility: overhead and gantry cranes, including semigantry, cantilever gantry, wall cranes, storage bridge cranes, and others having the same fundamental characteristics. [§1926.1438(a)]

(b) **Overhead and gantry cranes that are not permanently installed in a facility.** [§1926.1438(b)]

(1) *This paragraph applies to the following* equipment when used in construction and not permanently installed in a facility: Overhead and gantry cranes, overhead/bridge cranes, semigantry, cantilever gantry, wall cranes, storage bridge cranes, launching gantry cranes, and similar equipment having the same fundamental characteristics, irrespective of whether it travels on tracks, wheels, or other means. [§1926.1438(b)(1)]

(2) *The following requirements apply* to equipment identified in paragraph (b)(1) of this section: [§1926.1438(b)(2)]

(i) *Sections 1926.1400 through* 1926.1414; §§1926.1417 through 1926.1425; Sec. 1926.1426(d), §§1926.1427 through 1926.1434; Sec. 1926.1437, Sec. 1926.1439, and Sec. 1926.1441. [§1926.1438(b)(2)(i)]

(ii) *The following portions of Sec. 1910.179:* [§1926.1438(b)(2)(ii)]

[A] Paragraphs (b)(5),(6),(7); (e)(1),(3),(5),(6); (f)(1),(4); (g); (h)(1),(3); (k); and (n) of Sec. 1910.179. [§1926.1438(b)(2)(ii)[A]]

[B] The definitions in Sec. 1910.179(a) except for "hoist" and "load." For those words, the definitions in Sec. 1926.1401 apply. [§1926.1438(b)(2)(ii)[B]]

[C] Section 1910.179(b)(2), but only where the equipment identified in paragraph (b)(1) of this section (Sec. 1926.1438) was manufactured before September 19, 2001. [§1926.1438(b)(2)(ii)[C]]

(iii) *For equipment manufactured on or after* September 19, 2001, the following sections of ASME B30.2-2005 (incorporated by reference, *see* Sec. 1926.6) apply: 2-1.3.1; 2-1.3.2; 2-1.4.1; 2-1.6; 2-1.7.2; 2-1.8.2; 2-1.9.1; 2-1.9.2; 2-1.11; 2-1.12.2; 2-1.13.7; 2-1.14.2; 2-1.14.3; 2-1.14.5; 2-1.15.; 2-2.2.2; 2-3.2.1.1. In addition, 2-3.5 applies, except in 2-3.5.1(b), "29 CFR 1910.147" is substituted for "ANSI Z244.1." [§1926.1438(b)(2)(iii)]

§1926.1439
Dedicated pile drivers

(a) **The provisions of subpart CC** apply to dedicated pile drivers, except as specified in this section. [§1926.1439(a)]

(b) **Section 1926.1416(d)(3)** (Anti two-blocking device) does not apply. [§1926.1439(b)]

(c) **Section 1926.1416(e)(4)** (Load weighing and similar devices) applies only to dedicated pile drivers manufactured after November 8, 2011. [§1926.1439(c)]

(d) **In Sec. 1926.1433,** only §§1926.1433(d) and (e) apply to dedicated pile drivers. [§1926.1439(d)]

§1926.1440
Sideboom cranes

(a) **The provisions of this standard apply,** except Sec. 1926.1402 (Ground conditions), Sec. 1926.1415 (Safety devices), Sec. 1926.1416 (Operational aids), and Sec. 1926.1427 (Operator qualification and certification). [§1926.1440(a)]

(b) **Section 1926.1426** (Free fall and controlled load lowering) applies, except Sec. 1926.1426(a)(2)(i). Sideboom cranes in which the boom is designed to free fall (live boom) are permitted only if manufactured prior to November 8, 2010. [§1926.1440(b)]

(c) **Sideboom cranes mounted** on wheel or crawler tractors must meet all of the following requirements of ASME B30.14-2004 (incorporated by reference, *see* Sec. 1926.6): [§1926.1440(c)]

(1) *Section 14-1.1 ("Load Ratings").* [§1926.1440(c)(1)]

(2) *Section 14-1.3* ("Side Boom Tractor Travel"). [§1926.1440(c)(2)]

(3) *Section 14-1.5* ("Ropes and Reeving Accessories"). [§1926.1440(c)(3)]

(4) *Section 14-1.7.1 ("Booms").* [§1926.1440(c)(4)]

(5) *Section 14-1.7.2* ("General Requirements — Exhaust Gases"). [§1926.1440(c)(5)]

(6) *Section 14-1.7.3* ("General Requirements — Stabilizers (Wheel-Type Side Boom Tractors)"). [§1926.1440(c)(6)]

(7) *Section 14-1.7.4* ("General Requirements — Welded Construction"). [§1926.1440(c)(7)]

(8) *Section 14-1.7.6* ("General Requirements — Clutch and Brake Protection"). [§1926.1440(c)(8)]

(9) *Section 14-2.2.2* ("Testing — Rated Load Test"), except that it applies only to equipment that has been altered or modified. [§1926.1440(c)(9)]

(10) *In section 14-3.1.2* ("Operator Qualifications"), paragraph (a), except the phrase "When required by law." [§1926.1440(c)(10)]

(11) *In section 14-3.1.3* ("Operating Practices"), paragraphs (e), (f)(1) — (f)(4), (f)(6), (f)(7), (h), and (i). [§1926.1440(c)(11)]

(12) *In section 14-3.2.3* ("Moving the Load"), paragraphs (j), (l), and (m). [§1926.1440(c)(12)]

§1926.1441
Equipment with a rated hoisting/lifting capacity of 2,000 pounds or less

The following paragraphs of this section specify requirements for employers using equipment with a maximum rated hoisting/lifting capacity of 2,000 pounds or less.

(a) **The employer using this equipment must comply** with the following provisions of this subpart: §1926.1400 (Scope); §1926.1401 (Definitions); Sec. 1926.1402 (Ground conditions); Sec. 1926.1403 (Assembly/disassembly — selection of manufacturer or employer procedures); Sec. 1926.1406 (Assembly/disassembly — employer procedures); §§1926.1407 through 1926.1411 (Power line safety); Sec. 1926.1412(c) (*Post-assembly*); §§1926.1413 through 1926.1414 (Wire rope); Sec. 1926.1418 (Authority to stop operation); §§1926.1419 through 1926.1422 (Signals); Sec. 1926.1423 (Fall protection); Sec. 1926.1425 (Keeping clear of the load) (except for Sec. 1926.1425(c)(3) (qualified rigger)); Sec. 1926.1426 (Free fall and controlled load lowering); Sec. 1926.1432 (Multiple crane/derrick lifts — supplemental requirements); Sec. 1926.1434 (Equipment modifications); Sec. 1926.1435 (Tower cranes); §1926.1436 (Derricks); Sec. 1926.1437 (Floating cranes/derricks and land cranes/derricks on barges); Sec. 1926.1438 (Overhead & gantry cranes). [§1926.1441(a)]

(b) **Assembly/disassembly.** [§1926.1441(b)]

(1) *In addition to compliance with §§1926.1403* (Assembly/disassembly — selection of manufacturer or employer procedures) and 1926.1406 (Assembly/disassembly — employer procedures), the employer must also comply with Sec. 1926.1441(b)(2)-(3). [§1926.1441(b)(1)]

(2) *Components and configuration.* The employer must ensure that: [§1926.1441(b)(2)]

(i) *The selection of components,* and the configuration of the equipment, that affect the capacity or safe operation of the equipment complies with either the: [§1926.1441(b)(2)(i)]

[A] Manufacturer instructions, recommendations, limitations, and specifications. When these documents and information are unavailable, a registered professional engineer familiar with the type of equipment involved must approve, in writing, the selection and configuration of components; or [§1926.1441(b)(2)(i)[A]]

[B] Approved modifications that meet the requirements of Sec. 1926.1434 (Equipment modifications). [§1926.1441(b)(2)(i)[B]]

(ii) *Post-assembly inspection.* Upon completion of assembly, the equipment is inspected to ensure that it is in compliance with paragraph (b)(2)(i) of this section (*see* Sec. 1926.1412(c) for post-assembly inspection requirements). [§1926.1441(b)(2)(ii)]

(3) *Manufacturer prohibitions.* The employer must comply with applicable manufacturer prohibitions. [§1926.1441(b)(3)]

(c) **Operation — procedures.** [§1926.1441(c)]

(1) *The employer must comply with all manufacturer* procedures applicable to the operational functions of the equipment, including its use with attachments. [§1926.1441(c)(1)]

(2) *Unavailable operation procedures.* The employer must: [§1926.1441(c)(2)]

(i) *When the manufacturer's procedures are unavailable,* develop, and ensure compliance with, all procedures necessary for the safe operation of the equipment and attachments. [§1926.1441(c)(2)(i)]

(ii) *Ensure that procedures for the operational* controls are developed by a qualified person. [§1926.1441(c)(2)(ii)]

(iii) *Ensure that procedures related* to the capacity of the equipment are developed and signed by a registered professional engineer familiar with the equipment. [§1926.1441(c)(2)(iii)]

(3) *Accessibility.* The employer must ensure that: [§1926.1441(c)(3)]

(i) *The load chart is available* to the operator at the control station; [§1926.1441(c)(3)(i)]

(ii) *Procedures applicable to the operation* of the equipment, recommended operating speeds, special hazard warnings, instructions, and operator's manual are readily available for use by the operator. [§1926.1441(c)(3)(ii)]

(iii) *When rated capacities are available* at the control station only in electronic form and a failure occurs that makes the rated capacities inaccessible, the operator immediately ceases operations or follows safe shut-down procedures until the rated capacities (in electronic or other form) are available. [§1926.1441(c)(3)(iii)]

(d) **Safety devices and operational aids.** [§1926.1441(d)]

(1) *The employer must ensure that safety* devices and operational aids that are part of the original equipment are maintained in accordance with manufacturer procedures. [§1926.1441(d)(1)]

(2) *Anti two-blocking.* The employer must ensure that equipment covered by this section manufactured more than one year after November 8, 2010 have either an anti two-block device that meets the requirements of Sec. 1926.1416(d)(3), or is designed so that, in the event of a two-block situation, no damage or load failure will occur (for example, by using a power unit that stalls in response to a two-block situation). [§1926.1441(d)(2)]

(e) **Operator qualifications.** The employer must train each operator, prior to operating the equipment, on the safe operation of the type of equipment the operator will be using. [§1926.1441(e)]

(f) **Signal person qualifications.** The employer must train each signal person in the proper use of signals applicable to the use of the equipment. [§1926.1441(f)]

(g) *[Reserved.]* [§1926.1441(g)]

(h) **Inspections.** The employer must ensure that equipment is inspected in accordance with manufacturer procedures. [§1926.1441(h)]

(i) *[Reserved.]* [§1926.1441(i)]

(j) **Hoisting personnel.** The employer must ensure that equipment covered by this section is not used to hoist personnel. [§1926.1441(j)]

(k) **Design.** The employer must ensure that the equipment is designed by a qualified engineer. [§1926.1441(k)]

§1926.1442
Severability

Should a court of competent jurisdiction hold any provision(s) of subpart CC to be invalid, such action shall not affect any other provision of the subpart.

Subpart CC Appendix A
Standard Hand Signals

STOP –With arm extended horizontally to the side, palm down, arm is swung back and forth.

EMERGENCY STOP – With both arms extended horizontally to the side, palms down, arms are swung back and forth.

HOIST – With upper arm extended to the side, forearm and index finger pointing straight up, hand and finger make small circles.

RAISE BOOM – With arm extended horizontally to the side, thumb points up with other fingers closed.

SWING – With arm extended horizontally, index finger points in direction that boom is to swing.

RETRACT TELESCOPING BOOM – With hands to the front at waist level, thumbs point at each other with other fingers closed.

RAISE THE BOOM AND LOWER THE LOAD – With arm extended horizontally to the side and thumb pointing up, fingers open and close while load movement is desired.

DOG EVERYTHING – Hands held together at waist level.

LOWER – With arm and index finger pointing down, hand and finger make small circles.

LOWER BOOM – With arm extended horizontally to the side, thumb points down with other fingers closed.

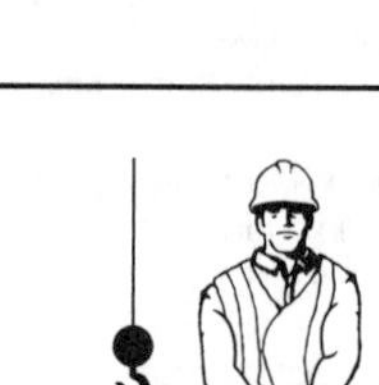

EXTEND TELESCOPING BOOM – With hands to the front at waist level, thumbs point outward with other fingers closed.

TRAVEL/TOWER TRAVEL With all fingers pointing up, arm is extended horizontally out and back to make a pushing motion in the direction of travel.

LOWER THE BOOM AND RAISE THE LOAD – With arm extended horizontally to the side and thumb pointing down, fingers open and close while load movement is desired.

MOVE SLOWLY – A hand is placed in front of the hand that is giving the action signal.

USE AUXILIARY HOIST *(whipline)* – With arm bent at elbow and forearm vertical, elbow is tapped with other hand. Then regular signal is used to indicate desired action.

CRAWLER CRANE TRAVEL, BOTH TRACKS – Rotate fists around each other in front of body; direction of rotation away from body indicates travel forward; rotation towards body indicates travel backward.

USE MAIN HOIST – A hand taps on top of the head. Then regular signal is given to indicate desired action.

CRAWLER CRANE TRAVEL, ONE TRACK – Indicate track to be locked by raising fist on that side. Rotate other fist in front of body in direction that other track is to travel.

TROLLEY TRAVEL – With palm up, fingers closed and thumb pointing in direction of motion, hand is jerked horizontally in direction trolley is to travel.

Subpart CC Appendix B

Assembly/Disassembly: Sample Procedures for Minimizing the Risk of Unintended Dangerous Boom Movement

1. **Section 1926.1404(f)(1) provides** that when pins (or similar devices) are being removed, employees must not be under the boom, jib, or other components, except where the requirements of Sec. 1926.1404(f)(2) are met. The exception in Sec. 1926.1404(f)(2) applies when the employer demonstrates that site constraints require one or more employees to be under the boom, jib, or other components when pins (or similar devices) are being removed. In such a situation, the A/D director must implement procedures that minimize the risk of unintended dangerous movement and minimize the duration and extent of exposure under the boom.

   The following scenario is an example of how the exception applies: A boom cannot be disassembled on the ground because of aboveground piping (as might be found, for example, in an oil refinery) that precludes lowering the boom to the ground. The boom must therefore be disassembled in the air, and the employees who remove the pins must perform that work from an aerial lift whose base is positioned on one side (the near side) of the boom. To gain access to the pins on the far side, the aerial lift basket must move under the boom, since, due to lack of room, the aerial lift cannot be repositioned on the far side. Due to lack of room, the aerial lift cannot be repositioned on the far side, so the aerial basket must move under the boom to gain access to the pins on the far side.

   To minimize the risk of unintended dangerous movement while the pins are removed, the A/D director uses an assist crane that is rigged to support the boom section that is being detached, using particular care to ensure that the section end that is near the employee(s) removing the pins is well supported. The duration and extent of exposure is minimized by removing the far side pins first, moving the aerial lift basket as soon as possible to the near side so that the employees are no longer under the boom, and then removing the near side pins.

2. **Section 1926.1404(h)(6)(i) provides** that, during assembly/disassembly, the center of gravity of the load must be identified if that is necessary for the method used for maintaining stability. Section 1926.1404(h)(6)(ii) states that, where there is insufficient information to accurately identify the center of gravity, measures designed to prevent unintended dangerous movement resulting from an inaccurate identification of the center of gravity must be used.

   An example of the application of Sec. 1926.1404(h)(6)(ii) is as follows: The boom is assembled by lowering boom sections sequentially into place using an assist crane. The A/D director's plan is to keep the boom sections stable while they are lowered into place by attaching the assist crane hoist line above the center of gravity of each section. However, in assembling the non-symmetrical top section of the boom, the A/D director is not able to determine where to attach the assist crane hoist line so that it is above the center of gravity. In this situation, before raising the section, all personnel are kept clear of the section and the section is first raised a few inches to determine whether it tips when raised (if it did tip, it would indicate it is not rigged over the center of gravity). If this occurs, the hoist line is repositioned and the procedure repeated (with employees kept clear of the section while it is raised) until the A/D director determines that it is rigged over the center of gravity and can be moved into place without dangerous movement.

Subpart CC Appendix C

Operator Certification: Written Examination: Technical Knowledge Criteria

This appendix contains information for employers, accredited testing organizations, auditors and government entities developing criteria for a written examination to test an individual's technical knowledge relating to the operation of cranes.

(a) **General technical information.**

- (1) *The functions and limitations of the crane* and attachments.
- (2) *Wire rope:*
  - (i) *Background information necessary* to understand the inspection and removal from service criteria in Sec. 1926.1413 and Sec. 1926.1414.
  - (ii) *Capacity and when multi-part rope is needed.*
  - (iii) *Relationship between line* pull and safe working load.
  - (iv) *How to determine the manufacturer's recommended* rope for the crane.
- (3) *Rigging devices and their use, such as:*
  - (i) *Slings.*
  - (ii) *Spreaders.*
  - (iii) *Lifting beams.*
  - (iv) *Wire rope fittings,* such as clips, shackles and wedge sockets.
  - (v) *Saddles (softeners).*
  - (vi) *Clamps (beams).*
- (4) *The technical limitations of protective* measures against electrical hazards:
  - (i) *Grounding.*
  - (ii) *Proximity warning devices.*
  - (iii) *Insulated links.*
  - (iv) *Boom cages.*
  - (v) *Proximity to electric power* lines, radii, and microwave structures.
- (5) *The effects of load share* and load transfer in multi-crane lifts.
- (6) *Basic crane terms.*
- (7) *The basics of machine power flow systems.*
  - (i) *Mechanical.*
  - (ii) *Electrical.*
  - (iii) *Pneumatic.*
  - (iv) *Hydraulic.*
  - (v) *Combination.*
- (8) *The significance of the instruments and gauge readings.*
- (9) *The effects of thermal expansion* and contraction in hydraulic cylinders.
- (10) *Background information necessary* to understand the requirements of pre-operation and inspection.
- (11) *How to use the safety devices* and operational aids required under Sec. 1926.1415 and Sec. 1926.1416.
- (12) *The difference between duty-cycle* and lifting operations.
- (13) *How to calculate net capacity* for every possible configuration of the equipment using the manufacturer's load chart.
- (14) *How to use manufacturer-approved attachments* and their effect on the equipment.
- (15) *How to obtain dimensions,* weight, and center of gravity of the load.
- (16) *The effects of dynamic loading from:*
  - (i) *Wind.*
  - (ii) *Stopping and starting.*

(iii) *Impact loading.*

(iv) *Moving with the load.*

(17) *The effect of side loading.*

(18) *The principles of backward stability.*

(b) Site information.

(1) *How to identify the suitability of the supporting* ground/surface to support the expected loads of the operation. Elements include:

(i) *Weaknesses below the surface* (such as voids, tanks, loose fill).

(ii) *Weaknesses on the surface* (such as retaining walls, slopes, excavations, depressions).

(2) *Proper use of mats,* blocking/cribbing, outriggers, stabilizers, or crawlers.

(3) *Identification of site hazards* such as power lines, piping, and traffic.

(4) *How to review operation plans* with supervisors and other workers (such as the signal person), including how to determine working height, boom length, load radius, and travel clearance.

(5) *How to determine if there is adequate* room for extension of crawlers or outriggers/stabilizers and counterweights.

(c) Operations.

(1) *How to pick,* carry, swing and place the load smoothly and safely on rubber tires and on outriggers/stabilizers or crawlers (where applicable).

(2) *How to communicate at the site with supervisors,* the crew and the signal person.

(3) *Proper procedures and methods* of reeving wire ropes and methods of reeving multiple-part lines and selecting the proper load block and/or ball.

(4) *How to react to changes in conditions* that affect the safe operation of the equipment.

(5) *How to shut down and secure* the equipment properly when leaving it unattended.

(6) *Know how to apply the manufacturer's* specifications for operating in various weather conditions, and understand how environmental conditions affect the safe operation of the equipment.

(7) *How to properly level the equipment.*

(8) *How to verify the weight of the load* and rigging prior to initiating the lift.

(9) *How to determine where the load is to be picked* up and placed and how to verify the radii.

(10) *Know basic rigging procedures.*

(11) *How to carry out the shift inspection* required in this subpart.

(12) *Know that the following operations* require specific procedures and skill levels:

(i) *Multi-crane lifts.*

(ii) *Hoisting personnel.*

(iii) *Clamshell/dragline operations.*

(iv) *Pile driving and extracting.*

(v) *Concrete operations,* including poured-in-place and tilt-up.

(vi) *Demolition operations.*

(vii) *Operations on water.*

(viii) *Magnet operations.*

(ix) *Multi-drum operations.*

(13) *Know the proper procedures* for operating safely under the following conditions:

(i) *Traveling with suspended loads.*

(ii) *Approaching a two-block condition.*

(iii) *Operating near power lines.*

(iv) *Hoisting personnel.*

(v) *Using other than full outrigger/crawler* or stabilizer extensions.

(vi) *Lifting loads from beneath* the surface of the water.

(vii) *Using various approved counterweight* configurations.

(viii) *Handling loads out of the operator's* vision ("operating in the blind").

(ix) *Using electronic communication systems* for signal communication.

(14) *Know the proper procedures* for load control and the use of hand-held tag lines.

(15) *Know the emergency response procedure for:*

(i) *Fires.*

(ii) *Power line contact.*

(iii) *Loss of stability.*

(iv) *Control malfunction.*

(v) *Two-blocking.*

(vi) *Overload.*

(vii) *Carrier or travel malfunction.*

(16) *Know how to properly use* outriggers and stabilizers in accordance with manufacturer specifications.

(d) Use of load charts.

(1) *Know the terminology necessary* to use load charts.

(2) *Know how to ensure that the load* chart is the appropriate chart for the equipment in its particular configuration and application.

(3) *Know how to use load* charts. This includes knowing:

(i) *The operational limitations of load* charts and footnotes.

(ii) *How to relate the chart to the configuration* of the crane, crawlers, or outriggers/stabilizers extended or retracted, jib erected or offset, and various counterweight configurations.

(iii) *The difference between structural* capacity and capacity limited by stability.

(iv) *What is included in capacity ratings.*

(v) *The range diagram and its relationship* to the load chart.

(vi) *The work area chart* and its relationship to the load chart.

(vii) *Where to find and how to use the "parts-of-line" information.*

(4) *Know how to use the load* chart together with the load indicators and/or load moment devices.

§1926.1500
Scope

This subpart applies only to employers engaged in demolition work covered by §1926.856 and §1926.858, and underground construction work covered by §1926.800. This subpart applies in lieu of §1926 subpart CC. [§1926.1500)]

§1926.1501
Cranes and derricks

(a) **General requirements.** [§1926.1501(a)]

(1) *The employer shall comply* with the manufacturer's specifications and limitations applicable to the operation of any and all cranes and derricks. Where manufacturer's specifications are not available, the limitations assigned to the equipment shall be based on the determinations of a qualified engineer competent in this field and such determinations will be appropriately documented and recorded. Attachments used with cranes shall not exceed the capacity, rating, or scope recommended by the manufacturer. [§1926.1501(a)(1)]

(2) *Rated load capacities,* and recommended operating speeds, special hazard warnings, or instruction, shall be conspicuously posted on all equipment. Instructions or warnings shall be visible to the operator while he is at his control station. [§1926.1501(a)(2)]

(3) *[Reserved]* [§1926.1501(a)(3)]

(4) *Hand signals to crane and derrick operators* shall be those prescribed by the applicable ANSI standard for the type of crane in use. An illustration of the signals shall be posted at the job site. [§1926.1501(a)(4)]

(5) *The employer shall designate a competent person* who shall inspect all machinery and equipment prior to each use, and during use, to make sure it is in safe operating condition. Any deficiencies shall be repaired, or defective parts replaced, before continued use. [§1926.1501(a)(5)]

(6) *A thorough, annual inspection* of the hoisting machinery shall be made by a competent person, or by a government or private agency recognized by the U.S. Department of Labor. The employer shall maintain a record of the dates and results of inspections for each hoisting machine and piece of equipment. [§1926.1501(a)(6)]

(7) *Wire rope shall be taken out of service* when any of the following conditions exist: [§1926.1501(a)(7)]

(i) *In running ropes,* six randomly distributed broken wires in one lay or three broken wires in one strand in one lay; [§1926.1501(a)(7)(i)]

(ii) *Wear of one-third the original diameter* of outside individual wires. Kinking, crushing, bird caging, or any other damage resulting in distortion of the rope structure; [§1926.1501(a)(7)(ii)]

(iii) *Evidence of any heat damage from any cause;* [§1926.1501(a)(7)(iii)]

(iv) *Reductions from nominal diameter* of more than one-sixty-fourth inch for diameters up to and including five-sixteenths inch, one-thirty-second inch for diameters three-eighths inch to and including one-half inch, three-sixty-fourths inch for diameters nine-sixteenths inch to and including three-fourths inch, one-sixteenth inch for diameters seven-eighths inch to 1 1/8 inches inclusive, three-thirty-seconds inch for diameters 1 1/4 to 1 1/2 inches inclusive; [§1926.1501(a)(7)(iv)]

(v) *In standing ropes,* more than two broken wires in one lay in sections beyond end connections or more than one broken wire at an end connection. [§1926.1501(a)(7)(v)]

(vi) *Wire rope safety factors* shall be in accordance with American National Standards Institute B30.5-1968 or SAE J959-1966. [§1926.1501(a)(7)(vi)]

(8) *Belts, gears, shafts, pulleys, sprockets, spindles,* drums, fly wheels, chains, or other reciprocating, rotating, or other moving parts or equipment shall be guarded if such parts are exposed to contact by employees, or otherwise create a hazard. Guarding shall meet the requirements of the American National Standards Institute B15.1-1958 Rev., Safety Code for Mechanical Power Transmission Apparatus. [§1926.1501(a)(8)]

(9) *Accessible areas within the swing radius* of the rear of the rotating superstructure of the crane, either permanently or temporarily mounted, shall be barricaded in such a manner as to prevent an employee from being struck or crushed by the crane. [§1926.1501(a)(9)]

(10) *All exhaust pipes shall be guarded* or insulated in areas where contact by employees is possible in the performance of normal duties. [§1926.1501(a)(10)]

(11) *Whenever internal combustion engine powered equipment* exhausts in enclosed spaces, tests shall be made and recorded to see that employees are not exposed to unsafe concentrations of toxic gases or oxygen deficient atmospheres. [§1926.1501(a)(11)]

(12) *All windows in cabs shall be of safety glass,* or equivalent, that introduces no visible distortion that will interfere with the safe operation of the machine. [§1926.1501(a)(12)]

(13)(i) *Where necessary for rigging or service requirements,* a ladder, or steps, shall be provided to give access to a cab roof. [§1926.1501(a)(13)(i)]

(ii) *Guardrails, handholds, and steps* shall be provided on cranes for easy access to the car and cab, conforming to American National Standards Institute B30.5. [§1926.1501(a)(13)(ii)]

(iii) *Platforms and walkways shall have anti-skid surfaces.* [§1926.1501(a)(13)(iii)]

(14) *Fuel tank filler pipe* shall be located in such a position, or protected in such manner, as to not allow spill or overflow to run onto the engine, exhaust, or electrical equipment of any machine being fueled. [§1926.1501(a)(14)]

(i) *An accessible fire extinguisher* of 5BC rating, or higher, shall be available at all operator stations or cabs of equipment. [§1926.1501(a)(14)(i)]

(ii) *All fuels shall be transported, stored, and handled* to meet the rules of subpart F of this part. When fuel is transported by vehicles on public highways, Department of Transportation rules contained in 49 CFR Parts 177 and 393 concerning such vehicular transportation are considered applicable. [§1926.1501(a)(14)(ii)]

(15) *Except where electrical distribution and transmission lines* have been deenergized and visibly grounded at point of work or where insulating barriers, not a part of or an attachment to the equipment or machinery, have been erected to prevent physical contact with the lines, equipment or machines shall be operated proximate to power lines only in accordance with the following: [§1926.1501(a)(15)]

(i) *For lines rated 50 kV. or below,* minimum clearance between the lines and any part of the crane or load shall be 10 feet; [§1926.1501(a)(15)(i)]

(ii) *For lines rated over 50 kV.,* minimum clearance between the lines and any part of the crane or load shall be 10 feet plus 0.4 inch for each 1 kV. over 50 kV., or twice the length of the line insulator, but never less than 10 feet; [§1926.1501(a)(15)(ii)]

(iii) *In transit with no load and boom lowered,* the equipment clearance shall be a minimum of 4 feet for voltages less than 50 kV., and 10 feet for voltages over 50 kV., up to and including 345 kV., and 16 feet for voltages up to and including 750 kV. [§1926.1501(a)(15)(iii)]

(iv) *A person shall be designated to observe clearance* of the equipment and give timely warning for all operations where it is difficult for the operator to maintain the desired clearance by visual means; [§1926.1501(a)(15)(iv)]

(v) *Cage-type boom guards, insulating links, or proximity warning devices* may be used on cranes, but the use of such devices shall not alter the requirements of any other regulation of this part even if such device is required by law or regulation; [§1926.1501(a)(15)(v)]

(vi) *Any overhead wire* shall be considered to be an energized line unless and until the person owning such line or the electrical utility authorities indicate that it is not an energized line and it has been visibly grounded; [§1926.1501(a)(15)(vi)]

(vii) *Prior to work near transmitter towers* where an electrical charge can be induced in the equipment or materials being handled, the transmitter shall be de-energized or tests shall be made to determine if electrical charge is induced on the crane. The following precautions shall be taken when necessary to dissipate induced voltages: [§1926.1501(a)(15)(vii)]

[A] The equipment shall be provided with an electrical ground directly to the upper rotating structure supporting the boom; and [§1926.1501(a)(15)(vii)[A]]

[B] Ground jumper cables shall be attached to materials being handled by boom equipment when electrical charge is induced while working near energized transmitters. Crews shall be provided with nonconductive poles having large alligator clips or other similar protection to attach the ground cable to the load. [§1926.1501(a)(15)(vii)[B]]

[C] Combustible and flammable materials shall be removed from the immediate area prior to operations. [§1926.1501(a)(15)(vii)[C]]

(16) *No modifications or additions which affect the capacity* or safe operation of the equipment shall be made by the employer without the manufacturer's written approval. If such modifications or changes are made, the capacity, operation, and maintenance instruction plates, tags, or decals, shall be changed accordingly. In no case shall the original safety factor of the equipment be reduced. [§1926.1501(a)(16)]

(17) *The employer shall comply* with Power Crane and Shovel Association Mobile Hydraulic Crane Standard No. 2. [§1926.1501(a)(17)]

(18) *Sideboom cranes mounted* on wheel or crawler tractors shall meet the requirements of SAE J743a-1964. [§1926.1501(a)(18)]

(19) *All employees shall be kept clear* of loads about to be lifted and of suspended loads. [§1926.1501(a)(19)]

(b) Crawler, locomotive, and truck cranes. [§1926.1501(b)]

(1) *All jibs shall have positive stops* to prevent their movement of more than 5° above the straight line of the jib and boom on conventional type crane booms. The use of cable type belly slings does not constitute compliance with this rule. [§1926.1501(b)(1)]

(2) *All crawler, truck, or locomotive cranes* in use shall meet the applicable requirements for design, inspection, construction, testing, maintenance and operation as prescribed in the ANSI B30.5-1968, Safety Code for Crawler, Locomotive and Truck Cranes. However, the written, dated, and signed inspection reports and records of the monthly inspection of critical items prescribed in section 5-2.1.5 of the ANSI B30.5-1968 standard are not required. Instead, the employer shall prepare a certification record which includes the date the crane items were inspected; the signature of the person who inspected the crane items; and a serial number, or other identifier, for the crane inspected. The most recent certification record shall be maintained on file until a new one is prepared. [§1926.1501(b)(2)]

(c) Hammerhead tower cranes. [§1926.1501(c)]

(1) *Adequate clearance shall be maintained* between moving and rotating structures of the crane and fixed objects to allow the passage of employees without harm. [§1926.1501(c)(1)]

(2) *Each employee required to perform duties* on the horizontal boom of hammerhead tower cranes shall be protected against falling by guardrails or by a personal fall arrest system in conformance with subpart M of this part. [§1926.1501(c)(2)]

(3) *Buffers shall be provided* at both ends of travel of the trolley. [§1926.1501(c)(3)]

(4) *Cranes mounted on rail tracks* shall be equipped with limit switches limiting the travel of the crane on the track and stops or buffers at each end of the tracks. [§1926.1501(c)(4)]

(5) *All hammerhead tower cranes in use* shall meet the applicable requirements for design, construction, installation, testing, maintenance, inspection, and operation as prescribed by the manufacturer. [§1926.1501(c)(5)]

(d) Overhead and gantry cranes. [§1926.1501(d)]

(1) *The rated load of the crane* shall be plainly marked on each side of the crane, and if the crane has more than one hoisting unit, each hoist shall have its rated load marked on it or its load block, and this marking shall be clearly legible from the ground or floor. [§1926.1501(d)(1)]

(2) *Bridge trucks shall be equipped* with sweeps which extend below the top of the rail and project in front of the truck wheels. [§1926.1501(d)(2)]

(3) *Except for floor-operated cranes,* a gong or other effective audible warning signal shall be provided for each crane equipped with a power traveling mechanism. [§1926.1501(d)(3)]

(4) *All overhead and gantry cranes* in use shall meet the applicable requirements for design, construction, installation, testing, maintenance, inspection, and operation as prescribed in the ANSI B30.2.0-1967, Safety Code for Overhead and Gantry Cranes. [§1926.1501(d)(4)]

(e) Derricks. All derricks in use shall meet the applicable requirements for design, construction, installation, inspection, testing, maintenance, and operation as prescribed in American National Standards Institute B30.6-1969, Safety Code for Derricks. [§1926.1501(e)]

(f) Floating cranes and derricks. [§1926.1501(f)]

(1) *Mobile cranes mounted on barges.* [§1926.1501(f)(1)]

(i) *When a mobile crane is mounted* on a barge, the rated load of the crane shall not exceed the original capacity specified by the manufacturer. [§1926.1501(f)(1)(i)]

(ii) *A load rating chart,* with clearly legible letters and figures, shall be provided with each crane, and securely fixed at a location easily visible to the operator. [§1926.1501(f)(1)(ii)]

(iii) *When load ratings are reduced* to stay within the limits for list of the barge with a crane mounted on it, a new load rating chart shall be provided. [§1926.1501(f)(1)(iii)]

(iv) *Mobile cranes on barges* shall be positively secured. [§1926.1501(f)(1)(iv)]

(2) *Permanently mounted floating cranes and derricks.* [§1926.1501(f)(2)]

(i) *When cranes and derricks are permanently installed* on a barge, the capacity and limitations of use shall be based on competent design criteria. [§1926.1501(f)(2)(i)]

(ii) *A load rating chart with clearly legible letters* and figures shall be provided and securely fixed at a location easily visible to the operator. [§1926.1501(f)(2)(ii)]

(iii) *Floating cranes and floating derricks* in use shall meet the applicable requirements for design, construction, installation, testing, maintenance, and operation as prescribed by the manufacturer. [§1926.1501(f)(2)(iii)]

(3) *Protection of employees working on barges.* The employer shall comply with the applicable requirements for protection of employees working onboard marine vessels specified in §1926.605. [§1926.1501(f)(3)]

(g) Crane or derrick suspended personnel platforms. [§1926.1501(g)]

(1) *Scope, application and definitions.* [§1926.1501(g)(1)]

(i) *Scope and application.* This standard applies to the design, construction, testing, use and maintenance of personnel platforms, and the hoisting of personnel platforms on the load lines of cranes or derricks. [§1926.1501(g)(1)(i)]

(ii) *Definitions.* For the purposes of this paragraph (g), the following definitions apply: [§1926.1501(g)(1)(ii)]

[A] **Failure** means load refusal, breakage, or separation of components. [§1926.1501(g)(1)(ii)[A]]

[B] **Hoist (or hoisting)** means all crane or derrick functions such as lowering, lifting, swinging, booming in and out or up and down, or suspending a personnel platform. [§1926.1501(g)(1)(ii)[B]]

[C] **Load refusal** means the point where the ultimate strength is exceeded. [§1926.1501(g)(1)(ii)[C]]

[D] **Maximum intended load** means the total load of all employees, tools, materials, and other loads reasonably anticipated to be applied to a personnel platform or personnel platform component at any one time. [§1926.1501(g)(1)(ii)[D]]

[E] **Runway** means a firm, level surface designed, prepared and designated as a path of travel for the weight and configuration of the crane being used to lift and travel with the crane suspended platform. An existing surface may be used as long as it meets these criteria. [§1926.1501(g)(1)(ii)[E]]

(2) *General requirements.* The use of a crane or derrick to hoist employees on a personnel platform is prohibited, except when the erection, use, and dismantling of conventional means of reaching the worksite, such as a personnel hoist, ladder, stairway, aerial lift, elevating work platform or scaffold, would be more hazardous, or is not possible because of structural design or worksite conditions. [§1926.1501(g)(2)]

(3) *Cranes and derricks.* [§1926.1501(g)(3)]

(i) *Operational criteria.* [§1926.1501(g)(3)(i)]

[A] Hoisting of the personnel platform shall be performed in a slow, controlled, cautious manner with no sudden movements of the crane or derrick, or the platform. [§1926.1501(g)(3)(i)[A]]

[B] *Load lines shall be capable of supporting,* without failure, at least seven times the maximum intended load, except that where rotation resistant rope is used, the lines shall be capable of supporting without failure, at least ten times the maximum intended load. The required design factor is achieved by taking the current safety factor of 3.5 (required under paragraph (b)(2)) of this section and applying the 50 percent derating of the crane capacity which is required by paragraph (g)(3)(i)(F) of this section. [§1926.1501(g)(3)(i)[B]]

[C] *Load and boom hoist drum brakes, swing brakes,* and locking devices such as pawls or dogs shall be engaged when the occupied personnel platform is in a stationary working position. [§1926.1501(g)(3)(i)[C]]

[D] *The crane shall be uniformly level* within one percent of level grade and located on firm footing. Cranes equipped with outriggers shall have them all fully deployed following manufacturer's specifications, insofar as applicable, when hoisting employees. [§1926.1501(g)(3)(i)[D]]

[E] *The total weight of the loaded personnel platform* and related rigging shall not exceed 50 percent of the rated capacity for the radius and configuration of the crane or derrick. [§1926.1501(g)(3)(i)[E]]

[F] *The use of machines having live booms* (booms in which lowering is controlled by a brake without aid from other devices which slow the lowering speeds) is prohibited. [§1926.1501(g)(3)(i)[F]]

(ii) *Instruments and components.* [§1926.1501(g)(3)(ii)]

[A] *Cranes and derricks with variable angle booms* shall be equipped with a boom angle indicator, readily visible to the operator. [§1926.1501(g)(3)(ii)[A]]

[B] *Cranes with telescoping booms shall be equipped* with a device to indicate clearly to the operator, at all times, the boom's extended length, or an accurate determination of the load radius to be used during the lift shall be made prior to hoisting personnel. [§1926.1501(g)(3)(ii)[B]]

[C] *A positive acting device* shall be used which prevents contact between the load block or overhaul ball and the boom tip (anti-two-blocking device), or a system shall be used which deactivates the hoisting action before damage occurs in the event of a two-blocking situation (two block damage prevention feature). [§1926.1501(g)(3)(ii)[C]]

[D] *The load line hoist drum* shall have a system or device on the power train, other than the load hoist brake, which regulates the lowering rate of speed of the hoist mechanism (controlled load lowering.) Free fall is prohibited. [§1926.1501(g)(3)(ii)[D]]

(4) *Personnel platforms.* [§1926.1501(g)(4)]

(i) *Design criteria.* [§1926.1501(g)(4)(i)]

[A] *The personnel platform and suspension system* shall be designed by a qualified engineer or a qualified person competent in structural design. [§1926.1501(g)(4)(i)[A]]

[B] *The suspension system shall be designed* to minimize tipping of the platform due to movement of employees occupying the platform. [§1926.1501(g)(4)(i)[B]]

[C] *The personnel platform itself,* except the guardrail system and personal fall arrest system anchorages, shall be capable of supporting, without failure, its own weight and at least five times the maximum intended load. Criteria for guardrail systems and personal fall arrest system anchorages are contained in Subpart M of this Part. [§1926.1501(g)(4)(i)[C]]

(ii) *Platform specifications.* [§1926.1501(g)(4)(ii)]

[A] *Each personnel platform shall be equipped* with a guardrail system which meets the requirements of subpart M, and, shall be enclosed at least from the toeboard to midrail with either solid construction or expanded metal having openings no greater than 1/2 inch (1.27 cm). [§1926.1501(g)(4)(ii)[A]]

[B] *A grab rail shall be installed* inside the entire perimeter of the personnel platform. [§1926.1501(g)(4)(ii)[B]]

[C] *Access gates, if installed,* shall not swing outward during hoisting. [§1926.1501(g)(4)(ii)[C]]

[D] *Access gates,* including sliding or folding gates, shall be equipped with a restraining device to prevent accidental opening. [§1926.1501(g)(4)(ii)[D]]

[E] *Headroom shall be provided* which allows employees to stand upright in the platform. [§1926.1501(g)(4)(ii)[E]]

[F] *In addition to the use of hard hats,* employees shall be protected by overhead protection on the personnel platform when employees are exposed to falling objects. [§1926.1501(g)(4)(ii)[F]]

[G] *All rough edges exposed to contact* by employees shall be surfaced or smoothed in order to prevent injury to employees from punctures or lacerations. [§1926.1501(g)(4)(ii)[G]]

[H] *All welding of the personnel platform* and its components shall be performed by a qualified welder familiar with the weld grades, types and material specified in the platform design. [§1926.1501(g)(4)(ii)[H]]

[I] *The personnel platform shall be conspicuously posted* with a plate or other permanent marking which indicates the weight of the platform and its rated load capacity or maximum intended load. [§1926.1501(g)(4)(ii)[I]]

(iii) *Personnel platform loading.* [§1926.1501(g)(4)(iii)]

[A] *The personnel platform shall not be loaded* in excess of its rated load capacity. When a personnel platform does not have a rated load capacity then the personnel platform shall not be loaded in excess of its maximum intended load. [§1926.1501(g)(4)(iii)[A]]

[B] *The number of employees occupying* the personnel platform shall not exceed the number required for the work being performed. [§1926.1501(g)(4)(iii)[B]]

[C] *Personnel platforms shall be used* only for employees, their tools, and the materials necessary to do their work, and shall not be used to hoist only materials or tools when not hoisting personnel. [§1926.1501(g)(4)(iii)[C]]

[D] *Materials and tools for use* during a personnel lift shall be secured to prevent displacement. [§1926.1501(g)(4)(iii)[D]]

[E] *Materials and tools for use* during a personnel lift shall be evenly distributed within the confines of the platform while the platform is suspended. [§1926.1501(g)(4)(iii)[E]]

(iv) *Rigging.* [§1926.1501(g)(4)(iv)]

[A] *When a wire rope bridle is used* to connect the personnel platform to the load line, each bridle leg shall be connected to a master link or shackle in such a manner to ensure that the load is evenly divided among the bridle legs. [§1926.1501(g)(4)(iv)[A]]

[B] *Hooks on overhaul ball assemblies, lower load blocks,* or other attachment assemblies shall be of a type that can be closed and locked, eliminating the hook throat opening. Alternatively, an alloy anchor type shackle with a bolt, nut and retaining pin may be used. [§1926.1501(g)(4)(iv)[B]]

[C] *Wire rope, shackles, rings, master links,* and other rigging hardware must be capable of supporting, without failure, at least five times the maximum intended load applied or transmitted to that component. Where rotation resistant rope is used, the slings shall be capable of supporting without failure at least ten times the maximum intended load. [§1926.1501(g)(4)(iv)[C]]

[D] *All eyes in wire rope slings* shall be fabricated with thimbles. [§1926.1501(g)(4)(iv)[D]]

[E] *Bridles and associated rigging* for attaching the personnel platform to the hoist line shall be used only for the platform and the necessary employees, their tools and the materials necessary to do their work, and shall not be used for any other purpose when not hoisting personnel. [§1926.1501(g)(4)(iv)[E]]

(5) *Trial lift, inspection, and proof testing.* [§1926.1501(g)(5)]

(i) *A trial lift with the unoccupied personnel platform* loaded at least to the anticipated liftweight shall be made from ground level, or any other location where employees will enter the platform, to each location at which the personnel platform is to be hoisted and positioned. This trial lift shall be performed immediately prior to placing personnel on the platform. The operator shall determine that all systems, controls and safety devices are activated and functioning properly; that no interferences exist; and that all configurations necessary to reach those work locations will allow the operator to remain under the 50 percent limit of the hoist's rated capacity. Materials and tools to be used during the actual lift can be loaded in the platform, as provided in paragraphs (g)(4)(iii) (D), and (E) of this section for the trial lift. A single trial lift may be performed at one time for all locations that are to be reached from a single set up position. [§1926.1501(g)(5)(i)]

(ii) *The trial lift shall be repeated* prior to hoisting employees whenever the crane or derrick is moved and set up in a new location or returned to a previously used location. Additionally, the trial lift shall be repeated when the lift route is changed unless the operator determines that the route change is not significant (i.e. the route change would not affect the safety of hoisted employees.) [§1926.1501(g)(5)(ii)]

(iii) *After the trial lift, and just prior to hoisting personnel,* the platform shall be hoisted a few inches and inspected to ensure that it is secure and properly balanced. Employees shall not be hoisted unless the following conditions are determined to exist: [§1926.1501(g)(5)(iii)]

[A] Hoist ropes shall be free of kinks; [§1926.1501(g)(5)(iii)[A]]

[B] Multiple part lines shall not be twisted around each other; [§1926.1501(g)(5)(iii)[B]]

[C] The primary attachment shall be centered over the platform; and [§1926.1501(g)(5)(iii)[C]]

[D] The hoisting system shall be inspected if the load rope is slack to ensure all ropes are properly stated on drums and in sheaves. [§1926.1501(g)(5)(iii)[D]]

(iv) *A visual inspection of the crane or derrick,* rigging, personnel platform, and the crane or derrick base support or ground shall be conducted by a competent person immediately after the trial lift to determine whether the testing has exposed any defect or produced any adverse effect upon any component or structure. [§1926.1501(g)(5)(iv)]

(v) *Any defects found during inspections* which create a safety hazard shall be corrected before hoisting personnel. [§1926.1501(g)(5)(v)]

(vi) *At each job site, prior to hoisting employees* on the personnel platform, and after any repair or modification, the platform and rigging shall be proof tested to 125 percent of the platform's rated capacity by holding it in a suspended position for five minutes with the test load evenly distributed on the platform (this may be done concurrently with the trial lift). After prooftesting, a competent person shall inspect the platform and rigging. Any deficiencies found shall be corrected and another proof test shall be conducted. Personnel hoisting shall not be conducted until the proof testing requirements are satisfied. [§1926.1501(g)(5)(vi)]

(6) *Work practices.* [§1926.1501(g)(6)]

(i) *Employees shall keep all parts of the body* inside the platform during raising, lowering, and positioning. This provision does not apply to an occupant of the platform performing the duties of a signal person. [§1926.1501(g)(6)(i)]

(ii) *Before employees exit or enter a hoisted personnel platform* that is not landed, the platform shall be secured to the structure where the work is to be performed, unless securing to the structure creates an unsafe situation. [§1926.1501(g)(6)(ii)]

(iii) *Tag lines shall be used* unless their use creates an unsafe condition. [§1926.1501(g)(6)(iii)]

(iv) *The crane or derrick operator* shall remain at the controls at all times when the crane engine is running and the platform is occupied. [§1926.1501(g)(6)(iv)]

(v) *Hoisting of employees shall be promptly discontinued* upon indication of any dangerous weather conditions or other impending danger. [§1926.1501(g)(6)(v)]

(vi) *Employees being hoisted shall remain* in continuous sight of and in direct communication with the operator or signal person. In those situations where direct visual contact with the operator is not possible, and the use of a signal person would create a greater hazard for that person, direct communication alone such as by radio may be used. [§1926.1501(g)(6)(vi)]

(vii) *Except over water,* employees occupying the personnel platform shall use a body belt/harness system with lanyard appropriately attached to the lower load block or overhaul ball, or to a structural member within the personnel platform capable of supporting a fall impact for employees using the anchorage. When working over water, the requirements of §1926.106 shall apply. [§1926.1501(g)(6)(vii)]

(viii) *No lifts shall be made* on another of the crane's or derrick's loadlines while personnel are suspended on a platform. [§1926.1501(g)(6)(viii)]

(7) *Traveling.* [§1926.1501(g)(7)]

(i) *Hoisting of employees while the crane is traveling* is prohibited, except for portal, tower and locomotive cranes, or where the employer demonstrates that there is no less hazardous way to perform the work. [§1926.1501(g)(7)(i)]

(ii) *Under any circumstances* where a crane would travel while hoisting personnel, the employer shall implement the following procedures to safeguard employees: [§1926.1501(g)(7)(ii)]

[A] Crane travel shall be restricted to a fixed track or runway; [§1926.1501(g)(7)(ii)[A]]

[B] Travel shall be limited to the load radius of the boom used during the lift; and [§1926.1501(g)(7)(ii)[B]]

[C] The boom must be parallel to the direction of travel. [§1926.1501(g)(7)(ii)[C]]

[D] A complete trial run shall be performed to test the route of travel before employees are allowed to occupy the platform. This trial run can be performed at the same time as the trial lift required by paragraph (g)(5)(i) of this section which tests the route of the lift. [§1926.1501(g)(7)(ii)[D]]

[E] If travel is done with a rubber tired-carrier, the condition and air pressure of the tires shall be checked. The chart capacity for lifts on rubber shall be used for application of the 50 percent reduction of rated capacity. Notwithstanding paragraph (g)(3)(i)[E] of this section, outriggers may be partially retracted as necessary for travel. [§1926.1501(g)(7)(ii)[E]]

(8) *Pre-lift meeting.* [§1926.1501(g)(8)]

(i) *A meeting attended by the crane or derrick operator,* signal person(s) (if necessary for the lift), employee(s) to be lifted, and the person responsible for the task to be performed shall be held to review the appropriate requirements of paragraph (g) of this section and the procedures to be followed. [§1926.1501(g)(8)(i)]

(ii) *This meeting shall be held* prior to the trial lift at each new work location, and shall be repeated for any employees newly assigned to the operation. [§1926.1501(g)(8)(ii)]

[44 FR 8577, Feb. 9, 1979; 44 FR 20940, Apr. 6, 1979, as amended at 52 FR 36382, Sept. 28, 1987; 53 FR 29139, Aug. 2, 1988; 54 FR 15406, Apr. 18, 1989; 54 FR 24334, June 7, 1989; 58 FR 35183, June 30, 1993; 59 FR 40730, Aug. 9, 1994; 61 FR 5510, Feb. 13, 1996]

1926 Subpart DD

Authority for 1926 Subpart DD

Authority:Section 3704 of the Contract Work Hours and Safety Standards Act (40 U.S.C. 3701); Sections 4, 6, and 8 of the Occupational Safety and Health Act of 1970 (29 U.S.C. 653, 655, 657); Secretary of Labor's Order Nos. 12–71 (36 FR 8754), 8–76 (41 FR 25059), or 9–83 (49 FR 35736), and 5–2007 (72 FR 31159).

General Duty Clause

The Williams-Steiger Occupational Safety and Health Act of 1970

An Act

To assure safe and healthful working conditions for working men and women; by authorizing enforcement of the standards developed under the Act; by assisting and encouraging the States in their efforts to assure safe and healthful working conditions; by providing for research, information, education, and training in the field of occupational safety and health; and for other purposes.

Be it enacted by the Senate and House of Representatives of the United States of America in Congress assembled, that this Act may be cited as the "Occupational Safety and Health Act of 1970".

Two important sections of this act are:

5. Duties

(a) Each employer

(1) *shall furnish to each of his employees* employment and a place of employment which are free from recognized hazards that are causing or are likely to cause death or serious physical harm to his employees;

(2) *shall comply with* occupational safety and health standards promulgated under this Act.

(b) Each employee shall comply with occupational safety and health standards and all rules, regulations, and orders issued pursuant to this Act which are applicable to his own actions and conduct.

What is OSHA's General Duty Clause?

Section 5(a)(1) of the **Williams-Steiger Occupational Safety and Health Act of 1970** has become known as **"The General Duty Clause"**. It is a **catch all** for citations if OSHA identifies unsafe conditions to which a regulation does not exist.

In practice, OSHA, court precedent, and the review commission have established that if the following elements are present, a **"general duty clause"** citation may be issued.

1. The employers failed to keep the workplace free of a hazard to which employees of that employer were exposed.
2. The hazard was recognized. (Examples might include: through your safety personnel, employees, organization, trade organization or industry customs.)
3. The hazard was causing or was likely to cause death or serious physical harm.
4. There was a feasible and useful method to correct the hazard.

11. Judicial Review

(c) (1) *No person shall discharge* or in any manner discriminate against any employee because such employee has filed any complaint or instituted or caused to be instituted any proceeding under or related to this Act or has testified or is about to testify in any such proceeding or because of the exercise by such employee on behalf of himself or others of any right afforded by this Act.

(2) *Any employee who believes that he has been discharged* or otherwise discriminated against by any person in violation of this subsection may, within thirty days after such violation occurs, file a complaint with the Secretary alleging such discrimination. Upon receipt of such complaint, the Secretary shall cause such investigation to be made as he deems appropriate. If upon such investigation, the Secretary determines that the provisions of this subsection have been violated, he shall bring an action in any appropriate United States district court against such person. In any such action the United States district courts shall have jurisdiction, for cause shown to restrain violations of paragraph (1) of this subsection and order all appropriate relief including rehiring or reinstatement of the employee to his former position with back pay.

(3) *Within 90 days of the receipt of a complaint* filed under this subsection the Secretary shall notify the complainant of his determination under paragraph 2 of this subsection.

OSHA's Citation Policy on Multi-Employer Worksite Inspections[1]

Employers must not create conditions that violate OSHA standards or make a workplace unsafe. On multi-employer worksites (in all industry sectors), more than one employer may be citable for a hazardous condition that violates an OSHA standard.

OSHA classifies employers into one or more of four categories — the creating, exposing, correcting, and controlling employers — to determine if a citation will be issued.

The Creating Employer: an employer who causes a hazardous condition that violates an OSHA standard. An employer who creates the hazard is citable even if the only employees exposed in the workplace are those who work for other employers.

The Exposing Employer: an employer whose own employees are exposed to the hazard.

If the exposing employer created the violation, he/she is citable for the violation as a creating employer.

If the violation was created by another employer, the exposing employer is citable if he/she
(1) knew of the hazardous condition or failed to exercise reasonable diligence to discover the condition, and
(2) failed to take steps to protect his/her employees.

If the exposing employer has the authority to correct the hazard, he/she must do so.

If he/she lacks the authority to correct the hazard, he/she is citable if he/she fails to do each of the following:
(1) ask the creating and/or controlling employer to correct the hazard
(2) inform his/her employees of the hazard, and
(3) take reasonable alternative protective measures.

Note: In some circumstances, the employer is citable for failing to remove his/her employees from the job to avoid the hazard.

The Correcting Employer: an employer who is responsible for correcting a hazard on the exposing employer's worksite, usually occurring while the correcting employer is installing and/or maintaining safety/health equipment. The correcting employer must exercise reasonable care in preventing and discovering violations and meet his/her obligation of correcting the hazard.

The Controlling Employer: an employer who has general supervisory authority over the worksite, including the power to correct safety and health violations or requiring others to correct them. A controlling employer must exercise reasonable care to prevent and detect violations on the site.

1. This information is found in OSHA Instruction CPL 2-0.124. See www.oshacfr.com for the full text of this document.

How to Fill Out the 300 Log

OMB Disclosure Statement: Public reporting burden for this collection of information is estimated to average 14 minutes per response, including time to review the instructions, search and gather the data needed, and complete and review the collection of information. Persons are not required to respond to the collection of information unless it displays a currently valid OMB control number. If you have any comments about these estimates or any other aspects of this data collection, contact: US Department of Labor, OSHA Office of Statistical Analysis, Room N-3644, 200 Constitution Avenue, NW, Washington, DC 20210. Do not send the completed forms to this office. See www.mancomm.com for updates or revisions.

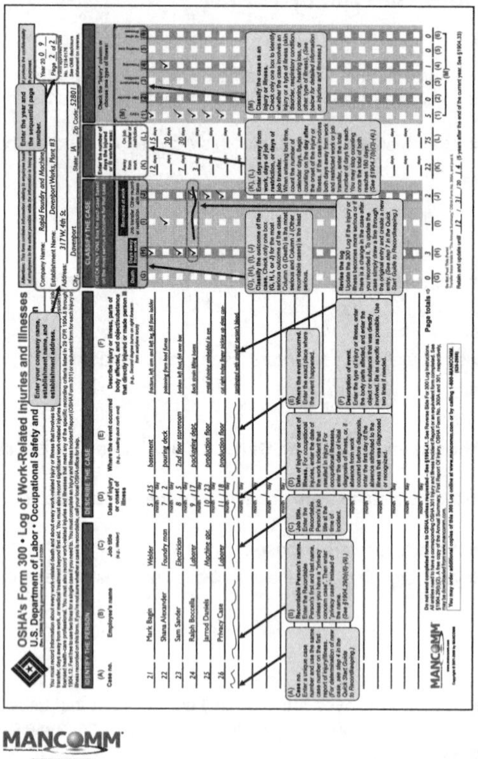

MANCOMM

www.mancomm.com

Key Code: BXA

U.S. Department of Labor OSHA Form No. 300 LOG

(Rev. 01/2009) Reflecting government revisions as of 01/2004

Step by Step

Quick Start Guide

1. Are You Required to Maintain a 300 Log?
Companies must keep records of all recordable occupational injuries and illnesses for each single business establishment that is expected to be in operation for one year or longer, unless the company-wide count of Recordable People (see Recordable People in Definitions section below) was 10 or fewer at all times during the last calendar year or your company is listed as a low hazard industry. *(See §1904.46 for definition of establishment. See §1904.1-.3 for information on partial exceptions. See Table §1904 Appendix A to Subpart B for the list of low hazard industries. See §1904.30(a) for multiple business establishments. See §1904.31 for definition of Recordable People.)*

2. Is this Person Recordable?
You record on the 300 Log the recordable injuries and illnesses of all Recordable People, *(see Recordable People in Definitions Section below). (See §1904.31 for information on Recordable People.)*

3. Is this Case Work-Related?
An injury or illness is considered work-related if an event or exposure in the workplace caused or contributed to the incident or significantly aggravated a preexisting condition. Injuries and illnesses that happen in the workplace are presumed to be work-related, unless there is an exception. *(See §1904.5(b)(2) for exceptions.)* The work environment includes not only the facility, but other areas where the company has sent at least one Recordable Person to work or be present. *(See §1904.5 for determination of work-relatedness. See §1904.5 (b)(1) for work environment. See §1904.5 (b)(4) for the definition of "significantly aggravated".)*

4. Is this a New Case?
A new case is when a Recordable Person suffers a new injury or illness. It is also a new case when a Recordable Person has fully recovered from the previous injury or illness, and a new event or exposure in the workplace causes the signs or symptoms to reappear. *See Step 7, if it is not a new case. (See §1904.6 for determination of new cases.)*

5. Is this Case Recordable?
A case is recordable on the 300 Log when it involves any of the following work-related incidents: death *(§1904.7(b)(2))*; days away from work *(§1904.7(b)(3))*; restricted work or transfer to another job *(§1904.7(b)(4))*; medical treatment, *(see Medical Treatment Exceptions in Definitions section below) (§1904.7(b)(5))*; loss of consciousness *(§1904.7(b)(6))*; significant diagnosis by a healthcare professional *(§1904.7(b)(7))*; needlestick or sharps injuries *(§1904.8)*; medical removal due to a specific OSHA Standard *(§1904.9)*; occupational hearing loss *(§1904.10)*; or tuberculosis *(§1904.11.)*

If you answered "yes" to all of the steps 1 through 5, record the case.

6. How to Record this Case:
Fill out a 301 Incident Report and enter the case on the OSHA 300 Log within seven (7) calendar days. A 301 Incident Report or equivalent form must be filled out for each entry on the 300 Log. Enter the days of away from work and the days of restricted work or job transfer on the 300 Log, do not count the first day of the incident. Restricted work activity is when either the company or a healthcare professional places a work restriction on a Recordable Person or keeps a Recordable Person from performing the same duties and working the full workday that was scheduled before the incident. *(See §1904.29 for information on forms. See §1904.29(b)(4) for equivalent forms. See §1904.7(b)(4) for restricted work.)*

7. How to Update this Case: (when required)
If there are changes to the case during a five year period, you must update the original 300 Log entry to show these changes. Also enter on the original 300 Log any newly discovered recordable injuries or illnesses. *(See §1904.33(b) for information on updating records.)*

8. How to Complete, Post, and Retain Records:
At the end of the calendar year, total each column on the 300 Log, and then use this information to fill out a Form 300A, Summary of Work-Related Injuries and Illnesses. The 300A must be signed and certified by the highest-ranking company official at that facility or his/her immediate supervisor. Post the 300A Summary in that facility from February 1 to April 30 of the following year. Keep all 301 Reports, 300 Logs, and 300A Summaries for 5 years. *(See §1904.32 for information on annual completion of 300 Log and 300A Summary. See §1904.32(b)(5)-(6) for information on posting the 300A Summary. See §1904.33(a) for information on retention.)*

References and Definitions

Fault *(§1904.0)*
Cases listed on the 300 Log are not necessarily eligible for worker's compensation; nor does it mean that the company or Recordable Person was at fault or that an OSHA standard was violated.

Location *(§1904.35(b)(2)(iii); 1904.40)*
You may prepare and maintain this 300 Log at another facility; however, you must be able to show the OSHA Inspector these records within four business hours. When a worker or former worker requests these records, you must give them the records, with some limitations, by the end of the next business day.

First Report of Injury/Illness *(§1904.31)*
You must complete a 301 Incident Report or equivalent form for each recordable case entered on the 300 Log. Many employers use an insurance form instead of the OSHA 301 Incident Report, or supplement an insurance form by adding any additional information required by OSHA.

Recordable People *(§1904.31)*
Recordable people are workers on your payroll and anyone else you supervise on a day-to-day basis who are not on your payroll.

Medical Treatment Exceptions
(are not considered recordable)
(§1904.7(b)(5), §1904.7(b)(5)(ii))
1. A visit to a doctor or health-care professional only for observation or counseling; or
2. Diagnostic procedures such as x-rays, blood tests, and taking prescription medications used only for diagnostic purposes; or
3. The following procedures:
   - Taking non-prescription medication at non-prescription strength;
   - Receiving a tetanus shot;
   - Cleaning, flushing, or soaking wounds on the surface of the skin; using wound coverings, like bandages, Band-Aids™, gauze pads, etc., or using SteriStrips™ or butterfly bandages;
   - Using hot or cold therapy; using any non-rigid means of support, such as elastic bandages, wraps, non-rigid back belts, etc.; using temporary immobilization devices while transporting an accident victim (splints, slings, neck collars, or back boards);
   - Drilling of a fingernail or toenail to relieve pressure, or draining fluids from blisters;
   - Using eye patches; removing foreign bodies from the eye using only irrigation or a cotton swab;
   - Removing splinters or foreign material from areas other than the eye by irrigation, tweezers, cotton swabs, or other simple means;
   - Using finger guards; using massages; or drinking fluids to relieve heat stress.

MANCOMM

Full-size versions available by calling toll-free 1-800-MANCOMM (1-800-626-2666), or you may order online at www.mancomm.com.

OSHA's Form 300 • Log of Work-Related Injuries and Illnesses

U.S. Department of Labor • Occupational Safety and Health Administration

(Rev. 01/2009) Reflecting government revisions as of 01/2004

Attention: *This form contains information relating to employee health and must be used in a manner that protects the confidentiality of employees to the extent possible while the information is being used for occupational safety and health purposes.*

You must record information about every work-related death and about every work-related injury or illness that involves loss of consciousness, restricted work activity or job transfer, days away from work, or medical treatment beyond first aid. You must also record significant work-related injuries and illnesses that are diagnosed by a physician or licensed health-care professional. You must also record work-related injuries and illnesses that meet any of the specific recording criteria listed in 29 CFR 1904.8 through 1904.12. Feel free to use two lines for a single case if you need to. You must complete an Injury and Illness Incident Report (OSHA Form 301) or equivalent form for each injury or illness recorded on this form. If you're not sure whether a case is recordable, call your local OSHA office for help.

Company Name: ____________ Year 20___

Establishment Name: ____________ Page ___ of ___

Address: ____________

City: ____________ State: ______ Zip Code: ______

Form approved OMB No. 1218-0176

See OMB disclosure statement on reverse.

IDENTIFY THE PERSON			DESCRIBE THE CASE			CLASSIFY THE CASE: CHECK ONLY ONE box for each case based on the most serious outcome for that case:				Enter the number of days the injured or ill worker was:		Check the "Injury" column or choose one type of illness: (M)					
(A) Case no.	(B) Employee's name	(C) Job title (e.g., Welder)	(D) Date of injury or onset of illness	(E) Where the event occurred (e.g., Loading dock north end)	(F) Describe injury or illness, parts of body affected, and object/substance that directly injured or made person ill (e.g., Second degree burns on right forearm from acetylene torch)	Death (G)	Days away from work (H)	Remained at work: Job transfer or restriction (I)	Remained at work: Other record-able cases (J)	Away from work (K)	On job transfer or restriction (L)	Injury (1)	Skin disorder (2)	Respiratory condition (3)	Poisoning (4)	Hearing loss (5)	All other illnesses (6)
			month / day							___ days	___ days						

Page totals ⇨ (G) (H) (I) (J) (K) (L) (1) (2) (3) (4) (5) (6) (M)

Do Not Post This Form! Transfer these totals to "The Annual Summary," OSHA Form No. 300A, and post it.

Retain and update until ____/____/20____ (5 years after the end of the current year. See §1904.33)

Do not send completed forms to OSHA unless requested – See §1904.41. See Reverse Side For 300 Log Instructions. All entries need to have a corresponding OSHA 301 Injury and Illness Incident Report or an equivalent form completed. See §1904.29(b)(2). A free copy of the Annual Summary, First Report Of Injury, OSHA Form No. 300A and 301, respectively, may be downloaded from www.mancomm.com.

You may order additional copies of this 300 Log online at www.mancomm.com or by calling 1-800-MANCOMM. (626-2666)

MANCOMM

www.mancomm.com

Full-size versions available by calling toll-free 1-800-MANCOMM (1-800-626-2666), or you may order online at www.mancomm.com.

OSHA's Form 300A (Rev. 01/2004)

Summary of Work-Related Injuries and Illnesses

Year 20__ __

U.S. Department of Labor
Occupational Safety and Health Administration

Form approved OMB no. 1218-0176

All establishments covered by Part 1904 must complete this Summary page, even if no work-related injuries or illnesses occurred during the year. Remember to review the Log to verify that the entries are complete and accurate before completing this summary.

Using the Log, count the individual entries you made for each category. Then write the totals below, making sure you've added the entries from every page of the Log. If you had no cases, write "0."

Employees, former employees, and their representatives have the right to review the OSHA Form 300 in its entirety. They also have limited access to the OSHA Form 301 or its equivalent. See 29 CFR Part 1904.35, in OSHA's recordkeeping rule, for further details on the access provisions for these forms.

Number of Cases

Total number of deaths	Total number of cases with days away from work	Total number of cases with job transfer or restriction	Total number of other recordable cases
______	______	______	______
(G)	(H)	(I)	(J)

Number of Days

Total number of days away from work	Total number of days of job transfer or restriction
______	______
(K)	(L)

Injury and Illness Types

Total number of . . .
(M)

(1) Injuries	______	(4) Poisonings	______
(2) Skin disorders	______	(5) Hearing Losses	______
(3) Respiratory conditions	______	(6) All other illnesses	______

Post this Summary page from February 1 to April 30 of the year following the year covered by the form.

Public reporting burden for this collection of information is estimated to average 50 minutes per response, including time to review the instructions, search and gather the data needed, and complete and review the collection of information. Persons are not required to respond to the collection of information unless it displays a currently valid OMB control number. If you have any comments about these estimates or any other aspects of this data collection, contact: US Department of Labor, OSHA Office of Statistical Analysis, Room N-3644, 200 Constitution Avenue, NW, Washington, DC 20210. Do not send the completed forms to this office.

Establishment information

Your establishment name ______

Street ______

City ______ State ______ ZIP ______

Industry description *(e.g., Manufacture of motor truck trailers)* ______

Standard Industrial Classification (SIC), if known *(e.g., 3715)* __ __ __ __

OR

North American Industrial Classification (NAICS), if known *(e.g., 336212)* __ __ __ __ __ __

Employment information
(If you don't have these figures, see the Worksheet on the back of this page to estimate.)

Annual average number of employees ______

Total hours worked by all employees last year ______

Sign here

Knowingly falsifying this document may result in a fine.

I certify that I have examined this document and that to the best of my knowledge the entries are true, accurate, and complete.

______ Company executive ______ Title

( ) - Phone / / Date

A free copy of the Annual Summary, First Report of Injury — OSHA Form No. 300A and 301, respectively, may be downloaded from www.mancomm.com.

OSHA's Form 301

Injury and Illness Incident Report

Attention: This form contains information relating to employee health and must be used in a manner that protects the confidentiality of employees to the extent possible while the information is being used for occupational safety and health purposes.

U.S. Department of Labor
Occupational Safety and Health Administration

Form approved OMB no. 1218-0176

This *Injury and Illness Incident Report* is one of the first forms you must fill out when a recordable work-related injury or illness has occurred. Together with the *Log of Work-Related Injuries and Illnesses* and the accompanying *Summary*, these forms help the employer and OSHA develop a picture of the extent and severity of work-related incidents.

Within 7 calendar days after you receive information that a recordable work-related injury or illness has occurred, you must fill out this form or an equivalent. Some state workers' compensation, insurance, or other reports may be acceptable substitutes. To be considered an equivalent form, any substitute must contain all the information asked for on this form.

According to Public Law 91-596 and 29 CFR 1904, OSHA's recordkeeping rule, you must keep this form on file for 5 years following the year to which it pertains.

If you need additional copies of this form, you may photocopy and use as many as you need.

Completed by ______

Title ______

Phone ( ) - ______ Date __/__/__

Information about the employee

1) Full name ______

2) Street ______

City ______ State ______ ZIP ______

3) Date of birth __/__/__

4) Date hired __/__/__

5) ☐ Male
☐ Female

Information about the physician or other health care professional

6) Name of physician or other health care professional ______

7) If treatment was given away from the worksite, where was it given?

Facility ______

Street ______

City ______ State ______ ZIP ______

8) Was employee treated in an emergency room?
☐ Yes
☐ No

9) Was employee hospitalized overnight as an in-patient?
☐ Yes
☐ No

Information about the case

10) Case number from the Log ______ *(Transfer the case number from the Log after you record the case.)*

11) Date of injury or illness __/__/__

12) Time employee began work ______ AM / PM

13) Time of event ______ AM / PM ☐ Check if time cannot be determined

14) ***What was the employee doing just before the incident occurred?*** Describe the activity, as well as the tools, equipment, or material the employee was using. Be specific. *Examples:* "climbing a ladder while carrying roofing materials"; "spraying chlorine from hand sprayer"; "daily computer key-entry."

15) ***What happened?*** Tell us how the injury occurred. *Examples:* "When ladder slipped on wet floor, worker fell 20 feet"; "Worker was sprayed with chlorine when gasket broke during replacement"; "Worker developed soreness in wrist over time."

16) ***What was the injury or illness?*** Tell us the part of the body that was affected and how it was affected; be more specific than "hurt," "pain," or "sore." *Examples:* "strained back"; "chemical burn, hand"; "carpal tunnel syndrome."

17) ***What object or substance directly harmed the employee?*** *Examples:* "concrete floor"; "chlorine"; "radial arm saw." *If this question does not apply to the incident, leave it blank.*

18) ***If the employee died, when did death occur?*** Date of death __/__/__

Public reporting burden for this collection of information is estimated to average 22 minutes per response, including time for reviewing instructions, searching existing data sources, gathering and maintaining the data needed, and completing and reviewing the collection of information. Persons are not required to respond to the collection of information unless it displays a current valid OMB control number. If you have any comments about this estimate or any other aspects of this data collection, including suggestions for reducing this burden, contact: US Department of Labor, OSHA Office of Statistical Analysis, Room N-3644, 200 Constitution Avenue, NW, Washington, DC 20210. Do not send the completed forms to this office.

A free copy of the Annual Summary, First Report of Injury — OSHA Form No. 300A and 301, respectively, may be downloaded from www.mancomm.com.

Instructions For 300A

At the end of the year, OSHA requires you to enter the average number of employees and the total hours worked by your employees on the summary. If you don't have these figures, you can use the information on this page to estimate the numbers you will need to enter on the Summary page at the end of the year.

How to figure the total hours worked by all employees:

Include hours worked by salaried, hourly, part-time, and seasonal workers, as well as hours worked by other workers subject to day to day supervision by your establishment (e.g., temporary help services workers).

Do not include vacation, sick leave, holidays, or any other non-work time, even if employees were paid for it. If your establishment keeps records of only the hours paid or if you have employees who are not paid by the hour, please estimate the hours that the employees actually worked.

If this number isn't available, you can use this optional worksheet to estimate it.

Optional Worksheet

	______	**Find** the number of full-time employees in your establishment for the year.
X	______	**Multiply** by the number of work hours for a full-time employee in a year.
	______	This is the number of full-time hours worked.
+	______	**Add** the number of any overtime hours as well as the hours worked by other employees. (part-time, temporary, seasonal)
	______	**Round** the answer to the next highest whole number. Write the rounded number in the blank marked *Total hours worked by all employees last year.*

How to figure the average number of employees who worked for your establishment during the year:

❶ **Add** the total number of employees your establishment paid in all pay periods during the year. Include all employees: full-time, part-time, temporary, seasonal, salaried, and hourly.

The number of employees paid in all pay periods = ❶ ______

❷ **Count** the number of pay periods your establishment had during the year. Be sure to include any pay periods when you had no employees.

The number of pay periods during the year = ❷ ______

❸ **Divide** the number of employees by the number of pay periods.

❸ ❶/❷ = ______

❹ **Round the answer** to the next highest whole number. Write the rounded number in the blank marked *Annual average number of employees.*

The number rounded = ❹ ______

For example, Acme Construction figured its average employment this way:

For pay period...	Acme paid this number of employees...
1	10
2	0
3	15
4	30
5	40
▼	▼
24	20
25	15
26	+10
	830

Number of employees paid = 830 ❶

Number of pay periods = 26 ❷

$\frac{830}{26}$ = 31.92 ❸

31.92 rounds to 32 ❹

32 is the annual average number of employees

MANCOMM
Mangan Communications, Inc.
Changing The Complex Into Compliance®
PHONE: 1-800-MANCOMM (626-2666)
FAX: 1-563-323-0804 WEB: www.mancomm.com

Calculating Injury and Illness Incidence Rates

What is an incidence rate?

An incidence rate is the number of recordable injuries and illnesses occurring among a given number of full-time workers (usually 100 full-time workers) over a given period of time (usually one year). To evaluate your firm's injury and illness experience over time or to compare your firm's experience with that of your industry as a whole, you need to compute your incidence rate. Because a specific number of workers and a specific period of time are involved, these rates can help you identify problems in your workplace and/or progress you may have made in preventing work-related injuries and illnesses.

How do you calculate an incidence rate?

You can compute an occupational injury and illness incidence rate for all recordable cases or for cases that involved days away from work for your firm quickly and easily. The formula requires that you follow instructions in paragraph (a) below for the total recordable cases or those in paragraph (b) for cases that involved days away from work, *and* for both rates the instructions in paragraph (c).

(a) *To find out the total number of recordable injuries and illnesses that occurred during the year,* count the number of line entries on your OSHA Form 300, or refer to the OSHA Form 300A and sum the entries for columns (G), (H), (I), and (J).

(b) *To find out the number of injuries and illnesses that involved days away from work,* count the number of line entries on your OSHA Form 300 that received a check mark in column (H), or refer to the entry for column (H) on the OSHA Form 300A.

(c) *The number of hours all employees actually worked during the year.* Refer to OSHA Form 300A and optional worksheet to calculate this number.

You can compute the incidence rate for all recordable cases of injuries and illnesses using the following formula:

Total number of injuries and illnesses X 200,000 ÷ Number of hours worked by all employees = Total recordable case rate

(The 200,000 figure in the formula represents the number of hours 100 employees working 40 hours per week, 50 weeks per year would work, and provides the standard base for calculating incidence rates.)

You can compute the incidence rate for recordable cases involving days away from work, days of restricted work activity or job transfer (DART) using the following formula:

(Number of entries in column H + Number of entries in column I) X 200,000 ÷ Number of hours worked by all employees = DART incidence rate

You can use the same formula to calculate incidence rates for other variables such as cases involving restricted work activity (column (I) on Form 300A), cases involving skin disorders (column (M-2) on Form 300A), etc. Just substitute the appropriate total for these cases, from Form 300A, into the formula in place of the total number of injuries and illnesses.

What can I compare my incidence rate to?

The Bureau of Labor Statistics (BLS) conducts a survey of occupational injuries and illnesses each year and publishes incidence rate data by various classifications (e.g., by industry, by employer size, etc.). You can obtain these published data at www.bls.gov/iif or by calling a BLS Regional Office.

Worksheet

Total number of injuries and illnesses				Number of hours worked by all employees		Total recordable case rate
☐	X	200,000	÷	☐	=	☐

Number of entries in Column H + Column I				Number of hours worked by all employees		DART incidence rate
☐	X	200,000	÷	☐	=	☐

SIC Division Structure[1]

Division A: Agriculture, Forestry, and Fishing

Major Group 01: Agricultural Production - Crops
Industry Group 011: Cash Grains
Industry Group 013: Field Crops, Except Cash Grains
Industry Group 016: Vegetables and Melons
Industry Group 017: Fruits and Tree Nuts
Industry Group 018: Horticultural Specialties
Industry Group 019: General Farms, Primarily Crop

Major Group 02: Agriculture Production Livestock and Animal Specialties
Industry Group 021: Livestock, Except Dairy and Poultry
Industry Group 024: Dairy Farms
Industry Group 025: Poultry and Eggs
Industry Group 027: Animal Specialties
Industry Group 029: General Farms, Primarily Livestock and Animal

Major Group 07: Agricultural Services
Industry Group 071: Soil Preparation Services
Industry Group 072: Crop Services
Industry Group 074: Veterinary Services
Industry Group 075: Animal Services, Except Veterinary
Industry Group 076: Farm Labor and Management Services
Industry Group 078: Landscape and Horticultural Services

Major Group 08: Forestry
Industry Group 081: Timber Tracts
Industry Group 083: Forest Nurseries and Gathering of Forest
Industry Group 085: Forestry Services

Major Group 09: Fishing, Hunting, and Trapping
Industry Group 091: Commercial Fishing
Industry Group 092: Fish Hatcheries and Preserves
Industry Group 097: Hunting and Trapping, and Game Propagation

Division B: Mining

Major Group 10: Metal Mining
Industry Group 101: Iron Ores
Industry Group 102: Copper Ores
Industry Group 103: Lead and Zinc Ores
Industry Group 104: Gold and Silver Ores
Industry Group 106: Ferroalloy Ores, Except Vanadium
Industry Group 108: Metal Mining Services
Industry Group 109: Miscellaneous Metal Ores

Major Group 12: Coal Mining
Industry Group 122: Bituminous Coal and Lignite Mining
Industry Group 123: Anthracite Mining
Industry Group 124: Coal Mining Services

Major Group 13: Oil and Gas Extraction
Industry Group 131: Crude Petroleum and Natural Gas
Industry Group 132: Natural Gas Liquids
Industry Group 138: Oil and Gas Field Services

Major Group 14: Mining and Quarrying of Nonmetallic Minerals, Except Fuels
Industry Group 141: Dimension Stone
Industry Group 142: Crushed and Broken Stone, Including Riprap
Industry Group 144: Sand and Gravel
Industry Group 145: Clay, Ceramic, and Refractory Minerals
Industry Group 147: Chemical and Fertilizer Mineral Mining
Industry Group 148: Nonmetallic Minerals Services, Except Fuels
Industry Group 149: Miscellaneous Nonmetallic Minerals, Except Fuels

Division C: Construction

Major Group 15: Building Construction General Contractors and Operative Builders
Industry Group 152: General Building Contractors-residential
Industry Group 153: Operative Builders
Industry Group 154: General Building Contractors-nonresidential

Major Group 16: Heavy Construction Other than Building Construction Contractors
Industry Group 161: Highway and Street Construction, Except Elevated Highways
Industry Group 162: Heavy Construction, Except Highway and Street

Major Group 17: Construction Special Trade Contractors
Industry Group 171: Plumbing, Heating and Air-conditioning
Industry Group 172: Painting and Paper Hanging
Industry Group 173: Electrical Work
Industry Group 174: Masonry, Stonework, Tile Setting, and Plastering
Industry Group 175: Carpentry and Floor Work
Industry Group 176: Roofing, Siding, and Sheet Metal Work
Industry Group 177: Concrete Work
Industry Group 178: Water Well Drilling
Industry Group 179: Miscellaneous Special Trade Contractors

Division D: Manufacturing

Major Group 20: Food and Kindred Products
Industry Group 201: Meat Products
Industry Group 202: Dairy Products
Industry Group 203: Canned, Frozen, and Preserved Fruits, Vegetables, and Food Specialties
Industry Group 204: Grain Mill Products
Industry Group 205: Bakery Products
Industry Group 206: Sugar and Confectionery Products
Industry Group 207: Fats and Oils
Industry Group 208: Beverages
Industry Group 209: Miscellaneous Food Preparations and Kindred

Major Group 21: Tobacco Products
Industry Group 211: Cigarettes
Industry Group 212: Cigars
Industry Group 213: Chewing and Smoking Tobacco and Snuff
Industry Group 214: Tobacco Stemming and Redrying

Major Group 22: Textile Mill Products
Industry Group 221: Broadwoven Fabric Mills, Cotton
Industry Group 222: Broadwoven Fabric Mills, Manmade Fiber and Silk
Industry Group 223: Broadwoven Fabric Mills, Wool (Including Dyeing and Finishing)
Industry Group 224: Narrow Fabric and Other Smallwares Mills
Industry Group 225: Knitting Mills
Industry Group 226: Dyeing and Finishing Textiles, Except Wool Fabrics
Industry Group 227: Carpets and Rugs
Industry Group 228: Yarn and Thread Mills
Industry Group 229: Miscellaneous Textile Goods

Major Group 23: Apparel and Other Finished Products Made from Fabrics and Similar Materials
Industry Group 231: Men's and Boys' Suits, Coats, and Overcoats
Industry Group 232: Men's and Boys' Furnishings, Work Clothing
Industry Group 233: Women's, Misses', and Juniors' Outerwear, and Allied Garments
Industry Group 234: Women's, Misses', Children's, and Infants' Underwear and Nightwear
Industry Group 235: Hats, Caps, and Millinery
Industry Group 236: Girls', Children's, and Infants' Outerwear
Industry Group 237: Fur Goods
Industry Group 238: Miscellaneous Apparel and Accessories
Industry Group 239: Miscellaneous Fabricated Textile Products

Major Group 24: Lumber and Wood Products, Except Furniture
Industry Group 241: Logging
Industry Group 242: Sawmills and Planing Mills
Industry Group 243: Millwork, Veneer, Plywood, and Structural Wood
Industry Group 244: Wood Containers
Industry Group 245: Wood Buildings and Mobile Homes
Industry Group 249: Miscellaneous Wood Products

Major Group 25: Furniture and Fixtures
Industry Group 251: Household Furniture
Industry Group 252: Office Furniture
Industry Group 253: Public Building and Related Furniture
Industry Group 254: Partitions, Shelving, Lockers, and Office and Store Fixtures
Industry Group 259: Miscellaneous Furniture and Fixtures

Major Group 26: Paper and Allied Products
Industry Group 261: Pulp Mills
Industry Group 262: Paper Mills
Industry Group 263: Paperboard Mills
Industry Group 265: Paperboard Containers and Boxes
Industry Group 267: Converted Paper and Paperboard Products

Major Group 27: Printing, Publishing, and Allied Industries
Industry Group 271: Newspapers: Publishing, Or Publishing and Printing
Industry Group 272: Periodicals: Publishing, Or Publishing and Printing
Industry Group 273: Books
Industry Group 274: Miscellaneous Publishing
Industry Group 275: Commercial Printing
Industry Group 276: Manifold Business Forms
Industry Group 277: Greeting Cards
Industry Group 278: Blankbooks, Looseleaf Binders, and Bookbinding
Industry Group 279: Service Industries for the Printing Trade

Major Group 28: Chemicals and Allied Products
Industry Group 281: Industrial Inorganic Chemicals
Industry Group 282: Plastics Materials and Synthetic Resins, Synthetic
Industry Group 283: Drugs
Industry Group 284: Soap, Detergents, and Cleaning Preparations
Industry Group 285: Paints, Varnishes, Lacquers, Enamels, and Allied Products
Industry Group 286: Industrial Organic Chemicals
Industry Group 287: Agricultural Chemicals
Industry Group 289: Miscellaneous Chemical Products

Major Group 29: Petroleum Refining and Related Industries
Industry Group 291: Petroleum Refining
Industry Group 295: Asphalt Paving and Roofing Materials
Industry Group 299: Miscellaneous Products of Petroleum and Coal

Major Group 30: Rubber and Miscellaneous Plastics Products
Industry Group 301: Tires and Inner Tubes
Industry Group 302: Rubber and Plastics Footwear
Industry Group 305: Gaskets, Packing, and Sealing Devices and Rubber
Industry Group 306: Fabricated Rubber Products, Not Elsewhere Classified
Industry Group 308: Miscellaneous Plastics Products

Major Group 31: Leather and Leather Products
Industry Group 311: Leather Tanning and Finishing
Industry Group 313: Boot and Shoe Cut Stock and Findings
Industry Group 314: Footwear, Except Rubber
Industry Group 315: Leather Gloves and Mittens
Industry Group 316: Luggage
Industry Group 317: Handbags and Other Personal Leather Goods
Industry Group 319: Leather Goods, Not Elsewhere Classified

Major Group 32: Stone, Clay, Glass, and Concrete Products
Industry Group 321: Flat Glass
Industry Group 322: Glass and Glassware, Pressed Or Blown
Industry Group 323: Glass Products, Made of Purchased Glass
Industry Group 324: Cement, Hydraulic
Industry Group 325: Structural Clay Products
Industry Group 326: Pottery and Related Products
Industry Group 327: Concrete, Gypsum, and Plaster Products
Industry Group 328: Cut Stone and Stone Products
Industry Group 329: Abrasive, Asbestos, and Miscellaneous

Major Group 33: Primary Metal Industries
Industry Group 331: Steel Works, Blast Furnaces, and Rolling and Finishing Mills
Industry Group 332: Iron and Steel Foundries
Industry Group 333: Primary Smelting and Refining of Nonferrous
Industry Group 334: Secondary Smelting and Refining of Nonferrous
Industry Group 335: Rolling, Drawing, and Extruding of Nonferrous
Industry Group 336: Nonferrous Foundries (Castings)
Industry Group 339: Miscellaneous Primary Metal Products

Major Group 34: Fabricated Metal Products, Except Machinery and Transportation Equipment
Industry Group 341: Metal Cans and Shipping Containers
Industry Group 342: Cutlery, Handtools, and General Hardware
Industry Group 343: Heating Equipment, Except Electric and Warm Air
Industry Group 344: Fabricated Structural Metal Products
Industry Group 345: Screw Machine Products, and Bolts, Nuts, Screws, Rivets, and Washers
Industry Group 346: Metal Forgings and Stampings
Industry Group 347: Coating, Engraving, and Allied Services
Industry Group 348: Ordnance and Accessories, Except Vehicles and Guided Missles
Industry Group 349: Miscellaneous Fabricated Metal Products

Major Group 35: Industrial and Commercial Machinery and Computer Equipment
Industry Group 351: Engines and Turbines
Industry Group 352: Farm and Garden Machinery and Equipment
Industry Group 353: Construction, Mining, and Materials Handling
Industry Group 354: Metalworking Machinery and Equipment
Industry Group 355: Special Industry Machinery, Except Metalworking
Industry Group 356: General Industrial Machinery and Equipment
Industry Group 357: Computer and Office Equipment
Industry Group 358: Refrigeration and Service Industry Machinery
Industry Group 359: Miscellaneous Industrial and Commercial

Major Group 36: Electronic and Other Electrical Equipment and Components, Except Computer Equipment
Industry Group 361: Electric Transmission and Distribution Equipment
Industry Group 362: Electrical Industrial Apparatus
Industry Group 363: Household Appliances
Industry Group 364: Electric Lighting and Wiring Equipment
Industry Group 365: Household Audio and Video Equipment, and Audio
Industry Group 366: Communications Equipment
Industry Group 367: Electronic Components and Accessories
Industry Group 369: Miscellaneous Electrical Machinery, Equipment and Supplies

1. *Standard Industrial Classification Manual.* Springfield, Va.: Government Printing Office, 1987, pp. 425-443.

Major Group 37: Transportation Equipment
Industry Group 371: Motor Vehicles and Motor Vehicle Equipment
Industry Group 372: Aircraft and Parts
Industry Group 373: Ship and Boat Building and Repairing
Industry Group 374: Railroad Equipment
Industry Group 375: Motorcycles, Bicycles, and Parts
Industry Group 376: Guided Missiles and Space Vehicles and Parts
Industry Group 379: Miscellaneous Transportation Equipment

Major Group 38: Measuring, Analyzing, and Controlling Instruments; Photographic, Medical and Optical Goods; Watches and Clocks
Industry Group 381: Search, Detection, Navigation, Guidance
Industry Group 382: Laboratory Apparatus and Analytical, Optical
Industry Group 384: Surgical, Medical, and Dental Instruments and Supplies
Industry Group 385: Ophthalmic Goods
Industry Group 386: Photographic Equipment and Supplies
Industry Group 387: Watches, Clocks, Clockwork Operated Devices and Parts

Major Group 39: Miscellaneous Manufacturing Industries
Industry Group 391: Jewelry, Silverware, and Plated Ware
Industry Group 393: Musical Instruments
Industry Group 394: Dolls, Toys, Games and Sporting and Athletic
Industry Group 395: Pens, Pencils, and Other Artists Materials
Industry Group 396: Costume Jewelry, Costume Novelties, Buttons
Industry Group 399: Miscellaneous Manufacturing Industries

Division E: Transportation, Communications, Electric, Gas, and Sanitary Services

Major Group 40: Railroad Transportation
Industry Group 401: Railroads

Major Group 41: Local and Suburban Transit and Interurban Highway Passenger Transportation
Industry Group 411: Local and Suburban Passenger Transportation
Industry Group 412: Taxicabs
Industry Group 413: Intercity and Rural Bus Transportation
Industry Group 414: Bus Charter Service
Industry Group 415: School Buses
Industry Group 417: Terminal and Service Facilities For Motor Vehicle

Major Group 42: Motor Freight Transportation and Warehousing
Industry Group 421: Trucking and Courier Services, Except Air
Industry Group 422: Public Warehousing and Storage
Industry Group 423: Terminal and Joint Terminal Maintenance

Major Group 43: United States Postal Service
Industry Group 431: United States Postal Service

Major Group 44: Water Transportation
Industry Group 441: Deep Sea Foreign Transportation of Freight
Industry Group 442: Deep Sea Domestic Transportation of Freight
Industry Group 443: Freight Transportation On The Great Lakes-st.
Industry Group 444: Water Transportation of Freight, Not Elsewhere Classified
Industry Group 448: Water Transportation of Passengers
Industry Group 449: Services Incidental To Water Transportation

Major Group 45: Transportation By Air
Industry Group 451: Air Transportation, Scheduled, and Air Courier
Industry Group 452: Air Transportation, Nonscheduled
Industry Group 458: Airports, Flying Fields, and Airport Terminal

Major Group 46: Pipelines, Except Natural Gas
Industry Group 461: Pipelines, Except Natural Gas

Major Group 47: Transportation Services
Industry Group 472: Arrangement of Passenger Transportation
Industry Group 473: Arrangement of Transportation of Freight and Cargo
Industry Group 474: Rental of Railroad Cars
Industry Group 478: Miscellaneous Services Incidental to Transportation Services

Major Group 48: Communications
Industry Group 481: Telephone Communications
Industry Group 482: Telegraph and Other Message Communications
Industry Group 483: Radio and Television Broadcasting Stations
Industry Group 484: Cable and Other Pay Television Services
Industry Group 489: Communications Services, Not Elsewhere Classified

Major Group 49: Electric, Gas, and Sanitary Services
Industry Group 491: Electric Services
Industry Group 492: Gas Production and Distribution
Industry Group 493: Combination Electric and Gas, and Other Utility
Industry Group 494: Water Supply
Industry Group 495: Sanitary Services
Industry Group 496: Steam and Air-conditioning Supply
Industry Group 497: Irrigation Systems

Division F: Wholesale Trade

Major Group 50: Wholesale Trade-durable Goods
Industry Group 501: Motor Vehicles and Motor Vehicle Parts and Supplies
Industry Group 502: Furniture and Home Furnishings
Industry Group 503: Lumber and Other Construction Materials
Industry Group 504: Professional and Commercial Equipment and Supplies
Industry Group 505: Metals and Minerals, Except Petroleum
Industry Group 506: Electrical Goods
Industry Group 507: Hardware, and Plumbing and Heating Equipment
Industry Group 508: Machinery, Equipment, and Supplies
Industry Group 509: Miscellaneous Durable Goods

Major Group 51: Wholesale Trade-non-durable Goods
Industry Group 511: Paper and Paper Products
Industry Group 512: Drugs, Drug Proprietaries, and Druggists' Sundries
Industry Group 513: Apparel, Piece Goods, and Notions
Industry Group 514: Groceries and Related Products
Industry Group 515: Farm-product Raw Materials
Industry Group 516: Chemicals and Allied Products
Industry Group 517: Petroleum and Petroleum Products
Industry Group 518: Beer, Wine, and Distilled Alcoholic Beverages
Industry Group 519: Miscellaneous Non-durable Goods

Division G: Retail Trade

Major Group 52: Building Materials, Hardware, Garden Supply, and Mobile Home Dealers
Industry Group 521: Lumber and Other Building Materials Dealers
Industry Group 523: Paint, Glass and Wallpaper Stores
Industry Group 525: Hardware Stores
Industry Group 526: Retail Nurseries, Lawn and Garden Supply Stores
Industry Group 527: Mobile Home Dealers

Major Group 53: General Merchandise Stores
Industry Group 531: Department Stores
Industry Group 533: Variety Stores
Industry Group 539: Miscellaneous General Merchandise Stores

Major Group 54: Food Stores
Industry Group 541: Grocery Stores
Industry Group 542: Meat and Fish (Seafood) Markets, Including Freezer Provisions
Industry Group 543: Fruit and Vegetable Markets
Industry Group 544: Candy, Nut, and Confectionery Stores
Industry Group 545: Dairy Products Stores
Industry Group 546: Retail Bakeries
Industry Group 549: Miscellaneous Food Stores

Major Group 55: Automotive Dealers and Gasoline Service Stations
Industry Group 551: Motor Vehicle Dealers (New and Used)
Industry Group 552: Motor Vehicle Dealers (Used Only)
Industry Group 553: Auto and Home Supply Stores
Industry Group 554: Gasoline Service Stations
Industry Group 555: Boat Dealers
Industry Group 556: Recreational Vehicle Dealers
Industry Group 557: Motorcycle Dealers
Industry Group 559: Automotive Dealers, Not Elsewhere Classified

Major Group 56: Apparel and Accessory Stores
Industry Group 561: Men's and Boys' Clothing and Accessory Stores
Industry Group 562: Women's Clothing Stores
Industry Group 563: Women's Accessory and Specialty Stores
Industry Group 564: Children's and Infants' Wear Stores
Industry Group 565: Family Clothing Stores
Industry Group 566: Shoe Stores
Industry Group 569: Miscellaneous Apparel and Accessory Stores

Major Group 57: Home Furniture, Furnishings, and Equipment Stores
Industry Group 571: Home Furniture and Furnishings Stores
Industry Group 572: Household Appliance Stores
Industry Group 573: Radio, Television, Consumer Electronics, and Music Stores

Major Group 58: Eating and Drinking Places
Industry Group 581: Eating and Drinking Places

Major Group 59: Miscellaneous Retail
Industry Group 591: Drug Stores and Proprietary Stores
Industry Group 592: Liquor Stores
Industry Group 593: Used Merchandise Stores
Industry Group 594: Miscellaneous Shopping Goods Stores
Industry Group 596: Nonstore Retailers
Industry Group 598: Fuel Dealers
Industry Group 599: Retail Stores, Not Elsewhere Classified

Division H: Finance, Insurance, and Real Estate

Major Group 60: Depository Institutions
Industry Group 601: Central Reserve Depository Institutions
Industry Group 602: Commercial Banks
Industry Group 603: Savings Institutions
Industry Group 606: Credit Unions
Industry Group 608: Foreign Banking and Branches and Agencies
Industry Group 609: Functions Related To Depository Banking

Major Group 61: Non-depository Credit Institutions
Industry Group 611: Federal and Federally-sponsored Credit Agencies
Industry Group 614: Personal Credit Institutions
Industry Group 615: Business Credit Institutions
Industry Group 616: Mortgage Bankers and Brokers

Major Group 62: Security and Commodity Brokers, Dealers, Exchanges, and Services
Industry Group 621: Security Brokers, Dealers, and Flotation
Industry Group 622: Commodity Contracts Brokers and Dealers
Industry Group 623: Security and Commodity Exchanges
Industry Group 628: Services Allied With The Exchange of Securities

Major Group 63: Insurance Carriers
Industry Group 631: Life Insurance
Industry Group 632: Accident and Health Insurance and Medical
Industry Group 633: Fire, Marine, and Casualty Insurance
Industry Group 635: Surety Insurance
Industry Group 636: Title Insurance
Industry Group 637: Pension, Health, and Welfare Funds
Industry Group 639: Insurance Carriers, Not Elsewhere Classified

Major Group 64: Insurance Agents, Brokers, and Service
Industry Group 641: Insurance Agents, Brokers, and Service

Major Group 65: Real Estate
Industry Group 651: Real Estate Operators (except Developers) and Lessors
Industry Group 653: Real Estate Agents and Managers
Industry Group 654: Title Abstract Offices
Industry Group 655: Land Subdividers and Developers

Major Group 67: Holding and Other Investment offices
Industry Group 671: Holding Offices
Industry Group 672: Investment Offices
Industry Group 673: Trusts
Industry Group 679: Miscellaneous Investing

Division I: Services

Major Group 70: Hotels, Rooming Houses, Camps, and Other Lodging Places
Industry Group 701: Hotels and Motels
Industry Group 702: Rooming and Boarding Houses
Industry Group 703: Camps and Recreational Vehicle Parks
Industry Group 704: Organization Hotels and Lodging Houses, on Membership Basis

Major Group 72: Personal Services
Industry Group 721: Laundry, Cleaning, and Garment Services
Industry Group 722: Photographic Studios, Portrait
Industry Group 723: Beauty Shops
Industry Group 724: Barber Shops
Industry Group 725: Shoe Repair Shops and Shoeshine Parlors
Industry Group 726: Funeral Service and Crematories
Industry Group 729: Miscellaneous Personal Services

Major Group 73: Business Services
Industry Group 731: Advertising
Industry Group 732: Consumer Credit Reporting Agencies, Mercantile
Industry Group 733: Mailing, Reproduction, Commercial Art and Graphic Design
Industry Group 734: Services To Dwellings and Other Buildings
Industry Group 735: Miscellaneous Equipment Rental and Leasing
Industry Group 736: Personnel Supply Services
Industry Group 737: Computer Programming, Data Processing
Industry Group 738: Miscellaneous Business Services

Major Group 75: Automotive Repair, Services, and Parking
Industry Group 751: Automotive Rental and Leasing, Without Drivers
Industry Group 752: Automobile Parking
Industry Group 753: Automotive Repair Shops
Industry Group 754: Automotive Services, Except Repair

Major Group 76: Miscellaneous Repair Services
Industry Group 762: Electrical Repair Shops
Industry Group 763: Watch, Clock, and Jewelry Repair
Industry Group 764: Reupholstery and Furniture Repair
Industry Group 769: Miscellaneous Repair Shops and Related Services

Major Group 78: Motion Pictures
Industry Group 781: Motion Picture Production and Allied Services

Industry Group 782: Motion Picture Distribution and Allied Services
Industry Group 783: Motion Picture Theaters
Industry Group 784: Video Tape Rental

Major Group 79: Amusement and Recreation Services
Industry Group 791: Dance Studios, Schools, and Halls
Industry Group 792: Theatrical Producers (except Motion Picture)
Industry Group 793: Bowling Centers
Industry Group 794: Commercial Sports
Industry Group 799: Miscellaneous Amusement and Recreation

Major Group 80: Health Services
Industry Group 801: Offices and Clinics of Doctors of Medicine
Industry Group 802: Offices and Clinics of Dentists
Industry Group 803: Offices and Clinics of Doctors of Osteopathy
Industry Group 804: Offices and Clinics of Other Health Practitioners
Industry Group 805: Nursing and Personal Care Facilities
Industry Group 806: Hospitals
Industry Group 807: Medical and Dental Laboratories
Industry Group 808: Home Health Care Services
Industry Group 809: Miscellaneous Health and Allied Services, Not Elsewhere Classified

Major Group 81: Legal Services
Industry Group 811: Legal Services

Major Group 82: Educational Services
Industry Group 821: Elementary and Secondary Schools
Industry Group 822: Colleges, Universities, Professional Schools
Industry Group 823: Libraries
Industry Group 824: Vocational Schools
Industry Group 829: Schools and Educational Services, Not Elsewhere Classified

Major Group 83: Social Services
Industry Group 832: Individual and Family Social Services
Industry Group 833: Job Training and Vocational Rehabilitation
Industry Group 835: Child Day Care Services
Industry Group 836: Residential Care
Industry Group 839: Social Services, Not Elsewhere Classified

Major Group 84: Museums, Art Galleries, and Botanical and Zoological Gardens
Industry Group 841: Museums and Art Galleries
Industry Group 842: Arboreta and Botanical Or Zoological Gardens

Major Group 86: Membership Organizations
Industry Group 861: Business Associations
Industry Group 862: Professional Membership Organizations
Industry Group 863: Labor Unions and Similar Labor Organizations
Industry Group 864: Civic, Social, and Fraternal Associations
Industry Group 865: Political Organizations
Industry Group 866: Religious Organizations
Industry Group 869: Membership Organizations, Not Elsewhere Classified

Major Group 87: Engineering, Accounting, Research, Management, and Related Services
Industry Group 871: Engineering, Architectural, and Surveying
Industry Group 872: Accounting, Auditing, and Bookkeeping Services
Industry Group 873: Research, Development, and Testing Services
Industry Group 874: Management and Public Relations Services

Major Group 88: Private Households
Industry Group 881: Private Households

Major Group 89: Miscellaneous Services
Industry Group 899: Miscellaneous Services

Division J: Public Administration

Major Group 91: Executive, Legislative, and General Government, Except Finance
Industry Group 911: Executive Offices
Industry Group 912: Legislative Bodies
Industry Group 913: Executive and Legislative Offices Combined
Industry Group 919: General Government, Not Elsewhere Classified

Major Group 92: Justice, Public Order, and Safety
Industry Group 921: Courts
Industry Group 922: Public Order and Safety

Major Group 93: Public Finance, Taxation, and Monetary Policy
Industry Group 931: Public Finance, Taxation, and Monetary Policy

Major Group 94: Administration of Human Resource Programs
Industry Group 941: Administration of Educational Programs
Industry Group 943: Administration of Public Health Programs
Industry Group 944: Administration of Social, Human Resource
Industry Group 945: Administration of Veteran's Affairs, Except Health and Insurance

Major Group 95: Administration of Environmental Quality and Housing Programs
Industry Group 951: Administration of Environmental Quality
Industry Group 953: Administration of Housing and Urban

Major Group 96: Administration of Economic Programs
Industry Group 961: Administration of General Economic Programs
Industry Group 962: Regulation and Administration of Transportation
Industry Group 963: Regulation and Admin. of Communications, Electric, Gas, & Other Utilities
Industry Group 964: Regulation of Agricultural Marketing and Commodities
Industry Group 965: Regulation, Licensing, & Inspection of Miscellaneous Commercial Sectors
Industry Group 966: Space Research and Technology

Major Group 97: National Security and International Affairs
Industry Group 971: National Security
Industry Group 972: International Affairs

Major Group 99: Nonclassifiable Establishments
Industry Group 999: Nonclassifiable Establishments

NAICS Codes

11 Agriculture, Forestry, Fishing, and Hunting
111 Crop Production
112 Animal Production
113 Forestry and Logging
114 Fishing, Hunting and Trapping
115 Support Activities for Agriculture and Forestry

21 Mining
211 Oil and Gas Extraction
212 Mining (except Oil and Gas)
213 Support Activities for Mining

22 Utilities
221 Utilities

23 Construction
236 Construction of Buildings
237 Heavy and Civil Engineering Construction
238 Specialty Trade Contractors

31-33 Manufacturing
311 Food Manufacturing
312 Beverage and Tobacco Product Manufacturing
313 Textile Mills
314 Textile Product Mills
315 Apparel Manufacturing
316 Leather and Allied Product Manufacturing
321 Wood Product Manufacturing
322 Paper Manufacturing
323 Printing and Related Support Activities
324 Petroleum and Coal Products Manufacturing
325 Chemical Manufacturing
326 Plastics and Rubber Products Manufacturing
327 Nonmetallic Mineral Product Manufacturing
331 Primary Metal Manufacturing
332 Fabricated Metal Product Manufacturing
333 Machinery Manufacturing
334 Computer and Electronic Product Manufacturing
335 Electrical Equipment, Appliance, and Component Manufacturing
336 Transportation Equipment Manufacturing
337 Furniture and Related Product Manufacturing
339 Miscellaneous Manufacturing

42 Wholesale Trade
423 Merchant Wholesalers, Durable Goods
424 Merchant Wholesalers, Nondurable Goods
425 Wholesale Electronic Markets and Agents and Brokers

44-45 Retail Trade
441 Motor Vehicle and Parts Dealers
442 Furniture and Home Furnishings Stores
443 Electronics and Appliance Stores
444 Building Material and Garden Equipment and Supplies Dealers
445 Food and Beverage Stores
446 Health and Personal Care Stores
447 Gasoline Stations
448 Clothing and Clothing Accessories Stores
451 Sporting Goods, Hobby, Book, and Music Stores
452 General Merchandise Stores
453 Miscellaneous Store Retailers
454 Nonstore Retailers

48-49 Transportation and Warehousing
481 Air Transportation
482 Rail Transportation
483 Water Transportation
484 Truck Transportation
485 Transit and Ground Passenger Transportation
486 Pipeline Transportation
487 Scenic and Sightseeing Transportation
488 Support Activities for Transportation
491 Postal Service
492 Couriers and Messengers
493 Warehousing and Storage

51 Information
511 Publishing Industries (Except Internet)
512 Motion Picture and Sound Recording Industries
515 Broadcasting (Except Internet)
516 Internet Publishing and Broadcasting
517 Telecommunications
518 Internet Service Providers, Web Search Portals, and Data Processing Services
519 Other Information Services

52 Finance and Insurance
521 Monetary Authorities — Central Bank
522 Credit Intermediation and Related Activities
523 Securities, Commodity Contracts, and Other Financial Investments and Related Activities
524 Insurance Carriers and Related Activities
525 Funds, Trusts, and Other Financial Vehicles

53 Real Estate and Rental and Leasing
531 Real Estate
532 Rental and Leasing Services
533 Lessors of Nonfinancial Intangible Assets (Except Copyrighted Works)

54 Professional, Scientific and Technical Services
541 Professional, Scientific, and Technical Services

55 Management of Companies and Enterprises
551 Management of Companies and Enterprises

56 Administrative and Support and Waste Management and Remediation Services
561 Administrative and Support Services
562 Waste Management and Remediation Services

61 Educational Services
611 Educational Services

62 Health Care and Social Assistance
621 Ambulatory Health Care Services
622 Hospitals
623 Nursing and Residential Care Facilities
624 Social Assistance

71 Arts, Entertainment and Recreation
711 Performing Arts, Spectator Sports, and Related Industries
712 Museums, Historical Sites, and Similar Institutions
713 Amusement, Gambling, and Recreation Industries

72 Accommodation and Food Services
721 Accommodation
722 Food Services and Drinking Places

81 Other Services (Except Public Administration)
811 Repair and Maintenance
812 Personal and Laundry Services
813 Religious, Grantmaking, Civic, Professional, and Similar Organizations
814 Private Households

92 Public Administration
921 Executive, Legislative, and Other General Government Support
922 Justice, Public Order, and Safety Activities
923 Administration of Human Resource Programs
924 Administration of Environmental Quality Programs
925 Administration of Housing Programs, Urban Planning, and Community Development
926 Administration of Economic Programs
927 Space Research and Technology
928 National Security and International Affairs

Incidence rates[1] of nonfatal occupational injuries and illnesses by selected industries and case types, 2010

NAICS Code[3]	Industry[2]	Total Recordable Cases
	Private industry[6]	3.5
	Goods-producing[6]	4.2
	Natural resources and mining[6,7]	3.7
11	**Agriculture, forestry, fishing and hunting[6]**	4.8
111	Crop production[6,8]	4.6
112	Animal production[6,8]	5.2
113	Forestry and logging	3.6
115	Support activities for agriculture and forestry	4.9
21	**Mining[7]**	2.3
211	Oil and gas extraction	1.2
212	Mining (except oil and gas)[9]	3.1
213	Support activities for mining	2.2
	Construction	4.0
23	**Construction**	4.0
236	Construction of buildings	3.5
237	Heavy and civil engineering construction	3.8
238	Specialty trade contractors	4.3
	Manufacturing	4.4
31-33	**Manufacturing**	4.4
311	Food manufacturing	5.8
312	Beverage and tobacco product manufacturing	6.4
313	Textile mills	3.3
314	Textile product mills[8]	3.1
315	Apparel manufacturing[8]	3.0
316	Leather and allied product manufacturing	5.0
321	Wood product manufacturing	6.2
322	Paper manufacturing	3.4
323	Printing and related support activities	3.2
324	Petroleum and coal products manufacturing	1.5
325	Chemical manufacturing	2.4
326	Plastics and rubber products manufacturing[8]	5.1
327	Nonmetallic mineral product manufacturing	6.0
331	Primary metal manufacturing	6.3
332	Fabricated metal product manufacturing	5.5
333	Machinery manufacturing[8]	4.4
334	Computer and electronic product manufacturing	1.5
335	Electrical equipment, appliance, and component manufacturing	3.3
336	Transportation equipment manufacturing[8]	5.5
337	Furniture and related product manufacturing[8]	5.2
339	Miscellaneous manufacturing	3.0
	Service-providing	3.4
	Trade, transportation, and utilities[10]	4.1
42	**Wholesale trade**	3.4
423	Merchant wholesalers, durable goods	3.1
424	Merchant wholesalers, nondurable goods	4.5
44-45	**Retail trade**	4.1
441	Motor vehicle and parts dealers	3.8
442	Furniture and home furnishings stores	4.2
443	Electronics and appliance stores	1.6
444	Building material, garden equipment and supplies dealers	5.1
445	Food and beverage stores	5.1
446	Health and personal care stores	2.8
447	Gasoline stations	2.9
448	Clothing and clothing accessories stores	2.7
451	Sporting goods, hobby, book, and music stores	3.2
452	General merchandise stores	5.0
453	Miscellaneous store retailers	3.1
454	Nonstore retailers	3.5
48-49	**Transportation and warehousing[10]**	5.2
481	Air transportation	8.1
482	Rail transportation[10]	2.0
483	Water transportation	3.2
484	Truck transportation	5.0
485	Transit and ground passenger transportation	5.3
486	Pipeline transportation	2.5
487	Scenic and sightseeing transportation	4.0
488	Support activities for transportation	3.8
492	Couriers and messengers	7.2
493	Warehousing and storage	5.9
22	**Utilities**	3.1
221	Utilities	3.1
	Information	1.8
51	**Information**	1.8
511	Publishing industries (except Internet)	1.2
512	Motion picture and sound recording industries	2.3
515	Broadcasting (except Internet)	2.8
517	Telecommunications[8]	2.2
519	Other information services[8]	.7
	Financial activities	1.3
52	**Finance and insurance**	.8
521	Monetary authorities - central bank	1.5
522	Credit intermediation and related activities	.9
523	Securities, commodity contracts, and other financial investments and related activities	.3
524	Insurance carriers and related activities	.8
525	Funds, trusts, and other financial vehicles	1.2
53	**Real estate and rental and leasing**	2.9
531	Real estate[8]	2.7
532	Rental and leasing services	3.3
533	Lessors of nonfinancial intangible assets (except copyrighted works)	1.0
	Professional and business services	1.7
54	**Professional, scientific, and technical services**	1.0
541	Professional, scientific, and technical services[8]	1.0
55	**Management of companies and enterprises**	1.5
56	**Administrative, support, waste management and remediation services**	2.8
561	Administrative and support services[8]	2.7
562	Waste management and remediation services	4.5
	Education and health services	4.8
61	**Educational services**	2.2
611	Educational services	2.2
62	**Health care and social assistance**	5.2
621	Ambulatory health care services	2.8
622	Hospitals	7.0
623	Nursing and residential care facilities	8.3
624	Social assistance	3.5
	Leisure and hospitality	3.9
71	**Arts, entertainment, and recreation**	4.8
711	Performing arts, spectator sports, and related industries	6.7
712	Museums, historical sites, and similar institutions	4.7
713	Amusement, gambling, and recreation industries	4.3
72	**Accommodation and food services**	3.7
721	Accommodation	5.3
722	Food services and drinking places	3.3
	Other services	2.7
81	**Other services, except public administration**	2.7
811	Repair and maintenance	3.3
812	Personal and laundry services	2.3
813	Religious, grantmaking, civic, professional, and similar organizations	2.6

[1] The incidence rates represent the number of injuries and illnesses per 100 full-time workers and were calculated as:
(N/EH) x 200,000, where
N = number of injuries and illnesses
EH = total hours worked by all employees during the calendar year
200,000 = base for 100 equivalent full-time workers (working 40 hours per week, 50 weeks per year)

[2] Totals include data for industries not shown separately.

[3] North American Industry Classification System — United States, 2007

[4] Employment is expressed as an annual average and is derived primarily from the BLS-Quarterly Census of Employment and Wages (QCEW) program.

[5] Days-away-from-work cases include those that result in days away from work, some of which also included job transfer or restriction.

[6] Excludes farms with fewer than 11 employees.

[7] Data for Mining (Sector 21 in the North American Industry Classification System– United States, 2007) include establishments not governed by the Mine Safety and Health Administration rules and reporting, such as those in Oil and Gas Extraction and related support activities. Data for mining operators in coal, metal, and nonmetal mining are provided to BLS by the Mine Safety and Health Administration, U.S. Department of Labor. Independent mining contractors are excluded from the coal, metal, and nonmetal mining industries. These data do not reflect the changes the Occupational Safety and Health Administration made to its recordkeeping requirements effective January 1, 2002; therefore, estimates for these industries are not comparable to estimates in other industries.

[8] Industry scope changed in 2009

[9] Data for mining operators in this industry are provided to BLS by the Mine Safety and Health Administration, U.S. Department of Labor. Independent mining contractors are excluded. These data do not reflect the changes the Occupational Safety and Health Administration made to its recordkeeping requirements effective January 1, 2002; therefore, estimates for these industries are not comparable to estimates in other industries.

[10] Data for employers in railroad transportation are provided to BLS by the Federal Railroad Administration, U.S. Department of Transportation.

NOTE: Because of rounding, components may not add to totals. Dash indicates data do not meet publication guidelines.
SOURCE: Bureau of Labor Statistics, U.S. Department of Labor

Most Common Standards Cited for Construction (29 CFR 1926), October 2010 through September 2011

Standard	#Cited	Description
1926.0501	7191	Duty to have fall protection.
1926.0451	7062	General requirements.
1926.1053	3283	Ladders.
1926.0503	1986	Training requirements.
1926.0100	1423	Head protection.
1926.0020	1369	General safety and health provisions.
1926.0453	1343	Aerial lifts.
1926.0102	1213	Eye and face protection.
1926.0651	1173	Specific Excavation Requirements.
1926.0405	1157	"Wiring methods, components, and equipment for general use."
1926.0404	936	Wiring design and protection.
1926.0652	902	Requirements for protective systems.
1926.0454	871	Training requirements.
1926.0502	753	Fall protection systems criteria and practices.
1926.0021	636	Safety training and education.
1926.0403	611	General requirements.
1926.0452	555	Additional requirements applicable to specific types of scaffolds.
1926.1101	528	Asbestos
1926.0062	518	Lead
1926.0416	515	General requirements.
1926.1060	511	Training requirements.
1926.1052	501	Stairways.
1926.0095	389	Criteria for personal protective equipment.
1926.0300	281	General requirements.
1926.0150	233	Fire protection.
1926.0701	227	General requirements
1926.0760	226	Fall protection.
1926.0602	222	Material handling equipment.
1926.0304	208	Woodworking tools.
1926.0025	206	Housekeeping.
1926.1051	192	General requirements.
1926.0350	185	Gas welding and cutting.
1926.0251	135	Rigging equipment for material handling.
1926.0050	129	Medical services and first aid.
1926.0850	111	Preparatory operations.
1926.0152	107	Flammable and combustible liquids.
1926.0200	106	Accident prevention signs and tags.
1926.0028	84	Personal protective equipment.
1926.0351	74	Arc welding and cutting.
1926.0307	70	Mechanical power-transmission apparatus.
1926.0153	66	Liquefied petroleum gas (LP-Gas).
1926.1412	66	Inspections
1926.0106	62	Working over or near water.
1926.0051	52	Sanitation.
1926.0600	52	Equipment.
1926.0034	51	Means of egress.
1926.0252	51	Disposal of waste materials.
1926.0601	51	Motor vehicles.
1926.0352	49	Fire prevention.
1926.0302	48	Power-operated hand tools.
1926.0550	48	Cranes and derricks.
1926.0055	43	"Gases, vapors, fumes, dusts, and mists."
1926.0303	41	Abrasive wheels and tools.
1926.0761	34	Training.
1926.0052	30	Occupational noise exposure.
1926.1417	27	Operation
1926.0201	25	Signaling.
1926.1428	25	Signal person qualifications
1926.1425	24	Keeping clear of the load
1926.0250	23	General requirements for storage.
1926.0706	21	Requirements for masonry construction.
1926.1126	19	Chromium (VI)
1926.0054	18	Nonionizing radiation.
1926.0605	18	Marine operations and equipment.
1926.0800	18	Underground Construction
1926.0417	17	Lockout and tagging of circuits.
1926.1402	17	Ground conditions
1926.1408	17	Power line safety (up to 350 kV) — equipment operations
1926.0552	16	"Material hoists, personnel hoists, and elevators."
1926.1404	15	Assembly/Disassembly
1926.1430	14	Training
1926.0859	13	Mechanical demolition.
1926.0301	12	Hand tools.
1926.0852	12	Chutes.
1926.0059	11	Hazard Communication.
1926.0096	11	Occupational foot protection.
1926.0151	11	Fire prevention
1926.1424	11	Work area control
1926.0752	10	"Site layout, site-specific erection plan and construction sequence."
1926.0754	10	Structural steel assembly.
1926.0856	10	"Removal of walls, floors, and material with equipment."
1926.1431	10	Hoisting personnel
1926.0353	9	"Ventilation and protection in welding, cutting, and heating."
1926.0703	9	Requirements for cast-in-place Concrete.
1926.0753	9	Hoisting and rigging.
1926.0755	9	Column anchorage.
1926.1127	9	Cadmium
1926.1413	8	Wire rope — inspection
1926.0101	7	Hearing protection.
1926.1437	7	Floating cranes/derricks and land cranes/derricks on barges
1926.0056	6	Illumination.
1926.0306	6	Air receivers.
1926.0104	5	"Safety belts, lifelines, and lanyards."
1926.0202	5	Barricades.
1926.0951	5	Tools and protective equipment.
1926.0035	4	Employee emergency action plans.
1926.0057	4	Ventilation.
1926.0154	4	Temporary heating devices.
1926.1416	4	Operational aids
1926.1427	4	Operator qualification and certification
1926.0026	3	Illumination.
1926.0105	3	Safety nets.
1926.0305	3	"Jacks-lever and ratchet, screw, and hydraulic."
1926.0432	3	Environmental deterioration of equipment.
1926.0851	3	"Stairs, passageways, and ladders."
1926.0858	3	Removal of steel construction.
1926.0900	3	General provisions.
1926.0950	3	General requirements.
1926.0955	3	Overhead lines.
1926.1414	3	Wire rope — selection and installation criteria
1926.1419	3	Signals — general requirements
1926.1422	3	Signals — hand signal chart
1926.1433	3	Design, construction and testing
1926.0702	2	Requirements for equipment and tools.
1926.0759	2	Falling object protection.
1926.0910	2	Inspection after blasting.
1926.0956	2	Underground lines.
1926.1000	2	Rollover protective structures (ROPS) for material handling equipment.
1926.1415	2	Safety devices
1926.1501	2	Cranes and derricks
1926.0103	1	Respiratory protection.
1926.0553	1	Base-mounted drum hoists.
1926.0554	1	Overhead hoists.
1926.0555	1	Conveyors.
1926.0603	1	Pile driving equipment.
1926.0700	1	"Scope, application, and definitions applicable to this subpart."
1926.0756	1	Beams and columns.
1926.0757	1	Open web steel joists.
1926.0758	1	Systems-engineered metal buildings.
1926.0854	1	"Removal of walls, masonry sections, and chimneys."
1926.0855	1	Manual removal of floors.
1926.1400	1	Scope
1926.1405	1	Disassembly
1926.1418	1	Authority to stop operation
1926.1423	1	Fall protection
1926.1426	1	Free fall and controlled load lowering

Sharps Injury Log[1]

For Period Ending: _____ / _____ / _________

Company Name: ____________________

Date Entered:	Date Incident Occurred & Time Incident Occured:	Type and Brand of Device Involved:	Department or Work Area Where Exposure Incident Occurred:	How Incident Occurred:
___/___/___ Month Day Year	___/___/___ ___:___ Month Day Year Hour Minute AM PM			
___/___/___ Month Day Year	___/___/___ ___:___ Month Day Year Hour Minute AM PM			
___/___/___ Month Day Year	___/___/___ ___:___ Month Day Year Hour Minute AM PM			
___/___/___ Month Day Year	___/___/___ ___:___ Month Day Year Hour Minute AM PM			
___/___/___ Month Day Year	___/___/___ ___:___ Month Day Year Hour Minute AM PM			
___/___/___ Month Day Year	___/___/___ ___:___ Month Day Year Hour Minute AM PM			
___/___/___ Month Day Year	___/___/___ ___:___ Month Day Year Hour Minute AM PM			
___/___/___ Month Day Year	___/___/___ ___:___ Month Day Year Hour Minute AM PM			
___/___/___ Month Day Year	___/___/___ ___:___ Month Day Year Hour Minute AM PM			
___/___/___ Month Day Year	___/___/___ ___:___ Month Day Year Hour Minute AM PM			
___/___/___ Month Day Year	___/___/___ ___:___ Month Day Year Hour Minute AM PM			

- **Retain until** _____ / _____ / _____ (5 years after the end of the current year - see §1904.44)
- You are required to maintain this log if the requirement to maintain a 300 log applies to you. See Part 1904.

[1] **Referred to in §1910.1030(h)(5)**

Download a printable version of this form free of charge at www.oshacfr.com.

Download a printable version of this form free of charge at www.oshacfr.com.

Instruction CPL 2-0.124

Multi-Employer Citation Policy

I. **Purpose.** This Directive clarifies the Agency's multi-employer citation policy and suspends Chapter III. C. 6. of OSHA's Field Inspection Reference Manual (FIRM).

II. **Scope.** OSHA-Wide.

III. **Suspension.** Chapter III. Paragraph C. 6. of the FIRM (CPL 2.103) is suspended and replaced by this Directive.

IV. **References.** OSHA Instructions: CPL 02-00.103; OSHA Field Inspection Reference Manual (FIRM), September 26, 1994. ADM 08-0.1C, OSHA Electronic Directive System, December 19,1997.

V. **Action Information**

A. *Responsible Office.* Directorate of Construction.

B. *Action Offices.* National, Regional, and Area Offices.

C. *Information Offices.* State Plan Offices, Consultation Project Offices

VI. **Federal Program Change.** This Directive describes a Federal Program Change for which State adoption is not required. However, the States shall respond via the two-way memorandum to the Regional Office as soon as the State's intent regarding the multi-employer citation policy is known, but no later than 60 calendar days after the date of transmittal from the Directorate of Federal-State Operations.

VII. **Force and Effect of Revised Policy.** The revised policy provided in this Directive is in full force and effect from the date of its issuance. It is an official Agency policy to be implemented OSHA-wide.

VIII. **Changes in Web Version of FIRM.** A note will be included at appropriate places in the FIRM as it appears on the Web indicating the suspension of Chapter III paragraph 6. C. and its replacement by this Directive, and a hypertext link will be provided connecting viewers with this Directive.

IX. **Background.** OSHA's Field Inspection Reference Manual (FIRM) of September 26, 1994 (CPL 2.103), states at Chapter III, paragraph 6. C., the Agency's citation policy for multi-employer worksites. The Agency has determined that this policy needs clarification. This directive describes the revised policy.

A. *Continuation of Basic Policy.* This revision continues OSHA's existing policy for issuing citations on multi-employer worksites. However, it gives clearer and more detailed guidance than did the earlier description of the policy in the FIRM, including new examples explaining when citations should and should not be issued to exposing, creating, correcting, and controlling employers. These examples, which address common situations and provide general policy guidance, are not intended to be exclusive. In all cases, the decision on whether to issue citations should be based on all of the relevant facts revealed by the inspection or investigation.

B. *No Changes in Employer Duties.* This revision neither imposes new duties on employers nor detracts from their existing duties under the OSH Act. Those duties continue to arise from the employers' statutory duty to comply with OSHA standards and their duty to exercise reasonable diligence to determine whether violations of those standards exist.

X. **Multi-employer Worksite Policy.** The following is the multi-employer citation policy:

A. *Multi-employer Worksites.* On multi-employer worksites (in all industry sectors), more than one employer may be citable for a hazardous condition that violates an OSHA standard. A two-step process must be followed in determining whether more than one employer is to be cited.

1. *Step One.* The first step is to determine whether the employer is a creating, exposing, correcting, or controlling employer. The definitions in paragraphs (B)-(E) below explain and give examples of each. Remember that an employer may have multiple roles (see paragraph H). Once you determine the role of the employer, go to Step Two to determine if a citation is appropriate. (Note: Only exposing employers can be cited for General Duty Clause violations.)

2. *Step Two.* If the employer falls into one of these categories, it has obligations with respect to OSHA requirements. Step Two is to determine if the employer's actions were sufficient to meet those obligations. The extent of the actions required of employers varies based on which category applies. Note that the extent of the measures that a controlling employer must take to satisfy its duty to exercise reasonable care to prevent and detect violations is less than what is required of an employer with respect to protecting its own employees.

B. *The Creating Employer*

1. *Step 1: Definition:* The employer that caused a hazardous condition that violates an OSHA standard.

2. *Step 2: Actions Taken:* Employers must not create violative conditions. An employer that does so is citable even if the only employees exposed are those of other employers at the site.

a. *Example 1:* Employer Host operates a factory. It contracts with Company S to service machinery. Host fails to cover drums of a chemical despite S's repeated requests that it do so. This results in airborne levels of the chemical that exceed the Permissible Exposure Limit.
Analysis:
Step 1: Host is a creating employer because it caused employees of S to be exposed to the air contaminant above the PEL.
Step 2: Host failed to implement measures to prevent the accumulation of the air contaminant. It could have met its OSHA obligation by implementing the simple engineering control of covering the drums. Having failed to implement a feasible engineering control to meet the PEL, Host is citable for the hazard.

b. *Example 2:* Employer M hoists materials onto Floor 8, damaging perimeter guardrails. Neither its own employees nor employees of other employers are exposed to the hazard. It takes effective steps to keep all employees, including those of other employers, away from the unprotected edge and informs the controlling employer of the problem. Employer M lacks authority to fix the guardrails itself.
Analysis:
Step 1: Employer M is a creating employer because it caused a hazardous condition by damaging the guardrails.
Step 2: While it lacked the authority to fix the guardrails, it took immediate and effective steps to keep all employees away from the hazard and notified the controlling employer of the hazard. Employer M is not citable since it took effective measures to prevent employee exposure to the fall hazard.

C. *The Exposing Employer*

1. *Step 1: Definition:* An employer whose own employees are exposed to the hazard. See Chapter III, section (C)(1)(b) for a discussion of what constitutes exposure.

2. *Step 2: Actions Taken:* If the exposing employer created the violation, it is citable for the violation as a creating employer. If the violation was created by another employer, the exposing employer is citable if it (1) knew of the hazardous condition or failed to exercise reasonable diligence to discover the condition, and (2) failed to take steps consistent with its authority to protect is employees. If the exposing employer has authority to correct the hazard, it must do so. If the exposing employer lacks the authority to correct the hazard, it is citable if it fails to do each of the following: (1) ask the creating and/or controlling employer to correct the hazard; (2) inform its employees of the hazard; and (3) take reasonable alternative protective measures. In extreme circumstances (e.g., imminent danger situations), the exposing employer is citable for failing to remove its employees from the job to avoid the hazard.

a. *Example 3:* Employer Sub S is responsible for inspecting and cleaning a work area in Plant P around a large, permanent hole at the end of each day. An OSHA standard requires guardrails. There are no guardrails around the hole and Sub S employees do not use personal fall protection, although it would be feasible to do so. Sub S has no authority to install guardrails. However, it did ask Employer P, which operates the plant, to install them. P refused to install guardrails.
Analysis:
Step 1: Sub S is an exposing employer because its employees are exposed to the fall hazard.
Step 2: While Sub S has no authority to install guardrails, it is required to comply with OSHA requirements to the extent feasible. It must take steps to protect its employees and ask the employer that controls the hazard - Employer P - to correct it. Although Sub S asked for guardrails, since the hazard was not corrected, Sub S was responsible for taking reasonable alternative protective steps, such as providing personal fall protection. Because that was not done, Sub S is citable for the violation.

b. *Example 4:* Unprotected rebar on either side of an access ramp presents an impalement hazard. Sub E, an electrical subcontractor, does not have the authority to cover the rebar. However, several times Sub E asked the general contractor, Employer GC, to cover the rebar. In the meantime, Sub E instructed its employees to use a different access route that avoided most of the uncovered rebar and required them to keep as far from the rebar as possible.

Analysis:
Step 1: Since Sub E employees were still exposed to some unprotected rebar, Sub E is an exposing employer.
Step 2: Sub E made a good faith effort to get the general contractor to correct the hazard and took feasible measures within its control to protect its employees. Sub E is not citable for the rebar hazard.

D. *The Correcting Employer*

1. *Step 1: Definition:* An employer who is engaged in a common undertaking, on the same worksite, as the exposing employer and is responsible for correcting a hazard. This usually occurs where an employer is given the responsibility of installing and/or maintaining particular safety/health equipment or devices.
2. *Step 2: Actions Taken:* The correcting employer must exercise reasonable care in preventing and discovering violations and meet its obligations of correcting the hazard.
   a. *Example 5:* Employer C, a carpentry contractor, is hired to erect and maintain guardrails throughout a large, 15-story project. Work is proceeding on all floors. C inspects all floors in the morning and again in the afternoon each day. It also inspects areas where material is delivered to the perimeter once the material vendor is finished delivering material to that area. Other subcontractors are required to report damaged/missing guardrails to the general contractor, who forwards those reports to C. C repairs damaged guardrails immediately after finding them and immediately after they are reported. On this project few instances of damaged guardrails have occurred other than where material has been delivered. Shortly after the afternoon inspection of Floor 6, workers moving equipment accidentally damage a guardrail in one area. No one tells C of the damage and C has not seen it. An OSHA inspection occurs at the beginning of the next day, prior to the morning inspection of Floor 6. None of C's own employees are exposed to the hazard, but other employees are exposed.
   Analysis:
   Step 1: C is a correcting employer since it is responsible for erecting and maintaining fall protection equipment.
   Step 2: The steps C implemented to discover and correct damaged guardrails were reasonable in light of the amount of activity and size of the project. It exercised reasonable care in preventing and discovering violations; it is not citable for the damaged guardrail since it could not reasonably have known of the violation.

E. *The Controlling Employer*

1. *Step 1: Definition:* An employer who has general supervisory authority over the worksite, including the power to correct safety and health violations itself or require others to correct them. Control can be established by contract or, in the absence of explicit contractual provisions, by the exercise of control in practice. Descriptions and examples of different kinds of controlling employers are given below.
2. *Step 2: Actions Taken:* A controlling employer must exercise reasonable care to prevent and detect violations on the site. The extent of the measures that a controlling employer must implement to satisfy this duty of reasonable care is less than what is required of an employer with respect to protecting its own employees. This means that the controlling employer is not normally required to inspect for hazards as frequently or to have the same level of knowledge of the applicable standards or of trade expertise as the employer it has hired.
3. *Factors Relating to Reasonable Care Standard.* Factors that affect how frequently and closely a controlling employer must inspect to meet its standard of reasonable care include:
   a. *The scale of the project;*
   b. *The nature and pace of the work,* including the frequency with which the number or types of hazards change as the work progresses;
   c. *How much the controlling employer knows both* about the safety history and safety practices of the employer it controls and about that employer's level of expertise.
   d. *More frequent inspections* are normally needed if the controlling employer knows that the other employer has a history of non-compliance. Greater inspection frequency may also be needed, especially at the beginning of the project, if the controlling employer had never before worked with this other employer and does not know its compliance history.
   e. *Less frequent inspections* may be appropriate where the controlling employer sees strong indications that the other employer has implemented effective safety and health efforts. The most important indicator of an effective safety and health effort by the other employer is a consistently high level of compliance. Other indicators include the use of an effective, graduated system of enforcement for non-compliance with safety and health requirements coupled with regular jobsite safety meetings and safety training.
4. *Evaluating Reasonable Care.* In evaluating whether a controlling employer has exercised reasonable care in preventing and discovering violations, consider questions such as whether the controlling employer:
   a. *Conducted periodic inspections* of appropriate frequency (frequency should be based on the factors listed in G.3.);
   b. *Implemented an effective system* for promptly correcting hazards;
   c. *Enforces the other employer's compliance* with safety and health requirements with an effective, graduated system of enforcement and follow-up inspections.
5. *Types of Controlling Employers*
   a. *Control Established by Contract.* In this case, **the Employer Has a Specific Contract Right to Control Safety**: To be a controlling employer, the employer must itself be able to prevent or correct a violation or to require another employer to prevent or correct the violation. One source of this ability is explicit contract authority. This can take the form of a specific contract right to require another employer to adhere to safety and health requirements and to correct violations the controlling employer discovers.
      (1) *Example 6:* Employer GH contracts with Employer S to do sandblasting at GH's plant. Some of the work is regularly scheduled maintenance and so is general industry work; other parts of the project involve new work and are considered construction. Respiratory protection is required. Further, the contract explicitly requires S to comply with safety and health requirements. Under the contract GH has the right to take various actions against S for failing to meet contract requirements, including the right to have non-compliance corrected by using other workers and back-charging for that work. S is one of two employers under contract with GH at the work site, where a total of five employees work. All work is done within an existing building. The number and types of hazards involved in S's work do not significantly change as the work progresses. Further, GH has worked with S over the course of several years. S provides periodic and other safety and health training and uses a graduated system of enforcement of safety and health rules. S has consistently had a high level of compliance at its previous jobs and at this site. GH monitors S by a combination of weekly inspections, telephone discussions and a weekly review of S's own inspection reports. GH has a system of graduated enforcement that it has applied to S for the few safety and health violations that had been committed by S in the past few years. Further, due to respirator equipment problems S violates respiratory protection requirements two days before GH's next scheduled inspection of S. The next day there is an OSHA inspection. There is no notation of the equipment problems in S's inspection reports to GH and S made no mention of it in its telephone discussions.
      Analysis:
      Step 1: GH is a controlling employer because it has general supervisory authority over the worksite, including contractual authority to correct safety and health violations.
      Step 2: GH has taken reasonable steps to try to make sure that S meets safety and health requirements. Its inspection frequency is appropriate in light of the low number of workers at the site, lack of significant changes in the nature of the work and types of hazards involved, GH's knowledge of S's history of compliance and its effective safety and health efforts on this job. GH has exercised reasonable care and is not citable for this condition.
      (2) *Example 7:* Employer GC contracts with Employer P to do painting work. GC has the same contract authority over P as Employer GH had in Example 6. GC has never before worked with P. GC conducts inspections that are sufficiently frequent in light of the factors listed above in (G)(3). Further, during a number of its inspections, GC finds that P has violated fall protection requirements. It points the violations out to P during each inspection but takes no further actions.

Analysis:

Step 1: GC is a controlling employer since it has general supervisory authority over the site, including a contractual right of control over P.

Step 2: GC took adequate steps to meet its obligation to discover violations. However, it failed to take reasonable steps to require P to correct hazards since it lacked a graduated system of enforcement. A citation to GC for the fall protection violations is appropriate.

(3) *Example 8:* Employer GC contracts with Sub E, an electrical subcontractor. GC has full contract authority over Sub E, as in Example 6. Sub E installs an electric panel box exposed to the weather and implements an assured equipment grounding conductor program, as required under the contract. It fails to connect a grounding wire inside the box to one of the outlets. This incomplete ground is not apparent from a visual inspection. Further, GC inspects the site with a frequency appropriate for the site in light of the factors discussed above in (G)(3). It saw the panel box but did not test the outlets to determine if they were all grounded because Sub E represents that it is doing all of the required tests on all receptacles. GC knows that Sub E has implemented an effective safety and health program. From previous experience it also knows Sub E is familiar with the applicable safety requirements and is technically competent. GC had asked Sub E if the electrical equipment is OK for use and was assured that it is.

Analysis:

Step 1: GC is a controlling employer since it has general supervisory authority over the site, including a contractual right of control over Sub E.

Step 2: GC exercised reasonable care. It had determined that Sub E had technical expertise, safety knowledge and had implemented safe work practices. It conducted inspections with appropriate frequency. It also made some basic inquiries into the safety of the electrical equipment. Under these circumstances GC was not obligated to test the outlets itself to determine if they were all grounded. It is not citable for the grounding violation.

b. *Control Established by a Combination of Other Contract Rights:* Where there is no explicit contract provision granting the right to control safety, or where the contract says the employer does not have such a right, an employer may still be a controlling employer. The ability of an employer to control safety in this circumstance can result from a combination of contractual rights that, together, give it broad responsibility at the site involving almost all aspects of the job. Its responsibility is broad enough so that its contractual authority necessarily involves safety. The authority to resolve disputes between subcontractors, set schedules and determine construction sequencing are particularly significant because they are likely to affect safety. (Note: citations should only be issued in this type of case after consulting with the Regional Solicitor's office).

(1) *Example 9:* Construction manager M is contractually obligated to: set schedules and construction sequencing, require subcontractors to meet contract specifications, negotiate with trades, resolve disputes between subcontractors, direct work and make purchasing decisions, which affect safety. However, the contract states that M does not have a right to require compliance with safety and health requirements. Further, Subcontractor S asks M to alter the schedule so that S would not have to start work until Subcontractor G has completed installing guardrails. M is contractually responsible for deciding whether to approve S's request.

Analysis:

Step 1: Even though its contract states that M does not have authority over safety, the combination of rights actually given in the contract provides broad responsibility over the site and results in the ability of M to direct actions that necessarily affect safety. For example, M's contractual obligation to determine whether to approve S's request to alter the schedule has direct safety implications. M's decision relates directly to whether S's employees will be protected from a fall hazard. M is a controlling employer.

Step 2: In this example, if M refused to alter the schedule, it would be citable for the fall hazard violation.

(2) *Example 10:* Employer ML's contractual authority is limited to reporting on subcontractors' contract compliance to owner/developer O and making contract payments. Although it reports on the extent to which the subcontractors are complying with safety and health infractions to O, ML does not exercise any control over safety at the site.

Analysis:

Step 1: ML is not a controlling employer because these contractual rights are insufficient to confer control over the subcontractors and ML did not exercise control over safety. Reporting safety and health infractions to another entity does not, by itself (or in combination with these very limited contract rights), constitute an exercise of control over safety.

Step 2: Since it is not a controlling employer it had no duty under the OSH Act to exercise reasonable care with respect to enforcing the subcontractors' compliance with safety; there is therefore no need to go to Step 2.

c. *Architects and Engineers:* Architects, engineers, and other entities are controlling employers only if the breadth of their involvement in a construction project is sufficient to bring them within the parameters discussed above.

(1) *Example 11:* Architect A contracts with owner O to prepare contract drawings and specifications, inspect the work, report to O on contract compliance, and to certify completion of work. A has no authority or means to enforce compliance, no authority to approve/reject work and does not exercise any other authority at the site, although it does call the general contractor's attention to observed hazards noted during its inspections.

Analysis:

Step 1: A's responsibilities are very limited in light of the numerous other administrative responsibilities necessary to complete the project. It is little more than a supplier of architectural services and conduit of information to O. Its responsibilities are insufficient to confer control over the subcontractors and it did not exercise control over safety. The responsibilities it does have are insufficient to make it a controlling employer. Merely pointing out safety violations did not make it a controlling employer. Note: In a circumstance such as this it is likely that broad control over the project rests with another entity.

Step 2: Since A is not a controlling employer it had no duty under the OSH Act to exercise reasonable care with respect to enforcing the subcontractors' compliance with safety; there is therefore no need to go to Step 2.

(2) *Example 12:* Engineering firm E has the same contract authority and functions as in Example 9.

Analysis:

Step 1: Under the facts in Example 9, E would be considered a controlling employer.

Step 2: The same type of analysis described in Example 9 for Step 2 would apply here to determine if E should be cited.

d. *Control Without Explicit Contractual Authority.* Even where an employer has no explicit contract rights with respect to safety, an employer can still be a controlling employer if, in actual practice, it exercises broad control over subcontractors at the site (see Example 9). Note: Citations should only be issued in this type of case after consulting with the Regional Solicitor's office.

(1) *Example 13:* Construction manager MM does not have explicit contractual authority to require subcontractors to comply with safety requirements, nor does it explicitly have broad contractual authority at the site. However, it exercises control over most aspects of the subcontractors' work anyway, including aspects that relate to safety.

Analysis:

Step 1: MM would be considered a controlling employer since it exercises control over most aspects of the subcontractor's work, including safety aspects.

Step 2: The same type of analysis on reasonable care described in the examples in (G)(5)(a) would apply to determine if a citation should be issued to this type of controlling employer.

F. *Multiple Roles*

1. *A creating, correcting or controlling employer* will often also be an exposing employer. Consider whether the employer is an exposing employer before evaluating its status with respect to these other roles.
2. *Exposing, creating and controlling employers* can also be correcting employers if they are authorized to correct the hazard.

· It's The Law! Mandatory Posting ·

You Have a Right to a Safe and Healthful Workplace.

All covered employers are required to display and keep displayed, a poster prepared by the Department of Labor*† informing employees of the protections of the Occupational Safety and Health Act P.L. 91-596, December 29, 1970 and its amendments. The poster must be displayed in a conspicuous place where employees and applicants for employment can see it. The new Plain Language poster (OSHA 3165) replaces OSHA's currently required workplace poster (OSHA 2203). As supplies of OSHA 2203 diminish, the new workplace poster will be phased in to take its place. Employers do not need to replace current 2203 posters. The OSHA 2203 poster will continue to be in compliance with OSHA regulations.

†(States with State Plans may have their own poster.)

*(Federal Government Agencies must use the Federal Agency Poster.)

Job Safety and Health

It's the law!

OSHA

Occupational Safety and Health Administration
U.S. Department of Labor

EMPLOYEES:

- You have the right to notify your employer or OSHA about workplace hazards. You may ask OSHA to keep your name confidential.
- You have the right to request an OSHA inspection if you believe that there are unsafe and unhealthful conditions in your workplace. You or your representative may participatte in that inspection.
- You can file a complaint with OSHA within 30 days of retaliation or discrimination by your employer for making safety and health complaints or for exercising your rights under the *OSH Act*.
- You have the right to see OSHA citations issued to your employer. Your employer must post the citations at or near the place of the alleged violations.
- Your employer must correct workplace hazards by the date indicated on the citation and must certify that these hazards have been reduced or eliminated.
- You have the right to copies of your medical records and records of your exposures to toxic and harmful substances or conditions.
- Your employer must post this notice in your workplace.
- You must comply with all occupatoinal safety and health standards issued under the *OSH Act* that apply to your own actions and conduct on the job.

EMPLOYERS:

- You must furnish your employees a place of employment free from recognized hazards.
- You must comply with the occupational safety and health standards issued under the *OSH Act*.

Free assistance in identifying and correcting hazards or complying with standards is available to employers, without citation or penalty, through OSHA-supported consultation programs in each state.

This free poster available from OSHA –
The Best Resource for Safety and Health

1-800-321-OSHA
www.osha.gov

OSHA 3165-12-06R

Safety and Health Management Guidelines Issuance of Voluntary Guidelines - 54:3904-3916

(a) **General.**

(1) *Employers are advised and encouraged* to institute and maintain in their establishments a program which provides systematic policies, procedures, and practices that are adequate to recognize and protect their employees from occupational safety and health hazards.

(2) *An effective program* includes provisions for the systematic identification, evaluation, and prevention or control of general workplace hazards, specific job hazards, and potential hazards which may arise from foreseeable conditions.

(3) *Although compliance with the law,* including specific OSHA standards, is an important objective, an effective program looks beyond specific requirements of law to address all hazards. It will seek to prevent injuries and illnesses, whether or not compliance is at issue.

(4) *The extent to which the program* is described in writing is less important than how effective it is in practice. As the size of a worksite or the complexity of a hazardous operation increases, however, the need for written guidance increases to ensure clear communications of policies and priorities and consistent and fair application of rules.

(b) **Major Elements.** An effective occupational safety and health program will include the following four elements. To implement these elements, it will include the actions described in paragraph (c).

(1) *Management commitment* and employee involvement are complementary. Management commitment provides the motivating force and the resources for organizing and controlling activities within an organization. In an effective program, management regards workers safety and health as a fundamental value of the organization and applies its commitment to safety and health protection with as much vigor as to other organizational purposes. Employee involvement provides the means through which workers develop and/or express their own commitment to safety and health protection, for themselves and for their fellow workers.

(2) *Worksite analysis* involves a variety of worksite examinations, to identify not only existing hazards but also conditions and operations in which changes might occur to create hazards. Unawareness of a hazard which stems from failure to examine the worksite is a sure sign that safety and health policies and/or practices are ineffective. Effective management actively analyzes the work and worksite, to anticipate and prevent harmful occurrences.

(3) *Hazard prevention and controls* are triggered by a determination that a hazard or potential hazard exists. Where feasible, hazards are prevented by effective design of the jobsite or job. Where it is not feasible to eliminate them, they are controlled to prevent unsafe and unhealthful exposure. Elimination or controls is accomplished in a timely manner, once a hazard or potential hazard is recognized.

(4) *Safety and health training* addresses the safety and health responsibilities of all personnel concerned with the site, whether salaried or hourly. It is often most effective when incorporated into other training about performance requirements and job practices. Its complexity depends on the size and complexity of the worksite, and the nature of the hazards and potential hazards at the site.

(c) **Recommended Actions**

(1) *Management Commitment and Employee Involvement.*

(i) *State clearly a worksite policy* on safe and healthful work and working conditions, so that all personnel with responsibility at the site and personnel at other locations with responsibility for the site understand the priority of safety and health protection in relation to other organizational values.

(ii) *Establish and communicate* a clear goal for the safety and health program and objectives for meeting that goal, so that all members of the organization understand the results desired and the measures planned for achieving them.

(iii) *Provide visible top management* involvement in implementing the program, so that all will understand that management's commitment is serious.

(iv) *Provide for and encourage* employee involvement in the structure and operation of the program and in decisions that affect their safety and health, so that they will commit their insight and energy to achieving the safety and health program's goal and objectives.

(v) *Assign and communicate* responsibility for all aspects of the program so that managers, supervisors, and employees in all parts of the organization know what performance is expected of them.

(vi) *Provide adequate authority* and resources to responsible parties, so that assigned responsibilities can be met.

(vii) *Hold managers, supervisors, and employees* accountable for meeting their responsibilities, so that essential tasks will be performed.

(viii) *Review program operations* at least annually to evaluate their success in meeting the goal and objectives, so that deficiencies can be identified and the program and/or the objectives can be revised when they do not meet the goal of effective safety and health protection.

(2) *Worksite Analysis.*

(i) *So that all hazards are identified:*

[A] Conduct comprehensive baseline worksite surveys for safety and health and periodic comprehensive update surveys;

[B] Analyze planned and new facilities, processes, materials, and equipment; and

[C] Perform routine job hazard analyses.

(ii) *Provide for regular site safety* and health inspection, so that new or previously missed hazards and failures in hazard controls are identified.

(iii) *So that employee insight and experience* in safety and health protection may be utilized and employee concerns may be addressed, provide a reliable system for employees, without fear of reprisal, to notify management personnel about conditions that appear hazardous and to receive timely and appropriate responses; and encourage employees to use the system.

(iv) *Provide for investigation of accidents* and "near miss" incidents, so that their causes and means for their prevention are identified.

(v) *Analyze injury and illness trends* over time, so that patterns with common causes can be identified and prevented.

(3) *Hazard Prevention and Control.*

(i) *So that all current and potential hazards,* however detected, are corrected or controlled in a timely manner, established procedures for that purpose, using the following measures:

[A] Engineering techniques where feasible and appropriate;

[B] Procedures for safe work which are understood and followed by all affected parties, as a result of training, positive reinforcement, correction of unsafe performance, and, if necessary, enforcement through a clearly communicated disciplinary system;

[C] Provision of personal protective equipment; and

[D] Administrative controls, such as reducing the duration of exposure.

(ii) *Provide for facility and equipment maintenance,* so that hazardous breakdown is prevented.

(iii) *Plan and prepare for emergencies,* and conduct training and drills as needed, so that the response of all parties to emergencies will be "second nature."

(iv) *Establish a medical program* which includes availability of first aid on site and of physician and emergency medical care nearby, so that harm will be minimized if any injury or illness does occur.

(4) *Safety and Health Training.*

(i) *Ensure that all employees* understand the hazards to which they may be exposed and how to prevent harm to themselves and others from exposure to these hazards, so that employees accept and follow established safety and health protections.

(ii) *So that supervisors* will carry out their safety and health responsibilities effectively, ensure that they understand those responsibilities and the reasons for them, including:

[A] Analyzing the work under their supervision to identify unrecognized potential hazards;

[B] Maintaining physical protections in their work areas; and

[C] Reinforcing employee training on the nature of potential hazards in their work and on needed protective measures, through continual performance feedback and, if necessary, through enforcement of safe work practices.

(iii) *Ensure that managers* understand their safety and health responsibilities, as described under (c)(1), "Management Commitment and Employee Involvement," so that the managers will effectively carry out those responsibilities.

U.S. Department of Labor

Occupational Safety and Health Administration
Washington, D.C. 20210

OSHA Instruction STD 1-11.6A February 19, 1979 Office of Compliance Programming

SUBJECT: Disparity Between 29 CFR 1910.178(c)(2)(vii) and Table N-1, 29 CFR 1910.178(c)(2)

A. Purpose. This instruction resolves a disparity between the context of 29 CFR 1910.178 (c)(2)(vii) and Table N-1, Summary Table On Use Of Industrial Trucks In Various Locations, 29 CFR 1910.178(c)(2).

B. Scope. This instruction applies OSHA-wide.

C. Cancellation. OSHA Instruction STD 1-11.6, July 18, 1978, Office of Compliance Programming, (previously OSHA Program Directive #100-104) is canceled.

D. Action. For compliance purposes, OSHA permits the use of truck types referenced in 29 CFR 1910.178(c)(1), with certain limitations, as specified in the following paragraphs:

1. *Truck types indicated* in 29 CFR 1910.178(c)(2)(vii) as DY, EE and EX are allowed in all Class II, Division 2, locations.
2. *Truck types designated* as DS, ES, GS or LPS may be used in the following Class II, Division 2 locations, rooms and areas containing only closed spouting and conveyors, closed bins or hoppers, or machines and equipment from which appreciable quantities of dust would escape only under abnormal operating conditions; rooms or areas into which explosive or ignitable concentrations of suspended dust might be communicated only under abnormal operating conditions; rooms or areas where the formation of explosive or ignitable concentrations of suspended dust is prevented by the operation of effective dust control equipment; warehouses and shipping rooms where dust producing materials are stored or handled only in bags or containers and other similar locations.

E. Federal Program Changes. This instruction describes a Federal program change which affects State programs. Each Regional Administrator shall:

1. *Ensure that this change* is forwarded to each State designee.
2. *Explain the technical content* of the change to the State designee as requested.
3. *Ensure that State designees* are asked to acknowledge receipt of the plan change in writing, within 30 days of notification, to the Regional Administrator. This acknowledgment should include a description either of the State's plan to implement the change or of the reasons why the change should not apply to that State.
4. *Review policies,* instructions and guidelines issued by the State to determine that this change has been communicated to State program personnel. Routine monitoring activities (accompanied inspections, spotcheck visits, and case file reviews) shall also be used to determine if this change has been implemented in actual performance.

F. Background. In NFPA No. 505-1969, the source standard for 29 CFR 1910.178(b), (c) and (d), paragraphs 206b and 209a state that approved power-operated trucks designated as DS, ES, GS or LPS may be used in Class II, Division 2, hazardous locations, if permitted by the authority having jurisdiction. These acronyms are earmarked by asterisks in NFPA Table 1 to indicate "Permitted with approval of the authority having jurisdiction." Although OSHA, by 29 CFR 1910.178(c)(2)(vii), chose not to permit these designated industrial trucks for use in all Class II, Division 2, locations, their acronyms, nevertheless, were inadvertently left in the Table when the NFPA Table 1 became OSHA Table N-1. Moreover, the first sentence of 29 CFR 1910.178(c)(2) states: "For OSHA Instruction STD 1-11.6A February 19, 1979 Office of Compliance Programming specific areas of use see Table N-1 which tabulates the information contained in this section." This statement is obviously inconsistent with the provisions of 29 CFR 1910.178(c)(2)(vii).

Irving Weisblatt Acting Director, Federal Compliance and State Programs

DISTRIBUTION: National, Regional and Area Offices All Compliance Officers Compliance Supervisors State Designees HEW-1 NIOSH Regional Program Directors

Material Approved for Incorporation by Reference

Revised as of July 1, 2005

The Director of the Federal Register has approved under 5 U.S.C. 552(a) and 1 CFR Part 51 the incorporation by reference of the following publications. This list contains only those incorporations by reference effective as of July 1, 2004. Incorporations by reference found within a regulation are effective upon the effective date of that regulation.

29 CFR (Part 1926)

Occupational Safety and Health Administration, Department of Labor

(Copies of the documents listed in this table are available through the Technical Data Center, U.S. Department of Labor, Washington, D.C., and through Regional Offices of the Occupational Safety and Health Administration. For a complete listing of these addresses, see the end of this table.)

Material	29 CFR
American Conference of Governmental Industrial Hygienists P.O. Box 1937, Cincinnati, OH 45201	
ACGHI Threshold Limit Values (1970)	1926.55
American National Standards Institute 25 West 43rd Street, Fourth Floor, New York, NY 10036; Telephone: (212) 642-4900	
ANSI A10.3-70 Safety Requirements for Explosive-Actuated Fastening Tools	1926.302
ANSI A10.4-63 Safety Requirements for Workmen's Hoists	1926.552
ANSI A10.5-69 Safety Requirements for Material Hoists	1926.552
ANSI A11.1-65 (R 70) Practices for Industrial Lighting (IES RP7-1965)	1926.56
ANSI A14.1-69 Safety Code for Portable Wood Ladders	1926.450; 1926.451
ANSI A14.2-56 Safety Code for Portable Metal Ladders	1926.450; 1926.451
ANSI A14.3-56 Safety Code for Fixed Ladders	1926.450
ANSI A17.1-65 Safety Code for Elevators, Dumbwaiters, and Moving Walks	1926.552
ANSI A17.1a-67 Supplement to A17.1-65	1926.552
ANSI A17.1b-68 Supplement to A17.1-65 and A17.1a-67	1926.552
ANSI A17.1c-69 Supplement to A17.1-65, A17.1a-67, and A17.1b-68	1926.552
ANSI A17.1d-70 Supplement to A17.1-65, A17.1a-67, A17.1b-68, and A17.1c-69	1926.552
ANSI A17.2-60 Practice for the Inspection of Elevators (Inspector's Manual)	1926.552
ANSI A17.2a-65 Addenda to A17.2-60	1926.552
ANSI A17.2B-67 Supplement to A17.2-60	1926.552
ANSI A92.2-69 Vehicle Mounted Elevating and Rotating Work Platforms	1926.451; 1926.453
ANSI A120.1-70 Safety Code for Power-Operated Platforms Used for Exterior Building Maintenance	1926.451
ANSI B7.1-70 Safety Code for the Use, Care, and Protection of Abrasive Wheels	1926.303
ANSI B15.1-53 (R 58) Safety Code for Mechanical Power-Transmission Apparatus	1926.300; 1926.550
ANSI B20.1-57 Safety Code for Conveyors, Cableways, and Related Equipment	1926.555
ANSI B30.2.0-67 Safety Code for Overhead and Gantry Cranes (Partial Revision of B30.2-43)	1926.550
ANSI B30.5-68 Safety Code for Crawler, Locomotive, and Truck Cranes (Partial Revision of B30.2-43)	1926.550
ANSI B30.6-69 Safety Code for Derricks (Partial Revision of B30.2-43)	1926.550
ANSI B56.1-69 Safety Standards for Powered Industrial Trucks (ISO R1074)	1926.602
ANSI D6.1-71 Manual on Uniform Traffic Control Devices for Streets and Highways	1926.200; 1926.201; 1926.202
ANSI J6.1-50 (R 71) Rubber Insulating Line Hose	1926.951
ANSI J6.2-50 (R 71) Rubber Insulating Hoods	1926.951
ANSI J6.4-71 Rubber Insulating Blankets	1926.951
ANSI J6.5-71 Rubber Insulating Sleeves	1926.951
ANSI J6.6-71 Rubber Insulating Gloves	1926.951
ANSI J6.7-35 Rubber Matting for Use Around Electrical Apparatus	1926.951
ANSI O1.1-61 Safety Code for Woodworking Machinery	1926.304
ANSI Z35.1-68 Specifications for Accident Prevention Signs	1926.200
ANSI Z35.2-68 Specifications for Accident Prevention Tags	1926.200
ANSI Z49.1-67 Safety in Welding and Cutting	1926.350
ANSI Z87.1-68 Practice for Occupational and Educational Eye and Face Protection (Partial Revision of Z2.1-59)	1926.102
ANSI Z89.1-69 Safety Requirements for Industrial Head Protection	1926.100
ANSI Z89.2-71 Safety Requirements for Industrial Protective Helmets for Electrical Workers, Class B	1926.100; 1926.951
American Society of Agricultural Engineers (ASAE) 2950 Niles Rd., P.O. Box 229, St. Joseph, MI 49085	
ASAE R313 Soil Cone Penetrometer, 1971	1926.1002
American Society of Mechanical Engineers Three Park Avenue, New York, NY 10016-5990; Order inquiries: 22 Law Drive, P.O. Box 2900, Fairfield, NJ 07007; Telephone: 1-800-843-2763	
Power Boilers (Sec. I), 1968	1926.603
Pressure Vessels (Sec. VIII), 1968	1926.603
American Society for Testing and Materials 100 Barr Harbor Drive, West Conshohocken, PA 19428-2959; Telephone: (610) 832-9585; Fax: (610) 832-9555	
ASTM A 370-68 Methods and Definitions for Mechanical Testing of Steel Products	1926.1001
ASTM B 117-64 (50 Hour Test)	1926.959
ASTM D 56-69 Test for Flash Point by the Tag Closed Tester	1926.155
ASTM D 93-69 Test for Flash Point by the Pensky-Martens Closed Tester	1926.155
ASTM D 323-58 (R-68) Test for Vapor Pressure of Petroleum Products (Reid Method)	1926.155
American Traffic Safety Services Association 15 Riverside Parkway, Suite 100, Fredericksburg, VA 22406-1022	
Part VI of the Manual on Uniform Traffic Control Devices, Millenium Edition, December 2000	1926.200(g); 1926.201; 1926.202
California, State of	
Construction Safety Orders, Department of Industrial Relations: Division 5, Labor Code, Section 6312	1926.1000
Commerce, Department of	
National Institute of Standards and Technology (formerly National Bureau of Standards) (Dept. of Commerce) PS 1-66, American Plywood Association (1966); PS 20-79, American Softwood Lumber Association (1970)	1926.451
General Services Administration	
QQ-P-416, Federal Specification Plating Cadmium (Electrodeposited)	1926.104

Institute of Makers of Explosives 1420 Lexington Ave., New York, NY 10017	
Publication No. 2, June 5, 1964, American Table of Distances for Storage of Explosives	1926.914
Publication No. 20, March 1968, Radio Frequency Energy — A Potential Hazard in the Use of Electrical Blasting Caps	1926.900
Interior, Department of Bureau of Reclamation (Dept. of Interior) Safety and Health Regulations for Construction, Part II (Sept. 1971)	1926.1000
National Fire Protection Association 1 Batterymarch Park, Quincy, MA 02269-9101; Telephone: (800) 344-3555	
NFPA 10A-1970 Recommended Good Practice for the Maintenance and Use of Portable Fire Extinguishers	1926.150
NFPA 13-1969 Standards for the Installation of Sprinkler Systems	1926.152
NFPA 30-1969 Flammable and Combustible Liquid Code, Chapters III and IV	1926.152
NFPA 70-1971 National Electrical Code (ANSI C1-1971)	1926.151; 1926.351; 1926.400; 1926.401; 1926.404; 1926.803
NFPA 80-1970 Standards for Fire Doors and Windows	1926.152
NFPA 251-1969 Standard Methods of Fire Tests of Building Construction and Materials	1926.151; 1926.155
NFPA 385-1966 Recommended Regulatory Standards for Tank Vehicles and Flammable and Combustible Liquids	1926.152
Power Crane and Shovel Association	
Std. No. 1-1968 Mobile Crane and Excavator Standards	1926.602
Std. No. 2-1968 Mobile Hydraulic Crane Standards	1926.550; 1926.602
Std. No. 3-1969 Mobile Hydraulic Excavator Standards	1926.602
Society of Automotive Engineers 400 Commonwealth Dr., Warrendale, PA 15096; Telephone: (412) 776-4841	
SAE Handbook-1970, pages 1088-1103 (Nomenclature and Description)	1926.602
SAE Handbook-1971	1926.1001; 1926.1002
SAE J166-1971 Minimum Performance Criteria for Brake Systems for Off-Highway Trucks and Wagons	1926.602
SAE J167-1970 Protective Frame with Overhead Protection — Test Procedures and Performance Requirements	1926.1003
SAE J168 (July 1970) Protective Enclosures — Test Procedures and Performance Requirements	1926.1002
SAE J236-1971 Minimum Performance Criteria for Brake Systems for Rubber-Tire Self-Propelled Graders	1926.602
SAE J237-1971 Minimum Performance Criteria for Brake Systems for Off-Highway Rubber-Tired Front End Loaders and Dozers	1926.602
SAE J319b-1971 Minimum Performance Criteria for Brake Systems for Off-Highway Rubber-Tired Self-Propelled Scrapers	1926.602
SAE J320a-1971 Minimum Performance Criteria for Roll-Over Protective Structures for Rubber-Tired Self-Propelled Scrapers	1926.1001
SAE J321a-1970 Fenders for Pneumatic-Tired Earthmoving Haulage Equipment	1926.602
SAE J333a-1970 Operation Protection for Wheel-Type Agricultural and Industry Tractors	1926.602; 1926.1002
SAE J334a-1970 Protective Frame Test Procedures and Performance Requirements	1926.1002
SAE J386-1969 Seat Belt for Construction Equipment	1926.602
SAE J394-1971 Minimum Performance Criteria for Roll-Over Protective Structures for Rubber-Tired Front End Loaders and Rubber-Tired Dozers	1926.1001
SAE J395-1971 Minimum Performance Criteria for Roll-Protective Structures for Crawler Tractors and Crawler-Type Loaders	1926.1001
SAE J396-1971 Minimum Performance Criteria for Roll-Over Protective Structures for Motor Graders	1926.1001
SAE J397-1969 Critical Zone Characteristics and Dimensions for Operators of Construction and Industrial Machinery	1926.1001
SAE J743a-1964 Tractor Mounted Side Boom	1926.550
SAE J959-1966 Lifting Crane Wire-Rope Strength Factors	1926.550
U.S. Army Corp of Engineers EM-385-1 (Mar. 1967), General Safety Requirements	1926.1000

Addresses

Technical Data Center: Frances Perkins Department of Labor Building, Room N2625, 200 Constitution Ave., NW, Washington, D.C. 20210.

Boston Regional Office-Region I: Regional Administrator, U.S. Department of Labor-OSHA, JFK Federal Building, Rm. E-340, Boston, MA 02203.

New York Regional Office-Region II: Regional Administrator, U.S. Department of Labor-OSHA, 201 Varick St., Room 670, New York, NY 10014.

Philadelphia Regional Office-Region III: Regional Administrator, U.S. Department of Labor-OSHA, Gateway Bldg., Suite 2100, 3535 Market St., Philadelphia, PA 19104.

Atlanta Regional Office-Region IV: Regional Administrator, U.S. Department of Labor-OSHA, 1375 Peachtree St., NE, Suite 587, Atlanta, GA 30367.

Chicago Regional Office-Region V: Regional Administrator, U.S. Department of Labor-OSHA, 32nd Fl., Room 3244, 230 S. Dearborn St., Chicago, IL 60604.

Dallas Regional Office-Region VI: Regional Administrator, U.S. Department of Labor-OSHA, 555 Griffin St., Room 602, Dallas, TX 75202.

Kansas City Regional Office-Region VII: Regional Administrator, U.S. Department of Labor-OSHA, 1100 Main St., Kansas City, MO 64106.

Denver Regional Office-Region VIII: Regional Administrator, U.S. Department of Labor-OSHA, 1999 Broadway, Suite 1690, Denver, CO 80202-5716.

San Francisco Regional Office-Region IX: Regional Administrator, U.S. Department of Labor-OSHA, 71 Stevenson St., Suite 420, San Francisco, CA 94105.

Seattle Regional Office-Region X: Regional Administrator, U.S. Department of Labor-OSHA, 1111 Third Avenue, Suite 715, Seattle, WA 91801-3212.

States with Approved Plans - State Office Directory

*State Plan for Public Employees Only

Alaska Dept. of Labor and Workforce Development
P.O. Box 111149
1111 W. 8th Street, Room 304
Juneau, Alaska 99811-1149
(907) 465-2700
Fax: (907) 465-2784

Industrial Commission of Arizona
800 W. Washington
Phoenix, AZ 85007-2922
(602) 542-5795
Fax: (602) 542-1614

California Dept. of Industrial Relations
455 Golden Gate Avenue
10th Floor
San Francisco, CA 94102
(415) 703-5050 Fax: (415) 703-5059

Connecticut Dept. of Labor (OSHA)*
200 Folly Brook Blvd.
Wethersfield, Connecticut 06109
(860) 263-6505
Fax: (860) 263-6529

Hawaii Dept. of Labor and Industrial Relations
830 Punchbowl Street, Suite 321
Honolulu, HI 96813
(808) 586-8842
Fax: (808) 586-9099

Illinois Dept. of Labor*
900 South Spring Street
Springfield, IL 62702
(217) 782-6206
Fax: (217) 782-0596

Indiana Dept. of Labor
State Office Building
402 West Washington Street,
Room W195
Indianapolis, Indiana 46204-2751
(317) 232-2655 Fax: (317) 233-3790

Iowa Division of Labor
1000 E. Grand Avenue
Des Moines, Iowa 50319-0209
(515) 242-5870
Fax: (515) 281-7995

Kentucky Deptartment of Labor
1047 U.S. Highway 127 South,
Suite 4
Frankfort, Kentucky 40601
(502) 564-3070
Fax: (502) 564-5387

Maryland Division of Labor and Industry
Dept. of Labor, Licensing and Regulation
1100 N. Eutaw St., Rm. 613
Baltimore, MD 21201-2206
(410) 767-2241 Fax: (410) 767-2986

Michigan Dept. of Energy, Labor and Economic Growth
MIOSHA
P.O. Box 30643
7150 Harris Drive
Lansing, MI 48909-8143
(517) 322-1817, Fax: (517) 322-1775

Minnesota Dept. of Labor and Industry
443 Lafayette Road N.
St. Paul, Minnesota 55155
(651) 284-5050
Fax: (651) 284-5721

Nevada Division of Industrial Relations
400 West King Street, Suite 400
Carson City, Nevada 89703
(775) 684-7260
Fax: (775) 687-6305

New Jersey Dept. of Labor and Workforce Development*
1 John Fitch Plaza
P.O. Box 110
Trenton, New Jersey 08625-0386
(609) 633-3896
Fax: (609) 292-3749

New Mexico Environment Department
1190 St. Francis Drive, Suite N4050
P.O. Box 5469
Santa Fe, New Mexico 87502
(505) 827-2855 Fax: (505) 827-2836

New York Dept. of Labor*
State Office Campus
Building - 12, Room 158
Albany, NY 12240
(518) 457-1263
Fax: (518) 457-5545

North Carolina Dept. of Labor
1101 Mail Service Center
Raleigh, NC 27699-1101
(919) 733-0359
Fax: (919) 733-6197

Oregon Occupational Safety & Health Division
Dept. of Consumer & Business Services
350 Winter St., NE, Room 430
P.O. Box 14480
Salem, Oregon 97309-0405
(503) 378-3272 Fax: (503) 947-7461

Puerto Rico Dept. of Labor and Human Resources
Prudencio Rivera Martínez Building
21 Floor
505 Muñoz Rivera Avenue
Hato Rey, Puerto Rico 00918
(787) 754-2119 Fax: (787) 753-9550

South Carolina Dept. of Labor, Licensing, and Regulation
Synergy Business Park,
Kingstree Building
110 Centerview Drive, PO Box 11329
Columbia, SC 29211
(803) 896-4300 Fax: (803) 896-4393

Tennessee Dept. of Labor and Workforce Development
220 French Landing Drive
Nashville, Tennessee
37243-1002
(615) 741-2582 Fax: (615) 741-5078

Utah Labor Commission (OSHA)
160 East 300 South, 3rd Floor
P.O. Box 146650
Salt Lake City, Utah 84114-6650
(801) 530-6848
Fax: (801) 530-6390

Vermont Dept. of Labor and Industry
5 Green Mountain Drive
P.O. Box 488
Montpelier, Vermont
05601-0488
(802) 828-4301
Fax: (802) 888-4022

Virgin Islands Dept. of Labor*
3012 Golden Rock
Christiansted, St. Croix, VI 00840
(340) 773-1994
Fax: (340) 773-1858

Virginia Dept. of Labor and Industry
Powers-Taylor Building
13 South 13th Street
Richmond, Virginia 23219-4101
(804) 371-2327 Fax: (804) 371-6524

Washington Dept. of Labor and Industries
General Administration Building
PO Box 44001
Olympia, Washington 98504-4001
(360) 902-4200
Fax: (360) 902-4202

Wyoming Dept. of Employment
Workers' Safety and Compensation Division
1510 E. Pershing Blvd., West Wing
Cheyenne, Wyoming 82002
(307) 777-7786 Fax: (307) 777-3646

U.S. Dept. of Labor Occupational Safety and Health Administration - Regional Offices

REGION 1

JFK Federal Building,
Room E340
Boston, Massachusetts 02203
(617) 565-9860
FAX: (617) 565-9827
Area Offices:
| Connecticut | Massachusetts | Maine | New Hampshire | Rhode Island | Vermont |

REGION 2

201 Varick Street,
Room 670
New York, New York 10014
(212) 337-2378
FAX: (212) 337-2371
Area Offices:
| New Jersey | New York | Puerto Rico | Virgin Islands |

REGION 3

The Curtis Center-Suite 740 West,
170 S. Independence Mall West
Philadelphia, PA 19106-3309
(215) 861-4900
FAX: (215) 861-4904
Area Offices:
| District of Columbia | Delaware | Maryland | Pennsylvania | Virginia | West Virginia |

REGION 4

61 Forsyth Street, SW Rm. 6T50
Atlanta, Georgia 30303
(404) 562-2300
FAX: (404) 562-2295
Area Offices:
| Alabama | Florida | Georgia | Kentucky | Mississippi | North Carolina | South Carolina | Tennessee |

REGION 5

230 South Dearborn Street,
Room 3244
Chicago, Illinois 60604
(312) 353-2220
FAX: (312) 353-7774
Area Offices:
| Illinois | Indiana | Michigan | Minnesota | Ohio | Wisconsin |

REGION 6

525 Griffin Street, Room 602
Dallas, Texas 75202
(972) 850-4145
FAX: (972) 850-4149
Area Offices:
| Arkansas | Louisiana | New Mexico | Oklahoma | Texas |

REGION 7

2 Pershing Square
2300 Main Street, Suite 1010
Kansas City, Missouri 64108
(816) 283-8745
FAX: (816) 283-0547
Area Offices:
| Iowa | Kansas | Missouri | Nebraska |

REGION 8

1999 Broadway, Suite 1690
Denver, Colorado 80202
(720) 264-6550
FAX: (720) 264-6585
Area Offices:
| Colorado | Montana | North Dakota | South Dakota | Utah | Wyoming |

REGION 9

90 7th Street, Suite 18100
San Francisco, California 94103
(415) 625-2547
FAX: (415) 625-2534
Area Offices:
| Arizona | California | Hawaii | Nevada | Guam | American Samoa |

REGION 10

1111 Third Avenue, Suite 715
Seattle, Washington 98101-3212
(206) 553-5930
FAX: (206) 553-6499
Area Offices:
| Alaska | Idaho | Oregon | Washington |

National Offices

Directorate of Cooperative and State Programs
U.S. Department of Labor
Directorate of Cooperative and State Programs
(OSHA) - Room: N3700
200 Constitution Ave. NW
Washington, D.C. 20210
(202) 693-2200

Office of State Programs
U.S. Department of Labor
Office of State Programs
(OSHA) - Room: N3700
200 Constitution Ave. NW
Washington, D.C. 20210
(202) 693-2244

STATUS OF STATE OCCUPATIONAL SAFETY AND HEALTH PLANS

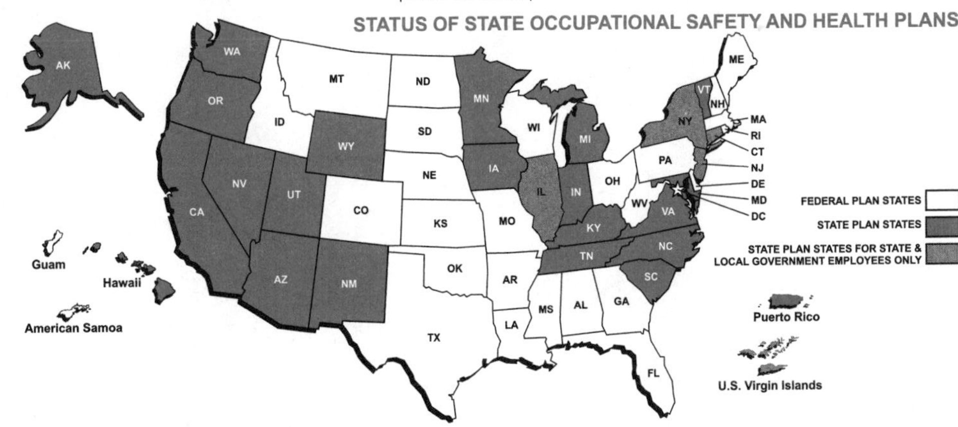

Notes

Letters of Interpretation

The letters of interpretation in this section are actual letters that OSHA sent in response to letters received from people who were confused about the meaning of the safety regulations. MANCOMM chose what we felt to be the most pertinent information from these letters and summarized it into a few sentences.

You will only find this one-of-a-kind letters of interpretation section in this CFR, available to you from MANCOMM.

Interpretations by topic **Pg. #**

§1926.701

U.S. Department of Labor Occupational Safety and Health Administration Washington, D.C. 20210

May 29, 1997

MEMORANDUM FOR: CONSTRUCTION COORDINATORS

FROM: RUSSELL B. SWANSON, DIRECTOR
Directorate of Construction

SUBJECT: Rebar Caps

The January 15, 1997, memo addressing the small mushroom style plastic rebar caps, commonly used in construction, was issued to provide information on studies conducted by the California Division of Occupational Safety and Health (Cal/OSHA) showing the general ineffectiveness of rebar caps as impalement protection under requirements in 29 CFR 1926.701(b). A number of offices as well as many in the construction community interpreted the memo as a new policy statement and a general ban on the use of small plastic rebar caps for any purpose. There is no change in OSHA policy nor is there a ban on the general use of the small plastic rebar caps as recommended by their manufacturer.

The standard, 29 CFR 1926.701(b), states: "all protruding reinforcing steel, onto and into which employees could fall, shall be guarded to eliminate the hazard of impalement" The key words are "to eliminate the hazard of impalement." Exposure to impalement is always a consideration when employees are working above rebar or other sharp protrusions. The critical element when evaluating any job activity is the recognition or identification of impalement hazards and the exposure to employees. As you know, construction activities constantly change and contractors must remain aware of and provide protection from or alternate work practices to eliminate impalement hazards.

When employees are working at any height above exposed rebar, fall protection/prevention is the first line of defense against impalement. Fall protection/prevention is also applicable when the rebar is below grade, e.g., footings or other excavations, where a fall into a trench would present an impalement hazard. When work is at grade, impalement exposure is dependent upon numerous situations and conditions; proximity of rebar to worker, height of rebar, e.g., working around rebar that is 3-6 feet high would not likely pose an impalement hazard. Rebar caps/covers are appropriate to prevent cuts, abrasions or other minor injuries when working at grade and there is no impalement hazard.

Please ensure that all compliance personnel, local contractors and labor organizations are informed of this clarification and they understand that 1926.701(b) addresses the hazard of impalement and not the use of rebar caps.

If you have any other questions or need further clarification, contact Tony Brown at (202) 219-8136.

Rebar caps ineffective as impalement protection.

"… the general ineffectiveness of rebar caps as impalement protection … [reinforces that] fall protection/prevention is the first line of defense against impalement. … Rebar caps/covers are appropriate to prevent cuts, abrasions or other minor injuries when working at grade and there is no impalement hazard."

32-002

Square Brackets
These brackets indicate where our safety professionals have added their own words in order to make the interpretations more clear.

Online Reference Code
If you would like to read the interpretations letter in its entirety, access www.oshacfr.com and select its five-digit code.

LOI
Letters of Interpretation

§1910.134(c), (e)

U.S. Department of Labor Occupational Safety and Health Administration Washington, D.C. 20210

October 5, 1998

Mr. William R. Jaggi
Trans World Airlines, Inc.
Ground Operations Center
P.O. Box 10007
St. Louis, MO 63145

Dear Mr. Jaggi:

This is in response to your letter of March 31, addressed to Ms. Bonnie Friedman, in the Occupational Safety and Health Administration's (OSHA's) Office of Public Affairs. We apologize for the long delay of this response. Enforcement of this standard will begin on the compliance date, October 5, 1998.

In your letter, you requested clarification on several respiratory protection issues. There were several statements listed that you sought to determine if they were true or not. The statements are responded to in the same order as given in your letter.

Please note that in your letter, you used the term "medical surveillance." The respiratory protection standard 1910.134 uses the term "medical evaluation."

1. **Statement: Use of the disposable paper type dust respirators does not require medical surveillance.**

   Response: Non-voluntary use of a filtering facepiece (dust mask/disposable paper type dust respirator) requires that the employer establish and implement a written respiratory protection program with worksite-specific procedures. The respiratory protection program must include the medical evaluation of employees.

   Voluntary use of a filtering facepiece respirator does not require medical evaluation. The employer needs only to ensure that the dust masks are not dirty or contaminated, that their use does not interfere with the employee's ability to work safely, and provide a copy of Appendix D to each voluntary wearer.

2. **Statement: Use of an air supplied positive pressure face mask does not require medical surveillance unless intended for rescue or emergency purposes.**

3. **Statement:** ... **surveillan...**

   Response: ... shall prov... before the... that use a... hoods mu... injuries a... pressure r... breathing... the body.

Please note that o... **Questions.** This ...

Medical evaluation for dust mask use.

"**Non-voluntary use** of a filtering facepiece (dust mask/disposable paper type dust respirator) requires that the employer establish and implement a written respiratory protection program with worksite-specific procedures [and] ... must include the medical evaluation of employees.

"**Voluntary use** ... does not require medical evaluation. The employer needs only to ensure that the dust masks are not dirty or contaminated, that their use does not interfere with the employee's ability to work safely, and provide a copy of Appendix D [Information for employees using respirators when not required under the standard] to each voluntary wearer."

32-019

§§1926.32; 1910.12

U.S. Department of Labor Occupational Safety and Health Administration Washington, D.C. 20210

August 11, 1994

MEMORANDUM FOR: REGIONAL ADMINISTRATORS

FROM: JAMES W. STANLEY
Deputy Assistant Secretary

SUBJECT: Construction vs. Maintenance

OSHA's regulations define "construction work" as "construction, alteration, and/or repair, including painting and decorating." 29 CFR 1926.32(g) and 29 CFR 1910.12(b). They further provide that OSHA's construction industry standards apply "to every employment and place of employment of every employee engaged in construction work." **Id.** at 1910.12(a). In interpreting definitional provisions in these regulations, the Occupational Safety and Health Review Commission and the courts have looked to similar definitional provisions in other federal laws and regulations. For example, OSHA's regulations make specific reference to definitions in the Davis-Bacon Act and regulations promulgated under that Act. The Davis-Bacon Act regulations issued by the Department of Labor broadly define construction work or activity as follows:

(i) The terms "building" and "work" generally include construction activity as distinguished from manufacturing, furnishing of materials, or servicing and maintenance work. The terms include, without limitation, buildings, structures, and improvements of all types, such as bridges, dams, plants, highways, parkways, streets, subways, tunnels, sewers, mains, power lines, pumping stations, heavy generators, railways,... excavating,....

(j) The terms construction, completion, or repair mean the following:

(1) all types of work done on a particular building or work at the site thereof...... 29 CFR 5.2(i) and (j).

In order for work to be construction work, the employer need not itself be a construction company. **See, e.g., New England Telephone & Telegraph Co.,** 4 OSHC 1838, 1939 (1976), **vacated on other grounds sub nom. New England Telephone & Telegraph Co. vs. Secretary of Labor** 589 F.2d 81 (1st Cir. 1978).

Further, construction work is not limited to new construction. It includes the repair of existing facilities. The replacement of structures and their components is also considered construction work. For example, in **Pacific Gas & Electric Co.,** 2 OSHC 1962 (1975), the Review Commission held that the replacement of a wooden utility pole is covered by the construction industry standards. The utility had argued that the replacement of the pole was "maintenance work," rather than "construction work." The Review Commission, however, concluded that pole replacement is "improvement" and, therefore, construction work. Similarly, construction work is typically performed outdoors, rather than at a manufacturing plant. This factor too is ... 910 F2d 1333 (6t...

There is no speci... "maintenance", "r... keeping a structu... anticipated fashio...

Construction vs. maintenance.

"... determinations of whether a contractor is engaged in maintenance operations rather than construction activities must be made on a case-by-case basis ... where an activity cannot be easily classified as construction or maintenance ... the activity should be classified so as to allow application of the more protective 1910 or 1926 standard, depending on the hazard."

32-003

§1910.134(g)(1)(i)-(iii)

U.S. Department of Labor Occupational Safety and Health Administration Washington, D.C. 20210

October 11, 1984

Mr. Mathew C. Kurzius
IBEW, Local 1673
235 Columbia Street
Dunellen, N.J. 08812

Dear Mr. Kurzius:

This is in response to your letter of September 29, 1984 concerning facial hair and the wearing of respirators. We are providing the following answers to your questions.

1. A copy of the pertinent section of the respirator standard that applies, [29 CFR 1910.134(g)(1)(i-iii)], is enclosed. It states that respirators shall not be worn when conditions prevent a good face seal. Such conditions may be a growth of beard, sideburns, a skull cap that projects under the facepiece, or temple pieces on glasses. This regulation does not ban facial hair on respirator users, per se, from the workplace.

However, when a respirator must be worn to protect employees from airborne contaminants, it has to fit correctly, and this will require the wearer's face to be clean-shaven where the respirator seals against it.

OSHA requires respirators to be used when they are necessary to protect employees against overexposure to air contaminants. When administrative or engineering controls have not kept workplace exposure to air, contaminants within OSHA's established permissible limits, then appropriate respirators must be worn by the exposed employees. The standard ([1910.134(g)(1)(i-iii)]) only applies to those employees who need the protection of a tight-fitting facepiece respirator, either routinely or in emergencies, because of such overexposure.

It does not matter if hair is allowed to grow on other areas of the face if it does not protrude under the respirator seal. Accordingly, mustaches, sideburns, and small goatees that are trimmed so that no hair underlies the seal of the respirator present no hazard and do not violate [1910.134(g)(1)(i)(A)].

2. The use of a self-contained breathing apparatus (SCBA), such as the Scott Air Pac, is not acceptable for bearded employees under emergency conditions. Since the SCBA is used in unknown concentrations for unspecified lengths of time, maximum protection must be achieved when the SCBAs are worn. The beard growth can significantly reduce the service life of the air cylinder on the SCBA which could restrict the performance in the emergency operation. The SCBA wearer can "overbreathe" when moderately heavy to heavy workloads are performed. If there is a leak caused by the beard, the air contaminant could be pulled inside the facepiece. Furthermore, the beard can interfere with the sealing of the exhalation valve and shortening the service life of the air supply. For emergency use, there is an escape hood with a continuous flow of air and a fifteen-minute service life which usually can be worn by bearded employees. Respirators of this type that have been approved by the National Institute for Occupational Safety and Health are available on the market.

3. The employer would be in violation of [1910.134(g)(1)(i)(A)] if a bearded employee wore a SCBA under a true emergency situation.

We hope this info...

Sincerely,

Cathie M. Manni...
Assistant Region...
for Technical Sup...

Facial hair and respirators.

"... mustaches, sideburns, and small goatees that are trimmed so that no hair underlies the seal of the respirator present no hazard and do not violate ...

"The use of a self-contained breathing apparatus (SCBA), such as the Scott Air Pac, is not acceptable for bearded employees ..."

32-007

§§1926.59; 1910.1200

U.S. Department of Labor Occupational Safety and Health Administration Washington, D.C. 20210

May 16, 1990

Mr. Kenneth E. Krauska
President
National Association of Plumbing-Heating-Cooling Contractors
Post Office Box 6808
Falls Church, Virginia 22046-1148

Dear Mr. Krauska:

Thank you for your letter of March 15, regarding the availability of material safety data sheets (MSDSs) on construction sites, as required under the Occupational Safety and Health Administration's (OSHA) Hazard Communication Standard (HCS), 29 CFR 1926.59. I apologize for the delay in this response.

As you are aware, the HCS "Multi-employer workplaces" provision at 29 CFR 1926.59 (e)(2) states that employers who use or store hazardous chemicals at a worksite in such a way that the employees of other employers may be exposed are required to incorporate additional elements into their hazard communication program. On a multi-employer worksite, each employer's hazard communication program must specify the method that the employer will use to provide the other employers with copies of his MSDSs, or the method he will use to make them available at a central location. The number of MSDSs that a contractor must have depends on the method used on the site to provide information to other employers.

If MSDSs are not maintained at a central location, each contractor must provide MSDSs to the other employers for each hazardous chemical the other employers' employees may be exposed to while working. Although a contractor does not actually have to physically give another employer the MSDS, he must make it available at the worksite for review. In some circumstances, a contractor may prefer to distribute copies of MSDS to other employers rather than provide continual access to those MSDSs.

If the contractor chooses to rely on a method of sharing his MSDSs at a central location, where a copy of all the employers' MSDSs will be kept, the method must also provide for access to them at all times employees are wo... location does not... would normally b... available because...

Each contractor h... hazardous chemic... maintain his MSD... be cited under 19... OSHA's complia... are set forth in th... of these sections...

The intent of OSI... chemical substan... employees, and th... The ultimate goal... healthful workpla... behind any and al...

Availability of MSDSs on multi-employer construction sites.

"... employers who use or store hazardous chemicals at a worksite in such a way that the employees of other employers may be exposed are required to incorporate additional elements into their hazard communication program. ... [the] program must specify the method that the employer will use to provide ... MSDSs to the other employers for each hazardous chemical the other employers' employees may be exposed to while working. Although a contractor does not actually have to physically give another employer the MSDS, he must make it available at the worksite for review. ... If the contractor chooses to rely on a method of sharing his MSDSs at a central location ... the method must also provide for access to them at all times employees are working on the site."

32-004

§1926.65

U.S. Department of Labor Occupational Safety and Health Administration
Washington, D.C. 20210

November 22, 1994

Ms. Jackie H. Ward
ENTERGY Gulf States Utilities
Post Office Box 2951
Beaumont, Texas 77704

Dear Ms. Ward:

Thank you for your letter of September 2, forwarded to the Occupational Safety and Health Administration's (OSHA's) Directorate of Compliance Programs from our Area Office in Baton Rouge, Louisiana, concerning the use of computer-based training to satisfy OSHA training requirements. In your letter, you ask a series of questions requesting clarification on whether the use of computer-based training is sufficient to comply with the minimum training requirements for initial employee training and retraining, in particular with regard to the number of hours of training required. We assume that your primary interest is in the training requirements of OSHA's Hazardous Waste Operations and Emergency Response Standard (HAZWOPER, 29 CFR 1910.120 and 1926.65), although you questions are also relevant to the training requirements of other OSHA standards. Each of the questions in your letter are answered in turn below.

1. What is OSHA's position on computer-based training programs for cognitive training?

In OSHA's view, self-paced, interactive computer-based training can serve as a valuable training tool in the context of an overall training program. However, use of computer-based training by itself would not be sufficient to meet the intent of most of OSHA's training requirements, in particular those of HAZWOPER. Our position on this matter is essentially the same as our policy on the use of training videos, since the two approaches have similar shortcomings. OSHA urges employers to be wary of relying solely on generic, "packaged" training programs in meeting their training requirements. For example, training under HAZWOPER includes site-specific elements and should also, to some degree, be tailored to workers' assigned duties.

Safety and health training involves the presentation of technical material to audiences that typically have not had formal education in technical or scientific disciplines, such as in areas of chemistry or physiology. In an effective training program, it is critical that trainees have the opportunity to ask questions where material is unfamiliar to them. In a computer-based program, this requirements may be providing a telephone hotline so that trainees will have direct access to a qualified trainer.

Equally important is the use of hands-on training and exercises to provide trainees with an opportunity to become familiar with equipment and safe practices in a non-hazardous setting. Many industrial operations, and in particular hazardous waste operations, can involve many complex and hazardous tasks. It is imperative that employees be able to perform such tasks safely. Traditional, hands-on training is the preferred m... hands-on training... first, to ensure th... workers have ma... program is likely...

Thus, OSHA beli... and health trainin... supplemented by...

Computer-based training must be supplemented with hands-on training.

"... computer-based training programs can be used ... provided that the program is supplemented by the opportunity for trainees to ask questions ... and provides trainees with sufficient hands-on experience. ... computer-based training will not relieve employers of their obligation to ensure that employees receive the minimum required amount of training ..."

32-001

§1926.100

U.S. Department of Labor Occupational Safety and Health Administration
Washington, D.C. 20210

July 22, 1992

Artie Scruggs
Business Manager
International Brotherhood
of Electrical Workers
P.O. Box 2796
Kokomo, Indiana 46904-2796

Dear Mr. Scruggs:

This is in response to your May 14 letter requesting an interpretation of an Occupational Safety and Health (OSHA) standard addressing the use of head protection at construction sites. I apologize for the delay in responding to your inquiry.

As you know, sections 1926.100(a) and (b) require employees to be protected by protective helmets that meet the specifications of the American National Standards Institute (ANSI) standard Z89.1, 1969, Safety Requirements for Industrial Head Protection. Because ANSI only tests and certifies hard hats to be worn with the bill forward, hard hats worn with the bill to the rear would not be considered reliable protection and would not meet the requirements of 29 CFR 1926.100(a) and (b) unless the hard hat manufacturer certifies that this practice meets the ANSI Z89.1- 1969 requirements.

If we can be of any further assistance, please contact Mr. Roy F. Gurnham or Mr. Dale R. Cavanaugh of my staff in the Office of Construction and Maritime Compliance Assistance at (202) 523- 8136.

Sincerely,

Patricia K. Clark, Director
Directorate of Compliance Programs

Hard hats should be worn with the bill forward.

"Because ANSI only tests and certifies hard hats to be worn with the bill forward, hard hats worn with the bill to the rear would not be considered reliable protection and would not meet the requirements of 29 CFR 1926.100(a) and (b) unless the hard hat manufacturer certifies that this practice meets the ANSI Z89.1-1969 requirements."

32-015

§1926.95; .100

U.S. Department of Labor Occupational Safety and Health Administration
Washington, D.C. 20210

December 31, 1998

MEMORANDUM FOR: Mr. John Jones
Region V - Indiana OSHA

FROM: Russell B. Swanson, Director
Directorate of Construction

SUBJECT: Western Hard Hats

This is in response to your e-mail requesting an interpretation regarding 29 CFR 1926.100, Head Protection. You asked if western hard hats [hard hats with brims] comply with the requirements of this standard. More specifically, you had concerns that the hat presents a hazard because hot items could fall into the brim and get caught or the brim could be struck by something and would cause the hat to easily fall off the employee's head.

Section 1926.100(a) provides that a hard hat must be used to protect against head injury from impact, or from falling or flying objects, or from electrical shock and [electrical] burns. Section .100(b) provides that a helmet for protection against impact and penetration of falling and flying objects must meet the requirements of ANSI Z89.1-1969. Therefore, as long as a helmet meets the 1969 ANSI standard, the helmet is sufficient for purposes of protecting against impact and penetration hazards of falling and flying objects.

A western hard hat complies with the 29 CFR 1926.100 standard for protection against falling and flying objects as long as it meets the requirements of ANSI Z89.1-1969. The impact testing requirements of ANSI Z89.1-1986 are even more specific than the 1969 version of the standard. Since the revised standard is, in effect, more rigorous, our position is that a hard hat that meets the criteria of the revised standard also meets the 1926.100 requirement for falling and flying object protection. The possibility that an object might strike the brim and cause the helmet to come off does not alter the fact that a helmet meeting the ANSI standard meets the section .100(a) requirement for falling object impact protection.

If there is a danger of head injury from something other than falling and flying objects, the employer is obligated under .100(a) to provide head protection against that hazard as well. There is no reference to the ANSI standard in 1926.100 with respect to protecting against these other impact hazards. If these other impact hazards are present, the employer must ensure that the head protection will protect against them. Because of the wide variety and circumstances of these other types of hazards, we cannot say, as a general matter, w... provided by a we...

Section 1926.100... such hazards, an ... equipment). Ther... an employer to pr... hazard. In such a ... general rule, a bri... provide protectio...

Western hard hats acceptable if ANSI requirements met.

"A western hard hat complies with the 29 CFR 1926.100 standard for protection against falling and flying objects as long as it meets the requirements of ANSI Z89.1-1969. ... The possibility that an object might strike the brim and cause the helmet to come off does not alter the fact that a helmet meeting the ANSI standard meets the section [1926].100(a) requirement for falling object impact protection."

32-014

§1926.150

U.S. Department of Labor Occupational Safety and Health Administration
Washington, D.C. 20210

May 26, 1995

Mr. Frank Schreiber
Kulback's Construction Inc.
6363 Transit Road
Depew, New York 14043

Dear Schreiber:

This is in response to your letter of March 7 regarding the size of fire extinguisher for construction equipment.

Per our conversation of May 23, the Occupational Safety and Health Administration's (OSHA) fire protection and prevention standard (29 CFR 1926.150 - Subpart F) does not state any specific weight or size requirements for fire extinguishers on construction equipment. As part of a fire prevention plan, a hazard assessment should be performed in order to determine potential hazards that exist and the types of fire fighting equipment that would be necessary.

We appreciate your interest in employee safety and health. If you require further assistance, please contact Dale Cavanaugh in the Office of Construction and Maritime Compliance Assistance on 202-219-8136.

Sincerely,

Raymond E. Donnelly, Director
Office of General Industry Compliance

Enclosures

Weight or size requirements for fire extinguishers on construction equipment.

"... the Occupational Safety and Health Administration's (OSHA) fire protection and prevention standard ... does not state any specific weight or size requirements for fire extinguishers on construction equipment. ... a hazard assessment should be performed ... to determine potential hazards that exist and the types of fire fighting equipment that would be necessary."

32-017

§1926.152(b)(4)(v)

U.S. Department of Labor — **Occupational Safety and Health Administration**
Washington, D.C. 20210

November 18, 1976

MEMORANDUM FOR: Robert C. Tice-Regional Administrator

ATTN: C. R. Holder-ARA for Technical Support

SUBJECT: Clarification of Difference Between 29 CFR 1910.106(d)(4)(iii) and 29 CFR 1926.152(b)(4)(v)

This is in response to your request of June 15, 1976, concerning a clarification of the two subject standards for Dr. C. Ray Asfahl of the University of Arkansas.

Class 1, Division 1, locations are where hazardous concentrations of flammable gases or vapors exist continuously, intermittently, or periodically under normal conditions.

Class 1, Division 2, locations are where volatile flammable liquids or flammable gases are handled, processed or used, but in which the hazardous liquids, vapors or gases will normally be confined within closed containers or closed systems from which they can escape only in case of accidental rupture or breakdown of such containers or systems, or in case of abnormal operation of equipment and in which hazardous concentrations of gases or vapors are normally prevented by positive mechanical ventilation, but which might become hazardous through failure or abnormal operation of the ventilating equipment.

The work environment found in the construction industry, being of a temporary and dynamic state, warrants the inside storage rooms for flammable and combustible liquids be classified as Class 1 Division 1 locations for electrical wiring and equipment. Storage areas are used for handling, transferring of liquids, spills, open and ruptured containers. These are normal conditions for the storage areas in construction because the areas are subject to traffic from employees of other employers, and are more difficult to control than the storage areas in general industry which usually are permanent facilities with positive mechanical ventilation systems.

The two standards have been written to be applied to their respective industries. The intent of both standards is to prevent a fire or an explosion from flammable gases or vapor which may accumulate in an inside storage room for flammable and combustible liquids.

Richard P. Wilson
Acting Director,
Federal Complian

Construction industry storage rooms are Class 1 Division 1 locations for electrical wiring and equipment.

"The work environment found in the construction industry, being of a temporary and dynamic state, warrants the inside storage rooms for flammable and combustible liquids be classified as Class 1 Division 1 locations for electrical wiring and equipment ... because the areas are subject to traffic from employees of other employers, and are more difficult to control than the storage areas in general industry which usually are permanent facilities with positive mechanical ventilation systems."

32-016

§1926.350(a)(10)

U.S. Department of Labor — **Occupational Safety and Health Administration**
Washington, D.C. 20210

December 31, 1998

Steven M. Dineen
Manager, Safety & Loss Control
Montgomery KONE Inc.
One Montgomery Court
Moline, Illinois 61265

Re: 29 CFR 1926.350(a)(10); definition of "in storage;" gas cylinders; intermittent use

Dear Mr. Dineen:

This is in response to a series of letters that you have written to OSHA beginning April 7, 1994, and most recently on July 16, 1996, requesting an interpretation of the gas cylinder storage requirements in 29 CFR 1926.350(a)(10). Please accept our sincere apologies for the inordinate and inexcusable delay in providing you with this interpretation. We have made changes that, we trust, will prevent such a delay from occurring in the future.

Section 1926.350(a)(10) requires that oxygen cylinders that are in storage be separated from fuel-gas cylinders and combustible materials. It specifies that the cylinders either be separated by a minimum of 20 feet or have a non-combustible fire wall (with a fire resistance rating of one half hour) at least five feet high. However, the standard does not specify when cylinders are considered to be in storage. You asked if the storage requirements apply to gas cylinders that are in "intermittent use."

When a gas cylinder is considered to be "in storage"
We consider a cylinder to be in storage when it is reasonably anticipated that gas will not be drawn from the cylinder within 24 hours (overnight hours included). At that point the storage requirements must be met. In contrast, i
hours, the cylinde
not apply. Wheth
whether specific
expected to be re

So, with respect t
anticipates that g
considered to be i
cylinders when th
use from being k
metal or other sou

Why 24 hours?
We believe that it
will be drawn fro
prevent the sprea
begins. Both stor
storage area some
be transported fro
requirements. Als
storage.

Storage requirements for gas cylinders.

"Section 1926.350(a)(10) requires that oxygen cylinders that are in storage be separated from fuel-gas cylinders and combustible materials. It specifies that the cylinders either be separated by a minimum of 20 feet or have a non-combustible fire wall (with a fire resistance rating of one half hour) at least five feet high. ... We consider a cylinder to be in storage when it is reasonably anticipated that gas will not be drawn from the cylinder within 24 hours (overnight hours included). At that point the storage requirements must be met. ... 'reasonably anticipated' ... is based on whether specific welding or cutting work is planned for that period and the number of gas cylinders expected to be required to do that work. ... The storage requirements cannot be met if an oxygen cylinder and an acetylene cylinder are on a welding cart ... proper storage in addition to the separation/fire wall requirement ... [includes] installing a valve protector cap, meeting storage area requirements, and properly securing the cylinders."

32-013

§1926.302(a); .403(a); .449; Subpart K

U.S. Department of Labor — **Occupational Safety and Health Administration**
Washington, D.C. 20210

January 28, 1994

Mr. Paul Weiss
Regional Sales Manager
Wyco Tool Company
P.O. Box 185
Augusta, WV 26704

Dear Mr. Weiss:

This is in response to your December 7 letter requesting an interpretation of Occupational Safety and Health Administration (OSHA) standards addressing hand tools used in construction.

With regard to whether all electric tools need to be tested by a qualified national testing laboratory and be listed and labeled, please be advised that 29 CFR 1926.403(a) requires all electric equipment, including double insulated and grounded hand tools, to be approved. OSHA defines "approved" at 1926.449 as being accepted, or certified, or listed, or labeled, or otherwise determined to be safe by a qualified testing laboratory.

With regard to the provisions addressing the grounding of electric hand tools, please be advised that 1926.302(a) requires that an electric hand tool be either double insulated **or** be equipped with a grounding conductor with a three-prong plug in accordance with Subpart K of part 1926. Therefore, a non-grounded, double-insulated electric hand tool with associated two-prong plug which has been listed by a nationally recognized testing laboratory would comply with 1926.302(a) and 1926.403(a).

If we can be of any further assistance, please contact me or Mr. Dale Cavanaugh at (202) 219-8136.

Sincerely,

Roy F. Gurnham, P.E., J.D.
Director
Office of Construction and Maritime
Compliance Assistance

All electric tools must be approved.

"... 29 CFR 1926.403(a) requires all electric equipment, including double insulated and grounded hand tools, to be approved. ... With regard to the provisions addressing the grounding of electric hand tools ... 1926.302(a) requires that an electric hand tool be either double insulated **or** be equipped with a grounding conductor with a three-prong plug ... a non-grounded, double-insulated electric hand tool with associated two-prong plug which has been listed by a nationally recognized testing laboratory would comply ..."

32-009

§1926.403(a), (e); .405(g)(2)(iii); .416(e)(1)

U.S. Department of Labor — **Occupational Safety and Health Administration**
Washington, D.C. 20210

December 16, 1998

Mr. Dennis Vance
Safety Specialist
711 Low Gap Road
Princeton, WV 24740

Re: 1926.403(e); 1926.405(g)(2)(iii); 1926.416(e)(1).

Dear Mr. Vance:

This is in response to your letter of March 20, and your follow-up letter of October 9, concerning the use of electrical tape to repair minor damage (abrasions and cuts of limited depth) on the outer jacket of an extension cord. We apologize for the lateness of this reply.

Generally, electrical tape may be used to cover superficial damage to cord jackets
You ask whether there is any prohibition against putting electrical tape over these kinds of abrasions and nicks when there is no damage beyond the jacket -- the conductors have not been scraped or exposed and the insulation inside the jacket has not been displaced or compressed.

Section 1926.416(e)(1) provides that "worn or frayed electrical cords or cables shall not be used." Superficial nicks or abrasions -- those that only slightly penetrate the outer jacket of a flexible cord, and do not permit the cord to bend more in that area than in the rest of the cord -- do not normally render a cord "worn or frayed." Therefore, there is no need to repair or replace such a cord.

Recommendation against taping
While taping these incidental abrasions and cuts does not necessarily violate any OSHA standard, we recommend that employers not tape this type of damage for two reasons. First, Section 1926.403(a) requires that "all electrical conductors and equipment shall be approved." This standard precludes the use of approved electrical conductors and equipment if their characteristics are significantly altered. Applying electrical tape that is too thick or applying too much of it could change the cord's original flexibility and lead to internal damage. Second, the depth of the abrasions and cuts cannot be monitored to see if they get worse without removing the tape.

It should also be kept in mind that the heavy duty extension cords commonly used on construction sites are designed to withstand a hostile environment. Damage to an extension cord that is bad enough to consider taping may have caused damage beyond the jacket.

Tape may not be used to repair significant damage to cord jackets
Repair or replacement of a flexible cord (depending on its gauge) is required when the outer jacket is deeply penetrated (enough to cause that part of the cord to bend more than the undamaged part) or
penetrated compl
the standard proh
1926.403(a) requ
of materials and c
characteristics of
particular concern
damage to the co
usually will not d
flexibility charact

Electrical tape may not be used to repair a cord.

"Section 1926.403(a) requires that 'all electrical conductors and equipment shall be approved.' This standard precludes the use of approved electrical conductors and equipment if their characteristics are significantly altered. ... Taped repairs ... will not duplicate the cord's original characteristics ... neither the jacket's strength nor flexibility characteristics will be restored. Therefore, tape repairs of the jacket may not be used ..."

32-005

§1926.450(b); .451(b); .452(e), (p)(4); Sub. L App. A

U.S. Department of Labor Occupational Safety and Health Administration Washington, D.C. 20210

February 16, 2001

Mr. Robert Harrell
President, Safety Management Services
4012 Santa Nella Place
San Diego, CA 92130-2291

Dear Mr. Harrell:

This is in response to your letter dated July 7, 2000, in which you requested our response to several questions or issues regarding wooden planks used in scaffold platforms. Your questions are restated and responded to as follows:

Question (1):
Which lumber grading agencies will OSHA accept as a qualified grading agency for inspecting and stamping scaffold grade lumber?

Response:
OSHA does not require that scaffold planks be graded and stamped by a qualified grading agency. However, employers must ensure that scaffolds are designed, erected, used and maintained in compliance with the provisions of Part 1926 Subpart L, Scaffolds. This includes selecting wooden planking that will meet the design specifications of the scaffold and loads imposed on it.

Appendix A to Subpart L is a non-mandatory guide that may be used to help select lumber for scaffold planks. Under this appendix, when calculating lumber strength for purposes of complying with Subpart L, the qualified person may base his/her calculations using the grading rules established by a recognized lumber grading association, or an independent lumber grading inspection agency, and certified by the Board of Review, American Lumber Standard Committee, as set forth in the American Softwood Lumber Standard (ASLS) of the U.S. Department of Commerce. For your convenience, the current list of certified agencies is enclosed.

Question (2):
Why doesn't OSHA require a grading stamp on the scaffold plank?

Response:
This issue was co... August 30, 1996, ... marking requirem... the consumer. Als... security or safety...

As you pointed o...

Question (3):
Do OSHA Compl... grade plank base...

Response:
Compliance offic... Instead, they are ... with the requirem...

Grades and stamps for scaffold planks.

"OSHA does not require that scaffold planks be graded and stamped by a qualified grading agency. However, employers must ensure that scaffolds are designed, erected, used and maintained in compliance with the provisions of Part 1926 Subpart L, Scaffolds. This includes selecting wooden planking that will meet the design specifications of the scaffold and loads imposed on it. Appendix A to Subpart L is a non-mandatory guide that may be used to help select lumber for scaffold planks. ...The qualified person who designed the scaffold system would be expected to be well informed as to the technical aspects associated with planking, including those that pertain to the strength and structural integrity of individual scaffold components."

32-008

§1926.502(d)(16); .1050(b); .1053(a)(18)-(19)

U.S. Department of Labor Occupational Safety and Health Administration Washington, D.C. 20210

January 13, 2000

Mr. Peter G. Chaney
Mechanical Contractors Association of America, Inc.
1385 Piccard Drive
Rockville, MD 20850-4340

RE: [29 CFR 1926] Subpart X

Dear Mr. Chaney:

This is in response to your May 26, 1999, letter in which you ask for clarification of several issues relating to the use of fall protection when working from ladders during construction work. You specifically ask if OSHA has any requirements for the use of fall protection when working from ladders at heights greater than six feet. We apologize for the lateness of this response.

Fixed ladders: fall protection must be provided for employees climbing or working on fixed ladders above 24 feet.
29 CFR 1926.1053(a)(19) states that fall protection must be provided whenever the length of climb on a fixed ladder equals or exceeds 24 feet. A fixed ladder is "a ladder that cannot be readily moved or carried because it is an integral part of a building or structure" (§1926.1050(b)). Also, even if the length of climb is less than 24 feet, under §1926.1053(a)(18), cages, wells, ladder safety devices, or self-retracting lifelines must be provided where the top of the fixed ladder is greater than 24 feet above lower levels.

Portable ladders: fall protection is not required for employees climbing or working on portable ladders.
Neither the ladder standard (29 CFR 1926, subpart X) nor the fall protection standard (29 CFR 1926, subpart M) requires fall protection for workers while working on portable ladders.

You note that a number of general contractors in Georgia "are attempting to require personal fall arrest systems for their subcontractors working on ladders 6 feet or higher." Although the OSHA standards do not require fall protection for workers on fixed ladders below 24 feet or on portable ladders, we encourage employers to provide additional protection.

Personal Fall Arrest Systems can be set up to limit arrested falls to less then 15 feet.
In your letter you assert that personal fall arrest systems will not arrest a fall from an elevation lower than 15 feet. When an... or less. Using a s... system were anch...

Several factors m... system to work -- ... §1926.502(d)(16)... below. A 6 foot la... about 9 1/2 feet b... at a point that is n... to assure that no ...

Second, under §1... not more than six... attachment point ...

Fall protection for ladders in construction.

"Fixed ladders: fall protection must be provided for employees climbing or working on fixed ladders above 24 feet. ... Also, even if the length of the climb is less than 24 feet, under §1926.1053(a)(18), cages, wells, ladder safety devices, or self-retracting lifelines must be provided where the top of the fixed ladder is greater than 24 feet above lower levels.

"Portable ladders: fall protection is not required for employees climbing or working on portable ladders. ...

"Although the OSHA standards do not require fall protection for workers on fixed ladders below 24 feet or on portable ladders, we encourage employers to provide additional protection."

32-010

§1926.451(g)(1)(vii); .452(w); .453

U.S. Department of Labor Occupational Safety and Health Administration Washington, D.C. 20210

July 21, 1998

Dennis Vance
Safety Specialist
Safety Services
711 Low Gap Road
Princeton, WV 24740

Dear Mr. Vance:

RE: 1926.451(g); 1926.452(w);1926.453; 1926.502(d); CPL 2-1.23; ANSI A92.2; ANSI A92.6; scissors lift.

This is in response to your letter of May 27 to the Occupational Safety and Health Administration (OSHA) in which you asked whether an employee working from a scissors lift equipped with guardrails needed to wear a "safety harness." You further expressed concern for the confusion created by OSHA's scaffold standard and its directive, CPL 2.112 (sic). Please accept our apology for the delay in responding to this inquiry and for any previously submitted requests.

We recognize that there is confusion regarding scissors lifts and the appropriate standards governing such equipment. The confusion stems from the way OSHA's directive is worded. In CPL 2-1.23, titled "Inspection Procedures for Enforcing Subpart L, Scaffolds Used in Construction - 29 CFR 1926.450-454," dated January 7, 1997, the statement in paragraph K.9.b states that scissors lifts are addressed by 1926.453, Aerial Lifts, and not by 1926.452(w), Mobile Scaffolds. The directive should have more clearly indicated that only aerial lifts meeting the design and construction of the American National Standards Institute (ANSI) A92.2, Vehicle Mounted Elevating and Rotating Work Platforms, are addressed by 1926.453 since the coverage of that section is specifically limited to such lifts. All other types of mobile lifts would be covered by the specific requirements at 1926.452(w) and/or the general requirements of 1926.451. Please note, however, that if an employer is in full compliance with the requirements of the relevant document of the ANSI A92 series, OSHA would consider that compliance as providing an appropriate degree of safety for employees.

In regards to your... worker need only... However, if the g... platform, an addi... provision found i... section, each emp... systems."

If you require any... Directorate of Co... 200 Constitution ...

Sincerely,

Russell B. Swans...
Directorate of Co...

Standards concerning scissors lifts.

"... the statement ... that scissors lifts are addressed by 1926.453, Aerial Lifts, and not by 1926.452(w), Mobile Scaffolds ... should have more clearly indicated that only aerial lifts meeting the design and construction of the ... (ANSI) A92.2, Vehicle Mounted Elevating and Rotating Work Platforms, are addressed by 1926.453 ... All other types of mobile lifts would be covered by the specific requirements at 1926.452(w) and/or the general requirements of 1926.451. ... [W]hen working from an elevated scissors lift ... a worker need only be protected from falling by a properly designed and maintained guardrail system. However, if the guardrail system is less than adequate, or the worker leaves the safety of the work platform, an additional fall protection device would be required."

32-020

§1926.701

U.S. Department of Labor Occupational Safety and Health Administration Washington, D.C. 20210

May 29, 1997

MEMORANDUM FOR: CONSTRUCTION COORDINATORS

FROM: RUSSELL B. SWANSON, DIRECTOR
Directorate of Construction

SUBJECT: Rebar Caps

The January 15, 1997, memo addressing the small mushroom style plastic rebar caps, commonly used in construction, was issued to provide information on studies conducted by the California Division of Occupational Safety and Health (Cal/OSHA) showing the general ineffectiveness of rebar caps as impalement protection under requirements in 29 CFR 1926.701(b). A number of offices as well as many in the construction community interpreted the memo as a new policy statement and a general ban on the use of small plastic rebar caps for any purpose. There is no change in OSHA policy nor is there a ban on the general use of the small plastic rebar caps as recommended by their manufacturer.

The standard, 29 CFR 1926.701(b), states: "all protruding reinforcing steel, onto and into which employees could fall, shall be guarded to eliminate the hazard of impalement" The key words are "to eliminate the hazard of impalement." Exposure to impalement is always a consideration when employees are working above rebar or other sharp protrusions. The critical element when evaluating any job activity is the recognition or identification of impalement hazards and the exposure to employees. As you know, construction activities constantly change and contractors must remain aware of and provide protection from or alternate work practices to eliminate impalement hazards.

When employees are working at any height above exposed rebar, fall protection/prevention is the first line of defense against impalement. Fall protection/prevention is also applicable when the rebar is below grade, e.g., footings or other excavations, where a fall into a trench would present an impalement hazard. When work is at grade, impalement exposure is dependent upon numerous situations and conditions; proximity of rebar to worker, height of rebar, e.g., working around rebar that is 3-6 feet high would not likely pose an impalement hazard. Rebar caps/covers are appropriate to prevent cuts, abrasions or other minor injuries when working at grade and there is no impalement hazard.

Please ensure that all compliance personnel, local contractors and labor organizations are informed of this clarification and they understand that 1926.701(b) addresses the hazard of impalement and not the use of rebar caps.

If you have any other questions or need further clarification, contact Tony Brown at (202) 219-8136.

Rebar caps ineffective as impalement protection.

"... the general ineffectiveness of rebar caps as impalement protection ... [reinforces that] fall protection/prevention is the first line of defense against impalement. ... Rebar caps/covers are appropriate to prevent cuts, abrasions or other minor injuries when working at grade and there is no impalement hazard."

32-002

LOI Letters of Interpretation

§1926.1101

U.S. Department of Labor Occupational Safety and Health Administration
Washington, D.C. 20210

June 21, 1993

Mr. Pasquale Navarro
Vice President of Operations
UNISERVE, Inc.
55(th) Street & A.V.R.R.
Pittsburgh, Pennsylvania 15201

Dear Mr. Navarro:

This is in response to your letter of March 16, concerning procedures for leaving from and returning to an asbestos removal area when recyclable protective uniforms are being used.

You requested that we advise you whether the procedures you recommended violate any provisions of Occupational Safety and Health Administration (OSHA) standards. The procedures you recommended are as follows:

The employee comes out of the asbestos containment area, removes his/her contaminated uniform in the equipment room while wearing proper respiratory protection, showers in the shower area, enters the clean room, removes his/her respiratory protection, dons clean clothes, and then goes to lunch or some other destination. When the employee returns, he/she removes his/her clean clothes, dons the proper respiratory protection, enters the equipment room and puts on the same uniform he/she wore in the morning, re-enters the containment area and continues working.

Your recommended practice of reusing a uniform that is contaminated with asbestos is prohibited by the standard. Provision [29 CFR 1926.1101(j)(1)(iii)(B)] requires the employer to ensure that employees remove their protective clothing in the equipment room upon leaving the containment area and deposit the clothing in labeled impermeable bags or containers; whereas provision [29 CFR 1926.1101(j)(1)(ii)(c)] requires the employer to ensure that employees put on protective clothing before leaving the clean room for the containment area. Since protective clothing must be put on in the clean room and cannot be taken beyond the equipment room after it has been worn in the asbestos containment area, the reuse of recyclable protective clothing until it has been properly laundered is prohibited.

In order to make your recommendations conform with the standard you must change them to indicate that when the employee returns, he/she is to don a clean uniform in the clean room before re-entering the containment area to continue working.

We appreciate the opportunity to clarify this matter for you.

Sincerely,

Ruth McCully, Director
Office of Health

Reuse of protective clothing before laundering is prohibited.

"… [the] practice of reusing a uniform that is contaminated with asbestos is prohibited … Since protective clothing must be put on in the clean room and cannot be taken beyond the equipment room after it has been worn in the asbestos containment area, the reuse of recyclable protective clothing until it has been properly laundered is prohibited."

32-011

Index

D

E

F

G

H

I

J

L

M

N

O

P

R

S

T

U

V

W

X

How to Fill Out the 300 Log

OMB Disclosure Statement: Public reporting burden for this collection of information is estimated to average 14 minutes per response, including time to review the instructions, search and gather the data needed, and complete and review the collection of information. Persons are not required to respond to the collection of information unless it displays a currently valid OMB control number. If you have any comments about these estimates or any other aspects of this data collection, contact: US Department of Labor, OSHA Office of Statistical Analysis, Room N-3644, 200 Constitution Avenue, NW, Washington, DC 20210. Do not send the completed forms to this office.
See www.mancomm.com for updates or revisions.

OSHA's Form 300 • Log of Work-Related Injuries and Illnesses

U.S. Department of Labor • Occupational Safety and Health Administration

(Rev. 01/2009) Reflecting government revisions as of 01/2004

You must record information about every work-related death and about every work-related injury or illness that involves loss of consciousness, restricted work activity or job transfer, days away from work, or medical treatment beyond first aid. You must also record significant work-related injuries and illnesses that are diagnosed by a physician or licensed health-care professional. You must also record work-related injuries and illnesses that meet any of the specific recording criteria listed in 29 CFR 1904.8 through 1904.12. Feel free to use two lines for a single case if you need to. You must complete an Injury and Illness Incident Report (OSHA Form 301) or equivalent form for each injury or illness recorded on this form. If you're not sure whether a case is recordable, call your local OSHA office for help.

Attention: This form contains information relating to employee health and must be used in a manner that protects the confidentiality of employees to the extent possible while the information is being used for occupational safety and health purposes.

Year 20 0 9
Page 2 of 2

Form approved OMB No. 1218-0176
See OMB disclosure statement on reverse.

Company Name: *Rapid Foundry and Machine*
Establishment Name: *Davenport Works, Plant #3*
Address: *317 W. 4th St.*
City: *Davenport* State: *IA* Zip Code: *52801*

Enter your company name, establishment name, and establishment address.

Enter the year and the sequential page number.

IDENTIFY THE PERSON			DESCRIBE THE CASE			CLASSIFY THE CASE											
(A) Case no.	(B) Employee's name	(C) Job title (e.g., Welder)	(D) Date of injury or onset of illness (month/day)	(E) Where the event occurred (e.g., Loading dock north end)	(F) Describe injury or illness, parts of body affected, and object/substance that directly injured or made person ill (e.g., Second degree burns on right forearm from acetylene torch)	(G) Death	(H) Days away from work	(I) Remained at work: Job transfer or restriction	(J) Remained at work: Other recordable cases	(K) Away from work (days)	(L) On job transfer or restriction (days)	(M)(1) Injury	(2) Skin disorder	(3) Respiratory condition	(4) Poisoning	(5) Hearing loss	(6) All other illnesses
21	Mark Bagin	Welder	5 / 25	basement	fracture, left arm and left leg, fell from ladder		✓			12	15	✓					
22	Shana Alexander	Foundry man	7 / 2	pouring deck	poisoning from lead fumes			✓			30				✓		
23	Sam Sander	Electrician	8 / 5	2nd floor storeroom	broken left foot, fell over box		✓			7	30	✓					
24	Ralph Boccella	Laborer	9 / 17	packaging dept.	Back strain lifting boxes		✓ (crossed out)		✓	3		✓					
25	Jarrod Daniels	Machine opr.	10 / 23	production floor	metal shaving embedded in eye				✓			✓					
26	Privacy Case	Laborer	11 / 18	production floor	cut right index finger picking up glass contaminated with another person's blood.				✓			✓					
					Page totals ⇨	0	3	1	2	22	75	5	0	0	1	0	0

Check ONLY ONE box for each case based on the most serious outcome for that case. Enter the number of days the injured or ill worker was away from work or on job transfer or restriction. Check the "Injury" column or choose one type of illness.

(A) Case no. Enter a unique case number and use the same case number on the first report of injury/illness. *(For determination of new case, see step 4 in the Quick Start Guide to Recordkeeping.)*

(B) Recordable Person's name. Enter the Recordable Person's first and last name, unless you have a "privacy concern case". Then enter "*privacy case*" instead of the name. *(See §1904.29(b)(6)-(9).)*

(C) Job title. Enter the Recordable Person's job title at the time of incident.

(D) Date of injury or onset of illness. For occupational injuries, enter the date of the work incident that resulted in injury. For occupational illnesses, enter the date of initial diagnosis of illness, or, if absence from work occurred before diagnosis, enter the first day of the absence attributed to the illness that was diagnosed or recognized.

(E) Where the event occurred. Enter the exact place where the event happened.

(F) Description of event. Enter the type of injury or illness, enter the body parts affected, and enter the object or substance that was directly involved. Be as specific as possible. Use two lines if needed.

(G), (H), (I), (J) Classify the outcome of the case. Check only one box **(G, H, I, or J)** for the most serious outcome of the case. Column G (Death) is the most serious and Column J (Other recordable cases) is the least serious.

Revise the log Update the 300 Log if the injury or illness becomes more serious or if there is a change in the case after you recorded it. To update the case simply draw a line through the original entry and create a new entry. *(See step 7 in the Quick Start Guide to Recordkeeping.)*

(K), (L) Enter days away from work, days of job restriction, or days of job transfer. When entering lost time, count the number of calendar days. Begin counting on the **day after** the onset of the injury or illness. If the case involves both days away from work and restricted work or job transfer, enter the total number of days for each. You may stop counting once the total of both reaches 180 days. *(See §1904.7(b)(3)-(4).)*

(M) Classify the case as an injury or illness. Check only one box to identify whether the case involves an injury or a type of illness (skin disorder, respiratory condition, poisoning, hearing loss, or other type of illness). *(See below for detailed information on injuries and illnesses.)*

MANCOMM® www.mancomm.com

Do not send completed forms to OSHA unless requested – See §1904.41. See Reverse Side For 300 Log Instructions. All entries need to have a corresponding OSHA 301 Injury and Illness Incident Report or an equivalent form completed. See §1904.29(b)(2). A free copy of the Annual Summary, First Report Of Injury, OSHA Form No. 300A and 301, respectively, may be downloaded from www.mancomm.com.

You may order additional copies of this 300 Log online at www.mancomm.com or by calling 1-800-MANCOMM. (626-2666)

Do Not Post This Form!
Transfer these totals to "The Annual Summary," OSHA Form No. 300A, and post it.

Retain and update until 12 / 31 / 20 14 (5 years after the end of the current year. See §1904.33)

(M)
Injury is any wound or damage to the body resulting from an incident in the workplace. *Examples:* Cut, puncture, laceration, abrasion, fracture, bruise, contusion, chipped tooth, amputation, insect bite, electrocution, or a thermal, chemical, electrical, or radiation burn. Sprain and strain injuries to muscles, joints, and connective tissues are injuries when they result from a slip, trip, fall, or other similar incidents.

Illnesses include:

Skin diseases or disorders that are caused by work exposure to chemicals, plants, or other substances. *Examples:* contact dermatitis, eczema, or rash caused by primary irritants and sensitizers or poisonous plants; oil acne; friction blisters; chrome, ulcers or inflammation of the skin;

Respiratory conditions are illnesses associated with breathing hazardous biological agents, chemicals, dust, gases, vapors, or fumes at work. *Examples:* silicosis, asbestosis, pneumonitis, pharyngitis, rhinitis or acute congestion; farmer's lung, beryllium disease, tuberculosis, occupational asthma, reactive airways dysfunction syndrome (RADS), chronic obstructive pulmonary disease (COPD), hypersensitivity pneumonitis, toxic inhalation injury, such as metal fume fever, chronic obstructive bronchitis, and other pneumoconioses;

Poisoning includes disorders caused by abnormal concentrations of toxic substances in blood, other tissues, other bodily fluids, or the breath from ingestion or absorption of toxic substances into the body. *Examples:* poisoning by lead, mercury, cadmium, arsenic, or other metals; poisoning by carbon monoxide, hydrogen sulfide, or other gases; poisoning by benzene, benzol, carbon tetrachloride, or other organic solvents; poisoning by insecticide sprays, such as parathion or lead arsenate; poisoning by other chemicals, such as formaldehyde;

Hearing loss is when an audiogram shows a change in hearing from the baseline audiogram of an average of 10 dB or more in either ear at 2000, 3000, and 4000 hertz and the Recordable Person's total hearing loss is 25 decibels (dB) or more above audiometric zero (also averaged at 2000, 3000, and 4000 hertz) in the same ear(s);

All other illnesses. *Examples:* heatstroke, sunstroke, heat exhaustion, heat stress, and other effects of environmental heat; freezing, frostbite, and other effects of exposure to low temperatures; decompression sickness; effects of ionizing radiation (isotopes, x-rays, radium); effects of non-ionizing radiation (welding flash, ultra-violet rays, lasers); anthrax; bloodborne pathogenic diseases, such as AIDS, HIV, hepatitis B or hepatitis C; brucellosis; malignant or benign tumors; histoplasmosis; or coccidioidomycosis.

www.mancomm.com

6 88550 50652 6

Key Code: BXA

U.S. Department of Labor OSHA Form No.

(Rev. 01/2009) Reflecting government revisions as of 01/2004

Step by Step

Quick Start Guide

1. Are You Required to Maintain a 300 Log?

Companies must keep records of all recordable occupational injuries and illnesses for each single business establishment that is expected to be in operation for one year or longer, unless the company-wide count of Recordable People (see Recordable People in Definitions section below) was 10 or fewer at all times during the last calendar year or your company is listed as a low hazard industry. *(See §1904.46 for definition of establishment. See §1904.1-.3 for information on partial exceptions. See Table §1904 Appendix A to Subpart B for the list of low hazard industries. See §1904.30(a) for multiple business establishments. See §1904.31 for definition of Recordable People.)*

2. Is this Person Recordable?

You record on the 300 Log the recordable injuries and illnesses of all Recordable People, *(see Recordable People in Definitions Section below). (See §1904.31 for information on Recordable People.)*

3. Is this Case Work-Related?

An injury or illness is considered work-related if an event or exposure in the workplace caused or contributed to the incident or significantly aggravated a preexisting condition. Injuries and illnesses that happen in the workplace are presumed to be work-related, unless there is an exception. *(See §1904.5(b)(2) for exceptions.)* The work environment includes not only the facility, but other areas where the company has sent at least one Recordable Person to work or be present. *(See §1904.5 for determination of work-relatedness. See §1904.5 (b)(1) for work environment. See §1904.5 (b)(4) for the definition of "significantly aggravated".)*

4. Is this a New Case?

A new case is when a Recordable Person suffers a new injury or illness. It is also a new case when a Recordable Person has fully recovered from the previous injury or illness, and a new event or exposure in the workplace causes the signs or symptoms to reappear. *See Step 7, if it is not a new case. (See §1904.6 for determination of new cases.)*

5. Is this Case Recordable?

A case is recordable on the 300 Log when it involves any of the following work-related incidents: death *(§1904.7(b)(2));* days away from work *(§1904.7(b)(3));* restricted work or transfer to another job *(§1904.7(b)(4));* medical treatment, *(see Medical Treatment Exceptions in Definitions section below) (§1904.7(b)(5));* loss of consciousness *(§1904.7(b)(6));* significant diagnosis by a healthcare professional *(§1904.7(b)(7));* needlestick or sharps injuries *(§1904.8);* medical removal due to a specific OSHA Standard *(§1904.9);* occupational hearing loss *(§1904.10);* or tuberculosis *(§1904.11.)*

If you answered "yes" to all of the steps 1 through 5, record the case.

6. How to Record this Case:

Fill out a 301 Incident Report and enter the case on the OSHA 300 Log within seven (7) calendar days. A 301 Incident Report or equivalent form must be filled out for each entry on the 300 Log. Enter the days of away from work and the days of restricted work or job transfer on the 300 Log, do not count the first day of the incident. Restricted work activity is when either the company or a healthcare professional places a work restriction on a Recordable Person or keeps a Recordable Person from performing the same duties and working the full workday that was scheduled before the incident. *(See §1904.29 for information on forms. See §1904.29(b)(4) for equivalent forms. See §1904.7(b)(4) for restricted work.)*

7. How to Update this Case: (when required)

If there are changes to the case during a five year period, you must update the original 300 Log entry to show these changes. Also enter on the original 300 Log any newly discovered recordable injuries or illnesses. *(See §1904.33(b) for information on updating records.)*

8. How to Complete, Post, and Retain Records:

At the end of the calendar year, total each column on the 300 Log, and then use this information to fill out a Form 300A, Summary of Work-Related Injuries and Illnesses. The 300A must be signed and certified by the highest-ranking company official at that facility or his/her immediate supervisor. Post the 300A Summary in that facility from February 1 to April 30 of the following year. Keep all 301 Reports, 300 Logs, and 300A Summaries for 5 years. *(See §1904.32 for information on annual completion of 300 Log and 300A Summary. See §1904.32(b)(5)-(6) for information on posting the 300A Summary. See §1904.33(a) for information on retention.)*

References and Definitions

Fault *(§1904.0)*
Cases listed on the 300 Log are not necessarily eligible for worker's compensation; nor does it mean that the company or Recordable Person was at fault or that an OSHA standard was violated.

Location *(§1904.35(b)(2)(iii); 1904.40)*
You may prepare and maintain this 300 Log at another facility; however, you must be able to show the OSHA Inspector these records within four business hours. When a worker or former worker requests these records, you must give them the records, with some limitations, by the end of the next business day.

First Report of Injury/Illness *(§1904.31)*
You must complete a 301 Incident Report or equivalent form for each recordable case entered on the 300 Log. Many employers use an insurance form instead of the OSHA 301 Incident Report, or supplement an insurance form by adding any additional information required by OSHA.

Recordable People *(§1904.31)*
Recordable people are workers on your payroll and anyone else you supervise on a day-to-day basis who are not on your payroll.

Medical Treatment Exceptions
(are not considered recordable)
(§1904.7(b)(5), §1904.7(b)(5)(ii))

1. A visit to a doctor or health-care professional only for observation or counseling; or
2. Diagnostic procedures such as x-rays, blood tests, and taking prescription medications used only for diagnostic purposes; or
3. The following procedures:
   - Taking non-prescription medication at non-prescription strength;
   - Receiving a tetanus shot;
   - Cleaning, flushing, or soaking wounds on the surface of the skin; using wound coverings, like bandages, Band-Aids™, gauze pads, etc., or using SteriStrips™ or butterfly bandages;
   - Using hot or cold therapy; using any non-rigid means of support, such as elastic bandages, wraps, non-rigid back belts, etc.; using temporary immobilization devices while transporting an accident victim (splints, slings, neck collars, or back boards);
   - Drilling of a fingernail or toenail to relieve pressure, or draining fluids from blisters;
   - Using eye patches; removing foreign bodies from the eye using only irrigation or a cotton swab;
   - Removing splinters or foreign material from areas other than the eye by irrigation, tweezers, cotton swabs, or other simple means;
   - Using finger guards; using massages; or drinking fluids to relieve heat stress.